ENGINEERED MATERIALS HANDBOOK™

Volume 1

COMPOSITES

Prepared under the direction of the
ASM INTERNATIONAL Handbook Committee

Theodore J. Reinhart, Technical Chairman

Cyril A. Dostal, Senior Editor
Mara S. Woods, Technical Editor
Heather J. Frissell, Editorial Supervisor
Alice W. Ronke, Assistant Editor
Diane M. Jenkins, Word Processing Specialist
Karen Lynn O'Keefe, Word Processing Specialist
Kelly L. Pilarczyk, Word Processing Specialist

Robert L. Stedfeld, Director of Reference Publications
Kathleen M. Mills, Manager of Editorial Operations

Editorial Assistance
J. Harold Johnson
Robert T. Kiepura
Dorene A. Humphries

METALS PARK, OHIO 44073

First printing, November 1987
Second printing, May 1988
Third printing, August 1989
Fourth printing, January 1993

Engineered Materials Handbook is a collective effort involving hundreds of technical specialists. It brings together in one book a wealth of information from world-wide sources to help scientists, engineers, and technicians solve current and long-range problems.

Great care is taken in the compilation and production of this volume, but it should be made clear that no warranties, express or implied, are given in connection with the accuracy or completeness of this publication, and no responsibility can be taken for any claims that may arise.

Nothing contained in the Engineered Materials Handbook shall be construed as a grant of any right of manufacture, sale, use, or reproduction, in connection with any method, process, apparatus, product, composition, or system, whether or not covered by letters patent, copyright, or trademark, and nothing contained in the Engineered Materials Handbook shall be construed as a defense against any alleged infringement of letters patent, copyright, or trademark, or as a defense against liability for such infringement.

Comments, criticisms, and suggestions are invited, and should be forwarded to ASM INTERNATIONAL.

Library of Congress Cataloging-in-Publication Data

ASM INTERNATIONAL

Engineered materials handbook

Includes bibliographies and indexes.
Contents: v. 1. Composites.

1. Materials—Handbooks, manuals, etc. I. ASM INTERNATIONAL. Handbook Committee.
TA403.E497   1987   620.1′1   87-19265
ISBN 0-87170-279-7 (v. 1)
SAN 204-7586

Printed in the United States of America

Foreword

Publication of Volume 1, *Composites,* of ASM INTERNATIONAL's new *Engineered Materials Handbook* is a signal event from many aspects.

It is a major step in fulfilling ASM INTERNATIONAL's commitment to expand its scope to include all materials of interest in the industries it serves, and new manufacturing areas as well. Future volumes in the *Engineered Materials Handbook* series will cover engineering plastics, ceramics, and other high technology materials.

It represents another major expansion of the ASM INTERNATIONAL handbooks publishing program, which began 60 years ago with a small looseleaf binder of metals properties data and other information. From this beginning the *Metals Handbook* has grown to the present 17-volume Ninth Edition, acknowledged worldwide as the definitive reference source for the metals and metalworking industry. With this inaugural volume, the *Engineered Materials Handbook* takes its place alongside the *Metals Handbook,* as a new series for the new materials.

It is the largest, most complete, most up-to-date single volume of in-depth engineering information on composite materials ever made available to the public. Its contents include articles on every essential topic pertaining to the use of composites: properties and forms of the basic fibers and matrix materials of which composites are made, as well as of the composite materials themselves; analysis and design of composite materials and of the structures made from them; testing of composites; manufacturing and fabrication processes; quality control; failure analysis; and applications and experience.

We are pleased to extend thanks and congratulations on behalf of ASM IN-TERNATIONAL to Technical Chairman Ted Reinhart and the volume's 18 Section Chairmen and Co-Chairmen, for the outstanding job they have done in recruiting an author list that includes many of the best-known authorities in the composites field. Our gratitude is also due these authors and the many reviewers who so generously donated their time and efforts to make this book a useful and authoritative reference. In addition, we express our appreciation to the ASM INTERNATIONAL editorial staff, for their hard work, diligence, and enthusiasm in beginning this new series at the same level of excellence so well established by the *Metals Handbook.*

William G. Wood
President,
ASM INTERNATIONAL

Edward L. Langer
Managing Director,
ASM INTERNATIONAL

Policy on Units of Measure

By a resolution of its Board of Trustees, ASM INTERNATIONAL has adopted the practice of publishing data in both metric and customary U.S. units of measure. In preparing this Handbook, the editors have attempted to present data primarily in metric units based on Système International d'Unités (SI), with secondary mention of the corresponding values in customary U.S. units. The decision to use SI as the primary system of units was based on the aforementioned resolution of the Board of Trustees, the widespread use of metric units throughout the world, and the expectation that the use of metric units in the United States will increase substantially during the anticipated lifetime of this Handbook.

For the most part, numerical engineering data in the text and in tables are presented in SI-based units with the customary U.S. equivalents in parentheses (text) or adjoining columns (tables). For example, pressure, stress, and strength are shown both in SI units, which are pascals (Pa) with a suitable prefix, and in customary U.S. units, which are pounds per square inch (psi). To save space, large values of psi have been converted to kips per square inch (ksi), where 1 kip = 1000 lb. Some strictly scientific data are presented in SI units only. For example, fatigue crack growth rates have been given only in millimeters per cycle (mm/cycle).

To clarify some illustrations that depict machine parts described in the text, only one set of dimensions is presented on artwork. References in the accompanying text to dimensions in the illustrations are presented in both SI-based and customary U.S. units. On graphs and charts, grids correspond to SI-based units, which appear along the left and bottom axes; where appropriate, corresponding customary U.S. units appear along the top and right axes.

Both nanometers (nm) and angstrom units (Å) are used (1 nm = 10 Å), the former as a unit of measure in x-ray crystallography, and the latter to measure light wavelengths. Data obtained according to standardized test methods for which the standard recommends a particular system of units are presented in the units of that system. Wherever feasible, equivalent units are also presented.

Conversions and rounding have been done in accordance with ASTM Standard E 380, with careful attention to the number of significant digits in the original data. For example, an annealing temperature of 1570 °F contains three significant digits. In this instance, the equivalent temperature would be given as 855 °C; the exact conversion to 854.44 °C would not be appropriate. For an invariant physical phenomenon that occurs at a precise temperature (such as the melting of pure silver), it would be appropriate to report the temperature as 961.93 °C or 1763.5 °F. In many instances (especially in tables and data compilations), temperature values in °C and °F are alternatives rather than conversions.

The policy on units of measure in this Handbook contains several exceptions to strict conformance to ASTM E 380; in each instance, the exception has been made to improve the clarity of the Handbook. The most notable exception is the use of $MPa\sqrt{m}$ rather than $MN \cdot m^{-3/2}$ or $MPa \cdot m^{0.5}$ as the SI unit of measure for fracture toughness. Other examples of such exceptions are the use of "L" rather than "l" as the abbreviation for liter and the use of g/cm^3 rather than kg/m^3 as the unit of measure for density (mass per unit volume).

SI practice requires that only one virgule (diagonal) appear in units formed by combination of several basic units. Therefore, all of the units preceding the virgule are in the numerator and all units following the virgule are in the denominator of the expression; no parentheses are required to prevent ambiguity.

Preface

This Volume focuses primarily on what are called "advanced composites." This is now perhaps a misnomer, since many of them have been in production use in the aviation industry for over 20 years. The phrase is shorthand nomenclature for what are, for the most part, thermoset polymer matrix materials, reinforced with continuous fibers, and highly engineered to the point where their properties approach the theoretical limits attainable from their particular combination of constituent materials. They are used mostly for primary structural applications.

This definition is offered to set them off from other composites which are less highly engineered but perhaps equally important in some nonaerospace areas of industry. These other composites, often referred to as "reinforced plastics," are most often used for secondary structure or components. They receive only minor attention in this Volume, but will be covered in more detail in Volume 2, *Engineering Plastics,* of this series.

In addition, the final section of Volume 1 affords a state-of-the-art overview of another type of composites which fit the definition offered above except for utilizing metal, carbon, or ceramic matrix materials, to better adapt them to high-temperature structural or component use.

Development of so-called advanced composites has for the most part been driven by the need for reduced weight and increased performance properties in aircraft, military, and space vehicle applications. Consequently, this development has taken place mostly in the aerospace community, much of it under Air Force, Department of Defense, and NASA sponsorship. Since cost was not a primary driver in this development, transfer of composites technology into commercial and consumer goods manufacturing was at first slow, confined mostly to areas like sporting goods, where cost is similarly secondary to performance.

However, the past five years have seen dramatic changes in this technology flow. Increased need in commercial and consumer sectors for stronger, lighter, more energy-efficient materials has accelerated interest in the available aerospace-developed advanced composites technology. At the same time, greatly increased aerospace use of advanced composites has decreased their material cost. And development of automated manufacturing methods has decreased their potential long-term-use procressing cost. These concurrent trends greatly increase the feasibility of using advanced composites in high-volume, nonaerospace production. As these benefits become a reality, they should, in turn, result in significant cost reductions for aerospace use of composites.

Another exciting trend is the inception of a back-flow of advanced composites technology from the commercial and consumer sector to aerospace, perhaps best exemplified by currently renewed aerospace interest in thermoplastic matrix composites, which owe most of their present state of development to automotive and other nonaerospace use. Continuous fiber reinforced thermoplastic matrix composites offer attractive possibilities for manufacturing cost reductions as well as increased service durability compared to thermoset matrix composites.

The substance of these trends is evident in the contents of this Volume. It affords extensive coverage of all phases of advanced composites technology, with particular emphasis on engineering properties and manufacturing. It is designed to be useful to engineers and other technical people, from novices to experts in composite materials, engaged in selecting, designing, evaluating, and applying advanced composite materials in a wide range of uses.

Theodore J. Reinhart
Technical Chairman

v

Authors and Reviewers

Ric Abbot
Beech Aircraft Corporation
Norman J. Abbott
Albany International Research Company
Frances L. Abrams
Air Force Wright Aeronautical Laboratories
D.F. Adams
University of Wyoming
R.C. Adams
Lockheed-Georgia Company
Richard G. Adams
J.P. Stevens & Company, Inc.
Robert D. Adams
University of Bristol (U.K.)
Norman R. Adsit
Rohr Industries, Inc.
Sam C. Aker
Sam C. Aker Inc.
James Allen
Fiberite Corporation
Ronald Allred
PDA Engineering
Maurice F. Amateau
The Pennsylvania State University
Roger A. Anderson
Enduro Fiberglass Systems
Jeanne M. Anglin
Boeing Commercial Airplane Company
Todd Ashton
Bentley-Harris Manufacturing Company
Roger Bacon
Amoco Performance Products, Inc.
L.L. Bahney
The Timken Company
J.A. Kim Bailie
Lockheed Missiles & Space Company Inc.
Richard C. Barr
Federal Mogul Corporation
W.D. Bascom
University of Utah
Robert J. Basso
Century Design Inc.
H. Dean Batha
Fiber Materials, Inc.
D.L. Baty
Babcock & Wilcox
Ronald S. Bauer
Shell Development Company
B.H. Beal
Grumman Aircraft Systems
Peter Beardmore
Ford Motor Company
Paul F. Becher
Oak Ridge National Laboratory

Anne R. Beck
Northrop Corporation
Donald Beckley
U.S. Polymeric Corporation
Scott W. Beckwith
Hercules Aerospace Company
David R. Beeler
Air Force Wright Aeronautical
Laboratories
Ron Bellew
Grumman Aircraft Systems
William P. Benjamin
Northrop Corporation
George Bennett
Mississippi State University
H.W. Bergmann
Deutsche-Forschungs- und Versuchsanstalt
für Luft- und Raumfahrt e.V.
Charles W. Bert
The University of Oklahoma
Roger L. Blaine
E.I. Du Pont de Nemours & Company,
Inc.
H. Wayland Blake
Martin Marietta Energy Systems Inc.
Paul Blanchard
General Electric Company
M.J. Bodnar
Organic Materials Branch
U.S. Army Armament Research and
Development Center
James A. Boldt
Northrop Corporation, Aircraft Division
Thomas Boljan
GTE Laboratories, Inc.
David J. Boll
Hercules Aerospace Company
Hans Borstell
Grumman Aircraft Systems
Robert Boudreau
Borden Chemical Division
Gordon Bourland
LTV Aerospace and Defense Company
Frank W. Bradish
Premix/E.M.S., Inc.
Richard A. Brand
General Dynamics Convair Division
Jürgen Brandt
Messerschmitt-Bölkow-Blohm GmbH
James P. Brazel
General Electric Company
Re-entry Systems Department
Walter V. Breitigam
Shell Development Company

F.T. Brewen
TRW Inc.
Hal F. Brinson
Virginia Polytechnic Institute and
State University
Gary G. Brown
Aerojet General Corporation
Gordon L. Brown, Jr.
Morrison Molded Fiber Glass Company
Richard T. Brown
Atlantic Research Corporation
William W. Brown, Jr.
Edler Industries, Inc.
Charles E. Browning
Air Force Wright Aeronautical
Laboratories
M.D. Brunner
Northrop Advanced Systems Division
Michael E. Buck
Avco Specialty Materials Division
Textron Inc.
Dwight A. Burford
Colorado School of Mines
Martin Burg
Composite Market Reports, Inc.
Mary T. Burgess
U.S. Army Materials Technology
Laboratory
Patrick D. Burke
Material Concepts, Inc.
Curtis V. Burkland
Amercom, Inc.
Robert Burns
Fiber Materials, Inc.
Steve Burpo
McDonnell Aircraft Company
Jose L. Camahort
Lockheed Missiles & Space Company Inc.
F.C. Campbell
McDonnell Aircraft Company
R.B. Cantley
Martin Marietta Aerospace
Michael F. Card
NASA Langley Research Center
Bernie Carpenter
Colorado School of Mines
H.E. Carroll
Lockheed-Georgia Company
J.R. Carroll, Jr.
Lockheed-Georgia Company
David Carver
Lockheed Missiles & Space Company Inc.
Kerry Casler
Consultant

Chris Chamis
NASA Lewis Research Center

Jai P. Chanani
Northrop Corporation, Aircraft Division

Tom Chance
LTV Aerospace and Defense Company

Benjamin T.A. Chang
Shell Development Company

G.B. Chapman II
Chrysler Corporation

M.A. Chaudhari
Ciba-Geigy Corporation

Robert G. Cheatham
United Technologies Corporation

Tsu-Wei Chou
University of Delaware

Eric Christiansen
Eagle Engineering, Inc.

Gail A. Clark
Consultant

Dave Clavadetscher
Premix, Inc.

Linda L. Clements
San Jose State University

William G. Colclough, Jr.
Fiberite Corporation

Jack L. Cook
Arco Chemical Company

Michael W. Cook
Technology Marketing Inc.

John Corden
Hexcel Corporation

James Cornie
Massachusetts Institute of Technology

Tony Cosentino
Grumman Aircraft Systems

Jeanne L. Courter
American Cyanamid Company

James Crawford
Techniweave, Inc.

Eugene R. Crilly
Rockwell International Corporation

Jane M. Crosby
LNP Corporation

Thomas Culkin
Lumonics Inc.

William D. Cumming
HITCO Woven Structures Division

James L. Cupp
Hexcel Corporation

M.J. Cusick
Colorado School of Mines

Peter Cyr
Sikorsky Aircraft
United Technologies Corporation

Daniel P. Dalenberg
Fiberite Corporation

George W. Dalley
Lockheed-California Company

Brian Damkroger
Colorado School of Mines

C.V. Darragh
The Timken Company

S.J. Dastin
Grumman Aircraft Systems

John E. Daunt
General Electric Company

H.O. Davis
Kaiser Aerotech

Peter Delvigs
NASA Lewis Research Center

Thomas W. DeMint
General Dynamics Corporation

Douglas L. Denton
Battelle—Columbus Division

L.E. DeShields
E.I. Du Pont de Nemours & Company, Inc.

H. Benson Dexter
NASA Langley Research Center

Anthony T. DiBenedetto
The University of Connecticut

James A. DiCarlo
NASA Lewis Research Center

Louis Dick
Lockheed Missiles & Space Company
Inc.

Russell J. Diefendorf
Rensselaer Polytechnic Institute

R.J. Dieterle
Amoco Chemicals Company

Peter R. DiGiovanni
Raytheon Company

Gail DiSalvo
Ciba-Geigy Corporation

B.B. Djordjevic
Martin Marietta Corporation

John F. Dockum, Jr.
PPG Industries, Inc.

Frank W. Doherty
Structurelite Plastics Corporation

F.S. Dominguez
Hercules Aerospace Company

Louis C. Dorworth
HITCO

Dave Douglas
General Electric Company

Norris F. Dow
Materials Sciences Corporation

Paul Doyle
U.S. Army Materials Technology
Laboratory

G.D. Drewry, Sr.
Hercules Inc.

Richard K. Dropek
Hercules Aerospace Company

Lawrence T. Drzal
Michigan State University

Charles D. Dudgeon
Ashland Chemical Company

Sidney G. Dunbar
Owens-Corning Fiberglas Corporation

Sunil Dutta
NASA Lewis Research Center

George J. Dvorak
Rensselaer Polytechnic Institute

John M. Edinger
E.I. Du Pont de Nemours & Company, Inc.

Coby Emshwiller
Sikorsky Aircraft
United Technologies Corporation

James Enos
Dow Chemical Company

Joseph N. Epel
Consultant

Richard H. Ericksen
Sandia National Laboratories

Diana M. Essock
General Electric Company

Glenn W. Ewald
Goldsworthy Engineering, Inc.

A.A. Fabry
Rockwell International Corporation

C.H. Fairbank, Jr.
Owens-Corning Fiberglas Corporation

G. Fantozzi
INSA, Lyon

Robert Farris
Shell Development Company

Frank Fechek
Air Force Wright Aeronautical
Laboratories

Mattison K. Ferber
University of Illinois, Urbana-Champaign

J.C. Finnell
McDonnell Douglas Corporation

A.L. Flescher
Grumman Corporation

John V. Foltz
Naval Surface Weapons Center

Charles R. Foreman
LTV Aircraft Products Group
Military Aircraft Division

Robert Forrer
Gerber Garment Technology, Inc.
Aerospace Division

Marvin Foston
Lockheed-Georgia Company

Kenneth A. Fouts
Teledyne Ryan Aeronautical

Bruce R. Fox
Douglas Aircraft Company

Richard B. Freeman
The Budd Company

K. Friedrich
Technische Universität Hamburg-Harburg

R. Fuquen
The Timken Company

Frank D. Gac
Los Alamos National Laboratory

Burton A. Gale
Consultant

James P. Gallivan
U.S. Army Materials Technology
Laboratory

Samuel P. Garbo
Sikorsky Aircraft
United Technologies Corporation

Hugh Gibbs
E.I. Du Pont de Nemours & Company, Inc.

John W. Gillespie, Jr.
University of Delaware

David M. Goddard
Material Concepts, Inc.

Gabriel D. Golam
Grumman Aircraft Systems

Bruce Goldstone
United Technologies Corporation
W. Brandt Goldsworthy
Goldsworthy Engineering, Inc.
Keith Goode
Composite Automation Equipment, Inc.
Kenneth N. Goodno
The Boeing Company
N.R. Gordon
Battelle Pacific N.W.
Rex B. Gosnell
King Mar Laboratories
Cape Composites, Inc.
D.C. Greene
BASF Structural Materials Inc.
Jacob Greenspan
U.S. Army Materials Technology
Laboratory
Glenn C. Grimes
Lockheed-California Company
Ray A. Grove
Boeing Commercial Airplane Company
Charles Gulotta
McDonnell Douglas Helicopter Company
Harry C. Haaxma
Haaxma & Associates
Deborah K. Hadad
Lockheed Missiles & Space Company Inc.
H. Thomas Hahn
The Pennsylvania State University
T.F.W. Hall
Northrop Aircraft Division
William B. Hall
University of Mississippi
R.G. Hallmark
Lockheed Missiles & Space Company Inc.
C.L. Hamermesh
Consultant
Gary E. Hansen
Hercules Aerospace Company
Niel W. Hansen
Hercules Aerospace Company
Steven F. Hanson
Composite Tooling International, Inc.
Gerald H. Hardesty
Amoco Chemicals Company
Foster L. Harding
Manville Service Corporation
Billy D. Harmon
Fiberite Corporation
John G. Harms
Dana Engine Products Division
William C. Harrigan, Jr.
DWA Composite Specialties, Inc.
L.J. Hart-Smith
Douglas Aircraft Company
James A. Harvey
The University of Dayton
Martin T. Harvey
ICI Americas Inc.
Zvi Hashin
Tel Aviv University
Mohamed Hashish
Flow Industries

Michael Henderson
Northrop Aircraft Division
Edmund G. Henneke II
Virginia Polytechnic Institute and State
University
Steve Henz
Northrop Aircraft Division
Paul Hergenrother
NASA Langley Research Center
David G. Hess
Northrop Advanced Systems Division
Robert Higgins
Technicut
William Hillig
General Electric Company
Jed Hilzinger
Alcoa Defense Systems, Inc.
Richard J. Hinrichs
Applied Polymer Technology Inc.
Kenneth E. Hofer
L.J. Broutman & Associates, Ltd.
Charles Holland
Rohr Industries, Inc.
Ed Hooper
Beech Aircraft Corporation
Lawrence C. Hopper
BASF Structural Materials Inc.
Arch Horner
Silmar
Ray E. Horton
Boeing Commercial Airplane Company
Keith Howe
Premix/E.M.S., Inc.
Rodney R. Howe
The Boeing Company
Maurice A.H. Howes
IIT Research Institute
J.P. Hrusovsky
The Timken Company
W. Donald Humphrey
Brunswick Corporation
Edward A. Humphreys
Materials Sciences Corporation
B.J. Hunter
Bell Helicopter Textron Inc.
Ramon L. Hurd
Huck Manufacturing Company
Michael Hyer
The University of Maryland
Hatsuo Ishida
Case Western Reserve University
Keith Jacobs
The Ironsides Company
Herb Jahnle
The Budd Company
Sushil Jain
General Motors Corporation
Arthur James
Lockheed-California Company
Carl F. Johnson
Ford Motor Company
Donald D. Johnson
Minnesota Mining & Manufacturing
Company

T.F. Jonas
Ciba-Geigy Corporation
John H. Jones
Boeing Commercial Airplane Company
R.J. Jones
TRW Inc.
Robert M. Jones
Virginia Polytechnic Institute and State
University
Raymond J. Juergens
McDonnell Douglas Corporation
Reid June
Boeing Commercial Airplane Company
D.H. Kaelble
Arroyo Polymer Center
Michael Kallaur
Freeman Chemical Corporation
B.E. Kaminski
Bell Helicopter Textron Inc.
Howard Katzman
The Aerospace Corporation
Manfred Katz
E.I. Du Pont de Nemours & Company, Inc.
A.A. Kays
Lockheed-Georgia Company
John J. Kibler
Materials Sciences Corporation
Ran Y. Kim
The University of Dayton Research
Institute
Donald E. Kizer
Material Concepts, Inc.
Howard S. Kliger
H.S. Kliger & Associates, Inc.
Stanley L. Klima
Grumman Aerospace Corporation
Ronald C. Knight
LTV Aerospace and Defense Corporation
Frank K. Ko
Drexel University
John R. Koenig
Southern Research Institute
Ron Kollmansberger
Alcoa Defense Systems, Inc.
Mark M. Konarski
U.S. Polymeric, HITCO
Jim Korican
Sundstrand Corporation
D. Korinke
Composite Automation Engineering
Peter Krag
Colorado School of Mines
Peter Kulas
Bonded Technology, Inc.
Walter L. Lachman
Materials International, Inc.
Paul A. Lagace
Massachusetts Institute of Technology
G.D. Lahoti
The Timken Company
Arthur H. Landrock
Organic Materials Branch
U.S. Army Armament Research and
Development Center

Lawrence A. Lang
Composites Manufacturing Engineering Inc.
Paul R. Langston
E.I. Du Pont de Nemours & Company, Inc.
Bruce Lanning
Colorado School of Mines
Richard C. Laramee
Morton Thiokol, Inc.
A.T. Laskaris
Avco Specialty Materials Division
Textron Inc.
J.A. Laverick
The Timken Company
A.F. Lawrence
MTS Systems Corporation
Irv Leichtle
Rockwell International Corporation
Arthur W. Leissa
The Ohio State University
G.H. Lemon
General Dynamics Corporation
Fort Worth Division
James C. Leslie
Advanced Composite Products &
Technology, Inc.
Edward J. Lesniak, Jr.
Chrysler Motors
Albert P. Levitt
U.S. Army Materials Technology
Laboratory
H. Walter Lewis
The Ingersoll Milling Machine Company
J.R. Lewis
Consultant
Stephen Lill
The Pennsylvania State University
John Lincoln
Air Force Wright Aeronautical Laboratories
Monroe Lindler
Shakespeare Company
Donald R. Linsenmann
E.I. Du Pont de Nemours & Company, Inc.
Stephen Liu
The Pennsylvania State University
Robert G. Loewy
Rensselaer Polytechnic Institute
Brian M. Louie
McClellan Air Force Base
John J. Lucas
Sikorsky Aircraft
United Technologies Corporation
James T. Luxon
GMI Engineering & Management Institute
David Maas
Advanced Composites Products, Inc.
Wesley C. Mace
W.C. Mace & Associates
F. Paul Magin, III
Hercules Aerospace Company
C. Lynn Mahoney
Hysol, Aerospace & Industrial Products
Division
Doug Malewicki
Aerovisions, Inc.

P.K. Mallick
University of Michigan, Dearborn
Leonard J. Marchinski
Leonard Associates, Inc.
Dan R. Marshall
Rockwell International Corporation
Jeffrey D. Martin
Pultrusion Technology, Inc.
Thierry N. Massard
Commissariat à l'Energie Atomique
John E. Masters
American Cyanamid Company
Louis Maus
Rockwell International Corporation
Paul Maxwell
Colorado School of Mines
Clayton A. May
Arroyo Research & Consulting Corporation
Lawrence E. McAllister
Allied Bendix Aerospace
John E. McCarty
Boeing Military Airplane Company
W.T. McCarvill
Hercules Aerospace Company
R.R. McClain
Interez, Inc.
John J. McCluskey
Owens-Corning Fiberglas Technical
Center
E.S. McCormick
General Electric Company
Louis R. McCreight
The Aerospace Corporation
John A. McElman
Avco Specialty Materials Division
Textron Inc.
T.W. McGann
Rockwell International Corporation
F.J. McGarry
Massachusetts Institute of Technology
Earl R. McGhee
McDonnell Aircraft Company
Patrick J. McGill
Technology Marketing Inc.
R.E. McGraugh
McDonnell Aircraft Company
Lee McKague
General Dynamics Corporation
L.E. Roy Meade
Lockheed-Georgia Company
Arnold Meeuwsen
General Electric Company
Brent Meredith
Northrop Corporation
Frederick J. Meyer
Ford Motor Company
Robert A. Meyer
The Aerospace Corporation
Michael J. Michno
Amoco Performance Products, Inc.
Leonard R. Migliore
Spectra-Physics
Alan Miller
Boeing Commercial Airplane Company

David M. Miller
Owens-Corning Fiberglas Corporation
Viorica E. Mindroiu
Ford Motor Company
Jovan Moacanin
California Institute of Technology
E.P. Moenning
Douglas Aircraft Company
Walter R. Mohn
Arco Chemical Company
Monib M. Monib
E.I. Du Pont de Nemours & Company, Inc.
Warren Moore
Boeing of Canada, Ltd.
Leonard Mordfin
National Bureau of Standards
Neal A. Mumford
Morton Thiokol, Inc.
Ashok K. Munjal
Aerojet Strategic Propulsion Company
Guy C. Murphy
General Electric Company
Meade Murphy
Douglas Aircraft Company
Allan D. Murray
Ford Motor Company
Frederick A. Myers
Armco Inc.
John Nardone
U.S. Army Plastics Technical Evaluation
Center
Donald Neal
U.S. Army Materials Technology
Laboratory
George Nielson
Boeing Commercial Airplane Company
James S. Noland
Consultant
Mark Norris
Bell Helicopter Textron Inc.
Bryan R. Noton
Battelle Columbus Laboratories
Ralph J. Nuismer
Hercules Aerospace Company
John Nunes
Materials Technology Laboratory
Kevin Obrachta
McDonnell Douglas Helicopter Company
T. Kevin O'Brien
NASA Langley Research Center
Francis Olix
CertainTeed Corporation
Howard Olson
Georgia Institute of Technology
Mark M. Opeka
Naval Surface Weapons Center
John F. Osterndorf
Organic Materials Branch
U.S. Army Armament Research and
Development Center
D.B. Owen
Southern Methodist University
N.J. Pagano
Air Force Wright Aeronautical Laboratories

R.J. Palmer
Douglas Aircraft Company
Russ Pancio
McDonnell Douglas Corporation
Stanley J. Paprocki
Material Concepts, Inc.
Robert T. Parker
Boeing Commercial Airplane Company
Emanuel Parzen
Texas A&M University
Marina R. Pascucci
GTE Laboratories Inc.
Ruth H. Pater
NASA Langley Research Center
Jocelyn M. Patterson
U.S. Air Force Materials Laboratory
John J. Pawlak
Chempure Corporation
Harry E. Pebly
Organic Materials Branch
U.S. Army Armament Research and
Development Center
S.T. Peters
Westinghouse Electric Company
Donald W. Petrasek
NASA Lewis Research Center
Joseph L. Phillips
Huck Manufacturing Company
James Phipps
Koppers Company, Inc.
S. Leigh Pheonix
Cornell University
M.R. Piggott
University of Toronto
Jack J. Pigliacampi
E.I. Du Pont de Nemours & Company, Inc.
Pat C. Pinoli
Lockheed Missiles & Space Company Inc.
Paul F. Pirrung
Air Force Wright Aeronautical Laboratories
Edwin P. Plueddemann
Dow Corning Corporation
Robin Podder
Northrop Aircraft Division
Frederick J. Policelli
Hercules Aerospace Company
Peter Popper
E.I. Du Pont de Nemours & Company, Inc.
John R. Porter
Rockwell International Corporation
William F. Prater
Martin Marietta Aerospace
John D. Pratt
Monogram Aerospace Fasteners
Roger Prescott
Great Lakes Carbon Corporation
Howard L. Price
Promatec Inc.
Frank S. Principe
E.I. Du Pont de Nemours & Company, Inc.
G. Pritchard
Kingston Polytechnic
John T. Quinlivan
Boeing Commercial Airplane Company

Henry J. Rack
Clemson University
N. Raghupathi
PPG Industries, Inc.
R.L. Ramkumar
Northrop Aircraft Corporation
Christopher Ramsay
Colorado School of Mines
Kenneth W. Ranger
Northrop Advanced Systems Division
Robert L. Rapson
Air Force Wright Aeronautical Laboratories
Stanley L. Ream
Amada Engineering & Service Company,
Inc.
John C. Reindl
General Motors Corporation
Harry S. Reinert, Jr.
Air Force Wright Aeronautical
Laboratories
Theodore J. Reinhart
Air Force Wright Aeronautical
Laboratories
Dan A. Reynolds
LTV Aerospace and Defense Company
Roy W. Rice
W.R. Grace & Company
Heinz Richter
Messerschmitt-Bölkow-Blohm GmbH
Dennis Riggs
R.J. Reynolds Tobacco Company
Richard W. Roberts
General Dynamics, Convair Division
A.B. Robertson
Ausimont U.S.A., Inc.
Richard E. Robertson
University of Michigan
Craig Robinson
Prince Manufacturing, Inc.
Peter Robinson
Olin Corporation
John W. Roman
Grumman Aircraft Systems
James C. Romine
E.I. Du Pont de Nemours & Company,
Inc.
B. Walter Rosen
Materials Sciences Corporation
Ernest P. Rossa
Fiberite Corporation
Charles R. Rowe
Naval Surface Weapons Center
Paul A. Roy
Ferro Corporation
R.G. Rudness
Martin Marietta Energy Systems, Inc.
Ward D. Rummel
Martin Marietta Aerospace
F.J. Russo
Naval Air Rework Facility
Victor N. Saffire
Reinhold Industries
Robert E. Sanders
Bell Helicopter Textron, Inc.

Gerald L. Sauer
BASF Structural Materials Inc.
Marvin E. Sauers
Amoco Performance Products, Inc.
Al Saurwein
Power Jet Inc.
E. Scala
Cortland Cable Company, Inc.
Weldon Scardino
Consultant
Centerville, OH
R.A. Schapery
Texas A&M University
Randy Scherer
Rogerson Aircraft Corporation
Cecil W. Schneider
Lockheed-Georgia Company
Sam Schneider
Rohr Industries
T. Schoenberg
Avco Specialty Materials Division
Textron Inc.
Scott Schoultz
Martin Marietta Corporation
W.J. Schultz
Minnesota Mining & Manufacturing
Company
Fred Schwab
Group 4 Associates, Inc.
Daniel A. Scola
United Technologies Research Center
Tom E. Scott
Michigan Technological University
Frederic D. Seaman
IIT Research Institute
Joel Self
Colorado School of Mines
George Sendeckyj
Air Force Wright Aeronautical
Laboratories
Tito T. Serafini
TRW Inc.
Ronald A. Servais
University of Dayton
P.T.B. Shaffer
Advanced Refractory Technology
Ray W. Shannon
Ray Shannon & Associates
Gordon Sharpe
Fortafil Fibers, Inc.
R.W. Sheatsley
Consultant
James E. Sheehan
GA Technologies Inc.
Don L. Sheldon
Electronics Metal Finishing Corporation
Wei Shih
The B.F. Goodrich Company
Vicki Sholtes
Boeing Vertol Company
Peter Shyprykevich
Grumman Aerospace Corporation
A.K. Sibal
Lockheed-Georgia Company

Mike Sibborn
Boeing Commercial Airplane Company
D.R. Sidwell
Ford Aerospace & Communications
Corporation
Robert Signorelli
NASA Lewis Research Center
Steven Singer
Ontario Die Company of America
Clyde Sipes
Lockheed Missiles & Space Company
Inc.
Steve Slaughter
Scaled Composites, Inc.
Adolph E. Slobodzinski
U.S. Army Plastics Technical Evaluation
Center
Brian W. Smith
Boeing Commercial Airplane Company
C.S. Smith
Admiralty Research Establishment
K.R. Smith
Consultant
W.S. Smith
E.I. Du Pont de Nemours & Company,
Inc.
Joseph Soderquist
Federal Aeronautics Administration
Harold G. Sowman
Minnesota Mining & Manufacturing
Company
J.F. Sproull
PPG Industries, Inc.
W.J. Spry
Consultant
Youngstown, NY
Eric B. Stark
Shell Development Company
James H. Starnes
NASA Langley Research Center
Gary L. Steckel
The Aerospace Corporation
Ed Stencel
Northrop Corporation
Robert H. Stone
Lockheed-California Company
E.R. Stover
The B.F. Goodrich Company
Sidney Street
Northrop Advanced Systems Division
James Strife
United Technologies Research Center
George Strumpf
Boeing Commercial Airplane Company
Pat Strumpff
Air Force Wright Aeronautical
Laboratories
Joseph E. Sumerak
Pultrusion Technology, Inc.
John Summerscales
Plymouth Polytechnic
United Kingdom
C.T. Sun
Purdue University

Raymond J. Suplinskas
Avco Specialty Materials Division
Textron Inc.
Stephen R. Swanson
The University of Utah
Saad Taha
Rockwell International Corporation
Eugene L. Talbot
Hercules Inc. (retired)
Todd Taricco
Thermal Equipment Corporation
Arthur Taverna
Avco Systems
T.N. Tiegs
Oak Ridge National Laboratory
John E. Theberge
LNP Corporation
Paul Thill
Rockwell International Corporation
Dale Thompson
Clairmont Enterprises, Inc.
Gaylen A. Thurston
NASA Langley Research Center
Milton W. Toaz
J.P. Industries, Inc.
Ronald Tobin
Borg-Warner Chemicals
Frederick Todt
Battelle Columbus Laboratories
M.E. Tohlen
LTV Aerospace & Defense Company
Peter Tomblin
DeHavilland Aircraft of Canada
Victor Tomporowski
Branson Sonic Power Company
Robert D. Torczyner
Lockheed Missiles & Space Company Inc
Frank T. Traceski
U.S. Army Materials Technology
Laboratory
Aurora A. Turla
Boeing Commercial Airplane Company
K.T. Turner
Grumman Aircraft Systems
Trevor Turner
Krauss Maffei Corporation
William R. Tyrrell
Branson Sonic Power Company
D.T. Uhl
General Dynamics Corporation
Fort Worth Division
Gary F. Van Aller
Bell Helicopter Textron Inc.
Mark Vangel
U.S. Army Materials Technology
Laboratory
H. Gary Van Schooneveld
Xerkon Company
Alex Vary
NASA Lewis Research Center
Thomas Vasilos
University of Lowell
John D. Venables
Martin Marietta Laboratories

Jozef Venner
BASF Structural Materials Inc.
Albert A. Vicario
Hercules Aerospace Company
Jack R. Vinson
University of Delaware
Ted K. Vogt
Rockwell International Corporation
H.F. Volk
LTV Aerospace and Defense Company
W.G. Wallis
Lockheed-California Company
A.S.D. Wang
Drexel University
Dick Warnock
McClellan Air Force Base
James C. Watson
PPG Industries, Inc.
F. Wawner
University of Virginia
John C. Weidner
Hercules Aerospace Company
Olin Weiss
General Dynamics Corporation
James A. Welbourn
Northrop Advanced Systems Division
Robert Wendt
Colorado School of Mines
R.G. White
The University, Southampton
United Kingdom
Robin Whitehead
Northrop Corporation
James M. Whitney
Air Force Wright Aeronautical
Laboratories
Mark A. Wigginton
Owens-Corning Fiberglas Corporation
T. Wilkinson
Ashland Chemical Company
Jerry G. Williams
Conoco Inc.
John Williams
Naval Air Development Center
Lynn A. Williams III
Ingersoll Milling Machine Company
Brian A. Wilson
Aerojet Strategic Propulsion Company
Dale W. Wilson
University of Delaware
Henry Wilson
Bell Helicopter Textron Inc.
Robert D. Wilson
Boeing Commercial Airplane Company
Steven Witschen
Owens-Corning Fiberglas Corporation
E.R. Witter
General Motors Corporation
E.P. Woo
Dow Chemical Company
Carl Zweben
General Electric Company

Contents

Introduction to Composites

Chairman: Theodore J. Reinhart, Air Force Wright Aeronautical Laboratories

Glossary of Terms

Harry E. Pebly, Army Armament Research & Development Center

A

A-basis. The "A" mechanical property value is the value above which at least 99% of the population of values is expected to fall, with a confidence of 95%. Also called A-allowable. See also *B-basis*, *S-basis*, and *typical-basis*.

abhesive. A material that resists adhesion. A film or coating applied to surfaces to prevent sticking, heat sealing, and so on, for example, a parting agent or mold release agent.

ablation. The degradation, decomposition, and erosion of a material caused by high temperature, pressure, time, percent oxidizing species, and velocity of gas flow. A controlled loss of material to protect the underlying structure.

ablative plastic. A material that absorbs heat (with a low material loss and char rate) through a decomposition process (pyrolysis) that takes place at or near the surface exposed to the heat.

ABL bottle. An internal pressure test vessel about 460 mm (18 in.) in diameter and 610 mm (24 in.) long used to determine the quality and properties of the filament-wound material in the vessel.

absorption. The penetration into the mass of one substance by another. The process whereby energy is dissipated within a specimen placed in a field of radiant energy. The capillary or cellular attraction of adherend surfaces to draw off the liquid adhesive film into the substrate.

accelerated test. A test procedure in which conditions are increased in magnitude to reduce the time required to obtain a result. To reproduce in a short time the deteriorating effect obtained under normal service conditions.

accelerator. A material that, when mixed with a catalyst or a resin, will speed up the chemical reaction between the catalyst and the resin (either in polymerizing of resins or vulcanization of rubbers). Also called promoter.

acoustic emission. A measure of integrity of a material, as determined by sound emission when a material is stressed. Ideally, emissions can be correlated with defects and/or incipient failure.

acrylic plastic. A thermoplastic polymer made by the polymerization of esters of acrylic acid and its derivatives. Its full name is polymethyl methacrylate. See also *polymethyl methacrylate*.

activation. The (usually) chemical process of making a surface more receptive to bonding to a coating or an encapsulating material.

activator. See *accelerator*.

addition polymerization. A chemical reaction in which simple molecules (monomers) are added to each other to form long-chain molecules (polymers) without forming by-products.

additive. Any substance added to another substance, usually to improve properties, such as plasticizers, initiators, light stabilizers, and flame retardants. See also *filler*.

adherend. A body that is held to another body, usually by an adhesive. A detail or part prepared for bonding.

adhesion. The state in which two surfaces are held together at an interface by mechanical or chemical forces or interlocking action or both.

adhesion, mechanical. See *mechanical adhesion*.

adhesion promoter. A coating applied to a substrate before it is coated with an adhesive, to improve the adhesion of the plastic. Also called primer.

adhesive. A substance capable of holding two materials together by surface attachment. Adhesive can be in film, liquid, or paste form.

adhesive, anaerobic. See *anaerobic adhesive*.

adhesive, cold-setting. See *cold-setting adhesive*.

adhesive, contact. See *contact adhesive*.

adhesive failure. Rupture of an adhesive bond such that the separation appears to be at the adhesive-adherend interface.

adhesive film. A synthetic resin adhesive, with or without a film carrier fabric, usually of the thermosetting type, in the form of a thin film of resin, used under heat and pressure as an interleaf in the production of bonded structures.

adhesive, gap-filling. See *gap-filling adhesive*.

adhesive, heat-activated. See *heat-activated adhesive*.

adhesive, heat-sealing. See *heat-sealing adhesive*.

adhesive, hot-melt. See *hot-melt adhesive*.

adhesive, hot-setting. See *hot-setting adhesive*.

adhesive, intermediate temperature setting. See *intermediate temperature setting adhesive*.

adhesive joint. The location at which two adherends or substrates are held together with a layer of adhesive. The general area of contact for a bonded structure.

adhesive, pressure-sensitive. See *pressure-sensitive adhesive*.

adhesive strength. Strength of the bond between an adhesive and an adherend.

adhesive, structural. See *structural adhesive*.

admixture. The addition and homogeneous dispersion of discrete components, before cure.

adsorption. The adhesion of the molecules of gases, dissolved substances, or liquids in more or less concentrated form, to the surfaces of solids or liquids with which they are in contact. A concentration of a substance at a surface or interface of another substance.

afterbake. See *postcure*.

aggregate. A hard, coarse material usually of mineral origin used with an epoxy binder (or other resin) in plastic tools. Also used in flooring or as a surface medium.

aging. The effect on materials of exposure to an environment for an interval of time. The process of exposing materials to an environment for an interval of time.

air-bubble void. Air entrapment within and between the plies of reinforcement or within a bondline or encapsulated area; localized, noninterconnected, spherical in shape.

air vent. Small outlet, to prevent entrapment of gases in a molding or tooling fixture.

alkyd plastic. Thermoset plastic based on resins composed principally of polymeric esters, in which the recurring ester groups are an integral part of the main polymer chain, and in which ester groups occur in most cross-links that may be present between chains.

allotropy. The existence of a substance and especially an element in two or more forms (as of crystals). See also *graphite*.

alloy. In plastics, a blend of polymers or copolymers with other polymers or elastomers under selected conditions; for example, styrene-acrylonitrile. Also called polymer blend. In metals, a substance having metallic properties and being composed of two or more chemical elements of which at least one is a metal.

allyl plastic. A thermoset plastic based on resins made by addition polymerization of monomers containing allyl groups; for example, diallyl phthalate (DAP).

alternating stress. A stress varying between two maximum values which are equal but with opposite signs, according to a law determined in terms of the time.

alternating stress amplitude. A test parameter of a dynamic fatigue test: one-half the algebraic difference between the maximum and minimum stress in one cycle.

ambient. The surrounding environmental conditions, such as pressure, temperature, or relative humidity.

amorphous plastic (amorphous phase). A plastic that has no crystalline component. There is no order or pattern to the distribution of the molecules.

anaerobic adhesive. An adhesive that cures only in the absence of air after being confined between assembled parts.

anelasticity. A characteristic exhibited by certain materials in which strain is a function of both stress and time, such that while no permanent deformations are involved, a finite time is required to establish equilibrium between stress and strain in both the loading and unloading directions.

angle-ply laminate. A laminate having fibers of adjacent plies oriented at alternating angles.

angle wrap. Tape fabric wrapped on a starter dam mandrel at an angle to the centerline.

anisotropic. Not isotropic. Exhibiting different properties when tested along axes in different directions. See also *anisotropy of laminates*.

anisotropic laminate. One in which the properties are different in different directions.

anisotropy of laminates. The difference of the properties along the directions parallel to the length or width of the lamination planes and perpendicular to the lamination.

annealing. In plastics, heating to a temperature at which the molecules have significant mobility, permitting them to reorient to a configuration having less residual stress.

antioxidant. A substance that, when added in small quantities to the resin during mixing, prevents its oxidative degradation and contributes to the maintenance of its properties.

antistatic agents. Agents that, when added to a molding material or applied to the surface of the molded object, make it less conducting, thus hindering the fixation of dust or the build-up of electrical charge.

aramid. A type of highly oriented organic material derived from polyamide (nylon) but incorporating aromatic ring structure. Used primarily as a high-strength high-modulus fiber. Kevlar and Nomex are examples of aramids.

arc resistance. Ability to withstand exposure to an electric voltage. The total time in seconds that an intermittent arc may play across a plastic surface without rendering the surface conductive.

areal weight. The weight of fiber per unit area (width × length) of tape or fabric.

aromatic. Unsaturated hydrocarbon with one or more benzene ring structures in the molecule.

artificial weathering. The exposure of plastics to cyclic laboratory conditions, consisting of high and low temperatures, high and low relative humidities, and ultraviolet radiant energy, with or without direct water spray and moving air (wind), in an attempt to produce changes in their properties similar to those observed in long-term continuous exposure outdoors. The laboratory exposure conditions are usually intensified beyond those encountered in actual outdoor exposure, in an attempt to achieve an accelerated effect. Also called accelerated aging.

ash content. Proportion of the solid residue remaining after a reinforcing substance has been incinerated (charred or intensely heated).

aspect ratio. The ratio of length to diameter of a fiber.

assembly time. The time interval between the spreading of the adhesive on the adherend and the application of pressure and/or heat to the assembly.

A-stage. An early stage in the polymerization reaction of certain thermosetting resins (especially phenolic) in which the material, after application to the reinforcement, is still soluble in certain liquids and is fusible. Also called resole. See also *B-stage* and *C-stage*.

attenuation. The diminution of vibrations or energy over time or distance. The process of making thin and slender, as applied to the formation of fiber from molten glass.

Audrey. The trade name of some equipment used for dynamic dielectric analysis (DDA).

autoclave. A closed vessel for conducting and completing a chemical reaction or other operation, such as cooling, under pressure and heat.

autoclave molding. A process in which, after lay-up, winding, or wrapping, an entire assembly is placed in a heated autoclave, usually at 340 to 1380 kPa (50 to 200 psi). Additional pressure permits higher density and improved removal of volatiles from the resin. Lay-up is usually vacuum bagged with a bleeder and release cloth.

automatic mold. A mold for injection or compression molding that repeatedly goes through the entire cycle, including ejection, without human assistance.

automatic press. A hydraulic press for compression molding or an injection machine that operates continuously, being controlled mechanically, electrically, hydraulically, or by a combination of any of these methods.

axial strain. The linear strain in a plane parallel to the longitudinal axis of the specimen.

axial winding. In filament-wound reinforced plastics, a winding with the filaments parallel or at a small angle to the axis (0° helix angle). See also *polar winding*.

B

back pressure. Resistance of a material, because of its viscosity, to continued flow when mold is closing.

bagging. Applying an impermeable layer of film over an uncured part and sealing the edges so that a vacuum can be drawn.

bag molding. A process in which the consolidation of the material in the mold is effected by the application of fluid or gas pressure through a flexible membrane.

bag side. The side of the part that is cured against the vacuum bag.

balanced construction. Equal parts of warp and fill in fiber fabric. Construction in which reactions to tension and compression loads result in extension or compression deformations only and in which flexural loads produce pure bending of equal magnitude in axial and lateral directions.

balanced design. In filament-wound reinforced plastics, a winding pattern so designed that the stresses in all filaments are equal.

balanced-in-plane contour. In a filament-wound part, a head contour in which the filaments are oriented within a plane and the radii of curvature are adjusted to balance the stresses along the filaments with the pressure loading.

balanced laminate. A composite laminate in which all laminae at angles other than 0° and 90° occur only in ± pairs (not necessarily adjacent) and are symmetrical around the centerline. See also *symmetrical laminate*.

balanced twist. An arrangement of twists in a combination of two or more strands that does not cause kinking or twisting on themselves when the yarn produced is held in the form of an open loop.

band density. In filament winding, the quantity of fiberglass reinforcement per inch of band width, expressed as strands (or filaments) per inch.

band thickness. In filament winding, the thickness of the reinforcement as it is applied to the mandrel.

band width. In filament winding, the width of the reinforcement as it is applied to the mandrel.

Barcol hardness. A hardness value obtained by measuring the resistance to penetration of a sharp steel point under a spring load. The instrument, called the Barcol impressor, gives a direct reading on a 0 to 100 scale. The hardness value is often used as a measure of the degree of cure of a plastic.

bare glass. Glass, such as yarns, rovings, and fabrics, from which the sizing or finish has been removed. Also, such glass before the application of sizing or finish.

barrier coat. An exterior coating applied to a composite wound structure to provide protection.

barrier film. The layer of film used to permit removal of air and volatiles from a composite lay-up during cure while minimizing resin loss.

batch. In general, a quantity of material formed during the same process or in one continuous process and having identical characteristics throughout. Also called a lot.

batt. Felted fabrics. Structures built by the interlocking action of compressing fibers, without spinning, weaving, or knitting.

B-basis. The "B" mechanical property value is the value above which at least 90% of the population of values is expected to fall, with a confidence of 95%. See also *A-basis*, *S-basis*, and *typical-basis*.

bearing area. The diameter of the hole times the thickness of the material. The cross-section area of the bearing load member on the sample.

bearing strain. The ratio of the deformation of the bearing hole, in the direction of the applied force, to the pin diameter. Also, the

stretch or deformation strain for a sample under bearing load.

bearing strength. The maximum bearing stress that can be sustained. Also, the bearing stress at that point on the stress-strain curve where the tangent is equal to the bearing stress divided by $n\%$ of the bearing hole diameter.

bearing stress. The applied load in pounds divided by the bearing area. Maximum bearing stress is the maximum load in pounds sustained by the specimen during the test, divided by the original bearing area.

bending-twisting coupling. A property of certain classes of laminates that exhibit twisting curvatures when subjected to bending moments.

bias fabric. Warp and fill fibers at an angle to the length of the fabric.

biaxial load. A loading condition in which a laminate is stressed in two different directions in its plane. A loading condition of a pressure vessel under internal pressure and with unrestrained ends.

biaxial winding. In filament winding, a type of winding in which the helical band is laid in sequence, side by side, with crossover of the fibers eliminated.

bidirectional laminate. A reinforced plastic laminate with the fibers oriented in two directions in its plane. A cross laminate. See also *unidirectional laminate*.

billet. A small ingot of nonferrous metal.

binder. The resin or cementing constituent (of a plastic compound) that holds the other components together. The agent applied to fiber mat or preforms to bond the fibers before laminating or molding.

bismaleimide (BMI). A type of polyimide that cures by an addition rather than a condensation reaction, thus avoiding problems with volatiles formation, and which is produced by a vinyl-type polymerization of a prepolymer terminated with two maleimide groups. Intermediate in temperature capability between epoxy and polyimide.

bladder. An elastomeric lining for the containment of hydroproof or hydroburst pressurization medium in filament-wound structures.

blanket. Fiber or fabric plies that have been laid up in a complete assembly and placed on or in the mold all at one time (flexible bag process). Also, the form of bag in which the edges are sealed against the mold.

bleeder cloth. A woven or nonwoven layer of material used in the manufacture of composite parts to allow the escape of excess gas and resin during cure. The bleeder cloth is removed after the curing process and is not part of the final composite.

bleeding. The removal of excess resin from a laminate during cure. The diffusion of color

out of a plastic part into the surrounding surface or part.

bleedout. The excess liquid resin that migrates to the surface of a winding. Primarily occurs in filament winding.

bloom. A visible local exudation or finish change on the surface of a plastic. Bloom can be caused by a lubricant or plasticizer or by atmospheric contamination.

BMC. See *bulk molding compound*.

BMI. See *bismaleimide*.

body putty. A pastelike mixture of plastic resin (polyester or epoxy) and talc used in repair of metal surfaces, such as auto bodies.

bond strength. The amount of adhesion between bonded surfaces. The stress required to separate a layer of material from the base to which it is bonded, as measured by load/bond area. See also *peel strength*.

boron fiber. A fiber produced by vapor deposition of elemental boron, usually onto a tungsten filament core, to impart strength and stiffness.

braiding. Weaving of fibers into a tubular shape instead of a flat fabric, as for graphite fiber reinforced golf club shafts.

branched polymer. In molecular structure of polymers, a main chain with attached side chains, in contrast to a linear polymer.

breaking extension. The elongation necessary to cause rupture of a test specimen. The tensile strain at the moment of rupture.

breaking factor. The breaking load divided by the original width of a test specimen, expressed in lb/in.

breaking length. A measure of the breaking strength of yarn. The length of a specimen whose weight is equal to the breaking load.

breakout. Fiber separation or break on surface plies at drilled or machined edges.

breather. A loosely woven material that serves as a continuous vacuum path over a part but is not in contact with the resin.

breathing. The opening and closing of a mold to allow gas to escape early in the molding cycle. Also called degassing; sometimes called bumping in phenolic molding.

bridging. Condition in which fibers do not move into or conform to radii and corners during molding, resulting in voids and dimensional control problems.

broad goods. Fiber woven to form fabric up to 1270 mm (50 in.) wide. It may or may not be impregnated with resin and is usually furnished in rolls of 25 to 140 kg (50 to 300 lb).

B-stage. An intermediate stage in the reaction of certain thermosetting resins in which the material softens when heated and is plastic and fusible but may not entirely dissolve or fuse. Also called resistol. The resin in an uncured prepreg or premix is usually in this stage. See also *A-stage* and *C-stage*.

buckling (composite). A mode of failure generally characterized by an unstable lateral material deflection due to compressive action on the structural element involved.

bulk density. The density of a molding material in loose form (granular, nodular, and so forth), expressed as a ratio of weight to volume.

bulk factor. The ratio of the volume of a raw molding compound or powdered plastic to the volume of the finished solid piece produced therefrom. The ratio of the density of the solid plastic object to the apparent or bulk density of the loose molding powder.

bulk modulus. The ratio of the hydrostatic pressure to the volume strain.

bulk molding compound (BMC). Thermosetting resin mixed with strand reinforcement, fillers, and so on, into a viscous compound for compression or injection molding. See also *premix* and *sheet molding compound*.

bundle. A general term for a collection of essentially parallel filaments or fibers.

burned. Showing evidence of thermal decomposition or charring through some discoloration, distortion, destruction, or conversion of the surface of the plastic, sometimes to a carbonaceous char.

burst strength (bursting strength). Measure of the ability of a material to withstand internal hydrostatic or gas dynamic pressure without rupture. Hydraulic pressure required to burst a vessel of given thickness.

bushing. An electrically heated alloy container encased in insulating material, used for melting and feeding glass in the forming of individual fibers or filaments. Also, special extra heavy load-carrying short cylinder inserted in bolt or pin holes.

butt joint. A type of edge joint in which the edge faces of the two adherends are at right angles to the other faces of the adherends.

C

calender. To produce a smooth finish and a desired dimensional thickness for sheet material by passing it between sets of pressure rollers.

carbon. The element that provides the backbone for all organic polymers. Graphite is a more ordered form of carbon. Diamond is the densest crystalline form of carbon.

carbon-carbon. A composite material consisting of carbon or graphite fibers in a carbon or graphite matrix.

carbon fiber. Fiber produced by the pyrolysis of organic precursor fibers, such as rayon,

polyacrylonitrile (PAN), and pitch, in an inert environment. The term is often used interchangeably with the term graphite; however, carbon fibers and graphite fibers differ. The basic differences lie in the temperature at which the fibers are made and heat treated, and in the amount of elemental carbon produced. Carbon fibers typically are carbonized in the region of 1315 °C (2400 °F) and assay at 93 to 95% carbon, while graphite fibers are graphitized at 1900 to 2480 °C (3450 to 4500 °F) and assay at more than 99% elemental carbon. See also *pyrolysis (of fibers)*.

carbonization. The process of pyrolyzation in an inert atmosphere at temperatures ranging from 800 to 1600 °C (1470 to 2910 °F) and higher, usually at about 1315 °C (2400 °F). Range is influenced by precursor, individual manufacturer's process, and properties desired.

catalyst. A substance that changes the rate of a chemical reaction without itself undergoing permanent change in composition or becoming a part of the molecular structure of the product. A substance that markedly speeds up the cure of a compound when added in minor quantity as compared to the amounts of primary reactants. See also *accelerator*, *curing agent*, *hardener*, *inhibitor*, and *promoter*.

catastrophic failures. Totally unpredictable failures of a mechanical, thermal, or electrical nature.

catenary. A measure of the difference in length of the strands in a specified length of roving as a result of unequal tension. The tendency of some strands in a taut horizontal roving to sag more than the others.

caul plates. Smooth metal plates, free of surface defects, the same size and shape as a composite lay-up, used immediately in contact with the lay-up during the curing process to transmit normal pressure and temperature, and to provide a smooth surface on the finished laminate.

cavity. The space inside a mold in which a resin or molding compound is poured or injected. The female portion of a mold. That portion of the mold that encloses the molded article (often referred to as the die). Depending on the number of such depressions, molds are designated as single cavity or multiple cavity.

cell. In honeycomb core, a cell is a single honeycomb unit, usually in a hexagonal shape.

cell size. The diameter of an inscribed circle within a cell of honeycomb core.

centrifugal casting. A production technique for fabricating cylindrical composites, such as pipe, in which composite material is positioned inside a hollow mandrel designed to be heated and rotated as resin is cured.

ceramic. A rigid, frequently brittle material made from clay and other inorganic, nonmetallic substances and fabricated into articles by sintering, that is, cold molding followed by fusion of the part at high temperature.

cermet. Composite materials consisting of two constituents, one being either an oxide, carbide, boride, or similar inorganic compound, and the other a metallic binder.

C-glass. A glass with a soda-lime-borosilicate composition that is used for its chemical stability in corrosive environments.

chain length. The length of the stretched linear macromolecule, most often expressed by the number of identical links.

chalking. Dry, chalklike appearance of deposit on the surface of a plastic.

Charpy impact test. A test for shock loading in which a centrally notched sample bar is held at both ends and broken by striking the back face in the same plane as the notch.

charring. The heating of a composite in air to reduce the polymer matrix to ash, allowing the fiber content to be determined by weight.

chemical vapor deposited (CVD) carbon. Carbon deposited on a substrate by pyrolysis of a hydrocarbon, such as methane

chemical vapor deposition (CVD). Process used in manufacture of several composite reinforcements, especially boron and silicon carbide, in which desired reinforcement material is deposited from vapor phase onto a continuous core, for example, boron on tungsten wire (core).

chromatography. See *thin-layer chromatography*.

circuit. In filament winding, one complete traverse of a winding band from one arbitrary point along the winding path to another point on a plane through the starting point and perpendicular to the axis.

circuit board. A sheet of insulating material laminated to foil that is etched to produce a circuit pattern on one or both sides. Also called printed circuit board or printed wiring board.

circumferential ("circ") winding. In filament wound reinforced plastics, a winding with the filaments essentially perpendicular to the axis (90° or level winding).

clamping pressure. In injection molding and transfer molding, the pressure that is applied to the mold to keep it closed in opposition to the fluid pressure of the compressed molding material.

closure. The complete coverage of a mandrel with one layer (two plies) of fiber. When the last tape circuit that completes mandrel coverage lays down adjacent to the first without gaps or overlaps, the wind pattern is said to have "closed."

cloth. See *woven fabric* and *nonwoven fabric.*

co-curing. The act of curing a composite laminate and simultaneously bonding it to some other prepared surface, or curing together an inner and outer tube of similar or dissimilar fiber-resin combination after each has been wound or wrapped separately. See also *secondary bonding.*

coefficient of elasticity. The reciprocal of Young's modulus in a tension test. See also *compliance.*

coefficient of expansion. A measure of the change in length or volume of an object, specifically measured by the increase in length or volume of an object per unit length or volume.

coefficient of friction. A measure of the resistance to sliding of one surface in contact with another surface.

coefficient of thermal expansion (CTE). The change in length or volume per unit length or volume produced by a 1° rise in temperature.

cohesion. The propensity of a single substance to adhere to itself. The internal attraction of molecular particles toward each other. The ability to resist partition of itself. The force holding a single substance together.

cohesive failure. Failure of an adhesive joint occurring primarily in an adhesive layer.

cohesive strength. Intrinsic strength of an adhesive.

coin test. Using a coin to tap a laminate in different spots, listening for a change in sound, which would indicate the presence of a defect. A surprisingly accurate test in the hands of experienced personnel.

coke. Carbonaceous residue resulting from the pyrolysis of pitch.

cold flow. The distortion that takes place in materials under continuous load at temperatures within the working range of the material without a phase or chemical change. See also *creep.*

cold-setting adhesive. A synthetic resin adhesive capable of hardening at normal room temperature in the presence of a hardener.

collet. A rigid, lateral container for the mold-forming material. A dam, a restriction box. The drive wheel that pulls glass fibers from the bushing. A forming tube is placed on the collet, and a package of strand is wound up on the tube. A metal band, ferrule, collar, or flange, often used to hold a tool or workpiece.

collimated. Rendered parallel.

collimated roving. Roving that has been made using a special process (usually parallel wound), so that the strands are more parallel than in standard roving.

colloidal. A state of suspension in a liquid medium in which extremely small particles are suspended and dispersed but not dissolved.

compaction. The application of a temporary vacuum bag and vacuum to remove trapped air and compact the lay-up.

compatibility. The ability of two or more substances combined with one another to form a homogeneous composition of useful plastic properties; for example, the suitability of a sizing or finish for use with certain general resin types. Nonreactivity or negligible reactivity between materials in contact.

complex dielectric constant. The vectorial sum of the dielectric constant and the loss factor.

complex shear modulus. The vectorial sum of the shear modulus and the loss modulus.

complex Young's modulus. The vectorial sum of Young's modulus and the loss modulus. Analogous to the complex dielectric constant.

compliance. Tensile compliance: the reciprocal of Young's modulus. Shear compliance: the reciprocal of shear modulus. Also, a term used in the evaluation of stiffness and deflection.

composite material. A combination of two or more materials (reinforcing elements, fillers, and composite matrix binder), differing in form or composition on a macroscale. The constituents retain their identities; that is, they do not dissolve or merge completely into one another although they act in concert. Normally, the components can be physically identified and exhibit an interface between one another.

compound. The intimate admixture of a polymer with other ingredients, such as fillers, softeners, plasticizers, reinforcement, catalysts, pigments, or dyes. A thermoset compound usually contains all the ingredients necessary for the finished product, while a thermoplastic compound may require subsequent addition of pigments, blowing agents, and so forth.

compression molding. A mold that is open when the material is introduced and that shapes the material by the pressure of closing and by heat.

compressive modulus. Ratio of compressive stress to compressive strain below the proportional limit. Theoretically equal to Young's modulus determined from tensile experiments.

compressive strength. The ability of a material to resist a force that tends to crush or buckle. The maximum compressive load sustained by a specimen divided by the original cross-sectional area of the specimen.

compressive stress. The normal stress caused by forces directed toward the plane on which they act.

condensation polymerization. A chemical reaction in which two or more molecules combine, with the separation of water or some other simple substance. If a polymer is formed, the process is called polycondensation. See also *polymerization.*

conditioning. Subjecting a material to a prescribed environmental and/or stress history before testing.

conductivity. Reciprocal of volume resistivity. The electrical or thermal conductance of a unit cube of any material (conductivity per unit volume).

consolidation. In metal matrix or thermoplastic composites, a processing step in which fiber and matrix are compressed by one of several methods to reduce voids and achieve desired density.

constituent. In general, an element of a larger grouping. In advanced composites, the principal constituents are the fibers and the matrix.

contact adhesive. An adhesive that is apparently dry to the touch and which will adhere to itself simultaneously upon contact. An adhesive applied to both adherends and allowed to become dry, which develops a bond when the adherends are brought together without sustained pressure.

contact molding. A process for molding reinforced plastics in which reinforcement and resin are placed on a mold. Cure is either at room temperature using a catalyst-promoter system or by heating in an oven, without additional pressure.

contact pressure resins. Liquid resins that thicken or polymerize on heating, and, when used for bonding laminates, require little or no pressure.

contaminant. An impurity or foreign substance present in a material or environment that affects one or more properties of the material, particularly adhesion.

continuous filament yarn. Yarn formed by twisting two or more continuous filaments into a single, continuous strand.

copolymer. A long-chain molecule formed by the reaction of two or more dissimilar monomers. See also *polymer.*

core. The central member, usually foam or honeycomb, of a sandwich construction to which the faces of the sandwich are attached or bonded. The central member of a plywood assembly. A channel in a mold for circulation of heat transfer media. A device on which prepreg is wound.

core crush. A collapse, distortion, or compression of the core.

core depression. A localized indentation or gouge in the core.

cored mold. A mold incorporating passages for electrical heating elements, steam, or water.

core separation. A partial or complete breaking of the core node bond.

core splicing. The joining of segments of a core by bonding, or by overlapping each segment and then driving them together.

corrosion resistance. The ability of a material to withstand contact with ambient natural factors or those of a particular artificially created atmosphere, without degradation or change in properties. For metals, this could be pitting or rusting; for organic materials, it could be crazing.

count. For fabric, number of warp and filling yarns per inch in woven cloth. For yarn, size based on relation of length and weight.

coupling agent. Any chemical substance designed to react with both the reinforcement and matrix phases of a composite material to form or promote a stronger bond at the interface.

coupon. Usually, a specimen for a specific test, as a tensile coupon.

crack. An actual separation of material, visible on opposite surfaces of the part, and extending through the thickness. A fracture.

crack growth. Rate of propagation of a crack through a material due to a static or dynamic applied load.

crazing. Region of ultrafine cracks, which may extend in a network on or under the surface of a resin or plastic material. May appear as a white band. Often found in a filament-wound pressure vessel or bottle.

creel. A device for holding the required number of roving balls (spools) or supply packages in desired position for unwinding onto the next processing step, that is, weaving, braiding, or filament winding.

creep. The change in dimension of a material under load over a period of time, not including the initial instantaneous elastic deformation. (Creep at room temperature is called cold flow.) The time-dependent part of strain resulting from an applied stress.

creep, rate of. The slope of the creep-time curve at a given time. Deflection with time under a given static load.

crimp. The waviness of a fiber or fabric, which determines the capacity of fibers to cohere under light pressure. Measured by the number of crimps or waves per unit length.

critical length. The minimum fiber length required for shear loading to its ultimate strength by the matrix.

critical longitudinal stress. Applied to fibers, the longitudinal stress necessary to cause internal slippage and separation of a spun yarn. The stress necessary to overcome the interfiber friction developed as a result of twist.

critical strain. The strain at the yield point.

cross-linking. Applied to polymer molecules, the setting-up of chemical links between the molecular chains. When extensive, as in most thermosetting resins, cross-linking makes one infusible supermolecule of all the chains.

cross-linking, degree of. The fraction of cross-linked polymeric units in the entire system.

cross-ply laminate. A laminate with plies usually oriented at 0° and 90° only.

crosswise direction. Crosswise refers to the cutting of specimens and to the application of load. For rods and tubes, crosswise is any direction perpendicular to the long axis. For other shapes or materials that are stronger in one direction than in another, crosswise is the direction that is weaker. For materials that are equally strong in both directions, crosswise is an arbitrarily designated direction at right angles to the lengthwise direction.

crystalline plastic. A polymeric material having an internal structure in which the atoms are arranged in an orderly three-dimensional configuration.

C-scan. The back-and-forth scanning of a specimen with ultrasonics. A nondestructive testing technique for finding voids, delaminations, defects in fiber distribution, and so forth.

C-stage. The final stage in the reaction of certain thermosetting resins in which the material is practically insoluble and infusible. Sometimes referred to as resite. The resin in a fully cured thermoset molding is in this stage. See also *A-stage* and *B-stage*.

cure. To irreversibly change the properties of a thermosetting resin by chemical reaction, that is, condensation, ring closure, or addition. Cure may be accomplished by addition of curing (cross-linking) agents, with or without heat and pressure.

cure cycle. The time/temperature/pressure cycle used to cure a thermosetting resin system or prepreg.

cure monitoring, electrical. Use of electrical techniques to detect changes in the electrical properties and/or mobility of the resin molecules during cure. A measuring of resin cure.

cure stress. A residual internal stress produced during the curing cycle of composite structures. Normally, these stresses originate when different components of a wet lay-up have different thermal coefficients of expansion.

curing agent. A catalytic or reactive agent that, when added to a resin, causes polymerization. Also called hardener.

CVD carbon. See *chemical vapor deposited (CVD) carbon*.

D

dam. Boundary support or ridge used to prevent excessive edge bleeding or resin runout of a laminate and to prevent crowning of the bag during cure.

damage tolerance. A design measure of crack growth rate. Cracks in damage tolerant designed structures are not permitted to grow to critical size during expected service life.

damping. The decay with time of the amplitude of free vibrations of a specimen. See also *hysteresis* and *attenuation*.

daylight. The distance, in the open position, between the moving and fixed tables or the platens of a hydraulic press. In the case of a multiplaten press, daylight is the distance between adjacent platens. Daylight provides space for removal of the molded part from the mold.

debond. A deliberate separation of a bonded joint or interface, usually for repair or rework purposes. Also, an unbonded or nonadhered region; a separation at the fiber-matrix interface due to strain incompatibility. In the United Kingdom, the term often refers to accidental damage. See also *disbond* and *delamination*.

debulking. Compacting of a thick laminate under moderate heat and pressure and/or vacuum to remove most of the air, to ensure seating on the tool, and to prevent wrinkles.

deep-draw mold. A mold having a core that is long in relation to the wall thickness.

deflashing. A finishing technique used to remove the flash (excess, unwanted material) on a plastic molding.

deflection temperature under load. The temperature at which a simple cantilever beam deflects a given amount under load. Formerly called heat distortion temperature.

deformation under load. The dimensional change of a material under load for a specified time following the instantaneous elastic deformation caused by the initial application of the load. See also *cold flow* and *creep*.

degassing. See *breathing*.

degradation. A deleterious change in the chemical structure, physical properties, or appearance of a plastic.

degree of polymerization. Number of structural units, or mers, in the average polymer molecule in a sample measure of molecular weight.

delamination. Separation of the layers of material in a laminate, either local or covering a wide area. Can occur in the cure or subsequent life.

denier. A yarn and filament numbering system in which the yarn number is numerically equal to the weight in grams of 9000 meters.

Used for continuous filaments. The lower the denier, the finer the yarn.

densification process. Consolidation of a loose or bulky material.

deposition. The process of applying a material to a base by means of vacuum, electrical, chemical, screening, or vapor methods, often with the assistance of a temperature and pressure container.

design allowables. Statistically defined (by a test program) material property allowable strengths, usually referring to stress or strain. See also *A-basis*, *B-basis*, *S-basis*, and *typical basis*.

desizing. The process of eliminating sizing, which is generally starch, from gray (also greige) goods before applying special finishes or bleaches (for yarn such as glass or cotton). Also, removing lubricant size following weaving of a cloth.

desorption. A process in which an absorbed material is released from another material. Desorption is the reverse of absorption, adsorption, or both.

devitrification. The formation of crystals (seeds) in a glass melt, usually occurring when the melt is too cold. These crystals can appear as defects in glass fibers.

D-glass. A high boron content glass made especially for laminates requiring a precisely controlled dielectric constant.

dielectric. A nonconductor of electricity. The ability of a material to resist the flow of an electrical current.

dielectric constant. The ratio of the capacitance of an assembly of two electrodes separated solely by a plastic insulating material to its capacitance when the electrodes are separated by air. See also *complex dielectric constant*.

dielectric curing. The curing of a synthetic thermosetting resin by the passage of an electric charge (produced from a high frequency generator) through the resin.

dielectric heating. The heating of materials by dielectric loss in a high-frequency electrostatic field.

dielectric loss. A loss of energy evidenced by the rise in heat of a dielectric placed in an alternating electric field.

dielectric monitoring. A means of tracking the cure of thermosets by changes in their electrical properties during material processing.

dielectric strength. The property of an insulating material that enables it to withstand electric stress. The average potential per unit thickness at which failure of the dielectric material occurs.

dielectrometry. Use of electrical techniques to measure the changes in loss factor (dissipation) and in capacitance during cure of the resin in a laminate.

differential scanning calorimetry (DSC). Measurement of the energy absorbed (endotherm) or produced (exotherm) as a resin system is cured. Also detects loss of solvents and other volatiles.

differential thermal analysis (DTA). An experimental analysis technique in which a specimen and a control are heated simultaneously and the difference in their temperatures is monitored. The difference in temperature provides information on relative heat capacities, presence of solvents, changes in structure (that is, phase changes, such as melting of one component in a resin system), and chemical reactions. See also *differential scanning calorimetry*.

dimensional stability. Ability of a plastic part to retain the precise shape to which it was molded, cast, or otherwise fabricated.

disbond. An area within a bonded interface between two adherends in which an adhesion failure or separation has occurred. Also, colloquially, an area of separation between two laminae in the finished laminate (in this case, the term delamination is normally preferred). See also *debond*.

displacement angle. In filament winding, the advancement distance of the winding ribbon on the equator after one complete circuit.

dissipation factor, electrical. See *electrical dissipation factor*.

distortion. In fabric, the displacement of fill fiber from the 90° angle (right angle) relative to the warp fiber. In a laminate, the displacement of the fibers (especially at radii), relative to their idealized location, due to motion during lay-up and cure.

doctor blade or bar. A straight piece of material used to spread resin, as in application of a thin film of resin for use in hot melt prepregging or for use as an adhesive film. Also called paste metering blade.

doily. In filament winding, the planar reinforcement applied to a local area between windings to provide extra strength in an area where a cut-out is to be made, for example, port openings. Usually placed at the knuckle joints of cylinder to dome.

dome. In filament winding, the portion of a cylindrical container that forms the spherical or elliptical shell ends of the container.

doubler. In filament winding, a local area with extra wound or fabric reinforcement, wound integrally with the part, or wound separately and fastened to the part.

doubler. See *tabs*.

draft. The taper or slope of the vertical surfaces of a mold designed to facilitate removal of molded parts.

draft angle. The angle of a taper on a mandrel or mold that facilitates removal of the finished part.

drape. The ability of a fabric or prepreg to conform to a contoured surface.

drawn fiber. Fiber with a certain amount of orientation imparted by the drawing process by which it was formed.

dry laminate. A laminate containing insufficient resin for complete bonding of the reinforcement. See also *resin-starved area*.

dry lay-up. Construction of a laminate by the layering of preimpregnated reinforcement (partly cured resin) in a female mold or on a male mold, usually followed by bag molding or autoclave molding.

dry winding. A term used to describe filament winding using preimpregnated roving, as differentiated from wet winding, where unimpregnated roving is pulled through a resin bath just before being wound onto a mandrel. See also *wet winding*.

DSC. See *differential scanning calorimetry*.

DTA. See *differential thermal analysis*.

ductility. The amount of plastic strain that a material can withstand before fracture. Also, the ability of a material to deform plastically before fracturing.

dwell. A pause in the application of pressure or temperature to a mold, made just before it is completely closed, to allow the escape of gas from the molding material. In filament winding, the time that the traverse mechanism is stationary while the mandrel continues to rotate to the appropriate point for the traverse to begin a new pass. In a standard autoclave cure cycle, an intermediate step in which the resin matrix is held at a temperature below the cure temperature for a specified period of time sufficient to produce a desired degree of staging. Used primarily to control resin flow.

dynamic modulus. The ratio of stress to strain under vibratory conditions (calculated from data obtained from either free or forced vibration tests, in shear, compression, or elongation).

E

edge distance ratio. The distance from the center of the bearing hole to the edge of the specimen in the direction of the principal stress, divided by the diameter of the hole.

edge joint. A joint made by bonding the edge faces of two adherends.

E-glass. A family of glasses with a calcium aluminoborosilicate composition and a maximum alkali content of 2.0%. A general-purpose fiber that is most often used in reinforced plastics, and is suitable for electrical laminates because of its high resistivity. Also called electric glass.

elastic deformation. The part of the total strain in a stressed body that disappears upon removal of the stress.

elasticity. That property of materials by virtue of which they tend to recover their original size and shape after removal of a force causing deformation. See also *viscoelasticity*.

elastic limit. The greatest stress a material is capable of sustaining without permanent strain remaining after the complete release of the stress. A material is said to have passed its elastic limit when the load is sufficient to initiate plastic, or nonrecoverable, deformation.

elastic recovery. The fraction of a given deformation that behaves elastically. A perfectly elastic material has an elastic recovery of 1; a perfectly plastic material has an elastic recovery of 0.

elastomer. A material that substantially recovers its original shape and size at room temperature after removal of a deforming force.

elastomeric tooling. A tooling system that uses the thermal expansion of rubber materials to form composite parts during cure.

electrical dissipation factor. The ratio of the power loss in a dielectric material to the total power transmitted through it; thus, the imperfection of the dielectric. Equal to the tangent of the loss angle.

electroformed molds. A mold made by electroplating metal on the reverse pattern of the cavity. Molten steel may then be sprayed on the back of the mold to increase its strength.

elongation. Deformation caused by stretching. The fractional increase in length of a material stressed in tension. (When expressed as percentage of the original gage length, it is called percentage elongation.)

elongation at break. Elongation recorded at the moment of rupture of the specimen, often expressed as a percentage of the original length.

encapsulation. The enclosure of an item in plastic. Sometimes used specifically in reference to the enclosure of capacitors or circuit board modules.

end. A strand of roving consisting of a given number of filaments gathered together. The group of filaments is considered an "end" or strand before twisting, a "yarn" after twist has been applied. An individual warp yarn, thread, fiber, or roving.

end count. An exact number of ends supplied on a ball of roving.

endurance limit. See *fatigue limit*.

environment. The aggregate of all conditions (such as contamination, temperature, humidity, radiation, magnetic and electric fields, shock, and vibration) that externally influence the performance of an item.

environmental stress cracking (ESC). The susceptibility of a thermoplastic resin to crack or craze when in the presence of surface active agents or other environments.

epichlorohydrin. The basic epoxidizing resin intermediate in the production of epoxy resins. It contains an epoxy group and is highly reactive with polyhydric phenols such as bisphenol A.

epoxide. Compound containing the oxirane structure, a three-member ring containing two carbon atoms and one oxygen atom. The most important members are ethylene oxide and propylene oxide.

epoxy plastic. A polymerizable thermoset polymer containing one or more epoxide groups and curable by reaction with amines, alcohols, phenols, carboxylic acids, acid anhydrides, and mercaptans. An important matrix resin in composites and structural adhesive.

equator. In filament winding, the line in a pressure vessel described by the junction of the cylindrical portion and the end dome. Also called tangent line or point.

ESC. See *environmental stress cracking*.

even tension. The process whereby each end of roving is kept in the same degree of tension as the other ends making up that ball of roving. See also *catenary*.

exotherm. The liberation or evolution of heat during the curing of a plastic product.

extend. To add fillers or low-cost materials in an economy producing endeavor. To add inert materials to improve void-filling characteristics and reduce crazing.

extenders. Low-cost materials used to dilute or extend high-cost resins without extensive lessening of properties. See also *filler*.

extensibility. The ability of a material to extend or elongate upon application of sufficient force, expressed as percent of the original length.

extensional-bending coupling. A property of certain classes of laminates that exhibit bending curvatures when subjected to extensional loading.

extensional-shear coupling. A property of certain classes of laminates that exhibit shear strains when subjected to extensional loading.

extensometer. A mechanical or optical device for measuring linear strain due to mechanical stress.

F

fabricating (fabrication). The manufacture of products from molded parts, rods, tubes, sheeting, extrusions, or other form by appropriate operations, such as punching, cutting, drilling, and tapping. Fabrication includes fastening parts together or to other parts by mechanical devices, adhesives, heat sealing, welding, or other means.

fabric fill face. That side of the woven fabric where the greatest number of the yarns are perpendicular to the selvage.

fabric, nonwoven. See *nonwoven fabric*.

fabric prepreg batch. Prepreg containing fabric from one fabric batch, impregnated with one batch of resin in one continuous operation.

fabric warp face. That side of the woven fabric where the greatest number of the yarns are parallel to the selvage.

fabric, woven. See *woven fabric*.

fairing. A member or structure, the primary function of which is to streamline the flow of a fluid by producing a smooth outline and to reduce drag, as in aircraft frames and boat hulls.

fatigue. The failure or decay of mechanical properties after repeated applications of stress. Fatigue tests give information on the ability of a material to resist the development of cracks, which eventually bring about failure as a result of a large number of cycles.

fatigue life. The number of cycles of deformation required to bring about failures of the test specimen under a given set of oscillating conditions (stresses or strains).

fatigue limit. The stress level below which a material can be stressed cyclically for an infinite number of times without failure.

fatigue ratio. The ratio of fatigue strength to tensile strength. Mean stress and alternating stress must be stated.

fatigue strength. The maximum cyclical stress a material can withstand for a given number of cycles before failure occurs. The residual strength after being subjected to fatigue.

faying surface. The surfaces of materials in contact with each other and joined or about to be joined together.

felt. A fibrous material made up of interlocked fibers by mechanical or chemical action, moisture, or heat. Made from fibers such as asbestos, cotton, glass, and so forth. See also *batt*.

fiber. A general term used to refer to filamentary materials. Often, fiber is used synonymously with filament. It is a general term for a filament with a finite length that is at least 100 times its diameter, which is typically 0.10 to 0.13 mm (0.004 to 0.005 in.). In most cases it is prepared by drawing from a molten bath, spinning, or deposition on a substrate. A whisker, on the other hand, is a short single-crystal fiber or filament made from a wide variety of materials, with diameters ranging from 1 to 25 μm (40 to 1400 μin.) and aspect ratios (a measure of length) between 100 and 15 000. Fibers can be continuous or specific short lengths (discontinuous), normally no less than 3.2 mm (1/8 in.).

fiber content. The amount of fiber present in a composite. This is usually expressed as a percentage volume fraction or weight fraction of the composite.

fiber count. The number of fibers per unit width of ply present in a specified section of a composite.

fiber diameter. The measurement (expressed in hundred thousandths) of the diameter of individual filaments.

fiber direction. The orientation or alignment of the longitudinal axis of the fiber with respect to a stated reference axis.

fiberglass. An individual filament made by drawing molten glass. A continuous filament is a glass fiber of great or indefinite length. A staple fiber is a glass fiber of relatively short length, generally less than 430 mm (17 in.), the length related to the forming or spinning process used.

fiberglass reinforcement. Major material used to reinforce plastic. Available as mat, roving, fabric, and so forth, it is incorporated into both thermosets and thermoplastics.

fiber pattern. Visible fibers on the surface of laminates or molding. The thread size and weave of glass cloth.

fiber-reinforced plastic (FRP). A general term for a composite that is reinforced with cloth, mat, strands, or any other fiber form.

fiber show. Strands or bundles of fibers that are not covered by plastic and that are at or above the surface of a composite.

fiber wash. Splaying out of woven or nonwoven fibers from the general reinforcement direction. Fibers are carried along with bleeding resin during cure.

filament. The smallest unit of a fibrous material. The basic units formed during drawing and spinning, which are gathered into strands of fiber for use in composites. Filaments usually are of extreme length and very small diameter, usually less than 25 μm (1 mil). Normally filaments are not used individually. Some textile filaments can function as a yarn when they are of sufficient strength and flexibility.

filament winding. A process for fabricating a composite structure in which continuous reinforcements (filament, wire, yarn, tape, or other), either previously impregnated with a matrix material or impregnated during the winding, are placed over a rotating and re-movable form or mandrel in a prescribed way to meet certain stress conditions. Generally the shape is a surface of revolution and may or may not include end closures. When the required number of layers is applied, the wound form is cured and the mandrel removed.

fill. Yarn oriented at right angles to the warp in a woven fabric.

filler. A relatively inert substance added to a material to alter its physical, mechanical, thermal, electrical, and other properties or to lower cost or density. Sometimes the term is used specifically to mean particulate additives. See also *inert filler*.

fillet. A rounded filling or adhesive that fills the corner or angle where two adherends are joined.

filling yarn. The transverse threads or fibers in a woven fabric. Those fibers running perpendicular to the warp. Also called weft.

film adhesive. A synthetic resin adhesive, usually of the thermosetting type, in the form of a thin, dry film of resin with or without a paper or glass carrier.

finish. A mixture of materials for treating glass or other fibers. It contains a coupling agent to improve the bond of resin to the fiber, and usually includes a lubricant to prevent abrasion, as well as a binder to promote strand integrity. With graphite or other filaments, it may perform any or all of the above functions.

first-order transition. A change of state associated with crystallization or melting in a polymer.

flame resistance. Ability of a material to extinguish flame once the source of heat is removed. See also *self-extinguishing resin*.

flame retardants. Certain chemicals that are used to reduce or eliminate the tendency of a resin to burn.

flammability. Measure of the extent to which a material will support combustion.

flash. That portion of the charge which flows from or is extruded from the mold cavity during the molding. Extra plastic attached to a molding along the parting line, which must be removed before the part is considered finished.

flexibilizer. An additive that makes a finished plastic more flexible or tough. See also *plasticizer*.

flexible molds. Molds made of rubber or elastomeric plastics, used for casting plastics. They can be stretched to remove cured pieces with undercuts.

flexural modulus. The ratio, within the elastic limit, of the applied stress on a test specimen in flexure to the corresponding strain in the outermost fibers of the specimen.

flexural strength. The maximum stress that can be borne by the surface fibers in a beam in bending. The flexural strength is the unit resistance to the maximum load before failure by bending, usually expressed in force per unit area.

flow. The movement of resin under pressure, allowing it to fill all parts of a mold. The gradual but continuous distortion of a material under continued load, usually at high temperatures; also called creep.

flow line. A mark on a molded piece made by the meeting of two flow fronts during molding. Also called striae, weld mark, or weld line.

flow marks. Wavy surface appearance of an object molded from thermoplastic resins, caused by improper flow of the resin into the mold.

fluted core. An integrally woven reinforcement material consisting of ribs between two skins in a unitized sandwich construction.

foamed plastics. Resins in sponge form, flexible or rigid, with cells closed or interconnected and density over a range from that of the solid parent resin to 0.030 g/cm^3. Compressive strength of rigid foams is fair, making them useful as core materials for sandwich constructions. Also, a chemical cellular plastic, the structure of which is produced by gases generated from the chemical interaction of its constituents.

foaming agent. Chemicals added to plastics and rubbers that generate inert gases on heating, causing the resin to assume a cellular structure.

foam-in-place. Refers to the deposition of foams when the foaming machine must be brought to the work that is "in place," as opposed to bringing the work to the foaming machine. Also, foam mixed in a container and poured into a mold, where it rises to fill the cavity.

force. The male half of the mold that enters the cavity, exerting pressure on the resin and causing it to flow. Also called punch.

FP fiber. Polycrystalline alumina fiber (Al_2O_3). A ceramic fiber useful for high-temperature (1370 to 1650 °C, or 2500 to 3000 °F) composites.

fracture. The separation of a body. Defined both as rupture of the surface without complete separation of laminate and as complete separation of a body because of external or internal forces.

fracture stress. The true, normal stress on the minimum cross-sectional area at the beginning of fracture.

fracture toughness. A measure of the damage tolerance of a material containing initial flaws or cracks. Used in aircraft structural design and analysis.

free-radical polymerization. A type of polymerization in which the propagating species is a long-chain free radical initiated by the introduction of free radicals from thermal or photochemical decomposition.

free wall. The portion of a honeycomb cell wall that is not connected to another cell.

friction, coefficient of. See *coefficient of friction*.

FRP. See *fiber-reinforced plastic*.

fungus resistance. The resistance of a material to attack by fungi in conditions promoting their growth.

fuzz. Accumulation of short broken filaments after passing glass strands, yarns, or rovings over a contact point. Often weighted and used as an inverse measure of abrasion resistance.

G

gage length. Length over which deformation is measured, for a tensile or compressive test specimen. The deformation over the gage length divided by the gage length determines the strain.

gap. In filament winding, the space between successive windings, which windings are usually intended to lay next to each other. Separations between fibers within a filament winding band. The distance between adjacent plies in a lay-up of unidirectional tape materials.

gap-filling adhesive. An adhesive subject to low shrinkage in setting, used as sealant.

gel. The initial jellylike solid phase that develops during the formation of a resin from a liquid. A semisolid system consisting of a network of solid aggregates in which liquid is held.

gelation. The point in a resin cure when the resin viscosity has increased to a point such that it barely moves when probed with a sharp instrument.

gelation time. That interval of time, in connection with the use of synthetic thermosetting resins, extending from the introduction of a catalyst into a liquid adhesive system until the start of gel formation. Also, the time under application of load for a resin to reach a solid state.

gel coat. A quick setting resin applied to the surface of a mold and gelled before lay-up. The gel coat becomes an integral part of the finished laminate, and is usually used to improve surface appearance and bonding.

gel permeation chromatography (GPC). A form of liquid chromatography in which the polymer molecules are separated by their ability or inability to penetrate the material in the separation column.

gel point. The stage at which a liquid begins to exhibit pseudoelastic properties. This stage may be conveniently observed from the inflection point on a viscosity time plot.

geodesic. The shortest distance between two points on a surface.

geodesic isotensoid. Constant stress level in any given filament at all points in its path.

geodesic-isotensoid contour. In filament wound reinforced plastic pressure vessels, a dome contour in which the filaments are placed on geodesic paths so that the filaments will exhibit uniform tensions throughout their length under pressure loading.

geodesic ovaloid. A contour for end domes, the fibers forming a geodesic line: the shortest distance between two points on a surface of revolution. The forces exerted by the filaments are proportioned to meet hoop and meridional stresses at any point.

glass. An inorganic product of fusion that has cooled to a rigid condition without crystallizing. Glass is typically hard and relatively brittle, and has a conchoidal fracture.

glass cloth. Conventionally woven glass fiber material. See also *scrim*.

glass fiber. A fiber spun from an inorganic product of fusion that has cooled to a rigid condition without crystallizing.

glass filament. A form of glass that has been drawn to a small diameter and extreme length. Most filaments are less than 0.15 mm (0.005 in.) in diameter.

glass filament bushing. The unit through which molten glass is drawn in making glass filaments.

glass finish. A material applied to the surface of a glass reinforcement to improve the bond between the glass and the plastic resin matrix.

glass flake. Thin, irregularly shaped flakes of glass, typically made by shattering a thin-walled tube of glass.

glass former. An oxide that forms a glass easily. Also, one which contributes to the network of silica glass when added to it.

glass, percent by volume. The product of the specific gravity of a laminate and the percent glass by weight, divided by the specific gravity of the glass.

glass stress. In a filament wound part, usually a pressure vessel, the stress calculated using the load and the cross-sectional area of the reinforcement only.

glass transition. The reversible change in an amorphous polymer or in amorphous regions of a partially crystalline polymer from, or to, a viscous or rubbery condition to, or from, a hard and relatively brittle one.

glass transition temperature (T_g). The approximate midpoint of the temperature range over which the glass transition takes place; glass and silica fiber exhibit a phase change at approximately 955 °C (1750 °F) and carbon/graphite fibers at 2205 to 2760 °C (4000 to 5000 °F). The temperature at which increased molecular mobility results in significant changes in the properties of a cured resin system. Also, the inflection point on a plot of modulus versus temperature. The measured value of T_g depends to some extent on the method of test.

graphite. The crystalline allotropic form of carbon.

graphite fiber. A fiber made from a precursor by oxidation, carbonization, and graphitization process (which provides a graphitic structure). See also *carbon fiber*.

graphitization. The process of pyrolyzation in an inert atmosphere at temperatures in excess of 1925 °C (3500 °F), usually as high as 2480 °C (4500 °F), and sometimes as high as 9750 °C (5400 °F), converting carbon to its crystalline allotropic form. Temperature depends on precursor and properties desired.

green strength. The ability of the material (such as a urethane elastomer), while not completely cured, to undergo removal from the mold and handling without tearing or permanent distortion.

greige, gray goods. Any fabric before finishing, as well as any yarn or fiber before bleaching or dyeing; therefore, fabric with no finish or size.

H

hand. The softness of a piece of fabric, as determined by the touch (individual judgment).

hand lay-up. The process of placing (and working) successive plies of reinforcing material or resin-impregnated reinforcement in position on a mold by hand.

handling life. The out-of-refrigeration time over which a material retains its handleability.

hardener. A substance or mixture added to a plastic composition to promote or control the curing action by taking part in it.

hardness. The resistance to surface indentation usually measured by the depth of penetration (or arbitrary units related to the depth of penetration) of a blunt point under a given load using a particular instrument according to a prescribed procedure. See also *Barcol hardness*, *Mohs hardness*, *Rockwell hardness*, and *Shore hardness*.

harness satin. Weaving pattern producing a satin appearance. "Eight-harness" means the warp tow crosses over seven fill tows and under the eighth (repeatedly).

heat-activated adhesive. A dry adhesive that is rendered tacky or fluid by application of heat, or heat and pressure, to the assembly.

heat build-up. The rise in temperature in a part resulting from the dissipation of applied strain energy as heat or from applied mold cure heat. See also *hysteresis*.

heat cleaned. A condition in which glass or other fibers are exposed to elevated temperatures to remove preliminary sizings or binders not compatible with the resin system to be applied.

heat distortion point. The temperature at which a standard test bar deflects a specified amount under a stated load. Now called deflection temperature.

heat resistance. The property or ability of plastics and elastomers to resist the deteriorating effects of elevated temperatures.

heat sealing. A method of joining plastic films by simultaneous application of heat and pressure to areas in contact.

heat-sealing adhesive. A thermoplastic film adhesive that is melted between the adherend surfaces by heat application to one or both of the adjacent adherend surfaces.

heat sink. A contrivance for the absorption or transfer of heat away from a critical element or part. Bulk graphite is often used as a heat sink.

heat treating. Term used to cover annealing, hardening, tempering, and so forth.

helical winding. In filament wound items, a winding in which a filament band advances along a helical path, not necessarily at a constant angle except in the case of a cylinder.

heterogeneous. Descriptive term for a material consisting of dissimilar constituents separately identifiable. A medium consisting of regions of unlike properties separated by internal boundaries. Note that not all nonhomogeneous materials are necessarily heterogeneous.

hexa. Shortened form of hexamethylenetetramine, a source of reactive methylene for curing novolacs.

high-frequency heating. The heating of materials by dielectric loss in a high-frequency electrostatic field. The material is exposed between electrodes and is heated quickly and uniformly by absorption of energy from the electrical field.

high-pressure laminates. Laminates molded and cured at pressures not lower than 6.9 MPa (1.0 ksi), and more commonly in the range of 8.3 to 13.8 MPa (1.2 to 2.0 ksi).

high-pressure spot. See *resin-starved area*.

HIP. See *hot isostatic pressing*.

homogeneous. Descriptive term for a material of uniform composition throughout. A medium that has no internal physical boundaries. A material whose properties are con-

stant at every point, that is, constant with respect to spatial coordinates (but not necessarily with respect to directional coordinates).

honeycomb. Manufactured product of resin-impregnated sheet material (paper, glass fabric, and so on) or metal foil, formed into hexagonal-shaped cells. Used as a core material in sandwich constructions. See also *sandwich constructions*.

hoop stress. The circumferential stress in a material of cylindrical form subjected to internal or external pressure.

hot isostatic pressing. A process for fabricating certain metal matrix composites. A preform is consolidated under fluid pressure (usually an inert gas) at high temperature and pressure in a pressure vessel.

hot-melt adhesive. An adhesive that is applied in a molten state and forms a bond after cooling to a solid state. A bonding agent that achieves a solid state and resultant strength by cooling, as contrasted with other adhesives, which achieve the solid state through evaporation of solvents or chemical cure. A thermoplastic resin that functions as an adhesive when melted between substrates and cooled.

hot-setting adhesive. An adhesive that requires a temperature at or above 100 °C (212 °F) to set.

hot working. Any form of mechanical deformation processing carried out on a metal or alloy above its recrystallization temperature but below its melting point.

hybrid. A composite laminate consisting of laminae of two or more composite material systems. A combination of two or more different fibers, such as carbon and glass or carbon and aramid, into a structure. Tapes, fabrics, and other forms may be combined; usually only the fibers differ. See also *interply hybrid* and *intraply hybrid*.

hydraulic press. A press in which the molding force is created by the pressure exerted by a fluid.

hydromechanical press. A press in which the molding forces are created partly by a mechanical system and partly by an hydraulic system.

hydrophilic. Capable of absorbing water. Easily wetted by water.

hydrophobic. Capable of repelling water. Poorly wetted by water.

hygroscopic. Capable of adsorbing and retaining atmospheric moisture.

hygrothermal effect. Change in properties due to moisture absorption and temperature change.

hysteresis. The energy absorbed in a complete cycle of loading and unloading. This energy

is converted from mechanical to frictional energy (heat).

I

ignition loss. The difference in weight before and after burning. As with glass, the burning off of the binder or size.

impact strength. The ability of a material to withstand shock loading. The work done in fracturing a test specimen in a specified manner under shock loading.

impact test. Measure of the energy necessary to fracture a standard notched bar by an impulse load. See also *Izod impact test*, *reverse impact test*, and *Charpy impact test*.

impregnate. In reinforced plastics, to saturate the reinforcement with a resin.

impregnated fabric. A fabric impregnated with a synthetic resin. See also *prepreg*.

inclusion. A physical and mechanical discontinuity occurring within a material or part, usually consisting of solid, encapsulated foreign material. Inclusions are often capable of transmitting some structural stresses and energy fields, but in a noticeably different degree from the parent material. See also *voids*.

inert filler. A material added to a plastic to alter the end-item properties through physical rather than chemical means.

infrared. Part of the electromagnetic spectrum between the visible light range and the radar range. Radiant heat is in this range, and infrared heaters are frequently used in the thermoforming and curing of plastics and composites. Infrared analysis is used for identification of polymer constituents.

inhibitor. A substance that retards a chemical reaction. Also used in certain types of monomers and resins to prolong storage life.

initial modulus. The slope of the initial straight portion of a stress-strain or load-elongation curve. See also *Young's modulus*.

initial strain. The strain produced in a specimen by given loading conditions before creep occurs.

initial (instantaneous) stress. The stress produced by force in a specimen before stress relaxation occurs.

initiator. Peroxides used as sources of free radicals. They are used in free-radical polymerizations, for curing thermosetting resins, as cross-linking agents for elastomers and polyethylene, and for polymer modification.

injection molding. Method of forming a plastic to the desired shape by forcing the heat-softened plastic into a relatively cool cavity under pressure.

inorganic pigments. Natural or synthetic metallic oxides, sulfides, and other salts that impart heat and light stability, weathering resistance, color, and migration resistance to plastics.

insert. An integral part of a plastic molding consisting of metal or other material that may be molded or pressed into position after the molding is completed.

insulation resistance. The electrical resistance between two conductors or systems of conductors separated only by insulating material. The ratio of the applied voltage to the total current between two electrodes in contact with a specified insulator. The electrical resistance of an insulating material to a direct voltage.

insulator. A material of such low electrical conductivity that the flow of current through it can usually be neglected. Similarly, a material of low thermal conductivity, such as that used to insulate structural shells.

integral composite structure. Composite structure in which several structural elements, which would conventionally be assembled together by bonding or mechanical fasteners after separate fabrication, are instead laid up and cured as a single, complex, continuous structure, for example, spars, ribs, and one stiffened cover of a wing box fabricated as a single integral part. The term is sometimes applied more loosely to any composite structure not assembled by mechanical fasteners. All or some parts of the assembly may be co-cured.

integrally heated. A term referring to tooling that is self-heating, through use of electrical heaters such as cal rods. Most hydroclave tooling is integrally heated. Some autoclave tooling is integrally heated to compensate for thick sections, to provide high heat-up rates, or to permit processing at a higher temperature than is otherwise possible with the autoclave.

integral skin foam. Urethane foam with a cellular core structure and a relatively nonporous skin.

interface. The boundary or surface between two different, physically distinguishable media. On fibers, the contact area between fibers and sizing or finish. In a laminate, the contact area between the reinforcement and the laminating resin.

interference fits. A joint or mating of two parts in which the male part has an external dimension larger than the internal dimension of the mating female part. Distension of the female by the male creates a stress, which supplies the bonding force for the joint.

interlaminar. Descriptive term pertaining to an object (for example, voids), event (for example, fracture), or potential field (for example, shear stress) referenced as existing or occurring between two or more adjacent laminae.

interlaminar shear. Shearing force tending to produce a relative displacement between two laminae in a laminate along the plane of their interface.

intermediate temperature setting adhesive. An adhesive that sets in the temperature range from 30 to 100 °C (87 to 211 °F).

interphase. The boundary region between a bulk resin or polymer and an adherend in which the polymer has a high degree of orientation to the adherend on a molecular basis. It plays a major role in the load transfer process between the bulk of the adhesive and the adherend or the fiber and the laminate matrix resin.

interply hybrid. A composite in which adjacent laminae are composed of different materials.

intralaminar. Descriptive term pertaining to an object (for example, voids), event (for example, fracture), or potential field (for example, temperature gradient) existing entirely within a single lamina without reference to any adjacent laminae.

intraply hybrid. A composite in which different materials are used within a specific layer or band.

irradiation. As applied to plastics, the bombardment with a variety of subatomic particles, usually alpha-, beta-, or gamma-rays. Used to initiate polymerization and copolymerization of plastics and in some cases to bring about changes in the physical properties of a plastic.

irreversible. Not capable of redissolving or remelting. Chemical reactions that proceed in a single direction and are not capable of reversal (as applied to thermosetting resins).

isocyanate plastics. Plastics based on resins made by the condensation of organic isocyanates with other compounds. Generally reacted with polyols on a polyester or polyether backbone molecule, with the reactants being joined through the formation of the urethane linkage. See also *polyurethane* and *urethane plastics*.

isostatic pressing. Pressing powder under a gas or liquid so that pressure is transmitted equally in all directions, for example, in sintering.

isotropic. Having uniform properties in all directions. The measured properties of an isotropic material are independent of the axis of testing.

Izod impact test. A test for shock loading in which a notched specimen bar is held at one end and broken by striking, and the energy absorbed is measured.

J

joint, adhesive. See *adhesive joint*.

joint, butt. See *butt joint*.

joint, edge. See *edge joint*.

joint, lap. See *lap joint*.

joint, scarf. See *scarf joint*.

K

kerf. The width of a cut made by a saw blade, torch, water jet, laser beam, and so forth.

Kevlar. An organic polymer composed of aromatic polyamides having a para-type orientation (parallel chain extending bonds from each aromatic nucleus).

knitted fabrics. Fabrics produced by interlooping chains of yarn.

***K* factor.** The coefficient of thermal conductivity. The amount of heat that passes through a unit cube of material in a given time when the difference in temperature of two opposite faces is 1°.

knuckle area. The area of transition between sections of different geometry in a filament-wound part, for example, where the skirt joins the cylinder of the pressure vessel. Also called Y-joint.

L

lamina. A single ply or layer in a laminate made up of a series of layers (organic composite). A flat or curved surface containing unidirectional fibers or woven fibers embedded in a matrix (metal matrix composite).

laminae. Plural of lamina.

laminate. To unite laminae with a bonding material, usually with pressure and heat (normally used with reference to flat sheets, but also rods and tubes). A product made by such bonding. See also *bidirectional laminate* and *unidirectional laminate*.

laminate coordinates. A reference coordinate system (used to describe the properties of a laminate), generally in the direction of principal axes, when they exist.

laminate orientation. The configuration of a cross-plied composite laminate with regard to the angles of cross-plying, the number of laminae at each angle, and the exact sequence of the lamina lay-up.

laminate ply. One fabric-resin or fiber-resin layer of a product that is bonded to adjacent layers in the curing process.

lap. In filament winding, the amount of overlay between successive windings, usually intended to minimize gapping. In bonding, the distance one adherend covers another adherend.

lap joint. A joint made by placing one adherend partly over another and bonding the overlapped portions.

lattice pattern. A pattern of filament winding with a fixed arrangement of open voids.

lay-up. The reinforcing material placed in position in the mold. The process of placing the reinforcing material in position in the mold. The resin-impregnated reinforcement. A description of the component materials, geometry, and so forth, of a laminate.

L-direction. The ribbon direction, that is, the direction of the continuous sheets of honeycomb.

level winding. See *circumferential winding.*

linear expansion. The increase of a given dimension, measured by the expansion or contraction of a specimen or component subject to a thermal gradient or changing temperature. See also *coefficient of thermal expansion.*

liner. In a filament-wound pressure vessel, the continuous, usually flexible coating on the inside surface of the vessel, used to protect the laminate from chemical attack or to prevent leakage under stress.

liquid crystal polymer. A newer thermoplastic polymer that is melt processable and develops high orientation in molding, with resultant tensile strength and high-temperature capability that is notably improved. First commercial availability was as an aromatic polyester. With or without fiber reinforcement.

liquid metal infiltration. Process for immersion of metal fibers in a molten metal bath to achieve a metal matrix composite; for example, graphite fibers in molten aluminum.

liquid shim. Material used to position components in an assembly where dimensional alignment is critical. For example, epoxy adhesive is introduced into gaps after the assembly is placed in the desired configuration.

load-deflection curve. A curve in which the increasing tension, compression, or flexural loads are plotted on the ordinate axis and the deflections caused by those loads are plotted on the abscissa axis.

longos. Low angle helical or longitudinal windings.

loop tenacity. The tenacity or strength value obtained by pulling two loops, as two links in a chain, against each other in order to demonstrate the susceptibility that a fibrous material has for cutting or crushing itself; loop strength.

loss factor. The product of the dissipation factor and the dielectric constant of a dielectric material.

loss modulus. A damping term describing the dissipation of energy into heat when a material is deformed.

loss on ignition. Weight loss, usually expressed as percent of total, after burning off an organic sizing from glass fibers, or an organic resin from a glass fiber laminate.

loss tangent. See *electrical dissipation factor.*

lot. A specific amount of material produced at one time using the same process and the same conditions of manufacture, and offered for sale as a unit quantity.

low-pressure laminates. In general, laminates molded and cured in the range of pressures from 2760 kPa (400 psi) down to and including pressure obtained by the mere contact of the plies.

lubricant. A material added to most sizings to improve the handling and processing properties of textile strands, especially during weaving.

M

macerate. To chop or shred fabric for use as a filler for a molding resin.

macro. In relation to composites, denotes the gross properties of a composite as a structural element but does not consider the individual properties or identity of the constituents.

mandrel. The core tool around which resin-impregnated paper, fabric, or fiber is wound to form pipes, tubes, or structural shell shapes.

mat. A fibrous material for reinforced plastic consisting of randomly oriented chopped filaments, short fibers (with or without a carrier fabric), or swirled filaments loosely held together with a binder. Available in blankets of various widths, weights, and lengths. Also, a sheet formed by filament winding a single-hoop ply of fiber on a mandrel, cutting across its width and laying out a flat sheet.

matched metal molding. A reinforced plastics manufacturing process in which matching male and female metal molds are used (similar to compression molding) to form the part, with time, pressure, and heat.

matrix. The essentially homogeneous resin or polymer material in which the fiber system of a composite is imbedded. Both thermoplastic and thermoset resins may be used, as well as metals, ceramics, and glasses.

mechanical adhesion. Adhesion between surfaces in which the adhesive holds the parts together by interlocking action.

mechanical properties. The properties of a material, such as compressive and tensile strengths, and modulus, that are associated with elastic and inelastic reaction when force is applied. The individual relationship between stress and strain.

melt. A charge of molten metal. See also *liquid metal infiltration.*

mer. The repeating structural unit of any polymer.

mesophase. An intermediate phase in the formation of carbon from a pitch precursor. This is a liquid crystal phase in the form of microspheres, which upon prolonged heating

above 400 °C (750 °F) coalesce, solidify, and form regions of extended order. Heating to above 2000 °C (3630 °F) leads to the formation of graphite structure.

metallic fiber. Manufactured fiber composed of metal, plastic-coated metal, metal-coated plastic, or a core completely covered by metal.

M-glass. A high beryllia (BeO_2) content glass designed especially for high modulus of elasticity.

micro. In relation to composites, denotes the properties of the constituents, that is, matrix, reinforcement, and interface only, and their effects on the composite properties.

microcracking. Cracks formed in composites when thermal stresses locally exceed the strength of the matrix. Since most microcracks do not penetrate the reinforcing fibers, microcracks in a cross-plied tape laminate or in a laminate made from cloth prepreg are usually limited to the thickness of a single ply.

microstructure. A structure with heterogeneities that can be seen through a microscope.

mil. The unit used in measuring the diameter of glass fiber strands, wire, and so forth (1 mil = 0.001 in.).

milled fiber. Continuous glass strands hammer milled into very short glass fibers. Useful as inexpensive filler or anticrazing reinforcing fillers for adhesives.

modulus, initial. See *initial modulus.*

modulus of elasticity. The ratio of the stress or load applied to the strain or deformation produced in a material that is elastically deformed. If a tensile strength of 13.8 MPa (2.0 ksi) results in an elongation of 1%, the modulus of elasticity is 13.8 MPa (2.0 ksi) divided by 0.01, or 1380 MPa (200 ksi). Also called Young's modulus. See also *offset modulus* and *secant modulus.*

modulus, offset. See *offset modulus.*

modulus of resilience. The energy that can be absorbed per unit volume without creating a permanent distortion. Calculated by integrating the stress-strain curve from zero to the elastic limit and dividing by the original volume of the specimen.

modulus of rigidity. The ratio of stress to strain within the elastic region for shear or torsional stress. Also called shear modulus or torsional modulus.

modulus of rupture, in bending. The maximum tensile or compressive stress value (whichever causes failure) in the extreme fiber of a beam loaded to failure in bending.

modulus of rupture, in torsion. The maximum shear stress in the extreme fiber of a member of circular cross section loaded to failure in torsion.

modulus, secant. See *secant modulus.*

modulus, tangent. See *tangent modulus.*

Mohs hardness. A measure of the scratch resistance of a material. The higher the number, the greater the scratch resistance (No. 10 being termed diamond).

moisture absorption. The pickup of water vapor from air by a material. It relates only to vapor withdrawn from the air by a material and must be distinguished from water absorption, which is the gain in weight due to the take-up of water by immersion.

moisture content. The amount of moisture in a material determined under prescribed conditions and expressed as a percentage of the mass of the moist specimen, that is, the mass of the dry substance plus the moisture present.

moisture equilibrium. The condition reached by a sample when it no longer takes up moisture from, or gives up moisture to, the surrounding environment.

moisture vapor transmission. A rate at which water vapor passes through a material at a specified temperature and relative humidity $(g/mil/24 \ h/100 \ in.^2)$.

mold. The cavity or matrix into or on which the plastic composition is placed and from which it takes form. To shape plastic parts or finished articles by heat and pressure. The assembly of all the parts that function collectively in the molding process.

molded edge. An edge that is not physically altered after molding for use in final form, and particularly one that does not have fiber ends along its length.

molded net. Description of a molded part that requires no additional processing to meet dimensional requirements.

molding. The forming of a polymer or composite into a solid mass of prescribed shape and size by the application of pressure and heat for given times. Sometimes used to denote the finished part.

molding cycle. The period of time required for the complete sequence of operations on a molding press to produce one set of moldings. The operations necessary to produce a set of moldings without reference to the total time taken.

molding powder or compound. Plastic material in varying stages of pellets or granulation, and consisting of resin, filler, pigments, reinforcements, plasticizers, and other ingredients, ready for use in the molding operation.

molding pressure. The pressure applied to the ram of an injection machine or compression or transfer press to force the softened plastic to fill the mold cavities completely.

mold-release agent. A lubricant, liquid, or powder (often silicone oils and waxes), used to prevent sticking of molded articles in the cavity.

mold shrinkage. The immediate shrinkage that a molded part undergoes when it is removed from a mold and cooled to room temperature. The difference in dimensions, expressed in inches per inch, between a molding and the mold cavity in which it was molded (at normal-temperature measurement). The incremental difference between the dimensions of the molding and the mold from which it was made, expressed as a percentage of the mold dimensions.

mold surface. The side of a laminate that faced the mold (tool) during cure in an autoclave or hydroclave.

molecular weight. The sum of the atomic weights of all the atoms in a molecule. A measure of the chain length for the molecules that make up the polymer.

monofilament. A single fiber or filament of indefinite length, strong enough to function as a yarn in commercial textile operations.

monolayer. The basic laminate unit from which cross-plied or other laminate types are constructed. Also, a "single" layer of atoms or molecules adsorbed on or applied to a surface.

monomer. A single molecule that can react with like or unlike molecules to form a polymer. The smallest repeating structure of a polymer (mer). For additional polymers, this represents the original unpolymerized compound.

morphology. The overall form of a polymer structure, that is, crystallinity, branching, molecular weight, and so on.

multicircuit winding. In filament winding, a winding that requires more than one circuit before the band repeats by laying adjacent to the first band.

multifilament yarn. A large number (500 to 2000) of fine, continuous filaments (often 5 to 100 individual filaments) usually with some twist in the yarn to facilitate handling.

MVT. See *moisture vapor transmission.*

N

NDE. See *nondestructive evaluation.*

NDI. See *nondestructive inspection.*

NDT. See *nondestructive testing.*

neat resin. Resin to which nothing (additives, reinforcements, and so on) has been added.

necking. The localized reduction in cross section that may occur in a material under tensile stress.

needled mat. A mat formed of strands cut to a short length, then felted together in a needle loom, with or without a carrier.

nesting. In reinforced plastics, the placing of plies of fabric so that the yarns of one ply lie in the valleys between the yarns of the adjacent ply (nested cloth).

netting analysis. The analysis of filament-wound structures that assumes the stresses induced in the structure are carried entirely by the filaments, and the strength of the resin is neglected; and assumes also that the filaments possess no bending or shearing stiffness, and carry only the axial tensile loads.

node. The connected portion of adjacent ribbons of honeycomb.

NOL ring. A parallel filament- or tape-wound hoop test specimen developed by the Naval Ordnance Laboratory (NOL), (now the Naval Surface Weapons Laboratory), for measuring various mechanical strength properties of the material, such as tension and compression, by testing the entire ring or segments of it. Also known as a parallel fiber reinforced ring.

nominal stress. The stress at a point calculated on the net cross section without taking into consideration the effect on stress of geometric discontinuities, such as holes, grooves, fillets, and so forth. The calculation is made by simple elastic theory.

nominal value. A value assigned for the purpose of a convenient designation. A nominal value exists in name only. It is often an average number with a tolerance so as to fit together with adjacent parts.

nondestructive evaluation (NDE). Broadly considered synonymous with nondestructive inspection (NDI). More specifically, the analysis of NDI findings to determine whether the material will be acceptable for its function.

nondestructive inspection (NDI). A process or procedure, such as ultrasonic or radiographic inspection, for determining the quality or characteristics of a material, part, or assembly, without permanently altering the subject or its properties. Used to find internal anomalies in a structure without degrading its properties.

nondestructive testing (NDT). Broadly considered synonymous with nondestructive inspection (NDI).

nonhygroscopic. Lacking the property of absorbing and retaining an appreciable quantity of moisture (water vapor) from the air.

nonwoven fabric. A planar textile structure produced by loosely compressing together fibers, yarns, rovings, and so forth, with or without a scrim cloth carrier. Accomplished by mechanical, chemical, thermal, or solvent means and combinations thereof.

normal stress. The stress component that is perpendicular to the plane on which the forces act.

notched specimen. A test specimen that has been deliberately cut or notched, usually in a V-shape, to induce and locate point of failure.

notch factor. Ratio of the resilience determined on a plain specimen to the resilience determined on a notched specimen.

notch sensitivity. The extent to which the sensitivity of a material to fracture is increased by the presence of a surface nonhomogeneity, such as a notch, a sudden change in section, a crack, or a scratch. Low notch sensitivity is usually associated with ductile materials, and high notch sensitivity is usually associated with brittle materials.

novolac. A linear thermoplastic B-staged phenolic resin, which, in the presence of methylene or other cross-linking groups, reacts to form a thermoset phenolic.

nylon. The generic name for all synthetic polyamides.

nylon plastics. Plastics based on a resin composed principally of a long-chain synthetic polymeric amide that has recurring amide groups as an integral part of the main polymer chain. Numerical designations (nylon 6, nylon 66, and so on) refer to the monomeric amides of which they are made. Characterized by great toughness and elasticity.

offset modulus. The ratio of the offset yield stress to the extension at the offset point.

offset yield strength. The stress at which the strain exceeds by a specific amount (the offset) an extension of the initial approximately linear proportional portion of the stress-strain curve. It is expressed in force per unit area.

olefin. A group of unsaturated hydrocarbons of the general formula C_nH_{2n} named after the corresponding paraffins by the addition of "ene" or "ylene" to the root, for example, ethylene, propylene, and pentene.

open-cell foam. Foamed or cellular material with cells that are generally interconnected. Closed cell refers to cells that are not interconnected.

orange peel. An uneven surface somewhat resembling that of an orange peel; said of injection moldings that have unintentionally ragged surfaces.

organic. Matter originating in plant or animal life or composed of chemicals of hydrocarbon origin, either natural or synthetic.

orientation. The alignment of the crystalline structure in polymeric materials in order to produce a highly aligned structure. Orientation can be accomplished by cold drawing or stretching in fabrication.

oriented materials. Materials, particularly amorphous polymers and composites, whose molecules and/or macroconstituents are aligned in a specific way. Oriented materials are anisotropic. Orientation can generally be divided into two classes, uniaxial and biaxial.

orthotropic. Having three mutually perpendicular planes of elastic symmetry.

out time. The time a prepreg is exposed to ambient temperature, namely, the total amount of time the prepreg is out of the freezer. The primary effects of out time are to decrease the drape and tack of the prepreg while also allowing it to absorb moisture from the air.

ovaloid. A surface of revolution symmetrical about the polar axis that forms the end closure for a filament-wound cylinder.

oven dry. The condition of a material that has been heated under prescribed conditions of temperature and humidity until there is no further significant change in its mass.

overlay sheet. A nonwoven fibrous mat (of glass, synthetic fiber, and so forth) used as the top layer in a cloth or mat lay-up, to provide a smoother finish, minimize the appearance of the fibrous pattern, or permit machining or grinding to a precise dimension. Also called surfacing mat.

oxidation. In carbon/graphite fiber processing, the step of reacting the precursor polymer (rayon, PAN, or pitch) with oxygen, resulting in stabilization of the structure for the hot stretching operation. In general usage, oxidation refers to any chemical reaction in which electrons are transferred.

P

package. Yarn, roving, and so forth in the form of units capable of being unwound and suitable for handling, storing, shipping, and use.

PAN. See *polyacrylonitrile.*

parallel laminate. A laminate of woven fabric in which the plies are aligned in the same position as originally aligned in the fabric roll. A series of flat or curved cloth-resin layers stacked uniformly on top of each other.

particulate composite. Material consisting of one or more constituents suspended in a matrix of another material. These particles are either metallic or nonmetallic.

parting agent. See *mold release agent.*

parting line. A mark on a molded piece where the sections of a mold have met in closing.

PAS. See *polyarylsulfone.*

PBI. See *polybenzimidazole.*

PEEK. See *polyether etherketone.*

peel ply. A layer of open-weave material, usually fiberglass or heat-set nylon, applied directly to the surface of a prepreg lay-up. The peel ply is removed from the cured laminate immediately before bonding operations, leaving a clean, resin-rich surface that needs no further preparation for bonding, other than application of a primer where one is required.

peel strength. Adhesive bond strength, as in pounds per inch of width, obtained by a stress applied in a peeling mode.

permanence. The property of a plastic that describes its resistance to appreciable changes in characteristics with time and environment.

permanent set. The deformation remaining after a specimen has been stressed a prescribed amount in tension, compression, or shear for a definite time period and released for a definite time period. For creep tests, the residual unrecoverable deformation after the load causing the creep has been removed for a substantial and definite period of time. Also, the increase in length, expressed as a percentage of the original length, by which an elastic material fails to return to original length after being stressed for a standard period of time.

permeability. The passage or diffusion (or rate of passage) of a gas, vapor, liquid, or solid through a barrier without physically or chemically affecting it.

pH. The measure of the acidity or alkalinity of a substance, neutrality being at pH 7. Acid solutions are less than 7, alkaline solutions are more than 7.

phenolic (phenolic resin). A thermosetting resin produced by the condensation of an aromatic alcohol with an aldehyde, particularly of phenol with formaldehyde. Used in high-temperature applications with various fillers and reinforcements.

phenylsilane resins. Thermosetting copolymers of silicone and phenolic resins. Furnished in solution form.

physical catalyst. Radiant energy capable of promoting or modifying a chemical reaction.

PI. See *polyimide.*

PIC. See *pressure-impregnation-carbonization.*

pick count. The number of filling yarns per inch of woven fabric.

pin holes. Small cavities that penetrate the surface of a cured part.

pit. A small, regular or irregular crater in the surface of a plastic, usually of a width approximately the same order of magnitude as its depth.

pitch. A high molecular weight material left as a residue from the destructive distillation of coal and petroleum products. Pitches are used as base materials for the manufacture of certain high-modulus carbon fibers and as matrix precursors for carbon-carbon composites.

plain weave. A weaving pattern in which the warp and fill fibers alternate; that is, the repeat pattern is warp/fill/warp/fill, and so on. Both faces of a plain weave are identical. Properties are significantly reduced relative to a weaving pattern with fewer crossovers.

planar. Lying essentially in a single plane.

planar helix winding. A winding in which the filament path on each dome lies on a plane that intersects the dome, while a helical path over the cylindrical section is connected to the dome paths.

planar winding. A winding in which the filament path lies on a plane that intersects the winding surface. See also *polar winding*.

plastic. A material that contains as an essential ingredient an organic polymer of large molecular weight, hardeners, fillers, reinforcements, and so forth; is solid in its finished state; and, at some stage in its manufacture or its processing into finished articles, can be shaped by flow. Made of plastic. A plastic may be either thermoplastic or thermoset.

plastic deformation. Change in dimensions of an object under load that is not recovered when the load is removed, as opposed to elastic deformation.

plastic flow. Deformation under the action of a sustained hot or cold force. Flow of semisolids in the molding of plastics.

plasticizer. A material incorporated in a plastic to increase its workability and flexibility or distensibility. Normally used in thermoplastics. A lower molecular weight material added to an epoxy to reduce stiffness and brittleness, thereby resulting in a lower glass transition temperature for the polymer.

plastic memory. The tendency of a thermoplastic material that has been stretched while hot to return to its unstretched shape upon being reheated.

platens. The mounting plates of a press, to which the entire mold assembly is bolted.

plied yarn. Yarn made by collecting two or more single yarns. Normally, the yarns are twisted together, though sometimes they are collected without twist.

ply. In general, fabrics or felts consisting of one or more layers (laminates, and so forth). The layers that make up a stack. Yarn resulting from twisting operations (three-ply yarn, and so forth). A single layer of prepreg. A single pass in filament winding (two plies forming one layer).

PMR polyimides. A novel class of high temperature resistant polymers. PMR represents *in situ* polymerization of monomer reactants.

Poisson's ratio. The ratio of the change in lateral width per unit width to change in axial length per unit length caused by the axial stretching or stressing of a material. The ratio of transverse strain to the corresponding axial strain below the proportional limit.

polar winding. A winding in which the filament path passes tangent to the polar opening at one end of the chamber and tangent to the opposite side of the polar opening at the other end. A one-circuit pattern is inherent in the system.

polyacrylonitrile (PAN). Used as a base material or precursor in the manufacture of certain carbon fibers.

polyamide. A thermoplastic polymer in which the structural units are linked by amide or thio-amide groupings (repeated nitrogen and hydrogen groupings). Many polyamides are fiber forming.

polyamideimide. A polymer containing both amide (nylon) and imide (as in polyimide) groups; properties combine the benefits and disadvantages of both.

polyamide plastic. See *nylon plastics*.

polyarylsulfone (PAS). A high temperature resistant thermoplastic (T_g = 275 °C, or 527 °F). The term is also occasionally used to describe the family of resins which includes polysulfone and polyethersulfone.

polybenzimidazole (PBI). A condensation polymer of diphenyl isophthalate and 3,3'-diaminobenzidine. Extremely high-temperature resistant. Available as adhesive and fiber.

polycarbonate resin. A thermoplastic polymer derived from the direct reaction between aromatic and aliphatic dihydroxy compounds with phosgene or by the ester exchange reaction with appropriate phosgene-derived precursors. Highest impact resistance of any transparent plastic.

polycondensation. See *condensation polymerization*.

polyesters, thermoplastic. See *thermoplastic polyesters*.

polyesters, thermosetting. See *thermosetting polyesters*.

polyether etherketone (PEEK). A linear aromatic crystalline thermoplastic. A composite with a PEEK matrix may have a continuous-use temperature as high as 250 °C (480 °F).

polyetherimide. An amorphous polymer with good thermal properties for a thermoplastic. Reported T_g of 215 °C (419 °F) and continuous-use temperature of about 170 °C (338 °F).

polyimide (PI). A polymer produced by reacting an aromatic dianhydride with an aromatic diamine. It is a highly heat-resistant resin $\geq$ 315 °C (600 °F). Similar to a polyamide, differing only in the number of hydrogen

molecules contained in the groupings. Suitable for use as a binder or adhesive. May be either thermoplastic or thermoset.

polymer. A high molecular weight organic compound, natural or synthetic, whose structure can be represented by a repeated small unit, the mer, for example, polyethylene, rubber, and cellulose. Synthetic polymers are formed by addition or condensation polymerization of monomers. Some polymers are elastomers, some are plastics, and some are fibers. When two or more dissimilar monomers are involved, the product is called a copolymer. The chain lengths of commercial thermoplastics vary from near a thousand to over one hundred thousand repeating units. Thermosetting polymers approach infinity after curing, but their resin precursors, often called prepolymers, may be relatively short— 6 to 100 repeating units—before curing. The lengths of polymer chains, usually measured by molecular weight, have very significant effects on the performance properties of plastics and profound effects on processibility.

polymerization. A chemical reaction in which the molecules of a monomer are linked together to form large molecules whose molecular weight is a multiple of that of the original substance. When two or more monomers are involved, the process is called copolymerization.

polymer matrix. The resin portion of a reinforced or filled plastic.

polymethyl methacrylate. A thermoplastic polymer synthesized from methyl methacrylate. It is a transparent solid with exceptional optical properties; available in the form of sheets, granules, solutions, and emulsions. Used as facing material in certain composite constructions. See also *acrylic plastic*.

polyphenylene sulfide (PPS). A high-temperature thermoplastic useful primarily as a molding compound. Optimum properties depend on slightly cross-linking the resin. Known for chemical resistance.

polypropylene. A tough, lightweight, thermoplastic made by the polymerization of high-purity propylene gas in the presence of an organometallic catalyst at relatively low pressures and temperatures.

polysulfide. A synthetic polymer containing sulfur and carbon linkages, produced from organic dihalides and sodium polysulfide. Material is elastomeric in nature, resistant to light, oil, and solvents, and impermeable to gases.

polysulfone. A high temperature resistant thermoplastic polymer with the sulfone linkage, with a T_g of 190 °C (375 °F).

polyurethane. A thermosetting resin prepared by the reaction of diisocyanates with polyols, polyamides, alkyd polymers, and polyether polymers. See also *isocyanate plastics* and *urethane plastics*.

porosity. A condition of trapped pockets of air, gas, or vacuum within a solid material. Usually expressed as a percentage of the total nonsolid volume to the total volume (solid plus nonsolid) of a unit quantity of material.

postcure. Additional elevated-temperature cure, usually without pressure, to improve final properties and/or complete the cure, or decrease the percentage of volatiles in the compound. In certain resins, complete cure and ultimate mechanical properties are attained only by exposure of the cured resin to higher temperatures than those of curing.

postforming. The forming, bending, or shaping of fully cured, C-staged thermoset laminates that have been heated to make them flexible. On cooling, the formed laminate retains the contours and shape of the mold over which it has been formed.

pot life. The length of time that a catalyzed thermosetting resin system retains a viscosity low enough to be used in processing. Also called working life.

power factor. The cosine of the angle between voltage applied and the current resulting. Measurements are usually made at millioncycle frequencies.

PPS. See *polyphenylene sulfide*.

precure. The full or partial setting of a synthetic resin or adhesive in a joint before the clamping operation is complete or before pressure is applied.

precursor. For carbon or graphite fiber, the rayon, PAN or pitch fibers from which carbon and graphite fibers are derived.

prefit. A process for checking the fit of mating detail parts in an assembly prior to adhesive bonding, to ensure proper bond lines. Mechanically fastened structures are sometimes prefitted to establish shimming requirements.

preform. A preshaped fibrous reinforcement formed by distribution of chopped fibers or cloth by air, water flotation, or vacuum over the surface of a perforated screen to the approximate contour and thickness desired in the finished part. Also, a preshaped fibrous reinforcement of mat or cloth formed to the desired shape on a mandrel or mock-up before being placed in a mold press.

preform binder. A resin applied to the chopped strands of a preform, usually during its formation, and cured so that the preform will retain its shape and can be handled.

pregel. An unintentional, extra layer of cured resin on part of the surface of a reinforced plastic. Not related to gel coat.

preheating. The heating of a compound before molding or casting, to facilitate the operation or reduce the molding cycle.

preimpregnation. The practice of mixing resin and reinforcement and effecting partial cure before use or shipment to the user. See also *prepreg*.

premix. A molding compound prepared prior to and apart from the molding operations and containing all components required for molding: resin, reinforcement, fillers, catalysts, release agents, and other ingredients.

premolding. The lay-up and partial cure at an intermediate cure temperature of a laminated or chopped-fiber detail part to stabilize its configuration for handling and assembly with other parts for final cure.

preply. A composite material lamina in the raw-material stage, ready to be fabricated into a finished laminate. The lamina is usually combined with other raw laminae before fabrication. A preply includes a fiber system that is placed in position relative to all or part of the required matrix material to constitute the finished lamina. An organic matrix preply is called a prepreg. Metal matrix preplies include green tape, flame-sprayed tape, and consolidated monolayers.

prepolymer. A chemical intermediate whose molecular weight is between that of the monomer or monomers and the final polymer or resin.

prepreg. Either ready-to-mold material in sheet form or ready-to-wind material in roving form, which may be cloth, mat, unidirectional fiber, or paper impregnated with resin and stored for use. The resin is partially cured to a B-stage and supplied to the fabricator, who lays up the finished shape and completes the cure with heat and pressure. The two distinct types of prepreg available are (1) commercial prepregs, where the roving is coated with a hot melt or solvent system to produce a specific product to meet specific customer requirements; and (2) wet prepreg, where the basic resin is installed without solvents or preservatives but has limited room-temperature shelf life.

press clave. A simulated autoclave made by using the platens of a press to seal the ends of an open chamber, providing both the force required to prevent loss of the pressurizing medium and the heat required to cure the laminate inside.

pressure bag molding. A process for molding reinforced plastics in which a tailored, flexible bag is placed over the contact lay-up on the mold, sealed, and clamped in place. Fluid pressure, usually provided by compressed air or water, is placed against the bag, and the part is cured.

pressure-impregnation-carbonization (PIC). A densification process for carbon-carbon composites involving pitch impregnation and carbonization under high temperature and isostatic pressure conditions. This process is carried out in hot isostatic press (HIP) equipment.

pressure intensifier. A layer of flexible material (usually a high-temperature rubber) used to ensure the application of sufficient pressure to a location, such as a radius, in a lay-up being cured.

pressure-sensitive adhesive. A viscoelastic material that, in solvent-free form, remains permanently tacky. Such material will adhere instantaneously to most solid surfaces with the application of very light pressure.

primer. A coating applied to a surface, before the application of an adhesive, lacquer, enamel, and so forth, to improve the adhesion performance or load-carrying ability of the bond.

printed wiring board. A completely processed conductor pattern, usually formed on a stiff, flat base (laminated plastic). It serves as a means of electrical interconnection and physical attachment for printed circuits. Also called printed circuit board.

processing window. The range of processing conditions, such as stock (melt) temperature, pressure, shear rate, and so on, within which a particular grade of plastic can be fabricated with optimum or acceptable properties by a particular fabricating process, such as extrusion, injection molding, sheet molding, and so forth. The processing window for a particular plastic can vary significantly with design of the part and the mold, with the fabricating machinery used, and with the severity of the end-use stresses.

promoter. A chemical, itself a feeble catalyst, that greatly increases the activity of a given catalyst. See also *accelerator*.

proof. To test a component or system at its peak operating load or pressure.

proof pressure. The test pressure that pressurized components shall sustain without detrimental deformation or damage. The proof pressure test is used to give evidence of satisfactory workmanship and material quality.

proportional limit. The greatest stress which a material is capable of sustaining without deviation from proportionality of stress and strain (Hooke's law). It is expressed in force per unit area. See also *elastic limit*.

prototype. A model suitable for use in complete evaluation of form, design, performance, and material processing.

puckers. Areas on prepreg materials where material has locally blistered from the separator film or release paper.

pulp molding. The process by which a resin-impregnated pulp material is preformed by application of a vacuum and subsequently is oven cured or molded.

pultrusion. A continuous process for manufacturing composites that have a constant cross-sectional shape. The process consists of pulling a fiber-reinforcing material through a resin impregnation bath and through a shaping die, where the resin is subsequently cured.

pyrolysis. With respect to fibers, the thermal process by which organic precursor fiber materials, such as rayon, polyacrylonitrile (PAN), and pitch, are chemically changed into carbon fiber by the action of heat in an inert atmosphere. Pyrolysis temperatures can range from 800 to 2800 °C (1470 to 5070 °F), depending on the precursor. Higher processing graphitization temperatures of 1900 to 3000 °C (3450 to 5430 °F) generally lead to higher modulus carbon fibers, usually referred to as graphite fibers. During the pyrolysis process, molecules containing oxygen, hydrogen, and nitrogen are driven from the precursor fiber, leaving continuous chains of carbon.

Q

quasi-isotropic laminate. A laminate approximating isotropy by orientation of plies in several or more directions.

R

random pattern. A winding with no fixed pattern. If a large number of circuits is required for the pattern to repeat, a random pattern is approached. A winding in which the filaments do not lie in an even pattern.

reaction injection molding (RIM). A process for molding polyurethane, epoxy, and other liquid chemical systems. Mixing of two to four components in the proper chemical ratio is accomplished by a high-pressure impingement-type mixing head, from which the mixed material is delivered into the mold at low pressure, where it reacts (cures).

reinforced plastics. Molded, formed, filament-wound, tape-wrapped, or shaped plastic parts consisting of resins to which reinforcing fibers, mats, fabrics, and so forth, have been added before the forming operation to provide some strength properties greatly superior to those of the base resin.

reinforced reaction injection molding (RRIM). A reaction injection molding with a reinforcement added. See also *reaction injection molding.*

reinforcement. A strong material bonded into a matrix to improve its mechanical properties. Reinforcements are usually long fibers, chopped fibers, whiskers, particulates, and so forth. The term should not be used synonymously with filler.

relaxation time. The time required for a stress under a sustained constant strain to diminish by a stated fraction of its initial value.

relaxed stress. The initial stress minus the remaining stress at a given time during a stress-relaxation test.

release agent. A material that is applied in a thin film to the surface of a mold to keep the resin from bonding to the mold. See also

mold release agent. Also called parting agent.

release film. An impermeable layer of film that does not bond to the resin being cured. See also *separator.*

residual gas analysis (RGA). The study of residual gases in vacuum systems using mass spectometry.

residual strain. The strain associated with residual stress.

residual stress. The stress existing in a body at rest, in equilibrium, at uniform temperature, and not subjected to external forces. Often caused by the forming and curing process.

resilience. The ratio of energy returned, on recovery from deformation, to the work input required to produce the deformation (usually expressed as a percentage). The ability to regain an original shape quickly after being strained or distorted.

resin. A solid or pseudosolid organic material, usually of high molecular weight, that exhibits a tendency to flow when subjected to stress. It usually has a softening or melting range, and fractures conchoidally. Most resins are polymers. In reinforced plastics, the material used to bind together the reinforcement material; the matrix. See also *polymer.*

resin content. The amount of resin in a laminate expressed as either a percentage of total weight or total volume.

resin pocket. An apparent accumulation of excess resin in a small, localized section visible on cut edges of molded surfaces, or internal to the structure and nonvisible. See also *resin-rich area.*

resin-rich area. Localized area filled with resin and lacking reinforcing material. See also *resin pocket.*

resin-starved area. Localized area of insufficient resin, usually identified by low gloss, dry spots, or fiber showing on the surface.

resin system. A mixture of resin and ingredients such as catalyst, initiator, diluents, and so forth, required for the intended processing and final product.

resin transfer molding (RTM). A process whereby catalyzed resin is transferred or injected into an enclosed mold in which the fiberglass reinforcement has been placed.

resistivity. The ability of a material to resist passage of electrical current either through its bulk or on a surface.

reverse helical winding. In filament winding, as the fiber delivery arm traverses one circuit, a continuous helix is laid down, reversing direction at the polar ends. In contrast to biaxial, compact, or sequential winding. The fibers cross each other at definite equators, the number depending on the helix. The minimum region of crossover is three.

reverse impact test. A test in which one side of a sheet of material is struck by a pendulum or falling object, and the reverse side is inspected for damage.

RGA. See *residual gas analysis.*

rheology. The study of the flow of materials, particularly plastic flow of solids and the flow of non-Newtonian liquids. The science treating the deformation and flow of matter.

rib. A reinforcing member designed into a plastic part to provide lateral, horizontal, hoop, or other structural support.

RIM. See *reaction injection molding.*

rise time. In urethane foam molding, the time between the pouring of the urethane mix and the completion of foaming.

Rockwell hardness. A value derived from the increase in depth of an impression as the load on an indenter is increased from a fixed minimum value to a higher value and then returned to the minimum value. Indenters for the Rockwell test include steel balls of several specific diameters and a diamond cone penetrator having an included angle of 120° with a spherical tip having a radius of 0.2 mm (0.0070 in.). Rockwell hardness numbers are always quoted with a prefix representing the Rockwell scale corresponding to a given combination of load and indenter, for example, HRC 30.

room-temperature curing adhesive. An adhesive that sets (to handling strength) within an hour at temperatures from 20 to 30 °C (68 to 86 °F) and later reaches full strength without heating.

room-temperature vulcanizing (RTV). Vulcanization or curing at room temperature by chemical reaction; usually applies to silicones and other rubbers.

roving. A number of yarns, strands, tows, or ends collected into a parallel bundle with little or no twist.

roving ball. The supply package offered to the winder, consisting of a number of ends or strands wound to a given outside diameter onto a length of cardboard tube. Usually designated by either fiber weight or length in yards.

roving cloth. A textile fabric, coarse in nature, woven from rovings.

RRIM. See *reinforced reaction injection molding.*

RTM. See *resin transfer molding.*

RTV. See *room-temperature vulcanizing.*

rubber. Cross-linked polymers with glass transition temperature below room temperature, which exhibit highly elastic deformation and have high elongation.

rupture. A cleavage or break resulting from physical stress. Work of rupture. The integral of the stress-strain curve between the origin and the point of rupture.

S

sandwich constructions. Panels composed of a lightweight core material, such as honeycomb, foamed plastic, and so forth, to which two relatively thin, dense, high-strength or high-stiffness faces or skins are adhered.

satin. A type of finish having a satin or velvety appearance, specified for plastics or composites.

satin weave. See *harness satin.*

S-basis. The S-basis property allowable is the minimum value specified by the appropriate federal, military, Society of Automotive Engineers, American Society for Testing and Materials, or other recognized and approved specifications for the material.

SBS. See *short beam shear.*

scarf joint. A joint made by cutting away similar angular segments on two adherends and bonding the adherends with the cut areas fitted together. See also *lap joint.*

scrim. A low-cost reinforcing fabric made from continuous filament yarn in an open-mesh construction. Used in the processing of tape or other B-stage material to facilitate handling. Also used as a carrier of adhesive, to be used in secondary bonding.

sealant. A material applied to a joint in paste or liquid form that hardens or cures in place, forming a seal against gas or liquid entry.

secant modulus. Idealized Young's modulus derived from a secant drawn between the origin and any point on a nonlinear stress-strain curve. On materials whose modulus changes with stress, the secant modulus is the average of the zero applied stress point and the maximum stress point being considered. See also *tangent modulus.*

secondary bonding. The joining together, by the process of adhesive bonding, of two or more already cured composite parts, during which the only chemical or thermal reaction occurring is the curing of the adhesive itself.

secondary structure. In aircraft and aerospace applications, a structure that is not critical to flight safety.

self-extinguishing resin. A resin formulation that will burn in the presence of a flame but will extinguish itself within a specified time after the flame is removed.

self-skinning foam. A urethane foam that produces a tough outer surface over a foam core upon curing.

selvage. The woven-edge portion of a fabric parallel to the warp.

semicrystalline. In plastics, materials that exhibit localized crystallinity. See also *crystalline plastic.*

separator. A permeable layer that also acts as a release film. Porous Teflon-coated fiber-glass is an example. Often placed between lay-up and bleeder to facilitate bleeder system removal from laminate after cure.

set. The irrecoverable or permanent deformation or creep after complete release of the force producing the deformation.

set up. To harden, as in curing of a polymer resin.

S-glass. A magnesium aluminosilicate composition that is especially designed to provide very high tensile strength glass filaments. S-glass and S-2 glass fibers have the same glass composition but different finishes (coatings). S-glass is made to more demanding specifications, and S-2 is considered the commercial grade.

shear. An action or stress resulting from applied forces that causes or tends to cause two contiguous parts of a body to slide relative to each other in a direction parallel to their plane of contact. In interlaminar shear, the plane of contact is composed primarily of resin. See also *shear strength* and *shear stress.*

shear edge. The cutoff edge of the mold.

shear modulus. The ratio of shearing stress to shearing strain within the proportional limit of the material.

shear strain. The tangent of the angular change, caused by a force between two lines originally perpendicular to each other through a point in a body. Also called angular strain.

shear strength. The maximum shear stress that a material is capable of sustaining. Shear strength is calculated from the maximum load during a shear or torsion test and is based on the original cross-sectional area of the specimen.

shear stress. The component of stress tangent to the plane on which the forces act.

sheet molding compound (SMC). A composite of fibers, usually a polyester resin, and pigments, fillers, and other additives that have been compounded and processed into sheet form to facilitate handling in the molding operation.

shelf life. The length of time a material, substance, product, or reagent can be stored under specified environmental conditions and continue to meet all applicable specification requirements and/or remain suitable for its intended function.

shell tooling. A mold or bonding fixture consisting of a contoured surface shell supported by a substructure to provide dimensional stability.

shoe. A device for gathering filaments into a strand, in glass fiber forming.

Shore hardness. A measure of the resistance of material to indentation by a spring-loaded indenter. The higher the number, the greater the resistance. Normally used for rubber materials.

short beam shear (SBS). A flexural test of a specimen having a low test span-to-thickness ratio (for example, 4:1), such that failure is primarily in shear.

short shot. Injection of insufficient material to fill the mold.

shot capacity. The maximum weight of material an injection machine can provide from one forward motion of the ram, screw, or plunger.

shrinkage. The relative change in dimension from the length measured on the mold when it is cold to the length of the molded object 24 h after it has been taken out of the mold.

silicon carbide. Reinforcement, in whisker, particulate, and fine or large fiber, that has application as metal matrix reinforcement because of its high strength and modulus, density equal to that of aluminum, and comparatively low cost. As a whisker or particulate, it gives the composite isotropic properties and is easily machined.

silicone plastics. Plastics based on resins in which the main polymer chain consists of alternating silicon and oxygen atoms, with carbon-containing side groups. Derived from silica (sand) and methyl chlorides and furnished in different molecular weights, including liquids and solid resins and elastomers.

single-circuit winding. A winding in which the filament path makes a complete traverse of the chamber, after which the following traverse lies immediately adjacent to the previous one.

sink mark. A shallow depression or dimple on the surface of an injection-molded part due to collapsing of the surface following local internal shrinkage after the gate seals. An incipient short shot.

sintering. The bonding of powders by solid-state diffusion, resulting in the absence of a separate bonding phase. The process is generally accompanied by an increase in strength, ductility, and, occasionally, density.

size. Any treatment consisting of starch, gelatin, oil, wax, or other suitable ingredient applied to yarn or fibers at the time of formation to protect the surface and aid the process of handling and fabrication or to control the fiber characteristics. The treatment contains ingredients that provide surface lubricity and binding action but, unlike a finish, contains no coupling agent. Before final fabrication into a composite, the size is usually removed by heat cleaning, and a finish is applied.

sizing content. The percent of the total strand weight made up by the sizing; usually determined by burning off or dissolving the organic sizing; known as loss on ignition.

skein. A continuous filament, strand, yarn, or roving, wound up to some measurable length and usually used to measure various physical properties.

skin. The relatively dense material that may form the surface of a cellular plastic or of a sandwich.

skirt. The extension of a motorcase from the tangency plane, used for interstage connections, usually wound or laid up as an integral part of the case.

slenderness ratio. The unsupported effective length of a uniform column divided by the least radius of gyration of the cross-sectional area.

slip angle. The angle at which a tensioned fiber will slide off a filament-wound dome. If the difference between the wind angle and the geodesic angle is less than the slip angle, fiber will not slide off the dome. Slip angles for different fiber-resin systems vary and must be determined experimentally.

sliver. A number of staple or continuous-filament fibers aligned in a continuous strand without twist. Pronounced "slyver." See also *strand*.

slurry preforming. Method of preparing reinforced plastic preforms by wet processing techniques similar to those used in the pulp molding industry. For example, glass fibers suspended in water are passed through a screen that passes the water but retains the fibers in the form of a mat.

SMC. See *sheet molding compound*.

S-N diagram. A plot of stress (*S*) against the number of cycles to failure (*N*) in fatigue testing. A log scale is normally used for *N*. For *S*, a linear scale is often used, but sometimes a log scale is used here, too. Also, a representation of the number of alternating stress cycles a material can sustain without failure at various maximum stresses.

softening range. The range of temperatures in which a plastic changes from a rigid to a soft state. Actual values will depend on the test method. Sometimes erroneously referred to as softening point.

solvation. The process of swelling, gelling, or dissolving a resin by a solvent or plasticizer.

specific gravity. The density (mass per unit volume) of any material divided by that of water at a standard temperature.

specific heat. The quantity of heat required to raise the temperature of a unit mass of a substance 1° under specified conditions.

specific properties. Material properties divided by the material density.

SPF. See *superplastic forming*.

splay. A fanlike surface defect near the gate on a part.

splice. The joining of two ends of glass fiber yarn or strand, usually by means of an air-drying adhesive.

sprayed metal molds. Molds made by spraying molten metal onto a master until a shell of predetermined thickness is achieved. The shell is then removed and backed up with plaster, cement, casting resin, or other suitable material. Used primarily as a mold in the sheet forming process.

spray-up. Technique in which a spray gun is used as an applicator tool. In reinforced plastics, for example, fibrous glass and resin can be simultaneously deposited in a mold. In essence, roving is fed through a chopper and ejected into a resin stream that is directed at the mold by either of two spray systems. In foamed plastics, fast-reacting urethane foams or epoxy foams are fed in liquid streams to the gun and sprayed on the surface. On contact, the liquid starts to foam.

spring constant. The number of pounds required to compress a spring or specimen 25 mm (1 in.) in a prescribed test procedure.

sprue. A single hole through which thermoset molding compounds are injected directly into the mold cavity.

spun roving. A heavy low-cost glass or aramid fiber strand consisting of filaments that are continuous but doubled back on themselves.

stabilization. In carbon fiber forming, the process used to render the carbon fiber precursor infusible prior to carbonization.

stacking sequence. A description of a laminate that details the ply orientations and their sequence in the laminate.

staging. Heating a premixed resin system, such as in a prepreg, until the chemical reaction (curing) starts, but stopping the reaction before the gel point is reached. Staging is often used to reduce resin flow in subsequent press molding operations.

standard deviation. A measure of dispersion of data from the average. The root mean square of the individual deviation from the average.

staple fibers. Fibers of spinnable length manufactured directly or by cutting continuous filaments to short lengths (usually 12.7 to 50 mm, or 1/2 to 2 in. long; 1 to 5 denier).

starved area. An area in a plastic part that has an insufficient amount of resin to wet out the reinforcement completely. This condition may be due to improper wetting, impregnation, or resin flow; excessive molding pressure; or improper bleeder cloth thickness.

starved joint. An adhesive joint that has been deprived of the proper film thickness of

adhesive due to insufficient adhesive spreading or to the application of excessive pressure during the lamination process.

static fatigue. Failure of a part under continued static load. Analogous to creep rupture failure in metals testing, but often the result of aging accelerated by stress.

static modulus. The ratio of stress to strain under static conditions. It is calculated from static stress-strain tests, in shear, compression, or tension. Expressed in force per unit area.

static stress. A stress in which the force is constant or slowly increasing with time, for example, test of failure without shock.

stiffness. A measure of modulus. The relationship of load and deformation. The ratio between the applied stress and resulting strain. A term often used when the relationship of stress to strain does not conform to the definition of Young's modulus. See also *stress-strain*.

stops. Metal pieces inserted between die halves. Used to control the thickness of a press-molded part. Not a recommended practice, because the resin will receive less pressure, which can result in voids.

storage life. The period of time during which a liquid resin, packaged adhesive, or prepreg can be stored under specified temperature conditions and remain suitable for use. Also called shelf life.

strain. Elastic deformation due to stress. Measured as the change in length per unit of length in a given direction, and expressed in percentage or mm/mm (in./in.).

strain, axial. See *axial strain*.

strain gage. Device to measure strain in a stressed material based on the change in electrical resistance.

strain, initial. See *initial strain*.

strain relaxation. Reduction in internal strain over time. Similar molecular processes occur as in creep, except that the body is constrained.

strain, residual. See *residual strain*.

strain, shear. See *shear strain*.

strain, transverse. See *transverse strain*.

strain, true. See *true strain*.

strand. Normally an untwisted bundle or assembly of continuous filaments used as a unit, including slivers, tows, ends, yarn, and so forth. Sometimes a single fiber or filament is called a strand.

strand count. The number of strands in a plied yarn. The number of strands in a roving.

strand integrity. The degree to which the individual filaments making up the strand or end are held together by the applied sizing.

strand tensile test. A tensile test of a single resin-impregnated strand of any fiber.

strength, compressive. See *compressive strength*.

strength, flexural. See *flexural strength*.

strength, shear. See *shear strength*.

strength, tensile. See *tensile strength*.

strength, wet. See *wet strength*.

strength, yield. See *yield strength*.

stress. The internal force per unit area that resists a change in size or shape of a body. Expressed in force per unit area.

stress concentration. On a macromechanical level, the magnification of the level of an applied stress in the region of a notch, void, hole, or inclusion.

stress-concentration factor. The ratio of the maximum stress in the region of a stress concentrator, such as a hole, to the stress in a similar strained area without a stress concentrator.

stress corrosion. Preferential attack of areas under stress in a corrosive environment, where such an environment alone would not have caused corrosion.

stress crack. External or internal cracks in a plastic caused by tensile stresses less than that of its short-time mechanical strength, frequently accelerated by the environment to which the plastic is exposed. The stresses that cause cracking may be present internally or externally or may be combinations of these stresses. See also *crazing*.

stress cracking. The failure of a material by cracking or crazing some time after it has been placed under load. Time-to-failure may range from minutes to years. Causes include molded-in stresses, postfabrication shrinkage or warpage, and hostile environment.

stress, fracture. See *fracture stress*.

stress, initial (instantaneous). See *initial (instantaneous) stress*.

stress, nominal. See *nominal stress*.

stress, normal. See *normal stress*.

stress relaxation. The decrease in stress under sustained, constant strain. Also called stress decay.

stress, relaxed. See *relaxed stress*.

stress, residual. See *residual stress*.

stress, shear. See *shear stress*.

stress-strain. Stiffness at a given strain.

stress-strain curve. Simultaneous readings of load and deformation, converted to stress and strain, plotted as ordinates and abscissae, respectively, to obtain a stress-strain diagram.

stress, tensile. See *tensile stress*.

stress, torsional. See *torsional stress*.

stress, true. See *true stress*.

structural adhesive. Adhesive used for transferring required loads between adherends exposed to service environments typical for the structure involved.

structural bond. A bond that joins basic load-bearing parts of an assembly. The load may be either static or dynamic.

superplastic forming (SPF). A strain rate sensitive metal forming process that uses characteristics of materials exhibiting high elongation-to-failure.

surface preparation. Physical and/or chemical preparation of an adherend to make it suitable for adhesive bonding.

surface resistivity (electrical). The surface resistivity of a material is the ratio of the potential gradient parallel to the current along its surface to the current per unit width of surface. Surface resistivity is numerically equal to the surface resistance between opposite sides of a square of any size when the current flow is uniform.

surface treatment. A material (size or finish) applied to fibrous material during the forming operation or in subsequent processes. For carbon fiber surface treatment, the process used to enhance bonding capability of fiber to resin.

surfacing mat. A very thin mat, usually 180 to 510 μm (7 to 20 mil) thick, of highly filamentized fiberglass, used primarily to produce a smooth surface on a reinforced plastic laminate, or for precise machining or grinding.

symmetrical laminate. A composite laminate in which the sequence of plies below the laminate midplane is a mirror image of the stacking sequence above the midplane.

syntactic foams. Composites made by mixing hollow microspheres of glass, epoxy, phenolic, and so forth, into fluid resins (with additives and curing agents) to form a moldable, curable, lightweight, fluid mass; as opposed to foamed plastic, in which the cells are formed by gas bubbles released in the liquid plastic by either chemical or mechanical action.

T

tabs. Extra lengths of composite or other material at the ends of a tensile specimen to promote failure away from the grips.

tack. Stickiness of an adhesive or filament reinforced resin prepreg material.

tack range. The period of time in which an adhesive will remain in the tacky-dry condition after application to the adherend, and under specified conditions of temperature and humidity.

tangent modulus. The slope of the line at a predefined point on a static stress-strain curve, expressed in force per unit area per unit strain. This is the tangent modulus at that point in shear, tension, or compression, as the case may be. See also *secant modulus*.

tape. Unidirectional prepreg fabricated in widths up to 305 mm (12 in.) for carbon and 75 mm (3 in.) for boron. Woven broad goods carbon and glass tapes up to 1250 or 1500 mm (50 or 60 in.) wide are available commercially.

tape wrapped. Fabric tape is heated and wrapped onto a rotating mandrel and subsequently cooled to firm the surface for the next tape layer application.

template. A pattern used as a guide for cutting and laying plies.

tenacity. The term generally used in yarn manufacture and textile engineering to denote the strength of a yarn or of a filament of a given size. Numerically, it is the grams of breaking force per denier unit of yarn or filament size. Grams per denier is expressed as gpd.

tensile modulus. See *Young's modulus*.

tensile strength. The maximum load or force per unit cross-sectional area, within the gage length, of the specimen. The pulling stress required to break a given specimen.

tensile strength, ultimate. See *ultimate tensile strength*.

tensile stress. The normal stress caused by forces directed away from the plane on which they act.

tenth-scale vessel. A filament wound material test vessel based on a one-tenth subscale of the prototype.

terpolymer. A polymeric system that contains three monomeric units.

tex. A unit for expressing linear density equal to the mass or weight in grams of 1000 meters of filament, fiber, yarn, or other textile strand.

textile fibers. Fibers or filaments that can be processed into yarn or made into a fabric by interlacing in a variety of methods, including weaving, knitting, and braiding.

T_g. See *glass transition temperature*.

TGA. See *thermogravimetric analysis*.

thermal conductivity. Ability of a material to conduct heat. The physical constant for the quantity of heat that passes through a unit cube of a substance in unit time when the difference in temperature of two faces is 1°.

thermal endurance. The time at a selected temperature for a material or system of materials to deteriorate to some predetermined level of electrical, mechanical, or chemical performance under prescribed conditions of test.

thermal expansion molding. A process in which elastomeric tooling details are constrained within a rigid frame to generate consolidation pressure by thermal expansion during the curing cycle of the autoclave molding process.

thermal stress cracking. Crazing and cracking of some thermoplastic resins, resulting from overexposure to elevated temperatures. See also *stress cracking*.

thermoforming. Forming a thermoplastic material after heating it to the point where it is soft enough to be formed without cracking or breaking reinforcing fibers.

thermogravimetric analysis (TGA). The study of the mass of a material under various conditions of temperature and pressure.

thermoplastic. Capable of being repeatedly softened by an increase of temperature and hardened by a decrease in temperature. Applicable to those materials whose change upon heating is substantially physical rather than chemical and that in the softened stage can be shaped by flow into articles by molding or extrusion.

thermoplastic polyesters. A class of thermoplastic polymers in which the repeating units are joined by ester groups. The two important types are (1) polyethylene terephthalate (PET), which is widely used as film, fiber, and soda bottles; and (2) polybutylene terephthalate (PBT), primarily a molding compound.

thermoset. A plastic that, when cured by application of heat or chemical means, changes into a substantially infusible and insoluble material.

thermosetting polyesters. A class of resins produced by dissolving unsaturated, generally linear, alkyd resins in a vinyl-type active monomer such as styrene, methyl styrene, or diallyl phthalate. Cure is effected through vinyl polymerization using peroxide catalysts and promoters or heat to accelerate the reaction. The two important commercial types are (1) liquid resins that are cross-linked with styrene and used either as impregnants for glass or carbon fiber reinforcements in laminates, filament-wound structures, and other built-up constructions, or as binders for chopped-fiber reinforcements in molding compounds, such as sheet molding compound (SMC), bulk molding compound (BMC), and thick molding compound (TMC); and (2) liquid or solid resins crosslinked with other esters in chopped-fiber and mineral-filled molding compounds, for example, alkyd and diallyl phthalate.

thin-layer chromatography (TLC). A micro type of chromatography in which a thin layer of special absorbent is applied to a glass plate, a drop of a solution of the material being investigated is applied to an edge, and that side of the plate is then dipped in an appropriate solvent. As the solvent travels up the thin layer, it selectively separates the molecules present in the material being investigated.

thixotropic (thixotropy). Concerning materials that are gel-like at rest but fluid when agitated. Having high static shear strength and low dynamic shear strength at the same time. To lose viscosity under stress.

thread. See *fiber*.

thread count. The number of yarns (threads) per inch in either the lengthwise (warp) or crosswise (fill or weft) direction of woven fabrics.

TLC. See *thin-layer chromatography*.

tolerance. The guaranteed maximum deviation from the specified nominal value of a component characteristic at standard or stated environmental conditions.

tooling resin. Resins that have applications as tooling aids, coreboxes, prototypes, hammer forms, stretch forms, foundry patterns, and so forth. Epoxy and silicone are common examples.

tool side. The side of the part that is cured against the tool (mold or mandrel).

torsion. Twisting stress.

torsional stress. The shear stress on a transverse cross section caused by a twisting action.

toughness. A property of a material for absorbing work. The actual work per unit volume or unit mass of material that is required to rupture it. Toughness is proportional to the area under the load-elongation curve from the origin to the breaking point.

tow. An untwisted bundle of continuous filaments. Commonly used in referring to manmade fibers, particularly carbon and graphite, but also glass and aramid. A tow designated as 140K has 140 000 filaments.

tracer. A fiber, tow, or yarn added to a prepreg for verifying fiber alignment and, in the case of woven materials, for distinguishing warp fibers from fill fibers.

transfer molding. Method of molding thermosetting materials in which the plastic is first softened by heat and pressure in a transfer chamber and then forced by high pressure through suitable sprues, runners, and gates into the closed mold for final shaping and curing.

transition, first order. See *first-order transition*.

transition temperature. The temperature at which the properties of a material change. Depending on the material, the transition change may or may not be reversible.

transversely isotropic. Term describing a material exhibiting a special case of orthotropy in which properties are identical in two orthotropic dimensions but not the third. Having identical properties in both transverse directions but not in the longitudinal direction.

transverse strain. The linear strain in a plane perpendicular to the loading axis of a specimen.

true strain. The natural logarithm of the ratio of gage length at the moment of observation to the original gage length for a body subjected to an axial force.

true stress. The stress along the axis calculated on the actual cross section at the time of observation instead of the original cross-sectional area. Applicable to tension and compression testing.

turns per inch (tpi). A measure of the amount of twist produced in a yarn, tow, or roving during its processing history. Also, the lead rate of a hoop layer at a specified band width. See also *twist*.

twist. The spiral turns about its axis per unit of length in a yarn or other textile strand. Twist may be expressed as turns per inch (TPI), and so forth. S and Z refer to direction of twist, in reference to whether the twist direction conforms to the middle-section slope of the particular letter.

twist, balanced. See *balanced twist*.

typical-basis. The typical property value is an average value. No statistical assurance is associated with this basis.

U

ultimate elongation. The elongation at rupture.

ultimate tensile strength. The ultimate or final (highest) stress sustained by a specimen in a tension test. Rupture and ultimate stress may or may not be the same.

ultrasonic testing. A nondestructive test applied to materials for the purpose of locating internal flaws or structural discontinuities by the use of high-frequency reflection or attenuation (ultrasonic beam).

ultraviolet (UV). Zone of invisible radiations beyond the violet end of the spectrum of visible radiations. Since UV wavelengths are shorter than visible wavelengths, their photons have more energy, enough to initiate some chemical reactions and to degrade most plastics, particularly aramids.

ultraviolet (UV) stabilizer. Any chemical compound that, when admixed with a thermoplastic resin, selectively absorbs UV rays.

unbond. An area within a bonded interface between two adherends in which the intended bonding action failed to take place, or where two layers of prepreg in a cured component do not adhere to each other. Also used to denote specific areas deliberately prevented from bonding in order to simulate a defective bond, such as in the generation of quality standards specimens.

undercure. A condition of the molded article resulting from the allowance of too little time and/or temperature or pressure for adequate hardening of the molding.

undercut. A protuberance or indentation that impedes the withdrawal of a molded part from a two-piece, rigid mold. Any such protuberance or indentation, depending on the design of the mold.

uniaxial load. A condition whereby a material is stressed in only one direction along the axis or centerline of component parts.

unidirectional laminate. A reinforced plastic laminate in which substantially all of the fibers are oriented in the same direction.

unsaturated compounds. Any compound having more than one bond between two adjacent atoms, usually carbon atoms, and capable of adding other atoms at that point to reduce it to a single bond.

unsymmetric laminate. A laminate having an arbitrary stacking sequence without midplane symmetry.

urethane plastics. Plastics based on resins made by condensation of organic isocyanates with compounds or resins that contain hydroxyl groups. The resin is furnished as two component liquid monomers or prepolymers that are mixed in the field immediately before application. A great variety of materials are available, depending upon the monomers used in the prepolymers, polyols, and the type of diisocyanate employed. Extremely abrasion and impact resistant. See also isocyanate plastics and polyurethane.

UV. See *ultraviolet*.

V

vacuum bag molding. A process in which a sheet of flexible transparent material plus bleeder cloth and release film are placed over the lay-up on the mold and sealed at the edges. A vacuum is applied between the sheet and the lay-up. The entrapped air is mechanically worked out of the lay-up and removed by the vacuum, and the part is cured with temperature, pressure, and time. Also called bag molding.

vacuum hot pressing (VHP). A method of processing materials (especially powders) at elevated temperatures and consolidation pressures, and low atmospheric pressures.

vapor-liquid-solid (VLS) process. A process utilizing vapor feed gases and a liquid catalyst, and producing solid crystalline whisker growth. Used to produce silicon carbide whiskers.

veil. An ultrathin mat similar to a surface mat, often composed of organic fibers as well as glass fibers.

vent. A small hole or shallow channel in a mold that allows air or gas to exit as the molding material enters.

vent cloth. A layer or layers of open-weave cloth used to provide a path for vacuum to "reach" the area over a laminate being cured, such that volatiles and air can be removed. Also causes the pressure differential that results in application of pressure to the part being cured. Also called breather cloth.

venting. In autoclave curing of a part or assembly, turning off the vacuum source and venting the vacuum bag to the atmosphere. The pressure on the part is then the difference between pressure in the clave and atmospheric pressure.

vermiculite. A granular material mixed with resin to form a filler of relatively high compressive strength.

VHP. See *vacuum hot pressing*.

vinyl esters. A class of thermosetting resins containing esters of acrylic and/or methacrylic acids, many of which have been made from epoxy resin. Cure is accomplished as with unsaturated polyesters by copolymerization with other vinyl monomers, such as styrene.

virgin filament. An individual filament that has not been in contact with any other fiber or any other hard material.

viscoelasticity. A property involving a combination of elastic and viscous behavior in the application of which a material is considered to combine the features of a perfectly elastic solid and a perfect fluid. Phenomenon of time-dependent, in addition to elastic, deformation (or recovery) in response to load.

viscosity. The property of resistance to flow exhibited within the body of a material, expressed in terms of relationship between applied shearing stress and resulting rate of strain in shear. Viscosity is usually taken to mean Newtonian viscosity, in which case the ratio of shearing stress to the rate of shearing strain is constant. In non-Newtonian behavior, which is the usual case with plastics, the ratio varies with the shearing stress. Such ratios are often called the apparent viscosities at the corresponding shearing stresses. Viscosity is measured in terms of flow in Pa · s (P), with water as the base standard (value of 1.0). The higher the number, the less flow.

VLS process. See *vapor-liquid-solid process*.

void content. Volume percentage of voids, usually less than 1% in a properly cured composite. The experimental determination is indirect, that is, calculated from the measured density of a cured laminate and the "theoretical" density of the starting material.

voids. Air or gas that has been trapped and cured into a laminate. Porosity is an aggregation of microvoids. Voids are essentially incapable of transmitting structural stresses or nonradiative energy fields.

volatile content. The percent of volatiles that are driven off as a vapor from a plastic or an impregnated reinforcement.

volatiles. Materials, such as water and alcohol, in a sizing or a resin formulation, that are capable of being driven off as a vapor at room temperature or at a slightly elevated temperature.

volume fraction. Fraction of a constituent material based on its volume.

volume resistance. The volume resistance between two electrodes in contact with or embedded in a specimen is the ratio between the direct voltage applied to them and that portion of the current between them that is distributed through the volume of the specimen. Also, the electrical resistance between opposite faces of a 1-cm (0.40 in.) cube of insulating material. Also called specific insulation resistance.

vulcanization. A chemical reaction in which a rubber is cured by reaction with sulfur or other suitable agents.

W

wafer. A reinforcement for motorcase port openings.

warp. The yarn running lengthwise in a woven fabric. A group of yarns in long lengths and approximately parallel. A change in dimension of a cured laminate from its original molded shape.

water absorption. Ratio of the weight of water absorbed by a material to the weight of the dry material.

water jet. Water emitted from a nozzle under high pressure (70 to 410 MPa, or 10 to 60 ksi or higher). Useful for cutting organic composites.

weathering. Exposure of plastics to the outdoor environment.

weathering, artificial. See *artificial weathering*.

weave. The particular manner in which a fabric is formed by interlacing yarns. Usually assigned a style number.

weeping. Slow leakage manifested by the appearance of water on a surface.

weft. The transverse threads or fibers in a woven fabric. Those fibers running perpendicular to the warp. Also called fill, filling yarn, or woof.

weld line. The mark visible on a finished part made by the meeting of two flow fronts of plastic material during molding. Also called weld mark or flow line.

weld mark. See *flow line*.

wet installation. A bolted joint in which sealant is applied to the head and shank of the fastener so that after assembly a seal is provided between the fastener and the elements being joined.

wet lay-up. A method of making a reinforced product by applying the resin system as a liquid when the reinforcement is put in place.

wet-out. The condition of an impregnated roving or yarn in which substantially all voids between the sized strands and filaments are filled with resin.

wet strength. The strength of an organic matrix composite when the matrix resin is saturated with absorbed moisture, or is at a defined percentage of absorbed moisture less than saturation. (Saturation is an equilibrium condition in which the net rate of absorption under prescribed conditions falls essentially to zero).

wetting. The spreading, and sometimes absorption, of a fluid on or into a surface.

wet winding. In filament winding, the process of winding glass on a mandrel in which the strand is impregnated with resin just before contact with the mandrel. See also *dry winding*.

whisker. A short single crystal fiber or filament used as a reinforcement in a matrix. Whisker diameters range from 1 to 25 μm (40 to 980 μin.), with aspect ratios between 100 and 15 000.

wind angle. The angular measure in degrees between the direction parallel to the filaments and an established reference. In filament-wound structures it is the convention to measure the wind angle with reference to the centerline through the polar bosses, that is, the axis of rotation.

winding pattern. The total number of individual circuits required for a winding path to begin repeating by laying down immediately adjacent to the initial circuit. A regularly recurring pattern of the filament path after a certain number of mandrel revolutions, leading eventually to the complete coverage of the mandrel.

winding tension. In filament winding or tape wrapping, the amount of tension on the reinforcement as it makes contact with the mandrel.

woof. See *weft*.

work hardening. Increase in resistance to further deformation with continuing distortion.

Hardening and strengthening of a metal or alloy caused by the strain energy absorbed from prior deformation.

working life. The period of time during which a liquid resin or adhesive, after mixing with catalyst, solvent, or other compounding ingredients, remains usable. See also *gelation time* and *pot life*.

woven fabric. A material (usually a planar structure) constructed by interlacing yarns, fibers, or filaments, to form such fabric patterns as plain, harness satin, or leno weaves.

woven roving. A heavy glass fiber fabric made by weaving roving or yarn bundles.

wrinkle. A surface imperfection in laminated plastics that has the appearance of a crease or fold in one or more outer sheets of the paper, fabric, or other base, which has been pressed in. Also occurs in vacuum bag molding when the bag is improperly placed, causing a crease.

X

x-axis. In composite laminates, an axis in the plane of the laminate which is used as the 0° reference for designating the angle of a lamina.

xy-plane. In composite laminates, the reference plane parallel to the plane of the laminate.

Y

yarn. An assemblage of twisted filaments, fibers, or strands, either natural or manufactured, to form a continuous length that is suitable for use in weaving or interweaving into textile materials.

yarn bundle. See *bundle*.

yarn, plied. See *plied yarn*.

y-axis. In composite laminates, the axis in the plane of the laminate that is perpendicular to the *x*-axis. Contrast with *x-axis*.

yield point. The first stress in a material, less than the maximum attainable stress, at which the strain increases at a higher rate than the stress. The point at which permanent deformation of a stressed specimen begins to take place. Only materials that exhibit yielding have a yield point.

yield strength. The stress at the yield point. The stress at which a material exhibits a specified limiting deviation from the proportionality of stress to strain. The lowest stress at which a material undergoes plastic deformation. Below this stress, the material is elastic; above it, the material is viscous. Often defined as the stress needed to produce a specified amount of plastic deformation (usually a 0.2% change in length).

Young's modulus. The ratio of normal stress to corresponding strain for tensile or com-

pressive stresses less than the proportional limit of the material. See also *modulus of elasticity*.

Z

z-axis. In composite laminates, the reference axis normal to the plane of the laminate.

zero bleed. A laminate fabrication procedure that does not allow loss of resin during cure. Also describes prepreg made with the amount of resin desired in the final part, such that no resin has to be removed during cure.

ACKNOWLEDGMENTS

While a number of individuals provided helpful comments and additions to the glossary, special thanks are given to Richard C. Laramee (Morton Thiokol, Inc.), Brian A. Wilson (Aerojet Strategic Propulsion Company), Eugene R. Crilly (Rockwell International Corporation), and Charles David Himmelblau (Lockheed Missiles & Space Company, Inc.).

SELECTED REFERENCES

- *Advanced Composites Design Guide*, Air Force Wright Aeronautical Laboratories, July 1983
- Chemical Propulsion Information Agency, Glossary, Applied Physics Laboratory, Johns Hopkins University, Joint Army-Navy-NASA-Air Force, to be published
- *Engineers' Guide to Composite Materials*, American Society for Metals, 1987
- *Glossary of Plastics Terms; A Consensus*, PLASTEC Note 14, Plastics Technical Evaluation Center, Dec 1966
- *Glossary to the Science of Composites*, C.M. Bower, Ed., Monsanto Company-Washington University, for Advanced Research Projects Agency, Department of Defense, 1965-1967
- A.H. Landrock, *Adhesives Technology Handbook*, Noyes Publications, 1985
- B. LeWark, Sr., "*Composites Glossary*," Society of Manufacturing Engineers, Jan 1985
- G. Lubin, Ed., *Handbook of Composites*, Van Nostrand Reinhold, 1982
- *Military Handbook 17A, Plastics for Aerospace Vehicles, Part 1, Reinforced Plastics*, Department of Defense, Jan, 1971
- "Standard Definitions of Terms Relating to Adhesives," D 907, *Annual Book of ASTM Standards*, American Society for Testing and Materials
- "Standard Definitions of Terms Relating to High-Modulus Reinforcing Fibers and Their Composites," D 3878, *Annual Book of ASTM Standards*, American Society for Testing and Materials

Introduction to Composites

Theodore J. Reinhart, Air Force Wright Aeronautical Laboratories
Linda L. Clements, San Jose State University

A COMPOSITE MATERIAL can be defined as a macroscopic combination of two or more distinct materials, having a recognizable interface between them. However, because composites are usually used for their structural properties, the definition can be restricted to include only those materials that contain a reinforcement (such as fibers or particles) supported by a binder (matrix) material.

Thus, composites typically have a discontinuous fiber or particle phase that is stiffer and stronger than the continuous matrix phase. In order to provide reinforcement, there generally must be a substantial volume fraction ($\sim$10% or more) of the discontinuous phase. There are, however, exceptions that may still be considered composites, such as rubber-modified polymers, where the discontinuous phase is more compliant and more ductile than the polymer, resulting in improved toughness.

Composites can be divided into classes in various manners. One simple classification scheme is to separate them according to reinforcement forms—particulate-reinforced, fiber-reinforced, or laminar composites. Fiber-reinforced composites can be further divided into those containing discontinuous or continuous fibers.

A reinforcement is considered to be a "particle" if all of its dimensions are roughly equal. Thus, particulate-reinforced composites include those reinforced by spheres, rods, flakes, and many other shapes of roughly equal axes. There are also materials, usually polymers, that contain particles that extend rather than reinforce the material. These are generally referred to as "filled" systems. Because filler particles are included for the purpose of cost reduction rather than reinforcement, these composites are not generally considered to be particulate composites. Nonetheless, in some cases the filler will also reinforce the matrix material. The same may be true for particles added for nonstructural purposes such as fire resistance, control of shrinkage, and increased thermal conductivity.

Fiber-reinforced composites contain reinforcements having lengths much greater than their cross-sectional dimensions. Such a composite is considered to be a discontinuous fiber or short fiber composite if its properties vary with fiber length. On the other hand, when the length of the fiber is such that any further increase in length does not, for example, further increase the elastic modulus of the composite, the composite is considered to be continuous fiber reinforced. Most continuous fiber (or continuous filament) composites, in fact, contain fibers that are comparable in length to the overall dimensions of the composite part.

Laminar composites are those composed of two (or more) layers with two of their dimensions being much larger than their third. Complicating the definition of a composite as having both continuous and discontinuous phases is the fact that in a laminar composite, neither of these phases may be regarded as truly continuous in three dimensions.

With some few specific exceptions, only "high-performance" composites will be considered in this Volume. These are composites that have superior performance compared to conventional structural materials such as steel and aluminum alloys. Thus, the emphasis will be on continuous fiber reinforced composites, although the principles will often be applicable to other types of composites as well. Furthermore, continuous fiber reinforced composites will generally be referred to as simply fiber-reinforced composites, and, in some cases, as merely fiber composites or composites. In addition, composites with organic (resin) matrices will be emphasized throughout this Volume, both because such composites are the most commonly used and because of the significant dissimilarities between organic-matrix composites and those made with metal, ceramic, and carbon matrices.

Composite materials were developed because no single, homogeneous structural material could be found that had all of the desired attributes for a given application. Fiber-reinforced composites were developed in response to demands of the aerospace community, which is under constant pressure for materials development in order to achieve improved performance. Aluminum alloys, which provide high strength and fairly high stiffness at low weight, have provided good performance and have been the main materials used in aircraft structures over the years. However, both corrosion and fatigue in aluminum alloys have produced problems that have been very costly to remedy. World War II promoted a need for materials with improved structural properties. In response, fiber-reinforced composites were developed, and by the end of the war, fiberglass-reinforced plastics had been used successfully in filament-wound rocket motors and in various other structural applications. These materials were put into broader use in the 1950s, and initially seemed to be the only viable approach available for the elimination of corrosion and crack formation in high-performance structures. Although more recent developments in metallic materials have led to some solutions to these problems, fiber-reinforced composites still provide other substantial benefits to designers and manufacturers.

Inexpensive fiberglass composites are used today in a wide variety of applications, from consumer products, such as the fiberglass boat shown in Fig. 1, to aerospace. More advanced fiber-reinforced composites, however, have been limited in their commercial use because of high materials cost, lack of widely distributed property and processing data bases, and the absence of rapid and efficient manufacturing techniques. However, fiber-reinforced composites have been developed and widely applied in aerospace applications to satisfy requirements for enhanced performance and reduced maintenance costs. In large commercial aircraft they have found application because of the weight considerations that were highlighted by the energy crisis of the 1970s.

Fiber composites offer many superior properties. Almost all high-strength/high-stiffness materials fail because of the propagation of flaws. A fiber of such a material is inherently stronger than the bulk form because the size of a flaw is limited by the small diameter of the fiber. In addition, if equal volumes of fibrous and bulk material are compared, it is found that even if a flaw does produce failure in a fiber, it will not propagate to fail the entire assemblage of fibers, as would happen in the bulk material. Furthermore, preferred orientation may be used (as in aramid and carbon fibers) to increase the lengthwise modulus, and perhaps strength, well above isotropic values. When this material is also lightweight, there is a tremendous potential

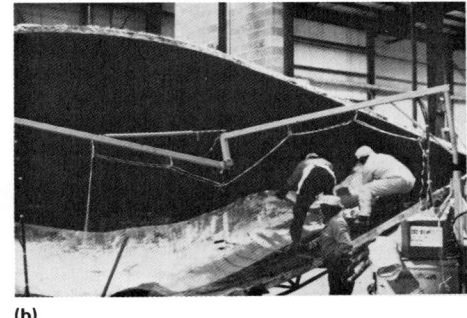

(a)

(b)

Fig. 1 (a) Eight-meter (27-ft) sailing sloop representing a fiberglass mat and roving reinforced resin matrix composite. (b) Fabrication of fiberglass hull by lay-up process. Courtesy of Pearson Yachts Corporation

advantage in strength-to-weight and/or stiffness-to-weight ratios over conventional materials. These desirable fiber properties can be converted to practical application when the fibers are embedded in a matrix that binds them together, transfers load to and between the fibers, and protects them from environments and handling. In addition, fiber-reinforced composites are ideally suited to anisotropic loading situations where weight is critical. The high strengths and moduli of these composites can be tailored to the high load direction(s), with little material wasted on needless reinforcement.

Plots of the specific tensile strength (strength/density) versus specific tensile modulus (modulus/density) for various fiber-reinforced composites are shown in Fig. 2(a)

and 2(b); the former gives fiber direction values for 65 vol% unidirectional composites, while the latter shows in-plane data for quasi-isotropic composites, in which the reinforcement is approximately isotropic in-plane. In both cases, the strengths and moduli are based on intrinsic fiber values, as taken from manufacturers' literature. These values for fibers will produce good estimates of composite modulus values, but may be off by a considerable amount—sometimes a factor of two or more—for strengths. (This discrepancy between the strength value estimated from fiber strengths and the actual value for the composite is due to factors such as processing damage, increased statistical likelihood of flaws in a larger part, and incorrect fiber orientation.) In spite of the approximate nature of the strength

data, the plots offer an excellent explanation for the wide use of fiber-reinforced composites. Compared to conventional structural materials, the improvements in specific properties can be tremendous.

Glass fiber reinforced organic matrix composites are the most familiar and widely used, and have had extensive application in industrial, consumer, military (Fig. 3), and aerospace markets. Carbon fiber reinforced resin matrix composites are by far the most commonly applied advanced (nonfiberglass) composites for a number of reasons. They offer extremely high specific properties, high quality materials that are readily available, reproducible material forms, increasingly favorable cost projections, and comparative ease of manufacture. Composites reinforced with aramid, other organics, and boron fibers, and with silicon-carbide, alumina, and other ceramic fibers are also used.

Recent technology has provided a variety of reinforcing fibers and matrices that can be combined to form composites having a wide range of very exceptional properties. In many instances the sheer number of available material combinations can make selection of materials for evaluation a difficult and almost overwhelming task. In addition, once a material is selected, the choice of an optimal fabrication process can be very complex. Simplifying these tasks is one of the purposes of this Volume.

This introduction will briefly outline the basic materials, design considerations, material forms, and fabrication processes used in the production of high-performance continuous fi-

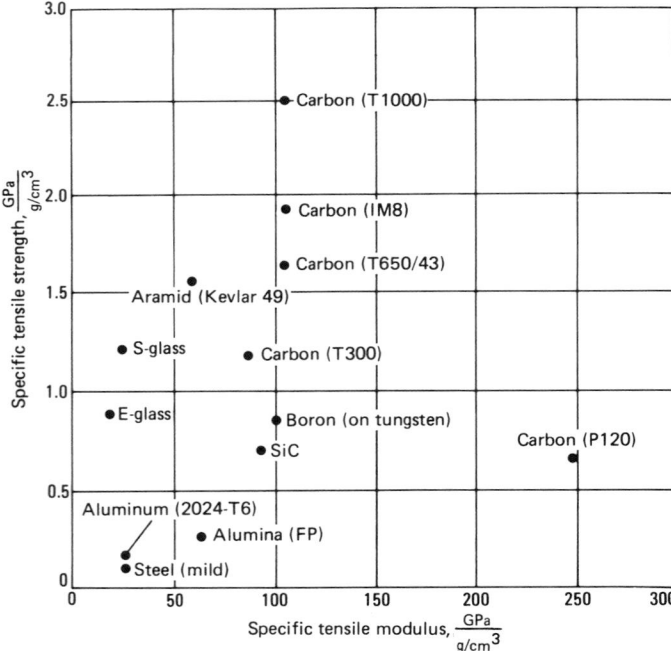

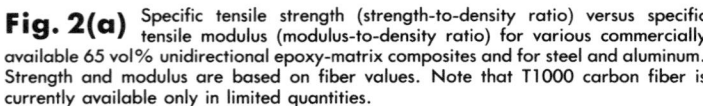

Fig. 2(a) Specific tensile strength (strength-to-density ratio) versus specific tensile modulus (modulus-to-density ratio) for various commercially available 65 vol% unidirectional epoxy-matrix composites and for steel and aluminum. Strength and modulus are based on fiber values. Note that T1000 carbon fiber is currently available only in limited quantities.

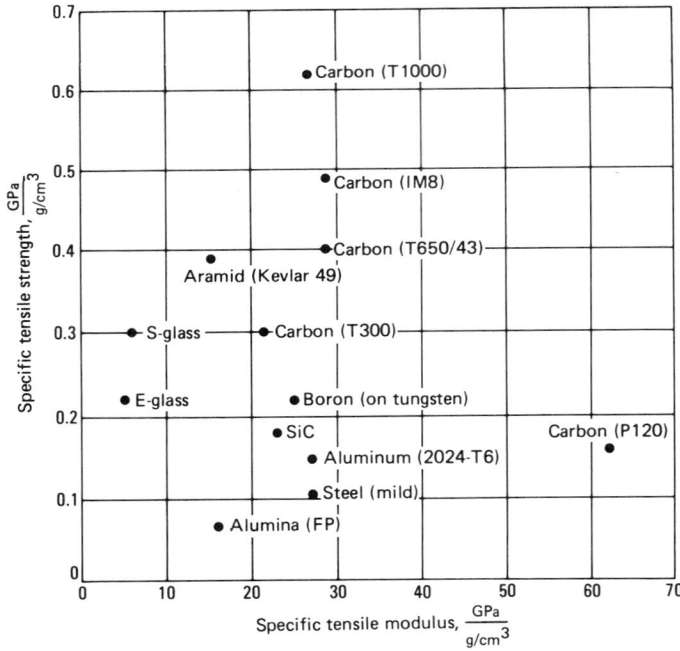

Fig. 2(b) Specific tensile strength (strength-to-density ratio) versus specific tensile modulus (modulus-to-density ratio) for various commercially available 65 vol% quasi-isotropic epoxy matrix composites and for steel and aluminum. Strength and modulus are based on fiber values. Note that T1000 carbon fiber is currently available only in limited quantities.

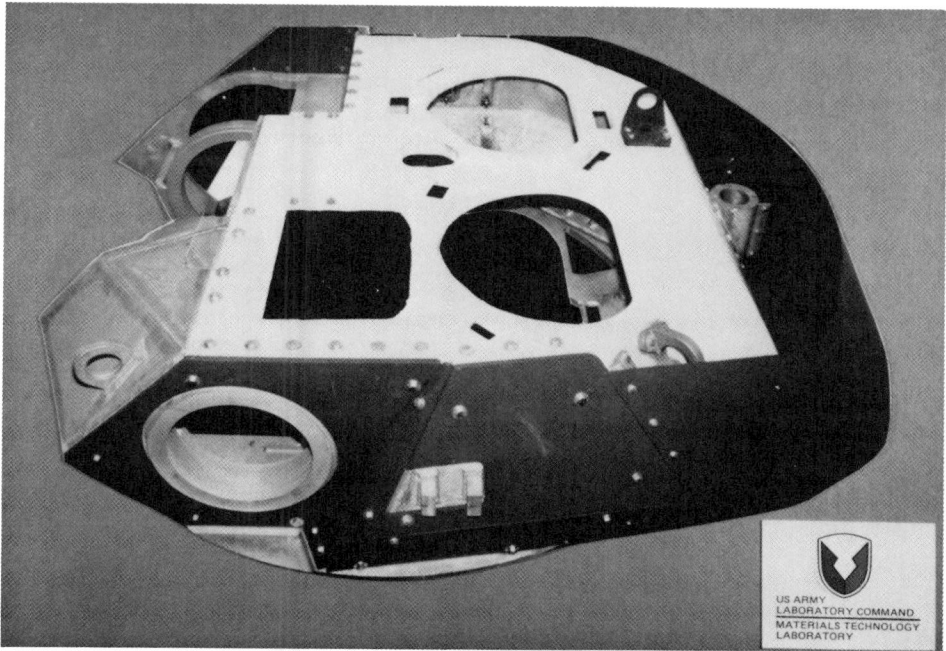

Fig. 3 Developmental-stage armored vehicle turret of S-2 glass woven roving fabric prepregged with polyester resin and vacuum-bag cured. Thickness: 45 mm (1¾ in.). Offers equal structural performance and ballistic protection, 16% weight savings over aluminum, and solution to a corrosion problem. Courtesy of U.S. Army Materials Technology Laboratory

ber reinforced organic matrix composite components.

Fibers

In a continuous fiber reinforced composite, the fibers provide virtually all of the load-carrying characteristics of the composite, the most important of which are strength and stiffness. (These fibers may also be referred to as filaments. In this case, a group of filaments is known as a fiber or fiber bundle.) The multiple fibers in a composite make it a very redundant material because the failure of even several fibers results in the redistribution of load onto other fibers rather than a catastrophic failure of the part.

Glass Fibers. Initial scientific and engineering understanding of fiber-reinforced organic matrix composites was based on studies of glass fiber reinforced composites. Both continuous and discontinuous glass fiber reinforced composites have found extensive application, ranging from nonstructural, low-performance uses such as panels in aircraft and appliances to such high-performance applications as rocket motor cases and pressure vessels. The reasons for the widespread usage of glass fibers in composites both in the past and in the present include competetive price, availability, good handleability, ease of processing, high strength, and other acceptable properties. Furthermore, the advent of highly efficient silane coupling agents, which are very compatible with either polyester or epoxy matrices, provided a strong and much needed boost in property translation

and in environmental durability. The sensitivity of the glass fiber surface to attack by moisture, however, still poses problems, as does stress-rupture failure under long-term loading at stress levels exceeding the threshold failure stresses.

The glass fiber most commonly used is known as E-glass, a calcium aluminoborosilicate glass having a useful balance of mechanical, chemical, and electrical properties, at very moderate cost. As in the manufacture of all glass fibers, E-glass is drawn from a melt of the appropriate composition. Typical strength and stiffness levels for the individual filaments are about 3450 MPa (500 ksi), tensile strength, and 72.4 GPa (10.5×10^6 psi), Young's modulus (Ref 1, 2).

Much effort has been expended to develop high-performance glass fibers, but the only kinds commercially available in the U.S. are the S-glass and S-2 glass fibers by Owens-Corning. Both S-glass and S-2 glass are magnesium aluminosilicates having a higher alumina content than E-glass. Both filament strength and stiffness are increased over those of E-glass fiber, to about 4600 MPa (670 ksi), tensile strength, and 85.5 GPa (12.4×10^6 psi), Young's modulus (Ref 2). Because of the property improvements, S-glass and S-2 glass are replacing E-glass in many applications.

Although these glass fibers have found extensive application in aircraft and in filament-wound rocket motor cases and pressure vessels, in many cases they are being supplanted by aramid or carbon fiber reinforced composites. The experience gained with high-performance

glass, however, was the starting point in the effort to develop stronger and stiffer reinforcing fibers to meet military aerospace needs.

For specialized applications, high-silica and quartz fibers are also used. High-silica fibers are made from typical fiberglass fibers by leaching out the impurities and then inducing crystallinity by heat treatment. High-silica fibers contain 95 to 99% silica, rather than the approximately 65% silica found in fiberglass. The density of high-silica glass is much less than that of fiberglass; 1.74 versus about 2.5 g/cm^3. Quartz fibers, on the other hand, are drawn from mineral quartz, which is mined in Brazil and made into precursor rods in France. Quartz fibers contain 99.95% silica but have a density of 2.25 g/cm^3, only somewhat less than that of fiberglass. Because both high-silica glass, with a tensile strength of 180 MPa (26 ksi), and quartz, at 900 MPa (130 ksi), are much weaker than fiberglass and also very expensive, they are used only when their superior temperature resistance and/or electrical signal transparency are critical. Applications include ablatives, thermal barriers, antenna windows, and radomes (Ref 3).

Carbon Fibers. Although the search for high-performance reinforcing fibers was highly successful, the limited demand outside the military aerospace industry did not permit the cost reductions that would have resulted from more extensive use. As a result, widespread industrial applications for the variety of new materials progressed very slowly in all but specialty applications where higher costs could be justified. Factors that changed this situation were the extensive use of carbon fiber reinforced composites in recreational equipment and the increased cost of energy in the early 1970s. The promise of commercial quantities of carbon fiber materials from a number of sources at attractive prices created a resurgence of interest in advanced composites in the general aerospace industry. Today, a large number of companies throughout the world are offering competitively priced carbon fibers, and many have initiated research and development programs aimed at placing advanced composite materials in widespread commercial use. Nearly 90% of the Voyager aircraft, shown in Fig. 4, was made with graphite fibers.

Currently, carbon fibers are the best known and most widely used reinforcing fibers in advanced composites. Although there are many reasons for this situation, two factors predominate. First, the manufacturing technology for carbon fibers, although complex, is more amenable to large-scale production than are those of many of the other advanced fibers. Second, carbon fibers have very useful engineering properties that, for the most part, can be readily translated into usable composite physical and mechanical properties (Ref 4).

Carbon fibers are available from a number of domestic and foreign manufacturers in a wide range of forms having an even wider range of mechanical properties. The variety of available

Fig. 4 The Voyager aircraft, of which nearly 90% is made from graphite fibers. Courtesy of Hercules Aerospace Company

carbon fiber properties, when translated into composite form, is evident from the data of Fig. 2a and 2b.

The earliest commercially available carbon fibers were produced by thermal decomposition of rayon precursor materials. The process involved highly controlled steps of heat treatment and tension to form the appropriately ordered carbon structure. Although a few carbon fibers are still made from these materials, rayon has been largely supplanted as a precursor by polyacrylonitrile (PAN). Polyacrylonitrile precursors produce much more economical fibers because the carbon yield is higher and because PAN-based fibers do not intrinsically require a final high-temperature "graphitization" step (Ref 5).

The patent literature abounds with the wide range of precursor materials that can be used to make carbon fibers. A great hope for the production of low-cost, high-performance carbon fibers was the use of either petroleum- or coal-based pitch as an inexpensive precursor. Pitch-based fibers having elastic moduli of up to 830 GPa (120 $\times$ 10^6 psi) are commercially available and are being used in several specialized applications. However, while the precursor material itself is intrinsically inexpensive, the purification and manufacturing processes are not. In addition, the strengths of pitch-based fibers tend to be relatively poor, so

they have never seriously challenged PAN-based fibers in structural aerospace applications (Ref 6).

Recent breakthroughs in PAN-based fiber technology have further strengthened the status of these fibers in the marketplace. In commercial processing, it is found that at low values of modulus, PAN-based fiber strengths increase with increases in modulus, but at higher values of modulus, the strengths reverse and begin to decrease (Ref 5). Thus, until about 1982, high-modulus PAN-based fibers tended to have low strengths. However, through refinements in the PAN precursor and more exacting control of the carbon fiber manufacturing process, higher-modulus PAN-based fibers have shown dramatic improvements in strength. Furthermore, PAN-based fibers having intermediate modulus values of about 280 to 450 GPa (40 to 65 $\times$ 10^6 psi), combined with astonishingly high strengths of up to 4140 to 5170 MPa (600 to 750 ksi), are now commercially available, and an intermediate modulus fiber with a strength of 6890 MPa (1000 ksi) is available in limited quantities. Because carbon fibers display linear stress-strain behavior to failure, the increase in strength also means an increase in the elongation to failure. The commercial fibers thus display elongations of up to 2%, which means that they exceed the strain capabilities of

conventional organic matrices. Thus, the high-strength PAN fibers have helped fuel the search for new matrices having higher strain capabilities, as will be described in the section "Matrices" in this article.

Aramid Fibers. Aramid is a generic term for a class of aromatic polyamide fibers introduced commercially during the early 1960s. Currently, commercially available versions of these fibers are produced by manufacturers in the Netherlands and Japan, and by E.I. Du Pont de Nemours & Company, Inc., in the U.S. These high-performance fibers are all variations of poly para-phenyleneterephthalamide. Before their introduction, it was not widely suspected that organic polymer fibers having such high levels of strength (up to 3450 MPa, or 500 ksi) and stiffness (132 GPa, or 19 $\times$ 10^6 psi) were possible.

Aramid fiber is unusual in that it is technically a thermoplastic polymer (like nylon), but rather than melting when heated, it decomposes before reaching its projected melting temperature. With polymerization, it forms rigid, rod-like molecules that cannot be drawn, as textile fiber molecules can, from a melt, but must instead be spun from a liquid crystalline solution in dilute sulfuric acid. The polymerization and manufacturing processes for aramid fibers are complex and exacting and involve many aggressive chemical species (Ref 7).

The high strength of aramid fiber, combined with a fiber modulus considerably higher than S-glass, gave it early application in filament-wound rocket motor cases, gas pressure vessels, and lightly loaded secondary structures on fixed-wing commercial aircraft and helicopters. The fiber shows linear tensile stress-strain behavior to failure, but unlike inorganic fibers, is surprisingly damage tolerant. However, it also displays far lower strength in compression than carbon and other inorganic fibers, a tendency to absorb water, and relatively poor adhesion to matrix resins. Low compressive strength hampers its competition with carbon fibers, and the new high-strength carbon fibers have caused it to be supplanted in many applications that it had earlier taken over from other fibers. Nevertheless, because of properties such as its high specific strength, low density, and toughness, significant markets still remain. Recent improvements include increased tensile strength (up to about 4500 MPa, or 650 ksi) and improved adhesion between the fiber and both epoxy and polyester resins. A new aramid fiber with a fiber modulus of about 186 GPa (27 $\times$ 10^6 psi) has recently been introduced by Du Pont.

Other Organic Fibers. The introduction of aramid fibers led to research into other rigid-rod fiber-forming polymers. This research has revealed several other classes of polymer fibers with properties that are highly competitive with those of carbon fibers. For example, polybenzimidazole (PBI), competes with aramid fiber because of its superior chemical and solvent resistance, but is significantly more expensive (Ref 8). For the most part, these fibers are presently available only in experimental quantities, and production may yet be years away.

Allied Chemical Corporation has introduced a high-performance polyethylene fiber. At the molecular level, it is composed of "extended chain" crystallites of the zigzag polyethylene chain. It is lighter, and offers better abrasion resistance and flexural-fatigue resistance than aramid fibers, at a similar cost. However, the fiber is limited to use below its melting point of 150 °C (300 °F). Promising applications include ballistic protection such as lightweight helmets and sailcloth (Ref 8). Only limited property data exist for the fiber, and adhesion to matrix materials is a definite problem. Only time and possible improvements will determine whether this fiber finds a place in the high-performance composites market.

Boron fibers were the first high-performance reinforcement available for use in advanced composites. Developed and first marketed in the early 1960s, these high strength, high modulus fibers found application in composite structural components on the U.S. Air Force F-15 and the U.S. Navy F-14 aircraft. Because these aircraft are still in production and the high costs of change-over are unacceptable, boron fibers are still in production and being used today, even though carbon fibers are now available with equivalent or better properties at a significantly lower price. Boron-epoxy composites have been used in the sporting goods industry and boron fibers have been used in metal matrix composites (MMC) because of their excellent mechanical properties, thermal stability, and reduced reactivity with the matrix (compared to carbon fibers). In order to prevent reaction between the boron fiber and the molten metal matrix during MMC production, boron fibers were, until recently, available with silicon carbide or boron carbide coatings that acted as diffusion barriers. They are no longer commercially available because silicon carbide fibers are now replacing boron in metal matrix composites (Ref 9).

Boron fibers are produced as a rather large monofilament fiber or "wire" (100 to 200 μm, or 4 to 8 mils) by chemical vapor deposition (CVD) of boron onto a tungsten or pyrolyzed carbon substrate. The resulting fibers have excellent strength (3450 MPa, or 500 ksi) and stiffness (400 GPa, or 58×10^6 psi). Because of their large fiber diameters, they form composites having extremely high compressive strengths. However, because both the precursor gases and the manufacturing process are inherently expensive, boron fibers cannot be expected to compete with carbon fibers on the basis of cost alone (Ref 10).

Continuous silicon carbide (SiC) fibers are being produced as large-diameter monofilaments or fine multifilament yarns. Monofilament SiC is produced by CVD in a manner similar to the method used in the production of boron fibers. Silicon carbide multifilaments are spun from a polycarbosilane precursor and then pyrolyzed to yield silicon carbide fibers (Ref 11).

Silicon carbide is also available as discontinuous fibers and as whiskers. Whiskers are produced by several methods, including a process described as the "vapor feed gases, liquid catalyst, solid crystalline whisker growth" (VLS) method. Silicon carbide whiskers are also being produced by pyrolysis of rice hulls (Ref 12).

The precursor materials for producing SiC fibers by CVD are much less expensive than those required to produce boron fibers. Thus, SiC fibers are inherently more economical than boron, although this decreased cost cannot be fully realized until the volume usage of SiC fibers increases considerably from current levels. The properties of SiC fibers are generally as good or better than those of boron. In particular, they are quite stable in molten metal during MMC production, and are stable in use to about 1200 °C (2200 °F). Silicon carbide fibers are therefore replacing boron in metal matrix composites, and, in both fiber and whisker forms, are being used to toughen ceramic matrix composites. They are also beginning to be used in organic matrices. Silicon carbide fibers are currently being used commercially for tennis rackets, as shown in Fig. 5, and are being considered for high-temperature components in automotive engines (Ref 9, 10).

Aluminum oxide fibers, known as alumina fibers, can be produced in monocrystalline and polycrystalline form as whiskers, discontinuous fibers, and continuous filaments. The fibers are produced by dry spinning from various solutions, followed by heat treatment. They are then coated with silica to increase wettability with molten metal for MMC appli-

Fig. 5 Tennis rackets consisting of graphite fibers (92%) and silicon carbide fibers (8%) in an epoxy resin matrix. Courtesy of Wilson Sporting Goods

cations. This coating also increases the fiber strength by approximately 20%, but it is dissolved during MMC fabrication, thus returning the fiber to its original strength value. Continuous single-crystalline fibers that are 50 to 200 μm (2 to 8 mil) in diameter may be produced by a melt-drawing process that yields a high-modulus but costly fiber (Ref 9).

Alumina fibers are being used or considered for use in organic matrix, ceramic matrix, and metal matrix composites. They are of particular interest in metal matrices such as aluminum because of their chemical and thermal stability in molten metal. Alumina fiber reinforced MMCs are being considered for engine components and for bearing materials. Because the fibers are not electrically conductive and resist thermal erosion and laser damage, alumina in organic matrices has potential applications in radar-transparent structures, circuit boards, antenna supports, radomes, and nose cones. Alumina fibers are also being used commercially in tennis rackets (Ref 9).

Matrices

As already mentioned, the purpose of the composite matrix is to bind the fibers together by virtue of its cohesive and adhesive characteristics, to transfer load to and between fibers, and to protect them from environments and handling. The matrix is the "weak link" in the composite, especially because resins do not presently exist that allow utilization of the stresses that the fibers are able to withstand. Thus, when the composite is under load, resins may microcrack and craze, form larger cracks through coalescence of microcracks, debond from the fiber surface, and generally break down at composite strains far lower than desired. Nonetheless, the matrix resin provides many essential functions; in addition to those mentioned above, the matrix keeps the reinforcing fibers in the proper orientation and position so that they can carry the intended loads, distributes the loads more or less evenly among the fibers, provides resistance to crack propagation and damage, and provides all of the interlaminar shear strength of the composite. Furthermore, the matrix generally determines the overall service temperature limitations of the composite, and may also control its environmental resistance.

Matrices for Commercial Applications

Polyester and vinyl ester resins are the most widely used of all matrix materials. They are utilized mainly in commercial, industrial, and transportation applications, including chemically resistant piping and reactors, truck cabs and bodies, appliances, bathtubs and showers, and automobile hoods, decks, and doors.

The very large number of resin formulations, curing agents, fillers, and other components provide a tremendous range of possible properties. The resulting resin systems include such types as general purpose, chemically resistant, and heat resistant, with general-purpose polyester and vinyl ester resin composites being the most widely used by far (Ref 12).

The development of highly effective silane coupling agents for glass fibers allowed the fabrication of glass fiber reinforced polyester and vinyl ester composites having excellent mechanical properties and acceptable environmental durability. These enhanced characteristics have been the major factors in the widespread use of these composites today.

The problems of attaining adequate adhesion to carbon and aramid fibers have discouraged the development of applications for polyester or vinyl ester composites that use these fibers. Although there are applications of high-performance fiberglass composites in military and aerospace structures, the relatively poor properties of advanced composites of polyester and vinyl ester resins when used with other fibers, combined with the comparatively large cure shrinkage of these resins, have generally restricted such composites to lower-performance applications (Ref 13).

Other Resins. When property requirements justify the additional costs, epoxies and other resins are used in commercial applications, including high-performance sporting goods (such as tennis rackets and fishing rods), piping for chemical processing plants, and printed circuit boards.

Matrices for Aerospace

Epoxy resins are presently used far more than all other matrices in advanced composite materials for structural aerospace applications. Although epoxies are sensitive to moisture in both their cured and uncured states, they are generally superior to polyesters in resisting moisture and other environmental influences, and offer better mechanical properties. Even though the elongation to failure of most cured epoxies is relatively low, for many applications epoxies provide an almost unbeatable combination of handling characteristics, processing flexibility, composite mechanical properties, and acceptable cost. Recent developments include modified epoxy resin formulations that have definitely improved elongation capabilities. In addition, a substantial data base exists for epoxy resins because both the U.S. Air Force and the U.S. Navy have been flying aircraft with epoxy-matrix structural components since 1972, and the in-service experience with these components has been very satisfactory.

Moisture absorption decreases the glass transition temperature (T_g) of an epoxy resin. To avoid subjecting the resins to temperatures equal to or higher than this so-called wet T_g, epoxy resins are presently limited to a maximum service temperature of about 120 °C (250 °F) for highly loaded applications and even lower temperatures (80 to 105 °C, or 180 to 220 °F) for toughened epoxy resins. Although this limit is conservative for some ap-

Fig. 6 A prototype aircraft access door using carbon fiber-polyphenylene sulfide woven sheeting prepreg to achieve substantial weight reduction (compared to metal construction) and cost savings (compared to thermoset composite fabrication). Courtesy of Phillips 66 Company, Advanced Composites Division

plications, its imposition has generally avoided serious thermal-performance difficulties. Considerable effort continues to be expended to develop epoxy resins that will perform satisfactorily at higher temperatures when wet. However, progress in increasing the 120 °C (250 °F) limit has been slow.

Bismaleimide resins (BMIs) possess many of the same desirable features as do epoxies, such as fair handleability, relative ease of processing, and excellent composite properties. They are superior to epoxies in maximum hot/wet use temperature, extending the safe in-service temperature to 205 to 220 °C (400 to 430 °F) or higher. They are available from a number of suppliers. Unfortunately, BMIs also tend to display the same deficiencies (or worse) as do epoxies: They have an even lower elongation to failure and are quite brittle. Damage tolerance is generally comparable to commercial aerospace epoxy resins. Work is continuing on developing these materials, and potentially improved formulations are now being evaluated by many aerospace users.

Polyimide resins are available with a maximum hot/wet in-service temperature of 260 °C (500 °F) and above. Unlike the previously mentioned resins, these cure by a condensation reaction that releases volatiles during cure. This poses a problem because the released volatiles produce voids in the resulting composite. Substantial effort has been made to reduce this problem, and there are currently several polyimide resins in which the final cure occurs by an addition reaction that does not release volatiles. These resins will produce good-quality, low void content composite parts. Unfortunately, like BMIs, polyimides are quite brittle.

Other Thermosetting Resins. The attempt to produce improved thermosetting resins is ongoing, with major efforts focusing on hot/wet performance and/or impact resistance of epoxies and BMIs. Other resins are constantly in development, and some are in commercial use for specialized applications. Phenolic resins, for example, have been used for years in applications requiring very high heat resistance (Ref 14). These resins also have good dielectric properties, combined with dimensional and thermal stability. Unfortunately, they also cure by a condensation reaction, giving off water as a by-product and producing a voidy laminate. However, they also produce low smoke and less toxic by-products upon combustion, and are therefore often used in such applications as aircraft interior panels where combustion requirements justify the lower properties.

Thermoplastic Resins. The dual goal of improving both hot/wet properties and impact resistance of composite matrices have led to the development, and limited use, of new high-temperature thermoplastic resin matrices. These materials are very different from the commodity thermoplastics (such as polyethylene, polyvinyl chloride, and polystyrene) that are commonly used as plastic bags, plastic piping, and plastic tableware. The commodity thermoplastics exhibit very little resistance to elevated temperatures; the new thermoplastics exhibit resistance that can be superior to that of epoxy.

Thermoplastic matrix materials are tougher and offer the potential of improved hot/wet resistance. Because of their high strains to failure, they also are the only matrices currently available that allow, at least theoretically, the

Table 1 Typical hot/wet in-service temperatures of selected organic matrix materials

Material	Temperature
Epoxy	120 °C (250 °F)
Bismaleimide	220 °C (430 °F)
Polyimide	260 °C (500 °F)
PEEK	$\geqq$120 °C (250 °F)
Polyamideimide	190 °C (375 °F)

new intermediate modulus, high-strength (and strain) carbon fibers to use their full strain potential in the composite. These materials include such resins as polyether etherketone (PEEK), polyphenylene sulfide (PPS), polyetherimide (PEI) (all of which are intended to maintain thermoplastic character in the final composite) and others, such as polyamideimide (PAI), which is originally molded as a thermoplastic but is then postcured in the final composite to produce partial thermosetting characteristics (and thus improved subsequent temperature resistance). The prototype aircraft access door shown in Fig. 6 uses PPS resin in woven sheeting.

Thermoplastic matrices do not absorb any significant amount of water, but organic solvent resistance is an area of concern for the noncrystalline thermoplastics. Because thermoplastic matrices have an unlimited shelf life before molding (unlike thermosetting matrices), and because they can potentially be remolded by the application of heat and pressure, thermoplastic matrix composites also offer the possibility of lower-cost fabrication. However, cross-linking or thermal degradation with repeated temperature cycling is a concern with most of the systems. The fabrication procedures necessary for the low-cost manufacture of thermoplastic matrix composites are being extensively studied. There is also a major ongoing effort to determine and understand the mechanical properties obtained from the various techniques currently available for fabricating thermoplastic matrix composites.

Table 1 compares the maximum hot/wet in-service temperatures of selected resins.

Designing With Composites

Because the Sections ''Composite Materials Analysis and Design'' and ''Composite Structures Analysis and Design'' in this Volume contain detailed information on design considerations for composites, only a few major factors will be discussed here. Obviously, any design must carefully match the requisite properties with the candidate composite material, along with cost considerations. Once the optimum, or best available, material is chosen, the designer must be concerned with any additional limitations that material selection might impose on the capabilities of the design. As already discussed, common areas of concern include hot/wet properties, in-service temperature, and

impact resistance. General limitations that must be considered in composite design include the relatively low strength and stiffness in the out-of-plane direction and often poor shear properties. Numerous efforts are being made, with some success, to improve these properties. However, if not corrected and/or accounted for in the design, these factors can cause difficulties, such as delamination under compressive loading or inadequate out-of-plane load-carrying capabilities.

Material Forms

Continuous reinforcing fibers are available in many product forms, ranging from monofilaments (for fibers such as boron and SiC) to multifilament fiber bundles, and from unidirectional ribbons to single-layer fabrics and multilayer fabric mats. If organic matrices are available in the neat (unreinforced) form, they are generally mixed from the individual components if the matrix is a thermoset, or are available as sheet, powders, or pellets, if the matrix is a thermoplastic. The reinforcing fibers and matrix resins may also be combined into many different nonfinal material or product forms that are designed for subsequent use with specific fabrication processes. In the case of continuous fibers, these combinations of unidirectional fiber ribbons or woven fabrics with resin are called prepregs.

Using prepregs rather than in-line impregnation of the fibers during the final composite fabrication process can offer significant advantages. Prepregs can have very precisely controlled fiber/resin ratios, highly controlled tack and drape (in the case of thermoset matrices), controlled resin flow during the cure process, and, in some processes, better control of fiber angle and placement. Prepreg materials can be produced and stored, normally under refrigeration for thermosetting matrices, and then used in processes ranging from hand lay-up to highly automated filament winding or tape laying. Processes such as pultrusion and braiding can also use prepreg forms instead of in-line resin impregnation. While the latter may be lower in initial cost, it may be prohibitive for some resin systems (such as thermoplastics), and parameters such as fiber/resin ratio may not be as easily controlled as is the case with a prepreg.

Discontinuous fiber reinforced product forms include sheet molding compounds (SMC), bulk molding compounds (BMC), injection molding compounds, and dry preforms fabricated for use in resin transfer molding (RTM) processes. Many other forms of reinforcement exist, primarily in fiberglass materials. Both continuous and discontinuous mats, with and without binder materials, are available. Of course, composites reinforced primarily with discontinuous fibers have lower mechanical properties than those with continuous fibers. This is because all of the loads in discontinuous fiber composites must be carried

by the matrix in shear from fiber length to fiber length (shear lag). In addition, fiber volume in discontinuous fiber composites is normally quite a bit lower than is typical in continuous fiber composites.

Current Composite Fabrication

A host of processes exist for the fabrication of composite components. Fiber-reinforced composites used in most high-performance applications are laminated with unidirectional (or fabric) layers at discrete angles to one another (such as in plywood), thereby distributing the in-plane load in several directions (Ref 15). This also facilitates the use of mechanical fasteners where necessary, as well as the repair of service-induced structural damage. A variety of fiber placement processes are used to achieve the desired combination of orientations. The two most common are hand or machine lay-up and filament winding. In lay-up, material that is usually in prepreg form is cut and laid up, layer by layer, to produce a laminate of the desired thickness, number of plies, and ply orientations. In filament winding, a fiber bundle or ribbon is impregnated with resin and wound upon a mandrel to produce a simple shape, such as a tube or pressure vessel, or a more complex shape, such as a wind turbine or helicopter rotor blade. As mentioned before, filament winding may use wet (or melted) resin or prepreg (Ref 15).

The fiber placement process is followed by some type of cure or molding process. This may simply involve holding at room temperature until cure completion, for a low cure temperature thermoset matrix. However, for applications involving elevated-temperature service, or for thermoplastics, there must be an elevated temperature cure or molding process. Filament-wound parts may be cured at elevated temperature in an open oven; in some cases, consolidation and surface finish may be improved by applying an external female mold or vacuum bag. Lay-ups are most commonly consolidated by applying both heat and pressure in an autoclave, but they may also be molded, pressed, or vacuum-bag cured. There are also special fabrication processes, such as pultrusion, that combine fiber placement, consolidation, and elevated-temperature cure in one continuous operation (Ref 14).

Composite fiber placement fabrication procedures can be labor intensive, so all major composite airframe manufacturers are developing and/or using automatic fabrication equipment. Such equipment is often used for composite components that have a relatively large area and reasonable production rate. Two methods predominate. One involves laying up the plies with tape that is 76 mm or 152 mm (3 or 6 in.) wide. Large tape-laying machines are computer controlled, include gantry robot system, and are equipped with a specially designed tape-dispensing head. Another method involves the cutout of entire plies from unidirectional broad goods (up to 1.5 m, or 60 in. wide) using laser, water-jet, or reciprocating-knife cutters. Cutout ply patterns are transferred to the tool and laid up by automatic equipment with specially designed pick-up and lay-down heads. After any intermediate debulking or consolidation steps, the lay-up is cured under heat and pressure in a press, vacuum bag/oven system, self-contained heated tool, or in an autoclave.

Very sophisticated, numerically controlled equipment is presently available for the lay-up process. High prepreg lay-down rates and fairly low scrap rates are claimed. However, most of the existing automatic lay-up processes are quite limited in their ability to produce satisfactory components having complex curvatures.

To select the best composite fabrication process, the designer generally chooses the process that will provide an acceptable-quality component for the lowest cost. In evaluating cost and quality, however, tooling cost, production rate, materials cost, desired part finish, and many other factors must be considered. Only after all the pertinent factors have been weighed can the fabrication method (or the material) be selected.

Composite Fabrication— the Future

Although aircraft manufacture is not the only area in which composite materials will produce significant changes in the future, it does illustrate the overall industrial future of composites. Because they provide structural efficiency at lower weights than equivalent metallic structures, advanced composites are rapidly emerging as the primary material for use in near-term and next-generation aircraft structures. These viable engineering materials are expected to revolutionize the entire airframe industry within the next 5 to 10 years. At the present time, the technology of advanced composites has already advanced to the point where all newly emerging military aircraft systems have a number of composite components in production. Based on trends to date, next-generation aircraft could have up to 65% of its structural weight made of advanced composites. However, except for advanced military aircraft, present production applications are usually limited to empennage and secondary structural assemblies, and are justified on the basis of cost savings through overall weight reduction. For next-generation aircraft, however, high-volume production should make use of composites that are more directly competitive with metals because a composite part will be the same price or lower than an equivalent metallic structure. Thus, to expand the existing production base and prepare for this projected high-volume, cost-competitive production, airframe manufacturers must reevaluate and significantly upgrade their facilities and technology if they are to remain competitive.

To meet these projected production requirements on a cost-competitive basis, airframe manufacturers have begun analyzing needs and installing new equipment required for an orderly transition to substantial use of this new material form. The facility requirements for producing advanced composites differ markedly from those of yesterday's aircraft plant, which primarily produced metallic structures. With the anticipated tremendous increase in the use of composites, it is evident that the aircraft production facility of the future will be quite different from those built in the past. Not only will processes and equipment be very different, but labor needs and manpower skills will also change significantly. Because the move to next-generation advanced composite aircraft will deal with more and larger structures, the manufacturing methods and assembly flow being used today to fabricate small composite assemblies will also have to be changed. As major portions of aircraft are produced from advanced composite materials, the hand labor or semiautomated approaches currently used will become costly and less and less efficient. In fact, in most cases, these approaches will become virtually unusable because of size limitations, required production rates, and cost.

Most experts feel that a step-by-step approach to the development of a completely integrated composite manufacturing plant is necessary to make such a system practical and economically viable in the near future. Because of the major costs involved, many organizations have categorized various functions of the factory for assessment purposes to determine what each requires for further development. From extensive cost studies, it is evident that the areas requiring cost-reducing automation at the earliest possible time are the labor-intensive "clean room" composite fabrication processes. These processes, which normally use hand-labor operations, include dispensing or lay-out of prepreg, cutting of the material, transferring cut material to individual work stations, laying up the actual structural assembly (pickup, transfer, and indexing the plies into the tool), removing prepreg scrap, and inspecting each operation. Not only are these steps expensive, as they are currently implemented, but they also can be low-quality operations. Even though automatic operations are used today for certain large-scale fabrication processes, considerable improvements and advances in existing operations are required. This area will be one of the major challenges, and major opportunities, for the manufacturers of the next-generation aircraft.

This introductory article was intended as a primer to introduce the reader to composite technology. Both current and future fiber and matrix materials, design concerns, material forms, and fabrication processes have been described briefly. For more specific and detailed data and guidelines on these topics, the reader should consult the technical sections in this Volume. However, this introduction should have clarified that composites offer tremendous opportunities—as well as impressive challenges—for the designer and manufacturer of high-performance components.

REFERENCES

1. I. Ahmad and B.R. Noton, Ed., *Advanced Fibers and Composites for Elevated Temperatures*, American Institute of Mining, Metallurgical, and Petroleum Engineers, 1980
2. Charles E. Knox, Fiberglass Reinforcement, in *Handbook of Composite Materials*, G. Lubin, Ed., Van Nostrand Reinhold, 1982, p 136-159
3. H. Shulock and R.R. Saffadi, High Silica and Quartz, in *Handbook of Composite Materials*, G. Lubin, Ed., Van Nostrand Reinhold, 1982, p 160-170
4. *Summary of the Eighth Refractory Composites Working Group Meeting, Vol III*, D.R. James and L.N. Hjelm, Ed., 1964
5. D.M. Riggs, R.J. Shuford, and R.W. Lewis, Graphite Fibers and Composites, in *Handbook of Composite Materials*, G. Lubin, Ed., Van Nostrand Reinhold, 1982, p 196-271
6. L.J. Broutman and R.H. Krock, in *Composite Materials, Vol 4, Metal Matrix Composites*, K.G. Kreider, Ed., Academic Press, 1974
7. C.C. Chiao and T.T. Chiao, Aramid Fibers and Composites, in *Handbook of Composite Materials*, G. Lubin, Ed., Van Nostrand Reinhold, 1982, p 272-317
8. M.S. Reisch, *Chem. Eng. News*, Vol 65, (No. 5), 1987, p 9-14
9. S. Swindlehurst, "Inorganic Fibers: A Brief Overview," to be published, 1987
10. H.E. DeBolt, Boron and Other High-Strength, High-Modulus, Low-Density Filamentary Reinforcing Agents, in *Handbook of Composite Materials*, G. Lubin, Ed., Van Nostrand Reinhold, 1982, p 171-195
11. J.W. Weeton, Design Guide: Fiber-Metal Matrix Composites, *Mach. Des.*, Feb 1969, p 141-156
12. I.H. Updegraff, Unsaturated Polyester Resins, in *Handbook of Composite Materials*, G. Lubin, Ed., Van Nostrand Reinhold, 1982, p 17-37
13. J.W. Weeton and E. Scala, Ed., *Composites: State of the Art*, American Institute of Mining, Metallurgical, and Petroleum Engineers, 1974
14. M.M. Schwartz, *Composite Materials Handbook*, McGraw-Hill, 1984
15. M.M. Schwartz, Ed., *Fabrication of Composite Materials: Source Book*, American Society for Metals, 1985

General Use Considerations

Bryan R. Noton, Battelle Memorial Institute

WHEN COMPARING COMPOSITE materials to metals in the materials selection stage of the design process, it is essential to compare the benefits of the composite technologies with those of the most advanced or emerging metal alloys, manufacturing methods, and design configurations. For example, composite structures designed to operate at moderate temperatures may have to compete with structures of aluminum-lithium alloys that may be manufactured using existing facilities and investments. Similarly, for higher-temperature applications, such as structures adjacent to jet engines, it may be necessary to conduct trade-off comparisons with titanium parts manufactured by superplastic forming. The latter process also provides benefits such as parts integration and flexibility of geometry selection, frequently cited as advantages of fibrous composite materials. When designing molded composite parts for high-performance applications, it is important to conduct trade-off studies among the available and emerging net shape manufacturing technologies. These technologies also offer efficient utilization of material by keeping the required machining to a minimum.

Comparison Between Composites and Metals

Although many factors, including material form, will significantly influence any design and manufacturing guidelines, some general differences between metals and composites may make the latter appear to be the more attractive choice. Literature and data from each fiber manufacturer compare specific fibers, material forms, resins, specimen types, and test methods with other fibers and metals. With this information, the potential user of composites can not only make reliable comparisons but can also understand the underlying assumptions on which they are based. Differences between composites and metals are as follows:

- Unidirectional aramid and carbon fiber reinforced epoxies provide a specific tensile strength (ratio of material strength to density) that is approximately four to six times greater than that of steel or aluminum
- Unidirectional carbon fiber reinforced epoxies provide a specific modulus (ratio of material stiffness to density) that is approximately 3½ to 5 times greater than that of steel or aluminum. Aramid falls between carbon and glass fiber reinforced epoxies
- Comparing efficiently designed structural elements, the fatigue endurance limit for aramid and carbon fiber reinforced epoxies may approach 60% of the ultimate tensile strength. For aluminum and steel, this value is considerably lower
- Because of the properties listed above, aramid, carbon, and hybrid fiber reinforced plastics can provide structures that are 25 to 45% lighter than aluminum structures designed to the same functional requirements
- Impact energy values for aramid-epoxy composites are significantly higher than those for carbon fibers and aerospace aluminum alloys
- Because fiber-reinforced plastics can be designed with excellent structural damping features, they are less noisy (important for automotive applications) and provide lower vibration transmission than metals
- When used for bearings, carbon fiber reinforced nylon (not requiring lubrication) provides friction coefficient and wear characteristics comparable to those of lubricated steel
- Fibrous composites are more versatile than metals and can be tailored to meet performance needs and complex design requirements, such as aeroelastic loading on wings and the vertical and horizontal stabilizers of aircraft. As evidenced by "Voyager," design requirements such as very high aspect ratio lifting surfaces sometimes cannot be satisfied by metal alloys within the critical weight and fuel conservation limitations
- The properties mentioned above can be balanced with cost by hybridization (mixing different fibers in a given composite to attain an optimum combination of properties)
- Corrosion and other attributes of fibrous composites will contribute to reduced life-cycle cost
- Composite parts can eliminate joints/fasteners, providing part simplification and integrated design

Perceptions and Problems

Although current applications of high-performance fibrous composites in many products and complex engineering systems are impressive, some misconceptions and problems may retard the acceptance of these materials. Among these factors can be included:

- Material cost is a growth inhibitor for some high-volume production applications. (Even a low cost $7/kg ($3/lb) for some composites is considered unacceptable in certain applications)
- Misconceptions about the lack of durability of plastic materials (for example the "plastic toy syndrome")
- Combustion and smoke liberation characteristics can inhibit acceptance
- Expertise is required for successful application of composites, and all industrial disciplines must be combined to achieve a cost-effective application
- Metals are really "moving targets" because new developments are occurring in many branches of metallurgy and manufacturing. Aluminum-lithium alloys, net shape technologies, and superplastic formed titanium are but a few of these emerging and important technologies. Therefore, trade-offs between the latest technologies for both metals and composites must be made carefully
- Supplier involvement in engineering and development is frequently required and may pose problems concerning protection of proprietary information on the product or system under development
- Manufacturing processes for composites are complex, must take place in special environments, and can be costly to control
- Unique assembly processes are required, for example, high-temperature ovens and autoclaves, and integrally heated tools. Special coating systems may be required for some applications
- Secondary processing methods are lacking; primary processes were developed first because of the overwhelming potential of properties promised by fibrous composites
- Porosity is sometimes a problem in applications such as deep-sea submersibles and

automobile body panels because of blow-through of the paint

- Special precautions are often necessary to prevent occurrences that would cause part rejection, such as composite dust being introduced during priming for adhesive application
- Managers who have made large equipment investments in metals may be reluctant to acquire new and expensive equipment for emerging composite technologies
- Some worker dermatitis reactions can be expected. Occupational Safety and Health Administration/Environmental Protection Agency regulations related to chemical constituents are management concerns

Advantages of Composites

The high performance and unique production applications of the numerous structural fibrous composite materials and forms now available to designers are proof of their many advantages. In some cases, however, advantages and benefits are not attainable with a single fiber type; rather, a combination of several fibers, or hybrids, may be necessary to achieve the required design properties, and these must be simultaneously balanced against acquisition, operations, and maintenance costs. The designer can make efficient and innovative improvements only by applying rigorous theoretical tools, ensuring strong interaction between the design and manufacturing processes, and integrating all disciplines. Among the benefits of high-performance fibrous composites are these:

- Because of their multifunctional aspect, composites are able to meet diverse design requirements
- Weight savings are significant, frequently ranging from 25 to 50% of the weight of conventional metallic designs
- The high torsional stiffness requirements of various vehicles, particularly high-speed aircraft, can be satisfied
- Corrosion resistance is outstanding
- Fatigue and fracture attributes are numerous
- Impact and damage tolerance characteristics are excellent
- Improved dent resistance is normally achieved (composite panels do not damage as easily as thin-gage sheet metals)
- Flexibility in selection and changing of styling and product aesthetic considerations is an important feature. Frequent styling modifications in response to changing customer needs can be made with limited investment
- Thermoplastics have rapid process cycles, making them attractive for high-volume commercial applications that traditionally have been the domain of sheet metals, and furthermore, thermoplastics can be reformed

- Like metals, thermoplastics have indefinite shelf lives
- Low thermal expansion can be achieved, but will vary significantly with the matrix material selected, the fiber types used, and their orientation (the coefficient of linear expansion for sheet molding compound can be close to that of steel, simplifying design and assembly in products requiring many material types
- Manufacturing and assembly are simplified because of part integration (joint/fastener reduction), which can reduce engineering, purchasing, and follow-up costs
- Inside/outside assembly is possible, which can reduce tooling costs
- Composites tooling is frequently two to five times cheaper than for metals, which significantly reduces amortization costs
- Lower freight costs are common

Advantages of composites for specific applications cover a wide range.

The higher strength and stiffness properties of fiber-reinforced polymers allows rotational components, such as in mechanical energy storage devices, to function significantly faster than those made of steel alloys; thus more energy per unit weight is stored. Composite components also have benign, or noncatastrophic, failure modes.

The improved torsional stiffness for drive shafts that is provided by a high-modulus fiber-reinforced composite implies high whirling speeds and a reduced number of intermediate bearings and supporting structural elements. The overall part count and manufacturing and assembly costs are thereby reduced.

The impact properties, including fracture and residual strengths, that can be achieved with aramid fibers are excellent. Furthermore, aramid fibers can be combined with carbon and glass fibers to optimize energy absorption and other design requirements, including cost.

Reduced energy dissipation and improved strength and stiffness properties can be achieved, which, for example, allow carbon fiber reinforced plastic tennis rackets to produce higher ball speed.

Corrosion Resistance. To achieve the properties and characteristics required to meet certain design objectives, such as resistance to a corrosive environment, metals such as carbon steel frequently require a polymeric liner with expensive pretreatment and bonding processes. With composites, this problem can be overcome by means of multifunctional designs.

The improved weatherability of composites in a marine environment, as well as their corrosion resistance and durability, reduce the downtime for maintenance to a fraction of that for wood and metal structures. For example, the income lost for fishing trawlers is reduced significantly. However, ultraviolet effects make a finish or paint on the composite necessary.

The dimensional stability of composite structural components is important for applications such as in x-ray equipment, machine tool elements, micrometers, calipers, and robot arms (improving positioning accuracy).

Excellent damping features can be designed into composites. Acoustical and mechanical vibrations can decay in 10% less time than in conventional metals.

The improved friction and wear properties of carbon fiber reinforced polymers are outstanding. Bearings represent one important application. The coefficient of friction of a carbon composite on steel is approximately 40% of that of lubricated steel on steel.

Heat sink properties of carbon-carbon composites designed for brakes, combined with dimensional stability, light weight, and friction characteristics, have resulted in their acceptance for high-performance aircraft and racing cars.

Thermal Expansion. Composite materials can be tailored to comply with a broad range of thermal expansion design requirements and to minimize thermal stresses. Aramid composites have a unique coefficient of thermal expansion (negative) that must be recognized and accommodated.

Improved insulation is provided by composites because they require less space and fewer parts, and they provide significantly better heat loss control than metals. Furthermore, composites minimize potential corrosion problems precipitated by some refrigerated liquids.

Electrical conductivity can be designed into composite laminates. For example, carbon fibers can be nickel coated.

The electronic transparency of glass fiber reinforced polymers, combined with their light weight and ability to form complex compound curvatures, has prompted their extensive use in airborne warning and control radar system housings, or radomes.

Electrical Waveguides. Because of their ability to minimize losses and achieve high-phase stability, carbon fiber reinforced plastics are used in electrical conductors for microwave frequencies in satellite missions. Metals are heavier and are subject to contour distortions precipitated by thermal gradients.

Parts consolidation frequently results in smaller mechanical and electrical devices, which is a design advantage. Furthermore, with some thin-gage contoured structures, no additional stiffeners may be required.

Close tolerances, along with contour accuracy and a tailored surface finish, can be achieved without machining, The cost and design problems associated with tolerances are reduced by parts consolidation, which minimizes the number of part interfaces.

Composite prototype parts and assemblies can be produced quickly and more economically than with metals. Furthermore, changes can be made easier than with sheet metals or other forms of metal.

Material waste is reduced because composite parts and structures are frequently built up to shape rather than machined to the required configuration, as is common with metals that are machined from blocker forgings or trimmed from sheet metals. The utilization rate (purchased material versus material in final product) for composites is therefore much higher, except in net shape manufacturing technologies, such as precision forgings and powder metallurgy.

Design cascading effects, or impact of the properties and characteristics of composites on an overall design, frequently become more important than the advantages of the individual component under study. For example, connecting rods may permit elimination of a balance shaft, transmission shafts may require fewer bearings, and smaller leaf-springs may enable a larger automobile trunk or a lower hood silhouette.

Portability. Because composite structures are lightweight, many products can be carried by hand or divided into segments that are easily transported.

Selection and Evaluation of Composites

Richard C. Laramee, Morton Thiokol, Inc.

SELECTING A COMPOSITE MATERIAL to use for a component of an assembly begins by matching customer requirements, manufacturing capabilities, and raw material characteristics of the suppliers' products.

Customer requirements for a product include general size, shape, weight, finish, cost, volume, environment, loads, performance, fit, function, quality, and maintainability levels. Manufacturing capabilities focus on fabrication space, facilities, equipment, and labor force; engineering design, analysis, test, and processing experience; quality inspection; material specifications and management make or buy decisions, and procurement policies. Considerations for suppliers' raw material or semifinished products include current technology for properties, dimensions, and performance; manufacturing lead time and cost; quality level; and general manufacturing processes. Each of these areas has a large tolerance band and can impose limitations on the general manufacturing operations.

The design engineer must meld the customer requirements with his own manufacturing capabilities and suppliers' products into a material system design that provides component performance, quality, maintainability, cost, and weight in a competitive time period. Selection of a material for composite fabrication should take into consideration the many resins, fillers, and fabrics that can be combined for optimization. The design process also requires that the material meet dimensional and property requirements, consistently satisfy quality requirements for acceptable parts, and operate within the mean time between failure (MTBF) requirement. The anisotropic properties of composite materials require an unusual amount of coordination between designer and fabricator to ensure that requirements are satisfied in a finished component.

In the material selection process, the single most important consideration is the composite material requirements (physical, mechanical, thermal, and electrical). These requirements may automatically eliminate some materials and point the way to the material family best suited to the design.

Table 1 depicts the evaluation of Material A in terms of its properties, of which four (tensile strength, density, coefficient of thermal expansion, and dielectric constant) have minimum required values that must be attainable.

Each selection parameter has a rating on a scale of 1 to 10. A perfect material for the application of interest would have a total summary score of 100 (10 parameters times rating of 10). Low numbers reflect low rating of the material.

As Table 1 shows, Material A cannot meet the required minimum property values for tensile strength and service temperature, which means automatic rejection for this product application.

An additional consideration is the priority attached to the selection parameters. Ideally, all parameters are equally important, but often a few factors are of high-priority consideration. Low numbers reflect low priority for the material characteristics. Thus, if Material A were subjected to the total ranking (rating times priority), the value of 48.7% of the ideal material would also eliminate Material A. Other materials more suited for the product design would have to be evaluated. As many as 5 to 30 material families may be considered in Table 1 for the design application, with the top candidates (from 3 to 20) subjected to a final selection matrix (Table 2).

The matrix shown in Table 2 summarizes the decision-making concepts behind material selection. After the component of the assembly and its design, manufacturing, and performance requirements are identified, the selection pa-

Table 1 Material properties selection matrix guidelines

Selection parameters (Material A)	Evaluation rating (1-10)	×	Priority factor (1-10)	=	Total ranking
Tensile strength at room temperature(a)	0(b)		10		0
Compressive strength at temperature	2(c)		10		20
Elastic, tensile, and compressive moduli	2(c)		10		20
Density(a)	10(c)		10		100
Thermal conductivity(a)	5		5		25
Coefficient of thermal expansion(a)	10		5		50
Specific heat	10		5		50
Service temperature(a)	0(b)		10		0
Volume resistivity	5		5		25
Dielectric constant at temperature(a)	10(c)		10		100
Scoring	Material A / Ideal material 100 max		8.0 average max		390 (48.7% of ideal material)(d) / 800 max ranking

(a) Minimum required value of property must be available to be acceptable. (b) Unacceptable. (c) Acceptable. (d) Material unacceptable because tensile strength and service temperature do not meet the minimum required property values.

Table 2 Material selection matrix guidelines

Selection parameters (Material B)	Evaluation rating (1-10)	×	Priority factor (1-10)	=	Total ranking
Good visual appearance	9		6		54
Compatibility in material assembly	5		7		35
Ease of fabrication/good production rate	9		8		72
Low weight/compact shape	10		10		100
Low cost	10		10		100
Good quality	10		8		80
Good maintainability/excellent performance	5		8		40
Good properties at temperature (mechanical, thermal, physical, and electrical)	6		8		48
Availability of process facilities	0		6		0
Material availability	8		7		56
Scoring	Material B / Ideal material 100 max		7.8 average max		585 (75% of ideal material)(a) / 780 max ranking

(a) Other materials need to be considered to meet component requirements.

rameters for Material B application can be determined. The application scenario requires that facilities be in place for the material fabrication process, and the component material be available to meet production quantities and schedules. The component must meet the minimum required tensile strength at the service temperature, be pleasing in appearance and functional after the final finishing operation, and have properties and dimensional tolerances that are compatible with the other interfacing assembly components.

Manufacturing considerations include ease of fabricating the part in volume after the production line is in place and inspected, at a production rate geared to the product sales program plan. Also, parts fabricated from the material should not hinder manufacturing processes and should meet quality assurance inspections of its visual characteristics and tag-end mechanical properties, and radiographic or ultrasonic evaluations. Dimensional tolerances should be acceptable to the specification and drawing requirements, and the reject rate should be very low.

Design and management objectives for the material include low weight, compact shape, and low net cost, including the tooling and facilities set-up charges. Achieving good maintainability with long MTBF and excellent performance in the customer's application represent management objectives toward keeping a successful and long-term production program.

As in Table 1, each selection parameter in Table 2 has a rating judgment from 1 to 10. A perfect material for the application would have a total summary score of 100 (10 parameters times 10 rating). Material B scores well in meeting customer requirements for visual appearance (9), easy fabrication (9), low weight (10), low cost (10), and good quality considerations (10), but fails to score well in compatibility in the assembly operation (5), maintainability (5), material properties (6), availability of process facilities (0), and material availability (8).

Each of the selection parameters is also rated in terms of priority. The material component requires good properties (8) and quality (8), long service life (8), reasonable lead times for raw materials (7), acceptable availability of process facilities (6), and acceptable final appearance (6). Some leeway is allowed on compatibility of the material to the other components of the assembly (7), while requiring a good learning curve for ease of fabrication and

good production rate (8). The priority factors are low weight, compact shape, and low net component cost (10).

The ideal material evaluation would have a maximum rating of 100 multiplied by an average weighting factor of 7.8, for a best total ranking of 780 points. In terms of total ranking, Material B lacks good properties at temperature (item 8), available facilities (item 9), good compatibility with interfacing parts (item 2), and good maintainability/performance (item 7). Its total ranking is 75% of an ideal material. The low total ranking, lack of good properties, and lack of process facilities justify elimination of Material B as a candidate. The conclusion is that other material systems must be considered.

In most selection situations, about ten materials are evaluated simultaneously. The best material is the one that has the highest number of points and meets all minimum requirements for the ten factors in the matrix. However, a good goal for final selection is to achieve 90% of the ideal material evaluation rating (100 points) and total ranking (780 points) and show no 0 rating anywhere in the matrix.

Definitions of the selection parameters identified in Table 2 are:

- *Good visual appearance*: The raw material, when processed through the primary and secondary forming processes and the finishing and joining process, must exhibit a surface that is pleasing to the eye and hand and projects a functional image of durability and toughness after final product assembly.
- *Compatibility in material assembly*: The finished material component, in the final product assembly, must interface with the other components in terms of dimensional tolerances, surface toughness, joining, long-term wear, and load-carrying ability, to satisfy the function of the product.
- *Ease of fabrication/good production rate*: The raw material, when processed through the primary and secondary forming processes (filament winding, curing, and post curing) must produce consistently good parts (with a low number of rejects) in a timely manner, without expensive equipment and facility modifications. The fabrication should not be labor intensive. In addition, the raw material must be easy on the tooling equipment and facilities, thus extending their operational life.
- *Low weight/compact shape*: The raw material must be capable of being processed into

a low total volume (cubic inches) component performing several functions or replacing several parts and exhibiting a low density.
- *Low cost*: The total cost of raw material, and forming, finishing, and joining processes, should be as low as possible for the final material component.
- *Good quality*: The finished material component must be free of surface and subsurface defects (cracks, delaminations, inclusions, and voids), must exhibit the required test properties from tag ends parted off the component, and must meet the drawing specification dimensions and tolerances.
- *Good maintainability/excellent performance*: The finished material component in the product assembly must have a good track record of functioning in a similar design application or testing environment, with a long service life between failures, and must provide operating characteristics that meet or exceed the customer requirements.
- *Good properties (mechanical, thermal, physical, and electrical) at temperature*: In the operating environment of the product, the finished material component must meet all the minimum requirements for such characteristics as strength, coefficient of thermal expansion, density, and volume resistivity.
- *Availability of process facilities*: All processing facilities for the raw material, for forming, finishing, joining, and assembly operations, including buildings, equipment, and tooling; and also personnel—trained engineers, manufacturing supervisors, and lead operators—must be in place at the production site or capable of being quickly ordered and delivered, or provided by local or regional suppliers. Training time for personnel must also be factored in.
- *Material availability*: The raw material specified must be capable of timely delivery by at least two local or regional suppliers, at a reasonable cost, with a reliable, constant quality. In addition, suppliers must be able to provide technical assistance for fabrication and finishing operation problems and to develop new or improved products as appropriate.

For any given material selection problem, other selection parameters may be needed to better define application of the material to a specific design.

Guide to General Information Sources

Frank T. Traceski, Department of the Army, Materials Technology Laboratory

THIS ARTICLE is a directory of information sources on composite materials technology, including processing, properties, and government/industry standardization documents. All technical documents described here are unclassified, although some are distribution limited or subject to export control. Organizations that do provide documents that are classified, subject to export control, or otherwise restricted in distribution have been included in the following listings; the reader is responsible for complying with applicable laws and regulations in order to obtain restricted documents. This guide does not include the numerous scientific and engineering journals and books that deal with composites technology.

American Carbon Society (ACS)

The ACS offers conferences and publications dealing with the chemistry, physics, and scientific aspects of materials ranging from carbons and graphites to organic crystals and polymers.

Contact: American Carbon Society, The Stackpole Corporation, St. Marys, PA 15857, (814) 781-8410.

American Ceramic Society (ACS)

The ACS publishes two monthly periodicals and a bimonthly abstract, and provides scientific and technical information to scientists and engineers involved in the glass, cements, refractories, ceramic-metal systems, nuclear ceramics, electronics, white wares, and structural clay products industries.

Contact: American Ceramic Society, 65 Ceramic Drive, Columbus, OH 43214, (614) 268-8645.

American Society for Testing and Materials (ASTM)

ASTM Standards. ASTM Committee D-30 on High Modulus Fibers and Their Composites and ASTM Committee D-20 on Plastics both prepare standards pertaining to composite materials and test methods. The D-30 standards encompass both polymeric and metallic reinforced composite materials. The D-30 and D-20 standards are published in the Annual Book of ASTM Standards.

ASTM Special Technical Publications (STPs). Approximately 30 STPs covering various aspects of composites technology have been published thus far.

Contact: American Society for Testing and Materials, 1916 Race Street, Philadelphia, PA 19103, (215) 299-5585.

ASM International

Information on composites is available from ASM International, a technical society for materials, through its reference books, conferences, and videocourses. It also produces a monthly publication, in both print and database form, which abstracts and indexes articles on composites. In addition, ASM is assembling, in conjunction with Materials Sciences Corporation, a composites properties database for Military Handbook 17B. The database also will be available from ASM as personal computer software.

Contact: ASM International, Metals Park, OH 44073, (216) 338-5151.

Chemical Propulsion Information Agency (CPIA)

The CPIA is a Department of Defense Information Analysis Center operated by The Johns Hopkins University Applied Physics Laboratory in accordance with DoD Instruction 5100.45, "Centers for Analysis of Scientific and Technical Information." The CPIA provides technical and administrative support to the Joint Army-Navy-NASA-Air Force (JAN-NAF) Interagency Propulsion Committee. It maintains a very broad data base on composites for military and aerospace applications.

Contact: Chemical Propulsion Information Agency, The Johns Hopkins University, Applied Physics Laboratory, Johns Hopkins Road, Laurel, MD 20707, (301) 953-5000.

Department of Defense (DoD)

Government/Industry Standardization Documents. As defined by DoD 4120.3-M, standardization documents include specifications, standards, handbooks, and related engineering documents used in engineering design, development, manufacturing, maintenance, and supply management. Standardization documents pertaining to composite materials, processes, and tests methods are prepared by three primary standardization organizations: military standardization (SD-1) activities, ASTM Committees D-30 and D-20, and the Society of Automotive Engineers, Aerospace Materials Division, Composites Committee. The DoD *Index of Specifications and Standards* (DoDISS) is the primary information source for these three groups of standardization documents. It consists of three volumes: an alphabetical index, a numerical index, and a Federal Supply Class (FSC) listing. Standardization documents related to composites can be found under the Composites Technology (CMPS) standardization area listing in the FSC volume. The establishment of a new CMPS area has been approved by DoD, but at this time the DoDISS does not contain a CMPS listing of documents.

Military Handbook 17 is the primary military standardization document that encompasses engineering properties of polymer matrix composites. It is updated continuously by a government-industry working group led by the U.S. Army Materials Technology Laboratory (AMTL) in Watertown, MA.

Military Standard 1944 contains military and industry specifications and standards for polymer matrix composite materials, processes, and test methods.

Contact: Commanding Officer, Naval Publications and Forms Center, 5801 Tabor Avenue, Philadelphia, PA 19120, (215) 697-3321.

Engineering Index (EI)

The EI, a monthly publication, compiles bibliographic citations and abstracts covering the technological literature in all engineering disciplines. The literature covered is found in journals, technical reports, books, and conference proceedings.

Contact: Engineering Information, Inc., 345 E. 47th Street, New York, NY 10017-2387, (800) 221-1044.

Fiber Society (FS)

The FS serves scientists and engineers involved in the research of fibers, fibrous materials, and fiber products.

Contact: Fiber Society, Box 625, Princeton, NJ 08540, (609) 924-3150.

Fiberglass Fabrication Association (FFA)

The FFA disseminates information to companies that fabricate fiberglass products.

Contact: Fiberglass Fabrication Association, 1010 Wisconsin Avenue N.W., Suite 630, Washington, DC 20007, (202) 544-0262.

International Aerospace Abstracts (IAA)

The IAA covers the literature in the field of aeronautics and space science and technology. Periodicals, books, conference proceedings, and translations are abstracted, indexed, and published twice monthly. A cumulative index is prepared annually. IAA complements NASA's STAR abstract journal (see the section on NASA in this article).

Contact: American Institute of Aeronautics and Astronautics, Inc., 555 W. 57th Street, New York, NY 10019, (212) 247-6500.

Materials Research Society (MRS)

The MRS promotes interaction among researchers at universities and in industry. It provides short courses, conferences, and tutorial lectures.

Contact: Materials Research Society, 110 Materials Research Laboratory, University Park, PA 16802, (814) 865-3424.

Metal Matrix Composites Information Analysis Center (MMCIAC)

The MMCIAC was established by the DoD in 1980 to provide scientific and technical information analysis services on metal matrix composite (MMC) materials technology to government agencies and the private sector. This center establishes and maintains an MMC properties data base for designers concerned with MMC applications.

Contact: Metal Matrix Composites Information Analysis Center, Kaman Tempo, 816 State Street, P.O. Drawer QQ, Santa Barbara, CA 93102-1479, (805) 963-6497.

National Aeronautics and Space Administration (NASA)

The Scientific and Technical Aerospace Reports (STAR) is a major component of the comprehensive NASA information system. The STAR abstract journal, published twice monthly, includes abstracts on composite materials.

Contact: NASA Scientific and Technical Information Facility, ATTN: Registration Activity, P.O. Box 8757, Baltimore-Washington International Airport, MD 21240, (301) 859-5300.

National Institute of Ceramic Engineers (NICE)

The NICE is a professional organization for ceramics engineers. It confers accreditation on educational programs.

Contact: National Institute of Ceramic Engineers, 65 Ceramic Drive, Columbus, OH 43214, (614) 268-8645.

Plastics Institute of America (PIA)

The PIA is a cooperative venture of companies in the plastics industry to support education and research in the plastic fields. It conducts a graduate-level program of education in plastics in cooperation with over 100 universities and colleges.

Contact: Plastics Institute of America, Stevens Institute of Technology, Castle Point Station, Hoboken, NJ 07030, (201) 420-5553.

The Refractories Institute (TRI)

TRI serves producers and suppliers in the refractory industry and publishes a product directory. It supports research and awards scholarships.

Contact: The Refractories Institute, 3760 One Oliver Plaza, Pittsburgh, PA 15222, (412) 281-6787.

National Technical Information Service (NTIS)

The NTIS serves as a clearinghouse for reports funded and issued by government agencies. NTIS publishes *Government Reports Announcements & Index* twice monthly, which abstracts and provides acquisition information for reports on advanced materials and technologies. NTIS also publishes weekly newsletters in 27 subject categories, one newsletter being *Materials Science*. The newsletters summarize unclassified federally funded research as it is made available to the public and provides abstracts of reports. NTIS also offers Selected

Research in Microfiche (SRIM) on an automatic biweekly basis and a master catalog of published searches, which lists more than 3000 bibliographies.

Contact: National Technical Information Service, Springfield, VA 22161, (703) 487-4600.

Suppliers of Advanced Composite Materials Association (SACMA)

SACMA is a trade association formed in 1985 to address technical problems in the development of composite materials, such as standardization of test methods, specifications, certification procedures, and export control. It disseminates relevant business, marketing, and technical information on advanced composite materials.

Contact: Suppliers of Advanced Composite Materials Association, 1600 Wilson Boulevard, Suite 1008, Arlington, VA 22209, (703) 841-1556.

Society of Automotive Engineers (SAE)

The Composites Committee of the SAE Aerospace Materials Division prepares Aerospace Material Specifications (AMS) for composite materials. These specifications are listed in the annual SAE/AMS index under the AMS Series 3000 documents for nonmetallic materials.

Contact: Society of Automotive Engineers, 400 Commonwealth Drive, Warrendale, PA 15096, (412) 776-4841.

Society for the Advancement of Material and Process Engineering (SAMPE)

SAMPE is a technical society that publishes a journal, a quarterly, and symposia and technical conference proceedings on a variety of processes and materials, including composites.

Contact: Society for the Advancement of Material and Process Engineering, 843 West Glentana, P.O. Box 2459, Covina, CA 91722, (818) 331-0616.

Society of Plastics Engineers (SPE)

SPE is a technical society and an information source for a wide variety of publications related to composites technology.

Contact: Society of Plastics Engineers, 14 Fairfield Drive, Brookfield, CT 06805, (203) 775-0471.

Society of the Plastics Industry (SPI)

The Reinforced Plastics/Composites Institute of SPI is an information source for reinforced plastics and composites technology. The proceedings of its annual meetings have a wealth of technological and commercial information.

Contact: Society of the Plastics Industry, Inc., The Reinforced Plastics/Composites Institute, 355 Lexington Avenue, New York, NY 10017, (212) 503-0600.

Other Sources

Periodic conferences, symposia, and seminars related to composites technology include:

- DoD/NASA conferences on fibrous composites in structural design
- High-temperature plastic laminate evaluation (high temple) workshops, coordinated by a DoD/NASA steering group. Contact: Wright-Patterson Air Force Base, OH 45433, (513) 257-1110
- Technology Transfer Society (TTS), advanced composites seminars. Contact: TTS Conferences, P.O. Box 3608, 3420 Kashiwa Street, Suite 2000, Torrance, CA 90510-3608, (213) 534-3922
- Advanced composites working group of the American Ceramic Society, Ceramic-Metal Systems Division, which is co-ordinated by DoD/NASA. Contact: NASA Langley Research Center, Mail Stop 387, Hampton, VA 23665, (804) 865-3131

SELECTED REFERENCES

- *AMS Index Aerospace Material Specifications,* Society of Automotive Engineers, Jan 1984
- *Annual Book of ASTM Standards,* Vol 08.01, *Plastics (I): C177-D1600,* American Society for Testing and Materials, 1985
- *Annual Book of ASTM Standards,* Vol 08.02, *Plastics (II): D1601-D3099,* American Society for Testing and Materials, 1985
- *Annual Book of ASTM Standards,* Vol 08.03, *Plastics (III): D3100-latest,* American Society for Testing and Materials, 1985
- *Annual Book of ASTM Standards,* Vol 15.03, *Space Simulation; Aerospace Materials; High Modulus Fibers and Composites,* American Society for Testing and Materials, 1985
- K.I. Clayton, "High Temperature Plastic Laminate Evaluation (High Temple)," AFWAL-TR-84-4190, Air Force Wright Aeronautical Laboratories, March 1985
- Defense Standardization and Specification Program Policies, Procedures and Instructions, in *Defense Standardization Manual DoD 4120.3-M,* Aug 1978
- *Department of Defense Index of Specifications and Standards,* July 1985
- "A Guide for Private Industry," Naval Publications and Forms Center, Dec 1983
- *Military Handbook 17,* 1985
- Millitary Standard 1944, "Polymer Matrix Composites," 10 June 1985
- J.D. Oetting, "Proceedings of the Seventh Conference on Fibrous Composites in Structural Design," AFWAL-TR-85-3094, Air Force Wright Aeronautical Laboratories, June 1985
- *Scientific and Technical Aerospace Reports (STAR),* National Aeronautics and Space Administration

Properties of Constituent Materials

Co-Chairmen: W.S. Smith, E.I. Du Pont de Nemours & Co., Inc.
Carl Zweben, General Electric Company

Introduction

THIS SECTION is concerned with the major matrix resins and reinforcing fibers used in composites, with emphasis on the properties of the individual components of composites and how these properties vary from fiber to fiber and resin to resin. Knowledge of constituent material properties is important in defining a composite structure for a given application and in understanding how that structure will respond to the various stimuli likely to be imposed on it. In some cases, fiber properties will be the most important in defining performance of the composite, as exemplified by tensile strength and stiffness of a unidirectional laminate in the fiber direction. In other cases, the matrix resin properties determine the important properties, such as maximum upper use temperature of a composite. On the other hand, sometimes fiber and matrix contribute to the composite response in direct proportion to their respective volume fractions. Composite dielectric constant and water absorption are examples.

The composites revolution owes its success to the fortuitous, independent development of resins and structurally efficient reinforcing fibers. Although composites are now widely used materials, the technology is at a relatively early stage of development, and it is continuing to evolve. Of the many fibers that have been developed, the most important at this time are E-glass, S-2 glass, para-aramid, silicon carbide, boron, alumina, fused silica, alumina-boria-silica and carbon/graphite.

The most widely used, economical reinforcement in polymer matrix composites is E-glass. When high stiffness and strength are required, graphite and para-aramid fibers are the preferred materials. Boron is also an excellent structural fiber, but its high cost restricts its use. Alumina, alumina-boria-silica, and fused silica are used primarily for their unique physical properties, such as low dielectric constants and high melting points. At this time, silicon carbide fibers are used mainly for metal and ceramic matrix composites, rather than for polymer matrix composites.

There are many types of graphite fibers. They are derived mainly from polyacrilonitrile and petroleum and coal pitch precursor materials. Graphite fiber technology is continuing to develop rapidly, and major property improvements and cost reductions are possible. Because other fibers are still in the laboratory stage, it seems likely that developments in these areas can be expected as well.

The polymer matrix resins considered in this Section include both thermosetting and thermoplastic types, with emphasis on the former because they account for more than 80% of all matrices in reinforced plastics and essentially all matrices used in advanced composites. The most widely used thermosetting resins are the polyesters, which are most often combined with E-glass. This combination accounts for the bulk of the fiber-reinforced plastics (FRP) market. Polyesters offer a combination of low cost, versatility in many processes, and reasonably good property performance unmatched by any other resin type. The most common orthophthalic types and the premium vinyl ester and bisphenol A fumarate types are discussed in this Section.

For more demanding structural uses, epoxy resins are the preferred candidates. Although the amount of epoxies used in reinforced plastics is small in comparison to the volume of polyester used, epoxy use dominates the more demanding aircraft/aerospace structural applications as well as the printed wiring board market. Epoxy resin performance is highly dependent on the particular curing agent chosen for a specific resin; therefore curing agents and chemistry are discussed in this Section.

Polyimide resins are more expensive and less widely used than polyesters or epoxies, but are preferred when optimum thermal stability at high temperature is required. Although polyimides may be thermosetting or thermoplastic, most composite applications use the thermosetting types, which are fully covered in this Section. Also covered are the addition-type bismaleimides.

The use of high-performance thermoplastics as matrices in continuous fiber reinforced composites is currently an area characterized by very low use but very high interest. The term high performance is used loosely to categorize resins with relatively high-temperature capability, good resistance to moisture and solvents, and good mechanical and toughness performance; these criteria eliminate most of the thermoplastics now available and leave only a few premium resins as candidates for the more demanding structural and semistructural applications. Most commercially available thermoplastics are available in chopped fiber reinforced grades; in fact, these short-fiber grades account for 15 to 20% of the FRP market and are the most commonly used fiber-reinforced thermoplastics. With a few exceptions these resins are not covered in this Section.

Glass Fibers

David M. Miller, Owens-Corning Fiberglas Corporation

GLASS FIBERS are unique materials that exhibit the familiar bulk glass properties of hardness, transparency, resistance to chemical attack, stability, and inertness, as well as fiber properties of strength, flexibility, lightness of weight, and processability. Glass is made by fusing silicates with soda or with potash, lime, or various metallic oxides. The molten mass is then cooled rapidly to prevent crystallization. All glass compositions described in this article are derived from silica and the silicates.

Two basic processes are used to manufacture continuous glass filaments: marble melt and direct melt. In the marble melt process, glass marbles are first produced by melting the appropriate mixture of raw materials and forming marbles, which are usually 2 to 3 cm (0.8 to 1.2 in.) in diameter. These marbles are then remelted (at the same or different location) and formed into the glass fiber product. This older technology is more common in Europe than in the United States today. In the direct melt process, the raw materials are melted and formed directly into the glass fiber product. Direct melt is the primary process used in continuous glass fiber manufacture today.

Commercial continuous glass fibers are formed by extruding molten glass through an orifice that is usually 0.793 to 3.175 mm (0.0312 to 0.125 in.) in diameter, and then rapidly drawing this glass to a fine diameter. Typical fiber diameters range from 3 to 20 μm (118 to 787 μin.). Mechanical winders pull the fibers at lineal velocities of up to 61 m/s (200 ft/s). The orifices through which the molten glass passes for fiberizing are in the base of a refractory metal bushing typically made of a platinum alloy. Most bushings carry 204 orifices or some multiple thereof. The individual filaments are combined into a strand, which is the basic building block for glass fiber products. Glass filaments are highly abrasive to each other. To minimize an abrasion-related degradation of filament strength, "size" coatings are applied before the strand is gathered. The size may be temporary, as in the form of a starch-oil emulsion that is subsequently removed by heating and replaced with a glass-to-resin coupling agent known as a finish. On the other hand, the size may be a compatible treatment that performs several necessary functions during the subsequent forming operation and which, during impregnation, acts as a coupling agent to the resin being reinforced.

Glass Fiber Types

There are several glass fiber types with different compositions that reflect the chemistry needed to provide the specific chemical and/or physical properties required (Ref 1).

E-glass is a family of glasses with a calcium aluminoborosilicate composition and a maximum alkali content of 2.0%. E-glasses are used as a general-purpose fiber when strength and high electrical resistivity are required.

S-glass has a magnesium aluminosilicate composition, which demonstrates high strength and is therefore used in applications where very high tensile strength is required. S-glass and S-2 glass fibers have the same glass composition (Ref 2), but different coatings. More stringent quality control procedures are necessary with S-glass to meet military specifications.

C-glass has a soda-lime-borosilicate composition that is used for its chemical stability in corrosive environments. Therefore it is often used in composites that contact or contain acidic materials.

Glass Fiber Chemical Components

When discussing the chemical composition of a specific glass type, the range of each of its oxide components must be presented. This is necessary because each manufacturer, and even different manufacturing plants for the same company, may use slightly different compositions for the same glass. These variations result from differences in the available glass batch (raw materials), or in the melting and forming processes, or from different environmental constraints at the manufacturing site. These compositional fluctuations do not significantly alter the physical and chemical properties of the glass type. It should be mentioned that while the compositions may vary from location to location, very tight control is maintained within a given production facility to achieve consistency in glass composition and maximize production efficiencies. Table 1 presents the oxide components and their ranges for the three types of glass fibers that have been produced and used in composites.

Glass Fiber Properties

Some glass fiber properties, such as tensile strength, Young's modulus, and chemical durability, are measured on the fibers directly. Other properties, such as relative permittivity, dissipation factor, dielectric strength, volume/surface resistivities, and thermal expansion, are measured on glass that has been formed into a bulk patty or block sample and annealed (heat treated) to relieve forming stresses. Properties such as density and refractive index are measured on both fibers and bulk samples, in annealed or unannealed form.

Table 1 Compositional ranges for glass fibers used in composite materials

	E-glass range, %(a)	S-glass range, %(b)	C-glass range, %(c)
Silicon dioxide	52-56	65	64-68
Aluminum oxide	12-16	25	3-5
Boric oxide	5-10	...	4-6
Sodium oxide and potassium oxide	0-2	...	7-10
Magnesium oxide	0-5	10	2-4
Calcium oxide	16-25	...	11-15
Barium oxide	...	...	0-1
Zinc oxide	...	...	...
Titanium dioxide	0-1.5	...	...
Zirconium oxide	...	...	...
Iron oxide	0-0.8	...	0-0.8
Iron	0-1	...	...

(a) Ref 3-7. (b) Ref 4-6, 8. (c) Ref 3, 8

Table 2 Typical properties for the glass fiber types

Material	Density, bulk annealed, g/cm³	Tensile strength at −190 °C (−310 °F) MPa	ksi	at 23 °C (72 °F) MPa	ksi	at 371 °C (700 °F) MPa	ksi	at 538 °C (1000 °F) MPa	ksi	Young's modulus of elasticity at 538 °C (1000 °F) GPa	10⁶ psi	Elongation, %
E-glass	2.62	5310	770	3445	500	2620	380	1725	250	81.3	11.8	4.88
S-glass	2.50	8275	1200	4585	665	4445	645	2415	350	88.9	12.9	5.7
C-glass	2.56	5380	780	3310	480	...	...	...	...	...	...	4.8

Material	Chemical resistance (percent weight loss) in H₂O 24 h	186 h	10% HCl 24 h	168 h	10% H₂SO₄ 24 h	168 h	1% Na₂CO₃ 24 h	168 h	10% NaOH 168 h	Relative permittivity at 1 MHz	at 60 Hz	Dissipation factor at 1 MHz	at 60 Hz
E-glass	0.7	0.9	42	43	39	42	2.1	2.1	20	6.6	6.7	0.0025	0.0034
S-glass	0.5	0.7	3.8	5.1	4.1	5.7	2.0	2.1	66	5.3	5.4	0.0034	0.0129
C-glass	1.1	2.9	4.1	7.5	2.2	4.9	24	31	...	6.9	...	0.0085	...

Material	Volume resistivity, Ω·m	Surface resistivity, Ω	Dielectric strength kV/cm	V/mil	Viscosity softening point °C	°F	Viscosity annealing point °C	°F	Viscosity strain point °C	°F	Thermal expansion 10⁻⁶/K(a)	Specific heat at 23 °C (72 °F) kJ/kg·K (Btu/lb·°F)	at 200 °C (392 °F) kJ/kg·K (Btu/lb·°F)	Refractive index, bulk annealed
E-glass	0.402×10^{15}	0.42×10^{16}	103	262	846	1555	657	1215	615	1140	5.4	0.810 (0.193)	1.03 (0.247)	1.562
S-glass	0.905×10^{13}	0.886×10^{13}	130	330	970	1778	810	1490	760	1400	1.6	0.737 (0.176)	...	1.525
C-glass	...	...	...	...	750	1382	588	1090	552	1025	6.3	0.787 (0.188)	0.90 (0.215)	1.537

(a) From −30 °C (−20 °F) to 250 °C (480 °F)

Just as the composition of glass fibers may vary slightly, so may their physical properties. The products of any given supplier, however, have constant properties. For the purposes of this article, the properties presented are representative of the compositional ranges discussed. Also, unless otherwise noted, the data presented have been collected by Owens-Corning Fiberglas (Ref 9). The data provided in Table 2 correspond to the following brief overview of glass fiber properties.

The physical properties of glass fibers are discussed below.

Density of glass fibers is most commonly reported as a bulk annealed value. The fiber density is actually less than the bulk annealed value; for E-glass, this difference is approximately 0.04 g/cm³ at room temperature. The bulk densities for glass fibers used in composites range from approximately 2.50 g/cm³ for C-glass to 2.62 g/cm³ for E-glass. Reference 10, ASTM C 693, is used for density determinations.

Tensile strength of glass fibers is usually reported as the pristine single-filament or the multifilament strand measured in air at room temperature. The respective strand strengths will normally be 20 to 40% lower than the values reported in Table 2 because of surface defects introduced during the strand-forming process.

Moisture has a detrimental effect on the pristine strength of glass. This is best illustrated by measuring the pristine single-filament strength at liquid nitrogen temperatures where moisture influence has been minimized. The result is an increase of 50 to 100% in strength over a measurement at room temperature in 50% relative humidity air. The loss in strength of glass exposed to moisture while under an external load is known as static fatigue (Ref 11, 12).

The pristine strength of glass fibers will likewise decrease as the fibers are exposed to increasing temperature. While there is no general rule for the magnitude of this decrease, E-glass and S-glass fibers have been found to retain approximately 50% of their pristine room-temperature strength at 538 °C (1000 °F).

The Young's modulus of elasticity of unannealed silicate glass fibers ranges from about 69 GPa (10.0 × 10⁶ psi) to 85 GPa (12.4 × 10⁶ psi). As the fiber is heated, the modulus gradually increases. E-glass fibers that have been annealed to compact their atomic structure will increase in Young's modulus from 72 GPa (10.5 × 10⁶ psi) to 84.7 GPa (12.3 × 10⁶ psi) (Ref 12).

The Poisson's ratio does not vary much with composition in silicate glasses (Ref 13). For most glasses, it falls between 0.15 and 0.26. The Poisson's ratio for E-glasses is 0.22 ± 0.02 and is reported not to change with temperature when measured up to 510 °C (950 °F) (see Ref 14).

The chemical resistance of glass fibers to the corrosive and leaching actions of acids, bases, and water is expressed as a percent weight loss. The lower this value, the more resistant the glass is to the corrosive solution. The test procedure involves subjecting a given weight of 10 μm (390 μin.) diam bare glass fibers, without binders or sizes, to a known volume of corrosive solution held at 96 °C (205 °F). The fibers are held in the solution for the time desired and then are removed, washed, dried, and weighed to determine the weight loss. The results reported are for 24-h (1 day) and 168-h (1 week) exposures. As Table 2 shows, the chemical resistance of glass fibers depends on the composition of the fiber, the corrosive solution, and the exposure time.

The electrical properties in Table 2 were measured on annealed bulk glass samples according to the testing procedures cited in Ref 15-17.

The relative permittivity is the ratio of the capacitance of a system with the specimen as the dielectric to the capacitance of the system with a vacuum as the dielectric. Capacitance is the ability of the material to store an electrical charge. Permittivity values are affected by test frequency, temperature, voltage, relative humidity, water immersion, and weathering.

The dissipation factor of a dielectric is the ratio of the parallel reactance to the parallel resistance, or the tangent of the loss angle, which is usually called the loss tangent. It is also the reciprocal of the quality factor, and when the values are small, tangent of the loss angle is essentially equal to the power factor, or sine of the loss angle. The power factor is the ratio of power in watts dissipated in the dielectric to the effective volt-amperes. The dissipation factor is dimensionless.

In almost every electrical application, a low value for the dissipation factor is desired. This reduces the internal heating of the material and keeps signal distortion low. The dissipation factor is generally measured simultaneously with permittivity measurements, and is greatly influenced by frequency, humidity, temperature, and water immersion.

The loss factor, or loss index, as it is sometimes called, is occasionally confused with the dissipation factor, or loss tangent. The loss factor is simply the product of the dissipation factor and permittivity and is proportional to the energy loss in the dielectric.

The dielectric breakdown voltage, usually expressed in kilovolts, is the voltage at which electrical failure occurs under prescribed test conditions in an electrical insulating material that is placed between two electrodes. When the thickness of the insulating material between the electrodes can be accurately measured, the

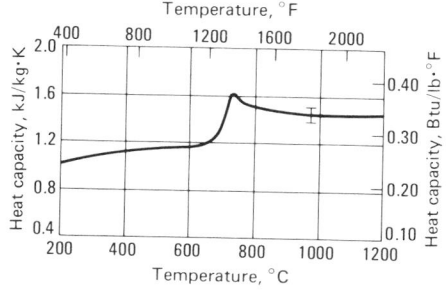

Fig. 1 Specific heat of E-glass

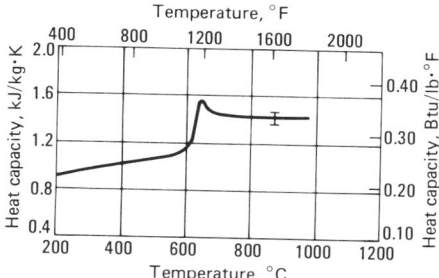

Fig. 2 Specific heat of C-glass

ratio of the dielectric breakdown voltage to the specimen thickness can be expressed as the dielectric strength in kV/cm.

Breakdown voltages are influenced by electrode geometry, specimen thickness (because dielectric strength varies approximately as the reciprocal of the square root of the thickness), temperature, voltage application time, voltage waveform, frequency, surrounding medium, relative humidity, water immersion, and directionality in laminated and inhomogeneous plastics.

The thermal properties of interest are considered below.

The viscosity of a glass varies continuously with temperature, increasing as the temperature decreases. Several reference viscosity points have been established and are defined below.

The softening point is the temperature at which a glass fiber of uniform diameter elongates at a specific rate under its own weight when measured by ASTM C 338 (Ref 18). The softening point is sometimes defined as the temperature at which glass will deform under its own weight; it occurs at a viscosity of approximately $10^{6.6}$ Pa · s ($10^{7.6}$ P).

The annealing point is the temperature corresponding to either a specific rate of elongation of a glass fiber when measured by ASTM C 336 (Ref 19), or a specific rate of midpoint deflection of a glass beam when measured by ASTM C 598 (Ref 20). At the annealing point of glass, internal stresses are substantially relieved in a matter of minutes. The viscosity at the annealing point is approximately 10^{12} Pa · s (10^{13} P).

The strain point is measured following ASTM C 336 or C 598 as described above for annealing point. At the strain point of glass, internal stresses are substantially relieved in a matter of hours. The viscosity at the strain point is approximately $10^{13.5}$ Pa · s ($10^{14.5}$ P).

The mean coefficient of thermal expansion over the temperature range from −30 to 250 °C (−22 to 482 °F) is provided in Table 2. The coefficient is expressed in units of 10^{-6}/K. The expansion measurements were made on bulk annealed bars using ASTM D 696 (Ref 21).

The specific heat data was determined using high-temperature differential scanning calorimetry techniques. The specific heat versus temperature curves for E-glass and C-glass are shown in Fig. 1 and 2. The shape of these curves is typical for the glasses in Table 1, except for their widely varying glass transition temperatures. In general, the average specific heat values can be represented as follows: 0.94 kJ/kg · K (0.224 Btu/lb · °F) at 200 °C (390 °F), 1.12 kJ/kg · K (0.267 Btu/lb · °F) just below the transition point, and 1.40 kJ/kg · K (0.333 Btu/lb °F) in the liquid state above the transition. These values are accurate to about 5%. Above the transition temperature, no further increase in specific heat was observed. It should be noted that the transition temperature is nearly identical to the annealing temperature of bulk glass.

Thermal conductivity characteristics in glasses differ considerably from those found in crystalline materials. For glasses, the conductivity is lower than that of the corresponding crystalline materials. Also, the conductivity of glasses drops steadily with temperature and reaches very low values, near absolute zero. For crystals, the conductivity continues to rise with decreasing temperature until very low temperatures are reached (Ref 22).

One of the main difficulties in attempting to present thermal conductivity data for glasses of various compositions is the wide divergence in results among investigators for materials which are normally identical (Ref 23). Results will vary by as much as a factor of two.

In general it is found that fused silica (silicon dioxide) glass and the alkali and alkaline earth silicate glasses have relatively similar conductivities at room temperature, whereas conductivities of borosilicate and glass that contain lead and barium are somewhat lower. It has been found that at near room temperature, the thermal conductivity for glasses ranges from 0.55 W/m · K (3.8 Btu · in./h · ft² · °F) for lead silicate (80% lead oxide, 20% silicon dioxide) to 1.4 W/m · K (9.6 Btu · in./h · ft² · °F) for fused silica glass (Ref 24).

In addition, property coefficients have been developed for predicting thermal conductivity from the percentage weight compositions of component oxides making up the glass (Ref 24). Using this calculation, it is found that the thermal conductivity of E-glass is approximately 1.3 W/m · K (8.9 Btu · in./h · ft² · °F), and that of C-glass is 1.1 W/m · K (7.9 Btu · in./h · ft² · °F) at near room temperature.

Optical properties. Refractive index is measured on annealed glass samples using standard oil immersion techniques and monochromatic sodium D light at 25 °C (77 °F). While the unannealed fiber refractive index is not presented in Table 2, it generally will range from approximately 0.0030 to 0.0060 below that of the annealed glass sample.

Radiation properties. E-glass fibers have excellent resistance to all types of nuclear radiation. Alpha and beta radiation have almost no effect, while gamma radiation and neutron bombardment produce a 5 to 10% decrease in tensile strength, a less than 1% decrease in density, and a slight discoloration of fibers. This data was true of 10^{20} NVT neutrons or gamma radiation up to 10^5 J/g. It is expected that C-glass, S-glass, and S-2 glass would also be suitable for use in areas of high radiation.

Glass fibers resist radiation because the glass is amorphous, and the radiation does not distort the atomic ordering. Glass can also absorb a few percent of foreign material and maintain the same properties to a reasonable degree. Also, because the individual fibers have a small diameter, the heat of atomic distortion is easily transferred to a surface for dispersion.

It should be noted that E-glass and C-glass are not recommended for use inside atomic reactors because of their high boron content. S-glass and S-2 glass should be suitable.

Even though glass fibers are ideal for use in areas of high irradiation, glass fiber products may not be because most have an organic binder, size, or coating of some kind. These organics are more easily degraded by all types of radiation than are the glass fibers. The use of a product will be determined by such factors as resistance of organic compounds to nuclear radiation, requirement of organic compounds with product in place, and the effect of environment and mechanical forces on the product.

Because quite a wide variety of organic products are used in diverse radiation environments, it is usually necessary to try out most products in simulated conditions to determine whether the organics will be satisfactory.

ACKNOWLEDGMENT

Special thanks are due to T.W. Church, D.A. Hofmann, N.T. Huff, J.B. Langensiepen, R.M. Potter, and T.R. Washer for their help in collecting the chemical and physical property data necessary for this review.

REFERENCES

1. P.F. Aubourg and W.W. Wolf, ''Glass Fibers—Glass Composition Research,'' Paper presented at Glass Division Meeting, American Ceramic Society, Grossinger, NY, Oct 1984
2. R.W. Fulmer, ''S-2 Glass Fiber Bridges a Gap in the Reinforcement Spectrum,'' Paper presented at the National SAMPE Symposium, Society for the Advancement of Material and Process Engineering,

May 1980

3. D.E. Campbell and H.E. Hagy, Glass and Glass-Ceramics, in *Handbook of Material Science*, Vol II, CRC Press, 1975, p 323-338

4. K.L. Loewenstein, *The Manufacturing Technology of Continuous Glass Fibres*," Elsevier, 1973, p 28-30

5. W.W. Wolf, "The Glass Fiber Industry — The Reason for the Use of Certain Chemical Compositions," Seminar presented at University of Illinois, Urbana, IL, Oct 1982

6. W.W. Wolf and S.L. Mikesell, Glass Fibers, in *Encyclopedia of Materials Science and Engineering*, 1st ed., 1986

7. "Roving, Glass, Fibrous (For Prepreg Tape and Roving, Filament Winding, and Pultrusion Applications)," MIL-R-60346C, U.S. Department of Defense, June 1981

8. J.G. Mohr and W.P. Rowe, *Fiber Glass*, Van Nostrand Reinhold Company, 1978, p 207

9. *Textile Fibers for Industry*, Publication 5-700-8285-B, Owens-Corning Fiberglas

10. "Standard Test Method for Density of Glass by Buoyancy," C 693, *Annual Book of ASTM Standards*, American Society for Testing and Materials

11. P.K. Gupta, Examination of the Tensile Strength of E-Glass Fiber in the Context of Slow Crack Growth, *Fracture Mechanics of Ceramics*, Vol 5, 1983, p 291

12. R.E. Lowrie, Glass Fibers for High-Strength Composites, in *Modern Composites Materials*, Addison-Wesley, 1967

13. J.R. Hutchins, III and R.W. Harrington, Glass, in *Encyclopedia of Chemical Technology*, Vol 10, 2nd ed., p 533-604

14. R.T. Brannan, *J. Am. Ceram. Soc.*, Vol 36, 1953, p 230-231

15. "Standard Test Methods for A-C Loss Characteristics and Permittivity (Dielectric Constant) of Solid Electrical Insulating Materials," D 150, *Annual Book of ASTM Standards*, American Society for Testing and Materials

16. "Standard Test Methods for D-C Resistance or Conductance of Insulating Materials," D 257, *Annual Book of ASTM Standards*, American Society for Testing and Materials

17. "Standard Test Method for Dielectric Breakdown Voltage and Dielectric Strength of Solid Electrical Insulating Materials at Commercial Power Frequencies," D 149, *Annual Book of ASTM Standards*, American Society for Testing and Materials

18. "Standard Test Method for Softening Point of Glass," C 338, *Annual Book of ASTM Standards*, American Society for Testing and Materials

19. "Standard Test Method for Annealing Point and Strain Point of Glass by Fiber Elongation," C 336, *Annual Book of ASTM Standards*, American Society for Testing and Materials

20. "Standard Test Method for Annealing Point and Strain Point of Glass by Beam Bending," C 598, *Annual Book of ASTM Standards*, American Society for Testing and Materials

21. "Standard Test Method for Coefficient of Linear Thermal Expansion of Plastics, D 696, *Annual Book of ASTM Standards*, American Society for Testing and Materials

22. C. Kittel, *Phys. Rev.*, Vol 75, 1949, p 972

23. C.L. Babcock, Symposium on Heat Transfer Phenomena in Glass, *J. Am. Ceram. Soc.*, Vol 44 (No. 7), July 1961

24. E.H. Ratcliffe, Thermal Conductivities of Glass Between −150 and 100 °C, *Glass Technol.*, Vol 4 (No. 4), Aug 1963

Carbon/Graphite Fibers

Russell J. Diefendorf, Rensselaer Polytechnic Institute

THE ELEMENT CARBON has two low-density allotropes, graphite and diamond, both of which have strong covalent bonding between the carbon atoms. Graphite has a hexagonal structure in which the strong sp^2 bonding in the hexagonal-layer planes generates the highest absolute modulus, highest specific modulus (modulus/density), and highest theoretical tensile strength of all known materials (see Tables 1 and 2). However, the weak, dispersive bonding between planes produces a low shear modulus and cross-plane Young's modulus that is detrimental to fiber properties (Ref 2). Diamond, with a cubic crystallographic structure, possesses the next highest absolute and specific modulus and does not suffer from a low shear modulus, as does graphite (Ref 3). Diamond or diamondlike carbon fibers have not yet been made, but could have great applicability, especially for compressively loaded structures.

This article deals with carbon and graphite fibers that are based on the graphene- (hexagonal-) layer networks present in graphite. Graphene is the accepted term used by the International Committee for the Characterization and Terminology of Carbon. If these graphene-layer planes stack with three-dimensional order, the material is defined as graphite (Ref 4). Because the bonding between planes is weak, disorder frequently occurs as rotation and/or translation such that only the two-dimensional ordering in the layers is present. This material is defined as carbon (Ref 4). In earlier literature, two-dimensionally ordered structures were referred to as turbostratic

graphite. Historical usage continues, and the term graphite fibers often is applied, improperly, to carbon fibers, which only have two-dimensional ordering.

Carbon and graphite fibers offer the highest modulus and highest strength of all reinforcing fibers. The fibers do not suffer from stress corrosion or stress rupture failures at room temperature, as glass and organic polymer fibers do. At high temperatures, the strength and modulus are outstanding compared to other materials. Finally, aggressive development of new processes promises significant improvements in the performance/cost ratio.

Carbon Fiber Processes

Carbon fibers have been made inadvertently from natural cellulosic fibers such as cotton or linen for thousands of years. However, it was Thomas Edison who, in 1878, purposely took cotton fibers and later, bamboo, and converted them into carbon in his quest for incandescent lamp filaments (Ref 5). Interest in carbon fibers was renewed in the late 1950s when synthetic rayons in textile forms were carbonized to produce carbon fibers for high-temperature

missile applications (Ref 6-8). All these fibers had low elastic moduli ($\leq$50 GPa, or 7 $\times$ 10^6 psi).

All continuous carbon fibers produced to date have started with organic precursors that were subsequently converted to carbon fibers. Discontinuous carbon whiskers have been produced by vapor-liquid-solid (VLS) growth from an iron catalyst and a hydrocarbon gas (Ref 9, 10). High-modulus carbon fibers ($\geq$200 GPa, or 30 $\times$ 10^6 psi) require that the stiff graphene layers be aligned approximately parallel to the fiber axis. (The low shear modulus between the planes significantly decreases fiber stiffness for off-axis layers.) Commercial processes develop this orientation by plastic deformation. This preferred orientation, which can be introduced in the precursor fibers, will be at least partially preserved upon conversion to carbon (Ref 11-14) and also may be introduced into the carbon fiber by high-temperature deformation (Ref 15, 16).

Three different precursor materials are used at present to produce carbon fibers: rayon, polyacrylonitrile (PAN), and isotropic and liquid crystalline pitches. Rayon and isotropic

Table 1 Theoretical properties of graphite and diamond

	Tensile modulus		Tensile strength		Density, g/cm^3	Specific modulus	Specific strength
	GPa	10^6 psi	GPa	10^6 psi			
Graphite(a) 1020		150	150	20	2.26	451	66
Diamond 20		3	90	15	3.51	177	25

(a) Parallel to basal plane. Source: Ref 1-3

Table 2 Properties of carbon fibers: polyacrylonitrile precursor fibers

Properties	Standard grades				New grades				
	Low modulus		High modulus		Low modulus		Intermediate modulus		High modulus
Axial									
Tensile modulus, GPa (10^6 psi). 230	(30)	390	(55)	230	(35)	270	(40)	320	(45)
Tensile strength, GPa (10^6 psi) 3.3	(0.48)	2.4	(0.35)	4.5	(0.65)	5.3-6.8	(0.77-0.99)	5.5	(0.80)
Elongation at break, %. 1.4	...	0.6	...	2.0	...	2.0-2.5	...	1.8	...
Thermal conductivity, W/m · K (Btu · in./h · ft^2 · °F) 8.5	(59)	70	(490)	7	(50)				
Electrical resistivity, μΩ · m (μΩ · cm) 18	(1800)	9.5	(950)	18	(1800)				
Coefficient of thermal expansion at 21 °C (70 °F), 10^{-6}/K . . . . −0.7	...	−0.5	...						
Transverse									
Tensile modulus, GPa (10^6 psi). 40	(6)	21	(3)						
Coefficient of thermal expansion at 50 °C (120 °F), 10^{-6}/K 10	(2)	7	(1)						
Bulk									
Density, g/cm^3. . . . . 1.76	...	1.9	...	1.8	...	1.8	...	1.8	...
Filament diameter, μm (μin.) 7-8	(280-310)	7	(280)	5-6	(200-240)	6	(240)	4	(160)
Carbon assay, % 92-97	...	100	...	92-97	...	96	...	96	...

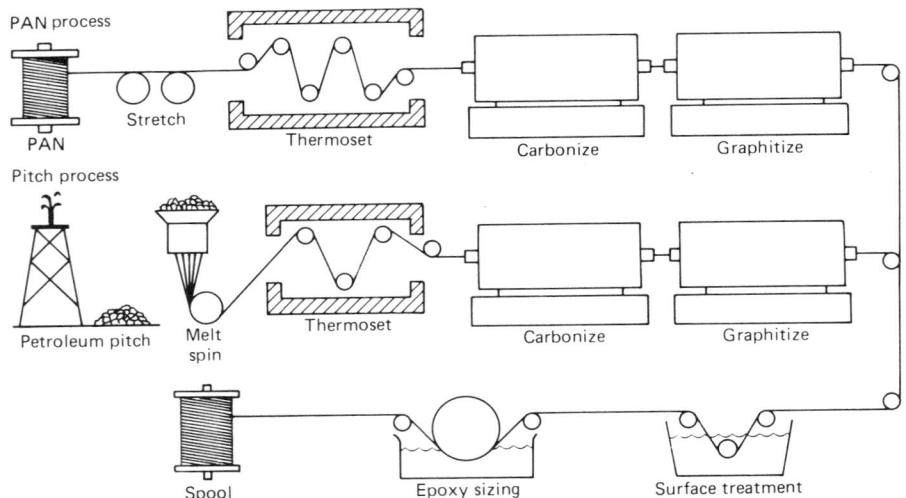

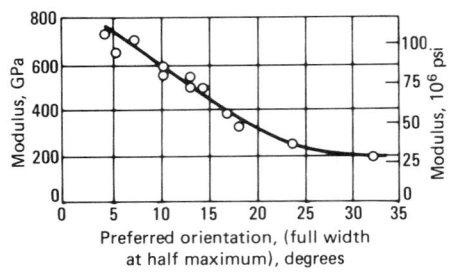

Fig. 1 The processing sequence for polyacrylonitrile (PAN) and mesophase pitch–based precursor fibers shows the similarities for the two processes. Highly oriented polymer chains are obtained in PAN by hot stretching, while high orientation in pitch is a natural consequence of the mesophase (liquid crystalline) order.

Fig. 3 The modulus of a carbon fiber is determined by the preferred orientation, microstructure, and elastic constants. The relationship between modulus and preferred orientation for a pitch-based carbon fiber is shown.

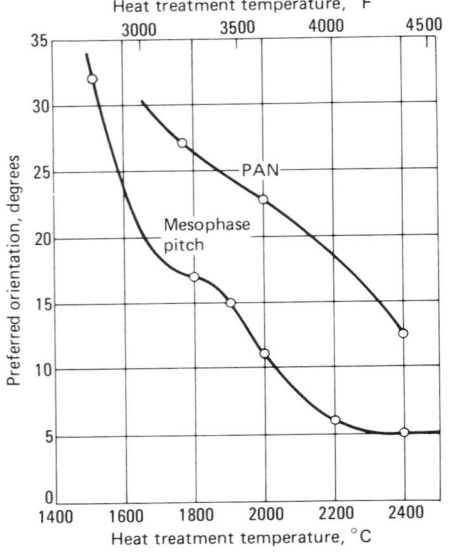

Fig. 2 The preferred orientation of the graphene planes is determined by the heat treatment temperature and the precursor type. Source: Ref 2, 23

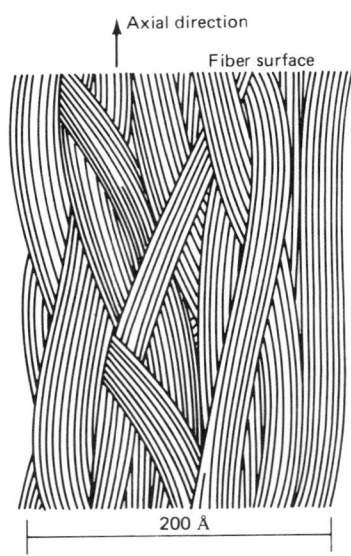

Fig. 4 The undulating ribbon structure of the graphene layers for a PAN-based carbon fiber with a 400 GPa (60 × 10⁶ psi) modulus. The ribbons at the surface have lower amplitude than in the core. There are about 20 graphene layers in the ribbons in the core and about 30 near the surface.

pitch precursors are used to produce low-modulus carbon fibers ($\leq$50 GPa, or 7 × 10⁶ psi) (Ref 6, 16-19). Both rayon-based and isotropic pitch-based carbon fibers can be strained at high temperature to increase fiber modulus but this process is not used commercially at present (Ref 15, 16). Higher modulus carbon fibers ($\geq$200 GPa, or 30 × 10⁶ psi) are made from PAN or liquid crystalline (mesophase) pitch precursors (Fig. 1). In both cases, an oriented precursor fiber is spun, slightly oxidized to thermoset the fibers, and carbonized to temperatures above 800 °C (1400 °F) to produce a carbon fiber (Ref 11-14, 20). The fiber modulus increases with heat treatment to temperatures from 1000 to 3000 °C (1830 to 5430 °F), although not uniformly with temper-

ature. The exact relation depends on the precursor (Ref 2, 21). Fiber strength usually maximizes at an intermediate temperature (1500 °C, or 2730 °F) for PAN and some pitch precursor fibers, but continuously increases for most mesophase pitch precursor fibers (Ref 2, 21).

Carbon Fiber Microstructures

The axial preferred orientation of the graphene layers determines the modulus of the fiber, while both axial and radial textures, as well as flaws, affect fiber strength. The orientation of the graphene layers at the fiber surface affects wetting and the strength of the interfacial bond to the matrix. The following sections describe axial and radial textures, illustrate a model of a three-dimensional carbon fiber, and describe the consequences of these microstructures on fiber performance.

Axial Structure. An overall measure of preferred orientation of graphene planes with respect to the fiber axis is obtained by x-ray diffraction (Ref 22). The full width at half maximum of the graphene-layer diffraction (0002) is useful for describing preferred orientation. The effect of heat treatment temperature on the preferred orientation of PAN and mesophase pitch-based carbon fibers is shown in Fig. 2. The trend for PAN is similar to that for mesophase pitch, but the curve is shifted about 400 °C (750 °F) higher. The relation between preferred orientation and fiber modulus for the mesophase pitch-based fiber is illustrated in Fig. 3. For a carbon fiber with a modulus of 220 GPa (30 × 10⁶ psi), about two-thirds of the graphene layers are aligned within about 15° of the fiber axis. Correspondingly, the orientation improves for a 400-GPa (60 × 10⁶-psi) fiber such that two-thirds of the layers fall within about 6° of the fiber axis. A PAN-based fiber shows similar behavior, except that it is

more difficult to obtain preferred orientations below 10° and a modulus greater than 400 GPa (60 × 10⁶ psi), whereas at least one mesophase pitch-based fiber has a modulus of 827 GPa (120 × 10⁶ psi). Transmission electron microscopy (TEM) has revealed, at lattice resolution, that the structure looks like wrinkled ribbons (Ref 23-25), as shown in Fig. 4. Lower-resolution TEM shows that for pitch-based fibers of 220 GPa (30 × 10⁶ psi), ribbons are typically about 16 layer planes thick, essentially continuous, and parallel to the fiber axis (Ref 26-30). The amplitude of the undulation is greater than the wavelength. As modulus is increased, the ribbons thicken and the amplitude of the undulation decreases.

Radial Structure. Radial texturing increases with axial preferred orientation. In contrast to axial structure, radial structure depends substantially on precursor type and processing (Ref 23, 26-30). Mesophase pitch-based fibers can have the graphene planes showing radial,

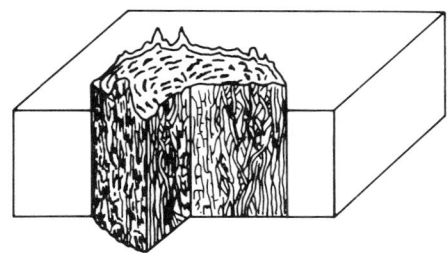

Fig. 5 A 400-GPa (60 × 10⁶ psi) PAN-based fiber. Source: Ref 26

onion-skin, or highly contorted planes with no overall transverse preferred orientation (Ref 31). However, most commercial pitch-based fibers have a radial preferred orientation. PAN-based fibers can have contorted graphene planes, or onion-skin structures. Low-modulus PAN-based fibers (≤350 GPa, or 50 × 10⁶ psi) now have a highly contorted transverse graphene-layer structure; higher-modulus fibers have a thin, highly oriented, onion-skin surface layer 50 to 100 × 10⁻⁹ m (1950 to 3950 × 10⁻⁹ in.) thick about a randomly oriented, contorted layer core (Ref 32).

Three-Dimensional Structure. A schematic model of a 400-GPa (60 × 10⁶-psi) PAN-based fiber is shown in three dimensions in Fig. 5. The axial orientation is higher at the surface than in the core. The ribbons are randomly, radially oriented in the core, with an onion-skin orientation at the fiber surface. A gradient in axial preferred orientation from skin to core produces a compressive axial stress at the surface after cool-down from processing temperatures. This stress makes the fiber insensitive to surface abrasion. Lower-modulus fibers do not have the preferred orientation gradient and beneficial residual stress.

Consequences of Structure on Properties. The increase in radial texture with increasing fiber modulus has some deleterious consequences (Ref 33). First, the onion skin that develops on the surface of a higher-modulus PAN-based fiber gives poor bonding to the matrix. The essentially basal plane char-

acter of very high modulus fibers is difficult to wet. As the fiber modulus increases, the surface becomes smoother, thereby decreasing mechanical interlocking, and the highly oriented surface layer becomes weak in shear. Second, microcracking occurs with the fiber upon cool-down from the processing temperature because of the anisotropy of thermal expansion both parallel and perpendicular to the graphene planes. The fracture surface of the fibers becomes rougher, especially when a strong radial texture develops, because microcracking causes decoupling between the undulating ribbons. Also, the fiber strength decreases. For high-modulus PAN-based fibers (≥400 GPa, or 60 × 10⁶ psi), the onion-skin structure and the axial gradient in preferred orientation place the fiber surface in axial and hoop compression and in radial tension, upon cool-down. These compressive surface stresses and the onion-skin microstructure protect the fiber from strength degradation due to surface abrasion. Other fibers are sensitive to surface abrasion. Finally, as better orientation develops, the interaction or tangling between ribbons decreases, and the shear modulus drops, making the fiber susceptible to compressive microbuckling at low compressive loads. Very highly oriented carbon and organic fibers both suffer from this problem.

Fiber Properties. Commercially available high-modulus carbon fibers are available from a number of manufacturers in an array of yarns and tows with differing moduli, strengths, cross-sectional areas and shapes, twists, plies, and number of fiber ends. They may be purchased in continuous lengths, or chopped to dimension. The diversity of physical properties is one of the benefits of carbon fibers, but is also a problem because complete evaluation is expensive. Certain generalizations can be made about the types of fibers available. First, fibers produced from mesophase pitch-based and PAN-based fibers have quite different combinations of properties and, hence, applications. At present, they cannot be considered interchangeable. Second, there are generally four classes of carbon fiber, based on moduli: low, intermediate, high, or ultrahigh. Unfortunately,

these categories are not like the specifications for alloys. Although fibers made by different manufacturers may be similar, they may not behave identically in all respects. Subtle differences in precursor types and carbon fiber processes can significantly affect the behavior of the carbon fiber in a composite. Moreover, improvements in fibers from different manufacturers have tended to splinter the categories. Third, the fiber selections that make up a particular manufacturer's grade, but have different numbers of fibers, are usually, but not always, based on the same precursor. In addition, while the fibers may be considered to have the same properties, the fiber count or twist may affect composite properties. Finally, because it is not realistic to list all the different manufacturers' fiber properties, general properties for each major category of carbon fibers are presented in Tables 2 to 4.

Bulk Properties. The carbon content of low-modulus PAN-based carbon fibers (≤300 GPa, or 45 × 10⁶ psi) is significantly below 100% because of retained nitrogen. High-modulus PAN-, rayon-, and pitch-based carbon fibers are essentially 100% carbon.

The density of carbon fibers, which increases with increasing modulus, varies from 80 to 94% and 86 to 100% of theoretical for PAN- and pitch-based fibers, respectively. The fractional density for rayon-based fibers is about 68% and increases to as high as 94% for higher-modulus fibers.

Finally, the diameters of present reinforcement grade carbon fibers are in the range of 4 to 10 μm (160 to 390 μin.). The trend has been toward smaller fiber diameters to attain improved tensile strength and processing speeds. However, the compressive and transverse properties of composites made with smaller-diameter fibers do not increase proportionally with tensile properties, and can contribute to failure in compression caused by buckling at lower loads.

Axial Properties. Of the commercially available carbon fibers, those made from PAN precursor provide the highest strain at failure, but only at lower moduli (≤300 GPa, or 45 ×

Table 3 Properties of carbon fibers: mesophase pitch precursor fibers

| Properties | Standard grades | | | | | | New grades | | | |
	Low modulus		High modulus		Very high modulus		Low modulus		High modulus	
Axial										
Tensile modulus, GPa (10⁶ psi)	160	(25)	380	(55)	725	(110)	225	(35)	380	(55)
Tensile strength, GPa (10⁶ psi)	1.4	(0.20)	1.7	(0.25)	2.2	(0.32)	3.1	(0.45)	3.1	(0.45)
Elongation at break, %	0.9	...	0.4	...	0.3	...	...	...	...	...
Thermal conductivity, W/m · K (Btu · in./h · ft² · °F)	...	...	100	(690)	520	(3600)				
Electrical resistivity, μΩ · m (μΩ · cm)	13	(1300)	7.5	(750)	2.5	(250)	...	...	...	...
Coefficient of thermal expansion at 21 °C (70 °F), 10⁻⁶/K	...	...	−0.9	...	−1.6	...	...	...	...	...
Transverse										
Tensile modulus, GPa (10⁶ psi)	...	...	21	(3)	...	...	...	...	...	...
Coefficient of thermal expansion at 50 °C (120 °F), 10⁻⁶/K	...	...	7.8	...	...	...	...	...	...	...
Bulk										
Density, g/cm³	1.9	...	2.0	...	2.15	...	2.05	...	2.15	...
Filament diameter, μm (μin.)	11	(430)	10	(390)	10	(390)	8	(310)	8	(310)
Carbon assay, %	97+	...	99+	...	99+	...	99+	...	99+	...

Note: The data in this table was obtained from technical data sheets from Great Lakes Carbon Corp., BASF Structural Materials Inc., Amoco Corp., Hercules Corp., Fiber Materials Inc., Stackpole Fibers Co., Polycarbon Inc., and Ashland Petroleum Co.

Table 4 Properties of carbon fibers: low-modulus rayon and isotropic pitch precursor fibers

Properties	Rayon precursor		Isotropic pitch precursor	
Axial				
Tensile modulus, GPa (10^6 psi)	41	6	55	8
Tensile strength, GPa (10^6 psi)	1.0	(0.15)	0.7	0.10
Elongation at break, %	2.5	...	1.4	...
Electrical resistivity, $\mu\Omega \cdot$ m ($\mu\Omega \cdot$ cm)	20	(2000)	30	(3000)
Bulk				
Density, g/cm^3	1.6	...	1.6	...
Filament diameter, μm (μin.)	8.5	(330)	10	(390)
Carbon assay, %	99	...	98	...

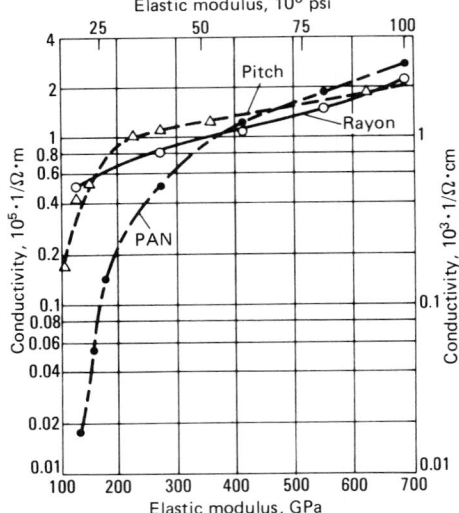

Fig. 6 The conductivity of carbon fibers increases with fiber modulus (heat treatment temperature). These data have been collected by laboratory heat treatments of fibers and may differ from conductivity measured on commercially produced fibers

10^6 psi). In general, at least 1.5% strain at failure is required for secondary composite structure and 2% or better for primary structure. The newer PAN precursor grades of carbon fiber achieve 2% strain at failure and can provide slightly higher stiffness as well. The newer grades also provide much higher strength and specific strength, which is attractive for nonstiffness-critical structures such as pressure vessels. The advantages of mesophase pitch-based fibers are that they can be made in ultrahigh-modulus grades ($\geq$600 GPa, or 90 × 10^6 psi), which are desirable for stiffness-critical but lightly loaded structures.

Carbon fibers increase in modulus by about 10% when highly loaded (Ref 34). However, the fibers show complete elastic recovery upon unloading (Ref 35) and do not appear to degrade because of mechanical fatigue or by stress rupture at temperatures below 2000 °C (3650 °F), as glass or aramid fibers do at or near room temperature (Ref 36-38). High-temperature mechanical properties depend on the maximum heat treatment temperature and the inherent mechanical properties of graphite. Measurements performed above the heat treatment temperature of the fiber will cause changes in microstructure and properties. The moduli and strength of carbon fibers are essentially constant up to temperatures of 1000 °C (1830 °F) for fibers given a higher heat treatment temperature. There is a drop of about 30% in modulus at 1900 °C (3450 °F). Above 2200 °C (3990 °F), creep becomes appreciable (Ref 39).

The coefficients of thermal expansion (CTEs) are slightly negative and become more negative with increasing modulus. This contraction, which is due to layer bending, can be combined with the positive coefficient of the matrix to produce a material with a near-zero CTE over a temperature range of several hundred degrees. At higher temperatures, (>700 °C, or 1290 °F), the CTE for all carbon fibers is positive. Zero-CTE metal matrix composites require the more negative CTE and very high modulus fibers ($\geq$650 GPa, or 95 × 10^6 psi) that are only available in mesophase pitch-based carbon fibers. Both high-modulus PAN- or pitch-based fibers can be used with zero-CTE resin matrix composites because of the lower matrix modulus.

Thermal conductivity and electrical resistivity depend on preferred orientation, crystallite size, and purity (Ref 40). All three increase with increasing heat treatment temperatures. Thermal conductivity can equal or exceed that of copper in the very high modulus, commercially available pitch-based fibers. Thermal conductivities that are several times higher have been made in the laboratory and should become available in the future. The electrical conductivity of carbon fibers varies with precursor type and heat treatment temperature, and is 1/50 or less that of copper for commercial 230-GPa (35 × 10^6-psi) fibers (Fig. 6). The fiber precursor is important for low heat treatment temperature, but to a lesser extent for high-temperature heat treatment. The relatively high conductivity of carbon fibers can cause electrical failures. Free-floating carbon fibers, which could be produced during composite manufacture, can be a problem in shorting out electrical equipment, and must be protected against.

Transverse Properties. While property measurements parallel to the fiber axis are easy to make, transverse property measurements on an 8-μm (315-μin.) fiber are nearly impossible. Values in Table 2 were derived from composite measurements and theory (Ref 41, 42). As can be seen, the fibers are anisotropic with a relatively low transverse modulus and a positive transverse CTE.

Interfacial Bonding

Carbon fibers are not wet by molten metals and are difficult to wet with resins, especially the higher-modulus fibers (Ref 43, 44). Surface treatments, which increase the number of active chemical groups and sometimes roughen the fiber surface, have been developed for the more mature resin matrix composites. Carbon fibers are commonly shipped with an epoxy size, which usually prevents fiber abrasion, improves handling, and provides an epoxy matrix compatible interface. Special surface treatments and sizes have been developed for other resins, especially polyimides and thermoplastics, and should be specified. Fiber and matrix interfacial bond strengths approach the strength of the neat matrix resin for lower-modulus carbon fibers (Ref 45). Higher-modulus PAN-based fibers ($\geq$400 GPa, or 60 × 10^6 psi) show substantially lower interfacial bond strengths (Ref 45). Failure in high-modulus fiber occurs in its surface layer,

in much the same way failure occurs within the fiber for aramids (Ref 46).

Environmental Interaction

Carbon fibers are not affected by moisture, atmosphere, solvents, bases, and weak acids at room temperature (Ref 47). However, oxidation becomes a problem at elevated temperature. For low-modulus PAN-based fibers and high-modulus PAN- or pitch-based fibers, the threshold for oxidation for extended operating times is 350 °C (660 °F) and 450 °C (840 °F), respectively (Ref 48). Oxidation is catalytic at these low temperatures, and somewhat improved oxidation resistance can be expected with higher-purity fibers and resins in the future (Ref 49).

Carbon fibers react with molten aluminum and titanium and must be protected by a barrier coating (Ref 50). While carbon does not react with nickel, a small amount of solution/dissolution can occur (even in the solid state), which degrades the fibers (Ref 51-53). Reactions occur with most oxides in the 1200 to 1500 °C (2190 to 2730 °F) range, and in the 1600 to 1700 °C (2910 to 3090 °F) range for zirconia (ZrO_2), hafnia (HfO_2), and thoria (ThO_2) (Ref 54).

REFERENCES

1. O.L. Blakslee, D.G. Procter, E.J. Seldin, G.B. Spence, and T. Weng, Elastic Constants of Compression Annealed Pyrolytic Graphite, *J. Appl. Phys.*, Vol 41, 1970, p 3373

2. G.D. D'Abate and R.J. Diefendorf, The Effect of Heat on the Structure and Prop-

erties of Mesophase Precursor Carbon Fibers, in *Proceedings of the 17th Biennial Conference on Carbon*, American Carbon Society, 1985, p 390
3. R. Berman, *The Properties of Diamond*, J.E. Field, Ed., Academic Press, 1979
4. *Carbon*, Vol 20 (No. 5), 1982, p 445-449
5. T. Edison, U.S. Patent 223,898, Jan 1880
6. C.E. Ford and C.V. Mitchell, U.S. Patent 3,107,152, Oct 1963
7. R. Bacon and M.M. Tang, Carbonization of Cellulose Fibers I, *Carbon*, Vol 2, 1964, p 211
8. R. Bacon and M.M. Tang, Carbonization of Cellulose Fibers II, *Carbon*, Vol 2, 1964, p 220
9. T.V. Hughes and C.R. Chambers, British Patent 405,480, June 1889
10. A. Oberlin, M. Endo, and T. Koyama, Filamentous Growth of Carbon Through Benzene Decomposition, *J. Cryst. Growth*, Vol 32, 1976, p 335
11. W. Johnston, L.N. Phillips, and W. Watt, British Patent 1,110,791, April and Dec 1964
12. W. Johnston, L.N. Phillips, and W. Watt, U.S. Patent 3,412,062, Nov 1968
13. L.S. Singer, Netherlands Patent 239490, April 1972
14. L.S. Singer, U.S. Patent 4,005,183, Jan 1977
15. R. Bacon and W.A. Schalamon, High Strength and High Modulus Carbon Fibers, in *Proceedings of the Eighth Biennial Conference on Carbon*, American Carbon Committee, 1967
16. H. Hawthorne, Structure and Properties of Strain-Graphitized Glassy Carbon Fibers, in *Proceedings of the First International Conference on Carbon Fibers*, Plastics Industry, 1971, p 81
17. W. Soltes, U.S. Patent 3,011,981, 1961
18. W. Abbott, U.S. Patent 3,053,775, 1962
19. S. Otani, The Fundamental of MP Carbon Fiber, *Carbon*, Vol 3, 1965, p 213
20. R.J. Diefendorf and D.M. Riggs, U.S. Patent 4,208,267, June 1980
21. R. Prescott and A. Standage, High Elastic Modulus Carbon Fibre, *Nature*, Vol 211, 1966, p 169
22. W. Ruland, The Relationship Between Preferred Orientation and Young's Modulus of Carbon Fibers, *Polymer Preprints*, Vol 9 (No. 2), Polymer Chemistry Division, American Chemical Society, 1968, p 1368
23. C.W. LeMaistre and R.J. Diefendorf, The Origin of Structure in Carbonized PAN Fibers, *SAMPE Q.*, Vol 4, 1973, p 1
24. R. Perret and W. Ruland, The Microstructure of PAN-Base Carbon Fibers, *J. Appl. Crystallogr.*, Vol 3, 1970, p 525
25. S.C. Bennett and D.J. Johnson, Structural Characterization of a High Modulus Carbon Fibre by High-Resolution Electron Microscopy and Electron Diffraction, *Car-*

bon, Vol 14, 1976, p 117
26. E.W. Tokarsky and R.J. Diefendorf, High Performance Carbon Fibers, *Polym. Eng. Sci.*, Vol 15 (No. 3), 1975, p 150
27. R.J. Diefendorf and E.W. Tokarsky, "The Relationships of Structure to Properties in Graphite Fibers, Part I," AFML-TR-72-133, Air Force Materials Laboratory, 1971
28. R.J. Diefendorf and E.W. Tokarsky, "The Relationships of Structure to Properties in Graphite Fibers, Part II," AFML-TR-72-133, Air Force Materials Laboratory, 1973
29. R.J. Diefendorf and E.W. Tokarsky, "The Relationships of Structure to Properties in Graphite Fibers, Part III," AFML-TR-72-133, Air Force Materials Laboratory, 1975
30. R.J. Diefendorf and E.W. Tokarsky, "The Relationships of Structure to Properties in Graphite Fibers, Part IV," AFML-TR-72-133, Air Force Materials Laboratory, 1975
31. D.M. Riggs, "The Characterization and Kinetic Mechanism of Mesophase Formation in High Molecular Weight Carbonaceous Materials and Its Relationship to the Concepts of Thermal and Catalytic Cracking," Ph.D. thesis, Rensselaer Polytechnic Institute, 1979
32. B.L. Butler and R.J. Diefendorf, Graphite Filament Structure, in *Proceedings of the Tenth Carbon Composite Technology Symposium*, American Society of Mechanical Engineers, 1970, p 109
33. K.J. Chen, C.W. LeMaistre, J.H. Wang, and R.J. Diefendorf, The Consequences of Residual Stress in High Modulus Carbon Fibers on Composite Performance, in *Polymer Preprints*, American Chemical Society, Vol 22 (No. 2), 1981, p 212
34. C.P. Beetz, Strain Induced Stiffening in Carbon Fibers, in *Proceedings of the 15th Biennial Conference on Carbon*, American Carbon Society, 1981, p 302
35. R.J. Diefendorf, unpublished data
36. J. Awerback and H.T. Hahn, Fatigue and Proof Testing of Unidirectional Graphite/ Epoxy Composites, in *Fatigue of Filamentary Composite Materials*, STP 636, American Society for Testing and Materials, 1977, p 248
37. T.T. Chiao, C.C. Chiao, and R.J. Sherry, Lifetimes of Fiber Composites Under Sustained Tensile Loading, in *Proceedings of the 1977 International Conference on Fracture Mechanics and Technology*, 1977
38. R.J. Diefendorf, unpublished data
39. L.A. Feldman, High Temperature Creep Effects in Carbon Yarns and Composites, in *Proceedings of the 17th Biennial Conference on Carbon*, American Carbon Society, 1985, p 393
40. A.A. Bright and L.S. Singer, Electronic and Structural Characteristics of Carbon Fibers from Mesophase Pitch, in *Proceed-*

ings of the 13th Biennial Conference on Carbon, American Carbon Society, 1977, p 100
41. J.H. Helmer and R.J. Diefendorf, Transverse Properties of Carbon and Kevlar Fiber Composites, in *Proceedings of the 16th Biennial Conference on Carbon*, American Carbon Society, 1983, p 511
42. I.M. Kowalski, Determining the Transverse Modulus of Carbon Fibers, *SAMPE J.*, Vol 22 (No. 4), 1986, p 38
43. B.L. Butler, C.W. LeMaistre, and R.J. Diefendorf, The Structure of High Modulus Graphite Fibers, in *Eighth Annual Conference Proceedings*, Society of the Plastics Industry, 1973, p 21-C
44. J.B. Donnet and P. Ehrberger, Carbon Fibre in Polymer Reinforcement, *Carbon*, Vol 15, 1977, p 143
45. J.C. Goan and S.P. Prosen, *Interfacial Bonding in Graphite-Resin Composites*, STP 452, American Society for Testing and Materials, 1969, p 3
46. L.T. Drzal, Adhesion of Graphite Fibers in Epoxy Matrices: I. Role of Fiber Surface Treatment, *J. Adhesion*, Vol 16, 1982, p 1-30
47. N.C.W. Judd, The Chemical Resistance of Carbon Fibres and a Carbon Fibre/ Polyester Composite, in *Proceedings of the First International Conference on Carbon Fibres*, Plastics Institute, 1971, p 258
48. D.W. McKee and V.J. Mimeault, Surface Properties of Carbon Fibers, in *Chemistry and Physics of Carbon*, Vol 8, Marcel Dekker, 1973, p 235
49. D.W. McKee, The Catalysed Gasification Reactions of Carbon, in *Chemistry and Physics of Carbon*, Vol 17, Marcel Dekker, 1981, p 1
50. W. Meyerer, D. Kizer, and H. Paul, Versatility of Graphite-Aluminum Composites, in *Proceedings of the 1978 International Conference on Composite Materials*, American Institute of Mining, Metallurgical, and Petroleum Engineers, 1978, p 141
51. P.W. Jackson and J.R. Marjoram, Recrystallization of Nickel-Coated Carbon Fibers, *Nature*, Vol 218, 1968, p 83
52. R.V. Sara, Fabrication and Properties of Graphite-Fiber Nickel-Matrix Composites, in *Proceedings 14th SAMPE Symposium*, Society for the Advancement of Material and Process Engineering, 1968, p II-4a-4
53. S. Sarian, Elevated Temperature of Carbon-Fibre, Nickel-Matrix Composites, Morphological and Mechanical Property Degradation, *J. Mater. Sci.*, Vol 8, 1973, p 251-260
54. R.M. Bushong, "Improved Graphite Materials for High-Temperature Aerospace Use, Vol II, Development of Graphite-Refractory Composites Technical Documentary Report," AFML ML-TDR-64-125, Air Force Materials Laboratory

Organic Fibers

Jack J. Pigliacampi, E.I. Du Pont de Nemours & Company, Inc.

THE FIRST ORGANIC FIBER with a high enough tensile modulus and strength to be used as a reinforcement in advanced composites was an aramid or aromatic polyamide fiber introduced in the early 1970s. The technology advances that led to these new levels of fiber tensile properties have been reported by many authors (Ref 1-3). This article will focus on the properties of the Kevlar aramid fibers, produced by Du Pont, which are the predominant organic reinforcing fibers. More recently, related aramid fibers are becoming available, specifically, Twaron by the Enka Corporation (Netherlands) and HM-50 by Teijin (Japan). Also, a high-strength high-modulus polyethylene fiber is being introduced by Allied Corporation (U.S.).

Para-Aramid Fiber

The chemical composition of Kevlar aramid fiber is poly para-phenyleneterephthalamide. This fiber is known as PPD-T because it is made from the condensation reaction of para-phenylene diamine and terephthaloyl chloride (Fig. 1). The aromatic ring structure contributes high thermal stability, while the para configuration leads to stiff, rigid molecules that contribute high strength and high modulus.

Para-aramid fibers belong to a class of materials known as liquid crystalline polymers. Because these polymers are very rigid and rodlike, in solution they can aggregate to form ordered domains in parallel arrays (Ref 4). This is shown in Fig. 2 and contrasted to more conventional, flexible polymers, which in solution can bend and entangle, forming random coils.

When PPD-T solutions are extruded through a spinneret and drawn through an air gap during fiber manufacture, the liquid crystalline domains can orient and align in the direction of flow (Ref 5-7), as shown in Fig. 3. With PPD-T, there is an exceptional degree of alignment of long, straight polymer chains parallel to the fiber axis. This structure, which has been analyzed and characterized extensively (Ref 8-11), is anisotropic and gives higher strength and modulus in the fiber longitudinal direction than in the axial direction. It is also fibrillar, as shown in Fig. 4, which has a profound effect on fiber properties and failure mechanisms, as will be discussed.

Material Properties

Key representative properties of para-aramid (p-aramid) fibers are given in Table 1. Kevlar 49 is the dominant form used today in structural composites because of its higher modulus. Kevlar 29 is used in composites when higher toughness, damage tolerance, or ballistic stopping performance is desired. An ultra-high-modulus fiber, Kevlar 149, is also available. Kevlar aramid fibers are available in a wide range of yarn or tow counts and in a variety of short fiber forms.

Tensile modulus of p-aramid is a function of the molecular orientation. As a spun fiber, Kevlar 29 has a modulus of 62 GPa (9 × 10⁶ psi). In composite form, it has a modulus that is slightly higher than that of E-glass (83 versus 69 GPa, or 12 versus 10 × 10⁶ psi). Heat treatment under tension increases crystalline orientation, and the resulting fiber, Kevlar 49, has a modulus of 131 GPa (19 × 10⁶ psi). Kevlar 149, a p-aramid with an even higher modulus (186 GPa, or 27 × 10⁶ psi), is now available. This modulus level approaches the theoretical maximum predicted for p-aramid (Ref 18).

The tensile strength of p-aramid fiber is in the range of 3.6 to 4.1 GPa (0.525 to 0.600 × 10⁶ psi). This is more than twice the strength of conventional organic fibers such as Nylon 66, and is 50% greater than E-glass strength. It

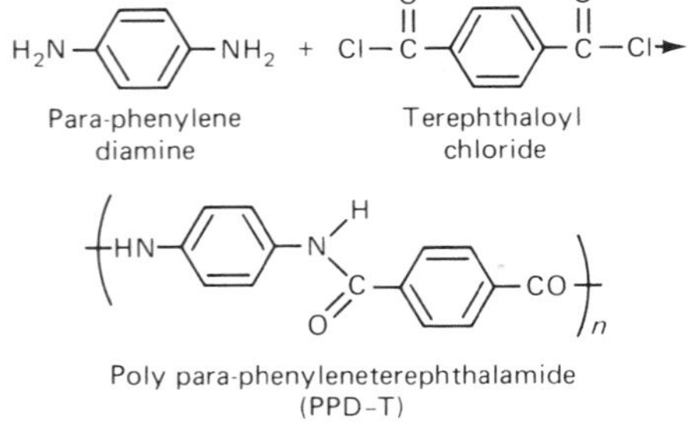

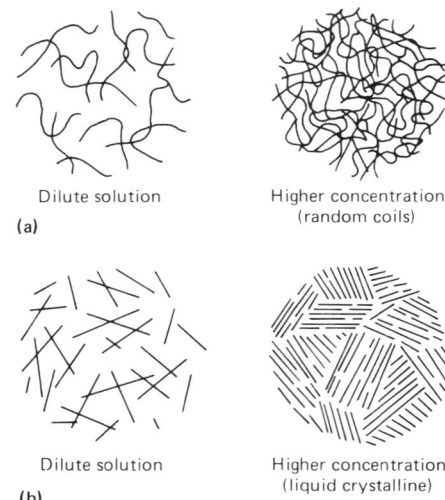

Fig. 1 Chemical structure of para-aramid

Para-phenylene diamine + Terephthaloyl chloride → Poly para-phenyleneterephthalamide (PPD-T)

Fig. 2 Polymer states in solution. (a) Flexible molecules. (b) Rigid molecules

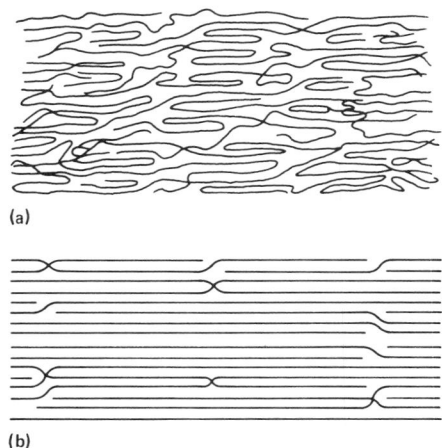

Fig. 3 Polymer chain orientation. (a) Conventional organic, characterized by chain folds, misalignment, and crystalline and amorphous regions. (b) Para-aramid, characterized by long, straigh t chains without folds, parallel to the fiber axis, crystalline

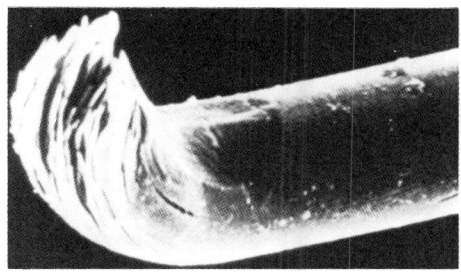

(a)

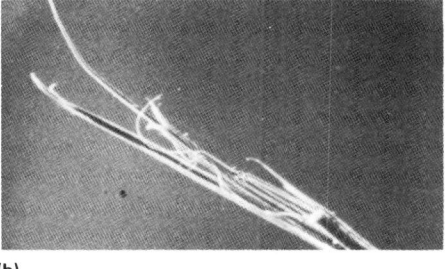

(b)

Fig. 4 Fibrillar structure of Kevlar aramid fibers. (a) Loop break. (b) Tensile failure

Table 1 Properties of para-aramid fibers

Material	Density, g/cm³	Filament diameter μm	μin.	Tensile modulus(a) GPa	10⁶ psi	Tensile strength(a) GPa	10⁶ psi	Tensile elongation, %	Available yarn count, No. filaments
Kevlar 29 (high toughness) 1.44		12	470	83	12	3.6 2.8(b)	0.525 0.400(b)	4.0	134-10 000
Kevlar 49 (high modulus) 1.44		12	470	131	19	3.6-4.1	0.525-0.600	2.8	134-5000
Kevlar 149 (ultra-high modulus) . . 1.47		12	470	186	27	3.4	0.500	2.0	134-1000

(a) ASTM D 2343, impregnated strand (Ref 12). (b) ASTM D 885, unimpregnated strand (Ref 13). Source: Ref 14-17

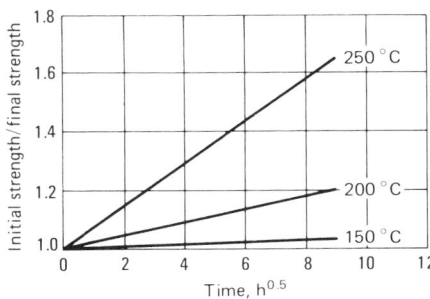

Fig. 6 Room-temperature tensile strength retention after air aging of Kevlar 49 aramid yarn at various temperatures

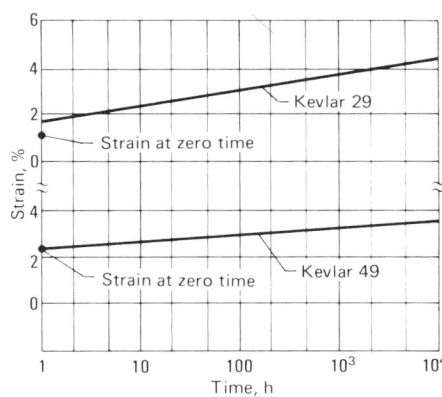

Fig. 8 Creep of Kevlar 29 and Kevlar 49 aramid yarns at 50% of ultimate strength

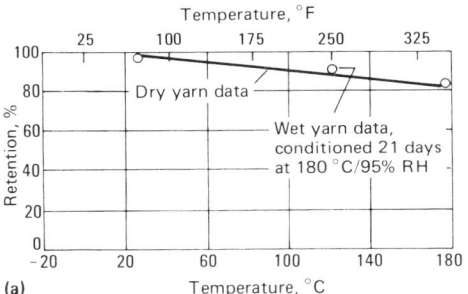

(a)

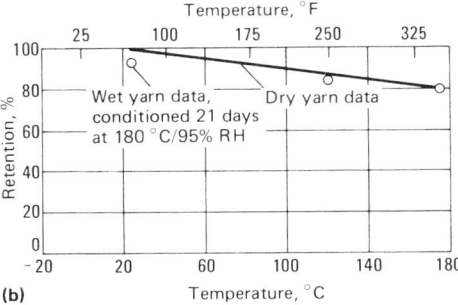

(b)

Fig. 5 Effect of temperature on tensile strength and modulus of dry and wet Kevlar 49 aramid yarn. (a) Tensile strength. (b) Tensile modulus

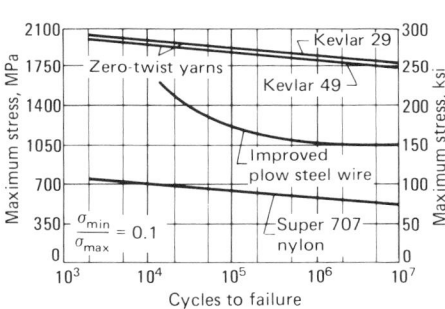

Fig. 7 Comparative tension-tension fatigue for yarns and wire

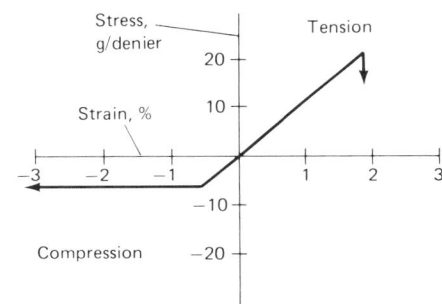

Fig. 9 Para-aramid stress-strain behavior in tension and compression

is believed that tensile failure initiates at fibril ends and propagates via shear failure between the fibrils.

Tensile Properties in Hot/Wet Conditions. High modulus *p*-aramid yarns show a linear decrease of both tensile strength and modulus when tested at elevated temperatures in air (Fig. 5). More than 80% of these prop-

erties are retained at 180 °C (355 °F). Figure 6 shows retention of room temperature yarn strength after long exposures at elevated temperatures. More than 80% of strength is retained after 81 h at 200 °C (390 °F).

At room temperature, the effect of moisture on tensile properties is <5% (Ref 19). At elevated temperature, the effect of moisture appears to be reversible: Yarn conditioned for 21 days at 180 °C (355 °F)/95% relative humidity tested at high temperature had essentially the same behavior as dry yarn (Fig. 5). Tests of yarn while immersed in hot water suggest a loss of 10% due to the water alone (Ref 19, 20).

Creep and Fatigue. Para-aramid is resistant to dynamic and static fatigue (Fig. 7 and 8). Creep rate is low and similar to that of fiberglass, but unlike glass, *p*-aramid is less susceptible to creep rupture (Ref 21).

Compressive Properties. Although *p*-aramid fiber responds elastically in tension, it exhibits nonlinear, ductile behavior under com-

pression. At a compression strain of 0.3 to 0.5%, a yield is observed (Fig. 9). This corresponds to the formation of structural defects known as kink bands (Fig. 10), which are related to compressive buckling of *p*-aramid molecules. As a result of this compression behavior, the use of *p*-aramid fibers in applications that are subject to high strain compressive or flexural loads is limited.

Toughness. Para-aramid fiber is noted for its toughness and general damage tolerance characteristics (Ref 23). In part, this is related directly to conventional tensile toughness, or the area under the stress-strain curve. Toughness is also related to composite impact resistance and ballistic stopping power (Ref 24, 25). The *p*-aramid fibrillar structure and compressive behavior contribute to composites that are less notch sensitive (Ref 26) and that fail in a

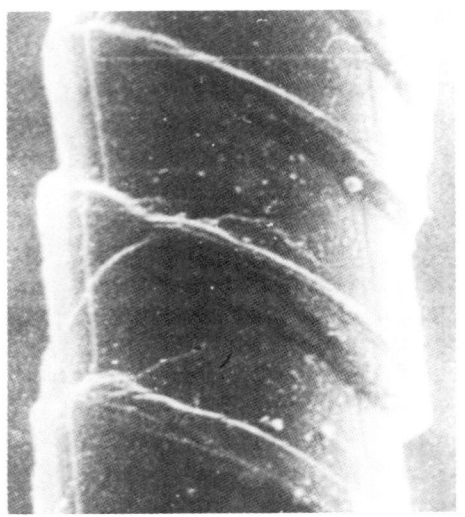

Fig. 10 Kink bands from severe compression. Source: Ref 22

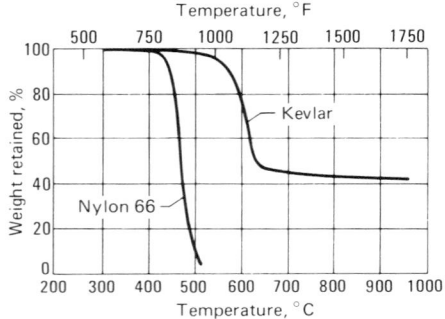

Fig. 11 Thermal stability; thermogravimetric analysis done at 20 °C (70 °F)/min in nitrogen

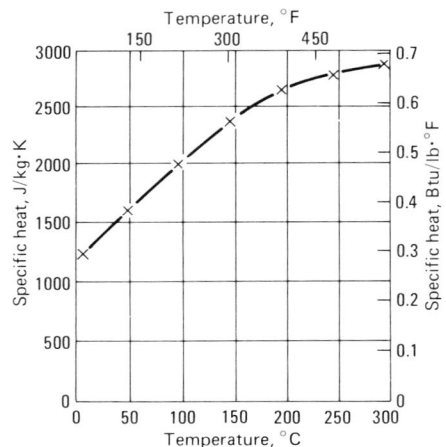

Fig. 12 Effect of temperature on the specific heat of Kevlar 49 aramid

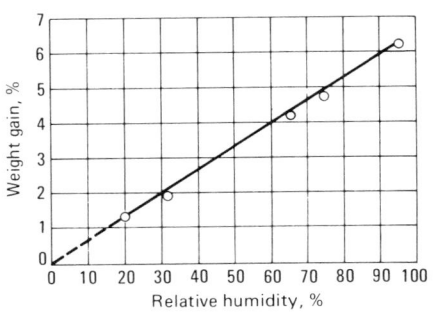

Fig. 13 Equilibrium moisture content versus relative humidity for Kevlar 49 at room temperature

Table 2 Effect of electron radiation on Kevlar 49

Using resonant transformer and filament wrapped in aluminum foil over dry ice, conditions of exposure were 1 Mrad every 13.4 s, 0.5 mA, 2 MV, and 30-cm (10-in.) distance.

| | Single-filament properties | | | | |
| | Tenacity | | Tensile modulus | | Elongation, |
Mrad exposure	MPa	ksi	GPa	10⁶ psi	%
0	2860	415	128	18.6	2.4
100	2940	426	130	18.8	2.4
200	3010	436	133	19.3	2.5

Table 3 Properties of HM-50 aramid fiber

Density, g/cm³	1.39
Filament diam, μm (μin.)	12 (470)
Tensile modulus(a), GPa (10⁶ psi)	81 (12)
Tensile strength(a), GPa (10⁶ psi)	3.1 (0.445)
Tensile elongation, %	4.4
Available yarn counts, No. filaments	134-1000

(a) ASTM D 885, unimpregnated yarn (Ref 12)

Table 4 Properties of Spectra polyethylene fibers

	Spectra 900	Spectra 1000
Density g/cm³	0.97	0.97
Filament diameter μm (μin.)	38 (1500)	27 (1060)
Tensile modulus GPa (10⁶ psi)	117 (17)	172 (25)
Tensile strength GPa (10⁶ psi)	2.6 (0.380)	2.9-3.3 (0.430-0.480)
Tensile elongation %	3.5	0.7
Available yarn counts No. filaments	60-120	60-120

ductile, nonbrittle or noncatastrophic manner, as opposed to glass and carbon (Ref 27).

Thermal Properties. The aromatic chemical structure of *p*-aramid imparts a high degree of thermal stability (Fig. 11). Fibers from PPD-T do not have a literal melting point or a glass transition temperature (T_g) (estimated at ≥375 °C, or 710 °F), as normally observed with other synthetic polymers. They decompose in air at 425 °C (800 °F) and are inherently flame resistant (oxygen index of 0.29). They have utility over a broad temperature range of about −200 to 200 °C (−330 to 390 °F), but are not generally used long-term at temperatures above 150 °C (300 °F) because of oxidation (Fig. 6). Para-aramid fiber has a slightly negative longitudinal coefficient of thermal expansion of -2×10^{-6}/K and a positive transverse expansion of 60×10^{-6}/K.

Para-aramid fiber has a low thermal conductivity that varies by about an order of magnitude in the longitudinal versus transverse direction. Combustion heat is about 35 MJ/kg (15 000 Btu/lb). The specific heat versus temperature is shown in Fig. 12.

Electrical and Optical Properties. Para-aramid is an electrical insulator. Its dielectric constant of ~4.0, measured at 10^6 Hz, is lower than that of fiberglass and about the same as that of quartz. The index of refraction of *p*-aramid fiber is 2.0 parallel to the fiber axis and 1.6 perpendicular to the fiber axis. E-glass is 1.55 in both directions.

Environmental Behavior. Para-aramid fiber has an equilibrium moisture content that is determined by the relative humidity, as shown in Fig. 13. At 60% RH, equilibrium moisture of Kevlar 49 is about 4%. Kevlar 149 ultra-high-modulus fiber has a corresponding value of about 1.5%.

Para-aramid fiber can be chemically degraded by strong acids and bases. It is resistant to most other solvents and chemicals (Ref 14). Ultraviolet radiation also can degrade *p*-aramid. The degree of degradation depends on material thickness because *p*-aramid is self-screening. In polymeric composites, strength loss of *p*-aramid has not been observed (Ref 28).

Para-aramid is resistant to electronic radiation, as shown in Table 2. Vacuum outgassing has been performed on *p*-aramid yarn that had been predried at 125 °C (257 °F) for 24 h under a vacuum of 1.3×10^{-4} Pa (10^{-6} torr). With a colorless deposit of 0.02 wt% micro volatile condensible materials (VCM), there was a 2.28% weight loss.

Other Fibers

The Enka Corporation is developing a newer *p*-aramid fiber, whose properties (Ref 29) are essentially the same as those previously described. Another recently introduced aramid copolymer (Ref 30) has properties (Table 3) that are similar to as-spun *p*-aramid fiber, but with some significant differences: The copolymer has a slightly lower density (~3%); its strength and modulus are similar to those of Kevlar 29, but its modulus is about 38% lower than that of Kevlar 49 *p*-aramid; its oxidative and hydrolytic stability are superior to *p*-aramid; moisture regain is similar to that of Kevlar 149; and high-temperature fiber shrinkage is higher than for *p*-aramid.

Allied Corporation has developed a high-strength high-modulus polyethylene fiber for tensile applications such as rope and cord. It may also have potential use in polymer matrix composites (Ref 31). Room-temperature properties of this fiber are shown in Table 4. While this fiber exhibits superior mechanical proper-

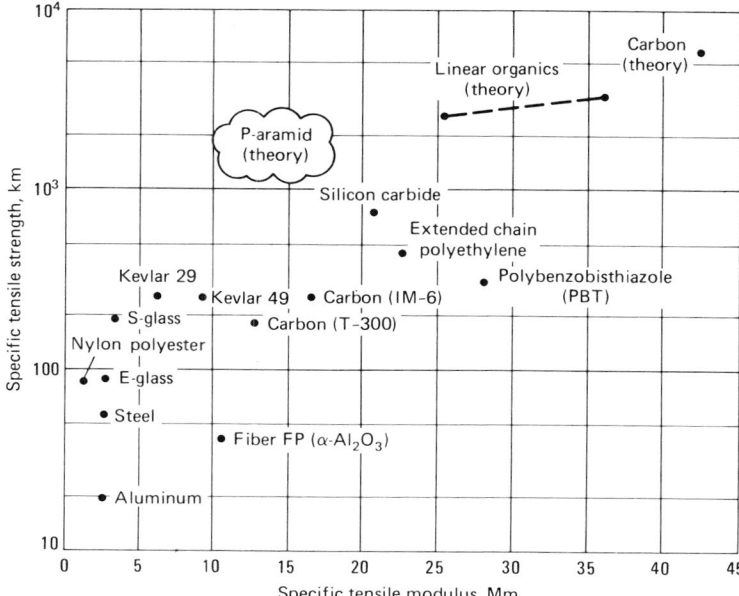

Fig. 14 Comparative specific strength and modulus

ties at ambient temperatures, its properties decrease rapidly with increasing temperature because of the relatively low melting point of the polymer.

As composite technology advances and manufacturing techniques become more cost effective, composite use will expand rapidly. Organic reinforcing fibers will play a major role in the expansion because they can be molecularly tailored for improved properties to meet specific application requirements. Figure 14 shows the density considered tensile properties of a wide range of commercial and experimental materials, as well as theoretical values of organic and carbon fibers, indicating the potential for future advances.

REFERENCES

1. M. Jaffe and R.S. Jones, chapter 9 in *High Technology Fibers*, Vol III, Part A, M. Lewin, Ed., Marcel Dekker, 1985
2. J. Preston, Aramid Fibers, in *Kirk-Othmer Encyclopedia of Chemical Technology*, Third Edition, Vol 4, Wiley-Interscience, 1978
3. W.B. Black and J. Preston, *High Modulus Wholly Aromatic Fibers*, Marcel Dekker, 1973
4. P.J. Flory, Molecular Theory of Liquid Crystals, in *Advances in Polymer Science*, Vol 59, Springer-Verlag, 1984, p 1-36
5. P.W. Morgan, *Macromolecules*, Vol 10, 1977, p 1381
6. S.L. Kwolek, P.W. Morgan, J.R. Schaefgen, and L.W. Gulrich, *Macromolecules*, Vol 10, 1977, p 1390
7. H. Blades, U.S. Patent 3,767,756, 1973

8. D. Tanner, J.A. Fitzgerald, W.F. Knoff, and J.J. Pigliacampi, "Aramid Fiber Structure/Property Relationships and Their Applications to Industrial Materials," Paper presented at Japan Fiber Society Meeting, Aug 1985
9. M.G. Northolt, *Eur. Polym. J.*, Vol 10, 1974, p 799
10. J.W. Ballou, *Polym. Preparation*, Vol 17, 1976, p 75
11. R. Hagege, M. Jarrin, and M. Sotton, *J. Microsc.*, Vol 115, 1979, p 65
12. "Standard Test Method for Tensile Properties of Glass Fiber Strands, Yarns, and Rovings Used in Reinforced Plastics," D 2343, *Annual Book of ASTM Standards*, American Society for Testing and Materials
13. "Standard Methods of Testing Tire Cords, Tire Cord Fabrics, and Industrial Filament Yarns Made From Man-Made Organic-Base Fibers," D 885, *Annual Book of ASTM Standards*, American Society for Testing and Materials
14. "'Kevlar' 49 Aramid", E.I. Du Pont de Nemours & Company, Inc.
15. "Characteristics and Uses of 'Kevlar' 49 Aramid High Modulus Organic Fiber," Technical Bulletin, E.I. Du Pont de Nemours & Company, Inc.
16. "Characteristics and Uses of 'Kevlar' 29 Aramid," Technical Bulletin, E.I. Du Pont de Nemours & Company, Inc.
17. R.E. Wilfong and J. Zimmerman, Strength and Durability Characteristics of "Kevlar" Aramid Fiber, *J. Appl. Polym. Sci.*, 31st Applied Polymer Symposium, John Wiley & Sons, 1977, p 1-21

18. E.E. Magat, *Philos. Trans. R. Soc.*, (London) A, Vol 294 (No. 463), 1980
19. W.S. Smith, Environmental Effects on Aramid Composites, in *Proceedings of Society of Plastics Engineers Conference*, Dec 1979
20. N.J. Abbott, *et al.*, *Some Mechanical Properties of "Kevlar" and Other Heat Resistant, Nonflammable Fibers, Yarns and Fabrics*, AFML-TR-74-65, Part III, Air Force Materials Laboratory, March, 1975
21. P.G. Riewald, *et al.*, Strength and Durability Characteristics of Ropes and Cables from "Kevlar" Aramid Fibers, in *Oceans '77 Conference Record*, Third Combined Conference sponsored by Marine Technology Society and Institute of Electrical and Electronics Engineers, Oct 1977
22. M.G. Dobb, D.J. Johnson, and B.P. Saville, *Polymer*, Vol 22 (No. 7), 1981, p 960
23. M.W. Wardle, Designing Composite Structure for Toughness, in *Technical Symposium V, Design and Use of "Kevlar" Aramid Fiber in Composite Structures*, 1984
24. M.W. Wardle and E.W. Tokarsky, Drop Wings Testing of Laminates Reinforced with "Kevlar" Aramid Fibers, E-Glass and Graphite, in *Composites Technology Review*, American Society for Testing and Materials, 1983
25. L.H. Miner, "Fragmentation Resistance of Aramid Fabrics and Their Composites," Paper presented at Symposium on Vulnerability and Survivability, San Diego, March 1985
26. C. Zweben, "Fracture of 'Kevlar' 49, E-Glass and Graphite Composites," Paper presented at ASTM Symposium on Fracture Mechanics of Composites, American Society for Testing and Materials, Gaithersburg, MD, Sept 1974
27. J.D. Cronkhite, Design of Helicopter Composite Structures for Crashworthiness, in *Technical Symposium V, Design and Use of "Kevlar" Aramid Fiber in Composite Structures*, E.I. Du Pont de Nemours & Co., Inc., April 1984
28. R.H. Stone, "Flight Service Evaluation of 'Kevlar' 49/Epoxy Composite Panels in Wide Body Commercial Transport Aircraft," NASA Contractor Report 159231, National Aeronautics and Space Administration, March 1980
29. Herbert Blumberg, "The Future of Newly Developed Fibers," Paper presented at Eurofabric 1983, Mainz, West Germany, Sept 1986
30. "High Tenacity Aramid Fibre HM-50", Technical Brochure, Teijin Ltd.
31. R.C. Wincklhofer, "'Spectra' 900 Ultra-High Strength From Polyethylene," Paper presented at State of the Art Symposium, Clemson University, Greenville, SC, Feb 1985

Boron and Silicon Carbide Fibers

T. Schoenberg, Avco Specialty Materials, Textron, Inc.

BOTH BORON AND SILICON CARBIDE monofilament fibers are continuous fibers that are produced at Avco by chemical vapor deposition (CVD) on substrate wires that are drawn through glass reactor tubes. Boron fiber is presently produced in production quantities, predominantly as a reinforcement for epoxy matrix structures used in aerospace and sporting goods applications. Silicon carbide (SiC) monofilament has found greatest application, in development quantities, as a reinforcement for metal structures, most notably aluminum and titanium alloys.

The intent of this article is to provide some important information about these two fiber types. References 1 to 5 give current sources for further study of various aspects of these fibers and their applications in composite structures.

Boron Fibers

Boron fiber is a continuous monofilament produced in two nominal diameters, approximately 100 μm (4000 μin.) and approximately 140 μm (5600 μin.). The fiber is typified by high strength, high stiffness (modulus of elasticity), and low density. Properties of the fiber, listed in Table 1, are compared with typical properties of other fibers, including SiC.

Almost all of the boron fiber produced is used to form a boron-epoxy preimpregnated tape or prepreg, a product form in which an array of fibers is coated with an epoxy resin and backed on one side with a light fiberglass fabric. The tacky prepreg is then sold to users for ply lay-ups and part fabrication.

Figure 1 illustrates the boron fabrication process. A tungsten substrate wire, typically about 12.5 μm (500 μin.) in diameter, is continuously drawn through a vertical glass reactor. Both ends of the reactor tube are sealed by a shallow pool of mercury, which acts also as an electrical contact to the fiber. Power is applied across the ends of the reactor tube such that the tungsten becomes incandescent. A mixture of boron trichloride (BCl_3) vapors and hydrogen (H_2) gas is admitted to the reactor tube, and the hydrogen reduction of the BCl_3 to elemental boron on the surface of the hot substrate takes place. The fiber diameter increases as the substrate travels through the tube, and the diameter of the fiber emerging from the bottom electrode and mercury seal is a function of the fiber throughput rate, other reactor parameters being constant.

The boron fiber is wound onto a take-up spool, with a core diameter of about 200 mm (8 in.). High-quality boron fiber can typically be formed in a loop to a radius of 6.4 mm (0.25 in.). For a 100-μm (4000-μin.) fiber, this corresponds to a filament surface stress of about 6.9 GPa (1.0 × 10^6 psi).

During the course of the deposition reaction, the tungsten substrate wire is also completely reacted to form a mixture of tungsten borides, expanding in diameter to about 17.5 μm (700 μin.). The internal stress states in the fiber set-up from this core expansion and simultaneous boron mantle deposition are key factors in determining the ultimate tensile strength of the fiber.

The typical surface appearance of the boron fiber is that of a corncob structure (Fig. 2). The deposited boron is of the polycrystalline beta-rhombahedral form. The crystalline size, of the order of 20 Å, is so small that it is considered to be amorphous.

The average tensile strength of high-quality boron fiber is the statistical result of many individual fiber tests; a typical histogram depicting these results is shown in Fig. 3. Note that a comparatively low-strength "tail" is present and is representative of the fiber properties. Its presence is related to various circumstances that can lead to premature failure of an individual fiber segment during a tensile test. Statistically, for example, a finite chance always exists for the localized formation of oversized crystallite caused by a microtemperature perturbation during the deposition process, for the inclusion in the fiber of a microscopic dust particle picked up in the top mercury electrode, for the presence of a substrate flaw, or for any number of other occurrences that introduce stress concentrators into the fiber and cause the reduced strength test results shown in the histogram. Note that the high-strength end of the histogram drops off much more sharply than the low-strength tail. This right-hand side of the histogram represents tests of completely unflawed fiber segments where the ultimate strength is controlled by the normal boron surface strength and then, ultimately, by the internal stresses at the core-mantle interface. Much higher test values for boron fibers of up to 6.9 GPa (1.0 × 10^6 psi) can result from smoothing the surface by chemical etching or, especially, by removing this fiber core. However, neither practice is routine in boron filament processing.

Silicon Carbide Fibers

SiC continuous monofilament is produced with a 140 μm (5600 μin.) diam and is typified by high strength, high stiffness (modulus of elasticity), and low density. Properties of this fiber are included in Table 1, and a typical histogram of strength properties is shown in Fig. 4.

Almost all of the SiC monofilament fiber presently produced is used in development programs for reinforcing metallic alloys of aluminum and titanium. However, it has been used successfully to reinforce organic and ceramic matrices.

Three variations of SiC fiber being produced by one manufacturer are SCS-2, which is ori-

Table 1 Comparative properties of reinforcing fibers

Fiber	Density, g/cm³	Average tensile strength(a)		Modulus of elasticity		Approx cost	
		GPa	10^6 psi	GPa	10^6 psi	$/kg	$/lb
Boron, 100 μm (4000 μin.)	2.57	3.6	0.52	400	60	700	320
Boron, 140 μm (5600 μin.).	2.49	3.6	0.52	400	60	700	320
Carbon, AS-4	1.75	3.1	0.45	221	32.1	65	30
E-glass	2.54	3.4	0.49	69	10	5.5	2.5
Aramid	1.44	3.6	0.52	124	18.0	45	20
SiC	3.0	3.9	0.57	400	60	~220(b)	~100

(a) Based on room temperature measurements at 25-mm (1-in.) gage length (b) Projected cost in production quantities

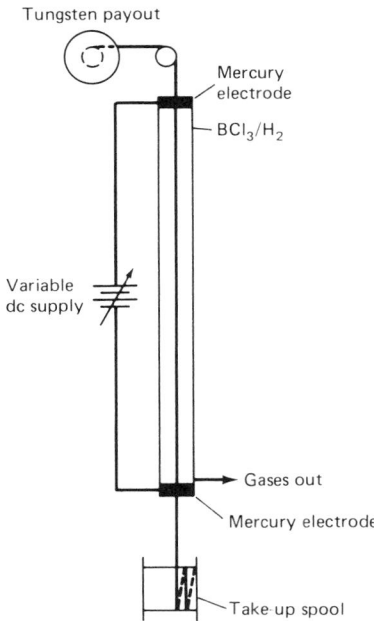

Fig. 1 Boron filament reactor

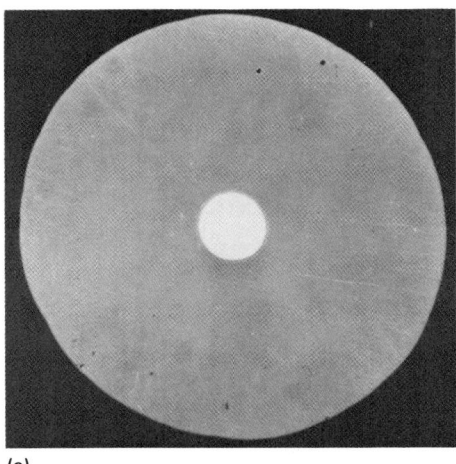

Fig. 2 Photomicrographs of boron fiber. (a) Filament 100-μm (4000-μin.) cross section with 17.5-μm (700-μin.) core. 560×. (b) Magnification of boron filament surface. 110×

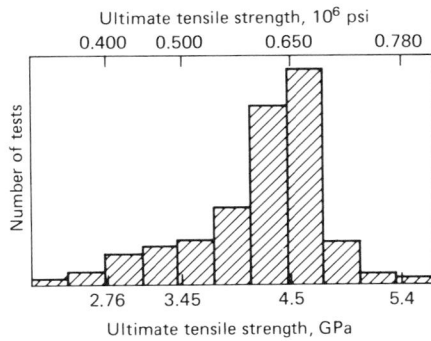

Fig. 3 Histogram of boron fiber tensile strengths

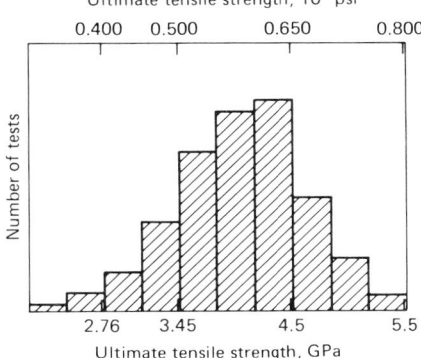

Fig. 4 Histogram of CVD SiC fiber tensile strengths

ented toward reinforcing aluminum alloys, SCS-6, which is used to reinforce titanium alloys, and SCS-8, which is used to reinforce an aluminum alloy structure when the composite must display higher transverse properties than can be obtained by using SCS-2.

The fiber is produced in a CVD process similar to that used for boron and is formed from the reaction of hydrogen with a mixture of chlorinated alkyl silanes at the surface of a resistively heated substrate. The substrate currently being used for SiC fiber is a spun carbon monofilament that has a 33 μm (1300 μin.) diam and is coated with a thin layer of pyrolitic graphite. The raw material for the substrate is coal tar pitch.

One of the main attractions of SiC fiber is its potentially low cost. Projections for volume production levels that are roughly equivalent to present boron production rates indicate that a cost of under $220/kg ($100/lb) is achievable, which is lower than the cost of boron by a factor of three. The explanation for this cost reduction resides in the lower cost of the substrate and raw materials for SiC production, along with a higher reaction and deposition rate than is achievable with boron.

There is a major difference between using fibers to reinforce metallic structures and using them to reinforce organic or ceramic matrices. In organic composites, the organic phase is used to coat the fiber surface, and the bonds at the fiber-matrix interface are essentially physical in character. In ceramic composites, a relatively weak bond at the fiber-matrix interface is actually desirable in order to achieve improved fracture toughness, which is the principal reason for reinforcing the ceramic structure. In metallic matrices, however, a very

strong diffusional bond at the atomic level is required for the reinforcing fiber to impart its strength properties to the metal. Therefore, a high degree of compatibility at the fiber-metal interface is essential for the benefit of the reinforcement properties to be exhibited in the composite. However, if the degree of affinity between the fiber and matrix is too high, the reaction between the metal and the fiber at the interface will result in a severe degradation of fiber properties such that a strong composite will not result.

The above comments introduce the reason that the chemistry and morphology of the surface region of a fiber used to reinforce a metal must be carefully controlled or tailored for optimum compatibility with both the specific metallic alloy and the fabrication method used to form the consolidated composite structure. In general, a fiber being used to consolidate a particular metal structure will probably not be a monolithic multipurpose fiber, but will more likely be a specialized material having a surface composition specifically aimed at matrix compatibility. For SiC fiber, the significance of this is displayed in the three SCS fibers types, which differ only in the gradations of surface composition, or in the several outermost micrometers of the fiber radius. In all three cases, the bulk fiber consists of polycrystalline β-SiC of much larger crystalline size than is the case with boron fibers. It is possible, however, to form β-SiC with a finer grain than is possible

with the bulk of the SiC monofilament. Thus, for each fiber type in the bulk, the Si/C atomic ratio is 1/1, but each of these fibers is produced with a carbon-rich zone near the surface. The differences among the three fibers reside in the respective values of the Si/C atomic ratio and the fineness of the grain structure as a function of fiber radius in this several-micrometer-thick surface zone.

REFERENCES

1. H.E. DeBolt, Boron and Other High Strength, High Modulus, Low-Density Filamentary Reinforcing Agents, in *Handbook of Composites*, G. Lubin, Ed., Van Nostrand Reinhold, 1982
2. V.J. Krukonis, Chemical Vapor Deposition of Boron Filaments, in *Boron and Refractory Borides*, V.S. Matkovich, Ed., Springer, 1977
3. R.J. Suplinskas and J.V. Marzik, Boron and Silicon Carbide, in *Handbook of Fillers for Plastics*, H. Katz, Ed., Van Nostrand Reinhold, 1987
4. A.M. Tsirlin, Boron Filaments, in *Strong Fibres*, W. Watt and B.V. Perov, Ed., North-Holland, 1985
5. J. Nunes, in *Failure Mechanisms in High Performance Materials*, J.E. Early, T.R. Shives, and J.H. Smith, Ed., Cambridge University Press, 1984, p 138-146

Ceramic Fibers

D.D. Johnson and H.G. Sowman, Minnesota Mining & Manufacturing Company

THE FIRST MAN-MADE CERAMIC FIBERS were colored glass threads used for decorative purposes on artware. In the 1930s, processes for manufacturing glass fibers were significantly improved. Subsequently, glass fibers became useful for the reinforcement of synthetic resins and polymers. In the late 1940s, all-silica ceramic fibers were developed by leaching fiberglass and reheating the leached fibers to densify them. These fibers were not considered reinforcing fibers but rather high-temperature insulating fibers.

In the early 1950s, a new fiber, derived from kaolin (a form of clay) and other mineral sources, was first known and used as an insulating fiber. The term ceramic fiber has been identified historically with that development. Ceramic fibers are very short staple fibers, which have proved to be suitable for reinforcement.

Continuous and discontinuous ceramic fibers having oxide, nitride, and carbide compositions are the focus of this article. Asbestos, a naturally occurring fibrous material used for over 2000 years, would generally be considered with this group for comparison purposes. However, because its use is being abandoned in the U.S. and several other countries due to its health hazards, it will not be discussed here.

Remarkable progress has been made since 1965 in the development of many types of ceramic fibers. Information compiled in Ref 1 and 2 provides an excellent starting point for investigation of this development. The European Patent Office has identified companies associated with this technology development, summarized their activities, and listed foreign and U.S. patents (Ref 3).

The development of ceramic fibers has been spurred on by the need for high-temperature reinforcing fibers for aerospace applications. Progress in the development and production of new continuous polycrystalline oxide and carbide fibers has been especially significant. Bracke *et al.* (Ref 3) have stated that "Japanese organizations are only involved to a small extent in the development of polycrystalline oxide fibers, most work being done in Europe and the U.S.A." However, the continuous silicon carbide (SiC) based fiber Nicalon is

made by a Japanese firm using controlled pyrolysis of spun polymeric precursors and has been publicized widely, and a continuous polycrystalline oxide fiber discussed later in this article is known to be available. In addition, the announcement by Toshiba Monofrax (Ref 4) of Japan that it was exporting fiber technology on the preparation of a ceramic fiber having 72% aluminum oxide (Al_2O_3) and 28% silicon dioxide (SiO_2) to Kennecott Copper for transfer to Sohio Engineered Materials (formerly Carborundum) would indicate more activity in Japan than suggested by Bracke *et al.*

Discussion of the special types of fibers and whiskers produced in the past can be found in the literature. This article focuses on the commercially available continuous and discontinuous oxide and nonoxide fibers that are listed in Tables 1 to 3. The data provided in Tables 4 to 7 illustrate the wide range of properties obtainable with fibers and whiskers.

Continuous Oxide Fibers

Aluminum Oxide Fibers. Important features of Al_2O_3 fibers, especially of dense alumina fibers in the alpha (α) crystalline form, include high modulus of elasticity, high melting point, and exceptional resistance to corrosive environments. Properties of alumina fibers are shown in Table 4.

Fiber FP is the only essentially all-Al_2O_3 continuous polycrystalline fiber available from a commercial source. It has a rough surface and polycrystalline grains that average 0.5 μm (20 μin.) in size. An alternate form consists of the as-made Fiber FP overcoated with a layer of silica. This composite fiber has a smoother surface and, at 1.9 GPa (0.275 $\times$ 10^6 psi), is significantly stronger than the uncoated fiber, at 1.38 GPa (0.200 $\times$ 10^6 psi) measured on a 0.6-cm (¼-in.) gage length. Several U.S. patents relate to an alumina fiber of this type (Ref 38-41).

Alumina-Silica and Alumina-Boria-Silica Fibers. Ceramic fibers of the alumina-silica system, including those containing significant quantities of boria, are probably the most advanced of all continuous polycrystalline fibers available from commercial sources, especially with regard to handling qualities. Some are available as continuous filament yarns, which can be braided and woven like

Table 1 Commercially available continuous oxide fibers

Composition, wt%	Identification	Company	Forms(a)
Al_2O_3, > 99	Fiber FP	E.I. Du Pont de Nemours & Co., Inc.	C,Y,F
Al_2O_3, 85 SiO_2, 15	High performance alumina fiber	Sumitomo Chemical Co. Ltd., Japan; distributed by Avco Specialty Materials, Textron, Inc.	C,Y,F
Al_2O_3, 80 SiO_2, 20	Long alumina fiber	Denki Kagaku Kogyo K.K., with Nichibi Co. Ltd.	C,Y
Al_2O_3, 70 B_2O_3, 2 SiO_2, 28	Nextel 440 and Nextel 480	Minnesota Mining & Manufacturing Co.	C,R,Y,F
Al_2O_3, 62 B_2O_3, 14 SiO_2, 24	Nextel 312	Minnesota Mining & Manufacturing Co.	C,R,Y,F
SiO_2, 99.95	Astroquartz II	Distributed by J.P. Stevens & Co. Inc.	C,R,Y,F
SiO_2, 98 rem, 2	Refrasil	Hitco Materials Div., Armco Inc.	Y,F
SiO_2, 98 rem, 2	Siltemp	Ametek, Inc.	F
$ZrO2$, 68 SiO_2, 32	Nextel Z-11	Minnesota Mining & Manufacturing Co.	C,R,Y,F
(a) C, continuous; Y, yarn; F, fabric; R, roving			

fiberglass; they are identified in Tables 1, 2, and 4. Recent information (Ref 14) discloses that another alumina-silica fiber in both continuous and discontinuous forms is available for sampling.

High-performance alumina fiber is available, at least in experimental quantities, from Sumitomo Chemical Company. Because the composition reportedly is 15% SiO_2 and 85% Al_2O_3, it is being considered in this article as an alumina-silica fiber. The data shown in Table 4 for this fiber are supplemented by data obtained from Ref 42. The Sumitomo fibers were prepared in continuous multifilament form by spinning a polymerized organoaluminum compound that was blended with a compound containing silicon. A U.S. patent provides details on the fabrication of a high (85%) alumina fiber with the properties of the Sumitomo fiber (Ref 43). In this process, precursor fibers were hydrolyzed and fired under controlled conditions. The Sumitomo fiber, described as a reinforcement-grade fiber with excellent mechanical properties and thermal stability, is considered suitable for reinforcing resins, ceramics, and especially light metals (Ref 42).

Nextel ceramic fiber, most readily available as Nextel 312, contains 62% Al_2O_3, 14% boric oxide (B_2O_3), and 24% SiO_2. It can be obtained as roving or yarn in several deniers (for example, 600, 900, and 1800) or as fabric of selected weaves.

Chemical properties of Nextel 312 ceramic fibers may be strongly influenced by microstructure. When these fibers are going to be exposed to corrosive atmospheres, to coatings with coupling agents followed by high-temperature heating, or to hot, humid environments (95 °C, or 200 °F, with 100% relative humidity) for extended periods of time, they should be preheat-treated to a minimum of 900 to 940 °C (1650 to 1725 °F) so that each fiber is heated at this temperature for an hour or more. This ensures that the major portion of the composition is crystallized into a more corrosion-resistant species than the unreacted components in an amorphous structure.

Nextel 440 and 480 are newer ceramic fibers, both of which contain 70% Al_2O_3, 2% B_2O_3, and 28% SiO_2. Not only do these fibers have a higher modulus of elasticity, but they also survive a higher temperature exposure than Nextel 312 fibers before significant degradation of properties occurs. Nextel 440 fibers are essentially gamma (γ) or eta (η) Al_2O_3 with amorphous SiO_2; Nextel 480 fibers are mullite, which is recognized for its ability to strengthen ceramics and its corrosion resistance, even to hydrofluoric acid.

All Nextel ceramic fibers are transparent, unless colored internally by modifying components, and have the appearance of glass fibers. Crystallites in these fibers are very small, which contributes to a glossy surface appearance and exceptional handling qualities.

Table 2 Commercially available discontinuous oxide fibers

Composition, wt%	Identification	Company	Forms(a)
Al_2O_3, 95 SiO_2, 5 . . . . Saffil		Imperial Chemical Industries, PLC Ltd., England; distributed by Babcock & Wilcox	D,B,M,Ch
Al_2O_3, 72 SiO_2, 28 . . . . Fibermax		Sohio Engineered Materials (formerly Carborundum)	D,B
Al_2O_3, 70 B_2O_3, 2 SiO_2, 28 . . . . Nextel 440 and Nextel 480 Ultrafiber		Minnesota Mining & Manufacturing Co.	D,M,Ch
Al_2O_3, 60-68 B_2O_3, 4-9 SiO_2, 23-32 . . . . Staple Fiber		Nichias Corp.	D,B
Al_2O_3, 62 B_2O_3, 14 SiO_2, 24 . . . . Nextel 312 Ultrafiber		Minnesota Mining & Manufacturing Co.	D,M,Ch
Al_2O_3, 52 SiO_2, 48 . . . . Fiberfrax		Sohio Engineered Materials	D,B,M,F
Al_2O_3, 49-50 SiO_2, 50-51 . . . . Innswool		A.P. Green Refractories	D,B,M
Al_2O_3, 52-55 SiO_2, 41-44 . . . . Cer-wool		Combustion Engineering	D,B,M
Al_2O_3, 47 SiO_2, 53 . . . . Cerafiber		Manville Corp.	D,B,M
Al_2O_3, 42.5 Cr_2O_3, 2.5 SiO_2, 55 . . . . Cerachrome		Manville Corp.	D,B,M
Al_2O_3, 40 SiO_2, 50 CaO, 5 MgO, 3.5 TiO_2, 1.5 . . . . Cerawool		Manville Corp.	D,B,M
ZrO_2, 92 Y_2O_3, 8 . . . . Zircar		Zircar Products, Inc.	D,B,M,F

(a) D, discontinuous; B, bulk; M, mat or blanket; Ch, chopped; F, fabric

Table 3 Commercially available carbide and nitride fibers

Composition	Identification	Company	Forms(a)
SiC . . . . Nicalon		Nippon Carbon Co. Ltd.	C,Y,Ch,F,M,R
Si-Ti-C . . . . Tyranno		Ube Industries, Ltd.	C
Si-Zr-C . . . . Tyranno		Ube Industries Ltd.	C
SiC on C Core . . . . CVD SiC		Tokai Carbon Co., Ltd.; distributed by Avco Specialty Materials, Textron, Inc.	C
SiC . . . . Tokawhisker		Tokai Carbon Co., Ltd.; distributed by Avco Specialty Materials, Textron, Inc.	D
SiC . . . . Silar		Arco Metals Co.	D
SiC . . . . Tateho		Tateho Chemical Industry Co., Ltd.; distributed by ICD Group Inc.	D
Si_3N_4 . . . . Tateho		Tateho Chemical Industry Co., Ltd.; distributed by ICD	D

(a) C, continuous; Y, yarn; Ch, chopped; F, fabric; M, mat or blanket; R, roving; D, discontinuous

Zirconia-Silica. The composition of the Nextel Z-11 ceramic fiber is 68% zirconium oxide (ZrO_2) and 32% SiO_2, which is approximately equivalent to a mole ratio of 1 ZrO_2: 1 SiO_2. Its microstructure consists of microcrystalline tetragonal ZrO_2 and amorphous SiO_2. This continuous specialty fiber can be special ordered in roving, yarn, textile, bulk, or chopped form. Its reinforcement properties are modest, with a modulus of elasticity of about 76 GPa (11×10^6 psi). As a fabric, it has excellent mechanical durability. Its most outstanding feature is its resistance to flame penetration, a requirement for firewall construction.

Fused-Silica Fiber. Astroquartz II fibers are continuous filaments of pure fused silica, specifically 99.9% SiO_2. The fibers are chemically stable, water insoluble, and nonhygroscopic. They can be used at much higher temperatures (1050 °C, or 1920 °F) than either E-glass or S-glass fibers. With a tensile strength of 3.45 GPa (0.500×10^6 psi), Astroquartz II fibers have a higher strength-to-weight ratio than other ceramic fibers. Properties are shown in Table 4.

Leached-Glass Fibers. The two commercially available continuous leached-glass fibers are Refrasil and Siltemp. These fibers are sold primarily as insulating rather than as reinforcing fibers because of their modest mechanical properties; they are compared in Table 4 with other available continuous oxide fibers.

Table 4 Properties of continuous oxide fibers

Fiber	Composition, wt% Al₂O₃	B₂O₃	SiO₂	ZrO₂	rem	Density, g/cm³	Average diameter μm	μin.	Tensile strength GPa	10⁶ psi	Tensile modulus of elasticity GPa	10⁶ psi	Use temperature °C	°F
Fiber FP(a) (Ref 5)	>99	...	...	...	...	3.95	20	790	1.38	0.200	379	55	1320	2410
Sumitomo Alumina (Ref 6)	85	...	15	...	...	3.2	17	670	1.45	0.210	193	28	1250	2280
Nextel 440 (Ref 7)	70	2	28	...	...	3.05	11	430	2.07	0.300	193	28	1430	2605
Nextel 480 (Ref 8)	70	2	28	...	...	3.1	11	430	2.24	0.325	207-241	30-35	1430	2605
Nextel 312 (Ref 9)	62	14	24	...	...	2.7	11	430	1.72	0.250	155	22	1200	2190
Nextel Z-11 (Ref 10)	...	...	32	68	...	3.7	14	550	1.31	0.190	76	11	1000	1830
Astroquartz II (Ref 11)	...	...	99.95	...	...	2.2	9	350	3.45	0.500	69	10	1050	1920
Refrasil (Ref 12)	...	...	97.9	...	2.1	2.1	...	...	0.21-0.41	0.03-0.06	72	10.5	1095	2000
Siltemp (Ref 13)	...	...	98	...	2	2.2	...	...	...	...	...	...	...	...
Denki (Ref 14)	80	...	20	...	...	...	10	390	...	...	...	...	...	...

Fiber	Melt or liquidus temperature °C	°F	Coefficient of thermal expansion, 10⁻⁶/K	Dielectric constant	Resistivity Ω·m at 20 °C	Ω·cm at 68 °F	Refractive index
Fiber FP(a)	2045	3710	...	...	...	...	...
Sumitomo Alumina	...	...	8.8(b)	...	10¹¹	10¹³	1.65
Nextel 440	>1800	>3270	5	5.8	...	...	1.617
Nextel 480	>1800	>3270	...	5.8	...	...	>1.617
Nextel 312	1800	3270	3.5	5	...	...	1.572
Nextel Z-11	2000	3630	...	...	...	...	1.75
Astroquartz II	1700	3090	0.54	3.8	10¹⁶	10¹⁸	1.459
Refrasil	>1760	>3200	...	...	...	...	...
Siltemp	>1760	>3200	...	...	...	...	...
Denki	...	...	...	...	...	...	...

(a) Compressive strength is 6.9 GPa (1.0 × 10⁶ psi). (b) At 200-400 °C (390-750 °F)

Table 5 Properties of discontinuous oxide fibers

Fiber	Composition, wt% Al₂O₃	B₂O₃	CaO	Cr₂O₃	Fe₂O₃	MgO	SiO₂	TiO₂	Y₂O₃	ZrO₂	rem	Density, g/cm³	Average diameter μm	μin.
Cerachrome (Ref 15)	42.5	...	...	2.5	...	...	55	...	...	...	0.2	...	3.5	138
Cerafiber (Ref 16)	47	...	...	...	...	...	52.8	...	...	...	0.2	2.65	3.5	138
Cerawool (Ref 17)	40	...	5	...	...	3.5	50	1.5	...	...	...	2.54	3.5	138
Cer-wool (Ref 18)	52-55	...	...	...	0.1-0.2	...	41-44	0.1-0.2	...	...	1-2	...	3.0	118
Fiberfrax (Ref 19)	51.9	...	...	...	...	...	47.9	...	...	...	0.1	2.73	2-3	79-118
Fibermax (Ref 20)	72	...	...	...	...	...	28	...	...	...	...	3.0	2-3.5	79-138
Innswool (Ref 21)	49-50	...	...	...	...	...	50-51	...	...	...	<0.5	...	3-5	118-197
Kaowool (Ref 22)	45	...	...	...	...	...	53	...	...	...	2	2.56	2.8	110
Nextel 312 Ultrafiber (Ref 23)	62	14	...	...	...	...	24	...	...	...	...	2.75	3.5	138
Nextel 440 Ultrafiber (Ref 24)	70	2	...	...	...	...	28	...	...	...	...	3.1	3.3	130
Nichias (Ref 25)	60-68	4-9	...	...	...	...	23-32	...	...	...	...	...	10.5	413
Saffil RF Grade (Ref 26)	96-97	...	...	...	...	...	3-4	...	...	...	...	3.3	3.0	118
Saffil RG Grade (Ref 26)	96-97	...	...	...	...	...	3-4	...	...	...	...	3.3-3.5	3.0	118
Zircar (Ref 27)	...	...	...	...	...	...	...	...	8	92	...	5.6-5.9	4-6	157-236

Fiber	Tensile strength GPa	10⁶ psi	Tensile modulus of elasticity GPa	10⁶ psi	Use temperature °C	°F	Melt or liquidus temperature °C	°F	Specific heat J/kg·K at 1000 °C	Btu/lb·°F at 1830 °F
Cerachrome	...	...	...	...	1425	2600	>1760	>3200	1148	0.2741
Cerafiber	...	...	...	...	1315	2400	>1760	>3200	1148	0.2741
Cerawool	...	...	...	...	875	1610	>1648	>3000	1148	0.2741
Cer-wool	...	...	...	...	...	...	...	...	...	...
Fiberfrax	1.90	0.276	100	14.6	1260	2300	1790	3255	1130	0.2698
Fibermax	1.03	0.150	150	22	1650	3000	1890	3435	...	...
Innswool	...	...	...	...	1235	2255	1760	3200	...	...
Kaowool	1.13	0.165	84	12.2	1260	2300	1760	3200	1088	0.2598
Nextel 312 Ultrafiber	1.72	0.250	152	22	1200	2190	1800	3270	...	...
Nextel 440 Ultrafiber	1.31	0.190	207-241	30-35	1430	2605	>1800	>3270	...	...
Nichias	1.79	0.260	...	...	...	...	...	...	...	...
Saffil RF Grade	2.0	0.290	310	45	1600	2910	>2000	>3630	...	...
Saffil RG Grade	1.0-2.0	0.145-0.290	297	43	1600	2910	>2000	>3630	...	...
Zircar	...	...	...	...	2200	3990	2600	4710	...	...

Discontinuous Oxide Fibers

The ceramic fibers developed primarily for insulating purposes in the early 1950s were derived from clay or slag (by-products of the steel industry) and were generally made into low-cost fibrous wools or batts. They are commercially available throughout the world from a variety of sources. Table 5 compares the fibrous wools and battings consisting essentially of Al₂O₃ and SiO₂ that are prepared by the leading manufacturers of their types. Historically, such battings have contained a significant amount of particles, or shot, from the manufacturing process, which is an undesirable additive for reinforcing or insulating applications. These discontinuous fibers are generally not used to

Table 6 Properties of continuous silicon carbide fibers

Fiber	Composition, wt% Si	O	C	Ti	H	N	rem	Crystalline species	Density, g/cm³	Average diameter μm	μin.	Tensile strength GPa	10⁶ psi	Tensile modulus of elasticity GPa	10⁶ psi
Nicalon SiC (Ref 28, 29)....	54.3	11.8	30.0	...	...	...	3.9	β-SiC	2.55	10-15(a)	390-590	2.5-3.2	0.36-0.47	180-200	26-29
Tyranno Si-Ti-C (Ref 30).......	44.2	12.3	24.5	11.0	0.6	3.4	4.0	β-Sic + TiC	2.3	10-15	390-590	1.99	0.286	117	17
Avco CVD SiC (Ref 31).......	...	...	...	...	...	...	...	β-SiC on carbon core	...	~ 140, with 33-μm core	~ 5510, with 1300-μin. core	>3.4(b)	>0.50	428	62

Fiber	Use temperature °C	°F	Coefficient of thermal expansion, 10⁻⁶/K	Dielectric constant	Resistivity Ω · m at 20 °C	Ω · m at 68 °F	Chemical resistance % wt reduction in 24 h at 80 °C (175 °F) 6N HCl	18N H₂SO₄	7N HNO₃	30% Na OH aqueous
Nicalon SiC	1200(c, d)	2190	3.1(e)	6-8	10³	10¹	< 1%	< 1%	< 1%	< 1%
Tyranno	1300	2370	...	...	...	...	...	...	...	...
Avco CVD SiC	1100 (Ref 32)	2010	4.9	...	...	...	...	...	...	...
	1200 (Avco)	2190								
	>1400 (Ref 33)	2550								

(a) Round cross section. (b) See Ref 33. (c) Substantial loss of strength. (d) Loss in strength above ~ 1000 °C (1830 °F), Ref 32. (e) Along fiber axis, 0-200 °C (32-390 °F)

Table 7 Properties of discontinuous silicon carbide and silicon nitride whiskers

Fiber	Composition	Max free carbon, wt%	Crystalline species	Whisker content, %	Particle content, %	Density, g/cm³	Average diameter μm	μin.	Predominant length μm	μin.
Silar SC-9	SiC	0.10	α	80-90	10-20	3.2	0.6	24	10-80	390-3150
Silar SC-10 (Ref 34)	SiC	0.20	α	70-80	20-30	3.2	6.6	260	10-80	390-3150
Tokawhisker (Ref 35)	SiC	Negligible	β	...	< 1	3.19	0.1-0.5	4-20	30-100	1200-3950
Tateho Sic (Ref 36)	SiC	...	β	...	...	3.18	0.05-1.5	2-59	20-200	790-7900
Tateho Si₃N₄ (Ref 37)	Si₃N₄	...	α	...	...	3.18	0.1-1.6	4-63	20-200	790-7900

Fiber	Surface area m²/kg	ft²/lb	Tensile strength GPa	10⁶ psi	Tensile modulus of elasticity GPa	10⁶ psi	Use temperature stability to °C	°F	Electrical conductivity
Silar SC-9	3000	14 600	6.9(a)	1.0	690	100	1760(b)	3200	...
Silar SC-10	3000	14 600	6.9(a)	1.0	690	100	1760(b)	3200	...
Tokawhisker	...	...	3-14	0.44-2.03	400-700	58-101	1600 (in air)	2910	...
Tateho SiC.	...	...	...	...	...	...	...	...	Conductive
Tateho Si₃N₄	...	...	...	...	...	...	...	...	Nonconductive

(a) Estimates. (b) Atmospheric environment not stated

reinforce organic materials because of their short average length; long or continuous fibers are preferred for reinforcing plastics in order to carry a greater portion of the load. Although continuous fibers provide the highest degree of reinforcement, short fibers can be used effectively to reinforce metals, provided they are well bonded to the matrix. Kaowool discontinuous fibers have been used successfully in an aluminum matrix for automotive pistons developed by the Toyota Motor Corporation.

During the last 10 to 15 years a new family of discontinuous oxide fibers made from chemicals or sol-gels has emerged. Fibermax, Nextel ultrafibers, and saffil fibers, which are compared in Table 6, generally contain very small amounts of shot and have improved mechanical properties, making them candidates for the reinforcement of metals and ceramics. Also available commercially is Zircar, which offers a high use temperature, as shown in Table 5.

Nonoxide Fibers

Carbide and nitride fibers that are prepared by pyrolysis of polymers or by chemical vapor deposition (CVD) represent another class of ceramic fibers whose properties differ significantly from those of the oxides. Silicon carbide fibers are available commercially in both continuous and discontinuous (whisker) forms. Carbide mixtures, such as SiC with titanium carbide (TiC) or zirconium carbide (ZrC), have been described (Ref 30). The commercial availability is not known. Although continuous boron nitride (BN) fibers have been prepared (Ref 44), they are not commercially available and are therefore not represented in Tables 6 and 7. Discontinuous SiC and silicon nitride (Si₃N₄) fibers can be obtained in whisker form. A considerable amount of developmental work has been devoted to their use in inorganic composites with ceramic, metal, and glass matrices. The

commercial availability of both continuous and discontinuous carbide and nitride fibers is summarized in Table 3.

Continuous SiC fibers are either monolithic or bicomponent types. The monolithic, or single structure, type includes those known as Nicalon (Ref 45-47).

Modified SiC fibers that are reported to have higher temperature stability were developed with titanium doping (Ref 48, 29). Tyranno fibers are commercial modifications consisting of either silicon, titanium, and carbon, or silicon, zirconium, and carbon.

Silicon carbide fibers that are prepared by chemical vapor deposition of SiC on a carbon core are also available commercially (see Table 3). These bicomponent fibers are reported to have significantly higher strength and modulus of elasticity (Ref 33) than continuous SiC fibers produced by polymer pyrolysis. An improved fiber of this type developed by Avco has a protective CVD SiC coating called SCS that consists of carbon enriched with silicon at the surface (Ref 49, 50).

Properties of these continuous SiC fibers are given in Table 6. Where possible, data were obtained from manufacturers' published information.

The principal advantage of carbide fibers over oxide fibers is their superior mechanical properties, especially higher modulus of elasticity and axial compression in the case of large-diameter fibers. Both carbide and nitride fibers offer additional property choices, such as relative resistance to a reaction with the glass, metal, or other ceramic matrices used in reinforced composites. Its mechanical properties make SiC an especially promising candidate for reinforcement applications. Although SiC has superior oxidation resistance when compared to other nonoxide filaments, the most readily available SiC fibers are still affected adversely in corrosive environments. Therefore, their potential will probably be realized in reinforcement applications shielded from corrosive exposure or in applications that require only short periods of high-temperature exposure.

A study of SiC oxidation showed that it occurs at relatively low temperatures (1100 °C, or 2010 °F) (Ref 51) and that although the oxide layer on SiC is protective, impurities, changes in atmospheric conditions, and even fabrication processes could result in severe corrosion. Because fibers and whiskers generally have small diameters, these corrosion effects could be dramatically more important to them than to macroscale SiC ceramic bodies.

The degradation of Nicalon SiC fibers was studied in various environments by Mah *et al.* (Ref 52). Regardless of conditions, strength decreased after fibers were exposed to 1200 °C (2190 °F) or above for 2 h in argon, a 6.5 × 10^{-3} Pa (5 × 10^{-5} torr) vacuum, and air. The degradation was believed to be associated with β-SiC grain growth as well as carbon monoxide evaporation. Nevertheless, the authors expressed the opinion that in a composite, thermal stability might be entirely different and that research in that area should be promising.

In a study of the thermal stability of SiC ceramic fibers, it was found that both standard-grade and ceramic-grade fibers were affected differently with respect to compositional change by the presence of a nitrogen or by oxidizing conditions, but showed similar strength loss from their as-received conditions (Ref 54). When two types of Nicalon fibers were evaluated for creep and structural characterization, it was found that strength fell off at temperatures above 1000 °C (1830 °F) because of microcrystallization of the fiber structure (Ref 55).

Discontinuous Silicon Carbide and Silicon Nitride Whiskers. The term whisker is usually applied to single crystals having such a high degree of growth anisotropy that they have fibrous or fibrillar characteristics. Compared to polycrystalline continuous or discontinuous fibers, they commonly have exceptionally high tensile strength and high modulus of elasticity.

Silicon carbide whiskers can be prepared by chemical reaction processes (Ref 53) or by rice hull pyrolysis (Ref 32). Commercial products include Silar, Tokawhisker, and Tateho. Tateho Si_3N_4 whiskers are also available. Manufacturers' published information is the basis for the data presented in Table 7.

REFERENCES

1. L.R. McCreight, H.W. Rauch, Sr., and W.H. Sutton, *Ceramic and Graphite Fibers and Whiskers: A Survey of the Technology*, Academic Press, 1965
2. H.W. Rauch, Sr., W.H. Sutton, and L.R. McCreight, *Ceramic Fibers and Fibrous Composite Materials*, Academic Press, 1968
3. P. Bracke, H. Schurmans, and J. Verhoest, *Inorganic Fibers and Composite Materials: A Survey of Recent Developments*, Pergamon Press, 1984
4. Short-Fiber of Alumina Capable of Resisting 1600 °C Is Produced, *Jpn. Econ. J.*, 5 Oct 1982
5. A.K. Dhingra, Alumina Fiber FP, *Philos. Trans. R. Soc. (London)* A294, 1980, p 411-417
6. "Alumina Fiber Composite Data Sheet," Avco Specialty Materials, Textron Inc.
7. "Nextel® 440 Ceramic Fiber," Technical Bulletin, Minnesota Mining & Manufacturing Company
8. "Nextel® 480 Ceramic Fibers," Technical Bulletin, Minnesota Mining & Manufacturing Company
9. "Nextel® 312 Ceramic Fiber," Technical Bulletin, Minnesota Mining & Manufacturing Company
10. "Nextel® Z-11 Ceramic Fiber," Technical Bulletin, Minnesota Mining & Manufacturing Company
11. "ASTROQUARTZ® II Products," J.P. Stevens & Company, Inc.
12. H.G. Sowman and D.D. Johnson, Ceramic Oxide Fibers, *Ceram. Eng. Sci. Proc.*, Vol 6 (No. 9-10), Conference on Raw Materials for Advanced and Engineered Ceramics, 1985
13. Data from product brochures, Ametek, Inc., Haveg Division
14. *Technocrat*, Vol 18 (No. 4), April 1985, and chemical analysis data, Minnesota Mining & Manufacturing Company
15. "CERACHROME®," Manville Corporation
16. "CERAFIBER®," Manville Corporation
17. "CERAWOOL®," Manville Corporation
18. Data Sheet, Combustion Engineering, Inc.
19. Data from product literature, Carborundum
20. "FIBERMAX®," Toshiba Monofrax Company, Ltd.
21. "INNSWOOL®," A.P. Green Refractories Company
22. "KAOWOOL®," Babcock and Wilcox
23. "Nextel® 312 Ultrafiber," Technical Bulletin, Minnesota Mining & Manufacturing Company
24. "Nextel® 440 Ultrafiber," Technical Bulletin, Minnesota Mining & Manufacturing Company
25. Data from Nichias Corporation
26. "SAFFIL® Alumina Fibre," Imperial Chemical Industries PLC
27. "ZIRCAR® Fibrous Ceramics," Zircar Products, Inc.
28. "New Product Information," Dow Corning Corporation, 1983
29. K.J. Wynne and R.W. Rice, Ceramics Via Polymer Pyrolysis, *Ann. Rev. Mater. Sci.*, Vol 14, 1984, p 297-334
30. Data Sheet, Ube Industries, Ltd.
31. "Silicon Carbide Composite Materials," Avco Specialty Materials, Textron Inc.
32. C.H. Andersson and R. Warren, Silicon Carbide Fibers and Their Potential for Use in Composite Materials, Part 1, *Composites*, Vol 15 (No. 1), Jan 1984, p 16-23
33. J.A. DiCarlo, Fibers for Structurally Reliable Metal and Ceramic Composites, *J. Met.*, Vol 37 (No. 6), June 1985, p 44-49
34. "SILAR® Silicon Carbide Whiskers," Avco Metals Company, Silag Operation
35. "TOKAWHISKER® Silicon Carbide Whiskers," Tokai Carbon Company, Ltd.
36. Published Data, Tateho Chemical Industries Company, Ltd.
37. Published Data, Tateho Chemical Industries Company, Ltd.
38. R.F. Tietz, Process of Strengthening Polycrystalline Refractory Oxide Fibers, U.S. Patent 3,837,891
39. A.K. Dhingra, Polycrystalline Alumina Fibers as Reinforcement in Magnesium Matrix, U.S. Patent 4,036,599
40. B. D'Ambrosio, Continuous Filaments and Yarns, U.S. Patent 3,853,688
41. J.R. Green, Product and Process, U.S. Patent 3,849,181
42. Y. Abe, S. Horikiri, K. Fujimura, and E. Ichiki, High-Performance Alumina Fiber and Alumina/Aluminum Composites, in *Progress and Science and Engineering of Composites*, Proceedings of the Fourth International Conference on Composite Materials, ICCM-IV, Japan Society for Composite Materials, Oct 1982
43. S. Horikiri, K. Tsuji, Y. Abe, A Fukui, and E. Ichiki, Process for Producing Alumina or Alumina-Silica Fiber, U.S. Patent 4,101,615
44. R.Y. Lin, J. Economy, H.H. Murty, and R. Ohnsorg, Preparation and Characterization of High Strength, High Modulus Continuous Boron Nitride Fibers, in *Applied Polymer Symposium No. 29*, John Wiley & Sons, 1976, p 175-188
45. S. Yajima, H. Kayano, K. Okamura, M. Omori, J. Hayashi, T. Matsuzawa, and K. Akutsu, Elevated Temperature Strength of Continuous SiC Fibers, *Am. Ceram. Soc. Bull.*, Vol 55 (No. 12), 1976, p 1065-1066

46. S. Yajima, Special Heat-Resisting Materials from Organometallic Polymers, *Am. Ceram. Soc. Bull.*, Vol 62 (No. 8), 1983, p 893-903

47. S. Yajima, K. Okamura, J. Hayashi, and M. Omori, Synthesis of Continuous SiC Fibers With High Tensile Strength, *J. Am. Ceram. Soc.*, Vol 59 (No. 7-8), 1976, p 324-327

48. S. Yajima, T. Iwai, T. Yamamura, K. Okamura, and Y. Hasegawa, Synthesis of a Polytitanocarbosilane and Its Conversion Into Inorganic Compounds, *J. Mater. Sci.*, Vol 16, 1981, p 1349-1355

49. S.R. Nutt and F.E. Warner, Silicon Carbide Filaments: Microstructure, *J. Mater. Sci.*, Vol 20, 1985, p 1953-1960

50. F.W. Wawner, A.Y. Teng, and S.R. Nutt, Microstructural Characterization of SiC (SCS) Filaments, *SAMPE Q.*, April 1983, p 39-45

51. B.O. Yavuz and L.L. Hench, Low Temperature Oxidation of SiC, *Ceram. Eng. Sci. Proc.*, 1982, p 596-600

52. T. Mah, N.L. Hecht, D.E. McCullum, J.R. Hoenignian, H.M. Kim, A.P. Katz, and H.A. Lipsitt, Thermal Stability of SiC Fibers (NICALON®), *J. Mater. Sci.*, Vol 19, 1984, p 1191-1201

53. J.J. Petrovic, J.V. Milewski, D.L. Rohr, and F.D. Gac, Tensile Mechanical Properties of SiC Whiskers, *J. Mater. Sci.*, Vol 20, 1985, p 1167-1177

54. T.J. Clark, M. Jaffe, J. Rabe, and N.R. Langley, Thermal Stability Characterization of SiC Ceramic Fibers:1, Mechanical Property and Chemical Structure Effects, in *Proceedings of the 10th Annual Conference on Composites and Advanced Ceramic Materials*, The American Ceramic Society, 1986, p 901-909

55. G. Simon and A.R. Bunsell, Creep Behavior and Structural Characterization at High Temperatures of Nicalon SiC Fibres, *J. Mater. Sci.*, Vol 19, 1984, p 3658-3670

SELECTED REFERENCES

- I.B. Cutler, Production of SiC From Rice Hulls, U.S. Patent 3,754,076, 1970

- H.S. Katz and J.V. Milewski, Ed., *Handbook of Fillers and Reinforcements for Plastics*, Van Nostrand & Reinhold, 1978

- J.G. Lee and I.B. Cutler, Formation of SiC From Rice Hulls, *Amer. Ceram. Soc. Bull.*, Vol 54, 1975, p 195-198

- J.V. Milewski, J.G. Sandstrom, and W.S. Brown, Production of SiC from Rice Hulls, in *Silicon Carbide*, R.C. Marshall, G.W. Faust, and C.E. Ryan, Ed., University of South Carolina Press, 1974, p 634-639

Epoxy Resins

Clayton A. May, Arroyo Research & Consulting Corporation

EPOXY RESINS are used extensively in composite materials for a variety of demanding structural applications. They are the most versatile of the commercially available matrices. Unsaturated polyesters may cost less, and polyimides may perform better at elevated temperatures, but epoxy resins have a broad range of physical properties, mechanical capabilities, and processing conditions that makes them invaluable by comparison. Excellent sources are available to those wishing to study these materials in detail (Ref 1-5).

Depending on the chemical structures of the resin and the curing agent, the availability of numerous modifying reactants, and the conditions of cure, it is possible to obtain toughness, chemical and solvent resistance, mechanical responses ranging from extreme flexibility to high strength and hardness, resistance to creep and fatigue, excellent adhesion to most fibers, heat resistance, and excellent electrical properties. No by-products are formed during the cure. Shrinkage that is due to cure is low. The uncured resins have a variety of physical forms, ranging from low-viscosity liquids to nontacky solids, which, combined with a large selection of curing agents, affords the composite fabricator a wide range of processing conditions.

The chemistry involved in the use and application of epoxy resins is the key to their outstanding performance. All epoxy resins contain the epoxide, oxirane, or ethoxylene group:

where R represents the point of attachment to the remainder of the resin molecule. The epoxide function is usually a 1,2- or α-epoxide that appears in the form:

$$CH_2 - CH - CH_2 -$$

called the glycidyl group, which is attached to the remainder of the molecule by an oxygen, nitrogen, or carboxyl linkage, hence, the terms glycidyl ether, glycidyl amine, or glycidyl ester.

Curing of the resin results from the reaction of the oxirane group with compounds that contain reactive hydrogen atoms:

where R′—H is:

R′NH₂ (primary amine) R′R″NH (secondary amine) R′ C—OH (carboxylic acid)

R′SH (mercaptan) or phenol —OH

Two other chemical reactions must be considered in the curing of epoxy resins: the reaction with carboxylic acid anhydrides, and catalysis by acids and bases. The carboxylic acid anhydrides, similar to the carboxylic acids, yield ester cross-links. Acid and base catalysis results in homopolymerization of the epoxide to a polyether:

where n can be any number of ether units, depending on the catalyst and the reaction conditions. The effects of the chemical structure of the curing agents and resins on processing and properties will be considered in the section "Curing Agents and Curing Reactions" in this article. Many epoxy resins and curing agents present toxicological problems in the uncured state. Accordingly, these materials should be handled with care, and the manufacturer's warnings should be heeded.

Resin Manufacture and Products

Two basic processes are used in the manufacture of epoxy resins: the reaction of epichlorohydrin with compounds containing reactive hydrogen atoms, such as phenols, or amines, and the peracid epoxidation of olefins. The former is by far the most common.

The primary resin in epoxy technology is the diglycidyl ether of bisphenol A (DGEBPA) and its higher homologs. Resin synthesis results from the reaction of epichlorohydrin, an intermediate in the synthetic glycerine process, with bisphenol A (BPA) in the presence of alkali (Fig. 1). Bisphenol A is also a product of the petroleum industry where it is formed by the condensation of acetone with phenol. As the value of n is increased (for DGEBPA, where n = 0), the resin progresses from a viscous liquid to a solid with a high softening point. The properties of commercial grades of BPA epoxy resins are shown in Table 1. The higher homologs are manufactured by controlling the ratio of epichlorohydrin to BPA. The higher the BPA concentration during the reaction, the higher the molecular weight of the resin. The higher homologs are also manufactured by the reaction of BPA with DGEBPA using Lewis base catalysis. This latter reaction is important to resin formulators because the higher homologs can also be formed in situ in a composite prepreg material, during either its manufacture or the subsequent cure. Brominated epoxy resins of the general structure shown in Fig. 2 are also formed by this reaction using tetrabromobisphenol A and are important in the electrical laminate field, as will be discussed in the section "Electrical Laminates" in this article. The values of x and y and the bromine content of the resin is varied by controlling the BPA tetrabromo-BPA ratio.

The second most important epoxy resin manufacturing process is the peracid epoxidation of olefins (Ref 6-8). Current manufacturing technology involves the use of peracetic acid as the epoxidizing agent (Fig. 3) (Ref 9, 10).

The glycidyl ethers of various novolac resins are the second most important class of epoxy resins. They are manufactured by the reaction of epichlorohydrin with various phenolic novolacs, resulting in the chemical structures shown

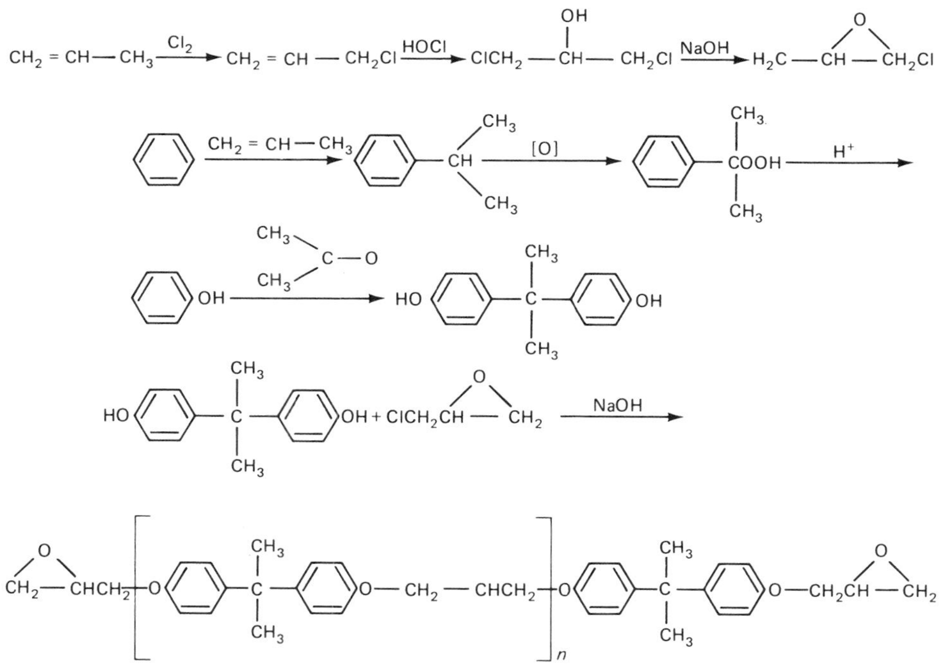

Fig. 1 Epoxy resin synthesis

12-14) indicate low moisture absorption, good hot/wet strength, and improved fracture toughness.

Three glycidyl amines are also commercially available; their structures are shown in Fig 6. Tetraglycidylmethylenedianiline (TGMDA) is the most widely used epoxy resin in the fabrication of carbon fiber reinforced hardware for the aerospace industry. Triglycidyl p-aminophenol (TGAP) is used extensively in epoxy adhesive formulations. Only one glycidyl ester, diglycidyl o-phthalate, is used in fiber-reinforced matrices, primarily as a reactive diluent that reduces matrix viscosity without appreciably affecting elevated-temperature performance.

Finally, a number of monofunctional epoxy diluents are available, which formulators can use to lower matrix viscosity and increase tack. Because of their monofunctionality, elevated-temperature performance drops rapidly with increasing concentration, so care should be exercised in their use. Typical commercial products are butyl glycidyl ether (BGE), allyl glycidyl ether (AGE), phenyl glycidyl ether (PGE), cresyl glycidyl ethers (CGE), and mixed C_{9-12} carboxylic acid glycidyl esters. Structures of these diluents are shown in Fig 7.

in Fig. 4. The glycidyl novolacs are characterized by better elevated-temperature performance than the BPA-based resins.

Figure 5 gives the chemical structures of other commercially available epoxy resins. Butylene glycol diglycidyl ether (BGDGE) is a low-viscosity difunctional epoxide useful as a diluent. Because it is bifunctional, the reduction of elevated-temperature performance is minimal. The diglycidyl ether of bisphenol F (DGEBPF) epoxy has lower viscosity than the BPA epoxies, with minimal loss in the glass transition temperature (T_g). The next two

products in Fig. 5 are among the most common of the peracid epoxies and are usually used with anhydride curing agents. VCDO also is useful as a diluent. The triglycidyl ether of triphenyl methane (TGETPM), a relative newcomer to the epoxy family, is said to give high T_g's, thermal oxidative stability, and excellent moisture resistance (Ref 11). Resorcinol diglycidyl ether (RDE) has good elevated-temperature performance but use is limited because of its toxicity. The last two products shown are newcomers. Although their full potential has not yet been determined, preliminary data (Ref

Curing Agents and Curing Reactions

Three chemical reactions are of major importance to the curing of epoxy composite matrices: the amine/epoxide reaction, the anhydride/epoxide reaction, and the Lewis acid-catalyzed epoxide homopolymerization. A simplified version of each reaction mechanism is shown in Fig. 8 to give a general picture of the chemistry. The most common curing agents are the amines, in which each of the amino hydrogens reacts with an epoxide group (Fig. 8a). Depending on the number of amino hydrogens on the curing agent, their reactivity, the number of epoxide groups per resin molecule, and the supporting structures of each, a wide variety of mechanical properties can be obtained and various laminating processes can be used.

The chemistry of curing with anhydrides is more complex. It was initially proposed that the anhydride reacted with an alcohol to form a half-acid ester, which in turn reacted with the epoxide to form a second ester linkage, resulting in a new hydroxyl. The reaction continued in this manner until all of the reactive groups were consumed (Fig. 8b). This mechanism was based on the logic that some hydroxyl was

Table 1 Properties of commercial grades of BPA epoxy resins

Average mol wt	Average wpe(a)	Approximate value of n	Viscosity at 25 °C (80 °F)	Softening point(b) °C	°F
350	182	0	80	...	...
380	188	0.15	140	...	...
600	310	0.9	Semisolid	40	105
900	475	2.0	Solid	70	160
1400	900	3.7	Solid	100	212
2900	1750	9.0	Solid	130	265
3750	3200	11.9	Solid	150	300

(a) Weight per epoxide, that is, grams of resins needed to provide 1 molar equivalent of epoxide. Also referred to as EEW (epoxide equivalent weight) and EMM (epoxy molar mass). All three items are interchangeable. (b) Softening point by Durran's mercury method

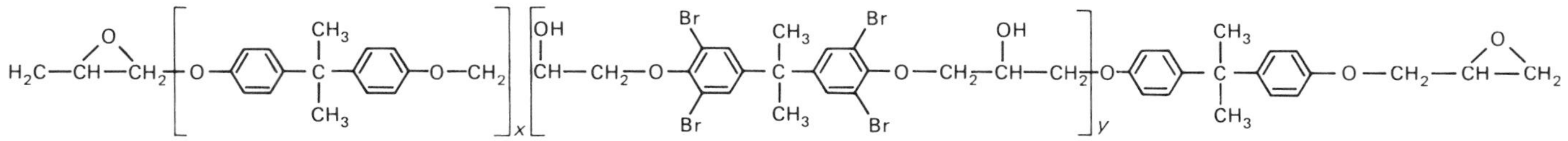

Fig. 2 Brominated epoxy resins

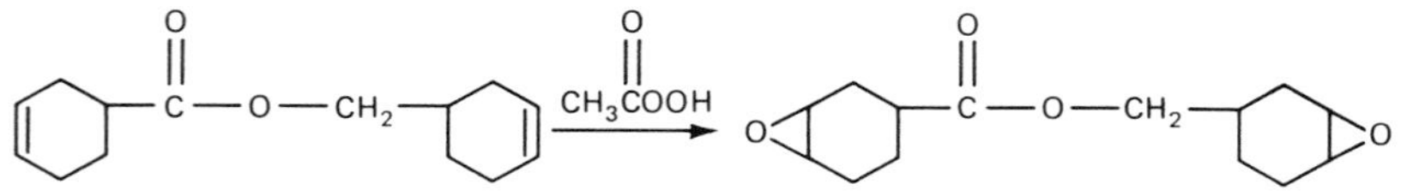

Fig. 3 Peracid epoxidation process

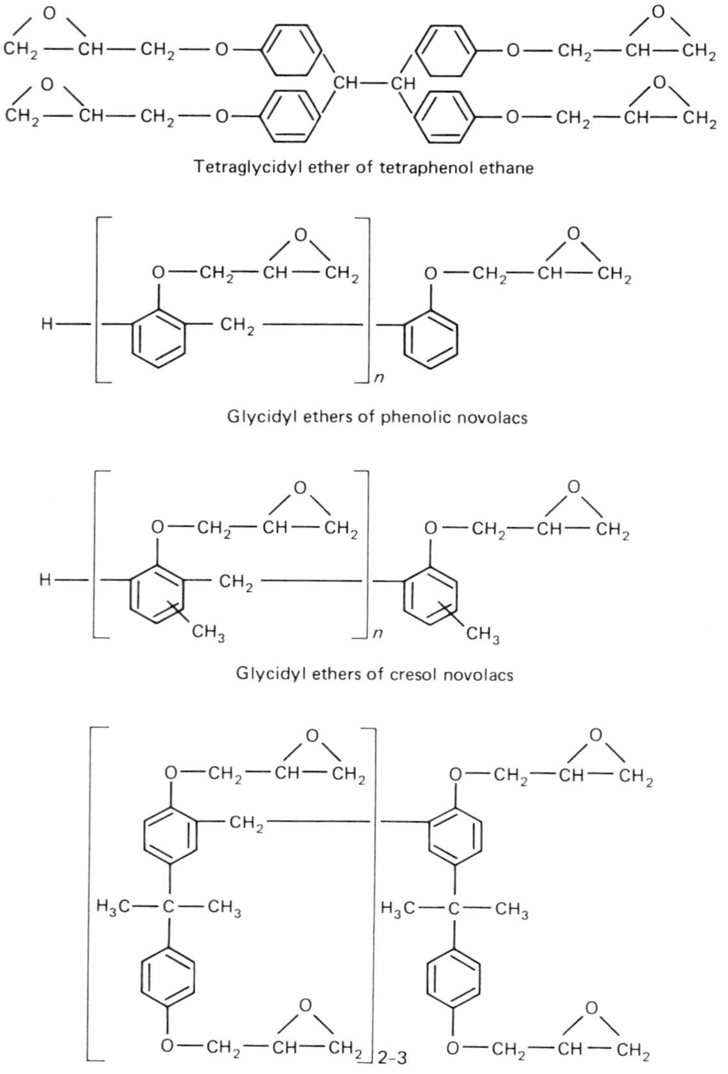

Tetraglycidyl ether of tetraphenol ethane

Glycidyl ethers of phenolic novolacs

Glycidyl ethers of cresol novolacs

Glycidyl ethers of bisphenol A novolacs

Fig. 4 Epoxy novolacs

always present, either in the resin itself or as an impurity, such as water (Ref 15). Optimum properties for this system resulted at an anhydride/epoxide ratio of 0.85/1, indicating that a side reaction was occurring: Probably epoxide homopolymerization was catalyzed by the acid, giving rise to some polyether formation.

It was later found that Lewis acids and bases act as catalysts for this reaction. In this system the anhydride/epoxide stoichiometry was found to be 1/1. This leads to a second mechanism in which the anhydride reacts with the catalyst to form a carboxyl anion, and the anionic process continues to the completion of the cure (Fig. 8c) (Ref 16). This latter curing agent system is now preferred since it leads to somewhat better elevated-temperature properties in the cured composites. It should be noted that in anhydride cures, one anhydride group, not carbonyl, reacts with each epoxide. This is important in calculating resin/curing agent ratios for anhydride systems. Because

stoichiometry is highly important to successful lamination with epoxies, the methods for calculating curing-agent concentrations for amines and anhydrides are shown in Table 2.

The third curing process to consider is epoxide homopolymerization, referred to earlier as polyether formation. Lewis acid catalysis is most prevalent. It proceeds by a cationic mechanism (Fig. 8d) (Ref 17).

Curing agents important to the fabrication of epoxy resin composites are shown in Table 3. Although others are available, including a broad range of proprietary products, the list presented gives an overall picture of the scope and handling characteristics of materials available to the laminator. The table, developed from commercial literature, is based on the diglycidyl ether of BPA as the epoxy resin. Included in the table are the recommended concentrations, typical cure and postcure schedules, pot life (working life) of the catalyzed resin, and maximum heat distortion temperature (an assessment of expected elevated-temperature performance). The left-hand column is subdivided into curing-condition classes and chemical types, which in general dictate the areas of composite application. The lamination processes for each class of curing agent will be described in the section "Laminating Processes" in this article. It should be remembered that epoxy resins other than DGEBPA are available and afford an even broader range of mechanical and processing properties than those shown in Table 3. Thus, a higher functionality epoxy may be used for better elevated-temperature performance (Fig. 4 to 6); a monofunctional diluent (Fig. 7) or a cycloaliphatic (Fig. 5) may be used to lower processing viscosity, and so forth.

The aliphatic amines and their derivatives (Fig. 9) are recommended for ambient-temperature curing. Consequently, they would be expected to have a limited pot life or shelf life. Typical applications include wet lay-up laminating operations such as tank linings, wet filament winding, patching (repairs), tooling, certain air frames and radomes, and electrical insulation. Pure chemical compounds such as diethylenetriamine (DTA) and triethylenetetramine (TETA), when properly postcured, display strengths above 100 °C (212 °F). However, poorer elevated-temperature performance should be expected with room-temperature curing, which may be required of some lamination processes because of hardware size. Although compounds of this type are low in cost and viscosity, they are toxic, have high vapor pressures, and can cause surface blushing on a laminate when exposed to the atmosphere. To overcome these disadvantages, they are often used in the form of an epoxy resin adduct (Fig. 5), at a sacrifice of the lower viscosity. As shown in Table 3, equivalent performance and cure characteristics can be expected from products of this type. The polyamides are condensation products

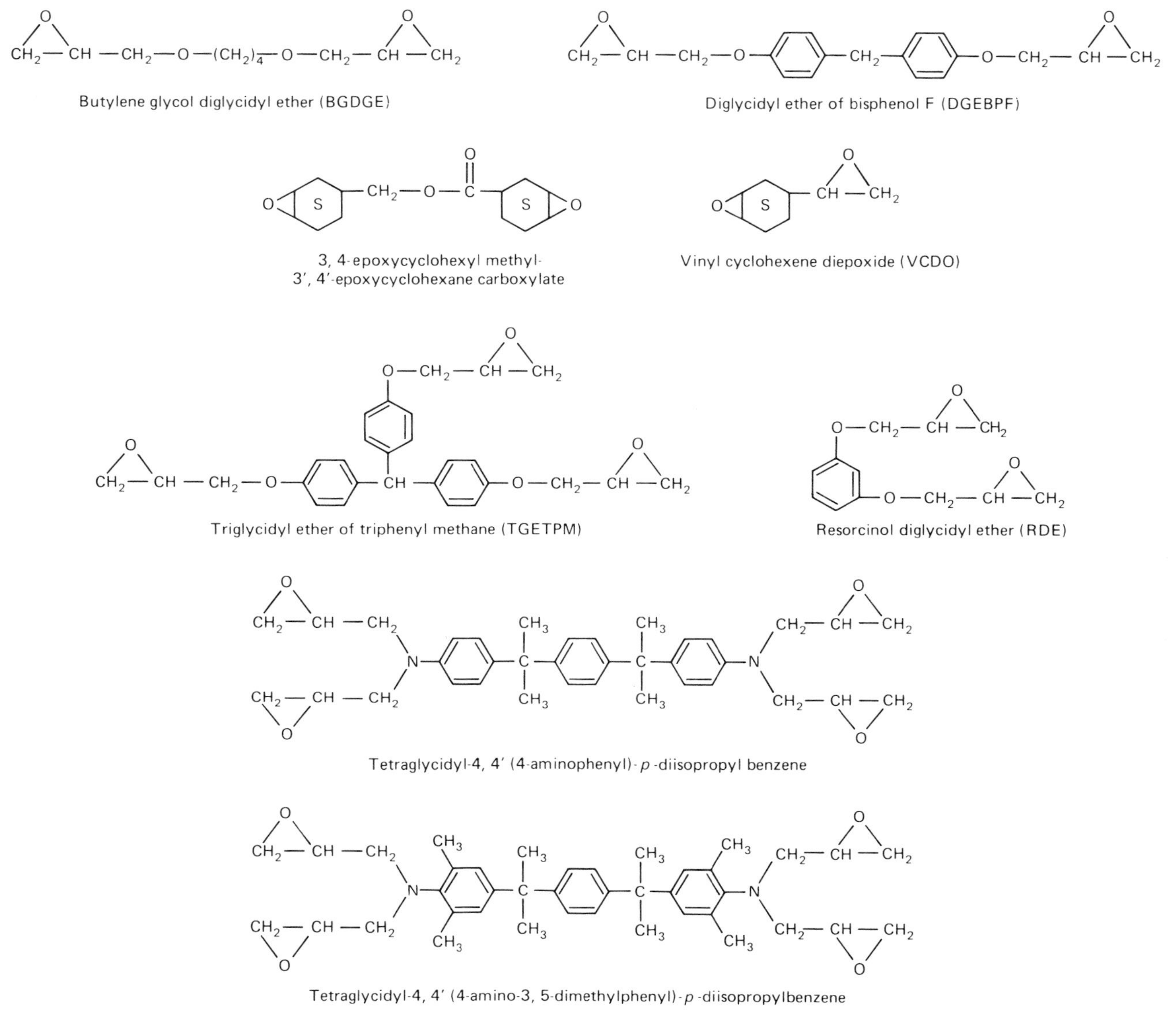

Fig. 5 Other commercial epoxy resins

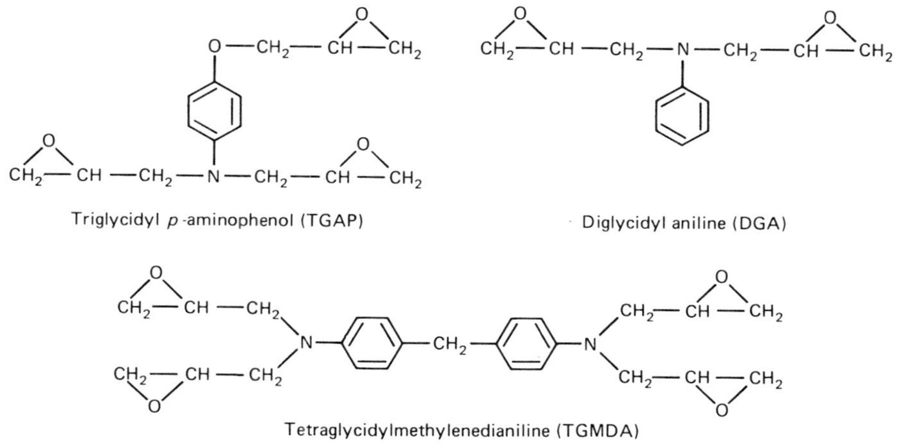

Fig. 6 Glycidyl amines

of a C_{36} dimerized fatty acid and an aliphatic amine, such as DTA. Here, elevated-temperature performance and mix viscosity are sacrificed to gain toughness, moisture resistance, pot life, and a much lower toxicity.

The moderate temperature cured cycloaliphatic and tertiary/primary aliphatic amines offer a compromise between the room-temperature curing agents and the higher-temperature curing aromatic amines. They can be cured under milder conditions than the aromatic amines, but their elevated-temperature performance, solvent resistance, and chemical resistance are generally superior to those obtained from the ambient temperature cured aliphatic amines. The exception, as noted in Table 3, is *n*-aminoethylpiperazine (AEP). Although the heat distortion temperature is low, AEP yields significantly better impact strengths. Some unique

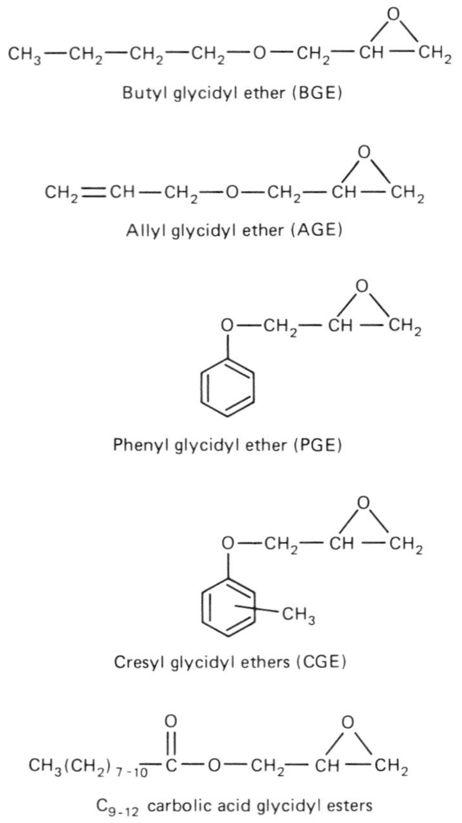

Fig. 7 Monofunctional epoxy diluents

Table 2 Methods for calculating the curing agent concentration for amine- and anhydride-cured systems

Amine-cured systems(a)

$$\frac{\left(\dfrac{\text{Molecular wt of amine}}{\text{No. of available hydrogens}}\right)}{\text{Wt per epoxide}} \times 100 = \text{parts by weight of amine to be used with 100 parts by weight of resin}$$

Anhydride-cured systems

$$\frac{\left(\dfrac{\text{Molecular wt of anhydride}}{\text{No. of anhydride groups}}\right)}{\text{Wt per epoxide}} \times 100 = \text{parts by weight of anhydride to be used with 100 parts by weight of resin}$$

(a) When tertiary amino groups are present, equation does not apply. Proper concentration is best found by experimentation.

combinations of performance and curing characteristics are found in this class of materials. Typical structures are shown in Fig. 10.

The aromatic amines are widely used in composite fabrication in both wet and dry lay-up applications for filament winding, electricals, piping, tooling, and whenever maximum chemical resistance is needed. They provide moderate viscosities at room temperature with the liquid resins, long pot life, excellent chemical resistance and electrical properties, and good elevated-temperature performance. Higher-temperature cures and longer cure times are required to obtain these advantages. m-Phenylenediamine (MPDA) is the best known and provides the lowest viscosities in the catalyzed resin systems. The aromatic amine 4,4'-methylenedianiline (MDA) provides a slightly longer pot life, but requires longer cures and has a higher processing viscosity. Both products are solids and must be melted before being blended with the resin. This drawback led to the development of aromatic amine eutectics, which are liquid at room temperature. The result is easier mixing but with a slight penalty in elevated-temperature performance. Although 4,4'-diaminodiphenylsulfone (DDS) has a much higher melting temperature and a slower reaction time and is normally used in conjunction with a Lewis acid accelerator (BF₃-monoethyl-amine complex), it provides the best elevated-temperature performance and elevated-temperature resistance of the three. Other resins, such as the novolacs (Fig. 4), are used with this curing agent to further enhance elevated-temperature performance. Combinations of the TGMDA resin (Fig. 6) and DDS are widely used in the aerospace industry for carbon fiber composites. Unfortunately, although T_g's in excess of 200 °C (390 °F) have been observed in this latter system, the presence of moisture can lower the T_g to about 130 °C (265 °F). The chemical structures of these amines, with the exception of the amine eutectics, are shown in Fig. 11a.

The acid anhydrides are characterized by a long pot life, better heat aging in air at elevated temperatures, and better electrical properties. Their composite applications are generally the same as those of the aromatic amines, and they are chosen to impart specific properties to the laminating system. Cures normally involve an added accelerator, such as benzyldimethylamine (BDMA), for reasons discussed above. Because anhydrides are hygroscopic materials, extended exposure to the atmosphere results in hydrolysis of the anhydride to the parent acid. This changes the resin/curing agent stoichiometry and can drastically reduce the elevated-temperature performance. The dibasic acid anhydrides with cycloaliphatic ring structures, such as hexahydrophthalic anhydride (HHPA) and methyltetrahydrophthalic anhydride (MTHPA), perform equally well at elevated temperatures (heat distortion temperatures ~130 to 140 °C, or 265 to 285 °F), with the exception of nadic methyl anhydride (NMA), which is preferred for elevated-temperature applications. It reacts further on postcuring to give a cured resin system with increased cross-link density, as the result of a reverse Diels-Alder reaction. It is the preferred curing agent for the epoxy novolacs in elevated-temperature applications.

All three anhydrides afford low viscosity in the laminating resin system. Hexahydrophthalic anhydride has excellent electrical properties and gives almost colorless laminates with superior weathering and antiyellowing properties. The anhydride MTHPA also has excellent electrical properties, while providing low viscosity at a moderate cost. Dodecenylsuccinic anhydride (DDSA) affords low viscosity and ease of handling. Although the heat distortion temperature is lowered by this curing agent, good ambient-temperature mechanical performance should be expected.

Chlorendic anhydride (CA) imparts fire retardancy, but is difficult to use alone because it is highly reactive and, accordingly, has limited pot life. The product is solid at room temperature but blends with other anhydrides to produce eutectics that can be handled at lower temperatures. Anhydrides of higher functionality cause handling problems because of their solid nature and poorer solubility, but higher heat distortion temperatures are attainable. Typical examples are benzophenonetetracarboxylic acid dianhydride (BTDA), pyromellitic dianhydride (PMDA), and trimellitic anhydride (TMA). Because of the handling problems, the use of these products is somewhat limited. Chemical structures of the various anhydride curing agents are shown in Fig. 12.

The principal Lewis acid used in epoxy composites is boron trifluoride in the form of its monoethylamine complex (BF₃-MEA). This curing agent is characterized by a long pot life and a high T_g. Samples of DGEBPA catalyzed with BF₃-MEA, although showing considerably increased viscosity, are still usable after storage for 6 months at room temperature.

Dicyanamide (DICY) is a true latent catalyst for epoxy resin curing. In admixture with DGEBPA, it has demonstrated a room-temperature storage life in excess of 4 years with little or no change in viscosity. This curing agent is used only for dry lay-up laminating. Dicyanamide systems find their major use in circuit board materials requiring good electrical properties, water resistance, and retention of strength at elevated temperatures (Ref 21). Cures can be conducted in the range of 120 to 175 °C (250 to 350 °F) but are very slow at the lower temperatures. As a result, it is common practice to add accelerators such as BDMA and mono and dichlorophenyl substituted ureas to these systems. The chemical structure and cure mechanism of DICY have been the subject of numerous studies (Ref 22, 23). Recent efforts to polymerize monoepoxides with DICY (Ref

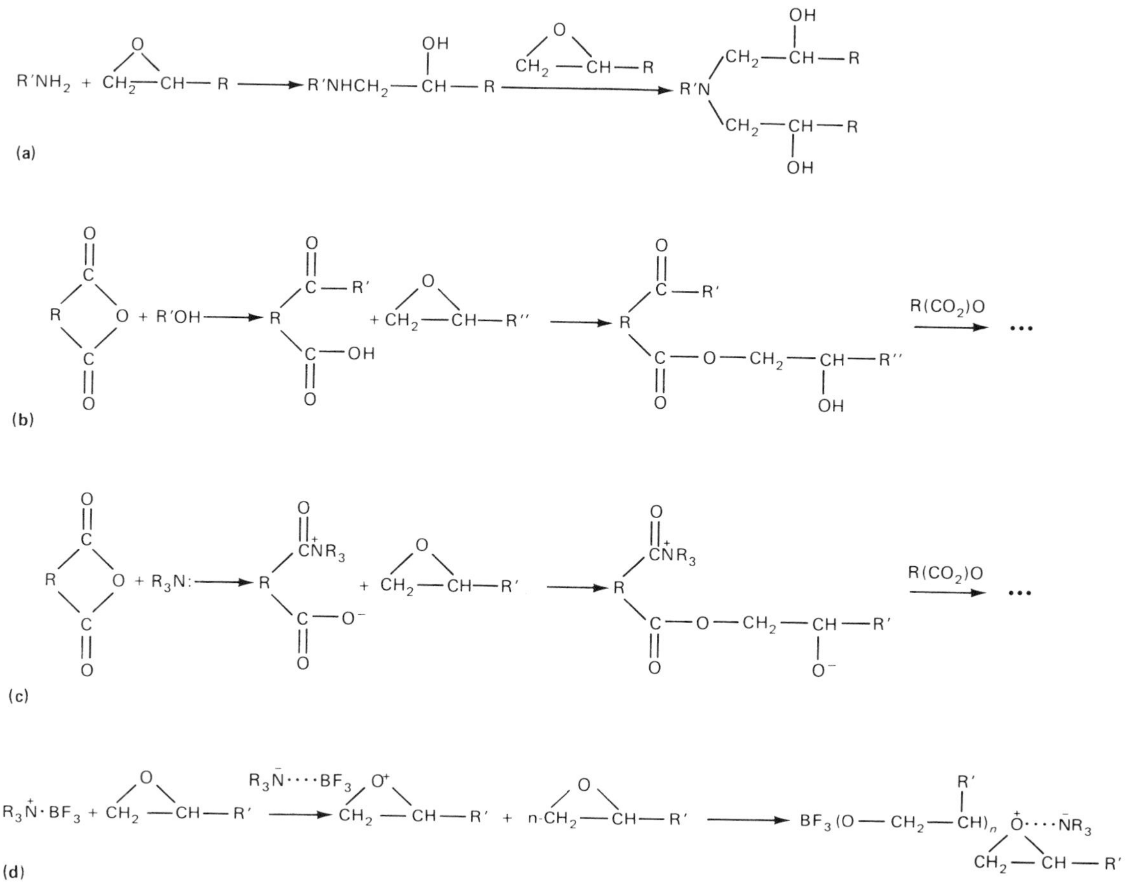

Fig. 8 Epoxy curing reactions. (a) Amine/epoxide reaction. (b) Anhydride/epoxide reaction with alcohol. (c) Anhydride/epoxide reaction with catalysts of Lewis acids and bases. (d) Lewis acid-catalyzed epoxide homopolymerization

24) have led to a better understanding of the main reactions occurring during cure.

Laminating Processes

The two basic procedures for fabricating epoxy composites are the wet lay-up and the dry lay-up methods. In wet lay-up laminating, a liquid low-viscosity resin impregnates the reinforcement, either before or during its placement on the tool. The hardware being formed is shaped to its final configuration while the resin is in the uncured state. In the dry lay-up process, the reinforcement is impregnated with the matrix resin, either in solution form or as hot melt, to produce a prepreg sheet. The hot-melt process has the advantage of not requiring the added expense of solvent removal. The final prepregs are supplied either as dry stock, commonly used by the electrical circuit board industry, or as a sticky or high-tack material, preferred by aerospace manufacturers. The prepreg is cut and placed on a tool in the desired configuration before cure. Heat and pressure are used to further shape and densify the fiber-resin combination as it cures.

The most common wet lay-up procedure involves hand lay-up or spray-up of the resin-reinforcement combination, followed by a con-

tact, vacuum bag, pressure bag, compression molding (matched metal die), or autoclave cure cycle. Other processes involving liquid resins include filament winding, reaction injection molding (RIM), resin transfer molding (RTM), and pultrusion.

Although there are many versions of the hand lay-up procedure, basically the resin and the reinforcement are placed in a mold, with the resin being introduced by either pouring or spraying it onto the fibers. The assembly is then worked with squeegees or rollers to densify the structure and remove much of the entrapped air. The laminate can be left in this form to cure at either room or elevated temperatures; however, better, denser products can be obtained by enclosing the lay-up in a vacuum or pressure bag. Vacuum bags apply additional pressure to the laminate (100 kPa, or 15 psig) and aid in the removal of entrapped air. Pressure bags also involve the use of vacuum but are more complex, applying additional pressure to the assembly through an elastomeric pressure bag or bladder contained in a clamshell cover, which fits over the mold. However, only mild pressures can be applied with this type of tooling, which must conform to state and local boiler codes. If more pressure is needed, compression or autoclave molding can be used. In the

compression process, a matched metal die is used to provide accurate control of the resin-fiber ratio, accurate part size, and precise surfaces on both sides of the laminate. Autoclave processing is a pressure cooker operation which usually involves the use of a vacuum bag in conjunction with the pressure applied by the autoclave gas. Autoclave processes are usually limited to pressures of about 1400 kPa (200 psig) because of safety code regulations. Epoxy resins, because of their excellent adhesive and wetting characteristics, do not require high laminating pressures.

Wet filament winding involves passing a continuous fiber roving or tape through a resin bath, then through a die that controls the resin content, and finally onto a rotating mandrel. The fiber strands can be laid in precise geometric patterns, permitting the mechanical strengths to be accurately predetermined. A dry winding process is also used, in which a prepreg tape or roving is wound onto a mandrel in a similar fashion. A dense structure is ensured by heating the prepreg in order to melt the resin formulation at the point of contact with the mandrel.

Reaction injection molding involves bringing together two reactive streams of resin by means of high-pressure impingement mixing. The

Table 3 Typical epoxy resin curing agents for composite fabrication

Curing agent type and typical products	Concentration, phr(a)	Typical cure Time, h	Typical cure days	Temperature °C	Temperature °F	Post cure(b) Time, h	Post cure(b) Temperature °C	Post cure(b) Temperature °F	Pot life at 25 °C, h	Heat distortion temperature(c) °C	Heat distortion temperature(c) °F
Aliphatic amines and derivatives, room-temperature cure:											
Diethylenetriamine (DTA)	12	...	7	25	80	1	200	390	1/4-1/2	124	255
Triethylenetetramine (TETA)	14	...	7	25	80	1	200	390	1/4-1/2	123	255
Polyamides (Versamides)(d)	30-50	...	7	25	80	None	...	...	2-3	55	130
Amine adducts(e)	26	...	4	25	80	1	150	300	1/4-1/2	120	250
Aliphatic/cycloaliphatic amines, moderate-temperature cure:											
Diethylaminopropylamine (DEAPA)	8	0.5	...	115	240	None	...	...	3-4	100	212
Bis (*p*-aminocyclohexyl) methane (PACM-20)	29	1.0	...	150	300	3	150	300	1.5	149	300
Isophorone diamine (IPD)	23	1.0	...	100	212	3	150	300	1.0	146	295
n-aminoethylpiperazine (AEP)	20	1.0	...	150	300	3	150	300	1/4-1/2	110	230
2-ethyl-4-methyl imidazole (EMI-24)	10	8.0	...	60	140	None	...	...	4-6	110	230
Same	4	4.0	...	60	140	2	150	300	20+	160	320
Aromatic amines, elevated-temperature cure:											
m-phenylenediamine (MPDA)	14	2.0 + 2.0	...	80(175)/150	300	2	150	300	5-6	150	300
4,4′-methylenedianiline (MDA)	28	2.0 + 2.0	...	80(175)/150	300	2	150	300	5-6	160	320
4,4′-diaminodiphenylsulfone (DDS) + 1phr BF₃ − MEA	30(f)	2.0 + 2.0	...	125(257)/200	390	2	200	390	(g)	175	350
Aromatic amine eutectics(h)	20	2.0 + 2.0	...	80(175)/150	300	2	150	300	6-8	145	290
Carboxylic acid anhydrides, elevated-temperature cure:											
Hexahydrophthalic anhydride (HHPA) + 1phr BDMA(i)	78	3.0 + 1.0	...	90(195)/150	300	3	200	390	24	132	270
Nadic methyl anhydride (NMA) − 1phr BDMA	90	3.0 + 1.0	...	120(250)/150	300	3	200	390	60-80	144	290
Chlorendic anhydride (CA)(j)	117	Gel + 4.0	...	25(80)/150	300	3	200	390	(j)	197	385
Dodecenylsuccinic anhydride (DDSA) + 1phr BDMA	134	4.0 + 1.0	...	90(195)/150	300	3	200	390	120	74	165
Methyltetrahydrophthalic anhydride (MTHPA) + 1phr BDMA	80	1.0 + 1.0	...	100(212)/150	300	4	150	300	24	130	265
Catalytic Lewis acids, elevated-temperature cure:											
Boron trifluoride monoethylamine (BF₃-MEA)	3	3.0 + 4.0	...	120(250)/200	390	4	200	390	>250	174	345
Latent curing agents, elevated-temperature cure:											
Dicyanamide (DICY)	6	1.0	...	175	350	1.0	175	350	∞	135	275(k)

(a) Parts per 100 parts resin by weight for the DGEBPA-WPE = 189. (b) Used to obtain heat distortion temperature. Post cure should be made at or above HDT. (c) By ASTM D 648 (Ref 18). (d) Based on polyamide (Versamide) resin; other grades are available. (e) Based on curing agent U, Shell Chemical Co. Numerous proprietary types are available. (f) Less than stoichiometric (33 phr) because BF₃-MEA also assists curing. (g) Mixture is too viscous for this measurement. Prepreg useful life is 10 to 20 days. (h) Data on curing agent Z, Shell Chemical Co. Many others available. (i) BDMA = benzyldimethylamine. (j)Chlorendic anhydride seldom used alone or at stoichiometry (203 phr) due to fast gel time. (k) T_g as measured by thermal mechanical analysis on a laminate. Source: Ref 19, 20

mixed stream passes directly into the mold, where heat and pressure cure the part in 2 to 5 min. The reinforcement is placed in the mold in the desired pattern before injection of the resin.

In resin transfer molding, a transfer cylinder and ram are used to inject a molding compound directly into a mold. Thus, this process is not used with continuous strand fiber reinforcement. In the case of reinforced plastics, how-ever, the molding compounds contain chopped fiber strands and are considered part of the composites family.

Pultrusion is a continuous lamination process useful for fabricating linear stock in the form of

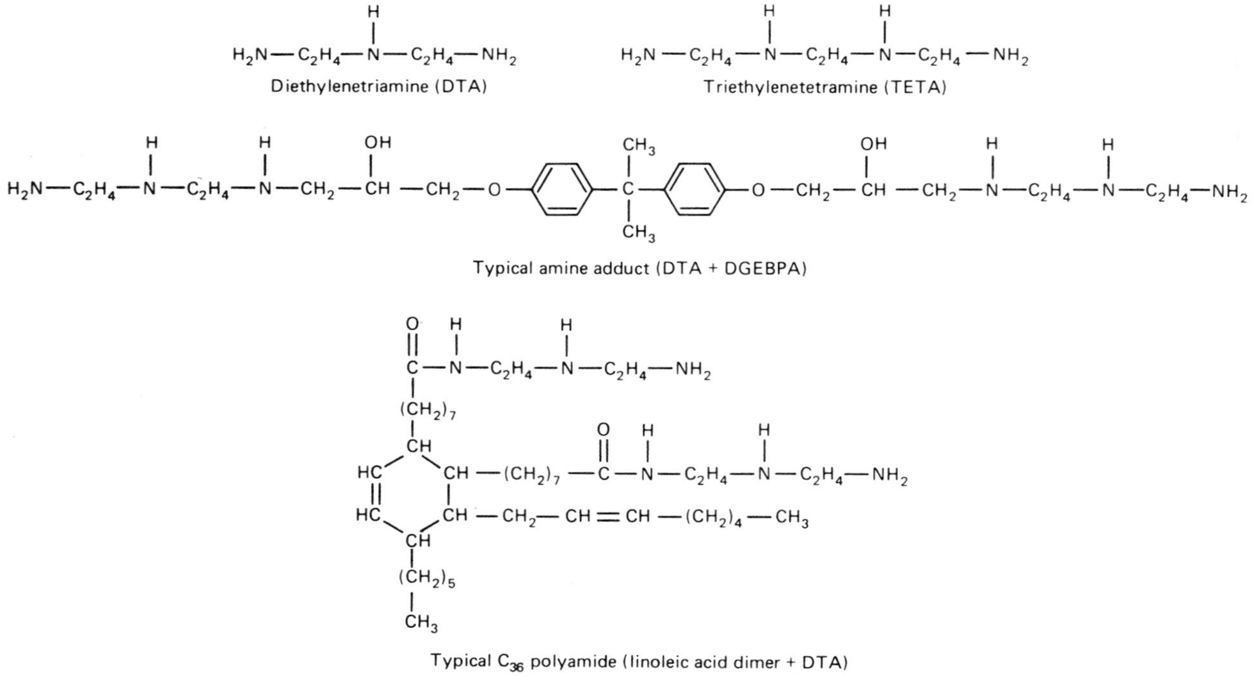

Diethylenetriamine (DTA)

Triethylenetetramine (TETA)

Typical amine adduct (DTA + DGEBPA)

Typical C₃₆ polyamide (linoleic acid dimer + DTA)

Fig. 9 Aliphatic amine curing agents

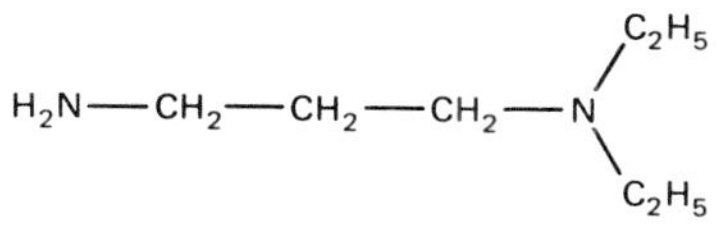

Diethylaminopropylamine (DEAPA)

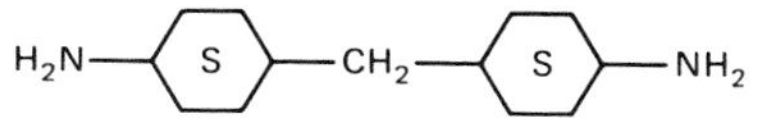

Bis (*p*-aminocyclohexyl) methane (PACM-20)

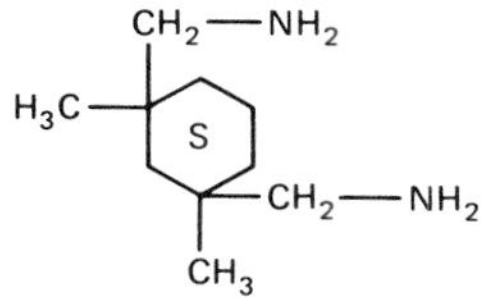

Isophorone diamine (IPD)

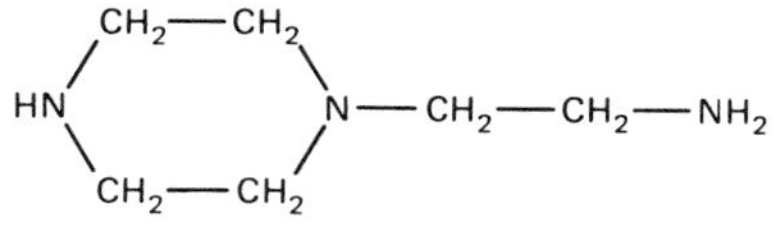

n-aminoethylpiperazine (AEP)

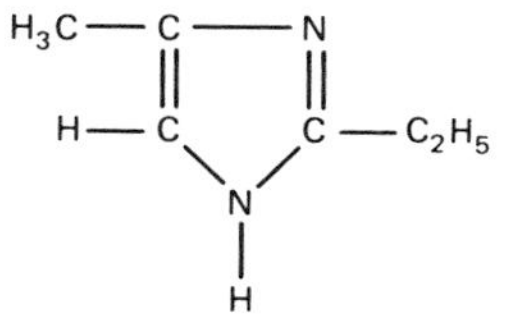

2-ethyl-4-methyl imidazole (EMI-24)

Fig. 10 Moderate-temperature curing agents

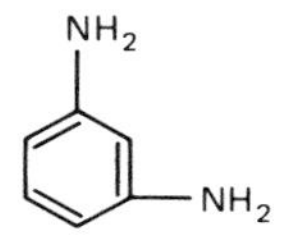

Metaphenylenediamine (MPDA)

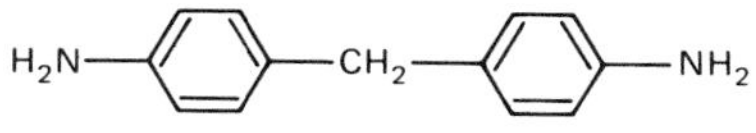

Methylenedianiline (MDA)

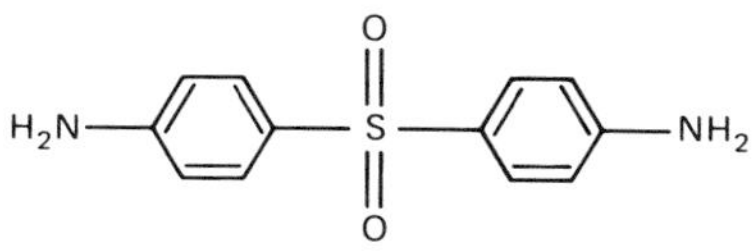

4, 4′-diaminodiphenylsulfone (DDS)

Fig. 11 Aromatic amine curing agents

bars, tubes, rods, and other cross-sectioned shapes, such as Js, Ls, Ts, Zs, hats, and channels. The reinforcing fiber is fed continuously through a trough of liquid resin and then through a die that controls the resin fiber volume. It then passes through a heated die that shapes and cures the laminate into the desired configuration. The laminated stock is thus available in any desired length. The cure rate of the resin determines the throughput rate of the process.

Although the processes for dry lay-up laminating are more limited in variety, large volumes of dry prepregs are used by the electronics and aerospace industries. Electrical laminates, principally circuit board materials, are made from nontacky, truly dry prepregs. The prepregs, along with copper sheets (which are later etched into electrical circuits), are cut, stacked, and laminated in multiplaten flat presses. Heat and pressure cause the resin to melt, flow, densify, and finally cure before the etch processing of the colaminated copper sheets.

Aerospace composites usually require prepregs that have tack and drape characteristics, because of the complex curvatures of the hardware being fabricated. The prepreg is placed by hand on the tool, layer by layer. The tack of the prepreg holds the layers in place until the lay-up is complete. Drape or flexibility of the material allows each layer to conform to the tool shape. Curing is accomplished as described above, typically with a vacuum bag in an autoclave. However, alternative, simpler methods are now being considered, such as processes that involve near-net resin content prepregs and no bleeding of excess resin, using vacuum bags in ovens or self-contained tools that heat, cure, and shape the hardware without the need for an autoclave.

Laminate Properties

Most of the property data on laminates made from epoxy resin matrices involve the use of glass fibers as the reinforcement. Although the newer fibers, particularly polyaramid and carbon, are major focal points of current research and development, the variety of epoxy matrices used with the newer fibers has been limited. Accordingly, in describing the effect of the chemical structures of the resins and curing agents on material and processing properties, the focus will mainly be on epoxy resins in glass fiber reinforced composites. Using the same fiber and fiber form (weave) insofar as is possible permits examination of the effects of resins and curing agents independent of the reinforcement. The effect of fiber type will be discussed in the section "Effect of Fiber Reinforcements" in this article, after the criteria for matrix selection have been established.

In epoxy resin laminates, and with other matrices as well, the fiber usually dominates the mechanical response of a composite. This, of course, depends on the test procedure, but it is readily evident from the simple rule-of-mixtures equation:

$$\sigma_c = V_f \sigma_f + V_m \sigma_m$$

where σ_c is the overall strength of the composite, V_f is the volume fraction of the fiber, σ_f is the tensile strength of the fiber, V_m is the volume fraction of the matrix, and σ_m is the shear strength of the matrix. Since fiber tensile strength is approximately 20 to 100 times the shear strength of the resin, depending on fiber type, the resin will contribute less than 10% of the composite strength in any fiber-dominated test mode. Since composite hardware is designed to take advantage of the fiber strength along two and sometimes all three axes, the mechanical properties of the resin usually have little effect on the overall strength of the composite, other than from the load transfer characteristics and the strength of the interphase.

This is evident from the data in Table 4, where the strengths of a series of glass cloth laminates, prepared from DGEBPA using several of the curing agents discussed in Table 3, are compared at essentially the same fiber volume. The flexural and compressive strengths of all five resin systems at room temperature are essentially the same. Because these are fiber-dominant test procedures, the pure resin mechanical properties are masked, as would be expected from the rule of mixtures. Considering the shear measurements, which are a resin-dominated property, the results are more consistent in magnitude with what would be expected from resin properties. The fact that all three resin systems give similar shear values indicates that the three matrix systems shown here vary little in this property. However, if the data at 150 °C (300 °F) are considered, pronounced differences are observed. This clearly reflects a matrix resin property, the T_g, and is compatible with the data shown in Table 3. Above the T_g, the resin properties contributing to the overall strength of the composite would be expected to decrease by at least an order of magnitude, making it a poor load transfer medium. Thus, three key points are the importance of T_g to elevated-temperature performance, the utility of the rule of mixtures, and the fact that criteria other than pure resin mechanical performance must be considered in matrix resin selection.

One additional point should be noted regarding the data in Table 4. Although T_g is important to elevated-temperature performance, a high T_g is not obtained without paying a price. It has been demonstrated that internal stresses

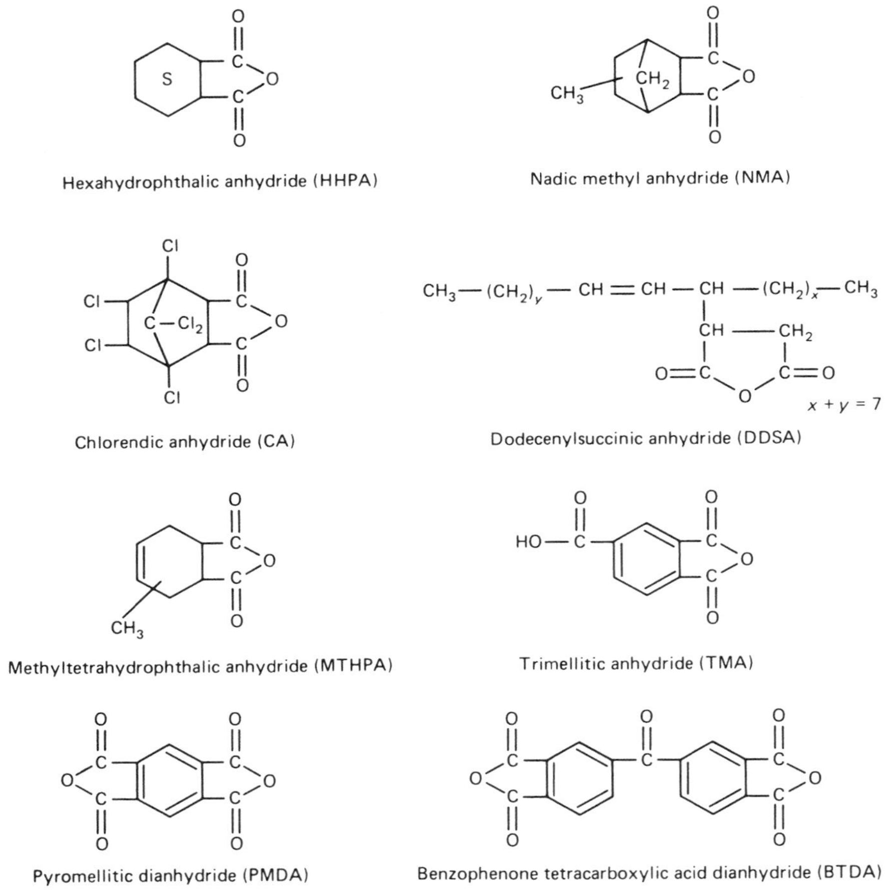

Fig. 12 Anhydride curing agents

Tables 6 to 8 contain additional data on epoxy resin based composites that should be of value in the matrix selection process. In addition to information on the curing agents discussed earlier (Table 3), data have been included on other epoxy resins. Table 6 gives the strength characteristics resulting from a number of wet lay-up laminating systems. Dry lay-up laminating systems are presented in Table 7, along with data on some higher-temperature performance resins. Several matrix systems show useful strengths in the 260 °C (500 °F) range. Epoxy resins, when properly cured with a carefully chosen curing agent, display excellent strength retention on aging at elevated temperatures. The diglycidyl ether of BPA, while not considered a high-temperature performance resin, shows excellent strength retention, even after aging for 200 h at 330 °C (630 °F), when cured with NMA (Ref 19).

Table 8 demonstrates another principle in epoxy resin selection. Previous studies showed that the water absorption characteristics of cured epoxy resins correlate with the hydrocarbon or nonpolar content of the backbone molecule (Ref 29). The data in Table 8 (Ref 11) confirm this conclusion because the triphenylmethane based epoxy resin displays much better wet strength retention than the more polar glycidyl amine based resin in the carbon fiber laminates shown. Although not yet verified by composite data, the tetraglycidyl diisopropylbenzene resins shown in Fig. 5, which have a substantial hydrocarbon backbone, should prove to be interesting, especially when combined with curing agents of similar structure (Ref 30, 31).

Electrical Laminates. Approximately 12% of all epoxy resins are used in electrical applications, mainly in the form of fiber-reinforced products. Since most electrical laminates are flat, the circuit board manufacturers prefer dry lay-up prepregs for ease of handling, ready application of the copper foil face sheets required for subsequent circuitry etching, and high production rates in multiplaten steam-heated presses. The epoxy resins used in the manufacture of the dry lay-up laminates are based primarily on the diglycidyl ethers of BPA of the lowest molecular weight grade that affords a solid resin (where $n = 2$ to 3) and

are generated within a glassy polymer as it cools from the cure temperature below T_g (Ref 26). Thus, the higher the matrix T_g, the greater the stress build-up in a laminate when cooled to room temperature. The fiber-resin interface is consequently more highly stressed, and room-temperature composite performance would be expected to suffer. The data in Table 4 confirm this: The higher T_g aromatic amine cured epoxies have room-temperature flexural properties that are lower than those of the anhydride systems.

Another factor to consider in epoxy resin lamination is the degree of cure. This effect is

illustrated in Table 5, which shows the physical properties of a series of wet lay-up laminates based on aliphatic amines, comparing a room-temperature cure to a moderate-heat cure. The heat-cured laminates are superior, emphasizing the point that optimum properties in epoxy composites are normally achieved with fully cured matrix systems. The excellent strengths shown for the diethylaminopropylamine-cured (DEAPA-cured) laminate is not unexpected, inasmuch as this curing agent is known for its adhesive properties, and the stronger fiber-resin interface does result in a somewhat stronger laminate.

Table 4 Properties of glass fabric laminates prepared from the diglycidyl ether of BPA using various curing agents

Curing agent	Cure cycle, min/ °C	(°F)	Post cure, h/ °C	(°F)	Resin contents, wt%	Flexural strength(a) at 25 °C (80 °F) MPa	ksi	at 150 °C (300 °F) MPa	ksi	Edgewise compressive strength(b) at 25 °C (80 °F) MPa	ksi	at 150 °C (300 °F) MPa	ksi	Interlaminar shear strength(c) at 25 °C (80 °F) MPa	ksi	at 150 °C (300 °F) MPa	ksi
MPDA	60/145	(290)	1/200	(390)	29	554	80	339	49	386	56	208	30	13.1	1.9	13.1	1.9
MDA	60/145	(290)	1/200	(390)	31	567	82	374	54	381	55	138	20	16.6	2.4	10.4	1.5
DDS/BF₃·MEA	60/170	(340)	2/200	(390)	23	498	72	249	36	346	56	125	18	15.2	2.2	13.8	2.0
HHPA/BDMA	10/120 58/180	(250) + (355)	1/200	(390)	29	581	84	69.2	10	386	56	41.5	6	19.4	2.8	2.1	0.30
NMA/DMP-30(d)...	45/120 20/165	(250) + (330)	2/200 2/245	(390) + (470)	29	623	90	104	15	388	56	41.5	6	15.2	2.2	3.5	0.50

(a) ASTM D 790 (Ref 25). (b) Federal Test Method Standard 406-1021. (c) Federal Test Method Standard 406-1042. (d) Tris-dimethylaminomethyl phenol

various brominated analogs thereof, as discussed in the section ''Resin Manufacture and Products'' in this article. The diglycidyl ether of tetrabromobisphenol A (BrDGEBPA) is a semisolid resin that is also used in fire retardant grade laminates.

Dicyanamide is the most commonly used curing agent for electrical laminating, and prepregs manufactured with it have been studied extensively. It derives its latency (shelf life) from insolubility in the resins. Although soluble in the prepregging varnish, it recrystallizes from the resin as the solvent is evaporated to form the dry prepreg. However, care must be exercised during prepreg manufacture; any undissolved or recrystallized DICY in the prepreg varnish may cause problems, such as erratic flow during the lamination process. Further, if it comes into contact with the copper face sheet, brown spots or voids may appear in the laminates (Ref 21). Manufacturers of epoxy resins have had considerable experience with this curing agent and will assist formulators and prepreg manufacturers as needed.

Electrical laminates are manufactured in accordance with standards set forth by the National Electrical Manufacturers Association (NEMA). Useful epoxy formulations and their consensus flexural properties are shown in Table 9 for the various NEMA grades of electrical laminates. A variety of resins are used, depending on the degrees of elevated-temperature performance and fire retardancy required.

Effect of Fiber Reinforcements. Epoxy resins, because of their excellent adhesive properties, bond easily to most fibers. Because they form a strong interphase, laminates made with these matrices reflect their fiber properties to a large extent. The three fibers most commonly used are glass, polyaramid, and carbon. In recent years, carbon fiber reinforced epoxy matrix composites have been studied extensively, because of the structural efficiency of carbon fibers in military and civil aircraft and in space structures. Consequently, these studies have received considerable DoD and NASA funding. The epoxy resin matrix systems receiving most of the attention are based on TGDMA. Usually DDS is used as the curing agent, either with or without a borontrifluoride amine salt as the catalyst. Many of these systems are also modified with small amounts of a second epoxy resin added to achieve special effects during processing. The three fibers in a TGMDA/DDS epoxy matrix are compared with various metals in Table 10. Boron fibers also play a minor but important role in epoxy matrix composites. Although expensive, these fibers combine high specific modulus with high specific strength and are used when high stiffness is required.

Carbon fibers are the most versatile and have the best balance of properties as an epoxy matrix reinforcement. Compared to glass and polyaramid fibers, they are superior in tensile,

flexural, compressive, and fatigue (tension-tension) properties. Polyaramid fibers are the lowest in density, which leads to a specific strength similar to that of some of the high-strength carbon fibers, but they have poor compressive and flexural properties because of lack of stiffness. A further advantage of carbon

fibers over the other fibers is in impact strength. Glass fibers are the least expensive and have shear properties equal to those of carbon fibers, but they are the densest of the three most popular fibers and have the lowest specific moduli. Boron fibers have an interesting combination of strength and stiffness but

Table 5 Effect of degree of cure on room-temperature strength(a)

Resin/curing agent system(b)	Cure schedule	Flexural strength(c) Ultimate MPa	ksi	Modulus GPa	10^6 psi
DGEBPA/TETA...............	7 days at 25 °C (80 °F)	370	53	14.2	2.1
	2 h at 110 °C (230 °F)	492	71	20.6	3.0
DGEBPA/DEAPA	2 h at 110 °C (230 °F)	574	83	24.9	3.6
DGEBPA/BGE/TETA(d)..........	7 days at 25 °C (80 °F)	327	47	18.9	2.7
	2 h at 110 °C (230 °F)	532	77	19.4	2.8

(a) On 181-style, E-glass fabric, Volan A finish. (b) See Fig. 5 for chemical structures, Table 3 for more details. (c) Normalized to 67 wt% glass. (d) Resin system contains approximately 12.5% butyl glycidyl ether

Table 6 Properties of wet lay-up heat-cured laminates prepared using various curing agents(a)

Epoxy resin	Curing agent(b) and concentration(c)	Test temperature °C	°F	Normalized flexural properties(d) Ultimate MPa	ksi	Modulus GPa	10^6 psi
DGEBPA	MPDA(14)	25	80	627	90.5	24.8	3.6
		150	300	434	63	21.3	3.1
DGEBPA	MDA(28)	25	80	526	76	19.4	2.8
		150	300	304	44	15.9	2.3
DGEBPA	Aromatic amine eutectic(e)	25	80	556	82	21.8	3.2
DGEBPA	DDS(20) BF₃-MEA(1)	25	80	471	68	20.8	3.0
		127	260	394	57	18.0	2.6
DGEBPA	NMA(90) BDMA(1)	25	80	476	69	20.5	3.0
		150	300	241	35	16.3	2.3
		260	500	78	11.3	7.2	1.0
DGEBPA	HHPA(78) BDMA(1)	25	80	491	71	20.1	2.9
DGEBPA	Chlorendic anhydride(117)	25	80	391	57	23.5	3.4
TGMDA	HHPA(110)	25	80	510	74	...	...
		170	340	225	32.5	...	...

(a) Data on 181-style, E-glass cloth, Volan A finish. (b) Structural formulas appear in Fig. 1-8, chemical names in Table 3. (c) Parts per 100 parts resin by weight. (d) Normalized to 67 wt% glass. (e) Many products of this type are commercially available. Data presented based on Curing Agent Z, Shell Chemical Company. (f) Data on N,N′-tetraglycidylmethylenedianiline, glass fabric type not specified (Ref 27). Source: Ref 18

Table 7 Properties of dry lay-up epoxy laminates(a)

Curing agent	Resin	Test temperature °C	°F	Flexural strength(b) Ultimate MPa	ksi	Modulus GPa	10^6 psi
DICY	DGEBPA, $n = 2 - 3$	25	77	541	78	23.8	3.4
		150	302	58	8.4	...	...
MPDA	DGEBPA	25	77	512	74	19.1	2.8
		150	302	136	20	...	...
NMA/BPMA	DGEBPA	25	77	476	69	20.4	3.0
		150	302	241	35	16.3	2.3
		260	500	78	11.5	7.2	1.0
NMA/BDMA	Tetraglycidyl ether of tetraphenol ethane(d)	25	77	491	71	20.8	3.0
		150	302	339	49	19.4	2.8
		260	500	86.5	12.5	13.1	1.9
CA	DGEBPA, $n = 2 - 3$	25	77	373	54	23.2	3.3
		150	302	28.9	4	3.8	0.6
CA	Tetraglycidyl ether of tetraphenol ethane(d)	25	77	543	78.5	21.8	3.1
		150	302	123	18	10.1	1.5
BF₃ · MEA	Tetraglycidyl ether of tetraphenol ethane(c)(d)/ DGEBPA, 1:1	25	77	601	87	24.6	3.6
		150	302	377	54	23.1	3.4
BF₃ · MEA	Tetraglycidyl ether of tetraphenol ethane(d)	25	77	524	75.5	22.9	3.3
		260	500	218	31.5	18.8	2.7
DDS/BF₃ · MEA	Tetraglycidyl ether of tetraphenol ethane(d)	25	77	443	64	22.6	3.3
		260	500	179	26	18.9	2.7
DDS/BF₃ · MEA	O-cresol novolac glycidyl ether(e)	25	77	587	85.5	24.7	3.6
		150	302	515	77	22.7	3.3
		225	437	278	40	19.6	2.8

(a) Primarily using 181-style, E-glass cloth, Volan A finish. (b) Normalized to 67 vol% glass fiber. (c) Data on 30-style, E-glass cloth. (d) See Fig. 1 for chemical structure. (e) See Fig. 1 for chemical structure, $n = ~4$. (Ref 28). Source: Ref 18

Table 8   Wet and dry properties of carbon fiber laminates(a)

Resin system	Test temperature °C	°F	Test condition	0° flexural properties Ultimate MPa	ksi	Modulus GPa	10^6 psi
TGETPM(b)/DGEBPA, 3:1/DDS........	25	80	Dry	1850	267	149	21.6
	93	200	Dry	1460	211	140	20.3
	93	200	Wet	1360	197	170	24.5
	149	300	Dry	1320	191	138	19.9
	149	300	Wet	990	143	163	23.6
	205	400	Dry	1120	162	137	19.8
	205	400	Wet	650	93.5	162	23.4
TGMDA/DGEBPA, 3:1/DDS	25	80	Dry	1640	237	128	18.5
	93	200	Dry	1440	208	127	18.4
	93	200	Wet	1160	168	132	19.1
	149	300	Dry	1260	182	125	18.0
	149	300	Wet	720	104	129	18.6
	205	400	Dry	800	115	112	16.2
	205	400	Wet	370	54	113	16.4

(a) Unidirectional carbon fiber laminates. (AS4-6K) normalized to 62 vol % fiber. Source: Ref 11. (b) See Fig. 2 for chemical structure.

Table 9   General formulations and properties of epoxy electrical laminates(a)

NEMA grade	Typical formulation Epoxy resin, pbw(b)	Curing agents, pbw(b)	Test temperature °C	°F	Average flexural properties Strength MPa	ksi	Modulus GPa	10^6 psi
G-10	DGEBPA, n = 2 – 3(100)	DICY(3)/CAT.(0.2)(c)	25	80	520	75	26.3	3.8
			150	300	5.2	7.5	21.4	3.1
G-11	DGEBPA(100)	DDS(30)/BF₃ – MEA(1.0)	25	80	450	65	20.0	2.9
			150	300	330	48	17.3	2.5
FR-3	DGEBPA, n = 2 – 3(100)	CA(45)	25	80	400	58	25.0	3.6
			150	300	31	4.5	4.1	0.6
FR-4	BrDGEBPA n = ~2(100)(d)	DICY(3)/CAT.(0.2)(c)	25	80	470	68	18.8	2.7
			150	300	93	13.5	0.5	0.5
FR-5	DGEBPA(60) BrDGEBPA(40)(e)	DDS(20)/BF₃ – MEA(1)	25	80	578	80	27.7	4.0
			150	300	277	40	23.5	3.4

(a) Mechanical properties are highly dependent on glass cloth weave and type. Properties and formulations are a consensus based on the technical literature. (b) pbw = parts by weight. (c) Typically, 0.1 to 0.4% of either BDMA or 2-methylimidazole is used for this purpose CAT., catalyst. (d) Solid brominated diglycidyl ether of bisphenol A. 18 to 20% bromine. (e) Brominated DGEBPA. 44 to 48% bromine. Source: Ref 18, 20, 30-33

Table 10   Property comparisons between an epoxy resin with various reinforcements and various metals

Material	Density, g/cm³	Unidirectional strength Tensile MPa	ksi	Compressive MPa	ksi	Unidirectional tensile modulus GPa	10^6 psi
Carbon(AS4)	1.55	1482	215	1227	178	145	21.0
Carbon(HMS)	1.63	1276	185	1020	148	207	30.0
S-glass	1.99	1751	254	496	72	59	8.6
E-glass	1.99	1103	169	490	71	52	7.6
Aramid	1.38	1310	190	290	42	83	12.0
Aluminum (7075-T6)	2.76	572 MPa (83 ksi)				69	10.0
Titanium (6Al-4V)	4.42	1103 MPa (160 ksi)				114	16.5
Steel (4130)...................	8.0	1241-1379 MPa (180-220 ksi)				207	30.0

Source: Ref 34

are more expensive. A broad range of carbon fibers that exceed boron in specific strength or modulus is commercially available.

Principles for Epoxy Matrix Selection

From the data presented throughout this article, it is obvious that many epoxy resin curing agent combinations are available as composite matrix materials. Because fiber properties dominate the mechanical strength in most hardware designs, matrix selection is difficult and sometimes arbitrary. The mechanical advantages of one resin system over another often disappear when compared in composite form. Yet the matrix is undoubtedly important. It acts as the load transfer medium, which permits the use of the high strengths of various fibers. This is a remarkable synergism; the 0° flexural properties of a bundle of unidirectional fibers are zero, but when used in conjunction with the proper matrix, the strength of the composite in properly designed hardware greatly exceeds that of the pure matrix resin and can come close to that of the fiber.

In selecting an epoxy composite matrix, there is no surer way than to make it and break it. This, however, is time consuming and expensive. Thus, the general suggestions below

are offered as starting points to aid in the selection process.

From the standpoint of composite strength, often several matrices will do the job. Thus, processing and processability become dominant factors in matrix selection, assuming other properties to be equal. There are many laminating processes from which to choose, depending on hardware geometry and performance needs. The matrix that is most compatible with the process should be the one chosen. For example, if ambient-temperature wet lay-up is to be used, working time and viscosity are primary considerations. In autoclave processing, cure rate and viscosity as a function of temperature are important; the matrix resin must retain sufficient viscosity so that it will not be drained from the fibers during heating, yet not cure at a rate so fast that autoclave pressure cannot be applied to the part before it gels and sets. Resins used in matched metal die processes must flow and cure rapidly. Process economics dictate that tooling must be recycled as rapidly as possible. Stand-alone tooling may require special flow characteristics so that dense laminates can be made at low pressures.

Each lamination process has its particular matrix requirements. Careful study of these needs is one of the key factors in successful epoxy resin composite hardware fabrication. As experience with these materials grows, selection of the proper material processing characteristics will become easier.

Without question, T_g is an excellent selection criterion for elevated-temperature performance. However, selection of a matrix with a T_g that exceeds service requirements, just to be safe (a common fault of designers), is not recommended. Doing so may impact cure schedules by requiring higher temperatures and longer cure times to achieve complete cure, which can result in thermal strains and a lower overall performance of the composite structure. Composites are used because of their excellent strength-to-weight ratios. In brittle high-temperature matrix systems, thermal strains can lead to microcracking and a variety of secondary problems. Therefore, it is best to select materials based on actual elevated-temperature needs.

It is important to consider the effects of moisture on the resultant hardware. Water exerts a plasticizing action on composite matrices and may actually appear to improve mechanical performance of the structure under ambient test conditions. However, because of this plasticizing action, elevated-temperature performance can be lowered drastically. Data on the mechanical properties and lowering of the T_g of water-saturated resin castings can be most useful in the matrix selection process. The more hydrocarbon-like and the less polar the molecular structure of a matrix, the better the resistance to moisture absorption is likely to be (see Table 8). It is possible, however, to go too far with this kind of thinking. All hydrocarbon low-polarity polymers have excellent water resistance, but their fiber-bonding characteristics are poor, resulting in low composite strengths.

The matrix should be considered only in the fully cured state. Occasionally a partially cured composite matrix seems to offer better flexibility or toughness. This can cause disaster in the long run. The use environment can cause unreacted chemical groups to undergo secondary reactions that may drastically alter the mechanical performance of a composite. It is true that ambient temperature cured structures may never achieve full cure, but this is unavoidable. In these cases, if there is any question as to results, extensive hardware testing may be a costly but necessary prerequisite to acceptance. Thermal analysis and dynamic mechanical property testing can be helpful during this part of the selection process.

Designing for so-called toughness in composite structures is a popular new phrase in matrix selection; everyone wants tougher resins. This new demand undoubtedly resulted from the realization that matrix tensile elongation characteristics in carbon fiber composites were often exceeded by those of the fiber. In the search for elevated-temperature performance, the fact that high cross-link density in the matrix can also lead to more brittle materials was probably overlooked in earlier efforts. The fracture energy characteristics of a matrix are useful in screening for toughness, but caution should be exercised. Discussions have revealed that matrix resin/composite fracture energy translations are useful to only about 20 J/m (5 ft · lbf/ft) (Ref 37). Above this value, a much greater increase in matrix fracture energy is required to improve composite toughness. It should be recognized that the term toughness lacks precise mechanical definition and means different things to different people, depending on their needs.

The adaptability of the matrix to the tooling required for hardware fabrication is often overlooked. On occasion, the design engineer conceives of an outstanding piece of hardware, the manufacturing engineer provides an excellent tool on which to make it, but the prepreg matrix system—if it is used according to vendor data and meets the mechanical requirements when cured—can be anything that happens to be in the shop. This often results in disaster on the production floor. The role of the matrix at this preliminary stage of fabrication must be kept in mind. Because of the versatility of the product, there are probably several epoxy resin curing agent combinations that will give good mechanical results. Tack and drape must be adequate to form the part on the tool surface. The matrix viscosity must not drop drastically during the cure, leaving starved areas in the cured composite. Sometimes the cure cycle should be changed from that recommended by the manufacturer to one that is more compatible with the tooling. Consideration of the tooling/matrix interactions is an important part of the matrix selection process.

Ultimately, the long-term effects of environmental exposures on the matrix dictate hardware longevity. Extensive data are available from manufacturers on epoxy resin heat stability; fatigue properties; creep characteristics; electrical properties; resistance to solvents, corrosive chemicals, thermal shock, stain, abrasion, and ultraviolet light; and other properties which may be important to a particular application. Many times information of this type, which can finalize the selection process, is free for the asking from resin manufacturers.

Finally, once the selection has been made, it is important to note how the material ages in storage and on the shop floor, whether overstaged material can be processed, how to make this decision, and how to alter the curing process accordingly. Physical and chemical test procedures are available to help determine whether the correct selection has been made (Ref 38).

REFERENCES

1. C.A. May and Y. Tanaka, Ed., *Epoxy Resins, Chemistry and Technology*, Marcel Dekker, 1973
2. H. Lee and K. Neville, *Handbook of Epoxy Resins*, McGraw-Hill, 1967
3. J. Schrade, *The Epoxy Resins*, McGraw Hill, 1967
4. A.M. Paquin, *Epoxyverbindungen und Epoxyharze*, Springer-Verlag, 1958
5. I. Skeist, *Epoxy Resins*, Reinhold, 1958
6. D. Swern, *Chem. Rev.*, Vol 45 (No. 1), 1949
7. D. Swern, chapter 7 in *Organic Reactions*, R. Adams, Ed., Vol 7, John Wiley & Sons, 1953
8. D. Swern, in *Encyclopedia of Polymer Science and Technology*, H.F. Mark and N.G. Gaylord, Ed., Vol 6, Interscience, 1967, p 83
9. R.J. Gall and F.P. Greenspan, *Ind. Eng. Chem.*, Vol 47, 1955, p 147
10. F.P. Greenspan and R.J. Gall, *J. Am. Oil Chem. Soc.*, Vol 33, 1956, p 391
11. "TACTIC 742 High Temperature Epoxy Resin," Technical Bulletin 296-662-86, Dow Chemical Company
12. "EPON HPT Resin 1071," Technical Bulletin SC: 875-86, Shell Chemical Company
13. "EPON HPT Resin 1072," Technical Bulletin, SC: 876-86, Shell Chemical Company
14. R.S. Bauer, Paper presented at the 31st International Symposium, Society for the Advancement of Material and Process Engineering, Las Vegas, April 1986
15. W. Fisch and W.J. Hofmann, *J. Polym. Sci.*, Vol 12, 1954, p 497
16. R.F. Fischer, *J. Polym. Sci.*, Vol 44, 1960, p 155
17. Y. Tanaka and T.F. Mika, in *Epoxy Resins, Chemistry and Technology*, C.A. May and Y. Tanaka, Ed., Marcel Dekker, 1973, p 198-199
18. "Standard Test Method for Deflection Temperature of Plastics Under Flexural Load," D 648, *Annual Book of ASTM Standards*, American Society for Testing and Materials
19. "EPON Resins for Fiberglass Reinforced Plastics," Technical Bulletin SC: 227-85, Shell Chemical Company
20. "General Guide, Formulating with Epoxy Resins," Technical Bulletin 296-346-583, Dow Chemical Company
21. "Formulating with Dow Epoxy Resins, Electrical Laminates," Technical Bulletin 296-364-484, Dow Chemical Company
22. H.H. Levine, Paper 41, *American Chemical Society, Division of Organic Coatings and Plastics Chemistry*, 1964
23. T.F. Saunders *et al.*, *American Chemical Society, Division of Organic Coatings and Plastics Chemistry*, 1966
24. S.A. Zahir, in *Proceedings of the Sixth International Conference on Organic Coatings Science and Technology*, July 1980, p 745-779
25. "Standard Test Method for Flexural Properties of Unreinforced and Reinforced Plastics and Electrical Insulating Materials," D 790, *Annual Book of ASTM Standards*, American Society for Testing and Materials
26. H. Dannenberg, *Plast. Eng.*, Vol 21 (No 7), 1965, p 7
27. "Araldite MY-720, A Medium Viscosity Multifunctional Epoxy Resin," Product Data Bulletin CR 801A5M32, Ciba-Geigy Corporation
28. "EPON Resin DPS-64," Technical Bulletin SC: 735-84, Shell Chemical Company
29. C.J. Busso, H.A. Newey, and H.V. Holler, AFML-TR-69-328, Air Force Materials Laboratory, Jan 1970
30. "EPON HPT Curing Agent 1061," Technical Bulletin SC: 878-86, Shell Chemical Company
31. "EPON HPT Curing Agent 1062," Technical Bulletin SC: 879-86, Shell Chemical Company
32. "Electrical Laminates, NEMA G-10," Technical Bulletin 296-395-484, Dow Chemical Company
33. "Electrical Laminates, NEMA G-11," Technical Bulletin 296-397-484, Dow Chemical Company
34. "Electrical Laminates, NEMA FR-4," Technical Bulletin 296-396-484, Dow Chemical Company
35. "Electrical Laminates, NEMA FR-5," Technical Bulletin 296-398-484, Dow Chemical Company
36. "Hercules Composite Structures," Technical Bulletin, Hercules, Inc.
37. N. Johnston, NASA-Langley Research Center, and D. Hunston, National Bureau of Standards, private communications
38. C.A. May, Ed., Epoxy Resins, *Chemistry and Technology*, 2nd ed., Marcel Dekker, to be published

Polyimide Resins

D.A. Scola, United Technologies Research Center

POLYIMIDE MATERIALS can be categorized by their temperature capabilities into those with an upper limit of 232 °C (450 °F) for extended time periods, and those capable of extended use up to 316 °C (600 °F). Bismaleimides and some condensation polyimides such as Avimid K-III belong in the former category, while those materials such as PMR-15, LARC TPI, and Avimid N (NR-150B2) belong in the latter. In terms of chemistry, there are two general types of commercial polyimides: thermoplastic polyimides, derived from a condensation reaction between anhydrides or anhydride derivatives and diamines, and cross-linked polyimides, derived from an addition reaction between unsaturated groups of the preformed imide monomers or oligomers. The imide monomers or oligomers are also derived from the typical condensation reaction to form the imide group, but polymer formation stems from the addition reaction. For completeness, the chemistry of both types will be described in this article. Polyimide types and their upper temperature capabilities are shown in Table 1. A comparison of common properties is provided in Table 2. Source information on all commercial data sheets used for all tables is provided in an appendix to this article.

Chemistry of Polyimides

Descriptions of both condensation-type and addition-type polyimides follow.

Condensation-type polyimides, discovered in 1908 by T.M. Bogert and R.R. Renshaw (Ref 20) and made practical by W.M. Edwards and I.M. Robertson (Ref 21), are derived from polyamic acids by either chemical or thermal treatment over a temperature range from room temperature to 300 °C (570 °F). The polyamic acids are produced by a series of step growth reactions at room temperature from a dianhydride or dianhydride derivative and a diamine. The general reaction for polyimide formation is illustrated in Fig. 1. The structures of several commercially available thermoplastic polyimides are shown in Fig. 2.

Addition-type polyimides are discussed below.

Bismaleimides can be converted to homopolymers, copolymers, or terpolymers and used as cross-linking agents. The first successful homopolymerization of bismaleimides into thermoset polymers was accomplished by F. Grundschober and J. Sambeth in 1968 (Ref 22).

Table 1 Commercial types of polyimides used in structural composites

Type	Upper temperature capability °C	°F
Condensation		
Monsanto Skybond 700, 703	316	600
DuPont NR-150B2 (Avimid N)	316	600
LARC TPI	300	572
Avimid K-III	225	432
Ultem	200	400
Kapton	316	600
Addition		
PMR-15 (Reverse Diels-Alder nadic end-capped)	316	600
LARC 160 (Reverse Diels-Alder nadic end-capped)	316	600
Thermid 600 (Acetylene end-capped)	288	550
BMIs (Bismaleimides, maleimide end-capped)	232	450

Table 2 Physical and mechanical properties of polyimides

Material	Density, g/cm³	Tensile strength MPa	ksi	Tensile modulus GPa	10⁶ psi	Flexural strength MPa	ksi	Flexural modulus GPa	10⁶ psi
Avimid K-III (Ref 1, 2)	1.31	102	15.0	3.8	0.55	...	...	...	...
Skybond 701 (Ref 3)	1.35	69	10.0	4.1	0.60	...	...	...	...
PMR-15 (Ref 4-6, 19, 59)	1.32	38.6	5.6	3.9	0.57	176	25.5	4.0	0.58
NR-150B2 (Ref 7-9)	1.40	110	16.0	4.1	0.60	...	...	...	...
Thermid MC-600 (Ref 10-12)	1.34	83	12.0	4.1	0.60	145	21.0	4.5	0.66
UpJohn 2080 (Ref 13)	1.40	120	17.1	1.3	0.19	117	17.0	3.4	0.48
BMIs (Ref 14, 15, 18, 24–26, 32–58)	1.22-1.30	41-82	6-12	4.1-4.8	0.60-0.70	76-145	11-21	3.4-4.8	0.50-0.70
Ultem 1000 (Ref 16)	1.27	104	15.2	3.0	0.43	145	21	3.4	0.48
Torlon 4203 (Ref 17)	1.38	186	27.0	4.4	0.64	211	30.7	4.5	0.66

Material	Izod impact strength, notched J/m	ft · lbf/in.	Strain-to-failure, %	Glass transition temperature, T_g °C	°F	Fracture toughness, G_{Ic} J/m²	in. · lb/in.²
Avimid K-III (Ref 1, 2)	...	...	14	250	482	1900	11.0
Skybond 701 (Ref 3)	53.4	1.0	1.00	330	626	...	...
PMR-15 (Ref 4-6, 19, 59)	53.4	1.0	1.5	340	644	280	1.57
NR-150B2 (Ref 7-9)	42.7	0.8	6.0	340	644	2400	13.4
Thermid MC-600 (Ref 10-12)	48	0.9	1.5	320	608	...	...
UpJohn 2080 (Ref 13)	37.4	0.7	10.0	280	536	...	...
BMIs (Ref 14, 15, 18, 24–26, 32–58)	...	...	1.3-2.3	230-290	446	34-260	0.19-1.45
Ultem 1000 (Ref 16)	53.4	1.0	60	210	426	...	...
Torlon 4203 (Ref 17)	133.5	2.5	20	267	512	3900	21.9

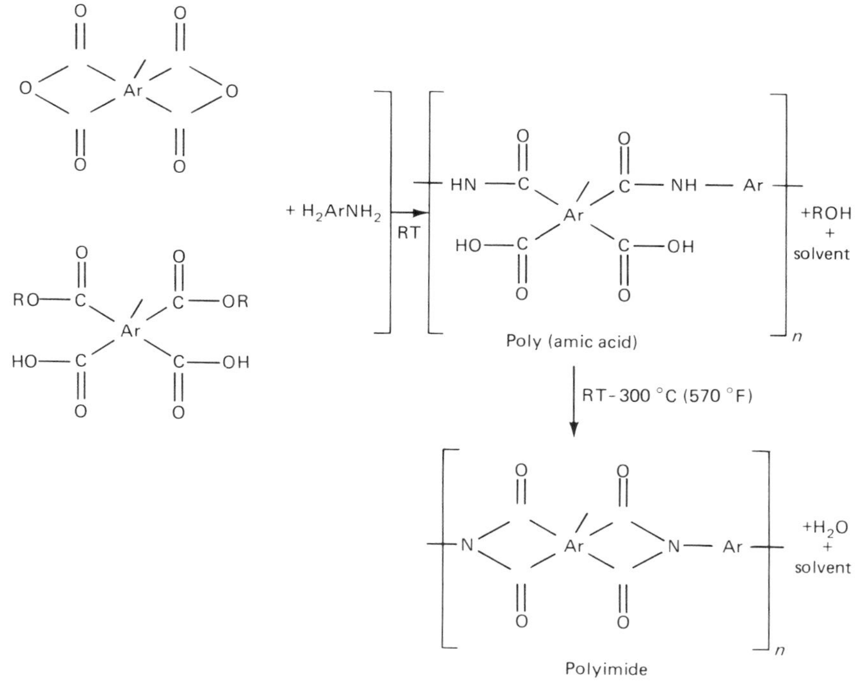

Fig. 1 Condensation polyimides. Solvents can be NMP, DMF, alcohols, and diglyme. RT, room temperature

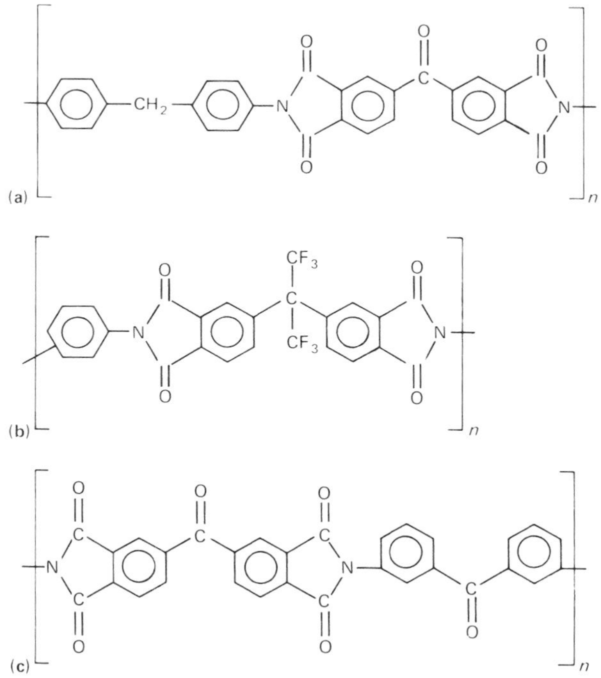

Fig. 2 Condensation polyimide resins. (a) Monsanto Kapton Skybond 701 (contains cross-linking via amide); thermoset. (b) DuPont NR-150 B2; thermoplastic. (c) LARC TPI

bismaleimides or mixtures of bismaleimides (Ref 23) (homopolymerization) (Fig. 4); bismaleimide or bismaleimide amine oligomer mixtures (Ref 23, 24) (copolymerization) (Fig. 4, 5); bismaleimide-olefin monomer mixtures and/or oligomers (Ref 25) (copolymerization) (Fig. 6); and bismaleimide, 0,0'-dicyanobisphenol A mixtures, called BT resins, and modifications of these systems (Ref 26) (polymer blending) (Fig. 7). There is presently no evidence for a reaction between the bismaleimide and 0,0'-dicyanobisphenol A in the BT resin systems.

More recent bismaleimides include the product derived from the copolymerization of a bismaleimide with m-aminobenzoic acid hydrazide (Ref 24) (Fig. 5) and the cross-linked product from copolymerization of bismaleimide with a diolefin (Ref 25) (Fig. 6).

In contrast to the condensation polyimides and other addition-type polyimides, the bismaleimides undergo polymerization by reaction of the maleimide double bond with another unsaturated system or by the Michael addition of nucleophylic species at relatively low temperature ($<$250 °C or 480 °F) without the evolution of volatile by-products. For this reason, bismaleimides are capable of epoxylike processing. The epoxylike processing characteristics of BMIs showing autoclave cure at low pressure and postcure requirements are defined in Table 3.

The Reverse Diels-Alder (RDA) polyimides (Ref 27, 28), represented by PMR-15, not only undergo the amidation and imidization reactions to form a low molecular weight oligomer, typical of condensation polyimides, but also undergo an irreversible Diels-Alder reaction leading to a high molecular weight cross-linked polyimide. Processing of this resin is further complicated by isomerization of the endo nadic end-capped imide oligomer to the exoisomer. These latter two reactions distinguish the RDA addition-type polyimides from the condensation polyimides.

The first process step is amide formation, which occurs between room temperature and 150 °C (300 °F) (Fig. 8). In the second step, imidization occurs over a temperature range of 150 to $\geq$ 250 °C (300 to $\geq$ 480 °F). Simultaneously, in the temperature range of 175-260 °C (350-500 °F), isomerization of the endoisomer to the exoisomer occurs (Fig. 9) (Ref 29, 30). Finally, in the fourth step, an irreversible Diels-Alder reaction of the endo/exo-oligomers occurs, yielding the reactive intermediates cyclopentadiene and bismaleimide oligomers (Fig. 10). These recombine to form a stable cross-linked polyimide. The critical steps in the processing of this polyimide are the removal of residual by-products due to the imidization reaction and removal of low molecular weight components formed during the various process steps. Properties of these materials are described in the section "Constituent Material Properties" in this article.

It should be noted that the cross-linking

Since then, several publications on homopolymerization of bismaleimides have appeared in the literature. However, practical application of bismaleimides to polymeric systems required copolymerization with amines or nucleophylic monomers by means of the Michael addition reaction, or with olefinic or acetylenic monomers, illustrated in Fig. 3.

Commercial bismaleimides are marketed essentially in four distinct forms, represented by the types discussed below, but with molecular modifications depending on the supplier:

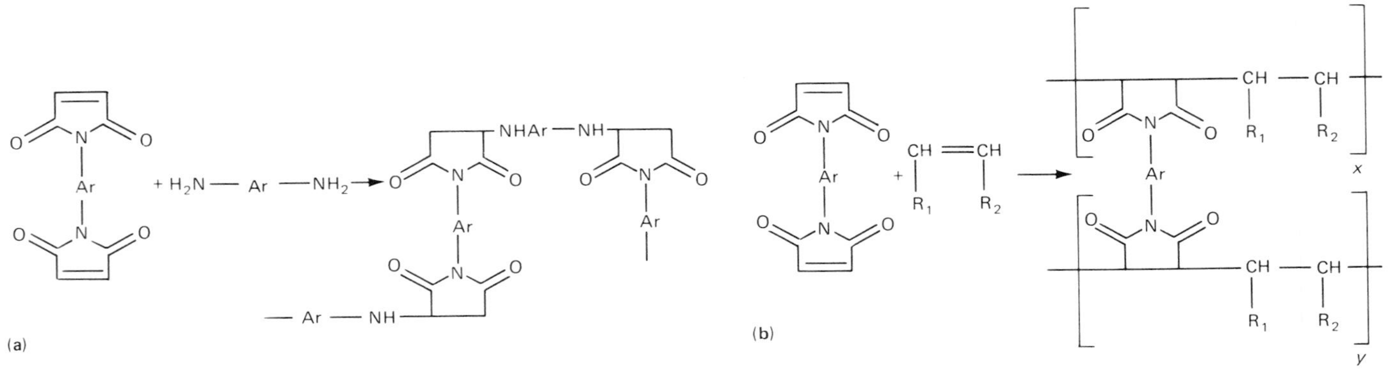

Fig. 3 Bismaleimide addition-type polyimides. (a) Michael addition BMIs (amine cross-linked polybismaleimide). (b) Bismaleimide-olefin copolymers

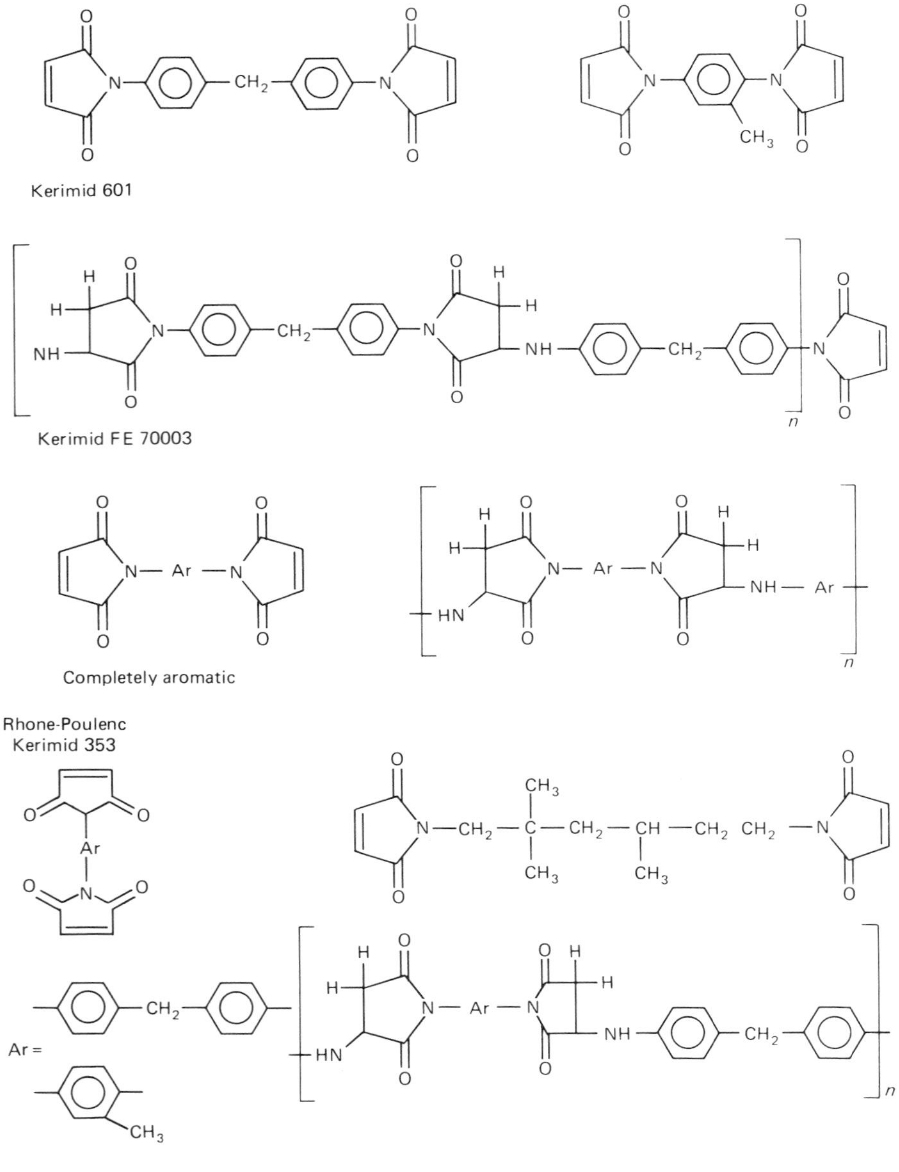

Kerimid 601

Kerimid FE 70003

Completely aromatic

Rhone-Poulenc Kerimid 353

Ar =

Fig. 4 Commercial bismaleimides

mechanism for PMR-15 and other related addition polyimides, such as LARC-160 and PN-modified PMR polyimide, remains controversial. There are at least three mechanisms that have been proposed in the literature. The cross-linking mechanism of PMR-15 described in this article represents just one of these mechanisms.

The acetylene end-capped polyimide Thermid 600 (Ref 31) can be processed from monomers of Thermid AL-600 (Fig. 11) in a manner similar to that of RDA polyimide PMR-15 and condensation polyimides in their initial stages of oligomer imide formation, or it can be processed from preformed imide or isoimide oligomers (Fig. 12).

If the PMR approach is used, the monomers combine to form amic acid oligomers (with the removal of ethanol and/or water and solvent) over the temperature range of 40 to 150 °C (105 to 300 °F). The second step involves formation of the acetylene-terminated Thermid 600 oligomer imide, with the elimination of water or ethanol and solvent over the temperature range of 150 to 250 °C (300 to 480 °F). Simultaneously, during the latter stages of imide formation, the acetylene end-cap undergoes homopolymerization to form a poly-(eneimide) over the temperature range of 175 to 350 °C (350 to 660 °F). As with other polyimides, the critical steps are elimination of the by-product off-gases and solvent, and consolidation during the low-viscosity point or before the gelation point without trapping volatiles evolved during the cross-linking reaction. With Thermid 600, pressure is applied at 200 °C (390 °F), and maintained up to 316 °C (600 °F). The cross-linking reaction is illustrated in Fig. 13.

Constituent Materials Properties

Descriptions of various BMI resins, PMR-15 polyimide, Thermid MC-600, and IP-600 resins follow.

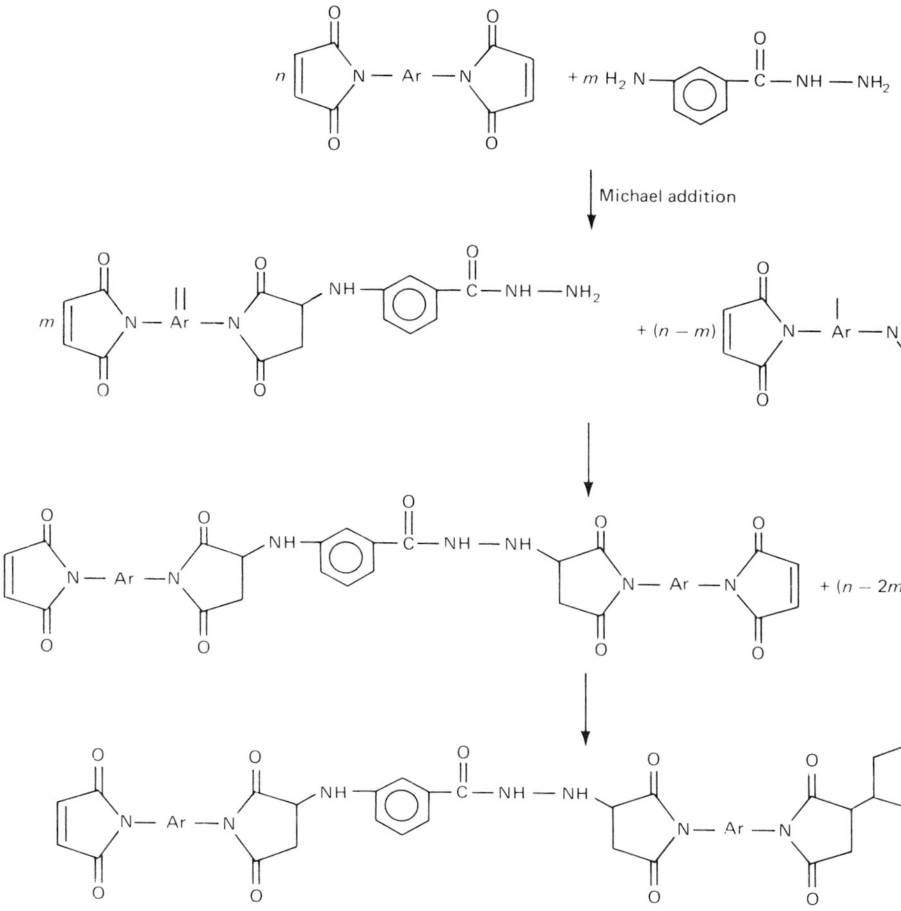

Fig. 5 Chemistry of Compimide resins

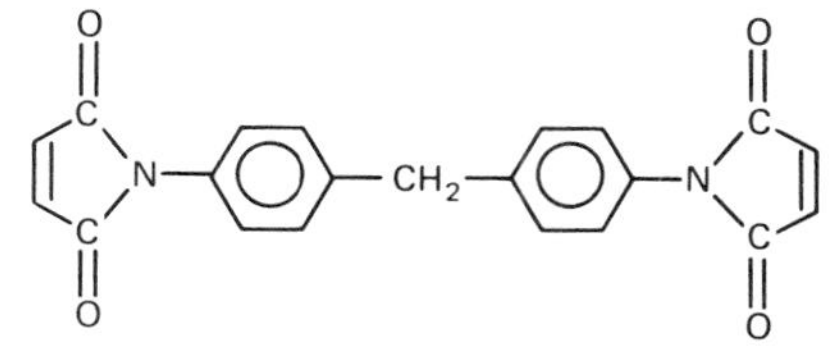

4, 4'- bismaleimido-diphenylmethane (component A)

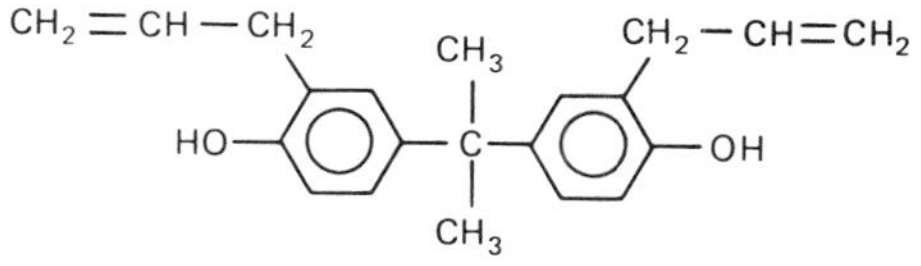

0, 0'-diallyl bisphenol A (component B)

Fig. 6 Two-component BMI-olefin resins

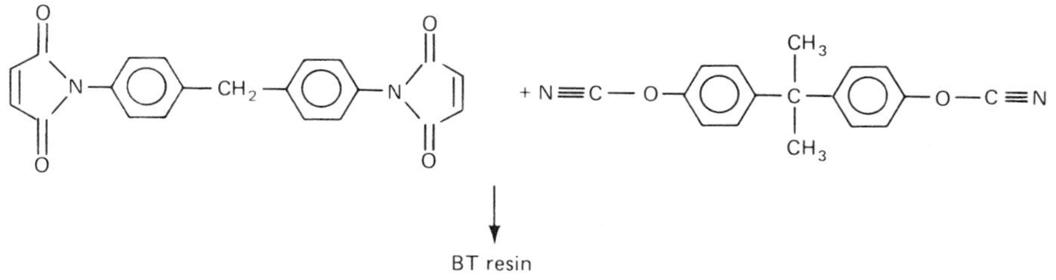

BT resin

Fig. 7 Bismaleimide-triazine resins

Table 3 Processing parameters of bismaleimide resins

Step	Parameters
Autoclave cure	80 °C/1 h at 1.2-2.5 °C/min (175 °F/1 h at 2.2-4.5 °F/min); 520-690 kPa (75-100 psi)
	177 °C/4 h at 1.2-2.5 °C/min (350 °F/4 at 2.2-4.5 °F/min); 520-690 kPa (75-100 psi)
	204 °C/4 h; 520-690 kPa (75-100 psi)
Postcure	220-260 °C/4-24 h (freestanding) (430-500 °F/4-24 h)
Cool-down	1.5 °C/min to 50 °C (2.7 °F/min to 120 °F)

Bismaleimide Rhone-Poulenc Kerimid
601 and Kerimid 353 (also known as Compimide 353), are multicomponent systems consisting of 4,4'-bismaleimido-diphenylmethane (BMI-4,4'-MDA); oligomer of BMI-4,4'-MDA and 4,4'-MDA; 1,4-bismaleimido-2-methylbenzene (BMI-1,4-tolyl); and other

components, illustrated in Fig. 4. The constituent properties of these materials are listed in Table 4.

The electrical and magnetic properties of these materials in the neat form have not been

determined. Some electrical, thermal, and magnetic properties have been determined in glass cloth laminates.

Bismaleimide Matrimid 5292 (Ref 28). The chemical composition of this two-comp-

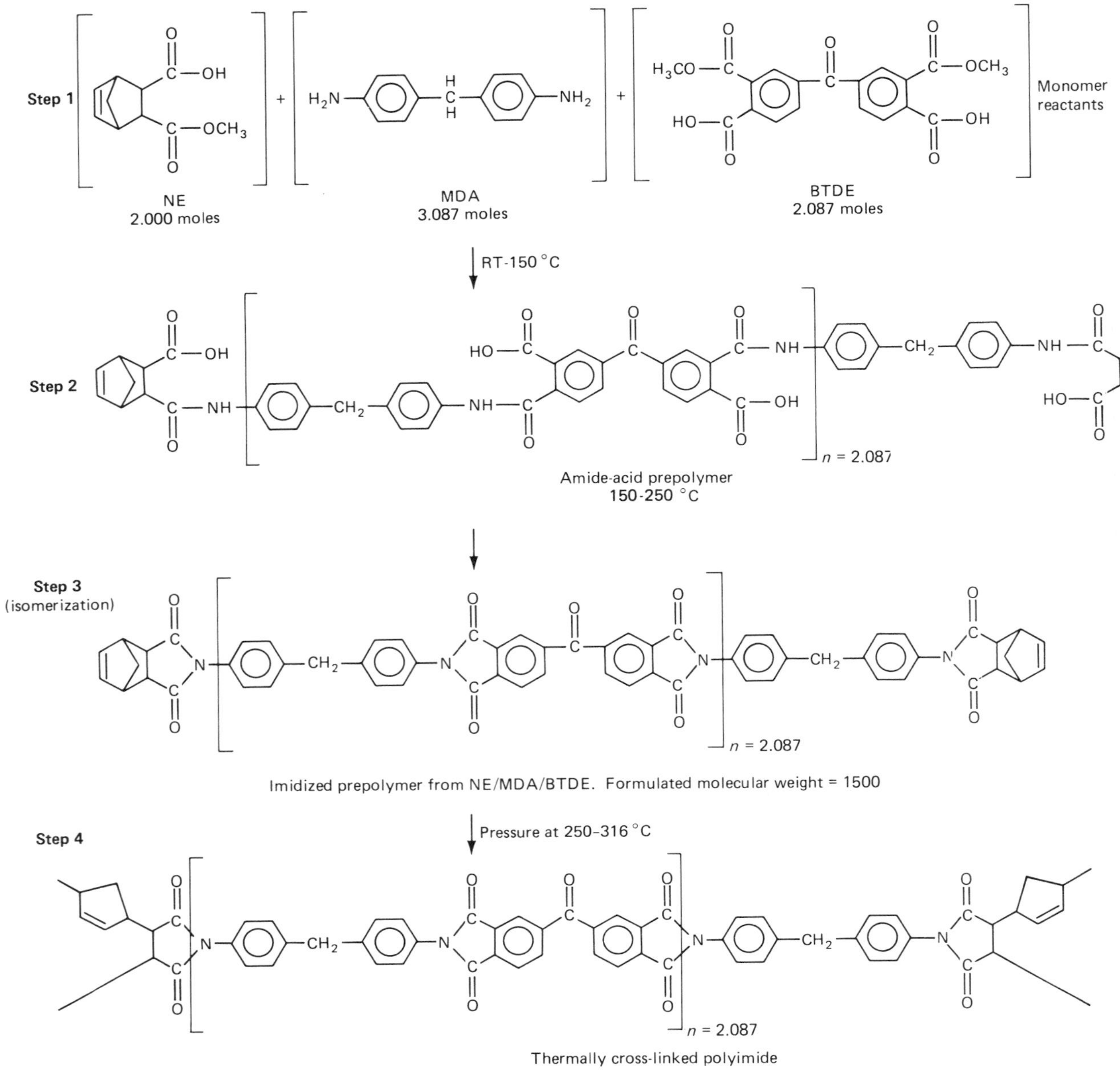

Fig. 8 General reaction scheme for PMR-15 reverse Diels-Alder polyimide. NE, monomethyl ester of nadic anhydride; MDA, 4,4'-methylene dianiline; BTDE, diethyl ester of 3,3',4-4'-benzophenone tetracarboxylic acid dianhydride

onent bismaleimide, formerly known as XU-292, consists of 4,4'-bismaleimido-diphenylmethane, component A, and 0,0'-diallyl bisphenol A, component B, in a weight ratio of 100/85 for System 1, or 113/85 for System 2, as illustrated in Fig. 6. Only properties of System 1 will be described, because those of System 2 are very similar. The constituent properties of this material are listed in Table 5.

Compimide Bismaleimides. Boots/Technochemie Compimide resins are Michael addition reaction polymerization products of 4,4'-bismaleimido-diphenylmethane (4,4'-MDA-BMI) or eutectic mixtures of bismaleimides. The molar proportions of bismaleimide and m-aminobenzoic acid are con-

trolled to produce a resin with various properties, such as melt viscosity, reactivity, and solubility, which is available in the Compimide resin forms H795, 796, 800, and 183. The reaction sequence and products are shown in Fig. 5 (Ref 24). A catalyst such as diazobicyclooctane (0.2-0.5 wt%) or 2-methylimidazole (0.2 to 0.5 wt%) is recommended to decrease the reaction process time for these materials. These four basic resins, formulated with processing additives and toughening agents such as 4,4'-bis (2-propenylphenoxy) diphenylsulfone or 4,4'-bis (2-methoxy-4-propenylphenoxy) diphenylsulfone, form the basis of a series of Compimide resins. Compimide H795 exhibits a lower viscosity

after aging for 2 h at 110 °C (230 °F) of 1200-5750 mPa · s or cP than Compimide 183 (3500-16000 mPa·s) and also reacts at an initial lower temperature than Compimide 183 (185 versus 195 °C, or 365 versus 380 °F). The heat of polymerization, ΔH reaction, for both materials is essentially the same, at 260 J/g (0.10 Btu/lb). Physical, thermal, and mechanical properties of Compimide resins H795, 796, and 183 and modified (toughened) Compimide 353 and 796 are listed in Table 6.

The bismaleimide-triazine (BT) (Ref 26) resins are reaction products of a bismaleimide and 0,0'-dicyanobisphenol A, illustrated in Fig. 7. The triazine form of the resin is derived from a trimerization of the

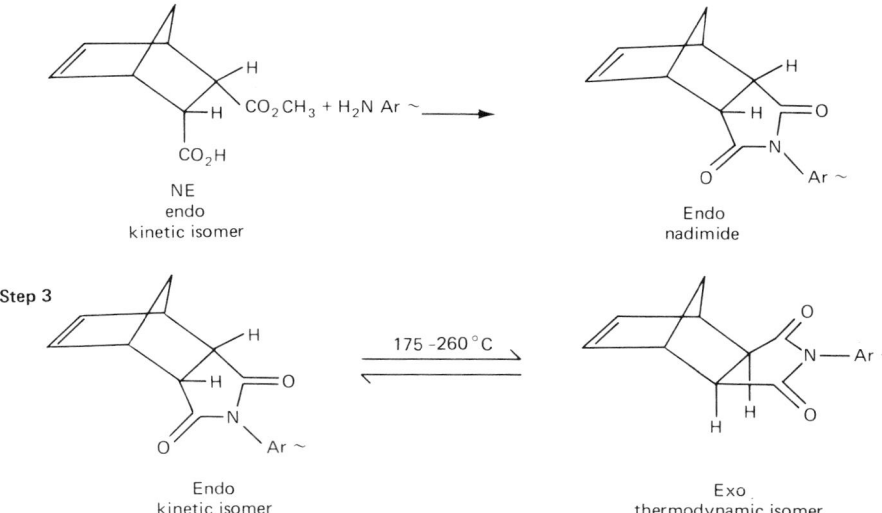

Fig. 9 Isomerization during PMR-15 polymerization

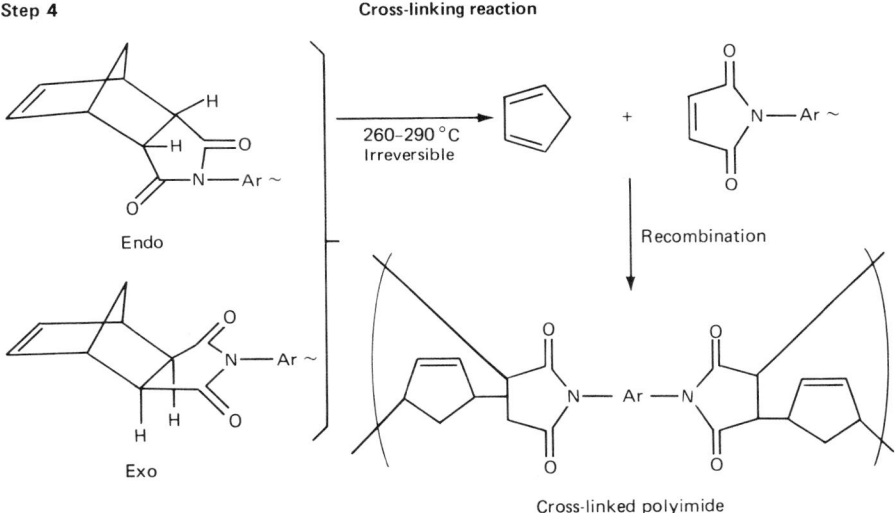

Fig. 10 Reverse Diels-Alder during cure

the molecular weight of each segment of the oligomer, is shown in Fig. 14. The theoretical empirical formula for this oligomer, where $n = 2.087$, is $C_{93.61}H_{59.392}N_{6.174}O_{14.435}$, yielding a molecular weight (MW) of 1501.56. This prepolymer, when heated to between 275 and 316 °C (530 and 600 °F), undergoes a reverse Diels-Alder reaction to yield a bismaleimide oligomer and cyclopentadiene or cyclopentadiene-associated BMI oligomers, which recombine to yield a cross-linked polyimide with the idealized structure shown in Fig. 10. The constituent material properties are listed in Table 11.

PMR-15 is extremely resistant to most organic solvents, including aliphatic hydrocarbons, ether, ketones, aromatic hydrocarbons, and chlorinated aliphatic and aromatic hydrocarbons. It is hydrolyzed in strong acids and strong bases, including aqueous hydrazine solutions, at elevated temperatures.

Thermid MC-600 and IP-600 Resins. In terms of chemical composition (Ref 10, 31, 62), the empirical formula for these acetylene terminated imide oligomers is $C_{68}H_{34}N_4O_{12}$. The molecular structures of these imide oligomers are shown in Fig. 12. The preformed imide oligomer is called MC-600, and the preformed isoimide oligomer is called IP-600. These materials differ in their flow properties in the temperature range between 225 to 250 °C (440 to 480 °F), with the isoimide oligomer reported to have better flow properties in this temperature range than the imide oligomer MC-600. Both materials yield similar cross-linked polymer materials because the isoimide is reportedly easier to process and undergoes thermal rearrangement to the imide form during its melt phase (225 to 250 °C or 440 to 480 °F). The constituent properties of these materials are listed in Table 12.

Thermid 600 is resistant to all polar and nonpolar organic solvents but is attacked by strong alkali and strong acids and is decomposed by hydrazine hydrate in water at 163 °C (325 °F) under pressure.

dicyanate bisphenol A component of the resin mixture (Ref 47). The BT resins are mixtures of BMI-4,4'-MDA and dicyanate bisphenol A in ratios of from 10/90 to 60/40. The BT resins can be used as a basic resin for many applications or can be further modified chemically for specific applications. The constituent properties of the BT resin systems are listed in Table 7.

Bismaleimide 5245C resin is a multicomponent system consisting of an epoxy resin, 0,0'-dicyanobisphenol A; BMI-4,4'-MDA (4,4'-bismaleimido-diphenylmethane); and a catalyst, dialkyltin dilaurate. The constituent properties are shown in Table 8.

Other commercial bismaleimide and modified bismaleimide resins are available. The constituent material properties available for these resin systems are listed in Tables 9 and 10. With the exceptions of Dayton

Research Institute's BPA-BMI and Rhone Poulenc's Kerimid 70003, these materials are formulated bismaleimides containing various components, such as epoxy resins, unsaturated alkylaromatic compounds, and catalysts, required for impregnating cloth and tape goods, controlling the cure process, and optimizing cured laminate properties.

PMR-15 Polyimide. The prepolymer composition of this material consists of three monomers: monomethylester of 5-nornbornene-2, 3-dicarboxylic acid; dimethyl ester of 3,3', 4,4'-benzophenone tetracarboxylic acid; and 4,4'-methylene dianiline, in the mole ratio of 2:2.087:3.087 (Ref 10). These monomers undergo a reaction to yield PMR-15 polyimide (Fig. 8).

The idealized structure of the imide oligomer resulting from thermal treatment of this monomer mixture (free of solvent), as well as

REFERENCES

1. R.C. Boyce, T.P. Gannett, H.H. Gibbs, and A.R. Wedgewood, in *32nd International SAMPE Symposium and Exhibition*, Society for the Advancement of Material and Process Engineering, Vol 32, 1987, p 167-184
2. "Avimid K-III," Data Sheet, E.I. Du Pont de Nemours & Company, Inc., 5 Feb 1987
3. Monsanto Company, Plastics Division, private communication
4. P.J. Cavana and W.F. Winters, "PMR-15 Polyimide/Graphite Fiber Composite Fan Blade," NASA CR-135113, National Aeronautics and Space Administration, Feb 1976
5. "PMR-15 Hexcel," Data Sheet F670, Hexcel Corporation, April 1986

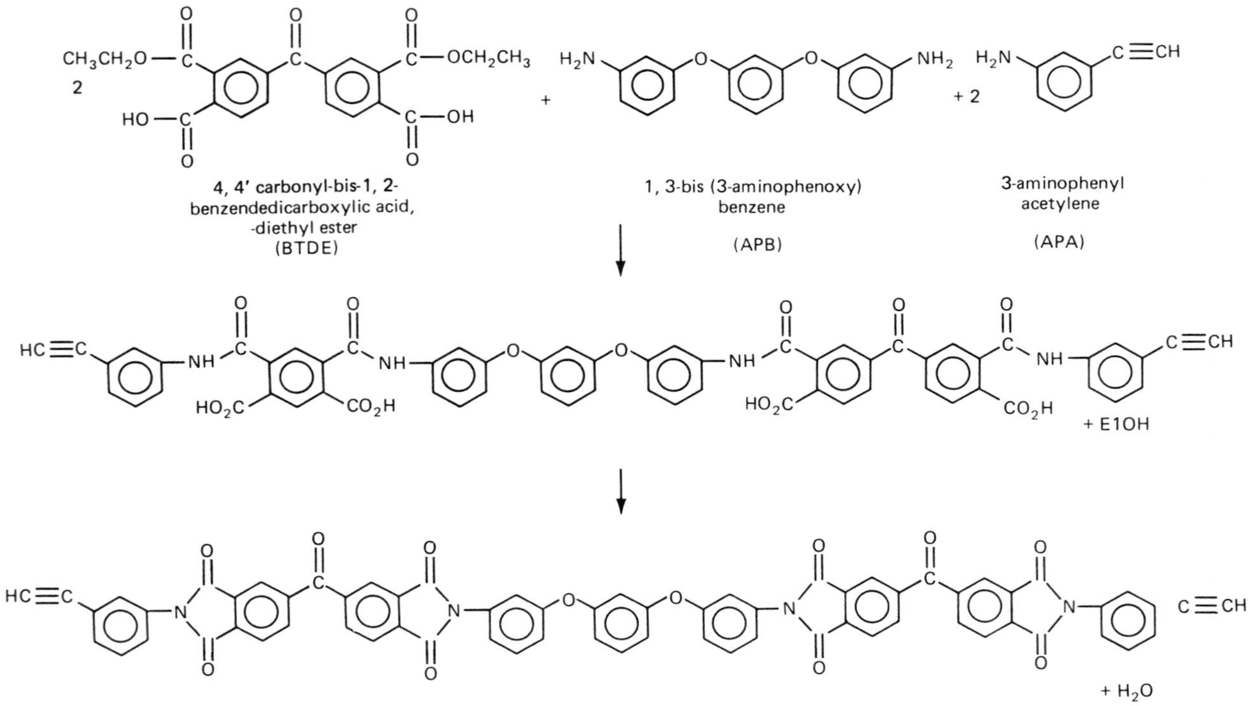

Fig. 11 PMR acetylene end-capped Thermid AL-600

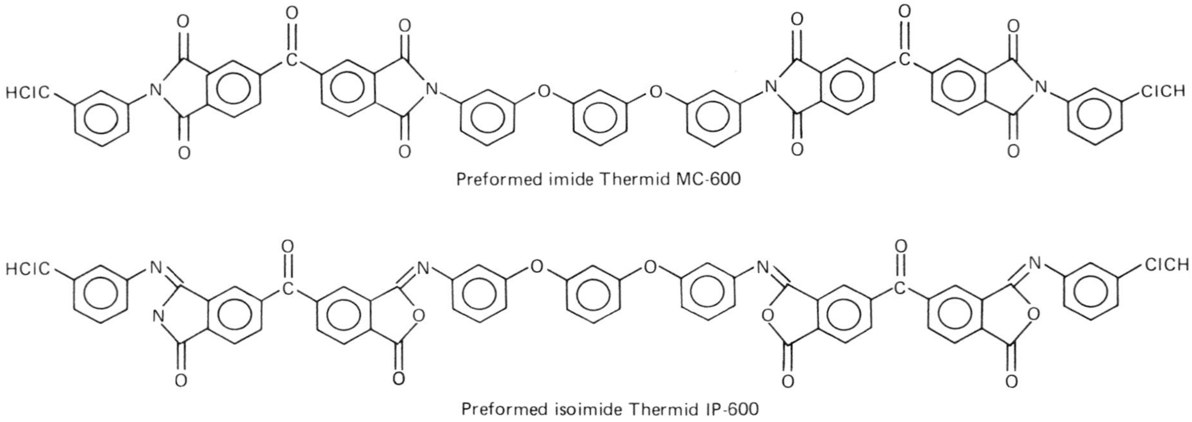

Fig. 12 Acetylene end-capped oligomer imides

6. "PMR-15," Ferro Data Sheet (PI-2237), Ferro Corporation

7. H. Gibbs, Section 2-D, in *Proceedings of the 28th Annual Technical Conference*, Reinforced Plastics/Composites Institute, Society of the Plastics Industry, 1973

8. H. Gibbs, Section 11-D, in *Proceedings of the 29th Annual Technical Conference*, Reinforced Plastics/Composites Institute, Society of the Plastics Industry, 1974

9. "NR-150B2 (Avimid N)," DuPont NR-150B2 Data Sheet, E.I. Du Pont de Nemours & Company, April, 1976

10. A.L. Landis and A.B. Naselow, in *14th National SAMPE Technical Conference*, Society for the Advancement of Material and Process Engineering, Vol 14, 1982, p 236-242

11. N. Bilow, *American Chemical Society Symposium Series 132*, C.A. May, Ed., 1980, p 139-150

12. "Thermid MC-600," Data Sheet 26283, National Starch and Chemical Corporation, Feb 1985

13. "UpJohn 2080 Polyimide Resin," Data Sheet Report 5, UpJohn Company, Oct 1975

14. "Bismaleimide 353, Kerimid 353," Rhodia Technical Data Sheet, Rhodia, Incorporated, Feb 1975

15. "Bismaleimide 353," Boots-Technochemie Compimide C353 Data Sheet, The Boots Company, March 1986

16. "Ultem 1000," Commercial Data Sheet ULT-301, General Electric Company

17. "Torlon 4203," Commercial Data Sheet AMOCO Chemical Corporation, July 1974

18. D.A. Scola, unpublished research

19. D.A. Scola and D.J. Parker, in *Proceedings of 43rd Annual Technical Conference and Exhibition*, Society of Plastics Engineers, Vol 31, 1985, p 399-400

20. T.M. Bogert and R.R. Renshaw, *American Chemical Society Symposium Series 132*, C.A. May, Ed., Vol 30, 1908, p 1140

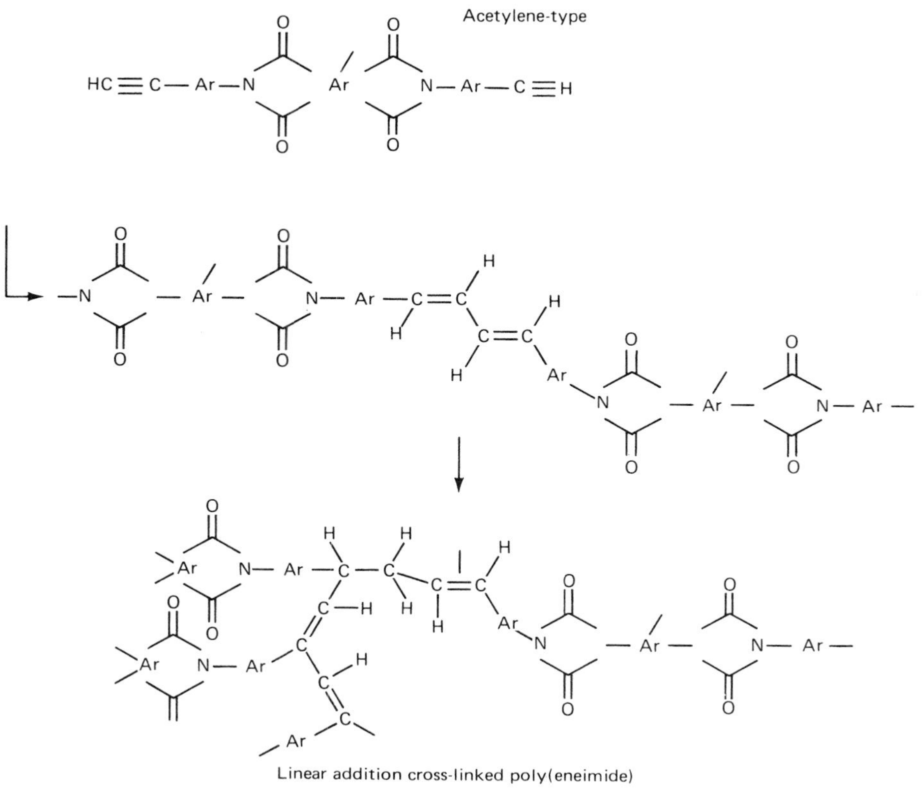

Fig. 13 Acetylene addition polyimides

32. Les Polyaminobismaleimides Dans les Circuits en Prémis Multicouches, *J. d'Etudes*, April 1978

33. M. Bargain and P. Grogean, U.S. Patent 3,526,764, Feb 1971

34. M. Bargain and P. Grogean, U.S. Patent 3,658,764, April 1972

35. C.L. Leung, T.T. Liao, and R.J. Dynes, in *28th National SAMPE Symposium and Exhibition*, Society for the Advancement of Material and Process Engineering, Vol 28, April 1983, p 818-823

36. H.D. Stenzenberger, P. Konig, M. Herzog, W. Roemer, M.S. Canning, and S. Pierce, in *18th International SAMPE Technical Conference*, Society for the Advancement of Material and Process Engineering, Vol 18, 1986, p 500-509

37. R.T. Alverez and F.P. Darmory, in *20th National SAMPE Symposium and Exhibition*, Society for the Advancement of Material and Process Engineering, Vol 20, 1975, p 253-269

38. "Matrimid 5292" Data Sheet, Ciba-Geigy Corporation, 1985

39. H.D. Stenzenberger, M. Herzog, W. Roemer, R. Scheiblich, N.J. Reeves, and S. Pierce, in *29th National SAMPE Symposium and Exhibition*, Society for the Advancement of Material and Process Engineering, Vol 29, 1984, p 1043-1059

40. "C795," Boots-Technochemie Compimide C353 Data Sheet, The Boots Company, March 1986

41. C.L. Segal, H.D. Stenzenberger, M. Herzog, W. Roemer, S. Pierce, and M.S. Canning, in *17th National SAMPE Technical Conference*, Society for the Advancement of Material and Process Engineering, Vol 17, 1985, p 147-160

42. "C800," Boots-Technochemie Compimide C353 Data Sheet, The Boots Company, March 1986

43. H.D. Stenzenberger, M. Herzog, W. Roemer, R. Scheiblich, S. Pierce, M. Canning, K. Fear, in *30th National SAMPE Symposium and Exhibition*, Society for the Advancement of Material and Process Engineering, Vol 30, 1985, p 1568-1601

44. "C183," Boots-Technochemie Compi-

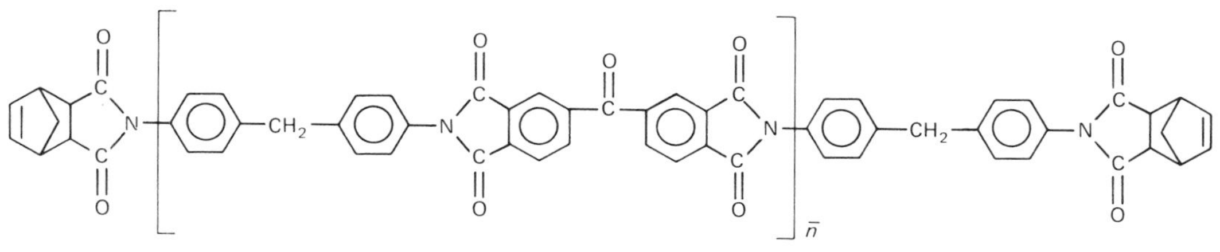

21. W.M. Edwards and I.M. Robertson, U.S. Patent 2,710,853, 1955

22. F. Grundschober and J. Sambeth, U.S. Patent 3,380,964, April 1968

23. F. Grundschober, U.S. Patent 3,533,996, Oct 1970

24. H.D. Stenzenberger, U.S. Patent 4,211,860, 1980, U.S. Patent 4,211,861, 1980, U.S. Patent 4,303,799, 1981

25. J.S. King, M. Chandhari, and S. Zhair, in *29th National SAMPE Symposium and Exhibition*, Society for the Advancement of Material and Process Engineering, Vol 29, 1984, p 1034-1041

26. S. Ayano, *Chem. Econ. Eng. Rev.*, Vol 10 (No. 3), (No. 115), March 1978, p 1-8

27. H.R. Lubowitz, U.S. Patent 3,528,590, Sept 1970

28. T.T. Serafini, P. Delvigs, and G.R. Lightsey, *J. Appl. Polym. Sci.*, Vol 16 (No. 4), 1972, p 905-916; also U.S. Patent 3,765,149, July 1973

29. P.R. Young and A.C. Chang, *J. Heterocyclic Chem.*, Vol 20, 1983, p 177

30. D.A. Scola, in *Proceedings of Second International Polyimide Conference*, Society of Plastic Engineers, 1985, p 247-252

31. A.L. Landis, N. Bilow, R.H. Boschan, R.E. Lawrence, and T.J. Aponyi, *Polymer Preprint*, American Chemical Society, Vol 15 (No. 2), 1974, p 537-541

$C_9H_8NO_2$
MW 162.154

16
$C_{30}H_{16}N_2O_5$
MW 484.442

$C_{22}H_{18}NO_2$
MW 328.364

Fig. 14 Idealized PMR-15 structure

Table 4 Constituent material properties of Keramid 601 and 353 bismaleimides

Property	601(a)	353(b)
Density, g/cm^3 . . . . 1.30	1.30	...
Melting point, °C (°F). 80	80	70-125
	(177)	(158-298)
T_g, °C (°F). 290	290	285
	(554)	(545)
Morphology. . . . . Amorphous	Amorphous	Amorphous
Gel time at 170 °C (338 °F), min	...	30
Melt viscosity, Pa · s (cP).	...	0.115-0.130
		(115-130)
Tensile strength (dry)		
at RT, MPa (ksi) 63.4	63.4	...
	(9.2)(c)	
at 200 °C (392 °F), MPa (ksi) 42.0	42.0	...
	(6.1)	
Tensile strain-to-failure at RT, % 3.1	3.1	...
at 200 °C (392 °F), % 4.9	4.9	...
Flexural strength (dry)		
at RT, MPa (ksi) 150	150	60 (8.7)
	(21.7)(d)	
at 250 °C (482 °F), MPa (ksi)	...	50 (7.3)
at 250 °C (482 °F)	...	...
after 210 h at 250 °C (482 °F), MPa (ksi) 103 (15.0)	103 (15.0)	
Flexural strength (dry), at 250 °C (482 °F), after		
1650 h at 250 °C (482 °F), MPa (ksi) 82 (11.9)	82 (11.9)	...
Flexural modulus (dry)		
at RT, GPa (10^6 psi)	...	5.6 (0.81)
at 250 °C (478 °F), GPa (10^6 psi)	...	3.4 (0.49)
Flexural elongation (dry)		
at RT, %	...	1.8
at 250 °C (478 °F), %	...	1.2
Fracture toughness at RT		
K_{Ic}, MPa$\sqrt{}$m (ksi$\sqrt{}$in.) 382	382	...
	(348)(e)	
G_{Ic}, J/m^2 (in. · lbf/in.^{2}). 34	34	30 (0.17)
	(0.19)(e)	
Thermogravimetric wt loss under nitrogen . . . . 1 wt% loss up	1 wt% loss up to 400 °C (750 °F)	...
Moisture absorption(f)		
at RT, 24 h, wt% 0.3	0.3	...
at 100 °C (212 °F), 2 h, wt%. 1.0	1.0	...
Coefficient of thermal expansion (CTE) 10^{-6}/K . . . . 61(f)	61(f)	...

RT, room temperature. (a) Ref 32-35. (b) Ref 14, 15, 36, 37. (c) Ref 35. (d) Ref 33. (e) Ref 19. (f) Ref 32, 33

Table 5 Constituent material properties of Matrimid 5292 bismaleimide

Property	5292
Density, g/cm^3. . . . . 1.23	1.23
Melting point	
component A, 150-160 °C (302-320 °F)	...
component B, amber liquid, viscosity at	
25 °C (77 °F), mPa · s 12 000-20 000	12 000-20 000
T_g (TMA method), °C (°F) 273 (523)	273 (523)
Morphology. . . . . Amorphous, cross-linked	Amorphous, cross-linked
Tensile strength (dry)	
at RT, MPa (ksi) 82 (11.9)	82 (11.9)
at 150 °C (300 °F), MPa (ksi) 51 (7.4)	51 (7.4)
at 204 °C (400 °F), MPa (ksi) 40 (5.8)	40 (5.8)
Tensile modulus (dry)	
at RT, GPa (10^6 psi). 4.3	4.3
	(0.62)
at 150 °C (300 °F), GPa (10^6 psi) 2.4	2.4
	(0.35)
at 204 °C (400 °F), GPa (10^6 psi) 2.0	2.0
	(0.29)
Tensile strain-to-failure	
at RT, % 2.3	2.3
at 150 °C (300 °F), % 2.6	2.6
at 204 °C (400 °F), % 2.3	2.3
Tensile strength (wet)	
at RT, MPa (ksi) 66 (9.6)	66 (9.6)
at 150 °C (300 °F), MPa (ksi) 30 (4.3)	30 (4.3)
Tensile modulus (wet)	
at RT, GPa (10^6 psi). 3.8	3.8
	(0.55)
at 150 °C (300 °F), GPa (10^6 psi) 1.9	1.9
	(0.27)
Tensile strain-to-failure	
at RT, % 2.1	2.1
at 150 °C (300 °F), % 1.95	1.95
Flexural strength (dry) at RT, MPa (ksi) 167	167
	(24.2)
Flexural modulus (dry) at	
RT, GPa (10^6 psi). 4.0 (0.59)	4.0 (0.59)
T_g (by DMA)	
(dry), °C (°F) 295 (563)	295 (563)
(wet), °C (°F) 305 (581)	305 (581)
Moisture absorption (Ref 25), wt% 1.40	1.40
Fracture toughness (compact tension),	
J/m^2 (in. · lbf/in.^{2}) 170 (0.97)	170 (0.97)
	216
	(1.22)(a)

Properties taken from commercial data sheet and technical literature. (a) Ref 19
Source: Ref 25, 38

mide C353 Data Sheet, The Boots Company, March 1986

45. "C796," Boots-Technochemie Compimide C353 Data Sheet, The Boots Company, March 1986

46. H.D. Stenzenberger, M. Herzog, W. Roemer, R. Scheiblich, S. Pierce, M. Canning, and K. Fear, in *31st National SAMPE Symposium and Exhibition*, Society for the Advancement of Material and Process Engineering, Vol 31, 1986, p 920

47. E. Gigat and R. Putter, *Angew. Chem. Int. Ed.*, Vol 6 (No. 3), 1967, p 206-215

48. "BT Resins," Commercial Data Brochure, 2nd Ed., Mitsubishi Gas Chemical Company, Inc., June 1980

49. "BMI 5245C," Rigidite 5245 Data Bulletin, Narmco Materials, April 1986

50. S.W. Street, U.S. Patent 4,351,932, Sept 28, 1982

51. J.A. Harvey, R.P. Chartoff, and J.M. Butler, in *Proceedings of 44th Annual Technical Conference and Exhibition*, Society of Plastics Engineers, Vol 32, 1985, p 1311-1315

52. "BMI 70003 Kerimid 70003," Rhone-Poulenc Leaflet, Rhone-Poulenc Company, Nov 1983

53. "Fiberite X-86," Document 86-1, Fiberite Corporation, July 1985

54. B. Stern, American Cyanamide Company, private communication, April 1984

55. "Hysol EA-9102," Hysol High Temperature Resins, Data Sheets, Hysol Division, Dexter Corporation

56. "Hysol EA-9655," Hysol High Temperature Resins, Data Sheets, Hysol Division, Dexter Corporation

57. "Hexcel F650," Hexcel Data Sheet F650, Hexcel Corporation, May 1985

58. "Fiberite 987A," Document 87-1, Fiberite Corporation, Nov 1984, rev. July 1985

59. Data Sheet CPI-2237, Ferro Corporation

60. Technical Data Sheet F-767, Hexcel Corporation, April 1986

61. D.A. Scola, in *22nd National SAMPE Symposium and Exhibition*, Society for the Advancement of Material and Process Engineering, Vol 22, 1977, p 238

62. R.H. Boschan, A.L. Landis, and A.A. Castillo, AFML-TR-77-149, Air Force Materials Laboratory, Aug 1977

63. "Thermid MC-600," Data Sheet 26283, National Starch and Chemical Corporation, Feb 1983

64. "Thermid IP-600," Data Sheet 02885, National Starch and Chemical Corporation, Aug 1985

65. D.J. Capo and J.E. Schoenberg, in *18th International SAMPE Technical Conference*, Society for the Advancement of Material and Process Engineering, Vol 18, 1986, p 710-721

Table 6 Properties of Compimide bismaleimide resins

Property	C795 (Ref 24, 39, 40)	C800 (Ref 24, 39, 40, 42)	C183 (Ref 24, 39, 43)	C796 (Ref 24, 36, 43, 45)	C353/TM-122 (Ref 24, 43, 46)(a)(b)	C796/TM-122 (Ref 24, 43, 46)(a)(b)
Density, g/cm³	1.32	...	1.31	...	...	...
T_g (dry), °C (°F)	290 (554)	290 (554)	250 (482)	...	...	...
Morphology	Amorphous	Amorphous	Amorphous	Amorphous	Amorphous	Amorphous
Flexural strength (dry), at RT, MPa (ksi)	110 (15.9)	92 (13.2)	106 (15.4)	76 (11.0)	110 (15.9)	114 (16.5)
at RT after isothermal aging for 500 h at 200 °C (392 °F) in circulating air, MPa (ksi)	100 (14.5)	... ...	... ...	... ...	... ...	... ...
at RT after isothermal aging for 500 h at 250 °C (478 °F) in circulating air, MPa (ksi)	108 (15.6)	... ...	... ...	... ...	... ...	... ...
at RT after isothermal aging for 500 h at 250 °C (478 °F) in circulating air and then exposed to 200 °C (392 °F), MPa (ksi)	65 (9.4)	... ...	58 (8.4)	... ...	... ...	... ...
at 200 °C (392 °F) after isothermal aging for 500 h at 200 °C (392 °F) in circulating air, MPa (ksi)	71 (10.2)	... ...	... ...	... ...	... ...	... ...
at 200 °C (392 °F) after isothermal aging for 500 h at 250 °C (478 °F) in circulating air, MPa (ksi)	62 (8.9)	46 (6.6)	... ...	31 (4.5)	64 (9.3)	73 (10.5)
at 250 °C (478 °F) after isothermal aging for 500 h at 250 °C (478 °F) in circulating air, MPa (ksi)	41 (5.9)	... ...	... ...	... ...	... ...	... ...
Flexural modulus (dry) at RT, GPa (10^6 psi)	5.5 (0.79)	3.9 (0.56)	4.1 (0.59)	4.6 (0.66)	3.7 (0.53)	3.9 (0.56)
at RT after isothermal aging for 500 h at 200 °C (392 °F) in circulating air, GPa (10^6 psi)	5.5 (0.79)	... ...	... ...	... ...	... ...	... ...
at RT after isothermal aging for 500 h at 250 °C (478 °F) in circulating air, GPa (10^6 psi)	5.3 (0.77)	... ...	... ...	... ...	... ...	... ...
at 200 °C (392 °F) after isothermal aging for 500 h at 200 °C (392 °F) in circulating air, GPa (10^6 psi)	4.7 (0.68)	... ...	3.2 (0.47)	... ...	... ...	... ...
at 200 °C (392 °F) after isothermal aging for 500 h at 250 °C (478 °F), GPa (10^6 psi)	3.4 (0.49)	2.1 (0.30)	... ...	3.0 (0.43)	2.5 (0.36)	2.62 (0.38)
at 250 °C (478 °F) after isothermal aging for 500 h at 250 °C (478 °F) in circulating air, GPa (10^6 psi)	4.5 (0.65)	... ...	... ...	... ...	... ...	... ...
Flexural elongation (dry) at RT, %	2.4	...	2.6	1.7	2.7	3.0
at 200 °C (392 °F), %	...	...	1.8	...	...	...
at 250 °C (478 °F), %	2.2	...	...	1.0	2.6	3.0
Tensile strength (dry) at RT, MPa (ksi)	89 (13)	... ...	... ...	... ...	... ...	... ...
Tensile strain-to-failure (dry), at RT, %	2.2	...	...	...	...	...
(wet), at 250 °C (478 °F), %	2.0	...	...	...	...	...
Fracture toughness, G_{Ic} J/m² (in. · lbf/in.²)	40 (0.23)	160 (0.80)	180 (1.0)	63 (0.36)	389 (2.2)	230 (1.27)
Gel time at 170 °C (338 °F), min	25	50	45	>30	...	...
Complex viscosity at 110 °C (230 °F), Pa·s (cP)	0.4-2.8 (400-2800)	0.2-2.5 (220-2500)	2.0-8.0 (2000-8000)	0.4-3.5 (400-3500)	... ...	... ...
Heat of polymerization, J/g (Btu/lb)	265 (0.10)	260 (0.10)	260 (0.10)	>200 (>0.85)	...	...
Melting range, °C (°F)	110-120 (230-248)	...	100-140 (212-284)	...	...	...
CTE to 250 °C (478 °F) 10^{-6}/K	73.4	...	66	...	...	...
Moisture absorption, wt%	4.85 (Ref 39)	...	...	...	...	...

(a) Wt ratio C353/TM-122, 76/24; TM-122 is 4,4′-bis(2-propenylphenoxy) diphenylsulfone. (b) Wt ratio 796/TM-122 70/30 TM-122 is 4,4′-bis(2-propenylphenoxy) diphenylsulfone

Table 7 Constituent material properties of BT resins

Material	Density, g/cm³	Gel time at 168 °C (330 °F), min	Melting range for uncured resin °C	°F	Viscosity at 160 °C (320 °F) Pa · s	cP	Tensile strength (dry) at RT MPa	ksi	Tensile modulus (dry) at RT GPa	10^6 psi	Tensile strain-to-failure, %
BT 2600(a)	1.31	2	60-70	140-158	0.015	15	60	8.7	...	...	...
BT 2160(b)	1.23	9	< 20	< 68	...	...	31	4.47	4.1	.60	1.0
BT 2460T(c)	1.25	...	< 20	< 68	...	...	...	...	...	...	...
BT 2164(d)	...	12	...	...	0.010	10	85	12.3	3.7	.53	3.6
BT 3209(e)	1.24	...	...	liquid	...	...	...	...	3.7	.54	...

Material	T_g °C	°F	Use time at 200 °C (400 °F), h	Moisture absorption, wt%	CTE, 10^{-6}/K	Dielectric constant, 1 MHz	Dissipation factor, 1 MHz	Surface resistance, Ω	Volume resistivity Ω · cm	Ω · in.	Thermal conductivity W/m · K	kcal/m/h · °C
BT 2600(a)	310	590	20 000	0.11	72	4.15	0.0020	1×10^{14}	1×10^{15}	5×10^{14}	...	...
BT 2160(b)	252	485	...	...	...	3.1	0.0030	...	...	...	...	...
BT 2460T(c)	257	496	...	...	...	...	...	...	...	...	...	...
BT 2164(d)	205	401	...	...	...	...	...	...	...	...	...	...
BT 3209(e)	236	459	...	...	...	...	...	...	...	...	0.207	0.178

(a) Weight ratio of BMI–4,4′–MDA to dicyanate bisphenol A is 60/40. (b) Weight ratio of BMI–4,4′–MDA to dicyanate bisphenol A is 10/90. (c) Weight ratio of BMI–4,4′–MDA to dicyanate bisphenol A is 40/60.
(d) Weight ratio of BMI–4,4′–MDA to dicyanate bisphenol A is 10/90. (e) Weight ratio of BMI–4,4′–MDA to dicyanate bisphenol A is 20/80.
Source: Ref 26 and 47

Table 8 Constituent material properties of modified bismaleimide resin, 5245C

Property	5245C (Ref 48)	Property	5245C (Ref 49)
Density, g/cm^3	1.25	Flexural modulus (dry)	
T_g, °C (°F)	220 (428)	at RT, GPa (10^6 psi)	3.3 (0.49)
Tensile strength (dry) at RT, MPa (ksi)	84 (12.2)	at 93 °C (200 °F), GPa (10^6 psi)	3.1 (0.46)
Tensile modulus (dry) at RT, GPa (10^6 psi)	33 (0.48)	at 130 °C (270 °F), GPa (10^6 psi)	2.7 (0.40)
Tensile strain-to-failure (dry) at RT, %	2.9	Flexural modulus (wet)(a)	
Flexural strength (dry)		at 93 °C (200 °F), GPa (10^6 psi)	2.9 (0.43)
at RT, MPa (ksi)	145 (21.0)	at 130 °C (270 °F), GPa (10^6 psi)	2.7 (0.39)
at 93 °C (200 °F), MPa (ksi)	115 (16.7)	Izod impact strength, unnotched, J/m (ft · lbf/in.)	410 (7.7)
at 132 °C (270 °F), MPa (ksi)	107 (15.5)	Fracture toughness (Ref 19), G_{Ic}, J/m^2	
Flexural strength (wet)(a)		(in. · lbf/in.^{2})	67 (0.38)
at 93 °C (200 °F), MPa (ksi)	96 (14.0)	Moisture absorption (72-h water boil), %	1.7
at 130 °C (200 °F), MPa (ksi)	83 (12.1)	CTE, 10^{-6}/K (Ref 18)	72

(a) Wet condition, 40-h water boil, specimen held 5 min at test temperature before loading

Table 9 Constituent properties of bismaleimide resins

Material	Density, g/cm^3	Uncured melting or softening temperature °C	Uncured melting or softening temperature °F	Tensile strength (dry) at RT MPa	Tensile strength (dry) at RT ksi	Tensile strength (dry) at 232 °C (450 °F) MPa	Tensile strength (dry) at 232 °C (450 °F) ksi	Tensile strain-to-failure (dry) at RT, %	Tensile modulus (dry) at RT GPa	Tensile modulus (dry) at RT 10^6 psi	Flexural strength (dry) at RT MPa	Flexural strength (dry) at RT ksi
Hexcel F178	...	...	...	56	8.14	...	...	1.3	...	...	...	...
Narmco 5250-2	1.24	...	...	62	9.0	...	...	1.7	2.7	0.39	138	20
U.S. Polymeric V378A (Ref 50)	1.26	...	...	78	11.3	44	6.4	6.6	...	...	...	...
Univ. Dayton BPA-BMI (Ref 51)	1.26	70	155	48	7.0	...	...	1.5	3.4	0.50	...	...
Kerimid 70003 (Ref 52)	1.25	...	...	...	...	...	...	2.2	...	...	...	...

Material	Flexural modulus (dry) at RT GPa	Flexural modulus (dry) at RT 10^6 psi	Fracture toughness at RT J/m^2	Fracture toughness at RT in. · lbf/in.^{2}	T_g (dry) °C	T_g (dry) °F	T_g (wet) °C	T_g (wet) °F	CTE, 10^{-6}/K	Moisture absorbtion, wt%
Hexcel F178	...	...	29.4	0.17	260	500	140	284	...	3.7
Narmco 5250-2	2.9	0.43	100	0.56	321	610	...	...	...	3.3 (steam 96 h)
U.S. Polymeric V378A	...	...	...	...	...	...	...	...	...	1.7
Univ. Dayton BPA-BMI	...	...	...	...	280	536	...	...	...	1.7
Kerimid 70003	...	...	82	0.46	320 (Ref 19)	620	...	...	21	1.7 (100 h BW)

Table 10 Constituent properties of bismaleimide resins

Property	Fiberite X-86 (Ref 53)	Cycom 3100 (Ref 54)	Hysol EA9102 (Ref 55)	Hysol EA9655 (Ref 56)	Hexcel F650 (Ref 57)	Fiberite 987A (Ref 58)
Density, g/cm^3	1.22	...	...	...	...	1.27
Uncured melting or softening temperature, °C (°F)	70-125 (158-257)	...	...	...	...	...
Tensile strength (dry)						
at RT, MPa (ksi)	...	...	58.5 (8.5)	52 (7.5)	49.3 (7.15)	...
at 177 °C (350 °F), MPa (ksi)	...	...	43 (6.3)	27 (3.9)	...	...
Tensile strength (wet), at RT, MPa (ksi)	...	...	...	39 (5.7)	...	...
Tensile strain-to-failure (dry)						
at RT, %	...	2.1	2.2	1.6	...	1.2
at 177 °C (350 °F), %	...	2.2	3.3	2.2	...	...
Tensile strain-to-failure (wet), at 177 °C (350 °F), %	...	3.0	...	...	...	...
Tensile modulus (dry), at RT, GPa (10^6 psi)	...	...	...	...	...	4.2 (0.61)
Flexural strength (dry)						
at RT, MPa (ksi)	130 (18.9)	117 (17)	...	...	...	121 (17.6)
at 177 °C (350 °F), MPa (ksi)	85 (12.3)	93 (13.5)	...	...	...	53.7 (7.8)
at 232 °C (450 °F), MPa (ksi)	72 (10.5)	...	...	...	...	41 (6.0)
Flexural strength (wet), at 177 °C (350 °F), MPa (ksi)	46 (6.7)	58 (8.5)	...	...	...	38 (5.5)
Flexural modulus (dry)						
at RT, GPa (10^6 psi)	4.6 (0.66)	...	...	...	...	4.3 (0.62)
at 177 °C (350 °F), GPa (10^6 psi)	3.0 (0.44)	...	...	...	...	2.2 (0.32)
at 232 °C (450 °F), GPa (10^6 psi)	2.6 (0.38)	...	...	...	...	1.8 (0.26)
Flexural modulus (wet), at 177 °C (350 °F), GPa (10^6 psi)	1.7 (0.25)	...	...	...	...	1.6 (0.23)
Fracture toughness at RT, G_{Ic}, J/m^2 (in. · lbf/in.^{2})	...	...	...	...	67 (0.38)	...
T_g (dry) °C (°F)	290 (554)	300 (527)	298 (568)	253 (489)	>316 (>600)	320 (608)
T_g (wet) °C (°F)	...	...	210 (410)	118 (244)	...	...
Moisture absorption, wt%	4.4(a)	...	...	...	4.3	2.97(a)

(a) Equilibrium water boil

Table 11 Constituent properties of PMR-15 polyimide

Property	PMR-15 polyimide
Density, g/cm³	1.32(a)
T_g after 316 °C (600 °F), 16-h postcure °C (°F)	340 (662)(a)
Morphology	Amorphous, cross-linked
Tensile strength (dry) at RT, MPa (ksi)	38.6 (5.6)(b)
Tensile modulus (dry) at RT, GPa (10⁶)	39 (0.57)
Tensile strain-to-failure, %	1.1
Flexural strength (dry)	
at RT, MPa (ksi)	176 (25.5)(c)
at 288 °C (550 °F), MPa (ksi)	73 (10.7)(c)
at 316 °C (600 °F), MPa (ksi)	72 (10.4)(c)
at 343 °C (650 °F), MPa (ksi)	52 (7.6)(c)
Flexural modulus (dry)	
at RT, GPa (10⁶ psi)	4.0 (0.58)(c)
at 288 °C (550 °F), GPa (10⁶ psi)	2.3 (0.34)(c)
at 316 °C (600 °F), GPa (10⁶ psi)	1.9 (0.27)(c)
at 343 °C (650 °F), GPa (10⁶ psi)	1.8 (0.26)(c)
Fracture toughness	
K_{Ic}, MPa $\sqrt{m}$ (ksi $\sqrt{in.}$)	1110 (1010)(d) 648 (590)(e)
G_{Ic}, J/m² (ft · lbf/in.²)	280 (1.6)(d) 94 (0.52)(e)
Izod impact strength, notched, J/m (ft · lbf/in.)	53.37 (1.0)
Moisture absorption, equilibrium moisture absorption(e)	
(95% RH, 71 °C, or 160 °F), wt%	4.2
T_g, °C (°F)(a)	
after 316 °C (1-h) cure	320 (608)
after 316 °C (16-h) cure	340 (662)
Wt% loss after 1000 h at 288 °C (550 °F)(f)	
in flowing air (100 cc/min)	0.3
after 2000 h	0.8
after 3000 h	2.0
CTE(e), 10⁻⁶/K	14

(a) Ref 18. (b) Ref 4. (c) Ref 59. (d) Ref 19. (e) Ref 60. (f) Ref 61

Table 12 Constituent properties of Thermid resins

Property	Thermid MC-600(a)	Thermid IP-600(b)
Density, g/cm³	1.37	1.34
Moisture absorption (24-h water boil), wt%	1.24	1.18
Uncured melting range, °C (°F)	190-210 (374-410)	149-171 (300-340)
T_g, °C (°F),		
after 371 °C (700 °F), 8-h postcure	320 (608)(c)	300 (572)
after 371 °C (700 °F), 16-h postcure	350 (662)	350 (662)
Morphology	Amorphous cross-linked	Amorphous cross-linked
Heat of reaction, J/g (Btu/lb)	335 (0.15)	335 (0.15)
Tensile strength (dry)		
at RT, MPa (ksi)	82.7 (12.0)(c)	58 (8.5)
at 316 °C (600 °F), MPa (ksi)	...	28 (4.01)
Tensile modulus (dry)		
at RT, GPa (10⁶ psi)	4.1 (0.60)(c)	5.0 (0.73)
at 316 °C (600 °F), GPa (10⁶ psi)	...	1.2 (0.18)
Tensile strain-to-failure		
at RT, %	1.5(c)	1.2
at 316 °C (600 °F), %	...	4.2
Compressive strength (dry) at RT, MPa (ksi)	172 (25)(d)	...
Flexural strength (dry)		
at RT, MPa (ksi)(e)	146.6 (21)	106 (15.3)
after 1000 h, MPa (ksi)	92 (13.4)	...
Flexural strength (dry)		
at 316 °C (600 °F), MPa (ksi)(e)	29 (4.2)	44 (6.4)
after 1000 h, MPa (ksi)	18 (2.6)	...
Flexural modulus (dry), at RT, GPa (10⁶ psi)(f)	4.6 (0.66)	...
Izod impact, notched, J/m (ft · lbf/in.)(c)	80 (1.5)	...
CTE, 10⁻⁶/K(g)	35-50	...
Wt% loss		
after 500 h at 316 °C (600 °F), %	2.89	...
after 1000 h at 316 °C (600 °F), %	4.04	...
Dielectric constant, 1 MHz	3.496	...
Dissipation factor, 1 MHz	0.0096	...

(a) Ref 10, 11, 63. (b) Ref 64. (c) Ref 10, 11. (d) Ref 31, 65. (e) Ref 31, 63, 64. (f) Ref 31. (g) Ref 65

Polyester Resins

Charles D. Dudgeon, Ashland Chemical Company

MORE THAN 450 000 t OF POLYES-TER RESIN were consumed in the United States in 1985, for a wide variety of uses. While much of this volume was used for general-purpose applications, there has been significant growth in the last few years because of the specially designed formulations that are required in high-performance composite applications. Without describing the large number of products currently available, this article does indicate the range of properties that can be obtained with polyester resins.

Polyester Resin Chemistry

Polyesters are macromolecules that are prepared by the condensation polymerization of dibasic acids or anhydrides with dihydric alcohols (Ref 1). Unsaturated polyesters, commonly referred to as polyester resins, are the group of polyesters in which the dibasic acid or anhydride is partially or completely composed of a 1,2-ethylenically unsaturated material, such as maleic anhydride or fumaric acid. In most cases this polymer is dissolved in a reactive vinyl monomer, such as styrene, vinyl toluene, diallyl phthalate, or methyl methacrylate, to give a solution that will typically have a viscosity in the range of 0.2 to 2 Pa · s (200 to 2000 cP). The addition of heat and a free radical initiator, such as an organic peroxide, will result in a cross-linking reaction between the unsaturated polymer and the unsaturated monomer, converting the low-viscosity solution into a three-dimensional thermoset plastic (Ref 2). Cross-linking can also be accomplished at room temperature through the use of peroxides and suitable activators. Polyester resins fall into the following generic classifications, which are identified by a key feature of their chemical composition.

Ortho resins are prepared using blends of phthalic anhydride and maleic anhydride or fumaric acid. The original family of polyester resins, orthophthalics, is still widely used and continues to dominate in many applications, although these resins have limited thermal stability, chemical resistance, and processibility.

Iso resins are prepared using blends of isophthalic acid and maleic anhydride or fu-maric acids. Although higher in cost than the corresponding ortho resins, isophthalics are of higher quality, having better thermal resistance, mechanical properties, and chemical resistance. These resins are especially responsive to changes in glycols.

Bisphenol A (BPA) fumarates are prepared by the reaction of propoxylated or ethoxylated BPA with fumaric acid. The aromatic BPA group imparts a higher degree of hardness and rigidity and an improved thermal performance. Because of their good chemical resistance and thermal stability, BPA fumarates are used almost exclusively in high-performance applications.

Chlorendics are prepared using a blend of chlorendic anhydride or chlorendic (HET) acid (Ref 3) and maleic anhydride or fumaric acid. These materials show excellent chemical resistance and some fire retardancy because of the presence of chlorine.

Vinyl ester is the common name for an unsaturated resin prepared by the reaction of a monofunctional unsaturated acid, such as methacrylic or acrylic acid, with a bisphenol diepoxide. The resulting polymer, which contains unsaturated sites only in the terminal positions, is mixed with an unsaturated monomer, such as styrene. At this point, the appearance, handling properties, and cure are the same as for conventional polyester resins; because of this, vinyl esters are used in many of the same applications. Although they are higher in cost than polyesters, vinyl esters exhibit exceptional mechanical and chemical performance characteristics.

Within each of these five classifications, specific polyesters from rigid to flexible can be formulated (Ref 4) by varying the type of dihydric alcohol used or by varying the fumaric/saturated acid ratio, as shown in Table 1.

Mechanical Properties

Mechanical properties are often the critical factor in the selection of a polyester resin for a specific application. Table 2 lists the common test methods of the American Society for Testing and Materials that are used to characterize the mechanical properties of polyester resins.

Table 1 Influence of acid ratio on flexural modulus

Polyester flexural modulus	Total unsaturated acids, mol	Total saturated acids, mol
Rigid	1.0	1.0
Resilient	0.7	1.3
Very resilient	0.5	1.5
Flexible	0.3	1.7
Very flexible	0.2	1.8

Source: Ref 4

Table 2 ASTM Procedures for characterizing mechanical properties of polyester resins

Properties	ASTM Test Method
Tensile strength, modulus, and % elongation	D 638
Flexural strength and modulus	D 790
Compressive strength, modulus, and % compression on break	D 695
Izod impact	D 256
Heat distortion	D 648
Barcol hardness	D 2583

Representative examples of clear casting data are shown in Table 3. All of the polyesters contained styrene, which was dominant on a molar basis. This dominance tended to mask some of the differences between the five classifications, but those differences are noted here. The isophthalic resin showed higher tensile and flexural properties than the orthophthalic resin. This may be because isophthalics usually form more linear, higher molecular weight polymers than orthophthalics. The BPA fumarate and chlorendic, on the other hand, were much more rigid, and in clear castings this imparted a brittle behavior with low tensile elongation and strength. The vinyl ester, because of its bisphenol diepoxide content, showed excellent tensile and flexural properties, and a high elongation.

While it is impossible to review in this article the effects that dihydric alcohols, acids, levels of unsaturation, monomer types and amounts, and cure temperatures have on mechanical properties, Ref 5 provides an excellent review. In general, it was found that increasing the chain length of the dihydric alcohol increased

Table 3 Clear casting mechanical properties

Material	Barcol hardness	Tensile strength		Tensile modulus		Elongation, %	Flexural strength		Flexural modulus		Compressive strength		Heat deflection temperature	
		MPa	ksi	10^{-2} Pa	10^{-5} psi		MPa	ksi	10^{-2} Pa	10^{-5} psi	MPa	ksi	°C	°F
Orthophthalic		55	8	34.5	5.0	2.1	80	12	34.5	5.0	...	...	80	175
Isophthalic 40		75	11	33.8	4.9	3.3	130	19	35.9	5.2	120	17	90	195
BPA fumarate 34		40	6	28.3	4.1	1.4	110	16	33.8	4.9	100	15	130	265
Chlorendic 40		20	3	33.8	4.9	...	120	17	39.3	5.7	100	15	140	285
Vinyl ester 35		80	12	35.9	5.2	4.0	140	20	37.2	5.4	...	...	100	212

Table 4 Mechanical properties of fiberglass-polyester resin composites

Material	Glass content, wt%	Barcol hardness	Tensile strength		Tensile modulus		Elongation, %	Flexural strength		Flexural modulus		Compressive strength		Izod impact	
			MPa	ksi	10^{-3} Pa	10^{-6} psi		MPa	ksi	10^{-3} Pa	10^{-6} psi	MPa	ksi	J/mm	ft·lbf/in.
Orthophthalic 40		...	150	22	5.5	0.8	1.7	220	32	6.9	1.0	...	...	...	...
Isophthalic 40		45	190	28	11.7	1.7	2.0	240	35	7.6	1.1	210	30	0.57	10.7
BPA fumarate 40		40	120	18	11.0	1.6	1.2	160	23	9.0	1.3	180	26	0.64	12
Chlorendic 40		40	140	20	9.7	1.4	1.4	190	28	9.7	1.4	120	18	0.37	7
Vinyl ester 40		...	160	23	11.0	1.6	...	220	32	9.0	1.3	210	30	...	...

Table 5 Effect of glass content on mechanical properties

Material	Glass content, wt%	Flexural strength		Flexural modulus		Tensile strength		Tensile modulus		Compressive strength	
		MPa	ksi	10^{-2} Pa	10^{-5} psi	MPa	ksi	10^{-2} Pa	10^{-5} psi	MPa	ksi
Orthophthalic	30	170	25	55	8.0	140	20	48	7.0	...	...
	40	220	32	69.0	10.0	150	22	55	8.0	...	...
Isophthalic	30	190	28	55	8.0	150	22	82.7	12.0	...	...
	40	240	35	75.8	11.0	190	28	117	17.0	210	30
BPA fumarate	25	120	17	51	7.4	80	12	75.8	11.0	170	24
	35	150	22	82.7	12.0	100	14	103	15.0	170	24
	40	160	23	89.6	13.0	120	18	110	16.0	180	26
Chlorendic	24	120	17	59	8.5	80	11	75.8	11.0	140	21
	34	160	23	68.9	10.0	120	18	96.5	14.0	120	18
	40	190	28	96.5	14.0	140	20	96.5	14.0	120	18
Vinyl ester	25	110	16	54	7.9	86.2	12.5	69.6	10.1	180	26.5
	35	260	37.3	95.2	13.8	153.4	22.25	108	15.6	230	34
	40	220	32	88.9	12.9	160	23	110	15.9	210	30

Table 6 The effect of glass type and amount on mechanical properties

Type of glass fiber reinforcement	Glass content, wt%	Density, g/cm³	Tensile strength		Tensile modulus		Elongation, %	Flexural strength		Flexural modulus		Compressive strength	
			MPa	ksi	10^{-3} Pa	10^{-6} psi		MPa	ksi	10^{-3} Pa	10^{-6} psi	MPa	ksi
Neat cured resin	0	1.22	59	8.6	5.40	0.783	2.0	88	12.8	3.90	0.565	156	22.6
Chopped strand mat	30	1.50	117	17.0	10.80	1.566	3.5	197	28.6	9.784	1.419	147	21.3
Chopped strand mat	50	1.70	288	41.8	16.70	2.422	3.5	197	28.6	14.49	2.102	160	23.2
Roving fabric	60	1.76	314	45.5	19.50	2.828	3.6	317	46.0	15.00	2.175	192	27.8
Woven glass fabric	70	1.88	331	48.0	25.86	3.750	3.4	403	58.4	17.38	2.520	280	40.6
Unidirectional roving fabric	70	1.96	611	88.6	32.54	4.720	2.8	403	58.4	29.44	4.270	216	31.3

Source: Ref 2

the flexibility of the cross-linked resin. The same result occurred with saturated acids. Aromatic groups, in either the dihydric alcohol or acid component, promoted increased stiffness and hardness.

Using a reinforcing fiber to produce a polyester composite dramatically improved both the tensile and flexural characteristics. Table 4 gives the mechanical properties for the five representative examples shown in Table 3. In Tables 5 and 6, the properties obtained were dependent upon the amount and type of glass fiber used. The last two entries in Table 6 show the influence of fiberglass orientation. Both contained 70 wt% glass fiber, but the unidirectional composite showed much higher tensile properties and flexural modulus when tested in the glass fiber direction. The properties of unidirectional composites are, of course, anisotropic, and mechanical properties measured transverse to the glass direction will approach those observed for clear castings. Polyester composites employing unidirectional reinforcements are commercially produced by the pultrusion or filament-winding process. They can be used in structural applications where strength or stiffness is required in only one direction. A comparison of glass fiber reinforced polyester composites to metals at room temperature can be seen in Table 7.

Using different reinforcing fibers can also affect mechanical properties. While E-glass is the most commonly used reinforcement of polyester resins, S-glass, aramid, and carbon fibers can also be used. Table 8 compares orthophthalic polyester to vinyl ester with E-glass, S-glass, and aramid fiber reinforcements. While the tensile strength of the

Table 7 Comparative properties of structural materials at room temperature

Material	Glass content, wt%	Density, g/cm³	Flexural strength Pa	Flexural strength 10⁻³ psi	Flexural modulus 10⁻³ Pa	Flexural modulus 10⁻⁶ psi	Tensile strength Pa	Tensile strength 10⁻³ psi	Tensile modulus 10⁻³ Pa	Tensile modulus 10⁻⁶ psi
Unidirectional glass roving, rod and bar	70	2.0	690	100	40	6	690	100	40	6
Unidirectional glass roving, sheet	50	1.6	205	30	15	2	140	20	12	1.8
Chopped strand mat	30	1.5-1.7	110-190	16-28	6.9-8.3	1.0-1.2	60-120	9-18	6-12	0.8-1.8
Aluminum sheet	...	2.8	140	20	70	10	40-190	6-27	70	10
Stainless steel	8.0	8.0	210-240	30-35	190	28	210-240	30-35	190	28
Low-carbon steel	...	8.0	190	28	210	30	200-230	29-33	210	30

Material	Elongation, %	Compressive strength Pa	Compressive strength 10⁻³ psi	Impact strength J/mm	Impact strength ft·lbf/in.	Thermal conductivity W/m·k	Thermal conductivity Btu·in./h·ft²·°F	Specific heat kJ/kg·K	Specific heat Btu/lb °F	Coefficient of thermal expansion, 10⁻⁶/K
Unidirectional glass roving, rod and bar	1.5	410	60	2.6	49	0.70	5	1.0	0.24	5
Unidirectional glass roving, sheet	1.5	140	20	0.96	18	0.60	4	1.2	0.28	9
Chopped strand mat	1-1.2	100-170	15-25	0.2-0.64	4-12	0.20-0.25	1.2-1.6	1.3-1.4	0.31-0.34	20-35
Aluminum sheet	30-40	...	...	...	...	120-230	810-1620	0.92-0.96	0.22-0.23	20-25
Stainless steel	40-50	210	30	0.45-0.59	8.5-11	14-25	96-185	0.50	0.12	15-20
Low-carbon steel	38-50	190	28	...	...	35-65	260-460	0.42-0.46	0.10-0.11	10-15

Table 8 Effect of reinforcement on mechanical properties

Material	Tensile strength MPa	Tensile strength ksi	Tensile modulus GPa	Tensile modulus 10⁶ psi	Elongation, %
E-glass-ortho polyester					
Ambient temperature	157	22.8	11.0	1.59	1.7
50 °C (125 °F)	148	21.5	8.41	1.22	2.4
65 °C (150 °F)	140	20.3	7.31	1.06	2.6
S-glass-ortho polyester					
Ambient temperature	159	23.0	10.3	1.49	1.9
50 °C (125 °F)	165	24.0	8.55	1.24	2.6
65 °C (150 °F)	157	22.8	7.38	1.07	2.6
Aramid-ortho polyester					
Ambient temperature	212	30.7	12.5	1.82	2.0
50 °C (125 °F)	208	30.2	11.5	1.67	2.1
65 °C (150 °F)	200	29.0	9.52	1.38	2.4
E-glass-vinyl ester					
Ambient temperature	206	29.9	12.6	1.83	2.1
50 °C (125 °F)	192	27.9	11.4	1.66	2.2
65 °C (150 °F)	201	29.2	11.6	1.68	2.4
S-glass-vinyl ester					
Ambient temperature	198	28.7	11.4	1.66	2.2
50 °C (125 °F)	172	25.0	9.6	1.39	2.3
65 °C (150 °F)	199	28.8	10.3	1.49	2.5
Aramid-vinyl ester					
Ambient temperature	189	27.4	12.1	1.76	1.8
50 °C (125 °F)	221	32.0	11.7	1.70	2.2
65 °C (150 °F)	218	31.6	11.7	1.69	2.2

orthophthalic polyester was much improved with aramid, no great difference was seen with the vinyl ester.

Inorganic fillers are commonly used in polyester resin compositions. While they do improve stiffness, as shown by an increase in modulus (see Table 9), they exhibit little effect on other strength characteristics and are used mainly to reduce cost.

Mechanical properties at elevated temperatures show the expected difference between four general classifications of polyester resins (Fig. 1). The extreme rigidity and high glass transition temperature, T_g, for BPA fumarate or chlorendic polyesters resulted in a high flexural strength retention up to 120 °C (250 °F). The thermal performance of vinyl esters, combined with their excellent mechanical properties and toughness, explains why they are often chosen for structural polyester resin composites.

Vinyl esters also showed performance advantages over isophthalic polyesters (Fig. 2) in fatigue studies (Ref 6). The advantage of vinyl ester was also shown at elevated temperatures (Ref 7). At 105 °C (220 °F), vinyl ester and isophthalic polyester composites (60% glass)

Table 9 Effect of filler and glass fiber reinforcement on mechanical properties of polyester resins

Reinforcement material	Flexural strength MPa	Flexural strength ksi	Flexural modulus 10⁻² Pa	Flexural modulus 10⁻⁵ psi	Tensile strength MPa	Tensile strength ksi	Tensile modulus 10⁻² Pa	Tensile modulus 10⁻⁵ psi	Elongation, %	Barcol hardness	Heat deflection temperature(a) °C	Heat deflection temperature(a) °F
Neat resin casting (A material)	129	18.7	36.0	5.22	70.0	10.1	35.0	5.07	3.0	30	58	135
80 pph A; 20 pph CaCO₃ (B material)	109	15.8	43.0	6.23	52	7.5	56.0	8.12	1.26	45	67	150
74 pph B; 26 pph 2.5-cm-long chopped strand (C material)	183	26.5	61.0	8.84	116	16.8	96.94	14.06	1.72	45	>260	>500

(a) For 250 μm (10 mil) deflection at 1.82 MPa (0.264 ksi). Source: Ref 2

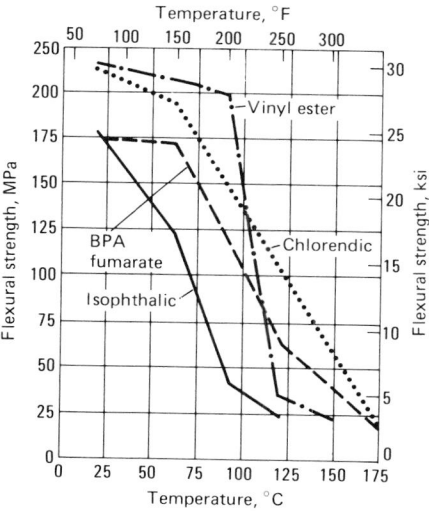

Fig. 1 Flexural strength versus temperature, glass-polyester composites of 40% glass

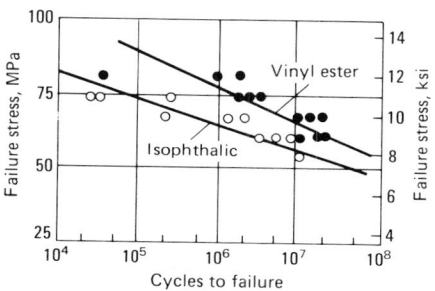

Fig. 2 Flexural fatigue strength

were cycled to a stress level of 60 to 70 MPa (9 to 10 ksi). After 200 000 cycles, the drop in flexural modulus was only 5% for the vinyl ester, compared to 12% for the isophthalic polyester.

Thermal and Oxidative Stability

Polyester resins are commonly used in elevated-temperature applications, especially in the electrical and corrosion resistance areas. It has been found that when polyester resin composites, as well as most organic polymers, are heated, the polymer begins to dissociate chemically. This decomposition will be accelerated by auto-oxidative reactions leading to the formation of oxidized fragments, the major ones being benzaldehyde, benzene, and acetaldehyde. The temperature at which this decomposition occurs and the type of fragment produced depends on the structure of the polyester used. Regardless of the polymer composition, at temperatures in the range of 300 °C (570 °F) the cured polyester resin will undergo spontaneous decomposition. This is a characteristic of

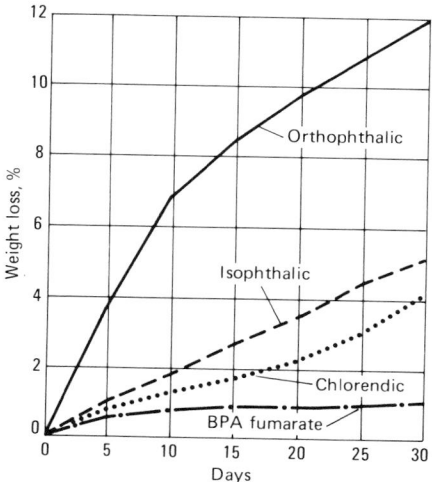

Fig. 3 Thermal stability of glass-polyester composites at 180 °C (355 °F)

vinyl polymers, which is caused by their depolymerization to form monomeric species.

Orthophthalic resins have shown very poor performance at elevated temperatures, while the very rigid, highly aromatic BPA fumarates have shown very good stability (Fig. 3). However, the orthophthalics have performed admirably in many applications and are typically favored when temperature requirements are not excessive, because of their much lower cost.

Monomer type also plays an important role in the thermal performance of a polyester resin. A polyester resin in vinyl toluene, for example, showed superior thermal performance when compared to the same resin in styrene. This has been attributed to the stronger copolymer bonds formed with the fumarate unsaturation. The retention of flexural strength for an isophthalic polyester when aged at 200 °C (390 °F) (Fig. 4) showed improvement with vinyl toluene versus styrene. While vinyl toluene versions of an isophthalic polyester and a BPA fumarate showed comparable performance at 200 °C (390 °F), the BPA fumarate had much better flexural strength retention at 220 °C (430 °F) (Fig. 5), and 240 °C (460 °F) (Fig. 6).

Chemical Resistance

Polyester resins have been used for many years in applications requiring resistance to chemical attack. Numerous applications for corrosion-resistant tanks, pipes, ducts, and liners can be found in the chemical process and pulp/paper industries. Different polyester resin classifications are used in corrosion-resistant applications depending upon the specific chemical environment to be contained. Table 10 shows the representative performance of four commonly used polyester resin classifications.

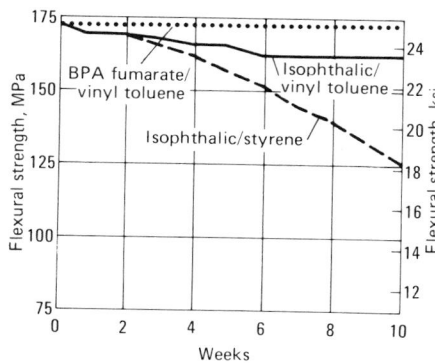

Fig. 4 Flexural strength retention of glass-polyester composite when aged at 200 °C (390 °F), tested at room temperature

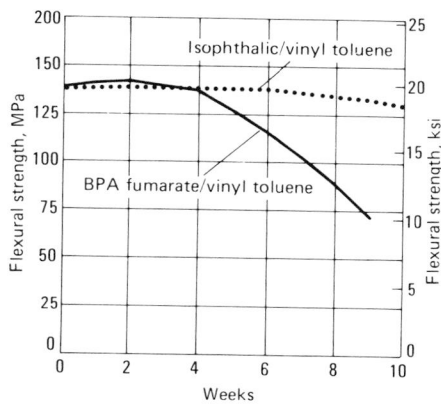

Fig. 5 Flexural strength retention of glass-polyester composite when aged at 220 °C (430 °F), tested at room temperature

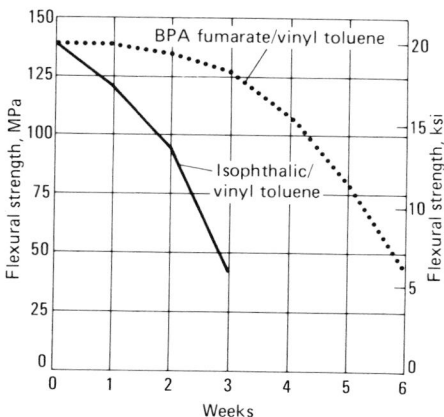

Fig. 6 Flexural strength retention of glass-polyester composite when aged at 240 °C (465 °F), tested at room temperature

Chlorendics are typically chosen for strong acid environments, especially at elevated temperatures, while BPA fumarate is better in strong basic solutions. Using glass fiber does not improve the corrosion resistance of polyester resins and, in many cases, actually reduces

Table 10   Corrosion resistance of glass fiber-polyester resin composites

Resin	75% H_2SO_4 80 °C (175 °F)	15% NaOH 65 °C (150 °F)	5¼% NaOCl 65 °C (150 °F)	Xylene ambient	Deionized water 100 °C (210 °F)	Seawater 80 °C (180 °F)
Isophthalic	–	–	–	+	–	–
Chlorendic	+	–	–	+	+	+
BPA fumarate	–	+	–	+	–	+
Vinyl ester	–	–	+	–	–	+

Table 11   Effect of methyl methacrylate on the gloss retention of weathered polyester panels

Components	Panel number 1	2	3	4	5
Polyester, %	75	75	60	60	60
Methyl methacrylate, %	25	. . .	20	10	0
Styrene, %	0	25	20	30	40
Gloss retention, %	5.2	48	83.5	56.5	11.8

Source: Ref 9

Table 12   Electrical properties of glass-polyester composites

Volume resistivity, $\Omega \cdot m$
50% relative humidity . . . . 10^{10}-10^{12}

Dielectric strength, V/μm (V/mil)
Short-time, 3.2 mm (1/8 in.) 13.6-16.5   345-420
Step-by-step,
3.2-mm (1/8-in.) increments . . 10.8-15.4   275-390

Dielectric constant
60 Hz 5.3-7.3
1 kHz 4.68
1 MHz 5.2-6.4

Dissipation factor
60 Hz 0.011-0.041
1 MHz 0.008-0.022

Arc resistance 120-200

performance. This is especially true in hydrofluoric acid or strong caustic environments because these chemicals can attack and dissolve glass. In this and other special cases, reinforcing materials, such as carbon fibers, may be useful (Ref 8).

Ultraviolet (UV) Resistance

Polyester resins can be formulated to exhibit exceptional UV resistance and have been used in many outdoor applications. They can survive exposure to the elements for periods exceeding 30 years, although some discoloration and loss of strength will occur. The onset of surface degradation is marked by a yellow discolora-

tion, which becomes progressively darker as erosion and surface stress crazing occur. In translucent systems, this UV radiation causes yellowing of the composite as a whole, although the color is usually more intense on the surface. These negative characteristics can be effectively eliminated by proper selection of polyester ingredients. Styrene and other aromatic vinyl monomer derivatives are more susceptible to oxidative degradation and are usually supplemented with more resistant acrylate or methacrylate monomers in cross-linked composition. Of the acrylate monomers, methyl methacrylate (MMA) is the most common. Because MMA does not polymerize very well with the fumarate/maleate unsaturation in polyester resins, using MMA as the only monomer usually results in a cured resin with poor properties. It has been found that when MMA is copolymerized with styrene, the cured polyesters have superior durability, color retention, and resistance to fiber erosion (Ref 9).

The improved UV resistance with MMA is evident in Table 11, which compares four polyester resins composed of identical polymers and four different monomer systems. These resins were used to prepare glass fiber reinforced panels that were exposed to 5 years of outdoor weathering. The gloss retention for styrene/MMA monomer blends was much greater than for either monomer alone. The refractive index of MMA is also lower than for styrene, allowing the formulation of polyester resins with a refractive index matched to the glass fibers. This, combined with improved UV resistance, has resulted in the use of MMA polyesters to prepare glass reinforced transparent building panels that can be used in green-

Table 13   Electrical properties of isophthalic polyester(a) 3.2-mm (1/8-in.) laminates with various fillers

Material	Dielectric strength short time V/μm	V/mil	Volume resistivity, 10^{-13} $\Omega \cdot m$	Dielectric constant, 1 MHz	Dissipation factor, 1 MHz	Dielectric constant, 1 kHz	Dissipation factor, 1 kHz	Dielectric constant, 60 Hz	Dissipation factor, 60 Hz	Arc resistance Avg	Max	Min	Track resistance, V	Dielectric breakdown short time, kV	Dielectric breakdown step-by-step, kV
Calcium carbonate	15.0	380	7.8	4.10	0.007	4.18	0.005	4.19	0.003	157	181	140	840	58	61
Gypsum CaSO4	14.4	365	2.1	3.69	0.011	4.04	0.023	4.19	0.027	153	184	141	840	70	55
Alumina trihydrate	15.4	390	2.6	3.67	0.009	3.81	0.010	3.89	0.011	183.5	184	183	860	67	51
Clay	14.4	365	6.4	4.08	0.018	4.61	0.040	5.10	0.057	182.5	183	182	840	59	57

(a) Vinyl toluene monomer

Table 14   Electrical properties of BPA fumarate polyester(a) 3.2-mm (1/8-in.) laminates with various fillers

Material	Dielectric strength short time V/μm	V/mil	Volume resistivity, 10^{-13} $\Omega \cdot m$	Dielectric constant, 1 MHz	Dissipation factor, 1 MHz	Dielectric constant, 1 kHz	Dissipation factor, 1 kHz	Dielectric constant, 60 Hz	Dissipation factor, 60 Hz	Arc resistance Avg	Max	Min	Track resistance, V	Dielectric breakdown short time, kV	Dielectric breakdown step-by-step, kV
Calcium carbonate	6.1	155	1.6	3.94	0.005	4.00	0.004	4.03	0.004	140	143	133	840	58	52
Gypsum CaSO4	5.9	150	3.3	3.72	0.009	4.03	0.024	4.24	0.029	144	151	137	820	50	40
Alumina trihydrate	11.8	300	3.3	3.64	0.008	3.81	0.015	3.93	0.025	182	184	181	820	55	52
Clay	12.6	320	3.5	4.08	0.023	4.68	0.043	5.11	0.053	183	184	181	840	61	43

(a) Vinyl toluene monomer

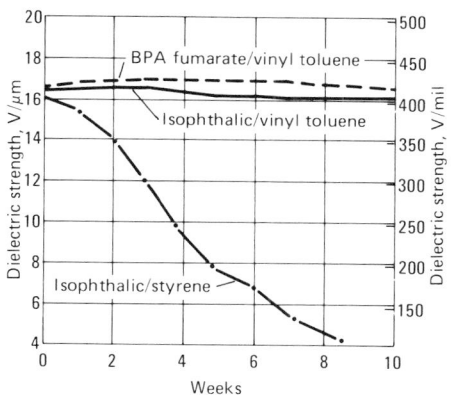

Fig. 7 Dielectric strength retention of glass-polyester composite when aged at 200 °C (390 °F), tested at room temperature

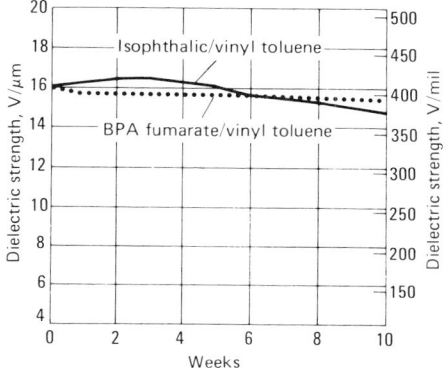

Fig. 8 Dielectric strength retention of glass-polyester composite when aged at 220 °C (430 °F), tested at room temperature

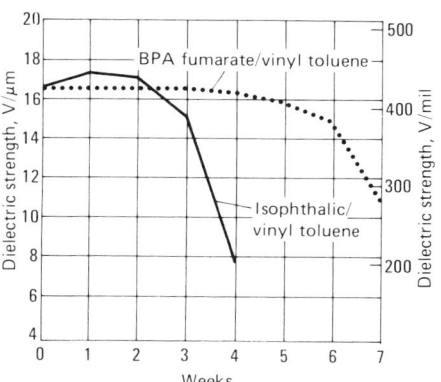

Fig. 9 Dielectric strength retention of glass-polyester composite when aged at 240 °C (465 °F), tested at room temperature

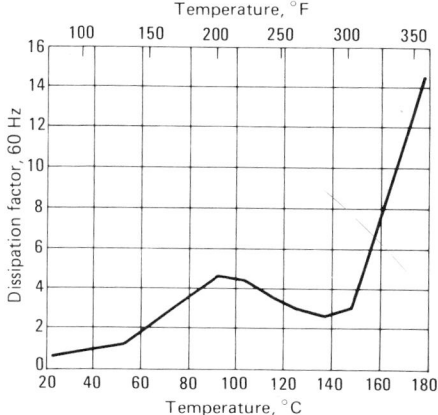

Fig. 10 Isophthalic polyester at 2.4 V/μm (60 V/mil)

Table 15 Performance of selected polyester composites in fire tests

	System I	System II	System III
Material			
Resin................	100(a)	100(b)	100(c)
Alumina trihydrate.......	100	100	100
Antimony oxide.........	...	5	...
Ferrous oxide..........	...	...	5
Test method and property			
ASTM E 162			
Flame-spread index....	75	7	7
ASTM E 84			
Flame spread........	64	23	25
Smoke emission.......	608	270	268
NBS chamber			
Flaming mode			
Max density........	203	433	264
90-s density........	2.5	18	11
240-s density.......	162	245	128
Nonflaming mode			
Max density........	481	400	350
90-s density........	1	1	5
240-s density.......	16	45	50

(a) Orthophthalic resin. (b) HET acid resin A, 26% Cl. (c) HET acid resin B, 26% Cl. Source: Ref 12

and a BPA fumarate as well as the influence of filler type.

Because many electrical applications require elevated-temperature performance, polyester composites must have good thermal stability. Thermal stability and electrical performance at elevated temperatures are directly related, as can be seen by comparing the retention of dielectric strength at 200 °C (390 °F), shown in Fig. 7, with the retention of flexural strength at 200 °C (390 °F), shown in Fig. 4. As with flexural strength, vinyl toluene outperformed a styrene-based polyester. At temperatures above 200 °C (390 °F), BPA fumarate outperformed isophthalic polyester when both were used with vinyl toluene (Fig. 8, 9).

Large electrical equipment, such as high-voltage motors or generators, often operate at elevated temperatures. The electrical property of greatest concern in this application is the dissipation factor, especially the dissipation factor versus the temperature. Polyester resins can be formulated for a low dissipation factor at elevated temperatures (Fig. 10). They are used as electrical varnishes at continuous-use temperatures up to 180 °C (355 °F).

house, skylight, and other applications. The effect of radiation can be further reduced by using UV-absorbing chemical additives, such as the substituted benzophenones or benzotriazoles.

Electrical Properties

Almost all organic polymers have medium to excellent electrical properties, and a wide range

of thermoset and thermoplastic materials are used in the electrical and electronics industries. Applications in which polyester resins have been used include the insulation of motor windings, encapsulation of electrical components, fabrication of printed circuit boards, high-voltage stand-off insulators, switch boxes, and miscellaneous equipment for high-voltage line work. Generalized properties for polyester resins are shown in Table 12. Tables 13 and 14 show specific data on an isophthalic polyester

Table 16 Flame spread and smoke emission characteristics of unfilled and filled polyester systems(a)

Component	Property value									
Unfilled system										
Resin...................100(b)	100(c)	100(c)	100(d)	100(e)	100(e)	100(f)	100(f)	100(g)	100(g)	
Filled system										
Antimony oxide............ ...	...	5	...	...	5	...	5	...	5	
Flame spread............ 350	74	25	...	72	33	67	23	69	23	
Smoke emission.......... 1100	780	450	...	747	731	980	1043	837	838	
Alumina trihydrate.......... 100	100	100	100	100	100	100	100	100	100	
Antimony oxide............ ...	5	...	...	5	...	5	...	5	...	
Ferrous oxide............. ...	...	5	5	...	5	...	5	...	5	
Flame spread............ 64	23	20	25	28	28	18	25	20	35	
Smoke emission.......... 608	270	242	168	364	260	450	400	761	620	

(a) Glass-reinforced laminates 3.2-mm (1/8-in.) thick, with 30% glass, tested using ASTM E 4-76. (b) Orthophthalic resin. (c) HET acid resin A, 26% Cl. (d) HET acid resin B, 26% Cl. (e) Dibromotetrahydrophthalic resin, 18% Br. (f) Dibromoneopentyl glycol resin, 18% Br. (g) Tetrabromophthalic resin, 18% Br. Source: Ref 12

Flame-Retardant Polyester Resins

All organic materials, including polyesters, will burn in the presence of a flame. In many applications, polyester resins are required to have some degree of resistance to burning, which can be accomplished by using either a filler or a specially formulated flame-retardant polyester resin, depending on the degree of resistance required.

Incorporating halogen into a polyester resin has been found to be an effective way of improving flame retardance. This can be done by using a halogenated dibasic acid, such as chlorendic anhydride or tetrabromophthalic anhydride, or a halogenated dihydric alcohol, such as dibromoneopentyl glycol or tetrabromo bisphenol A. Halogen can also be incorporated into a preformed polyester by the direct bromination of tetrahydrophthalic anhydride containing resins (Ref 10). At equivalent concentrations, bromine is much more effective than chlorine. It has also been found that additives such as antimony oxides and ferrous oxide act as synergists with halogenated polyesters, improving their performance (Ref 11).

Burning rate and smoke generation can be measured using the Steiner Tunnel Test (ASTM E 84). In this test, a gas burner is placed at one end of a 53-cm by 7.6-m (21-in. by 25-ft) section. The distance of flame travel and the amount of smoke liberated are measured (by the obscuration of a photoelectric beam). These are compared to two standards: red oak board, which is given a rating of 100 for flame spread and smoke generation, and asbestos cement, which is given a rating of 0. Smoke generation can also be measured with an NBS smoke chamber, which by means of a photoelectric cell measures smoke buildup in a closed chamber. The sample can be burned either with or without a direct flame. When NBS smoke chamber testing is used, it is often common to use ASTM E 162, Flame Spread Index, to measure this variable.

A comparison of various resins by these test methods (Table 15) showed the improvement in flame retardance obtained with a halogenated versus an orthophthalic resin. Ferrous oxide also showed a reduction in smoke in the NBS chamber compared to antimony oxide.

Table 16 compares various halogenated polyester resin composites. The improvement of an orthophthalic resin by the incorporation of alumina trihydrate (ATH) is dramatic (flame spread is reduced from 350 to 64), although halogenated resins reached this level of performance without filler. Halogenated resins with a synergist showed only a slight reduction in flame spread when ATH was added, but smoke emission was greatly reduced. In the relationship of chlorine to bromine, 26% chlorine was comparable to 18% bromine.

REFERENCES

1. W.H. Carothers, Studies on Polymerization and Ring Formation. Part I. An Introduction to the General Theory of Condensation Polymers, *J. Am. Chem. Soc.*, Vol 51, 1929, p 2548
2. M. Grayson and D. Eckroth, Ed., *Encyclopedia of Chemical Technology*, 3rd ed., Vol 18, John Wiley & Sons, 1982, p 575
3. P. Robitschek and C.T. Bean, Flame-Resistant Polyesters from Hexochlorocyclopentadiene, *Ind. Eng. Chem.*, Vol 46, 1954, p 1628
4. E.N. Doyle, *The Development and Use of Polyester Products*, McGraw Hill, 1969, p 258
5. H.V. Boenig, *Unsaturated Polyesters: Structure and Properties*, Elsevier, 1964
6. B. Das, H.S. Loveless, and S.J. Morris, Effects of Structural Resins and Chopped Fiber Lengths on the Mechanical and Surface Properties of SMC Composites, in *36th Annual Conference of the Reinforced Plastics/Composites Institute*, The Society of the Plastics Industry, Inc., 1981
7. P.K. Mallick, Fatigue Characteristics of High Glass Content SMC Materials, in *37th Annual Technical Conference*, Society of Plastics Engineers, Inc., 1979, p 589
8. H.S. Kliger and E.R. Barker, A Comparative Study of the Corrosion Resistance of Carbon and Glass Fibers, in *39th Annual Conference of the Reinforced Plastics/Composites Institute*, The Society of the Plastics Industry, Inc., 1984
9. A.L. Smith and J.R. Lowry, Long Term Durability of Acrylic Polyesters Versus 100% Acrylic Resins in Glass Reinforced Constructions, in *15th Annual Conference of the Reinforced Plastics/Composites Institute*, The Society of the Plastics Industry, Inc., 1960
10. U. Toggweiler and F.F. Roselli, Halogenated Polyester Compositions and Process for Preparing the Same, U.S. Patent 3,536,782, 1970
11. E. Dorfman, W.T. Schwartz, Jr., and R.R. Hindersinn, Fire-Retardant Unsaturated Polyesters, U.S. Patent 4,013,815, 1977
12. J.E. Selley and P.W. Vaccarella, Controlling Flammability and Smoke Emissions in Reinforced Polyesters, *Plast. Eng.*, Vol 35, 1979, p 43

Thermoplastic Resins

Rex B. Gosnell, King Mar Laboratories, Cape Composites, Inc.

THERMOPLASTIC RESINS that are potentially useful as matrices for advanced composites for aerospace applications are assessed in this article in terms of fiber impregnation, fabrication, and in-service performance. Future developments in thermoplastic resin prepregs are also considered. This article discusses the use of resins as the matrix binder for advanced fibrous composites, and the similarities and differences in the behavior and processing of thermosets and thermoplastics. Emphasis is on the development of tougher materials for reduced weight, lower component costs, and higher damage tolerance, the latter of which tightens design allowables.

When dealing with a thermosetting polymer, the process engineer must realize that the polymerization is dynamic, that both its course and the final product performance are a function of the thermal history and environment of the living, growing molecule, and that the processing must be consistent with these factors. In contrast, a promising feature of a thermoplastic resin is that, ideally, the size and shape of the polymer are basically unchanged in the engineer's hands. Of course, this is not entirely true in practice; but the goal is to produce the thermoplastic polymer under carefully monitored and controlled conditions so that the polymer will be the same in every batch. No further chemistry occurs during fabrication of the composite part by the engineer.

In this article the emphasis is placed on this finished type of thermoplastic, although some information is presented on the intentional alteration of molecular weight or branching by the use of reactive plasticizers or reactive chain terminal functionality. These approaches help offset some of the shortcomings of thermoplastic resins, but they result in both a loss of constancy in processing and significantly increased costs.

Cost Elements

The oil crisis that began in 1973 and raised the price of fuel was a main impetus for increasing the fuel efficiency of airframes by manufacturers. This activity evolved into an effort not only to replace metal structures with composites but also to develop basic aircraft designs that would exploit the unique advantages of composite structures. An early assessment of potential savings from the use of composites in aircraft is shown in Table 1.

As the oil crisis eased, the emphasis on cost saving shifted from increasing fuel efficiency to decreasing manufacturing costs. The use of thermoplastic resins in composites offers significant advantages in cost savings in both these areas. Weight reduction remains important, however.

Emergence of Damage Tolerance Requirements

The initial work on composites in the late 1970s involved many aerospace companies and the National Aeronautics and Space Administration (NASA). One important conclusion that became evident during the course of these efforts was that the state-of-the-art 175 °C (350 °F) curing epoxy systems failed to provide the level of damage tolerance required for the extensive use of composites in primary structures on military and commercial aircraft.

The 175 °C (350 °F) curing epoxy systems and some similar prepregs, however, were the only products available for use on commercial aircraft as late as 1981. Although this type of system offered an attractive combination of properties, such as processability and high strength, commercial airplane builders felt that the reduction in mechanical properties after damage from runway stones, dropped tools, and bird impacts was of sufficient concern to limit their use. The U.S. military had similar concerns regarding damage due to hostile action. Although the use of aramid and glass with carbon fiber resulted in hybrid composites having a somewhat higher damage tolerance, the prime contractors sought tougher resins and carbon fibers that would have higher strain-to-failure properties in addition to all the properties of the 175 °C (350 °F) curing epoxy resins.

Development of Tests. As new, tougher resins began to appear, various tests were used to evaluate their toughness and/or damage tolerance. To promote systematic evaluation, NASA and industry representatives selected and standardized a set of five common tests for the characterization of fiber-resin composites (Ref 1-7). These include the edge-delamination and double-cantilever-beam tests for interlaminar fracture toughness (G_{Ic}) and the open-hole-tension, open-hole-compression, and compression-after-impact tests for damage tolerance.

Each test has its own set of problems. For example, failures in the double-cantilever-beam test seldom occur solely by mode I (peel). In the testing process, different laboratories used composite specimens with different resin contents, fiber types, dimensions; additionally, a variety of impact stress levels were used. After extensive refinement of the tests, reliable data have begun to emerge.

Testing will remain an important factor in assessing the quality of thermoplastic matrix composites. In the past, inadequate testing has been a serious shortcoming in realizing the complete potential of composite materials. Improved toughness of thermoplastic composites may help offset edge-effect problems, such as notch effects, encountered in high-modulus materials.

Supplemental tests will probably be required to assess thermoplastic systems, particularly as a guide to design in the areas of temperature versus creep and environmental stress cracking. Because of the unusual behavior of some of these linear molecules, designers will require tests to ensure service life and reliability. The difference in failure behavior in the stressed and unstressed conditions of many thermoplastics is

Table 1 Estimated cost savings from the use of composites in aircraft over a 15-year life span

Composite weight per aircraft		Weight saved per aircraft		Fuel saved per aircraft		Dollars saved per aircraft
kg	lb	kg	lb	L	gal	
1800	4000	450	1000	1 130 000	300 000	320 000

Table 2 Interlaminar fracture toughness of carbon fiber-thermoset resins as determined by double-cantilever-beam specimen

Material(a)	Interlaminar fracture toughness, G_{Ic}	
	J/m²	ft·lbf/ft²
5208	80-90	5-6
3502	120-150	8-10
3501-6	150-214	10-15
1504	95-123	7-8
2220-1	256	18
914	220-250	15-17
BP907	324-397	22-27
HST-7	543	37
V378A(b)	72-86	5-6
5245	134-141	9-10

(a) Composites fabricated at 175 °C (350 °F). (b) Postcured at 205 °C (400 °F)

Table 3 Comparison of residual compression strength after impact at 6650 J/m (1500 ft·lbf/ft)

Material	Compression strength	
	MPa	ksi
AS4/BP907	275	40
AS4/3502	210	30.1
AS4/3501-6	165	24
T300/914	130	18.7
CHS/5245	175	25.3
C12/5245	225	32.2
AS4/5245	210	30.2
CHS/1504	180	26.1
T700/1504	215	30.9
C6/HST-7	270	39.3
AS4/2501	250	36
AS6/6376	260	38
AS4/8551	310	45
XASn/PEEK APC-1	285	41.6
AS4/PEEK APC-2	275	40

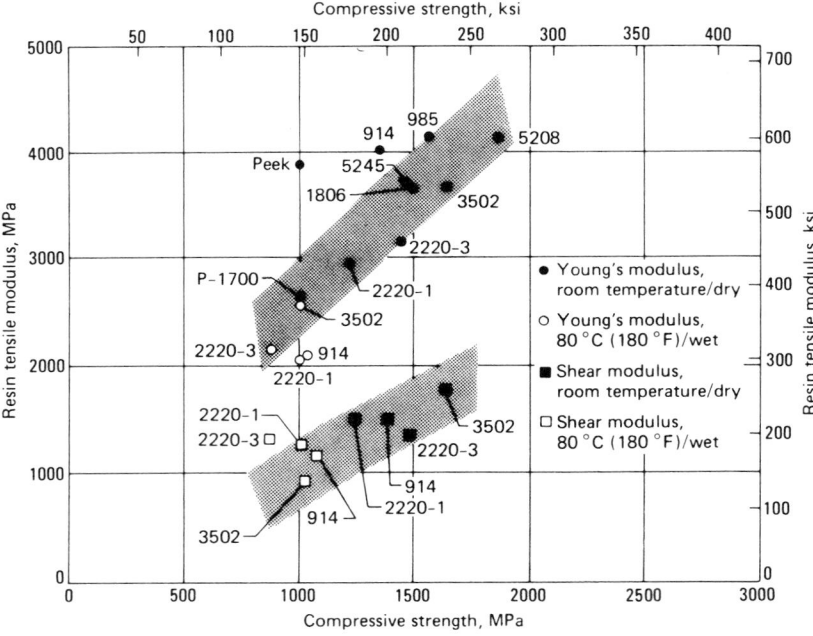

Fig. 1 Resin modulus versus composite 0° compressive strength (experimental values). Courtesy of NASA Langley Research Center

dramatic, especially when the environmental effects of solvents and surface-active agents are imposed on the material.

Furthermore, crystallinity of a linear thermoplastic must be considered. In a thermoset, this property has never been of concern. Changes in the crystalline nature of a linear thermoplastic, however, caused by time, temperature history, stress, and chemical factors, can profoundly affect its mechanical behavior, both during processing and in service.

Consideration of factors such as crystallinity suggest that thermoplastic composites do not offer ready answers to the known and documented difficulties with thermosets. To think so would be natural because much of the current effort with thermoplastic matrix composites has been an extrapolation of experience with thermosetting systems. However, all that would be accomplished would be the exchange of one set of problems for a new set of unknown or underestimated problems. The G_{Ic} value will certainly be a widely used criterion in the evaluation of thermoplastic-based composites, as will be other test values, such as compression strength after impact.

G_{Ic} values for carbon fiber composites from several thermosetting systems are presented in Table 2. Originally, a G_{Ic} value of about 700 J/m² (50 ft · lbf/ft²) was considered an acceptable level, although none of the systems available then, as shown in Table 2, reached that level. As it became apparent that the use of G_{Ic} values alone was not the best way to evaluate toughness or damage tolerance in composites, mode II (shear) and even mode III (scissor shear) values were also considered important. Concurrent with G_{Ic} testing by the double-cantilever-beam method, the edge-delamination test was designed in an attempt to determine true G_{Ic} values and compensate for the effects of mode II and mode III failures. However, this test also has shortcomings, such as complex failure modes.

The G_{Ic} testing drew some criticism from those who felt that toughness or damage tolerance should be evaluated on the basis of a test that more closely approximated actual conditions. Therefore, tests were designed to measure the retention of mechanical properties after specified impact, frequently while under load. For example, a test developed by NASA Langley involved striking a quasi-isotropic 48-ply panel preloaded in compression with a 12.7-mm (0.50-in.) aluminum ball traveling at a velocity of about 90 m/s (300 ft/s). In a different test selected by a major aircraft manufacturer, a similar specimen was impacted with a falling weight, and the resulting residual compressive strength was determined.

The goal was a residual compressive strength of 360 MPa (50 ksi) after impacting the composite specimen at a stress level of 6650 J/m (1500 ft · lbf/ft). Table 3 lists the residual compression strengths for some thermosetting and thermoplastic systems after the designated impact of 6650 J/m (1500 ft · lbf/ft).

Many different composite systems have been evaluated on the basis of the tests described above. A comparison of the performance of toughened composites, including thermoplastics, based on NASA standard damage tolerance tests has been published (Ref 1).

Thermoplastic Suitability for High-Modulus Composites

As work progressed in developing tougher resins and refining the testing of the resulting composites, some relationships between neat resin properties and composite performance were established. The relationship between resin modulus and 0° compressive strength is indicated in Fig. 1. Note that the thermoplastics P-1700 and PEEK may not have been the best composites, possibly because of poor fiber bundle wet-out and improper interface. Figure 2 presents data on the relationship between resin modulus and interlaminar fracture toughness. The thermoplastics perform fairly well, suggesting that in some cases incomplete bundle wet-out is not critical to good performance

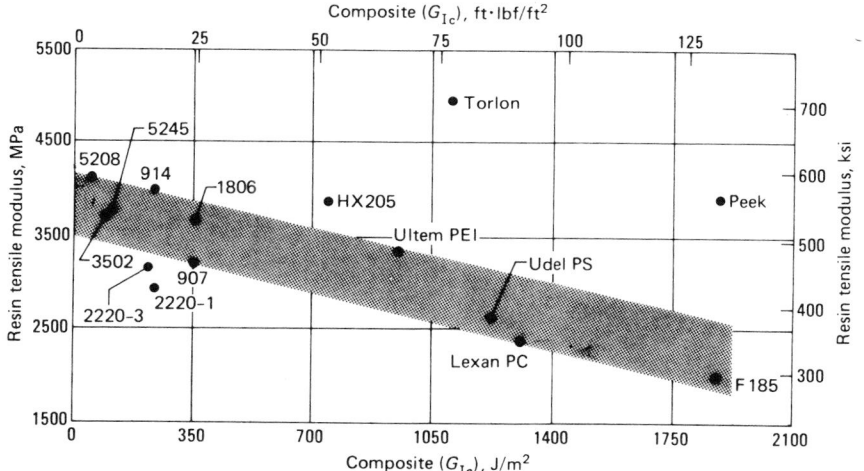

Fig. 2 Resin modulus versus composite interlaminar fracture toughness at room temperature/dry conditions. Courtesy of NASA Langley Research Center

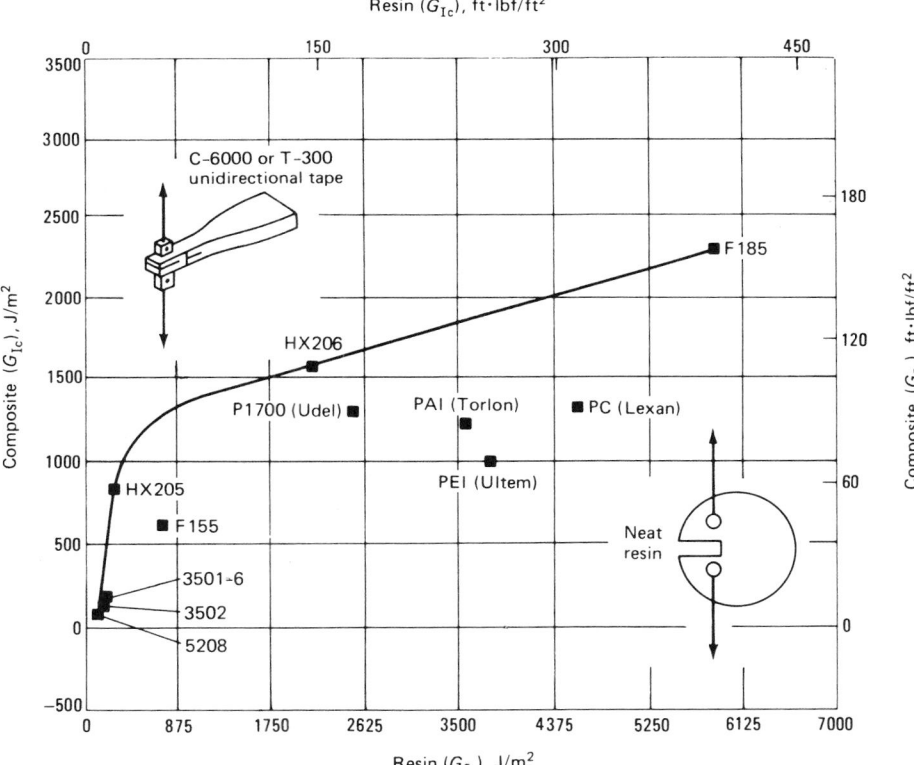

Fig. 3 Empirical relationship between interlaminar fracture toughness (G_{Ic}) values of resins and composites. Courtesy of NASA Langley Research Center

in interlaminar fracture toughness. Figure 3 shows a relationship between resin fracture energy and composite fracture energy. Figure 4 presents data on strain and fracture toughness.

Based on the information above, the following recommendations can be made for the major constituent properties required to obtain improved damage tolerance in high-modulus composites:

Impact strength
- High fiber strain
- High resin shear strength
- Moderate interlaminar fracture toughness

Delamination resistance
- Moderate resin fracture toughness
- Good fiber-resin relationship

Compression strength
- High 0° compressive strength
- High resin tensile and shear modulus

These guidelines should help polymer chemists and formulators of thermoplastic matrix resins design systems with the desired chemical, rheological, and mechanical characteristics. Because it was felt that the toughened thermosets could be further improved, thermoplastics began to receive more attention early in 1984. Few of the toughened thermosets had met the combined requirements of damage tolerance and hot/wet compression strength. In addition, manufacturing cost reduction, a potential offered by thermoplastics, had become a more important incentive than weight reduction. Although some early projections of production cost savings were quite optimistic, it was the initial programs that projected the potential advantages of lower costs using thermoplastic systems in production of composite parts. With updated qualifications, these conclusions are still widely accepted today.

Composites for use on aircraft must demonstrate resistance to aircraft fluids, lubricants, and solvents, especially while under stress. Many linear amorphous thermoplastics, such as Udel and Ultem, are susceptible to attack by hydraulic fluid and paint strippers (methylene chloride); semicrystalline polymers, such as Ryton and PEEK, show significantly better resistance. A recent report (Ref 4) indicated that PEEK was resistant to hydraulic fluid but exhibited an equilibrium weight gain of 15% when exposed to methylene chloride. This disclosure is in contrast to other reports that PEEK is resistant to methylene chloride.

The potential of thermoplastic prepregs was recognized from the early studies. The chemistry involved in the production of thermoplastics provides prepregs with indefinite shelf life, which gives the fabricator better quality assurance and avoids the storage/refrigeration problem associated with thermosetting prepregs. Thermoplastic polymers offer high toughness, believed to translate into improved damage tolerance. For example, some of the interlaminar fracture toughness values in Table 4 are excellent.

An overview of the differences between thermosets and thermoplastics as prepreg materials is presented in Table 5. Despite some criticisms of thermoplastic prepreg material, this class of resin appears certain to be used to some extent in aerospace composites: By the year 1990, approximately 3 to 5% of aerospace prepregs are expected to be from thermoplastic resins. The success of new aluminum alloys will have a significant effect on the volume. Beyond 1990, growth in the use of thermoplastic prepreg will depend on a number of factors, particularly the degree of success realized in thermoplastic prepreg automated fabrication. Also of great importance will be the degree of success in attaining performance requirements, including damage tolerance, en-

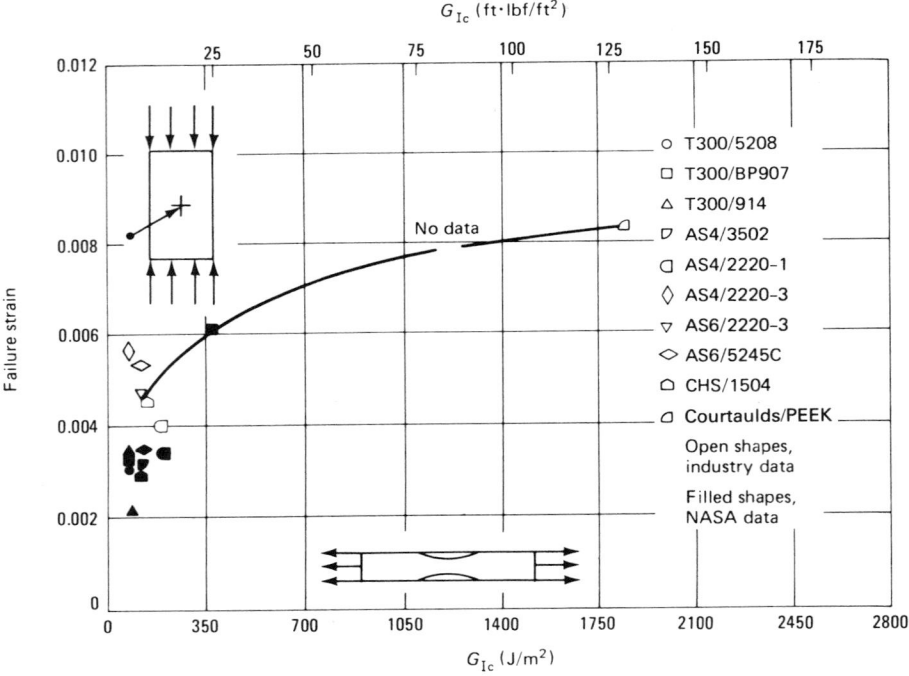

Fig. 4 Compression after impact failure strain versus interlaminar fracture toughness after impact at 4450 J/m (1000 ft · lbf/ft). Courtesy of NASA Langley Research Center

Table 4 Interlaminar fracture toughness of thermoplastic composites as determined by the double-cantilever-beam test

Material	Process conditions		Interlaminar fracture toughness, G_{Ic}	
	°C/kPa	°F/psi	J/m²	ft·lbf/ft²
Polysulfone (Udel)	340/4150	650/600	1175	80
Polyetherimide (Ultem)	400/4850	750/700	950	65
Polyamideimide (Torlon)	340/1400	650/200	1050	70
Polyphenylene sulfide (Ryton)	340/1400	650/200	720	50
Polyether etherketone (PEEK)	400/700	750/100	1600	110

Table 5 Thermoset and thermoplastic trade-offs for commercial aircraft composites

Property	Thermosets	Thermoplastics
Resin cost	Low	Low to high
Prepregability	Excellent	Poor (new methods such as emulsions could change this)
Prepreg tack/drape	Excellent	None (revised lay-up techniques are required)
Volatile-free prepreg	Good	Good to excellent
Prepreg shelf life and out-time	Poor	Excellent
Prepreg quality assurance	Fair	Excellent
Prepreg cost	Good	High (new methods needed)
Composite processing	Slow	Slow (unless automated processes are developed)
Shrinkage	Moderate	Low
Composite mechanical properties	Good (room for improvement in damage tolerance)	Good (more data and experience needed)
Interlaminar fracture toughness	Low	High
Resistance to fluids/solvents	Good	Poor to good
Resistance to creep	Good	Currently not known
Crystallinity problems	None	Yes

vironmental integrity, and solvent and creep resistance.

Chemical Structure and Properties of High-Performance Thermoplastics

Many definitions of thermoplastics include two principal descriptors: being linear and being repeatedly meltable (or reprocessable). Both features have important ramifications in the use of thermoplastics as matrix resins for composites. Because they are linear, rheological and solubility behavior must be taken into consideration if these resins are to be used with reinforcing fiber. Also, for aircraft applications, appropriate structural or processing modifications must be made in order for the resin to resist solvents such as hydraulic fluid, fuel, and methylene chloride used in paint strippers.

The most effective means of imparting solvent resistance is to establish order (crystallinity) in the polymer structure. The intermolecular forces of crystallinity offset the solvating strength of liquids. Reproducible control of crystallinity then becomes important because crystallinity also profoundly affects modulus. Consequently, cool-down times significantly affect the degree of crystallinity, which is largely a function of the polymer organizing time in the repeating pattern of the crystal. If a polymer is cooled rapidly through the temperature range where solidification occurs, the viscosity of the melt increases so rapidly that the mobility decreases to such a degree that extensive organization into an ordered crystalline structure cannot occur. As a result, there is little or no crystallinity in the polymer. In contrast, if the cooling is slow, the mobility changes slowly, and the polymer molecule has time to undergo the movement required to form the ordered structure that is resulting from forces of free energy of the forming crystal lattice structure. The importance of this to thermoplastic composite fabrication is clear: The cooling rate is critical, and all factors that affect the cooling rate (such as thickness of the composite) are significant.

Describing thermoplastics as being repeatedly meltable requires qualification. The integrity of the linear molecule in the molten state is subject to disruption by mechanical as well as thermal factors. Shear, for example, can break molecular bonds and reduce the properties of the plastic. Similarly, thermal exposure degrades the properties of the polymer. For example, long high-temperature exposures can drastically decrease polymer integrity. Thermoplastic materials such as acrylonitrile-butadiene-styrene (ABS) and polycarbonate lose their good impact properties if they are extruded in an injection machine when the screw barrel is too hot or the residence time in the screw is too long.

The fact that the molecule of a thermoplastic is linear results in increased elongation before failure. The molecules tend to disentangle, orient, and draw out. There is a large area under the stress-strain curve which is a direct function of the work to failure. In other words, thermoplastic molecules are tougher than the rigid cross-linked network of the thermosets. The elongations of high-performance plastics being considered for advanced composite matrix resins range from 30 to 100%, compared to 1 to 2% for thermosets such as 5208.

The elongation and toughness of a thermoplastic are modified by the degree of crystallinity. The higher the degree of crystallinity, the higher the modulus and the higher the resistance to solvents and water. Some trade-offs between solvent resistance and mechanical properties are required to establish the extent of crystallinity. Other behavioral phenomena with linear molecules cause some concern when thermoplastics are to be used in fiber-reinforced composites. For example, these linear molecules display a memory of their chain configuration in the melt; this reduces the required entanglement of polymers in adjacent plies in the lamination process. Also, many thermoplastics become brittle when exposed to temperatures approaching the glass transition temperature (T_g), which is attributed to reduction of molecular freedom in the amorphous regions. These amorphous regions also cause creep under load; the linear molecules can move. The boundary between these amorphous regions and crystalline regions represents a discontinuity where failure or microcracking may initiate. This is also true for crystalline-crystalline boundaries. The discontiguities seem to be particularly susceptible to failure in the presence of solvents, especially when under stress. By contrast, none of the foregoing mechanisms apply to thermosets.

The expectations for thermoplastics are optimistic, but the idiosyncrasies of their molecular structures must be dealt with in order to use them successfully in advanced composites. The National Research Council formed a National Material Advisory Board Committee to study some of the problems as well as the strengths of thermoplastic composites as structural components. A report is expected to be available in 1987.

Figure 5 shows the chemical structures of several engineering thermoplastics potentially useful as matrix systems for advanced composites. Other pertinent data are given in Table 6. Because there are several different forms of these thermoplastic resins, the properties listed vary from one form to another.

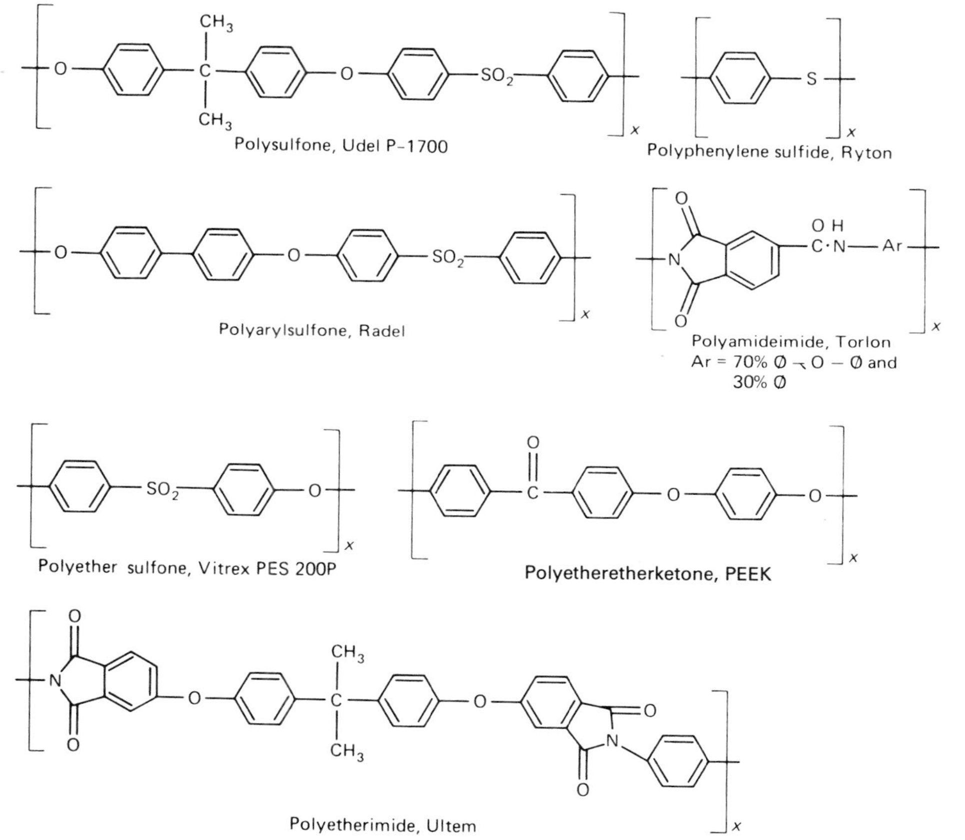

Problems of Impregnation

In both thermoplastic and thermoset prepreg material, each fiber bundle must be completely wetted with resin. In the case of thermoplastic resins, attempting to impregnate from the melt is difficult. Most thermoplastics have number average molecular weights of at least 20 000, and the melt viscosities of molecules of that size are so high that their movement between fibers makes reasonable process times virtually impossible. Increasing the temperature lowers the viscosity, but can cause decomposition before sufficiently low viscosities are realized. It is not realistic to assume that wetting of the fiber bundle will be completed during the fabrication process; the flow dynamics in that situation do not permit sufficient permeation of the bundle by the molten resin. (See Table 7 for typical melt viscosities.) The difficulty of melt impregnation of thermoplastics has been demonstrated by several unsuccessful attempts to build prepreg machines for this purpose.

Thermoplastics, depending on their chemical structure, molecular weight, and thermal his-

Fig. 5 Chemical structures of high-performance thermoplastics

Table 6 Properties of thermoplastics

Material	Supplier	Cost, $/lb	T_g °C	T_g °F	T_m °C	T_m °F	Tensile Strength MPa	Tensile Strength ksi	Tensile Modulus MPa	Tensile Modulus ksi	Fracture toughness, G_{Ic} J/m²	Fracture toughness, G_{Ic} ft·lbf/ft²
Udel P-1700	Union Carbide	4.50	190	375	None	...	76	11	2200	320	3200	220
Radel	Union Carbide	...	220	430	None	...	...	...	...	...	5500	380
Vitrx PES 200P	ICI	...	220	430	None	...	83	12	2410	350	2600	180
Ultem	General Electric	6.80	220	430	None	...	110	16	3300	480	3700	250
Torlon	Amoco	22.00	275	530	None	...	193	28	4800	700	3400	230
Ryton	Phillips	...	85	185	285	545	65	9.5(a)	3800	550(a)	210(b)	15
PEEK	ICI	28.00	143	290	343	650	...	...	...	...	...	...
Avimid K-II	Du Pont	...	277	530	None	...	110	16	2850	415	14000	960
CM-X	Ausimont U.S.A.	100.00	...	...	327	620	37	5.5(b)	3800	550	(b)	...

(a) 15% crystallinity. (b) High crystallinity

Table 7 Melt viscosity of thermoplastics

Material	Temperature °C	°F	Viscosity(a) Pa·s	P
Udel P-1700	240	460	10^5	10^6
	300	570	10^4	10^5
	400	750	10^3	10^4
Ultem	305	580	10^5	10^6
Ryton	313	595	10^4	10^5
	328	620	10^3	10^4
Torlon............	350	660	10^6	10^7
Peek	400	750	10^3	10^4
For reference:				
Molasses..........	25	80	10^2	10^3
5208	100	212	10^{-1}	10^0

(a) Viscosity at low shear rates

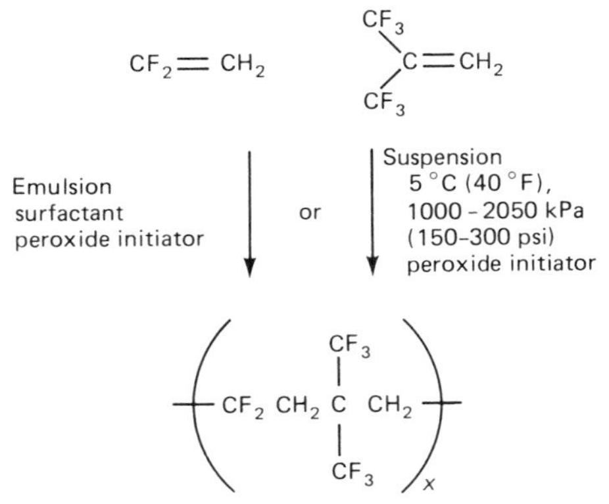

Fig. 6 Preparation of CM-X

tory, undergo various degrees of shear thinning at higher temperatures; they are not Newtonian in their flow/shear behavior. This shear thinning can result in a significant reduction in viscosity, by several orders of magnitude. Thus, a large investment in equipment is required to perform high-shear impregnation. This approach is said to be used in the impregnation of carbon fiber with Peek by forcing the fiber and resin through a die at high temperatures under conditions that create high shear rates, possibly supplementing this with laser heating. The chemical behavior of Peek limits its use for other methods of impregnation; it is insoluble, and solvent methods are unworkable.

The shear-thinning method of impregnation might be suitable for other thermoplastics, but problems of incomplete wetting could be encountered. No reports on the use of this method with other resins were found. Even if complete wetting of the fiber bundle could be achieved by the high-shear method, another difficulty with thermoplastics would become a problem: The inherently high orientation effects of the flow cause all the molecules to lie primarily in the same direction as the fiber, with undesirable effects on interlaminar shear strength and transverse properties.

Melt impregnation processes must be conducted with complete respect for the thermal endurance of the thermoplastic material, which can otherwise undergo loss of mechanical properties. This seems to be a common tendency of higher molecular weight linear polymers. Attempts to decrease the viscosities of higher molecular weight resins (such as those shown in Table 7) by increasing the impregnation temperature do involve risk.

There is much room for creativity in resolving problems associated with melt impregnation of thermoplastics. One modification consists of stacking a film layer over spread tow, heating above the melt or softening temperature, and forcing the resin into the fiber with nip rolls or platen pressure. This approach does not resolve problems previously mentioned, however, and tow bundle wetting is poor. This approach has also been used to prepare small laminates from several experimental thermoplastics; stacks of alternating unidirectional material (or, more commonly, fabric) and films of the thermoplastic polymer are laid up and then pressed.

Another modification of melt impregnation consists of intermingling fibers of the thermoplastic polymer and the carbon fiber. This mixed fiber rope is then melted or softened, and effective impregnation is achieved depending on the degree of randomness of the intermingling of the thermoplastic fiber throughout the system. The spinning process to form the thermoplastic fiber does represent an added cost. The concept has interesting possibilities, although some of the disadvantages of molecular orientation remain. Drawn, or spun, fiber owes its unique strength to its highly oriented nature, which would not be lost while melting among the carbon fibers. Nevertheless, this approach could be useful with highly crystalline polymers and polymers that are otherwise insoluble. It seems to be particularly advantageous in the pultrusion of carbon fiber reinforced shapes, for example, sucker rods.

The use of polyethyleneterephthalate or polybutyleneterephthalate has some cost advantages that may outweigh the cost of fiberization. In addition, it is believed by many in this field that the temperature requirements for thermoplastic composites may be too high and that thermoplastic matrix composites with a lower temperature capability may be suitable in some applications, such as commercial aircraft, which do not need the higher temperature capability in many components. Because such materials would require solvent resistance, polymer chemical structure would have to be considered. The questions of creep and embrittlement need to be resolved, but the list of polymers currently popular for thermoplastic composites may include materials with unnecessary temperature capability. If this is true, the use of lower-cost fibers in the intermingling approach to impregnation may be possible.

A variation of the fiber intermingling method is the intermingling of finely powdered resin with the fiber tow. Resin powder having a particle size that is the same or smaller than the diameter of the fiber is powder coated onto the spread tow, and the polymer is then melted to form prepreg. Alternately, fabric can be powder coated but not melted, which forms a drapeable ply with no tack. One company has reported making such a product using Nylon 12 and Nylon 6. The fiber or fabric is appropriately sized and passed through a fluidized bed of powdered resin having a particle size ranging from 8 to 20 μm (310 to 790 μin.). There is a cost problem in the production of powders with this particle size. Of course, the polyamides would not be an appropriate choice for moisture or solvent resistance, but the approach is interesting.

One way to make powdered polymers with a very small particle size is to synthesize the resin as an emulsion. A resin emulsion system could then be dispersed into the tow as a variation of the intermingling of solid powder with fiber. Sufficiently powerful surfactants are available to produce forming-polymer particle sizes that are truly colloidal. The use of such colloidal dispersions, or nearly colloidal dispersions, to impregnate carbon fiber does not appear to present any insurmountable difficulties. Concern about the possible presence of surfactant in the final composite can be resolved by the use of available volatile surfactants, which could be evolved during the melt step of the prepreg formation. If the continuous phase in the emulsion is water, attractive possibilities are presented in terms of environmental and health considerations. There are also advantages in the handling of carbon fiber tow inasmuch as deionized water has been shown to be a useful agent in such fiber handling.

In the late 1970s, one chemical company examined a one-to-one, head-to-tail copolymer of hexafluoroisobutylene and vinylidene fluoride, designated CM-X (Fig. 6). This polymer was produced as an emulsion and could be used as such for impregnation. A major aircraft company examined the system, successfully prepared prepreg from the aqueous emulsion with the use of a volatile surfactant, and judged the resulting composites of great potential. The polymer CM-X is a high-modulus resin with high crystallinity and solvent resistance. Table 8 compares its properties to those of epoxy and polysulfone. The emulsion approach is a viable resolution to the apparently contradictory problem of solvent impregnation and solvent resistance. Undoubtedly, the candidate thermoplastic polymers would have to be synthesized by emulsion methods.

Regarding the use of fiber, powder, and emulsion as impregnation forms, it should be noted that none of these methods has the problem of high viscosity in the melt. The thermoplastic molecule is small enough to exist in these three states of subdivision and to avoid problems such as flow restriction between molecules and molecular entanglement, which cause high melt viscosities. These methods can also provide solvent resistance.

Table 8   Comparative properties of CM-X, epoxy, and polysulfone

	Density, g/cm³	Melt temperature	Processing temperature		Tensile strength		Tensile modulus		Izod impact strength		Hardness	Heat distortion temperature at 1800 kPa (264 psi)		Water absorption, %	Creep modulus at 170 °C (340 °F)		Continuous service temperature		Flammability limiting oxygen index	Solvent resistance
			°C	°F	MPa	ksi	MPa	ksi	ft·lbf/in.	J/m		°C	°F		kPa	psi	°C	°F		
Epoxy(a)	1.20	Decomposes	175	350	41	6	4500	650	0.2	10.7	M120	160	320	0.1-0.4	1400	200	204	400	<29	Fair
Polysulfone(b) . .	1.24	Amorphous	315	600	70	10.2	2500	360	1.3	69.4	M69	174	345	0.3	950	140	204	400	31	Attacked
CM-X(b)	1.88	327 °C (620 °F)	340	650	38	5.5	3800	550	0.4	21.4	R-116	220	428	0.01	9150	1330	288	550	60	Resistant

(a) Thermoset. (b) Thermoplastic.

Table 9   Government-funded thermoplastic programs (1985)

Agency	Contractor	Contract number	Funding, thousands of $/year	Comment
NASA Langley	Aerotherm/Acurex	NAS1-16808	100	Thermoplastic polyimides as tough solvent-resistant matrix resins
NASA Langley	Hercules, Inc.	NAS1-17918	139	Interfacial adhesion studies
NASA Langley	UTRC	NAS1-16914	80	Mechanical properties of matrix resins
NASA Langley	Boeing Aerospace Co.	NAS1-17432	68	Comparative properties of advanced thermoplastic polyimides
NASA Langley	AM Tech Industries	NAS1-17858	36	Thermoplastic prepreg
NASA Langley	Arroyo Research	Pending	100	Resin and prepreg preparation
U.S. Navy	Boeing Aerospace Co.	N62269-83-C-0262	80	Development and optimization of modified polysulfone
U.S. Navy	Boeing Aerospace Co.	N00019-84-C-0134	180	Nadic-terminated polysulfone
U.S. Navy	Boeing Aerospace Co.	N62269-84-R-0282	80	Thermoplastics for repair

Much of the work that has been conducted on impregnation with thermoplastics has been done with solvents, despite the fact that if a polymer is soluble, it cannot have solvent resistance. Thermoplastics usually exhibit limited solubility; at higher solids content, high solution viscosities result, causing the same impregnation problems as with hot melts. As a result, multiple passes from a low-solids low-viscosity solution are usually required to obtain adequate resin pick-up and good wetting of the fiber bundle. Sometimes the solvent required to dissolve a particular resin may fail to comply with environmental restrictions; present health, safety, and odor problems; or be difficult to remove from the polymer (affinity). In the latter case, traces of solvent left in the prepreg, either intentionally or unintentionally, lead to undesirable porosity in the composite; attempts have been made to leave some solvent in thermoplastic prepreg in the expectation of improving processing by improving flow. A newer method that shows promise for the preparation of thermoplastic prepreg involves the use of an effective plasticizer to lower the melt temperature and viscosity. The plasticizer must be carefully chosen to provide the proper degree of behavior alteration and still be removable during a heated posttreatment of the prepreg. A possible way to avoid any problem of removal would be to use a reactive plasticizer that would become a part of the polymer matrix.

As part of the impregnation process, the interface is important in obtaining high translation of fiber properties in the thermoplastic composite. The interface problem, along with other problems, is being studied in thermoplastic programs funded by NASA and the U.S. Navy (Table 9). It seems clear that the sizings used on carbon fibers for thermosetting prepreg would not be appropriate for thermoplastic prepreg. Sizings such as epoxy resins are consistent for the thermosets and are probably washed off to become part of the new polymerizing resin, which would not be true of the thermoplastics.

Fabrication With Thermoplastic Prepreg

Thermoplastic composite production is similar in many respects to sheet metal forming processes. Consequently, it should be possible to borrow manufacturing technology from the metal-forming industry.

Several approaches to fabrication have been considered (Ref 6). Basically, the fabricator would be supplied with one or more of the following forms manufactured by the prepregger: thin single-ply fabric or unidirectional material on a roll (probably having a large diameter), flat sheets of single-ply fabric or unidirectional material, or flat sheets of multiple-ply fabric or unidirectional material (including preoriented plies). The fabricator would heat these precut shapes either in flat form or supported with a minimal frame. The prepreg would be heated above its softening temperature in an oven or other unit and then removed to a hard tool, perhaps a type of matched tool, which would be held at a temperature slightly below the flow temperature. The tool would then be closed, and the appropriate pressure applied for cure. Soon thereafter, the tool would be opened, and the formed part (now slightly cooled and handle-able) would be removed from the tool for trim and subsequent operations.

This simplistic description could be modified for the fabrication of parts of varying thickness by stacking plies before the first heat-up. Plies could be held in place by spot welding or ironing. Because the composite would be formed in a net flow manner, uniform thickness would be required in the prepreg ply.

Attractive cost reductions to the fabricator would result from:

- Reduced capital investment
- Reduced refrigeration
- Reduced handling of refrigerated rolls
- Elimination of in-and-out transfers and record keeping
- Elimination of warm-up time from refrigeration
- Elimination of quality control to follow aging of prepreg
- Reduced tool preparation between parts
- Elimination of bagging procedures and bagging materials
- Reduced cure times, heat-ups, and cooldowns
- Elimination of autoclave problems (such as excessive energy, fires, venting, leaks, and pumps)
- Reduced shipping costs

Under the first thermoplastic composite study funded by the U.S. Navy, the contractor reported optimistically that the extra cost of tooling could be recovered by these cost reductions, when manufacturing as few as ten of the parts on an aircraft that appear in numerous counts. Such parts seem to be particularly attractive for thermoplastic fabrication, for example, a seat part of an interior window panel.

Automated lay-up processes are believed by many to be the logical approach to labor costs in thermoplastic prepreg fabrication. However, aircraft manufacturers, unlike automotive manufacturers, may be reluctant to make the large capital investment in equipment required for sophisticated thermoplastic composite production. Therefore, some preppregers have postponed entrance into the thermoplastic prepreg business.

If the fabricator chooses to use preplied thermoplastic stock, the capital investment problem shifts to the prepregger. The production of preplied and oriented prepreg will require some expensive automated equipment.

The position of the prepregger then is like that of the sheet metal producer, with fabricators certain to want similar credibility, quality assurance, and traceability. Thermoplastic materials have the potential to be much better in those respects than thermoset prepregs.

In addition, it is likely that the prepregger would be required to supply a variety of widths, thicknesses, ply counts, ply orientations and so on. Intermediate processors might then spring up, especially if the prepreggers did not have the capability of producing these variations economically by programming them into their computer controlled production equipment. The transverse integrity of unidirectional thermoplastic prepreg would be an advantage in the production of such variations in form.

Intermediate processors might also find an attractive market in such preformed shapes as hat sections, beams, and other varied cross-sectional lengths. Skins with integral stiffeners will be evolving in thermoplastic components, and the stiffeners can be bonded or thermoplastically welded to the skins. Thermoplastic composites do not appear to be as attractive for honeycomb structures, although thermoplastic skins would be feasible.

REFERENCES

1. J.G. Williams, T.K. O'Brien, and A.J. Chapman III, Conference Publication 2321, National Aeronautics and Space Administration, 1984
2. S. Oken and J.J. Hoggatt, AFWAL-TR-80-3023, Air Force Wright Aeronautical Laboratories, 1980
3. M.G. Maximovich, *SAMPE Ser.*, Vol 19, Society for the Advancement of Material and Process Engineering, 1974, p 262
4. E.J. Stober, J.C. Seferis, and J.D. Keenan, *Polymer*, Vol 25, 1984, p 1845
5. P.E. McMahon and L. Ying, Contractor Report 3607, National Aeronautics and Space Administration, 1982
6. G.R. Griffiths *et al.*, *SAMPE J.*, Vol 20 (No. 32), 1984
7. "Standard Tests for Toughened Resin Composites," Reference Publication, Revised Edition, National Aeronautics and Space Administration, 1983, p 1092

SECTION 3

Constituent Material Forms

Chairman: John C. Weidner, Hercules Aerospace Company

Introduction

THIS SECTION addresses those types of raw materials that are available for use in fiber-reinforced polymeric matrix composites. It offers the engineer a shopping list that complements the Sections in this Volume that focus on design, manufacturing, and material properties. A key advantage of working with composite materials is the opportunity to integrate material properties, design, and manufacturing technique so that the end product—a completed structure—is optimized from both a performance and an economics standpoint.

The individual articles within this Section are written to give an understanding of the composite raw materials available today and how they are processed. Because there is a great deal of flexibility in the manufacture of composite material forms, the engineer is encouraged to be creative when selecting a particular material form. If a particular form fits a particular design and manufacturing technique but is not listed here, he is encouraged to ask for what he wants.

Most of the forms listed are the result of such requests. For example, the tow sizes available for glass and carbon fibers resulted from particular needs. In the first case, large bundles of glass were needed to feed choppers to make sheet molding compounds. After the request was made and the potential market was found to be significant, large multiple-strand bundles in center-pull packages were developed. In the case of carbon fibers, the need for woven fabrics to form complex shapes brought a request for smaller tows. The resulting 1000-, 3000-, and 6000-filament tows are now used to produce woven goods.

In another example, resin systems reinforced with glass, carbon, or boron fibers were initially low-temperature (80 °C, or 180 °F) performing systems. As composites moved from low-performance structures to high-performance primary structures, the temperature requirement increased to 175 °C (350 °F). The resin system that resulted from the need for higher-temperature resins performs at 175 °C (350 °F) and above.

Similar examples can be cited for prepreg tape width, resin content, and thickness. The point is that constituent raw-material forms can be adjusted to meet specific requirements. This is important to remember when considering the overall system for producing a composite structure.

Cost is always a key consideration in material selection. Some guidelines to keep in mind are:

- *Fiber:* the fewer filaments per tow, the higher the cost
- *Resin:* the lower the temperature performance, the lower the cost
- *Prepreg:* the wider the product (tape or fabric), the lower the cost

The proper selection of material form to fit with the structure and process design is critical. Using continuous reinforcement where discontinuous will do the job can increase the cost of the final structure to the point where it is non-competitive. The same logic applies to tape versus fabric, carbon versus glass, glass versus aramid, etc.

Glass Fibers

James C. Watson and N. Raghupathi, PPG Industries, Inc.

GLASS FIBER USE dates back to the ancient Egyptians, who used coarse fibers drawn from heat-softened glass to manufacture vessels by wrapping the fibers around clay mandrels and heating them to fuse the glass into a solid container. The clay mandrel was removed after the glass cooled. In the mid-1600s, English physicist Robert Hooke described experiments with glass filaments, and in the next century, French scientist René Antoine Ferchault de Réaumur predicted that fine glass fibers would be pliable enough to weave into fabric. Glass fabric was made in 1893 by Edward Libbey, the founder of Libbey Glass Company; a dress made from this fabric was exhibited at the Columbian Exposition held that year in Chicago. It was not until 1931 that a commercial process for the production of glass fibers was demonstrated by the Owens Illinois Glass Company. By 1936 the company had developed a process for continuous glass filament production. Owens Illinois merged with Corning Glass Works in 1938 to form Owens-Corning Fiberglas. This marked the beginning of the modern fiberglass industry. In the late 1940s and early 1950s, Owens-Corning granted licenses to Pittsburgh Plate Glass (now PPG), Libbey-Owens-Ford, St. Gobain in France, and Pilkington in England (Ref 1-3).

Glass Composition

Glass is an amorphous material obtained from the molten (melt) state by cooling the liquid at a rate such that no ordered regions (known as crystals) are formed. Chemically, glass is essentially composed of a silica network. However, pure silica, or quartz, requires very high temperatures before it can be melted and drawn into fibers. Therefore, other chemical components are added to decrease the glass viscosity to levels suitable for melting, homogenizing, removal of gaseous inclusions, and fiberizing. The physical properties of the resultant glass are altered to varying degrees by the type and amount of modifiers.

Although a number of glass compositions have been developed, only a few are used commercially to create continuous glass fibers. The four main glasses used are high alkali (A-glass), electrical grade (E-glass), a modified E-glass that is chemically resistant (ECR-glass), and high strength (S-glass). High-alkali glass, essentially consisting of soda-lime-silica, is used in applications such as windows and containers. The fiber form of this composition is used in applications requiring good reinforcing properties coupled with good chemical resistance. For applications requiring excellent electrical properties as well as durability, an E-glass composition is used. This glass is chemically a calcium alumino-borosilicate material with a low alkali oxide content. E-glass fiber is the material most widely used as a reinforcing medium for plastic as well as for textile fiberglass product applications. ECR-glass fibers are used in applications that require good electrical properties, coupled with better chemical resistance. S-glass fibers are used in those applications requiring higher tensile strength and higher thermal stability. S-glass is a magnesium alumino-silicate glass that essentially contains no boron oxide. The representative chemical compositions of these four glasses are given in Table 1.

In terms of mechanical properties, S-glass fiber has the highest tensile strength and elastic modulus of these four glasses. Its elevated temperature property retention is also better than that of any of the other glass fibers. However, it is expensive and is therefore used in those applications where cost-performance benefits can be justified. The inherent properties of the four glass fibers are given in Table 2.

Most S-glass fiber is used in the aircraft/aerospace industry. ECR-glass and S-glass fibers show excellent resistance to moisture over a period of time, compared to A-glass. E-glass is predominantly used as a reinforcement in printed circuit boards because of its excellent electrical properties, superior dimensional stability, good moisture resistance, and lower cost. The same attributes also make it the preferred reinforcement in the plastics industry, as mentioned above.

Many other glass compositions have been formulated for specific applications, two of which are minor commercial entities and several of which have not been commercialized. The commercial compositions are chemical-resistant C-glass and high-strength high-modulus R-glass. The noncommercialized ones are high-modulus glass compositions (110 GPa, or 16×10^6 psi) based on beryllium oxide, lead glass (for radiation protection), D-glass (for its low dielectric constant), and lithium-oxide-based glass (for x-ray transparency).

Glass Melting and Forming

The glass melting process begins with the weighing and blending of the appropriate raw

Table 1 Glass composition

Glass type	Silica	Alumina	Calcium oxide	Magnesia	Boron oxide	Soda	Calcium fluoride	Total minor oxides
E-glass	54	14	20.5	0.5	8	1	1	1
A-glass	72	1	8	4	...	14	...	1
ECR-glass	61	11	22	3	...	0.6	...	2.4
S-glass	64	25	...	10	...	0.3	...	0.7

Table 2 Inherent properties of glass fibers

	Specific gravity	Tensile strength MPa	ksi	Tensile modulus GPa	10⁶ psi	Coefficient of thermal expansion, 10⁻⁶/K	Dielectric constant(a)	Liquidus temperature °C	°F
E-glass	2.58	3450	500	72.5	10.5	5.0	6.3	1065	1950
A-glass	2.50	3040	440	69.0	10.0	8.6	6.9	996	1825
ECR-glass	2.62	3625	525	72.5	10.5	5.0	6.5	1204	2200
S-glass	2.48	4590	665	86.0	12.5	5.6	5.1	1454	2650

(a) At 20 °C (72 °F) and 1 MHz. Source: Ref 4

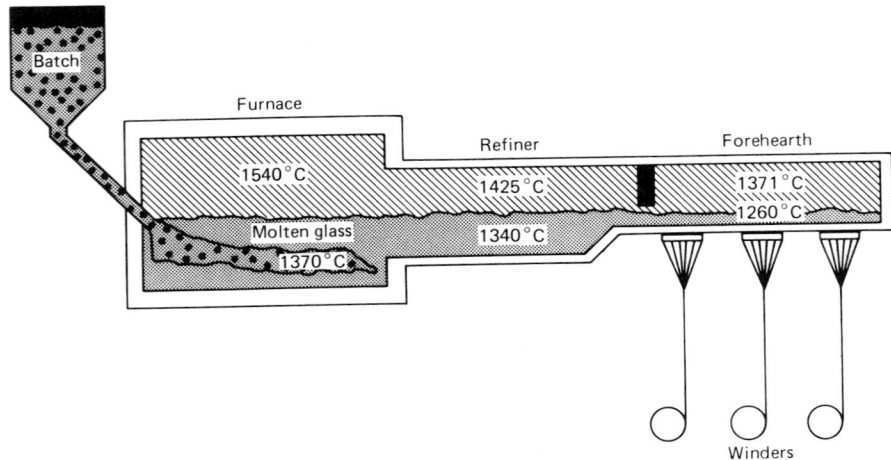

Fig. 1 Glass melting

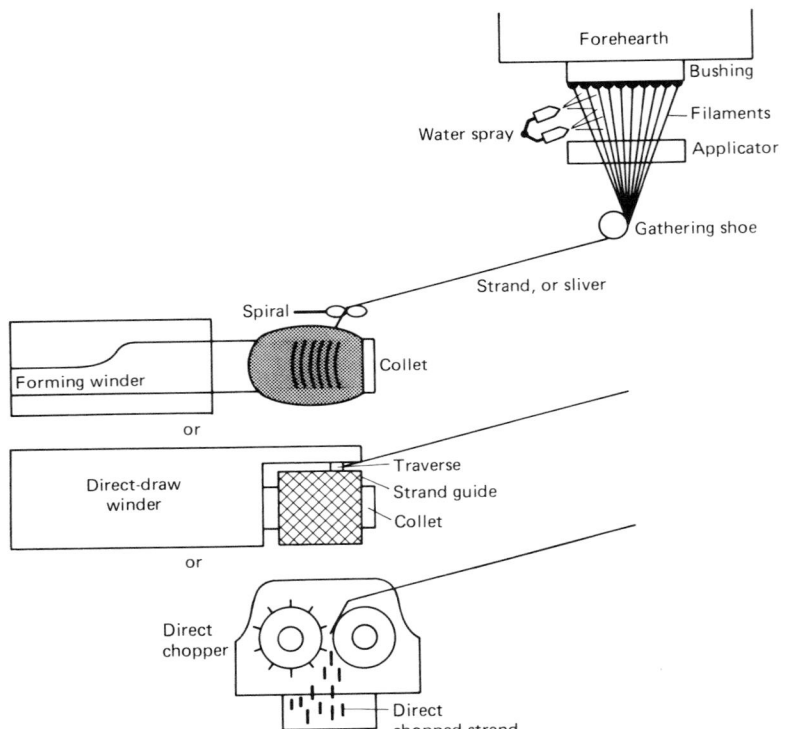

Fig. 2 Fiberglass forming process

bushing with a large number of holes or tips (400 to 8000, in typical production). The bushing is heated electrically, and the heat is controlled very precisely to maintain a constant glass viscosity. The fibers are drawn down and cooled rapidly as they exit the bushing. A sizing is then applied to the surface of the fibers by passing them over an applicator that continually rotates through the sizing bath to maintain a thin film through which the glass filaments pass. It is this step, in addition to the original glass composition, which primarily differentiates one fiberglass product from another. The components of the sizing impart strand integrity, lubricity, resin compatibility, and adhesion properties to the final product, thus tailoring the fiber properties to the specific end-use requirements. After the sizing is applied, the filaments are gathered into a strand before approaching the take-up device. If smaller bundles of filaments (split strands) are required, multiple gathering devices (often called shoes) are used.

The attenuation rate, and therefore the final filament diameter, is controlled by the take-up device. Fiber diameter is also impacted by bushing temperature, glass viscosity, and the pressure head over the bushing. The most widely used take-up device is the forming winder, which employs a rotating collet and a traverse mechanism to distribute the strand in a random manner as the forming package grows in diameter. This facilitates strand removal from the package in subsequent processing steps, such as roving or chopping. The forming packages are dried and transferred to the specific fabrication area for conversion into the finished fiberglass roving, mat, chopped strand, or other product. In recent years, processes have been developed to produce finished roving or chopped products directly during forming, thus leading to the term direct draw roving or direct chopped strand. Special winders and choppers designed to perform in the wet-forming environment are used in these cases (Fig. 2).

Filament Diameter Nomenclature. It is standard practice in the fiberglass industry to refer to a specific filament diameter by a specific alphabet designation, as listed in Table 3. Fine fibers, which are used in textile applications, range from D through G. One reason for using fine fibers is to provide enough flexibility to the yarn to enable it to be processed in high-speed twisting and weaving operations. Conventional plastics reinforcement, however, uses filament diameters that range from G to T.

Fabrication Process

Once the continuous glass fibers have been produced they must be converted into a suitable form for their intended composite application. The major finished forms are continuous roving, woven roving, fiberglass mat, chopped strand, and yarns for textile applications.

materials. In modern fiberglass plants, this process is highly automated, with computerized weighing units and enclosed material transport systems. The individual components are weighed and delivered to a blending station, where the batch ingredients are thoroughly mixed before being transported to the furnace.

Fiberglass furnaces generally are divided into three distinct sections (Fig. 1). Batch is delivered into the furnace section for melting, removal of gaseous inclusions, and homogenization. Then, the molten glass flows into the refiner section, where the temperature of the

glass is lowered from 1370 °C (2500 °F) to about 1260 °C (2300 ° The molten glass next goes to the forehearth section, located directly above the fiber-forming stations. The temperatures throughout this process are determined by the viscosity characteristics of the particular glass. In addition, the physical layout of the furnace can vary widely, depending on the space constraints of the plant.

The conversion of molten glass in the forehearth into continuous glass fibers is basically an attenuation process (Fig. 2). The molten glass flows through a platinum-rhodium alloy

Table 3 Filament diameter nomenclature

Alphabet	Filament diameter	
	μm	10^{-4} in.
AA	0.8 - 1.2	0.3 - 0.5
A	1.2 - 2.5	0.5 - 1.0
B	2.5 - 3.8	1.0 - 1.5
C	3.8 - 5.0	1.5 - 2.0
D	5.0 - 6.4	2.0 - 2.5
E	6.4 - 7.6	2.5 - 3.0
F	7.6 - 9.0	3.0 - 3.5
G	9.0 - 10.2	3.5 - 4.0
H	10.2 - 11.4	4.0 - 4.5
J	11.4 - 12.7	4.5 - 5.0
K	12.7 - 14.0	5.0 - 5.5
L	14.0 - 15.2	5.5 - 6.0
M	15.2 - 16.5	6.0 - 6.5
N	16.5 - 17.8	6.5 - 7.0
P	17.8 - 19.0	7.0 - 7.5
Q	19.0 - 20.3	7.5 - 8.0
R	20.3 - 21.6	8.0 - 8.5
S	21.6 - 22.9	8.5 - 9.0
T	22.9 - 24.1	9.0 - 9.5
U	24.1 - 25.4	9.5 - 10

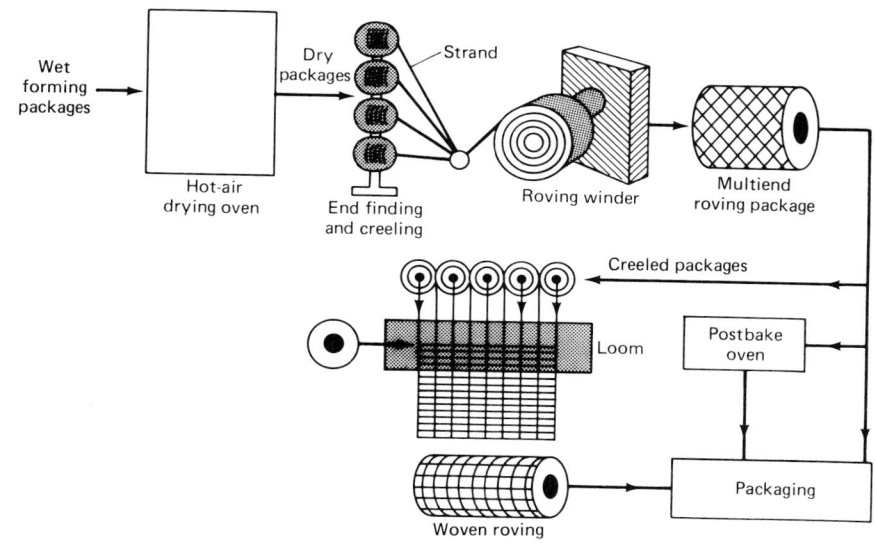

Fig. 3 Multiend roving processproduction

Fiberglass roving is produced by collecting a bundle of strands into a single large strand, which is wound into a stable, cylindrical package. This is called a multi-end roving process. The process begins by placing a number of oven-dried forming packages into a creel. The ends are then gathered together under tension and collected on a precision roving winder that has a constant traverse-to-winding ratio, called the waywind. This ratio has a significant effect on package stability, strand characteristics, and ease of payout in subsequent operations. The yield (yd/lb) of the finished roving is determined by the number of input ends and the yield of the input strand, or sliver. Final package weight and dimensions can be made to vary widely, depending upon the required end-use. Figure 3 shows the entire process.

Rovings are used in many applications. When used in a spray-up fabrication process, the roving is chopped with an air-powered gun that propels the chopped-glass strands to a mold while simultaneously applying resin and catalyst in the correct ratio. This process is commonly used for bath tubs, shower stalls, and many marine applications. In another important process, the production of sheet molding compound (SMC), the roving is chopped onto a bed of formulated polyester resin and compacted into a sheet, which thickens with time. This sheet is then placed in a press and molded into parts. Many fiber-reinforced plastic automotive body panels are made by this process.

Filament winding and pultrusion are processes that use single-end rovings in continuous form. Applications include pipes, tanks, leaf springs, and many other structural composites. In these processes the roving is passed through a liquid resin bath and then shaped into a part by winding the resin-impregnated roving onto a mandrel or by pulling it through a heated die. Because of a property called catenary (the

presence of some strands that have a tendency to sag within a bundle of strands), multiend rovings sometimes do not process efficiently. Catenary is caused by uneven tension in the roving process, which results in poor strand integrity. While providing desirable entanglement for transverse strength in pultrusion, the looser ends in the roving may eventually cause loops and breakouts in close-tolerance orifices, making reinforced plastics processing difficult. Consequently, the fiberglass manufacturing process of direct forming single-end rovings was developed, by using very large bushings and a precision winder specially designed to operate in the severe forming environment. No subsequent step other than drying is required. Single-end rovings have become the preferred product for many filament-winding and pultrusion applications.

Woven roving is produced by weaving fiberglass rovings into a fabric form. This yields a coarse product that is used in many hand lay-up and panel molding processes to produce fiber-reinforced plastics. Many weave configurations are available, depending upon the requirements of the laminate. Plain or twill weaves provide strength in both directions, while a unidirectionally stitched or knitted fabric provides strength primarily in one dimension. Many novel fabrics are currently available, including biaxial, double bias, and triaxial weaves for special applications.

Fiberglass mats may be produced as either continuous- or chopped-strand mats. A chopped-strand mat is formed by randomly depositing chopped fibers onto a belt or chain and binding them with a chemical binder, usually a thermoplastic resin with a styrene solubility ranging from low to high, depending on the application. For example, hand lay-up processes used to make corrosion-resistant liners or boat hulls require high solubility, whereas closed-mold processes such as cold

press or compression molding require low solubility to prevent washing in the mold during curing. Continuous-strand mat is formed in a similar manner, but without chopping, and usually less binder is required because of increased mechanical entanglement, which provides some inherent integrity. Continuous-strand mat may be used in closed-mold processes and as a supplemental product in unidirectional processes such as pultrusion, where some transverse strength is required.

A number of specialty mats are also produced. Surfacing veil made with C-glass is used to make corrosion-resistant liners for pipes and tanks. Surfacing veils made from other glass compositions are used to provide a smooth finished surface in some applications. Glass tissue is used in some vinyl flooring products.

Combinations of fiberglass mat and woven roving have been developed for specific products in recent years. In many lay-up processes the laminate is constructed from alternate layers of fiberglass mat and woven roving. Fiberglass producers thus began to provide products that make this process more efficient. The appropriate weights of fibergass mat (usually chopped-strand mat) and woven roving are either bound together with a chemical binder or mechanically knit or stitched together. This product can then be used as a significant labor saver by the fabricators.

Chopped-Strand Products. Chopped strands are produced by two major processes. In the first process, dried forming packages are used as a glass source. A number of strand ends are fed into a chopper, which chops them into the correct length, typically 3.2 to 12.7 mm (1/8 to 1/2 in.). The product is then screened to remove fuzz and contamination, and boxed for shipment (Fig. 4). The second process, used in recent years to produce many chopped-strand products, is the direct-chop process, in which large bushings are used in forming, and the

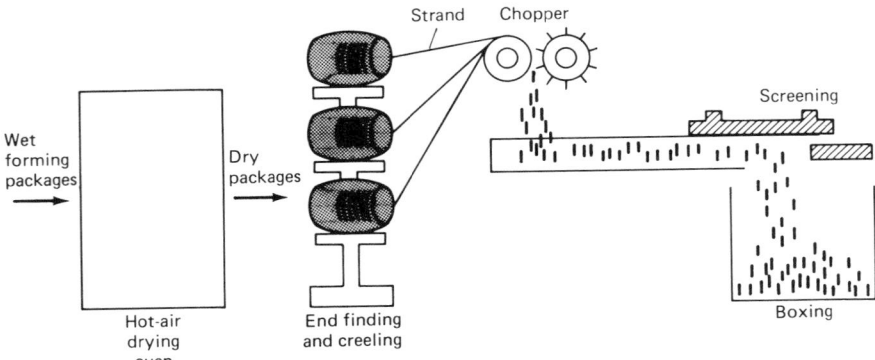

Fig. 4 Chopped-strand production

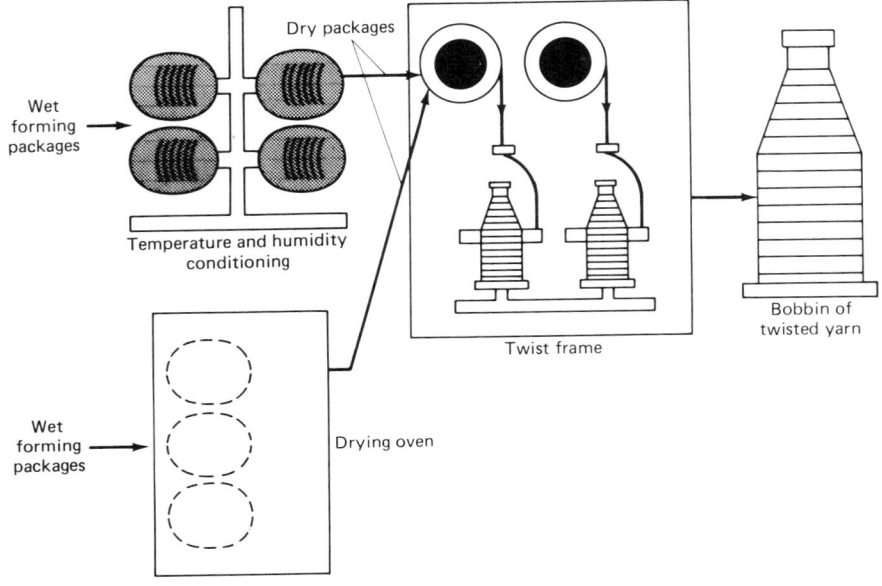

Fig. 5 Twisting

ing resin is added on-line, followed by drying and curing, to form the fiberglass paper. This paper is then combined with the appropriate resin system to form roofing shingles.

Textile Yarns. Fine-fiber strands of yarn from the forming operation are air dried on the forming tubes to provide sufficient integrity to undergo a twisting operation. Twist provides additional integrity to yarn before it is subjected to the weaving process, a typical twist consisting of up to one turn per inch. The twisting operation is shown in Fig. 5. In many instances heavier yarns are needed for the weaving operation. This is normally accomplished by twisting together two or more single strands, followed by a plying operation. Plying essentially involves retwisting the twisted strands in the opposite direction from the original twist. The two types of twist normally used are known as S and Z, which indicate the direction in which the twisting is done. Usually, two or more strands twisted together with an S twist are plied with a Z twist in order to give a balanced yarn. Thus, the yarn properties, such as strength, bundle diameter, and yield, can be manipulated by the twisting and plying operations.

Yarn Nomenclature. Fiberglass yarns are designated by nomenclatures consisting of both alphabets and numbers. For instance, in ECG 75 2/4:

- The first letter specifies the glass composition, in this case, E-glass
- The second letter specifies the filament type (staple, continuous, texturized), in this case, continuous
- The third letter specifies the filament diameter, in this case, G
- The next series of numbers represents the basic strand yield in terms of 1/100th of the yield; in this case, 75 means 7500 yd/lb.
- The fraction represents the number of strands twisted together (numerator) to form a single end and the number of such ends plied together (denominator) to form the final yarn. In the above case, 2/4 means two basic strands are twisted together to form a single end, and four such ends are plied together (usually in the opposite direction) to form the final yarn

For details on commercially available fabrics, refer to product brochures from various weavers as well as to Ref 4.

Fiberglass Fabric. Fiberglass yarns are converted to fabric form by conventional weaving operations. Looms of various kinds are used in the industry, but the air jet loom is the most popular.

The major characteristics of a fabric include its style or weave pattern, fabric count, and the construction of warp yarn and fill yarn. Together, these characteristics determine fabric properties such as drapability and performance in the final composite. The fabric count identifies the number of warp and fill yarns per inch.

strands are chopped in a wet state directly after sizing is applied. The wet, chopped strands are then transported to an area where they are dried, screened, and packaged. The direct-chop process has provided the industry with a wide variety of chopped reinforcements for compounding with resins.

Chopped glass is widely used as a reinforcement in the injection molding industry. The glass and resin may be dry blended or extrusion compounded in a preliminary step before molding, or the glass may be fed directly into the molding machine with the plastic resin. Hundreds of different parts for many applications are made in this manner. Chopped glass may also be used as a reinforcement in some thermosetting applications, such as bulk molding compounds.

Milled fibers are prepared by hammer milling chopped or sawed continuous strand glass fibers, followed by chemically sizing for some specific applications and by screening to length. Fiber lengths typically vary from particulates to screen opening dimensions for the reported nominal length (0.79 to 6.4 mm, or 1/32 to 1/4 in.). As such, milled fibers have a relatively low apsect ratio (length to diameter). They provide some increased stiffness and dimensional stability to plastics, but minimal strength. Their use is primarily in phenolics, reaction injection molded urethanes, fluorocarbons, and potting compounds.

Fiberglass Paper. Chopped strands of 25- to 50-mm (1- to 2-in.) length are usually used in making thin fiberglass mat, known as fiberglass paper, which is the reinforcing element for fiberglass roofing shingles. In this process, chopped fibers are dispersed in water to form a dilute solution. The fiberglass strands filamentize during the mixing and dispersion process. The solution is pumped onto a continuously moving chain, where most of the water is removed by vacuum, leaving behind a uniformly distributed, thin fiberglass mat. A bind-

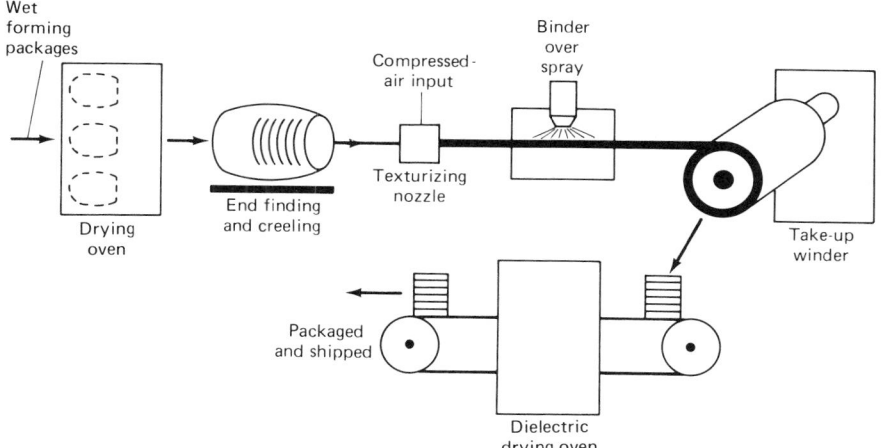

Fig. 6 Texturizing

that impinges on its surface to make the yarn "fluffy" (Fig. 6). The air jet causes the surface filaments to break at random, giving the yarn a bulkier appearance. The extent to which this occurs can be controlled by the velocity of the air jet and the yarn feed rate. The texturizing process allows the resin-to-glass ratio to be increased in the final composite. One of the major applications of texturized yarns is as an asbestos replacement.

Carded Glass Fibers. Carding is a process that makes a staple fiberglass yarn from continuous yarn. The continuous yarn is chopped into 38- to 50-mm (1.5- to 2.0-in.) lengths, then aligned in one direction in a mat form, and finally converted to a staple yarn. The yarn produced by this process can absorb much more resin than texturized yarn. Carded glass fibers are also used as an asbestos replacement in friction applications, such as automotive brake lining.

REFERENCES

1. K.L. Loewenstein, *The Manufacturing Technology of Continuous Glass Fibers*, Elsevier, 1983
2. H.F. Mark, S.M. Atlas, and E. Cernia, Ed., *Man-Made Fibers*, Vol 3, John Wiley & Sons, 1968
3. C.B. Chapman, *Fibres*, Butterworths, 1974
4. G. Lubin, *Handbook of Composites*, Van Nostrand Reinhold, 1982

Warp yarns run parallel to the machine direction, and fill yarns are perpendicular.

There are basically four weave patterns: plain, basket, twill, and satin. Plain weave is the simplest form, in which one warp yarn interlaces over and under one fill yarn. Basket weave has two or more warp yarns interlacing over and under two or more fill yarns. Twill weave has one or more warp yarns floating over at least two fill yarns. Satin weave (crowfoot) consists of one warp yarn interfacing over three and under one fill yarn, to give an irregular pattern in the fabric. The eight harness satin weave is a special case, in which one warp yarn interlaces over seven and under one fill yarn to give an irregular pattern. In fabricating a composite part, the satin weave gives the best conformity to complex contours, followed in descending order by twill, basket, and plain weaves.

Texturized Yarn. Texturizing is a process in which the textile yarn is subjected to an air jet

Carbon Fibers

Niel W. Hansen, Hercules Aerospace Company

CARBON FIBERS, though known since Thomas Edison's development of the incandescent light in the 1870s, were not made in large quantities until the late 1960s. At that time it was found that carbonizing several fibrous materials resulted in a continuous fiber with relatively low density and high Young's modulus of elasticity. This fiber could then be used much as glass fiber had been used: to provide a continuous reinforcement in various resin systems for the fabrication of structural components.

Initial interest was in the aerospace field, where the advantages of low weight and high strength/stiffness are most obvious. Substantial broadening of the application base has taken place since then to include recreational sports equipment as well as industrial and commercial products. The price of carbon fibers has dropped dramatically, and its mechanical properties have increased. In the early 1970s, the cost of carbon fibers exceeded $220/kg ($100/lb). Carbon fibers produced in the 1980s possess more impressive properties and sell for as low as $9/kg ($44/lb). As prices continue to drop and mechanical property values to rise, the number of applications for continuous filament carbon fibers will undoubtedly grow.

Carbon Fiber Feedstocks

Continuous filament carbon fibers are produced by pyrolyzing, or decomposing by heating, fibers that contain enough carbon so that the resultant carbon fiber is both physically and economically attractive. Carbon fibers retain the physical shape and surface texture of the precursor fibers from which they have been made. Commercial quantities of carbon fibers are derived from three major feedstock or precursor sources: rayon, polyacrylonitrile, and petroleum pitch.

Rayon precursors, which are derived from cellulosic materials, were one of the earliest precursors used to make carbon fibers. Their advantages were that they were well characterized and readily available; their most significant disadvantage was a relatively high weight loss, or low conversion yield to carbon fiber. Typically, only 25% of the initial fiber mass remains after carbonization, which means that carbon fiber made from these materials is compar-

atively more expensive than carbon fibers made from other materials.

Polyacrylonitrile (PAN) precursors are the basis for the majority of carbon fibers commercially available today. They provide a carbon fiber conversion yield that is 50 to 55%. These precursors can be thermally rearranged before thermal decomposition, which allows them to be oxidized and stabilized before the carbon fiber conversion process, while maintaining the same filamentary configuration. The chemical composition of PAN precursors defines the thermal characteristics that the material displays throughout the oxidation/stabilization portion of the conversion process. These thermal characteristics influence the processing sequences that are used to convert PAN precursors to carbon fiber. Carbon fiber based on a PAN precursor generally has a higher tensile strength than a fiber based on any other precursor. This is due to a lack of surface defects, which act as stress concentrators and, hence, reduce tensile strength.

Pitch precursors based on petroleum asphalt, coal tar, and polyvinyl chloride can also be used to produce carbon fiber. Pitches are relatively low in cost and high in carbon yield. Their most significant drawback is nonuniformity from batch to batch. One way to alleviate this problem is to preprocess the pitch into a mesophase pitch. Although this reduces overall carbon yield, it does improve both the properties and processibility of the resultant carbon fiber.

Carbon Fiber Conversion Processes

Each precursor category requires different processing techniques. There are several processes that are used by different manufacturers using the same type of precursor, but generally, the precursor-to-carbon-fiber conversion process follows this sequence: stabilization, carbonization, graphitization (optional), surface treatments, application of sizings or finishes, and spooling.

Stabilization is carried out at temperatures <400 °C (<750 °F) in various atmospheres. The fibers are often stressed during this stage of the process to improve the orientation of the

molecular structure and increase carbon fiber strength and modulus. All fiber-handling equipment is designed to minimize any damage that could occur in transit from one sequence to the next throughout the conversion process.

Carbonization is accomplished at temperatures from 800 to 1200 °C (1470 to 2190 °F). The stabilized fibers are pyrolyzed in inert environments to reduce their impurity levels and increase their crystallinity. Fibers may be shrunk or stretched in this step. As in the stabilization step, any increase in orientation generally increases tensile modulus. If this orientation increase does not produce surface defects, tensile strength will increase also.

Graphitization is an additional pyrolysis step that is in excess of 2000 °C (3630 °F) in inert environments. This step also reduces the level of impurities and stimulates crystal growth. With all precursor categories, the higher the process temperatures that are used in carbonization and graphitization, the higher the modulus of the resultant fibers.

Surface Treatments. Various materials can be applied to the carbonized/graphitized

Fig. 1 Carbon fiber spools

Table 1 Typical mechanical property values of commercially available carbon fibers

Product name	Manufacturer	Precursor type	Density, g/cm³	Tensile strength		Tensile modulus	
				GPa	10⁶ psi	GPa	10⁶ psi
AS-4	Hercules, Inc.	PAN	1.78	4.0	0.580	231	33.5
AS-6	Hercules, Inc.	PAN	1.82	4.5	0.652	245	35.5
IM-6	Hercules, Inc.	PAN	1.74	4.8	0.696	296	42.9
T300	Union Carbide/Toray	PAN	1.75	3.31	0.480	228	32.1
T500	Union Carbide/Toray	PAN	1.78	3.65	0.530	234	33.6
T700	Toray	PAN	1.80	4.48	0.650	248	36.0
T-40	Toray	PAN	1.74	4.50	0.652	296	42.9
Celion	Celanese/ToHo	PAN	1.77	3.55	0.515	234	34.0
Celion ST	Celanese/ToHo	PAN	1.78	4.34	0.630	234	39.0
XAS	Grafil/Hysol	PAN	1.84	3.45	0.500	234	34.0
HMS-4	Hercules, Inc.	PAN	1.78	3.10	0.450	338	49.0
PAN 50	Toray	PAN	1.81	2.41	0.355	393	57.0
HMS	Grafil/Hysol	PAN	1.91	1.52	0.220	341	49.4
G-50	Celanese/ToHo	PAN	1.78	2.48	0.360	359	52.0
GY-70	Celanese	PAN	1.96	1.52	0.220	483	70.0
P-55	Union Carbide	Pitch	2.0	1.73	0.250	379	55.0
P-75	Union Carbide	Pitch	2.0	2.07	0.300	517	75.0
P-100	Union Carbide	Pitch	2.15	2.24	0.325	724	100
HMG-50	Hitco/OCF	Rayon	1.9	2.07	0.300	345	50.0
Thornel 75	Union Carbide	Rayon	1.9	2.52	0.365	517	75.0

Table 2 Specific tensile strength and modulus of carbon fiber relative to other reinforcements(a)

Fiber	Tensile strength/density		Tensile modulus/density	
	10⁷ cm	10⁶ in.	10⁹ cm	10⁹ in.
AS-4	2.25	8.86	1.29	0.508
IM-6	2.76	10.9	1.70	0.669
E-glass	1.33	5.24	0.28	0.11
S-glass	1.73	6.81	0.32	0.13
Kevlar 49	2.50	9.84	0.90	0.35
Boron	1.50	5.91	1.60	0.63
P-75	1.04	4.09	2.58	1.02

(a) Data derived in epoxy matrix

Table 3 Graphite fiber tensile modulus versus bulk resistivity

Fiber	Bulk dc resistivity ohm·m × 10⁻⁴	Tensile modulus	
		GPa	10⁶ psi
Carbon			
AS-4	15	231	33.5
T300	18	230	33.4
Graphite			
HMS-4	7.5	341	49.9
P-75	1.8	517	75.0

fibers to control interaction between the fibers and the reinforced matrix. For example, if epoxy resins are the matrix to be reinforced, the composite properties may benefit in laminate shear performance from an increase of hydroxyl and amine groups on the fiber surface.

Sizings (coatings) and finishes are often applied to fiber bundles (tows) to improve their handling characteristics. They must be formulated to adhere to but not to interfere with the performance of the composite.

Spooling, or winding, of tows onto a carrier tube is the final step in the conversion process. The tows must be tight enough to form a stable package, yet not so tight as to impede their removal from the package or to be damaged in the spooling process. Variously sized spooled tows are shown in Fig. 1.

Fiber Properties

The names of some commercially available carbon fibers and their manufacturers are identified in Table 1. Data are provided to give an overview of the wide range of properties attainable with carbon fibers and to show that within each precursor group, the mechanical properties can be broadly tailored. Table 2 shows the relative specific performance of carbon fibers compared to that of other reinforcements. Specific properties are mechanical properties that are normalized by the specific gravity of the material. The relationship between tensile modulus and fiber bulk resistivity is shown in Table 3.

SELECTED REFERENCES

- J.B. Barr, S. Chwastiak, R. Didchenko, I.C. Lewis, R.T. Lewis, and L.S. Singer, in *Applied Polymer Symposia No. 29,* 1976, p 161
- J.-B. Donnet, *International Fiber Science and Technology: Carbon Fibers,* Marcel Dekker, Vol 3, 1984
- Federal Republic of Germany Patent 2, 027, 384, 1970
- Federal Republic of Germany Patent 2, 161, 532, 1972
- Japanese Patent, 6902510
- J.W. Johnson and W. Watt, *Applied Polymer Symposia No. 9,* 1969, p 215
- J.W. Johnson and W. Watt, Polymer reprints, Atlantic City Meeting, American Chemical Society, 1968
- W. Johnson, in *Proceedings of the Third Conference on Ind. Carbon and Graphite,* Society of Chemical Industry, 1971, p 447
- G. Lubin, *Handbook of Composites,* Van Nostrand Reinhold, 1982, p 172
- R. Moreton, W. Watt, and W. Johnson, *Nature,* Vol 213, 1967, p 690
- A. Shindo, Y. Nakanishi, and I. Sema, in *Applied Polymer Symposia No. 9,* 1969, p 271
- A. Shindo, Report 317, Osaka Industrial Research Institute, 1961
- W. Watt, *Nature,* Vol 236, 1972, p 10

Aramid Fibers

Jack J. Pigliacampi, E.I. Du Pont de Nemours & Company, Inc.

THE PREDOMINANT ORGANIC REIN-FORCING FIBER used in advanced composites since the early 1970s has been an aramid, or aromatic polyamide, known as Kevlar. Its chemical composition and material properties are thoroughly described in the article "Organic Fibers" in this Volume.

Numerous aramid fiber forms have been developed that support commercial applications in a variety of composite components, ballistic products, tires, ropes and cables, asbestos replacement, and protective apparel. The major fiber forms are continuous filament yarns, rovings, and woven fabrics, and discontinuous staple and spun yarns, fabrics, and pulp. Aramid is also available as textured yarn, needle-punched felts, spunlaced sheets, and wet-laid papers.

Continuous Filament

Continuous filament aramid fiber is manufactured by extruding a solution of polymer through a spinneret. Continuous filaments are the most widely used fiber form in composite applications. Aramid filament can be converted into a wide range of intermediate forms for specific composite applications or fabrication processes.

Yarns and Rovings. Kevlar 49 is a high-modulus aramid fiber that is available in six yarn sizes and two roving sizes, as shown in Table 1. Yarns are multifilament products that are spun directly during fiber manufacture and range from a very fine, 25-filament yarn to 1000-filament yarns. Rovings are produced by combining ends of yarns in a process similar to that used to produce glass fibers (see the article "Glass Fibers" in this Section of the Volume). For example, four ends of 1140-denier yarn are combined to make 4560-denier roving. Denier, a textile unit of size, is the weight in grams of 9000 meters of yarn or roving (1 denier = 1.111×10^{-7} kg/m). Table 2 lists the yarn and roving sizes of Kevlar 29, which has a lower tensile modulus than Kevlar 49 and is used extensively in ballistic armor, asbestos replacement, and certain composites when greater damage tolerance is desired. Kevlar 149, with a tensile modulus 25 to 40% higher than Kevlar 49, is available as 1140-, 1420- and 2130-denier yarn.

Because aramid yarns and rovings are relatively flexible and nonbrittle, they can be processed in most conventional textile operations, such as twisting, weaving, knitting, carding, and felting. Yarns and rovings are used in the filament winding, prepreg tape, and pultrusion processes (Ref 4, 5). Applications include missile cases, pressure vessels, sporting goods, cables, and tension members.

Fabrics and Woven Rovings. Conventional woven fabric is the principal aramid form used in composites. Of the wide range of fabric weights and constructions available, those most commonly used are identified in Table 3 (Ref 1). Many of these aramid fabrics of Kevlar were designed and constructed to be the volume equivalent to the same style number of fiberglass fabric. Generally, fabrics made of a very fine size of aramid yarn are thin, lightweight, and relatively costly, and are used when ultralight weight, thinness, and surface smoothness are critical. Fabrics are available from weavers worldwide (Ref 1). Plain, basket, crowfoot, and satin weave patterns are available. Generally, crowfoot and satin weaves are recommended when a high degree of mold conform-

Table 1 Kevlar 49 yarn and roving sizes

Denier	Yield		Number of filaments
	m/kg	yd/lb	
55	163636	81175	25
195	46155	22895	134
380	23684	11749	267
1140	7895	3916	768
1420	6388	3144	1000
2130	4225	2097	1000
4560	1973	980	3072
7100	1268	630	5000

Note: All Kevlar 49 yarns have approximately 1.5 denier units/filament, with the exception of the 2130-denier product and the 55-denier product, which have a denier/filament ratio of approximately 2.1 and 2.25, respectively. Source: Ref 1, 2

Table 2 Kevlar 29 yarn and roving sizes

Denier	Yield		Number of filaments
	m/kg	yd/lb	
200	45000	22320	134
400	22500	11160	267
1000	9000	4464	1000
1500	6000	2976	4000
9000	1000	497	4000
15000	600	298	10000

Source: Ref 3

Table 3 Kevlar 49 fabric and woven roving specifications

Style no.	Weave	Basis weight g/m²	Basis weight oz/yd²	Fabric construction ends/cm	Fabric construction ends/in.	Yarn denier	Fabric thickness mm	Fabric thickness 10⁻³ in.
Light weight								
166(a)	Plain	30.6	0.9	37 × 37	94 × 94	55	0.04	1.5
199(a)	Plain	61.13	1.8	24 × 24	60 × 60	55	0.05	2
120	Plain	61.1	1.8	13 × 13	34 × 34	195	0.11	4.5
220	Plain	74.7	2.2	9 × 9	22 × 22	380	0.11	4.5
Medium weight								
181	8-harness satin	169.8	5.0	20 × 20	50 × 50	380	0.23	9
281	Plain	169.8	5.0	7 × 7	17 × 17	1140	0.25	10
285	Crowfoot	169.8	5.0	7 × 7	17 × 17	1140	0.25	10
328	Plain	230.9	6.8	7 × 7	17 × 17	1420	0.33	13
335	Crowfoot	230.9	6.8	7 × 7	17 × 17	1420	0.30	12
500	Plain	169.8	5.0	5 × 5	13 × 13	1420	0.28	11
Unidirectional								
143	Crowfoot	190.2	5.6	39 × 8	100 × 20	380 × 195	0.25	10
243	Crowfoot	227.5	6.7	15 × 7	38 × 18	1140 × 380	0.33	13
Woven roving								
1050	4 × 4 basket	356.6	10.5	11 × 11	28 × 28	1420	0.46	18
1033	8 × 8 basket	509.4	15.0	16 × 16	40 × 40	1420	0.66	26
1350	4 × 4 basket	458.5	13.5	10 × 9	26 × 22	2130	0.64	25

(a) Only available on special order; custom fabric will be woven to specifications.

ability is required. Heavy, woven roving fabrics are also available and are used in marine applications where hand lay-up is appropriate, and for ballistic fabrics and aircraft cargo liners. Unidirectional fabrics are used when maximum properties are desired in one direction. Applications for high-modulus aramid fabrics include commercial aircraft and helicopter secondary composite parts, particularly facings of honeycomb core constructions, boat hulls, electrical and electronic parts, ballistic systems, and coated fabrics.

A wide range of fabrics made from Kevlar 29 yarn are also available. Composite applications are primarily in ballistic armor systems (Ref 6) and asbestos replacement. Because of its inherent toughness, products made with aramid require care and special tooling in cutting and machining operations (Ref 7).

Textured aramid can be processed through a high-velocity air jet to attain filament loops in the continuous filament yarn. This produces a bulkier yarn that has more air space between the filaments and a drier, less slick, tactile characteristic. The yarn is used in asbestos replacement to give a composite a higher resin-to-aramid ratio and in protective apparel, to achieve superior textile aesthetics.

Discontinuous Filament Forms

Although continuous filament forms dominate composite applications, the use of aramid in discontinuous or short fiber forms is rapidly increasing. One reason for the increase is that the inherent toughness and fibrillar nature of aramid allows the creation of fiber forms not readily available with other reinforcing fibers. Discontinuous aramid fibers are discussed in the article "Other Discontinuous Fibers" in this section of the Volume.

Staple and Spun Yarns. Staple or short aramid fiber is available in crimped or uncrimped form in lengths ranging from 6.4 to 100 mm (0.25 to 4.0 in.) (Ref 8). Crimped versions that are 25 mm (1.0 in.) or longer are used to make spun yarns on conventional cotton, woolen, or worsted system equipment. Spun yarns formed in this way are not as strong or as stiff as continuous aramid filament yarns but are bulkier, pick up more resin, and have tactile characteristics similar to cotton or wool. They are used in asbestos replacement (for example, clutch facings) and as sewing thread. The shorter aramid fibers, crimped and uncrimped, are used to reinforce thermoset, thermoplastic, and elastomeric resins (Ref 9). Applications include automotive and truck brake and clutch linings, gaskets, electrical parts, and wear-resistant thermoplastic parts. Selection of a crimped or uncrimped form usually depends on the resin-fiber mixing method or on equipment used for the specific application. Special mixing methods and equipment are usually necessary to achieve uniform dispersion of the aramid fiber (Ref 10-12).

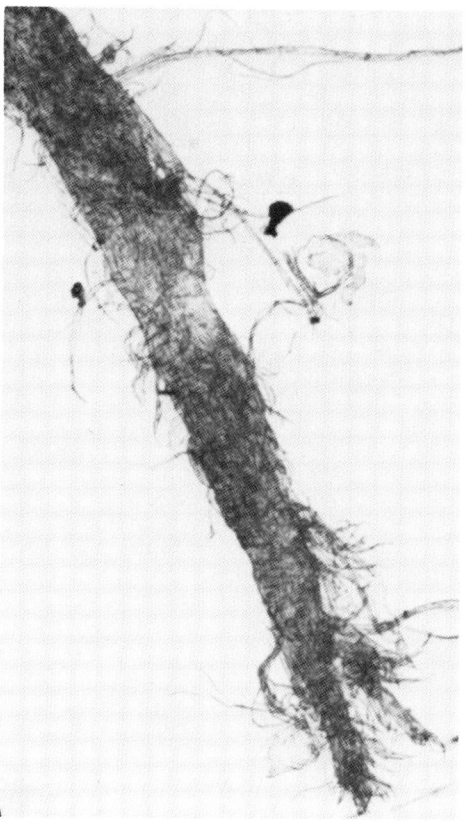

Fig. 1 Aramid pulp

Fabrics and Felts. Woven and knit fabrics made from aramid spun yarns are available. They are used in asbestos replacement and in protective apparel because of their resistance to cutting, puncture, or thermal exposure. Aramid staple is also processed into needle-punched felts, which are used in asbestos replacement, ballistic armor, and marine laminates.

Pulp. Kevlar is available in a unique short-fiber form known as pulp (Fig. 1). It is a very short fiber (2 to 4 mm, or 0.08 to 0.16 in.) with many attached fibrils. These fibrils are complex in that they are curled, branched, and often ribbonlike. They are a direct result of the inherent fibrillar structure of this fiber. The large surface area (40 times standard fiber) and high aspect ratio of the fibrils (greater than 100) can provide very efficient reinforcement. In general, pulp is more easily mixed into resin formulations than is staple fiber and is now used extensively in replacing asbestos in gaskets, friction products, sealants, caulks, and coatings (Ref 13-15).

Spunlaced Sheets. Aramid staple fibers can be processed into lightweight, nonwoven sheet structures. One of the available forms is spunlaced sheet, in which webs of staple fibers are entangled by high-pressure water jets. No binder resin is used. These sheets are low-density (0.008 to 0.16 g/cm^3), lightweight (0.02 to 0.07 kg/m^2, or 0.5 to 2.0 oz/yd^2)

structures that are very drapable and readily impregnated. Applications include surfacing veil, printed circuit boards, and fire-blocking layers in aircraft seating.

Papers. The dominant aramid paper used in advanced composites in honeycomb sandwich constructions is made from Nomex aramid fiber (Ref 16). It is chemically related to Kevlar, but its tenacity and modulus are considerably lower, and are more like those of conventional textile fibers. A range of Nomex papers in varying thicknesses and weights are available. Honeycomb cores of Nomex are also available; densities range from 0.24 to 0.14 g/cm^3 (Ref 17).

Aramid chopped fibers and pulp can be processed into wet-laid papers on conventional fourdrinier machines. Applications include asbestos replacement, such as in gasketing and automatic transmissions (Ref 13). Straight, uncrimped fibers can be used to maximize stiffness and mechanical properties of the wet-laid papers. These thin, lightweight papers are readily impregnated and can be cost competitive with expensive, lightweight, thin, continuous filament fabrics. Composites that use these papers are being developed for printed circuit boards, aerospace, and industrial applications. A proprietary all-Kevlar paper is available.

REFERENCES

1. "Kevlar 49 Aramid Data Manual," E.I. Du Pont de Nemours & Co., Inc.
2. "Characteristics and Uses of Kevlar 49 Aramid High Modulus Organic Fiber," Technical Bulletin, E.I. Du Pont de Nemours & Co., Inc.
3. "Characteristics and Uses of Kevlar 29 Aramid," Technical Bulletin, E.I. Du Pont de Nemours & Co., Inc.
4. "Filament Winding With Yarns or Rovings of Kevlar 49 Aramid Fiber," Technical Bulletin, E.I. Du Pont de Nemours & Co., Inc.
5. "Kevlar 49 Aramid for Pultrusion," Preliminary Information Memo No. 351, E.I. Du Pont de Nemours & Co., Inc.
6. "A Guide to Designing and Preparing Ballistic Projection of Kevlar Aramid," Technical Bulletin, E.I. Du Pont de Nemours & Co., Inc.
7. "A Guide to Cutting and Machining Kevlar Aramid," Technical Bulletin, E.I. Du Pont de Nemours & Co., Inc.
8. "Spun Yarns of Kevlar Type 970 Aramid Staple," Technical Bulletin, E.I. Du Pont de Nemours & Co., Inc.
9. E. Galli, The Use of Short-Length (Chopped) Kevlar Fibers in Friction Products and Thermoplastics, *Plast. Compd.*, Nov/Dec, 1981
10. "Processing Staple Kevlar Fiber and Kevlar Pulp by Littleford Mixing Technology," Technical Bulletin, Littleford Brothers, Inc., Jan 1982

11. "Eirich Processing Equipment for Kevlar Aramid Fibre and Pulp," Technical Bulletin, Eirich Machines, Ltd., Feb 1983
12. "Guide to Mixing and Processing Kevlar Aramid Fiber and Pulp," Technical Bulletin, E.I. Du Pont de Nemours & Co., Inc.
13. "Kevlar Aramid for Asbestos Replacement," Technical Bulletin, E.I. Du Pont de Nemours & Co., Inc.
14. "Gasket Sheeting With Kevlar Aramid Fiber," Technical Bulletin, E.I. Du Pont de Nemours & Co., Inc.
15. "Manufacturing Guide for Gasket Sheeting Reinforced With Kevlar Aramid Pulp," Technical Bulletin, E.I. Du Pont de Nemours & Co., Inc.
16. H.Y. Loken, Du Pont Aramids for Advanced Composites, in *Proceedings of Industrial Composites Seminar, SAMPE Midwest Chapter*, Society for the Advancement of Material and Process Engineering, March 1982
17. "Design and Fabrication Techniques for Honeycomb of Nomex Aramid Sandwich Structures," Technical Bulletin, E.I. Du Pont de Nemours & Co., Inc.

Other Continuous Fibers

Willard D. Bascom, University of Utah

THE CONTINUOUS FIBER REINFORCE-MENTS most often used in present-day commercial composite materials are glass, carbon, synthetic organic, and silicon carbide fibers. However, there are applications for which other continuous fibers are appropriate. Natural fibers have been used in structures since ancient times, and specialty ceramic and metallic fibers are used today in high-temperature applications. Some specialty fibers are in the early stages of development so their real potential has yet to be determined. This article identifies these less commonly used fibers, both old and new, and describes their properties.

Natural Fibers

The Egyptians and Romans, and possibly earlier civilizations, used a straw-reinforced brick for construction in semiarid Mediterranean regions. The matrix was a red mud, and the bricks were either sun dried or kiln fired. Biblical accounts of brick making suggest the straws were at least a few inches in length (longer than typical chopped fiber) and at a fiber volume of about 20%. Various civilizations have used and continue to use natural fibers, such as flax, jute, and cotton, as reinforcement in mud, clay, and cement structures.

Flax-reinforced organic matrix composites came close to being used in the aerospace industry during World War II. The British were faced with being cut off from a primary source of bauxite ore in France needed to produce aluminum for the Spitfire fighter. To demonstrate the feasibility of using composite material as a direct substitute for aluminum, a fuselage was built with flax-reinforced phenolic resin (Ref 1). A prepreg tape, made by a wet-roller dip-coating process, was used to fabricate the fuselage section from the forward main-spar frame to the joint at the tail section. The composite members included frame sections, longerons, and skin sections. The prototype structure passed all static load tests. However, the project was terminated when alternate bauxite sources became available, and it was over 30 years before fiber-reinforced polymer composites were being used extensively on military aircraft.

Nonetheless, the use of natural fiber reinforcements, primarily flax, linen, and silk, had not disappeared from the composite materials field in the intervening years. A number of projects have been undertaken using natural fiber, polyester, and epoxy matrix composites for housing construction, primarily in developing countries. It was hoped that it would be cheaper to import the matrix resin and fabricate the laminates with locally available reinforcements than to import prefabricated fiberglass structures, but none of these projects progressed beyond the development and demonstration stage. However, some of these same countries are now exporters of carbon, aramid, and glass-reinforced composite materials and finished products, because of low labor costs.

The use of natural fibers as reinforcement in cement and polyester resins has been studied, but a major problem in their use is their susceptibility to fungal and insect attack and degradation by moisture (Ref 2, 3).

Some key properties of natural fibers are compared with corresponding properties of glass, carbon, and aramid fibers in Table 1. Clearly, in tensile strength and modulus, the natural fibers are inferior. However, they exhibit significantly better elongation, which affords better composite damage tolerance. Low cost and damage tolerance make natural fibers attractive for housing construction with low load requirements.

Synthetic Fiber Reinforcement

Continuous fibers other than glass, aramid, and carbon are generally high-cost materials and are therefore used in applications where this cost is offset by their unique thermal and/or mechanical properties.

Boron (B) filaments deposited on tungsten (W) or carbon (C) have been used in organic matrix composites, but their current applications are primarily in metal matrix composites (MMCs). The microstructure and chemical composition of the boron deposit differs significantly for the two substrates: The tungsten wire is converted to the borides W_2B_5 and WB_4, whereas there is no significant reaction of the boron with carbon fiber substrate. The B-W fiber structure is nodular, which results in a relatively rough surface. The microstructure of the B-C fiber is a dense nucleation of boron from the carbon surface so that the outer surface of the deposit is relatively smooth. Because of volume changes during deposition and reaction, both fiber types have residual thermal stress, compressive stress at the surface, and tensile stress at the midplane between surface and core. The outer surface of these fibers is boric oxide, which offers oxidation protection up to 500 °C (930 °F) but decomposes at higher temperatures, with a resulting dramatic loss in properties (essentially, complete loss in tensile strength at 600 °C, or 1110 °F). The high reactivity of boron with most metals prohibits its use in MMCs unless it is given a protective overlay of silicon carbide, boron carbide, or boron nitride. The key properties of boron fibers are presented in Table 2.

High-Alumina Fiber. Although alumina silicate fibers are generally classified as glass fibers, those with a high alumina content ($\alpha Al_2O_3 > 99\%$) have special properties and applications. These alumina fibers are produced

Table 1 Properties of natural fibers and typical synthetic fibers

Fiber	Density, g/cm³	Tensile strength		Tensile modulus		Elongation at failure, %
		GPa	10⁶ psi	GPa	10⁶ psi	
Hemp	1.52	0.92	0.13	70	10	1.7
Jute	1.52	0.86	0.12	60	9	2.0
Flax	1.52	0.84	0.12	100	15	1.8
Cotton	1.52	0.2-0.8	0.03-0.12	27	4	6-12
Silk	1.34	0.6	0.09	10	2	18-20
S-glass	2.50	4.6	0.67	84	12	2-5
Carbon (type I)	1.90	2.0	0.29	380	55	1-2
Aramid	1.44	2.8	0.41	133	20	2-4

Soures: Ref 4-6

Table 2 Properties of boron and alumina fibers

Fiber	Density, g/cm³	Diameter		Tensile strength		Tensile modulus	
		µm	µin.	GPa	10⁶ psi	GPa	10⁶ psi
Boron-tungsten............	2.6	100-200	3950-7850	5.5-7.0	0.80-1.0	400	58
Boron-carbon.............	2.3	100-200	3950-7850	5.0	0.73	400	58
α-alumina(a)	3.95	20	790	14(b)	2.0	390	57
				19(c)	2.8	390	57

(a) Slurry-spun continuous fiber. (b) Uncoated. (c) Silicon carbide-coated.

Table 3 Physical properties of metallic wires

Material	Specific gravity	Melting point		Tensile strength		Young's modulus of elasticity		Coefficient of thermal expansion, 10⁻⁶/K
		°C	°F	MPa	ksi	GPa	10⁶ psi	
Aluminum.................	2.71	660	1220	290	40	68.9	10.0	23.6
Beryllium	1.85	1350	2460	1100	160	310	45.0	11.6
Copper	8.90	1083	1980	413	60	124	18.0	16.5
Tungsten	19.3	3410	6170	2890	130	345	50.0	4.6
Austenitic stainless steel	7.9	1539	2800	2390	350	200	29.0	8.5
Molybdenum	10.2	2625	4750	2200	320	331	48.0	...

Source: Ref 9

by a continuous slurry spinning process of alumina particles in an aqueous medium with organic polymer stabilizers and viscosity control agents. The spun fiber is dried at a low temperature and then fired to form a dense, high-temperature stable α-alumina fiber. The morphology is polycrystalline, with a grain size of ~0.5 µm (20 µin.). Tensile strengths are limited by surface flaws, primarily grain boundaries. Therefore, a silica coating is generally applied to heal surface flaws (Table 2). The principal feature of these fibers is their excellent strength and modulus retention, which is nearly 100% (up to 1000 °C, or 1830 °F). Their primary application is in metal matrix composites and ceramic fiber-ceramic matrix composites.

Silicon Nitride Fibers. Recent advances in silicon nitride (Si_3N_4) chemistry may lead to the development of continuous Si_3N_4 fibers (Ref 7). Conventional Si_3N_4 ceramics are produced by high-temperature reactions of silicon with nitrogen. Coatings are produced by reaction bonding, and small components are produced by hot-pressing or sintering techniques. Catalysts have been found for the pyrolysis of polysilizane to essentially pure Si_3N_5 (Ref 8). Work is underway on the development of solution spinning of the polymer and catalyst to form a precursor fiber that, in principle, could be thermally converted into a Si_3N_4 fiber in much the same way that polyacrylonitrile is converted into carbon fiber. The potential properties of a continuous Si_3N_4 fiber are very attractive, based on the properties of the current ceramic forms: use temperatures of 1500 °C (2730 °F), tensile strength and modulus of 1000 MPa (145 ksi) and 300 GPa (45 × 10⁶ psi), low coefficient of thermal expansion, electrical nonconductivity, and wear resistance. Achieving these properties will depend on the densification and crystal structure developed during pyrolysis.

Metallic Wires

The continuous reinforcements in many MMCs are metal wires. The properties of the more widely used wires are listed in Table 3. Beryllium stands out for its low-density and high-tensile properties. However, its use is hampered by its toxicity as a lung irritant and carcinogen, although these dangers have been somewhat exaggerated because of poor handling practices during World War II. Awareness of the toxicity problem, implementation of protective measures, and medical surveillance of personnel can reduce the toxicity hazard to a minimum.

Aside from the properties inherent to the wire reinforcement, the properties of a composite are highly dependent on the matrix metal and the processing methods. This technology has been thoroughly documented. References 10 and 11 give two excellent reviews.

REFERENCES

1. W.I.L.F. Bishop, High-Performance Fibers—An Overview, in *Advanced Composites*, I.K. Partridge, Ed., Elsevier, to be published
2. D.G. Swift, in *Composite Structures*, I.H. Marshall, Ed., Applied Science, 1981, p 602
3. K.G. Satyanarayana, A.G. Kulkarni, K. Sukumaran, S.G.K. Pillai, K.A. Cherion, and P.K. Rohatgi, in *Composite Structures*, I.H. Marshall, Ed., Applied Science, 1981, p 618
4. S.A. Wainwright, W.D. Biggs, J.D. Curry, and J.M. Gosline, *Mechanical Design in Organisms*, John Wiley & Sons, 1976
5. A. Kelly, *Strong Solids*, Clarendon Press, 1973
6. M.B. Bever, *Encyclopedia of Materials Science and Engineering*, Vol 3, MIT Press, 1986, p 1729
7. W. Verbeck, *Ger. Offen.*, Vol 2 (No. 218), 1973, p 960
8. S.R. Jones, A New Silicon Nitride Ceramic, *Chem. Week*, Vol 138 (No. 17), 23 April, 1986, p 15
9. J.R. Vinson and T.W. Chou, *Composite Materials and Their Use in Structures*, Elsevier, 1975
10. K.G. Kreider, Metal Matrix Composites, in *Composite Materials*, Vol 4, Academic Press, 1974
11. T.W. Chou, A. Kelly, and A. Okura, *Composites*, Vol 16 (No. 187), 1985

Other Discontinuous Fibers

Willard D. Bascom, University of Utah

SHORT FIBER REINFORCED COMPOS-ITES, with fiber lengths of 0.5 to 6 mm (0.020 to 0.25 in.) or less, account for the largest share, at approximately 60%, of the fiber-reinforced composite market. This market includes organic polymers with particulate fillers or chopped-glass, aramid, and carbon fibers. In most of these materials, the reinforcement functions either as an extender (filler) for the more expensive matrix polymer or to impart stiffness and thermal stability, with some sacrifice in tensile strength and strain-to-failure. In the past 5 to 10 years, there has been a growing trend to increase the reinforcing efficiency of the fibers to the point that the mechanical properties of short fiber reinforced composites approach at least 90% of the modulus and 50% of the tensile strength of continuous fiber reinforced composites.

One advantage of short over continuous reinforcements is rapid, low-cost processibility by compression or injection molding, or by extrusion. Another advantage is the ability to mold intricate component configurations that cannot be fabricated using continuous fiber reinforcements.

High performance in discontinuous fiber composites depends on the alignment, uniform distribution, and length of the fiber and on good adhesion between the fibers and the matrix. The two basic approaches to controlling short fiber alignment and distribution are mold design and processing conditions in injection molding or extrusion, and controlled fiber alignment in mat forms that are subsequently impregnated with a matrix resin and then compression molded. The basic concepts for predicting short fiber reinforced composite properties are given below, followed by a brief discussion of the processing methods already mentioned.

Properties

The material properties of discontinuous fibers, which are essentially the same as those of the corresponding continuous fiber, are discussed in other articles. In some instances, discontinuous fibers do not have a corresponding continuous form (for example, single-crystal whiskers and ceramic fibers). Key properties of short fibers and whiskers are listed in Table 1. It is evident that much higher tensile

strengths are attainable from single-crystal whiskers than from the corresponding ceramic form. These high-performance fibers are used in metal or ceramic matrices for high-temperature or ultrahigh modulus applications.

Until recently, the full potential of discontinuous fiber reinforcements has not been utilized for essentially two reasons. First, discontinuous fibers are often shorter than the critical length required to realize the full reinforcing capability of the fiber. Second, the fibers must be closely aligned in the principal stress direction. These restraints can be at least partially relieved by advanced processing methods.

The reinforcing efficiency of a fiber depends, in part, on stress transfer from the matrix into the fiber. If the fibers are too short, that is, less than the critical length, L_c, the full strength of the fiber cannot be realized. The average critical length is:

$$L_c = \frac{\sigma_f d}{2\tau_c} \qquad \text{(Eq 1)}$$

where σ_f is the fiber strength, d is the fiber diameter, and τ is the shear stress at the fiber-resin boundary (Ref 2). At fiber lengths of L_c, the average fiber stress is just equal to fiber strength, so that for maximum reinforcing efficiency in discontinuous fiber reinforced composites, the fiber length must be $>L_c$. Note that L_c depends on τ_c, the fiber-matrix adhesive strength, or the shear yield strength of the matrix, which limits stress transfer to the fiber, making the reinforcement dependent on the polymer matrix. The critical lengths for carbon and glass fibers in various polymers are listed in

Table 2, and were determined in single-fiber tests (Ref 3-7). In well-bonded systems, such as carbon or glass fibers in epoxy polymers, the critical lengths are low, but if adhesion is poor, as with some carbon fibers in thermoplastics, the critical length is much higher.

Fiber alignment is another critically important factor for reinforcement efficiency. Even for fully aligned short fibers, the tensile strength of composites is less than 50% of that obtainable from continuous reinforcement. In Fig. 1, data from L. Kacir et al. (Ref 8) are plotted for average fiber lengths of 3.1 mm (0.125 in.), 6.3 mm (0.25 in.), and 12.5 mm (0.50 in.), along with the theoretical curve for continuous reinforcement and data for continuous fibers from S.W. Tsai (Ref 9), and from Ref 10. Both systems used E-glass fibers in an epoxy matrix. The reinforcement and matrix resins were not identical, but had comparable properties. In the case of the short fibers, the fiber volume fraction, V_f, was 0.5, and for the continuous fiber, it was 0.66. This difference in V_f would increase the discontinuous fiber strength by no more than 20% (Ref 2). The major reason for the lower property values of the discontinuous fiber reinforced composites is not entirely clear. Certainly, one contributing factor is their high concentration of fiber ends, which act as sites for stress concentration.

The materials tested by Kacir et al. were fiber mats (see below), in which orientation of the fiber was carefully controlled and the fiber was uniformly distributed to prevent any fiber-rich or resin-rich regions. In more commercially typical discontinuous fiber reinforced composites, the fiber length is much shorter,

Table 1   Properties of short fibers and whiskers

Material	Density g/cm³	Tensile strength		Tensile modulus	
		GPa	10⁶ psi	GPa	10⁶ psi
Alumina					
Whiskers	4.0	10-20	1-3	700-1500	100-220
Sintered fibers	<4.0	0.2-0.7	0.030-0.10	140-300	20-40
Boron, thermally formed fibers	2.3	2.75	0.400	400	60
Boron nitride, fibers	1.8-2.0	0.3-1.4	0.045-0.20	28-80	4-10
Carbon					
Whiskers	>2.0	...	...	700	100
Fibers	1.8-2.0	2-3	0.30-0.45	230-550	35-80
Silicon nitride, whiskers	3.2	5-7	0.75-1.0	350-380	50-55

Adapted from Ref 1

the fibers are not as highly aligned, and the distribution is not fully uniform.

The strength of discontinuous fiber composites is given for ideally aligned and uniform distribution by:

$$\sigma_c = \sigma_f^c V_f \left(1 - \frac{L_c}{2_L}\right) + \sigma_m (1 - V_f) \qquad \text{(Eq 2)}$$

where σ_f^c is the fracture strength of the fiber and σ_m is the stress in the matrix when the composite fails. This equation points up the importance in discontinuous fiber reinforced composites of high fiber strength, volume fraction, and fiber critical length. The bond strength between fiber and matrix does influence composite properties by its effect on critical length (Eq 1).

If the fiber length, L, is less than the critical length, L_c, then the equation must be modified to reflect the fact that maximum stress transfer into the fiber is less than the fiber breaking stress. Equation 2 then takes the form:

$$\sigma_c = \frac{L\tau_c}{d}(V_f) + \sigma_m (1 - V_f) \qquad \text{(Eq 3)}$$

It should be noted that this discussion oversimplifies the effects of fiber properties and orientation on the strength of discontinuous fiber reinforced composites. For more comprehensive discussions, see Ref 11 and 12.

The tensile modulus of short fiber reinforced composites can closely match that of continuous fiber reinforced composites (see Fig. 2),

Table 2 Critical lengths (L_c) and critical aspect ratios (L_c/d)

Fiber	Matrix	L_c mm	L_c in.	L_c/d mm	L_c/d in.	Reference
E-glass(a)	Polypropylene	1.78	0.0700	140	6	4
E-glass(a)	Epoxy	0.43	0.017(b)	34(b)	1.3	3
E-glass(a)	Polyester	1.27	0.0500(c)	100(c)	4	3
Carbon (AS1)(d)	Epoxy	0.33	0.013	47	1.9	5
Carbon (AS4)(d)	Epoxy	0.42	0.017	60	2.4	6
Carbon (AS4)(d)	Polycarbonate	0.74	0.030	106	4.17	7
Carbon (XAS)(e)	Polycarbonate	0.35	0.014	50	2	7
Carbon (XAS)(e)	Epoxy	0.36	0.014	51	2	7

(a) Proprietary sizing. (b) At 40 °C (105 °F). (c) At 50 °C (120 °F). (d) Hercules Inc. (e) Hysol-Grafil Ltd.

provided that the fibers are aligned in the direction of the applied stress.

Processing

The major methods for manufacturing discontinuous fiber organic polymer composite structures are compression molding, injection molding, and extrusion. All these processes require premixing of the fiber and the polymer. This mixing process is critically important to uniform dispersion of the fiber, is usually accomplished by shear mixing, and is sometimes aided by dispersing agents or by heating to lower the polymer viscosity. Some breakdown in fiber length occurs and reduces reinforcing efficiency, but this loss must be weighed against the development of weak resin-rich or fiber-rich regions, if fiber dispersion is not uniform.

Table 2 shows that the critical lengths of carbon fibers are generally less than for glass fiber. This difference means that fiber breakage during processing has a less severe effect on reinforcement efficiency for carbon fiber than for glass fiber. For this reason, and because of their high modulus, the high-strength fibers such as carbon, silicon carbide, and aramid are attractive for short fiber reinforced composite applications.

The fiber-polymer mix for thermosetting compounds is generally in two basic forms, bulk molding compounds (BMC) and sheet molding compounds (SMC). The essential difference is a fiber length of <1.0 mm (0.040 in.) for BMC and 2 to 6 mm (0.080 to 0.25 in.) for SMC. Both contain 14 to 40% fiber. In the case of SMC, the mixing process is deliberately less aggressive than for BMC, in order to preserve fiber length.

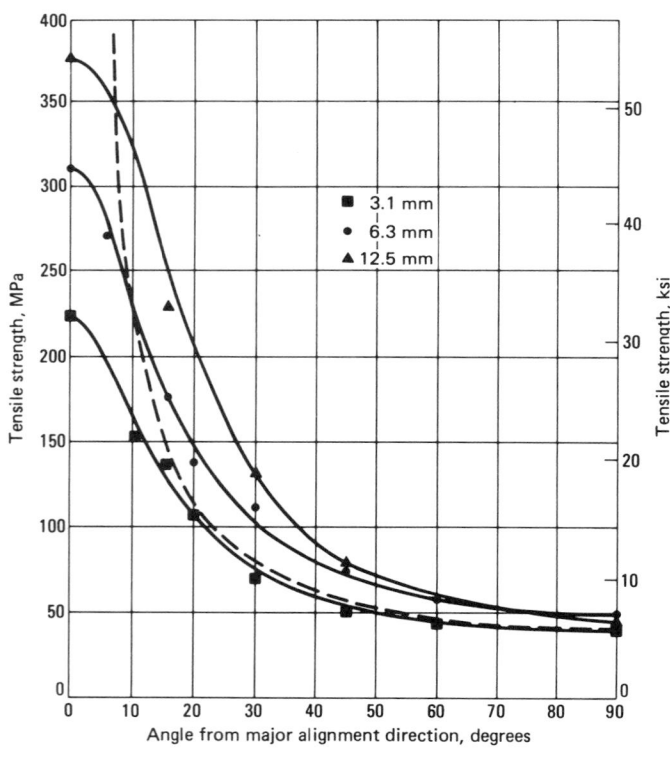

Fig. 1 Effect of fiber orientation on composite tensile strength. E-glass fiber in epoxy matrix with average fiber lengths of 3.1 mm (0.125 in.), 6.3 mm (0.25 in.), and 12.5 mm (0.50 in.). Dashed line is theoretical curve for continuous fiber. Source: Ref 8

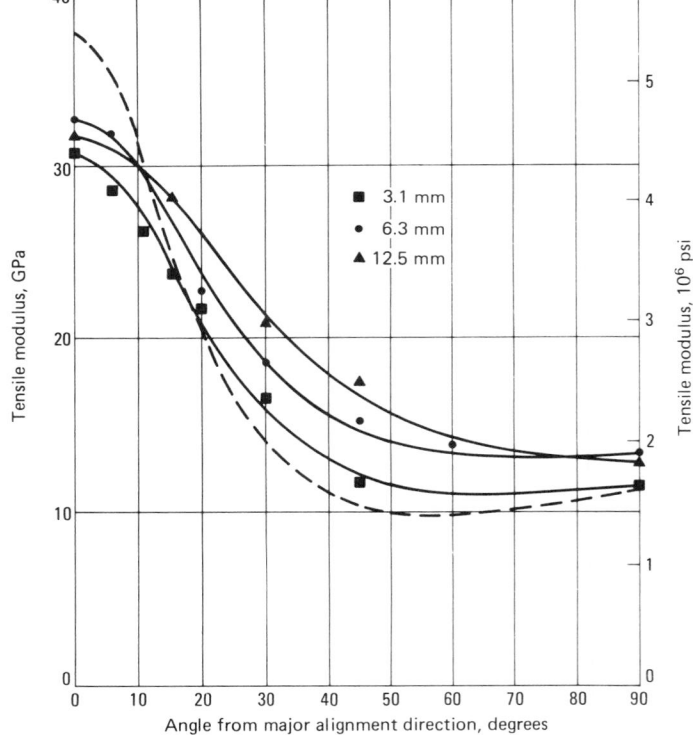

Fig. 2 Effect of fiber orientation on composite modulus, E-glass fibers in epoxy matrix. Source: Ref 8

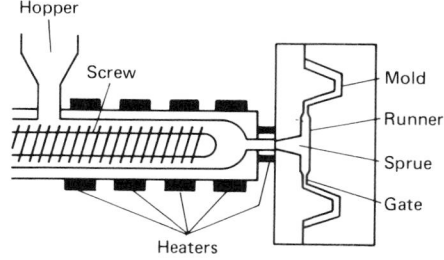

Fig. 3 Schematic of injection molding equipment. Source: Ref 13

Sheet molding compounds are generally used in compression molding processes, which, compared to injection molding or extrusion, are less expensive but are slow and produce more scrap because of mold overflow. Fiber orientation is established in the SMC raw material before molding because there is little opportunity for flow and fiber alignment in the compression cycle.

Injection molding is the most widely used process for discontinuous fiber reinforced thermoplastic composites. Figure 3 illustrates the essential configuration. The molding compound, usually in the form of extruded, chopped pellets, is introduced into a reciprocating screw molder, which provides additional mixing in the melt but also further reduces fiber length. Molten mix is forced into the mold through the sprues in one or more gates. Once the mold is filled, a back pressure is maintained (the packing cycle) to compensate for shrinkage of the molding during cool-down and solidification. Heat transfer to the metal mold causes the material to solidify rapidly so that the mold can be opened and the part ejected automatically by ejection pins. Inserts can be prepositioned in the mold and incorporated into the part during molding, thereby obviating expensive machining or other fabrication steps. It is this rapid production rate and minimal machining requirement that has made injection molding so attractive. The major disadvantage is the short fiber length of the BMC, usually less than the critical length, which results in low reinforcement efficiency. This difficulty can be offset by using matrix polymers that are inherently strong so that the principal function of the reinforcement is to provide stiffness; taking advantage of the aramid, carbon, or silicon carbide fibers, which have higher moduli than the more widely used glass fibers; and using sophisticated premix methods, molding techniques, and mold design to preserve fiber length and optimize fiber alignment in the finished part. There have been some major advances in mold design using finite-element and finite-difference analysis (Ref 14). These analyses consider the rheological behavior of a melt through the gates and in the mold itself. Usually the mold is treated like a network of relatively simple geometries that are analyzed separately, using viscous and viscoelastic

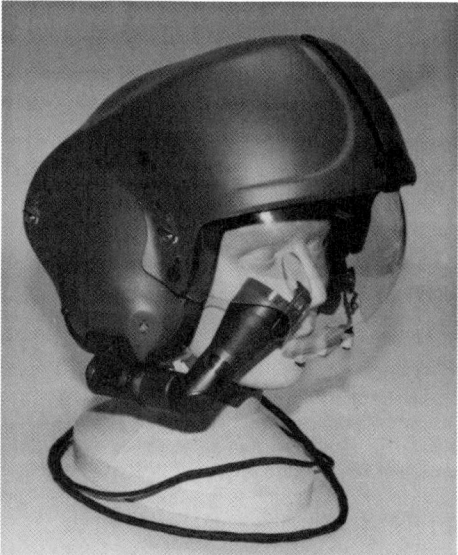

Fig. 4 Injection molded helicopter pilot helmet. Courtesy of LNP Corporation

constitutive equations. The solutions are then coupled to predict flow patterns.

Injection molding can be used to produce highly sophisticated structures, such as the helicopter pilot helmet shell shown in Fig. 4. Within the shell molding is a complete electronic package (for gun-firing control), visor tracks, nut plates, adjusting knobs, coil supports, trip levers, and other components. Injection molding is not limited to thermoplastics. Reinforced reaction injection molding (RRIM) of thermosets is widely used and has the advantage of a polymer matrix with a much lower viscosity than thermoplastic melts; thus lower injection pressures are required and fiber damage is reduced. Polyurethanes, polyesters, and epoxies, all specifically designed for rapid cure, are the principal polymers used in RRIM technology.

Mats of highly aligned fibers with fiber lengths of 5 to 10 mm (0.20 to 0.40 in.) can be formed by suspending the fibers in a moderate-viscosity water-soluble medium such as glycerol (Ref 15, 16). The dispersion is pumped through nozzles onto a screen. The flow rate and tube nozzle design are such that a convergent flow is produced, and the fibers deposit on the screen with a high degree of alignment. In one such arrangement, the dispersion is pumped through a tube along the axis of a rotating cylinder. Convergent flow nozzles spaced along the tube spray the dispersion onto a screen lining on the inside of the cylinder. Centrifugal force carries the glycerol away as the mat is formed. The mat is washed in place with water, then with a water-soluble organic solvent such as acetone. The mat is dried and impregnated with a matrix resin, usually an epoxy or polyester. This prepreg can then be fabricated into relatively complex parts by matched-die molding.

This alignment process is described in Ref 8, which presents data on the effect of fiber orientation on composite mechanical properties. Some of this data is summarized in Fig. 1 and 2.

Profile extrusion of short fiber composites is in some respects a combination of injection molding and the alignment orientation process. The melt is forced under high shear stresses through dies to form rods or tubular structures. Because of the high shear rates (100/s), there is actually a reduction in effective viscosity that presumably is due to plug flow. Variations in channel cross section produce extensional flow fields that cause fibers to move past each other (or reorient) and create shear stresses in the matrix. Die configuration can be designed for convergent flow to orient the fiber in the flow direction, or for extensional flow for transverse orientation (Ref 17). This technology lends itself to a rapid production rate (fast cool-down of the melt), but is somewhat limited in the type of structure that can be formed.

REFERENCES

1. M.O.W. Richardson, in *Polymer Engineering Composites*, M.O.W. Richardson, Ed., Applied Science Publishers, 1977, p 31
2. A. Kelly, *Strong Solids*, Clarendon Press, 1973, p 172
3. M. Miwa, T. Ohsawa, and K. Tahara, *J. Appl. Polym. Sci.*, Vol 25, 1980, p 795
4. W.A. Fraser, F.H. Achker, and A.T. DiBenedetto, 30th Annual Technical Conference, Section 22-A, Reinforced Plastics/Composites Institute, Society of the Plastics Industry, 1975, p 1
5. L.T. Drzal, M.J. Rich, and P.F. Lloyd, *J. Adhesion*, Vol 16, 1982, p 1
6. W.D. Bascom and R.M. Jensen, *J. Adhesion*, Vol 19, 1986, p 219
7. W.D. Bascom, unpublished results
8. L. Kacir, M. Narkis, and O. Ishai, *Polym. Eng. Sci.*, Vol 15, 1975, p 525, 532; Vol 17, 1977, p 234; and Vol 18, 1978, p 45
9. S.W. Tsai, NASA Report CR-224, National Aeronautics and Space Administration
10. A. Kelly, *Strong Solids*, Clarendon Press, 1973, p 218
11. M.J. Folkes, *Short Fibre Reinforced Thermoplastics*, Research Studies Press, John Wiley & Sons, 1982
12. C.W. Bert and R.A. Kline, *Polym. Composites*, Vol 6, 1985, p 133
13. R.P. Sheldon, *Composite Polymeric Materials*, Applied Science Publishers, 1982, p 144
14. M.R. Kamal and P.G. Lafleur, *Polym. Eng. Sci.*, Vol 22, 1982, p 1067
15. N.J. Parratt, *Composites*, Vol 1, 1969, p 25
16. G.E.G. Bagg, M.E.N. Evans, and A.W.H. Pryde, *Composites*, Vol 1, 1969
17. L.A. Goettler, *Polym. Composites*, Vol 5, 1984, p 60

Fiber Sizing

Willard D. Bascom, University of Utah

SIZINGS ARE AN ESSENTIAL FACTOR in fibrous composite technology. They are critical in composite manufacturing and can have both negative and positive effects on composite properties. Unfortunately, the development of sizing agents is too often an afterthought in composite materials technology. This can be changed through concentrated efforts to develop sizings that not only protect the fiber during processing but also enhance composite properties.

In the composite industry, the term sizing has come to mean any surface coating applied to a reinforcement to protect it from damage during processing, aid in processing, or improve the mechanical properties of the composite. This article reviews the primary functions and major types of sizing agents and the theory behind the effect of these agents on composite properties. The focus will be on fiber reinforced polymer matrix composites.

Other terms used synonymously for sizing include finishing agent, which comes from the textile industry and refers to fiber coatings that render flexibility, drape, and special features such as fire retardance to fabrics (Ref 1). This term still finds use in fibrous composite nomenclature, especially for woven glass or carbon fiber products. Sometimes sizing is referred to as a coupling agent when it is designed to enhance composite mechanical properties or durability. As discussed in the section "Theory" in this article, this term implies the presence of mechanisms such as chemical bonding that may not in fact be operative. A better, less specific, term would be adhesion promoter.

Surface treatment is sometimes confused with sizing, especially in carbon fiber reinforcement technology (Ref 2). Surface treatment always involves a chemical modification of the nascent fiber surface that alters the surface chemical composition, such as oxidation, acid-base reactions, and etching, without a deliberate coating of the surface. The distinction between sizing and surface treatment is fairly clear in the case of carbon fibers, but is less clear in the case of boron fibers that are treated chemically to form a boron carbide or boron nitride coating (Ref 3). For the purposes of this article, a sizing is a deliberate coating of the reinforcement, which may incidentally react chemically with the surface; a surface treatment is a deliberate chemical modification of the reinforcement, which may incidentally result in the formation of a coating.

Functions and Types of Sizing Agents

Table 1 defines the major functions of sizing agents. Many commercial sizings are formulated to be multifunctional. For example, a glass fiber sizing may contain a film-forming polymer to produce a uniform protective coating and an organofunctional silane to promote adhesion.

Sizings may be a "necessary evil" in that they are needed in one stage of processing but interfere with subsequent processing and/or adversely affect composite mechanical properties or durability. For example, in the carbon fiber industry, sizings must be applied to the fiber tow (which may consist of 12 000 filaments or more) to prevent the individual filaments from contact damage between themselves or with eyelets or guides during weaving or prepregging. However, this same sizing may bond the filaments together and prevent uniform impregnation of the tows by the matrix. This problem can be avoided by applying the sizing in a diluted solution when the tow is spread apart. Glass fibers are usually sized just after they are drawn from the melt but before they are gathered into a tow. Carbon fibers must be spread into a flat ribbon on a spreader bar to allow the sizing solution to flow between the filaments.

A sizing may adversely affect the mechanical properties of the composite. For example, a sizing that holds the filaments in a bundle so that the strand (tow) can be chopped for discontinuous fiber composites hinders later efforts to disperse the fibers during injection molding or extrusion. In developing sizings, the formulator is often faced with a "no-win" situation and can only hope that a sizing does minimal damage to composite performance.

Chemical Composition

Sizings designed to protect the reinforcement during processing must coat the surface uniformly. For this reason, polymers that are widely used in the coating industry because of their good film-forming ability (Ref 4) are also used as sizing agents. Typical examples are starch and starch derivatives, the vinyl polymers, and the phenoxys. The choice is dictated by a number of considerations: compatibility

Table 1 Sizing classifications and functions

Type	Purpose	Example	Remarks
Film-forming organics and polymers	To protect the reinforcement during processing	Polyvinyl alcohol (PVA), polyvinyl acetate (PVAc)	The polymer is formulated to wet-spread to form a uniform coating that is applied to aid processing but later may be removed by washing or heat cleaning, for example, fugitive sizing.
Adhesion promoters	To improve composite mechanical properties and/or moisture resistance	Silane coupling agents	Principally used on inorganic reinforcement, for example, glass fiber. See the section "Theory" in this article.
Interlayer	To enhance composite properties by creating an interphase between matrix and reinforcement	Elastomeric coating	Not in commercial use. See the section "Theory" in this article.
Chemical modifiers	React to form protective coating	Silicon carbide on boron fibers	

with the matrix polymer, the level of protection required (for example, weaving is more severe on continuous fibers than prepregging of unidirectional tape), pliability or drape of the sized tow or cloth (for example, a stiff, "boardy" fabric is difficult to process), and cost. Sizings are usually applied at a level of 1.0 wt% or less, making it necessary to remove and dispose of large volumes of solvent. Environmental pollution and cost considerations mandate that the sizing be applied from aqueous media, which requires that it be soluble in water or able to be applied as a water-based emulsion.

Ideally, a sizing should be chemically compatible with the matrix polymer and should not adversely affect the mechanical properties of the interphase between reinforcement and matrix. If these requirements cannot be met, the sizing may be removed by washing or heating before processing the reinforcement into the final composite form. However, these manipulations usually either damage the fiber or leave residues that may prevent good bonding between reinforcement and matrix. Nevertheless, these fugitive sizings are still used, especially for woven reinforcements.

There are a variety of film-forming polymers that are compatible with the more widely used polyester and epoxy matrix resins. However, there are very few sizings that can be used with the newer high-temperature matrix polymers, such as the bismaleimides and polyimides, or with the tough thermoplastic matrices, such as polyphenylene sulfide or polyether etherketone. One approach to developing sizing for these newer matrix materials is to use the polymers themselves as the sizing. However, they usually do not have the wetting and spreading behavior necessary to form a uniform coating. Developing sizings for these new matrix polymers, especially for carbon fiber reinforced composites, is essential in order to realize their full potential.

The moisture resistance of composites is very frequently compromised by the sizing, especially at elevated temperatures. An inherent characteristic of film-forming polymers is that they are highly polar because of polarizable hydroxyl, carboxylic, or ester moieties. Therefore, they readily absorb moisture, and unless they are fully reacted with the matrix resin to reduce their polarity, they will attract water into the interphase region and reduce the bond strength between reinforcement and matrix.

Moisture resistance is a concern with all composite materials, but is especially critical for glass fiber reinforced polymer composites because the glass fiber itself is hydrophilic. This problem has been overcome by changes in its chemical composition and, more pertinent to this article, through the use of adhesion promoters, or coupling agents. The most widely used coupling agents are organofunctional alkoxysilanes (Ref 5), although chromium complexes (Ref 6) and titanates (Ref 7) are also used. Various theories have been developed to explain how these agents work. Generally, they

are not good film formers, and therefore are not in themselves good sizings. Instead, they are usually applied in combination with a film-forming polymer. A typical sizing for glass fibers used in an epoxy matrix would be an epoxy-compatible film former and a silane coupling agent.

The third classification of sizings in Table 1 are the interlayer materials designed to develop specific mechanical properties in the interphase region between reinforcement and matrix. The chemistry of these sizings is discussed in the next section.

Theory

The use of sizings as protective coatings is largely empirical, with relatively little documentation. As already mentioned, polymers that are used in the coating industry because of their good wetting and spreading properties (good film formers) are sizing candidates because it is essential to obtain a continuous coating on each reinforcing filament for maximum protection during processing and interaction with the matrix.

In distinct contrast to the film-forming sizings, the adhesion promoters—especially the silanes—have been extensively studied. The reason for this interest is primarily that these agents do in fact improve the moisture resistance of glass fiber reinforced polymer composites. To understand how they work would provide valuable insight into the mechanisms that affect composite moisture durability. However, despite all the studies of silanes on glass (and other) surfaces, the mechanisms involved in their protection of the glass-polymer interface are not well understood.

The early concept of the silanes forming a simple chemical coupling between fiber and matrix (Fig. 1) has been largely discredited. The silanes readily form three-dimensional polysiloxane networks through hydrolysis and condensation of the alkoxy groups (Fig. 2). This polymerization is acid-base catalyzed, and the silanes are frequently applied from acid solution. Indeed, one of the more widely used silanes, especially for epoxy matrix polymers, is the aminopropyl triethoxysilane $NH_2CH_2CH_2CH_2Si(OC_2H_5)_3$, which is a strong base. It seems reasonable, and has been demonstrated experimentally (Ref 8, 9), that these silanes form a polymeric network on solid substrates that, as in the case of glass, has an occasional chemical attachment to the surface. It has been suggested that this is a relatively open network that is easily penetrated by the molecules of the matrix polymer (Ref 8) so that an entanglement of the polymer networks is formed in the interphase region between matrix and reinforcement (Fig. 3). This polymer network formation does not, in itself, explain how the silanes protect the boundary from attack by water. It does, however, present a more realistic picture of the adsorbed silane film.

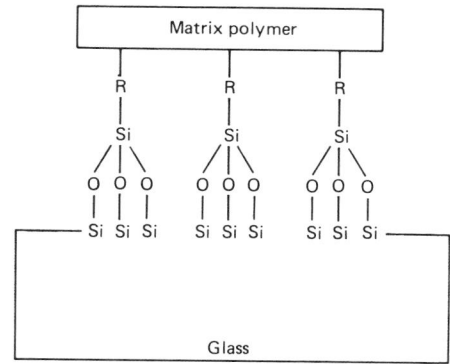

Fig. 1 Idealized "coupling" of matrix and glass by organofunctional silane

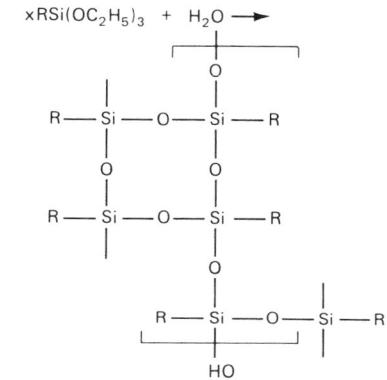

Fig. 2 Hydrolysis of organofunctional silanes to polysiloxane networks

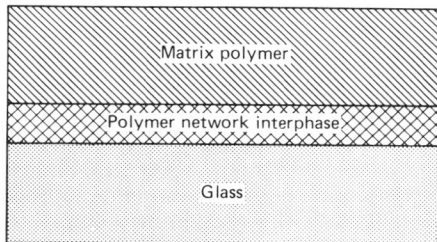

Fig. 3 Silane and matrix interphase polymer network

In commercial practice, the silanes are often applied with a film-forming polymer. Presumably, the coating polymer becomes entangled in the silane network along with the matrix polymer. The composition of this complex interphase is critical to understanding the moisture durability of composite materials. The possible interpenetration formation of silane and epoxy molecules is a subject of recent study (Ref 10).

The interlayer is included in this article because it is the ultimate sizing: a coating that aids processing and also enhances composite mechanical properties. The concept of an interlayer has been the subject of a number of studies (Ref 11, 12), notably the work by Subramanian *et al.* (Ref 13). In all of this work, evidence has been presented that coatings can in fact increase specific mechanical properties, especially impact strength. In most cases, however, the effect is marginal or is achieved at the expense of some other property. None of these interlayer coatings have been reduced to practice. Therefore, there is no data base from which to access their actual effectiveness.

Recently, it has been found that very thin 50 to 70 nm (500 to 700 Å) coatings of butadiene-acrylonitrile elastomers on particulate and fibrous reinforcement in epoxy matrix polymers significantly increase the fracture energy and, in the case of the continuous fiber composite, the delamination resistance (Ref 14-16). These elastomers react with the epoxy to form low-modulus interphase coatings. Moreover, because they are thin, the stiffness and the elevated-temperature resistance of the composite are not seriously reduced. Thick elastomer coatings decouple the matrix from the reinforcement, which could explain why attempts to develop elastomeric interlayers in the past have not been especially successful. The interlayer concept holds the most promise for protecting the reinforcement during processing and enhancing composite properties, but a considerable amount of research is needed to develop commercially viable sizings.

REFERENCES

1. A. Rose and E. Rose, Ed., *The Condensed Chemical Dictionary*, 6th ed., Reinhold, 1956, p 497
2. J.B. Donnet and R.C. Bansal, *Carbon Fibers*, Marcel Dekker, 1985
3. M. Basche, Interfacial Stability of Silicon Carbide Coated Boron Filament Reinforced Metals, in *Interfaces in Composites*, STP 452, American Society for Testing and Materials, 1968, p 130
4. R.R. Meyers and J.S. Long, Ed., *Film Forming Compositions*, Parts I and II, Marcel Dekker, 1968
5. E.P. Plueddeman, Interfaces in Polymer Matrix Composites, in *Composite Materials*, Vol 6, L.J. Broutman and R.H. Krock, Ed., Academic Press, 1975
6. A. Rose and E. Rose, Ed., *The Condensed Chemical Dictionary*, 6th ed., Reinhold, 1956, p 1219
7. S.J. Monte, G. Sugarman, and D.J. Seeman, "Titanate Coupling Agents—Current Applications," Paper presented at Rubber Division Meeting, American Chemical Society, May 1977, p 40
8. W.D. Bascom, Structure of Silane Adhesion Promoter Films on Glass and Metal Surfaces, *Macromolecules*, Vol 5, 1972, p 792
9. H. Ishida and Y. Suzuki, Hydrolysis and Condensation of Aminosilane Coupling Agents in High Concentration Aqueous Solutions: A Simulation of Silane Interphase, in *Composite Interfaces*, H. Ishida and J.L. Koenig, Ed., North-Holland, 1986
10. K. Hoh, H. Ishida, and J.L. Koenig, The Diffusion of Epoxy Resin Into a Silane Coupling Agent Interphase, in *Composite Interfaces*, H. Ishida and J.L. Koenig, Ed., Elsevier, 1986, p 251
11. L.J. Broutman and B.D. Agarwal, A Theoretical Study of the Effect of an Interfacial Layer on the Properties of Composites, *Polym. Eng. Sci.*, Vol 14, 1974, p 581
12. J.H. Cramner, G.C. Tesoro, and D.R. Uhlmann, Chemical Modification of Carbon Fiber Surfaces With Organic Polymer Coatings, *Ind. Eng. Chem. Prod. Res. Dev.*, Vol 21, 1982, p 185
13. R.V. Subramanian, J.J. Jakubowski, and F.B. Williams, Interfacial Aspects of Polymer Coating by Electrodeposition, *J. Adhesion*, Vol 9, 1978, p 185
14. J. Kawamoto, F.J. McGarry, and J.F. Mandell, "Impact Resistance of Rubber Modified Carbon Fiber Composites," PPST-85-5, Massachusetts Institute of Technology, 1985
15. M. Tse, Ph.D. thesis, Massachusetts Institute of Technology
16. F.J. McGarry, "Rubber in Crosslinked Glassy Polymers," Paper presented at the 129th Meeting, Rubber Division, American Chemical Society, New York, 1986

Unidirectional and Two-Directional Fabrics

William D. Cumming, HITCO Woven Structures Division

WOVEN MATERIALS, in laminate form, are currently displacing more traditional structural forms primarily because of the availability of fibers (such as carbon and aramid) whose enhanced mechanical properties in composite form surpass the property values of corresponding hardware in aluminum or steel on a strength-to-weight basis.

Woven broad goods, considered to be intermediate forms, present these fibers in a more convenient format for the design engineer, resin coater, and hardware fabricator. The many variations of properties made possible by combining different yarns and weaves allow the structural engineer a wide range of laminate properties. The designer should understand the operation of weaving hardware and textile design details in order to select the best fabric style.

Fabric Design

The fabric pattern, often called the construction, is an x, y coordinate system. The y-axis represents warp yarns and is the long axis of the fabric roll (typically 30 to 150 m, or 100 to 500 ft). The x-axis is the fill direction, that is, the roll width (typically 910 to 3050 mm, or 36 to 120 in.). Basic fabric weaves are few in number, but combinations of different types and sizes of yarns with different warp/fill counts allow for hundreds of variations.

The most common weave construction used for everything from cotton shirts to fiberglass stadium canopies is the plain weave, shown in Fig. 1. The essential construction requires only four weaving yarns: two warp and two fill. This basic unit is called the pattern repeat. Plain weave, which is the most highly interlaced, is therefore the tightest of the basic fabric designs and most resistant to in-plane shear movement. Basket weave, a variation of plain weave, has warp and fill yarns that are paired: two up and two down. The satin weaves represent a family of constructions with a minimum of interlacing. In these, the weft yarns periodically skip, or float, over several warp yarns, as shown in Fig.

2. The satin weave repeat is x yarns long and the float length is $x - 1$ yarns; that is, there is only one interlacing point per pattern repeat per yarn. The floating yarns that are not being woven into the fabric create considerable looseness or suppleness. The satin weave produces a construction with low resistance to shear distortion and is thus easily molded (draped) over compound curves, such as an aircraft wingroot area. This is one reason that satin weaves are preferred for many aerospace applications. Satin weaves can be produced as standard four-, five-, or eight-harness forms. As the number of harnesses increases, so do the float lengths and the degree of looseness and sleaziness, making the fabric more difficult to control during handling operations. Textile fabrics generally exhibit greater tensile strength in plain weaves, but greater tear strength in satin weaves. This distinction fades in the composites field.

The ultimate laminate mechanical properties are obtained from unidirectional-style fabric (Fig. 3), where the carrier properties essentially vanish in the laminate form. The higher the yarn interlacing (for a given-size yarn), the fewer the number of yarns that can be woven per unit length. The necessary separation between yarns reduces the number that can be packed together. This is the reason for the higher yarn count (yarns/in.) that is possible in unidirectional material and its better physical properties. Unidirectional material has the most "unbalanced" weave and is usually reserved for special applications involving hardware with axial symmetry (such as a carbon fiber reinforced shuttle motor case) fabricated using a tape-wrapping operation.

A weave construction known as locking leno (Fig. 4), which is used only in special areas of the fabric, such as the selvage, is woven on a shuttleless loom. The gripping action of the intertwining leno yarns anchors or locks the open selvage edges produced on rapier looms. The leno weave helps prevent selvage unraveling during subsequent handling operations, but is unsatisfactory for obtaining good laminate

physical properties. However, it has found applications where a very open (but stable) weave is desired.

The textile designer is concerned with only a few fabric parameters: type of fiber, type of yarn, weave style, yarn count, and areal weight. Standard methods for measuring such parameters are well documented in Ref 1.

Fiber finish is vital to the weaver because it helps lubricate and protect the fiber as it is exposed to the sometimes harsh weaving oper-

Fig. 1 Plain weave yarn interlacing

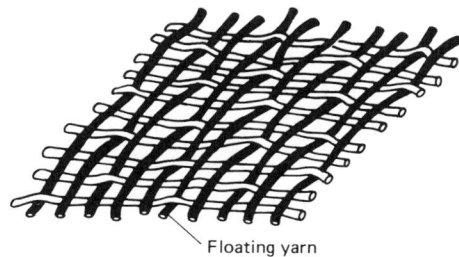

Floating yarn

Fig. 2 Five-harness satin weave interlacing

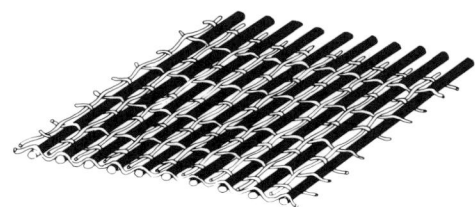

Fig. 3 Unidirectional weave

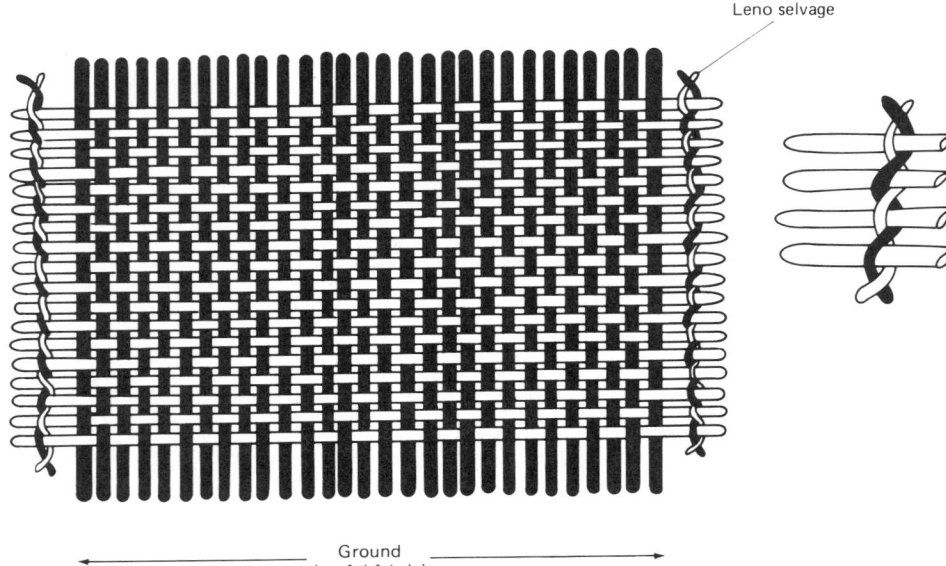

Fig. 4 Full-width plain weave with leno selvage

Fig. 5 Curved tape (19.1 mm, or 3/4 in.); wide weaving (140 μm, or 5.6 mil); boron wire warp (75 μm, or 3.0 mil); and copper ribbon fill

Table 1 Typical fabric styles and composite properties

Typical fabric weaves

Weave	Yarns/in. Warp	Fill	Weight kg/m²	oz/yd²	Thickness at 25 kPa (3.4 psi) mm	in.	Yarn (carbon)
Eight-harness satin	24 ×	23	0.370	10.9	0.46	0.018	Thornel 300 3K
Eight-harness satin	24 ×	23	0.370	10.9	0.48	0.019	Celion 3K
Plain	12½ ×	12½	0.190	5.6	0.30	0.012	Thornel 300 3K, Kevlar aramid tracers
Five-harness satin	24 ×	24	0.125	3.7	0.20	0.008	Thornel 300 1K
CFS	24 ×	12	0.20	6.0	0.23	0.009	Celion 3K warp 150 l/o glass fill
Plain	11½ ×	11½	0.19	5.7	0.25	0.010	Magnamite AS-4 3K
Five-harness satin	11 ×	11	0.370	10.9	0.50	0.020	Magnamite AS-4 6K
Plain	8 ×	8	0.525	15.5	0.81	0.032	Celion 12K
Eight-harness satin	10½ ×	10½	0.755	22.2	1.0	0.040	Thornel 300 15K
Plain	10 ×	10	0.345	10.2	0.48	0.019	75 l/o glass warp, Grafil E/XA-S 12K fill
8HS	21 ×	21	0.393	11.6	0.38	0.015	HITEX 3K

Typical composite properties (balanced weave)

Tensile strength, MPa (ksi)	620-690 (90-100)
Tensile modulus, GPa (10⁶ psi)	69-76 (10-11)
Flexural strength, MPa (ksi)	690-900 (100-130)
Flexural modulus, GPa (10⁶ psi)	62-69 (9-10)
Compressive strength, MPa (ksi)	620-690 (90-100)
Compressive modulus, GPa (10⁶ psi)	62-69 (9-10)
Short beam shear strength, kPa (psi)	55-69 (8-10)
Specific gravity	1.6

ation. The quality of the woven fabric is often determined wholly by the type and quality of the fiber finish. The finish of choice, however, is usually dictated by end-use and resin chemistry. Epoxy resin laminates require a compatible finish, which itself may be an epoxy, although that is not the best weaving finish available.

The verification of quality is an important aspect of the aerospace materials business. Quality is usually governed by military specification as part of the purchasing requirements.

Typical quality defects, such as missing or broken warp or fill yarns, fabric misorientation (pucker), and misweaves in the pattern due to equipment failure or foreign material on the fabric, are documented in Ref 1 and 2.

Weave construction is the realm of the textile engineer, but fabric mechanical properties and how they translate into the laminate are concerns of the composite design engineer. Maximum directional properties for the minimum material (thickness) are attained with unidirectional-style material. The more usual goal of

balanced properties requires two-directional styles. The fiber obviously dominates those properties carried by the fabric into a structural composite. Fiber properties are covered in the Section "Properties of Constituent Materials" in this Volume. Another source of fiber properties is manufacturer data sheets, which may also provide data on the fabric forms of fibers. This is typical of the fiberglass industry, which has these well-established fabric styles and categories:

Fabric	Areal wt kg/m²	oz/yd²
Light weight	0.10-0.35	3-10
Intermediate weight	0.35-0.70	10-20
Heavy weight	0.50-1.0	15-30

Fabric	Thickness μm	mil
Light weight	25-125	1-5
Intermediate weight	125-250	5-10
Heavy weight	250-500	10-20

The newer carbon and aramid fiber industries are somewhat oriented to custom design, but as the aerospace market matures, a few fabric constructions may become standards. Table 1 provides a sampling of styles that have found uses, along with corresponding order-of-magnitude epoxy resin composite properties.

The following data illustrate the relative market importance of various aerospace textile intermediate forms:

Woven	90 + %
Filament winding	5%
Braided	<1%
Knit	<1%
Prepregs	(a)

(a) Depends on prepreg manufacturer and its market segment/product line

Special Fabric Applications. Combining fibers with very different properties, such as carbon warp with glass fill, can provide a fabric with good longitudinal strength/stiffness values, as well as transverse (fill direction) toughness and impact resistance. The ability to hy-

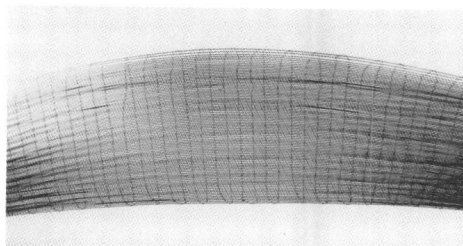

Fig. 6 Boron-copper tape with continuous reentrant fill, no selvage

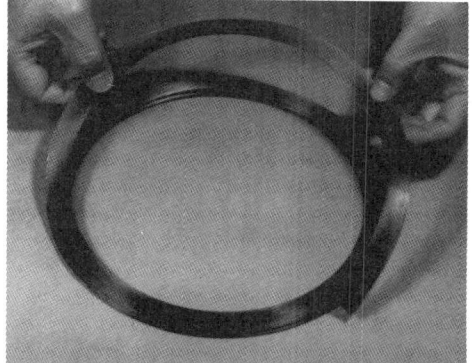

Fig. 7 Woven curved tape

Fig. 8 Fly-shuttle loom, weaving unidirectional fabric

Fig. 9 Rapier loom, weaving plain-weave carbon fabric

bridize the fabric allows the composite designer the freedom to build hardware with different and perhaps conflicting requirements without major compromises. It is also possible to "teach" the looms new tricks, particularly in three-directional weaving, but interesting modifications are even possible for two-directional fabric. As shown in Fig. 5 to 7, the loom has the capability of weaving an endless helix using boron warp and copper fill. The final hardware is to be an aluminum ring for use in high-speed rotation. The design goal is to hoop-reinforce for burst strength and to have the fabric mimic the hardware geometry.

Textile Equipment

As already discussed, yarns may be woven in two basic formats: unidirectional and the more conventional two-directional. Traditional textile fly-shuttle looms are capable of weaving most of the newer fibers (Fig. 8). Carbon fiber is the major exception to this rule because it is too fragile to withstand the whipping action of the shuttle. The shuttleless, rapier-type loom is the preferred equipment for carbon fibers (Fig. 9). Because the operation of this loom is typical for most looms, a description follows.

Warp yarns move from the creel supply magazine, shown in Fig. 10, toward the back of the loom and are taken up on the roll near the floor. Each individual warp yarn (>500 in this case) is captured in an eyelet integral to a series of heddles (identified in Fig. 9) and then passes through the comblike reed that is behind the rapier (Fig. 11). The fabric appears immediately in front of the reed, where the weaving action forms the fabric. Transverse fill yarns are placed in front of the reed by the in-and-out motion (picking) of the rapier. The rapier grips the fill yarn, pulls it through an open shed, and releases it in front of the reed, which is used to compact the fill yarns by pivoting forward, thereby plowing the fill yarn into the open shed. The shed then closes around the fill yarn.

The weave pattern is determined by the up/down heddle position, which in turn is controlled by the geared head motion or dobby head (upper right, Fig. 9), which reads an endless punched tape programmed with the weave pattern. A simple plain weave using 3000 filament aerospace grade carbon fiber to produce a 110-cm (42-in.) wide, open but fringed selvage fabric is shown in Fig. 12. Rapier looms may also produce an open or tucked-in selvage. Aramid locking-leno pattern yarns are woven in the selvage with aramid tracer yarns to verify layer-to-layer registration (witness proper ply orientation) in a multi-ply laminate.

Figure 13 shows a unidirectional fabric that was woven on the fly-shuttle-type loom depicted in Fig. 8. The bobbin (far left) is a small fill supply package that is inserted in and carried piggyback by the shuttle (center). The projectilelike shuttle is thrown rapidly from one

Fig. 14 Shuttle emerging from shed

Fig. 10 Creel warp supply spools

Fig. 11 Rapier within shed during filling sequence

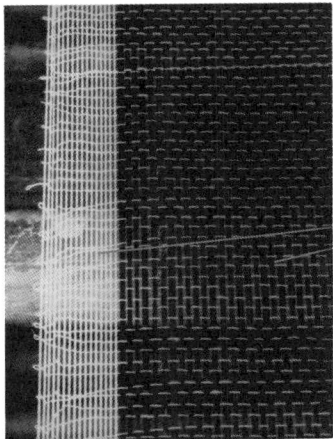

Fig. 15 Fly-shuttle loom selvage with "uni-type" fill

side of the loom to the other. The shuttle trails fill yarn behind it as it moves in front of the reed (Fig. 14). The fill yarn may be a very light rayon, nylon, or polyester yarn, all of which easily survive the rapid back and forth shuttle movements that might damage the more fragile carbon yarn. Note the relatively small size of the carrier fill yarn (Fig. 15), compared to the warp yarns, which will be the major load-carrying yarns in the laminate. The shuttle loom also produces a closed selvage (continuous fill), regardless of the weave style. This is the type of selvage commonly seen on ordinary textile fabric. Only the fly-shuttle loom can produce tubular woven fabrics.

REFERENCES

1. *Annual Book of ASTM Standards*, Vol 07.01, Textiles — Yarns, Fabrics, General Test Methods, Vol 07.02, Textiles — Fibers, Zippers, American Society for Testing and Materials
2. "Textile Test Methods," Federal Specification 191a, 1978

Fig. 12 Open-selvage plain-weave carbon (aerospace grade fiber) fabric produced on rapier-type loom

Fig. 13 Flying shuttle, trailing fill yarn

Multidirectionally Reinforced Fabrics and Preforms

F. Paul Magin, III, Hercules Aerospace Company

EARLY APPLICATIONS OF COMPOSITE MATERIALS, such as boat hulls and sports car bodies, involved moderate and simple mechanical loads that allowed the use of randomly oriented chopped fibers or two-directionally oriented fabric reinforcement. With the emergence of carbon-carbon composites and the resultant increased use of composite materials for high-temperature aerospace applications, the anisotropic mechanical property of these materials presented a design problem: Although mechanical properties were satisfactory in the two directions containing reinforcement fabric, the mechanical properties in the third direction were matrix dominated and typically more than an order of magnitude less than in the reinforced direction. This problem was especially critical in applications involving high thermal stresses, such as carbon-carbon composite ballistic reentry nose tips and solid rocket throats. The obvious solution was to add fiber reinforcements in the third direction, and additional directions when necessary, to provide composite materials with isotropism approaching that of metals.

Early work in multidirectional reinforcement, performed in the late 1960s, emphasized the development of geometric principals for a variety of fiber orientations, ranging from orthogonal, 3-directional to 11-directional reinforcement.

By the mid 1970s, it became apparent that the high cost of hand-assembled preforms (a preshaped fibrous reinforcement) would severely limit the application of multidirectionally reinforced composites to a very few aerospace applications. As a result, two trends developed in the field. For the most part, development activities began to de-emphasize the more complicated reinforcement geometries and concentrate on three-directional (3-D) reinforcements, seeking to narrow the program scope and to optimize the 3-D configuration for specific applications, such as reentry nose tips and solid rocket motor throats. This optimization included fiber selection, heat treatment variations, and weave balance.

With the scope of multidirectional reinforcement development thus reduced, it became possible to design and fabricate semiautomated equipment to reduce preform costs while increasing quality by eliminating the human error potential associated with hand assembly. As a result of these activities, semiautomated weavers were developed by the late 1970s and used to produce propulsion hardware for strategic weapon systems. At the present time, fully automated computer-controlled 3-D weavers are operational. This equipment and its capabilities and limitations will be described in the section "Three-Directional Weaving Machines" in this article. For further information, see the articles "Braiding" and "Fiber Preforms and Resin Injection" in this Volume.

Reinforcement Materials

There are few limits on the composition of reinforcement fibers that can be woven into 3-D preforms; if a material can be made into a fiber, it can probably be woven into some type of 3-D preform. Fibers that have been woven into 3-D preforms include carbon/graphite, glass, silica, alumina, aluminosilicates, silicon carbide, cotton, and aramid. When design requirements necessitate, it is possible to weave a 3-D preform with a combination of fibers, as illustrated by the silica/carbon cylinder in Fig. 1.

Generally, the only limitation to fiber selection is the combination of brittle fibers and small yarn bend radii, the latter being caused by either weave geometry or the yarn delivery system of the automated weaving machine. This is particularly true of carbon and graphite fibers, which account for about 90% of all 3-D woven preforms. High-modulus graphite fiber is particularly prone to fracture during preform construction.

Weave Geometry

There are more than 20 varieties of multidirectionally reinforced preforms. Because describing them all is beyond the intended scope of this article, the three variations of 3-D preforms that are most widely used and best characterized will be described.

Fig. 1 Cylinder showing the combined use of silica and carbon fibers

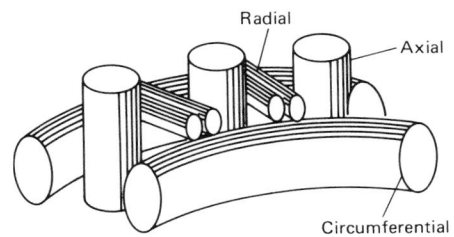

Fig. 2 Geometry of 3-D polar weave preform

Polar weave 3-D preforms have reinforcement yarns in the circumferential, radial, and axial (longitudinal) directions, as shown in Fig. 2. Preforms of this geometry normally contain 50 vol% fibers that can be introduced equally in the three directions. Some variation in relative yarn distribution can be accomplished when a specific application requires unbalanced properties. For example, if high-hoop tensile strength is required, additional fibers can be added in the circumferential direction, at some sacrifice of radial and longitudinal properties.

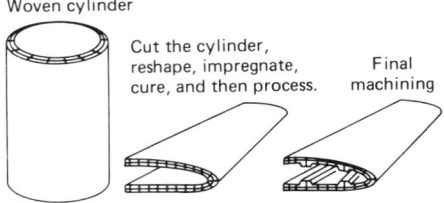

Fig. 3 Deformation of 3-D cylindrical preform to form a leading edge

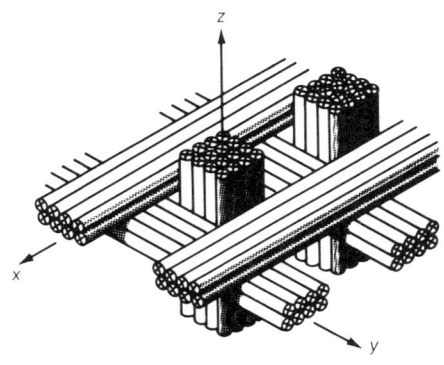

Fig. 4 Geometry of 3-D orthogonal weave preform

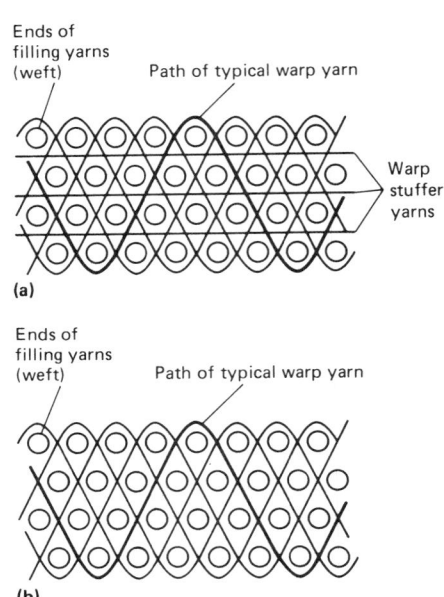

Fig. 5 Geometry of angle-interlock fabric (a) with and (b) without added stuffer yarns

Although originally developed as 305- to 510-mm diam (12- to 20-in. diam) thick-walled cylinders, polar weave 3-D preforms are presently fabricated in a number of body-of-revolution shapes, including cylinders, cylinders/cones, and convergent/divergent sections. When nonaxisymmetric shapes are needed, such as leading edges or conic/rectangular transitions, a two-step process is used. First, a preform of appropriate geometry is woven. Next, the preform is placed in a metal die, deformed into the required shape, and impregnated with a suitable resinous material to ensure geometric stability during the remainder of the densification process, as shown in Fig. 3. It may or may not be necessary to slit the preform before deformation; although the leading-edge example in Fig. 3 required slitting, a simpler deformation from conic to conic/rectangular would not.

Present size limitations for 3-D polar weave cylindrical preforms are an approximate maximum of 210 cm (84 in.) in diameter and 130 cm (50 in.) in length. Wall thicknesses vary from 6.4 to 200 mm (0.25 to 8 in.). Inside diameters are limited to 75 mm (3 in.) because of space requirements for the weaving mechanism.

Although yarn spacing will not be discussed in specific quantitative terms because of the large number of variables, some general observations are appropriate. In a polar weave 3-D preform, the angular spacing of the radial fibers is a critical factor. The number and diameter of the radial yarn bundles remains constant as they move from the inside diameter to the outside diameter of the part, forming a pie-shaped "corridor" of increasing width. The volume percentage of radial fibers is therefore lower at the outside diameter than at the inside diameter. To account for this and to maintain a uniform fiber volume throughout the preform, the fiber bundle size of both the circumferential and longitudinal yarns is increased.

Orthogonal weave 3-D preforms have reinforcement yarns arranged in an orthogonal (Cartesian) geometry, with all yarns intersecting at 90° angles, as shown in Fig. 4. Typical yarn content varies from 45 to 55%, similar to that of a polar weave preform. These fibers can be introduced uniformly in each of the three directions to provide isotropic properties or in unbalanced amounts when design considerations require anisotropic properties. Unlike polar weave preforms, they are rarely woven to

near-net configuration. Instead, they are woven as blocks, and parts are machined to the requisite size and shape.

In regard to yarn spacing, it is worth noting that orthogonally woven 3-D preforms generally have a much finer unit cell size than their polar weave counterparts, resulting in superior mechanical properties and erosion resistance after densification into a composite.

Angle interlock, also known as warp interlock, is a multilayered fabric in which the warp yarns travel from one surface of the fabric to the other, holding together up to eight layers of fabric, thus creating a thick, two-directional (2-D) fabric, as shown in Fig. 5(a). When higher in-plane strength is needed, additional stuffer yarns can be added to create a quasi 3-D fabric, as shown in Fig. 5(b).

Although angle interlock fabric is economical to produce on commercially available weaving equipment, its use has been limited. Because it is unavailable in closed shapes, its use when closed cones/cylinders are required necessitates the use of joints and their attendant plane of weakness.

Stitched fabric and needled felt can be marginally included in a summary of 3-D preforms and fabrics. Although both materials have reinforcement fibers in all three directions, the amount of fiber in the cross-ply direction is frequently such a small fraction of the total fiber volume that the cross-ply mechanical properties are only slightly better than the matrix-dominated properties of 2-D materials.

Three-Directional Weaving Machines

There are nearly as many varieties of 3-D weaving machines as there are weaving designs, but this discussion focuses on three machines developed by Aerospatiale in Bordeaux, France, and presently operating at the Hercules Aerospace Company facility in Bacchus, UT.

The three machines, designated TL600, TL1000, and TL1250, were designed primarily to manufacture 3-D polar weave preforms for solid rocket throats, small- to medium-exit

cones, and medium- to large-exit cones, respectively. The designation TL refers to the French words *tissage* and *laçage*, which are derived from the terms to weave and to lace and describe the two steps used to produce a preform (Fig. 6). Their size limitations are presented in Table 1. Although designed for the cited applications, they have adequate versatility to produce a wide variety of shapes and sizes. Because this is an emerging technology, products have been highly customized, and prospective users should not expect to find standard or stocked items.

The fabrication process begins with the insertion of fine steel wires into a series of pierced aluminum plates to form a weaving network, as shown in Fig. 7. The relative position of this network within the weaving machine is shown in Fig. 8. Actual weaving begins with the simultaneous placement of radial and circumferential yarns, as shown in Fig. 9. As the entire network assembly rotates, circumferential yarns are continuously laid down at the weaving site by a tubular yarn delivery system that includes yarn-tensioning devices and missing-yarn sensors. At each pie-shaped corridor formed by the wire network, a radial knitting needle traverses the corridor, captures a radial yarn at the inside of the port, and returns to the outside of the port, where it makes a locking stitch that prevents movement of the radial yarn during subsequent operations, such as reshaping in a steel die. Circumferential and radial yarns are woven into the wire network until the preform is of sufficient length.

The final step of the fabrication process, called lacing, consists of replacing the network wires with yarn bundles using a lacing needle with a drilled hole at its point and an opening eye. The lacing needle is electronically posi-

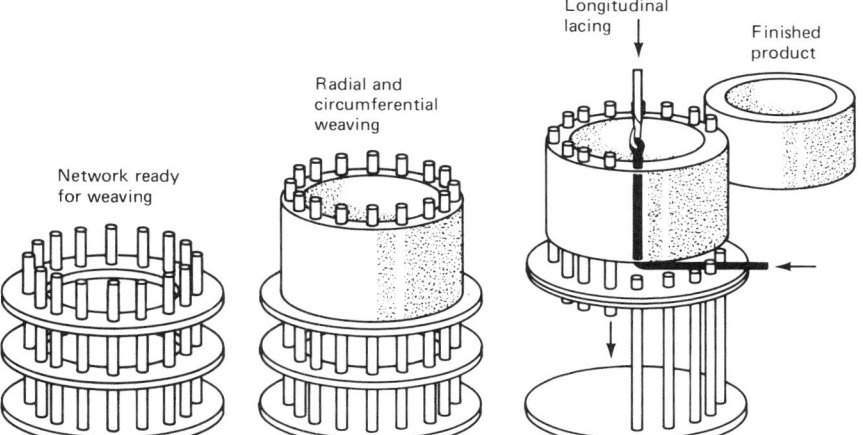

Fig. 6 Summary of 3-D polar weaving process

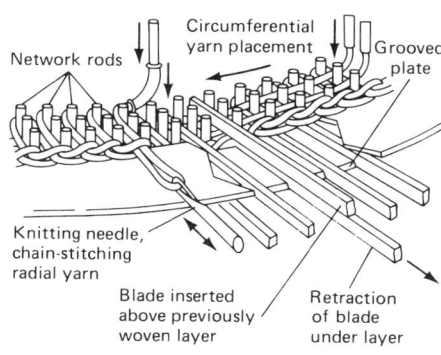

Fig. 9 Radial and circumferential weaving

Table 1   Weaving machine size limitations for 3-D polar preforms

Weaving Machine	Min inside diam		Max outside diam		Wall thickness, min/max	
	mm	in.	mm	in.	mm	in.
TL600	56	2.2	610	24	3.8/150	0.15/6.0
TL1000	100	4.0	1000	40	3.8/150	0.15/6.0
TL1250	250	10.0	1250	50	3.8/150	0.15/6.0

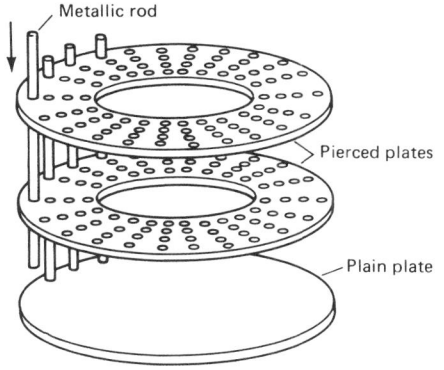

Fig. 7 Network construction

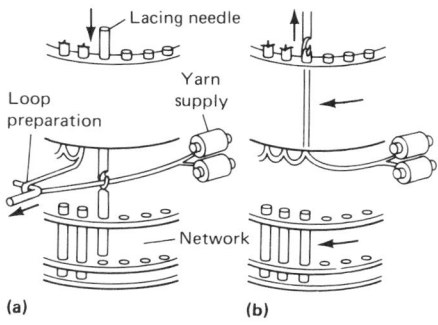

Fig. 10 The lacing operation, showing the replacement of a steel network rod with a yarn bundle: (a) the needle pushes the rod out of the preform and hooks a fiber bundle, and (b) the needle pulls the fiber bundle up through the preform, network, and preform index to the next lacing position.

Fig. 8 The TL1250 weaving machine, showing pierced plates, wires, and a partially woven exit cone

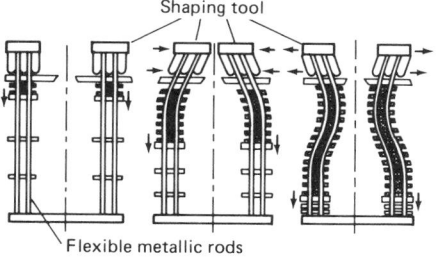

Fig. 11 Weaving of noncylindrical preforms using a shaping tool

tioned so that a rod end enters the drilled hole at the point of the needle. The needle is then thrust downward, pushing the rod until the rod is free of the preform, as shown in Fig. 10. The eye of the lacing needle then opens and captures a yarn loop, which has been prepared by a loop preparation mechanism. The needle then pulls along the captured yarn loop as it moves upward through the preform until it has cleared the top. This step is repeated until every wire in the network has been replaced with a yarn bundle, which completes the fabrication operation. The preform is then ready for densification into a composite material. Weaving noncylindrical preforms is accomplished by de-

forming the wire network with a shaping tool, as shown in Fig. 11.

Testing

Consisting as it does of geometrically arranged dry fibers only, a 3-D preform is not normally subjected to the extensive mechanical

testing used on densified composite materials. Dimensional inspection and weighing are used to calculate bulk density to ensure that the desired fiber volume has been achieved. Radiographic and computer-aided tomography are used to confirm a proper yarn geometry and to detect low- or high-density areas that may be indicative of missing yarns or metallic inclusions, such as broken lacing needles, rod ends, or wear particles from the weaving machine.

Wet Lay-up Resins

William T. McCarvill, Hercules Aerospace Company

WET LAY-UP using hand or spray techniques is one of the oldest and simplest methods of combining a fiber reinforcement with a solidifying resin to form a structure (Ref 1, 2). It is suitable for continuous unidirectional or chopped fibers, and for mat or woven fabric forms.

In the hand lay-up technique, fiber is hand-positioned in or on a mold, and liquid resin is poured onto the fiber. Cure occurs at room temperature with no applied pressure. Tools roll the blend to work out air bubbles and ensure complete fiber wet-out. Unfortunately, this can also result in fiber damage. The proper level of handling to maintain adequate properties requires at least semiskilled workers.

Large, complex structures or small production runs are economically served by this method. Mold costs are typically low. Very short runs can be made from plaster or wood molds, whereas longer runs usually require more durable materials. Expensive molds made from steel are rarely used. Applications include boats, chemical tanks, ducts, swimming pools, and bathroom fixtures. Fiber orientation and resin content cannot be accurately controlled. Consequently, wet lay-up is not used in strength/weight-critical applications, such as those found in the aerospace industry. Glass fibers are used primarily, because of their low cost and effectiveness. Higher-cost, higher-performance aramid and graphite fibers cannot be used in a way that justifies their cost.

The spray lay-up technique is faster than hand lay-up, but produces a less-uniform product. It consists of feeding a stream of chopped fibers into a spray of liquid resin in a mold cavity. The direction of the fibers is random, as opposed to the mats or woven fabrics that can be used in hand lay-up. Automated spray processes have been developed that improve uniformity within a part and from part to part, and increase production rates.

Because of the large size or low cost of parts fabricated by these methods, the part must achieve its properties at room temperature. To ensure adequate handling, wet-out, and fiber impregnation, the matrix must be liquid at room temperature. A sufficient time interval must exist during part lay-up to permit the entire part to be laid up before solidification. This time interval is called gel time. After lay-up of the part is completed, the resin must solidify or cure in a reasonable time at room temperature. Gelling and curing occur by means of chemical reactions that convert the monomers or liquid reactants to a solid. This conversion results in the liberation of heat. Too rapid a conversion will cause an exotherm in which the part and the mold rapidly increase in temperature; high temperatures can be reached that will cause damage to the part and mold, and in some instances, fires. Too slow a conversion will result in uneconomical production rates or an uncured part. The resin chemistry must be properly balanced to achieve the middle ground. Because both temperature and humidity affect the conversion rate, the actual resin formulation must be changed to meet actual use conditions.

If the part is cured at room temperature, the maximum operating temperature will be somewhat above room temperature. Additional temperature capability can be achieved with alternate resin formulations that require curing at elevated temperatures in an oven or during self-heating during cure. Hand lay-up techniques are not effective in completely removing entrapped air or compacting the fibers. The resulting voids and uneven fiber placement degrade the physical properties. Spray lay-up techniques produce better wetting of the fiber and fewer voids in the part. Composite parts made by this method are usually lighter and more corrosion resistant than the same structures made from metals. A complex part that consists of a number of small pieces can be made without the fastening and joining steps that would be required using other materials.

The hand lay-up technique requires precautions to prevent exposure of plant personnel to the resins. Skin contact with liquid resin components must be avoided through the use of gloves and protective clothing. Many chemical components used in resins are toxic or are irritants, while other components are highly volatile and flammable. Fresh air masks or adequate ventilation can prevent inhalation of these components. Contaminated clothing and tools must be cleaned before they are used. Once the part is cured, this hazard is reduced or eliminated. Improperly controlled reactions can cause exotherms that result in plant fires.

Resin Types

Polyesters and vinyl esters form the foremost family of resins used for wet lay-up.

Polyesters are based on reacting a mixture of acid anhydrides, acids, unsaturated anhydrides, and alcohols (Fig. 1). The nature of the acid or anhydride constituent exerts an influence on the properties of the final part. Table 1 lists various components and their applications. The resin, as prepared, is too viscous to use. Diluents that participate in the chemical reaction, such as styrene, vinyl toluene, and diallyl

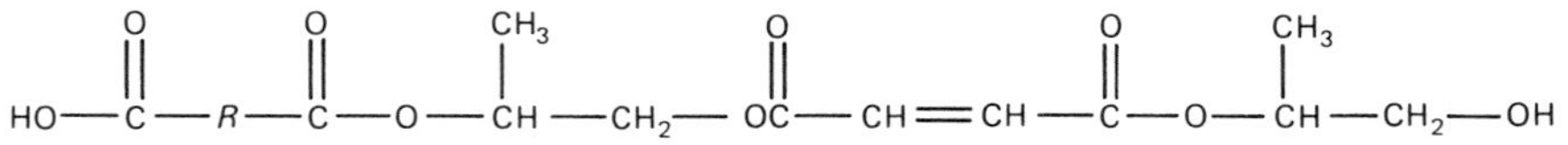

Fig. 1 Reaction of a mixture of acid anhydrides, acids, unsaturated anhydrides, and alcohols to form polyester. R, acid or anhydride constituent

Table 1 Components of polyester resins

Anhydride(a)	Diluent	Use
Orthophthalic	Styrene	General low cost
Isophthalic	Styrene	Better mechanical
Isophthalic	Vinyl toluene	Better mechanical, less volatile
Orthophthalic	Diallylphthalate	Improved electrical, less volatile
Isophthalic	Methyl methacrylate	Outdoor exposure
Tetrabromophthalic	Styrene	Fire retardant
Isophthalic & bisphenol A	Styrene	Corrosion resistance

(a) In addition to maleic or fumaric anhydride

Table 2 Partial list of suppliers of unsaturated esters and related compounds

Company	Product
Apogee Product Division,	
MIT Chemicals...................	Activators
Lucidol Division,	
Penwalt Corp.	Catalysts
U.S. Peroxygen	Catalysts
Plastics Division,	
American Cyanamid Co.............	Polyester resin
Kopper's Company Inc.	Polyester resin
Reichhold Chemicals Inc.	Polyester resin
ICI Americas	Polyester resin
U.S.S. Chemicals..................	Polyester resin
Ashland Chemical Co.	Polyester resin

Table 3 Typical polyester formulations

% MEKP (0.5% cobalt naphthenate)	Gel time at 30 °C (86 °F), min
2.0	9.0
1.0	18.5
0.75	25.0
0.5	38.5

% BPO (0.2% dimethyl aniline)	Gel time at 30 °C (86 °F), min
2.0	4.5
1.0	7.5
0.5	12.0
0.25	21.0

Table 4 Catalyst-promoter-inhibitor systems for room temperature cure polyester resins

Application or end use	System, %	Gel time starting at room temperature, min	Approximate time at 21-24 °C (70-75 °F) for development of Barcol Hardness = 35
Gel coats	MEKP-1.5(a) Cobalt naphthenate-0.4(b) (assessory promoters usually omitted because of tendency to discolor)	30 (high filler content)	6-8 (can proceed with lay-up over gel coat in 30-45 min)
For normal lay-up resins	MEKP-1.0 Cobalt naphthenate-0.4	32	6-8
For fast-cure resins	MEKP-1.0 Cobalt naphthenate-0.4 Dimethyl aniline-0.1	16	2-2.5
For fast-cure resins	MEKP-1.0 Cobalt naphthenate-0.4 Quaternary ammonium salt-0.1	15	2-2.5
Alternate room-temperature cure	Cyclohexanone peroxide(c)-1.0 Cobalt naphthenate-0.4	30	~6-8
Alternate room-temperature cure	Bis-I-hydroxy cyclohexyl peroxide(c)-1.0	30	~6-8
Alternate room-temperature cure	Benzoyl peroxide-1.0 Dimethyl aniline-0.1	20	2
Effect of inhibitor................	MEKP-1.0 Cobalt naphthanate-0.4 Hydroquinone-0.1	∞	∞

(a) Percentages based on 100 parts polyester resin. (b) Concentration of cobalt metal, 6%. (c) Peroxides costlier than MEKP.

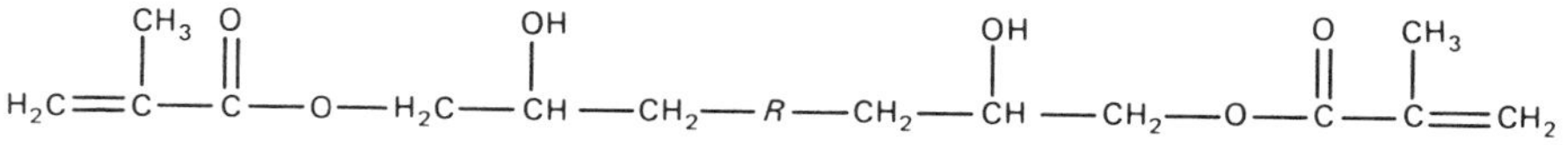

Fig. 2 Reaction of an unsaturated acid or anhydride with an epoxy to form vinyl ester. *R* is derived from the epoxy component

phthalate, promote a usable flow. Additives are also used to prevent running and bleeding of the resin on vertical surfaces after lay-up. A list of suppliers of polyester resins is provided in Table 2. These types of resins are called thermoset polyesters to differentiate them from a much different type of family called thermoplastic polyesters.

The chemical reaction that causes the hardening of the liquid mixture is a free-radical polymerization initiated by organic peroxides. The most common catalysts used are methyl ethyl ketone peroxide (MEKP) and benzoyl peroxide (BPO). In order to work at room temperature, they require an activator or accelerator. A typical formulation for hand lay-up is shown in Table 3. A wide range of gel times can be achieved by varying the amount of the peroxide and activator. A small change in activator quantity will exert a profound change in rate of solidification, as shown in Table 4. Amounts of catalyst should be carefully weighed. In no case should the activator be mixed with the catalyst because a violent reaction may occur. Incoming ingredients should be tested on a small scale to ensure proper reactivity before large amounts are mixed. In many cases the activator is mixed with the liquid resin, and the peroxide or catalyst is added by the fabricator.

Several resin-handling systems can be used to combine resin, activator, and catalyst safely. In the two-pot system, twice the needed catalyst is combined with half the resin, and twice the activator is combined with the other half of the resin. The two components are sprayed together and mix on the way to the mold. In the one-part system, after the resin is combined with the desired level of activator, pure catalyst is added in special spray guns designed to mix the two streams internally.

Vinyl esters shrink less and are more chemically resistant than the polyester family. They are prepared by reaction of an unsaturated acid or anhydride with an epoxy (Fig. 2). As in the case of polyesters, this resin is too viscous to be easily handled, and the same types of reactive diluents are used. The curing mechanism, free-radical polymerization catalyzed by an organic peroxide, also is the same. (Vinyl ester suppliers include Koppers Company, Ashland Chemical Company, Dow Chemical Company, ICI Americas, and Interplastic Corporation.)

Again, an activator is used to start the curing at room temperature. In order to more closely control gel time, a retarder can be used to delay the onset of polymerization and gelation. A typical vinyl ester formulation is shown in Table 5, and the effect of a retarder on gel time is shown in Table 6.

The parts produced by laying up against a mold have an uneven appearance caused by the

Table 5 Typical vinyl ester formulation

Percentages based on 100 parts vinyl ester resin

Vinyl ester, %	100%
Promoter (6% cobalt naphthenate), %	0.2-0.5%
Activator (100% dimethyl aniline), %	0.0-0.2%
Catalyst, (9% MEKP), %	0.9-2.5%

Source: Ref 3

Table 6 Delayed gel times for Derakane 411-45 resin

MEKP, wt%	Material — Cobalt naphthenate, wt%	2,4 pentanedione, wt%	Typical gel time, min	Typical peak time, min	Typical exotherm °C	Typical exotherm °F
1.0	0.25	0.0	21	37	40	108
1.0	0.25	0.05...................	23	39	60	138
1.0	0.25	0.1	60	74	55	132
1.0	0.25	0.2	180	191	70	153
1.0	0.25	0.3	265	280	60	147

Source: Ref 3

Table 7   Epoxy resins and curatives for wet lay-up

Company	Product
Shell Chemical Co.	Epoxy resins, curatives
Dow Chemical Co.	Epoxy resins, curatives
Ciba-Geigy Corp.	Epoxy resins, curatives
Interez Inc.	Epoxy resins, curatives
Reichhold Chemicals Inc.	Curatives
Applied Plastics Co., Inc.	Curatives
Pacific Anchor Chemicals	Curatives, activators

Table 8   Epon resin/curing agent system selection guide for fiberglass-reinforced plastics

Diluted Epon resins(a)	Curing agents	Recommended curing agent concentration(b)	Comments on curing agent
Epon resins 813, 815, and 8132, unmodified Epon resins 825, 826, and 828, and Araldite GY 6010, GY 6008, GY 6005	Epon Curing Agent U	20-25	Formulations with Epon curing agent U are useful for tooling or patching. Short pot life is a distinctive feature. The systems cure quickly, and strong composites are produced at room temperature or under heat lamps at 70–90 °C (160–200 °F).
	Epon Curing Agents V-15, V-25, V-40, V-50 (polyamides)	33-133	Polyamide systems provide longer pot life but longer cures than do those above. V-15 is extremely viscous; V-25 cures at room temperature or elevated temperatures; V-40 has lower viscosity than V-25. They are less toxic than the corresponding unmodified polyamines.
	Diethylenetriamine (DETA)	8-12	These systems provide good room-temperature properties, tensile strength, chemical resistance, short pot life, and cure at moderate temperatures.
	Triethylenetetramine (TETA)	11-14	

(a) Containing reactive diluents. (b) Parts per 100 parts of resin; recommended range given is calculated for use with liquid Epon resins, such as Epon resin 815 or 828; when solid Epon resins are used, curing agent level must be adjusted in accordance with the higher epoxide equivalent weight of the solid resins. Source: Ref 4

mixture of resin and fiber. Liquid resin containing pigment can be applied to the mold surface and allowed to gel and partially cure before laying on the fiber and resin that will actually make up the structural element. This coating, called the gel coat, provides the part with a smooth and uniform surface, but it is not used in all industrial applications.

Epoxies are a chemical family that is least used in the hand lay-up technique. The epoxide group consists of a cyclic ether that can be hardened or cured in a number of ways. Suppliers of epoxy resins and curatives used in room-temperature fabrication are shown in Table 7, and a typical formulation is presented in Table 8. Room-temperature cure systems are based upon amine curing agents. With reference to Table 8, the low viscosity of diluted resins permits rapid wet-out of tightly woven glass cloth and higher glass content (60-70%), which results in greater room-temperature strength than higher-viscosity resins. The unmodified resins are widely used in wet lay-up. Epon resins 825 or 826 are preferred when moderate viscosity handling characteristics are needed, such as in filament winding.

The hardening reaction is not catalytic, as it is for the vinyl ester and polyester families. The curative must be used in an amount that is sufficient to react with each epoxide group. This reaction can be speeded up by using catalysts. In general, epoxy resins are more expensive than the other resin families and are slower to react. Advantages of epoxy systems over the vinyl ester and polyester systems are higher strength, higher modulus, fewer

volatiles, excellent adhesion to reinforcement, lower shrinkage, and better chemical resistance. Disadvantages are more brittleness, properties lowered by water exposure, and long curing schedules.

Future Trends

Hand and spray lay-up techniques are mature technologies. Their advantages lie in the room-temperature low-pressure cure of low-viscosity impregnating resins. The properties achieved are not outstanding when compared to the careful lay-down of fibers in filament winding and prepreg lay-up. However, they are adequate for many large non-performance-critical applications. They will continue to be the preferred methods for small to moderate production runs of medium to large structures.

The vinyl ester and polyester families will continue to be the workhorses of the industry because of their low cost and well-developed production technologies. Epoxies will be used when their special performance characteristics justify their higher cost.

REFERENCES

1. J.G. Mohr, Ed., *SPI Handbook of Technology and Engineering of Reinforced Plastics Composites*. R.E. Krieger, 1981
2. C. Wittman, G.D. Shook, in *Handbook of Composites*, G. Lubin, Ed., Van Nostrand Reinhold, 1982, p 321-367
3. Publication 296-315-85, Dow Chemical Company, 1985
4. Technical Bulletin SC:227-85, Shell Chemical Company, 1985

Filament-Winding Resins

William T. McCarvill, Hercules Aerospace Company

FILAMENT WINDING is the process in which continuous strands or filaments of fiber are wound on a supporting form or mandrel (Ref 1, 2). Its best use is for making tube- and pipe-shaped objects, such as chemical storage tanks, corrosion-resistant and lightweight piping, liners for smokestacks, golf club shafts, aerospace missiles, and drive shafts. The continuous laying down of fibers in this process lends it to automation. A variety of fibers (glass, aramid, and graphite) can be used, depending on cost and the needed level of performance. Blends of fibers are also used to optimize cost/performance. In general, glass is the least expensive, but has the lowest performance level based on its density. Carbon fibers are the most expensive but have the highest performance level.

Processing

The manufacturing method, performance level, and cost of the wound tube are also controlled by the matrix or resin component (Ref 3). Resins can be applied to the fiber in a number of ways. The fiber can be dipped in a bath of liquid resin during the process of winding. The resin must have a low viscosity in order to flow easily into the fiber bundle and impregnate it with resin. The application of tension to the fiber helps to compact the winding and reduce the number of voids. Viscosities of 1 to 10 Pa · s (1000 to 10 000 cP) are usually acceptable. These can be achieved by the proper selection of resin components for a suitable room-temperature flow. It is also possible to heat the resin mixture to reduce the viscosity, although this usually shortens the working life of the resin. Elevated temperatures used to reduce viscosity will also shorten pot life, as the higher temperature will accelerate the chemical reactions that build viscosity. Pot life is the time that the resin mixture can be used until its viscosity becomes too high to impregnate the fiber bundle.

Important factors to consider in processing are the pot life, gel time, and cure time of the resin. In order for the part being formed to become a load-bearing structure, the liquid resin must chemically react or cure into a strong solid. This is achieved chemically by cross-linking the resin, thereby forming a three-

dimensional structure. This solidification process begins the moment the resin components are mixed. As the chemical reaction occurs, the mixture progressively builds higher viscosities, going from a liquid state to a state where no flow occurs. This is called the gel point, at which the full physical properties are not yet developed. Further exposure to higher temperatures or additional time is required to achieve maximum properties. As a general rule, the temperature of cure is somewhat below the maximum temperature capability of the composite. Room temperature cured resins will perform at or slightly above room temperature.

Wet-winding advantages include good fiber wet-out and consolidation of resin-wet fiber bundles. The amount of resin required can be mixed and used as needed, and a wide variety of resin types to fit the application can be used. Disadvantages include the need to inventory and mix chemicals, as well as inability to control precisely the resin content in the fiber.

Fiber bundles with resin already applied, known as prepreg tow, can be used for winding, bypassing the need for the resin bath. This also eliminates the need to measure and mix the resin and to impregnate the dry fiber. Resin content is usually more tightly controlled with this method than with wet winding. The tow must be kept at a low temperature to prevent premature resin reaction prior to warming and use. Fiber lay-down and compaction are poorer than in wet winding. After winding and before cure, the composite must be pressure compacted to reduce voids and delaminations between layers of winding. The resins used for prepregging are somewhat more viscous than wet-winding resins. The fiber bundles tend to stay where they are laid down and do not slip in regions of sharp curvature. The primary advantage lies in eliminating the need for resin mixing and content control during the winding process.

The least-used technique is winding the dry fibers to the desired form and enclosing the form in a sealed mold. The mold is evacuated, and liquid resin is injected. The composite is cured in the mold, yielding very smooth inner and outer surfaces that conform to the tolerances designed into the mold. Resins used for this technique must have very low viscosities in order to penetrate the fiber form fully with no

dry areas. A slow viscosity build-up and long gel time are also required to maximize fiber wet-out and encapsulation by resin.

Resins

The term resin is applied to a wide variety of chemically reactive compounds that solidify into load-bearing materials. The primary type of chemistry is based on thermoset formation. The reaction builds a three-dimensional chemical bonding. Because of the three-dimensional structure, the cured material never melts or flows, but will soften somewhat as the temperature is increased, and will even ignite or burn at sufficiently high temperatures. At some temperature, the material will lose stiffness to a degree that the composite will not carry the designed loads. Called the glass transition or heat distortion temperature, it is a function of the type of resin system used and the cure temperature.

Epoxy resins have the widest range of properties of the resins used for filament winding (Ref 4, 5). Compositions can be cured at room temperature to give tough, chemically resistant composites. Other compositions require elevated cure temperatures and may be used to make missile bodies with high-pressure capabilities and excellent resistance to temperature. The epoxies tend to be more expensive than other available resins, but are used where their unique properties are needed.

Some commercially available epoxies are shown in Table 1. Aromatic groups found in bisphenol A, bisphenol F, and novolac epoxies give high load-bearing and temperature capabilities. Aliphatic groups give low viscosity, relatively low temperature performance, and high flexibility. Various members of the families can be mixed together to give the precise properties needed. Epoxies are resistant to all common solvents and to mild acids and bases. It is this flexibility that has contributed to the widespread high-volume use of epoxies over the last 20 to 30 years.

A curative is needed to co-react with the epoxy group. An enormous number of curatives have been used, including the amines, anhydrides, and acids shown in Table 2. In filament winding, a popular class is based on amine (NH_2)-containing curatives. The type of amine

Table 1 Selected epoxy resins

Resin	Supplier	Formula
DER 332	Dow Chemical Co.	
EPON 826	Shell Chemical Co.	
EPI-REZ 509	Interez Inc.	
Araldite GY 6008		
(Diglycidyl ether of bisphenol A)	Ciba-Geigy Corp.	
EPN 1139	Ciba-Geigy Corp.	
DEN 431		
(Polyglycidyl ether of		
phenol-formaldehyde novolac)	Dow Chemical Co.	
EPI-REZ 5022	Interez Inc.	
RD-2		
(Diglycidyl ether of butanediol)	Ciba-Geigy Corp.	
Tonox 60/40		
40% *m*-phenylenediamine		
60% 4,4′-methylenedianiline	UniRoyal	

Table 2 Selected curatives for epoxies

Curative	Supplier	Formula
DEH 20		
(Diethylenetriamine)	Dow Chemical Co.	
Hardener HT 972	Ciba-Geigy Corp.	
MDA		
(Methylenedianiline)	Ciba-Geigy Corp.	
Hardener HT 976		
(4,4′-diaminodiphenyl sulfone)	Ciba-Geigy Corp.	
Phthalic anhydride	Allied Chemical	
DICY		
(Dicyandiamide)	Pacific Anchor Chemicals	
T403		
(Aliphatic polyether triamine)	Jefferson Chemical	

Table 3   Accelerators for epoxy curing

Accelerator	Supplier	Formula
DICY (Dicyandiamide)	Pacific Anchor Chemicals	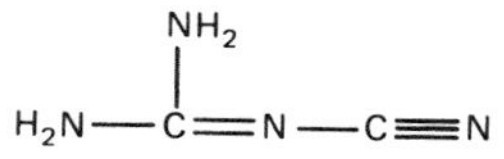
BF$_3$-MEA (Boron trifluoride-monoethyl amine)	Harshaw/Filtrol	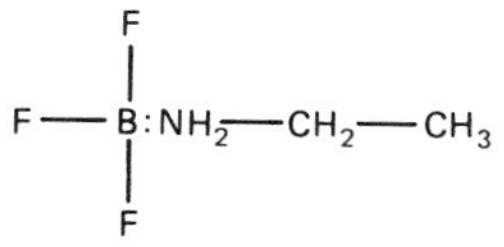
EMI-24 (2-ethyl-4-methyl imidazole	BASF Air Products	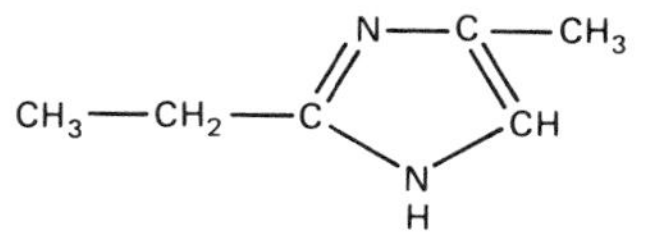

Table 4   Comparison of resin formulations

High-temperature formulation
  Components, parts by wt
    EPON 826 resin 100
    RD-2 diluent 25
    Tonox 60/40 curative 20
  Cure cycle . . . . 3 h at 60 °C (140 °F) and
                                       2 h at 120 °C (250 °F)
  Heat distortion temperature, °C (°F) 121   (250)
  Failure stress 50   (7.54)
Room-temperature formulation
  Components
    DER 332 resin 100
    Jeffamine T-403 curative 45
  Cure cycle . . . . 48 h at 20 °C (70 °F)
  Heat distortion temperature, °C (°F) 62   (140)
  Failure stress, MPa (ksi) 60   (8.85)

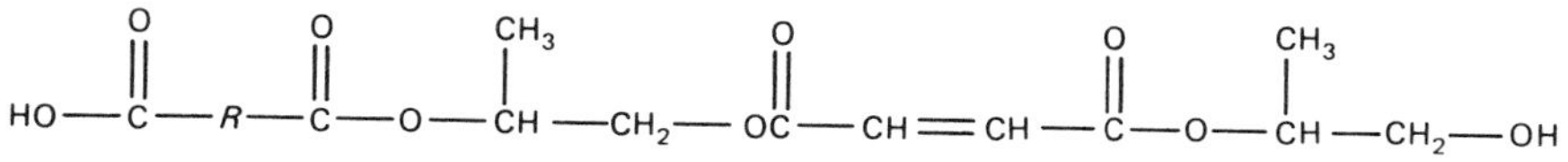

Fig. 1 Typical polyester. *R*, adipic acid, isophthalic, orthophthalic, or phthalic anhydride

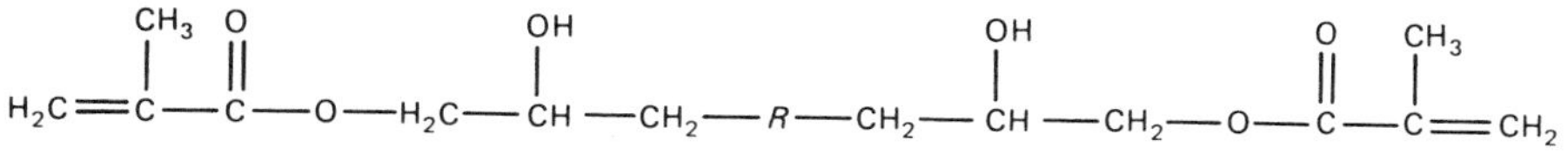

Fig. 2 Typical vinyl ester. *R* is derived from the epoxy component

Table 5   Typical vinyl ester formulation

Vinyl ester, % 100
Promoter (5% cobalt naphthanate), % . . . . .0.2 to 0.5
Catalyst (50% methyl ethyl ketone peroxide), % . .9 to 2.5
Activator (100% dimethylaniline), % . . . . .0.0 to 0.2

Source: Ref 7

Table 6   Gel time variation for Derakane 411-45 resin

	Material		
MEKP, wt%	Cobalt naphthanate, wt%	2,4 pentandione, wt%	Gel time, min
1.0	0.25	0.0	21
1.0	0.25	0.05	23
1.0	0.25	0.1	60
1.0	0.25	0.2	180
1.0	0.25	0.3	265

Source: Ref 7

used also controls the rate of curing and final properties. Aromatic amines are slow curing, but give high performance with aromatic epoxies. Aliphatic amines provide room-temperature cures with poor elevated-temperature capability. It is common to mix curatives in the same way that epoxies are blended, to achieve the desired properties and cure rate.

Catalysts described in Table 3 are used to accelerate the rate of cure. A wide range of flexibilizers and modifiers are also available to tailor end properties.

Anhydride curing agents represent the second most popular class. They give better thermal stability than amines, but can be more difficult to handle because of their selective solubilities in epoxies and tendency to volatilize during cure.

Table 4 compares a typical winding resin for high-temperature performance with a room-temperature cure system. The mixture of epoxies and amine curatives designed for high-temperature use needs an elevated-temperature cure to achieve the desired level of performance (Ref 6).

Composites being made from epoxy resin formulations must be rotated until gelation is achieved. If the pipe or tube is not rotated, resin will tend to flow to the lowest point, which results in a nonuniform distribution. Elevated-temperature cure cycles are based on the resin chemistry, mass, and geometry of the part. Heat-up rates and time at a given temperature must be controlled to completely cure all portions of the part.

Unsaturated polyesters and vinyl esters are similar enough chemically to be considered to be in the same family. Polyesters are less expensive than vinyl esters, but vinyl esters have somewhat better chemical resistance. Both are much less expensive than epoxy-based resins and generally cure more rapidly. They are both well suited for economical production of glass fiber based composites. Both are also styrene or vinyl toluene diluted, free-radical initiated, liquid thermosetting resins.

Polyesters are based on reacting an unsaturated acid anhydride, such as maleic anhydride, and another anhydride with a polyol. Various types of anhydrides and polyols will change the properties of the cured composite, although over a much narrower range than with epoxies. The polyester itself is too viscous for use in winding, and must be diluted with another low-viscosity unsaturated compound. The unsaturated bonds left in the polyester are the basis for further reaction and cross-linking (Fig. 1).

Vinyl esters are prepared from an unsaturated acid and an epoxy (Fig. 2). Again, the unsaturation reacts with the diluent to provide the polymerization and cross-linking. A typical vinyl ester formulation is shown in Table 5. Peroxides are compounds that break down thermally or by catalytic action to create free radicals that initiate and propagate polymerization. The temperature and rate at which polymerization will occur are dependent on the type of peroxides and activators used. Activators speed up the rate of cure; retarders extend the pot life and gel time. Examples of the flexibility to control gel times are shown in Table 6. Methyl ethyl ketone peroxide (MEKP) is the source of free radicals in the system,

Table 7 Derakane vinyl ester resins

Product name	Type	Heat distortion temperature °C	°F	Resin/ styrene ratio	Applications
Derakane 411-45	Bisphenol A epoxy	100	215	55/45	General
Derakane 510N	Brominated bisphenol A epoxy	120	250	64/36	Flame retardant
Derakane 8084	Flexibilized bisphenol A epoxy	80	180	60/40	Toughened
Derakane 470-36	Epoxy novolac	140-150	290-300	64/36	High temperatures

Source: Ref 8

Table 8 Types of polyester resin

Type	Anhydride	Glycol
General purpose	Orthophthalic, maleic	Ethylene, diethylene, or propylene
Corrosion resistant	Isophthalic, maleic, bisphenol A, chlorendic	Propylene
Flame resistant	Brominated tetra hydropththalic, tetrabromo- pththalic, chlorendic anhydride	Dibromoneopentyl glycol

Source: Ref 8

Table 9 High-temperature thermoplastics

Product	Producer	Chemistry	Glass transition temperature °C	°F	Melting point °C	°F	Form
Avimid K	Du Pont	Polymide	210	410	...	None	Amorphous
Ryton PAS	Phillips	Polyarylene sulfide	145	293	170	345	Crystalline
Torlon	Amoco	Polyamideimide	335	638	...	None	Amorphous
Ultem	General Electric	Polyetherimide	215	422	...	None	Amorphous
PEEK	ICI	Polyether etherketone	180	360	370	700	Crystalline

cobalt naphthanate acts as a promoter, and the 2,4 pentandione controls the gel time.

Various vinyl ester and polyester compound types are shown in Tables 7 and 8. The activator should be mixed into the resin before adding the peroxide. Mixing it directly with the peroxide may cause a fire or explosion. Because of the small amounts of catalyst and activator used to cure the unsaturated resin, care must be taken to weigh them accurately. Small variations in amount will greatly change the rate of reaction and can result in an uncured or too rapidly cured system.

The ambient conditions will also change the gel time and polymerization rate. If manufacturing conditions are not uniform, changes in formulation will be required. This situation is most acute for room-temperature cure systems. Formulations designed for elevated-temperature cures are more stable and less hazardous to use.

Future Trends

Thermoset-based resin systems will continue to be the workhorse of the filament-winding industry because of their ease of use, long history, familiarity, and low material cost. Thermoplastics are linear polymers that soften and flow at elevated temperatures. They are already fully reacted, in contrast to the liquids that become thermosets. With thermoplastics, subjecting them for a brief time to temperatures above the melting point (with application of pressure) is sufficient to melt or fuse them. Some high-temperature thermoplastics are shown in Table 9.

Several types of prepreg tow that use carbon fiber reinforcements are becoming available. These materials are now aimed at the high-performance markets because of their current high costs. Thermosets have low viscosities, impregnate tows easily, and can be wet-wound. Thermoplastics have very high viscosities at their flow temperatures. Special heating and impregnating equipment is needed to ensure full fiber wet-out and impregnation at the high temperature needed to melt the thermoplastic. Once made, prepreg tape can be briefly heated and pressed during the winding operation. No further heating is needed to cure the composite, as in the case of thermoset resins. Because the resin is fully reacted, no toxic chemicals are used or liberated during cure. There is also no danger from exothermic runaway caused by improper mixing or curing of reactive ingredients, and the storage life of the prepreg is unlimited, even at room temperature. As prepreg processes for fiber impregnation and automated rapid winding and fusion are developed, thermoplastics will begin to replace thermosets in some markets.

REFERENCES

1. J.G. Mohr, Ed., *SPI Handbook of Technology and Engineering of Reinforced Plastics Composites*, R.E. Keiger, 1981
2. Publication 5-PL-13199-A, Owens-Corning Fiberglas Corporation, 1986
3. A.M. Shibley, in *Handbook of Composites*, G. Lubin, Ed., Van Nostrand Reinhold, 1982, p 448-477
4. H. Lee and K. Neville, *Handbook of Epoxy Resins*, McGraw-Hill, 1967
5. N.L. Hancox, *Some Properties of Epoxy Resins Used For Filament Winding and Preimpregnating Fiber*, AERE-R-10966, United Kingdom Atomic Energy Authority, 1983
6. L.S. Penn and T.T. Chiao, in *Handbook of Composites*, G. Lubin, Ed., Van Nostrand Reinhold, 1982, p 57-88
7. Publication 296-315-85, Dow Chemical Company, 1985
8. Publication 296-623-85, Dow Chemical Company, 1985

Prepreg Resins

William T. McCarvill, Hercules Aerospace Company

FIBERS THAT ARE PREIMPREGNATED with matrix resin in the uncured state are known as prepregs. Either continuous or chopped fiber prepregs are supplied to a part fabricator to be laminated or molded. The laid-up part is then subjected to heat and pressure to cure (chemically react) the resin. Prepreg has become an article of commerce because it frees the end user from having to develop resin formulations and impregnate fiber. The composite material can be bought with resin content, resin type, and fiber type already made to order. Handling characteristics such as curing time and temperature can be controlled to precise levels to meet user requirements. Prepregs can be divided into at least two classes: those suitable for aerospace applications, and those to be used in lower-performance molding compounds. Aerospace applications demand high-performance, high-quality composites and moldings. The lower-performance applications employ sheet molding prepregs for automotive components and appliance housings. The two general classes differ widely in composition, handling, part manufacture, and use.

Aerospace Applications

The reinforcement in prepregs supplied for aerospace application is typically carbon fiber. The high strength and stiffness of carbon, coupled with its low density, result in composites with a higher performance/weight ratio than is possible in either metals or composites using glass fiber. The prepreg consists of resin-impregnated fiber in either uniaxial or woven form. Typical properties of a graphite prepreg tape consisting of amine-cured epoxy resin reinforced with unidirectional graphite fibers are shown in Table 1.

The part fabricator requires a prepreg with tack, drape, and a certain tack life and out time. Tack is the tendency of two plies or layers to adhere sufficiently to allow laying up of complex parts yet allow a clean strip-back if layers are applied incorrectly. Too low an adhesion level will allow layers to slip, while too aggressive a level will not allow repositioning. Drape is the ability of the prepreg to bend and conform to mold curvature. Tack life refers to the amount of time that the prepreg can be at room temperature and still retain enough tack for

lay-up. Out time is the total amount of time that the prepreg can be left at room temperature before curing and still make a good part. Fabricators must consider all of these handling characteristics, in addition to the cured properties generated by certain resin and fiber combinations.

Epoxy Resins. A resin chemistry that satisfies both manufacturing and composite property requirements is based on epoxy resins with a latent curative system. The cure system will be slow, at room temperature, to prevent reactions that reduce tack, drape, and out time, but sufficiently rapid at elevated temperatures to permit reasonably short curing times. Even the most latent systems in use today do not completely eliminate room-temperature reaction. After the fiber is impregnated with resin, it is stored and shipped at low temperatures. The material is allowed to warm to room temperatures for lay-up. Typically, a 1-year storage life at −20 °C (0 °F) is provided.

The epoxide group is well known and is a mature technology (Ref 2, 3, 4). A wide range of epoxy-containing ingredients are available, as well as a wide range of curing agents and catalysts. A partial list is shown in Table 2. Resins with different viscosities, amounts of reactive groups, and structures are available. Additives that change the uncured resin viscosity, reduce brittleness, or impart some other

property are available. Aromatic backbones and high functionality give a strong high-temperature highly cross-linked matrix that is usually brittle. Aliphatic epoxies and low functionality usually result in matrices with higher elongation, lower temperature capability, and higher toughness. The primary resin for aerospace application is $N,N,N'N'$-tetraglycidyl-4, 4'-methylenebisbenzenamine. When reacted with the appropriate curative, it yields a hard resin with temperature capabilities of about 190 to 205 °C (375 to 400 °F).

Curatives. The epoxide group can react chemically with other molecules to form a three-dimensional network. This chemical reaction changes the liquid resin into a load-bearing solid. Curing agents can include amines, anhydrides, acids, and many others. Two commonly used amine curatives for prepreg resins are shown in Table 3. Other curatives are too reactive with the epoxies at low and room temperatures, resulting in an unacceptable reduction of storage and use life. Dicyandiamide appears to decompose at elevated temperatures of 145 to 154 °C (290 to 310 °F) to yield other nitrogen-containing species, which cause the curing reaction to occur. The curative 4,4'-diaminodiphenylsulfone may or may not completely dissolve in the epoxy resins. Insolubility contributes to its latent curing behavior. Curatives are usually mixed with the epoxy

Table 1 Typical graphite prepreg tape properties

Typical composite properties		
Tensile strength at 25 °C (77 °F), MPa (ksi)	1600	(230)
Tensile modulus at 25 °C (77 °F), GPa (10^6 psi)	140	(20.0)
Flexural strength at 25 °C (77 °F), MPa (ksi)	1800	(260)
Flexural strength at 175 °C (350 °F), MPa (ksi)	1250	(180)
Flexural modulus at 25 °C (77 °F), GPa (10^6 psi)	120	(17.5)
Flexural modulus at 175 °C (350 °F), GPa (10^6 psi)	117	(17.0)
Short beam shear strength at 25 °C (77 °F), MPa (ksi)	120	(17.5)
Short beam shear strength at 175 °C (350 °F), MPa (ksi)	65	(9.5)
Fiber vol%	62	
Cured-ply thickness, μm (mil)	132	(5.2)
Typical prepreg characteristics		
Fiber areal weight, g/m² (oz/yd²)	149	(4.4)
Standard width, mm (in.)	300	(12)
Approximate yield, kg/m (ft/lb)	28.1	(18.9)
Resin content, % by weight	42 ± 3, 35 ± 3	
Gel time at 175 °C (350 °F), min	6-12	
Volatile content, max % by weight	1	
Out time at room temperature, min days	10	
Shelf life at −20 °C (0 °F), mo	12	

Note: Tensile and flexural strength and modulus measured in 0° direction. Source: Ref 1

Table 2 Epoxy resins used in aerospace prepregs

Name	Supplier	Formula
Araldite MY 0510 (Triglycidyl p-aminophenol	Ciba-Geigy Corp.	[chemical structure]
Araldite MY 720 (N,N,N',N'-tetraglycidyl-4',4'-methylenebisbenzenamine)	Ciba-Geigy Corp.	[chemical structure]
EPON 826 (Diglycidyl ether of bisphenol A)	Shell Chemical Co.	[chemical structure]
DER 330 (Diglycidyl ether of bisphenol A)	Dow Chemical Co.	
Epiclon 830 (Diglycidyl ether of bisphenol F)	Dinippon	[chemical structure]
Araldite ECN 1235 (Epoxy novolac)	Ciba-Geigy Corp.	[chemical structure]

content on a 1:1 chemical basis. In the calculation of amounts, each hydrogen of the amine group is considered to react with one epoxy group. The actual mix ratio may be varied to optimize desired properties.

Catalysts are employed to accelerate the latent curatives in order to achieve a complete cure in a shorter time. Boron trifluoride (BF_3) can be rendered latent by complexing with nitrogen-containing compounds, such as monoethylamine. Other amines can be used to adjust latency.

The types of amines and epoxies used for prepregs react slowly at room temperature, and elevated temperatures are needed for complete cure and attainment of ultimate properties. However, as a general rule, the temperature-use capability of the cured resin is slightly above the actual curing temperature. Therefore, a room-temperature system will operate at room temperature within a range; a formulation cured at 175 °C (350 °F) will perform at or somewhat above that temperature.

The actual mixture of epoxy resin or resins, curative, and catalyst that is formulated and

Table 3 Curatives for prepreg resins

Name	Supplier	Formula
Hardener HT 976 (4,4'-diaminodiphenylsulfone)	Ciba-Geigy Corp.	[chemical structure]
DICY (Dicyandiamide)	Pacific Anchor Chemical Co.	[chemical structure]

blended is designed to meet end-use, prepregging, handling, and storage requirements. Low-viscosity resin components are used to reduce overall viscosity and aid in prepreg manufacture and flow during curing. Aromatic epoxies that differ in epoxy functionality are used to vary the cross-link density. Highly cross-linked aromatic resins employed in aerospace prepregs make a very strong composite that is nonetheless brittle; flexibilizers or tougheners may be added to decrease brittleness. The amount and type of epoxy resins is varied to maximize

desired composite properties as well as improve processing and handling characteristics. A typical formulation is shown in Table 4.

Cure cycles for resin formulations are determined empirically. A given cure cycle may have several hold steps on the temperature rise to the maximum cure temperature. Hold steps at a given temperature allow resin flow to ensure a void-free part. They also prevent runaway temperature increases caused by rapid rates of reaction and cross-linking. The programmed heat-up cycle allows reactive groups to be

Table 4 Prepreg resin formulation

Components, parts by wt
  DGEBA epoxy 100
  DDS curative 36
  BF₃-MEA accelerator 0.5
Cure cycle . . . . 24 h at 120 °C (250 °F) and
                                  4 h at 175 °C (350 °F)
Gel time for 30-g (1-oz) mass, h
  at 100 °C (212 °F) 3
  at 120 °C (250 °F) 1.93
Heat distortion temperature after cure, °C (°F). . .190 375

Source: Ref 3

Table 5 Prepreg suppliers

Company	Prepreg products
American Cyanamid Co.	
Engineered Materials Dept.	thermoset
Amoco Performance Products Inc . . .	thermoplastic
Ciba-Geigy Corp.	
Composite Materials Dept.	thermoset
Ferro Corp.	
Composites Div	thermoset
Fiberite Corp.	thermoset/thermoplastic
Hercules Inc.	
Aerospace Div	thermoset
Hexcel Corp.	thermoset
Dexter Corp.	
Hysol Div.	thermoset
Narmco Materials Inc.	thermoset
Phillips Chemical Co.	
Engineering Plastics	thermoplastic
U.S. Polymeric Inc.	
Hitco Materials Div.	thermoset

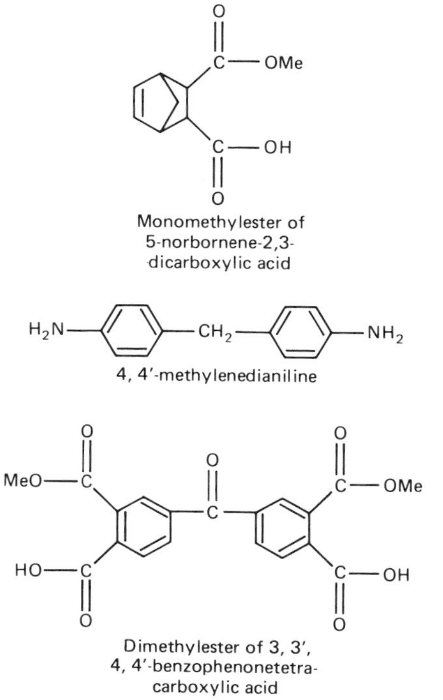

consumed at a rate that permits removal of the heat of reaction. Subsequent heating to a higher temperature (and higher polymerization rate) is safe, because the amount of reactive material has been reduced. Conversely, hold steps can build viscosity through a reaction that reduces resin flow when the temperature is increased. Cure cycles recommended by the prepreg supplier are designed for specific composite thicknesses and resin formulations. Significant departures from customary lay-ups and cure schedules should be reviewed for safety as well as for achievement of ultimate properties.

Another epoxy application for prepregs is in circuit boards. Glass is used for this application because carbon fiber is electrically conductive. The combination of glass and epoxy resin yields good electrical properties, in addition to good temperature capabilities and ease of epoxy chemistry processing.

A partial list of companies supplying prepreg materials is shown in Table 5.

High-temperature epoxy resins have a maximum continuous-use temperature of 205 to 230 °C (400 to 450 °F). Temperature spikes up to 290 °C (550 °F) can be tolerated by some formulations based on epoxy novolacs. The amine-cured epoxy resins are also affected by water. Because of the hydroxyl groups generated during cure, water is absorbed readily and acts like a plasticizer. Water exposure can reduce the operating temperature by 56 °C (100 °F). Although much effort has been spent to synthesize polymers that possess adequate thermal and water resistance, very few have proved

Fig. 1 Chemical structure of PMR-15 and its monomers

to be economically feasible. However, two chemical approaches appear able to solve the problem: addition polyimides and condensation polyimides (Ref 5).

Addition polyimides are based on the reaction of bismaleimide molecules with aromatic amines or dienes. These systems possess temperature capabilities in excess of 260 °C (500 °F). Addition polyimides are similar to epoxies in that no volatiles are released by the polymerization reaction. In order to achieve the high-temperature properties, autoclave cures at 175 °C (350 °F) are followed by postcures at 315 to 340 °C (600 to 650 °F) in a free-standing oven. The primary drawback to addition polyimides is their extreme brittleness, which can be compensated for, somewhat, but usually at the loss of temperature capability or water resistance.

Condensation polyimides have temperature capabilities exceeding 315 °C (600 °F) and are not as brittle as the addition polyimides. A PMR polyimide solution (*in situ* polymerization of monomer reactants) consists of a mixture of monomers dissolved in solvent. The monomers are latent, as in the case of prepreg epoxy resins, but react at high temperatures. The monomers are listed in Fig. 1. The presence of solvent, as well as the liberation of volatiles during polymerization, can create voids in the

Table 6 Typical vinyl ester formulation

Component	Parts by wt
Vinyl ester resin	40
Styrene	30
Polyvinyl acetate	10
Calcium carbonate	150
AS P400 clay	17
Zinc stearate	4
t-Butyl perbenzoate	0.5

Source: Ref 9

cured parts. The cure cycle and devolatilization are critical for producing high-quality parts. A typical cure cycle (in this case, for PMR-15, Ref 6) would be:

- Apply 17 kPa (5 in. Hg) vacuum at room temperature
- Heat to 250 °C (485 °F) at 15 °C (30 °F) per min
- Hold at 250 °C (485 °F) for 30 min
- Apply a minimum 81 kPa (24 in. Hg) vacuum at 1400 kPa (200 psi) pressure
- Heat to 325 °C (620 °F) at 20 °C (40 °F) per min
- Hold at 325 °C (620 °F) for 3 h
- Cool down at 3 °C (5 °F) per min to 65 °C (150 °F) under full vacuum and pressure
- Postcure for 6 h at 330 °C (625 °F) in an air-circulating oven

Glass-phenolic and carbon-phenolic prepregs are used in specialty flame-resistant and ablative applications (Ref 7, 8), including aircraft interior panels and exit cones for rocket motors. Phenolic resins have excellent heat resistance, but their drawbacks include brittleness and the need for press curing at high pressures to suppress volatiles generated during the cross-linking reaction.

Lower Performance Applications

Sheet molding compounds (SMCs), used in applications where the high performance and high cost of carbon prepregs are not justified, consist of continuous or chopped fibers and a polyester or vinyl ester resin. The formulation includes inorganic filler, thixotrope, catalyst, release agent, and pigments. Processing conditions usually are 1 to 3 min in a press at 90 to 150 °C (200 to 300 °F). A typical vinyl ester formulation is shown in Table 6.

Sheet molding compound is manufactured by calendering a strip that contains the fiber and resin. Because of the latent nature of the catalyst, molding compounds can be stored and shipped at ambient temperature.

The two types of resin families that are used—polyester and vinyl ester—are similar in

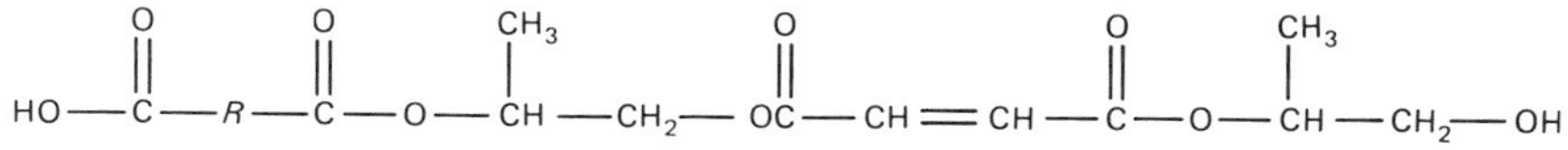

Fig. 2 Reaction of glycols and dibasic acids and anhydrides to form polyester resin. *R*, acid constituent

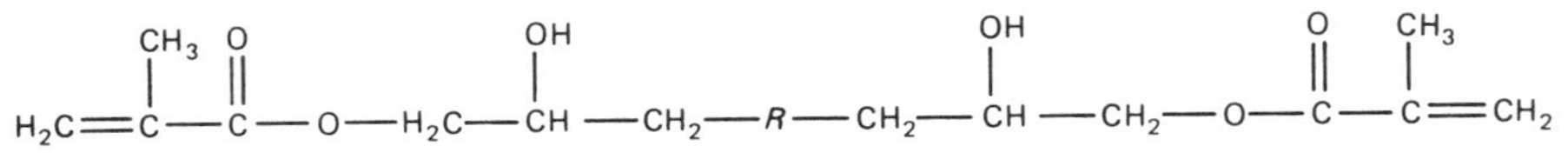

Fig. 3 Reaction of unsaturated carboxylic acids and epoxies to form vinyl ester resin. R is derived from the epoxy component

Table 7 Gel times of a vinyl ester resin with various initiators

Initiator	Gel time, min(a) at		
	25 °C (80 °F)	80 °C (175 °F)	120 °C (250 °F)
2% benzoylperoxide (BPO) (50%) + 0.1% dimethylaniline (DMA)	10	...	...
1.5% methy ethyl ketone peroxide (MEKP) (50%) + 3% Co-octoate (1%) + 0.1% DMA	14	...	...
1.5% MEKP + 3% Co-octoate (1%) + 0.015% DMA	34	...	...
2% BPO (50%) + 0.01% DMA	117	...	...
2% MEKP (50%)	700	6	2
2% MEKP (50%)	...	25	15
1% BPO (50%)	...	25	15
1% cumylhydroperoxide	...	32	10
1% t-butylperbenzoate	...	120	6
1% t-butylcumylperoxide	...	360	9

(a) 5g (140 oz.), isothermally. Source: Ref 10

Table 8 Subjective assessment of thermoset (TS) and thermoplastic (TP) processing

Aspect	Advantage	Reason
Prepreg formation		
Viscosity	TS	Lower
Solvents	TS	Greater choice
Hand	TS	More flexible
Tack	TS	Prepolymer variable
Storage	TP	Not reactive
Quality control	TP	Fewer variables
Composite fabrication		
Lay-up	TS	Ease of handling
Degassing	TP	Fewer volatiles
Temperature changes	TP	Fewer
Maximum temperature	TS	Lower
Pressure changes	TP	Fewer
Maximum pressure	TS	Lower
Cycle time	TP	Lower
Postcure cycle	TP	Not required
Repair	TP	Remelt
Post forming	TP	Remelt

Source: Ref 11

chemistry. A typical polyester resin is formed by the reaction of glycols and dibasic acids and anhydrides (Fig. 2). Other unsaturated compounds, such as styrene, are added to cross link and act as a diluent. Curing is accomplished by peroxide-initiated free-radical polymerization.

Vinyl ester resins are manufactured by reacting unsaturated carboxylic acids and epoxies (Fig. 3).

As in the case of polyesters, styrene or vinyl toluene is used as a diluent for processing and chemical reactions. Vinyl esters are tougher, shrink less, and are more resistant to chemical attack than polyesters.

For both families, the composition of the resin affects the physical properties of a final part. The type of chemical backbone and diluent will allow weather resistance, chemical resistance, impact properties, and flow during processing to be tailored to fit the application. The curing time and storage life is altered by the amount and type of catalyst, accelerator, and inhibitors, as shown in Table 7. Suppliers of SMCs include: Applied Composites, Budd Company, Interplastic Corporation, Polymer Engineering Inc., Poly Ply Inc., and Premix Inc.

Future Trends

Thermoplastic resins are materials that are already fully polymerized in a linear fashion, as opposed to the three-dimensional nature of thermosets. Their high-temperature properties are achieved through backbone stiffness, whereas thermosets achieve their properties through their high degree of cross-linking.

The advantages and disadvantages of thermosets and thermoplastics are summarized in

Table 9 Selected high-temperature thermoplastics

Polymer	Symbol	Glass transition temperature		Melting point	
		°C	°F	°C	°F
Polybutylene terephthalate	PBT	40	105	228	440
Polyethylene terephthalate	PET	80	175	265	510
Polysulfone	PS	190	375	(a)	
Polyphenylene sulfide	PPS	93	200	288	550
Polyether sulfone	PES	230	445	(a)	
Polyether etherketone	PEEK	143	290	340	645
Polyimides	PI	280	535	(a)	
Polyetherimide	PEI	210	410	(a)	

(a) Amorphous polymer. Source: Ref 11

Table 8. Table 9 identifies several high-temperature thermoplastics. The primary advantage of thermoplastics lies in the fact that they are already fully reacted. Subjecting them for a brief time to a temperature above their softening point will cause them to melt and fuse. In the case of thermosets, a long time is needed to react the molecules and achieve the fully cured state. An epoxy system requires vacuum degassing, heating under vacuum, and curing under vacuum and high temperature for a period exceeding 6 h. Thermoplastics require heat and pressure for a brief time at the softening or melting point, followed by pressure cooling below the melting point.

REFERENCES

1. Product Data for Magnamite graphite prepreg tape, AS/3501-6, No. 844-1, Hercules Inc.
2. H. Lee, K. Neville, *Handbook of Epoxy Resins*, McGraw-Hill, 1967
3. L.S. Penn and T.T. Chiao, in *Handbook of Composites*, G. Lubin, Ed., Van Nostrand Reinhold, 1982, p 57-88
4. P.F. Bruins, *Epoxy Resin Technology*, Wiley-Interscience, 1968
5. K.L. Mittal, Ed., *Polyimides*, Vol 1, Plenum, 1984
6. Product Data Bulletin, U.S. Polymeric, Hitco Material Division
7. A. Knop and L.A. Pilato, *Phenolic Resins*, Springer-Verlag, 1985
8. K.L. Forsdyke, G. Lawrence, R.M. Mayer, and I. Patter, The Use of Phenolic Resins for Load Bearing Structures, in *Engineering With Composites*, Society for the Advancement of Material and Process Engineering, 1983
9. M.B. Launikitis, in *Handbook of Composites*, G. Lubin, Ed., Van Nostrand Reinhold, 1982, p 38-49
10. U. Gruber, in *Recent Adv. Material Res., Proc. Silver Jubilee Comm. Sem. Indian Inst. Technol.*, C.M. Srivastava and A.A. Balkema, Ed., 1982, p 289-306
11. J.D. Muzzy and A.O. Kays, Thermoplastic vs. Thermosetting Structural Composites, *Polym. Compos.*, Vol 5 (No. 3), 1984, p 169-172

Unidirectional Tape Prepregs

Fred S. Dominguez, Hercules Aerospace Company

PREPREG TAPES of continuous fiber reinforcement in uncured matrix resin are one of the most widely used forms of composite materials for structural applications. Tapes offer the designer advantages in the areas of economics and translation of fiber properties. To select the proper prepreg form for an application, the designer must be familiar with the available product forms. Prepreg tape is a collimated series of fiber-reinforcing tows impregnated with a matrix resin. Before tape is manufactured from fiber, the fiber is usually found in a spooled form with widely varying tow size, weight per unit length (denier), and filament size, as shown in Table 1.

Tape Manufacture and Product Forms

The fiber is typically converted into a prepreg by bringing a number of spooled tows into a collimated form, as shown in Fig. 1. The prepregging operation consists of heating a matrix resin to obtain low viscosity and creating a well-dispersed fiber-resin mass. The amount of fiber is controlled by the number of tows brought into the prepreg line, and the resin can be cast onto the substrate paper either on the prepreg line or in a separate filming operation to obtain the desired fiber-resin ratio. The prepreg is calendered to obtain a uniform thickness and to close fiber gaps before being wound on a core. Substrate paper is ordinarily left between layers of tape. The paper can be any releasing film but is typically a calendered paper coated with a nontransferable, cured silicone coating. This substrate paper is available in many forms, which can be matched to the tack of the matrix resin and to the user's needs.

The finished product, a thin sheet of fiber-reinforced resin, is usually wound on a cardboard core and interleafed with paper, as described above. Figure 2 shows a typical spool of graphite-epoxy prepreg tape, which is available in a wide variety of widths, thicknesses, and package sizes. Table 2 shows typical ranges of product dimensions.

Using narrow prepreg tape (75 mm, or 3 in.) usually results in a minimal material loss of 7 to 10%. Narrow tape is ideally suited for a very expensive material, such as boron-epoxy. However, using narrow tape increases labor costs, which must be balanced against material costs.

Tape Properties

Mechanical. Reinforcement fibers by their nature are anisotropic (reinforcing primarily in one direction). Consequently, unidirectional tapes reinforce primarily in the 0° direction of the reinforcing fibers. Other structural properties also vary depending on fiber direction.

Tapes offer the best translation of fiber properties because the fibers are not crimped or distorted as in fabric prepregs. Significant differences exist between tape and fabric mechanical properties. Figures 3 and 4 show typical tensile property translation differences between tape and fabric prepregs.

Physical properties that are critical to the selection of a material form include tack, flow, gel time, and drape.

Tack, which is the measure of the adhesion of the tape to tool surfaces and to other prepregs, is an adhesion characteristic that is con-

Table 1 Fiber bundle dimensions

Material	Yield/tow		Filament size	
	m/kg	yd/lb	μm	μin.
Graphite (1000 to 12000 filaments per tow)	300-1200	150-600	5-10	200-390
Fiberglass (2450-12240 filaments per tow)	490-2400	245-1200	4-13	160-510
Aramid (800-3200 filaments per tow)	2000-7850	980-3900	12	470

Table 2 Tape dimensions

Parameter	Typical range
Thickness, mm (in. × 10⁻³)	0.08-0.25 (3-10)
Resin content, %	28-45 nominal ± 2
Dry fiber areal weight, g/m² (oz/ft²) . .	30-300 (0.10-1.0)
Width, mm (in.).	25-1525 (1-60)
Package size, kg (lb)	4.5-225 (10-500)

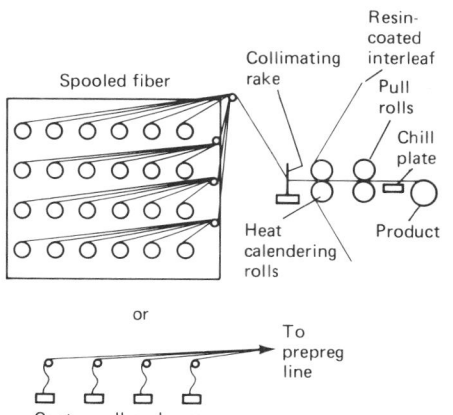

Fig. 1 Typical prepreg machine

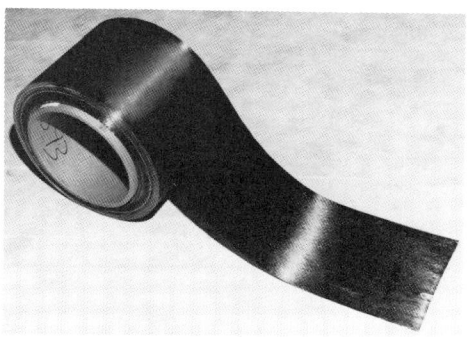

Fig. 2 Spool of graphite-epoxy tape

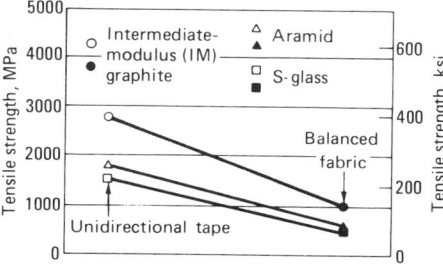

Fig. 3 Tensile strength comparison—fiber-epoxy tape versus fabric

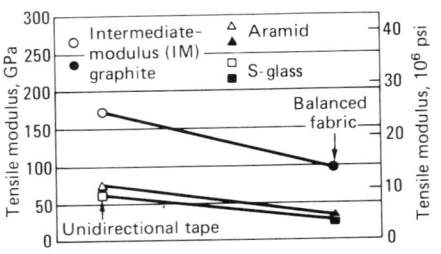

Fig. 4 Tensile modulus comparison—fiber-epoxy tape versus fabric

trolled in order to facilitate lay-up operations. It is affected by the apparent viscosity of the resin, which may be affected by inert volatile content, advancement of the matrix cure, or lay-up room temperature and humidity. Sometimes tack can be increased by increasing resin and volatile contents, by retarding prepreg advancement, or by increasing lay-up room temperature or humidity. Resin formulations can be reapportioned or new additives can be blended into the resin to increase tack.

Prepregs with excessive tack generally are difficult to handle without disrupting resin distribution and fiber orientation or causing a roping (fiber bundling) of the reinforcements. Constituents are not reproducible because undetermined amounts of resin are removed when the release film or backing is separated from the prepreg. In general, all the disadvantages of wet lay-up systems are inherent to overly tacky prepregs.

Prepregs with no tack are either excessively advanced, have exceeded their normal storage life, or are inherently low in tack. Such materials cannot attain adequate cured properties and should be discarded. Exceptions are silicones and some polyimides, which can only be prepared with no tack. Lay-ups with these materials are limited to those situations where lower mechanical properties can be tolerated in exchange for improved heat resistance or electrical properties. A lack of tack in thermoplastic prepregs does not interfere with their consolidation, provided that they can be heated to the melting point of the polymer during processing.

The tack qualities should be adequate to allow the prepreg to adhere to prepared molding surfaces or preceding plies for a lay-up, but light enough to part from the backing film without loss of resin. Tack qualities can be specified to require the prepreg to remain ad-

hered to the backing until a predetermined force is applied to peel it off.

Tack requirements can also be modified to suit plant fabrication conditions, provided that cured laminates will not be adversely affected. Local temperature and humidity sensitivities are minimized by air conditioning. Heavy tack can be made manageable by reducing the temperature, and drier tack can be improved by judicious use of hot-air guns or tackifying adhesives.

Flow is the measure of the amount of resin squeezed from a specimen as it cures (under heat and pressure) between press platens. Flow measurement indicates the capability of the resin to fuse successive plies in a laminate and to bleed out volatiles and reaction gases. Flow can be an indicator of prepreg age or advancement. It is often desirable to optimize resin content and viscosity to attain adequate flows. In some cases, prepreg flow can be controlled by adding thickening or thixotropic additives to the resin.

Gel time, the measure of the time a specimen remains between heated platens until the resin gels or reaches a very high viscosity stage (Ref 1), can be an indicator of the degree of prepreg advancement. The useful life of prepregs is limited by the amount of staging or advancement. Most prepregs are formulated to attain a useful life of 10 days or more at standard conditions. Life can be prolonged by cold storage, but each time the prepreg is brought to thermal equilibrium at lay-up room temperatures, useful life is shortened. Gel time measurements are used as quality control verifications (Ref 1).

Drape is the measure of the formability of a material around contours, which is critical to fabrication costs. Tape drapability is typically measured by the ability of a prepreg to be formed around a small-radius rod. The pass/fail criterion for drape is the ability to undergo this forming without incurring fiber damage. This measurement translates to the ability of fabrication personnel to form the prepreg to complex tools. Of the physical properties mentioned, drape is one property where tapes differ from other prepreg forms. Tapes are typically less drapable than fabric forms of prepreg, and this difference must be considered when specifying a prepreg form for manufacture.

It is essential that prepregs for structural applications be staged to desirable tack and drape qualities. The combination of manageable tack and drape is sometimes best attained

from woven satin fabric-reinforced prepregs. Cross-plied or multiplied prepregs are sometimes used to provide transverse strengths for lay-ups of broad goods. The term broad goods refers to wide prepreg tape (>305 mm, or >12 in.) that consists of one or more plies of tape oriented at 0° or off-axis to each other.

During the fabrication or lay-down process, temperature and humidity sensitivities are usually minimized in air-conditioned and pressurized clean rooms. The pressure for clean rooms is maintained by filtered air that is kept at positive gage pressure by blowers. The pressure is just high enough to prevent airborne contaminants in the surrounding atmosphere from entering the clean room (Ref 1).

Applications

Unidirectional tape manufacturing processes fall into three major categories: hand lay-up, machine-cut patterns that are laid-up by hand, and automatic machine lay-up.

Hand Lay-up. Historically, tapes have primarily been used in hand lay-up applications in which the operator cuts lengths of tape (usually 305 mm, or 12 in.) and places them on the tool surface in the desired ply orientation. Although this method uses one of the lower-cost forms of reinforcement and has a low facility investment, it results in a high material scrap rate, fabrication time/cost, and operator-to-operator part variability. The scrap factor on this type of operation can exceed 50%, depending on part complexity and size.

Auxiliary processing aids should be used extensively to expedite the lay-up operation and to use molds and tools more efficiently. It is customary to presize the laid-up ply before it is applied to the mold. Usually, an auxiliary backing is fixed in position on the lay-up tool, which is sometimes equipped with vacuum ports to anchor the backings. Plies are oriented to within ±1° using tape-laying heads, or, manually, using straight edges, drafting machine dividing heads (Ref 2) or ruled lines on the table (Ref 2).

Indexes or polyester film templates also can be used to reduce the lay-up times on molds. The presized plies are first laid up and oriented on the templates. When the mold is available for the lay-up, the plies are positioned on them and transferred. Positioning is achieved by using the references used for indexing. Reference posts for the templates are sometimes located on the mold; corresponding holes in the

Table 3 Mechanical properties versus fiber orientation

Material	Tensile strength, 0° MPa	ksi	Tensile modulus, 0° GPa	10^6 psi	Tensile strength, 90° MPa	ksi	Tensile modulus, 90° GPa	10^6 psi	Compressive strength, 0° MPa	10^6 psi	Compressive modulus, 0° GPa	10^6 psi	Compressive strength, 90° MPa	ksi	Coefficient of longitudinal thermal expansion, 10^{-6}/K	Coefficient of transverse thermal expansion, 10^{-6}/K
E-glass . . .	1104	160	39	6	36	5.2	10	2	600	85	32	4.6	138	20	5.4	36
Aramid . . .	1310	190	83	12	39	5.7	5.6	0.81	286	41	73	11	138	20	−2.3 to −4.0(a)	35(b)
Graphite . .	1725	250	159	23	42	6.1	10.9	1.58	1366	198	138	20	230	35	0.045	20.2

(a) −79 to +100°C, (−110 to +212 °F) 0° = fiber direction. (b) −195 to +120°C, (−320 to +250°F), 90° = perpendicular to fiber. Source: Ref 1

Fig. 5 Gerber cutting machine

Table 4 Factors affecting prepreg form selection

10 = highest cost or worst case; 1 = lowest cost or best case

Material form and fabrication process	Facility cost	Production rate	Importance of operator's skill	Part complexity possible	Part reproducibility	Material cost	Material use efficiency
Unidirectional tape							
Hand lay-up	1	10	10	5	10	3	7
Machine-cut,							
hand lay-up	5	5	5	5	5	3	4
Machine lay-down	10	1	1	5	1	6	2
Multidirectional tape							
Hand lay-up	1	5	9	7	8	8	5
Machine-cut,							
hand lay-up	5	3	7	7	4	8	4
Fabrics							
Hand lay-up	1	10	8	1	8	5	7
Machine-cut,							
hand lay-up	5	5	4	1	4	5	7
Towpreg	8	5	7	3	5	2	3

Fig. 6 Automatic tape laying equipment on flo tooling

Fig. 7 Automatic tape laying equipment on contoured tooling. Courtesy Ingersoll-Rand Company

templates fit exactly over the posts. In some cases, the templates are shaped so that they fit only one way in the mold. The plies are rubbed out from the templates onto the mold, the mold is removed, the bleeder systems are laid up, and the assemblies are bagged and cured.

Machine-Cut Patterns. More recent technology uses machine-cut patterns that are then laid up by hand. This method of manufacture involves a higher facility cost but increases part fabrication output and reduces operator error in lay-up. The right-sized pattern can be automatically cut in one or more ply thicknesses using wider tapes of up to 1500 mm (60 in.), which are potentially more economical to fabricate.

The cut is normally done on a pattern-cutting table, where up to eight plies of material are laid up. Various templates are located on top of the lay-up, and the most economical arrangement is determined by matching templates. The patterns are then cut and and stored until required. Cutting of plies can be done by laser, water jet, or high-speed blades. The machine-cut method is often used in modern composites shops and is best suited for broad goods and wide tapes. A typical cutting machine is shown in Fig. 5.

Automatic Machine Lay-Up. Numerically controlled automatic tape-laying machines, especially in the aerospace industry, are now programmed to lay down plies of tape in the quasi-isotropic patterns required by most design applications. In addition to being able to lay down a part in a short time and with reduced scrappage, robotics also lend consistency to lay-down pressures and ply-to-ply separations. These advantages are rapidly causing the aerospace industry to switch from hand lay-up operations. Automatic tape layers (ATL) are evolving from being able to handle

only limited tape widths and simple tool contours to being able to fabricate large, heavily contoured parts. Figure 6 shows a typically flat tool lay-up, while Fig. 7 shows a tape layer on a severely contoured tool.

Future Trends. Table 4 shows advantages and disadvantages of typical material forms in areas of interest to designers and manufacturers, who must consider factors other than mechanical performance when choosing a material form or a manufacturing scheme. Factors such as part reproducibility can be significant in composite use.

The advance of ATL technology has appreciably extended the product life of unidirectional tapes. Although potentially more economical forms of prepreg/composite manufacture, such as filament winding and towpregs, are becoming more widely used, the use of tape in a robotic lay-down operation will result in part reproducibility and lower cost. Aerospace companies in particular are investing significant facility capital to purchase and upgrade ATL equipment. This indicates their confidence that this form of part manufacture will continue to play an important role in composite part manufacturing technology.

REFERENCES

1. B.D. Agarwol and L.J. Broutman, *Analysis and Performance of Fiber Composites*, John Wiley & Sons, 1980
2. G. Lubin, *Handbook of Composites*, Van Nostrand Reinhold, 1982

SELECTED REFERENCES

- *Carbon and Graphite Fibers, Manufacture and Applications*, Noyes Data Corporation, 1980
- D. Clark, N.J. Wadsworth, and W. Watt, *Carbon Fibers: Their Place in Modern Technology*, The Plastics Institute, 1974

Multidirectional Tape Prepregs

Fred S. Dominguez, Hercules Aerospace Company

FIBROUS REINFORCEMENTS by nature reinforce primarily in the 0° direction, parallel to the longitudinal axis of the fiber. The consensus is that the discrepancy in directional properties is due to the presence of flaws in the fiber. Properties in the 0° direction are maximized because of the small cross-sectional dimensions of the fiber. The stress-strain response of an orthotropic (unidirectional) ply is characterized by high modulus of elasticity, strength, and elongation parallel to the fibers, whereas the corresponding values in the transverse direction are relatively low. When a number of plies are laminated at several orientations, the stress-strain relationship will be intermediate to the longitudinal and transverse relations. As the number of oriented plies is increased, the isotropic strength is approached asymptotically. Four ply directions are sufficient, and a 0°/90°/±45° laminate can be selected for isotropic simulation (Ref 1). This is one limitation to the designer using a unidirectionally reinforced matrix material. One solution to this problem is the use of multidirectionally reinforced tapes. However, woven goods or fabrics, even though multidirectionally reinforced, do not yield the same translation of mechanical properties as unidirectional plies of tape.

Product Forms

Multidirectional tapes can be manufactured with multiple plies of unidirectional tape oriented to the designer's choice. These tapes are available in the same widths and package sizes as unidirectional tape, with varying thickness. Up to four or five plies of tape, with each ply typically being 0.125 mm (0.005 in.), can be plied together in various orientations to yield a multidirectionally reinforcing tape. Figure 1 depicts the difference between unidirectional and multidirectional tapes.

By using a preplied quasi-isotropic prepreg, the fabricator can avoid a substantial lay-up cost. However, preplied prepregs are typically more costly than unidirectional prepregs because of the additional work necessary to ply the tape.

Tape Properties

The performance of multioriented prepregs can be accurately predicted from test data that have been generated on these configurations. Tables 1 and 2 show typical mechanical prop-

Table 1 Comparative strength/weight versus material form

Material(a)	Strength, 0° MPa	ksi	Strength, 0°/±45°/90° MPa	ksi	Density, g/cm³	Strength/density, 0° 10⁶ cm	10⁶ in.	Strength/density, 0°/±45°/90° 10⁶ cm	10⁶ in.
Graphite, high-strength, low modulus	2.2	0.32	0.73	0.11	1.55	14.3	5.63	4.8	1.9
Graphite, high-strength, intermediate modulus	2.4	0.35	0.80	0.12	1.52	...	...	...	...
Graphite, low-strength, high modulus	1.2	0.17	0.43	0.06	1.63	15.1	5.94	2.7	1.1
S-glass	1.8	0.26	0.76	0.11	1.99	9.2	3.6	3.9	1.5
E-glass	0.82	0.12	0.52	0.075	1.99	4.2	1.7	2.7	1.1
Aramid	1.5	0.22	0.39	0.057	1.36	10.9	4.29	2.9	1.1
Aluminum	...	0.41	0.059	...	2.77	...	1.5	0.59	...
Steel	...	2.1	0.30	...	8.00	...	2.6	1.0	...

(a) In epoxy-resin matrix

Table 2 Comparative stiffness/weight versus material form

Material(a)	Stiffness, 0° MPa	ksi	Stiffness, 0°/±45°/90° MPa	ksi	Density, g/cm³	Stiffness/density, 0° 10⁶ cm	10⁶ in.	Stiffness/density, 0°/±45°/90° 10⁶ cm	10⁶ in.
Graphite, high-strength, low modulus	0.15	0.022	0.046	0.0067	1.55	0.98	0.39	0.30	0.12
Graphite, high-strength, intermediate modulus	0.17	0.025	0.065	0.0094	1.52	1.14	0.45	0.43	0.17
Graphite, low-strength, high modulus	0.20	0.029	0.052	0.0075	1.63	1.25	0.49	0.33	0.13
S-glass	0.055	0.0080	0.025	0.0036	1.99	0.28	0.11	0.13	0.051
E-glass	0.041	0.0059	0.018	0.0026	1.99	0.21	0.083	0.09	0.035
Aramid	0.073	0.011	0.025	0.0036	1.36	0.59	0.23	0.19	0.075
Aluminum	...	0.069	0.010	...	2.77	...	0.25	0.098	...
Steel	...	0.19	0.028	...	8.00	...	0.24	0.094	...

(a) In epoxy-resin matrix

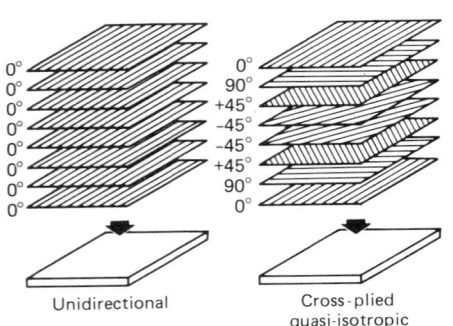

Fig. 1 Unidirectional versus quasi-isotropic lay-ups

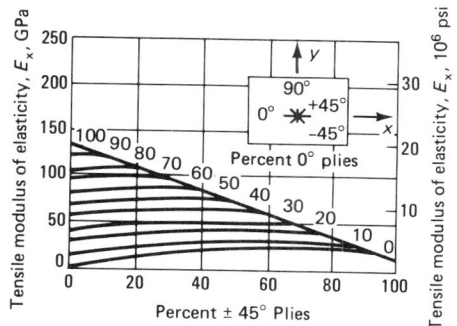

Fig. 2 Tensile modulus of elasticity of carbon-epoxy laminates at room temperature

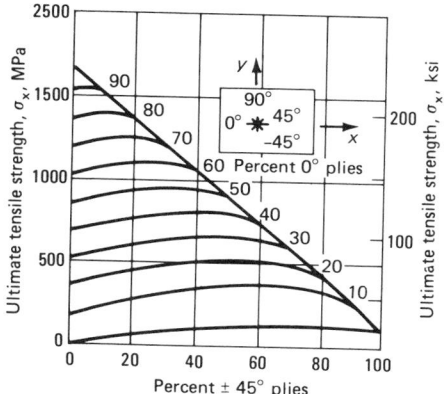

Fig. 3 Ultimate tensile strength of carbon-epoxy laminates at room temperature

erty data for these lay-ups compared to other structural materials.

Cross-plied tapes offer controlled anisotropy, that is, properties can be varied and modified in selected directions, but these tapes are generally more expensive than unidirectional tapes because of the additional manufacturing steps.

This disadvantage is often overcome, however, by the cost savings from using a preplied tape in part lay-up.

Properties are controlled by the number of plies of tape oriented in critical directions. Figures 2 and 3 show typical changes in tensile properties when ply orientation is changed.

Applications

Multidirectional tapes are ordinarily used in lay-up applications that require a repeated fiber orientation but neither exceptional drapability nor formability of material. Multiplied goods are typically thicker and less pliable than unidirectional tapes. Typical applications where unidirectional broad goods are used either in hand-cut or machine-cut patterns represent good applications for multioriented tapes.

REFERENCES

1. G. Lubin, *Handbook of Composites*, Van Nostrand Reinhold, 1982
2. B.D. Agarwol and L.J. Broutman, *Analysis and Performance of Fiber Composites*, John Wiley & Sons, 1980

Woven Fabric Prepregs

Fred S. Dominguez, Hercules Aerospace Company

WOVEN FABRIC PREPREGS are one of the most widely used fiber reinforced resin forms. Fabrics typically offer flexibility in fabrication technique, but at a higher cost than other prepreg forms. The designer must consider these and other factors before selecting a prepreg form for structural application.

Fabric Construction

Fibers can be woven into many different types of weave patterns, widths, and thicknesses. The warp yarns, or ends, lie in the lengthwise (machine) direction of the fabric, whereas the filling yarns, or picks, lie crosswise, at right angles to the warp yarn. Fabric construction is specified by the number of warp yarns per centimeter of fabric width and the number of filling yarns per centimeter in the lengthwise direction. Therefore, fabric weight, thickness, and breaking strength are proportional to the number and types of warp and filling yarns used in weaving.

A variety of weave patterns can be used to interlace the warp and filling yarns to form a stable fabric (see Fig. 1). The weave pattern controls the handling characteristics of a fabric and, to some degree, the properties of a product that uses it as reinforcement. Some applications require that all fabric construction variables be specifically designed so that the desired performance criteria can be met.

The plain weave, which interlaces one warp yarn over and under one filling yarn, demonstrates the greatest degree of stability with respect to yarn slippage and fabric distortion; yarn count and content, however, also contribute to fabric stability.

The basket weave has two or more warp yarns that interlace over and under two or more filling yarns. Although the basket weave is less stable than the plain weave, it is more pliable and will conform more readily to simple contours.

The twill weave interlaces one or more warp yarns over one and under two or more filling yarns in a regular pattern. This produces either a straight or a broken diagonal line in the fabric, which consequently has greater pliability and better drapability than either plain-woven or basket-woven fabric.

A crowfoot satin weave has one warp yarn interlacing over three and under one filling yarn in an irregular pattern, resulting in a pliable fabric capable of conforming to complex or compound contours.

The 8-end satin weave has one warp yarn interlacing over seven and under one filling yarn in an irregular pattern, which yields a pliable fabric that will readily conform to compound contours. Since this weave pattern allows a comparatively high yarn count per centimeter and fewer fiber distortions, it translates into better strength properties in all directions than a tighter weave, such as the plain weave. A variation of the 8-end satin weave is the 5-end satin weave.

Fabrics woven with heavy warp yarns and fine filling yarns in either the crowfoot or long-shaft (such as the 8-end) satin weave patterns are called unidirectional fabrics. These fabrics are characterized by a high strength contribution to composites in the heavy-yarn

Table 1 Graphite fabric forms

Weave	Construction tows/cm	tows/in.	Weight g/m²	oz/yd²	Thickness(a) mm	mil
Plain	4.5 × 4.5	11 × 11	193	5.7	0.18	7.2
8-end satin	8.5 × 8.5	22 × 22	370	10.9	0.34	13.2
5-end satin	4.3 × 4.3	11 × 11	370	10.9	0.34	13.2
Crowfoot satin	4.1 × 4.1	10 × 10	185	5.5	0.17	6.7

(a) Cured ply thickness at 62 vol% fiber for high-strength, low-modulus fiber-epoxy prepreg

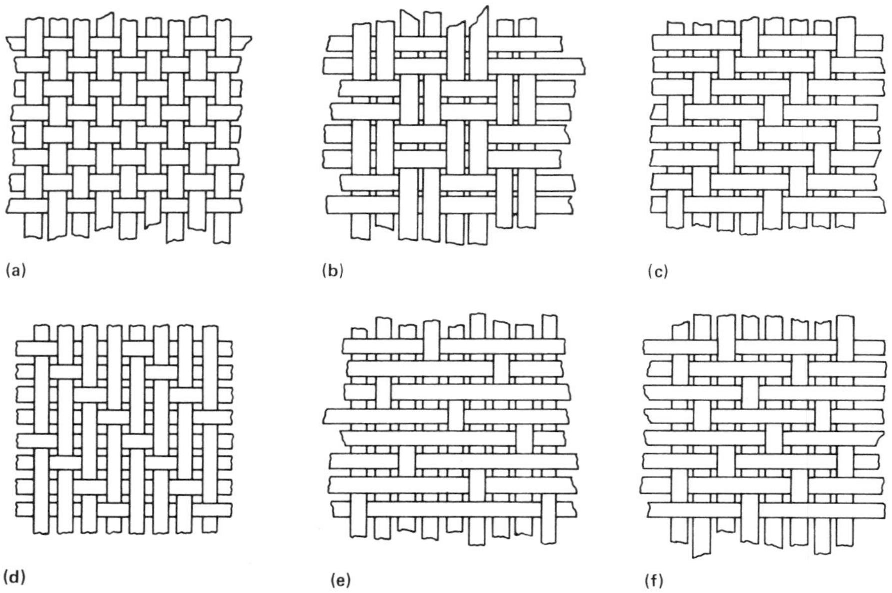

Fig. 1 Fabric construction forms. (a) Plain weave. (b) Basket weave. (c) Twill. (d) Crowfoot satin. (e) 8-end satin. (f) 5-end satin

Table 2   Properties of graphite, aramid and hybrid fabric composites compared to 0°/90° laminates made from unidirectional layers (data normalized to 65 vol% fiber)

Ratio of aramid to graphite fiber	Tensile modulus, 0°/90°		Tensile modulus, fabric		Fabric efficiency, %	Tensile strength, 0°/90°		Tensile strength, fabric		Fabric efficiency, %	Compressive strength, 0°/90°		Compressive strength, fabric		Fabric efficiency, %
	GPa	10⁶ psi	GPa	10⁶ psi		MPa	ksi	MPa	ksi		MPa	ksi	MPa	ksi	
100/0	36.5	5.29	35.8	5.19	98	579	84.0	544	78.9	94	165	23.9	152	22.0	92
50/50	55.1	7.99	48.2	6.99	87	572	83.0	400	58.0	70	407	59.0	227	32.9	56
25/75	69.6	10.1	57.2	8.30	82	661	95.9	434	62.9	66	641	93.0	317	45.0	49
0/100	72.3	10.5	59.9	8.69	83	730	105	434	62.9	59	965	140	558	80.9	58

Source: Ref 2

(warp) direction. Table 1 shows typical graphite weave patterns.

Nonwoven unidirectional fabrics can be produced by chemically bonding the warp and filling yarns rather than interlacing them. Although the chemical bonding contributes to the stability of these nonwoven products, they tend to be somewhat firm and therefore do not readily conform to complex or compound contours (Ref 1).

The handling characteristics of a fabric are determined by the yarn count and the weave pattern holding the yarns together. If the weave pattern is too tight, the fabric will not conform to various contours and will not accept resin, resulting in a weak composite. On the other hand, if the weave pattern is too open or loose, the fabric will not contain sufficient fiber to attain its maximum possible strength and will be easily distorted, precluding the alignment of the fibers with preferred strength axes.

Because precise fiber orientation is necessary for optimum translation of fiber properties, it is critical that fabrics be aligned properly on the tool surface. To achieve this, tracer yarns can be woven into the fabric, particularly a graphite fabric. The tracer provides a very fine (small denier) fiber of contrasting color that can be used by the fabricator to verify fiber direction in the warp and fill directions. This tracer is also used by the prepregger to verify correct orientation of warp and fill tows during prepreg operations. In addition, if the tracer contains x-ray detectable pigment, it can be used to verify ply orientation and count in cured laminates.

Table 3   Impact resistance of hybrid composites

Hybrid composite(a), wt%	Izod impact strength, unnotched J/m	ft·lbf/in.
Graphite, 100%	1495	28
Graphite, 75%; aramid, 25%	1815	34
Graphite, 50%; aramid, 50%	2349	44
Aramid, 100%	2562	48
Graphite, 100%	1495	28
Graphite, 75%; glass, 25%	2349	44
Graphite, 50%; glass, 50%	2989	56
Glass, 100%	3843	72

(a) With epoxy matrix. Source: Ref 1

Fabric Prepreg Forms

Fabrics can be prepregged using either a hot-melt or a solvent-coating process. The hot-melt process uses a machine similar to that used for fabricating unidirectional tape. Resin can be applied to the fabric either by using prefilmed substrate paper, a "knife over roll," or a similar coating mechanism. Solvent coating is typically accomplished by immersing the fabric into a bath containing 20 to 50% of a solvent and resin mixture and then drying the fabric in a one-pass or multipass former coater. The two techniques generate different characteristics in the prepreg:

Hot melt

- Less drape and lower tack, due to higher resin viscosity, which in turn is due to lack of residual solvent in the prepreg
- Better hot/wet mechanical properties, less flow, and longer gel time, due to the absence of volatiles
- Higher cost, due to slower process speed and higher resin scrappage

Solvent coating

- Better drape, due to lower resin viscosity and, usually, higher tack

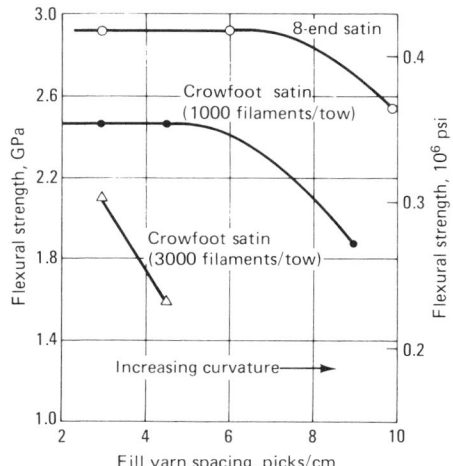

Fig. 2 Influence of fabric construction on graphite composite flexural strength (data normalized to 100 vol% warp fiber)

- Residual solvent of 1 to 2%, which incurs longer gel time, higher flow, and lower hot/wet mechanical properties
- Lower cost, due to increased process speed and reduced resin waste

Hybrids

Hybrid fabrics are those woven from several different types of fibers, in contrast to fabrics woven from a single type of fiber. Table 2 shows typical properties of one hybrid graphite/aramid fabric. Combining fiber reinforcements allows the designer considerable flexibility. Among the reasons for the hybridization of graphite composites are (1) adding another fiber to a predominantly graphite composite in order to overcome the inherent disadvantages of graphite, (2) adding graphite fibers to a predominantly nongraphite composite or structure in order to take advantage of the benefits of graphite, and (3) producing a lower-cost structure. Normally, the impact resistance of graphite fiber composites can be improved by adding high-strength fibers with a greater strain-to-failure ratio than graphite. Several energy-absorbing mechanisms that have been proposed include interlacing resin layers to absorb energy and using fillers to stop cracks. Table 3 shows typical hybrid graphite/aramid and graphite/glass impact properties.

Graphite hybrid composites can be fabricated using conventional techniques and can be com-

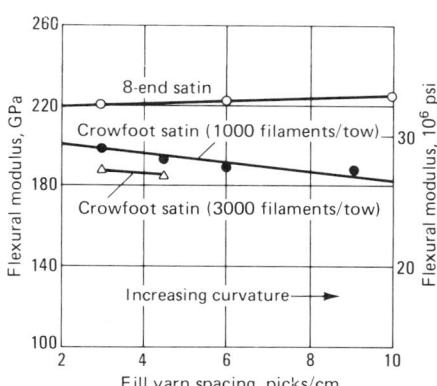

Fig. 3 Influence of fabric construction on graphite composite flexural modulus (data normalized to 100 vol% warp fiber)

Table 4   Effect of weave pattern on fiberglass composite mechanical properties

Fabric style(a)	Plies, number	Resin content, wt%	Thickness mm	Thickness in.	Flexural strength MPa	Flexural strength ksi	Flexural modulus GPa	Flexural modulus 10^6 psi	Compressive strength MPa	Compressive strength ksi	Tensile strength MPa	Tensile strength ksi
7628	18	37.1	3.15	0.124	371	53.8	26.8	3.89	177	25.7	317	45.9
7628l(b)	18	36.7	3.07	0.121	584	84.7	23.5	3.41	393	57.0	408	59.2
16-149	12	36.5	3.05	0.120	436	63.2	26.3	3.81	331	48.0	405	58.7
7781(c)	12	37.6	3.05	0.120	600	87.0	22.3	3.24	443	64.3	414	60.0

(a) Each material is polyester compatible. (b) 5-end satin weave version of style 7781. (c) 8-end satin weave. Source: Ref 1

bined with boron, glass, aramid fibers, or metals in a laminated structure. Hybrids usually have the same matrix and can be fabricated by the co-curing process.

Fabric Mechanical Properties

Generally, the longer the float, which is the portion of a warp or a filling yarn that extends unbound over two or more yarns lying 90° to it, the higher the composite strength. As shown in Table 4, extending the length of the float, which reduces the interlacing frequency, increases the composite strength for fiberglass composites as well as for other fibrous reinforcements. The transition from a plain weave to a crowfoot weave also results in improved composite mechanical properties. Figures 2 and 3 show how weave style and construction can influence graphite composite properties.

Fabrication Techniques

Woven fabric prepregs can be darted to conform to convoluted shapes of low-stressed items. Darting is the practice of slitting the prepregs at locations where folds would nor-mally occur in a lay-up; the excess material at those locations is removed completely, and the remaining edges butted together. As an alternative, the prepreg can be slit where a crease would normally form, and the excess material can overlap, provided that it is wrinkle free. When the former method is used, an additional ply may be required to compensate for the weak butt joints in the lay-up. Darting is not recommended for highly stressed, lightweight construction; on those occasions, prepregs should be cut in predetermined patterns such that joints in successive plies do not coincide. Overlapping joints must be deliberately placed and joint widths must be controlled. Usually, patterns for precutting the prepregs allow for 13-mm (0.5-in.) overlaps on the lay-up.

When woven fabric reinforcements are laid up on convoluted shapes, weave patterns become distorted and the fiber directions change. Orientations of 0°, ±60° or 0°, ±45°, 90° are used to compensate for undetermined deficiencies. These plying sequences provide reinforcements for laminate plane quasi-isotropic properties. However, ply alignments of heavily draped lay-ups of fabric-reinforced prepregs are difficult to control. Colored tracer fibers, woven into the fabrics, simplify the lay-up and inspection of the composite (Ref 1).

Structural composites that are required to resist loads must be designed to obtain reproducible properties. Their shapes should permit the plies to be oriented in predetermined directions. Whether the lay-up is produced manually or is automated, the principles behind the lay-up techniques are similar. When the structural shapes permit, the most reproducible properties are developed by laying up plies that are cut to size and then applied to transfer films. These transfer films, or polyester film templates, are indexed to specified ply locations and orientations with respect to the mold. Plies that are laid up on templates are transferred to molds without additional distortion; after the plies are laid up and transferred, the templates are removed.

Anisotropies of a fabric-reinforced prepreg in one ply are corrected with equal but opposite anisotropy in adjacent plies. The symmetry achieved by making these corrections is important to avoid distortions in the cured laminates. Other corrections are sometimes made by cross-plying compensating misalignments to attain orthotropy (Ref 3).

REFERENCES

1. G. Lubin, *Handbook of Composites*, Van Nostrand Reinhold, 1982
2. C. Zweben and J.C. Norman, "Kevlar" 49/"Thornel" 300 Hybrid Fabric Composites for Aerospace Applications, *SAMPE Q.*, July 1976
3. B.D. Agarwol, L.J. Broutman, *Analysis and Performance of Fiber Composites*, John Wiley & Sons, 1980

SELECTED REFERENCE

- D. Clark, N.J. Wadsworth, and W. Watt, *Carbon Fibers: Their Place in Modern Technology*, Plastics Institute of America, 1974

Prepreg Tow

Fred S. Dominguez, Hercules Aerospace Company

ONE OF THE NEWER PREPREG FORMS is a towpreg, which is either a single tow or a strand of fiber that has been impregnated with matrix resin. The impregnated fiber is typically wound on a cardboard core before being packaged for shipment. Because a towpreg is potentially the lowest-cost form of prepreg, it is of significant interest to designers. It also lends itself to potentially low-cost manufacturing schemes, such as filament winding. Towpreg is being considered by filament winders as a way to combine the advantages of low-cost part manufacture and high-performance matrix resins. The fibers that are typically used are shown in Table 1.

Towpreg Manufacture and Forms

Most towpregs are converted in a solvent-coating process (Fig. 1) in which base resin is first dissolved in a mix containing 20 to 50% solvent and resin. The dry fiber is then routed through the solvent-resin mix and dried in a tower consisting of one or more heated zones. Resin content is controlled either by using metering rolls after impregnation or by adjusting the solvent-resin ratio. This drying step reduces volatiles and advances the resin so that the towpreg will not adhere to itself during unspooling in part manufacture. Towpregs can also be manufactured in a hot-melt operation by filming resin on substrate paper, impregnating strands between two layers of filmed paper, and then advancing the resin to an intermediate point between freshly mixed and cured (B-staging) on a prepreg line. However, this tends to result in a higher-cost towpreg.

Table 2 shows typical form parameters that a manufacturing shop might specify. A designer must evaluate the size and complexity of the part being designed before selecting material parameters. Resin content will determine part mechanical performance and thickness by determining fiber volume, assuming that little or no resin is lost in the curing process. Tow width, which is important in establishing ply thickness and gap coverage, can be modified during lay-down. Package size can be important to manufacturing personnel, especially when more than one spool is used in the manufacturing process. In such cases, manufacturing personnel often try to match the sizes of spools that are used, to minimize spool doffs (changes) and splices in the manufactured part.

To determine the mechanical properties of a towpreg, it can be tested by a single-strand type of test or by winding tows on a drum to specified thicknesses and then laying up laminates from this wind. Mechanical properties of towpregs are comparable to those of tapes, if they are cured under autoclave conditions. Filament-wound structures that are not autoclave cured will typically have higher void contents than autoclave-cured parts.

Applications

Currently, the two basic uses for towpregs are as a filler in hard-to-form areas and in joints of structural components such as I-beams (Fig. 2) and as a replacement for low-performance filament-winding resins in filament-winding

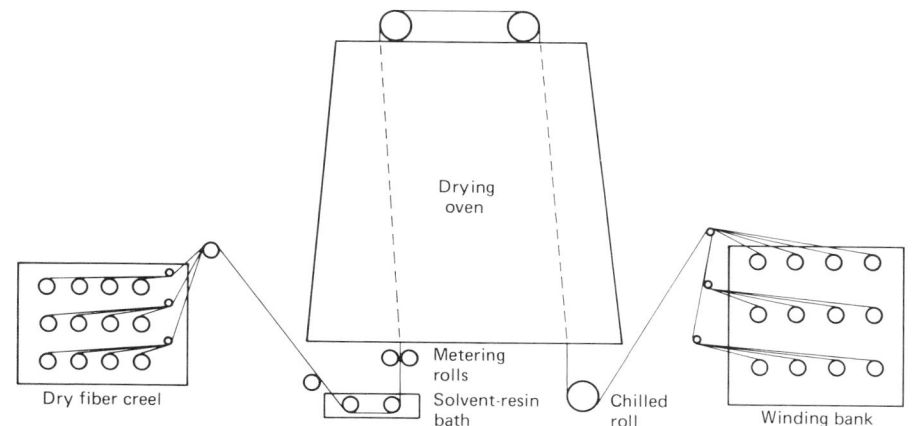

Fig. 1 Typical towpreg manufacturing process

Table 1 Fiber tow characteristics

Before impregnation

Material	Yield/tow m/kg	Yield/tow yd/lb	Filament size μm	Filament size μin.
Graphite (1000-12 000 filaments/tow)	300-1200	150-600	5-10	200-390
Fiberglass (2450-12 240 filaments/tow)	490-2400	245-1200	4-13	160-510
Aramid (800-3200 filaments/tow)	2000-7850	980-3900	12	470

Table 2 Towpreg form parameters

Parameter	Typical range
Strand weight per length, g/m (lb/yd)	0.74-1.48 (0.0015-0.0030)
Resin content, %	28-45
Tow width, cm (in.)	0.16-0.64 (0.06-0.25)
Package size, kg (lb)	0.25-4.5 (0.5-10)

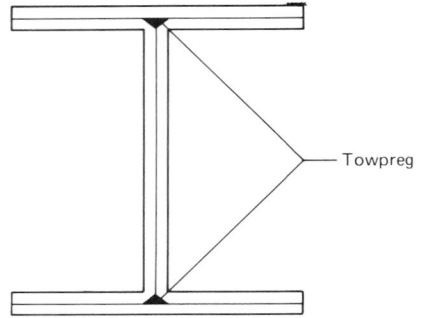

Fig. 2 Towpreg used as filler in an I-beam

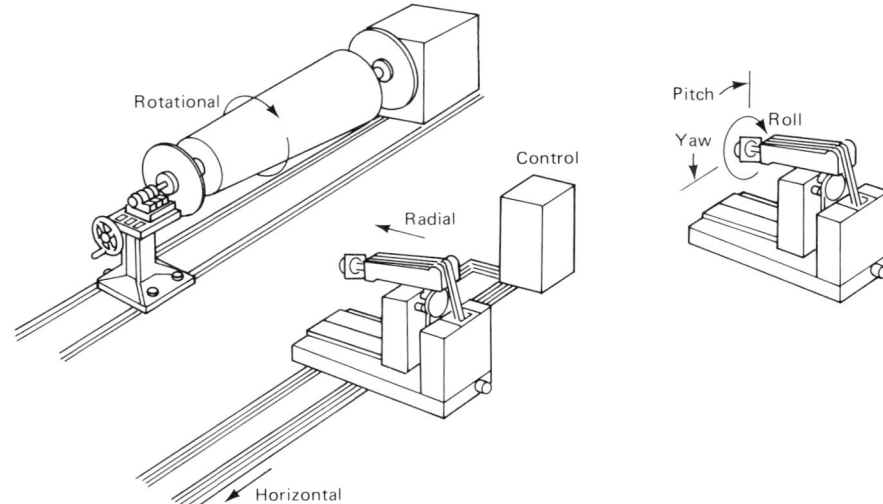

Fig. 3 Six-axis winding machine

operations. Using a towpreg as a filler material in areas where tape or fabric prepregs will not lay down involves hand lay-up.

Most of the development in towpreg technology has been in the area of winding, particularly using a graphite-epoxy towpreg. The six-axis winding machine shown in Fig. 3 is one of the newest pieces of equipment being evaluated for towpreg use. This machine unspools the towpreg bundles and collimates them into a band of prepregs before laying down a unified band. The band of prepreg can be laid into complex cylindrical or nongeodesic forms, as shown in Fig. 4. This technology has the potential of making significant inroads into complex low-cost aerospace-grade part manufacture and may revolutionize the amount of composites and types of techniques used in aircraft fuselage manufacture.

REFERENCES

1. B.D. Agarwol and L.J. Broutman, *Analysis and Performance of Fiber Composites*, John Wiley & Sons, 1980
2. G. Lubin, *Handbook of Composites*, Van Nostrand Reinhold, 1982
3. "Carbon and Graphite Fibers, Manufacture and Applications," Noyes Data Corporation, 1980
4. D. Clark, N.J. Wadsworth, and W. Watt, "Carbon Fibers: Their Place in Modern Technology," The Plastics Institute, 1974

Fig. 4 Complex structure wound with towpreg on six-axis winding machine

Recycling Carbon Fiber Scrap

Heinz Richter and Jürgen Brandt, Messerschmitt-Bölkow-Blohm GmbH

RECOVERING CARBON FIBERS from scrap that accrues in processing carbon fiber reinforced plastic (CFRP) is a technologically new aspect of plastics recycling. In the manufacture of CFRP components, continuous filaments in the form of woven and nonwoven fabrics are predominantly used. These fibrous materials are usually impregnated with epoxy resins, and the resultant prepregs are subsequently processed. The prepregs are primarily available in web form, which is cut to size and then arranged according to the desired configuration. The cutting and trimming process yields considerable quantities of scrap, which currently are discarded.

Recycling CFRP scrap is economically interesting because of the high cost of carbon fibers. In principle, recovering glass fibers from reinforced plastic is also possible, but is not as economically appealing because of the low fiber cost.

Potential Applications of Reprocessed Fibers

Reprocessing can be aimed at recovering discontinuous carbon fibers of defined length from cut and trimmed scrap. Recently developed processes for producing unidirectional, short-fiber nonwoven fabric allow the use of short fibers in the manufacture of high-performance composite materials. This is the optimal approach to utilizing the properties of short fibers in composite materials and obtaining high fiber content. Fiber lengths from 2 to 4 mm (0.075 to 0.15 in.), used in the alignment process, are reflected in the reprocessing effort.

Another application of short fibers is in compression and injection molding processes in which fiber direction is arranged randomly. Injection molded materials exhibit substantially lower mechanical properties than composite materials that are reinforced with carbon fibers in order to provide high strength and stiffness levels and low density.

Currently, carbon fiber reinforced thermoplastics are being considered for industrial applications. Fiber recycling is potentially interesting because the requirements for fiber length distribution are less exacting than they are for short-fiber prepregs.

Reprocessing

The scrap that accrues in CFRP processing, in order of increasing recovery costs, consists of sections of fibers, fabrics, prepregs, and cured laminates. Most scrap consists of prepreg sections, which can be reprocessed in two basic operations: removing the resin and/or the sizing, and cutting the remaining fibers and/or fabrics.

Because the matrix materials currently used for CFRP are predominantly epoxy resins, studies have been conducted for that type of material. The resin content of prepregs is usually about 40 wt% and the sizing on the fibers is usually 1 to 2 wt%. As epoxy resin is in an uncured condition in a prepreg and is therefore soluble in common organic solvents such as acetone or methyl ethyl ketone, the resin can be washed out with a solvent without difficulty. Apart from complete resin removal, it is essential that the fiber surface not undergo any change because that would affect resin-fiber adhesion. It can be proved by strength determination tests that resin removal with solvents has no adverse effects.

Another approach to resin removal is thermal degradation. In normal atmosphere, carbon fibers are destroyed by oxidation at temperatures above ~400 °C (~750 °F). In tests on scrap fabrics and prepregs, the resin could be removed at temperatures below 400 °C (750 °F) in normal atmosphere, and at higher temperatures under inert gas. In both cases, however, thermal treatment leads to a change of the fiber surface, causing lower resin-fiber adhesion and thus a lower strength level. The fiber content of CFRP is commonly determined by oxidative resin removal with sulfuric acid and hydrogen peroxide, which also can be used to remove the resin from the prepregs. This technique has not been pursued because the acid requirement is relatively high, and the handling of acids necessitates extensive precautions in terms of equipment and process technologies. Currently, no economical way to remove epoxy resins from cured laminates is known.

The essential aspect of reusing carbon fibers in the form of aligned, short-fiber nonwoven materials is the production of short fibers with a defined length of ~3 mm (~0.10 in.). Longer lengths would affect the fiber orientation, which is obtained hydrodynamically. The lower degree of alignment would cause a reduction in mechanical properties and more frequent fiber breakage in the compression molding process. Shorter fibers, too, would lead to lower strength levels because of insufficient length/diameter ratios. Therefore, to obtain the optimum reinforcing effect, fibers must be produced with as uniform a cutting length as possible.

Commercially available cutting equipment consists of fiber cutters, cutting mills, and granulators. However, none of these are suitable for cutting the recovered carbon fibers discussed in this article. Because fiber and fabric scraps are randomly oriented, the cutting operation should be carried out in two directions, which is not possible with conventional fiber cutters. Although cutting mills and special-purpose granulators meet this requirement, they do not produce the necessary uniformity of cutting lengths, as could be demonstrated. The cutting could be satisfactorily achieved with a newly developed cutting device into which scrap fibers are fed via a conveyor belt and passed through a gap between two rotary cutters. The first rotor carries cutting discs that cut the scrap fibers in the conveying direction, while the second rotor has longitudinal knives that cut transverse to that direction. Knife distances of 3 mm (0.10 in.) on both rotors theoretically should result in fiber lengths between 0 and 4.2 mm (0.15 in.). Experimental investigations found that about 60% of the fibers had a length of 2.5 to 3.5 mm (0.10 to 0.14 in.), and only 20% were shorter than 1.5 mm (0.06 in.). These fiber lengths are sufficient for the alignment process and the desired laminate properties. Reusing scrap carbon fibers appears to be economically promising; at the present stage of investigation, converting the scrap to short fibers accounts for only 20% of the cost of comparable new fibers.

Fiber-Reinforced Plastic With Aligned Discontinuous Fibers

Alignment techniques have been developed for eventual use with high grade fiber materials

Fig. 1 Facility for producing aligned short fiber based nonwoven materials. Courtesy of Verpreg GmbH

The alignment technique is based on the hydrodynamic principle; that is, fibers are aligned by a controlled flow of a fiber-liquid slurry. A defined fiber suspension is prepared in a storage tank and then passed through a special feed channel onto a rotating vacuum drum filter (Fig. 1). The carrier liquid is sucked off by the vacuum in the drum, while the fibers are retained on the drum surface, which is covered with a filter cloth. Fiber alignment occurs while the fiber suspension flows through the channel and passes from the channel to the drum. Fiber sheets are formed on the drum surface in the rotating direction by the alignment of fibers. Upon removal of the fiber sheets from the drum, the carrier liquid is washed out, and the fibers are fixed. The dimensions of the resulting nonwoven sheets depend on the size of the drum, and their thickness can be varied by manipulating feed of the fiber suspension. The preferred fiber lengths for optimum alignment range from 2 to 4 mm (0.075 to 0.15 in.).

To produce laminates, the matrix materials must first be introduced into the short-fiber sheets by one of two methods. The first method begins with the preparation of a resin solution with a low-boiling solvent, which is then used to impregnate the fiber sheets. The solvent is subsequently evaporated by heat treatment in a recirculated-air or vacuum-type drying cabinet. The second method introduces a resin film into the short-fiber sheets under pressure and heat. The two methods are applicable to both thermosetting and thermoplastic resin systems. In a prepreg with a thermosetting matrix, the resin-hardener system is present in an uncured condition. During processing, the resin is cured under pressure and heat. In thermoplastic prepregs, the resin is present in its final state. Applying the appropriate temperature results in a processable melt. After cooling, this melt resumes the original state of the material. Processing cycles of only a few minutes are realistic.

now obtainable only as whiskers. Because of their very high cost, such fibers have not yet been used. Developmental work has been based primarily on short carbon fibers, which are potentially the most interesting reinforcement for high-performance composites. The alignment technique used with these fibers is also applicable to short glass fibers and blends of short glass and carbon fibers. Other types of short fibers also can be aligned, but a good orientation effect is possible only if the fibers are straight and have no split or frayed ends from the cutting operation.

The matrix systems used for continuous fiber reinforced composites can also be used for materials reinforced with unidirectional short fibers. The current standard resins for CFRP are epoxy resins in different modifications. For higher temperature ranges, thermosetting resins based on polyimides are available, with a thermal stability that is clearly higher than that of epoxy resins. Recent developments with CFRP are aimed at replacing thermosetting resins with thermoplastic resins, which offer advantages in terms of material properties and processing technologies.

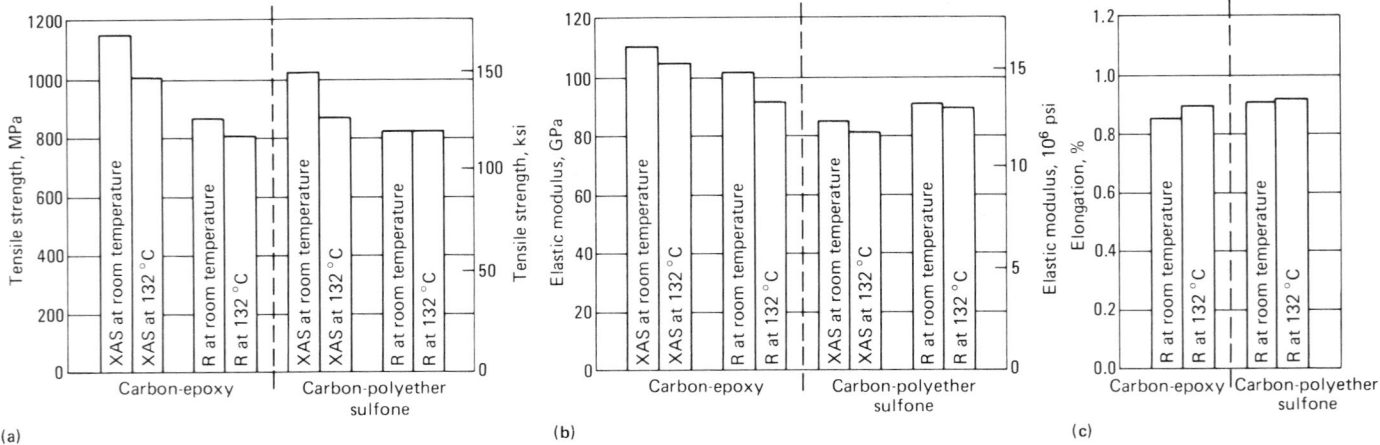

Fig. 2 Tensile properties of unidirectional short fiber based CFRP, 0° direction. (a) Tensile strength. (b) Elastic modulus. (c) Elongation. R: Specimens with recycled fibers

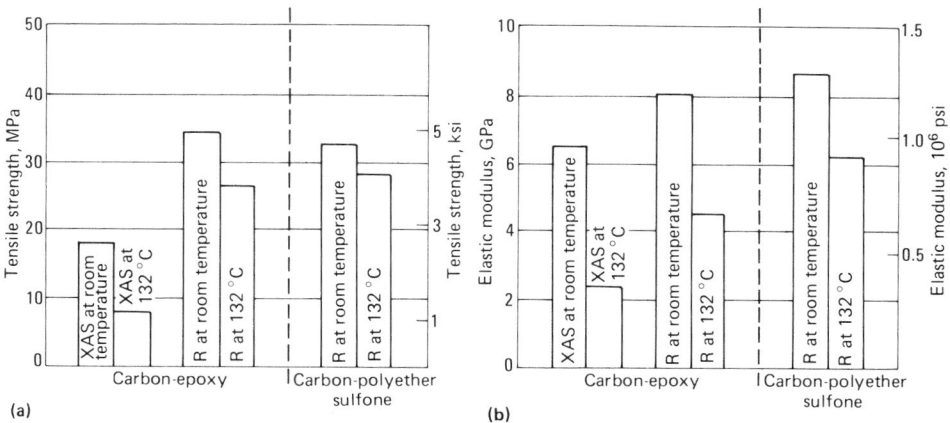

Fig. 3 Tensile properties of unidirectional short fiber based CFRP, 90° direction. (a) Tensile strength. (b) Elastic modulus. R: Specimens with recycled fibers

An evaluation of mechanical properties was performed on unidirectional laminates with aligned short carbon fibers, of the high tensile (HT) grade, with a tensile strength of about 3000 MPa (440 ksi) and an elastic modulus of about 220 GPa (30 × 10⁶ psi). The unidirectional, short fiber based laminates were prepared using 3-mm (0.10 in.) long new fibers, as well as reprocessed fibers. An epoxy resin based prepreg system with a 1 h cure at 180 °C (355 °F) and a polyether sulfone system with a 320 °C (610 °F) processing temperature were used as matrix materials. The epoxy resin laminates were consolidated under vacuum and at a pressure of 1 MPa (10 bars) and cured under the above-mentioned conditions. The polyether sulfone laminates were produced in normal atmosphere at a pressure of 3 MPa (30 bars). In all cases, the laminate thickness was about 2 mm (0.075 in.). The dimensions of the test specimens and the test procedure were selected in accordance with the specifications observed for CFRP in the aviation industry. The most significant properties are shown in Fig. 2 to 5. Aligning short fibers results in composites with mechanical properties that are below those of long fiber reinforced materials, but considerably above those of short fiber reinforced injection and compression molded composites.

Potential Applications of Discontinuous Fiber Reinforced Composites

Although the processing behavior of composites having aligned short fibers is equal to that of continuous fiber reinforced materials, short-fiber reinforcement does offer advantages. First, prepregs with discontinuous fibers have good draping properties because before and during the molding operation the fibers can slide relative to each other, enabling them to be used to make complex three-dimensional components. Second, hybrid materials, that is, composites containing two types of fibers with differing properties in the matrix, are a possibility. Because the short-fiber sheets are obtained from a fiber suspension, various types of fibers can be mixed in any composition, resulting in composites with a wide range of properties. Third, low-cost short carbon fibers can be obtained by reusing scrap from CFRP processing.

These possibilities and the good mechanical properties of aligned, short fiber composites make them suitable for many structural applications, such as:

- Compression moldings with complex geometries that require high draping capabilities of prepregs
- Components with small radii of curvature in different directions and with varying section thicknesses
- Compact, thick-walled components
- Components with high impact resistance gained by using fiber-blend composites

Figures 6 and 7 illustrate two typical components that were prepared within the test programs. The sine wave beam shown in Fig. 6 was laid up mainly from ±45° cross plies. The web and flanges were manufactured together in one step; the tight radius between the two shows the excellent draping properties of the material. A section of a window frame for potential use in transport aircraft is shown in Fig. 7. Essentially, this frame is made from a ring of aligned discontinuous fibers with a few additional skin layers. Using a special vacuum drum, the technology based on the application of short fibers enables the

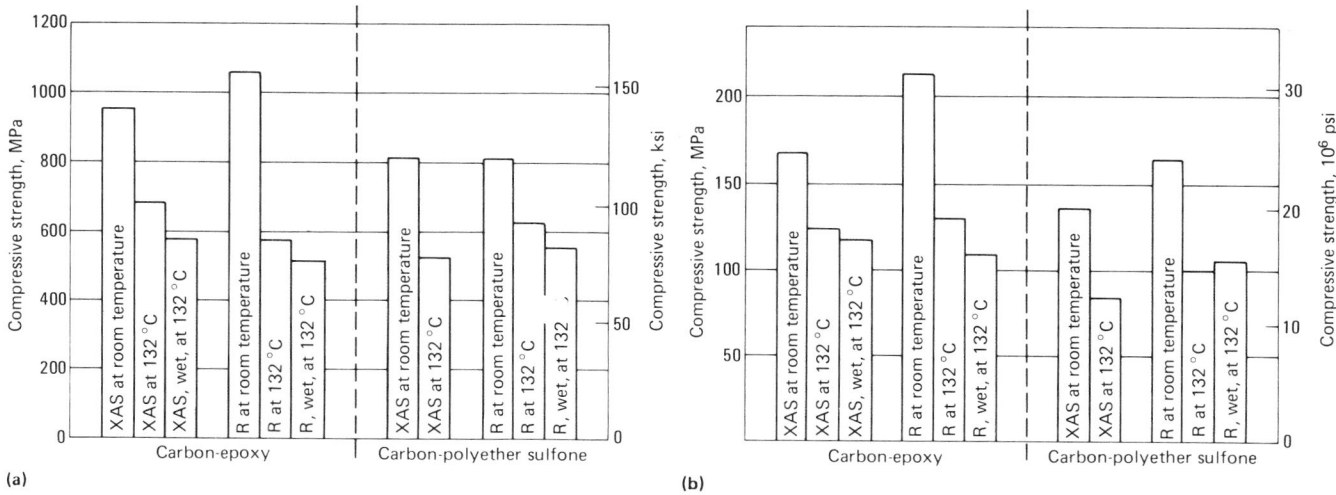

Fig. 4 Compression properties of unidirectional short fiber based CFRP, (a) in the 0° direction, (b) in the 90° direction. R: specimens with recycled fibers. Wet: moisture saturated at 70 °C (160 °F) and 75% RH

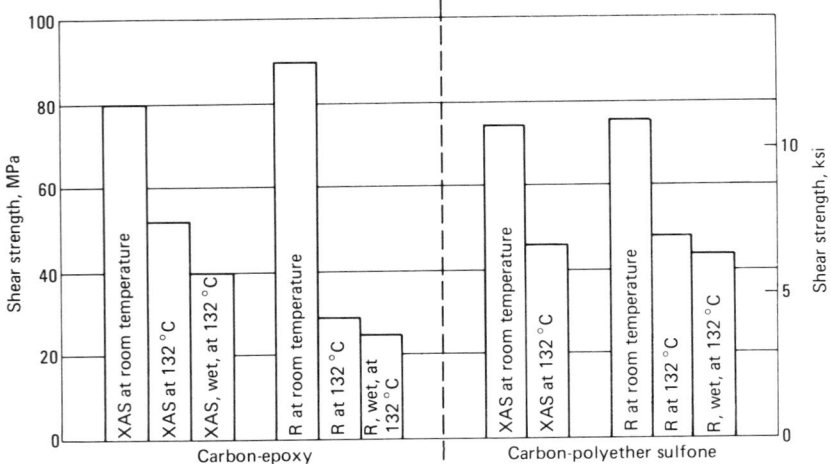

Fig. 5 Interlaminar shear strength of unidirectional short fiber based CFRP. R: specimens with recycled fibers

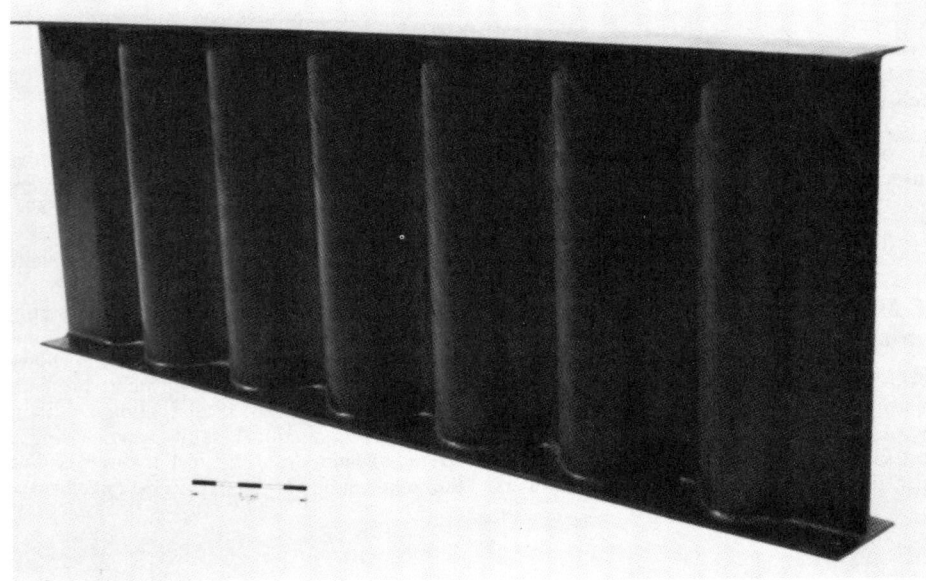

Fig. 6 Sine wave beam serving as a shear and stiffening member in aircraft structures

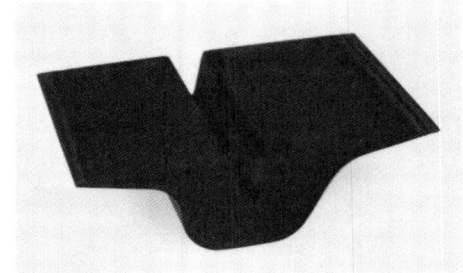

Fig. 7 Segment of a window frame

manufacture of annular prepregs so that the unidirectional core does not have to be assembled from a large number of individual layers.

SELECTED REFERENCES

- H. Richter, Fibre-Reinforced Composites With Unidirectional Short Fibres, *Kunstst.*, Vol 75 (No. 2), 1985, p 80-83

- H. Richter, Recovery of Carbon Fibres From Scrap Accruing in CRP Processing, *Kunstst.*, Vol 75 (No. 5), 1985, p 266-267

- K. Friedrich, K. Schulte, G. Horstenkamp, and T.W. Chou, Fatigue Behavior of Aligned Short Carbon-Fiber Reinforced Polyimide and Polyethersulfone Composites, *J. Mater. Sci.*, Vol 20, 1985, p 3353-3364

- K. Schulte, G. Horstenkamp, and K. Friedrich, "Mechanical Properties of Aligned Short Carbon Fiber Reinforced PES and PI Laminates," Technical Translation ESA-TT-898, European Space Agency, March 1985, translated from Report FB-84-24, Deutsche Forschungs-und Versuchsanstalt für Luft- und Raumfahrt, 1984

Sheet Molding Compounds

J.J. McCluskey and Frank W. Doherty*, Owens Corning Fiberglas Corporation

SHEET MOLDING COMPOUND (SMC) refers to both a material and a process for producing glass fiber reinforced polyester resin items. The material is basically composed of a filled, thermosetting resin and a chopped or continuous strand reinforcement of glass fiber. The uncomplicated SMC processing machine (Fig. 1) produces molding compound in sheet form that is not unlike that of rolled steel. The size of the machine is designated by the width of the sheet it produces. Machine manufacturers generally offer a range of sizes from 0.61 to 1.52 m (2 to 5 ft), the most common being 1.22 m (4 ft).

The process starts in the paste reservoir (below the chopper in Fig. 1), which meters a specified amount of resin filler paste onto a plastic carrier film. The paste consists of several ingredients, which can be changed to fit the particular needs of specific processing conditions and applications. The carrier film passes under a chopper, which cuts glass roving into 25.4-mm (1-in.) lengths. After the glass falls to the resin bed another carrier film with another layer of paste is added on top, sandwiching the glass between the two layers.

When the paste is first mixed and put in the SMC machine, it has the consistency of pancake batter. After maturation, when the thickening agents have had the opportunity to react, the material attains the consistency of heavy putty or caulking compound. Once matured, all carrier film is removed, the SMC material is cut into charges, and the charges are placed in matched metal die molds made of machined steel. A high-tonnage hydraulic press then applies molding pressure. The application of heat and pressure causes the SMC to flow to all areas of the mold. Heat from the mold, normally 149 °C (300 °F), also activates the catalyst in the material, and cure or cross-linking takes place. The part is then removed from the mold.

A number of advantages can be credited to the SMC compression molding process:

- High-volume production
- Excellent part reproducibility
- Low labor requirement per unit produced
- Minimum material scrap
- Excellent design flexibility (from simple to very complex shapes)
- Parts consolidation
- Weight reduction

Material Components

With an unsaturated polyester resin system as its base, the resin paste incorporates other materials for desirable processing and molding characteristics and optimum physical and mechanical properties. Glass fiber reinforcements improve the performance of polyester by upgrading mechanical strength, impact resistance, stiffness, and dimensional stability. Other additives are catalysts, fillers, thickeners, mold release agents, pigments, thermoplastic polymers, polyethylene powders, flame retardants, and ultraviolet absorbers, all of which are mixed by the SMC manufacturer to exact proportions for specific resin paste formulations. Some ingredients, such as release agents and thermoplastic syrups, can be added by the resin supplier. Each additive provides important properties to the SMC, either during the processing and molding steps or in the finished parts.

The catalyst initiates the chemical reaction (copolymerization) of the unsaturated polyester and monomer ingredients from a liquid to a solid state. This is the primary purpose of a catalyst. Heat from the mold causes the catalyst to decompose, which activates the monomer and polyester to form cross-linked thermosetting polymers.

Catalysts are only a small part of an SMC resin formulation. Generally, the addition of 0.3 to 1.5 wt% of catalytic agents will adequately promote the cross-linking reaction. Organic peroxides are the principal catalysts used for SMC resin pastes. The temperature at which the curing process is to be carried out usually determines the selection of a catalyst. For any given catalyst-resin system there is an optimum temperature at which peroxide decomposition

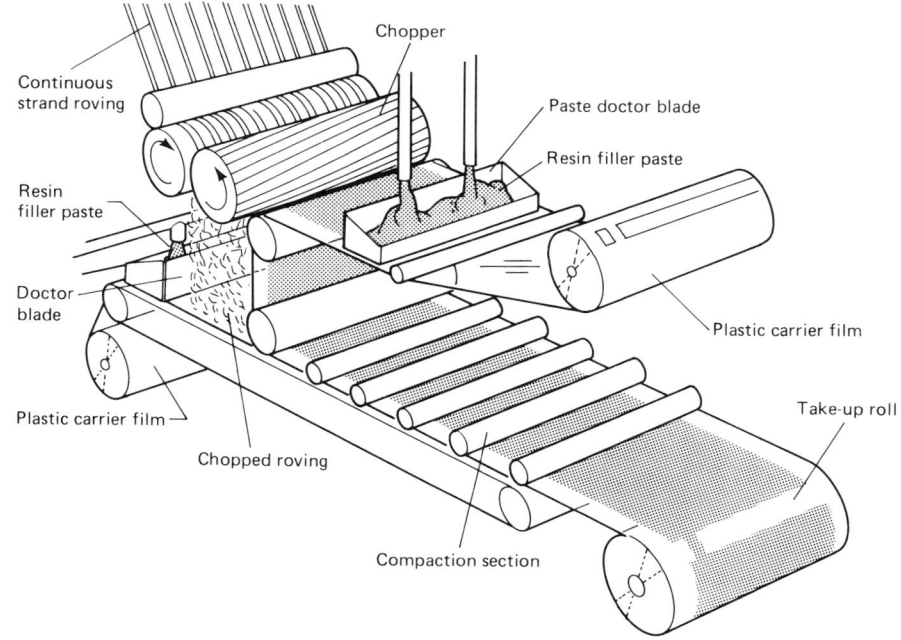

Fig. 1 Sheet molding compound processing machine

Labels in figure:
Continuous strand roving — Chopper — Paste doctor blade — Resin filler paste — Resin filler paste — Doctor blade — Plastic carrier film — Plastic carrier film — Chopped roving — Compaction section — Take-up roll

*Now with Structurelite Plastics Corporation

initiates the monomer-resin polymerization process. Since SMC is usually molded at temperatures of 132 to 165 °C (270 to 330 °F), catalysts that are the most effective as polymerization initiators over this temperature range are the ones used most often.

Fillers enhance the appearance of molded parts, promote flow of the glass reinforcement during the molding cycle, and reduce the overall cost of the compound. Commonly used fillers include calcium carbonate, hydrated alumina, and clay. Calcium carbonates are readily available and can be added to polyester resin in large amounts, while still maintaining a processable paste. They assist in reducing shrinkage of the molded parts and in distributing glass reinforcement for better strength uniformity. Hydrated alumina fillers are incorporated in SMC formulations to provide flame retardancy while maintaining good electrical properties. They are used in most electrical and appliance applications and in some construction applications where material requirements call for specific Underwriters' Laboratories (UL) standards established for flame spread, burning, and smoke density. Kaolin clays are sometimes combined with calcium carbonates or hydrated aluminas. When they represent 10 to 20% of the total filler weight, the clays serve to control paste viscosity, promote flow, and improve resistance to cracking in molded parts.

Thickeners include calcium and magnesium oxides and hydroxides. They initiate the reaction that transforms the mixture of SMC ingredients into a handleable, reproducible molding material. Usually 1 to 3% of the SMC resin formulation is thickener. It is the final ingredient added to the resin mix, and it begins the chemical thickening process immediately.

The thickening reaction must:

- Be slow enough to allow wet-out and impregnation of the glass reinforcement
- Be fast enough to allow the handling required by molding operations, as soon as possible after the impregnation step, in order to keep storage inventories low
- Give a viscosity at molding temperatures that is low enough to permit sufficient flow to fill out the mold at reasonable molding pressures
- Give a viscosity at molding temperatures that is high enough to carry the glass reinforcement along with the resin paste as it flows into the mold
- Level off in the moldable range to give a long shelf life

A typical thickening curve is shown in Fig. 2.

Release agents are common components of SMC formulations. They are selected on the basis of their melting points being just below that of the molding temperature. In theory, the release agent at the molding compound-mold surface interface melts upon contact and forms a barrier against adhesion.

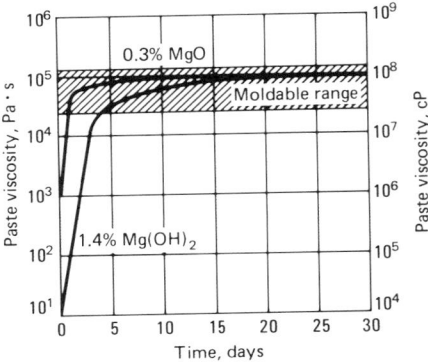

Fig. 2 SMC paste thickening curve

Commonly used internal release agents include zinc stearate, calcium stearate, and stearic acid. Zinc stearate has a melting point of 133 °C (272 °F) and can be used at molding temperatures up to 155 °C (310 °F). Calcium stearate, with a higher melting point of 150 °C (302 °F) can be used at molding temperatures up to 165 °C (330 °F). Stearic acid should be used only if molding temperatures are below 127 °C (260 °F).

Mold release agents must be used at the lowest concentration possible to do an adequate job, which normally is a concentration less than 2 wt% of the total compound. Excessive amounts can reduce mechanical strength, cause objectionable cosmetic appearance on the molded part surface, and affect paint and/or bond adhesion characteristics.

Pigments are supplied as either dry powders or paste dispersions. Two advantages of paste dispersion pigments are that there are fewer agglomerates in the SMC resin paste and that they can be added at lower concentrations than dry powders. Pigment concentration generally is 1 to 5 wt% of the resin paste. Pigments can affect the cure time and shelf life stability of SMC systems and may accelerate or inhibit the reactivity of the catalyst-resin system. Thus, preevaluation of the reactivity of a specific pigment is essential.

Thermoplastic polymers are combined with polyester resins to achieve low polymerization shrinkage for many SMC applications. Shrinkage is primarily controlled by varying the polyester/thermoplastic ratio. It is possible to attain near-zero shrinkage in molded parts when thermoplastic polymers are added to polyester resins at concentrations of 40 wt% of the total resin system.

There are a number of thermoplastic additives that are compatible with polyester resins developed for SMC low-shrink and low-profile systems. Among those in use are acrylics, polyvinyl acetate, styrene copolymers, polyvinyl chloride (PVC), PVC copolymers, cellulose acetate butyrate, polycaprolactones, thermoplastic polyester, and polyethylene powder.

Flame Retardants. As a filler, hydrated alumina compounds normally satisfy most UL

Table 1 Typical mechanical properties for sheet molding compounds with 15 to 30 wt% glass fiber

Tensile modulus, GPa (10^6 psi)	11-17 (1.6-2.5)
Tensile strength, MPa (ksi)	55-138 (8-20)
Elongation at failure, %	0.3-1.5
Compressive strength, MPa (ksi)	103-206 (15-30)
Flexural modulus, GPa (10^6 psi)	96-138 (1.4-2.0)
Flexural strength, MPa (ksi)	124-207 (18-30)
Izod impact strength, notched, J/m (ft·lbf/in.)	430-1176 (8-22)
Dielectric strength, kV/cm (V/mil)	120-160 (300-400)
Heat distortion, °C (°F)	205-260 (400-500)
Continuous heat resistance, °C (°F)	150-205 (300-400)
Thermal conductivity, W/m·K (Btu·in./h·ft^2·°F)	0.19-0.25 (1.3-1.7)
UL flammability class, V	945
Rockwell hardness number	H50-H112
Specific gravity	1.7-2.1
Density, g/cm^3	1.7-2.1
Coefficient of thermal expansion, 10^{-6}/K	15-22

requirements, but more stringent flammability classifications necessitate the use of flame retardant additives. These additives are used in conjunction with hydrated alumina fillers and halogenated polyester resins to provide maximum retardancy performance. Flame retardant additives recommended for SMC are antimony trioxides, tris (2,3,-dibromopropyl) phosphates, chlorinated paraffins, and zinc borates. Two of these additives are often combined at 1:1 ratios to offer more selective properties. About 3 to 5% of the SMC formulation consists of these additives.

Ultraviolet (UV) absorbers can be added to SMC resin blends when the molded parts need to withstand extended exposure to sunlight. Generally, SMC resins are stabilized with approximately 0.1 to 0.25% of UV absorber of the benzotriazole or benzophenone type.

Physical Properties

The typical SMC properties shown in Table 1 represent a broad range of data on composites having 15 to 30% glass reinforcements. Many of these properties were tested on specific SMC formulations to better define their suitability for various end-product applications. Standard test methods, developed by the American Society for Testing and Materials (ASTM) and others, were used. Mechanical properties were obtained on both dry (as-molded) and wet (2-h submersion in boiling water) test specimens. Additional properties normally reported on dry specimens include Izod impact (notched and unnotched), Barcol hardness, water absorption, and heat distortion.

It should be noted that most standardized laboratory tests are, at best, a simplification or approximation of what may happen to a finished part in use. The shape and dimensions of the test specimens and the procedure by which they are molded practically never duplicate those of any end product.

Mixing Techniques for SMC Resin Pastes

The three types of resin paste mixing techniques for an SMC operation are batch, batch/continuous, and continuous.

Batch mixing is an economical method adequate for preparing small amounts of resin paste for short production runs or for evaluating an experimental formulation. All raw materials can be mixed in a single mixing unit. The disadvantages are:

- Material efficiencies are low (generally considered to be 85% or less); the resin paste often becomes too thick for good wet-out of the glass
- Batch-to-batch variations of the resin paste can lead to SMC inconsistencies
- Additional manpower is required to make the paste and deliver it to the SMC machine

The SMC supplier generally can justify a more automated resin-mixing system if large quantities of material are to be produced.

Batch/continuous mixing has these primary advantages over the batch system:

- More reproducible resin paste thickening
- Higher material efficiencies
- Less manpower

This mixing technique normally employs two tanks, A and B, one for holding the thickenable resin bath mix (A) and the other for holding the thickener material or component in a nonthickenable resin mix (B). A metering pump or cylinder is used to proportion each side of the paste system and simultaneously pump the paste through a static or dynamic mixer to the paste doctor (metering) blades. Batch-mixing equipment for each paste side is still required in this system, but the combined A and B resin paste delivered to the machine has marked improvement in reproducibility.

Continuous mixing eliminates a separate resin paste mixing facility, which is its biggest advantage over the batch/continuous mixing system. Predetermined amounts of the liquid ingredients are individually pumped into a continuous mixer. The dry ingredients can be preblended or fed individually by automatic-metering equipment into the same continuous mixer. The amount of thickened resin paste in the continuous mixer can be kept to a minimum level for delivery to the SMC machine. Better control of the resin mix is accomplished because reproducible pumping and metering rates are used. The continuous-mixing system also provides higher material efficiencies than either the batch or batch/continuous system. A change of resin formulation of the normal clean-up operation wastes only a small amount of material. Only the head of the mixer requires cleaning (flushing) with a suitable solvent, whereas in the batch/continuous system, all lines must

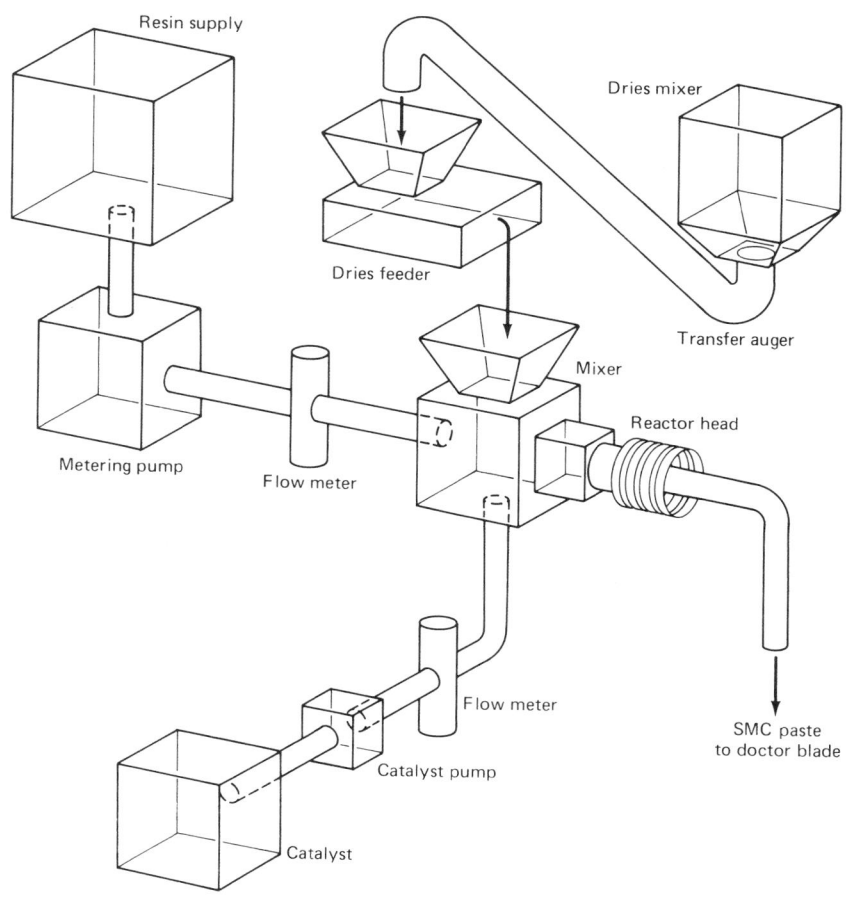

Fig. 3 Continuous SMC paste mixing system

be flushed, which involves more cleaning time and greater material waste. Water jacketing of the continuous-mixing head in an automatic system offers temperature control, thereby maintaining a more reproducible resin paste. The temperature of the resin mix is critical in viscosity control, with 32 °C (90 °F) at the time the thickening ingredient is added considered optimum. The continuous-mixing system (Fig. 3) is best for long runs. The length of set-up time makes short runs of various formulations impractical.

SMC Machines

The two most common types of SMC machines are the continuous-belt and the beltless machines. Each type has unique design features, although the functional operations of both are very similar (see Fig. 1). Both types can handle chopped roving or chopped and continuous strand mat glass reinforcements.

Paste Metering. There are two adjustable paste doctor blades (Fig. 4) on the SMC continuous-belt machine, which meter a predetermined thickness of resin paste onto upper and lower carrier films of (usually untreated) polyethylene that is 0.05-mm (0.002-in.) thick.

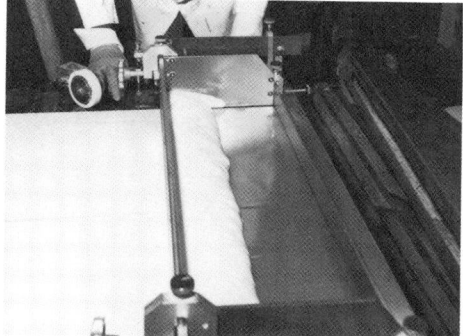

Fig. 4 Adjustable paste metering blades

Paste doctor blades are adjustable for both product width and thickness. They are rigid blades positioned vertically to the belt and are beveled to a fine edge on the side of the belt away from the paste flow. The bevel makes the metering of the paste less sensitive to viscosity and temperature variations.

The amount of resin paste in the final SMC product is determined by the height adjustment of the paste doctor blades. The speed of the machine and viscosity of the resin formulation have been found to have little or no effect on

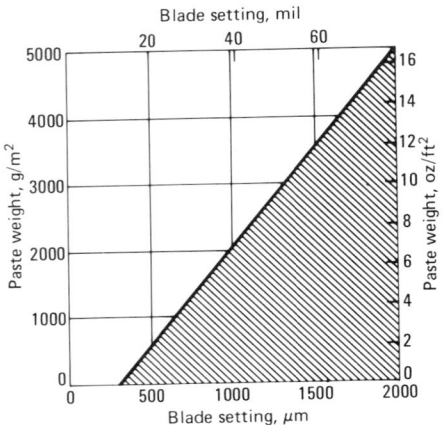

Fig. 5 Paste metering blade setting versus resin paste weight for typical SMC formulation

the paste distribution. Figure 5, which shows the relationship of paste weight to paste doctor blade settings, is representative of the settings for each of the two paste doctor blades when a typical SMC resin paste formulation (1.5:1 filler/ratio) is used.

Compaction. After the materials are brought together, they go through a compaction section to help push the resin paste through the glass to ensure wet-out of the glass. The two designs commonly used for compaction sections are serrated steel rollers and a chain belt.

Serrated Steel Rollers. This type of compaction system consists of three series of serrated steel rollers. The first series initiates the wet-out process, the second more intimately compacts the partially wet-out sheet, and the third completes the wet-out. There are a number of roller designs that do an adequate job of compacting the glass-resin sheet, but one particularly popular design uses grooved, spiral-cut roller flutes to force the resin paste into the glass pack with some lateral movement. Usually the rollers are pneumatically adjustable so that they can be set for maximum pressures (equal at each end) without causing shifting of the fibers or build-up of resin paste.

Chain Belt. New machines can be equipped with one or two dual wire mesh belt compaction modules. The two-dual module unit supports the sheet of SMC on both sides with a flexible wire mesh as it follows a sinusoidal path through the unit. The two modules have different mesh sizes in order to increase the range of materials that can be compounded, with the upstream module having a coarser mesh than the downstream module. The reasoning behind this is that a coarse mesh forces the resin paste to the center of the glass layer, and the fine mesh spreads the paste evenly through the layers. The compaction modules can also impart an oscillating motion to the movement of the upper belt to aid wet-out.

Take-up. SMC production machines are normally equipped with a dual turret take-up system for continuous operation, although festooning into 1350- to 1800-kg (3000- to 4000-lb) boxes is becoming more popular. The speed of the turrets is controlled by a weighted dancer roll that automatically adjusts to the correct tension as the SMC roll diameter increases.

Electronic devices can be used to provide torque control on the wind-up mandrel.

When a full roll of SMC is ready, the sheet is cut and transferred to the second wind-up turret. The full roll is taped to prevent unwinding, and a vapor barrier sleeve is applied. The sleeve film can be any material, such as aluminum foil or monomer resistant film, that can contain the styrene monomer in the SMC and prevent UV light or moisture contamination. SMC rolls may vary in weight, but for shelf life stability and/or shipping in cardboard boxes, a weight of about 450 kg (1000 lb) per roll is considered the maximum. SMC rolls for in-house storage and subsequent molding are suspended on racks that are mounted on wheels or designed for transporting by forklift vehicles.

Maturation Room Environments. A common practice among SMC processors is to condition their products in a temperature-controlled environment, known as a maturation room, to provide a uniform, reproducible viscosity for sheet molding. Maturation rooms are usually maintained at temperatures in the range of 29 to 32 °C (85 to 90 °F). Storage times may vary from 1 to 7 days, depending on the resin formulations. Maturation of most SMC formulations requires approximately 3 days.

Output and Feed Requirements. A single SMC machine can satisfy the material requirements of many molding presses. At 100% efficiency and continuous operation, one 1.2-m (4-ft) SMC machine can produce 11×10^9 g (25×10^6 lb) of molding compound per year.

Bulk Molding Compounds

William G. Colclough, Jr. and Daniel P. Dalenberg, Fiberite Corporation

BULK MOLDING COMPOUNDS, or fiber-reinforced thermoset molding compounds, represent the original engineering plastic materials. Phenolic molding compounds reinforced with cellulose fibers were developed by L. Baekeland early in this century. Fiberglass reinforcement and polyester, melamine, and epoxy resin systems extended the range of thermoset material capabilities. More recent developments of carbon and aramid fibers and of polyimide and silicone matrices offer further enhancements of the engineering properties that are available in thermoset molding compounds.

Thermoset materials reinforced with long fibers offer excellent engineering properties in a moldable material. Bulk molding compounds can be molded into a variety of complex shapes by methods that can be readily automated for high-volume production. At the same time, their engineering properties can approach those attainable with continuous fiber reinforced composites.

Formulation. There is a general similarity in the many formulations of bulk molding compounds. However, because of the wide variety of reinforcements, thermoset resins, and additives available, a multitude of compounds can be tailored for a variety of uses. Most compounds consist of:

- Resin matrix, with the necessary hardeners, catalysts, and plasticizers to make a moldable composition capable of curing
- Fiber reinforcement
- Additives, such as colorants, lubricants for mold release, and others for special properties such as flame retardancy, dimensional stability, and crack resistance

Processing

Just as there are many formulations, there are many ways to compound and process the requisite raw materials into a bulk molding compound. These methods influence the final physical properties. Minimizing damage to the fiber reinforcement during processing is a key factor in maintaining optimum-strength properties.

Compound Preparation. One of the most common processing methods is use of the sigma blade mixer. With this method, fiber reinforcement is charged to the mixer in pre-chopped form. Resin matrices are added in liquid form, such as polyester resins dissolved in styrene, or a varnish solution of resin (such as a phenolic) dissolved in methanol. With polyester bulk molding compound, the material is ready to mold when it is discharged from the mixer. With material made from a varnish, the solvent has to be volatilized and the resin advanced to a moldable viscosity.

Other methods of manufacture are used to obtain a longer chop length or a higher loading of reinforcement. One method is to pull reinforcement from a roving form through a dip tank of resin varnish, evaporate the solvent, and chop to length. Prepreg fabric can also be diced or chopped, with fabric used as the reinforcement. The resultant fabric is in a macerated form.

Molding Methods. Bulk molding compounds are processed by compression, transfer, and injection molding. Compression molding is used for large parts and wherever strength is critical to a molded part. This molding method causes the least amount of damage to reinforcing fibers, and in some cases the orientation of the fibers can be predetermined.

A particular use of transfer molding is to mold inserts into the part, by prepositioning them in the mold prior to the transfer operation. Transfer molding is also used to ensure accurate dimensions in the end part and reduce clean-up of molded flash.

Injection molding is capable of the highest degree of automation and lowest processing cost. Special techniques are needed to feed the bulky materials into the injection machine. This process is not widely used for bulk molding compounds because of the poor flow characteristics of many compounds.

Properties

A discussion of the effects of fiber type and length and matrix type on thermoset bulk molding compounds follows.

Effect of Fiber Type. The translation of fiber properties to bulk molding compounds that use the same matrix, fiber length, and percentage of fiber but different types of fiberglass, carbon, and aramid fibers, is shown in Table 1. Specific gravity, as might be expected, translates well with the fiber properties. Tensile and flexural strengths, and flexural modulus especially, also closely follow fiber properties. However, compressive-strength properties show a distinct pattern of higher values produced by fiberglass reinforcement, followed by polyacrylonitrile- (PAN-) based carbon fiber materials. Aramid fiber compounds have lower compressive-strength values. PAN-based carbon fibers, being brittle, demonstrate the lowest Izod impact strength. Glass fibers have higher elongation values than aramid fibers, but are notch sensitive. The toughness of aramid fibers is demonstrated by their higher impact values;

Table 1 Properties of molded composite materials

Fiber type	Material Fiber product code	Resin type	Specific gravity, g/cm³	Fiber wt%	vol%	Tensile strength MPa	ksi	Flexural strength MPa	ksi	Flexural modulus GPa	10⁶ psi	Compressive strength MPa	ksi	Impact strength J/mm	ft · lbf/in.
Fiberglass	Type E	Epoxy	1.88	63	46	190	27	470	68	28	4.1	290	42	1.6	30
Fiberglass	Type S-2	Epoxy	1.85	63	46	210	30	430	62	30	4.3	260	38	1.7	32
PAN-based Carbon	High-strength	Epoxy	1.48	58	49	140	20	330	48	38	5.5	190	28	0.55	10
PAN-based Carbon	High-modulus	Epoxy	1.51	58	48	170	25	340	50	55	8.0	210	30	0.70	13
Aramid	Kevlar 49	Epoxy	1.34	53	49	160	23	290	42	21	3.0	150	22	1.8	34
Aramid	Kevlar 29	Epoxy	1.33	53	49	110	16	270	39	19	2.8	130	19	2.1	40

Table 2 Effect of chop length on properties of composite materials

Fiber type	Material Fiber product code	Resin type	Chop length mm	Chop length in.	Specific gravity, g/cm³	Fiber wt%	Fiber vol%	Tensile strength MPa	Tensile strength ksi	Flexural strength MPa	Flexural strength ksi	Flexural modulus GPa	Flexural modulus 10⁶ psi	Compressive strength MPa	Compressive strength ksi	Impact strength J/mm	Impact strength ft · lbf/in.
Fiberglass	Type E	Epoxy	6.4	¼	1.88	63	46	120	17	270	39	25	3.6	190	27	0.75	14
Fiberglass	Type E	Epoxy	12.7	½	1.88	63	46	190	27	470	68	28	4.1	290	42	1.6	30
Fiberglass	Type E	Epoxy	31.8	1¼	1.88	63	46	310	45	760	110	32	4.6	290	42	2.8	53
Fiberglass	Type E	Phenolic	12.7	½	1.78	56	34	110	16	240	35	21	3.0	240	35	1.1	20
Fiberglass	Type E	Phenolic	25.4	1	1.78	56	34	120	18	280	40	24	3.5	260	37	1.6	30
Fiberglass	Type E	Polyimide	6.4	¼	1.90	63	47	100	15	250	36	19	2.8	230	34	0.40	7
Fiberglass	Type E	Polyimide	12.7	½	1.95	63	47	140	21	260	37	21	3.1	220	32	1.2	22
PAN-based carbon	High strength	Epoxy	12.7	½	1.48	53	49	140	20	330	48	38	5.5	190	28	0.55	10
PAN-based carbon	High Strength	Epoxy	50.8	2	1.44	53	49	160	23	470	68	38	5.5	220	32	1.0	18

Table 3 Effect of resin matrices on properties of composite materials

Fiber type	Material Fiber product code	Resin type	Specific gravity, g/cm³	Fiber wt%	Fiber vol%	Tensile strength MPa	Tensile strength ksi	Flexural strength MPa	Flexural strength ksi	Flexural modulus GPa	Flexural modulus 10⁶ psi	Compressive strength MPa	Compressive strength ksi	Impact strength J/mm	Impact strength ft · lbf/in.
Fiberglass	Type E	Epoxy	1.88	63	46	190	27	470	68	28	4.1	290	42	1.6	30
Fiberglass	Type E	Polyimide	1.95	63	47	140	21	260	37	21	3.1	220	32	1.2	22
Fiberglass	Type E	Phenolic	1.78	56	34	110	16	240	35	21	3.0	340	35	1.1	20
Fiberglass	Type E	Polyester	1 98	55	39	80	12	170	25	17	2.5	180	26	0.8	15
Fiberglass	Type E	Silicone	2.02	46	34	30	4	70	10	14	2.0	80	11	0.25	5

the difficulty encountered in cutting aramid fibers is demonstrated by this property.

Effect of Fiber Length. Generally, the longer the fiber reinforcement, the higher the physical properties. Typical lengths range from 3.2 to 50 mm (⅛ to 2 in.); with 3.2 mm (⅛ in.), 6.4 mm (¼ in.); and 12.7 mm (½ in.) predominating. The tensile, flexural, and particularly the impact strength properties show great improvement with increasing fiber length. Modulus values also show improvement with increasing length, but to a lesser degree than strength values.

The values for strength and modulus properties listed in Table 2 were obtained from compression molded test specimens. Partial orientation of the fiber reinforcement is obtained by mold geometry and resin flow. Orientation of the fiber reinforcement can be used with long fiber reinforced bulk molding compounds to enhance performance in molded parts.

Effect of Matrix. Epoxy is used whenever the best structural performance is required, and polyimide is used when higher temperature performance is a criterion. Phenolics are used for their good heat resistant and flame retardant properties. Polyester is used in electrical applications that require high arc track resistance. Silicone does not offer good structural properties, but is used in applications where temperatures requiring continuous exposure up to 300 °C (570 °F) are needed.

The comparison in Table 3 of properties of bulk molding compounds reinforced with 12.7-mm (½-in.) chopped fiberglass measured at room temperature, shows that an epoxy matrix yields the best structural properties. Polyimide and phenolics are close in their

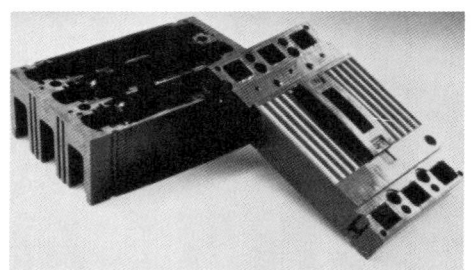

Fig. 1 High amperage breaker components (glass-phenolic)

Fig. 2 Tactical rocket nozzle (glass-phenolic)

performance at room temperature. The slight advantage of polyimide improves with increasing temperature. Polyester offers good performance in electrical applications. Silicones, although unimpressive at room temperature, maintain their mechanical properties at temperatures up to 300 °C (570 °F).

Markets

Established markets for long fiber reinforced bulk molding compounds are electrical, ordnance, aerospace, industrial, sporting goods, and automotive.

Many electrical applications require good structural as well as electrical insulation properties. Impact strength is particularly important in high-amperage switch and breaker housings (see Fig. 1). Long glass fiber reinforced phenolic, melamine, and polyester materials are used for this type of application. Other applications using long glass fiber reinforcements include connectors, commutators,

and insulators. Carbon fiber reinforced thermoset molding compounds are candidates for electrically conductive applications that require high physical property values. Nickel-coated graphite fibers, which offer improved conductivity, are now available.

Ordnance. The use of high-impact glass fiber reinforced phenolic molding compounds for military rifle armguard components is presently established. Heat resistance is required because of high barrel temperatures. Small tactical weapon components also use long fiber reinforced phenolic materials. Phenolic is selected for its ablative properties (see the rocket nozzle in Fig. 2). Glass, carbon, and specialty silica fibers are all being used as reinforcements. Fiber-reinforced phenolic molding compounds allow more design flexibility and lower unit costs than biased tape lay-ups, which are also used in ablative applications.

Aerospace. High-strength fiberglass-reinforced epoxy molding compound was selected for use on the wing section of a drone aircraft. Reproducible structural performance in a large molding was the primary criterion. The necessity for heat resistance up to 175 °C (350 °F) required the use of a high-temperature epoxy matrix.

Industrial. Present applications for long fiber reinforced epoxy molding compounds include pump component parts for chemical and food handling (Fig. 3). Fiberglass-reinforced epoxy is used in chemical pump parts because of its excellent chemical resistance and high strength. Fiberglass-reinforced epoxies approved by the Federal Drug Administration and U.S. Department of Agriculture are available to the food processing industry. Pump rotors molded to dimensions are being used in place of stainless steel. Rotor inserts that are overmolded with rubber are replacing stainless steel inserts because of improved adhesion to the rubber overmold. Other industrial applications for long fiber thermoset molding compounds include insulators for welding equipment and end caps for copiers.

Sporting Goods. Carbon fiber reinforced epoxy molding compound is being used in a line of molded golf club heads (Fig. 4). Light weight, high modulus, and high impact resistance were requirements for this application. Hybrid glass and graphite materials are being considered to improve impact resistance and lower material costs.

Automotive. Air-pump compressor vanes are a potential application for long fiber reinforced thermoset molding compounds. The molded part would replace a punched-glass

Fig. 3 Chemical pump component (glass-epoxy)

Fig. 4 Molded graphite driver head (graphite-epoxy)

laminate, which sometimes fails by delaminating. High strength, low thermal expansion, and temperature resistance up to 205 °C (400 °F) are criteria for this application.

SELECTED REFERENCES

- "Composite Materials" and "Molding Compounds," Fiberite Corporation, 1983
- D.P. Dalenberg, "Long Fiber Thermoset Molding Compounds," Paper presented at Plastic Fillers and Reinforcements, Regional Technical Conference, Atlanta, Society of Plastics Engineers, Oct 1984
- H.S. Katz, chapter 29, in *Handbook of Reinforced Plastics*, Reinhold, 1978
- J.G. Mohr, chapter 26, in *Handbook of Reinforced Plastics*, Reinhold, 1978
- R.E. Shepler, M.K. Towne, and D.K. Saylor, *Carbon-Fiber Composites*, Penton, 1979
- D.L.G. Sturgeon and R.I. Lacy, chapter 27, in *Handbook of Reinforced Plastics*, Reinhold, 1978
- "S-2 Glass Fiber High Performance/Low Cost Reinforcements," Owens-Corning Fiberglas Corporation, 1981
- G. Lubin, Ed., *Handbook of Composites*, Van Nostrand Reinholt, 1982

Injection Molding Compounds

Frederick J. Meyer, Ford Motor Company

INJECTION MOLDING COMPOUNDS are thermoplastic or thermosetting materials and their composites, which are specifically formulated for the injection molding process. This process requires materials capable of being fed into a molding machine, transported to accumulate pressure, injected through channels, and made to flow into a small opening in the mold. The process may cause major changes in both the physical and chemical properties of the molding compound. Because of their resistance to flow, neither high molecular weight resins nor long reinforcing fibers, or flakes, can be effectively manipulated through the molding process. Consequently, parts produced from molding compounds represent a compromise between optimum physical properties and the essential ability to flow under pressure. This compromise is offset by the ability to produce three-dimensional products with holes, ribs, and bosses, often without secondary operations or direct labor.

While the flow process has the potential to physically change the molding compound, these changes are not always negative. Improved homogeneity, better reinforcement wetting, and higher physical properties may be achieved if the response of the molding compound to the injection process conditions is well understood. Where this understanding is absent, reinforcements may be broken, and polymers may be either degraded or prematurely cured. This article focuses on enhancing the general understanding of the forces that these compounds encounter in the injection molding process, rather than on the thousands of commercially available molding compounds. Table 1 itemizes the more common thermoplastic and thermosetting molding compounds and indicates where, within the four subprocesses of injection molding, specific formulations demand special care or attention.

The injection process (Fig. 1) may be divided into four specific zones in which the molding compound properties may be changed. These zones, which correspond to the four injection molding subprocesses, are:

- Feeding the molding compound into the molding machine
- Transporting and melting the compound, while developing pressure

- Injecting the molding compound through runners and gates
- Flowing the molding compound material into the mold cavity

Within each of these areas, several forces may combine to affect one molding compound in a manner entirely different from the way they affect another. The uniformity of a thermoplastic may be improved by the combination of shearing forces and heat, while the fibrous reinforcement of a thermoset molding compound may be broken, or the polymers may be prematurely cured. An extremely high shear rate may be beneficial to an unfilled polymer blend or alloy, but similar conditions might cause another molding compound to flow unevenly.

Plastic resins obtained directly from a chemical reaction rarely exhibit the properties that are essential to meet the flow requirements for feeding into injection machines. Therefore, a separate compounding operation is used to convert the form of a resin while also introducing the stabilizers, lubricants, reinforcements, fillers, and even the pigments that will constitute the final injection molding compound formulation. During this operation, the molding compound resin may be exposed to heat, shearing forces, and ambient moisture. Most "virgin" molding compounds carry a history of these events. The nominal molecular weight distribution, length of the reinforcement (if one is used), and moisture content may vary as a result.

In most cases, the chemical effects are slight and would require precision analytical instruments to detect. It is wise, however, to monitor the consistency of the average molecular weight of the "as received" molding compound, using ASTM D 1238 (Ref 1), ASTM D 3123 (Ref 2), or gel permeation chromatography.

Because fiber length may be dramatically reduced during compounding (Ref 3), some compounders of glass or carbon fiber composites coat the strands of fibers with coupling agents or molten/fluid ingredients and chop them into pellets of the desired length to ensure a specific fiber length in the "after compounding" product.

Moisture will usually interfere with thermoset curing reactions. Compounding, unless done in controlled environments, may introduce extreme moisture content variations between the extremes of dry winter periods and humid summer days. Thermoplastic molding compounds are often produced as extruded strands and quenched in water. The moisture content of thermoplastics produced in this manner may vary, based on the efficiency of the compounder's driers and the moisture-sealing effectiveness of the shipping containers. Thermoplastic molding compounds should be dried before feeding. At the very least, moisture can affect surface quality, cause steam erosion of the mold, or interfere with the curing reaction. At worst, moisture can actually interfere with polymerization, and hence diminish the molecular weight of certain condensation thermoplastics, such as nylon, polycarbonate, or polyester, or consume vital cross-linking sites in thermoset compounds such as urethanes. Hopper driers and routine quality assurance tests are usually all that is required to ensure that compounding has not already changed the chemistry of the plastic.

Feeding thermoplastic, thermoset, or composite molding compounds is greatly facilitated when they are in the form of dry, free-flowing pellets. However, if this is not the case, the feeding process may cause variations in both the base polymer and any reinforcements. Typically, the molder will elect to integrate some or all of the compounding operations into the feeding and transport processes. The injection of liquid pigments, cointroduction of glass fibers and thermoplastic pellets, and building of a bulk molding compound at the press are examples of this dual-process integration. While economically feasible, compounding integration during the molding process should not be attempted without a solid understanding of the underlying effects on a molding compound.

Molding compounds that are sticky, rubbery, or powdery present feeding problems that often require specialized prefeeders. Force-feeding units may be employed, although the potential improvement in mixing may be outweighed by fiber length attrition in reinforced molding compounds. There are mechanical auger-feeder devices and so-called strap feeders that are designed to introduce a ribbon of molding compound (usually rubber) directly into the feed throat of the injection molding machine.

Table 1 Common thermoplastic and thermoset molding compounds

Base polymer	Principal applications	Feeding	Transporting	Injecting	Flowing
Thermoplastics					
ABS	Furniture, cabinets, containers, trim	B	H I	K L N O	Q
Acetal	Clock gears, miniature engineered parts	B	F H I J	O	P S
Acrylic	Automobile light lenses, plastic glazing	B	H I	K L N O	Q
Cellulose	Esters trim, moldings, screwdrivers	B	H I	K L N O	Q R
Polycarbonate	Auto bumpers, traffic lights, lenses	B C	F H I J	K L N O	Q
Polyester	Appliance parts, pump and electrical housings	B C	H I J	O	P S
Polyethylene	Houseware, food storage, dunnage	B	I	O	P
Fluoroplastics	Corrosion/solvent-resistant parts	B C	F I J	K L N O	Q
Polyimide	Aerospace items, electrical insulators	B C	H I J	K L N O	Q
Ionomer	Bumper rub strips, golf ball covers	B	H I	K L N O	P
Nylon	Auto parts, bearing retainers, appliances	B C	H I J	O	P S
Polyphenylene oxide and alloys	Auto instrument panels	B	H I (J)	K L N O	Q R
Polypropylene	Battery cases, auto parts, containers	B	H I	O	P S
Polystyrene	Toys, advertising displays, picture frames	B	H I	K L N O	Q
Polysulphone	Camera cases, aircraft parts, connectors	B C	H I J	K L N O	Q
Polyvinylchloride	Soft steering wheels, trim items	B	F H I J	K L N O	Q R
Thermosets					
Alkyd	Switches, motor housings, pot/pan handles	A B	G H I J	K L M N O	Q R S T
Allyl	Electrical connectors, circuit boards	A B	G H I J	K L M N O	Q R S T
Epoxy	Electrical insulators, electronic cases	B E	G H I J	K L M N O	Q R S T
Polyester	Automotive structural parts	D E	G I J	K L M N O	P Q R S T
Polyimide	Aircraft components, aerospace parts	A D E	G H I J	K L M N O	P Q R S T
Melamine	Dinnerware, microwave cookware	B	G H I J	K L M N O	Q R S T
Phenolic	Distributor caps, plastic ash trays	A B	G H I J	K L M N O	Q R S T
Urethane	Automotive body panels, bumpers	A D E	G I	L M O	Q R T
Vinylester	Composite car/truck springs, wheels	B D E	G I J	K L M N O	P Q R S T

(a) Feeding
A—Moisture may chemically react to degrade the polymer base
B—Drying is recommended to avoid splay in molded product
C—Drying is essential to prevent molecular weight attrition
D—Drying may volatilize monomers essential to the curing reaction
E—Fiber reinforcement breakage may occur during force feeding

Transporting
F—Overheating may cause explosive depolymerization
G—Overheating may cause premature curing of (thermoset) compound
H—Venting is recommended to remove volatiles and reduce splay
I—Fiber reinforcement lengths may be severely reduced
J—Overheating may produce chemical changes in the base polymer

Injecting
K—Fast injection of the molding compound can lead to serious overheating
L—Filled or reinforced compounds will exhibit much higher viscosity
M—Open runners required as material can cure in closed channels
N—Melt fracture may occur with high injection speeds
O—Fiber-filler orientation will occur if molding compound is reinforced

Flowing
P—Major sink marks may develop if part sections are thick, not uniform
Q—Weak knitlines may develop if compound packing pressure is low
R—Large mold vents recommended to allow volatiles to escape
S—Hot molds required to promote cure, crystalline growth
T—Curing reaction may produce peak exotherm leading to degradation

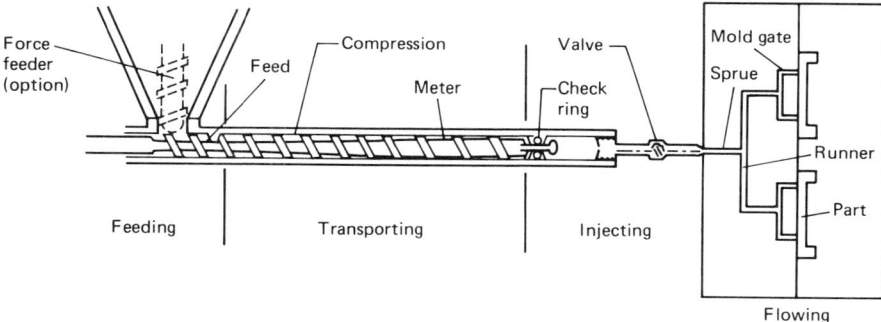

Fig. 1 The tortuous path of injection molding compounds

Auxiliary feeders generally add no heat to the molding compound.

Transporting the molding compound forward from the feed section to the injection section often provides high shearing forces, heat, and extensive mixing, which can be good or bad depending on the polymer. In normal processing of thermosets, the transport zone is designed only to densify the molding compound by removing entrapped gases. For thermoplastics, the densification process requires heat. A pressure gradient, required to degas both thermoplastics and thermosets, is pro-

duced by use of variable flight depth (tapered) screws to provide pumping, compression, decompression, and mixing functions. The design of these screws will vary greatly from thermoplastic to thermoset, from crystalline to amorphous polymers and from inherently solid to potentially volatile materials (see Fig. 2). The description of an injection molding screw is complex and involves two key factors: the length and the compression ratio.

Short screws are recommended for heat-sensitive thermosetting injection molding compounds. Longer screws are needed for thermo-

plastic molding compounds that must be both compressed and melted. The compression ratio is determined by dividing the "open volume" in the feed zone of the screw by the volume at the end of the screw. Thermoplastic molding compounds require compression ratios in the range of 2.0 to 3.0 or more. Thermoset molding compounds rarely require compression ratios above 1.5. Exactly how the compression is achieved is also of great importance to the ultimate properties of the molding compound. A heat-sensitive thermoset should not be compressed quickly during the transport process, as reinforcing fiber lengths will be broken and/or excessive heat will be created. Similarly, a thermoplastic molding compound should not be compressed quickly unless the intent is to create a hot spot of extremely high work energy. The specification of the screw profile of the injection molding machine is just as vital as the specification of its compression ratio and length.

Thermoplastic molding compounds must be gradually melted, compressed, degassed (if volatile), sheared (if crystalline and amorphous), and accumulated before injection. This must be accomplished with minimal thermal gradient, at the lowest possible temperature, and with minimal heat history. Otherwise, the part cooling time will be extended, part shrink-

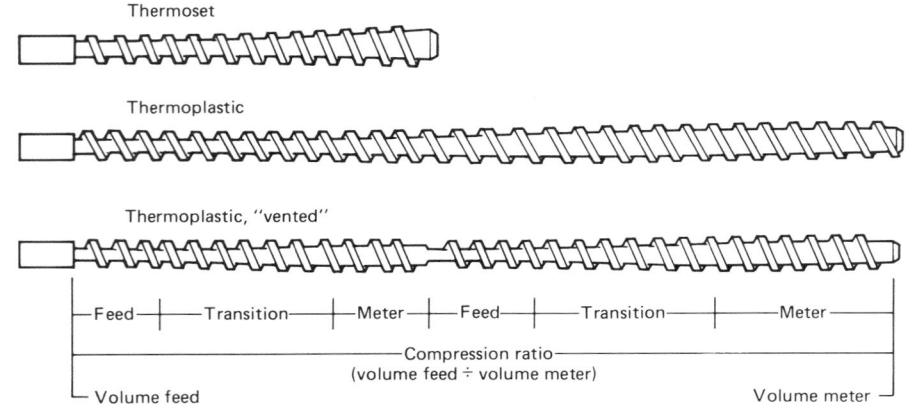

Fig. 2 Injection molding compound screws

age may vary, and the potential for polymer degradation may be increased, although degradation is not always the result. Reinforced thermoplastic molding compounds of improved strength may be obtained if sufficient time is allowed for the polymer to wet the reinforcement surface (Ref 4). The proper compression ratio for an injection molding screw will be slightly greater than the bulk density of the as-received molding compound divided by the density of the molten metal (melt).

Crystalline thermoplastics, such as acetal, nylon, polypropylene, and polyester, should be processed with screws containing a meter zone (no further compression) at the end of the profile. The meter zone is really a high shear zone where the crystalline regions can be more effectively melted or fluxed. Amorphous thermoplastics, such as ABS, styrene, polycarbonate, and acrylic, require little, if any, meter zone (Ref 5) because it would create useless shear and heat that could reduce molecular weight, which in turn could reduce impact strength.

Volatile materials must be removed from any molding compound to prevent either internal voids or splay (a fanlike surface defect near the gate) on the part. With thermoplastic molding compounds, a two-stage "vented" screw may be employed, as shown in Fig. 2. Only thermally stable molding compounds can safely employ two-stage injection molding screws. The orderly transport of molding compound is interrupted where the second screw stage begins. Thermally unstable molding compounds may accumulate, and the increased residence of time and heat will degrade them at this vent point. Attaching a vacuum pump to this vent point will further reduce the volatile content of the molding compound.

Thermoset molding compounds are rarely vented because their formulations usually include a proportion of potentially volatile, low molecular weight monomers that are needed later in the molding process to cross-link the resin. Venting could seriously deplete these vital ingredients. The objective of the transport process in the injection molding of thermosets

is to preserve both the physical and chemical integrity of the compound, while providing mixing and densification. Low compression ratios are required to prevent the reduction of reinforcing fiber length and to avoid the mechanical heat. Any hot spots could prematurely initiate the curing reaction. For this reason, it is standard operating procedure to cool the barrel of an injection molding machine that is used for thermosets, whereas the barrel is always heated for use with thermoplastics.

In either thermoplastic or thermoset injection molding, the end of the transport process is the accumulation of a volumetrically predetermined amount (shot) of compound, under pressure, somewhere in front of the screw. In some thermoplastic molding techniques, such as molding of structural foam, this shot is held in an accumulator. Normally, however, the shot is accumulated between the end of the screw and the end of the barrel. The screw must be allowed to slide back out of the barrel, under some controlled back pressure, to allow for this accumulation. In the subsequent injection subprocess, the travel of the screw is reversed, as the shot is forced forward.

A check ring, ball check, or some other mechanical valve must be placed at the tip of the screw to prevent the molding compound from flowing back down the screw during the subsequent injection. The resistance to flow through these devices can be so difficult that the polymer and any reinforcement will be degraded. The development of the free-flow check ring concept minimizes this difficulty (Ref 6). Another mechanical device, the nozzle valve, is needed to contain the molding compound within the barrel of the injection machine as the shot is being accumulated. As with the check ring, nozzle valves can be areas of extremely high shear, work energy, and heat. In some cases, nozzle valves can be eliminated by assuming that the tail end of the prior shot acts as a mechanical plug. This is effective only if the back pressure is low and the viscosity of the molding compound is high. Otherwise, the nozzle will "drool" after the molded part is removed.

Injecting thermoset molding compounds is done directly into the mold cavity through a single hole known as a sprue. The potential thermal instability of thermoset molding compounds usually precludes the use of "close" manifold systems used to support multicavity thermoplastic injection molding. When molding thermoplastics, however, the molding compounds will flow from the sprue into a system of "runners" leading to the gate of the mold. Injection pressures of up to 100 MPa (15 ksi) are employed to cause the molding compound to flow, although higher pressures are available on special injection machines equipped with boosters. The compound must flow if the mold is to be filled, and molding compounds are formulated specifically for this moment. If the flow is inadequate under the available conditions, an internal lubricant or processing aid may be used by the molding compound formulators to ensure that the molding compound flows smoothly without melt fracture or jetting (Ref 7), that is, the formation of disorderly ropelike patterns.

Jetting can occur as a result of fill rate in materials without these additives; it is a process phenomenon rather than a material condition. As the molding compound is injected from the nozzle through the sprue and then forward toward the mold, the narrowness of the runners exerts high shearing forces. In some cases, the presence of additional shear is beneficial because it lowers the apparent viscosity in those molding compounds that exhibit shear thinning. Under these high-shear conditions, fibrous and flake reinforcements will align parallel to the runner walls. Random orientation will become flow orientation. As the compound reaches the entrance, or gate, of the mold, it will begin to flow into the cavity unless there is a gate valve restriction. Gate valves may be used in thermoplastic molding compound applications to avoid the scrap loss associated with the runner system. In some cases, however, these restrictions may create conditions of high shear and heat.

Flowing from the gate into the mold cavity proceeds as in the runner system. If the molding compound is reinforced and the gates are smaller than the fiber lengths, the reinforcements will break (Ref 4). If the formulator, compounder, and molder have all done their jobs, the molding compound should flow in an even front, or curtain. Experimenting with short shots, shown in Fig. 3, is a valuable analytical tool. Successively increasing the size of a short shot should produce a progressively more complete product with good surface finish. The presence of splay, bubbles, pits, or roughness indicates that something is wrong in the preceding feed, transport, and injection subprocesses. There are two notable exceptions to this rule. Rib sink, or a reverse vein appearance over a rib, is normal in a short shot. The molding compound has not been "packed out". If the molding compound is a structural foam, the front will not be smooth,

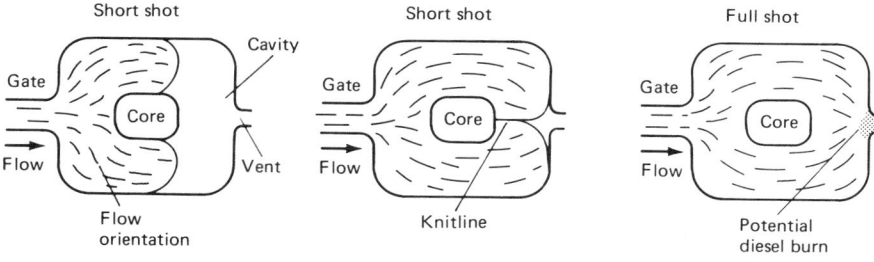

Fig. 3 Flowing molding compounds into molds

as in a properly delivered unfoamed short shot, because under the tremendous injection pressure it will tend to explode into the unpressurized mold.

In order to inject the molding compound into a mold, the air within the mold must be adequately vented. Usually this is accomplished by determining the last area to be filled and providing a narrow vent slot from this point to the outside edge of the mold. When the mold fills to a point that is inaccessible to the parting line plane, an undersized ejector pin or flat-sided pin may be utilized. If either of these actions is not taken or if the vents clog with processing oils or volatile additives, the force of injection may locally ignite the molding compound. Known as diesel burn, the ignition is caused by compressing the combustible oxygen in the mold. The molding compound within a burn area is chemically destroyed, and adjacent areas also may be affected.

Knitlines occur in any part that is molded in a cavity with more than one gate, or in any part where the molding compound is forced to flow around a pin or core. In cases of high shear stress, reinforcements such as talc, mica, and glass may physically protrude from the molding compound front. If knitlines are formed under these conditions, they will have greatly reduced strength. Although knitlines may be nearly invisible to the eye, they are areas of very high stress, fiber orientation, and potential weakness. To maximize the strength of a knitline, it is necessary to arrange the mold gating such that the melt fronts join before the completion of the shot. The molding compound fronts may thereby "scuff" together with sufficient force to achieve a higher-strength knit.

At the end of the flow process, the mold cavity will be filled. Depending upon whether the molding compound is thermoplastic or thermoset, it will begin to cool or cure. The cold molds used for crystalline thermoplastics must be held above the minimum crystalline growth temperature of the molding compound. Crystallinity initiators and so-called nucleated resins accelerate the rate of crystal growth but generally do not effectively lower the minimum crystal growth temperature. When thermoplastics crystallize, they shrink. If the mold is too cold to permit crystallization, the molded part will crystallize at some later time/temperature schedule, which may cause warpage or distortion. Thermoset molding compounds must

cure. To initiate their chemical cross-linking reactions, it is necessary to heat the mold above the initiation temperature of the curing reaction. Once initiated, thermoset reactions liberate exothermic heat. At its peak, the heat produced in this manner may raise the molded temperature above the polymer degradation threshold. Thick thermoset molding compound products may present extreme difficulties both in the ability to uniformly initiate the curing process and, once initiated, to control the heat generated during the curing process.

REFERENCES

1. "Standard Test Method for Flow Rates of Thermoplastics by Extrusion Plastometer," D 1238, *Annual Book of ASTM Standards*, American Society for Testing and Materials

2. "Standard Test Method for Spiral Flow of Low-Pressure Thermosetting Molding Compounds," D 3123, *Annual Book of ASTM Standards*, American Society for Testing and Materials

3. R.A. Schweizer, "Glass Fiber Length Degradation in Thermoplastics Processing," Paper presented at the 36th Annual Conference, Reinforced Plastics/Composites Institute, 1981

4. S. Newman, and F.J. Meyer, Mica Composites of Improved Strength, *Polymer Composites*, Vol 1 (No. 1), Sept 1980, p 37

5. Zero Meter Screws, *Plast. World*, March 1982, p 34

6. A.J. Keeney, "Free Flow Check Ring and Zero Metering Screw," Paper presented at the 37th Annual Conference, Reinforced Plastics/Composites Institute, 1982

7. K. Oda, *et al.*, Jetting Phenomena in Injection Mold Filling, *Polym. Eng. and Sci.*, Vol 16 (No. 8), Aug 1976, p 585

Resin Transfer Molding Materials

Eric B. Stark and Walter V. Breitigam, Shell Development Company

RESIN TRANSFER MOLDING (RTM) is becoming a popular and effective fabrication technique for producing fiber-reinforced high-performance composites. In the RTM process, the resin system is transferred at low viscosity using low pressure into a closed mold containing all the appropriate reinforcements and inserts.

This molding technique offers many potential processing, cost, and design advantages over other common composite fabrication methods (Ref 1-13), including the easy manufacture of complex shapes without high-cost tooling. In many cases, RTM represents the only method available for manufacturing certain complex structures, such as those requiring hand tooling of both inside mold line (IML) and outside mold line (OML) surfaces. This is especially true for stitched lay-ups.

Another advantage of RTM is the ability to build larger net shapes. A large percentage of the cost of producing composite parts comes in the assembly (bonding and fastening) of large numbers of smaller detail parts after they have been fabricated. Since RTM is not limited to the size of an autoclave or by pressure application, novel tooling approaches can be used to produce not only complex contours, but large, complete structures. A further benefit of RTM is that worker exposure to chemical environments can be greatly reduced, compared to other systems of composite manufacturing.

Examples of typical parts are categorized in Table 1 (Ref 3, 4, 6-11, 14, 15). Parts consolidation can be easily accomplished by placing inserts or fittings, along with the reinforcement, into the mold in a single operation. Reinforcements can be "preformed" into the appropriate lay-up sequence and shape, chemically or mechanically bonded to maintain integrity, and placed into the mold as single or multiple units. Composites with high fiber volumes (> 60 vol%) can be made in this fashion. Resin transfer molding also allows relatively fast cycle times, highly reproducible part dimensions, good surface definition and appearance, good quality control, low clamping pressure, easily learned operator skills, and low capital investment, as opposed to other techniques, such as hand lay-up (prepreg) (Ref 1, 2, 4-13, 15). Resin transfer molding can also be highly automated to reduce production costs (Ref 15).

Materials Selection

Material considerations for RTM include tooling, resin pumping/dispensing equipment, mold releases and cleaner, resin selection, and reinforcements. Before considering specific material issues, general design guidelines must be considered. Some of these guidelines are outlined in Table 2 in the form of general questions that must be answered for a number of specific areas.

Mold design and construction are the most critical factors in the successful use of the RTM process (Ref 1, 4, 6, 9-11, 16, 17). The mold itself may be broken down into five major areas: the injection port(s), the air vent(s), the guide pins, the mold cavity, and the gasket. The injection port(s) and air vent(s) provide resin access to the mold and a means for removing volatiles and trapped air from the part. The guide pins ensure the proper alignment of the mold halves. The mold cavity imparts the desired shape to the part, while the gasket seals the mold and restricts resin flow out of it. Two other important considerations in mold design and construction are surface finish of the mold cavity and temperature control.

Before tooling material selection, all the issues in Table 2 must be addressed. The mold must maintain its mechanical integrity under the temperature, chemical, and pressure conditions required to make the part. The mold must also maintain its dimensional stability and stay within specified tolerances. The typical materials for tooling are metals, ceramics, and poly-

Table 1 Typical parts currently manufactured using RTM

Use	Part
Industrial	Solar collectors
	Electrostatic precipitator plates
	Fan blades
	Business machine cabinetry
	Water tanks
Recreational	Canoe paddles
	Television antennae
	Snowmobiles
Construction	Seating
	Bathtubs
	Roof sections
Aerospace	Airplane wing ribs
	Cockpit hatch covers
	Airplane escape doors
Automobile	Crash members
	Leaf springs
	Car bodies
	Bus shelters

Source: Ref 3, 4, 6-11, 14, 15

Table 2 Design guidelines for use in RTM materials selection

General question	Specific areas of concern
Mold	
How many parts are required from the process in a given time period?	Primarily, tooling and pumping/dispensing Secondarily, release, cleaners, resin
Is design life a consideration?	Tooling, pumping/dispensing, cleaners, release agents
What are the dimensional requirements?	Tooling, cleaners, release agents
What are the strength requirements?	Tooling, cleaners, release agents
What are the surface finish requirements?	Tooling, cleaners, release agents, resin system, reinforcement
Part	
What are the performance requirements of the part?	Reinforcement, resin system
Production	
What are the shop environmental requirements?	Resin system, cleaners, release agents, tooling, pumping/dispensing
What are the capabilities of personnel?	All
What are the cost objectives?	All

Source: Ref 16

Table 3 Typical characteristics and applications of RTM tooling materials

Tooling material	Characteristics	Applications
Metals	Excellent mechanical strength, chemical resistance, dimensional stability, durability, surface finish; difficult to handle because of weight; expensive to machine; requires durable mold	High volume, elevated temperatures (50 to 600 °C, or 120 to 1110 °F)
Ceramics	Excellent mechanical and chemical properties, dimensional stability; problems with surface finish, cracking, low thermal conductivity, manufacturing to close tolerance specifications	Extremely high temperatures (over 1000 °C, or 1830 °F)
Composites	Easy to manufacture to tolerance specifications; low density; low processing cost; excellent mechanical and chemical properties; shorter life cycle than metals; difficult to maintain surface finish	Low to intermediate temperatures (room temperature to 180 °C, or 355 °F); low to intermediate volume

Source: Ref 4, 6, 8-11, 16, 17

Table 4 Application areas generally associated with generic resin types, based on performance characteristics

Resin type	Applications	Typical neat resin properties
Polyester	Consumer products, tanks, pipes, pressure vessels, automotive structures	Tensile strength of 3.4 to 90 MPa (0.5 to 13 ksi); compressive strength of 90 to 210 MPa (13 to 30 ksi); up to 120 °C (250 °F) continuous use; low viscosity; fast reaction; can be catalyzed; high shrinkage
Vinyl ester	Consumer products, pipes, ducts, stacks, automotive structures, flooring, linings	Tensile strength of 60 to 90 MPa (9 to 13 ksi); elongation of 2-6%; up to 120 °C (250 °F) continuous use; low viscosity; fast reaction; can be catalyzed; intermediate shrinkage
Polybutadiene	Resin modifiers, coatings, adhesives, potting compounds	Good chemical resistance; up to 120 °C (250 °F) continuous use; high viscosity; fast reaction; can be catalyzed; low moisture pick-up
Epoxy	Adhesives, tooling, electronics, aerospace and automotive structures	Tensile strength of 55 to 130 MPa (8 to 19 ksi); excellent chemical resistance; up to 175 °C (350 °F) continuous use; high viscosity; can be catalyzed; intermediate reaction; low shrinkage
Polyimide	Primary and secondary aerospace structures in high-temperature areas, electronics	Tensile strength of 55 to 120 MPa (8 to 17 ksi); up to 315 °C (600 °F) continuous use; high viscosity; can be catalyzed; slow reaction; reaction by-products; microcracking
Bismaleimide	Similar to polyimide	Similar to polyimide, except that continuous use only up to 230 °C (450 °F); no reaction by-product
Low-performance thermoplastic	Automotive panels, appliance housings, gears, bearings, fixtures, consumer products	Amorphous or semicrystalline; high toughness; up to 120 °C (250 °F) continuous use; high processing temperatures and pressures; high viscosity
Engineering-grade thermoplastic	Automotive and aerospace structures	High toughness; up to 230 °C (450 °F) continuous use; high processing temperatures and pressures; high viscosity; amorphous or semicrystalline

Source: Ref 18

meric composites. Each of these material types has been used successfully in RTM, with the material choice depending on the performance requirements and the durability required to achieve the economically necessary product volume (that is, about 3000 parts for composite tooling and about 20 000 parts for metal tooling) (Ref 1, 4, 6, 9-11, 16, 17). Typical characteristics and applications for these tooling materials appear in Table 3.

The type of mold release and mold cleaner to be used as processing aids also must be considered. Neither the mold release agent nor the mold cleaner should affect the quality of the part or mold surface. Mold release agents take many forms, such as waxes; silicon and silicone sprays; and polyvinyl chloride, polyester, or polyimide films. The selection depends on the type of resin, the molding conditions, and the tooling material used. Gel coats are also often used to achieve good surface characteristics. Compatibility with part materials is the key factor. Generally, mold cleaners are common, environmentally acceptable solvents for the resin system and/or mold release agent being used. Two examples are acetone and methylene chloride (Ref 9, 11, 16-18).

Pumping/Dispensing Equipment. Pumps, rams, and air lines can all be utilized to apply the necessary force to transfer the resin into the closed mold. In addition, a vacuum is often desirable to remove air from the mold cavity before and during resin injection. Multicomponent resin systems require accurate metering, mixing heads, and/or static mixers to bring the components together and provide adequate mixing. Temperature control is important for reproducibility. Once the resin system is mixed, a probe must be designed in consort with the injection port(s) on the mold to permit resin flow into the mold. Finally, a flushing system is necessary to prevent resin gelation in the transfer system. Materials for construction of this type of equipment are described in the literature (Ref 4, 10, 16).

Resin Selection. Although a resin system selection should be based primarily on the performance requirements of the end-use application, there are several key characteristics specific to the selection of a resin system. The term resin system in this context will refer to everything needed to make the system suitable for the desired processing and for the end-use application. This includes the resin, curing agent, catalysts, fillers, pigments, promoters, and inhibitors. The available choices for a resin system tend to fall into certain application/performance areas and resin types/characteristics. These are outlined in Table 4. Although there are exceptions, the initial resin system selection can be based on these generic categories. The systems include both thermoset and thermoplastic polymers, which cover a broad range of applications (see parts listed in Table 1).

In general, resin systems that are most suitable for the RTM process have a long pot life (at least 2 h) and low viscosity at the temperature used to transfer the resin (1 Pa · s, or 1000 cP, or less), a short gel time at the curing temperature (less than 1 h), and low levels of outgases, volatiles, and cure by-products (Ref 1, 3, 4, 6, 7, 12, 13, 15-17, 19, 20). A long pot life allows the resin system to completely fill large, complex parts with high reinforcement

Table 5 Advantages and disadvantages of various RTM reinforcement forms

Reinforcement form	Advantages/disadvantages
Continuous strand mat	Good formability, wash resistance, high bulk factor, high part fill-out, uses glass fibers
Woven roving/fabric	High strength (biaxial), good formability
Unidirectional roving/fabric	High strength (unidirectional), high stiffness, good formability
Chopped-strand mat	Low formability, low wash resistance, low cost, high bulk factor, uses glass fibers
Preforms	Highly complex forms possible, little forming/handling necessary, high initial cost
Veils/surfacing mats	Good surface quality, wear resistance

Source: Ref 4, 6, 7, 12, 14, 16, 17, 21

Table 6 Breakdown of costs of RTM versus other molding techniques

	Resin transfer molding	Sheet molding compounds	Injection molding
Process operation			
Production volume	5000-10 000/press	25 000/press	30 000/press
Fixed assets	Moderate	High	High
Labor	High	Moderate	Moderate
Skill dependency	Considerable	Very low	Lowest
Operation	Movements/intersections	Flowing, neat	Flowing, neat
Inspection/control			
Raw materials	Yes	Compounds for degradation	Compounds for degradation
Products	Visual with attention	Visual, easy	Visual, easy
Finishing	Trim flash, and so on	Very little	Very little
Product			
Complexity	Preform limit	Yes	Best
Size	Big parts for low investment	Big parts if flat	Not very big parts
Tolerance	Good	Very good	Very good
Surface appearance	Gel-coated	Very good	Very good
Voids/wrinkles	Occasional	Extremely rare	Least
Reproducibility	Skill dependent	Yes	Yes
Cores/inserts	Possible	Not easy	Possible
Strength	Moderate	Best	Very good
Material usage			
Raw-material cost	Lowest	Highest	High
Handling/applying	Skill dependent	Easy	Automatic feed
Inventory	Raw materials	Dependent on number of types	Dependent on number of types
Precision	Skill dependent	Very good	Automatic feed
Waste	<3%	Very low	Attention runner
Scrap	Skill dependent	Cuts reusable	Low
Reinforcement flexibility	Yes	No	No
Mold			
Initial cost	Moderate	Very high	Very high
Cycle life	3000-4000 parts	Years	Years
Handling	With care	Metal	Metal
Preparation	In-factory	Special shop	Special shop
Maintenance	In-factory	Special machine shop/equipment	Special machine shop/equipment

Source: Ref 5

Table 7 Breakdown of costs of RTM versus hand lay-up

	Resin transfer molding	Hand lay-up
Product weight, kg (lb)	30 (65)	33 (75)
Production rate, pieces/month	1000	1000
Direct laborers	14	30
Materials cost		
Resin	28.4%	27.4%
Glass fibers	27.7%(a)	26.4%
Others	0.3%	2.7%
Subtotal	56.4%	56.5%
Depreciation cost		
Mold	9.0%(b)	3.0%(c)
Equipment	1.8%	0.7%
Subtotal	10.8%	3.7%
Scrap	1.6%	0.0%
Manufacturing cost	14.4%	39.6%
Fuel cost	0.0%	0.2%
Subtotal	16.0%	39.8%
Total	83.2%	100.0%

(a) Including the cost of auxiliary materials, fuel, manufacturing, depreciation of preformer and preforming screen. (b) life of the mold is assumed to be 2000 pieces. (c) life of the mold is assumed to be 200 pieces. Source: Ref 2

volume before resin gelation. This is one of the major advantages of RTM. The low viscosity is desirable because it permits the liquid to fill all areas of the mold quickly, with little resistance or disturbance of material placement in the mold. In addition, good fiber wet-out can be more easily achieved at low resin viscosities. The controllable gel and cure times are needed to aid part reproducibility and to minimize the mold cycle time, which is the most constrained process step. Finally, low outgassing, volatility, and by-product formation all help minimize void formation in the part during production. Resin systems providing these processing characteristics as well as the necessary performance properties make RTM an extremely cost-competitive processing technique (Ref 1, 3, 4, 6, 7, 12, 13, 15-17, 19, 20).

Reinforcement Selection. As with the resin system, the selection of the appropriate reinforcement (including veils and surface mats) is primarily governed by the performance and cost requirements of the end-use application. However, there are several important mechanical, processing, and fiber characteristics that also influence the choice of reinforcement. These attributes are the physical form of the reinforcement, the base fiber material, the sizing (if any), and the type of stitching (Ref 4, 6, 7, 12, 14, 16, 17, 21). Some common fiber reinforcement forms and their advantages and disadvantages are given in Table 5.

The predominant fiber materials are boron, aramid, ceramic, glass, and graphite, the latter two being the most common. Graphite provides the best property performance with respect to its weight, and is used in applications such as aerospace parts, where reduced weight and high-performance characteristics are dominant factors. However, it is also more costly than glass or aramid. Glass is often used in parts with lower cost and property performance requirements, such as automotive, industrial, and consumer products. Boron exhibits high-performance properties that exceed those of most graphite fibers, but it is very expensive to manufacture and has a greater density, which results in a heavier part. Ceramic fibers are useful primarily for very high temperature applications, and although they provide excellent mechanical properties, are very brittle and relatively expensive. Neither boron nor ceramic fibers are used much at this time, but as temperature and property requirements increase, particularly in aerospace applications, they may become more important (Ref 16, 18).

The effect of fiber sizing on composite properties is becoming widely recognized as an important issue. As higher-temperature/performance resins and fibers are developed, there is a greater need for well-understood interfacial characteristics. Sizing is intended to improve the handling of the fiber bundles as well as to provide a bond between the reinforcement and the matrix, thereby enhancing physical properties. Therefore, the use of sizing and the type of sizing to be used become important to the overall performance of the composite part. Typically, most commercial fiber sizings are epoxy or epoxy-silanes, but others are available that are designed for the chemical characteristics of the resin system of interest.

Chemical and physical interactions between the fiber, interface, and matrix must be addressed in order to judge the necessity and effectiveness of sizing in a specific application (Ref 6, 18).

Finally, the manner in which preforms and fabrics are made and held together is also important, particularly in high-volume production. The most common method used is stitching, which is intended to maintain orientation of the individual fiber tows or to keep reinforcement plies together while handling. Benefits include better interlaminar shear properties, damage tolerance, and fiber alignment. The concerns here arise primarily from the type and amount of material used for the stitching. For graphite fabrics, polyester stitching is commonly used. Other techniques for holding fibers or fiber forms together include adhesives or other types of chemical bonding and physical attachments, such as staples or braiding. Compatibility of the resin-fiber combination is the key to good performance in composites that are made by the RTM process (Ref 4, 6, 7, 21).

Cost Analysis

Once the materials have been specified with respect to their properties, compatibility, and cost, the process to produce the required parts must be cost-competitive with other available fabrication techniques. The economics of RTM versus other molding techniques and hand lay-up are shown in Tables 6 and 7 (Ref 5 and 2). As can be seen in Table 7, the cost advantages of RTM over hand lay-up are predominantly in the areas of part fabrication and labor. These factors more than offset the increased mold and equipment depreciation costs of RTM. However, when compared to other molding techniques that have much lower manufacturing and labor depreciation costs than the hand lay-up technique, RTM becomes less economical with respect to these factors (Table 6). The advantages of RTM compared to other molding techniques are the types of parts that can be manufactured and the generally lower tooling costs, although RTM "clamshell" tools are more expensive than single-sided autoclave tools. Improvements in the RTM process that can be made in order to further reduce part cost include the use of automation, assembly lines, multiple tools, mechanical assistance for mold handling, and better resin systems (Ref 2, 4, 5, 8-11, 13, 15).

REFERENCES

1. N.E. Michaels, J. Laven, and J. Bauer, RTM—The Right Choice for the 80's, Section 15-B, *Proceedings of the 37th Annual Conference*, Society of the Plastics Industry, 1982, p 1-16
2. K. Tabei, H. Kittaka, S. Yoshimura, and M. Hori, Resin Injection Process—Its Productivity and Cost Effectiveness, Section 24-B, *Proceedings of the 38th Annual Conference*, Society of the Plastics Industry, 1983, p 1-5
3. I. Sayama, I. Nomura, K. Tabei, and S. Gotoh, New Applications of Resin Injection Molding Process in Japan, Section 15-E, *Proceedings of the 36th Annual Conference*, Society of the Plastics Industry, 1981, p 1-5
4. J. Raymer and D. Clarke, Evaluation of Economic and Design Considerations for Production of Business Machine Cabinetry with the Resin Injection (RTM) Process, Section 15-A, *Proceedings of the 36th Annual Conference*, Society of the Plastics Industry, 1981, p 1-5
5. G.C. Grigoropoulos, Technoeconomical Criteria for Selecting Pressure Molding Processes, Section 18-C, *Proceedings of the 40th Annual Conference*, Society of the Plastics Industry, 1985, p 1-8
6. P.W. Vaccarella, RTM: A Proven Molding Process, Section 24-A, *Proceedings of the 38th Annual Conference*, Society of the Plastics Industry, 1983, p 1-5
7. M. Ware, Thermal Expansion Resin Transfer Molding (TERTM)—A Manufacturing Process for RP Sandwich Core Structures, Section 18-A, *Proceedings of the 40th Annual Conference*, Society of the Plastics Industry, 1985, p 1-4
8. M.H. Naitove, RTM Goes Mechanized to Make Large Parts, *Plast. Technol.*, Vol 28 (No. 6), June 1982, p 79-82
9. H. El-Amin, In Resin Transfer Molding: Tooling is the Key to Success, *Plast. Technol.*, Vol 27 (No. 12), Nov 1981, p 97-99
10. M. Hartung, Resin Transfer Molding: How It's Done on a Larger Scale, *Plast. Technol.*, Vol 25 (No. 3), March 1979, p 73-77
11. R.S. Morrison, Resin Transfer Molding of Fiber Glass Preform Reinforced Polyester Resin, Section 15-D, *Proceedings of the 36th Annual Conference*, Society of the Plastics Industry, 1981, p 1-7
12. S.G. Dunbar and T.E. Griffith, Reinforcement Selection for Resin Transfer Molding, Section 15-B, *Proceedings of the 36th Annual Conference*, Society of the Plastics Industry, 1981, p 1-3
13. P.W. Vaccarella, New Resin Transfer Molding (RTM) Resins for Corrosion Resistance and Fire Retardance, Section 8-E, *Proceedings of the 36th Annual Conference*, Society of the Plastics Industry, 1981, p 1-7
14. R.R. Lacovara and J.T. Woehr, Feasibility of Structural Coring in the RTM Process, Section 15-C, *Proceedings of the 36th Annual Conference*, Society of the Plastics Industry, 1981, p 1-4
15. C. Lodge, "RTM Process Invades High-Volume Markets", *Plast. World*, June 1985, p 79-82
16. "Resin Transfer Molding", Technical Bulletin, Owens-Corning Fiberglas Corporation, 1980
17. D.R. Sayers and R.D. Howard, The Potential for Mass Production with Resin Transfer Molding Using New Methacrylate Based Resins, Section 18-B, *Proceedings of the 40th Annual Conference*, Society of the Plastics Industry, 1985, p 1-5
18. G. Lubin, Ed., *Handbook of Composites*, Van Nostrand Reinhold, 1982
19. P. LaFontaine, L.-P. Hebert, and R. Gauvin, Material Characterization for the Molding of Resin Transfer Molding, Section 17-C, *Proceedings of the 39th Annual Conference*, Society of the Plastics Industry, 1984, p 1-4
20. P. Emrich and R.M. Riddell, Resin Reactivity and Its Effect on Physical Properties of RTM Molded Parts, Section 24-D, *Proceedings of the 38th Annual Conference*, Society of the Plastics Industry, 1983, p 1-6
21. K. Tabei, H. Kimure, and S. Gotoh, The Supply of Preforms for the Use of FRP Manufacturers, Section 17-E, *Proceedings of the 39th Annual Conference*, Society of the Plastics Industry, 1984, p 1-2

Composite Materials Analysis and Design

Chairman: B. Walter Rosen, Materials Sciences Corporation

Introduction

FULL UTILIZATION of the advantages of composite materials requires an understanding of the analytical tools and methodologies employed in the design process. The following article, "Overview of Composite Materials Analysis and Design," summarizes these design concepts and the characteristics of composite structures, and identifies the available analytical and experimental methods of evaluation.

The topics of unidirectional composites, laminates, test methods, and structural analysis and design all must be considered and understood. The articles in this Section focus on the first two topics. The article "Computer Programs for Structural Analysis" is included because it covers programs applicable to unidirectional composites and laminates. In addition, this article covers programs for structural analysis, which is covered in depth in the Section "Composite Structures Analysis and Design" in this Volume.

The Overview is followed by the article "Design Requirements," in which the complex interactions of properties, performance, fabrication methods, and costs that influence material selection are described. This is followed by articles on the properties of basic unidirectional fiber composites.

Unidirectional composite properties must be considered in detail to characterize this fundamental building block upon which the technology of fiber-reinforced composites is based. These are all "effective properties," that is, properties relating statistical averages of state variables over representative volume elements sufficiently large to characterize the composite. Properties to be understood include:

- Elastic moduli
- Coefficients of thermal expansion
- Moisture swelling coefficients
- Viscoelastic properties
- Thermal and electrical conductivities
- Tensile, compressive, and shear strengths
- Strength under combined stresses
- Fatigue, creep, and impact failure characteristics

The viscoelastic properties are examined further and related to experimental data in the article "Damping Properties Analysis of Composites."

The important issue of laminate analysis is treated next. The objective of laminate analysis is to develop methods for calculating overall laminate properties, given the unidirectional composite ply properties. The elements of this objective include:

- Stress-strain relationships for membrane and bending response
- Thermal and moisture effects
- Inelastic behavior
- Strength and failure
- Interlaminar stresses
- Fracture
- Fatigue
- Impact and damage tolerance

These topics reflect the complexity involved in the understanding of the characteristics of a structural laminate. When the large number of geometric variables associated with laminate construction is added to this list of properties to be considered, it becomes apparent that analytical methods are required. In general, a combination of analysis with limited experimentation provides the desirable approach to defining the laminate properties needed for structural analysis and design.

The basics of lamination theory are presented in the article "Properties Analysis of Laminates." This is followed by a series of articles that deal with important special aspects of laminate response, with emphasis on failure.

Given the understanding of the relationships between ply properties and laminate properties, it is then possible to address the statistical aspects of material property definition. These issues are discussed in the article "Utilization of Test Data to Obtain Design Properties."

At the present state of development of methods for analysis and design of fiber composites, extensive use has been made of computer methods. Reliable computer methods for composites are reviewed and specific codes for basic material and laminate properties are presented in the final two articles.

This Section demonstrates that rational methods do exist for analysis and design of composite materials and structures. The use of fiber-reinforced composites, in combination with each other and with other materials, will continue to grow by applying these methods. Improvements in properties, the generation of unique combinations of properties, and simplification of fabrication processes resulting from the use of fiber-reinforced composites, will further their use as prominent materials of construction.

Overview of Composite Materials Analysis and Design

B. Walter Rosen and Norris F. Dow, Materials Sciences Corporation

COMBINING MATERIALS to compensate for the shortcomings of one, such as the use of straw in mud bricks of ancient times, or to capitalize on the advantages of another, as in today's composites, has a long history. Roman engineers used the properties of stone as the aggregate in concrete to build structures that survived the Roman empire. Unfortunately, the concept died with the Romans and was not rediscovered until the advent of Portland cement. The Japanese combined hard but brittle materials with ductile ones to make laminated swords with properties respectable by today's standards. Both classes of composites continue to be developed, and the mud brick has evolved into reinforced plastic, with growing applications in fields where cost is the prime concern. Where performance is the criterion, as in aerospace structures, fiber composites predominate.

This article covers the characteristics of composite materials, their performance potentials, the reasons for their use, and the general principles involved in their utilization.

Fiber reinforcement materials are available in a wide variety of forms:

- Natural fibers (jute and sisal), formerly used for economy, but now generally supplanted by synthetics with better properties and lower costs
- Synthetic organic fibers, both thermoplastic (such as nylon, polyester, and polypropylene), and thermosetting (such as the aramids), which offer low densities and high strengths but low stiffnesses. The range of application is limited because of their low stiffness
- Synthetic inorganic fibers (such as glass, boron, carbon, aluminum oxide, and silicon carbide), of which glass use far outstrips the others, primarily because its cost is much lower

The wide range of properties available from reinforcements is indicated by the values in Table 1. The densities range from 1.36 g/cm^3 for the polyester to 3.96 g/cm^3 for the aluminum oxide. The strength variation is from 1100

MPa (160 ksi) for the polyester to 4130 MPa (600 ksi) for the S-glass. The stiffnesses cover the range from 13.8 GPa (2.0×10^6 psi) to more than 345 GPa (50×10^6 psi). One or more types of fibers may be combined with any of a number of matrix materials. Thus, the composite designer has a far wider selection of mechanical properties than are available in conventional structural materials.

Almost all of these reinforcement materials (except polyester) are characterized by linear stress-strain relationships over their entire tensile load range, as shown in Fig. 1. As illustrated, the very high modulus fibers are limited to 1% strain or less; aramid and the glasses can accommodate strains of 3 to 4%. In composites, however, the same linear range of response as in the fibers alone may not be achieved, because of nonlinear responses of the matrix material.

Matrix materials cover the range from polymers to metals to ceramics. The polymers are characterized by low densities, relatively low strengths, a nonlinear stress-strain relationship (Fig. 2), and relatively high strains-to-failure. Polymeric matrix composites can be manufactured more readily and can incorporate higher volume fractions of the reinforcing fibers than composites with metal or ceramic matrices. For these reasons, polymers are the most generally used matrix materials. Accordingly, this article emphasizes the technology of polymeric matrix composites including methods of design, analysis, and evaluation to ex-

plain ways to minimize their shortcomings and maximize their advantages.

Polymeric matrix materials may be either thermoplastic or thermosetting. Thermoplastic materials, which can be heated and formed repeatedly, are primarily used for injection molding, or for thermoformable graphite lami-

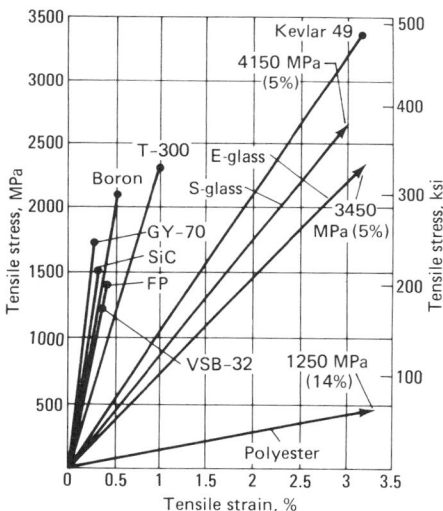

Fig. 1 Stress-strain relationships for various fibers. Generally, the stiffer the fiber, the smaller the strain at failure. Only the polyester stress-strain curve deviates appreciably from a straight line below the failure stress.

Table 1 Properties of fiber materials

Material	Density, g/cm^3	Longitudinal Young's modulus		Tensile strength	
		GPa	10^6 psi	MPa	ksi
Polyester	1.36	13.8	2.0	1100	160
E-glass	2.52	72.3	10.5	3450	500
S-glass	2.49	85.4	12.4	4130	600
Kevlar 49	1.44	124	18.0	2760	400
T-300	1.72	218	31.6	2240	325
VSB-32	1.99	379	55.0	1210	175
FP	3.96	379	55.0	1380	200
Boron	2.35	455	66.0	2070	300
Silicon Carbide	3.19	483	70.0	1520	220
GY-70	1.97	531	77.0	1720	250

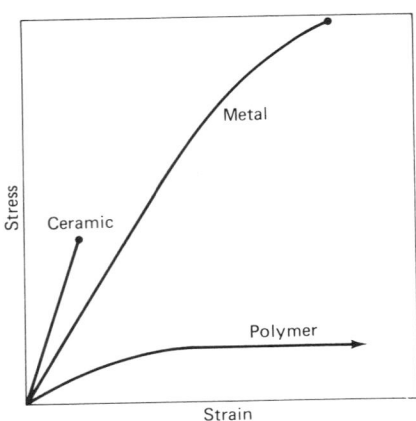

Fig. 2 Characteristic stress-strain curves for three classes of matrix materials. The ceramics have linear stress-strain curves to failure and low failure strains. Curves for metals have an appreciable linear portion, yield before failure, and have many times the failure strains of ceramics. Polymers yield at low stresses and elongate even more than metals at failure.

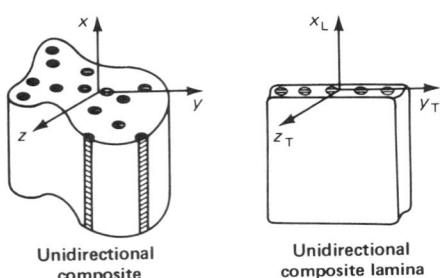

Fig. 3 Basic fiber composite elements and axis systems. For a lamina, the letters L and T are used as subscripts to identify the longitudinal and transverse directions.

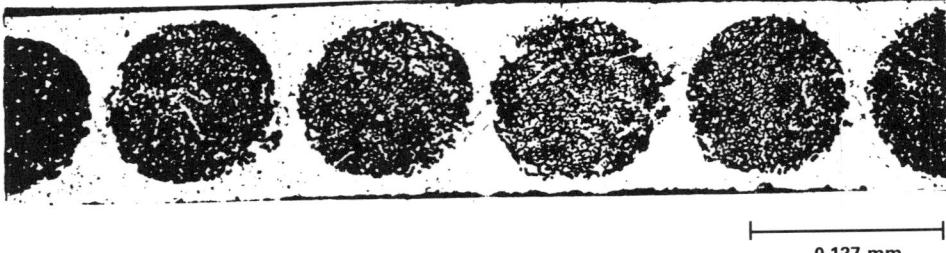

Fig. 4 Lamina cross section perpendicular to the fibers showing discrete yarns made up of a multiplicity of filaments

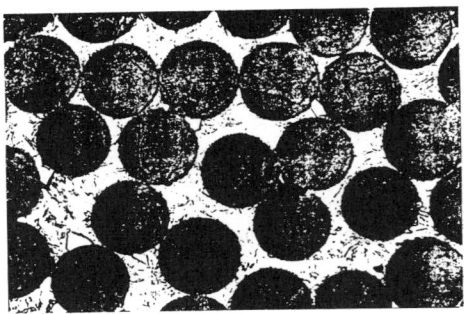

Fig. 5 Typical random dispersion of glass fibers in a lamina cross section

Composite Materials

The fundamental building block upon which fiber-reinforced composites are based is an element of a unidirectional array of fibers in a surrounding matrix, as illustrated in Fig. 3. In the element or lamina, the basic fiber unit is generally a multifilament yarn containing hundreds of individual filaments. These filament bundles may retain their individual identity, as shown in Fig. 4 or, more commonly, they may disperse and intermingle to form a random but more-or-less uniform distribution of filaments throughout the cross section (Fig. 5).

Figure 5 shows the typical disarray, with irregular gaps and some contiguity, encountered in a composite. The particular cross section shown is of glass filaments; carbon fibers would also be random in distribution and have some irregularity in fiber cross section. Similarly shaped irregularities are evident in the crystal structure of metals at the appropriate magnification level. In both metals and composites, a kind of orderly disorder exists, which varies with the magnification.

The implications of changes with magnification levels are important to understand the development of the analysis of the mechanics of composites. Consider, for example, the silicon carbide fibers embedded in a pyrolytic graphite matrix shown in Fig. 6. At 10 000×, the individual constituents are identifiable, and it appears appropriate to consider the properties of

the local constituents. Thus, the fibers, for example, would be treated as brittle materials having a statistical strength distribution and some definable geometry. This constituent information defines a characteristic unit cell at the microscopic level. At 1000×, a large number of these unit cells are seen. Collectively, they define a typical region, called a representative volume element (RVE). The properties of this element define the characteristics of the material. It is at the level of the RVE, corresponding to fiber bundles and their associated matrix, that the effective properties of a composite material can be defined. These are the desired material properties that relate average values of the state variables. From the RVE, the properties can be translated from the micro to the macro level. The properties of the assemblage of elements are volume-averaged properties that vary from point to point (where the point may be chosen at the center of the RVE), with the variation defining the statistical variability on the macro scale (200×). These assemblages are still of very small dimensions compared to the overall material, and, importantly, are small compared to the characteristic dimension over which any of the average stress or strain variables would change significantly.

Terminology. Mechanical properties, such as stiffness, and expansion coefficient, are values relating state variables in a material, such as stress, strain, and temperature. If the properties vary from point to point, as at the RVE level in the composite, the material is identified as heterogeneous. Because the average values of the state variables within an RVE are often a concern, the term effective properties will be used. This concept is illustrated in Fig. 7, in which a heterogeneous material subjected to a uniform stress, σ_0, in the y direction, has an internal stress distribution that varies from point to point in the x direction by approximately an average value, $\bar{\sigma}_y$, which is equivalent to the uniform stress in a homogeneous material.

If the properties vary with direction, the material is identified as anisotropic, of which there are many types. Those that exhibit particular symmetries have special names, such as orthotropic and cubic. Figure 8, in which the distance from the origin in any direction is a measure of the material effective stiffness,

nates. The resins employed in injection molding, usually with chopped-fiber reinforcements, include acrylics, nylon, vinyls (polyvinyl chloride, vinyl acetate, vinylidene chloride), polyethylenes, polypropylene, and acrylonitrile-butadiene-styrene (ABS). The resins used for laminates include acrylics, polysulfone or polyether sulfone, and polyether etherketone. These thermoplastic resins are chosen primarily for composite applications in which economy of manufacture, especially for moldability to complex shapes, is at a premium. Thermosetting resins, which are generally the choice for structural composites, include: polyesters, for low cost and high resistance to chemical degradation; vinyl esters, for strength combined with ease of processing; and epoxies, for highest strength. Other thermosetting resins may be employed for their special properties, such as butadiene for improved electrical properties and polyimides and silicones for use at elevated temperatures. Thermoplastics may be used at temperatures up to 205 °C (400 °F). Because of its superior structural properties, epoxy is considered the baseline material for fiber composites.

(a)

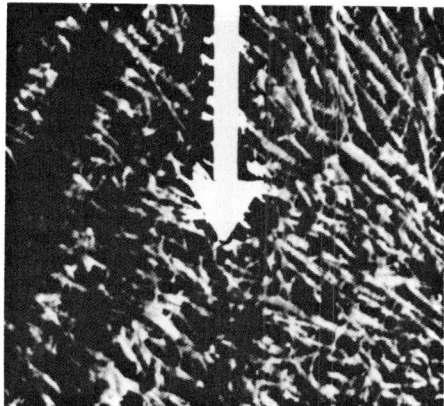

(b)

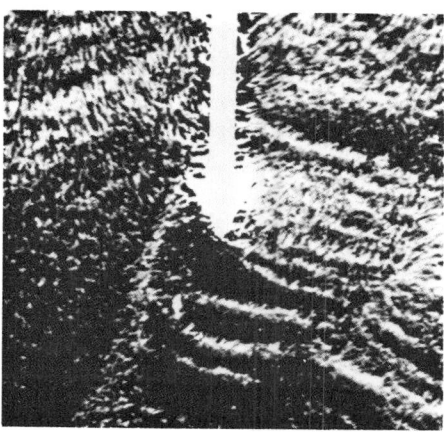

(c)

Fig. 6 Scanning electron microscope views of a composite at various magnifications showing the blending of the nonuniform constituents evident at 10 000× into the uniformly distributed assemblage of the composite as a whole. (a) 10 000×. (b) 1000×. (c) 200×

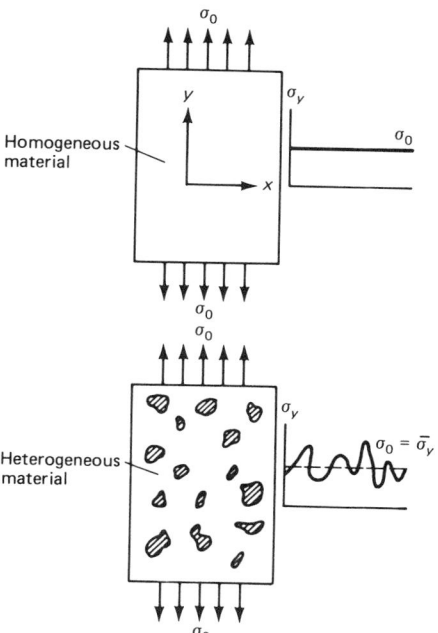

Fig. 7 Averaging of nonuniform stresses in heterogeneous materials for equivalence to uniform stress in a homogeneous material

shows that both a unidirectional carbon fiber composite and a biaxial fiberglass fabric are anisotropic, but in very different ways, while the aluminum is isotropic. The fiber composite

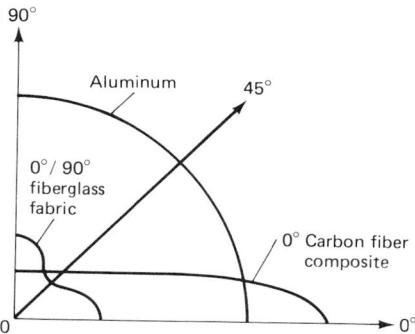

Fig. 8 The directional nature of composites, indicated by the radial distance from the origin at any angle

shows a maximum stiffness in the fiber direction, 0°, and a minimum at 90°. The fabric shows the characteristic minimum stiffness in the bias (45°) direction of woven biaxial fabrics. The aluminum is uniformly stiff in all directions.

Properties. A unidirectionally fiber-reinforced composite behaves like a homogeneous anisotropic material. In general, the distribution of fibers over a given cross section is adequately random so that transversely they provide no preferential direction of reinforcement. Thus, the unidirectional composite is effectively transversely isotropic. It can be shown that the effective elastic properties of a transversely isotropic material can be fully characterized by five elastic constants, as illustrated in Fig. 9 (note that the transverse, or TT, plane is isotropic) and defined as:

● A longitudinal Young's modulus, E_L, and associated Poisson's ratio, ν_L, where E_L is defined by the inverse ratio of longitudinal strain, ϵ_L, to simple, uniform, unidirectional stress applied in the fiber direction, σ_L:

$$E_L = \frac{\sigma_L}{\epsilon_L} \qquad (Eq\ 1)$$

● ν_{LT} is defined by the negative of the ratio of the associated transverse strain, ϵ_T, to the longitudinal strain:

$$\nu_{LT} = \nu_L = -\frac{\epsilon_T}{\epsilon_L} \qquad (Eq\ 2)$$

● A longitudinal shear modulus, G_{LT}, defined by the inverse ratio of shear strain, γ_{LT}, to pure shear stress in a longitudinal plane, τ_{LT}:

$$G_{LT} = G_L = \frac{\tau_{LT}}{\gamma_{LT}} \qquad (Eq\ 3)$$

● A transverse shear modulus, G_{TT}, similar to G_{LT} but for pure shear stress, τ_{TT}, and strain, γ_{TT}, in the transverse plane:

$$G_{TT} = G_T = \frac{\tau_{TT}}{\gamma_{TT}} \qquad (Eq\ 4)$$

● A transverse bulk modulus, K_{TT}, defined for the case of equal transverse stresses, σ_T, which produce equal transverse strains, ϵ_T (in the absence of longitudinal strains):

$$K_{TT} = K = \frac{\sigma_T}{2\epsilon_T} \qquad (Eq\ 5)$$

With these five basic elastic constants evaluated, any other desired constant may be calculated. For example, the transverse Young's modulus, E_T, which is the same in all directions in the TT plane, is:

$$E_T = \frac{4KG_T}{K + \left[1 + \dfrac{4K\nu_L^2}{E_L}\right]G_T} \qquad (Eq\ 6)$$

Defining the basic elastic constants in terms of loading conditions makes possible the measurement of the elastic properties needed for design and analysis. This empirical approach is necessary and sufficient for metals as well as for the constituent properties in composites. Composites have an element of complexity because their constituents may be combined in various ways. Therefore, determining the properties of the many combinations for all feasible cases becomes a practical impossibility. Accordingly, analytical methods have been devel-

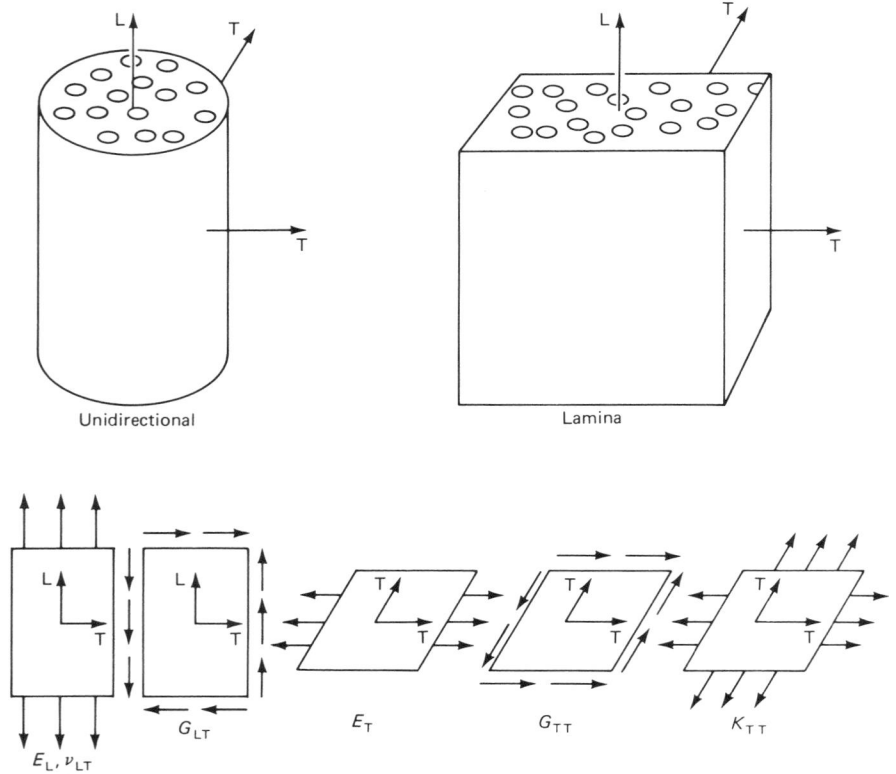

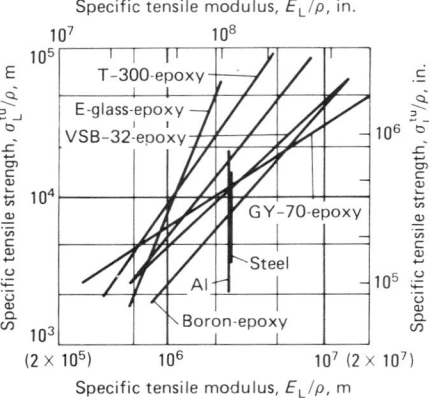

Fig. 10 Approximate ranges of values of strength/ weight and stiffness/weight ratios for structural metals and for composites of various reinforcements. Maximum values for the composite are for unidirectional reinforcements in all cases.

Fig. 9 Loading conditions for the evaluation of basic elastic properties

oped to permit the derivation of predicted composite properties from the properties of their constituents.

Before considering the technology behind performance potentials of fiber properties such as those displayed in Table 1, Table 2 and Fig. 10 should be reviewed because they summarize typical properties attainable in the current state of the art. Table 2 presents data on the mechanical properties of typical fiber-epoxy composites and of reinforcements covering the currently available stiffness range of 72.3 to 530 GPa (10.5 to 77.0 × 10⁶ psi). New fibers continually become available. Included for comparison are properties for the aluminum alloys 2024-T3 and 7075-T6 and the chrome-molybdenum steel 4130. Table 2 reveals the

characteristic effectiveness for all composites of the use of fiber properties as reinforcements in the fiber direction, and the general ineffectiveness of reinforcement transverse to the fibers.

Disparities in properties are being successfully accommodated, as evidenced by the ever widening range of fiber reinforced composite applications. A proper perspective for this success can be attributed to Galileo, who defined an important measure of a material as the longest length of a uniform bar of the material that could be hung vertically from a height without breaking. This measure of "structural efficiency" is simply the strength/density ratio of the material, in the units of Table 2 equal to σ_L^{tu}/ρ, in meters (feet). A similar measure of

structural efficiency when deflection and strength are concerned is the modulus/density ratio from Table 2.

For generalized comparisons of materials, the simple measures σ_L^{tu}/ρ and E_L/ρ are adequate to portray structural composite weight savings potentials when loading is uniaxial. A comparison of this kind is presented in Fig. 10, where specific tensile strength σ_L^{tu}/ρ and specific modulus E_L/ρ values are plotted for a wide range of composites and structural metals. For aluminum and steel, the range of strengths available through changes in heat treatment, alloy content, and so forth, are represented by the range of values of σ_L^{tu}/ρ; that is, the moduli are not affected by such changes. For the composites, however, both strength and stiffness may be substantially altered by changes in reinforcement configuration and volume fraction. Hence, the composite curves cover a range of values of both σ_L^{tu}/ρ and E_L/ρ.

The maximum values of σ_L^{tu}/ρ and E_L/ρ represent the substantial weight savings potentials provided by fiber composites — factors of two to three in most cases, for specific strength or modulus or both. Unfortunately, the practical attainment of these potentials is complicated by less favorable factors, primarily low transverse properties. To provide improved trans-

Table 2   Properties of typical fiber-epoxy composites and structural metals

Material	Density, g/cm³	Young's modulus of elasticity E_L GPa	E_L 10⁶ psi	E_T GPa	E_T 10⁶ psi	Shear modulus, G_{LT} GPa	G_{LT} 10⁶ psi	Poisson's ratio, ν_{LT}	Ultimate tensile strength σ_L^{tu} MPa	σ_L^{tu} ksi	σ_T^{tu} MPa	σ_T^{tu} ksi	Ultimate compressive strength σ_L^{cu} MPa	σ_L^{cu} ksi	σ_T^{cu} MPa	σ_T^{cu} ksi	Ultimate shear strength τ_{LT}^{su} MPa	τ_{LT}^{su} ksi
Unidirectional composites ($V_f = 0.6$)																		
E-glass	1.94	45	6.5	12	1.7	4.4	0.64	0.25	1000	150	34	5	550	80	140	20	40	6
Kevlar 49 ..	1.30	76	11.0	5.5	0.8	2.1	0.3	0.34	1380	200	28	4	280	40	140	20	55	8
T-300	1.47	132	19.2	10.3	1.5	6.5	0.95	0.25	1240	180	45	6.5	830	120	140	20	62	9
VSB-32....	1.63	229	33.2	6.9	1.0	5.5	0.8	0.25	1170	170	41	6	690	100	140	20	680	11
Boron	1.86	274	39.8	15	2.2	52	7.5	0.25	1310	190	34	5	2480	360	310	45	100	15
GY-70.....	1.61	320	46.4	5.5	0.8	4.1	0.6	0.25	690	100	41	6	620	90	140	20	96	14
Metals																		
2024-T3 ...	2.77	72.3	10.5	72.3	10.5	27.6	4.0	0.31	462	67	455	66	345	50	345	50	276	40
7075-T6 ...	2.80	71.0	10.3	71.0	10.3	27.6	4.0	0.31	544	79	530	77	475	69	475	69	324	47
4130	7.84	207	30.0	207	30.0	82.7	12.0	0.25	655	95	655	95	1100	160	1100	160	380	55

verse properties, composites commonly are used in laminates, such as plywood, with individual layers at different ply orientations. Thus, the laminate gains enhanced transverse properties at the expense of some of the potential inherent in the unidirectional material. This loss of potential is evidenced by the lowest values indicated by the curves in Fig. 10.

Other Reinforcement Forms. In addition to continuous filaments, which offer the greatest potential for performance, other forms of reinforcement include:

- Chopped fibers, usually used in molding compounds, for ease in the formation of complex shapes
- Continuous-strand rovings, or multiple-filament bundles, for economy of manufacture of thick sections
- Mats, that is, random or semioriented arrays of fibers of varying lengths, for economical reinforcement in bulk, and for ease in the formation of complex shapes when performance requirements are less important than economy
- Woven fabrics of various configurations, to provide a compromise between the maximum performance attainable with unidirectional fibers and the ease of handling and formability of chopped fibers and mats

In many manufacturing processes, reinforcements are preimpregnated with resin to form a partially cured composite, known as a prepreg. This form can be converted to the desired end shape during the final curing step of the fabrication process.

Fabrication processes for fiber-reinforced composites usually employ a die mold or mandrel to establish the desired shape, a method of fitting the fibers and resin to this shape, and a method of applying proper curing conditions (temperature and pressure as a function of time, along with provisions for removing excess resin and volatiles). Options include:

- Contact molding, in which the prepregs are placed in an open mold and cured. When a flexible bag is used to cover the composite, and pressure or vacuum and heat are applied to speed the cure and improve the quality of the part, the process is called bag molding. Further increases in pressure can be achieved by using an autoclave. Contact molding is generally used when the production quantities are too small to justify the cost of closed dies
- Compression and injection molding, in which closed, matched dies (male and female) are used and the composite is either placed into the mold before it is closed (compression molding) or injected into the closed mold (injection molding). With the addition of heat and pressure, parts of uniform quality can be produced
- Filament winding, in which filaments are wrapped around a mandrel to form a part.

Sophisticated filament winding machines have been developed to wind complex shapes with precisely oriented reinforcements
- Pultrusion, a continuous, extrusionlike process, which efficiently produces low-cost, accurate, finished shapes of constant cross section
- Braiding, a variant of molding or filament winding, in which a braid of filaments replaces a single filament in the process, in order to gain transverse strength, but at the expense of some longitudinal properties in some applications

This variety of fabrication processes allows the designer the freedom to tailor fiber composite characteristics to meet specific requirements by precisely placing any number of fibers in specific locations and directions. For composites, the design process becomes a complete cycle, from selection of constituents to final configuration, as discussed in the following section.

Design Cycle for Composite Structures

The first step in approaching any structural design problem is to establish the governing design requirements, not only the functional or property requirements (mechanical, thermal, electrical, and chemical), but also the economic and performance objectives. To meet these requirements, the designer must select a material, a configuration, and a process for manufacture. This approach is essentially the same for composites as for conventional materials, except that composites require that the material be designed along with the structure; thus, more steps are required in the design cycle.

The first step is the selection of composite constituents and their volume fractions. These selections define the unidirectional composite, which is the basic element of the composite structural material. Generally, however, multidirectional properties are required; thus, the next step is to provide these by designing a suitable multidirectional lay-up of the basic unidirectional element. It is only at this stage that the composite design cycle reaches the same starting point for design that exists with isotropic metals. From this stage, the designer proceeds to determining the configuration of each of the structural parts and the overall structural design configuration.

With composites, the possibilities of improving the design by design cycle iteration are increased by the added steps in the design cycle. Thus, by performing structural efficiency analyses during the final, structural design step in the cycle, guidelines may be generated to change constituent properties and laminate configurations, leading to improvements in performance of the composite structure.

In designs for commercial applications, the motivation for using composites

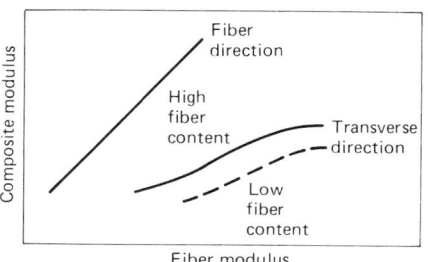

Fig. 11 Characteristic relationships between fiber and composite stiffness are directly proportional in the fiber direction, but much less than proportional transversely.

focuses on longevity and low manufacturing costs rather than improved strength and stiffness and attendant weight reductions. The design process can minimize fabrication costs by integrating structural elements, thereby reducing part count. The ability to design the material as well as the structure gives the designer new flexibility, as suggested in Fig. 11. The usual constraints and requirements of the design process are amenable to treatment in both the structure and material. The technology of structural design is long established; that of material design is new but sufficiently advanced so that if it is understood adequately, it may be applied with confidence.

Service Characteristics

Most in-service experience with composites has been established by marine applications. While composite performance in this area has been exemplary in structural integrity, ease of maintenance, and long service life, much of this durability derives from what is considered over-design in the aircraft industry. Service characteristics for aircraft applications must be more thoroughly evaluated.

Durability of fiber composites in aircraft is influenced by factors addressed quantitatively in design, such as fatigue loadings and environment scenarios, and by service factors such as damage tolerance and inspection and repair in the field.

Fatigue characteristics of composites are different in many respects from those of metals. For simple tension of unidirectionally reinforced composites, fatigue limits are generally higher percentages of ultimate tensile strengths. Thus, composite durability might be expected to be superior to that of metals. However, for more complex loadings or reinforcement configurations, the superiority diminishes. In particular, compressive or fully reversed tension-compression load cycles tend to develop more damage in composites than in metals. Intralaminar cracks and interlaminar defects often grow from points of stress concentration at some interior layer where they are not evident to visual inspection.

Environment effects upon the durability of composites further complicate the analysis of fatigue lifetimes. Temperature cycles between sunny tropical landing fields and subzero conditions at high altitudes induce thermal stresses due to differential expansion of fibers and matrix in the composite and differential expansion of adjoining structural components. Absorbed moisture in the composite, particularly differences in the level of absorption in matrix versus that in fibers, induces expansion similar to that accompanying temperature increases. Both temperature and moisture increases in the material affect its properties. Allowances for these interacting effects must be made in the design and analysis of the composite.

Damage Tolerance. As a general rule, the greater the ductility of a material, the greater its impact tolerance, and fiber composites lack ductility. The greatest concern is the inadvertent, inevitable, in-service damage. The stone bouncing up from the runway, the wrench dropped on the wing, or the bird or hailstone that collides with the composite structure can cause lateral-impact damage, to which composites are particularly susceptible. Lateral impact on a fiber composite can induce a crazing, cracking type of damage, which is usually most severe on the inner surface and often undetectable on the outer surface. Inner surface damage may be accompanied by internal disbonds and local separation of plies (delamination). Much research has been directed toward alleviating this problem.

Test, Inspection, and Repair. Determining the adequacy of a composite at any stage of its service life is more complex than for metals. The fact that the composite material itself is designed and manufactured requires that it be inspected and qualified. When it is in service, complexity increases because damage may develop internally or externally.

Quality control of the manufactured material can be implemented through both destructive and nondestructive testing. The destructive method involves cutting specimens from the material and identifying them as being representative within that structure; it is employed in usually simple tests of strength (flexure or short beam shear) or quality (volume fraction reinforcement or void content). Nondestructive testing or evaluation (NDT or NDE) methods transmit nondestructive electrical, ultrasonic, or x-ray signals through the material. Automated scanning equipment maps the area in question and identifies cracks, voids, and disbonds.

Repair techniques have been developed for various types of damage encountered in service. The usual approach is to replace the damaged material and bond or bolt metal or composite reinforcing plates to adjacent surfaces to bridge the damaged area.

Design Requirements

Frank S. Principe, Monib M. Monib, and Donald R. Linsenmann,
E.I. Du Pont de Nemours & Company, Inc.

DESIGN REQUIREMENTS, the criteria against which designs are checked, are application specific. Only professionals in their respective industries can address the specific design requirements of end products. This article will focus on the key design concerns associated with the use of polymeric matrix (thermoset and thermoplastic resin) in composite materials. Although many design requirements for metals are appropriate for composites, some changes and additions are required.

This article will discuss new design considerations that may arise from inherent composite material properties or result from fabrication processes in the course of satisfying part design requirements. The required design process for a successful program will be described, and an example of a typical advanced polymeric composite part development program will be given.

Design Considerations

Designing a part that optimizes the overall functionality of a system is the ideal role of the designer. Functionality is a measure of performance capabilities versus cost, weight, and geometric envelope. To assess the functionality of composite structures, the designer must consider total system cost, anisotropic behavior of materials, material property data base, environmental effects, damage tolerance, manufacturing and quality control, assembly, and inspection and repair. Although most of these also represent concerns in structures of conventional materials, such as metals, the importance of these topics is particularly emphasized here because the functionality of composite structures is highly sensitive to them.

Total system cost, rather than just raw material cost, must be considered when comparing costs of metallic versus composite structures. Designing replacement parts one for one (metallic to composite) is almost certain to be a more expensive alternative because composite raw material costs can be from 10 to 100 times greater than the cost of metallic materials, and, in some industries the manufacture of composite parts is labor intensive. Significant system cost reductions, however, are possible by re-

ducing the number of joining steps, which reduces the requirements for manufacturing, assembly, inventory, inspection, and machining (for example, of bolt holes). Fewer joints can also improve dimensional control (near-net shape), thereby reducing rejection and rework rates for the overall structure. Additional savings can be realized by lessening either total structural weight or the inertia of rotating and translating parts, thereby making the machinery (aircraft, rocket, or automobile, for example) more fuel efficient. Decreased weight can also result in improved performance, such as increased range, speed, or payload. Therefore, when designing composite replacement parts, the functionality of the total system must be considered in terms of part consolidation.

Anisotropy. Many structural materials generally have homogeneous and isotropic properties. This implies that the strength, stiffness, and coefficients of thermal and moisture expansion of the material are equal in all directions and at all locations. Advanced polymeric composite structures that incorporate continuously oriented fibers laid up in plies can be radically anisotropic in nature; that is, they exhibit different properties along different axes. Strength, stiffness, coefficient of thermal expansion (CTE), and coefficient of moisture expansion can vary by more than 10 times in different directions. For instance, in the through-thickness direction, strength and stiffness properties are significantly less than in-plane properties, while the coefficients of thermal and moisture expansion are greater in the through-thickness direction.

Because of these stiffness and strength differences, the designer must have a thorough understanding of the stiffness requirements and anticipated load directions and magnitudes, to ensure proper fiber alignment. In contrast to metallic materials, the stiffness and strength of laminates can be engineered to meet a wide variety of needs. Differences in coefficients of thermal and moisture expansion between joined parts are also a concern because of the large stresses that result from temperature and moisture variation.

The anisotropic properties inherent in composite structures are the key to developing

highly efficient structures. The designer must have training in and access to computerized modeling techniques, such as finite-element analysis or specialized programs that predict the performance of anisotropic composite structures, as well as a thorough knowledge of the limitations imposed by the chosen manufacturing process.

Design Data Base. Obtaining accurate and reliable material property values is one of the most important steps toward achieving a functional design, although the process can be expensive and time consuming. Material properties and allowables for unidirectional composite materials under room-temperature and standard conditions can be obtained from material suppliers and the sources cited in Ref 1 and 2.

Using laminated plate theory, the designer can combine properties and the orientation of each ply in a predetermined stacking sequence to predict the overall performance characteristics for the laminate. Laminated plate theory works well in most cases, but care must be taken to recognize its limitations.

Test coupons made with the proposed raw materials must be evaluated to establish and verify the true properties and allowables for a given lay-up or joint design before it can be used to manufacture a part. The designer must compare this data with the prediction of the model to verify the analysis.

Determining laminate properties for the effects of moisture ingress, outgassing, and oxidation can be very costly and time consuming because not all accelerated testing procedures provide consistently accurate predictions of final properties. In some cases, properties are reduced because of long-term exposure. The designer must consider to what degree the properties of the laminate will change throughout its life cycle and analyze it at each critical point.

To establish design allowables such as modulus of elasticity and strength, a statistical study of the proposed laminate structure must be conducted. The results of these studies provide A-basis, B-basis, S-basis, and typical-basis design allowables. An A-basis allowable is the value that 99% of the population of values should exceed, with 95% confidence. A B-basis allowable is the value that 90% of the values should exceed, with 95% confidence. S-basis is the minimum expected value, while the typical-basis is the average value expected.

Environmental Effects. As discussed in the preceding material on design data base, environmental effects, including heat, cold, moisture, ultraviolet light, and acids, can, over time, degrade mechanical properties to varying degrees, depending on the fiber-resin system. A common environmental effect of concern to the designer is a hot/wet condition. Under an extreme hot/wet condition, the elasticity and strength values of the composite can easily be cut in half. This condition, which mostly affects the matrix material, is a concern when matrix performance is important, such as with compressive loads. Testing under the expected environmental conditions is required.

One of the benefits of composites use in the industrial market is their resistance to corrosion. However, testing is required to determine the effects of long-term exposure to chemicals and environment on composite properties. For example, in an automobile application, a composite structure must be compatible with, or protected from, water, oil, gasoline, battery acid, brake fluid, transmission fluid, and coolant. Structures must be able to withstand vibrations, ultraviolet (UV) light, rain, hail, road salt, and temperature extremes. These conditions, when they are long term, translate to stringent performance criteria for the material designer.

Damage tolerance is a measure of the ability of a structure to function adequately after a flaw has been introduced. Flaws include scratches, dents, holes, voids, delaminations, and cracks, each of which can result from improper manufacturing or handling techniques. The designer must anticipate that not all flaws will be detected. Furthermore, holes for fasteners or step changes in ply lay-up, although planned, are analogous to flaws because they provide high stress-concentration areas, where failure can start because of a generally low tolerance for strain among composites.

A basic design concern is to ensure, through analysis and testing, the existence of alternate load paths so that the composite structure can function adequately even with flaws. Toughened matrix materials, such as thermoplastics, have demonstrated high resistance to impact damage. Stones kicked up from a runway or tools dropped by mechanics onto airplane fuselage skins are major sources of concern since possible damage can go undetected between flights. Composite structures, which do not always show visible damage from impact, can nonetheless sustain internal damage that reduces performance.

Residual stresses must also be considered in designing composite parts. When metallic parts are heated or cooled in an unconstrained state, dimensional changes will occur, but internal stresses will not necessarily result. Composites, on the other hand, can be designed to be dimensionally stable, with a low coefficient of thermal expansion, yet significant increases in internal stresses can result. Composite parts have been known to fracture during cool-down in the cure cycle. These residual stresses are superimposed on the external loads that are incurred during handling and operation.

Analysis programs for modeling flaw growth rates and residual strength and stiffness should only be used to guide design development, not to provide exact solutions. In addition to visual inspection, measuring compressive strength after impact is a common test technique for quantifying the damage tolerance of a structure. The key impact damage tolerance concerns are:

- Range of anticipated impact energies
- Frequency of impacts
- Criticality of part to system performance
- Ability to be inspected and repaired
- Residual strength/alternate load paths
- Shape of projectile

Manufacturing and Quality Control. Materials should be characterized before and after a part is manufactured. Because thermosetting resin based composites have limited shelf lives, they require constant refrigeration before use in order to minimize polymerization. Properties of these raw materials can change, even when stored properly. Maintaining good records and monitoring material life cycle are essential; composite materials should be requalified before being used in production. In selection of materials, the availability of the specific manufacturing equipment required for handling and curing must be considered. Cure cycle requirements vary significantly from one material to another. The cure cycle must also be fast enough to suit production needs. The cured production laminate can be fabricated to a greater length than required, then cut down to the proper length. The excess pieces can be used for quality control testing.

Part warpage, usually caused by poor mold design or an inappropriate curing cycle, can also occur when asymmetric stacking sequences for the laminate are used, or when the sheet is cured in tight-bend corners. Using symmetric and balanced laminates will balance out residual curing stresses, and molds with tight bend corners can be fabricated specifically to compensate for expected warpage.

Aircraft and space-related programs often require the composite parts manufacturer to use serial numbers based on a parts identification system, in which key material properties are documented and full life cycle histories are maintained and certified for each composite material used. Life cycle history records describe details of the production processes and cure cycle used during fabrication, identify the personnel who made the specific part, and give information on all environmental changes. These documentation requirements, though te-dious and costly, are essential to a quick determination of the source and extent of problems that might arise both before and after a part is in service.

Other certification requirements focus on performance. The Federal Aviation Administration, for example, has long had an extensive testing and "prove-out" procedure to qualify new parts and aircraft for flight worthiness. The amount of testing required for a new composite material or application may be greater than that required for a metallic part. Because qualification testing may significantly increase developmental costs and lead time, a designer should be aware of the full implications of specifying a new material system for any given part.

Assembly refers to the joining and handling of composite structures. Joints are used to transfer a load from one part to another. Designing and manufacturing reliable, economical joints are key challenges in making a functional system because of anticipated high loads and stress concentrations and the brittle nature of composite laminates. Most thermoset matrix laminates have limited yielding capability. When the stresses exceed the load limit, the laminate develops microcracks or can fail catastrophically, unlike most metals, which exhibit yield and thereby redistribute the stresses. Thus, there is an especially high concern for the optimum placement and design of joints to ensure reliable operations.

The two basic types of joints currently being used are mechanically fastened joints and adhesively bonded joints. The first step in making a mechanically fastened joint (bolted or riveted) is to drill a hole through the laminate; the stress-concentration factor at a drilled hole can be as great as six to eight, depending on the stacking sequence of the laminate and the materials used.

Mechanically fastened joints typically require compressive preload to provide an efficient load transfer path. The amount of compressive preload applied by the fasteners is important because the through-thickness strength of the composite may not be sufficient to prevent crushing, which will significantly increase the damaged area. Viscoelastic effects, such as creep, can reduce the fastener preload over an extended time period. Because creep rates are not well established for composites, they should be found by testing under the expected environmental conditions. Additional loads are imposed on fasteners and laminates by differential coefficients of thermal and moisture expansion and by stiffness mismatches between the parts being joined.

Adhesively bonded joints are made by gluing together two or more parts (for example, a lap joint). Like mechanical fasteners, properly designed adhesive joints should have minimal differentials for coefficients of thermal and moisture expansion and gradual stiffness transitions. In addition, the cure temperature of the adhesive must not exceed the temperature limits of the materials being bonded. Adhesives themselves can be adversely affected by environmental conditions such as temperature, moisture, and chemicals. Laminate surface preparation must be thorough and must meet all specifications to ensure proper bonding. Additionally, the proper adhesive must be selected for the expected loading conditions within the joint, because some adhesives perform better in shear and others in tension modes.

Handling composite structures includes moving and storing. Unlike isotropic metallic structures, anisotropic composites are primarily designed to meet stiffness and strength requirements in specific directions. Therefore, imposing loads in an unexpected direction or magnitude should be avoided when moving them to preclude premature failure of the part before it is even in service. When handling large structures, it is good practice to have attachment points for lifting and moving the parts. Of course, dropping or bumping the composite structure could result in damaged areas.

Storage surfaces should be clean to avoid scratching composite surfaces. Some composite materials are sensitive to environmental conditions, such as humidity and UV light, even after curing. Any special storage requirements should be thoroughly understood to prevent degraded performance.

Inspection and Repair. Not all damaged areas in composites are visible to the naked eye. Furthermore, those that are visible cannot be quantified by visual techniques. It is essential to locate and ascertain the extent of damage to determine whether repairs can or should be attempted.

Nondestructive evaluation (NDE) techniques can be used to inspect composite parts. To ensure the functionality of a part, the inspection schedule and NDE techniques required to recertify the parts must be specified in the design process. The recommended repair procedures and expected restored property values also should be documented. Recommended repair techniques can vary depending on the location of the work. Some composite structures must be repaired in the field, while others must be serviced in a specialized facility. Analysis and testing are required to verify the adequacy of the restored values.

Design Process

The standard design/development process for parts made from metals has long been established as a linear, compartmentalized process. That is, rarely do personnel representing the customer, marketing, analysis, design, manufacturing, quality control, and material suppliers team up to work simultaneously on a problem. This linear approach would not be an effective process for designing composite structures because the part configurations, material specifications, and means of manufacture are generated simultaneously. Thus, as an

alternate approach to the design process, an interdisciplinary three-phase team concept has been used effectively.

During the first phase of the team approach, functionality requirements, such as cost, size, weight, performance, loading specifications, and environment are defined, and preliminary designs are developed. In the second phase, new materials, joint designs, and manufacturing techniques specified in the first phase are evaluated and verified. Additionally, detailed finite-element analyses of critical areas may be conducted. In the third phase, a full-scale prototype is fabricated and evaluated for cost, performance capabilities, and effects from environmental factors. If all specifications are satisfied, production can begin. As composite technology matures and standards for materials, designs, manufacturing, and inspection prove to be reliable, the team approach could evolve into a more linear form.

Application Example

In this section a typical application for composite structural materials will be discussed. In Ref 3, the source of this example, the design concerns for developing a dimensionally stable and stiffness-critical support structure strut are discussed. Typical applications for composite supports with low thermal expansion include medical lasers, silicon chip processing equipment, and satellite antennas. Although a composite strut is a relatively simple component, it does serve to illustrate the way many of the concerns already discussed can affect part design. Although larger and more complex structures have been developed in the aircraft/aerospace industry, fully discussing their design concerns is beyond the scope of this article. More information on other commercialized composite structures can be found in Ref 1, 2, and 4.

The given design requirements for this structural strut were:

- $0 \pm 0.09 \times 10^{-6}$/K at -20 to 18 °C (-10 to 65 °F), coefficient of thermal expansion (CTE)
- 150 cm (60 in.) long, 3.0 cm (1.2 in.) outside diameter, 0.38-cm (0.15-in.) thick walls
- 110 GPa (15.5 × 10^6 psi), elastic modulus, axial direction

- Invar end fitting attachment points at both ends of strut
- Minimum weight

These specifications for low CTE, high stiffness, and light weight cannot be met using standard metallic materials, but can be met with the proper use of advanced carbon fiber-epoxy resin composite materials. Therefore, in performance-driven projects such as this, the cost of developing and producing a composite material part may not be the highest-priority design concern. Since CTE and elastic modulus requirements were the critical factors, the primary design emphasis was placed on choice of materials, ply lay-up, and manufacturing techniques. The manufacturing techniques were concurrently developed to minimize parameters such as fiber misalignment and percentage void content that could adversely affect tube performance. In addition, special inspection techniques had to be developed to certify the overall CTE of the tube assembly, including the Invar end fittings. This application exemplifies the interdisciplinary team approach required to develop a functional composite part.

For the analytical design, 12 different material systems and a family of ply orientations were investigated, using traditional laminate analysis. Because ply orientation, laminate modulus, and CTE are interrelated, plots were used to find the material and ply orientation that met both the CTE and minimum elastic modulus requirements. Generally, two families of symmetric ply orientations are used for zero CTE applications: the shallow-angled $+/-$ lay-up and the $(0_x, \pm \theta)$, where x is a variable number of plies and θ is a variable ply angle. Coefficient of thermal expansion and elastic modulus data were compiled for both of these families of laminates, using combinations of four different materials. A low-modulus fiber (T300), an intermediate-modulus fiber (P-55), and two high-modulus fibers (P-75 and GY-70) were used in the design study (Table 1). For the laminates studied, each material was used in the 0° direction, with lower-modulus fibers used in the angle plies.

The plots of CTE and elastic modulus in Fig. 1 and 2 show that a wide range of mechanical characteristics were available by using different materials and ply orientations. For the current tube design, the traditional flat plate

Table 1 Input laminate material properties for the design study

Materials	Elastic modulus(a) GPa	Elastic modulus(a) 10^6 psi	Elastic modulus(b) GPa	Elastic modulus(b) 10^6 psi	Shear modulus GPa	Shear modulus 10^6 psi	Coefficient of thermal expansion(a) 10^{-6}/K	Coefficient of thermal expansion(b) 10^{-6}/K
T300/934	148	21.5	9.7	1.4	4.6	0.66	0.018	20.7
P-55/934	228	33	6.9	1.0	4.1	0.6	−0.9	27
P-75/934	283	41	6.6	0.95	6.6	0.95	−1.3	27
GY-70/934	296	43	7.6	1.1	2.0	0.29	−0.13	30.2

(a) In fiber direction, 0°. (b) In transverse direction, 90°. Source: Ref 3

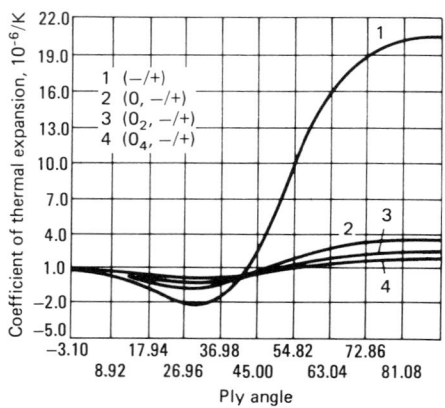

Fig. 1 Longitudinal CTE for a typical graphite laminate. Source: Ref 3

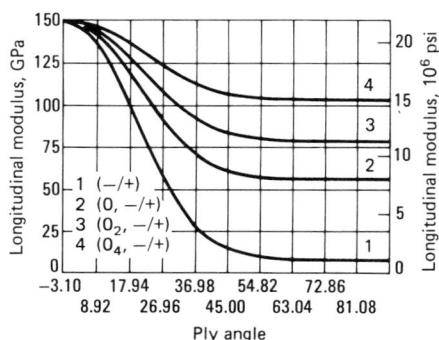

Fig. 2 Longitudinal modulus for a typical graphite laminate. Source: Ref 3

$(0_x, \pm \theta)$ lay-up did not successfully meet the modulus requirements. The dilemma was that as higher-modulus fibers were used to meet the modulus requirements, the slightly negative CTE of those fibers began to dominate. Thus, laminates that met the required modulus had a much too negative CTE.

The advantages of the shallow-angled laminates became apparent after examining the tube requirements. The tube was predominantly an axial member with few or no requirements in its circumferential direction. For instance, the circumferential modulus of the laminate bears little significance to the function of the tube, and there was only a small torsion load applied during installation. Thus, a laminate with $(+/-)$ angle plies primarily in the axial direction of the tube could be used. This design illustrates the utility of composites at meeting design requirements by placing the fibers in the preferred direction. Another advantage of shallow-angled laminates for this design is the ability to use lower-modulus fibers whose values more closely resemble those of the matrix. This causes lower residual stresses to develop during cure. In addition, lower-modulus fibers can have a greater product consistency than higher-modulus fibers, thus reducing CTE scatter. Also, because of less traumatic differences in fiber and matrix moduli, shallow-angled laminates are less susceptible to microcracking from temperature excursions and moisture absorption.

As a result of the design study, several promising laminates were identified. Laminate testing was performed on several zero CTE design configurations. The prime reason behind the testing was to ensure that the mechanical performance of the laminates met expectations. Additionally, laminate theory seldom matches test results to the tolerances required for the CTE of the tubes. Tests were performed using approximately 0.11-cm (0.045-in.) thick tubular laminates with a 1.59-cm (0.625-in.) inside diameter. Once the mechanicals were met, the laminates were tested for CTE using 10-cm (4-in.) samples. Based on those test results, full-scale specimens were made using T300/934 $\pm$ 3° and T300/934 $\pm$ 5° laminates.

The selected fabrication process involved roll-wrapping of precut patterns onto steel mandrels. This technique was selected because it produces a tubular part of the highest and most consistent quality. During the lay-up process, excess resin was removed with woven nylon tape by heat compacting. If a higher fiber volume was required, additional resin was bled during cure. Consistent fiber volumes were difficult to achieve, but fortunately, because of the fiber dominance in the axial direction of the tube, the changes in fiber volume did not greatly affect CTE. The Invar end fittings were adhesively bonded into the composite tubes because mechanically fastening them would have added cost and weight, caused higher stress-concentration factors, and had the potential of loosening over a period of time.

In the CTE design, the fiber orientation was a critical parameter, controlled by assembling the patterns on a template of polyester film marked with the proper angles. Deviation from the pattern in fiber placement is typically $\pm$0.08 cm ($\pm$0.03 in.). This small misalignment resulted in an insignificant ¼° variation in fiber direction over the full tube length. This tolerance was verified experimentally. Final testing of full-scale tubes with Invar end fittings showed that the $\pm$3° design had lower CTE but still met the 110 GPa (15.5 × 10^6 psi) elastic modulus criteria.

REFERENCES

1. G. Lubin, Ed., *Handbook of Composites*, Van Nostrand Reinhold, 1982
2. M.M. Schwartz, *Composite Materials Handbook*, McGraw-Hill, 1984
3. R.J. Carter and A.J. Brzezinski, Design and Fabrication, and CTE Testing of a Low Coefficient of Thermal Expansion Tube, *SAMPE J.*, Vol 22 (No. 2), March/April 1986
4. K.T. Kedward and J.M. Whitney, *Composite Design Guide*, Vol 5, *Design Studies*, University of Delaware, 1984

Analysis of Material Properties

B. Walter Rosen, Materials Sciences Corporation
Zvi Hashin, Tel Aviv University

WITH INCREASING USE of high-strength and high-stiffness fibers in materials designed to yield a desired set of properties, new interest has arisen in the relationships between the mechanical and physical properties of composites and those of their constituents. A study of the property relationships facilitates analysis of the performance of structures using these heterogeneous materials and provides guidelines for the development of improved materials.

The entire design process has been greatly affected by inclusion of the material design phase in the structural design process. In the preliminary design, the materials considered will usually include many that are still experimental, for which property data are not available. Thus, preliminary material selection may be based on analytically predicted properties. The analytical methods used are the result of studies of the relationship between the effective properties of composites and their constituents (studies which, although not actually conducted at a microscopic level, are frequently described by the term micromechanics). When the relationships between the overall or average response of a composite and the properties of its constituents are understood, the nonhomogeneous composite can be represented by an effective homogeneous (and, usually, anisotropic) material. The properties of this homogeneous material are the effective properties of the composite; that is, they are the properties which give the average values of the state variables in the composite. When the effective properties of a unidirectional composite have been determined, the material may be viewed as a homogeneous anisotropic material for many aspects of the design process. The evaluation of these effective properties is the major topic of this article.

Fiber Composites: Physical Properties

A unidirectional fiber composite (UDC) consists of aligned continuous fibers that are embedded in a matrix (Fig. 1). Fibers currently used are glass, carbon, graphite, and boron; typical matrices are polymeric, such as epoxy, and light metallic, primarily aluminum alloys. The physical properties of a UDC as measured by means of laboratory specimens will be called effective properties. A typical specimen is a flat coupon containing many fibers. The effective physical properties are functions of both fiber and matrix physical properties, of their volume fractions, and perhaps also of statistical parameters associated with fiber distribution. The fibers usually have circular cross sections with little variability of diameters. A UDC is clearly anisotropic, because properties in the fiber direction are different from properties transverse to the fibers. The effective properties that will be considered include elasticity, thermal expansion coefficients, moisture swelling coefficients, static and dynamic viscoelastic properties, conductivity, and moisture diffusivity.

Traditionally, material properties have been obtained experimentally and have been compiled in handbooks. Such an approach is not practical for fiber composites because of their large variety: There are many kinds of carbon and graphite fibers with different anisotropic properties, there are many kinds of matrix materials with different properties (Ref 1), and there are different environmental effects on those matrix properties. Because experimental determination of all of the effective anisotropic

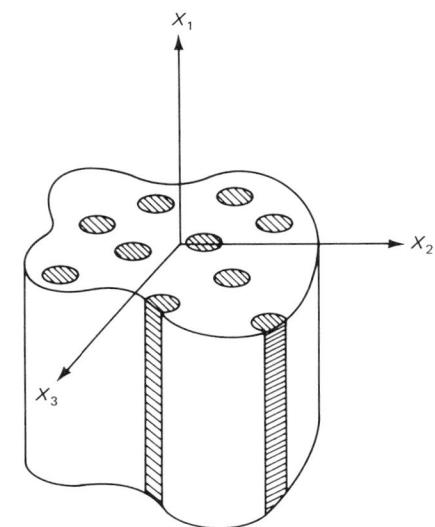

Fig. 1 Properties of unidirectional composites

properties of interest is impossible, analytical procedures must be developed (on the basis of fiber and matrix properties, volume fractions, and perhaps fiber distribution) to determine those properties. The task of experimentation is to check the validity of the analytical procedures. Thus, the properties are determined from the point of view of structural mechanics, not of materials science. Indeed, composite materials are complicated structures, not materials in the classical sense.

A variety of analytical methods that may be used to determine various properties of a UDC are described in the following sections. Details of derivations may be found in Ref 2 and 3. Reference 4 is also helpful.

Elastic Properties. The elastic properties of a material are a measure of its stiffness and are used to determine the deformations produced by loads. In a UDC, the stiffness is provided by the fibers; the matrix prevents lateral deflection of the fibers. An illustration may be obtained by comparing a bundle of stiff fibers with a UDC containing the same amount of fibers embedded in a polymeric matrix. If the stiffness of the polymer is neglected, the bundle and the UDC will have the same stiffness for a tensile load; in this case, the bundle functions like a rope. However, if compressive load is applied to the bundle in the fiber direction, the bundle will buckle at once; its stiffness for this load is zero. This is in contrast to the UDC, which has the same stiffness for compressive and tensile loads, because the matrix prevents fiber buckling until high values of load are applied. Similarly, the bundle has no transverse tensile stiffness, because the fibers will separate at once. This again will be prevented in the UDC by the matrix.

The elastic properties of a UDC are functions of the elastic properties of fibers and matrix and of their relative volumes in the composite material. Clearly, the stiffness in the fiber direction is much greater than the stiffness transverse to the fibers. If a load is applied in the fiber direction, it is carried primarily by the fibers, which deform very little and constrain the matrix to small deformation. On the other hand, in the direction normal to the fibers, the matrix is a continuous load carrying structure and the fibers move with the deforming matrix, not significantly impeding deformation. Therefore, the stiffness in the direction transverse to the fibers is much less than the stiffness in the fiber direction, making the material highly anisotropic.

For engineering purposes, it is necessary to determine Young's modulus in the fiber direction (large), Young's modulus transverse to the fibers (small), shear modulus along the fibers, and shear modulus in the plane transverse to the fibers, as well as various Poisson's ratios. This can be done in terms of simple analytical expressions.

Elastic properties of homogeneous materials are defined by relations between homogeneous (constant) stress and strain. Because of the various symmetries, there are 21 independent elastic moduli or compliances in the most general case.

The fundamental property of a fiber composite or any other composite material is statistical homogeneity. This implies that the properties of a sufficiently large sample element containing many fibers are the same as those of the entire specimen. Because the fibers are usually randomly placed, there is no preferred direction in the transverse $x_2 x_3$ plane, which implies that the UDC is statistically transversely isotropic.

Experimental determination of the properties of homogeneous materials is based on induction of homogeneous states of stress and strain in suitable specimens. The mathematical interpretation is the application of suitable boundary conditions in terms of tractions or displacements that produce homogeneous states of stress and strain, or so-called homogeneous boundary conditions. Examples are simple tension, pure shear, and hydrostatic loading. An experimenter would naturally think to apply the same homogeneous boundary conditions to composite specimens. In this case, however, the states of stress and strain in the specimen are no longer homogeneous, but highly complex. The variations of stress and strain on any plane through the composite material are random; nothing specific distinguishes the variation on one plane from that on another. Such stress and strain fields are called statistically homogeneous. They consist essentially of constant averages with superimposed random noise and are produced in geometrically statistically homogeneous specimens subjected to homogeneous boundary conditions. Consequently, effective elastic properties are defined by relations between average stress and average strain.

A typical transverse section of a UDC shows random fiber placement; hence, the material is statistically transversely isotropic. Its effective elastic stress-strain relations have the form:

$$\bar{\sigma}_{11} = n^* \bar{\epsilon}_{11} + \ell^* \bar{\epsilon}_{22} + \ell^* \bar{\epsilon}_{33}$$

$$\bar{\sigma}_{22} = \ell^* \bar{\epsilon}_{11} + (k^* + G_T^*) \bar{\epsilon}_{22} + (k^* - G_T^*) \bar{\epsilon}_{33}$$

$$\bar{\sigma}_{33} = \ell^* \bar{\epsilon}_{11} + (k^* - G_T^*) \bar{\epsilon}_{22} + (k^* + G_T^*) \bar{\epsilon}_{33}$$

$$(\text{Eq 1a})$$

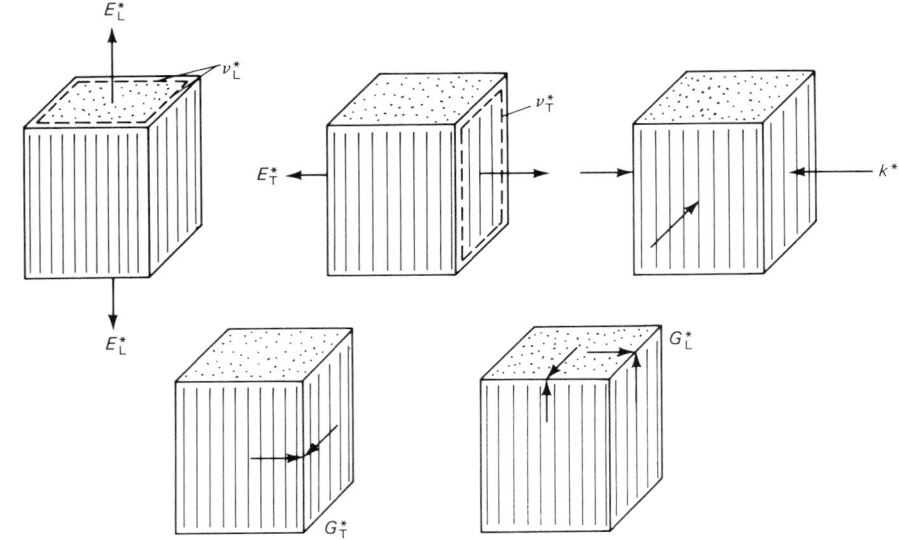

Fig. 2 Basic loadings associated with effective elastic properties

$$\bar{\sigma}_{12} = 2G_L^* \bar{\epsilon}_{12}$$

$$\bar{\sigma}_{23} = 2G_T^* \bar{\epsilon}_{23} \qquad (\text{Eq 1b})$$

$$\bar{\sigma}_{13} = 2G_L^* \bar{\epsilon}_{13}$$

with inverse

$$\bar{\epsilon}_{11} = \frac{\bar{\sigma}_{11}}{E_L^*} - \frac{\nu_L^*}{E_L^*} \bar{\sigma}_{22} - \frac{\nu_L^*}{E_L^*} \bar{\sigma}_{33}$$

$$\bar{\epsilon}_{22} = -\frac{\nu_L^*}{E_L^*} \bar{\sigma}_{11} + \frac{\bar{\sigma}_{22}}{E_T^*} - \frac{\nu_T^*}{E_T^*} \bar{\sigma}_{33} \qquad (\text{Eq 1c})$$

$$\bar{\epsilon}_{33} = -\frac{\nu_L^*}{E_L^*} \bar{\sigma}_{11} - \frac{\nu_T^*}{E_T^*} \bar{\sigma}_{22} + \frac{\bar{\sigma}_{33}}{E_T^*}$$

where * denotes effective property relating values of state property, n^* is C_{11}^*, ℓ^* is C_{12}^*, E_L^* is the longitudinal Young's modulus in fiber direction, ν_L^* is the associated longitudinal Poisson's ratio, E_T^* is the transverse Young's modulus, normal to fibers, ν_T^* is the associated transverse Poisson's ratio (in transverse plane), G_T^* is the transverse shear modulus, G_L^* is the longitudinal shear modulus, and k^* is the transverse bulk modulus. Figure 2 illustrates the loadings associated with these properties. The Poisson's ratio ν_L^* is an abbreviated notation for ν_{LT}^*, which defines the lateral strain in the transverse direction due to a stress in the fiber direction. Similarly, the Poisson's ratio ν_T^* is an abbreviated notation for ν_{TT}^*, which defines strain in transverse direction 3 due to stress in transverse direction 2, or vice versa. There is also a Poisson's ratio ν_{TL}^*, which defines strain in the longitudinal direction due to stress in the transverse direction, but it is seldom used and does not enter into the stress-strain relations presented here. Its value is given by $\nu_{TL}^* = \nu_{LT}^* E_T^*/E_L^*$. All of these Poisson's ratios are illustrated in Fig. 2.

The longitudinal shear modulus G_L^* is an abbreviation for $G_{LT}^* = G_{TL}^*$, which is associated with shear acting on perpendicular longitudinal and transverse planes. Similarly, the transverse shear modulus G_T^* is an abbreviation for G_{TT}^*, associated with shear or transverse perpendicular planes.

The effective modulus k^* is obtained by subjecting a specimen to the average state of strain:

$$\bar{\epsilon}_{22} = \bar{\epsilon}_{33}$$

all others vanish, in which case it follows that:

$$(\bar{\sigma}_{22} + \bar{\sigma}_{33}) = 2k^*(\bar{\epsilon}_{22} + \bar{\epsilon}_{33})$$

Unlike the other properties listed, k^* is of little engineering significance but is of considerable analytical importance.

Only five of the properties in Eq 1 are independent. The most important interrelations of properties are:

$$n^* = E_L^* + 4k^* \nu_L^{*2} \qquad (\text{Eq 2a})$$

$$\ell^* = 2k^* \nu_L^* \qquad (\text{Eq 2b})$$

$$4/E_T^* = 1/G_T^* + 1/k^* + 4\nu_L^{*2}/E_L^* \qquad (\text{Eq 2c})$$

$$2/(1 - \nu_T^*) = 1 + k^*/(1 + 4k^*\nu_L^{*2}/E_L^*)G_T^* \qquad (\text{Eq 2d})$$

$$G_T^* = E_T^*/2(1 + \nu_T^*) \qquad (\text{Eq 2e})$$

Computation of effective elastic moduli is a difficult problem in elasticity theory. It is first necessary to assume a suitable arrangement of

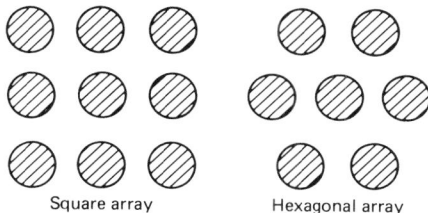

Fig. 3 Models for numerical computation of properties

Square array   Hexagonal array

fibers and thus a geometrical model of the UDC. Suitable homogeneous boundary conditions are then applied to fiber-reinforced specimens. For example, to compute k^*, it is convenient to apply displacement boundary conditions for which there is no external longitudinal deformation and for which the plane deformation in the transverse plane is isotropic, preserving the shape of the cross section. To find the associated average stress, however, it is necessary to determine in detail the elastic displacement fields in matrix and fibers. These displacements must satisfy the differential equation of elasticity theory in matrix and fibers, the displacement and traction continuity conditions at fiber-matrix interfaces, and the external boundary conditions. Once these displacements are known, the strain fields are computed by differentiation, the stress fields are found from the local Hooke's laws, and then the stress average, which is necessarily proportional to the strain average, is computed. Then, $2k^*$ is the coefficient of proportionality.

In view of the difficulty of the problem, only a few simple models permit exact analysis. One kind of model is periodic arrays of identical circular fibers, for example, square and hexagonal periodic arrays (Fig. 3). These models are analyzed by numerical finite-difference or finite-element procedures. It is necessary in each case to identify a suitable repeating element of the fiber composite and to express its boundary conditions on the basis of symmetry requirements in terms of the external boundary conditions (see, for example, Ref 2). The hexagonal array was apparently analyzed first in Ref 5 and the square array in Ref 6. It should, however, be noted that the square array is not a suitable model for most UDC analyses, because it is tetragonal, but not transversely isotropic. Further discussion on periodic arrays is available in Ref 7.

The only existing model that permits exact analytical determination of effective elastic moduli is the composite cylinder assemblage (CCA), introduced in Ref 8. To construct the model, one might imagine a collection of composite cylinders, each consisting of a circular fiber core and a concentric matrix shell. The sizes of outer radii b_n of the cylinders may be chosen at will. The size of fiber core radii a_n is restricted by the requirement that in each cylinder the ratio a_n/b_n be the same, which also implies that matrix and fiber volume fractions

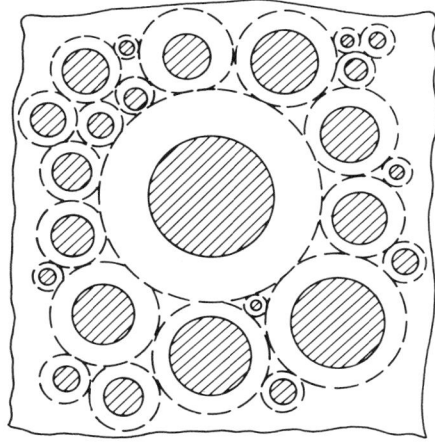

Fig. 4 Composite cylinder assemblage (CCA)

are the same in each composite cylinder. It may be shown that for various loadings of interest, each composite cylinder behaves as some equivalent homogeneous cylinder. A hypothetical homogeneous cylindrical specimen is assigned these equivalent properties and is progressively filled out with composite cylinders. Because the radii of the cylinders can be arbitrarily small, the remaining volume can be made arbitrarily small. In the limit, the properties of the assemblage converge to the properties of one composite cylinder. The construction of a CCA is shown in Fig. 4. A desirable feature of the model is the randomness of fiber placement; an undesirable feature is the large variation of fiber sizes. It will be shown, however, that the latter is not of serious concern.

Analysis of the CCA gives closed-form results for the effective properties k^*, E_L^*, ν_L^*, n^*, ℓ^*, and G_L^* and close bounds for the properties G_T^*, E_T^*, and ν_T^*. Such results are listed below for isotropic fibers, with the necessary modifications for transversely isotropic fibers. Details of derivation are given in Ref 2 and 9.

$$k^* = \frac{K_m(K_f + G_m)V_m + K_f(K_m + G_m)V_f}{(K_f + G_m)V_m + (K_m + G_m)V_f}$$

(Eq 3)

$$= K_m + \frac{V_f}{1/(K_f - K_m) + V_m/(K_m + G_m)}$$

$$E_L^* = E_m V_m + E_f V_f + \frac{4(\nu_f - \nu_m)^2 V_m V_f}{V_m/K_f + V_f/K_m + 1/G_m}$$

(Eq 4a)

$$\cong E_m V_m + E_f V_f$$

(Eq 4b)

where K_m and K_f are the bulk moduli of elasticity of matrix and fiber, and V_m and V_f are the volume fractions of matrix and fiber. The last is an excellent approximation for all UDCs.

$$\nu_L^* = \nu_m V_m + \nu_f V_f +$$
$$\frac{(\nu_f - \nu_m)(1/K_m - 1/K_f)V_m V_f}{V_m/K_f + V_f/K_m + 1/G_m}$$

(Eq 5)

$$G_L^* = G_m\left[\frac{G_m V_m = G_f(1 + V_f)}{G_m(1 + V_f) + G_f V_m}\right]$$

(Eq 6)

$$= G_m + \frac{V_f}{1/(G_f - G_m) + V_m/2G_m}$$

As indicated previously, the result for G_T^* is a pair of bounds on the actual value. One or the other of these bounds is recommended, depending on the ratios of the constituent properties (to compute the resulting E_T^* and ν_T^*, Eq 2c and 2d are used). When $G_f > G_m$ and $K_f > K_m$, the upper bound is recommended:

$$G_{T(+)}^* \sim G_T^* = G_m$$

$$\left[1 + \frac{(1 + \beta_m)V_f}{\rho - [1 + 3\beta_m^2 V_m^2/(\alpha V_f^3 + 1)]V_f}\right]$$

(Eq 7)

where

$$\alpha = \frac{\beta_m - \gamma\beta_f}{1 + \gamma\beta_f} \qquad \rho = \frac{\gamma + \beta_m}{\gamma - 1}$$

$$\beta_m = \frac{1}{3 - 4\nu_m} \qquad \beta_f = \frac{1}{3 - 4\nu_f}$$

and $\gamma = \dfrac{G_f}{G_m}$

When $G_f < G_m$ and $K_f < K_m$, the lower bound is recommended:

$$G_{T(-)}^* \sim G_T^* = G_m$$

$$\left[1 + \frac{(1 + \beta_m)V_f}{\rho - [1 + 3\beta_m^2 V_m^2/(\alpha V_f^3 - \beta_m)]V_f}\right]$$ (Eq 8)

The result (Eq 8) is of interest for metal matrix composites consisting of carbon and graphite fibers in an aluminum matrix because in that case, the elastic properties G_m and K_m of the matrix are larger than the G_{Tf} and K_f of the fibers. Note that K_m in Eq 3 is the isotropic plane-strain bulk modulus.

For transversely isotropic fibers, there are the following modifications (Ref 2, 9):

For k^* — k_f is the fiber transverse bulk modulus

For E_L^*, ν_L^* — $E_f = E_{Lf}$
$\nu_f = \nu_{Lf}$
k_f as above   (Modification 1)

For G_L^* — $G_f = G_{Lf}$

For G_T^* — $G_f = G_{Tf}$
$\beta_f = k_f/(k_f + 2G_{Tf})$

Table 1   Elastic properties of graphite-aluminum composite; comparison of analytical and experimental results

Fiber	E_L GPa	10^6 psi	E_T GPa	10^6 psi	K GPa	10^6 psi	G_T GPa	10^6 psi	G_L GPa	10^6 psi	ν_L	ν_T
T50 graphite	388	56.3	7.17	1.04	7.03	1.02	2.41	0.350	6.79	0.985	0.23	0.486

Matrix	E GPa	10^6 psi	G GPa	10^6 psi	ν
Aluminum 201	71.0	10.3	26.7	3.87	0.33

Composite	E_L^* GPa	10^6 psi	E_T^* GPa	10^6 psi	G_T^* GPa	10^6 psi	G_L^* GPa	10^6 psi	ν_T^*
30% T50-70% Al									
Experimental	160	23.2	29.6	4.30	10.3	1.50	18.5	2.69	0.43
Analytical	66	24.1	32.0	4.64	10.4	1.51	18.6	2.70	0.41

A rational approximate evaluation of G_T^* of the CCA model is developed (Ref 3) by assuming that any composite cylinder behaves as if it were embedded in the effective medium with effective shear modulus G_T^*.

Numerical analysis of the effective elastic properties of the hexagonal-array model reveals that their values are extremely close to those predicted by the CCA model, as given by the preceding equations. The simple analytical results given here predict effective elastic properties with sufficient engineering accuracy. They are of considerable practical importance for two main reasons: They permit easy determination of effective properties for a variety of matrix properties, fiber properties, and volume fractions, and they constitute the only approach known today for experimental determination of carbon fiber properties.

With respect to the first reason, it is to be noted that the variety of matrix properties to be considered arises not only from the choice of different materials but also, and perhaps more important, from environmental changes of properties of a specified matrix. Significant examples are property changes due to temperature, moisture absorption, fatigue damage accumulation, and radiation. Thus, environmental changes of composite stiffness can be determined by measurement of environmental matrix stiffness changes and computation of such changes for the composite. This avoids a great deal of expensive experimentation.

With respect to the second reason, it should be recalled that a carbon or graphite fiber has five independent elastic properties (Eq 2a to 2e) and that its diameter is of the order of 0.01 mm (0.0004 in.). It is thus impossible to measure directly the elastic properties except for E_L in the fiber direction. A remaining alternative is to measure the elastic modulus of the matrix and the five effective elastic moduli of the composite and then to compute the five fiber elastic properties from Eq 3 to 7 with Modification 1. The current standard method for doing this is to prepare a flat UDC specimen and measure the effective elastic properties by ultrasonic wave propagation. The principle of the method is to measure various wave velocities in various directions relative to the fiber direction. Because wave velocities are defined in terms of elastic moduli and density, their measurement determines the elastic moduli. Table 1 lists the anisotropic elastic properties of a carbon fiber having an axial modulus of 350 GPa (50×10^6 psi). Its properties were determined in this fashion by using carbon-epoxy specimens. The fiber properties obtained were then used to compute the elastic properties of a UDC consisting of an aluminum matrix and the carbon fiber. Also shown is the comparison between measured and computed composite elastic properties, based on the results (Eq 3 to 6, Eq 8) of Ref 10. The agreement is excellent.

The foregoing results are of sufficient accuracy and reliability to provide a predictive tool for evaluation of effective elastic properties of fiber composites. Various approximate treatments given in the literature are at least as complex and are less reliable than those presented here. Attempts have also been made to devise empirical or semiempirical equations for effective elastic properties, but in view of the complexity of the problem, this is not a helpful approach.

For purposes of laminate analysis, it is important to consider the plane-stress version of the effective stress-strain relations. Let x_3 be the normal to the plane of a thin, unidirectionally reinforced lamina. The plane-stress condition is defined by:

$$\bar{\sigma}_{33} = \bar{\sigma}_{13} = \bar{\sigma}_{23} = 0 \qquad \text{(Eq 9)}$$

Then, from Eq 1:

$$\bar{\epsilon}_{11} = \frac{\bar{\sigma}_{11}}{E_L^*} - \frac{\nu_L^*}{E_L^*} \bar{\sigma}_{22}$$

$$\bar{\epsilon}_{22} = -\frac{\nu_L^*}{E_L^*} \bar{\sigma}_{11} + \frac{\bar{\sigma}_{22}}{E_T^*} \qquad \text{(Eq 10)}$$

$$2\bar{\epsilon}_{12} = \frac{\bar{\sigma}_{12}}{G_L^*}$$

The inversion of Eq 10 is:

$$\bar{\sigma}_{11} = C_{11}^* \bar{\epsilon}_{11} + C_{12}^* \bar{\epsilon}_{22}$$

$$\bar{\sigma}_{22} = C_{12}^* \bar{\epsilon}_{11} + C_{22}^* \bar{\epsilon}_{22} \qquad \text{(Eq 11)}$$

$$\bar{\sigma}_{12} = 2G_L^* \bar{\epsilon}_{12}$$

where

$$C_{11}^* = \frac{E_L^*}{1 - \nu_L^{*2} E_T^*/E_L^*}$$

$$C_{12}^* = \frac{\nu_L^* E_T^*}{1 - \nu_L^{*2} E_T^*/E_L^*} \qquad \text{(Eq 12)}$$

$$C_{22}^* = \frac{E_T^*}{1 - \nu_L^{*2} E_T^*/E_L^*}$$

The value of ν_L^* for polymeric matrix composites at the usual 60 vol% of fiber is of the order of 0.25, while the ratio E_T^*/E_L^* is of the order of 0.1 to 0.2. Consequently, the denominator is usually practically equal to unity; hence, the approximation:

$$C_{11}^* \cong E_L^*$$

$$C_{12}^* \cong \nu_L^* E_T^* \qquad \text{(Eq 13)}$$

$$C_{22}^* \cong E_T^*$$

Thermal Expansion and Moisture Swelling. The elastic behavior of composite materials discussed in the previous section was concerned with deformations produced by stresses, thus by loads. Deformations are also produced by temperature changes and absorption of moisture, thus by environmental changes. The two phenomena are similar and therefore will be discussed together. A change of temperature in a free body produces thermal strains, while moisture absorption produces swelling strains. The relevant physical parameters to quantify these phenomena are coefficients of thermal expansion (CTEs) and coefficients of swelling.

Fibers have significantly smaller CTEs than do polymeric matrices. The CTE of glass fibers is 5.0×10^{-6}/K while a typical epoxy value is 54×10^{-6}/K. Carbon and graphite fibers are anisotropic in thermal expansion. The CTEs in the fiber direction are usually extremely small, either positive or negative, of the order of 0.9×10^{-6}/K. It follows that a UDC will have very small CTEs in the fiber direction because the fibers will restrain matrix expansion. On the other hand, transverse CTEs will be much larger because the fibers move with the expanding matrix and thus provide less restraint to matrix expansion.

These phenomena are of considerable practical importance, particularly for laminates made of unidirectionally reinforced layers. When such a laminate is heated, the free expansion of any layer is prevented by the adjacent laminae because the fiber directions in all layers are different. This causes internal stresses that could be considerable. To compute these stresses, it is necessary to know the CTEs of the layers. Procedures to determine the CTEs in terms of elastic properties and the CTEs of component fibers and matrix are discussed in this section.

A physically similar situation arises in moisture swelling. Polymers absorb moisture in a wet environment and consequently swell. When this swelling is restrained, stresses are produced. If a UDC is placed in a wet environment, the matrix absorbs moisture but, with the exception of aramid, the fibers do not. Therefore, the fibers restrain—literally prevent—swelling in the fiber direction but not in the direction transverse to the fibers. Thus, the UDC is highly anisotropic with respect to moisture swelling.

When a laminate absorbs moisture, the same phenomenon occurs as with heating. Because free swelling of the layers cannot take place, internal stresses develop, which can be computed if the UDC coefficients of swelling are known. Procedures for computing those coefficients are also discussed in this section.

The CTEs of homogeneous solids are defined by the strains produced by unit change of temperature in bodies that are free of load and thus free of stress. The main engineering importance of thermal expansion is in the stresses produced when temperature changes occur under conditions of constraint and free expansion is prevented.

The CTEs of a composite are defined by the average strains produced by unit temperature change. It should be noted, however, that in this case, free thermal expansion cannot take place on the microscale, because the difference in constituent CTEs will produce microstresses when the temperature changes. Thus, failure of a composite material can be caused by changes in temperature. The CTEs of a UDC are indispensable for stress analysis of laminates subjected to temperature changes.

Consider a free cylindrical specimen of a UDC under uniform temperature change, $\Delta\Phi$, implying that the boundary temperature is changed from Φ to $\Phi + \Delta\Phi$. When a composite body is subjected to uniform boundary temperature, the temperature will also be uniform throughout the constituents. On the other hand, the stresses and strains in the phases are nonuniform and complex. The stress-strain relations of Eq 1 in the presence of temperature change then assume the form:

$$\bar{\epsilon}_{11} = \frac{\bar{\sigma}_{11}}{E_L^*} - \frac{\nu_L^*}{E_L^*}\bar{\sigma}_{22} - \frac{\nu_L^*}{E_L^*}\bar{\sigma}_{33} + \alpha_L^*\Delta T$$

$$\bar{\epsilon}_{22} = -\frac{\nu_L^*}{E_L^*}\bar{\sigma}_{11} + \frac{\bar{\sigma}_{22}}{E_T^*} - \frac{\nu_T^*}{E_T^*}\bar{\sigma}_{33} + \alpha_T^*\Delta T$$

$$\bar{\epsilon}_{33} = -\frac{\nu_L^*}{E_L^*}\sigma_{11} - \frac{\nu_T^*}{E_T^*}\bar{\sigma}_{22} + \frac{\bar{\sigma}_{33}}{E_T^*} + \alpha_T^*\Delta T$$

(Eq 14)

where α_L^* is the effective axial expansion coefficient, and α_T^* is the effective transverse expansion coefficient.

This definition would indicate that to compute α_{ij}^*, it would be necessary to compute the detailed strain fields in the two phases of a composite subjected to uniform temperature rise. Fortunately, however, this is not necessary; it has been found (Ref 11) that there is a unique mathematical relation between the effective CTEs and the effective elastic properties of a two-phase composite (for the general relations, see Ref 2 and 12). The present discussion is confined to specific cases of interest for a UDC. When fibers and matrix are isotropic:

$$\alpha_L^* = \alpha_m + \frac{\alpha_f - \alpha_m}{1/K_f - 1/K_m}\left[\frac{3(1 - 2\nu_L^*)}{E_L^*} - \frac{1}{K_m}\right]$$

$$\alpha_T^* = \alpha_m + \frac{\alpha_f - \alpha_m}{1/K_f - 1/K_m}$$

$$\left[\frac{3}{2k^*} - \frac{3(1 - 2\nu_L^*)\nu_L^*}{E_L^*} - \frac{1}{K_m}\right] \quad (Eq\ 15)$$

where α_m and α_f are the matrix and fiber isotropic expansion coefficients, K_m and K_f are the matrix and fiber three-dimensional bulk modulus, and E_L^*, ν_L^*, and k^* are the effective axial Young's modulus, axial Poisson's ratio, and transverse bulk modulus. These equations are suitable for glass-epoxy and boron-epoxy or boron-aluminum composites. They have also been derived in Ref 13 and 14.

For carbon and graphite fibers, it is necessary to consider the case of transversely isotropic fibers whose elastic and thermal expansion behavior is characterized by five independent elastic constants and two independent thermal expansion coefficients. This complicates the results considerably (Ref 7, 9):

$$\alpha_L^* = \alpha_m + (\alpha_{Lf} - \alpha_m)[P_{11}(S_{11}^* - S_{11}) +$$
$$2P_{12}(S_{12}^* - S_{12})] +$$
$$2(\alpha_{Tf} - \alpha_m)[P_{12}(S_{11}^* - S_{11}) +$$
$$(P_{22} + P_{23})(S_{12}^* - S_{12})]$$

$$\alpha_T^* = \alpha_m + (\alpha_{Lf} - \alpha_m)\{P_{11}(S_{12}^* - S_{12}) +$$
$$P_{12}[S_{22}^* + S_{23}^* - (S_{22} + S_{23})]\} +$$
$$2(\alpha_{Tf} - \alpha_m)\{P_{12}(S_{12}^* - S_{12}) +$$
$$\frac{1}{2}(P_{22} + P_{23})[S_{22}^* + S_{23}^* - (S_{22} + S_{23})]\} \quad (Eq\ 16)$$

where

$$P_{11} = (\Delta S_{22} + \Delta S_{23})/D$$

$$P_{12} = -\Delta S_{12}/D$$

$$P_{22} + P_{23} = \Delta S_{11}/D$$

$$D = \Delta S_{11}(\Delta S_{22} + \Delta S_{23}) - 2\Delta S_{12}^2$$

$$\Delta S_{11} = 1/E_{Lf} - S_{11}$$

$$\Delta S_{12} = -\nu_{Lf}/E_{Lf} - S_{12}$$

$$\Delta S_{22} + \Delta S_{23} = \frac{1}{2}(1/k_f + 4\nu_{Lf}^2/E_{Lf}) - (S_{22} + S_{23})$$

$$S_{11} = 1/E_m$$

$$S_{12} = -\nu_m/E_m$$

$$S_{22} + S_{23} = (1 - \nu_m)/E_m$$

$$S_{11}^* = 1/E_L^*$$

$$S_{12}^* = -\nu_L^*/E_L^*$$

and

$$S_{22}^* + S_{23}^* = \frac{1}{2}(1/k^* + 4\nu_L^{*2}/E_L^*)$$

Frequently, fiber and matrix CTEs are functions of temperature. It is not difficult to show that Eq 15 and 16 remain valid for temperature-dependent properties if elastic properties are taken at final temperature and CTEs are taken as secant values between that temperature and the stress-free temperature.

To evaluate the CTEs from Eq 15 or Eq 16, it is necessary to know the effective elastic properties k^*, E_L^*, and ν_L^*. These may be taken as the values predicted by the CCA model as given by Eq 3, 4a, 4b, and 5 with Modification 1 when the fibers are transversely isotropic. Comparison of the values thus obtained with a numerical analysis performed for a hexagonal array of carbon fibers (with an axial modulus of 210 GPa) in epoxy (Ref 15) shows that the results obtained are numerically indistinguishable.

The CTEs of carbon and graphite fibers can be measured directly only in the fiber direction, thus obtaining α_L. Direct measurement of transverse α_T is not possible because of the minute fiber diameters. Again, the only possible way to evaluate α_T is to measure α_L^* and α_T^* experimentally for a specified composite and then compute α_T from Eq 16. If the equations are written in the form:

$$\alpha_L^* = \alpha_m + (\alpha_{Lf} - \alpha_m)\,a_{11} + (\alpha_{Tf} - \alpha_m)\,a_{12}$$

(Eq 17)

$$\alpha_T^* = \alpha_m + (\alpha_{Lf} - \alpha_m)\,a_{21} + (\alpha_{Tf} - \alpha_m)\,a_{22}$$

it follows that:

$$\alpha_{Lf} = \alpha_m + \frac{(\alpha_L^* - \alpha_m)a_{22} - (\alpha_T^* - \alpha_m)a_{12}}{a_{11}a_{22} - a_{12}a_{21}}$$

$$\text{(Eq 18)}$$

$$\alpha_{Tf} = \alpha_m + \frac{(\alpha_T^* - \alpha_m)a_{11} - (\alpha_L^* - \alpha_m)a_{21}}{a_{11}a_{22} - a_{12}a_{21}}$$

Figure 5 shows typical plots of the effective CTEs of graphite-epoxy. For 50 vol% fiber, the axial expansion coefficient, α_L^*, is practically equal to that of the fibers. Because many carbon and graphite fibers have small negative axial expansion coefficients, many composites with such fibers have small negative or practically vanishing axial expansion coefficients. It is also interesting that for small fiber volume fractions, the transverse CTE becomes larger than the matrix CTE, which is the maximum constituent CTE.

With respect to deformation due to moisture, when a body that absorbs moisture is placed in a wet environment, moisture will diffuse through the external boundary. The internal moisture concentration, C, is defined by the mass of moisture accumulated per unit volume and is initially a function of space and time. Ultimately, the moisture concentration will stabilize and become time independent. The time-dependent stage is called the transient stage, and the ultimate time-independent stage is called the stationary stage. There is a complete mathematical analogy between the equations describing heat conduction and moisture diffusion, in both the transient and stationary stages. Thus, in the latter stage, the concentration C satisfies the Laplace equation.

When a composite with a polymeric matrix is placed in a wet environment, the matrix will begin to absorb moisture. However, the moisture absorption of most fibers used in practice is negligible. Aramid fibers alone do absorb significant amounts of moisture when exposed to high humidity. The total moisture absorbed by an aramid-epoxy composite, however, may not be substantially greater than that absorbed by other epoxy composites.

In most composites, the fibers act as barriers to moisture diffusion, analogous to perfect insulators in heat conduction. After sufficient time has elapsed, the matrix will be in an equilibrium moisture state with uniform concentration on the boundary. After a long time in this stage, the moisture concentration throughout the matrix will also be uniform, and thus the same as the boundary concentration. It is customary to define the specific moisture concentration, c, by:

$$c = C/\rho \qquad \text{(Eq 19)}$$

Thus, c is the mass of moisture per unit mass, a nondimensional number. The swelling strains due to moisture are functions of c and, to a first approximation, are given by:

$$\epsilon_{ij} = \beta_{ij} c \qquad \text{(Eq 20)}$$

where β_{ij} are the swelling coefficients.

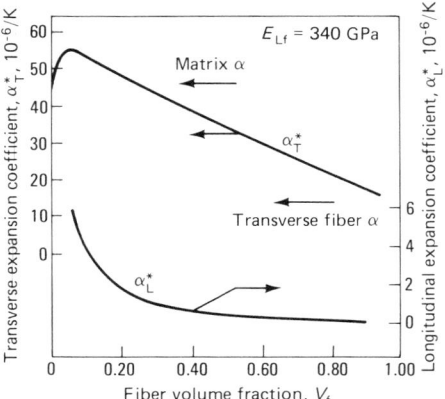

Fig. 5 Effect of fiber volume fraction on thermal expansion for representative carbon-epoxy composite

If there are also mechanical stresses and strains, the swelling strains are superimposed on the latter, which is exactly analogous to the thermoelastic stress-strain relations of an isotropic material where β replaces the CTE and c replaces the temperature. The swelling coefficients of most fibers are zero since their moisture absorption is negligible. (Aramid fiber swelling coefficients are unknown.) It follows that moisture swelling of a UDC is mathematically analogous to thermal expansion of such a composite with vanishing fiber expansion coefficients. Therefore, all of the results previously given for thermal expansion can be transcribed to moisture swelling. The effective swelling coefficients, β_{ij}^*, are defined by the average strains produced in a free sample subjected to uniform unit change of specific moisture concentration in the matrix. Thus, these coefficients follow directly from the equations for CTEs.

Other aspects of moisture absorption and transient and steady state are discussed in Ref 16 and 17, which also contain survey articles on this subject.

With respect to the important question of simultaneous moisture swelling and temperature rise, often called hygrothermal behavior, the simplest approach is to assume that thermal expansion strains and moisture swelling strains can be superimposed. Thus, for a free specimen:

$$\bar{\epsilon}_{11} = \alpha_L^* \Delta T + \beta_L^* c$$

$$\text{(Eq 21)}$$

$$\bar{\epsilon}_{22} = \bar{\epsilon}_{33} = \alpha_T^* \Delta T + \beta_T^* c$$

In this situation, the matrix elastic properties in Eq 15 and 16 may be functions of the end temperature and the equilibrium moisture concentration, c; this dependence must be known in order to evaluate α_L^*, α_T^*, β_L^*, and β_T^* in Eq 21.

Viscoelastic Properties. All polymers exhibit time dependence. This manifests itself by the increase with time of deformations under

constant load, which is called creep, and, conversely, by the decrease with time of stresses under deformation constraints, which is called relaxation. Another important effect of time dependence is the damping of vibrations due to energy dissipation in the polymeric matrix. The significance of all these phenomena increases with rise in temperature.

The effects described are of considerable engineering importance for fiber-reinforced composite structures, because stresses and deformations determined on the basis of elastic analysis may change considerably with time because of polymeric matrix time dependence. Vibration damping is a beneficial effect that is of particular significance for the higher vibration modes, which often become negligible because of damping.

The simplest description of time dependence is linear viscoelasticity. Viscoelastic behavior of polymers manifests itself primarily in shear and is negligible for isotropic stress and strain. This implies the elastic stress-strain relation:

$$\sigma_{11} + \sigma_{22} + \sigma_{33} = 3K(\epsilon_{11} + \epsilon_{22} + \epsilon_{33}) \quad \text{(Eq 22)}$$

where K, the three-dimensional bulk modulus, remains valid for polymers. When a polymeric specimen is subjected to shear strain ϵ_{12}°, which does not vary with time, the stress needed to maintain this shear strain is given by:

$$\sigma_{12}(t) = 2G(t)\epsilon_{12}^\circ \qquad \text{(Eq 23)}$$

where $G(t)$ is defined as the shear relaxation modulus. When a specimen is subjected to shear stress σ_{12}° constant in time, the resulting shear strain is given by:

$$\epsilon_{12}(t) = \frac{1}{2} g(t)\sigma_{12}^\circ \qquad \text{(Eq 24)}$$

and $g(t)$ is defined as the shear creep compliance.

Typical variations of relaxation modulus and creep compliance with time are shown in Fig. 6. These material properties change significantly with temperature: The relaxation modulus decreases, and the creep compliance increases. This implies that stiffness decreases with temperature. The initial values of these properties at time-zero (at the beginning of deformation) are denoted $G(o)$ and $g(o)$ and are the elastic properties of the matrix. Note that $G(o)g(o) = 1$. If the applied shear strain is an arbitrary function of time, commencing at time zero, Eq 23 is replaced by:

$$\sigma_{12}(t) = 2G(t)\epsilon_{12}(0) + 2\int_o^t G(t - t')\frac{d\epsilon_{12}}{dt'}\,dt'$$

$$\text{(Eq 25)}$$

Similarly, for applied shear stress, which is a function of time, Eq 24 is replaced by:

$$\epsilon_{12}(t) = \frac{1}{2}g(t)\sigma_{12}(0) + \frac{1}{2}\int_o^t g(t - t')\frac{d\sigma_{12}}{dt'}\,dt'$$

$$\text{(Eq 26)}$$

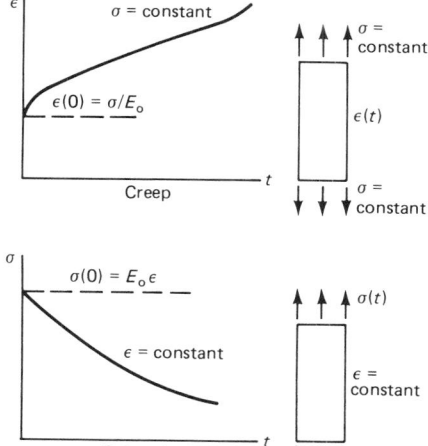

Fig. 6 Viscoelasticity

The viscoelastic counterpart of the Young's modulus is obtained by subjecting a cylindrical specimen to axial strain ϵ_{11} constant in space and time. Then:

$$\sigma_{11}(t) = E(t)\epsilon^{\circ}_{11} \qquad \text{(Eq 27)}$$

and $E(t)$ is the Young's relaxation modulus. If the specimen is subjected to axial stress σ°_{11} constant in space and time, then:

$$\epsilon_{11}(t) = e(t)\sigma^{\circ}_{11} \qquad \text{(Eq 28)}$$

and $e(t)$ is the Young's creep compliance (Fig. 6).

Obviously $E(t)$ is related to K and $G(t)$, and $e(t)$ is related to k and $g(t)$. The relations are not simple (Ref 18). To write the viscoelastic stress-strain relations for general states of three-dimensional stress and strain, it is customary to separate stress and strain into isotropic (or hydrostatic) and deviatoric parts. Thus:

$$\sigma_{ij} = \sigma\delta_{ij} + s_{ij} \qquad \sigma = \frac{1}{3}(\sigma_{11} + \sigma_{22} + \sigma_{33})$$
$$\text{(Eq 29)}$$

$$\epsilon_{ij} = \epsilon\delta_{ij} + e_{ij} \qquad \epsilon = \frac{1}{3}(\epsilon_{11} + \epsilon_{22} + \epsilon_{33})$$

where ij represents range indices varying from 1 to 3. Then:

$$\sigma(t) = 3K\epsilon(t)$$
$$S_{ij}(t) = 2G(t)e_{ij}(0) + 2\int_o^t G(t - t')\frac{\partial e_{ij}}{\partial t'}dt' \quad \text{(Eq 30)}$$
$$e_{ij}(t) = \frac{1}{2}g(t)s_{ij}(0) + \frac{1}{2}\int_o^t g(t - t')\frac{\partial s_{ij}}{\partial t'}dt'$$

where S_{ij} represents the deviatoric stress components.

The problem is to evaluate the effective viscoelastic properties of a UDC in terms of polymeric matrix viscoelastic properties and the elastic properties of the fibers. (It is assumed that the fibers themselves do not experience any time-dependent effect.) This problem has been resolved in general fashion in Ref 19 and 20, which show that the Laplace transforms of the effective relaxation moduli and creep compliances of a composite can be written in terms of expressions for effective elastic moduli in which elastic matrix properties are replaced by Laplace transforms of viscoelastic matrix properties. The main difficulty is the inversion of these Laplace transforms into the time domain. Details and applications of such procedures are given in Ref 2. Only some simple illustrative cases are presented here.

Detailed analysis shows that the viscoelastic effect in a UDC is significant only for axial shear, transverse shear, and transverse uniaxial stress. For any of average strains $\bar{\epsilon}_{22}$, $\bar{\epsilon}_{23}$, and $\bar{\epsilon}_{12}$ constant in time, thus stress relaxation, the time-dependent stress responses will be, respectively:

$$\bar{\sigma}_{22}(t) = E_T^*(t)\bar{\epsilon}_{22}$$
$$\bar{\sigma}_{23}(t) = 2G_T^*(t)\bar{\epsilon}_{23} \qquad \text{(Eq 31)}$$
$$\bar{\sigma}_{12}(t) = 2G_L^*(t)\bar{\epsilon}_{12}$$

For any of stresses $\bar{\sigma}_{22}$, $\bar{\sigma}_{23}$, and $\bar{\sigma}_{12}$ constant in time, thus creep deformation, the time-dependent strain responses will be, respectively:

$$\bar{\epsilon}_{22}(t) = e_T^*(t)\bar{\sigma}_{22}$$
$$\bar{\epsilon}_{23}(t) = \frac{1}{2}g_T^*(t)\bar{\sigma}_{23} \qquad \text{(Eq 32)}$$
$$\bar{\epsilon}_{12}(t) = \frac{1}{2}g_L^*(t)\bar{\sigma}_{12}$$

where the material properties in Eq 31 are effective relaxation moduli and the properties in Eq 32 are associated effective creep functions. All other effective properties may be considered elastic. This implies in particular that if a fiber composite is subjected to stress $\bar{\sigma}_{11}(t)$ in the fiber direction, then:

$$\bar{\sigma}_{11}(t) \cong E_L^*\bar{\epsilon}_{11}(t)$$
$$\bar{\epsilon}_{22}(t) = \bar{\epsilon}_{33}(t) \sim -\nu_L^*\bar{\epsilon}_{11}(t) \qquad \text{(Eq 33)}$$

where E_L^* and ν_L^* are the elastic results (Eq 4 and 5) with matrix properties taken as initial (elastic) matrix properties. Similar considerations apply to the relaxation modulus k^*.

The simplest case of the viscoelastic properties entering into Eq 31 and 32 is the relaxation modulus $G_L^*(t)$ and its associated creep compliance, $g_L^*(t)$. A simple result has been obtained for fibers that are infinitely more rigid than the matrix (Ref 2). In this case, the elastic result (Eq 6) reduces to:

$$G_L^* = G_m\frac{1 + V_f}{1 - V_f}$$

For a viscoelastic matrix, the corresponding results are:

$$G_L^*(t) = G_m(t)\frac{1 + V_f}{1 - V_f}$$
$$\text{(Eq 34)}$$
$$g_L^*(t) = g_m(t)\frac{1 - V_f}{1 + V_f}$$

This result is an acceptable approximation for glass fibers in a polymeric matrix and an excellent approximation for boron fibers in a polymeric matrix. However, it is not applicable to carbon or graphite fibers in a polymeric matrix, since the axial shear modulus of these fibers is not large enough in relation to the matrix shear modulus to justify rigid-fiber approximation. In this case, it is necessary to use the correspondence principle previously mentioned. Laplace transform inversion can be carried out by representing the matrix in terms of a viscoelastic spring-dashpot model (see Ref 2 and 19). The situation for transverse shear is more complicated and involves complicated Laplace transform inversions (Ref 20).

All polymeric matrix viscoelastic properties, such as creep and relaxation functions, are significantly temperature dependent. If this temperature dependence is known, all of the results given in this section can be obtained for the temperature of interest by introducing into the results matrix properties at that temperature.

At elevated temperatures, the viscoelastic behavior of the polymeric matrix may become nonlinear. In this event, the UDC will also be nonlinearly viscoelastic and all the results given are not valid. The problem of analytical determination of nonlinear properties is, of course, much more difficult than the linear problem. Some results are given in Ref 21.

Conduction and Moisture Diffusion. The conductivity of a UDC implies both thermal and electrical conductivity. Because all of the conductivity problems are governed by similar equations, the results obtained also apply to dielectric and magnetic properties and to steady-state moisture diffusion. The various

Table 2   Equivalent physical quantities

Physical subject	T	$H = -\Delta T$	D	μ_{ij}	ξ_{ij}
Thermal conduction	Temperature	Temperature gradient	Heat flux	Heat conductivities	Resistivities
Electric conduction	Electric potential	Electric field intensity	Current density	Electric conductivities	Resistivities
Electrostatics	Electric potential	Electric field intensity	Electric induction, electric displacement	Dielectric constants, permittivities	
Magnetostatics	Magnetic potential	Magnetic field intensity	Magnetic induction	Magnetic permeabilities	
Moisture diffusion	Concentration	Moisture gradient	Moisture flux	Diffusivities	

equivalent physical quantities in these different areas are listed in Table 2. The language of thermal conductivity will be used in the following discussion.

Let $T(x)$ be a steady-state temperature field in a homogeneous body. The temperature gradient is given by:

$$H_i = -\frac{\partial T}{\partial x_i} \qquad \text{(Eq 35)}$$

and the heat flux vector by:

$$D_i = \mu_{ij} H_j \qquad \text{(Eq 36)}$$

where μ_{ij} is the conductivity tensor. The inverse of Eq 36 is:

$$H_i = \xi_{ij} D_j \qquad \text{(Eq 37)}$$

where ξ_{ij} is the resistivity tensor. In an isotropic material:

$$D_i = \mu\, H_i$$
$$H_i = \xi\, D_i \qquad \text{(Eq 38)}$$
$$\xi = 1/\mu$$

For a transversely isotropic material, such as carbon and graphite fibers with x_1, the axis of transverse isotropy (fiber axis):

$$D_1 = \mu_L H_1$$
$$D_2 = \mu_T H_2 \qquad \text{(Eq 39)}$$
$$D_3 = \mu_T H_3$$

where μ_L is longitudinal conductivity and μ_T is transverse conductivity.

A simple, common example is heat conduction through a slab whose faces are maintained at different temperatures, T and $T + \Delta T$. Then:

$$H_1 = -\frac{\Delta T}{h}$$
$$D_1 = \mu_{11} H_1$$

where h is the slab thickness and μ_{11} is the conductivity coefficient normal to the faces of the slab.

Because a UDC is statistically transversely isotropic, it has two different effective conductivities, μ_L^* in the fiber direction and μ_T^* transverse to the fibers. The effective constituent relations are analogous to Eq 39:

$$\bar{D}_1 = \mu_L^* \bar{H}_1$$
$$\bar{D}_2 = \mu_T^* \bar{H}_2 \qquad \text{(Eq 40)}$$
$$\bar{D}_3 = \mu_T^* \bar{H}_3$$

where overbars denote averages over the representative volume element (RVE).

It can be shown (Ref 2) that for isotropic matrix and fibers, the axial conductivity μ_L^* is given by:

$$\mu_L^* = \mu_m V_m + \mu_f V_f \qquad \text{(Eq 41)}$$

and for transversely isotropic fibers:

$$\mu_L^* = \mu_m V_m + \mu_{Lf} V_f \qquad \text{(Eq 42)}$$

where μ_{Lf} is the longitudinal conductivity of the fibers. The results (Eq 41 and 42) are valid for any fiber distribution and any fiber cross section.

The problem of transverse conductivity is mathematically analogous to the problem of longitudinal shearing (Ref 2). It follows that all results for the effective longitudinal shear modulus, G_L^*, can be interpreted as results for transverse effective conductivity, μ_T^*. In particular, for the CCA model:

$$\mu_T^* = \mu_m \left[\frac{\mu_m V_m + \mu_f (1 + V_f)}{\mu_m (1 + V_f) + \mu_f V_m} \right]$$
$$\text{(Eq 43)}$$
$$= \mu_m + \frac{V_f}{1/(\mu_f - \mu_m) + V_m/2V_m}$$

These results are for isotropic fibers. For carbon and graphite fibers, μ_f should be replaced by the transverse conductivity of the fibers, μ_{Tf} (Ref 9).

As in the elastic case, there is reason to believe that Eq 43 represents with great accuracy all cases of circular fibers that are randomly distributed and are not in contact. The reason, again, is the numerical coincidence of hexagonal array numerical analysis results with the number predicted by Eq 43.

To interpret the results for moisture diffusion, the quantity μ_m is interpreted as the diffusivity of the matrix. Because moisture absorption of fibers is negligible, μ_f is set equal to zero. The results are then:

$$\mu_L^* = \mu_m V_m$$
$$\mu_T^* = \mu_m \left[\frac{V_m}{1 + V_f} \right] \qquad \text{(Eq 44)}$$

which leads to the interesting relation:

$$\mu_L^*/\mu_T^* = 1 + V_f \qquad \text{(Eq 45)}$$

Strength and Failure

The mathematical treatment of the relationships between the strength of a composite and the properties of its constituents is less developed than the analyses for the other physical property relationships already discussed. One of the reasons is that failure is likely to initiate in a local region because of the influence of the local values of constituent properties and the geometry in that region. Thus, both the high degree of variability of local geometry (that is, relative locations of adjacent fibers) and the high degree of variability of the local strength of the fibers contribute to the onset of initial localized damage within the composite. This dependence upon local characteristics makes the analysis of the composite failure mechanisms much more complex.

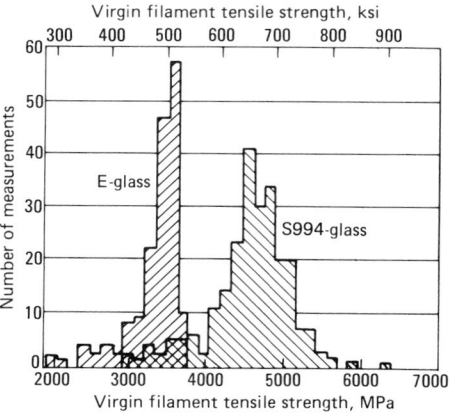

Virgin filament tensile strength, ksi

Fig. 7 Single-filament tensile strength of glass using gage length of 70 mm (2.75 in.) and strain rate of 0.062 mm/mm/min

The strength of a fiber composite clearly depends on the orientation of the applied load with respect to the direction in which the fibers are oriented, as well as on whether the applied load is tensile or compressive. The following sections describe failure mechanisms and composite constituent property relationships for each of the principal loading conditions.

Axial Tensile Strength. One of the most attractive properties of advanced fiber-reinforced composites is their high tensile strength. The simplest model for the tensile failure of a UDC subjected to a tensile load in the fiber direction is based on the elasticity solution of uniform axial strain throughout the composite. Generally, the fibers have a lower strain-to-failure value than the matrix, and composite fracture occurs at the failure strain of the fibers alone. This results in a composite tensile strength, σ_L, given by:

$$\sigma_L = V_f \sigma_f + V_m \sigma_m \qquad \text{(Eq 46)}$$

where σ_f is the fiber tensile strength and σ_m is the stress in the matrix at a strain equal to the fiber failure strain, σ_f/E_f.

The problem with this approach is the variability of fiber strength. Nonuniform strength is characteristic of most current high-strength filaments. This is illustrated in Fig. 7, which shows strength distributions for single filaments of two different types of commercial glass fibers. This statistical distribution of single-filament strength is generally considered to result from a distribution of imperfections along the length of these brittle fibers. There are two important consequences of a wide distribution of individual fiber strengths (the word fiber is used to denote a single filament, usually of the order of 10 μm, or 400 μin., in diameter). First, all fibers will not be stressed to their maximum value at the same time. Thus, the strength of a group of fibers will not equal the sum of the strengths of the individual fibers, nor will it equal the mean strength of these fibers. Second, those fibers which break first during the loading process will cause perturbations of the stress field in the vicinity of the break, resulting in localized high fiber-matrix interface shear stresses. These shear stresses transfer the load across the interface and also introduce stress concentrations into adjacent, unbroken fibers.

At each local fiber break, the stress in the vicinity of the broken fiber changes so that the axial stress in the fiber vanishes at the fiber break and gradually builds back up along the fiber length to its undisturbed stress value. The general form of the local stress pattern in the fiber is shown in Fig. 8; the interfacial stress is a maximum close to the fiber break and decays rapidly along the length of the fiber. As a result, the axial fiber stress builds up to its initial undisturbed value over a relatively short dimension.

As a result of this stress distribution, several possible failure events may occur. The shear stresses may cause a crack to progress along the interface. If the interface is weak, such propagation can be extensive, and the strength of the composite material may differ only slightly from that of a bundle of unbonded fibers. This undesirable mode of failure can be prevented by the attainment of a strong fiber-matrix interface or by the use of a soft, ductile matrix that permits redistribution of the high shear stresses. When the bond strength is high enough to prevent interface failure, the local stress concentrations may cause the fiber break to propagate through the matrix, to and through adjacent fibers. Alternatively, the stress concentration in adjacent fibers may cause one or more to break before the intermediate matrix fails. If such a crack or such fiber breaks continue to propagate, the strength of the composite may be no greater than that of the weakest fiber. This failure mode is defined as a weakest-link failure. If the matrix and interface properties have sufficient strength and toughness to prevent or arrest these failure mechanisms, continued load increases will produce new fiber failures at other locations in the material, resulting in an accumulation of dispersed internal damage as the loading continues.

All of these effects can be expected to occur before material failure; that is, local fractures will propagate for some distance along the fibers and normal to the fibers. These fractures will initiate and grow at various points within the composite. Increasing the load will produce a statistical accumulation of dispersed damage regions until enough regions interact to provide a weak surface, resulting in composite tensile failure. On the basis of experimental observations of dispersed fiber fractures at relatively low loads and of irregular failure surfaces showing combinations of broken-fiber clusters and axial shear failure planes, this complex failure mechanism appears to represent the practical case in most, if not all, composites of interest.

Contemporary filaments exhibit very high tensile strengths. When they are incorporated

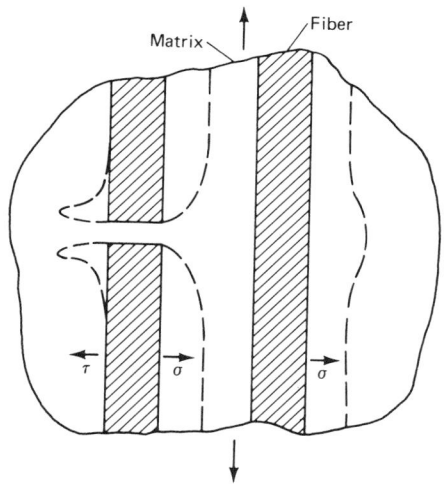

Fig. 8 Perturbation of stresses in vicinity of broken fiber end

into a composite material, the degree to which their high strength can be utilized depends on matrix characteristics, which may vary with the service and environmental history of the composite material. Furthermore, new matrix materials are constantly under development to improve composite resistance to temperature, impact, and so forth, and to lower manufacturing costs. The influence of these changes on composite tensile strength must be understood.

Fiber tensile strength is generally reported as the average strength of a group of fibers of a particular type. Actually, the strength is rarely determined by testing single filaments and obtaining a numerical average of their strength values. Usually, a bundle or yarn of such fibers is impregnated with a polymer and loaded to failure. The average fiber strength is then defined by the maximum load divided by the cross-sectional area of the fibers alone. It is important to recognize the difference between these two techniques for measuring fiber strength. At a constant gage length, the strength of single filaments will exhibit a certain amount of dispersion because of fiber imperfections. The probability of finding an imperfection of given severity increases with gage length; hence, the average strength of the fiber can be expected to, and indeed does, decrease with increasing gage length. An example of this effect is shown in Fig. 9. The question of which value of fiber strength influences composite strength can be resolved only by determining the composite failure mechanism. It will be shown that a characteristic length of the fiber that is effective within the composite can be defined.

Nonetheless, the prevalent technique for defining fiber strength in the industry is to impregnate a strand of fibers with a low-quality matrix material and to use the failure load to define average fiber strength (Ref 22). Fiber volume fraction and fiber alignment do not

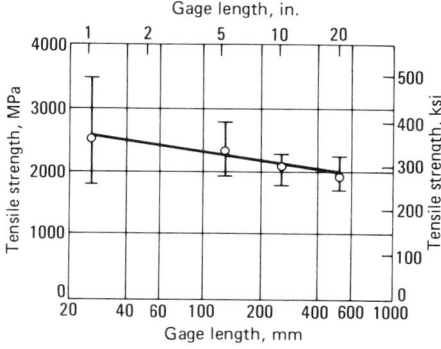

Fig. 9 Variation of mean and range of individual fiber strength values. Boron fiber data
source: Ref 1

necessarily duplicate those of the actual composites of interest.

A Weibull distribution, $F(\sigma)$, has been found to provide a good fit to experimental data of the type shown in Fig. 9. The cumulative distribution function for this case can be expressed in the form:

$$F(\sigma) = 1 - \exp(-\alpha L \sigma^\beta) \qquad \text{(Eq 47a)}$$

and the Weibull strength distribution function is then:

$$f(\sigma) = L\alpha\beta\sigma^{\beta - 1} \exp(-L\alpha\sigma^\beta) \qquad \text{(Eq 47b)}$$

where α and β are two parameters to be fit to the experimental fiber tensile strength data, and L is the fiber length. For the Weibull distribution of Eq 47b, the mean strength, $\bar{\sigma}$, can be found to be:

$$\bar{\sigma} = (\alpha L)^{-1/\beta} \Gamma\left(1 + \frac{1}{\beta}\right) \qquad \text{(Eq 48)}$$

where Γ is the standard gamma function. The standard deviation, s, is:

$$s = (\alpha L)^{-1/\beta} \left[\Gamma\left(1 + \frac{2}{\beta}\right) - \Gamma^2\left(1 + \frac{1}{\beta}\right)\right] \qquad \text{(Eq 49)}$$

Thus, the coefficient of variation, c.v., for this distribution is given by:

$$\text{c.v.} = \frac{s}{\bar{\sigma}} = \frac{\Gamma\left(1 + \frac{2}{\beta}\right) - \Gamma^2\left(1 + \frac{1}{\beta}\right)}{\Gamma\left(1 + \frac{1}{\beta}\right)} \qquad \text{(Eq 50)}$$

It is now possible to assign some physical significance to the parameters of the distribution function of Eq 47b. First, the coefficient of variation is seen to be a function only of the parameter β. Indeed, Eq 50 can be approximated to within 3% for $0.05 \leq \text{c.v.} \leq 0.50$, by:

$$\text{c.v.} = \beta^{-0.92} \qquad \text{(Eq 51)}$$

Thus, β is essentially an inverse measure of c.v., and for practical fibers:

$$\beta \approx \frac{1}{\text{c.v.}} \qquad \text{(Eq 52)}$$

Also, because for $\beta > 1$ the gamma function is close to unity, for a unit length:

$$\bar{\sigma}_1 \approx \alpha^{-1/\beta} \qquad\qquad \text{(Eq 53)}$$

Thus, the quantity $\alpha^{-1/\beta}$ may be viewed as a reference stress level.

With these definitions, Eq 48 can be plotted in the form shown in Fig. 10, where the tensile strength of single fibers of length L is shown as a function of some normalized fiber length for different values of β. The upper curve, where β is infinity, applies to a set of fibers having identical strength ($s = 0$). For this case, there is no length dependence. For β of 10, which is approximately a 12% c.v., an order-of-magnitude change in the fiber length produces about a 20% drop in its average strength; for β of 4, an order-of-magnitude change in length produces almost a 50% drop in strength.

Weakest-link failure occurs when a UDC is loaded in axial tension and scattered fiber breaks occur through the material at various stress levels. One of these fiber breaks may trigger a stress wave or initiate a crack in the matrix, resulting in localized stress concentrations, which cause fracture of one or more adjacent fibers. In turn, the failure of these fibers may result in additional stress waves or matrix cracks, leading to overall failure. This produces a catastrophic mode of failure which is associated with the occurrence of one, or a small number of, isolated fiber breaks and is referred to as the weakest-link mode of failure.

The lowest stress at which this type of failure can occur is the stress at which the first fiber will break. The expressions for the expected value of the weakest element in a statistical population (Ref 23) have been applied by C. Zweben (Ref 24) to determine the expected stress at which the first fiber will break. Assuming that the fiber strength is characterized by a Weibull distribution, the expected first fiber break will occur at a stress:

$$\sigma_w = \left(\frac{\beta - 1}{NL\alpha\beta}\right)^{1/\beta} \qquad\qquad \text{(Eq 54)}$$

where N is the number of fibers in the material. Thus, Eq 54 provides an estimate of the failure stress associated with the weakest-link mode.

It should be noted that the occurrence of the first fiber break is a necessary, but not a sufficient, condition for failure. In general, the occurrence of a single fiber break rarely produces catastrophic failure; however, in very small samples containing fibers having a narrow dispersion of strength, it is a possible failure mechanism. For practical materials in realistic structures, the calculated weakest-link failure stress is small, and failure usually cannot be expected to occur in this mode.

Cumulative Weakening Failure. If the weakest-link failure mode does not occur, it is possible to continue loading the composite until, with increasing stress, fibers will begin to break randomly throughout the material. When a fiber breaks, there is a redistribution of stress in the vicinity of the fracture site (Fig. 10). This stress perturbation is the origin of important mechanisms involved in composite failure.

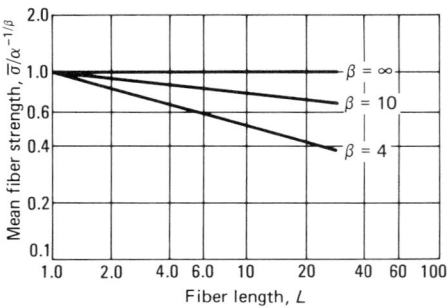

Fig. 10 Strength-length variation of fiber strengths following Weibull statistical distribution functions

When a fiber breaks, the broken surfaces displace axially, inducing stresses in the matrix and large shear stresses at the fiber-matrix interface. The interface shear stress acting on the broken fiber localizes the axial fiber dimension over which the stress in the broken fiber is greatly reduced. Were it not for some form of interfacial shear stress, a broken fiber would be unable to carry any load and the composite would be, in effect, a bundle of separate fibers, from the standpoint of resisting axial tensile loading.

An important function of the matrix is to localize the reduction of fiber stress when a fiber breaks. The axial dimension over which the axial fiber stress is significantly reduced is referred to as the ineffective length, δ, which is a significant length parameter in the failure of fiber-reinforced composite materials. The magnitude of δ depends on the stress distribution in the region of the fiber break. This distribution is complex and is influenced by fiber and matrix elastic properties as well as by any inelastic phenomena, such as debonding, matrix fracture, or yield that may occur. Also, the definition of δ is somewhat arbitrary because the stress in the broken fiber is a continuously varying quantity that asymptotically approaches the average stress in unbroken fibers. An estimate of ineffective length is obtained for an elastic fiber and a perfectly plastic matrix as:

$$\delta = \frac{\sigma_f}{4\tau_y} d_f \qquad\qquad \text{(Eq 55)}$$

where δ is the ineffective length, σ_f is the fiber tensile strength, τ_y is the matrix shear yield strength, and d_f is the fiber diameter.

The concept of representing this variable stress field and a fiber composite material having distributed fractures by an assemblage of elements of length δ was introduced by B.W. Rosen (Ref 25). In this model, shown in Fig. 11, the composite may be considered to be a chain of layers of dimension equal to the ineffective length. Any fiber which fractures within this layer will be unable to transmit a load across the layer. The applied load at that cross section is then assumed to be distributed among the unbroken fibers in each layer. (The effective load concentrations, which would in-

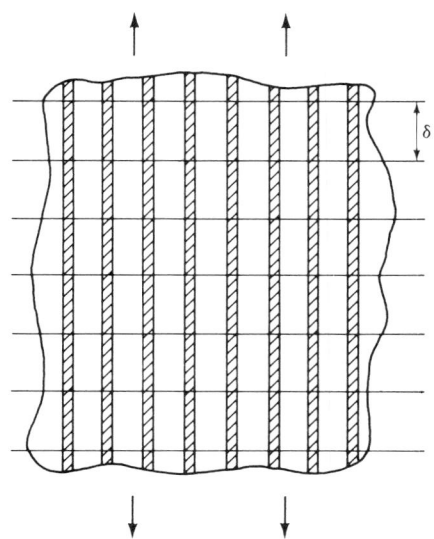

Fig. 11 Geometry of composite for statistical tensile failure model

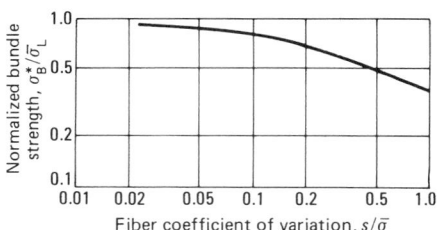

Fig. 12 Variation of fiber bundle strength with fiber coefficient of variation. Source: Ref 26

troduce a nonuniform redistribution of these loads, are not considered initially.) A segment of a fiber within one of these layers may be considered to be a link in the chain that constitutes an individual fiber. Each layer of the composite is then a bundle of such links, and the composite itself is a series of such bundles (Fig. 12).

The filaments in each bundle act in parallel, and the strength of a bundle of filaments whose members are not of uniform strength is not equal to the mean fiber strength. Following the work of B.D. Coleman, Fig. 12 shows the variation of bundle strength (again a statistical characterization, so that the bundle strength is in this case measured by the mode of the strength) normalized with respect to the mean strength of individual fibers of the same length as the bundles. This is plotted as a function of the coefficient of variation of the fiber population; the general result is that the average strength of the bundle is somewhat less than the average strength of the fibers, and is a decreasing function of the coefficient of variation of the fibers.

Treatment of a fiber as a chain of links is appropriate to the hypothesis that fracture is due to local imperfections. The links may be considered to have a statistical strength distribution equivalent to the statistical flaw distribution along the fibers. The validity of such a model for the fibers is demonstrated by the length dependence of fiber strength. For this model, it is necessary to define the link dimension, δ; the probability of failure of fiber elements of that length; and the statistical strength distribution of the assemblage. This analysis leads to the cumulative-weakening mode of failure. The definition of ineffective length and the determination of the link strength distribution is treated in Ref 25. When these are known, the relationship of the strength of the assemblage to the strength of the elements, or links, can be treated by the methods of Ref 23. The result for fibers having a strength distribution of the form given in Eq 47a is given in Ref 25 as:

$$\sigma^* = (\alpha\delta\beta e)^{-1/\beta} \qquad \text{(Eq 56)}$$

where σ^* is the statistical mode of the composite tensile strength based on fiber area.

As pointed out previously, the cumulative-weakening model represents the varying stress near a fiber break by a step function in stress. The model also neglects the possibility of failures involving parts of more than one layer. More importantly, the overstress in unbroken fibers adjacent to the broken fibers has not been treated. This stress concentration increases the probability of failure for these adjacent elements, and creates the probability of propagation of fiber breaks. This combination of variable fiber strength and variable fiber stress can be expected to lead to a growth in both the number of damaged regions and the size of a given damaged region.

In the situation described, the possibility exists that one damaged group may propagate, causing failure, or that the cumulative effect of many smaller damaged groups will weaken a cross section, causing failure.

Fiber break propagation failure was proposed as a possibility by C. Zweben (Ref 24). The effects of stress perturbations on fibers adjacent to broken ones are of significance. When a fiber breaks, equilibrium requires that the net load on the cross section containing the broken fiber be unchanged. Therefore, the average stress in the remaining fibers must increase. Because of the matrix, the stress redistribution is highly nonuniform. The shear stress that arises in the matrix when a fiber breaks results in localized increases of average stress in the fibers surrounding the break. To differentiate this increase in the average stress over a fiber cross section from the increase at a point, the term load concentration is used for the former and the conventional term stress concentration for the latter.

The load concentration in the fibers adjacent to a broken one increases the probability that one or more of them will break. When such an event occurs, the load concentration in neighboring fibers intensifies, increasing the probability of additional fiber breaks, and so on. From this description, it is not difficult to

identify the propagation of fiber breaks as a mechanism of failure. The probability of occurrence of this mode of failure increases with the average fiber stress because of the increasing number of scattered fiber breaks and the increasing stress level in overstressed fibers.

The fiber-break propagation mode of failure was studied by C. Zweben (Ref 24), who proposed that the occurrence of the first fracture of an overstressed fiber could be used as a measure of the tendency for the fiber breaks to propagate and, hence, as a failure criterion for this mode, at least for small volumes of material. The effects of load concentrations on fiber break propagation in three-dimensional unidirectional composites, as well as on cumulative weakening failures, was treated in Ref 27. In Ref 28, C. Zweben reviewed experimental data available for various fiber-matrix systems to support the contention that the first multiple break is a lower bound to strength. Although the first multiple break criterion may provide good correlation with experimental data for small specimens and may be a lower bound on the stress associated with fiber break propagation, it gives low stresses for large volumes of materials, which appears to conflict with practical experience with composites. However, no reliable data appear to be available to shed light on the influence of sample material size on composite tensile strength.

The approximate model of Ref 27 for including effects of load concentrations in the cumulative weakening model was also of limited success. The resulting mathematical expression for composite strength is a sequence in which each term corresponds to a group of broken fibers of increasing size. A large number of terms is required for convergence. Continuing investigations of procedures for reflecting such statistics have been undertaken by D.G. Harlow and S.L. Phoenix (Ref 29), who have treated the condition probabilities that result from sequential breaks of adjacent fibers in order to obtain manageable expressions for failure probabilities. Such approaches require assumptions regarding the redistributions of load from the broken fibers. Some of these load-sharing rules have been explored. The failure mode remains a cumulative weakening of the material.

Cumulative group mode failure is the fourth basic mode of failure to be considered. As multiple broken fiber groups grow, the magnitude of the local axial shear stress increases, and axial cracking can occur. This effect has been treated in the failure model (Ref 30) formulated to incorporate the following three effects, which were deemed to be of importance in the tensile failure of high-strength fibrous composites:

- The variability of fiber strength will result in distributed fiber fractures at stress levels well below the composite strength
- Load concentrations in fibers adjacent to broken fibers will influence the growth in

size of the crack regions to include additional fibers
- High shear stresses will cause matrix shear failure or interfacial debonding, which will serve to arrest the propagating crack

Thus, as the stress level increases from that at which fiber breaks are initiated toward that at which the composite fails, the material will have distributed groups of broken fibers. Each group will have an ineffective length that increases with group size and, after matrix failure, with stress. This situation may be viewed as a generalization of the cumulative weakening model of Ref 25, in which the effect of the isolated breaks was modeled by a chain-of-bundles model such as that used in Ref 23.

In practical materials, the problem is complicated by the presence of bundles of various sizes; that is, both the number of broken fibers in a bundle and the ineffective length of that bundle vary. Thus, the basic problems of defining the required input information for the analysis of the chain-of-bundles model are of increased complexity. The size of the basic element must first be defined, and then the probability of failure of that element can be determined.

At stress levels above those required to cause some number of isolated breaks in the composite, there is an increasing probability of occurrence of multiple adjacent breaks as a result of stress concentrations. Thus, at moderate stress levels, there will usually be a nonnegligible probability of existence of a crack containing n broken fibers, for many values of n. For each crack size n, there is a different elastic ineffective length and also different values of both shear load concentration and fiber load concentration factors. Thus, for different-sized cracks, there are different stress levels at which matrix failure initiates and different distances over which it propagates.

The statistical problem represented by the situation described is exceedingly complex. Various authors have taken different approaches in an attempt to obtain a reliable model. D.G. Harlow and S.L. Phoenix (Ref 29) have pursued development of the statistics by focusing on two-dimensional materials and specified load-sharing rules. B.W. Rosen and C. Zweben (Ref 30) have pursued an approximate model based on the definition of a characteristic group size determined by an estimate of the size at which growth would be arrested by axial shear cracks. S.B. Batdorf (Ref 31) has used simplifying approximations of the statistical functions that result in a propagation failure mode having a defined critical crack size.

Each of these models has severe limitations with regard to the quantitative prediction of tensile strength. However, the models show the important effect of variability of fiber strength and matrix stress-strain characteristics on composite tensile strength.

Fracture Mechanics. In the foregoing discussion of composite failure mechanics, a number of basic modes of failure have been described, including those associated with propagation effects. However, in the discussion of the analytical treatment of these modes, no mention has been made of classical fracture mechanics. Because a large, well-developed body of knowledge exists relating to the failure of homogeneous materials, the possibility of applying classical fracture mechanics techniques to analysis of the failure of composite materials merits examination. A basic principle of classical fracture mechanics is that a crack will grow when the energy required to extend a crack a given amount is equal to the change in strain energy in the body resulting from that crack advance. This condition has a number of implications for the analysis of the modes of failure.

First, consider the weakest-link mode of failure, in which a single fiber break results in a catastrophic crack propagation. If failure results from a crack that propagates in a continuous manner through both phases, fiber and matrix, the fracture mechanics approach can reasonably be expected to describe the process, although consideration of propagation through the two phases separately may be necessary. Also, when the crack size becomes large with respect to fiber diameter and interfiber spacing distance, it seems reasonable to expect that the material can be adequately treated as a homogeneous, anisotropic material.

In the case of the fiber break propagation mode, the onset of unstable crack growth is governed by fiber load concentrations and the statistical aspects of material strength. This mode of failure need not result from a crack propagating through the matrix. Thus, the total energy of fracture of the composite has no relation to the conditions precipitating fiber break propagation. However, it is reasonable to expect that when the crack grows appreciably, the form of the damage region will become stabilized and will advance through the material without significant change. If this should occur, it is reasonable to believe that increments of strain energy could be related to the energy expended in extending the damaged region. This type of energy balance, which may be valid when the crack size is large with respect to fiber diameter and spacing, would not be expected to be applicable at the early stages of instability, when the crack is small and the effects of heterogeneity are important.

The remaining modes of failure are associated with failure due to an accumulation of weakened areas, which is not directly related to crack propagation. Therefore, it seems reasonable to assume that classical fracture mechanics has no relevance in this case.

A multiplicity of internal planes of weakness creates the possibility for various failure modes in composites. It appears that the heterogeneity must be considered in the development of failure criteria. After an understanding of failure modes is obtained, it may be possible to formulate effective fracture mechanics param-

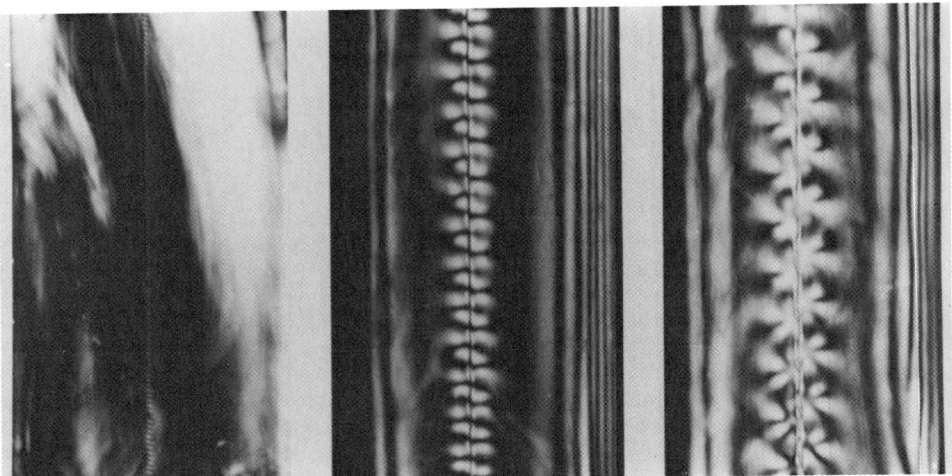

Fig. 13 Photoelastic stress pattern for three individual E-glass fibers embedded in epoxy matrix. Source: Ref 33

eters for some composites under some loading conditions not yet known.

In summary, although uniaxial tensile strength is one of the most attractive properties of contemporary advanced fiber composites, it is generally viewed in a rudimentary fashion. Thus, manufacturers of fibers frequently measure the strength of their product by using yarns impregnated with low-temperature curing resin systems in composites having properties that are not necessarily those of the composites of interest. The ratio of the measured tensile strength of a sound, unidirectional composite to the tensile strength of this nonuniform impregnated strand is identified as a measure of the efficiency of the composite. The acceptance of this procedure apparently stems from the fact that, for current test specimen dimensions and materials, this ratio approaches unity.

When improved materials are developed, they can be expected to be treated by improved methods. The statistical methods presented here provide the basis for such a treatment. Although these methods are not yet adequate for definitive quantitative prediction, they can be useful for assessment of the relative merits of different materials. The rule of mixtures need no longer be a substitute for an understanding of material behavior. Assessment of merit should use both cumulative- and propagation-type models to define both the relative level of strength and the expected mode of failure.

For preliminary estimates, cumulative weakening (Eq 56) can be used with an inelastic ineffective length (Eq 55) to provide an overestimate of the actual value and an indication of the role of the properties of each constituent.

Axial Compressive Strength. For compressive loads applied parallel to the fibers of a UDC, both strength and stability failures must be considered. Microbuckling was proposed as a failure mechanism by N.F. Dow (Ref 32), with the suggestion that small wavelength microinstability of the fibers occurs in a fashion

analogous to the buckling of a column on an elastic foundation. It has been demonstrated that this will occur even for a brittle material, such as glass. For example, Fig. 13 (Ref 33) shows three specimens of a single E-glass filament embedded in a block of epoxy that has been cooled to produce a compressive strain. Each specimen contains a fiber of 0.08 mm (0.003 in.) diameter. The photoelastic stress pattern in the epoxy emphasizes the repetitive nature of the deformation pattern for each fiber and supports the contention that the deformations result from an elastic instability.

Analyses of this instability were performed independently in Ref 34 and 35. The analyses approximate the problem by treating a layered two-dimensional medium (Fig. 14). The model consists of plates of thickness h separated by a matrix of dimension $2c$. Each fiber is subjected to a compressive load, P, and the fiber length is given by the dimension L. Two possibilities are considered for the instability failure mode. First, the fibers may buckle in opposite directions in adjacent fibers, as shown on the left portion of the figure, and the so-called extension mode occurs. This mode receives its name from the fact that the major deformation of the matrix material is an extension in the direction perpendicular to the fibers. The analysis treats the fibers as stiff relative to the matrix so that shear deformations in the fiber can be neglected relative to those in the matrix.

The second possibility is shown on the right portion of the figure, where adjacent fibers buckle in the same wavelength and in phase with one another, so that the deformation of the matrix material between adjacent fibers is primarily a shear deformation; hence, the shear mode designation for this potential mode.

The buckling stress for these modes has been evaluated by the energy method, in which the composite stressed to the buckling load is considered and the strain energy in this compressed but straight deformation pattern is then

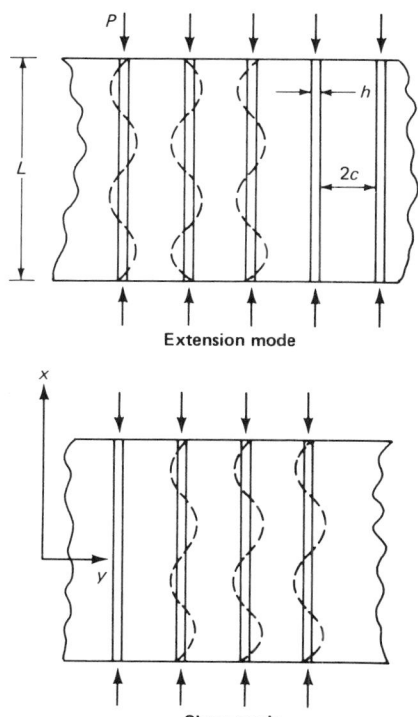

Fig. 14 Analytical model for compressive strength of fibrous composite

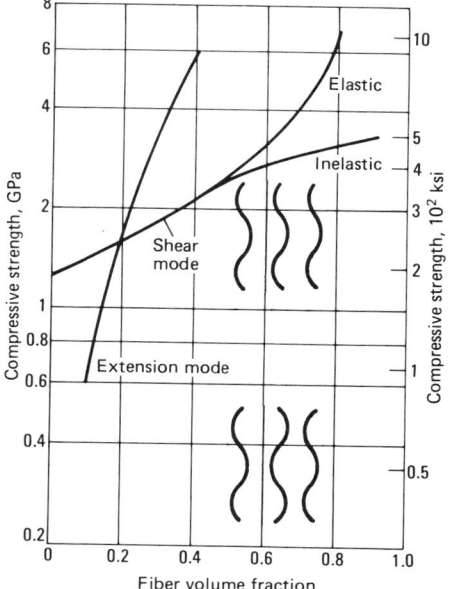

Fig. 15 Compressive strength of glass-reinforced epoxy composites

compared to a deformation pattern following an assumed buckling shape under the same load. Thus, a change in the strain energy of the composite consisting of the strain-energy change in the fiber and in the matrix can be compared to the change in the potential energy associated with the shortening of the distance between the applied loads at the end of the fibers. The condition for instability is given by equating the strain-energy change to the work done by the external loads during buckling. Buckling is found to occur at a relatively short wavelength (10 to 100 fiber diameters).

The result for the compressive strength, σ_c, for the extension mode is given by:

$$\sigma_c = 2V_f [(V_f E_m E_f)/3(1 - V_f)]^{1/2} \qquad \text{(Eq 57)}$$

The result for the shear mode is given by:

$$\sigma_c = G_m/(1 - V_f) \qquad \text{(Eq 58)}$$

These results are plotted in Fig. 15 for E-glass fibers embedded in an epoxy matrix. The compressive strength of the composite is plotted as a function of the fiber volume fraction, V_f. The two curves represent the two failure modes considered. For the low fiber volume fractions, the extension mode is the lower stress; for high fiber volume fractions, the shear mode predominates. The compressive strength of reasonable glass-reinforced plastic containing fiber volume fractions of 0.6 to 0.7 ranges from 3100 to 4150 MPa (450 to 600 ksi). Values of this magnitude do not appear to have been measured for any realistic specimens. However, achievement of a

strength of 3450 MPa (500 ksi) in a composite of this type would require an average shortening greater than 5%. For the epoxy materials used, such a shortening would decrease the effective shear stiffness of the matrix material because the proportional limit of the matrix would be exceeded. Hence, the analysis must be modified to consider inelastic deformation of the matrix material. A first simple approximation of this can be provided by replacing the matrix modulus in the formulas previously shown with a reduced modulus. As an example, a reduced modulus is assumed that varies linearly for the epoxy from its elastic value at 1% strain to a zero value at 5% strain. The result of this assumption is the curve labeled inelastic in Fig. 15 where, for high fiber volume fractions, the strength is bounded, and although higher than any results obtained to date, the values are not unreasonably high.

A more general result can be obtained by modeling the matrix as an elastic, perfectly plastic material. For this matrix, the secant modulus value at each axial strain value may be assumed to govern the instability. These assumptions (Ref 33) yield the following result, in place of Eq 58 for the dominant shear mode:

$$\sigma = \left[\frac{V_f E_f \sigma_y}{3(1 - V_f)} \right]^{1/2} \qquad \text{(Eq 59)}$$

where σ_y is the matrix yield stress level.

For the generally dominant shear mode, the elastic results of Eq 58 are independent of the fiber modulus; yet the compressive strength of boron-epoxy is much greater than that of glass-epoxy composites. One hypothesis to explain this is that use of the stiffer boron fibers yields lower matrix strains and less of a strength

reduction due to inelastic effects. Thus, the results of Eq 59 show a ratio of $\sqrt{6}$ or 2.4 for the relative strength of boron compared to that of glass fibers in the same matrix.

All of the foregoing analytical results indicate that compressive strength is independent of fiber diameter. However, different diameter fibers may yield different compressive strengths for composites because large-diameter fibers such as boron (~0.13 mm, or 0.005 in. diam) are better collimated than small-diameter fibers such as glass (~0.01 mm. or 0.004 in. diam).

For small-diameter fibers such as aramid and carbon, local out-of-straightness can introduce matrix shear stresses, cause fiber-matrix debonding, and produce lower instability stress levels. The effects of fiber eccentricities were treated in Ref 36, and those of interface separation were discussed in Ref 36 and 8. Both effects were shown to be capable of causing large decreases in the calculated compressive strength. However, it is difficult to define a realistic quantitative measure of the extent of interface debonding or of fiber eccentricity.

Carbon and aramid fibers are anisotropic and have extremely low axial shear moduli. As a result, the elastic buckling stress in the shear mode is reduced. When this was studied in Ref 36, the following result was obtained:

$$\sigma_c = \frac{G_m}{1 - V_f(1 - G_m/G_{Lf})}$$

where G_{Lf} is the fiber longitudinal shear modulus. This result reduces to that of Eq 58 for high fiber shear moduli.

Another failure mechanism that has been observed (Ref 36) for oriented polymeric fibers is a kink-band formation at a specific angle to the direction of the compressive stress. A typical formation of slip planes in aramid fibers is illustrated in Fig. 16. The formation of kink-bands is mainly attributed to the fibrillar structure of the highly anisotropic fiber and its poor shear strength. Breakup of the fiber into small-diameter fibrils results in the degradation of shear stiffness and hence compressive strength.

The results of the analyses of compressive strength indicate that for the elastic case, the matrix Young's modulus is the dominant parameter. For the inelastic case, however, there are strength limitations that depend on both the fiber modulus and the matrix strength. The nature of changes made in matrix properties to improve the compressive strength of composites of a given fiber depends on the base of reference. In some cases, performance is limited by a matrix yield stress at a given fiber modulus; for other cases, compressive strength could be increased by improving the matrix modulus. The analytical results provide guidance for determination of reasonable changes in matrix properties to yield improved composite compressive strength.

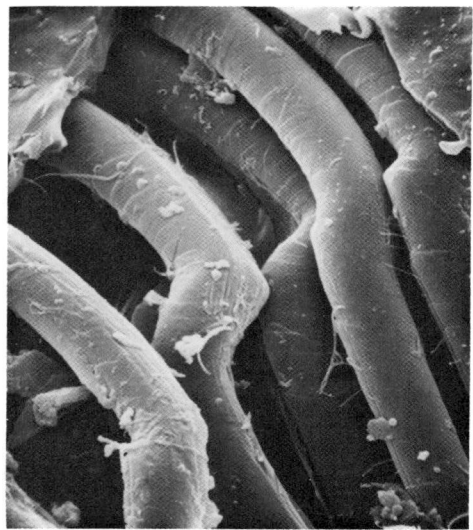

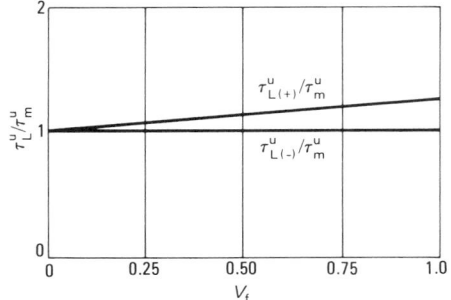

Fig. 17 Bounds on limit load τ_L^u shear stress applied in plane parallel to fibers of uniaxial fibrous composite (normalized for matrix shear strength τ_m^u)

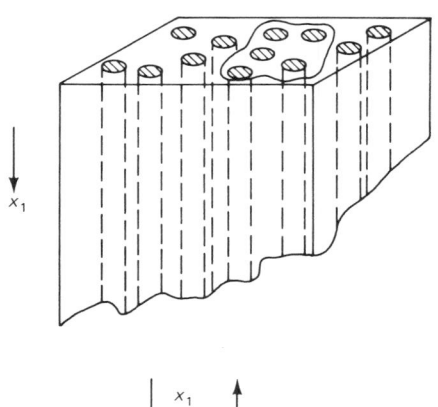

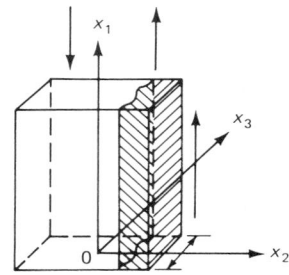

Fig. 18 Axial shear failure mode

Fig. 16 Local fiber failure mechanisms resulting from compression of aramid-epoxy composites

Matrix Mode Strength. The remaining failure modes of interest are transverse tension and compression, and axial shear. For each of these loading conditions, material failure can occur without fracture of the fibers; hence the terminology matrix dominated or matrix modes of failure.

Micromechanical analyses of these failure modes are complex because, unlike the axial failure modes previously treated, the critical stress states for these matrix modes are in the matrix, are highly nonuniform, and depend on local details of the geometry. As a result, the most fruitful approaches would appear to be those that consider average states of stress rather than local details.

Under transverse tensile stress, failure may occur within the matrix or along the fiber-matrix interface. Composite transverse tensile strengths will probably not be significantly greater than matrix tensile strength. Indeed, the addition of fibers may weaken the matrix in this direction because of local stress concentrations, weak interfaces, and so forth.

For transverse compression, failure may occur by shearing along a surface through the matrix parallel to the fiber axes, in a fashion somewhat similar to the compressive failure of a homogeneous material. Thus, there are two types of shearing stresses of interest for matrix-dominated failures: in a plane that contains the filaments, and in a plane normal to the filaments. In the first case, the filaments provide little reinforcement to the composite and the shear strength depends on the shear strength of the matrix material. In the second case, some reinforcement may occur and, at high volume fractions of filaments, may be substantial. Because the analysis shows that reinforcement does not take place in planes or surfaces parallel to the filaments, these planes may be consid-

ered planes of shear weakness. Surfaces of shear weakness do indeed exist in filamentary composites. It is important to recognize that filaments provide little resistance to shearing in any surfaces parallel to them.

The approach to the shear failure analysis is to consider that a uniaxial fibrous composite consists of strong and stiff fibers embedded in a matrix that is characterized by its initial elastic modulus and by a maximum stress level. Accordingly, the matrix is idealized so that its stress-strain relation is that of an elastic, perfectly plastic material. For homogeneous materials, the existence of this plastic region generally signifies the possibility of unbounded structural deformations beyond some limiting load.

For the composite, the theorems of limit analysis of plasticity (Ref 36, 37) may be used to obtain upper and lower bounds of a composite limit load (Ref 38). This is defined as the load at which the matrix yield stress permits composite deformation to increase with no increase in load. This limit load has been defined as composite failure and may be considered an approximation of the strength of a composite having a ductile matrix. It is assumed that the filaments are elastic-brittle and that the matrix is elastic-perfectly plastic and obeys the von Mises yield criterion.

The lamina analyzed consists of a matrix containing a uniaxial set of filaments. The reinforcements are assumed to be in the CCA array configuration described in Ref 8 so that all filaments are considered surrounded by circular-cylindrical surfaces such that the ratios of filament radii to surrounding circular-binder radii are the same for all cylinders. A single cylinder, consisting of a filament and associated concentric matrix shell, will be referred to as a composite cylinder.

The procedure is to select "admissible" stress and velocity fields for construction of the lower and upper bounds, respectively (details are presented in Ref 39). An admissible stress field must satisfy the equilibrium equations everywhere and the traction boundary conditions where specified. In the approach used, a uniform stress field was used as the admissible

stress field in all cases. A kinematically admissible velocity field is a continuous field (with certain permissible exceptions) that satisfies the displacement and velocity boundary conditions. For this approach, the elastic displacement fields of Ref 8 were used as admissible velocity fields to obtain the upper bounds for the two shear strengths previously described. The results obtained from this bounding procedure for axial shear are presented in Fig. 17. The upper bound is plotted as a function of V_f, as shown by the solid curve, coinciding with the lower bound at $V_f = 0$ and approaching the value $4/\pi$ as $V_f \rightarrow 1$. From Fig. 17 it appears that the maximum possible increase in in-plane shear strength due to the filamentary reinforcements is approximately 27%. Figure 18 illustrates this mode of failure.

For transverse shear, the bounds are plotted in Fig. 19 as a function of V_f. In this case, the upper bound approaches infinity (because of the assumption of rigid filaments) as $V_f \rightarrow 1$. The implication here is that reinforcement against transverse shear may in fact take place, particularly in compression.

Strength Under Combined Stress. The developments presented in the previous three sections demonstrate that the analytical determination of the ultimate stresses of a UDC in terms of microstructure and constituent strength characteristics is extremely complex. The qualitative treatments discussed provide valuable insights into the problem of strength and guidelines for the development of improved materials, but design values of the ultimate stresses must be determined experimentally. Even such experimental information, however, is not sufficient for characterization of the strength of a UDC because in most structural applications

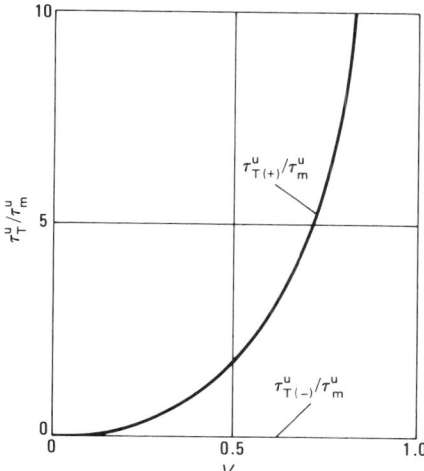

Fig. 19 Bounds on V_f limit load τ_T^u for shear stress applied in plane normal to fibers of uniaxial fibrous composite (normalized for matrix shear strength τ_m^u)

the material is subjected to states of combined stress. The reason is that the most common structural use of a UDC is in the form of unidirectionally reinforced laminae that are joined to form a laminate. As will be shown subsequently, the state of stress in such a lamina is at least plane; thus, it has three components: σ_{11} in the fiber direction, σ_{22} transverse to the fibers, and σ_{12} in axial shear. Therefore, the strength of a UDC must be determined under a state of combined stresses. In view of the extreme difficulty of analytical determination of strength for a single stress component, resolution of the problem for combined stresses by microstructural analysis is impractical. On the other hand, resolution of the problem by experimentation is also impractical, because the number of tests required to develop the full failure surface would be enormous, and some of the combined stress tests are difficult to perform. The remaining alternative is to establish the ultimate stress values for single-stress components by experiment and construct analytical failure criteria in terms of these values by global considerations.

The failure of a UDC under single-stress components has been shown to involve a number of different failure mechanisms. This knowledge and quantitative experimental data for single-stress components will now be used to formulate practical failure criteria for combined stresses. For simplicity, the discussion will deal with the case of plane stress. Results for more complicated stress systems will be presented subsequently. The stresses considered are averages over a representative volume element. The fundamental assumption is that there exists a failure criterion of the form:

$$F(\sigma_{11}, \sigma_{22}, \sigma_{12}) = 1 \qquad \text{(Eq 60)}$$

which characterizes failure of the UDC. A convenient description of Eq 60 is a surface in stress space with coordinates σ_{11}, σ_{22}, and σ_{12} (Fig. 20). Because all failure stresses are finite, the surface must be closed. Each point in stress space represents a state of stress. Points inside the surface indicate no failure, points on the surface indicate failure, and points outside the surface are not realizable stress states. A failure surface can also be expressed in terms of strain. The stress and strain failure surfaces must be connected by the stress-strain relations of the UDC.

The most primitive failure criterion is the assumption that failure will occur whenever any single stress component reaches its ultimate value, regardless of the values of the other stress components. This maximum-stress failure criterion is not realistic because it disregards the combined effects of stresses on failure. Although simple to use in practice, such a criterion is nonconservative because it overestimates the strength of the material under combined stress. A similar possibility is the assumption that failure occurs whenever any strain component reaches its ultimate value regardless of the others. This maximum-strain criterion is also frequently used, although it is open to similar criticism.

The usual approach to construction of a failure criterion is to assume that it can be represented by a quadratic polynomial in the stresses. The coefficients of the polynomial should be expressed in terms of simple experimental information such as ultimate stresses under individual load components. All the various failure criteria that have been proposed attempt to use these basic test data to define a surface of the form given in Eq 60. No fundamental significance should be attached to the choice of a quadratic form. This should be regarded as the simplest choice that can adequately describe the experimental data. Apparently, such an approach has been simultaneously initiated in Ref 40 and 41. A general quadratic failure criterion including linear terms is presented in Ref 42. A comprehensive general discussion of quadratic failure criteria is given in Ref 43. All the quadratic failure criteria referred to are smooth, which implies that the failure criterion is represented in terms of a single polynomial. For example, in the case of plane stress, the general quadratic version of Eq 60 would be:

$$F_{11}\sigma_{11}^2 + F_{22}\sigma_{22}^2 + F_{66}\sigma_{12}^2 + 2F_{12}\sigma_{11}\sigma_{22} +$$
$$2F_{16}\sigma_{11}\sigma_{12} + 2F_{26}\sigma_{22}\sigma_{12} +$$
$$F_1\sigma_{11} + F_2\sigma_{22} + F_6\sigma_{12} = 1$$

The material has different strengths in uniaxial, longitudinal, and transverse tensions and compressions, but evidently the shear strength is not affected by the sign of the shear stress. It follows that all powers of shear stress in the failure criterion must be even and therefore all the terms containing a shear stress to power one must be rejected. Consequently, the criterion simplifies to:

$$F_{11}\sigma_{11}^2 + F_{22}\sigma_{22}^2 + F_{66}\sigma_{12}^2 + 2F_{12}\sigma_{11}\sigma_{22} +$$
$$F_1\sigma_{11} + F_2\sigma_{22} = 1 \qquad \text{(Eq 61)}$$

Let the ultimate stresses for single stress components be denoted:

σ_L^{tu} = tensile ultimate, fiber direction
σ_L^{cu} = compressive ultimate, fiber direction

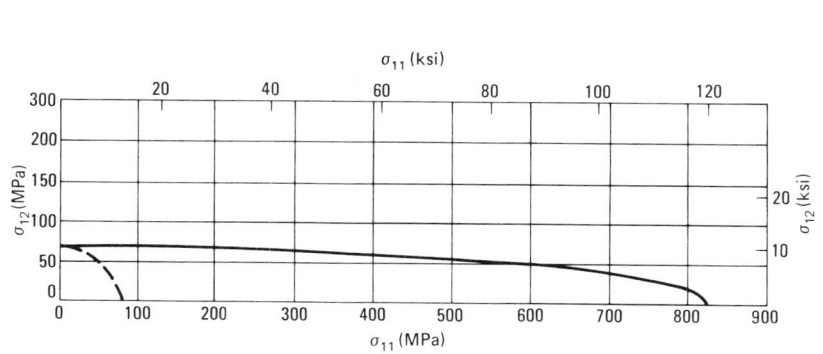

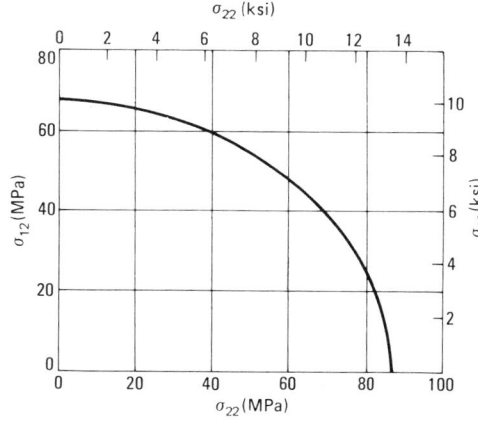

Fig. 20 Failure surfaces in stress space for graphite-epoxy, 60 vol% fibers

σ_T^{tu} = tensile ultimate, transverse to fibers
σ_T^{cu} = compressive ultimate, transverse to fibers
τ_L^u = ultimate longitudinal shear stress

These values determine the points on the σ_{11}, σ_{22}, and σ_{12} axes through which the surface (Eq 61) must pass. It follows that:

$$F_{11} = \frac{1}{\sigma_L^{tu}\sigma_L^{cu}} \qquad F_{22} = \frac{1}{\sigma_T^{tu}\sigma_T^{cu}}$$

$$F_1 = \frac{1}{\sigma_L^{tu}} - \frac{1}{\sigma_L^{cu}} \qquad F_2 = \frac{1}{\sigma_T^{tu}} - \frac{1}{\sigma_T^{cu}} \qquad \text{(Eq 62)}$$

$$F_{66} = \frac{1}{(\tau_L^u)^2}$$

However, F_{12} cannot be determined in terms of the ultimate stresses, and therefore failure tests must be performed under biaxial stress to find this coefficient. Frequently, this coefficient is established by relating Eq 61 to the Mises-Hencky yield criterion for isotropic materials. This yields:

$$F_{12} = -\frac{1}{2}(F_{11}F_{22})^{1/2}$$

which is the two-dimensional version of the Tsai-Wu criterion (Ref 42). Its implementation raises several problems: First, the coefficient F_{12} must be determined by biaxial testing. Tension or compression values must be selected for both σ_{11} and σ_{22} for these tests. It is reasonable to expect that F_{12} will have four different values corresponding to these choices. This contradicts the premise of description of the failure criterion by a single polynomial. Second, because the coefficients are expressed in terms of both tensile and compressive strengths, it follows that in a biaxial tension-tension test, the failure will depend on the compressive ultimate stresses. This is not satisfactory from a physical point of view. Finally, the failure criterion ignores the different failure modes of a UDC and does not specify how the composite fails. In view of the diverse failure mechanisms previously discussed, this third problem is particularly severe. Identification of the different failure modes of a UDC can be used to lead to physically more realistic and, at the same time, simpler failure criteria (Ref 44). Testing of a polymer matrix UDC reveals that tensile stress in the fiber direction produces a jagged, irregular failure surface, while tensile stress transverse to the fibers produces a smooth, straight failure surface (Fig. 21). The underlying microstructural phenomena are as follows: Under tension in the fiber direction, fibers break progressively at random places, leading to small, transverse cracks and to longitudinal cracks along the fibers (Fig. 22); the load-carrying capacity in the fiber direction deteriorates because of the transverse cracks until a sufficient number coalesce to result in catastrophic failure.

Such failure will be identified as the tensile fiber mode. Because the carrying-capacity de-

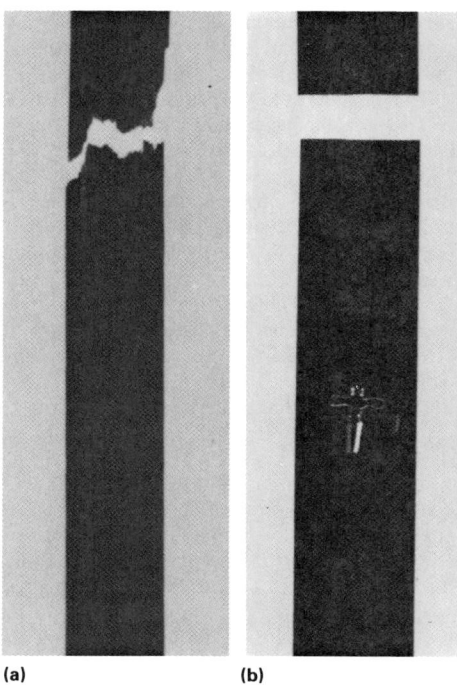

Fig. 21 Typical failures of [0°] boron-epoxy laminates subjected to tension loading. (a) Longitudinal. (b) Transverse

terioration in this mode is due to transverse cracks, and because transverse stress σ_{22} has no effect on such cracks (although it may affect load transfer after cracking), the plane tensile fiber mode is assumed to depend only on the stresses σ_{11} and σ_{12}.

In the case of compressive σ_{11}, failure is due to fiber buckling, as discussed previously. Because the fibers buckle in the shear mode, it may again be assumed that the transverse stress, σ_{22}, has little effect on the compressive failure. Thus, in this compressive fiber mode, failure again depends primarily on σ_{11} and also on σ_{12}. The dependence on σ_{12} is not known, and arguments may be offered for and against including it in the failure criterion.

In the case of tension transverse to the fibers, failure occurs by a sudden crack in the fiber direction (Fig. 23). The phenomenon is so abrupt that its initiation and extension are extremely difficult to detect. This failure mode is called the tensile matrix mode. Because stress in the fiber direction has no effect on a crack in the fiber direction, the axial stress does not enter into this failure mode, which is thus dependent only on σ_{22} and σ_{12}.

For compressive stress transverse to the fibers, failure occurs on some plane parallel to the fibers, not necessarily normal to the direction of σ_{22}. This is called the compressive matrix mode and is produced by normal stress and shear stress on the failure plane. Again, the stress σ_{11} does not affect this failure.

Each of the failure modes described can be modeled separately by a quadratic polyno-

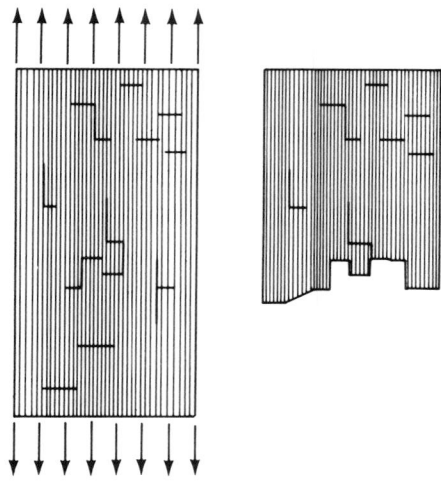

Fig. 22 Tensile fiber failure mode

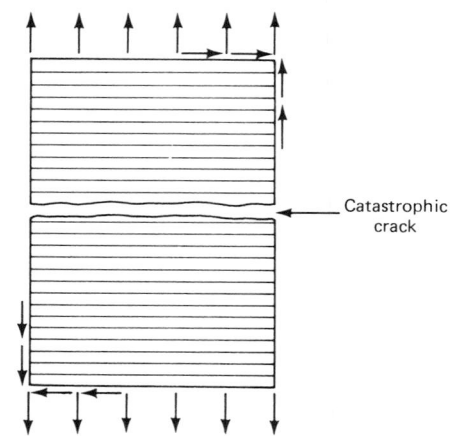

Fig. 23 Tensile matrix failure mode

mial (for details, see Ref 44, in which the general case of three-dimensional stress is considered). This results in four individual failure criteria.

The choice of stress components included in each of these criteria, and the particular mathematical form used, are subjects that are not yet fully resolved. The following appear to be a reasonable set of criteria that satisfy the desire to deal separately with different modes of failure:

Fiber Modes

Tensile

$$\left(\frac{\sigma_{11}}{\sigma_L^{tu}}\right)^2 + \left(\frac{\sigma_{12}}{\tau_L^u}\right)^2 = 1 \qquad \text{(Eq 63a)}$$

Compressive

$$\left(\frac{\sigma_{11}}{\sigma_L^{cu}}\right)^2 + \left(\frac{\sigma_{12}}{\tau_L^u}\right)^2 = 1 \qquad \text{(Eq 63b)}$$

Matrix Modes

Tensile

$$\left(\frac{\sigma_{22}}{\sigma_T^{tu}}\right)^2 + \left(\frac{\sigma_{12}}{\tau_L^u}\right)^2 = 1 \qquad \text{(Eq 64a)}$$

Compressive

$$\left(\frac{\sigma_{22}}{2\tau_T^u}\right)^2 + \left[\left(\frac{\sigma_T^{cu}}{2\tau_T^u}\right)^2 - 1\right]\frac{\sigma_{22}}{\sigma_T^{cu}} + \left(\frac{\sigma_{12}}{\tau_L^u}\right)^2 = 1$$

$$\text{(Eq 64b)}$$

Note that σ_T^{cu} in Eq 64b should be taken with its absolute value. The ultimate transverse shear, $\sigma_{23} = \tau_T^u$, is difficult to measure. However, a reasonable approximation for this quantity is the ultimate shear stress for the polymeric matrix. For any given state of stress, one of each of Eq 63 and 64 is chosen according to the signs of σ_{11} and σ_{22}. The stress components are introduced into the appropriate pair, and whichever criterion is satisfied first is the operative criterion.

The advantages of Eq 63 and 64 are:

● The failure criteria are expressed in terms of single-component ultimate stresses. No biaxial test results are needed
● The failure mode is identified by the criterion that is satisfied first

The last feature is of fundamental importance for analysis of fiber-reinforced composite structural elements, because it permits identification of the nature of initial damage. Moreover, in conjunction with a finite-element analysis, it is possible to identify the nature of the failure in elements, modify their stiffnesses accordingly, proceed with the analysis to predict new failure, and so forth.

Fatigue Failure. Understanding the response of composites to repeated load applications is of primary importance, because aeronautical structures are subjected to many load cycles during their lifetime. Local areas of damage may be initiated by cyclic loading. Such localized damage may also result from deficiencies in the fabrication process or unanticipated events in service, such as foreign object damage. All of these causes may be expected to propagate further damage as a result of subsequent cyclic loading. This may lead to a loss in structural performance and eventually to failure.

For metal structures, fatigue problems have been approached with some success by use of the methodology of fracture mechanics. In metals fatigue, failure consists of the initiation and propagation of one dominant crack. Over the years, a fracture mechanics design procedure for fatigue of metal aeronautical structures has emerged that is based on periodic inspection and monitoring of visible cracks (Ref 45). There are significant differences between the fatigue failure processes in metals and those in fiber composites. This requires new design

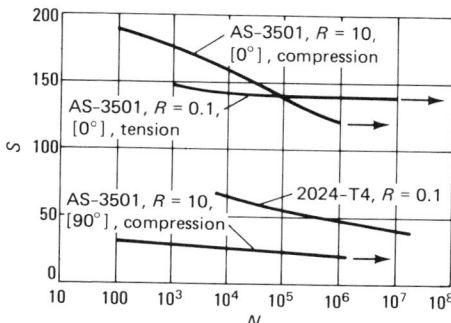

Fig. 24 Typical *S-N* curves showing effect of different loadings on carbon-epoxy composites

procedures for fiber composite structures, but these have not yet been established in definitive form.

It is emphasized again that in the design process of a UDC structure, both the laminate configuration and the constituents of the UDC are design variables. Clearly, to obtain the necessary laminate fatigue data by testing all possible candidate laminates for all loading conditions of interest is prohibitively expensive. Therefore, either the number of alternatives to be examined must be severely limited, or an analytical procedure must be developed. Again, it appears desirable to assess the fatigue characteristics of a laminate for general loading conditions in terms of the fatigue characteristics of the laminae and the laminate geometry. The fatigue response of the UDC would be determined experimentally. Therefore, the present discussion is concerned with UDC fatigue failure. As in the case of static failure, the point of view is adopted that the complexities of the microstructural failure process prohibit quantitative analytical determination of fatigue failure in terms of the microcharacteristics of the composite. The observed nature of microfailures is of great qualitative importance, however, primarily for the detection and description of different failure modes.

Single cyclic stress involves testing components in the traditional fatigue test in which a specimen is subjected to uniform sinusoidal cyclic stress (Fig. 24). The mean, σ^m, and alternating, σ^a, parts of the cyclic stress are defined by:

$$\sigma^m = \frac{1}{2}(\sigma_{max} + \sigma_{min})$$
$$\sigma^a = \frac{1}{2}(\sigma_{max} - \sigma_{min}) \qquad \text{(Eq 65)}$$

and the stress amplitude ratio by:

$$R = \sigma_{min}/\sigma_{max} \qquad \text{(Eq 66)}$$

The most common fatigue characterization is a plot of number of cycles to failure—also called lifetime—versus the maximum ampli-

tude in a constant sinusoidal cycling. This is known as the *S-N* curve and is usually presented for stress ratios, *R*, of either 0 or 1. Because lifetimes may attain millions of cycles and most of the fatigue damage develops toward the end of the lifetime, it is convenient to plot log *N* versus amplitude and sometimes versus log amplitude. In the examples of *S-N* curves for cyclic stress of carbon-epoxy shown in Fig. 24, the tension-tension curve ($R = 0.1$) illustrates the characteristic insensitivity of the UDC to repeated tensile loadings in the fiber direction. This is in contrast to the axial compression results ($R = 10$), which demonstrate significant fatigue degradation. For both loading conditions, however, the maximum stress values are substantially greater than those obtained for a good aluminum alloy. The figure also contains a representative result for compressive cyclic loading transverse to the fibers. In this mode, the *S-N* curve is flat but substantially below all the other curves.

An intrinsic characteristic of fatigue failure is large scatter in lifetime data, which implies that repeated tests of the same cyclic loading produce different lifetimes. This scatter is due to the complexity of the developing microdamage under cyclic loading, which is different in detail for every specimen tested. Such scatter is inherent to fatigue for all materials and must be considered in the analytical methodology.

Because most of the experience in fatigue failure has been accumulated for metals, some fundamental differences between fatigue failure of metals and composites should be mentioned. Metals are polycrystalline aggregates. During stress cycling, microdamage develops in the form of microcracks and plastifications of single crystals, the latter being known as persistent slip bands. At some stage of the cycling, a crack develops that grows with continued cycling until the specimen fails. These two stages are known as initiation and propagation. At present, the fraction of lifetime spent in initiation cannot be predicted. In the propagation stage, however, the growth of the crack with cycling can be predicted reasonably well (Ref 46). The use of a crack growth law, in addition to a measured or assumed size of the existing crack, forms the basis for design methods for fatigue of metals. These methods take into account possible crack locations and characteristics of available inspection methods to define safe inspection periods.

For a UDC, the nature of damage growth is quite different. The microstructure consists of a polymeric matrix containing stiff, strong fibers. The material is highly anisotropic, and cracks propagate easily in the matrix along surfaces parallel to the fibers. When a unidirectional specimen is cycled in tension-tension in the fiber direction, damage accumulates in the form of random fiber ruptures that are the source of small cracks in the matrix along the fibers (Fig. 22). More and more cracks develop, both perpendicular and parallel to the fibers, with additional cycling until some of these coalesce to produce catastrophic failure. The failure surface is jagged and irregular, as suggested in Fig. 22 and illustrated in Fig. 23. The failure mode is similar to the static tensile fiber mode discussed previously. For some fibers, failure may occur by propagation of one sudden transverse crack. Observations of this mode are rare.

This damage accumulation mechanism of fatigue failure is called the tensile-tensile fiber mode. Its main characteristics are that failure occurs because of fiber rupture, and that the entire lifetime consists of damage initiation.

When the cycling is other than tension-tension, the situation is less clear. For tension-compression cycling, the same phenomena might be expected because of the tension component. The matter of the failure mechanism in compression-compression cycling is an open question. Fibers might buckle at a much lower cyclic load than for a static buckling load because of deterioration of the matrix shear modulus due to cycling and/or due to the opening of longitudinal cracks at fiber-matrix interfaces. Careful experimentation is needed to uncover the governing mechanisms.

As in the static case, the second primary failure mode consists of a sudden cracking in a plane along the fibers and is called the matrix mode, shown in Fig. 23 and illustrated for the tension-tension loading of a boron-epoxy composite in Fig. 21. Pending thorough experimental investigation, it seems reasonable to expect that for tension-compression and compression-compression cycling, the directionality of the fibers will lead to the same kind of plane matrix failure mode.

The main characteristics of the matrix failure mode are that failure occurs by plane cracking along fibers, and that the crack propagates suddenly; thus, the entire fatigue life consists of initiation.

The *S-N* curve describes the relation between maximum cyclic stress amplitude and lifetime for a single cyclic stress component having a specified stress ratio. It has been introduced in the investigation of fatigue of metal structures where there are many cases of a single-component cyclic stress. For a given stress ratio, *R*, of a single stress component, the *S-N* curve defines the lifetime, *N*, as a function of the maximum applied stress, *S*. Another useful format for presentation of fatigue data is the Goodman diagram (Fig. 25), which presents various combinations of uniaxial mean and alternating stress for failure at a specified lifetime. From Eq 65 and 66, it is clear that constant values of *R* are represented by straight lines through the origin. Thus, different regimes of fatigue loading can be located, as shown in Fig. 25. This format is convenient for interpolation for different values of stress ratio. Representative uniaxial fatigue data for several UDC materials are presented in Fig. 26.

In practical metal structures, fatigue failure can be expected to occur first in the vicinity of a discontinuity that causes a stress concentration. Appropriate *S-N* curves may be of value

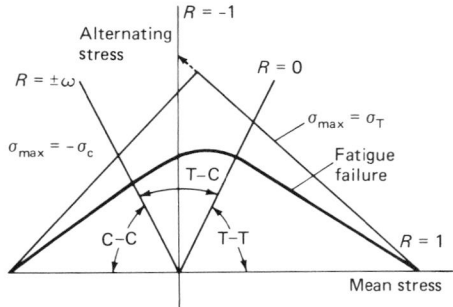

Fig. 25 Characteristics of Goodman diagram

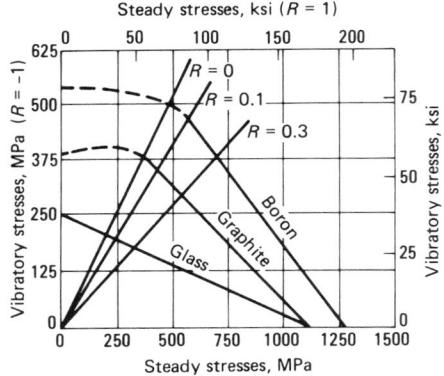

Fig. 26 Goodman diagram for representative unidirectional composites, 10^7 cycles to fracture

for predicting the onset of such failure. Following such an initial crack, the dominant effect may become the subsequent growth of that crack. As indicated previously, metal fatigue failures are ascribed to fracture following a stable crack growth. Thus, design methods require the definition of an appropriate crack-growth law. It has been customary in metal fatigue to use notched specimens for this purpose.

In a plain specimen, the stress in the test section is uniform and equal to the applied stress. Thus, a dominant crack may start at any place in the test section. In a notched specimen, there is significant magnification (stress concentration) of applied stress at the notch, where the state of stress is locally uniaxial tension. Thus, a fatigue crack will develop and grow at this location. The notched specimen, however, does not provide useful information for the fatigue characteristics of a UDC, because it initiates different kinds of fatigue failure. A common case is the propagation of a crack along the fibers, and not a transverse crack, as in metals. This situation is much more complicated in a laminate, where a notch produces a complicated state of three-dimensional stress that must be found numerically by the use of complex computer programs. Furthermore, as previously noted, the simplest case of interest for static loading of a UDC is plane stress with

three stress components. This is of practical interest because a planar stress state exists in a lamina away from the free edges of the laminate. Therefore, the *S-N* curve for a single-stress component of a UDC is by itself of limited importance. The conclusions are that for a UDC, the required fatigue information is for failure under plane cyclic stress, and that notched specimens are not helpful for determination of fatigue characteristics.

Combined cyclic stress requires the introduction of the concept of a fatigue failure criterion for such a stress state. This is a generalization of the failure criterion for combined static stress. In the general case of three-dimensional cyclic stress, each stress component has a mean component, σ_{ij}^m, and an alternating component, σ_{ij}^a, as in Eq 65. In addition, the stress cycles are not necessarily in phase and may have phase lags, λ_{ij}. If the lifetime, *N*, under such stress cycling is defined by the quantities cited, then the relation among all of these is written:

$$F(\sigma_{ij}^m, \sigma_{ij}^a, \lambda_{ij}, N) = 1 \qquad \text{(Eq 67)}$$

Nothing seems to be known about such a general relation.

Even if it is assumed that there are no phase lags (this is the case when the applied cyclic loads are in phase and the material is linear elastic), establishment of a general failure criterion in terms of mean and alternating stresses is a formidable problem. The only existing work in the literature seems to be for the case of one stress component and is purely empirical. In this case, it is customary to plot the experimentally obtained pairs of σ_{ij}^m and σ_{ij}^a that lead to the same lifetime, in the form of the Goodman diagram.

To simplify matters further, it is assumed that all stress components have the same ratio, *R*. Thus, from Eq 65 and 66,

$$\sigma_{ij}^{min}/\sigma_{ij}^{max} = R$$
$$\sigma_{ij}^m = \sigma_{ij}^{max}(1 + R) \qquad \text{(Eq 68)}$$
$$\sigma_{ij}^a = \sigma_{ij}^{max}(1 - R)$$

Inserting this into Eq 67 yields:

$$F(\sigma_{ij}^{max}, R, N) = 1 \qquad \text{(Eq 69)}$$

The case described occurs in practice when all cyclic loads are in phase and cycle with the same ratio, *R*.

Just as for static failure criteria, it is convenient to represent Eq 69 as a surface in stress space, where in the present case the stress variables are σ_{ij}^{max} (which will be called simply σ_{ij}). If *R* has some fixed value, then Eq 69 may be interpreted as a family of surfaces, each of which is associated with a different *N* (Fig. 27). Each surface is the locus of all cyclic stress states that produce fatigue failure after the same lifetime, *N*. This description ignores scatter; therefore, *N* must be interpreted in some mean sense. The static failure criterion is a special

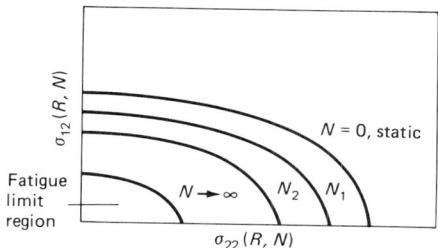

Fig. 27 Fatigue failure surface family

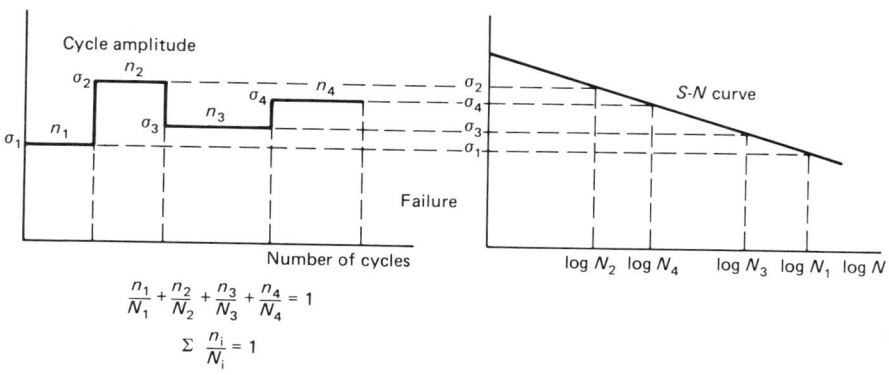

$$\frac{n_1}{N_1} + \frac{n_2}{N_2} + \frac{n_3}{N_3} + \frac{n_4}{N_4} = 1$$

$$\Sigma \frac{n_i}{N_i} = 1$$

Fig. 28 Palmgren-Miner Rule

case, $N = 0$. At the other extreme, there are stress states of such low amplitudes that fatigue failure does not occur. This defines the fatigue (or endurance) limit region in Fig. 27, which is a generalization of the usual fatigue (or endurance) limit of an S-N curve. The family of surfaces is for a fixed R; a different family will be associated with a different R.

As in the static case, the failure surfaces will be described in terms of quadratics in the stress amplitudes σ_{ij}. Because the surfaces vary with N and R, the coefficients of the quadratics must also be functions of N and R. This approach can be applied to plane cyclic stress, as in laminates.

Residual strength must be considered in connection with repeated applications of stresses that are not large enough to produce static failure but eventually can cause failure. This concept suggests a continual deterioration in material strength. The residual strength, σ_r, is defined as the static strength after n cycles have elapsed. A specimen is first cycled for n cycles with specified σ_{min} and σ_{max} or, equivalently, specified σ_{max} (amplitude) and R (Eq 65 and 66). It is then removed from the testing machine and failed statically. The residual strength is clearly a decreasing function of n, because increasing n will increase the internal damage and therefore decrease the strength. It is usually argued that fatigue failure will occur when the residual strength becomes equal to σ_{max}, implying that then the last quarter cycle is equivalent to a static failure experiment. Thus, if N is the lifetime for the cyclic stress under consideration, we have:

$$\sigma_r = \sigma_r (\sigma_{max}, R, n) \qquad \text{(Eq 70)}$$

The relationship in Eq 70 can only be obtained experimentally. Furthermore, it is not a well-defined relationship because both σ_r and N are scattered (thus random) variables. The subject has been considered in a number of papers, notably Ref 47 to 49. The statistical relation between σ_r and N has been obtained on the basis of various assumptions. The relation involves several parameters that must be found in terms of a considerable amount of test data.

All of the work done on residual strength seems to be confined to a single stress component. The problem of residual strength for combined states of stress does not appear to have received attention in the literature.

Cumulative damage represents another central problem of fatigue failure. Most fatigue testing is conducted for constant-amplitude cycling resulting in the S-N curve. Constant-amplitude cycling, however, rarely occurs in practice. As an example, during takeoff, maneuvering, cruising, and landing, the wing of an airplane is subjected to different aerodynamic loads. Also, variable weather conditions result in gust loads of different intensities. Thus, the loads vary significantly in time and the wing is subjected to a program of cyclic loading. The problem that arises is to predict the lifetime of a structural member that is subjected to a cyclic loading program of variable amplitude. This has become known as the cumulative-damage problem.

There is little hope that the cumulative-damage problem can be treated on the basis of the microstructure of a UDC, because the failure process of damage accumulation that terminates in catastrophic failure is much too complex and unpredictable to be handled by analytical means. At present, the only viable approach seems to consist of lifetime prediction under complex cyclic load in terms of lifetime test data for simple cyclic loading, for example, in terms of the usual S-N curves for specimens. This point of view has been used in numerous theoretical and experimental investigations of the problem in the context of metal fatigue (Ref 45, 49).

The simplest and best-known procedure for lifetime evaluation is an empirical formula that has become known as Miner's Rule, or the Palmgren-Miner Rule, referred to here as the PM rule. Consider a cyclic-loading program where the amplitude varies in jumps from one constant value to another. Let the number of cycles applied at amplitude σ_i be n_i and let N_i (σ_i) be the lifetime under constant-amplitude cycling as defined by the S-N curve. The PM rule asserts that:

$$\sum_i \frac{n_i}{N_i (\sigma_i)} = \frac{n_1}{N_1} + \frac{n_2}{N_2} + \dots = 1 \qquad \text{(Eq 71)}$$

This defines the last n_i, which terminates at failure; thus, the total lifetime is:

$$L = \sum_i n_i \qquad \text{(Eq 72)}$$

The procedure is illustrated in Fig. 28.

Experience with metal fatigue has shown that the simple PM rule is not reliable. The major problem is that because it is a linear superimposition rule, it cannot take into account a sequence of loading effects. For example, if the loading is two-stage, that is, first n_1 cycles at amplitude σ_1 and then n_2 cycles to failure at amplitude $\sigma_2 > \sigma_1$, the total lifetime is $n_1 + n_2$ cycles. If the sequence of loading were reversed and σ_2 were first applied for n_2 cycles, the PM rule would predict the same total lifetime, $n_1 + n_2$. Experience with metal fatigue shows, however, that the sequence of loading does significantly affect the lifetime. Usually, the lifetime is larger for low-high than for high-low two-stage cycling.

Many alternative cumulative-damage approaches have been proposed for metals (Ref 45, 49), usually on an empirical or semiempirical basis. A general cumulative-damage theory, based on the concept of damage curves and an equivalent cyclic-loading postulate, has been developed in Ref 51. A cumulative-damage theory based on residual-strength degradation has been proposed in Ref 52. A definitive theory, however, is not available at present, either for metals or for composites. Consequently, many engineers still continue to use the simple PM rule, despite its severe limitations.

Most cumulative-damage work is concerned with one-component cyclic stress. As previously emphasized, this is not of great interest for UDC studies because the layers in a laminate are at least in states of plane cyclic stress. It is necessary, therefore, to construct a cumulative-damage theory for UDC materials under combined states of cyclic stress.

In conclusion, although the subject of fatigue failure of fiber-reinforced composites is still in a developing stage, it is clear that fatigue failure mechanisms of composites differ significantly from those of metals. Furthermore, the widely used carbon fiber-polymeric composites have demonstrated excellent fatigue characteristics when loaded in the fiber direction.

Much of the current research on fatigue of composites is concerned with laminates rather than unidirectional specimens. In view of the enormous diversity of candidate laminates, theories relating laminate fatigue failure to unidirectional layers are clearly needed; therefore, the unidirectional material merits more attention. Many interesting and important fatigue studies have already been performed (Ref 53-59).

REFERENCES

1. R.J. Palmer, "Investigation of the Effect of Resin Material on Impact Damage to Graphite/Epoxy Composites," NASA CR-165677, National Aeronautics and Space Administration, March 1981
2. Z. Hashin, "Theory of Fiber Reinforced Materials," NASA CR-1974, National Aeronautics and Space Administration, 1972
3. R.M. Christensen, *Mechanics of Composite Materials*, Wiley-Interscience, 1979
4. R. Hill, Theory of Mechanical Properties of Fiber Strengthened Materials, I, Elastic Behavior, *J. Mech. Phys. Solids*, Vol 12, 1964, p 199-212
5. G. Pickett, AFML TR-65-220, 1965; also, Elastic Moduli of Fiber Reinforced Plastic Composites, in *Fundamental Aspects of Fiber Reinforced Plastic Composites*, R.T. Schwartz and H.S. Schwartz, Ed., Wiley-Interscience, 1968, p 13-27
6. D.F. Adams, D.R. Doner, and R.L. Thomas, AFML TR-67-96, 1967. Also, *J. Compos. Mater.*, Vol 1, 1967, p 4-17, p 152-165
7. G.P. Sendeckyj, Elastic Behavior of Composites, in *Mechanics of Composite Materials*, Vol II, G.P. Sendeckyj, Ed., Academic Press, 1974
8. Z. Hashin and B.W. Rosen, The Elastic Moduli of Fiber Reinforced Materials, *J. Appl. Mech.*, Vol 31, 1964, p 223-232
9. Z. Hashin, Analysis of Properties of Fiber Composites With Anisotropic Constituents, *J. Appl. Mech.*, Vol 46, 1979, p 543-550
10. G.V. Blessing and W.L. Elban, Aluminum Matrix Composite Elasticity Measured Ultrasonically, *J. Appl. Mech.*, Vol 48, 1981, p 965-966
11. V.M. Levin, On the Coefficients of Thermal Expansion of Heterogeneous Materials, in *Mechanics of Solids*, Vol 2, 1967, p 58-61
12. B.W. Rosen and Z. Hashin, Effective Thermal Expansion Coefficients and Specific Heats of Composite Materials, *Int. J. Eng. Sci.*, Vol 8, 1970, p 157-173
13. R.A. Schapery, Thermal Expansion Coefficients of Composite Materials Based on Energy Principles, *J. Compos. Mater.*, Vol 2, 1968, p 380ff
14. B.W. Rosen, "Thermal Expansion Coefficients of Composite Materials," Ph.D. thesis, University of Pennsylvania, 1968
15. T. Ishikawa, K. Koyama, and S. Kobayashi, Thermal Expansion Coefficients of Unidirectional Composites, *J. Compos. Mater.*, Vol 12, 1978, p 153-168
16. G.S. Springer, Environmental Effects on Epoxy Matrix Composites, in *Fifth Conference on Composite Materials: Testing and Design*, STP 674, S.W. Tsai, Ed., American Society for Testing and Materials, 1979, p 291-311
17. S.W. Tsai and H.T. Hahn, *Introduction to Composite Materials*, Technomic, 1980
18. R.M. Christensen, *Theory of Viscoelasticity*, Academic Press, 1971
19. Z. Hashin, Viscoelastic Behavior of Heterogeneous Media, *J. Appl. Mech.*, Vol 32, 1965, p 630-636
20. Z. Hashin, Viscoelastic Fiber Reinforced Materials, *AIAA J.*, Vol 4, 1966, p 1411-1417
21. R.A. Schapery, Viscoelastic Behavior of Composites, in *Mechanics of Composite Materials*, Vol II, G.P. Sendeckyj, Ed., Academic Press, 1974
22. "Standard Test Method for Tensile Properties of Glass Fiber Strands, Yarns, and Rovings Used in Reinforced Plastics," ASTM D 2343, *Annual Book of ASTM Standards*, American Society for Testing and Materials
23. D.E. Gucer and J. Gurland, Comparison of the Statistics of Two Fracture Modes, *J. Mech. Phys. Solids*, 1962, p 365-373
24. C. Zweben, Tensile Failure Analysis of Fibrous Composites, *AIAA J.*, Vol 6, 1968, p 2325-2331
25. B.W. Rosen, Tensile Failure of Fibrous Composites, *AIAA J.*, Vol 2, 1964, p 1982-1991
26. B.D. Coleman, On the Strength of Classical Fibres and Fibre Bundles, *J. Mech. Phys. Solids*, Vol 7, 1958
27. C. Zweben and B.W. Rosen, A Statistical Theory of Material Strength With Application to Composite Materials, *J. Mech. Phys. Solids*, Vol 18, 1970, p 180-206
28. C. Zweben, A Bounding Approach to the Strength of Composite Materials, *Eng. Fract. Mech.*, Vol 4, 1972, p 1-8
29. D.G. Harlow and S.L. Phoenix, "The Chain-of-Bundles Probability Model for the Strength of Fibrous Materials, I, Analysis and Conjectures, II, A Numerical Study of Convergence, *J. Compos. Mater.*, Vol 12, 1978, p 195-214, p 300-313
30. B.W. Rosen and C.H. Zweben, "Tensile Failure Criteria for Fiber Composite Materials," NASA CR-2057, National Aeronautics and Space Administration, Aug 1972
31. S.B. Batdorf, "Fracture Statistics of Brittle Materials With Intergranular Cracks," Paper presented at the Third International Conference on Structural Mechanics in Reactor Technology, London, Sept 1975
32. N.F. Dow and I.J. Gruntfest, "Determination of Most Needed Potentially Possible Improvements in Materials for Ballistic and Space Vehicles," GE-TIS 60SD389, June 1960
33. B.W. Rosen, Strength of Uniaxial Fibrous Composites, in *Mechanics of Composite Materials*, Pergamon Press, 1970
34. B.W. Rosen, Mechanics of Composite Strengthening, in *Fiber Composite Materials*, American Society for Metals, 1965
35. H. Schuerch, Prediction of Compressive Strength in Uniaxial Boron Fiber-Metal Matrix Composite Materials, *AIAA J.*, Vol 4, 1966
36. C.H. Chen and S. Cheng, Mechanical Properties of Fiber Reinforced Composites, *J. Compos. Mater.*, Vol 1, 1967, p 30
37. D.C. Drucker, H.J. Greenberg, and W. Prager, The Safety Factor of an Elastic-Plastic Body in Plane Strain, *J. Appl. Mech.*, Vol 18, 1951, p 371
38. W.T. Koiter, General Theorems for Elastic-Plastic Solids, chapter IV in *Progress in Solid Mechanics*, I.N. Sneddon and R. Hill, Ed., North-Holland, 1960
39. L.S. Shu and B.W. Rosen, Strength of Fiber Reinforced Composites by Limit Analysis Method, *J. Compos. Mater.*, Vol 1, 1967, p 366
40. S.W. Tsai, "Strength Characteristics of Composite Materials," NASA CR-224, National Aeronautics and Space Administration, 1965
41. E.K. Ashkenazi, *Zavod. Lab.*, Vol 30, 1964, p 285-287, also, *Mekh. Polim*, Vol 1, 1965, p 60-70
42. S.W. Tsai and E.M. Wu, A General Theory of Strength for Anisotropic Materials, *J. Compos. Mater.*, Vol 5, 1971, p 58
43. E.M. Wu, Phenomenological Anisotropic Failure Criterion, in *Mechanics of Composite Materials*, G.P. Sendeckyj, Ed., Academic Press, 1974
44. Z. Hashin, Failure Criteria for Unidirectional Fiber Composites, *J. Appl. Mech.*, Vol 47, 1980, p 329-334
45. W.S. Pellini, *Principles of Structural Integrity Technology*, Office of Naval Research, 1976
46. N.E. Frost, K.J. Marsh, and L.P. Pook, *Metal Fatigue*, Oxford University Press, 1974
47. J.C. Halpin, K.L. Jerina, and T.A. Johnson, *Analysis of the Test Methods for High Modulus Fibers and Composites*, STP 521, 1973, p 5-64
48. J.N. Yang, Fatigue and Residual Strength Degradation for Graphite/Epoxy Composites and Tension-Compression Cyclic Loadings, *J. Compos. Mater.*, Vol 12, 1978, p 19-39
49. J.N. Yang, R.K. Miller, and C.T. Sun, Effect of High Load on Statistical Fatigue of Unnotched Graphite/Epoxy Laminates, *J. Compos. Mater.*, Vol 14, 1980, p 82-94
50. H.L. Leve, Cumulative Damage Theories, in *Metal Fatigue: Theory and Design*, A.F. Madayag, Ed., John Wiley & Sons, 1969

51. Z. Hashin and A. Rotem, A Cumulative Damage Theory of Fatigue Failure, *Mater. Sci. Eng.*, Vol 34, 1978, p 147-160

52. J.N. Yang and D.L. Jones, Load Sequence Effects on the Fatigue of Unnotched Composite Materials, in *Fatigue of Fibrous Composite Materials*, STP 723, American Society for Testing and Materials, 1981

53. *Composite Materials: Testing and Design*, STP 497, American Society for Testing and Materials, 1971

54. *Composite Materials: Testing and Design*, STP 546, American Society for Testing and Materials, 1974

55. *Fatigue of Composite Materials*, STP 569, American Society for Testing and Materials, 1975

56. *Composite Materials: Testing and Design*, STP 617, American Society for Testing and Materials, 1977

57. *Fatigue of Filamentary Composite Materials*, STP 636, American Society for Testing and Materials, 1977

58. *Composite Materials: Testing and Design*, STP 674, American Society for Testing and Materials, 1979

59. *Fatigue of Fibrous Composite Materials*, STP 723, American Society for Testing and Materials, 1981

Damping Properties Analysis of Composites

Robert D. Adams, University of Bristol, United Kingdom

HIGH-PERFORMANCE COMPOSITE materials that use high-strength high-stiffness fibers embedded in a suitable matrix are being increasingly applied to aircraft, space, and propulsion systems. Their use complements the already extensive use of low-performance composite materials in general industrial and shipbuilding applications. The distinguishing features of these new materials are their high specific strength and stiffness, together with a directionality of properties. The first two properties mean that significant weight reductions may be achieved. The directionality factor complicates structural design procedures but, at the same time, leads to a new degree of freedom in that the strength and stiffness of the material may be tailored to the amount and in the direction required for a specific application.

The matrices most commonly used are resin types, although metal-, rubber-, ceramic-, and carbon-based materials are also used. The resins available may be divided into thermosetting and thermoplastic systems. Of the former group, by far the most important types are the polyesters and epoxides. Polyesters are low-cost general-purpose resins that cure at room temperature, while epoxides generally require both temperature and pressure during the cure cycle. The fiber-reinforced epoxy composite is often supplied in sheet or tape form that has been preimpregnated with partially cured resin, but which retains sufficient flexibility to facilitate molding. Thermally stable polymers have been developed for continuous operation at up to 300 °C (570 °F). Thermoplastics such as nylon and polypropylene may be reinforced with fibrous material, especially glass, but one of the most exciting recent developments is polyether etherketone (PEEK). This material, in addition to polyphenylene sulfide and polyetherimide, offers greatly improved impact resistance and, in some cases, the potential of improved hot/wet temperature resistance. Because of their high strains to failure, they are the only matrices currently available that allow the new carbon fibers to use their full strain potential in the composite.

In the past, the damping capacity of conventional engineering materials has not generally provided sufficient energy dissipation to limit resonant or near-resonant amplitudes of vibration. The position has been further aggravated by the development of high-strength alloys of aluminum and titanium, which generally have lower damping than those provided by their weaker counterparts. Conventional structures have many additional sources of energy dissipation, such as bolted and riveted joints, lubricated bearings, and so on. In space applications, because of the absence of a surrounding fluid or gas, aerodynamic damping is essentially zero, thus removing an important source of energy dissipation, especially in thin-sheet structures. However, when using composite materials, it is usually necessary to use adhesively bonded joints, because bolts and rivets tend to pull out. This seriously reduces structural damping, which makes material damping far more important. This situation can be alleviated in fiber-reinforced materials by making a suitable choice of components so that the damping derives essentially from the matrix and fiber-matrix interface. It is therefore more important to understand the mechanisms of damping in composites and to appreciate their significance than is the case for metallic materials.

The main sources of internal damping in a composite material arise from microplastic or viscoelastic phenomena associated with the matrix and from relative slipping at the interface between the matrix and the reinforcement. Thus, excluding the contribution from any cracks and debonds, the internal damping of the composite will be influenced by:

- The properties and relative proportions of matrix and reinforcement in the composite (the latter is usually represented by the volume fraction of the reinforcement, V_f)
- The size of the inclusions
- The orientation of the reinforcing material to the loading axis
- The surface treatment of the reinforcement

In addition, loading and environmental factors, such as amplitude, frequency, and temperature, may affect damping.

To cover the dynamic properties of all composite materials is beyond the limited scope of this article. Only those advanced composites used in stress-bearing situations in modern engineering will be discussed. Unreinforced polymers will not be covered here, except as a component of the composite, nor will composites with randomly oriented fibers, nor those containing nonfibrous reinforcements. The damping and moduli of short-fiber composites were recently discussed by S.A. Suarez *et al.* (Ref 1). Because modulus and damping are interrelated, both are considered.

The vibration properties of concern are the damping and the dynamic modulus, which are defined in Fig. 1. When subjected to a stress cycle, all materials show a nonsingular relationship between stress and strain. The modulus is given by the mean slope of the stress-strain loop. For most materials, there is little ambiguity in this definition, because the loop is almost indistinguishable from a straight line.

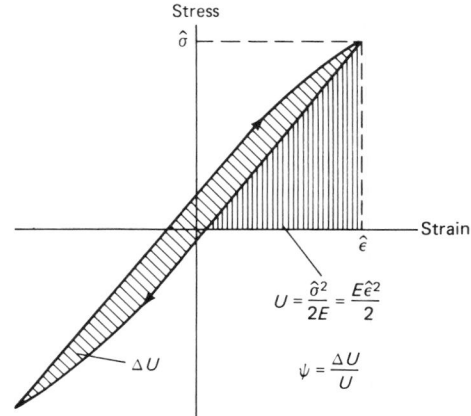

Fig. 1 Definition of specific damping capacity, Ψ. ΔU, energy dissipated per cycle, U, maximum stored energy

The area, ΔU, of the loop represents the work done against "internal friction" and is the amount of energy dissipated during the cycle.

It can be seen from Fig. 1 that the maximum strain energy stored per unit volume in the cycle is $U = \hat{\sigma}^2/2E = E\hat{\epsilon}^2/2$. The specific damping capacity, Ψ, of the material is defined as:

$$\Psi = \frac{2\pi}{Q} = 2\delta = 2\pi\eta = 4\pi C =$$

$$2\pi \left(\frac{f_2 - f_1}{f_n}\right)_{3\ dB} \qquad \text{(Eq 1)}$$

where Q is the quality (amplification) factor, δ is the logarithmic decrement, η is the loss factor, C is the proportion of critical damping, f_n is the natural frequency (Hz), and f_1 and f_2 are the half-power (3 dB) points.

Finally, it must be emphasized that just as there are many possible sources of energy dissipation in structures, most apparatuses for measuring damping need some method of applying the cyclic loads, measuring the dynamic response, and locating the specimen in some frame of reference. Many authors have not been sufficiently careful in making such measurements with metals and composites. Whereas thermoplastics often have substantial amounts of damping, structural metals and composites do not. It is therefore crucially important that any damping data used should be from an impeccable source. All things being equal, the lowest damping values quoted for a given material are likely to be the nearest to the truth. Not only should damping measurements be carefully made, but the apparatus must be purged of extraneous losses by calibrating it with some metal of very low damping properties (such as high-strength titanium or aluminum alloys).

Unidirectional Composites

The basic building block of layered composite structures is a single lamina of unidirectionally reinforced material. All the fibers are considered to be parallel and lying in the direction of the major axis of the specimen.

Longitudinal shear involves the twisting of a bar of an aligned composite. Thus, the longitudinal shear modulus of carbon fiber reinforced plastic (CFRP) or glass fiber reinforced plastic (GFRP) is principally a function of the matrix shear modulus, the fiber shear modulus, and the volume fraction of fibers. None of the existing micromechanics theories accurately fits the experimental data (Ref 2), but the numerical prediction of D.F. Adams and D.R. Doner (Ref 3) gives good agreement. From the type of curve shown in Fig. 2, it is possible to determine the fiber volume fraction, V_f, if the matrix shear modulus, G_m, is known, and vice versa.

For longitudinal shear loading, it can be shown that for viscoelastic materials (Ref 4):

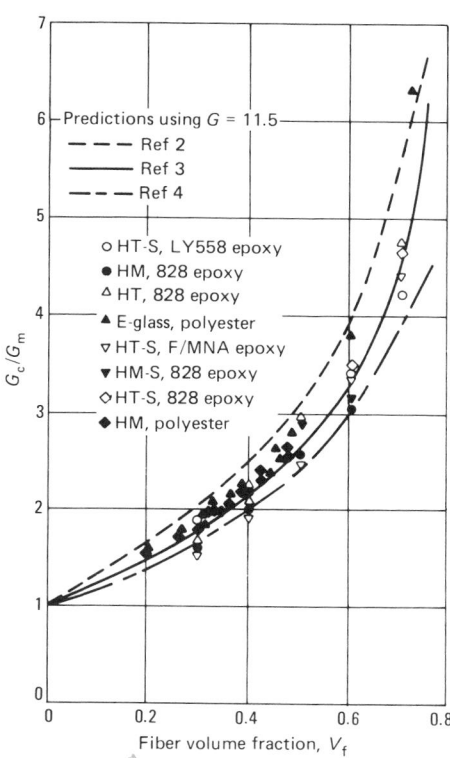

Fig. 2 Variation of reduced composite longitudinal shear modulus (G_{LT}/G_m) with fiber volume fraction (V_f)

$$\Psi_{LT} =$$

$$\frac{\Psi_m(1 - V_f)\ |(G + 1)^2 + V_f\ (G - 1)^2|}{[G(1 + V_f) + 1 - V_f]\ [G(1 - V_f) + 1 + V_f]}$$

$$\text{(Eq 2)}$$

where f and m represent fiber and matrix, V_f is the fiber volume fraction, G is the ratio of the shear modulus of the fiber to that of the matrix, and Ψ is the specific damping capacity. For many fiber-matrix combinations, G is of the order of 10, which leads to a composite specific damping capacity that is not influenced very much by the fiber volume fraction. This relationship is shown in Fig. 3 for a typical carbon fiber reinforced composite, along with an alternative solution proposed by R.D. Adams and D.G.C. Bacon (Ref 2). To allow a comparison of these predictions with the variety of experimental data, the damping values have been nondimensionalized. To explain the lower damping measured in their torsional tests, R.D. Adams and D.G.C. Bacon cited effects that were due to fiber misalignment and dilatational strains in the materials that contribute little to the damping but add significantly to the stored strain energy. Thus, if the damping capacity or shear modulus of the matrix is known, it is possible to estimate the damping and shear modulus of a composite with a given volume

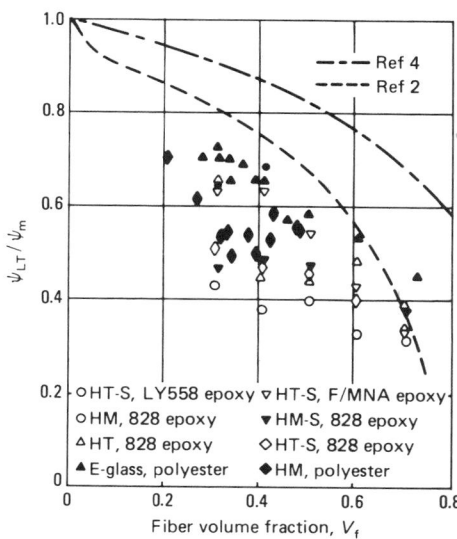

Fig. 3 Variation of the ratio of longitudinal shear damping (Ψ_{LT}) to the matrix damping (Ψ_m) with volume fraction (V_f)

fraction. Alternatively, by using curves such as those in Fig. 3, or the similar ones given by R.G. Ni and R.D. Adams (Ref 5), it is possible to estimate the properties of a composite with a given volume fraction if those at some other volume fraction are known.

Longitudinal Tension/Compression. The longitudinal Young's modulus, E_L, (the tensile modulus in the direction of the fibers in a unidirectional composite), is given by the rule of mixtures and is:

$$E_L = E_f V_f + E_m (1 - V_f)$$

where E_f and E_m are the fiber and matrix Young's moduli, and V_f is the fiber volume fraction. This rule is well obeyed experimentally, as shown in Fig. 4, and may be used as a check on any parameter, provided the others are known. This relationship was derived for normal axial loading (tension or compression), but also applies in flexure, provided that shear effects can be neglected. The scatter of the experimental results in Fig. 4 is mainly due to fiber misalignment, which is usually worse at low volume fractions. Other errors can be due to incorrect assessment of the volume fraction. It is also possible to predict the damping capacity of a unidirectional material when it is stressed in the fiber direction by using the rule of mixtures and assuming that all the energy dissipation occurs in the matrix. This gives the equation:

$$\Psi_L = \Psi_m (1 - V_f) E_m/E_L$$

where E is Young's modulus and L represents longitudinal tensile/compressive properties of the composite.

However, it is found that this expression considerably underestimates the measured value of Ψ_L, even when considerable effort has been made to eliminate extraneous losses (Fig.

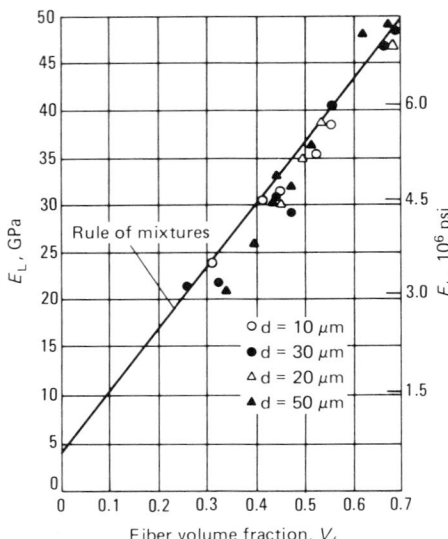

Fig. 4 Longitudinal Young's modulus (E_L) against fiber volume fraction (V_f) for different glass fibers

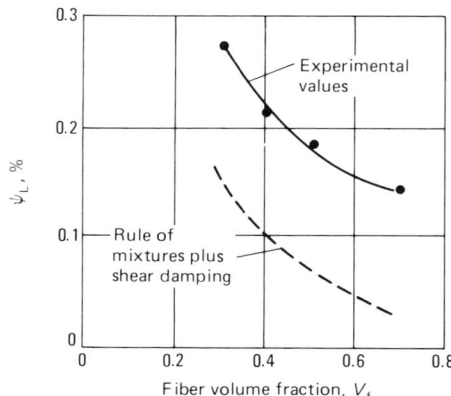

Fig. 5 Variation of flexural damping (Ψ_L) with fiber volume fraction (V_f) for HT-S carbon fiber in epoxy resin

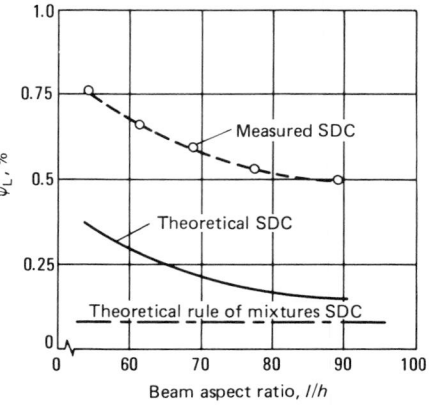

Fig. 6 Variation of flexural damping (Ψ_L) with aspect ratio l/h (l, length; h, thickness) for high-modulus carbon fiber in DX209 epoxy resin $V_f = 0.5$. SDC, shear damping contribution

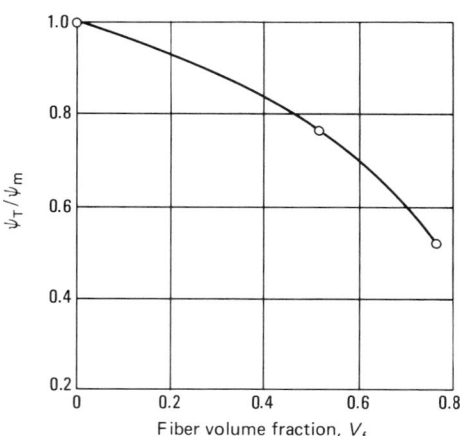

Fig. 7 Variation of ratio of transverse damping (Ψ_T) to matrix damping (Ψ_m) with fiber volume fraction (V_f). Results from GFRP specimens in flexure

5). Basically, there are several factors contributing to the discrepancy. First, the smaller the fiber diameter, the larger the surface area of fiber per unit volume. R.D. Adams and D.F. Short (Ref 6) showed that for 10-, 20-, 30-, and 50-μm (390-, 790-, 1200-, and 1950-μin.) diam glass fibers in polyester resin, there was a consistent increase in Ψ_L with reduction in fiber diameter. Second, the problem of misalignment is not insignificant, as is shown below for angle-ply composites. Third, any structural imperfections, such as cracks and debonds, lead to interfacial rubbing and, hence, to additional losses. Finally, the unidirectional lamina is very often loaded in flexure rather than uniform in-plane tension or compression. Thus, although the effect of shear is negligible in stiffness measurements, this is less true for damping because shear damping is essentially that of the matrix, and Ψ_{LT} is of the order of 50 to 100 times larger than Ψ_L. Although only a small percentage of the energy is stored in shear, it can make a substantial contribution to the total predicted value of damping. Figure 6 shows that as the aspect ratio of a beam was reduced from 90 to 50, the shear damping contribution was increased. Further, by subtracting the shear damping from the experimental values, the effect of aspect ratio is essentially eliminated. The difference remaining between the rule of mixtures prediction and the "experimental minus shear" values was mainly due to the combination of misalignment, internal flaws, and fiber diameter. It has been suggested that the discrepancy can be explained by the composite being modeled such that it considers damping of the fibers. Unfortunately, this is unlikely to be a realistic solution because the damping of the fibers is extremely small. R.D. Adams (Ref 7) has directly measured the longitudinal shear damping of a variety of

single carbon fibers and found values of the order of 0.13% specific damping capacity. This gives a loss factor of the order of 2×10^{-4}. In tension/compression, the graphite microfibrils that make up the carbon fiber structure will be preferentially stressed in their strong direction, with much less interfacial slipping than might occur in torsion. Thus, the longitudinal damping (tension/compression) ought to be at least an order of magnitude lower than that measured in torsion, giving a loss factor of approximately 2×10^{-5}. There is, therefore, no way in which damping of this level can reasonably be used to explain the discrepancies in the micromechanics models. On the other hand, it is known that aramid fibers possess quite high damping levels, even in tension/compression, and might therefore offer an exception to the above generalization.

Transverse Tension/Compression. In the transverse direction, damping is, as in shear, very heavily matrix dependent. Again, there is no reliable micromechanics theory for predicting Ψ_T. Experiments covering a wide variety of fibers, from E-glass to high-modulus carbon, showed that transverse damping is largely independent of both fiber type and surface treatment. Volume fraction, like longitudinal shear, does have a significant effect on Ψ_T. Some experimental results to illustrate this point are given in Fig. 7. Figure 8 shows the transverse Young's modulus, E_T, which, like G_{LT} (Fig. 2), is seen to increase markedly with volume fraction. The theoretical curve is based on the expression proposed by S.W. Tsai and H.T. Halpin (Ref 8), and the values are for a glass composite for which it has been assumed that the longitudinal and transverse moduli of the glass fibers are identical (that is, the fiber is isotropic). This cannot be assumed for carbon fibers, which are highly anisotropic.

General Comments on the Simple Loading of Laminae. While the various micromechanics theories, including those proposed by S. Chang and C.W. Bert (Ref 9), are sufficiently accurate for predicting moduli, they

are generally poor at predicting damping. This is because the various theories do not contain some of the important factors (such as microcracks, misalignment, and surface area) that contribute to the damping of unidirectional materials while having little effect on the moduli. Without the development of some very complex models, it is unlikely that the situation can be changed in the near future. The only safe feature is that, with the possible exception of aramid fibers, there is essentially no damping in the fibers themselves.

R.G. Ni and R.D. Adams (Ref 5) used a combination of micromechanics and experimental results to produce a series of predictive curves for the variation of the unidirectional moduli and damping values with fiber volume fraction. They also showed the importance of using the correct volume fraction for this basic data. Thus, when predictions of the damping for laminated plates are being made, it is important to know the volume fraction of both

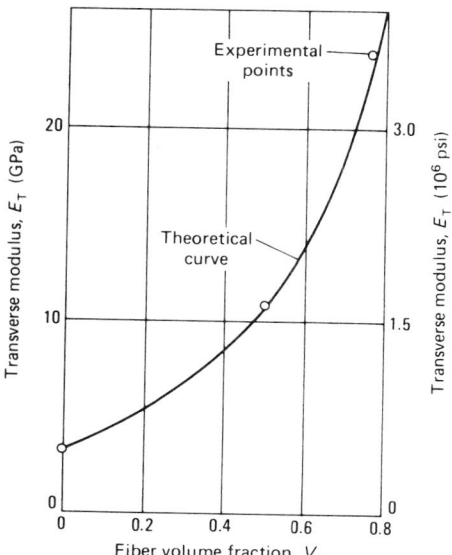

Fig. 8 Variation of the transverse modulus (E_T) with fiber volume fraction (V_f) of GFRP in flexure. $E_{fT} = 70$ GPa (10×10^6 psi). $E_m = 3.21$ GPa (0.466×10^6 psi)

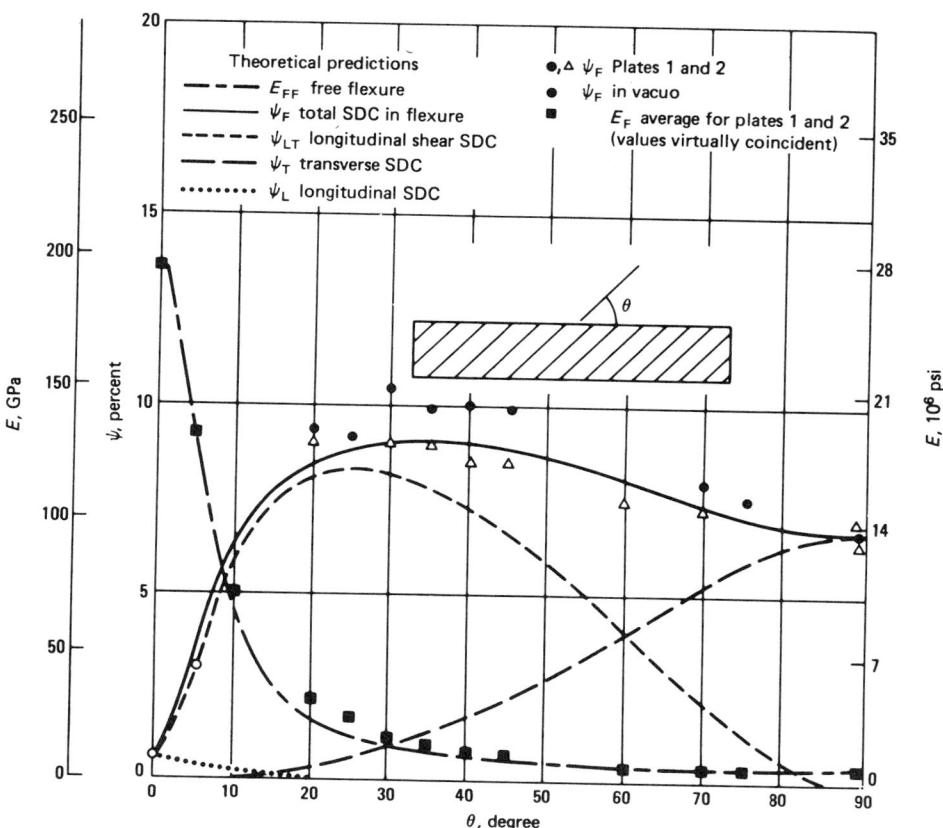

Fig. 9 Variation of flexural Young's modulus (E) and damping (Ψ) with fiber orientation (θ) for high-modulus carbon fiber in DX209 epoxy resin. $V_f = 0.5$

the plates and the unidirectional material used in making the prediction. Guides were given for converting data, and an example showed the errors that can occur if the corrected data are not used. It is suggested that the Ni and Adams approach is far more suitable in practical terms than trying to evolve increasingly complex micromechanics models. A further practical point is that it is difficult to make representative pure resin (matrix) specimens, the results of which are necessary for any micromechanics prediction. This is because the resins used for making preimpregnated fiber sheets or tapes (prepregs) contain volatiles that are difficult to remove from the bulk without creating bubbles or causing other chemical changes in the cured blocks.

Off-Axis Loading. When the specimen axis, and thus the direction of loading, is at an angle, θ, to the fiber direction in a unidirectional composite, an off-axis situation exists.

R.D. Adams and D.G.C. Bacon (Ref 10) derived closed-form expressions for the damping, Ψ_θ, of a unidirectional beam with fibers at an angle to that of the specimen axis. Figure 9 shows the theoretical and experimental values of Ψ_θ for a CFRP beam, together with the separate theoretical contributions from stresses in the L, T, and LT directions. Figure 9 also shows the separate contributions from direct stresses in the direction of the fibers, Ψ_L, transverse to the fibers, Ψ_T, and in shear, Ψ_{LT}. The theoretical prediction and experimental measurements of the variation of Young's modulus, E_θ, with angle, are also shown in Fig. 9. Excellent agreement between theory and experiment is shown for both modulus and damping.

R.D. Adams and D.G.C. Bacon showed that, for a carbon composite in which $E_L \geqslant$

G_{LT}, $E_L \geqslant E_{LT}$, $\Psi_L \ll \Psi_T$, and $\Psi_L \ll \Psi_{LT}$, then to a very good approximation over the range $5° < \theta < 90°$:

$$\Psi_\theta = \frac{1}{s_{11}} \left[\frac{\Psi_T}{E_T} \sin^4\theta + \frac{\Psi_{LT}}{G_{LT}} \sin^2\theta \cos^2\theta \right]$$

where S_{11} is the compliance in the direction of the specimen axis.

Beams Cut From Laminated Plates

In practice, structures made from composites contain a series of layers of unidirectional fibers such that each layer has some predetermined orientation with respect to the defined dimensions of the structure (Fig. 10). The orientations and transverse dispositions of the fibers depend on the loads to be carried (strength) and the deflections that can be tolerated (stiffness). For any arrangement of layers, it is now possible to predict not only structural strength and stiffness, but also 'inherent' damping. Laminated plate theory is used to evaluate the contributions to damping made by each layer. Beams are a special case of plates, but are often treated separately because the theory of vibrating beams is much easier than that of plates. In an

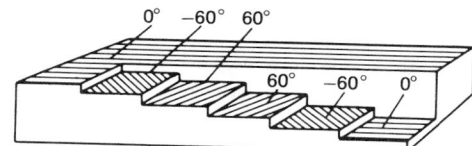

Fig. 10 Lamina stacking arrangement for (0°, −60°, 60°), laminate. The suffix s indicates symmetry of stacking about the midplane.

article of this length, the theory can only be outlined. A fuller treatment is given by R.G. Ni and R.D. Adams (Ref 11).

The constitutive equation relating stresses, σ, and strains, ϵ, in the kth lamina is (using standard notation for composites):

$$\begin{bmatrix} \sigma_1 \\ \sigma_2 \\ \sigma_6 \end{bmatrix}_k = \begin{bmatrix} Q_{11} & Q_{12} & Q_{16} \\ Q_{12} & Q_{22} & Q_{26} \\ Q_{16} & Q_{26} & Q_{66} \end{bmatrix}_k \begin{bmatrix} \epsilon_1 \\ \epsilon_2 \\ \epsilon_6 \end{bmatrix}_k$$

where the values Q_{ij}^k are the stiffness matrix components in the specimen system of axes 1, 2, 3 of the kth lamina, and are obtained from the values in the axes related to the fiber direction x, y, z by using the appropriate geometric transformation. For a beam specimen, the stresses σ_2 and σ_6 (transverse and interlaminar shear) can generally be neglected in comparison with σ_1 although M.M. Wallace and C.W. Bert cite cases where this may not always be so (Ref 12).

With the appropriate geometric transformation, these stresses can be converted from the specimen axes to the fiber directions. It is then possible to calculate the stresses in the fiber direction σ_x (that is, σ_L), normal to it σ_y (that is, σ_T), and the shear components σ_{xy} (that is, σ_{LT}). The total energy stored in the x (or L) direction, Z_L, for example, can then be calculated, and the energy dissipation in this layer and in this direction can then be given by:

$$\Delta Z_L = \Psi_L \cdot Z_L$$

For the beam, the overall specific damping capacity Ψ_{ov} is then given by the total energy dissipated divided by the total energy stored:

$$\Psi_{ov} = \frac{\Sigma \Delta Z}{\Sigma Z} = \frac{\Psi_L Z_L + \Psi_T Z_T + \Psi_{LT} Z_{LT}}{Z_L + Z_T + Z_{LT}} \quad \text{(Eq 3)}$$

If the elastic moduli and damping coefficients are known for the unidirectional material, it is possible to calculate the overall damping of a beam. R.G. Ni and R.D. Adams (Ref 11) gave the theory for generally laminated beams and obtained excellent agreement with the measured results.

Whereas specimens with all the layers at θ will twist as they are bent, the twisting can be restrained internally by using several layers at $\pm\theta$. The damping contributions can again be assessed, and the measured values accounted for (Ref 10, 11). Figure 11 shows theoretical predictions and experimental measurements for the modulus and damping of a series of CFRP beams made with ten layers of high-modulus carbon fibers in epoxy resin, alternately at $\pm\theta$. Note that the modulus is higher than that of the off-axis specimens because of the internal restraint, while the damping is generally lower.

More generally laminated composites, as shown in Fig. 10, are commonly used in practice. Fortunately, the same method as that just described can be used to predict damping. Figure 12 shows the excellent agreement between theory and experiment for the variation of damping (and stiffness) with θ of a symmetrical, high-modulus, graphite fiber reinforced epoxy plate. Beam specimens were cut at angles from $-90°$ to $+90°$ relative to the fiber direction in the outer layer of this $(0°, -60°, +60°)_s$ plate.

Laminated Plates

Fiber-reinforced plates of various shapes and different boundary conditions (free, clamped, and hinged) commonly occur in practice. Designers need to be able to predict the stiffness parameters and damping values of such plates for conditions such as aeroelasticity, acoustic fatigue, and so on. Much attention has been devoted to stiffness predictions, but very little to damping. The objective here is to develop a suitable mathematical model that can be used to

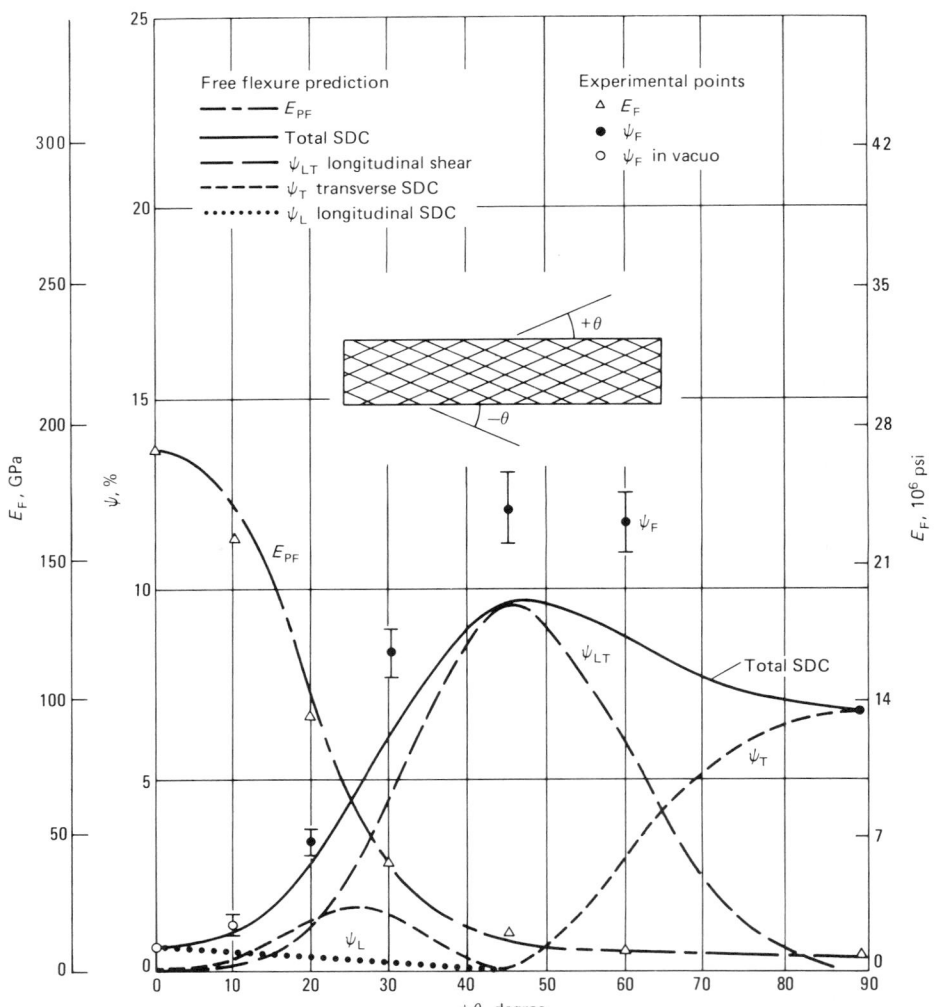

Fig. 11 Variation of flexural Young's modulus (E_F) and damping (Ψ) with ply angle $\pm\theta$ for high-modulus carbon fiber in DX209 epoxy resin. $V_f = 0.5$

predict the damping values of plates laminated from fibers of various types at various orientations. Such is the mathematical complexity of the equation of motion of plates (even those made from isotropic materials) that closed-form solutions exist only for special cases, such as hinged (simply supported) rectangular plates and circular plates (involving Bessel functions). The solution is therefore best obtained using finite-element techniques, which can readily accommodate different shapes, thicknesses, and boundary conditions. Some examples are given by P. Cawley and R.D. Adams (Ref 13).

All the plates discussed here are midplane symmetric, which eliminates bending-stretching coupling. It is, however, possible to include this effect in the analysis if asymmetrical laminates are used.

The first ten modes of vibration of a typical plate can be adequately described by using a coarse finite-element mesh with six elements per side ($6 \times 6 = 36$ elements for a rectangular plate). The essence of the technique is first to determine the values of strain energy stored because of the stresses relative to the fiber axes of each layer of each element. Use of modulus parameters determined from unidirectional bars makes it possible to determine the total energy stored in each layer of each element. These are then summed through the thickness to give the energy stored in each element (related to the strains and the mean elasticity matrix for the element). It is then possible to use standard finite-element programs and avoid the mathematical complication of working in terms of the standard plate equations. This approach provides the stiffness of the plate, the maximum strain energy, U, stored in any given mode of vibration, the natural frequencies, and the mode shape. The energy dissipated in an element of unit width and length situated in the kth layer can also now be determined. This is done by transforming the stresses and strains to the fiber directions and using the damping properties of

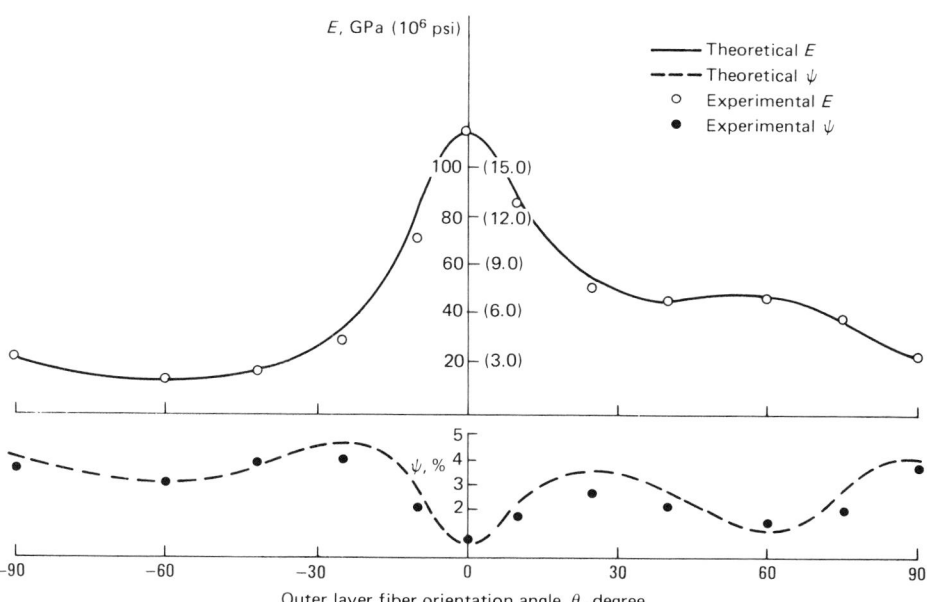

Fig. 12 Variation of flexural modulus (E) and damping (Ψ) with outer layer fiber orientation angle (θ) for $(0°, -60°, 60°)_s$ made from high-modulus carbon fibers in DX210 epoxy resin

Number	Frequency, Hz	Mode shape	SDC, %
1	58.10 (68.88)	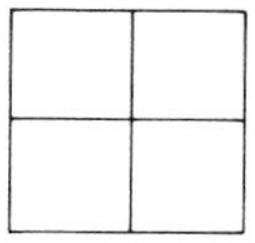	7.80 (6.65)
2	213.31 (218.9)	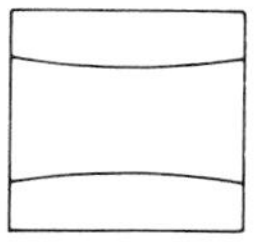	0.91 (1.05)
3	243.47 (251.2)	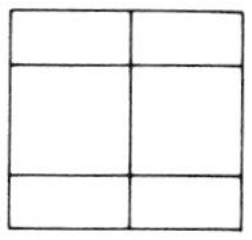	2.50 (2.6)
4	302.51 (305.4)	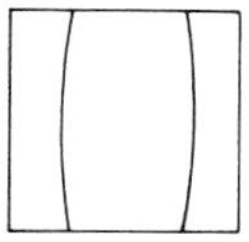	0.60 (0.92)
5	324.16 (323.5)	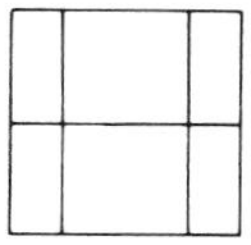	1.51 (1.7)
6	441.62 (452.5)	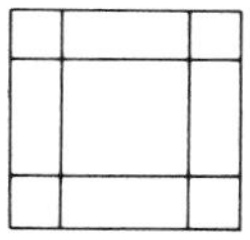	2.74 (3.0)

Outer layer fiber direction $\longrightarrow$

Fig. 13 Natural frequency and damping of various modes of an eight-layer $(0°, 90°, 0°, 90°)_s$ CFRP plate (plate 1). Experimental values in parentheses

$0°$ bars. The energy dissipated in the element in the kth layer is integrated over the whole area of the plate, and the contributions of each layer are summed to give ΔU, the total energy dissipated in the plate. The overall specific damping capacity, Ψ_{ov}, is then given by $\Psi_{ov} = \Delta U/U$. Alternatively, the damping can first be summed through the thickness of the damped element to give a damped element stiffness matrix. This can then be treated by standard finite element techniques (Ref 14).

It is useful to express in mathematical terms the technique described above. The maximum strain energy, U, is obtained as for an undamped system as follows:

$$U = \frac{1}{2} \int_V |\epsilon_{ij}|^T |\sigma_{ij}| \, dV \qquad \text{(Eq 4)}$$

where ϵ_{ij} and σ_{ij} are the strains and stresses related to the fiber direction, and V refers to the volume.

This equation may be reduced to a standard form as:

$$U = \frac{1}{2} |\delta|^T [K] |\delta| \qquad \text{(Eq 5)}$$

where $|\delta|$ is the nodal point displacement matrix. Here, five degrees of freedom for each nodal point and eight nodal points for each element are used, and $[K]$ is the stiffness matrix. In the evaluation of the maximum strain energy, U, the Young's modulus of $0°$ and $90°$ unidirectional fiber-reinforced beams, E_L, E_T, and the shear modulus of a $0°$ unidirectional rod, G_{LT}, are used. Now:

$$\Delta U = \int_V \delta(\Delta U) \, dV \qquad \text{(Eq 6)}$$

where $\delta(\Delta U)$ is the energy dissipated in each element, and is defined as:

$$\delta(\Delta U) = \delta(\Delta U_1) + \delta(\Delta U_2) + \delta(\Delta U_{23}) + \\ \delta(\Delta U_{13}) + \delta(\Delta U_{12})$$

$$\delta(\Delta U_1) = \frac{1}{2}\Psi_L \, \epsilon_{11} \, \sigma_{11}, \; \delta(\Delta U_2) = \frac{1}{2}\Psi_T \, \epsilon_{22} \, \sigma_{22}$$

and

$$\delta(\Delta U_{23}) = \frac{1}{2}\Psi_{TT}\, \epsilon_{23} \, \sigma_{23} \; \delta(\Delta U_{13}) = \frac{1}{2}\Psi_{LT} \, \epsilon_{13} \, \sigma_{13}$$

$$\delta(\Delta U_{12}) = \frac{1}{2}\Psi_{LT} \, \epsilon_{12} \, \sigma_{12}$$

where subscript 1 denotes the fiber direction, while 2 and 3 denote the two directions transverse to the direction of the fibers, and Ψ_L, Ψ_T and Ψ_{LT} are the associated damping capacities that are also obtained from tests on unidirectional beams.

We may now reduce Eq 7 to matrix form as:

$$\Delta U = \frac{1}{2} \int_V |\epsilon_{ij}|^T [\Psi] |\sigma_{ij}| \, dV \qquad \text{(Eq 7)}$$

where:

$$[\Psi] = \begin{bmatrix} \Psi_L & 0 & 0 & 0 & 0 \\ 0 & \Psi_T & 0 & 0 & 0 \\ 0 & 0 & \Psi_{TT} & 0 & 0 \\ 0 & 0 & 0 & \Psi_{LT} & 0 \\ 0 & 0 & 0 & 0 & \Psi_{LT} \end{bmatrix}$$

Using the same method as with Eq 4, Eq 7 may be reduced to:

$$\Delta U = \frac{1}{2} |\delta|^T [K_d] |\delta| \qquad \text{(Eq 8)}$$

where $|\delta|$ is the same matrix as in Eq 4 and was obtained from the finite element results. The stiffness matrix of the damped system is $[K_d]$, and it may be evaluated separately. D.X. Lin, R.G. Ni, and R.D. Adams described this method in much more detail (Ref 14).

Some results are given for theoretical predictions and experimental measurements on sev-

eral plates made from glass or high-modulus carbon fibers in DX210 epoxy resin. The plates were made of 8 or 12 layers of preimpregnated fiber to give different laminate orientations; details of the plates used are given in Table 1. The material properties used in the theoretical prediction are given in Table 2. All the values in this table were established either by using beam specimens cut from a unidirectional plate (longitudinal and transverse damping and Young's moduli) or cylindrical specimens (for measuring the shear modulus and damping in torsion). It should be noted that the value of the torsional damping of a bar with fibers at $90°$ to the axis, Ψ_{23}, is not important in the prediction, because changing it from 6% to 15% gave no significant difference to the overall theoretical results. In the prediction, Ψ_{23} is taken as the same value as Ψ_{12}, which is the value of torsional damping of a unidirectional rod in longitudinal shear. Because of variations in the fiber volume fraction of the plates, the material properties used in the theoretical prediction were each corrected from a standard set given

Table 1   Plate data

Plate number	Material	No. of layers	Density g/cm³	V_f	Ply orientation
1	CFRP(a)	8	1.446	0.342	(0°, 90°, 0°, 90°)$_s$(b)
2	CFRP	12	1.636	0.618	(0°, −60°, 60°, 0°, −60°, 60°)$_s$
3	GFRP	8	1.813	0.451	(0°, 90°, 0°, 90°)$_s$
4	GFRP	12	2.003	0.592	(0°, −60°, 60°, 0°, −60°, 60°)$_s$

(a) Using high-modulus carbon fiber. (b) *s* represents midplane symmetric.

Table 2   Moduli and damping values for materials used in the plates

Material	E_L GPa	E_L 10⁶psi	E_T GPa	E_T 10⁶psi	G_{LT} GPa	G_{LT} 10⁶psi	ψ_L, %	ψ_T, %	ψ_{LT}, %	ν_1, ν_2	V_f
HMS-DX210(a)	172.1	30.0	7.20	1.04	3.76	0.55	0.45	4.22	7.05	0.3	0.50
Glass-DX210	37.87	5.49	10.90	1.58	4.91	0.71	0.87	5.05	6.91	0.3	0.50
DX210-BF₃400	3.21	0.47	3.21	0.47	1.20	0.17	6.54	6.54	6.68	0.34	0

(a) HMS, high-modulus carbon fiber

Number	Frequency, Hz	Mode shape	SDC, %
1	165.17 (156.6)	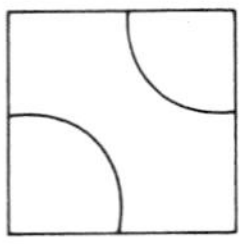	1.44 (1.40)
2	279.14 (272.0)	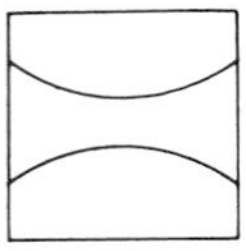	0.93 (0.88)
3	387.8 (372.3)	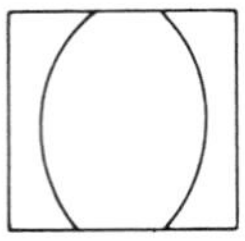	0.63 (0.65)
4	432.57 (407.8)	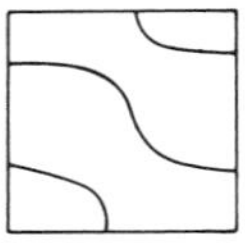	1.23 (1.26)
5	511.43 (486.1)	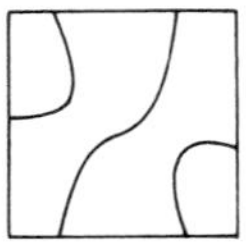	0.98 (0.99)
6	800.37 (779)	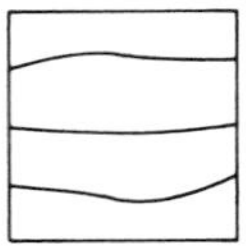	0.92 —

Outer layer fiber direction ⟶

Fig. 14 Natural frequency and damping of various modes of a 12-layer (0°, −60°, 60°, 0°, −60°, 60°)$_s$ CFRP plate (plate 2). Experimental values in parentheses

Number	Frequency, Hz	Mode shape	SDC, %
1	66.42 (62.2)	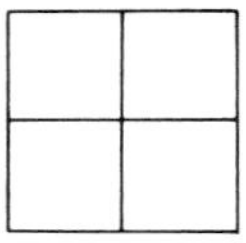	7.16 (6.7)
2	131.62 (131.4)	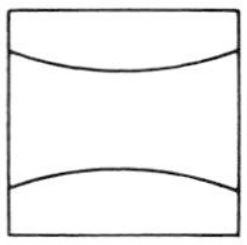	2.47 (2.8)
3	164.46 (159.2)	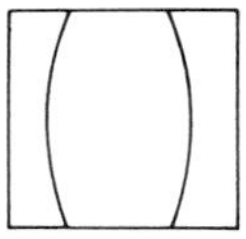	1.62 (1.9)
4	189.79 (180.5)	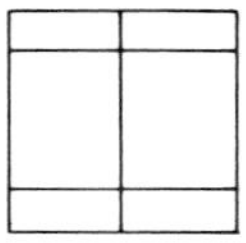	4.87 (4.9)
5	208.87 (200.05)	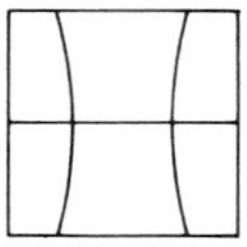	3.73 (3.2)
6	347.16 (326.7)	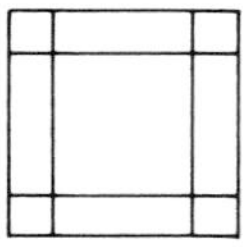	5.09 (5.8)

Outer layer fiber direction ⟶

Fig. 15 Natural frequency and damping of various modes of an eight-layer (0°, 90°, 0°, 90°)$_s$ GFRP plate (plate 3). Experimental values in parentheses

Number	Frequency, Hz	Mode shape	SDC, %
1	108.17 (90.4)	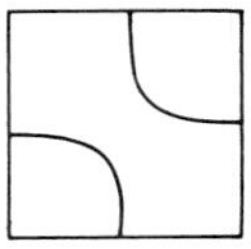	3.74 (4.4)
2	168.64 (144.7)	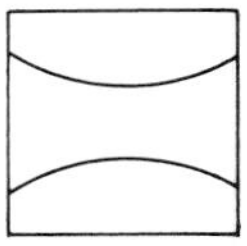	2.81 (3.5)
3	218.64 (222.3)	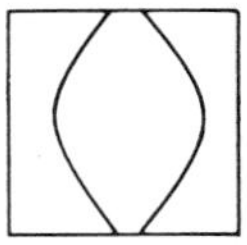	1.90 (2.6)
4	280.15 (264.1)	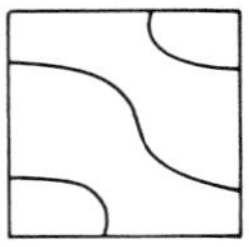	3.40 (3.4)
5	301.00 (281.1)	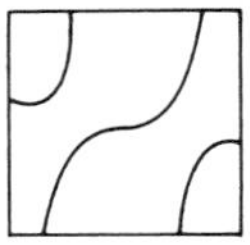	2.84 (3.0)
6	505.15 (492.6)	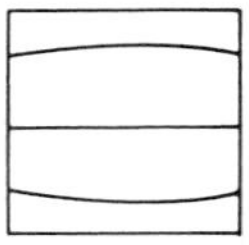	2.76 (2.7)

Outer layer fiber direction ⟶

Fig. 16 Natural frequency and damping of various modes of a 12-layer (0°, −60°, +60°, 0°, −60°, +60°)$_s$ GFRP plate (plate 4). Experimental values in parentheses

diction and experimental results of CFRP plates for various fiber orientations. On the whole, there is good agreement between the predicted and measured values. Mode 6 in plate 3 could not be obtained experimentally because the input energy from the transient technique used for measuring the frequency and damping (Ref 15) was insufficient. Figures 15 and 16 give the results for GFRP plates in free-free vibration. All show good agreement between prediction and measurement.

The effect of air damping and the additional energy dissipation associated with the supports affect the results of the very low damping modes, such as mode 4 of plate 1, mode 4 of plate 3, and so on. These are essentially beam modes in which the large majority of the strain energy is stored in tension/compression in the fibers and not in matrix tension or shear. However, the results for all the plates used are satisfactory, even when the specimens have imperfections, such as slight variations in thickness and the nominal angle of the fibers (±2° to ±3° error). It can be said that the more twisting

for 50 vol%, using the method of R.G. Ni and R.D. Adams (Ref 11). The plates were vibrated in the free-free condition (with all the edges freely supported). Although hinged or clamped

edges can be readily incorporated into the finite element model, they are not easy to reproduce in an experiment. Figures 13 and 14 show, for the first six free-free modes, the theoretical pre-

there is, the higher the damping. For instance, for an eight-layer cross-ply (0°/90°) GFRP plate (Fig. 15), the two beam-type modes, that is, modes 2 and 3, appear to be similar, but the relationship of the nodal lines to the outer fiber direction means that the higher mode has much less damping than the lower one. The other modes of vibration of this plate all involve much more plate twisting, and hence matrix shear, than do modes 2 and 3, and so the damping is higher.

Design Considerations for Plates. It is important for designers to realize the significance of these results, which show that for all the plates, the damping values are different for each mode. For instance, for the all-0° square GFRP plate in Fig. 17, the damping of the first mode was over 14 times that of the sixth mode. Also, it should be noted that some modes may have much less damping than others, especially when most of the fibers are in one direction. If such a low-damping mode has its natural frequency close to or in the frequency range of any excitation, it may well lead to excessive motion and cause fatigue, noise radiation, component malfunction, and so on.

D.X. Lin, R.G. Ni, and R.D. Adams (Ref 14) showed how the prediction of natural frequencies and damping values could be simplified for geometrically similar structures by using a series of design charts. First, the damping is a function only of the mode shape. Thus, for a series of square plates with a given fiber orientation, the damping value of each mode is constant, irrespective of the dimensions of the plate. It should be noted that the damping of epoxy and similar high-performance resins is not frequency dependent in the way that is normally associated with thermoplastics, except near the glass transition temperature, which would normally be outside the design range, even though aerospace structures have a wide temperature range of operation. Furthermore, the natural frequencies can all be scaled from a series of simple numbers, or read off a design chart. Figure 17 shows such a chart for an all-0° GFRP plate that has a volume fraction of 0.50. For a given mode, the natural frequency of the ith mode, f_i, is given by $f_i = k_i h/a^2$, where h is the plate thickness and a is its side length. Thus, it is possible, by using charts such as those in Fig. 17, to determine quickly and accurately the natural frequency and damping of any of the first six modes of a square plate with that particular fiber arrangement.

Because the volume fraction can also change, it is necessary to construct a further series of graphs to allow for this. Figure 18 shows the variation of natural frequency and damping of an all-0° GFRP plate with volume fraction. Again, the damping will not change with plate dimensions (h and a), although it does decrease as the volume fraction increases. Figure 18 is based on a plate for which $h/a^2 = 0.032$ m^{-1}. Now, because:

$$f_i = k_i h/a^2$$

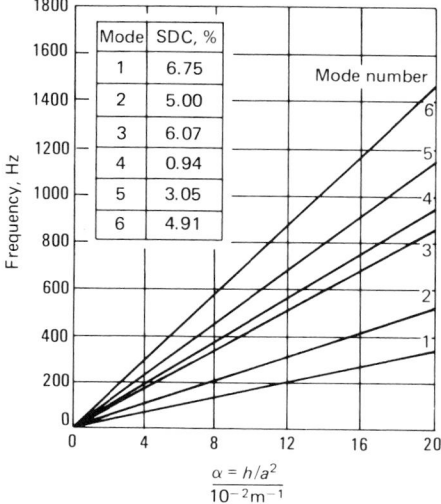

Mode	SDC, %
1	6.75
2	5.00
3	6.07
4	0.94
5	3.05
6	4.91

Fig. 17 Variation of natural frequencies and damping with the ratio of thickness to the area of a square plate (all 0°) GFRP. $V_f = 0.50$

k_i must be some function of *inter alia*, the volume fraction. Thus:

$$\frac{f_i}{f_i{}'} = \frac{h}{h'}\left(\frac{a'}{a}\right)^2$$

and by cross correlating from charts such as those given in Fig. 17 and 18, it is possible to predict the damping and frequency of a given mode as h, a, and V_f vary (Ref 14).

Woven Fibrous Composites

Woven fiber reinforced plastics are becoming increasingly important as they have the following advantages over laminates made from individual layers of unidirectional material:

- Improved formability and drape
- Bidirectional reinforcement in a single layer
- Improved impact resistance
- Balanced properties in the fabric plane

The woven composite is formed by interlacing two sets of threads, the warp and the weft, in a wide variety of weaves and balances.

Figure 19 shows the damping results obtained by cutting a series of beams at various angles, θ, from a 16-ply plate of CFRP. The fibers were woven to a balanced five-harness satin weave pattern in which one weft thread was interwoven with every fifth warp thread. This weave is the most widely used in laminates, as it gives higher mechanical properties than do plain and twill weaves, because of reduced crimping. The damping increases from about 1% at θ = 0° and 90° to a maximum of about 6% at θ = 45°. R.G. Ni and R.D. Adams' prediction (Ref 11) for a 0°/90° cross-ply made from unidirectional laminae is also shown on Fig. 19, and is in reasonable agree-

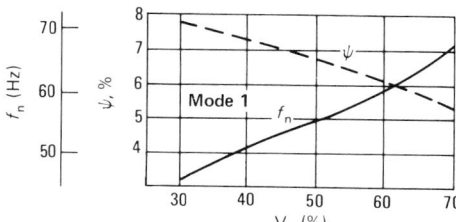
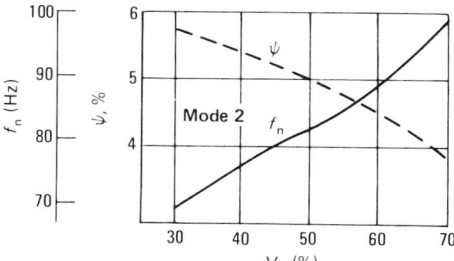
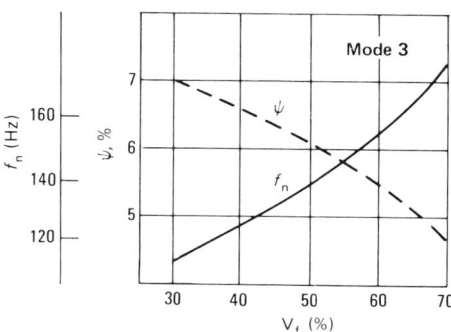

Fig. 18 Variation of natural frequencies (f_n) and damping (Ψ) of a GFRP (all 0°) plate with fiber volume fraction (V_f). $\alpha = h/a^2 = 0.32$ m^{-1}

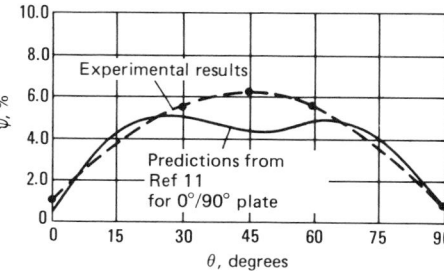

Fig. 19 Variation of specific damping capacity (Ψ) for a series of beams cut at various angles (θ) from a woven 16-ply CFRP plate

ment with the experimental data for the woven material. The damping of the woven material is modified because the twisting of the specimens is restrained internally by the perpendicular arrangement of the warp and weft threads.

While further work is necessary to characterize the various weaves, it appears that the results will, at first approximation, be similar to those for laminates made from unidirectional laminae with the same in-plane fiber orientations.

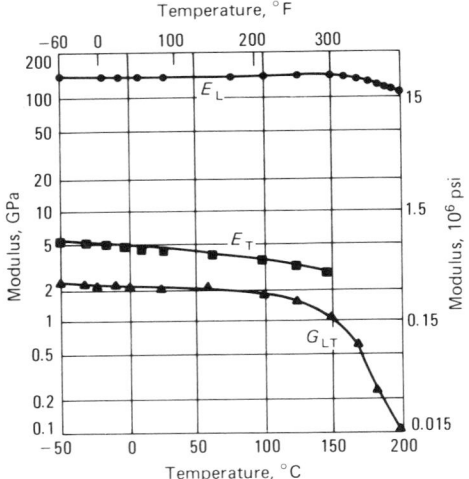

Fig. 20 Variation of longitudinal modulus (E_L), transverse modulus (E_T), and longitudinal shear modulus (G_{LT}) with temperature for high-modulus carbon fibers in DX209 epoxy resin. $V_f = 0.5$

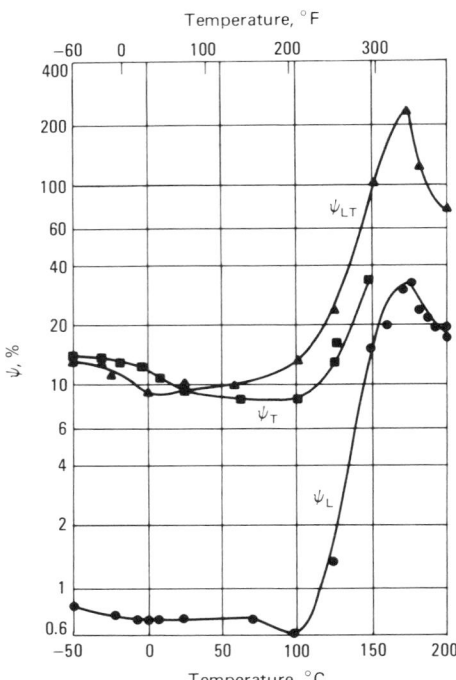

Fig. 21 Variation of longitudinal damping (Ψ_L), transverse damping (Ψ_T), and longitudinal shear damping (Ψ_{LT}) with temperature for high-modulus carbon fibers in DX209 epoxy resin. $V_f = 0.5$

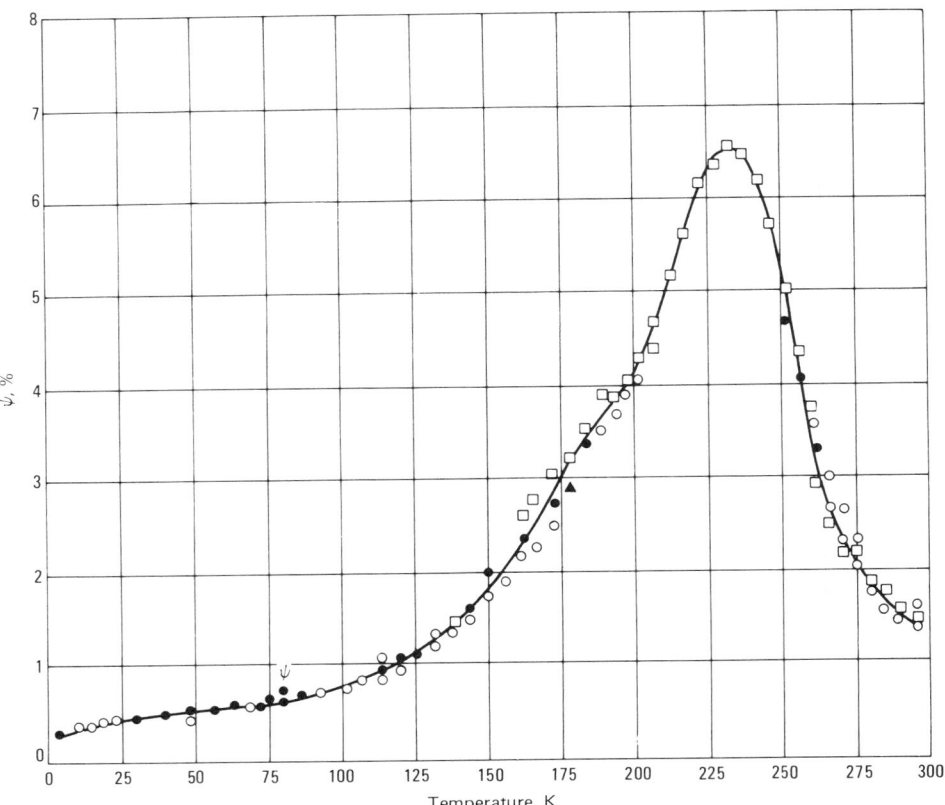

Fig. 22 Variation of specific damping capacity (Ψ) with temperature for a glass cloth-epoxy specimen

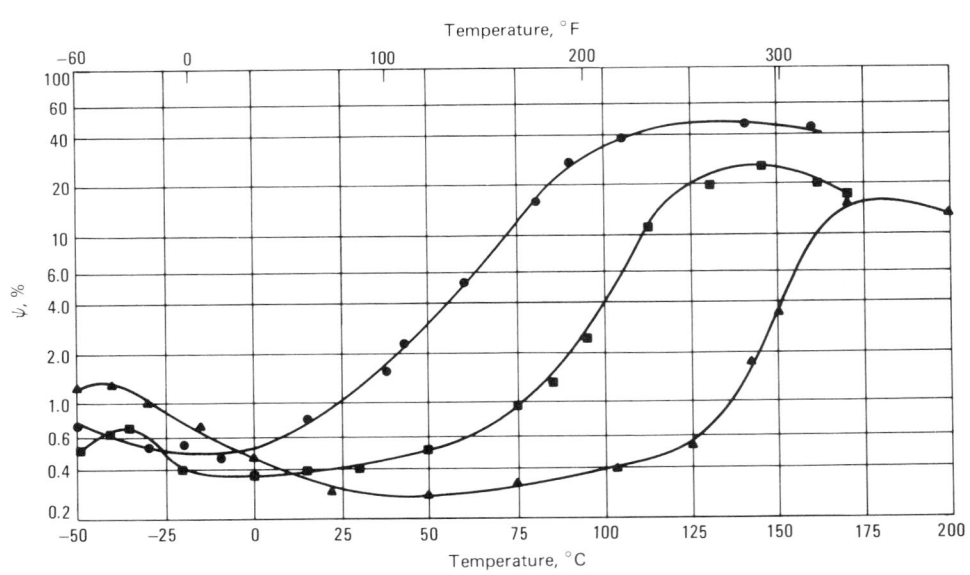

Fig. 23 Variation of specific damping capacity (Ψ) with temperature for 0° unidirectional composite made from Epikote flexibilized resin. $V_f = 0.5$

Sandwich Laminates

To maximize the stiffness of GFRP laminates, it is common to add thin skins of CFRP. R.G. Ni, D.X. Lin, and R.D. Adams (Ref 16) made mathematical predictions of the dynamic properties of such hybrid laminates. They obtained excellent agreement between their experimental results and their theoretical predictions for the damping and moduli of plates, and of beams cut from these plates. These authors also showed how to maximize both the stiffness of a laminate from the ratio of the amounts of glass and carbon, and their relative costs.

The theoretical analysis showed that the effect of the core material on the flexural modulus and damping of this type of hybrid is generally not great. This allows some freedom in choos-

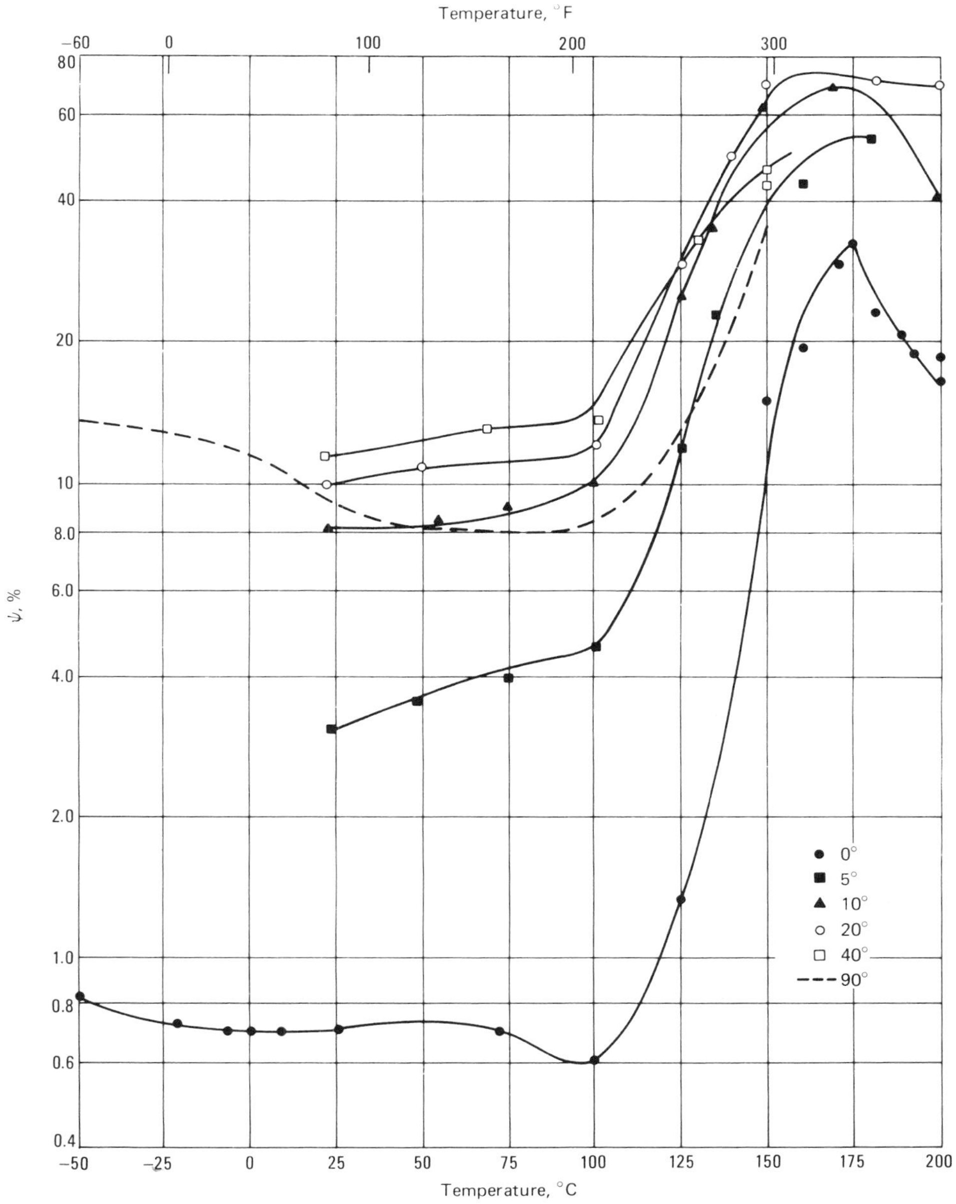

Temperature, °F

ψ, %

Temperature, °C

Fig. 24 The effect of fiber orientation on the variation of specific damping capacity (Ψ) with temperature for high-modulus carbon fibers in DX209 epoxy resin. $V_f = 0.5$

Legend (Fig. 24):
- ● 0°
- ■ 5°
- ▲ 10°
- ○ 20°
- □ 40°
- - - - 90°

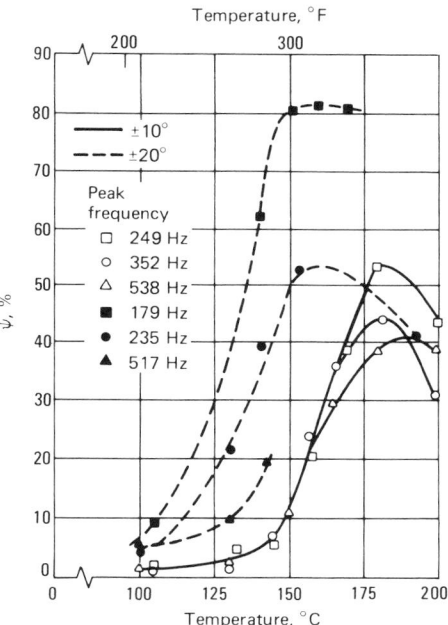

Fig. 25 Variation of specific damping capacity (Ψ) with temperature for $\pm 10°$ and $\pm 20°$ angle-plies made from high-modulus carbon fibers in DX209 epoxy resin. $V_f = 0.5$

Legend (Fig. 25):
- —— $\pm 10°$
- - - - $\pm 20°$

Peak frequency
- □ 249 Hz
- ○ 352 Hz
- △ 538 Hz
- ■ 179 Hz
- ● 235 Hz
- ▲ 517 Hz

ing the orientation of the GFRP core, and even in the selection of core materials.

Effect of Temperature

Many composites are based on polymeric matrices, usually epoxy resins, for which there are temperature-dependent damping capacities and moduli. While the influence of frequency on damping and modulus is not as strong for epoxies as it is for the high-damping polymers, as used in constrained-layer and similar damping treatments, it is not negligible. Figure 20 shows the change of E_L, E_T, and G_{LT} of a unidirectional CFRP composite over the range of -50 to $+200$ °C (-60 to $+390$ °F). (The

matrix material was DX209 epoxy resin cured for 2 h at 180 °C (355 °F) and postcured for 6 h at 150 °C (300 °F).) A logarithmic scale was used, and it can be seen that the two matrix-dependent moduli, E_T and G_{LT}, were significantly reduced at temperatures above 150 °C (300 °F). Indeed, the transverse specimen (90° orientation) could not be tested at temperatures above 150 °C, as it sagged under its own weight. In contrast, the 0° modulus was essentially unaffected until the matrix became shear soft, at which point the deformation became more by shear than by bending and fiber deformation. Figure 21 shows damping on a logarithmic scale and the much higher damping levels that are available in shear and transverse

loading than in longitudinal tension/compression. The Ψ_L damping is due not only to increased matrix damping, according to the rule of mixtures, but also to the enhanced shear deformation referred to above. The damping peak, at about 180 °C (360 °F), represents classical viscoelastic behavior. Testing beyond 200 °C (390 °F) was impossible because of charring.

At lower temperatures, the β relaxation phenomenon comes into effect. This is illustrated in Fig. 22 for a cryogenic grade, woven glass fiber reinforced epoxy material.

To achieve a wide range of resin properties, a standard resin was modified by the addition of a flexibilizer. By varying the proportions of resin to flexibilizer, precondensates with different glass transition temperatures were formed. Shell Epikote 828 was used as the standard resin and flexibilized by the addition of Epikote 871 in the proportions 1:1 and 2:1 (828/871) by weight to give FO (pure 828), F50 (50% flexibilizer), and F33 (33% flexibilizer). The resin was made into prepreg with type II (high-tensile strength) carbon fiber and laminates prepared from it. Figure 23 shows that increasing the flexibilizer content increases the damping and decreases the glass transition temperature.

Figure 24 shows the combined effect of fiber orientation and temperature on the flexural damping of a series of beams cut at various angles from a unidirectional plate (type II carbon fibers in DX209 epoxy resin). The damping results of specimens with angles of 0° to 40° are presented in Fig. 24; specimens from

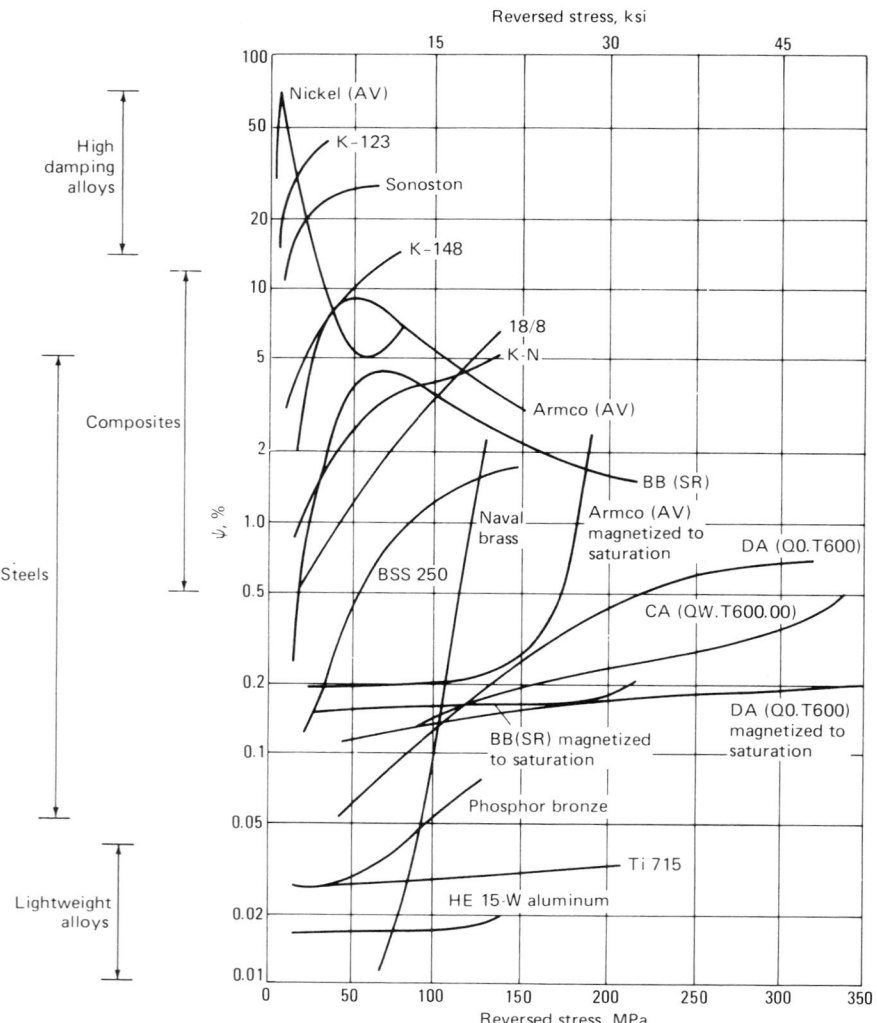

Fig. 26 Specific damping capacity (Ψ) versus stress for a range of ferrous and nonferrous metals. Nickel (AV), annealed in vacuum. K-123, K-148, K-N, grades of cast iron. 18/8, stainless steel. Armco (AV), low-carbon iron, annealed in vacuum. BB(SR), 0.12% carbon steel, stress relieved. BSS 250, Naval brass, and phosphor bronze are copper-based alloys. CA, DA, high-carbon steels. HE 15-W, Duralumin aluminum alloy. Ti 715, titanium alloy

where the fiber angle is near 0°, there is only a small reduction at high temperatures, but at larger fiber angles (20 to 90°), the modulus can decrease by more than an order of magnitude.

Relationship Between Damping and Strength

If improving damping properties of a laminate at no detriment to its mechanical properties is an objective, it is interesting to examine the differences between ±15° angle-plies and 0°/90° cross-plies. In flexure, the strengths and moduli are almost identical for the range of fibers, whereas the damping of the angle-plies is double that of the cross-plies. In torsion, the shear moduli of angle-plies are largely dependent on the fiber modulus and are much larger than the cross-ply shear moduli, the values of which are nominally independent of fiber modulus. The damping of angle-plies is, in this latter case, less than half that of the cross-plies.

Thus, different lamination geometries and fiber moduli can be arranged to give some common properties between laminates while having very different properties in other modes. In design, this gives greater flexibility to cater for strength and stiffness in one direction, with optional properties in others, according to requirements, than can possibly be achieved with isotropic materials. The damping properties of laminates can now be added to these design parameters (Ref 17).

Composites Versus Metals

To put the damping of composites in context, a comparison should be made with the damping of metals. Figure 26 shows the variation of damping with cyclic stress amplitude for a range of common structural metals. The metallic specimens were tested in axial vibration (tension-compression) using the apparatus described by R.D. Adams and A.L. Percival (Ref 18); more details of the results are given in Ref 19 and 20. Composites provide slightly higher damping than steels, but significantly less than conventional high-damping alloys. On the other hand, low-weight high-strength alloys such as aluminum and titanium give extremely low damping; values of less than 0.01% SDC have been reported (Ref 21).

REFERENCES

1. S.A. Suarez, R.F. Gibson, C.T. Sun, and S.K. Chaturvedi, The Influence of Fiber Length and Fiber Orientation on Damping and Stiffness of Polymer Composite Materials, *Exp. Mech.*, Vol 26, 1986, p 175-184
2. R.D. Adams and D.G.C. Bacon, The Dynamic Properties of Unidirectional Fibre Reinforced Composites in Flexure and Torsion, *J. Compos. Mater.*, Vol 7, 1973, p 53-67

50° to 70° (not shown) showed very little difference in behavior. The frequency of vibration used in measuring the glass transition temperature varied from 324 Hz for the 0° specimen to 120 Hz for the 40° specimen. There was a reduction of the peak temperature of about 10 °C over the range of fiber angles of 0° to 20°.

The damping properties near the relaxation peaks for a range of frequencies are given for ±10° and ±20° specimens (Fig. 25). In both cases, there is a reduction of the damping peak with increase of frequency. A comparison of traces with approximately the same frequency at the peak (that is, 249 Hz, ±10° and 235 Hz, ±20°), showed that the damping was almost identical, at about 53% SDC, but that the peak temperature for the ±20° specimen was nearly 20 °C (36 °F) below that of the ±10° specimen.

The contribution of shear damping for the ±20° specimen is much larger than is that for the ±10° specimen (Fig. 11), and the longitudinal tensile component is almost negligible. However, due to the fairly high flexural modulus of the ±10° specimen, $E_{\pm 10}$, and its relatively low torsion modulus, $G_{\pm 10}$, there will be shear deformation in flexure. The difference between the two types of shears is one of direction; the flexural shear is denoted by σ_{zx} where z is perpendicular to the plane of the laminate, and the shear that is due to the fiber angle is denoted by σ_{xy}. For 0° specimens, σ_{zx} and σ_{xy} lie in the plane of symmetry, and the effect will be identical, but for more complex laminates involving adjacent laminae at different angles, the result is not as obvious.

The effect of temperature on flexural modulus depends to a large extent on the lay-up;

3. D.F. Adams and D.R. Doner, Longitudinal Shear Loading of a Unidirectional Composite, *J. Compos. Mater.*, Vol 1, 1967, p 4-17

4. Z. Hashin, Complex Moduli of Viscoelastic Composites, II, Fibre Reinforced Materials, *Int. J. Solids Struct.*, Vol 6, 1970, p 797-804

5. R.G. Ni and R.D. Adams, A Rational Method for Obtaining the Dynamic Mechanical Properties of Laminae for Predicting the Stiffness and Damping of Laminated Plates and Beams, *Composites*, Vol 15, 1984, p 193-199

6. R.D. Adams and D.F. Short, The Effect of Fibre Diameter on the Dynamic Properties of Glass-Fibre-Reinforced Polyester Resin, *J. Phys. D. Appl. Phys.*, Vol 6, 1973, p 1032-1039

7. R.D. Adams, The Dynamic Longitudinal Shear Modulus and Damping of Carbon Fibres, *J. Phys. D., Appl. Phys.*, Vol 8, 1975, p 738-748

8. S.W. Tsai and H.T. Halpin, *Introduction to Composite Materials*, Technomic Publishing Co., Westport, CT, 1980

9. S. Chang and C.W. Bert, Analysis of Damping for Filamentary Composite Materials, *Composite Materials in Engineering Design*, B.R. Noton, Ed., American Society for Metals, 1973, p 51-62

10. R.D. Adams and D.G.C. Bacon, Effect of Fibre Orientation and Laminate Geometry on the Dynamic Properties of CFRP, *J. Compos. Mater.*, Vol 7, 1973, p 402-428

11. R.G. Ni and R.D. Adams, The Damping and Dynamic Moduli of Symmetric Laminated Composite Beams — Theoretical and Experimental Results, *J. Compos. Mater.*, Vol 18, 1984, p 104-121

12. M.M. Wallace and C.W. Bert, Transfer-Matrix Analysis of Dynamic Response of Composite-Material Structural Elements With Material Damping, *Shock & Vib. Bull. 50*, Part 3, Sept 1980, p 27-38

13. P. Cawley and R.D. Adams, The Predicted and Experimental Natural Modes of Free-Free CFRP Plates, *J. Compos. Mater.*, Vol 12, 1978, p 336-347

14. D.X. Lin, R.G. Ni, and R.D. Adams, Prediction and Measurement of the Vibrational Damping Parameters of Carbon and Glass Fibre-Reinforced Plastics Plates, *J. Compos. Mater.*, Vol 18, 1984, p 132-152

15. D.X. Lin and R.D. Adams, Determination of the Damping Properties of Structures by Transient Testing Using Zoom-FFT., *J. Phys. E., Sci. Instrum.*, Vol 18, 1985, p 161-165

16. R.G. Ni, D.X. Lin, and R.D. Adams, The Dynamic Properties of Carbon-Glass Fibre Sandwich-Laminated Composites: Theoretical, Experimental and Economic Considerations, *Composites*, Vol 15, 1984, p 297-304

17. R.D. Adams and D.G.C. Bacon, The Effect of Fibre Modulus and Surface Treatment on the Modulus, Damping, and Strength of Carbon-Fibre-Reinforced Plastics, *J. Phys. D., Appl. Phys.*, Vol 7, 1974, p 7-23

18. R.D. Adams and A.L. Percival, Measurement of the Strain-Dependent Damping of Metals in Axial Vibration, *J. Phys. D., Appl. Phys.*, Vol 2, 1969, p 1693-1704

19. R.D. Adams, The Damping Characteristics of Certain Steels, Cast Irons, and Other Metals, *J. Sound and Vibr.*, Vol 23, 1972, p 199-216

20. R.D. Adams, Damping of Ferromagnetic Materials at Direct Stress Levels Below the Fatigue Limit, *J. Phys. D., Appl. Phys.*, Vol 5, 1972, p 1877-1889

21. G.A. Cottell, K.M. Entwistle, and F.C. Thompson, The Measurement of the Damping Capacity of Metals in Torsional Vibration, *J. Inst. Metals*, Vol 74, 1948, p 373-424

Properties Analysis of Laminates

Edward A. Humphreys and B. Walter Rosen, Materials Sciences Corporation

THE PROPERTIES of unidirectional composite (UDC) materials are quite different from those of conventional, metallic materials. The primary difference, from an analytical viewpoint, results from the material anisotropy. UDC materials typically have exceptional properties in the direction of the reinforcing fibers, but poor to mediocre properties perpendicular (transverse) to the fibers. Thus, with the exception of one-dimensionally loaded members (for example, truss members), UDC materials would be expected to perform poorly compared to conventional materials. The problem then is how to obtain maximum advantage from the exceptional fiber directional properties while minimizing the effects of the low transverse properties. One obvious solution is to use the approach taken in the manufacture of plywood.

Plywood consists of layers, or plies, of wood bonded together, with the wood grain in each ply oriented perpendicular to the adjacent plies. With this orientation of the plies of wood, the lesser properties of the wood perpendicular to the grain are augmented by the superior properties in the direction of the wood grain. At the same time, however, the superior properties in the grain direction cannot be fully utilized because of the perpendicular plies.

The bonding of individual UDC plies is used to form laminates. The plies, or laminae, are oriented such that the effective properties of the laminate match the loading environment. Laminate effective material properties are tailored to meet performance requirements through the use of lamination theory.

Lamination theory can be considered a form of structural analysis but can also be used when a structural material is being designed. This adds another level of effort to the design process but allows the structural material to be tailored to match the loadings. Thus, if a 2:1 biaxial loading environment is prescribed, the structural laminate can be designed for a 2:1 strength. The amount of material is thereby minimized in a way that is not possible with conventional materials.

This article first presents a treatment of UDC stress-strain relations in the forms appropriate for analysis of thin plies of material. The analysis is then developed for an assemblage of plies (laminate), and finally the stress-strain relations for a thin, laminated plate are developed for the case of plate membrane forces and bending moments.

For purposes of structural analysis, it is desirable to represent a laminate by a set of effective stiffnesses, just as a homogeneous plate is defined by its extensional and bending stiffnesses. Accordingly, the calculation of these laminate mechanical properties is defined and illustrated in this article. Also, the analysis is expanded to include treatment of laminate expansion resulting from temperature and moisture changes.

Next, calculations of temperature and moisture distributions through the thickness of a laminate are discussed. With this information and the definition of applied loads or deformations, the stresses in each ply can be calculated. Procedures for doing this are presented.

The sections just described deal with what may be classified as the nondestructive response of the laminate to external load and environment. The remaining sections address laminate failure resulting from static, cyclic, and impact loading conditions.

Lamina Stress-Strain Relations

A laminate is composed of unidirectionally reinforced laminae oriented in various directions. Each lamina is the type of UDC discussed in the article "Overview of Composite Materials Analysis and Design" in this Volume. The elastic stress-strain relations of the lamina will be expressed in this article in matrix form using a different notation for elastic properties. With x_1 in the fiber direction, x_2 transverse to the fibers in the plane of the lamina, and x_3 normal to the plane of the lamina, these material properties define the lamina properties:

$$
\begin{aligned}
E_1 &= E_L^* & \nu_{12} &= \nu_L^* \\
E_2 &= E_3 = E_T^* & \nu_{23} &= \nu_T^* \\
G_{12} &= G_L^* & G_{23} &= G_T^*
\end{aligned}
\qquad \text{(Eq 1)}
$$

where E represents Young's modulus of elasticity, ν is Poisson's ratio, G is shear modulus, L is longitudinal, T is transverse, and * represents the effective properties that were found in the article "Overview of Composite Materials Analysis and Design" in this Volume.

Furthermore, the laminae, at this point, are treated as effective, homogeneous, transversely isotropic materials, and the strains are written without overbars. Thus,

$$
\begin{Bmatrix} \epsilon_{11} \\ \epsilon_{22} \\ \epsilon_{33} \\ 2\epsilon_{23} \\ 2\epsilon_{13} \\ 2\epsilon_{12} \end{Bmatrix} =
\begin{bmatrix}
\dfrac{1}{E_1} & \dfrac{-\nu_{12}}{E_1} & \dfrac{-\nu_{12}}{E_1} & 0 & 0 & 0 \\[6pt]
\dfrac{-\nu_{12}}{E_1} & \dfrac{1}{E_2} & \dfrac{-\nu_{23}}{E_2} & 0 & 0 & 0 \\[6pt]
\dfrac{-\nu_{12}}{E_1} & \dfrac{-\nu_{23}}{E_2} & \dfrac{1}{E_2} & 0 & 0 & 0 \\[6pt]
0 & 0 & 0 & \dfrac{1}{G_{23}} & 0 & 0 \\[6pt]
0 & 0 & 0 & 0 & \dfrac{1}{G_{12}} & 0 \\[6pt]
0 & 0 & 0 & 0 & 0 & \dfrac{1}{G_{12}}
\end{bmatrix}
\begin{Bmatrix} \sigma_{11} \\ \sigma_{22} \\ \sigma_{33} \\ \sigma_{23} \\ \sigma_{13} \\ \sigma_{12} \end{Bmatrix}
\qquad \text{(Eq 2)}
$$

where ϵ is strain and σ is stress. Account has been taken of the symmetry relations:

$$
\frac{\nu_{21}}{E_2} = \frac{\nu_{12}}{E_1} = \frac{\nu_{31}}{E_3} = \frac{\nu_{13}}{E_1} \qquad \frac{\nu_{23}}{E_2} = \frac{\nu_{32}}{E_3}
$$

It has been common practice in the analysis of laminates to use engineering shear strains rather than tensor shear strains. Thus, the factor of 2 has been introduced into the stress-strain relations of Eq 2.

The most important state of stress in a lamina is plane, that is:

$$
\sigma_{13} = \sigma_{23} = \sigma_{33} = 0 \qquad \text{(Eq 3)}
$$

because it occurs for both in-plane loading and bending at a sufficient distance from the laminate edges. In this case, Eq 2 reduces to:

$$
\begin{Bmatrix} \epsilon_{11} \\ \epsilon_{22} \\ 2\epsilon_{12} \end{Bmatrix} =
\begin{bmatrix}
\dfrac{1}{E_1} & \dfrac{-\nu_{12}}{E_1} & 0 \\[6pt]
\dfrac{-\nu_{12}}{E_1} & \dfrac{1}{E_2} & 0 \\[6pt]
0 & 0 & \dfrac{1}{G_{12}}
\end{bmatrix}
\begin{Bmatrix} \sigma_{11} \\ \sigma_{22} \\ \sigma_{12} \end{Bmatrix}
\qquad \text{(Eq 4)}
$$

which may be written:

$$\{\epsilon_\ell\} = [S]\{\sigma_\ell\}$$

where ℓ identifies lamina coordinates, and $[S]$, the compliance matrix, relates the stress and strain components in the principal material directions.

The three relations in Eq 4 relate the in-plane strain components to the three in-plane stress components. For this plane stress state, the three additional strains can be found to be:

$$\epsilon_{23} = \epsilon_{13} = 0$$

$$\epsilon_{33} = -\sigma_{11}\frac{\nu_{13}}{E_1} - \sigma_{22}\frac{\nu_{23}}{E_2}$$

and thus the complete state of stress and strain is determined.

The relations in Eq 4 can be inverted to yield:

$$\{\sigma_\ell\} = [S]^{-1}\{\epsilon_\ell\} \tag{Eq 5}$$

or

$$\{\sigma_\ell\} = [Q]\{\epsilon_\ell\}$$

where the matrix $[Q]$ is defined as the inverse of the compliance matrix and is known as the reduced lamina stiffness matrix. Its terms can be shown to be given by:

$$[Q] = \begin{bmatrix} Q_{11} & Q_{12} & 0 \\ Q_{12} & Q_{22} & 0 \\ 0 & 0 & Q_{66} \end{bmatrix}$$

$$Q_{11} = \frac{E_1}{\dfrac{E_1 - \nu_{12}^2 E_2}{E_1}}$$

$$Q_{22} = \frac{E_2}{\dfrac{E_1 - \nu_{12}^2 E_2}{E_1}} \tag{Eq 6}$$

$$Q_{12} = \frac{\nu_{12}E_2}{\dfrac{E_1 - \nu_{12}^2 E_2}{E_1}}$$

$$Q_{66} = G_{12}$$

The notation used for the $[Q]$ matrix follows from the simplified engineering representation of the fourth-rank stiffness tensor. Each pair of subscripts of the stiffness components is replaced by a single subscript according to the scheme:

$$11 \rightarrow 1 \quad 22 \rightarrow 2 \quad 33 \rightarrow 3$$
$$23 \rightarrow 4 \quad 31 \rightarrow 5 \quad 12 \rightarrow 6$$

The reduced stiffness and compliance matrices, Eq 6 and 4, relate stresses and strains in the principal material directions of the material. To define the material response in directions other than these material coordinates, transformation relations must be developed for the material stiffnesses.

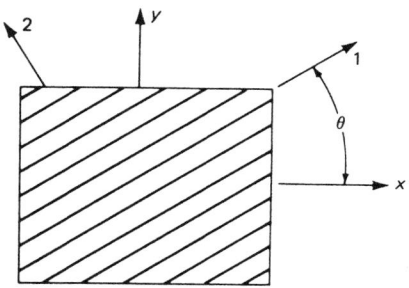

Fig. 1 Coordinate systems; 1, 2, principal material coordinates; *x, y*, laminate or arbitrary coordinates

In Fig. 1, two sets of coordinate systems are shown. The 1-2 coordinate system corresponds to the principal material directions for a lamina, while the *x-y* coordinates are arbitrary and are related to the 1-2 coordinates through a rotation about the axis out of the plane of the figure. The angle, θ, is defined as the rotation from the arbitrary *x-y* system to the material 1-2 system (θ is positive for a counterclockwise rotation).

The transformation of stresses from the 1-2 system to the *x-y* system follows the rules for transformation of tensor components. Thus:

$$\begin{Bmatrix} \sigma_{xx} \\ \sigma_{yy} \\ \sigma_{xy} \end{Bmatrix} = \begin{bmatrix} m^2 & n^2 & -2mn \\ n^2 & m^2 & 2mn \\ mn & -mn & m^2-n^2 \end{bmatrix} \begin{Bmatrix} \sigma_{11} \\ \sigma_{22} \\ \sigma_{12} \end{Bmatrix} \tag{Eq 7}$$

or

$$\{\sigma_x\} = [\theta]\{\sigma_\ell\}$$

where $m = \cos\theta$ and $n = \sin\theta$. In these relations, the subscript *x* is used to refer to the laminate coordinate system.

The same transformation matrix $[\theta]$ can also be used for the tensor strain components. However, since the engineering shear strains have been used, a different transformation matrix is required. Thus:

$$\begin{Bmatrix} \epsilon_{xx} \\ \epsilon_{yy} \\ 2\epsilon_{xy} \end{Bmatrix} = \begin{bmatrix} m^2 & n^2 & -mn \\ n^2 & m^2 & mn \\ 2mn & -2mn & m^2-n^2 \end{bmatrix} \begin{Bmatrix} \epsilon_{11} \\ \epsilon_{22} \\ 2\epsilon_{12} \end{Bmatrix} \tag{Eq 8}$$

or

$$\{\epsilon_x\} = [\psi]\{\epsilon_\ell\}$$

where $[\psi]$ is now the transformation matrix.

Given the transformations for stress and strain to arbitrary coordinate systems, the relations between stress and strain in the laminate system can be determined. Substituting Eq 7 and 8 into Eq 5 yields:

$$\{\sigma_x\} = [\bar{Q}]\{\epsilon_x\} \tag{Eq 9}$$

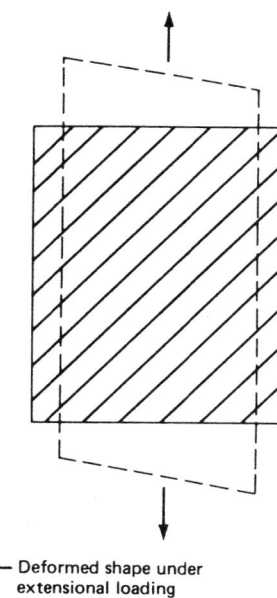

- - - Deformed shape under extensional loading

Fig. 2 Extensional-shear coupling

The reduced-stiffness matrix, $[\bar{Q}]$, relates the stress and strain components in the laminate coordinate system. Here:

$$[\bar{Q}] = [\theta][Q][\psi]^{-1} \tag{Eq 10}$$

The terms within $[\bar{Q}]$ are defined by the appropriate matrix multiplication to be:

$$\bar{Q}_{11} = Q_{11}m^4 + Q_{22}n^4 + 2m^2n^2(Q_{12} + 2Q_{66})$$

$$\bar{Q}_{12} = m^2n^2(Q_{11} + Q_{22} - 4Q_{66}) + (m^4 + n^4)Q_{12}$$

$$\bar{Q}_{16} = [Q_{11}m^2 - Q_{22}n^2 - (Q_{12} + 2Q_{66})(m^2 - n^2)]mn$$

$$\bar{Q}_{22} = Q_{11}n^4 + Q_{22}m^4 + 2m^2n^2(Q_{12} + 2Q_{66})$$

$$\bar{Q}_{26} = [Q_{11}n^2 - Q_{22}m^2 + (Q_{12} + 2Q_{66})(m^2 - n^2)]mn$$

$$\bar{Q}_{66} = (Q_{11} + Q_{22} - 2Q_{12})m^2n^2 + Q_{66}(m^2 - n^2)^2$$

$$\bar{Q}_{21} = \bar{Q}_{12}$$

$$\bar{Q}_{61} = \bar{Q}_{16} \quad \bar{Q}_{62} = \bar{Q}_{26} \tag{Eq 11}$$

where the subscript 6 has been retained in keeping with the discussion following Eq 6. Thus:

$$[\bar{Q}] = \begin{bmatrix} \bar{Q}_{11} & \bar{Q}_{12} & \bar{Q}_{16} \\ \bar{Q}_{21} & \bar{Q}_{22} & \bar{Q}_{26} \\ \bar{Q}_{16} & \bar{Q}_{26} & \bar{Q}_{66} \end{bmatrix} \tag{Eq 12}$$

A feature of the $[\bar{Q}]$ matrix that is immediately noticed as being dissimilar to previous constitutive relations is that $[\bar{Q}]$ is fully populated. The additional terms that have appeared in $[\bar{Q}]$, namely, $\bar{Q}_{16}$ and $\bar{Q}_{26}$, relate shear strains to extensional loading and vice versa. This effect of a shear strain resulting from an extensional stress is shown in Fig. 2.

Referring to Eq 11, each of the extensional-shear coupling terms can be seen to contain a sine-cosine multiplier term. Obviously, if either $\sin \theta$ or $\cos \theta$ is zero, the extensional-shear coupling terms are zero. For the product $\sin \theta \cos \theta$ to be zero, the angle θ must be $0°$ or $90°$. Physically, this means that the fibers are either parallel or perpendicular to the loading direction. For this case, extensional shear coupling does not occur in an orthotropic material, because the loadings are in the principal material directions (Eq 4).

The procedure used to develop the transformed stiffness matrix can also be used to find a transformed compliance matrix. Thus:

$$\{\epsilon_\ell\} = [S]\{\sigma_\ell\}$$
$$\{\epsilon_x\} = [\psi][S][\theta]^{-1}\{\sigma_x\} \qquad \text{(Eq 13)}$$
$$\{\epsilon_x\} = [\bar{S}]\{\sigma_x\}$$

The relations between the terms in $[\bar{S}]$ and $[S]$ are developed by a procedure identical to that for the relations between $[\bar{Q}]$ and $[Q]$.

With the stress-strain relations now defined in the laminate coordinate system, lamina stiffnesses can also be defined in this system. Thus, expanding the last of Eq 13 yields:

$$\begin{Bmatrix} \epsilon_{xx} \\ \epsilon_{yy} \\ 2\epsilon_{xy} \end{Bmatrix} = \begin{bmatrix} \bar{S}_{11} & \bar{S}_{12} & \bar{S}_{16} \\ \bar{S}_{21} & \bar{S}_{22} & \bar{S}_{26} \\ \bar{S}_{16} & \bar{S}_{26} & \bar{S}_{66} \end{bmatrix} \begin{Bmatrix} \sigma_{xx} \\ \sigma_{yy} \\ \sigma_{xy} \end{Bmatrix}$$

The engineering constants for the material can be defined by specifying the conditions for an experiment. Thus, the ratio $\sigma_{xx}/\epsilon_{xx}$, for $\sigma_{yy} = \sigma_{xy} = 0$, is the Young's modulus in the x direction. For this same stress state, $-\epsilon_{yy}/\epsilon_{xx}$ is the Poisson's ratio. In this way, the lamina stiffnesses in the coordinate system of Eq 13 are found to be:

$$E_x = \frac{1}{\bar{S}_{11}}$$

$$E_y = \frac{1}{\bar{S}_{22}}$$

$$G_{xy} = \frac{1}{\bar{S}_{66}} \qquad \text{(Eq 14)}$$

$$\nu_{xy} = -\frac{\bar{S}_{21}}{\bar{S}_{11}} = -\frac{\bar{S}_{12}}{\bar{S}_{11}}$$

It is sometimes desirable to obtain elastic constants directly from the reduced stiffnesses, $[\bar{Q}]$, by utilizing Eq 9. In the general case where the $\bar{Q}_{ij}$ matrix is fully populated, this can be accomplished by using Eq 14 and the solution for $\bar{S}_{ij}$ as functions of $\bar{Q}_{ij}$ obtained from the inverse relationship of the two matrices. An-

other approach is to evaluate extensional properties for the case of zero shear strain, $\epsilon_{xy} = 0$, as opposed to the previous case of zero shear stress, $\sigma_{xy} = 0$. For single stress states and zero shear strain it is found that, in terms of the transformed stiffness matrix terms (Eq 12), the elastic constants are:

$$E_x = \bar{Q}_{11} - \frac{\bar{Q}_{12}^2}{\bar{Q}_{22}}$$

$$E_y = \bar{Q}_{22} - \frac{\bar{Q}_{12}^2}{\bar{Q}_{11}} \qquad \text{(Eq 15)}$$

$$\nu_{xy} = \frac{\bar{Q}_{12}}{\bar{Q}_{22}} = \frac{\bar{Q}_{21}}{\bar{Q}_{22}}$$

Also:

$$G_{xy} = \bar{Q}_{66}$$

Referring to the terms in the $[\bar{Q}]$ matrix (Eq 11) and the stiffness relations (Eq 15), it can be seen that, in general, the elastic constants in an arbitrary coordinate system are functions of all of the elastic constants in the principal material directions as well as the angle of rotation.

The variation of elastic modulus E_x with angle of rotation is shown in Fig. 3 for a typical graphite-epoxy material. For demonstration purposes, two different shear moduli have been used in generating the figure. The differences between the two curves demonstrate the effect of the principal material shear modulus on the transformed extensional stiffness. The two curves are identical at $\theta = 0°$ and $\theta = 90°$. This is to be expected because at these angles, the extensional stiffness, E_x, is simply E_1 or E_2. Between the two end points, substantial differences exist. For the smaller shear modulus, from approximately $50°$ to just less than $90°$, the extensional stiffness is less than the E_2 value. This is a most interesting result, indicating that for these angles, the material stiffness is governed more strongly by the principal material shear modulus than by the transverse extensional stiffness.

The curves of Fig. 3 can also be used to determine the modulus E_y, simply by reversing the angle scale. Thus, the values shown at $0°$ correspond to E_y at $90°$ and the values at $90°$ correspond to E_y at $0°$.

With the transformed stress-strain relations, it is now possible to develop an analysis for an assemblage of plies, that is, a laminate.

Lamination Theory

The development of procedures to evaluate stresses and deformations of laminates is crucially dependent on the fact that the thickness of laminates is so much smaller than the in-plane dimensions. Typical thickness values for individual plies range between 0.15 and 0.25 mm (0.005 and 0.010 in.). Consequently, aerospace laminates using from 8 to 50 plies are generally still thin plates and therefore can be analyzed on

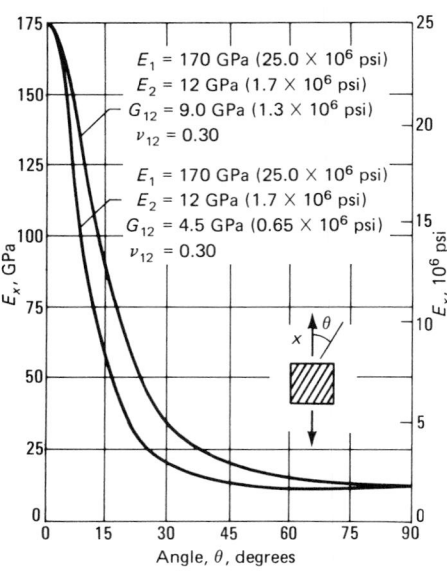

Fig. 3 Variation of E_x with angle of rotation and G_{12} for typical graphite-epoxy materials

the basis of the usual simplifications of thin-plate theory.

Analysis of isotropic thin plates is an old, established field in which in-plane loading and bending are usually analyzed separately. The former is described by plane stress elastic theory, and the latter is described by classical plate-bending theory. This separation is possible because the two loadings are uncoupled; when both occur, the result is given by superposition. In the case of anisotropic laminates, in-plane loading and bending are generally coupled and must be treated together. It is only for symmetric laminae stacking sequences that uncoupling occurs. Consequently, lamination theory will be developed first for the general case, and simplifications will then be introduced.

The classical assumptions of thin-plate theory are:

- The thickness of the plate is much smaller than the in-plane dimensions
- The strains in the deformed plate are small compared to unity
- Normals to the undeformed plate surface remain normal to the deformed plate surface
- Vertical deflection does not vary through the thickness
- Stress normal to the plate surface is negligible

On the basis of the second, third, and fourth assumptions, the displacement field in the plate, u, can be expressed as:

$$u_z = u_z°(x,y)$$

$$u_x = u_x°(x,y) - z\frac{\partial u_z}{\partial x} \qquad \text{(Eq 16)}$$

$$u_y = u_y°(x,y) - z\frac{\partial u_z}{\partial y}$$

with the x-y-z coordinate system defined as in Fig. 4. These relations (Eq 16) indicate that the in-plane displacements consist of a midsurface displacement, designated by the superscript °, plus a linear variation through the thickness. The two partial derivatives are simply bending rotations of the midsurface. The use of the fourth assumption prescribes that u_z not vary through the thickness. The z coordinate is measured from the midsurface of the plate. This is a natural convention for a laminate that is symmetrically stacked with reference to the midsurface but an arbitrary one when this is not the case.

The linear strain displacement relations are:

$$\epsilon_{xx} = \frac{\partial u_x}{\partial x}$$

$$\epsilon_{yy} = \frac{\partial u_y}{\partial y} \qquad \text{(Eq 17)}$$

$$\epsilon_{xy} = \frac{1}{2}\left(\frac{\partial u_x}{\partial y} + \frac{\partial u_y}{\partial x}\right)$$

Performing the required partial differentiations yields:

$$\epsilon_{xx} = \epsilon^{\circ}_{xx} + z\,\kappa_{xx}$$
$$\epsilon_{yy} = \epsilon^{\circ}_{yy} + z\,\kappa_{yy} \qquad \text{(Eq 18)}$$
$$2\epsilon_{xy} = 2\epsilon^{\circ}_{xy} + 2z\,\kappa_{xy}$$

or

$$\{\epsilon_x\} = \{\epsilon^{\circ}\} + z\{\kappa\}$$

where

$$\{\epsilon^{\circ}\} = \begin{matrix} \dfrac{\partial u^{\circ}_x}{\partial x} \\[2mm] \dfrac{\partial u^{\circ}_y}{\partial y} \\[2mm] \left(\dfrac{\partial u^{\circ}_x}{\partial y} + \dfrac{\partial u^{\circ}_y}{\partial x}\right) \end{matrix} \qquad \text{(Eq 19)}$$

and

$$\{\kappa\} \equiv \begin{matrix} \kappa_{xx} \\[2mm] \kappa_{yy} \\[2mm] 2\kappa_{xy} \end{matrix} = \begin{matrix} -\dfrac{\partial^2 u_z}{\partial x^2} \\[2mm] -\dfrac{\partial^2 u_z}{\partial y^2} \\[2mm] -2\dfrac{\partial^2 u_z}{\partial x \partial y} \end{matrix} \qquad \text{(Eq 20)}$$

Thus, the strain at any point in the plate is defined as the sum of a midsurface strain, $\{\epsilon^{\circ}\}$, and a curvature, $\{\kappa\}$, multiplied by the distance from the midsurface.

For convenience, as is usual in plate theory, stress and moment resultants will be used rather than stresses for the remainder of the develop-

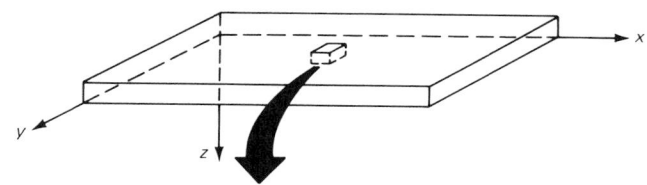

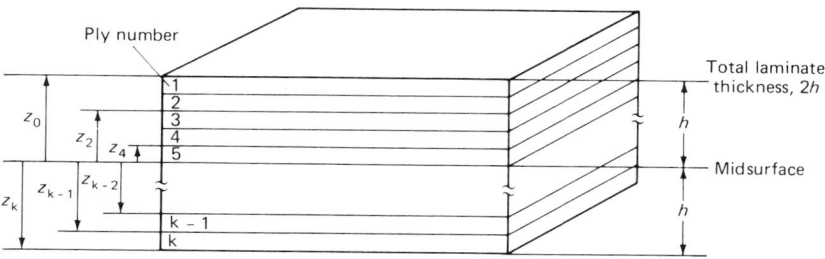

Fig. 4 Laminate construction

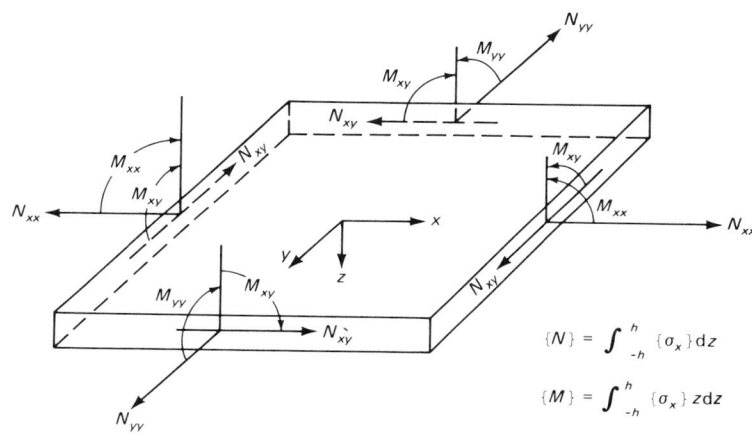

$$\{N\} = \int_{-h}^{h} \{\sigma_x\}\,dz$$

$$\{M\} = \int_{-h}^{h} \{\sigma_x\}\,z\,dz$$

Fig. 5 Stress and moment resultants

ment (Fig. 5). The stress resultants, N, are defined as:

$$\{N\} = \begin{Bmatrix} N_{xx} \\ N_{yy} \\ N_{xy} \end{Bmatrix} = \int_{-h}^{h} \{\sigma_x\}dz \qquad \text{(Eq 21)}$$

and the moment resultants, M, are defined as:

$$\{M\} = \begin{Bmatrix} M_{xx} \\ M_{yy} \\ M_{xy} \end{Bmatrix} = \int_{-h}^{h} \{\sigma_x\}z\,dz \qquad \text{(Eq 22)}$$

where the integrations are carried out over the plate thickness, $2h$.

Noting Eq 9 and 18, relations between the stress and moment resultants and the midplane strains and curvatures can be written as:

$$\{N\} = \int_{-h}^{h} \{\sigma_x\}\,dz = \int_{-h}^{h} [\bar{Q}]\,(\{\epsilon^{\circ}\} + z\{\kappa\})\,dz \qquad \text{(Eq 23)}$$

$$\{M\} = \int_{-h}^{h} \{\sigma_x\}\,z\,dz = \int_{-h}^{h} [\bar{Q}]\,(\{\epsilon^{\circ}\} + z\{\kappa\})\,z\,dz \qquad \text{(Eq 24)}$$

Since the transformed lamina stiffness matrices are constant within each lamina and the midplane strains and curvatures are constant with respect to the z coordinate, the integrals in Eq 23 and 24 can be replaced by summations, Σ, over the total number of plies.

Introducing three matrices equivalent to the needed summations, the relations can be written as:

$$\{N\} = [A]\{\epsilon^{\circ}\} + [B]\{\kappa\}$$
$$\{M\} = [B]\{\epsilon^{\circ}\} + [D]\{\kappa\}$$

or

$$\begin{Bmatrix} N \\ \hline M \end{Bmatrix} = \begin{bmatrix} A & B \\ \hline B & D \end{bmatrix} \begin{Bmatrix} \epsilon^{\circ} \\ \kappa \end{Bmatrix} \qquad \text{(Eq 25)}$$

where the stiffness matrix is composed of the following 3 by 3 matrices:

$$[A] = \sum_{i=1}^{K} [\bar{Q}]^i (z_i - z_{i-1})$$

$$[B] = \frac{1}{2} \sum_{i=1}^{K} [\bar{Q}]^i (z_i^2 - z_{i-1}^2) \qquad \text{(Eq 26)}$$

$$[D] = \frac{1}{3} \sum_{i=1}^{K} [\bar{Q}]^i (z_i^3 - z_{i-1}^3)$$

where K is the total number of plies, z_i is defined as in Fig. 4, and superscript i denotes a property of the ith ply. Note that $z_i - z_{i-1}$ is equal to the ply thickness. Here the reduced lamina stiffnesses for the ith ply are found from Eq 11 using the principal ply properties and orientation angle of each ply in turn. Thus, the constitutive relations for a laminate in terms of stress and moment resultants have been developed.

In examining the relations (Eq 25), the first interesting feature revealed is a coupling between bending and extension. The $[B]$ matrices relate stress resultants to bending curvatures and moment resultants to midplane strains. Thus, for a general laminate, the application of a stress resultant produces curvatures, and application of a bending moment produces extensional strain. The feature is known as bending-extensional coupling.

In typical structural laminates, bending-extensional coupling is eliminated by proper specification of the stacking sequence. Stacking sequence refers to the order in which the various plies are put together. From the relations for computing the $[B]$ matrix (Eq 26), it can be seen that if the plies are arranged in an even z function (symmetric about the midplane), the $[B]$ matrix is eliminated. Thus, if the laminate is designed with identically oriented plies at equal distances from the midsurface, bending-extensional coupling is eliminated.

Other forms of coupling are inherent in the relations (Eq 25). To examine these couplings, it is necessary to evaluate individual terms in the $[A]$, $[B]$, and $[D]$ matrices. The most general form of these matrices, combined as in Eq 25, is:

$$\begin{bmatrix} A_{11} & A_{12} & A_{16} & B_{11} & B_{12} & B_{16} \\ A_{21} & A_{22} & A_{26} & B_{21} & B_{22} & B_{26} \\ A_{16} & A_{26} & A_{66} & B_{16} & B_{26} & B_{66} \\ B_{11} & B_{12} & B_{16} & D_{11} & D_{12} & D_{16} \\ B_{21} & B_{22} & B_{26} & D_{21} & D_{22} & D_{26} \\ B_{16} & B_{26} & B_{66} & D_{16} & D_{26} & D_{66} \end{bmatrix}$$

In general, the $[A]$, $[B]$, and $[D]$ matrices are fully populated. Thus, there is coupling between membrane extension and membrane shear (A_{16}, A_{26}) and between bending and twisting (D_{16}, D_{26}). Both of these forms of coupling can be eliminated by judicious ply

orientation and stacking sequence selection, but in some cases this selection may be impractical. Extensional-shear coupling can be eliminated by specifying a balanced construction. Balance indicates that for every $+\theta$ ply, there is a $-\theta$ ply. These plies do not have to be adjacent to satisfy this requirement.

When bending-twisting coupling is undesirable, it can be eliminated, but only by using a unidirectional or cross-ply construction. Unidirectional construction implies that all layers have the same orientation, which is aligned with the loading direction. A cross-ply laminate has plies oriented at 0° and 90° only, again oriented with the loading direction. It is also possible to limit the effect of bending-twisting coupling by other means. If a symmetric laminate is constructed with many plies, and plies with the same angular orientation are not grouped together, the magnitude of the D_{16} and D_{26} terms will be reduced with respect to the other terms in the $[D]$ matrix. Thus, while the bending-twisting coupling is not eliminated, its effect is reduced. In certain applications, such coupling may be desirable.

Laminate Properties

The previous section presented the development of the relations between midsurface strains and curvatures and membrane stress and moment resultants. The elastic stiffnesses in these relationships are functions of the ply elastic constants and the ply orientations and arrangement or stacking sequence. In the present section, these results will be used to calculate plate bending and extensional stiffnesses suitable for use in structural analysis. The effects of orientation variables on plate properties will also be discussed.

For practical structures it is necessary to understand not only the mechanical loading conditions but also the effects of temperature changes on laminate behavior. Thus, the thermal expansion characteristics of laminates will be presented. Further, for polymeric matrix composites, high moisture content has been found to cause dimensional changes. This effect is treated in this section to define effective swelling coefficients.

In general, as shown previously, laminates exhibit coupling between bending and extension. Elimination of this coupling was shown to be possible through the specification of midplane symmetric construction. This leads to a natural division of laminates into symmetric and nonsymmetric categories.

Before these two laminate categories are discussed, typical laminate notation will be described. A shorthand notation has been devised and is in common use. The basis for the notation comes from the fact that structural laminates typically have repeating groups of plies and pairs of plies at $+\theta$ and $-\theta$ angles. Using these groupings, the laminate construction

$$[0°/0°/45°/-45°/90°90°/-45°45°0°0°]$$

can be specified as:

$$[0_2°/\pm45°/90°]_s$$

The numerical subscript indicates the number of adjacent identical plies. Adjacent balanced angle plies are lumped together, and the subscript s indicates that the pattern is repeated in reverse order, forming a symmetric laminate. A multiplier subscript is also used to denote multiple groups of plies. Thus, the laminate

$$[45°/-45°/0°/0°/45°/-45°/0°/0°/0°/0°/-45°/45°/0°/0°/-45°/45°]$$

is specified as:

$$[\pm45°/0_2°]_{2s}$$

When the central ply of a symmetric laminate is not repeated, this is denoted by an overbar. Thus the laminate

$$[45°/-45°/0°/0°/90°/0°/0°/-45°/45°]$$

is specified as:

$$[\pm45°/0_2°/\overline{90°}]_s$$

Additional conventions can be designated, but those shown here are usually sufficient.

Symmetric Laminates. Because a symmetric laminate does not exhibit coupling between extension and bending, the design of composites is considerably simplified because symmetric constructions behave somewhat similarly to conventional materials. Recalling Eq 25 and noting that for this case the $[B]$ matrix is zero, the relations can be rewritten as:

$$\begin{Bmatrix} N_{xx} \\ N_{yy} \\ N_{xy} \end{Bmatrix} = \begin{bmatrix} A_{11} & A_{12} & A_{16} \\ A_{12} & A_{22} & A_{26} \\ A_{16} & A_{26} & A_{66} \end{bmatrix} \begin{Bmatrix} \epsilon_{xx}^\circ \\ \epsilon_{yy}^\circ \\ 2\epsilon_{xy}^\circ \end{Bmatrix}$$

and $\qquad\qquad$ (Eq 27)

$$\begin{Bmatrix} M_{xx} \\ M_{yy} \\ M_{xy} \end{Bmatrix} = \begin{bmatrix} D_{11} & D_{12} & D_{16} \\ D_{12} & D_{22} & D_{26} \\ D_{16} & D_{26} & D_{66} \end{bmatrix} \begin{Bmatrix} \kappa_{xx} \\ \kappa_{yy} \\ 2\kappa_{xy} \end{Bmatrix}$$

Since the extensional and bending behavior are uncoupled, effective laminate elastic constants can be readily determined. Inverting the stress resultant midplane strain relations yields:

$$\{\epsilon^\circ\} = [A]^{-1} \{N\} = [a] \{N\}$$

from which the laminate elastic constants are seen to be:

$$E_x = \frac{1}{2ha_{11}} \qquad G_{xy} = \frac{1}{2ha_{66}}$$

$$E_y = \frac{1}{2ha_{22}} \qquad \nu_{xy} = -\frac{a_{12}}{a_{11}}$$

(Eq 28)

where the divisor $2h$ corresponds to the laminate thickness.

Noting that the $[A]$ matrix consists of $[\bar{Q}]$ matrices from each layer in the laminate, it is obvious that the laminate elastic properties are functions of the angular orientation of the plies. This is illustrated in Fig. 6 for a typical high-modulus graphite-epoxy system. The lamina properties for this material are listed in Table 1. The laminae are oriented in $\pm\theta$ pairs in a symmetric, balanced construction, creating what is called an angle-ply laminate.

Figure 6 indicates a variation in extensional modulus similar to that shown for the off-axis unidirectional material (Fig. 3). The differences between an off-axis material and an angle-ply laminate are due to the effects of extensional-shear coupling. In an off-axis material, extensional loadings produce shear deformations as well as extensional deformations. In an angle-ply laminate, the shear deformations are eliminated through internal constraints. Since the shear strains are eliminated, the material exhibits higher effective stiffnesses.

Two other features are noteworthy in Fig. 6, namely, the variation of shear modulus and Poisson's ratio. The shear modulus is equal to the unidirectional value for $\theta = 0°$ and $\theta = 90°$; it rises sharply and reaches a maximum at $\theta = 45°$. The peak at 45° can be explained by noting that shear is equivalent to a combined state of equal tension and compression loads oriented at 45°. Thus, the shear loading on a $[\pm 45°]$ laminate is equivalent to tensile and compressive loading on a $[0°/90°]_s$ laminate. Effectively, the fibers are aligned with the loading; hence, the large shear stiffness.

An even more interesting effect is seen in the variation of Poisson's ratio. The peak value in this example is greater than 1.5. In an isotropic material, this would be impossible. In an orthotropic material, the isotropic restrictions do not hold, and a Poisson's ratio greater than one is valid and realistic. In fact, large Poisson's ratios are typical for laminates constructed of UDC materials with the plies oriented at approximately $\pm 30°$.

This effect can readily be explained by referring to Fig. 7, which shows two separate loadings on a $\pm 30°$ lamina. When a positive N_{xx} load is applied, the deformed shape contains a positive ϵ_{xx} and a negative ϵ_{xy}. In the $[\pm 30°]_s$ laminate, the shear strain is constrained by the presence of the $-30°$ plies, causing an internal shear load, which eliminates the shear strain. The large Poisson's ratio is a result of this internal shear load. As shown in the second part of Fig. 7, a compressive N_{yy} load promotes a positive ϵ_{xy}. Because of material symmetries, the application of a positive

Table 1 Properties of a high-modulus graphite-epoxy lamina

E_1 = 170 GPa (25.0 × 10⁶ psi)
E_2 = 12 GPa (1.7 × 10⁶ psi)
G_{12} = 4.5 GPa (0.65 × 10⁶ psi)
ν_{12} = 0.30
ρ = 0.056
α_1 = −0.54 × 10⁻⁶/K
α_2 = 35.1 × 10⁻⁶/K
σ_L^{tu} = 758 MPa (110.0 ksi)
σ_T^{tu} = 28 MPa (4.0 ksi)
σ_L^{u} = 62 MPa (9.0 ksi)
σ_L^{cu} = 758 MPa (110.0 ksi)
σ_T^{cu} = 138 MPa (20.0 ksi)
V_f = 0.6

Note: Ply thickness, 0.13 mm (0.0052 in.)

N_{xy} would produce a compressive ϵ_{yy}. Therefore, the positive internal shear load required to constrain the shear strains developed by the positive N_{xx} in the first part of Fig. 7 produces a compressive ϵ_{yy}. Thus, the Poisson-induced strain is made up of two components: First, the applied tensile N_{xx} produces a compressive ϵ_{yy}, and, second, the induced positive N_{xy} produces a compressive ϵ_{yy}. Thus, the large Poisson's ratio is simply a function of extensional-shear coupling in the individual laminae.

Because of the infinite variability of the angular orientation of the individual laminae, presumably a laminate could be constructed having a stiffness which behaves isotropically in the plane of the laminate by using a large number of plies having small, equal differences in their orientation. It can be shown that a symmetric, quasi-isotropic laminate can also be constructed with as few as six plies, three plies above and three below the midplane. The simplest quasi-isotropic laminate is $[0°/\pm 60°]_s$. A general rule for describing a quasi-isotropic laminate states that the angles between plies are equal to π/N, where N is an integer greater than or equal to 3, and the number of plies at each orientation is identical, in a symmetric laminate. For plies of a given material, all such quasi-isotropic laminates will have the same elastic properties, regardless of the value of N.

As was stated, a quasi-isotropic laminate has in-plane stiffnesses which follow isotropic relationships. Thus:

$$E_x = E_y = E_\theta$$

where the subscript θ indicates any arbitrary angle. Additionally,

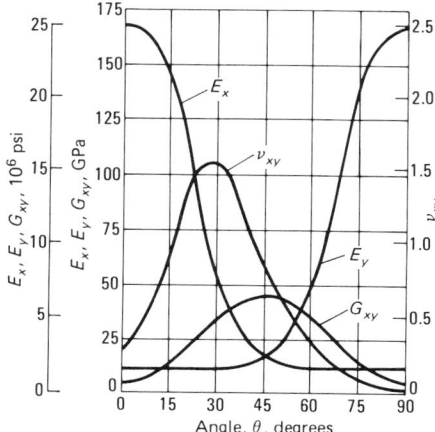

Fig. 6 Laminate elastic constants for high-modulus graphite-epoxy system, $[\pm\theta]_s$

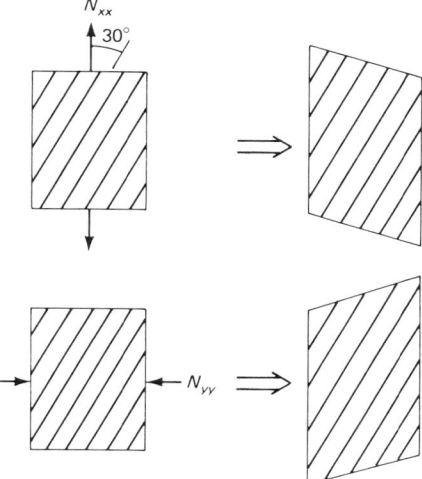

Fig. 7 Extensional-shear coupling and Poisson's ratio

$$G_{xy} = \frac{E_x}{2(1 + \nu_{xy})}$$

There are two items which must be remembered with respect to quasi-isotropic laminates. First and foremost, only the elastic in-plane properties are isotropic; the strength properties, in general, will vary with direction. The second item is that two equal moduli, $E_x = E_y$, do not necessarily indicate quasi-isotropy. This second item is graphically demonstrated in Table 2

Table 2 Elastic properties of laminates

| | $E_x = E_y$ | | ν_{xy} | G_{xy} | |
	GPa	10⁶ psi		GPa	10⁶ psi
$[0°/90°]_s$	92.46	13.41	0.038	4.5	0.65
$[\pm 45°]_s$	16.4	2.38	0.829	44.5	6.46
$[0°/90°/+45°/-45°/90°/0°]_s$	75.64	10.97	0.213	17.9	2.59

Note: See Table 1 for lamina properties.

where three laminates each have identical E_x and E_y moduli.

The first two laminates in Table 2 are actually the same. If one rotates the $[0°/90°]_s$ laminate 45°, it is easily seen that it becomes a $[\pm 45°]_s$ laminate. Note that the extensional moduli of these laminates are not the same and that the shear modulus in each laminate is not related to the extensional moduli and Poisson's ratio. For these laminates, the π/N relation has not been satisfied and they are not quasi-isotropic.

The third laminate has plies oriented at 45° to each other, but there are not equal numbers of plies at each angle. This laminate is also not quasi-isotropic. This can be verified by computing a shear modulus using the isotropic relation.

The discussion of symmetric laminates has thus far centered on membrane behavior. It has been shown that symmetric laminates can be constructed that are very well behaved in the membrane sense. The bending behavior of symmetric laminates is considerably more complex, primarily due to the arrangement of the plies throughout the thickness of the laminate.

Unsymmetric Laminates. The use of laminates that are not midplane symmetric introduces some fundamental difficulties for the designer and analyst, the most important being how one defines the membrane and bending stiffnesses of such a material. Because this type of laminate bends when subjected to membrane loading, how is an extensional modulus defined? Conversely, since the material extends when subjected to a bending load, how is a bending stiffness defined? Basically, there are two approaches to defining these elastic constants. A membrane stiffness can be derived either with zero curvatures or with zero bending moments, and a bending stiffness can be defined with either zero membrane strains or zero membrane forces. In both cases, the stiffness defined will be greater when the bending-extensional coupling is restrained. The nature of the problem can sometimes help the designer determine which approach to use.

In order to define stiffnesses of unsymmetric laminates with constrained bending-extensional coupling, it is necessary to use zero curvatures for extension and zero midplane strains for bending. Writing the full constitutive relations:

$$\{N\} = [A]\{\epsilon°\} + [B]\{\kappa\}$$
$$\{M\} = [B]\{\epsilon°\} + [D]\{\kappa\} \qquad \text{(Eq 29)}$$

the extensional stiffness can be found by substituting $\kappa = 0$; thus:

$$\{N\} = [A]\{\epsilon°\} \qquad \text{(Eq 30)}$$

and bending stiffness can be found by substituting $\{\epsilon°\} = 0$.

$$\{M\} = [D]\{\kappa\} \qquad \text{(Eq 31)}$$

The relations shown in Eq 30 and 31 are identical to those used in symmetric laminates; therefore, the constrained stiffnesses are identical to those of symmetric laminates. However, moment resultants are required to develop the zero curvatures in Eq 30, and stress resultants are required to develop the zero midplane strains in Eq 31.

The moment resultants that are developed because of the prescribed zero curvatures can easily be determined. Solving Eq 30 for the resulting midplane strains yields:

$$\{\epsilon°\} = [A]^{-1}\{N\} \qquad \text{(Eq 32)}$$

Substituting these into Eq 29, while noting that the curvatures are zero, yields:

$$\{M\} = [B]\{\epsilon°\}$$

or $\qquad \text{(Eq 33)}$

$$\{M\} = [B] [A]^{-1}\{N\}$$

Stress resultants developed because of curvature restraint can be found similarly. Thus:

$$\{\kappa\} = [D]^{-1}\{M\}$$
$$\{N\} = [B]\{\kappa\} \qquad \text{(Eq 34)}$$

or

$$\{N\} = [B] [D]^{-1}\{\kappa\} \qquad \text{(Eq 35)}$$

Defining stiffnesses without constraining the bending-extensional coupling requires more effort. Referring to Eq 29, the extensional stiffnesses can be found by specifying that $\{M\} = 0$. Therefore:

$$\{0\} = [B]\{\epsilon°\} + [D]\{\kappa\} \qquad \text{(Eq 36)}$$

The curvatures can then be found to be:

$$\{\kappa\} = -[D]^{-1}[B]\{\epsilon°\} \qquad \text{(Eq 37)}$$

Substituting these into the first set of relations in Eq 29 yields:

$$\{N\} = [A]\{\epsilon°\} + [B] (-[D]^{-1}[B]\{\epsilon°\})$$

or

$$\{N\} = ([A] - [B] [D]^{-1}[B])\{\epsilon°\} \qquad \text{(Eq 38)}$$

or

$$\{N\} = [A*]\{\epsilon°\}$$

The effective laminate extensional stiffnesses can now be found using the relations in Eq 28 and the effective stiffness matrix $[A*]$.

An effective bending stiffness matrix can be determined similarly by specifying $\{N\} = 0$. Thus:

$$\{0\} = [A]\{\epsilon°\} + [B]\{\kappa\} \qquad \text{(Eq 39)}$$

Now the midplane strains are seen to be:

$$\{\epsilon°\} = -[A]^{-1}[B]\{\kappa\} \qquad \text{(Eq 40)}$$

and the second of relations in Eq 29 is rewritten as:

$$\{M\} = (-[B] [A]^{-1} [B] + [D]) \{\kappa\}$$

or $\qquad \text{(Eq 41)}$

$$\{M\} = [D*] \{\kappa\}$$

yielding an effective bending stiffness matrix, $[D*]$.

Thermal Expansion. As the use of composite materials becomes more common, they are subjected to mechanical and environmental loading conditions that are increasingly severe, and with the advent of high-temperature resin systems, the range of temperatures over which the composite system can be used has increased. Thus, it is necessary to understand the response of laminates to temperature and moisture, as well as to applied loads. Previously, laminate extensional and bending stiffnesses were determined; in this section, laminate conductivities and expansivities will be defined.

The presence of stresses induced by free thermal expansion is new to the designer and analyst of conventional materials. The mechanism which produces such stresses is qualitatively described in Fig. 8, which shows that free laminae and bonded laminae (laminate) expand differently. In the direction shown, the 0° plies will develop tensile stress, while the

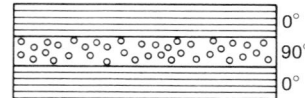

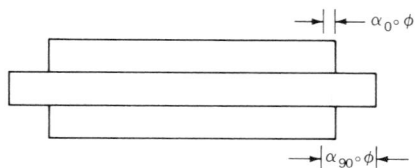

Heating of free laminae

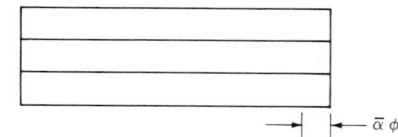

Heating of laminate

Fig. 8 Thermally induced loads

90° ply will develop compressive stress because of the differences between the laminae and laminate thermal expansion coefficients. This effect is similar to heating a conventional material and providing constraints such that thermal expansion strains cannot develop.

To determine quantitatively the laminate thermal expansion coefficients and thermally induced stresses, it is necessary to begin at the ply level. The thermoelastic relations for strain in the principal material directions are:

$$\{\epsilon_\ell\} = \{\epsilon_\ell^M\} + \{\alpha_\ell\}\Delta T$$

or   (Eq 42)

$$\{\epsilon_\ell\} = \{\epsilon_\ell^M\} + \{\epsilon_\ell^T\}$$

The vector $\{\alpha_\ell\}$ represents the free thermal expansion coefficients of a ply. The individual components are:

$$\{\alpha_\ell\} = \begin{Bmatrix} \alpha_1 \\ \alpha_2 \\ 0 \end{Bmatrix}$$ (Eq 43)

and ΔT represents a change in temperature. The thermal strains, $\{\alpha_\ell\}\Delta T$, are the lamina free thermal expansions. They produce no stress in an unconstrained lamina. The thermal expansion coefficients, α_1 and α_2, are the effective thermal expansion coefficients α_L^* and α_T^*, respectively, of the unidirectional composite.

Substituting for mechanical strain terms in Eq 42 and inverting, yields:

$$\{\sigma_\ell\} = [Q]\{\epsilon_\ell\} - \{\Gamma_\ell\}\Delta T$$

where   (Eq 44)

$$\{\Gamma_\ell\} = [Q]\{\alpha_\ell\}$$

The components in the thermal stress coefficient vector, $\{\Gamma_\ell\}$, can be shown, by carrying out the indicated multiplication, to be:

$$\{\Gamma_\ell\} = \begin{Bmatrix} \dfrac{E_1\alpha_1 + \nu_{12}E_2\alpha_2}{\Delta} \\[2ex] \dfrac{E_2\alpha_2 + \nu_{12}E_1\alpha_1}{\Delta} \\[2ex] 0 \end{Bmatrix}$$ (Eq 45)

where

$$\Delta = 1 - \frac{E_2}{E_1}\nu_{12}^2$$

The vector $\{\Gamma_\ell\}\Delta T$ physically represents a correction to the stress vector, which results from the full constraint of the free thermal strains in a lamina.

Both the thermal expansion vector, $\{\alpha_\ell\}\Delta T$, and thermal stress vector, $\{\Gamma_\ell\}\Delta T$, can be transformed to an arbitrary coordinate system using the relations developed for stress and strain transformation (Eq 7 and 8).

With transformed thermal expansion and stress vectors, it is possible to develop thermoelastic laminate relations. Following directly the development of Eq 21 to 25, the membrane relations are:

$$\{N\} = [A]\{\epsilon^\circ\} + [B]\{\kappa\} + \{N^T\}$$ (Eq 46)

where

$$\{N^T\} = -\int_{-h}^{h} \{\Gamma_x\}\Delta T\, dz$$ (Eq 47)

Similarly, the bending relations are

$$\{M\} = [B]\{\epsilon^\circ\} + [D]\{\kappa\} + \{M^T\}$$ (Eq 48)

where

$$\{M^T\} = \int_{-h}^{h} \{\Gamma_x\}\Delta T\, zdz$$ (Eq 49)

The integral relations for the thermal stress resultant vector, $\{N^T\}$, and thermal moment resultant vector, $\{M^T\}$, can be evaluated only when the change in temperature variation through the thickness of the laminate is known. For uniform temperature change through the thickness, the term ΔT is constant and can be factored out of the integration, yielding:

$$\{N^T\} = \Delta T \sum_{i=1}^{K} \{\Gamma_x\}^i (z_i - z_{i-1})$$

$$\{M^T\} = \frac{1}{2}\Delta T \sum_{i=1}^{K} \{\Gamma_x\}^i (z_i^2 - z_{i-1}^2)$$

With the relations in Eq 46 and 48, it is possible to determine effective laminate coefficients of thermal expansion and thermal curvature. These quantities are the extension and curvature changes resulting from a uniform temperature distribution.

Noting that for free thermal effects $\{N\} = \{M\} = 0$, and defining a free thermal expansion vector as

$$\{\alpha_x\} = \{\epsilon^\circ\}\frac{1}{\Delta T}$$ (Eq 50)

and a free thermal curvature vector as

$$\{\delta_x\} = \{\kappa\}\frac{1}{\Delta T}$$ (Eq 51)

the relations in Eq 46 and 48 can be solved. After suitable matrix manipulations, the following expressions for thermal expansion, $\{\alpha_x\}$, and thermal curvature, $\{\delta_x\}$, are found:

$$\{\alpha_x\} = \frac{1}{\Delta T}[L_1]^{-1}([B][D]^{-1}\{M^T\} - \{N^T\})$$ (Eq 52)

$$\{\delta_x\} = \frac{1}{\Delta T}[L_2]^{-1}([B][A]^{-1}\{N^T\} - \{M^T\})$$ (Eq 53)

where

$$[L_1] = [A] - [B][D]^{-1}[B]$$

$$[L_2] = [D] - [B][A]^{-1}[B]$$

The relations in Eq 52 and 53 are complicated expressions containing the $[A]$, $[B]$, and $[D]$ matrices. In many practical cases, the laminates of interest are symmetric, which simplifies the expressions considerably. For symmetric laminates, the bending-extensional coupling vanishes (that is, $[B] = 0$), and the relations are reduced to the form:

$$\{\alpha_x\} = -\frac{1}{\Delta T}[A]^{-1}\{N^T\}$$

$$\{\delta_x\} = -\frac{1}{\Delta T}[D]^{-1}\{M^T\}$$ (Eq 54)

Examination of the relations for $\{M^T\}$, Eq 49, shows that the symmetry that eliminates the $[B]$ matrix also eliminates the $\{M^T\}$ vector. Thus:

$$\{\delta_x\} = \{0\}$$ (Eq 55)

and no curvatures occur because of uniform temperature changes in symmetric laminates.

The variation of the longitudinal thermal expansion coefficient for a symmetric angleply laminate is shown in Fig. 9 to illustrate the effect of laminae orientation. At $\theta = 0°$, the term α_x is simply the axial lamina coefficient of thermal expansion; at $\theta = 90°$, α_x equals the lamina transverse thermal expansion coefficient. An interesting feature of the curve is the large negative value of α_x in the region of 30°. In Fig. 6, the value of Poisson's ratio also behaves peculiarly in the region around 30°. The odd variation of both the coefficient of thermal expansion and Poisson's ratio stems from the magnitude and sign of the shear-extensional coupling in the individual lamina.

Previously, classes of laminates were shown to have isotropic stiffnesses in the plane of the laminate. Similarly, laminates that are isotropic in thermal expansion within the plane of the laminate can be specified. The requirements for thermal expansion isotropy are considerably less restrictive than those for elastic constants. In fact, any laminate that has two identical, orthogonal thermal expansion coefficients and a zero-shear thermal expansion coefficient is isotropic in thermal expansion. Thus, $[0°/90°]_s$ and $[\pm 45°]_s$ laminates are isotropic in thermal expansion even though they are not quasi-isotropic for elastic stiffnesses.

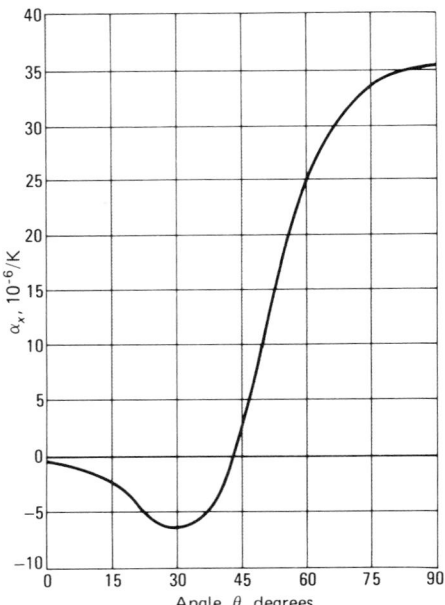

Fig. 9 Thermal expansion coefficients for high-modulus graphite-epoxy system, $[\pm\theta]_s$

Laminates that are isotropic in thermal expansion have thermal expansions of the form:

$$\{\alpha_x\} = \begin{Bmatrix} \alpha_x \\ \alpha_y \\ \alpha_{xy} \end{Bmatrix} = \begin{Bmatrix} \alpha^* \\ \alpha^* \\ 0 \end{Bmatrix} \qquad \text{(Eq 56)}$$

where the term α^* can be shown to be a function of lamina properties only, as follows:

$$\alpha^* = \alpha_1 + \frac{(\alpha_2 - \alpha_1)(1 + \nu_{12})}{1 + 2\nu_{12} + \dfrac{E_1}{E_2}}$$

Thus, for a given ply material, all laminates that are isotropic in thermal expansion have identical expansion coefficients.

Moisture Expansion. Resin matrix composites are said to be hygroelastic when the matrix absorbs and desorbs moisture from and to the environment. The primary effect of moisture sorption is a volumetric change in the laminae. When a lamina absorbs moisture, it expands; when moisture is lost, the lamina contracts. Thus, the effect is similar to thermal expansion.

In a lamina, a free moisture expansion coefficient vector can be defined as:

$$\{\epsilon_\ell\} = \{\beta_\ell\}\Delta M \qquad \text{(Eq 57)}$$

where

$$\{\beta_\ell\} = \begin{Bmatrix} \beta_1 \\ \beta_2 \\ 0 \end{Bmatrix}$$

and ΔM is moisture change by weight. Noting that the relations in Eq 57 are identical to those of thermal expansion, with $\{\beta_\ell\}$ substituted for $\{\alpha_\ell\}$ and ΔM substituted for ΔT, it can easily be seen that all of the relations developed for thermal effects can be used for moisture effects, as long as the substitutions mentioned are performed.

The procedure required to determine moisture gains or losses will be described in the section "Stresses Due to Temperature and Moisture" in this article. The analytical procedures detailed here allow for the prediction of moisture-induced stresses when the percent moisture change is known for the laminate. As with the thermoelastic analysis, only elastic effects have been treated.

Conductivity. The conductivity (thermal or moisture) of a laminate in the direction normal to the surface is equal to the transverse conductivity of a unidirectional fiber composite. This follows from the fact that normal conductivity for all plies is identical and unaffected by ply orientation.

The in-plane conductivities will be required for certain problems involving spatial variations of temperature and moisture. For a given uniform state of moisture in a laminate, the effective thermal conductivities in the x and y directions can be obtained by methods entirely analogous to those used previously for stiffnesses:

$$\mu_x = \frac{1}{2h} \sum_{i=1}^{K} (\mu_1 m^2 + \mu_2 n^2)\, t_{\text{ply}}^i \qquad \text{(Eq 58)}$$

$$\mu_y = \frac{1}{2h} \sum_{i=1}^{K} (\mu_1 n^2 + \mu_2 m^2)\, t_{\text{ply}}^i \qquad \text{(Eq 59)}$$

where μ_1 is the conductivity in the fiber direction, μ_2 is the conductivity transverse to the fibers, m is $\cos\theta^i$, n is $\sin\theta^i$, θ^i is the orientation of ply i, t_{ply} is the thickness of ply i, K is the total number of plies, and $2h$ is the laminate thickness. The results apply to symmetric as well as to asymmetric laminates. The results for moisture conductivity are identical.

Thermal and Hygroscopic Analysis

This section is concerned with the distribution of temperature and moisture through the thickness of a laminate. The mathematical descriptions of these two phenomena are identical and the physical effects are similar. Some of these aspects have already been discussed.

A free lamina undergoes stress-free deformation due to temperature change or moisture swelling. In a laminate, stress-free deformation is constrained by adjacent layers, producing internal stresses. In addition to the swelling-induced stresses, temperature and moisture content also affect the properties of the mate-

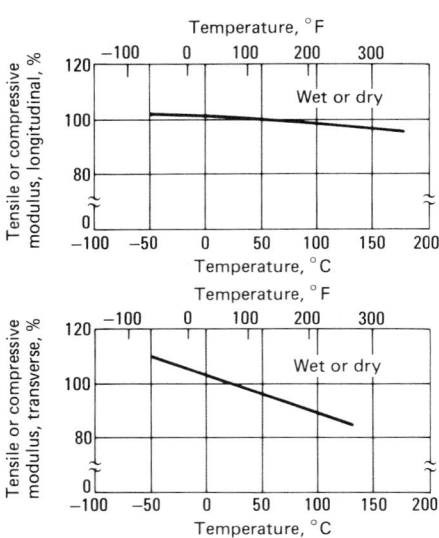

Fig. 10 Effects of moisture and temperature on lamina stiffness measured at room temperature

rial. These effects are primarily related to matrix-dominated strength properties. Figure 10 (Ref 1) indicates that for the typical graphite-epoxy system, moisture has little or no effect on stiffness. Figures 11 and 12 (Ref 1) show a large variation in strength, with the exception of axial tension. In each of these figures, the presence of moisture is seen to decrease the elevated-temperature strengths of the material while neither detrimentally affecting nor increasing the lower-temperature strengths.

The principal strength-degrading effect is related to a change in the glass transition temperature of the matrix material. As moisture is absorbed, the temperature at which the matrix changes from a glassy state to a viscous state decreases. Thus, the elevated-temperature strength properties decrease with increasing moisture. Limited data suggest, however, that this process is reversible. When the moisture content of the composite is decreased, the glass transition temperature increases and the original strength properties return.

In Fig. 10, 11, and 12 the term wet indicates that the material is fully saturated. This condition corresponds to a state of equilibrium with the environment at which the relative humidity is nearly 100%. The equilibrium moisture content at full saturation for typical epoxy composite systems ranges from about 1.0 to 2.5% weight gain.

All of these considerations also apply in the case of temperature rise in the sense that the matrix and therefore the laminae lose stiffness and strength when the temperature rises. Therefore, this effect is primarily important for matrix-dominated properties.

The differential equation governing time-dependent moisture sorption of an orthotropic homogeneous material is given by:

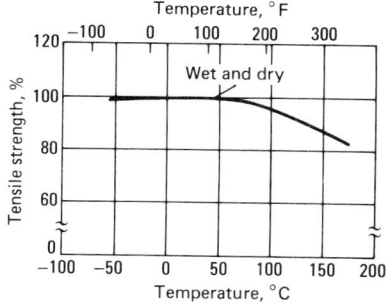

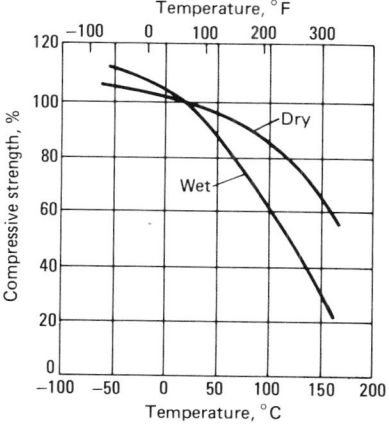

Fig. 11 Effects of moisture and temperature on lamina longitudinal strength, measured at room temperature

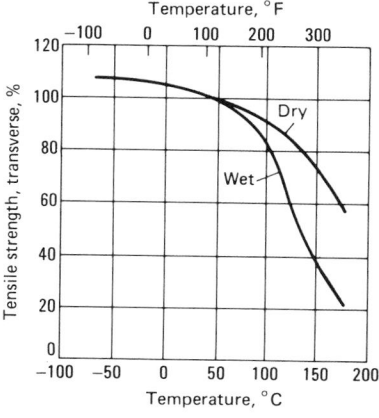

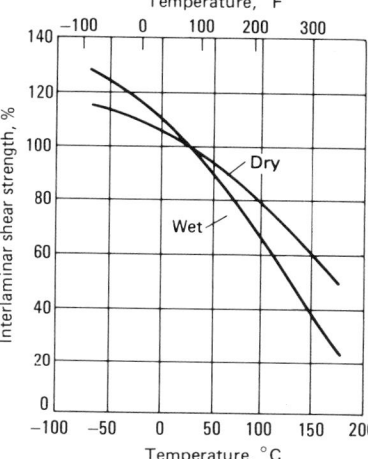

Fig. 12 Effects of moisture and temperature on lamina matrix-dominated strengths, measured at room temperature

$$D_1 \frac{\partial^2 c}{\partial x_1^2} + D_2 \frac{\partial^2 c}{\partial x_2^2} + D_3 \frac{\partial^2 c}{\partial x_3^2} = \frac{\partial c}{\partial t} \qquad \text{(Eq 60)}$$

where t is time, x_1, x_2, x_3 are coordinates in principal material directions, c is the specific moisture concentration, and D_1, D_2, and D_3 are moisture diffusivity coefficients. Equation 60 is based on Fick's law of moisture diffusion and is entirely analogous to the equation governing time-dependent heat conduction, with temperature,ϕ, replacing concentration, c, and thermal conductivities μ_1, μ_2, and μ_3 replacing the moisture diffusivities. For a transversely isotropic lamina, with x_1 in the fiber direction, x_2 in the transverse direction, and $x_3 = z$ in the direction normal to the lamina:

$$\begin{aligned} D_1 &= D_L \\ D_2 &= D_3 = D_T \end{aligned} \qquad \text{(Eq 61)}$$

These quantities are analogous to the thermal conductivities of a UDC.

An important special case is one-dimensional diffusion or conduction through the thickness of a lamina. In this case, Eq 60 reduces to:

$$D_T \frac{\partial^2 c}{\partial z^2} = \frac{\partial c}{\partial t} \qquad \text{(Eq 62)}$$

This equation also applies to moisture diffusion or thermal conduction through a laminate, in the direction normal to its laminae planes, since all laminae are homogeneous in the z direction with equal diffusion coefficients, $D_T = D_z$.

Equation 62 is applicable to the important problem of time-dependent moisture diffusion through a laminate where the two faces are in different moisture environments. After enough time has elapsed, the concentration settles down to a time-independent (so-called stationary) state. In this state, since c is no longer time-dependent, Eq 62 simplifies to:

$$\frac{d^2 c}{dz^2} = 0$$

Thus, c is a linear function of z, and if the laminate faces are in environments with constant saturation concentrations, c_1 and c_2, then:

$$c = \frac{1}{2} [(c_2 - c_1)z/h + c_2 + c_1] \qquad \text{(Eq 63)}$$

where the laminate thickness is $2h$, and z originates in the middle surface. In the important case where $c_2 = c_1$, Eq 63 reduces to:

$$c = c_1 = \text{constant} \qquad \text{(Eq 64)}$$

The foregoing discussion of moisture also applies to heat conduction.

Solutions to the time-dependent problem, Eq 62, are readily available, and considerable work has been performed in the area of moisture absorption (Ref 2). The most interesting feature of the solutions to Eq 62 relates to the magnitude of the coefficient D_z, which is a measure of the speed of moisture diffusion. In typical epoxy matrix systems, D is of the order of 645×10^{-8} mm^2/s (1×10^{-8} in.^{2}/s) to 645×10^{-10} mm^2/s (1×10^{-10} in.^{2}/s). The diffusion coefficient is sufficiently small that full saturation of a resin matrix composite may require months or years, even when subjected to 100% relative humidity.

The approach typically taken for design purposes is to assume a worst-case scenario. If the material is assumed to be fully saturated, reduced allowable strengths can be computed. This is a conservative approach, because typical service environments do not generate full saturation. It is used because it allows for inclusion of the effects of moisture in a relatively simple fashion. As the design data base and analytical methodologies mature, the development of more physically realistic methods can be expected. At present, however, the conservative approach is appropriate.

In the case of heat conduction, the time required to achieve the stationary state, the analogue of saturation, is extremely rapid. Thus, the transient time-dependent state is generally of little practical importance for laminates.

Laminate Stress Analysis

The physical properties defined in the section "Laminate Properties" in this article enable any laminate to be represented by an equivalent homogeneous anisotropic plate or shell element for the purpose of structural analysis. The results will be the definition of stress resultants, bending moments, temperature, and moisture at any point on the surface defining the plate. (Temperature and moisture distributions through the thickness of the plate will also be defined when they exist.) With this definition of the local values of the state variables, a laminate analysis is next performed to determine the state of stress in each lamina in order to assess margins for each critical design condition.

Stresses Due to Mechanical Loads. To determine stresses in the individual plies, the laminate midplane strain and curvature vectors must be used. Writing the laminate constitutive relations as:

$$\left\{ \frac{N}{M} \right\} = \left[\begin{array}{c|c} A & B \\ \hline B & D \end{array} \right] \left\{ \frac{\epsilon^\circ}{\kappa} \right\} \qquad \text{(Eq 65)}$$

shows that a simple inversion will yield the required relations for $\{\epsilon^\circ\}$ and $\{\kappa\}$. Thus:

$$\left\{\frac{\epsilon^\circ}{\kappa}\right\} = \left[\frac{A \mid B}{B \mid D}\right]^{-1} \left\{\frac{N}{M}\right\} \qquad \text{(Eq 66)}$$

Given the strain and curvature vectors, the total strain in the laminate can be written as:

$$\{\epsilon_x\} = \{\epsilon^\circ\} + z\{\kappa\} \qquad \text{(Eq 67)}$$

The strains at any point through the laminate thickness are now given as the superposition of the midplane strains and the curvatures multiplied by the distance from the midplane. Thus, the strain field at the center of ply i in a laminate can be seen to be:

$$\{\epsilon_x\}^i = \{\epsilon^\circ\} + \frac{1}{2}\{\kappa\}\,(z^i + z^{i+1}) \qquad \text{(Eq 68)}$$

where the term

$$\frac{1}{2}\,(z^i + z^{i+1})$$

corresponds to the distance from the midplane to the center of ply i. Curvature-induced strains at a point through the laminate thickness can be defined simply by specifying the distance to the point in question from the midplane.

The strains defined in Eq 68 correspond to the arbitrary laminate coordinate system. These strains can be transformed into the principal material coordinates for this ply by using transformations developed previously. Thus:

$$\{\epsilon_\ell\}^i = [\theta^i]^{-1}\{\epsilon_x\}^i \qquad \text{(Eq 69)}$$

where the superscript i indicates which ply and therefore which angle of orientation is to be used. Because the orientations of the various plies may differ, it is necessary to use the transformation matrix corresponding to the proper ply orientation.

With the strains in the principal material coordinates defined, stresses in these same coordinates are written using the lamina reduced stiffness matrix (Eq 5). Thus:

$$\{\sigma_\ell\}^i = [Q^i]\{\epsilon_\ell\}^i \qquad \text{(Eq 70)}$$

Again, it is important that the stiffness matrix used correspond to the correct ply, as each ply may be of a different material.

The stresses in the principal material coordinates can be determined without the use of principal material strains. Using the strains defined in the laminate coordinates (Eq 68) and the transformed lamina stiffness matrix (Eq 9,

11, 12), stresses in the laminate coordination system are written as:

$$\{\sigma_x\}^i = [\bar{Q}^i]\{\epsilon_x\}^i \qquad \text{(Eq 71)}$$

and these stresses are then transformed to the principal material coordinates. Thus:

$$\{\sigma_\ell\}^i = [\theta^i]^{-1}\{\sigma_x\}^i \qquad \text{(Eq 72)}$$

Reviewing these relations, it can be seen that for the case of symmetric laminates and membrane loading, the curvature vector is zero. This implies that the laminate coordinate strains are identical in each ply and equal to the midplane strains. The differing angular orientation of the various plies will promote different stress and strain fields in the principal material coordinates of each ply.

Stresses Due to Temperature and Moisture. When the equations for the thermoelastic response of composite laminates were developed, it was indicated that thermal loading in laminates can cause stresses within the plies even when the laminate is allowed to expand freely. The stresses are induced because of a mismatch in thermal expansion coefficients between plies oriented in different directions. Either the mechanical stresses of the preceding section or the thermomechanical stresses to be calculated in this section can be used to evaluate laminate strength.

To determine the magnitude of thermally induced stresses, the thermoelastic constitutive relations (Eq 46, 48) are required. Since free thermal effects require that $\{N\} = \{M\} = 0$, these relations are written as:

$$\left\{\frac{0}{0}\right\} = \left[\frac{A \mid B}{B \mid D}\right]\left\{\frac{\epsilon^\circ}{\kappa}\right\} + \left\{\frac{N^T}{M^T}\right\} \qquad \text{(Eq 73)}$$

Inverting these relations yields the free thermal strain and curvature vectors for the laminate:

$$-\left[\frac{A \mid B}{B \mid D}\right]^{-1}\left\{\frac{N^T}{M^T}\right\}\left\{\frac{\epsilon^\circ}{\kappa}\right\} \qquad \text{(Eq 74)}$$

Proceeding as before, the strain field in any ply is written as:

$$\{\epsilon_x\}^i = \{\epsilon^\circ\} + z^i\{\kappa\} \qquad \text{(Eq 75)}$$

Stresses in the laminate coordinates are written as:

$$\{\sigma_x\}^i = [\bar{Q}^i]\{\epsilon_x\}^i - \{\Gamma_x\}^i\,\Delta T^i \qquad \text{(Eq 76)}$$

which can then be transformed to the principal material coordinates. Thus:

$$\{\sigma_\ell\}^i = [\theta^i]^{-1}\{\sigma_x\}^i \qquad \text{(Eq 77)}$$

As was shown for mechanically induced loadings, the stresses can be found in another fashion. Transforming the strains of Eq 75 directly yields:

$$\{\epsilon_\ell\}^i = [\theta^i]^{-1}\{\epsilon_x\}^i \qquad \text{(Eq 78)}$$

Recalling Eq 44, stresses in the principal material coordinates are written as:

$$\{\sigma_\ell\}^i = [Q^i]\{\epsilon_\ell\}^i + \{\Gamma_\ell\}^i\,\Delta T^i \qquad \text{(Eq 79)}$$

Some interesting physical interpretations can be obtained by restricting the discussion to uniform-temperature fields and symmetric laminates. In the presence of these restrictions, the coupling matrix, $[B]$, and the thermal moment resultant vector, $\{M^T\}$, vanish, and:

$$\{\epsilon^\circ\} = \{\alpha_x\}\,\Delta T$$

and

$$\{\kappa\} = \{0\}$$

The strains in the laminate coordinates are identical in each ply. Thus:

$$\{\epsilon_x\}^i = \{\epsilon^\circ\} = \{\alpha_x\}\,\Delta T \qquad \text{(Eq 80)}$$

For this case, it can be shown that:

$$\{\sigma_x\}^i = [\bar{Q}^i]\,(\{\alpha\} - \{\alpha_x\}^i)\,\Delta T \qquad \text{(Eq 81)}$$

These relations indicate that stresses induced by the free thermal expansion of a laminate are related to the differences between the laminate and the ply thermal expansion vectors. Thus, the stresses are proportional to the difference between the amount the ply would freely expand and the amount the laminate will allow it to expand.

A further simplification can be found if the laminate under investigation is isotropic in thermal expansion. It can be shown that if this class of laminates is subjected to a uniform temperature change, the stresses in the principal material coordinates are identical in every ply. The stress vector can be shown to be:

$$\{\sigma_\ell\} = \frac{E_{11}\,(\alpha_{22} - \alpha_{11})\Delta T}{1 + 2\nu_{12} + \dfrac{E_{11}}{E_{22}}} \left\{\begin{array}{c} 1 \\ -1 \\ 0 \end{array}\right\} \qquad \text{(Eq 82)}$$

where the transverse direction stress is equal and opposite to the fiber direction stress. This is important because unidirectional transverse material strengths are typically an order of magnitude lower than fiber direction strengths.

A similar development can be generated for moisture-induced stresses. All of the results of this section apply when the moisture swelling coefficients, $\{\beta_1\}$, are substituted for the thermal expansion coefficients, $\{\alpha_1\}$.

Netting Analysis. Another approach to the calculation of ply stresses is sometimes used for membrane loading of laminates. This procedure, called netting analysis, treats the laminate as a net; that is, all loads are carried in the fibers while the matrix material is present only to hold the geometric position of the fibers.

Since only the fibers are assumed to be loaded in this analytical model, stress-strain relations in the principal material directions can be written as:

$$\sigma_{11} = E_1 \epsilon_{11}$$

or   (Eq 83)

$$\epsilon_{11} = \frac{1}{E_1} \sigma_{11}$$

and

$$E_2 = G_{12} = \sigma_{22} = \sigma_{12} = 0$$

The laminate stiffnesses predicted by means of a netting analysis will be smaller than those predicted using lamination theory, because of exclusion of the transverse and shear stiffnesses from the formulation. This effect is demonstrated in Table 3 for a quasi-isotropic laminate consisting of high-modulus graphite-epoxy. The stiffness properties predicted using a netting analysis are on the order of 10% smaller than those predicted using lamination theory. Experimental work has consistently shown that lamination theory predictions are more realistic.

Although the stiffness predictions using netting analyses are of limited value, such an analysis can be used as an approximation of the response of a composite in which the matrix has been damaged. In this sense it may be considered a worst-case analysis and is frequently used to predict ultimate strengths of composite laminates.

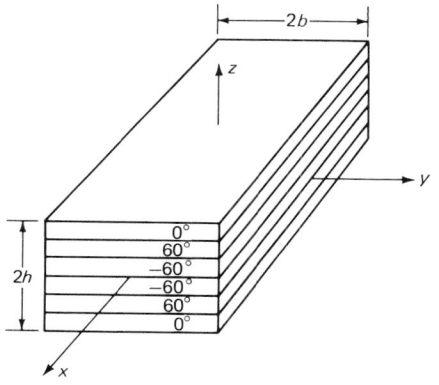

Fig. 13 Laminate construction (interlaminar stress example)

Table 3 Laminate elastic constants

Analysis	E_x		E_y		G_{xy}		
	GPa	10^6 psi	GPa	10^6 psi	GPa	10^6 psi	ν_{xy}
Lamination theory	64.9	9.42	64.9	9.42	24.5	3.55	0.325
Netting analysis	57.4	8.33	57.4	8.33	21.6	3.13	0.333

Note: Unit thickness: high-modulus graphite-epoxy

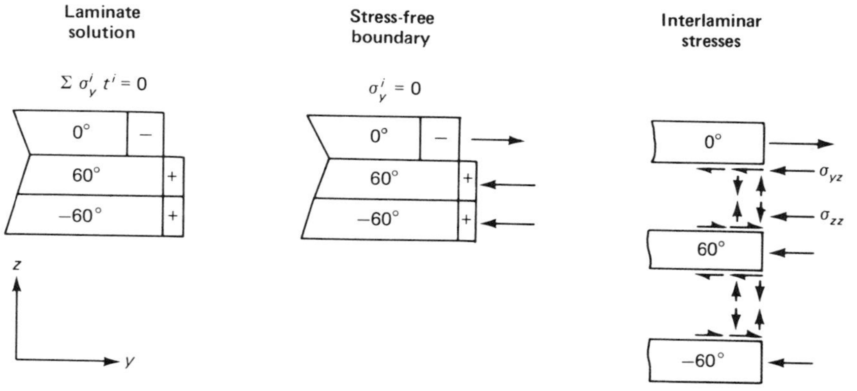

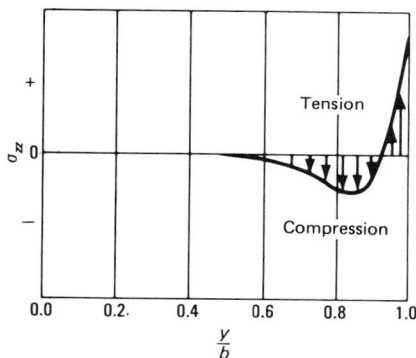

Fig. 14 Equilibrium and boundary conditions, y direction. t^i, thickness of ith ply

Interlaminar Stresses. The analytical procedures that have been developed in this article can be used to predict stresses within each lamina in a laminate. The stresses predicted are planar because of the assumed state of plane stress. There are cases for which the assumption of plane stress is not valid and a three-dimensional stress analysis is required.

An example of such a case exists at certain free edges in laminates where stress-free boundary conditions must be imposed. When the laminate in Fig. 13 is subjected to a membrane loading in the x direction, Poisson's ratio mismatches between the $0°$ plies and the $\pm 60°$ plies will introduce a stress in the y direction in each ply. Additionally, extensional-shear coupling effects will generate equal but opposite shear stresses in the $60°$ and $-60°$ plies. In Fig. 13, all σ_y and σ_{xy} stresses must vanish at $y = \pm b$ to satisfy the stress-free boundary conditions. This cannot be accomplished with lamination theory; a three-dimensional analysis is required.

Using equilibrium and symmetry considerations, the other components of stress which arise at the free edge can be visualized. In Fig. 14, lamination theory stresses are shown for the uniaxial loading discussed above; it can be seen that the laminate analysis satisfies equilibrium in the y direction but not the conditions at the free boundary. Thus, additional forces are required to satisfy the boundary conditions. For example, in the $0°$ plies, a compressive σ_y is predicted using lamination theory. This does not satisfy the boundary condition locally. A second, self-equilibrating set of edge stresses may be regarded as a superposed stress field that brings the lamination theory stresses to the necessary boundary values. Thus, an interlaminar shear stress, σ_{yz}, develops between the $0°$

and $60°$ plies. This introduces an x-moment disequilibrium even though the y-direction force equilibrium has been satisfied. To achieve x-moment equilibrium, a couple is needed. The required couple develops as an interlaminar normal stress distribution which, in Fig. 15, is seen to be compressive at $y = b$. For a couple to develop, the normal stress must change sign as the distance from the free edge increases.

In Fig. 16, a similar situation exists with respect to the in-plane shear stresses, σ_{xy}. At the free edge, $y = b$, these stresses must vanish. Therefore, the other component of interlaminar shear stress, σ_{xz}, develops, and the stress-free boundary conditions are satisfied. In each case it is seen that lamination theory satisfies strain compatibility and equilibrium, while the interlaminar stresses satisfy the stress-free boundary conditions.

The sign of the interlaminar normal stress at the free edge can be of great importance. As

Fig. 15 Interlaminar normal stresses and moment equilibrium

can be inferred from previous discussions, a large Poisson's ratio mismatch between plies produces large interlaminar normal stresses. If these stresses are tensile, the laminate may split between plies, or delaminate, under membrane loading. Obviously, if the laminate is designed such that expected loadings produce only compressive interlaminar normal stresses at the free edges, delamination will not occur.

The free-edge problem occurs wherever ply stresses must become zero to satisfy boundary conditions. Thus, any free edge, including holes, may introduce interlaminar stresses. The consequences of these stresses obviously need to be evaluated for structural applications, particularly where fatigue loadings are present. It can become extremely complex to provide quantitative analytical treatment of free-edge stresses in all but the most simple geometric shapes. In most cases, the possibility of delaminations due to the effects of free edge stresses is tested experimentally.

Nonlinear Stress Analysis. All of the preceding material in this article has related to laminae that behave in a linear elastic fashion. However, composites can behave in a nonlinear fashion because of such factors as internal damage or because of nonlinear stress-strain behavior of the polymeric matrix. Matrix nonlinearity (or microcracking) can result in laminae that have nonlinear stress-strain curves for transverse stress or axial shear stress. When this situation exists, the previously presented elastic laminate stress analysis must be replaced by a nonlinear analysis. A convenient procedure for doing this is found in Ref 3.

This subject is of importance primarily for such specialized applications as dimensionally stable space structures. Hence, further treatment is beyond the scope of this article.

Strength and Failure

Methods for stress analysis of laminates subjected to mechanical loads, temperature changes, and moisture absorption were described in the previous section. The main reason for performing a stress analysis is to determine or assess the strength of a laminate. The problem is of primary practical significance, because design without knowledge of strength characteristics is impossible. Such information for laminates cannot be obtained on a purely experimental basis, because the number of candidate laminates is immense, and each of them is an anisotropic structural element. Thus, in each case it would be necessary to determine a set of strengths and to incorporate them into a failure criterion for a laminate under combined load. The situation is much more complex than for a UDC where the one-dimensional strengths can be obtained experimentally and only one failure criterion under combined stress is needed. In the present case, however, the one-dimensional strengths of a laminate are by themselves of great variety, because they depend on stacking sequence and layer orienta-

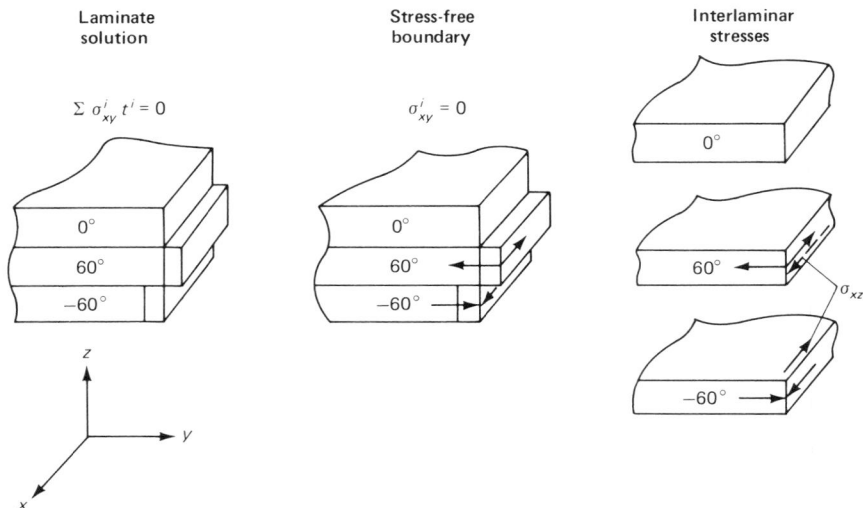

Fig. 16 Equilibrium and boundary conditions, x direction

tions. To complicate matters further, both in-plane and bending loads must be considered.

This situation clearly requires analytical criteria for strengths of laminates, which, unfortunately, are not yet available. Consequently, the approaches and methods discussed will be both quantitative and qualitative.

The case of a perfect laminate, which does not contain defects of any kind, is considered first. When such a laminate is loaded in its plane, a simple stress distribution results, consisting of plane stress in each layer and a complex three-dimensional stress distribution at the edges that involves interlaminar normal and shear stress. Some of these stress components may become large at the edges; theoretically, they may become infinite. This at once indicates several modes of initial failure. Away from the edges, a laminate may fail either in what may be called a fiber mode, implying rupture or buckling of fibers, or in what may be called a matrix mode, implying longitudinal cracking parallel to the fibers. Since the state of stress is planar, there are no other failure modes. The advantage of the criteria to be developed for these modes is that they predict not only failure but also the mode. At the edges, the interlaminar stresses may produce delamination as the initial failure. The major difficulties of analytical strength definition are determining failures subsequent to the initial failure and providing a criterion for when these will produce ultimate failure of the laminate.

Initial Failure. For a laminate that is loaded by in-plane forces and/or bending moments, the stresses in the layers at a sufficient distance from the edges can be obtained by the methods developed in the section "Laminate Stress Analysis" in this article. If there is no external bending, if the membrane forces are constant along the edges, and if the laminate is balanced and symmetric, the stresses in the kth layer are constant and planar. With reference to the material axes of the laminae, fiber direction x_1

and transverse direction x_2, they are written σ_{11}^k, σ_{22}^k, and σ_{12}^k. Failure will occur when either one of the following failure criteria is satisfied:

Fiber mode

$$\left(\frac{\sigma_{11}^k}{\sigma_L^u}\right)^2 + \left(\frac{\sigma_{12}^k}{\tau_L^u}\right)^2 = 1$$

Matrix mode

$$\left(\frac{\sigma_{22}^k}{\sigma_L^u}\right)^2 + \left(\frac{\sigma_{12}^k}{\tau_L^u}\right)^2 = 1 \qquad \text{(Eq 84)}$$

where σ_L^u is ultimate longitudinal stress and τ_L^u is ultimate longitudinal shear stress. It is necessary to distinguish between tensile and compressive normal stresses. Thus:

$$\sigma_L^u \begin{cases} \sigma_L^{tu} & \sigma_{11} > 0 \qquad \text{(a)} \\ \\ \sigma_L^{cu} & \sigma_{11} < 0 \qquad \text{(b)} \end{cases} \text{(Eq 85)}$$

$$\sigma_T^u \begin{cases} \sigma_T^{tu} & \sigma_{22} > 0 \qquad \text{(a)} \\ \\ \sigma_T^{cu} & \sigma_{22} < 0 \qquad \text{(b)} \end{cases} \text{(Eq 86)}$$

where σ_T^u is ultimate transverse stress, σ_L^{tu} and σ_T^{tu} are ultimate tensile stress, longitudinal and transverse; and σ_L^{cu} and σ_T^{cu} are ultimate compressive stress, longitudinal and transverse. Sometimes the simplistic maximum-stress criterion is used to assess the initiation of failure. For the kth lamina, the maximum-stress criterion assumes the form:

$$-\sigma_L^{cu} \leq \sigma_{11}^k \leq \sigma_L^{tu}$$
$$-\sigma_T^{cu} \leq \sigma_{22}^k \leq \sigma_T^{tu} \qquad \text{(Eq 87)}$$
$$|\sigma_{12}|^k \leq \tau_L^u$$

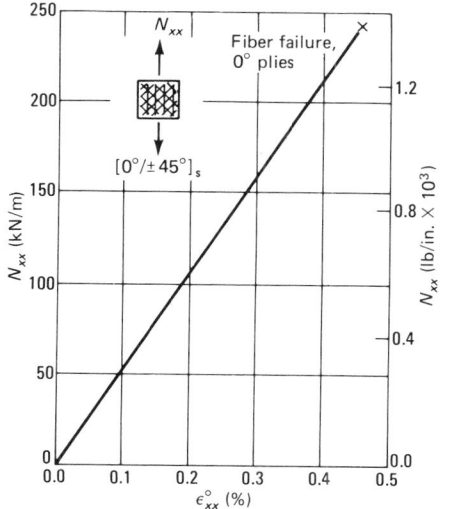

Fig. 17 Load-strain response, tensile N_{xx} loading

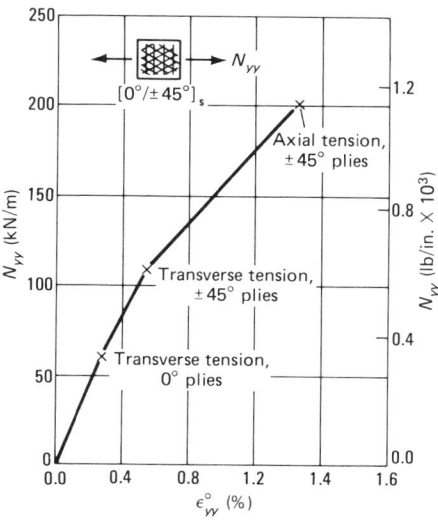

Fig. 18 Load-strain response, tensile N_{yy} loading

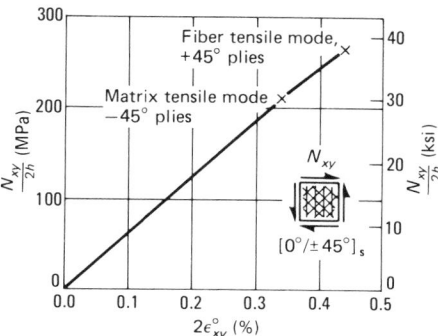

Fig. 19 Stress-strain response, N_{xy} loading

The maximum-stress criterion typically overestimates the strength of a lamina, because the interaction of stress components is neglected. Some stress interaction is obtained when the maximum-strain criterion is used. This involves replacing the stresses by ϵ_{11}, ϵ_{22}, and ϵ_{12} and replacing the ultimate stresses by the corresponding ultimate strains.

The failure criteria (Eq 84) must be used for each layer. The layer for which the criteria are satisfied by the lowest external load will define the load that produces the initial laminate failure, while also identifying the layer that fails and the nature of the failure, (that is, fiber failure or cracking along fibers). This is called first-ply failure.

Bending occurs when there are external bending and/or twisting moments or when the laminate is not symmetric. In these cases, the stresses σ_{11}^k, σ_{22}^k, and σ_{12}^k in a layer are linear in x_3. Consequently, the stresses assume their maximum and minimum values at the layer interfaces; thus, the failure criteria (Eq 84) must be examined for each layer at these locations.

The situation is much more complicated for the edge stresses, where the state of stress is three-dimensional and must be obtained by numerical analysis in each case. The major problem is that there are edge stress singularities, that is, some interlaminar stresses become theoretically infinite. Numerical methods cannot uncover the nature of a singularity, but there are analytical treatments (Ref 4) that do this. The major problem is how to assess the implication for failure of such edge stress singularities, a problem as yet unresolved. The situation is reminiscent of fracture mechanics in the sense that stresses at a crack tip are theoretically infinite, thus singular. Fracture mechanics copes with this difficulty in terms of a criterion for crack propagation that is based either on the amount of energy required to open

a crack or, equivalently, on the value of the stress intensity factor. Similar considerations seem to apply for edge singularities, but the situation is much more complicated, since a crack initiating at the edge will propagate between anisotropic layers. It appears, therefore, that the problem of edge failure must be relegated to experimentation for the time being.

Subsequent Failures. In many cases, laminates have considerable strength remaining after first-ply failure; the difficult and important problem arises of analytically determining subsequent failures. As noted, this problem is far from resolved, and it is an active area of research in composite materials. A simplistic, well known approach to this problem is as follows: When initial failure takes place, the failure may occur in the fiber or in the matrix mode. In the first case, the stiffness in the fiber direction, E_L, is reduced; in the second case, the elastic properties, E_T and G_L, are reduced. These elastic properties are not reduced to zero, because the initial failure produces cracks in the layer and the uncracked regions remain bonded to adjacent layers. Although a precise estimate of these stiffness reductions is not yet available, the simplest (but drastic) assumption is that E_L reduces to zero in the fiber mode and that E_T and G_L reduce to zero in the matrix mode. Since a laminate will usually not survive a fiber-mode failure, the progressive-failure cases of interest are initial and subsequent failures in the matrix mode. If it is assumed that, for failure of a lamina group in the matrix mode, E_T and G_L of that group can be set equal to zero, eventually the basic assumptions of netting analysis result. For netting analysis, the ultimate load is defined by the state at which E_T and G_L vanish in all laminae.

To demonstrate these effects, some specific examples of membrane loadings will be considered. The unidirectional lamina properties used in the examples are listed in Table 1.

The predicted load-strain response of a $(0°/\pm 45°)_s$ laminate subjected to three different membrane-loading environments is shown in Fig. 17, 18, and 19. For simplicity, failure analyses for these examples have been performed using the maximum-stress failure criterion.

The laminate loading in Fig. 17 consists of uniaxial tension in the direction of the 0° fibers, or the laminate x direction. The laminate response is seen to be linear to the load where the 0° fibers fail in axial tension. Because the failure mode is fiber breakage, no additional load-carrying capability exists.

In Fig. 18, the laminate loading consists of uniaxial tension perpendicular to the 0° fiber direction, or in the laminate y direction. The laminate response in this example is not linear to first fiber failure. Two distinct knees exist in the load-strain response. The first slope discontinuity corresponds to a transverse tensile failure in the 0° fibers. At this point, the lamina properties for the 0° plies are modified to account for the damage. Thus, for the 0° plies, the moduli E_2 and G_{12} are set to zero, and laminate properties are reformulated.

Subjecting this damaged laminate to additional loading then yields transverse tensile failure in both the 45° and −45° plies, resulting in the second knee in the load-strain curve. When a new laminate is again formulated with all plies having moduli E_2 and G_{12} set to zero, additional loading produces fiber compressive mode failure in the 0° plies (followed immediately by fiber tensile mode failure in the 45° and −45° plies), which corresponds to the ultimate strength under N_{yy} loading.

The final membrane-loading state considered is an applied N_{xy} (Fig. 19). The response under this shear loading produces a single matrix-mode failure before ultimate or fiber-mode failure.

Examination of the three figures reveals major differences in laminate response under different loading states, simply because the ply stress states are very different from each other. This is apparent in Table 4, in which the states of stress in each ply of the laminate are listed at first-ply failure under the three membrane loadings applied. The stresses listed in Table 4

Table 4 Ply stresses at first-ply failure, [0°/±45°]$_s$ laminate

Loading MPa (ksi)	0° plies						45° plies						−45° plies					
	σ_1		σ_2		σ_{12}		σ_1		σ_2		σ_{12}		σ_1		σ_2		σ_{12}	
	MPa	ksi	MPa	ksi	MPa	ksi	MPa	ksi	MPa	ksi	MPa	ksi	MPa	ksi	MPa	ksi	MPa	ksi
$\frac{N_{xx}}{2h} = 308$ (44.7)	760	110.0(a)	−25	−3.6	0.0	0.0	88.3	12.8	7.58	1.10	−35.4	−5.14	88.3	12.8	7.58	1.10	35.4	5.14
$\frac{N_{yy}}{2h} = 75.2$ (10.9)	−140	−19.9	30	4.0(a)	0.0	0.0	150	22.4	13.4	1.95	15.4	2.23	154	22.4	13.4	1.95	−15.4	−2.23
$\frac{N_{xy}}{2h} = 208$ (30.2)	0.0	0.0	0.0	0.0	30.0	4.35	570	82.2	−28	−4.0	0.0	0.0	567	−82.3	28	4.0(a)	0.0	0.0

(a) Critical stress value

demonstrate that at the three different first-ply failure loads, different stress components in different plies are critical. When subjected to N_{xx} loading, the fiber-direction stress in the 0° plies reaches the lamina strength value first, hence promoting failure. The other ply stresses are not critical at this load. Under N_{yy} loading, the first stress component to reach the lamina strength is the transverse stress in the 0° plies. Shear loading produces a critical stress in the −45° plies in transverse tension. Thus, the critical stresses depend on loading direction.

The differences between first-ply failure and ultimate failure are shown in Fig. 20, in which the laminate failure surface for combined in-plane extensional stresses is shown for the laminate discussed above. The dashed curve defines the combination of applied laminate stresses for which the first-ply failure occurs in the matrix mode. The solid curve defines the combined laminate stress states, which result in failure of a ply in the fiber mode. Note that for most positive values of N_{yy}, a matrix-mode failure precedes the first fiber-mode failure. Conversely, for negative values of N_{yy}, the first-ply failure is always a fiber-mode failure. The large region between these two surfaces indicates the potentially limiting effect of low transverse and shear strengths of the individual plies and hence the limiting effect of the matrix properties.

The related failure surfaces for combined laminate membrane extensional and shear stress are presented in Fig. 21, which shows that for large membrane shear stresses, the first-ply failure is in the matrix mode. Again, the potential for performance improvement through changes in matrix properties appears to be worth considering.

The failure surfaces of Fig. 20 and 21 illustrate the importance of knowing the difference between the loads causing first-ply failure and ultimate failure. It is first necessary to determine whether there is additional load capacity in the laminate beyond the point of first-ply failure. A necessary condition for this is stability of the laminate (that is, it must be able to carry loads) after the loss of laminae stiffnesses associated with matrix-mode failure. For example, an angle-ply laminate ($\pm\theta$) experiencing matrix-mode failure when subjected to an axial load will be unable to carry that load. However, any laminate having fibers oriented in three or

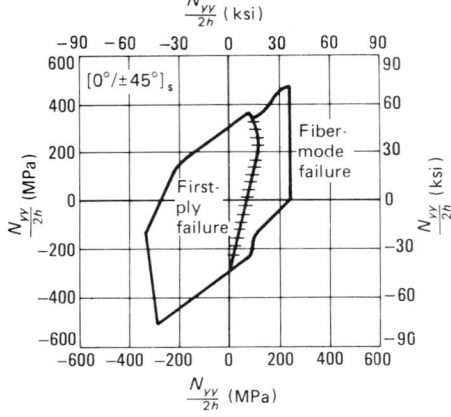

Fig. 20 Laminate failure surface for combined in-plane extensional stress states

more directions can generally carry load after matrix-mode failure. An estimate of load-carrying capability can be obtained by using netting analysis.

In Table 5, laminate analysis predictions for first-ply failure strength and ultimate failure strength are shown, as well as netting analysis predictions for ultimate failure strength. In both the N_{yy} and N_{xy} loadings, the netting-analysis strength predictions are considerably larger than the first-ply failure laminate analysis predictions, indicating that considerable strength remains beyond first-ply failure. The netting analysis, however, predicts a lower strength than lamination theory for N_{xx} loadings, indicating that first-ply failure corresponds to ultimate failure. Because of the trends of the type indicated in Table 5, netting analysis can be a useful tool in composite strength predictions.

Strength. A conventional definition of material strength is the maximum static stress, or combination of stresses, that can be carried by the material. For example, in a simple uniaxial tensile test, strength is the highest stress achieved before the material breaks. In a metal, other points on the material stress-strain curve have also been found to be significant in considering the maximum stress to be applied to a material; for example, the proportional-limit stress and the material 0.2% offset yield stress are valuable measures of material strength.

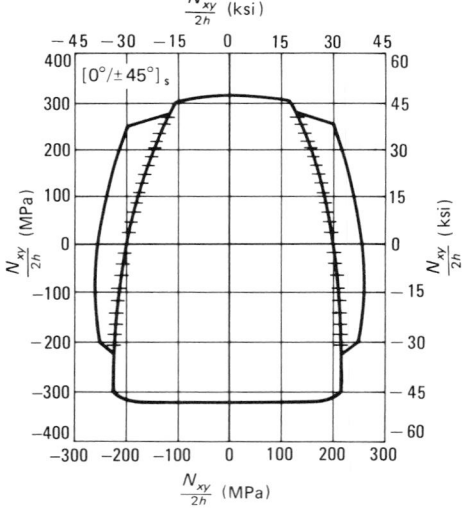

Fig. 21 Laminate failure surface for combined in-plane extensional and shear stress states

Similarly, in composites, several characteristic stress levels should be considered in the evaluation of strength. The primary stress levels of interest are the stress at which first-ply failure occurs and the maximum static stress (stress resultant divided by laminate thickness) that the laminate can carry.

In the calculation of first-ply failure, residual thermal stresses must be considered. Residual stresses are the ply stresses induced in a laminate as a result of fabrication processes. Typical resin matrix composites are formed under a combination of elevated temperature and pressure, which promotes matrix curing. When the laminate is subsequently cooled to room temperature, significant residual processing stresses develop. How to include these stresses in the failure analysis presents a problem. The rationale for including them is obvious: the stresses exist after processing and can therefore be expected to influence the occurrence of first-ply failure. However, there is also a rationale for not including them. Because all resin matrix materials exhibit viscoelastic, or time-dependent, effects, a significant portion of the residual processing stresses can be assumed to dissipate through stress relaxation. Addition-

Table 5 Failure loads and modes, [0°/±45°]$_s$ laminate

Loading type	First-ply failure load MPa	ksi	Failure mode	Ultimate failure load MPa	ksi	Failure mode	Ultimate load MPa	ksi	Failure mode
				Laminate analysis			Netting analysis		
$\frac{N_{xx}}{2h}$ 308		44.7	Axial tension, 0° plies	308	44.7	Axial tension, 0° plies	250	36.5	Axial tension, 0° plies
$\frac{N_{yy}}{2h}$ 75.2		10.9	Transverse tension, 0° plies	250	36.6	Axial tension, ±45° plies; axial compression, 0° plies	250	36.5	Axial tension, ±45° plies; axial compression, 0° plies
$\frac{N_{xy}}{2h}$ 208		30.2	Transverse tension, −45° plies	270	38.7	Axial tension, 45° plies	250	36.5	Axial tension, 45° plies; axial compression, −45° plies

ally, the processing stresses may be reduced through the introduction of transverse matrix microcracking. A limited amount of cracking may occur without significantly affecting the laminate elastic properties. This problem is complicated by the difficulty of measuring the residual stresses in a laminate and of observing first-ply failure during a laminate test.

From an analytical point of view, when laminae properties are known, it is possible to calculate a laminate stress-strain curve (including or excluding residual thermal stress effects) and determine first-ply failure, subsequent ply failures, and maximum stress at failure. From an experimental point of view, it is possible to measure a laminate stress-strain curve and determine proportional-limit stress and maximum stress. The existence of these different characteristic stress levels in laminated composite materials must be recognized, just as both a yield stress and an ultimate stress are known to characterize metallic materials. These stress levels for composites, as for metals, must be used with different factors of safety. At present, the proper action seems to be to use analytical strength predictions (both first-ply failure and ultimate failure) in the preliminary design phase, and experimental strength values for final design.

When experimental strength data are used for laminates, the problem of combined loading conditions arises again. The difficulties of combined-load testing preclude the purely experimental approach; the uncertainties of failure mechanism cloud the choice of analytical interaction criteria. A reasonable approach appears to be represented by the following sequence: analytical calculation of ultimate strength via netting analysis, or any other straightforward method; use of the analytical result to define the shape of the interaction curve; use of experimental data for single-stress components to define amplitudes of the interaction curve; and confirmation via limited combined-stress testing for a condition of practical interest.

It is common practice in the aerospace industry to neglect the residual thermal stresses in the calculations of ply failure. Data to support this approach do not appear to be available. However, at the present time, damage tolerance requirements limit allowable strain levels in laminates to a range of 3000 to 4000 μm/m.

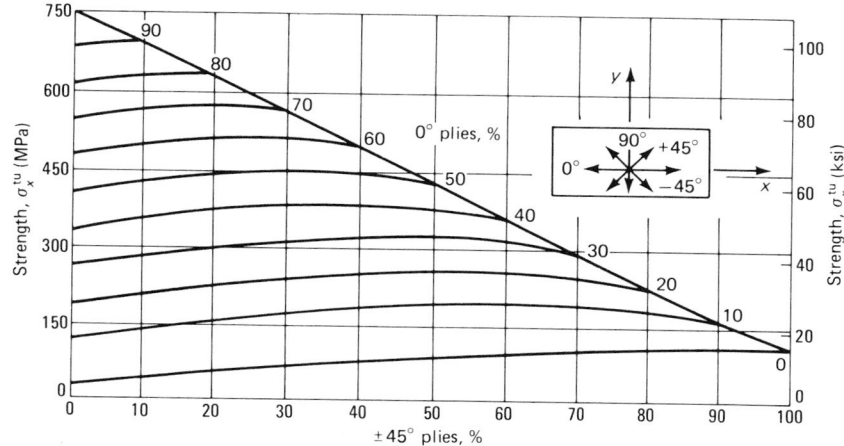

Fig. 22 Tensile strength of [0°$_i$/±45°$_j$/90°$_k$]$_s$ family of high-modulus graphite-epoxy laminates at room temperature

This becomes the dominant design restriction and eliminates the need to resolve the effects of residual thermal stresses. As materials of improved impact resistance are developed, first-ply failure calculations will become more significant. It appears that thermal stresses should be included in first-ply failure calculations unless their omission were to create an improved degree of conservatism. Careful stress-strain testing of laminate coupons should be used to define design allowables.

Whichever stress values are used, design charts in the form of so-called carpet plots are of great value for the selection of the appropriate laminate. Figure 22 presents a representative plot of this type for the axial tensile strength of laminates having various proportions of plies oriented at 0°, ±45°, and 90°. Appropriate strength data suitable for preliminary design can be found for various materials in Ref 1 and 5.

Fracture. The presence in a structure of a hole or other discontinuity introduces local stress concentrations which, if high, can result in initial localized failure. In metals, a common consequence of these effects is the formation of a small crack at the region of stress concentration. When there is a combination of high stress intensity and low resistance of the metal to crack propagation, the result is failure due to

cracking. This failure mechanism is generally termed fracture. The mathematical technology that has been developed to treat this problem uses fracture mechanics methods and is called structural-integrity technology (Ref 1).

For composite laminates, the problem is complicated by the heterogeneity of the material, both at the microscale level and on a layer-to-layer basis. The complexity of this problem is illustrated by the laminate analysis of a plate with a hole. In a symmetric, balanced laminate having a circular hole and subjected to in-plane axial tension, the plate may be regarded as an orthotropic plate whose properties are the effective laminate moduli defined in the section "Laminate Properties" in this article. Solutions for stress concentrations in such a plate are available (Ref 6). The stresses obtained from such a solution define local values of the stress resultants, which can be treated as the stress resultants applied to the actual laminate. In general, at any point in the plate this will define a combined loading with all three stress resultants being nonzero. Following the laminate stress analysis methods, the stress state in each ply can then be found, thereby defining the variation of ply stresses throughout the plate.

Sample results obtained by the above procedure are presented in Fig. 23. The tangential

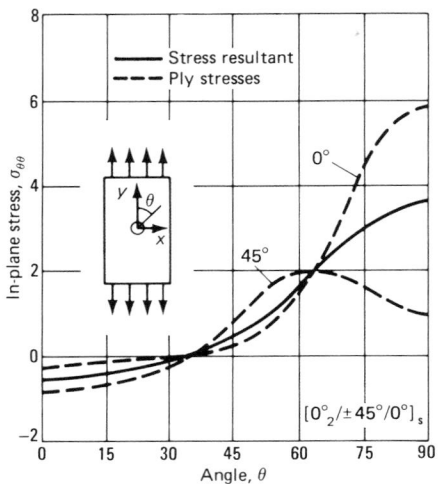

Fig. 23 Laminate theory calculations for stresses around a hole

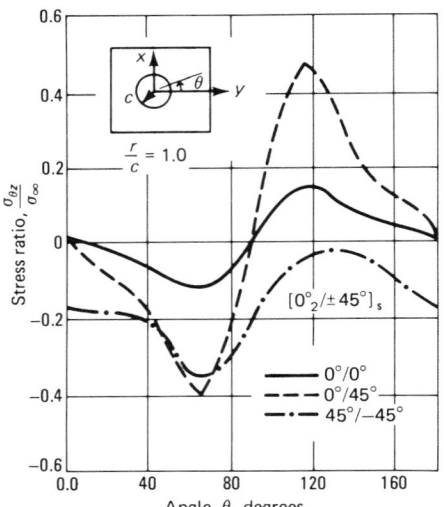

Fig. 24 Interlaminar shear stress at a hole

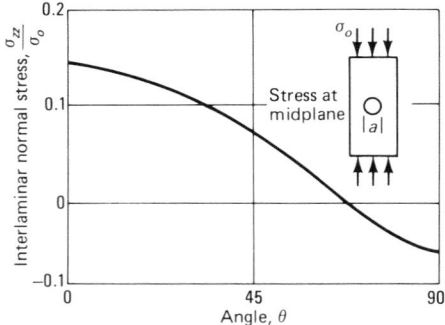

Fig. 25 Interlaminar normal stress at a hole

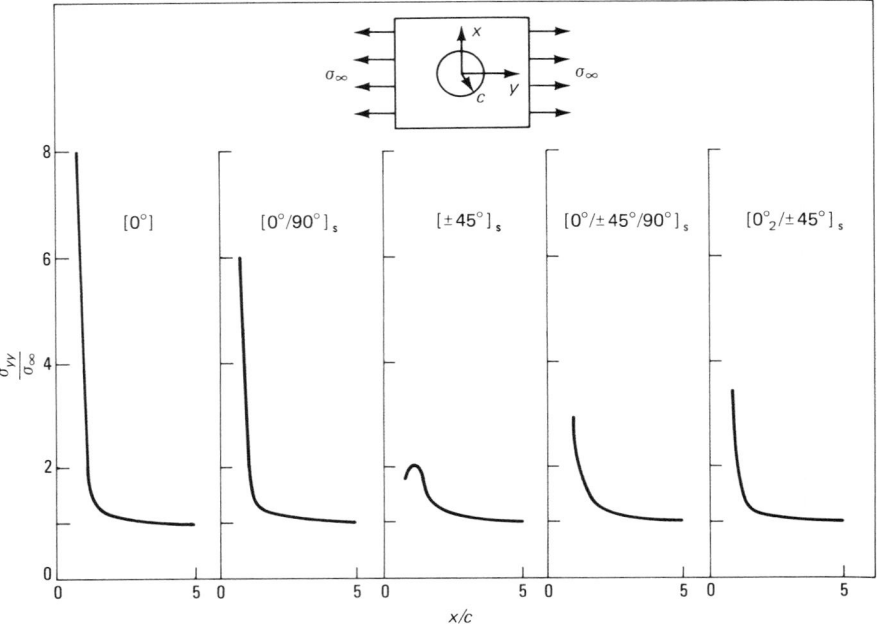

Fig. 26 Stress-concentration factors for a circular hole in a homogeneous, orthotropic, infinite plate

stress around the periphery of the hole is plotted. The solid line represents the stress from the orthotropic plate analysis, which is equivalent to the average value of the in-plane ply stresses. At the edge of the hole, these average tangential stresses are the only nonzero components of the stress resultants; elsewhere in the plate all three stress resultants generally will be nonzero. The dotted lines are the tangential stresses in the individual plies computed from laminate theory. In general, all the stress components in each ply will be nonzero, even at the edge of the hole. This is the same effect as treated in the discussion of interlaminar stress: The boundary conditions are satisfied in laminate analysis only by the average stresses.

In Fig. 23, the lamina stresses at the edge of the hole will not be correct, and interlaminar stresses will exist. These interlaminar shear stresses are shown in Fig. 24 at the various ply interfaces; the interlaminar normal stresses are shown in Fig. 25 at the midsurface interface. These interlaminar stresses are normalized with respect to the average applied stress. The magnitudes of these stresses are substantial.

Associated with the interlaminar stresses is a modification of the in-plane stress state in each ply. Thus, as described for the free-edge problem, the only nonzero in-plane stress component in each ply at the edge of the hole is the tangential stress, $\sigma_{\theta\theta}$. This combination of in-plane and interlaminar stresses varies strongly with the coordinate. The variation takes place near the hole, where the average in-plane stress components are also varying. This complex state of stress can cause various failure mechanisms. Clearly, interlaminar disbonding is a possibility at various locations at or near the edge of the hole. Intraply matrix-mode failure in any ply is another possibility. At elevated stress levels, it is reasonable to expect diverse local matrix failures. This complex problem can be treated by the method described below.

Using the laminate properties methods, the effective in-plane laminate stiffnesses, E_x, E_y, and G_{xy}, may be calculated for any laminate. With these properties specified, a balanced symmetric laminate may be regarded, for purposes of structural stress analysis, as a homogeneous, orthotropic plate. Orthotropic elasticity theory may be used for evaluating stresses around a hole in such a plate (Ref 6). Examples of the resulting stress concentrations are shown in Fig. 26 for various carbon-epoxy laminates. The laminae orientation combinations influence both the magnitude and the shape of the stress variation in the vicinity of the hole. The high stresses at the edge of the hole may initiate fracture.

If the laminate fails as a brittle material, fracture will be initiated when the maximum tensile stress at the edge of the hole equals the strength of the unnotched material. In a tensile coupon with a hole, as in Fig. 26, the failure will occur at the minimum cross section and will initiate at the edge of the hole, where the stress concentration is at a maximum. The stress-concentration factor of interest is one that includes the finite-width effect, illustrated in Fig. 27 for an isotropic plate with a central, circular hole. Stress distributions are shown for various ratios of hole diameter, a, to plate width, W.

The basic stress-concentration factor for this problem is the ratio of the axial stress at the edge of the hole ($x = a/2$ and $y = 0$) to the applied axial stress, σ_{∞}. For small holes in an isotropic plate, this factor is equal to 3. For large hole sizes, the average stress at the minimum section, σ_n, is higher than the applied stress, σ_{∞}, and is given by the relation:

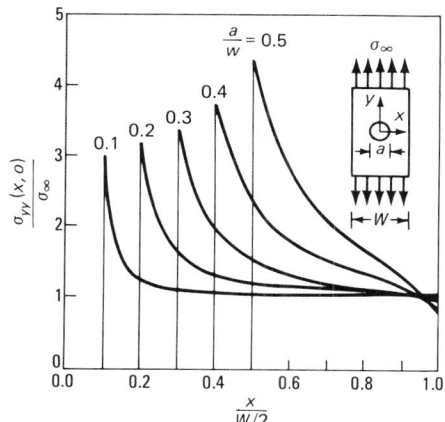

Fig. 27 Stress distribution around a hole in a finite-width plate

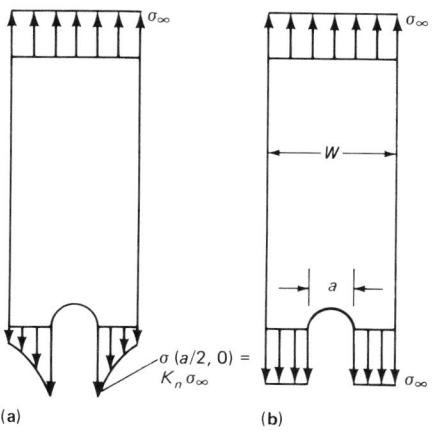

Fig. 28 Idealized limit cases for failure. (a) Notch-sensitive. (b) Notch-insensitive

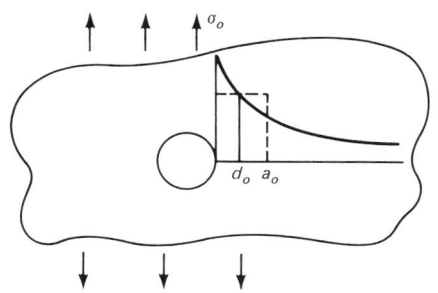

a_o Parameter for average-stress theory
d_o Parameter for point stress theory

Fig. 29 Representation of characteristic dimensions for semiempirical fracture theories

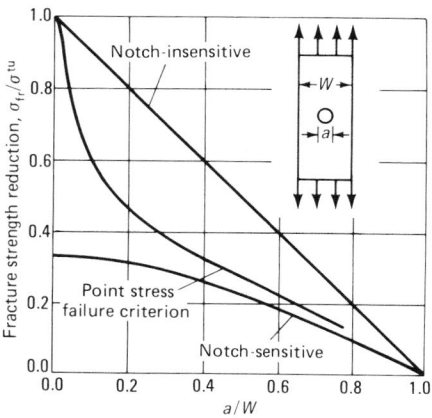

Fig. 30 Approximate-failure theories

$$\sigma_n = \frac{\sigma_\infty}{\left(1 - \dfrac{a}{W}\right)} \qquad \text{(Eq 88)}$$

The net section stress-concentration factor, K_n, is the ratio of the maximum stress to this average stress. Thus:

$$K_n = \frac{\sigma\left(\dfrac{a}{2}, 0\right)}{\dfrac{\sigma_\infty}{1 - \dfrac{a}{W}}} \qquad \text{(Eq 89)}$$

Laminate fracture for the elastic-brittle case will occur at stress σ_{fr}, that is:

$$\sigma_{fr} = \frac{\sigma^{tu}}{K_n} \qquad \text{(Eq 90)}$$

A material which fails in this fashion is called a notch-sensitive material. In contrast, a ductile material will yield locally, alleviating the stress-concentration effect. In the extreme, this will result in a uniform stress distribution at the net section, as shown in Fig. 28. For this case, failure will occur when the average net section stress equals the material strength, or:

$$\sigma_{fr} = \sigma^{tu}\left(1 - \frac{a}{W}\right) \qquad \text{(Eq 91)}$$

A material that fails in this way is known as a notch-insensitive material. Practical laminates may be expected to fail in a fashion which is neither of these two extreme cases.

Various matrix damage effects are expected to occur at the maximum-stress locations. These localized damages reduce the material stiffness and diminish and spread the stress-concentration effects. As a result, although fracture will occur because of stress concentration, it can be expected to occur because of lesser stress-concentration factors than those

calculated from orthotropic elastic stress analysis of the notched laminate. Semiempirical methods for evaluating this reduction in stress concentration have been proposed.

The point stress theory (Ref 7) proposes that the elastic-stress distribution curve (Fig. 26) be used but that the stress concentration be selected at a distance, d_o, from the edge of the hole. Thus, the numerator of Eq 89 should be evaluated at the point $x = a/2 + d_o$. The characteristic length d_o must be evaluated experimentally. The average-stress theory (Ref 7) takes a similar approach by proposing that the elastic-stress distribution be averaged over a distance, a_o, to obtain the stress concentration. Thus:

$$K_n = \frac{\displaystyle\int_{a/2}^{(a/2)+a_o} \sigma_y \, d_x}{\displaystyle\int_{a/2}^{W/2} \sigma_y \, d_x} \qquad \text{(Eq 92)}$$

Again, the characteristic dimension, a_o, must be found experimentally. The two approaches are illustrated in Fig. 29. For both methods, the resulting stress concentration is used in Eq 90 to define the fracture stress.

Representative results are plotted in Fig. 30 to illustrate the differences associated with the

different types of material behavior. The ratio of strength of the plate with a hole, σ_f, to that of an unnotched laminate, σ^{tu}, is plotted as a function of hole size. For the notch-insensitive material, this course clearly shows a linear decrease in failure stress with hole size due to the reduction in net cross-sectional area. The lower curve, for a notch-sensitive material, shows an immediate drop in strength with the introduction of even a small hole due to the local stress-concentration effect. The semiempirical theories (both have similar effects) show a rapid but more modest drop due to stress concentration. Experimental data for laminates support the use of either of the semiempirical theories for fracture of a laminate with a hole.

REFERENCES

1. *Advanced Composites Design Guide*, 3rd ed., North American Rockwell Corporation, AFML F33615-71-C-1362, Air Force Materials Laboratory, Jan 1973
2. C. Shen and G.S. Springer, Moisture Absorption and Desorption of Composite Materials, *J. Compos. Mater.*, Vol 10, 1976, p 1
3. Z. Hashin, D. Bagchi, and B.W. Rosen, "Nonlinear Behavior of Fiber Composite Laminates," NASA CR-2313, National Aeronautics and Space Administration, April 1974
4. S.S. Wang and I. Choi, Boundary Layer Thermal Stresses in Angle-Ply Composite Laminates, in *Modern Developments in Composite Materials and Structures*, J.R. Vinson, Ed., American Society of Mechanical Engineers, 1979, p 315-342
5. *Plastics for Aerospace Vehicles, Part I, Reinforced Plastics*, MIL-HDBK-17A, Department of Defense, Jan 1971
6. G.N. Savin, "Stress Distribution Around Holes," NASA TT-F-607, National Aeronautics and Space Administration, Nov 1970
7. J.M. Whitney and R.J. Nuismer, Stress Fracture Criteria for Laminated Composites Containing Stress Concentrations, *J. Compos. Mater.*, Vol 8, 1974, p 253-265

Strength, Failure, and Fatigue Analysis of Laminates

A.S.D. Wang, Drexel University

DIMENSIONAL SCALE is an important factor in the failure analysis of engineering materials and structures. It is commonly thought that an engineering material is a homogeneous substance from which large-sized structures or structural components are fabricated. Consequently, it is assumed a given material has a set of elastic and strength properties, and that these properties are not affected by any extrinsic geometric factors. Thus, one common requirement in structural design and analysis is to determine the elastic and strength properties of the material, and local material failure is then determined solely from its strength properties. Only an overall structural failure would involve additional influencing factors of geometric origin.

When viewed at the fiber-matrix scale, fiber-reinforced composites are seen as a complex structure rather than a basic homogeneous substance. Depending on the dimensional scale of interest, a composite laminate is an even more complex structure. The host of geometric factors that affect laminate failure behavior are shown in Table 1. Note that the physical sizes of these factors span a wide range of dimensional scale. While some of the factors are well defined geometrically, many are essentially probabilistic in nature.

For this reason, failure analysis of laminates is not a commonly used term. Here, the selection of a proper dimensional scale for analysis is of utmost importance. For instance, microscopic failures involve factors of size of the order of the fiber diameter; sublaminate failures involve additional factors of size on the order of the laminating layer thickness; and global failure adds factors relating to the overall structural configuration of the laminate.

The typical approach to laminate failure analysis has been to regard the laminate as a quasi-structure fabricated by one or more laminating ply materials, such as the unidirectional ply. Each ply material is assumed to be a homogeneous medium, even though it is heterogeneous at the fiber-matrix scale. In this sense, each material ply is assumed to possess a given set of elastic properties; and the laminate is a structure consisting of many distinct plies. Thus, one concept of laminate failure is based on the individual ply failure being governed by its strength properties. Another concept is based on local fracture mechanisms that initiate and propagate in the laminate. In this case, determination of material fracture resistance relative to the considered fracture mechanisms is required and replaces the ply strength properties. In this view, laminate failure then occurs at some critical or unstable fracture propagation state.

This article reviews current methods of laminate failure analysis for both these concepts. Because of the exclusion of the effects of material heterogeneity at the fiber-matrix scale, the analysis methods reviewed in this article are generally limited to laminate failures caused by statically applied loads and, to some extent, by low-frequency cyclically applied loads.

Laminate Stress Analysis Approaches

Composite laminates used in structural applications are most efficient in the form of a plate, a curved panel, or a shell. In these structures, principal stresses develop in the plane of the laminate, while out-of-plane stresses are gener-ally secondary, except near locations of abrupt geometric change, such as near a small hole, a cut-out, or free edges. Near these locations, a localized three-dimensional stress state is present.

The classical laminated plate theory (Ref 1) is normally used to analyze laminate stress fields that are free from local stress concentrations. Solution methods based on this theory provide only the in-plane stresses in each lamina of the laminate. These lamina stress solutions are incorporated in laminate failure analysis procedures developed on the basis of the ply failure concept.

Table 1 Geometric variables and size factors (fiber-reinforced composite laminates)

Variables	~ Size
Fiber diam	$d = 5\text{-}10 \ \mu m$
Fiber volume fraction	...
Microdefects/voids	$1\text{-}10d$
Defects/voids content.	...
Ply thickness	$2t = 20\text{-}100d$
Number of plies	...
Ply angle orientation/sequence	...
Laminate thickness	$h = 10\text{-}100t$
Laminate defects.	$1\text{-}10h$
Laminate plane dimensions	$> 100h$

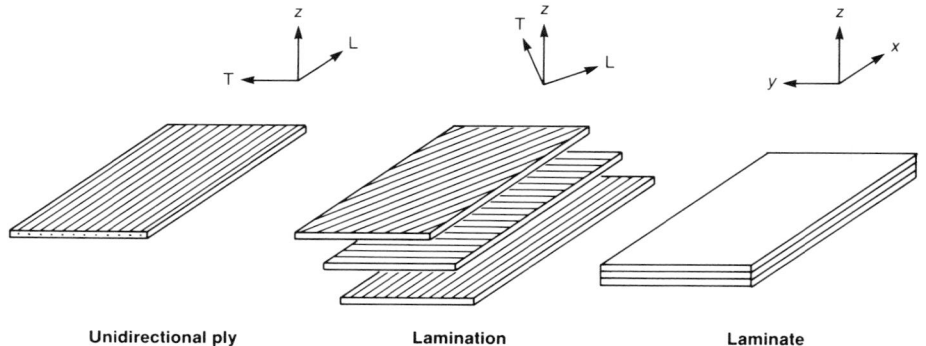

Fig. 1 Material (ply) and global (laminate) reference systems

Unidirectional ply   Lamination   Laminate

Near locations of abrupt geometric change, however, the highly concentrated stress field requires a three-dimensional analysis. This is commonly done by means of the theory of anisotropic elasticity (Ref 2) applied to layered media. Such analyses provide both the intraply and the interply stress components. Because of the sharp stress concentrations in all directions, local failure modes are usually complicated. Typically, multiple intraply cracks and interply delamination occur during the course of the applied load. Fracture mechanics approaches, describing crack initiation and propagation mode, have been employed to treat laminate failure under these conditions.

Hence, depending on the requirement of the analysis, the basic lamina can be characterized either as a two-dimensional or a three-dimensional elastic solid.

For the purpose of clarity, let the basic laminating ply be the unidirectionally fiber-reinforced ply, as depicted in Fig. 1. Two reference frames will be used here. The (L,T,z) frame refers to the principal material directions of the ply, with L in the longitudinal fiber direction, T transverse to the fiber direction, and z normal to the ply plane. The (x,y,z) frame refers to the global coordinates of the laminate. Normally, the x-y plane is the midplane of the laminate. The orientation of a ply with respect to the laminate is defined by the angle θ, rotating counterclockwise from L to x.

The assumption of ply material homogeneity requires that the thickness, t, of the unidirectional ply be substantially larger than the fiber diameter, d (for example, most carbon fiber reinforced epoxy resin systems have a d/t ratio on the order of 1/30). The assumption of ply elasticity provides the constitutive relations in the form of the generalized Hooke's law. Thus, if the ply is characterized as a three-dimensional body, the stress-strain relations, expressed in contracted notation, are given (by Ref 1) as:

$$\sigma_i = C_{ij} \epsilon_j \qquad \text{(Eq 1)}$$

where $i,j = 1, 2, 3, 4, 5, 6$.

If the ply is characterized as a two-dimensional body, assuming plane stress in the x-y plane, Eq 1 reduces to:

$$\sigma_i = Q_{ij} \epsilon_j \qquad \text{(Eq 2)}$$

where $i,j = 1, 2, 6$ and Q_{ij} is the reduced stiffness, related to the stiffness C_{ij} in the form:

$$Q_{ij} = C_{ij} - \frac{(C_{i3}C_{j3})}{C_{33}} \qquad \text{(Eq 3)}$$

where $i,j = 1, 2, 6$.

Both C_{ij} and Q_{ij} are symmetric matrices that can be expressed in either of the two reference frames. By referring to the ply reference frame, (L,T,z), material symmetry can be observed whenever it is appropriate. Typically, the unidirectional ply is assumed to possess either an orthogonal symmetry with respect to (L,T,z), or a transversely isotropic symmetry with the

T-z plane being the plane of isotropy. In some cases, monoclinic symmetry with respect to the z = 0 plane has also been assumed (Ref 1).

Material symmetry assumption determines the number of independent elastic constants required in C_{ij} or Q_{ij}. For a three-dimensional body of monoclinic symmetry, for example, it requires 13 constants in C_{ij}, for orthotropic symmetry, it requires nine constants, and for transversely isotropic symmetry, it requires five constants. For a two-dimensional body, orthotropy and transverse isotropy coincide; only four independent constants are required in Q_{ij}. Table 2 lists the engineering constants required for orthotropic and transversely isotropic plies under two- and three-dimensional assumptions. The mathematical relationships between the engineering constants and the constants in the stiffness matrices C_{ij} or Q_{ij} are found in textbooks of composite materials (for example, Ref 1).

Because most laminates are fabricated at an elevated temperature, residual thermal strains and stresses are present in the laminating plies after they are cooled down to the ambient temperature. To incorporate the thermal strains, the material ply is assumed to expand linearly with increasing temperature. The stress-strain relations in Eq 1 and 2 take the respective forms:

$$\sigma_i = C_{ij}(\epsilon_j - \alpha_i \Delta T) \qquad \text{(Eq 4)}$$

where $i,j = 1, 2, 3, 4, 5, 6$, and

$$\sigma_i = Q_{ij}(\epsilon_j - \alpha_i \Delta T) \qquad \text{(Eq 5)}$$

where $i,j = 1, 2, 6$ and where α_i are the thermal expansion coefficients of the ply and ΔT is the temperature difference. Table 2 lists the coefficients of thermal expansion required for orthotropic and transversely isotropic plies.

Eq 5 provides the ply constitutive relations for the two-dimensional laminated plate theory, while Eq 4 provides the same for three-dimensional ply elasticity formulations. The required elastic constants in each case are obtained either by direct experiments using the unidirectional ply material (Ref 3) or by a micromechanics analysis based on the exact fiber and matrix microstructure of the unidirectional ply (as seen in Ref 4, for example).

Ply Strength Property Characterization

The concept of ply failures for laminate failure analysis requires the determination of the strength constants of the unidirectional ply. In theory, the ply strength constants can be determined from a micromechanics analysis of the exact fiber-matrix structure, given the strength properties of the constituents. In practice, however, ply strength constants are frequently determined experimentally along with the ply elastic constants (as seen in Ref 3, for example).

The apparent strength properties of composites are known to vary with the loading method

Table 2 Engineering elastic constants (orthotropic and transversely isotropic plies)

Ply symmetry	Three-dimensional assumption	Two-dimensional assumption
Orthotropy	$E_L\ E_T\ E_z$	$E_L\ E_T$
	$\nu_{LT}\ \nu_{Lz}\ \nu_{Tz}$	ν_{LT}
	$G_{LT}\ G_{Lz}\ G_{Tz}$	G_{LT}
	$\alpha_L\ \alpha_T\ \alpha_z$	$\alpha_L\ \alpha_T$
Transverse isotropy	$E_L\ E_T = E_z$	$E_L\ E_T$
	$\nu_{Tz}\ \nu_{LT} = \nu_{Lz}$	ν_{LT}
	$G_{LT} = G_{Lz}$	G_{LT}
	$G_{Tz} = 2(1 + \nu_{Tz})/E_T$	
	$\alpha_L\ \alpha_T = \alpha_z$	$\alpha_L\ \alpha_T$

and the associated failure mode. The factors determining failure modes are related to the fiber-matrix microstructure of the ply. The latter, however, is disregarded in the material homogeneity assumption. For these reasons, the apparent ply strength constants can exhibit a wide range of scatter for a given unidirectional ply system. When used in the context of ply failure analysis, only the average values are used.

Then, for a two-dimensional unidirectional ply, the following five strength constants are typically determined.

Uniaxial Tension. The limiting tensile stress in the fiber direction, denoted by X_t, is defined as the ply axial tensile strength. This quantity is routinely determined by testing unidirectional laminate specimens under simple tension. Experience has shown that X_t can have a wide range of experimental values for a given ply system (Ref 3). Generally, the spread of the X_t values is not symmetric about its mean value. A skewed distribution, such as the Weibull function, is usually required to describe the scatter. The physical reasons for the skewness of the scatter are attributed to the various ply failure modes that occur during the complicated ply failure process. In the context of ply strength, a fixed value is assigned to X_t. It is usually the mean or the characteristic value of the strength distribution. In any event, X_t is a fiber-controlled variable.

Uniaxial Compression. The limiting compressive stress in the fiber direction, denoted by X_c, is defined as the ply axial compressive strength. It is usually determined by testing unidirectional laminate specimens under simple compression. In this case, an effort must be made to circumvent global buckling of a test-piece (Ref 3). Even so, the specimen is still capable of failing in a number of possible modes, including fiber microbuckling, fiber-matrix interfacial debonding, and kinkband formation. As expected, X_c is also a variable with large scatter, but unlike X_t, it is essentially a matrix-controlled variable.

Transverse Tension. The limiting tensile stress transverse to the fibers, denoted by Y_t, is defined as the ply transverse tensile strength. This variable is sensitive to the fiber-matrix interfacial bonding strength and to material defects such as voids in the matrix, broken fibers, and fiber-matrix disbonds. These microanomalies cause tensile crack propagation and crack coalescence in the matrix phase of the material, and significantly lower the apparent ply transverse strength. As can be expected, Y_t can be extremely scattered. Because of the very low values of Y_t, designers usually avoid loading transverse to the ply. Even so, this property plays a dominant role in initiating ply failures in multidirectional laminates.

Transverse Compression. This variable is denoted by Y_c. It can be determined, at least in principle, by a simple compression applied in the plane of the ply and transverse to the fibers. Such a loading state exists, for instance, in the extreme compressive surface layer of a sandwiched beam subject to pure bending. Failure modes under this or other simulated loading states are sensitive to factors such as the thickness of the compressed surface layer, interfacial bonding between the compressed surface layer and its substrate material, and so forth. Because these factors can determine the likelihood of delamination and subsequent buckling of the compressed layer, the apparent value of Y_c can vary significantly, depending on the test method. However, for most polymer-based composites, the mean value of Y_c is several times greater than Y_t. Hence, its practical role in causing laminate failure is relatively insignificant.

In-Plane Shear. This strength property, denoted by S, is determined by subjecting the unidirectional ply to a pure plane shear in the L-T plane. In principle, such a pure shear state can be created by torsion of a thin-walled tube, with the fibers oriented either in the longitudinal or the circumferential direction of the tube. A biaxial test of a flat, unidirectional laminate with equal tension and compression applied 45° from the fiber orientation can also create a pure shear state in the L-T plane. But these test methods, aside from being difficult to perform, also create many undesirable effects and can cause failure of the testpiece in a way unassociated with a pure shear. In some cases, an indirect method is used, for example, by testing a $[\pm45°]_s$ laminate specimen under axial tension. In such tests, a nonlinear shear stress-strain relation with excessively large shear strain has been observed (Ref 3). Shear strength derived from such a test can also be complicated by other extrinsic effects.

Clearly, the strength constants of a ply as discussed above are not a well-defined ply property. Even within the context of ply failure analysis, one must exercise caution as well as engineering judgment in their use.

Ply Failure Criteria Based on Stress at a Point

As has been mentioned, the first concept of laminate failure is to assume that one or more of the individual plies fail before the laminate does. Failure of any one ply is determined by the stresses in that ply, which satisfy a certain chosen criterion. Several schools of thought have been followed by numerous analysts in the field to formulate a ply failure theory that is appropriate for the material used. There are 50 such theories available in the open literature, according to a recent survey by M.N. Nahas (Ref 5). The majority of these theories is based on the assumption of failure at a point. Specifically, whenever the stress state at a point in the ply satisfies a certain critical condition, the material at that point fails. When this happens, the entire ply fails.

As for the critical condition for failure, it is usually formulated on the basis of some physical or purely mathematical reasoning. Only two commonly used ply failure criteria and their use in laminate failure analysis are described in this article.

Maximum Stress Criterion. When referring to the L-T plane of the ply, the stress state $[\sigma_1, \sigma_2, \sigma_6]$ at a point will cause failure if one or more of the following conditions are satisfied:

$$\sigma_1 \geq X_t \text{ or } \sigma_1 \leq X_c \qquad \sigma_2 \geq Y_t \text{ or } \sigma_2 \leq Y_c$$

$$|\sigma_6| \geq S \qquad\qquad \text{(Eq 6)}$$

Note that X_c and Y_c are negative, while X_t, Y_t, and S are positive.

Individually, these five conditions apply in the mean to the five test cases discussed previously in the section "Ply Strength Property Characterization" in this article. However, when the stresses σ_1, σ_2, and σ_6 exist simultaneously at the point, failure may still occur even if none of the five conditions is satisfied. This is known as the stress interaction effect, which enhances or retards material failure. Thus, the maximum stress criterion (Eq 6) can be used primarily for plies in which only one stress component is predominant.

The strength potential theory assumes that some energylike potential exists in the material. When the level of the potential reaches a certain limit, the material fails. This potential is assumed to depend solely on the current state of stress at a point, consistent with the point failure concept. Now, let the stress state at the point be denoted by σ_i, which can be either in a two-dimensional ($i = 1, 2, 6$) or three-dimensional ($i = 1, 2, 3, 4, 5, 6$) state. The desired form for the strength potential $f(\sigma_i)$ must be such that failure of the material is universally predictable, given the state of stress σ_i. Assuming that $f(\sigma_i)$ can be expanded in a power series, and that only the first few terms in the expansion are essential:

$$f(\sigma_i) = F_o + F_i\sigma_i + F_{ij}\sigma_i\sigma_j \qquad \text{(Eq 7)}$$

where $i,j = 1, \ldots, 6$.

Assume also that the potential $f(\sigma_i)$ is positive definite and it is null at zero stress. It then follows that $F_0 = 0$. Hence, failure of material can now be defined when the following general condition is satisfied:

$$F_i\sigma_i + F_{ij}\sigma_i\sigma_j = 1 \qquad \text{(Eq 8)}$$

where $i,j = 1, \ldots, 6$ and where F_i and F_{ij} are constants to be determined for the material considered.

The criterion in Eq 8 was first proposed by S.W. Tsai and E.M. Wu (Ref 6), who showed that the form of Eq 8 must be invariant with respect to coordinate transformation. In fact, F_i is a second rank tensor, and F_{ij} is a fourth rank tensor (both are in contracted notation). Furthermore, criterion (Eq 8) can be applied to both two- and three-dimensional stress states, provided that the necessary strength constants, F_i and F_{ij}, can be uniquely determined for a given material.

By virtue of material symmetry, it can be shown that there are three independent constants in F_i and nine in F_{ij} if a three-dimensional stress state is considered for an orthotropic ply. In a two-dimensional stress state, there are two constants in F_i and four in F_{ij}, for a total of six for an orthotropic or transversely isotropic ply. These strength constants must be determined experimentally, but not all of the constants in the three-dimensional case can be determined uniquely. Even in the two-dimensional case, only five of the six strength constants have so far been obtained.

The strength tests discussed in the section "Ply Strength Property Characterization" in this article can provide five independent conditions for determining the constants of two-dimensional orthotropic plies, yielding:

$$F_1 = \frac{1}{X_t} + \frac{1}{X_c} \qquad F_2 = \frac{1}{Y_t} + \frac{1}{Y_c} \qquad F_{11} = -\frac{1}{(X_t X_c)}$$

$$F_{22} = -\frac{1}{(Y_t Y_c)} \qquad F_{66} = \frac{1}{S^2} \qquad \text{(Eq 9)}$$

The sixth constant, F_{12}, requires a combined loading test. However, from the finiteness constraint imposed on the stress state, Eq 8 can only describe the surface of an ellipsoid. Hence, the following inequality must be imposed on the strength constants under two-dimensional stress conditions:

$$F_{11}F_{22} > (F_{12})^2 \qquad \text{(Eq 10)}$$

or

$$-\sqrt{F_{11}F_{22}} < F_{12} < \sqrt{F_{11}F_{22}} \qquad \text{(Eq 11)}$$

In practice, the value of F_{12} is often estimated within the bound of Eq 11. Some experiments using an off-axis unidirectional specimen loaded in tension suggested that the influence of F_{12} on failure is insignificant (Ref 7).

Progressive Ply Failures in Laminates

From laminate stress analysis, the ply that first fails under a given load program can be determined according to an appropriately selected ply failure criterion. After the first ply fails, the laminate may still be able to carry a higher load. This, then, requires additional analysis of subsequent ply failures.

One simple approach is to assume that the failed ply can no longer carry any load; the load which it carried before failure is redistributed to the remaining plies. The failed ply is then deleted from the laminate, and a new round of laminate stress calculation and ply failure analysis is conducted under the last applied load level. If no additional ply fails under this load, the load can be increased to a higher level until failure occurs in one or more plies. The procedure can be repeated as many times as are required until all plies are failed by the applied load.

This procedure, known as the ply discount method (Ref 8), establishes not only the ultimate laminate load but also a nonlinear or piecewise linear relationship between the applied laminate load and laminate deformation.

One obvious weakness of the ply discount method is that it disregards the ability of load redistribution through ply interface bonding. Specifically, when one ply fails, the failure is localized by its adjacent plies through their high stiffness constraint and high ply interface bonding strength. When this happens, the failed ply can continue to carry load in other parts of the laminate. In fact, this very ply, which failed first, is capable of sustaining multiple failures throughout the loading process.

Several modifications of the ply discount methods have been suggested in order to retain the aforementioned load redistribution capacity. One such method is to treat the failed ply as an elastic/plastic material (Ref 9, 10). Essentially, first-ply failure is defined as the transition from elastic to plastic deformation. Postfailure plastic deformation then provides the mechanisms for load redistribution.

This scheme, however, may not describe the true mechanism of load redistribution that occurs in the laminate. Indeed, the analysis methods of progressive ply failure are essentially incapable of describing the actual failure and failure modes that develop in the laminates. As the laminate is a highly redundant structure, load redistribution mechanisms exist both at the fiber-matrix interface scale and the ply-to-ply interface scale. Both types of mechanisms control the initiation and progression of failure and failure modes. However, the fiber-matrix interface mechanisms are discarded by virtue of the ply homogeneity assumption, and the ply-to-ply interface failure mechanisms are discarded through the independently applied ply failure criterion.

Three-Dimensional Laminate Stress Analysis

Three-dimensional stress analysis of laminates can be performed based on the theory of anisotropic elasticity (Ref 2), which provides both the in-plane stresses (σ_1, σ_2, σ_6) and the out-of-plane stresses (σ_3, σ_4, σ_5) in the individual plies. The out-of-plane stress components, known as the interlaminar stresses, are the result of the ply interface load-transfer

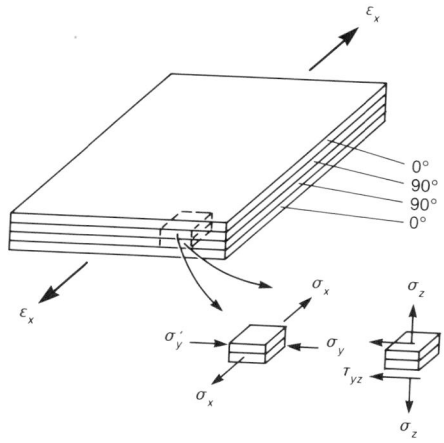

Fig. 2 A $(0°/90°)_s$ laminate under uniaxial tension

mechanisms. Hence, by proper accounting for the effects of these stresses, failure modes stemming from interply load-transfer can be treated under a number of conditions. The celebrated works on laminate free-edge stress analysis (Ref 11-14) are classical examples. In this section, the free-edge problem will be used to illustrate some of the peculiarities of the elastic stress fields in composite laminates.

Consider the $[0°/90°]_s$ laminate loaded in uniaxial tension, shown in Fig. 2. Assume the laminate is long and has a uniform width. Then, excluding the local effects at the far ends of the laminate, the stress field is in a state of generalized plane strain (Ref 11). The laminate displacement field (u_1, u_2, u_3) can be expressed in the general form:

$$u_1(x,y,z) = (\epsilon_0)x + U(y,z) \qquad u_2(x,y,z) = V(y,z)$$
$$u_3(x,y,z) = W(y,z) \qquad\qquad (Eq\ 12)$$

where e_0 is the applied uniaxial strain, and (U,V,W) are the generalized displacements, which are continuous functions of (y,z) only.

The laminate strain field, ϵ_i ($i = 1, ..., 6$), is derived from Eq 12, and for each ply of the laminate, the ply stress field, σ_i ($i = 1, ..., 6$), is given by the generalized Hooke's law, Eq 1 or Eq 4. The displacement functions (U,V,W) are determined by solving the stress equilibrium equations of each of the plies, which must simultaneously satisfy the traction-free surface and the ply interface matching conditions in the (y,z) domain. The matching conditions are derived by requiring displacement continuity and stress equilibrium at each ply interface, assuming perfect bonding.

This problem was originally treated by R.B. Pipes and N.J. Pagano (Ref 11) using a finite-difference solution scheme (Ref 11), and later by others using finite elements (Ref 13, 14).

Figure 2 depicts both the dominant in-plane and the interlaminar stresses acting on the inner 90° plies of a $[0°/90°]_s$ laminate. The tensile σ_x is due directly to the applied strain ϵ_0 and the compressive σ_y is caused by the Poisson's ratio

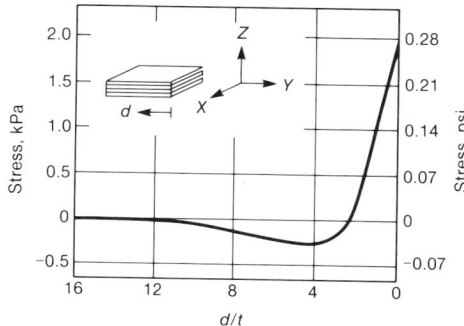

Fig. 3 Distribution of the interlaminar normal stress σ_z on the laminate midplane of the $(0°/90°)_s$ laminate

mismatch between the 0° ply and the 90° ply. These two stress components exist uniformly throughout the interior of the plies, except near the free edges of the laminate, where the interlaminar normal stress σ_z and the shear stress τ_{yz} also exist. In many instances, these latter interlaminar stresses cause laminate failure.

The behavior of the interlaminar edge stresses can best be illustrated by some specific numerical examples. Figure 3 shows the σ_z distribution on the laminate midplane (90°/90° interface), calculated according to a finite-element model for the $[0°/90°]_s$ laminate made of a T300/934 graphite-epoxy ply system (Ref 12). Here, σ_z is normalized with respect to the applied axial strain $\epsilon_0 = 1\mu\epsilon$ and is plotted against the normalized distance from the free edge, d/t, where t is half the ply thickness. It is seen that σ_z does not exist in the laminate interior, but develops gradually from a small compression to a sharply rising tension as it approaches the free edge. The sharp concentration of σ_z near the free edge is often blamed for interply failure in the form of free-edge delamination (Ref 11).

But, in this example, the value of the tensile σ_z, near the free edge is about 2 kPa/$\mu\epsilon$ (0.3 psi/$\mu\epsilon$) while the tensile σ_x existing in the plane of the 90° ply (normal to the fiber direction) is more than 12 kPa/$\mu\epsilon$ (1.8 psi/$\mu\epsilon$). Thus, tension failure of the 90° ply caused by σ_x is more likely to occur than a delamination of the laminate through midplane caused by σ_z.

In another example, the $[\pm 25°/90°]_s$ laminate made of the same ply system was considered (Ref 13). Here, a tensile σ_z also develops near the free edge on the laminate midplane when the laminate is under uniaxial tension (Fig. 4). It can be seen that the maximum value of σ_z at the free edge is nearly 20 kPa/$\mu\epsilon$ (2.9 psi/$\mu\epsilon$), compared to the tensile σ_x of about 12 kPa/$\mu\epsilon$ (1.8 psi/$\mu\epsilon$) in the 90° ply. Hence, in this case, a midplane delamination caused by the tensile σ_z could occur as the first laminate failure.

Actually, σ_z is not the only interlaminar edge stress that can cause delamination, nor is σ_z acting on the laminate midplane necessarily the

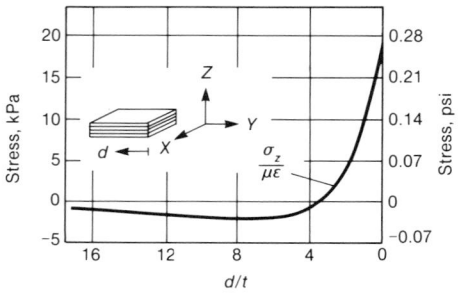

Fig. 4 Distribution of the interlaminar normal stress σ_z on the laminate midplane of the $(\pm 25°/90°)_s$ laminate

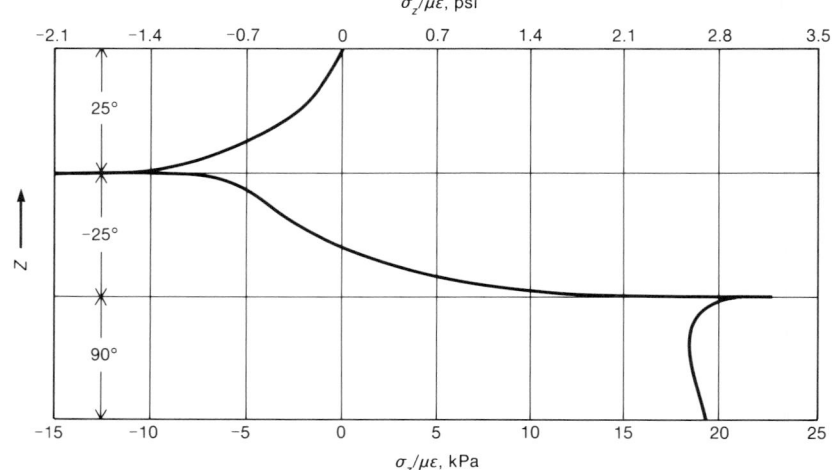

Fig. 5 Through-thickness distribution of the interlaminar normal stress σ_z near the laminate free edge of the $(\pm 25°/90°)_s$ laminate

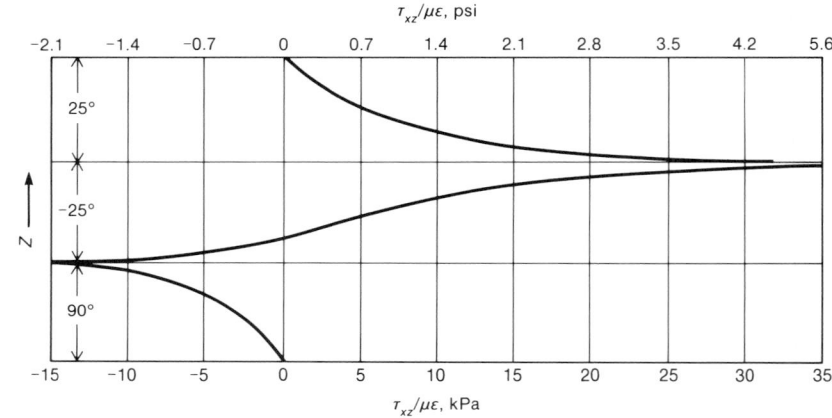

Fig. 6 Through-thickness distribution of the interlaminar shear stress τ_{xz} near the laminate free edge of the $(\pm 25°/90°)_s$ laminate

largest valued. Figure 5 shows the thickness distribution of σ_z near the free edge, which indicates that σ_z develops into a compressive stress in the 25° ply and appears to be singular on the 25°/$-$25° interface. It then changes sharply into a tensile stress within the thickness of the $-$25° ply, and becomes unbounded again on the $-$25°/90° interface. However, it returns to be finite-valued once inside the 90° ply, and remains nearly uniform throughout the thickness of the 90° ply. From these results, we see that the through-thickness variation of σ_z at the free edge is rather violent. At the points where the 25°/$-$25° and the $-$25°/90° interfaces intersect the free-edge boundary, the stress field is singular according to exact ply elasticity solutions (Ref 15, 16).

Similarly, Fig. 6 shows the thickness distribution of the interlaminar shear stress τ_{xz} near the free edge. In this case, τ_{xz} also becomes unbounded on the 25°/$-$25° and the $-$25°/90° interfaces. It is probable that the interlaminar shear stress also contributes to free-edge delamination failure (the other interlaminar shear stress, τ_{yz}, also intensifies on these ply interfaces, but it is relatively insignificant compared to the problem with either σ_z or τ_{xz}.

In the two examples above, interlaminar stresses are shown to play an important role in initiating laminate failures near geometrical discontinuities, such as free edges, holes, and other internal or external flaws of large proportion. Because of their unusual distribution patterns, these stresses are not always easily computed. Often, a truly three-dimensional field analysis incorporating singular stress behavior is required (Ref 16).

Laminate Failure Based on Failure Modes

Assume that the three-dimensional laminate stress field can be accurately calculated for any given laminate situation, including regions of severe stress concentration. Appropriate criteria are then needed to determine the exact condition under which a certain mode of failure will first initiate. Once the first failure is initiated, whether it is in the plane of a ply or in the interface between plies, a new stress

concentration will develop near the failure location. This local failure may either propagate in the same mode, or a new failure mode may evolve. In this sense, failure in a laminate can both propagate in space and accumulate in time. Clearly, the stationary ply failure criteria are inadequate to treat such evolving failure processes. A new predictive approach then becomes necessary.

Recently, an energy method based on the concept of brittle fracture and the assumption of effective material flaws has been employed to describe the initiation and propagation of matrix-dominant cracks in laminates (Ref 17-19). The essential feature of the method is that it simulates the evolving state of matrix cracking as a function of the applied load or load history. Because matrix cracks are primarily due to the weakness of the matrix and the weak fiber-matrix interface bonding, they normally occur before the breakage of any fibers. Therefore, by matrix-cracking analysis alone, the energy

method does not generally predict the ultimate strength of the laminate.

The initiation and propagation of matrix cracks can occur both at the fiber-matrix interface level and the ply-to-ply interface level. Consequently, they are classified as two basic modes. The former is known as the intralaminar cracking mode and the latter is the interlaminar cracking mode.

Intralaminar cracking is caused primarily by the in-plane stresses (σ_1, σ_2, σ_6) in a lamina, in which the lamina breaks apart because of the tensile stress component normal to the fibers. Figure 7 shows a $[0°/90°]_s$ laminate loaded in uniaxial tension. A number of lamina separations are seen to have formed in the inner 90° layer. Macroscopically, each of these separations is a crack formed parallel to the fibers and normal to the applied tension. Hence, these are commonly called transverse cracks.

Interlaminar cracking is caused mainly by the interlaminar stresses (σ_3, σ_4, σ_5) associated

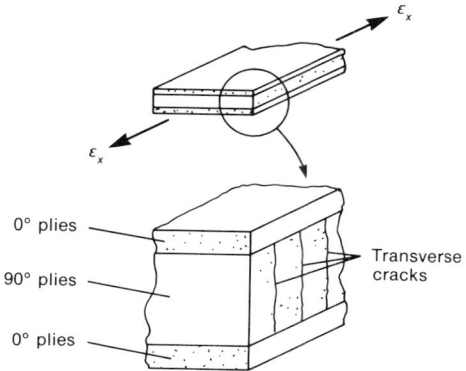

Fig. 7 Transverse cracks in a $(0°/90°)_s$ laminate under uniaxial tension

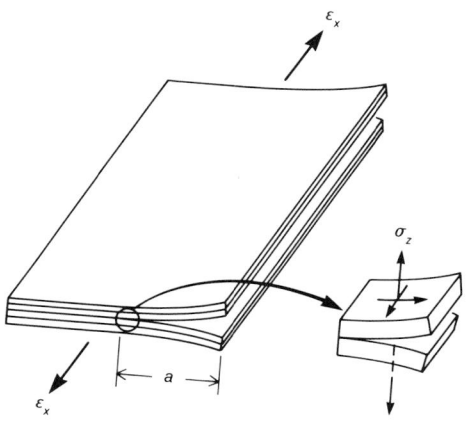

Fig. 8 Schematic view of free-edge delamination in a laminate under uniaxial tension

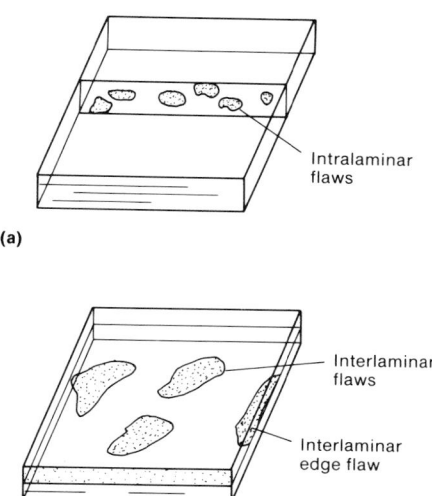

Fig. 9 View of (a) effective intralaminar flaws. (b) Effective interlaminar flaws

with the thickness effect of the laminate. As it was discussed in the section "Three-Dimensional Laminate Stress Analysis" in this article, interlaminar stresses tend to concentrate near locations of geometric discontinuity, such as the laminate free edge. These localized interlaminar stresses can then initiate and propagate a plane crack parallel to the ply interface. Such a plane crack is commonly termed a delamination. Figure 8 depicts free-edge delamination in a laminate under uniaxial tension.

Of course, the intralaminar stresses and the interlaminar stresses are not two independent stress groups. They are, in fact, coupled through the equilibrium relations of the solid element on which they act (Fig. 8). Therefore, when a matrix crack occurred in one mode, a transfer of stress from one group to the other takes place. This may cause the matrix crack to change into the other mode. Indeed, intralaminar and interlaminar cracks often occur concurrently in laminates, resulting in a complex network of cracks long before the actual rupture of the laminate.

A brief discussion follows on an analytical model (energy method) as it is applied to simulate intralaminar and interlaminar cracking that occur independently rather than concurrently.

The energy method for matrix cracking basically incorporates the method of ply elasticity and the concept of effective material flaws into the classical theory of fracture mechanics. The theory of ply elasticity, which is based on the continuum assumption, is employed for the purpose of three-dimensional laminate stress field analysis. Similarly, elastic fracture mechanics criteria are used to describe the propagation behavior of matrix cracks. However, in order to render a rational prediction for matrix crack initiation, the concept of effective material flaws is first introduced.

In this concept, two basic types of effective flaws are postulated. First, for the unidirectional ply, there exists a known distribution of effective intralaminar flaws, lying in the ply thickness direction, as shown in Fig. 9a. These flaws effect intralaminar cracking. Second, in

each ply-to-ply interface, there exists a known distribution of effective interfacial flaws, lying parallel to the ply interface, as shown in Fig. 9b. These flaws effect interlaminar cracking.

At this point, the effective flaws remain a hypothetical quantity, though they are considered inherent properties of the basic ply material system. It is postulated that the presence of the effective flaws does not effect any change in the elastic constitutive properties of the ply. But, at some critical laminate loading, one (or more) of the flaws can be transformed into a matrix crack that is measurable at the phenomenological scale.

The condition governing the transformation from a particular effective flaw to a detectable matrix crack is provided by a criterion of the theory of brittle fracture (Ref 20). Because the size and the location of a particular flaw are known, it is possible to perform an elastic stress analysis by treating the flaw as a crack; the crack-tip stress intensity factor $K(a_0, \sigma_0)$ or the strain energy release rate $G(a_0, \sigma_0)$ can be calculated in terms of the initial crack (flaw) size a_0 and the applied laminate load σ_0. The condition for a_0 to propagate and to become detectable is given, for example, by:

$$G(a, \sigma_0) = G_c \quad \text{at } a = a_0 \quad \text{(Eq 13)}$$

where G_c is the material fracture toughness against the considered mode of matrix crack propagation. The stability of propagation is determined by the functional dependence of G on the crack size, a. Namely, if $dG/da > 0$ at $a = a_0$, the propagation is unstable; if $dG/da < 0$, the propagation is stable, that is, there is no crack growth.

Interlaminar Cracking, Free-Edge Delamination. The energy approach outlined above has been applied to analyze the initiation and propagation of free-edge delamination (Ref 17, 18). In order to illustrate the specific procedures involved in the energy analysis, let us consider the $[\pm 25°/90°]_s$ tension coupon discussed in the section "Three-Dimensional Laminate Stress Analysis" in this article.

When this coupon is loaded in uniaxial tension ϵ_0, significant interlaminar stresses develop near the free edges. In fact, these interlaminar stresses are large enough to cause free-edge delamination as the first failure mode in the laminate. Of course, the question is at what critical ϵ_0 and in which ply interface will the delamination initiate. Also, the question of delamination stability must be answered.

Interface Flaw. To answer these questions, we begin with the assumed distribution of interfacial (effective) flaws. In particular, we identify in each ply interface and along the laminate free edge the most dominant effective flaw having a uniformly elongated shape with a width a_0 (Fig. 9b). This interfacial flaw is capable of transforming into an interfacial crack as soon as the laminate stress field reaches the critical level.

Energy Release Rate Curve. Second, the most dominant edge flaw is treated as a crack of the same size, a_0, and the crack tip strain energy release rate $G(a_0, \epsilon_0)$ is calculated. The calculation can be rendered most effective by a numerical procedure, such as that outlined in the Appendix.

The strain energy release rate, G, when calculated numerically, can be expressed in explicit terms of the crack size a, and the applied laminate strain ϵ_0.

$$G(a, \epsilon_0) = C_\epsilon(a) \, 2t \, \epsilon_0^2 \quad \text{(Eq 14)}$$

where $C_\epsilon(a)$ is a coefficient function explicitly depending on a, but implicitly depending on the location of a ply interface, the ply stacking sequence of the laminate, and the ply elastic constants (see details in Appendix A).

If the residual thermal stresses due to laminate curing are significant and their effects are to be included in the calculation of G, it then is necessary to define the laminate stress-free temperature, T_0. Usually, T_0 is approximately

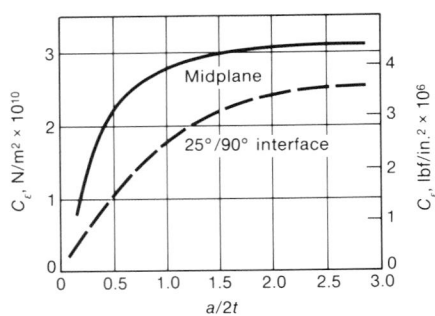

Fig. 10 Strain energy release rate coefficients, C_ϵ, for interfacial cracking in the 90°/90° and the −25°/90° interfaces of the (±25°/90°)$_s$ laminate

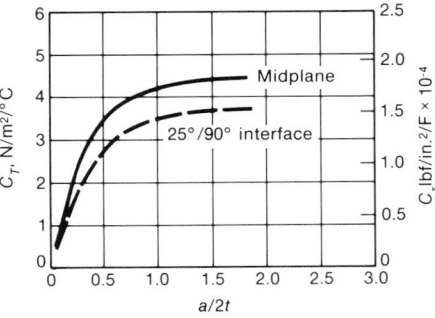

Fig. 11 Strain energy release rate coefficients, C_T, for interfacial cracking in the 90°/90° and the −25°/90° interfaces of the (±25°/90°)$_s$ laminate

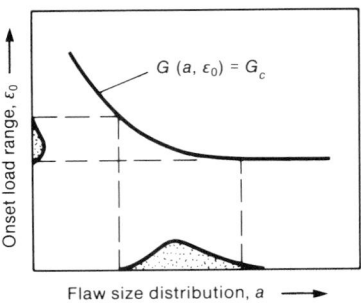

Fig. 12 Fracture relationship between the critical load range (ϵ_0) and the effective flaw size distribution $f(a)$

the same as the laminate curing temperature. Thus, at the ambient temperature, T, the laminate is exposed to a uniform thermal loading, $\Delta T = T - T_0$. In such a case, the strain energy release rate due to the combined loading of ϵ_0 and ΔT can be expressed as:

$$G(a, \epsilon_0, \Delta T) = [\sqrt{C_\epsilon}\, \epsilon_0 - \sqrt{C_T}\, \Delta T]^2\, (2t) \quad \text{(Eq 15)}$$

where C_T is another coefficient function expressed in terms of a. Note that a minus sign associated with the ΔT term is only a convention, because the value of ΔT is always negative.

From the section "Three-Dimensional Laminate Stress Analysis" in this article, it was seen that in free-edge stress analysis, the most probable delamination interfaces in the [±25°/90°]$_s$ laminate are the midplane and the −25°/90° interface. However, the exact delaminating interface could not be determined based on edge stresses alone. The following discussion attempts to resolve the problem by an energy analysis.

Figure 10 shows the computed strain energy release rate coefficient functions C_ϵ for interfacial cracking in the midplane (90°/90° interface) and the −25°/90° interface. Note that the computed C_ϵ curve for delamination in the midplane contains exclusively mode I energy, as it should, while the C_ϵ curve for delamination in the −25°/90° interface contains both mode I and mode III energies. The ratio of mode I energy to mode III energy in the latter case is about 0.8. Similarly Fig. 11 shows the strain energy release rate coefficients C_T for delamination in each of the same two ply interfaces.

It is seen that the characteristics of all the energy curves are similar. Namely, the curve rises sharply from 0 at $a = 0$ and reaches a limiting value at $a = a_m$. The physical meaning of a_m is not exactly known. However, the free-edge effect may be interpreted as fully developed when a delamination reaches the size a_m. The size of a_m is approximately of the order of a couple of ply thicknesses for most laminate situations (Ref 18).

Determining the Exact Delaminating Interface. A comparison of magnitudes of the strain

energy release rate curves for the [±25°/90°]$_s$ laminate shown in Fig. 10 and 11, shows that edge delamination in the laminate midplane would release more strain energy than delamination in the −25°/90° interface. Whether the midplane (mode I) or the −25°/90° interface (mixed modes) will delaminate, however, is determined by the fracture criterion of Eq 13, for which the material fracture toughness G_c and the effective flaw size a_0 are yet to be specified.

Laboratory experience with graphite-epoxy laminates has shown that G_c measured for mode I delamination is generally lower than that measured for mixed-mode delamination. In fact, mixed-mode G_c has been found to increase with the amount of shearing mode (Ref 21-23). Assuming this is the case in the present example, it is then possible to conclude that the laminate midplane will delaminate first. Hence, the C_ϵ and C_T curves for the midplane and G_c for mode I should be used in Eq 13.

Determining the Critical Laminate Strain. With the interface and mode of delamination identified, the last information needed in using Eq 13 is the size of a_0. Because a_0 is a hypothetical quantity, however, it is essentially a random variable. Yet, given the functional relationship $G(a)$ as expressed in Eq 14 or 15 and the values for ΔT and G_c, a one-to-one relationship between ϵ_0 and a can be obtained from Eq 13 (Fig. 12). If all the possible values of a_0 are represented by some probability function, $f(a_0)$, then there is a corresponding range of ϵ_0 values for which Eq 13 is satisfied. While the exact size, a_0, of the most dominant effective flaw is not known, the minimum critical ϵ_0, which is associated with all values of $a_0 > a_m$, can now be determined from Fig. 12.

As mentioned before, the value of a_m is on the order of one or two ply thicknesses, while the actual initial edge delamination observable at the macroscale is on the order of the laminate thickness. Hence, delamination onset can be predicted by the minimum critical strain associated with $a_0 > a_m$. Thus, from Eq 13 and in conjunction with Eq 15, the predicted minimum

critical strain for initial delamination propagation is given by:

$$(\epsilon_0)_{cr} = \frac{[\sqrt{(G_c/2t)} + \sqrt{C_T}\, \Delta T]}{\sqrt{C_\epsilon}} \quad \text{at } a > a_m \quad \text{(Eq 16)}$$

where the coefficients C_ϵ and C_T are taken from the calculated energy curves for the midplane.

To determine the propagation stability, the functional form of C_ϵ and C_T for all $a > a_m$ is examined. Accordingly, it is concluded that the stability behavior of free-edge delamination is neutral. In reality, however, the propagation is generally stable, with the degree of stability depending on whether other modes of matrix cracking develop subsequent to delamination initiation.

Intralaminar Cracking, Transverse Cracks. The energy analysis procedures outlined above for the free-edge delamination problem have also been applied to simulate intralaminar cracking in laminates (Ref 19). Here, the [0°/90°]$_s$ graphite-epoxy tensile coupon serves as an example. As was discussed earlier, free-edge effect is relatively unimportant when the coupon is under uniaxial tension. Hence, an analysis of transverse cracking in the 90° layer can be developed without regard to the influence of free-edge stresses.

Initiation and Growth Characteristics. At the macroscopic scale, the formation of a transverse crack is simply a sudden separation of the 90° layer. When this happens, it often gives off acoustic emissions. At the microscopic scale, however, the exact nature of the event is unclear. It may be postulated again that the crack is caused initially by the coalescence of those material microflaws which lie in the thickness direction in the 90° layer. The effect of this coalescence is represented by the propagation of an effective flaw (Fig. 9a) in the thickness direction. The crack, however, is unable to propagate through the 0° ply and is either arrested at or blunted to propagate along the 0°/90° interface. The latter then becomes a localized delamination. In either case, the propagation is not catastrophic. This allows an increase of the applied tension, which, in turn, can cause another transverse crack to form at another location. In this manner, a series of

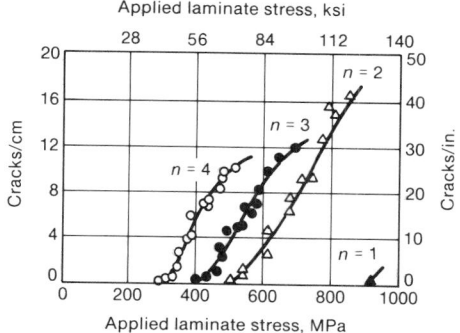

Fig. 13 Experimental crack density versus load plots for a family of T300-934 $(0°/90°_n/0°)$ laminates under uniaxial tension, 2 specimens each. $n = 1, 2, 3, 4$

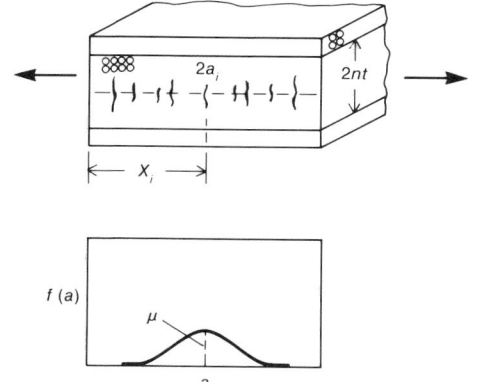

Fig. 14 Idealized effective intralaminar flaw distribution in the 90° layer

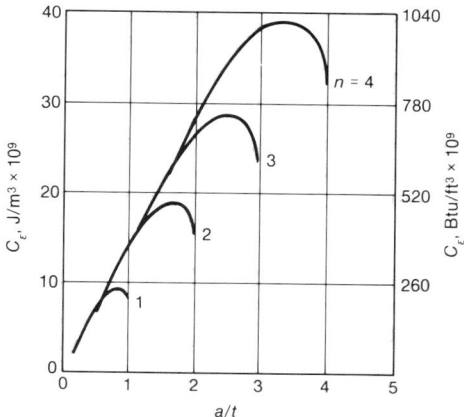

Fig. 15 Strain energy release rate coefficients, C_ϵ, for transverse cracking in $(0°/90°_n/0°)$ laminates. $n = 1, 2, 3, 4$

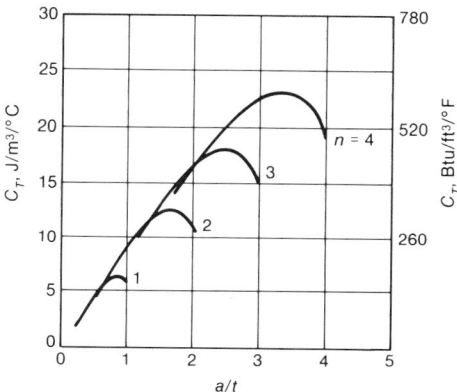

Fig. 16 Strain energy release rate coefficients, C_T, for transverse cracking in $(0°/90°_n/0°)$ laminates. $n = 1, 2, 3, 4$

transverse cracks can be formed along the length of the coupon as a function of the tension loading.

Figure 13 shows a plot for the number of transverse cracks (per cm length) versus applied tension relationships, obtained by testing a family of $[0°/90°_n/0°]$ coupons ($n = 1, 2, 3, 4$) made of the T300/934 graphite-epoxy system (Ref 13). From the plot, it can be seen that for the $n = 1$ coupon transverse cracking did not occur until the load had reached the critical level, causing fiber breaking in the 0° plies. For the other three coupons, however, each yielded a cumulative crack development curve that is characteristically distinct from the others. These results clearly indicate that the occurrence of any one particular crack in a given coupon is essentially probabilistic in nature, but that the developmental character of the cumulative cracks as a whole seems to follow a rather deterministic rule.

Effective Flaws. Now, for the sake of simplicity, it may be assumed that the effective intralaminar flaws are one-dimensional, being oriented in the ply thickness direction, as shown in Fig. 14. The linear size of an individual flaw is denoted by $2a$, and its location is denoted by x. Then, for the unidirectional ply (thickness $2t$), the discrete random variables $(a_i, i = 1, M)$ and $(x_i, i = 1, M)$ characterize the size and the location distributions of the effective intralaminar flaws. At this point, the exact values for (a_i) and (x_i) are not known, but are assumed to be inherent ply properties. As such, these can be determined by some suitable experimental measurements, as will be discussed.

When two or more 90° plies are grouped together, such as in the $(0°/90°_n/0°)$ coupons with $n > 1$, the effective flaw size distribution in the grouped plies will be different from that of the single ply, although their spacing distribution may be assumed to be the same. Here, a volumetric rule (Ref 24) can be used to express the flaw size distribution, $(a_{i,n})$, in n-plies in terms of the flaw sizes, (a_i), of the single ply:

$$a_{i,n} = a_i(n)^{2/\alpha} \qquad \text{(Eq 17)}$$

where $i = 1, M$, and α is an arbitrary constant related to the distributional characteristics of (a_i) and the particular volumetric rule used in deriving the relationship in Eq 17 (see, for example, Ref 24).

Onset of the First Crack. The $(0°/90°_n/0°)$ coupon is now considered under the applied tensile strain ϵ_0 as shown in Fig. 14. Each of the flaws is capable of propagating into a transverse crack. The propagation of any one of the flaws, say the ith flaw a_i, is governed by Eq 13. There, the strain energy release rate $G(a_i, \epsilon_0, \Delta T)$ associated with the ith flaw must first be calculated. In addition, the fracture toughness G_c must also be determined *a priori*. It is noted that, for the present problem, transverse cracking is essentially in mode I. Hence, G_{Ic} should be determined first and then used in Eq 13.

The calculation of $G(a_i, \epsilon_0, \Delta T)$ can be performed by the finite-element procedures outlined in the Appendix to this article. The calculated G is then expressed in the form of Eq 15. Figures 15 and 16 show, respectively, the $C_\epsilon(a)$ and $C_T(a)$ functions for $n = 1, 2, 3, 4$. Here, the unique character of the C_ϵ or C_T curves is worth noting. It is seen that both C_ϵ and C_T increase sharply from $a = 0$ to reach their respective maxima, $C_{\epsilon,max}$ and $C_{T,max}$; these maxima occur at about three-fourths the thickness of the 90° layer. After this point, they both decrease toward the 0°/90° interface at $a = nt$. This behavior is consistent with the observed fact that transverse cracking can be initially unstable (that is, in giving off acoustic emissions) but is immediately arrested at the 0°/90° interfaces.

Now, for the first crack to form, let the largest of (a_i) be denoted by a_{max}. Then, the critical strain (ϵ_0) for the onset of the first crack is given by Eq 16 for $a = a_{max}$ and $G_c = G_{Tc}$.

Actually, because each (a_i) must be smaller than the half-thickness of the 90° layer, nt, and in all likelihood, a_{max} is only slightly less than nt, a minimum $(\epsilon_0)_{cr}$ can be found by substituting $C_{\epsilon,max}$ and $C_{T,max}$ in Eq 16.

Shear Lag Effect. Assume that the transverse crack is arrested at the 0°/90° interfaces. Then,

the local tensile load formerly carried by the unbroken 90° layer is now transferred to the adjacent 0° plies. If the 0°/90° interface bonding is strong, a slip of the interface (delamination) is not possible. Then, a localized interlaminar shear stress τ_{xz} is developed near the transverse crack termini, as shown in Fig. 17. This interlaminar shear stress decays exponentially a small distance away from the transverse crack, and at the same time, the *in-situ* tensile stress σ_x in the 90° layer regains its far-field strength. This local load transfer zone, known as the shear lag zone, is proportional to the thickness of the 90° layer, $2nt$. Hence, the thicker the 90° layer, the larger the shear lag zone. For the present problem, one side of the shear lag zone is about $12nt$.

If there is an effective flaw located near a transverse crack (Fig. 18) the flaw is under the influence of the shear lag casted by the transverse crack. The degree of influence depends on their relative distance, s. Specifically, if the size of this flaw is $2a$, and the associated strain energy release rate is $G(a, \epsilon_0, \Delta T, s)$, then the shear lag effect on the strain energy release rate

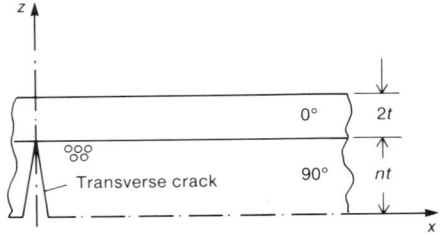

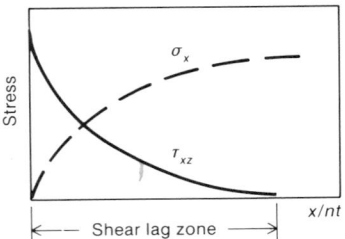

Fig. 17 Shear lag zone due to a transverse crack in the $(0°/90°)_s$ laminate

can be expressed by the factor, $R(s)$, defined by:

$$R(s) = \frac{G(a,\epsilon_0,\Delta T,s)}{G(a,\epsilon_0,\Delta T)} \qquad \text{(Eq 18)}$$

where $G(a,\epsilon_0,\Delta T)$ is calculated without the influence of shear lag. It may be noted that the numerical range of $R(s)$ is between 0 and 1, as shown in Fig. 18.

If a flaw is situated between two consecutive cracks, then it is under the shear lag effect from both cracks. The associated strain energy release rate, G^*, is given by:

$$G^*(a,\epsilon_0,\Delta T) = R(s_l)\, G(a,\epsilon_0,\Delta T)\, R(s_r) \qquad \text{(Eq 19)}$$

where s_l and s_r are distances from the flaw to the left crack and the right crack, respectively.

Multiple Cracks. After the formation of the first crack from the largest flaw in (a_i), subsequent cracks can form from the remaining flaws at laminate strain appropriately higher than $(\epsilon_0)_{tr}$. A Monte Carlo search routine is then commenced to determine the next flaw that yields the highest strain energy release rate G^*, with the shear lag effect casted by any existing crack or cracks included. The laminate strain corresponding to the next crack, which should be higher than $(\epsilon_0)_{cr}$, is determined by using G in Eq 16.

Successive searches for the next most energetic flaws follow, and the entire load sequence of transverse cracks is simulated. In essence, this search procedure mimics the transverse cracking process as it would occur naturally.

Determining the Effective Flaw Distribution. The difficulty in the above simulation procedure lies in the fact that the effective flaws are hypothetical quantities, and that they can only be determined indirectly at the macroscale. Appropriate experiments in which distributed intralaminar cracks occur must be devised.

In Ref 24, the effective flaw distribution in

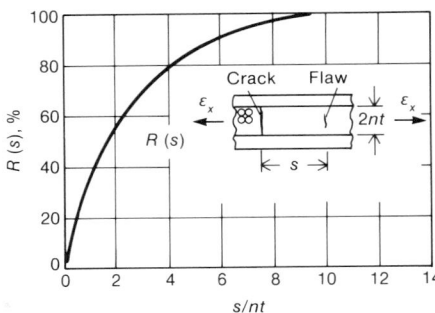

Fig. 18 Strain energy release rate retention factor, $R(s)$

$(90°_2)$ was determined by testing $(0°_2/90°)_s$ tension coupons. The shaded band in Fig. 19 was formed by test data obtained from four specimens, plotted in terms of crack density versus applied tension (average laminate stress). This band, statistically speaking, resembles a cumulative formation behavior of the transverse cracks as a function of the applied tension. The band possesses a certain position relative to the stress scale, a certain characteristic curvature, and an asymptotic value in the crack density scale.

This band is used to determine the flaw distribution in the $(90°_2)$ layer. To do so, a random number generation scheme is employed. First, the number of flaws, M, in a unit length of the 90° layer is assumed. Second, a set of M random values is generated in the interval of $(0,1)$. These M random values are assigned to be (x_i), which are the locations of the flaws along the unit length of the 90° layer. Third, the sizes of the M flaws (a_i) are described by a Weibull cumulative function:

$$F(a) = 1 - \exp\left[-\left(\frac{a}{\beta}\right)^\alpha\right] \qquad \text{(Eq 20)}$$

with the parameters α and β so far being unknown.

A new set of M random values is again generated in the interval $(0,1)$. These values are then assigned to (F_i), corresponding to the values of $F(a)$ at $a = a_i$. Thus, for appropriately assumed values of α and β, one determines from Eq 20 the flaw size a_i for each assigned value of F_i.

With the flaw distribution (sizes and locations) now characterized, though the values of α, β, and M are still assumed, a simulation of the transverse cracking process can be performed as outlined previously. The correct choices of α, β, and M must be ones that closely simulate the experimental band shown in Fig. 19. Generally, α primarily affects the curvature of the band, β shifts the band along the stress scale, and M determines the asymptotic value of the band on the crack density scale (Ref 24). Figure 19 shows the crack density development relations for four simulated specimens, along with a set of appropriately chosen α, β, and M values.

With α, β, and M chosen from the above experiment, flaw size distribution in any number of grouped 90° plies can be found using Eq 17. For example, the band in Fig. 20 represents the experimental results from four $(0°_2/90°_2)_s$ coupons, while the simulated results for four samples of the same coupons are shown by scattered dots. Figure 21 shows a similar comparison for four $[0°_2/90°_4]$ coupons. In both cases, the basic flaw distribution found from the $(0°_2/90°_2)_s$ coupons was used in conjunction with Eq 17 in the simulation.

Of course, the uniqueness of the determined flaw distribution in the basic ply cannot be proved. Specifically, the values of α, β, and M determined experimentally could assume different sets of values for the same set of experiments. This difficulty, it is felt, will remain as long as an exact analysis of the cracking mechanisms at the fiber-matrix scale is unavailable.

Fatigue Failure Concepts

Material damage or local failure can accumulate in time under sustained loading. This can happen under creep conditions in which the load amplitude is held constant in time, or under fatigue conditions in which the load amplitude varies with time. The two central problems in fatigue are, first, to predict the useful lifetime of the material under the projected applied load, and, second, to predict the safe load limit that can be applied to the material after a certain lifetime. To predict both, it is often desirable to describe damage accumulation accurately as it evolves with time and load.

These problems have been considered for many years for the more homogeneous metallic materials, and during the past decade, some of the analysis approaches developed primarily for fatigue in metals have been applied to composites. In this context, this article will briefly describe two general approaches that have traditionally been followed to treat fatigue failure in metals and are increasingly being applied to treat fatigue failure in composites. These approaches follow essentially the same failure concepts adopted in the static failure cases.

The point-stress failure approach is an extension of the point-stress failure concept of the static loading case, which assumes that failure can be defined by the stress state at one point without regard to the actual failure mode or modes. Thus, let σ_i ($i = 1, \ldots, 6$) be the stress state at a point, with the stresses σ_i oscillating between the maximum and the minimum amplitudes σ_{imax} and σ_{imin}. Then, the problem of fatigue failure may be approached from two different views, fatigue life and residual strength.

Fatigue Life. This view concerns failure at a point at which the oscillating stress state satisfies a certain mathematical criterion. For instance, one of the commonly proposed criteria for fatigue failure in metals is expressed as:

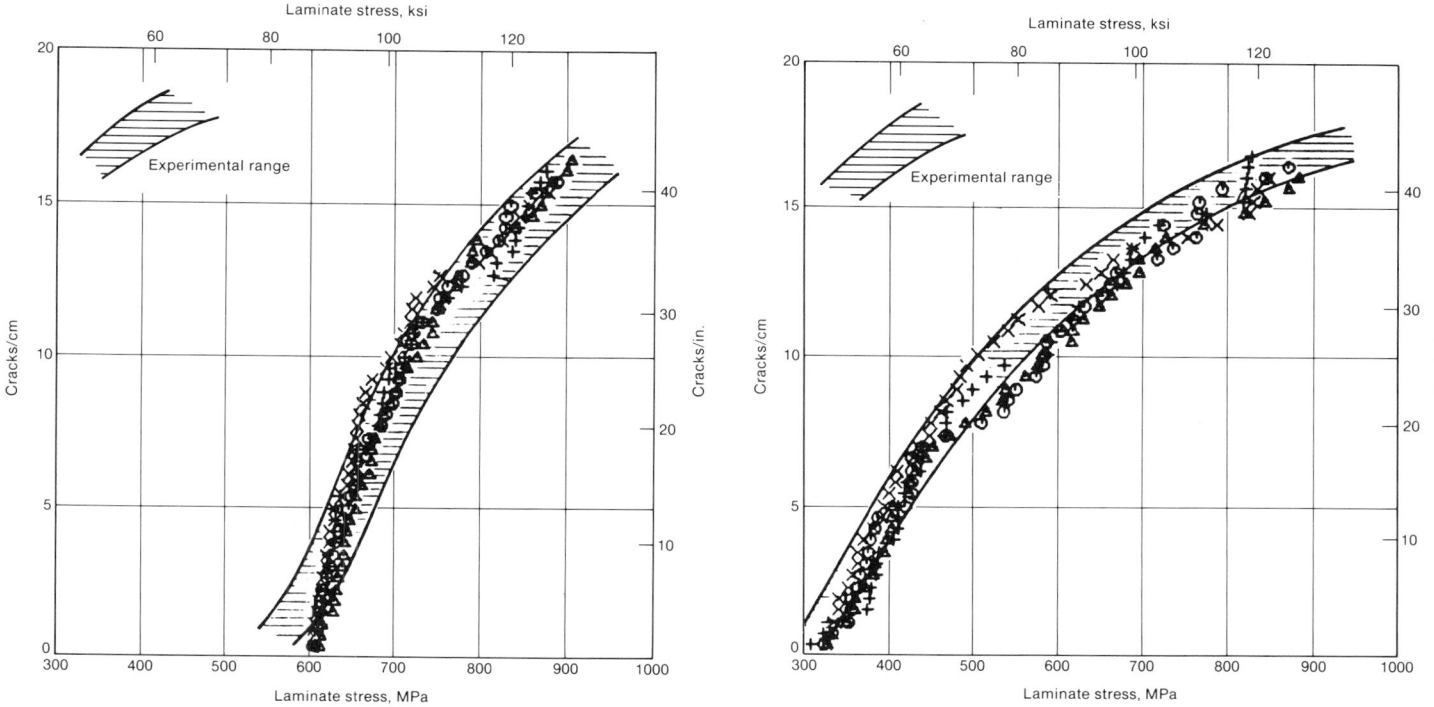

Fig. 19 Experimental (shaded) and simulated (points) transverse crack density versus laminate tension relations. Data from four $(0^\circ_2/90^\circ)_s$ coupons

Fig. 20 Experimental (shaded) and simulated (points) transverse crack density versus laminate tension relations. Data from four $(0^\circ_2/90^\circ_2)_s$ coupons

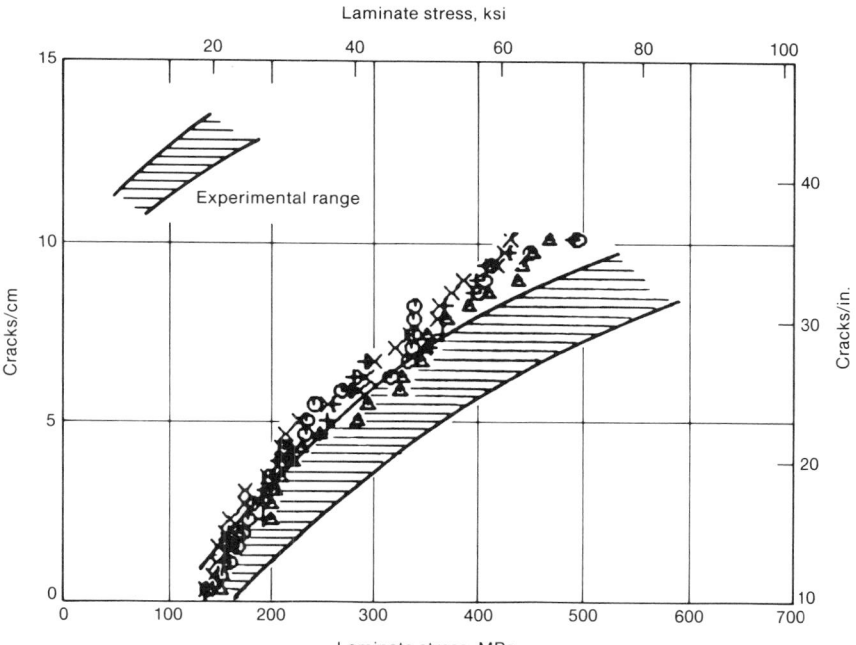

Fig. 21 Experimental (shaded) and simulated (points) transverse crack density versus laminate tension relations. Data from four $(0^\circ_2/90^\circ_4)_s$ coupons

$$F(\sigma_{i\max}R,N) = 1 \qquad \text{(Eq 21)}$$

where R is the stress amplitude ratio, $\sigma_{i\min}/\sigma_{i\max}$, and N is the number of cycles required to cause failure.

The criterion in Eq 21 is based strictly on phenomenological supposition. Usually, the explicit form of $F(\sigma_{i\max}, R,N)$ is determined empirically by fitting a large body of experimental data. Hence, one common approach is

to conduct extensive fatigue experiments under some constant stress amplitude, S, and the stress ratio, R. Both S and R are varied widely as major influencing parameters. Essentially, for each fixed S and R value, the experiment provides the fatigue cycle N at which material failure occurs. When a large body of such *S-N-R* data is collected, an empirical formula relating these three variables can be established. It is noted that the stress state σ_i at any point can be theoretically related to S and R through an elastic stress analysis, and in some cases where the data exhibit large scatter, a statistical analysis is required.

A fatigue failure model based on Eq 21 can be used to predict the fatigue lifetime, N, given the loading combination of S (or σ_i at the point) and R. However, in practice this model has been primarily applied to fatigue involving uniform uniaxial stress states. In the case of multidirectional loading or stress states involving a sharp gradient (for example, stress concentration), the model evaluation procedure can become extremely cumbersome as well as impractical to apply.

Residual Strength. The alternative view is to consider the strength of the material after a prescribed number of fatigue cycles. The material at the point of consideration is presumably damaged but not totally failed. Hence, some of the material strength that is assumed to remain is still enough to resist the applied load. In this case, an empirical expression for the residual strength may be expressed as:

$$\sigma_{k,r} = f_k(\sigma_{i\max}R,n); \quad n < N \qquad \text{(Eq 22)}$$

where n is the number of the lapsed cycles, N is the lifetime under constant $\sigma_{i\max}$ and R, and the subscript k refers to the various strength components of the material.

For a fixed R value, f_k represents the degradation of the kth material strength as a function of fatigue cycles under the applied load. In a sense, the strength degradation is a measure of the material damage incurred during the fatigue cycling. Obviously, the form of f_k must conform to the conditions that f_k is the static strength of the material for no cycling ($n = 0$) and that fatigue failure occurs when the degrading strength f_k becomes equal to the maximum stress amplitude $\sigma_{k\max}$ at the point when $n = N$. In principle, the explicit form of f_k in Eq 22 can be determined from the above conditions, in conjunction with a set of properly generated experimental static strength and fatigue S-N-R data. Again, such an empirical procedure is primarily for simple loading cases involving uniform stress states.

Crack Propagation Approach. This approach is concerned with the prediction of the slow growth of a dominant crack in the material due to cyclic loading. The approach incorporates the concepts of fracture mechanics and thus, it can follow the actual mode of failure. Let a be the size of a linear crack in the material and K be the stress intensity factor at the crack tip associated with the applied stress σ. Then, $K_{\max}$ and $K_{\min}$ are the stress intensity factors corresponding to the maximum stress amplitude $S_{\max}$ and the minimum stress amplitude $S_{\min}$, respectively. For a given fatigue loading condition of $S_{\max}$ and R, the size of the crack can slowly grow as a function of the fatigue cycles. The problem is then reduced to properly describing crack growth under the prescribed fatigue loading. For example, one proposition for the rate of crack growth in metals is expressed in the form:

$$\frac{\mathrm{d}a}{\mathrm{d}n} = F(\Delta K, R) \qquad \text{(Eq 23)}$$

where $\Delta K = K_{\max}$ and $K_{\min}$. Because ΔK includes the crack size factor and the loading factor, which may be a function of the fatigue cycle n, Eq 23 can be integrated when the explicit form of F is known.

For some structural metals, the experimental relationship between the growth rate $\mathrm{d}a/\mathrm{d}n$ and ΔK with a fixed R value has been known to maintain a power relationship during the period of stable crack growth. This leads to the well-known Paris formula (Ref 25):

$$\frac{\mathrm{d}a}{\mathrm{d}n} = \alpha(\Delta K)^\beta \qquad \text{(Eq 24)}$$

where the constants α and β are assumed to be material related. They are therefore determined by fitting a body of fatigue crack growth test data.

It should be noted that the crack growth rate rule, such as Eq 24, is strictly empirical. Its correctness can be adjudicated only by experimental comparison.

Fatigue Failure in Composite Laminates. Many empirical formulations based on similar assumptions (as in Eq 21, 22, or 24) have been developed and applied to composite laminates. Generally, methods based on point-stress failure are applied to laminates under uniform stress conditions (Ref 26). In these cases, the stress state σ_i in the individual plies is analyzed using either the laminated plate theory or ply elasticity, whichever is appropriate. Ply failure is then defined according to criteria based on Eq 21 for fatigue lifetime prediction or on Eq 22 for residual strength prediction. Clearly, the model does not take into consideration any of the possible failure modes, nor does it include any of the possible interfacial stress redistribution mechanisms that exist among the different plies.

The crack propagation methods, on the other hand, are applied to laminates with a known dominant through-thickness crack or a man-made through-thickness notch (Ref 27). Usually, the laminate itself is treated as a homogeneous plate in which a severe stress concentration is present near the crack tip. The assumption is that fatigue stressing will result in stable crack growth from the crack tip, as in the case of metallic materials. If this were the case, a crack growth rate rule, similar to Eq 24, could be devised to predict the crack growth behavior.

All these methods, however, are severely restricted in their use because of their empirical nature and because failure in composites is actually a time process in which many local failures occur and interact. As discussed in the section "Laminate Failure Based on Failure Modes" in this article, local failures in the form of small matrix cracks can be propagated stably and/or arrested in time. The process allows for the formation of a distribution of sublaminate cracks, and the manner in which these cracks form and interact depends largely on the local reinforcement constraints. Thus, the actual result of fatigue cycling is often in the form of distributed sublaminate cracks rather than the growth of a single dominant crack, as in the case of metal fatigue.

Recently, several analysis methods have been advanced which attempt to relate sublaminate matrix crack accumulation to fatigue load history. Because the problem of sublaminate cracking remains extremely difficult to analyze rationally and rigorously, these methods are essentially in their early developmental stages. However, two such methods are briefly described below.

Fatigue Models Based on Damage Accumulation

There are two ways to bring damage accumulation into a predictive fatigue model. The first is to identify the exact damage modes (for example, local matrix cracks); follow their growth progression in space and in time; and account for the interactions among the applied loading, the details of the laminate internal geometry, and the damages themselves. This proposition clearly requires a complete description of the various damage mechanisms at the sublaminate level. Although such an approach is highly desirable, it is intractable, at least when the damage modes are complex.

The second approach is to represent damage accumulation indirectly in the predictive model by some globally measurable quantity, provided that quantity is uniquely related to the damage development process. Indeed, experiments on composite laminates have shown that development of sublaminate damages can be associated with a measurable decrease of stiffness in the laminate continuously in time and under load. Thus, an empirical model describing laminate stiffness loss in terms of the applied fatigue loading can usually be established from a body of test data. Such a model can be established without any prior analysis of the damage mechanisms themselves. However, the extent of damages must be evaluated from the amount of stiffness loss, and this requirement is difficult to do when the damage modes are complex.

Nevertheless, these approaches represent a first step toward the inclusion of damage accumulation in the predictive fatigue failure models for composite laminates.

Stiffness Reduction Model. Given the physical modes of the laminate damages that are induced by the applied fatigue loading, the extent of the damages can be represented quantitatively by the parameter D. Let E_0 be the initial laminate stiffness and $E(n)$ be the stiffness after n cycles under the fatigue load. A loss of the stiffness will result if a damage $D(n)$ has occurred during the n cycles:

$$E(n) - E_0 = g(D) \qquad \text{(Eq 25)}$$

where the function $g(D)$ is established either by some micromechanics analysis or through an experimental correlation. Usually, it is through the latter.

From Eq 25, we can express the rate of damage in terms of the rate of stiffness loss:

$$\frac{\mathrm{d}D}{\mathrm{d}n} = \left(\frac{\partial g}{\partial D}\right)^{-1}\left(\frac{\mathrm{d}E}{\mathrm{d}n}\right) \qquad \text{(Eq 26)}$$

where $\mathrm{d}E/\mathrm{d}n$ may depend on several factors. For instance, a recent study by A. Poursartip et al. (Ref 28, 29) on carbon-epoxy laminates suggests the following relationship:

$$\frac{\mathrm{d}E}{\mathrm{d}n} = f(\Delta\sigma, R, D) \qquad \text{(Eq 27)}$$

The explicit form for $f(\Delta\sigma, R, D)$ and the empirical parameter associated with the form

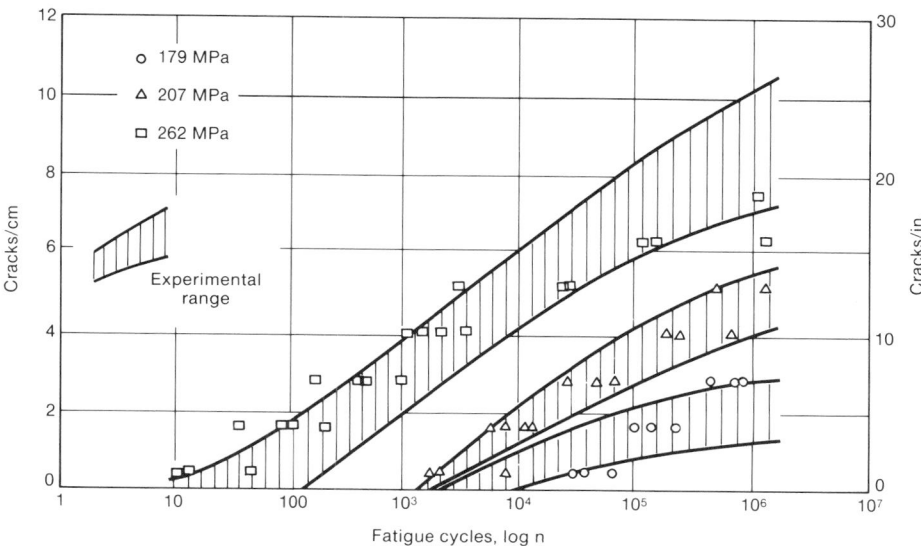

Fig. 22 Experimental (shaded) and simulated (points) transverse crack density versus fatigue cycle relations. Data from four $(0°_2/90°_3)_s$ coupons under each of three load amplitudes

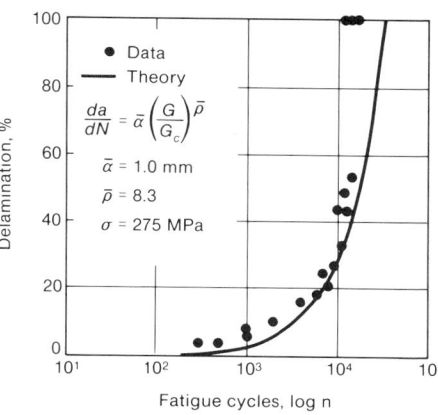

Fig. 23 Experimental (points) and simulated (solid line) delamination size (percent coupon width) versus fatigue cycle relations

Appendix: Finite-Element Procedure

The finite-element routine is formulated to compute the elastic stress field in a long, finite-width, symmetrically stacked laminate that is subjected to in-plane loading. Because of the long-length assumption, the solution domain can be reduced to a transverse laminate cross section. The routine is then formulated on the basis of generalized plane strain. A line crack of finite size, or several similar cracks, may be introduced into the solution domain. In such cases, the routine can compute the crack tip strain energy release rate as well as the elastic stresses. Provisions can be made such that the crack is propagated incrementally in the prescribed direction, with the resulting stress field and the associated strain energy release rate computed accordingly. In short, the routine is aimed at specifically treating the free-edge delamination problem in symmetric laminate coupons.

Finite-Element Formulation

Figure 24 shows the geometrical representation of a long, finite-width, symmetrically stacked laminate coupon. For definiteness, let the coupon be of infinite length, of width $2b$, and of thickness $2h$. Let all the laminating plies be of the same ply thickness, $2t$. Also, let the coupon be loaded by a uniform axial strain ϵ_x. Then, excluding the local effect at the far ends, the solution for the displacements, u, v, w, of the laminate will have the following general form (Ref 11):

$$u(x,y,z) = \epsilon_x\, x\, +\, U(y,z)$$
$$v(x,y,z) = V(y,z)$$
$$w(x,y,z) = W(y,z) \qquad \text{(Eq 29)}$$

are established by experiment in which the specific damage modes must be induced.

Model Based on Exact Damage Modes. When the damage modes in the laminate are relatively simple, an analysis based on the exact damage modes is possible. In the section "Laminate Failure Based on Failure Modes" in this article it was shown that sublaminate matrix failures in the form of intralaminar or interlaminar cracks could be simulated as a load-dependent process by means of fracture mechanics methods, with the assumption of effective material flaw. Similarly, when these cracking processes occur under certain fatigue loading conditions, they can be simulated as time-under-load processes by the same fracture mechanics/effective flaw approach.

In fact, the process of 90° ply transverse cracks in $(0°/90°)_s$ laminates subjected to cyclic tension has been successfully simulated by A.S.D. Wang, S.C. Lei, *et al.* (Ref 19, 24). The simulation procedure is similar to that described in the static loading case in the above-cited section in this article. In this case, the growth of each individual effective flaw, a_i, is treated as a small crack undergoing stable growth under the applied fatigue load. The growth is stopped as it reaches the 0°/90° interface, at which time a full transverse crack is formed. To simulate the time sequence of multiple transverse cracks, the Monte-Carlo search procedure outlined in the above-cited section is then followed.

As for the rate of individual flaw growth, it is assumed to depend only on the crack tip strain energy release rate G at the maximum stress amplitude:

$$\frac{da}{dn} = \alpha \left[\frac{G(\sigma_{\max}, a, \Delta T, R)}{G_c} \right]^\rho \qquad \text{(Eq 28)}$$

where α and ρ are empirical constants assumed to be material related only. The magnitude of α

and ρ are determined by experiments that simulate the mode I crack extension.

Equation 28 can be integrated to yield the required number of cycles when the considered flaw growth has reached the 0°/90° interface. Thus, a crack density versus fatigue cycle relationship is obtained for a set of given $S_{\max}$ and R. Figure 22 compares the simulated and experimental results for transverse cracks in a graphite-epoxy $(0°_2/90°_3)_s$ laminate under three levels of cyclic tension (Ref 24). Note that the simulation provides not only the mean behavior of the transverse cracking process but also its scatter, by virtue of the randomly generated effective flaws.

As reported in a related article (Ref 30), free-edge delamination in a laminate coupon under cyclic tension has also been simulated. The delamination is represented by a single interface crack whose growth rate is again governed by the rule stated in Eq 28. Here, the empirical constants α and ρ may assume different values because delamination growth is generally controlled by mixed mode under fatigue loading, rather than by mode I under static loading. Figure 23 compares the simulated and experimental results of free-edge delamination growth in a graphite-epoxy $(\pm 25°/90°)_s$ laminate under cyclic tension (Ref 28). It should be noted in this case that the predicted result provides only the mean behavior of the growth.

In both these examples, damage accumulation (that is, matrix cracking) is explicitly described as a time-under-load process. The considered matrix cracking modes are but two of the simplest, separate cases. Even so, the crack growth criteria used have not met the required scrutinization accorded other analysis methods, such as those developed for metals. Further development, on both a theoretical and experimental basis, is clearly needed.

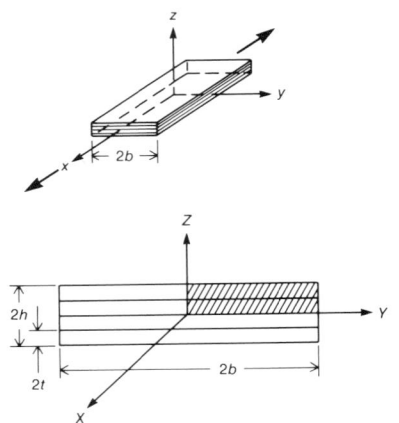

Fig. 24 A long finite-width symmetric laminate under uniaxial loading

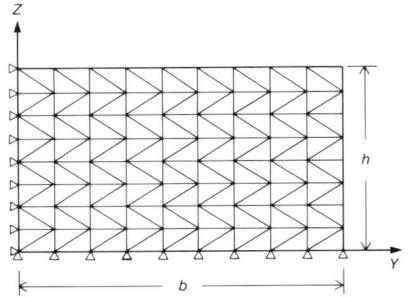

Fig. 25 Finite element discretization of the laminate solution domain

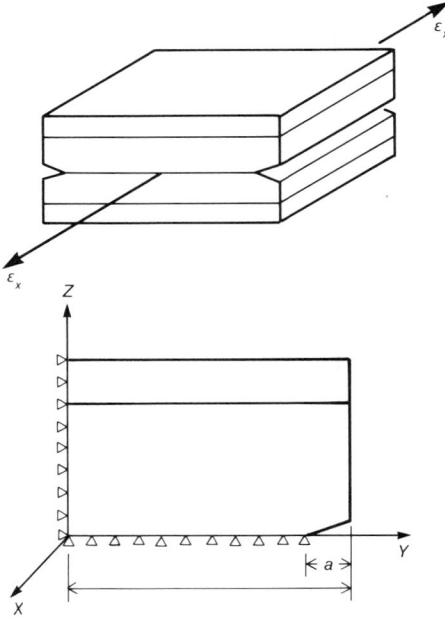

Fig. 26 Finite-element model for free-edge delamination in the midplane

where U, V, and W are functions of y and z only. In particular, U represents the warping of the laminate cross section. Because of the assumed lamination symmetry, the solution domain can further be reduced to the first quadrant of the cross section, as shown by the shaded area in Fig. 24. Then, the following boundary conditions for the functions U, V, and W are assumed:

$$U(0,z) = V(0,z) = 0$$

$$W(y,0) = 0 \qquad \text{(Eq 30)}$$

Now, let the solution domain be made discrete by a network of finite elements as shown in Fig. 25. Here, the constant strain triangular element is used. For a typical element, the strains in the element are determined by the displacement components of its three nodes. Because the coupon is prestrained by ϵ_x, the element strains are given by:

$$\epsilon_i = \epsilon_x$$

$$\epsilon_{m+1} = B_{mi}\bar{o}_i \quad m = 1,5 \quad i = 1,9 \qquad \text{(Eq 31)}$$

where $\bar{o}_i$ are the nine nodal values of U, V, and W functions and B is a 5×9 matrix representing the differentiation operator in the y-z domain. The values of B for the considered element depend only on the coordinates of its three nodes.

The repeated indices appearing in Eq 31 and all subsequent equations imply a summation over the integer range indicated.

Let the element material constitutive relations be described by generalized Hooke's law:

$$\sigma_i = C_{ij}(\epsilon_j - a_j\Delta T) \quad i,j = 1,6 \qquad \text{(Eq 32)}$$

where the thermal effect by ΔT is included.

By a virtual work procedure, the element stiffness matrix is obtained with the form:

$$k_{ij} = (B_{im}D_{mn}B_{nj})A \quad i,j = 1,9 \quad m,n = 1,5 \qquad \text{(Eq 33)}$$

where A is the element area, and:

$$D_{mn} = C_{(m+1)(n+1)} \quad m,n = 1,5 \qquad \text{(Eq 34)}$$

Corresponding to the nine nodal displacement components, $\bar{o}_i$, are nine nodal force components, f_i, $i = 1, 9$. The relation between the nodal forces and the nodal displacements is given by:

$$f_i = r_i + k_{ij}\bar{o}_j \quad i,j = 1,9 \qquad \text{(Eq 35)}$$

where r_i represents the corresponding nodal forces due to the prestrain ϵ_x and the thermal loading ΔT; r_i is calculated by:

$$r_i = (B_{im}D_{mn}\epsilon_n^* + B_{in}\sigma_n^*)A$$

$$i = 1,9 \quad m,n = 1,5 \qquad \text{(Eq 36)}$$

with

$$\epsilon_n^* = -a_{(n+1)}\Delta T$$

$$a_n^* = C_{1(n+1)}[\epsilon_x - a_i\Delta T] \quad n = 1,5 \qquad \text{(Eq 37)}$$

Let the solution field be a network of N elements with M nodes. Then, there will be N sets of nine equations obtained by using Eq 35. However, a consolidation of these nine times N equations will result in a set of three times M equations for the entire finite-element system in the form:

$$F_p + R_p = K_{pq}\bar{\sigma}_q^* \quad p,q = 1,3M \qquad \text{(Eq 38)}$$

where F_p are the system nodal forces, R_p are the system nodal forces due to the prestrain ϵ_x and the thermal loading ΔT, K_{pq} is the system stiffness matrix, and $\bar{\sigma}_q^*$ are the system nodal values of U, V, and W.

The system of equations in Eq 38 can be solved when one of the following nodal conditions are prescribed: all the nodal forces, F_p, are prescribed; all the nodal displacements, $\bar{\sigma}_p^*$, are prescribed; and a mixed nodal force and nodal displacement condition is prescribed. The specification of the nodal forces and/or nodal displacements must, of course, be consistent with the appropriate boundary conditions.

Once the system of equations is solved, the individual element strains are calculated from the solutions for $\bar{\sigma}_q^*$ through the element Eq 31, and the element stresses are calculated using Eq 32, giving:

$$\sigma_l = C_{lj}(\epsilon_j - a_j\Delta T) \ j = 1,6$$

$$\sigma_{(m+1)} = D_{mn}[\epsilon_n^* + \epsilon_{(n+1)}] + \sigma_n^*$$

$$m,n = 1,5 \qquad \text{(Eq 39)}$$

If the applied laminate load is specified as an averaged value of the laminate axial stress $\bar{\sigma}_x$ instead of the uniform strain ϵ_x, then the term must be treated as an unknown; it is to be determined by satisfying the condition that the sum of all the nodal forces in the x-direction must be equal to $\bar{\sigma}_x A$.

Crack-Closure Scheme for Energy Release Rate

As has been mentioned earlier, the solution domain can admit the presence of a line crack, and the crack can be propagated incrementally in some prescribed direction. In the case of free-edge delamination occurring in one of the ply interfaces, the solution model may be represented by one of two possible configurations. Figure 26 shows the configuration for delamination in the laminate midplane, while Fig. 27 shows the configuration for delamination in an interface away from the laminate midplane. In the latter, there will be two identical delaminating interfaces due to laminate symmetry, one above the midplane and one below the midplane.

In either case, a starter crack of size a is introduced in the intended interface, and near the free edge. This starter crack represents the effective interface flaw. Under the applied strain ϵ_x for instance, the crack is allowed to propagate incrementally in a stable manner along the interface in which the crack is situated. The finite-element scheme for incremental crack propagation is illustrated in Fig. 28, in which the crack tip originally at node c is now propagated to node d with the extension of Δa. As the node c is opened, the crack tip opening

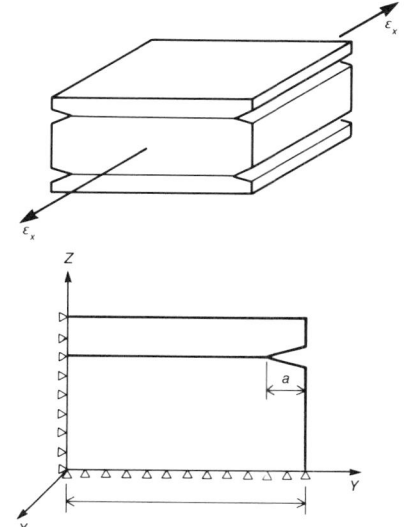

Fig. 27 Finite-element model for free-edge delamination in an off-midplane interface

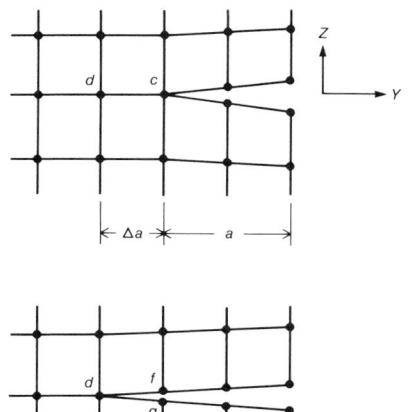

Fig. 28 Incremental crack extension in the finite-element solution domain

is represented by the pair of nodes f and g. Because the finite-element procedure calculates the displacement components for all nodes, including nodes f and g, the crack tip opening due to the incremental extension from node c to node d can thus be determined.

Accordingly, let Δ_i, $i = 1,3$, denote the crack tip opening displacement components between nodes f and g. Then Δ_i can be expressed in terms of the applied strain ϵ_x:

$$\Delta_i = (\Delta_\epsilon)_i \, \epsilon_x \quad i = 1,3 \tag{Eq 40}$$

If ΔT is included as an independently applied loading, then Δ_i can be expressed as:

$$\Delta_i = (\Delta_\epsilon)_i \, \epsilon_x + (\Delta_T)_i \quad i = 1,3 \tag{Eq 41}$$

The terms of Δ_ϵ and Δ_T in the above two equations are the influence coefficients for Δ due to a unit of ϵ_x and ΔT, respectively.

Now, if the nodes f and g are closed again by force and returned to the position originally occupied by node c, the components of the required nodal force can also be calculated by the finite-element procedure, provided the adjusted nodal conditions are prescribed. Then, let F_i, $i = 1,3$, denote the required nodal force components, whose numerical solutions can also be expressed in terms of the applied loads, ϵ_x and ΔT, in the form:

$$F_i = (F_\epsilon)_i \, \epsilon_x + (F_T)_i \, \Delta T \quad i = 1,3 \tag{Eq 42}$$

where F_ϵ and F_T are the influence coefficients for F due to a unit of ϵ_x and ΔT, respectively.

The work done, ΔW, during the incremental crack closure is then expressed as:

$$\Delta W = \frac{[F_i \Delta_i]}{2} \quad i = 1,3 \tag{Eq 43}$$

On the basis of elastic deformation assumption, the work done during the incremental crack closure is assumed to be equivalent to the strain energy released during the same incremental crack extension (Eq 34). Thus, it follows that the crack tip strain energy release rate, G, may be approximated by:

$$G(\epsilon_x, \Delta T, a) = (\Delta W / \Delta a) t^* \tag{Eq 44}$$

where t^* is the linear scale between the physical model and the finite-element model. Substituting Eq 41 to 43 into Eq 44, G can be expressed in explicit terms of ϵ_x and ΔT:

$$G(\epsilon_x, \Delta T, a) = [C_\epsilon \, \epsilon_x^2 + C_T \, \Delta T^2 + C_{\epsilon T} \, \epsilon_x \, \Delta T] t^* \tag{Eq 45}$$

where the coefficients C_ϵ, C_T and $C_{\epsilon T}$ in Eq 45 denote:

$$C_\epsilon = [(F_\epsilon)_i \, (\Delta_\epsilon)_i]/(2\Delta a)$$

$$C_T = [(F_T)_i \, (\Delta_T)_i]/(2\Delta a) \quad i = 1,3$$

$$C_{\epsilon T} = [(F_\epsilon)_i \, (\Delta_T)_i + (F_T)_i \, (\Delta_\epsilon)_i]/(2\Delta a) \tag{Eq 46}$$

It is noted that the coefficient C_ϵ is associated with the loading ϵ_x only, while C_T is associated with ΔT only. If each load produces individually admissible crack openings, then it can be shown that the coefficient $C_{\epsilon T}$ can be expressed as:

$$C_{\epsilon T} = \pm 2\sqrt{(C_\epsilon C_T)} \tag{Eq 47}$$

Consequently, Eq 45 takes the following simple form:

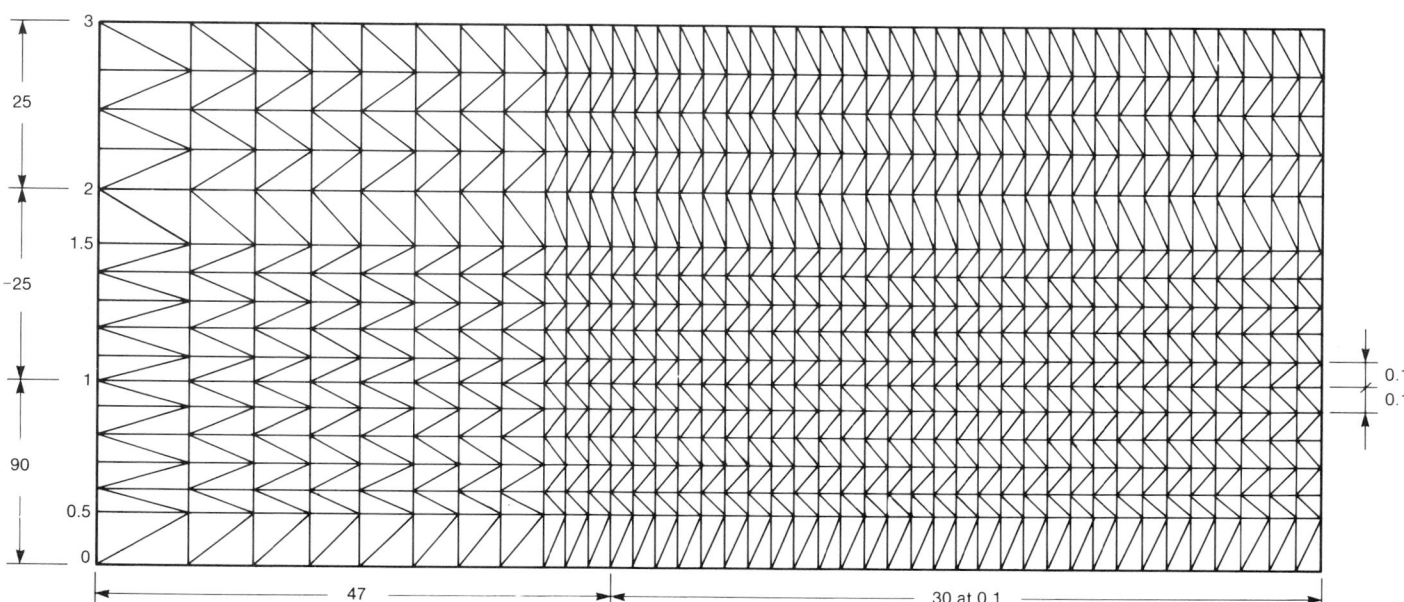

Fig. 29 Finite element grid used to compute $C_\epsilon(a)$ for the $(\pm 25°/90°)_s$ laminate having delamination growth in $-25°/90°$ interface

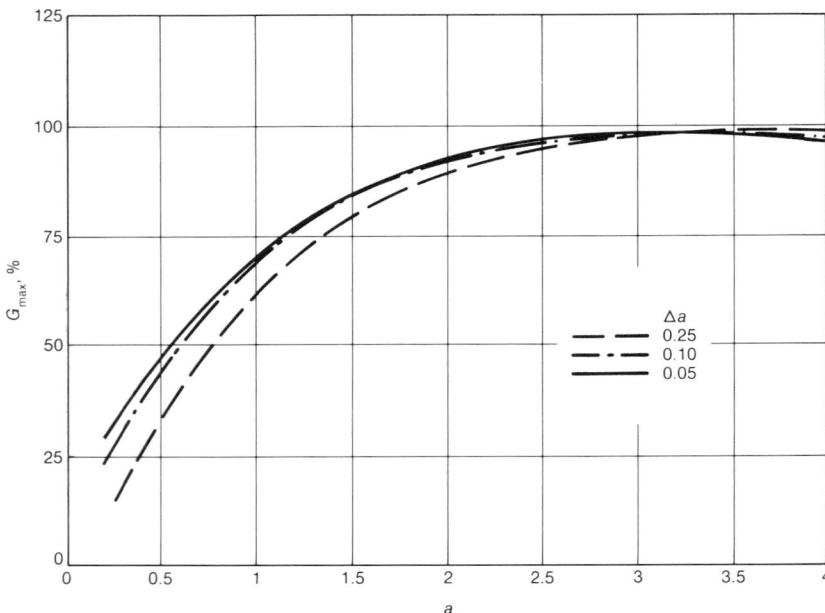

Fig. 30 The effect of $\Delta a/a$ on computed energy release rate by the finite-element technique

$$G(\epsilon_x, \Delta T, a) = [\sqrt{C_\epsilon}\, \epsilon_x \pm \sqrt{C_T}\, \Delta T]^2 t^* \qquad \text{(Eq 48)}$$

The ambiguous sign in Eq 47 and 48 is chosen in order to be consistent with the actual sign of ΔT. Because in most practical cases $\Delta T\, (= T - T_0)$ is negative, a minus sign is taken in Eq 47 and 48.

It is noted from these equations that G, so computed by the crack closure technique, contains the sum of three scalars: $G = G_{\mathrm{I}} + G_{\mathrm{II}} + G_{\mathrm{III}}$. Each is related to a physical mode of the crack propagation, namely:

$$G_{\mathrm{I}} = (F_3 \Delta_3 / 2\Delta a) t^*$$

$$G_{\mathrm{II}} = (F_2 \Delta_2 / 2\Delta a) t^*$$

$$G_{\mathrm{III}} = (F_1 \Delta_1 / 2\Delta a) t^* \qquad \text{(Eq 49)}$$

The foregoing discussions pertained to the general procedure of the finite-element crack closure scheme for calculating the strain energy release rate. The next section will discuss some specific details in computing the coefficients C_ϵ and C_T.

Specifics of the Finite-Element Computation

Ply Constitutive Equations. The basic unidirectional ply may be treated as an elastic homogeneous medium having one plane of symmetry, namely, the x-y plane. The material stiffness matrix C_{ij} appearing in Eq 32 will then have 13 independent material constants (see, for example, Ref 1). For the example problems presented in this article, an orthotropic symmetry was assumed for the material ply, with the

principal axes being L, T, and z. In this case, C_{ij} will contain only nine independent material constants. Similarly, there are three independent thermal expansion coefficients, α_{L}, α_{T}, and α_z. These are listed in Table 3 for the AS-3501-06 graphite-epoxy system in which two of its constants ($\nu_{\mathrm{T}z}$ and $G_{\mathrm{T}z}$) are estimated values.

Finite-Element Grid. The deployment of the finite elements in the solution domain can affect the accuracy of the computed results. This is especially true because the elastic stress field contains singularities of various strengths. Hence, it is necessary to employ smaller elements near the singular points. Even so, the numerical routine cannot determine the singularities themselves, but can provide an approximate solution for the stresses near the singular points.

Figure 29 shows the finite-element grid being used for the $(\pm 25^\circ/90^\circ)_s$ laminate. The solution domain has a rectangular area of 3×50, so that each ply has a thickness of unity in the model. Because the actual thickness of the material ply is $2t$, the linear scale factor t^* appearing in Eq 48 is $t^* = 2t$. Also, it is known that the free-edge effect extends only a few ply thicknesses into the interior. Thus, a very dense grid is also deployed near the edge region. As for the width of the solution domain of 50, it is considered large compared to its thickness of 3. In this way, the boundary effect near the z-axis will not interfere with that near the free edge or that of a propagating delamination.

The grid for the $(\pm 25^\circ/90^\circ)_s$ laminate shown in Fig. 29 allows the $-25^\circ/90^\circ$ interface to be examined for edge delamination. Note the smaller elements deployed near both sides of the intended interface. Material layers situated

Table 3 Material constants for AS-3501-06 unidirectional ply

Constants	SI Unit
E_{LL}, GPa (10^6 psi)	140.0 (20.4)
E_{TT}, E_{zz}, GPa (10^6 psi)	11.0 (1.6)
ν_{LT}, $\nu_{\mathrm{L}z}$	0.29
$\nu_{\mathrm{T}z}$	0.3
G_{LT}, $G_{\mathrm{L}z}$, GPa (10^6 psi)	5.5 (0.8)
$G_{\mathrm{T}z}$, GPa (10^6 psi)	5.5 (0.8)
α_{T}, α_z, 10^{-6}/K	28.8
α_{L}, 10^{-6}/K	0.36

away from this interface, however, are represented by much larger elements. In this particular example, the network contains 697 nodes, and the resulting solution field has 2091 degrees of freedom. This is a relatively small problem for a high-speed machine.

Effect of $\Delta a/a$ Ratio. The computed energy release rate G is approximated according to Eq 44. Its accuracy is clearly dependent on the value of Δa, but because of the finite-element crack closure scheme, Δa is determined by the size of the elements near the crack tip. The relative effect of Δa diminishes, however, as the crack size a becomes large, or the ratio of $\Delta a/a$ becomes smaller.

Figure 30 shows the computed G in terms of $G_{\max}$ for the $(\pm 25^\circ/90^\circ)_s$ laminate with the intended delamination in the $-25^\circ/90^\circ$ interface. Here, G is plotted against the delamination size a for three values of Δa. It is seen that when a is small, the effect of the size of Δa is significant. As a becomes larger, the effect of Δa diminishes. In the present case, it seems that G can be computed with good accuracy when $\Delta a/a$ is smaller than $\frac{1}{10}$. No additional effort was made to evaluate the accuracy of the computed G against an exact solution, however.

REFERENCES

1. S.W. Tsai and H.T. Hahn, *Introduction to Composite Materials*, Technomic, 1980
2. S.G. Lekhnitsky, *Theory of Elasticity of An Anisotropic Elastic Body*, Holden-Day, 1963
3. J.M. Whitney, I.M. Daniel, and R.B. Pipes, *Experimental Mechanics of Fiber Reinforced Composite Materials*, Society for Experimental Stress Analysis, 1982
4. Z. Hashin, "Theory of Fiber Reinforced Materials," NASA-CR-1974, National Aeronautics and Space Administration, 1970
5. M.N. Nahas, Survey of Failure/Post-Failure Theories of Laminated Fiber Reinforced Composites, *J. Compos. Technol.*, Vol 8, 1986, p 138-153
6. S.W. Tsai and E.M. Wu, A General Theory of Strength for Anisotropic Materials, *J. Compos. Mater.*, Vol 5, 1971, p 58-80
7. R.B. Pipes and B.W. Cole, On the Off-

Axis Strength Test for Anisotropic Materials, *J. Compos. Mater.*, Vol 7, 1973, p 102-108

8. K.D. Chiu, Ultimate Strength of Laminated Composites, *J. Compos. Mater.*, Vol 3, 1969, p 578-582

9. H.T. Hahn and S.W. Tsai, On the Behavior of Composite Laminates After Initial Failure, *J. Compos. Mater.*, Vol 8, 1974, p 280-305

10. P.H. Petit and M.E. Waddoups, A Method of Predicting the Nonlinear Behavior of Laminated Composites, *J. Compos. Mater.*, Vol 3, 1969, p 2-19

11. R.B. Pipes and N.J. Pagano, Interlaminar Stresses in Composite Laminates Under Uniaxial Extension, *J. Compos. Mater.*, Vol 4, 1970, p 538-548

12. A.S.D. Wang and F.W. Crossman, Some New Results on Edge Effects in Symmetric Composite Laminate, *J. Compos. Mater.*, Vol 11, 1977, p 92-102

13. F.W. Crossman and A.S.D. Wang, The Dependence of Transverse Cracking and Delamination on Ply Thickness in Graphite-Epoxy Laminates, in *Damages in Composite Materials*, STP 775, American Society for Testing and Materials, 1982, p 118-139

14. S.S. Wang, and I. Choi, "The Interface Crack Behavior of Dissimilar Anisotropic Composites Under Mixed-Mode Loading," *J. Appl. Mech. (Trans. ASME)*, Vol 50, 1983, p 178-183

15. F. Delale, A.S.D. Wang, and W. Binienda, On the Strain Energy Release Rate for a Crack at the Interface of Two Bonded Materials, in *Advances in Aerospace Sciences and Engineering*, ASME AD-08, 1984, p 113-120

16. A.S.D. Wang, N.N. Kishore, and C.A. Li, On Crack Development in Graphite-Epoxy $(0_2/90_n)_s$ Laminates Under Uniaxial Tension, *J. Compos. Sci. Technol.*, Vol 23, 1985, p 1-31

17. A.S.D. Wang and F.W. Crossman, Initiation and Growth of Transverse Cracks and Edge Delamination in Composite Laminates, Part 1, An Energy Method, *J. Compos. Mater.*, Vol 14, 1980, p 71-87

18. F.W. Crossman, W.J. Warren, A.S.D. Wang, and G.E. Law, Initiation and Growth of Transverse Cracks and Edge Delamination in Composite Laminates, Part 2, Experimental Correlation, *J. Compos. Mater.*, Vol 14, 1980, p 88-106

19. A.S.D. Wang, P.C. Chou, and S.C. Lei, A Stochastic Model for the Growth of Matrix Cracks in Composite Laminates, *J. Compos. Mater.*, Vol 18, 1984, p 239-254

20. A.A. Griffith, The Phenomena of Rupture and Flow in Solids, *Phil. Trans. R. Soc.* (London) A, Vol 221, 1920, p 163-198

21. D.J. Wilkins, J.R. Eisenmann, R.A. Camin, W.S. Margolis, and R.A. Benson, Characterizing Delamination Growth in Graphite-Epoxy, in *Damages in Composite Materials*, STP 775, American Society for Testing and Materials, 1982, p 168-183

22. A.S.D. Wang, N.N. Kishore, and W.W. Feng, On Mixed-Mode Fracture in Off-Axis Unidirectional Graphite-Epoxy Composites, in *Progress in Science and Engineering of Composites*, Vol 1, Japan Society for Composite Materials, 1982, p 599-606

23. A.J. Russell and K.N. Street, Moisture and Temperature Effects on the Mixed-Mode Delamination Fracture of Unidirectional Graphite-Epoxy, in *Delamination and Debonding of Materials*, STP 876, American Society for Testing and Materials, 1985, p 349-370

24. S.C. Lei, "A Stochastic Model for the Damage Growth During the Transverse Cracking Process in Composite Laminates," Ph.D thesis, Drexel University, 1986

25. P.C. Paris and F. Erdogan, A Critical Analysis of Crack Propagation Laws, *J. Basic Eng.*, ASME Trans., D, Vol 85 (No. 3), 1963

26. J.C. Halpin, K.L. Jerina, and T.A. Johnson, Characterization of Composites for the Purpose of Reliability Evaluation, in *Fatigue of Fibrous Composite Materials*, STP 723, American Society for Testing and Materials, 1973, p 5-64

27. J.A. Awerbuch and M.S. Madhukar, "Notched Strength of Composite Laminates: Predictions and Experiments—A Review," *J. Reinf. Plast. Compos.*, Vol 4, 1985, p 3-159

28. A. Poursartip, M.F. Ashby, and P.W.R. Beaumont, The Fatigue Damage Mechanics of a Carbon Fibre Composite Laminate: I. Development of the Model, *J. Compos. Sci. Technol.*, Vol 25, 1986, p 193-218

29. A. Poursartip, M.F. Ashby, and P.W.R. Beaumont, The Fatigue Damage Mechanics of a Carbon Fibre Composite Laminate: II. Life Prediction, *J. Compos. Sci. Technol.*, Vol 25, 1986, p 283-299

30. A.S.D. Wang, M. Slomiana, and R.B. Bucinell, Delamination Crack Growth in Composite Laminates, STP-876, American Society for Testing and Materials, 1985, p 135-167

Fracture Analysis of Laminates

James M. Whitney, Air Force Wright Aeronautical Laboratories

LAMINATED COMPOSITES containing through-thickness discontinuities in the form of sharp cracks and circular holes have received considerable attention because of their importance in design. Models have been proposed for notched composites in which details of the damage region adjacent to the notch are included (Ref 1-5). These models are useful for gaining insight into the mechanisms associated with failure of notched composites. Two simplified approaches are of practical interest for predicting the notched strength of laminated composites. One approach uses concepts of linear-elastic fracture mechanics (LEFM), while the second approach is based on the stress distribution adjacent to the notch. These approaches are generally applicable to laminates containing a number of 0° plies parallel to the load direction (that is, filament-dominated laminates). Before these simplified models are considered, stress concentrations and stresses developed adjacent to various through-thickness discontinuities must be reviewed briefly.

Stress Concentrations

Consider a laminated plate constructed such that the in-plane behavior can be characterized by a homogeneous orthotropic analysis. The plate contains an elliptical hole with the major and minor axes parallel to the x and y material axes, respectively. The plate is of infinite extent with a remote uniaxial stress, σ_y^∞, applied parallel to the y-axis (Fig. 1). In terms of laminated-plate theory, the resultant far-field force is denoted by N_y^∞, yielding the average stress:

$$\sigma_y^\infty = N_y^\infty/h = \text{constant}$$

where h is plate thickness of the composite. In terms of strength reduction due to the elliptical hole, the effective normal stress along the major axis is of particular interest, that is, the normal stress is:

$$\sigma_y(x,0) = N_y(x,0)/h \qquad \text{(Eq 1)}$$

where $x \geq a$. The precise function corresponding to Eq 1 can be determined using a complex variable technique (Ref 6). Although the solution can be expressed in closed form, it is a complicated function of complex variables.

The stress-concentration factor for an infinite plate, K_t^∞, is also of interest and can be expressed in a simplified form (Ref 6):

$$K_t^\infty = \frac{\sigma_y(a,0)}{\sigma_y^\infty} = \frac{N_y(a,0)}{N_y^\infty} = 1 + n\left(\frac{a}{b}\right) \qquad \text{(Eq 2)}$$

where a and b are the length of the major and minor semiaxes of the ellipse, respectively, and:

$$n = \sqrt{2\left(\frac{E_y}{E_x} - \nu_{xy} + \frac{E_y}{2G_{xy}}\right)}$$

where E_x and E_y are the effective Young's modulus in the x- and y-directions, ν_{xy} is the major Poisson's ratio, and G_{xy} is the in-plane shear modulus of the laminate. For an isotropic material, $n = 2$ and Eq 2 becomes:

$$K_t^\infty = 1 + 2\left(\frac{a}{b}\right)$$

The stress field in the vicinity of a circular hole of radius R in an orthotropic laminate under uniaxial loading, σ_y^∞, can be determined from the solution to the elliptical hole as the

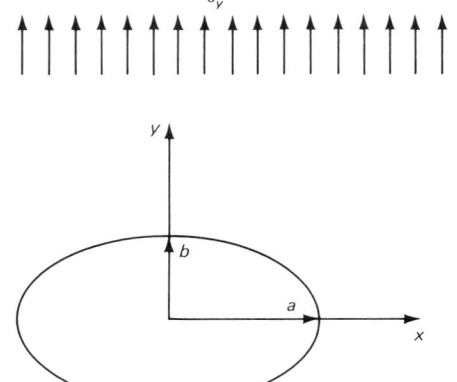

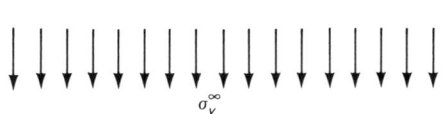

Fig. 1 Infinite plate containing an elliptical hole under uniform far-field tension

special case where $a = b = R$. Again, the effective normal stress, $\sigma_y(x,0)$, is a function of complex variables. This function can be approximated, however, by the relationship (Ref 7):

$$\sigma_y(x,0) = \frac{\sigma_y^\infty}{2}\left\{2 + \left(\frac{R}{x}\right)^2 + 3\left(\frac{R}{x}\right)^4 - \right.$$
$$\left. (K_t^\infty-3)\left[5\left(\frac{R}{x}\right)^6 - 7\left(\frac{R}{x}\right)^8\right]\right\} \qquad \text{(Eq 3)}$$

where $x \geq R$. For a circular hole, $a = b$ in Eq 2 and:

$$K_T^\infty = 1 + n \qquad \text{(Eq 4)}$$

For the isotropic case, $n = 2$, Eq 4 reduces to the classical value $K_t^\infty = 3$, and Eq 3 reduces to the form:

$$\sigma_y(x,0) = \frac{\sigma_y^\infty}{2}\left[2 + \left(\frac{R}{x}\right)^2 + 3\left(\frac{R}{x}\right)^4\right]$$

where $x \geq R$, which is an exact solution for the isotropic case.

The accuracy of Eq 3 is illustrated in Fig. 2, which compares it to the exact elasticity solution for a laminate with $K_t^\infty = 3.852$. Excellent agreement between the approximate and exact solution is obtained.

The stress in the vicinity of a center crack of length $2c$ in an orthotropic laminate under uniaxial loading, σ_y^∞, can be derived as a special case of an elliptical hole with $a = c$ and $b \rightarrow 0$. A cursory examination of Eq 2 reveals that the stress-concentration factor becomes unbounded, yielding a stress singularity at the crack tip. For this case the exact expression for the normal stress ahead of the crack is given by (Ref 6):

$$\sigma_y(x,0) = \frac{\sigma_y^\infty x}{\sqrt{x^2-c^2}} \qquad \text{(Eq 5)}$$

where $x > c$. Note that Eq 5 is independent of material properties, for either orthotropic or isotropic laminates.

Because of the stress singularity at the crack tip, the concept of a stress-concentration factor is replaced by a stress-intensity factor, which forms the basis of classical fracture mechanics. Uniaxial tension is referred to as mode I, and the notation K_I^∞ is often used to denote the mode

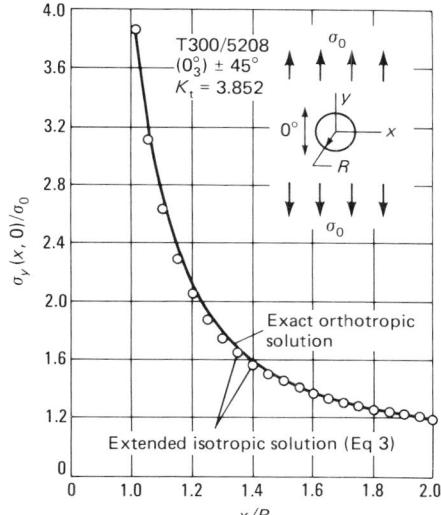

Fig. 2 Comparison of stress distribution solutions adjacent to a hole in an infinite orthotropic plate

I stress-intensity factor. Eq 5 can also be written:

$$\sigma_y(x,0) = \frac{K_I^\infty}{\sqrt{\pi c(x^2 - c^2)}} \; , \; x > c \qquad \text{(Eq 6)}$$

where:

$$K_I^\infty = \sigma_y^\infty \sqrt{\pi c} \qquad \text{(Eq 6b)}$$

For values of x close to the crack tip, Eq 6 can be approximated by the relationship (Ref 8):

$$\sigma_y(x,0) = \frac{K_I^\infty}{\sqrt{2\pi(x-c)}} \qquad \text{(Eq 6c)}$$

Fracture Mechanics Criteria

It has been shown that the strength of a laminated composite containing a sharp center crack of length $2c$ subjected to a uniaxial load can be predicted by LEFM methods (Ref 9-11). In particular:

$$K_{Ic}^\infty = \sigma_N^\infty \sqrt{\pi c} \qquad \text{(Eq 7)}$$

where σ_N^∞ is the notched strength for a laminate of infinite width and K_{Ic}^∞ is the mode I critical stress-intensity factor. The parameter K_{Ic}^∞ is a function of c for small crack lengths and asymptotically approaches a constant value for large crack lengths. The concept of a plastic zone has been utilized in metals (Ref 12) to explain the increasing value of K_{Ic}^∞ with increasing crack length. Although most polymeric matrix composites fail in a brittle manner, a damage zone does develop that is analogous to the plastic zone. Using this concept in conjunction with Eq 7 yields the relationship:

$$K_c = \sigma_N^\infty \sqrt{\pi(c + c_0)} \qquad \text{(Eq 8)}$$

where K_c is the fracture toughness and c_0 is an inherent flaw size. The term inherent flaw size is used because the unnotched strength, σ_0, of a composite laminate is given by Eq 8 for the case of vanishing c, that is:

$$K_c = \sigma_0 \sqrt{\pi c_0} \qquad \text{(Eq 9)}$$

It should be noted that c_0 does not refer physically to an inherent crack, but refers to a cracklike damage region that develops before ultimate failure. From a cursory examination of Eq 7 and 8, it can be seen that:

$$K_c = \lim_{c \to \infty} K_{Ic}^\infty$$

Combining Eq 8 and 9 yields the failure criterion:

$$\sigma_N^\infty = \sigma_0 \sqrt{1 - \xi_1} \qquad \text{(Eq 10)}$$

where:

$$\xi_1 = \frac{c}{c + c_0} \qquad \text{(Eq 11)}$$

Although LEFM methods can be used in conjunction with composite materials, they must be applied cautiously to strength predictions. In particular, unlike homogeneous metals, common failure processes in laminated composites do not involve the development of through-thickness cracks. Thus, LEFM information is valid only for assessing strength reduction when a sharp discontinuity is inflicted on a composite material from an external source, for example, impact damage.

Experimental data (Ref 10, 11, 13) have shown that the tensile strength of laminated composites containing a circular hole depends on the hole size. Such a phenomenon obviously cannot be predicted by a classical stress-concentration factor; that is, a failure criterion of the form:

$$\sigma_N^\infty = \frac{\sigma_0}{K_t^\infty} \qquad \text{(Eq 12)}$$

is independent of hole size. M.E. Waddoups *et al.* used LEFM methods to explain the hole size effect (Ref 13). In particular, they assumed that a damage region that developed adjacent to the hole, perpendicular to the load direction, could be modeled as through-thickness cracks of length a, as illustrated in Fig. 3. A solution for this problem has been developed by O.L. Bowie (Ref 14) for an infinite isotropic plate, with the result:

$$K_c = \sigma_N^\infty \sqrt{\pi a} \; f(a/R) \qquad \text{(Eq 13)}$$

where R is the hole radius. The function $f(a/R)$ has been tabulated in Ref 15. It is assumed that a is independent of hole radius. For a vanishing hole radius, $a/R \to \infty$, $f(a/R)$ approaches unity. Thus:

$$K_c = \sigma_0 \sqrt{\pi a} \qquad \text{(Eq 14)}$$

Combining Eq 13 and 14 results in:

$$\sigma_N^\infty = \frac{\sigma_0}{f(a/R)} \qquad \text{(Eq 15)}$$

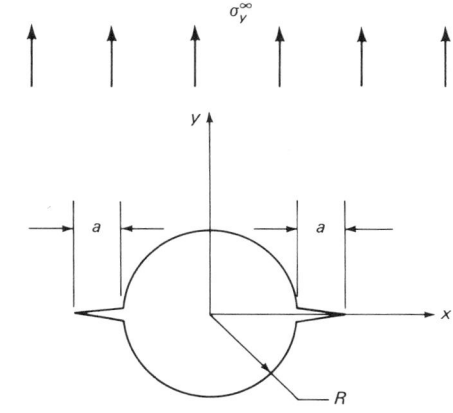

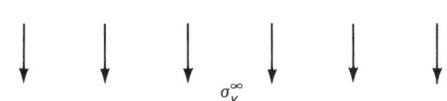

Fig. 3 Fracture mechanics model for circular hole

Because a is assumed to be independent of hole radius, Eq 15 predicts the notch strength to be a function of absolute hole size. It should be noted that $f(0) = 3.37$. Thus, for increasing hole size, the strength reduction drops below the stress-concentration factor; that is:

$$\sigma_N^\infty = \frac{\sigma_0}{3.37} \; , \; a/R \to 0$$

while, for an isotropic material ($K_t = 3$), Eq 12 leads to the result:

$$\sigma_N^\infty = \frac{\sigma}{3}$$

Despite the simplicity of the fracture mechanics failure criterion for circular holes, as displayed by Eq 15, numerical difficulties are encountered because the function $f(a/R)$ has not been tabulated for orthotropic materials.

Stress Fracture Criteria

An alternate approach to LEFM methods for predicting uniaxial notch strength, as proposed in Ref 10 and 11, is based on the stress distribution adjacent to the notch, perpendicular to the load. In this approach the explanation of the hole size effect is based on the difference that exists in the normal stress distribution ahead of the hole for different hole sizes, as shown for an isotropic material in Fig. 4. Although all hole sizes have the same stress-concentration factor, the normal stress perturbation from a uniform stress state can be seen to be considerably more concentrated near the hole boundary in the smaller hole. Thus, intuitively, the plate containing the smaller hole might be expected to be the stronger of the two, because it offers a greater opportunity to redis-

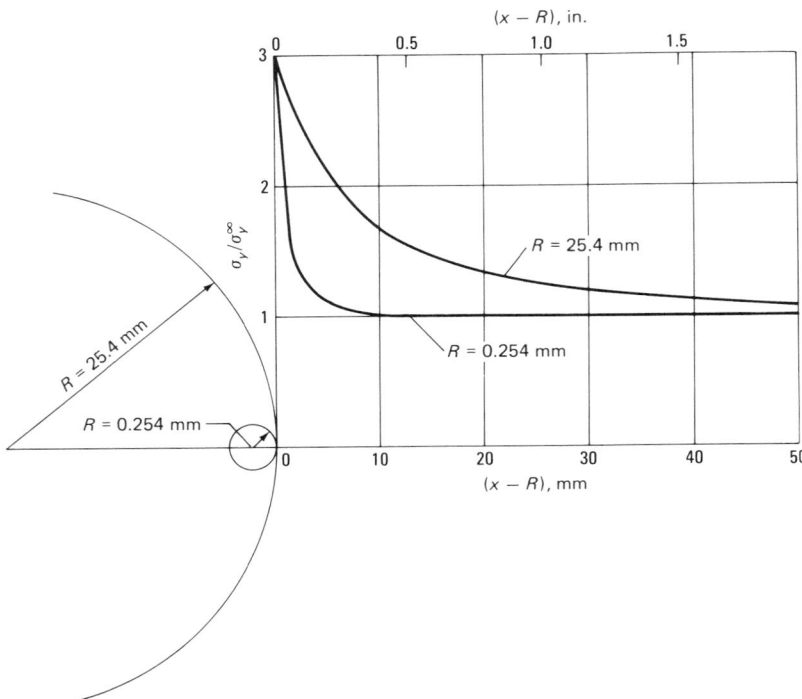

Fig. 4 Effect of hole radius on adjacent normal stress distribution in an isotropic material

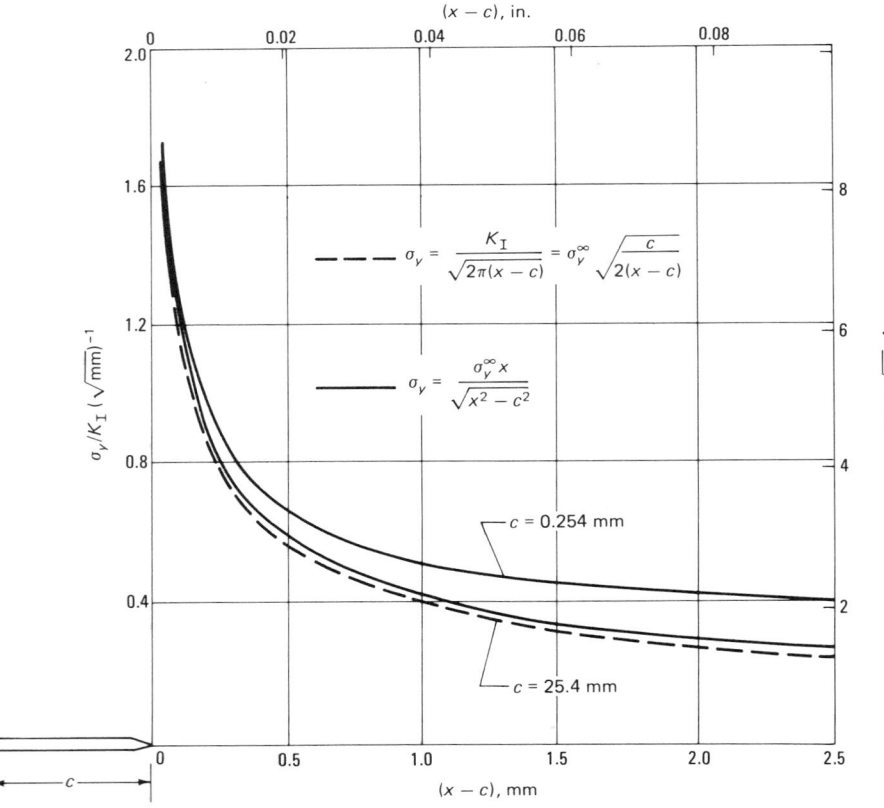

Fig. 5 Comparison of approximate and exact solutions for stress distribution ahead of a center crack

tribute high stresses. Through-thickness cracks were also considered in Ref 10 and 11, where the crack size effect was explained by considering the exact elasticity solution for the normal stress ahead of a crack rather than the singular term of the asymptotic expansion (Fig. 5). Two stress criteria were proposed: point stress and average stress.

Point stress criterion assumes that failure occurs when the stress, σ_y, at some fixed distance, d_0, ahead of the notch becomes equal to the unnotched tensile strength, σ_0, of the material. For a circular hole, failure occurs when:

$$\sigma_y(x,0)|_{x=R+d_0} = \sigma_0 \qquad \text{(Eq 16)}$$

Using the approximate stress distribution for a circular hole in an infinite orthotropic plate, Eq 3, in conjunction with Eq 16, yields the failure criterion:

$$\sigma_N^\infty = \frac{2\sigma_0}{[2 + \xi_2^2 + 3\xi_2^4 - (K_t^\infty-3)(5\xi_2^6-7\xi_2^8)]} \text{(Eq 17)}$$

where:

$$\xi_2 = \frac{R}{R + d_0}$$

Note that for increasing hole size, ξ_2 approaches unity, and the strength reduction as given by Eq 17 reduces to the stress-concentration factor criterion, Eq 12.

Considering the center-cracked geometry of Fig. 5, the exact anisotropic elasticity solution for the normal stress ahead of a crack of length $2c$ in an infinite orthotropic plate under uniform axial tension, σ_y^∞, is given by Eq 5. Replacing R with c in Eq 16 and combining the result with Eq 5 yields the failure criterion:

$$\sigma_N^\infty = \sigma_0 \sqrt{1 - \xi_3^2} \qquad \text{(Eq 18)}$$

where:

$$\xi_3 = \frac{c}{c + d_0}$$

Average stress criterion, proposed in Ref 10 and 11, assumes that failure occurs when the average stress value of σ_y over some fixed distance, a_0, ahead of the notch first reaches the unnotched tensile strength of the material; that is, for the circular hole, failure occurs when:

$$\frac{1}{a_0} \int_R^{R+a_0} \sigma_y(x,0)\, dx = \sigma_0 \qquad \text{(Eq 19)}$$

Using Eq 19 in conjunction with Eq 3 yields the strength reduction relationship:

$$\sigma_N^\infty = \frac{2\sigma_0(1 - \xi_4)}{[2 - \xi_4^2 - \xi_4^4 + (K_t-3)(\xi_4^6-\xi_4^8)]} \qquad \text{(Eq 20)}$$

where:

$$\xi_4 = \frac{R}{R = a_0}$$

Again, for increasing hole size, Eq 20 reduces to the stress-concentration factor failure criterion, Eq 12.

Replacing R with c in Eq 19 and combining the result with Eq 5 yields the failure criterion:

$$\sigma_N^\infty = \sigma_0 \sqrt{\frac{1 - \xi_5}{1 + \xi_5}} \qquad \text{(Eq 21)}$$

where:

$$\xi_5 = \frac{c}{c + a_0}$$

Crack Size Effect. The predicted crack size effect on the measured value of the critical stress-intensity factor, K_{Ic}, can be observed by writing Eq 18 and 21 in the form:

$$K_{Ic} = \sigma_0 \sqrt{\pi c (1 - \xi_3^2)} \qquad \text{(Eq 22)}$$

$$K_{Ic} = \sigma_0 \sqrt{\pi c \frac{(1 - \xi_5)}{(1 + \xi_5)}} \qquad \text{(Eq 23)}$$

respectively. These results are obtained by using the relationship between K_{Ic} and σ_N^∞ as given by Eq 7. For increasing crack lengths, K_{Ic} approaches a constant value in both Eq 22 and 23, yielding a predicted fracture toughness given by the relationships:

$$K_{Ic} = \sigma_0 \sqrt{2\pi d_0} \qquad \text{(Eq 24)}$$

$$K_{Ic} = \sigma_0 \sqrt{\pi \frac{a_0}{2}} \qquad \text{(Eq 25)}$$

respectively.

By using the inherent flaw model as given by Eq 9 in conjunction with Eq 24 and 25, a relationship is obtained between c_0 and the stress fracture parameters d_0 and a_0. In particular:

$$c_0 = 2d_0 = \frac{a_0}{2} \qquad \text{(Eq 26)}$$

By applying the inherent flaw model in the form of Eq 8 in conjunction with Eq 23 and 26, Eq 25 is obtained. Thus, the average stress criterion in conjunction with the inherent flaw model yields a fracture toughness that is independent of crack size. The same procedure in conjunction with the point stress criterion yields the relationship:

$$K_c = \sigma_0 \sqrt{2\pi d_0 \left[1 + \frac{cd_0}{(c + d_0)^2} \right]} \qquad \text{(Eq 27)}$$

which is not independent of crack length. Numerical results, however, show K_c to be relatively constant for practical values of c.

Three-Parameter Point Stress Criterion. R.F. Karlak found that d_0 and a_0 are not independent of hole size for circular holes (Ref 16). This observation led to a three-parameter point stress criterion developed by R.B. Pipes *et al.* (Ref 17, 18). In particular, they assumed:

$$d_0 = \frac{a}{K} \left(\frac{a}{L_0} \right)^{m-1} \qquad \text{(Eq 28)}$$

where a is the half-notch length, L_0 is a unit reference length, k is the notch sensitivity factor, and m is an exponential constant. Note that k and m are dimensionless.

By combining Eq 28 with the point stress criterion, Eq 16, and the approximate stress distribution for a circular hole, Eq 3, Eq 17 is obtained with:

$$\xi_2 = \frac{k}{[k + (R/R_0)^{m-1}]} \qquad \text{(Eq 29)}$$

For values of $0 \le m < 1$, $\xi_2 \to 1$ as $R \to \infty$, and Eq 17 reduces to the stress-concentration factor failure criterion, Eq 12. For $m = 1$, a cursory examination of Eq 29 reveals that ξ_2 is independent of hole size, which leads to a notch strength independent of hole size. For $m > 1$, $\xi_2 \to 0$ as $R \to \infty$, and $\sigma_N^\infty \to \sigma_0$. Thus, from physical considerations, $0 \le m < 1$.

Other Notch Geometries. Although results presented here are limited to circular holes and center cracks, the point stress and average stress criteria can be extended to any notch geometry in which the stress distribution adjacent to the notch is known. These criteria are most effective when used to predict the failure of notched laminates that are fiber dominated (where failure mode is characterized by fiber breakage rather than matrix fracture). It should also be noted that, to date, experimental data on notched composite laminates have been limited primarily to straight cracks and circular holes.

Experimental Data

Experimental procedures for generating data on laminated composites containing circular holes and through-thickness center cracks, including data reduction and comparison to the theories previously presented, will now be discussed.

Center Crack Specimen. Although a number of test specimen geometries, including the three-point bend test (Ref 19) and the side notch (Ref 20), have been used to determine the strength reduction of a composite laminate containing a through-thickness crack, the center notch specimen is the most widely used. A straight-sided tensile coupon is used for this test (Fig. 6). The cracks are typically formed by drilling a 0.254-mm (0.01-in.) diameter pilot hole and then using 0.125-mm (0.005-in.) diameter diamond wire to complete the crack (Ref 11).

The far-field failure stress, σ_N (failure load divided by unnotched cross-sectional area), is obtained for notched specimens. These values can be adjusted to obtain σ_N^∞ from the relationship (Ref 15):

$$K_{Ic}^\infty = Y_1(2c/W) \, \sigma_N \sqrt{\pi c} \qquad \text{(Eq 30)}$$

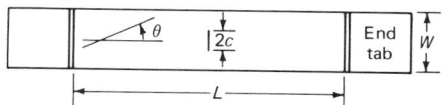

Fig. 6 Center crack specimen geometry

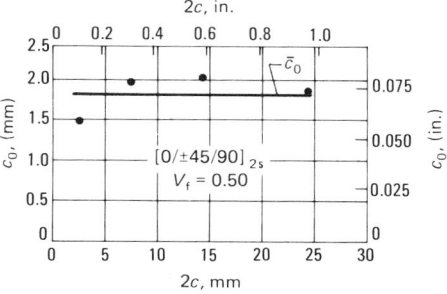

Fig. 7 Variation of inherent flaw size, c_0, with crack length. Source: Ref 11

where Y_1 is the finite-width correction factor for isotropic materials and W is width. Combining Eq 7 and 30 results in:

$$\sigma_N^\infty = Y_1(2c/W) \, \sigma_N \qquad \text{(Eq 31)}$$

The function Y_1 can be approximated by the expression (Ref 21):

$$Y_1(2c/W) = 1 + 0.128(2c/W) - 0.288(2c/W)^2 + 1.52(2c/W)^3 \qquad \text{(Eq 32)}$$

This relationship is accurate within 0.5% for $2c/W \le 0.7$. H.J. Konish (Ref 22) showed that the isotropic finite-width correction factor yields satisfactory results for most orthotropic plates of practical interest, provided that $0.3 \ge 2c/W$.

An unnotched tensile test yields σ_0, which can be used in conjunction with σ_N^∞ and Eq 10 to determine c_0. Because K_{Ic} is not constant for many crack sizes of interest, it is necessary to use Eq 10 to determine σ_N^∞ for other crack sizes. An average value of c_0, denoted by $\bar{c}_0$, can be obtained from Eq 10 and 11 in conjunction with experimentally measured notch strengths, σ_{Ni}, associated with number, n, of crack sizes, c_i. Thus:

$$\bar{c}_0 = \frac{1}{n} \sum_{i=1}^{n} \frac{c_i}{\left[\left(\dfrac{\sigma_0}{\sigma_{Ni}^\infty} \right)^2 - 1 \right]} \qquad \text{(Eq 33)}$$

where σ_{Ni}^∞ can be determined from σ_{Ni} by applying Eq 31 and 32. Results of this procedure are shown in Fig. 7 for quasi-isotropic fiberglass. Because of data scatter, variation in c_0 with crack size is anticipated. For crack sizes of the same order as laminate thickness, free-edge effects and local heterogeneities often dominate the fracture process. In such cases c_0

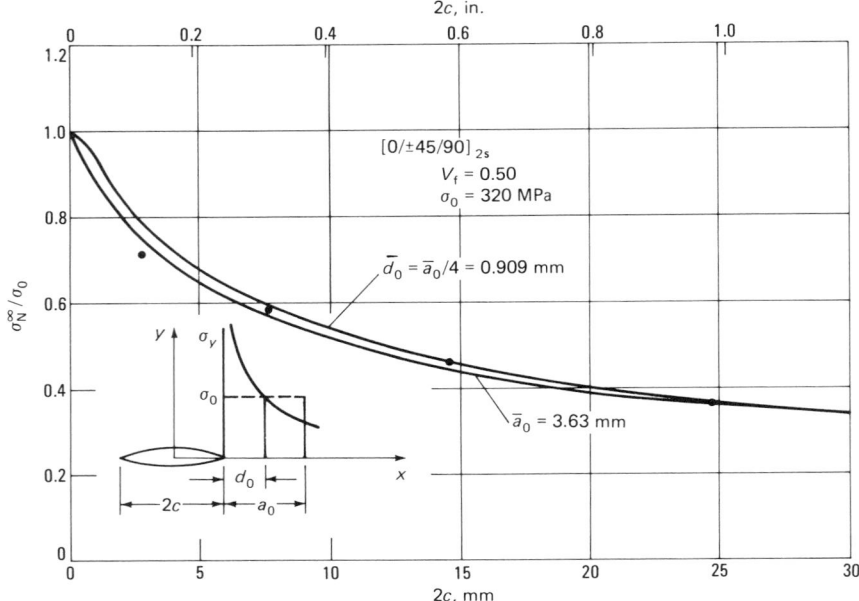

Fig. 8 Comparison between stress fracture criteria and experimental data for center notch

may be substantially different than it is for cracks that are large compared to laminate thickness.

The stress fracture criteria require a value of σ_N^∞ for one crack length and a measurement of σ_0. Equations 18 and 21 can then be used to determine d_0 and a_0, respectively. If the notch strength is determined for a number of crack sizes, average values of these parameters, denoted by d_0 and a_0, can be determined from Eq 33 in conjunction with Eq 26, which yields:

$$\bar{d}_0 = \frac{\bar{c}_0}{2}, \ \bar{a}_0 = 2\bar{c}_0$$

Theory and experiment are compared in Fig. 8 to 10 (Ref 11) for the data in Fig. 7. Each data point represents an average of three specimens. Note that in Fig. 8 K_{Ic} does increase with crack size, as predicted by the fracture criteria. As previously shown (Eq 27), the point stress criterion does not yield a constant value of K_c for all crack sizes. The results in Fig. 10 illustrate, however, that the departure from a constant value of K_c is not of practical significance.

Circular Hole Specimen. As for the center crack, a straight-sided tensile coupon is utilized for a circular hole (Fig. 11). The holes are formed by drilling mechanically or ultrasonically. A backing material should be used in

conjunction with mechanical drilling procedures to prevent delamination of the bottom plies as the drill bit exits.

As for the center crack, the notched strength, σ_N (failure load divided by unnotched cross-sectional area), is obtained for specimens containing circular holes. For applying the fracture mechanics approach to the problem of circular hole size effect in an isotropic plate (Fig. 3), the following relationship can be used (Ref 23):

$$K_c = \sigma_N \sqrt{\pi a} \ F(2R/W, 2b/W) \qquad \text{(Eq 34)}$$

where $b = R + a$. The function f(2R/W, 2b/W) is displayed graphically in Ref 23. Equation 13 in conjunction with Eq 34 yields the following results for an isotropic material:

$$\sigma_N^\infty = Y_2(a/R, 2R/W, 2b/W) \ \sigma_N \qquad \text{(Eq 35)}$$

where

$$Y_2(a/R, 2R/W, 2b/W) = \frac{\text{f}(2R/W, 2b/W)}{\text{f}(a/R)}$$

For orthotropic materials, it is possible that Y_2 will also be a function of elastic properties.

For isotropic materials:

$$K_t = Y_3(2R/W) \ K_t^\infty$$

For $2R/W \leq \frac{1}{3}$, Y_3 can be accurately approximated by the expression (Ref 24):

$$Y_3(2R/W) = \frac{2 + (1 - 2R/W)^3}{3(1 - 2R/W)}$$

If it is assumed that the stress-concentration factor is of primary importance in determining strength reduction in a circular hole, then:

$$\sigma_N^\infty = Y_3(2R/W) \ \sigma_N \qquad \text{(Eq 36)}$$

Implicit in Eq 36, in conjunction with the stress failure criterion, is the assumption that

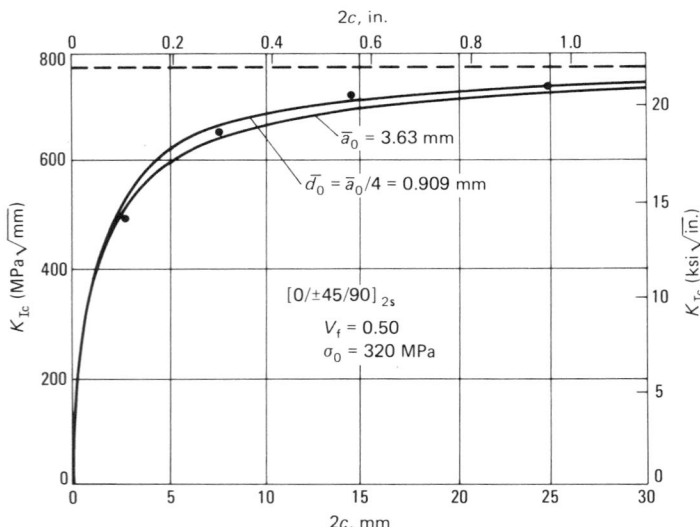

Fig. 9 Comparison between stress fracture criterion for critical stress-intensity factor and experimental data

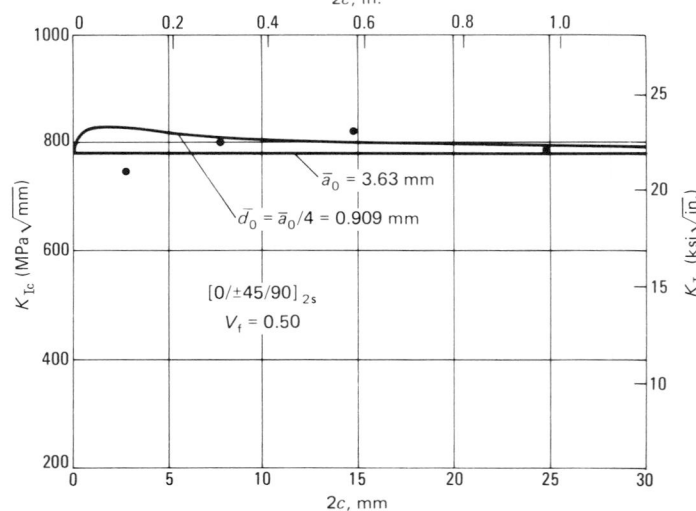

Fig. 10 Comparison between stress fracture criteria for fracture toughness and experimental data

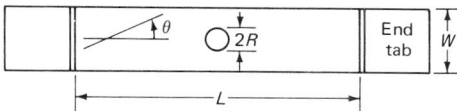

Fig. 11 Circular-hole specimen geometry

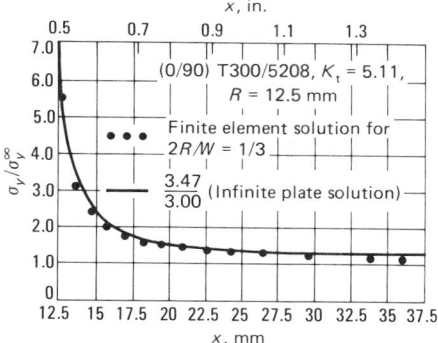

Fig. 12 Comparison of normal stress distribution across the ligament of an anisotropic finite-width plate containing a circular hole

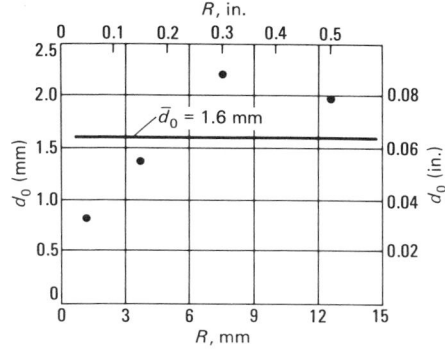

Fig. 13 Variation of damage zone parameter with hole size

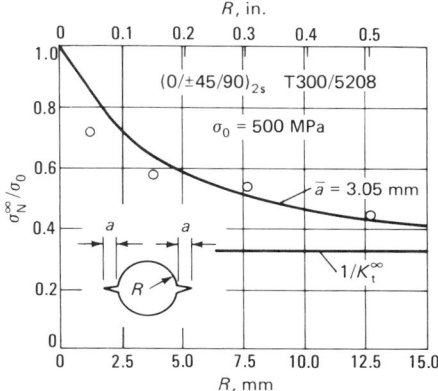

Fig. 14 Comparison between experimental data and fracture mechanics criterion for circular hole

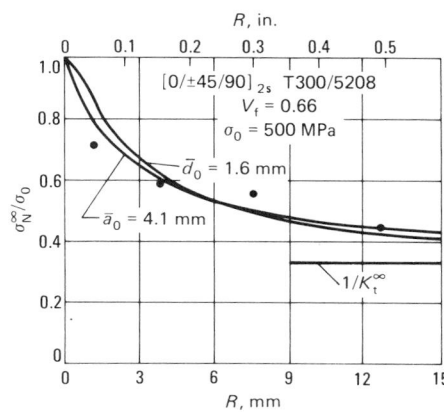

Fig. 15 Comparison between experimental data and stress fracture criteria for circular hole

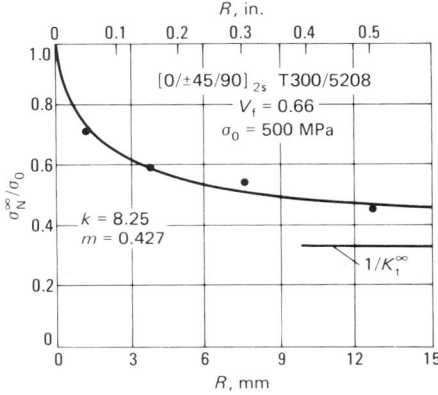

Fig. 16 Comparison between experimental data and three-parameter stress fracture criterion for circular hole

the normal stress distribution adjacent to a circular hole in a finite-width orthotropic plate is equal to the stress distribution in an infinite plate multiplied by the isotropic finite-width correction factor. In Ref 11, a constant-stress finite-element program was used to generate the normal stress ahead of a hole in a finite-width plate of $2R/W = \frac{1}{3}$ for a number of laminated composites. These results were plotted against the normal stress distribution obtained by multiplying the infinite plate stress by the isotropic finite-width correction factor, Y_3. A typical result is shown in Fig. 12 for a hole in a [0/90]s graphite-epoxy plate, where the comparison is good.

Both the fracture mechanics approach and the stress fracture criteria require a measurement of the unnotched strength, σ_0, and the notched strength for one hole size to determine the damage zone parameters a, a_0, and d_0. As for the center crack, these parameters can be determined for a number of hole sizes such that theory and experiment coincide, and an average value can then be used to develop an interpolation curve for any hole size. Notched strength for at least two hole sizes must be determined if the three-parameter point stress model is used. The parameters k and m can be determined from a log-log plot of d_0 versus R. In particular, because L_0 is of unit value, Eq 28 can be written in the log-log linear form:

$$\log d_0 = m \log R - \log k \qquad \text{(Eq 37)}$$

Then m will be the slope of the line and k can be determined from the y intercept. If the minimum two hole sizes are used, an exact fit to Eq 37 can be obtained. A standard least squares fit can be used in conjunction with a number of hole sizes to determine k and m. Because Eq

15, 17, and 20 cannot be solved explicitly for the damage zone parameters, an iteration process is necessary in conjunction with each measured hole size strength reduction to match theory and experiment.

The variation of d_0 with hole size is shown in Fig. 13 for a quasi-isotropic graphite-epoxy laminate (Ref 11). As with the center crack, it is anticipated that hole sizes of the same order as laminate thickness could produce a fracture process that is dominated by free-edge effects and hole heterogeneities, resulting in a substantially different value of the damage zone parameters.

Theory and experiment for the graphite-epoxy laminate of Fig. 13 are compared in Fig. 14 to 16. The results in Fig. 14 (Ref 25) are for the fracture mechanics criterion (Eq 15), while the results in Fig. 15 are for the point stress and average stress criteria. The three-parameter point stress criterion is illustrated in Fig. 16. The bars over the damage zone parameters denote average values. The data reduction procedure used in conjunction with Fig. 14 is based on Eq 36 rather than the more complex expression given by Eq 35. In this procedure, as for the stress fracture criteria, it is assumed that failure is dominated by the stress-concentration factor. This procedure was also used in

Ref 13. All data points are based on an average of three specimens.

REFERENCES

1. S.V. Kulkarni, B.W. Rosen, and C. Zweben, Load Concentration Factors for Circular Holes in Composite Laminates, *J. Compos. Mater.*, Vol 7 (No. 3), 1973, p 387-393
2. C. Zweben, Fracture Mechanics and Composite Materials: A Critical Analysis, in *Analysis of the Test Methods for High Modulus Fibers and Composites*, STP 521, American Society for Testing and Materials, 1973, p 65-97
3. S.S. Wang, J.F. Mandell, and F.J. McGarry, Three-Dimensional Solution for a Through-Thickness Crack With Crack Tip Damage in a Cross-Plied Laminate, in *Fracture Mechanics of Composites*, STP 593, American Society for Testing and Materials, 1975, p 61-85
4. R.J. Nuismer and G.E. Brown, Progressive Failure of Notched Composite Lami-

nates Using Finite Elements, in *Advances in Engineering Science*, NASA CP-2001, Vol 1, National Aeronautics and Space Administration, 1976, p 183-192

5. S.C. Chun, O. Orringer, and J.H. Rainey, Post-Failure Behavior of Laminates: II, Stress Concentration, *J. Compos. Mater.*, Vol 11 (No. 1), 1977, p 71-78

6. S.G. Lekhnitskii, *Anisotropic Plates*, S.W. Tsai and T. Cheron, Trans., Gordon and Breach Science Publishers, 1968

7. H.J. Konish and J.M. Whitney, Approximate Stresses in an Orthotropic Plate Containing a Circular Hole, *J. Compos. Mater.*, Vol 9 (No. 2), 1975, p 157-166

8. G.C. Sih, P.C. Paris, and G.R. Irwin, On Cracks in Rectilinearly Anisotropic Bodies, *Int. J. Fract. Mech.*, Vol 1 (No. 2), 1965, p 189-203

9. H.J. Konish and T.A. Cruse, Jr., Determination of Fracture Strength in Orthotropic Graphite/Epoxy Laminates, in *Composite Reliability*, STP 580, American Society for Testing and Materials, 1975, p 490-503

10. J.M. Whitney and R.J. Nuismer, Stress Fracture Criteria for Laminated Composites, *J. Compos. Mater.*, Vol 8 (No. 3), 1974, p 253-265

11. R.J. Nuismer and J.M. Whitney, Uniaxial Failure of Composite Laminates Containing Stress Concentrations, *Fracture Mechanics of Composites*, STP 593, American Society for Testing and Materials, 1975, p 117-142

12. F.A. McClintock and G.R. Irwin, Plasticity Aspects of Fracture Mechanics, in *Fracture Toughness Testing and Its Applications*, STP 381, American Society for Testing and Materials, 1984, p 84-113

13. M.E. Waddoups, J.R. Eisenmann, and B.E. Kaminski, Macroscopic Fracture Mechanics of Advanced Composite Materials, *J. Compos. Mater.*, Vol 5 (No. 4), 1971, p 446-454

14. O.L. Bowie, An Analysis of an Infinite Plate Containing Radial Cracks Originating From the Boundary of an Internal Circular Hole, *J. Math. Phys.*, Vol 35 (No. 1), 1956, p 60-71

15. P.C. Paris and G.C. Sih, Stress Analysis of Cracks, in *Fracture Toughness Testing and Its Applications*, STP 381, American Society for Testing and Materials, 1964, p 30-81

16. R.F. Karlak, Hole Effects in a Related Series of Symmetrical Laminates, in *Proceedings of Failure Modes in Composites, IV*, The Metallurgical Society, 1977, p 105-117

17. R.B. Pipes, R.C. Wetherhold, and J.W. Gillespie, Jr., Notched Strength of Composite Materials, *J. Compos. Mater.*, Vol 12 (No. 2), 1979, p 148-160

18. R.B. Pipes, J.W. Gillespie, Jr., and R.C. Wetherhold, Superposition of the Notched Strength of Composite Laminates, *Polym. Eng. Sci.*, Vol 19 (No. 16), 1979, p 1151-1155

19. H.J. Konish, J.L. Swedlow, and T.A. Cruse, Experimental Investigation of Fracture in an Advanced Fiber Composite, *J. Compos. Mater.*, Vol 6 (No. 1), 1972, p 114-125

20. J.F. Mandell, S.S. Wang, and F. McGarry, The Extension of Crack Tip Damage Zones in Fiber Reinforced Plastic Laminates, *J. Compos. Mater.*, Vol 9 (No. 3), 1975, p 266-287

21. W.F. Brown and J.E. Srawley, *Plane Strain Crack Toughness Testing of High Strength Metallic Materials*, STP 410, American Society for Testing and Materials, 1965, p 30-83

22. H.J. Konish, Jr., Mode I Stress Intensity Factors for Symmetrically-Cracked Orthotropic Strips, in *Fracture Mechanics of Composites*, STP 593, American Society for Testing and Materials, 1975, p 99-116

23. H. Toda, P.C. Paris, and G.R. Irwin, *The Stress Analysis of Cracks Handbook*, Del Research Corporation, 1973, p 199

24. R.E. Peterson, *Stress Concentration Factors*, Wiley-Interscience, 1974, p 110-111

25. J.M. Whitney, I.M. Daniel, and R.B. Pipes, *Experimental Mechanics of Fiber-Reinforced Composite Materials*, rev. ed., Society for Experimental Mechanics, 1984

Damage Tolerance of Composites

Ray E. Horton, Boeing Commercial Airplane Company
John E. McCarty, Boeing Military Airplane Company

DAMAGE TOLERANCE in the aircraft industry is defined as the ability of a structure to tolerate a reasonable level of damage or defects that might be encountered during manufacture or while in service, without jeopardizing aircraft safety. Thus, safety is the primary goal of damage tolerance. Other important considerations, however, are that damage tolerance must be achieved with maximum structural efficiency (minimum weight) and with minimum manufacturing, maintenance, and supportability costs.

Safety. The primary structure must be sufficiently damage tolerant to ensure that the aircraft will not fail catastrophically during its operational life. There are three aspects of damage tolerance safety: Damage or defects that might be undetected should be able to be sustained for the life of the aircraft; detectable damage also should be able to be sustained for a selected period of time before its detection; and, in the case of in-flight damage, the aircraft should be able to complete the flight safely.

Structural efficiency. A design approach associated with achieving the above damage tolerance requirements is to keep normal operational stresses in the structure low. Then, if damage does occur, the structure will still have sufficient residual strength to resist failure. However, the structural efficiency problem is that unless the structure is designed for stability, low design stresses are commonly achieved with sacrifice of minimum weight. A goal then is to achieve damage tolerance while still keeping design stresses at the highest possible level consistent with residual strength requirements, thus achieving a minimum-weight structure.

The economics of damage tolerance has two aspects. The first is related to the manufacturability or, for the buyer, procurement cost. The material and design configuration should preferably be such that high processing or fabrication costs are not required to produce an acceptable damage-tolerant product. In the past, for example, special care was required for the surface treatment of high-strength steels because of extreme notch sensitivity. Close control of the interstitial content of titanium alloys for cryogenic applications was another undesirable situation.

The second aspect of damage tolerance economics is related to maintenance. Composites have a definite advantage over metals in incidences of corrosion or fatigue. This lower cost-of-maintenance advantage can be lost, however, if the composite structure does not have a reasonable tolerance to foreign-object damage that is routinely encountered during ground handling operations, or if the structure is simply too expensive to repair when damage does occur.

This balance between damage tolerance, structural efficiency, and low-cost fabrication and maintenance has been accomplished for most metal structures. The safety record for metal aircraft structures has been very good. This safety level must also be maintained for composite materials, which have become a material design option for primary aircraft structures. To accomplish this, the inherent differences between the damage tolerance behavior of composites and that of metal structures must be recognized and accounted for in the design process.

Comparison of Composites and Metals

When considering damage tolerance, the physical characteristics of composites and of metals are significantly different. Metals are largely isotropic while composites are very anisotropic. In-plane strength and stiffness are usually high and directionally variable, depending on the orientation of the reinforcing fiber. Properties that do not benefit from this reinforcement, at least for the polymeric matrix laminates, are comparatively low in strength and stiffness. An important example of these is the transverse, that is, through-the-thickness tensile strength. The low transverse tensile strength of a representative composite laminate compared to that of aluminum is illustrated in Fig. 1. A similar situation occurs in a typical laminate for the interlaminar or transverse shear strength versus the shear capability in the reinforced plane.

Metals, at least those commonly used in aircraft structures, also have reasonable ductility. When they reach some load level, they continue to elongate or compress considerably without picking up more load and without failure. This ductile yielding has two important benefits. It provides for local load relief by distributing excess load to adjacent material or structure. Consequently, ductile metals have a great capacity to provide relief of stress concentrations when statically loaded. Second, the ductility of metals provides great energy-absorbing capability, as is indicated by the area under a stress-strain curve. As a result, when impacted, a metal structure will typically deform but not actually fracture.

In contrast, composites are relatively brittle. A comparison of typical tensile stress-strain curves for the two materials is shown in Fig. 2. The brittleness of the composite is reflected in its poor ability to tolerate stress concentrations, as shown in Fig. 3. The characteristically brittle composite material has poor ability to resist impact damage without extensive internal matrix fracture. The fracture paths are coincident with the low-strength matrix-dominated planes previously discussed. This can be seen in the cross section of an impact-damaged laminate (Fig. 4). Diagonal shear cracks have occurred in the matrix between the transverse oriented fibers, that is, the fibers oriented normal to the plane of Fig. 4. Extensive delamination has also taken place in the matrix between the plies.

In most instances of low-velocity impact, that is, those other than ballistic, the damage is

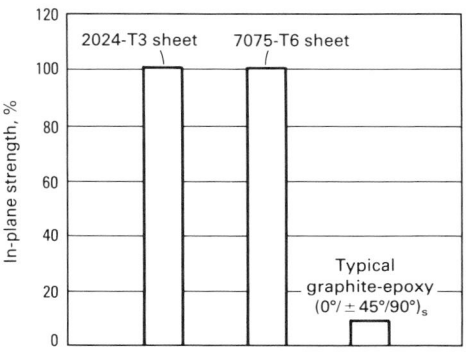

Fig. 1 Comparison of transverse in-plane tensile strength for metal versus composite

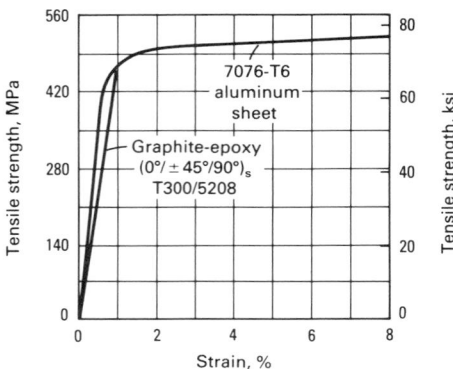

Fig. 2 Comparative tension stress-strain curves for 7075-T6 aluminum versus graphite-epoxy isotropic lay-up

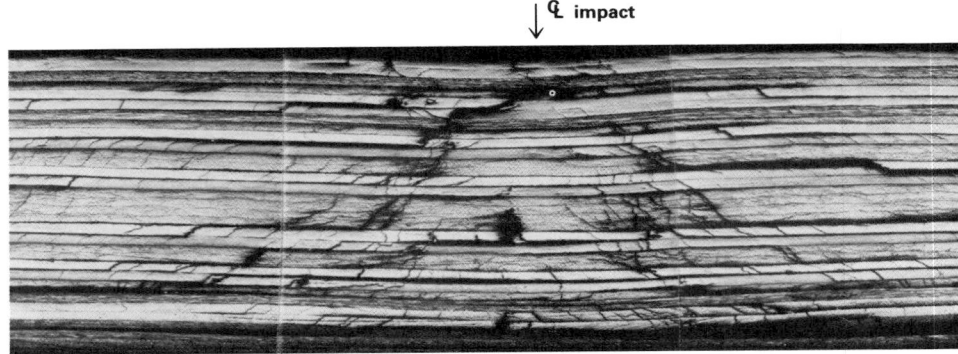

Fig. 4 Cross section of a composite laminate showing damage due to impact. Material: AS6/2220-3. Impact: 40 J (350 in. · lbf)

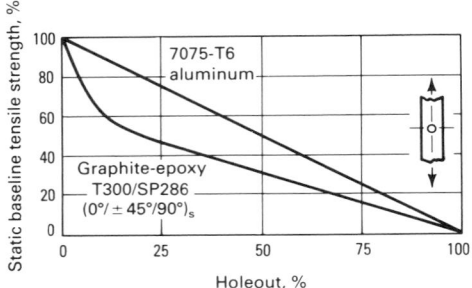

Fig. 3 Comparison of static strength resistance to stress concentrations for 7075-T6 aluminum versus graphite-epoxy

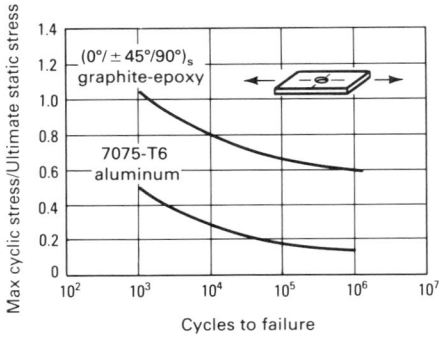

Fig. 5 Comparative fatigue performance of graphite-epoxy versus aluminum. $R = 1.0$; $K_t = 3.0$

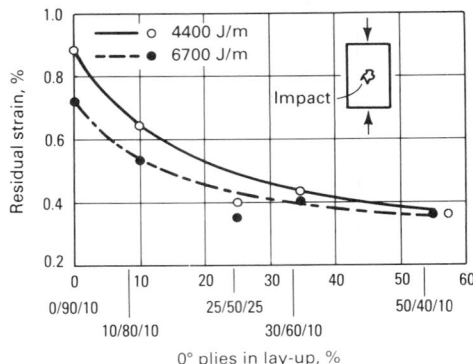

Fig. 6 Effect of ply lay-up on residual compression strain after impact (%0°/%±45°/%90° plies)

confined to the matrix, and little fiber damage occurs. Therefore, the in-plane tension strength of the laminate may not be seriously degraded. However, even with impact levels that leave little indication of damage on the surface, the matrix damage may be significant, and therefore its ability to stabilize the fibers in compression may be seriously degraded. Because of this, tolerance to impact damage is often the critical design consideration and compression is the critical loading mode.

The response of damaged composites to cyclic loading is also significantly different from that of metals. In contrast to the poor composite static strength when it has damage or defects, the ability of composites to withstand cyclic loading is far superior to that of metals. The comparison of the normalized notched specimen fatigue response of a common aircraft metal, 7075-T6 aluminum, and a composite laminate is shown in Fig. 5. The fatigue strength of the composite is much higher relative to its static or residual strength. The static or residual strength requirement for structures is typically much higher than the fatigue requirement. Therefore, because the fatigue threshold of composites is a high percentage of their static or damaged residual strength, they are usually not fatigue critical. In metal structures, fatigue is typically a critical design consideration.

Other properties of composites affect their damage tolerance characteristics. One factor is ply orientation. Impacted laminates with stiff ply orientations, that is, a higher percentage of plies oriented parallel to the loading direction, typically fail at lower strains than those with softer orientations (Fig. 6). Structural weight, however, is more significantly related to stress capability than to strain. Thus, a laminate with a higher modulus may be able to operate to a higher stress even though its strain capability may be less. Each specific case must be carefully examined. The potential advantage of using a soft laminate, especially in damage-prone areas or in areas where there are stress concentrations, should be considered. For example, a soft laminate might be used effectively for an outer skin, whereas the interior stiffeners, which are exposed to less damage, might advantageously use a stiffer and hence a more weight-efficient but less damage-tolerant lay-up.

Other properties of composites, as opposed to metals, also contribute to damage-tolerant response. While Poisson's ratio for metals is typically near 0.3, the Poisson's ratio for composites may vary widely, depending on the particular combination of ply lay-up orientations. Consequently, Poisson's ratios may be unbalanced between particular sets of plies in a given laminate. These imbalances create residual stresses that may add to damage-induced stresses and thereby increase the damage extent or the tendency for damage growth (Ref 1, 2). These stresses can be either normal to the laminate plane or in shear. A plot of normal stresses at the ply interfaces of a laminate example is shown in Fig. 7. As an example of the significance of the effect, quasi-isotropic laminates (0°/±45°/90° ply stacking) with various stacking sequences were tensile tested in Ref 3. Variations in through-the-thickness Poisson's ratios ranged from 0.85 to −0.30. The strength of the laminates varied from 410 to 620 MPa (60 to 90 ksi), with delamination being a prevalent failure mode. The delamination initiation and its effect on residual strength correlated with the magnitude of calculated normal stresses. Similarly, the effect of stacking sequence on the tensile strength of laminates with holes (analogous to defects) was examined (Ref 4). Laminates with stacking sequences giving high interlaminar shear or normal stresses failed at 10 to 20% lower loads than alternate lay-ups. Again, prefailure delamination was prevalent, with the failure mode being strongly controlled by the lay-up. It follows that the laminate lay-up sequence and orientation significantly influence the residual strength of damaged composites. Laminate lay-ups that have high interlaminar stress gradients should be avoided.

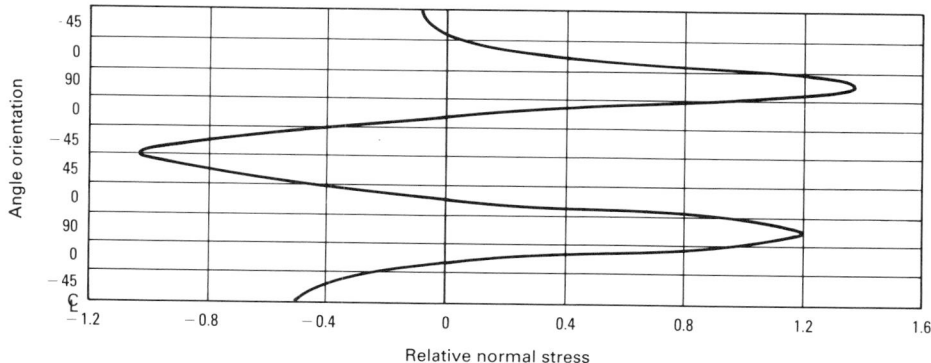

Fig. 7 Relative normal stresses in an axial-loaded laminate due to Poisson's effects

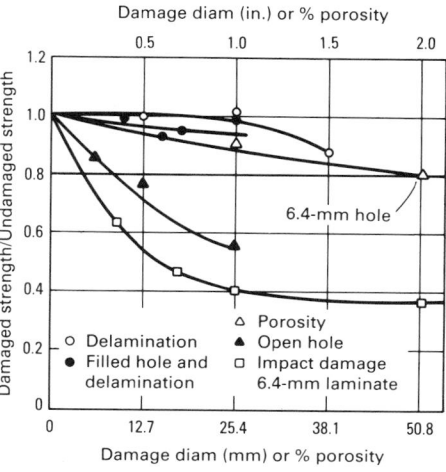

Fig. 8 Relative severity of defect damage on static compression strength

Defect Sensitivity

Although through-the-thickness cracks and cuts typically associated with tensile loading must be considered relative to damage tolerance of composites, their importance is not as dominant as with metals. The in-plane fibers act as effective barriers to resist the initiation of a through-the-thickness crack or subsequent crack growth. They also effectively redistribute load at the crack tip. Thus, if a cut or puncture occurs, it causes a strength reduction similar to that caused by a hole of comparative width (Ref 5). There is little additional sharp notch effect. Consequently, the effect of small cracks and cuts is usually accounted for by the allowance for fastener holes. Large cuts usually occur on exposed surfaces, and, in contrast to the non-visible damage that may result from low-velocity impact, are readily detectable during normal inspection activities. Through-the-thickness crack growth will not occur unless the damage and/or stresses are very large. Growth tends to be in an interlaminar mode. The through-the-thickness damage growth thresholds and subsequent damage growth rates can be predicted reasonably well using a fracture mechanics approach (Ref 6).

As noted, damage affecting compression strength is typically more critical than that affecting strength in tension. A comparison of the relative severity of defects or damage commonly contributing to premature compression failure is given for static loading in Fig. 8 and for fatigue or cyclic loading in Fig. 9. Of the defects or damage shown, impact is readily seen as being most critical. There is another important aspect when damage occurs. Delaminations, flawed holes, and porosity usually occur during processing. Consequently, they are subject to detection during manufacturing quality control inspection. As a result, their severity can be controlled within safe limits. Impact, although it can occur during manufacture, more likely occurs in service where, unless it is visible on the surface, it may remain undetected either for a long period of time or for the life of the aircraft.

The concern with impact damage is made apparent in Fig. 8. Note that a laminate can lose 60 to 65% of its undamaged static strength with impact damage that is essentially nonvisible. The loss in cyclic-load capability is also greatest from impact (Fig. 9).

Cyclic Loading

The fatigue curve for composites is relatively flat compared to the curve for metals. For a constant-amplitude loading condition of $R = 10$, that is, compression-compression, the 10^6 cycle endurance condition typically corresponds to a maximum cyclic stress (in compression) of 60 to 65% of the undamaged static ultimate (Fig. 10). The fatigue curve is much higher for spectrum-type loading, which is more representative of typical aircraft loading conditions. A typical transport loading sequence is shown in Fig. 10. The reason for the improvement in the spectrum load condition is that, for composites, only the high-level loads cause significant fatigue damage or damage growth. Consequently, typical aircraft loading is relatively benign compared to constant amplitude load cycling because of the few high or peak load cycles encountered. This is in contrast to metals, for which the large number of low-stress cycles is the primary contributor to crack growth, and the high-stress cycles may actually be beneficial because of their retardation effect. Because composites are only sensitive to the high-level loads, this property gives the advantage of being able to truncate spectrums drastically in simulating lifetimes of cyclic loading. Ironically, however, a large number of loading lifetimes may be required to obtain an exceedance confidence because of the characteristic flat cyclic loading curve of the material and the associated data scatter inherent in time-to-failure data.

When cyclic tests are conducted at ambient temperatures, experience has shown that, within reasonable limits, results are relatively insensitive to cyclic rate. For example, test results on panels tested at 5 Hz have correlated

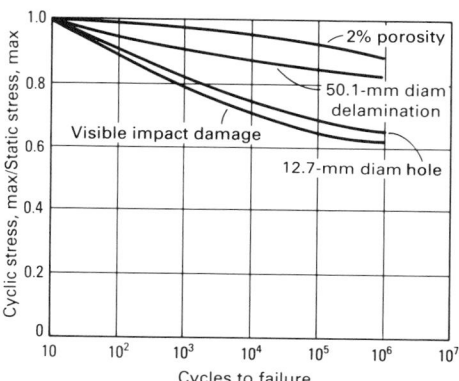

Fig. 9 Relative severity of defect damage on compression fatigue strength. $R = 10.0$

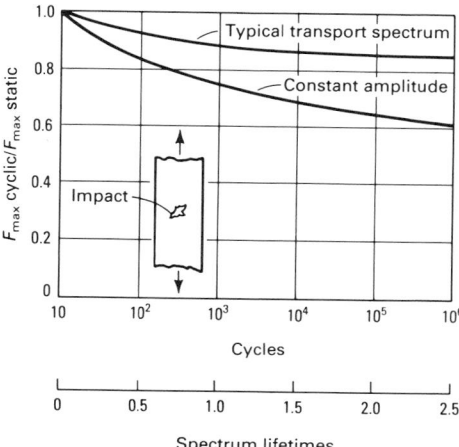

Fig. 10 Typical cyclic load response, constant amplitude versus spectrum loading. $R = 10.0$

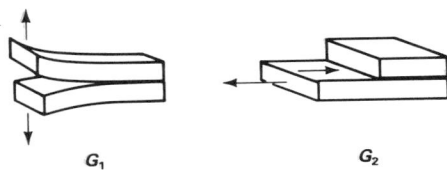

Fig. 11 G_1 and G_2 loading modes

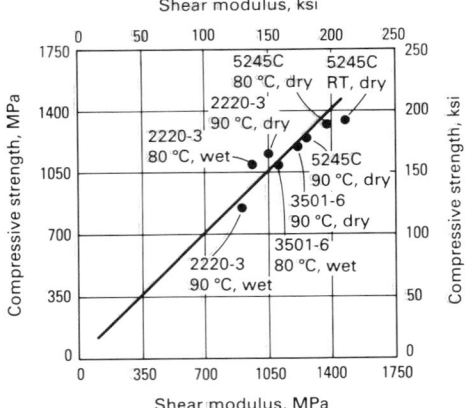

Fig. 12 Influence of resin shear modulus on compressive strength

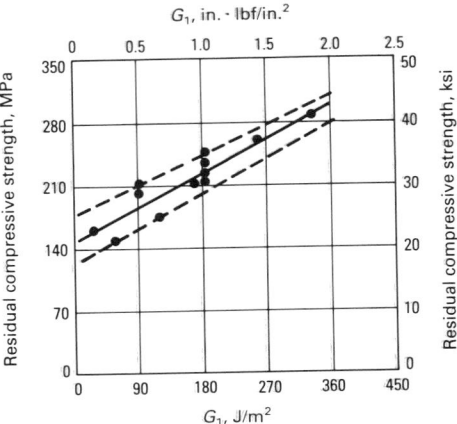

Fig. 13 Relation of G_1 to residual strength after impact of 6700 J/m (1500 in.·lbf/in.)

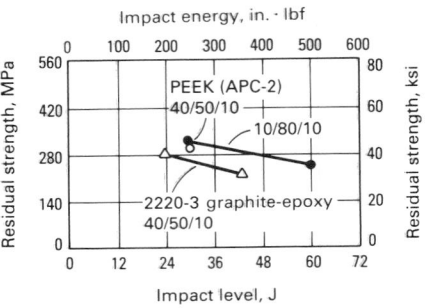

Fig. 14 Residual compressive strength after impact, AS6/2220-3 versus AS4/APC-2 (PEEK) for 6.4- × 125- × 250-mm (0.25- × 5.0- × 10.0-in.) laminate, with 13-mm (0.50-in.) diam impactor tup

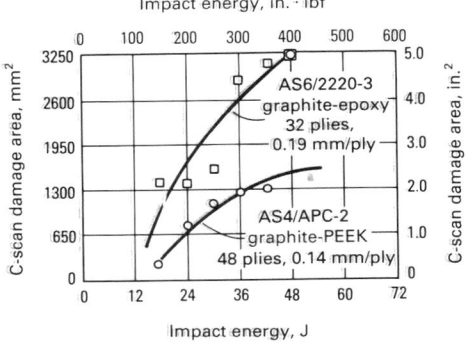

Fig. 15 Influence of material on impact damage size. C-scan measurements on outer periphery of damage

well with results from panels tested at 25 cycles per minute (~0.5 Hz). The latter simulated a loading rate that might be applied to spectrum cycling of a large full-scale test component. The 5 Hz rate is considered a maximum rate for generic composite small coupon cyclic load testing.

Material Effects

The ability of composite structures to tolerate impact damage is largely dependent on the material properties and structural configuration of the composite. Each plays a meaningful role in composite design. Those material properties involving the resin matrix are the most significant and include elongation of the material, or the area under the stress-strain curve (Ref 7). This is indicative of its energy absorption capability. It is also related to its interlaminar fracture toughness (G_{1c}), as indicated by energy release rate properties. This is, most importantly, G_1, but also is G_2. Parameters G_1 and G_2 represent the ability of the resin to resist delamination in the tensile and shear modes respectively, as shown in Fig. 11.

The resin must also have sufficient stiffness, as indicated by its shear or axial modulus, to stabilize the fibers in compression. This is illustrated in Fig. 12, where the shear moduli of the resins for several laminates with the same fibers are related to the compressive strength of the laminate. This property is equally important to stabilize fibers in damaged composites. This is evidenced by the results of tests on impacted specimens tested at ambient versus elevated

temperatures and dry versus wet conditioning. While the higher temperatures and moisture act to plasticize the resin, that is, increase toughness, this is offset by modulus loss, and, as a net effect, the residual compression strength is reduced.

The effect of the resin toughness G_1 on impact damage residual strength is shown in Fig. 13. The materials have the same fiber reinforcement. The resin toughness and the laminate damage tolerance are directly related. A more specific comparison is made between a graphite-epoxy thermoset system and a tougher, thermoplastic material in Fig. 14. The tougher thermoplastic material has a much higher residual strength after impact. The G_1 of the thermoplastic material is approximately 1050 J/m² (6.0 in. · lbf/in.²), whereas the G_1 of the graphite-epoxy is approximately 180 J/m² (1.0 in. · lbf/in.²).

The toughness of the material also has a significant effect on the size of the damaged area as related to a specific impact energy level. A comparison is shown for graphite-epoxy and the tougher thermoplastic system in Fig. 15. Because it is less tough, the graphite-epoxy has a much larger damage area, as indicated by ultrasonic C-scan, a measurement of the extent of internal delaminations. Although the thermoplastic laminate was impacted with a greater energy level, its internal damage both in terms of transverse shear cracking and delaminations is much less.

Some investigations have been conducted on the effect of variation in fiber properties on impact tolerance. In general, laminates made with fabric-style reinforcement have had better resistance to impact than laminate constructions using unidirectional tape. However, differences between the various graphite fiber tape materials seem small. Some studies have been made of composites constructed with hybrid fiber construction, that is, composites in which some percentage of the graphite fibers were replaced by fibers with higher elongation, such as E-glass or aramid (Ref 8, 9). Studies in both Ref

8 and 9 showed improvement in compression residual strength after impact, using the hybrid approach. However, basic undamaged properties, that is, strength and stiffness, were usually reduced.

The use of hybrid designs has been shown to enhance the tension damage tolerance, that is, residual tension strength of a composite structure. The use of glass fibers has been particularly effective. Because of the modulus differences between graphite and glass, the glass can absorb more elongation and thus pick up load from the graphite fibers after their failure from defects or damage. The load transfer to the glass fibers arrests damage growth. This damage arrest capability is especially beneficial for tolerance to penetration damage.

Configuration Effects. Considerable work has been done on the effects of design configuration on the damage tolerance of composites (Ref 10-14). Two different types of structures were evaluated in these studies. One was typical of a multirib transport wing, while the other was representative of a multispar fighter wing. A comparison of the two designs, which are considerably different, is shown in Fig. 16. The multirib design has a relatively low modulus skin (10/80/10 lay-up) with unidirectional reinforcement concentrated under the co-cured stiffeners. The multispar design has a mono-

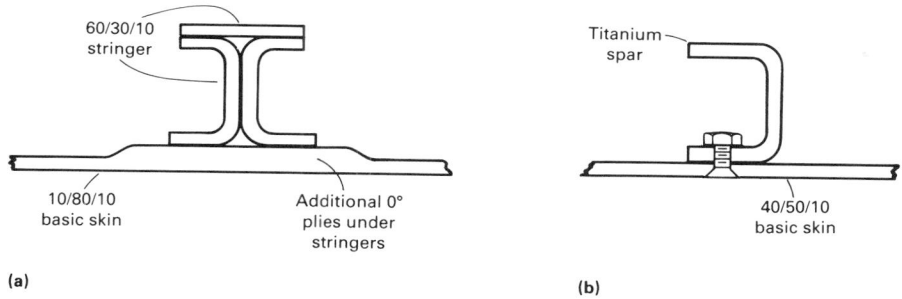

(a)

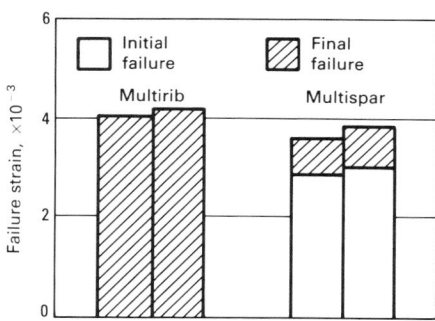

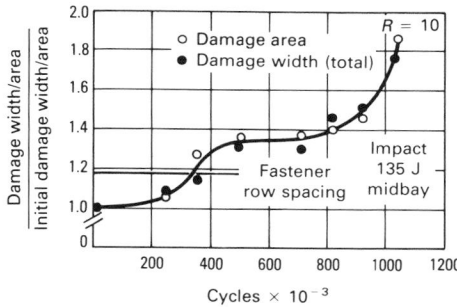

(b)

Fig. 16 Comparison of (a) multirib and (b) multispar test panel configurations

Fig. 17 Comparison of residual strength of impact damaged multirib versus multispar panels at 135 J (100 ft · lbf) impact

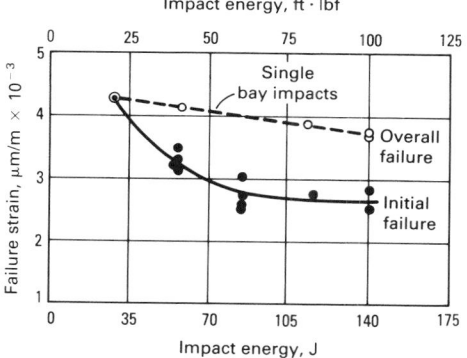

Fig. 18 Influence of fastener row arrest on the static failure response of impact-damaged panels

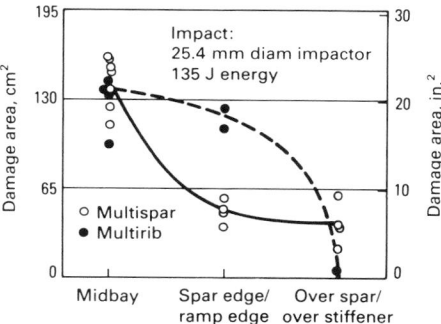

Fig. 19 Influence of stiffener fastener rows on cyclic load damage growth

Fig. 20 Design configuration effect on damage size

lithic stiffened higher modulus skin design (40/50/10 lay-up) with the stiffeners, that is, spars, mechanically fastened. The two designs provide a meaningful comparison because panels were fabricated with the same material and were evaluated against the same damage tolerance specification requirements.

Results of testing panels with these two configurations are shown in Fig. 17. Two significant differences are evident. First, the final failure strain for the multirib configuration panels was somewhat higher than the multispar panels. The multirib panels were specifically designed to be damage tolerant, while damage tolerance was not given special emphasis for the multispar design. The greater-depth transport wing is also more amenable to some damage tolerance features. A thinner fighter-type wing employing the same features would sacrifice considerable structural efficiency. For instance, the multirib design incorporates a soft, that is, low-stiffness skin with a predominance of ±45° plies. The soft skin is damage tolerant at high strains. The 45° plies act to redistribute load from damaged to undamaged areas. The primary axial load-carrying reinforcement, that is, 0° fibers, is concentrated at the stiffeners where the thicker mass is more damage resistant, and much of the reinforcement is subsurface and therefore is not exposed to impact. While this approach is attractive for a thick transport wing, it is much less attractive for a thin fighter wing, where it is more

efficient to keep primary load-carrying material near the outer surface for greater effectiveness in bending.

The second difference is that the multirib design failure was a single-stage catastrophic event, while for the multispar design, the failure was in two stages. In the latter case, at the initial failure, the damage popped locally (that is, it propagated to the spar fastener line). It was then arrested, until, at a higher load, ultimate failure occurred. The phenomenon is shown as a function of the imposed impact energy level in Fig. 18.

Note that at some lower impact energy level, the failure energy, or strain, to cause initial damage propagation may be sufficiently high that the damage will not arrest. Tests have indicated, however, that this strain is higher than the final strain at which panels impacted at higher energy levels will fail.

The fastener rows attaching the spar in the multispar design were effective in arresting damage growth during cycle loading. Figure 19 shows the resulting pattern of growth. The growth retardation is similar to the initial failure arrest noted for the static tests. Lateral propagation of the damage arrests at the fasteners and then may proceed along the fastener line until it grows through and failure follows.

Another significant difference in the response of the two designs to impact damage can be noted in Fig. 20. While the damage areas at the midbay and over the stiffener/spar locations were similar, there was considerable difference in damage at the edge of the stiffener/spar locations. The damage area near the edge of the

multirib stiffener was much larger. There was also a difference in the damage location criticality. The hard-skin multispar panels consistently failed at the lowest strain when impacted on the skin, midway between spars. The edge-of-stiffener damage was most critical for the soft-skin multirib panels.

Results of the stiffened panel tests were compared to tests on small, unstiffened panels (Fig. 21). A parametric correction was made to account for variation in the panel thickness. However, the small-panel test results were still much lower than were those for the larger panels. It was concluded that the small-panel tests were good for comparing materials. This is shown for tests involving a comparison of small and larger panels using comparative sets of materials in Fig. 22. The small panels do, however, give conservative results as related to an actual structural design.

Impact damage is normally in the form of resin shear fracture and interlaminar delamination. Consequently, when the laminate is loaded in compression, the fibers are poorly stabilized and buckle prematurely. Some work has been done to evaluate the effectiveness of through-the-thickness stitching to reduce the size of the damage, that is, of interply delamination. A stitched three-stringer panel that was evaluated is shown in Fig. 23. Two stitching patterns were evaluated: One was a single-row wide-spaced pattern, and the other was a close-spaced pattern, defined as having 6.4 mm

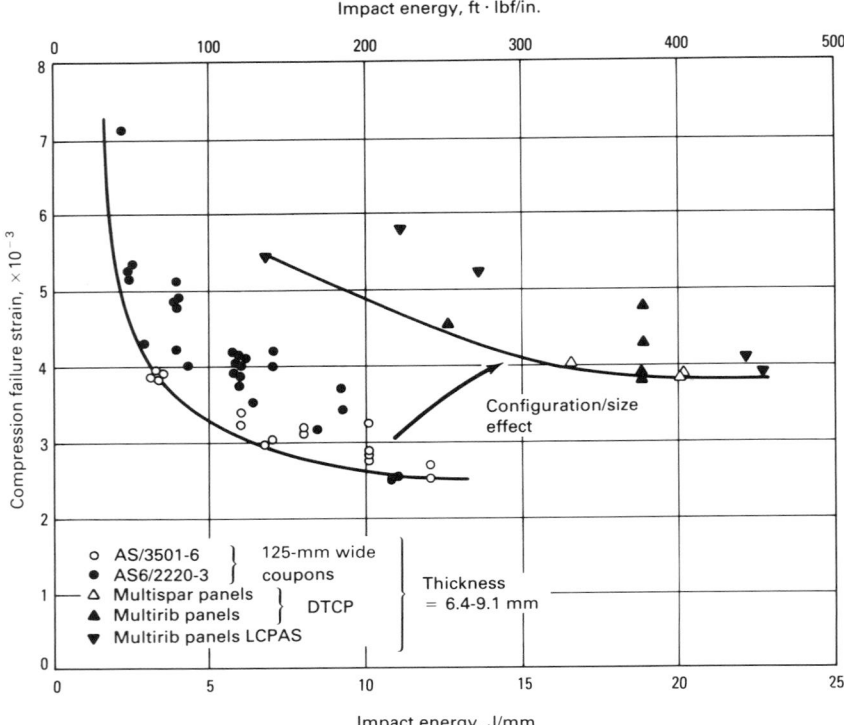

Fig. 21 Effect of test panel size/configuration on residual strength

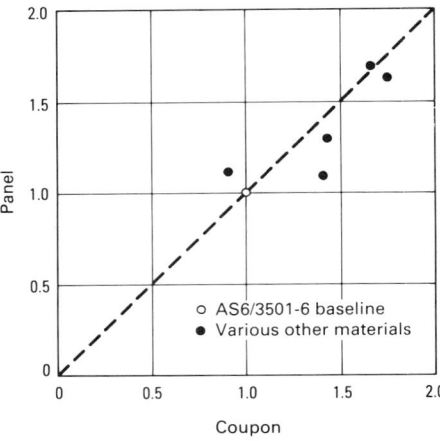

Fig. 22 Normalized panel strength compared with normalized coupon strength after impact

(0.25 in.) between stitched rows and a 6.4-mm (0.25-in.) stitch length. Test results are shown in Fig. 24. Little or no benefit was realized from the single-row wide-spaced approach. An approximate strength increase of 15% was obtained using the stitching that was closely spaced. Because stitching to this extent is expensive, the merit of its use must be evaluated on a case-by-case basis.

Test Methods

There are several standard test methods that can be used to advantage for determining damage tolerance characteristics of laminates or properties of the resin interface. The materials compression strength response to impact damage or other defects can be evaluated directly with the compression specimen developed by the National Aeronautics and Space Administration and included in its compilation of test methods (Ref 15). The test set-up is shown in Fig. 25. The specimen is nominally 6.4 mm (0.25 in.) thick by 125 mm (5.0 in.) wide and 250 mm (10.0 in.) high. Impacted specimens are initially 180 mm (7.0 in.) wide, impacted in a standard support fixture with a 13-mm (0.5-in.) diam impacting tup and then trimmed. The specimen use should be restricted to damage that is a maximum of 50 mm (2.0 in.) in width. The test is valuable for determining the comparative damage tolerance of resin systems or a basic laminate. It may be quite conservative as an indication of the damage tolerance of the actual structure, as noted previously.

There are several methods of measuring fracture toughness of the interlaminar matrix that give an indication of damage tolerance. These properties are most significantly G_1, resistance to the tensile opening mode; G_2, resistance to delamination in shear; and $G_{1,2}$, a combination or mixed mode.

The tensile opening mode characteristic, G_1, is most appropriately determined by the double cantilever beam test method. The test set-up is shown in Fig. 26. Either static or cyclic loading can be applied to determine the onset and rate of delamination growth. Details of the test are given in Ref 16-19. A standard for the specimen configuration and test procedure is included in Ref 15.

The test to determine delamination resistance in the pure shear mode, G_2, is the most difficult. This is in part due to the difficulty in obtaining pure shear without inducing a significant tensile component. Because the resins are typically much weaker in tension, it is difficult to avoid failure in the tensile mode or at least have the tensile component strongly influence the test value. A flexure test using a specimen with an end delamination was used with considerable success (Ref 20). The difficulty of obtaining satisfactory results is noted in the reference by the statement, "The G_2 values for all the materials tested were nearly an order of magnitude greater than their interlaminar tension, G_1, values." A simple two-point supported beam was used in the above tests. Some success has also been obtained using a cantilever beam specimen (Ref 21).

Although the G_2 forces may be a significant factor in damage-related failure because of the much greater G_2 capability of the resin, it is probable that G_2 will rarely be the most important parameter associated with impact damage failure. For example, although the deflections caused by the impact event cause large shear stresses, these can normally be resolved into tension components that cause interply failure long before the critical shear value is reached. However, because the interaction of the shear and tensile stresses is considered important, testing in which both modes of stress are involved is considered perhaps most representative of the actual failure case. Various mixed-mode test methods have been developed to represent this condition. Most noted are tensile tests involving laminates with ply stacking sequences that create especially high interlaminar stresses. These high interlaminar stresses and their influence in causing early delamination at the free edge of tensile specimens were noted in Ref 1 and 22. It was subsequently proposed that this characteristic be incorporated in a test method to evaluate edge delamination effects. These include a ±30/90 lay-up (Ref 23) and a ±25/90 lay-up (Ref 24). Both methods have been thoroughly analyzed and have shown good agreement between theoretical and experimental results. An additional investigation of the ±30/90 specimen was made in Ref 25. A standard specimen configuration and procedure involving this test is included in Ref 15. Other mixed-mode methods of testing include use of a 90° center-notched tension coupon and a cracked lap shear specimen. These are discussed and evaluated in Ref 25 and 18, respectively.

Damage Tolerance Requirements

A similar approach has been used in establishing damage tolerance requirements for both civil/commercial and U.S. Air Force aircraft. For civil/commercial aircraft, the requirements

Fig. 23 Stitched test panel

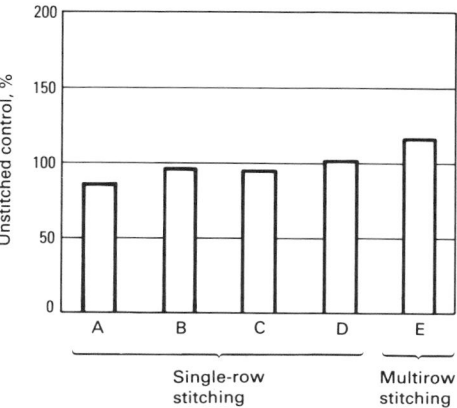

Fig. 24 Effect of stitching on the residual strength of impact-damaged coupons and panels. A, 25- × 305-mm (1- × 12-in.) coupon; B, 150- × 280-mm (6- × 11-in.) coupon; C, three-stringer multispar panel; D, three-stringer multirib panel; E, two three-stringer multirib panels

are established in the Federal Aviation Regulation (FAR) 25 (Ref 26) and in the Advisory Guideline Circular AC-107A (Ref 27). The latter is not a mandatory requirement, but provides both industry and the Federal Aviation Administration (FAA) with guidelines for composite structure compliance with FARs.

The key guideline element relative to impact damage is as follows: "It should be shown that impact damage that can be realistically expected from manufacturing and service, but not more than the established threshold of detectability for the selected inspection procedure, will not reduce the structural strength below ultimate load capability. This can be shown by analysis supported by test evidence, or by tests at the coupon, element, or subcomponent level."

The key elements of the guideline for impact damage relate to two areas: the detectability of the damage, including the method used to detect

it, and the residual strength requirement to carry ultimate load after damage that is not detectable by the inspection plan and methods designated.

The U.S. Air Force has an established damage tolerance criterion for metal structures (Ref 28). Recent updating of structural requirements has placed this damage tolerance criterion in a new, general, structural military aircraft specification (Ref 29). To bring this latest requirements document up to date for coverage of composite durability and damage tolerance, draft requirements for composites were written. The key elements of the requirements define nondetectable flaws and damage (which must be considered to be present in the structure) and corresponding design load requirements.

The key statement with regard to flaw/damage definition is: "Flaws considered critical should be defined in terms of size for use in the

validation process. The selected sizes should be related to and supported by the inspection plan for both manufacturing and in-service flaw/damage. The selection should also be related to the likelihood of occurrence, for example, the susceptibility of a particular structural location to foreign object damage of a certain intensity and frequency of occurrence, etc.

"In the absence of a comprehensive inspection demonstration program, the flaw/damage types and sizes given in Table 1 should be assumed."

Again, detectability of the damage is a key consideration. However, in this case a maximum boundary or default sizes are provided.

The residual strength load requirements, given in Ref 29, are that the structure must sustain the maximum internal member load, P_{xx}, that will occur once in 20 lifetimes. In cases in which P_{xx} is determined to be less than the design limit load, the design limit load shall be the required residual strength load level. P_{xx} need not be greater than 1.2 times the maximum load in one lifetime, if greater than design limit load. The designated residual strength is required after a period of two lifetimes of unrepaired service use.

With regard to the assumption of flaws and damage, the two specifications (Ref 27 and 29) differ significantly only in that the Air Force specification (Ref 29) defines an upper boundary for required impact energy and detectability, while the civil/commercial (Ref 27) guideline does not.

The cut-off limits of the Air Force requirements (Ref 29) are illustrated in Fig. 27, in which dent depth (that is, damage visibility) is associated with laminate thickness for a fixed impact energy level. In reality, it is a more complex function of the material properties and structural details.

To establish the limits shown in Fig. 27, it was decided that a dent depth of 2.5 mm (0.1 in.) provided a reasonable damage indication

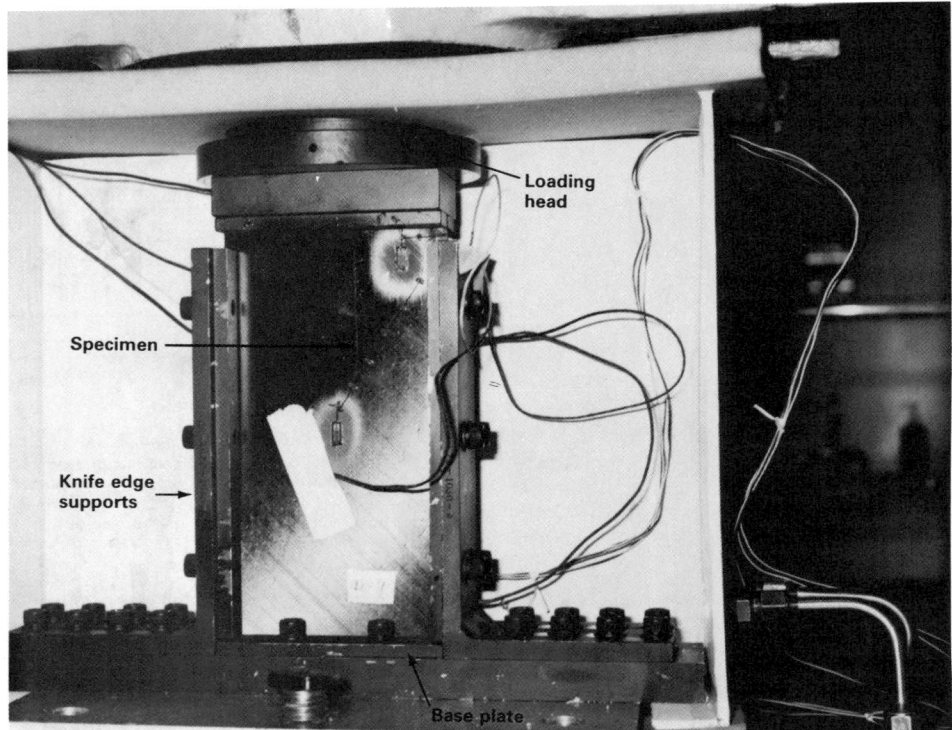

Fig. 25 Test set-up for compression testing on small 25- × 50-mm (5- × 10-in.) laminate panels

Fig. 26 Test set-up for G_1 testing

for visible detection. The 135 J (100 ft · lbf) energy level for the maximum impact energy level was established as a reasonable upper boundary for a remote-impact occurrence.

An illustration similar to Fig. 27 can be visualized for the civil/commercial requirements where an upper boundary or detectability limit is defined by the selected inspection procedure. The difference, however, is that the right boundary defining an energy cut-off is removed; consequently, it can be visualized that in the case of very damage-resistant structures, for example, the thick root area of a wing skin, very large impact energies may be involved to be visible if that level of detectability is required. In such a case, it might be argued that the occurrence of such damage would be obvious or at least sufficiently rare to warrant negotiation of the requirement.

The most significant difference between the U.S. Air Force damage tolerance requirements (Ref 29) and those in the FAA guideline circular (Ref 27) concerns the requirement for residual strength. The FAA circular requires a residual strength equal to ultimate load, while the Air Force requirement for nondetectable damage is established at the highest once-in-20-lifetimes load or limit, whichever is greatest, but not to exceed 1.2 times limit. This authorization of less than an ultimate load requirement by the Air Force is identical to that for metals in MIL-A-83444 (Ref 28). It reflects several considerations, including the highly improbable simultaneous occurrence of worst-case conditions. The 135 J (100 ft · lbf) impact is considered a rogue occurrence. Also, it is considered likely that the occurrence of an impact event of that magnitude will be obvious.

Table 1   Assumed nondetectable flaw/damage

Type	Size
Scratches	A surface scratch that is 100 mm (4.0 in.) long and 0.50 mm (0.02 in.) deep
Delamination	An interply delamination that has an area equivalent to a 50-mm (2.0-in.) diameter circle with dimensions most critical to its location (a)
Impact Damage	Damage caused by the impact of a 25-mm (1.0-in.) diameter hemispherical impactor with 135 J (100 ft · lbf) of kinetic energy, or with that kinetic energy required to cause a dent 2.5 mm (0.10 in.) deep, whichever is least

(a) For limited access areas such as the interior of the wing, the contractor shall have the option of proposing an inspection procedure before closeout, which will allow the assumed damage area size or imposed impact intensity to be reduced.

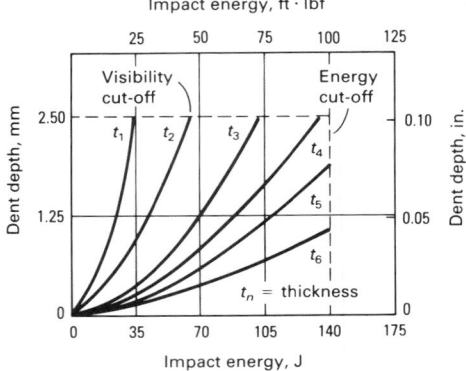

Fig. 27 Damage tolerance impact energy and damage visibility requirement cut-off limits designated by the Air Force. Source: Ref 29

It certainly must be considered a possibility then that the damage would be reported, inspected, and a repair or use-as-is decision made. It is also highly probable that any such damage would not occur in the most highly loaded location on the aircraft and that the aircraft would not be subjected to a once-in-20-lifetimes (maximum) loading scenario.

REFERENCES

1. N.J. Pagano and R.B. Pipes, The Influence of Stacking Sequence on Laminate Strength, *J. Compos. Mater.*, Vol 5, 1971
2. R.B. Pipes and N.J. Pagano, Interlaminar Stresses in Composite Laminates—An Approximate Elastic Solution, *J. Appl. Mech. (Trans. ASME)*, Vol 41, Sept 1974
3. J.G. Bjeletich, F.W. Crossman, and W.J. Warren, "The Influence of Stacking Sequence on Failure Modes in Quasi-Isotropic Graphite-Epoxy Laminates," Paper presented at TMS-AIME/ASM Joint Composite Materials Committee Symposium,

Chicago, Oct 1977

4. I.M. Daniel, R.E. Rolands, and J.B. Whiteside, Effects of Material and Stacking Sequence on Behavior of Composite Plates With Holes, *Exp. Mech.*, Vol 14, Jan 1974

5. M. Ashizawa, Improving Damage Tolerance of Laminated Composites Through the Use of New Tough Resins, in *Proceedings of the Sixth Conference on Fibrous Composites in Structural Design*; also, AMMRC MS 83-2, Army Materials and Mechanics Research Center, Nov 1983

6. W.I. Griffith, M.F. Kanninen, and E.F. Rybicki, A Fracture Mechanics Approach to the Analysis of Graphite/Epoxy Laminated Precracked Tension Panels, in *Nondestructive Evaluation and Flow Criticality for Composite Materials*, STP 696, American Society for Testing and Materials, 1979

7. R.J. Palmer, "Resin Properties to Improve Impact in Composites," Paper presented at the Fifth DOD/NASA Conference on Fibrous Composites in Structural Design, Department of Defense/National Aeronautics and Space Administration, New Orleans, Jan 1987

8. P.W.R. Beaumont, P.G. Riewald, and C. Zweben, Methods for Improving the Impact Resistance of Composite Materials, in *Foreign Object Impact Damage to Composites*, STP 568, American Society for Testing and Materials, 1975

9. G. Doney, G.R. Sidey, and J. Hutchings, *Impact Properties of Carbon Fibre/Kevlar 49 Fibre Hybrid Composites*, IPC Science and Technology Press, LTD., 1975

10. J.M. Hopper, E. Demuts, and G. Miliziano, "Damage Tolerant Design Demonstration," Paper presented at the AIAA/ASME/ASCE/AHS 25th Structures, Structural Dynamics and Materials Conference, Palm Springs, California, May 1984

11. E. Demuts and R.E. Horton, "Damage Tolerant Composite Design Development," Paper presented at the Third International Conference on Composite Structures, Paisley College of Technology, Paisley, Scotland, Sept 1985

12. J.E. McCarty and R.E. Horton, "Damage Tolerance of Composites," Paper presented at the 15th Congress, International Council of the Aeronautical Sciences, London, Sept 1986

13. R.S. Whitehead, and E. Demuts, "Damage Tolerance Qualification of Composite Structures," Paper presented at the U.S. Air Force ASIP Conference, Dayton, Ohio, Nov 1985

14. E. Demuts, R.S. Whitehead, and R.B. Deo, "Assessment of Damage Tolerance in Composites," Paper presented at the Structural Impact and Crashworthiness Conference, Imperial College, London, July 1984

15. "Standard Tests for Toughened Resin Composites," NASA Reference Publication 1092, National Aeronautics and Space Administration, 1982

16. D.J. Wilkins, J.R. Eisenmann, R.A. Camin, W.J. Margolis, and R.A. Benson, Characterizing Delamination Growth in Graphite-Epoxy, in *Damage in Composite Materials: Basic Mechanisms, Accumulation, Tolerance and Characterization*, STP 775, American Society for Testing and Materials, 1982

17. J.M. Whitney, F.E. Browning, and W. Hoogsteden, A Double Cantilever Beam Test for Characterizing Mode I Delamination of Composite Materials, *J. Rein. Plast. Compos.*, Vol 1 (No. 4), 1982

18. R.C. Shah, G. Miliziano, and A.V. Viswanathan, "Interlaminar Fracture Characteristics of Tougher Thermoset Materials," Paper presented at the AIAA/ASME/ASCE/AHS 26th Structures, Structural Dynamics and Materials Conference, Orlando, Florida, April 1985

19. F.X. de Charentenay, J.M. Harry, Y.J. Prel, and M.L. Benzeggagh, Characterizing the Effect of Delamination Defect by Mode 1 Delamination Test, in *Effects of Defects in Composite Materials*, STP 836, American Society for Testing and Materials, 1984

20. G.B. Murri and T.K. O'Brien, "Interlaminar G_{110} Evaluation of Toughened-Resin Matrix Composites Using the End-Notched Flexure Test," Paper presented at AIAA/ASME/ASCE/AHS 26th Structures, Structural Dynamics and Materials Conference, Orlando, Florida, April 1985

21. L.B. Ilcewicz, P.E. Keary, and J.J. Trostle, Interlaminar Fracture Toughness Testing of Composite Mode I and Mode II DCB Specimens, *J. Polymer Compos.*, June 1987

22. R.B. Pipes, B.E. Kaminski, and N.J. Pagano, "Influence of the Free Edge Upon the Strength of Angle-Ply Laminates," ASTM Special Technical Publication 521, American Society for Testing and Materials, 1973

23. T.K. O'Brien, N.J. Johnson, D.H. Morris, and R.A. Simonds, A Simple Test for the Interlaminar Fracture Toughness of Composites, *SAMPE J.*, July/August 1982

24. G.E. Law, A Mixed Mode Fraction Analysis of ($\pm 25/90n$)s Graphite/Epoxy Composite Laminates, in *Effects of Defects in Composite Materials*, STP 836, American Society for Testing and Materials, 1984

25. J.M. Whitney, and C.E. Browning, Materials Characterization of Matrix Dominated Failure Modes, in *Effects of Defects in Composite Materials*, STP 836, American Society for Testing and Materials, 1984

26. U.S. Federal Aviation Regulation (FAR) 25

27. "Composite Aircraft Structure Advisory Circular AC-107-A," U.S. Department of Transportation, Federal Aviation Administration

28. "Airplane Damage Tolerance Requirements," MIL-A-83444, U.S. Air Force, July 1974

29. "General Specification for Aircraft Structures," MIL-A-87221, U.S. Air Force, Feb 1985

Computer Programs for Structural Analysis

Richard T. Brown, Atlantic Research Corporation

IN A SURVEY of analysis methods suitable for composite materials, 46 mainframe computer-based structural analysis programs were reviewed. Codes that did not treat surfaces of revolution were not evaluated. The major problems addressed by all of the programs were:

- Structural response of laminated and multi-directionally reinforced composites
- Changes in material properties with temperature, moisture, and ablative decomposition
- Thin-shelled, thick-shelled, and/or plate structures
- Thermal-, pressure-, traction-, deformation-, and vibration-induced load states
- Failure modes

Secondary, but not trivial, concerns addressed by the more general methods were

Table 1 Programs evaluated

Program	Main application
ABAQUS	Marine structures
ADINA	Nonlinear stress analysis
AGGIE	General nonlinear analysis
AMGO 54 & AMGO 72 (pstr/pstn)	Rocket nozzle analysis
ANSYS REVISION 4.0+	General structural analysis
ARGUS	General structural analysis
ASAAS II	Nose tips
ASKA	Rocket nozzle analysis—Europe
BOPACE 3d version 6	Liquid rocket nozzle analysis
BOSOR 4	Shell analysis
BOSOR 5	Shell buckling
CASSE	Thermal process analysis
CHAMPION 3D	Composite rocket motor cases
DCAP	Micromechanics
DIAL	General structural analysis
DOASIS	Nose tip analysis
EASE2	General structural analysis
FASOR	Shell buckling
GIFTS 5	Graph oriented fin el system
LAMPS-A (FE Analy of LAM)	Composite plates and shells
MARC	General nonlinear analysis
MHOST	General nonlinear analysis
MIPAC (Micro Process Analy Code)	Micromechanics and processing
NASTRAN	General structural analysis
ND PROP	Micromechanics
NEPSAP	General structural analysis
NIFDI	Axisymmetric joints
NISA	General structural analysis
NONSAP	General nonlinear analysis
PAC 78	Composites analysis
PAFEC	General structural analysis
PATCHES-III	Rocket nozzle analysis
PATRAN-G	General stress analysis
PLANS (Plas Lrg Defl Ana Nln St)	Nonlinear structural analysis
SAAS 3M	Graphite nose tip analysis
SAAS III	Rocket nozzle analysis
SAP 4-5	General structural analysis
SAP 6 (Linear), 7 (Nonlinear)	General composite analysis
SPAR (Struc Perf Analy and Redsign)	Aircraft and trusses
STAGSC-1	Shell analysis
STARDYNE	General structural analysis
SUPERB	General structural analysis
TASS	Rocket nozzle analysis
TEXGAP84 2d-3d texcap for contact	Rocket case/grain analysis
TEXLESP (Texas Large Elas Stn Pg)	Rocket case/grain interface
TSAAS	Nuclear

Table 2 Definition of terms

Symbol	Definition
adhsn	Contact element with adhesion model
ani	Anisotropic laminate model
aniso	Anisotropic material properties
axi	Surface of revolution
axi. press	Axisymmetric pressure loading
bimod	Different properties in tension and compression
body	Body forces
char	Coupling with material decomposition
defects	Defect modeling capability
fail	Failure criteria for composites
fidif	Finite difference
fluid	Aeroelastic coupling
frac	Fracture (crack tip) element
fric	Contact element with spring and/or friction
harmc	Harmonic series input to simulate a local load
1d	Truss and beam geometry
invo	Involute laminate model
lin	Linear elasticity
lrgrot	Large rotation, small strains
lrgstn	Large strains and rotations
micro	Coupling with micromechanics model
moving	Internal contour coupled with surface recession
nonlin	Various kinds of nonlinear material behavior
opn	Open and closing contact element
optim	Composite optimization
ortho	Orthotropic material properties
pore press	Pore pressure loading
post	Postprocessing of results into graphics
pre	Preprocessor for grid
preload	Preload on interfaces
pstn	Plane strain model
pstr	Plane stress model
pt	Point loading
re-mesh	Automatic mesh regeneration in high-stress area
rubr	Incompressible material model
smlstn	Small strain displacements
spcl elmts	Special elements
stfn	Stiffener (rib and beam) element
subrtn	Subroutine
thermal	Thermal loading or heat transfer calculation
trans.shr	Transverse shear laminate calculation
3d	Full three-dimensional model
ud	Unidirectional properties
xnb	x Number of nodes in a brick element
xnq	x Number of nodes in a quadrilateral element

material (constitutive law) nonlinearity, geometric (large-deformation) nonlinearity, structural instability (buckling), and fracture mechanics. Some of the codes surveyed covered specialized topics, such as micromechanics, process modeling, fastener and other contact problems, and aero/fluid elastic coupling.

Program Evaluation

Table 1 summarizes the programs reviewed. The information was collected and sorted using the data base module of the Appleworks personal computer program. Copies of the data base (PRODOS operating system only) are available in ASCII or Appleworks formats. Information sources included user manuals or reports, vendor literature, source codes, and Ref 1 to 11. The IN-RAM data base program required extensive use of abbreviations to condense 70 criteria and 30 categories into a limit of 45K bytes on the Apple IIC. Table 2 explains the terms and abbreviations. Code descriptions in the survey format were textual. Many of the sources cited did not identify the program version. Because features are constantly being updated, some may not be correctly identified, and the ease-of-use assessment is admittedly subjective.

Program evaluations are listed in Table 3. An item identified as being unknown should be understood to mean, "I do not know about this feature, but this type of code should be able to do this." It is possible to use the criteria-matching capability of the data base to find programs that contain specific features. Key words, such as "micromechanics," can be used, as well as combinations, such as "mate-

rial model: contains aniso-y, and loading types: contains it-y or axi.press-y." The words "yes," "no," and "unknown" are represented by "y," "n," and "u" or "unk," respectively.

It should be noted that many of the mainframe programs have versions suitable for use on mini- or microcomputer systems. The number of such programs is rapidly expanding, and Ref 12 is a good source of vendors through 1986.

Laminated Composite Analysis

The programs surveyed in Table 3 use finite-element or finite-difference numerical methods. The structural response of laminated plate structures can be assessed using the closed-form solutions of Classical Laminate Analysis. While the actual matrix manipulations are too tedious for hand calculations, their implementation on programmable calculators and personal computers has provided a means for rapid design analysis and stacking laminate optimization.

Commercial and public domain laminate analysis programs are available and several are listed in Table 4. While each program contains special features, all are generally based on the same laminate analysis methodology. The next article in this Volume, "Software for Composite Materials Analysis," treats the CLASS code in detail and is representative of this type of program. The ready availability of these calculation tools in a variety of calculator and computer languages has facilitated rational composite material

design. This is a significant improvement over the netting analysis and carpet plot methods of the previous decade.

ACKNOWLEDGMENTS

Portions of this survey were funded by Failure Analysis Associates, Palo Alto, CA, and SNIA-BPD, Rome, Italy.

REFERENCES

1. E.J. Jeter and J.H. Hildreth, "Structural Computer Code Evaluation," Vol I and II, AFRPL-TR-76-68, Air Force Rocket Propulsion Laboratories, Nov 1976
2. D. Ehrenpreis, "Structural Analytical and Assessment Tasks, AFRPL Nozzle Programs," AFRPL-TR-79-85, Air Force Rocket Propulsion Laboratories, March 1980
3. R. Brown and J. Nachlas, Criteria for Structural Code Selection, in *Composite Structures*, I.H. Marshal, Ed., Applied Science Publishers, 1985
4. W.D. Pilkey, K. Salzalski, and J. Shaeffer, *Structural Mechanics Computer Programs*, University of Virginia Press, 1974
5. H.H. Fong, "An Evaluation of Eight U.S. General Purpose Finite Element Computer Programs," AIAA-82-0699, American Institute of Aeronautics and Astronautics, 1982
6. O.H. Griffin, Jr., Evaluation of Finite-Element Software Packages for Stress Analysis of Laminated Composites, *Compos. Technol. Rev.*, Vol 4 (No. 4), 1982, p 136-141
7. A.K. Noor, Survey of Computer Programs

Table 3 Program evaluations

Method: ABAQUS   Source: Hibbitt, Karlsson & Sorensen
Reference: draft user's manual (10/81)   Availability: public-n,lease-y,buy-u
Validation: 39 sample problems   Machines: unk
Main appl: marine structures   Features: pre-y,post-y,fail-u,optim-u
Programming: sub-struc,user subs   Ease of use: free form,texgap style
Maintenance: vendor-frequent updates   Language: unk
Analysis: static-y,vibration-y,buckling-y   Machine dependency: unk
Material model: iso-y,ortho-y,aniso-y   Lamination: ud-y,ortho-y,ani-y,invo-n
Dependency: lin-y,rubr-y,bimod-y,nonlin-y   Coupling: micro-y,fluid-y,thermal-y
Geometry: 1d-y,axi-y,pstr-y,pstn-y,3d-y   Disp type: smlstn-y,lrgrot-y,lrgstn-y
Contact elements: opn-y,fric-y,adhsn-y   Spcl elmts: frac-n,stfn-y,trans.shr-y
Element library: fidif-n,4nq-y,8nq-y,12nq-n,8nb-y,20nb-y,32nb-n,shell-y,wedge-n
Loading types: pt-y,harmc-n,axi.press-y,pore press-y,body-y,preload-y,thermal-y
Moving boundary: y Auto re-mesh: y Step loading: y Char and decomp: n Defects: n

Method: ADINA   Source: MIT
Reference: Composites Tech Revw (Griffen)   Availability: public-n,lease-y,buy-y
Validation: unk   Machines: CDC, IBM, Univac, VAX
Main appl: nonlinear stress analysis   Features: pre-y,post-y,fail-n,optim-n
Programming: pre-post from Adina Eng.   Ease of use: unk
Maintenance: MIT   Language: unk
Analysis: static-y,vibration-y,buckling-y   Machine dependency: unk
Material model: iso-y,ortho-y,aniso-y   Lamination: ud-y,ortho-n,ani-y,invo-n
Dependency: lin-y,rubr-n,bimod-u,nonlin-y   Coupling: micro-n,fluid-n,thermal-n
Geometry: 1d-y,axi-y,pstr-y,pstn-y,3d-y   Disp type: smlstn-y,lrgrot-y,lrgstn-u
Contact elements: opn-n,fric-n,adhsn-n   Spcl elmts: frac-n,stfn-n,trans.shr-n
Element library: fidif-n,4nq-y,8nq-y,12nq-n,8nb-y,20nb-y,32nb-n,shell-y,wedge-n
Loading types: pt-y,harmc-n,axi.press-y,pore press-n,body-y,preload-n,thermal-u
Moving boundary: n Auto-re-mesh: n Step loading: u Char and decomp: n Defects: n

Method: AGGIE   Source: Cosmic, Texas A&M
Reference: Cosmic (1984)   Availability: public-y,lease-n,buy-n
Validation: unk   Machines: IBM 360
Main appl: general nonlinear analysis   Features: pre-n,post-y,fail-n,optim-n
Programming: moderate size,ltd.substruc   Ease of use: unk
Maintenance: unk (developed in 1977)   Language: FORTRAN IV
Analysis: static-y,vibration-y,buckling-y   Machine dependency: unk
Material model: iso-y,ortho-y,aniso-n   Lamination: ud-n,ortho-n,ani-n,invo-n
Dependency: lin-y,rubr-y,bimod-n,nonlin-y   Coupling: micro-n,fluid-n,thermal-n
Geometry: 1d-n,axi-y,pstr-n,pstn-n,3d-y   Disp type: smlstn-y,lrgrot-y,lrgstn-y
Contact elements: opn-n,fric-n,adhsn-n   Spcl elmts: frac-n,stfn-n,trans.shr-n
Element library: fidif-n,4nq-y,8nq-y,12nq-n,8nb-y,20nb-y,32nb-n,shell-n,wedge-n
Loading types: pt-y,harmc-n,axi.press-n,pore press-n,body-n,preload-n,thermal-u
Moving boundary: n Auto re-mesh: n Step loading: y Char and decomp: n Defects: n

Method: AMGO 54 and AMGO 72 (pstr-pstn)   Source: Rohm & Haas
Reference: ASME Therm Strc Analys Surv 72   Availability: public-u,lease-u,buy-u
Validation: Becker and Parr   Machines: IBM 360,Univac 1108,CDC 660
Main appl: rocket nozzle analysis   Features: pre-y,post-y,fail-n,optim-n
Programming: unk   Ease of use: unk
Maintenance: none   Language: FORTRAN IV
Analysis: static-y,vibration-n,buckling-n   Machine dependency: unk
Material model: iso-y,ortho-y,aniso-n   Lamination: ud-n,ortho-n,ani-n,invo-n
Dependency: lin-y,rubr-n,bimod-n,nonlin-n   Coupling: micro-n,fluid-n,thermal-n
Geometry: 1d-n,axi-y,pstr-y,pstn-y,3d-n   Disp type: smlstn-y,lrgrot-n,lrgstn-n
Contact elements: opn-n,fric-n,adhsn-n   Spcl elmts: frac-n,stfn-n,trans.shr-n
Element library: fidif-n,4nq-y,8nq-y,12nq-n,8nb-y,20nb-y,32nb-n,shell-n,wedge-n
Loading types: pt-n,harmc-n,axi.press-y,pore press-n,body-n,preload-n,thermal-y
Moving boundary: n Auto re-mesh: n Step loading: n Char and decomp: n Defects: n

(continued)

Table 3 (continued)

Method: ANSYS revision 4.0+
Reference: user's manual (2/1/82)
Validation: 126 test cases
Main appl: general structural analysis
Programming: unk
Maintenance: Swanson Analysis
Analysis: static-y,vibration-y,buckling-y
Material model: iso-y,ortho-y,aniso-y
Dependency: lin-y,rubr-n,bimod-y,nonlin-y
Geometry: 1d-y,axi-y,pstr-y,pstn-y,3d-y
Contact elements: opn-y,fric-y,adhsn-n
Element library: fidif-y,4nq-y,8nq-n,12nq-n,8nb-n,20nb-n,32nb-n,shell-y,wedge-n
Loading types: pt-y,harmc-y,axi.press-y,pore press-y,body-y,preload-y,thermal-y
Moving boundary: n Auto re-mesh: y Step loading: y Char and decomp: n Defects: n

Source: Swanson Analysis System
Availability: public-n,lease-y,buy-n
Machines: VAX,Cyber,Univac,IBM,Cray
Features: pre-y,post-y,fail-y,optim-y
Ease of use: moderate (free field)
Language: FORTRAN
Machine dependency: unk
Lamination: ud-y,ortho-y,ani-n,invo-n
Coupling: micro-n,fluid-y,thermal-y
Disp type: smlstn-y,lrgrot-y,lrgstn-n
Spcl elmts: frac-y,stfn-y,trans.shr-n

Method: ARGUS
Reference: sales pamphlet, Comp Tech Revw
Validation: unk
Main appl: general structural analysis
Programming: NEPSAP derivative,user subs
Maintenance: Merlin
Analysis: static-y,vibration-y,buckling-y
Material model: iso-y,ortho-y,aniso-y
Dependency: lin-y,rubr-y,bimod-n,nonlin-y
Geometry: 1d-n,axi-y,pstr-y,pstn-y,3d-y
Contact elements: opn-y,fric-y,adhsn-n
Element library: fidif-n,4nq-n,8nq-y,12nq-y,8nb-y,20nb-y,32nb-n,shell-y,wedge-n
Loading types: pt-y,harmc-y,axi.press-y,pore press-n,body-y,preload-y,thermal-y
Moving boundary: n Auto re-mesh: n Step loading: y Char and decomp: n Defects: n

Source: Merlin
Availability: public-n,lease-y,buy-y
Machines: CDC,Univac
Features: pre-y,post-y,fail-y,optim-y
Ease of use: unk (nepsap like)
Language: FORTRAN
Machine dependency: unk
Lamination: ud-y,ortho-y,ani-y,invo-y
Coupling: micro-n,fluid-n,thermal-n
Disp type: smlstn-y,lrgrot-y,lrgstn-y
Spcl elmts: frac-n,stfn-n,trans.shr-y

Method: ASAAS II
Reference: PDA rept tr-4370-00-85 (9/77)
Validation: 6 sample problems
Main appl: nose tips
Programming: harmonic analysis
Maintenance: no
Analysis: static-y,vibration-n,buckling-n
Material model: iso-y,ortho-y,aniso-y
Dependency: lin-y,rubr-n,bimod-y,nonlin-y
Geometry: 1d-n,axi-y,pstr-y,pstn-y,3d-n
Contact elements: opn-n,fric-n,adhsn-n
Element library: fidif-n,4nq-n,8nq-n,12nq-n,8nb-n,20nb-n,32nb-n,shell-n,wedge-n
Loading types: pt-n,harmc-y,axi.press-y,pore press-n,body-n,preload-n,thermal-y
Moving boundary: y Auto re-mesh: n Step loading: n Char and decomp: n Defects: n

Source: NSWC, White Oak, and PDA Engineering
Availability: public-y,lease-n,buy-n
Machines: CDC 6600,Univac 1108
Features: pre-y,post-y,fail-n,optim-n
Ease of use: similar to SAAS
Language: FORTRAN (version unk)
Machine dependency: Univac
Lamination: ud-n,ortho-n,ani-n,invo-n
Coupling: micro-n,fluid-n,thermal-y
Disp type: smlstn-y,lrgrot-n,lrgstn-n
Spcl elmts: frac-m,stfn-n,trans.shr-n

Method: ASKA
Reference: Composites Tech Revw (Griffen)
Validation: unk

Main appl: rocket nozzle analysis-Europe
Programming: sub rtn calls from main
Maintenance: none
Analysis: static-y,vibration-n,buckling-n
Material model: iso-y,ortho-y,aniso-n
Dependency: lin-y,rubr-n,bimod-n,nonlin-y
Geometry: 1d-n,axi-y,pstr-n,pstn-y,3d-n
Contact elements: opn-n,fric-n,adhsn-n
Element library: fidif-n,4nq-y,8nq-n,12nq-n,8nb-n,20nb-n,32nb-n,shell-n,wedge-n
Loading types: pt-n,harmc-n,axi-press-n,pore press-n,body-y,preload-n,thermal-y
Moving boundary: n Auto re-mesh: n Step loading: n Char and decomp: n Defects: n

Source: Inst for Statics and Dyn (ISD)
Availability: public-n,lease-y,buy-y
Machines: CDC 600,Univac 1108,IBM 360
Features: pre-y,post-y,fail-y,optim-n
Ease of use: easy (SAASlike)
Language: FORTRAN 77
Machine dependency: unk
Lamination: ud-n,ortho-n,ani-n,invo-n
Coupling: micro-n,fluid-n,thermal-n
Disp type: smlstn-y,lrgrot-u,lrgstn-n
Spcl elmts: frac-n,stfn-n,trans.shr-n

Method: BOPACE 3d version 6
Reference: Cosmic (1984)
Validation: unk
Main appl: liquid rocket nozzle analysis
Programming: unk
Maintenance: Boeing
Analysis: static-y,vibration-n,buckling-n
Material model: iso-y,ortho-u,aniso-u
Dependency: lin-y,rubr-n,bimod-n,nonlin-y
Geometry: 1d-y,axi-y,pstr-u,pstn-u,3d-y
Contact elements: opn-n,fric-n,adhsn-n
Element library: fidif-n,4nq-n,8nq-n,12nq-n,8nb-n,20nb-n,32nb-n,shell-n,wedge-y
Loading types: pt-y,harmc-n,axi.press-n,pore press-n,body-n,preload-n,thermal-y
Moving boundary: n Auto re-mesh: n Step loading: y Char and decomp: n Defects: n

Source: Cosmic, Boeing Aerospace
Availability: public-y,lease-n,buy-n
Machines: Univac 1100
Features: pre-y,post-n,fail-n,optim-n
Ease of use: unk
Language: FORTRAN IV
Machine dependency: unk
Lamination: ud-u, ortho-u,ani-u,invo-u
Coupling: micro-n,fluid-n,thermal-n
Disp type: smlstn-y,lrgrot-y,lrgstn-y
Spcl elmts: frac-n,stfn-n,trans.shr-n

Method: BOSOR 4
Reference: Lockheed Missiles & Space Company, Inc. (LMSC) D243605 (3/72) AD748639
Validation: 6 sample cases + many pubs
Main appl: shell analysis

Source: LMSC
Availability: public-y,lease-n,buy-n

Machines: VAX,CDC,IBM,UNIVAC
Features: pre-n,post-n,fail-n,optim-n

Method: BOSOR 4 (continued)
Programming: unk
Maintenance: LMSC
Analysis: static-y,vibration-y,buckling-y
Material model: iso-y,ortho-y,aniso-n
Dependency: lin-y,rubr-n,bimod-n,nonlin-n
Geometry: 1d-n,axi-y,pstr-n,pstn-n,3d-n
Contact elements: opn-n,fric-n,adhsn-n
Element library: fidif-y,4nq-n,8nq-n,12nq-n,8nb-n,20nb-n,32nb-n,shell-n,wedge-n
Loading types: pt-y,harmc-y,axi.press-y,pore press-n,body-n,preload-n,thermal-y
Moving boundary: n Auto re-mesh: n Step loading: n Char and decomp: n Defects: y

Ease of use: easy (field oriented)
Language: FORTRAN IV
Machine dependency: unk
Lamination: ud-y,ortho-y,ani-n,invo-n
Coupling: micro-n,fluid-n,thermal-n
Disp type: smlstn-y,lrgrot-y,lrgstn-y
Spcl elmts: frac-n,stfn-y,trans.shr-n

Method: BOSOR 5
Reference: LMSC d407166 (12/74)
Validation: 5 sample cases

Main appl: shell buckling
Programming: unk
Maintenance: LMSC
Analysis: static-n,vibration-n,buckling-y
Material model: iso-y,ortho-y,aniso-n
Dependency: lin-y,rubr-n,bimod-n,nonlin-n
Geometry: 1d-n,axi-y,pstr-n,pstn-n,3d-n
Contact elements: opn-n,fric-n,adhsn-n
Element library: fidif-y,4nq-n,8nq-n,12nq-n,8nb-n,20nb-n,32nb-n,shell-n,wedge-n
Loading types: pt-y,harmc-y,axi.press-n,pore press-n,body-y,preload-n,thermal-y
Moving boundary: n Auto-re-mesh: n Step loading: y Char and decomp: n Defects: n

Source: LMSC
Availability: public-n,lease-n,buy-n
Machines: Univac 1108,CDC 6600, VAX,IBM
Features: pre-n,post-n,fail-n,optim-n
Ease of use: easy (fixed field)
Language: FORTRAN IV
Machine dependency: unk
Lamination: ud-y,ortho-y,ani-n,invo-n
Coupling: micro-n,fluid-n,thermal-n
Disp type: smlstn-y,lrgrot-y,lrgstn-y
Spcl elmts: frac-n,stfn-y,trans.shr-n

Method: CASSE
Reference: Pacific Numerix sales pamphlet
Validation: 4 example cases
Main appl: thermal process analysis
Programming: modular system
Maintenance: unk
Analysis: static-y,vibration-n,buckling-n
Material model: iso-y,ortho-y,aniso-n
Dependency: lin-y,rubr-n,bimod-n,nonlin-n
Geometry: 1d-y,axi-y,pstr-n,pstn-n,3d-y
Contact elements: opn-n,fric-n,adhsn-n
Element library: fidif-n,4nq-y,8nq-u,12nq-u,8nb-u,20nb-u,32nb-u,shell-u,wedge-u
Loading types: pt-y,harmc-n,axi.press-y,pore press-y,body-y,preload-n,thermal-y
Moving boundary: y Auto re-mesh: u Step loading: y Char and decomp: y Defects: n

Source: Pacific Numerix
Availability: public-n,lease-n,buy-y
Machines: unk
Features: pre-n,post-y,fail-u,optim-y
Ease of use: unk
Language: unk
Machine dependency: unk
Lamination: ud-y,ortho-y,ani-n,invo-n
Coupling: micro-n,fluid-y,thermal-y
Disp type: smlstn-y,lrgrot-n,lrgstn-n
Spcl elmts: frac-n,stfn-n,trans.shr-n

Method: CHAMPION 3D
Reference: NWC tech memo 5294 (3/84)
Validation: 3 sample cases
Main appl: composite rocket motor cases
Programming: unk
Maintenance: unk
Analysis: static-y,vibration-n,buckling-y
Material model: iso-y,ortho-y,aniso-y
Dependency: lin-y,rubr-y,bimod-n,nonlin-y
Geometry: 1d-n,axi-y,pstr-n,pstn-n,3d-y
Contact elements: opn-n,fric-n,adhsn-n
Element library: fidif-n,4nq-n,8nq-y,12nq-n,8nb-n,20nb-y,32nb-n,shell-n,wedge-n
Loading types: pt-y,harmc-n,axi.press-y,pore press-n,body-y,preload-n,thermal-y
Moving boundary: n Auto re-mesh: n Step loading: n Char and decomp: y Defects: y

Source: Naval Weapons Ctr, NASA (NWC) MSFC
Availability: public-y,lease-n,buy-n
Machines: unk
Features: pre-n,post-n,fail-n,optim-n
Ease of use: poor (NASTRAN quality)
Language: unk
Machine dependency: unk
Lamination: ud-n,ortho-n,ani-n,invo-n
Coupling: micro-n,fluid-n,thermal-n
Disp type: smlstn-y,lrgrot-y,lrgstn-y
Spcl elmts: frac-n,stfn-n,trans.shr-n

Method: DCAP
Reference: MSC tfr/7510 (7/29/75)
Validation: 1 sample ge 3d case + Brown
Main appl: micromechanics
Programming: two sub rtns (uni and weaks)
Maintenance: none
Analysis: static-y,vibration-n,buckling-n
Material model: iso-y,ortho-y,aniso-n
Dependency: lin-y,rubr-n,bimod-n,nonlin-n
Geometry: 1d-n,axi-n,pstr-n,pstn-n,3d-n
Contact elements: opn-n,fric-n,adhsn-n
Element library: fidif-n,4nq-n,8nq-n,12nq-n,8nb-y,20nb-n,32nb-n,shell-n,wedge-n
Loading types: pt-n,harmc-n,axi.press-n,pore press-n,body-n,preload-n,thermal-y
Moving boundary: n Auto re-mesh: n Step loading: n Char and decomp: y Defects: n

Source: NSWC, Edwards Air Force Base
Availability: public-y,lease-n,buy-n
Machines: IBM,VAX,CDC
Features: pre-n,post-y,fail-y,optim-n
Ease of use: hard (name list input)
Language: FORTRAN IV
Machine dependency: unk
Lamination: ud-y,ortho-n,ani-n,invo-n
Coupling: micro-n,fluid-n,thermal-n
Disp type: smlstn-y,lrgrot-n,lrgstn-n
Spcl elmts: frac-n,stfn-n,trans.shr-n

Method: DIAL
Reference: srvy of nonln strc prbs (noor)
Validation: unk
Main appl: general structural analysis
Programming: modular structure, user subs
Maintenance: LMSC
Analysis: static-y,vibration-y,buckling-y

Source: LMSC
Availability: public-n,lease-u,buy-u
Machines: Univac,CDC
Features: pre-y,post-y,fail-n,optim-n
Ease of use: interactive
Language: FORTRAN
Machine dependency: Univac,CDC

(continued)

Table 3 (continued)

Method: DIAL (continued)
Material model: iso-y,ortho-y,aniso-y
Dependency: lin-y,rubr-y,bimod-u,nonlin-y
Geometry: 1d-y,axi-y,pstr-y,pstn-y,3d-y
Contact elements: opn-y,fric-y,adhsn-n
Element library: fidif-n,4nq-y,8nq-y,12nq-y,8nb-y,20nb-y,32nb-y,shell-y,wedge-u
Loading types: pt-y,harmc-y,axi.press-y,pore press-y,body-y,preload-y,thermal-y
Moving boundary: n Auto re-mesh: n Step loading: n Char and decomp: n Defects: n
Lamination: ud-n,ortho-y,ani-u,invo-u
Coupling: micro-n,fluid-n,thermal-n
Disp type: smlstn-y,lrgrot-y,lrgstn-u
Spcl elmts: frac-n,stfn-y,trans.shr-u

Method: DOASIS
Reference: AFML tr-75-32 (1975)
Validation: unk
Main appl: nose tip analysis
Programming: unk
Maintenance: none
Analysis: static-y,vibration-n,buckling-n
Material model: iso-y,ortho-y,aniso-y
Dependency: lin-y,rubr-n,bimod-y,nonlin-y
Geometry: 1d-n,axi-y,pstr-y,pstn-y,3d-n
Contact elements: opn-n,fric-n,adhsn-n
Element library: fidif-n,4nq-y,8nq-y,12nq-n,8nb-n,20nb-n,32nb-n,shell-n,wedge-n
Loading types: pt-n,harmc-n,axi.press-n,pore press-n,body-n,preload-n,thermal-y
Moving boundary: y Auto re-mesh: n Step loading: y Char and decomp: n defects: n
Source: Weiler Research
Availability: public-y,lease-n,buy-n
Machines: unk
Features: pre-y,post-y,fail-n,optim-n
Ease of use: unk
Language: unk
Machine dependency: unk
Lamination: ud-n,ortho-n,ani-n,invo-n
Coupling: micro-n,fluid-n,thermal-n
Disp type: smlstn-y,lrgrot-n,lrgstn-n
Spcl elmts: frac-n,stfn-n,trans.shr-n

Method: EASE2
Reference: computer sys for analy (3/78)
Validation: unk
Main appl: general structural analysis
Programming: unk
Maintenance: Engineering Analysis Corp
Analysis: static-y,vibration-y,buckling-n
Material model: iso-y,ortho-n,aniso-n
Dependency: lin-y,rubr-n,bimod-n,nonlin-n
Geometry: 1d-y,axi-n,pstr-n,pstn-y,3d-y
Contact elements: opn-n,fric-n,adhsn-n
Element library: fidif-n,4nq-y,8nq-n,12nq-n,8nb-n,20nb-n,32nb-n,shell-y,wedge-n
Loading types: pt-y,harmc-n,axi.press-y,pore press-n,body-y,preload-n,thermal-y
Moving boundary: n Auto re-mesh: n Step loading: n Char and decomp: n Defects: n
Source: Engineering Analysis Corp
Availability: public-n,lease-y,buy-n
Machines: Cyber
Features: pre-y,post-y,fail-n,optim-n
Ease of use: unk
Language: unk
Machine dependency: unk
Lamination: ud-n,ortho-n,ani-n,invo-n
Coupling: micro-n,fluid-n,thermal-n
Disp type: smlstn-y,lrgrot-n,lrgstn-n
Spcl elmts: frac-n,stfn-n,trans.shr-n

Method: FASOR
Reference: user's manual version 1.2
Validation: 4 sample cases
Main appl: shell buckling
Programming: field method (invar imbedng)
Maintenance: Structures Research Assoc
Analysis: static-y,vibration-y,buckling-y
Material model: iso-y,ortho-y,aniso-y
Dependency: lin-y,rubr-y,bimod-n,nonlin-n
Geometry: 1d-n,axi-y,pstr-n,pstn-n,3d-n
Contact elements: opn-n,fric-n,adhsn-n
Element library: fidif-y,4nq-n,8nq-n,12nq-n,8nb-n,20nb-n,32nb-n,shell-n,wedge-n
Loading types: pt-n,harmc-n,axi.press-y,pore press-n,body-y,preload-n,thermal-y
Moving boundary: n Auto re-mesh: n Step loading: y Char and decomp: n defects: n
Source: Cybernet, Struc Research Assoc
Availability: public-n,lease-y,buy-y
Machines: CDC
Features: pre-n,post-y,fail-n,optim-n
Ease of use: hard (name list input)
Language: unk
Machine dependency: unk
Lamination: ud-y,ortho-y,ani-y,invo-n
Coupling: micro-n,fluid-n,thermal-n
Disp type: smlstn-y,lrgrot-y,lrgstn-y
Spcl elmts: frac-n,stfn-n,trans.shr-y

Method: GIFTS 5
Reference: adv gifts 5 user wkshop pamp
Validation: unk
Main appl: graph oriented fin el system
Programming: unk
Maintenance: unk
Analysis: static-y,vibration-y,buckling-u
Material model: iso-y,ortho-y,aniso-u
Dependency: lin-y,rubr-u,bimod-u,nonlin-y
Geometry: 1d-y,axi-y,pstr-u,pstn-u,3d-y
Contact elements: opn-u,fric-u,adhsn-u
Element library: fidif-u,4nq-y,8nq-u,12nq-u,8nb-y,20nb-u,32nb-u,shell-u,wedge-u
Loading types: pt-u,harmc-u,axi.press-y pore press-u, body-u, preload-u,thermal-y
Moving boundary: u Auto re-mesh: u Step loading: u Char and decomp: u Defects: u
Source: Univ. of Arizona, Int Gr Eng Lab
Availability: public-n,lease-y,buy-n
Machines: Eclipse-s230,Sel 32/27
Features: pre-y,post-y,fail-n,optim-n
Ease of use: unk
Language: unk
Machine dependency: unk
Lamination: ud-u,ortho-u,ani-u,invo-u
Coupling: micro-n,fluid-n,thermal-n
Disp type: smlstn-y,lrgrot-u,lrgstn-u
Spcl elmts: frac-u,stfn-u,trans.shr-u

Method: LAMPS-A (finite element analysis of laminate shell)
Reference: Cosmic (1984)
Validation: unk
Main appl: composite plates and shells
Programming: unk
Maintenance: Cosmic
Analysis: static-y,vibration-n,buckling-n
Material model: iso-y,ortho-y,aniso-y
Dependency: lin-y,rubr-n,bimod-n,nonlin-n
Geometry: 1d-n,axi-n,pstr-n,pstn-n,3d-y
Contact elements: opn-n,fric-n,adhsn-n
Source: Cosmic
Availability: public-y,lease-n,buy-n
Machines: IBM 360
Features: pre-n,post-n,fail-n,optim-n
Ease of use: unk
Language: FORTRAN IV
Machine dependency: unk
Lamination: ud-y,ortho-y,ani-n,invo-n
Coupling: micro-n,fluid-n,thermal-n
Disp type: smlstn-y,lrgrot-n,lrgstn-n
Spcl elmts: frac-n,stfn-n,trans.shr-y

Method: LAMPS-A (continued)
Element library: fidif-n,4nq-n,8nq-n,12nq-n,8nb-n,20nb-n,32nb-n,shell-y,wedge-n
Loading types: pt-y,harmc-y,axi.press-y,pore press-n,body-n,preload-n,thermal-u
Moving boundary: n Auto re-mesh: n Step loading: n Char and decomp: n Defects: n

Method: MARC
Reference: MARC user's manual (6/11/74)
Validation: extensive
Main appl: general nonlinear analysis
Programming: user subs,prob substructng
Maintenance: MARC
Analysis: static-y,vibration-y,buckling-y
Material model: iso-y,ortho-y,aniso-y
Dependency: lin-y,rubr-y,bimod-y,nonlin-y
Geometry: 1d-y,axi-y,pstr-y,pstn-y,3d-y
Contact elements: opn-y,fric-y,adhsn-n
Element library: fidif-y,4nq-y,8nq-y,12nq-y,8nb-y,20nb-y,32nb-n,shell-y,wedge-y
Loading types: pt-y,harmc-y,axi.press-y,pore press-n,body-y,preload-y,thermal-y
Moving boundary: n Auto re-mesh: n Step loading: y Char and decomp: n Defects: y
Source: MARC Corp
Availability: public-n,lease-y,buy-y
Machines: CDC,IBM,Univac,VAX, Prime
Features: pre-y,post-y,fail-y,optim-n
Ease of use: hard (NASTRANlike)
Language: FORTRAN 77
Machine dependency: unk
Lamination: ud-n,ortho-y,ani-y,invo-n
Coupling: micro-n,fluid-y,thermal-y
Disp type: smlstn-y,lrgrot-y,lrgstn-y
Spcl elmts: frac-y,stfn-y,trans.shr-n

Method: MHOST
Reference: user's manual version 2.1
Validation: 10 sample problems
Main appl: general nonlinear analysis
Programming: MARC deriv,multi iter proc
Maintenance: MARC
Analysis: static-y,vibration-y,buckling-y
Material model: iso-y,ortho-y,aniso-y
Dependency: lin-y,rubr-n,bimod-y,nonlin-y
Geometry: 1d-y,axi-y,pstr-y,pstn-y,3d-y
Contact elements: opn-n,fric-n,adhsn-n
Element library: fidif-n,4nq-y,8nq-n,12nq-n,8nb-y,20nb-n,32nb-n,shell-y,wedge-y
Loading types: pt-y,harmc-y,axi.press-y,pore press-y,body-y,preload-n,thermal-y
Moving boundary: n Auto re-mesh: n Step loading: y Char and decomp: y Defects: n
Source: NASA Lewis Research Center
Availability: public-y,lease-n,buy-n
Machines: Prime,VAX,CDC,Cray,IBM
Features: pre-n,post-y,fail-n,optim-n
Ease of use: medium,key words input
Language: FORTRAN 77
Machine dependency: none (double pre)
Lamination: ud-y,ortho-y,ani-y,invo-n
Coupling: micro-n,fluid-n,thermal-n
Disp type: smlstn-y,lrgrot-y,lrgstn-y
Spcl elmts: frac-n,stfn-y,trans.shr-y

Method: MIPAC
Reference: AFWAL TR-81-4118
Validation: APIC/RN Test Program
Main appl: micromechanics and processing
Programming: sub rtn calls from main
Maintenance: none
Analysis: static-y,vibration-n,buckling-n
Material model: iso-y,ortho-y,aniso-n
Dependency: lin-y,rubr-n,bimod-n,nonlin-n
Geometry: 1d-y,axi-n,pstr-n,pstn-n,3d-y
Contact elements: opn-y,fric-y,adhsn-n
Element library: fidif-n,4nq-n,8nq-n,12nq-n,8nb-n,20nb-n,32nb-n,shell-n,wedge-n
Loading types: pt-n,harmc-n,axi.press-n,pore press-n,body-n,preload-n,thermal-y
Moving boundary: n Auto re-mesh: n Step loading: n Char and decomp: y Defects: n
Source: SAI, Air Force Materials Laboratory (AFML)
Availability: public-y,lease-n,buy-n
Machines: CDC 170/175
Features: pre-y,post-y,fail-y,optim-n
Ease of use: unk
Language: FORTRAN IV
Machine dependency: CDC
Lamination: ud-n,ortho-y,ani-n,invo-n
Coupling: micro-y,fluid-y,thermal-y
Disp type: smlstn-y,lrgrot-n,lrgstn-n
Spcl elmts: frac-n,stfn-n,trans.shr-n

Method: NASTRAN
Reference: MSC manual (2/78) primer (8/84)
Validation: extensive validation
Main appl: general structural analysis
Programming: direct matrix abstraction
Maintenance: MSC (excellent)
Analysis: static-y,vibration-y,buckling-y
Material model: iso-y,ortho-y,aniso-y
Dependency: lin-y,rubr-n,bimod-y,nonlin-y
Geometry: 1d-y,axi-y,pstr-y,pstn-y,3d-y
Contact elements: opn-y,fric-y,adhsn-n
Element library: fidif-n,4nq-y,8nq-y,12nq-y,8nb-y,20nb-y,32nb-y,shell-y,wedge-y
Loading types: pt-y,harmc-y,axi.press-y,pore press-y,body-y,preload-y,thermal-y
Moving boundary: n Auto re-mesh: n Step loading: y Char and decomp: n Defects: n
Source: MSC, Cosmic (ltd version)
Availability: public-y,lease-y,buy-y
Machines: CDC,IBM,Univac,VAX, Cray
Features: pre-y,post-y,fail-y,optim-n
Ease of use: hard (fixed field)
Language: FORTRAN IV (97–99%)
Machine dependency: 1–3% assembler
Lamination: ud-y,ortho-y,ani-y,invo-n
Coupling: micro-n,fluid-n,thermal-y
Disp type: smlstn-y,lrgrot-y,lrgstn-n
Spcl elmts: frac-n,stfn-y,trans.shr-y

Method: ND PROP
Reference: User's guide and Pagano's notes
Validation: 1 sample problem
Main appl: micromechanics
Programming: CDC machine calls
Maintenance: AFML
Analysis: static-y,vibration-n,buckling-n
Material model: iso-y,ortho-y,aniso-y
Dependency: lin-y,rubr-n,bimod-n,nonlin-n
Geometry: 1d-n,axi-n,pstr-n,pstn-n,3d-y
Contact elements: opn-n,fric-n,adhsn-n
Element library: fidif-n,4nq-n,8nq-n,12nq-n,8nb-n,20nb-n,32nb-n,shell-n,wedge-n
Loading types: pt-y,harmc-n,axi.press-n,pore press-n,body-n,preload-n,thermal-y
Moving boundary: n Auto re-mesh: n Step loading: n Char and decomp: n Defects: n
Source: AFML (Pagano)
Availability: public-y,lease-n,buy-n
Machines: CDC,VAX
Features: pre-n,post-n,fail-n,optim-y
Ease of use: hard (name list input)
Language: FORTRAN IV
Machine dependency: CDC
Lamination: ud-y,ortho-n,ani-n,invo-n
Coupling: micro-n,fluid-n,thermal-n
Disp type: smlstn-y,lrgrot-n,lrgstn-n
Spcl elmts: frac-n,stfn-n,trans.shr-n

(continued)

Table 3 (continued)

Method: NEPSAP
Reference: LMSC-D556041 (10/76)
Validation: unk
Main appl: general structural analysis
Programming: unk
Maintenance: LMSC (ARC has NWC source)
Analysis: static-y,vibration-y,buckling-y
Material model: iso-y,ortho-y,aniso-y
Dependency: lin-y,rubr-y,bimod-n,nonlin-y
Geometry: 1d-y,axi-y,pstr-y,pstn-y,3d-y
Contact elements: opn-n,fric-n,adhsn-n
Element library: fidif-n,4nq-n,8nq-y,12nq-y,8nb-y,20nb-y,32nb-n,shell-y,wedge-n
Loading types: pt-y,harmc-n,axi.press-y,pore press-n,body-y,preload-n,thermal-y
Moving boundary: n Auto re-mesh: n Step loading: y Char and decomp: n Defects: n

Source: LMSC, NWC (limited version)
Availability: public-y,lease-n,buy-n
Machines: CDC
Features: pre-y,post-y,fail-n,optim-n
Ease of use: unk
Language: FORTRAN
Machine dependency: unk
Lamination: ud-n,ortho-n,ani-n,invo-n
Coupling: micro-n,fluid-n,thermal-n
Disp type: smlstn-y,lrgrot-y,lrgstn-y
Spcl elmts: frac-n,stfn-n,trans.shr-n

Method: NIFDI
Reference: NIFDI manual TR-9204 (10/76)
Validation: 10 sample cases
Main appl: axisymmetric joints
Programming: sub rtn calls from main
Maintenance: PDA (none recently)
Analysis: static-y,vibration-n,buckling-n
Material model: iso-y,ortho-y,aniso-n
Dependency: lin-y,rubr-n,bimod-y,nonlin-y
Geometry: 1d-y,axi-y,pstr-y,pstn-y,3d-n
Contact elements: opn-y,fric-y,adhsn-n
Element library: fidif-n,4nq-y,8nq-y,12nq-y,8nb-n,20nb-n,32nb-n,shell-n,wedge-n
Loading types: pt-n,harmc-n,axi.press-y,pore press-y,body-y,preload-y,thermal-y
Moving boundary: n Auto re-mesh: n Step loading: y Char and decomp: n Defects: n

Soruce: PDA Engineering
Availability: public-n,lease-y,buy-y
Machines: Univac 1108,CDC
Features: pre-n,post-n,fail-n,optim-n
Ease of use: easy (SAASlike)
Language: FORTRAN IV
Machine dependency: unk
Lamination: ud-n,ortho-n,ani-n,invo-n
Coupling: micro-n,fluid-n,thermal-n
Disp type: smlstn-y,lrgrot-n,lrgstn-n
Spcl elmts: frac-n,stfn-n,trans.shr-n

Method: NISA
Reference: EMRC sales pamphlet
Validation: unk
Main appl: general structural analysis
Programming: unk
Maintenance: EMRC
Analysis: static-y,vibration-y,buckling-y
Material model: iso-y,ortho-y,aniso-y
Dependency: lin-y,rubr-n,bimod-n,nonlin-y
Geometry: 1d-y,axi-y,pstr-y,pstn-y,3d-y
Contact elements: opn-n,fric-n,adhsn-n
Element library: fidif-n,4nq-y,8nq-y,12nq-y,8nb-y,20nb-y,32nb-y,shell-y,wedge-n
Loading types: pt-y,harmc-y,axi.press-y,pore press-n,body-y,preload-n,thermal-y
Moving boundary: n Auto re-mesh: n Step loading: y Char and decomp: n Defects: n

Source: Eng Mechanics Research Corp
Availability: public-n,lease-y,buy-y
Machines: Cray, Cyber
Features: pre-y,post-y,fail-n,optim-y
Ease of use: unk
Language: FORTRAN
Machine dependency: unk
Lamination: ud-y,ortho-y,ani-y,invo-u
Coupling: micro-n,fluid-n,thermal-n
Disp type: smlstn-y,lrgrot-y,lrgstn-y
Spcl elmts: frac-y,stfn-y,trans.shr-n

Method: NONSAP
Reference: Composites Tech Revw (Griffen)
Validation: unk
Main appl: general nonlinear analysis
Programming: unk
Maintenance: unk
Analysis: static-y,vibration-y,buckling-n
Material model: iso-y,ortho-y,aniso-u
Dependency: lin-y,rubr-n,bimod-n,nonlin-y
Geometry: 1d-n,axi-y,pstr-y,pstn-y,3d-y
Contact elements: opn-n,fric-n,adhsn-n
Element library: fidif-n,4nq-y,8nq-y,12nq-y,8nb-y,20nb-y,32nb-n,shell-y,wedge-n
Loading types: pt-y,harmc-n,axi.press-n,pore press-n,body-n,preload-n,thermal-y
Moving boundary: n Auto re-mesh: n Step loading: y Char and decomp: n: Defects: n

Source: Univ of CA, Berkeley
Availability: public-y,lease-n,buy-n
Machines: unk
Features: pre-n,post-n,fail-n,optim-n
Ease of use: unk
Language: FORTRAN 77
Machine dependency: unk
Lamination: ud-n,ortho-n,ani-n,invo-n
Coupling: micro-n,fluid-n,thermal-n
Disp type: smlstn-y,lrgrot-y,lrgstn-y
Spcl elmts: frac-n,stfn-n,trans.shr-n

Method: PAC 78
Reference: Composites Tech Revw (Griffen)
Validation: unk
Main appl: composites analysis
Programming: unk
Maintenance: Duke Univ
Analysis: static-y,vibration-n,buckling-n
Material model: iso-y,ortho-y,aniso-y
Dependency: lin-y,rubr-n,bimod-n,nonlin-y
Geometry: 1d-n,axi-n,pstr-n,pstn-n,3d-y
Contact elements: opn-n,fric-n,adhsn-n
Element library: fidif-n,4nq-y,8nq-y,12nq-y,8nb-y,20nb-y,32nb-n,shell-n,wedge-n
Loading types: pt-y,harmc-n,axi.press-n,pore press-n,body-n,preload-n,thermal-u
Moving boundary: n Auto re-mesh: n Step loading: n Char and decomp: n defects: n

Source: Duke Univ
Availability: public-u,lease-u,buy-u
Machines: IBM 360/175
Features: pre-n,post-n,fail-n,optim-n
Ease of use: unk
Language: unk
Machine dependency: unk
Lamination: ud-y,ortho-y,ani-y,invo-n
Coupling: micro-y,fluid-n,thermal-n
Disp type: smlstn-y,lrgrot-n,lrgstn-n
Spcl elmts: frac-n,stfn-n,trans.shr-n

Method: PAFEC
Reference: Composites Tech Revw (Griffen)
Validation: unk
Main appl: general structural analysis
Programming: unk
Maintenance: PAFEC

Source: PAFEC Ltd (England)
Availability: public-n,lease-y,buy-y
Machines: CDC,VAX,Prime
Features: pre-y,post-y,fail-n,optim-n
Ease of use: unk
Language: unk

Method: PAFEC (continued)
Analysis: static-y,vibration-u,buckling-u
Material model: iso-y,ortho-y,aniso-y
Dependency: lin-y,rubr-u,bimod-u,nonlin-u
Geometry: 1d-n,axi-y,pstr-y,pstn-y,3d-y
Contact elements: opn-u,fric-u,adhsn-u
Element library: fidif-n,4nq-n,8nq-y,12nq-y,8nb-y,20nb-y,32nb-y,shell-y,wedge-y
Loading types: pt-y,harmc-y,axi.press-y,pore press-n,body-y,preload-u,thermal-y
Moving boundary: n Auto re-mesh: n Step loading: u Char and decomp: n defects: n

Machine dependency: unk
Lamination: ud-y,ortho-y,ani-u,invo-u
Coupling: micro-y,fluid-y,thermal-u
Disp type: smlstn-y,lrgrot-u,lrgstn-u
Spcl elmts: frac-u,stfn-y,trans.shr-y

Method: PATCHES-III

Reference: users manual (5/77) tr 1045-00
Validation: 4 sample cases + rpl involute

Main appl: rocket nozzle analysis
Programming: unk,PATRAN compatible
Maintenance: PDA
Analysis: static-y,vibration-n,buckling-n
Material model: iso-y,ortho-y,aniso-y
Dependency: lin-y,rubr-n,bimod-y,nonlin-y
Geometry: 1d-n,axi-n,pstr-n,pstn-n,3d-y
Contact elements: opn-n,fric-n,adhsn-n
Element library: fidif-n,4nq-n,8nq-y,12nq-y,8nb-n,32nb-n,64nb-y,shell-n,wedge-n
Loading types: pt-y,harmc-n,axi.press-y,pore press-n,body-y,preload-n,thermal-y
Moving boundary: n Auto re-mesh: n Step loading: n Char and decomp: n Defects: n

Source: PDA Engineering,AFRPL,
    NADC,ONR (ltd vers)
Availability: public-y,lease-y,buy-y
Machines: VAX,Prime,CDC,IBM,
    Univac
Features: pre-y,post-y,fail-y,optim-n
Ease of use: hard (NASTRANlike)
Language: unk
Machine dependency: unk
Lamination: ud-y,ortho-y,ani-y,invo-y
Coupling: micro-y,fluid-n,thermal-y
Disp type: smlstn-y,lrgrot-n,lrgstn-n
Spcl elmts: frac-y,stfn-n,trans.shr-n

Method: PATRAN-G

Reference: user's guide, vol 1,2 (1980)
Validation: 5 example cases
Main appl: general stress analysis
Programming: coupled to many stress progs
Maintenance: PDA Engineering
Analysis: static-y,vibration-u,buckling-u
Material model: iso-y,ortho-y,aniso-n
Dependency: lin-y,rubr-y,bimod-n,nonlin-n
Geometry: 1d-y,axi-y,pstr-y,pstn-y,3d-y
Contact elements: opn-n,fric-n,adhsn-n
Element library: fidif-n,4nq-y,8nq-y,12nq-y,8nb-y,20nb-y,32nb-n,shell-y,wedge-y
Loading types: pt-y,harmc-n,axi.press-n,pore press-n,body-n,preload-n,thermal-y
Moving boundary: n Auto re-mesh: n Step loading: n Char and decomp: n Defects: n

Source: PDA Engineering (ANSYS
    analy)
Availability: public-n,lease-y,buy-y
Machines: VAX,CDC,Prime,IBM
Features: pre-y,post-y,fail-n,optim-n
Ease of use: easy (menu driven)
Language: unk
Machine dependency: unk
Lamination: ud-n,ortho-n,ani-u,invo-u
Coupling: micro-n,fluid-n,thermal-y
Disp type: smlstn-y,lrgrot-n,lrgstn-n
Spcl elmts: frac-n,stfn-n,trans.shr-n

Method: PLANS
Reference: Cosmic (1984)
Validation: unk
Main appl: nonlinear structural analysis
Programming: 5 separate programs
Maintenance: Grumman
Analysis: static-y,vibration-n,buckling-n
Material model: iso-y,ortho-y,aniso-n
Dependency: lin-y,rubr-n,bimod-n,nonlin-y
Geometry: 1d-y,axi-y,pstr-y,pstn-y,3d-y
Contact elements: opn-n,fric-n,adhsn-n
Element library: fidif-n,4nq-n,8nq-y,12nq-y,8nb-y,20nb-y,32nb-n,shell-y,wedge-n
Loading types: pt-y,harmc-n,axi.press-y,pore press-n,body-n,preload-n,thermal-y
Moving boundary: n Auto re-mesh: n Step loading: y Char and decomp: n Defects: n

Source: Cosmic, Grumman Aerospace
Availability: public-y,lease-n,buy-n
Machines: CDC,Cyber 175,IBM 360
Features: pre-y,post-n,fail-n,optim-n
Ease of use: unk
Language: FORTRAN IV
Machine dependency: unk
Lamination: ud-y,ortho-y,ani-n,invo-n
Coupling: micro-n,fluid-n,thermal-n
Disp type: smlstn-y,lrgrot-y,lrgstn-y
Spcl elmts: frac-n,stfn-y,trans.shr-y

Method: SAAS 3M
Reference: AIAA 75-769,Nelson,SMU PhD
Validation: Jortner ATJ-S data
Main appl: graphite nosetip analysis
Programming: works with JNM data and Pfail
Maintenance: none
Analysis: static-y,vibration-n,buckling-n
Material model: iso-y,ortho-y, aniso-n
Dependency: lin-y,rubr-n,bimod-y,nonlin-y
Geometry: 1d-n,axi-y,pstr-y,pstn-y,3d-n
Contact elements: opn-n,fric-n,adhsn-n
Element library: fidif-n,4nq-y,8nq-n,12nq-n,8nb-n,20nb-n,32nb-n,shell-n,wedge-n
Loading types: pt-n,harmc-n,axi.press-y,pore press-y,body-y,preload-n,thermal-y
Moving boundary: n Auto re-mesh: n Step loading: n Char and decomp: n Defects: n

Source: Bob Jones Univ, VPI
Availability: public-y,lease-n,buy-n
Machines: IBM,Univac
Features: pre-n,post-n,fail-n,optim-n
Ease of use: easy (SAASlike)
Language: FORTRAN IV
Machine dependency: dp reqd for IBM
Lamination: ud-n,ortho-n,ani-n,invo-n
Coupling: micro-n,fluid-n,thermal-n
Disp type: smlstn-y,lrgrot-n,lrgstn-n
Spcl elmts: frac-n,stfn-n,trans.shr-n

Method: SAAS III

Reference: Aerospace TR-0059-1 (6/22/71)
Validation: 8 smpl problems + NWC,
    ARC,AFML
Main appl: rocket nozzle analysis
Programming: subrtn calls from main
Maintenance: none
Analysis: static-y,vibration-n,buckling-n
Material model: iso-y,ortho-y,aniso-n

Source: Atlantic Research Corporation
    and others
Availability: public-y,lease-n,buy-n
Machines: IBM 360,CDC 6600,Univac
    1108
Features: pre-y,post-y,fail-n,optim-n
Ease of use: simple and easy
Language: FORTRAN IV
Machine dependency: IBM 360
Lamination: ud-n,ortho-n,ani-n,invo-n

(continued)

Table 3 (continued)

Method: SAAS III (continued)
Dependency: lin-y,rubr-n,bimod-y,nonlin-y
Geometry: 1d-n,axi-y,pstr-y,pstn-y,3d-n
Contact elements: opn-n,fric-n,adhsn-n
Element library: fidif-n,4nq-n,8nq-n,12nq-n,8nb-n,20nb-n,32nb-n,shell-n,wedge-n
Loading types: pt-n,harmc-n,axi.press-n,pore press-y,body-y,preload-n,thermal-y
Moving boundary: n Auto re-mesh: n Step loading: n Char and decomp: n Defects: n
Coupling: micro-n,fluid-n,thermal-n
Disp type: smlstn-y,lrgrot-n,lrgstn-n
Spcl elmts: frac-n,stfn-n,trans.shr-n

Method: SAP 4-5
Reference: Composites Tech Revw (Griffen)
Validation: unk
Main appl: general structural analysis
Programming: unk
Maintenance: Computer Software Inc
Analysis: static-y,vibration-y,buckling-n
Material model: iso-y,ortho-y,aniso-n
Dependency: lin-y,rubr-n,bimod-n,nonlin-n
Geometry: 1d-n,axi-n,pstr-n,pstn-n,3d-y
Contact elements: opn-n,fric-n,adhsn-n
Element library: fidif-n,4nq-n,8nq-n,12nq-n,8nb-y,20nb-n,32nb-n,shell-n,wedge-n
Loading types: pt-y,harmc-n,axi.press-n,pore press-u,body-y,preload-n,thermal-y
Moving boundary: n Auto re-mesh: n Step loading: n Char and decomp: n Defects: n
Source: Computer Software Inc
Availability: public-n,lease-y,buy-y
Machines: CDC,IBM,Univac,VAX
Features: pre-n,post-n,fail-n,optim-n
Ease of use: unk
Language: FORTRAN IV
Machine dependency: unk
Lamination: ud-n,ortho-n,ani-n,invo-n
Coupling: micro-n,fluid-n,thermal-n
Disp type: smlstn-y,lrgrot-n,lrgstn-n
Spcl elmts: frac-n,stfn-n,trans.shr-n

Method: SAP 6 (linear), 7 (nonlinear)
Reference: Composites Tech Revw (Griffen)
Validation: unk
Main appl: general composite analysis
Programming: NONSAP derivative
Maintenance: Computer Software Inc
Analysis: static-y,vibration-y,buckling-n
Material model: iso-y,ortho-y,aniso-u
Dependency: lin-y,rubr-n,bimod-n,nonlin-y
Geometry: 1d-n,axi-y,pstr-y,pstn-y,3d-y
Contact elements: opn-n,fric-n,adhsn-n
Element library: fidif-n,4nq-y,8nq-y,12nq-n,8nb-y,20nb-y,32nb-n,shell-y,wedge-n
Loading types: pt-y,harmc-n,axi.press-y,pore press-u,body-y,preload-n,thermal-y
Moving boundary: n Auto re-mesh: n Step loading: y Char and decomp: n Defects: n
Source: Computer Software Inc
Availability: public-n,lease-y,buy-y
Machines: CDC,IBM,Univac,VAX
Features: pre-y,post-y,fail-y,optim-n
Ease of use: unk
Language: FORTRAN 77
Machine dependency: unk
Lamination: ud-n,ortho-y,ani-u,invo-n
Coupling: micro-n,fluid-n,thermal-n
Disp type: smlstn-y,lrgrot-y,lrgstn-y
Spcl elmts: frac-n,stfn-n,trans.shr-y

Method: SPAR (Struc Perf Analy and Redsign)
Reference: NASA tech memo 78701/83181
Validation: unk
Main appl: aircraft and trusses
Programming: 25 analysis modules
Maintenance: Engineering Info Systems
Analysis: static-y,vibration-y,buckling-y
Material model: iso-y,ortho-y,aniso-y
Dependency: lin-y,rubr-n,bimod-n,nonlin-n
Geometry: 1d-y,axi-y,pstr-n,pstn-n,3d-y
Contact elements: opn-n,fric-n,adhsn-n
Element library: fidif-n,4nq-y,8nq-y,12nq-n,8nb-n,20nb-n,32nb-n,shell-y,wedge-n
Loading types: pt-y,harmc-n,axi.press-n,pore press-n,body-n,preload-n,thermal-y
Moving boundary: n Auto re-mesh: n Step loading: y Char and decomp: n Defects: n
Source: NASA Langley, Engn Info Sys
Availability: public-y,lease-y,buy-n
Machines: CDC,Cyber,Univac,Prime,VAX
Features: pre-n,post-y,fail-n,optim-n
Ease of use: hard (NASTRANlike)
Language: FORTRAN IV (94%)
Machine dependency: assembler (6%)
Lamination: ud-y,ortho-y,ani-y,invo-n
Coupling: micro-n,fluid-n,thermal-n
Disp type: smlstn-y,lrgrot-u,lrgstn-u
Spcl elmts: frac-n,stfn-y,trans.shr-n

Method: STAGSC-1
Reference: Cosmic (1984)
Validation: unk
Main appl: shell analysis
Programming: substructured and user def subs
Maintenance: LMSC
Analysis: static-y,vibration-y,buckling-y
Material model: iso-y,ortho-n,aniso-n
Dependency: lin-y,rubr-n,bimod-n,nonlin-y
Geometry: 1d-y,axi-y,pstr-n,pstn-n,3d-n
Contact elements: opn-n,fric-n,adhsn-n
Element library: fidif-n,4nq-y,8nq-n,12nq-n,8nb-n,20nb-n,32nb-n,shell-y,wedge-n
Loading types: pt-y,harmc-y,axi.press-y,pore press-n,body-y,preload-y,thermal-y
Moving boundary: n Auto re-mesh: n Step loading: y Char and decomp: n Defects: n
Source: Cosmic, LMSC
Availability: public-n,lease-y,buy-n
Machines: Univac,CDC,VAX
Features: pre-n,post-n,fail-n,optim-n
Ease of use: hard
Language: FORTRAN IV (97%)
Machine dependency: 1-3% assembler
Lamination: ud-y,ortho-y,ani-y,invo-n
Coupling: micro-n,fluid-n,thermal-n
Disp type: smlstn-y,lrgrot-y,lrgstn-y
Spcl elmts: frac-n,stfn-n,trans.shr-n

Method: STARDYNE
Reference: sales pamphlet
Validation: unk
Main appl: general structural analysis
Programming: PATRAP, Unistruc compatible
Maintenance: SDC
Analysis: static-y,vibration-y,buckling-y
Material model: iso-y,ortho-u,aniso-u
Dependency: lin-y,rubr-n,bimod-n,nonlin-n
Geometry: 1d-y,axi-y,pstr-n,pstn-n,3d-y
Contact elements: opn-y,fric-n,adhsn-n
Element library: fidif-n,4nq-y,8nq-y,12nq-u,8nb-y,20nb-u,32nb-u,shell-y,wedge-y
Source: System Development Corp
Availability: public-n,lease-y,buy-y
Machines: VAX,Cyber,Cray
Features: pre-y,post-y,fail-y,optim-n
Ease of use: unk
Language: FORTRAN
Machine dependency: unk
Lamination: ud-u,ortho-u,ani-u,invo-u
Coupling: micro-n,fluid-y,thermal-n
Disp type: smlstn-y,lrgrot-y,lrgstn-y
Spcl elmts: frac-n,stfn-y,trans.shr-n

Method: STARDYNE (continued)
Loading types: pt-y,harmc-y,axi.press-y,pore press-n,body-y,preload-y,thermal-y
Moving boundary: n Auto re-mesh: n Step loading: y Char and decomp: n Defects: n

Method: SUPERB
Reference: computer sys for analy (3/78)
Validation: unk
Main appl: general structural analysis
Programming: unk
Maintenance: SDRC
Analysis: static-y,vibration-y,buckling-n
Material model: iso-y,ortho-y,aniso-n
Dependency: lin-y,rubr-n,bimod-n,nonlin-n
Geometry: 1d-y,axi-y,pstr-y,pstn-y,3d-y
Contact elements: opn-n,fric-n,adhsn-n
Element library: fidif-n,4nq-y,8nq-y,12nq-n,8nb-y,20nb-y,32nb-n,shell-n,wedge-n
Loading types: pt-y,harmc-n,axi.press-y,pore press-n,body-y,preload-y,thermal-y
Moving boundary: n Auto re-mesh: n Step loading: y Char and decomp: n Defects: n
Source: Structural Dynamics Research
Availability: public-n,lease-y,buy-y
Machines: Cyber,IBM,Univac
Features: pre-y,post-y,fail-y,optim-n
Ease of use: unk
Language: unk
Machine dependency: unk
Lamination: ud-n,ortho-n,ani-n,invo-n
Coupling: micro-n,fluid-n,thermal-y
Disp type: smlstn-y,lrgrot-n,lrgstn-n
Spcl elmts: frac-n,stfn-n,trans.shr-n

Method: TASS
Reference: Thml-Mech Stres Anl (Brown 74)
Validation: Cook and Anderson test cases
Main appl: rocket nozzle analysis
Programming: unk
Maintenance: Thiokol
Analysis: static-y,vibration-n,buckling-n
Material model: iso-y,ortho-y,aniso-n
Dependency: lin-y,rubr-y,bimod-y,nonlin-y
Geometry: 1d-n,axi-y,pstr-y,pstn-y,3d-n
Contact elements: opn-n,fric-n,adhsn-n
Element library: fidif-y,4nq-y,8nq-y,12nq-y,8nb-n,20nb-n,32nb-n,shell-n,wedge-n
Loading types: pt-n,harmc-y,axi.press-y,pore press-n,body-n,preload-n,thermal-y
Moving boundary: n Auto re-mesh: n Step loading: n Char and decomp: n Defects: n
Source: Morton Thiokol/Wasatch Div
Availability: public-n,lease-n,buy-n
Machines: IBM
Features: pre-y,post-y,fail-y,optim-n
Ease of use: easy (free field)
Language: FORTRAN IV
Machine dependency: unk
Lamination: ud-n,ortho-n,ani-n,invo-n
Coupling: micro-n,fluid-n,thermal-n
Disp type: smlstn-y,lrgrot-y,lrgstn-y
Spcl elmts: frac-n,stfn-n,trans.shr-n

Method: TEXGAP84 2d-3d Texcap for contact
Reference: AFRPL TR 84-070/TR 78-86,87
Validation: 6 sample cases
Main appl: rocket case/grain analysis
Programming: substructuring
Maintenance: 84-Anatech International
Analysis: static-y,vibration-n,buckling-n
Material model: iso-y,ortho-y,aniso-y
Dependency: lin-y,rubr-y,bimod-n,nonlin-n
Geometry: 1d-n,axi-y,pstr-y,pstn-y,3d-y
Contact elements: opn-y,fric-y,adhsn-n
Element library: fidif-n,4nq-y,8nq-n,12nq-n,8nb-n,20nb-n,32nb-n,shell-n,wedge-y
Loading types: pt-y,harmc-n,axi.press-y,pore press-n,body-y,preload-y,thermal-y
Moving boundary: n Auto re-mesh: y Step loading: n Char and decomp: n Defects: n
Source: AFRPL/MKPB Edwards AFB
Availability: public-y,lease-n,buy-n
Machines: IBM,VAX,CDC,Prime,Cray,Univ.
Features: pre-y,post-y,fail-n,optim-n
Ease of use: moderate (free field)
Language: FORTRAN 77
Machine dependency: many compiler prb
Lamination: ud-n,ortho-n,ani-n,invo-n
Coupling: micro-n,fluid-n,thermal-y
Disp type: smlstn-y,lrgrot-n,lrgstn-n
Spcl elmts: frac-y,stfn-n,trans.shr-n

Method: TEXLESP (Texas Large Elas Stn Pg)
Reference: TEXLESP newsletter (1/85)
Validation: 2 sample problems
Main appl: rocket case/grain interface
Programming: unk
Maintenance: Mechanism Software, Inc.
Analysis: static-y, vibration-n,buckling-n
Material model: iso-y,ortho-y,aniso-y
Dependency: lin-y,rubr-y,bimod-n,nonlin-u
Geometry: 1d-n,axi-y,pstr-y,pstn-y,3d-y
Contact elements: opn-n,fric-n,adhsn-n
Element library: fidif-n,4nq-y,8nq-n,12nq-n,8nb-n,20nb-n,32nb-n,shell-n,wedge-n
Loading types: pt-n,harmc-n,axi.press-n,pore press-n,body-n,preload-n,thermal-y
Moving boundary: n Auto re-mesh: y Step loading: y Char and decomp: n defects: n
Source: AFRPL/MKPB Edwards AFB
Availability: public-y,lease-n,buy-n
Machines: IBM,VAX,CDC
Features: pre-n,post-n,fail-n,optim-n
Ease of use: medium (TEXGAPlike)
Language: FORTRAN V
Machine dependency: unk
Lamination: ud-n,ortho-n,ani-n,invo-n
Coupling: micro-n,fluid-n,thermal-n
Disp type: smlstn-y,lrgrot-y,lrgstn-y
Spcl elmts: frac-y,stfn-n,trans.shr-n

Method: TSAAS
Reference: 1a-5599-ms (5/74)
Validation: 2 sample problems
Main appl: nuclear
Programming: subrtn calls from main
Maintenance:
Analysis: static-y,vibration-n,buckling-n
Material model: iso-y,ortho-y,aniso-n
Dependency: lin-y,rubr-n,bimod-y,nonlin-y
Geometry: 1d-n,axi-y,pstr-y,pstn-y,3d-n
Contact elements: opn-n,fric-n,adhsn-n
Element library: fidif-n,4nq-y,8nq-n,12nq-n,8nb-n,20nb-n,32nb-n,shell-n,wedge-n
Loading types: pt-n,harmc-n,axi.press-y,pore press-y,body-y,preload-n,thermal-y
Moving boundary: n Auto re-mesh: n Step loading: n Char and decomp: y defects: n
Source: Los Alamos Laboratory
Availability: public-y,lease-n,buy-n
Machines: unk
Features: pre-y,post-y,fail-n,optim-n
Ease of use: simple and easy
Language: FORTRAN (version unk)
Machine dependency: unk
Lamination: ud-n,ortho-n,ani-n,invo-n
Coupling: micro-n,fluid-n,thermal-y
Disp type: smlstn-y,lrgrot-n,lrgstn-n
Spcl elmts: frac-n,stfn-n,trans.shr-n

Table 4 Laminate analysis programs

Program	Special feature	Source	Ref
CLASS	HP 41C calculator program; UNI subroutine calculates fiber-matrix bundle properties using the upper bound formulation of the composite cylinders assemblage model	Materials Sciences Corp.	(a)
Digital algorithm for laminate analysis	FORTRAN program for general laminate analyses	University of Dayton	13
Personal computer based laminate analysis programs	BASIC language, with versions for TI-59, Apple II, Timex 1000, TRS-80, and Sharp PC-1500	Air Force Wright Aeronautical Laboratories (AFWAL)	13
Laminate weight optimization	BASIC language; minimum ply thickness optimization using the Sharp PC-1500	AFWAL	14
Laminate ranking	Macintosh spread sheet based stacking-sequence ranking program	Think Composites	15
CLA2D	FORTRAN and BASIC languages using Sharp PC-1500; laminate analyses with mixed-constraint (line load and deformation) boundary conditions	Atlantic Research Corp.	16
CLAP	Microsoft BASIC Program for IBM PC, includes plotting of failure envelopes	Adtech Systems Research	17

(a) Refer to the next article in this Volume, "Software for Composite Materials Analysis."

for Solution of Non-linear Structural and Solid Mechanics Problems, *Computers and Structures*, Vol 13, 1981

8. *Cosmic Software Catalog*, University of Georgia, 1984
9. Thermal Structural Analysis Programs, in *Proceedings of the 27th Petroleum-Mechanical Engineering Conference*, American Society of Mechanical Engineers, 1972
10. W.L. Hufferd, "Composite Design and Analysis Code, 5th Monthly Contract Status Report," Document CSD

2845-CFR-5, Air Force Rocket Propulsion Laboratories, Jan 1985
11. R.J. Melosh and T.J. Dwyer, "Computer Systems for Analysis of Structures," MARC Analysis Research Corporation, 1978
12. *CAE Computer-Aided Engineering*, Vol 5 (No. 12), Dec 1986
13. S.R. Soni, "A Digital Algorithm for Composite Laminate Analyses—Fortran," AFWAL-TR-81-4073, Air Force Wright Aeronautical Laboratories, Oct 1983
14. G.V. Flanagan and W.J. Park, "Com-

posite Laminate Weight Optimization on the Sharp PC-1500 Pocket Computer," AFWAL-TR-83-4095, Air Force Wright Aeronautical Laboratories, Aug 1983
15. S.W. Tsai, *Composites Design 1986*, Think Composites, 1986
16. M. Buczek, "A Mixed Boundary Condition Laminate Analysis," unpublished report, Atlantic Research Corporation
17. *Composite Laminate Analysis Program*, Adtech Systems Research Systems Inc., June 1986

Software for Composite Materials Analysis

John J. Kibler, Materials Sciences Corporation

THE ANALYSIS OF LAMINATED COMPOSITE MATERIALS can be reduced to relatively simple, user-friendly, desk top computer codes. The ability to compute composite material properties and to estimate strength levels provides a very powerful tool for the designer who is ready to optimize the material at the same time that its structural dimensions are being formulated. The optimization process requires that the engineer allow the material properties to be a variable in the design procedure and determine the structural dimensions that best apply to each postulated material. Before the advent of single-user microcomputers, the optimization process could be long and tedious, with the result that few material combinations were considered. Currently available codes for microcomputers have brought the synthesis of material properties to the fingertips of the engineer. Only one such code and its capabilities is described in this article. Other codes do exist.

The code presented here also allows the user to conduct point stress analyses of given laminated materials, which then show both the critical layers and failure modes for a given loading system. The code predicts the elastic constants, thermal expansions, strengths, and internal stress states of a laminated composite plate. The user may start the analysis by inputting fiber and matrix properties, or by supplying layer properties. Layers can be allowed to fail in a matrix mode and loading can be continued such that the process of failing the laminate can be determined.

The CLASS code was originally written as a teaching aid to demonstrate the computation of both layer and laminate properties of composite materials. It was intended for engineers with very little or no computer experience, and is therefore entirely user interactive and menu driven. Its popularity in a classroom environment led to developing strength predictions and a stress analysis capability within the code. With the increased popularity of single-user microcomputers, the code is being used increasingly by practicing engineers who wish to be able to evaluate a number of materials at their desk.

A significant feature of the code is that the constituent fiber, matrix, and layer properties can be saved in a disk file for future use, and therefore only have to be input into the code once. Subsequent analyses can call upon the library of properties stored on disk to construct composite properties.

A summary of the features of the code follows. Details of the underlying theory can be found in Ref 1 to 4. The code computes layer properties by combining fiber and matrix properties and applying the composite cylinder assemblage model for determining the properties of representative elements of fiber and matrix. Laminate properties are computed using classical laminated plate theory, which can be found in detail in Ref 1 and 2. Strengths of the computed laminates are estimated using any one of several popular failure criteria.

CLASS Microcomputer Program Description

Before specific code features are described, several generic features should be discussed: The code checks input properties to the farthest extent possible and will not store properties that are not thermodynamically admissible. To be thermodynamically admissible, the properties must satisfy the expressions found in Ref 2. These expressions arise from certain physical requirements, such as nonnegative strain energy. In addition, because any stored properties may be used in the stress analysis section, the code requires that some strength value be input for every direction of every constituent. The user may supply any input strength except 0. The code has extensive error checking capabilities to prevent locking the system because of improper input. Therefore, the user can try any possibility during an analysis. The following points apply to all areas of the code and should be understood before conducting analyses.

Fiber and Matrix Constituents. All constituents are stored as transversely isotropic materials. One hundred each of fiber, matrix, and layer materials can be stored in the Fibers, Matrix, and Layers files, respectively. Several initial constituents are supplied, but the user must review the supplied properties to ensure that current properties are reflected.

The user can supply new constituents by selecting appropriate options in the code. When supplying new constituents, the user must provide values for the following properties: axial, transverse, and axial shear moduli; axial Poisson's ratio (axial load, transverse strain); transverse Poisson's ratio (transverse plane, ν_{TT}); axial and transverse coefficient of thermal expansion (CTE); axial tensile and compressive strengths; transverse tensile and compressive strengths; axial shear strength; and density.

For a typical isotropic matrix, the axial and transverse properties are identical, even though the code requires both as input. Axial refers to the direction in which the fibers lie, and transverse is at 90° to the fiber axis. The code can also handle anisotropic matrices.

Each constituent has an identifying number and a user-supplied title to identify the material being stored. When layer properties are calculated, a layer name is constructed from the fiber name, matrix name, and fiber volume fraction. The user can accept the supplied name, or edit the name to a more appropriate value.

Each input constituent must be defined by supplying all of the above thermoelastic and strength properties. A constituent can be made isotropic by simply supplying a 0 input value to the appropriate constants. The program determines whether the supplied constituent properties are thermodynamically admissible, and will not allow the user to store properties that are not thermodynamically admissible. (See Ref 2, p 41-44, for definitions of the functions that must be met for a transversely isotropic material to be thermodynamically admissible.)

Layer Analysis. The code permits storage of 100 layer materials within any one layer file. These layers can be input layers that the user supplies based on unidirectional-layer data, or they can be computed layer properties. The computed layer properties are obtained by combining a fiber and a matrix, using the composite cylinder assemblage model for representing the layer.

The same transversely isotropic properties are required for the layers as for the fiber and matrix. When layer properties are computed, the layer strengths are estimated by the program to supply initial strengths for the layer. The user must review these estimated strengths and compare them with what may be available for similar layer materials. The strength values are only provided as estimates.

There are currently no really good strength-calculating procedures that can compute all of the required strengths for the unidirectional material. This is simply because the transverse strength of a layer is very dependent upon the bond between the fiber and the matrix, the matrix ability to deform plastically, and residual stresses that may be locked into the matrix because of processing conditions. Fiber direction, tensile, and even compressive strengths can be estimated with reasonable accuracy. However, matrix-dominated strengths require some experimental definition in order to account for the process history, and so forth.

Computation of Layer Properties. At the user's option, layer properties can be constructed using combinations of the fiber and matrix stored in the constituent files. A fiber number, matrix number, and fiber volume fraction must be specified for the layer. The code combines the fiber and matrix using the composite cylinder assemblage model to predict all of the thermoelastic properties for the layer. Strengths are estimated, using the fiber and matrix strengths as guidelines. The user should review the layer properties before constructing a laminate from the layer.

After computing layer properties, the code allows a repeated layer property analysis, with the currently computed layer being saved in the Layers file. Thus, parametric analyses can be conducted by generating a number of layers at a given time.

Laminate analysis follows the general laminated plate theory found in Ref 1 and 2. The user can specify a different material, layer thickness, and layer orientation for each layer in the laminate, if desired. The code can model laminates containing 200 layers, and the laminate construction need not be balanced about 0°, nor need it be symmetric about the centerline of the laminate. That is, fully general laminate constructions can be handled, with full coupling of all loading and expansion terms. Examples of CLASS output are provided in an Appendix to this article.

Specification of Laminate. A shorthand notation has been programmed to allow specification of the laminate construction with text that closely follows industry standard specifications.

The user constructs a laminate by defining the layer material, layer thickness, and layer orientation. Because the code allows fully general laminates, each layer can be specified with these three parameters. In most cases, once the layer material and layer thickness are defined, only the layer orientation is changed. If the material type or layer thickness is not specified, then the code assumes that the

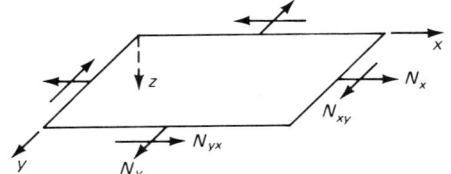

Fig. 1 In-plane forces on a flat laminate

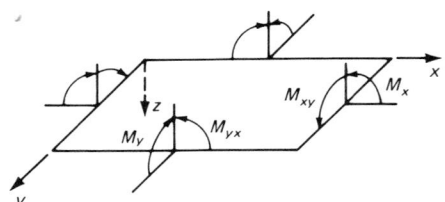

Fig. 2 Moments on a flat laminate

current layer is the same material and thickness as the previously defined layer. Thus, material type and layer thickness need only be specified initially and when changes occur within the laminate.

The maximum number of layers allowed is 200. If the construction is balanced and symmetric, then when loads are applied, the curvatures can be constrained, resulting in modeling one half of a laminate. In this way, a 400-layer laminate can be modeled.

One of the main strengths of the CLASS code is in the shorthand notation used to specify a laminate. The laminate is defined by a string of orientations separated by a slash. The material type for succeeding layers is placed ahead of a layer orientation and specified as /Mi/, and the thickness is specified as /Tj/. Thus, 0°/45°/−45°/90°/symmetric laminate of Material 3 from the Layers file, with layer thicknesses of 0.20 mm (0.007 in.) would be specified as M3/T0.007/0/45/−45/90/S. The /S signifies that the laminate is symmetric about the centerline.

Coordinates defining the laminate will not generally concern the user. When modeling a balanced and symmetric laminate, the top/bottom layers and direction of positive angle rotation from 0 for the laminate is of little or no importance. However, when analyzing an unbalanced or unsymmetric laminate, there are times when one must know which layer is "on top" and precisely what a positive angle for the laminate is.

Figures 1 to 4 describe the coordinates and angles used in the code. The figures are self-explanatory and should be consulted whenever complex laminates and loadings are being analyzed. Alternatively, several test loadings can be run to determine the directions that loads, especially bending moments, are being applied.

Thermoelastic Properties. The laminate properties are displayed after computations, and the initial display presents the "engineer-

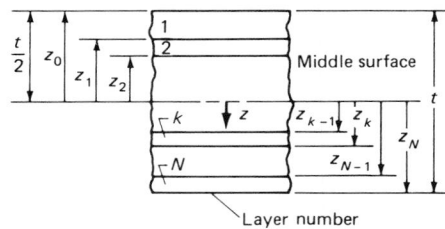

Fig. 3 Geometry of an *N*-layered laminate

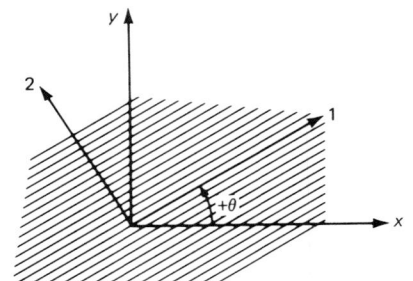

Fig. 4 Positive rotation of principal material axes from arbitrary *x*- and *y*-axes

ing" constants as defined by the centerline response of the laminate. These properties are presented regardless of whether or not the laminate possesses significant coupling terms in its stiffness matrix. Another screen presents the laminate A,B,D matrices. This screen should be consulted to determine whether there are coupling terms that exist for the laminate analyzed.

Also, the option exists for having the code write the A,B,D matrices to a disk file under a user-supplied name and with a user-defined title. This permits use of the code to develop input properties for a more extensive stress analysis code that would require the plate stiffness matrices as input.

Strength Computations. Once the laminate properties have been computed, the user can take that laminate into the failure and stress analysis section of the program. Initial strength calculations are based upon initial failure of a ply, using the ply strength properties stored in the Layers file.

Strength can be calculated using several failure criteria: maximum stress, maximum strain, plane stress interaction, and quadratic interaction.

The plane stress interaction criteria of Z. Hashin is an interaction that attempts to account for failure modes and interactions of the various stress components where applicable (see Ref 4). The Tsai-Wu failure criterion is a general stress interaction criterion that is used in industry (see Ref 5).

The merits of each criterion are the subject of many professional debates and will not be discussed here. The recommended approach has been to use the simplest criteria for parametric studies and for determining the strengths that may improve a laminate. The final analyses should include all available criteria so that the designer can be sure that the particular criteria are not being extremely unconservative for his particular loading conditions.

Strength is calculated one component at a time. The user is asked to define the loading condition to be applied and the temperature change to be applied as a uniform temperature change from the stress-free state on the laminate. A menu provides the standard loading conditions as options for computing strengths under single components of loading.

Generalized loading conditions can also be applied to the laminate to determine the laminate response under a given loading condition, as described in the section "Stress Analysis" below.

If a temperature is applied to the laminate, the code separates the thermal-induced stresses and strains from the load-induced values.

Stress Analysis. If the user applies general loading conditions to the laminate, then point stress analysis is performed. The user can define the temperature, as described above, along with the load and strain state to be applied to the laminate.

The code is capable of handling mixed loading conditions on a laminate. That is, for each loading direction, either the load or the strain in that direction must be specified. This permits application of a load in the *x*-direction while holding the strain in the *y*-direction to a fixed value. Care must be taken in interpreting the resulting laminate response.

As with the strength calculations, the thermal component of stress is held constant, and the code scales the load- and strain-induced components to the point where the first ply would fail under the given failure criteria.

In both this and the above cases, the code will show a screen of the laminate midplane response to the applied loading. At the user's option, the code will also show screens of layer stresses and strains in layer coordinates for each layer of the laminate.

The point stress analysis section of the code allows the user to develop a full understanding of the load-carrying capability of the laminate and to define the way the laminate will begin to fail under the given load state.

Layer Cracking. After the initial load is applied to a laminate and the first-ply failure loads are defined, the operator has an option to crack the most highly stressed plies (in matrix mode) and to reapply loads. This option allows analysis of the laminate after first-ply failure.

Several points should be made relative to this option. The code defines the ratio of strength in a given layer direction to the stress in that direction. The closer the ratio is to unity, the closer the layer is to failing due to that mode of failure. If the failure mode represents cracking of a layer in a matrix mode, then it may be possible that the laminate could carry additional load following the matrix cracking. Only the minimum stress ratio is listed in the output. The stress ratios relative to failing a layer in a fiber

failure mode and in a matrix failure mode are computed and stored within the program.

An option exists to allow the layers closest to failing in a matrix mode to crack, and to reanalyze the laminate to determine the next mode of failure. If the operator desires to let the layers fail in a matrix mode to determine if first-ply cracking is, or is not, the maximum load that the laminate can carry, then the following occurs.

The code examines the matrix modes of failure to define which ply will fail first in a matrix mode. Note that no regard is paid to fiber failure modes. The lowest stress ratio for matrix failure becomes the base ratio. The operator can then define how closely the other ratios should be examined to identify other plies to fail on that load increment. This permits the operator to allow cracking of all plies with close to the same stress ratio. Fiber direction properties are not affected.

Given a criterion for how many plies to crack, the code recomputes the laminate properties based on the cracked plies. The previous stress states are then applied to the laminate, and the load at which failure will occur is computed. If the new load is higher than in the initial case, the laminate will carry the load beyond first-ply failure. By following the strain response of the laminate, along with the loads for each successive ply failure, one can develop an estimate of the nonlinear stress-strain response of the laminate.

Cracking can be continued until all of the laminate plies are cracked in a matrix mode, at which time the analysis represents a netting analysis. The internal procedure for cracking plies is to divide the transverse modulus and shear modulus of the layer by a factor of 100 times the axial modulus. All other properties, including strength, are left the same as for the uncracked layer. Examining the layer stresses for plies that have been cracked will show that the code does not allow the layer to develop any transverse normal or in-plane shear stresses. Fiber direction stresses will develop.

Programmable Calculator Programs

The equations that need to be solved to determine the elastic constants and CTE of unidirectional layers and balanced and symmetric laminates are simple enough to be programmed on a hand-held calculator. Source codes are provided in Table 1.

Unidirectional Material Properties (UNI) is the name of a program that computes the elastic constants and CTE for a unidirectional fiber reinforced composite consisting of transversely isotropic constituents. The program calculations utilize the upper-bound formulations for the composite cylinder assemblage.

Program execution is initiated by keying in XEQ INPUT. This is the routine that reads in the fiber and matrix properties. Each required

input parameter is identified by a statement ending with a question mark. The keyboard is used to define the required data, and the R/S key is used to enter the data and continue execution of the program. The data below must be supplied.

- Sentinal for thermal expansion input:

  ALPH   Enter a 0 for no thermal expansion Enter a 1 to compute thermal expansion

- Matrix properties:

  EL   Young's modulus of elasticity in the axial direction

  ET   Young's modulus of elasticity in the transverse direction

  NU L   Poisson's ratio for stress in the axial direction

  NU T   Poisson's ratio in the transverse plane

  GL   Shear modulus in an axial plane

  AL L   CTE in the axial direction

  AL T   CTE in the transverse direction

- Fiber properties:

  EL   Young's modulus of elasticity in the axial direction

  ET   Young's modulus of elasticity in the transverse direction

  NU L   Poisson's ratio for stress in the axial direction

  NU T   Poisson's ratio in the transverse plane

  GL   Shear modulus in an axial plane

  AL L   CTE in the axial direction

  AL T   CTE in the transverse direction

Once the fiber and matrix properties are input, the program will automatically execute the UNI routine that requires the fiber volume fraction as input (**VF**).

The program output consists of the following fiber bundle properties:

E*L   Young's modulus of elasticity in the axial direction

E*T   Young's modulus of elasticity in the transverse direction

NU*L   Poisson's ratio for stress in the axial direction

NU*T   Poisson's ratio in the transverse plane

G*L   Shear modulus in an axial plane

AL*L   CTE in the axial direction

AL*T   CTE in the transverse direction

Successive cases may be run for various fiber volume fractions by keying in XEQ UNI. The program will use the fiber and matrix properties that were input previously.

A sample set of input properties and resulting unidirectional material properties are given in Table 2.

Laminate Material Properties (LAM) is the title of a program that computes the extensional and bending stiffnesses of laminates and is applicable only to balanced and symmetric laminates.

Table 1 UNI and LAM program source codes

UNI program listing

```
PRP "UNI"

01 ◆LBL ·UNI·
02 "FIB- VF=?"
03 PROMPT
04 STO 10
05 1
06 X⟨ ⟩Y
07 –
08 STO 00
09 RCL 02
10 2
11 /
12 RCL 04
13 1
14 +
15 /
16 STO 07
17 RCL 12
18 2
19 /
20 RCL 14
21 1
22 +
23 /
24 STO 17
25 1
26 RCL 04
27 –
28 RCL 01
29 *
30 2
31 *
32 RCL 03
33 X↑2
34 RCL 02
35 *
36 4
37 *
38 –
39 RCL 01
40 RCL 02
41 *
42 X⟨ ⟩Y
43 /
44 STO 06
45 1
46 RCL 14
47 –
48 RCL 11
49 *
50 2
51 *
52 RCL 13
53 X↑2
54 RCL 12
55 *
56 4
57 *
58 –
59 RCL 11
60 RCL 12
61 *
62 X⟨ ⟩Y
63 /
64 STO 16
65 RCL 06
66 –
67 1/X
68 RCL 06
69 RCL 07
70 +
71 1/X
72 RCL 00
73 *
74 +
75 RCL 10
76 X⟨ ⟩Y
77 /
78 RCL 06
79 +
80 STO 21
81 GTO 01

82 ◆LBL 02
83 RCL IND 20
84 RCL 00
85 *
86 10
87 ST+ 20
88 X⟨ ⟩Y
89 RCL IND 20
90 RCL 10
91 *
92 +
93 RTN

94 ◆LBL 01
95 RCL 00
96 RCL 16
97 /
98 RCL 10
99 RCL 06
100 /
101 +
102 RCL 07
103 1/X
104 +
105 RCL 00
106 X⟨ ⟩Y
107 /
108 RCL 10
109 *
110 STO 28
111 4
112 *
113 RCL 13
114 RCL 03
115 –
116 X↑2
117 *
118 STO 29
119 1
120 STO 20
121 XEQ 02
122 RCL 29
123 +
124 STO 22
125 RCL 13
126 RCL 03
127 –
128 RCL 06
129 1/X
130 RCL 16
131 1/X
132 –
133 *
134 ST* 28
135 3
136 STO 20
137 XEQ 02
138 RCL 28
139 +
140 STO 23
141 RCL 15
142 RCL 05
143 –
144 1/X
145 RCL 05
146 2
147 *
148 1/X
149 RCL 00
150 *
151 +
152 RCL 10
153 X⟨ ⟩Y
154 /
155 RCL 05
156 +
157 STO 24
158 RCL 06
159 ENTER↑
160 ENTER↑
161 RCL 07
162 2
163 *
164 +
165 /
166 STO 28
167 RCL 16
168 ENTER↑
169 ENTER↑
170 RCL 17
171 2
172 *
173 +
174 /
175 STO 29
176 RCL 17
177 RCL 07
178 /
179 STO 30
180 ENTER↑
181 ENTER↑
182 RCL 28
183 +
184 X⟨ ⟩Y
185 1
186 –
187 /
188 STO 31
189 RCL 30
190 RCL 29
191 *
192 ENTER↑
193 ENTER↑
194 RCL 28
195 X⟨ ⟩Y
196 –
197 X⟨ ⟩Y
198 1
199 +
200 /
201 RCL 10
202 3
203 Y↑X
204 *
205 1
206 +
207 RCL 00
208 X↑2
209 RCL 28
210 X↑2
211 *
212 3
213 *
214 X⟨ ⟩Y
215 /
216 1
217 +
218 RCL 10
219 *
220 RCL 31
221 X⟨ ⟩Y
222 –
223 RCL 28
224 1
225 +
226 RCL 10
227 *
228 X⟨ ⟩Y
229 /
230 1
231 +
232 RCL 07
233 *
234 STO 25
235 RCL 23
236 X↑2
237 RCL 21
238 *
239 4
240 *
241 RCL 22
242 /
243 1
244 +
245 RCL 25
246 *
247 RCL 21
248 +
249 RCL 21
250 RCL 25
251 *
252 4
253 *
254 X⟨ ⟩Y
255 /
256 STO 26
257 RCL 25
258 /
259 2
260 /
261 1
262 –
263 STO 27
264 FS? 02
265 GTO 04
266 GTO 06

267 ◆LBL 04
268 FIX 3
269 "COMPOSITE"
270 AVIEW
271 PSE
272 "E*L="
273 ARCL 22
274 PROMPT
275 "E*T="
276 ARCL 26
277 PROMPT
278 "NU*L="
279 ARCL 23
280 PROMPT
281 "NU*T="
282 ARCL 27
283 PROMPT
284 "G*L="
285 ARCL 24
286 PROMPT
287 FS? 02
288 GTO 03
289 "AL*L="
290 ARCL 28
291 PROMPT
292 "AL*T="
293 ARCL 29
294 PROMPT

295 ◆LBL 03
296 CLA
297 CLX
298 STOP

299 ◆LBL 06
300 RCL 22
301 1/X
302 RCL 01
303 1/X
304 –
305 STO 06
306 1
307 RCL 14
308 –
309 RCL 12
310 /
311 1
312 RCL 04
313 –
314 RCL 02
315 /
316 –
317 STO 07
318 RCL 23
319 RCL 22
320 /
321 RCL 03
322 RCL 01
323 /
324 –
325 STO 16
326 RCL 13
327 RCL 11
328 /
329 RCL 03
330 RCL 01
331 /
332 –
333 STO 17
334 RCL 11
335 1/X
336 RCL 01
337 1/X
338 –
339 STO 20
340 1
341 RCL 27
342 –
343 RCL 26
344 /
345 1
346 RCL 04
347 –
348 RCL 02
349 /
350 –
351 STO 30
352 RCL 20
353 RCL 07
354 *
355 RCL 17
356 X↑2
357 2
358 *
359 –
360 STO 31
361 RCL 06
362 RCL 07
363 *
364 RCL 16
365 RCL 17
366 *
367 2
368 *
369 –
370 RCL 18
371 RCL 08
372 –
373 *
374 RCL 06
375 RCL 17
376 *
377 RCL 16
378 CHS
379 RCL 20
380 *
381 +
382 RCL 19
383 RCL 09
384 –
385 *
386 2
387 *
388 +
389 RCL 31
390 /
391 RCL 08
392 +
393 STO 28
394 RCL 16
395 CHS
396 RCL 07
397 *
398 RCL 17
399 RCL 30
400 *
401 +
402 RCL 18
403 RCL 08
404 –
405 *
406 RCL 20
407 RCL 30
408 *
409 RCL 17
410 RCL 16
411 *
412 2
413 *
414 –
415 RCL 19
416 RCL 09
417 –
418 *
419 +
420 RCL 31
421 /
422 RCL 09
423 +
424 STO 29
425 GTO 04

426 ◆LBL "INPUT"
427 "UNI"
428 AVIEW
429 PSE
430 CF 02
431 1
432 "ALPHA 0 OR1?"
433 PROMPT
434 X=0?
435 SF 02
436 "MATRIX"
437 AVIEW
438 PSE
439 "MIX-"
440 ASTO 22
441 SF 01

442 ◆LBL 07
443 CLA
444 ARCL 22
445 "|- EL=?"
446 PROMPT
447 FS? 01
448 STO 01
449 STO 11
450 CLA
451 ARCL 22
452 "|- ET=?"
453 PROMPT
454 FS? 01
455 STO 02
456 STO 12
457 CLA
458 ARCL 22
459 "|- NU L=?"
460 PROMPT
461 FS? 01
462 STO 03
463 STO 13
464 CLA
465 ARCL 22
466 "|- NU T=?"
467 PROMPT
468 FS? 01
469 STO 04
470 STO 14
471 CLA
472 ARCL 22
473 "|- GL=?"
474 PROMPT
475 FS? 01
476 STO 05
477 STO 15
478 FS? 02
479 GTO 05
480 CLA
481 ARCL 22
482 "|- AL L=?"
483 PROMPT
484 FS? 01
485 STO 08
486 STO 18
487 CLA
488 ARCL 22
489 "|- AL T=?"
490 PROMPT
491 FS? 01
492 STO 09
493 STO 19

494 ◆LBL 05
495 FC? 01
496 GTO "UNI"
497 "FIBER"
498 AVIEW
499 PSE
500 "FIB-"
501 ASTO 22
502 CF 01
503 GTO 07
504 END
```

LAM program listing

```
PRP "LAM"

01 ◆LBL "LAM"
02 1
03 "NO. MTLS=? 1,2"
04 PROMPT
05 X⟨ ⟩Y?
06 SF 01
07 10.1
08 STO 00
09 CLA
10 "MTL. NO.=1"
11 AVIEW
12 PSE
13 GTO 02

14 ◆LBL 01
15 "MTL. NO.=2"
16 AVIEW
17 PSE

18 ◆LBL 02
19 CLA
20 "E1=?"
21 XEQ 03
22 "E2=?"
23 XEQ 03
24 "HU12=?"
25 XEQ 03
26 "G12=?"
27 XEQ 03
28 FS?C 01
29 GTO 01
30 GTO "STRT"

31 ◆LBL 03
32 PROMPT
33 STO IND 00
34 ISG 00
35 CLA
36 RTN

37 ◆LBL "STRT"
38 18.031
39 STO 00
40 CLX

41 ◆LBL 08
42 STO IND 00
43 ISG 00
44 GTO 08
45 ◆LBL 10
46 XEQ 11
47 STO 18
48 ST+ 19
49 RDN
50 STO 20
51 RDN
52 4
53 *
54 9
55 +
56 STO 09

57 ◆LBL 04
58 XEQ 05
59 STO 04
60 XEQ 05
61 STO 03
62 STO 00
63 XEQ 05
64 STO 02
65 ST* 00
66 XEQ 05
67 STO 01
68 ST/ 00
69 1
70 RCL 00
71 RCL 03
72 *
73 –
74 ST/ 01
75 ST/ 02
76 RCL 02
77 ST* 03
78 4
79 ST* 04
80 RCL 01
81 RCL 02
82 –
83 STO 05
84 2
85 ST/ 05
86 ST* 03
87 RCL 02
88 ST+ 01
89 9
90 STO 00
91 –1
92 STO 06
93 ST* 03
94 XEQ 06
95 ST* 04
96 XEQ 06
97 ST* 03
98 3
99 ST* 03
100 XEQ 06
101 ST* 04
102 3
103 ST/ 03
104 ST* 01
105 XEQ 06
106 RCL 20
107 2
108 *
109 COS
110 ST* 05
111 LASTX
112 2
113 *
114 COS
115 ST* 08
116 RCL 06
117 STO 01
118 STO 02
119 RCL 07
120 STO 03
121 RCL 09
122 STO 04
123 RCL 05
124 ST+ 01
125 ST– 02
126 RCL 08
127 ST+ 01
128 ST+ 02
129 ST– 03
130 ST– 04
131 4
132 STO 00
133 24
134 STO 09

135 ◆LBL 07
136 RCL 18
137 RCL IND 00
138 *
139 2
140 *
141 ST+ IND 09
142 DSE 09
143 DSE 00
144 GTO 07
145 RCL 19
146 RCL 18
147 2
148 /
149 –
150 X↑2
151 RCL 18
152 *
153 LASTX
154 3
155 Y↑X
156 12
157 /
158 +
159 2
160 *
161 STO 00
162 RCL 01
163 *
164 ST+ 25
165 RCL 00
166 RCL 02
167 *
168 ST+ 26
169 RCL 00
170 RCL 04
171 *
172 ST+ 27
173 GTO 10
174 RTN

175 ◆LBL 05
176 RCL IND 09
177 DSE 09
178 RTN

179 ◆LBL 06
180 RCL 01
181 RCL 03
182 +
183 RCL 04
184 +
185 8
186 /
187 STO IND 00
188 RCL 06
189 DSE 00
190 RTN
191 ◆LBL "ELA"
192 FIX 4
193 RCL 21
194 RCL 22
195 *
196 RCL 23
197 STO 03
198 X↑2
199 –
200 STO 01
201 STO 02
202 RCL 24
203 STO 04
204 RCL 19
205 2
206 *
207 ST/ 01
208 ST/ 02
209 ST/ 04
210 RCL 21
211 ST/ 02
212 RCL 22
213 ST/01
214 ST/ 03
215 "LAM. PROP."
216 PROMPT
217 1.1
218 STO 00
219 "EX ="
220 XEQ 09
221 "EY="
222 XEQ 09
223 "NUXY ="
224 XEQ 09
225 "GXY="
226 XEQ 09
227 25.1
228 STO 00
229 "D11="
230 XEQ 09
231 "D22="
232 XEQ 09
233 "D66="
234 XEQ 09
235 GTO 12

236 ◆ LBL 09
237 ARCL IND 00
238 PROMPT
239 CLA
240 ISG 00
241 RTN

242 ◆LBL 11
243 ISG 28
244 FIX 0
245 FIX 0
246 "LYR"
247 ARCL 28
248 "|- MTL.=?"
249 PROMPT
250 STO 29
251 "LYR"
252 ARCL 28
253 "|- THETA="
254 PROMPT
255 STO 30
256 "LYR"
257 ARCL 28
258 "|- THK.=?"
259 PROMPT
260 STO 31
261 RCL 29
262 RCL 30
263 RCL 31
264 RTN

265 ◆LBL 12
266 CLA
267 CLX
268 END
```

Note: These programs operate on Hewlett Packard 41C hand-held calculators.

Table 2 Sample UNI data

					Constituent properties						
	E_L		E_T				G_L		α_L	α_T	
	GPa	10^6 psi	GPa	10^6 psi	ν_L	ν_T	GPa	10^6 psi	10^{-6}/K	10^{-6}/K	
Fiber 467		67.7	9.10	1.32	0.41	0.45	17.6	2.55	0.263	12.6	
Matrix 10.1		1.46	10.1	1.46	0.13	0.13	4.45	0.646	4.12	4.12	

					Unidirectional material properties					
	E_L^*		E_T^*				G_L^*		α_L^*	α_T^*
V_f	GPa	10^6 psi	GPa	10^6 psi	ν_L^*	ν_T^*	GPa	10^6 psi	10^{-6}/K	10^{-6}/K
0.40	193	28.0	9.86	1.43	0.251	0.275	7.24	1.05	0.414	8.03
0.50	239	34.6	9.72	1.41	0.279	0.306	8.21	1.19	0.745	8.84
0.55	261	37.9	9.65	1.40	0.293	0.321	8.83	1.28	0.353	9.25
0.60	284	41.2	9.65	1.40	0.306	0.336	9.38	1.36	0.338	9.63
0.65	307	44.5	9.58	1.39	0.320	0.351	10.1	1.46	0.324	10.0

The program accepts layer material properties, orientations, and thicknesses. The output consists of the laminate elastic constants and bending stiffnesses. The in-plane properties computed are moduli, not plate stiffnesses. Plate stiffnesses can be obtained by multiplying the moduli by the plate thickness. The bending stiffnesses are the actual bending stiffnesses based upon the supplied total plate thickness.

Program execution is simplified through the use of descriptive information supplied on the calculator display. Once execution is initiated, that is, XEQ LAM, the program prompts the user for the following input:

● *Material Definition.* Number of materials to be used in the analysis. The program can use two material types. For each material, the user is then asked for elastic constants:

E1   Young's modulus of elasticity in fiber direction

E2   Young's modulus of elasticity in transverse direction

NU 12 Poisson's ratio for load in fiber direction and Poisson's strain in transverse direction

G12   In-plane shear modulus

● *Laminate Definition.* The following items are repeated until all layers of the laminate have been entered:

LYR J MATERIAL   Material (1,2) for current layer

LYR J THETA   Orientation for current layer

LYR J THICKNESS   Thickness of current layer

The laminate configuration in this program is limited to symmetric laminates. This was done to permit an arbitrary number of layers having any orientation and thickness. A description of the numbering system for the symmetric laminate follows.

The program assumes that Layer 1 lies directly above the centerline of the laminate and that it constructs a symmetric laminate by repeating the user input for the top half of the laminate as input for the bottom half. After the user has defined N layers, the laminate lay-up is:

Layer N
Layer $N-1$
.
.
.
Layer 3
Layer 2
Layer 1
----------centerline
Layer 1
Layer 2
Layer 3
.
.
.
Layer $N-1$
Layer N

Therefore, when N layers are entered, a symmetric laminate composed of $2N$ layers is analyzed.

The program will continue to ask for layer information until the procedure to compute

engineering constants is initiated, that is, XEQ ELA. This is entered when the layer material definition is requested and all of the layers in the laminate have been input. The program capabilities allow for unequal layer thicknesses as well as arbitrary layer orientations.

Program output can be obtained when the calculator displays LAM PROP. This is accomplished by pressing the R/S key to obtain each laminate engineering constant. Output consists of:

EX   Laminate Young's modulus of elasticity in the 0° direction.

EY   Laminate Young's modulus of elasticity in the 90° direction.

NUXY Laminate Poisson's ratio for load in the 0° direction and strain in the 90° direction.

GXY Laminate in-plane shear modulus.

D11   Laminate 0° direction bending stiffness.

D22   Laminate 90° direction bending stiffness.

D12   Laminate twisting stiffness.

Once output has been obtained, the user may analyze another laminate composed of the same materials that were input at the start of the program. This is accomplished by keying in XEQ STRT. It is not necessary to reinput material properties if they are to remain unchanged.

REFERENCES

1. Z. Hashin, and B.W. Rosen, with the collaboration of E.A. Humphreys, *Fiber Composites Analysis and Design, Vol 1, Composites Materials and Laminates*, CT-85/6, Department of Transportation/Federal Aviation Administration, 1983
2. R.M. Jones, *Mechanics of Composite Materials*, McGraw Hill, 1975
3. Z. Hashin, Analysis of Properties of Fiber Composites With Anisotropic Constituents, *J. Appl. Mech. (Trans. ASME)*, Vol 46, 1979, p 543
4. Z. Hashin, Failure Criteria for Unidirectional Fiber Composites, *J. Appl. Mech. (Trans. ASME)*, Vol 47, 1980, p 329-334
5. S.W. Tsai and E.M. Wu, A General Theory of Strength for Anisotropic Materials, *J. Compos. Mater.*, Vol 5, 1971, p 58

Appendix: Sample CLASS Output

```
MATERIALS SCIENCES CORPORATION

       J. J. KIBLER

COMPOSITE MATERIALS ANALYSIS SYSTEM

        options are -

  1 - CONSTITUENT/LAYER PROPERTIES
  2 - COMPUTE LAYER PROPERTIES
  3 - COMPUTE LAMINATE PROPERTIES
      includes elastic and
      strength calculations

      'ESC' to exit codes
```

The Primary Menu, above, allows paths to all main subroutines.

```
CONSTITUENT AND LAYER PROPERTIES

        options are:

1 = view/edit/input FIBER  properties
2 = view/edit/input MATRIX properties
3 = view/edit/input LAYER  properties

9 = initialize files (BE CAREFUL)
```

The Constituents section allows full mainte-nance of a constituent data base.

```
UNIDIRECTIONAL MATERIAL PROPERTIES

        options are:

1 = view INSTRUCTION Screen
2 = view LAYER names in layer file
3 = COMPUTE LAYER PROPERTIES

9 = initialize layer file (BE CAREFUL)
```

The above menu provides for computing layer properties from fiber and matrix constituents.

```
COMPOSITE LAMINATE ANALYSIS

       options are:

1 - Instruction screen
2 - View Names of LAYERS available

3 - COMPUTE LAMINATE PROPERTIES


    'ESC' for main menu

code written by: J. J. KIBLER
MATERIALS SCIENCES CORPORATION
```

The Laminate Menu leads to laminate property computations and the stress and failure analysis sections.

```
AVAILABLE FAILURE CRITERIA

1 - Maximum Stress
2 - Maximum Strain
3 - Plain Stress Interaction (Hashin)
4 - Quadratic Interaction (Tsai-Wu)

    'ESC' for Laminate menu

*** Default Criteria is ***

PLANE STRESS INTERACTION (Hashin)
```

The user can select any of four failure criteria for predicting laminate failure loads. The se-lected criteria may be changed for comparisons.

```
TOLERANCE TO LOWEST MATRIX STRESS RATIO

1 -  2 Percent of lowest Matrix Ratio
2 - 10      '       '       '       '
3 - 25      '       '       '       '
4 - all Plies ( Netting Analysis )

     'ESC' for LOADING  menu

*** Current Laminate Status ***

       0 Plies are Cracked.
```

If the user selects to allow plies to crack in a matrix failure mode, the code will then ask for a definition of how to select plies to be cracked.

```
        EXPLANATION OF LAMINATE DESCRIPTION

    This  code formulates a GENERAL LAMINATE using classical laminated plate
theory.   It is important to understand the  nomenclature  used in the code to
define the laminate being analyzed.    A shorthand description is used to
identify the laminate as follows:

    /T j/      following layers are thickness 'j' (default = .1)
    /xxx/      orientation of layer is 'xxx' degrees
       /S      at end of input means laminate is symmetric
       /C      at end means to continue on next line (max # = 7)
[../../..]N    repeats pattern in [..] N times
               maximum number of layers is 200

   Example:  M1/T.005/0/45/-45/90/S      ( 8 layers)
             M2.007/[0/60/-60]6/S       (36 layers)

The Following are equivalent ---
       a) M1/T.01/0/45/-45/0/45/-45/M2/0/60/-60/T.1/0/C
          0/T.01/-60/60/0/M1/-45/45/0/-45/45/0

       b) M1/T.01/[0/45/-45]2/M2/0/60/-60/T.1/0/S
```

Several HELP screens are available during operation. The definition of a laminate lay-up can handle very general laminate constructions. The screen below defines the parameters that can be used to define a laminate. Note that nested sets of square brackets are permissible.

```
                     LAMINATE THERMOELASTIC PROPERTIES
••••••••••••••••••••••••••••••••••••••••••••••••••••••••••••••••••••••••••••••••

MATERIAL IS : M1/T.007/0/45/-45/90/S

   LAMINATE LONGITUDINAL MODULUS ( 0)      (E xx) =   7.722D+06
   LAMINATE TRANSVERSE MODULUS   (90)      (E yy) =   7.722D+06
   LAMINATE SHEAR MODULUS  (in plane)      (G xy) =   2.934D+06
   LAMINATE MAJOR POISSON'S RATIO          (NU xy) =  0.3157

   LAMINATE LONGITUDINAL BENDING MODULUS   (D xx) =   1.888D+02
   LAMINATE TRANSVERSE BENDING MODULUS     (D yy) =   4.464D+01
   LAMINATE SHEAR BENDING MODULUS          (D xy) =   3.328D+01

   LAMINATE DENSITY                               =   5.680D-02

   LAMINATE LONGITUDINAL EXPANSION COEFFICIENT (α xx) =  1.030D-06
   LAMINATE TRANSVERSE EXPANSION COEFFICIENT   (α yy) =  1.030D-06
   LAMINATE SHEAR EXPANSION COEFFICIENT        (α xy) =  2.404D-19
   LAMINATE THERMAL CURVATURE COEFFICIENT      (k xx) = -1.668D-20
   LAMINATE THERMAL CURVATURE COEFFICIENT      (k yy) =  4.123D-19
   LAMINATE THERMAL CURVATURE COEFFICIENT      (k xy) = -1.414D-19
```

Once a laminate is specified, the code predicts the standard engineering thermoelastic constants. The resulting constants are summarized on a single screen.

```
              LAMINATE A, B, D STIFFNESS MATRICES
 ..............................................................

 MATERIAL IS : M1/T.007/0/45/-45/90/S
```

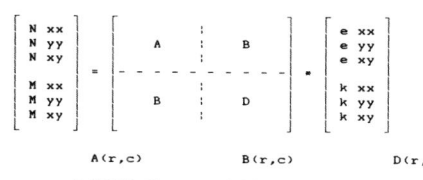

```
     ROW,COL          A(r,c)           B(r,c)            D(r,c)
      1-1          4.8027D+05        0.0000D+00        2.1036D+02
      1-2          1.5162D+05        2.1000D-11        3.3297D+01
      2-2          4.8027D+05       -4.0000D-11        5.3305D+01
      1-3          0.0000D+00        0.0000D+00        1.3088D+01
      2-3         -1.7000D-08        0.0000D+00        1.3088D+01
      3-3          1.6433D+05        3.0000D-11        3.6619D+01
```

The full A,B,D matrices are available for each laminate analyzed. If unbalanced or nonsymmetric laminates are modeled, then these matrices, rather than the engineering constants, should be utilized for determining laminate response.

```
           LAYER STRESSES and STRAINS  - Initial Loading.
 ..............................................................

 MATERIAL IS : M1/T.007/0/45/-45/90/S

 Results for analysis during  : GENERALIZED Loading Conditions
 Applied Temperature is        : -125
```

LYR	LOAD INDUCED STRESSES/STRAINS S11 e11	S22 e22	S12 e12	TEMPERATURE INDUCED STRESSES/STRAINS S11 e11	S22 e22	S12 e12
1	5.237D+04	1.204D+03	3.181D-12	-2.768D+03	2.768D+03	-1.919D-11
	2.562D-03	1.288D-04	5.054D-18	-1.772D-04	2.116D-03	-3.049D-17
2	2.802D+04	2.346D+03	-1.532D+03	-2.768D+03	2.768D+03	6.295D-12
	1.345D-03	1.345D-03	-2.433D-03	-1.772D-04	2.116D-03	1.000D-17
3	2.802D+04	2.346D+03	1.532D+03	-2.768D+03	2.768D+03	-6.295D-12
	1.345D-03	1.345D-03	2.433D-03	-1.772D-04	2.116D-03	-1.000D-17
4	3.661D+03	3.489D+03	1.848D-10	-2.768D+03	2.768D+03	1.896D-11
	1.288D-04	2.562D-03	2.936D-16	-1.772D-04	2.116D-03	3.012D-17
5	3.661D+03	3.489D+03	1.833D-10	-2.768D+03	2.768D+03	1.888D-11
	1.288D-04	2.562D-03	2.913D-16	-1.772D-04	2.116D-03	2.999D-17

At the user's option, layer stresses and strains in layer coordinates will be displayed. Again, the load- and temperature-induced responses are separated, and the response to the defined loading is the sum of the two.

```
       LAMINATE MID PLANE RESPONSE  - Initial Loading.
 ..............................................................

 MATERIAL IS : M1/T.007/0/45/-45/90/S
 Results for analysis during  : GENERALIZED Loading Conditions
 Applied Temperature is        : -125
```

```
              LAMINATE MID PLANE STRAINS AND CURVATURES

       APPLIED              LOAD INDUCED            TEMPERATURE INDUCED
       LOADING                response                 response

 N xx =  1.2500D+03    e xx =   2.5621D-03     e xx =  -1.2876D-04
 N yy =  4.5029D+02    e yy =   1.2876D-04     e yy =  -1.2876D-04
 N xy =  0.0000D+00    e xy =   1.3320D-17     e xy =  -3.0055D-17
 M xx =  0.0000D+00    k xx =   1.3731D-16     k xx =   2.0852D-18
 M yy =  0.0000D+00    k yy =  -1.0814D-15     k yy =  -5.1537D-17
 M xy =  0.0000D+00    k xy =   3.3742D-16     k xy =   1.7675D-17
```

When determining the laminate response to applied loading, the midplane extension and bending levels are determined and displayed. Applied load response is separated from the temperature response, and the net laminate response is the sum of the two.

```
           FAILURE RESULTS  - Initial Loading.
 ..............................................................

 MATERIAL IS : M1/T.007/0/45/-45/90/S
 Results for analysis during  : GENERALIZED Loading Conditions
 Failure loads predicted using : PLANE STRESS INTERACTION (Hashin) Criterion
 Applied Temperature is        : -125

 NOTE: Failure LOADS are predicted (not STRESSES).  Failure stresses are
       equal to failure loads only for a unit thickness laminate.
 ..............................................................
```

```
       CRITICAL LOADING

 N xx =  2.448D+03        LAYER  5 FAILS FIRST due to -
 N yy =  8.818D+02
 N xy =  0.0000D+00       MATRIX FAILURE (combined stresses)
 M xx =  0.0000D+00
 M yy =  0.0000D+00
 M xy =  0.0000D+00


                         `R' to See Failure Ratios
                         `C' to Crack Plies & Continue
                         `L' to Apply NEW loads
                         `ESC' to Exit to Laminate Menu
```

The applied mechanical load response is scaled by the program to determine when failure will occur. Temperature-induced stresses and strains are held constant; only the applied load response is scaled. The code also defines the first layer that will fail, and what the failure mode of that layer will be.

Testing

Chairman: Gordon Bourland, LTV Aerospace and Defense Company
Co-Chairman: Linda L. Clements, San Jose State University

Introduction

ADVANCED COMPOSITE materials with anisotropic properties created a need for new test specimens and test techniques. New tests were required to evaluate reinforcing fibers, characterize matrix materials, and determine lamina and laminate mechanical properties. This Section describes the tests that have been developed during the past three decades to define the chemical, physical, and mechanical properties of these engineered structural materials. It begins with coverage of tests on the constituent fiber and matrix materials and ends with techniques for full-scale testing.

Graphite and boron fibers with very high strengths and stiffnesses necessitated new tests for fiber strand and lamina properties. Organic matrix materials required new techniques for chemical fingerprinting and characterization. An alphabet soup of acronyms for analytical chemical techniques began to appear in aerospace industry specifications, such as

HPLC (high-performance liquid chromatography), DSC (differential scanning calorimetry), DMA (dynamic mechanical analysis), and IR (infrared spectroscopy). New types of test specimens were required to determine tension and compression properties of $0°$ lamina materials. Anisotropic laminate design required new methods for evaluating element specimens, subcomponents, and full-scale components for the aerospace and automotive industries.

Testing of composite materials is still a changing technology, particularly in determining compression and shear properties of $0°$ lamina with fibers having tensile strengths above 4.2 GPa (0.6 psi $\times$ 10^6) and in analytical chemical characterization of matrix materials, such as epoxies, bismaleimides, and polyimides. New complexities in matrix material characterization are to be expected with the introduction of high temperature resistant thermoplastic matrix materials with crystalline structures.

Tests for Reinforcement Fibers

Gary E. Hansen, Hercules Aerospace Company

REINFORCING FIBERS can be tested in the form of single filaments, tows, fabric, or unidirectional tape. While all reinforcements are made up of bundles of filaments, it is possible to isolate single filaments to determine fiber properties. Academically this is interesting, but such testing has not been representative of composite performance. Testing the single tow or bundle allows an evaluation of the reinforcing fiber in its simplest form and is valuable in across-the-band variability studies. Because many fibers are woven into fabrics for use in composites, physical tests may be used to characterize the fabrics, but it is recommended that mechanical tests also be performed on laminates of impregnated material. Unidirectional tapes are a convenient form for testing mechanical properties of fibers. Columnated tows of fiber are impregnated with an appropriate resin, and the resultant tape is cured with heat and pressure to form a laminate, which is subsequently machined into test specimens.

Chemical Tests

Surface characteristics and polymer analysis can be accomplished with x-ray photoelectron spectroscopy (XPS), generally regarded as an important and key technique. Also known as electron spectroscopy for chemical analysis (ESCA), this technique provides a total elemental analysis, with the exception of hydrogen and helium, of the top 10-200 Å (depending on the sample and instrumental conditions) of any solid surface that is vacuum stable or can be made vacuum stable by cooling. It also provides chemical bonding information. Of all the presently available instrumental techniques for surface analysis, XPS is generally regarded as being the most quantitative, the most readily interpretable, and the most informative with regard to chemical information. A relatively simple, straightforward, and quantitative technique, its advantages include nondestructive and surface sensitive (10-200 Å) characteristics and elemental sensitivity (parts per 1000) for all elements except hydrogen and helium. Although the method requires relatively sophisticated, expensive instrumentation, many universities, industrial research and development groups, and commercial service laboratories provide access to their instruments on a collaborative or fee-for-service basis.

The information content in a typical XPS spectrum is enormous. There are various hierarchies of spectral interpretation, including simple elemental analysis, detailed considerations of chemical shifts and chemical bonding nature in the surface region, and loss or relaxation structures that provide further information on the chemical nature of the surface. Many potential artifacts, often related to the preparation of samples, are also made available in an XPS experiment. The interpretation becomes progressively more difficult as one considers more complex surfaces, such as those of multicomponent and multiphase materials. Some disadvantages of XPS include the need for a large analysis area of several mm^2 (although an instrument being developed will have an analysis area of about a 150 μm, or 5900 μin., diameter), a high vacuum requirement (10^{-6} to 10^{-9} Pa, or 10^{-8} to 10^{-11} torr), lengthy test time ($\frac{1}{2}$ to 8 h per sample), and low resolution ($\sim$0.1 to 1.0 eV). Charging and energy referencing can also be a problem. It may be necessary to consult the literature and examine a large number of spectra to obtain practical experience with spectral interpretation and analysis for a wide range of sample types (see the Selected References for a summary of texts, monographs, and review articles generally regarded as basic to XPS).

Other instrumental surface analysis techniques that are available and widely used include Auger electron spectroscopy (AES) and secondary ion mass spectroscopy (SIMS). The Auger technique is actually much older in practical application than XPS. Because AES uses an electron beam, it is generally highly damaging to organic polymer surfaces, but, as a form of electron spectroscopy, it is complementary to XPS. Secondary ion mass spectroscopy is now beginning to be applied to polymer surfaces, but is not yet used routinely nor is it as easily interpreted as XPS. Secondary ion mass spectroscopy is undergoing extensive development and may prove to be useful for routine polymer surface analysis. Table 1 compares the features of numerous techniques.

The carbon assay is done according to ASTM C 571 (Ref 1) or ASTM C 831 (Ref 2) and basically involves combustion of the product to form carbon dioxide and an absorption train to collect and compute the original carbon content.

Sizing Content. High-strength fibers may have sizing applied to their surfaces to improve bonding and/or handling characteristics. Determining sizing content is very direct: The fibers are weighed, the sizing removed, and the fibers are weighed again. Typical methods include extraction with an appropriate solvent to achieve total solvation of the size, but no attack on the fiber, and pyrolysis at a temperature that will burn off the sizing but not affect the fiber. It should be noted that both methods are idealistic from a practical viewpoint. Extraction frequently does not solvate all of the sizing, while pyrolysis frequently results in fiber weight loss.

Physical Tests

Seven physical characteristics and the tests that measure them are described below.

Density is best measured by means of ASTM D 792 (Ref 3) using displacement, ASTM D 3800 (Ref 4) using Archimedes prin-

Table 1 Comparison of instrumental surface analysis techniques

Technique	Quantitative analysis	Structure	Electronic levels	Vibrational levels	Depth profiling	Spatial resolution	Adsorption energy	In situ fingerprint
Auger electron spectroscopy	+	+	+		+	+		
Low-energy electron diffraction		+						
X-ray photoelectron spectroscopy	+	+	+					
Electron energy loss spectroscopy		+		+	+			
Secondary ion mass spectroscopy	+				+	+		
Scanning electron microscopy						+		
Thermal desorption mass spectroscopy	+						+	
Fourier transform infrared spectroscopy		+		+				+

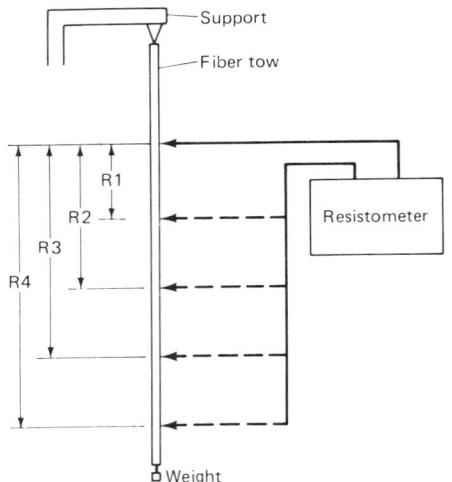

Fig. 1 Typical test setup for measuring electrical conductivity

ciple, or ASTM D 1505 (Ref 5) using a density gradient column. These are all well-established methods that can be used with confidence.

Weight per length (yield). A simple test is accomplished by weighing a known length of fiber. It is typically reported in g/μm (lb/μin.).

Filament diameter can be measured in two ways, direct and calculated, both of which are limited to round cross sections. The direct method is to view the fiber under a microscope or by microprojection per ASTM D 578 (Ref 7). Alternatively, the average filament diameter can be calculated from the following equation:

$$D = \frac{4WPL^{1/2}}{\pi \cdot \rho_f \cdot k}$$

where D is the filament diameter, WPL is the weight per length, ρ_f is fiber density, and k is the number of filaments/tow. Other methods of measurement include cross-sectional microscopy and image analysis with a vibroscope. Irregular cross sections are microphotographed and measured planimetrically.

Electrical conductivity can be measured by means of a simple resistance probe. A test setup similar to Fig. 1 may be used. The weight is hung to preclude wrinkling of the fiber. The resistance is measured at 25-, 50-, 75-, and 100-mm (1-, 2-, 3-, and 4-in.) intervals. Resistance versus distance plotted, and slope of line indicates the resistivity of the fiber. Because the resistivity is also a function of the cross-sectional area, the necessary filament data, that is, the number of filaments per tow and the filament diameter, should also be reported.

Thermal Expansion. Many thermal expansion procedures exist, but none give consistent, reproducible data when measuring fibers.

The number of twists per unit length of a carbon fiber tow is determined by a twist test. The sequence is:

- Any frayed surface fiber is removed from the carbon fiber to be tested
- Spool to be sampled is placed on spool holder
- Fiber is unspooled, while being kept from twisting, and locked in the fixed clamp at the end of the cutting board
- The free clamp is attached at the 1220-mm (48-in.) cutting edge, but sample is not cut from spool
- A fine, pointed, polished stylus is inserted into the center of the sample at the opposite 1220-mm (48-in.) cutting edge, where the free clamp is located
- The stylus is drawn down the sample, splitting the tow to the free clamp (making sure the movable clamp does not rotate)
- The stylus is backed off approximately 25 mm (1 in.) from the free clamp and the number of fiber rotations to the free clamp is observed. The twist per mm (in.) (tpmm, or tpi) equals the number of rotations of fiber/1220 mm (48 in.)
- Results are reported to two significant digits. Example: 1.5 rotations/1220 mm (48 in.) = 0.0012 tpmm (0.031 tpi)

Fabric Weave. When fibers are woven into fabrics, the physical properties listed below are measured.

Pick count, which is defined as the number of tows/mm (in.) in both the warp and fill directions. This is done by simply measuring the dimensions of an undistorted fabric and removing and counting each thread. Data should be reported as the number of warp threads/mm (in.) and number of fill threads/mm (in.).

Fabric areal weight is measured by weighing a piece of undistorted fabric and measuring its surface area. The data is reported as g/m² (oz/yd²).

Tensile strength of dry fabric may be measured using ASTM D 579 (Ref 7); however, it would be better to measure the tensile strength of an impregnated laminate.

Mechanical Tests

It is important to note that mechanical properties of fibers are very test dependent. For example, Table 2 shows the difference in tensile strength of a typical fiber tested as a filament, tow, and laminate, all reduced to 100% fiber volume. This data emphasizes the importance of identifying the test method when reporting test data.

Single-filament tensile strength can be determined using ASTM D 3379 (Ref 8), which can be summarized as a random selection of single filaments made from the material to be tested. Filaments are centerline-mounted on special slotted tabs. The tabs are gripped so that the test specimen is aligned axially in the jaws of a constant-speed movable-crosshead test machine. The filaments are then stressed to failure at a constant strain rate. For this test method, filament cross-sectional areas are determined by planimeter measurements of a representative number of filament cross sections as displayed on highly magnified photomicrographs. Alternative methods of area determination use optical gages, an image-splitting microscope, a linear weight-density method, and others.

Tensile strength and Young's modulus of elasticity are calculated from the load elongation records and the cross-sectional area measurements.

The specimen is shown in Fig. 2. Note that a system compliance adjustment may be necessary for single-filament tensile modulus.

Tow tensile testing. Using ASTM D 4018 (Ref 9) or an equivalent is recommended. This is summarized as finding the tensile properties of continuous filament carbon and graphite yarns, strands, rovings, and tows by the tensile loading to failure of the resin-impregnated fiber forms. This technique loses accuracy as the filament count increases. Strain and Young's modulus are measured by extensometer.

The purpose of using impregnating resin is to provide the fiber forms, when cured, with enough mechanical strength to produce a rigid test specimen capable of sustaining uniform loading of the individual filaments in the specimen.

To minimize the effect of the impregnating resin on the tensile properties of the fiber forms, the resin should be compatible with the fiber, the resin content in the cured specimen should be limited to the minimum amount required to produce a useful test specimen, the individual filaments of the fiber forms should be well collimated, and the strain capability of the resin should be significantly greater than the strain capability of the filaments.

ASTM D 4018 Method I test specimens require a special cast-resin end tab and grip design to prevent grip slippage under high loads. Alternative methods of specimen mounting to end tabs are acceptable, provided that test specimens maintain axial alignment on the test machine centerline and that they do not slip in

Table 2   Effect of test method

	Nominal tensile strength					
	AS4		IM6		IM7	
Test	MPa	ksi	MPa	ksi	MPa	ksi
Filament	4100	595	4950	715	5400	780
Tow	4000	580	5050	730	5450	790
Laminate	3850	555	4300	625	4600	665

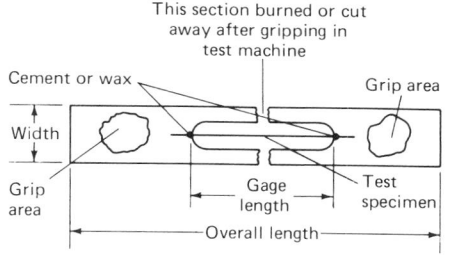

Fig. 2 Schematic showing typical specimen-mounting method

Table 3 Effect of resin on laminate properties

| Material | | Tensile strength | | Tensile modulus | | Short-beam shear | |
Fiber	Resin	MPa	ksi	GPa	10^6 psi	MPa	ksi
AS4	2220-3	3650	527	221	32.1	106	15.4
AS4	3501-6	3450	500	225	32.7	124	18.0
AS4	3502	3000	435	223	32.4	134	19.4
AS4	4502	3000	432	220	31.9	102	14.8

Note: Tensile data normalized to 100% fiber volume

the grips at high loads. ASTM D 4018 Method II test specimens require no special gripping mechanisms. Standard rubber-faced jaws should be adequate.

Laminate properties. The most generally representative procedure for measuring composite properties is to combine the fiber and resin and test the composite as a cured laminate because this is the form in which the materials are used. It is important to understand that laminate properties are a function of both fiber and resin properties. Table 3 shows the dependence of mechanical properties on the resins used. Another factor to consider is the fiber volume fraction of the laminate. A fiber volume of 55 to 65% has been found to allow consistent measurement of properties. Because tensile properties are fiber dominated, it is recommended that tensile strength and modulus be normalized to a constant fiber volume. A normalized fiber volume of 100% is common. This is done simply by using the following equation:

$$\text{Property (100\%)} = \frac{\text{property} \times 100}{\text{fiber volume (\%)}}$$

Laminate tensile testing should be conducted per ASTM D 3039 (Ref 10). Major concerns when conducting this test are:

- Verification of collimation of fibers. The specimens should be cut $0° \pm \frac{1}{4}°$ from the fiber direction
- Machining quality. Specimens should be cut with a diamond blade circular saw, 180-grit minimum, using a water coolant. Specimens should be inspected to verify that no specimen damage was caused by machining
- Surface quality. The panel should have a resin-rich surface to preclude fiber splitting. This can best be done by using an appropriate peel ply. Burlington peel ply 51789, style 52006, has been found to yield excellent results

The short-beam shear test is of little value to a designer because short-beam shear strength is not necessarily representative for a pure shear test. However, this test is an inexpensive, easily done measurement of the quality of the laminate and fiber interface. It should not be used to compare fiber-resin systems. For example, a true shear test run on the systems shown in Table 3 would show 2220-3 resin-AS4 fiber to have higher shear strength than 3501-6 or 3502 resin. The test, however, is sensitive to changes within a product; because it has been shown to be a very valuable quality control test to identify changes in fiber-resin interfacial quality, it should be used to monitor product changes.

The short-beam shear strength should be determined per ASTM D 2344 (Ref 11). This is a simple three-point bending test with a span-to-depth ratio of 4:1 and a nominal thickness of 2.03 mm (0.08 in.).

REFERENCES

1. "Standard Methods for Chemical Analysis of Carbon and Carbon-Ceramic Refractories," C 571, *Annual Book of ASTM Standards*, American Society for Testing and Materials
2. "Standard Test Methods for Residual Carbon, Apparent-Residual Carbon, and Apparent Carbon Yield in Coked Pitch-Containing Brick and Shapes," C 831, *Annual Book of ASTM Standards*, American Society for Testing and Materials
3. "Standard Test Methods for Specific Gravity and Density of Plastics by Displacement," D 792, *Annual Book of ASTM Standards*, American Society for Testing and Materials
4. "Standard Test Method for Density of High-Modulus Fibers," D 3800, *Annual Book of ASTM Standards*, American Society for Testing and Materials
5. "Standard Test Method for Density of Plastics by the Density-Gradient Technique," D 1505, *Annual Book of ASTM Standards*, American Society for Testing and Materials
6. "Standard Specification for Glass Fiber Yarns," D 578, *Annual Book of ASTM Standards*, American Society for Testing and Materials
7. "Standard Specification for Greige Woven Glass Fabrics," D 579, *Annual Book of ASTM Standards*, American Society for Testing and Materials
8. "Standard Test Method for Tensile Strength and Young's Modulus for High-Modulus Single-Filament Materials,"
   D 3379, *Annual Book of ASTM Standards*, American Society for Testing and Materials
9. "Standard Test Methods for Tensile Properties of Continuous Filament Carbon and Graphite Yarns, Strands, Rovings, and Tows," D 4018, *Annual Book of ASTM Standards*, American Society for Testing and Materials
10. "Standard Test Method for Tensile Properties of Fiber-Resin Composites," D 3039, *Annual Book of ASTM Standards*, American Society for Testing and Materials
11. "Standard Test Method for Apparent Interlaminar Shear Strength of Parallel Fiber Composites by Short-Beam Method," D 2344, *Annual Book of ASTM Standards*, American Society for Testing and Materials

SELECTED REFERENCES

- D. Briggs, Ed., *Handbook of X-Ray and Ultraviolet Photoelectron Spectroscopy*, Heyden and Son, 1977
- C.R. Brundle and A.D. Baker, Ed., *Electron Spectroscopy*, Vol 1-5, Academic Press, 1977-1983
- T.A. Carlson, *Photoelectron and Auger Spectroscopy*, Plenum Press, 1975
- D.T. Clark, ESCA Applied to Polymers, *Adv. Polym. Sci.*, Vol 24, 1977, p 125-188
- D.T. Clark and W.J. Feast, Application of ESCA to Studies of Structure and Bonding and Polymeric Systems, *J. Macromol. Sci. Rev., Macromol. Chem.*, Vol C12, 1975, p 191-286
- C.S. Fadley, Basic Concepts of XPS, in *Electron Spectroscopy*, Vol II, C.R. Brundle and A.D. Baker, Ed., Pergamon Press, 1978, p 1-156
- P.K. Ghosh, *Introduction to Photoelectron Spectroscopy*, Wiley-Interscience, 1983
- M.M. Millard, Fibers and Polymers, in *Industrial Applications of Surface Analyses*, American Chemical Society Symposium Series, L.A. Casper and C.J. Powell, Ed., Vol 199, American Chemical Society, 1982, p 143-202
- B.D. Ratner, Application of XPS to Biomedical Polymers: A Review, *Annual of Biomedical Engineering*, 1983
- K. Siegbahn and L. Karlsson, *Electron Spectroscopy for Atoms, Molecules, and*

Condensed Matter, North-Holland, 1984
- K. Siegbahn, C. Nordling, A. Fahlman, R. Nordberg, K. Hamrin, J. Hedman, G. Johansson, T. Bergmark, S.-E. Karlsson, I. Lindgren, and B. Lindberg, *ESCA—Atomic, Molecular and Solid State Structure Studied by Means of Electron Spectroscopy, Nova Acta Regiae Societatis Sci.*

Upsallensis, Series 4, Vol 20, 1967, p 1-282; also, U.S. Government Report A.D. 844 315, National Technical Information Service, Oct 1986
- K. Siegbahn, C. Nordling, G. Johansson, J. Hedmark, P.F. Heden, K. Hamrin, U. Gelius, T. Bergmark, L.O. Werme, R. Manne, and Y. Baer, *ESCA Applied to*

Free Molecules, North-Holland, 1969
- C.D. Wagner, W.M. Riggs, L.E. Davis, J.F. Moulder, and G.E. Muilenberg, *Handbook of X-Ray Photoelectron Spectroscopy*, Perkin-Elmer Corporation, 1979
- H. Windawi and F.F.-L. Ho, *Applied Electron Spectroscopy for Chemical Analysis*, John Wiley & Sons, 1982

Properties Tests for Matrix Resins

Lawrence C. Hopper and Gerald L. Sauer, BASF Structural Materials Inc.

A COMPOSITE IS COMPOSED of some form of fibrous reinforcement combined with a resin matrix. The thermal and mechanical performance of composites depends on the properties of the matrix resin as well as on the reinforcement. This article defines some of the basic materials used for thermoset and thermoplastic resin matrices and their most significant chemical and physical tests. The individual characteristics are identified along with the test methods normally used for their determination. Many of the recommended tests are directed at the chemical properties of the resin. A complete evaluation of physical properties must also be performed to ensure that controls are adequate. Properties can be varied rather widely by the selection of the matrix resins, curing agents, and modifiers. Therefore, it is important to understand what these properties are and what effect they have on the resin matrix, impregnated fiber, and processing of the product. All tests described in this article are relative to uncured components and uncured and cured matrix systems.

Thermosets

The term matrix refers only to the nonfiber component of a composite. The key function of the matrix is to aid in the handling and processing of the fibers while translating the properties of the fibrous component in the composite. The matrix must also perform in a desired temperature envelope specified by the end-use application. The thermoset resins most widely used for matrices are epoxies, bismaleimides, phenolics, polyesters, and polyimides. The properties of the uncured resins that are of most interest are the equivalent weight and viscosity, or rheological values. When the equivalent weight is known, it is possible to calculate the optimum amount of curing agent required for the resin. The viscosity of the resin is a convenient index for its handling and flow characteristics during cure. Other properties of more limited interest include softening point, melting point, molecular weight, molecular weight distribution, specific gravity, refractive index, and chlorine content.

Of the thermoset systems being considered, only polyesters and epoxies require the addition of another ingredient to facilitate polymerization. There are other ingredients in some systems, particularly epoxies and bismaleimides, which serve as modifiers and as such may or may not enter into the chemical reactions during curing.

Epoxies are among the most widely used resins because of their overall balance of properties. Their mechanical properties are superior to those of polyesters and phenolics, and they offer improvements in environmental/moisture resistance, fatigue resistance, interlaminar shear strengths, and ease of processing. They also provide a combination of cross-link density and elongation that yields a balance of suitable properties over a broad temperature range. Because of this versatility, epoxies are finding increased use as matrix materials.

There are a large number of epoxy resins commercially available. The diglycidyl ether of bisphenol A (DGEBA), epoxy phenol novolac, and tetraglycidyl-4,4′ diamino diphenyl methane (TGDDM) are the most commonly used epoxy resins for matrix formulation. Key property tests are listed in Table 1.

The number of bismaleimide (BMI) resins available commercially are limited. These resins have backbones similar to those of epoxies but contain different functional end groups. The most basic and widely used resin is the bismaleimide of methylene dianiline (bis S). Bismaleimide resins cure by means of addition polymerization with little or no evolution of volatiles. They are prime candidates for matrix resins because of their high-temperature resistance, environmental stability under hot/wet conditions, and superior smoke/toxicity properties, and epoxy-like processing. However, they tend to be more brittle than epoxies, but can be formulated to achieve various combinations of cross-link density and elongation that provide matrix resins with balanced properties. Bismaleimides are being used in applications that require temperature performance between that of state-of-the-art epoxies and polyimides. Key property tests are listed in Table 2.

Phenolic resins are available commercially in wide variety. The two main types are a single-stage resole and a two-stage novolac. The resole phenolic is the most widely used

Table 1 Epoxy properties and tests

Ingredient	Property	Test method
Epoxy resins	Epoxide content	Titration
	Viscosity/softening point	Viscometer/Duran or rheometer
	Residual chlorides	Titration
	Moisture content	Titration/Karl Fisher
	Molecular weight distribution	GPC
	Characterization	HPLC/infrared spectroscopy
Hardener (amine)	Amine content	Titration
	Purity	Melting point refractive index, HPLC
Catalyst	Purity	Melting point
	Cation	Atomic absorption
Modifier (inorganic)	Particle size	Sedigraph/particle sizer
	Moisture	Moisture analyzer/Karl Fisher
Modifier (organic)	Viscosity	Rheometer
	Reactivity	Titration

GPC, gel permeation chromatography; HPLC, High-performance liquid chromatography

Table 2 Bismaleimide properties and tests

Ingredient	Property	Test method
Bismaleimide resin	Viscosity	Rheometer
	Composition	HPLC/infrared spectroscopy
Modifier (organic)	Viscosity	Rheometer
	Molecular weight	GPC

Table 3 Phenolic properties and tests

Ingredient	Property	Test method
Phenolic resin	Phenol	Titration
	Molecular weight	GPC
	Characterization	HPLC/infrared spectroscopy
	Solids	Evaporation
Modifier (organic)	Viscosity	Rheometer
	Molecular weight	GPC

Table 4 Polyester properties and tests

Ingredient	Property	Test method
Polyester	Reactivity	Titration of peroxide
	Molecular weight	GPC
	Purity	HPLC
		H_2O determination

Table 5 Polyimide properties and tests

Ingredient	Property	Test method
Polyimide Resin	Ingredient ratio	HPLC/infrared titration
	Purity	HPLC
	Functional groups	Titration

because of its handling characteristics in the impregnated form. It cures by means of a condensation-type reaction in which water is formed as a by-product. The resultant matrix is highly cross-linked, but can be formulated with a wide variety of materials. The volatile by-products limit the use of phenolic for some composite applications. Phenolics have high heat and chemical resistance, good dielectric properties, dimensional and thermal stability, and surface hardness. They yield low smoke and toxicity properties after combustion, which is important for many applications. Key property tests are listed in Table 3.

Polyesters. Of the wide variety of polyester resins commercially available, those most commonly used are alkyd and unsaturated polyester. These resins can be combined with cross-linking monomers, such as styrene, to form a matrix. Both alkyd and unsaturated polyester matrices can be used for a wide variety of composite applications. Alkyd polyesters have good arc-track resistance and show outstanding retention of dielectric strengths up to 175 °C (350 °F). Their highly cross-linked resin structure gives good dimensional stability at elevated temperatures. However, the percentage strength retention at elevated temperature is not as great as for some of the other matrix resins. In general, polyesters yield moderate composite properties and have poor chemical and hydrolytic resistance compared to other materials. Higher levels of matrix shrinkage during cure are also a detriment for some applications. Because polyesters can be cured quickly by means of a free-radical process and generally are less costly, there is incentive for using them in composites when mechanical properties suit the application. Key property tests are listed in Table 4.

Polyimides. There are a limited number of polyimide resins available for use as composite matrices. These systems cure by means of a condensation reaction with evolution of a high level of volatiles. The two types of polyimide resins most widely used are PMR-15 and the family of pyromellitic dianhydride/oxydianiline (PMDA/ODA) polymers. These materials possess exceptionally high thermo-oxidative stability. However, the higher the oxidative stability, the more difficult they are to process. Polyimides are being used in composite applications that require temperature performance at temperatures (110 to 190 °C, or 230 to 375 °F) above

the capabilities of BMIs. Their use is limited because of the high volatile levels given off during cure. These volatiles pose severe processing problems for applications requiring large, thick, composite parts. Nevertheless, the commercial success of polyimides in a variety of high-technology applications can be attributed to resin matrices with a good balance of thermo-oxidative stability, high glass transition temperature, and processing parameters. Key property tests are listed in Table 5.

Cyanate Esters. Another thermoset resin that is currently finding increasing potential as a matrix material is the cyanate ester. These resins perform in the same temperature range as epoxy resins, but offer a different balance of properties. Cyanate esters provide different formulation opportunities because of their unique cure chemistry.

Formulations

Matrices or neat resin systems are mostly blends of ingredients that are added to impart a desired property to the prepreg and/or cured composite. Proper formulation of resin compounds requires knowledge of the molecular structure of the resins as well as their chemical/physical properties and curing reactions. The structure of the resin determines its physical and chemical properties. The number and location of the reactive sites determine the functionality and the cross-linking density. These properties establish the rigidity, thermal stability, solvent resistance, and so forth, of the cured system. Resin structure also determines the viscosity of the resin, an important factor. Resins are only building blocks in the development of a matrix formulation. A formulation will ultimately involve several resins combined with curing agents, catalysts, and modifiers (such as fillers, flow control agents, or flexibilizers). Each component contributes to the final properties of the prepreg and cured composite and represents a major variable. Once an application is known, a neat resin matrix can be formulated.

It is essential to identify key properties and define an acceptable range for each before assigning a suitable test to measure those properties. Tests are typically performed on the individual components as well as on the uncured final formulation in order to better understand and characterize the resin system. Various chemical and physical tests are run on all types of resin matrices to ensure overall consistency and correct ingredient amounts, func-

tional groups, molecular weight distribution, kinetics, and advancement levels.

To exemplify a neat resin matrix formulation, a typical epoxy matrix would contain epoxy A, epoxy B, a catalyst, and a curing agent. Most applications requiring a service temperature up to 120 °C (250 °F) would combine DGEBA (epoxy A) and/or DGEBA with epoxidized phenol novolac resins (epoxy B) with a dicyandiamide curing agent and tertiary amine salt catalyst.

An example of a current state-of-the-art epoxy matrix system suited to a service temperature up to 177 °C (350 °F) would have TGDDM as epoxy A, diaminodiphenylsulfone (DADPS) as the curing agent, and a salt of BF_3 as the catalyst. Elastomer modifiers or flexible resins would be added as needed to improve toughness. For both of these examples, high-performance liquid chromatography (HPLC), differential scanning calorimetry (DSC), infrared spectroscopy, and rheometric dynamic scanning (RDS) viscosity tests are run to characterize the resin and ensure the quality of the molecular weight distribution cure kinetics, functional groups, and rheological behavior for each batch of material. The chemical properties normally reported for these neat resin systems are those given in Tables 1 and 6. Chemical and physical property testing is performed on uncured neat resin systems prior to impregnation, and on impregnated systems after cure. During system development, some properties are also measured on cured neat resin.

A typical bismaleimide could contain a bismaleimide of methylene dianiline (bis S), divinyl benzene (DVB), and thermoplastic resin. This example is a formulated matrix for applications requiring service temperatures up to 230 °C (445 °F). The bis S resin provides high-temperature properties, the thermoplastic resin modifies the rheology, and the DVB provides both matrix toughness (elongation) and tack. Applicable property tests are listed in Table 7.

A typical phenolic matrix would combine resole phenolic, ethyl alcohol, polyamide resin, and a catalyst. Resole phenolics, available in an alcohol solution, can be used directly as a matrix material with an inorganic acid catalyst. Nylon 6,6 polyamide resin can be added to improve matrix toughness and flow characteristics. Applicable property tests are listed in Table 8.

A typical polyester matrix would have an unsaturated polyester resin, styrene monomer, hydroquinone, and a promoter. A large variety

Table 6   Epoxy matrix

Material	Property	Test method
Uncured resin	Composition	HPLC
		Infrared spectroscopy
		GPC
	Processibility	RDS viscosity
		Gel time
		Volatile content
		Moisture content
	Reactivity	DSC
		ARC
Cured neat resin	Completeness of cure	Glass transition
		DSC
	Solvent/H_2O resistance	Glass transition, wet
		Moisture weight gain
		Solvent weight gain
	Resin toughness	Cleavage (G_{Ic})
		Residual compression
		strength after impact
Uncured impregnated system	Characterization	HPLC
		Infrared spectroscopy
		DSC
	Processibility	Flow
		Gel
		RDS viscosity
		Volatiles
		Moisture
		Tack/drape
		Resin content
		Fiber weight
Cured impregnated system	Completeness of cure	DSC
		Glass transition
	Thermal properties	TGA
		Flammability
		Heat distribution temperature
	Electrical properties	Dielectric

ARC, accelerated rate calorimeter; TGA, thermogravimetric analysis

Table 7   Bismaleimide matrix

Material	Property	Test method
Uncured resin	Composition	HPLC
		Infrared spectroscopy
		GPC
		DSC
	Processibility	RDS viscosity
		Gel time
		Volatile content
Cured neat resin	Completeness of cure	Glass transition
	H_2O/solvent resistance	Glass transition, wet
		Moisture weight gain
		Solvent weight gain
	Resin toughness	Cleavage (G_{Ic})
Uncured impregnated system	Characterization	HPLC
		Infrared spectroscopy
		DSC
	Processibility	Resin content
		Fiber content
		Flow
		Gel
		RDS viscosity
		Volatiles
		Tack/drape
Cured impregnated system	Completeness of cure	Glass transition
		DSC
	Moisture resistance	Weight gain
	Thermal properties	Thermal conductivity
	Laminate properties	Ply thickness
		Fiber volume

of thermoset polyesters are commercially available and are already cut with styrene monomer. Inhibitors (hydroquinone) and promoters are added for gel and cure time control. Additional resins, such as isocyanates, can be added to improve toughness. Applicable property tests are listed in Table 9.

As discussed previously, only a limited number of polyimide resins are available for use as a matrix for composites. The complex processing technology required narrows the number of resins that would be useful for a matrix resin. One of the most commonly used resins is PMR-15, which is typically supplied in alcohol at 40-50% solids. Applicable property tests are listed in Table 10.

Prepreg Forms

All five families of matrix resins can be used to impregnate various fiber forms. The resin is then no longer in a low-viscosity stage, but has been advanced to a B-stage level of cure for better handling characteristics. The type of prepreg form is highly dependent on the end-use requirement. Four of the main forms of composite materials are unidirectional fiber tapes, woven fabrics, roving, and chopped mat. Fiber reinforcement may be in the form of strands that are composed of a number of very fine filaments. Strands can be gathered into continuous roving, chopped to provide mats, or twisted into yarns and woven into cloth. Fiber type can be glass, carbon, or aramid. Resin content in all forms must be tested because it is one of the critical parameters for composite performance. The resin content is dependent on the laminating pressure and is of major importance because, in conjunction with the fiber, it determines the strength of the ultimate laminate. Likewise, fiber areal weight is monitored because the amount of fiber present in a composite laminate affects the ultimate strength. Property tests for four prepreg forms are discussed below.

Unidirectional Tapes. Numerous fiber strands (tows) are creeled and spread while being combined with the resin matrix to produce a unidirectional tape prepreg. Once a tape is produced, the physical tests of resin content, fiber areal weight, volatile content, gel time, and flow can be run to characterize the system and ensure consistency. Chemical tests performed include HPLC, DSC, and infrared spectroscopy, as shown in Tables 6 through 10.

Woven fabrics provide more uniform but lower composite strength properties than do tapes, but are nonetheless used extensively because of the ease of fabrication and strength in both directions. Fabrics are produced in a broad range of styles, widths, and lengths. Woven fabrics can be impregnated either by means of a hot-melt or solution method, with solvent being subsequently removed. The physical and chemical tests used to characterize woven fabrics are the same as those used for tapes.

Continuous strand rovings lend themselves to automated fabrication techniques by means of winding. Rovings are typically impregnated in a solvent process and then wound on a core after the removal of solvent. Prepreg rovings typically provide better resin content control than that provided by wet winding. Physical tests include resin content, volatiles, band width, gel time, and yield (m/kg). Chemical tests are the same as those used for tapes and fabrics.

Chopped mat is a form that produces lower-cost structures with uniform strength characteristics. Composite strength properties

Table 8   Phenolic matrix

Material	Property	Test method
Uncured resin	Composition	HPLC
		Infrared spectroscopy
		GPC
	Processability	Solids
		Gel
		Volatile content
	Chemical activity	DSC
Cured neat resin	Completeness of cure	Glass transition
		Solvent extraction
Uncured impregnated system	Characterization	HPLC
		Infrared spectroscopy
		DSC
	Processability	Resin content
		Volatile content
		Flow
		Gel
		Tack/drape
		Fiber weight
Cured impregnated system	Completeness of cure	DSC
	Thermal properties	TGA
		Flammability
	Electrical properties	Dielectric

Table 9   Polyester matrix

Material	Property	Test method
Uncured resin	Composition	Infrared spectroscopy
		HPLC
	Processability	RDS viscosity
		Gel
		Volatile content
Cured neat resin	Completeness of cure	DSC
Uncured impregnated system...................	Characterization	HPLC
	Processability	Resin content
		Flow
		Gel
		Tack/drape
		Fiber weight
Cured impregnated system	Completeness of cure	DSC
	Laminate properties	Ply thickness
		Fiber volume
		Hardness

Table 10   Polyimide matrix

Material	Property	Test method
Uncured resin	Composition	HPLC
		Infrared spectroscopy
		GPC
	Processability	RDS viscosity
		Gel time
		Volatile content
Cured neat resin	Completeness of cure	Glass transition
		DSC
	H_2O/solvent resistance	Glass transition, wet
		Solvent weight gain
Uncured impregnated system	Characterization	HPLC
		Infrared spectroscopy
		DSC
	Processability	Resin content
		Fiber weight
		Flow
		Gel
		RDS viscosity
		Volatiles
		Tack/drape
Cured impregnated system	Completeness of cure	Glass transition
		DSC
	Moisture resistance	Weight gain
	Thermal properties	TGA
		Thermal conductivity
	Laminate properties	Ply thickness
		Fiber volume

are generally lower than those of woven fabrics. Physical tests should include resin content, weight (g/m^2), volatiles, gel time, and flow. Chemical tests should be run according to the applicable resin family (Tables 6 to 10).

Thermoplastics

In conventional applications of thermoplastics, the physical and chemical property tests listed below are usually offered by the resin supplier to inform the user of the behavior of the material.

- Tensile strength, yield
- Tensile strength, break
- Elongation, yield
- Elongation, break
- Tensile modulus
- Linear thermal expansion
- Electricals:
    dielectric strength
    dielectric constant
    volume resistivity
    dissipation factor
- Flexural strength, yield
- Flexural strength, break
- Compressive strength
- Izod impact, notched
- Hardness (Rockwell)
- Water absorption, 24-h
- Thermal conductivity
- Continuous service temperature
- Processing temperature or melting point
- Heat deflection temperature
- Solvent and chemical resistance
- Environmental resistance

Each thermoplastic polymer application has a need for some data from this list of tests. The processing method or the end-use usually requires specific information about the chemical and physical nature of the material.

Impact Strength. Some of these tests assume special importance and even criticality for fiber-reinforced thermoplastics. For instance, the total consideration of impact strength is significant in advanced composite applications. In fact, the behavior of thermoplastics in terms of their stress-strain relationship is the underlying reason for their consideration in impact-resistant and damage-tolerant applications. Looking at the stress-strain behavior of neat conventional thermosets, such as DADPS-cured epoxies, one sees low strain and a steep curve all the way to failure. In the case of thermoplastics, strain is usually greater, but a unique behavior is observed before failure.

This phenomenon is the special involvement and the changes in molecular orientation that are known as yield. At the yield point the average axis of molecular orientation begins to conform increasingly with the direction of the stress. Other terminology, such as draw, is sometimes used. There is usually a break in the stress-strain curve as it begins to flatten out and more strain is observed with a given increased stress. The result is that the giant molecules

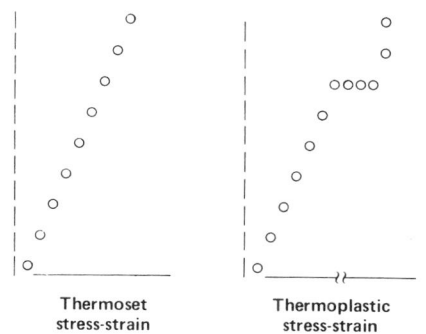

Fig. 1 Thermoset versus thermoplastic stress-strain behavior

begin to align and team up in their resistance to the implied stress. Frequently, there is a final increase in the slope of the curve just before ultimate failure (Fig. 1). The extent to which this orientation takes place varies from one linear thermoplastic to the next, but the effect is quite significant in the behavior of such materials in these fiber-reinforced systems. Even the smallest amount of the teaming-up effect imparts greatly improved impact resistance and damage tolerance. In thermoplastics, there is much more area under the stress-strain curve; this area is a direct function of the work to failure.

Conventional thermosets do not behave this way and can be thought of as rigid networks that have a more glassy failure mode with much less area under the stress-strain curve. Thermoplastic resins being considered today as matrices for composites have relatively high values for their modulus or low strain-stress ratios. The molecular orientation effects are small but nevertheless quite important in their effect on toughness. Values for Young's modulus approaching 3450 MPa (500 ksi) have been reported. Such numbers are comparable to those of the thermosets. When these new, stiffer, and tougher resins began to appear, the need developed for more discriminating testing to evaluate the toughness and damage tolerance of composites fabricated with them.

Toughness and Damage Tolerance. Both National Aeronautics and Space Administration (NASA) and industry representatives have selected and standardized a set of five common tests for characterizing the toughness of fiber-resin composites. As the development of new, tougher thermoplastic matrix resins is reduced to specific applications, these identified tests will become the standards for characterizing the new thermoplastic-based composites.

For interlaminar fracture toughness, the edge delamination and double cantilevered beam tests have been defined as appropriate. For damage tolerance, open-hole compression, open-hole tension, and compression after impact have been defined. These tests are described in Ref 1.

It should be noted that each test has its own set of problems. For example, failures in the double cantilevered beam test seldomly occur solely in mode I (peel). In the past, different laboratories have used composite specimens with different resin content and fiber types, as well as different specimen dimensions and impact stress levels. In spite of these problems, test refinements have been made and reliable test data have begun to emerge. See Table 11 for G_{Ic} values.

Creep and Environmental Resistance. Special supplemental tests will be needed to assess thermoplastic resins as advanced composite matrices and to provide design assurances in the areas of temperature versus creep and environmental-stress cracking. Because the linear nature of thermoplastics means a departure from the familiar behavior of thermosets, tests will be needed to provide assurance of service life and reliability. For example, the differences in the failure behavior of many thermoplastics in the stressed and unstressed condition are dramatic, especially when the environmental effects of agents such as solvents and surface-active organics are present.

Creep in thermoplastics will have to be appropriately addressed if these resins are to become reliable matrices for advanced composites. The rigid, random network of the thermosetting resin is, by nature, a significantly different system than the collection of amassed, large, linear molecules in a thermoplastic. Test procedures and a test data bank will be required.

Creep occurs in thermoplastics because, unlike thermoset molecules, the molecules of thermoplastics are not bonded together by primary (covalent) bonds. Thus, thermoplastic molecules can make, break, and remake associations with their molecular neighbors. Such behavior allows subtle changes in form and shape. Corresponding reversible changes in thermosets are not possible because the primary covalent chemical bond is the mainstay for retention of form and shape: Once the limit of load-carrying ability is reached, bonds break and the thermoset fails catastrophically. Break and remake are not possible. The integrity of thermoplastic molecules can be attributed to simple mechanical entanglement, but a great deal of this integrity is due to reversible associations of ordered molecules by van der Waals forces, dipole-dipole interactions, and crystalline lattice energy. Some of the new thermoplastics rely on a high crystalline content for stiffness and creep resistance. The polyphenylene sulfides, for example, exhibit some of the best of both worlds and seem to be tough and damage tolerant.

One of the most difficult test configurations that could be used to define creep behavior would be a static load on a $\pm 45°$, tensile, thermoplastic matrix coupon. This test would place a sustained load on the thermoplastic polymer, which would require the association forces within and between the polymer mole-

Table 11 Interlaminar fracture toughness values of thermoplastic resins

Material	J/m²	ft · lbf/ft²
Polyphenylene sulfide	720	50
Polyetherimide	950	65
Polyamideimide	1050	70
Polysulfone	1175	80
Polyether etherketone	1600	110

cules to have enough strength and integrity to resist slow deformation and maintain the spatial configuration in this scissoring plane. This situation, difficult enough for the thermoplastic polymer, would be made even more severe by the imposition of an elevated temperature. Unfortunately, this most difficult scenario for a test coupon can be a real-life load situation for an advanced composite structural component. Similarly, a beam in a static flexing load represents another very difficult real-life configuration in which retention of shape and form would be of critical importance. The influence of the composite designer can be significant in terms of selecting fiber configurations in the composite lay-up that would permit the fiber to handle this stress and let the thermoplastic resin synergistically contribute its best innate features.

Solvent and Chemical Resistance. Because the associative forces of thermoplastics are by nature completely different from the strong covalent chemical bond acting as the principal associative force in thermosets, the weaker forces of the thermoplastics are much more susceptible to disruption by solvents. For example, any solvation around a polar site would alter molecular spacing and upset van der Waals association. Similarly, the strongest of these forces, the crystalline lattice energy, would also be disrupted if a solvent were present with enough solvating ability to come into the interstices in the crystalline structure of the ordered polymer. Thermosets are not immune to such disruption of molecular associative forces, however. It is well documented that the intrusion of water will reduce the glass transition temperature of a typical epoxy system. In fact, water is an agent that also must be considered in thermoplastic matrices.

Resistance to hydraulic fluids, lubricants, fuel, and solvents, especially when under stress, is an important requirement of composites used on aircraft. Many amorphous thermoplastics, such as polyetherimide and polysulfone, are prone to attack by hydraulic fluid and paint strippers, such as methylene chloride. When the crystalline character of a polymer is increased, its resistance to solvents improves. Consequently, polymers such as polyphenylene sulfide and polyether etherketone show better solvent resistance. The long-term suitability of such crystalline polymers in contact with solvents is yet to be determined.

Tests to evaluate solvent resistance have usually been before-and-after assessments of a given mechanical property after immersing the thermoplastic resin in the solvent or agent. A quick, reliable test consists of determining the weight picked up after a period of immersion. Also, dimensional changes can be measured after exposure, especially on neat resins. As described above, solvent intrusion into polymer matrices disrupts the associative forces and consequently alters molecular volume. This can be detected by a simple measurement of sample dimensions and a calculation of volume change. Some thermoplastic resin suppliers provide some information on water with data on weight increase after a 24-h immersion in water. A more severe test is the 24-h water boil.

The intrusion of solvent or any other agent, including water, may also interfere with the associative or adhesive forces at the interface between the resin and the fiber. This is true for both thermosets and thermoplastics. Such interference may be more severe for thermoplastics because there is no reactive functionality to assist in forming special adhesion sites on the fiber. With thermoplastics, the surface-surface strength is more a problem of physical interactions and classical adhesion. Testing interlaminar shear or transverse mechanical properties can assess the detrimental effects of such solvent or water intrusion.

Thermal Expansion. In another volume change phenomenon, thermosets undergo a significant amount of volume shrinkage when the polymer is formed from its monomeric precursors. This change can be in the range of 3 to 5% of the volume of epoxies. In some addition-type unsaturated polymers, such as BMIs, it can be even higher and can become a source of difficulty by initiating microcracks. This behavior represents a potential advantage for thermoplastics in that such shrinkage is a possible source of microcracking in composite fabrication, especially in fabric or cross-plied laminates. Because the polymerization of thermoplastics is already over at the processing stage, the shrinkage caused by packing improvements on a molecular scale is minimal. The shrinkage that remains is due to thermal expansion, which is reported in the supplier's standard data as mold shrinkage or linear thermal expansion.

Viscosity measurements on thermoplastic candidates for composite matrices are all taken at elevated temperature and, consequently, are not routinely provided by the resin suppliers. The melt values reported in Table 12 give some information as to how the resin will flow under laminating conditions, but the real nature of flow is important in good composite fabrication. If the flow is merely a sintering, there will be a discontinuity between plies, and voids will most likely be present. In processing composites, a good rule of thumb is that some degree of edge flow or bleed will be visible. If this is not

Table 12 Melt values of thermoplastic resins

Material, temperature	Pa · s	P
Polyamideimide at 350 °C (662 °F)	10^6	10^7
Polysulfone at 300 °C (570 °F)	10^4	10^5
Polyetherimide at 305 °C (580 °F)	10^5	10^6
Polyphenylene sulfide at 313 °C (595 °F)	10^4	10^5
Polyether etherketone at 400 °C (750 °F)	10^3	10^4

the case, voids are likely to be present because of air entrapment. Some fabricators use unrealistic pressures to prepare study samples; resins that require such treatment will not be suitable for producing good quality composite structures. Brute force and the preparation of 75- × 75-mm (3- × 3-in.) test laminates are not suitable approaches for evaluating a resin for use in composite fabrication. In addition to the requirement for flow, there is a need for a workable viscosity for good tow wet-out during the initial prepreg process. The viscosity test values in Table 12 will change, usually downward, if thermal mistreatment of the polymer occurs during the prepreg or lamination processes, or during rework of the finished composite.

When melting a thermoplastic resin in the prepreg process or, more important, in the final fabrication step, all crystalline order is erased (except in unique liquid crystalline polymers). Subsequently, during cool-down the polymer reestablishes crystalline lattice order. The degree to which this order is reinstated is a function of the viscosity and mobility of the cooling system. If the cooling is fast, there is little time for the molecules to find their appropriate positions in the growing lattice structure. Many of the molecules will lose their mobility as the temperature is decreased and, being unable to move into the proper position to form the crystalline state, will be stranded in an amorphous condition. Conversely, if the cooling rate is slow, most of the molecules, or segments of molecules, will maintain freedom long enough to find their appropriate position in the crystalline state.

Characterization Methods. Some common characterization tests used for oligomers and polymers are not appropriate for the class of thermoplastics being considered for high-performance composite matrices because the test methods require that the material be in solution. If the thermoplastic composite is to have solvent resistance, it cannot be soluble in any solvent. This is the case for inherent viscosity, for example, because this measurement is made in solution. Similarly, HPLC cannot be used as a characterization method. In composite applications that do not require a high degree of solvent resistance, the methods

Table 13 Useful property data for a typical carbon-resin composite

Property	Typical (carbon)
Fiber density	1.80 g/cm³
Tow yield	1100 m/kg (565 yd/lb)
Cured resin density	1.2 g/cm³
Prepreg density	1.5 g/cm³
Prepreg thickness/ply	0.15 mm (0.0067 in.)
Resin film thickness	0.085 mm (0.0034 in.)
Resin film areal wt	0.099 kg/m² (2.946 oz/yd²)
Fiber areal wt	0.149 kg/m² (4.420 oz/yd²)
Prepreg resin content	40%
Tow count	0.17 tows/mm (4.3 tows/in.)
Fiber volume fraction	49%
Prepreg yield	4.00 m²/kg (19.55 ft²/lb)
Flow	5%
Cured ply thickness	0.15 mm (0.0061 in.)
Fiber vol% (cured)	53%

described might be used. With thermosets, HPLC has been used as a fingerprint and quality control tool. This analytical technique could be used for soluble thermoplastics.

Other methods, such as infrared spectroscopy, are more suitable for routine characterization or fingerprinting of thermoplastics. Before the advent of Fourier transform infrared (FTIR), infrared spectroscopy was difficult to use as a quantitative tool for measuring subtle changes in structure and concentration. Now that FTIR has been supplemented with improved surface reflectance methods, this technique has become very viable for characterizing and studying thermoplastic resins and is informative for finished thermoplastic composites. Further work is needed to exploit the capability of FTIR reflectance as a nondestructive method of characterizing composite components.

Perhaps the most important characterization methods for both thermoplastic and thermoset composites and composite precursors are gravimetry and mensuration. The values obtained are the criteria most critical to the outcome of successful composite components. Table 13 gives the type of data that are of vital significance to the prepregger, designer, and fabricator.

The basic mechanical performance and/or quality of polymer, fiber, prepreg, or composite is primarily determined by mechanical testing of a test coupon to attain the truest performance of the material. The coupon test method is thoroughly described in the article "Mechanical Properties and Environmental Exposure Tests" in this Section of the Volume.

ACKNOWLEDGMENT

The authors would like to thank Dr. Rex Gosnell for his contribution to the thermoplastics section of this article.

Mechanical Properties and Environmental Exposure Tests

N.R. Adsit, Rohr Industries, Inc.

SELECTION OF TEST METHODS used to determine the mechanical properties of composite materials must take environmental factors into account, since exposure to any environment affects these properties. The tests to be discussed are for coupons only. Data currently available indicate that the coupon test method is related to the form of the composite material. The more anisotropic the material, the more limited the method of testing. A method that works adequately for chopped fiber composites may not work for unidirectional materials. A method that works for unidirectional materials probably will work well for all materials, although it may not always be the most cost effective.

This discussion is limited to planar and molded composites, including chopped fiber and reinforced plastics. No attempt has been made to discuss test methods for braided or three-dimensional woven structures; since the latter tend to be isotropic, more acceptable test methods may be available. Materials to be discussed are those made from fibers of graphite, ceramics, aramids, metals, and glass. Matrices covered are those commonly identified as organic, either thermoset or thermoplastic. Specifically excluded from this discussion are metal matrix composites and ceramic matrix composites, for example, carbon-carbon. However, the principles used in testing the subject composites could apply equally to the excluded classes of materials, because the real criterion may be the behavior of the fiber and matrix rather than the actual material used in fabrication of the composite.

Environmental Exposures

Early in the development of high-performance carbon fiber composites, it was discovered that merely leaving specimens in the laboratory to await testing had an effect on measured properties. From this experience one must conclude that all exposures, intentional or not, may have an effect and should be documented. Environmental exposures that often occur during specimen preparation and testing are described below, and those that are planned are described in the section "Planned Exposures" in this article.

Exposures During Specimen Preparation

Cutting. There is currently no accepted standard method for specimen preparation, but there appears to be a consensus (Ref 1, 2) that the preparation method is critical. Therefore, a prudent investigator will document all steps in this process. After a panel is fabricated, the first exposure that a specimen experiences is when it is being cut from that panel. The most reliable cutting method is to use a diamond saw with a coolant. Although this process takes only a short time and the excess moisture is usually wiped off, the specimen is still wet and can pick up moisture. However, the amount of moisture characteristic of this process is acceptable for ambient conditions.

A second method is to cut specimens from panels using little or no coolant and then dress the edges to remove any damage. The dressing operation usually introduces minimal moisture. Diamond tooling of the type shown in Fig. 1 has been used for dressing operations. In some cases, investigators have polished the edges of specimens to remove any damage.

Any induced cracks or delaminations represent mechanical damage. This will be discussed in the section on "Planned Exposures." Recent data (Ref 3) show that unidirectional glass specimens cut in the transverse direction exhibited a 50% reduction in strength because of damage from cutting. Machining operations for specimens that are shaped or slotted or have holes also can induce damage and provide opportunities for moisture pickup.

Tab Bonding. In order to test many specimen configurations, it is necessary to attach tab material, usually by adhesive bonding. However, if an excessively elevated temperature is used in the bond cycle, it can affect the mate-

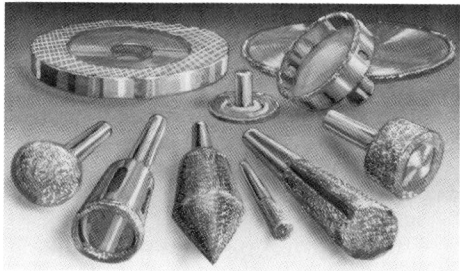

Fig. 1 Diamond tools. Courtesy of DoALL

rial. At the very least, the bond cycle will dry out the specimen. If the tab material is not the same as the specimen material, or if they have different coefficients of thermal expansion, residual stresses can occur, which can cause a stress riser at the end of the tab and induce failure at that location.

It is usually advisable to bond the tabs onto the panels before they are cut. Film adhesives with scrim are recommended because they control the bond line thickness. It also is necessary to support the center section of the panel when it is cut, to prevent buckling.

Instrumentation. As with tab installation, attaching instrumentation can dry out the specimen if an adhesive requiring elevated temperature is used. Before a strain gage or other sensor is installed, the surface must be lightly abraded. Care must be taken not to damage or change the dimensions of the specimen while cleaning the surface. Following manufacturers' recommendations for strain gage installation, personnel have used gages and adhesives successfully over a wide range of temperatures.

The strain gages should be large enough to measure average properties. Gages with a 6.4-mm (1/4-in.) active length have been used for tensile specimens, and those with a 3.2-mm (1/8-in.) length have been used for compression specimens. To minimize localized heating, it is recommended that the voltage on the bridge be kept low and the gage resistance be kept high;

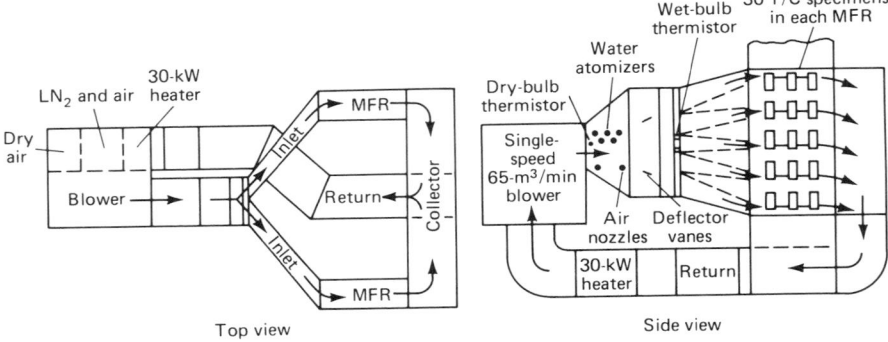

Fig. 3 Standard exposure chamber

Fig. 2 Environmental conditioning system. MFR, multiple-specimen fatigue rig. Source: Ref 6

350-ohm gages are commonly used for materials of low conductivity.

Laboratory Conditions. Holding specimens in the laboratory awaiting testing conditions them. If a specimen were heated to install tabs or instrumentation and then held in the laboratory awaiting testing, it could reabsorb moisture. Even more critical to the specimen is the situation where the laboratory environment is not controlled, which should be noted and reported. In addition, care should be exercised by laboratory personnel not to inflict damage (visible or nonvisible) to the specimens (Ref 4).

Planned Exposures

There are many environments to which materials may be exposed during their lives. It is important that test environment and techniques be as realistic as possible. Early in the development of graphite composites, specimens were put in boiling water for 24 h to simulate high-temperature moisture conditions, but it was found to be too harsh an exposure. Additionally, two very detailed studies (Ref 5, 6) show that neither short-time nor accelerated tests accurately predict long-term behavior. Three exposure techniques are discussed below: moisture, damage (impact), and thermal cycling.

Moisture Cycling (Atmospheric). One of the most elaborate systems used to condition a specimen to a realistic cycle for moisture is shown in Fig. 2. More conventional methods use salt in the chamber to control humidity while the specimen is being heated (Ref 7). The usual method of monitoring moisture content, however, is to weigh the specimens as a function of time. The specimen is then tested after it reaches equilibrium or a given moisture level. One problem with this method is that after such conditioning the specimen must be tested before the condition changes. Since the bonding of tabs and instrumentation at elevated temperatures will dry out the specimen, it is important that the specimen be ready for testing before exposure. This limits the use of strain gages because the adhesive used in the installation is itself sensitive to moisture. Therefore, extensometers are usually used. As tabs must be

held in place, testing for moisture content often uses flexural and short beam shear specimens, which do not need tabs. The data are then used to establish trends, which are superimposed upon tensile, compression, or other types of shear tests.

Damage (Impact). Flaws caused by damage can have a significant effect on the measured properties. Since specimens that have been damaged even without visible evidence can experience a reduction in strength of up to 50% (Ref 4), it is important that specimens that form a baseline in the testing process be free of defects. Ongoing work by the ASTM Committee D-30 (Ref 1) should result in fracture and damage test methods.

Thermal cycling of specimens can be carried out either in conjunction with moisture exposure or separately (Ref 5). Thermal cycling of specimens or panels has been carried out in a standard chamber (Fig. 3). Thermal cycles that have been used include −155 to 110 °C (−250 to 232 °F) for space applications of epoxy systems (Ref 8) and −18 to 230 °C (0 to 450 °F) for engine applications of polyimide systems (Ref 9). The cycle to be used should be tailored to the application.

Test Methods

After defining their common characteristics, this section will describe four mechanical property test methods: tensile, compression, flexural, and shear. Recommended configurations are discussed, as well as the proper use and the validity of the tests.

Definition of Data. Many composites are made from brittle fibers and behave like brittle materials, with notable exceptions, such as aramid fiber, which can exhibit nonlinear behavior. Failure is defined as the complete loss of load-carrying capability of the specimen. It should be recognized that other failure criteria, such as first ply failure, are also used.

The stress is defined in the classical sense, that is, P/A or MC/I, where P is the load, A_o is the original area, M is the moment, C is the half-thickness, and I is the moment of inertia. The area used in the calculation is the original

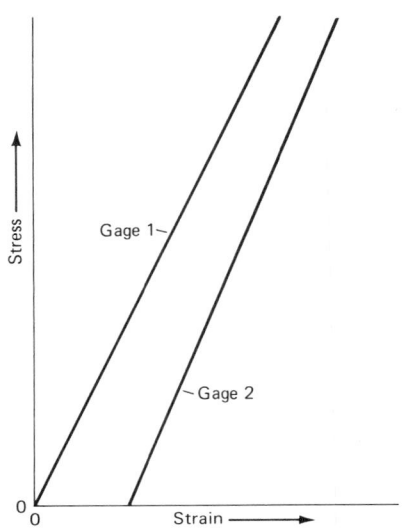

Fig. 4 Stress-strain curve

measured area. The P/A is used for axial loading, and the MC/I is used for bending of lamina only. The simple beam theory will not work for bending laminates; laminate theory must be used.

As the type of micrometer used with these composites can make a difference, a ball-type micrometer is recommended rather than the type with points or flat anvils. In some cases, the stress is calculated using nominal ply thickness, but this is not the recommended method.

The strain can be measured by a variety of methods. Since extensometers and strain gages give the same results, either can be used. There are extensometers that measure both axial and transverse strain; these systems can be used for both tensile and compression tests. However, there does not seem to be a reliable extensometer to measure shear strain for the shear specimens, so strain gages must be used in this case.

One of the most difficult values to obtain is the strain at failure. For brittle materials it is sometimes possible to use separable extensometers that can be left on the specimen to failure, but this is not usually recommended. Strain gages, which have the capacity to be used to failure for brittle materials, are another

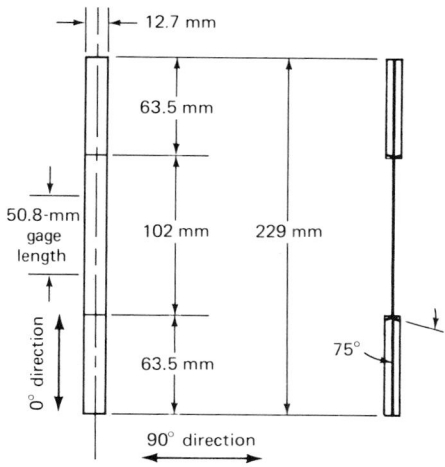

Fig. 5 Tensile test specimen. (1) Fiberglass tabs shall be positioned, both sides, two places; (2) Tabs shall be bonded with adhesives to graphite, dependent upon test temperature; (3) Specimen thickness shall not vary more than 0.076 mm (0.003 in.) from nominal; (4) Specimen edges shall be parallel to 0.076 mm (0.003 in.); (5) Specimen shall be used for unidirectional (0°) and bidirectional (0°/90°), graphite tape laminates. Unless otherwise specified, O.*XX* dimensions shall be ±0.76 mm (0.030 in.), O.*XXX* dimensions shall be ±0.25 mm (0.010 in.). Source: Ref 12

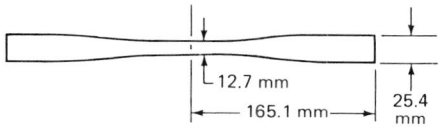

Fig. 6 Streamlined tensile specimen. Source: Ref 13

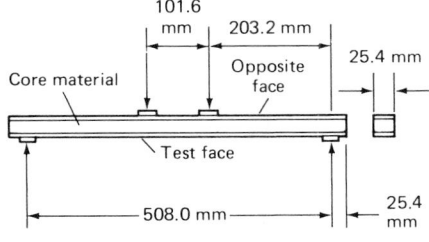

Load pads: 25.4 mm (1.0 in.) wide
Reaction pads: 38.1 mm (1.5 in.) wide

Fig. 7 Sandwich beam test specimen. Source: Ref 1

choice, but they can be damaged or become inoperative near the failure point. However, they still represent the best method for obtaining strain at failure if this value is required.

The application of strain gages is relatively straightforward. The type of gage and adhesive used should follow the manufacturer's recommendations. Gages and adhesives have been used on composites at temperatures from liquid helium (4.2 K) to 589 K (Ref 10, 11). Higher temperature systems are available and have been used on an experimental basis (Ref 10).

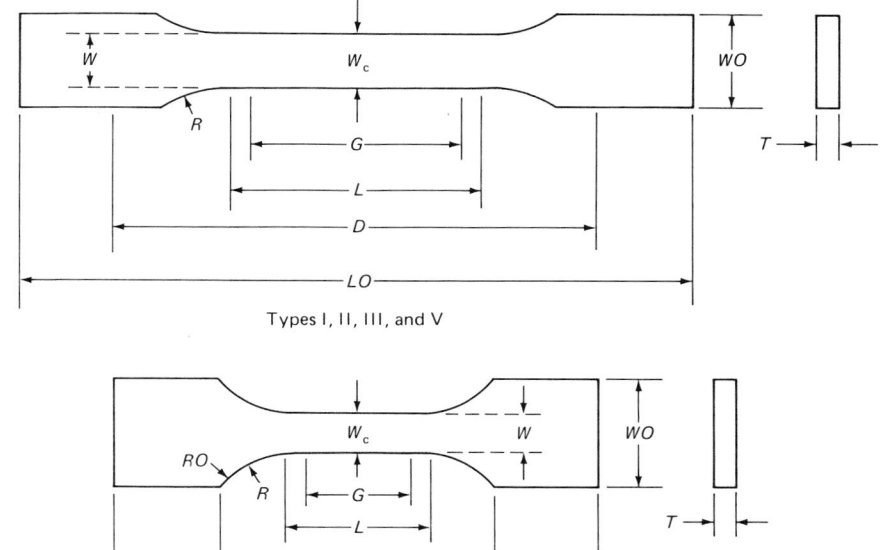

Types I, II, III, and V

Type IV

Specimen dimensions for thickness *T*, mm (in.) (a)

Dimensions (see drawings)	7 (0.28) or under		Over 7 to 14 (0.28 to 0.55)	4 (0.16) or under		Tolerances
	Type I	Type II	Type III	Type IV(b)	Type V(c)	
W-width of narrow section(d, e)	13 (0.50)	6 (0.25)	19 (0.75)	6 (0.25)	3.18 (0.125)	±0.5 (±0.02)(b, c)
L-length of narrow section	57 (2.25)	57 (2.25)	57 (2.25)	33 (1.30)	9.53 (0.375)	±0.5 (±0.02)(c)
WO-width overall, min(f)	19 (0.75)	19 (0.75)	29 (1.13)	19 (0.75)	9.53 (0.375)	±6 (±0.25)
LO-length overall, min(g)	165 (6.5)	183 (7.2)	246 (9.7)	115 (4.5)	63.5 (2.5)	no max (no max)
G-gage length(h)	50 (2.00)	50 (2.00)	50 (2.00)	... ...	7.62 (0.300)	±0.25 (±0.010)(c)
G-gage length(h)	... ...	... ...	... ...	25 (1.00)	... ...	±0.13 (±0.005)
D-distance between grips	115 (4.5)	135 (5.3)	115 (4.5)	64 (2.5)	25.4 (1.0)	±5 (±0.2)
R-radius of fillet	76 (3.00)	76 (3.00)	76 (3.00)	14 (0.56)	12.7 (0.5)	±1 (±0.04)(c)
RO-outer radius (Type IV)	... ...	... ...	... ...	25 (1.00)	... ...	±1 (±0.04)

(a) Thickness *T*, shall be 3.2 ± 0.4 mm (0.13 ± 0.02 in.) for all types of molded specimens, and for other Types I and II specimens where possible. If specimens are machined from sheets or plates, thickness, *T*, may be the thickness of the sheet or plate provided this does not exceed the range stated for the intended specimen type. For sheets of nominal thickness greater than 14 mm (0.55 in.) the specimens shall be machined to 14 ± 0.4 mm (0.55 ± 0.02 in.) in thickness, for use with the Type III specimen. For sheets of nominal thickness between 14 and 51 mm (0.55 and 2 in.) approximately equal amounts shall be machined from each surface. For thicker sheets both surfaces of the specimen shall be machined and the location of the specimen with reference to the original thickness of the sheet shall be noted. Tolerances on thickness less than 14 mm (0.55 in.) shall be those standard for the grade of material tested. (b) For the Type IV specimen, the internal width of the narrow section of the die shall be 6.00 ± 0.05 mm (0.250 ± 0.002 in.). The dimensions are essentially those of Die C in "Standard Test Methods for Rubber Properties in Tension," D 412, *Annual Book of ASTM Standards*. (c) The Type V specimen shall be machined or die cut to the dimensions shown, or molded in a mold whose cavity has these dimensions. The dimensions shall be *W* = 3.18 ± 0.03 mm (0.125 ± 0.001 in.), *L* = 9.53 ± 0.08 mm (0.375 ± 0.003 in.), *G* = 7.62 ± 0.02 mm (0.300 ± 0.001 in.), and *R* = 12.7 ± 0.08 mm (0.500 ± 0.003 in.). The other tolerances are those in the table. Supporting data on the introduction of the L specimen of Method D 1822 as the Type V specimen may be obtained from American Society for Testing and Materials (RR:D-20-1038). (d) The width at the center W_c shall be + 0.000 mm, − 0.10 mm (+0.000 in., −0.004 in.) compared with width *W* at other parts of the reduced section. Any reduction in *W* at the center shall be gradual, equal on each side, so that no abrupt changes in dimension result. (e) For molded specimens, a draft of not over 0.13 mm (0.005 in.) may be allowed for either Type I or II specimens 3.2 mm (0.13 in.) in thickness, and this should be taken into account when calculating the width of the specimen. Thus a typical section of a molded Type I specimen, having the maximum allowable draft, could be as follows:

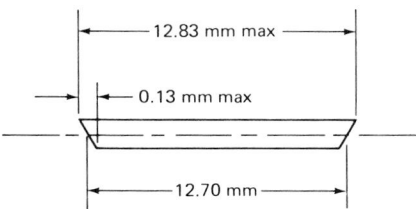

(f) Overall widths greater than the minimum indicated may be desirable for some materials in order to avoid breaking in the grips. (g) Overall lengths greater than the minimum indicated may be desirable either to avoid breaking in the grips or to satisfy special test requirements. (h) Test marks or initial extensometer span.

Fig. 8 Tension test specimens for sheet, plate, and molded plastics. Source: Ref 14

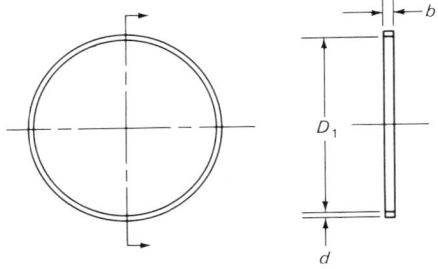

D_1 = 146.05 ± 0.051 mm (5.750 ± 0.002 in.)
b = 6.35 ± 0.127 mm (0.250 ± 0.005 in.)
Type A d = 1.52 ± 0.051 mm (0.060 ± 0.002 in.)
Type B d = 1.52 ± 0.0254 mm (0.060 ± 0.010 in.)

Fig. 9 Parallel fiber reinforced ring specimen. Source: Ref 16

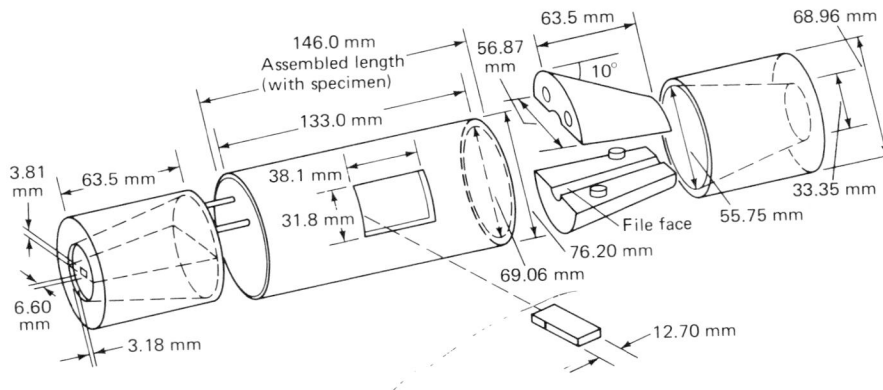

Fig. 11 Compression test fixture with conical wedges. Source: Ref 20

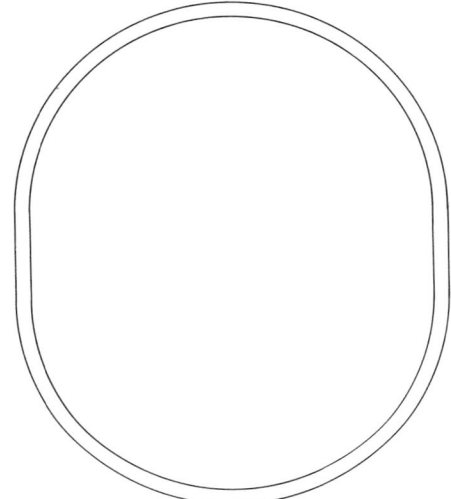

Fig. 10 Elongated parallel fiber reinforced ring specimen. Source: Ref 17

The modulus of elasticity and shear modulus usually reported are obtained from the initial straight line of the stress-strain curve, such as the one shown in Fig. 4. Other types of moduli are of interest, and, if the total stress-strain curve is provided, these other values can be calculated.

With a stress-strain curve, the yield stress can also be calculated if the criterion for yield is defined and the material is such that a yield stress exists. This offset is similar to that used for metal: 0.2% strain, and the calculation is the same.

Tensile Test. The measured strength of composites can be sensitive to the specimen configuration. In general, the more isotropic the specimen, the greater the number of configurations that can be used. For unidirectional specimens, the recommended configuration is shown in Fig. 5, which is the configuration defined in ASTM D 3039 (Ref 12). A recent round robin by the ASTM Committee D-30 (Ref 1) showed that the unidirectional straight-sided specimen gave higher values than the

streamlined (reduced-section) specimen (Fig. 6). The initial data comparing these two specimen configurations revealed comparable results, but it was determined that specimens had fibers in multiple directions (Ref 13). The cost of machining the steamlined or other reduced-section specimens (as defined in ASTM D 638, Ref 14), which is usually done using mechanical or computer-controlled tracing machines, does not seem to be justified even for fabrics or laminates. For nonunidirectional laminates, the width of the coupon shown in Fig. 5 is usually increased to at least 25 mm (1 in.). A wider specimen also is usually used for testing transverse (or 90°) unidirectional specimens.

For normal applications, the coupon is preferred, but for sandwich honeycomb structures, the sandwich beam test specimen is usually preferred (Fig. 7). Also, refer to ASTM C 273 (Ref 15).

The specimen shown in Fig. 8 is the recommended configuration for unreinforced resins and for chopped fiber composites.

Filament-winding applications sometimes require special specimens, such as the parallel fiber reinforced ring used by ASTM D 2290 (Ref 16) and shown in Fig. 9. This specimen is useful, but because the split diameters of the test fixture cause stress concentration at the edge, the tensile strength values are conservative. A better test (Ref 17) is to pressurize the ring or to make a parallel fiber reinforced ring with an elongated section, as shown in Fig. 10. A third approach, which seems to be preferred by the JANNAF committee (Ref 18), is to fabricate a small bottle and then pressure-test it. One size and test procedure is provided in ASTM D 2585 (Ref 19).

Tests conducted at elevated temperatures can use the same specimen configuration and gripping arrangement (Ref 4) as those conducted in ambient conditions but require the use of high-temperature adhesives to bond tabs to the specimen. At temperatures below ambient, the same specimen configuration can be used (Ref 5), but proper adhesives must be used, and any jaws or moving parts must be free of moisture, which could freeze and prevent the required motion for a valid test.

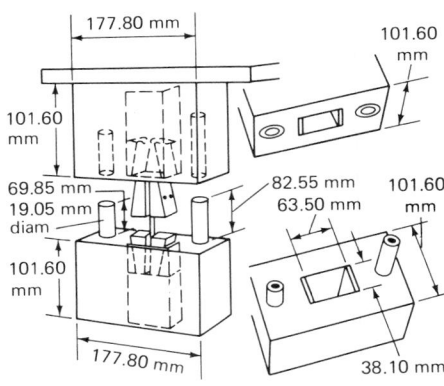

Fig. 12 Compression test fixture with pyramidal wedges. Source: Ref 21

Compression Test. The measured compression strength of composites is sensitive to the specimen configuration and the fixture used for loading. The specimen must first be constrained from buckling (general instability). Several years of development have shown that the three methods given in ASTM D 3410 (Ref 20), two of which are shown in Fig. 11 and 12, give similar measured strengths (Ref 21). These methods were developed for unidirectional composites. The method shown in Fig. 13 did not give as high a strength value as the other methods and therefore is not recommended for unidirectional materials.

As in the case of measuring tension, the sandwich beam test specimen is recommended for applications that use sandwich construction. Coupon specimens are recommended for all other applications.

For nonunidirectional laminates, a variety of fixtures can be used, all of which will result in equivalent strength values (Ref 1). Since several fixtures can be used, no standard has been established yet. Therefore, the three methods in ASTM D 3410 are recommended, since they will give acceptable results. A possible difficulty with using them is that the specimen is relatively small, and it is not generally possible to test specimens that have been intentionally

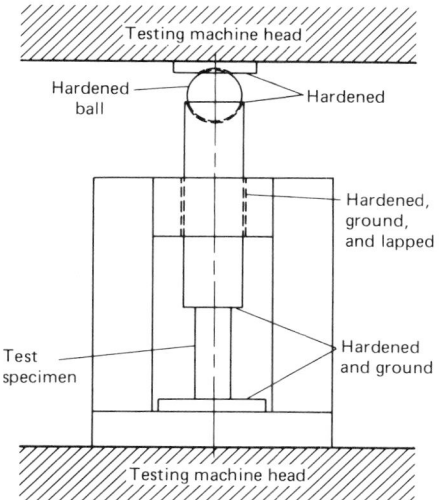

Fig. 13 Compression specimen and tool for plastics. Source: Ref 22

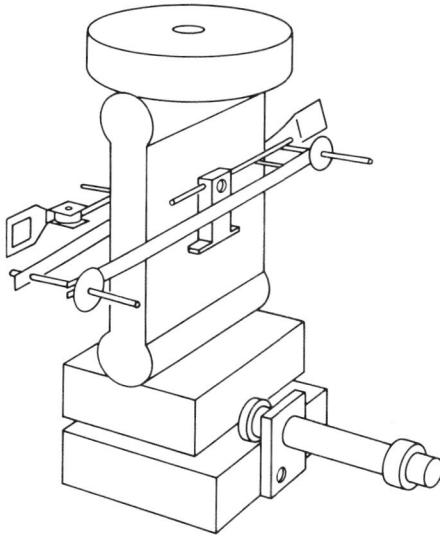

Fig. 14 Edgewise compression test for sandwich specimen. Source: Ref 23

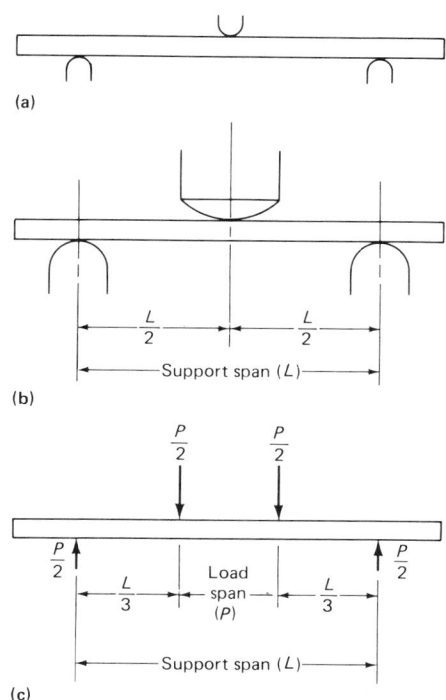

Fig. 15 Flexural test arrangement. (a) Minimum radius, 3.2 mm (⅛ in.). (b) Maximum radius supports, 1.5 × specimen depth; Maximum radius loading nose, 4 × specimen depth. (c) Loading diagram. Source: Ref 24

damaged. No recommendations can be made for this condition.

For chopped fiber reinforced specimens and reinforced plastic, the method outlined in ASTM D 695 (Ref 22) is recommended. The use of cylinders or rectangular prisms, shown in Fig. 14, is especially useful.

No special method for compression testing of filament-wound materials has been identified, although parallel fiber reinforced rings can be used with application of uniform inward pressure. In general, compression values should be found by one of the methods discussed in this section.

A final compression test uses the edge compression specimen described in ASTM C 364 (Ref 23) and shown in Fig. 14, although a test using this specimen is generally considered a subcomponent test rather than a coupon test.

Flexural Test. The one test method most quoted for flexural testing is ASTM D 790 (Ref 24). In general, flexural strength is not considered an intrinsic property; however, the test is inexpensive to run and is usually considered a good quality-control test (Ref 25). It is important that the actual conditions of the test be given, since many variations exist.

Laminates are usually tested with a span-to-depth ratio of 32 to 1 (Fig. 15). One notable exception is an aramid fiber composite, in which case a span-to-depth ratio of 16 to 1 is recommended for testing strength, and a span-to-depth ratio of 60 to 1 is recommended for testing modulus (Ref 26).

The calculations of strength and modulus depend upon thickness squared and cubed, respectively. Therefore, the measurements of thickness are critical. Ball-type micrometers are recommended for this. It is important to note the failure mode. If the specimens fail at the load points, modifications of the configuration must be made. Larger loading points can raise the reported strength values.

Simple beam theory used to reduce the data for lamina assumes equal tensile and compressive moduli for the material; this assumption is not necessarily true and leads to a significant difference between the properties determined from flexural tests and tension or compression tests. Also, there are two methods employed to determine flexural modulus: measurement of deflection at the midspan and measurement of strain on the outer fibers at the

Table 1 Shear test methods

Method name	ASTM designator	Type
Short beam shear (three-point)	D 2344	Interlaminar
Short beam shear (four-point)	. . .	Interlaminar
Notched shear	D 3846	Interlaminar
±45° tensile.	D 3518	In-plane
10° off-axis	. . .	In-plane
Double-V-notch (Iosipescu).	. . .	In-plane
Rail shear	D 4255	In-plane
Torsion tube	. . .	In-plane

midspan with a strain gage. In many cases, the deflection method must employ a shear correction.

ASTM D 790 was developed for nonreinforced and chopped fiber reinforced plastics. A series of recommendations on span-to-depth and thickness are given for using these materials. For the material with a shear-strength-to-tensile-strength ratio of less than 1 to 8, a span-to-depth ratio of 16 to 1 is satisfactory. This usually is the case for chopped fiber reinforced material.

Shear Test. There are a significant number of shear test specimens and methods; a partial list is shown in Table 1. All of these methods are used to some degree. The torsion tube is generally recognized as the truest shear test method. It has only limited use, but does give accurate results (Ref 1, 27). The problem with using a tube is the strain concentrations caused by the end attachments. Two tube designs are shown in Fig. 16, and there are others that could be used, but in all cases the ends must be such that the stress concentration is minimized.

The most used, and some say abused, shear test is the apparent interlaminar shear strength, ASTM D 2344 (Ref 28). This method does not measure modulus nor accurately measure the shear strength; however, it is a very useful quality-control test. Recent efforts (Ref 29) suggest that the four-point beam can be used to measure interlaminar shear strength accurately. However, the three-point method remains a standard because it is such a good measure of quality of the laminate, and so much data are based on it.

The standard notched shear specimen in ASTM D 3846 (Ref 30) is not recommended for any test. The ±45° tensile test, in ASTM D 3518 (Ref 31), is reported to measure the in-plane shear modulus; however, there is a biaxial state of stress, that is, tensile and shear, in this specimen that limits its usefulness. The same is true of the +10° off-axis specimen.

Currently the most recommended method uses some variation of the double-V-notch specimen. There are several fixtures for this coupon, but the method attributed to Iosipescu is the most used and is the one recommended (Fig. 17). This method is used to measure both modulus and strength and can be used for lamina, laminates, and short fiber reinforced plastics. If the notch is properly tailored, it can even be used for brittle materials (Ref 32).

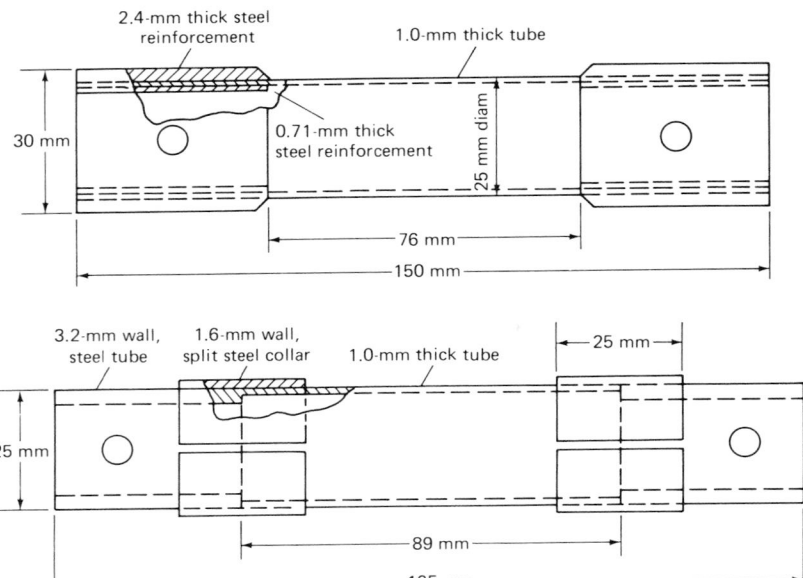

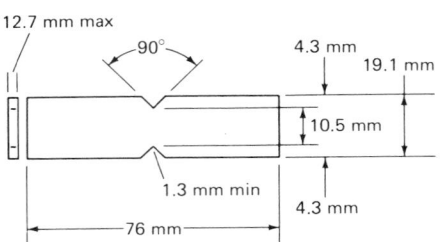

Fig. 16 Torsion tube shear configurations. Source: Ref 5

Fig. 17 Iosipescu shear test specimen. Source: Ref 32

The rail shear test (ASTM D 4255, Ref 33) measures the same property as the double-V-notch shear test but uses substantially more material. The rail shear test is subject to a great deal of variability and is only recommended in cases where it is not practical to use the double-V-notch test.

Documentation

It is important to document all aspects of the material and test. The report should contain:

- A complete description of the material, including code numbers, lot numbers, run numbers, and any physical test data
- A description of the technique used to fabricate the panels, tubes, and so forth, especially noting out-time and moisture conditions of the lay-up room
- The number and orientation of plies
- The specimen fabrication techniques, including cutting methods, moisture condition, and storage times
- The measured dimensions of each specimen
- The instrumentation and methods of attachment, including storage time

- All information relating to deliberate exposures
- The test environment, including temperature and humidity
- The type of machine, the grip, and the speed of the test
- All the calculated values of modulus, strength, strain-to-failure, and Poisson's ratio (if reported), including any test points that were discarded and the reasons for discarding them
- The type of failure

Remember that it is easier to record information than to reconstruct it later.

REFERENCES

1. ASTM Committee D-30, Subcommittee on "High Modulus Fibers and Their Composites," American Society for Testing and Materials
2. S.L. Channon, "Minutes of Workshop on Composite Materials Test Methods," Report M-81, Institute for Defense Analyses, April 1985
3. Unpublished data, Rohr Industries, Inc.
4. N.R. Adsit and J.P. Waszczak, Effect of Near Visual Damage on the Properties of Graphite/Epoxy, in *Fracture Mechanics of Composites*, STP 674, American Society for Testing and Materials, 1979, p 163-176
5. J.R. Kerr and J.F. Haskins, "Time-Temperature-Stress Capabilities of Composite Materials for Advanced Supersonic Technology Applications," GDC-MAP-80-001, General Dynamics Convair, 1980
6. P. Shyprykevich and W. Wolter, Effects of Extreme Aircraft Storage and Flight Environments on Graphite/Epoxy, in *Compos-*

ites for Extreme Environments, STP 768, N.R. Adsit, Ed., American Society for Testing and Materials, 1981, p 118-134
7. J.F. Haskins, R.D. Holmes, *et al.*, "Advanced Composites Design Data for Spacecraft Structural Applications," AFML-TR-79-4208, Air Force Materials Laboratory, 1979
8. D.R. Dunbar, A.R. Robertson, and R. Kerrison, "Graphite/Epoxy Booms for the Space Shuttle Remote Manipulator," presented at the International Conference on Composite Materials, Toronto, 1978
9. J.R. Goulding, Microcracking of PMR-15/Gr Laminates Due to Thermal Cycling, in *High Temple Workshop V*, AFWAL/MLSE, March 1985
10. N.R. Adsit, Some Experiences in Elevated Temperature Testing of Graphite-Reinforced Composite Materials, in *The Enigma of the Eighties: Environment, Economics, Energy*, 24th National SAMPE Symposium and Exhibition, Vol 24 (No. 1), Society for the Advancement of Material and Process Engineering, 1979, p 829-838
11. M.B. Kasen, Cryogenic Properties of Filament-Reinforced Composites: An Update, in *Cryogenics*, June 1981, p 323-339
12. "Standard Test Method for Tensile Properties of Fiber-Resin Composites," D 3039, *Annual Book of ASTM Standards*, American Society for Testing and Materials
13. D.W. Oplinger, B.S. Parker, K.R. Gandhi, R. Lamothe, and G. Foley, On the Streamlined Specimen for Tension Testing of Composite Materials, in *Recent Advances in Composites*, STP 864, J.R. Vinson and M. Taya, Ed., American Society for Testing and Materials, 1985, p 532-555
14. "Standard Test Method for Tensile Properties of Plastics," D 638, *Annual Book of ASTM Standards*, American Society for Testing and Materials
15. "Standard Method of Shear Test in Flatwise Plane of Flat Sandwich Constructions or Sandwich Cores," C 273, *Annual Book of ASTM Standards*, American Society for Testing and Materials
16. "Standard Test Method for Apparent Tensile Strength of Ring or Tubular Plastics and Reinforced Plastics by Split Disk Method," D 2290, *Annual Book of ASTM Standards*, American Society for Testing and Materials
17. L.L. Clements, and T.T. Chiao, Engineering Design Data for an Organic Fiber/Epoxy Composite, *Composites*, Vol 8 (No. 2), April 1977, p 87-92
18. Joint-Army-Navy-NASA-Air Force Interagency Propulsion Committee in conjunction with Laurel, MD, Chemical Propulsion Information Agency, private communication
19. "Standard Method for Preparation and Tension Testing of Filament-Wound Pres-

sure Vessels,'' D 2585, *Annual Book of ASTM Standards*, American Society for Testing and Materials

20. "Standard Test Method for Compressive Properties of Unidirectional or Crossply Fiber-Resin Composites,'' D 3410, *Annual Book of ASTM Standards*, American Society for Testing and Materials

21. N.R. Adsit, Compression Testing of Graphite/Epoxy, in *Compression Testing of Homogeneous Materials and Composites*, STP 808, R. Chait and R. Papirno, Ed., American Society for Testing and Materials, 1983, p 175-186

22. "Standard Test Method for Compressive Properties of Rigid Plastics,'' D 695, *Annual Book of ASTM Standards*, American Society for Testing and Materials

23. "Standard Test Method for Edgewise Compressive Strength of Flat Sandwich Constructions,'' C 364, *Annual Book of ASTM Standards*, American Society for Testing and Materials

24. "Standard Test Methods for Flexural Properties of Unreinforced and Reinforced Plastics and Electrical Insulating Materials, D 790, *Annual Book of ASTM Standards*, American Society for Testing and Materials

25. R.B. Pipes and R.A. Blake, Jr., Test Methods, in *Composites Design Encyclopedia*, Vol 6, University of Delaware, 1983, p 68-72

26. C. Zweben, W.S. Smith, and M.W. Wardle, Test Methods for Fiber Tensile Strength, Composite Flexure Modulus, and Properties of Fabric-Reinforced Laminates, in *Composite Materials: Testing and Design* (Fifth Conference), STP 674, S.W. Tsai, Ed., American Society for Testing and Materials, 1979, p 228-262

27. N.R. Adsit, Shear Testing of Advanced Composites, in *Proceedings of the Sixth St. Louis Symposium on Composite Materials in Engineering Design*, B.R. Norton, Ed., American Society for Metals, 1973, p 448-460

28. "Standard Test Method for Apparent Interlaminar Shear Strength of Parallel Fiber Composites by Short-Beam Method,'' D 2344, *Annual Book of ASTM Standards*, American Society for Testing and Materials

29. C.E. Browning, F.L. Abrams, and J.M. Whitney, A Four-Point Shear Test for Graphite/Epoxy Composites, in *Composite Materials: Quality Assurance and Processing*, STP 797, C.E. Browning, Ed., American Society for Testing and Materials, 1983, p 54-74

30. "Standard Test Method for In-Plane Shear Strength of Reinforced Plastics,'' D 3846, *Annual Book of ASTM Standards*, American Society for Testing and Materials

31. "Standard Practice for Inplane Shear Stress-Strain Response of Unidirectional Reinforced Plastics,'' D 3518, *Annual Book of ASTM Standards*, American Society for Testing and Materials, June 1985

32. D. Walrath and D.F. Adams, "Iosipescu Shear Properties of Graphite Fabric/Epoxy Composite Laminates,'' UWME-DR-501-103-1, University of Wyoming

33. "Standard Guide for Testing Inplane Shear Properties of Composite Laminates,'' D 4255, *Annual Book of ASTM Standards*, American Society for Testing and Materials

Statistical Analysis of Mechanical Properties

Donald Neal and Mark Vangel, U.S. Army Materials Technology Laboratory
Frederick Todt, Battelle Columbus Laboratories

STATISTICAL PROCEDURES for obtaining material basis values in order to evaluate composite material capabilities in structural designs are the subject of this article.

Unlike most traditional structural materials, which are homogeneous and isotropic, composite materials have extensive intrinsic statistical variability in many material properties. This variability, particularly important to strength properties, is due not only to inhomogeneity and anisotropy, but also to the basic brittleness of many matrices and most fibers and to the potential for property mismatch between the components. Because of this inherent statistical variability, careful statistical analysis of composite material properties is not only more important but is also more complex than for traditional structural materials.

A basis property is a stress value that is determined so that there is a specified probability of the strength exceeding this value with 95% confidence in the assertion. The survival probabilities are in the 99th percentile for the A-basis and 90th percentile for the B-basis. Only the B-basis value will be considered in this article.

The statistical procedures necessary for obtaining basis values from composite material strength data are described in flowchart form in Fig. 1. The sequence of operations is dependent on the batch size, the number of batches, the differences among the batches, and the identification of statistical models.

Military and aircraft industry requirements for the statistical evaluation of composite material properties guided the selection of methods shown in Fig. 1. Both the development of new statistics and the modification of current methods were necessary to meet the requirements. The k-sample Anderson-Darling test, improvements to the modified Lemon method, the extension and correction of the tolerance factors for Weibull basis properties, and the development of a nonparametric small sample procedure were required. A computer program written to perform the computations necessary to obtain B-basis values would be helpful. Following Fig. 1 should prevent incorrect use of the statistical procedures. An increase in either the number of batches or the number of test values within each batch would augment the power of the tests and the precision of the estimates.

In the single-batch case, the data is initially screened for outliers. This is required so that the user can recognize possible manufacturing or testing problems. Determining which statistical model is acceptable involves applying goodness-of-fit tests. In order of preference, the Weibull and normal statistical distributions are the best candidates for such modeling. If neither candidate meets the test requirements, nonparametric methods are employed, when appropriate.

The computation of basis properties for more than one batch of material requires applying a test for equality of populations. If equality can be assumed, then the data from the batches are pooled, and the basis value is determined using procedures outlined for the single-batch case. If there are three or more batches, and pooling is not possible because of differences among the batch populations, additional tests are required. For two batches, either additional batches or correction of the variability is necessary. The hypothesis of equality of variance among the batches is tested. If this assumption is tenable, then the modified Lemon method (Ref 1, 2) is employed to obtain the B-basis value. This method is based on a balanced one-way random-effects ANOVA model. It is important to be aware of inequality among the batch variances, although this inequality does not affect the use of the modified Lemon method.

The statistics chapter in *Military Handbook 17B* on composite material in aircraft applications (Ref 3) provides a more detailed descrip-

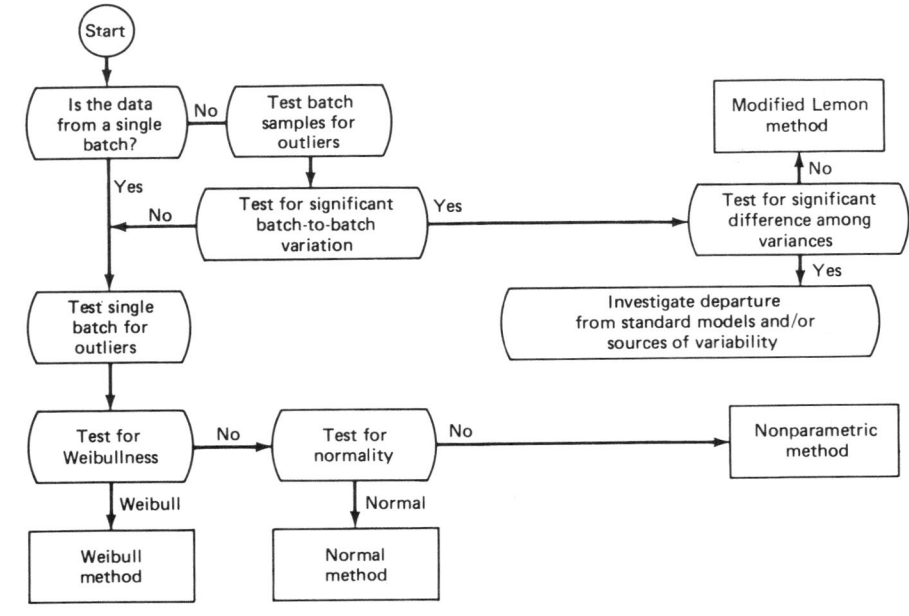

Fig. 1 Computational procedures for statistically based material properties

tion of the methods described in this article and a number of carefully chosen example problems.

Detecting Outliers

Before computing a B-basis value, it is necessary to screen the data carefully for outliers and determine which method of analysis is appropriate (see Fig. 1). If outlying data values are identified, they should be investigated. If a rationale is established for the existence of the outliers, it is suggested that they either be corrected or removed from the data set.

Exploratory data analysis techniques should be considered before any formal data analysis is attempted (Ref 4). These procedures, though qualitative, may indicate potential problems.

In searching for outliers, the first step is to compute the maximum normed residual (MNR) statistic (Ref 5). If $x_1, x_2, \ldots, x_n$ denote the data values in a sample of size n, the normed residuals are defined as:

$$r_i = \frac{x_i - \bar{X}}{S} \qquad \text{(Eq 1)}$$

where $i = 1, 2, \ldots, n$ and $\bar{X}$ and S are the sample mean and sample standard deviation, respectively ($\bar{X}$ and S can be computed by Eq 9). The normed residuals are scaled deviations from the "center" of the data. The MNR statistic is the maximum of the absolute values of the normed residuals:

$$MNR = \max_i |r_i| \qquad \text{(Eq 2)}$$

Next, the MNR statistic is compared to the critical value for the sample size n. If MNR is smaller than the critical value, then no outliers are detected. If MNR is larger than the critical value, the datum associated with the largest absolute normed residual is declared to be an outlier. When sample sizes are small, values which appear to be outliers should be investigated, even if they are not declared outliers by this test.

The critical values, CV, for this test are computed from the following formula (Ref 5):

$$CV = \frac{n-1}{\sqrt{n}} \sqrt{\frac{t^2}{n-2+t^2}} \qquad \text{(Eq 3)}$$

where t is the $1 - \alpha/(2n)$ quantile of the t distribution with $n - 2$ degrees of freedom (Table 1). A significance level of $\alpha = 0.05$ is suggested.

Single-Batch Analysis

If a B-basis value is to be obtained for a single batch of material or from pooled data from several batches, the following procedures should be applied. It is assumed that the examination for outliers has been completed. Therefore, the next step is to attempt to identify an appropriate model for the data using goodness-

Table 1 Critical values (CV) for the maximum normed residual outlier test

n	CV $\alpha = .01$	$\alpha = .05$	$\alpha = .10$
5	1.764	1.715	1.671
6	1.973	1.887	1.822
7	2.139	2.020	1.938
8	2.274	2.127	2.032
9	2.387	2.215	2.110
10	2.482	2.290	2.176
11	2.564	2.355	2.234
12	2.636	2.412	2.285
13	2.699	2.462	2.331
14	2.755	2.507	2.372
15	2.806	2.548	2.409
16	2.852	2.586	2.443
17	2.894	2.620	2.475
18	2.932	2.652	2.504
19	2.968	2.681	2.531
20	3.001	2.708	2.557
21	3.031	2.734	2.580
22	3.060	2.758	2.603
23	3.087	2.780	2.624
24	3.112	2.802	2.644
25	3.135	2.822	2.663
26	3.158	2.841	2.681
27	3.179	2.859	2.698
28	3.199	2.876	2.714
29	3.218	2.893	2.730
30	3.236	2.908	2.745

of-fit tests. The Weibull model with the cumulative distribution function

$$F_o(x) = 1 - e^{-(x/\alpha)^\beta} \qquad \text{(Eq 4)}$$

should be considered first (see Fig. 1). The estimates for the shape parameter, β, and scale parameter, α, are obtained by the maximum likelihood method (Ref 6). Figure 2 defines the computer code for obtaining these Weibull parameters.

The Anderson-Darling (AD) goodness-of-fit test statistic (Ref 6, 7) is suggested for identifying a model because it is sensitive to discrepancies in the tail regions. This test is based on the comparison of the cumulative distribution function of the data with the cumulative distribution function of the model.

For the Weibull distribution (Eq 4):

$$Z_{(i)} = \left(\frac{X_{(i)}}{\hat{\alpha}}\right)^\beta \qquad \text{(Eq 5)}$$

where $i = 1, \ldots, n$ and $\hat{\alpha}$ and $\hat{\beta}$ are the maximum likelihood estimates of α and β, the sample size is n, and the sample observations ordered from least to greatest are $x_{(1)}, \ldots, x_{(n)}$.

The AD test statistic is:

$$AD = \sum_{i=1}^{n} \frac{1-2i}{n} \{\ln[1 - \exp(-Z_{(i)})] - Z_{(n+1-i)}\} - n \qquad \text{(Eq 6)}$$

The observed significance level (OSL) is calculated as:

$$OSL = 1/\{1 + \exp[-0.10 + 1.24 \ln(AD^*) + 4.48\, AD^*]\} \qquad \text{(Eq 7)}$$

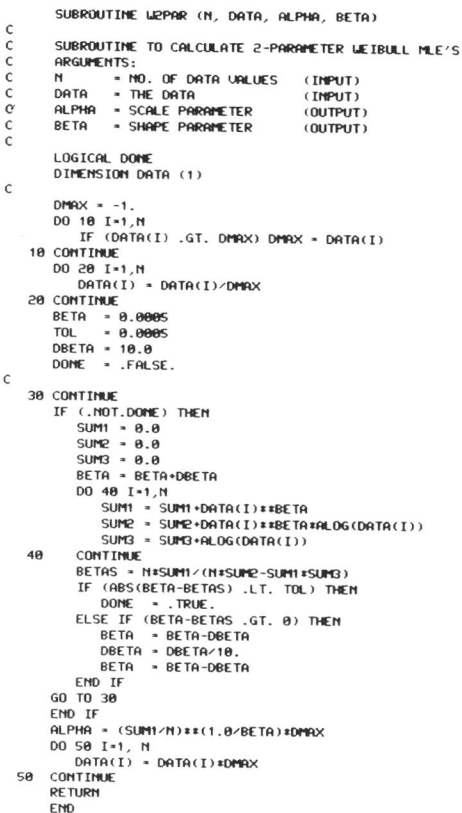

```
C
      SUBROUTINE W2PAR (N, DATA, ALPHA, BETA)
C
C     SUBROUTINE TO CALCULATE 2-PARAMETER WEIBULL MLE'S
C     ARGUMENTS:
C        N      = NO. OF DATA VALUES   (INPUT)
C        DATA   = THE DATA             (INPUT)
C        ALPHA  = SCALE PARAMETER      (OUTPUT)
C        BETA   = SHAPE PARAMETER      (OUTPUT)
C
      LOGICAL DONE
      DIMENSION DATA (1)
C
      DMAX = -1.
      DO 10 I=1,N
         IF (DATA(I) .GT. DMAX) DMAX = DATA(I)
10    CONTINUE
      DO 20 I=1,N
         DATA(I) = DATA(I)/DMAX
20    CONTINUE
      BETA = 0.0005
      TOL = 0.0005
      DBETA = 10.0
      DONE = .FALSE.
C
30    CONTINUE
      IF (.NOT.DONE) THEN
         SUM1 = 0.0
         SUM2 = 0.0
         SUM3 = 0.0
         BETA = BETA+DBETA
         DO 40 I=1,N
            SUM1 = SUM1+DATA(I)**BETA
            SUM2 = SUM2+DATA(I)**BETA*ALOG(DATA(I))
            SUM3 = SUM3+ALOG(DATA(I))
40       CONTINUE
         BETAS = N*SUM1/(N*SUM2-SUM1*SUM3)
         IF (ABS(BETA-BETAS) .LT. TOL) THEN
            DONE = .TRUE.
         ELSE IF (BETA-BETAS .GT. 0) THEN
            BETA = BETA-DBETA
            DBETA = DBETA/10.
            BETA = BETA-DBETA
         END IF
         GO TO 30
      END IF
      ALPHA = (SUM1/N)**(1.0/BETA)*DMAX
      DO 50 I=1, N
         DATA(I) = DATA(I)*DMAX
50    CONTINUE
      RETURN
      END
```

Fig. 2 Computer code for obtaining maximum likelihood estimates of Weibull parameters

where $AD^* = (1 + 0.2/\sqrt{n})AD$ (Ref 3, 8). The OSL measures the goodness of fit of the two-parameter Weibull distribution to the data. Specifically, the OSL is the probability of observing an AD statistic as extreme as the value calculated if the two-parameter Weibull distribution is in fact the correct model.

If the $OSL \leq 0.05$, one may conclude (at a 5% or less risk of being in error) that the data is not a sample from a two-parameter Weibull population. Otherwise, a two-parameter model may be used to calculate the B-basis value. If the Weibull model is rejected, the AD procedure is applied to the normal distribution. Although there is no simple "weakest-link" model consistent with the strength distribution of composite materials, there is substantial empirical evidence supporting the use of the Weibull distribution. It is therefore suggested that the Weibull model be considered first. Note that for small sample sizes, it is difficult to distinguish between Weibull and normal distributions.

The normal distribution is written as:

$$F_o(x) = 1/\sqrt{2\pi} \int_{-\infty}^{x} e^{-(t-\mu)^2/2\sigma^2}\, dt \qquad \text{(Eq 8)}$$

The parameters are the mean μ and standard deviation σ. These are estimated by the sample mean and sample standard deviation, respectively:

$$\bar{X} = \sum_{i=1}^{n} X_i/n \quad \text{and}$$

$$S = \sqrt{\sum_{i=1}^{n} (X_i - \bar{X})^2/(n-1)} \qquad \text{(Eq 9)}$$

The goodness-of-fit procedure for the normal distribution is similar to that for the Weibull distribution, except for the definitions of the $Z_{(i)}$, AD, and the OSL. Let:

$$Z_{(i)} = (X_{(i)} - \bar{X})/S \qquad \text{(Eq 10)}$$

where $i = 1, ..., n$. The AD test statistic is:

$$AD = \sum_{i=1}^{n} \left(\frac{1-2i}{n} \right) [\ln F_o(Z_{(i)}) +$$

$$\ln (1 - F_o(Z_{(n+1-i)}))] - n \qquad \text{(Eq 11)}$$

where F_o is the standard normal distribution function, and the OSL is calculated as (Ref 3, 9):

$$OSL = 1/\{1 + \exp[-0.48 + 0.78 \ln(AD^*) +$$

$$4.58(AD^*)]\} \qquad \text{(Eq 12)}$$

where $AD^* = (1 + 4/n - 25/n^2)AD$. The OSL measures the goodness-of-fit of a normal distribution to the data. Specifically, the OSL is the probability of observing an AD statistic that is as extreme as the value calculated, if the normal distribution is in fact the correct model.

If the $OSL \leq .05$, one may conclude (at a maximum 5% risk of being in error) that the data is not a sample from a normal population. Otherwise, a normal model may be used to calculate the B-basis value.

If neither distribution is accepted by the Anderson-Darling test, nonparametric methods (Ref 10, 11) should be used (see Fig. 1).

Weibull Method for Obtaining B-Basis Values. If the Weibull model has been accepted and the parameter estimates have been obtained, the B-basis value can then be determined from the following relationship (Ref 12, 13):

$$B = \hat{\alpha} [\ln(1/P_B)]^{1/\beta} \qquad \text{(Eq 13)}$$

where P_B is obtained from Table 2 as a function of the sample size n.

Normal Method for Obtaining B-Basis Values. The B value is calculated in terms of the sample mean $\bar{X}$ and standard deviation S by using the following formula (Ref 14):

$$B = \bar{X} - K_B S \qquad \text{(Eq 14)}$$

where K_B is obtained from Table 3 as a function of the sample size n.

Nonparametric Methods (Single Population). Nonparametric methods are suggested only if the hypotheses of normality and Weibullness are both rejected. The nonparametric procedures usually provide a more conservative B-basis value than procedures that assume a parametric model.

To compute a B-basis value when the sample size is greater than 28 (Ref 10), first the sample is ordered so that the observations are in order of increasing value. Then, using Table 4(a), the r value corresponding to the sample size n can be determined. For sample sizes between tabulated values, the r value associated with the largest sample size that is smaller than the actual n should be used. The B-basis value is the rth lowest observation in the ordered sample.

For example, in a sample of size $n = 30$, the lowest ($r = 1$) observation is the B-basis value. As another example, in a sample of size $n = 200$, the 15th ordered observation ($r = 15$) is the B-basis value.

For sample sizes less than 29, the procedure (Ref 11) to be followed is similar to the large sample method described above. Order the sample values from least to greatest and obtain values r and k from Table 4(b). For example, if n equals 10, then the sixth ordered value with a k of 2.137 is used in Eq 15. If $x_{(r)}$ denotes the rth ordered value and $x_{(1)}$ denotes the first ordered value, the B-basis value is:

$$B = X_{(r)} (X_{(1)}/X_{(r)})^k \qquad \text{(Eq 15)}$$

Procedures for Multiple Batches

When analyzing failure data from material produced in several batches, the extent of the batch-to-batch variation should be assessed first (see Fig. 1). Each batch should be examined for outliers at the 10% significance level (using $\alpha = 0.10$ in Eq 3). The k-sample Anderson-Darling statistic (Ref 3, 15) tests the hypothesis that each of the k batches is a sample from the same unspecified population. If the AD statistic indicates that this hypothesis is consistent with the data, the single-batch procedures may be used for the pooled batches. Otherwise, the modified Lemon method (Ref 1-3) should be employed.

The Anderson-Darling Test. If the data consist of two batches, the following method will determine whether pooling is justified. Let $x_1, ..., x_{n_A}$ denote a sample of size n_A from batch A and let $y_1, ..., y_{n_B}$ denote a sample of size n_B from batch B. Let $N = n_A + n_B$ represent the total number of data values, and $Z_{(1)}, ..., Z_{(N^*)}$ denote the distinct values of the pooled data, ordered from smallest to largest. If there are no ties in the data, $N = N^*$. Otherwise, $N^* < N$. Let h_i be the number of the ties at the value $Z_{(i)}$. The two-sample Anderson-Darling statistic (Ref 3, 16) is:

$$AD2 = \frac{1}{n_A n_B} \sum_{i=1}^{N^*} h_i \frac{(NF_i - n_A H_i)^2}{H_i(N - H_i)} \qquad \text{(Eq 16)}$$

where F_i is defined as the number of x's less than or equal to $Z_{(i)}$, and H_i is the number of x's and y's less than $Z_{(i)}$, plus one-half the number of x's and y's equal to $Z_{(i)}$. If $AD2 < 2.492 - 2.3126/N$, the batches may be pooled. If the data were in fact from a single batch, the test would reach a false conclusion 5% of the time.

If the data consists of k batches, $AD2$ is calculated k times, with the ith batch being batch A and the remaining $k-1$ batches pooled being batch B. These k statistics are denoted AD_i. The k sample Anderson-Darling statistic is given by:

Table 2 B-Basis factors (P_B) for the Weibull distribution

n	P_B	n	P_B	n	P_B	n	P_B
5	0.9987	15	0.9771	25	0.9636	55	0.9453
6	0.9969	16	0.9754	26	0.9626	60	0.9435
7	0.9948	17	0.9738	27	0.9616	65	0.9420
8	0.9929	18	0.9723	28	0.9607	70	0.9406
9	0.9895	19	0.9708	29	0.9599	75	0.9393
10	0.9875	20	0.9695	30	0.9590	80	0.9381
11	0.9852	21	0.9682	35	0.9553	85	0.9371
12	0.9830	22	0.9669	40	0.9522	90	0.9361
13	0.9809	23	0.9658	45	0.9495	95	0.9352
14	0.9790	24	0.9647	50	0.9473	100	0.9344

Table 3 B-Basis factors (K_B) for the normal distribution

n	K_B	n	K_B	n	K_B	n	K_B
1	...	11	2.275	21	1.905	35	1.732
2	20.581	12	2.210	22	1.886	40	1.697
3	6.155	13	2.155	23	1.869	45	1.669
4	4.162	14	2.109	24	1.853	50	1.646
5	3.407	15	2.068	25	1.838	55	1.626
6	3.006	16	2.033	26	1.824	60	1.609
7	2.755	17	2.002	27	1.811	70	1.581
8	2.582	18	1.974	28	1.799	80	1.559
9	2.454	19	1.949	29	1.788	90	1.542
10	2.355	20	1.926	30	1.777	100	1.527

$$ADk = \frac{1}{N(k-1)} \sum_{i=1}^{k} (N - n_i) \, AD_i \qquad \text{(Eq 17)}$$

where n_i is the size of the ith batch and N is the sum of the n_i's. If the ADk statistic does not exceed the critical value

$$CV = 1 + \frac{\left(1.25 - \dfrac{1.75}{\bar{n}}\right)}{(k-1)^{0.5}} + \frac{0.262}{(k-1)^{0.75}}$$

where $\bar{n} = \dfrac{1}{k} \sum_{i=1}^{k} n_i \qquad \text{(Eq 18)}$

then the batches may be pooled, with the same risk of a wrong decision as in the two-batch case.

Test for Equality of Variance. If the k sample Anderson-Darling test indicates that the batches cannot be pooled, the modified Lemon procedure is the only acceptable method currently available. This procedure is based on the assumption that the data represent a sample from a normal population with equal variance in each batch. The hypothesis of equal variance must be tested, and if this hypothesis is inconsistent with the data, there is no available procedure for obtaining a statistically valid material basis property. However, verification of the normality assumption does not appear to be necessary.

In the following equations, the jth specimen in the ith of k batches is denoted by x_{ij}. The size of the ith batch is n_i, and the pooled sample size is N. For equality of variance, first determine the batch mean and standard deviation, $\bar{X}_i$ and S_i, respectively:

$$\bar{X}_i = \sum_{j=1}^{n_i} x_{ij}/n_i$$

and

$$S_i = \sqrt{\frac{n_i \sum_{j=1}^{n_i} x_{ij}^2 - \left(\sum_{j=1}^{n_i} x_{ij}\right)^2}{n_i(n_i - 1)}} \qquad \text{(Eq 19)}$$

The equality of variance (EV) statistic may then be calculated as follows:

$$EV = \frac{1}{2} \sum_{i=1}^{k} (n_i - 1) Z_i^2 - \frac{1}{2}(N - k)\bar{Z}^2 \qquad \text{(Eq 20)}$$

where

$$Z_i = \ln(S_i^2)$$

and

$$\bar{Z} = \sum_{i=1}^{k} (n_i - 1) Z_i/(N - k)$$

Table 4(a) Ranks (r) of observations (n) for determining B-basis values for an unknown distribution

n	r	n	r	n	r
<29	... See Table 4(b)	129	... 8	227	... 16
29	... 1	142	... 9	239	... 17
46	... 2	154	...10	251	... 18
61	... 3	167	...11	263	... 19
76	... 4	179	...12	275	... 20
89	... 5	191	...13	298	... 22
103	... 6	203	...14	321	... 24
116	... 7	215	...15	345	... 26

If the value of the Lehmann equality of variance statistic (Ref 17) does not exceed the 95th percentile of a χ^2 distribution with $k - 1$ degrees of freedom (Tables 5 and 6), one may conclude with a 5% risk of error that the batch variances are equal.

Modified Lemon Method. If the assumption of equality of variance has not been contradicted, the modified Lemon procedure (Ref 1-3) should be used to obtain a statistically valid material basis property that takes into account variation among the batches.

The first step in this method is to compute the between-batch mean square (MSB) and the within-batch mean square (MSE). These values are defined as:

$$MSB = \sum_{i=1}^{k} (n_i \bar{X}_i^2 - N \bar{X}^2)/(k - 1) \qquad \text{(Eq 21)}$$

and

Table 4(b) B-basis values for small sample sizes

n	r	k
2	2	35.177
3	3	7.859
4	4	4.505
5	4	4.101
6	5	3.064
7	5	2.858
8	6	2.382
9	6	2.253
10	6	2.137
11	7	1.897
12	7	1.814
13	7	1.738
14	8	1.599
15	8	1.540
16	8	1.485
17	8	1.434
18	9	1.354
19	9	1.311
20	10	1.253
21	10	1.218
22	10	1.184
23	11	1.143
24	11	1.114
25	11	1.087
26	11	1.060
27	11	1.035
28	12	1.010

$$MSE = \left(\sum_{i=1}^{k} \sum_{j=1}^{n_i} x_{ij}^2 - \sum_{i=1}^{k} n_i \bar{X}_i^2\right)/(n - k) \qquad \text{(Eq 22)}$$

where

$$\bar{X} = \sum_{i=1}^{k} \sum_{j=1}^{n_i} x_{ij}/N \qquad \text{(Eq 22a)}$$

The within-batch variance, σ_w^2, is estimated by MSE. The between-batch variance, σ_b^2, is estimated by:

Table 5 The 95th percentile of the χ^2 distribution (q) associated with degrees of freedom (df)

df	q	df	q	df	q	df	q
1	3.842	11	19.681	21	32.678	35	49.811
2	5.995	12	21.030	22	33.933	40	55.753
3	7.817	13	22.367	23	35.178	45	61.652
4	9.492	14	23.691	24	36.421	50	67.501
5	11.073	15	25.000	25	37.660	55	73.308
6	12.596	16	26.301	26	38.894	60	79.078
7	14.070	17	27.593	27	40.119	70	90.528
8	15.512	18	28.877	28	41.344	80	101.876
9	16.925	19	30.148	29	42.565	90	113.143
10	18.311	20	31.416	30	43.782	100	124.340

Table 6 $C_{k,n}$ factors for modified Lemon method

For k batches of average size n, use the value at the intersection of row k with column n.

	3	4	5	6	7	8	9	≥10
3	2.0281	2.6230	3.0540	3.3810	3.6377	3.8448	4.0156	4.1585
4	2.3995	2.7834	3.0022	3.1400	3.2335	3.2999	3.3494	3.3875
5	2.4244	2.6059	2.6900	2.7365	2.7647	2.7836	2.7968	2.8067
6	2.2944	2.3708	2.4023	2.4187	2.4288	2.4354	2.4401	2.4436
7	2.1410	2.1751	2.1890	2.1967	2.2016	2.2049	2.2075	2.2093
8	2.0085	2.0265	2.0346	2.0393	2.0426	2.0449	2.0467	2.0481
9	1.9024	1.9140	1.9198	1.9235	1.9259	1.9278	1.9293	1.9305
≥10	1.8180	1.8269	1.8319	1.8346	1.8368	1.8384	1.8396	1.8406

Note: These factors provide upper confidence estimates of R with the confidence level chosen so that the B-basis estimates will, on average, be conservative when the population variance ratio is less than 10.

Table 7 The 95th percentile of the *t* distribution associated with *n* degrees of freedom

n	t	n	t	n	t	n	t
1	12.706	11	2.201	21	2.080	35	2.030
2	4.303	12	2.179	22	2.074	40	2.021
3	3.182	13	2.160	23	2.069	45	2.014
4	2.776	14	2.145	24	2.064	50	2.009
5	2.571	15	2.131	25	2.060	55	2.004
6	2.447	16	2.120	26	2.056	60	2.000
7	2.365	17	2.110	27	2.052	70	1.994
8	2.306	18	2.101	28	2.048	80	1.990
9	2.262	19	2.093	29	2.045	90	1.987
10	2.228	20	2.086	30	2.042	100	1.984

$$S_b^2 = (MSB - MSE)/n' \qquad \text{(Eq 23)}$$

where

$$n' = (N - n^*)/(k - 1)$$

and

$$n^* = \sum_{i=1}^{k^*} n_i^2/N$$

The ratio of between-batch to within-batch variances is next estimated as:

$$\hat{R} = \frac{\dfrac{MSB}{MSE} C_{kn} - 1}{n'} \qquad \text{(Eq 24)}$$

when C_{kn} is the value from Table 6 corresponding to k batches of average size n.

From these values, one next calculates the degrees of freedom as:

$$\lambda = \frac{(\hat{R} + 1)^2}{\dfrac{\left(\hat{R} + \dfrac{1}{n^*}\right)^2}{k^* - 1} + \dfrac{\left(\dfrac{n^* - 1}{n^*}\right)^2}{k^*(n^* - 1)}} \qquad \text{(Eq 25)}$$

where $k^* = N/n^*$.

The tolerance limit factor, T, as a function of k, R, and noncentrality parameter

$$\delta = 1.282 \sqrt{\frac{N(\hat{R} + 1)}{n^*\hat{R} + 1}} \qquad \text{(Eq 26)}$$

can be obtained from tables (Ref 2, 3). Tolerance limit factors can be approximated by the following formula:

$$
\begin{aligned}
T = {}& 1.282 \\
& + 1.282\,(t_{\lambda,0.95}/\delta) \\
& + 1.80\,(1/\lambda) \\
& - 1.85\,(1/\lambda^2) \\
& + 0.567\,(\delta/\lambda) \qquad \text{(Eq 27)} \\
& + 5.24\,(\delta/\lambda^2) \\
& - 1.08\,(\delta^2/\lambda^2) \\
& + 0.0166\,(\delta^3/\lambda^2) \\
& + 7.79\,(1/\lambda^4)
\end{aligned}
$$

where $t_{\lambda,0.95}$ is the 0.95 quantile of a t distribution with λ degrees of freedom (Table 7). This approximation is accurate to within 0.45% of the values in the tables. The B-basis value is then calculated according to the formula:

$$B = \bar{X} - T\sqrt{S_b^2 + MSE} \qquad \text{(Eq 28)}$$

Example. Hypothetical tensile strength measurements from nine batches of material:

1	2	3	4	5
61.3	66.5	66.0	61.9	68.9
68.5	64.7	72.7	68.0	65.0
62.5	64.9	67.1	63.3	70.9
66.0	65.2	67.7	74.6	65.4
66.6	70.3	65.7	66.2	66.5
64.8			68.2	64.9
69.5				69.1

6	7	8	9
75.8	72.8	71.9	68.7
75.2	75.0	71.0	76.3
71.5	66.3	69.5	76.6
69.6	69.5	69.5	66.2
66.1	71.9	72.6	72.4
		74.6	72.8
			109.6

A single outlier, 109.6, is determined ($MNR = 2.17$ with critical value 1.938). This value is investigated and replaced with a corrected value of 69.6. The k-sample Anderson-Darling test suggests that the data cannot be pooled ($ADk = 1.74$ with critical value 1.39). The test for equality of variance ($EV = 5.6$ with critical value 15.51) justifies the use of the modified Lemon method.

After correcting the value, the summary statistics for the data are:

Batch	n_i	$\bar{X}_i$	S_i
1	7	65.60	2.99
2	5	66.32	2.33
3	5	67.84	2.84
4	7	67.33	4.17
5	6	66.93	2.45
6	5	71.64	4.03
7	5	71.10	3.33
8	6	71.52	1.96
9	7	71.80	3.88

Preliminary calculations for the modified Lemon method give:

$SSW = 460.68$	$MSE = 10.47$
$SSB = 317.255$	$C_{9,6} = 2.4401$
$\bar{X} = 88.84$	$S_b^2 = 4.97$
$n^* = 6.0189$	$N = 53$
$k^* = 8.8056$	$n' = 5.8726$
$MSB = 39.66$	$\hat{R} = 1.404$

$$\lambda = 17.43$$
$$t_{17.43,0.95} = 1.737$$
$$\delta = 4.707$$

from Eq 27, $T = 2.014$. Finally, from Eq 28, the B-basis value is $B = 68.84 - 2.014 (3.929) = 60.93$. Note that this value is less than the smallest ordered value in the pooled sample. If pooling could have been justified, the nonparametric B-basis value would be the second smallest ordered value (61.9). Because of batch-to-batch variation, the modified Lemon method for this example results in a B-basis value less than that obtained from the single-batch nonparametric procedure.

REFERENCES

1. G.H. Lemon, Factors for One-Sided Tolerance Bounds for Balanced One-Way ANOVA Random Effects Model, *J. Am. Statis. Assoc.*, Vol 72, 1977, p 676-680
2. R.W. Mee and D.B. Owen, Improved Factors for One-Sided Tolerance Limits for Balanced One-Way ANOVA Random Model, *J. Am. Statis. Assoc.*, Vol 78, 1983, p 901-905
3. Military Handbook 17B, *Polymer Matrix Composites, Volume 1, Guidelines,* Naval Publications and Forms Center, 1988
4. F. Mosteller and J.W. Tukey, *Data Analysis and Regression*, Addison-Wesley, 1977
5. W. Stefansky, Rejecting Outliers in Factorial Designs, *Technometrics*, Vol 14, 1972, p 469-479
6. K.A. Brownlee, *Statistical Theory and Methodology in Science and Engineering*, John Wiley & Sons, 1965
7. T.W. Anderson and D.A. Darling, A Test of Goodness-of-Fit, *J. Am. Statis. Assoc.*, Vol 49, 1954, p 765-769
8. M.A. Stephens, Goodness of Fit for the Extreme Value Distribution, *Biometrika*, Vol 64, 1977, p 583-588
9. M.A. Stephens, EDF Statistics for Goodness of Fit and Some Comparisons, *J. Am. Statis. Assoc.*, Vol 69, 1974, p 730-737
10. W.J. Conover, *Practical Nonparametric Statistics*, John Wiley & Sons, 1980, p 111
11. D.L. Hanson and L.H. Koopmans, Toler-

ance Limits for the Class of Distributions With Increasing Hazard Rates, *Annals of Mathematical Statistics*, Vol 35, 1964, p 1561-1570

12. D. Neal and L. Spiridigliozzi, An Efficient Method for Determining "A" and "B" Allowables ARO 83-2, in *Proceedings of the Twenty-Eighth Conference on the Design of Experiments in Army Research, Development, and Testing*, Army Research Office, 1983, p 199-235

13. D.R. Thoman, L.S. Bain, and C.E. Antle, Maximum Likelihood Exact Confidence Intervals for Reliability and Tolerance Limits on the Weibull Distribution, *Technometrics*, Vol 12, 1970, p 363-371

14. D.B. Owen, A Survey of Properties and Applications of the Noncentral t-Distribution, *Technometrics*, Vol 10, 1968, p 445-478

15. F.W. Scholz and M.A. Stephens, K-Sample Anderson Darling Tests, *Journal of the American Statistical Association*, Vol 82, p 918, 1987

16. A.N. Pettit, A Two-Sample Anderson-Darling Rank Statistic, *Biometrika*, Vol 63, 1976, p 161-168

17. E.L. Lehmann, *Testing Statistical Hypotheses*, John Wiley & Sons, 1959, p 274-275

Generation of Design Allowables

Mark E. Tohlen, LTV Aerospace and Defense Company

LAMINA PROPERTY TEST RESULTS are not in a form that is useful to a stress engineer until the data are reduced, translated into design allowables, and then reported in a format that is easy to use. This article outlines the process of obtaining laminate design properties from lamina level tests, focusing on assumptions and basic models. Actual calculations are available in the article "Element and Subcomponent Tests" in this Volume. Two easily understood formats for displaying the results of design-allowable generation are tables, for lamina level properties, and carpet plots, for laminate level properties.

Basic Assumptions for Lamina Allowables

Generation of design allowables begins with lamina level tests. The first feature apparent in the raw data is its lack of measurements for through-the-thickness stiffness and, sometimes, strength. Development of basic allowables assumes that the out-of-plane properties do not influence the in-plane stiffness or strength of the laminate. Another feature is the change in stiffness and strength in transverse directions due to the ropelike characteristic of the fiber. This feature enables the tailoring of laminate properties by combining different lamina orientations. Another feature immediately apparent is the nonlinearity in some of the data. A useful design allowable requires the translation of this curved, plotted data into a linearized data set. Each of these features must be addressed in generating design allowables.

Material Homogeneity. Although composites are by definition inhomogeneous materials, laminate analysis very often proceeds on the assumption of lamina homogeneity (no fiber and matrix distinction). Figure 1, which depicts actual construction, shows tows (bundles of fibers) for a unidirectional laminate. One of the consequences of inhomogeneity on the lamina level is a thermal mismatch between the fiber and matrix. For a 175 °C (350 °F) cure of a graphite-epoxy thermoset material, this force imbalance may cause matrix strain as high as 0.010 mm/mm. However, current materials and curing processes reduce this

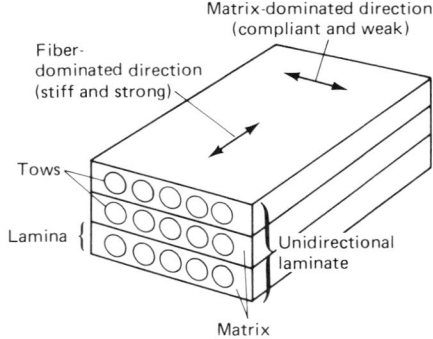

Fig. 1 Unidirectional ply properties as the basis for lamina allowables, with balanced thermal forces being force$_{tow}$ = −force$_{matrix}$

Table 1 Constituent dominance of orthotropic lamina properties

Dominant constituent	Lamina property	Lamina parameter
Fiber	Longitudinal	
	Stiffness	E_1
	Ultimate stress	σ_{11}
	Strain	ϵ_{11}
	Transverse	
	Minor Poisson's ratio	ν_{21}
Matrix	Transverse	
	Stiffness	E_2
	Ultimate stress	σ_{22}
	Strain	ϵ_{22}
	Longitudinal	
	Major Poisson's ratio	ν_{12}
	Shear	
	Stiffness	G_{12}
	Ultimate stress	τ_{12}
	Strain	γ_{12}

Table 2 Matrix content fluctuation in cured parts

Cured lamina condition(a)	Physical model	Constituent total		Constituent percents	
		Fiber	Matrix	Fiber	Matrix
Resin rich		Constant	High	Low	High
Nominal		Constant	Average	Average	Average
Resin poor.		Constant	Low	High	Low

(a) Cured parts usually contain a repeatable level of resin within certain limits.

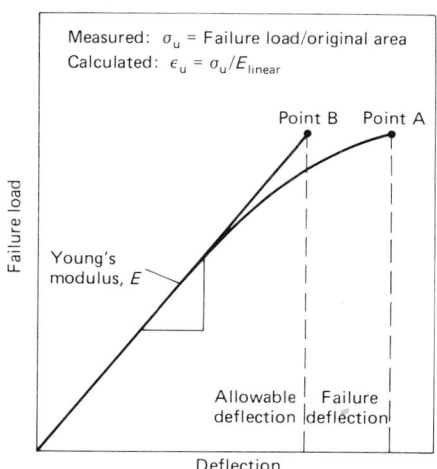

Measured: σ_u = Failure load/original area

Calculated: $\epsilon_u = \sigma_u / E_{linear}$

Fig. 2 Nonlinear load-deflection

Table 3 Linearization assumptions in stress-strain calculations

	Calculated	
Measured	Strain	Stress
Strain	Accurate	Conservative
Stress	Nonconservative	Accurate

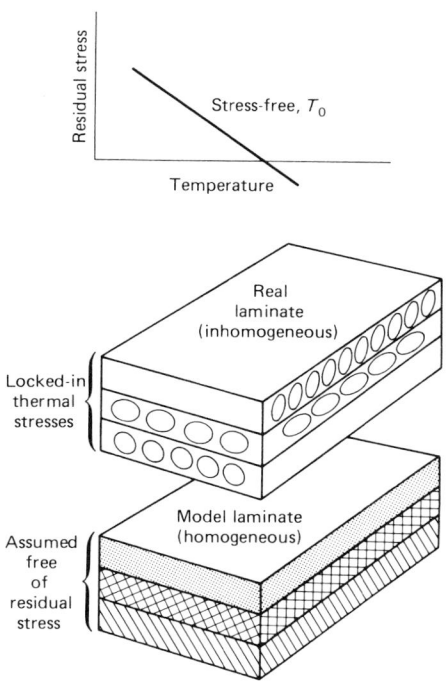

Fig. 3 Laminate analysis assumes homogeneous laminae without residual stresses

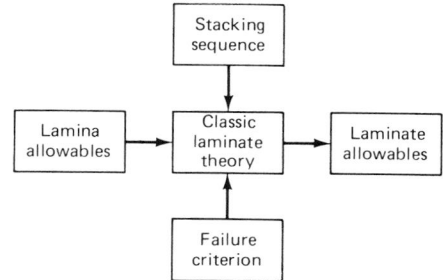

Fig. 4 Lamina data and failure criterion support CLP predictions

to a considerably lower and acceptable level. Therefore, this lamina or micromechanic mismatch is usually ignored when reducing data for design allowables. The mismatch or residual stress is assumed not to affect the stiffness of the material, nor its ability to strain uniformly. In fact, it is the uniform straining of the composite material that allows it to be modeled as homogeneous.

Material Orthotropy. Lamina properties exhibit large differences in directional properties. Lamina properties that are measured in the direction dominated by fiber properties reflect higher stiffness and strength. Measurements in directions that are transverse to the fiber exhibit matrix dominance; the lamina properties are compliant and weak. The unidirectional laminate shown in Fig. 1 emphasizes fiber and matrix dominance. The matrix, though weak, is still very critical to overall material performance, because it acts as the bonding agent, binding the fibers together, providing a shear path between fibers and giving lateral support. Dissimilarity between the fiber and resin mechanical properties means that lamina are only idealized as orthotropic. Through-the-thickness properties are not addressed here; rather, plane stress conditions are assumed. To aid in generating lamina allowables, Table 1 gives the lamina mechanical properties that are generally provided to a

design project. Fiber and matrix dominance categorization is given as an aid to understanding, but is not rigorous in its application. For instance, compressive properties are generally thought of as fiber dominated, but as matrix influences are actually known to be significant, the distinction between fiber and matrix dominance for 0° compression is obscure.

Material Variability. Because cured laminates tend to have a variable matrix content, the fiber and matrix load-carrying capability does become an issue when reducing test data to an allowables format. Table 2 illustrates the effect of variations in matrix content on a unidirectional lamina test specimen. The phenomenon of variable resin content occurs within laminates as well. As Table 2 indicates, the fiber content by total volume remains constant within much tighter tolerances on fiber areal weight; therefore, matrix content is typically normalized for fiber-dominated properties. The general rule is to use nominal ply thickness for fiber-dominated properties (assumes fibers carry the load) and as-measured thickness for matrix-dominated properties.

Material Linearity. The raw data are often plotted on a load-deflection curve. The data generally form a straight line up to useful strain levels, although some data do not, and are therefore nonlinear. The degree of nonlinearity is dependent on the specimen, load condition, and test environment. Because allowables development typically requires a linearization of

this test data, load-deflection curves (Fig. 2) must be "fitted" with a straight line. This fitting could be analytically determined, but is often obtained by strain measurement up to a given percentage of the failure strain. Linearizing the data means, computationally, that the material strain is equal to material stress divided by a constant modulus. Considering the real load-deflection curve reveals an ambiguity in that after linearization, the measured deflection (strain) may not equal the applied load (stress) divided by the linearized Young's modulus, E. As shown in Fig. 2 by Point A, a measured failure strain may not be the allowable strain for a given stress, as shown by Point B. Table 3 shows the results of material failure prediction. Parameter selection is dependent on the application and is therefore a design choice. One special comment is necessary. Load-deflection curves for the shear lamina properties are extremely nonlinear. Development of allowables for this characteristic of the shear data is again a design choice. Solutions have ranged from one secant modulus linear curve to useful strain levels with project-dependent design strain values, to multiple secant modulus curves with up to ultimate design strain values, to digitized curve representations. Factors such as level of computer support and structural use will determine the selection of a reasonable solution.

Probabilistic Nature of Material Properties. Because lamina properties vary from coupon to coupon, the randomness may be treated statistically. After a linearization has been applied to the data, strength and stiffnesses are calculated, with normalization applied when appropriate. Strength is often reported with a typical or mean value for test prediction, and as a statistically reduced value for design purposes. The statistical problems are complex; refer to the article "Statistical Analysis of Mechanical Properties" in this Volume. Usually, elastic moduli are reported as average values, and strength allowables are statistically reduced. Typical values are reported for stiffness, since it is not clear from an application viewpoint that a statistical reduction is conservative. An illustration of this is a structure with multiple load paths. The structure will distribute load based on stiffness. Statistically reduced stiffness alters the most probable load distribution. As a result, structural members might be subjected to a higher load than predicted, even though conservative analysis is intended.

Material Hygrothermal Elasticity. Composite materials have mechanical and physical properties that are temperature- and moisture-dependent. The environmental temperature at which the composite data is of interest is application-dependent. Design conditions caused by aerodynamic heating, engine exhaust, flight at altitude, and operations in areas of high humidity determine the hygrothermal conditions at which lamina data is required. The hygrothermal dependence of the composite material usually dictates the development of lamina data for multiple environmental conditions. Lamina allowables for these design conditions are generally displayed in a format similar to the one shown in Table 4. The data has been normalized, linearized, statistically reduced, and obtained for specific design environments.

Laminate Tests

Basic Assumptions. Laminate behavior can be predicted using the completed lamina table. As in the case of the unidirectional laminate, certain analysis assumptions have to be made. It has already been assumed that the individual lamina are homogeneous, with orthotropic properties. When several homogeneous plies of multiple orientations are laminated together, residual thermal stresses are again created because of the material orthotropy of the homogeneous lamina. For most material

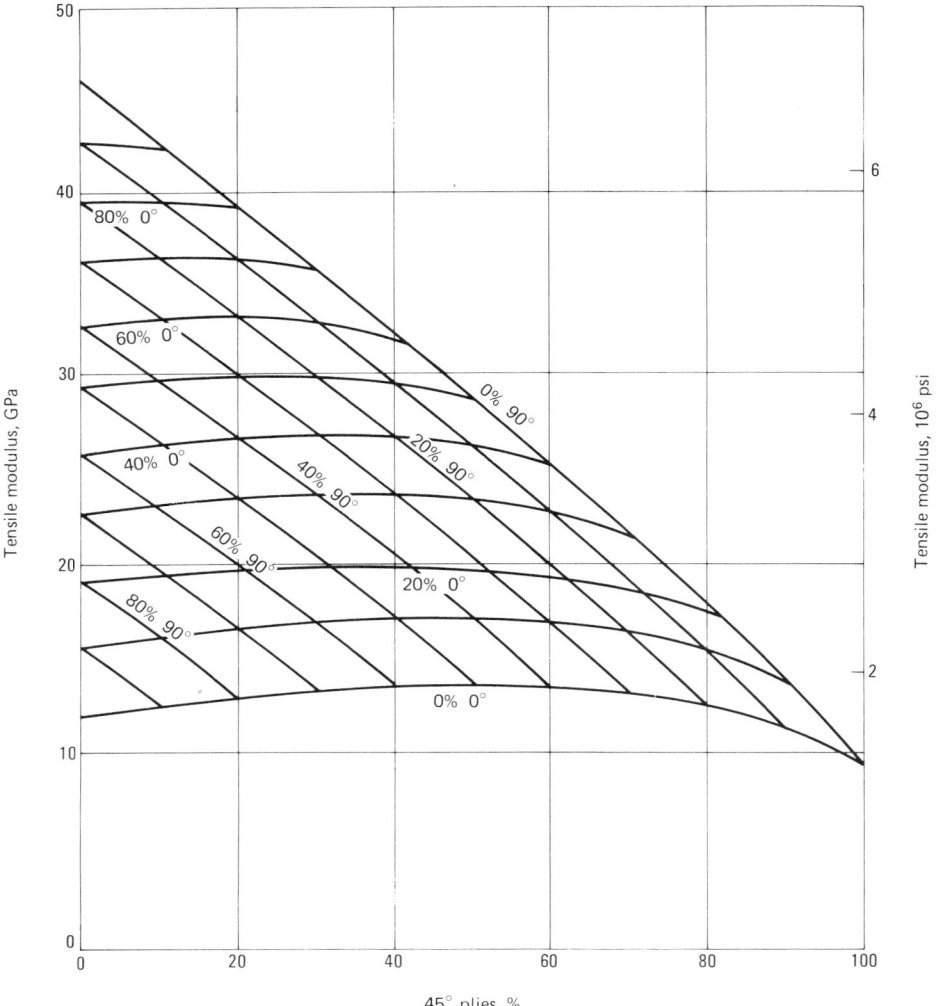

Fig. 5 Carpet plots illustrating tensile modulus of laminates

Table 4 Environmental effects on lamina properties

Data based on 49 ± 4 vol% fiber; ply thickness is 0.12 ± 0.10 mm (0.0048 ± 0.004 in.)

Conditioning and testing environment	Tensile modulus, E_{t1} GPa	10^6 psi	Ultimate tensile stress, σ_{tu1} MPa	ksi	Ultimate tensile strain, ϵ_{tu1} mm/mm	Poisson's ratio, ν_{12}	Compression modulus, K_{c1} GPa	10^6 psi	Ultimate compressive stress, σ_{cu1} MPa	ksi	Ultimate compressive strain, ϵ_{cu1} mm/mm	Tensile modulus, E_{t2} GPa	10^6 psi	Ultimate tensile stress, σ_{tu2} GPa	10^6 psi
−65/D, B-basis(a)	47.9	6.95	923.17	133.89	0.0193	0.294	56.53	8.198	1132.0	164.18	0.0200	15.0	2.18	13.8	2.00
RTD, typical-basis(a)	46.5	6.75	1377.9	199.84	0.0296	0.298	52.8	7.66	1249.9	181.27	0.0237	12.3	1.79	41.0	5.94
RTD, B-basis(b)	46.5	6.75	1185.5	171.94	0.0255	0.298	52.8	7.66	1026.4	148.86	0.0194	12.3	1.79	35.0	5.08
160 ETW, B-basis(b)	48.7	7.07	586.1	85.00	0.0120	0.307	59.4	8.61	601.4	87.22	0.0101	10.4	1.51	12.6	1.83
250 ETW, B-basis(b)	51.0	7.39	584.1	84.72	0.0115	0.355	55.0	7.98	309.4	44.87	0.0056	4.93	0.715	6.90	1.00

Conditioning and testing environment	Ultimate tensile strain, ϵ_{tu2} mm/mm	Compression modulus, K_{c2} GPa	10^6 psi	Ultimate compressive stress, σ_{cu2} MPa	ksi	Ultimate compressive strain, ϵ_{cu2} mm/mm	Shear modulus, G_{12} GPa	10^6 psi	Ultimate shear stress, τ_{su} MPa	ksi	Ultimate shear strain, γ_{12} mm/mm	Ultimate interlaminar shear strength, F_{ISU} MPa	ksi	Coefficient of thermal expansion, α α_1, 10^{-6}/K	α_2, 10^{-6}/K
−65/D, B-basis(a)	0.0009	16.3	2.37	207.1	30.04	0.0127	3.54	0.513	79.22	11.49	0.0224	79.230	11.491	5.4	29.5
RTD, typical-basis(a)	0.0033	13.9	2.01	206.2	29.90	0.0149	2.62	0.380	97.84	14.19	0.0373	65.84	9.549	5.4	29.5
RTD, B-basis(b)	0.0028	13.9	2.01	183.5	26.61	0.0132	2.62	0.380	90.05	13.06	0.0344	46.93	6.806	5.4	29.5
160 ETW, B-basis(b)	0.0012	10.6	1.54	98.94	14.35	0.0093	1.66	0.241	60.47	8.77	0.0364	...	...	5.4	29.5
250 ETW, B-basis(b)	0.0014	5.81	0.843	61.0	8.85	0.0105	0.834	0.121	44.1	6.40	0.0529	9.473	1.374	5.4	29.5

Note: 1, fiber direction; 2, matrix direction. (a) As fabricated. (b) Environmentally conditioned

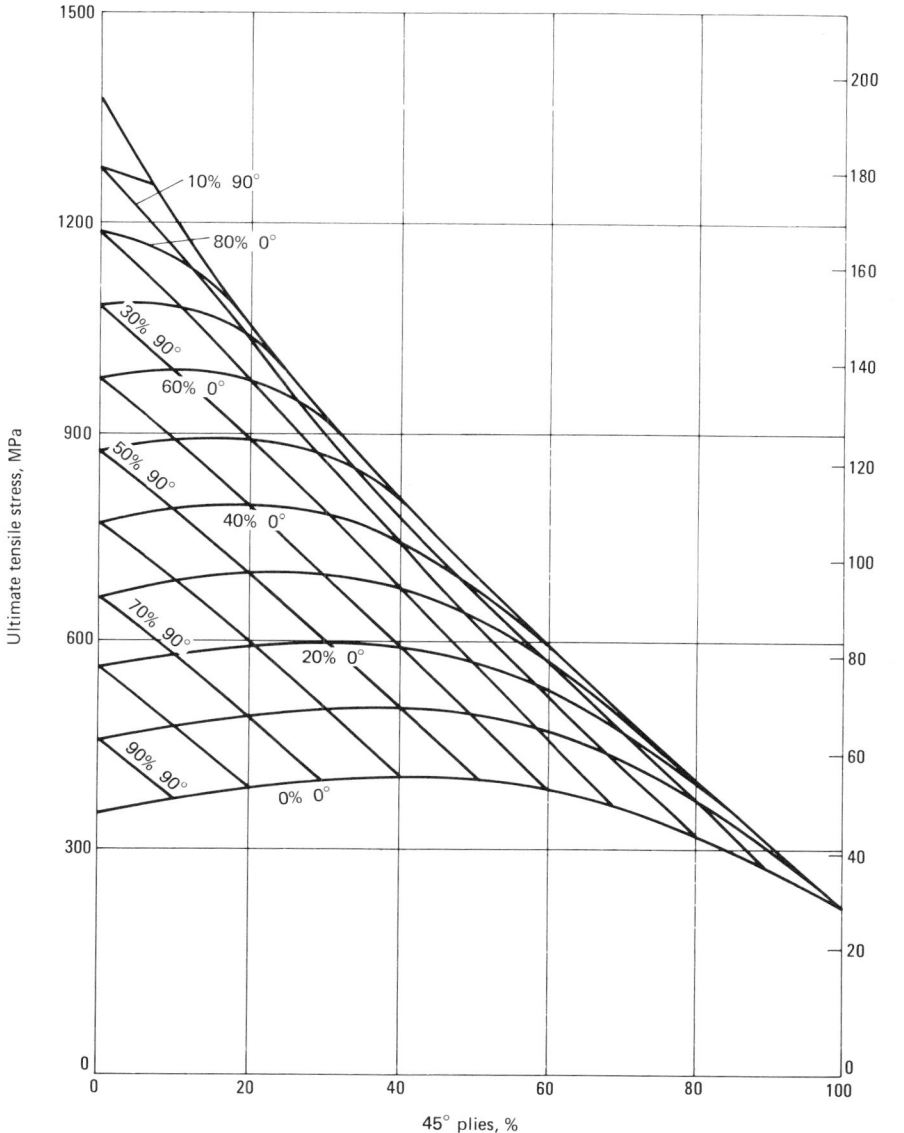

Fig. 6 Carpet plots illustrating tensile strength of laminates

Theoretical predictions are based on classical laminated plate (CLP) theory and failure theory. Figure 4 illustrates the way lamina properties and laminate stack descriptions contribute to CLP to make the laminate predictions. Lamina stiffness and laminate stacking sequence are used to calculate the laminate stiffness. Laminate strength predictions require lamina strength data and an interactive failure criterion. The important point is that from a finite number of lamina tests, an infinite number of laminate material behaviors can be predicted.

Carpet Plots

The ability to predict an infinite number of variations in laminates without sophisticated software could potentially constrain and confuse the design process. Some aerospace companies have restricted themselves to the use of 0°, ±45°, and 90° lamina orientations. This is becoming less true as a generalization with the advance in composite technologies, as exemplified by the aeroelastic tailoring of designs. Given specific angle orientations as a constraint and the assumption of a uniaxial load case, laminate data may be represented by carpet plots, which are maps of a laminate property versus the percentage of one of the lamina orientations. Figures 5 to 7 show the axial stiffness, axial strength, and Poisson's ratio for a 0°, 45°, and 90° laminate family. The carpet plot is for a given design temperature and for a certain design strength criterion, such as B-basis (90% confidence that 95% of the failures will exceed the value) for the ultimate properties. The carpet plots assume that the laminate is linear to failure (that is, no material nonlinearities, no ply cracks, and no extraneous failures such as free edge stresses). For some materials, carpet plots of strength are inaccurate at the 0% 0° ply laminates.

Carpet plots are simple to use. The designer interested in a particular laminate property would first select the appropriate plot for the design application. Stiffness properties are given in Fig. 5. For a laminate composed of 40% 0° fibers, 50% ±45° fibers, and 10% 90° fibers, the stiffness is read at 0.55 GPa (3.8 × 10⁶ psi). The procedure is to first find the percentage of ±45° plies and then follow that value vertically until it intercepts the correct percentage of 0° or 90° plies, both at the same point. Then, the ordinate value for the laminate stiffness can be read.

Laminate Complexities and Design Practices

Complex load cases allow the possibility for interaction of failure modes not found at the lamina level. Figure 8, which illustrates such a

design allowables, the laminate is generally assumed to be free of these thermal stresses. Figure 3 depicts the actual inhomogeneous character of each lamina and laminate, shows the assumed stress-free laminate, and provides a linearized sketch of the magnitude of residual stresses in the actual laminate as a function of temperature. The temperature at which no residual stress exists is called the stress-free temperature, which generally is not used in developing laminate allowables.

Laminate and Test Restrictions. Actual laminate tests are performed on balanced and symmetric stacking sequences when these are typical in the design cases. The exact consequences of not maintaining a balanced and symmetric stacking sequence are complex and will not be discussed here, except to say that the

lack of balance puts a skew into plate expansions, and lack of symmetry causes a twist or warp in plate expansions. Another laminate restriction applies to the stacking sequence. The stacks are selected so that no one orientation is clustered. The rules of clustering are not precise. Having no plies of like-orientation together is best. Two like-orientations together are commonly accepted. Four or more like-orientations together put the laminate at risk for maintaining structural integrity. Uniaxial tests are conducted at 0° and 90° to obtain tensile and compressive results. Shear tests of laminates require specialized fixtures such as rail shears or picture frames. The article "Mechanical Properties and Environmental Exposure Tests" in this Volume contains more information on these tests.

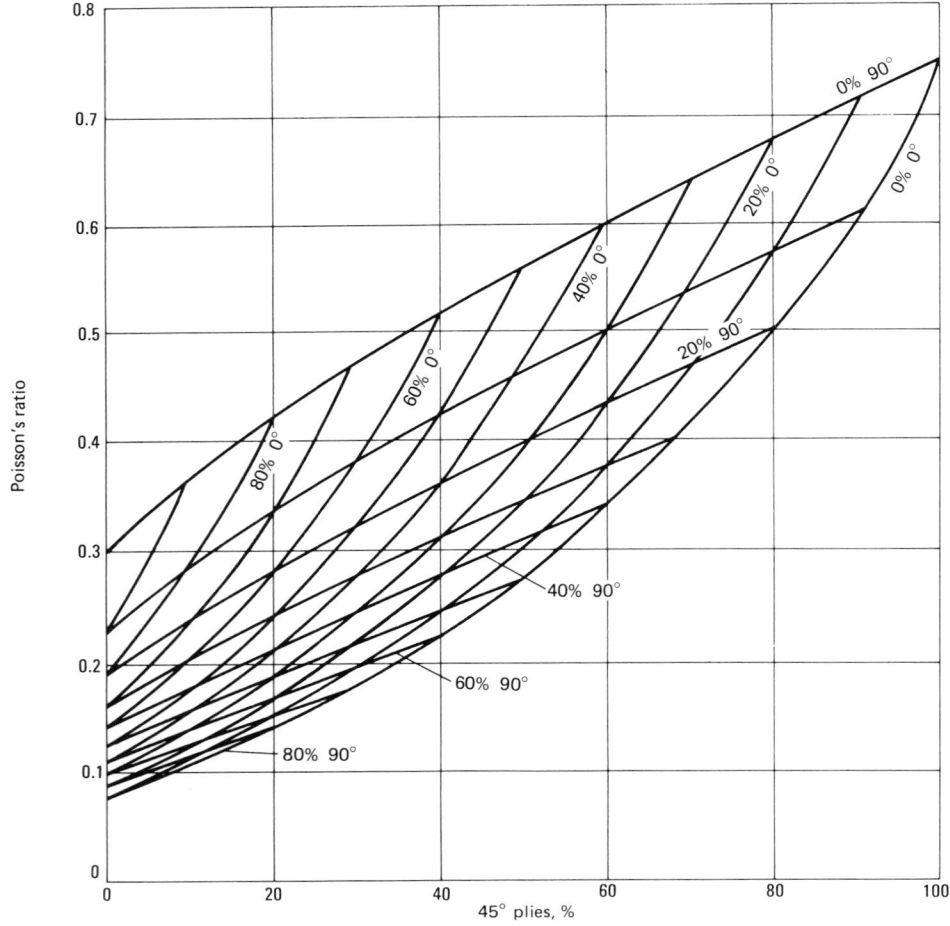

Fig. 7 Carpet plots illustrating Poisson ratio variation

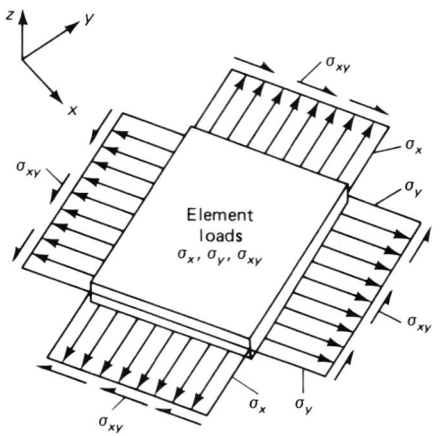

Fig. 8 Structural elements under three in-plane loads

case, shows a unit plate of laminate under a general load case with in-plane forces. It is important to understand that the predictions of these complex load cases are made with the same lamina allowables.

Laminate nonlinearities and peculiar failures are virtually eliminated when certain design rules of thumb are applied. One such rule is that each of the lamina directions of $0°$, $\pm45°$, and $90°$ should have at least 10% of the laminate plies in its direction. The fibers then guarantee a fiber-dominant (and therefore linear) stiffness for all loads, which also eliminates most concerns about viscoelastic effects. Another rule of thumb is to construct laminates with balanced and symmetric ply stacking sequences. A third rule is to avoid clustering like-orientations.

Element and Subcomponent Tests

G.C. Grimes,* K.W. Ranger, and M.D. Brunner, Northrop Advanced Systems Division

COMPOSITE STRUCTURAL TESTING is similar to metal structural testing in that it requires knowledge of design and analysis. The differences stem from the anisotropic behavior of composite materials under static, durability, and damage tolerance testing. A thorough experimental and analytical study of anisotropic behavior is necessary, not only in regard to the structure design but also the test specimen design, whether a specimen be coupon, element, or subcomponent. To facilitate such studies, materials and simple joint testing are described in this article followed by a discussion of element and subcomponent testing, using a typical fighter airframe as a model for the design requirements and related test verification. The article concludes with a discussion of durability and damage tolerance as they relate to composite structures.

Testing Philosophy and Criteria

The design philosophy used here is called material substitution; that is, a wing that is already designed in aluminum alloy, namely, a Northrop F-5E wing structure configuration, is the model (Ref 1). The design objective is to substitute a graphite-epoxy composite wing skin and related joints for the existing metal one. In material substitution, the graphite-epoxy laminate thickness and stiffness will closely duplicate those of the metal being replaced. Static and fatigue strengths of such composite structures usually exceed those of the metal. The result of material substitution will be a wing skin weight saving of 15 to 20%. Although the density of the graphite-epoxy material is 35% less than that of aluminum, added material in the joint areas reduces the total weight saving. Structural design criteria remain the same, with a few important exceptions.

The composite wing skin and joint design are in the area of or just outside the root rib (Fig. 1) at the 15%-of-chord front spar and the 39%-of-chord midspar, where the running loads differ by a factor of two. The composite skin and joints must be capable of sustaining design ultimate load and two lifetimes of realistic

*Now with Lockheed-California Company

fatigue loads. Both bonded and bolted joints are studied, and existing or redesigned metal substructure is used. In this 4.8%-of-chord airfoil thickness, the running load at the 15%-of-chord front spar is 1.58 MN/m (9000 lbf/in.), and the running load at the 39%-of-chord midspar is 3.15 MN/m (18 000 lbf/in.). Because torsion strength and stiffness are not critical, these properties can be somewhat lower than those of the aluminum alloy. Structural design requirements are summarized in Table 1.

A substantial material properties and allowables data base is necessary for composite structural design. Standard coupon testing of the material (ply and laminate) from which the structure is to be made is performed either to verify the allowables (minimal testing) or to develop allowables (substantial testing). In addition, representative elements in the form of small sections or pieces of the structure in critical and highly loaded areas (including joints and cutouts) are typically designed and tested as a quick way to start verifying the structural design without incurring costly design risks. Larger representative structural pieces, called subcomponents, are then designed and tested to verify manufacturing approaches and structural integrity. Full-size test components and/or a full-scale vehicle test may or may not be needed for design verification; vehicle economics, customer requirements, and funding affect such decisions. However, testing at the materials (coupon), element, and subcomponent levels is almost always needed to satisfy design structural integrity requirements. Testing at the generic material level typically includes coupon (ply and laminate) static testing, with fatigue testing being optional. Structural element and subcomponent static, durability, and damage tolerance tests are necessary to establish the required reliability and survivability of the airframe.

Composite Materials and Simple Joint Data

This section reviews ply and laminate testing and related data on pin and bolt bearing strength. The experimental methodology used to obtain such data is discussed, and data analysis procedures and test reporting are described.

Design with composite materials requires a knowledge of lamination theory and appropriate failure criteria as well as related analytical methods. This creates the need for experimental measurement of properties of plies used in making a laminate. By developing mean stiffness and B-basis allowable strength properties at the ply level, it is easy and economical to obtain accurate predictions of laminate mean stiffness and B-basis strength properties (unholed and holed). Bolted and bonded joint strengths can also be predicted with the appropriate analytical methods, using ply properties for both, and neat adhesive properties for the latter.

Ply and Laminate Properties. The orthotropic ply and its properties are, physically and analytically, the basic building blocks for anisotropic laminates and composite structural elements and subcomponents. Single-ply (lamina) properties are obtained experimentally from multi-ply, unidirectional laminate specimens with the same orientation. For tape laminates (all fibers aligned), the ply properties needed for design are the elastic constants E_{11}^t, E_{11}^c, E_{22}^t (or E_{22}^c), G_{12}, and ν_{12} and the related ultimate-strength values F_1^{tu} and ϵ_1^{tu}, F_1^{cu} and ϵ_1^{cu}, γ_{12}^u, F_2^{tu}, ϵ_2^{tu} and F_2^{cu} and ϵ_2^{cu}.

A typical tensile stress-strain curve for an AS1/3501-5/6 graphite-epoxy tape laminate are shown in Fig. 2, and a typical compression stress-strain curve for the same material is shown in Fig. 3. For obtaining graphite-epoxy tape ply shear properties, a $[\pm 45°]_{ns}$ laminate compression stress-strain curve is shown in Fig. 4; this curve is used to generate the shear-stress strain curve given in Fig. 5 in accordance with ASTM D 3518 (Ref 2). Figure 6 shows a $[90°]_{nt}$ laminate tensile stress-strain curve for graphite-epoxy tape, and Fig. 7 shows a $[90°]_{nt}$ compression stress-strain curve for the same material.

Ply properties are used with analysis to predict general laminate and structural strengths, stiffnesses, deflections, failure loads, and modes. Table 2 presents thermal, mechanical, and physical properties for the AS1/3501-5/6 tape graphite-epoxy material. Figures 8, 9, and 10 provide elastic-constant carpet plots for this material based on the room temperature, dry, and room temperature, wet,

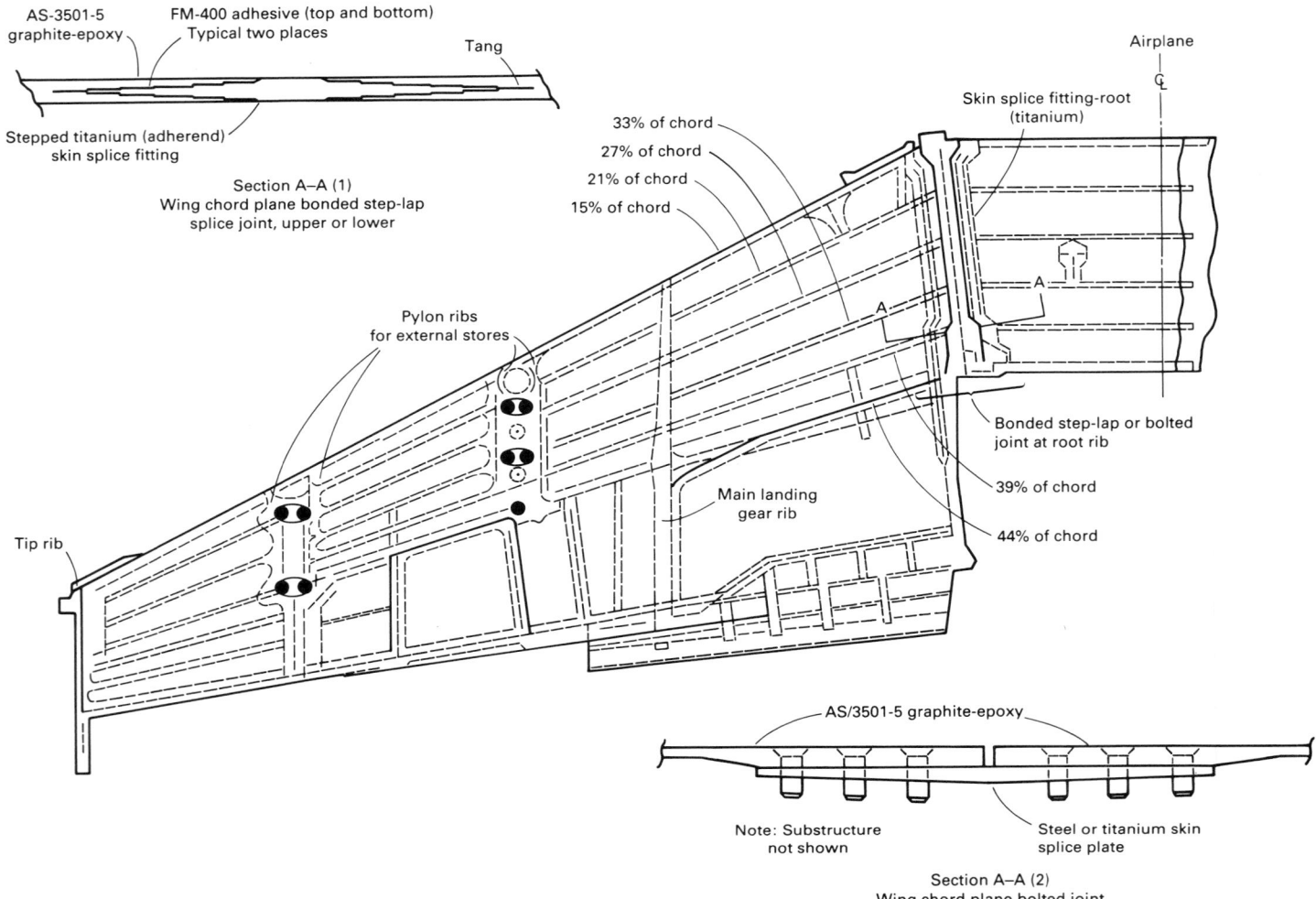

Fig. 1 F-5E wing structural configuration with bolted and bonded composite root rib joint concepts

Table 1   Design requirements for root rib joint, F5E wing composite skin

Design considerations	Bonded joint	Bolted joint
Design environment	−30 °CW to 120 °CW (−20 °FW to 250 °FW)	−30 °CW to 120 °CW (−20 °FW to 250 °FW)
Desired mode of failure	Bondline/matrix	Fiber failure
Specified joint location	Wing skin/root rib splice	Wing skin/root rib splice
Design load (static . . . . ultimate), standard- sized joints, 15% chord	1.6 MN/m (9000 lbf/in.)	1.6 MN/m (9000 lbf/in.)
Design load (static . . . . ultimate), large- scale joints, 39% chord	3.2 MN/m (18 000 lbf/in.)	1.6 MN/m 18 000 lbf/in.
Fatigue-design . . . . considerations	Survive 2 LT	Survive 2 LT
Restrictions imposed . . . . for realism	. . .	Countersunk fasteners
Geometry restrictions	25-mm (1.0-in.) wide standard and large scale	25-mm (1.0-in.) wide, standard; 88.9-mm (3.5-in.) wide, large-scale
Loading mode	Static tension and compression and tension-dominated and compression-dominated spectrum fatigue	Static tension and tension-dominated fatigue

LT, lifetimes

(RTD/RTW) ply properties given in Table 2. Elastic constants for any balanced and symmetric orientation laminate may be determined from Fig. 8 to 10.

Unnotched ply strength allowables are shown in Table 3. Carpet plots of notched (6.4-mm, or 0.25-in. diam hole) cross-plied balanced and symmetric laminate strength allowables are presented in Fig. 11 and 12 for tension and compression loading. These plots cover the RTD/RTW linear-behavior laminate range for balanced and symmetric laminates in which ±45° plies make up 20 to 80% of the total number of plies. Figures 13 to 18 represent the 6.4- and 4.8-mm (¼- and ³⁄₁₆-in.) diam open-hole (notched) geometries for several balanced and symmetric laminate orientations. These material properties were obtained from Ref 3 to 10.

Pin and Bolt Bearing Strength. Figures 19, 20, and 21 show double-shear pin (no torque-up) bearing strength for 4.8-, 6.4-, and

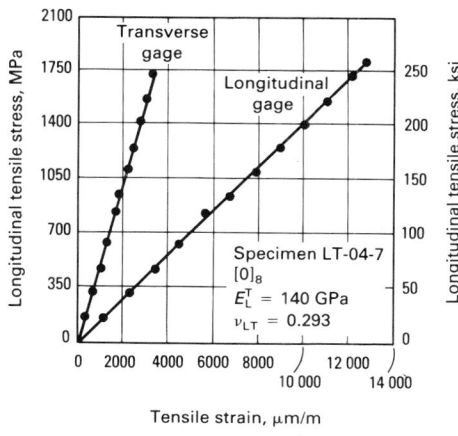

Fig. 2 Typical longitudinal tension stress-strain curves, AS1/3501-5/6 graphite-epoxy tape, room temperature, dry and room temperature, wet (RTD/RTW)

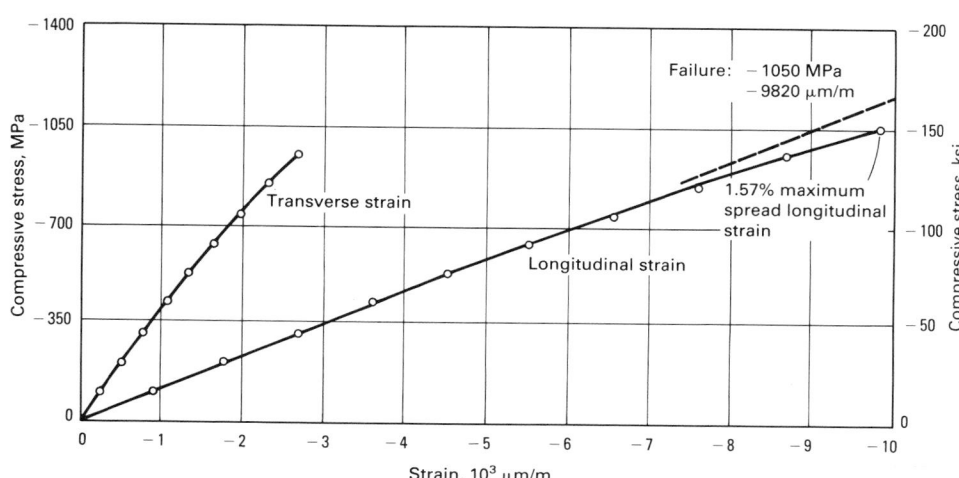

Fig. 3 Typical longitudinal compression stress-strain curves, AS1/3501-5/6 graphite-epoxy tape, RTD/RTW

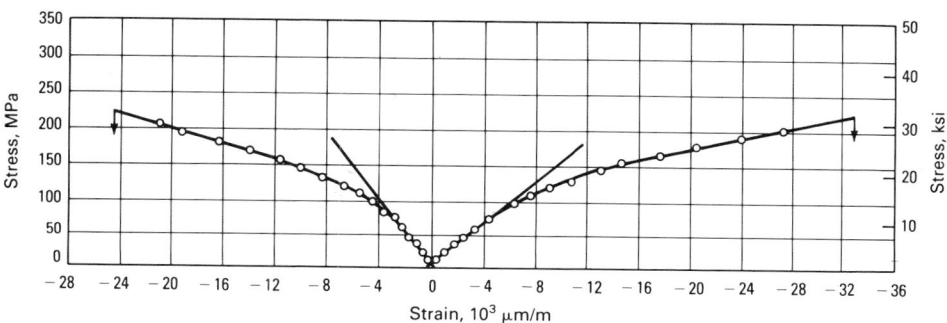

Fig. 4 Typical bias static compression stress-strain curves, AS1/3501-5/6 [±45°]ns graphite-epoxy tape laminate, RTD/RTW

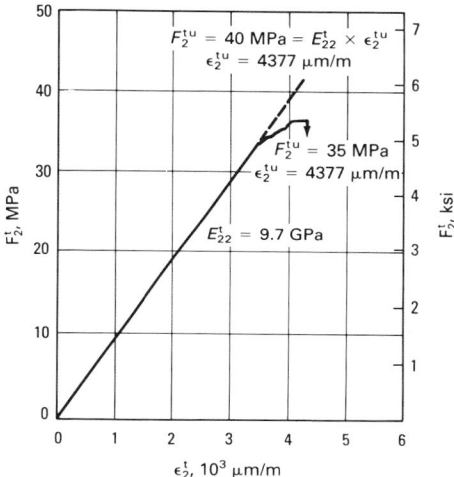

Fig. 6 Typical transverse static tension stress-strain curve, AS1/3501-5/6 [90°]nt tape, RTD/RTW. (Ref 3)

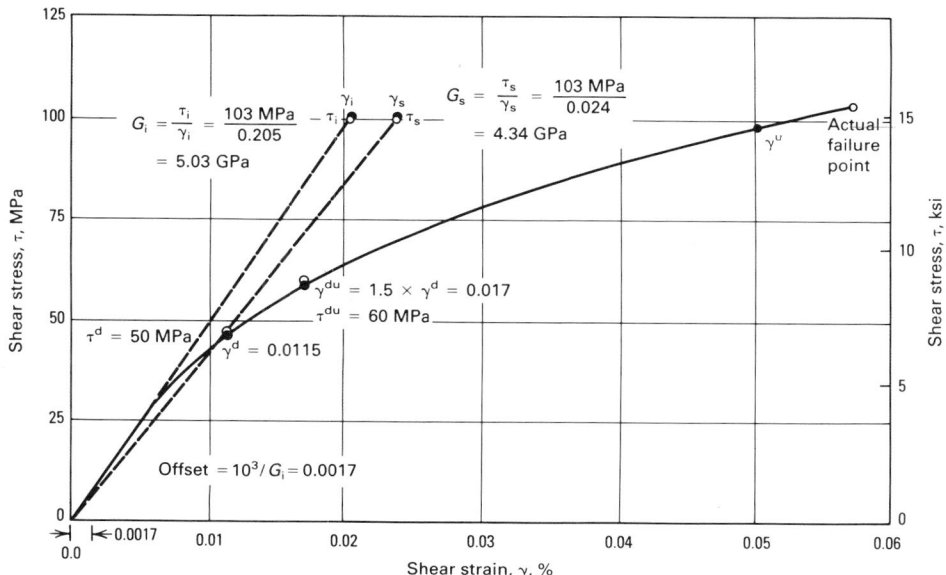

Fig. 5 Typical bias static compression shear stress-strain curve, AS1/3501-5/6 [±45°]ns graphite-epoxy tape, RTD/RTW. ETL platen-supported fixture. γ^u is the shear strain that gives a notched allowable equal to γ^{du}.

7.9-mm ($\frac{3}{16}$-, $\frac{1}{4}$-, and $\frac{5}{16}$-in.) diam holes in balanced and symmetric laminates with 20 to 80% ±45° plies. Figures 22 to 27 show carpet plots of bearing versus far-field strength (allowable gross section strain) for four pin diameters in six balanced and symmetric lami-

nates for this type of joint. When fastener torque-up is included, the clamped bolted joint ultimate strength is higher and can be determined from tests; this strength usually ranges from 1.0 to 2.4 times the pin-bearing strength for low torque-up values. However, pin/bolt load-deformation data show that pin-bearing ultimate strength and bolt-bearing yield strength are about the same and represent the end of nonlinearity in the load-displacement behavior.

Therefore, any empirical factors used on the analytical predictions to derive bolt-bearing ultimate strengths should not exceed the design factor of safety, usually 1.5 for aircraft structures. This prevents such bolted joints from being stressed into the nonlinear behavior range at design limit loads. Figure 28 shows the bolt- and pin-bearing set-ups, and Fig. 29 shows typical bolt-bearing and pin-bearing load-displacement curves. These illustrate that bolt-bearing yield strength is approximately the

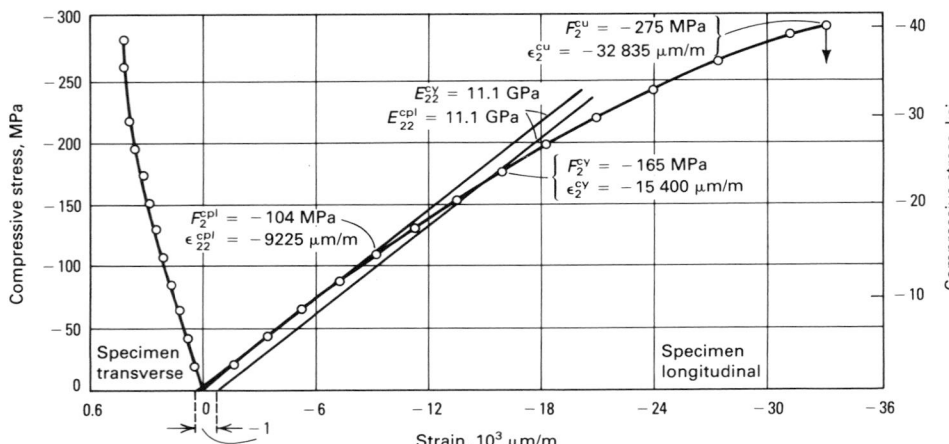

Fig. 7 Typical transverse static compression stress-strain curves, AS1/3501-5/6 [90°]ₙₜ, graphite-epoxy tape, RTD/RTW, ETL platen-supported fixture

same as pin-bearing ultimate strength and that the nonlinear portion of the bolt-bearing load-displacement curve goes much higher (than the end of linearity) to failure (Ref 11, 12).

Ply-Property Inputs to Analytical Methods. Generally, ply stiffness properties (Table 2) are used with a lamination theory computer program to predict laminate elastic properties, or they may be read directly from carpet plots (Fig. 8 to 10). To obtain strength values, both elastic and strength ply properties (Table 3) must be used, with the latter usually in terms of ultimate-strain values. Usually, a failure criterion such as maximum strain is combined with lamination theory to predict strengths. However, to predict strengths accurately, input property guidelines must be estab-

lished, and the appropriate strength values must be selected from the ply stress-strain curves (Fig. 2 to 7).

In laminate design, two areas must be considered: linear and nonlinear behavior. To make sure that uniaxially loaded laminates remain linear, a minimum of 10% of the total number of plies must be oriented in each principal ply direction, that is, the 0°, 45°, −45°, and 90° directions, with plies balanced (equal + and − orientations) and symmetric about the thickness centerline, such as [0°/±45°/90°]ₛ, an 8-ply (25/50/25) laminate representing (%0°/%±45°/%90°) plies relative to the primary material axis. In the work presented here, a minimum of 10% of the total number of laminate plies are oriented in the 0°, +45°, −45°, and 90° directions. Thus, the analytical strength

predictions cover only the 20 to 80% ±45° laminate range. This keeps the laminates in the linear-behavior range. Even so, some biaxial and shear loadings require special properties selections from the shear stress-strain curve (Fig. 5).

While unnotched-laminate strength values can be predicted, much of the structural design may require notched-laminate properties. The philosophy used here is that a 6.4-mm (¼-in.) diam hole represents a "rogue flaw," which inspection misses, or holes that may be drilled in a structure subsequent to its fabrication, for various reasons ranging from draining condensed water to attaching parts and equipment not included in the initial design. In strength data shown here, plain laminate strength for design purposes is assumed to be one with a 6.4-mm (¼-in.) diam hole. This and other open-hole laminate strength predictions, as well as those for pinned double-shear joints, can be predicted by the bolted-joints stress field model (BJSFM) computer program. This program, summarized in Ref 12 and developed in Ref 13, uses lamination theory, maximum-strain failure criterion, and variable stress concentration around the hole. It distinguishes between open (unloaded) holes and loaded holes. An empirically derived factor, R_c, is necessary and is obtained from plain and open-hole tension and compression data on laminate test coupons.

Ply mean value elastic properties are needed as input data for the BJSFM prediction technique. However, the B-basis values ϵ_1^{tu} and ϵ_1^{cu} are used in both the 1 (fiber) and 2 (matrix) directions to eliminate matrix-cracking laminate failure predictions. Some microcracking of the 90° and ±45° plies of fiber-dominated laminates may be acceptable (Ref 14); thus,

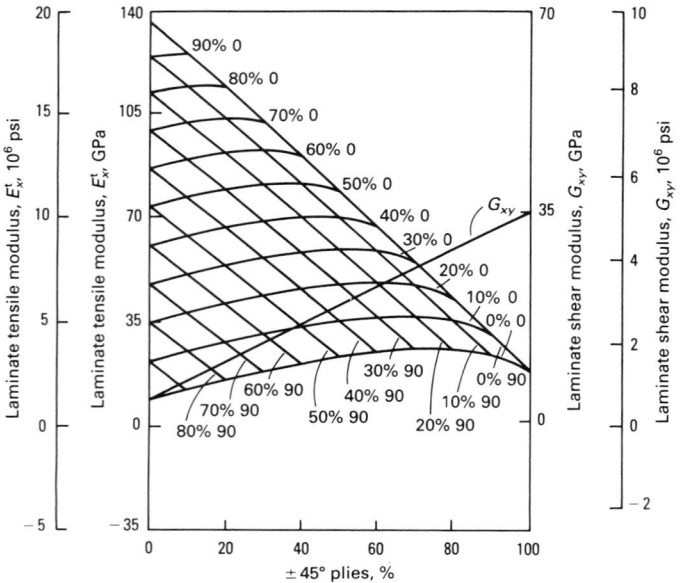

Fig. 8 Laminate tension and shear moduli, AS1/3501-5/6 laminate E_x^t and G_{xy} RTD/RTW

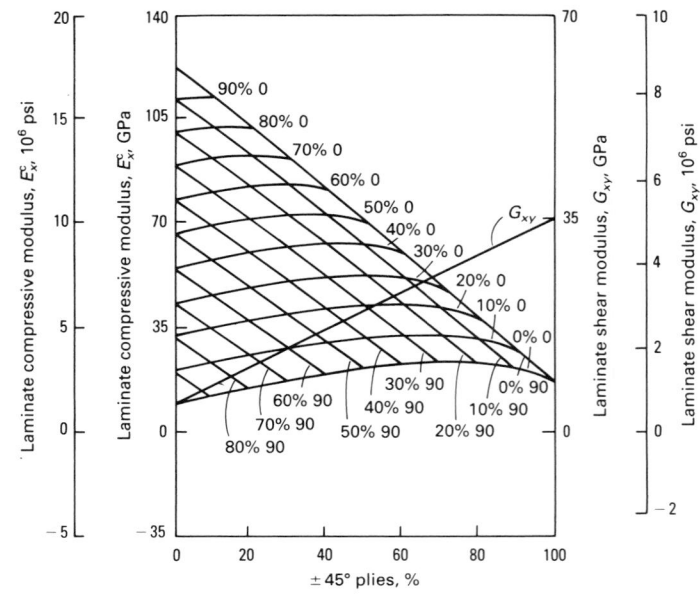

Fig. 9 Laminate compression and shear moduli, AS1/3501-5/6 laminate E_x^c and G_{xy}, RTD/RTW

Table 2 Elastic, physical, and thermal properties of AS1/3501-5/6 graphite-epoxy tape, unnotched ply
Nominal ply thickness = 0.140 mm (0.0055 in.); fiber volume = 60%

Mean moduli and Poisson's ratio values	Temperature/moisture condition					
	−55 °C (−67 °F) dry/wet		RT dry/wet		130 °C (265 °F)/wet	
	GPa	10^6 psi	GPa	10^6 psi	GPa	10^6 psi
Longitudinal tensile modulus, E_{11} 137	19.8	140	20.3	143	20.8	
Transverse tensile modulus, E_{22} 11	1.6	9	1.35	2	0.25	
Longitudinal compressive modulus, E_{11} 125	18.2	119	17.3	128	18.6	
Transverse compressive modulus, E_{22} 12	1.7	10	1.4	1	0.21	
In-plane shear modulus, G_{12} 5	0.7(a)	5	0.7(a)	3	0.5(a)	
Longitudinal Poisson's ratio (tension), ν_{12} 0.205		0.320		0.397		
Longitudinal Poisson's ratio (compression), ν_{12} 0.375		0.324		0.280		

Physical constants (typical or mean)	Temperature/moisture condition		
	−55 °C (−67 °F) dry/wet	RT dry/wet	130 °C (265 °F)/wet
Density (ρ), g/cm³.....................0.058		0.058	0.058
Longitudinal coefficient of thermal expansion (α_{11}), 10^{-6}/K 0.0		0.0	0.0
Transverse coefficient of thermal expansion (α_{22}), 10^{-6}/K 16.0		16.0	16.0

(a) Estimated from similar materials

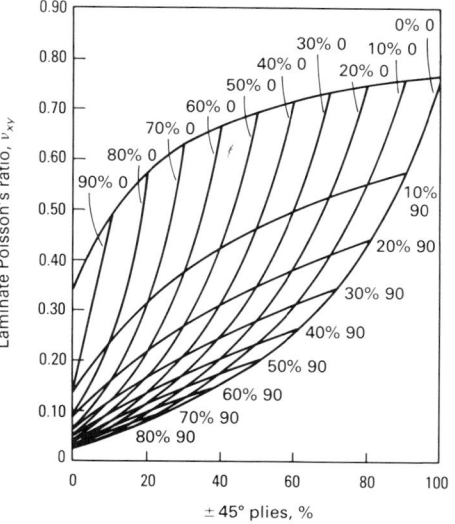

Fig. 10 Laminate Poisson's ratio, AS1/3501-5/6 laminate ν_{xy}, RTD/RTW

Table 3 Strength allowables for AS1/3501-5/6 graphite-epoxy tape, unnotched ply
Nominal ply thickness = 0.1397 mm (0.0055 in.); fiber volume = 60%

Type of stress allowable, B-basis ultimate and yield strength values	Temperature/moisture condition					
	−55 °C (−67 °F) dry/wet		RT dry/wet		130 °C (265 °F) wet	
	MPa	ksi	MPa	ksi	MPa	ksi
Longitudinal tensile ultimate stress, F_1^{tu}........................... 1434 (1413a)	208 (205a)	1544 (1572a)	224 (228a)	1524 (1448a)	221 (210a)	
Transverse tensile ultimate stress, F_2^{tu}....................... 29 (28a)	4.2 (4.1a)	40 (41a)	5.8 (6.0a)	8 (11a)	1.1 (1.6a)	
Longitudinal compressive ultimate stress, F_1^{cu} .... −1117 (−1207a)	−162 (−175a)	−910 (−965a)	−132 (−140a)	−496 (−517)	−72 (−75a)	
Transverse compressive ultimate stress, F_2^{cu} .... −270	−39.1	−196	−28.4	−56	−8.1	
Transverse compressive yield stress, F_2^{cy}	...	−138	−20.0	...	...	
In-plane shear ultimate stress, F_{12}^{su}	...	±90	±13.0	...	...	
In-plane shear yield stress, F_{12}^{sy} .... ±47.5	±6.9	±41	±6.0	±25	±3.6	

Type of strain allowable B-basis ultimate and yield strength values	Temperature/moisture condition		
	−55 °C (−67 °F) dry/wet	RT dry/wet	130 °C (265 ° F) wet
Longitudinal tensile ultimate strain, ϵ_1^{tu}..................... 10 330		11 260	10 110
Transverse tensile ultimate strain, ϵ_2^{tu}..................... 2563		4470	6285
Longitudinal compressive ultimate strain, ϵ_1^{cu}..................... −9610		−8090	−4059
Transverse compressive ultimate strain, ϵ_2^{cu}..................... −26 530		−28 580	−34 990
Transverse compressive yield strain, ϵ_2^{cy}.................... ...		−13 090	...
In-plane shear ultimate strain, γ_{12}^u..................... ...		±47 600	...
In-plane shear yield strain, γ_{12}^y..................... ...		±9760	...

(a) σ = Eε

first-ply matrix cracking does not necessarily denote laminate failure. The B-basis value for the shear ultimate strain, γ_{12}^u, is chosen to preclude shear failure predictions, because they do not occur when the loading is primarily uniaxial.

For such uniaxially loaded laminates with at least 10% of the plies in the 0°, +45°, −45°, and 90° directions, that is, 20 to 80% ±45° plies, such strength predictions are accurate.

For some biaxial or shear loading of these laminates as well as for uniaxially loaded laminates with greater than 80% ±45° plies, different input shear-strain allowables must be used. A value that is 1.5 times a defined yield point, such as γ_{12}^{du}, must be used (Fig. 5). For laminates with less than 20% ±45° plies, the linear approach accurately predicts the strength as long as the 90° plies do not exceed 10% of the total number of plies. Otherwise, for lami-

nates with either <20% or >80% ±45° plies, 1.5 times the matrix damage (defined as matrix cracking or yielding or fiber-matrix interface bond failure) threshold values should be used from the [90°]$_{nt}$ tension or compression ply stress-strain curves (Fig. 6 and 7). For tension on [90°]$_{nt}$ plies, the damage threshold is the ultimate value because this is essentially a linear stress-strain curve to failure (Fig. 6). For compression [90°]$_{nt}$ plies, the damage threshold

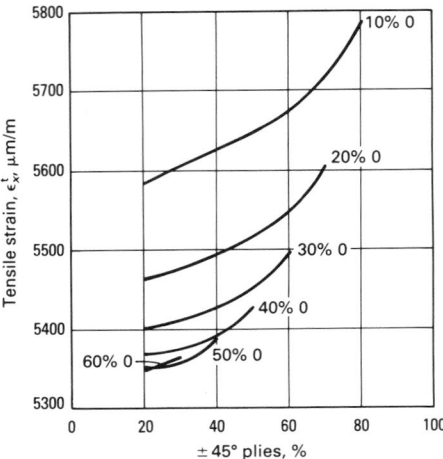

Fig. 11 Laminate tensile loading, 6.4-mm (0.25-in.) open-hole ultimate strain, AS1/3501-5/6 tape laminate ϵ_x^{tu}, RTD/RTW

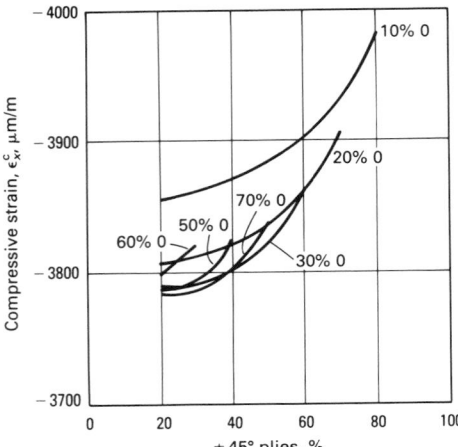

Fig. 12 Laminate compressive loading, 6.4-mm (0.25-in.) open-hole ultimate strain, AS1/3501-5/6 tape laminate ϵ_x^{cu}, RTD/RTW

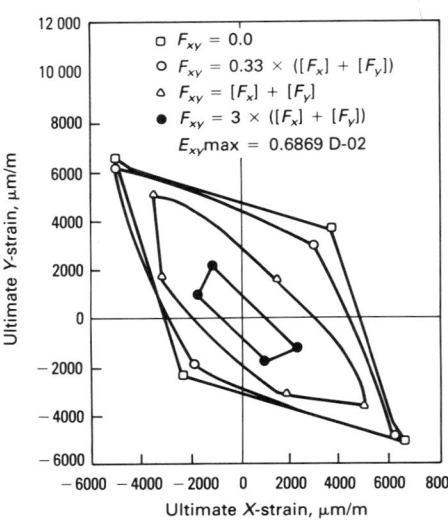

Fig. 13 Notched allowables, 6.4-mm (0.25-in.) open hole, AS1/3501-5/6 (10/80/10) tape laminate ϵ_x^u and ϵ_y^u, RTD/RTW

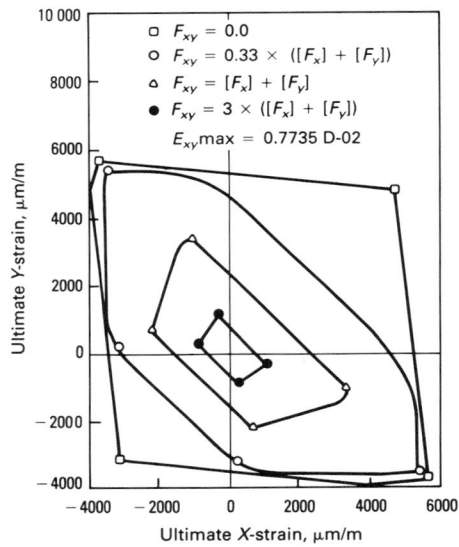

Fig. 14 Notched allowables, 6.4-mm (0.25-in.) open hole, AS1/3501-5/6 (25/50/25) tape laminate ϵ_x^u and ϵ_y^u, RTD/RTW

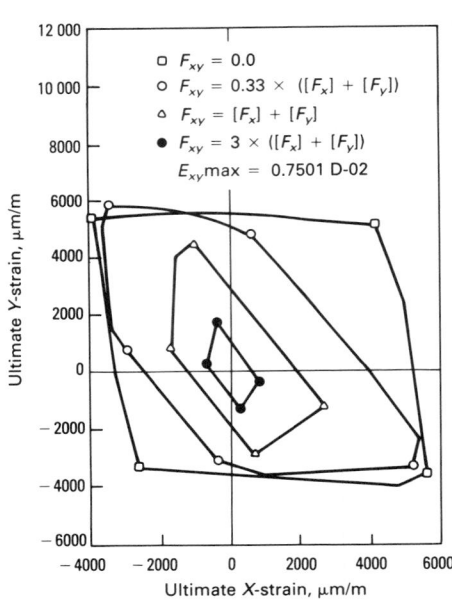

Fig. 15 Notched allowables, 6.4-mm (0.25-in.) open hole, AS1/3501-5/6 (40/50/10) tape laminate ϵ_x^u and ϵ_y^u, RTD/RTW

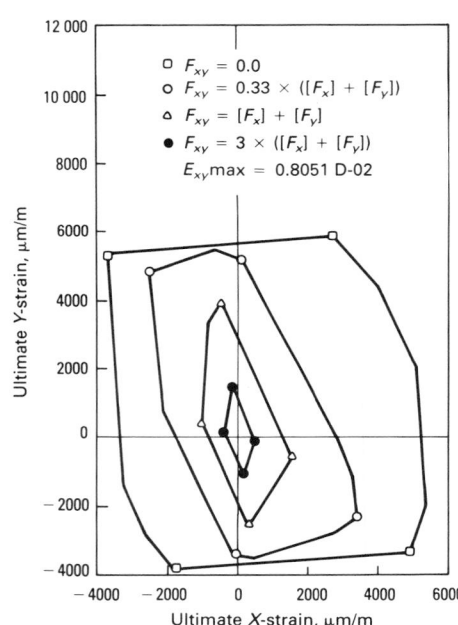

Fig. 16 Notched allowables, 6.4-mm (0.25-in.) open hole, AS1/3501-5/6 (60/30/10) tape laminate ϵ_x^u and ϵ_y^u, RTD/RTW

will be a defined yield point on the nonlinear stress-strain curve (Fig. 7) such as is defined on the shear stress-strain curve (Fig. 5).

Data Analysis and Reporting. Analyzing composite materials test data involves identifying and tabulating the related specifications, materials and processing records, inspection records, specimen preparation records, test data logs, and stress-strain curves. Complete traceability is necessary for individual specimens regarding their locations in the laminate; inspection and processing records; and material identification, including prepreg batch number and fiber lot number. Obtaining this level of traceability requires proper planning of the test program from the beginning, monitoring it in progress, and collecting all the documentation when data analysis begins. "Go/no-go" decisions must be made along the way, using acceptance levels based on test program needs and the capability of the materials, processing

facilities, instrumentation and test facilities, and people. Controlling and monitoring all these steps virtually assures that > 90% of the test data will be acceptable, thus minimizing retest requirements.

Correlation of experimental and analytical data is one of the basic parts of data analysis. For example, pin and bolt bearing joint test data are compared with BJSFM analytical predictions in Table 4. Load displacement (or strain) data (Fig. 29 and Table 4) show that pin-bearing ultimate values and finger-tight (1 N · m, or 10 in. · lbf, torque on 6.4-mm, or ¼-in. diam bolt) bolt-bearing yield values are approximately the same. These values, representing the end of nonlinearity,

are accurately predicted by the BJSFM analysis in Table 4.

Figure 30 illustrates the correlation of the bolt-bearing yield (BBY) value with the pin-bearing ultimate (PBU) values for the fiber-dominated (63/25/12) tape laminates. Plotting both PBU and BBY values as the ordinate against the BJSFM predictions as the abcissa (Fig. 31) shows a reasonable fit to a perfect-correlation (45°) line. Figure 31 covers (63/25/12) and (10/80/10) tape laminates and a

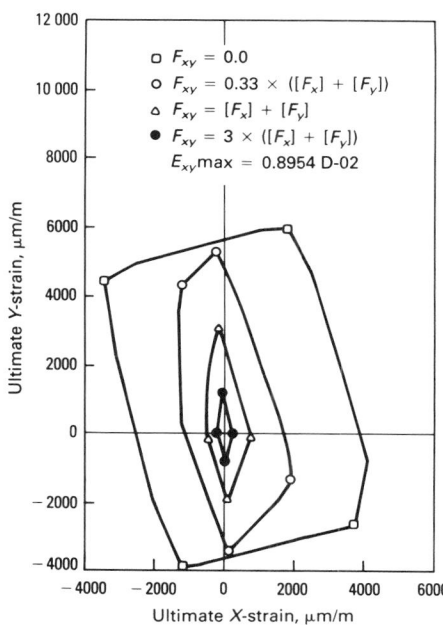

Fig. 17 Notched allowables, 6.4-mm (0.25-in.) open hole, AS1/3501-5/6 (80/10/10) tape laminate ϵ_x^u and ϵ_y^u, RTD/RTW

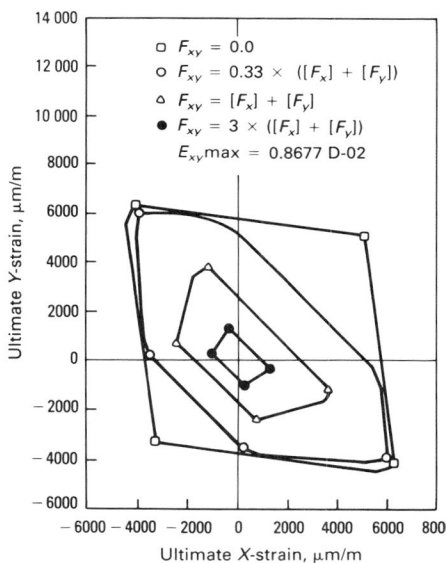

Fig. 18 Notched allowables, 4.8-mm (0.19-in.) open hole, AS1/3501-5/6 (25/50/25) tape laminate ϵ_x^u and ϵ_y^u, RTD/RTW

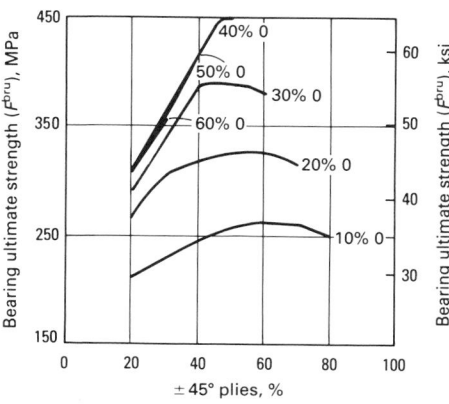

Fig. 19 AS1/3501-5/6 loaded hole (4.8-mm, or ³/₁₆-in.) tape laminate F^{bru}, RTD/RTW

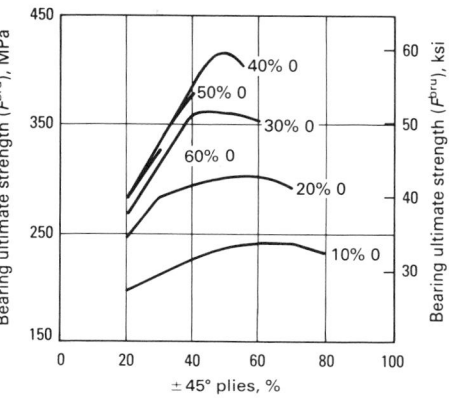

Fig. 20 AS1/3501-5/6 loaded hole (6.4-mm, or ¼-in.) tape laminate F^{bru}, RTD/RTW

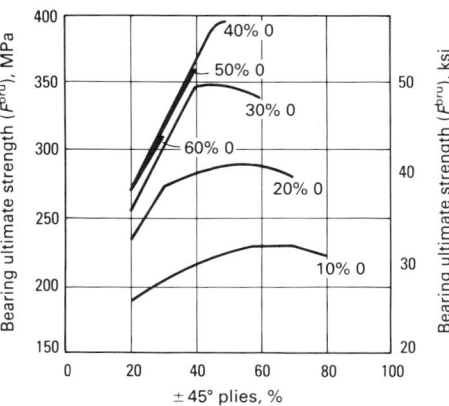

Fig. 21 AS1/3501-5/6 loaded hole (7.9-mm or ⁵/₁₆-in.) tape laminate F^{bru}, RTD/RTW

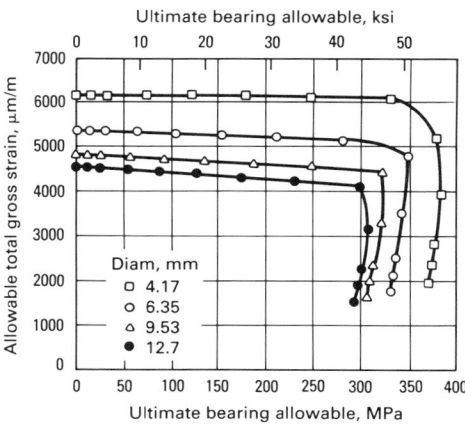

Fig. 22 AS1/3501-5/6 (25/50/25) tape laminate tension-loaded bolt-bearing bypass, ϵ_x^{tu} versus F^{bru}, RTD/RTW

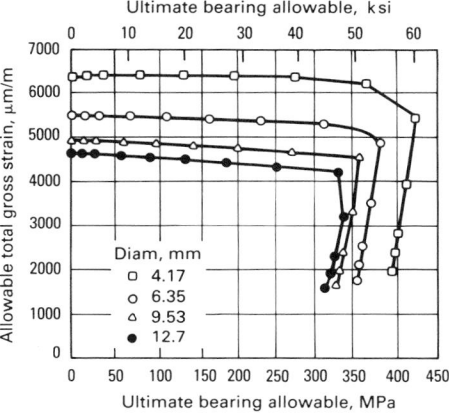

Fig. 23 AS1/3501-5/6 (30/60/10) tape laminate tension-loaded bolt-bearing bypass, ϵ_x^{tu} versus F^{bru}, RTD/RTW

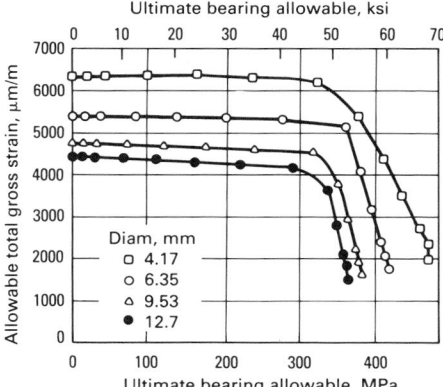

Fig. 24 AS1/3501-5/6 (40/50/10) tape laminate tension-loaded bolt-bearing bypass, ϵ_x^{tu} versus F^{bru}, RTD/RTW

(25/50/25) fabric laminate, giving good correlation of test data and analytical predictions. When bolt-bearing ultimate (BBU) strength is plotted against the BJSFM predictions, the BBU/BJSFM slopes result in numbers ranging from 1.0 to 2.4 (Fig. 32). These slopes represent the bump factors, which can be used with analytical predictions to derive BBU strength. While such bump factors could be established for each data point, there are apparently two data groupings: one covering the fiber-dominated (63/25/12) tape laminates

dry and wet and the (25/50/25) intermediate fabric laminates dry, all with an average bump factor of 1.2; and the other covering the matrix-dominated (10/80/10) tape laminates dry and wet and the intermediate (25/50/25) fabric laminates wet, all with an average bump factor of 2.2. However, these latter laminates should be limited to a bump factor

of 1.5 for aircraft structures to prevent exceeding their bearing yield point at design limit load.

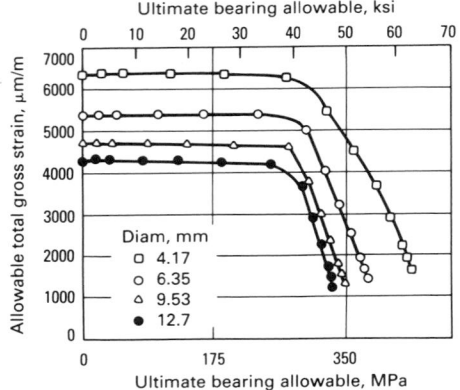

Fig. 25 AS1/3501-5/6 (50/40/10) tape laminate tension-loaded bolt-bearing bypass, ϵ_x^{tu} versus F^{bru}, RTD/RTW

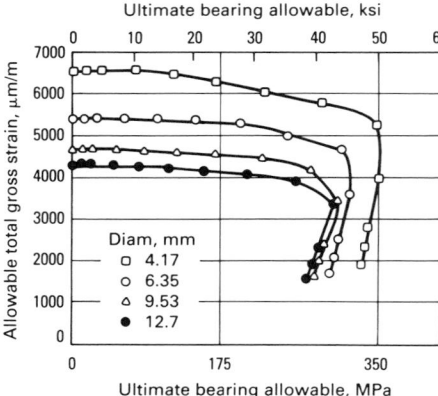

Fig. 27 AS1/3501-5/6 (25/75/0) tape laminate tension-loaded bolt-bearing bypass, ϵ_x^{tu} versus F^{bru}, RTD/RTW

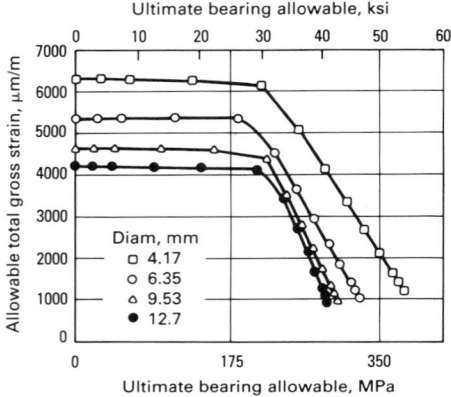

Fig. 26 AS1/3051-5/6 (60/30/10) tape laminate tension-loaded bolt-bearing bypass, ϵ_x^{tu} versus F^{bru}, RTD/RTW

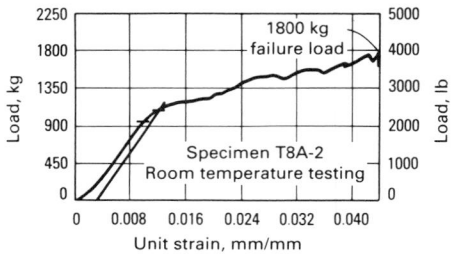

(a)

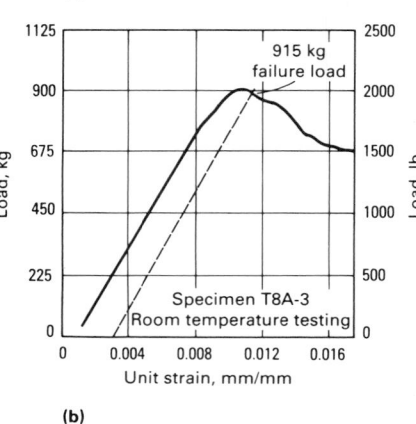

(b)

Fig. 29 Double-shear bolt- and pin-bearing test, AS1/3501-5/6 (10/80/10) tape laminates, load versus strain, RTD/RTW (Ref 12)

Structural-Element Testing

With the materials and simple joint data presented in the previous section, tests for structural elements and subcomponents can be designed based on the actual structural components being studied experimentally. These

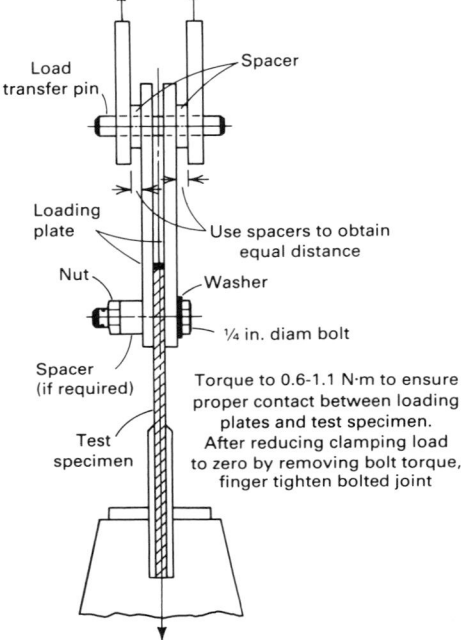

Fig. 28 Double-shear bolt- and pin-bearing test set-up

tests are used for allowable verification and for fulfillment of structural-integrity requirements.

This section discusses simple structural-element testing under in-plane unidirectional, multidirectional, and combined loadings as well as out-of-plane bending loadings. Simple joint elements under uniaxial loading are also included, as are environmental effects, experimental methodology, reporting, and data analysis.

In-Plane Loaded Elements. In many respects, simple element testing under uniaxial tension and compression loading resembles coupon testing. However, the specimens are usually larger and occasionally contain discon-

Table 4 Comparison of pin- and bolt-bearing test data with BJSFM analytical predictions

AS1/3501-5/6 graphite-epoxy laminates

Specimen	Environmental condition, RTD °CW(a)	°FW	Bearing stress								Ratio of thickness to diam(c)
			B-basis test data				BJSFM analysis		Thickness		
			Bolt-bearing yield or pin-bearing ultimate MPa	ksi	Bolt-bearing ultimate MPa	ksi	MPa	ksi	mm	in.	
3250	105	220	390	56	700	101	505	73.3	2.2	0.088	0.352
			230	33	440	64	383	55.6	2.2	0.088	0.352
3122, 3097	105(b)	220	430	63	...	...	505	73.3	2.2	0.088	0.352
			250	36	...	...	383	55.6	2.2	0.088	0.352
3251	105	220	480	69	910	132	374	54.5	2.2	0.088	0.352
			280	40	650	94	283	41.0	2.2	0.088	0.352
3916	105	220	320	46	430	63	435	63.1	2.8	0.112	0.448
			190	28	370	54	197	28.5	2.8	0.112	0.448

(a) 1 to 1.2% moisture content by weight. (b) interpolated to given temperatures. (c) diam = 6.4 mm (1/4 in.). (Tape joints: 3122 and 3097, $(0_3/\pm45/0_2/90)_s$, F.V. = 60%, V.V. = 0%, pinned. 3250, same as 3122 and 3097 except F.V. = 62.2%, bolted. 3251, $(\pm45_2/0/90/\pm45_2)_s$, F.V. = 59.5%, V.V. = 0.06%, bolted. Cloth joints: 3916, $(0/\pm45/90)_s$, F.V. = 61.9%, V.V. = 0%, bolted. Source: Ref 12

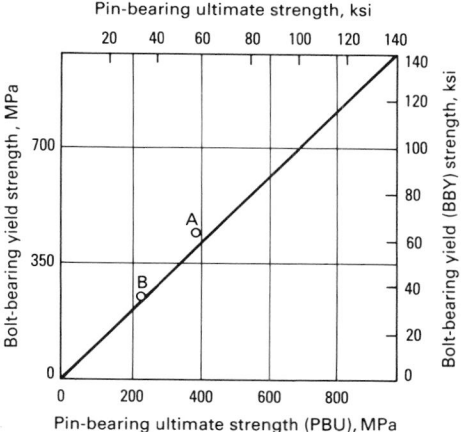

Fig. 30 Correlation of test double-shear bolt-bearing yield and pin-bearing ultimate strengths, for AS1/3501-5/6 (63/25/12) tape laminates, RTD/RTW. Point A = 3250 bolted (63/25/12) tape. Point B = 3122 and 3097 pinned (63/25/12) tape

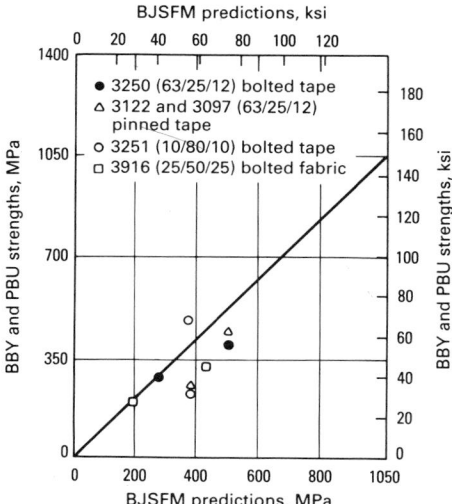

Fig. 31 Correlation of test double-shear bolt-bearing yield to BJSFM predicted strength for AS1/3501-5/6 laminates, RTD/RTW

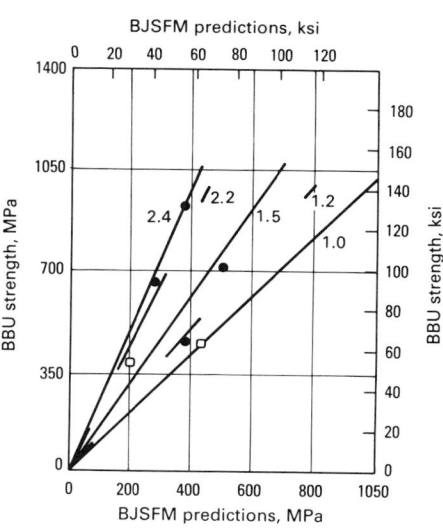

Fig. 32 Correlation of test double-shear bolt-bearing ultimate to BJSFM predicted strength for AS1/3501-5/6 laminates, RTD/RTW

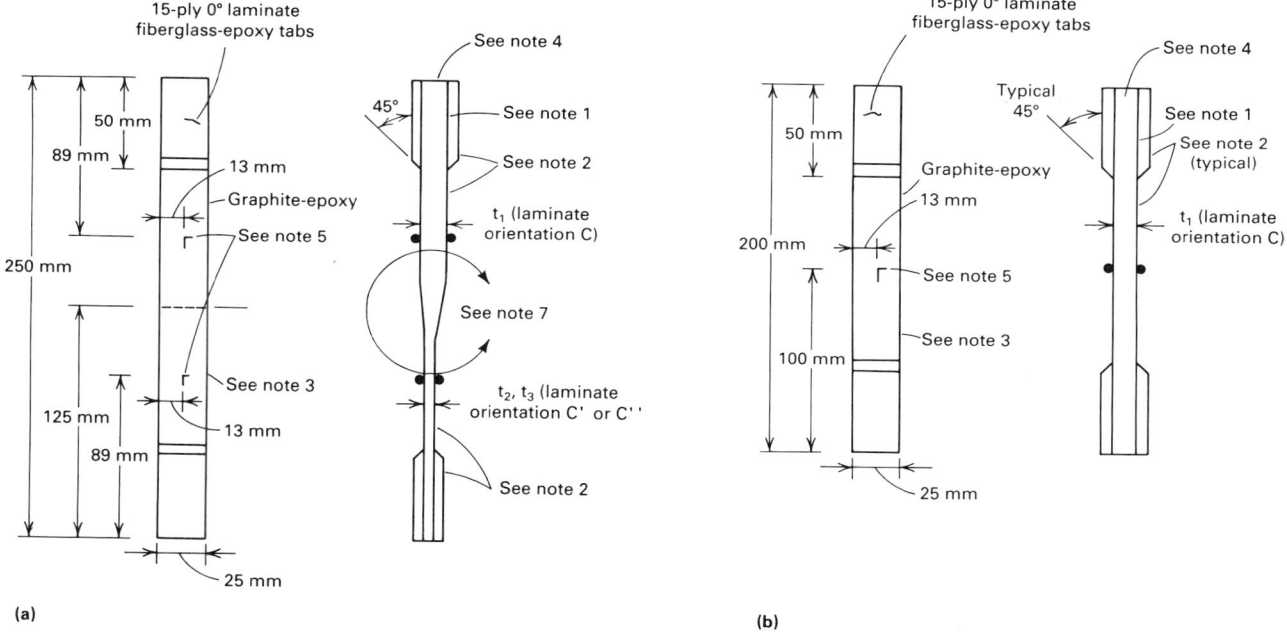

Fig. 33 Longitudinal or transverse tensile specimens for static and fatigue testing (a) Ply-drop-off specimen. (b) Plain specimen. (1) Bond glass-epoxy tabs with an epoxy film adhesive. (2) Specimen thickness shall not vary more than ±0.13 mm (0.005 in.) from nominal. (3) Specimen longitudinal edges shall be parallel to 0.13 mm (0.005 in.). (4) Top end and bottom end surfaces shall be flat and parallel to 0.25 mm (0.010 in.). (5) Strain gages shall be micromeasurements EA 03-250BF-350 or equivalent, located as shown. (6) Testing shall be in accordance with ASTM D 3039 (Ref 3). (7) See Fig. 34 and 35 for details.

tinuities such as holes, notches (slots), ply drop-offs, or joggles. These elements typically have more instrumentation than do simple coupons, and more effort is required to provide adequate design of the load introduction and test fixtures, because few standard methods are available at the element level. In this case, adequate design means that the load introduction and test fixture configuration will allow the specimen to be loaded to the required stress (or strain) level at failure. Such

design will also allow an acceptable failure mode to occur, that is, one that is expected in the full-sized structure (component and subcomponent).

Tension-Loaded Elements. Although much information on tension-loaded coupons is given in ASTM D 3039 (Ref 3) that may be useful in design and testing of tension elements, additional work on load-introduction tabs, consideration of discontinuities, and added instrumentation are usually necessary. In the lami-

nate evaluation longitudinal-tension specimens with and without a thickness discontinuity illustrated in Fig. 33, the discontinuity is a ply drop-off, but it could have been a center hole or a combination of both. Two types of ply drop-offs are described in Fig. 34 and 35, and material identification and laminate orientation are described in Table 5. Ply drop-off (PDO) 1 has two 0° plies dropped, while PDO 2 has four 45° plies dropped, both of which are located near the plan-form middle of the test

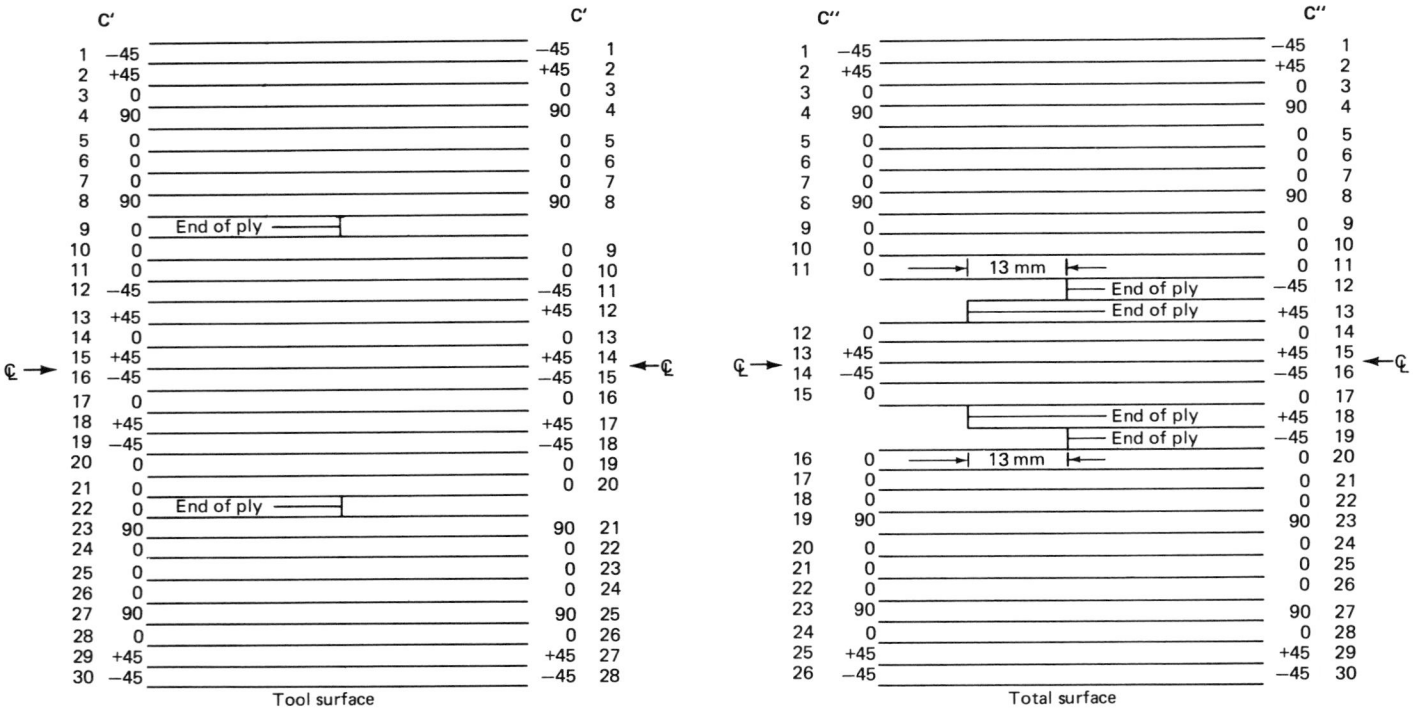

Fig. 34 Details of ply drop-off

Fig. 35 Details of ply drop off 2

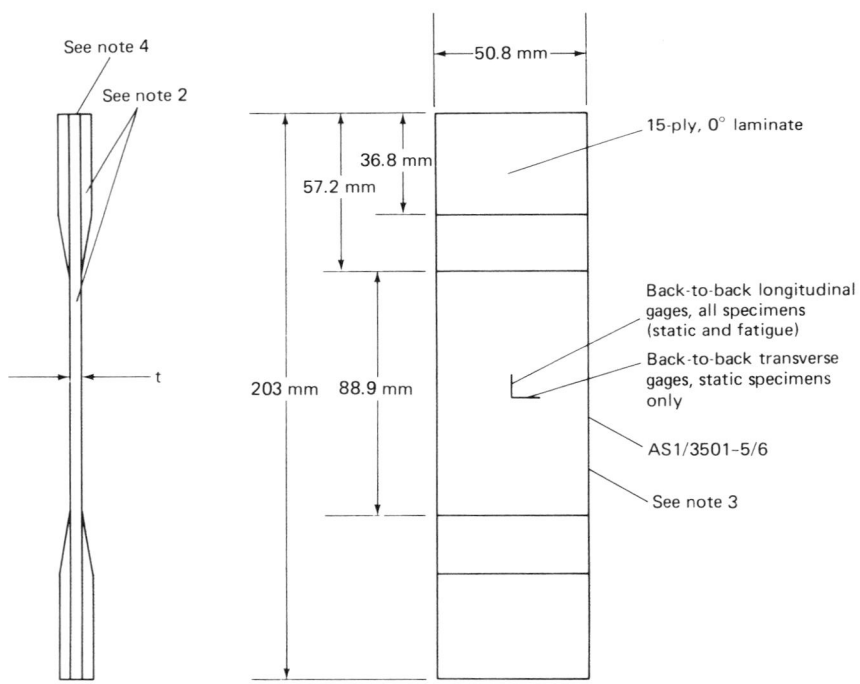

Fig. 36 Test specimen, plain laminate, static and fatigue. (1) Bond fiberglass-epoxy tabs with epoxy film adhesive. (2) Specimen thickness shall not vary more than ± 0.13 mm (0.005 in.) from nominal. (3) Specimen longitudinal edges shall be parallel to ± 0.13 mm (0.005 in.). (4) Top end and bottom end surfaces shall be flat and parallel to 0.025 mm (0.001 in.).

laminate in a balanced and symmetric fashion through the thickness. Effects of such discontinuities are covered in Ref 7 and 8. Small reductions in static compression strength and no reduction in fatigue strength are observed in these ply drop-off tests.

Design of the grip tabs that are bonded onto the ends of the tensile specimens for load introduction must be considered first. Because the orientation and material identification of the graphite-epoxy composite are given in Table 5, the laminate tensile strength can be calculated with lamination theory/failure-criterion equations and the ply properties given in the section "Composite Materials and Simple Joint Data" in this article.

Tabs for graphite-epoxy test specimens are made of woven glass-epoxy or suitable graphite-epoxy prepregs that have been premolded under suitable pressures and temperatures into solid laminate sheets. Unidirectional glass-epoxy and graphite-epoxy tape laminates have been successfully used as tab materials, but they are usually more expensive. Either $[0°/90°]_{nt}$ or $[\pm 45°]_{ns}$ glass cloth laminate is recommended, unless the specimen and tab width are greater than 50 mm (2 in.); then the same orientation as the test specimen should be used to prevent Poisson's ratio mismatch problems, which could cause premature or unacceptable failures. Graphite-epoxy or metal tabbing material such as titanium or steel may also be used when higher tab-to-specimen bonded-area strengths are required, but they are more expensive than the glass-epoxy materials. Also, machining and surface preparation costs for metals exceed those for fiberglass-epoxy or graphite-epoxy.

All the materials mentioned above perform well as load introduction grip tabs. Neither aluminum nor magnesium should be used, however, because of the large differences in the coefficients of thermal expansion (CTEs) com-

Table 5 Condensed and full lamination codes, AS1/3501–5/6 graphite-epoxy

Condensed code (%0°/%±45°/%90°)	Section ID	Full lamination sequence code	Number of plies
$[0_{16}/(\pm45)_5/90_4]_c$			
(53/33/14) C		$[\mp45/0/90/0_3/90/0_3/\mp45/0/\pm\overline{45}]_s$	30
$[0_{14}/(\pm45)_5/90_4]_c$			
(50/36/14) . . . . C'		$[\mp45/0/90/0_3/90/0_2/\mp45/0/\pm\overline{45}]_s$	28
$[0_{16}/(\pm45)_3/90_4]_c$			
(62/23/15) . . . . C''		$[\mp45/0/90/0_3/90/0_4/\pm\overline{45}]_s$	26

Ply drop-off 1 is symmetrical transition from section C to section C', and ply drop-off 2 is symmetrical transition from section C to section C''. All thickness change taken on one side only in making the laminates.

Fig. 38 Support fixture for compression static and fatigue test

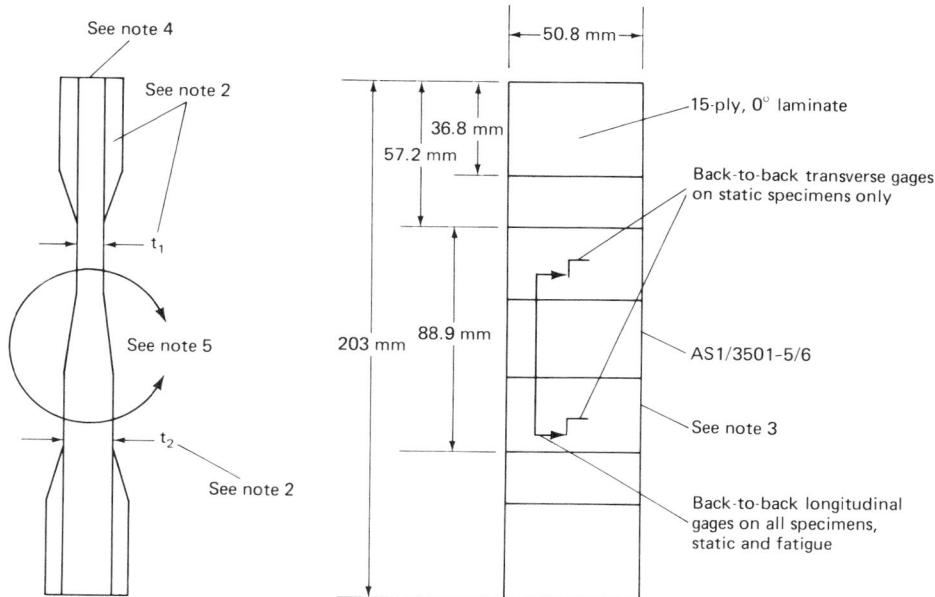

Fig. 37 Test specimen, ply drop-off, static and fatigue. (1) Bond fiberglass-epoxy tabs with epoxy film adhesive. (2) Specimen thickness shall not vary more than ±0.13 mm. (3) Specimen longitudinal edges shall be parallel to 0.13 mm (0.005 in.). (4) Top end and bottom end surfaces shall be flat and parallel to 0.025 mm (0.001 in.). (5) See Fig. 34 and 35 for details.

Fig. 39 Machine set-up for compression static and fatigue test

pared with those of the composite materials to which they would be bonded. At the element level, a small mismatch of CTEs between the tab and test material is acceptable, but when the specimen becomes larger, such a mismatch is unacceptable. In extreme CTE mismatch cases, even a tab area of 2600 mm² (4 in.²) may cause premature failure in the tab-to-specimen bonded area or in the specimen itself, leading to an unacceptable failure mode or location or both.

At 30 plies, the specimen thickness is 4.19 mm (0.165 in.). If the calculated tensile strength is 972 MPa (141 ksi), then the estimated load at failure is 103.2 kN (23.2 kip). Because the load is taken out in shear in two equal-sized bonded areas on each side of the specimen, each side must take 36.7 kN (8.25 kip) at specimen failure. The actual lap shear ultimate strength of the film adhesive should be divided by 2 to obtain an acceptable test-allowable value. Using lap-joint test ultimate strengths of 30 to 35 MPa (4 to 5 ksi) for most epoxy film adhesives and plastic behavior val-

ues above a stress level of 17 to 21 MPa (2.5 to 3.0 ksi), it can be calculated that the necessary tab length is 105 mm (4.12 in.) for a 25-mm (1-in.) wide specimen.

If load introduction grips of this size are not readily available or cannot be easily designed and fabricated, the standard 50-mm (2-in.) long grips may be used. If this is the case, however, higher quality materials, processing, and inspection of the tab bonding will be necessary because the bonded-area stress at failure will have increased to 28.4 MPa (4.12 ksi), well into the plastic range for most epoxy film adhesives. If the tabs, as received, are too long for the available test machine grips and the ends of the specimens are cut off to fit into the 50-mm (2-in.) long wedge or hydraulic grips, the bond joint stress at specimen failure will increase significantly. This will result in unacceptably low tab bonded area strength and test data reliability and some premature tab-to-specimen bond failures. However, under no circumstance should the test engineer try to test the 105-mm (4.12-in.) long tab specimens with

50-mm (2-in.) long test machine grips. Use of any gripping length shorter than the tab length will induce premature and unacceptable failure under the portion of the tab that is not gripped, for example, delamination in the bonded area or at the tab beveled ends or in the specimen laminate.

Compression-Loaded Elements. A typical compression-loaded plain laminate element is shown in Fig. 36, and one with ply drop-off defects is shown in Fig. 37. The clamped platen-supported fixture for these specimens is shown in Fig. 38. Load introduction is by bearing on the ends of the specimen, which is machined flat and parallel to within 0.025 mm (0.001 in.). The test set-up, which can be used in compression loading for both the static and fatigue test conditions, is shown in Fig. 39.

Because load introduction in these compression specimens is primarily by end bearing on the specimen and the tab, the bonded-area quality is not as critical as before. However, grinding the ends flat and parallel to ≤0.025 mm (≤0.001 in.) is critical, and both ends of

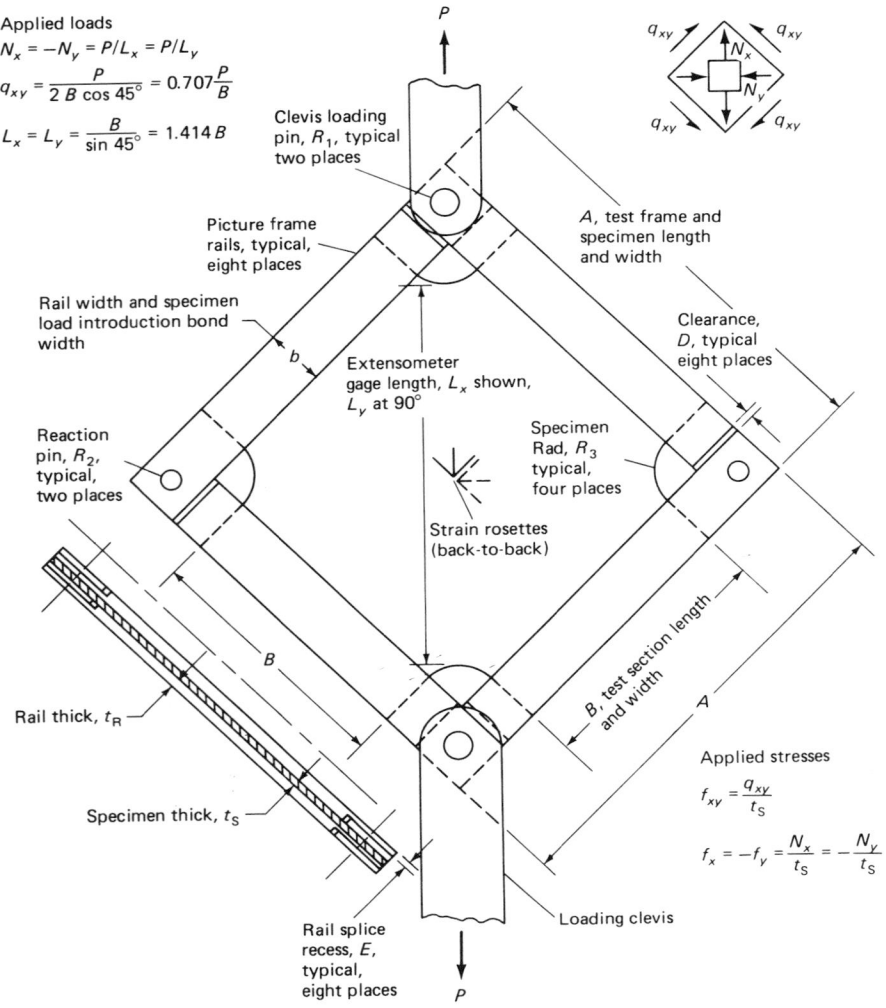

Applied loads

$N_x = -N_y = P/L_x = P/L_y$

$q_{xy} = \dfrac{P}{2B \cos 45°} = 0.707\dfrac{P}{B}$

$L_x = L_y = \dfrac{B}{\sin 45°} = 1.414B$

Clevis loading pin, R_1, typical two places

Picture frame rails, typical, eight places

Rail width and specimen load introduction bond width

A, test frame and specimen length and width

Clearance, D, typical eight places

Extensometer gage length, L_x shown, L_y at 90°

Reaction pin, R_2, typical, two places

Specimen Rad, R_3 typical, four places

Strain rosettes (back-to-back)

Rail thick, t_R

Specimen thick, t_S

B, test section length and width

Applied stresses

$f_{xy} = \dfrac{q_{xy}}{t_S}$

$f_x = -f_y = \dfrac{N_x}{t_S} = -\dfrac{N_y}{t_S}$

Loading clevis

Rail splice recess, E, typical, eight places

Fig. 40 Picture frame test set-up and specimen

every specimen should be inspected on a surface plate after machining. Those that do not meet this requirement should be rejected and reground.

Whether the bearing area is large enough to cause failure in the test section without end failure is determined as follows. The bearing area is 213 mm² (0.330 in.²). The $[0]_{nt}$ fiberglass-epoxy tabs are 15 plies at 0.25 mm (0.01 in.)/ply or 3.8 mm (0.15 in.)/side, or a 26-mm (0.30-in.) thickness for both sides. The fiberglass-epoxy bearing area is 387 mm² (0.60 in.²). However, the compression modulus of the graphite-epoxy laminate is 73.8 GPa (10.7 × 10⁶ psi) (Ref 5, 6). Assuming that the bearing load absorbed by each material is related to its compression modulus of elasticity times its bearing area ($E^c \cdot A^b$), the fraction of the load taken by the graphite-epoxy is:

$$\frac{(E^c A^b)_{GR\text{-}EP}}{(E^c A^b)_{GR\text{-}EP} + (E^c A^b)_{GL\text{-}EP}} = C_{GR\text{-}EP} \quad \text{(Eq 1)}$$

and

$$\frac{(E^c A^b)_{GL\text{-}EP}}{(E^c A^b)_{GR\text{-}EP} + (E^c A^b)_{GL\text{-}EP}} = C_{GL\text{-}EP} \quad \text{(Eq 2)}$$

Thus, the total load at specimen failure is

$$C_{GR\text{-}EP} \cdot P^u + C_{GL\text{-}EP} \cdot P^u = P^u \quad \text{(Eq 3)}$$

and

$$C_{GR\text{-}EP} + C_{GL\text{-}EP} = 1 \quad \text{(Eq 4)}$$

Calculating $E^c A^b$ shows: $(E^c A^b)_{GR\text{-}EP} = 14.7$ MN (3.53 × 10⁶ lbf) and $(E^c A^b)_{GL\text{-}EP} = 9.07$ MN (2.04 × 10⁶ lbf). Then:

$$C_{GR\text{-}EA} = \frac{3.53}{3.53 + 2.04} = 0.63$$

and

$$C_{GL\text{-}EP} = \frac{2.04}{3.53 + 2.04} = 1 - 0.63 = 0.37$$

That is, the graphite-epoxy takes 63% of the load (0.63 P^u), and the glass-epoxy takes 37% of the load (0.37 P^u). The failure load, based on

a −597.1 MPa (86.6 ksi) failure stress, is $P^u = -60.1$ kN (−14.04 kip).

The tab-to-specimen bonded area must now be checked for adequate strength. The effective bonded area (Fig. 36, 37) is 3740 mm² (5.8 in.²). The tab-to-specimen bondline stress at failure is 5.95 MPa (0.862 ksi). This assumes pure shear in these bonded areas, which is not literally true but is acceptable; as long as conservative film-adhesive ultimate strengths (30 to 35 MPa, or 4 to 5 ksi, for 175-°C, or 350-°F, cured epoxies) and related allowable strengths (15 to 20 MPa, or 2 to 3 ksi) are used, it keeps the failure modes in the test section. Thus, the bonded area has a margin of safety of 132%. It can be seen from the established bearing allowable strength values for graphite-epoxy tape and glass-epoxy cloth (Ref 15) that there are also large margins of safety in bearing. Therefore, failure will be in the test section of the specimens, that is, it will be a successful test, assuming proper specimen preparation, specimen quality, dimensional tolerances in machining (especially the ends), and support fixture installation.

Combined Loaded Elements. Combined in-plane loaded composite material test elements and methodology can be as simple as the rail-shear test detailed in ASTM D 4255 (Ref 16) for obtaining shear stress-strain curves to failure of simple, unnotched, unidirectional or multidirectional laminates. A picture frame test set-up (Fig. 40) is used for testing notched and unnotched multidirectional laminates in shear; with proper instrumentation (strain gages or deflectometers), this gives a shear stress-strain curve to failure. More complex testing includes biaxially loaded notched or unnotched multi-directional laminates (Fig. 41). Such tests are usually combinations of compression and compression or tension and compression loadings, each at 90° to the other.

Composite tubes are also used for studying biaxial loading in composites from simple pure shear to complex biaxial loading modes that are possible only with tube test specimens. Specialized tube test facilities are required. While precise, rigid mounting of the tubes in a test machine is usually necessary to impart torsion, longitudinal tension, and compression loadings individually or simultaneously, additional modes of loading may be desirable. Such modes include circumferential tension, obtained by applying internal fluid pressure, or circumferential compression, obtained by applying external fluid pressure. Typical tube test specimens with load introduction grips are shown in Fig. 42, and typical tube test specimens, including the internal and external pressure plugs and support fixtures, are shown in Fig. 43.

All of these test elements, both flat and tubular, require load introduction techniques that involve either friction gripping imparted by through-the-thickness mechanical clamping or hydraulic pressure, or gripping or mechanically fastening to bonded-on tabs, which introduce

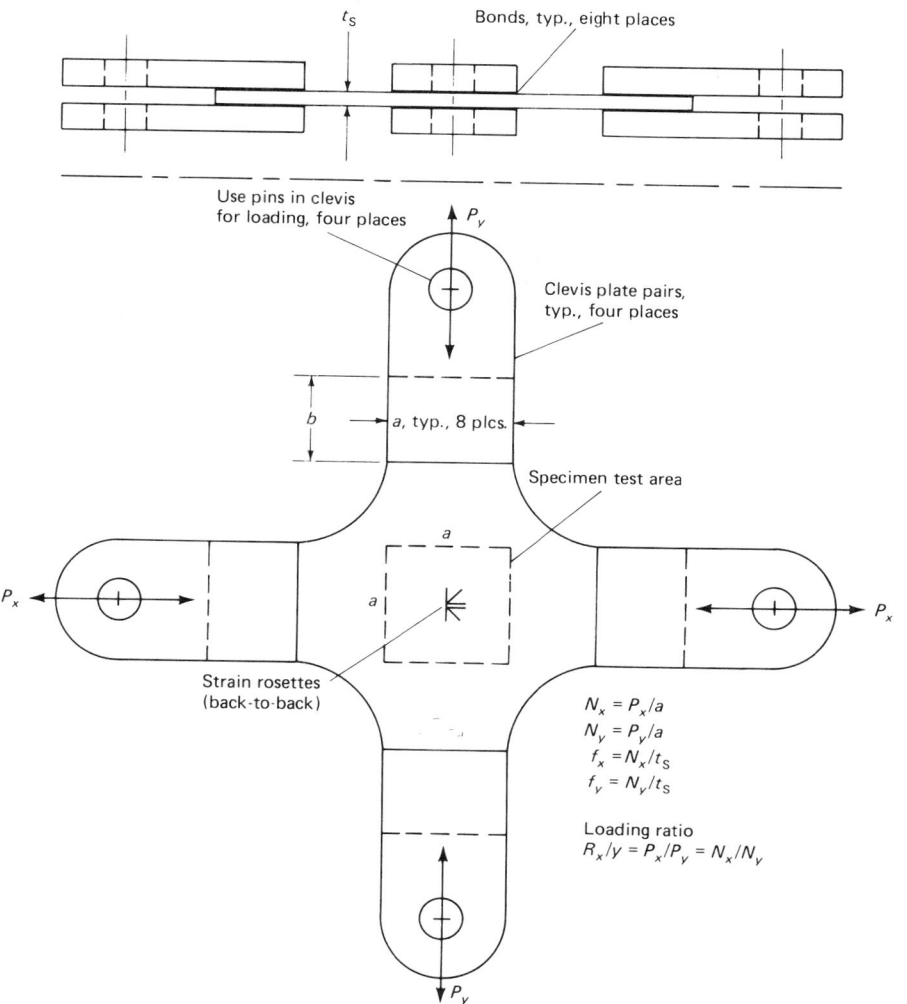

Fig. 41 Biaxially loaded flat element

Within the figure:

t_s

Bonds, typ., eight places

Use pins in clevis
for loading, four places

P_y

Clevis plate pairs,
typ., four places

b

a, typ., 8 plcs.

Specimen test area

P_x

P_x

Strain rosettes
(back-to-back)

$N_x = P_x/a$
$N_y = P_y/a$
$f_x = N_x/t_S$
$f_y = N_y/t_S$

Loading ratio
$R_x/y = P_x/P_y = N_x/N_y$

P_y

---

load by shear through the bonded area. Internal pressure loads on tubes that impart only circumferential tension loading are accommodated by O-ring seals on male end plugs, with the longitudinal force being resisted by the holding fixture or the test machine rather than by the tube. External pressure applied to obtain circumferential compression loadings requires an external pressure vessel of some kind using O-ring seals at the ends or a fluid-pressure cylindrical test tank such as those found at deep-submergence structural test facilities. When the latter technique is used, the test tube internal end plugs with O-ring seals must have a rod or tube structure inside the test tube to take the axial compression loads applied on the tube ends by the hydrostatic fluid pressure.

The best mechanical clamping and hydraulic friction grips may approach 15 MPa (2 ksi) friction shear-stress transfer; however, the load-introduction design for such gripping should not exceed 7 MPa (1 ksi) stress. Combined friction (or bonding) and bolt-bearing load in-troduction should not be used; the friction shear or the bonded-area shear transfers all the load until the friction or bonded area fails, where-upon the bolts are suddenly required to transfer all the load in bearing. Such joints should be designed as friction or bonded-area load transfer joints with clearance holes for clamping bolts, or should be designed to take the full bearing load in the holes (without failure) at specimen failure load. Bonded-joint load transfer is performed similarly to that of tensile coupons, assuming pure shear loading in the bonded area and using a 15 to 20 MPa (2 to 3 ksi) allowable bonded-area stress for epoxy film adhesives (cured at 175 °C or 350 °F).

Normal and Bending Loaded Elements. At the element level, these tests will be primarily in one category, that of beams. They will probably be three- or four-point loaded simple or single-point loaded cantilever beams, or uniformly distributed loaded (UDL) simple or cantilever beams. Occasionally, fixed-end beams are tested with either type of loading.

These beams could be solid laminates or stiff-ened sections (blade, hat, I) or sandwich beams with composite faces. Beam technology, in general, is well documented. Simple sandwich beams are discussed in Ref 17 and 18.

Testing composite and sandwich beam elements differs from testing isotropic material (metal) beam elements in that shear stresses (or strains) and deformations must be taken into consideration when designing, testing, and an-alyzing them. In addition, load and support-point softness should be considered in load introduction, as should the requirement for load pads to prevent local crushing, especially in sandwich beams with honeycomb or foam cores that are relatively weak in flatwise compression strength.

A four-point loaded simple sandwich beam element is shown in Fig. 44, with the applicable equations for calculating maximum face stress, core shear stress/bondline shear stress, and core shear modulus. Such elements are used to measure core or face properties or to perform quality assurance testing, and are tested as structural elements representing larger beam-bending sandwich structures. Generally, the sandwich flatwise tension strength and the flat-wise compression strength and modulus are required for complete analysis, in addition to similar sandwich core shear and face sheet properties. This is because of the need for face-wrinkling analysis plus intercellular-buckling analysis of honeycomb core sandwich specimens on the compression face (Ref 17). In addition, the core flatwise compression strength is used to calculate the load/support pad area necessary for use in testing the beam. Once the properties of all the sandwich constituents are known, the maximum load at failure can be calculated, and then the load pad size can be calculated. For failure load, P^u, on a four-point sandwich beam, each load pad must take P^u load. The compression stress on the core is then:

$$F_{TL}^{cu} = \frac{P^u}{2A_p} \qquad \text{(Eq 5)}$$

where A_p is the load area. Hence, the required load pad area for test is:

$$A_p = \frac{P^u}{2 \cdot \bar{F}_{TL}^{cu}} \cdot (SF) \qquad \text{(Eq 6)}$$

and $\bar{F}_{TL}^{cu}$ is the core allowable compression strength, while (SF) is the safety factor, usually 1.5 or greater.

Joint Elements. The joint element types in composite structures discussed in the following sections are the bolted and bonded joint config-urations used in the tests described in Ref 1 and 9.

Bolted-Joint Elements. The bolted-joint structural element shown in Fig. 45 represents a static and fatigue specimen that was used to evaluate joints for aircraft wing structures. The joint was designed to fail at 1.58 MN/m (9000 lbf/in.). The grip tabs on the graphite-epoxy composite end represent an effective bonded

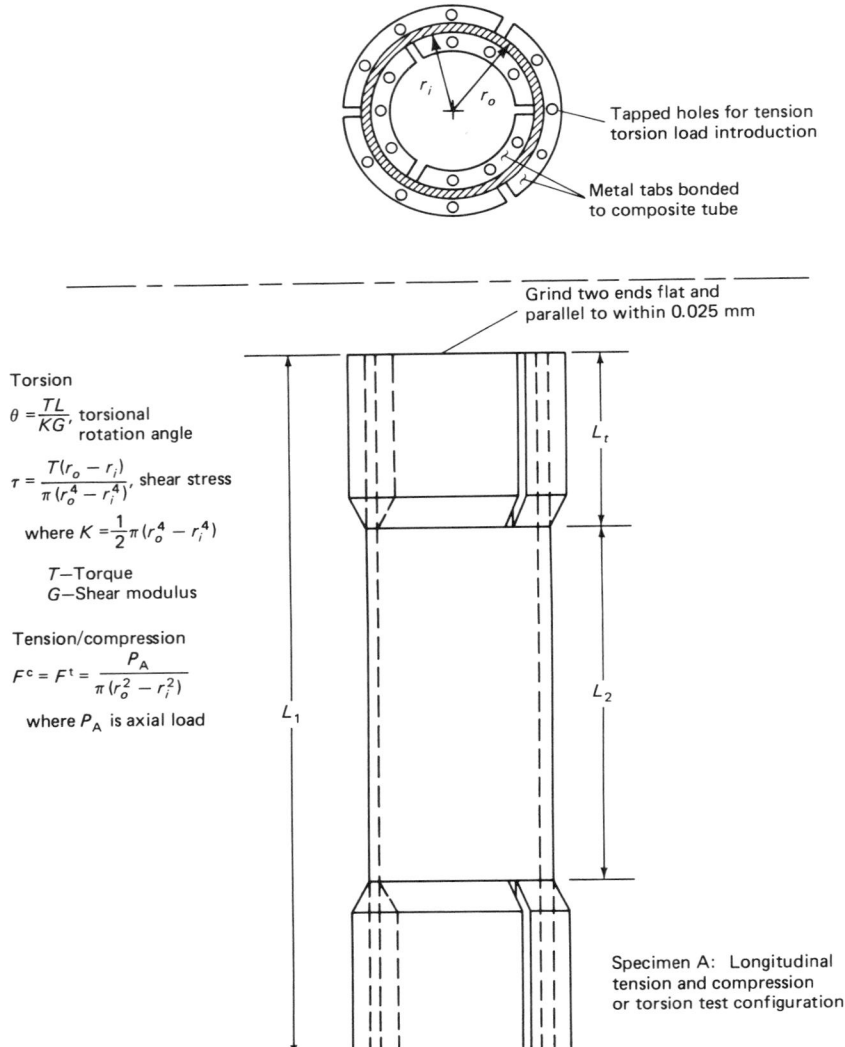

Tapped holes for tension torsion load introduction

Metal tabs bonded to composite tube

Grind two ends flat and parallel to within 0.025 mm

Torsion

$\theta = \dfrac{TL}{KG}$, torsional rotation angle

$\tau = \dfrac{T(r_o - r_i)}{\pi (r_o^4 - r_i^4)}$, shear stress

where $K = \dfrac{1}{2}\pi(r_o^4 - r_i^4)$

T—Torque
G—Shear modulus

Tension/compression

$F^c = F^t = \dfrac{P_A}{\pi(r_o^2 - r_i^2)}$

where P_A is axial load

Specimen A: Longitudinal tension and compression or torsion test configuration

Fig. 42 Tube test specimens, axial and torsion loading

area of 3.03×10^3 mm² (4.70 in.²). At 15 MPa (2 ksi), the bonded area allowable used above, the joint will be safe for at least 1.65 MN/m (9400 lbf/in.). The steel ends were tied directly with screws to the load plate on the test machine. Failure of the specimen was designed to be gross thin-section tension (away from the joint) or thick-section bolt bearing or both. Stress in the thin, unnotched, gross composite area (outside the joint build-up) is 418 MPa (60.6 ksi) at design failure, whereas the un-notched stress allowable is 1048 MPa (152 ksi).

Using the properties derived in the section "Composite Materials and Simple Joint Data" in this article, the bolted joint of Fig. 45 was analyzed with BJSFM at a net area through the first hole (from the taper) in a load/net area analysis (Fig. 46). The tension test data are slightly above the analytical bearing-failure values predicted by BJSFM. When the net tension analysis is compared with the test failure, a

stress-concentration factor of approximately 2 is observed for this joint. Although the most obvious specimen failure was net tension through the first bolt hole, subsequent inspection of the failed specimens showed the onset of bearing failure that correlates to a reasonable degree with the BJSFM predicted bearing (yield) failure. A finite-element analysis of this joint, assuming that half the load is taken by each of the two fasteners, accurately predicted the failure load and mode as net tension through the first hole from the thickness taper. Therefore, both modes of failure, net tension through the first hole from the taper and bolt bearing at the same location, are applicable.

Bonded-Joint Elements. A typical bonded-joint element design for the same wing joint is shown in Fig. 47. This three-step lap graphite-epoxy-to-titanium bonded joint is designed to fail at 1.58 MN/m (9000 lbf/in.). Basic thickness and tab design on the composite are the

same as on the thin section of the bolted joint, assuring that the tabs and the composite are strong enough. On the other end, the test machine grips the titanium directly (hydraulic or mechanical gripping of metals is valid for at least 20 MPa, or 3 ksi, in a friction-shear allowable). Total step-lap bonded area is 1.6×10^3 mm² (2.5 in.²). Average bondline stress at 1.58 MN/m (9000 lbf/in.) design ultimate load is 24.8 MPa (3.6 ksi). Stresses in some of the composite steps will be above the 418 MPa (60.6 ksi) calculated earlier for the design load, and may even be above the 1048 MPa (152 ksi) material design allowable. Thus, failure is intended to occur in either the bonded area or in one of the graphite-epoxy adherend steps.

The bondline stress of 24.8 MPa (3.6 ksi) represents a reasonable estimate of the mean strength of the FM-400 epoxy film adhesive tested in a 19.1-mm (¾-in.) double-overlap bonded joint (Fig. 48), which correlated well with the step-lap joint strength. The wet conditions shown in Fig. 48 are extreme (30 days at 75 °C, or 170 °F, 95% RH), that is, approximately 1.4% moisture content in the 27-ply laminate by weight. Dry laminate conditions represent approximately 0.3% moisture content by weight, and the design-wet conditions were subsequently determined to be approximately 0.85% moisture content by weight. Thus, the FM-400 adhesive mean strength for the design-wet condition is half-way between the wet and dry conditions shown in Fig. 48 at room temperature (20 °C, or 72 °F), that is, 24.5 MPa (3.55 ksi). Thus, the failure will be in the bonded area or in the laminate steps or both.

Test Methodology. Experimental techniques for structural-element testing are usually on a larger scale and more complex than those for coupon testing. Also, instrumentation is extensive to permit measuring more than laminate material properties; it is usually necessary or desirable to measure mechanical behavior under static monotonically increasing loads to failure and during fatigue loading. Results from such tests provide data that can be correlated more directly with subcomponent test results and that can be used to represent discrete portions of the subcomponent behavior.

Jigs and fixtures, such as those shown in Fig. 38 and 39 for the compression specimens shown in Fig. 36 and 37, are usually custom designed. The test fixture fully supports the specimen against buckling in compression between tabs and also provides a floating wedge support for the beveled area of the tabs. The flat part of the tabs is bolt clamp friction-gripped for load introduction in nearly pure shear in the tab-to-specimen bonded area. Additional fixture side support is necessary as shown in Fig. 39 for both static-compression and compression-compression fatigue loading.

Instrumentation may be as simple as that shown in Fig. 33 for tensile elements or as complex as that shown in Fig. 41 for biaxially loaded flat elements. Such instrumentation on elements is used to measure the material and

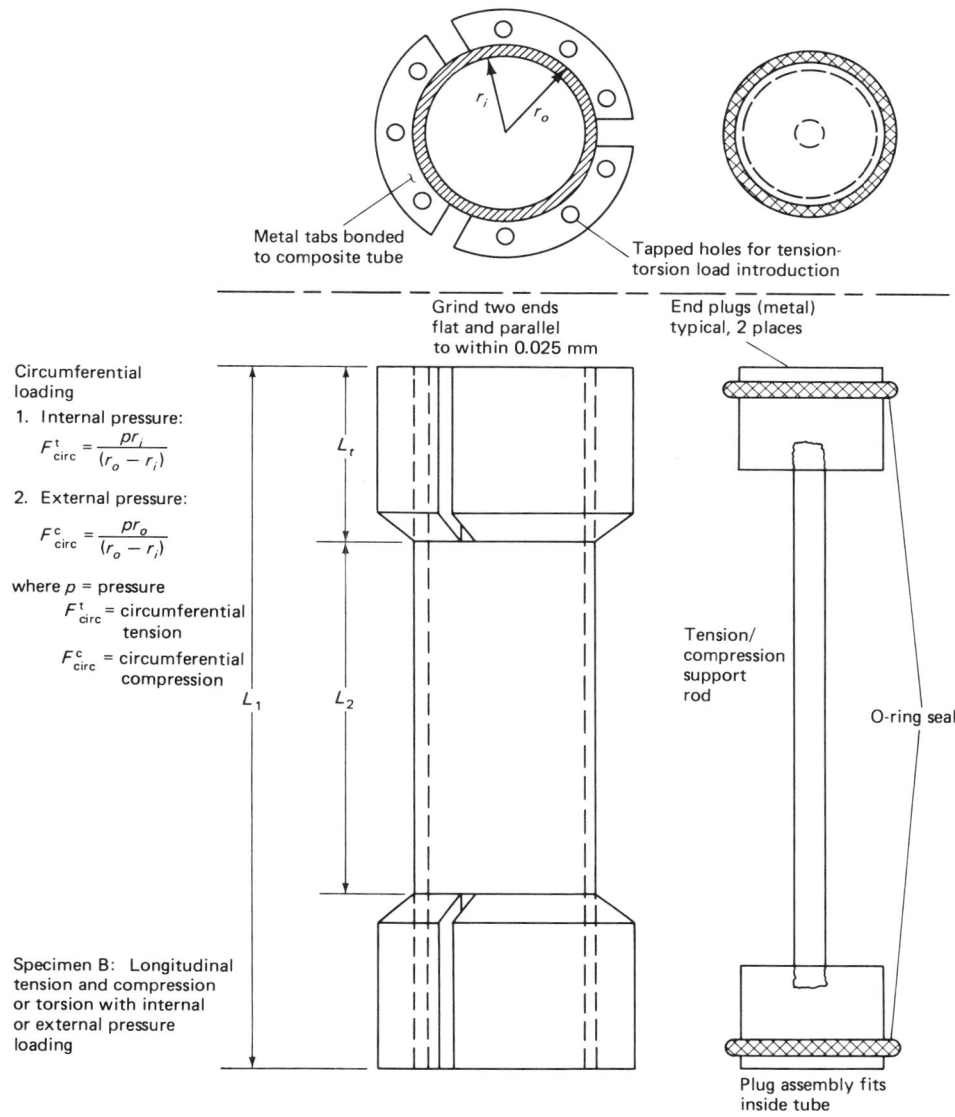

Circumferential
loading

1. Internal pressure:

$$F_{circ}^{t} = \frac{pr_i}{(r_o - r_i)}$$

2. External pressure:

$$F_{circ}^{c} = \frac{pr_o}{(r_o - r_i)}$$

where p = pressure

F_{circ}^{t} = circumferential
tension

F_{circ}^{c} = circumferential
compression

Specimen B: Longitudinal
tension and compression
or torsion with internal
or external pressure
loading

Fig. 43 Tube test specimens—axial, torsion, and circumferential loading

element behavior under loading, and will almost always include a number of back-to-back biaxial strain gages. It may also include extensometers and deflection gages, especially if the test objectives include measurement of deformation or buckling of critical elements.

Data Analysis and Reporting. Only key usable material property data are collected, reduced, and analyzed, as well as simple structural-element mechanical-behavior data. Load versus strain or load versus deflection data may represent not only material properties, but also important structural-element behavior under loading, such as linearity range, deformation at elastic limit, and load versus deformation behavior beyond the elastic range to failure. Failure modes and their classification and relation to material behavior become even more important than in coupons because they more

closely resemble selected discrete areas of failure in the subcomponent test. Visual and photographic studies of the failure modes are conducted, as well as failure surface studies with microscopic visual inspection, scanning electron microscope (SEM) photographs, and microscopic cross-sectional photomicrographs. Accurate analytical prediction of failure load and mechanical behavior of structural elements is also more important than at the coupon level because the structural-element behavior may represent a discrete part of the subcomponent and lead to the development of analytical methodology for the subcomponent.

For the two-bolt joint element shown in Fig. 45, good correlation is apparent when failure and mechanical-behavior analysis is compared with test data (Fig. 46, 49, 50). Figure 46 presents failure prediction analysis and test

results, Fig. 49 compares predicted with measured axial strain behavior between hole edges and joint laminate edge, and Fig. 50 compares predicted with measured axial stress distribution along the joint centerline through-the-hole sections. These studies were conducted at design-limit loading in the elastic-behavior range.

Occasionally, structural-element behavior must be compared to the variation in as-received material quality so as to remove material variability from the data analysis. In fiber-dominated structural elements such as the simple two-bolt joint shown in Fig. 46, the test results are usually normalized to a nominal fiber volume before further reduction and analysis so as to remove the normally induced material and process variables from the data analysis. However, in this instance, the relationship of ultimate bolted-joint strength to fiber volume appeared to be only casual at best (Fig. 51); such normalizing did little to reduce the scatter. Further investigation revealed that the bolted-joint strength was a function of the individual prepreg lot numbers used to make the joint panels. Joint specimen strength values grouped by prepreg lot number fell into discrete groups that correlated directly with the graphite fiber lot strength used in each prepreg lot (Fig. 52). After this fiber lot strength variation was taken into consideration, the scatter in bolted-joint strength data was approximately ±4% from the mean (Fig. 53). When such fiber strength variability is factored into the analysis, the significance of the data related to structural design can be determined.

The mean RTD static test results for this simple two-bolt joint element was 1.90 MN/m (10 870 lbf/in.). Statistical data reduction for allowables gave values somewhat above the 1.58-MN/m (9000-lbf/in.) design ultimate value previously established. Fatigue-loading maximum value was set early in the test program at 1.14 MN/m (6500 lbf/in.), representing the 5% over-design-limit load that the F-5E aircraft is expected to exceed occasionally. (Design limit load is 0.67 times design ultimate load, and the estimated B allowable is taken as 0.85 times the static test data mean.) It was also found that this fiber-dominated joint design was not sensitive to the test environment or the laminate moisture content. Thus, testing statically or in spectrum fatigue at −30 °C (−20 °F), 120 °C (250 °F), or at a spectrum variable temperature and moisture condition ranging from −30 °C (−20 °F) to 120 °C (250 °F), called mission profile temperature wet (MPTW), made no difference in residual mechanical behavior and strength. Figure 54, a comparison of the static and spectrum fatigue residual strengths for a number of fatigue variables at the MPTW environmental condition, shows no effect of these variables and thus no joint wear-out in a fatigue spectrum loading under these severe environmental conditions. In fact, a comparison of all the static and fatigue residual-strength test data from these tests (Ref 1 and 11) and other tests (Ref 19) shows that at

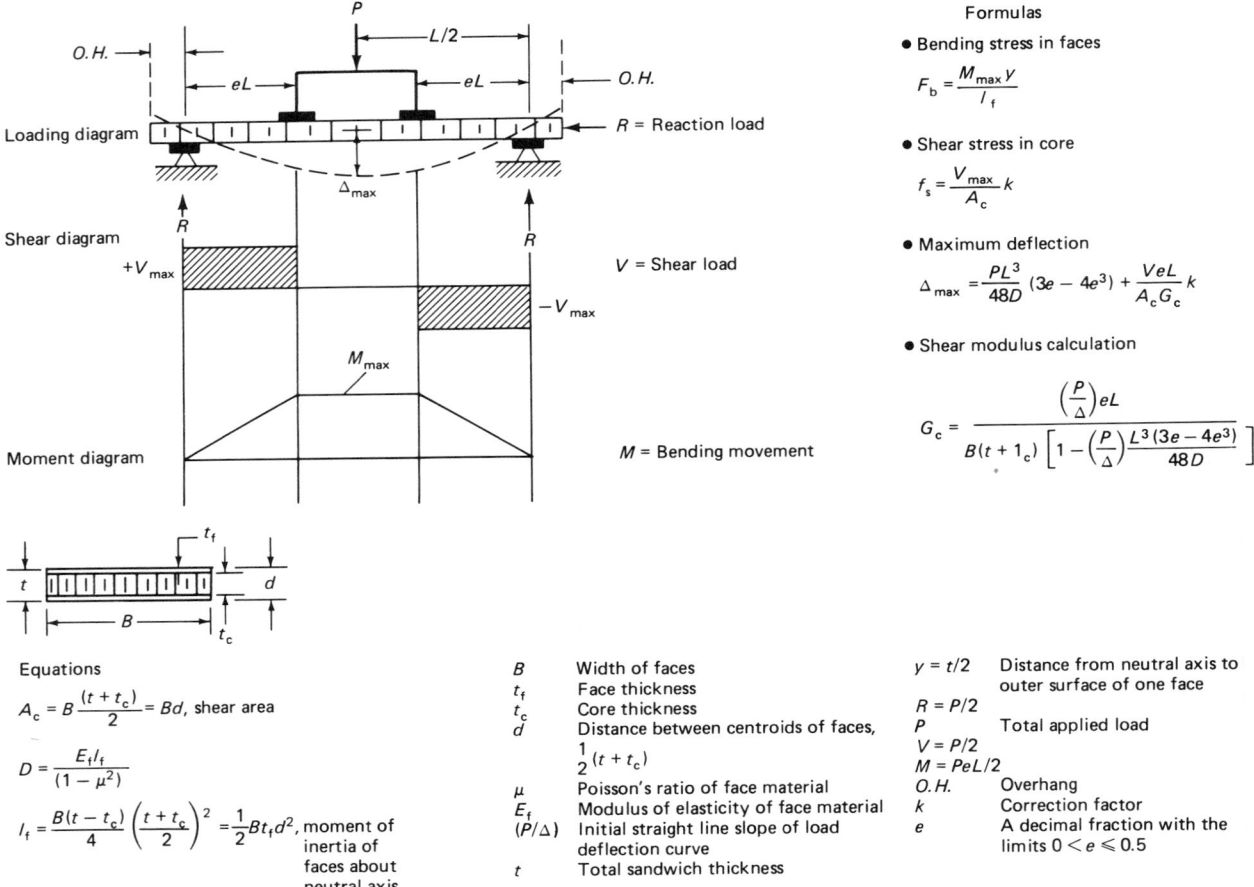

Formulas

- Bending stress in faces

$$F_b = \frac{M_{max} y}{I_f}$$

- Shear stress in core

$$f_s = \frac{V_{max}}{A_c} k$$

- Maximum deflection

$$\Delta_{max} = \frac{PL^3}{48D}(3e - 4e^3) + \frac{VeL}{A_c G_c} k$$

- Shear modulus calculation

$$G_c = \frac{\left(\frac{P}{\Delta}\right)eL}{B(t + 1_c)\left[1 - \left(\frac{P}{\Delta}\right)\frac{L^3(3e - 4e^3)}{48D}\right]}$$

Equations

$$A_c = B\frac{(t + t_c)}{2} = Bd, \text{ shear area}$$

$$D = \frac{E_f I_f}{(1 - \mu^2)}$$

$$I_f = \frac{B(t - t_c)}{4}\left(\frac{t + t_c}{2}\right)^2 = \frac{1}{2}Bt_f d^2, \text{ moment of inertia of faces about neutral axis}$$

B	Width of faces	$y = t/2$	Distance from neutral axis to outer surface of one face
t_f	Face thickness		
t_c	Core thickness	$R = P/2$	
d	Distance between centroids of faces, $\frac{1}{2}(t + t_c)$	P	Total applied load
		$V = P/2$	
μ	Poisson's ratio of face material	$M = PeL/2$	
E_f	Modulus of elasticity of face material	O.H.	Overhang
(P/Δ)	Initial straight line slope of load deflection curve	k	Correction factor
		e	A decimal fraction with the limits $0 < e \leqslant 0.5$
t	Total sandwich thickness		

Fig. 44 Sandwich beam four-point load test

a coefficient of variation of 10%, the safe B-basis allowable is greater than 80% of the mean test value when using many different prepreg and fiber lots (Fig. 55).

For the double-overlap three-step titanium-to-graphite-epoxy bonded joint, the average tension-test failure at RTD conditions was 1.90 MN/m (10 830 lbf/in.), with adhesive and prepreg processing-quality variables being extremely important. Every acceptable group of specimens was normalized to the mean strength of that fabrication lot before other variables were compared. The Hart-Smith step-lap bonded-joint computer program A4EG/F (Ref 20) was used in this analysis with the composite material properties given in the section "Composite Materials and Simple Joint Data" in this article, the adhesive properties given in Fig. 56 (Ref 21), and the *Military Handbook 5* properties for titanium alloy (Ti-6Al-4V) in the annealed condition (Ref 22). The results (Fig. 57) show that adhesive bonded area failure was the main mode of failure at all environmental and loading (static or fatigue) conditions under tension loading. Thus, the predicted strength of 1.98 MN/m (11 300 lbf/in.) is 4% above the measured static strength. The maximum fa-

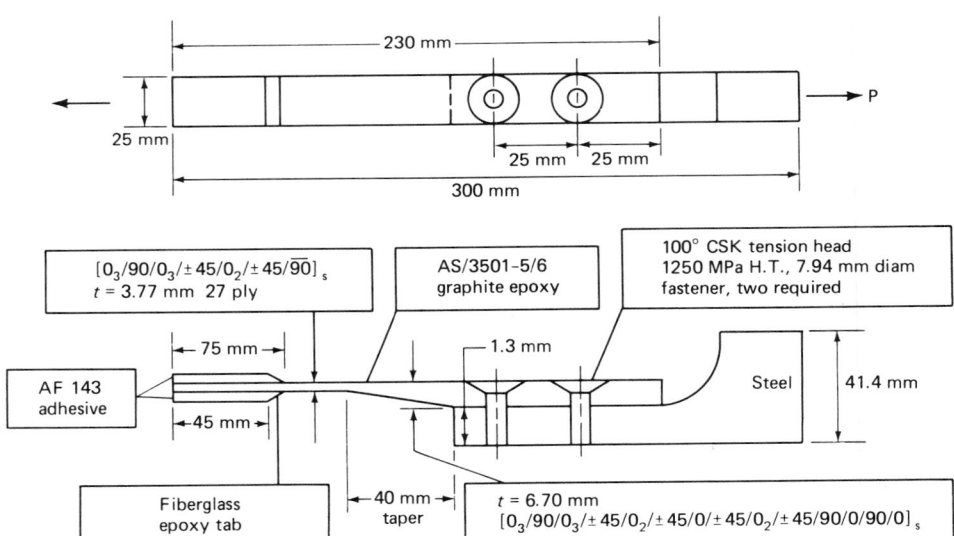

Fig. 45 Bolted-joint test specimen

tigue spectrum loading was calculated at 1.02 MN/m (5850 lbf/in.) by the same calculation used for bolted joints but with the addition of a 10% knockdown factor for the environmental

(moisture) degradation effects shown for the FM-400 adhesive in Fig. 48.

In the analysis (Fig. 57), the chosen maximum fatigue spectrum loading was in fact

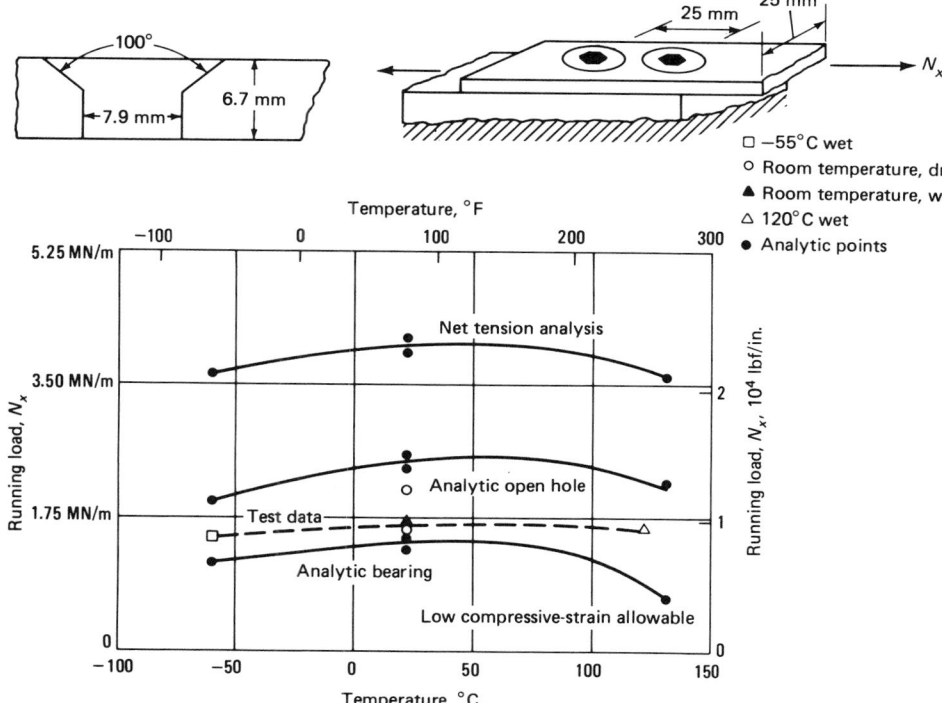

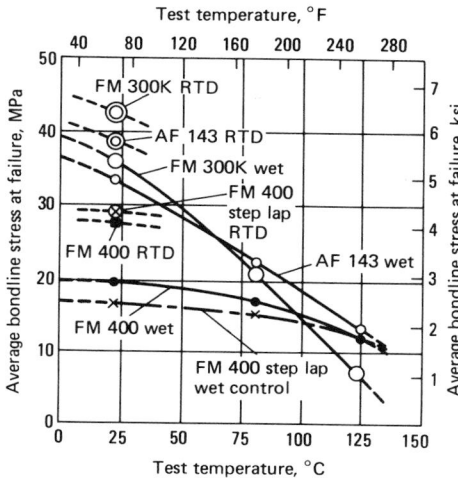

Fig. 48 Adhesive evaluation data

Fig. 46 Correlation between bolt-bearing data and several analytical models

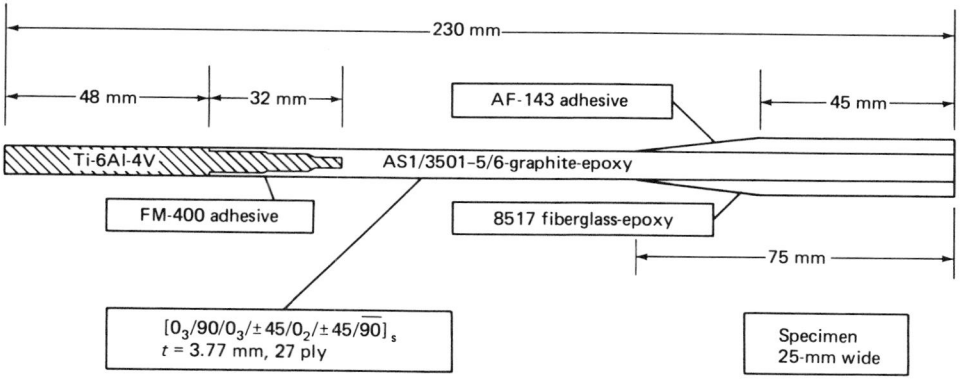

Fig. 47 Standard step-lap bonded-joint specimen

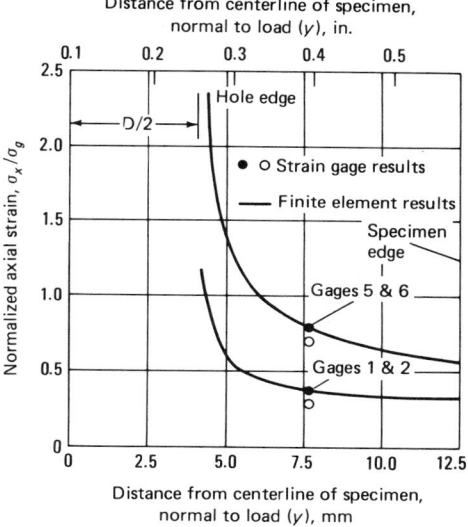

Fig. 49 Axial strain distribution of standard joint at hole sections, RTD

Subcomponent Testing

The purpose of subcomponent testing is to obtain a structural overview, in contrast with lower-level testing. The following sections treat in-plane loaded subcomponents, normal and bending loaded subcomponents, bolted and bonded joint components, experimental procedures, and test reporting.

In-Plane Loaded Subcomponents. Test equipment and instrumentation used for in-plane loaded subcomponents are described in the following sections, along with tension and compression testing and edgewise shear testing.

Test Equipment and Instrumentation. One philosophy that is used is to design the structure so that the joints are always statically stronger than the components located away

15% above the calculated adhesive bond area elastic limit of 0.89 MN/m (5080 lbf/in.) at RTD conditions. Thus, considerable wear-out was observed in the fatigue spectrum, along with many spectrum variable effects. Nonetheless, it was found that under all the testing conditions described for bolted joints, a safe design limit allowable for bonded joints was 0.68 MN/m (3900 lbf/in.), or 36% of the static mean value of 1.90 MN/m (10 830 lbf/in.), with the safe design ultimate allowable being 0.89 MN/m (5850 lbf/in.), or 54% of the static mean test value. While this is lower than the bolted-joint strength discussed previously and the design ultimate goal of 1.58 MN/m (9000 lbf/in.), the bonded joint is substantially

lighter than the bolted joint and, on a weight/strength basis, more efficient in transferring load. However, it is obvious that the joint will need to be strengthened for this design application. At a safe design ultimate loading of 0.89 MN/m (5850 lbf/in.), the joint develops a shear stress of 1.61 MPa (2.34 ksi) with the 31.8-mm (1.25-in.) overlap length (1.6 × 10³ mm², or 2.5-in.², bonded area). Therefore, a step-lap joint design of 48.3 to 50.8 mm (1.9 to 2.0 in.) in length would be required. This would also require increasing the laminate thickness and the number of steps. Even when such a design change is made, the bonded joint is still about 10% lighter than the bolted joint.

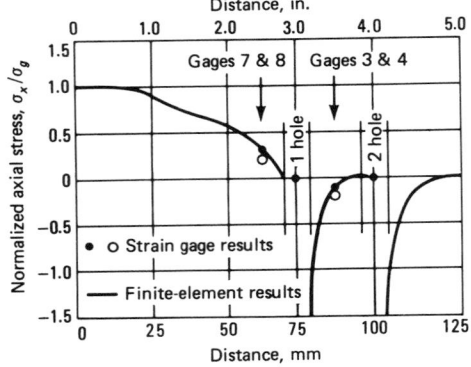

Fig. 50 Axial stress distribution along the centerline of standard joint through-hole sections, RTD

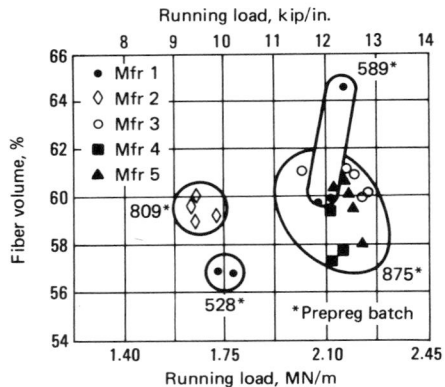

Fig. 51 Fiber volume versus quality control destructive-test data for various prepreg batches, standard bolted joint, RTD

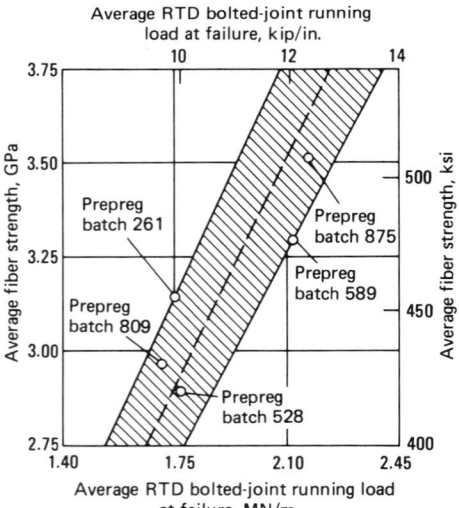

Fig. 52 Graphite fiber tensile strength versus quality control destructive-test data for various prepreg batches, standard bolted joint, RTD

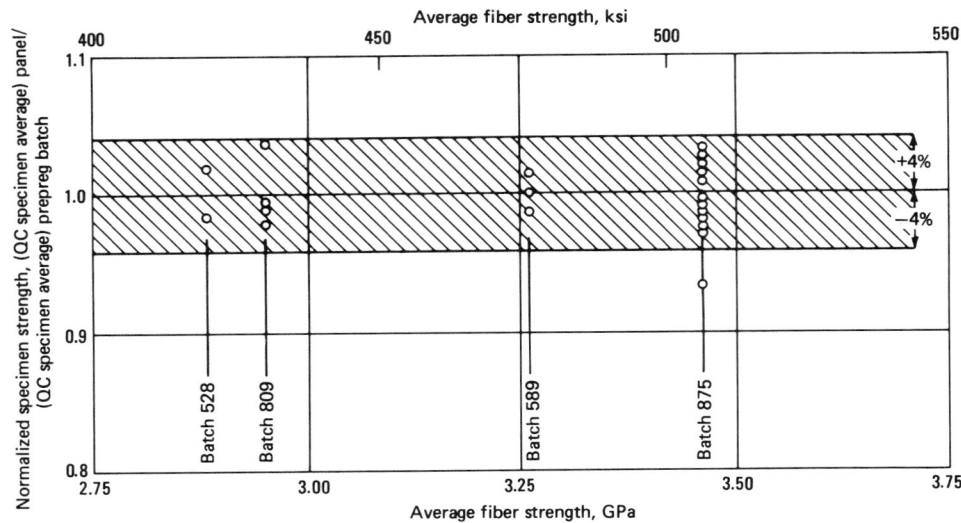

Fig. 53 Standard bolted-joint panel-strength scatter for each prepreg batch, RTD

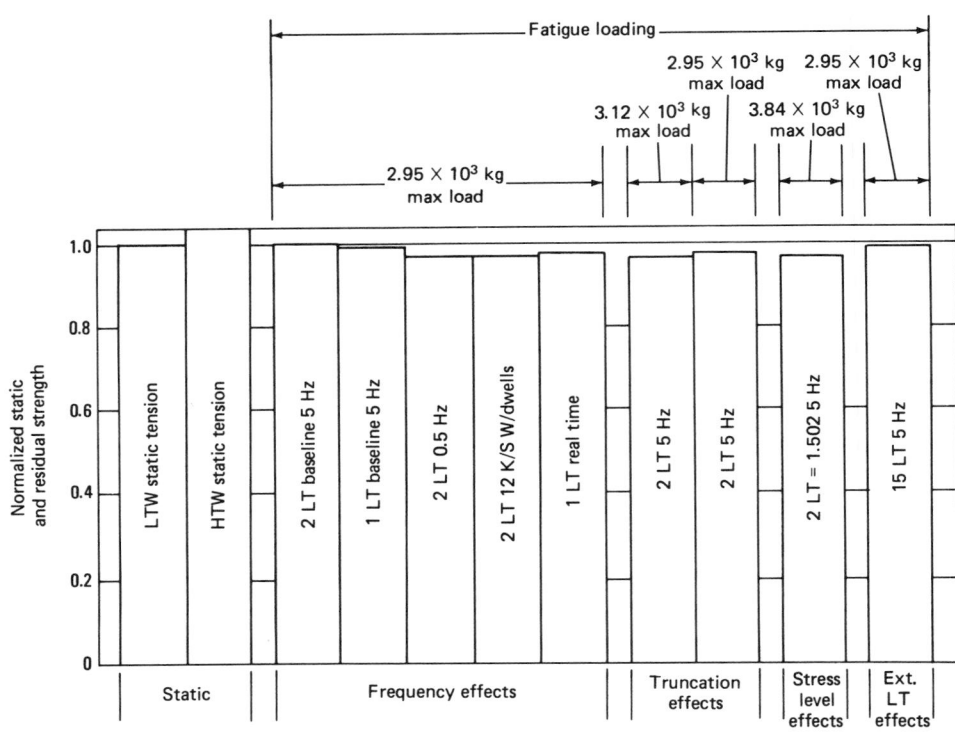

Fig. 54 Summary of static and residual-strength parameter effects on standard bolted-joint strength, MPTW. LT, lifetime; EXT LT, extended lifetime.

from the attachment. Where joints are structurally critical, subcomponents with joints are tested.

A typical in-plane tension-loaded or compression-loaded blade-stiffened composite subcomponent structure is shown in Fig. 58. This subcomponent and test set-up may be used for both static and fatigue testing, although the load introduction equipment and the instrumen-

tation may differ for the two testing modes. Servohydraulic load introduction equipment with load, strain, or deflection feedback and control may be used for either type of testing, but the power requirements and the test equipment cost may be relatively high. Electromechanical test equipment for deflection or strain-controlled fatigue testing requires less power and a lower financial investment. For either

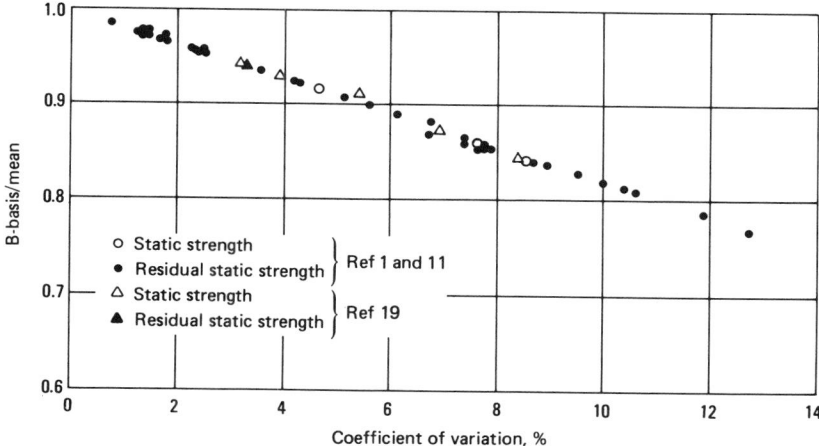

Fig. 55 Variability of standard bolted-joint strength, all environments

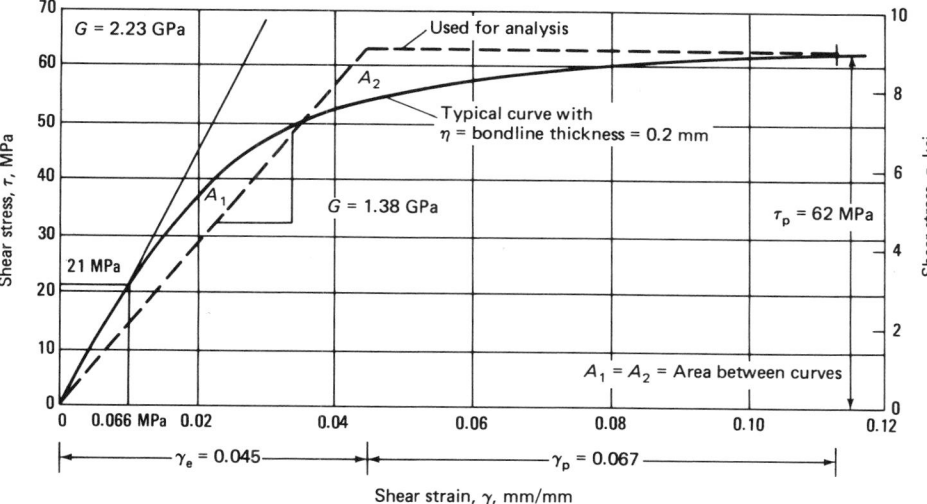

Fig. 56 Shear stress-strain curve for FM-400 adhesive film, RTD (Ref 21)

desirable for static residual-strength testing so that an accurate comparison can be made with the original static strength. Special instrumentation may be required for subcomponents that exhibit elastic buckling before reaching design ultimate load. Back-to-back biaxial gages or three-gage rosettes are preferred for nonbuckling compression testing, but are a necessity for compression buckling testing. A divergence of the strain readings of the back-to-back compression gages of more than ±10% from the mean strain usually indicates buckling.

Tension- and Compression-Loaded Subcomponents. Load introduction attachments to subcomponent ends and side supports are designed to simulate those used in the full-size structure, which may take tension, compression, or both types of loading. The blade-stiffened subcomponent shown in Fig. 58 is typical, but the structural configuration might also be a plain, solid laminate; a sandwich panel with composite face sheets; a double-skin corrugation; or other configuration. Environmental (temperature and humidity) conditioning may be done before installation in the test fixture or in the environmental chamber used in the testing. Such a chamber requires additional expense and design effort.

For a monotonically increasing load, strain, or deflection-controlled static test, the important properties to measure at critical points are load-strain and load-deflection behavior to failure. Also, the failure mode, location, and type should be recorded photographically.

In some cases, the design philosophy may require that one or more holes, impact damage points, sawcut edges, or blades, be put into the subcomponent before testing. Such simulation of "rogue flaw" damage may be used to demonstrate the ability of the structure to perform to design ultimate or limit load, or to sustain a certain number of lifetimes or cycles of durability (fatigue) testing without failure.

Edgewise Shear Loaded Subcomponents. Subcomponents loaded in edgewise shear can be tubular or box configurations loaded in torsion; picture frame flat configurations (unstiffened, stiffened, or sandwich); or fixed-end, cantilever double-shear configurations. The objective is to simulate a portion of a full-sized component such as an airplane wing structural box (a wing skin or spar web), a fuselage section segment, or a control or power transmission torque tube.

A typical torque tube subcomponent configuration is one that might be used for control surface actuation from a remote spot in the wing or fuselage. Optimum design of such a torque tube would seem to indicate a $[\pm\theta]_{ns}$ filament or tape machine winding or broad goods lay-up (hand or machine). However, such a lay-up gives Poisson's ratio values that are too high (0.6 to 0.8) and that create excessively high stress concentrations and interlaminar stresses at the load introduction locations. A tube lay-up orientation of (5/90/5) to (15/70/15) lowers the Poisson's ratio to ≤0.4

static or fatigue testing, computerized electronic monitoring/controlling is always the best method.

For coupon testing and some element testing, deflection or strain control is generally considered to be superior to load (stress) control because any nonlinear or viscoelastic behavior can be more easily and accurately monitored. This provides more information for design and analysis. However, for some element testing and most subcomponent testing, load (stress) control is preferred for static testing because it represents actual in-service loading conditions for aircraft structures. Certain deflection-limited structures may be the exception. When the design philosophy assumes that design limit and/or maximum fatigue loading (spectrum or constant cycle) will not cause the composite material in any area of the structure to exceed its proportional limit or yield point, it makes little difference in fatigue loading whether load, deflection, or strain control is used because

fatigue failure will be rare and little or no wear-out will occur. However, if the design philosophy allows design limit or maximum spectrum fatigue loading to exceed such proportional limit or yield point values, deflection or strain-controlled fatigue testing is preferred because the nonlinear or viscoelastic range of the structure must be more accurately charted and studied to ensure that the design fulfills its goals, including safety and design lifetime. In the latter case, there will probably be some fatigue failures and some wear-out after 1 lifetime of fatigue testing.

Instrumentation is usually greater for static than for fatigue testing, and to some extent, the static-test results will influence the location and type of instrumentation to be used in the fatigue testing. Certainly the highest stress (strain) and deflection areas observed in the static test need to be monitored at least periodically, if not continuously, in fatigue testing. For surviving subcomponents, added instrumentation may be

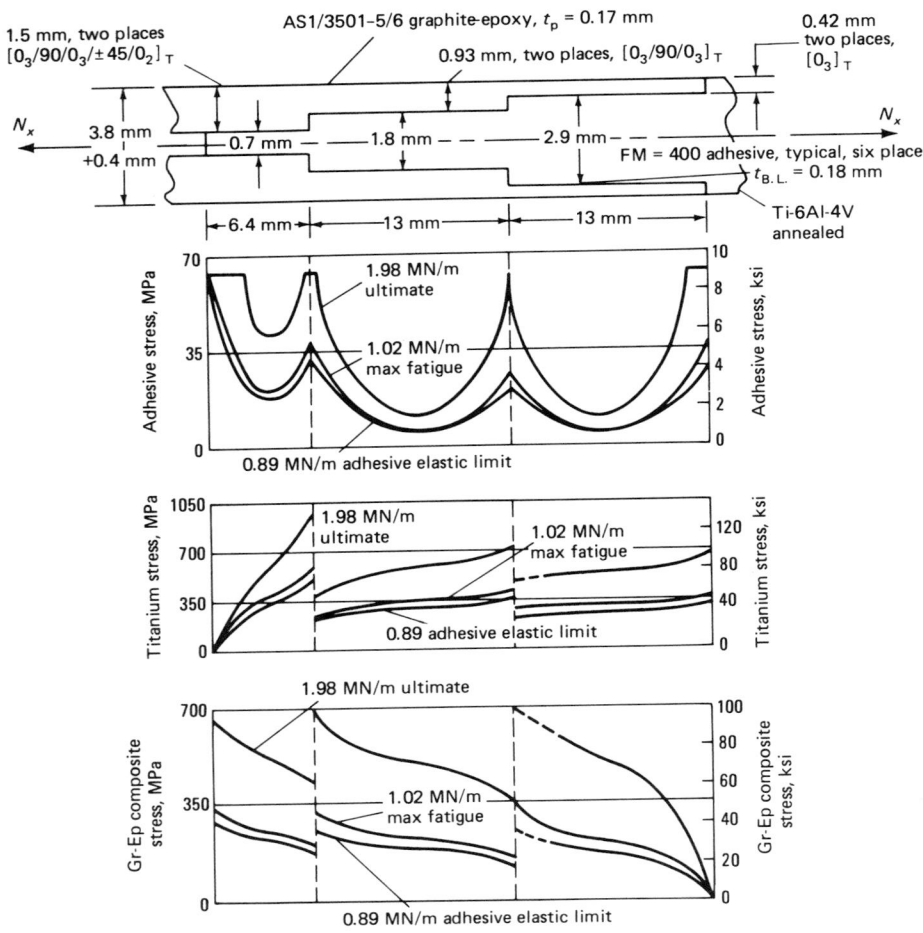

Fig. 57 Summary of standard step-lap bonded-joint analysis, RTD, elastic-plastic analysis program, A4EG (Ref 20)

calculations and measuring lateral deflection that is no less than $\frac{1}{10}$ (nor greater than 1) of laminate thickness, the critical buckling load can be expressed as:

$$\frac{a_i}{(P_{cr/P}) - 1} \qquad \text{(Eq 7)}$$

where δ is deflection, a_i is constant, P is applied load, and P_{cr} is critical buckling load, which may be rewritten as:

$$\left(\frac{\delta}{P}\right) P_{cr} - \delta = a_i \qquad \text{(Eq 8)}$$

A plot of δ/P versus δ will give a straight line near the critical buckling load, and the slope of the line is equal to $1/P_{cr}$. A Southwell plot along with a load-deflection curve for panel 7 (Fig. 61) shows good agreement between the analytically predicted buckling load and the Southwell plot experimentally determined buckling load. The equation for the critical buckling load is:

$$N_{xy_{cr}} = K \frac{D_{11}}{b^2} \qquad \text{(Eq 9)}$$

where D_{11} is the calculated flexural rigidity of the composite in the x-direction and K is the analytically predicted buckling constant. Table 7 shows the analytically predicted and theoretical buckling factors (K) for all the panels studied; a correlation plot of this data (Fig. 62) shows the excellent analytical/experimental agreement. Thus, the Southwell plot is observed to be a useful tool for determining the onset of buckling in picture frame shear tests.

A typical fixed-end cantilever double-shear subcomponent loaded so as to simulate a buckling or nonbuckling spar web is shown in Fig. 63. Much of what was said regarding picture frame shear subcomponents applies to cantilever beam subcomponents, especially if they are buckling subcomponents.

In a nonbuckling case (Fig. 63), the 0° or x-axis will be horizontal, that is, spanwise. Laminates with a general [0°/90°]$_{ns}$ orientation are not recommended for such structures because the shear load-strain results will be highly nonlinear for most of the loading region and a high Poisson's ratio (0.6 to 0.8) in the diagonal direction will create severe stress concentrations at the corners. In addition, such an orientation is not efficient in carrying shear loads. The most efficient orientation is the (0/100/0), that is, [±45°]$_{ns}$ plies. Even with the recommended orientations, some stress concentrations will occur at the corners as well as at the perimeter load introduction caps and stiffeners. Therefore, some reinforcement build-up as shown in Fig. 63 is recommended.

When obtaining stress and deflection equations from a beam handbook, one must get those for the shear stresses and the bending stresses and both the shear deflections and the

and likewise reduces the severity of the related stress concentration and interlaminar effects. The numbers in parentheses mean (%0°/% ± θ°/%90°) ply orientations relative to the longitudinal axis of this tube.

Use of local build-ups of the same material used in the tube permits successful end bolted joint load introduction; that is, flat-laminate bolted-joint technology applied around the circumference of the tube in this example shows a possible bonded-joint load introduction to the test tube that may be efficient, though somewhat more difficult to build (Fig. 59). Lay-up would be on a washout cast-material male tool with metal end plugs and a center ring that are co-cure bonded with adhesive onto the lay-up. At the ends and at the intermediate external fiberglass build-up locations, a woven material is used that can be co-cured with adhesive or laid up and cured as a secondary operation with adhesive. These build-up pads allow split-metal clamp bonding of the actuator to the tube for load introduction. It is assumed that the actuators also function as bearing-support points. A loading diagram of the actuator tube test is shown in Fig. 60; strain gage instrumentation and deflection gages will be needed to evaluate

the mechanical behavior under either static or fatigue loading. The spring stiffnesses for static loading must simulate what will be needed to model the actual control surface/wing structural component stiffnesses. If accelerated fatigue loads are run, some damping at the spring points may be necessary.

For testing edgewise shear in a square or rectangular picture frame test set-up, subcomponent test specimens may be plain, stiffened, or sandwich laminates, buckling or nonbuckling. A good example of a plain laminate subjected to buckling is presented by J.E. Ashton and T.S. Love in Ref. 23; such a configuration might be used to represent a spar web or a wing skin.

The 16-ply 5505 boron-epoxy laminates used by J.E. Ashton and T.S. Love (Table 6) were midplane symmetric and anisotropic in nature. The analytical methods used (Ref 23) were obtained by means of the principle of stationary potential energy, using the Ritz method. The experimentally determined critical buckling load was established by using five linear differential transformers located along the diagonal line between the loading corners (Fig. 61). Using the Southwell method (Ref 24) for these

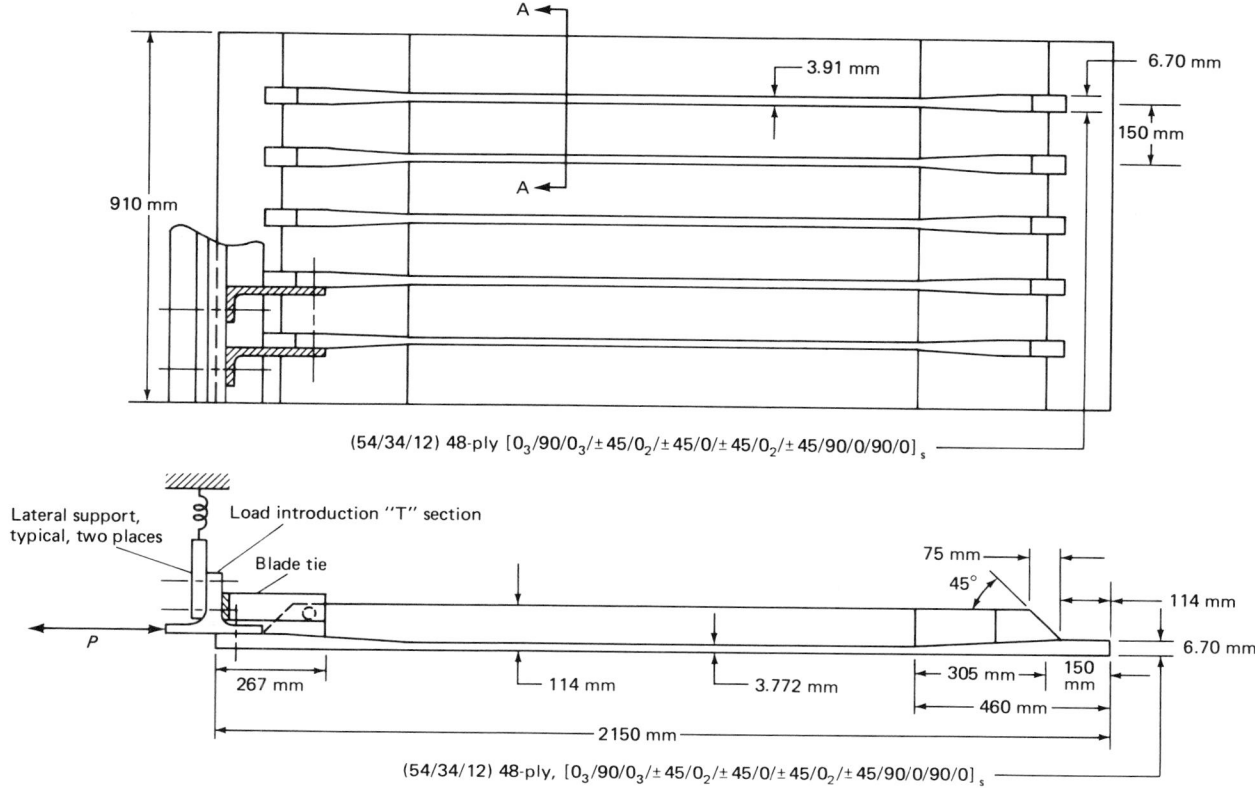

(54/34/12) 48-ply [0₃/90/0₃/±45/0₂/±45/0/±45/0₂/±45/90/0/90/0]ₛ

(54/34/12) 48-ply, [0₃/90/0₃/±45/0₂/±45/0/±45/0₂/±45/90/0/90/0]ₛ

Fig. 58 Blade-stiffened subcomponent test conceptual design, tension or compression, (AS1/3501-5/6 graphite-epoxy tape)

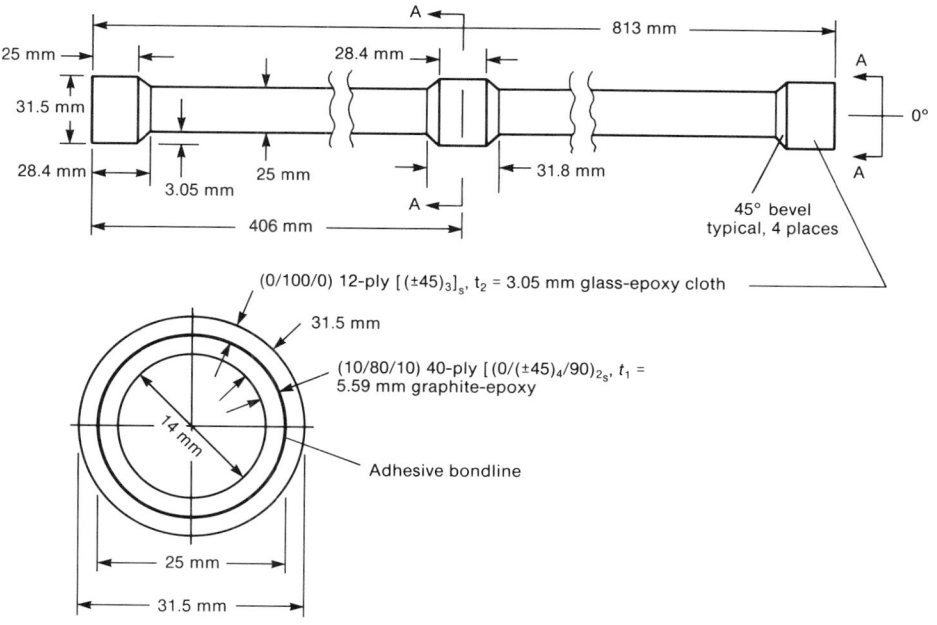

Fig. 59 Control surface actuator tube configuration subcomponent

bending deflections. The design of this test should simulate the actual spar web, caps, and stiffeners. Such designs are usually critical in one of these areas, preferably the cap. In some cases, it may be desirable to strengthen the cap so that the shear web fails first.

The instrumentation needed will be strain rosettes at the corners and middle of the shear webs, with a deflection gage at the end of the cantilever. In some cases, axial gages (spanwise) along the caps will be needed, and some axial gages running vertically along the stiffeners may be necessary.

Normal loaded subcomponents of aircraft structures that must be considered include integrally stiffened or sandwich wing skin structures that must take fuel tank pressure in addition to in-plane bending and torsion and vertical shear loadings; the latter are taken by the spar and rib webs or the full-depth honeycomb core that is sometimes used instead of discrete spar/rib shear webs. Concentrated direct normal tension loads on composite laminates utilize them in the worst possible way because their flatwise tension and peel strengths are low compared with in-plane laminate strengths. Thus, testing of wing subcomponents to determine integral blade-stiffener pull-off strength is used to evaluate the structural behavior of such fuel tank structures.

Figure 64 shows such a test in the form of a simple beam. The pull-off results can be translated into a running-load strength that can then be compared with the design running-load requirements. Design ultimate running-load requirements should never exceed 60% of their mean test values; or, given sufficient test data, they should not exceed 75% of the 85% cut-off B-basis allowable in the 95% confidence limit, 90% probability statistical value (normal,

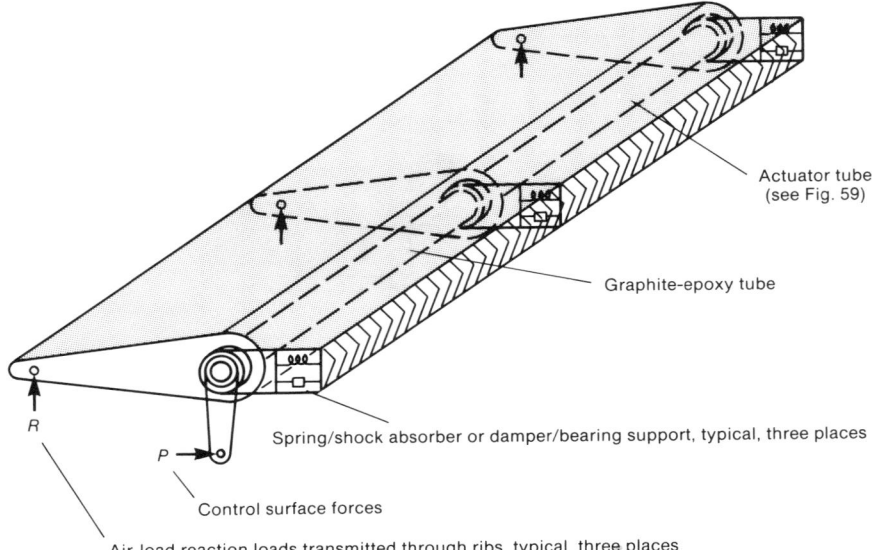

Fig. 60 Loading of control surface actuator tube subcomponent

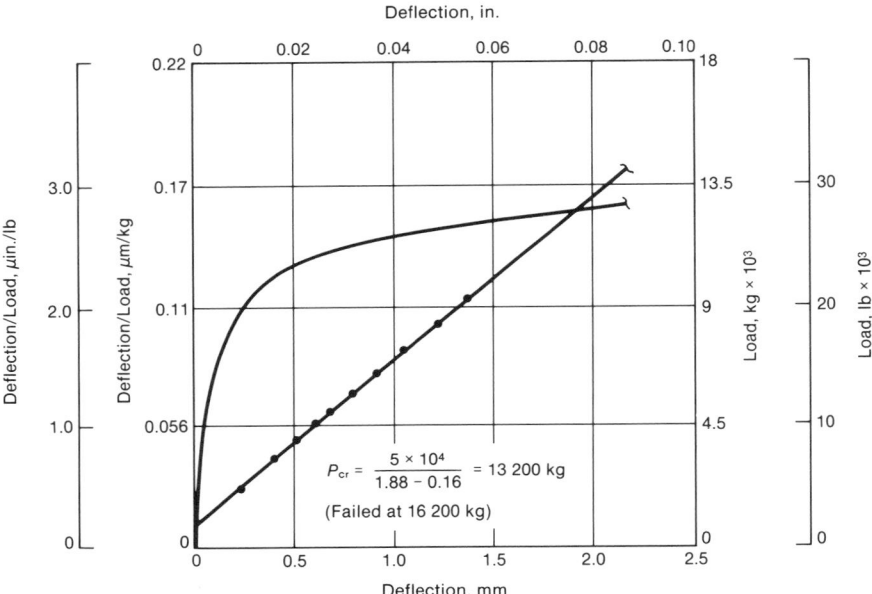

$$P_{cr} = \frac{5 \times 10^4}{1.88 - 0.16} = 13\ 200\ kg$$

(Failed at 16 200 kg)

Fig. 61 Southwell plot and load-deflection curve for boron-epoxy panel 7 (Ref 23)

long-normal, or Weibull statistics), or 85% of the mean value of the test data, whichever is higher (depending on design). Because most of these tests fail in the interlaminar tension and peel mode, the main data of interest are those for ultimate strength; however, strain and deflection data under loading may also be of interest for determining such things as the onset and progression of delamination.

However, if the blade-to-skin transition area must carry high running vertical shear loads as well as high (normal-to-the-skin) running loads due to fuel pressure, the test shown in Fig. 64 should have these shear loads applied while this

normal loading is applied. The matrix in the high-stressed area will be an isotropic material under combined tension and shear; thus, a Mohr's circle representation (Fig. 65) illustrates that the combination of the two stresses results in maximum principal and shear stresses that are higher than the individually applied tension and shear stresses. It is these higher maximum principal and shear stresses that should be compared to the allowable interlaminar tension and shear stresses, and such analysis should be compared to the analyzed subcomponent test results based on testing of the type shown in Fig. 64.

Bending Loaded Subcomponents. Many aircraft wing and empennage trailing-edge and control surface structures are made with bonded full-depth honeycomb sandwich using composite face sheets, spars, ribs, and close-out parts. The honeycomb core may be made of either corrosion-resistant aluminum, Nomex, glass-phenolic, or graphite-phenolic materials. For mild environments, lightly loaded parts that are not transverse-shear-stiffness critical may be made of Nomex honeycomb. Epoxy film adhesives requiring 170 to 690 kPa (25 to 100 psi) pressure and 120 to 175 °C (250 to 350 °F) curing temperatures are the best bonding materials. For the skins, spars, ribs, and close-outs, laminates precured at 690 kPa (100 psi) are the best, although co-curing at 69 to 280 kPa (10 to 40 psi) has been used successfully, especially for composite skins over honeycomb. Some reduction in the allowable results from this fabrication procedure due to the high porosity (4 to 15 vol%) and intercellular sagging that occurs. Although such co-cured skins may have allowables that are 60 to 90% of those of the high pressure precured skins, the manufacturing and tooling costs are reduced significantly.

The typical sandwich wedge cantilever beam test shown in Fig. 66 might be used to simulate a trailing-edge fixed or movable control surface structure. Load-strain and load-deflection behaviors in such structures are usually recorded and translated to the real structure behavior for comparison. In many cases, durability (fatigue) and damage tolerance testing must be performed on these subcomponents; thus, static behavior through design limit/maximum spectrum fatigue load must be known.

Such structures are usually cost-effective if a large number (>50) are to be built and if they have life-cycle use costs that are lower than those of built-up stiffened structures. In addition, they are usually lighter. Their disadvantages include water entry, weight pick-up, and possible aluminum core corrosion, if not properly designed and built. Also, small production quantities (<25) may increase part expense because of tooling, material costs, and learning-curve and development costs.

Combined In-Plane and Normal Loaded Subcomponents. Such structures can take many forms. For example, shell structures can simulate aircraft fuselages that must take in-plane bending and torsion loads and normal direction internal-pressure loads as well as out-of-plane point loads. Deep-submergence vehicle shells would present a similar loading picture, except that the normal pressure loading would be external instead of internal. Pressure bulkheads in aircraft and naval structures are another example.

A prime example of such a structure is a wing box that has a flexible-bladder fuel cell or is made to be an integral fuel tank. The combined bending and torsion-induced in-plane loading is also combined with internal box fuel pressure as well as point loads induced by

Table 6   Picture frame shear buckling study of 5505 boron-epoxy laminates

Material	Orientation	Thickness mm	in.	(%0°/%±45°/%90°) plies	Number of plies
Aluminum 1	...	3.07	0.121	...	...
Aluminum 2	...	2.24	0.088	...	...
5505 boron-epoxy composite		2.24	0.088	(50/0/50)	16
1	$[(0/90)_4]_s$	2.24	0.088	(50/0/50)	16
2	$[0]_{16T}$	2.21	0.087	(100/0/0)	16
3	$[0]_{16T}$	2.24	0.088	(100/0/0)	16
4	$[(0/90)_4]_s$	2.16	0.085	(50/0/50)	16
5	$[(90/\mp45/0)_2]_s$	2.13	0.084	(25/50/25)	16
6	$[(90/\mp45/0)_2]_s$	2.13	0.084	(25/50/25)	16
7	$[(\mp45)_4]_s$	2.18	0.086	(0/100/0)	16
8	$[(\pm45)_4]_s$	2.24	0.088	(0/100/0)	16
9	$[(0/\pm45/90)_2]_s$	2.29	0.090	(25/50/25)	16
10	$[(0/\pm45/90)_2]_s$	2.26	0.089	(25/50/25)	16
11	$[(\mp45)_4]_s$	2.24	0.088	(0/100/0)	16
12	$[(\mp45)_4]_s$	2.18	0.086	(0/100/0)	16
13	$[-45]_{16T}$	2.29	0.090	(0/100/0)	16
14	$[-45]_{16T}$	2.16	0.085	(0/100/0)	16

Note: Material properties for aluminum:
$E = 72$ GPa (10.5×10^6 psi), $\nu = 0.33$

For 5505 boron-epoxy:
$E_1 = 210$ GPa (31×10^6 psi) $E_2 = 19$ GPa (2.7×10^6 psi)
$\nu_{12} = 0.28$ $G_{12} = 5$ GPa (0.75×10^6 psi)

Source: Ref 23

Table 7   Analytical and experimental values for shear buckling factor, K

Panel	K (experimental)	K (theory)
Aluminum 1	91.1	95.2
Aluminum 2	88.5	95.2
1	53.2	54.4
2	13.5	13.4
3	13.8	13.4
4	49.4	54.4
5	137.8	137.8
6	143.0	137.8
7	109.5	118.5
8	131.9	149.3
9	63.0	65.3
10	70.0	65.3
11	111.2	118.5
12	125.5	118.5
13	48.8	37(a)
14	50.6	37(a)

(a) 44.4 with $E_2 = 23.5$ GPa (3.4×10^6 psi). Source: Ref 19

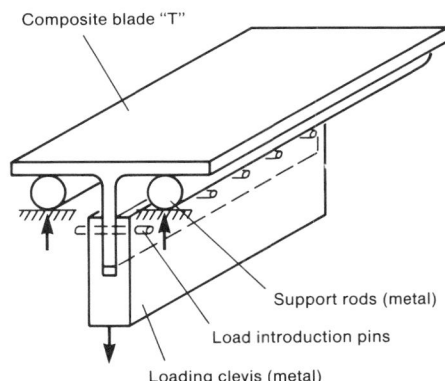

Fig. 62 Analytical/experimental correlation, shear buckling factor, K (Ref 23)

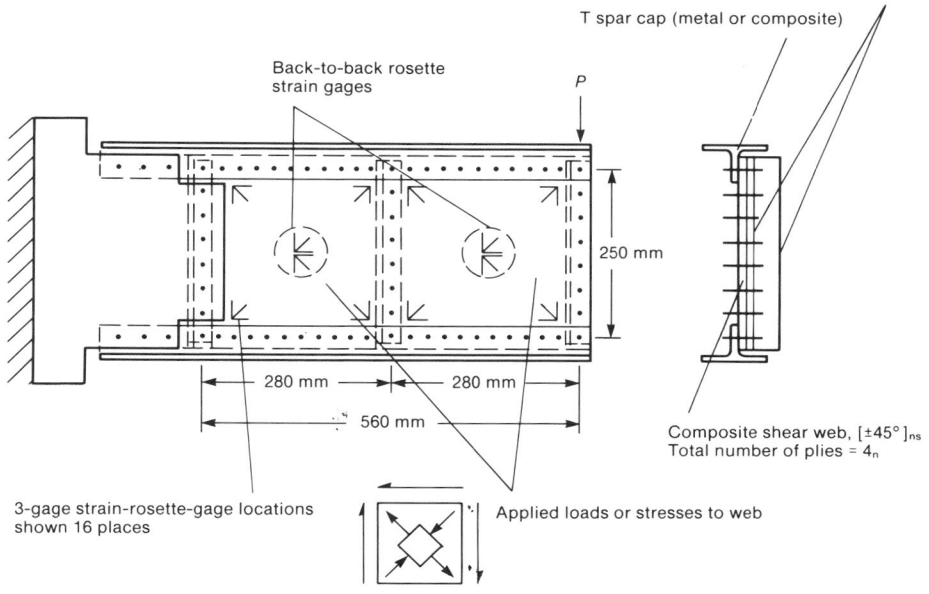

Fig. 63 Cantilever beam double-shear subcomponent

Fig. 64 Test configuration for normal loaded blade-stiffened "T" subcomponent pull-off

internal or external equipment or stores. Carrying fuel in aircraft wings is structurally more efficient than carrying it in the fuselage because it reduces the high gross-weight wing bending moments when flying, that is, it reduces the structural weight.

Design and testing of composite material, box beam subcomponents is detailed and complex. However, the results can provide confidence that a wing structure is going to work (or realization that it will not work), which is important in the design and development process.

Composite box beam design with balanced and symmetric quasi-isotropic skins, spar webs, and spar caps, which is covered in detail in Ref 25, might be done in a preliminary structure design (Fig. 67). The test box is usually a model of the aircraft wing structural box. Additional box design work is then required to cover the load introduction and reaction points as well as the box loading and support structure. Finite-element modeling and analysis of these test box structures as well as the loading and support structures is desirable because of their complexity and the need to accurately model the actual wing structural box. The support structure must be designed not to fail and to simulate the stiffnesses of the surrounding wing structure.

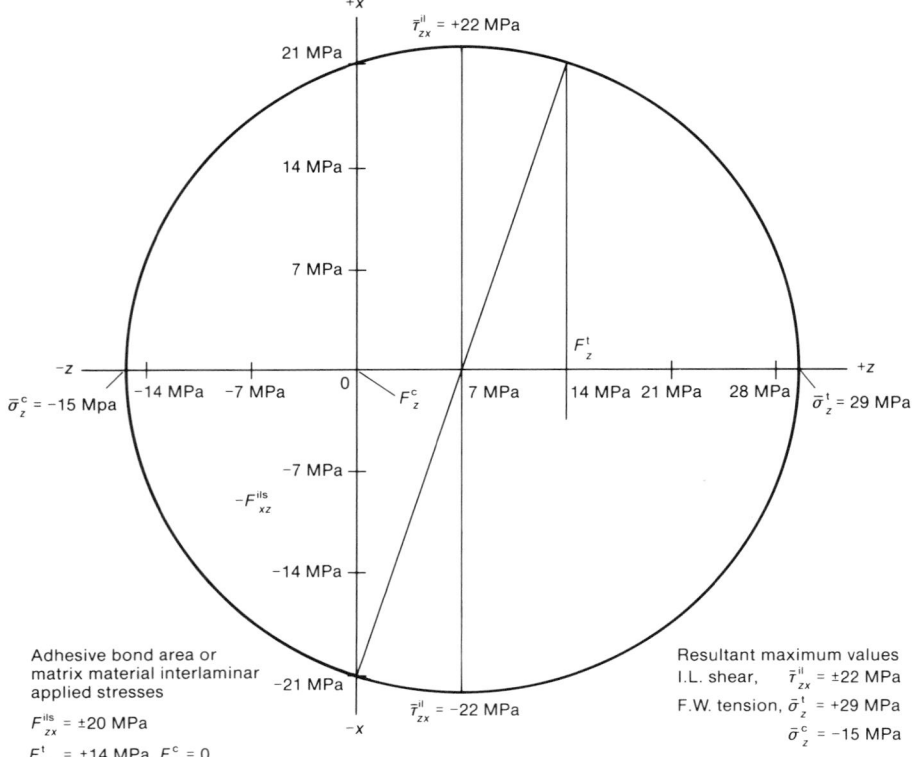

Fig. 65 Mohr's circle for interlaminar shear and tension stresses on matrix material or adhesive

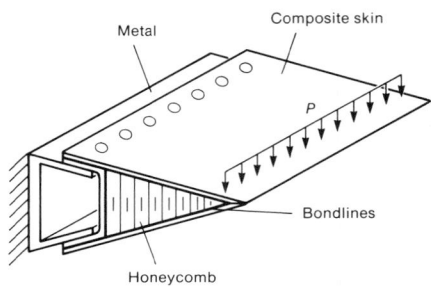

Fig. 66 Cantilever beam sandwich wedge subcomponent test

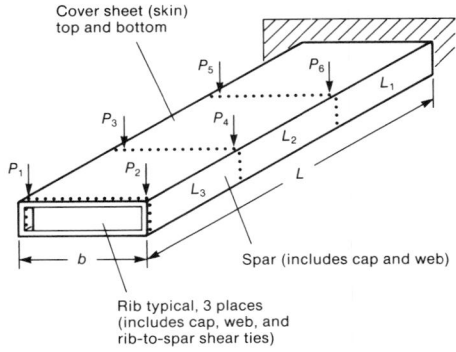

Fig. 67 Wing structure box beam model

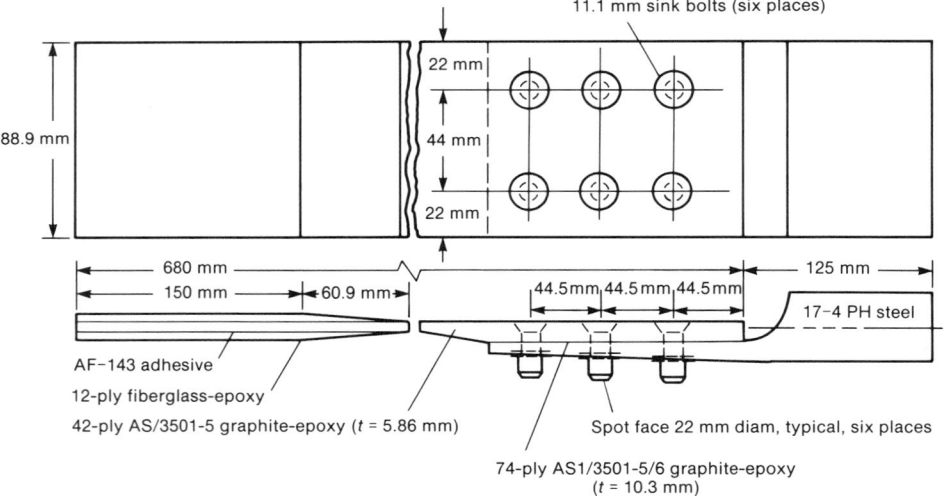

Fig. 68 Large-scale bolted-joint specimen

Box beam subcomponent tests combine elements and smaller subcomponents. Box beams are used to simulate wing structures closely, and the test results should relate directly to full-scale component and vehicle static, durability, or damage tolerance testing. These boxes may be without joints, or they may include them because they are closely simulating a real structure, especially if the joints are designed to fail first.

The box design in Ref 21 is set up for optimization using quasi-isotropic lay-ups, but the procedure could be modified and used for other lay-up orientations. The quasi-isotropic design lay-up with thin skins and heavy spar caps in Ref 21 is good for multispar wing structures with large multidirectional loadings (bending, torsion, chordwise, and internal pressure). Many wings are lightly loaded in torsion and chordwise bending and will probably require different orientations, as will a multirib wing structure configuration. Damage-tolerant designs may require heavy spar caps with a high percentage ($>50\%$) of 0° fibers with skins using a high percentage (70 to 100%) of $\pm 45°$ plies. Testing such boxes requires the use of reversible hydraulic cylinders and servocontrollers at the load introduction points, deflection gages at critical points, and strain gages at selected locations on the test box structure.

The box beam represents the final stage of the development cycle before full-scale component or vehicle testing. A successful box beam test provides considerable asssurance that structural-integrity goals will be met on the full-scale structure; or a failure will pinpoint the weakness, and the necessary redesign can be accomplished with confidence.

Bolted-Joint Subcomponents. The aluminum wing structure (Fig. 1) is used as a model for a direct-substitution composite skin and joint design, with titanium used as splice plates. Two design ultimate running-load intensities are taken from this wing design, one at 1.6 MN/m (9000 lbf/in.) (small-scale two-bolt joints) and one at 3.2 MN/m (18 000 lbf/in.) (large-scale six-bolt joints). The small-scale design and testing was covered previously in

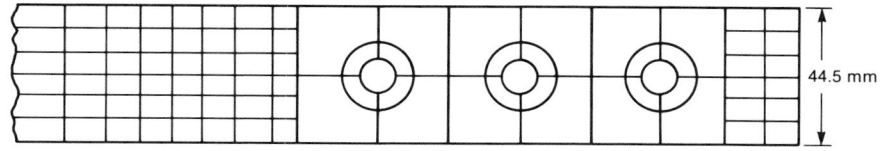

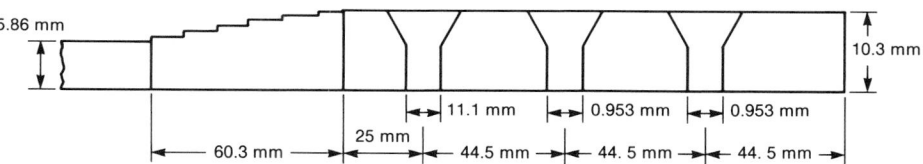

Fig. 69 Finite-element model for large-scale six-bolt specimen

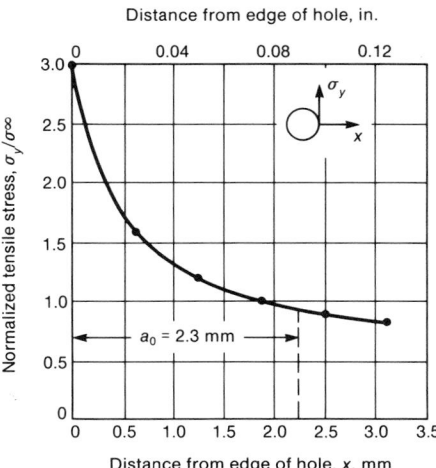

Fig. 70 Tensile-stress distribution at first hole section (next to taper in thickness) large-scale six-bolt joint, RTD

the sections "Composite Materials and Simple Joint Data" and "Structural Element Testing" of this article. The large-scale bolted-joint configuration is shown in Fig. 68, with the finite-element model shown in Fig. 69 and the resultant stress distribution around one of the holes depicted in Fig. 70. A stress concentration of 3.0 is shown at the hole edge, perpendicular to the loading, dropping to 1.0 at a characteristic distance of $a_0 = 2.3$ mm (0.09 in.) into the edge of the laminate hole. The tensile-stress distribution shown in Fig. 70 is across the two holes next to the tapered laminate area of the joint. With this stress distribution, the average stress criterion predicts an RTD joint strength of 3.2 MN/m (18 100 lbf/in.) of running load with a net tension mode of failure, which is about twice the standard-sized joint capacity, as desired. The RTD static test data on the large-scale joints show an average strength of 3.6 MN/m (20 453 lbf/in.), 13% above the predicted value. Using the properties previously derived, lamination theory, and maximum strain criterion, the predicted unnotched tension

and compression laminate strengths of the 74-ply laminate are 803 MPa (116.4 ksi) and −600 MPa (−87 ksi), respectively. Predicted failure mode is net tension through the holes next to the taper, with the effect of variations in load distribution among the bolts being insignificant.

Table 8 shows that the standard-sized two-bolt joint and the large-scale six-bolt joint compare favorably in experimental behavior at RTD conditions. The large-scale joint has 87% of the bearing strength and 91% of the net tension strength observed in the thick areas of the standard-sized joint. In the thin areas, the large-scale joint at failure has a strain value of 108% of that of the standard-sized joint. Joint stiffnesses in both the thin and the thick laminates of the two sizes of joints are similar. Also, bending in the thick bolt area is similar for the two joint sizes. Thus, no reduction in strength due to scale-up is apparent. However, on the basis of static unit load-carrying weight efficiency, the large-scale joint is 80% heavier than the standard joint.

Detailed design study with testing support may be able to reduce this.

Figure 71 compares the static and fatigue strength of the large-scale and standard-sized bolted joints, all data being normalized to the baseline (frequency effects) fatigue residual strength. All fatigue specimens survived and exhibited residual-strength values about the same as their original static strength values. All tests fell within a ±4% band of the baseline fatigue residual-strength values. Thus, the 70% of static strength established previously on the standard-sized joints was maintained for the large-scale joints; this fiber-dominated tensile property proved to be a good basis for static ultimate design and fatigue strength.

Bonded-Joint Subcomponents. A large-scale bonded joint designed to carry twice the running load (3.2 MN/m, or 18 000 lbf/in.) of the standard-sized joint is shown in Fig. 72. This corresponds to the higher-loaded second

Table 8 Results of comparative strain-gage study, large-scale versus standard bolted-joint tension tests, RTD

Laminate description	Laminate thickness(a) mm	in.	Bearing stress MPa	ksi	Running load at failure MN/m	lbf/in.	Net tension stress at failure MPa	ksi	Strain failure(b) 10^3 μm/m	E_1(c) GPa	10^6 psi	E_2(c) GPa	10^6 psi	$E_{average}$(c) GPa	10^6 psi	Average bending based on modulus of elasticity, %
Six-bolt large-scale joints [W = 88.9 mm (3.4961 in.)]																
3-specimen average for 42-ply solid laminate 6	0.2412		. . .	. . .	3.58	20 453	586	84.93	7.68	75	10.87	77	11.22	76	11.05	0.7
3-specimen average for 74-ply 1st-line 2-C's holes 11	0.4445		342	49.64	3.58	20 453	458	66.37	5.39	104	15.07	105	15.26	105	15.17	18.5
Two-bolt standard-sized joints [W = 25.4 mm (0.9986 in.)]																
3-specimen average for 27-ply solid laminate 4	0.1529		. . .	. . .	2.15	12 268	553	80.22	7.09	78	11.37	79	11.46	79	11.42	9.7
3-specimen average for 48-ply 1st C's hole 7	0.2756		392	56.90	2.15	12 268	504	73.13	6.68	107	15.55	103	14.92	105	15.24	15.0

(a) Basic laminate average test strength is 1147 MPa (166.37 ksi) in tension. (b) Based on differences between top and bottom strain readings. (c) E_1 is the top and bottom average for the primary modulus; E_2 is the top and bottom average for the secondary modulus; $E_{average}$ is the average of the primary (E_1) and secondary (E_2) modulus.

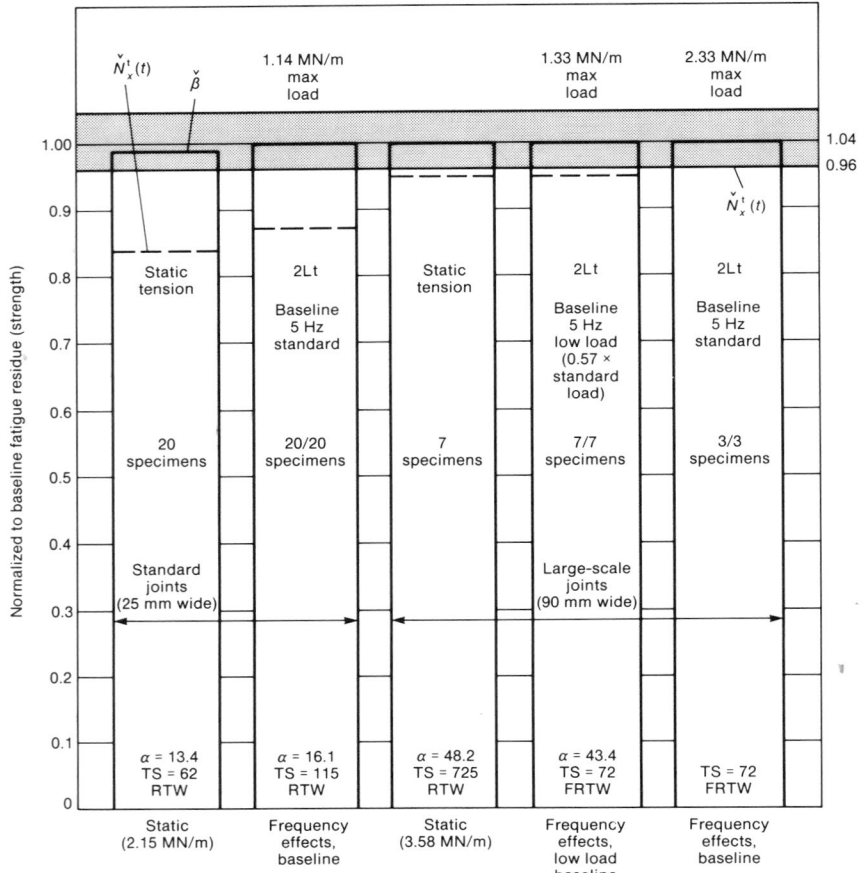

Fig. 71 Large-scale and standard joints, normalized static and fatigue residual strength, RTW

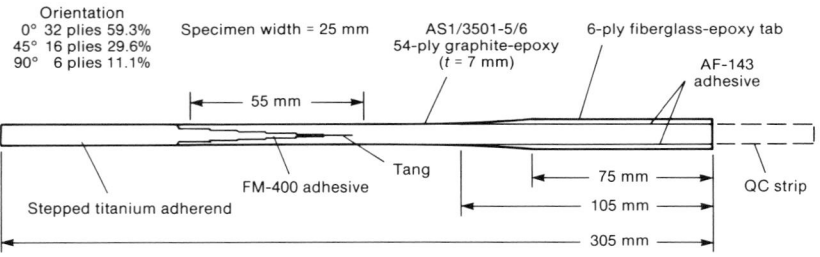

Fig. 72 Large-scale bonded joint

area of the wing structure chordwise joint presented in the section "Testing Philosophy and Criteria" in this article.

Eight cases of tension-loaded joints (Fig. 73) were studied using a trial-and-error Hart-Smith approach (Ref 17). Cases 2, 4, and 7 were chosen for further study and detailed design because they represent the most practical and efficient joints of those under study. Case 2 at 3.8 MN/m (21 960 lbf/in.) predicted failure load was the second highest, but had a much lighter and shorter titanium tab than did case 4, at 3.9 MN/m (22 229 lbf/in.). Case 7 at 3.5 MN/m (19 921 lbf/in.) failure load provided the highest running load at the adhesive shear-

stress elastic limit of 1.9 MN/m (10 615 lbf/in.), which was not much lower than it was for case 2. Cases 2 and 7 gave the same failure mode prediction as the standard-sized three-step lap joint analysis. Changing the titanium-tang (first-step) thickness and length on the large-scale joint design changed the failure mode but reduced the magnitude of the predicted failure load. Case 7 had the same configuration as case 2 except that the tang (first-step) thickness was reduced from 0.84 to 0.70 mm (0.0330 to 0.0275 in.), resulting in a 12.6% increase in the predicted adhesive elastic-limit load but a 9.3% reduction in the ultimate joint failure load. Both cases had the

same predicted failure mode. The case 7 joint may have had a higher fatigue strength than the case 2 joint if the maximum tension-dominated spectrum fatigue loadings had been at the same percentage of ultimate load and the same percentage above the elastic-limit load as were the standard joints during this testing.

The case 2 joint configuration was selected for final design, fabrication, and testing because it gave the highest ultimate load and the desired failure modes. The RTD mean experimental static strength of this joint is 4.08 MN/m (23 314 lbf/in.) running load, or about 6% above the analytical value, with primary failure occurring in the adhesive rather than in the step 1 titanium tang as predicted. This difference in failure mode is partially caused by the bondline normal stresses being added to the bondline shear stresses as a vector sum (Mohr's Circle), and partly due to the radius on the titanium at the step 1 (tang) intersection with step 2; the predictive technique used considers only bond-line shear stresses and not radius at the intersection. While these analytical limitations are not critical, they do show the need for experimental verification.

Using the Hart-Smith A4EG/F computerized analytical method (Ref 20), the case 2 large-scale six-step lap joint was analyzed with the input properties shown in Table 9, which were assembled from the properties of these materials given in the section "Composite Materials and Simple Joint Testing" in this article. The analytically predicted stresses at the six steps of the composite outer adherends are shown in Fig. 74, and the stresses in the titanium inner adherends are shown in Fig. 75. The maximum titanium tang stress is approximately 1030 MPa (150 ksi), close to its ultimate strength, while the maximum composite adherend stress of 690 MPa (100 ksi) at the other end of the joint on a (100/0/0) laminate is well below its B-basis allowable strength of 910 MPa (132 ksi). The maximum pure-adhesive shear strength of ±62 MPa (±9 ksi) has been exceeded at the two ends of the joint, at the end of steps 1 and 6 (Fig. 76), at the ultimate predicted running load of 3.85 MN/m (21 960 lbf/in.). Because perfectly elastic/perfectly plastic adhesive-shear stress-strain behavior is assumed in Ref 20, the adhesive also reaches the ±60 MPa (±9 ksi) shear stress at the end of the perfectly elastic strain range, but continues to strain substantially more into the plastic range before ultimate joint failure is reached at a shear ultimate strain value of 0.112 mm/mm. The lower plate in Fig. 74, 75, and 76 shows the stress distribution at the elastic-limit running load of 1.65 MN/m (9430 lbf/in.) as defined by the Hart-Smith perfectly elastic shear strain of 0.045 mm/mm (Table 9). It can be seen from Fig. 76 that the bonded-area elastic limit is reached at only one point, that is, at the end of the titanium tang (step 1).

The large-scale bonded-joint static and fatigue data are summarized in Table 10 and Fig. 77. The large drop (27%) in static strength for

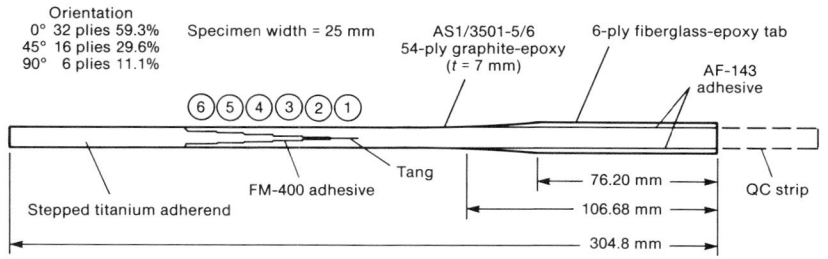

Joint design case	Thickness of 1st step (tang), mm	Step 1, mm	Step 2, mm	Step 3–5, mm	Step 6, mm	Predicted joint strength Elastic, MN/m	Predicted joint strength Ultimate, MN/m	Predicted failure At step:	Predicted failure Type
1	0.838	6.4	9.5	9.5	9.5	1.549	3.694	1	Adhesive
2	0.838	9.5	9.5	9.5	9.5	1.651	3.846	1	Ti
3	0.838	9.5	9.5	13.0	9.5	1.651	3.838	1	Ti
4	0.838	9.5	13.0	13.0	13.0	1.652	3.893	1	Ti
5	0.978	9.5	9.5	9.5	9.5	1.492	3.671	1	Adhesive
6	0.978	13.0	9.5	9.5	9.5	1.520	3.751	1	Adhesive
7	0.699	9.5	9.5	9.5	9.5	1.859	3.489	1	Ti
8	0.699	6.4	9.5	9.5	9.5	1.750	3.658	1	Ti

Fig. 73 Predicted RTD strengths for tension-loaded large-scale step-lap bonded joints

the wet condition compared to the dry condition under static tension loading is apparent, although this drop is not as large as that for the standard-scale joints. From wet static tension loading to baseline spectrum tension dominated fatigue (TDF) maximum spectrum loading at RTW conditions, applied running load drops 28%. All these TDF/RTW specimens failed in fatigue at less than 2 lifetimes, that is, at an average of 1.28 lifetimes.

By contrast, wet compression static strengths were slightly higher than the dry tension values and substantially above the wet tension values. In addition, the wet compression dominated fatigue (CDF) residual-strength values are about the same as the static values. No CDF specimens failed in fatigue after 2 lifetimes of testing. These results further demonstrate that wet conditions are significantly more severe in their effects on tension static and fatigue bonded-joint strength than in their effects on compression-loaded bonded joints. Exceeding the static adhesive-shear proportional limit stress is considerably more severe in TDF loading than it is in CDF loading because of the normal tensile stresses induced in the bondline in tensile-loaded joints, in contrast to the normal compression stresses in compression-loaded joints.

The foregoing data further verify the experimental observations that bonded-joint design limit load shear stresses should not exceed the adhesive bonded area proportional limit shear strength, and that design ultimate load shear strength should not exceed the adhesive bonded area ultimate load shear strength (suitably reduced statistically).

General Experimental Procedures for Subcomponents. Subcomponent tests usually represent a full-sized or scale-model segment of the real structure being evaluated. Load introduction and support structures must be carefully designed to simulate those of the real structure, yet must not fail before the subcomponent does. Loading should represent design static or durability (fatigue) load cases based on the internal loading calculated for the particular structural segment. Data to be recorded should be planned so that instrumentation can be selected and located in a manner that will maximize the useful information to be recorded. Strains and deflections at critical points in the structure should be measured, and the locations should be such that the onset and magnitude of expected or unexpected buckling can be charted. Load points at which noise occurs should be recorded, and failure modes should be observed and recorded.

Loading and recording equipment should be of a type that allows continuous reading such that constant monotonically increasing loading rates, or deflection or strain rates, at critical points can be maintained to failure. The familiar static loading in *n* lb increments, which requires stopping to read instrumentation at each increment, should not be used with composite structures for testing to failure. Preliminary instrumentation static checkout runs should not exceed 20% of predicted design ultimate failure load. For fatigue loading of subcomponents, the temperature increases of the structure at high-strain areas should be monitored; if increases exceed a delta of 3 °C (5 °F), either the temperature or the fatigue loading rate should be reduced.

Data and Test Reports. Load versus strain and load versus deflection data from static tests should be tabulated, and continuous plots of their behavior curves recorded to failure. Failure modes should be analyzed visually, microscopically, and with cross-sectional photomicrographs; selected critical failure surface areas should be studied with SEM photographs to determine the character, cause, and progression of the fracture at the microstructure level that leads to the macrostructural failure causing the subcomponent structural failure. Such study

Table 9   AS1/3501-5/6 properties input for computer program A4EG for analysis of six-step lap joint

Joint cure temperature parameters

$\alpha_{outer\ adherend} = 6 \times 10^{-6}/K$ $\alpha_{inner\ adherend} = 5.3 \times 10^{-6}/K$ $\Delta T = 125$ °C (225 °F)

Bondline (FM-400) RTD parameters

$\tau_{ultimate} = 62$ MPa (9 ksi) $\gamma_{ultimate} = 0.112$ mm/mm $\tau_{BL} = 0.178$ mm (0.007 in.) $\gamma_e = 0.045$ mm/mm

Composite outer adherend

Step 1	$E_{o1} = 100$ GPa	$(14.5 \times 10^6$ psi)	[67%/25%/8%]	$\tau = 6.32$ mm (0.249 in.)	48 plies	$\epsilon_{o1} = 0.011$
Step 2	$E_{o2} = 103$ GPa	$(15.0 \times 10^6$ psi)	[70%/20%/10%]	$\tau = 5.30$ mm (0.209 in.)	40 plies	$\epsilon_{o2} = 0.011$
Step 3	$E_{o3} = 94$ GPa	$(13.7 \times 10^6$ psi)	[62%/25%/13%]	$\tau = 4.22$ mm (0.166 in.)	32 plies	$\epsilon_{o3} = 0.011$
Step 4	$E_{o4} = 110$ GPa	$(15.9 \times 10^6$ psi)	[75%/17%/8%]	$\tau = 3.18$ mm (0.125 in.)	24 plies	$\epsilon_{o4} = 0.011$
Step 5	$E_{o5} = 117$ GPa	$(16.9 \times 10^6$ psi)	[75%/25%/0%]	$\tau = 2.11$ mm (0.083 in.)	16 plies	$\epsilon_{o5} = 0.011$
Step 6	$E_{o6} = 137$ GPa	$(19.97 \times 10^6$ psi)	[100%/0%/0%]	$\tau = 1.06$ mm (0.0416 in.)	8 plies	$\epsilon_{o6} = 0.011$

Titanium inner adherend

Step 1	$E_{i1} = 110$ GPa (16.0 $\times 10^6$ psi)	$\tau = 0.084$ mm (0.033 in.)	$\epsilon_{i1} = 0.010$
Step 2	$E_{i2} = 110$ GPa (16.0 $\times 10^6$ psi)	$\tau = 1.96$ mm (0.077 in.)	$\epsilon_{i2} = 0.010$
Step 3	$E_{i3} = 110$ GPa (16.0 $\times 10^6$ psi)	$\tau = 3.07$ mm (0.121 in.)	$\epsilon_{i3} = 0.010$
Step 4	$E_{i4} = 110$ GPa (16.0 $\times 10^6$ psi)	$\tau = 4.19$ mm (0.165 in.)	$\epsilon_{i4} = 0.010$
Step 5	$E_{i5} = 110$ GPa (16.0 $\times 10^6$ psi)	$\tau = 5.31$ mm (0.209 in.)	$\epsilon_{i5} = 0.010$
Step 6	$E_{i6} = 110$ GPa (16.0 $\times 10^6$ psi)	$\tau = 6.43$ mm (0.253 in.)	$\epsilon_{i6} = 0.010$

Source: Ref 20

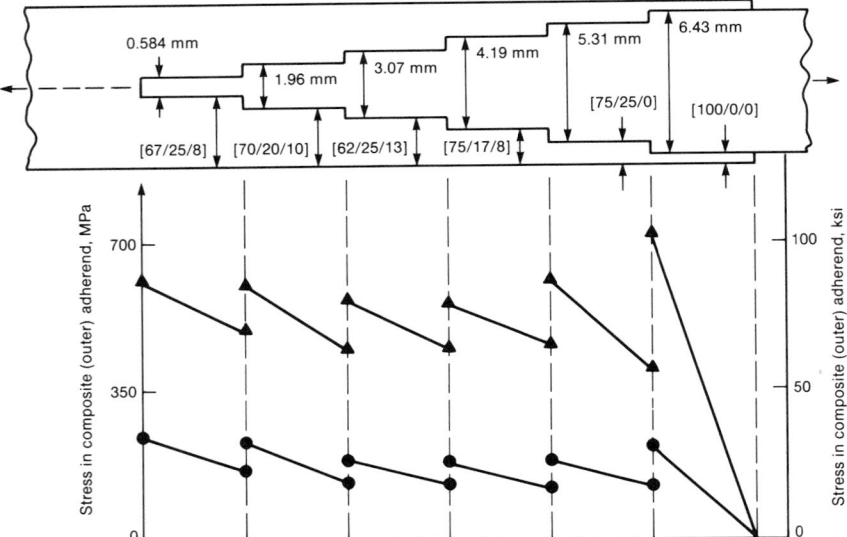

Fig. 74 Predicted stresses in composite (outer) adherend, six-step lap joint

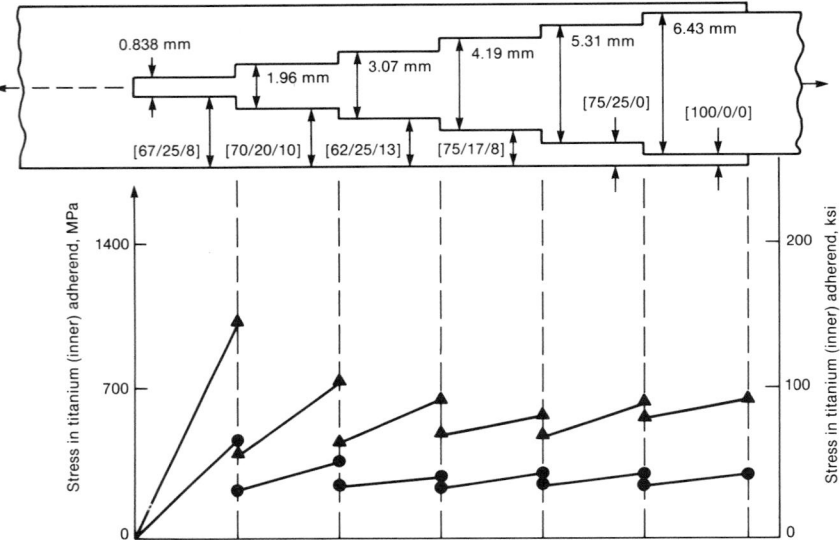

Fig. 75 Predicted stresses in titanium (inner) adherend, six-step lap joint

of the microstructure can provide information on material, processing, or design defects that might cause premature Euler buckling. Laminate lay-ups that have too many plies of the same orientation laid up together can also cause premature failure under certain loading conditions, such as hot/wet compression, which can be identified along with the microfailure mode.

Key points on the load-strain or load-deflection curves should be tabulated. These might be the end of linearity; any points of sudden load, strain, or deflection changes; and the initial straight-line slope, which gives the elastic-range structural stiffness. Recorded experimental data will need to be reported in detail, along with explanations of what data were obtained

and how, when, and where. In addition, the data should be compared with analytical predictions of stresses, strains, and deflections at various points along the load path, and the predicted and actual failure load and mode should be compared. Once the body of the report is complete, conclusions can be drawn and the meaning and importance of the test results detailed. These will include such things as whether the structure performed in the manner for which it was designed, along with discussion of the meaning of any discrepancies in behavior. The concluding remarks should also state whether the testing met its scope of work and whether the test results satisfied the test plan objectives.

Durability and Damage Tolerance Testing

This section discusses durability testing and requirements for composite structures, including laminates and joints, as well as composite structural damage tolerance testing and requirements for both elements and subcomponents of full-sized structures.

Durability Testing. Durability testing is generally understood to be fatigue testing, either constant cycle or lifetime-cyclic spectrum load testing. However, with polymer matrix composite materials, the effects of environmental exposure on static and dynamic behaviors must be considered. Thus, durability testing for composite material structures becomes a function of load cycling and environmental exposure. In aircraft structures, such durability testing has been accomplished in a complicated, sophisticated manner, using flight-by-flight real-time loading spectrums related to aircraft lifetime and, concurrently, environmental exposure to flight temperature and ground-based moisture environments. In addition, accelerated flight spectrum loading and accelerated moisture/temperature environments have been used to simulate real-time testing, with some successes and some failures in correlation.

The successes have been with fiber-dominated plain and open-hole laminates and bolted-joint structures ($\geq$25% to 40% 0° fibers) that are highly fatigue resistant and not sensitive to either fatigue-loading magnitude, frequency variations, or environmental exposure, assuming that the latter conditions do not exceed the wet glass transition temperature of the materials. For epoxy matrices, the worst environmental exposure condition that is safe usually ranges from a maximum of 80 °C (180 °F) wet to 120 °C (250 °F) wet, with the term wet meaning a material moisture content of 1.0 to 1.4 wt%. For instance, the epoxy-based materials discussed in this article are tested up to a maximum condition of 120 °C (250 °F) (1.4 wt% wet); however, study of the data shows that design with AS1/3501-5/6 composites should be limited to 105 °C (220 °F) (1.4% wet) conditions.

The failure to obtain correlations between accelerated and real-time durability testing has been related to matrix-dominated (< 25% to 40% 0° fibers) materials and adhesive-bonded joints. However, the test programs that show noncorrelation dynamically load adhesive-bonded joints to maximum stresses significantly above their static-yield or proportional limit shear-strength values. Because this puts the dynamic loading of the bonded joints into the nonlinear behavior range, the observed wear-out is understandable. Design limit or maximum spectrum loading stresses in matrix-dominated structures and adhesive-bonded joints should not exceed their mean proportional limit or yield values (tension, compression, or shear) at the required environmental conditions; their design ultimate loading stresses should not

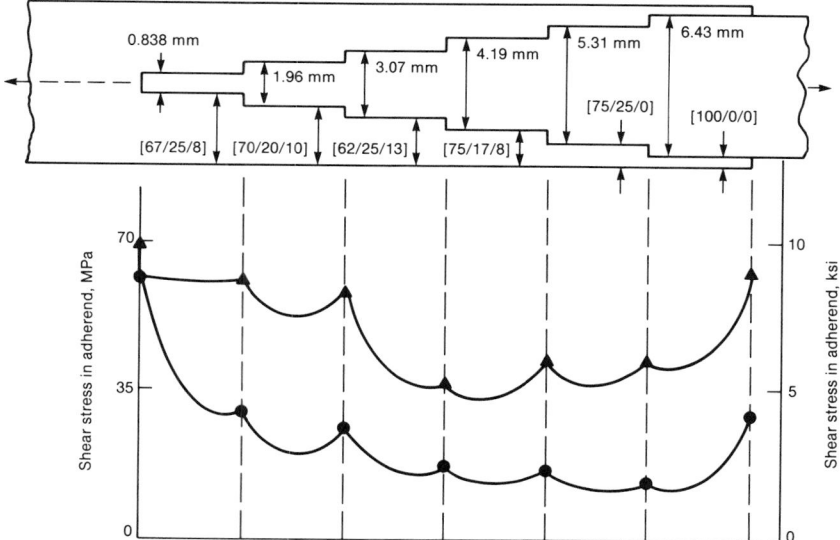

Fig. 76 Predicted shear stresses in the adhesive bondlines, six-step lap joint

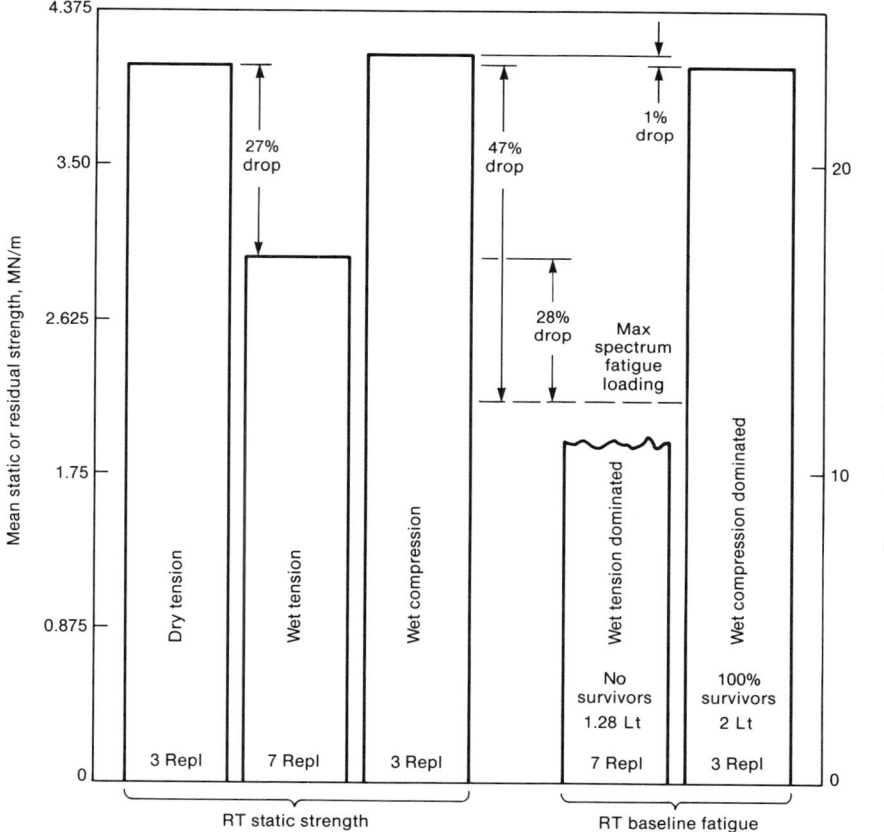

Fig. 77 Tests results for large-scale bonded step-lap joint. Repl, replacement

where overload (underdesign) conditions may occur.

For fiber-dominated (≥25 to 40% 0° fibers) plain and open-hole laminates and bolted joints, Fig. 55 shows that a safe static B-basis allowable is 85% of the mean test values, and that a safe durability B-basis allowable based on residual strength after 2 lifetimes of testing is 77% of the mean test value. For aircraft structures that use a static design safety factor of 1.5 (design ultimate load divided by design limit load), the durability design value of 1.05 times 0.67 times mean test value, (that is, 0.70 times mean test value) is safely conservative because the B-basis allowable is 0.77 times mean test value. This assumes that the materials and processes are controlled to a high level of quality and receive adequate postfabrication nondestructive inspection such as through-transmission ultrasonic inspection with C-scan readouts. This also allows for the one big variable observed in such composite structures, namely, the material lot-to-lot variation in fiber strength.

For matrix-dominated (<25 to 40% 0° fibers) plain and open-hole laminates and composite adherend-bonded joints, the picture is quite different. Figure 78 shows that the most significant strength reductions in composite adherend-bonded joints are static phenomena caused by moisture absorption. These joints failed within the adhesive (cohesive fracture), and sustained some interlaminar failure (matrix fracture) as well. Figure 78 also shows that the largest scatter in tension is at RTD test conditions, while in compression it is at 30 °C (−20 °F) wet conditions. Scatter in compression is the larger of the two, with the B-basis allowable being at 75% of the mean test value. The largest environmentally caused strength reductions occur in the 120 °C (250 °F) RTW condition, with compression loading being slightly worse than tension loading. Statically, the 120 °C (250 °F) wet B-basis allowable is 52% of the RTD joint mean strength in tension and 40% of it in compression.

At a coefficient of variation of 14%, the static B-basis minimum allowable is 0.75 times the mean test values (Fig. 79). For 2 lifetime durability residual strength, Fig. 79 shows that the B-basis minimum allowable is 0.65 times mean test value at a coefficient of variation of 22%.

Regarding the durability test data in terms of fatigue failure strength versus lifetimes, Fig. 80 shows the tension-dominated bonded-joint stress level and environmental effects in accelerated testing. There are significant reductions due to severe (wet and hot/wet) environmental exposures, and some wear-out. The same conclusions apply to the compression-dominated bonded-joint stress level and environmental-effects data in Fig. 81, except that the wear-out is more pronounced. The relative size of the bonded joints tested (scale effects) does not appear to be significant for tension-dominated

exceed their B-basis allowable ultimate strength at these conditions. However, many of the polymer matrix materials and most of the adhesives used in structural design do not have their mechanical-property stress-strain curves

adequately characterized at the needed environmental conditions. Consequently, this static and fatigue testing under design environmental conditions provides information that is useful in defining design limits in cases

Table 10 Summary of large-scale bonded-joint data on mechanical behavior

Test series	Environment	Load direction	Load rate	Lifetimes	Truncation(a)	Mean static or residual strength running load(b) MN/m	lbf/in.
76SRTD(T)	RTD	Tension	...	Static	...	...	...
76SRTW(T)	RTW	Tension	...	Static	...	...	...
76FRTW(T)	RTW	Tension	5Hz	2	9/2	2.17	12 400
76SRTW(C)	RTW	Compression	...	Static	...	...	...
76FRTW(C)	RTW	Compression	5Hz	2	9/2	2.17	12 400

Test series	Max spectrum running load(c) MN/m	lbf/in.	$\hat{\alpha}$ ($\hat{\alpha}_L$)	$\check{\beta}$ ($\check{\beta}_L$) MN/m	lbf/in.	$\check{\beta}$ ($\check{\beta}_L$) MN/m	lbf/in.	$\check{N}_x(t)$ (T) MN/m	lbf/in.	Number of specimens
76SRTD(T)	4.08	23 300	0.150(e)	...	...	...	...	...	...	3
76SRTW(T)	2.99	17 100	15.47	3.10	17 700	2.99	17 100	2.56	14 800	7
76FRTW(T)	...	...	(3.99)(f)	(1.42)(f)		(1.25)(f)		(0.70)(f)		7
76SRTW(C)		(1.28)(f)	0.190(e)	...	...	...	...	...	...	3
76FRTW(C)	4.08	23 300	0.125(e)	...	...	...	...	...	...	3

(a) 9/2 denotes 9g maximum load, the 1/2g-2g peaks truncated. (b) Weibull residual strength values are shown first with the fatigue lifetime values shown in parentheses where applicable. (c) Maximum spectrum loading $\approx 1.05 \times 0.67 \times$ [$\check{\beta}$ of static RTW tension strength]. (d) Scatter factor $\hat{\alpha}$. (e) Coefficient of variation. (f) Lifetimes

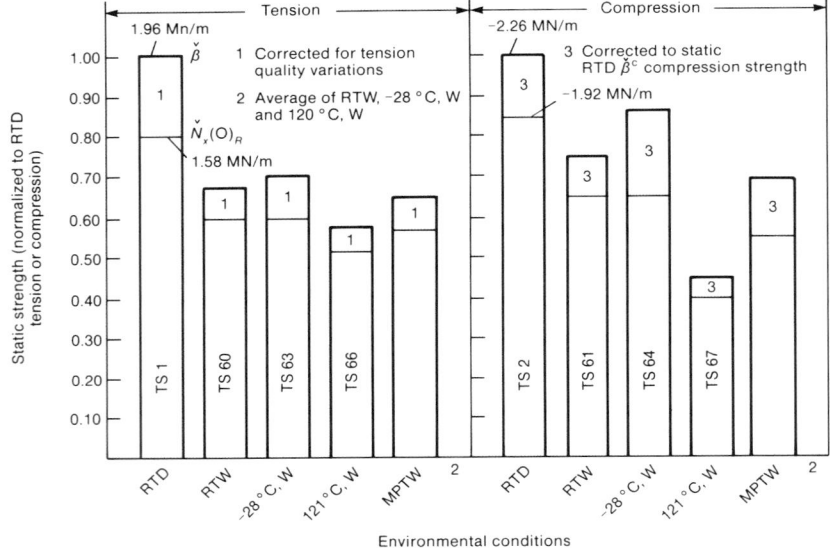

$\check{\beta}$ = 95% confidence β valve = Weibull mean

$\check{N}_x(0)_R$ = B-basis allowable = 95% confidence, 90% probability value via Weibull statistics

Fig. 78 Standard-sized, step-lap, bonded-joint, static-test results, normalized for environmental conditions

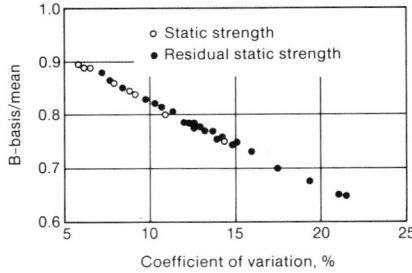

Fig. 79 B-basis/mean values versus coefficient of variation for bonded-joint static and residual strength

testing and is probably not significant for compression-dominated testing.

Comparison of accelerated and real-time tension-dominated durability testing (Fig. 82) shows the nonconservative nature of accelerated testing. While the reduction in the real-time versus accelerated strength is 10 to 20% at a given number of lifetimes, the lifetime is reduced by a factor of 4 to 5 at a given strength level, which is very significant.

In fighter aircraft, then, accelerated testing to 4 or 5 lifetimes under the worst environmental conditions would simulate 1 lifetime of real-time testing. Design requirements for fighter aircraft generally state that the durability testing must simulate 4 real lifetimes; therefore, accelerated testing would require at least 16 lifetimes. For a fighter aircraft designed for a 4000

flight hour lifetime, 4 lifetimes (16 000 h) of real-time testing give a minimum spectrum B-basis allowable load-to-mean static-strength ratio of 0.57 at the worst environmental condition studied (MPTW), as shown in Fig. 83. This compares with a ratio of 0.65 based on residual strength. The 0.57 ratio relates to a durability bondline B-basis allowable strength of 10 MPa (1.5 ksi), compared with the 0.65 ratio giving a durability bondline B-basis allowable strength of 15 MPa (2 ksi). These values represent approximately 30 and 40%, respectively, of the RTD mean static ultimate bondline strength (~35 MPa, or 5 ksi) for this three-step double-overlap joint. The larger six-step double-overlap joint studied would have the same durability requirements. It is assumed that for plain and open-hole matrix-dominated

laminates, durability behavior would be similar.

The same high-quality materials and process controls used in bolted-joint processing are used in the bonded-joint/matrix-dominated laminate processing, with postfabrication nondestructive inspection. Such materials and process variations are found to cause data scatter not exceeding $\pm 10\%$ of any of the mean values.

A comparison of bolted and bonded joints under durability conditions reveals that the former B-basis allowables are 77% of the mean RTD test values versus 30% for the latter. This means that bonded joints will require special design consideration and that the durability allowable, rather than the static allowable, will control the design. Static design shows bonded joints to be considerably more weight-efficient than bolted joints, but the weight added by the required durability design of the former will bring the weight efficiency of the two joints to about the same value.

Although fiber-dominated laminates are considerably more weight-efficient in load-carrying ability than are matrix-dominated laminates, the latter are sometimes needed, as for multidirectional loadings and damage tolerance requirements. It is assumed from this study that matrix-dominated laminate design is governed by durability strength, whereas fiber-dominated laminate design is governed by static strength.

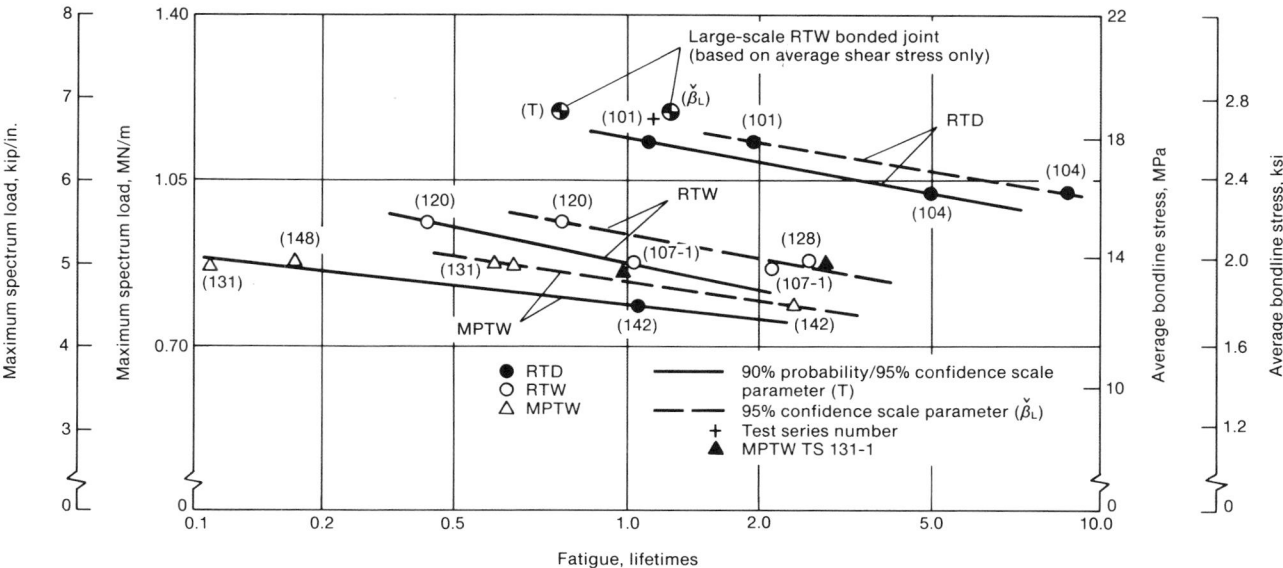

g. 80 Stress level and environmental effects of accelerated tension-dominated spectrum fatigue testing on bonded step-lap joints

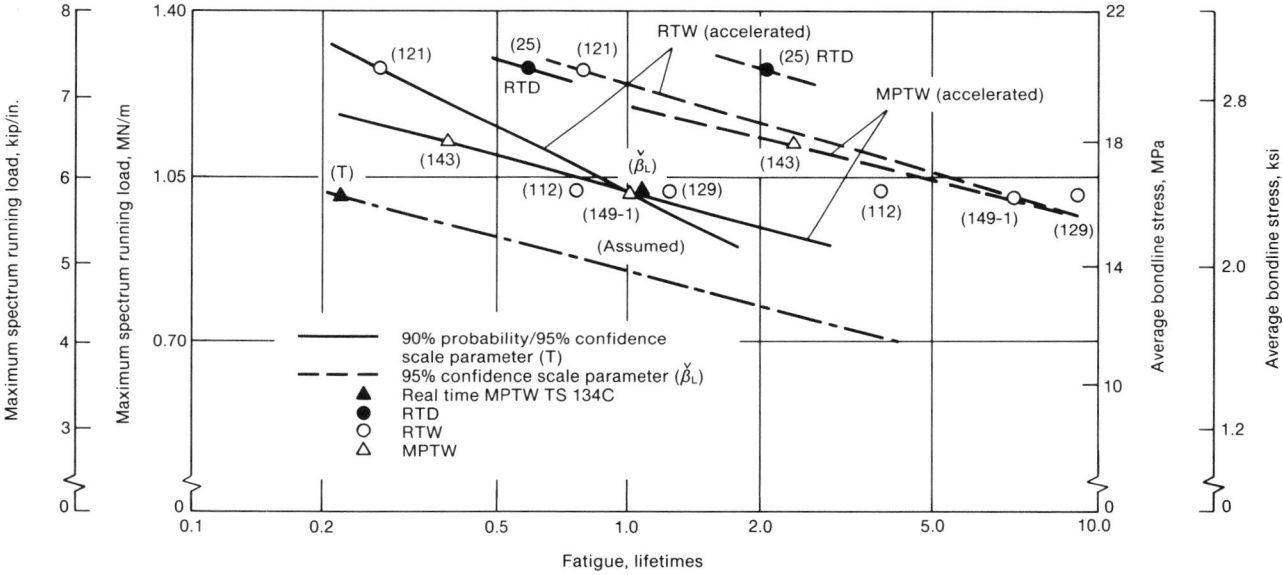

Fig. 81 Stress level and environmental effects of accelerated and real-time compression-dominated spectrum fatigue testing on bonded step-lap joints

Therefore, durability testing for structural-integrity verification of matrix-dominated laminates becomes necessary, as it is for structural integrity verification of bonded joints.

Damage Tolerance Testing. As with durability, damage tolerance design requirements and testing are significantly different for composites than they are for metals. With metals, damage tolerance is related to the rate of propagation of a crack of a given size and location, whereas in composites, it usually means relating impact damage resistance to design requirements. Some work has been done on interlaminar fracture toughness (Ref 26, 27) through the use of the double cantilever beam and the

edge delamination tests (Fig. 84) and their results relating to impact damage resistance. However, the exact meaning of such results has not yet been defined or correlated with impact test data.

Impact damage testing for the most part has been done on laminates and laminate-faced sandwich panels using a square support fixture with either simple support or fixed edges. The central impactor used has ranged from 12.7 to 25.4 mm (½ to 1 in.) in diameter and the impact energy has ranged from 3 to 140 J (2 to 100 ft · lbf). Dent depth damage allowance may range from 0.64 to 6.4 mm (0.025 to 0.25 in.), depending on the application.

Aircraft structure requirements vary considerably, depending on mission and lifetime requirements. Damage tolerance requirements for a typical fighter aircraft composite wing structure usually fall into two ranges. First, low-level impact damage requirements might necessitate an impact of 7 J (5 ft · lbf) with a 12.7-mm (½-in.) diam impactor. The resulting damage must not prevent the structure from carrying static design ultimate load or keep it from meeting durability requirements. High-level impact requirements on such a fighter might include a 140 J (100 ft · lbf) impact with a 25.4-mm (1-in.) diam impactor, or an impact energy level such that an impactor would not

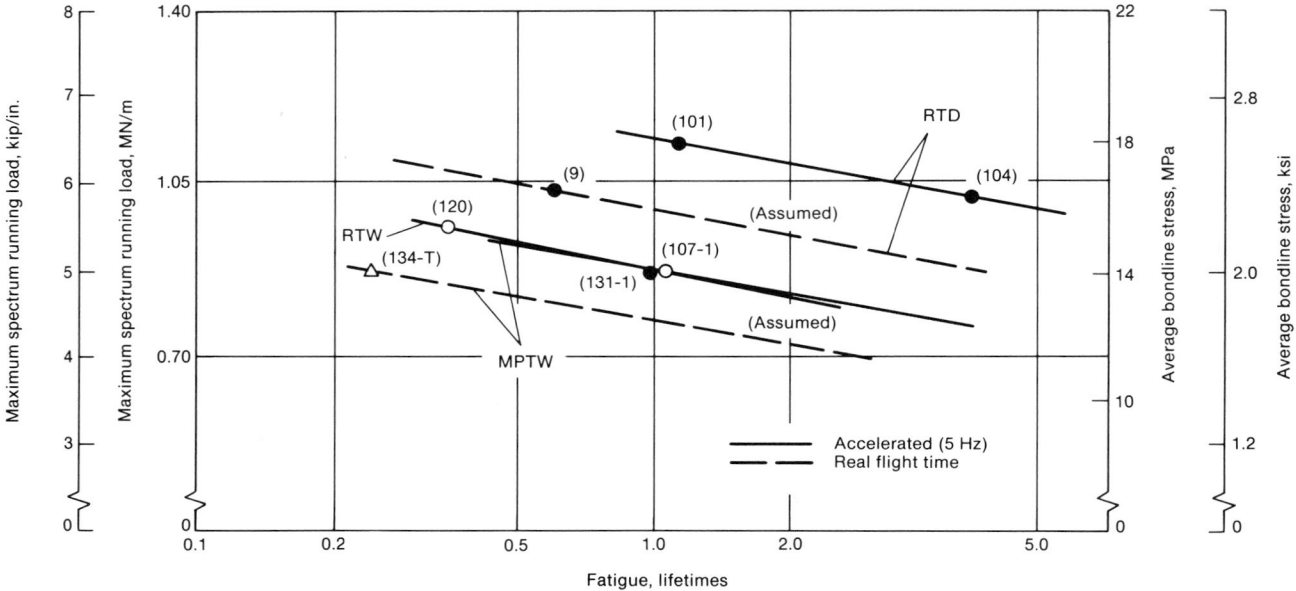

Fig. 82 Environmental effects of accelerated and real-time tension-dominated spectrum fatigue testing on step-lap bonded joints

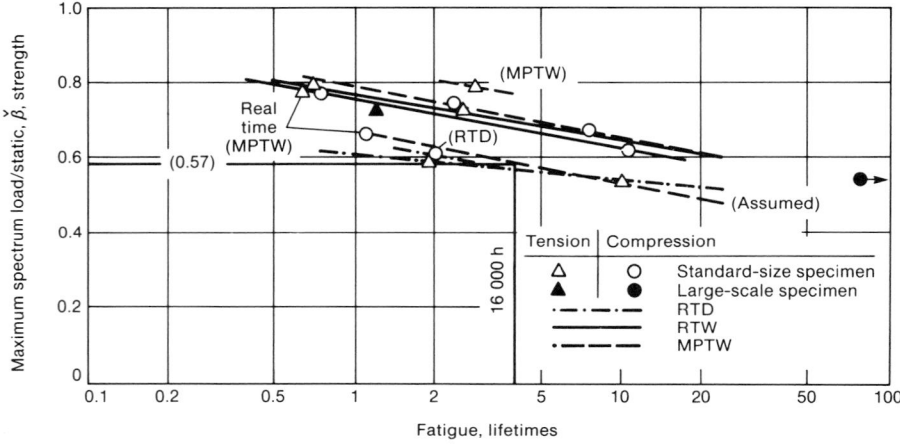

Fig. 83 Fatigue/static strength versus lifetimes for step-lap bonded joints

cause a dent deeper than 5.1 mm (0.20 in.). With this kind of damage, the aircraft structure would be expected to take static design limit or maximum spectrum durability load once and be safe for 10 h of flight before being repaired.

In-service delaminations are usually included under damage tolerance studies. For the fighter aircraft structure discussed, 3.2×10^3 mm^2 (5 in.^{2}) (64 mm, or 2.5 in. diam) delamination would not be expected to keep the structure from meeting all its static and durability design requirements, that is, requirements similar to those of the 7 J (5 ft · lbf) impact with a 12.7-mm (½-in.) diam impactor. A more serious sized delamination of 16.1×10^3 mm^2 (25 in.^{2}) would require that the structure be able to take static design limit load or maximum spectrum durability load once in 10 h of flying before being repaired.

Bolted and bonded joints in such composite structures are usually more damage resistant, but are more critical to structural integrity. Thus, such joints must meet the same damage tolerance requirements as the structure away from the joints.

Durability testing covers fatigue under severe environmental conditions and is a vehicle economic-life consideration. Fiber-dominated composite structures and bolted joints are not usually sensitive to fatigue spectrum or environmental variations within the design envelope. Degradation caused by severe environmental conditions on matrix-dominated composite structures or stress concentrations on fiber- or matrix-dominated structures show up primarily as static-strength reductions. Little or no wear-out is observed in fiber-dominated structures, but some wear-

out is observed in matrix-dominated structures, especially if the loads induce local stresses in the material above its proportional limit or yield strength. For the fighter aircraft considered, the structure is to be tested to 2 lifetimes (8000 h) of flight loading, but designed to be safe for 4 lifetimes (16 000 h) of flight loading. Accelerated durability testing of fiber-dominated composite structure and bolted joints is a feasible, faster, and less-expensive alternative to real-time testing. Real-time testing should be performed on matrix-dominated composite structures and bonded joints because accelerated durability testing is nonconservative. However, if such a structure is designed so that the composite and the adhesive bond do not exceed their proportional limit values, accelerated testing may be safely performed if a load multiplication factor of 1.1 is used in the durability testing. This is done by multiplying the fatigue spectrum load magnitudes by the 1.1 load multiplication factor.

Damage tolerance in composites also differs appreciably from that in metals. A typical requirement is survivability after repairable damage for 10 h of flight in which the loading does not exceed the design limit or maximum spectrum fatigue loading. It is expected that aircraft inspection will reveal the damage and that repair (field or depot) will be accomplished. Thus, such damage will consist of low-level impact not requiring repair; a high-level impact requiring repair; or a large, detectable delamination requiring repair.

An airframe must satisfy static, durability, and damage tolerance requirements to meet its structural-integrity and reliability goals.

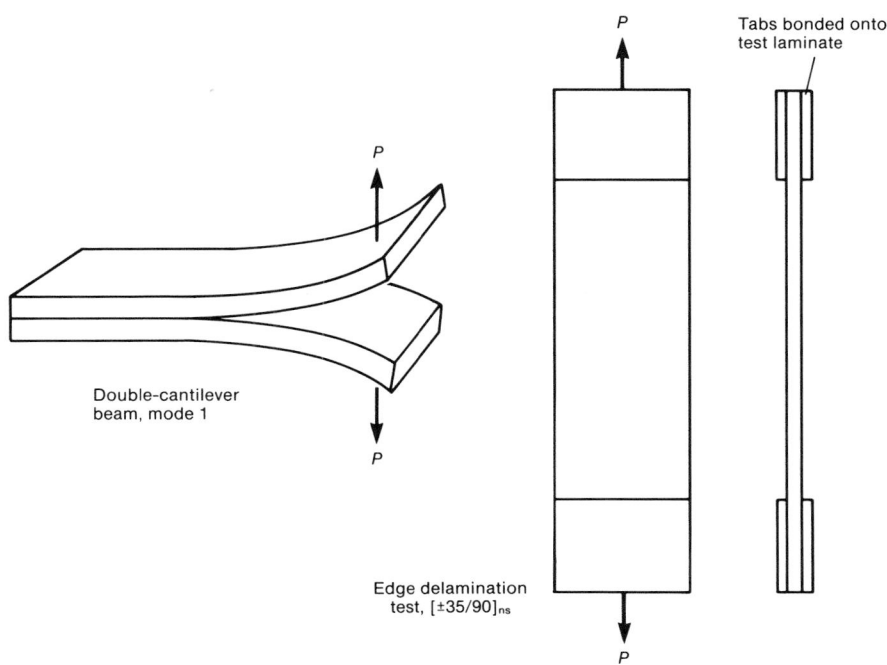

P

Tabs bonded onto
test laminate

P

Double-cantilever
beam, mode 1

P

Edge delamination
test, [±35/90]ns

P

Fig. 84 Specimens for interlaminar fracture toughness testing

REFERENCES

1. L.L. Jeans, G.C. Grimes, and H.P. Kan, Fatigue Sensitivity of Composite Structure for Fighter Aircraft, in *Proceedings of the AIAA/ASME/ASCE/AH 22nd Structures, Structural Dynamics, and Materials Conference*, 1981; also, *J. Aircr.*, 1982

2. "Standard Practice for Inplane Shear Stress-Strain Response of Unidirectional Reinforced Plastics," D 3518, *Annual Book of ASTM Standards*, American Society for Testing and Materials

3. "Standard Test Method for Tensile Properties of Fiber-Resin Composites, D 3039, *Annual Book of ASTM Standards*, American Society for Testing and Materials

4. J.D. Labor and R.M. Verette, Environmentally Controlled Fatigue Tests of Composite Box Beams with Built-in Flaws, in *Proceedings, of Conferences on Aircraft Composites*, American Institute of Aeronautics and Astronautics, March 1977; also, *J. Aircr.*, April 1978

5. G.C. Grimes and D.F. Adams, "Investigation of Compression Fatigue Properties of Advanced Composites," Final Technical Report NOR 79-17, Northrup Aircraft Division, Oct 1979

6. G.C. Grimes, Experimental Study of Compression-Compression Fatigue of Graphite/Epoxy Composites, in *Test Methods and Design Allowables for Fibrous Composites*, STP 734, American Society for Testing and Materials, 1981

7. G.C. Grimes, D.F. Adams, and E.G. Dusablon, "The Effects of Discontinuities on Compression Fatigue Properties of Advanced Composites," CN. N00019-79-C-0275 and 0276, Final Technical Report NOR 80-158, Northrup Aircraft Division, Oct 1980

8. G.C. Grimes and E.G. Dusablon, Study of the Static and Fatigue Compression Properties of Graphite/Epoxy Composites With Discontinuities Under Severe Environmental Exposures, in *Composites Materials: Testing and Design*, STP 787, I.M. Daniel, Ed., American Society for Testing and Materials, 1982

9. R.L. Ramkamar, G.C. Grimes, D.F. Adams, and E.G. Dusablon, "Effects of Materials and Process Defects on the Compression Properties of Advanced Composites," Technical Report NOR 82-103, Northrop Aircraft Division, May 1982

10. R.L. Ramkamar and D.F. Adams, "Compression Properties of Porous Laminates in the Presence of Ply Drop-offs and Fastener Holes," NOR 84-1, Northrop Aircraft Division, March 1984

11. G.C. Grimes *et al.*, "Composites and Joints," Seminar presented at Fiber Composites: Three Technical Seminars, L.J. Broutman & Associates, Ltd., Chicago, April 1986

12. S.J. Kong, Bolt Bearing Strengths of Graphite/Epoxy Composites, in *Proceedings of the AIAA/ASME/ASCE/AH 22nd Structures, Structural Dynamics, and Materials Conference*, April 1981

13. S. Garbo and M. Ogonowski, "Effects of Variances and Manufacturing Tolerances on the Design Strength and Life of Mechanically Fastened Joints," 3 Vol, AFWAL TR-81-3041, Air Force Wright Aeronautical Laboratories, April 1981

14. R.B. Deo, Post First-Ply Failure Fatigue Behavior of Composites, in *Proceedings of the AIAA/ASME/ASCE/AH 22nd Structures, Structural Dynamics, and Materials Conference*, April 1981

15. Military Handbook 17A, *Plastics for Aerospace Vehicles, Part I, Reinforced Plastics*, Department of Defense, Jan 1971

16. "Standard Guide for Testing Inplane Shear Properties of Composite Laminates," D 4255, *Annual Book of ASTM Standards*, American Society for Testing and Materials

17. Military Handbook 23A, Structural Sandwich Composites, Department of Defense, Dec 1968

18. G.C. Grimes, Honeycomb Structures, chapter 20, *Handbook of Adhesive Bonding*, C.V. Cagle *et al.*, Ed., McGraw-Hill, 1973

19. L.L. Jenks, R. Deo, G.C. Grimes, and R.S. Whitehead, Durability Certification of Fighter Aircraft Primary Composite Structure in *Aircraft Fatigue in the Eighties*, Document 1216, International Committee on Aeronautical Fatigue, Proceedings of the 11th ICAF Symposium, J.B. de Jonge and H.H. van der Linden, Ed., 1981

20. L.J. Hart-Smith, "Adhesive Bonded Scarf and Stepped-Lap Joints," NASA CR 112237, National Aeronautics and Space Administration, Jan 1973

21. "Reliability of Step Lap Bonded Joints," AFFDL-TR-75-26, Air Force Wright Aeronautical Laboratories April 1975

22. *Military Handbook 5D, Military Standardization Handbook, Metallic Materials and Elements for Aerospace Vehicle Structures*, Vol 1 and 2, Department of Defense, June 1983

23. J.E. Ashton and T.S. Lave, Shear Stability of Laminated Anisotropic Plates, in *Composite Materials: Testing and Design*, STP 460, American Society for Testing and Materials, 1969, p 352-361

24. R.V. Southwell, *Proceedings of the Royal Society*, Series A, Vol 135, 1932, p 601

25. A.S. Bicos and G.S. Springer, Design of a Composite Box Beam, *J. Compos. Mater.*, Vol 20, Jan 1986

26. H. Chai, "Bond Thickness Effect in Adhesive Joints and Its Significance for Mode I Interlaminar Fracture of Composites," *Composites Materials: Testing and Design (Seventh Conference)*, STP 893, J.M. Whitney, Ed., American Society for Testing and Materials, 1986

27. V.S. Avva, J.R. Vala, and M. Jeyoseelan, Effect of Impact and Fatigue Loads on the Strength of Graphite/Epoxy Composites, *Composite Materials: Testing and Design (Seventh Conference)*, STP 893, J.M. Whitney, Ed., American Society for Testing and Materials, 1986

Full-Scale Tests

John E. McCarty, Boeing Military Airplane Company

VEHICLE STRUCTURAL DESIGN requires a continuing assessment of structural functions to determine whether or not their requirements have been satisfied. First to be assessed are the initial functional requirements of the concept or configuration. Throughout the design process, as material selection, structural element identification, structural arrangement, manufacturing and quality assurance methods, and in-service maintenance are defined and their effects on performance are quantified, the design is reassessed. The successful incorporation of each design element is determined by the in-service performance of the structure.

Because development and production involve a substantial financial commitment, the expected in-service performance must be assessed before the structure enters its service environment. Therefore, tests are performed throughout the developmental cycle to establish a data base for the design and to evaluate the potential in-service performance of the selected elements. Full-scale testing of the completed structure (Fig. 1), or testing of large segments as a single unit (Fig. 2), is the major test in an extensive series. At their current stage of development, composite structures are very dependent on all levels of test in the validation process. This section describes the role of the full-scale test in assessing composite structural systems and qualifying them for in-service use.

The designer's first step in selecting his approach is to understand the need that the design is to fulfill. To define this specific need and ensure that the end product satisfies it, the designer must have a set of design requirements. The requirements usually do not define the structure itself, but rather its performance. Performance requirements are divided into two categories: those related to structural function (for example, the structural wing box, which transmits lift forces to the body for payload support) and those imposed to ensure safety and durability of the structural component. This article focuses on the second set of requirements. The full-scale test is one of the primary means of demonstrating how successfully a structure meets these structural performance requirements and is extremely important be-

cause it tests all combined relationships of the critical elements of the structure in the most realistic manner.

In airframe production, the full-scale structural test is most often performed on one of the first three or four airframes manufactured. This places the full-scale test either well into the production commitment or critical to initiation of the production commitment. Timing of this test in the overall schedule if an unexpected failure were to occur has the potential to impact both the cost of the airframe and the delivery schedule to the customer. If major redesign and/or tooling changes were required, costs would increase and the delivery schedule would slip, which could involve financial penalties and customer dissatisfaction.

The full-scale test enables the internal load distribution and the stress-strain level of each structural element to be imposed correctly. As it is the best representation of the true structural performance of the system, it must be a significant part of the certification or qualification process. It is imperative, therefore, that the purpose of each test element in the process be defined.

There are two basic approaches to full-scale testing for certification: the structure is certified by analysis and supported by test evidence, or the structure is certified by a successful full-scale test validation of the structure and the analysis. Regardless of which approach is taken, full-scale testing is usually required to satisfy the requirements of the agency confer-

Fig. 1 Commercial aircraft full-scale fatigue test

The use of full-scale tests must recognize the unique characteristics of composite structures and their response to the expected in-service conditions as simulated by these tests. Each test type described in the following sections is oriented to aircraft testing requirements, with emphasis on the special considerations needed for testing composite structures. The aircraft structure was selected as the basis for discussion because it is the most generic type of vehicle structure that uses composites.

Static Test

The full-scale static test is the most important test in qualifying composite structures because of their brittle nature, sensitivity to stress concentrations, and insensitivity to fatigue cycling. The parameters to consider when developing the basic requirements for the static test are the type of test article, the type and number of load conditions, the usage environment to be simulated, the load level, and the type and quantity of data to be obtained. The ability of the test data to meet certification requirements must be inherent in each of these static test requirements.

The full-scale test article, whether it is a major component or an entire airframe, is considered representative of the production structure, which implies that it has been fabricated according to production drawings using specification-controlled materials on production tooling and following fabrication and assembly specifications, and inspected according to production quality control requirements. This inspection level represents the detection, acceptance, or rejection of manufacturing anomalies as required by the specifications and directed by the material review board.

To ensure that test data are a useful part of the certification data base, additional special inspections are performed, which further establish the test data validity. For composites, ultrasonic and x-ray inspection procedures are required to detect processing flaws such as porosity and delamination. This same level of additional inspection is performed after each critical test sequence.

Material Considerations. The type of test article used depends on its material composition, function, size, structural configuration, and the degree to which it represents the entire airframe or its full-sized major components.

The material composition of the test article can affect the test procedures and the load level. Carbon-epoxy is the composite material that is currently most commonly used. Properties unique to carbon-epoxy materials, as compared to metals, are displayed in Table 1. Carbon-epoxy materials are known to be generally linear to failure in their response to monotonic loading. This brittle nature of composites must be considered in selecting testing levels for test setup and check-out, strain surveys, limit load

Fig. 2 Aircraft composite horizontal stabilizer full-scale test

ring qualification or certification. Depending on the approach, there may be some difference in the test steps or their sequence and in the data required from the full-scale test.

Typical full-scale tests are static, durability (fatigue), and damage tolerance, the last of which may not require a full-scale test. These full-scale tests are designed to address these questions about the structure:

- Is the analysis of the internal load distribution correct?
- Have there been any errors or omissions in design, manufacture, or quality assurance measures?
- Are there any unexpected deflections that impose functional constraints?

- Have composite structures that are sensitive to out-of-plane or through-the-thickness loads been correctly or adequately covered by analysis or lower-level testing?
- Are there any deflections that significantly alter the load path and increase the stress-strain level of a structural element (that is, are there large displacement effects)?
- Has the assessment of durability been correctly made for composites, metals, and the combined structure, particularly in the interface areas?
- Has the damage tolerance of the structure been correctly evaluated by test and analysis? For composite structures in particular, have the nonvisual flaw/damage effects on the structures been adequately assessed?

Table 1 Behavior of carbon-epoxy versus metals under various conditions

Condition	Carbon-epoxy behavior relative to metals
Stress-strain relationship	More linear strain to failure
Notch sensitivity	
Static	Greater sensitivity
Fatigue	Less sensitivity
Transverse properties	Weaker
Mechanical properties variability	Higher
Sensitivity to aircraft hygrothermal environment	Greater
Damage growth mechanism	In-plane delamination instead of through-the-thickness cracks

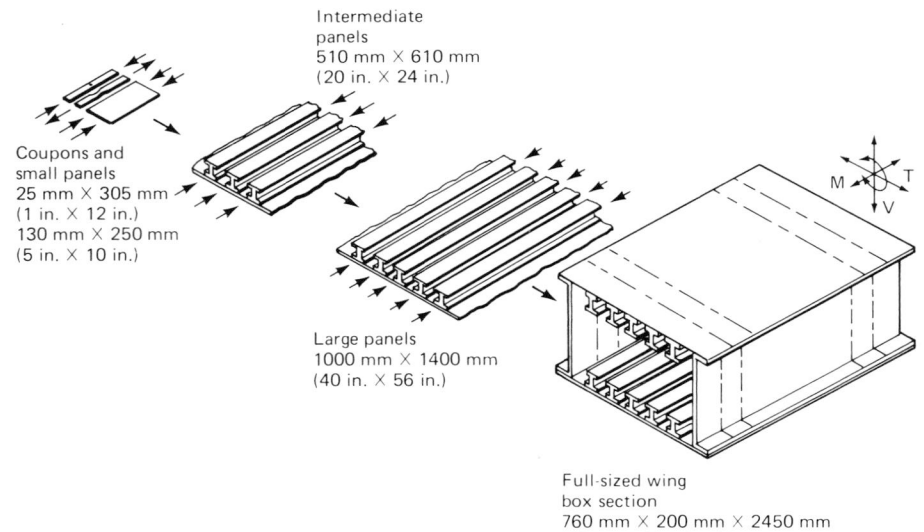

Fig. 3 Building-block testing approach

conditions, and the sequence of ultimate load conditions.

Moisture and Temperature Effects. Composites are also sensitive to temperature and can absorb moisture from in-service environmental exposure. As a full-scale test of even a small aircraft in a fully simulated environment would be very expensive and time consuming, alternative ways to determine environmental effects must be found. Although testing at room temperature and in ambient moisture is the most economical way to test a large structural system, it does not account sufficiently for environmental effects. The elements of the hot/wet (temperature and absorbed moisture) environmental test involve the same environmental considerations for both subsonic and supersonic flight environments. For flight and ground environments, only the temperature levels, time at temperature, and absorbed moisture content of the composite material will vary in magnitude.

The temperature profile of the structure as a function of heat input and its material and structural response must be accounted for by means of the analyses and tests required for certification. Changes in stress-strain caused by moisture absorption also must be accounted for by the same generic analysis and test sequences. The induced stress-strains are caused

by two levels of structural response. The first level is due to the physical compliance of the laminate, which is usually accounted for by ply-level properties and analysis at the laminate level or by laminate-level properties that include the effect of both moisture and temperature. The inclusion of this effect in the analysis of the structure can be accounted for by increased design stress-strain and by reduced allowables. The second level is introduced at the structural element or component level. It is induced by nonlinear temperature or moisture profiles imposed on the structure, the nonuniform coefficient of expansion of the composite elements due to different ply orientations, and the inherent structural redundancy of a complex airframe structure.

Because satisfying both moisture and temperature requirements is difficult at the full-scale level, most composite structures have required a "building-block" testing approach (Fig. 3), in which environmental effects are addressed at the analysis, coupon, structural element, subcomponent, component, and full-scale levels. The sums of these levels of analyses and tests must be consolidated in such a way as to validate the consideration of environmental effects on the composite structure. The methods that have satisfied this requirement at the full-scale level are to:

- Pick the maximum environmental factor (that is, allowable reduction factor) used to correct the room temperature allowables for the effect of environment, and factor up the critical load conditions by the reciprocal of this factor
- Pick the environmental factor used with the minimum margin of safety in the structure for each load condition being tested
- Pick the environmental factor from the elements of the structure that are loaded in such a way that the type of loading and/or the structural load paths are the most sensitive to either the magnitude or the rate of change in capability with change in the environment
- Pick the environmental factor from the elements of the structure that are loaded in such a way that the type of loading or the structural load paths are through the area of the structure where the effects of environment are the most severe or are the most poorly defined in the allowables used to establish that margin of safety
- Test to the ultimate load for each condition considered critical, use the strain gage results to extrapolate the strains that would be added by the environmental effects, and require that these extrapolated strains still show a positive margin of safety

All these methods are suitable for an all-composite structure. In structures containing metal structural elements, these methods would impose the penalty of either overdesigning the metal structure or necessitating two test articles: one with an augmented metal structure for the environmentally factored composite test and one with no factors on the metal being tested to the ultimate load level to validate the metal elements. The only method that avoids these penalties is the last one given in the list above. The selection of the environmental factors by any of the above scenarios is dependent on analysis. The effect on failure mode and location should be considered in the selection of test environmental factors.

Structure Size. There are two considerations in choosing the size of the test structure: the cost of testing to get the environmental information on "allowables" to the statistical level usually considered acceptable and the cost of testing a full-scale small component in a simulated environment. If the component is small enough that a full-scale environmental test would be less costly than using a large data base (that is, environmental allowables) and analysis to certify the structure, then testing in the environment is justified. Other considerations, such as type of structure (primary or secondary), type and complexity of loading, and structural configuration also play a role in the environmental conditioning required.

Critical Test Condition Selection. The steps in full-scale composite structure test procedure follow the same generic path as in a

Fig. 4 Use of formers for loading aircraft horizontal stabilizer

Fig. 5 Use of skin pads for loading aircraft elevator

metal structure test procedure. These steps start with a review of the analysis of the structure. The selection of the most critical load conditions for the structure is based on this analysis.

The critical conditions selected are usually not based solely on the minimum safety margins, although this is the main consideration. Additional considerations may include:

- Stability-critical structures that have low but not minimum margins of safety
- Combined loading conditions that are difficult to analyze
- Major structural joints or intersections that are difficult to analyze
- Areas where the building block approach to certification must be demonstrated, such as when there is concern for a failure mode change between testing in the simulated environment and testing in an ambient condition
- Areas where the environmental factors methods described above must be demonstrated

Load Application Alternatives. Once the critical conditions have been selected, the means of load application must be established to facilitate the most cost-effective method of simulating the real flight and ground loads application. The detail method of loading a composite structure needs careful consideration, because most composites have weak through-the-thickness strength and sensitivity to stress concentrations. Selection of a load application method must take into account composite structure sensitivity, the engineer's desire for the truest simulation of the load distribution, and the cost of the test set-up. Loading methods include: formers that contact the structure on the exterior where the surface is supported by substructure, direct fastening to the substructure by penetrating the surface panels,

and bonding or mechanically fastening to the surface panels.

Formers (Fig. 4) usually concentrate the loading more and do not give as good a representation of the distributed air load as the other methods. This method usually has a less complex loading set-up and is therefore often the least costly.

Direct attachment to the substructure is usually similar in loading simulation and cost to the former method. However, for both composite and metal structures, care must be taken in the method of attachment to the substructure. If the attachment can be made at the same fastener locations that are used for attaching the surface panels and substructure elements together, the procedure is relatively easy. However, if special access holes are needed, this method is less acceptable for both metals and composites. The effects of holes are particularly bad for composites because of their stress concentration sensitivity.

Direct surface attachment (Fig. 5) usually offers a better chance of uniform load representation. This closer representation of the real vehicle structural load usually involves a more complex test set-up. The application of the load directly to a composite surface must be done more carefully than to a metal surface because of stress concentration sensitivity and through-the-thickness weakness.

Loading set-up is now controlled mostly by computers, which control not only the loading applied by each hydraulic jack but also the rate of loading. They also check the displacement of the jack or the load cell to prevent overload of the specimen. Computer control of load application has allowed more complex test set-ups and load application in a truly representative manner.

Instrumentation Requirements. Once the load application method and test set-up have been selected, the instrumentation required to

collect data (for submittal as part of the qualification base for the structure) must be defined. The data display must:

- Ensure correct application and introduction of the load
- Allow monitoring of the testing in real time and protect the test article from being incorrectly loaded
- Validate that the loads applied produce not only the correct loading, but also the correct deflected shape
- Validate that the correct interaction of the applied load is being made in the correct ratio and magnitude for combined load conditions
- Provide a means to automatically terminate loading when loading errors are detected, precluding a human reaction to the error and avoiding loss of the test article
- Validate that the internal strain distribution is as predicted by analysis

The types of instrumentation that are often applied in full-scale testing are strain gages; deflection measurement indicators; stress coats; photostress, moiré fringe, and acoustic emission detectors; and accelerometers. All data are electrically recorded, and computers are used to enhance the on-line presented critical locations data.

Test Procedure Considerations. Establishing test procedures and test sequence requires that all test participants have defined responsibilities both in planning the test and during the test itself. As specific assignments will vary from company to company, they will not be detailed here except to state that they generally include test planning, test fixture design and set-up, test functional check-out, and test conducting and monitoring for continuance/discontinuance during a particular test loading. Deciding who will make the critical decision to stop loading or to proceed is the key to providing maximum protection of the test article during critical portions of the test.

Selecting and establishing the test sequence is generic to most large-scale tests and is particularly important to composite structures. The test sequence usually starts with a checkout of the test set-up, which involves functional testing of loading jacks and evener system, instrumentation, data recording, and real-time critical data displays. A simple loading case and low load levels are applied to ensure that loads are being introduced as expected. The unique features of each test may require more functional monitoring, data recording, and tracking procedures.

Following check-out of the test set-up, a strain and deflection survey is usually run to determine whether the strain distributions and deflections are as predicted. This survey checks the analysis of the structure and the test set-up again and is usually done at a loading level that will not affect the certification test results. For composite structures, the testing load level is usually in the range of 30 to 50% of the design ultimate load level. The load conditions applied are simple singular types used to determine whether the strains for these simple cases agree with the analysis. Then, low load levels of the selected critical certification loads are applied.

These results can then be extrapolated to the design ultimate load levels to determine whether the structure can sustain ultimate load as predicted. If these extrapolations continue to show positive margins of safety comparable to the analysis, testing can proceed. If the extrapolations show negative margins of safety or deviations from the expected load distribution, then testing is delayed until these problems are resolved. A serious problem arises when the deviations are small but critical and there is some risk of failure before reaching the required design load level. A review should be conducted before proceeding.

With the stress-strain at the proper level and the test anomalies resolved, testing can proceed through the design certification critical load tests. These tests run in sequence, with the conditions for which there is the highest confidence of success usually being run first and those with the highest risk of premature failure being run last. The first loading cases are often the most simple, which minimizes the loading complexity before running critical combined loading cases.

Ultimate Load Requirements. It is necessary at this point to discuss the type of load levels required by the qualifying or certifying agencies to meet their validation requirements. The static test load level requirement for Federal Aviation Administration (FAA) certification is based on experience the manufacturer and the FAA have had with a particular type of structure. To meet FAA requirements, conventional metal structures are usually tested only to the limit load. This may also be true for composite structures if the manufacturer has had experience with similar composite structures; otherwise, testing to the ultimate load level is required. Various Department of Defense (DoD) agencies require the ultimate load level for the static test in order to qualify most structures.

After completing the required testing, whether to limit or ultimate load, the manufacturer often picks the most critical condition and tests the article to destruction. This destruction test further validates not only the ability of the analysis to predict the load distribution but also the strength of the structure. If the destruction test failure load exceeds the required ultimate load, vehicle performance growth is warranted.

Test Results Correlation. The final step in the static test sequence is a review of the data obtained from the test and evaluation of its correlation with the stress analysis. The structure is also carefully inspected to determine whether damage that has occurred cannot be readily detected visually. This is of particular importance to composite structures because many of their failure modes and sequences are interlaminar and thus may not be visible.

Durability (Fatigue) Test

The effects of cyclic loading on current carbon-epoxy composites have generally been shown to be noncritical, due to the static load sensitivity of composite structures to stress concentration. In addition, the load level threshold at which composites become sensitive to cyclic loading is a very high percentage of their static failure load. Because this threshold is so high and most vehicles do not experience repeated loads that approach their ultimate loads, composite structures are not fatigue critical. Even if they were to experience loads near this threshold, or slightly above, there are so few of these high cycles in the spectrum life of the vehicle that no significant fatigue damage would occur that would affect structural capability. No industry acceptable-damage rule has been developed for fatigue of composite structures because of this noncritical factor of fatigue loading and the complexity of the fatigue mechanism for composite materials.

Some of the new "tougher" matrix resin systems being used in composites may have characteristics that make them more fatigue-sensitive than the current epoxy systems. If this is true, the importance of the full-scale fatigue test for composite structures will grow significantly, along with the need to develop an acceptable-damage rule.

Spectrum Loading Considerations. To date, cyclic testing of composite structures has been conducted to evaluate a metal structure used with a composite structure. Without the damage rule, validation of the durability of composite structures has generally used spectrum loading that represents a compromise between the spectrums most critical for composites and metals. This compromise, necessitated by the difference in response of composite and metal structures to the magnitude of repeated loads in the applied spectrum, is particularly valid at the full-scale test level. Composites are cyclic-sensitive to loads high in the spectrum, which produce the most fatigue damage and shortest test life. Both the high and low loads in the spectrum can damage a metal structure. However, the high-spectrum loads that damage composites produce a generally unconservative test life (longer than the in-service life) for metal components of the structure.

In general, full-scale cyclic testing has been limited to two to four lifetimes of spectrum loading, including a spectrum load enhancement factor. The flat stress versus cycles curve for composite materials would require, from a statistically significant point of view, consideration of the large scatter in repeated load life, in order to technically validate the fatigue performance of composite structures. This may be accomplished through a life and/or load factor.

Most of the test set-up and performance considerations for the full-scale cyclic composite structures test discussed in the previous section apply to the fatigue test. Only those requirements unique to the fatigue test will be discussed in the following two sections.

Testing and Inspection Requirements. Selecting the loads to be applied represents a compromise for both the composite and metal structural elements. It is very easy to apply a random sequence of spectrum loads using computers to control the loading jacks. Most full-scale fatigue tests, if not all, are spectrum loaded. The methods of loading, attachment of load fixtures, instrumentation, data recording, and check-out of the test set-up are all similar to those used in static tests. The fatigue test has the additional feature of inspection intervals throughout the test life. These inspections are conducted to determine whether any damage is progressing because of cyclic loading, to obtain fatigue performance of the structural details, and to catch a critical damage growth that could cause loss of the test article during load cycling.

Stiffness change of a composite structure has been found to be an indicator of fatigue damage. Therefore, stiffness checks are conducted at various times throughout the test, in a manner similar to the imposed inspection sequence. Because a significant stiffness change in a full-scale test article is very difficult to detect, nondestructive (primarily ultrasonic and x-ray) inspection methods are commonly used to detect damage and monitor its growth throughout the fatigue test.

Accounting for Environmental Effects. After selecting the spectrum loading for a composite structure, a decision must be made as to a way to apply environmental effects during the fatigue test, or account for them in the test results. The cost of either conditioning or applying a real environment for a full-scale test article may be prohibitive. Enhanced or factored spectrum loads are the easiest means of accounting for environmental effects. Although mechanically loading the full-scale structure in a manner that truly represents the effects of environmentally induced strains is very difficult, if not impossible, the factored spectrum load approach is currently the only real option. Modifying the test results by analysis will not be feasible until an acceptable-damage rule for composite structure fatigue is developed. With the full-scale fatigue test, as with the static test, a post-test inspection of the test article is very important to ensure that no fatigue damage has occurred.

Damage Tolerance Test

The damage tolerance test, like the static test, is a qualification requirement of both the FAA and the DoD. The fatigue test is required by the FAA only if the primary structure is not certified damage-tolerant. Although the load level required by the FAA and the DoD varies, both specify residual strength requirements, which vary with the flaw damage assumption, ability to inspect damage, type of in-service inspection used, and type of aircraft. The loading requirement must be carefully reviewed to establish the damage tolerance test residual strength requirements.

Testing composite structures for damage tolerance is particularly important, because it addresses the concerns associated with both the static and fatigue tests. As with the static test, the brittle nature and notch sensitivity of composites is a concern in the damage tolerance test. As in the fatigue test, the critical flaw or damage may be associated with either its initial state or its growth after cyclic loading. Because the full-scale damage tolerance test has many

other similarities with the static and fatigue tests, information on instrumentation, load application, loading control, data display and recording, test set-up and check-out, and test assignment responsibilities will not be repeated. Concerns unique to the damage tolerance test are discussed below.

The types of flaws or damage that are critical to current composite structures are penetrations, delaminations, and low-velocity impact damage. The following discussion will be limited to impact damage, since it is currently the most critical type of damage for composite structures and its concerns are representative.

Cycling and Inspection Requirements. The damaged full-scale structure should be subjected to cyclic loading that substantiates a statistically significant life. Because structural response to impact damage is unique to composites, the imposed spectrum can be tailored to the needs of the composite structure only. Full-scale flaw or damage cyclic testing of a metal structure is not regularly done; therefore, the spectrum imposed for the damage tolerance test of composite structures can be tailored to a composite-sensitive spectrum. The repeated inspections required for the fatigue test are also required for this test.

Accounting for environmental effects of temperature and absorbed moisture is a necessary part of the damage tolerance test. For the residual strength part of the test, the approaches noted for the static test can be applied here as well. The damage effect is usually so dominant that temperature and moisture do not have a strong effect on residual strength. However, the question of environmental effect during the cyclic test portion of the damage tolerance test is not easily addressed. Load enhancement of the spectrum, as suggested for the fatigue test, is currently the only option.

Residual Strength Test. Once the lifetime spectrum loading requirements have been completed, the required residual strength loads are applied. As in the static test, there will be more than one type of loading required, and a critical loading selection must be made. If the structure successfully passes these residual strength tests, one option is to load it to failure to further integrate its damage tolerance capability. As for the other full-scale tests, a detailed inspection is made before the destruction test. The resulting test data are reviewed and correlated with the analysis as part of the certification of the structure for damage tolerance requirements.

SELECTED REFERENCES

- "Composite Aircraft Structure Advisory Circular," AC-107A, Federal Aviation Administration, U.S. Department of Transportation
- R.W. Johnson, J.E. McCarty, and D.R. Wilson, "Damage Tolerance Testing for the Boeing 737 Graphite-Epoxy Horizontal Stabilizer," Paper presented at The Fifth Conference on Fibrous Composites in Structural Design, Department of Defense/National Aeronautics and Space Administration, New Orleans, LA, Jan 1981
- R.W. Johnson, J.E. McCarty, and D.R. Wilson, "737 Graphite-Epoxy Horizontal Stabilizer Certification," Paper No. 82-0745 presented at the 23rd SDM Conference, AIAA/ASME/ASCE/AHS, New Orleans, LA, May 1982
- J.E. McCarty and D.R. Wilson, "Advanced Composite Stabilizer for Boeing 737 Aircraft," Paper presented at The Sixth Conference on Fibrous Composites in Structural Design, Department of Defense/National Aeronautics and Space Administration, Jan 1983

SECTION 6

Forms and Properties of Composite Materials

Chairman: Richard C. Laramee, Morton Thiokol, Inc.

Introduction

TO AFFORD TRULY comprehensive coverage of the forms and properties of composite materials would be a staggering task. The results of such a study, appropriately supported by test data, would fill an entire volume the size of this one. Undertaking such a venture is well beyond the bounds of practicality, as would be keeping its results updated, in pace with present-day advances in composites development. Instead, the effort represented here is limited to a much more modest scope.

The approach taken here is to select, from the many available composite fiber-resin combinations manufactured by conventional processing methods, a few which are representative of the common, high-volume or special-interest types. Their important material properties are presented as an aid for the designer and analyst, and for process and test engineers. These selected materials are identified and briefly described, and the properties test data are presented in graphic form for the various fibers, resins, and fiber-resin combinations.

The availability of such information in this form should greatly simplify the task of developing safe, quality products which match customer requirements and incorporate adequate safety factors for the special loading environment (pressure, temperature, electromagnetic field, and other service conditions) of a specific application. Because the number of possible composite material combinations is very large, comparison of the properties of different fiber-resin composite types, to evaluate materials for a given application, is left to the reader.

The information presented here is as current as possible, from research of open-literature government and industrial reports and manuals, trade publications and research journals, material supplier laboratory test data brochures, and technical papers from universities and research institutes. The reader should bear in mind that a considerable body of information exists that is not in the open literature but is available to corporations and individuals with appropriate security clearance.

If the composite material required for a particular application is not covered in this Section, or if additional properties data are required, the references listed here for the composites selected will serve as a handy guide to the necessary engineering data, or to material suppliers who can provide them. If a special composite is required, a unique design propertry, or extensive test data for a property, statistical confidence level, design allowable curve, a test data base must be generated for it. The composites industry at present has only a 50-year history, and more years will be required before an in-depth, universal, material family characterization can be made available, which will be suitable for a larger number of application designs.

Materials and Properties Description

Richard C. Laramee, Morton Thiokol, Inc.

THE DESIGN AND ANALYSIS of aerospace and industrial composite components and assemblies require a detailed knowledge of material properties, which, in turn, are dependent on the manufacturing methods and machining-assembly operations used.

Composite components are in the form of fibers, resins, and fiber-resin combinations, as shown in Fig. 1. The fiber starts with single filaments grouped by 1000 to 12 000 into a fiber bundle, which can be chopped into short (3.2 to 50 mm, or ⅛ to 2 in.) fibers, or woven into fabric, or further combined into a fiber tow containing as many as 40 000 to 300 000 filaments. Fabrics and tows can be further processed into chopped 12.7- × 12.7-mm (½- × ½-in.) fabric squares, or 6.4- to 12.7-mm (¼- to ½-in.) chopped fiber tows, respectively. Typical fiber-resin applications, such as filament winding, tape wrapping, and molding, are shown in Fig. 2. Typical fabric weaves and fiber directions are shown in Fig. 3 for plain, leno, satin, and crowfoot patterns.

Figure 3 illustrates four types of fabric weave. In the plain weave, yarns are interlaced in an alternating fashion over and under every other yarn to provide maximum fabric stability and firmness and minimum yarn slippage. The thinnest and lightest-weight fabrics are also achieved by use of the plain weave. Fabrics as fine as 25.4 μm (1 mil) are being produced for the electrical industry for mica backing. It is the primary weave used in the coating industry.

The leno weave maintains uniformity of threads and minimizes distortion of threads where a relatively low number of threads is required. Grinding wheel reinforcements, lightweight membranes, and laminating fabrics use the leno weave to good advantage.

The eight-harness satin weave, a very pliable weave that conforms readily to intricately contoured planes, can also be woven with more threads per inch to achieve a high density. Satin weaves are used to best advantage in the reinforced-plastic field, especially for prepregs for aircraft and missiles. Satin weave fabrics require more threads per inch to retain stability and are generally produced in the medium and heavy weight range (200 to 610 g/m², or 6 to 18 oz/yd²).

In the crowfoot satin weave, one warp yarn weaves over three and under one fill yarn. It is used primarily in unidirectional fabrics for fishing rod reinforcement. Crowfoot weave fabrics also conform easily to contoured surfaces.

Resin forms are categorized as liquid, powder, or solid pellets and can be further combined with fiber or have fillers, such as carbon, graphite, calcium carbonate, clay, hydrated alumina or silica powder, added to them.

The fiber-resin forms in a majority of structural component applications are either filament-wound fiber bundles or tows, unidirectional fiber tape lay-down (tape wrapped), woven fabric, chopped fiber molding compound, or lay-up fabrics that are all either combined with a staged (partially cured) liquid resin or impregnated with a wet resin during the fabrication process.

The physical, mechanical, thermal, and electrical properties and in-service conditions of these forms have been obtained from a survey of technical literature and publications and corporate product engineering data sheets and reports. Where a multitude of data points exist for a given material property, multiple curves are drawn and the user can interpret the information based on the references.

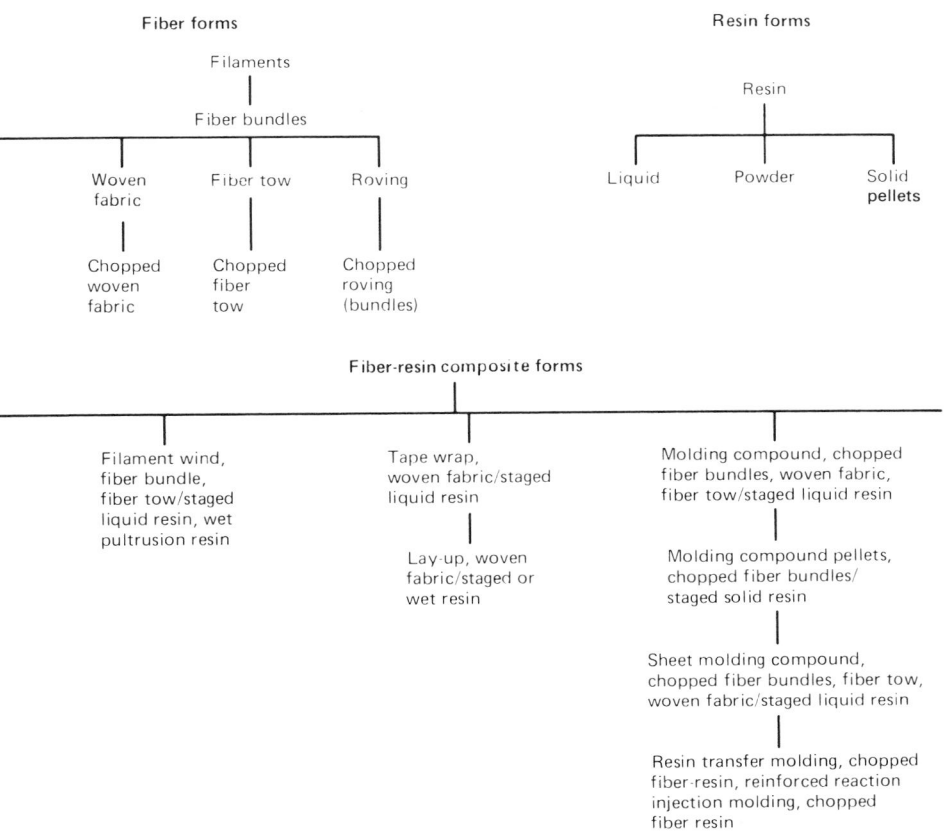

Fig. 1 Composite forms

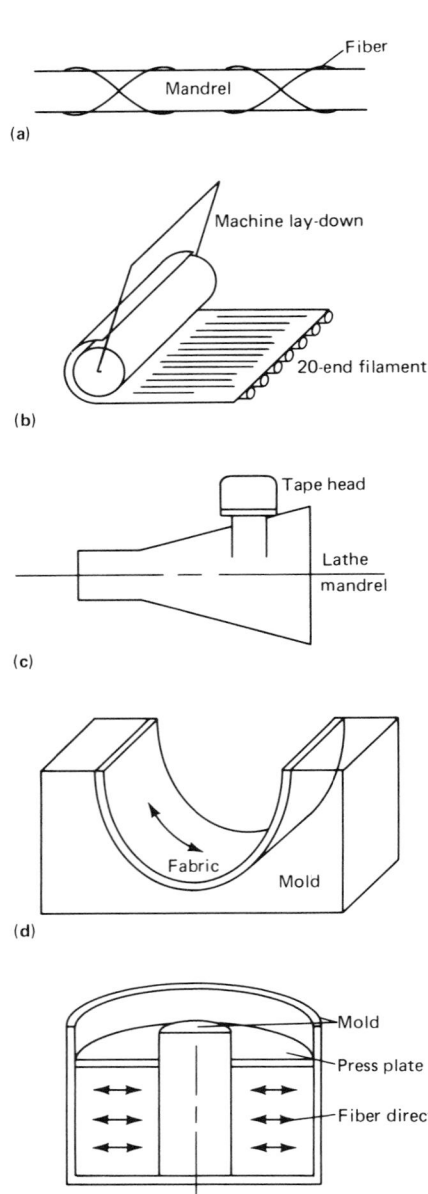

Fig. 2 Typical fiber-resin applications. (a) Filament winding of pipe or rocket motor case. (b) Unidirectional tape lay-down for aircraft panel. (c) Tape-wrapping cone for rocket motor nozzle. (d) Lay-up of fabric resin for boat hull. (e) Molding of hollow cylinder for rocket motor nozzle

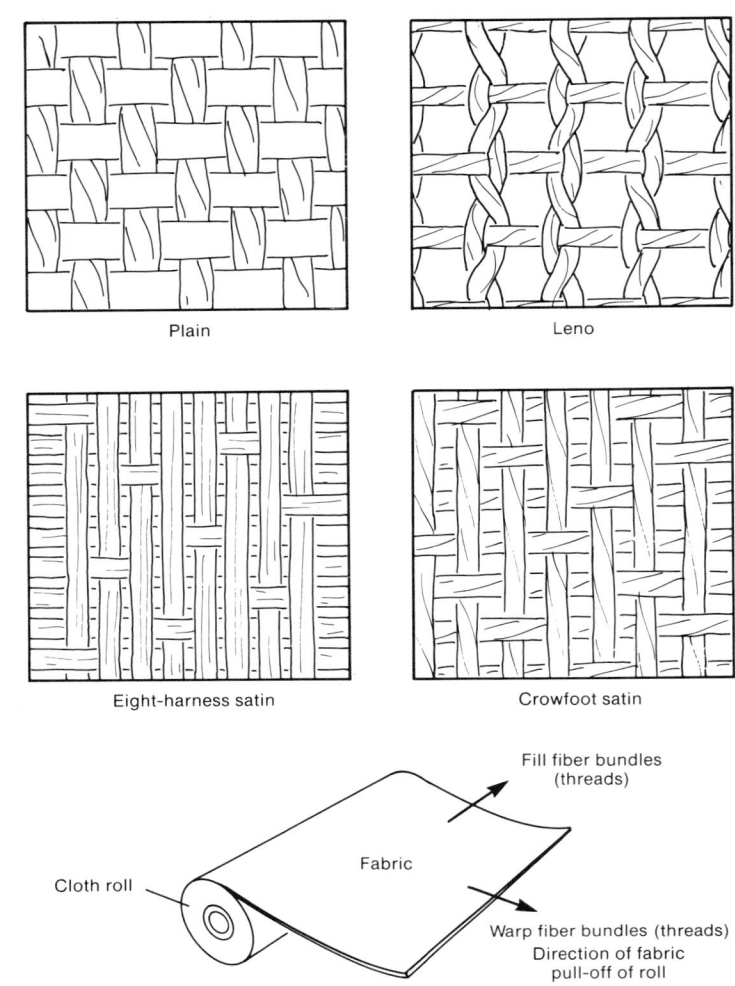

Fig. 3 Typical fabric weaves. Source: J.P. Stevens & Company, Inc.

Table 1 Fiber-resin composite properties

Physical	Mechanical	Thermal	Electrical	In-service conditions
Specific gravity	Tensile strength	Coefficient of	Dielectric constant	Service temperature
Density	Tensile modulus	thermal expansion	Dielectric strength	TGA
	Poisson's ratio	Thermal conductivity	Dissipation factor	Temperature
		Specific heat	Volume resistivity	allowed on all
	Compressive strength			standard loads
	Compressive modulus			Flammability
	Poisson's ratio			EMI/RFI protection
	Shear strength			
	Shear modulus			

An attempt was made to provide as many values as possible for the properties listed in Table 1 for each fiber-resin composite material. Whenever possible, property values were plotted versus temperature. Some unavailable properties await future material characterization. Table 1 primarily lists the key physical, mechanical, thermal, and electrical properties and in-service conditions, all of which will be covered in the articles in this Section of the Volume.

The following general statements on material properties versus temperature represent guidelines for composite material usage:

● Mechanical properties of fibers, resins, and fiber-resin composites decrease with temperature

● Most mechanical properties of fiber-resin composites are higher than those for resin properties but lower than those for fiber properties

● Most thermal properties from room temperature to 260 °C (500 °F) increase with temperature

● Carbon and graphite fibers increase the thermal and electrical conductivity of a fiber-resin composite material system

● Properties may be different in all three planes, depending on fiber configuration

● Factors that affect mechanical properties are: resin-to-fiber ratio, resin and fiber types, resin-fiber interface, type of composite processing (filament winding, tape wrapping, and molding), and type of composite cure (autoclave, hydroclave, vacuum bag molding, extrusion, and compression or injection molding)

In addition, these observations on material properties in respect to form of reinforcement should be considered:

- Maximum properties in load direction 1 are achieved by unidirectional lamination of continuous fiber reinforcement
- In a unidirectional laminate, mechanical properties in load directions 2 and 3 are much lower than in load direction 1, and are highly dependent on the matrix resin
- Bidirectional reinforcement can be achieved by cross-plying unidirectional tapes or broad goods, and by using woven fabric reinforcement
- Strength can be tailored to end-use requirements by directional placement of individ-

Table 2 Forms and constituents of fiber-resin composite materials

Fibers	No resin	Thermoplastic resins			Thermoset resins				
		Polyesters	Polyamides (nylon)	Polysulfones	Epoxies	Phenolics	Polyesters	Polyimides	Bismaleimides
No fiber . .	No	Yes	Yes	Yes	Yes	Yes	Yes	Yes	Yes
Glass	Yes	Yes Fiber, MC	Yes Fiber, MC	Yes Fiber, MC	Yes Fiber, MC Fabric, TW	Yes Fiber, MC Fabric, TW	Yes Fiber, MC	Yes Fiber, MC	No
Aramid . . .	Yes	No	No	No	Yes Fiber, FW	No	No	No	No
Quartz . . .	Yes	No	No	No	Yes Fiber, FW	No	No	No	No
Carbon . . .	Yes	Yes Fiber, MC	Yes Fiber, MC	Yes Fiber, MC	Yes Fiber, FW	Yes Fabric, TW	No	No	Yes Fiber, FW
Graphite . .	Yes	No	No	No	Yes Fiber, FW	Yes Fabric, TW	No	No	No

Note: MC, molding compound; FW, filament wind; TW, tape wrap

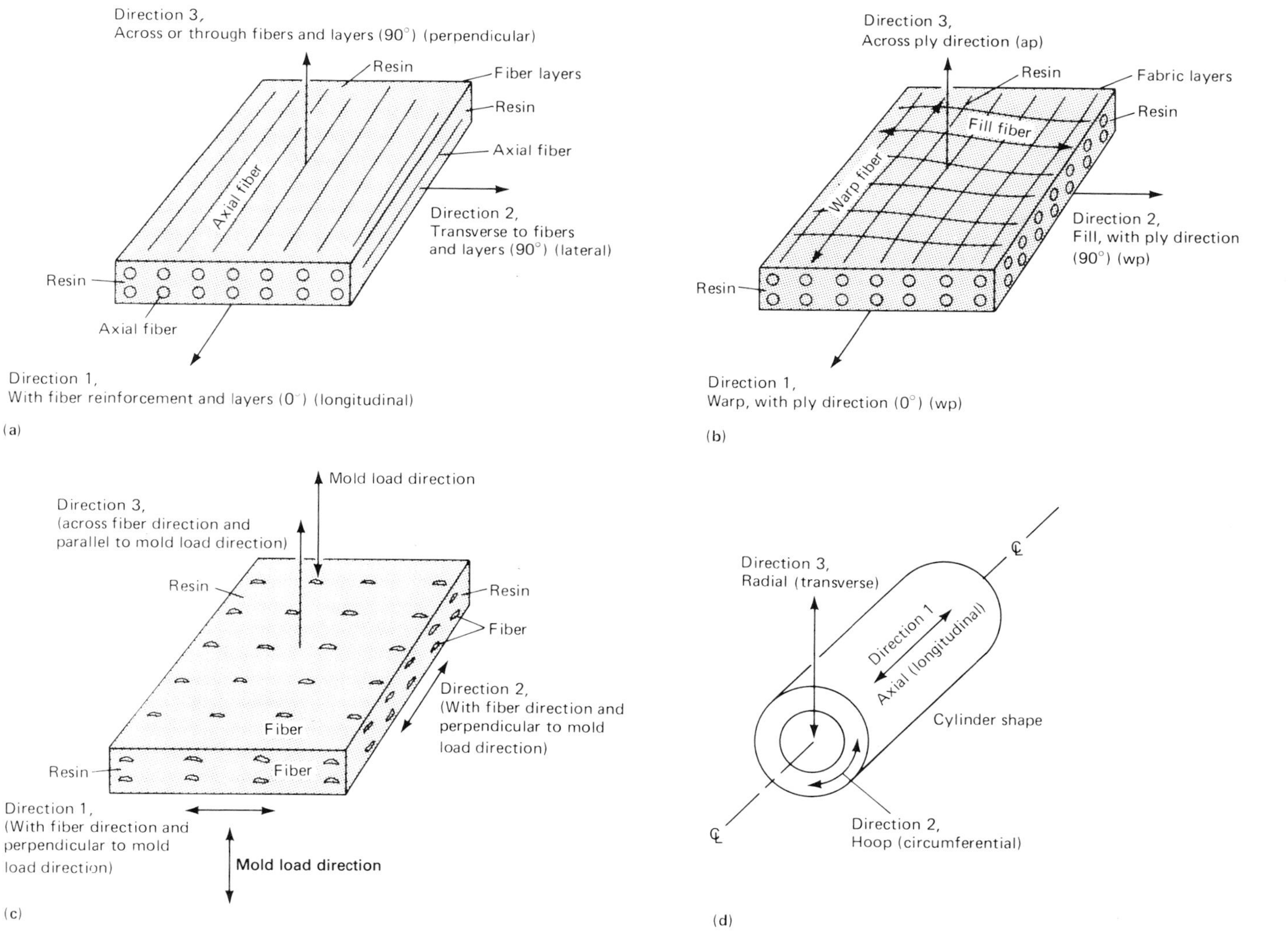

Fig. 4 Definitions of test axes for various composite types. (a) Unidirectional continuous fiber-resin composites (filament wind, tape lay-down, pultrusion). (b) Fabric-resin composites (tape wrap, lay-up). (c) Chopped fiber or fabric-resin molding compounds. (d) Shape components

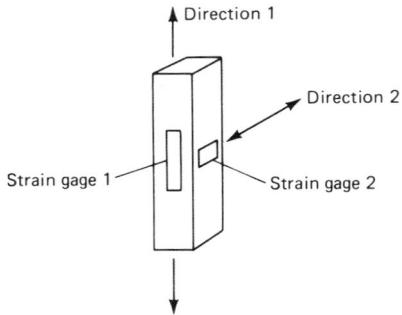

Fig. 5 Directions for Poisson's ratio defined

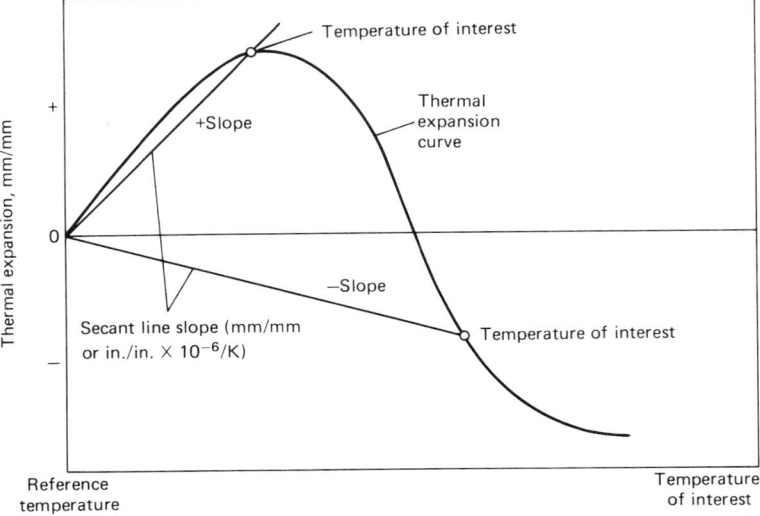

Fig. 6 Coefficient of thermal expansion is calculated as the slope of the secant line between the reference temperature (room temperature) and the temperature of interest. $10^{-6}/K \times \sfrac{5}{9} = \mu.in./in. \times °F$

ual plies of reinforcement, such as 0°/90°/60°/60°/90°/0°, or 0°/45°/45°/0°, and so forth

- Mechanical properties of discontinuous fiber reinforced composites (chopped-fiber or chopped-fabric molding compounds) are usually substantially lower than those of continuous filament reinforced composites
- Properties of composites made from molding compounds are generally omnidirectional in the plane of the part unless flow in molding causes directional orientation of the reinforcing fibers

Some data references for resin-fiber composites may not, at the present time, give fiber percentage by volume or weight, fiber length or diameter, resin type, supplier, or cure cycle. Despite the fact that complete data on these general properties is not available, the information given is intended to inform the engineer of the effects of fiber reinforcement. For individual applications of a fiber-resin composite, specific property data must be generated for design analysis, as well as to assist in manufacturing processes. More data references are available from raw material suppliers, the research and development laboratories of the user corporation, contract laboratories that service government and industry needs, and unclassified or nonproprietary documents. Unfortunately, competitive or classified information is not available to the general engineer. However, some basic material property data on a specific design application can be obtained by making a request to these information sources or by obtaining the necessary classification clearance, although designers will have to "flesh out" the data by property testing in their own test laboratories.

The combinations of major fibers and major resins covered in this Section of the Volume are listed in Table 2. Six reinforcement choices (no fiber, glass, quartz, aramid, carbon, and graphite), and nine resin choices (no resin; polyester, polyamide, and polysulfone thermoplastics; and epoxy, phenolic, polyester, polyimide, and bismaleimide thermosets) allow a possible matrix combination of 54 (6 × 9) choices. Each fiber or fabric can be chopped or continuous. All fiber-resin combinations list references that

can be researched for additional background information on a particular set of data points.

This Section of the Volume also contains descriptions and data that characterize each generic material according to its composition and method of manufacture. Because the composition and manufacture of virtually all materials are subject to change by their manufacturers, a composite may bear the same name for several years, but its mechanical and thermal properties may be modified appreciably.

Data sets for a particular material property may contain a single allowable curve. The curve and its extrapolation are based on knowledge of the material, combined with engineering judgment. No attempt was made to derive an analytical function through the data set.

To assess properties, these factors need to be considered:

- Three major axes of directionality of properties
- Type of fabrication for component
- Fiber or fabric type, length, and fiber percentage by weight or volume
- Fiber finish and strength; fabric weave pattern
- Number of filaments and twist in fiber bundle
- Resin type and manufacturer
- Resin-to-fiber ratio, percent of filler, percent of volatile content, percent of resin flow
- Test panel preparation and cure
- Test specimen preparation, test method, number of tests
- Test specimen design, test fixture, and load ratio
- Test specimen condition, dry or wet

Table 3   Criteria characteristics for flammable ratings

Rating test	Burn criteria
94 HB, horizontal test specimen	<38 mm (1.5 in.)/min burn for 3-13 mm (0.120-0.500 in.) thickness
	<75 mm (3.0 in.)/min burn for <3 mm (0.120 in.) thickness
	No burn >100 mm (4.0 in.) specimen length
94 V-0, vertical test specimen	<10 s burn after each of two F.A.(a)
	<50 s burn for 10 F.A.
	No total burn of specimen
	No dripping burn of cotton 305 mm (12 in.) below specimen
	<30 s after second F.A.
94 V-1, vertical test specimen	<30 s burn after each of two F.A.
	<25 s burn for 10 F.A.
	No total burn of specimen
	No dripping burn of cotton 305 mm (12 in.) below specimen
	<60 s glow after second F.A.
94 V-2, vertical test specimen	<30 s burn after each of two F.A.
	<290 s burn for 10 F.A.
	No total burn of specimen
	Brief dripping burn of cotton
	305 mm (12 in.) below specimen
	<60 s glow after second F.A.
95 V-5, vertical test specimen	<60 s burn after fifth F.A.
	No drip specimens

(a) F.A., flame application, 10 s

Axes Definitions, Symbols, and Special Property Calculations

Because composite materials can have different properties in each of the three directions, definitions of test axes are very important. Each test axis is defined as 1, 2, or 3 for unidirectional fiber, fabric, and fiber or cut-fabric molding compounds, as shown in Fig. 4. Directions 1 and 2 are usually either in the direction of the fiber reinforcement or transverse to the fiber, while direction 3 is perpendicular to the fiber or fabric layers, or across ply. Composite material shapes also have directional properties as illustrated in Fig. 4(d), with direction 1 being axial or longitudinal, direction 2 being hoop or circumferential, and direction 3 being radial or transverse. Figure 4(b), for example, indicates that ultimate tensile strength, with ply, could be expressed as σ_{tu_1}, σ_{tu_2}, or $\sigma_{tu_{12}}$. The Section "Abbreviations and Symbols" in this Volume includes commonly used symbols for material properties.

For Poisson's ratio, ν, numeric subscripts are used:

$$\nu_{12} = \frac{\text{Strain in direction 1}}{\text{Strain in direction 2}}$$

where direction 1 is tension loaded and direction 2 is perpendicular to load application, as shown in Fig. 5. Poisson's ratio, by convention, is expressed as a positive number even though the strain is actually a negative value because most materials shrink in the lateral direction when stretched in the longitudinal direction.

The coefficient of thermal expansion, α, is defined in Fig. 6 as the slope of the secant line from a reference temperature (20 °C, or 70 °F: no expansion) to the expansion or contraction at the temperature of interest.

Definitions of the electical properties used as a standard for comparison of resins, fibers, and resin-fiber composites are:

- *Dielectric:* An insulating medium between two conductors to store and release energy
- *Dielectric constant:* Pure number indicating capability to store energy per unit volume
- *Dielectric strength:* The voltage that an insulating material of a given thickness can withstand before breakdown occurs
- *Volume (electrical) resistivity:* The electrical resistance between opposite faces of a 1-cm (0.40-in.) cube of insulation material
- *Dissipation factor (power factor) or loss tangent:* Multiplier used with apparent power to determine how much of the supplied power is available for use, that is, a measure of power loss in the material
- *Electromagnetic interference/radiofrequency interference (EMI/RFI) protection:* Provided by a material that will conduct an electrical signal away from the component that needs shielding or dissipate a localized electrical strike over a large surface area. The best material has high electrical conductivity, low electrical resistivity or surface resistivity, or low dielectric strength. The introduction of carbon fibers to all resin composite materials will ensure a better EMI/RFI protective characteristic

The definition of flammability characteristics of plastic materials is given by Underwriters' Laboratory (UL94) rating tests for flammability of plastic materials. A typical test specimen for all ratings is 125 mm (5 in.) long by 12.7 mm (½ in.) wide by 12.7 mm (½ in.) thick. A Bunsen burner is applied at one end of a specimen that is held at the opposite horizontal or vertical end. Table 3 identifies the specifications for flammability ratings. Generally, a 94 V-0 rating is satisfactory, while 94 HB, 94 V-1, and 94 V-2 are less satisfactory. A 94 V-5 is the highest rating.

Fibers

ASM Committee on Forms and Properties of Composite Materials*

Glass Fiber

Glass fiber is manufactured from a melted glass batch into continuous filaments, and chopped fiber is cut from continuous filaments (Ref 1, 2). Continuous E-glass and S-glass fibers, made by Owens-Corning Fiberglas Corporation, contain 52 to 56% silicon dioxide (SiO_2) and 60 to 65% SiO_2, respectively, with the balance being oxides of calcium, aluminum, boron, sodium, potassium, and magnesium. The softening points of E- and S-glass fibers are 846 °C (1555 °F) and 970 °C

1778 °F), respectively. PPG Industries and Certainteed Corporation also manufacture glass fibers. Properties are found in Table 1.

Quartz Fiber

Astroquartz II is manufactured by J.P. Stevens, Inc. (Ref 3) as a continuous yarn of fused silica. Fiber composition is 99.9% silica (SiO_2). The fiber softens at 1300 °C (2372 °F) and starts to volatilize at 2000 °C (3632 °F). Properties are listed in Table 2.

Aramid Fiber

Aramid fiber is manufactured from a poly-p-phenyleneterephthalamide (PPTA) polymer into continuous Kevlar 49 and 149 filaments by DuPont (Ref 4, 5). Long-term service temperature in air is 160 °C (320 °F), while the decomposition temperature is 500 °C (930 °F). Total ash occurs at 700 °C (1292 °F). Table 3 lists properties.

Carbon Fiber

Continuous fiber (T300) is manufactured from polyacrylonitrile (PAN) polymer and carbonized at under 1650 °C (3000 °F), by Amoco Performance Products, Inc. (formerly Union Carbide), Amoco Chemicals (Ref 6). The carbon assay is 92% and the fiber is serviceable at temperatures up to its maximum processing temperature (1650 °C, or 3000 °F) in controlled environments. The sublimation temperature is over 3315 °C (6000 °F). Other manufacturers are Hercules, Inc., Celion Carbon Fibers, Division of BASF Structural Materials, Toray, and Courtaulds. Properties are listed in Table 4.

Table 1 Properties of glass fiber

Properties	E-glass	S-glass
Physical (fiber)		
Specific gravity	2.55	2.48
Density, g/cm^3 (lb/in.^{3})	2.55 (0.092)	2.49 (0.090)
Filament diameter, μm (mils)	8.9-20.3 (0.35-0.80)	5.3-9.9 (0.21-0.39)
Mechanical (fiber)		
Tensile strength at RT, MPa (ksi)	3450 (500)	4600 (665) (See Fig. 1)
Tensile modulus at RT, GPa (10^6 psi)	72.4 (10.5)	85.5 (12.4)
Elongation at break at RT, %	4.8	5.7
Thermal (bulk glass)		
CTE, 10^{-6}/K	5.0	2.9
Thermal conductivity, W/m · K (Btu/ft · h · °F)	0.87 (0.50)	...
Specific heat, kJ/kg · K (Btu/lb · °F)	0.825 (0.197)	0.737 (0.176)
Electrical (bulk glass)		
Dielectric constant at RT		
At 1 MHz	5.9-6.4	5.0-5.4
At 10 kHz	6.13	5.21
Dissipation factor at RT		
At 1 MHz	0.002-0.005	0.002
At 60 kHz	0.002-0.005	0.005
Dielectric strength at RT, 5 mm (190 mils) thick, kV/m (kV/in.)	10 250	260
Volume resistivity at RT, 500 V dc, Ω · m (Ω · cm)	10^{13}-10^{14}	10^{14}

RT, room temperature; CTE, coefficient of thermal expansion

Table 2 Properties of quartz fibers

Properties	Astroquartz II
Physical	
Specific gravity	2.20
Density, g/cm^3 (lb/in.^{3})	2.19 (0.079)
Filament diameter, μm (mils in.)	8.9 (0.35)
Mechanical	
Tensile strength at RT, MPa (ksi)	3450 (500)
Tensile modulus at RT, GPa (10^6 psi)	69.0 (10.0)
Elongation, %	5
Thermal (pure silica block)	
CTE, axial and lateral, 10^{-6}/K	...
Thermal conductivity	...
Specific heat from −20 to 500 °C (0° to 932 °F), kJ/kg · K (Btu/lb · °F)	0.96 (0.23)
Electrical (pure fused silica block)	
Electrical resistivity at RT, Ω · m	10^{16}
Dielectric constant at RT, 1 MHz	3.78

*Chairman: Richard C. Laramee, Morton Thiokol, Inc.; Richard G. Adams, J.P. Stevens & Company, Inc.; Roger Bacon, Michael J. Michno, Marvin E. Sauers, Amoco Performance Products, Inc.; Donald Beckley, U.S. Polymeric Corporation; Paul Blanchard, General Electric Company; Robert Boudreau, Borden Chemical Division; Douglas L. Denton, Battelle—Columbus Division; L.E. DeShields, Paul R. Langston, E.I. Du Pont de Nemours & Company, Inc.; William B. Hall, University of Mississippi; Gary E. Hansen, Hercules Aerospace Company; Keith Jacobs, The Ironsides Company; John R. Koenig, Southern Research Institute; Ernest P. Rossa, Fiberite Corporation; John E. Theberge, LNP Corporation

Table 3 Properties of aramid fibers

Properties	Kevlar 49	Kevlar 149
Physical		
Specific gravity	1.44	1.38
Density, g/cm³ (lb/in.³)	1.44 (0.052)	1.42 (0.050)
Filament diameter, μm (mils)	12 (0.47)	...
Mechanical		
Tensile strength, MPa (ksi)	See Fig. 2	1400 (203),0°, Dir.1
Tensile modulus, GPa (10⁶ psi)	See Fig. 3	105 (15.4),0°, Dir.1
Elongation at RT, %	2.5	1.3,0°, Dir.1
Thermal		
CTE, axial and lateral, 10⁻⁶/K	See Fig. 4	−2.68,0°, Dir.1; 120,90°, Dir. 2
Thermal conductivity	...	...
Specific heat	...	...
Electrical		
Dielectric constant	...	At 1 kHz, 4.14 dry; at 1 MHz, 3.90 dry
Dissipation factor	...	At 1 kHz, 0.0103 dry; at 1 MHz, 0.0142 dry

Table 4 Properties of carbon fibers

Properties	Amoco T300	Amoco	Toray	Hercules	FMI
Physical					
Specific gravity	1.77	2.18	1.82	1.88	1.80
Density, g/cm³ (lb/in.³)	1.77 (0.064)	2.18 (0.079)	1.82 (0.066)		
Filament diameter, μm (mils)	7.01 (0.276)	...	...	...	...
Mechanical					
Tensile strength at RT, MPa (ksi)	3650 (530)	2240 (325)	6900 (1000)	3790 (550)	5170 (750)
Tensile modulus at RT, GPa (10⁶ psi)	230 (33.5)	830 (120)	290 (42)	425 (62)	260 (38)
Elongation, %	1.4	...	2.4	0.75	1.97
Thermal					
CTE, 10⁻⁶/K	−0.54	...	...	...	...
Thermal conductivity, W/m · K (Btu/ft · h · °F)	8.7 (5.0)	...	...	...	...
Specific heat					
At RT, kJ/kg · K (Btu/lb · °F)	0.92 (0.22)	...	...	...	...
From 20-1480 °C mean (70-2700 °F mean)	1.7 (0.4)	...	...	...	...
Electrical					
Electrical resistivity, Ω · m	18.0 × 10⁻⁶	...	...	...	...

Table 5 Properties of graphite fibers

Properties	Hercules HMS PAN	Amoco WYB rayon	Amoco P-120 pitch	Amoco-rayon T-50	Amoco-rayon T-75
Physical					
Specific gravity	1.83	1.32	2.18	1.67	1.80
Density, g/cm³ (lb/in.³)	1.83 (0.066)	1.32 (0.048)	2.18 (0.079)	1.67 (0.060)	1.80 (0.065)
Filament diameter, μm (mils)	8.00 (0.315)	9.4 (0.37)	10 (0.4)	6.6 (0.26)	6.0 (0.24)
Mechanical					
Tensile strength, MPa (ksi)	2200 (320)	620 (90)	2240 (325)	2170 (315)	2620 (380)
Tensile modulus, GPa (10⁶ psi)	340 (50)	40 (6)	825 (120)	395 (57)	540 (78)
Elongation, %	0.58	1.5	0.27	0.60	0.50
Thermal					
CTE, 10⁻⁶/K					
Axial	−0.99	...	−1.62	...	...
Lateral	16.8	...	...	...	...
Thermal conductivity, axial, W/m · k (Btu/ft · h · °F)	104 (60.0)	...	609 (352)	118 (68)	156 (90)
Specific heat kJ/kg · k (Btu/lb · °F)	0.71 (0.17)	...	...	0.71 (0.17)	0.71 (0.17)
Electrical					
Electrical resistivity, Ω · m					
Axial	13 × 10⁻⁶	...	2.2 × 10⁻⁶	...	...
Lateral	10 × 10⁻⁶	...	...	...	...

Graphite Fiber

Continuous fiber (HMS) is manufactured from polyacrylonitrile polymer and graphitized at over 2205 °C (4000 °F) by Hercules, Inc. (Magna, UT) (Ref 7, 8). The carbon content is 99.7% and the fiber is serviceable at temperatures up to its maximum processing temperature (2205 °C, or 4000 °F) in controlled environments. The sublimation temperature is over 3315 °C (6000 °F). Other manufacturers are Amoco Performance Products, Celion Carbon Fibers (division of BASF), and Fiber Materials Inc. In addition, pitch-based polymer may be graphitized to form a graphite fiber by Amoco Performance Products. Rayon-based polymer also may be graphitized to form a graphite fiber, but this is an older technology no longer used to produce high-strength high-modulus fiber. Properties are listed in Table 5.

Figure 5 provides dry bundle tensile strength and modulus values versus temperature for a rayon-based graphite fiber polymer. The data (Ref 9) are not related to the current PAN-based HMS fibers produced by Hercules, Inc., but, rather, show a trend of fiber strength and modulus versus temperature. Impregnated strand-cured strength and modulus values are usually reported in supplier and research laboratory reports and data sheets.

REFERENCES

1. "S-2 Glass Fiber High Performance/Low Cost Reinforcements," Owens-Corning Fiberglas Corporation, 1981, 1984

2. "Textile Fiber Materials for Industry," Publications 1-GT-1375-A and 5-TOD-8285-C, Owens-Corning Fiberglas Corporation, 1961, 1980, 1985

3. "Astroquartz II," Product Data Sheet, J.P. Stevens Inc.

4. "Data Manual for Kevlar 49 Aramid," Textile Fibers Department, E.I. Du Pont de Nemours & Company, Inc., 1974, 1986

5. C.O. Pruneda *et al.*, The Impurities in Kevlar 49 Fibers, in *Proceedings of the Thirtieth National SAMPE Symposium*, Society for the Advancement of Material and Process Engineering, March 1985

6. "Thornel Carbon Fiber T-300-3K," Product Information Sheet F4311, Rev 9, Amoco Performance Products

7. "Magnamite Graphite Fiber Type HMS," Product Data Sheet 840-1, Hercules, Inc., Aug 1981

8. "General Composite Properties," Bulletin ACM-7, Hercules, Inc.

9. F.S. Inman, "Elevated Temperature Graphite Yarn Tensile Tests," Morton Thiokol, June 1975

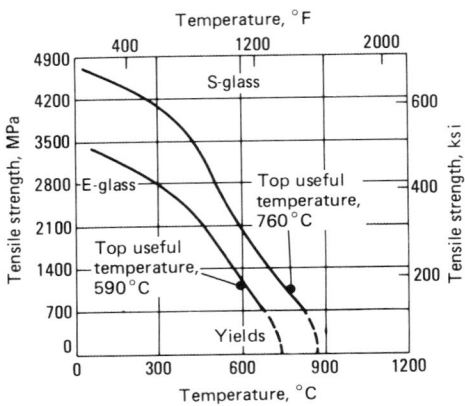

Fig. 1 Glass filament tensile strengths versus temperature. Comparative data for E, S, and S₂ glass, Publication ASP-6870, Owens-Corning Fiberglas Corporation

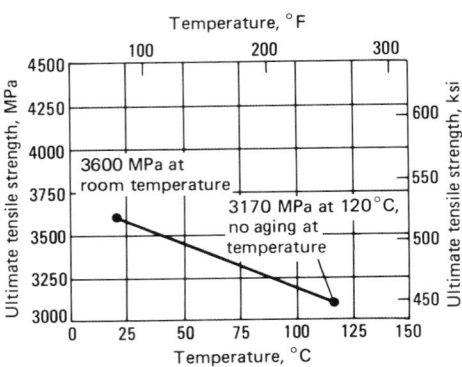

Fig. 2 Ultimate tensile strength versus temperature, for Kevlar 49 yarn. Based on strand test, resin impregnated

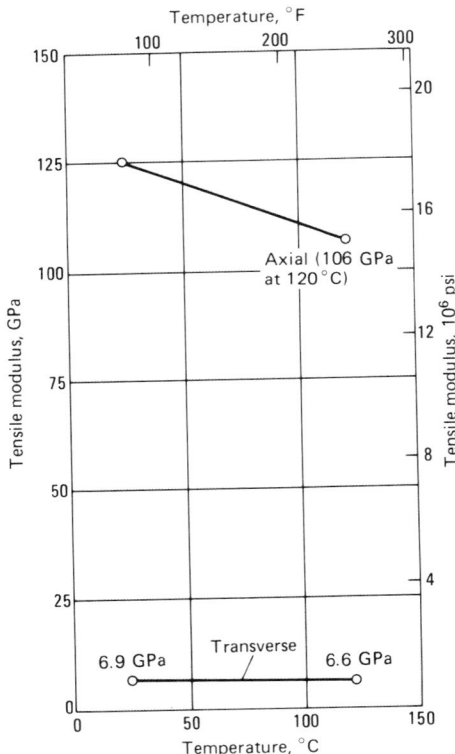

Fig. 3 Tensile modulus versus temperature for Kevlar 49 yarn. Based on strand test, resin impregnated

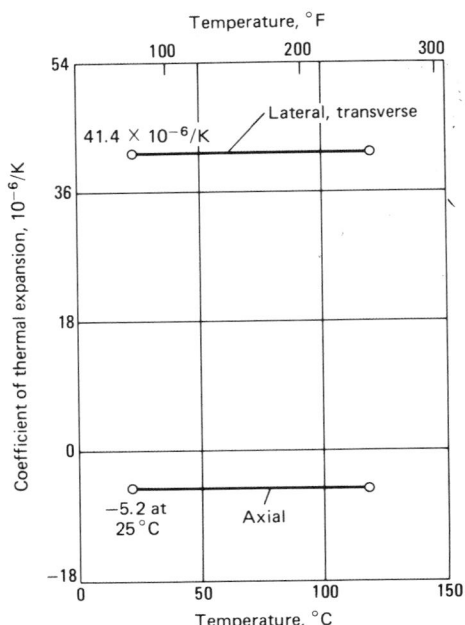

Fig. 4 Coefficient of thermal expansion versus temperature for Kevlar 49 yarn. Based on unidirectional composite data. $10^{-6}/K \times 5/9 = \mu in./in. \times °F$

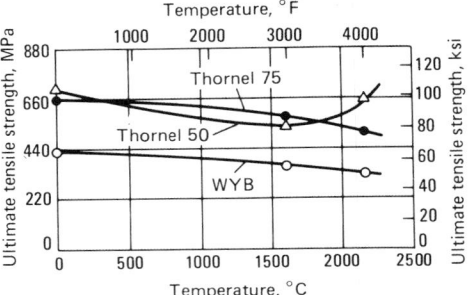

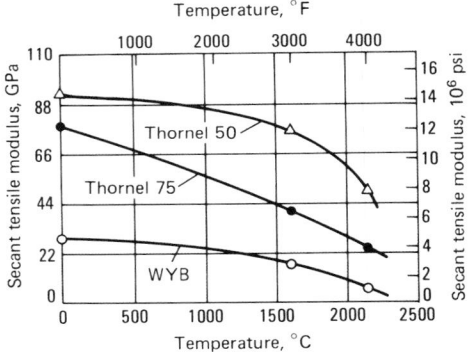

Fig. 5 Dry bundle tensile strength and modulus of graphite yarn (rayon-based polymer) of Union Carbide Fibers. Source: Ref 6

Thermoplastic Matrix Composites

ASM Committee on Forms and Properties of Composite Materials*

Polyester Thermoplastic Resin and Fiber-Resin Composites (Ref 1-11)

Three polyester resin systems—aromatic copolyesters (Xydar, Dart Industries), polybutylene terephthalate (PBT) (Valox, General Electric), and polyethylene terephthalate (PET) (Rynite, DuPont)—are used to describe the family of properties (Table 1). Short-length glass and carbon fiber reinforcements of up to 55% of the fiber-resin composite are used to improve or alter the properties in molding compounds. The Underwriters' Laboratory (U.L.) service temperature of the unfilled resins ranges from 120 to 240 °C (253 to 464 °F). Applications are electrical/electronic components, chemical processing and oil field equipment, aerospace and transportation vehicles, appliance and consumer products, furniture, and bottles for toiletries and food products.

Polyamide Thermoplastic Resin and Fiber-Resin Composites (Ref 12, 13)

Zytel polyamide (nylon) resins are manufactured by DuPont. Some variations of the nylon resin may also include alloys of polyolefin and other copolymers to tailor properties for specific uses. The melt point will vary with the different product families but will be between 212 to 270 °C (414 to 518 °F). Short-length carbon and glass fiber reinforcements of up to 43% of the fiber-resin composite are used to alter or improve the properties in molding compounds. The range of properties listed in Table 2 include unreinforced and reinforced polyamide resin systems.

It should be noted that because all nylons pick up moisture, final design properties and dimensions will change depending on the relative humidity.

Polysulfone Thermoplastic Resin and Fiber-Resin Composites (Ref 14-16)

Polysulfone resin is manufactured by Union Carbide and LNP Corporation. It exhibits the highest service temperature (150 to 205 °C, or 300 to 400 °F) of any melt-processable thermoplastic. The resin is strong, noted for high heat deflection temperatures, and is stable in the presence of moisture. Care must be taken not to expose material to excessive ultraviolet rays or organic solvents. Glass fiber reinforcements of up to 40% of the fiber-resin composite are used to alter properties. Property values listed in Table 3 include resin with and without glass reinforcement. Applications include medical instrumentation, food processing equipment, chemical processing equipment, camera and watch cases, automotive and aerospace components, and water purification devices.

Table 1   Properties of thermoplastic matrix composites

| Properties | Aromatic copolyester | | PBT | | PET | |
	Resin	40% glass fiber-resin	Resin	15-40% glass fiber-resin	Resin	30-45% glass fiber-resin
Physical						
Heat deflection temperature at 1800 kPa (264 psi), °C (°F)	355 (671)	...	55 (130)	205 (400)	...	225 (435)
U.L. in-service temperature rating, °C (°F)	240 (464)	...	120 (248)	140 (284)	140 (284)	150-180 (302-356)
Processing melt temperature, °C (°F)	400-450 (750-840)	...	270 (520)	...	290 (550)	...
Specific gravity	1.35	1.70	1.31	1.53	...	1.56-1.69
Density, g/cm³ (lb/in.³)	1.35 (0.049)	1.70 (0.061)	1.31 (0.047)	1.53 (0.055)	...	1.56-1.69 (0.056-0.061)
U.L. flammability	94 V-0	94 V-0	94 HB / 94 V-0	94 HB / 94 V-0	94 HB / 94 V-0	94 HB / 94 V-0

Mechanical

Tensile strength . . . . See Fig. 1
Tensile modulus . . . . See Fig. 2
Elongation . . . . See Fig. 3
Compressive strength . . . . See Fig. 4
Compressive modulus . . . . See Fig. 5
Shear strength . . . . See Fig. 6

Thermal

Coefficient of thermal expansion . . . . See Fig. 7
Thermal conductivity . . . . See Fig. 8
Specific heat . . . . See Fig. 9

Electrical

Dielectric constant . . . . See Fig. 10
Volume resistivity . . . . See Fig. 11
Dielectric strength . . . . See Fig. 12
Dissipation factor . . . . See Fig. 13

*Chairman: Richard C. Laramee, Morton Thiokol, Inc.; Richard G. Adams, J.P. Stevens & Co., Inc.; Roger Bacon, Michael J. Michno, Marvin E. Sauers, Amoco Performance Products, Inc.; Donald Beckley, U.S. Polymeric Corporation; Paul Blanchard, General Electric Company; Robert Boudreau, Borden Chemical Division; Douglas L. Denton, Battelle—Columbus Division; L.E. DeShields, Paul R. Langston, E.I. Du Pont de Nemours & Co., Inc.; William B. Hall, University of Mississippi; Gary E. Hansen, Hercules Aerospace Company; Keith Jacobs, The Ironsides Company; John R. Koenig, Southern Research Institute; Ernest P. Rossa, Fiberite Corporation; John E. Theberge, LNP Corporation; Steven Witschen, Owens-Corning Fiberglas Corporation

REFERENCES

1. *Modern Plastics Encyclopedia*, Vol 62 (No. 10A), McGraw-Hill, 1985-1986
2. "Xydar—High Performance Engineering Resins, SRT-500, FC-Series," Dartco Manufacturing Inc.
3. E. Galli, New Thermoplastics Perform at Very High Temperatures, *Plast. Des. Forum*, March-April 1985
4. "Valox Engineering Thermoplastic Properties Guide," Plastics Group, Composite Polymers Products Department, General Electric Company
5. "Designing With Plastics," *Design Handbook*, DuPont Engineering Plastics Module IV Rynite, Polymer Products Department, E.I. Du Pont de Nemours & Company, Inc.
6. Generic Thermoplastic Polyesters, 1986 Materials Reference Issue, *Mach. Des.*, 17 April, 1986
7. "Carbon Fiber Reinforced Thermoplastic Composites," Engineering Plastics, LNP Corporation
8. Material Selector 1985, *Mater. Eng.*, Dec 1984
9. *The International Plastics Selector, Extruding and Molding Guides*, Desk Top Data Bank, Cordura, 1977
10. "Zytel Nylon Resins," General Guide to Products and Properties, DuPont Engineering Plastics, Polymer Products Department, E.I. Du Pont de Nemours & Company, Inc.
11. "Carbon Fiber Reinforced Thermoplastic Composites," Engineering Plastics, LNP Corporation
12. "Nylon," *Modern Plastics Encyclopedia*, Vol 62 (No. 10A), McGraw-Hill, 1985-1986
13. H.R. Clausen, *Encyclopedia/Handbook of Materials, Parts and Finishes*, Technical Publishing, 1976
14. "Thermocomp—GF Series," LNP Corporation
15. "Udel-Polysulfone," Engineering Polymers Product Data, Union Carbide Corporation
16. Material Selector, *Mater. Eng.*, Dec 1985

Table 2 Properties of thermoplastic resin and fiber-resin composites

Properties	Nylon 66 Resin	Nylon 66 30% carbon fiber-resin	43% glass fiber-resin
Physical			
Specific gravity	1.14	1.28	1.51
Density, g/cm³ (lb/in.³)	1.14 (0.041)	1.28 (0.046)	1.51 (0.055)
U.L. electrical rating °C (°F)	130 max (266 max)	...	...
Deflection temperature at 1800 kPa, °C (°F)	90 (194)	...	...
U.L. flammability	94 V-2	...	...
Mechanical			
Tensile strength		See Fig. 14	
Tensile modulus		See Fig. 15	
Tensile elongation at break		See Fig. 16	
Compression strength		See Fig. 17	
Compression modulus		See Fig. 18	
Shear strength		See Fig. 19	
Thermal			
Coefficient of thermal expansion		See Fig. 20	
Thermal conductivity		See Fig. 21	
Specific heat		See Fig. 22	
Electrical			
Dielectric constant		See Fig. 23	
Volume resistivity		See Fig. 24	
Dielectric strength		See Fig. 25	
Dissipation factor		See Fig. 26	

Table 3 Properties of polysulfone resin and fiber-resin composites

Properties	Neat resin	30% glass fiber-resin	30% carbon fiber-resin
Physical			
Specific gravity	1.24	1.45	1.37
Density, g/cm³ (lb/in.³)	1.24 (0.045)	1.45 (0.052)	1.37 (0.049)
U.L. electrical rating, °C (°F)	160 (320)	...	...
Deflection temperature at 1800 kPa (264 psi), °C (°F)	175 (345)	185 (365)	185 (365)
In-service temperature, °C (°F)	150-205 (300/400)	...	...
U.L. flammability rating	94 V-0	94 V-0	94 V-0
Mechanical			
Tensile strength		See Fig. 27	
Tensile modulus		See Fig. 28	
Tensile elongation at break		See Fig. 29	
Compressive strength		See Fig. 30	
Shear strength		See Fig. 31	
Thermal			
Coefficient of thermal expansion		See Fig. 32	
Thermal conductivity		See Fig. 33	
Specific heat		See Fig. 34	
Electrical			
Dielectric constant		See Fig. 35	
Volume resistivity		See Fig. 36	
Dielectric strength		See Fig. 37	
Dissipation factor		See Fig. 38	

Polyester Resin—Thermoplastics

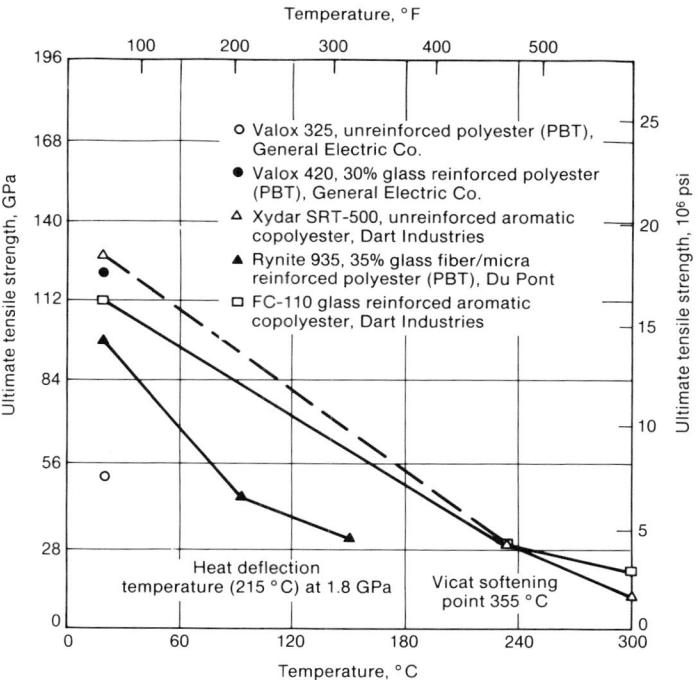

Fig. 1 Ultimate tensile strength versus temperature

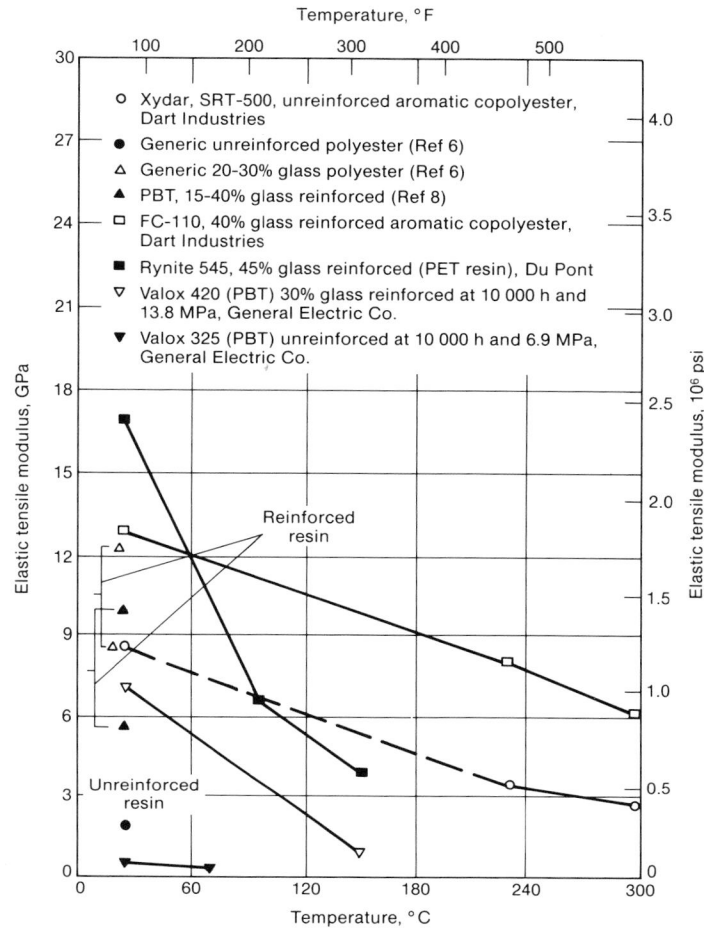

Fig. 2 Elastic tensile modulus versus temperature

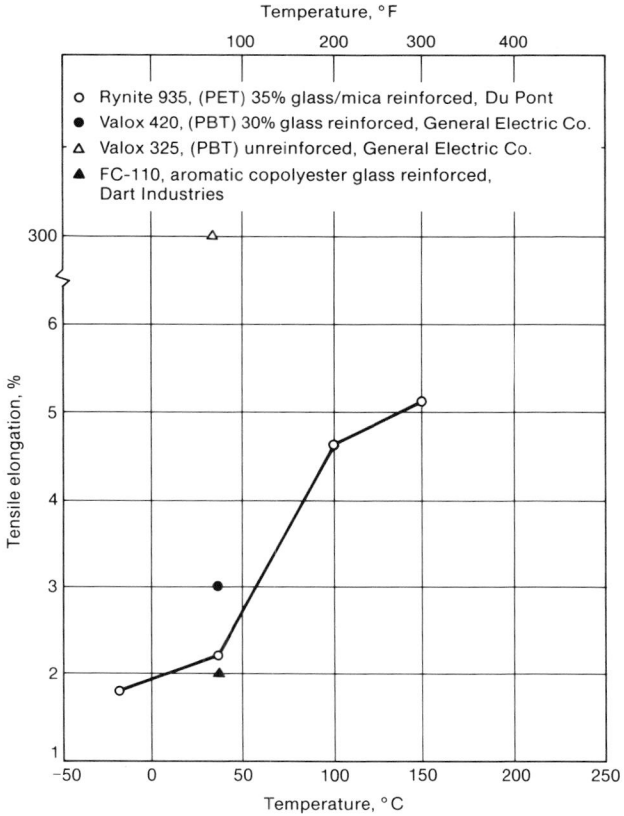

Fig. 3 Tensile elongation at break versus temperature

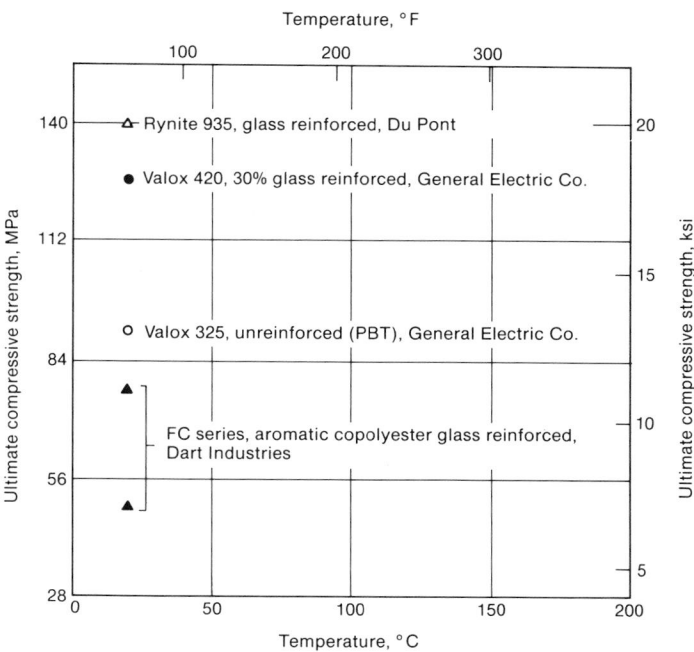

Fig. 4 Ultimate compressive strength versus temperature

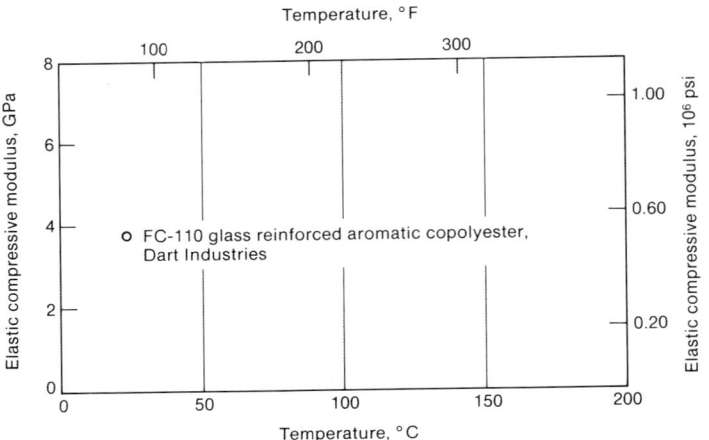

Fig. 5 Elastic compressive modulus versus temperature

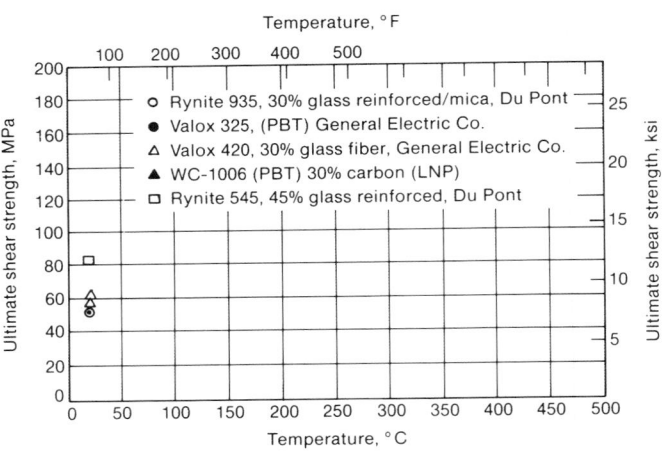

Fig. 6 Ultimate shear strength versus temperature

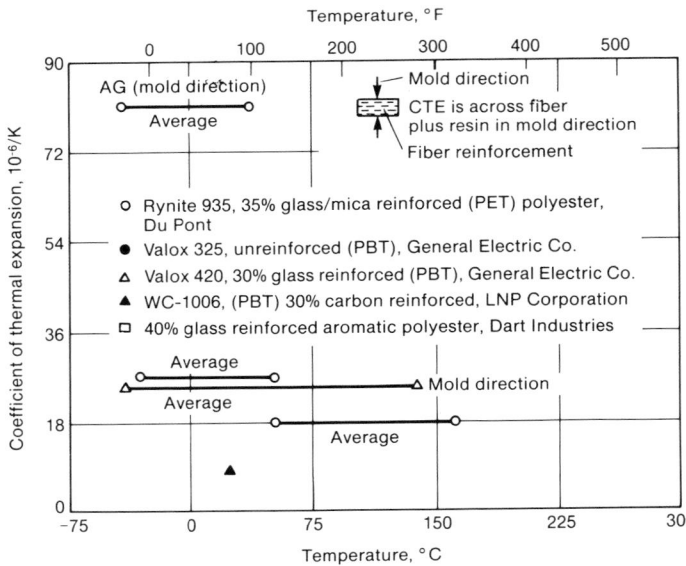

Fig. 7 Coefficient of thermal expansion versus temperature. $10^{-6}/K \times 5/9 =$ μin./in. $\times$ °F

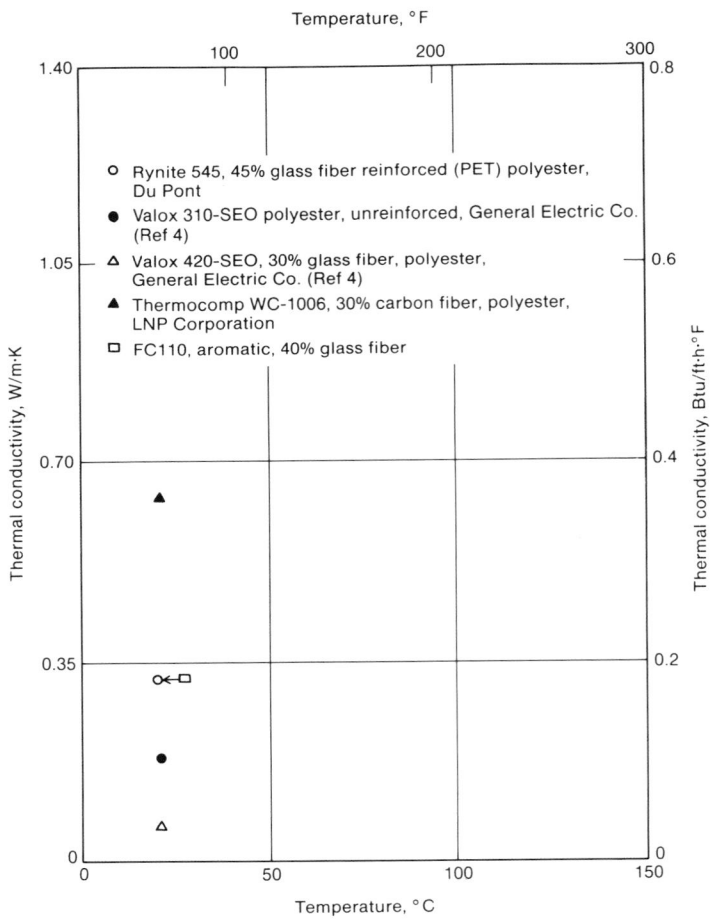

Fig. 8 Thermal conductivity versus temperature

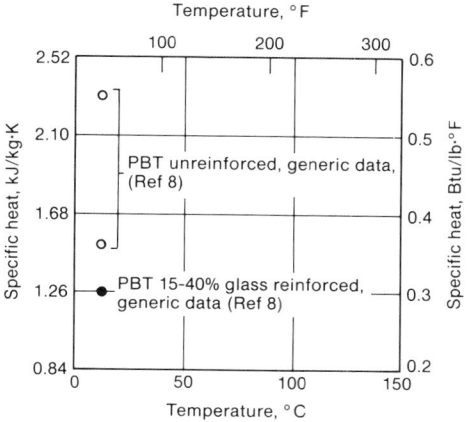

Fig. 9 Specific heat versus temperature

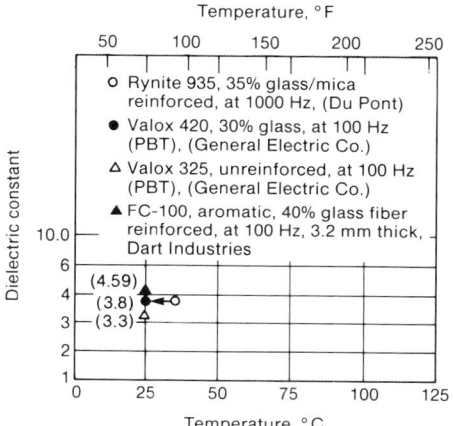

Fig. 10 Dielectric constant versus temperature

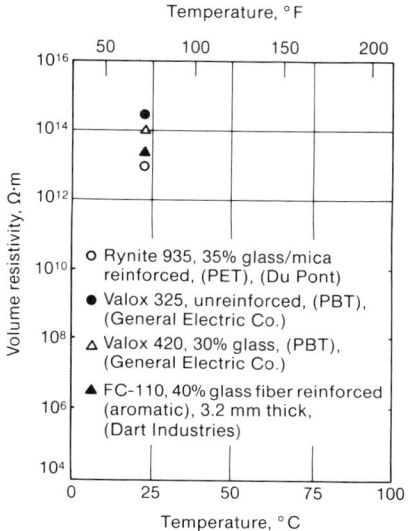

Fig. 11 Volume resistivity versus temperature

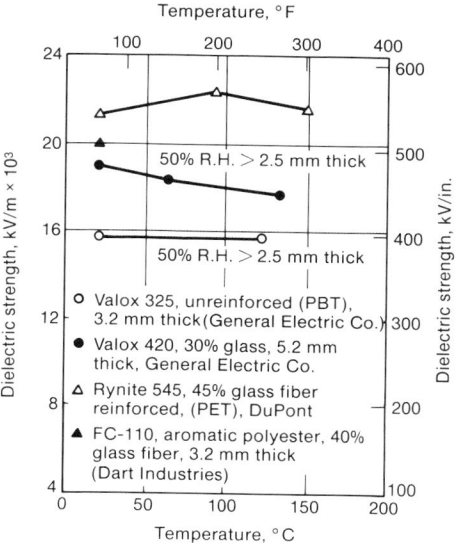

Fig. 12 Dielectric strength versus temperature

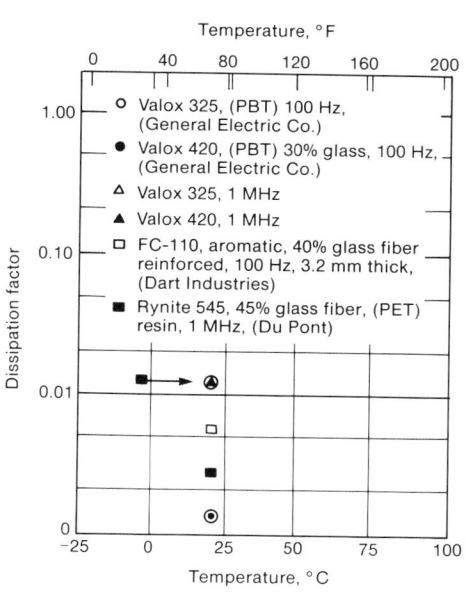

Fig. 13 Dissipation factor versus temperature

Polyamide Resin—Thermoplastic

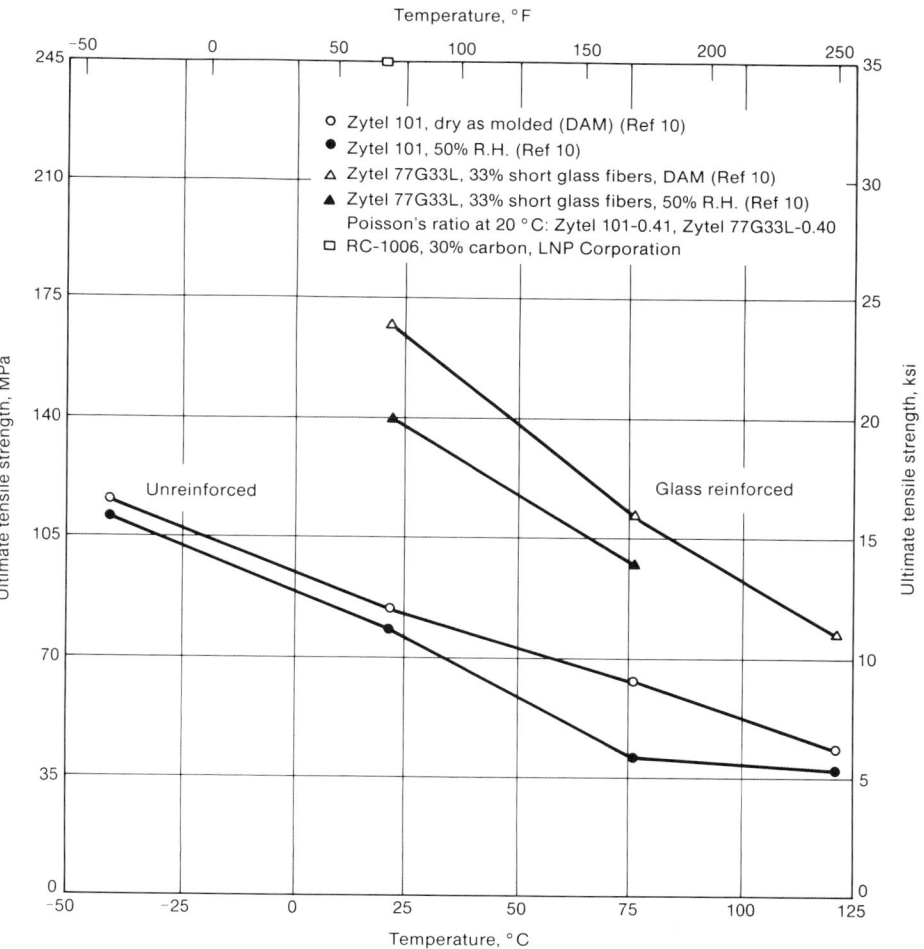

Fig. 14 Ultimate tensile strength versus temperature

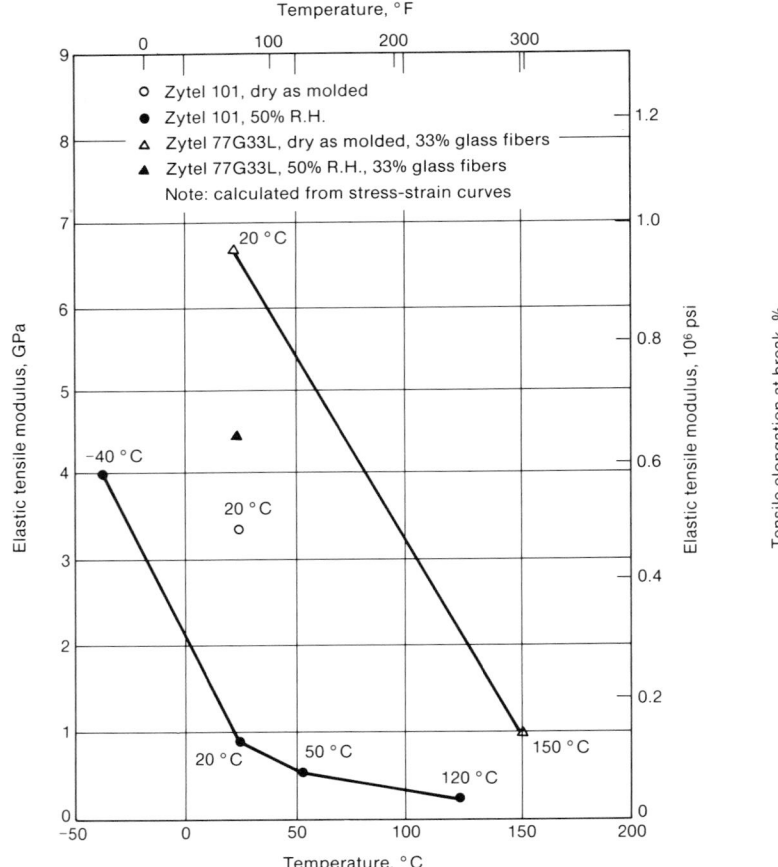

Fig. 15 Elastic tensile modulus versus temperature

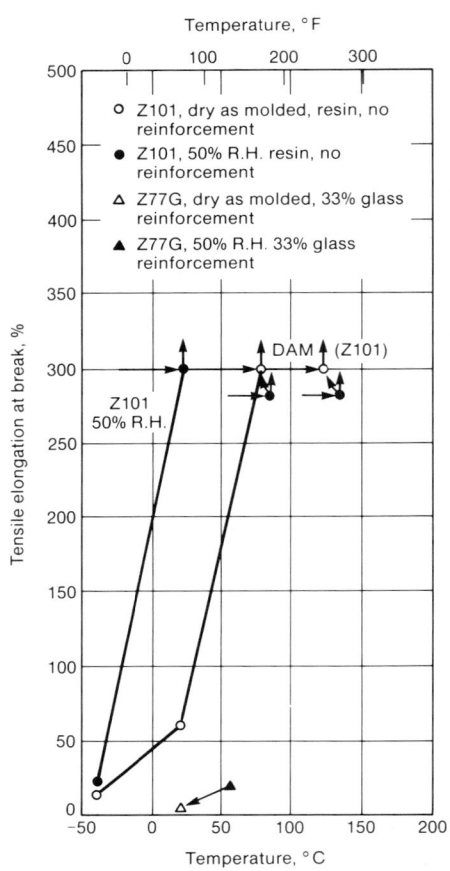

Fig. 16 Tensile elongation at break versus temperature

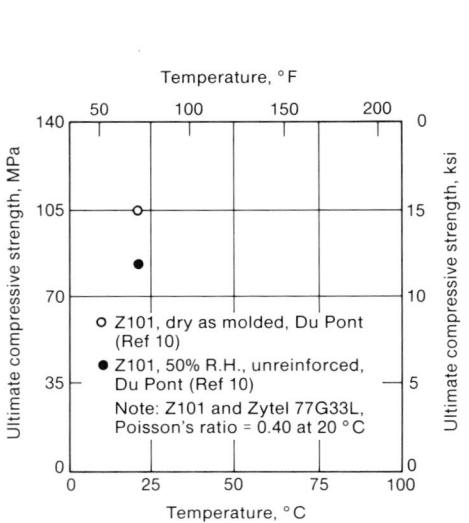

Fig. 17 Ultimate compressive strength versus temperature

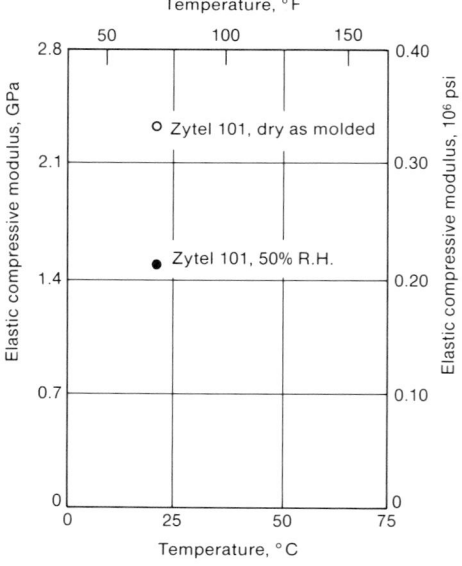

Fig. 18 Elastic compressive modulus versus temperature

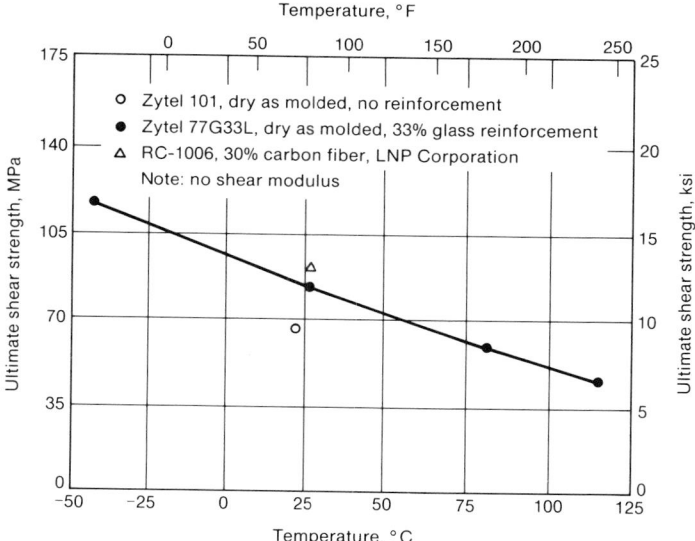

Fig. 19 Ultimate shear strength versus temperature

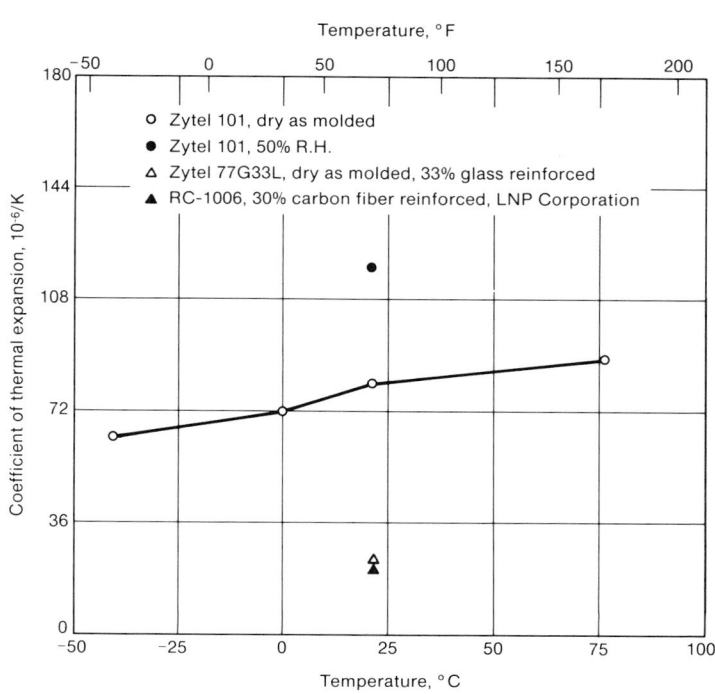

Fig. 20 Coefficient of thermal expansion versus temperature. $10^{-6}/\text{K} \times \frac{5}{9} = \mu\text{in./in.} \times \text{°F}$

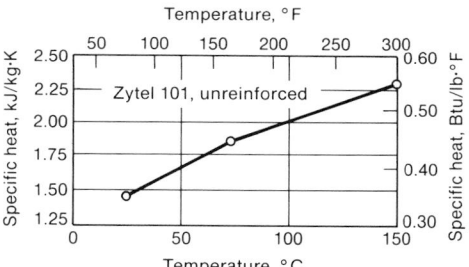

Fig. 22 Specific heat versus temperature

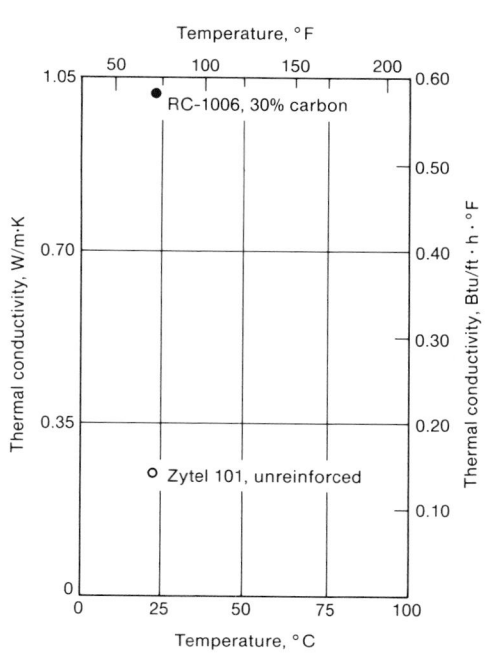

Fig. 21 Thermal conductivity versus temperature

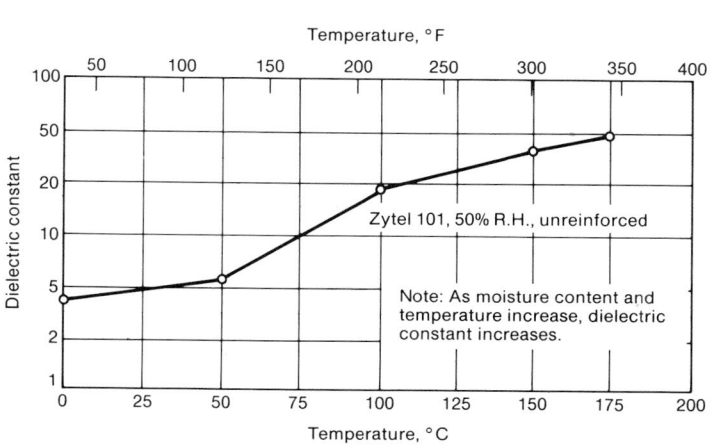

Fig. 23 Dielectric constant (100 Hz) versus temperature

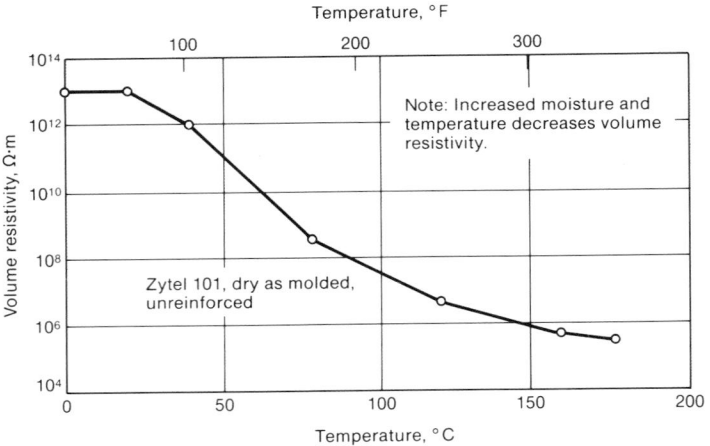

Fig. 24 Volume resistivity versus temperature

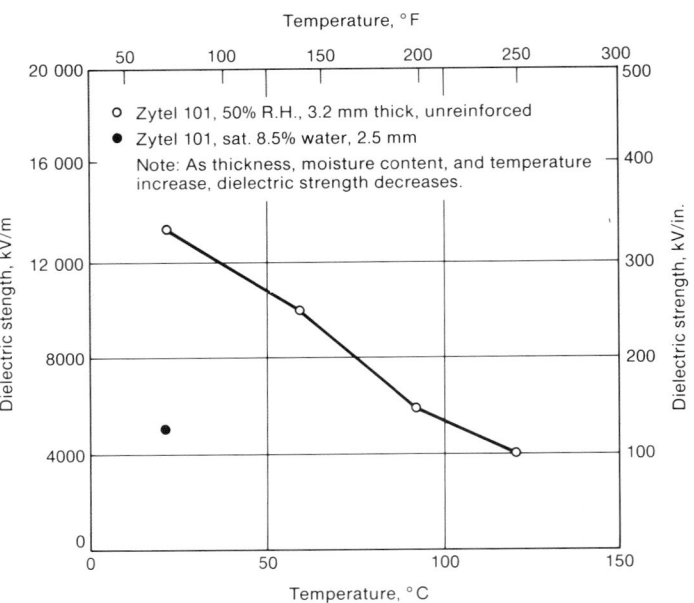

Fig. 25 Dielectric strength versus temperature

Polysulfone Resin—Thermoplastic

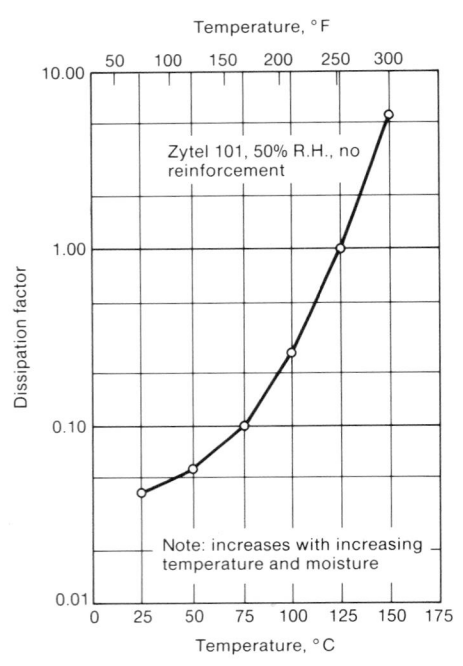

Fig. 26 Dissipation factor (100 Hz) versus temperature

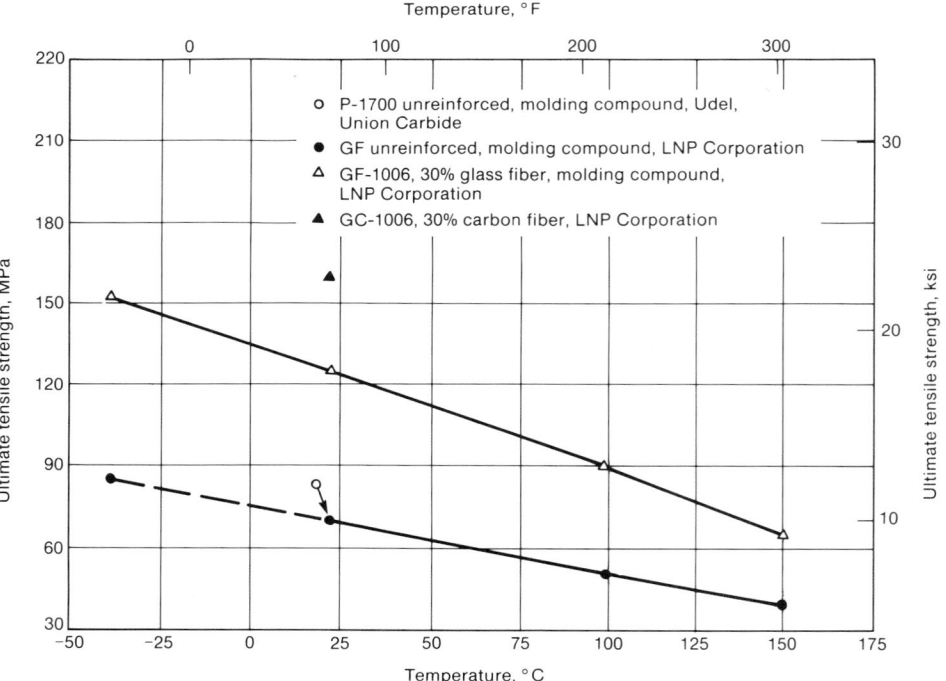

Fig. 27 Ultimate tensile strength versus temperature

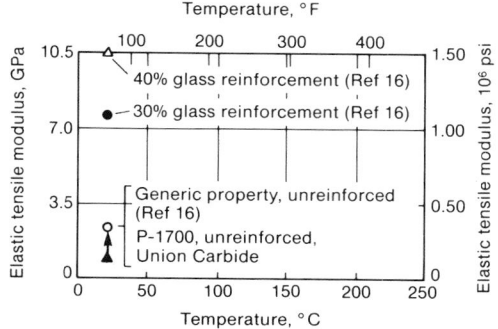

Fig. 28 Tensile elastic modulus versus temperature

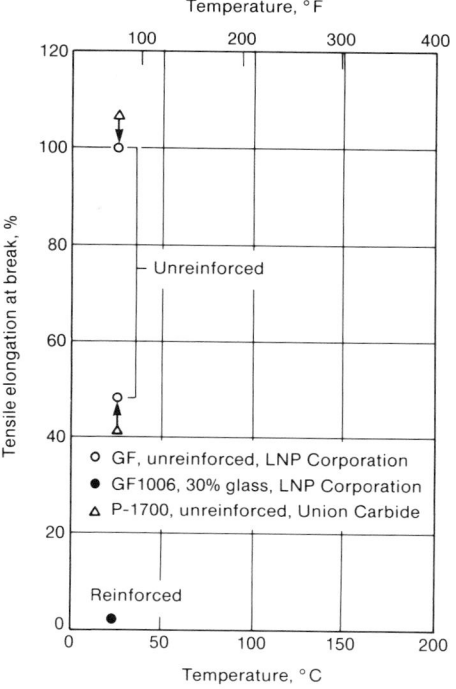

Fig. 29 Tensile elongation at break versus temperature

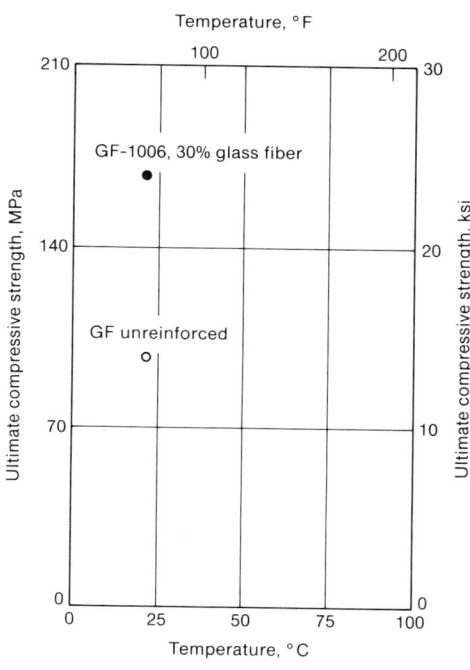

Fig. 30 Ultimate compressive strength versus temperature

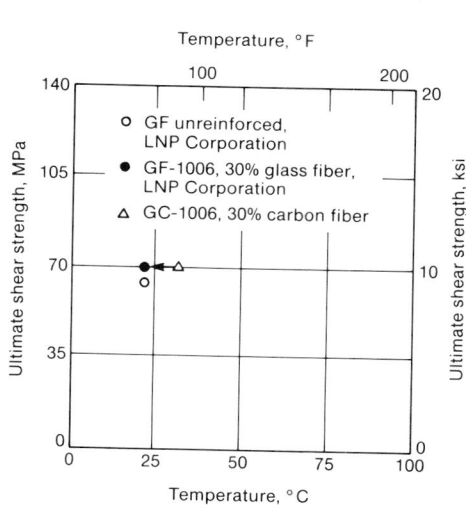

Fig. 31 Ultimate shear strength versus temperature

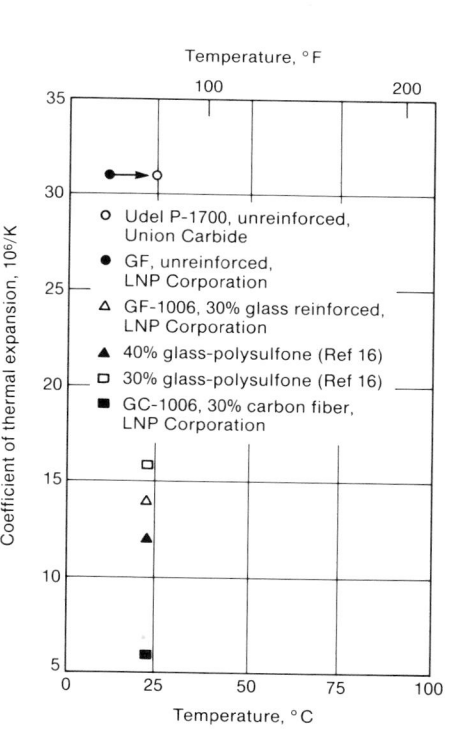

Fig. 32 Coefficient of thermal expansion versus temperature. $10^{-6}/K \times \frac{5}{9} = \mu in./in. \times °F$

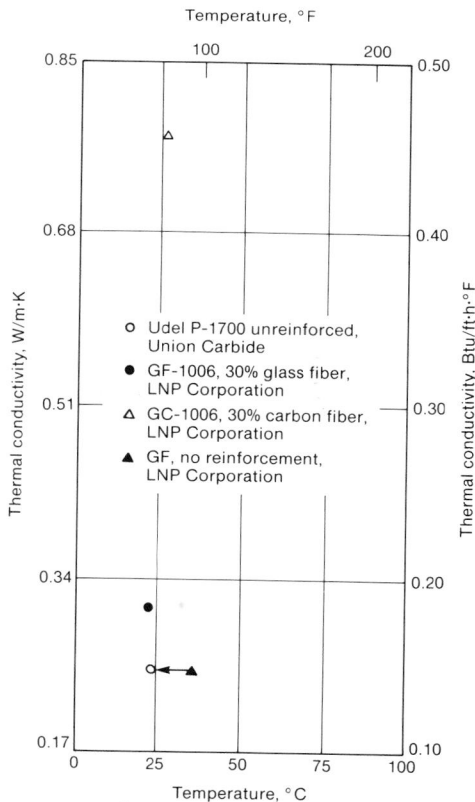

Fig. 33 Thermal conductivity versus temperature

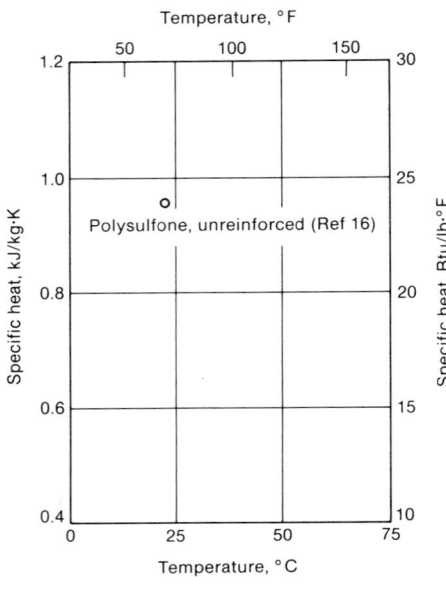

Fig. 34 Specific heat versus temperature

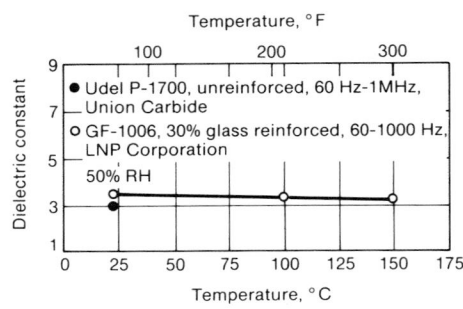

Fig. 35 Dielectric constant versus temperature

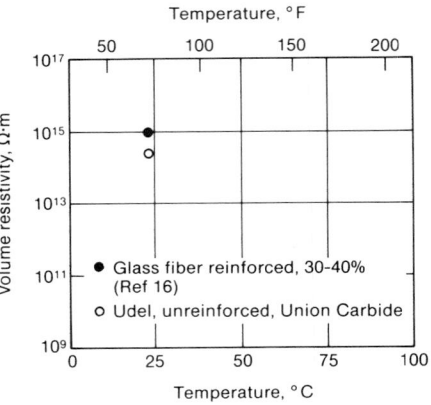

Fig. 36 Volume resistivity versus temperature

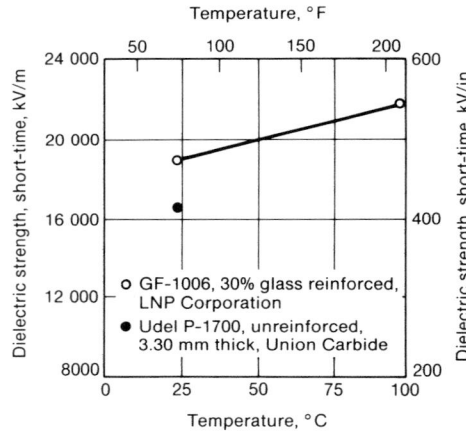

Fig. 37 Dielectric strength versus temperature

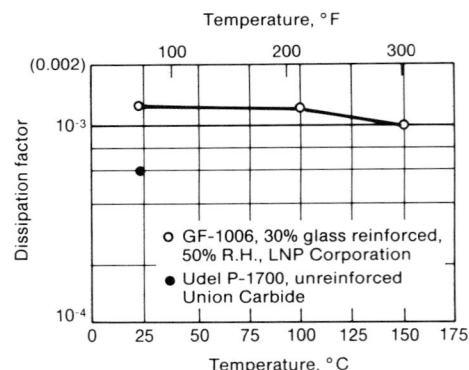

Fig. 38 Dissipation factor (60 Hz) versus temperature

High-Temperature Thermoset Matrix Composites

ASM Committee on Forms and Properties of Composite Materials*

Polyimide Thermoset Resin and Fiber-Resin Composites (Ref 1-10)

Polyimide thermoset resin reinforced with fillers and short fibers is used as molding compounds by DuPont, Fiberite, and Rhone-Poulenc. In addition, continuous fibers or fabric that is preimpregnated with DuPont resin polymer is used for lay-ups, filament winding, and tape wrapping. Resin polymer is popular for its high-temperature characteristics (315 °C, or 600 °F for long durations, and 480 °C, or 900 °F for short durations), low-temperature properties at −195 °C (−320 °F) for long duration, and low creep and deflection under load at temperature. In addition, its electrical properties are excellent. Properties information is given in Table 1. Major applications are in the aerospace and electrical industries.

Bismaleimide Thermoset Resin and Fiber-Resin Composites (Ref 11-14)

The bismaleimide thermoset resins, which may be cured at the lower epoxy cure temperature (175 °C, or 350 °F) and pressure (690 kPa, or 100 psi), do not have gaseous by-products, thus providing void-free parts. They also offer a new polymer system to combine with various fiber types for improved composite designs. Table 2 lists properties data. Suppliers of the resin system are HITCO/Bristol Composite Materials Division, British Petroleum North America Inc., Ciba-Geigy, and The Boots Company. Applications include aerospace wing skin ribs, helicopter firewalls, and printed wiring boards.

REFERENCES

1. 1985 Material Selector, *Mater. Eng.*, Dec 1984
2. 1986 Materials Reference Issue, *Mach. Des.*, April 1986
3. *Modern Plastics Encyclopedia*, Vol 62 (No. 10), McGraw Hill, 1985-1986

Table 1 Properties of polyimide thermoset resin and fiber-resin composites

Properties	Resin	50% glass fiber-resin
Physical		
Heat deflection temperature at 1820 kPa (264 psi), °C (°F)	305–360 (582–680)	350 (660)
UL rating, °C (°F)	... ...	... ...
In-service temperature, °C (°F)	260–370 (500–700)	260 (500)
Specific gravity	1.43	1.65
Density, g/cm³ (1b/in.³)	1.43 (0.052)	1.65 (0.060)
Mechanical		
Ultimate tensile strength	See Fig. 1	
Initial elastic tensile modulus	See Fig. 2	
Ultimate compressive strength	See Fig. 3	
Initial elastic compressive modulus	See Fig. 4	
Ultimate shear strength	See Fig. 5	
Thermal		
Coefficient of thermal expansion	See Fig. 6	
Thermal conductivity	See Fig. 7	
Specific heat	See Fig. 8	
Electrical		
Dielectric constant	See Fig. 9	
Volume resistivity	See Fig. 10	
Dielectric strength	See Fig. 11	
Dissipation factor	See Fig. 12	

Table 2 Properties of bismaleimide thermoset resin and fiber-resin composites

Properties	Neat resin	68.3 vol% T300 carbon fiber-resin	57.7 vol% E-glass fiber fabric-resin
Physical			
Specific gravity	1.23	1.60	2.0
Density (1b/in.³)	1.23 (0.044)	1.60 (0.058)	2.0 (0.072)
In-service temperature			
Short-term, °C (°F)	315 (600)	315 (600)	315 (600)
Long-term °C (°F)	230 (450)	230 (450)	230 (450)
Mechanical			
Ultimate tensile strength		See Fig. 13	
Elastic tensile modulus		See Fig. 14	
Ultimate compressive strength		See Fig. 15	
Elastic compressive modulus		See Fig. 16	
Ultimate shear strength		See Fig. 17	
Thermal			
Coefficient of thermal expansion		See Fig. 18	
Electrical			
None			

*Chairman: Richard C. Laramee, Morton Thiokol, Inc.; Richard G. Adams, J.P. Stevens & Company, Inc.; Roger Bacon, Michael J. Michno, Marvin E. Sauers, AMOCO Performance Products, Inc.; Donald Beckley, U.S. Polymeric Corporation; Paul Blanchard, General Electric Company; Robert Boudreau, Borden Chemical Division; Douglas L. Denton, Battelle—Columbus Division; L.E. DeShields, Paul R. Langston, E.I. Du Pont de Nemours & Company, Inc.; William B. Hall, University of Mississippi; Gary E. Hansen, Hercules Aerospace Company; Keith Jacobs, The Ironsides Company; John R. Koenig, Southern Research Institute; Ernest P. Rossa, Fiberite Corporation; John E. Theberge, LNP Corporation

4. "Thermid Polyimide Resin Product Sheets," National Starch and Chemical Corporation
5. "Vespel, Avimid Product Sheets," E.I. Du Pont de Nemours & Company, Inc.
6. "Polyimide Molding Compounds," Fiberite Corporation
7. "Kinel Polyimide Molding Compounds," Rhone-Poulenc Inc.
8. Astroquartz Polyimide, *Military Handbook 17A*, Part 1, Sept 1973
9. R.V. Wolff, and S.F. Monroe, Vacuum Cured Polyimide Laminate Properties, *SAMPE J.*, Aug-Sept 1969
10. W.T. Freeman and M.D. Campbell, Thermal Expansion Characteristics of Graphite Reinforced Composite Materials, *Composite Materials: Testing and Design (Second Conference)*, STP 497, American Society for Testing and Materials, 1972
11. "'Compimide,' A Family of Bismaleimide Resins," The Boots Company
12. M.S. Hsu, T.S. Chen, J.A. Parker, and A.H. Heimbuch, NASA New Bismaleimide Matrix Resins for Graphite Fiber Composites, *SAMPE J.*, July-Aug 1985
13. M. Chaudhari, T. Galvin, and J. King, "Characterization of Bismaleimide System XU-292," Ciba-Geigy; also, Matrimid 5292 System, *SAMPE J.*, July-Aug 1985
14. "V-378A Addition Cured Bismaleimide Matrix Resin," and "XV-388 Addition Cured Modified Bismaleimide Matrix Resin," Hitco

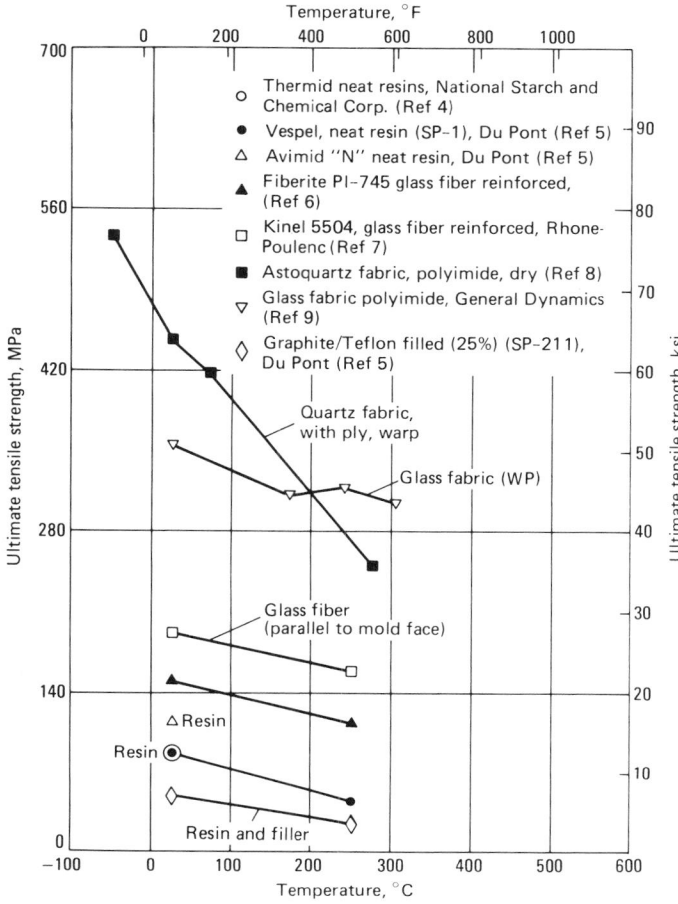

Fig. 1 Ultimate tensile strength versus temperature

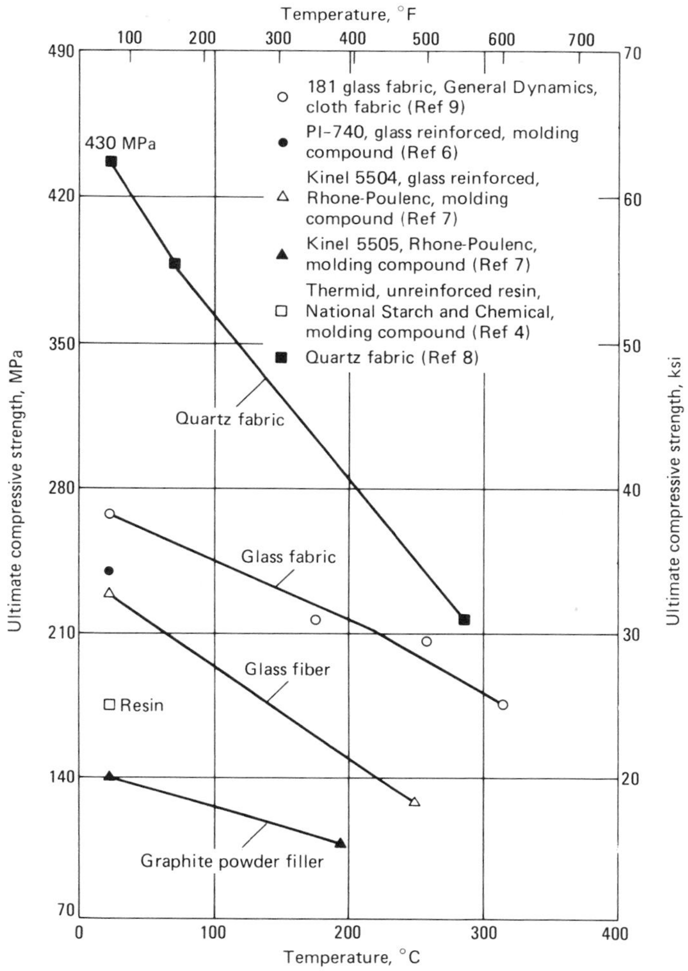

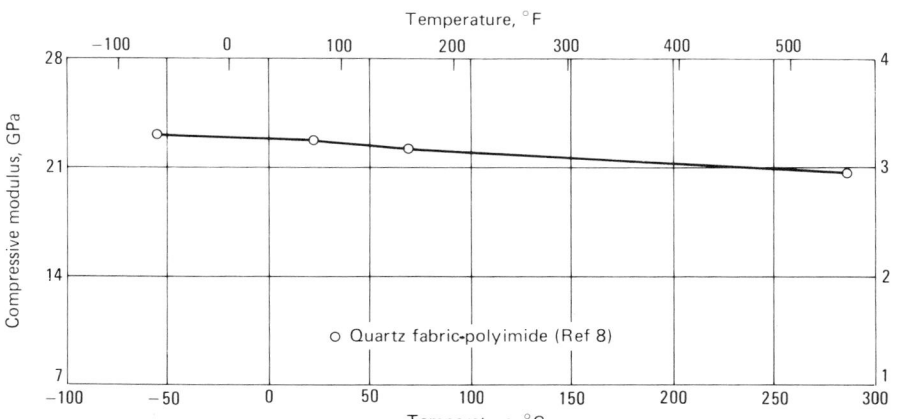

Fig. 3 Ultimate compressive strength versus temperature

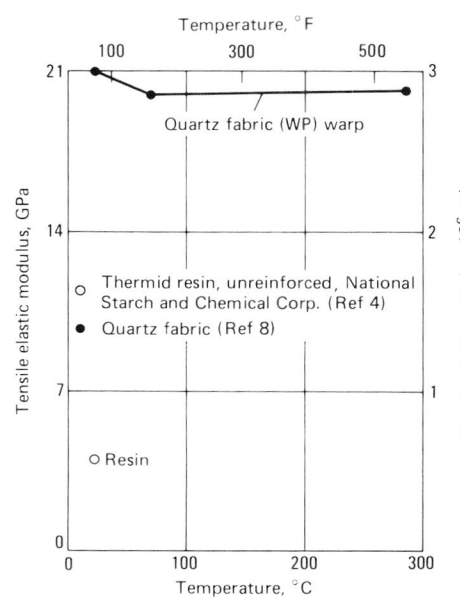

Fig. 2 Tensile elastic modulus versus temperature

Fig. 4 Initial elastic compressive modulus versus temperature

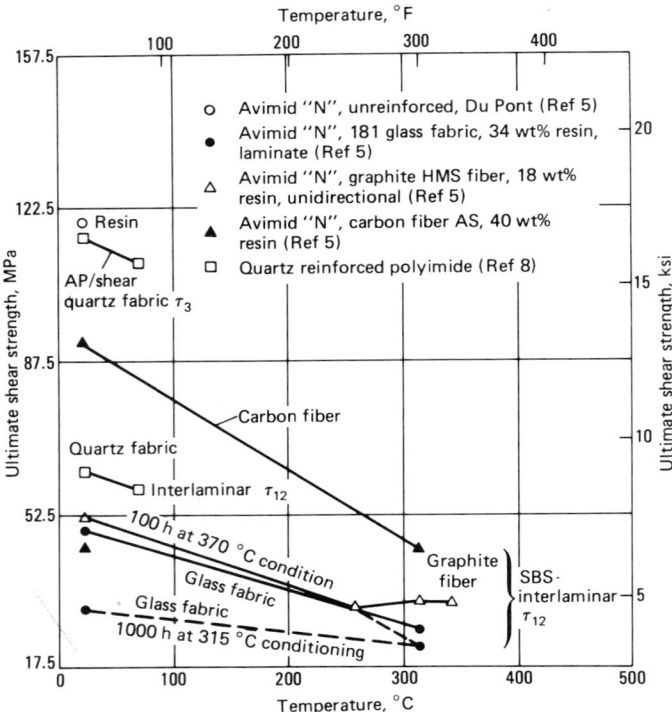

Fig. 5 Ultimate shear strength versus temperature

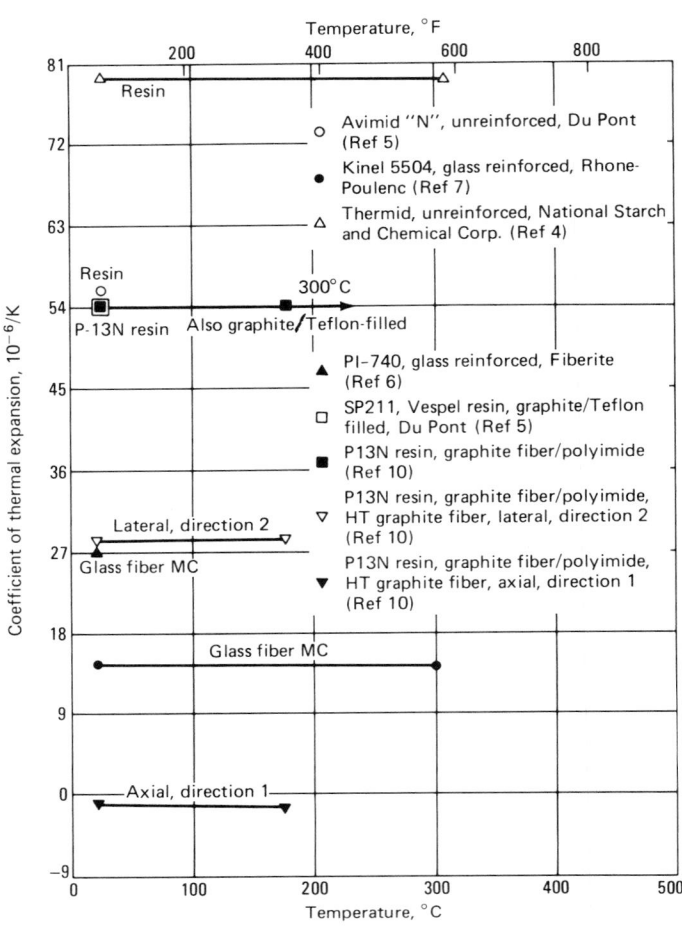

Fig. 6 Coefficient of thermal expansion versus temperature. $10^{-6}/K \times \frac{5}{9} = \mu in./in. \times °F$

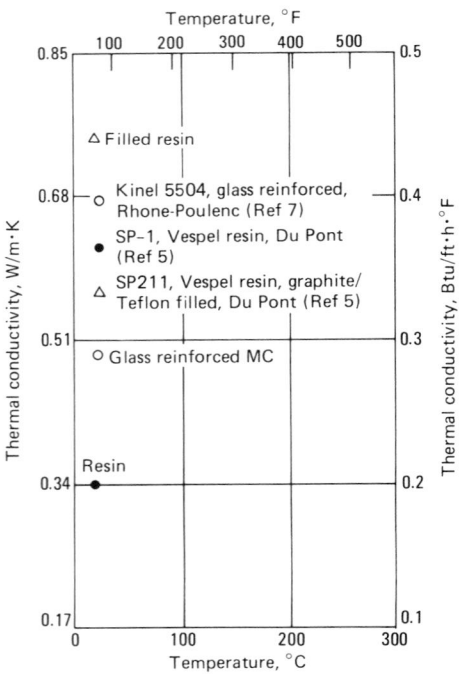

Fig. 7 Thermal conductivity versus temperature

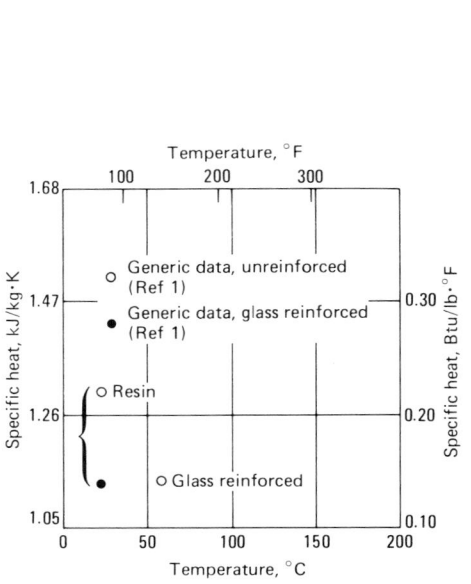

Fig. 8 Specific heat versus temperature

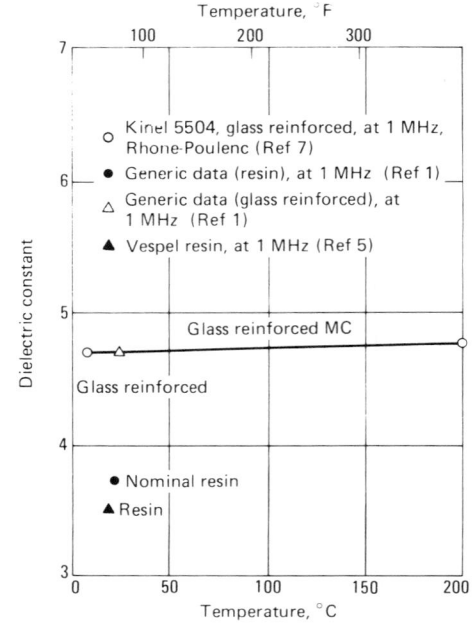

Fig. 9 Dielectric constant versus temperature

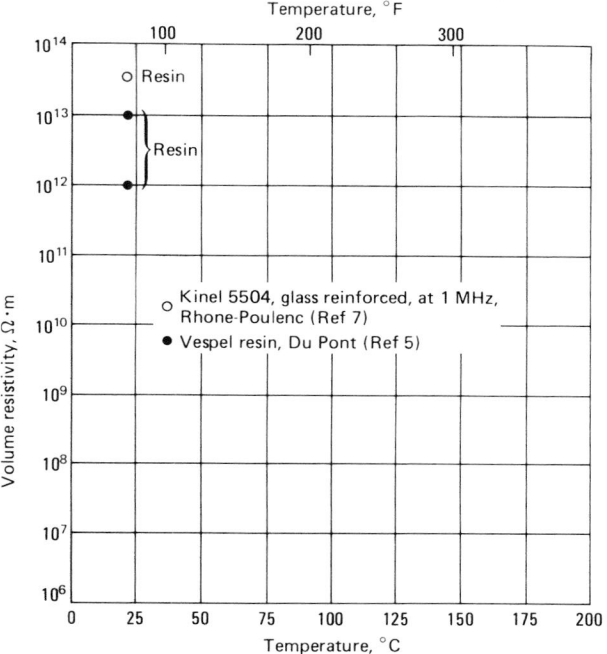

Fig. 10 Volume resistivity versus temperature

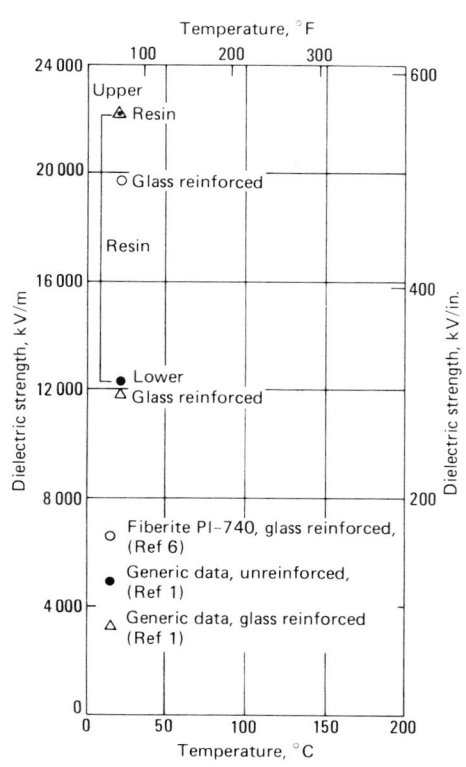

Fig. 11 Dielectric strength versus temperature

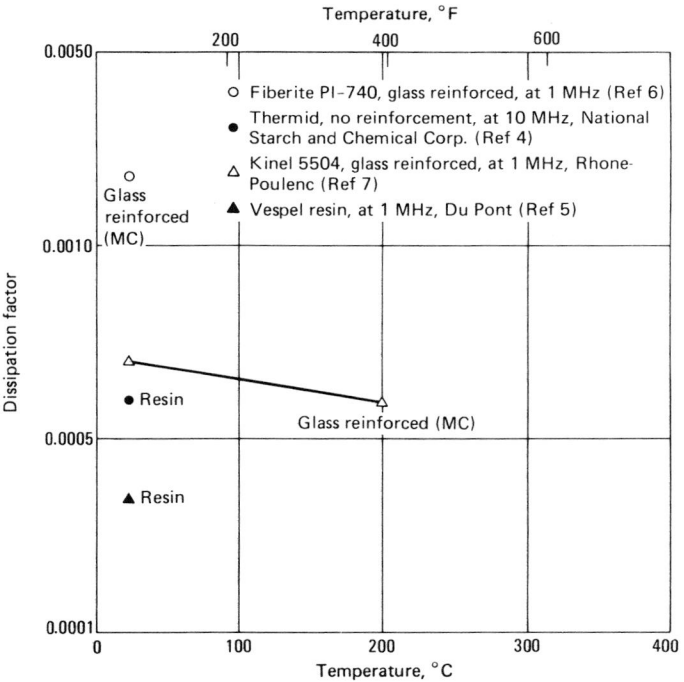

Fig. 12 Dissipation factor versus temperature

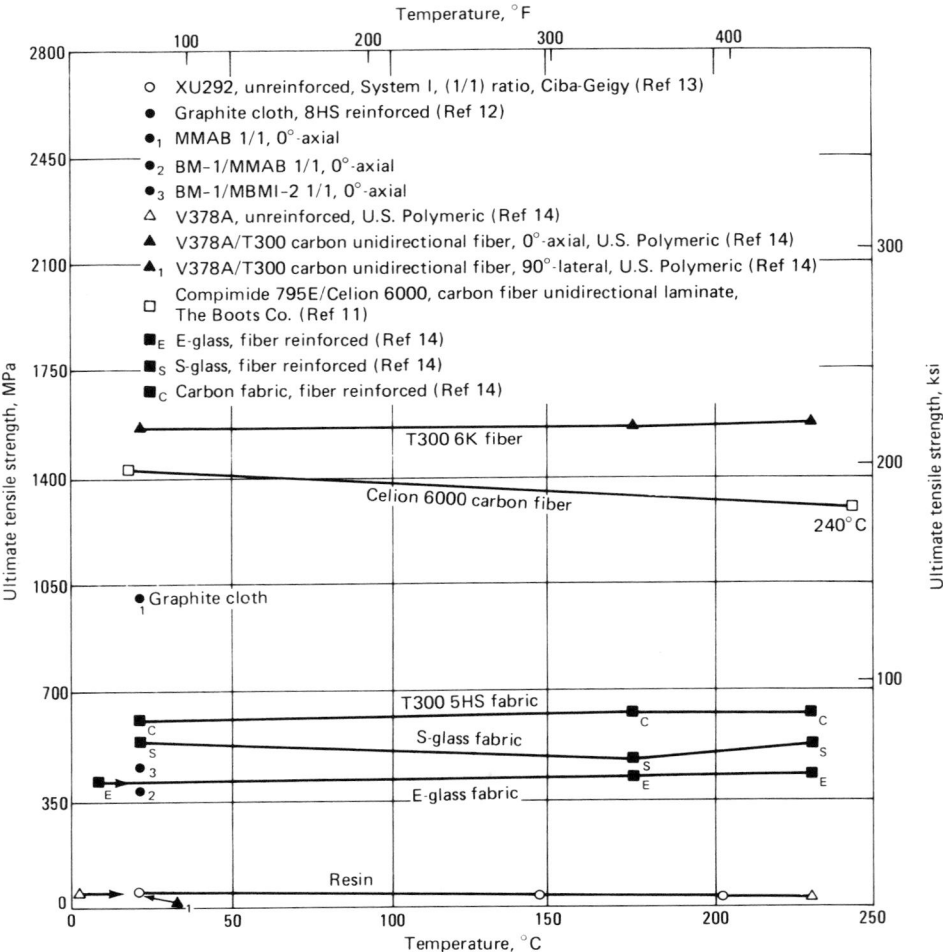

Fig. 13 Ultimate tensile strength versus temperature

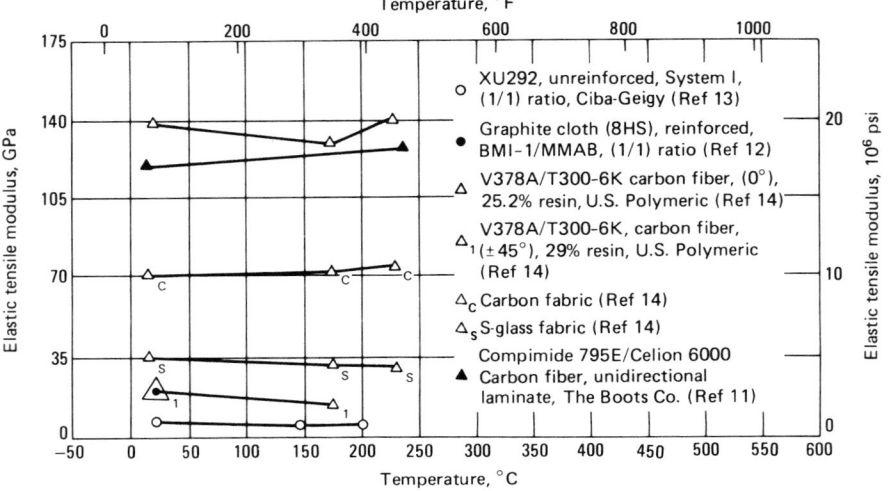

Fig. 14 Elastic tensile modulus versus temperature

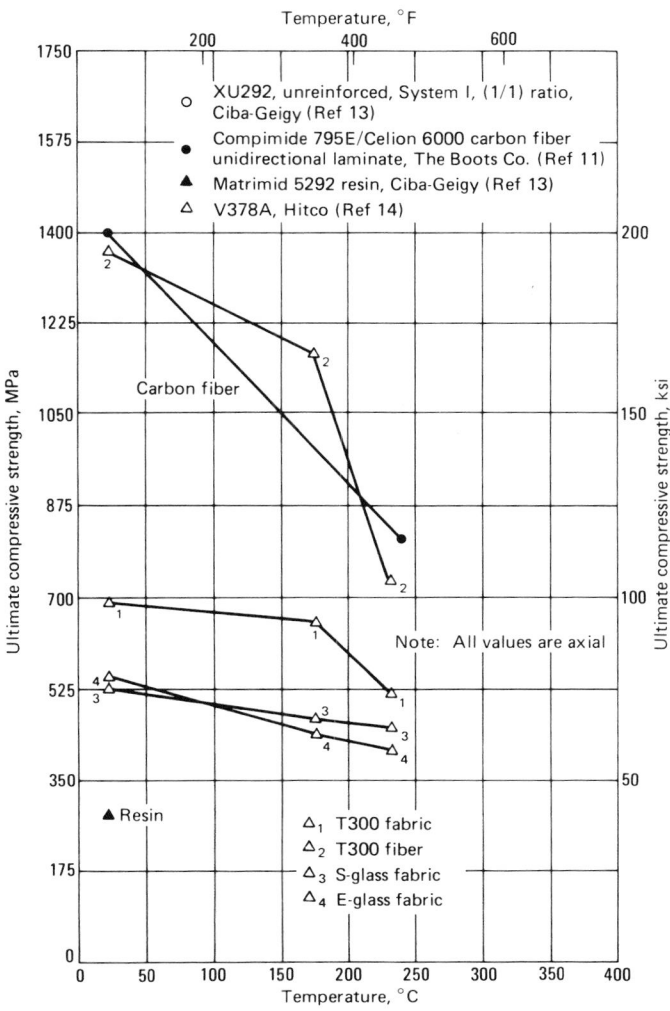

Fig. 15 Ultimate compressive strength versus temperature

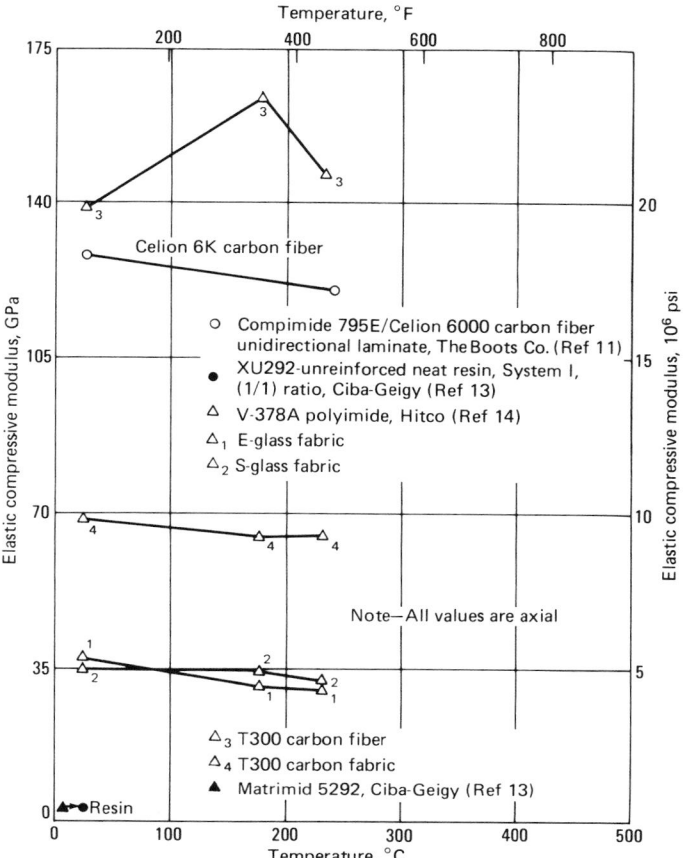

Fig. 16 Elastic compressive modulus versus temperature

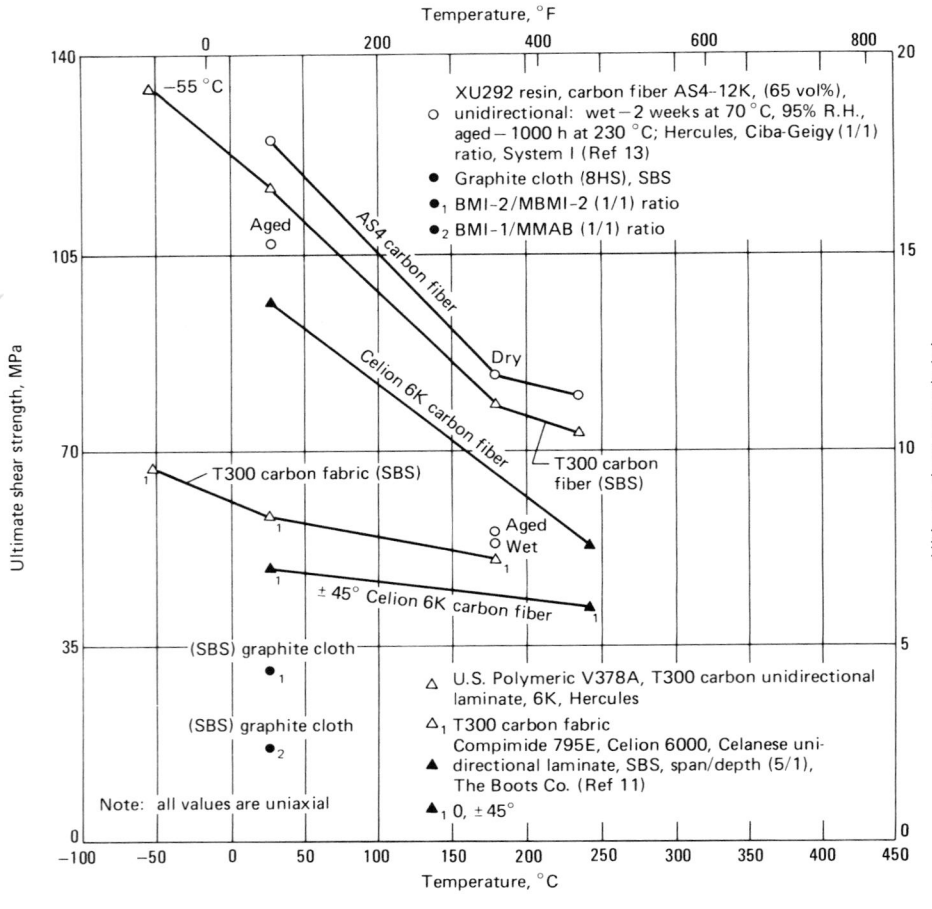

Fig. 17 Ultimate shear strength versus temperature

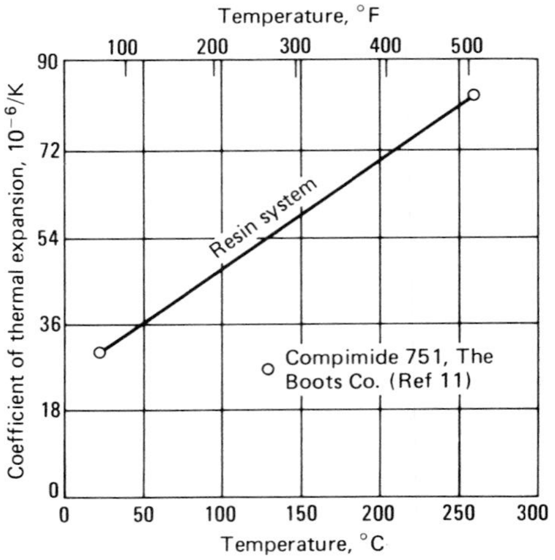

Temperature, °F

XU292 resin, carbon fiber AS4–12K, (65 vol%),
○ unidirectional: wet – 2 weeks at 70 °C, 95% R.H.,
aged – 1000 h at 230 °C; Hercules, Ciba-Geigy (1/1)
ratio, System I (Ref 13)
● Graphite cloth (8HS), SBS
●₁ BMI-2/MBMI-2 (1/1) ratio
●₂ BMI-1/MMAB (1/1) ratio

△ U.S. Polymeric V378A, T300 carbon unidirectional
laminate, 6K, Hercules
△₁ T300 carbon fabric
Compimide 795E, Celion 6000, Celanese uni-
▲ directional laminate, SBS, span/depth (5/1),
The Boots Co. (Ref 11)
▲₁ 0, ±45°

Note: all values are uniaxial

Fig. 18 Coefficient of thermal expansion versus temperature. $10^{-6}/K \times \frac{5}{9} = \mu in./in. \times °F$

Medium-Temperature Thermoset Matrix Composites

ASM Committee on Forms and Properties of Composite Materials*

PHENOLIC RESIN SYSTEMS are formulated from the reaction between phenol and formaldehyde. The two main resin types are resoles and novolacs. Two-stage phenolic resins (novolacs) are used for general-purpose molding compounds, while hybrids of the novolacs are used as impregnating resins with glass, carbon, and graphite cloth for tape wrapping or hand lay-up of aerospace components, rocket nozzle ablative, and insulation liners. Chopped-fiber molding compounds are used mostly in the automotive, appliance, and electrical component markets. General characteristics of these materials that make them suited for the above applications are high service temperatures, good electrical properties, excellent moldability and dimensional stability, and good moisture resistance.

The following sections of this article describe phenolic-resin molding compounds, with or without chopped glass fiber, and three types of tapes for tape wrapping and lay-ups, namely, glass fabric-phenolic tape, carbon fabric-phenolic tape, and graphite fabric-phenolic tape.

Table 1, a properties data sheet, gives specific mechanical, thermal, and electrical properties for each type of phenolic resin thermoset. Physical properties are identified in the table, but some test data and references are described in the following sections.

Phenolic Resin, With or Without Reinforcement Fibers. Phenolic resin thermoset characterization includes unfilled resin and filled resin systems. The fillers include glass fiber, wood flock, aluminum powder, rubber, cellulose fabric, minerals, cotton flock, and polyamide fiber.

Glass fabric reinforced phenolic resin is provided by at least four prepreg suppliers: U.S. Polymeric, Fiberite, Ferro, and Hexcel.

Table 1 Phenolic glass fiber-resin and fiber-resin properties

Physical and service properties	Phenolic resin	Molding compound glass fiber-resin	E-glass fabric-resin	Carbon fabric-resin	Graphite fabric-resin
Physical					
Specific gravity	1.28	1.95	1.90	1.45	1.42
Density, g/cm³(lb/in.³)	1.28 (0.046)	1.95 (0.070)	1.90 (0.069)	1.45 (0.052)	1.42 (0.051)
Service temperature, °C (°F)	150-230 (300-450)	150 (300)	538 (1000)	3038 (5500)	3038 (5500)
Heat deflection temperature at 1820 MPa (264 psi), °C (°F)	120-315 (250-600)	na	na	na	na
Flammability (U.L.)	94 V-1	94 V-0	94 V-0	...	...
Thermogravimetric analysis— stability temperature at 5-10% wt loss, °C (°F)	230 (450)	na	...	...	...
Reinforcement, wt%		35-45	63	55	55

	Phenolic with/without glass fiber		Glass fabric phenolic tape	Carbon fabric-phenolic tape	Graphite fabric-phenolic tape
Mechanical					
Ultimate tensile strength	See Fig. 1		See Fig. 14	See Fig. 26	See Fig. 35
Tensile modulus	See Fig. 2		See Fig. 15	See Fig. 27	See Fig. 36
Tensile strain to failure	See Fig. 3		See Fig. 16	...	na
Tensile Poisson's ratio	na		See Fig. 17	...	...
Ultimate compressive strength	See Fig. 4		See Fig. 18	See Fig. 28	See Fig. 37
Compressive modulus	na		See Fig. 19	See Fig. 29	See Fig. 38
Ultimate shear strength	na		See Fig. 20	See Fig. 30	See Fig. 39
Shear modulus	See Fig. 5		na	...	...
Thermal					
Coefficient of thermal expansion (ap) slow heating	See Fig. 6		See Fig. 21	See Fig. 31	See Fig. 40
Coefficient of thermal expansion (ap) high heating	...		...	...	...
Coefficient of thermal expansion (wp)	...		See Fig. 21	See Fig. 32	See Fig. 41
Specific heat	See Fig. 7		See Fig. 22	See Fig. 33	See Fig. 42
Thermal conductivity	See Fig. 8		See Fig. 23	See Fig. 34	See Fig. 43
Thermogravimetric analysis	See Fig. 9		...	na	na
Electrical					
Dielectric constant	See Fig. 10		See Fig. 24	na	na
Volume resistivity	See Fig. 11		na	na	na
Dielectric strength	See Fig. 12		na	na	na
Dissipation factor	See Fig. 13		See Fig. 25	na	na

na, not applicable

*Chairman: Richard C. Laramee, Morton Thiokol, Inc.; Richard G. Adams, J. P. Stevens & Company, Inc.; Roger Bacon, Michael J. Michno, Marvin E. Sauers, Amoco Performance Products, Inc.; Donald Beckley, U.S. Polymeric Corporation; Paul Blanchard, General Electric Company; Robert Boudreau, Borden Chemical Division; Douglas L. Denton, Battelle—Columbus Division; L. E. DeShields, Paul R. Langston, E. I. Du Pont de Nemours & Company, Inc.; William B. Hall, University of Mississippi; Gary E. Hansen, Hercules Aerospace Company; Keith Jacobs, The Ironsides Company; John R. Koenig, Southern Research Institute; Ernest P. Rossa, Fiberite Corporation; John E. Theberge, LNP Corporation; Steven Witschen, Owens-Corning Fiberglas Corporation

The glass fabric is supplied by weaving manufacturers such as Burlington Mills, Clark-Schwebel, and J.P. Stevens. The glass fiber is supplied by Owens-Corning or PPG Industries. Phenolic resin-based polymers are provided by the Borden Chemical Division and Ironsides Company. The material is used for lay-ups, tape wrapping, and molding compounds in the chopped cloth (12.7-mm × 12.7-mm, or ½-in. × ½-in.) version.

Carbon fabric reinforced phenolic resin is provided as FM5055 by U.S. Polymeric, and as MX4926 by Fiberite. The base fiber is a special variation of high-tenacity rayon tire cord reinforcement from Avtex. Other reinforcements are pitch fibers (Amoco-Union Carbide) and polyacrylonitrile (PAN) fibers (Amoco-Union Carbide, Hercules, and BASF-Celion). The base fiber is woven into plain, leno, and satin weaves by Burlington Mills and subsequently carbonized at below 1650 °C (3000 °F) by Polycarbon, Hitco, and Amoco-Union Carbide. Polyacrylonitrile fibers are processed in much the same way as is rayon, but in the base process the pitch fibers are carbon, which need no further carbonization after cloth weaving. The phenolic resin-based polymers are supplied by the Borden Chemical Division and Ironsides Company. The material is supplied either as a broad goods roll that is 965 to 1200 mm (38 to 48 in.) wide and approximately 20 to 45 kg (50 to 100 lb) per roll, or as a slit tape roll width. The raw material is used for hand lay-ups, tape wrapping, and molding compounds in the chopped cloth (12.7-mm × 12.7-mm, or ½-in. × ½-in.) version.

The tensile, compressive, and shear (strength and modulus) mechanical properties decrease in value with an increase in temperature.

The conductivity and specific heat thermal properties increase with a temperature increase. The coefficient of thermal expansion (with ply) has the same characteristics versus temperature, except for a decrease around 540 °C (1000 °F) due to the phenolic weight loss and conversion to a carbon char. The CTE (across ply), directly related to the phenolic resin, increases up to 400 °C (750 °F), then decreases to 815 °C (1500 °F), then increases again to 1926 °C (3500 °F).

No electrical properties are currently available because carbon fabric phenolic is too highly conductive for most electrical industry applications.

Graphite fabric reinforced phenolic resin is manufactured as FM5014 and FM5064 by U.S. Polymeric, and as MXG-175 and MX2630A by Fiberite. The fiber and resin manufacture is the same as the carbon fabric-phenolic resin, except that the rayon-based fabric is graphitized at over 2205 °C (4000 °F). All suppliers remain the same as well as the form and process use of the broad goods material. Graphitization at a higher temperature stabilizes the fabric from further thermal shrinkage at application temperatures of 3315 °C (6000 °F) in aerospace components and rocket nozzles, increases thermal conductivity, and lowers the strength levels.

A description of mechanical, thermal, and electrical properties would be much the same as it is for carbon fabric-phenolic resin, above.

REFERENCES

1. *Modern Plastics Encyclopedia*, Vol 62 (No. 10A), McGraw-Hill, 1985-1986
2. 1986 Materials Reference Issue, April 1986, *Mach. Des.*, Penton
3. Resinox SC1008 Phenolic Product Data Sheets, Plastics Division, Monsanto Chemical Company
4. Materials Selector 1985, *Mater. Eng.*, Penton, Dec 1984
5. "Reinforced Molding Compounds," Fiberite Corporation
6. "Plenco Molding Compounds," Plastics Engineering Company
7. *1970 Guide to Plastics, Modern Plastics Encyclopedia*
8. G. Lubin, Ed., *Handbook of Fiberglass and Advanced Plastics Composites*, Van Nostrand Reinhold, 1969
9. K. Boller, "Tensile and Compressive Strength of Reinforced Plastic Laminates After Rapid Heating," Wright Air Development Division Report WADD-TR-60-804, Aug 1960
10. F.R. O'Brien and S. Oglesby, Jr., "Investigation of Thermal Properties of Plastic Laminates," Wright Air Development Center Report WADD-TR-54-306, 1955
11. *Military Handbook 17, Plastics for Flight Vehicles*, Department of Defense, June 1955
12. *Military Handbook 17A, Plastics for Aerospace Vehicles*, Department of Defense, Jan 1971
13. K. Boller, "Strength Properties of Reinforced Plastic Laminates at Elevated Temperatures," Wright Air Development Center Report WADD-TR-59-569, July 1959
14. "Mechanical Properties of MXB-6001 Phenolic Resin Impregnated Glass Cloth," Fiberite Corporation, May 1965
15. "Mechanical Properties of FM-5042 Glass Fabric Reinforced Prepreg Employing a MIL-R-92299 Resin," U.S. Polymeric Chemical Inc., June 1964
16. R.R. Barnet, "Evaluation of Glass Fabric Reinforced Plastic Laminates," Navord Report 2669, U.S. Naval Ordnance Laboratory, Jan 1953
17. L. Holliday, Ed., *Composite Materials*, Elsevier, 1966
18. Research and Development Laboratory Report LWR 278149, Morton Thiokol, Inc., Aug 1977
19. "Elevated Temperature Properties of Carbonaceous and Silica Fabric Reinforced Phenolic Composites," Table III, AGC MF-540, Aerojet General Company, 1964
20. "Various Mechanical and Thermal Properties of Carb-I-Tex and Carbon Phenolic," SoRI, for Lockheed Propulsion Laboratories, 1971
21. "Mechanical Properties Test Data Package," FM5064, FM5055, FM5014, U.S. Polymeric
22. "Subscale Ring Shearout Load Data," Report TWR-9071, Morton Thiokol, Inc., 1975
23. "Subscale Ring Shearout Test," Report TWR-20393, Morton Thiokol, Inc., 1977
24. Data Sheet, MXG175, MX4926, Fiberite Corporation

Phenolic Resin and Fiber Reinforcement

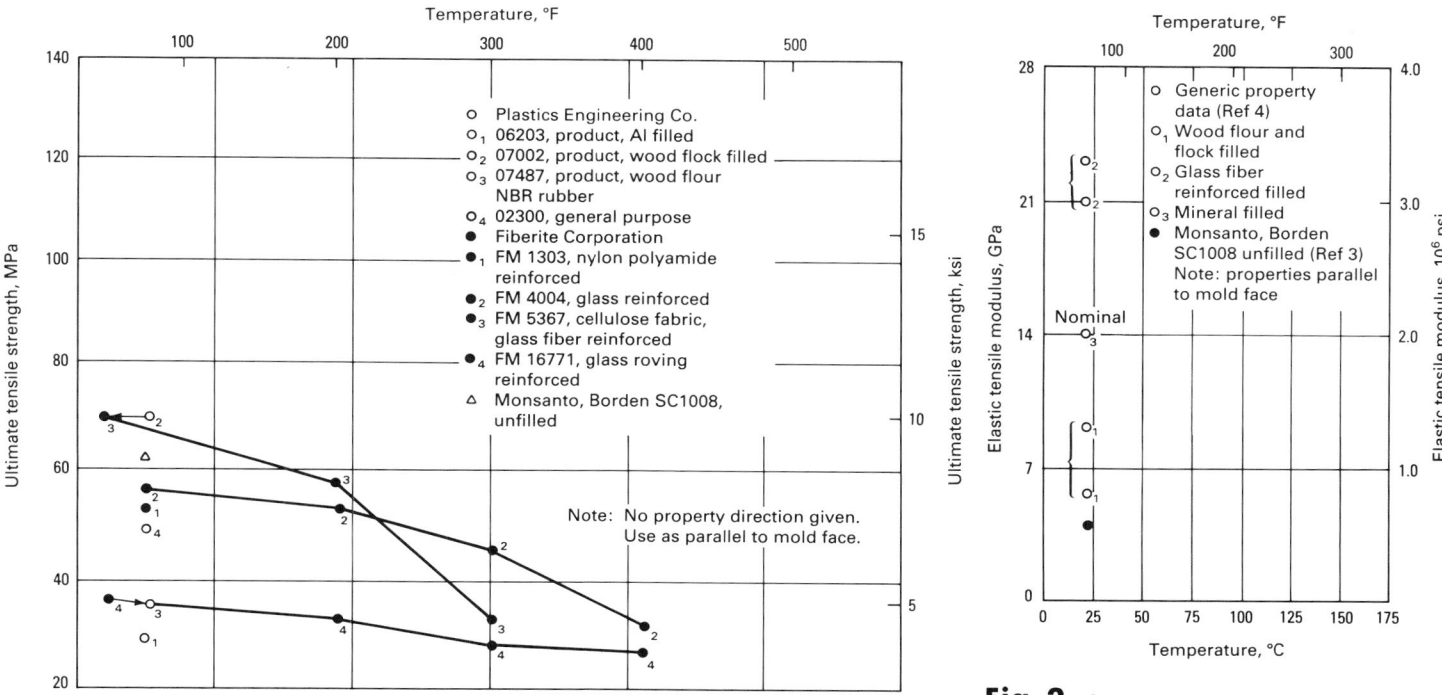

Fig. 1 Ultimate tensile strength versus temperature

Fig. 2 Elastic tensile modulus versus temperature

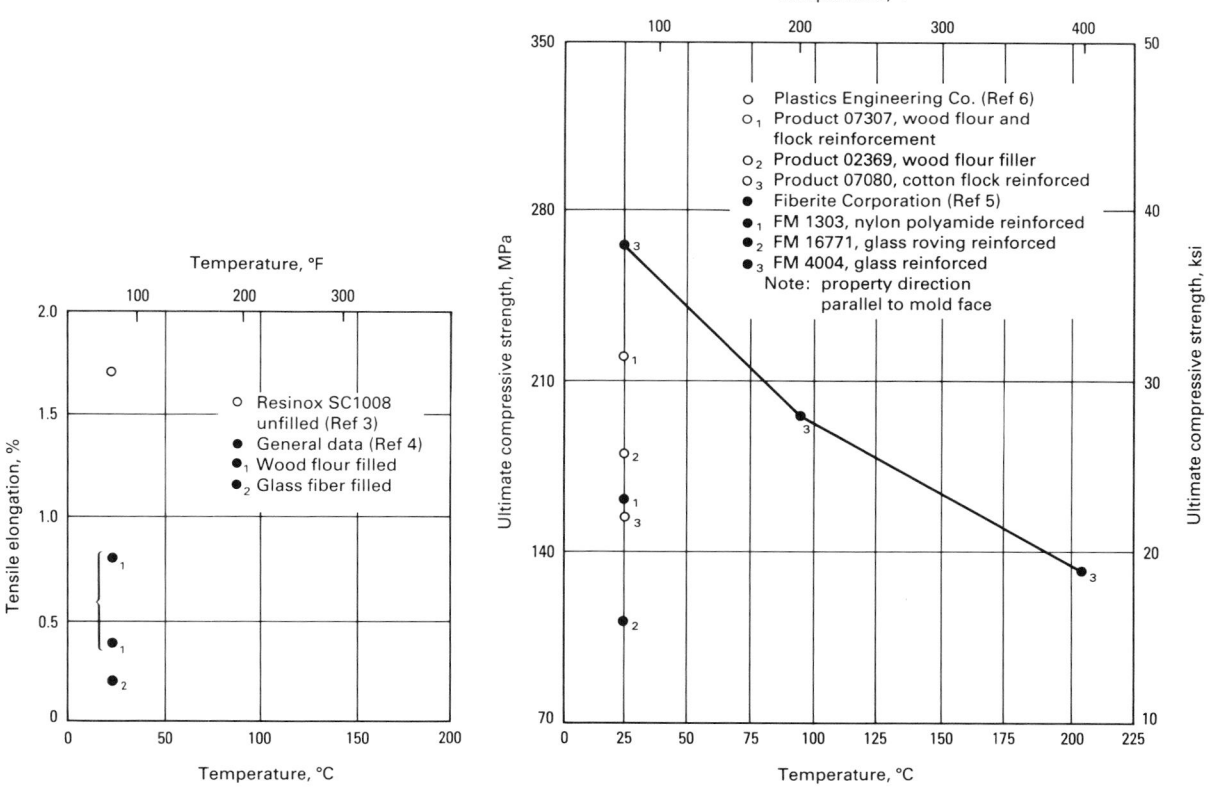

Fig. 3 Tensile elongation versus temperature

Fig. 4 Ultimate compressive strength versus temperature

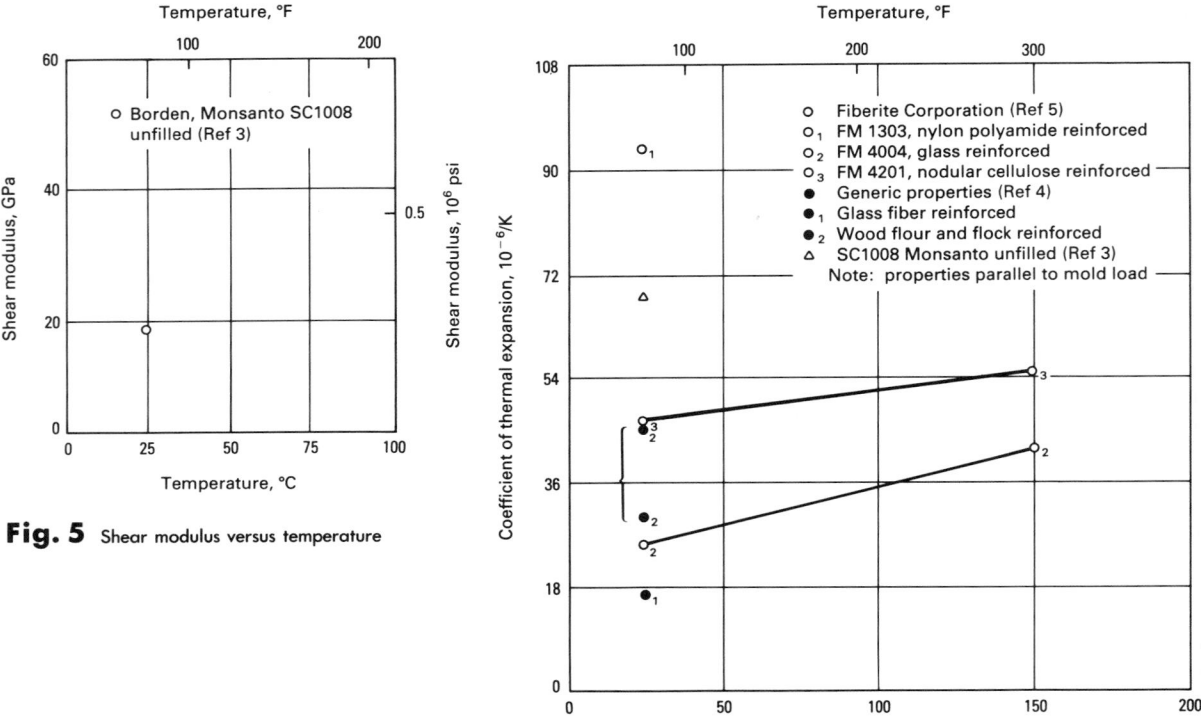

Fig. 5 Shear modulus versus temperature

Fig. 6 Coefficient of thermal expansion versus temperature. $10^{-6}/K \times 5/9 = \mu$in./in. $\times$ °F

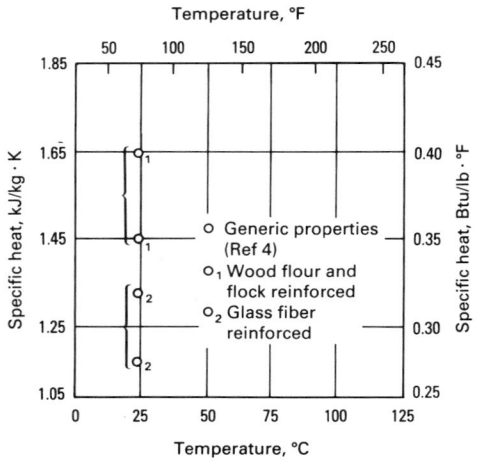

Fig. 7 Specific heat versus temperature

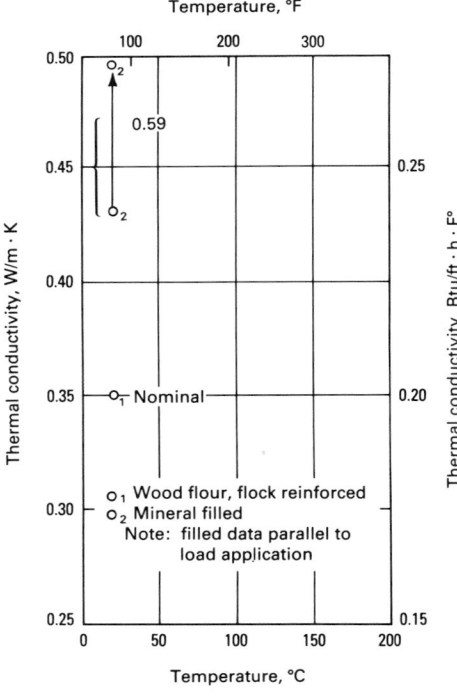

Fig. 8 Thermal conductivity versus temperature

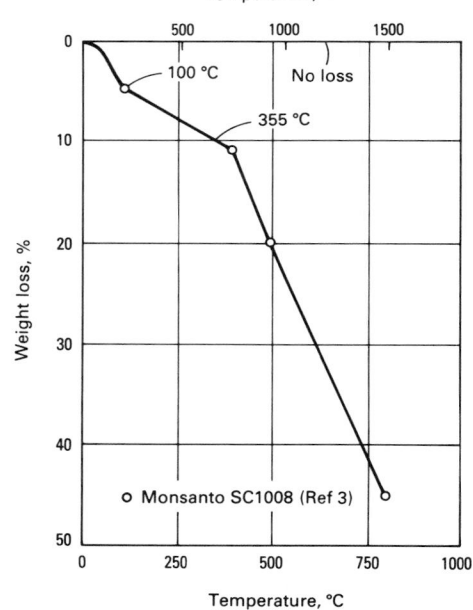

Fig. 9 Thermogravimetric analysis, weight loss percent versus temperature (in argon atmosphere)

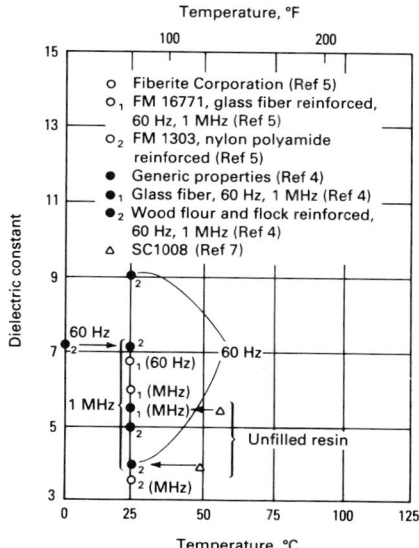

Fig. 10 Dielectric constant versus temperature

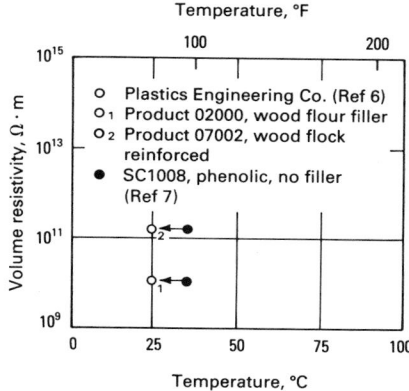

Fig. 11 Volume resistivity versus temperature

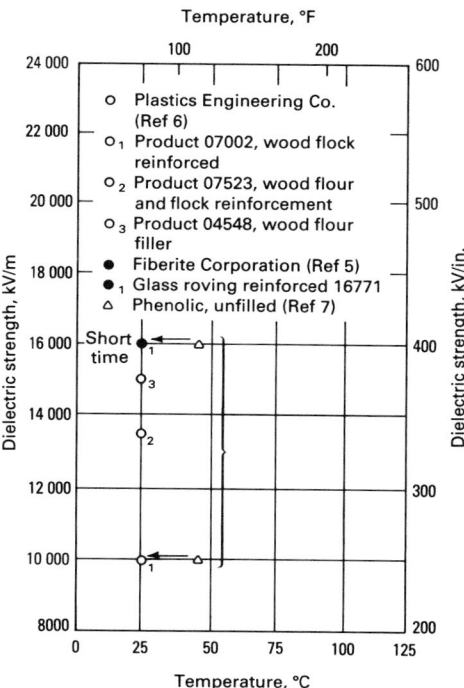

Fig. 12 Dielectric strength versus temperature

Glass Cloth Fabric-Phenolic Thermoset

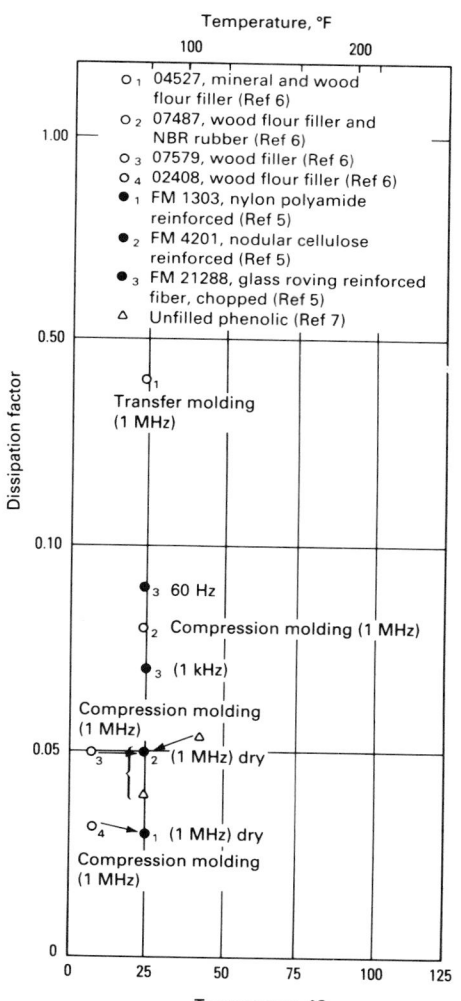

Fig. 13 Dissipation factor versus temperature

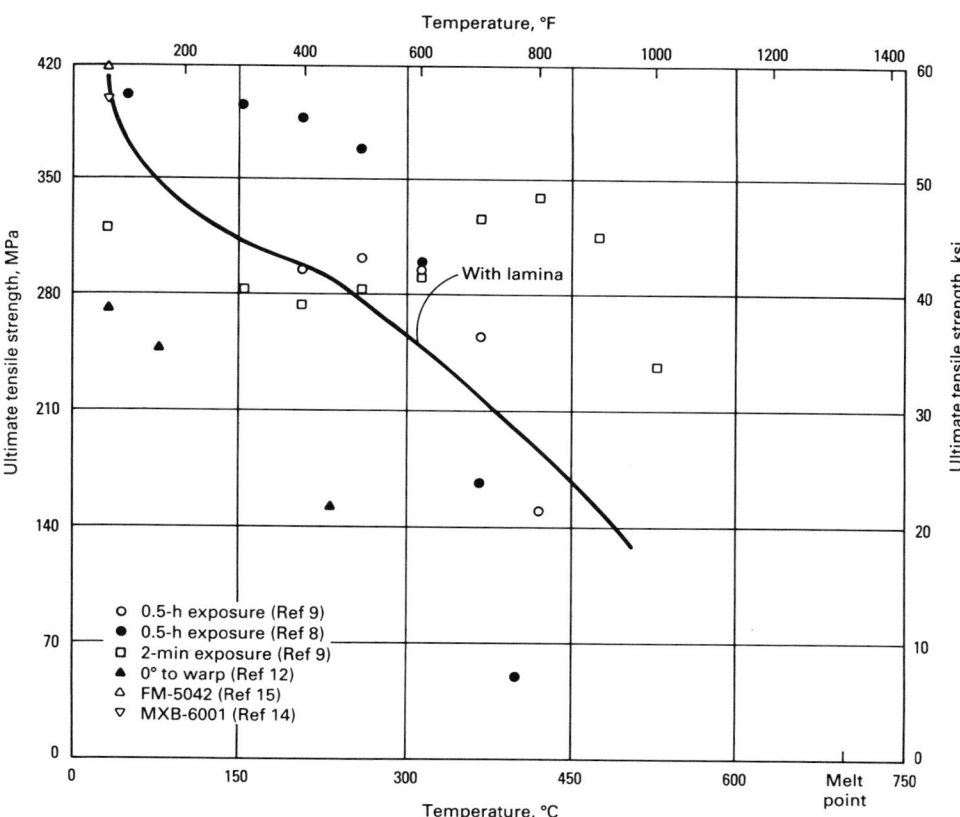

Fig. 14 Ultimate tensile strength versus temperature

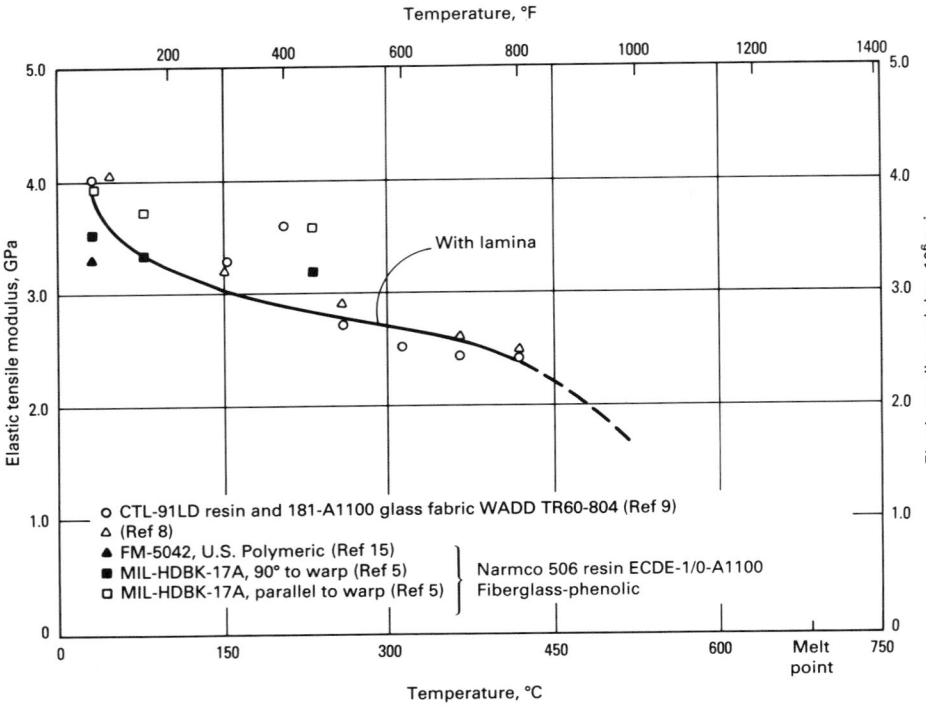

Fig. 15 Elastic tensile modulus versus temperature

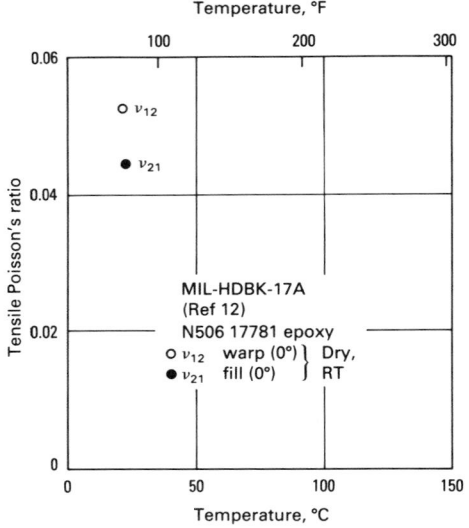

Fig. 17 Tensile Poisson's ratio versus temperature

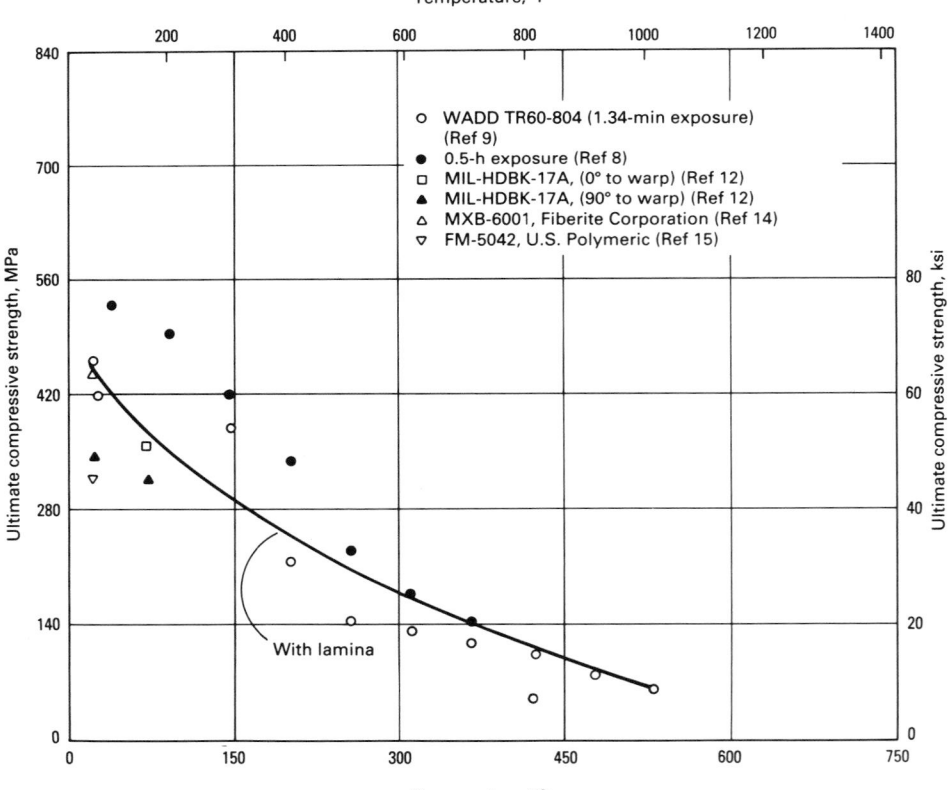

Fig. 16 Tensile elongation versus temperature

Fig. 18 Ultimate compressive strength versus temperature

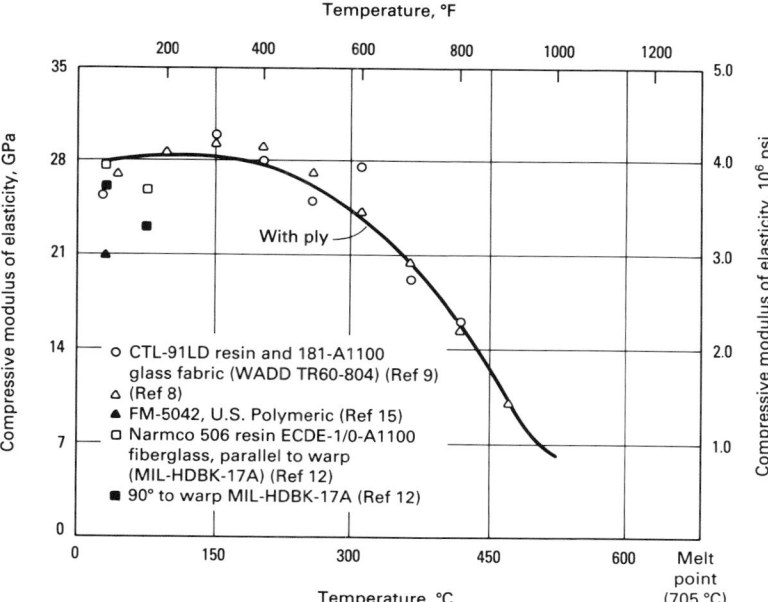

Fig. 19 Elastic compressive modulus versus temperature

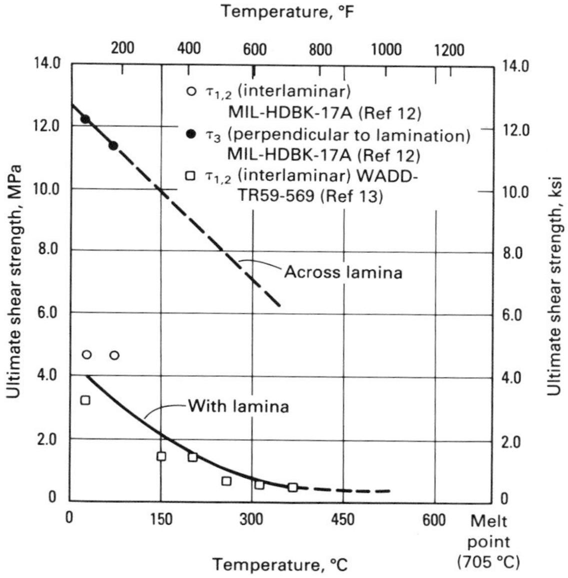

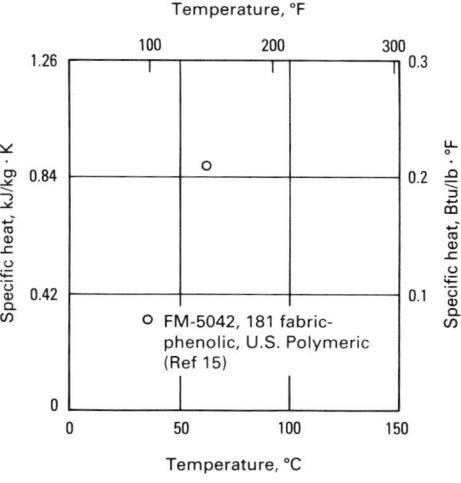

Fig. 20 Ultimate shear strength versus temperature

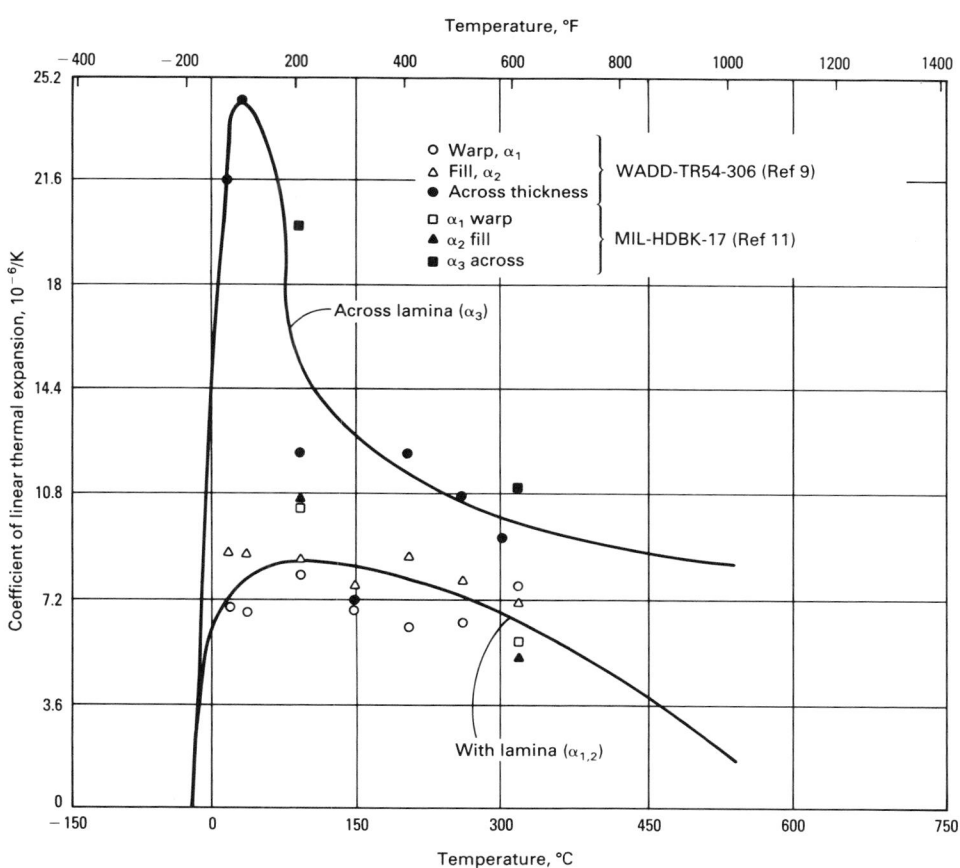

Fig. 21 Coefficient of thermal expansion versus temperature. 10^{-6}/K $\times \frac{5}{9} = \mu$in./in. $\times$ °F

Fig. 22 Specific heat versus temperature

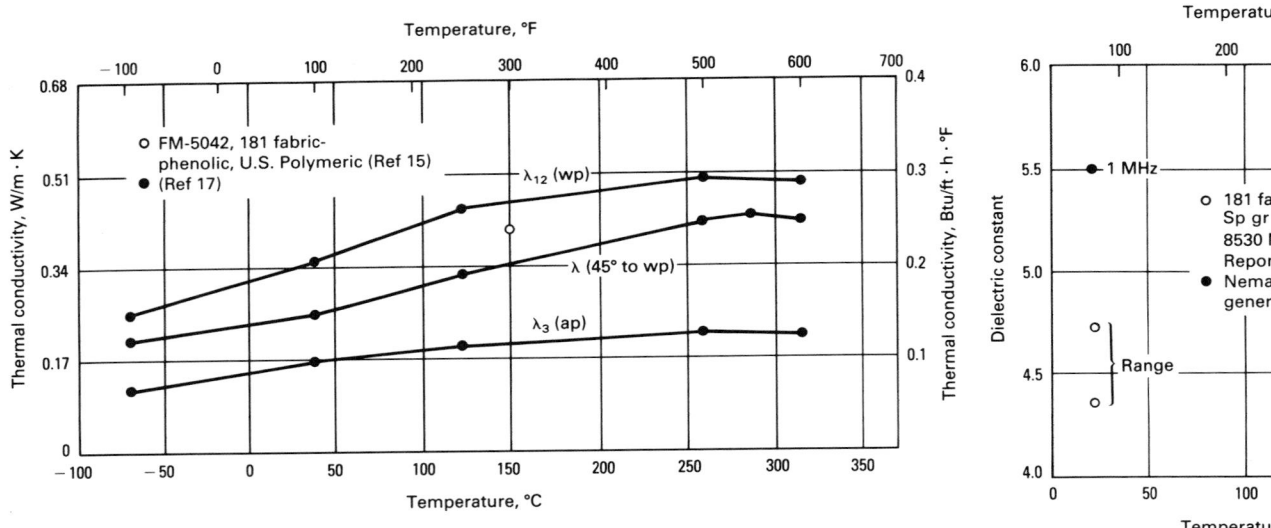

Fig. 23 Thermal conductivity versus temperature

Fig. 24 Dielectric constant versus temperature

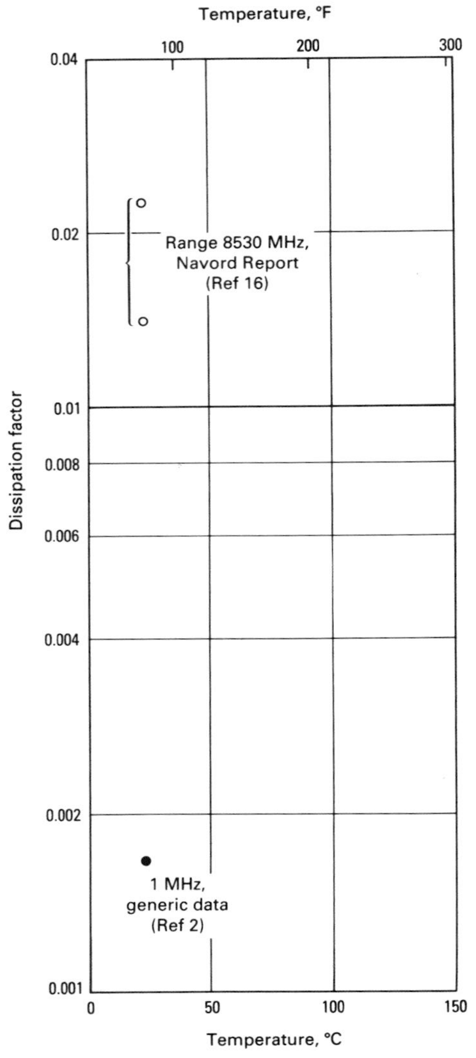

Carbon Fabric-Phenolic Thermoset

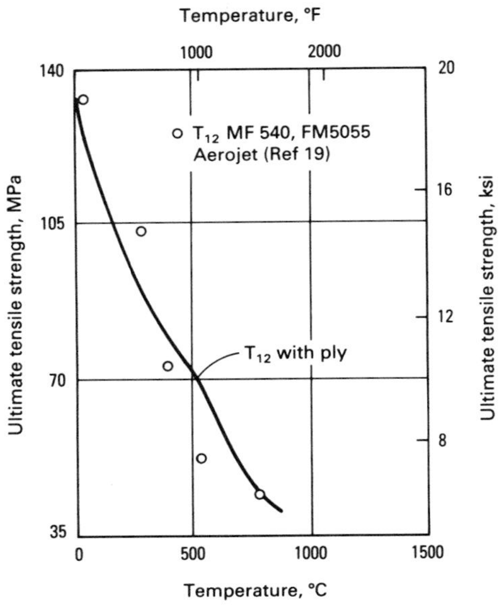

Fig. 25 Dissipation factor versus temperature

Fig. 26 Ultimate tensile strength versus temperature

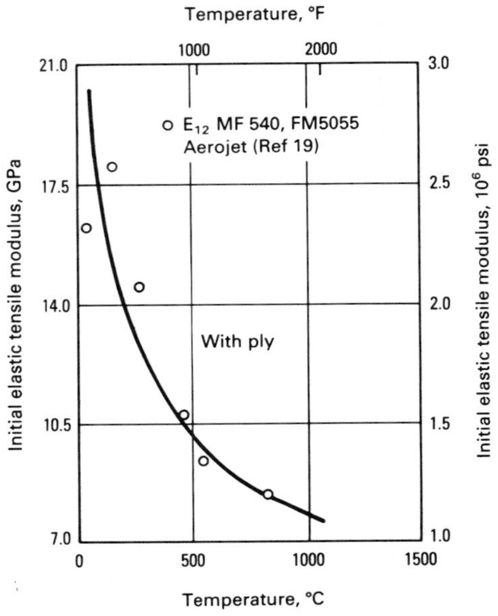

Fig. 27 Initial elastic tensile modulus versus temperature

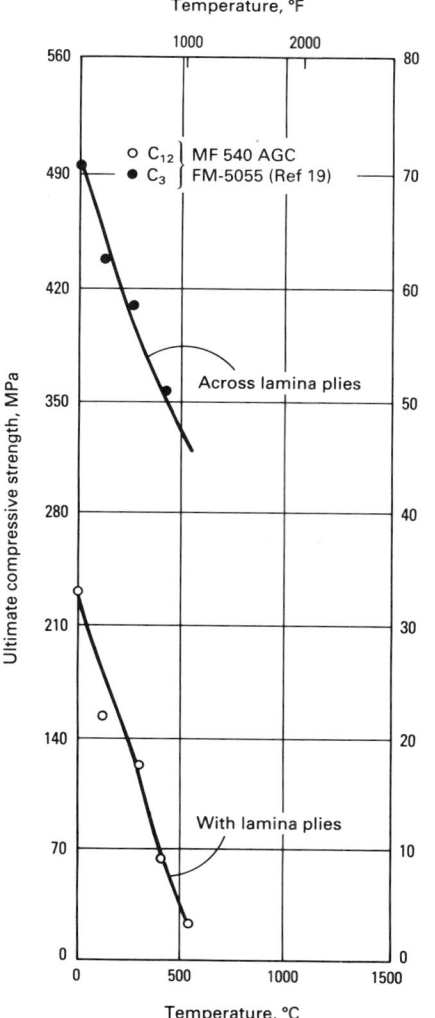

Fig. 28 Ultimate compressive strength versus temperature

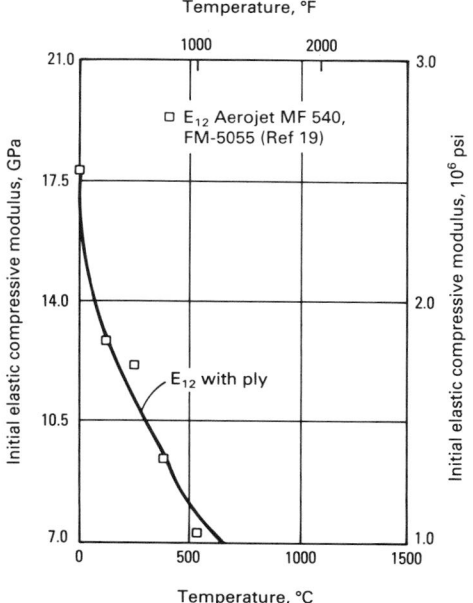

Fig. 29 Initial elastic compressive modulus versus temperature

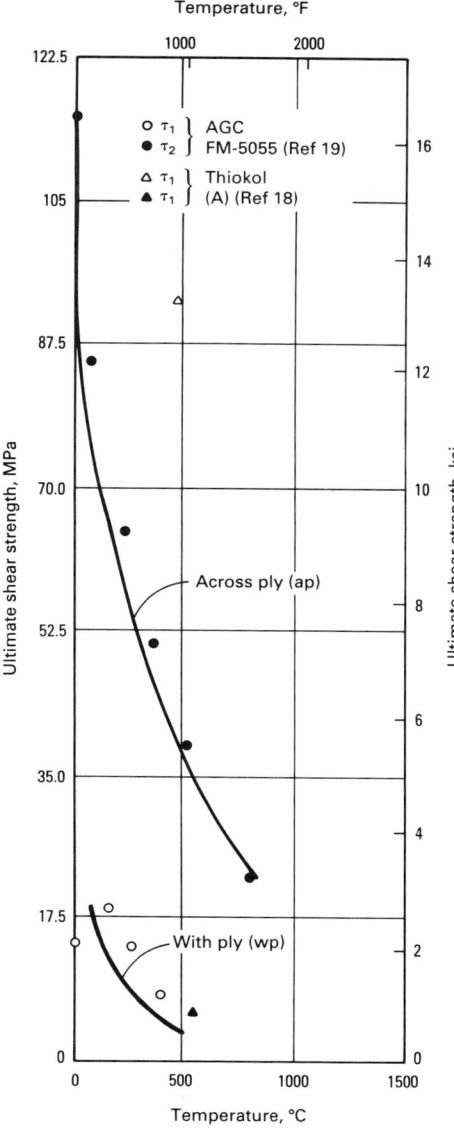

Fig. 30 Ultimate shear strength versus temperature

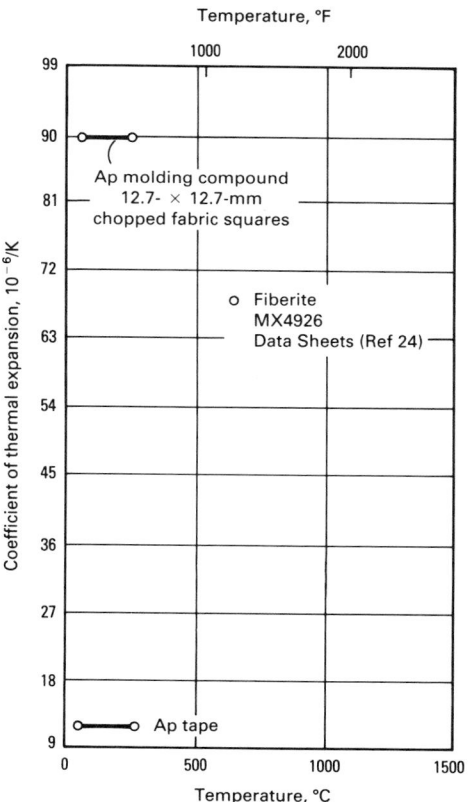

Fig. 31 Coefficient of thermal expansion versus temperature across ply—low heating rates. 10^{-6}/K × ⁵⁄₉ = μin./in. × °F

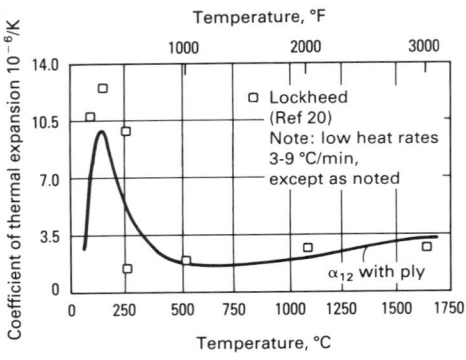

Fig. 32 Coefficient of thermal expansion versus temperature with ply—low heating rates. $10^{-6}/K \times 5/9 = \mu in./in. \times °F$

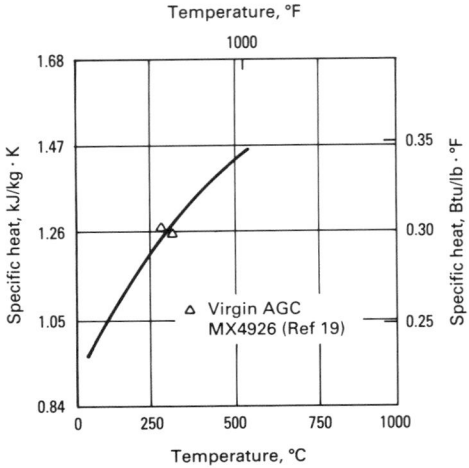

Fig. 33 Specific heat versus temperature

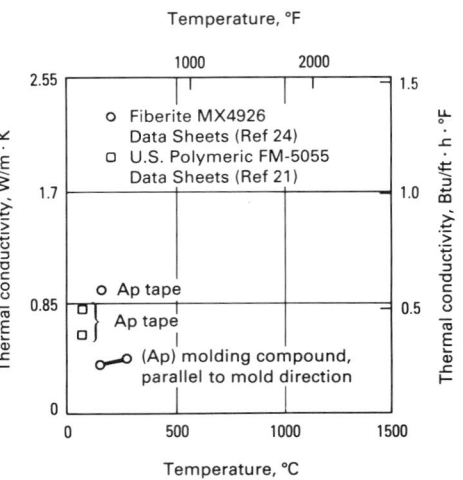

Fig. 34 Thermal conductivity versus temperature

Graphite Fabric-Phenolic Thermoset

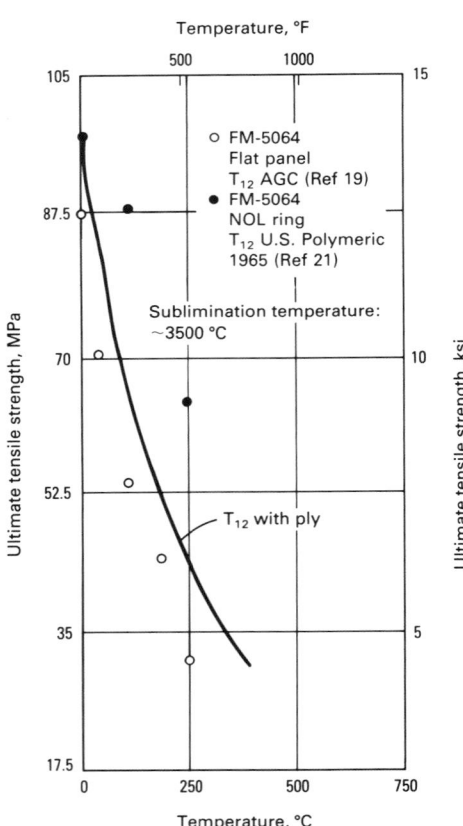

Fig. 35 Ultimate tensile strength versus temperature

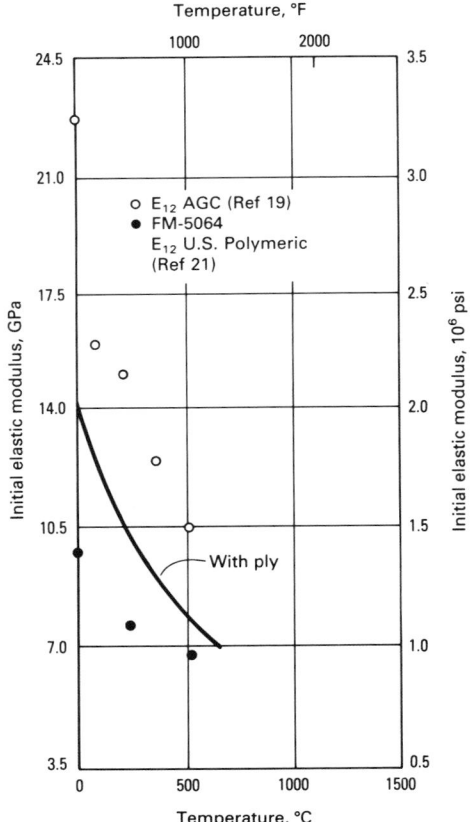

Fig. 36 Initial elastic tensile modulus versus temperature

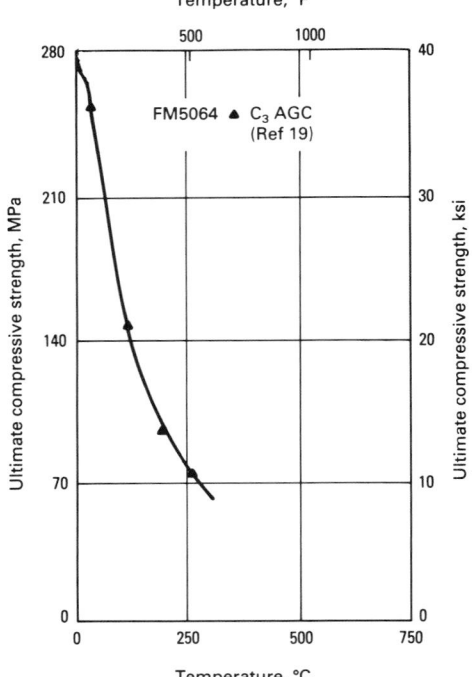

Fig. 37 Ultimate compressive strength versus temperature

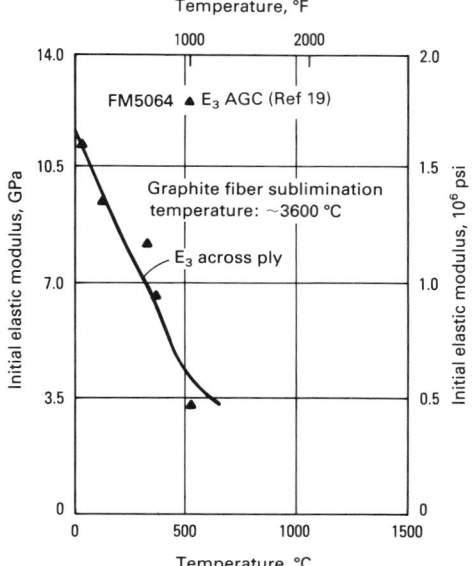

Fig. 38 Initial elastic modulus versus temperature

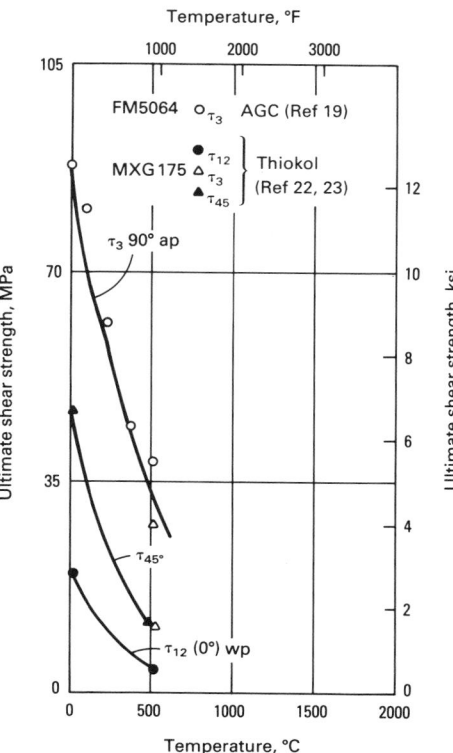

Fig. 39 Ultimate shear strength versus temperature

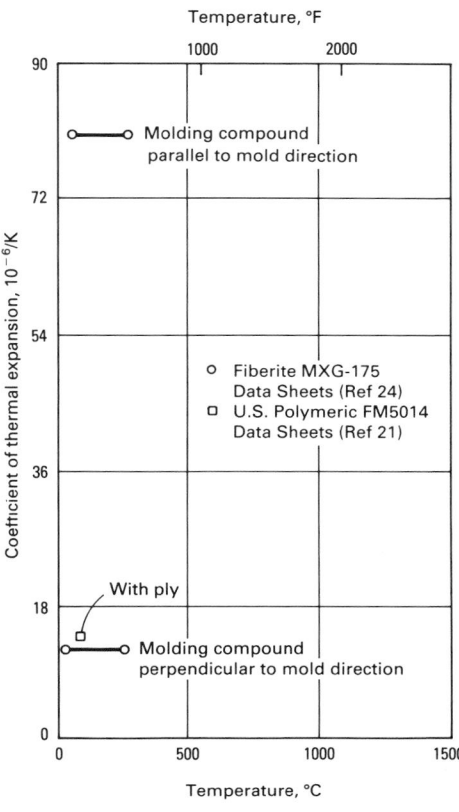

Fig. 40 Coefficient of thermal expansion versus temperature across ply. 10^{-6}/K $\times$ 5/9 = μin./in. $\times$ °F

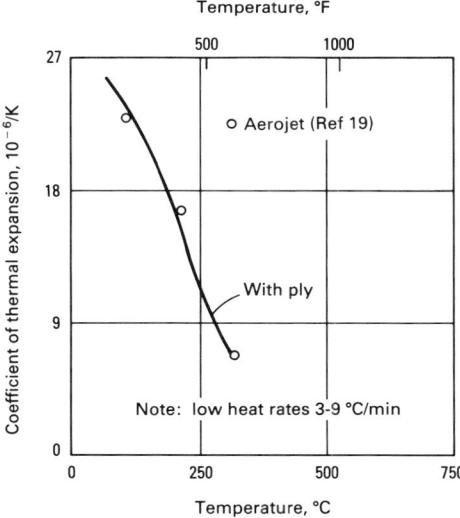

Fig. 41 Coefficient of thermal expansion versus temperature with ply. 10^{-6}/K $\times$ 5/9 = μin./in. $\times$ °F

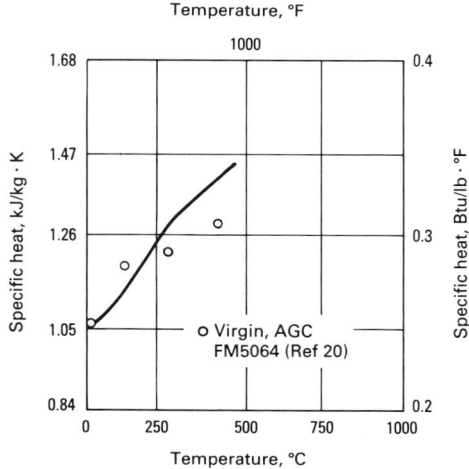

Fig. 42 Specific heat versus temperature

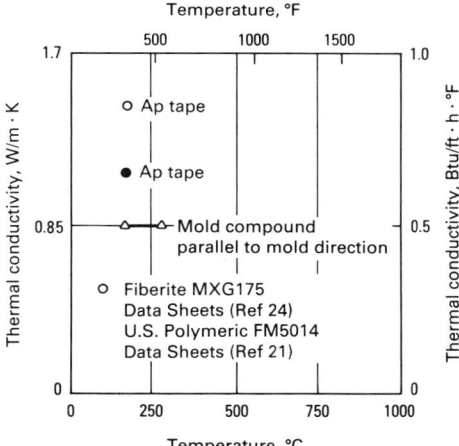

Fig. 43 Thermal conductivity versus temperature

Low-Temperature Thermoset Matrix Composites

ASM Committee on Forms and Properties of Composite Materials*

THERMOSET POLYESTER RESINS are generally produced from the reaction of an organic alcohol (glycol) with both a saturated (isophthalic) and an unsaturated (maleic or fumaric) organic acid. The polyester is then dissolved in a liquid reactive monomer such as styrene, and the solutions are sold as polyester resins. Some polyesters are supplied as pellets or granular solids, and some are premixed with glass fiber for bulk molding compounds (BMC) or sheet molding compounds (SMC). Polyester resins with fiber reinforcements can be formulated to provide different mechanical, thermal, electrical, and flammability properties. The manufacturers of thermoset polyesters include Aristech Chemical Company (formerly USS Chemicals), Reichhold Chemicals Inc., Ashland Chemical Company, Owens-Corning Fiberglas Corporation, Freeman Chemical Company, Koppers Chemical Company, Alpha Corporation, and Interplastic Corporation.

Because of their low cost, ease of processing, and good performance characteristics, unsaturated polyesters are the most extensively used type of thermoset resin. See Table 1 for properties. Unsaturated polyesters are generally combined with chopped, continuous, or woven glass fibers, as well as fillers and additives, to alter the properties to fit the application. Figure 1 depicts the effects of fabric type, and Fig. 2 illustrates four types of fabric weave. (See the article "Materials and Properties Description" in this Volume for more information on fabric weave.)

Thermoset polyesters are widely used in transportation, construction, electrical, and consumer products. In addition, the versatility of polyesters allows them to be used in a broad variety of processes. Through appropriate selection of the cross-linking initiator, these resins can be cured anywhere from room temperature to 175 °C (350 °F). Resin and glass fibers are combined at the "mold" in hand lay-up, spray-up, filament winding, pultrusion, and

resin transfer molding. Both BMC and SMC, as well as other molding compounds, are used as input materials for compression, injection, and transfer molding processes. Because the fibers are not "preplaced" in the later molding operations, fiber orientation caused by molding compound flow can produce variable anisotropy in the finished parts. Molding compounds are available from many manufacturers, including Budd Company; Premix Inc.; Durez Division, Occidental Chemical Company; Plastic Engineering Company; and Fiberite Corporation.

Properties of cast neat resin and molded glass fiber reinforced resin composites are presented in this article (Ref 1-10).

REFERENCES

1. "Industrial Glass Fabrics," Glass Fabrics Division, J.P. Stevens & Company, Inc.

Table 1 Properties of low-temperature thermoset matrix composites

Properties	Neat resin	10-40 wt% glass fiber-resin
Physical		
Specific gravity	1.2-1.4	1.6-1.9
Density, g/cm^3(lb/in^3)	1.2-1.4 (0.043-0.051)	1.6-1.9 (0.058-0.069)
In-service temperature °C (°F)	120-150 (250-300)	120-205 (250-400)
Heat deflection temperature at 1820 kPa (264 psi) °C (°F)	50-205 (120-400)	190-205 (375-400)
U.L. rated in-service temperature °C (°F)	180 (356)	...
U.L. flammability	...	...
Mechanical		
Ultimate tensile strength	See Fig. 3 to 6	
Tensile modulus	See Fig. 6, 7	
Tensile elongation	See Fig. 8	
Tensile Poisson's ratio	See Fig. 9	
Ultimate compressive strength	See Fig. 10, 11	
Compressive modulus	See Fig. 12	
Compressive elongation	See Fig. 13	
Ultimate shear strength	See Fig. 11, 14	
Thermal		
CTE	See Fig. 15	
Thermal conductivity	See Fig. 16	
Specific heat	See Fig. 17	
Electrical		
Dielectric constant	See Fig. 18, 19	
Volume resistivity	See Fig. 20	
Dielectric strength	See Fig. 21	
Dissipation factor	See Fig. 22, 23	

CTE, coefficient of thermal expansion

*Chairman: Richard C. Laramee, Morton Thiokol, Inc.; Richard G. Adams, J.P. Stevens & Company, Inc.; Roger Bacon, Michael J. Michno, Marvin E. Sauers, Amoco Performance Products, Inc.; Donald Beckley, U.S. Polymeric Corporation; Paul Blanchard, General Electric Company; Robert Boudreau, Borden Chemical Division; Douglas L. Denton, Battelle—Columbus Laboratories; L.E. DeShields, Paul R. Langston, E.I. Du Pont de Nemours & Company, Inc.; William B. Hall, University of Mississippi; Gary E. Hansen, Hercules Aerospace Company; Keith Jacobs, The Ironsides Company; John R. Koenig, Southern Research Institute; Ernest P. Rossa, Fiberite Corporation; John E. Theberge, LNP Corporation; Steven Witschen, Owens-Corning Fiberglas Corporation

2. D.L. Denton, "Mechanical Properties of an SMC-R 50 Composite," Owens-Corning Fiberglas Corporation, 1979
3. *Modern Plastics Encyclopedia*, Vol 62 (No. 10A), McGraw-Hill, 1985-1986
4. 1985 Materials Selector, *Mater. Eng.*, Dec 1984
5. 1986 Materials Reference Issue, *Mach.*

Des., April 1986
6. "Plenco Molding Compounds," Plastics Engineering Company
7. "Fiberite Reinforced Molding Materials," Fiberite Corporation
8. "Plastics for Aerospace Vehicles," *Military Handbook 17A*, Table 4.64, "8HS Fabric-27.4% Resin," Sept 1973, Depart-

ment of Defense, Sept 1973
9. "Evaluation of Glass Fabric Reinforced Plastic Laminates," Navord Report 2669, U.S. Naval Ordnance Laboratory, Jan 1953
10. S.S. Wang, D.P. Goetz, and H.T. Corten, *J. Compos. Mater.*, Vol 18 (No. 2), 1984

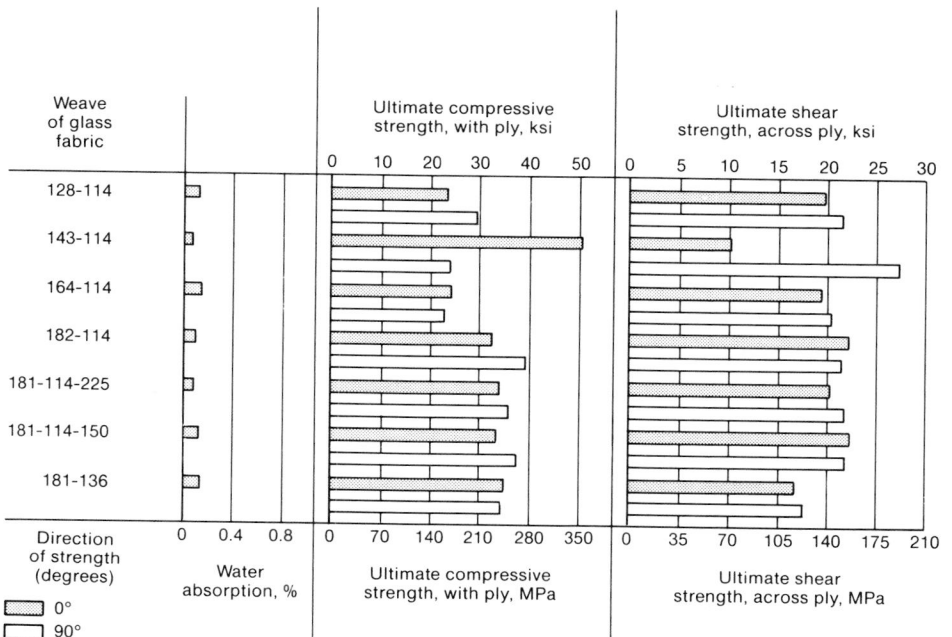

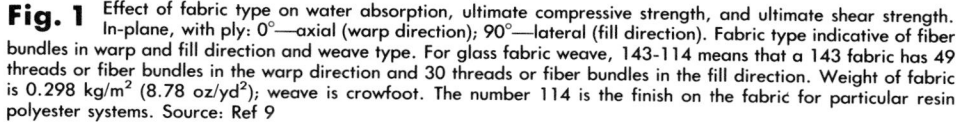

Fig. 1 Effect of fabric type on water absorption, ultimate compressive strength, and ultimate shear strength. In-plane, with ply: 0°—axial (warp direction); 90°—lateral (fill direction). Fabric type indicative of fiber bundles in warp and fill direction and weave type. For glass fabric weave, 143-114 means that a 143 fabric has 49 threads or fiber bundles in the warp direction and 30 threads or fiber bundles in the fill direction. Weight of fabric is 0.298 kg/m² (8.78 oz/yd²); weave is crowfoot. The number 114 is the finish on the fabric for particular resin polyester systems. Source: Ref 9

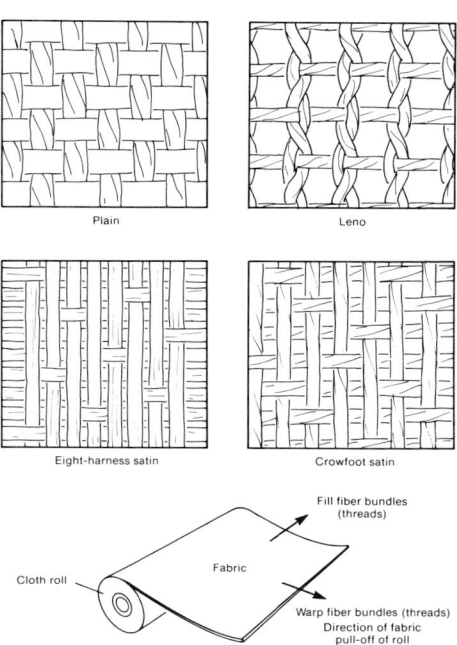

Fig. 2 Typical fabric weaves. Source: Ref 1

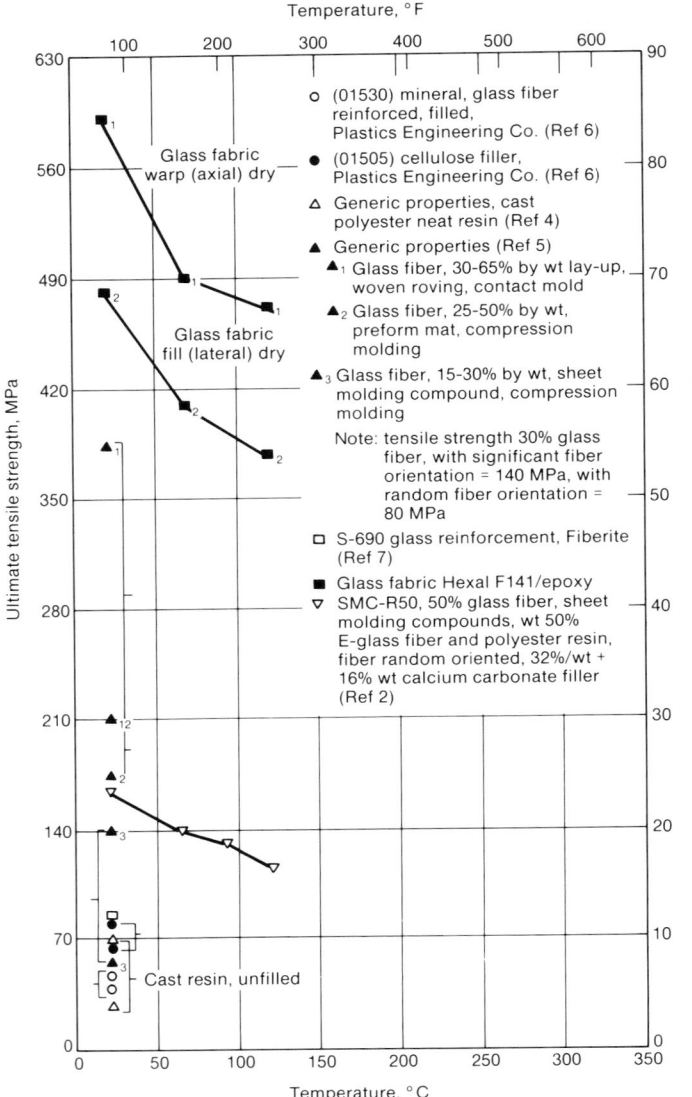

Key for Fig. 3:

- ○ (01530) mineral, glass fiber reinforced, filled, Plastics Engineering Co. (Ref 6)
- ● (01505) cellulose filler, Plastics Engineering Co. (Ref 6)
- △ Generic properties, cast polyester neat resin (Ref 4)
- ▲ Generic properties (Ref 5)
  - ▲₁ Glass fiber, 30-65% by wt lay-up, woven roving, contact mold
  - ▲₂ Glass fiber, 25-50% by wt, preform mat, compression molding
  - ▲₃ Glass fiber, 15-30% by wt, sheet molding compound, compression molding
  - Note: tensile strength 30% glass fiber, with significant fiber orientation = 140 MPa, with random fiber orientation = 80 MPa
- □ S-690 glass reinforcement, Fiberite (Ref 7)
- ■ Glass fabric Hexal F141/epoxy
- ▽ SMC-R50, 50% glass fiber, sheet molding compounds, wt 50% E-glass fiber and polyester resin, fiber random oriented, 32%/wt + 16% wt calcium carbonate filler (Ref 2)

Fig. 3 Ultimate tensile strength versus temperature. Note: Ranges are due to property differences among compression, transfer, and injection molding processes

Fig. 5 Effect of panel thickness on ultimate tensile stress and secant tensile elastic modulus at RT. In-plane, with ply: 0°—axial (warp); 90°—lateral (fill). Data presented versus thickness to show the efficiency of the glass fabric-polyester resin laminate bonding as determined by small specimen testing. There may be some deviation for material application. Source: Ref 9

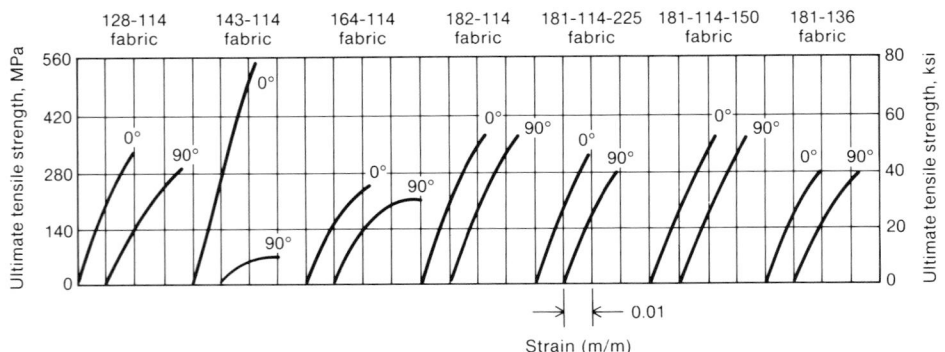

Fig. 4 Effect of glass fabric on tensile stress-strain data curves at RT. In-plane, with ply: 0°—axial (warp direction); 90°—lateral (fill direction). Fabric types indicate number of fiber bundles in the warp and fill direction and weave type. Source: Ref 9

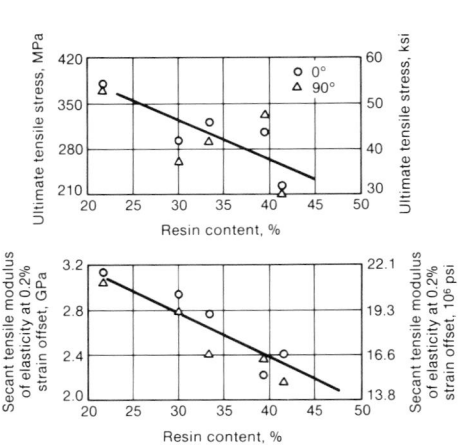

Fig. 6 Effect of resin content on ultimate tensile stress and secant tensile modulus. In-plane, with ply: 0°—axial (warp); 90°—lateral (fill). Source: Ref 9

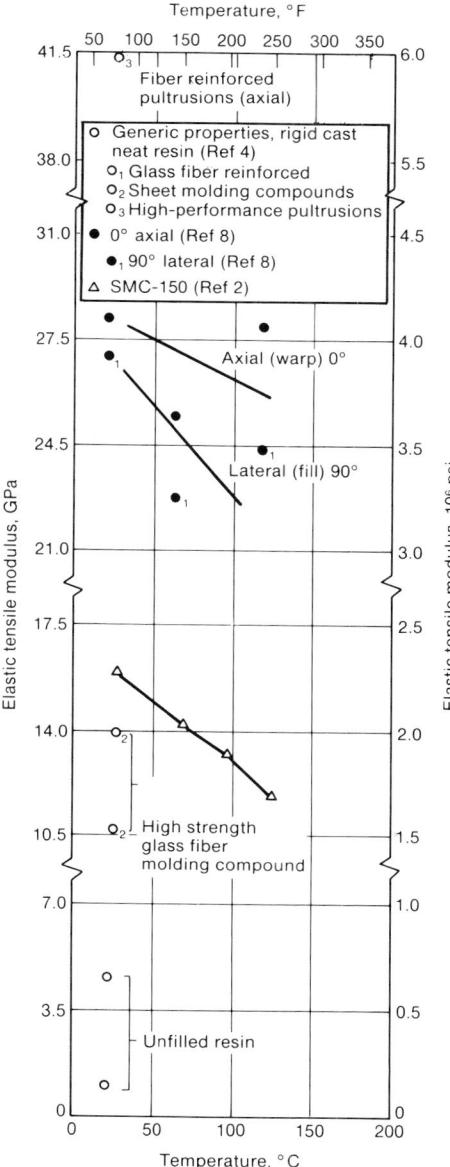

Fig. 7 Elastic tensile modulus versus temperature

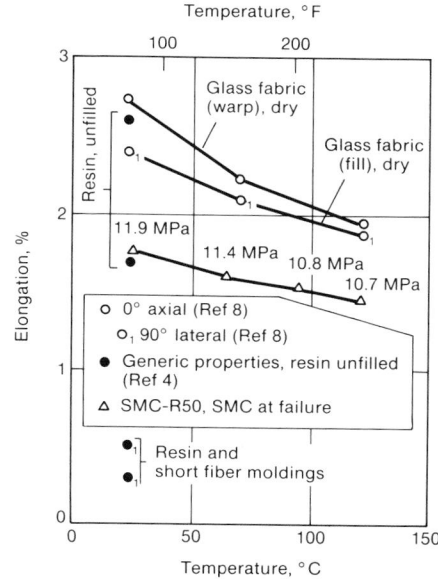

Fig. 8 Percent elongation versus temperature

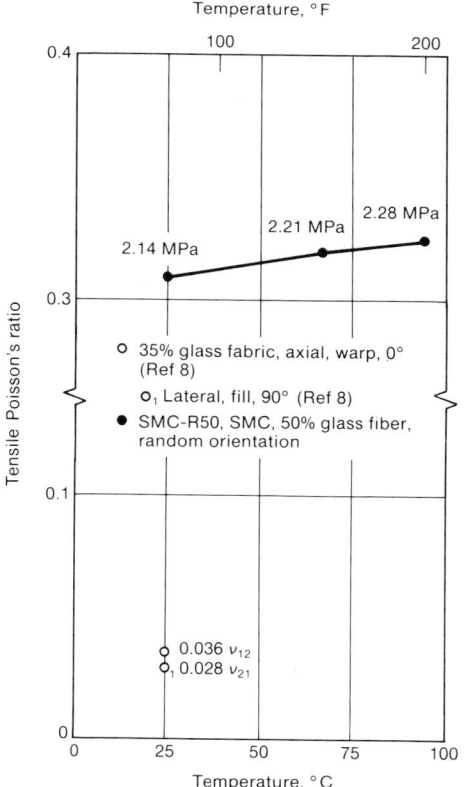

Fig. 9 Tensile Poisson's ratio versus temperature

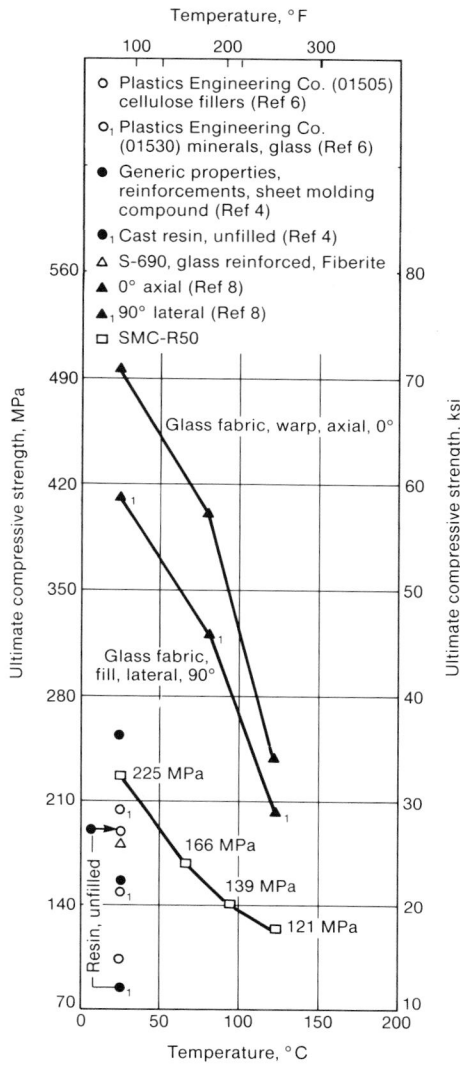

Fig. 10 Ultimate compressive strength versus temperature

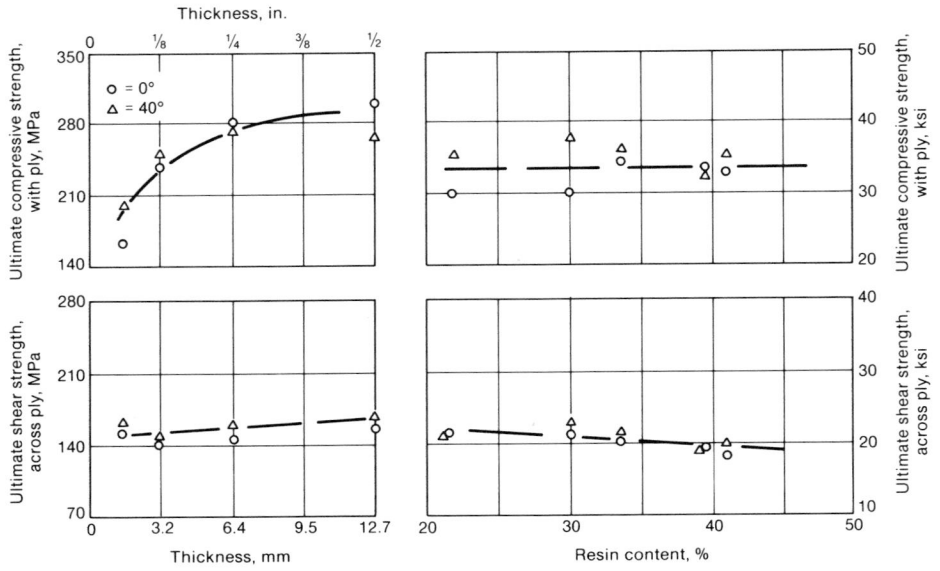

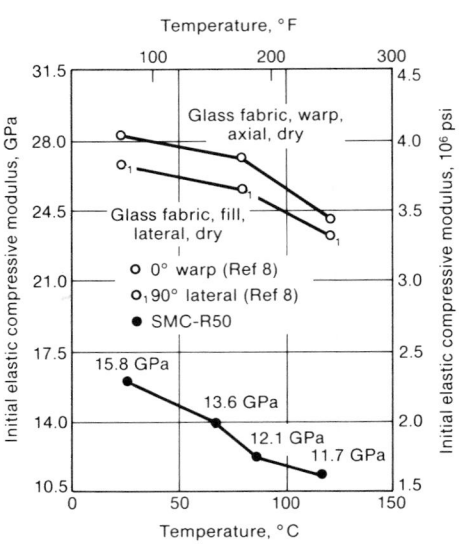

Fig. 11 Effect of panel thickness and resin content on ultimate compressive strength and shear strength. All types of fabric. In-plane, with ply: 0°—axial (warp); 90°—lateral (fill)

Fig. 12 Initial elastic compressive modulus versus temperature

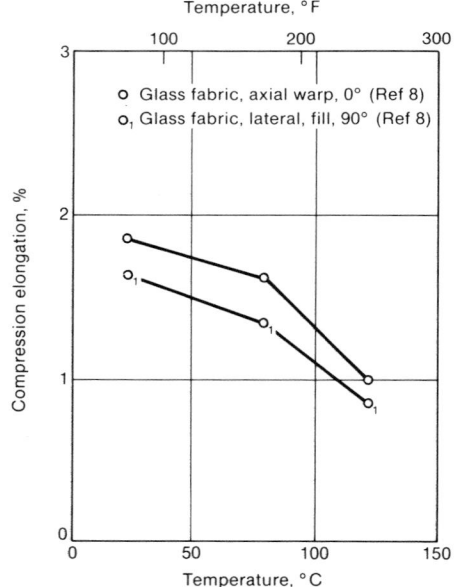

Fig. 13 Compression elongation percent versus temperature

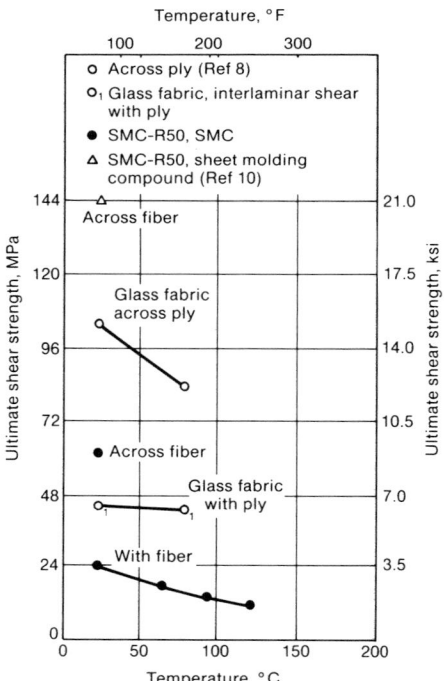

Fig. 14 Ultimate shear strength versus temperature

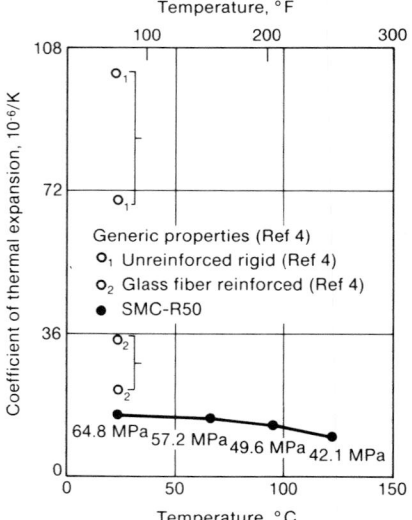

Fig. 15 Coefficient of thermal expansion versus temperature. 10^{-6}/K × $5/9$ = μin./in. × °F

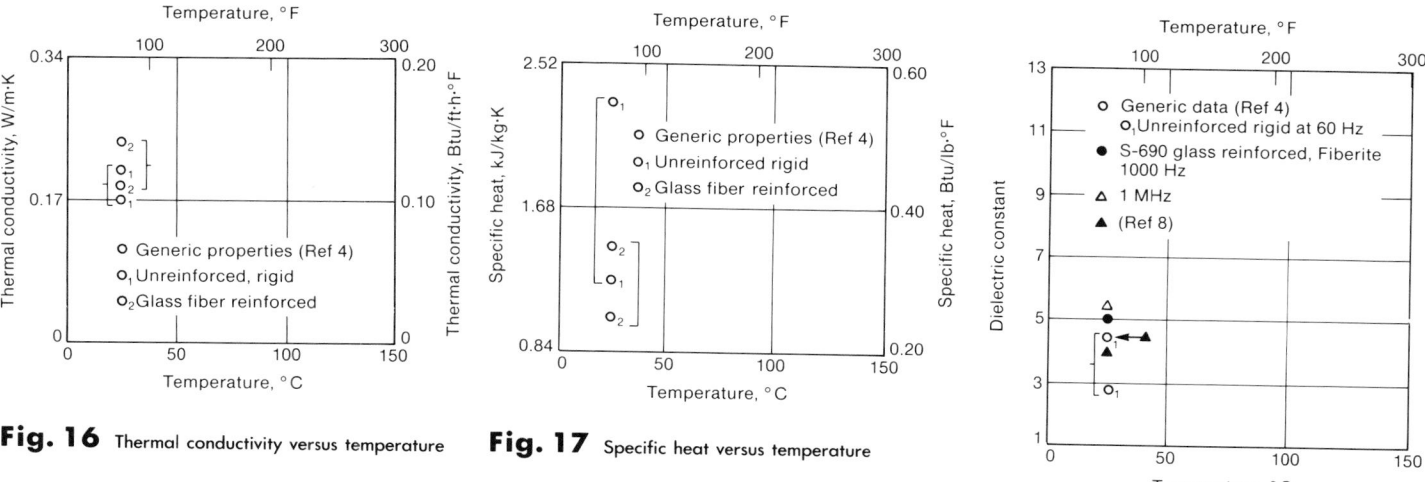

Fig. 16 Thermal conductivity versus temperature

Fig. 17 Specific heat versus temperature

Fig. 18 Dielectric constant versus temperature

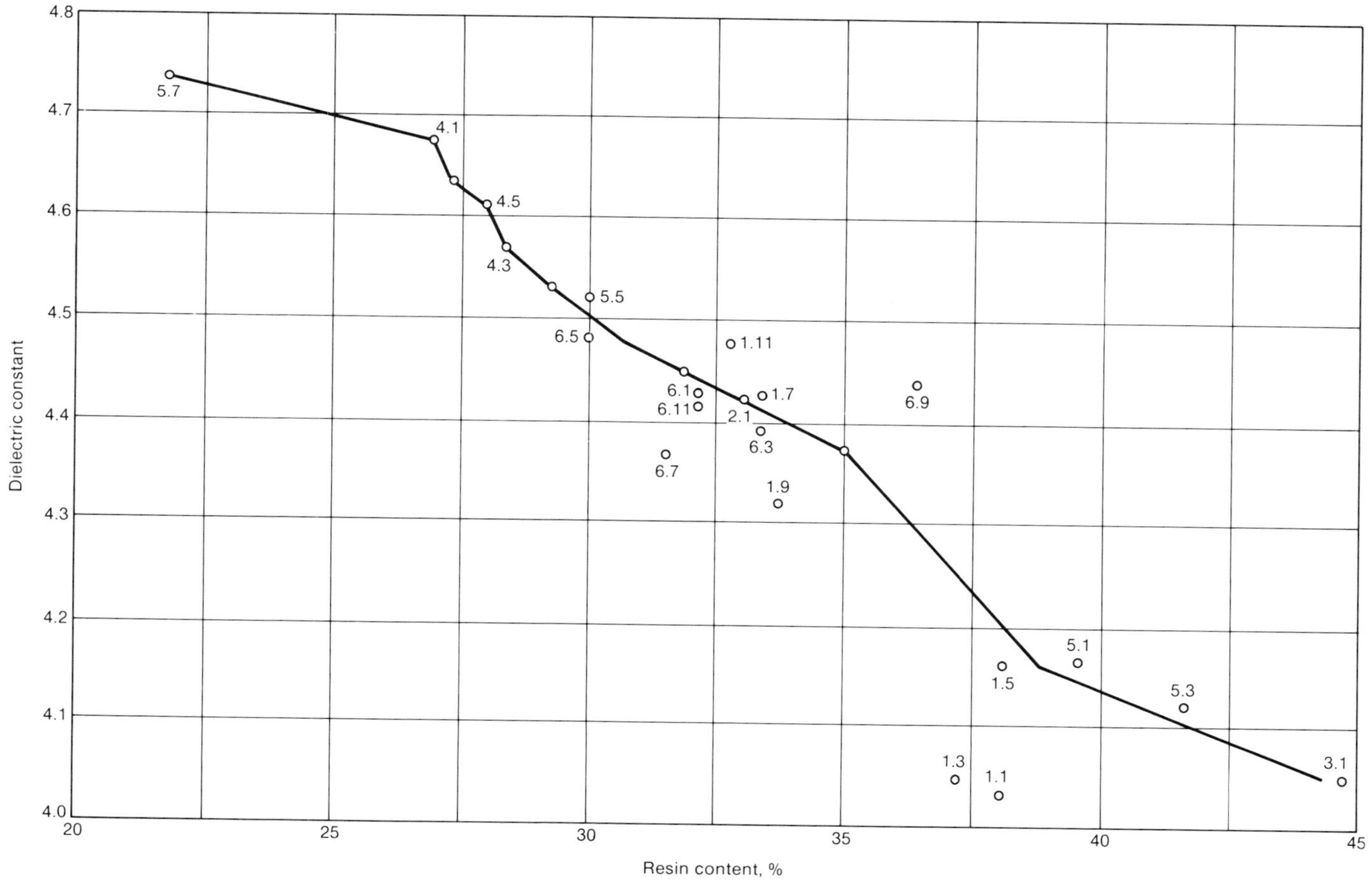

Fig. 19 Effect of resin content on dielectric constant. Glass fabric, all types. Note: Numbers on field of figure are identification of data sets. Source: Ref 9

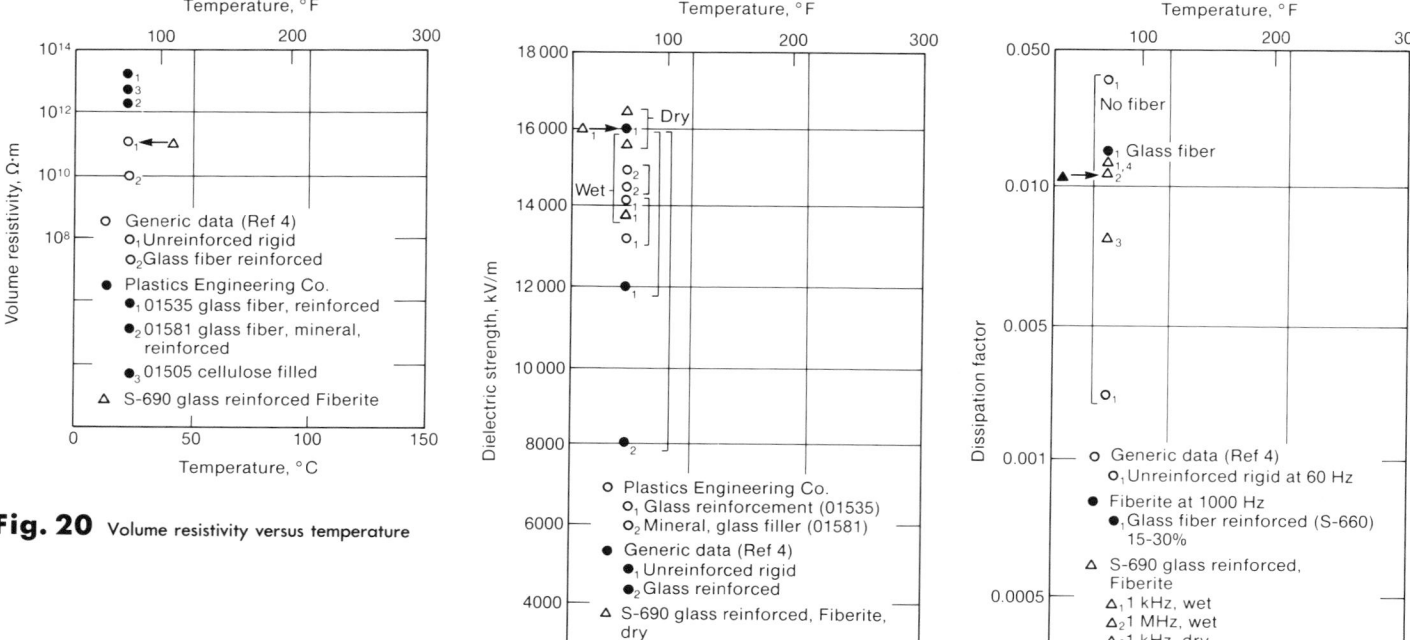

Fig. 20 Volume resistivity versus temperature

Fig. 21 Dielectric strength versus temperature

Fig. 22 Dissipation factor versus temperature

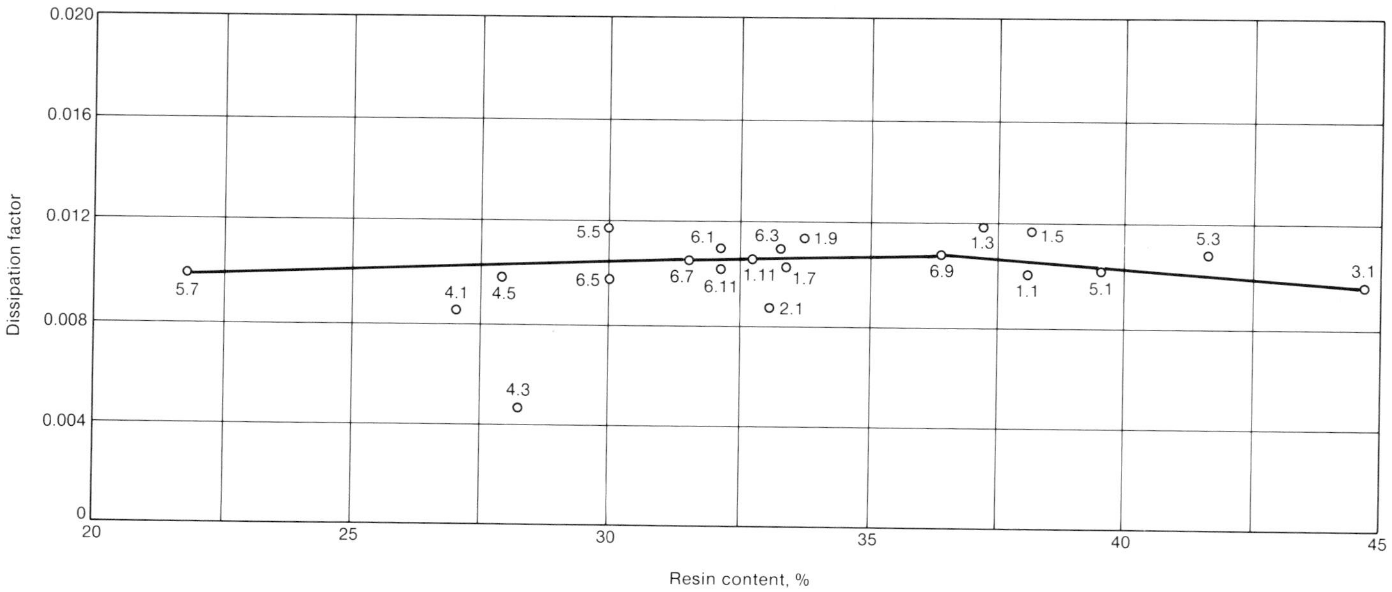

Fig. 23 Effect of resin content on dissipation factor (power factor). Glass fabric types of reinforcement. Note: Numbers on field of figure are identification of data sets. Source: Ref 9

High-Strength Medium-Temperature Thermoset Matrix Composites

ASM Committee on Forms and Properties of Composite Materials*

EPOXY RESIN SYSTEMS can be formulated into compounds, such as diglycidyl ether of bisphenol A (DGEBA); multifunctional epoxies, such as phenolic novolac; and aliphatic epoxies, such as cycloaliphatic. Temperatures up to 230 to 260 °C (450 to 500 °F) can be used for the latter two types of resin systems for short periods of time. At 540 °C (1000 °F), the hydrogen and oxygen have been driven off, leaving a weak carbonaceous char. Reinforced epoxy structures provide high strength-to-weight ratios and good thermal and electrical properties. Filament winding and machine or hand lay-up processes are used for advanced aircraft fuselages, wing and control surface panels; rocket motor cases; rocket nozzle structural shells; and commercial pressure vessels, tanks, and pipe. Glass fiber/fabric, carbon, graphite, quartz, and aramid fibers are used in molding compounds, hand lay-ups, and fiber/fabric prepreg composites for different applications to match pressure, temperature, service life, weight, and cost requirements (Ref 1-3).

Table 1 Epoxy resin and fiber-resin properties

Properties	Epoxy resin systems					Continuous fiber-fabric epoxy resin systems				
	DGEBA	Phenolic-novolac	Cyclo-aliphatic	Glass-resin fiber	Chopped glass fabric-resin	S-2 filament winding glass fiber-resin	Kevlar-resin	Carbon-resin	Graphite fiber-resin	Lay-up quartz fabric-resin
Physical										
Specific gravity	1.15	1.24	1.22	1.72	1.79	1.86	1.25	1.46	1.58	1.75
Density, g/cm³ (lb/in.³)	1.15 (0.042)	1.24 (0.045)	1.22 (0.044)	1.72 (0.062)	1.79 (0.065)	1.86 (0.067)	1.25 (0.045)	1.46 (0.053)	1.58 (0.057)	1.75 (0.063)
Service temperature, °C (°F) . . .	80-88 (175-190)	230 (450)	230-260 (450-500)	150 (300)	150 (300)	150 (300) for 1 to 2 minutes	150 (300)	150 (300)	150 (300)	150 (300)
Heat deflection temperature at 1820 kPa (264 psi) °C (°F) . . .	110 (230)	150-205 (300-400)	150-275 (300-525)	NA	NA	NA	NA	NA	NA	NA
TGA, stability temperature °C (°F) 0% wt loss (Ref 3)	200 (390)			NA						
Al-asbestos filled						NA	NA	NA	NA	NA
Flammability	NA	NA	NA	NA	NA	NA	NA	NA	NA	NA
Reinforcement volume.	NA	NA	NA	35	49.3	60	62.9	56	54	65 wt%
Mechanical										
Ultimate tensile strength		See Fig. 1			See Fig. 15	See Fig. 25	See Fig. 36	See Fig. 50	See Fig. 62	See Fig. 73
Tensile modulus.		See Fig. 2			See Fig. 16	See Fig. 26	See Fig. 37	See Fig. 51	See Fig. 63	See Fig. 74
Tensile elongation		See Fig. 3			NA	See Fig. 27	NA	See Fig. 52	See Fig. 64	NA
Tensile Poisson's ratio		NA			NA	See Fig. 28	See Fig. 38	See Fig. 53	See Fig. 65	NA
Ultimate compressive strength		See Fig. 4			See Fig. 17	See Fig. 29	See Fig. 39	See Fig. 54	See Fig. 66	See Fig. 75
Compressive modulus.		See Fig. 5			See Fig. 18	See Fig. 30	See Fig. 40	See Fig. 55	See Fig. 67	See Fig. 76
Compressive Poisson's ratio		NA			NA	NA	NA	NA	NA	NA
Ultimate shear strength		See Fig. 6			See Fig. 19	See Fig. 31	See Fig. 41	See Fig. 56	See Fig. 68	NA
Shear modulus.		NA			NA	See Fig. 32	See Fig. 42	See Fig. 57	NA	NA
Thermal										
CTE		See Fig. 7			NA	See Fig. 33	See Fig. 43	See Fig. 58	See Fig. 69	NA
Thermal conductivity		See Fig. 8			See Fig. 20	NA	See Fig. 44	See Fig. 59	See Fig. 70	NA
Specific heat		See Fig. 9			See Fig. 21	See Fig. 34	See Fig. 45	See Fig. 60	See Fig. 71	NA
TGA.		See Fig. 10			NA	NA	NA	NA	NA	NA
Electrical										
Dielectric constant		See Fig. 11			See Fig. 22	NA	See Fig. 46	NA	NA	See Fig. 77
Volume resistivity		See Fig. 12			See Fig. 23	See Fig. 35	See Fig. 47	See Fig. 61	See Fig. 72	NA
Dielectric strength		See Fig. 13			NA	NA	See Fig. 48	NA	NA	NA
Dissipation factor.		See Fig. 14			See Fig. 24	NA	See Fig. 49	NA	NA	See Fig. 78

*Chairman: Richard C. Laramee, Morton Thiokol, Inc.; Richard G. Adams, J.P. Stevens & Co., Inc.; Roger Bacon, Michael J. Michno, Marvin E. Sauers, Amoco Performance Products, Inc.; Donald Beckley, U.S. Polymeric Corporation; Paul Blanchard, General Electric Company; Robert Boudreau, Borden Chemical Division; Douglas L. Denton, Battelle—Columbus Division; L. E. DeShields, Paul R. Langston, E. I. Du Pont de Nemours & Company, Inc.; William B. Hall, University of Mississippi; Gary E. Hansen, Hercules Aerospace Company; Keith Jacobs, The Ironsides Company; John R. Koenig, Southern Research Institute; Ernest P. Rossa, Fiberite Corporation; John E. Theberge, LNP Corporation; Steven Witschen, Owens-Corning Fiberglas Corporation

This article focuses on these seven different fabric forms:

- Epoxy resin with and without E-glass fiber reinforcement
- Glass fabric reinforced epoxy resin
- S-glass fiber reinforced epoxy resin
- Quartz fabric reinforced epoxy resin
- Kevlar 49 fiber reinforced epoxy resin
- Carbon fiber reinforced (T300) epoxy resin
- Graphite fiber reinforced (HM) epoxy resin

Table 1, a properties data sheet, shows specific characteristics of each material, including specific gravity, useful temperatures, and reinforcement percentage. The remaining mechanical, thermal, and electrical properties are shown in the figures.

Epoxy Resin With and Without Reinforcement Fibers, and Molding Compounds. Epoxy resin characterization includes DGEBA, phenolic novolac, and aliphatic epoxies without fibers or fillers, as well as other resin systems with glass fiber reinforcement or asbestos fibers and aluminum filler.

Glass fabric reinforced epoxy resin, used as a lay-up, provides a structural material with a large variety of epoxies, numerous manufacturers, and at least two glass fibers (E and S) by Owens-Corning Fiberglas. Characterization includes mechanical, thermal, and electrical properties.

S-glass fiber reinforced epoxy resin is used as a filament winding prepreg to manufacture pressure vessels, tanks, tubes, and cones. S-glass is manufactured by Owens-Corning Fiberglas, and the resin is a hybrid epoxy resin available from several manufacturers. Characterization includes mechanical, thermal, and limited electrical properties.

Kevlar 49 fiber and fabric reinforced epoxy resin is used as a lay-up fabric for boats, aircraft surface panels, and electrical circuit boards. The fiber prepreg is also used in filament winding pressure vessels, tanks, tubes, and cone shapes. Kevlar 49 is manufactured by E.I. Du Pont de Nemours & Company, Inc., and the resins are hybrid epoxies from several manufacturers.

Carbon Fiber and Fabric Reinforced Epoxy Resin. The epoxy resin, supplied by Hercules, Inc., Fiberite, Inc., and other suppliers, is reinforced with carbon fibers supplied by Amoco Performance Products (formerly Union Carbide) (T300) and Hercules (AS4) to form unidirectional prepreg tapes suitable for filament winding of components such as pressure vessels, tanks, and pipes, as well as shapes of revolution. It is also used for aircraft external structural panels and space structures.

Graphite Fiber-Epoxy Resin Thermoset. The epoxy resins, primarily supplied by Hercules, Inc., are reinforced with HMS graphite fibers (Hercules, Inc.) and supplied as unidirectional fiber prepreg tapes suitable for filament winding of components or as fabric prepreg rolls suitable for hand lay-ups.

Quartz Fabric Reinforced Epoxy Resin. The CE-9000 epoxy resin system (Ferro Corp.) and Epon 828 resin (Shell Chemical) are reinforced with woven quartz (S/557 unidirectional fabric and S/581 eight-harness satin weave fabric). The supplier of the quartz is J.P. Stevens Company, Inc., Industrial Products Group.

REFERENCES

1. *Modern Plastics Encyclopedia 1985-1986*, Vol 62 (No. 10A), McGraw-Hill
2. Materials Selector, 1985, *Mater. Eng.*, Dec 1984, 1986; Materials Reference Issue, *Mach. Des.*, 17 April 1986
3. Epon 934 Product Data Sheet, Shell Chemical Company
4. Molding Compound Product Data Sheets, Fiberite Corporation
5. Epoxylite 5403 Product Data Sheets, Epoxylite Corporation
6. Materials Reference 1986, *Mach. Des.*, 17 April 1986
7. E-710 Data Sheet, U.S. Polymeric, 1965
8. "Thermal Conductivity for Nine Glass and Asbestos Reinforced Plastics," Research Paper FPL-36, U.S. Forest Service, Aug 1965
9. ANC-17, Plastics for Aircraft, June 1955
10. Short Beam Shear, Thiokol Laboratory Report, Morton Thiokol, Inc., Jan 1979
11. Short Beam Shear, Lockheed Laboratory Report, Lockheed Corporation, March 1979
12. M.L. White, "Design and Fabrication of an Improved Performance, Low Cost, Composite Main Rotor Blade for the Cobra Helicopter," Kaman Aerospace Corporation, 1977
13. W.T. Freeman and G.C. Kuebeler, Mechanical and Physical Properties of Advanced Composites, in *Composite Materials: Testing and Design, (Third Conference)*, STP 546, American Society for Testing and Materials, 1974, p 435-456
14. L. Holliday, Ed., *Composite Materials*, Materials Science Series, Elsevier
15. "Data Manual for Kevlar 49 Aramid," Textile Fibers Department, E.I. Du Pont de Nemours & Company, Inc., 1974, 1986
16. "AS4/3502 Graphite Prepreg Tape and Fabric Module," Hercules, Inc.
17. Yeow *et al.*, "On Time-Temperature Behaviour of Carbon-Epoxy," Fifth ASTM Conference Composite Materials, American Society for Testing and Materials
18. "Advanced Composite Development Studies of Graphite Epoxy," Lockheed Missiles and Space Company, Inc.
19. J.C. Ekvall and C.F. Griffin, Design Allowables for T300/5208 Graphite Epoxy Composite Materials, *J. Aircr.*, Vol 19 (No. 8), Aug 1982
20. D.N. Yates, R.D. Torczyner, and D.R. Sidwell, "Design and Development of Woven Structures Products for Aerospace Structures," Paper presented at AIAA/ASME/SAE 16th Structures, Structural Dynamics, and Materials Conference, Denver, May 1975
21. AS/3002, in *Advanced Composites Design Guide*, Vol IV, 3rd Ed., 1973
22. W.T. Freeman and G.C. Kuebeler, Mechanical and Physical Properties of Advanced Composites, in *Composite Materials: Testing and Design (Third Conference)*, STP 546, American Society for Testing and Materials, 1974
23. "Magnamite Graphite Fibers," Hercules, Inc.
24. HM-HMS Graphite Epoxy Product Data Sheets, Hercules, Inc.
25. D.M. Mazenko and R.J. Milligan, "Epoxy For Structural Composites," SAMPE Technical Conference, Society for the Advancement of Material and Process Engineering, Oct 1980
26. Astro Quartz II Data Product Sheets, J.P. Stevens

Epoxy Resin With and Without E-Glass Fiber Reinforcement (Ref 3-6)

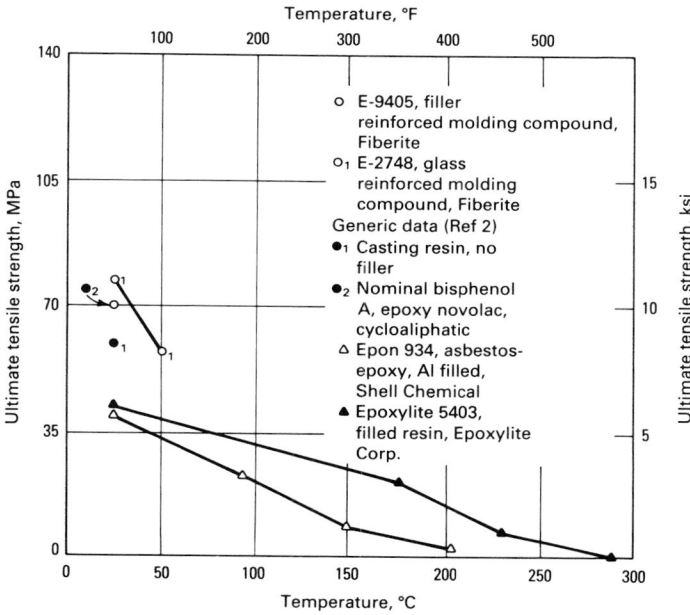

Fig. 1 Ultimate tensile strength versus temperature

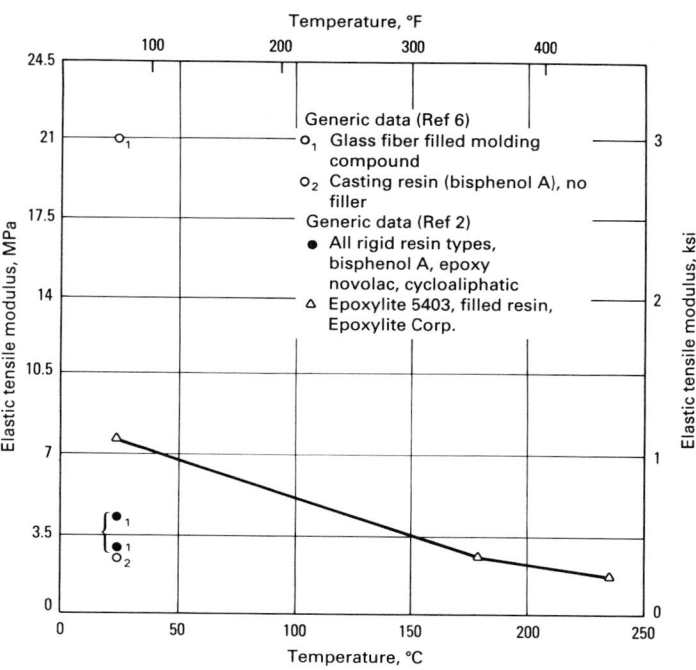

Fig. 2 Elastic tensile modulus versus temperature

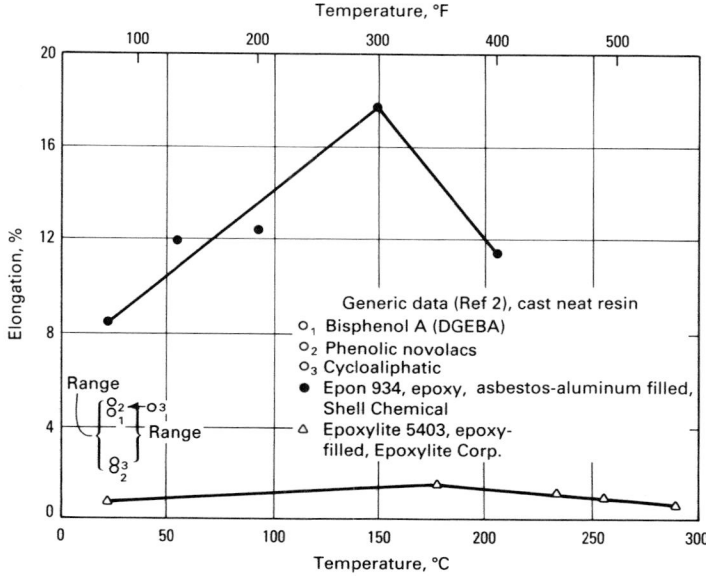

Fig. 3 Tensile elongation percent versus temperature

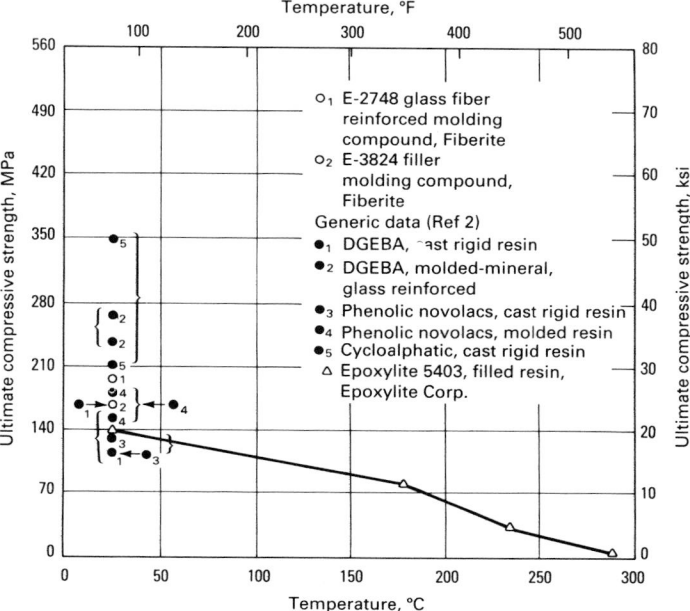

Fig. 4 Ultimate compressive strength versus temperature

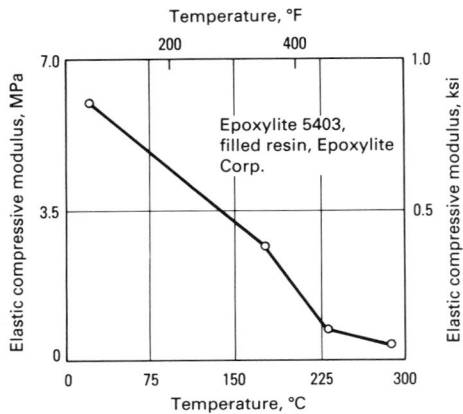

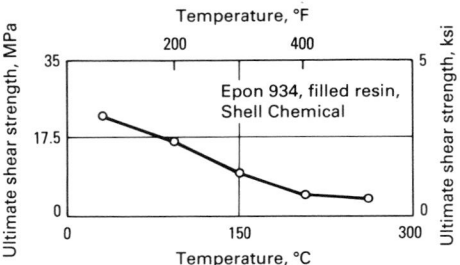

Fig. 6 Ultimate shear strength versus temperature

Fig. 5 Elastic compressive modulus versus temperature

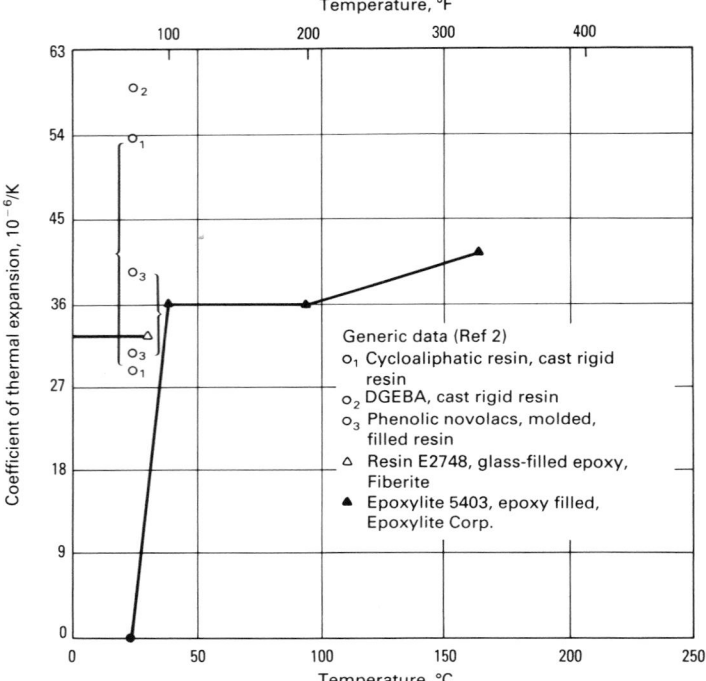

Fig. 7 Coefficient of thermal expansion versus temperature. $10^{-6}/K \times \frac{5}{9} = $ μin./in. × °F

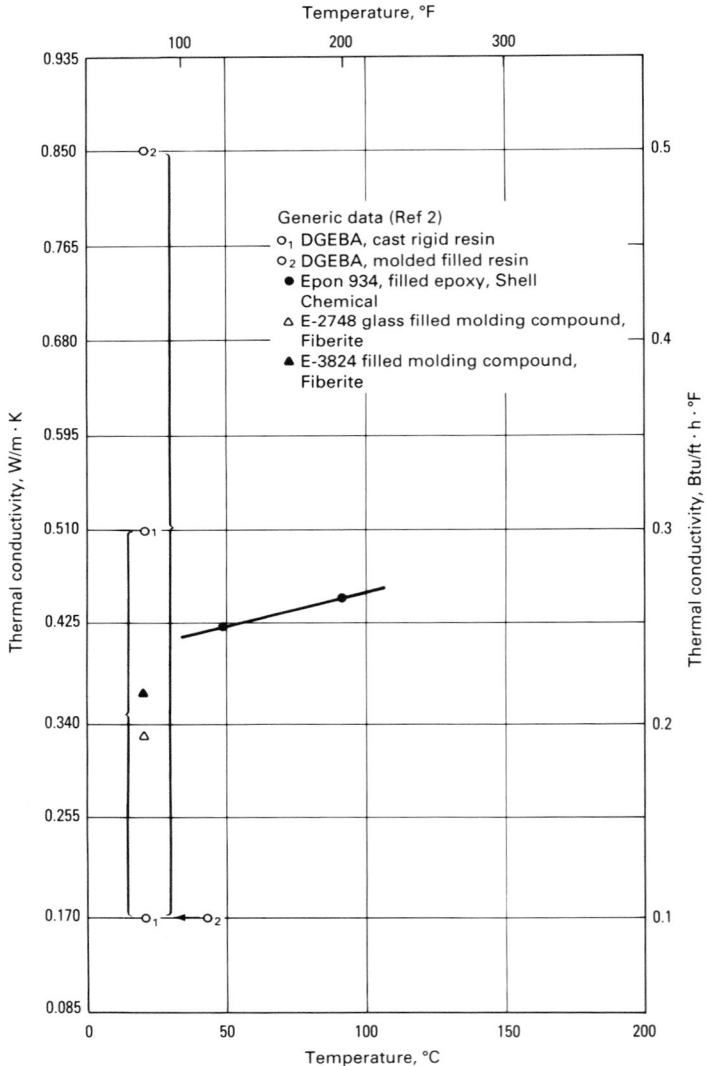

Fig. 8 Thermal conductivity versus temperature

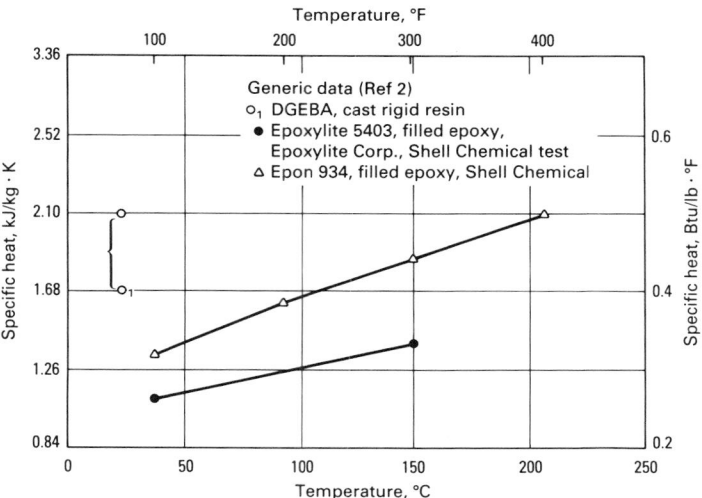

Fig. 9 Specific heat versus temperature

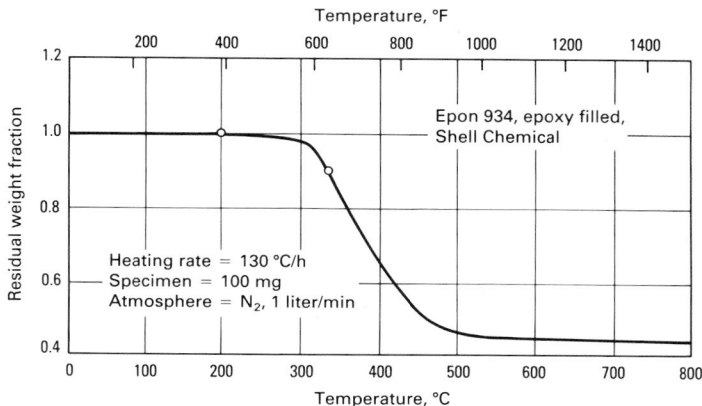

Fig. 10 Thermogravimetric analysis residual weight versus temperature

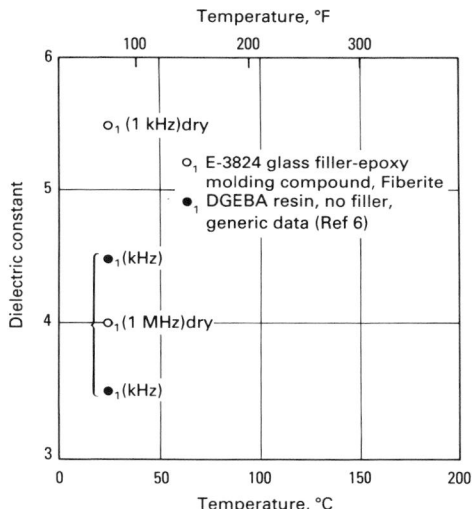

Fig. 11 Dielectric constant versus temperature

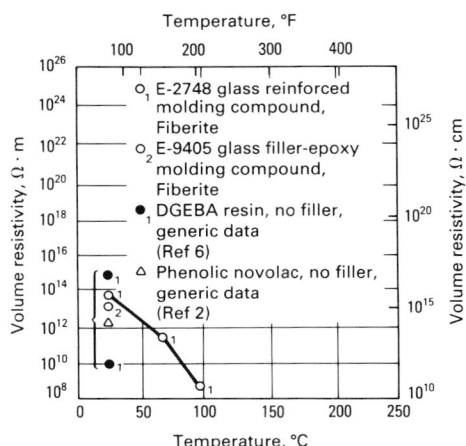

Fig. 12 Volume resistivity versus temperature

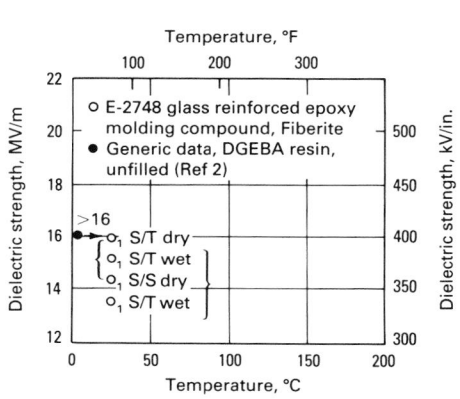

Fig. 13 Dielectric strength versus temperature

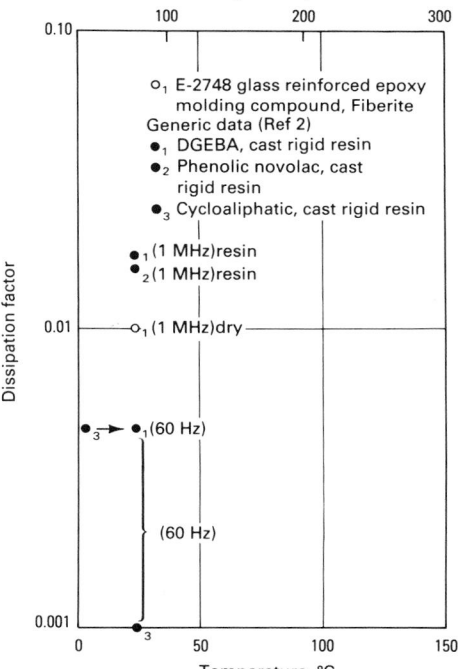

Fig. 14 Dissipation factor versus temperature

Glass Fabric Reinforced Epoxy Resin (Ref 7-11)

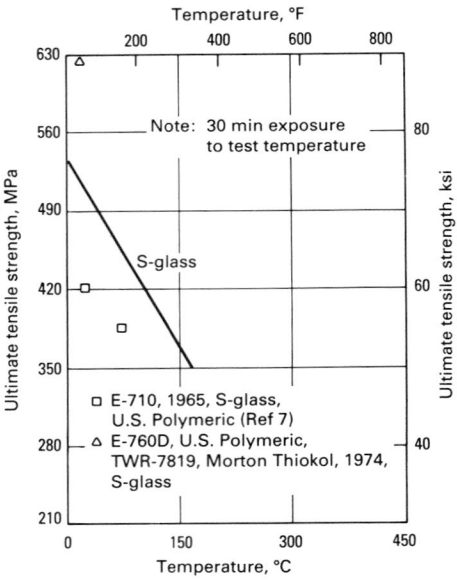

Fig. 15 Ultimate tensile strength versus temperature

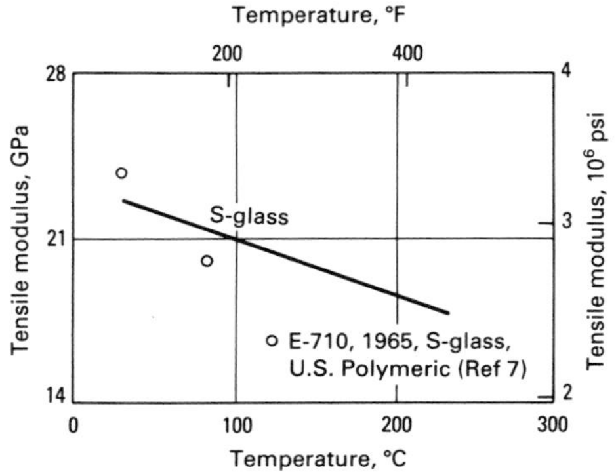

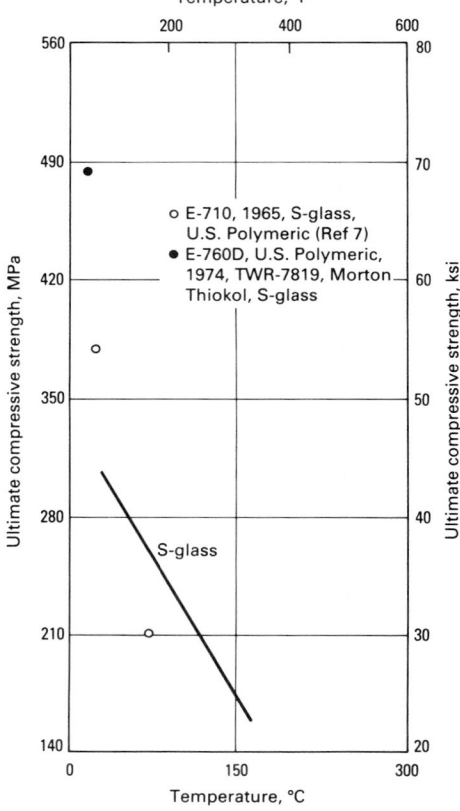

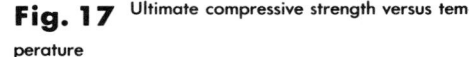

Fig. 16 Tensile modulus versus temperature

Fig. 17 Ultimate compressive strength versus temperature

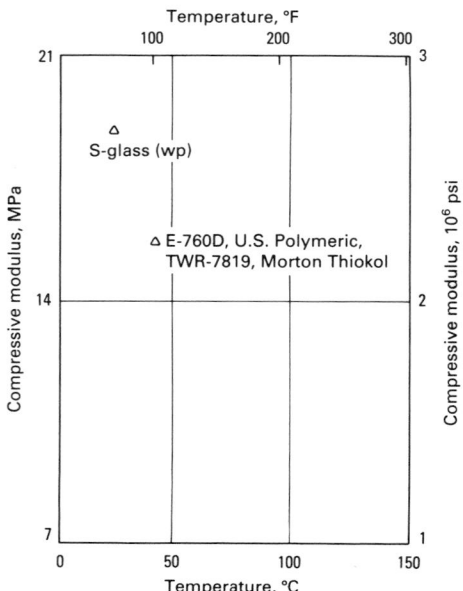

Fig. 18 Compressive modulus versus temperature

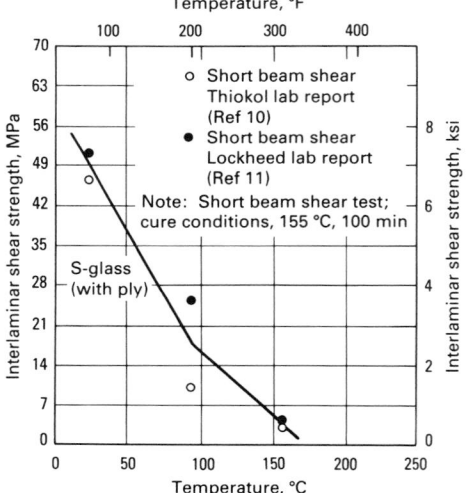

Fig. 19 Interlaminar shear versus temperature

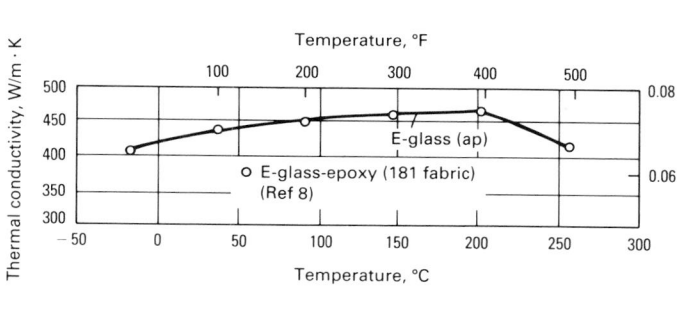

Fig. 20 Thermal conductivity versus temperature

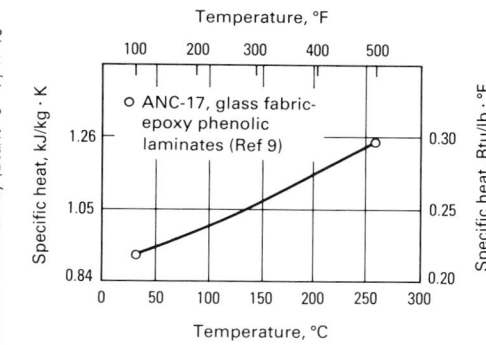

Fig. 21 Specific heat versus temperature

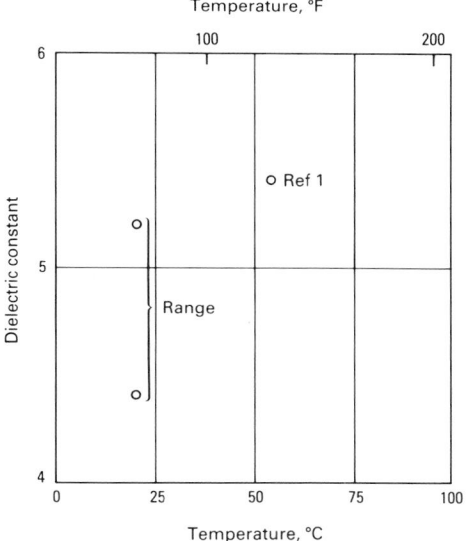

Fig. 22 Dielectric constant versus temperature

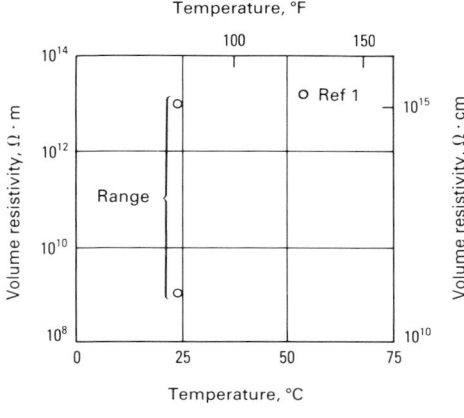

Fig. 23 Volume resistivity versus temperature

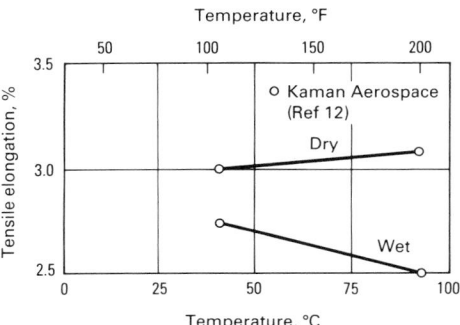

Fig. 24 Dissipation factor versus temperature

S-Glass Fiber Reinforced Epoxy Resin (Ref 12-14)

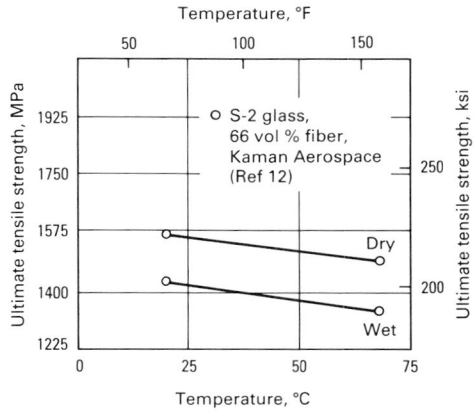

Fig. 25 Ultimate tensile strength versus temperature

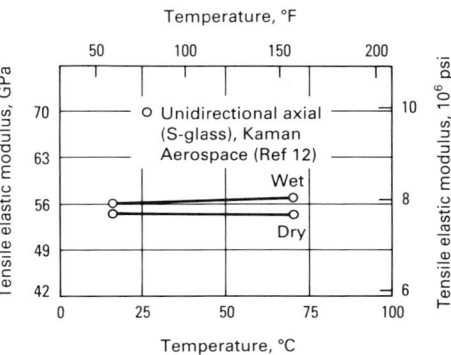

Fig. 26 Tensile elastic modulus versus temperature

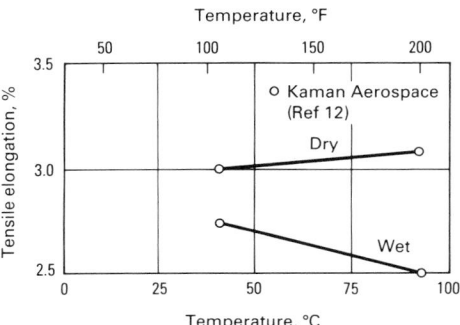

Fig. 27 Tensile elongation percent versus temperature

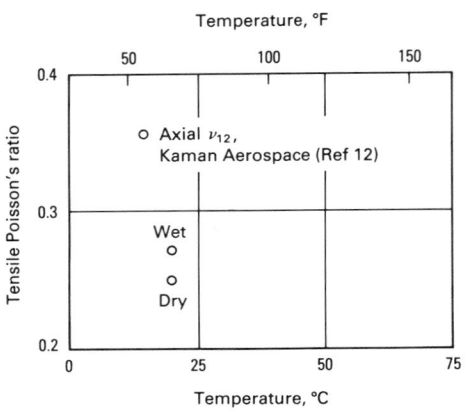

Fig. 28 Tensile Poisson's ratio versus temperature

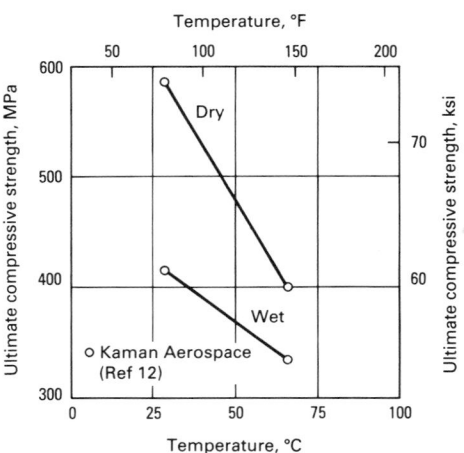

Fig. 29 Ultimate compressive strength versus temperature

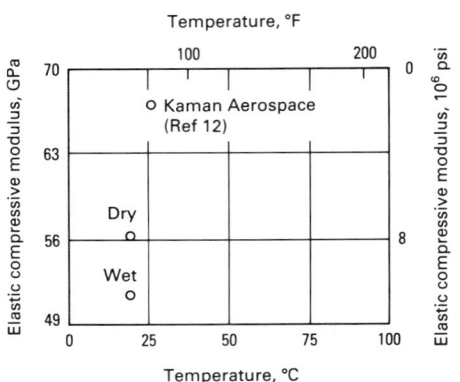

Fig. 30 Elastic compressive modulus versus temperature

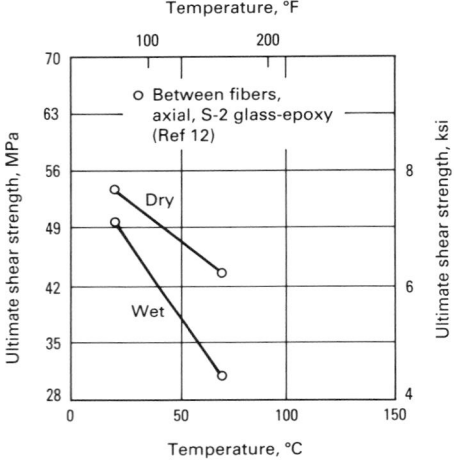

Fig. 31 Ultimate shear strength versus temperature

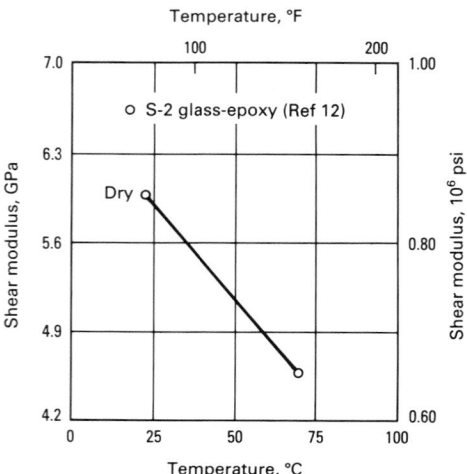

Fig. 32 Shear modulus versus temperature

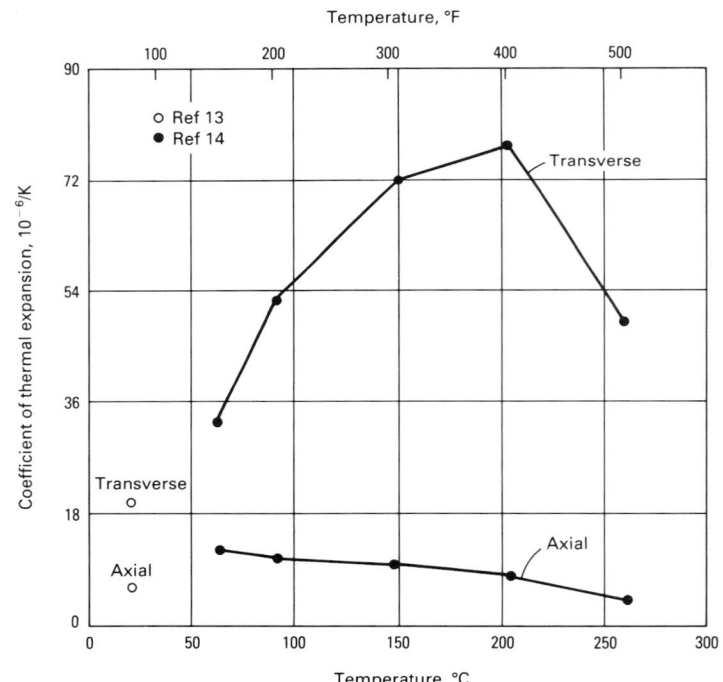

Fig. 33 Coefficient of thermal expansion versus temperature. $10^{-6}/\mathrm{K} \times \frac{5}{9} = \mu\mathrm{in./in.} \times °\mathrm{F}$

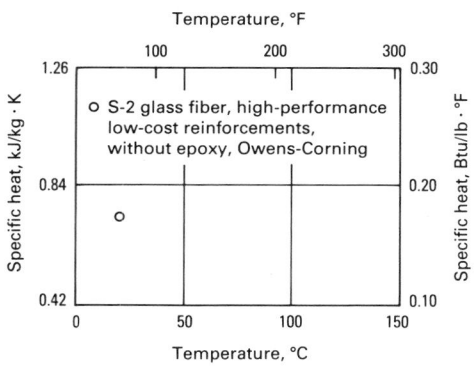

Fig. 34 Specific heat versus temperature

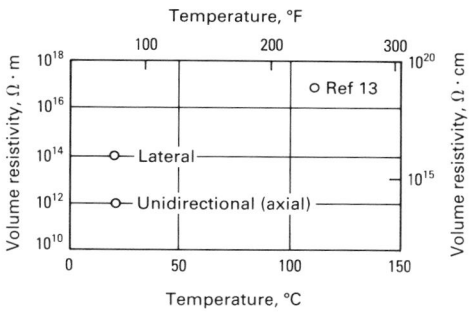

Fig. 35 Volume resistivity versus temperature

Kevlar 49 Fiber and Fabric Reinforced Epoxy Resin for Filament Winding (Ref 15)

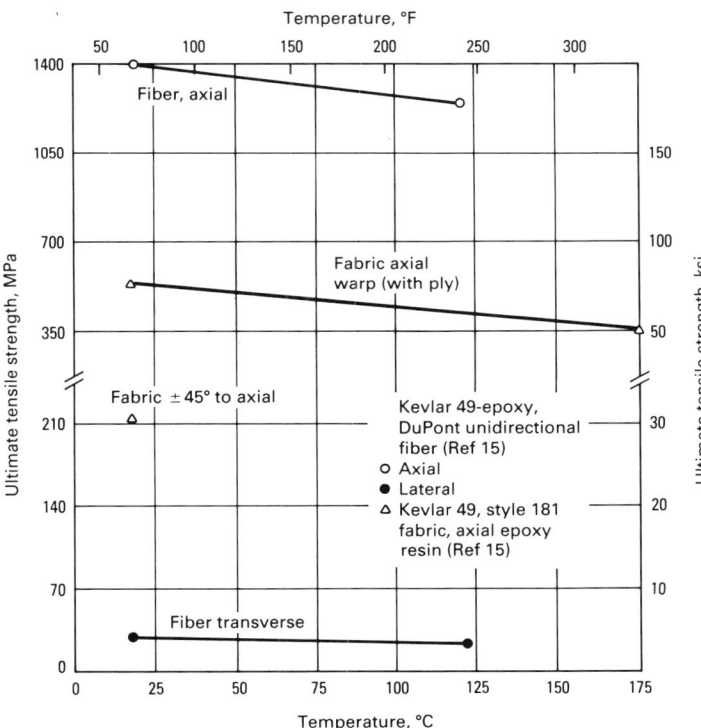

Fig. 36 Ultimate tensile strength versus temperature

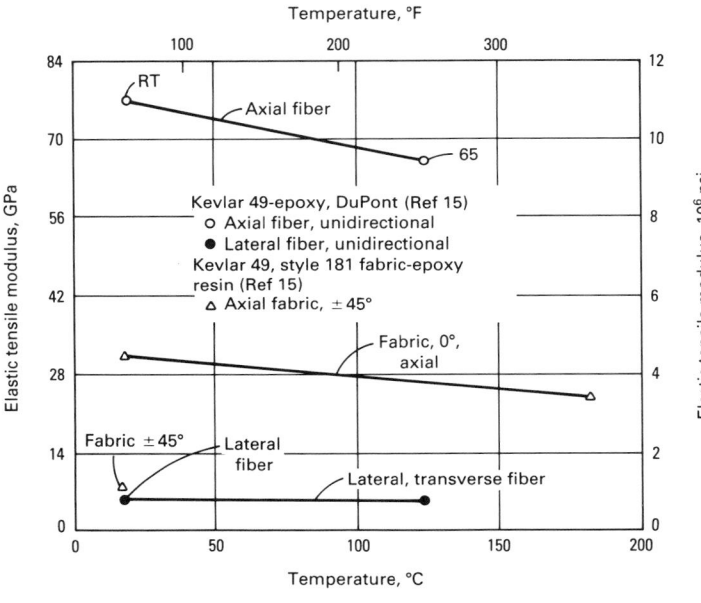

Fig. 37 Elastic tensile modulus versus temperature

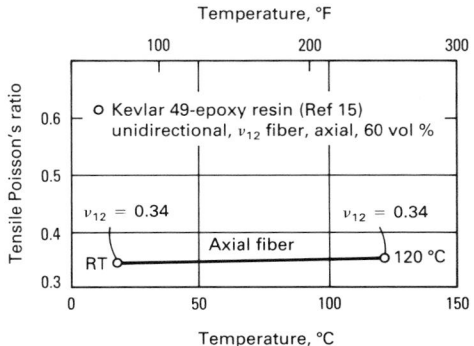

Fig. 38 Tensile Poisson's ratio versus temperature

Note: Tensile elongation at room temperature for axial fiber = 1.85%, for axial fabric = 1.78%, for transverse lateral fiber = 0.58%

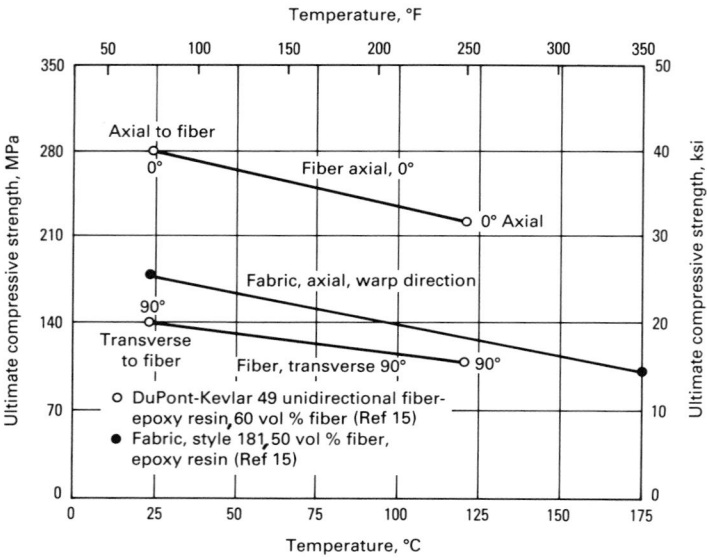

Fig. 39 Ultimate compressive strength versus temperature

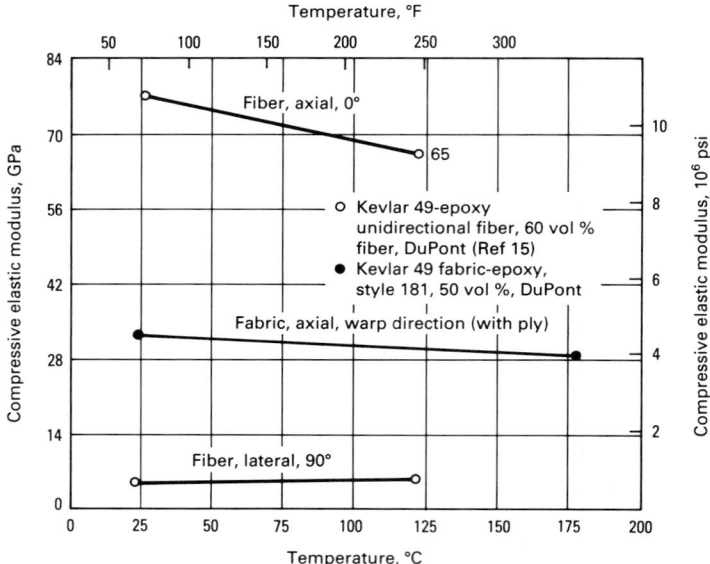

Fig. 40 Compressive elastic modulus versus temperature

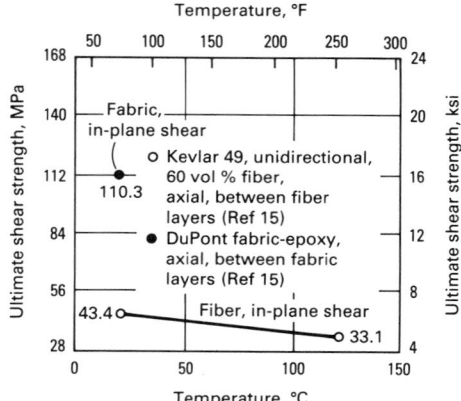

Fig. 41 Ultimate shear strength versus temperature

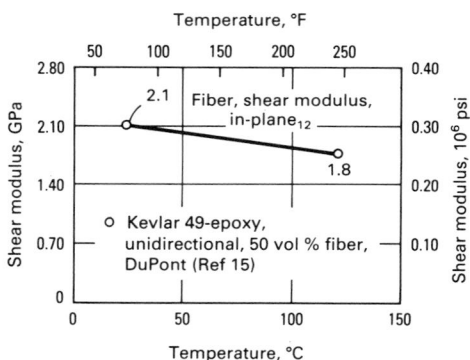

Fig. 42 Shear modulus versus temperature

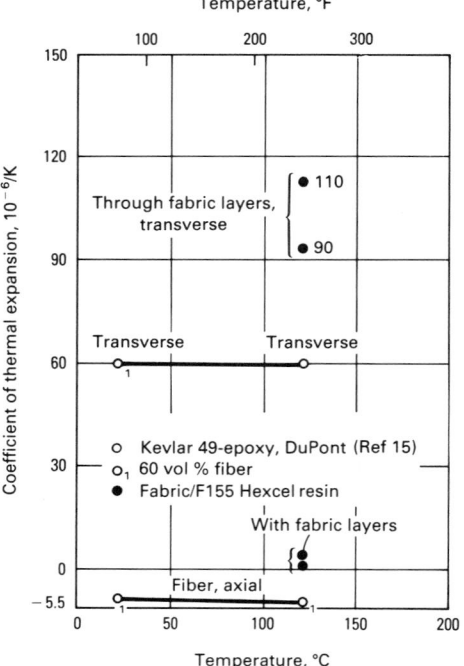

Fig. 43 Coefficient of thermal expansion versus temperature. $10^{-6}/K \times \frac{5}{9} = \mu in./in. \times °F$

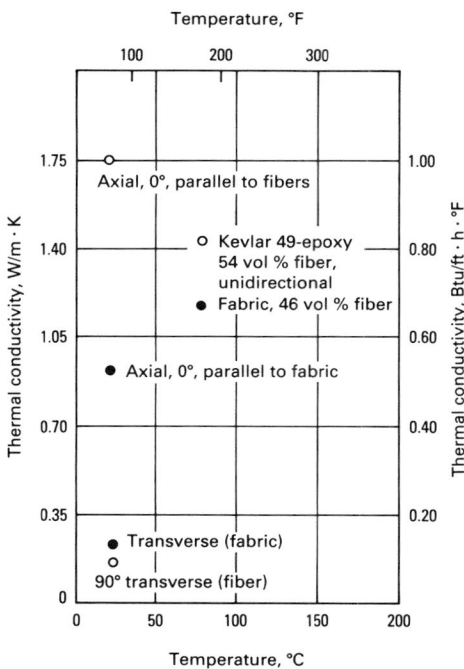

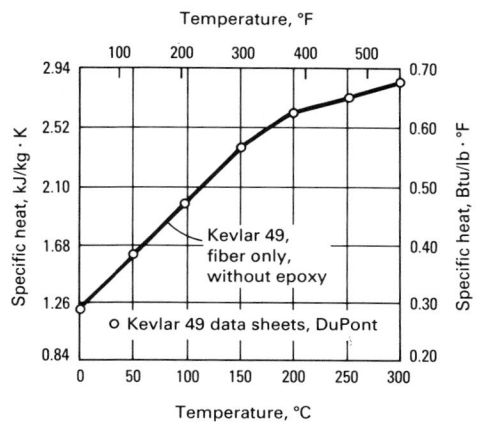

Fig. 45 Specific heat versus temperature

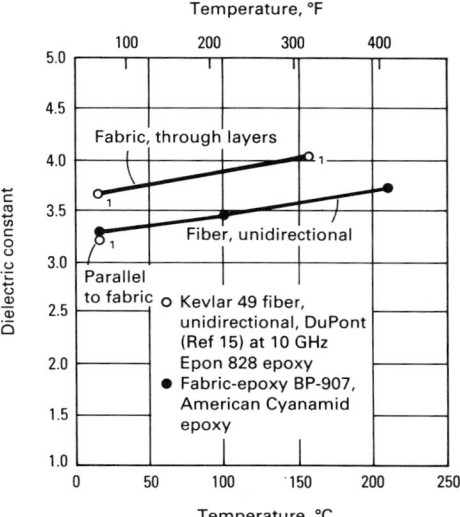

Fig. 46 Dielectric constant versus temperature

Fig. 44 Thermal conductivity versus temperature

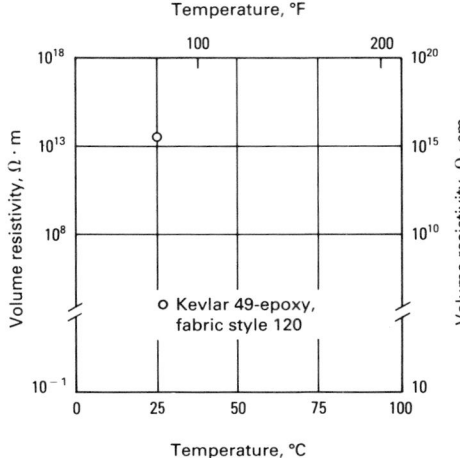

Fig. 47 Volume resistivity versus temperature

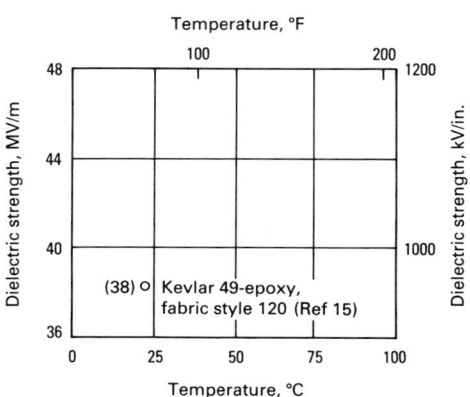

Fig. 48 Dielectric strength versus temperature

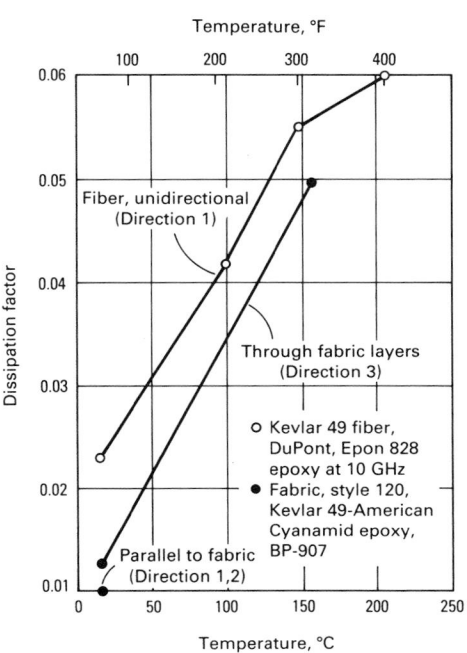

Fig. 49 Dissipation factor versus temperature

Carbon Fiber and Fabric Reinforced Epoxy Resin (Ref 16-22)

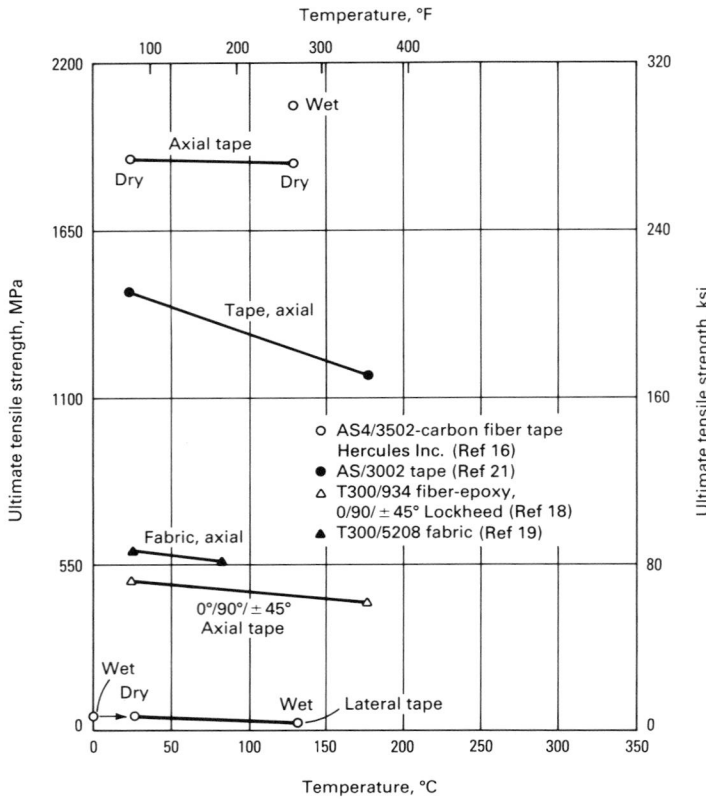

Fig. 50 Ultimate tensile strength versus temperature

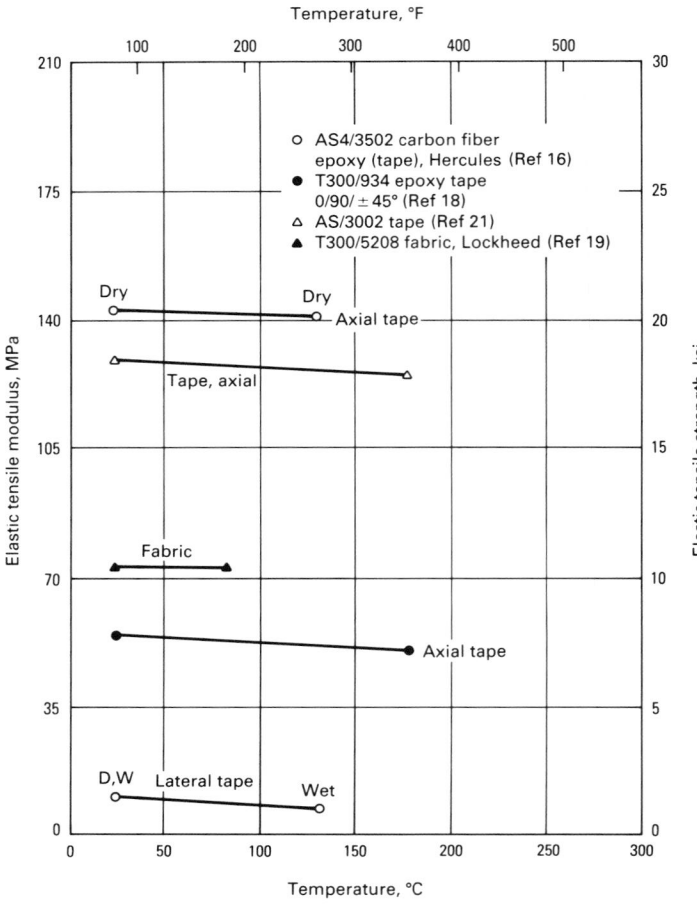

Fig. 51 Elastic tensile modulus versus temperature

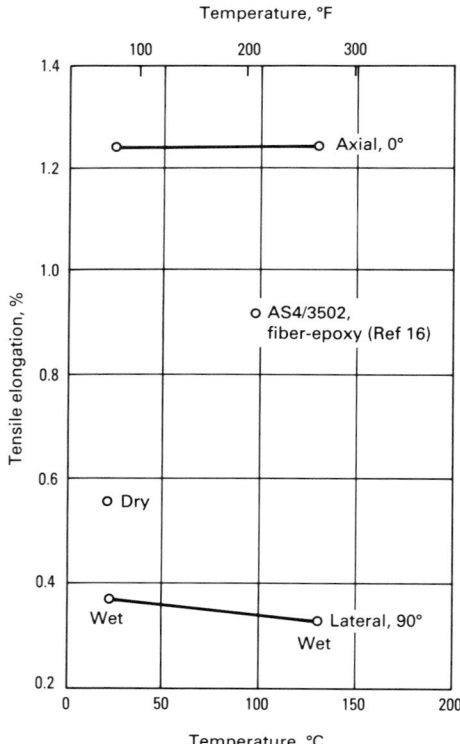

Fig. 52 Tensile elongation versus temperature

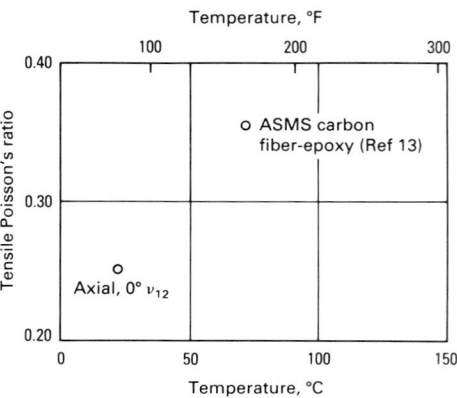

Fig. 53 Tensile Poisson's ratio versus temperature

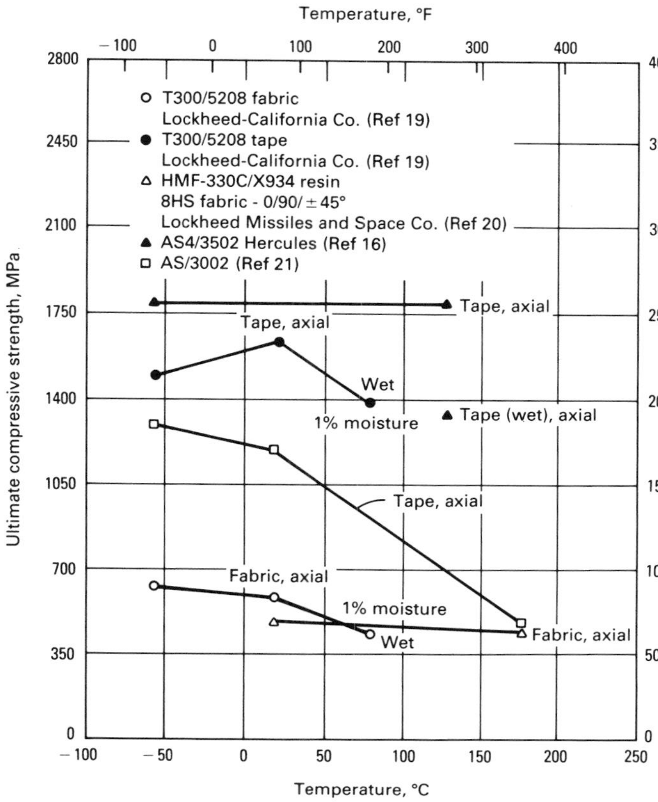

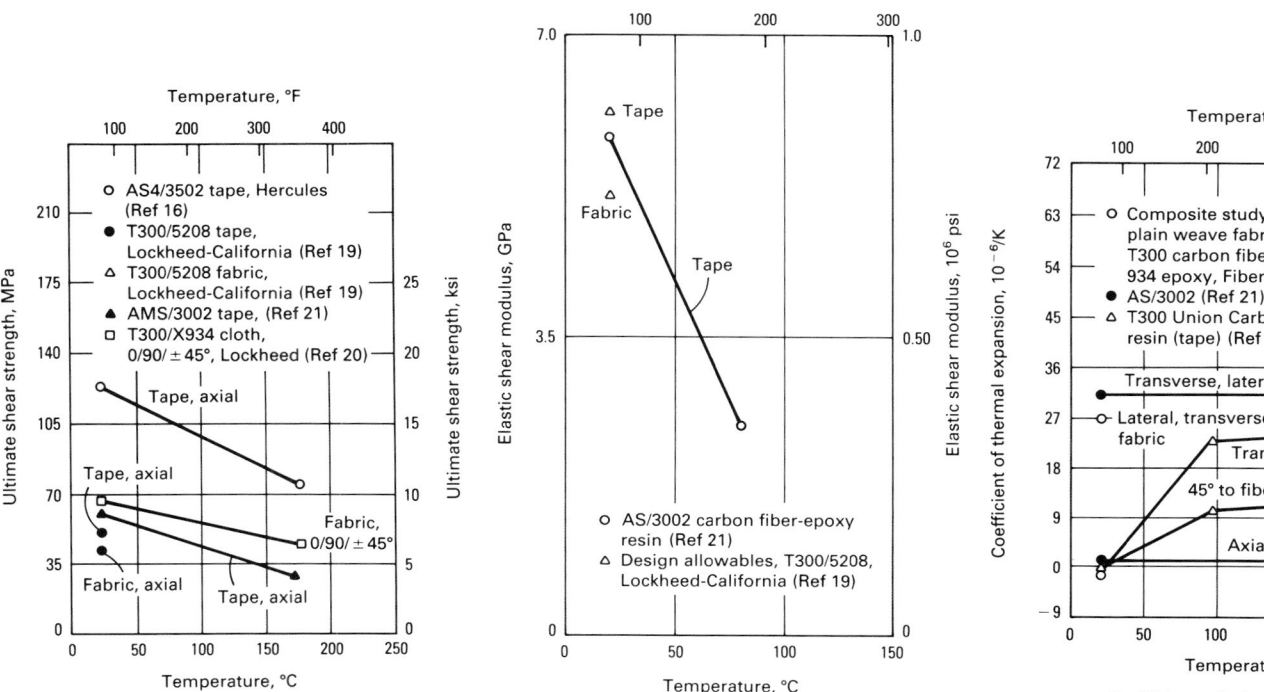

Fig. 54 Ultimate compressive strength versus temperature

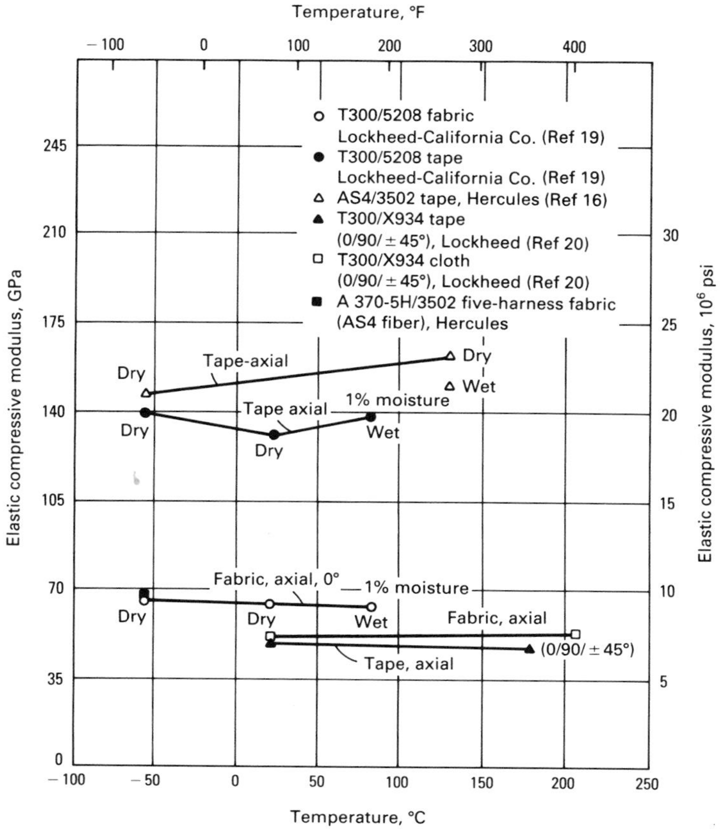

Fig. 55 Elastic compressive modulus versus temperature

Fig. 56 Ultimate shear strength versus temperature

Fig. 57 Elastic shear modulus versus temperature

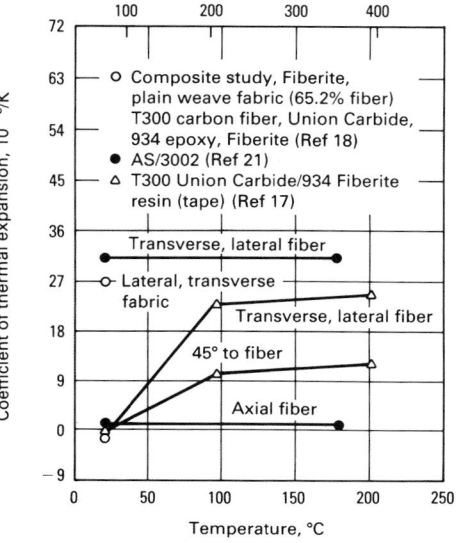

Fig. 58 Coefficient of thermal expansion versus temperature. $10^{-6}/K \times \frac{5}{9} = \mu in./in. \times °F$

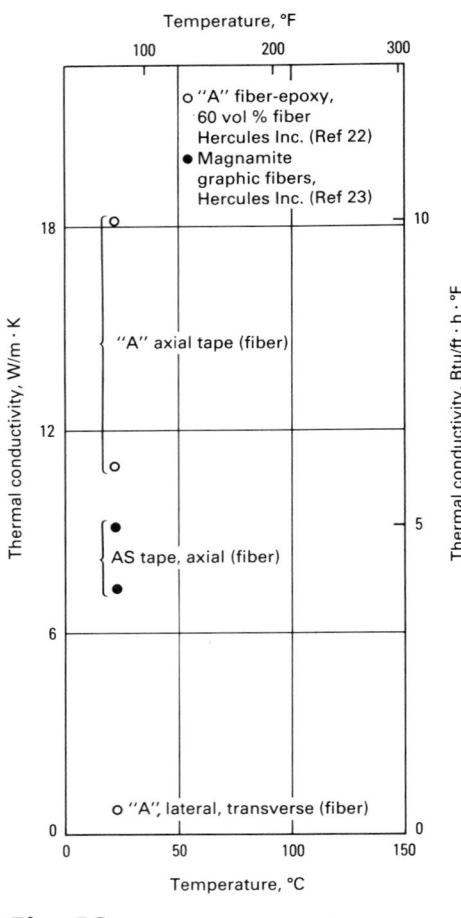

Fig. 59 Thermal conductivity versus temperature

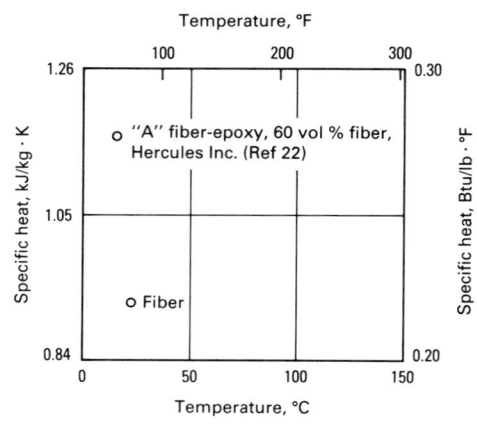

Fig. 60 Specific heat versus temperature

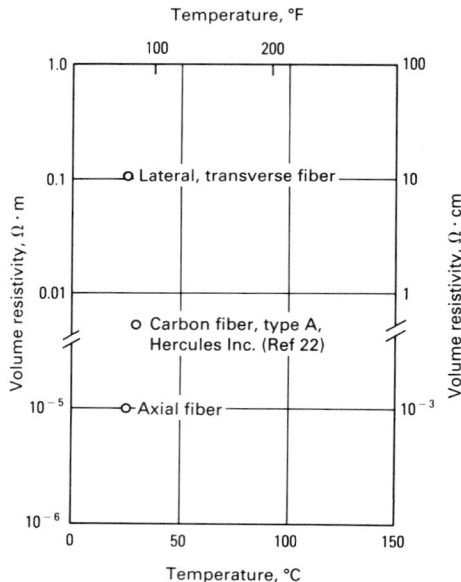

Fig. 61 Volume resistivity versus temperature

Graphite Fiber Reinforced Epoxy Resin (Ref 23-25)

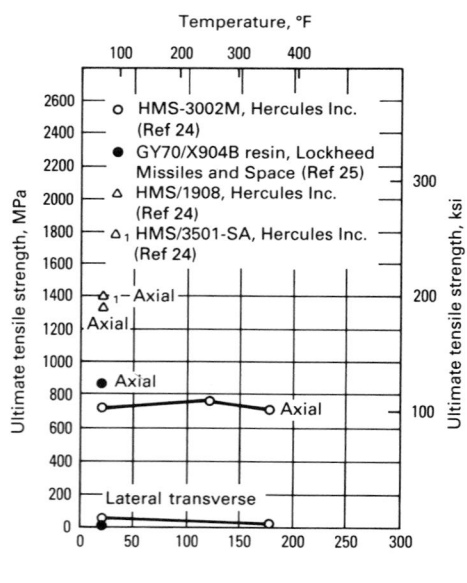

Fig. 62 Ultimate tensile strength versus temperature

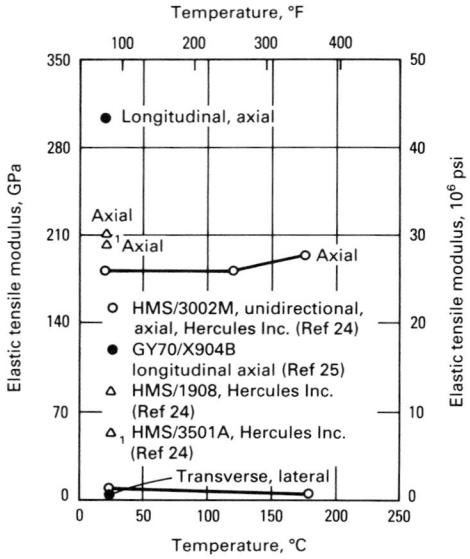

Fig. 63 Elastic tensile modulus versus temperature

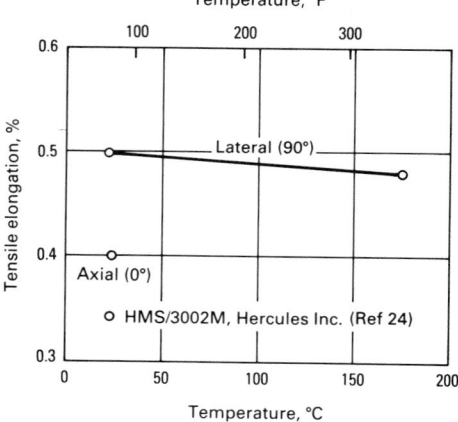

Fig. 64 Tensile elongation versus temperature

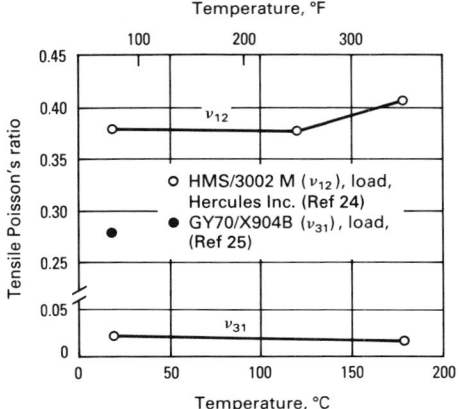

Fig. 65 Tensile Poisson's ratio versus temperature

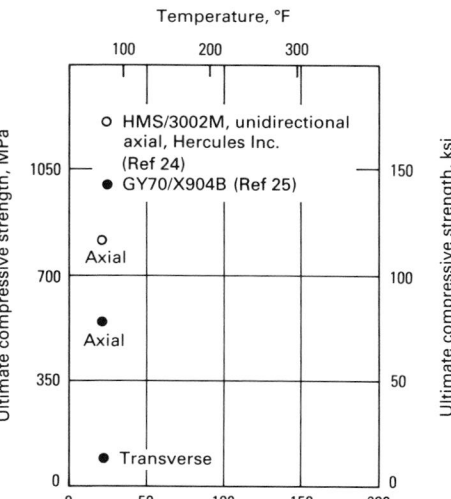

Fig. 66 Ultimate compressive strength versus temperature

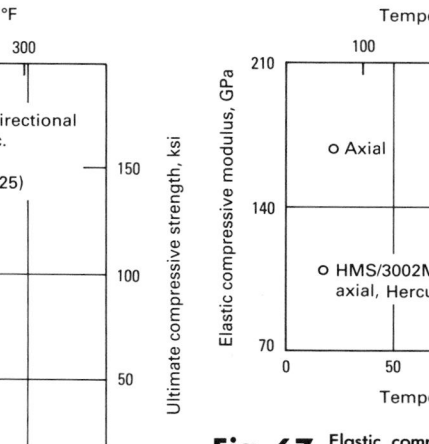

Fig. 67 Elastic compressive modulus versus temperature

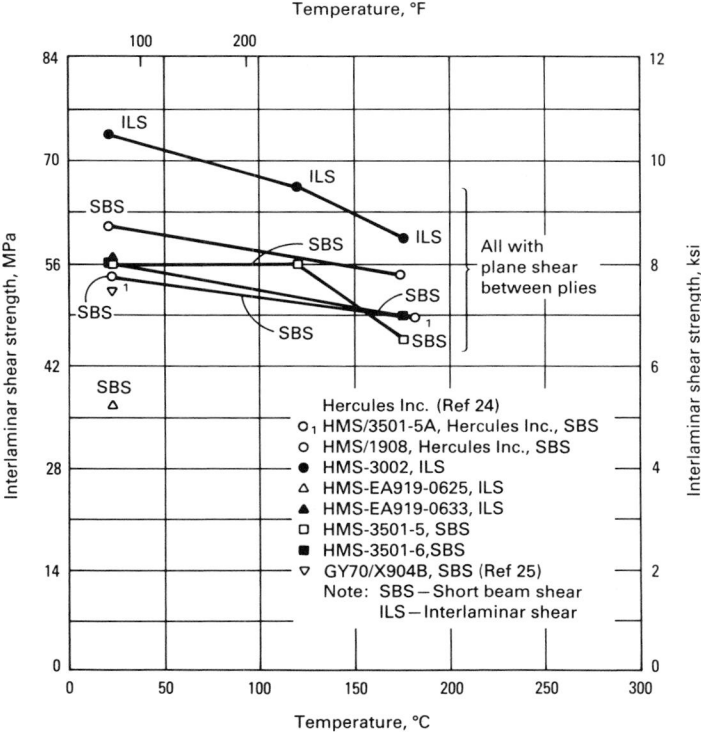

Fig. 68 Interlaminar shear versus temperature

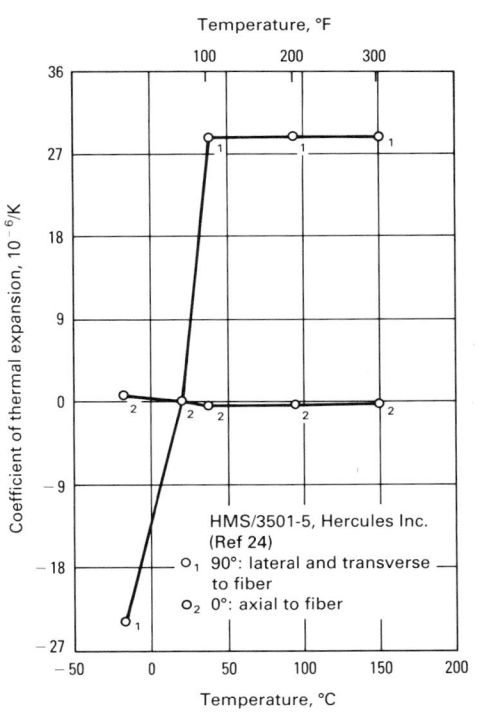

Fig. 69 Coefficient of thermal expansion versus temperature. $10^{-6}/K \times 5/9 = \mu in./in. \times °F$

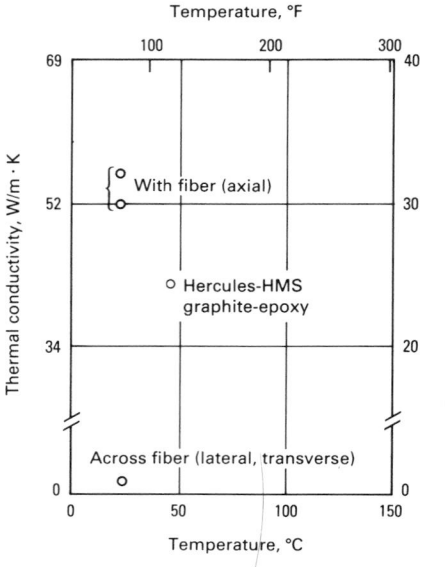

Fig. 70 Thermal conductivity versus temperature

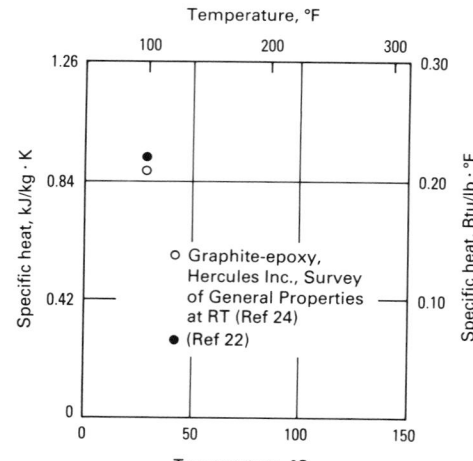

Fig. 71 Specific heat versus temperature

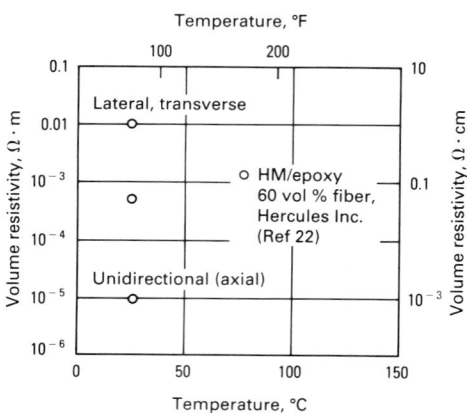

Fig. 72 Volume resistivity versus temperature

Quartz Fabric Reinforced Epoxy Resin (Ref 26)

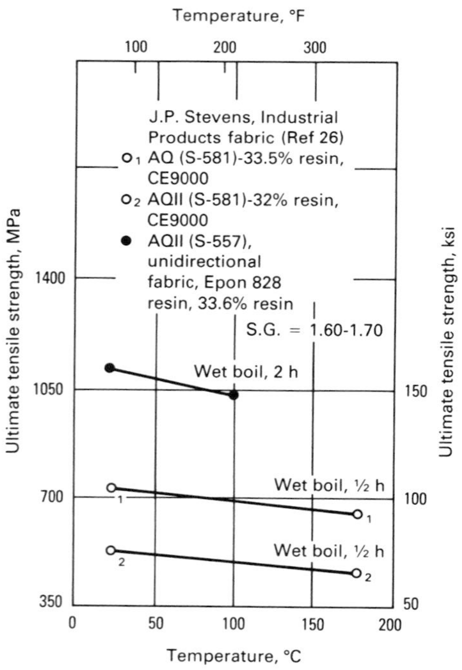

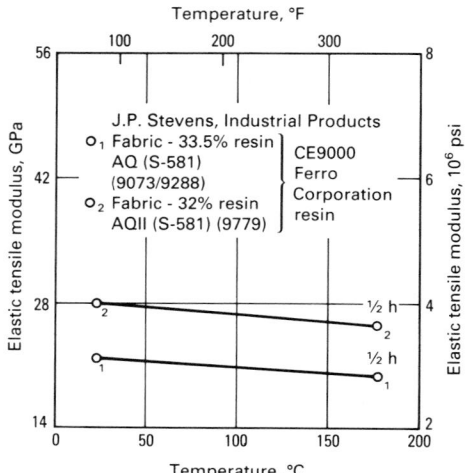

Fig. 74 Elastic tensile modulus versus temperature

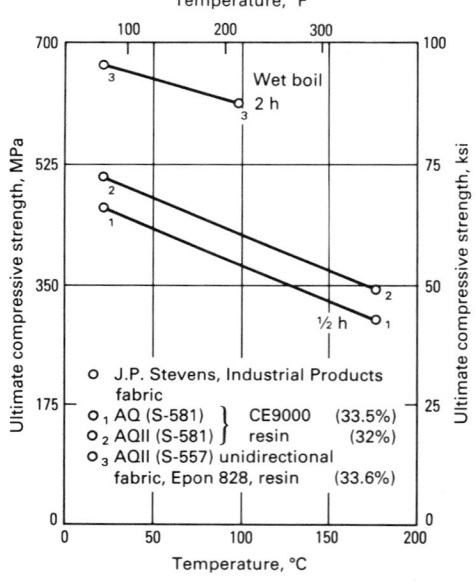

Fig. 75 Ultimate compressive strength versus temperature

Fig. 73 Ultimate tensile strength versus temperature

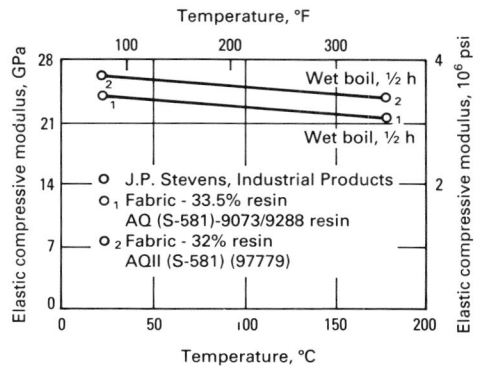

Fig. 76 Elastic compressive modulus versus temperature

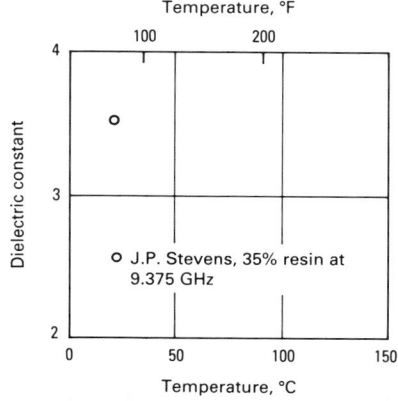

Fig. 77 Dielectric constant versus temperature

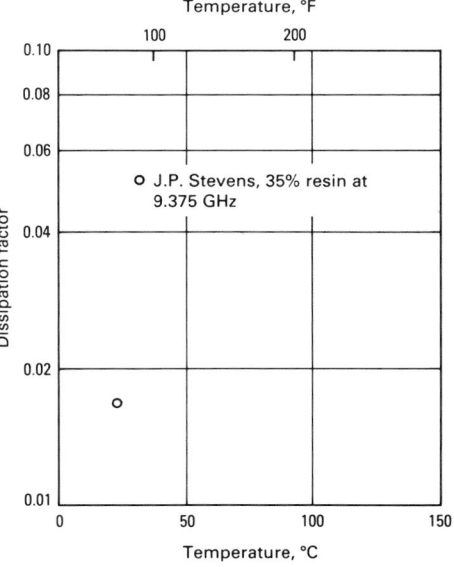

Fig. 78 Dissipation factor versus temperature

Composite Structures Analysis and Design

Chairman: Scott W. Beckwith, Hercules Aerospace Company

Introduction

APPLYING COMPOSITE materials to full-scale structures means using analysis and design techniques that are more encompassing and sophisticated than those required for metal applications. This introduction provides an overview of the interactive concepts that are important to the analysis and design of composite structures for a variety of structural applications:

- Cost drivers in the design and manufacturing processes
- Interfaces in design, tooling, and manufacture
- Static strength considerations and limitations
- Fatigue strength assessment
- Instability considerations
- Laminate ranking as a design tool for sizing laminates
- Analysis of composite structures
- Numerical design and analysis
- Joint design

In the design and analysis of practical composite structures, one factor that should not be overlooked is cost. The first article in this section describes the strong interaction between design, manufacture, and cost, with emphasis on approaches that are geared toward product life cycle. The designer is shown to have a strong influence on cost considerations during both the design and manufacturing phases.

The second article, on interfaces, discusses the problems associated with the communication link between the designer and those responsible for the actual tooling and manufacture of the product. Formerly, the design analyst was removed from problems associated with actual fabrication, but this is no longer the case, and this article defines the areas in which the analyst plays a key role.

Static strength and fatigue of composite structures are key elements of the two succeeding articles. Static strength is evaluated on three scales: micro, mini, and macro. The various zones, their limitations, and their input to the structural design are defined. Fatigue and multiple load-cycling effects, which are often dominant design drivers in aircraft composite structures, are discussed in depth.

Instability considerations in the design phase, using flat plate and cylindrically curved shell panels, are discussed in the next article. Buckling analysis for orthotropic materials is formulated in terms of the laminate properties necessary for structural design. Considerations during the postbuckling phase also are discussed.

The article on laminate sizing offers an approach for improving design efficiency by potentially optimizing the overall laminate design based on strength and stiffness considerations as well as lay-up and orientation. Two computer programs and subroutines available to the design analyst are shown to provide alternatives to conventional optimization methods.

The article on the analysis of structures describes the contributions made using classical laminated plate theory to determine ply and laminate stiffnesses for structural design input.

Numerical design and analysis is the subject of an article that summarizes the state of the art in composite analysis, with an in-depth discussion on closed-form and finite-element methods. The article covers analysis methods and their results and limitations, and then focuses on practical design applications. Several numerical computer codes are compared in terms of their contribution to composite structures analysis.

Because joints are a key area of composite structural design, joint design considerations, analysis techniques, and experimental verification for practical applications are presented in the final article of this Section.

A discussion of the computer programs available for macrostructural and microstructural analysis is provided in the article "Computer Programs for Structural Analysis" in the Section "Composite Materials Analysis and Design" in this Volume.

Cost Drivers in Design and Manufacture of Composite Structures

Bryan R. Noton, Battelle—Columbus Division

THE NEED TO REDUCE COSTS at all levels of product life cycle is becoming increasingly important. Qualitative and quantitative data on cost drivers applicable to manufacture, operation, and maintenance of products are essential. Cost must be considered throughout the design process because the reduction in the number of high-performance engineering products and systems has resulted in increased performance requirements, such as reduced weight, higher quality, and lower energy consumption, but at lower acquisition and ownership costs. Affordable performance of future products and systems is an important goal.

A realistic perspective on cost effectiveness is given by W.B. Goldsworthy (Ref 1), who states that in all too many cases it is merely a rationalization to justify the production of any overly expensive item. Goldsworthy stresses the need for "designing for economics versus property optimization, evolving toward better balance between economics and properties."

Improved performance of products, within cost limitations, depends upon engineering design excellence. Affordable performance depends upon both designers and manufacturing engineers recognizing cost drivers and controlling them in new designs, and improving manufacturing methods for existing products. Cost drivers can be controlled by the design-to-cost (DTC) process. Early identification of cost drivers and corrective action for existing and new products can also be facilitated by proficiency in the manufacturing-to-cost (MTC) process.

Cost drivers are more easily understood when related to appropriate categories of system development:

- Concept/performance requirements
- Design
- Material selection
- Manufacturing

Within these categories, cost drivers typical of a mechanical system are:

- Performance related
  Reduced weight
  Higher operating speeds
  Increased reliability
  Improved maintainability
- Design related
  Part count
  Nonstandardization
  Special tolerances
- Material related
  Cost
  Availability
  Utilization level
  Energy requirements
  Inventory
- Manufacturing related
  Cyclic production
  Small lot size
  Job shop environment
  Highly skilled labor
  Material removal
  Deburring/hand-finishing
  Hand fit-up
  Energy requirements
  Facilities
  Qualification
  Test, inspection, and evaluation

The individual designer seldom has the training or experience to conduct structural performance/manufacturing cost trade-off studies. However, today's designers are rated not only on ingenuity in meeting weight, reliability, and cost objectives, but also on achieving these objectives within schedule limitations. Designing to the lowest cost is now an important discipline. Design teams must be provided with the tools for identifying and documenting cost drivers, evaluating cost reduction methods in design and manufacture, and determining cost targets. Incentives must be provided for achieving cost targets against which their performance can be measured.

The designer has an important resource for determining cost: the cost estimator, whose experience is very significant in the final iteration of the design before production commit-

ment. However, it is sometimes difficult to evaluate an adequate number of design choices in order to select with confidence the lowest-cost design alternative, while still meeting scheduling requirements. The importance of reducing cost to achieve the structural revolution based on advanced composites was stressed by R.C. Forney (Ref 2). Recognizing that composite fabrication techniques and manufacturing equipment must improve, Forney discusses the necessity for conceptual changes in manufacturing to be closely allied to those in design. Vigorous interaction between design and manufacturing is therefore essential to achieve the lowest cost.

While cost reduction efforts are essential at all levels of the design process, the importance of the preliminary design phase, the "window of opportunity," needs to be stressed. Figure 1 illustrates how the leverage for achieving cost savings dramatically decreases as the development of a system progresses. The preliminary design phase provides the principal opportunity to achieve a low-cost design; innovative materials, design concepts, and manufacturing technologies can significantly impact cost. In the early design phase, the designer must:

- Consider cost as a primary design objective
- Identify cost drivers in early decisions
- Provide designers with meaningful cost data at the inception of the effort
- Increase the number of performance/cost trade-offs of alternative designs
- Determine realistic system mission costs
- Determine the cost of changes in design objectives and engineering solutions
- Improve interaction between design and other disciplines

Configuration selection often offers a spectacular opportunity to reduce costs. As Fig. 1 shows, at the early design phase only a small percentage of the program costs have been expended, yet decisions are made which influence 90 to 95% of the total cost, including operations and maintenance costs.

The ability to achieve low cost by means of design concepts that facilitate simplicity in fabrication and permit use of existing capital equipment is described in Ref 3. Detailed trade-off studies between metallic (as the baseline) and various composite designs are made. As the program progresses through detail design and production, it becomes increasingly difficult to reduce the cost by more than a few percent, even with new materials. As soon as the detail design phase has been entered, the majority of components considered for redesign (to utilize alternative advanced materials or manufacturing processes) must meet form, fit, and function requirements of the part being replaced. The process economics of thermoplastics versus thermosets for aerospace and automotive exterior panels are discussed by J.M. Margolis in Ref 4, including the influence of the detail design concept on panel fabrication economics. Figure 2 shows the cost impact of decisions as a function of their number. The major milestones throughout the development process, in this case, of an aircraft system committed to production, are indicated.

Material Cost Drivers

The following are cost drivers for high-performance composite materials and represent areas in which further research and development is required:

- Lamina (fiber, resin, and prepreg)
- Material identification and tracking
- Material storage and shelf-life requirements
- Improved three-dimensional properties
- Nonautomated production
- Autoclave costs (dedicated autoclaves designed for one cycle may alleviate problems)
- Finishing requirements
- Paint removal
- Inventory
- Improved standards
- Waste disposal of liquid/solids required for large structures
- Corrosion of lightning-protection materials
- Capital investment for large-assembly tooling
- Refurbishment of manufacturing and inspection tools
- Special fasteners
- Long lead-time for most types of equipment
- Large-area nondestructive inspection
- High personnel turnover
- High ratio of supervisors/labor for hand lay-up operations
- Training requirements for supportability
- Repairs (new and simpler concepts)
- Acquisition of experience for operations and maintenance personnel

Lot Size Considerations

Selection of a lot size has a significant impact on the economics of many types of manufacturing processes. Products are manufactured in

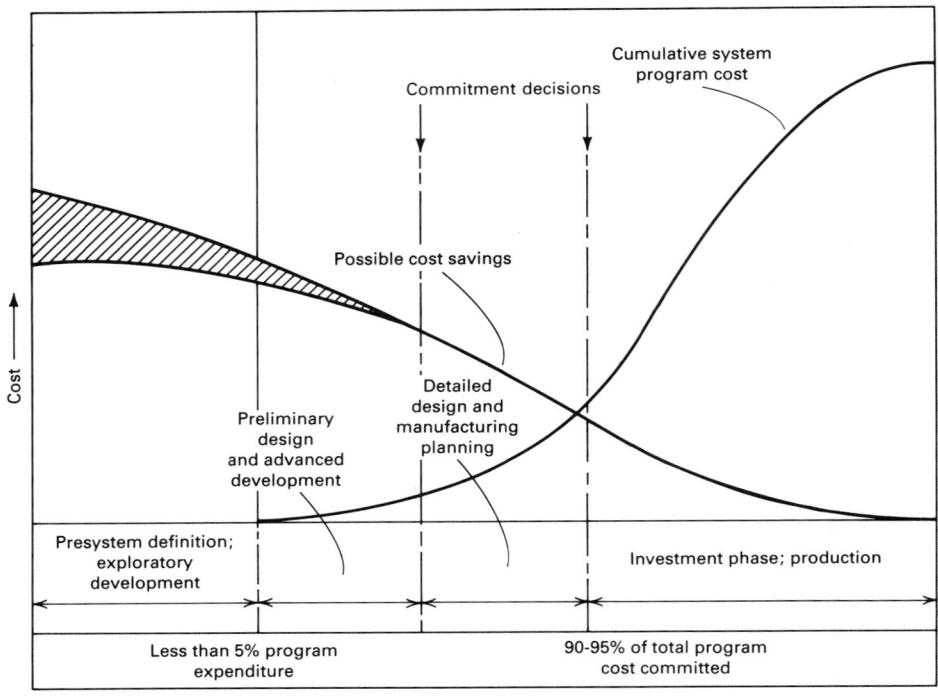

Fig. 1 Decreasing leverage for cost savings as program progresses

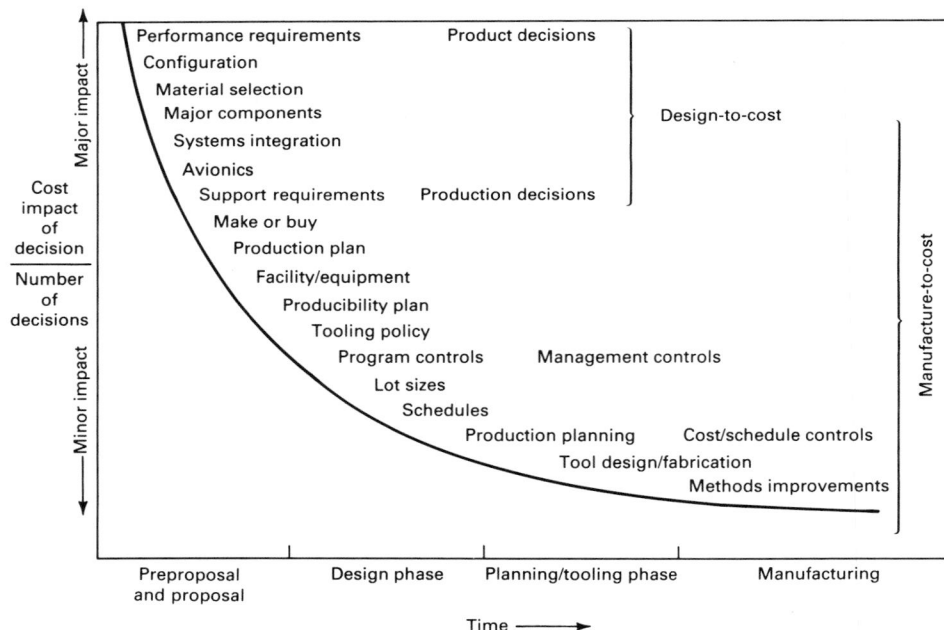

Fig. 2 Impact of decisions on cost

lots, batches, or blocks depending on customer schedule (inventory management), customer or manufacturer's requirement to maintain configuration control for a specified number of units, incremental funding, and the need to achieve the most efficient manufacturing process.

Although the product determines the type of line flow and lot method of manufacture, most methods fall into one or more of these categories: lot (batch) details/assemblies, lot/line details/assemblies, and line-type assemblies.

The following contingencies indicate the importance of lot size in manufacturing:

- New programs with large lots can result in excessive scrap rates due to engineering and tooling changes
- Inadequate allowance for scrap can result in shortages, schedule delays, and uneconomical runs
- Large lots are more cost effective because set-up time is spread over many parts
- Having parts available when required results in smooth running operations
- Efficient use of facilities and equipment depends upon well-planned lot sizes
- Large lot sizes help justify tooling and degree of automation
- Operators achieve a more favorable learning curve with large lot sizes
- A high degree of interchangeability is feasible with large lot sizes and production quantities

The prime factors in determining lot size are:

- Total number of completed deliverable products on order for projected firm business
- Contract delivery schedule
- Material availability
- Fiscal or incremental funding
- Configuration control requirements
- Necessity for an efficient learning curve and build up rate
- Average flow time for parts in manufacturing
- Set-up versus run-time
- Equipment loading and capacity
- Economical order quantity
- Size and part complexity
- Commonality of parts
- Transportation and storage of parts in process

There are, however, secondary factors that also influence lot size selection, including projected spares requirement, estimated scrapped or lost parts, and anticipated impact of change traffic.

Design Impact on Lot Size. When developing candidate hardware to meet design objectives, the design can influence the selection of the lot size in an effort to promote efficient manufacturing and cost minimization. Some considerations are to:

- Use standard common parts for efficient lot size
- Select the most cost-effective technology, for example, pultrusion versus shape from preimpregnated material
- Anticipate the number of changes affecting configuration and part number (this will limit lot size)
- Freeze the design effort as soon as large lots can be introduced
- Provide the designer with a simple method to evaluate set-up versus run-time to allow evaluation of cost of initial part and of large quantities.

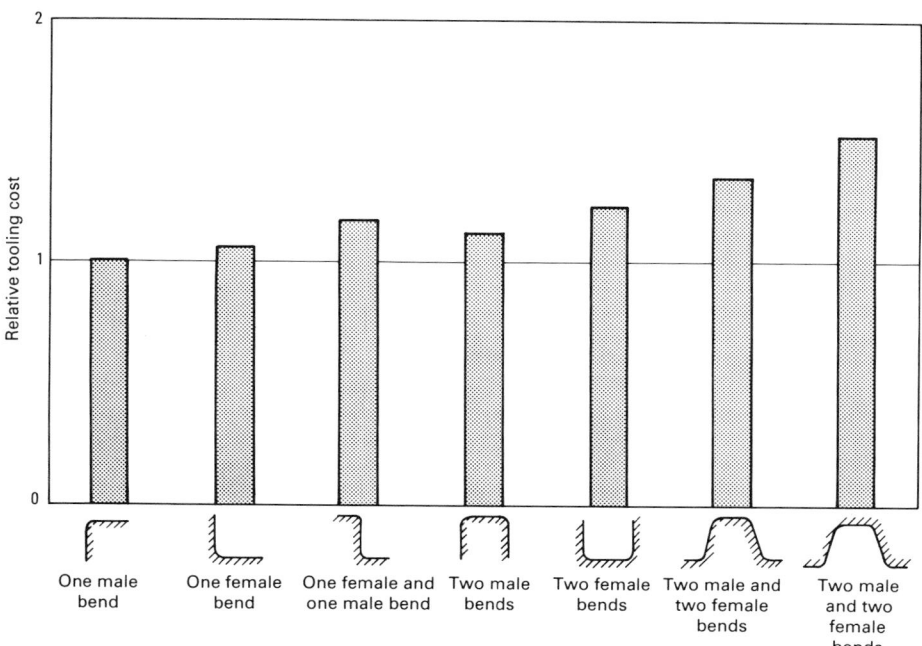

Fig. 3 Typical designer-oriented format showing effects of number of bends, shape, and tool types on tooling cost for carbon-epoxy stiffeners

Test, Inspection, and Evaluation Cost Drivers

Cost drivers for high-performance composite mechanical components and structural elements are the man-hours and equipment required for the various tasks in test, inspection, and evaluation (TI&E). However, considerable commonality exists in the cost drivers attributable to TI&E for several manufacturing technologies. There is also a definite correlation between manufacturing and TI&E cost drivers as a result of the design requirements established by the customer and the engineering operation.

TI&E costs can vary greatly according to the product being manufactured, and in some high-technology industries are significant, ranging from 10 to 30% of the manufacturing cost.

The cost of achieving the required quality varies, of course, according to the product. In airframe and engine manufacture, quality control costs may range from 4 to 5% of total sales, or 8 to 12% of manufacturing costs. On some high-performance aircraft components, such as constant-speed drives and fuel pumps, the inspection or quality control costs can approach 30% of manufacturing costs.

Following is the approximate allocation of the quality control or quality assurance costs for airframe and engine manufacture:

Cost	Airframe	Engines
Prevention	15%	20%
Detection	57%	65%
Others	28%	15%

These values indicate that more than half the quality costs are related to the review of finished or partially finished articles for defects. Increasing the size of allowable defects in composite parts by means of design changes will greatly reduce the cost of TI&E. For a 12-ply 180 × 180 cm (72 × 72 in.) carbon-epoxy composite skin panel, TI&E functions would include receiving; material quality; dimensional, in-process inspection; nondestructive evaluation; mechanical tests of coupons; and final acceptance tests.

Quality control or quality assurance costs, although necessary, obviously have a major impact on total cost. However, as with manufacturing and other costs, these costs can be alleviated if the cost drivers are identified and reduced. Detail designers, in particular, need guidance on the cost impact of design decisions in their paricular area. TI&E spans all phases of system development.

Data Methodologies and Formats

When presenting cost drivers and manufacturing man-hour data to designers, the terms that have been found to be useful (Ref 5) are cost driver effects (CDE) and cost-estimating data (CED). An objective of both the CDE and CED methodologies is to develop a simple approach for the use of formatted data during all design phases to achieve lower fabrication costs. The CDE approach also provides qualitative cost guidance for performing simple trade-offs to achieve lowest fabrication cost, while the CED approach provides the designer with the capability to perform trade-offs to achieve quantitative rough order of magnitude

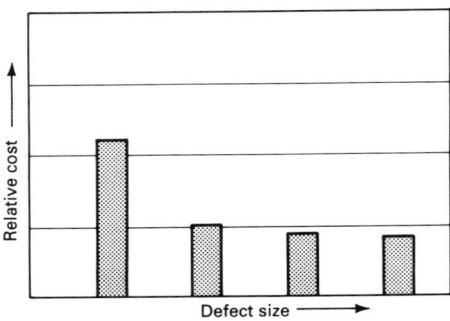

Fig. 4 Effect of maximum allowable defect size (desinger-determined) on test, inspection, and evaluation cost

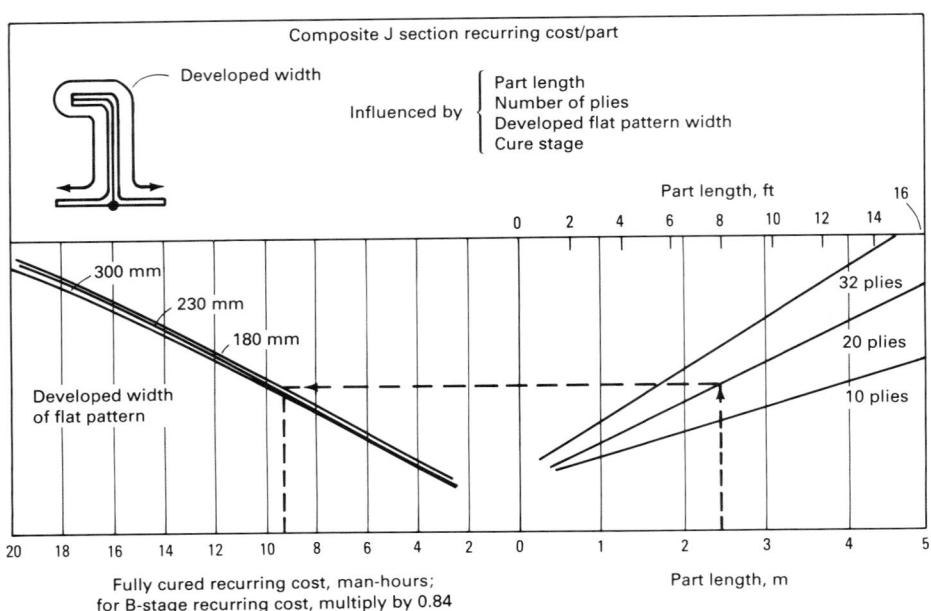

Fig. 5 Typical format showing recurring costs in man-hours for carbon-epoxy composite stiffeners

(ROM) estimated fabrication costs. These methodologies give the designer the cost guidance required for achieving lower manufacturing costs at all design phases. As shown in Fig. 3 and 4, cost driver effects achieve qualitative results, while Fig. 5 to 11 show that cost-estimating data provide quantitative results.

The CDE approach enables preliminary phase and production designers to identify the features that increase the manufacturing cost of the design, determine the relative effects of cost drivers over which they have control, and use cost data to perform simple trade-offs to achieve comparative costs for those configurations evaluated.

The CDE approach motivates designers. Low-cost designs can be realized provided full advantage is taken of the CDE data and the lower end of the cost range is used whenever possible, while still satisfying performance, reliability, and other design requirements. The CED approach provides preliminary and detail designers with the ability to estimate costs by using simplified charts giving manufacturing man-hour data.

Reference 6 provides a cost comparison of various composite manufacturing processes. Two such formats that are similar to the above categories are included as Fig. 12 and 13.

Composite Part Definitions. Designers find that subdividing mechanical components and structural parts of composites into various elements is useful (Ref 5) for determining the manufacturing man-hours required for trade-off studies. An effective approach has been to consider an element as a base part, a designer-influenced cost element, or a detailed or discrete part.

A base part is a detailed part in its simplest form, for example, a pultruded straight "Z" or angle section.

A designer-influenced cost element (DICE) is an element over which the designer has control and that adds cost because of the increased fabrication operations and tooling required over the standard manufacturing method. There are two distinct types of designer-influenced cost elements. One requires adding standard manufacturing operations when dealing with joggles,

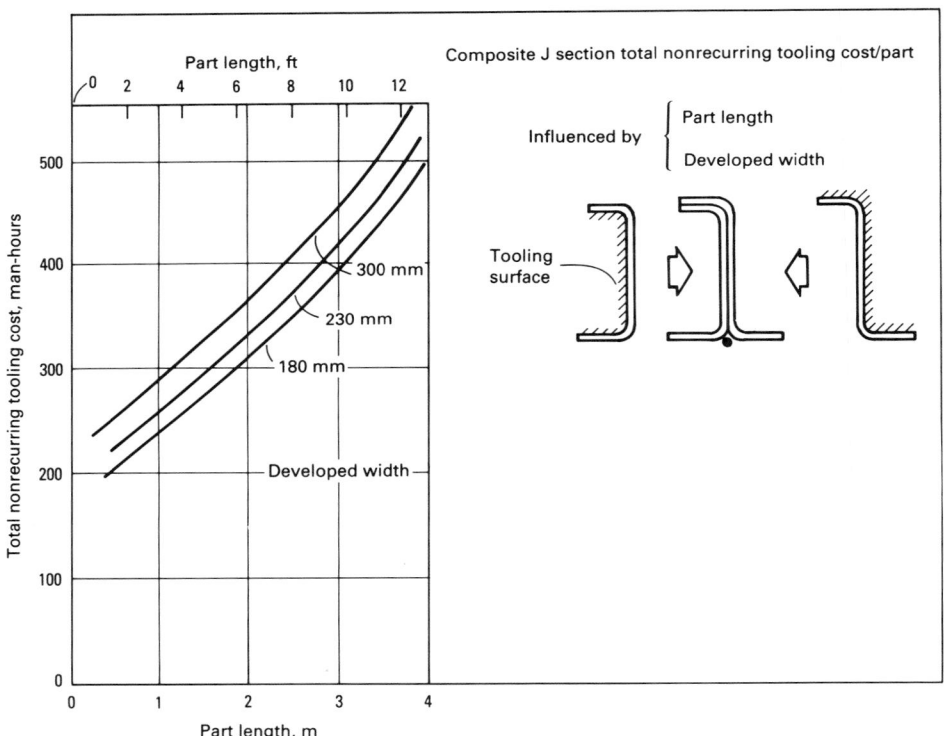

Fig. 6 Typical format showing nonrecurring tooling cost in man-hours for carbon-epoxy composite stiffeners

special lineal trim, bend radii, inserts, and beads. The other type involves manufacturing complexities, such as hybrid fibers, laminate transition, special tolerances, and special treatment. Examples of potential cost drivers in carbon-epoxy composite structures are part type and function, part size, fiber types and mix (hybrids), resin system, number of plies, curvature (single or compound), thickness transition, reinforcing doublers, curing method, automatic versus manual lamination, and quality assurance requirements.

A detailed or discrete part is a specific structural part that is likely to incorporate com-

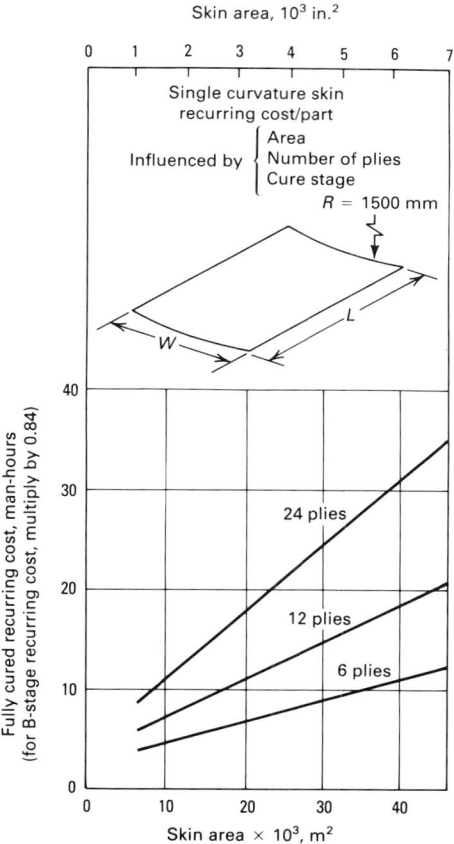

Fig. 7 Typical format showing recurring costs in man-hours for a carbon-epoxy panel

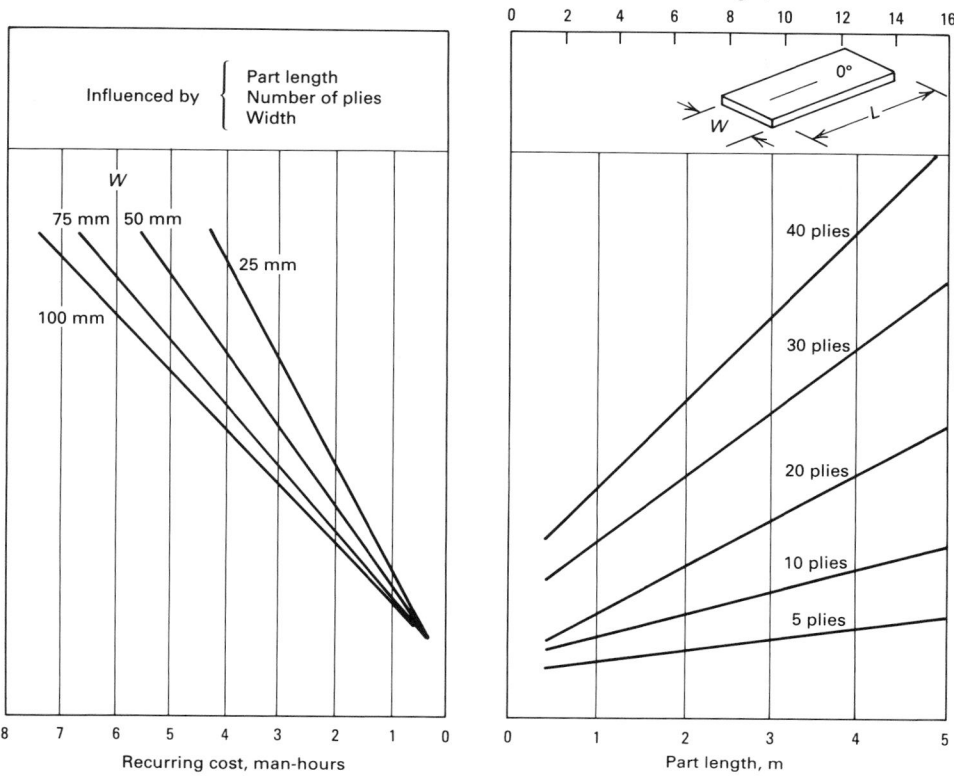

Fig. 8 Co-cured flat parts for carbon-epoxy strip plies. Recurring cost/part

plexities to meet the design objectives, for example, a base part plus DICE. A discrete part is one that is ready for assembly.

Production Definitions. The terms used to determine man-hours and costs for manufacturing parts are:

- *Set-up.* Time required to prepare the work place or machine for a fabrication or assembly operation by obtaining any necessary tooling, materials, and so forth
- *Run-time.* Actual time required to perform a given operation, or series of operations, necessary to fabricate or assemble a part or assembly, after the set-up has been completed
- *Set-up versus run-time.* When the set-up time for tools and equipment, in preparation for production run, is very high in relation to the run-time per part, it is often more economical to run as large a lot size as practical, so that the total cost per piece is lower, for example: cost per part = run-time per part + set-up/number of pieces

Defining Ground Rules. When comparing the cost of products or systems that are being developed by means of various techniques, a particularly important step is to establish a series of ground rules. These provide a com-

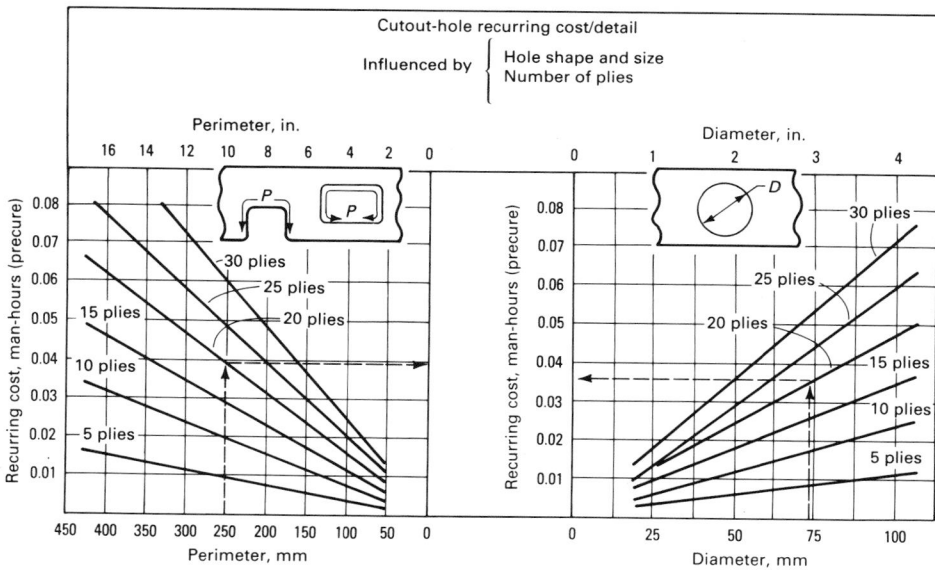

Fig. 9 Typical format showing designer-influenced cost element for cutouts in carbon-epoxy panels

mon base for promoting understanding, consistency, uniformity, and accuracy in generating and using manufacturing man-hour data, and are appropriately characterized as either general or detail ground rules.

General ground rules include composite discrete parts under study, general composite material types, manufacturing technologies, facilities, data generation (recurring costs), data generation (nonrecurring costs), and support function modifiers (for example, inspection). Detailed ground rules include specific material types, base part drawings and sketches used to develop cost data for formats, design tolerances, surface requirements, and estimating methods.

Designer-Oriented Format/Chart Criteria. In the development of cost data and the formats/charts for designer use, the information must emphasize cost drivers; be simple to use; focus on designer terminology; instill confidence; and be economical, accessible, and maintainable, which implies the need for computerization. The formats shown in Fig. 3 to 11 meet these criteria (Ref 5).

The importance of the microcomputer for cost analysis is demonstrated in Ref 7 for an automobile hood designed in sheet molding compound.

Cost Estimating. A number of cost-estimating models for composite parts and assemblies have been developed under contract for the U.S. Air Force. While much of this information is Department of Defense proprietary, Ref 8 provides a brief review of four of these cost models.

The following procedure has been used by designers to determine manufacturing costs in an effort to minimize costs (Ref 5):

- Develop engineering drawings for discrete parts and/or assemblies to isolate designer-influenced cost drivers in the manufacturing technologies to be analyzed
- Establish operational sequences, including equipment requirements
- Apply industrial engineering base standards to each operational step in order to determine set-up time (man-hours) and run-time (man-hours). (Applied standards are the sum total of all elements required to carry out each operation, plus set-up time.)
- Establish tooling required for each element of manufacturing operations
- Record total set-up time in man-hours for each element of manufacturing operations
- Record total run-time in man-hours for each element of manufacturing operations
- Amortize total set-up time in man-hours over the lot size and add to total run-time
- Establish the number of man-hours using company improvement/learning curves (no variances included, for example, personal fatigue and delay, clean-up, equipment downtime, and supervisory instruction)
- Determine both manufacturing man-hours for discrete parts and/or assemblies (using worksheet) and material costs and labor rates at projected time of manufacture

The learning curve (LC) is a generally accepted method for projecting labor or man-hour costs for a part or product over a specific production period. The theory is based on the premise that skill in performing a specific task increases as the task is repeated. The rate of increase in skill depends on the type of task.

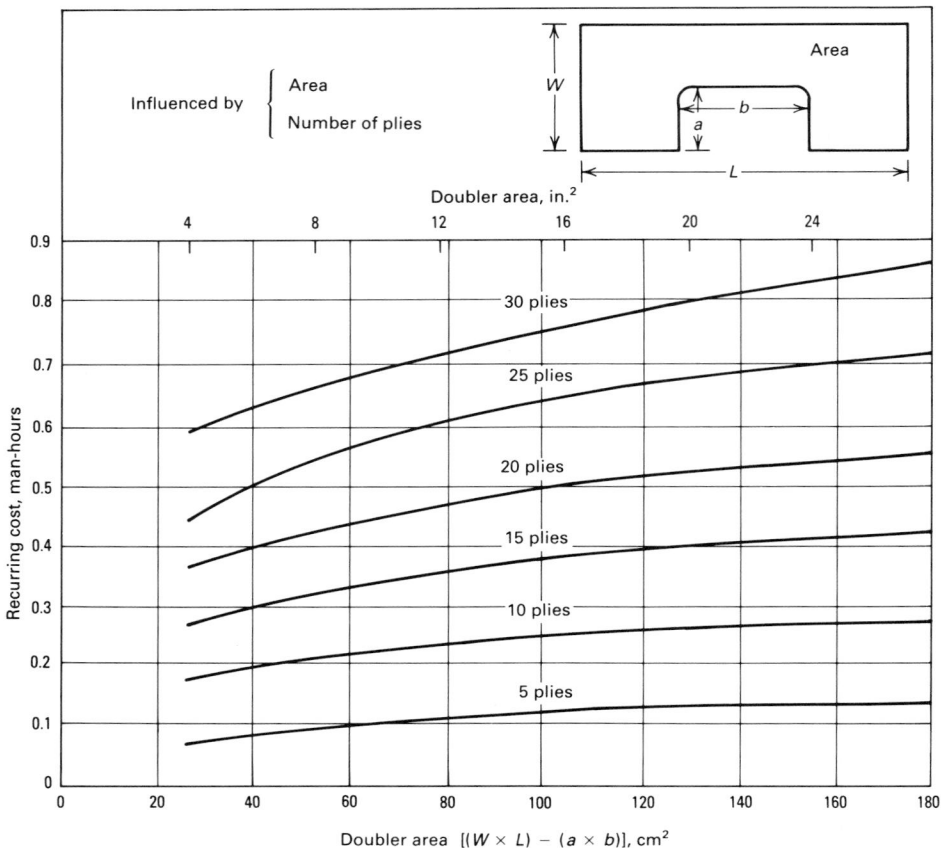

Fig. 10 Co-curing of carbon-epoxy reinforcing doubler for cutout. Recurring cost/detail

In general, a steep curve (65 to 70%) indicates that learning/skill is attained over an extended period of time. The task requires an extended learning period because it is complex or requires learning an excessive amount of hand-fitting or assembly techniques.

A flat curve (95 to 98%) is often obtained when a part or product is processed by a completely automated technology with a minimum of learning required (machines do not learn, they repeat a preprogrammed operation). Once any faults or deficiencies have been resolved and the automated manufacturing processes are in operation, learning, or any reduction in labor or machining time, is negligible. Therefore, automation tends to flatten a learning curve, while manual work steepens the slope.

The aerospace industry has found that the so-called standard man-hour, that point at which maximum learning has been achieved, generally occurs at the 200 to 250th part or product. From that point on, a learning curve on a log/log scale has a slope based on past history of similar parts or products. The curve is generally a modified LC, that is, the first 100 or so articles have a steeper LC (~75%), and the second 100 or so have a less steep curve (~80%). From the 200 to 250th article, the curve is generally at about 95% because learning has been completed. At this point,

reduction in man-hours will be achieved only by method studies, improved tooling, and cost-effective engineering redesign. A typical LC for an aerospace engineering system is shown in Fig 14. However, in actual practice, a projected learning curve is seldom achieved. Rather, a series of variations occur that must be constantly monitored to ensure that projected costs remain within the target values.

Selecting the Learning Curve Factor. Labor costs are normally collected by cost centers, each representing a different manufacturing technology, and are not traceable to individual parts or assemblies. Labor costs are for a production lot representing a mix of single-usage and multiple-usage parts/assemblies. From these data, learning curve slopes (%) are established for the various cost centers. When estimating the cost of parts/assemblies, the appropriate learning curve factor is selected by the LC% for the technology involved and the design quantity, regardless of the quantity of parts/assemblies per engineering system.

Factors and reasons for not achieving the projected or desired learning curve are:

- Engineering design (complexity of parts or products—too many cost drivers and design errors)

- Lack of adequate facilities/equipment for fabrication of the product or system
- Inadequate or insufficient tooling
- Introduction of engineering changes, which can decrease or increase cost
- Inadequate test, inspection, or evaluation
- Rate/production tooling not phased in timely manner
- Failure to conduct methods improvement and adequate trade-off studies during program
- Inadequate manpower/loading plan
- Poor scheduling (too rapid or too slow buildup to desired production rates/quantities)
- Parts or products not suitable for automation or division into simpler work tasks
- Poor planning, for example, inadequate breakdown of work tasks
- Poor supervision/management
- Poorly trained or underskilled operators
- Turnover of manpower (new/added untrained manpower)
- Poor estimating of manpower and skill requirements at start of program
- Labor problems

Designer's Worksheet. In determining the total program costs for both composite and other discrete parts and assemblies, designers have used the cost worksheet shown in Table 1 and described in Table 2.

Because experience and facilities vary across the industry, it is necessary for each company to determine its own learning curve factors. Examples of learning curve values are shown in Table 3.

Labor rate fluctuations can be handled in much the same way as material price variations. As labor rates increase, the need to design a part that can be manufactured with the least amount of hands-on labor will become more important. With computerized cost analysis systems, the designer can use projected labor rate values for the proposed time period of production in a trade-off study to determine whether the labor rate would cause a major problem in the cost of the product or system.

Determining the impact of the product quantity on learning curves must be factored into trade-off studies. Such information also has potential use for a computerized system. The data shown in the formats in the figures in this article are based on 200 units. A prototype development of about five products or systems would have higher values on the learning curve, whereas, at the other end of the scale, a very large production contract would display a much lower value. The impact of this learning curve value would be a major factor in management decisions.

It is sometimes important to determine the impact of lot release size, especially for lots of less than 25 units, depending on whether the process is computer aided or manual. Beyond 25 units, the impact of lot size is normally negligible for trade-off study purposes, but as the lot release size decreases below 25 units, its impact increases dramatically for the majority of processes.

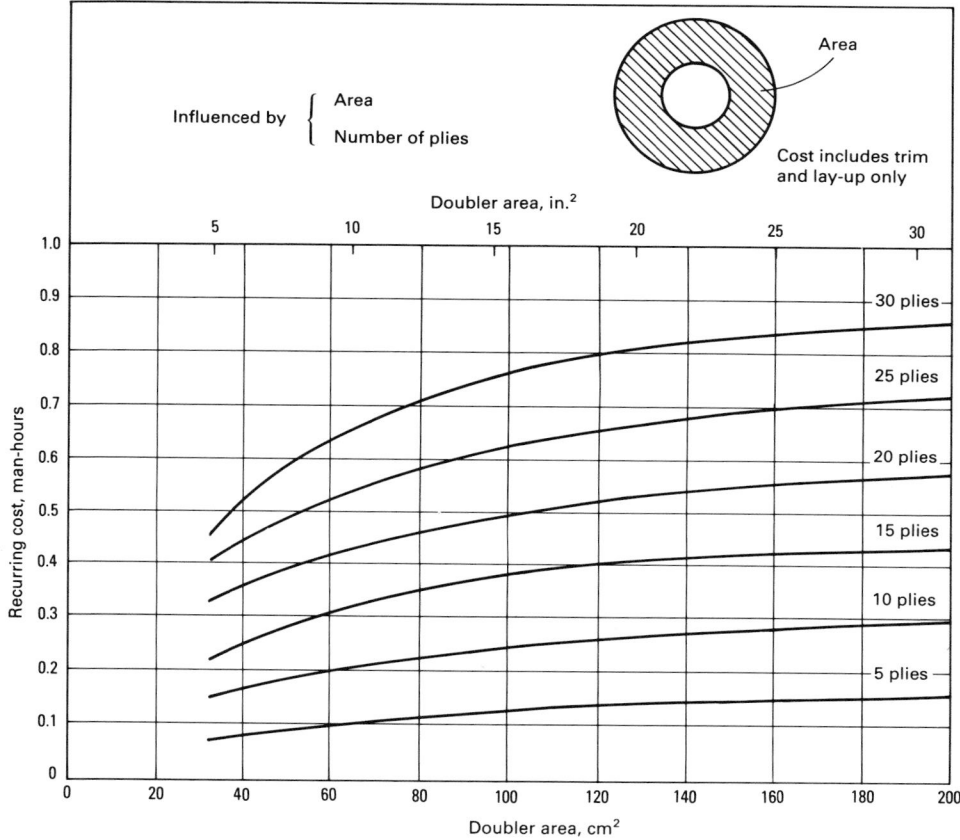

Fig. 11 Co-curing of carbon-epoxy reinforcing doubler. Recurring cost/detail

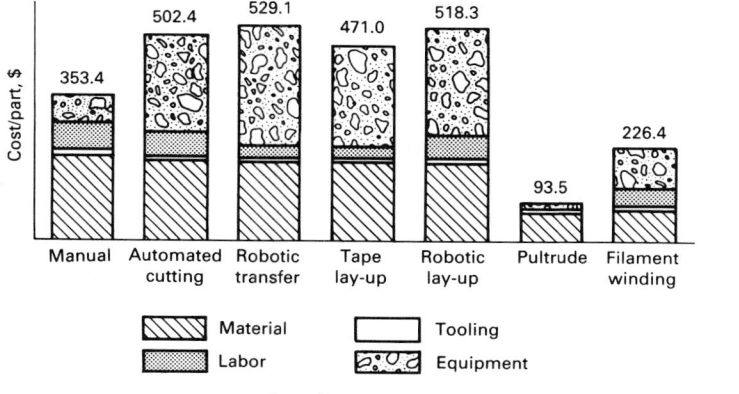

Fig. 12 Costs for a 0.40-m² (4-ft²) 24-ply part for an annual production of 2000 parts. Source: Ref 6

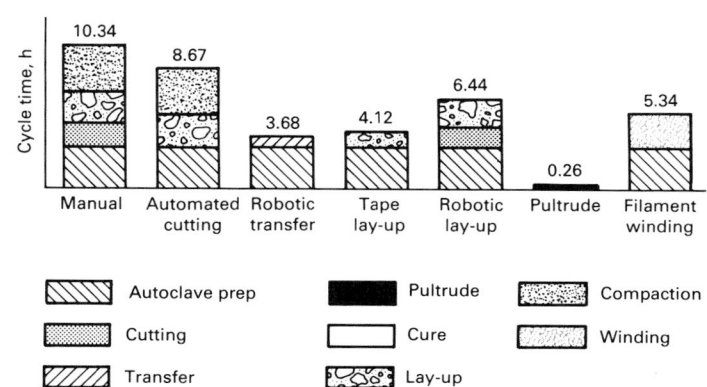

Fig. 13 Cycle times for a 0.4-m² (4-ft²) 24-ply part for a 100-part batch size. Source: Ref 6

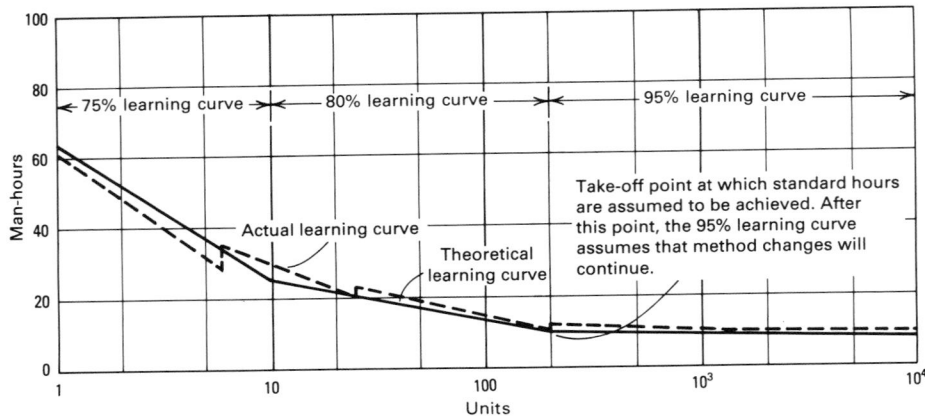

Fig. 14 Example of actual learning curve for typical aerospace system

- Determine test, inspection, and evaluation (TI&E) costs
- Determine total manufacturing costs, including materials and tooling
- Determine weight of each panel assembly
- Present manufacturing man-hours/costs and structural weight in summary tables and, if appropriate, on design charts that show structural weight on the ordinate versus manufacturing cost on the abscissa

The designer and management then select the optimum structure (discrete part, subassembly, or assembly), with respect to structural weight and other design factors such as those described above, and manufacturing costs.

The designer, having developed candidate structural configurations to meet all design requirements, then utilizes CDE and CED data. The following steps are typical of those taken to arrive at a lowest manufacturing cost design:

Step 1. Select materials that meet corrosion, elevated temperature, or other design requirements.

Step 2. Review the ground rules to determine the correct parameters for the discrete parts and assemblies to be analyzed.

Step 3. Record on the designer's worksheet the concept number, part number, description, labor rate, number of parts per system, and design quantity. One worksheet can be used for each part when conducting the trade-off between parts, or a separate worksheet can be used for each subassembly.

Procedure to Conduct Trade-off Studies

The designer needs to find the lowest-cost structural candidate that meets the design objectives. These goals may include:

- Strength and stiffness
- Minimum weight
- Performance at elevated temperature
- Fatigue strength
- Low maintenance
- Crashworthiness
- Corrosion resistance
- Damage tolerance
- Ease of repair

The designer can use the following procedure to conduct manufacturing cost trade-off studies:

- Develop concepts that, in the case of an airplane fuselage panel, require selecting or determining the material type, panel size, frame shape, number of frames, stringer shape, number of stringers, joining methods (for example, bonding versus mechanical), and candidate manufacturing methods for each discrete part in the assembly
- Determine manufacturing costs for parts in each configuration
- Determine assembly cost for each configuration

Table 1   Designer's cost worksheet

Page ———————

Design concept		Recurring cost (RC)										Nonrecurring cost (NRC)				Program cost		
		(L. LC + TI&E) LR = L$ + M$ = RC. P/AC. DQ = PRC										(NRC + TI&E) LR = PNRC				10 + 14 DQ = Cost/AC		
Part No.	Description	Labor MC/DG MH/PT (1)	Learning curve factor (2)	Labor TI&E MH/PT (3)	Labor rate $/MH (4)	Labor RC $/PT (5)	Material $/part (6)	REC. cost/ part $ (7)	Parts per AC (8)	Des. qty. (9)	Prog. RC $ (10)	NRC MC/DG MH (11)	NRC TI&E MH (12)	Labor rate $/MH (13)	Prog. NRC $ (14)	Prog. cost $ (15)	Des. qty. (16)	Cost/ AC $ (17)
Totals																		

Remarks _____

_____

_____

By: _____

Date: _____

Note: L, labor; LC, learning curve; LR, labor rate; DQ (or Des. qty.), design quantity, Prog; RC, program recurring cost; M, material; P, parts; PT, part; MH, man-hour; AC, aircraft; MC, material cost

Table 2   Use of worksheet for designers

Part number		Enter identification, if available.
Description		Enter brief description, for example, stiffener, zee, J section.
1	Manufacturing labor	Enter man-hours per part at required quantity.
2	Learning curve (LC) factor	Enter LC factor based on learning curve percentage and design quantity. Factor is likely to vary for each company.
3	TI&E labor	Enter recurring cost for TI&E (man-hours).
4	Labor rate	Enter current manufacturing labor rate, including direct labor fringe benefits and overhead charges.
5	Labor recurring costs (RC)	Enter the product of column 1 times column 2 plus column 3 times column 4.
6	Material cost	Based upon furnished data in company, enter material cost per part in dollars.
7	Recurring cost (RC) per part	Enter the total of columns 5 and 6.
8	Parts per system	Enter number of identical parts per system.
9	Design quantity	Enter number of systems to be procured.
10	Program recurring cost (RC)	Enter the product of column 7 times column 8 times column 9.
11	Nonrecurring tooling costs . . . . (NRTC) for part/assembly	Enter NRTC in man-hours.
12	NRTC for TI&E	Enter NRTC for TI&E in man-hours.
13	Labor rate	See column 3.
14	Program nonrecurring . . . . tooling costs (NRTC)	Enter the product of column 13 times the total of columns 11 and 12.
15	Program cost	Enter the sum of columns 10 and 14.
16	Design quantity	See column 9.
17	Cost per aircraft	Enter the quotient of column 15 divided by column 16.

Table 3   Typical learning curve values

Numerically controlled machining	95%
Manually operated machining	80%
Filament winding	85-90%
Pultrusion/wrap	85%
Adhesive bonding	
Assembly	75-80%
Curing	90%
Bench assembly	80%
Final assembly	80-85%

quirements, conduct methods value analyses, and conduct cost reduction/production readiness reviews. Where manufacturing systems are concerned, the information can be used to characterize design/manufacturing interaction; evaluate impact of engineering changes during system development; implement computer-aided design/computer-integrated manufacturing, construction cost trade-offs, and component ranking baselines; and plan upgraded, computer-integrated manufacturing facilities.

Step 4. Use company data or data from other sources for the CDE formats, CED formats for the manufacturing methods, and TI&E methods for the materials and parts analyzed in accordance with the ground rules.

Step 5. Develop CDE formats, as shown in Fig. 3 and 4, providing relative-cost information for the materials, parts, and assemblies being analyzed. These CDE formats are to provide qualitative information leading to the lowest cost.

Step 6. Study CED formats, as shown in Fig. 5 to 11, for the base parts and any required designer-influenced cost elements (DICE) using the required dimensions, for example, length for sheet metal stringers or area for panels. Note on the designer's worksheet the total labor man-hours per part (including applicable DICE) recorded on the cost worksheet for each discrete part in the assembly.

Step 7. Check for applicable DICE. The formats must indicate which DICE are applicable. In some cases DICE will be incorporated in the manufacturing methods for the base part.

Step 8. Apply the learning curve tables as required, based on the total design quantity.

Step 9. Determine the value (man-hours) for the nonrecurring tooling costs. Note the quantity of parts or assemblies for which these values are intended. Record the man-hours divided by the required quantity on the designer's worksheet.

Step 10. Record the current manufacturing labor rate, including direct labor fringe benefits and overhead charges, on the designer's worksheet.

Step 11. Using the same procedure as was used for manufacturing methods, determine TI&E manufacturing man-hours and record the TI&E recurring and nonrecurring tooling costs on the designer's worksheet.

Step 12. Insert the cost of materials based on data furnished by the company and enter material costs per part in dollars on the designer's worksheet.

Step 13. Consult instructions accompanying the designer's worksheet to determine aerospace vehicle program cost for the discrete part and assembly.

Step 14. Compare results from the designer worksheets for each part and/or subassembly and, if desired, enter on a diagram (graph), showing weight versus manufacturing cost, and compare each concept. In the case of an advanced system, such as a supersonic aircraft, management and the customer may elect to sacrifice some manufacturing cost for improved performance, but in the case of a low-speed aircraft, they may decide to sacrifice some performance to achieve a lower manufacturing cost.

Using this procedure, the designer can compare different design concepts, possibly using different materials, for example, sheet metal versus composites, or castings versus an injection molded part. With each trade-off conducted in accordance with the same ground rules, for example, lot sizes, design quantity, and so on, credible results can be achieved.

Cost Analysis Applications

The focus of this article on the analysis of manufacturing costs of candidate designs to meet the design objectives has shown that the analysis can be used in the design phase to guide manufacturing method selection, understand cost implications of new manufacturing processes, and estimate costs for group technology part families. Determination of manufacturing cost and cost drivers has other timely applications. Where manufacturing processes are concerned, the information can be used to select cost-effective routing and assembly alternatives, meet producibility re-

REFERENCES

1. W.B. Goldsworthy, "Transition to Volume Composite Component Manufacturing," Extension Course on Composite Materials Processing and Quality Assurance, University of California, Los Angeles, Sept 1981

2. R.C. Forney, "Advanced Composites—The Technical Linchpin in the Structural Revolution," Paper presented at the Fifth International Conference on Composite Materials, San Diego, July/Aug 1985

3. J. Van Hamersveld and L.D. Fogg, Producibility Aspects of Advanced Composites for an L-1011 Aileron, *SAMPE J.*, May/June 1976, p 6-13

4. J.M. Margolis, "Thermoplastics vs. Thermosets Process Economics Aerospace/Aircraft and Automotive Exterior Panels," Paper presented at the Conference on Advanced Composites—the Test Developments, American Society for Metals and Engineering Society, Detroit, Nov 1986

5. B.R. Noton, "Optimizing Aerospace Structures for Manufacturing Cost," Paper presented at the 13th Congress of the International Council of the Aeronautical Sciences, Seattle, Aug 1982

6. S. Krolewski and T. Gutowski, Effect of the Automation of Advanced Composite Fabrication Processes on Part Cost, in *Proceedings of 18th SAMPE Technical Conference*, Oct 1986, p 83-97

7. J.V. Busch, "Micro-Computer Based Cost Estimation for the SMC Process," Paper presented at the Conference on Advanced Composites—the Latest Developments, American Society for Metals and Engineering Society, Detroit, Nov 1986

8. V. Zaloom and C. Miller, A Review of Cost Estimating for Advanced Composite Materials Applications, *Engineering Costs and Production Economics*, Vol 7, Elsevier, 1982, p 81-86

Design/Tooling/Manufacturing Interfaces

Brian A. Wilson, Aerojet Strategic Propulsion Company

INTERFACE CONSIDERATIONS represent an age-old factor in the production of any manufactured part. This consideration is present from the simplest of consumer products to the most complicated aircraft, missile, and spacecraft hardware. An interface is any communication link between the various disciplines involved in producing a product for use in the marketplace. Good communication between the design, tooling, and manufacturing disciplines will minimize problems and provide a superior product. Quality assurance concerns, such as part inspectability, also warrant attention.

With the current emphasis on composites use in both aerospace and consumer hardware, the importance of interfaces has increased significantly. Because design, tooling, and manufacturing aspects are integral components of the composites manufacturing industry, the three disciplines must work together throughout the development phase of any product, from idea inception to its culmination as hardware. To make the best selections for any given structure, the designer should examine in detail the various available manufacturing processes described in the Section "Manufacturing Processes," as well as the machining and assembly methods covered in the Section "Machining, Assembly, and Assembly Forms" in this Volume. In many cases, the articles in these Sections contain design information, or information pertinent to design, that would be useful both in selecting the basic type of structure (for example, filament wound versus braided) and in actually designing it.

The qualitative nature of this article is due to the unavailability of precise data on interface effects between the three disciplines. However, the evidence of these effects is certainly apparent from the evolution of composites over the past 15 to 20 years. The amount of interface depends on the materials specified for a particular design and product. Although an interface, its tolerances, and its effects on the three primary disciplines cannot be properly estimated without this specification, this article does define particular concerns that should be considered in the development of any composite product.

Design

The following discussion of the effects of tooling and manufacturing on the design process is the focus of this article, although many of the issues could just as easily have been discussed in the subsequent section on effects of manufacturing and tooling on their complementary disciplines.

Tooling Effects. Composite tooling differs from the tooling used in metallic structures in that the final machined dimensions do not necessarily apply to the composite manufacturing process. Rather, the final dimensions and their tolerances are extremely dependent upon the type of tooling used and the thermal-expansion characteristics of each tool material. The majority of composite materials and structures are subjected to heat as a means of curing the part to its final dimensions. As tooling is associated with the product through its final phase of manufacture, the thermal history that is provided to the unit is provided to the tooling as well as to the part.

Two particular types of tooling are used, depending upon the manufacturing process. Many of the composite structures used for aircraft, spacecraft, and industrial applications are built on partially contoured, but essentially flat tooling. This type of tooling is used as a form or fixture for laminate lay-up, which is either done by hand or with the assistance of robotics, and is performed on a layer-by-layer basis. The material lay-up can be in the form of cloth, mat, or tape, all of which are usually preimpregnated with resin systems but can be impregnated after the lay-up process. The structure takes on the shape of the tooling to which it is applied, and the final dimensions of the part are the dimensions that will prevail at the ultimate gelation temperature of the resin system. This involves consideration of the dimensional size of the tool at the curing temperature.

The other form of tooling is designed for the filament-winding process, which is used primarily for large, strategic rocket motors, small tactical missiles, launch tubes, driveshafts, pressure vessels, and fuel tanks. The tools involved in filament winding, while different from a standard lay-up fixture, are nonetheless subject to the same criteria dictated by the dimensional properties of a specific material system in the thermal environment.

The variety of materials used for tools and fixtures depends on the individual part requirements and on the background and experience of the tool designers and manufacturing engineers. Table 1, which lists the coefficients of thermal expansion (CTE) for the major tool materials in use to date, shows that aluminum has the highest CTE and carbon has the lowest. For lay-up fixtures, the designer generally looks for a material with the least CTE, because any pressure that is required for void elimination and densification can be provided in the standard autoclave curing environment by applying a vacuum bag and external pressure to the part. For filament-wound applications, however, the designer must consider the dimensions of the tooling at the specific cure temperature. Plaster and sand were the primary tooling materials for large structures until recent times. Now, assembled metal stringers and skins are used to produce a mandrel with the precise internal dimensions of the structure to be wound onto it.

After the development of the net metal mandrel came a similar structure with significant

Table 1   Coefficients of thermal expansion for major tool materials

Material	Thermal expansion coefficient, 10^{-6}/K
Aluminum	22.5
Steel	12.1
Sand/polyvinyl alcohol	12.2
Sand/sodium silicate	11.5
Graphite prepreg	3.6
Glass prepreg	11.7-13.1
Ceramic	12.1-12.6

use of carbon fiber tooling materials. This type of tooling is used to manufacture a part that is ultimately cured in an oven. In an oven cure, composite materials are heated up first because they are external to the assembly. Curing, which takes place during the stages of oven ramp-up and hold, causes the composite part to shrink. The part then expands elastically to conform with the secondary mandrel tool expansion, and contracts in the same way. Final dimensions will thus be very close to the room-temperature dimensions of the tool. Some pressure will be put on the composite by tool expansion, but it is probable this will be limited in its densification and void removal effect because the resin may already be substantially cross linked when this expansion takes place. The tool must be designed to be capable of disassembly and removal through openings in the final structure or pressure vessel.

In the case of pure cylindrical structures, such as launch tubes, and full, open-ended rocket motorcases and drive shafts, the part must be removable from the tool by a push-off process. To accomplish this, the tool must have taper or, alternatively, the part must be cured by heating the tool from within. Steam, hot oil, or hot gas is passed through internal passages in the mandrel. The tool, now heated, expands to the cure temperature condition. The composite is thus put under pressure prior to curing and cross-linking. This means that the composite cures essentially at the expanded dimensions of the tool, and the shrinkage that takes place helps reduce the size of voids and densify the composite. Upon cooling, the tool contracts more than the composite, and the part is removed from the tool with comparative ease. These processes are discussed separately in order to emphasize the importance of tooling expansion in the initial design of the product.

Another tooling consideration is the residual stress factor that could be built up in the tool during its fabrication, and which could result in some shape change in the tooling upon heat-up. This factor can normally be eliminated by emphasizing temperature stability when designing tools. It may also be eliminated, to a degree, by a stress relief process that may take place during the initial thermal processing cycle of the tool.

A final important consideration, specific to the filament-winding process, is the surface condition that is required on the final part. The internal surface on a filament-wound part is always smooth because it conforms to the winding mandrel. However, the external surface, unless it is machined or cured to an external shell dimension, will have a rough, as-wound surface. Pressurized expansion of a filament-wound part during cure into an external shell is a sophisticated process that is just beginning to be used for filament-wound aircraft fuselages. The alternative in the past was to wind the part until it was slightly oversized and then machine or grind it to the final dimensions, or to provide an external sacrificial fiberglass wrap that could be ground to the final dimensions.

The effects of manufacturing considerations on the design process are categorized by two types of processing: one for laid-up parts and the other for filament-wound parts, with some additional considerations when either molding or pultrusion is the primary manufacturing process. Some considerations, however, are common to all manufacturing methods. Resin handling during fabrication is a concern because of the potential for formula variation in the mixing of the resin components. This can be due to operator error, the type of mixer used, or the efficiency and duration of the mixing process. Specification of the tolerances of the final formulations should be provided by the designer and discussed with the manufacturing personnel to ensure that the tolerances specified are obtainable with the available equipment. The possibility of error in formulation and of lack of uniformity in the mixing process should be evaluated during the design process for their effects on the ultimate strength of the product.

A further consideration for resin systems is the allowable tolerance band on temperature during a cure cyle. A ramp-up that is too fast will cause a surface skin curing and may ultimately inhibit resin flow through the composite structure during the balance of the cure cycle. Conversely, a slow ramp-up will result in a viscosity reduction before cross-linking of the resin system begins to take place, which can result in the resin being sufficiently fluid to drain from the part. This is especially important when using low-density hydrocarbon resin systems such as those in the polybutadiene family.

If the actual cure temperature sustained for a resin system is either too high or too low, it will also have a bearing on final part properties. Too high a resin temperature may cause the resin to become embrittled, and if the over-temperature is excessive, discoloration and charring of the product could result. Too low a resin temperature could cause inadequate curing, which may be internal to the part structure and not detectable by nondestructive inspection methods.

An additional consideration, common to all the manufacturing processes, that influences material selection is that of material handling. For example, resins in both the mixed state and prepregged materials have very precise specifications for the period of time during which they can be stored at room temperature; exposure to room-temperature conditions for either of these forms results in some degree of resin advancement. Without proper materials selection, this can be a factor during the manufacturing process as well, an important consideration for thick-walled structures. An additional consideration here is whether or not the fabrication process is interrupted overnight or over a weekend.

When hygroscopic systems and, in particular, aramid fiber are involved, the moisture environment normally inherent in manufacturing buildings becomes an important factor, such that many of these buildings are now being converted to a controlled moisture condition. If the manufacturing environment does contain moisture, safety factors must be incorporated in design to allow for degradation effects. This is particularly important in high-humidity areas of the United States.

In the flat laminate lay-up process, one of the primary considerations is the care with which autoclave parts are bagged for vacuum conditioning and pressure application. This should be evaluated by manufacturing and design engineering personnel during the design process so that bagging can take place without any wrinkling of the bag during vacuum application, which would cause a crease in the final part. Such a crease results in an unacceptable stress riser condition, which causes premature failure. Bleeder cloth materials, required during most autoclave curing to absorb excess resin from the composite structure, should also receive attention from both manufacturing and design engineering. Careful selection of the type of bleeder, which is very dependent upon the specified resin system, is required. If too much bleeder cloth is applied and the cloth is too open, the resulting composite part could be starved of resin in critical areas, resulting in high void content, strength degradation, and unacceptable surface conditions.

With respect to mechanical aspects of the manufacturing process, discussion between manufacturing and engineering personnel is vital. Tolerances of mat or tape placement can have a severe effect on mechanical properties of a composite structure. There is a limit to the tolerance that can be achieved in a hand lay-up process. There is also a labor cost factor when ultrahigh tolerances must be achieved in laminate layer placement. There is a stress discontinuity at every drop-off of a laminate layer; these are normally staggered to achieve a gradual reduction of thickness. Tolerances must be reasonable, and ply thickness, an important design factor, must be considered when positioning these layers. Ply thickness should be specified and checked in an "as-laid" condition. The thickness specified must be sufficient to achieve the proper distribution of fibers in the layer and result in the required structural thickness. Ply thickness can be severely affected by the amount of manual squeegeeing that takes place, which also affects the resin content. Furthermore, in tape-laying, the pressure applied by the lay-down roller of a tape machine can affect the ply thickness.

In the filament-winding process, there are a variety of machine control and software factors that affect the structural properties of the filament-wound part. One of the major factors is the ability of the machine to control wind tension in the fiber. The primary type of tensioner, borrowed from the textile industry and used for many years, has been the rotary spring-loaded plate-type tensioner, which uses two free-floating spindles. However, because this particular tensioner is limited to 150 to 200 mm (6 to 8 in.) travel, it will not maintain tension under some of the conditions of fiber path length change during lay-down. Microprocessor-controlled tensioners, which are able to wind up any slack that may develop in the fiber, have now been developed. These ''wind-back'' tensioners can be calibrated directly from the machine control console, and can provide a print-out of tension at any point during the winding process. Wind tension can be very important: Too much tension will squeeze the underlying layers, which may result in slack in some of the fibers in these layers; insufficient tension will not provide any squeeze at all, which can result in fiber looseness and a high external dimension on the product. Another problem is that when winding noncircular parts, any sharp edges will cause the tensioned fiber to ''cut in'' at these edges. The ability to vary tension during winding has not yet been realized. In addition, laying fiber across a wide, flat surface will result in a very low-tension condition. Thus, extreme care must be taken in the design and manufacture of airfoil-shaped parts, which incorporate both of these conditions. The addition of a pressure roller to the winding machine will help maintain tension over long, relatively flat dimensions. The effect of tension variation in delivered strength in a composite structure has never been properly defined. The type and accuracy of available tensioning systems should be reviewed by the design engineering group with the manufacturing group before defining design-allowable properties of the composite.

A further consideration is the effect of wind angle tolerance and pattern-closing errors on the strength and stiffness of properties designed into the composite. There is a limit to the accuracy with which wind angles can be stipulated and followed. Depending on the number of circuits needed to close a helical wind pattern, the slightest deviation (on the order of minutes of arc) in wind angle can result in a closing error of 12 to 14 mm (0.5 to 0.6 in.) or more.

The type of fiber-handling pulleys and fiber pay-out systems on the production filament-winding machines should also be reviewed prior to design. The materials used in these pulleys and pay-out systems will affect the quality of the delivered fiber into the product. For example, carbon fiber requires a minimum pulley diameter of 75 to 100 mm (3 to 4 in.) in order to minimize breakage of the high-modulus filaments. Aramid fiber normally requires a 60 to 70 RMS satin-finish roller or pulley in order to minimize the wear on the pulleys from the fiber and ultimately the abrasion of the fiber by a worn pulley. Fiber pay-out systems will have an effect on the variation in band width, amount of gapping, fiber band density and uniformity, and assurance that the fiber band will lay down normal to the surface of the wind mandrel. The design-allowable material properties should be factored to reflect the available options on the filament-winding machine.

A final consideration in the manufacturing effects on design is in the placement and accuracy of any attachment holes that are to be used in the composite structure. The accuracy of the manufacturing equipment should be considered to determine permissible tolerances on hole drilling. This involves not only hole diameter, but also the accuracy of hole-to-hole placement and the ability of the equipment to produce a hole that is normal to the composite surface. Misalignment of the hole pattern must be evaluated. The number of fibers that are cut during the drilling process also should be considered. Joint strength and joint design are important parts of composite structure design. Many studies on the effect of manufacturing misalignment, angled holes, and effects of hole-to-hole accuracy on overall joint strength have been conducted (Ref 1-16). Before finalizing joint design, sections of the joint should be fabricated and tested to determine shear and bearing strengths and the possibility of creep under bearing load. Because bonded joint strengths are improving, bonded joints are currently a viable alternative to mechanically bolted or pinned joints. However, the same degree of care should be taken with this design, and a full-scale joint coupon should be tested before the joint design is finalized.

Manufacturing

While the bulk of the interface requirements are those that come from manufacturing and tooling and affect design engineering, by the same token there are some developments in tooling and design that must be recognized by the manufacturing engineer, so that he can develop an effective and responsive process.

Tooling Effects. The initial concept of tool design for the composites industry was a hold-over from metal fabrication technology. Tools were usually massive, with welded or bolted construction, and were machined to very close tolerance. Because most of the forming work was done at room temperature, thermal expansion and thermal distortion were not a problem. With composites, the tool design engineer thinks in terms of low-mass rigid structures. The CTE must be designed to either match the composite part or have the necessary degree of expansion to accommodate the effects of pressure application on the composite during cure.

Composite fabrication permits the design engineer to use fewer parts in assembly, which results in more complex molds and fixtures for these parts. Where complexity of the tooling requires that it be assembled and disassembled with each use, lightweight material is an advantage. In considering lightweight tooling in the composite manufacturing process, the tool designer must communicate with the manufacturing engineer so that the tools will not only manufacture the product to the correct design, but will also be suitable for the environment of the total manufacturing process.

Design Effects. Virtually all of the interfaces between the design engineering and manufacturing operations have been discussed in the section ''Design'' in this article. The influence of design on manufacturing is primarily concerned with the tolerances imposed on the various manufacturing functions, such as ply thickness, wind angles, wind tension, band density, band width, void content, and drilled holes. Reviewing these factors before the design is created promotes realistic tolerances and allows the design engineer to use appropriate safety factors.

Tooling

The revolution in the use of composite materials for tooling to produce composite structures has only occurred in the last 5 years. Before that, composites were manufactured using metal tooling, with all the inherent problems previously described in this article. The move to composite tooling was permitted by the advent of carbon cloth and carbon mat prepregs specifically manufactured by prepreg suppliers for tool manufacturing purposes. These materials use the 230 GPa (34×10^6 psi) modulus and 2.4 GPa (0.35×10^6 psi) tensile strength carbon fibers, coupled with very stable free-flowing resin systems with high-temperature capability in the range of 150 to 205 °C (300 to 400 °F). It is important that the tool be leak tight and the resins be designed to flow freely within the laminate layers of the tool to seal off the tool surface against vacuum and pressure leakage.

Design Effects. The primary interface factor concerns the trade-off between simplicity of tool design and the ability of composite materials to provide part reduction. In the automotive, helicopter, and aircraft fields, significant part reduction has been achieved on products such as doors, covers, vents, canopies, and access ports. This overall part reduction can have an accompanying cost reduction benefit for the final structure through ease of assembly and minimized time flow in production. However, as more and more part reduction occurs, the design of the necessary tooling to produce the part may become unnecessarily complicated. With this trade-off, optimum conditions can be achieved only with full communication between the design engineer, materials engineer, and tool design engineer.

Manufacturing Effects. The primary effect of manufacturing on tool design is in the environment to which the tools are subjected during the fabrication of composite structures. The specific effects of moisture and temperature have already been described. One additional environmental factor is that of tool handling. Because of their light weight, composite tools are more prone to handling damage than their metal counterparts. Thus, repair and refurbishment of the tooling can become a factor in its use. The manufacturing engineer should review the tool design so that those areas subjected to significant handling stresses can be reinforced to minimize damage.

REFERENCES

1. M.E. Morgan and S.W. Beckwith, "Joint Design and Analysis of Composite Cases," Paper presented at AIAA Solid Rocket Lecture Series, AIAA Aerospace Sciences Meeting, American Institute of Aeronautics and Astronautics, Reno, Jan 1987

2. F.M. Norton, M.E. Hodgson, and S.W. Beckwith, "Pinned, Thick-Wall Composite Carbon/Epoxy Joint Behavior Subjected to Thermal and Humidity Exposure," Paper presented at the 31st International Symposium and Exhibition, Society for the Advancement of Material and Process Engineering, Las Vegas, April 1986

3. S.R. Swanson, J.S. Burns, C.L. Suplizio, and S.W. Beckwith, "Effect of Reinforcement Orientation on the Strength of Carbon/Epoxy Pinned Joints," Paper presented at the 31st International Symposium and Exhibition, Society for the Advancement of Material and Process Engineering, Las Vegas, April 1986

4. M.E. Morgan and S.W. Beckwith, "Bolt Torque Loading and Radial Gap Effects on Thick-Wall Composite Joint Strength," Paper presented at the 30th National Symposium and Exhibition, Society for the Advancement of Material and Process Engineering, Anaheim, March 1986

5. D.W. Wilson and R.B. Pipes, "Behavior of Composite Bolted Joints at Elevated Temperatures," NASA CR-159137, National Aeronautics and Space Administration, Sept 1979

6. D.W. Wilson *et al.*, Mechanical Characterization of PMR-15 Graphite/Polyimide Bolted Joints, in *Test Methods and Design Allowables for Fibrous Composites*, STP 734, C.C. Chamis, Ed., American Society for Testing and Materials, 1981

7. D.W. Wilson and R.B. Pipes, Analysis of the Shearout Failure Mode in Composite Bolted Joints, in *Proceedings of the International Conference on Composite Structures*, I.H. Marshal, Ed., Applied Science Publishers, 1981

8. J.L. York, D.W. Wilson, and R.B. Pipes, "Analysis of the Net Tension Failure Mode in Composite Bolted Joints, *J. Reinf. Plast. Compos.*, Vol 1, April 1982

9. Y. Tsujimoto and D.W. Wilson, Elasto-Plastic Failure Analysis of Composite Bolted Joints, *J. Compos. Mater.*, Vol 20 (No. 3), May 1986, p 236-252

10. D.W. Wilson and Y. Tsujimoto, On Phenomenological Failure Criteria for Composite Bolted Joints, *Compos. Sci. Technol.*, Vol 26 (No. 4), 1986, p 283-305

11. L.J. Hart-Smith, "Lessons Learned From the DC-10 Carbon-Epoxy Rudder Programs," Paper presented at SAE Aerospace Technical Conference and Exposition, SAE Transactions 86-1675, Society of Automotive Engineers, Long Beach, Oct 1986

12. L.J. Hart-Smith, Design and Analysis of Bolted and Riveted Joints in Fibrous Composite Structures, in *Joining Fibre-Reinforced Plastics*, F.L. Matthews, Ed., Elsevier, 1987, p 227-269

13. L.J. Hart-Smith, "Design of Repairable Advanced Composite Structures," Paper presented at SAE Aerospace Technical Conference and Exposition, SAE Transactions 85-1830, Society of Automotive Engineers Oct 1985

14. L.J. Hart-Smith, Bonded and Bolted Composite Joints, *J. Aircr.*, Vol 22, 1985, p 993-1000

15. W.D. Nelson, B.L. Bunin, and J. Hart-Smith, Critical Joints in Large Aircraft Structures, in *Proceedings of the Sixth Conference on Fibrous Composites in Structural Design*, Army Materials and Mechanics Research Center, MS 83-2, II-1 to II-38, New Orleans, Jan 1983

16. L.J. Hart-Smith, Mechanically Fastened Joints for Advanced Composites—Phenomenological Considerations and Simple Analyses, in *Proceedings of the Fourth Conference on Fibrous Composites in Structural Design*, San Diego, Nov 1978; also, in *Fibrous Composites in Structural Design*, E.M. Lenoe, D.W. Opplinger, and J.J. Burke, Ed., Plenum Press, 1980, p 543-574

Static Strength

Ralph J. Nuismer, Hercules Aerospace Company

PREDICTING FAILURE in composite structures is currently more of an art than a science. Although advances are being made at a rapid rate, they offer only general guidelines to the analyst, rather than a fixed set of procedures. At this time, no production computer codes are available to predict composite failure. However, adequate estimates of composite strength can be made in many cases, in spite of the difficulties, through a thorough knowledge of composite failure modes; careful consideration of how they are likely to interact with ultimate strength; and, whenever possible, testing.

Failure of a composite structure can take many forms. A part can fail cosmetically, it can fail to maintain a desired stiffness, or it can fail to support the loads for which it was designed. Only the latter two, closely related failure examples are addressed in this article.

There are two considerations in predicting the strength and stiffness loss of a composite structure. The first, stress analysis of the structure, is fairly straightforward, at least in the linear range of structural behavior. A number of simple models of composite structural elements, such as beams or plates, exist for predicting the stresses in these elements under load. For more complicated composite structures, a number of finite-element codes are available for performing the stress analysis. Composite stress analysis techniques are more fully addressed in the article "Analysis of Structures" in this Section of the Volume.

A complication in stress analysis that has important ramifications to strength prediction is that the analysis can be conducted on any of three scales:

- Microscale, where fibers and matrix are treated as separate elastic phases
- Miniscale, where each individual lamina is treated as a separate homogeneous, orthotropic elastic body
- Macroscale, where the entire composite laminate or structure is treated as a homogeneous, anisotropic elastic body

As might be expected, the latter two scales are emphasized for design purposes. Generally accepted models are available at these two levels for determining equivalent homogeneous elastic properties of laminae from properties of the fiber and matrix constituents or of the laminate from the properties of the individual laminae (refer to the Section "Composite Materials Analysis and Design" in this Volume for information on micromechanics and laminated plate theory). These elastic properties can then be input into the various analytical or numerical stress analysis models, enabling the stresses in the structure under the design loads to be determined. It should be noted here for later reference that the same models used to analytically convert individual fiber, matrix, and laminae elastic properties into homogeneous structural elastic properties can be reversed. That is, stresses in the individual fibers or matrix (microscale) or in the individual laminae (miniscale) can be obtained from the macrostress analysis models.

The second consideration, once stresses and strains are obtained from the structural stress analysis, is to determine their effect on the stiffness or strength of the composite structure. Although the stress analysis part of the problem may with some degree of accuracy be classified as a science, without generally accepted procedures or computer codes for predicting failure, the methods used are largely left to the experience and intuition of the designer involved.

Predicting composite structural strength is difficult for two major reasons. The first is that, unlike stiffness, which depends on "average" properties, the initiation of failure is highly affected by flaws that are distributed randomly and unpredictably throughout the structure. These flaws, a product of the manufacturing process, invariably cause regions of high stress, such as resin-rich areas, areas of high void content, contiguous fibers, and so forth. In general, these regions are too numerous to be readily characterized or modeled, yet are responsible for the onset of failure.

Second, the strength of composite structures is affected not only by the initiation of failure at flaw sites, but to a large extent by the progressive growth and accumulation of such microfailures, which result in stiffness changes and stress redistribution that ultimately lead to the inability of the structure to carry its design load. Because of the inherently inhomogeneous nature of composites, the progressive growth of microfailures can take an enormous number of different paths, both within and between laminae, depending on the unique geometric details and loading of the structure in question. As indicated previously, there are at present neither set procedures for analyzing this progressive failure process nor standard computer codes that include progressive failure as part of the analysis.

The result is that, on a practical level, the prediction of the strength of composite structures is still largely performed on an ad hoc basis using linear elastic stress analyses and rudimentary failure notions, which include initial matrix failure; first-fiber failure; elementary progressive failure ideas; and special approaches for stress concentrations, such as fracture mechanics or stress averaging concepts. These methods are typically supplemented by close observation of analog laboratory tests and, because of difficulties in scaling the results of such tests, full-scale tests, wherever possible.

In the remainder of this article, a more detailed discussion of the static strength of composite structures under both tension and compression loads is given. An attempt has been made to emphasize a practical, design-oriented approach throughout. Lamina strength, laminate strength, and stress concentrations and damage are discussed. Only continuous fiber composites are considered. Special topics in failure, such as buckling, fatigue loading, environmental effects, and dynamically applied loads are addressed in other articles in this Volume.

Lamina Strength

A composite lamina is a distinct layer in a composite laminate. It consists of an array of continuous parallel fibers embedded in a matrix material. Lamina strength is greatly dependent on how it is loaded. When loaded in the fiber direction (longitudinal load), the lamina is very strong because the failure mode involves fiber breakage or buckling. However, when loaded normal to the fiber direction (transverse load) or in shear, a different failure mode, typically involving only matrix failure, occurs, and the strength is more than an order of magnitude lower. Refer to the article "Basic Failure Modes of Continuous Fiber Composites" in this Volume, a subject important to the prediction of failure.

First, in the case of tensile loads in the fiber direction, failure has been demonstrated to occur progressively at the microlevel, with weaker fibers failing first in isolated locations of the lamina. As a fiber fails, the broken ends give rise to stress concentrations that result in the fiber either debonding from the matrix or yielding the matrix in shear along the fiber-matrix interface (Ref 1, 2). In addition, the unloading of the broken fiber in the vicinity of the break results in more load being taken up by the neighboring fibers. The stress concentration in the neighboring fibers depends on their proximity to the break, the matrix stiffness, and the damage the matrix suffers as a result of the break. Eventually, as the lamina tensile load increases, these concentrations result in clusters of multiple broken fibers forming at random locations. Finally, an instability results at a particular location, which results in total failure of the lamina.

This failure process has been modeled statistically with some success (Ref 2). However, at the present time the specific matrix properties necessary to optimize the delivered fiber strength are not fully understood. Furthermore, it is not possible at present to predict lamina tensile strength from a knowledge of fiber and matrix properties. Thus, tensile tests of unidirectional tows and laminates are routinely conducted to determine an average fiber strain or stress at ultimate failure. Unfortunately, the fiber strengths obtained from the two types of tests are frequently different because of the use of different resins and different gripping methods for load introduction. This may extend to the structure itself, where, for example, delivered fiber strength of a composite pressure vessel is often lower than that obtained from either tows or unidirectional laminates. For the purposes of design, an average fiber strength is assigned that is appropriate for the structure in question. This is best obtained, when possible, from a data base accumulated for the resin system and type of structure being designed.

If the loads in the fiber direction are compressive, the situation becomes even more difficult to assess. This is because many more failure modes are possible in compression than in tension, depending on the degree of support the lamina fibers receive (see the article "Basic Failure Modes of Continuous Fiber Composites" in this Volume). In essence, each fiber is a column, side supported by the surrounding matrix and contiguous fibers, and each lamina is a thin plate supported by surrounding laminae and perhaps by geometrical restraint features of the structure. Thus, failure under compression can occur by any of several modes: as gross structural buckling of the lamina; as compression/shear failure of fibers; as longitudinal splitting of the matrix followed by fiber instability; as fiber kinking or microbuckling; or as a

delamination of a portion of the lamina, which then buckles, causing ultimate failure of the remainder of the lamina because it can no longer support the entire load. Several of these modes, such as kinking or microbuckling, can take place either in the plane of the lamina or in the through-the-thickness direction. Furthermore, compression failure of a lamina is considerably more dependent on the matrix properties than is tensile failure. Thus, matrix toughness and processing affect compression strength considerably more than they affect tensile strength. As a result, the fiber direction compressive strength of a lamina is usually less than its tensile strength.

Although a number of micromechanics models of the basic failure mechanisms exist (Ref 3, 4), none is completely successful at predicting compressive failure of a laminate under all conditions, based on a knowledge of the fiber and matrix properties alone. Thus, as in the case of tension, the prediction of compressive fiber failure is usually based on an ultimate fiber stress or strain. This should be obtained as the average value at failure of a specimen manufactured with the same matrix and processing conditions as the final part, and tested under conditions as similar to the final design loading and support conditions as possible.

In the case of a lamina loaded in transverse tension or compression, or in shear, it is the matrix that typically fails first. In such cases, failure occurs either in the matrix itself, parallel to the fibers, or at the matrix-fiber interface (or combinations of these).

Typically, tensile transverse strength is considerably less than compressive strength, with the tensile failure mode being one of cleavage, while the compressive failure mode is typically one of shear, as exhibited by the approximately 45° angle of the failure surface to the lamina surface (through-the-thickness slip). In either case, obtaining the strength is relatively straightforward, with the use of a unidirectional coupon tested normal to the fiber direction being typical (refer to the Section "Testing" in this Volume). Obtaining the shear strength of a lamina is, however, less straightforward, with all tests using a flat laminate subjected to stress concentrations that cloud the interpretation of the test results (Ref 5). A cylindrical tube with reinforcement in only the circumferential (hoop) direction eliminates the stress-concentration problem, but is a more complicated and expensive test and involves possible differences in processing that can affect results. In spite of the difficulties, any of these tests can be used to obtain a shear strength estimate for the purposes of design and analysis.

When a lamina is loaded more generally, that is, with a combination of tension or compression in the fiber direction, tension or compression normal to the fiber direction, and shear, as is typical of laminae in a multidirectionally reinforced laminate, then a failure criterion that takes into account the effects of load interactions is necessary. Although noninteractive cri-

teria have been applied to such loading situations and have the advantage of simplicity, the interaction of load components has been well demonstrated to have a considerable effect on lamina failure. Interactive failure criteria can be divided into those that are phenomenological, such as the Tsai-Wu criterion (Ref 6), and those that are based on the physics of the failure process, such as that of Z. Hashin (Ref 7). Although the Tsai-Wu criterion enjoys a great deal of popularity, the physics of the failure process, including all possible failure modes, should never be forgotten when making strength predictions. Furthermore, extrapolation of any phenomenological criterion into regions where little or no data is available, such as combinations of longitudinal and transverse compression, should always be done with extreme caution.

Laminate Strength

Structural composite laminates are almost always multidirectionally reinforced to carry loads that are multiaxial in nature. That is, the typical laminate is made up of a number of laminae stacked together with their fibers oriented in a number of different directions. Although the loading is multiaxial, it is predominantly in plane in nature because of the poor load-carrying ability of composites in the thickness direction. Once the elastic structural stress analysis has been performed, the first step in laminate failure prediction is to resolve the structural stresses into the stresses in each lamina. This is done in terms of the lamina coordinates, that is, stresses parallel and normal to the fiber direction. This can be accomplished using laminated plate theory (Ref 8) and may or may not be done automatically by the structural analysis model or code being used.

With the stresses in each lamina known for a particular load level (and, therefore, for any load level through the use of scaling, which is possible because of the linear relationship that exists between load level and lamina stress), strength predictions can be made using a variety of methods. The simplest is known as first-ply failure. In this method, the stresses in each ply (lamina) are compared to any of the interactive lamina failure criteria already discussed, for example, the Tsai-Wu criterion (Ref 6). The strength of the structure is then taken to be the lowest load level at which any of the laminae satisfy the chosen failure criterion. The problem with this method is that the predicted failure is almost always a matrix failure in only one of the laminae. This implies that some longitudinal splitting of the matrix in that particular ply alone will occur, but that the laminate and the structure itself is nowhere near ultimate failure or even serious reduction of its stiffness. Thus, for most design purposes, the first-ply failure criterion of strength prediction is far too conservative for practical use.

A second simple method often used for predicting structural strength is ultimate fiber failure. In this approach, the longitudinal stresses or strains in the fibers of each ply are determined from the stress analysis and compared to the ultimate allowable values. The structural strength is then predicted to be the load at which the fibers in any lamina first equal the allowable value (tension or compression). The major difficulty here is that this criterion can be too optimistic in its predictions, overpredicting strength by a considerable margin. The reason for this lies in the progressive nature of composite failure.

The progressive growth of damage that leads to ultimate failure of laminated composites is reasonably well understood, although not totally predictable. Under tensile loading, for example, matrix cracks typically form parallel to the fibers in the individual laminae, where the transverse and shear stresses are at a maximum (in the sense of an interactive failure criterion). These cracks form at roughly periodic distances, forming a characteristic pattern that is dependent on the lay-up of the laminate (Ref 9, 10) as well as on the properties of the lamina. The matrix cracks extend through the entire lamina thickness, abutting the neighboring laminae. There, the stress concentrations from the crack tips can lead to progressive delamination of the cracked lamina and its neighbors. In addition, the stress concentration in the neighboring laminae can result in increased fiber failures in the delamination region.

Although the growth of damage under compression loads is not as well understood as in the tensile load case, it does occur here as well (Ref 11). However, the predominant microdamage preceding ultimate failure tends to be kink-band formation in the primary load-bearing fibers, rather than matrix cracking normal to the loading direction, because the matrix compressive strength is normally considerably higher than its tensile strength. These kink-bands can then lead to delaminations of the laminae in which they occur and the off-axis laminae. Depending on its toughness, splitting of the matrix can occur parallel to the fibers in the primary load-bearing laminae as well.

The consequence of the microdamage growth is that neither the redistribution in lamina load sharing or the stress concentrations that result from it is accounted for in the typical stress analysis conducted to predict structural strength. Hence, predictions of strength based on elastic analyses can be quite nonconservative because they can grossly underpredict the fiber stresses in the structure.

For example, consider a $(0°/90°)_s$ laminate loaded in tension in the 0° direction. The usual stress analysis of this laminate, based on the initial elastic stiffnesses of the 0° and 90° laminae, underpredicts the load carried by the fibers of the 0° laminae as ultimate failure nears. This is because of the matrix cracking that occurs in the 90° lamina before ultimate failure, which is not accounted for in the elastic analysis. The matrix cracking leads to both a reduced stiffness of the 90° lamina, resulting in more load being shifted to the 0° laminae, as well as stress concentrations in the 0° laminae fibers near the tips of the matrix cracks. The result is less strength than predicted.

Another simple example is that of a $(\pm30°)_s$ angle ply laminate tensile coupon. Here, the combination of an elastic analysis and fiber failure criterion grossly overpredicts the actual strength. This occurs because the actual failure mode is one of matrix cracking of both laminae, followed by a delamination that grows to join the matrix cracks of the two laminae. The laminate can then pull apart without breaking a single fiber.

The above examples serve to indicate the need to account for microdamage growth and accumulation when predicting the strength of composite structures. This is, of course, similar to the situation in metals, where accurate predictions of strength must account in general for plastic yielding. Such models in composites have been called progressive failure models. Unfortunately, at the present time progressive failure models are at the laboratory rather than production stage. No major finite-element codes generally available have a progressive failure capability. This is primarily due to the fact that progressive failure in composites, although fairly well understood qualitatively, has not yet been successfully modeled quantitatively. Although a lot of excellent work is currently being carried out in the area of quantitative modeling of microdamage modes, the modeling has typically been of one particular type of damage, such as fiber failure, or matrix cracking, or delamination growth. The important step that remains is to model the interaction of these individual modes. This is a very difficult step, however, being three dimensional in general, and one that has yet to receive the needed attention.

In spite of the fact that progressive-failure procedures are not yet fully developed, certain rudimentary concepts of progressive failure can be used to predict composite strength. For example, a number of researchers have used the idea of reducing laminae stiffnesses in the stress analysis to account for matrix failure. In early attempts at this approach, crude models were used in which certain stiffnesses were simply reduced or set to 0 (Ref 12-14) after matrix failure was predicted, using, for example, one of the interactive lamina failure criteria, such as the Tsai-Wu (Ref 6) or Hashin (Ref 7) criterion. More recently, elasticity models have been formulated for a matrix-cracked lamina that give the effective lamina stiffness as a function of the crack spacing (Ref 10) and crack spacing as a function of applied loads (Ref 15, 16). This approach allows stiffnesses of damaged plies to be lowered appropriately as the structural load increases, causing more load to be carried by other undamaged plies. Such modeling leads to an earlier prediction of failure by, say, an ultimate fiber strain criterion. It is cautioned here, however, that although models of this type are an improvement over linear-failure models, they still ignore other important effects, such as matrix-cracking interaction with delaminations, and damage stress-concentration effects on strength.

Other simple progressive-failure models seek to include the effect of delaminations on composite failure. For example, the modeling of delamination growth can be accomplished using the well-known principles of fracture mechanics (Ref 17). Then, stress analysis models can be used to obtain the effects of the delaminations on lamina stresses by modeling the delaminations as stress-free or sliding contact boundary conditions as appropriate. This allows the effects of the delaminations on stress redistribution to be determined and the strength of the structure to be estimated. The major difficulty here is that this generally becomes a three-dimensional stress analysis problem, which limits its practical use at the present time. However, in certain cases, such as axisymmetric problems involving cylindrical pressure vessels (Ref 18), two-dimensional stress solutions are valid, and practical results have been obtained.

Stress Concentrations and Damage

Stress concentrations occur in composites both by design and by accident. Examples of stress concentration by design are holes for mechanical fasteners, ply drop-offs to change laminate reinforcement and thickness needs along a part length, and any free edges where a composite ends. Examples of accidental stress concentrations include damage from impact, cuts, and abrasion as well as delaminations resulting from processing. The major difficulty in predicting strength reductions due to such concentrations is their typically three dimensional characteristics, such as free-edge effects in through-the-thickness holes, fastener cocking and bending effects, part-through cuts, and low-velocity impact damage. Adding to the difficulty in analyzing stress concentrations of an accidental nature is the fact that nondestructive evaluation (NDE) techniques are not advanced enough to give sufficient detail of the damage, such as broken fibers, or an accurate through-the-thickness mapping of damage (refer to the article "Destructive and Nondestructive Tests" in this Volume).

Approaches to predicting the failure of composite structures containing stress concentrations have typically tried to account for progressive damage growth with increasing load, in either direct or ad hoc ways. Examples of the latter approach are detailed in the article "Fracture Analysis of Laminates," where through-the-thickness stress concentrations are addressed. Therefore, stress concentrations of this type will not be further discussed here. Similar ad hoc methods have also been used to treat partial through-the-thickness damage, such as that from impact. In this case, the approach has been to model the damage as an equivalent through-the-thickness hole or crack (Ref 19, 20). It is very difficult, however, with present-day understanding and NDE techniques, to determine what the equivalent hole diameter or crack length should be.

In general, to predict structural strength in the presence of stress concentrations or damage, the designer needs to at least consider the potential for significant damage growth (progressive failure) prior to ultimate failure and to model its effect in some approximate way. Often, this can be done simply through the removal of material in the stress analysis model or the modeling of delaminations. For example, part-through cuts in the exterior hoop lamina of composite pressure vessels typically result in a peeling off of the band of cut fibers prior to ultimate burst. The strength loss due to the cut can then be modeled using an ultimate strain fiber failure criterion in combination with an axisymmetric finite-element model in which the cut band of fibers is eliminated (Ref 18).

The effect on strength of ply drop-offs can be estimated by comparing the predicted peak shear stresses with the interlaminar shear strength of the laminate. Under compression loading, an interlaminar failure may result in ultimate failure of the laminate because of the loss of lamina stability (Ref 21). Under tensile loading, however, interlaminar failure, although signifying the initiation of delaminations in the drop-off region, does not usually result in immediate failure. A strength estimate in tension can be obtained by predicting the onset of delamination (using an interlaminar stress failure criterion), the growth of delamination (using a fracture criterion), and the ultimate strength (using a fiber strain failure criterion in conjunction with a stress analysis of the delaminated part) (Ref 18).

The most difficult stress concentrations to analyze are those that result in stress gradients in the plane of the laminate as well as through the thickness. These include damage due to impact and mechanical fasteners of single lap-type design where fastener cocking becomes important. As indicated previously, the impact problem is complicated by the inability of present-day NDE techniques to adequately describe the damage present. However, even with an adequate description of the damage, this is a difficult problem to analyze because of its three-dimensional nature and the presence of irregular regions of broken fibers, matrix cracking, and delamination. At present, this problem has been inadequately treated in the literature, with most of the effort directed towards the damage caused by the impact event, as described in Ref 22 and 23. In contrast, very little effort has been directed toward modeling the residual strength after impact. Similarly, little

work has been done on the effects of through-the-thickness fastener effects on strength (Ref 24). At present, both the impact problem and the effects of fastener cocking are largely treated empirically.

REFERENCES

1. R.D. Jamison, "The Role of Microdamage in Tensile Failure of Graphite/Epoxy Laminates," Paper presented at the International Symposium on Composites: Materials and Engineering, University of Delaware, Sept 1984
2. D.G. Harlow and S.L. Phoenix, The Chain-of-Bundles Model for the Strength of Fibrous Materials I: Analysis and Conjectures, *J. Compos. Mater.*, Vol 12, 1978, p 195-214
3. L.B. Greszczuk, "On Failure Modes of Unidirectional Composites Under Compression Loading," in *Proceedings of the 2nd USA-USSR Symposium on Fracture of Composite Materials*, 9-12 March 1981, Lehigh University
4. A.G. Evans and W.F. Adler, Kinking as a Mode of Structural Degradation in Carbon Fiber Composites, *Acta Metall.*, Vol 26, 1978
5. H.E. Gascoigne, M.G. Abdallah, and D.S. Adams, Influence of Composite Layup on Shear Strain Distribution in the Iosipescu Shear Test Specimen, in *Proceedings of the 1986 Society of Experimental Mechanics, Spring Conference on Experimental Mechanics*, June 1986, p 314-328
6. S.W. Tasi and H.T. Hahn, *Introduction to Composite Materials*, Technomic, 1980
7. Z. Hashin, Failure Criteria for Unidirectional Fiber Composites, *J. Appl. Mech.*, Vol 47, 1980, p 329-334
8. R.M. Jones, *Mechanics of Composite Materials*, McGraw-Hill, 1975
9. K.L. Reifsnider and A. Talug, Characteristic Damage States in Composite Laminates, Research Workshop on Mechanics of Composite Materials, Duke University, Oct 1978, p 130-161
10. R.J. Nuismer and S.C. Tan, "Constitutive Relations of a Cracked Composite Lamina," *J. Compos. Mater.*, 1987
11. H.T. Hahn and J.G. Williams, "Compression Failure Mechanisms in Unidirectional Composites," NASA Technical Memorandum 85834, National Aeronautics and Space Administration, Aug 1984
12. P.H. Petit and M.E. Waddoups, A Method of Predicting the Nonlinear Behavior of Laminated Composites, *J. Compos. Mater.*, Vol 3, 1969, p 2-19
13. R.J. Nuismer and G.E. Brown, Progres-

sive Failure of Notched Composite Laminates Using Finite Elements, in *Advances in Engineering Science*, Vol 1, NASA CP-2001, National Aeronautics and Space Administration, p 183-192
14. R.J. Nuismer, Predicting the Performance and Failure of Multidirectional Polymeric Matrix Composite Laminates, in *Proceedings of the Third International Conference on Composite Materials*, Aug 1980, p 436-452
15. R.J. Nuismer and S.C. Tan, The Role of Matrix Cracking in the Continuum Constitutive Behavior of a Damaged Composite Ply, in *Mechanics of Composite Materials: Recent Advances*, Pergamon Press, 1983, p 437-448
16. D.L. Flaggs, Prediction of Tensile Matrix Failure in Composite Laminates, *J. Compos. Mater.*, Vol 19, 1985, p 29-50
17. E.F. Rybicki, D.W. Schueser, and J. Fox, An Energy Release Rate Approach for Stable Crack Growth in the Free-Edge Delamination Problem, *J. Compos. Mater.*, Vol 11, 1977
18. D.S. Adams, G.E. Colvin, C.L. Dodson, T.J. Itchkawich, J.W. Kordig, M.J. Messick, J.B. Goodro, and S.W. Beckwith, Fracture Control Document, WDI-FWC-11, Rev. 2, Hercules Aerospace Division, Contract No. 114010, Volumes I and II, January 1985
19. G.E. Husman, J.N. Whitney, and J.C. Halpin, "Residual Strength Characterization of Laminated Composites Subjected to Impact Loading," AFML-TR-73-309, Air Force Materials Laboratory, Feb 1974
20. N.R. Adsit and J.P. Waszczak, Effect of Near Visual Damage on the Properties of Graphite/Epoxy, in *Composite Materials: Testing and Design (Fifth Conference)*, STP 674, American Society for Testing and Materials, 1979, p 101-117
21. M.J. Messick, R.J. Nuismer, G.T. Jamison, and S.R. Graves, "Space Shuttle Filament Wound Case Compressive Strength Study: Part II—Analysis," AIAA/ASME/SAE/ASEE 22nd Joint Propulsion Conference, Huntsville, AL, June 1986, AIAA-86-1417
22. L.B. Greszczuk, Damage in Composite Materials Due to Low Velocity Impact, in *Impact Dynamics*, John Wiley & Sons, 1982
23. S.P. Joshi and C.T. Sun, Impact Induced Fracture in a Laminated Composite, *J. Compos. Mater.*, Vol 19, 1985, p 51-66
24. R.L. Ramkumar *et al.*, "Strength Analysis of Composite and Metallic Plates Bolted Together by a Single Fastener," AFWAL-TR-85-3064, Air Force Wright Aeronautical Laboratories, Aug 1985

Fatigue Strength

Ran Y. Kim, University of Dayton Research Institute

TESTS RESULTS obtained from small laboratory specimens cannot be directly applied to the design of composite structures without encountering some difficulties. Real-life structures are usually subjected to different loads and environments than are test coupons. The size, manufacturing process, state of stress, and other factors also influence the fatigue strength of structures. For these reasons, fatigue tests should be conducted on structural parts and, in some cases, complete structures. Tests on parts such as joints, beams, and columns provide valuable data for verification of design and fabrication methods, while component testing circumvents the costly and lengthy process of complete structure fatigue tests. However, because the results obtained from both types of tests apply only to the particular design being tested, and cannot be generalized, coupon-type fatigue tests can indeed be very useful for studying many variables during initial evaluations.

Knowledge of fatigue behavior at the laminate level is essential for understanding the fatigue life of a laminated composite structure. In fact, most published fatigue data have addressed laminates with or without a notch. Relatively little fatigue data exists on composite structures. Investigations have demonstrated that sufficient fatigue life for a composite structure is achieved if it is designed to satisfy static strength requirements (Ref 1, 2). Fatigue failure mechanisms, S-N relations and characterization, and statistical life prediction models of composite laminates are described in this article. Data on the fatigue behavior of bolted joints and helicopter blades made of composites also are presented.

Fatigue Failure

Composite materials exhibit very complex failure mechanisms under static and fatigue loading because of anisotropic characteristics of their strength and stiffness. Fatigue failure is usually accompanied by extensive damage that multiplies throughout specimen volume, in contrast to the localized formation of a predominant, single crack, as is common in isotropic, brittle materials. The four basic failure mechanisms in composites are layer cracking, delamination, fiber breakage, and fiber-matrix interfacial debonding. Any combination of these can cause fatigue damage that will result in reduced strength and stiffness. Both the type and degree of damage vary widely, depending upon material properties, laminations (including stacking sequence), and type of fatigue loading. It has been observed that, in general, damage development under fatigue loading is similar to that under static loading except that fatigue at a given stress level causes additional damage as fatigue cycles increase.

Layer Cracking. In multidirectional laminates under in-plane loading, failure from layer cracking usually occurs in succession from the weakest layer to the strongest. For example, successive transverse cracks in the off-axis layers of a $(0°/90°/\pm45°)_s$ graphite-epoxy laminate subjected to uniaxial tension are expected to occur as the load increases. The first crack occurs in the 90° layers; with increasing load, more cracks develop, but they are still confined to the 90° layers. As the load increases further, cracks occur in the adjacent 45° layers, appearing at the tips of the 90° cracks in most cases, and extending to the interface of the $+45°/-45°$ layers (see Fig. 1). Subsequently, the number of cracks increases with the load (Fig. 2) until final laminate failure takes place. However, some laminates reach a crack density limit, after which no new cracks occur before final failure, despite additional loading. The crack density limit for a given layer depends on its thickness and appears to be independent of laminate type (Ref 3-7).

During fatigue loading, more cracks occur in each layer, and reach a crack density limit, than during static loading. For instance, many cracks in the $-45°$ layers of the $(0°/90°/\pm45°)_s$ laminate occur during fatigue loading, whereas few or none occur during static loading. In addition, many axial cracks initiate at the tips of the transverse cracks and extend along the axial direction as fatigue cycles increase. The multiplication process of transverse cracks in the course of the fatigue cycle is shown in Fig. 3. The damage ratio is defined here as the ratio of

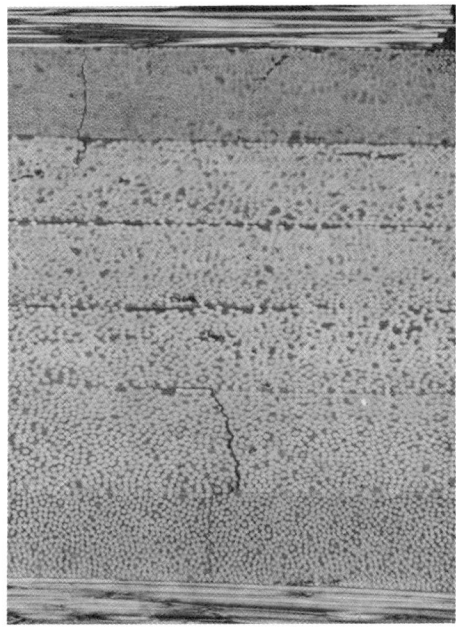 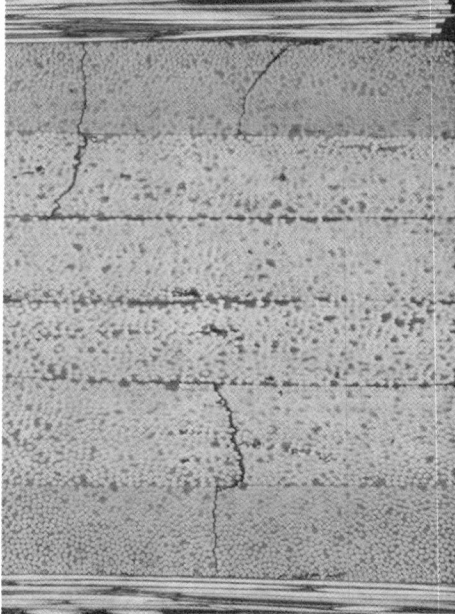

(a)　　　　　　　　　　　　　　　　(b)

Fig. 1 Photomicrographs showing crack patterns under static loading: $[0°/90°/\pm45°]_s$. (a) 440 MPa or 65 ksi. (b) 483 MPa or 70 ksi

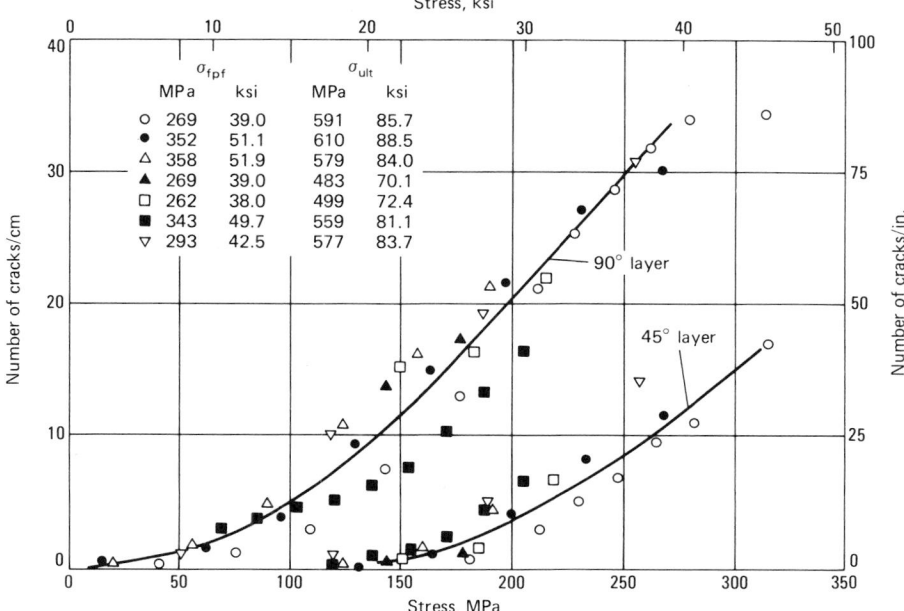

Fig. 2 Increase of number of cracks for [0°/90°/±45°]ₛ laminate subjected to static loading

the crack density at *n* cycles to the crack density at final failure. Cracks in all off-axis plies are grouped together. Most of the crack multiplica-tion occurs during the first 20% of the fatigue life. A significant amount of fatigue life re-mains after reaching the crack density limit.

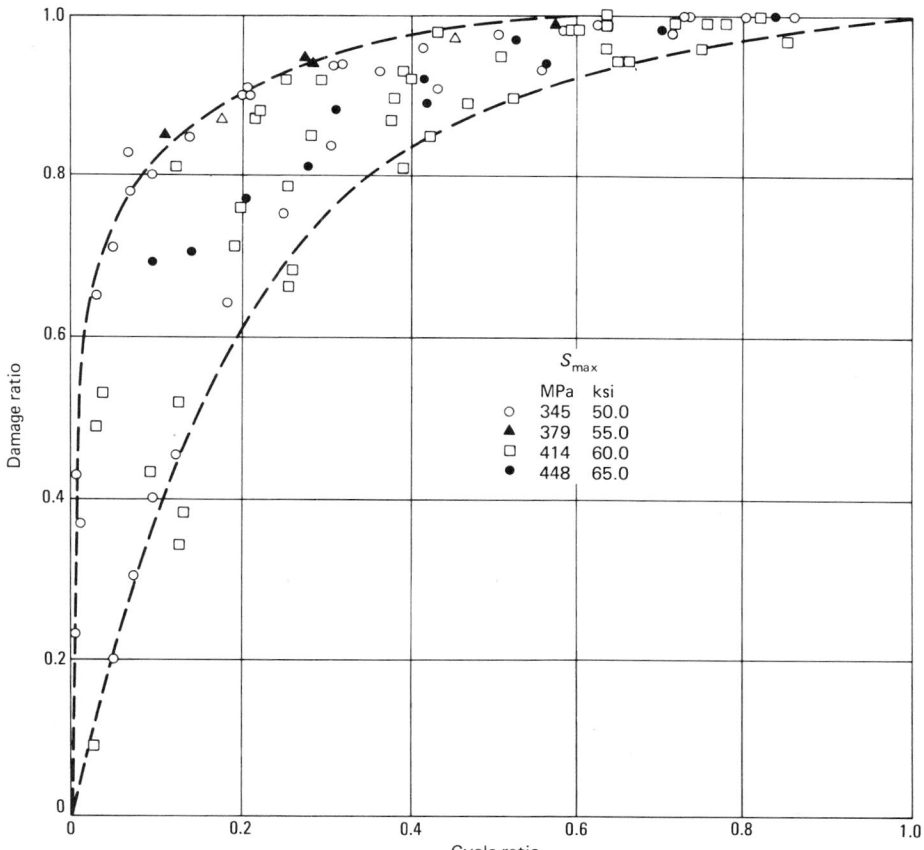

Fig. 3 Increase of number of cracks for [0°/90°/±45°]ₛ laminate subjected to fatigue loading

Cracks in the 90° layers of the $(0°/90°/\pm45°)_s$ laminate are found at fewer than one million cycles at a fatigue stress level of 170 MPa (25 ksi), which is less than the 280 MPa (40 ksi) stress level at which the first layer fails under static loading.

The 0° layers are also susceptible to cracking in the fiber direction because of the transverse stress resulting from a mismatch in Poisson's ratio between different layers. Because this transverse stress is usually small, axial cracking in 0° layers may not occur under static loading. However, under fatigue loading, axial cracking can occur, for example, in cross-ply laminates such as $(0°/90°)_s$.

Delamination. In composite laminates, free-edge delamination under in-plane axial loading is caused by interlaminar stresses that are highly localized around the free edge (Ref 8). The nature of interlaminar stresses with regard to their magnitude and sign can be accurately calculated using an analytical model (Ref 9). The magnitude and distribution of interlaminar stress components vary widely de-pending on the type of laminate, stacking se-quence, properties of the constituent materials, and type of loading. For a given laminate, the stacking sequence significantly influences free-edge delamination. As an example, consider a quasi-isotropic laminate with two stacking se-quences, $(0°/90°/\pm45°)_s$ and $(0°/\pm45°/90°)_s$. The $(0°/90°/\pm45°)_s$ laminate does not show any delamination under static tension, whereas the latter laminate shows extensive delamination under static tension. Figure 4 shows a pho-tomicrograph of a free edge and an x-ray radiograph of the specimen width showing delamination in the $(0°/\pm45°/90°)_s$ laminate un-der static tension. At about 350 MPa (50 ksi) of the applied tension, delamination occurred and extended almost instantly over the entire length of the specimen, taking irregular paths at the interfaces of the 90°/90°, 90°/−45°, and 45°/−45° plies. The change in the delamination path happened for the most part at the trans-verse crack tips. The delamination also grew continuously toward the middle of the specimen from both free edges as the load increased (Fig. 4). In fatigue, the onset of delamination oc-curred very early (and at fatigue stress levels that were smaller than static stress levels) and propagated rapidly toward the middle of the specimen width as the cycles increased. The lower bound of the delamination-free stress level for a predetermined life cycle is expected to vary, depending upon the type of laminate. In addition to interlaminar tensile stress, other factors, such as transverse cracking and in-terlaminar shear stress, appear to affect the onset and growth of delamination (Ref 10). The $(0°/90°/\pm45°)_s$ laminate, which has a compres-sive interlaminar normal stress, does not show any delamination under static tension, but un-der tensile fatigue, considerable delamination occurs at the interface between the ±45° and −45° plies. The interlaminar shear stress at the +45°/−45° interface is not large enough to

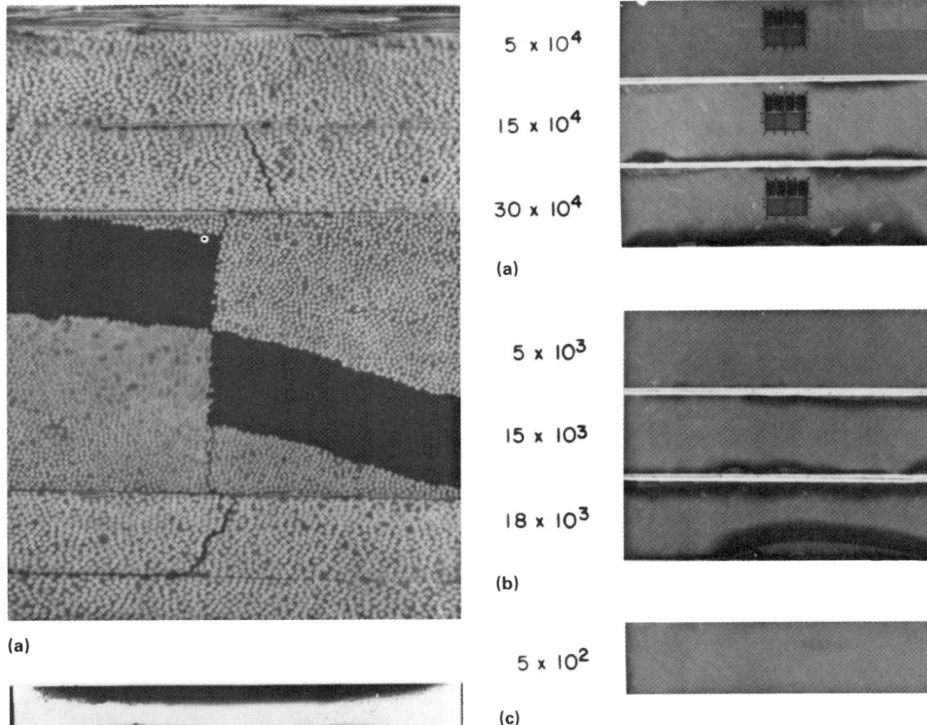

(a)

(b)

Fig. 4 Delamination in [0°/±45°/90°] graphite-epoxy subjected to static loading. (a) Micrograph of a free edge. 35× (b) An x-ray of the width. 0.2×

5 × 10⁴

15 × 10⁴

30 × 10⁴

(a)

5 × 10³

15 × 10³

18 × 10³

(b)

5 × 10²

(c)

Fig. 5 X-ray radiographs of delamination growth as a function of fatigue cycles for a [0°/90°/±45°]ₛ carbon fiber reinforced plastic T300-5208 laminate. (a) S_{max} = 345 MPa (50.0 ksi). N = 312 000 cycles. (b) S_{max} = 414 MPa (60.0 ksi). N = 19 400 cycles. (c) S_{max} = 483 MPa (70.0 ksi). N = 620 cycles

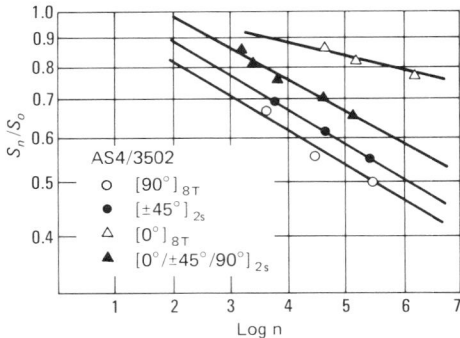

Fig. 6 S-N curves for various AS4-epoxy laminates

reach the interlaminar shear strength under static loading, but under fatigue loading, the shear stress becomes significant because of the high fatigue sensitivity of the epoxy matrix. Figure 5 shows delamination growth as a function of fatigue cycle for a (0°/90°/±45°)ₛ graphite-epoxy T300/5208 laminate. However, the growth of delamination in this case is much slower than is the case with interlaminar tension.

Fiber break and interface debonding differ widely, depending on the properties of the constituent materials and fiber defects. In most advanced composites, such as boron and graphite fiber with polymer matrix, the strain-to-failure is greater in the matrix than in the fibers. Therefore, fibers can break, because of defects or weakness, before the matrix fails. The crack created by a fiber break grows into the matrix, as load increases, along a path varying with matrix and interface properties. If bonding is strong, the crack grows into the matrix, resulting in a fairly smooth fracture surface across the section. With a weak bond, the crack is more likely to lead to interfacial debonding and extensive fiber pull-out. An intermediate bond shows irregular failure surfaces with some fiber pull-out. These failure mechanisms occur under static and fatigue loading, although fatigue failure depends on the sensitivity of the matrix, interface, and fiber.

Because in most advanced composites the matrix remains elastic until the composite fails, fatigue damage in the interface is negligible, except at the site of fiber breaks. Consequently, the modulus and strength of a unidirectional laminate subjected to fatigue loading along the fiber direction do not decrease until fracture is imminent.

S-N Relation

The fatigue behavior of a material is basically expressed in the form of a relation between applied maximum fatigue stress (S) and fatigue life (number of cycles to failure, N). In composite laminates, the S-N relation primarily depends on the constituent material properties (Ref 11, 12). Most advanced fibers are very insensitive to fatigue, and the resulting composites show good fatigue resistance. Figure 6 shows the S-N curves for a variety of laminates of AS 4/3502 graphite-epoxy. The ordinate represents the fatigue strength ratio, which is the ratio of fatigue stress to static strength. The laminates with matrix-dominant failure modes exhibit a smaller fatigue resistance than the laminates with fiber-dominant failure modes. The slope of the S-N curve for the [0°]₈T laminate is relatively flat because of the fatigue insensitivity of the graphite fiber. Conse-

quently, the slope of the S-N curve tends to increase as the content of the 0° ply decreases, as shown in Fig. 7 (Ref 13). The fatigue strength of a (±45°)ₛ laminate is frequently used to determine the longitudinal shear fatigue strength. The fatigue strength in compression-compression fatigue is slightly greater than that in tension-tension fatigue (Fig. 8) for a (0°/45°/90°/−45°)₂ₛ graphite-epoxy laminate. The static tensile and compressive strengths for this laminate are practically identical. The extensive damage incurred during tension-tension fatigue is mainly responsible for the reduction in tensile fatigue strength. The fatigue strength of multi-directional laminates is closely related to the 0° fatigue strength, as shown in Fig. 9, unless the laminate undergoes extensive damage during fatigue.

Mean Stress. A constant-amplitude fatigue loading may be regarded as a fully reversed cyclic loading superimposed upon a constant load. Consequently, the fatigue strength at a chosen cycle may be expressed in terms of alternating the stress, S_r, and mean stress, S_m. The values of S_m and S_r are given by the following equations:

$$S_m = (S_{max} + S_{min})/2$$

$$S_r = (S_{max} - S_{min})/2$$

where S_{max} and S_{min} are the maximum and minimum fatigue stresses, respectively.

Figure 10 shows the effect of mean stress on fatigue life in the form of a constant-life diagram (Ref 14). This diagram is practically symmetrical with respect to the alternating stress axis, S_r (R = −1), although the peaks of constant-life curves tend to lie slightly in the tension-dominant fatigue region. This indicates that the S-N behavior of the composite is independent of the type of fatigue loading (that is, tension-tension, tension-compression, or compression-compression) as long as the tensile and compressive strengths are equal. In tension-compression fatigue, the fatigue failure in the tension-dominant region (T > C) in Fig. 10 was accompanied by severe matrix damage. This effect on the S-N relation appears to be balanced out by the fact that the compressive stress is much more detrimental to the residual

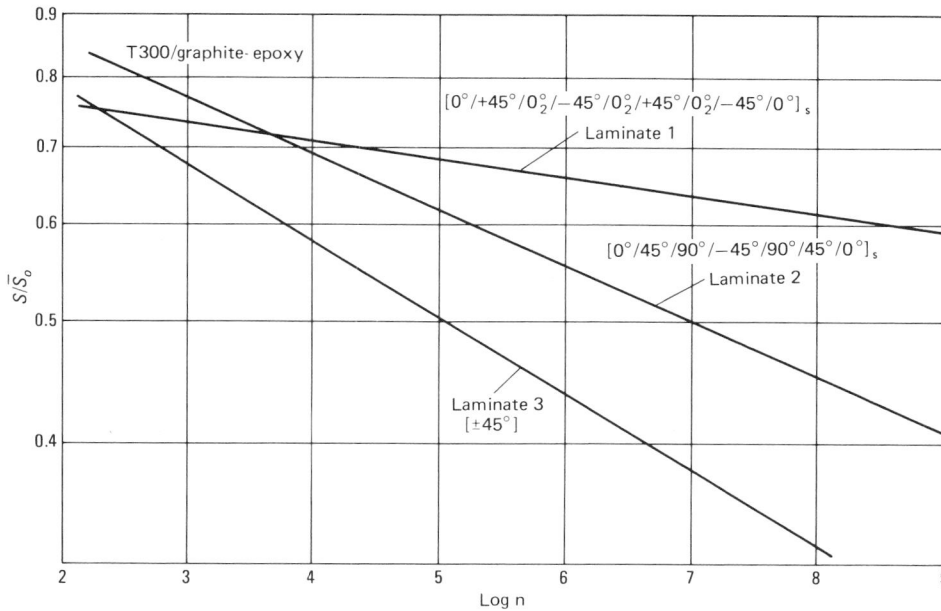

Fig. 7 Effect of 0° layer fraction on *S-N* relation

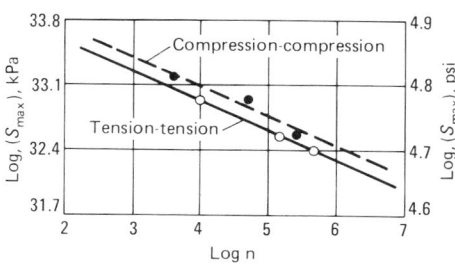

Fig. 8 Tension-tension and compression-compression fatigue tests of $[0°/45°/90°/-45°]_{2s}$ carbon fiber reinforced plastic T300/5208

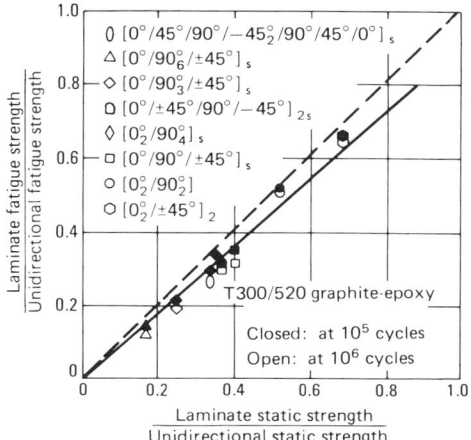

Fig. 9 Fatigue and static strengths normalized with respect to unidirectional tensile strengths

strength of damaged specimens than is the tensile stress. The peak of the constant-life curve for a $(0°/\pm 30°)_s$ graphite-epoxy lies in the positive mean stress region because of the lower static compressive strength (Ref 15).

Notch. With metals, the problem of stress concentration resulting from various notches (circular holes and cracks) is complicated by the many factors that influence behavior under cyclic stresses. In notched specimens, combined stresses are produced and vary widely in the region around the notch. Furthermore, any

damage that occurs during fatigue will change the stress distribution around the notch. The foregoing factors have made it difficult to predict the fatigue strengths of notched specimens from the fatigue strengths of the unnotched specimens. In contrast to metals, the reduction in fatigue strength resulting from the presence of a notch is found to be insignificant in most composite laminates (Ref 14, 16, 17). The value of fatigue notch factor (the ratio of the fatigue strength of an unnotched specimen to the fatigue strength of a notched specimen at

N cycles) is much smaller than the static stress-concentration factor, and is close to unity in

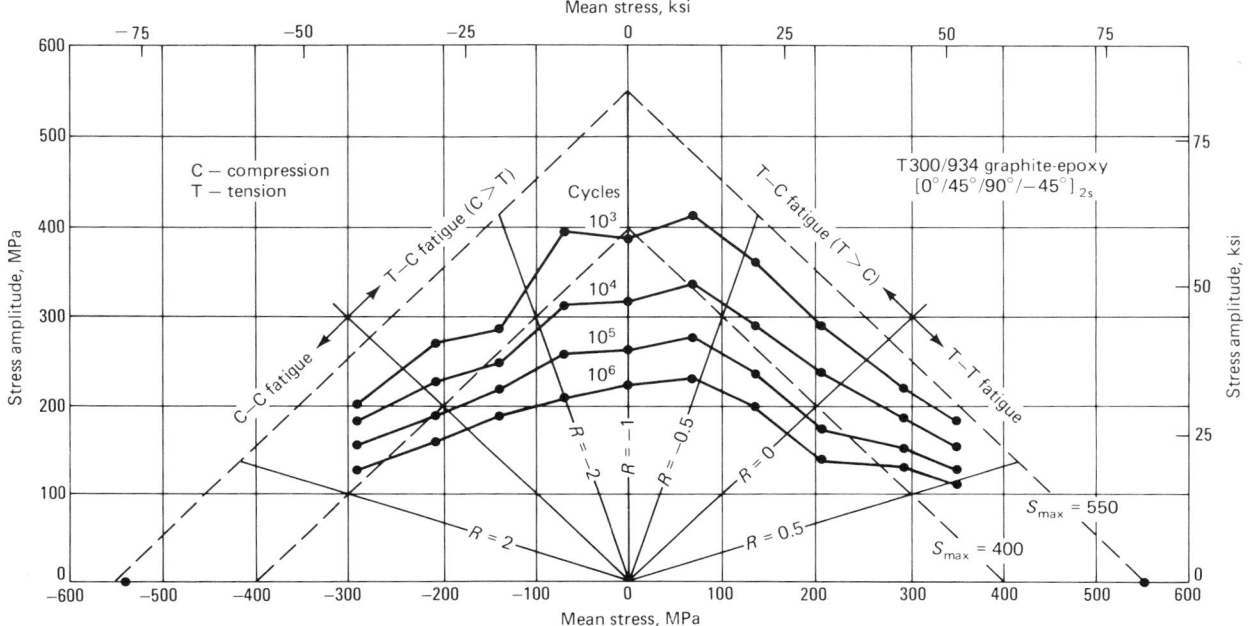

Fig. 10 Effect of mean stress on fatigue life of $[0°/45°/90°/-45°]_{2s}$ carbon fiber reinforced plastic T300/5208

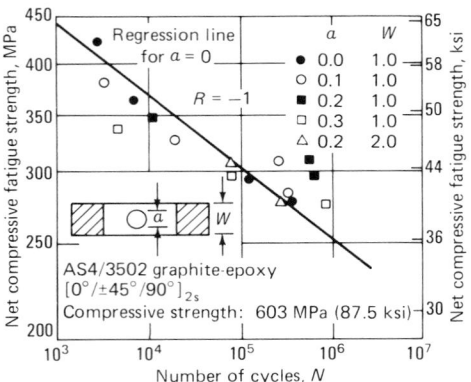

Fig. 11 Effect of circular hole on fatigue life

many cases. Figure 11 shows the net fatigue stress at the notch, instead of gross stress, versus the number of cycles to failure for unnotched and notched specimens under tension-compression fatigue with $R = -1$ (Ref 14). The various symbols in Fig. 11 represent the Weibull characteristic life estimated from at least five specimens for each stress level. The *S-N* relations for notched specimens are almost identical to those for unnotched specimens over the range of fatigue life considered, indicating no notch effect for the size considered in this laminate. The residual strength after fatigue loading is usually greater than the static strength. The excellent fatigue resistance of the notched specimen is mainly due to the fact that damage relaxes the stress concentration around the notch.

Bolted Joint Fatigue. Fatigue behavior of a bolted joint is very similar to the fatigue behavior of notched composites. The residual strength generally increases with the fatigue cycle toward the unnotched net section strength (Ref 18). This strength increase is mainly due to damage that occurs around fasteners, reducing the stress concentration. The residual strength of a bolted joint depends on the degree of fitness of the bolt. When a loose bolt in a hole is stressed back and forth, the hole elongates at an increasing rate, and failure is rapid. This is why it is important that all bolts in composites be fit tightly. A tight fit also promotes uniform sharing of load between multiple fasteners.

Bolt torque pressure has a significant effect on both static strength and fatigue strength (Ref 18, 19). Each strength is improved considerably, compared to the pin-bearing case, which has no torque pressure, as shown in Fig. 12 (Ref 18). High torque virtually eliminates hole elongation that is due to fatigue and also results in less temperature increase. With loose-fit bolts, a large increase in bolt temperature due to friction at high frequency is one of the major problems. This is associated with a decrease in mechanical properties because of temperature elevation.

Figures 13 and 14 show the effect of joint geometry in terms of different edge distance and width (Ref 18). Fatigue strength and static strength do not change with geometry variations in $[0_5^\circ/\pm45_2^\circ/90^\circ]_s$ laminates. However, variations of specimen width in $[0_{19}^\circ/\pm45_{38}^\circ/90_5^\circ]_s$ laminates have a pronounced effect on fatigue life, as shown in Fig. 14. Furthermore, a change of failure mode from the bearing to the net section occurs as the w/d ratio is reduced.

Fatigue strength of a bolted joint is also influenced by many other factors, such as type of joint (single or double), geometry of fastener head, interference fit, and so forth. Figure 15 shows the typical *S-N* relation of a bolted joint of graphite-epoxy laminate of $[0_5^\circ/\pm45_2^\circ/90^\circ]_s$ orientation with two different fatigue loadings. The fatigue life is indicated in terms of fatigue cycles required to produce a 0.50-mm (0.02-in.) elongation of the fastener hole. Tension-compression fatigue ($R = -1$) reduces fatigue strength of the bolted joint for most laminate types. The stacking sequence also influences fatigue strength.

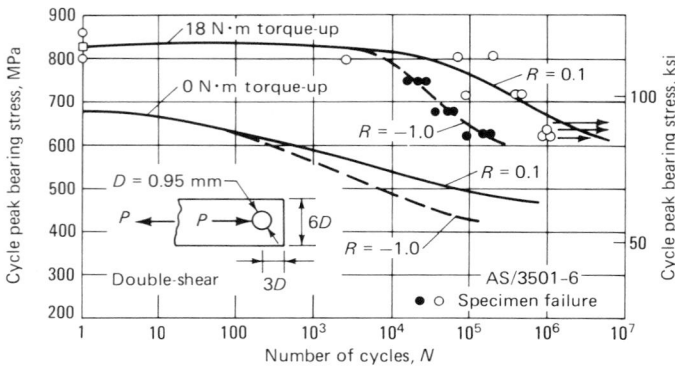

Fig. 12 Effect of torque-up on bolted joint fatigue life

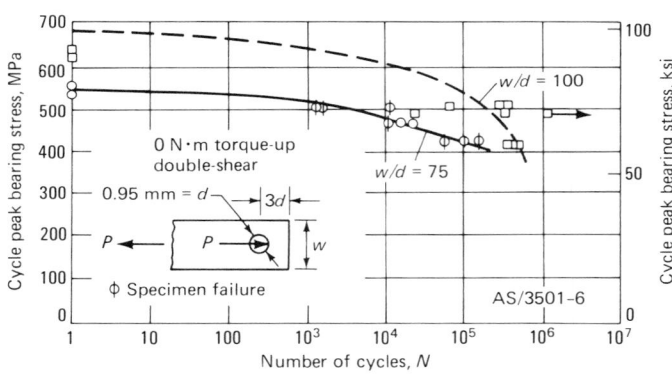

Fig. 14 Effect of geometry on bolted-joint fatigue life, $[0^\circ/\pm45^\circ/90^\circ]_s$ laminate

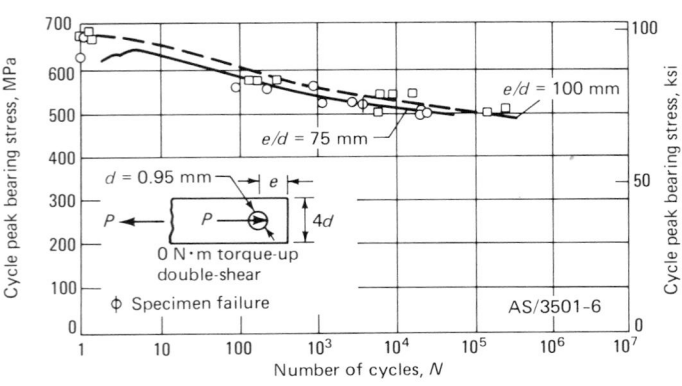

Fig. 13 Effect of geometry on bolted-joint fatigue life, $[0_5^\circ/\pm45_2^\circ/90^\circ]_s$ laminate

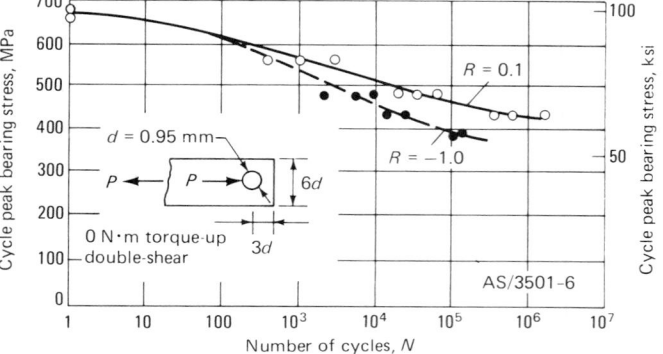

Fig. 15 Effect of *R* ratio on fatigue life

Helicopter Rotor Blade Fatigue. The helicopter rotor blade is the first large-volume use of a composite material as a primary airframe structure. During its service life, the helicopter experiences a severe high-cycle fatigue environment, typically on the order of 10^7 to 10^8 cycles. The blades show excellent performance with unlimited fatigue life when the composite materials are properly used in rotor blade design (Ref 20-22). The results of tests performed on the tail and main rotor blades of the Westland Sea King helicopter are reported in Ref 20. Blade construction is typically of a hollow D-spar composed largely of unidirectional glass-reinforced epoxy tape with internal and external woven wraps. A bolted joint is used at the root end with local reinforcement of the spar. The fatigue testing involved both structural part and complete (full-sized) structure tests. The structural part fatigue tests were performed for the four critical areas of the blade chosen for the test under a programmed series of loading blocks designed to correspond closely to actual flight conditions. The early design of tail rotor blade root end specimens indicated a weakness in the blade spar. Design and manufacturing changes were made at the root end to provide a new standard blade. After the fatigue testing at a fully factored, programmed load of two root end specimens, no failure occurred and unlimited fatigue life was shown.

Fatigue Data Analysis and Life Prediction

Weibull Distribution and Parameters Estimation. A random variable, X, has a Weibull distribution with shape parameter α and scale parameter β if its cumulative distribution function, $F_X(x)$, is:

$$F_X(x) = P_r\{X \le x\} = 1 - \exp\left[-\left(\frac{X}{\beta}\right)^\alpha\right] \quad \text{(Eq 1)}$$

where $x > 0$ and $F_X(x)$ represents the probability of the random number X being less than or equal to x-experimental. On the other hand, the distribution for the n-ordered data ($x_1 \le x_2, ..., \le x_n$) is given by the medium rank.

$$F_X(x_i) = 1 - \frac{i - 0.3}{n + 0.4} \quad \text{(Eq 2)}$$

The following relation exists:

$u = $ expected value of $X = \beta \cdot \Gamma\left(1 + \frac{1}{\alpha}\right)$

$\sigma^2 = $ variance of $X = $

$$\beta^2 \left\{\Gamma\left(1 + \frac{2}{\alpha}\right) - \Gamma^2\left(1 + \frac{1}{\alpha}\right)\right\} \quad \text{(Eq 3)}$$

where Γ is the gamma function.

A maximum-likelihood method is widely used for estimating Weibull parameters α and β:

Table 1 Percentage points l_γ^* such that $P_r\{\hat{\alpha} \ln (\hat{\beta}/\beta) < l_\gamma^*\} = \gamma$

$n \backslash \gamma$	0.90	0.95	0.98
5	0.772	1.107	1.582
6	0.666	0.939	1.291
7	0.598	0.829	1.120
8	0.547	0.751	1.003
9	0.507	0.691	0.917
10	0.475	0.644	0.851
12	0.425	0.572	0.752
14	0.389	0.520	0.681
16	0.360	0.480	0.627
18	0.338	0.447	0.584
20	0.318	0.421	0.549
22	0.302	0.398	0.519
24	0.288	0.379	0.494
26	0.276	0.362	0.472
28	0.265	0.347	0.453
30	0.256	0.334	0.435
40	0.222	0.285	0.371
50	0.195	0.253	0.328

Source: Ref 18

$$\frac{\sum\limits_{i=1}^{n} x_i^{\hat{\alpha}} \ln x_i}{\sum\limits_{i=1}^{n} x_i^{\hat{\alpha}}} - \frac{1}{\hat{\alpha}} - \frac{\sum\limits_{i=1}^{n} \ln x_i}{n} = 0 \quad \text{(Eq 4)}$$

and

$$\hat{\beta} = \left[\frac{1}{n} \sum\limits_{i=1}^{n} x_i^{\hat{\alpha}}\right]^{1/\hat{\alpha}} \quad \text{(Eq 5)}$$

An efficient iterative technique for obtaining the solution of $f(\hat{\alpha}) = 0$ is to use the Newton-Raphson method in which the $(j+1)$st successive approximation, $\hat{\alpha}_{j+1}$ to $\hat{\alpha}_j$, is given by:

$$\hat{\alpha}_{j+1} = \hat{\alpha}_j - f(\hat{\alpha}_j)/f'(\hat{\alpha}_j) \quad \text{(Eq 6)}$$

The maximum-likelihood method has the advantage in that the confidence intervals for α and β can be computed and it can be applied to a life-test model in which censoring is progressive, that is, a model in which a portion of the survivors is withdrawn from life test several times during the test.

The 95% confidence interval (one-sided) for β_i can be obtained from Table 1 (Ref 23) as follows:

$$P_r\{\hat{\alpha}\ln(\hat{\beta}_i/\beta_i) \le l_{0.95}^*\} = 0.95 \quad \text{(Eq 7)}$$

Let us denote the lower bound of 95% confidence interval for β_i by $\underline{\beta}_i$. Then:

$$\underline{\beta}_i = \hat{\beta}_i \exp\left[-(l_{0.95}^*/\hat{\alpha})\right] \quad \text{(Eq 8)}$$

where $\hat{\alpha}$ is the Weibull shape parameter estimated from fatigue data.

An A-allowable (or B-allowable) material fatigue design value is defined by the probabilistic statement that we can be 95% confident of the assertion that the probability of surviving the A-allowable value is at least 0.99 (or 0.90 for B-allowable). Therefore, A-allowable x_{Ai} under the stress level S_i, where $i = 1, 2, ..., m$, can be obtained by solving the following:

$$R(x_{Ai}) = \exp\left[-\left(\frac{x_{Ai}}{\underline{\beta}_i}\right)^{\hat{\alpha}}\right] = 0.99 \quad \text{(Eq 9)}$$

where $\underline{\beta}_i$ is the 95% confidence lower limit for β_i, that is:

$$x_{Ai} = \underline{\beta}_i[-\ln (0.99)]^{1/\hat{\alpha}} \quad \text{(Eq 10)}$$

Pooling Technique. To obtain the *S-N* relationship, m levels of stress are employed with n specimens tested at each level of stress. The shape parameter and scale parameter (characteristic fatigue life) can be obtained from the pooled fatigue data $x_{i1}, x_{i2}, ..., x_{in}$ (under its level of stress), where $i = 1, 2, ..., m$. By assuming that the shape parameter α for each stress level is equal to that for every other level, the pooled estimation of α can be obtained by normalizing the data, that is, by letting $Y_{ij} = X_{ij}/\beta_i$, for $i = 1, 2, ..., n$. The maximum-likelihood equations for the normalized data are:

$$\frac{\sum\limits_{i=1}^{m} \sum\limits_{j=1}^{n} y_{ij}^{\hat{\alpha}} \ln Y_{ij}}{\sum\limits_{i=1}^{m} \sum\limits_{j=1}^{n} y_{ij}^{\hat{\alpha}}} - \frac{1}{\hat{\alpha}} - \frac{\sum\limits_{i=1}^{m} \sum\limits_{j=1}^{n} \ln Y_{ij}}{n \cdot m} = 0 \quad \text{(Eq 11)}$$

and

$$\hat{\beta}_i = \left[\frac{1}{n} \sum\limits_{j=1}^{n} x_{ij}^{\hat{\alpha}}\right]^{1/\hat{\alpha}} \text{where } i = 1,2,...,m \quad \text{(Eq 12)}$$

This pooled estimate is more efficient than the one introduced in Eq 4 and 5. As before, the solutions of Eq 11 and 12 can be obtained by the iterative technique of the Newton-Raphson method. Table 2 (Ref 24) is useful in computing the confidence interval of β.

S-N Curve Characterization. Traditionally, the fatigue behavior of materials is expressed by the relationship between fatigue stress and fatigue life, that is, cycle to failure. Although no universally accepted rules are available for *S-N* characterization, the following discussion may be used as a general guide. Constant-amplitude fatigue tests are normally conducted under three to five different stress levels. The values of the extreme stress levels are approximately chosen so that the fatigue cycles under the extreme stress levels range between 10^3 and 10^6 cycles. The values of other intermediate stress levels are arbitrarily chosen. The number of specimens to be tested at each stress level depends on the size of scatter in the data and the availability of specimens and test time, but approximately four to ten specimens are typical for composite laminates. Large variation in fatigue life data is common in composites, and hence an *S-N* curve obtained from mean lifetimes only may be in error. In view of the foregoing variability in test results, a procedure developed for a more accurate evaluation of fatigue strength is discussed below.

Let us assume that the fatigue data follows a classical power law and a two-parameter Weibull distribution. The *S-N* curve takes the form:

$$KS^bN = 1 \quad \text{(Eq 13)}$$

Table 2   Percentage points, $I\gamma(a)$ such that P_r ($\hat{\alpha}$ ln ($\hat{\beta}/\beta$) < I_γ^*) = γ

m	$\gamma \backslash n$	5	6	7	8	9	10
3	0.90	0.655	0.578	0.533	0.488	0.453	0.420
3	0.95	0.875	0.768	0.701	0.639	0.589	0.548
3	0.98	1.116	0.997	0.905	0.816	0.751	0.710
4	0.90	0.656	0.571	0.521	0.480	0.445	0.422
4	0.95	0.849	0.746	0.679	0.628	0.581	0.555
4	0.98	1.094	0.955	0.858	0.781	0.736	0.700
5	0.90	0.648	0.570	0.513	0.466	0.436	0.416
5	0.95	0.836	0.737	0.660	0.606	0.569	0.537
5	0.98	1.063	0.940	0.845	0.764	0.710	0.671

(a) $I\gamma$ obtained from Ref 18

where K and b are parameters. The K and b can be estimated using the least squares linear regression, because Eq 13 becomes a straight line after logarithmic transformation. The fatigue life, N_i, for each stress level, S_i, can be replaced by Weibull fatigue characteristic life, $\hat{\beta}_i$. Taking natural logarithms, we obtain:

$$\ln\hat{\beta}_i = -b\ln S_i - \ln K \qquad \text{(Eq 14)}$$

The values of b and K can be easily determined by applying the least squares linear regression analysis.

Fatigue Life Prediction. Because composite laminates exhibit very complex failure processes, no one analytical model has been able to account for all possible processes. Consequently, statistical life prediction methodologies have frequently been adopted. This section describes a strength degradation model that is capable of predicting the statistical distribution of both fatigue life and residual strength (Ref 25, 26).

The residual strength, instead of a crack length, is chosen to describe the criticality of the damage because composite failure is characterized by a multitude of matrix cracks and fiber breaks rather than a single dominant crack growth.

The change of the residual strength, X, at a material age, L, is assumed to be:

$$\frac{dX}{dL} = -\frac{1}{c}X^{-c+1} \qquad \text{(Eq 15)}$$

Upon integration, Eq 15 yields the relation between the static strength, $X(o)$, and the residual strength, $X(L)$:

$$X^c(L) = X^c(o) - L \qquad \text{(Eq 16)}$$

If the material age, L, is related to test variables such as stress level, S, frequency, ω, stress ratio, R, and number of cycles incurred, then Eq 16 can be expressed as:

$$X^c(n) = X^c(o) - f(S,\omega,R)n \qquad \text{(Eq 17)}$$

For simplicity, the test frequency, ω, and stress ratio, R, will be fixed so that $f(S, \omega, R) = f(S)$; c is a constant.

The distribution of $X(o)$ is assumed to have a two-parameter Weibull distribution:

$$R_o(x) = P_r\{X(o) \geq x\} = \exp\left[-\left(\frac{x}{\beta_o}\right)^{\alpha_o}\right] \qquad \text{(Eq 18)}$$

in which α_o is the shape parameter and β_o is the scale parameter or the characteristic strength.

The distribution of $X(n)$ can be obtained from Eq 17 and 18 as follows:

$$R_n(x) = P_r\{X(n) \geq x\} =$$
$$\exp\left[-\left\{\left(\frac{x}{\beta_o}\right)^c + \frac{n}{\beta_o^c/f(S)}\right\}^{\alpha_o/c}\right] \qquad \text{(Eq 19)}$$

Let N denote the number of cycles at which fatigue failure occurs under the applied stress level, S. Then, at the moment of fatigue fracture, the relationship $X(n) = S$ if and only if $N = n$ holds, and the distribution of fatigue life, N, can be obtained from Eq 16 and 19 as:

$$R_N(n) = P_r\{N \geq n\} =$$
$$\exp\left[-\left\{\frac{n + (S^c/f(S))}{\beta_o^c/f(S)}\right\}^{\alpha_o/c}\right] \qquad \text{(Eq 20)}$$

In general, fatigue stresses are low, such that $(S/\beta_o)^c \ll 1$, and Eq 20 is reduced to:

$$R_N(n) = \exp\left[-\left\{\frac{n}{\beta_o^c/f(S)}\right\}^{\alpha_o/c}\right] \qquad \text{(Eq 21)}$$

The fatigue life, N, has a two-parameter Weibull distribution with the characteristic life $\beta_o^c/f(S)$ and the shape parameter α_o/c, which is independent of the stress level, S.

From the S-N curve relationship, $KS^bN = 1$, an expression for $f(S)$ is obtained as $f(S) = \beta_o^c KS^b$ because the characteristic life is $N =$

$1/KS^b = \beta_o^c/f(S)$). Equations 17, 20, and 21 can now be rewritten as:

$$X(n)^c = X(o)^c - \beta_o^c KS^b n \qquad \text{(Eq 22)}$$

$$R_n(x) = \exp\left[-\left\{\left(\frac{x}{\beta_o}\right)^c + \frac{n}{1/KS^b}\right\}^{\alpha_o/c}\right] \qquad \text{(Eq 23)}$$

and

$$R_N(n) = \exp\left[-\left(\frac{n}{1/KS^b}\right)^{\alpha_o/c}\right]$$

or $\exp\left[-\left(\frac{n}{N}\right)^{\alpha f}\right]$ \qquad (Eq 24)

where $c = \alpha_o/\alpha_f$.

For example, static strength and fatigue life data of graphite-epoxy laminates are listed in Tables 3 and 4. The fatigue test was performed under tension-tension fatigue loading with R = 0.1 at 10 Hz.

The Weibull parameters of static strength were estimated by Eq 4 and 5 and are given by:

$$\hat{\alpha}_o = 18.0 \text{ and } \hat{\beta}_o = 598.4 \text{ MPa (86.79 ksi)}$$

The pooled estimate $\hat{\alpha}$ of the common parameter α is obtained from fatigue data by Eq 11 and is given by:

$$\hat{\alpha} = 1.95$$

The estimate $\hat{\beta}_i$ of the characteristic life β_i is computed by Eq 12 with $\hat{\alpha} = 1.95$; the values are listed in Table 5.

The coefficient, c, in the residual strength degradation model is estimated by:

$$c = \hat{\alpha}_o/\hat{\alpha} = 18.0/1.95 = 9.231$$

Equation 14 was estimated by the least squares linear regression from the data given in Table 5, and the resulting equation is:

$$\ln \hat{\beta}_i = -17.33 \ln S_i + 114.985 \qquad \text{(Eq 25)}$$

From this equation, the constants b and K are obtained as:

Table 3   Static tensile strength data, graphite-epoxy T300/5208, [0°/90°/±45°]ₛ

MPa	ksi	MPa	ksi	MPa	ksi	MPa	ksi	MPa	ksi
496.7	72.04	...	...	528.2	76.61	532.2	77.19	533.4	77.36
547.2	79.36	554.9	80.48	562.1	81.52	565.2	81.97	566.8	82.20
566.9	82.22	568.9	82.51	574.7	83.35	578.5	83.90	586.0	84.99
590.9	85.70	591.5	85.79	595.9	86.42	597.2	86.61	598.3	86.77
606.5	87.96	610.5	88.54	612.7	88.86	617.8	89.60	621.7	90.17
630.5	91.44	632.5	91.73	636.9	92.37	643.5	93.33	...	...

Table 4   Fatigue data, graphite-epoxy T300/5208, [0°/90°/±45°]ₛ

Stress level MPa	ksi		Fatigue failure cycle, 10 Hz			
483	70.1	1 150	1 850	2 436	3 768	6,898
448	65.0	2 620	4 920	6 490	7 000	9 020
414	60.0	10 300	21 270	22 550	28 760	78 720
375	54.4	71 050	108 550	168 700	169 480	325 780
345	50.0	412 000	614 960	764 680	1 333 390	1 367 890

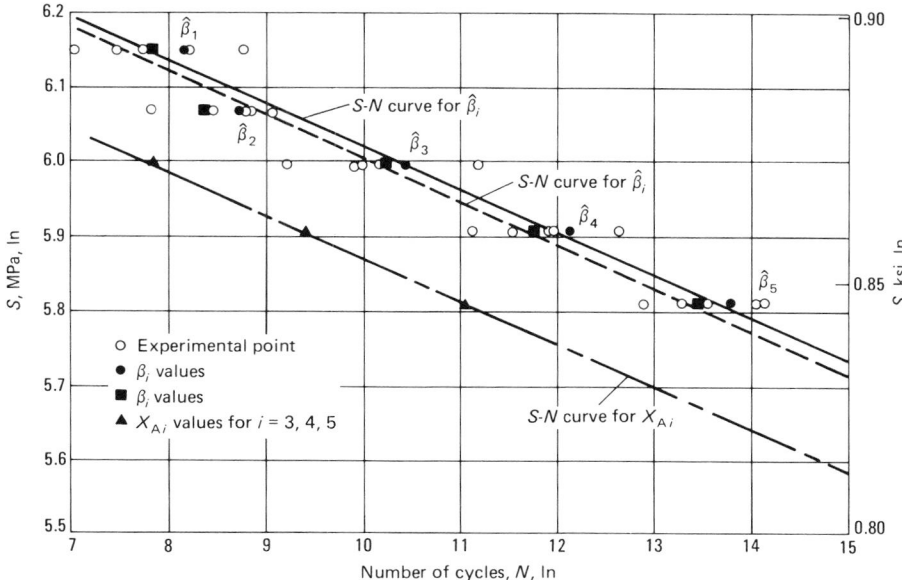

Fig. 16 *S-N* curves for T300/5208 [0°/90°/ ± 45°]ₛ graphite-epoxy

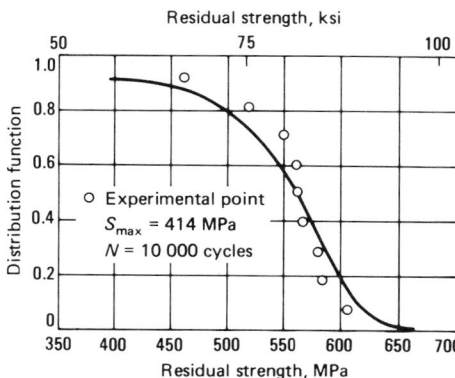

Fig. 17 Residual strength distribution for [0°/90°/ ± 45°]ₛ graphite-epoxy laminate after undertaking 10 000 fatigue cycles at 414 MPa (60 ksi)

Table 5 Maximum-likelihood estimates of β_i for $\hat\alpha$ = 1.95

Stress level, S_i		
MPa	ksi	β_i
483	70.1	3 779
448	65.0	6 364
414	60.0	39 834
379	54.4	188 780
345	50.0	974 263

b = 17.33 and K = $e^{-114.985}$ = 1.155 × 10^{-50}

The critical value $\ell_{0.95}^{*}$ = 0.836 was obtained from Table 2 (for m = 5 and n = 5) and the *S-N* curve relations for $N_i = \underline{\beta}_i$ and $N_i = X_{Ai}$ are obtained from Eq 8 and 10 as follows:

$$\ln \underline{\beta}_i = -17.33 \ln S_i + 114.555 \qquad \text{(Eq 26)}$$

and

$$\ln X_{Ai} = -17.33 \ln S_i + 112.195 \qquad \text{(Eq 27)}$$

The estimates $\hat\beta_i$ and the relations (Eq 25 to 27) are graphed in Fig. 16, with experimental points denoted by open circles.

The residual strength distribution is plotted for the same T300/5208 graphite-epoxy (0°/90°/±45°)ₛ laminate. Nine specimens were tested for residual strength after undertaking 10 000 cycles of tension-tension fatigue at the stress level of 414 MPa (60 ksi), and the results are given in Table 6. The theoretical distribution (Eq 23) is compared with the experimental data using Eq 2 in Fig. 17. The comparison is fairly good.

REFERENCES

1. R.A. Weinberger *et al.*, "U.S. Navy Certification of Composite Wings for the F-18 and Advanced Carrier Aircraft," AIAA

Table 6 Residual strength data, graphite-epoxy [0°/90°/ ±45°]ₛ

After N = 10 000 fatigue cycles at S = 414 MPa (60.0 ksi)

MPa	ksi	MPa	ksi	MPa	ksi
466.0	67.59	563.5	81.73	581.2	84.29
522.5	75.78	563.7	81.73	585.9	84.97
554.3	80.39	568.4	82.44	606.3	87.93

Paper 77-466, American Institute of Aeronautics and Astronautics, 1977

2. G.C. Grimes, "Investigation of Stress Levels Causing Significant Damage in Composites," AFML-TR-75-33, Air Force Materials Laboratory, 1975

3. R.Y. Kim, "Experimental Assessment of Static and Fatigue Damage of Graphite/Epoxy Laminates," *Advances in Composite Materials, Vol 2, Proceedings of the Third International Conference on Composite Materials*, A.R. Bunsell, Ed., Paris, 1980

4. K.L. Reifsnider, E.G. Henneke, II, and W.W. Stinchcomb, "Defect-Property Relationships in Composite Materials," Technical Report AFML-TR-76-81, Part IV, Air Force Materials Laboratory, 1979

5. A.S.D. Wang and F.W. Crossman, Initiation and Growth of Transverse Cracks and Edge Delamination in Composite Laminates, Part l. An Energy Method, *J. Compos. Mater.*, Supplement, 1980

6. K.W. Garrett and J.E. Bailey, Multiple Transverse Fracture in 90° Cross-Ply Laminates of a Glass Fiber-Reinforced Polyester, *J. Mater. Sci.*, Vol 12, 1977

7. A. Parvizi and J.E. Baily, On Multiple Transverse Cracking in Glass Fiber Epoxy Cross-Ply Laminates, *J. Mater. Sci.*, Vol 13, 1978

8. N.J. Pagano and R.B. Pipes, Some Observations on the Interlaminar Strength of Composite Laminates, *Int. J. Mech. Sci.*, Vol 15, 1973

9. N.J. Pagano and S.R. Soni, Global-Local Laminate Variational Model, *Int. J. Solids Struct.*, Vol 19 (No. 3), 1983

10. S.R. Soni and R.Y. Kim, Delamination of Composite Laminates Stimulated by Interlaminar Shear, in *Composite Materials: Testing and Design (Seventh Conference)*, STP 893, J.M. Whitney, Ed., American Society for Testing and Materials, 1986

11. H.T. Hahn, Fatigue Behavior and Life Prediction of Composite Laminates, in *Composite Materials: Testing and Design (Fifth Conference)*, STP 674, S. Tsai, Ed., American Society for Testing and Materials, 1979

12. M.J. Salkind, *Composite Materials: Testing and Design (Second Conference)*, STP 497, American Society for Testing and Materials, 1972, p 143

13. J.M. Whitney, "Fatigue Characterization of Composite Materials," AFWAL-TR-79-4111, Air Force Materials Laboratory, Oct 1979

14. R.Y. Kim, University of Dayton Research Institute, unpublished research, 1986

15. S.V. Ramani and D.P. Williams, Axial Fatigue of [0/ ± 30]₃ₛ Graphite/Epoxy, *Failure Modes in Composites*, Vol III, Feb 1976, p 115

16. T.R. Porter, Evaluation of Flawed Composite Structures Under Static and Cyclic Loading, in *Fatigue of Filamentary Composite Materials*, STP 636, R. Evans, Ed., American Society for Testing and Materials, 1977

17. R.W. Walter, R.W. Johnson, R.R. June, and J.E. McCarty, Design for Integrity in Long-Life Composite Aircraft Structures, in *Fatigue of Filamentary Composite Materials*, STP 636, R. Evans, Ed., American Society for Testing and Materials, 1977

18. S.P. Garbo and J.M. Ogonowski, "Effects

of Variances and Manufacturing Tolerances on the Design Strength and Life of Mechanically Fastened Composite Structures," AFWAL-TR-81-3041, Vol II, Air Force Wright Aeronautical Laboratories, April 1981

19. J.H. Crews, Jr., Bolt-Bearing Fatigue of a Graphite/Epoxy Laminate, in *Joining of Composite Materials*, STP 749, K. Kedward, Ed., American Society for Testing and Materials, 1981, p 131-144

20. A.J. Barnard, "Fatigue and Damage Propagation in Composite Rotor Blades," AGARD 288, North Atlantic Treaty Organization, April 1980

21. K. Brunsch, "Service Experience With GRC Helicopter Blades (BO-105)," AGARD 288, North Atlantic Treaty Organization, April 1980

22. E. Jarosch and A. Stepan, *J. Am. Helicopter Soc.*, Vol 15 (No. 1), Jan 1970

23. D.R. Thoman, L.J. Bain, and C.E. Antle, Inferences on the Parameters of the Weibull Distribution, *Technometrics*, Vol 11, 1979, p 257-269

24. W.J. Park, "Pooled Estimate of the Parameters on Weibull Distributions," AFML-TR-29-4112, Air Force Materials Laboratory, 1979

25. H.T. Hahn and R.Y. Kim, Fatigue Behavior of Composite Laminate, *J. Compos. Mater.*, Vol 10, April 1976

26. J.N. Yang, Reliability Prediction of Composites Under Proof Tests in Service, in *Composite Materials: Testing and Design (Fourth Conference)*, STP 617, J.G. Davis, Jr., Ed., American Society for Testing and Materials, March 1977, p 272-295

Instability Considerations

Arthur W. Leissa, Ohio State University

LAMINATED COMPOSITE flat plate and cylindrically curved shell panels are often used in modern structural applications when minimum weight is important. These structural elements are capable of withstanding compressive or shear stress loadings (in the plane of the plate, or tangent to the surface of a shell, at its boundaries) up to certain critical limits. At these critical limits, which are typically far below the material ultimate (or yield) stresses, instability that is characterized by a large decrease in the transverse stiffness of the element may arise. Specifically, if a very small static loading is applied perpendicular to the plate or shell surface when the element is at or near one of its critical limits, a large transverse displacement will result. The element is then said to "buckle." In a dynamic situation, as the in-plane loading approaches a critical limit, all of the natural frequencies of the element are reduced, and one of them approaches zero. Consequently, if the structural element is subjected to a transverse vibratory excitation while it is in or near its critical loading state, its vibratory response amplitude will probably be larger than it would be otherwise.

Buckling analysis for plates or shells fabricated from a laminated composite is considerably more complicated than it is for homogeneous, isotropic materials, such as ordinary metals. Theoretical analysis is complicated by the additional calculations that must be made to account for fiber and matrix material properties, fiber orientations, and stacking sequences. Moreover, if the plies are not symmetrically stacked, coupling will exist between bending and midplane stretching as transverse deflection takes place, which complicates the problem further. Reliable experimental results are also more difficult to realize, particularly because of increased difficulty in simulating desired edge constraints.

In the general case of an unsymmetric laminate, the in-plane stress resultants N_x, N_y, N_{xy} (forces per unit length) and moment resultants M_x, M_y, M_{xy} (moments per unit length) that occur during buckling are related to the midplane strains ϵ_x, ϵ_y, γ_{xy} and midplane curvature changes K_x, K_y, K_{xy} by six simultaneous equations, which may be written in matrix form as:

$$
\begin{bmatrix} N_x \\ N_y \\ N_{xy} \\ \hline M_x \\ M_y \\ M_{xy} \end{bmatrix} = \begin{bmatrix} A_{11} \ A_{12} \ A_{16} & B_{11} \ B_{12} \ B_{16} \\ A_{12} \ A_{22} \ A_{26} & B_{12} \ B_{22} \ B_{26} \\ A_{16} \ A_{26} \ A_{66} & B_{16} \ B_{26} \ B_{66} \\ \hline B_{11} \ B_{12} \ B_{16} & D_{11} \ D_{12} \ D_{16} \\ B_{12} \ B_{22} \ B_{26} & D_{12} \ D_{22} \ D_{26} \\ B_{16} \ B_{26} \ B_{66} & D_{16} \ D_{26} \ D_{66} \end{bmatrix} \begin{bmatrix} \epsilon_x \\ \epsilon_y \\ \gamma_{xy} \\ \hline -K_x \\ -K_y \\ -K_{xy} \end{bmatrix} \quad \text{(Eq 1)}
$$

Thus, the A_{ij} coefficients represent the stretching stiffnesses of a plate, the D_{ij} coefficients represent bending stiffnesses, and the B_{ij} coefficients indicate bending-stretching coupling. For a symmetric laminate, all B_{ij} are zero.

Although the study of plate and shell panel buckling is relatively recent for composite materials, compared with homogeneous, isotropic materials, a recent review of the relevant literature has uncovered approximately 400 references (Ref 1). This article is a very brief summary of some of the most important results included in Ref 1, which should be consulted for more detailed information. In addition, considerable information on the buckling of laminated composite plates may be found in Ref 2-7.

Orthotropic Plate Instability

Orthotropic plate theory may be used to determine critical buckling loads of rectangular composite plates for two important types of symmetric laminate configurations. The first is parallel fiber, in which all fibers are parallel to each other and to a set of plate edges. The second is cross-ply, in which fibers of adjacent plies are oriented at 90° to each other and parallel to the plate edges. For plates of these two configurations, not only is bending behavior uncoupled from stretching (all $B_{ij} = 0$, as in Eq 1), but twisting is also uncoupled ($D_{16} = D_{26} = 0$).

For orthotropic plates having all simply supported edges (that is, hinged or knife edged) that are subjected to uniaxial or biaxial in-plane stresses (σ_x = constant, σ_y = constant, and

$\tau_{xy} = 0$), the buckled mode shape may be represented as:

$$
w = C \sin \frac{m\pi x}{a} \sin \frac{n\pi y}{b} \quad \text{(Eq 2)}
$$

where m, n = 1, 2, 3..., and a and b are the plate dimensions in the x- and y-directions, respectively. The critical value of the buckling stress may be calculated from the exact, nondimensional formula:

$$
\frac{K_x}{\pi^2} = -\frac{\sigma_x h b^2}{\pi^2 D_{22}} =
$$

$$
\frac{\left(\dfrac{D_{11}}{D_{22}}\right)\left(\dfrac{b}{a}\right)^2 m^2 + 2\left(\dfrac{D_{12}}{D_{22}} + 2\dfrac{D_{66}}{D_{22}}\right)n^2 + \left(\dfrac{a}{b}\right)^2\left(\dfrac{n^2}{m}\right)n^2}{1 + \left(\dfrac{\sigma_y}{\sigma_x}\right)\left(\dfrac{a}{b}\right)^2\left(\dfrac{n}{m}\right)^2}
$$
$$\text{(Eq 3)}$$

where the negative sign shown is necessary because compressive stress, σ_x, is defined as being negative. A negative value of σ_y/σ_x is needed to denote that a tensile stress is acting in the y-direction at the same time that a compressive stress is acting in the x-direction. The total plate thickness is h.

Figure 1 shows plots of Eq 3 in the special situation of uniaxial loading ($\sigma_y/\sigma_x = 0$). The critical (lowest) value of the nondimensional buckling stress parameter K_x/π^2 is plotted versus the plate aspect ratio (a/b) for the case in which $(D_{12} + 2D_{66})D_{22} = 1$. For $D_{11}/D_{22} = 1$, the plate is isotropic, and for $0 < a/b \leq \sqrt{2}$, it buckles into a mode shape having one half-sine wave ($m = 1$) in the direction of loading, whereas for $\sqrt{2} \leq a/b \leq \sqrt{6}$ it has two half-sine waves ($m = 2$), and so on. In each case, the buckled mode shape for the critical load has only one half-wave in the y-direction ($n = 1$). The minimum value of K_x/π^2 in this case is 4, and occurs at $a/b = 1$, 2, 3, ... For a composite plate that is much stiffer in the direction of loading ($D_{11}/D_{22} = 10$), Fig. 1 shows not only that higher buckling loads are achieved (minimum values of $\kappa_x/\pi^2 = 8.324$), but also that the critical mode shapes have fewer longitudinal waves ($m = 1$ for $0 < a/b \leq 2.515$). Conversely, if the fibers lie primarily perpendicular to the direction of load-

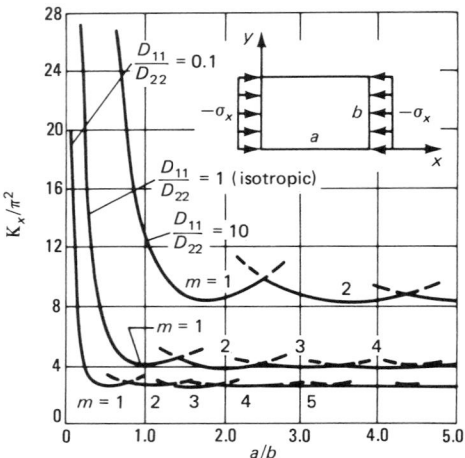

Fig. 1 Uniaxial buckling stress ($\sigma_y/\sigma_x = 0$) of SSSS plates with various D_{11}/D_{22}, for (D_{12} + $2D_{66})/D_{22} = 1$

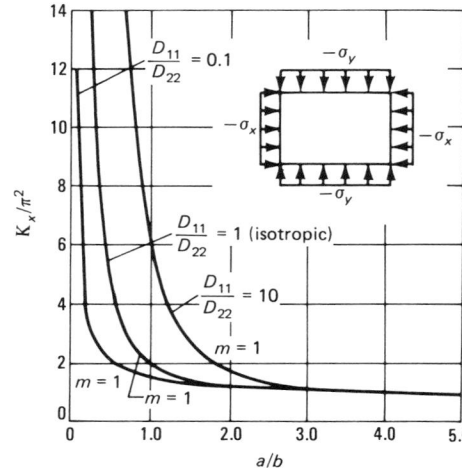

Fig. 2 Hydrostatic buckling stress ($\sigma_y/\sigma_x = 1$) of SSSS plates with various D_{11}/D_{22}, for (D_{12} = $2D_{66})/D_{22} = 1$

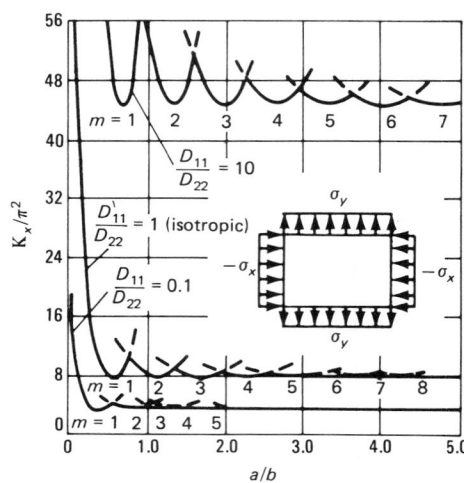

Fig. 3 Tension-compression buckling ($\sigma_y/\sigma_x = -1$) of SSSS plates with various D_{11}/D_{22}, for ($D_{12} + 2D_{66})/D_{22} = 1$

ing (for example, $D_{11}/D_{22} = 0.1$), the buckling load is lower and the plate is inclined to have more longitudinal waves in its critical mode.

Figures 2 and 3 show similar curves generated by Eq 3 for two situations involving biaxial loading. In the case of hydrostatic loading ($\sigma_y/\sigma_x = 1$, Fig. 2), all simply supported (SSSS) plate configurations have a critical mode shape with only one half-wave in each direction, and the minimum value of K_x/π^2 is 1. For tension-compression buckling ($\sigma_y/\sigma_x = -1$, Fig. 3), which corresponds to pure shear on planes making 45° angles with the plate edges, more half-waves are found in the direction of the compressive loading than for uniaxial loading, although the compressive stress required for buckling is higher. Plots for other biaxial loading combinations may be easily determined by means of Eq 3.

While Eq 3 is applicable only for simply supported plates, the effects of changing edge conditions may be reasonably estimated. Adding constraint at an edge increases the buckling stress, while taking it away decreases the buckling stress. Thus, for a uniaxially loaded, *isotropic* square plate, the value of $K_x/\pi^2 = 4$ is increased to 6.7, 7.7, or 10.2 if the loaded ends are clamped, the unloaded sides are clamped, or all edges are clamped, respectively, and its value is decreased to 1.5 if one of the unloaded sides is made free.

Anisotropic Plate Instability

Completely general anisotropic plate theory may be used to determine critical buckling loads of rectangular composite plates for two important types of symmetric laminate configurations. The first is parallel fiber or cross-ply, where the fibers are not parallel to the plate edges, but skewed. The second is angle-ply, where an odd number of adjacent plies are

oriented at alternating angles, $+\theta$, $-\theta$, $+\theta$, ..., with respect to the plate edges.

Figure 4, taken from the work of J.M. Housner and M. Stein (Ref 8), shows the variation of uniaxial buckling stress, σ_x, with orientation angle, $\pm\theta$, and aspect ratio, a/b, for graphite-epoxy angle-ply plates having a large number of layers. The buckling parameter contains E_1, which is the modulus of elasticity of a ply in the direction of its fibers, and which was approximately nine times as great as the transverse modulus, E_2, in this study. Data are given for two types of edge conditions: all sides either clamped or simply supported. Two important points are demonstrated in Fig. 4. First, the critical buckling stress is maximized for plies having $\pm\theta$ of approximately 45°. The resulting maximum may be more than twice as great, as in the case of parallel fiber plies ($\pm\theta = 0°$ or 90°). Second, a range of fiber orientations exists for which the buckling stress of the graphite-epoxy plates exceeds that of an aluminum plate of equal weight with the same planar dimensions ($a \times b$) and edge conditions. The latter data are shown as points on the right-hand ordinate of Fig. 4. Similar plots are seen in Fig. 5 for the case of shear stress loading, τ_{xy}. In this case the angle of optimum fiber orientation shifts to values exceeding 45° as a/b becomes greater than unity.

J. Crouzet-Pascal conducted optimization studies for the buckling of parallel fiber plates having skewed fibers (Ref 9). Figure 6 shows the optimum skew angle, θ, measured from the direction of loading, x, for uniaxially stressed glass-epoxy plates having SSSS edges. This is considered to be a composite material of moderate orthotropy ($E_1 = 54$ GPa, or 7.8 $\times$ 10^6 psi, $E_2 = 18$ GPa, or 2.6 $\times$ 10^6 psi, G_{12} = 4.5 GPa, or 0.65 $\times$ 10^6 psi, and $\nu_{12} =$ 0.25. The optimum value of fiber orientation is seen to vary between 30° and 55° for plates of moderate aspect ratio ($0.7 < a/b < 4$).

Unsymmetric Laminate Instability

For unsymmetrically laminated composite plates, coupling between bending and mid-plane stretching occurs in the buckled mode shape.

Two closed-form exact solutions are available for unsymmetrically laminated plates that are subjected to uniform, biaxial stresses (Ref 10). One solution is for cross-ply, and the other is for angle-ply plates. Both pertain to edges that are simply supported all around, but the former has *tangential* (S2) in-plane constraints at the boundaries, whereas the latter has *normal* (S3) constraints during the buckling process. In both cases the transverse displacement determining the buckled mode shape is given by Eq 2, but different forms for the in-plane displacements (u, v) are required. The critical buckling stress parameters may be determined from:

$$-\sigma_x h\alpha^2 - \sigma_y h\beta^2 =$$

$$C_{33} + \frac{2C_{12}C_{13}C_{23} - C_{11}C_{23}^2 - C_{22}C_{13}^2}{C_{11}C_{22} - C_{12}^2} \quad \text{(Eq 4)}$$

where $\alpha = m\pi/a$ and $\beta = n\pi/b$, and m and n are integers, as before. For the cross-ply plate, the constants C_{ij} are defined by:

$$C_{11} = A_{11}\alpha^2 + A_{66}\beta^2$$

$$C_{22} = A_{22}\beta^2 + A_{66}\alpha^2$$

$$C_{33} = D_{11}\alpha^4 + 2(D_{12} + 2D_{66})\alpha^2\beta^2 + D_{22}\beta^4$$

$$C_{12} = C_{21} = (A_{12} + A_{66})\alpha\beta \quad \text{(Eq 5)}$$

$$C_{13} = C_{31} = B_{11}\alpha^3 + (B_{12} + 2B_{66})\alpha\beta^2$$

$$C_{23} = C_{32} = (B_{12} + 2B_{66})\alpha^2\beta + B_{22}\beta^3$$

whereas for the angle-ply plate they are:

$$C_{11} = -(A_{11}\alpha^2 + A_{66}\beta^2)$$

$$C_{22} = -(A_{22}\beta^2 + A_{66}\alpha^2)$$

$$C_{33} = D_1\alpha^4 + 2(D_{12} + 2D_{66})\alpha^2\beta^2 + D_{22}\beta^4$$

$$C_{12} = C_{21} = -(A_{12} + A_{66})\alpha\beta$$ (Eq 6)

$$C_{13} = C_{31} = 3B_{16}\alpha^2\beta + B_{26}\beta^3$$

$$C_{23} = C_{32} = B_{16}\alpha^3 + 3B_{26}\alpha\beta^2$$

The coefficients A_{ij}, B_{ij}, and D_{ij} ($i, j = 1, 2, 6$) used in Eq 5 and 6 are the plate stiffnesses seen earlier in Eq 1. It is noted that for the cross-ply plate, $A_{16} = A_{26} = B_{16} = B_{26} = D_{16} = D_{26} = 0$, and for the angle-ply plate, $A_{16} = A_{26} = B_{11} = B_{22} = B_{12} = D_{16} = D_{26} = 0$, and Eq 5 and 6 reduce to Eq 3 when all $B_{ij} = 0$.

A typical plot of the nondimensional, uniaxial buckling stress arising from Eq 4 and 5 in the case of cross-ply plates is shown in Fig. 7. The curves plotted are for a high-modulus laminate having 2, 4, 6 and an infinite number of plies. For $N = \infty$, the bending-stretching coupling disappears and the plate behaves as if it were orthotropic and symmetrically laminated. The curves are seen to be similar in shape to those of Fig. 2. However, Fig. 7 shows how the critical buckling stress is dras-

tically reduced when a laminate is laid up unsymmetrically and is composed of only a few layers. For a square plate ($a/b = 1$) having only two plies, Fig. 7 indicates that the plate will buckle at a stress that is only about one-third as much as if it had a larger number of plies.

A similar plot is shown in Fig. 8 for the angle-ply buckling stress given by Eq 4 and 6. Here the nondimensional, uniaxial buckling stress is plotted versus fiber orientation angle, $\pm\theta$, for a high-modulus square plate, and is seen to be much lower for two or four layers than for $N = \infty$. The cusps in the curves indicate where the critical buckling mode shape changes from one to two half-waves in the direction of loading. The optimum value of θ is $\pm 45°$.

Shear Deformation Effects

Classical laminated plate theory is based upon the Kirchhoff hypothesis: Normals to the midplane of the undeformed plate remain straight and normal to the midplane during deformation. This assumption ignores the transverse shear deformation. Consideration of shear deformation results in added flexibility, which

becomes significant as the plate thickness increases relative to its length and width, and which may be considerably more important for laminates than for isotropic plates.

Figure 9 shows results obtained by J.M. Whitney (Ref 13) for uniaxially loaded, symmetrically laminated angle-ply ($\pm 45°$) plates ($E_1/E_2 = 40$, $G_{12}/E_2 = 0.6$, $G_{23}/E_2 = 0.5$, and $\nu_{12} = 0.25$) having all edges simply supported. Nondimensional buckling stress is plotted versus the length-to-thickness ratio, a/h, for square plates. The two curves shown are for classical plate theory and shear deformation theory. Shear deformation effects are seen to decrease the buckling stress considerably for plates with $a/h < 30$. This reduction in buckling stress has been observed by many researchers (see Ref 1).

Postbuckling Behavior of Plates

Although a theoretical and/or experimental study may establish the critical buckling stress of a plate where instability is found to occur, the plate will typically be capable of carrying considerable additional load before collapse or ultimate load is reached. In some cases this is as

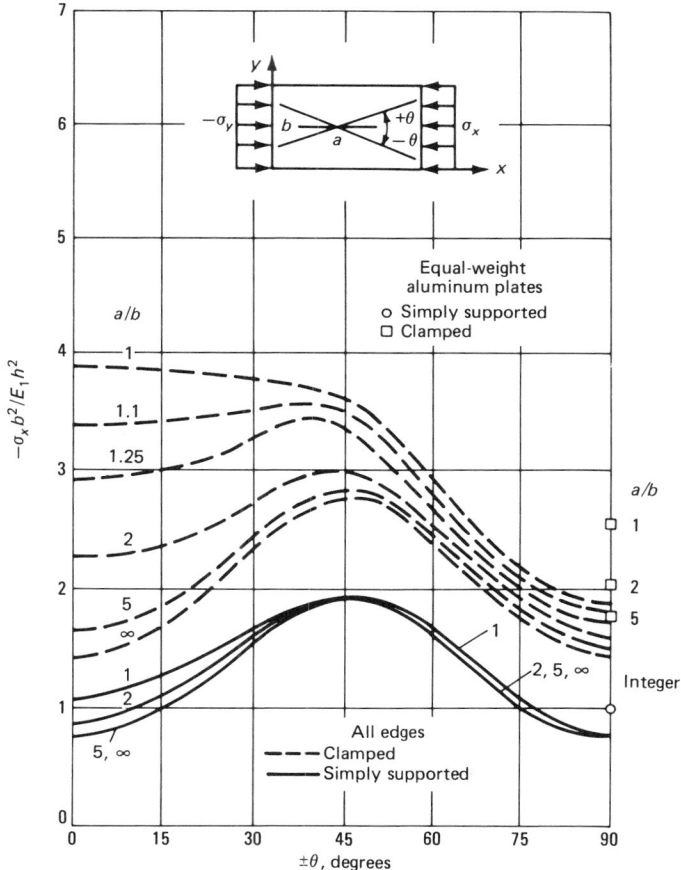

Fig. 4 Uniaxial buckling stress parameters for graphite-epoxy angle-ply plates. Source: Ref 8

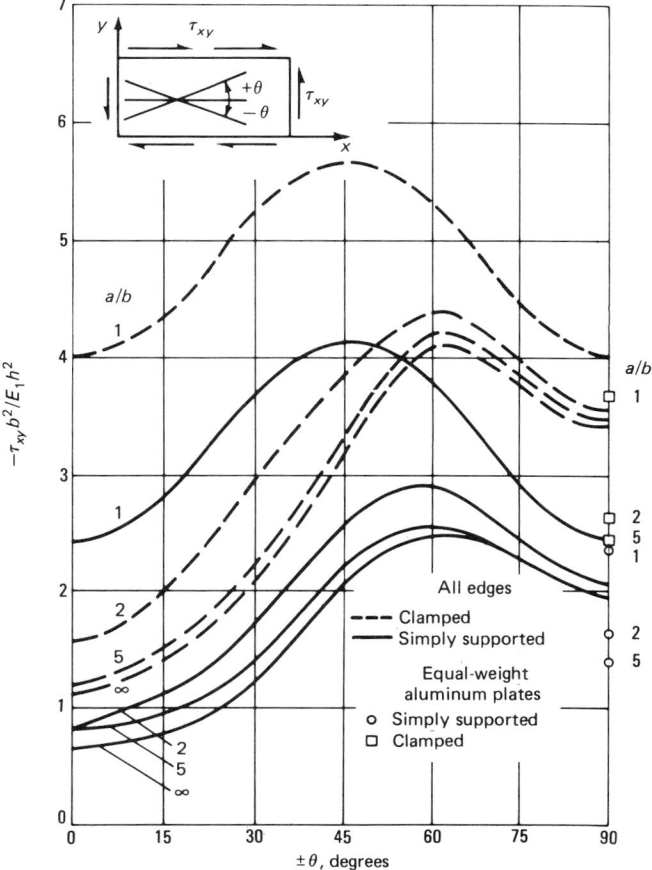

Fig. 5 Shear stress buckling parameters for graphite-epoxy angle-ply plates. Source: Ref 8

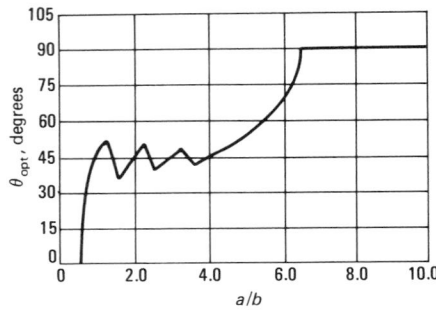

Fig. 6 Optimum material axis orientation versus aspect ratio for a uniaxially loaded SSSS plate (unidirectional, medium orthotropy laminate)

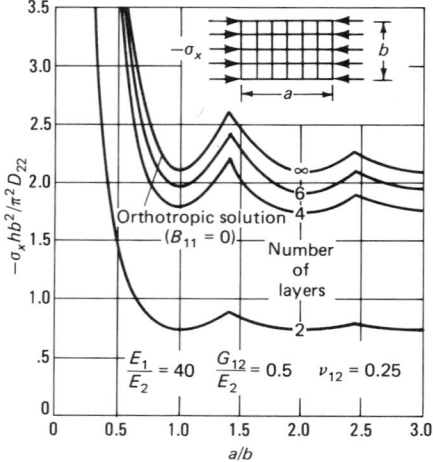

Fig. 7 Comparison of antisymmetrical and orthotropic solutions with varying plate aspect ratio a/b for uniaxially loaded cross-ply plates having S2 edge conditions. Source: Ref 11

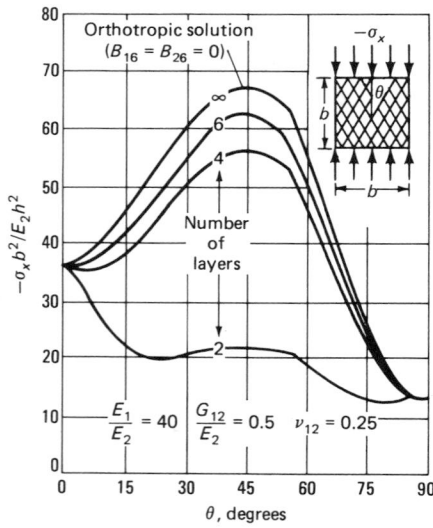

Fig. 8 Comparison of antisymmetrical and orthotropic solutions with varying lamination angle for uniaxially loaded angle-ply plates having S3 edge conditions ($a/b = 1$). Source: Ref 12

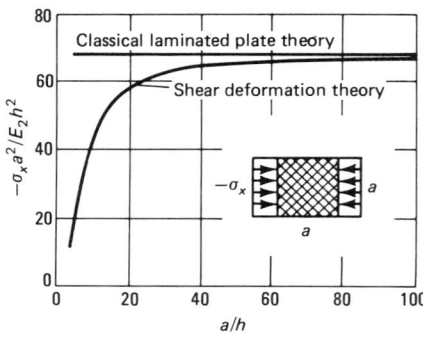

Fig. 9 Buckling of a uniaxially loaded, SSSS, $\pm45°$ angle-ply square plate having an infinite number of layers, with and without shear deformation

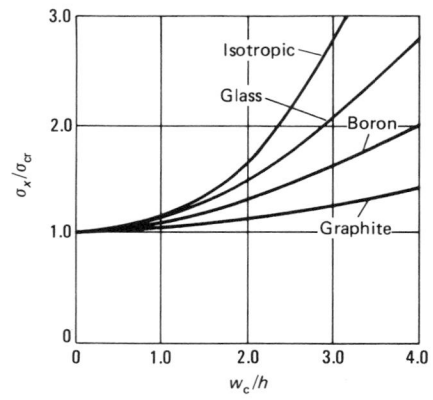

Fig. 10 Postbuckling uniaxial stress-deflection curves for isotropic and orthotropic SSSS plates ($a/b = 1$)

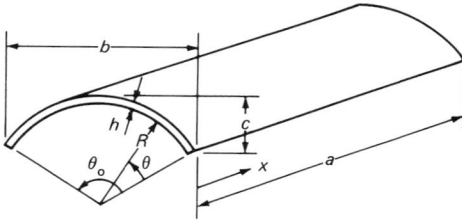

Fig. 11 Circular cylindrical shell panel

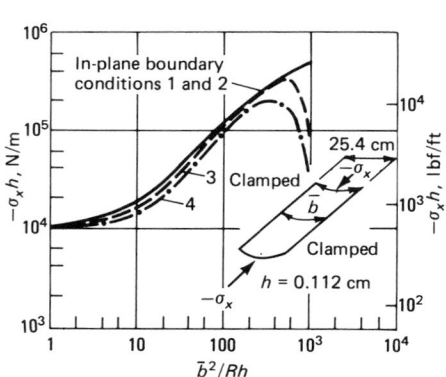

Fig. 12 Variation of $\sigma_x h$ with curvature parameter for shell panels. Source: Ref 15

much as several times the critical load, although this capability has been found to be less pronounced for laminated composite plates than for metal ones.

Figure 10 shows postbuckling deflection curves for uniaxially loaded, simply supported, parallel fiber, square plates made of glass, boron, or graphite fibers embedded in epoxy resin (Ref 7, 14). A curve is also shown for a homogeneous, isotropic plate. The variation of σ_x/σ_{cr} with w_c/h is depicted, where σ_{cr} is the critical uniaxial buckling stress and w_c is the plate deflection at its center. It may be observed that for a given percentage increase in uniaxial compressive stress beyond the buckling stress, the composite material plates all require greater deflection than the isotropic one, with graphite-epoxy needing the greatest increase in w_c/h. Ultimate failure occurs when excessive stress levels are reached within the materials, causing delamination and/or rupture.

Shell Panel Instability

A circular cylindrical shell panel with length a, radius R, platform width b, and thickness h is depicted in Fig. 11. The circumferential width, $\bar{b} = R\,\theta_o$, of its middle surface is determined by the subtending angle θ_o.

Buckling characteristics of shell panels may be considerably different from those of plates, even in the case of homogeneous, isotropic materials. For example, a shell panel subjected to uniaxial compressive stress ($-\sigma_x$) parallel to its straight edges may have a buckled mode shape with two or more half-waves in its circumferential (θ) direction corresponding to its critical (lowest) stress.

Figure 12 shows a representative set of curves for the uniaxial buckling load (per unit width) of long shell panels (large a/b) having their straight edges clamped. The shell comprises eight equal-thickness layers of boron-epoxy material having a $[0°/90°/+45°/-45°]_s$ stacking sequence and a total thickness of $h = 0.112$ cm (0.0440 in.). The curves show the effect of increasing shell curvature upon buckling load for four different types of in-plane constraint applied to the clamped edges. The out-of-plane edge conditions are those of zero displacement and circumferential slope change ($w = \partial w/\partial\theta = 0$) in each case. As the shell curvature, $1/R$, is increased in each case, the buckling load is seen to increase until a maximum value is reached.

Postbuckling Behavior of Plate Versus Shell Panels

A comparison of the postbuckling behaviors of plate and shell panels may be made from the representative load-deflection curves seen in Fig. 13. The curves are for boron-epoxy panels ($E_1/E_2 = 10$, $G_{12}/E_2 = 0.3$, $\nu_{12} = 0.3$) having unidirectional plies with fibers oriented at 45° with the longitudinal (x) axis. The edges are all

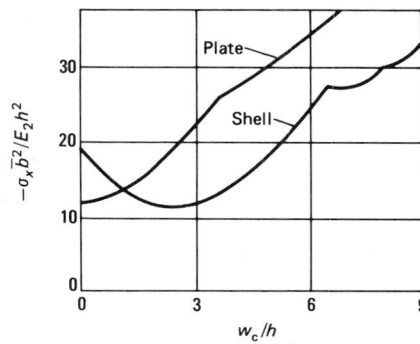

Fig. 13 Postbuckling load-deflection curves for SSSS boron-epoxy panels with unidirectional fibers at $\theta = 45°$. Source: Ref 16

simply supported, with $a/\bar{b} = 1$, and the panels are subjected to uniaxial compressive strength ($-\sigma_x$). The shell panel has a curvature parameter $\bar{b}^2/Rh = 25$.

Figure 13 shows that although the shell panel has a higher critical buckling stress than the plate, the compressive stress required to maintain equilibrium decreases with increasing center displacement (w_c) immediately after the critical stress is reached. Thus, if the applied axial stress is kept constant in the shell panel after the critical point is reached, a large transverse displacement (and bending stress) immediately ensues. Subsequently, if rupture has not occurred, the panel is capable of withstanding axial compressive stress. The one and two cusps seen in the curves for the plate and shell panels, respectively, are indicative of sudden changes in the deflected shapes (that is, points where load-deflection curves cross).

It should be noted that geometric imperfections in curved panels can cause collapse before reaching the theoretical bifurcation point.

REFERENCES

1. A.W. Leissa, "Buckling of Laminated Composite Plates and Shell Panels," AFWAL-TR-85-3069, Air Force Wright Aeronautical Laboratories, 1985
2. S.G. Lekhnitskii, *Anisotropic Plates*, 2nd ed., S.W. Tsai and T. Cheron, Trans., Gordon and Breach Science Publishers, 1968
3. S.A. Ambartsumyan, *Theory of Anisotropic Plates*, Technomic, 1970
4. J.E. Ashton and J.M. Whitney, *Theory of Laminated Plates*, Technomic, 1970
5. R.M. Jones, *Mechanics of Composite Materials*, Scripta, 1975
6. J.R. Vinson, T.W. Chou, *Composite Materials and Their Use in Structures*, Applied Science Publishers, 1975
7. C.-Y. Chia, *Nonlinear Analysis of Plates*, McGraw-Hill, 1980
8. J.M. Housner and M. Stein, "Numerical Analysis and Parametric Studies of the Buckling of Composite Orthotropic Compression and Shear Panels," NASA TN D-7996, National Aeronautics and Space Administration, Oct 1975
9. J. Crouzet-Pascal, Buckling Analysis of Laminated Composite Plates, *Fibre Sci. Technol.*, Vol 11, 1978, p 413-446
10. J.M. Whitney and A.W. Leissa, Analysis of Heterogeneous Anisotropic Plates, *J. Appl. Mech.* (Trans. ASME), Vol 36 (No. 2), 1969, p 261-266
11. R.M. Jones, Buckling and Vibration of Unsymmetrically Laminated Cross-Ply Rectangular Plates, *AIAA J.*, Vol 11 (No. 12), 1973, p 1626-1632
12. R.M. Jones, H.S. Morgan, and J.M. Whitney, Buckling and Vibration of Antisymmetrically Laminated Angle-Ply Rectangular Plates, *J. Appl. Mech.* (Trans. ASME), Vol 12, 1973, p 1143-1144
13. J.M. Whitney, The Effect of Transverse Shear Deformation on the Bending of Laminated Plates, *J. Compos. Mater.*, Vol 3, 1969, p 534-547
14. M.K. Prabhakara and C.Y. Chia, Post-Buckling Behaviour of Rectangular Orthotropic Plates, *J. Mech. Eng. Sci.*, Vol 15 (No. 1), 1973, p 25-33
15. A.V. Viswanathan, M. Tamekuni, and L.L. Baker, "Elastic Stability of Laminated Flat and Curved, Long Rectangular Plates Subjected to Combined Inplane Loads," NASA CR-2330, National Aeronautics and Space Administration, June 1974
16. Y. Zhang and F.L. Matthews, Postbuckling Behaviour of Curved Panels of Generally Layered Composite Materials, *Compos. Struct.*, Vol 1, 1983, p 115-135

Laminate Ranking as a Tool for Laminate Sizing

Thierry N. Massard, Commissariat à l'Energie Atomique
Jocelyn M. Patterson, U.S. Air Force Materials Laboratory

THE POTENTIAL OF LAMINATE RANK-ING as a design tool for determining an optimum multidirectional configuration and its required thickness is discussed in this article. A ranking method for the optimum design of symmetric laminates is compared to an optimization method. The ranking method, which offers many unique features not offered by the conventional optimization method, provides definitive laminates that are achievable in practice. This article discusses a ranking method that is limited to in-plane loading situations; the analysis is based on laminated plate theory and the quadratic failure criterion. A parameter sensitivity study indicates how final laminate size is influenced by factors such as the total number of plies in a sublaminate, the total number of angles in a sublaminate, and the angles selected as the basis for the laminate family.

Composite materials are unique and require equally unique methods of design. An important function of designing with composite materials is to determine the multidirectional ply composition of a laminate subjected to multiple loading conditions. A fundamental difference between isotropic and anisotropic design lies in the determination of the controlling load. For the isotropic design, the controlling load is obvious from the thickness required. For the anisotropic design, however, thickness alone, usually given as the number of plies, is not sufficient to determine the controlling load; ply composition also must be considered.

Having considered the multidirectional ply composition, the objective of composite structural design is to determine how thick that laminate must be in order to sustain loading conditions without failure, typically defined by either a stress level or deflection value. In this article, failure criteria constraints are based on a general quadratic failure criterion (Ref 1) evaluated for each ply in a laminate and here defined in terms of strain:

$$(G_{ij}\epsilon_i\epsilon_j)R^2 + (G_i\epsilon_i)R - 1 = 0 \qquad \text{(Eq 1)}$$

where $i, j = x, y, s$, which represent longitudinal, transverse, and shear components corresponding to the ply coordinates, x and y; G_{ij} and G_i are functions of the ply strength; and R is the strength/stress ratio:

$$|\sigma|_{max} = R|\sigma|_{applied} \text{ and } |\epsilon|_{max} = R|\epsilon|_{applied} \quad \text{(Eq 2)}$$

Laminate Ranking Notation. The laminate code convention created for laminate ranking is used for the sake of simplicity and states only the number of plies per angle. The ply orientations are not explicitly included in the notation because, after being specified initially, they are held constant during the ranking process. This notation can be illustrated by the following example of a sublaminate consisting of four angles:

$$[4\ 2\ 2\ 1] \text{ means } [(\theta 1)_4/(\theta 2)_2/(\theta 3)_2/(\theta 4)_1]_s \quad \text{(Eq 3)}$$

where s implies that the laminate is symmetric. Accordingly, for a sublaminate consisting of two or three angles, the same convention is used by simply setting the extra last digits to 0:

$$[2\ 2\ 0\ 0] \text{ means } [(\theta 1)_2/(\theta 2)_2]_s \quad \text{(Eq 4)}$$

It is important to note that laminate code indicates only the number of plies within a sublaminate, which can easily be converted into the ratio of ply groups within a laminate, but does not indicate the ply-angle orientations, for example, in Eq 4 above, θ_1 and θ_2. The specific ply-angle orientation should be given explicitly in the context of use, or, by default, can be assumed to correspond to $\pi/4$ laminate: $[0°/90°/45°/-45°]_s$. One special case is the quasi-isotropic laminate, which, as part of an eight-

ply family, assumes the form [2222], equivalent to $[0°_2/90°_2/45°_2/-45°_2]_s$.

The generation of families of sublaminates is defined by the total number of plies in a sublaminate and the total number and value of ply-angle orientations. Independent of the angles that are specified, the number of sublaminates in the family is systematically generated based on the total number of plies and the total number of angles. The sublaminates in each family can be unidirectional, bidirectional, tridirectional, or quadridirectional, and so on, depending on the number of angles selected. Tables 1, 2, and 3 show the makeup of each family in terms of the types of directionality for two-, three-, and four-angle families, respectively.

Table 4 describes the families and summarizes the information shown in Tables 1 to 3 by

Table 1 Number of different sublaminates as a function of the number of plies in a two-angle family

No. plies	Unidirectional	Bidirectional	Total
1	2	...	2
2	2	1	3
3	2	2	4
4	2	3	5
5	2	4	6
6	2	5	7
7	2	6	8
8	2	7	9
9	2	8	10
10	2	9	11

Table 2 Number of different sublaminates as a function of the number of plies in a three-angle family

No. plies	Unidirectional	Bidirectional	Tridirectional	Total
1	3	...	...	3
2	3	3	...	6
3	3	6	1	10
4	3	9	3	15
5	3	12	6	21
6	3	15	10	28
7	3	18	15	36
8	3	21	21	45
9	3	24	28	55
10	3	27	36	66

Table 3 Number of different sublaminates as a function of the number of plies in a four-angle family

No. plies	Unidirectional	Bidirectional	Tridirectional	Quadridirectional	Total
1	4	...	...	...	4
2	4	6	...	...	10
3	4	12	4	...	20
4	4	18	12	1	35
5	4	24	24	4	56
6	4	30	40	10	84
7	4	36	60	20	120
8	4	42	84	35	165
9	4	48	112	56	220
10	4	54	144	84	286

Table 4 Number of sublaminates in a family as a function of the number of ply-angle orientations and the number of plies within a sublaminate

No. plies	Family of two orientations	Family of three orientations	Family of four orientations
2	3	6	10
3	4	10	20
4	5	15	35
5	6	21	56
6	7	28	84
7	8	36	120
8	9	45	165
9	10	55	220
10	11	66	286
Total	65	285	1000

listing the total number of possible sublaminates in a family of a given number of orientations (between two and four) and total number of plies (between two and ten). For example, assuming six plies in each sublaminate and any four ply-angle orientations, a data base of 84 sublaminates can be generated, as shown in Table 5. In Table 6, a data base of 165 sublaminates is generated from the assumption of eight plies in each sublaminate and any four ply-angle orientations.

Selecting the most useful family would favor one that included an explicit, quasi-isotropic sublaminate, which generally represents the minimum performance of the composite material without claiming possible benefits of anisotropy and which is useful for comparing the performance of other sublaminates. To ensure that a quasi-isotropic sublaminate will be part of the family, the number of plies must be divisible by the number of angles, and the angles selected must be equally spaced. For example, π/4 laminates should consist of three angles, $[-60°/0°/60°]_s$, and six plies, and π/4 laminates should have four angles, $[-45°/0°/45°/90°]_s$, and eight plies. The data base shown in Table 6 for eight-ply π/4 sublaminates is in fact the most frequently recommended and used.

Although there are no limitations, sublaminates consisting of more than four angles are of little practical use because the quasi-isotropic characteristics attainable from four equally spaced angles are more than satisfactory. Similarly, there is no limit to the total number of plies in a sublaminate. More plies allow larger laminate families and a chance for a better laminate. As the number of plies in a sublaminate increases, however, it becomes difficult to optimize, costly to assemble, and difficult to conduct the ply drop-off procedure. Furthermore, for sublaminates having more than ten plies per angle, the four-digit code in Eq 3 is no longer valid.

Laminate Ranking Versus Optimization

Examples of both laminate ranking and optimizations are compared below.

Laminate Ranking Program. LamRank, for laminate ranking, is a FORTRAN program based on the process described by S.W. Tsai (Ref 1), but the approach described can be programmed by any user. LamRank, based on laminated plate theory, uses the quadratic failure criterion with a value of $-\frac{1}{2}$ for the strain interaction term. It considers residual stresses that are due to curing of multidirectional laminates and to changes in environmental conditions, but it is valid for in-plane loads only. The program is able to accommodate single and multiple loading conditions consisting of either simple or complex loads. Its data base is generated by the user and it can accommodate up to four different angles and ten total plies in a sublaminate. For each laminate in the data base, the program evaluates an effective strain invariant, the first-ply failure (FPF), and the last-ply failure (LPF). The LPF prediction is based on the matrix degradation model (Ref 1). All of these parameters are defined and described in detail in Ref 1. The data base then ranks the family of sublaminates for any of the following criteria, and also determines the number of multiples of the sublaminate needed.

Minimum Strain. To avoid any dependence on the definition of the axes orientation, this criterion evaluates a strain invariant value for the laminate based on all three components of strain:

$$|\epsilon|^2 = \epsilon_1^2 + \epsilon_2^2 + \epsilon_6^2/2 \tag{Eq 5}$$

Table 5 Summary of the 84 sublaminates that constitute the family in which each sublaminate has a total of six plies distributed among four possible angles

No.	Code	No. angles	Symmetry	No.	Code	No. angles	Symmetry	No.	Code	No. angles	Symmetry	No.	Code	No. angles	Symmetry
1	0006	1	Ortho	22	0330	3	Ortho	43	1230	4	Ortho	64	2400	4	Ortho
2	0015	2	Ortho	23	0402	3	Ortho	44	1302	4	Ortho	65	3003	4	Ortho
3	0024	2	Ortho	24	0411	3	Ortho	45	1311	4	Ortho	66	3012	4	Ortho
4	0033	2	Ortho	25	0420	2	Ortho	46	1320	3	Ortho	67	3021	3	Ortho
5	0042	2	Aniso	26	0501	2	Aniso	47	1401	3	Aniso	68	3030	2	Aniso
6	0051	2	Aniso	27	0510	2	Aniso	48	1410	3	Aniso	69	3102	3	Aniso
7	0060	1	Ortho	28	0600	1	Ortho	49	1500	2	Ortho	70	3111	4	Ortho
8	0105	2	Aniso	29	1005	2	Ortho	50	2004	2	Aniso	71	3120	3	Aniso
9	0114	3	Aniso	30	1014	3	Aniso	51	2013	3	Aniso	72	3201	3	Aniso
10	0123	3	Aniso	31	1023	3	Aniso	52	2022	3	Ortho	73	3210	3	Aniso
11	0132	3	Aniso	32	1032	3	Aniso	53	2031	3	Aniso	74	3300	2	Ortho
12	0141	3	Aniso	33	1041	3	Aniso	54	2040	3	Aniso	75	4002	2	Aniso
13	0150	2	Aniso	34	1050	2	Aniso	55	2103	3	Aniso	76	4011	3	Ortho
14	0204	2	Aniso	35	1104	3	Aniso	56	2112	4	Aniso	77	4020	2	Aniso
15	0213	3	Aniso	36	1113	4	Aniso	57	2121	4	Aniso	78	4101	3	Aniso
16	0222	3	Ortho	37	1122	4	Ortho	58	2130	3	Aniso	79	4110	3	Aniso
17	0231	3	Aniso	38	1131	4	Aniso	59	2202	3	Aniso	80	4200	2	Ortho
18	0240	2	Aniso	39	1140	3	Aniso	60	2211	4	Ortho	81	5001	2	Aniso
19	0303	2	Aniso	40	1203	3	Aniso	61	2220	3	Aniso	82	5010	2	Aniso
20	0312	3	Aniso	41	1212	4	Aniso	62	2301	3	Aniso	83	5100	2	Ortho
21	0321	3	Aniso	42	1221	4	Aniso	63	2310	3	Aniso	84	6000	1	Ortho

Table 6 Summary of the 165 sublaminates that constitute the family in which each sublaminate has a total of eight plies distributed among four possible angles

No.	Code	No. angles	Symmetry	No.	Code	No. angles	Symmetry	No.	Code	No. angles	Symmetry	No.	Code	No. angles	Symmetry
1	0008	1	Ortho	43	0701	2	Aniso	85	2033	3	Ortho	127	3320	3	Aniso
2	0017	2	Aniso	44	0710	2	Aniso	86	2042	3	Aniso	128	3401	3	Aniso
3	0026	2	Aniso	45	0800	1	Ortho	87	2051	3	Aniso	129	3410	3	Aniso
4	0035	2	Aniso	46	1007	2	Aniso	88	2060	2	Aniso	130	3500	2	Ortho
5	0044	2	Ortho	47	1016	3	Aniso	89	2105	3	Aniso	131	4004	2	Aniso
6	0053	2	Aniso	48	1025	3	Aniso	90	2114	4	Aniso	132	4013	3	Aniso
7	0062	2	Aniso	49	1034	3	Aniso	91	2123	4	Aniso	133	4022	2	Ortho
8	0071	2	Aniso	50	1043	3	Aniso	92	2132	4	Aniso	134	4031	3	Aniso
9	0080	1	Ortho	51	1052	3	Aniso	93	2141	4	Aniso	135	4040	2	Aniso
10	0107	2	Aniso	52	1061	3	Aniso	94	2150	3	Aniso	136	4103	3	Aniso
11	0116	3	Aniso	53	1070	2	Aniso	95	2204	3	Aniso	137	4112	4	Aniso
12	0125	3	Aniso	54	1106	3	Aniso	96	2213	4	Aniso	138	4121	4	Aniso
13	0134	3	Aniso	55	1115	4	Aniso	97	2222	4	Q-iso	139	4130	3	Aniso
14	0143	3	Aniso	56	1124	4	Aniso	98	2231	4	Aniso	140	4202	3	Aniso
15	0152	3	Aniso	57	1133	4	Ortho	99	2240	3	Aniso	141	4211	4	Aniso
16	0161	3	Aniso	58	1142	4	Aniso	100	2303	3	Aniso	142	4220	3	Aniso
17	0170	3	Aniso	59	1151	4	Aniso	101	2312	4	Aniso	143	4301	3	Aniso
18	0206	3	Aniso	60	1160	3	Aniso	102	2321	4	Aniso	144	4310	4	Aniso
19	0215	3	Aniso	61	1205	3	Aniso	103	2330	3	Aniso	145	4400	2	Ortho
20	0224	3	Aniso	62	1214	4	Aniso	104	2402	3	Aniso	146	5003	2	Aniso
21	0233	3	Ortho	63	1223	4	Aniso	105	2411	4	Ortho	147	5012	3	Aniso
22	0242	3	Aniso	64	1232	4	Aniso	106	2420	3	Aniso	148	5021	3	Aniso
23	0251	3	Aniso	65	1241	4	Aniso	107	2501	3	Aniso	149	5030	2	Aniso
24	0260	2	Aniso	66	1250	3	Aniso	108	2510	3	Aniso	150	5102	3	Aniso
25	0305	2	Aniso	67	1304	3	Aniso	109	2600	2	Ortho	151	5111	4	Aniso
26	0314	3	Aniso	68	1313	4	Aniso	110	3005	2	Aniso	152	5120	3	Aniso
27	0323	3	Aniso	69	1322	4	Ortho	111	3014	3	Aniso	153	5201	3	Aniso
28	0332	3	Aniso	70	1331	4	Aniso	112	3023	3	Aniso	154	5210	3	Aniso
29	0341	3	Aniso	71	1340	3	Aniso	113	3032	3	Aniso	155	5300	2	Ortho
30	0350	2	Aniso	72	1403	4	Aniso	114	3041	3	Aniso	156	6002	2	Aniso
31	0404	2	Aniso	73	1412	4	Aniso	115	3050	2	Aniso	157	6011	3	Aniso
32	0413	3	Aniso	74	1421	4	Aniso	116	3104	4	Aniso	158	6020	2	Aniso
33	0422	3	Aniso	75	1430	3	Aniso	117	3113	4	Aniso	159	6101	3	Aniso
34	0431	3	Aniso	76	1502	4	Aniso	118	3122	4	Ortho	160	6110	3	Aniso
35	0440	2	Aniso	77	1511	4	Ortho	119	3131	4	Aniso	161	6200	2	Ortho
36	0503	3	Aniso	78	1520	3	Aniso	120	3140	3	Aniso	162	7001	2	Aniso
37	0512	3	Aniso	79	1601	3	Aniso	121	3203	3	Aniso	163	7010	2	Aniso
38	0521	3	Aniso	80	1610	3	Aniso	122	3212	4	Aniso	164	7100	2	Ortho
39	0530	2	Aniso	81	1700	2	Ortho	123	3221	4	Aniso	165	8000	1	Ortho
40	0602	3	Aniso	82	2006	2	Aniso	124	3230	3	Aniso				
41	0611	3	Ortho	83	2015	3	Aniso	125	3302	3	Aniso				
42	0620	2	Aniso	84	2024	3	Aniso	126	3311	4	Ortho				

Maximum Strength Based on the FPF (analogous to a limit stress). The quadratic failure criterion is applied to each ply by considering the strains in the ply. Of all the plies in the laminate, the one with the lowest strength/stress ratio defines FPF.

Maximum Strength Based on the LPF (analogous to an ultimate stress). All the plies are considered degraded by matrix cracking, which is accounted for by a reduced matrix modulus. (The remaining failure mode is fiber failure). Using micromechanics, the reduced matrix modulus is related to the transverse and shear moduli. Using the modified material data, each ply is then reevaluated by laminated plate theory and the quadratic failure criterion, and the ply with the lowest strength/stress ratio defines LPF.

Maximum Strength Based on the Safety Factor Rule. Adherence to this rule is a means of guaranteeing that the design point never approaches the ultimate stress, within some margin. For example, when a safety factor is equal to 1.5, the ultimate stress will have a 50% margin over the design point. With composites, the actual margin is dependent on the laminate configuration. For example, for $[\pm 45°]_s$ under uniaxial tensile stress, the LPF stress is equal to

the FPF stress; conversely, for $[0°/90°/\pm 45°]_s$ under longitudinal tensile stress, the LPF stress can be four times the FPF stress. The safety factor rule requires that no plies have failed at the limit stress, and that the ultimate stress is typically taken to be 1.5 times the limit stress. If the actual ratio of LPF to FPF stresses is less than 1.5, a new design limit is defined as the LPF divided by 1.5. If the ratio of LPF to FPF stresses is more than 1.5, the ultimate stress is still only taken to be 1.5 times the limit stress. (Figure 1 depicts these two examples.) Instead of having to decide whether to design according to the FPF or LPF stress and having to consider their relative relationship, the designer can safely examine the design limit. Note that the value of 1.5 is arbitrary and that the same principles can be applied to force any consistent relationship between the limit and ultimate stresses.

In addition to displaying the absolute strength and stiffness of each laminate within the family, LamRank also displays the relative strength and stiffness over the quasi-isotropic laminate. This ratio is important for determining not only the advantage of directionality but also the relative performance over isotropic materials.

With the LamRank program, the designer chooses the number of total plies and the number and orientation of ply angles in the sublaminates, selects the material type, and inputs the loading and environmental conditions. Material selection is easy when using the data file that currently accommodates eight materials but can have more added. The constraints can be the FPF, the LPF, the safety factor rule, or any other rule based on criterion provided by the designer.

Optimization Program. Classic, a name adapted from Class to accommodate an initial condition, is a BASIC program that uses a nonlinear optimization algorithm proposed by G.V. Flanagan and W.J. Park (Ref 2). Classic represents one of many optimization programs and is presented here only as an example for comparison purposes. The theory behind Classic has been adapted to include new topics such as matrix degradation, which enables calculation of last-ply failure. The program is configured to accommodate up to ten loads and a maximum of ten different ply-angle orientations. The run time is about 20 seconds, for example, for the design problem of optimizing a laminate with four ply-angle orientations subjected to a single load. The program uses an

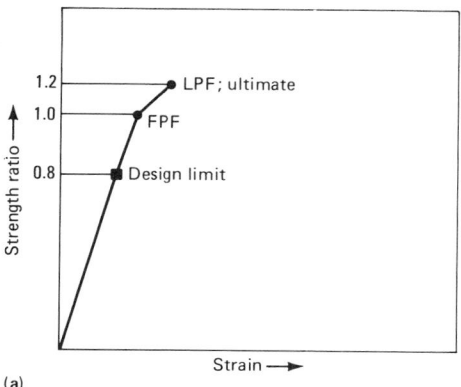

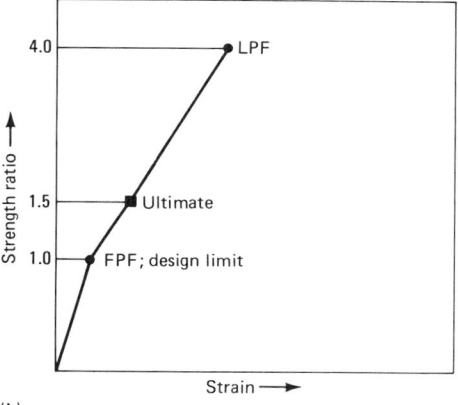

Fig. 1 Two types of strength ratio relationships explaining the safety factor rule relationship to FPF and LPF (in this case, SF = 1.5)

Table 7 Property data used by LamRank showing title block and corresponding T300-5208 graphite-epoxy values as a sample data block

Title block	NAME [SI] X,MPa(g) Fxy*(n) alph/x,E-6(u) E^lso,GPa(bb)	Ex,GPa(a) X',MPa(h) ho,E-6m(o) alph/y,E-6(v) X^lso,MPa(cc)	Ey,GPa(b) Y,MPa(i) Vf(p) beta/x(w) E^u/E^l(dd)	nu/x(c) Y',MPa(j) rho/ply(q) beta/y(x) X^u/X^l(ee)	Es,GPa(d) S,MPa(k) eta/y(r) eta/s(y)	Em,GPa(e) rho/m(l) a,Em(s) f,Ef(z)	T/cure(f) T/opr(m) b,eta(t) h,Xf(aa)
Data block	T3/N52[SI] 1500 −0.5 0.02 69.7	181 1500 125 22.5 325	10.3 40 0.7 0.0 0.916	0.28 246 1.6 0.6 1.56	7.17 68 0.500 0.400	3.4 1.2 0.3 0.004	122 22 0.1 0.004

(a) E_x, 0° elastic modulus. (b) E_y, 90° elastic modulus. (c) ν_x, Poisson's ratio. (d) E_s, shear modulus. (e) E_m, matrix modulus. (f) T_{cure}, cure temperature. (g) X, 0° tensile strength. (h) X', 0° compressive strength. (i) Y, 90° tensile strength. (j) Y' 90° compressive strength. (k) S, shear strength. (l) ρ_m, matrix density. (m) T_{opr}, operating temperature. (n) F_{xy}*, normalized interaction term. (o) h_o, unit ply thickness. (p) V_f, fiber volume fraction. (q) ρ_{ply}, ply density. (r) η_y, stress partitioning parameter for transverse stiffness. (s) a, E_m, hygrothermal exponent for matrix modulus. (t) b, η, hygrothermal exponent for stress partitioning parameter. (u) α_x, 0° coefficient of thermal expansion. (v) α_y, 90° coefficient of thermal expansion. (w) β_x, 0° moisture expansion. (x) β_y, 90° moisture expansion. (y) η_s, stress partitioning parameter for shear modulus. (z) f, E_f, hygrothermal modulus of fiber. (aa) h, X_f, hygrothermal exponent for fiber strength. (bb) E^Iso, quasi-isotropic stiffness. (cc) X^Iso, quasi-isotropic uniaxial strength. (dd) E^u/E^l, elastic modulus ultimate/elastic modulus limit. (ee) X^u/X^l, uniaxial strength ultimate/uniaxial strength limit

Table 8 Data for comparing LamRank and Classic

Test case	Load	LamRank	Total plies(a)	Initial laminate	Classic	Total plies
1	(−4,0,2)	...	...	[1 1 1 1]	[12 9 9 12]	84
		[1 0 4 3]	96 (80.5)	[1 0 4 3]	[5 20 0 16]	82
		[3 0 4 1]	96 (80.5)	[3 0 4 1]	[16 0 20 5]	82
2	(2,6,−4)	...	...	[1 1 1 1]	[0 24 3 58]	170
		[0 2 0 6]	112 (107.3)	[0 2 0 6]	[0 14 0 40]	108
3	(0,0,2)	...	...	[1 1 1 1]	[0 0 15 6]	42
		[0 0 5 3]	48 (41.3)	[0 0 5 3]	[0 0 14 7]	42
4	(−4,2,0)	...	...	[1 1 1 1]	[18 6 9 9]	84
		[2 0 3 3]	80 (76)	[2 0 3 3]	[11 0 14 14]	78
		[5 3 0 0]	96 (81)	[5 3 0 0]	[27 13 0 0]	80
5	(−4,2,0) and	...	...	[1 1 1 1]	[13 24 15 39]	182
	(2,6,−4)	[1 2 1 4]	176 (169)	[1 2 1 4]	[10 22 00 42]	168
		[2 2 0 4]	176	[2 2 0 4]	[25 20 0 40]	170
6	(4,2,0) and	...	...	[1 1 1 1]	[17 6 19 9]	84
	(0,0,2)	[2 0 3 3]	80 (78)	[2 0 3 3]	[11 0 14 14]	78
		[2 2 0 4]	96	[2 2 0 4]	[16 3 12 10]	82

(a) Practical (theoretical)

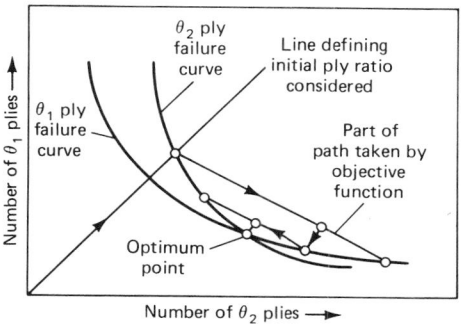

Fig. 2 Iterative process used by Classic

iterative process to converge toward the optimum solution, as shown in Fig. 2. It appears that because of local constraint function minimums, defined by the quadratic failure criterion for each ply, the solution given by Classic may depend upon the initial point.

LamRank and Classic Comparison. Test cases have been analyzed with a T300-5208 graphite-epoxy laminate consisting of four ply-angle orientations: 0°, 90°, 45°, −45°. Because only in-plane loads are considered, the strains are uniform across the laminate thickness and therefore the order of the angles is unimportant.

Table 7 summarizes the material data used for T300-5208. Table 8 is a summary of several test cases, showing the loading conditions and the recommended laminate configurations predicted by Classic and LamRank. Only the first few recommended laminates selected by LamRank are given. The first laminate obtained using Classic, starting with a quasi-isotropic initial laminate (one ply each of the above four angles) is shown. Other laminates listed in Table 8 were also obtained using Classic, but start with an initial laminate recommended by LamRank. Based on the comparisons of Lam-Rank and Classic in Table 8, several conclusions can be drawn.

First, Classic gives a real optimum solution if the starting point (initial laminate) of the optimization process resembles the optimum ratio of ply-angle thicknesses. If the initial laminate does not resemble the true optimum laminate, the final solution could be a local minimum. Indications of such a problem and troubleshooting hints can be found by looking at the length ratio for each ply-angle orientation. Unfortunately, there is no systematic way to avoid a local minimum.

Second, the solution given by Classic determines the number of plies per ply-angle group,

but not the stacking sequence. Therefore, the stacking sequence should be optimized before being used to manufacture parts. It is well known that interleaving the different ply-angle orientations gives greater toughness and less free-edge sensitivity. For instance, in the first test case shown in Table 8, the solution given by Classic is [5 20 0 16]. One equivalent configuration that could also be used is [1 4 0 3]₅ + [0 0 0 1].

LamRank gives a choice of several solutions, whereas Classic gives only one. Alternate solutions can be very useful if some nonexplicit constraints, such as manufacturing limits, prevent the use of the first solution.

Finally, LamRank provides an optimum sublaminate from the data base and the theoretical number of required plies based on the strength ratio analysis. In fact, the real construction has to be made from an integral number of sublaminates, because determining which plies in a sublaminate are the most essential cannot be done arbitrarily. For example, in the test cases considered in Table 8, the increment in sizing the laminate is 16 plies, where each sublaminate is made of 8 plies and the total laminate is symmetric. Therefore, even if the optimized sublaminate leads to an

Table 9   Rearrangement of laminate recommended by Classic for test cases shown in Table 8

Test case	Optimum laminate
1	$[1\ 4\ 0\ 3]_5 + [0\ 0\ 0\ 1]$
2	$[0\ 2\ 0\ 6]_6 + [0\ 2\ 0\ 4]$
3	$[0\ 0\ 5\ 3]_2 + [0\ 0\ 5\ 0]$
4	$[2\ 0\ 3\ 3]_4 + [3\ 0\ 2\ 2]$
5	$[1\ 2\ 1\ 4]_{10} + [0\ 2\ 0\ 2]$
6	$[2\ 0\ 3\ 3]_4 + [3\ 0\ 2\ 2]$

Table 10   Sequence of supplementary sublaminates investigated by LamRank

	No.	Code		No.	Code	No.	Code
One-ply sublaminate	1	0001		44	0202	87	0320
	2	0010		45	0211	88	0401
	3	0100		46	0220	89	0410
	4	1000		47	0301	90	0500
Two-ply sublaminate	5	0002		48	0310	91	1004
	6	0011		49	0400	92	1013
	7	0020		50	1003	93	1022
	8	0101		51	1012	94	1031
	9	0110		52	1021	95	1040
	10	0200		53	1030	96	1103
	11	1001		54	1102	97	1112
	12	1010		55	1111	98	1121
	13	1100		56	1120	99	1130
	14	2000		57	1201	100	1202
Three-ply sublaminate	15	0003		58	1210	101	1211
	16	0012		59	1300	102	1220
	17	0021		60	2002	103	1301
	18	0030		61	2011	104	1310
	19	0102		62	2020	105	1400
	20	0111		63	2101	106	2003
	21	0120		64	2110	107	2012
	22	0201		65	2200	108	2021
	23	0210		66	3001	109	2030
	24	0300		67	3010	110	2102
	25	1002		68	3100	111	2111
	26	1011		69	4000	112	2120
	27	1020		70	0005	113	2201
	28	1101		71	0014	114	2210
	29	1110		72	0023	115	2300
	30	1200		73	0032	116	3002
	31	2001		74	0041	117	3011
	32	2010	Five-ply sublaminate	75	0050	118	3020
	33	2100		76	0104	119	3101
	34	3000		77	0113	120	3110
Four-ply sublaminate	35	0004		78	0122	121	3200
	36	0013		79	0131	122	4001
	37	0022		80	0140	123	4010
	38	0031		81	0203	124	4100
	39	0040		82	0212	125	5000
	40	0103		83	0221		
	41	0112		84	0230		
	42	0121		85	0302		
	43	0130		86	0311		

efficient optimum solution, as is true for all test cases shown in Table 8, the round-off process (described below) unfortunately leads to a much thicker laminate than the optimum found by Classic.

It should be noted that the solutions from Classic can be rebuilt using the optimum sublaminate from LamRank, as shown in Table 9. Another method for achieving a similar result is to calculate the number of plies that should be removed from the LamRank solution. For instance, the optimum solution for test case 2 in Table 9 could also be stated as $[0\ 2\ 0\ 6]_7 - [0\ 0\ 0\ 2]$.

Round-Off Procedure. When designing with sublaminates, a round-off problem exists and must be rectified to achieve an optimum design. LamRank gives a solution in the form $[a\ b\ c\ d]$, where n equals $a + b + c + d$, the number of plies in each sublaminate; N is the total number of plies in the solution; and r, the ratio N/n, is the repeat index. If N is not a multiple of n, then the repeat index is $r = $ integer $(N/n) + 1$. This discrete sequence of sublaminates is a drawback of the laminate ranking method. The increment can become large, as the number of plies in the sublaminate increases. For instance, with an [eight-ply]$_s$ sublaminate, the increment is 16 plies, because the laminate is symmetric. To remove this round-off error and attain truly optimum laminate sizing, such as that found by Classic, a systematic round-off process to find the best supplementary sublaminate must be implemented for LamRank. The supplementary sublaminate should be designed using the same criterion as was used for the sublaminate, but could include other constraints as well, such as minimizing free-edge delamination.

Supplementary Sublaminates. In addition to ranking the best laminate configurations, LamRank indicates the total number of plies required for each configuration. This number, however, usually is not an integer number of sublaminates. If the LamRank solution is simplistically rounded to the next integer number of sublaminates, the result is many more plies than necessary and the improvement over the optimization solution is often sacrificed. To achieve optimum laminate sizing, a supplementary sublaminate, $[A\ B\ C\ D]$, is needed, for which the number of plies is less than the number of plies in the sublaminate, and such that the laminate has these criteria:

- $[a\ b\ c\ d]_{rs}$ does not meet the design criteria
- $[a\ b\ c\ d]_{(r+1)s}$ overshoots the design criteria
- $\{[a\ b\ c\ d]_r + [A\ B\ C\ D]\}_s$ or $\{[a\ b\ c\ d]_{r+1} - [A\ B\ C\ D]\}_s$ just meets the design criteria
- $[A\ B\ C\ D]$ has the minimum number of plies

The algorithm is brute force: The design must investigate all the possible one-ply sublaminates for each of the ply-angle orientations of the sublaminates $[a\ b\ c\ d]$ to see if any one meets the requirements. If not, two-ply laminates should be investigated, and so on, until a laminate that meets the design criteria is found. Of course, the number of possible supplementary sublaminates increases very quickly with the number of plies in the sublaminate.

In order to reduce the number of supplementary sublaminates to be ranked, Table 10 shows various supplementary sublaminates that are investigated by the round-off procedure. For a ten-ply sublaminate, the maximum number of supplementary sublaminates to be investigated is 250. Only five-ply sublaminates need to be investigated. It is easier to use supplementary sublaminates in positive and negative ways; that is, supplementary sublaminates can be added or dropped from the repeated sublaminates.

Round-Off Design Criteria. Consistent with the manner in which LamRank chooses the best sublaminate and the number of repetitions needed to make a laminate, three design criteria have been implemented into the round-off routine:

- *FPF criterion.* Plies are added or dropped until the strength ratio, defined as the ratio of strength to applied stress, at first-ply failure is equal to or greater than 1
- *LPF criterion.* Plies are added or dropped until the strength ratio at last-ply failure is equal to or greater than 1
- *Limit design criterion.* Plies are added or dropped until the following two conditions are met at the same time: The strength ratio at FPF is equal to or greater than 1, and the strength ratio at LPF is equal to or greater than the user-specified safety factor

The effectiveness of the round-off procedure is demonstrated by considering the first two test cases in Table 8. The output from LamRank for

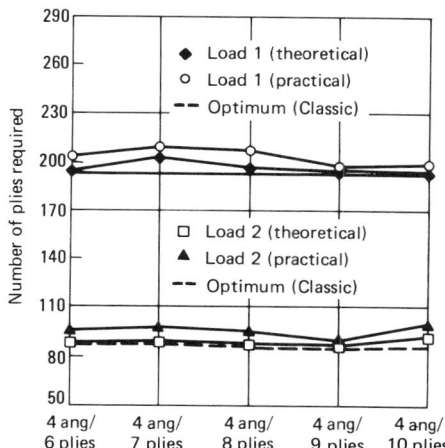

Fig. 3 Influence of number of plies in sublaminate on the design thickness of the laminate for two load cases: load 1 = (10,5,−2) MN/m or (30,15,−5) tonf/in.; and load 2 = (0,5,−2) MN/m or (0,15,−5) tonf/in.

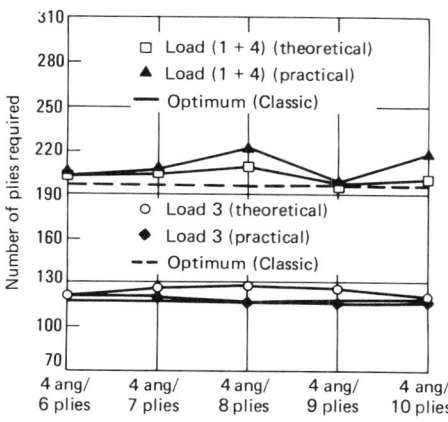

Fig. 4 Influence of number of plies in sublaminate on the design thickness of the laminate for two load cases: load 3 = (−3, 0, 4) MN/m or (−10, 0, 11) tonf/in.; and load 1 = (10,5,−2) MN/m or (30,15,−5) tonf/in. + load 4 = (0,10,−4) MN/m or (0,30,−11) tonf/in.

the top five laminates is shown in Tables 11(a) and 11(b) and 12(a) and 12(b) for the two test cases. Tables 11(a) and 12(a) show the original LamRank output defining the sublaminate and the number of repetitions required to satisfy the design criteria specified (in this case, the 1.5 rule). Tables 11(b) and 12(b) show the output after the round-off calculation, which refines the laminate as a number of repetitions of the sublaminate, plus a supplementary sublaminate.

For the first test case, Table 8 shows that the best laminate requires 96 plies, while the best Classic solution requires only 80 plies. Table 11(b) shows the answer from LamRank after the round-off procedure has been implemented. The best laminate requires 82 plies; 81 could not be achieved because of the symmetric lay-up. The optimum supplementary sublaminate predicted by Classic was also a one-ply laminate, [0 0 0 1].

Table 11(a) Further inquiry into test case 1 of Table 8 showing top five laminates recommended by LamRank, the number of sublaminates required based on the FPF criterion, and the theoretical number of plies required by different criteria

No.	Lay-up	r(a)	Lim/iso(b)	No. of plies required			Ratio of FPF/LPF
				Limit	FPF	LPF	
1	1403	6	0.83	118.99	80.02	79.33	1.01
2	3041	6	0.83	118.99	80.02	79.33	1.01
3	1304	6	0.82	119.66	80.22	79.77	1.01
4	4031	6	0.82	119.66	80.22	79.77	1.01
5	2213	6	1.07	91.7	84.56	61.13	1.38

(a) *r*, sublaminate repeat index. (b) Comparison with the limit strength of the quasi-isotropic laminate

Table 11(b) Number of sublaminates required and the supplementary sublaminate needed to exactly satisfy the FPF criterion for test case 1 of Table 8

No.	Lay-up	r(a)	Supplementary sublaminate	Strength ratio	Total number of plies
1	1403	5	1000	1.02	82
2	3041	5	1000	1.03	82
3	1304	5	1000	1.02	82
4	4031	5	1000	1.02	82
5	2213	5	200	1.01	84

(a) *r*, sublaminate repeat index

Table 12(a) Further inquiry into test case 2 of Table 8 showing top five laminates recommended by LamRank, the number of sublaminates required based on the FPF criterion, and the theoretical number of plies required by different criteria

No.	Lay-up	r(a)	Lim/iso(b)	No. of plies required			FPF/LPF
				Limit	FPF	LPF	
1	206	7	1.74	140.62	107.3	93.75	1.14
2	305	8	1.73	141.47	114.6	94.31	1.22
3	404	8	1.65	148.35	126.2	98.9	1.28
4	107	8	1.47	166.37	126.4	110.9	1.14
5	503	10	1.5	163.38	145.6	108.9	1.34

(a) *r*, sublaminate repeat index. (b) Comparison with the limit strength of the quasi-isotropic laminate

Table 12(b) Number of sublaminates required and the supplementary sublaminate needed to exactly satisfy the FPF criterion for test case 2 of Table 8

No.	Lay-up	r(a)	Supplementary sublaminate	Strength ratio	Total number of plies
1	206	6	204	1.003	108
2	305	7	101	1.009	116
3	404	7	404	1.014	128
4	107	7	300	1.017	118
5	503	9	1	1.015	146

(a) *r*, sublaminate repeat index

From the second test case, Table 8 shows that the best laminate has 112 plies for a required number of 107.3. Table 12(a) shows that after round-off the best laminate has 108 plies. Table 12(b) shows the supplementary sublaminate to be [0 2 0 4], identical to the one predicted by Classic in Table 9.

Parameter Sensitivity

This section discusses the importance of various parameters that the designer must con-

sider. Examples that use both single and multiple complex loads, which allow conclusions about the design of composite parts to be made, are given.

Number of Plies in Sublaminate. A sublaminate can vary from two to ten plies, in the LamRank program. The symmetric hypothesis of the design process implies that the thickness of the laminates varies from four to 20 plies, depending on the sublaminate size. A question arises as to whether a small or a large sublaminate should be used. Figures 3 and 4

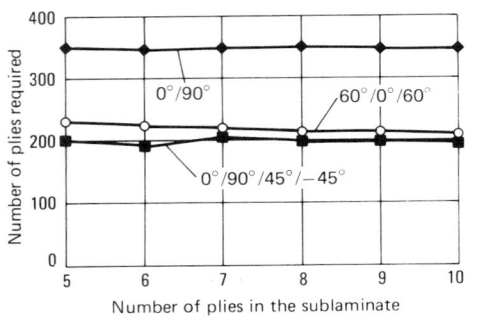

Fig. 5 Influence of the number of ply groups in sublaminate

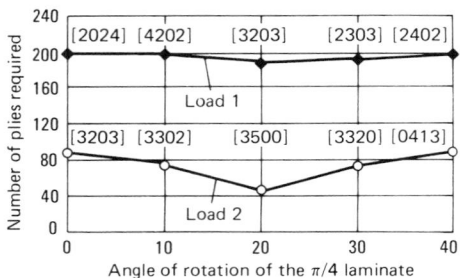

Fig. 6 Influence of ply-angle orientations: load 1 = (10,5,−2) MN/m or (30,15,−5) tonf/in.; and load 2 = (0,5,−2) MN/m, or (0,15,−5) tonf/in. [2222] sublaminate code

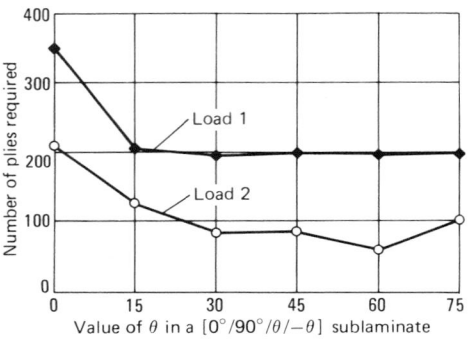

Fig. 7 Influence of θ [0°/90°/±θ°] sublaminate. Load 1 = (10,5,−2) MN/m or (30,15,−5) tonf/in.; and load 2 = (0,5,−2) MN/m, or (0,15,−5) tonf/in.

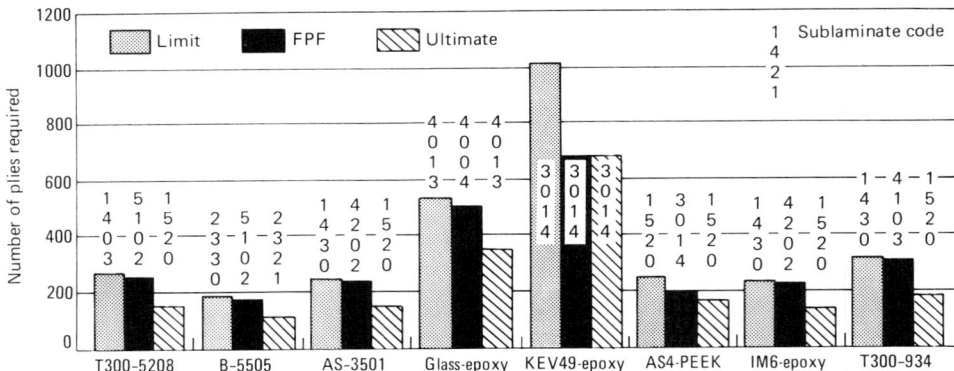

Fig. 8 Influence of material selection for a load of (5,−10,4) MN/m or (15,−30,11) tonf/in.

illustrate the influence of the number of plies in the sublaminate for typical loads. Note that in Fig. 4, a multiple loading condition is considered in the second test case (load 1 and load 4). The angles are fixed to four orientations, [0°/90°/45°/−45°]s. For comparison, the solution provided by Classic is also included in Fig. 3 and 4.

The theoretical number of plies in Fig. 3 and 4 refers to the exact amount computed by LamRank, while the practical number of plies required must be an integral number. Hence, a round-off process is needed to reconcile the often large disparity between the practical and theoretical number of plies. Figures 3 and 4

also confirm that the solution predicted by Classic, the optimization method, is very nearly approximated by the theoretical LamRank solution, the laminate ranking method.

Most important, it can be concluded from Fig. 3 and 4 that the number of plies in the sublaminate is not a critical parameter in the design process. Although the ratios between ply-angle orientations may be different (not explicitly indicated in Fig. 3 and 4), the required total number of plies in the laminate is constant. For practical reasons, such as round-off, edge effects, or manufacturing considerations, it might be better to work with smaller sublaminates.

Number of Orientations in Sublaminate. Sublaminate families can include ply angles of any orientation in the LamRank program. The number or orientations is limited, however, to four different angles. Figure 5 shows the influence of the number and selection of angles on the design requirements of two typical loads.

From Fig. 5 it can be observed that equivalent results are found using sublaminates consisting of three and four ply angles. Laminates consisting of sublaminates with only two angles, however, are severely penalized, presumably because of the inability to support shear loads. Again, consistent with the results of Fig. 3 and 4, it also appears that increasing the number of plies in the sublaminate has little effect on the total number of plies required for the laminate and cannot compensate for the lack of orientation.

Ply-Angle Orientation. The rigid-body rotation of a given family of sublaminates is an example that considers an eight-ply family of sublaminates with π/4 orientations. This set of ply-angle orientations is rotated between 0° and 40°, a form of rigid-body rotation of the given family of sublaminates.

From Fig. 6 it appears that the influence of the rigid-body rotation is small. It must be emphasized that the laminate design is very different from one point to another, as indicated in Fig. 6 by the laminate code of each data point, even though the total ply thickness does not vary. Knowing that a composite material is

by definition a highly oriented material, it can be foreseen that the choice of orientation is critical in the design process. Figure 6 shows, however, that it is not critical as long as a good optimization process is used to determine the right ratio between the different orientations.

Figure 7 shows the design for two typical load cases using the sublaminate [0°/90°/θ/−θ], where θ varies between 0° and 75°. From Fig. 7 it can be seen that for four different orientations (θ ≠0), the design is independent of θ. Again, although the total thickness of the laminate is constant, the stacking may be significantly different according to the value of θ (although not explicitly indicated in Fig. 7).

Material selection is probably the most delicate choice in design. In addition to the wide range of materials available on the market, technical and economic criteria have to be considered. Figure 8 shows the design thickness required to sustain a given complex load based on the FPF stress, the LPF stress, and the safety factor rule criteria for eight different materials.

Figure 8 shows that the final design depends on the criteria chosen by the designer. The three criteria often lead to significantly different sublaminates. The difference, between materials, of the ratio of thicknesses required by the three criteria reflects the way the different ratios of ply-angle orientations affect the relationship of FPF and LPF. For example, except for fiberglass and aramid systems, stacking for FPF is very different from stacking for the design limit criteria. This can be explained by pointing out that high-stiffness systems, such as graphite or boron, are more affected by degradation than are low-stiffness systems.

The large disparity in thickness among the different materials in Fig. 8 clearly shows the importance of choosing the right material. Stacking can be very different for different materials. The limit design is [1 4 0 3] for T300-5208 and [3 0 1 4] for Kevlar 49. Translating a laminate design from one material to another by some simple ratio that is inferred by comparing a presumably dominant property, such as longitudinal strength or stiffness, cannot be done. From this example it can be seen

Table 13   Cost of different materials, relative to cost for graphite-epoxy laminate

Material	Cost(a)
Graphite-epoxy (T300-5208)	1
Boron-epoxy (B4-5505)	5
Graphite-epoxy (AS-3501)	1
Glass-epoxy (E-glass-epoxy)	0.2
Kevlar49-epoxy	0.5
Graphite-thermoplastic AS4-APC2	2
Graphite-epoxy (IM6-epoxy)	1.5
Graphite-epoxy (T300-934)	1

(a) Normalized by cost of graphite-epoxy

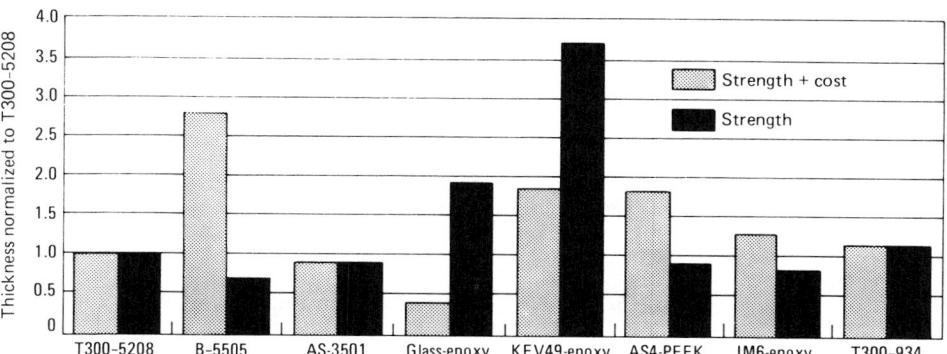

Fig. 9 Efficiency rating calculated as the cost (normalized to cost of T300-5208 graphite-epoxy shown in Table 13) per unit strength and the strength ratio at FPF (normalized to strength ratio of T300-5208 graphite-epoxy) as a function of material selection

that high compressive loads can be better supported by off-axis reinforcement with the aramid, while graphite is able to support the same type of load through on-axis reinforcement.

A sophisticated analysis can be built from this method. For instance, economic considerations can be included by making the cost-to-weight ratio of each material a design parameter. Table 13 shows typical product and manufacturing cost ratios based on the authors' experience and presented only to enable a qualitative comparison. In Figure 9, the limit design for the same load considered in Fig. 8

and the cost ratios shown in Table 14 are plotted as an efficiency value for each material. Figure 9 shows that the most cost-effective design would be to use glass-epoxy (ignoring weight). Boron is the most expensive design (ignoring stiffness). From Fig. 9, it can be observed that even though aramid is more economical than graphite, in this case, it is not as effective as he graphite designs.

REFERENCES

1. S.W. Tsai, *Composites Design—1986*, Think Composites, 1986
2. G.V. Flanagan and W.J. Park, "Composite Laminate Weight Optimization on the PC-1500 Pocket Computer," AFWAL TR-83-4095, Air Force Wright Aeronautical Laboratories, 1983

Analysis of Structures

S.R. Swanson, University of Utah

THE ANALYSIS OF COMPOSITE STRUCTURES has many features in common with that of more usual isotropic structures. However, the directional properties of stiffness and strength resulting from the fiber reinforcement must be accounted for in analysis, and this introduces certain new complications into the analysis procedure. Composites with continuous fibers are usually the most directional in properties, while short fiber reinforced composites may have very little orientation effect in properties. Thus, this article will emphasize composites with continuous fiber reinforcement, because this represents the most significant departure from isotropic structural analysis procedures.

A fundamental characteristic of fiber-reinforced composite structures is that both strength and stiffness are highly directional. As an example, the modulus of elasticity of a carbon-epoxy composite may be an order of magnitude higher in the fiber direction than in the direction transverse to the fiber. The strength properties may be even more directional. Thus, proper placement of the fibers is essential in design, and must be accounted for in analysis. In some designs the preponderance of fibers may be in a single load direction, but typically it is necessary to have at least some of the fibers in other directions to provide strength and stiffness in secondary directions. Thus, the usual analysis problem is to calculate the stresses and displacements in a composite that is made up of fiber reinforcement at various angles.

The procedure usually used to analyze composite structures is first to calculate the effective nonisotropic properties of the composite material by means of classical lamination theory (CLT). These overall effective properties are then used in conjunction with strength of materials or elasticity analysis procedures, often similar to those used in conventional isotropic analysis, to calculate the structural response and, particularly, the strains in the structure. Finally, the stresses in the individual plies or layers are then related to the strains.

Many typical analysis situations in composite structures involve stretching and/or bending of relatively thin laminates. The geometry may contain stress risers in the form of changes in thickness, fastener holes, or changes in shape.

Some of these problems can be treated by elementary strength-of-materials formulas. As an example, consider the case of direct tension of a laminated tensile member. The following steps illustrate the process of determining the stresses in the individual plies under the action of the tensile load. Assume that the lamination sequence is balanced and symmetric. This problem is very simple but illustrates the fundamental features of the analysis. The first step is to calculate the orthotropic extensional stiffnesses (moduli) for the tensile member by means of CLT. That is, the moduli of the individual plies are combined to form the overall moduli of the laminated material. The average stress in the tensile member is then determined by using the stress analysis formula $\sigma_x = P/A$, which for this case is independent of the material properties, where P is applied force and A is cross-sectional area. The strains are then calculated by using the average stress and the effective orthotropic moduli and Poisson's ratio. In this case, the formulas are:

$$\epsilon_x = \sigma_x/E_{xx} \quad \text{and} \quad \epsilon_y = -\nu_{xy}\epsilon_x \qquad \text{(Eq 1)}$$

Finally, the stresses in the individual plies are calculated from the strains by assuming that the plies are bonded together perfectly and therefore experience the same strain levels. It should be emphasized that the stresses in the individual plies may be quite different from the average stress, σ_x.

This simple analysis is part of an overall rational design and analysis approach for laminated structures. The final steps are to relate the calculated stresses for each ply to the design allowables for that material, and thus estimate the margin of safety for either matrix cracking or fiber failure by means of appropriate failure criteria.

Classical Lamination Theory

As already mentioned, the heart of the analysis for laminated structures is embodied in the classical lamination theory. The output of this theory is twofold: to obtain overall in-plane and bending stiffness properties from the individual ply properties, and to relate the individual ply stresses to structural strains and rotations. Be-

fore details are described, it should first be noted that the theory has been presented in a number of standard texts on composite materials (Ref 1-6). The notation used in most of these texts is fairly standard, and has been followed here. Second, the algebra involved is tedious enough that most users will prefer using a microcomputer to make the computations, although a programmable calculator could possibly be used. The basic equations are simple enough that any microcomputer and any programming language would be adequate. The article "Computer Programs for Structural Analysis" in this Volume should be useful. A relatively elaborate version of the basic theory with failure criteria included is available commercially (Ref 6).

The ply geometry of a typical continuous fiber structure can be described by the ply orientations, number of plies, and the stacking sequence, as shown in Fig. 1. This ply geometry can be written in a conventional shorthand notation. For example, a laminate composed of a 0°, +45°, −45°, 90°, 90° −45°, +45°, 0° ply sequence would be written as $[0°/\pm45°/90°]_s$ in standard notation. The s stands for symmetric, which means that the lay-up is symmetric about the midplane. If

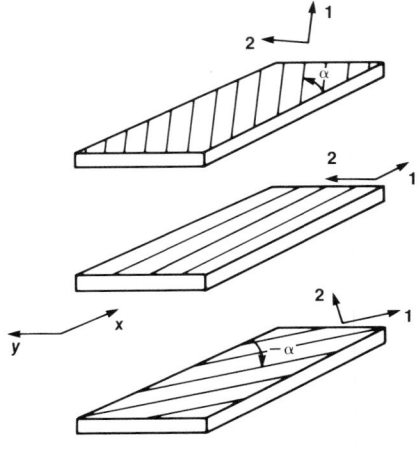

Fig. 1 Ply geometry for a $[\alpha/0/-\alpha]$ laminate

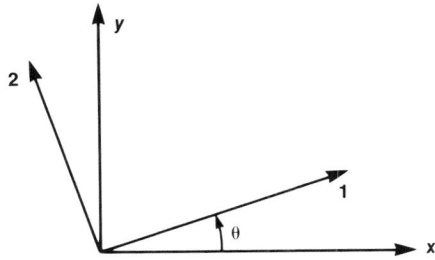

Fig. 2 Rotation of coordinate system from overall x,y,z to local 1,2,3 axes

Table 1   Typical in-plane stiffness properties of composite materials

Material	E_{11} GPa	E_{11} 10^6psi	E_{22} GPa	E_{22} 10^6psi	ν_{12}	G_{12} GPa	G_{12} 10^6psi	Ref
Carbon-epoxy prepreg (AS4/3501–6)	131.0	19.0	11.2	1.63	0.28	6.52	0.945	7, 8
Carbon-epoxy prepreg (T300/5208).	153	22.1	11.2	1.63	0.33	7.10	1.03	9
Boron-epoxy.	204	29.6	18.5	2.68	0.23	5.59	0.810	6
Kevlar 49-epoxy.	76.0	11.0	5.50	0.80	0.34	2.30	0.33	6
E-glass-epoxy.	38.6	5.60	8.27	1.20	0.26	4.14	0.60	6

symmetry is not used, a subscript t may be used to emphasize this fact. Repeated plies can be indicated by a subscript number, as in $[(0°_2/\pm 60°]_s$, for a 0°, 0°, +60°, −60° −60°, +60°, 0°, 0° laminate.

Transformation of Coordinates. An essential feature of laminate analysis is that the plies may have fiber directions oriented at an angle with respect to some overall body coordinate system. As shown in Fig. 2, we will use a 1,2,3 coordinate system to denote the fiber directions of a given ply, and an x,y,z overall coordinate system. The angle between these systems in the plane of the ply is θ (Fig. 2). The equations for relating components of stress in the two coordinate systems can be written in a shorthand notation as $(\bar{\sigma}) = [T]$ (σ) and $(\bar{\epsilon}) = [T] (\epsilon)$, where the $[T]$ matrix and its inverse are:

$$[T] = \begin{bmatrix} c^2 & s^2 & 2sc \\ s^2 & c^2 & -2sc \\ -sc & sc & c^2-s^2 \end{bmatrix}$$

$$[T^{-1}] = \begin{bmatrix} c^2 & s^2 & -2sc \\ s^2 & c^2 & 2sc \\ sc & -sc & c^2-s^2 \end{bmatrix}$$

(Eq 2)

where s and c stand for $\sin \theta$ and $\cos \theta$ respectively, and an overbar symbol is used to indicate the 1,2,3 fiber direction coordinate system. Note that the transformation equation must be used with the tensor shear strain, which is half of the engineering shear strain.

Ply Stress-Strain Law. The strain-stress relationship for a lamina can be written:

$$\begin{Bmatrix} \epsilon_1 \\ \epsilon_2 \\ \epsilon_3 \\ \gamma_{23} \\ \gamma_{31} \\ \gamma_{12} \end{Bmatrix} = \begin{bmatrix} 1/E_{11} & -\nu_{21}/E_{22} & -\nu_{31}/E_{33} & 0 & 0 & 0 \\ -\nu_{12}/E_{11} & 1/E_{22} & -\nu_{32}/E_{33} & 0 & 0 & 0 \\ -\nu_{13}/E_{11} & -\nu_{23}/E_{22} & 1/E_{33} & 0 & 0 & 0 \\ 0 & 0 & 0 & 1/G_{23} & 0 & 0 \\ 0 & 0 & 0 & 0 & 1/G_{31} & 0 \\ 0 & 0 & 0 & 0 & 0 & 1/G_{12} \end{bmatrix} \begin{Bmatrix} \sigma_1 \\ \sigma_2 \\ \sigma_3 \\ \tau_{23} \\ \tau_{31} \\ \tau_{12} \end{Bmatrix}$$

(Eq 3)

where the material property matrix in this form is termed the compliance matrix, S. Symmetry of the matrix implies relationships such as $\nu_{12}/E_{11} = \nu_{21}/E_{22}$ so that only nine material properties are required to fully characterize the linear behavior of a lamina or unidirectional

laminate. In many cases only the in-plane terms are needed. For plane stress, the in-plane terms are written:

$$\begin{Bmatrix} \epsilon_1 \\ \epsilon_2 \\ \gamma_{12} \end{Bmatrix} = \begin{bmatrix} S_{11} & S_{12} & 0 \\ S_{21} & S_{22} & 0 \\ 0 & 0 & S_{66} \end{bmatrix} \begin{Bmatrix} \sigma_1 \\ \sigma_2 \\ \tau_{12} \end{Bmatrix}$$

(Eq 4)

where $S_{12} = S_{21}$, and the compliance terms are as defined above. The usual notation for Poisson's ratio is that ν_{ij} refers to the negative of the strain in the j direction as a ratio to the strain in the i direction, because of the uniaxial stress in the i direction. However, this convention is sometimes reversed (Ref 6).

The inverse of the compliance matrix, S, is the stiffness matrix, Q, for the lamina, and for plane stress this is given by:

$$\begin{Bmatrix} \sigma_1 \\ \sigma_2 \\ \tau_{12} \end{Bmatrix} = \begin{bmatrix} Q_{11} & Q_{12} & 0 \\ Q_{21} & Q_{22} & 0 \\ 0 & 0 & Q_{66} \end{bmatrix} \begin{Bmatrix} \epsilon_1 \\ \epsilon_2 \\ \gamma_{12} \end{Bmatrix}$$

(Eq 5)

where again symmetry holds, so that $Q_{12} = Q_{21}$, and the terms are given explicitly as:

$$Q_{11} = E_{11}/(1-\nu_{12}\nu_{21})$$

$$Q_{22} = E_{22}/(1-\nu_{12}\nu_{21})$$

$$Q_{66} = G_{12}$$

$$Q_{12} = Q_{21} = \nu_{12}E_{22}/(1-\nu_{12}\nu_{21}) = \nu_{21}E_{11}/(1-\nu_{12}\nu_{21})$$

(Eq 6)

Typical in-plane elastic properties are given in Table 1 for various continuous fiber reinforced composite materials. The very large difference between the moduli in the fiber direction and those transverse to the fiber should be noted. Also, symmetry can be used to get the minor Poisson's ratio, so that for a typical carbon-epoxy prepreg, $\nu_{21} = (E_{22}/E_{11})$ $\nu_{12} \approx 0.02$.

It is necessary to be able to write the ply stress-strain law with respect to a set of x,y coordinates that are at an angle, θ, with respect to the fiber axes. This can be accomplished by using the rotation of coordinate axes equations given above, remembering that the rotation equations must be applied to tensor strains. As shown in Ref 1, this can be accomplished conveniently by means of the matrix, R, defined as:

$$\begin{Bmatrix} \epsilon_1 \\ \epsilon_2 \\ \gamma_{12} \end{Bmatrix} = [R] \begin{Bmatrix} \epsilon_1 \\ \epsilon_2 \\ \epsilon_{12} = \gamma_{12}/2 \end{Bmatrix} = \begin{bmatrix} 1 & 0 & 0 \\ 0 & 1 & 0 \\ 0 & 0 & 2 \end{bmatrix} \begin{Bmatrix} \epsilon_1 \\ \epsilon_2 \\ \epsilon_{12} \end{Bmatrix}$$

(Eq 7)

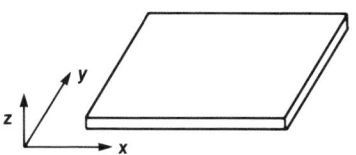

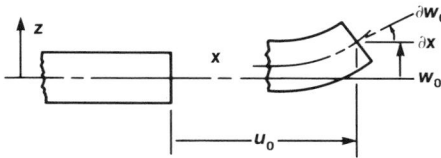

Fig. 3 Deformed plate geometry

Then, starting with the ply stress-strain law in fiber directions and substituting gives:

$$\begin{Bmatrix} \sigma_x \\ \sigma_y \\ \tau_{xy} \end{Bmatrix} = [\bar{Q}] \begin{Bmatrix} \epsilon_x \\ \epsilon_y \\ \gamma_{xy} \end{Bmatrix}$$

(Eq 8)

where $[\bar{Q}] = [T]^{-1} [Q] [R] [T] [R]^{-1}$. Multiplying this out gives the equations for the ply stress-strain law in an overall x,y coordinate system:

$$\bar{Q}_{11} = Q_{11} c^4 + 2(Q_{12} + 2Q_{66})s^2c^2 + Q_{22}s^4$$

$$\bar{Q}_{12} = \bar{Q}_{21} = (Q_{11}+Q_{22}-4Q_{66})s^2c^2 + Q_{12}(s^4+c^4)$$

$$\bar{Q}_{22} = Q_{11}s^4 + 2(Q_{12} + 2Q_{66})s^2c^2 + Q_{22}c^4$$

$$\bar{Q}_{16} = (Q_{11}-Q_{12}-2Q_{66})sc^3 + (Q_{12}-Q_{22}+2Q_{66})s^3c$$

$$\bar{Q}_{26} = (Q_{11}-Q_{12}-2Q_{66})s^3c + (Q_{12}-Q_{22}+2Q_{66})sc^3$$

$$\bar{Q}_{66} = (Q_{11}+Q_{22}-2Q_{12}-2Q_{66})s^2c^2 + Q_{66}(s^4+c^4)$$

(Eq 9)

Deformation and Strain. The usual assumptions of the deformation of thin laminated plates (and beams) are illustrated in Fig. 3. The plies are assumed to be perfectly bonded so that no slippage occurs between them. In the usual (simplest) theory, the Kirchhoff-Love hypothesis is invoked by assuming that normals to the centerline remain normal after deformation. This assumption neglects through-the-thickness shear deformation, which can be important in some applications of composites because of the relatively low values of the shear moduli. Further discussion of shear deformation will be given in the section "Plates and Shells" in this article. From the geometry of Fig. 3, the in-plane displacements u and v can be related to

Table 2 Analysis of $[0°_2/\pm45°]_s$ carbon-epoxy laminate under tension and bending

Matrix	Units	Matrix terms(a)					
		1,1	1,2	1,3	2,2	2,3	3,3
$\bar{Q}$, 0°	GPa (10^6 psi)	131.9 (19.13)	3.168 (.459)	0 (0)	11.32 (1.641)	0 (0)	13.03 (.945)
$\bar{Q}$, 45°	GPa (10^6 psi)	43.9 (6.37)	30.9 (4.48)	30.1 (4.37)	43.9 (6.37)	30.1 (4.37)	34.2 (4.96)
A	MN/m (10^3 lbf/in.)	92.9 (530)	17.98 (102.7)	0 (0)	29.2 (166.6)	0 (0)	21.5 (122.9)
B	N (lbf)	0 (0)	0 (0)	0 (0)	0 (0)	0 (0)	0 (0)
D	N · m (lbf · in.)	11.88 (105.2)	0.652 (5.77)	0.278 (2.46)	1.513 (13.39)	0.278 (2.46)	0.981 (8.68)
A^{-1}	10^{-9} m/N (10^{-6} in./lbf)	12.23 (2.14)	−7.54 (−1.320)	0 (0)	38.9 (6.82)	0 (0)	46.5 (8.14)
D^{-1}	10^{-2} (N · m)$^{-1}$ (10^{-3} lbf · in.)	8.64 (9.76)	−3.45 (−3.90)	−1.469 (−1.659)	71.1 (80.34)	−19.16 (−21.65)	107.8 (121.8)

(a) Matrices are symmetric; thickness/ply = 0.132 mm (0.0052 in.)

the centerline displacements u_0 and v_0 and the slopes by:

$$u = u_0 - z\frac{\partial w_0}{\partial x} \qquad v = v_0 - z\frac{\partial w_0}{\partial y} \qquad \text{(Eq 10)}$$

Using the usual strain-displacement relationships of linear elasticity given by:

$$\epsilon_x = \frac{\partial u}{\partial x} \qquad \epsilon_y = \frac{\partial v}{\partial y} \qquad \gamma_{xy} = \frac{\partial u}{\partial y} + \frac{\partial v}{\partial x} \qquad \text{(Eq 11)}$$

$$\begin{Bmatrix} \epsilon_x \\ \epsilon_y \\ \gamma_{xy} \end{Bmatrix} = \begin{Bmatrix} \epsilon_x^{\,0} \\ \epsilon_y^{\,0} \\ \gamma_{xy}^{\,0} \end{Bmatrix} + z \begin{Bmatrix} \kappa_x \\ \kappa_y \\ \kappa_{xy} \end{Bmatrix} \qquad \text{(Eq 12)}$$

where the matrices on the right are the centerline strains and curvatures, given by:

$$\begin{Bmatrix} \epsilon_x^{\,0} \\ \epsilon_y^{\,0} \\ \gamma_{xy}^{\,0} \end{Bmatrix} = \begin{Bmatrix} \partial u_0/\partial x \\ \partial v_0/\partial y \\ \partial u_0/\partial y + \partial v_0/\partial x \end{Bmatrix} \begin{Bmatrix} \kappa_x \\ \kappa_y \\ \kappa_{xy} \end{Bmatrix} = - \begin{Bmatrix} \partial^2 w_0/\partial x^2 \\ \partial^2 w_0/\partial y^2 \\ 2\partial^2 w_0/\partial x \partial y \end{Bmatrix}$$

$$\text{(Eq 13)}$$

Stress and Moment Resultants. The next step in the development is to relate the internal stresses (in overall x,y coordinates) to the external stress resultants [N] and moment resultants [M]. The term stress resultant refers to the stress integrated over the thickness of the laminate, and is thus the applied force per unit width. A similar interpretation can be given to the moment resultant, which is the applied moment per unit width. Using equilibrium, we equate the force and moment per unit width to the integral of the stress and stress times distance from the neutral axis, to get, for example:

$$N_x = \int_{-t/2}^{t/2} \sigma_x dz \qquad M_x = \int_{-t/2}^{t/2} \sigma_x z dz \qquad \text{(Eq 14)}$$

where the integral is taken over the total laminate thickness by summing the integrals over each ply. Substituting for the stress-strain law for each ply (in the overall global coordinate system) gives the classic relation between normal forces, moments, centerline strains, and curvature in the form:

$$\begin{Bmatrix} N \\ - \\ M \end{Bmatrix} = \begin{bmatrix} A & | & B \\ - & + & - \\ B & | & D \end{bmatrix} \begin{Bmatrix} \epsilon_0 \\ - \\ \kappa \end{Bmatrix} \qquad \text{(Eq 15)}$$

where the A, B, and D matrices are defined as:

$$A_{ij} = \sum_{k=1}^{n} (\bar{Q}_{ij})_k (z_k - z_{k-1}) \qquad \text{(Eq 16)}$$

$$B_{ij} = \frac{1}{2}\sum_{k=1}^{n} (\bar{Q}_{ij})_k (z_k^2 - z_{k-1}^2) \qquad \text{(Eq 17)}$$

$$D_{ij} = \frac{1}{3}\sum_{k=1}^{n} (\bar{Q}_{ij})_k (z_k^3 - z_{k-1}^3) \qquad \text{(Eq 18)}$$

Here the positions of the ply surfaces are denoted by z_k, n is the number of plies, and the $\bar{Q}_k$ is the stiffness in the x,y coordinate system of each ply. This relationship between the stress and moment resultants and the centerline strains and curvatures is the heart of laminate analysis.

Hygrothermal Behavior. It is straightforward to include moisture and thermal stress effects in the laminate analysis. Starting with the ply stress-strain law, but including thermal- and moisture-induced strains:

$$\begin{Bmatrix} \sigma_1 \\ \sigma_2 \\ \sigma_{12} \end{Bmatrix} = [Q] \begin{Bmatrix} \epsilon_1 - e_1 \\ \epsilon_2 - e_2 \\ \gamma_{12} \end{Bmatrix} = [Q] \begin{Bmatrix} \epsilon_1 \\ \epsilon_2 \\ \gamma_{12} \end{Bmatrix} - [Q] \begin{Bmatrix} e_1 \\ e_2 \\ 0 \end{Bmatrix}$$

$$\text{(Eq 19)}$$

where $e_i = \alpha_i \Delta T + \beta_i C$, α and β are thermal and moisture expansion coefficients, and ΔT and C are differences in temperature and moisture concentrations from cure conditions. Following the development given above for mechanical stresses alone gives:

$$[N] = [A] |\epsilon_0| + [B] |\kappa| - N^e$$
$$[M] = [B] |\epsilon_0| + [D] |\kappa| - M^e \qquad \text{(Eq 20)}$$

where the moisture and temperature terms are:

$$N^e = \sum_{k=1}^{n} ([T]^{-1}[Q] [e])_k (z_k - z_{k-1}) \qquad \text{(Eq 21)}$$

$$M^e = \sum_{k=1}^{n} ([T]^{-1}[Q] [e])_k (z_k^2 - z_{k-1}^2) \qquad \text{(Eq 22)}$$

Special Laminates. A number of special cases are of interest. If the laminate is symmetric with respect to the midplane, the B matrix will vanish. If the laminate is not symmetric, the extensional and bending responses will be coupled, so that, for example, bending stresses can result from uniform extensional strains. Another special case is termed a balanced laminate, in which the number of plies in a given θ direction is equal to the number in the −θ direction. In a balanced laminate, the A_{16} and A_{26} terms are zero, thus uncoupling the in-plane shear and extensional response.

The couplings cited above are features of composite structures that have no counterpart in isotropic metals, and can be taken advantage of to achieve unique characteristics. For example, the forward swept wings of the X-29 aircraft use coupling between bending and torsion to achieve stability, and the design could not function using metallic wings (Ref 10). The couplings available with composites have probably not been fully explored.

Another special category of laminates is termed quasi-isotropic, in that the in-plane stiffnesses are independent of orientation. This can be achieved by using laminates that include equal numbers of $[0°/\pm60°]$ or $[0°/\pm45°/90°]$ ply groups.

Applications

To illustrate the analysis process for laminates, consider a $[0°_2/\pm45°]_{2s}$ laminate under a direct tension load and also a bending moment. The properties of a carbon-epoxy material (Table 1) give the individual ply stiffness matrices as shown in the first two rows of Table 2, after transforming to the overall x,y coordinates that coincide with the 0° fibers. Substituting into Eq 15 to 17 gives the A and D matrices that are given in Table 2. The B matrix is identically zero because of the symmetry of the laminate. Inverting these matrices gives the centerline strains and curvatures, which then permit the strain to be calculated for each ply from Eq 12. Finally, the ply stresses can be calculated from the laminate strain, using the stress-strain relationship for the ply. It is usually desirable to transform these ply stresses into the ply fiber coordinate system directions, so that direct comparisons can be made with allowable stresses for the ply. The results of these computations for ply stresses are given in Tables 3 and 4 for the direct tension and bending loads, respectively. Superposition can be used to combine the direct tension and bending stresses.

It can be seen that the basic procedure of laminate analysis is straightforward. It can also be seen that the process is complicated enough to warrant programming on a computer.

Analysis of a Filament-Wound Driveshaft. Another application to be considered is that of torsion of a filament-wound automotive driveshaft. As explained in Ref 11, driveshafts are required to resist torsion and to have a sufficiently high natural frequency in beam bending. The torsion load can be carried effi-

Table 3 Ply stresses in [0°₂/±45°]ₛ carbon-epoxy laminate under direct tension load

Ply	Ply stress (a)					
	σ_1		σ_2		τ_{12}	
	MPa	ksi	MPa	ksi	MPa	ksi
−45°	31.7	4.59	3.40	0.493	12.88	1.868
45°	31.7	4.59	3.40	0.493	−12.88	−1.868
0°	158.9	23.0	−4.65	−0.675	0	
0°	158.9	23.0	−4.65	−0.675	0	

(a) $N_x = 10^5$ N/m (571 lbf/in.)

Table 4 Ply stresses in [0°₂/±45°]ₛ carbon-epoxy laminate under bending load

Ply	Ply stress (a)								
	z		σ_1		σ_2		τ_{12}		
	mm	in.	MPa	ksi	MPa	ksi	MPa	ksi	
−45°	0	0	0	0	0	0	0	0	
	0.132	0.0052	7.12	1.033	0.338	0.0490	1.040	0.1509	
45°	0.132	0.0052	2.13	0.309	0.654	0.095	−1.040	−0.1509	
	0.264	0.0104	4.26	0.617	1.308	0.1897	−2.08	−0.302	
0°	0.264	0.0104	29.8	4.32	−0.309	−0.044	−0.506	−0.073	
	0.396	0.0156	44.7	6.48	−0.463	−0.0672	−0.758	−0.110	
0°	0.396	0.0156	44.7	6.48	−0.463	−0.0672	−0.758	−0.110	
	0.528	0.0208	59.6	8.65	−0.618	−0.0896	−1.01	−0.147	

(a) $M_x = 10$ N · m/m (2.25 in. · lbf/in.); stresses have opposite sign in plies below midplane

ciently by fibers in the ±45° directions. Resistance to torsional buckling depends on the stiffness in both axial and hoop directions, while the natural frequency in bending depends on axial stiffness and inversely on mass per unit length. Reasons of economy suggest glass fibers in the ±45° and hoop directions. A hybrid with carbon fibers in the axial direction for increased stiffness relative to glass is illustrated in Ref 11.

The effective moduli in the axial and hoop directions can be obtained from the diagonal terms of the inverse of the A matrix; for example, if the x direction is taken to coincide with the axial direction of the tube, and letting $[A]^{-1} = [A']$, then:

$$\bar{E}_x = 1/(A'_{11}t) \qquad \text{(Eq 23)}$$

where t is the total wall thickness of the tube.

The calculation of the fiber stresses under the torsion load can be handled approximately by neglecting the variation of shear strain through-the-wall thickness. Thus, the equations given above for flat laminates can be employed, in conjunction with the equilibrium equation:

$$\text{Torque} = N_{xy} 2\pi \bar{R}^2 \qquad \text{(Eq 24)}$$

where $\bar{R}$ is a mean radius. As shown by comparison with the experiment in Ref 11, this approximate method is reasonably good for thin-walled shafts. An alternative method that includes the variation in strain through-the-wall thickness can easily be derived by using the usual deformation assumptions of strength of materials. Assuming $\gamma = r\theta$, and using equilibrium gives:

$$\text{Torque} = \int_{r_i}^{r_o} 2\pi r^2 \tau_{xy} dr =$$

$$\bar{\theta}\left[\sum_{k=1}^{n}(\bar{Q}_{66})_k(r_k^4 - r_{k-1}^4)\,\pi/2\right] = \bar{\theta}\bar{J}\bar{G} \qquad \text{(Eq 25)}$$

The term in brackets can be considered an effective value of JG. Solving the above for θ, the angle of twist per unit length, permits the stresses in each ply to be calculated in fiber coordinates as:

$$\begin{Bmatrix} \sigma_1 \\ \sigma_2 \\ \tau_{12} \end{Bmatrix} = ([T][\bar{Q}]_k \begin{Bmatrix} 0 \\ 0 \\ r\,\text{torque}/\bar{J}\bar{G} \end{Bmatrix} \qquad \text{(Eq 26)}$$

It may be noted that filament winding involves cross-overs and interspersing of the individual angle plies. This effect has apparently not been studied extensively, but is usually neglected (Ref 11).

An additional example of the use of classical lamination theory is to calculate the shear stress distribution in beams. Delamination caused by transverse shear stress is a possible failure mode in composites due to the relatively low interlaminar shear strength as compared to the tensile or compressive strength of the plies. The development of a formula for shear stress distribution in a composite beam closely follows that for isotropic beams, except that the stress distribution varies from ply to ply. Using, as usual, the axial equilibrium of a cut section of a beam gives (assuming a symmetric laminate):

$$\tau b \, dx = \int_z^{z_m} d\sigma dA \qquad \text{(Eq 27)}$$

Using the following one-dimensional stress relationships:

$$\sigma_x = (\bar{E})_k \, \epsilon_x$$

$$M_x = D_{xx} \, \kappa_x \qquad \text{(Eq 28)}$$

$$\epsilon_x = z\kappa_x$$

gives the expression for τ as:

$$\tau = \frac{V}{D_x b} \sum_{k=1}^{n} (\bar{E}_x b)_k \left(\frac{z_k^2 - z_{k-1}^2}{2}\right) \qquad \text{(Eq 29)}$$

This reduces to the usual shear stress distributions for isotropic beams.

The above examples indicate how classical lamination theory can be used in strength-of-materials stress analysis situations. There is an extensive literature concerning the analysis of composite structures using more sophisticated theories such as plates, shells, and anisotropic elasticity theory. Some brief highlights of this work are presented below to give an indication of more advanced treatments.

Plates and Shells. The analysis of anisotropic plates has been described in a number of references (Ref 2, 5, 12-14). The equilibrium equations relating the variations of the in-plane stress resultants N and the moment resultants M are of course the same as those for isotropic plates. A displacement formulation is obtained by substituting for N and M in terms of centerline strains and curvatures, Eq 15, and then substituting the usual strain-displacement relationships, Eq 13.

Considering only symmetric laminates eliminates the membrane-bending coupling, as the B matrix is then zero. Even with this simplification, the equations are complicated and are usually solved by approximate methods such as Ritz techniques, including finite-element methods. The assumption of specially orthotropic behavior in which A_{16}, A_{26}, D_{16}, and D_{26} vanish gives a considerable simplification, permitting standard Fourier series solutions (Ref 12, 13). As discussed previously, a balanced laminate will cause the A_{16} and A_{26} terms to vanish, but generally not the D_{16} and D_{26} terms. Cross-ply laminates are a special case and have been explored extensively (Ref 12, 13, 15).

It has been shown that transverse shear deformations are much more important in composite plates because of the lower transverse shear moduli relative to isotropic materials. Typical solutions demonstrate important effects at length-to-thickness ratios as high as 20 (Ref 15, 16). Plate theories have also been developed to improve the accuracy of the in-plane strains and stresses (Ref 17-19).

The analysis of laminated composite shells is similar to that of plates, but obviously is complicated by the shell curvature. The special case of cylindrical shells, and particularly that of specially orthotropic materials, has been explored in Ref 12, using Fourier series techniques.

The solution to problems involving thick-walled cylinders has been presented by S.G. Lekhnitskii (Ref 20), using three-dimensional elasticity theory. An interesting consequence of using typical laminate properties is that thick-walled cylinder effects are important at much higher ratios of mean radius-to-wall thickness than is the case with isotropic materials (Ref 7).

Stress Concentration at Cutouts. The solution to a number of important problems related to stress concentration in plates with holes is presented in Ref 14. The stress concentration may be quite different than it is in isotropic materials. As an example, the stress concentration at the edge of a circular hole in an orthotropic plate loaded by a uniform tension remote from the hole and in a direction coinciding with the principal material directions is given by:

$$K_t = 1 + \left\{ 2\left(\frac{\bar{E}_{11}}{\bar{E}_{22}} - \bar{\nu}_{12} \right) + \frac{\bar{E}_{11}}{\bar{G}_{11}} \right\}^{1/2} \quad \text{(Eq 30)}$$

or

$$K_t = 1 + \left\{ 2\left(\frac{A_{11} - A_{12}}{A_{22}} \right) + \frac{A_{11}^2 A_{22}}{A_{66}(A_{11}A_{22} - A_{12}^2)} \right\}^{1/2} \quad \text{(Eq 31)}$$

where $\bar{E}_{11}$ and A_{11} are in the direction of the applied load. For a quasi-isotropic laminate (or an isotropic material), the stress-concentration factor has a value of 3. For an orthotropic material with the properties of a carbon-epoxy lamina, the stress-concentration factor is 7.7. Thus, the stress concentration is seen to be sensitive to the orthotropic material properties. Other loadings are considered in Ref 14, including pressure on the interior of the hole, shear, and biaxial tension.

Free-Edge Effects. A surprising source of stress concentration exists at the lateral edge of laminates subjected to stress in the axial direction. Known as the free-edge effect, this phenomenon is caused by the transition in the stress state from the edge to the biaxial stresses within the interior of the laminate caused by the variation of Poisson's ratio from ply to ply. The result is a highly localized region of interlaminar normal and interlaminar shear stresses adjacent to the edge. The nature of these stresses has been investigated by a number of approximate methods, including finite-difference and finite-element methods (Ref 21-24) and by exact elasticity solutions (Ref 25, 26). It appears that the stresses are related to the stacking sequence of the plies, and that in practical cases the edge stress concentrations can cause early failure of the laminate.

REFERENCES

1. R.M. Jones, *Mechanics of Composite Materials*, McGraw-Hill, 1975
2. L.R. Calcote, *The Analysis of Laminated Composite Structures*, Van Nostrand Reinhold, 1969
3. S.W. Tsai and H.T. Hahn, *Introduction to Composite Materials*, Technomic, 1980
4. D. Hull, *An Introduction to Composite Materials*, Cambridge University Press, 1981
5. R.M. Christensen, *Mechanics of Composite Materials*, John Wiley & Sons, 1979
6. S.W. Tsai, *Composites Design 1986*, Think Composites, 1986
7. S.R. Swanson and A.P. Christoforou, Response of Quasi-Isotropic Carbon/Epoxy Laminates to Biaxial Stress, *J. Compos. Mater.*, Vol 20, 1986, p 457-471
8. S.R. Swanson, M. Messick, and G.R. Toombes, Comparison of Torsion Tube and Iosipescu In-Plane Shear Test Results for a Carbon Fibre-Reinforced Epoxy Composite, *Composites*, Vol 16, 1985, p 220-224
9. N.J. Pagano and H.T. Hahn, Evaluation of Composite Curing Stresses, in *Composite Materials: Testing and Design (Fourth Conference)*, STP 617, 1977, p 317-329
10. M.H. Shirk, T.J. Hertz, and T.A. Weisshaar, Aeroelastic Tailoring—Theory, Practice, and Promise, *J. Aircr.*, Vol 23, 1986, p 6-18
11. T.S. Brown and D.B. Rezin, Hybrid Composite Driveshaft Design Considerations, in *Modern Developments in Composite Materials and Structures*, American Society of Mechanical Engineers, 1979, p 173-187
12. J.R. Vinson and T.W. Chou, *Composite Materials and Their Use in Structures*, Applied Science Publishers, 1975
13. J.E. Ashton and J.M. Whitney, *Theory of Laminated Plates*, Technomic, 1970
14. S.G. Lekhnitskii, *Anisotropic Plates*, Gordon and Breach Science Publishers, 1968
15. J.M. Whitney, The Effect of Transverse Shear Deformation on the Bending of Laminated Plates, *J. Compos. Mater.*, 1969, p 534-547
16. N.J. Pagano, Exact Solutions for Composite Laminates in Cylindrical Bending, *J. Compos. Mater.*, 1969, p 398-411
17. K.H. Lo, R.M. Christensen, and E.M. Wu, A High-Order Theory of Plate Deformation, Part 1: Homogeneous Plates, and Part 2: Laminated Plates, *J. Appl. Mech. (Trans. ASME)*, Vol 44, 1977, p 663-676
18. J.N. Reddy, A Simple Higher-Order Theory for Laminated Composite Plates, *J. Appl. Mech. (Trans. ASME)*, Vol 51, 1984, p 745-752
19. A. Toledano and H. Murakami, A Composite Plate Theory for Arbitrary Laminate Configurations, *J. Appl. Mech. (Trans. ASME)*, Vol 54, 1987, p 181-189
20. S.G. Lekhnitskii, *Theory of Elasticity of an Anisotropic Elastic Body*, Holden-Day, 1963
21. R.B. Pipes and N.J. Pagano, Interlaminar Stresses in Composite Laminates Under Uniform Axial Extension, *J. Compos. Mater.*, Vol 4, 1970, p 538-548
22. E.F. Rybicki, Approximate Three-Dimensional Solutions for Symmetric Laminates Under In-Plane Loading, *J. Compos. Mater.*, Vol 5, 1971, p 354-360
23. J.D. Whitcomb and I.S. Raju, Superposition Method for Analysis of Free-Edge Stresses, *J. Compos. Mater.*, Vol 17, 1983, p 492-507
24. P.W. Hsu and C.T. Herakovich, Edge Effects in Angle-Ply Composite Laminates, *J. Compos. Mater.*, Vol 11, 1977, p 422-428
25. S.S. Wang and I. Choi, Boundary-Layer Effects in Composite Laminates: Part 1, Free-Edge Stress Singularities, *J. Appl. Mech. (Trans. ASME)*, Vol 49, 1982, p 541-548
26. S.S. Wang and I. Choi, "Boundary-Layer Effects in Composite Laminates: Part 2, Free-Edge Stress Solutions and Basic Characteristics," *J. Appl. Mech. (Trans. ASME)*, Vol 49, 1982, p 549-560

Numerical Design and Analysis of Structures

Richard K. Dropek, Hercules Aerospace Company

IN DESIGNING AND ANALYZING A STRUCTURE, the designer/analyst often finds that closed-form analysis methods are somewhat limited in defining the state of stress or strain. Analytical difficulties may arise in areas where geometric discontinuities and stress concentrations exist because of joints, cutouts, edge effects, complicated support conditions, and so forth. In addition, through-the-thickness stress gradients, multiple materials, and hydro-thermal effects may be difficult to characterize using closed-form techniques. The classical analytical methods have limited application for these and other classes of problems where local stress phenomena must be examined. In addition, large structures usually have a number of interrelated structural elements. Such large structural assemblies also preclude hand analysis methods. Fortunately, numerical analysis techniques, coupled with digital computers, have evolved and are capable of addressing both the local stress as well as the large assembly problems.

Numerical methods basically consist of subdividing the structural problem into convenient idealized elements. The solution of the individual element response to applied loads is easier to obtain than the analysis of the total structure as a continuum. The structural response then becomes the sum of the individual element responses to externally applied loads or displacements.

This article discusses the analytical principles and methods of the finite-difference and finite-element methods, which are the primary tools used today for numerical analysis of composite structures. Although detailed treatment of each of these topics is beyond the scope of this article, key steps and analytical equations will be introduced to provide a basis for understanding the example problems presented at the end of this article. The article "Computer Programs for Structural Analysis" in this Volume provides useful information as well.

Origins of Methods

The finite-difference (FD) approach evolved by replacing the governing differential equation of the continuum with its corresponding finite-difference equation. Because problems of elasticity usually require the solution of governing partial differential equations (PDE) having certain boundary conditions, representing the PDE using an FD approximation becomes particularly attractive. The drawback of closed-form elasticity analyses is that only the simpler boundary conditions can be easily accommodated. It was found that more complex boundary conditions could be analyzed using the FD approach. One of the first applications of the FD approach in elasticity was done by C. Runge (Ref 1), who used the method to solve torsional problems. Since that time, the method has been successfully applied to beams, columns, complex torsional elements, plates, and shells.

The finite-element (FE) method was an outgrowth of beam/frame networks and airframe structural analyses. Engineers had been using matrix analysis methods to analyze two-dimensional and three-dimensional beam/frame networks where each beam was considered a discrete element. The most widely used matrix analysis method for truss and frame systems was (and still is) the displacement method. In the displacement method, a system of algebraic equations of the form:

$$[k]\{u\} = \{R\}$$

is developed where $[K]$ is an $(n \cdot n)$ system stiffness matrix, $\{u\}$ is the generalized displacement vector, and $\{R\}$ is the $(n \cdot 1)$ system load vector. Based on a familiarity with the method of matrix structural analysis, engineers attempted to represent plate and shell continuums using beam networks and space trusses, respectively. These attempts to represent continuous media using beam approximations met with

limited success because of the gross approximations that had to be made.

From these attempts, a physical approach to the method of finite elements evolved. It became increasingly difficult to analyze the structural behavior of airframe structures during the advent of jet-powered aircraft and supersonic flight. Aircraft analysts needed stiffness data to perform structural dynamics calculations. The usual force matrix method, which was then being used to solve structural analyses, could not be easily applied to the new delta wing structures. The traditional force method solves for stresses within the structure, from which deflections are subsequently calculated in a secondary solution. Certain wing characteristics could not be incorporated into the flexibility matrix required for the solution using the force method. Thus, aircraft designers turned to the stiffness methods that were being used by civil engineers analyzing frame and truss networks. A group headed by M.J. Turner at Boeing first developed the concept of a triangular FE stiffness matrix based on an assumed displacement field. This approach was first documented by Turner et al. (Ref 2) in 1956. The approach was later put in more rigorous terms using the methods of variational calculus. This ultimately led to the FE method used today.

Basic Theory Development

The basic concepts and theories for both the FD and FE methods are presented below. Simple examples will be given for each method to illustrate their power and use. It must be emphasized that both numerical approaches only approximate the exact solution to a problem. However, the errors in approximation can be made sufficiently small so that an acceptable solution is obtained.

Finite-Difference Method. The approach in the FD method is to write the governing

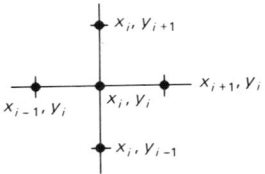

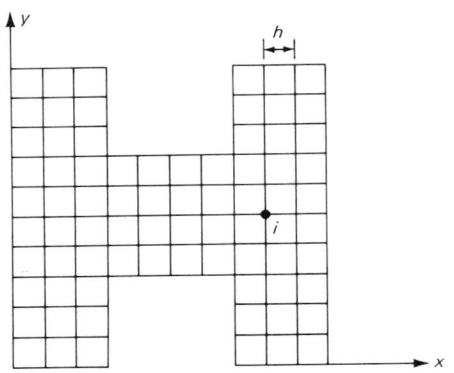

Fig. 1 Typical square net for finite-difference analysis

differential equation of the continuum by its FD approximation. The continuum is first divided into nodes. For ease of calculation, the nodes are often equally spaced into a linear or square net. However, hexagonal, triangular, as well as irregular net spacings have been used. Once the net type has been established, the approximating FD equations for the nodes can be developed.

The simplest 2-D net is the square net with spacing, h, in the x-y plane, as shown in Fig. 1. Consider the derivation of the FD equations for the node X_i. The nodes on either side of X_i are X_{i-2}, X_{i-1}, X_i, X_{i+1}, and X_{i+2}. A central difference scheme is usually employed such that the finite differences are centered about the reference node X_i, Y_i (Fig. 1). Recall the forward Taylor Series expansion of a function $f(x)$:

$$f(x + h) = f(x) + f'(x)h + \frac{f''(x)}{2!}h^2 + \ldots \quad \text{(Eq 1)}$$

where primes indicate differentiation. If the series is truncated after two terms, an approximation for the first derivative of x is obtained:

$$f(x + h) \cong f(x) + f'(x)h \quad \text{(Eq 2)}$$

Solving for the derivative, $f'(x)$, we have the first order forward difference.

First Order Forward Difference.

$$f'(x) \cong \frac{f(x + h) - f(x)}{h} \quad \text{(Eq 3)}$$

This is called the forward difference approximation to the derivative $f'(x)$. Suppose h is replaced by $-h$ in the Taylor Series. We then form the backward difference approximation to the derivative $f'(x)$.

First Order Backward Difference.

$$f'(x) \cong \frac{f(x) - f(x - h)}{h} \quad \text{(Eq 4)}$$

The central difference approximation is obtained by adding the forward and backward differences and then simplifying.

First Order Central Difference.

$$f'(x) \cong \frac{1}{2h}[f(x + h) - f(x - h)] \quad \text{(Eq 5)}$$

Retaining another term in the Taylor Series, we arrive at the forward and backward difference approximations to the second derivatives, as shown below.

Second Order Forward Difference.

$$f''(x) \cong \frac{2}{h^2}[f(x + h) - f(x) - f'(x)h] \quad \text{(Eq 6)}$$

Second Order Backward Difference.

$$f''(x) \cong \frac{2}{h^2}[f(x - h) - f(x) + f'(x)h] \quad \text{(Eq 7)}$$

When the forward difference is added to the backward difference, the central difference approximation to $f''(x)$ is obtained, as shown in Eq 8. Third order and higher derivative approximations can be obtained by including more terms in the Taylor Series expansion.

Second Order Central Difference.

$$f''(x) \cong \frac{1}{h^2}[f(x + h) - 2f(x) - f(x - h)] \quad \text{(Eq 8)}$$

The theory may be expanded to partial derivatives as shown in Appendix 1 at the end of this article.

Note that the partial and ordinary FD coefficients are identical (refer to Appendix 1). Detailed FD coefficient tables have been generated in the literature for forward, central, and backward difference schemes, including third order and higher ordinary and partial differential equations (Ref 3, 4). Determining which approximation to use (forward, central, or backward) depends on the problem at hand. Usually, central difference schemes are used within the net, with forward or backward FD equations used at the boundaries.

Once the ordinary or partial differential equation has been derived for a specific structural problem, it can then be represented by the appropriate form of the FD approximation. For example, the differential equations describing the deflection of a beam and torsion

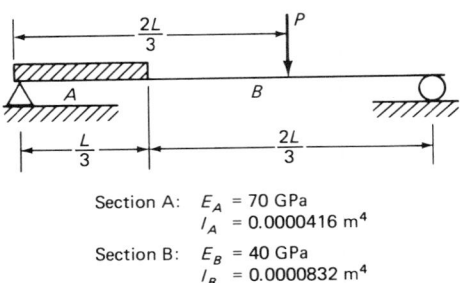

Section A: E_A = 70 GPa
I_A = 0.0000416 m^4

Section B: E_B = 40 GPa
I_B = 0.0000832 m^4

P = 900 kg L = 250 cm

Fig. 2 Solution of a simply supported beam using the finite-difference approach

of a rod are listed in Table 1, along with the corresponding FD approximations. The FD method is useful for composite beams and columns because it is able to handle sections with variable geometries and material properties. Additional beam, column, torsion, plate, and shell FD equations may be found in Ref 1, and 4 to 6.

A simple demonstration of the FD solution capability is the beam system shown in Fig. 2. The simply supported beam is composed of two materials and has two different cross sections. The problem is to solve for the deflection of the beam. The simple beam deflection equation may be used for the solution of the problem:

$$v'' = \frac{M}{EI} \quad \text{(Eq 9)}$$

where v is the deflection, v'' is the second derivative of deflection with respect to length, M is the moment, E is Young's modulus, and I is the beam moment of inertia. Using Table 1, the finite-difference form of Eq 9 is:

$$[v_{(x_i - h)} - 2v(x_i) + v(x_i + h)]\frac{1}{h^2} = \left[\frac{M}{EI}\right]_i \quad \text{(Eq 10)}$$

where h is the mesh spacing and i is the ith mesh point. If $h = L/9$ (equal spacing), then the following sets of equations are formed:

$$v_0 - 2v_1 + v_2 = \frac{M_1}{E_A I_A}(h^2)$$

$$v_1 - 2v_2 + v_3 = \frac{M_2}{E_A I_A}(h^2)$$

$$v_7 - 2v_8 + v_9 = \frac{M_8}{E_B I_B}(h^2) \quad \text{(Eq 11)}$$

Table 1   Example of finite-difference (FD) approximations for typical structural problems

Application	Differential equation	FD approximation
Beam flexure	$\dfrac{d^2v}{dx^2} = \dfrac{M}{EI}$	$\dfrac{1}{h^2}[v(x + h) - 2v(x) + v(x - h)] = \dfrac{M}{EI}$
Torsion	$\dfrac{\delta^2\phi}{\delta x^2} + \dfrac{\delta^2\phi}{\delta y^2} = -2G\,\theta$	$\dfrac{1}{h^2}[\theta(x + h,y) - \theta(x,y + h) + \phi(x - h,y) + \phi(x,y - h) - 4\phi(x,y)] = 2G\,\theta$

Note: u, deflection in x direction. v, deflection in y direction. M, moment. h, mesh spacing. P, force. E, Young's modulus. I, moment of inertia. ϕ, stress function. G, shear modulus. θ, angle of twist

Noting that the deflections at the supports are 0 ($v_0 = v_9 = 0$), Eq 11 may be written in matrix form:

$$\begin{bmatrix} -2 & 1 & 0 & 0 & 0 & 0 & 0 & 0 \\ 1 & -2 & 1 & 0 & 0 & 0 & 0 & 0 \\ 0 & 1 & -2 & 1 & 0 & 0 & 0 & 0 \\ 0 & 0 & 1 & -2 & 1 & 0 & 0 & 0 \\ 0 & 0 & 0 & 1 & -2 & 1 & 0 & 0 \\ 0 & 0 & 0 & 0 & 1 & -2 & 1 & 0 \\ 0 & 0 & 0 & 0 & 0 & 1 & -2 & 1 \\ 0 & 0 & 0 & 0 & 0 & 0 & 1 & -2 \end{bmatrix} \begin{Bmatrix} v_1 \\ v_2 \\ v_3 \\ v_4 \\ v_5 \\ v_6 \\ v_7 \\ v_8 \end{Bmatrix} = (h^2) \begin{Bmatrix} M_1/E_A I_A \\ M_2/E_A I_A \\ M_3/E_A I_A \\ M_4/E_B I_B \\ M_5/E_B I_B \\ M_6/E_B I_B \\ M_7/E_B I_B \\ M_8/E_B I_B \end{Bmatrix}$$

or in condensed form:

$$[K]\{v\} = (h^2)\{M/EI\} \qquad \text{(Eq 12)}$$

where $[K]$ is the $(8 \cdot 8)$ coefficient matrix. The moment, M, must be found at each mesh point before the problem can be solved. Equation 13 shows the moment equations for the beam:

$$M = \frac{P}{3}(x); \quad \left(0 \text{ to } \frac{2L}{3}\right)$$

$$M = \frac{P}{3}(x) - P\left(x - \frac{2L}{3}\right); \left(\frac{2L}{3} \text{ to } L\right) \qquad \text{(Eq 13)}$$

Solving for the moment at points 1 through 8, the terms $(M/EI)_{1\rightarrow 8}$ can be determined. After inverting the $(8 \cdot 8)$ coefficient matrix and multiplying by the terms $(M/EI)_{1\rightarrow 8}$, the deflections are found.

$$\begin{Bmatrix} v_1 \\ v_2 \\ v_3 \\ v_4 \\ v_5 \\ v_6 \\ v_7 \\ v_8 \end{Bmatrix} = \begin{matrix} -0.0249 \\ -0.0475 \\ -0.0655 \\ -0.0767 \\ -0.0798 \\ -0.0734 \\ -0.0554 \\ -0.0295 \end{matrix} \quad \begin{matrix} -0.0241 \\ -0.0460 \\ -0.0632 \\ -0.0739 \\ -0.0769 \\ -0.0691 \\ -0.0516 \\ -0.0272 \end{matrix}$$

| | Nine mesh point approximation (cm) | Exact analytical solution (cm) |

Note that the nine mesh point approximation is within about 5% of the exact solution. This relatively simple example shows how varying material properties and geometry are easily handled using the FD approach. The advantage of designing with composites is that the material properties can be varied throughout the part. This ability of the FD method to vary properties throughout the structure makes it particularly attractive for solving composite structures problems.

The finite-element method models a structure by subdividing it into "small" but discrete elements that are connected togther at the corner nodes. Three common element shapes used in representing structures are illustrated in Fig. 3, a two-noded rod, a three-noded plane stress triangle, and an eight-noded solid "brick" element. (Fig. 16 in the section "Examples" in this article shows an example of how eight-noded solid elements were used to model a bolted lug.) Once conveniently subdivided, the individual element stiffness matrices

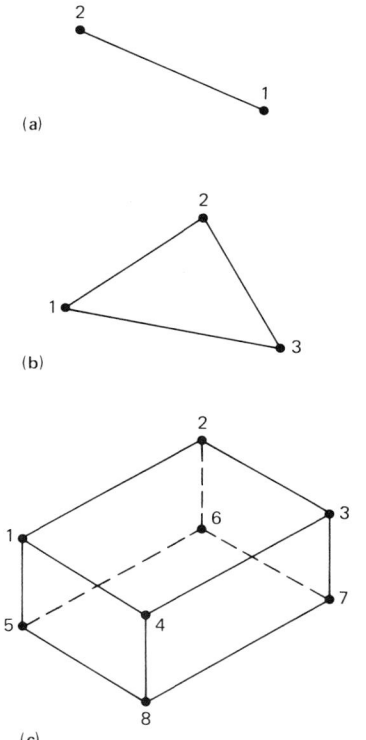

Fig. 3 Three simple-element types. (a) Two-noded rod. (b) Three-noded plane stress triangle. (c) Eight-noded solid "brick" element

are summed together to form the "global" structural matrix, which is then subjected to external loading to determine the total structural response. Thus, the foundation of the method is based upon the theory that the total structural response to loading is equivalent to the summed response of the individual elements.

The above discussion introduces the basic problem of the FE method, that is, determining element characteristics. An element must have the following four characteristics:

● The number of terms in the approximating polynomial displacement function must be at least equal to the degrees of freedom of the element
● The displacement function must provide for compatibility at the element interfaces
● The element must allow for constant strain and rigid body modes
● The displacement function must be differentiable

Development of various elements consumes a large part of the activity in the field of finite elements. Once the basic element stiffness matrix has been developed, a majority of the work is completed. The remainder of the solution becomes very repetitive and automatic. Researchers have developed a plethora of elements for structural mechanics, heat transfer, and fluid flow problems. Thus, the designer/analyst is not faced with developing element

formulations for each problem. However, it is essential that the engineer understand the method, how it works, and what approximations have been made in the development of a specific element. These details are usually included in the description of elements used in specific commercial codes. This overview will make no attempt at deriving detailed element formulations, but rather will outline the general procedures used in formulating an element stiffness matrix.

The element stiffness matrix relates the nodal forces to the resulting nodal displacements. The element stiffness matrix for an elasticity problem may be found using one of four techniques: direct method using statics (used in original derivation of triangular element), (Ref 2); virtual work; variational method; and weighted residuals. The weighted residuals method is very general and finds applications in the areas of thermal and fluid flow FE derivations. Elasticity problems have usually employed the variational method to compute the element stiffness matrix of which the principle of virtual work (the second method) is a specific minimization result. The virtual work principle provides a relatively straightforward approach to finite elements in solid mechanics. It will be used here to illustrate the way an element stiffness matrix may be derived.

The virtual work principle is stated in Eq 14. The equation states that the increment in internal strain energy, δU, must equal the work done by the surface force increment, δW_s, and the body force increment δW_b (Ref 7).

$$\delta W = \delta U + \delta W_s + \delta W_B = 0 \qquad \text{(Eq 14)}$$

The development of the element stiffness requires the evaluation of the three terms δU, δW_s, and δW_b. The FE approach is based on assuming a permissible element displacement function (usually a polynomial) that satisfies three conditions: the degrees of freedom associated with the element, compatibility, and differentiability. A simple derivation for the element stiffness equation using Eq 14 is given in Appendix 2. The approach in Appendix 2 is summarized in the following eight steps:

1. Select the element type that will be used to model the structure (rod, beam, plate, axisymmetric shell, and so forth).
2. Assume the functional form of the displacement field within the element (usually a polynomial where $[1 \times y \ldots]$ are polynomial terms and $\{a\}$ is a coefficient:

$$u = [1 \times y\ x^2\ xy \ldots] \begin{Bmatrix} a_0 \\ a_1 \\ a_2 \\ \cdot \\ \cdot \\ \cdot \\ a_i \end{Bmatrix} \qquad \text{(Eq 15)}$$

3. Solve for the unknown coefficients $\{a\}$ by defining the polynomial displacement equation (Eq 15) at the element nodes.

$$\{r\} = [G]\{a\} \qquad \text{(Eq 16)}$$

where $\{r\}$ defines the nodal displacements and $[G]$ defines the nodal x, y geometry: $[1\ x_i\ y_i\ x_i^2 \ldots]$.

4. Express the displacement field within the element in terms of the displacements defined at the nodes by combining Eq 15 and 16. This involves solving for $\{a\}$ in Eq 16 and substituting into Eq 15. This gives:

$$\{u\} = [N^{\text{T}}]\{r\} \qquad \text{(Eq 17)}$$

where

$$[N^{\text{T}}] = \{1 \times y\ xy \ldots\}[G]^{-1}$$

5. Introduce the strain displacement equations, thus determining the state of strain in the element corresponding to the assumed displacement field. For example, for two-dimensional displacements (small strain), strain is defined as:

$$\{\epsilon\} = \begin{Bmatrix} \delta/\delta x & 0 \\ 0 & \delta/\delta y \\ \delta/\delta y & \delta/\delta x \end{Bmatrix} \begin{Bmatrix} u\,(x,y) \\ v\,(x,y) \end{Bmatrix}$$

A substitution in the above equation for the displacements from Eq 17 and differentiating yields the B matrix:

$$\{\epsilon\} = [B]\{r\} \qquad \text{(Eq 18)}$$

See Appendix 2 for the exact definition of $[B]$.

6. Introduce the material properties of the element by writing the constitutive equations relating stress to strain. This introduces the property matrix $[D]$.

$$\{\sigma\} = [D]\{\epsilon - \epsilon_{\text{o}}\} \qquad \text{(Eq 19)}$$

7. Apply the calculus of variation principle to the element. For problems in solid mechanics, virtual work is often used:

$$\delta W = \delta U - \delta W_{\text{s}} - \delta W_{\text{B}} = 0$$

where the virtual work of the body forces δW_{B} and surface tractions δW_{s} must equal the element strain energy δU.

7a. Introduce the strain energy equation for the element to evaluate δU.

$$U = \int_v \{\epsilon\}^{\text{T}}\{\sigma\}\ dv$$

A substitution from Eq 19 and Eq 18 into U and use of the first variation δU yields the increment in internal strain energy:

$$\delta U = \delta\{r\}^{\text{T}}[K]_{\text{e}}\{r\}_{\text{e}} - \delta\{r\}^{\text{T}}[R]_{\text{o}}$$

where the element stiffness matrix is:

$$[K]_{\text{e}} = \int_v [B]^{\text{T}}[D][B]dv \qquad \text{(Eq 20)}$$

and the self-strain matrix is:

$$[R]_{\text{o}} = \int_v [B]^{\text{T}}[D]\{\epsilon_{\text{o}}\}dv$$

7b. Evaluate the virtual work term, δW_{s}, for surface forces. This may be expressed as:

$$\delta W_{\text{s}} = \delta\{r\}^{\text{T}}R_{\text{s}} \qquad \text{(Eq 21)}$$

where R_{s} is the surface force vector and $\delta\{r\}^{\text{T}}$ are the virtual nodal displacements.

7c. Evaluate the virtual work term, δW_{B}, for body forces. This may be expressed as:

$$\delta W_{\text{B}} = \delta\{r\}^{\text{T}}R_{\text{B}} \qquad \text{(Eq 22)}$$

where R_{B} is the body force vector and $\delta\{r\}^{\text{T}}$ is defined above.

8. Obtain the element stiffness equation by combining the expressions for δU, δW_{s}, and δW_{B} into Eq 14:

$$[K]_{\text{e}}\{r\}_{\text{e}} = R_{\text{o}} + R_{\text{s}} + R_{\text{B}} \qquad \text{(Eq 23)}$$

Equations 15 to 23 may be used in a simple example for the formulation of the stiffness equation for the axial stiffness (rod) element.

Let a rod element be represented below with nodal forces Fx_1 and Fx_2 acting at nodes 1 and 2, respectively.

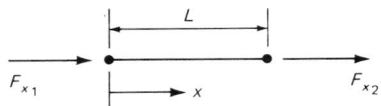

Assume a linear displacement function for the element; that is, assume displacements vary linearly within the element.

$$u(x) = [1\ x]\begin{Bmatrix} a_{\text{o}} \\ a, \end{Bmatrix}$$

Terms of the nodal displacements may be written in, as in Eq 17:

$$u(x) = \left(1 - \frac{x}{L}\ \frac{x}{L}\right)\begin{Bmatrix} u_1 \\ u_2 \end{Bmatrix} = [N]\{r\} \qquad \text{(Eq 24)}$$

Note that strain is defined for the element as $\epsilon = du/dx$. Substituting Eq 24 into the strain definition and differentiating gives:

$$\epsilon = \left[-\frac{1}{L}\ \frac{1}{L}\right]\begin{Bmatrix} u_1 \\ u_2 \end{Bmatrix} = [B]\{r\} \qquad \text{(Eq 25)}$$

The constitutive equation for an elastic axial stiffness rod involves only the elastic axial modulus and is expressed as:

$$\sigma = D(\epsilon - \epsilon_{\text{o}}) = E(\epsilon - \epsilon_{\text{o}})$$

From Eq 20, the element stiffness matrix is:

$$K_{\text{e}} = \int_v [B]^{\text{T}}[D][B]dv$$

Thus

$$K_{\text{e}} = \int_v \begin{Bmatrix} -\dfrac{1}{L} \\ \dfrac{1}{L} \end{Bmatrix} E \begin{Bmatrix} -\dfrac{1}{L}\ \dfrac{1}{L} \end{Bmatrix} dv = \frac{E}{L^2}\begin{bmatrix} 1 & -1 \\ -1 & 1 \end{bmatrix}\int_v dv$$

But $\int_v dv$ for the rod is simply AL (area times length). The element stiffness matrix becomes:

$$K_{\text{e}} = \frac{AE}{L}\begin{bmatrix} 1 & -1 \\ -1 & 1 \end{bmatrix}$$

Now assume there is no self-strain term, R_{o}, that is, no thermal or moisture strain effects. Also assume there are no body forces acting per unit volume, that is, $F_{\text{B}} = 0$. Thus, only the surface forces exist at nodes 1 and 2, Fx_1 and Fx_2. This then gives:

$$R_{\text{s}} = \begin{Bmatrix} Fx_1 \\ Fx_2 \end{Bmatrix}$$

and the element stiffness equation for the rod becomes:

$$\frac{AE}{L}\begin{bmatrix} 1 & -1 \\ -1 & 1 \end{bmatrix}\begin{Bmatrix} u_1 \\ u_2 \end{Bmatrix} = \begin{Bmatrix} Fx_1 \\ Fx_2 \end{Bmatrix} \qquad \text{(Eq 26)}$$

(assuming no self-strain or body forces). Note that distributed surface loads and point loads could have been included within the element but were not considered here for simplicity. This agrees with the typical PL/AE displacement solution for an axially loaded bar.

Solution Approach

The node or element characteristics have been defined for the FD and FE approaches. The individual node or element equations must now be assembled into a global system of equations describing the response of the entire structure. This assemblage results in a linear set of algebraic equations. The basic form of the equation for finite elements becomes:

$$[K]\{u\} = \{R\} \qquad \text{(Eq 27)}$$

where $[K]$ is an $(n \cdot n)$ stiffness matrix. The matrix $\{u\}$ is an $(n \cdot 1)$ column matrix, usually of the unknown deflections (for structural problems), while the $n \cdot 1$ column matrix $\{R\}$ contains the known applied loads. The FD equations take a similar $(n \cdot n)$ form as illustrated in the section "Finite-Difference Method" in this article. However, the $[K]$ matrix for the problem may represent stress function coefficients, deflection matrix coefficients, and so forth (see equations in Table 1).

Solution of the linear set of equations can be accomplished using a number of techniques, which include:

- Gaussian reduction
- Choleski decomposition
- Gauss-Seidel iteration

A discussion of these solution methods may be found in Ref 7. The first two methods are called direct methods. By direct application, they yield the exact solution of the linear system of equations within the error of computer round-off. The third method is called an iterative technique, whereby a ''suitable'' initial guess of the unknown vector $\{u\}$ is made. A judicious choice of an iterative algorithm produces subsequent vectors $\{u^i\}$, which, it is hoped, converge rapidly to the solution. The Gaussian elimination or Choleski direct methods are preferred over the iterative technique.

The Gaussian elimination method reduces the $[K]$ matrix to an upper triangular matrix of the form:

$$[K] = \begin{bmatrix} 0 \end{bmatrix} \qquad \text{(Eq 28)}$$

where the last equation has only one unknown and, thus, back substitution can be used for solution. The Choleski method is similar. If, in the case of most FE and FD problems, $[K]$ is positive, definite, and symmetric, then:

$$[K] = L\,L^{\mathrm{T}} \qquad \text{(Eq 29)}$$

where L is a lower triangular matrix, including the diagonal, and L^{T} is its transpose. Then Eq 27 can be written:

$$[L]\,[L^{\mathrm{T}}]\,\{u\} = \{f\}$$

let

$$[d] = [L]^{\mathrm{T}}\,\{u\}$$

then

$$[L]\,[d] = \{f\}$$

The values for $[d]$ are solved by using forward substitution. Then, knowing $[d]$ and $[L]^{\mathrm{T}}$, the values of $\{u\}$ are solved using back substitution.

The general solution approaches used to obtain the set of linear equations are discussed below for both the FD and FE methods.

Finite-Difference Solution. The solution of a beam using two different materials and two geometries using the FD approach was presented in the section ''Finite-Difference Method.'' This simple problem illustrated the basic elements used in the general solution approach to the FD method. The basic nine steps in the approach are listed below.

1. Establish the governing differential equation (for beams, rods, plates, and shells).
2. Establish the finite-difference representation of the governing differential equation (Table 1, Ref 4, 5).
3. Discretize the geometry into a mesh (usually with equal spacing and oriented in the global coordinate system).
4. ''Write out'' the finite-difference expression for each mesh point.
5. Define the local mesh point material properties and rotate into the global mesh system if required.
6. Assemble the finite-difference mesh equations into an algebraic set of linear equations: Form the $(n \cdot n)$ coefficient matrix; form the $(n \cdot 1)$ column matrix of unknown

values; form the $(n \cdot 1)$ column matrix of known values.
7. Apply the boundary conditions to the linear set of equations.
8. Solve the linear system of equations using methods previously outlined.
9. Postprocess the FD results (usually in the case of composites; this amounts to obtaining local ply stresses and strains).

Finite-Element Solution. The FE approach does not necessarily require the selection of the type of equation that will be used to model the structures. The commercially available codes have previously defined elements for which the individual element stiffness matrix has been solved. Thus, the user need only define the element type. If a specialized FE program is being written, the element stiffness matrix must be defined using methods outlined in the section ''Finite-Element Method'' in this article. Once the element is defined, the solution procedure follows steps similar to the FD approach:

1. Decide on required geometry to adequately model the problem: one-dimensional (rods, beams), two-dimensional (plane stress or strain), two-dimensional axisymmetric (shells and solids of revolution), or three-dimensional (shells and solids).
2. Select the element type and formulate the general element stiffness matrix if required (see the section ''Finite-Element Method''): Define the $[B]$ matrix relating element strains to nodal displacements; define the constitutive law matrix $[D]$; apply a numerical quadrature routine to evaluate the element stiffness integral, $\int_v [B]^{\mathrm{T}}\,[D]\,[B]\,dv$, over the element volume.
3. Discretize the geometry into a mesh: Define element node locations; refine mesh in areas of stress concentrations.
4. Define the local element properties: anisotropic, orthotropic, or isotropic
5. Calculate the local element stiffness matrices using steps 2, 3, and 4.
6. Transform the local element stiffness matrices into the global coordinate system.
7. Obtain the assemblage stiffness matrix by superimposing the global element stiffness matrices.
8. Define the applied boundary conditions to the system and formulate the nodal force vector, $\{f\}$; define slides, gaps, nodal fixes, displacements, loads, temperature, and moisture
9. Solve for the resulting linear set of algebraic equations using methods previously outlined: Obtain nodal deflections and solve for global system strains.
10. Solve for local element strains and stresses: Compute lamina strains and stresses and compare to failure criteria.

Steps 2, 5, 6, 7, and 9 are usually done internally to the selected FE code. Depending upon the code, step 10 may also be included. For composite analysis, the local strains and

stresses must be recovered for comparison with the selected lamina or laminate failure criteria. Typically, a lamina (ply) failure criteria is used, thus requiring the recovery of the local lamina strains and stresses. (Refer to the article ''Static Strength'' in this Section of the Volume for failure criteria.)

Method Selection

The analyst must decide which numerical analysis technique best suits the solution of the particular problem. Finite-difference techniques prove difficult when nonuniform meshes are required to define an irregular geometry or when a three-dimensional continuum must be represented. Furthermore, FD solutions are more easily solved when Dirichlet boundary conditions (BC), rather than Cauchy boundary conditions, are defined. Let the boundary be defined as S. The Dirichlet problem specifies the solution on the boundary.

$$\text{Dirichlet BC: } \phi = f(x,y,z) \text{ on } S$$

In the Cauchy problem, the outward normal derivatives are specified on the boundary.

$$\text{Cauchy BC: } \frac{\delta\phi}{\delta n} = g(x,y,z) \text{ on } S$$

Thus, the FE method lends itself better to the solution of problems having irregular geometry, nonuniform meshes, and Cauchy-type boundary conditions.

Finite-difference solutions have been used very successfully for beam, column, plate, and two-dimensional axisymmetric shell structural analyses. The ease with which geometry (thickness) and properties can vary using FD formulations for the above-mentioned structures makes the method particularly well suited for composite analysis. When irregular geometries or three-dimensional continuums are encountered, the FD method becomes cumbersome. The advantage of the FD approach is that once it has been set up, the solution of the problem is usually considerably faster than the corresponding FE analysis. This is generally true because a more sparsely populated matrix must be solved with the FD approach. If the analyst has a frequently used specific structure for which geometry and material property changes can be easily preprogrammed, the FD method would be a viable analysis approach.

The FE method can more easily accommodate irregular geometries and three-dimensional situations. This is because the basic geometry of the element can be ''warped'' into the desired shape to accommodate cutouts, thickness buildups, joints, irregular boundaries, and so on. The element geometry is an integral part of the stiffness formulation for each element, making changes in element size and shape trivial. In addition, the user has one-, two-, and three-dimensional elements from which to select. This provides the user with the capability to use different elements within the same analysis to simulate different substructures. Note that such mixing of different element types requires compatibility between the different elements. Some codes, such as ADINA, have been written with special compatible elements to overcome any difficulties.

Typical Problems and Corrective Actions

As with most analysis methods, whether they be thermal, structural, or flow, typical problems are encountered. The numerical analysis of composite structures is no different. The following sections list some of the more common problems that may occur. Some are unique to composites, while others are representative of problems in structural analysis in general. Ways to eliminate these problems are suggested. It is assumed that a problem has been adequately defined to the analyst and, hence, the discussion does not go into the problematics of program and problem definition. The following information can serve as a checklist during code selection, model representation, code execution, and results interpretation.

Problem 1: Poor code documentation or poor support from leasing/selling agency.

Suggested Corrective Action. Study the code documentation carefully before buying, leasing, or using an analysis code. Run test problems for which answers are known. Many document and support group shortcomings can be found using the test case approach. Software companies often allow running test cases free of charge.

Problem 2: A code has a limited element set.

Suggested Corrective Action. It is essential that the analysis code have at least both shell and solid three-dimensional orthotropic elements. In addition, two-dimensional solid, two-dimensional solid axisymmetric, and two-dimensional shell orthotropic elements are useful. The two-dimensional elements usually allow for less-costly analyses because fewer degrees of freedom are required for problem solution.

Problem 3: Inadequate definition of constitutive properties (stress/strain behavior).

Suggested Corrective Action. A reasonable set of constitutive properties is necessary for an accurate solution. The "garbage in, garbage out" principle applies here. The majority of composite codes (as of this printing) can handle only linear elastic properties. Thus, it is necessary to obtain the elastic moduli.

Moduli may be obtained in one of three ways: from direct laminate testing, from lamina testing in conjunction with a laminating constitutive theory (refer to the article "Analysis of Structures" in this Section of the Volume), and from the use of micromechanics to predict lamina properties in conjunction with a laminating constitutive theory to predict laminate behavior (refer to the article "Analysis of Material Properties" in this Volume for micromechanics analysis approach).

Sometimes each lamina layer is modeled as an element so that a laminating constitutive theory is not required. In such cases, only lamina data are needed. Note also that micromechanics is good for initial design or preliminary investigations, but should never be used to replace "real data." Test data should always be used for final design purposes.

The constitutive equation for orthotropic plate and shell analysis has been derived in the articles "Properties Analysis of Laminates" and "Analysis of Structures," the latter of which defines the A, B, and D matrix terms. The theory used for the derivation is commonly called classical lamination theory (CLT). The constitutive equation from CLT is restated as:

$$\left\{\frac{N}{M}\right\} = \left[\frac{A}{B}\Big|\frac{B}{D}\right]\left\{\frac{\epsilon_o}{\kappa}\right\} \qquad \text{(Eq 30)}$$

The terms N and M are the plate or shell line loads and moments, with ϵ_o and κ being the midplane strains and curvatures, respectively. The principal conclusion from the derivation is that eight basic material properties are required, along with the lay-up stacking sequence, to define the A, B, and D terms. The eight required properties are described in Table 2. Finite-element programs that use orthotropic plate or shell elements require either the ABD matrix for each element or the lay-up sequence and lamina properties for each element. The user must be careful to note the orientation of the local element coordinate system so that the lamina are input with the proper lay-up angles.

The constitutive equation for solid orthotropic elements is stated in the article "Analysis of Material Properties" in this Volume. The basic equation, in compliance matrix form, is written as:

$$\begin{Bmatrix} \epsilon_a \\ \epsilon_b \\ \epsilon_c \\ \gamma_{bc} \\ \gamma_{ac} \\ \gamma_{ab} \end{Bmatrix} = \begin{bmatrix} \frac{1}{E_{aa}} & \frac{-\nu_{ba}}{E_{bb}} & \frac{-\nu_{ca}}{E_{cc}} & 0 & 0 & 0 \\ \frac{-\nu_{ab}}{E_{aa}} & \frac{1}{E_{bb}} & \frac{-\nu_{ab}}{E_{cc}} & 0 & 0 & 0 \\ \frac{-\nu_{ac}}{E_{aa}} & \frac{-\nu_{bc}}{E_{bb}} & \frac{1}{E_{cc}} & 0 & 0 & 0 \\ 0 & 0 & 0 & \frac{1}{G_{bc}} & 0 & 0 \\ 0 & 0 & 0 & 0 & \frac{1}{G_{ac}} & 0 \\ 0 & 0 & 0 & 0 & 0 & \frac{1}{G_{ab}} \end{bmatrix} \begin{Bmatrix} \sigma_a \\ \sigma_b \\ \sigma_c \\ \tau_{bc} \\ \tau_{ac} \\ \tau_{ab} \end{Bmatrix}$$

$$\text{(Eq 31)}$$

Note that the matrix is symmetric and that there are only nine independent constants, which are as follows: elastic moduli Eaa, Ebb, and Ecc in the three orthogonal directions (tension or compression, depending on the problem); the Poisson's ratios, ν_{ij}, for transverse strain in the j directions caused by a stress in the i direction; and the shear moduli G_{ab}, G_{bc}, and G_{ac} in the a-b, b-c, and a-c planes, respectively.

Table 2 Elastic moduli and lamina properties required for orthotropic plate and shell analysis

Property	Units	Description
E_{11}............	GPa (10^6 psi)	Young's modulus in fiber direction for tension and compression
E_{22}............	GPa (10^6 psi)	Young's modulus transverse to fiber direction for tension and compression
G_{12}............	GPa (10^6 psi)	Shear in 1,2 plane
ν_{12}............	...	Poisson's ratio
α_{11}............	10^{-6}/K	CTE in fiber direction
α_{22}.............	10^{-6}/K	CTE transverse to fiber
β_{11}............	Strain/% RH	Coefficient of moisture expansion in fiber direction
β_{22}............	Strain/% RH	Coefficient of moisture expansion transverse to fiber

Equation 30 represents the individual lamina moduli. In contrast, Eq 31 represents the element orthotropic moduli located in the local element coordinate system (a, b, c) for the complete stacking sequence contained within the element.

Problem 4: Inadequate allowables and/or wrong failure criteria.

Suggested Corrective Action. Two-dimensional allowables should include axial and transverse tension and compression, in-plane and interlaminar shear, and often flatwise tension and compression. Three-dimensional laminates will require the above tests plus testing the third dimension in tension, compression, and shear. Stress or strain may be used for failure predictions, depending on selected failure criteria. Usually, a statistical A- or B-basis is used for allowables generation.

Problem 5. Selection of the wrong analysis or elements for problem solution.

Suggested Corrective Action. Unfortunately, there is no "cookbook" method to ensure use of the correct element or correct analysis. The element must have sufficient degrees of freedom to accurately predict structural behavior. The analysis type must include the desired loads and environmental conditions. Trouble begins when the analyst starts to simplify from a three-dimensional solid representation to simpler element forms: three-dimensional shell, two-dimensional axisymmetric, two-dimensional plane stress, and so forth. The simpler elements (and resulting simpler boundary conditions) must be carefully analyzed to ensure the structure is being correctly modeled. Experienced users should be consulted and asked to review the approach before the problem is submitted for analysis.

Problem 6: Incorrect boundary conditions (BCs).

Suggested Corrective Action. As mentioned in the problem above, improper BCs cause a large number of analysis errors. In fact, they are probably the single largest source of error in analysis. Incorrect BCs can cause structures to be too stiff or too flexible, or can cause stress risers where there are none or cause them to be underpredicted, and so on. The structure must be critically examined to determine reasonable BCs. Sometimes simple inspection is sufficient to establish reasonable BCs, but often simpler approximate models of the structure are analyzed first to determine whether the BCs are correct. Lastly, there is no substitute for experience; involve an experienced finite-element or finite-difference user in establishing the correct boundary conditions for the model.

Problem 7: Examination of the wrong lamina for failure (wrong location).

Suggested Corrective Action. Utilize postprocessing codes that have layer-by-layer *xy* plots and isoline plots for detailed stress and strain results. These plots can tell the analyst at a glance the location and magnitude of the peak values. It is also useful to have a code that summarizes the maximum and minimum values for each section of the structure.

Problem 8: Differential (or geometric) stiffness not used when large deflection occurred.

Suggested Corrective Action. Two-dimensional laminated composite structures are relatively flexible out of the plane of the laminate. For example, when a two-dimensional laminate is loaded in flexure, as in a leaf spring, large

deflections result. Such large deflection problems must be analyzed using geometric nonlinearity principles in which the deformed configuration must be used to write the equilibrium equations. This is often called the LaGrangian solution approach. Many codes have the capability to analyze large deflection problems, but the solution of such problems is more costly. The LaGrangian solution sequence must usually be requested during the analysis type definition. A single-step elastic analysis will usually be conducted first to determine whether deflections are large enough to require the LaGrangian solution technique.

Problem 9: Material properties not correctly input and/or not transformed into the global coordinate system.

Suggested Corrective Action. Care must be used in the input of material properties for the lamina and/or laminate. Mistakes include transposition of the moduli (E_{xx} for E_{yy}) or Poisson's ratio (ν_{xy} for ν_{yx}, and so on). Usually the properties are entered in the local element coordinate system. These properties must be transformed into the global system of stiffnesses prior to solution. Be sure the properties are correctly input into the local system and that the transformation angles between the local and the global system for each element are correctly determined. A preprocessor code should be used for the material transformations, making manual input unnecessary.

Problem 10: Results not checked by closed-form analysis where possible.

Suggested Corrective Action. Check all results by whatever closed-form analysis tech-

niques are applicable. Just because the solution converged, the user should not assume that the correct answer has been obtained. Be critical of the result and check the answer for reasonableness using strength of materials approximations, elasticity results, simple statics or dynamics results, and so forth. Prove that the results make sense. Sometimes an analysis of a similar but simpler structure is required to ensure that a complex problem is converging to a reasonable solution.

Representative Commercial Codes

Table 3 lists seven representative commercial codes available for composite structures analysis. The common feature of the codes is their ability to analyze orthotropic materials. These codes by no means represent a complete listing of all available composite analysis codes. Nevertheless, the table provides a starting point from which initial analyses investigations may begin. In addition, many of the codes in the table are recognized as industry standards, for example, MSC-NASTRAN, BOSOR, and PATRAN (solids modeler). The table lists eight basic code attributes (in the left-hand column) by which each code has been rated.

The interactive capability attribute is a qualitative measure of how easily pre- and postanalysis processing is done. The analysis types and the element types attributes are an abbreviated listing of capabilities particularly useful for composite structures analysis. Most of the listed codes have many more analysis and

Table 3   Numerical codes available for composite structures analysis

Attribute	ABAQUS F.E.	ADINA F.E.	ANSYS F.E.	BOSOR F.D.	MSC NASTRAN F.E.	STAGS F.E.	PATRAN-II COMPOSITES D&A MODULES F.E.
Interactive capability (mesh generator, material input, FE input)	Poor	Good (ADINA-IN)	Fair	Good (interactive BOSOR version)	Fair (FEMGEN)	Poor	Excellent
Typically used mesh generator	Internal code (fair-to-poor)	PATRAN, ADINA-IN	PATRAN, Code internal	Code internal	PATRAN	Code internal	PATRAN
Local lamina lay-up input	Yes (shell only)	None	Yes (triangular shell)	Yes	Yes (shell only)	Yes	Yes
Local-to-global transformation	Yes	None	Yes	Yes	Yes	Yes	Yes
Global-to-local lamina stress or strain recovery......................	Yes (shell only)	None	Limited	Limited	Yes (shell only)	Yes, stresses	Yes
Analysis types							
Static.........................	Yes	Yes	Yes	Yes	Yes	Yes	Yes
Dynamic......................	Modal, transient	Modal, transient	Modal, transient	Modal	Modal, transient	Modal, transient	No
Collapse......................	Yes	Yes	Yes	No	Yes	Yes	No
Bifurcation	Yes	Yes	Yes	Yes	Yes	Yes	No
Geometry nonlinearity	Yes	Yes	Yes	Yes	Yes	Yes	No
Hygrothermal	Yes	No	Yes	Yes	Yes	Yes	Yes
Orthotropic element types							
Three-dimensional plates	No	No	Yes	No	Yes	Yes	No
Two-dimensional axisymmetric shell ..	Yes	Yes	Yes	Yes	No	No	No
Three-dimensional shell.............	Yes	Yes	Yes	No	Yes	Yes	No
Two-dimensional solid	Yes	Yes	Yes	No	Yes	Yes	No
Two-dimensional axisymmetric solid ..	Yes	Yes	Yes	No	Yes	No	Yes
Three-dimensional solid.............	Yes	Yes	Yes	No	Yes	No	Yes
Code support	Hibbitt, Carlson & Solenson, Inc.	ADINA Engineering	Swanson Analysis Systems Inc.	David Bushnell	MacNeal-Schwendler Corp.	COSMIC	PDA

element capabilities than those given in the table.

One interesting observation emerges from the table: There is no one code that satisfies all of the attributes. Some of the codes have good pre/postprocessing capability but limited analysis capability (PATRAN-II); others excel in analysis capability but have poor pre/postprocessing capability (ABAQUS). To avoid these limitations, many commercial users have developed their own pre/postprocessors or added element modifications to obtain a more complete code. Thus, it is important to be able to obtain codes that have the capability for modification.

Three avenues of modification exist: being able to obtain source code (for example, ADINA, BOSOR), being able to access a neutral data file, or having a good working relationship with the corporation supporting the code. Having the source code allows the user to add elements, modify routines, and so forth. Access to a neutral data file allows the user to enhance pre/postprocessing capabilities. Modifications may also occur through formal requests to the supplier of the code. Each of these three options, as well as the code attributes, must be critically assessed by the user before leasing or purchasing a code.

Examples

Six analysis examples are presented to illustrate the usefulness and power of numerical methods in the solution of problems in composite structures. The examples were selected because they represent problems for which the numerical methods are particularly useful. The FE examples address joints, three-dimensional shells, and "compact" solids. The FD examples address shells and beams of varying materials and geometries. The examples also illustrate specific problems for which closed-form predictions would be difficult or impossible using composite materials. This is specifically true for the pinned- and bonded-joint analyses.

Example 1: Bonded Single Lap Joint Analysis With Metallic Fitting and Graphite-Epoxy Composite Tube. The problem presented is to determine the critical stresses in the bond region of the joint that are due to the mechanical and hygrothermal (nonmechanical) loads.

Structural Schematic and Laminate Definition. Figure 4 illustrates the axisymmetric bonded joint under study. The metal fitting is 7075-T73 aluminum. The composite laminate is composed of the following lay-up inside diameter (ID) to outside diameter (OD):

	-45°	0°2	90°	0°2	+45°
1	HMS-4	P75	HMS-4	P-75	HMS-4
2	0.0076	0.0305	0.0076	0.0305	0.0076

1, material; 2, thickness (cm)

A fiber volume fraction of 62% was used.

An even-weave glass cloth was used between the metal fitting and the composite tube to

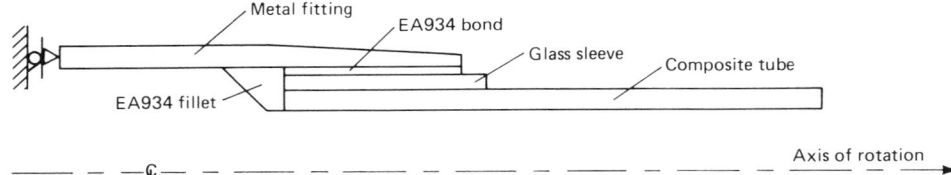

Fig. 4 Typical bond joint cross section for a tubular member

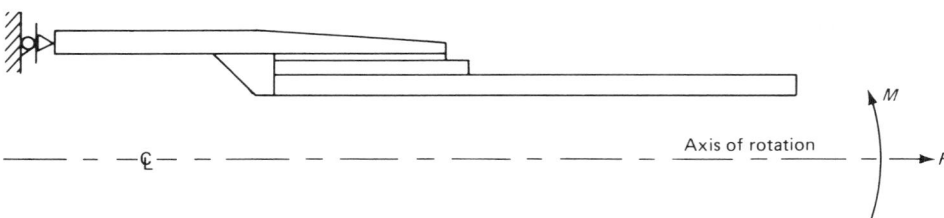

Fig. 5 Joint boundary conditions

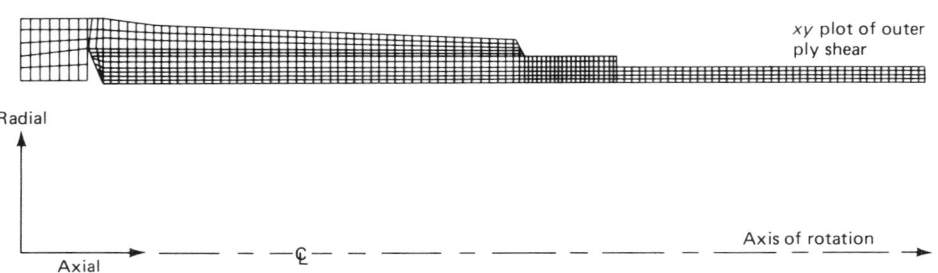

Fig. 6 Midfit joint finite element grid

reduce stress discontinuities and provide corrosion protection. The glass cloth was 0.50 mm (0.02 in.) thick. EA934 adhesive was used to bond the metal and composite together.

Selected Numerical Method. The FE method was used to analyze this problem. Isoparametric quadrilateral axisymmetric elements (QUAD 6) were used. This element type assumes linear strain in the element, thus providing for more accurate approximation of bending.

Simplifying Assumptions and Analytical Approach. The problem was modeled using axisymmetric elements in an asymmetric loading FE code. Loads were applied using terms of a Fourier series. Load terms were summed to produce the various load states on the joint. For example, the moment load was applied using the first term of the Fourier series, while the constant axial load, F, was applied using the 0th term of the series. Resulting strains and stresses were summed using the principle of superposition, because deflections were "small."

Each fiber reinforced plastic (FRP) composite layer of the structure was modeled as a discrete layer of finite elements so as to obtain

accurate interlaminar normal and shear stresses. Testing has shown these stresses to govern failure for many lap-joint designs.

Element (Mesh) Discretization and Boundary Conditions. Figures 5 and 6 show the boundary conditions and loads used to model the structure. The model was axially and circumferentially constrained at the end of the metallic fitting. Radial growth was unrestrained. Axial and moment mechanical loads were applied at the FRP composite end of the model. A steady-state change (ΔT) in thermal environment was also imposed on the joint. The discretized configuration is illustrated in Fig. 6.

Key Results. Figure 7 shows the interlaminar shear stress for the joint at the outermost composite ply interface. The peak stress occurs at the termination of the composite tube, as one might expect. The allowable stresses in the shear stress mode, stated in the figure, show that interlaminar shear drives the design. Using a maximum stress failure criterion, the margin of safety in the design under this load condition would be $MS = 0.08$. It should be noted that the shear stress was primarily due to the cold condition, not the applied mechanical loading.

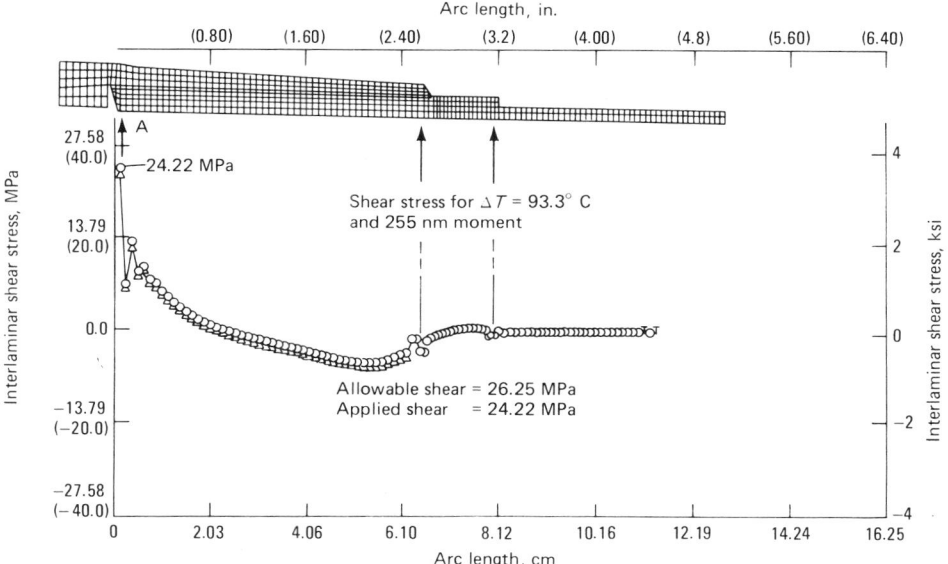

Fig. 7 Interlaminar shear

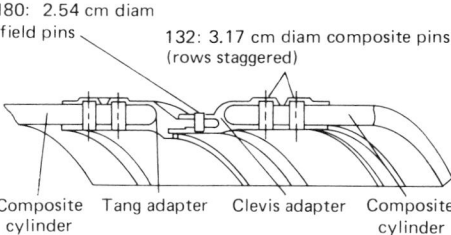

Fig. 8 Joint configuration

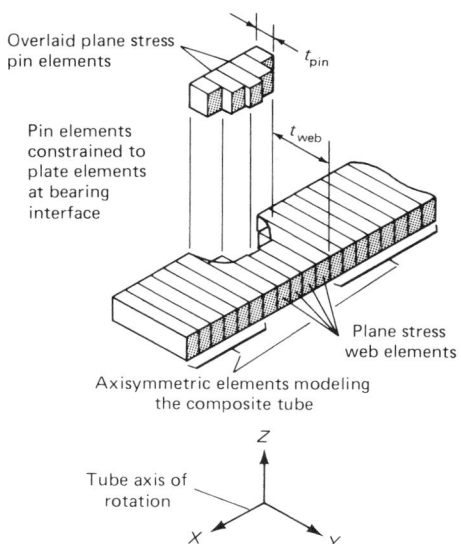

Fig. 9 Conceptual diagram of plane-stress overlays

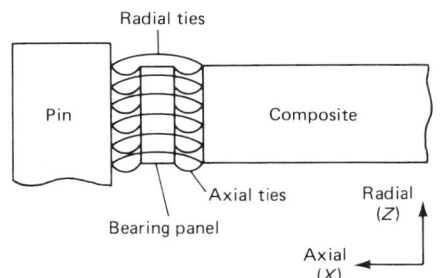

Fig. 10 Finite-element model, bearing element schematic

Comparison to Test Data and Conclusions. The joint described above had glass cloth bonded at room temperature after tube cure and used a scalloped aluminum fitting. Tube failure, not joint failure, was predicted and, in fact, actually occurred during tests.

However, earlier joint designs used a co-cured glass cloth and straight tapered (nonscalloped) fitting. Analyses of the earlier joint predicted outer tube ply shear failure due to thermal cool-down to occur at $\Delta T = -120$ °C (-190 °F). Acoustic events and inspection after cool-down indicated thermal cracking at $\Delta T \simeq -110$ °C (-170 °F). Sections cut from thermally cycled joints showed cracking initiating at the peak shear location point *A* of Fig. 7. Thus, model predictions showed good agreement with the experimental data. It should be noted that co-curing of the glass cloth added 6900 MPa (1 ksi) shear stress to the outer HMS ply, helping to precipitate the thermal failure.

Example 2: Clevis Joint Analysis With Metallic Fitting and Graphite-Epoxy Composite Tube. The problem is to determine the clevis joint strains and stiffness.

Structural Schematic and Laminate Definition. Figure 8 illustrates the joint with the D6AC steel adapter rings attached to a nonsymmetric laminate AS4 graphite-55A epoxy case. Two staggered rows of 31.8-mm (1.25-in.) diam Inconel 718 pins are also shown. The laminate lay-up in the joint is composed of 46% 0° material, 43% ±33.5° helical wraps, and 11% 90° hoop-wound material. The fiber volume fraction was 55%.

Selected Numerical Method. The FE method was used to analyze the joint. Isoparametric quadrilatrial axisymmetric and plane stress elements were used in the model. These elements

assumed a linear strain field that provided for a more accurate approximation of bending.

Simplifying Assumptions and Analytical Approach. A plane stress overlay technique was used to model the pins, while axisymmetric elements were used to model the composite tube and metal clevis joint. This modeling technique produces a quasi three-dimensional joint model (Ref 8). In practice, the pin and circumferential web section between pins are modeled with multiple plane stress finite elements of varying thickness. The element thicknesses are chosen to match *EI* and *EA* of the actual pin. Figure 9 shows the plane stress overlay elements as they relate to the composite failure. "Bearing" elements were also included in the model and are shown in Fig. 10. Bearing elements are a group of plane stress elements placed between the pin and bearing surface of the pin hole. A lower axial modulus is assigned to the bearing elements to replicate the localized lower compressive modulus in bearing (Ref 8).

Element Mesh Discretization and Boundary Conditions. Figure 11 illustrates the FE grid used for the study. The grid shows four pins being modeled using the plane stress overlay elements (two on each side of the clevis). Note also that the metal-to-metal tang and clevis pin was also modeled using plane stress elements. All other elements illustrated in Fig. 11 are axisymmetric.

The boundary conditions for the model are shown in Fig. 12. The boundary conditions included axisymmetric restraints (axial fix) on the left cylindrical member and an applied axial line load on the right member. Internal pressure was also applied. Note that slide and modified gap elements were used at all composite-metal interfaces.

Key Results. Key results are shown in Fig. 13 and 14. The dashed lines show the predicted circumferential strain and joint pin rotation, respectively, versus internal pressure. The circumferential strain was measured to ensure that the metal clevis was not overstraining. The pin rotation measurement was critical because it would establish whether the bearing elements were performing correctly, thereby producing the proper joint growth.

Comparison to Test Data and Conclusions. Actual test data are shown as solid lines in Fig. 13 and 14. The average circumferential strain

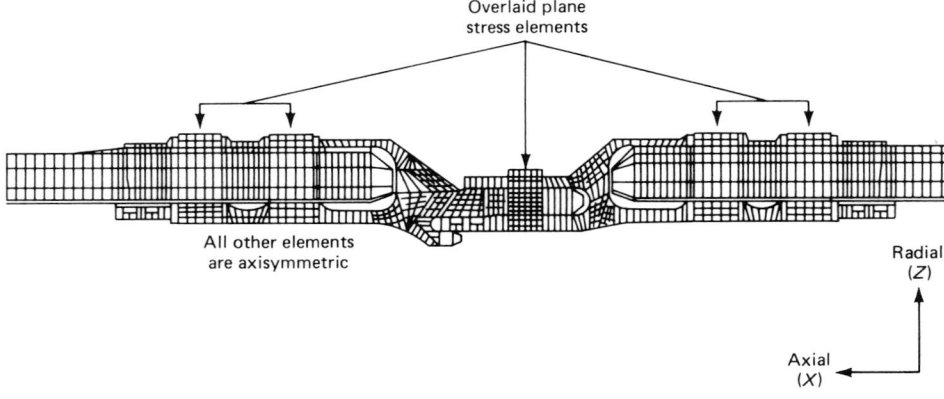

Fig. 11 Axisymmetric finite-element joint model

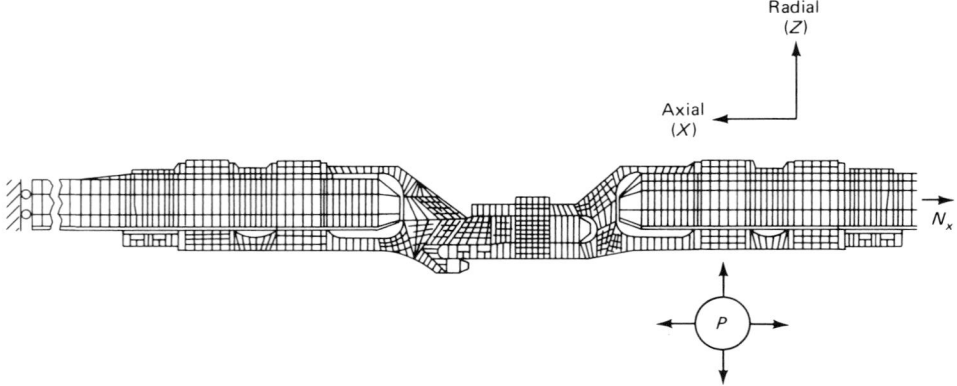

Fig. 12 Joint model boundary conditions

results are given in Fig. 13. The slope of the data to predictions agree to within 5%. The initial data nonlinearity is attributed to initial pin gap and permanent hole elongation. Fig. 14 shows the pin rotation data results. Again, the slope of the predicted curve is within 5% of the linear portion of the test data. The initial nonlinearity was caused by permanent hole elongation from previous pressure cycles. Thus, the FE model showed good correlation to the linear portion of the test data. Permanent hole elongation effects causing the initial nonlinearity could be included if the progressive gap increases were taken into account in the model.

Example 3: Three-Dimensional Pinned Lug Joint. The intent of the model is to find the stresses about a pinned lug for a vertically applied load condition.

Structural Schematic and Laminate Definition. The structural schematic in Fig. 15 shows a steel pivot pin passing through a three-dimensional graphite-epoxy composite clevis fitting. The clevis is restrained by graphite hoop overwraps passing over the plate. This fastens the clevis fitting to the cylindrical surface. The three-dimensional graphite-epoxy clevis and plate were machined from a single piece. The billet from which the clevis was machined was composed of fibers oriented in three orthogonal directions, x, y, and z. Fiber volume fractions were 18% along the y and z axes and 12% along the x axis. The material was Hercules fiber and resin IM6/55A. The total fiber volume fraction was 48%.

Selected Numerical Method. The FE method was chosen to analyze the three-dimensional graphite-epoxy clevis stresses. The selected codes included the PATRAN three-dimensional

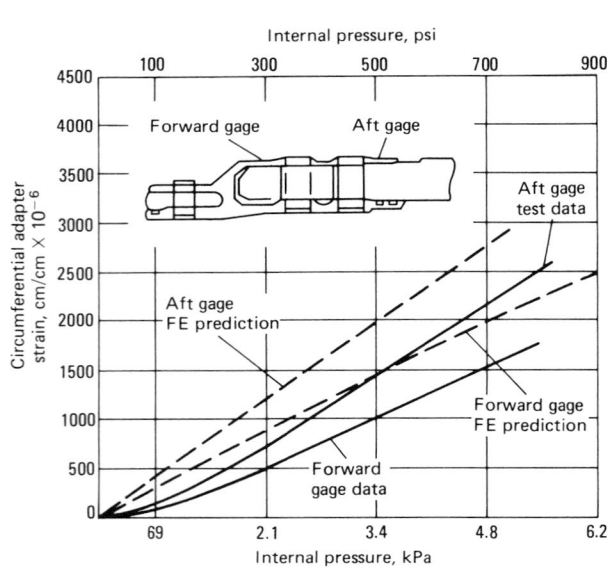

Fig. 13 Full-scale joint long wire gage results

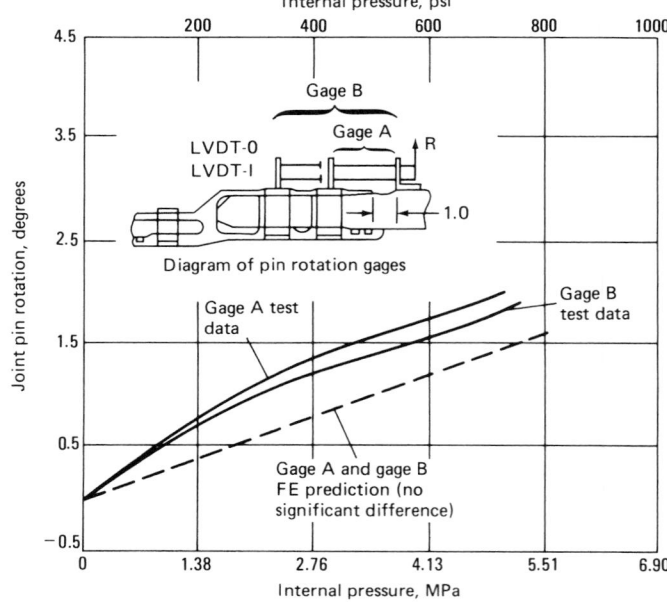

Fig. 14 Full-scale joint pin rotation gage results

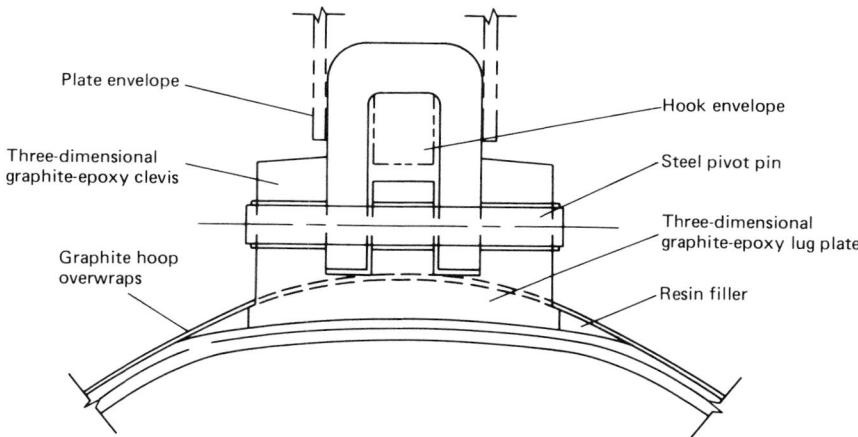

Fig. 15 Pinned lug joint schematic

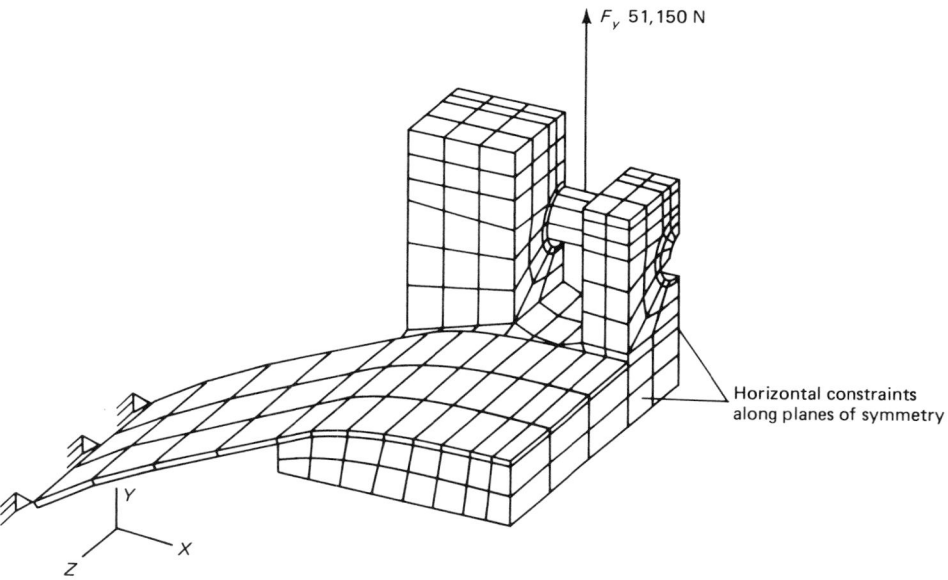

Fig. 16 3-D lug single quadrant grid and boundary conditions

solids model and the FE code ADINA. An ADINA translator was used to input PATRAN neutral file information into the ADINA input code format. An eight-noded hexahedral solid element was used in this study for the clevis/plate, steel pivot pin, and graphite-epoxy hoop overwraps.

Simplifying Assumptions and Analytical Approach. The three-dimensional analysis was performed for only one quadrant of the lug plate. Symmetry boundary conditions were assumed. Figure 16 illustrates the quadrant being modeled and shows the two surfaces for which symmetry was assumed. An approximation of the clevis pin was used. Only the upper half of the pin was modeled in order to achieve the proper surface loading in the clevis. This avoided the use of gap elements on the lower half of the pin. The pin was also given a small

hollow core to avoid problems with degenerate elements. The modulus of the pin was artificially increased to maintain the proper pin stiffness. These approximations were considered reasonable because it is the stresses in the three-dimensional clevis that are of interest, not the pin. The pin approximation was sufficient to accurately transmit the vertical load.

Element (Mesh) Discretization and Boundary Conditions. Figure 16 also illustrates the finite element grid and boundary conditions. The grid shows the three-dimensional graphite-epoxy tang and plate cone piece, pin, and graphite-epoxy hoop overwraps. A decoupled plate to cylinder condition was assumed. This was considered the worst case. One-quarter of the total vertical force was applied (5300 kg, or 11 600 lb) to the quarter model. The boundary at the

"left" side of the graphite-epoxy overwrap was assumed to be fixed.

Key results are shown in Fig. 17 and 18. Typical stress results are illustrated in Fig. 17 for the vertical, σy, stresses. This particular plot clearly shows the typical hole stress concentration, σy^{max}, for the tang. This peak stress was used for *MS* calculations in Fig. 18. Figure 18 shows both *MS* and deflections. Deflections are represented at a scale factor of 3. Margins are shown in the maximum y-axis stress. The allowable was 480 MPa (69.6 ksi). The calculated applied load with a 1.5 safety factor was 490 MPa (70.7 ksi). Thus, a negative margin of -0.015 was predicted.

Comparison to Test Data and Conclusions. No tests have been run on this design. However, the small negative margin would suggest that some redesign is required on the tang to further reduce the stress effects of the applied load. At least two redesign solutions might be considered. The percentage of y-axis fibers could be slightly increased, resulting in higher y-axis allowables. Also, the axial thickness (z-axis dimension) and pivot pin diameter could be increased to further distribute this vertical load and reduce the applied stress.

Example 4: Race Car Chassis Stiffness Determination (Three-Dimensional Shell Analysis). The problem is to find the torsional stiffness of the race car chassis about its axis of symmetry.

Structural Configuration and Laminate Definition. The chassis configuration and laminate lay-up are illustrated in Fig. 19 and 20. Figure 19 shows the stripped-down chassis, including engine, front suspension, roll bar, forward aluminum box, and composite chassis. The composite lay-up is illustrated in Fig. 19. This figure shows laminate types A, B, and C used in the bulkheads, as well as a generic outer-shell laminate. The basic chassis design used an aluminum core with HMS/3501-6 graphite-epoxy face sheets to form a honeycomb sandwich structure. Note that numerous local variations were made to the outer-shell laminate to accommodate local stress and deflection requirements. Thicknesses are not shown because they are still customer proprietary.

Selected Numerical Method. The FE method was again selected as the analysis tool. MSC-NASTRAN was used as the finite element code, and FEMGEN was used as the solids model to construct the grid. The NASTRAN shell element, QUAD 4, was used to model the composite portion of the structure.

Simplifying Assumptions and Analytical Approach. Half of the structure was modeled and asymmetric boundary conditions (234 fix) were applied to the plane of symmetry. This condition fixed the y translation, z translation, and rotation about the x axis.

Element (Mesh) Discretization and Boundary Conditions. Figure 20 shows the grid and boundary conditions used for the chassis analysis. The node at the junction of the four front suspension struts was completely fixed from all

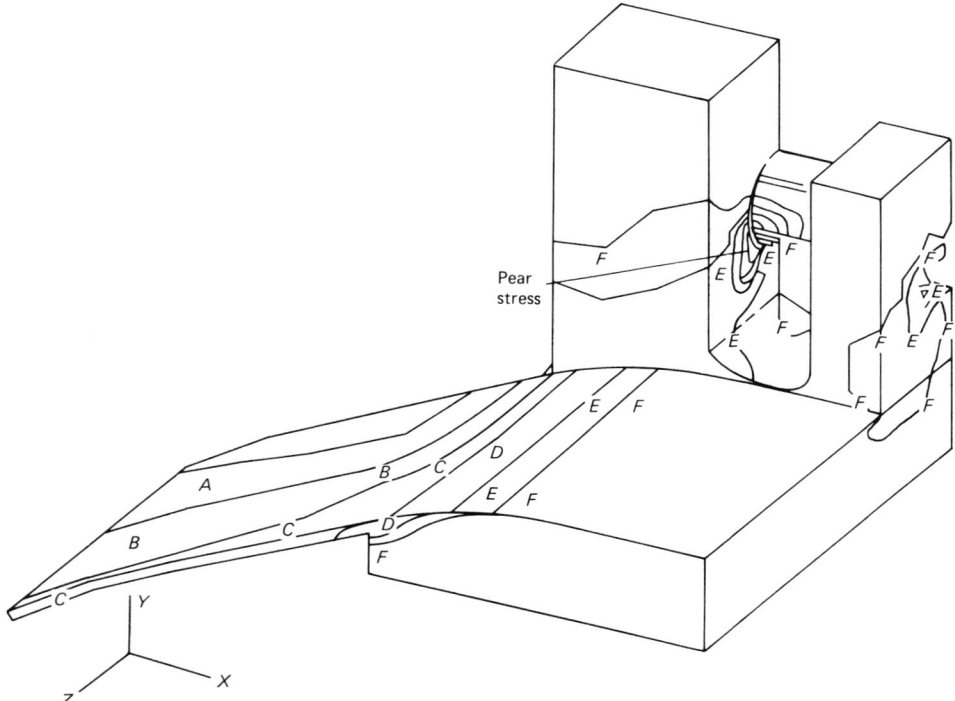

Fig. 17 σ_y isostress plot. A, 345 MPa (50 ksi). B, 275 MPa (40 ksi). C, 207 MPa (30 ksi). D, 138 MPa (20 ksi). E, 69 MPa (10 ksi). F, 35 MPa (5 ksi)

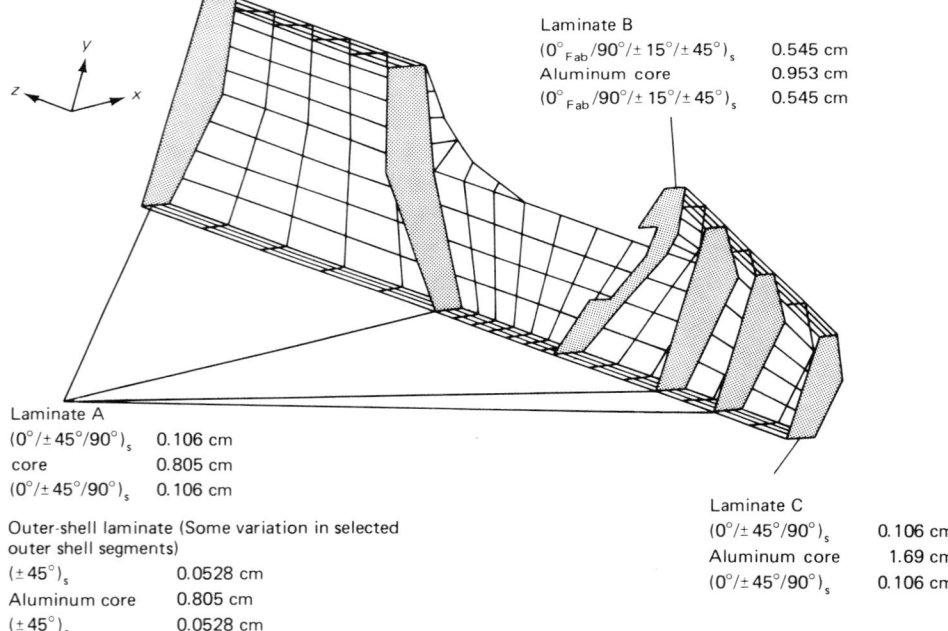

Laminate B
$(0^\circ_{Fab}/90^\circ/\pm 15^\circ/\pm 45^\circ)_s$ 0.545 cm
Aluminum core   0.953 cm
$(0^\circ_{Fab}/90^\circ/\pm 15^\circ/\pm 45^\circ)_s$ 0.545 cm

Laminate A
$(0^\circ/\pm 45^\circ/90^\circ)_s$ 0.106 cm
core   0.805 cm
$(0^\circ/\pm 45^\circ/90^\circ)_s$ 0.106 cm

Outer-shell laminate (Some variation in selected outer shell segments)
$(\pm 45^\circ)_s$ 0.0528 cm
Aluminum core   0.805 cm
$(\pm 45^\circ)_s$ 0.0528 cm

Laminate C
$(0^\circ/\pm 45^\circ/90^\circ)_s$ 0.106 cm
Aluminum core   1.69 cm
$(0^\circ/\pm 45^\circ/90^\circ)_s$ 0.106 cm

Fig. 19 Race car chassis configuration and laminate definition

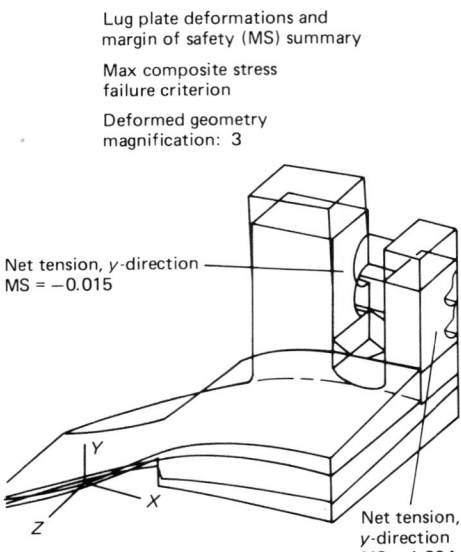

Lug plate deformations and margin of safety (MS) summary

Max composite stress failure criterion

Deformed geometry magnification: 3

Net tension, y-direction
MS = −0.015

Net tension, y-direction
MS = 1.024

Fig. 18 Lug plate deformations and selected margins of safety

translations and rotations. Torsional loading was applied at the rear bulkhead at the centroid of the motor mount bolts. A rigid body element was used at the centroid that contained node 900. The motor mount nodes 129 and 133 were attached to node 900 using a 12456 transformation; That is, the nodes were prevented from x and y translations and rotations about the x, y, z axes. Node 900 was fixed with the standard asymmetric boundary conditions (234 fix) and an additional "1" fix. This caused the rigid body element to rotate about node 900.

The torsional load of 9150 J (6750 ft · lbf) was applied at node 900 of the rigid body element. This torque was then transmitted to nodes 129 and 133 through the transformation constraint conditions. The 9150 J (6750 ft · lbf) torque represented one-half of the total required torque of the chassis, because only half of the chassis was modeled.

Key Results. The torsional stiffness was calculated using the rotation of the rigid body element about the z axis, using the following equation:

$$\frac{GJ}{L} = \frac{T}{\gamma} = \frac{\text{applied torque}}{\text{rotation}}$$

The computed chassis stiffness was 0.912 × 10^6 N · m/radian (8.07 × 10^6 lbf · in./ radian).

Comparison to Test Data and Conclusions. Chassis stiffness tests measured a torsional stiffness value of 0.845 × 10^6 N · m/radian (7.48 × 10^6 lbf · in./radian) (average of two tests). Thus, the agreement with predictions was within 8%.

Example 5: Externally Loaded Ring Stability Analysis. The problem is to determine the bifurcation buckling load for an externally pressurized I-section ring.

Structural Schematic and Laminate Definition. Figure 21 illustrates the ring being analyzed. The dimensions of the overall ring are also shown. An I section was used for the basic ring. The composite material for the ring was Hercules fiber IM6 in a 3501-6 epoxy matrix (V_f = 62%). The following table lists the various percentages of fibers for the flange and web. A balanced symmetric laminate was assumed.

Ply angle	Flange	Web
0°	60%	20%
90°	10%	20%
45°	30%	60%

Selected Numerical Method. The FD program BOSOR IV was used for analysis. The Branched Orthotropic Shells of Revolution (BOSOR) program is described in Ref 6.

Simplifying Assumptions and Analytical Approach. The ring was analyzed as an axisymmetric body. Thus, only one representative cross section needed to be modeled.

A bifurcation buckling analysis was conducted. Only the first Eigen value, λ, was extracted for each buckling mode. Because a unit external pressure (1 kPa, or 0.15 psi) was applied to the outer flange, the Eigen value represented the required pressure to cause bifurcation for the specified mode.

$$\rho = 1 \text{ kPa} \times \lambda$$

The minimum Eigen value represents the minimum pressure required to cause bifurcation. Usually the second, third, and higher Eigen values are extracted for a given mode to ensure that the minimum value has been found.

Element (Mesh) Discretization and Boundary Conditions. The FD mesh is shown in Fig. 22 along with the unit external pressure and axisymmetric boundary condition fix.

Key Results. Figures 23 and 24 show the first Eigen value deflection results for circumferential half-wave values of $N = 2$ and 3. The Eigen values for $\lambda = 2$, 3, and 4 are 3800, 9140, and 10 300, respectively. Clearly, the minimum bifurcation pressure is 3800 kPa for the mode 2 condition.

Comparison to Test Data and Conclusions. Experience from buckling experiments (Ref 6) has shown that a knockdown factor of 0.66 to 0.75 is usually required to correlate with the actual collapse pressure. Thus the estimated buckling pressure would be $P_c \approx 0.66 \cdot 3800$ kPa (551 psi) or $P_c \approx 2500$ kPa (363 psi). Buckling tests have not been conducted for correlation to the predicted value.

Example 6: Multiple Material and Geometry Beam Analysis. The intent is to size the member geometry (thicknesses) in Fig. 25 to optimize the stresses resulting from a 680-kg (1500-lb) tip load. Use the percentage material lay-ups shown in Fig. 25 for each segment. Also, find the tip deflection.

Structural Schematic and Lay-up Definition. A balanced, symmetric laminate was assumed for each material type. The lay-up percentages are shown in Fig. 25. A fiber volume fraction of 60% was used. An I-beam cross section was assumed for the beam sections. A constant web height of 88.9 mm (3.5 in.) and thickness of 1.5 mm (0.06 in.) were used.

Selected Numerical Method. An FD beam code, FINDIF, was used for the analysis. It is a Hercules, Inc., code used for quick analysis of beam structures having variable loading, geom-

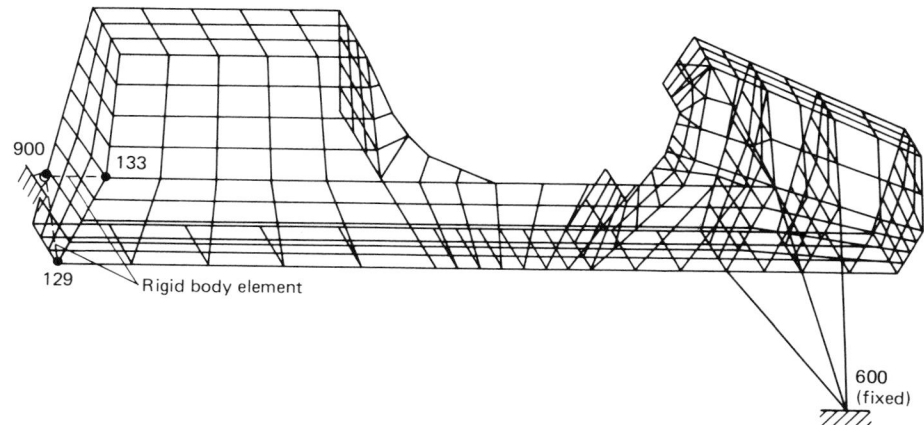

Fig. 20 Chassis grid discretization and boundary conditions. The motions of nodes 129 and 133 are fixed with respect to each other except in Z translation; they are also forced to rotate about node 900.

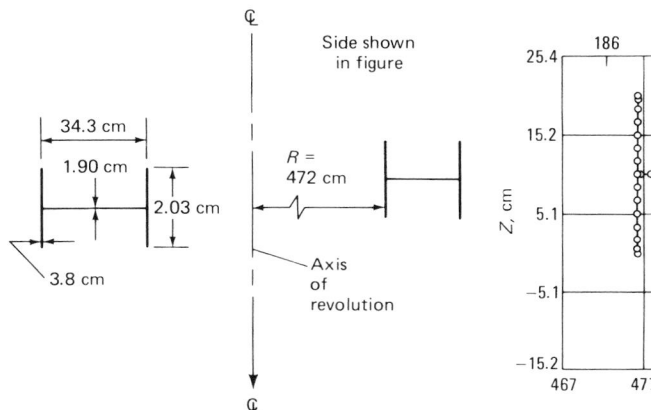

Fig. 21 Orthotropic I-section externally loaded ring

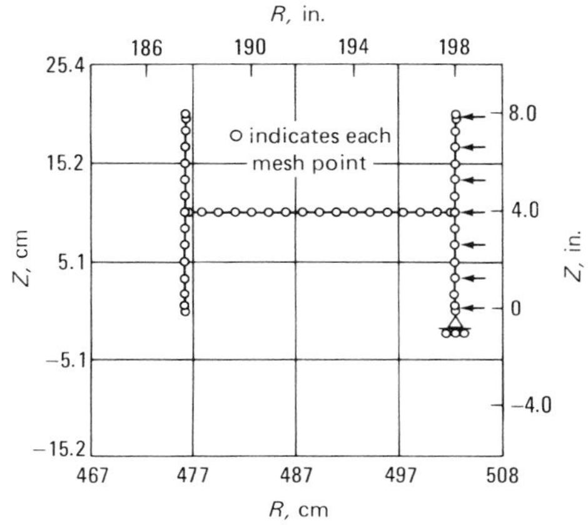

Fig. 22 Finite-difference mesh discretization and boundary conditions

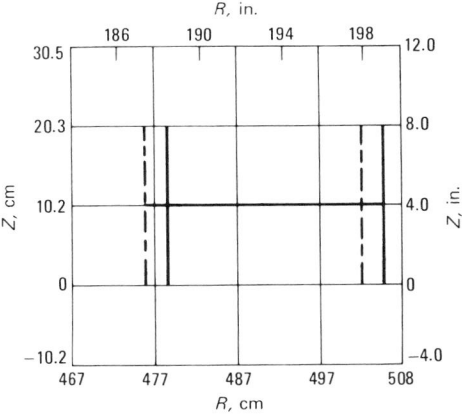

Fig. 23 Deflection plot for the first buckling mode for 2 circumferential half waves N = 2)

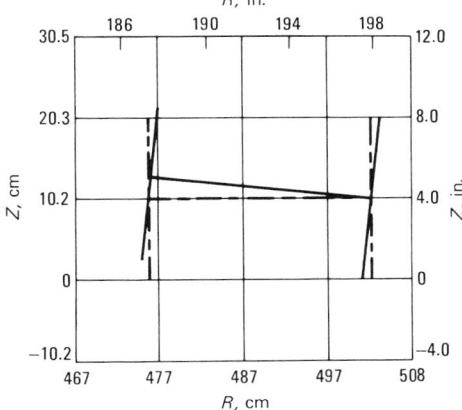

Fig. 24 Deflection plot of the first buckling mode for 3 circumferential half waves (N = 3)

etry, materials, and boundary/support conditions. Similar beam analysis codes are available in the industry (for example, STRUDLE).

Simplifying Assumptions and Analytical Approach. The objective of the analysis was to

optimize the member geometry to achieve a more uniform stress state along the beam. The width was held constant at 20 mm (0.75 in.) Only flange thicknesses were varied. A tapered flange thickness would be optimum, but this configuration often costs more to produce. Consequently, the geometry was adjusted only at the third points to achieve the required maximum stress.

Element (Mesh) Discretization and Boundary Condition. A total of 60 evenly spaced mesh points were used, 20 per segment. A clamped boundary condition was used on the left end of the beam; $\theta = 0$, $u = v = 0$.

Key Results. Figure 26 shows the maximum stress diagram for the beam. Note that the maximum stress is achieved at the clamped boundary and at the third points. The allowable stresses are indicated in Fig. 26. The resulting tip deflection was 550 mm (21.6 in.).

Comparison to Test Data and Conclusions. Test data were not available for this particular beam structure. This particular analysis was conducted to illustrate the FD capabilities for handling variable geometries and materials.

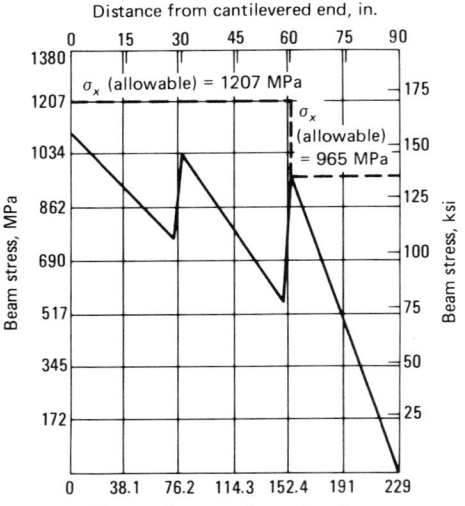

Fig. 25 Fiber reinforced plastic cantilever beam, with beam bending stresses along the cantilever beam length

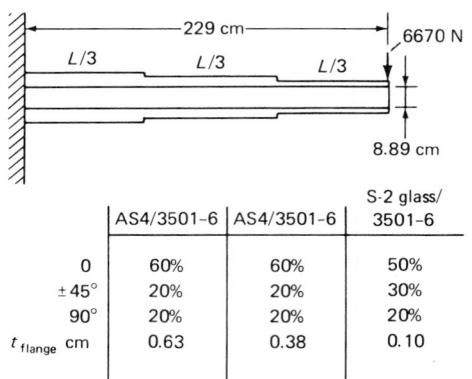

	AS4/3501-6	AS4/3501-6	S-2 glass/ 3501-6
0	60%	60%	50%
±45°	20%	20%	30%
90°	20%	20%	20%
t_{flange} cm	0.63	0.38	0.10

Fig. 26 Maximum axial laminate stress results

Appendix 1: Finite-Difference (FD) Formulations for Partial Differentials

The Taylor series expansion in two variables is expressed in Eq 32 and Eq 33 for both forward and backward differences.

Forward Taylor Expansion

$$f(x + h,y) = f(x,y) + f_x(x,y)h + f_{xx}(x,y)\frac{h^2}{2!} + \ldots \quad \text{(Eq 32)}$$

Backward Taylor Expansion

$$f(x - h,y) = f(x,y) - f_x(x,y)h + f_{xx}(x,y)\frac{h^2}{2!} + \ldots \quad \text{(Eq 33)}$$

The first and second FD approximations are found using methods identical to those used in determining the ordinary differential finite difference. The first order partial FD in x is determined by retaining the first two terms of the series.

Forward Difference Approximation

$$f_x(x,y) \cong \frac{f(x + h,y) - f(x,y)}{h} \quad \text{(Eq 34)}$$

Backward Difference

$$f_x(x,y) \cong \frac{f(x,y) - f(x - h,y)}{h} \quad \text{(Eq 35)}$$

As with the ordinary differential equation, the central difference is obtained by adding the forward and backward differences Eq, 34 and 35, and simplifying.

Central Difference

$$f_x(x,y) \cong \frac{1}{2h} [f(x + h,y) - f(x - h,y)] \quad \text{(Eq 36)}$$

If the third term of the series is retained, the second order partial FD in x is found. The central difference is given in Eq 37 below and was derived using steps identical to forming Eq 36.

Second Order Central Difference

$$f_{xx}(x,y) \cong \frac{1}{h^2} [f(x + h,y) - 2f(x,y) + f(x - h,y)] \quad \text{(Eq 37)}$$

Equations similar to Eq 34 to Eq 37 may be formulated for $f_y(x,y)$ and $f_{yy}(x,y)$, the first and second partial derivatives in y. These equations are shown without derivation in Eq 38 and 39.

First and Second Order Central Differences in y

$$f_y(x,y) \cong \left(\frac{1}{2h}\right) [f(x,y + h) - f(x,y - h)] \quad \text{(Eq 38)}$$

$$f_{yy}(x,y) \cong \frac{1}{h^2} [f(x,y + h) - 2f(x,y) + f(x,y - h)] \quad \text{(Eq 39)}$$

As an example, the partial difference approximation for the torsion equation may be derived. Table 1 shows the torsion equation to be:

$$\frac{\partial^2 \phi}{\partial x^2} + \frac{\partial^2 \phi}{\partial y^2} = -2G\theta$$

where ϕ is the stress function, G is the shear modulus, and θ is the angle of twist. The fiber difference terms are simply the sum of Eq 37 and Eq 39. Assuming uniform spacing, h, the FD equation is seen to be exactly that shown in Table 1.

Appendix 2: Finite-Element Stiffness Equation Using the Virtual Work Approach

The virtual work expression relates the increment in internal strain energy to the virtual work done by surface and body forces. This is expressed as:

$$\delta U = \delta W_s + \delta W_B \quad \text{(Eq 40)}$$

δU is the increment in internal strain energy. The terms δW_s and δW_B are the virtual work terms for surface and body forces, respectively. Each of these terms must be determined in order to develop the element stiffness equation.

The derivation begins by assuming a permissible displacement function for the element. The function must be differentiable, satisfy compatibility, and have the same degrees of freedom as those associated with the displacement of the element. Usually, a polynomial is used as the displacement function because it simply satisfies the above conditions. Thus, for two-dimensional elements, for example, the function in u for some location x, y within the element would be:

$$u(x,y) = [1 \times y \, x_2 \, xy \ldots] \begin{Bmatrix} a_0 \\ a_1 \\ a_2 \\ | \\ a_i \end{Bmatrix} \quad \text{(Eq 41)}$$

or more simply:

$$u(x,y) = \{f^T(x,y_2)\}\{a\}$$

where 1, x, y, x^2, xy, y^2, and so forth, are the polynomial terms written as $\{f^T\}$, and a_o, a_1, a_3, . . . are coefficients. Similarly, for deflections in y,

$$v(x,y) = \{[g^T(x,y)]\}\{b\}$$

Writing the displacement, u, at nodes $i, j, \ldots, n$ gives:

$$\begin{Bmatrix} u_i \\ u_j \\ | \\ u_n \end{Bmatrix} = \begin{bmatrix} 1 & x_i & y_i & x_i^2 & x_i y_i & \ldots \\ 1 & x_j & y_j & x_j^2 & x_j y_j & \ldots \\ & & \ldots & & & \\ 1 & x_n & y_n & x_n^2 & x_n y_n & \ldots \end{bmatrix} \begin{Bmatrix} a_0 \\ a_1 \\ | \\ a_n \end{Bmatrix}$$

or

or $\{r^{(u)}\} = [G^{(u)}]\{a\}$ (Eq 42)

using the notation in Ref 7. Likewise for v,

$$\{r^{(v)}\} = [G^{(v)}]\{b\} \quad \text{(Eq 43)}$$

$G_1^{(u)}$ and $G^{(v)}$ are usually identical because the same polynomial function is usually used to define both u and v displacements. As seen in Eq 42 and 43, $G^{(u)}$ and $G^{(v)}$ are defined at the specific nodal locations $(i, j, \ldots, r)$. Solving for Eq 42 and 43 for the coefficients $\{a\}$ and $\{b\}$ allows the assumed displacement coefficients, $\{a\}$ and $\{b\}$, to be written in terms of the nodal displacements. Continuing with the two-dimensional displacement example:

$$\begin{Bmatrix} u(x,y) \\ v(x,y) \end{Bmatrix} = \begin{bmatrix} f^T(x,y) & 0 \\ 0 & g^T(x,y) \end{bmatrix} \begin{bmatrix} G^{(u)} & 0 \\ 0 & G^{(v)} \end{bmatrix}^{-1} \begin{Bmatrix} r^{(u)} \\ r^{(v)} \end{Bmatrix} \quad \text{(Eq 44)}$$

or using $\{u\}$ as the generalized displacement vector:

$$\{u\} = [N^T] \{r\}$$

where

$$[N^T] = \begin{bmatrix} f^T & 0 \\ 0 & g^T \end{bmatrix} \begin{bmatrix} G^{(u)} & 0 \\ 0 & G^{(v)} \end{bmatrix}^{-1}$$

The displacements within the element have now been expressed in terms of the nodal displacements. Writing the strains in terms of the displacements (for the two-dimensional example) yields:

$$\{\epsilon\} = \begin{Bmatrix} \epsilon_x \\ \epsilon_y \\ \gamma_{xy} \end{Bmatrix} = \begin{bmatrix} \dfrac{\delta}{\delta x} & 0 \\ 0 & \dfrac{\delta}{\delta y} \\ \dfrac{\delta}{\delta y} & \dfrac{\delta}{\delta x} \end{bmatrix} \begin{Bmatrix} u(x,y) \\ v(x,y) \end{Bmatrix} \quad \text{(Eq 45)}$$

Substitute Eq 44 into Eq 45 and differentiate. Note that only f^T and g^T vary with x and y. $G^{(u)}$ and $G^{(v)}$ are defined at the nodes and are therefore constant.

$$\{\epsilon\} = \begin{bmatrix} \dfrac{\partial\{f^T\}}{\partial x} & 0 \\ 0 & \dfrac{\partial\{g^T\}}{\partial y} \\ \dfrac{\partial\{f^T\}}{\partial y} & \dfrac{\partial\{g^T\}}{\partial x} \end{bmatrix} \begin{bmatrix} G^{(u)} & 0 \\ 0 & G^{(v)} \end{bmatrix}^{-1} \begin{Bmatrix} r^{(u)} \\ r^{(v)} \end{Bmatrix}$$

or more simply:

$$\{\epsilon\} = [B] \{r\}$$

The variational strain increment is written as:

$$\delta\{\epsilon\} = [B] \, \delta\{r\} \quad \text{(Eq 46)}$$

Next define the stress-strain constitutive relationship. For a two-dimensional orthotropic material in plane stress:

$$\{\sigma\} = [D] \{\epsilon - \epsilon_o\} \quad \text{(Eq 47)}$$

where

$$[D] = \frac{1}{1-\nu_1\nu_2} \begin{bmatrix} E_1 & \nu_2 E_2 & 0 \\ \nu_1 E_1 & E_2 & 0 \\ 0 & 0 & G(1 - \nu_1\nu_2) \end{bmatrix}$$

The strain energy increment, δu, may now be calculated. Using the two-dimensional example, the variation in strain energy may be written as:

$$\delta u = \int_v (\sigma_x \delta\epsilon_x + \sigma_y \delta\epsilon_y + \tau_{xy}\delta\gamma_{xy}) \, dv$$

or, writing in vector form:

$$\delta u = \int_v \delta\{\epsilon\}^T\{\sigma\} \, dv$$

Substituting from Eq 46 and 47 gives:

$$\delta u = \int_v [B]^T \delta[r]^T [D]\{\epsilon - \epsilon_o\} \, dv \quad \text{(Eq 48)}$$

If $\{\epsilon\}$ is expressed in terms of nodal displacements $\{r\}$, the final expression for internal strain energy becomes:

$$\delta U = \delta\{r\}^T \left(\int_v [B]^T [D][B] \, dv\right)\{r\}$$
$$\quad - \delta\{r\}^T \int_v [B]^T [D]\{\epsilon_o\} \, dv \quad \text{(Eq 49)}$$

The surface and body force virtual work terms must also be obtained. As with displacements, it is convenient to write the forces in terms of

the nodal values. For surface forces this may be written as (two-dimensional example):

$$\rho(x,y) = \begin{bmatrix} p^{(u)} & 0 \\ 0 & p^{(v)} \end{bmatrix} \begin{Bmatrix} p_x \\ p_y \end{Bmatrix} \quad \text{(Eq 50)}$$

or more simply:

$$\rho(x,y) = P_p$$

where p_x and p_y are nodal-force values and $P^{(u)}$, $P^{(v)}$ are polynomial approximating functions. Likewise, for body forces:

$$f(x,y) = \begin{bmatrix} F^{(u)} & 0 \\ 0 & F^{(v)} \end{bmatrix} \begin{Bmatrix} f_x \\ f_y \end{Bmatrix} \quad \text{(Eq 51)}$$

where f_x and f_y are again nodal values and $F^{(u)}$ and $F^{(v)}$ are polynomial approximation functions. The virtual work due to surface traction is:

$$\delta W_s = \int_a^T \delta u(x,y) p(x,y) dA \quad \text{(Eq 52)}$$

However, δu has already been expressed in terms of nodal displacements in Eq 44. Thus, substituting Eq 45 and Eq 50 into Eq 52 gives:

$$\delta W_s = \int_A \delta r^T N^T P_p dA$$

Let

$$R_s = (\int_A N^T P dA)_p \quad \text{(Eq 53)}$$

Then

$$\delta W_s = \delta\{r\}^T R_s$$

Similarly for body forces:

$$\delta W_B = \int_v \delta U^T(x,y) f(x,y) dv \quad \text{(Eq 54)}$$

Substituting Eq 51 and Eq 44 into Eq 54 gives:

$$\delta W_B = \int_v \delta\{r\}^T N^T F f dv$$

Now let

$$R_B = (\int_v N^T F \, dv) f$$

Then

$$\delta W_B = \delta\{r\}^T R_B \quad \text{(Eq 55)}$$

The terms R_s and R_B are the nodal surface and body force vectors, respectively.

The final form of the element stiffness matrix is found by substituting Eq 49, 53, and 55 into Eq 40. This results in:

$$(\int_v [B]^T[D][B] dv)\{r\} =$$
$$\int_v [B]^T[D]\{\epsilon_o\} dv + R_s + R_B \quad \text{(Eq 56)}$$

where

$$\int_v [B]^T[D]\{\epsilon_o\} dv = R_o$$

is the initial strain force vector contribution (for example, thermal strain). The terms on the right-hand side of Eq 55 add to form the nodal-force vector. The vector $\{r\}$ consists the unknown nodal displacements. Thus, the element stiffness matrix is the remaining term:

$$[K]_e = \int_v [B]^T [D][B] dv \qquad \text{(Eq 57)}$$

The element stiffness (Eq 56) may then be expressed more simply as:

$$[K]_e \{r\}_e = R_o + R_s + R_B \qquad \text{(Eq 58)}$$

ACKNOWLEDGMENTS

The author would like to thank G.R. Toombes, R.E. Nelson, R.W. Nish, M.D. Nelson, and J.D. Hendrix for providing analytical support for the example problems.

REFERENCES

1. S.P. Timoshenko and J.N. Goodier, *Theory of Elasticity*, McGraw-Hill, 1934
2. M.J. Turner, R.W. Clough, H.C. Martin, and L.J. Topp, Stiffness and Deflection Analysis of Complex Structures, *J. Aeron. Sci.*, Vol 23 (No. 9), Sept 1956, p 805-824
3. S.J. Farlow, *Partial Differential Equations for Scientists and Engineers*, John Wiley & Sons, 1982
4. J.H. Faupel and F.E. Fisher, *Engineering Design*, 2nd ed., John Wiley & Sons, 1981
5. R.W. Hornbeck, *Numerical Methods*, Quantum, 1975
6. D. Bushnell, "Stress, Stability and Vibration of Complex Branches Shells of Revolution: Analysis and Users Manual for BOSOR 4," NASA CR-2116, National Aeronautics and Space Administration, Oct 1972
7. H.C. Martin and G.F. Carey, *Introduction to Finite Element Analysis*, McGraw-Hill, 1973
8. G.E. Colvin, Jr. and D.S. Adams, "A Finite Element Overlay Technique for Modeling Pinned Composite Joints," AIAA-86-1420, AIAA/ASME/SAE/ASEE 22nd Joint Propulsion Conference, June 1986

Joints

L.J. Hart-Smith, Douglas Aircraft Company

THE STRUCTURAL EFFICIENCY of a composite structure is established, with very few exceptions, by its joints, not by its basic structure. Joints can be manufacturing splices planned at predetermined locations in the structure, or unplanned repairs that could be needed anywhere in the structure. Consequently, unless a specific application needs no provision for repairs, or uses throw-away unrepairable components, the correct sequence for design is to first locate and size the joints, in fiber patterns optimized for that task, and then fill in the gaps in between.

This sequence is a marked departure from normal practice for conventional ductile metal alloys and is necessitated by the relative brittleness of fiber-reinforced composites. Yielding of ductile metals usually reduces the stress concentrations around bolt holes so that there is only a loss of area, with no stress concentration at ultimate load on the remaining (net) section at the joints. With composites, however, there is no relief at all from the elastic stress concentration if the holes or cutouts are large enough. Even for small holes in composite structures, the stress-concentration relief is far from complete, although the local disbonding (between the fibers and resin matrix and local intraply and interply splitting close to the hole edge) does locally alleviate the most severe stress concentrations.

The reason for emphasizing the importance of joints in the design of composite structures is that the availability of large computer optimization programs and the highly deficient treatment of residual thermal stresses within the resin* in most of the composite laminate theories have combined to create the illusion that optimized composite structures will necessarily be highly orthotropic and tailored precisely to match the load conditions and stiffness requirements. Were this actually true, the task of designing joints in composites would be much more difficult than it is now. The capability of bolted joints in such highly orthotropic materials is often unacceptably low. Hence, the lam-

inate can never be loaded to the levels suggested by lamination theory.

Fortunately, or unfortunately, depending on one's point of view, the strength of composite structures with both loaded and unloaded holes depends only slightly on the fiber pattern (for nearly quasi-isotropic laminates); the stress-concentration factor increases almost as rapidly as the unnotched strength for slightly orthotropic patterns. Indeed, throughout the range of fiber patterns surrounding the quasi-isotropic lay-up, the bearing strengths and gross-section strengths are almost constant, which simplifies the design process considerably. A valid case can often be made for a small amount of orthotropy, within the shaded area in Fig. 1, particularly when there is a preferred load direction or stiffness requirement that must be met. The farther a laminate pattern is outside the shaded area, the more likely it is to fail prematurely by through-the-thickness cracks parallel to the maximum concentration of fibers.

Even for those few truly non-strength-critical uses of highly orthotropic fiber patterns, such as space structures with zero coefficient of thermal expansion and one-shot missiles, there must be a transition to nearly quasi-isotropic patterns around any bolt holes; the structural efficiency of bolted joints in highly orthotropic laminates is known to be inadequate. Therefore, the material presented here is applicable to all sensible composite structures, although it deliberately excludes mechanical joints in highly orthotropic materials. Attention is confined to uniaxial membrane loading, because most of the relevant test data on bolted joints are similarly restricted and because the analysis methods for off-axis loading and applied bending moments are still being developed.

The design of joints in nearly quasi-isotropic composites is straightforward, although it is necessary to allow for nonlinearities in the material behavior; linearly elastic analyses are far too conservative. The analysis of adhesively bonded joints using elastic-plastic adhesive models has advanced to the stage at which it can legitimately be called a science. The design of straightforward bonded joints has been reduced to following a few procedures and obeying a few simple design refinements to prevent

premature failures due to induced peel loads. The design and analysis of the more complex stepped-lap bonded joints needed for much thicker and more highly loaded bonded structures is facilitated by the use of digital computer programs based on nonlinear continuum-mechanics solutions. Even the determination of the design load level for bonded joints is easy, regardless of the nominal applied loads. In no case should the strength of the joint be allowed to fall below that of the surrounding structure; otherwise, the bonded joint would have no damage tolerance and could act as a weak-link fuse. Fortunately, with the strong ductile adhesives typically used by the aerospace industry, the bond is inevitably stronger than the adherends for properly designed joints between thin members. Even for thicker structures, the bond can always be made stronger than the structure by using enough steps in the joint.

The design and analysis of bolted or riveted joints in fibrous composites, however, remains very much an art because of the need to rely on empirical correction factors in some form or other. Mechanically fastened joints differ from bonded composite joints in one further aspect: The presence of holes ensures that the joint

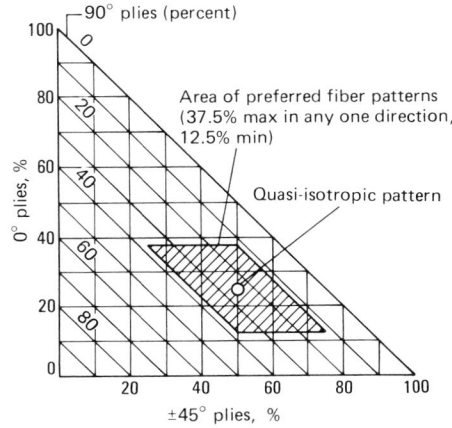

Fig. 1 Selection of lay-up pattern for fibrous composite laminates. All fibers in 0°, +45°, 90°, or −45° direction. Note: Lightly loaded minimum-gage structures tend to encompass a greater range of fiber patterns than indicated because of the unavailability of thinner plies.

---

*There are no terms in most theories to allow for separate residual thermal stresses in the fibers or matrix of the monolayer, which serves as the building block for cross-plied laminate theories. The omission is due to the artificial homogenization of distinctly two-phase composite materials into mathematically simpler one-phase models.

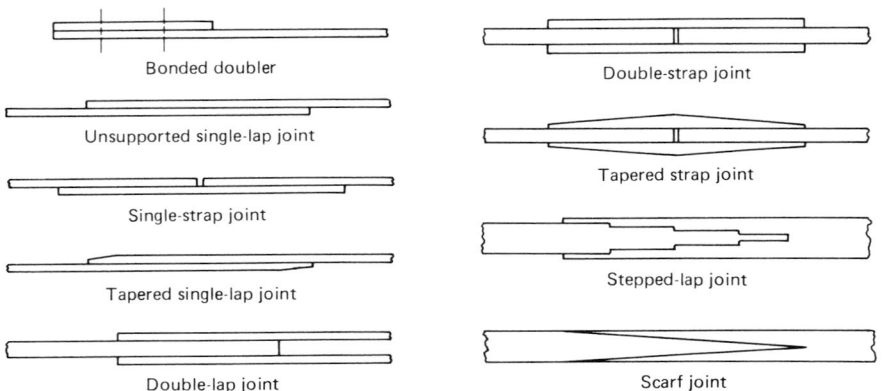

Fig. 2 Adhesively bonded joint types

strength can never exceed the local laminate strength. Indeed, after years of research and development, it appears that only the most carefully designed bolted composite joints will be even half as strong as the basic laminate. The simpler bolted joint configurations will attain no more than a third of the laminate strength. However, because the adhesively bonded repair of thick composite laminates is often impossible or impractical (Ref 1), there is a real need for bolted composite structures quite apart from the greater ease of assembly at mechanically fastened manufacturing breaks between subassemblies. A further problem with the design of bolted composite structural joints is that fibrous composites are so brittle that there is virtually no capability for redistributing load, as is afforded by yielding of ductile metals. Consequently, it is very important to calculate accurately the load sharing between fasteners and to identify the most critically loaded one.

Obviously, one can avoid the strength limitations of bolted or riveted composite joints by using such techniques as local pad-ups to thicken chordwise bolt seams on wing skins, for example, and glass softening strips in the skins over the spar caps. However, such an approach precludes the possibility of making repairs with mechanical fasteners throughout the remaining unprotected structure unless one is prepared to accept a substantial reduction in strength.

This article starts with a discussion of adhesively bonded joints, covering the keys to durability, the elastic-plastic mathematical model for the adhesive in shear, the simple design rules for thin bonded structures, the computer programs for the more highly loaded stepped-lap joints, and the two-dimensional effects associated with load redistribution around flaws and with damage tolerance. Three articles in this Volume, "Adhesives Selection," "Adhesives Specifications," and "Bonding Cure Considerations" provide additional information.

Mechanically fastened joints are then discussed, starting with the elastic-isotropic geo-

metric stress-concentration factors, the empirically established correlation factors to convert these elastic values to those observed in the composites at failure, the identification of optimum joint proportions for single-row joints, and the design and analysis of the stronger multirow joints with particular regard to the bearing-bypass interaction. The articles "Mechanical Fastener Selection" and "Blind Fastening" in this Volume should be consulted. As no single article can cover every aspect of so wide a subject, it is appropriate to cite Ref 2 and 3.

Fundamentals of Shear Load Transfer Through Adhesively Bonded Joints

Adhesively bonded joints can be strong in shear but are inevitably weak in peel, so the objective of good design practice is to arrange the joint to transfer the applied load in shear and to minimize any direct or induced peel stresses. The details of the design vary with the load intensity (and, hence, the thickness of the adherends) as shown in Fig. 2. The thinner members can be joined effectively by simple, uniformly thick overlaps, while thicker members require the more complex stepped-lap joints.

For each of the joints shown in Fig. 2, the potential shear strength of the bond—that is, the strength that the bond could have developed had the adherends not failed first—exceeds the direct strength of the adherends outside the joint, up to a determinable thickness. This characteristic is shown in Fig. 3, which also shows the loss of bond strength that is sometimes associated with flaws in or damage to the bond. Even with such degradation, the bond will be stronger than the members outside the joint, up to some lesser adherend thickness.

The key point of Fig. 3 is that for adherend thicknesses greater than that for which the

bond and member strengths are equal, there can be absolutely no tolerance with respect to flaws, porosity, or damage. The slightest imperfection would lead to catastrophic unzipping of the entire bond area if sufficient load were applied. That is why it is so important that bonded joint strengths must exceed those of the adherends, even to the point of exceeding the strength by at least 50% to permit the occurrence of minor manufacturing flaws or imperfections. Subject to that proviso, bonded joints between thin members have a remarkable insensitivity to very large local flaws, as explained in Ref 4. In this context, the term thin is adjusted to suit the complexity of the joint and refers to those sensible designs for which the nominally perfect bond is stronger than the members being joined. Such a design philosophy for bonded joints should always be followed, even when the nominal applied loads are less than the strength of the members. Otherwise, there will always be the possibility of a local flaw that is large enough to convert the nominally perfect bond into a weak-link fuse. A flaw in an underdesigned joint shares the characteristics of a through-crack in a metal sheet, as explained in Fig. 4, except that it is much harder to find. Figure 4 refers equally to metal and composite structures except that, for the latter, load redistribution around the bond flaw may also cause delaminations in the composite panel or possibly result in the unzipping of the bond.

Adhesive bonds must also be resistant to the environment in which they operate, which is typically thermal or chemical. This need has been demonstrated in the many in-service failures of secondary and, in at least two cases, primary bonded metal structures on U.S. aircraft made during the late 1960s and early 1970s. Those failures were not caused by poor design detailing; indeed, the most recent adhesives failures known to the author, which were due to mechanical overloading of bonded aircraft structures, occurred over half a century ago on glued wooden aircraft. There were also some structural failures on some wooden aircraft in the tropics during World War II due to a poor choice of glue, which was very sensitive to moisture. However, those joints were properly proportioned, and even the glue worked adequately in Europe.

Bonded joint failures in metal structures have occurred when the absorption of moisture by some adhesives on the surface of the adherends hydrolyzed and subsequently corroded the oxide surface of clad aluminum alloys. This subject was explored in depth during the U.S. Air Force Primary Adhesively Bonded Structure Technology (PABST) program (Ref 5-7). It is now well known that aluminum alloys must be anodized, in phosphoric or chromic acid, to create a stable, durable oxide surface, and that the surface must be promptly coated with a corrosion-inhibiting primer, usually BR-127 or Redux liquid. Using clad 7075 aluminum alloys should be avoided. Similarly, titanium alloys

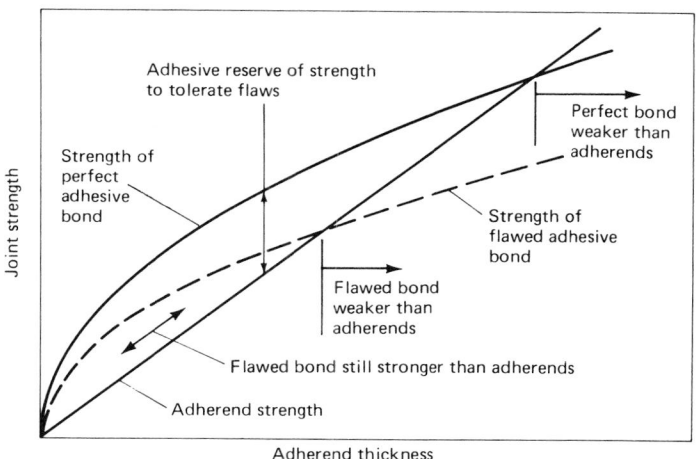

Fig. 3 Relative strength of adhesive and adherends as affected by bond flaws

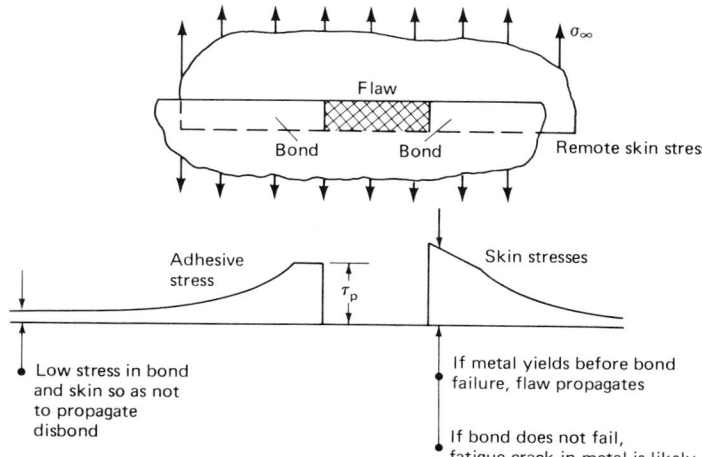

Fig. 4 Redistribution of load at flaws in bond

need appropriate surface preparations for reliable adhesive bonding, as do steels.

Somewhat surprisingly, the need for comparable attention to the preparation of fibrous composite surfaces for adhesive bonding has not received nearly as much publicity. Three articles in this Volume, "Adhesive Bonding Surface Preparation," "Dissimilar Material Separation," and "Faying Surface Sealing," contain useful information pertaining to bonded joints. The widespread use of inferior preparations, such as removing no more than a peel ply, or scuff sanding followed by solvent contamination, remains the norm. R.J. Schliekelmann, a pioneer of adhesive bonding of metal structures, has warned of the importance of this issue to composites (Ref 8). L.J. Hart-Smith *et al.* have strongly recommended light grit blasting as the best known treatment today (Ref 9). This view is shared by A.V. Pocius (Ref 10), who has also advocated mechanical abrasion with Scotchbrite pads. Promising work has also been done on composite surface preparation by so-called flash blasting, but no production applications are known yet.

The subsequent discussion in this article assumes that the durability of the surfaces to be bonded has been ensured by appropriate preparation. Otherwise, just as for metal bonding, no reliable life can be established for adhesively bonded composite structures.

The subject of environmental durability of the adhesive layer itself, rather than of the interface, is much more straightforward and can benefit from the massive amount of testing already done for metal bonding. The adhesive resin can be regarded as a well-behaved engineering material up to some service temperature, which depends on the amount of plasticizing additives as well as on the base resin. That temperature, called the glass transition temperature, can be reduced slightly by the absorption of moisture. Increasing the glass transition temperature for bonding on supersonic aircraft,

or near engines, has meant sacrificing most of the adhesive strength at lower temperatures by omitting the modification of the basic epoxy or phenolic resin by rubber, nylon, or vinyl additives. The analysis of adhesively bonded joints requires a nonlinear shear stress-strain curve for all adhesives, ductile or not, because even the brittle adhesives exhibit substantial nonlinear behavior at temperatures approaching their upper service limits. The strongest structural additives are suitable for most of the structure of subsonic aircraft, having an upper limit of about 70 °C (160 °F).

Even a small amount of moisture can be very harmful to bonded composites (Ref 11). Absorbed water in cured laminates must be removed by careful, gentle drying before any bonded repairs are performed. If such moisture is driven off too rapidly, it will delaminate the composite. Conversely, if it is not driven off completely, it will later react adversely with any uncured material (adhesive or resin) in the patch. Likewise, any moisture absorbed by the uncured resin (in a prepreg) or adhesive will prevent the proper curing of the material. Many uncured resins are hygroscopic. It is therefore very important that such materials be properly stored and subsequently thoroughly thawed out so that there is no opportunity for them to absorb the condensate formed when they are removed from the freezer.

Today, rational engineering design of bonded structures is based on the measured adhesive stress-strain relation in shear for a thin layer of adhesive between thick aluminum adherends. Given these stress-strain data for a range of operating temperatures, it is now possible to calculate the actual adhesive stress distributions within the bonded joints, at least in the short term. More work will be needed to characterize the time-dependent changes in internal load distribution under sustained loads. However, the lack of such information does not prevent the satisfactory completion of most designs.

After the surface preparation issue has been resolved, the real key to the durability of adhesively bonded joints is that the minimum adhesive shear stress in a joint needs to be restricted to prevent failure of the joint by creep rupture, every bit as much as the maximum stress needs to be restricted to prevent static failure. This issue is of tremendous importance in interpreting data from test coupons. The prime objective of designing bonded structural joints should be to ensure that the bond will never fail, while the objective of designing test coupons is to ensure that the adhesive will always fail, at as uniform a stress state as can be established. Unfortunately, therefore, bonded test coupons are in many ways totally unrepresentative of the behavior of real structural joints. In particular, a highly nonuniform stress distribution in the adhesive is necessary if a structural joint is to attain an adequate life.

Nonuniformity of Load Transfer Through Adhesive Bonds

The classical analysis by O. Volkersen (Ref 12) established in 1938 that the load transfer through adhesive bonds between uniformly thick adherends is not uniform but peaks at each end of the overlap, as shown in Fig 5. This nonuniformity results from the compatibility of deformations associated with the variation of direct stress, within the adherends, from one end of the bonded overlap to the other.

A few years later, M. Goland and E. Reissner analyzed the distribution of the peel stresses induced in the adhesive layer by the eccentricity in load path associated with single-lap joints (Ref 13). However, in determining the bending moments in the adherends at the ends of the overlap, they made a physically unrealistic and mathematically unnecessary simplifying assumption, thereby overestimating

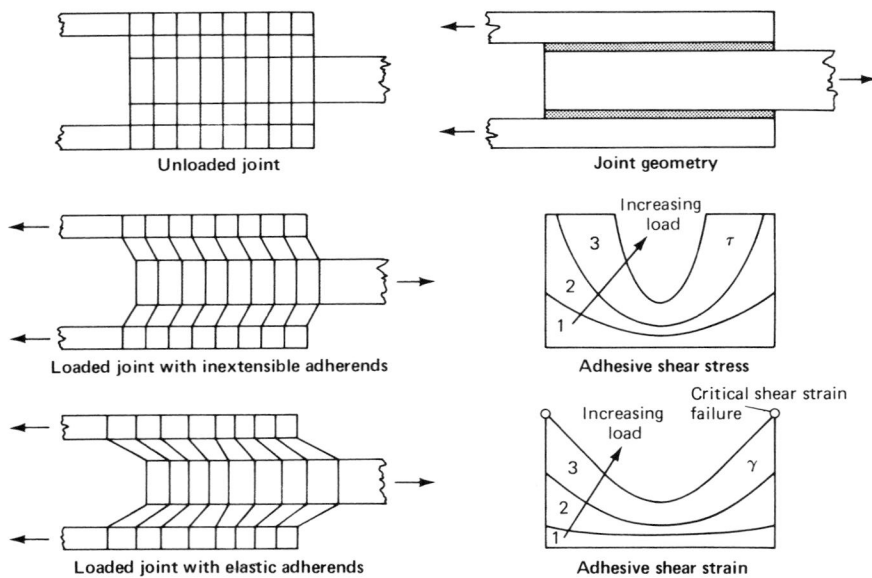

Fig. 5 Shearing of adhesive in balanced joints

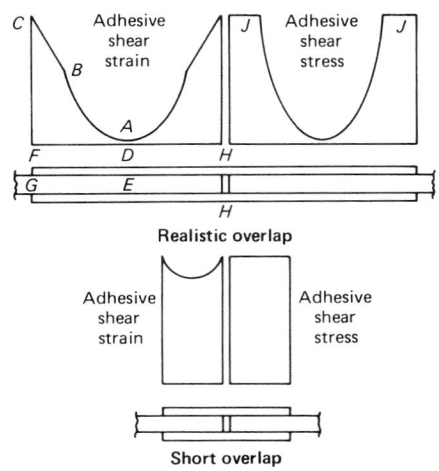

Fig. 6 Nonuniform stresses and strains in bonded joints

those moments, which act as boundary conditions on the adhesive stress distributions. The remainder of their derivation of the elastic adhesive stresses is analytically correct, but the predicted peak adhesive stresses in shear and peel are excessive. An improved estimate of that bending moment is found in Ref 14 and 15. Mention should also be made of the analyses published by N.A. de Bruyne (Ref 16) that have resulted from his work on Redux and its application in England during and after World War II.

L.J. Hart-Smith has built upon these pioneering investigations and added nonlinear adhesive behavior to the analysis and design of adhesively bonded joints in the form of an elastic-plastic adhesive model (Ref 17-19). Also, the A4E series of digital computer programs was developed for joints of various geometries, under contract to NASA Langley and the laboratories at Wright-Patterson Air Force Base. The origins of these programs are given in Ref 20 and 21.

The knowledge imparted by the precise analyses on which Fig. 5 is based makes it possible to understand the differences between the behavior of adhesive bonds in test coupons and structurally configured joints (see Fig. 6). The key difference is that for the short-overlap test coupon, the minimum adhesive shear stress and strain are nearly as high as the maximum values, while for the long-overlap structural joint, the minimum adhesive shear stress and strain can be made as low as desired by using a sufficiently long overlap. Consequently, the short-overlap test coupon is extremely sensitive to failure by creep rupture (which accumulates under both steady and cyclic loads) because there is no mechanism for restoring the adhesive to its original state when the load is removed. While there is creep in the adhesive at

the ends of the long overlap, between points F and G, where the stress (at J) is high, there can be none in the middle, between points D and E, if the stress (at A) is low enough. Consequently, the creep that does occur cannot accumulate because the stiff adherends push the adhesive back to its original position whenever the joint is unloaded. This memory, or anchor, in part of the adhesive is the key to a durable bonded structure. Without it, there can be no successfully bonded structure. Such recovery during unloading does not imply that the adhesive suffers no damage at all when loaded slightly beyond the knee in the stress-strain curve. However, that damage is limited and not catastrophic.

The question of just how low a minimum stress should be has yet to be resolved scientifically. However, during the PABST program, the minimum was set at 10% of the maximum, and environmental testing on both coupons and complete structures showed no adverse effects, even though premature failures were commonplace with the standard half-inch-overlap test coupons. The influence of minimum stress on the design overlap, for standard double-lap or double-strap bonded joints, is shown in Fig. 7. The width of the elastic trough is adjusted so that the minimum stress is 10% of the maximum. This value is reached when the elastic trough has a total length of $6/L$, where L is the exponent of the elastic adhesive shear stress distribution. To that elastic overlap, a sufficient plastic zone must be added at each end to transfer a load at least equal to the entire strength of the adherends, with the adhesive stressed to its maximum shear strength (for a particular environment). The maximum design overlap is normally associated with the highest service temperature for the bonded joint.

It was found during the PABST program that, for the thicknesses of aluminum alloy suitable for bonding on subsonic transport aircraft, the overlap could be calculated at approximately 30 times the central adherend thickness. In addition, because the modulus of cross-plied carbon-epoxy laminates within the shaded area in Fig. 1 is on the same order of magnitude as for aluminum alloys, a similar overlap-to-thickness ratio would also be satisfactory for such laminates.

Actually, the static strength of bonded joints between uniform adherends is quite insensitive to the precise (long) overlap, as shown in Fig. 8. Any longer overlap beyond point C would be superfluous. This insensitivity of the joint strength to the total bonded area is important in recognizing the folly of designing joints with the old notion that the bond strength is equal to the product of the bond area and some fictitious uniform "allowable" shear strength.

Another important point shown in Fig. 8 is that for all overlaps longer than the abrupt precipice in the upper diagram, the maximum strain in the adhesive is limited by the adherend strength to a value below that which would be needed to fail the adhesive. No such protection is afforded for short overlaps. This issue is explained more fully in Ref 22, where similar diagrams have been prepared for adherends of different thicknesses and adhesive properties appropriate for a range of thermal environments. It is shown there that if the adherends are too thick, the limit on the peak shear strain shown in Fig. 8 is removed. Therefore, a more complex stepped-lap joint is appropriate for adherends thicker than about 3.2 mm (1/8 in.). Also, it is found that the limiting strength of the joint is usually set by the lowest service temperature, while the design overlap is set by the highest service temperature, at which the adhesive is the softest.

Having established the design overlap for simple bonded joints, the elimination of adverse peel stresses will next be addressed.

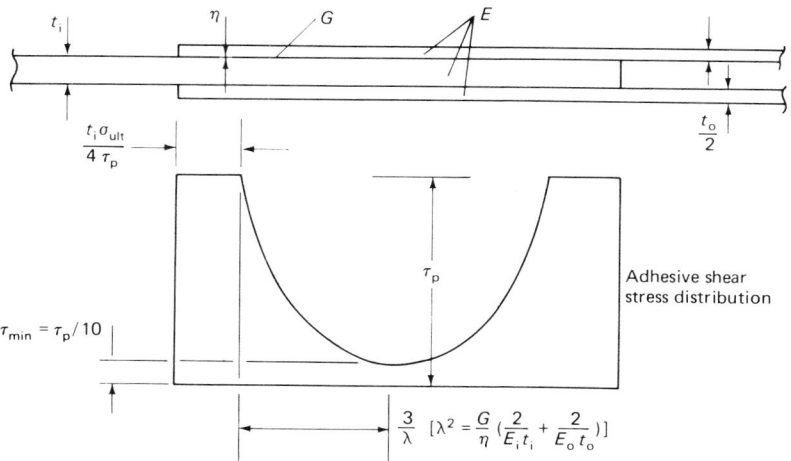

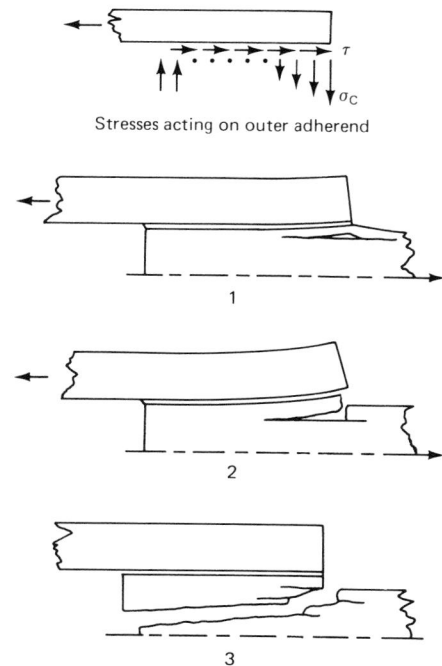

Stresses acting on outer adherend

Fig. 7 Design of double-lap bonded joints. Plastic zones long enough for ultimate load. Elastic trough wide enough to prevent creep at middle. Adequate strength must be verified

Fig. 9 Peel stress failure of thick composite joints, where 1, 2, and 3 indicate failure sequence

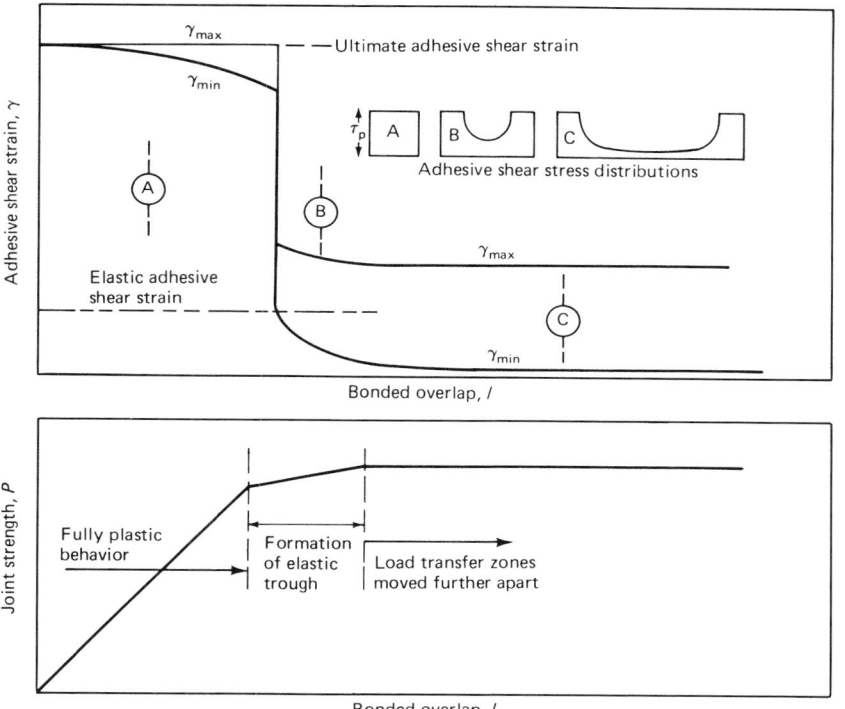

Fig. 8 Influence of overlap on maximum and minimum adhesive shear strains in bonded joints

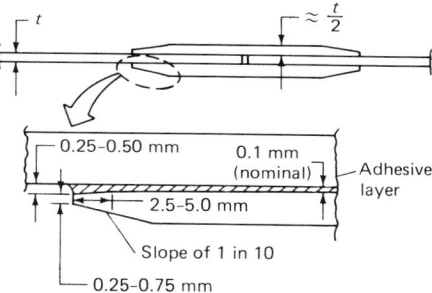

Fig. 10 Tapering of edges of splice plates to relieve adhesive peel stresses (slightly thicker tips permissible for aluminum)

These peel stresses occur for single-lap and single-strap joints having a primary eccentricity in load path and for double-lap and double-strap joints, as shown in Fig. 9, even though there is no obvious eccentricity in the seemingly balanced joints. While some have argued that it is more appropriate to modify the adhesive failure criteria to account for an interaction between shear and peel stresses, the author contends that the presence of any significant peel stresses necessarily detracts from the shear strength of the joint. Therefore, to improve structural effi-

ciency, those peel stresses should be removed from the structure by simple modifications in design detail rather than be included in a more complicated failure criterion. Such a philosophy also simplifies the analyses by separating the tasks of characterizing the adhesive stress components.

The simple design modifications that reduce the peel stresses to insignificance are shown in Fig. 10. The idea is to make the tips of the adherends thin and flexible so that only negligible peel stresses can develop. Reference 23

discusses the effects of variations in bond-line thickness like those shown in Fig. 10. The local thickening shown is beneficial, and, as could be expected, any pinch-off would be detrimental. Such local thickening of the adhesive layer must be used with caution with high-flow heat-cured adhesives, lest voids be created by capillary action. Additional adhesive or scrim fillers can be used, if necessary, to avoid any such problems.

The exact proportions in tapering the adherend or thickening the adhesive layer are not otherwise critical. If the overlap is long enough, it is impossible to overdo the peel-stress relief. This is explained by Fig. 11, which shows that the joint strength remains constant with varying amounts of tapering because the other end of the joint, where no peel stresses develop, is unchanged. The precise distribution of the shear stress transfer at the

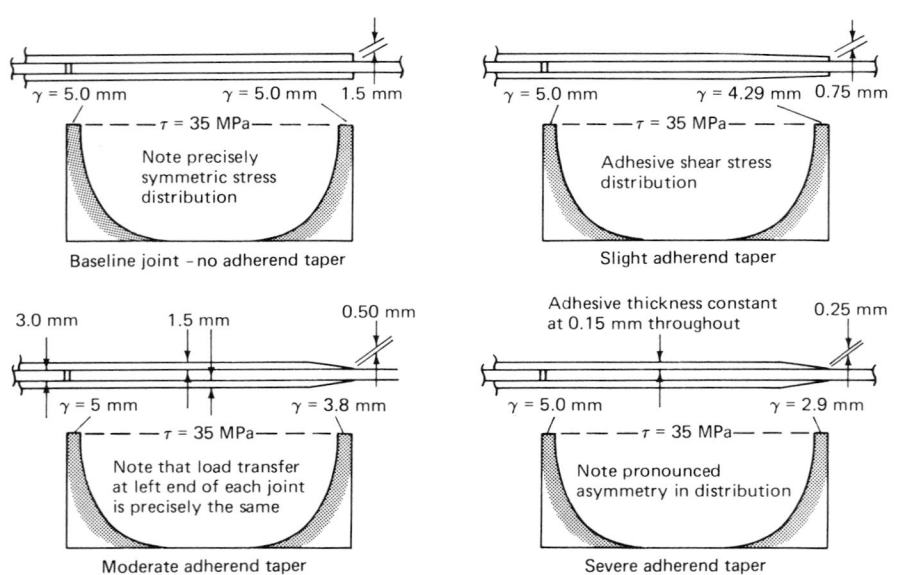

γ = 5.0 mm γ = 5.0 mm   1.5 mm

τ = 35 MPa

Note precisely symmetric stress distribution

Baseline joint - no adherend taper

γ = 5.0 mm γ = 4.29 mm   0.75 mm

τ = 35 MPa

Adhesive shear stress distribution

Slight adherend taper

3.0 mm   1.5 mm   0.50 mm

γ = 5 mm γ = 3.8 mm

τ = 35 MPa

Note that load transfer at left end of each joint is precisely the same

Moderate adherend taper

Adhesive thickness constant at 0.15 mm throughout   0.25 mm

γ = 5.0 mm γ = 2.9 mm

τ = 35 MPa

Note pronounced asymmetry in distribution

Severe adherend taper

Fig. 11 Insensitivity of adhesively bonded joint strength to modifications at one end of joint only. Adhesive strain at right end of joint decreases with more taper.

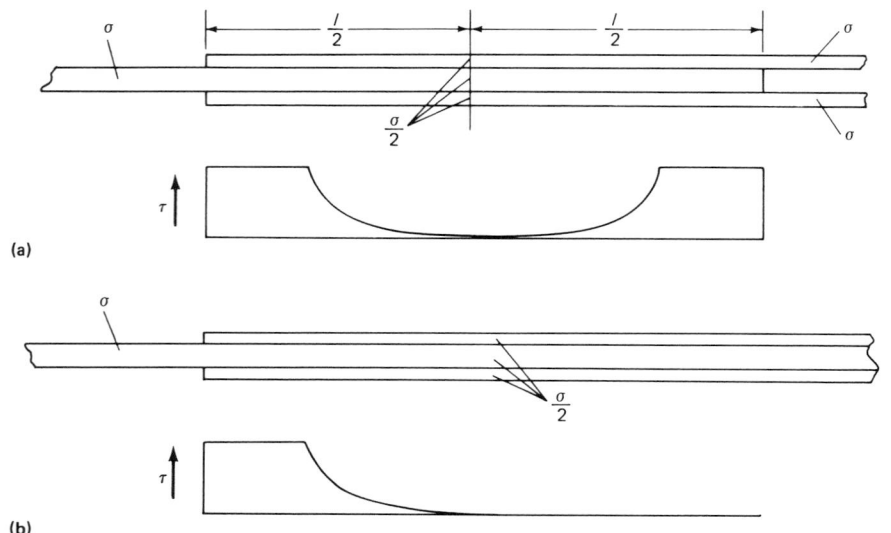

Fig. 12 Similarity of bonded stresses in joints and doublers. (a) Same adhesive stresses in each case. (b) Same maximum adhesive shear strain for same adherends and metal stresses

tapered end is modified, but the integral of those shear stresses is not. This insensitivity can also be deduced from the comparison of bonded joints and bonded doublers in Fig. 12. Compatibility of deformations for long overlaps requires that there be uniform strain at the middle of the joint and that consequently, for stiffness-balanced joints as shown, half the load must be transferred at each end of the joint even if the ends are not identical. For long-overlap bonded joints, it is fair to say that the adhesive at one end of the joint is unaware of the presence or absence of the other end of the joint. In other words, the adhesive stresses around the edges of bonded splices are the same as those around the periphery of wide-area doublers.

Elastic-Plastic Adhesive Shear Model

The linearly elastic analysis of bonded joints has been found to be far too conservative for the strong ductile adhesives used on subsonic transport aircraft. Of the possible nonlinear models that could have been proposed to characterize the actual adhesive behavior, only the simple elastic-plastic model has proved amenable to widespread application. This is because the mathematical simplicity permitted explicit closed-form solutions to be obtained for the simpler joints and those results facilitated comprehensive parametric studies. In addition, those same closed-form solutions apply to each step of the more complex and stronger stepped-lap joints. The elastic-plastic model in Fig. 13 is shown in comparison with an actual stress-strain curve, which is now customarily measured on thick-adherend test specimens using the Krieger KGR-I extensometer shown in Fig. 14.

The mathematical model at ultimate load has the same peak shear stress and strain as the actual characteristic, and the same strain energy (area under the curve). This was established as the appropriate model in Ref 25, in which the analysis for double-lap joints using the bilinear model shown in Fig. 13 reveals that the predicted joint strength would be the same for any two-straight-line adhesive model having the same strain energy. In other words, the only advantage of the bilinear model is that a single model would work for all load levels. The elastic-plastic model needs to be adjusted for less than ultimate loads, as shown in Fig. 13. Actually, it is usually sufficient to perform only two analyses: a linearly elastic one for limit load using the actual adhesive shear modulus, and an elastic-plastic model to predict the ultimate joint strength. The latter model is inappropriate for low load levels because the initial shear modulus is too low then, and the elastic strain energy is too high.

The adhesive stress-strain curves in shear vary with the operating environment, as shown in Fig. 15 for a typical ductile adhesive. While the individual properties, such as peak shear strength and strain-to-failure, vary greatly with the operating temperature, the areas under all three curves in Fig. 15 are quite similar. The ultimate strength of a long-overlap bonded joint between uniform adherends is shown by analysis to be defined by the strain energy of the adhesive layer in shear, not by any individual properties such as peak shear stress (Ref 25). The strain energy is proportional to the area under each of the curves in Fig. 15. Consequently, the strengths of realistically configured bonded joints are not very sensitive to the operating environment, provided that the temperature is kept below the glass transition temperature for each adhesive.

The difference between the behavior of ductile and brittle adhesives is characterized in Fig. 16. That difference is not as pronounced at the upper service temperatures, at which even the brittle adhesives are considerably ductile. It should also be noted that even at room temperature the brittle adhesive characteristic is significantly nonlinear. The reasons that both ductile and brittle adhesives have been developed are that ductile adhesives are typically limited to service environments no greater than about

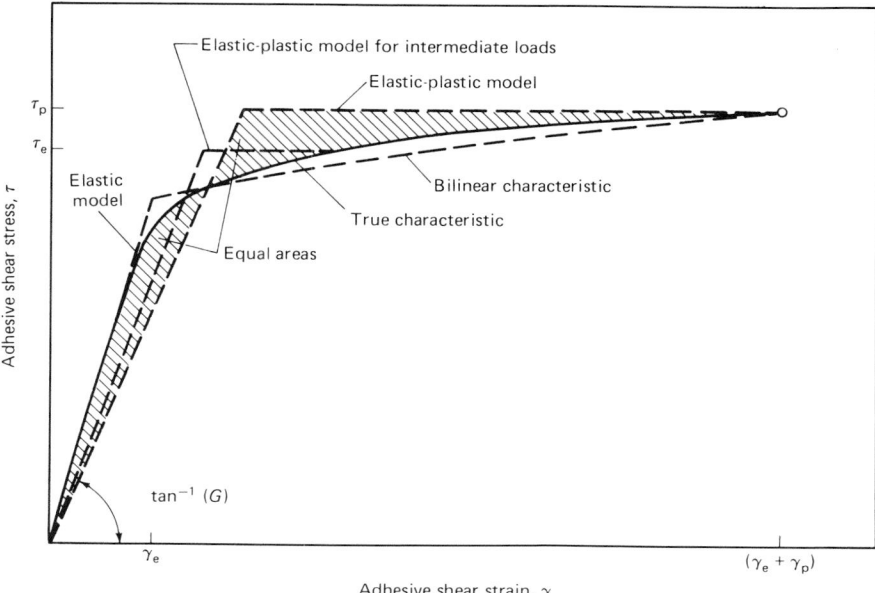

Fig. 13 Representations of adhesive nonlinear shear behavior

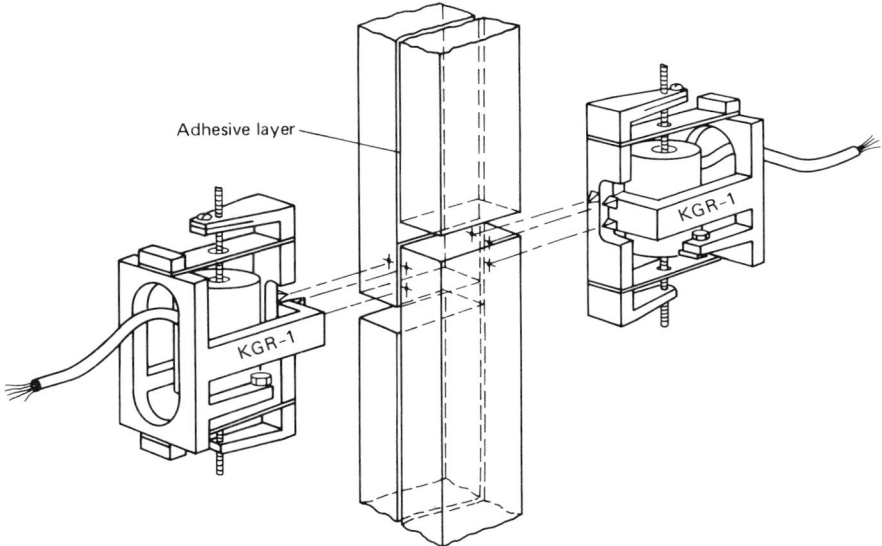

Fig. 14 KGR-1 extensometer and thick-adherend adhesive test specimen. Source: Ref 24

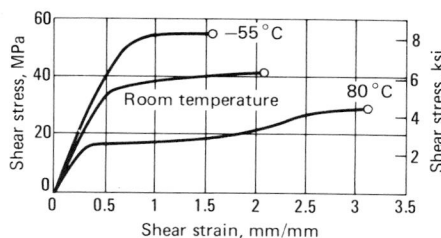

Fig. 15 Effect of temperature on adhesive stress-strain curves in shear. Nylon-epoxy adhesive (120 °C, or 250 °F cure)

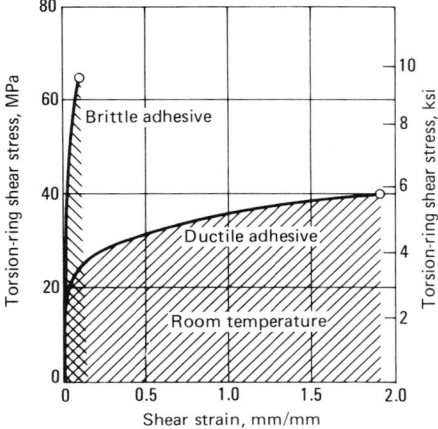

Fig. 16 Adhesive stress-strain curves in shear

70 °C (160 °F), and that there are some applications (in proximity to engines or on supersonic aircraft) where brittle adhesive bonding is still viable even if much of the strength has had to be sacrificed to attain much higher service temperatures.

Single-Lap Adhesively Bonded Joints

The eccentricity in load path inherent in unsupported single-lap joints decreases the joint strength below the level that could have been developed in the absence of the bending asso-

ciated with that eccentricity. That loss of joint strength is quantified in Fig. 17, which shows, for example, that for the standard ASTM D 1002 lap-shear test coupon, for which the abscissa is about 2, the structural efficiency is limited to no more than one-third. (The structural efficiency is defined as the ratio of the direct membrane stress outside the joint to the sum of the stretching and bending stresses at the ends of the bonded overlap.) However, Fig. 17 also points the way to alleviating the problem, that is, by increasing the overlap from the 8 to 1 ratio used on the test coupon to about 80 to 1 for structural joints. The joint can never be as

strong as the basic structure outside it. However, for long overlaps and thin adherends, the weakness is in the adherends rather than in the adhesive, and, in any case, the joint strength need only exceed the alternative, that is, riveting, which causes a significant loss of strength because of the holes.

A detailed analysis of single-lap bonded joints can be found in Ref 14. Naturally, if the joint is supported against out-of-plane rotations, the appropriate method of analysis would be to consider the actual joint as one side of a double-lap joint that was twice as thick. Normally, single-lap joints should not be considered for joints thicker than about 1.8 mm (0.07 in.) of aluminum alloy or its equivalent. The same peel-stress relief techniques shown in Fig. 10 are equally applicable to single-lap and single-strap joints, except that protection is needed at both ends of the bonded overlap instead of only at one.

Stepped-Lap Adhesively Bonded Joints

Composite laminates that are too thick and hence too strong to be joined by simple uniform lap-splice bonded joints can be bonded together successfully by stepped-lap joints of the type shown in Fig. 18. (Glued scarf joints, which work well for wood at a slope of about 1 in 20 at most, are not as attractive for advanced

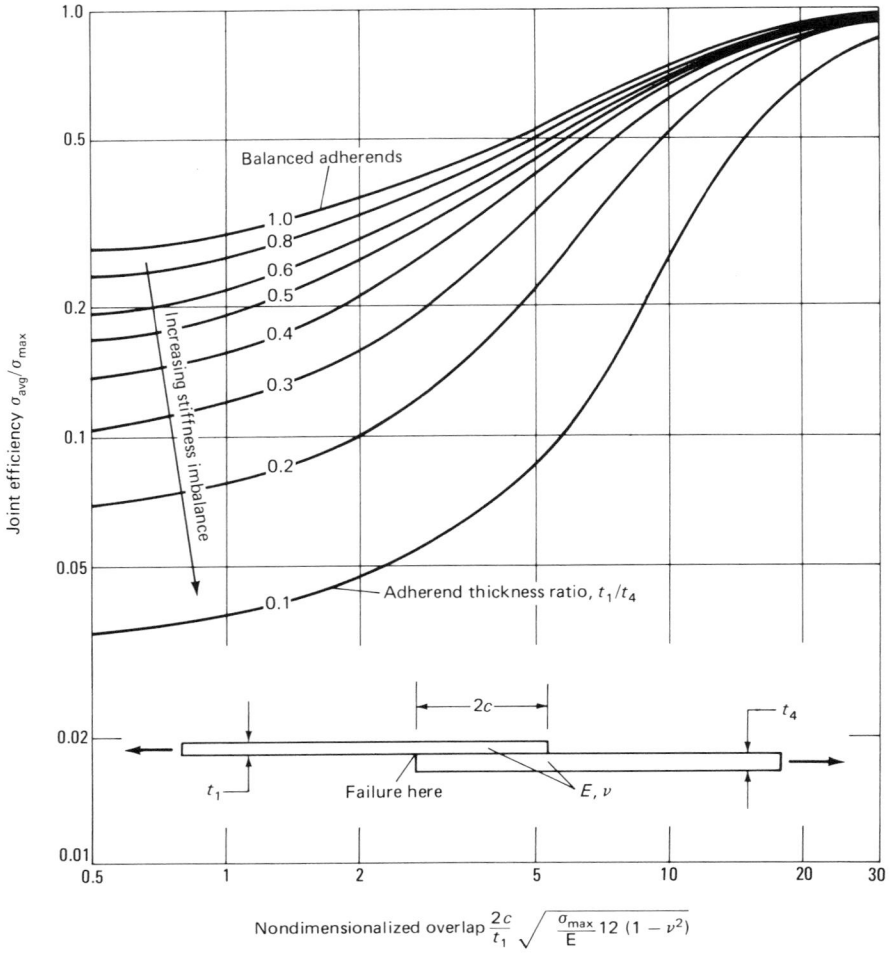

Fig. 17 Effect of adherend stiffness imbalance on adherend bending strength of single-lap bonded joints. Thinner adherend t_1 critical in combined bending and axial load at end of overlap

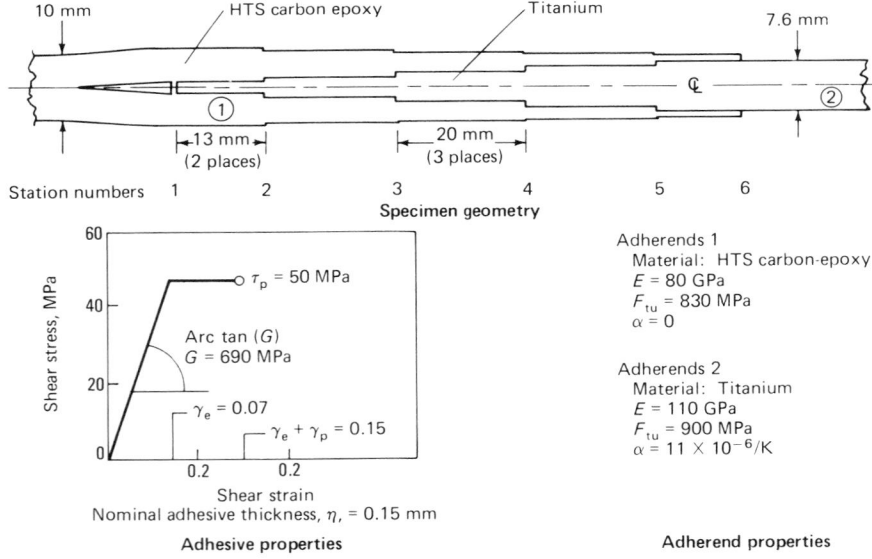

Fig. 18 Typical stepped-lap adhesively bonded joint

composites because the slope needs to be much shallower, at no more than about 1 in 50, resulting in scarf joints for thick laminates that are very large.)

Each step of the stepped-lap bonded joint is governed by the same differential equation that applies to double-lap joints. Consequently, there is a highly nonuniform shear stress distribution in the adhesive, with the load transfer concentrated at the ends of each step, as shown in Fig. 19.

The design and analysis of stepped-lap bonded joints have been made straightforward by the digital computer programs A4EG and A4EI, which are based on an elastic-plastic shear model (Ref 26, 27). The first program employs a uniform adhesive layer throughout the joint. The second program expanded on the first and incorporated variable adhesive properties so that the effect of flaws and defects could be investigated (worked examples of joints analyzed by these programs are found in Ref 4, 20, 21, 23, and 28). It should be noted that the A4EG and A4EI programs do not merely predict the strength of joints of specified geometries, but also serve as valuable tools for improving the initial designs.

Some rules of thumb for initial design are that the end steps must be neither too thick nor too long—0.76-mm (0.030-in.) thick titanium with a step length of 9.5 mm (0.375 in.) is typical—lest premature fatigue failures occur. Most of the other steps are usually 12.7 to 19.1 mm (0.50 to 0.75 in.) long, with step thicknesses no greater than 0.5 mm (0.02 in.) (and preferably much less) on each side of the joint, with one longer step near the middle of the overlap to provide creep resistance. The computer programs give enough detailed information about the internal stresses within the joint for the initial design to be improved upon by modifying design details. Any poorly proportioned step usually shows up rapidly. Once the proportions have been adjusted properly, the most powerful variable with which to increase the joint strength is the number of steps; merely increasing the bond area for the same number of steps is not effective. The strength continues to increase as the number of steps is increased to one per ply, and the number should always be sufficient to ensure that the bond strength (calculated by overriding the adherend strength limits while maintaining the same stiffnesses) exceeds the adherend strengths by at least 50%.

Load Redistributions With Flawed and Damaged Adhesively Bonded Joints

It is not sufficient to design bonded splices to a strength that is only adequate for the nominally applied loads unless there is no need for damage tolerance. If a bonded splice is everywhere weaker than the local strength of the adherends just outside the joint, any load redis-

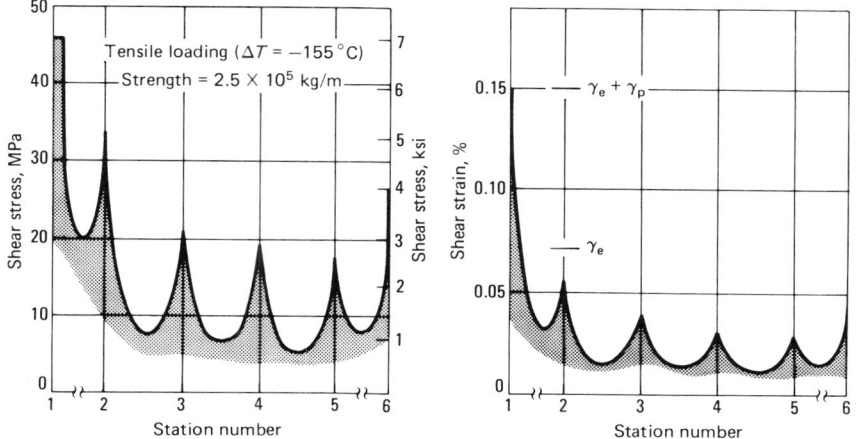

Fig. 19 Adhesive shear stresses and strains in unflawed joint

tribution associated with local flaws in the bond or damage to it could cause the remaining bond to unzip catastrophically. This phenomenon is explained in Fig. 4; the uniform remote stress in the adherends must be kept lower for larger flaws or damage lest the remaining bond become overloaded and fail just outside the ineffective bond area.

However, if the intact bond outside the weakened area were stronger than the adherends, no amount of load redistribution could ever fail the bond, as explained in Ref 4 and 20. Any subsequent failure would be transferred to the adherends, where it would become visible much sooner. The appropriate analysis for estimating the remaining life in the damaged bonded structure then pertains to the adherends and not to the adhesive. The same is true when structures so thick that they should not be bonded are bonded and then are protected by fail-safe mechanical fasteners. Any initial flaws or damage will unzip sufficiently to enable some of the fasteners to pick up the load through the defective area. From that point on,

the basic load transfer is redistributed, as explained in Ref 20 and 27. The disbond will grow no further and the subsequent life will be determined by the most severely loaded fastener through the area of defective bond.

A word of caution is in order about the current fashion of misapplying fracture mechanics theory (from cracked metal structures) to estimating the life of adhesively bonded structures. Such a malpractice is more likely to result in bonded structures inferior to those designed by classical nonlinear strength-of-materials approaches than it is to result in improvements. The key to designing bonded structures is that the nominally unflawed bond must never be weaker than the members being joined. If the bond is weaker, the adherends determine the structural life, whether there is a bond flaw or not. The notion that it is permissible to design weak-link fuse bonded joints and justify this practice by calculating a supposedly adequate finite fatigue life must be discouraged. The associated potential for instantaneous unzipping of the remaining bond

area would place an intolerable burden on inspection. Finding cracks in metal structures has been difficult enough, and they are much easier to detect.

Further, the two situations (flawed bonds and cracked metal) are not at all comparable. The basis of the damage tolerance approach to cracked metal structures is that there will be a relatively long period of slow, stable crack growth before a critical crack length is reached. That usually allows sufficient opportunity to detect small cracks and repair them. In adhesive bonds, on the other hand, the measured disbond rates have either been so rapid that the bonds would need to be inspected several times on each flight to ensure that it was safe to continue, or so slow that no growth could be detected. There has been no in-between behavior that would lend itself to metallic damage tolerance techniques. This difference is explained in Fig. 20 (from Ref 29). Properly designed bonded joints, on the other hand, exhibit remarkable tolerance to quite large bond flaws. Fracture mechanics analyses can be valuable in this context when used to calculate thresholds below which initial damage will not propagate rather than to calculate rates of damage growth.

Another beneficial use of fracture mechanics analysis of bonded and laminated composite structures is to identify design details and stacking sequences that should be avoided (see Ref 30). Special attention should be paid to the numerous publications on delaminations in composites by T.K. O'Brien at NASA Langley; one of these is included in Ref 30. The need for such an approach is accentuated by the current lack of capability in the standard composite laminate theories to account for residual thermal stresses in the resin, which is, after all, where the cracks and delaminations in fiber-reinforced composites occur. While a few of the published laminate theories do an adequate job of predicting the fiber-dominated strengths of composite laminates under arbitrary in-plane

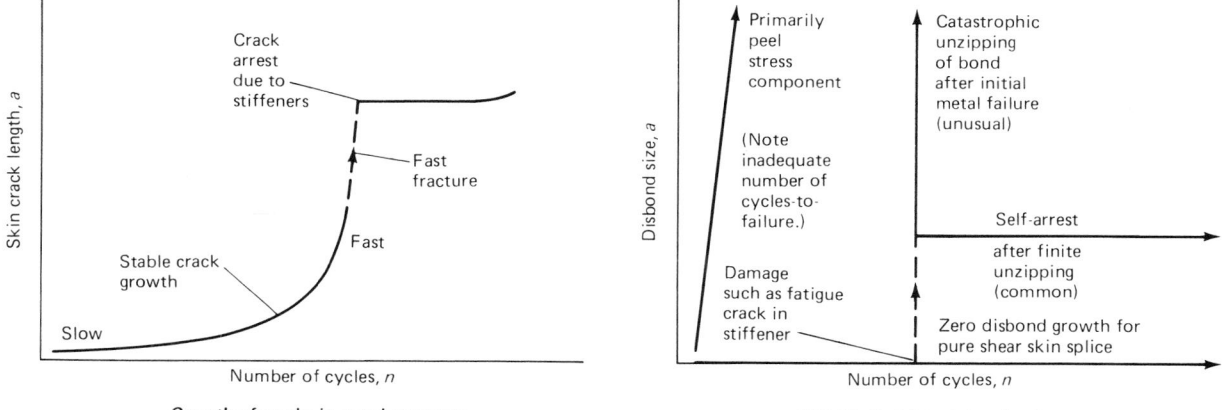

Fig. 20 Differences between growth of cracks in metal components and adhesive bonds

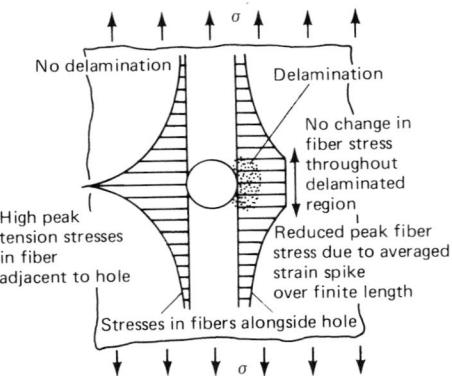

Fig. 21 Stress concentration relief in fibrous composites

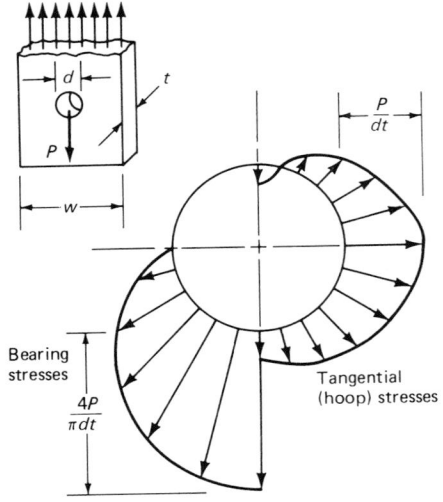

Fig. 22 Bearing and hoop stresses at bolt hole

loads, none of the published theories can do justice to resin-dominated strengths, and all are quite incapable of accounting for such things as stacking-sequence effects. Most laminate theories work only for uniaxial loading and are unacceptable for biaxial loads, although new and old cross-plied laminate theories that do work satisfactorily for biaxial in-plane loads are discussed in Ref 31 and 32. Such deficiencies in the laminate theories may be the reason that all bolted composite joint analyses have had to rely on substantial empirically determined correlation factors between theory and test.

Fundamentals of Shear Load Transfer Through Mechanical Fasteners

Perhaps the most important thing to understand about mechanically fastened joints in fibrous composite structures is that the eventual failure of the composite occurs long after the laminate has stopped behaving like the one-phase homogeneous engineering material on which it is usually modeled. While the fibers and the resin matrix are both essentially linear until failure, the microcracks and delaminations around bolt holes in composite laminates cause substantial internal load redistributions that are not accounted for in conventional mathematical models of bolted or riveted composite joints. Thus, there appears to be substantial nonlinear behavior associated with the normal rivet or bolt sizes used in composite structures. However, while there are indeed softened zones at the microlevel, as shown in Fig. 21, this softening is unlike the yielding associated with ductile metal alloys in similar circumstances.

However, there are striking similarities in the behavior at the macrolevel. For example, the residual-stress zone around cold-worked holes in metal structures leads to very substantial increases in the fatigue life. Likewise, any increase in the albeit much smaller softened zone around fasteners in composites causes an increase in the static strength. Either the careful installation of interference-fit fasteners or the

gentle fatiguing of bolted composite structures can increase the static tensile strength ever closer toward the unnotched net-section strength. The corresponding increase in compressive strength will not be as dramatic because it is dominated more by bearing stresses than by net-section stresses.

There is a very strong similarity between the effects of design details on the fatigue strength of bolted or riveted metal structures and on the static strength of composite structures. Perhaps the strongest analogy is associated with the desirability of restricting the bearing stress. As shown in Fig. 22, at the elastic level the peak tension stress alongside the loaded bolt hole in an isotropic panel is on the same order of magnitude as the average bearing stress, P/dt. Keeping the bearing stress low is the key to structurally efficient bolted composite joints, particularly for multirow joints, as explained in the section "Multirow Bolted Composite Joints" in this article.

There are also some curious juxtapositions of behavior between metallic and composite components. For example, in metal aircraft structures, a severe application of load early in the life of the aircraft will retard the subsequent growth of cracks by creating a larger plastic zone at the crack tips or likely crack sites. Conversely, if all bolted composite aircraft were subjected to five lifetimes of fatigue testing before delivery, their ultimate strength would be increased significantly, perhaps by as much as a factor of 2.

Because of the distinctly two-phase material behavior of fibrous composites around both loaded and unloaded holes (Fig. 21), there will be a continued need for a substantial empirically established correlation factor to reconcile test and theory for bolted composite joints. This can be done in a straightforward manner, as in the author's hypothesis (Ref 33-35, 27), in which the amount of stress-concentration relief is assumed to be proportional to the intensity of

the original elastic stress concentration. That hypothesis leads to an easily calculated residual stress concentration for other geometries that have not been tested. With a correlation factor determined from single-hole test specimens as a starting point, this method has been shown to be effective in predicting the strength of highly loaded multirow bolted composite joints (Ref 36). This method of analysis is presented here.

The literature on bolted composite joints contains another basic approach to this nonlinearity problem, usually referred to as the characteristic-length or characteristic-offset approach. The origins of this approach are the point stress and average-stress failure criteria (Ref 37). With that method, the linearly elastic analysis is presumed to be valid outside some empirically determined softened zone adjacent to the hole. The basic drawback to that approach, which can of course always be shown to be capable of explaining any test results one at a time, is that the so-called characteristic dimension varies considerably with bearing stress and that failure is being predicted at some place other than where it is known to occur. Nevertheless, both methods of analysis will continue to be used until it is possible to cover all joint geometries and bearing stress intensities with a single theory. At present, each approach covers some situations not covered by the other. Both have been used successfully in hardware applications, and both have led to increased understanding of stress concentrations around bolt holes in composite structures.

Single-Hole Bolted Composite Joints

The methods developed in Ref 33 and 34 for the analysis and design of bolted or riveted composite joints call for a major empirical correlation factor. Once that is accepted, it makes sense to use analyses for elastic isotropic materials, as reference points, because they are simple and widely available. Corrections for both the nonlinear behavior described in Fig. 21 and for orthotropy can be combined into a single factor, provided that the mode and location of failure do not change. Separate analyses are needed for bearing failures and for the tension-through-the-hole failure modes. The various modes of failure are illustrated in Fig. 23. Compression strengths tend to be higher and not as sensitive to stress concentrations because some of the load can be transmitted through the fastener instead of the entire load going around it.

The other major failure mode, shearout, does not occur within the shaded area of laminate patterns in Fig. 1. Outside that area, the widespread splitting accompanying shearout and the low load level at which it occurs (see, for example, Ref 33 and 38) discourage the installation of fasteners in such highly orthotropic laminates. A weakness in shearout cannot be corrected by adding more edge distance for the fasteners.

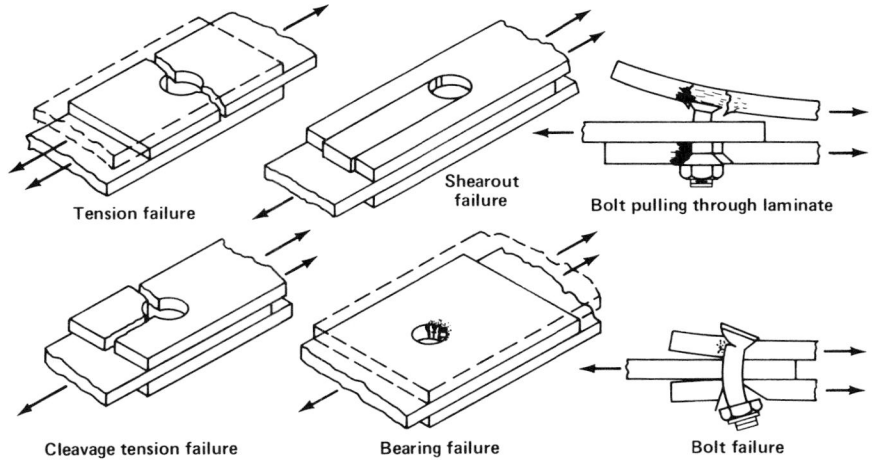

Fig. 23 Modes of failure for bolted joints in advanced composites

Tension failure

Shearout failure

Bolt pulling through laminate

Cleavage tension failure

Bearing failure

Bolt failure

is then considerable relief of the stress-concentration factors, as shown in Fig. 24, which also includes the efficiency of bolted joints in highly ductile and perfectly brittle materials. It is evident that fibrous composite behavior cannot be predicted on the basis of a minor perturbation from either the elastic or plastic analyses shown. (It should also be noted that the strength increase shown in Fig. 24 for roughly 6.5-mm (0.25-in.) diameter bolts in carbon-epoxy composites decreases asymptotically as the fastener diameter increases, and, for very large bolt holes or cutouts, the linearly elastic predictions would be expected to apply.)

The origin of this substantial stress-concentration relief for typical fastener sizes is explained in Fig. 25, in which the theoretical stress-concentration factors, K_{te}, are calculated according to Eq 1. The observed stress concentration at failure, K_{tc}, is calculated as:

$$K_{tc} = F_{tu}(w - d)t/P \qquad (Eq\ 3)$$

where P is the load at which the specimen failed and the numerator is the unnotched net-section strength. It was postulated in Ref 33 and 34 that the amount of (nonlinear) relief would be proportional to the intensity of the original (elastic) stress concentration. Thus, the effective stress-concentration factor experienced by the composite laminate at loaded and unloaded bolt holes is taken to be:

$$K_{tc} = 1 + C(K_{te} - 1) \qquad (Eq\ 4)$$

in which the correlation factor, C, varies with both the fiber pattern and the hole size. A value of 0 for C would indicate complete stress-concentration relief, as with ductile metals,

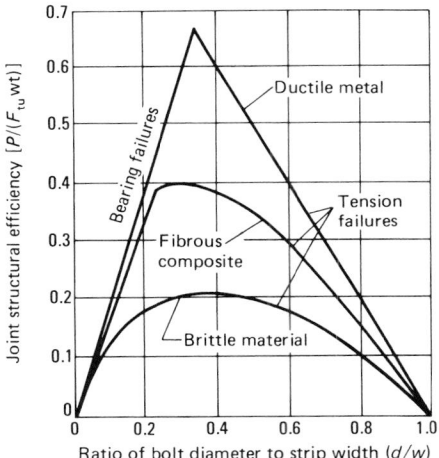

Fig. 24 Relationship between strengths of bolted joints in ductile, fibrous composite, and brittle materials

geometries. A major benefit of having explicit expressions for these stress concentrations is that they facilitate parametric studies and identification of optimum joint geometries. The maximum strength of brittle perfectly elastic strips loaded by a central bolt is 21% of the unnotched strength at an optimum w/d ratio of about 2.5. The bolts should be placed close together to minimize the peak hoop tension stress, which, as shown in Fig. 22, is on the same order as the average bearing stress.

Fortunately, at least for the small, typically 6.5-mm (0.25-in.) diameter fasteners used in most aircraft structures, the pessimistic outlook for bolt holes in window glass materials does not occur in fiber-reinforced composites because of the separate behavior of the two distinct constituents of such composites. There

References 33 and 34 contain formulas for the elastic-isotropic stress concentration associated with a loaded bolt hole of diameter, d, in a finite strip of width, w. With respect to the average net-section tension stress, the peak stress-concentration factor on the net section immediately adjacent to the hole is:

$$K_{te} = \frac{w}{d} + 1 - 1.5\ \frac{(w/d - 1)}{(w/d + 1)} \qquad (Eq\ 1)$$

provided that the edge distance, e, is adequate (that is, $e \geq w$). The cited references contain modifications for short edge distances. R.B. Heywood gives the corresponding expression for the stress-concentration factor at an unloaded hole in the middle of a strip (Ref 39):

$$K_{te} = 2 + (1 - d/w)^3 \qquad (Eq\ 2)$$

Reference 33 also contains formulas for stress conditions associated with holes in different

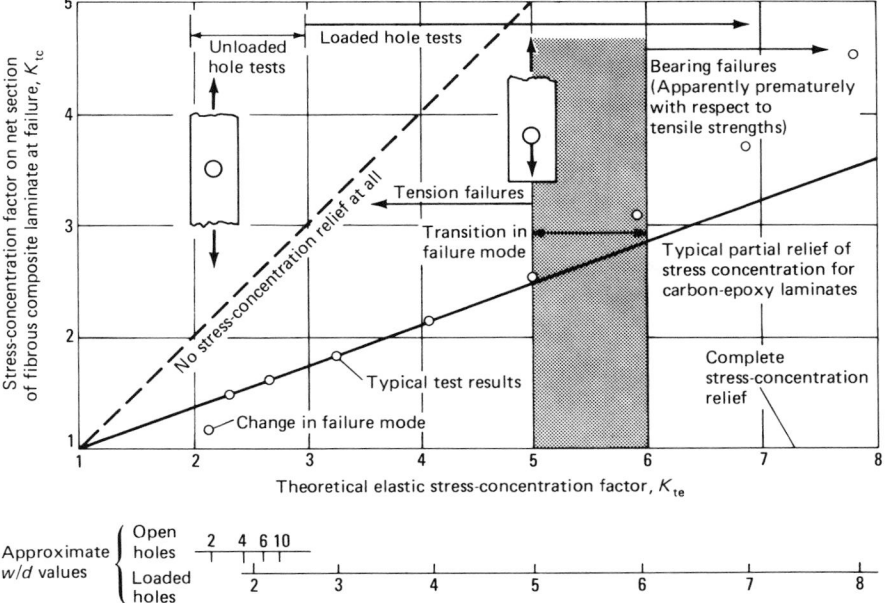

Fig. 25 Relation between stress-concentration factors observed at failure of fibrous composite laminates and predicted for perfectly elastic isotropic materials

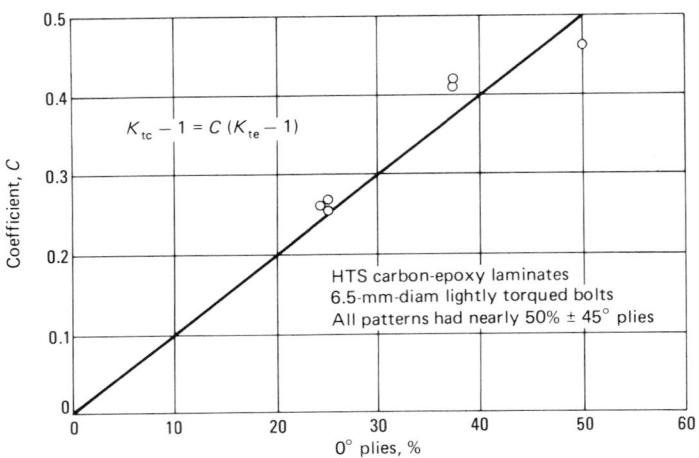

Fig. 26 Stress-concentration relief at bolt holes in composite laminates

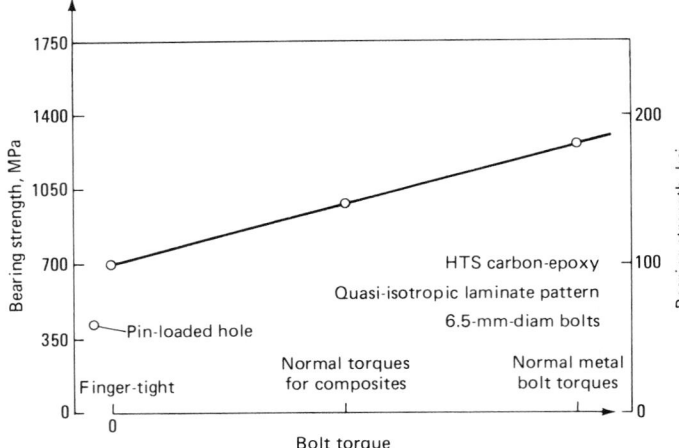

Fig. 27 Effect of bolt torque on bearing strength of fibrous composite laminates

while a value of 1 (for quasi-isotropic materials) would indicate no relief at all. As a useful aide-mémoire, the value of the C factor has been found to be close to 0.25 for 0.25-in. bolts in three different carbon-epoxy quasi-isotropic laminates, which fraction is the same as the percentage of 0° plies. Higher values have been deduced for orthotropic laminates. For 6.5-mm (0.25-in.) diameter fasteners in laminates within the shaded area in Fig. 1:

$$C \approx (\% \ 0° \text{ plies})/100 \qquad (\text{Eq } 5)$$

as shown in Fig. 26.

Absorbing the orthotropy factor into the single coefficient C can be justified only when the mode of failure does not change. Whether this effect of orthotropy were incorporated into an expanded abscissa or into a steepened slope of the line in Fig. 25, the result would be the same. This combination would be expected to become invalid for most of the fiber patterns outside the shaded area shown in Fig. 1 because of a predominance of shearout failures.

For small d/w (or large w/d) ratios, the laminates will fail at a lower load in bearing under the bolt rather than by tension through the hole. Therefore, a lower-bound cutoff is needed to cover this failure mode. This is shown on the left of the middle curve in Fig. 24. Such bearing failures also occur for ductile metal alloys, as shown to the left of the top curve in Fig. 24.

The optimum w/d ratio, when allowance is made for the nonlinear behavior of the composite around the bolt holes, is approximately 3 to 1 for 6.5-mm (0.25-in.) diameter bolts, being slightly higher for smaller bolts and lower for larger ones. These optimum ratios pertain to single-row joints, in which all of the load is transferred through a single row of fasteners; different values are given in the section "Multi-row Bolted Composite Joints" in this article. Also, whereas the maximum strength for ductile metals occurs at the intersection of the

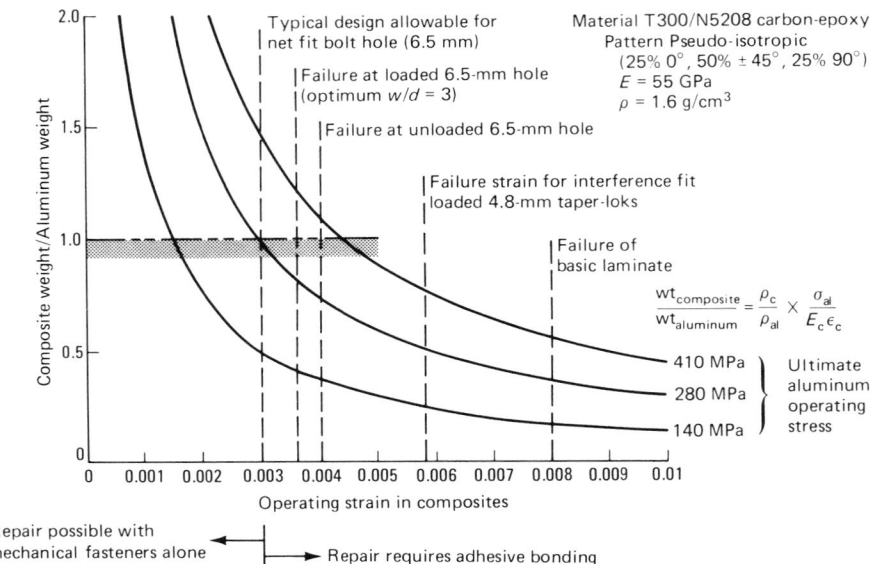

Fig. 28 Relative weights of aluminum and carbon-epoxy composite structures. Failure strains are less for loaded holes, for using statistical basis rather than average, and for larger holes. Reliance on benefit from interference fit fasteners requires absolutely no net or loose-fit bolts. Otherwise, static strength will be reduced by factor of 2. Multiple rows of bolts in uniformly thick members increase strength by only a few percent. Further strain limits for damage tolerance and impact resistance are not yet established.

bearing and tension strengths, the maximum efficiency for single-row bolted composite joints is developed by tensile failures of the net section, at a bearing stress that is typically only about three-fourths of the bearing stress the composite material could withstand at wider bolt spacings. The philosophy of simultaneous failures is definitely not applicable here.

The bearing strength of composite laminates is strongly influenced by the presence or absence of any through-the-thickness clamp-up, as shown in Fig. 27. There is a nearly 2 to 1 difference between the pin-loaded case, in which there is no clamp-up at all, and the finger-tight case, for which the bolt head and nut prevent any initially damaged composite material from unloading itself by deflecting

sideways. Because all the material is confined under these circumstances, the joint continues to carry on to higher loads, as shown. This improvement in strength is customarily accounted for in design. The even greater strengths achievable by torquing the bolts down tightly should not be relied upon for design purposes because it would be very difficult to detect a single undertorqued fastener that would substantially reduce the static strength of the composite structure. In metal structures, on the other hand, the loss of such clamp-up would merely reduce the fatigue life with no associated reduction in static strength. In any case, the improvement in bearing strength of composites due to additional bolt torque is often not realized because tension-through-the-hole

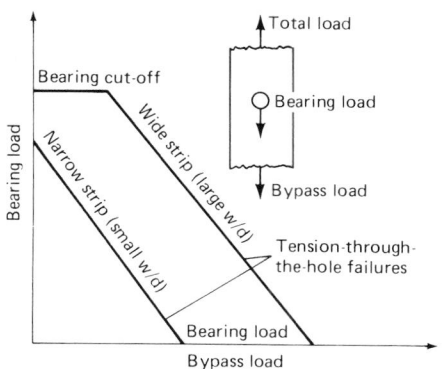

Fig. 29 Bearing-bypass load interaction for loaded bolts in advanced composites

strength may govern the design, particularly for large fasteners.

Multirow Bolted Composite Joints

The limited structural efficiency of bolted composite laminates containing single-row splices is not sufficient for them to compete on a weight basis with well-designed aluminum alloy structures. This is indicated by Fig. 28, which points to the need for operating at higher strain levels. However, this need has sometimes been misinterpreted and has been responded to by changing to structurally inferior patterns to acquire the increase in strain, only to have that goal nullified by an associated reduction in modulus. Interestingly, diagrams similar to Fig. 28, but prepared for mildly orthotropic laminate patterns, have shown that the increase in strength associated with additional 0° plies (those in the primary direction of loading) are almost nullified by the associated reduction in strain-to-failure because of the higher stress-concentration factors, K_{tc}. Likewise, the lower stress concentrations and higher strains-to-failure associated with softened laminates carry with them a balancing decrease in strength because of the lower modulus.

Tinkering with the fiber pattern usually cannot substantially enhance the strength of fibrous composite laminates. A well-known exception is the use of local softening strips and pad-ups, which can virtually eliminate the effect of stress concentrations with respect to the basic laminate. However, such an approach leaves the structure outside those locally protected areas with little, if any, damage tolerance (because of the higher operating strain permitted by the softening strips and pad-ups) and severely limits the opportunity to perform repairs; so the situations in which such an approach is practical are limited.

An alternative technique for improving the structural efficiency of bolted or riveted composite structures is to improve the joints themselves, by using more than one row of fasteners in conjunction with tapered splice plates. The analyses in Ref 33 and 34 pointed the way to

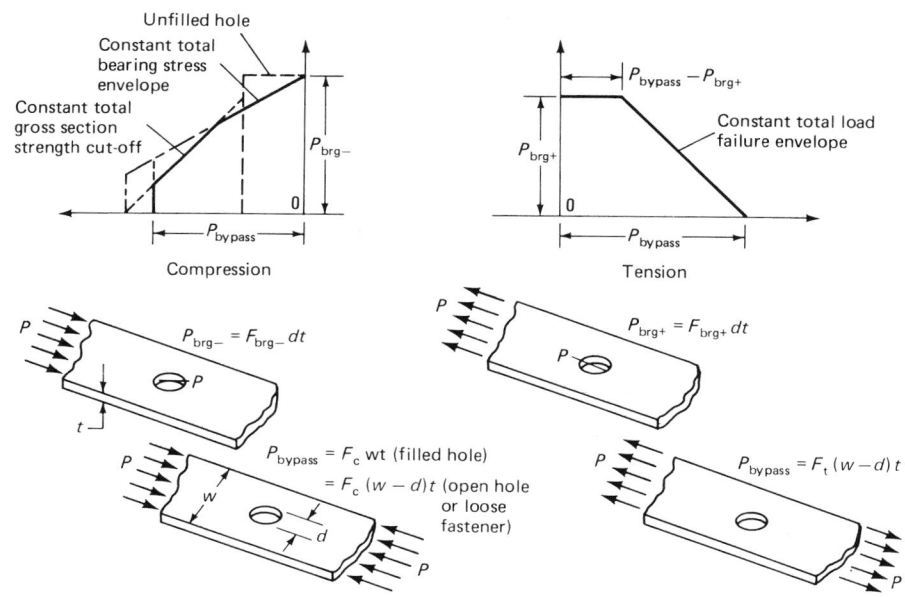

Fig. 30 Outer envelope of bearing-bypass load interactions

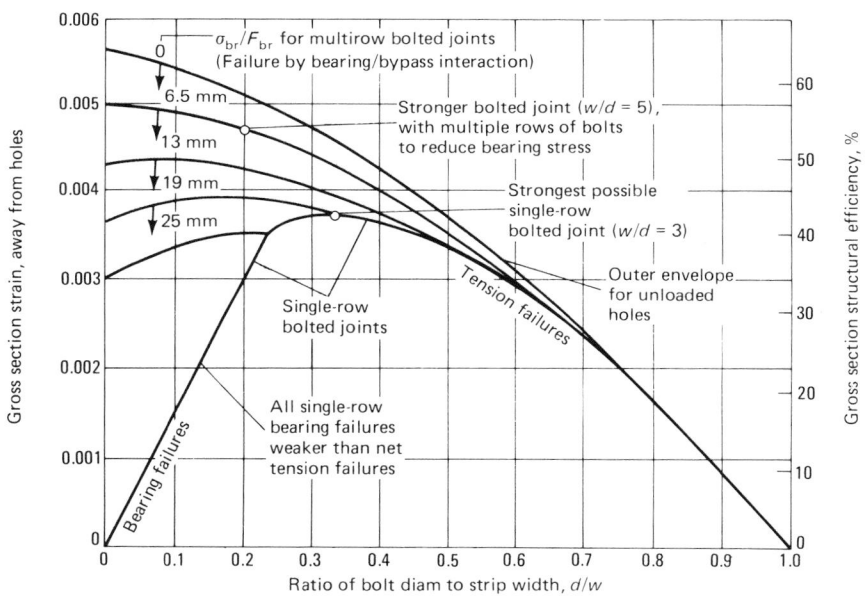

Fig. 31 Influence of bolted joint design on structural efficiency of carbon-epoxy composite structures

accomplishing this almost a decade ago, and considerable progress has since been made in the design and analysis of such joints. The validity of the methods has been confirmed by extensive testing of large, highly loaded bolted composite joints (Ref 36). The key to the analysis of multirow bolted composite joints is a formula for the bearing-bypass interaction. Such an interaction, for tensile loads, is shown in Fig. 29, where the terms bearing load and bypass load are defined. The bearing load is reacted at the particular bolt under consideration, while the bypass load, interrupted by the

bolt hole, passes by and is reacted elsewhere. Figure 30 also covers compressive loading, for which the bearing-bypass interaction is more complicated.

When the bearing load is high enough, there is a bearing-stress cutoff for sufficiently wide fastener spacing, but for closer spacings, the failure will be in the net section whether the load is all taken out on that fastener (pure bearing) or all reacted at other fasteners in the joint (pure bypass). The extremities of the interactions could be established by test if a theory were not available. The real key to Fig.

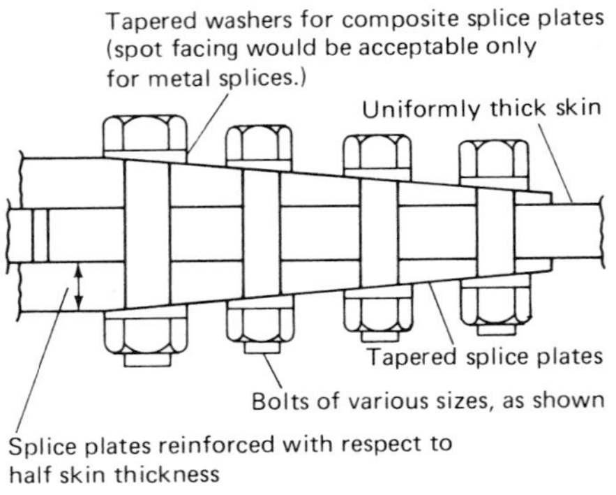

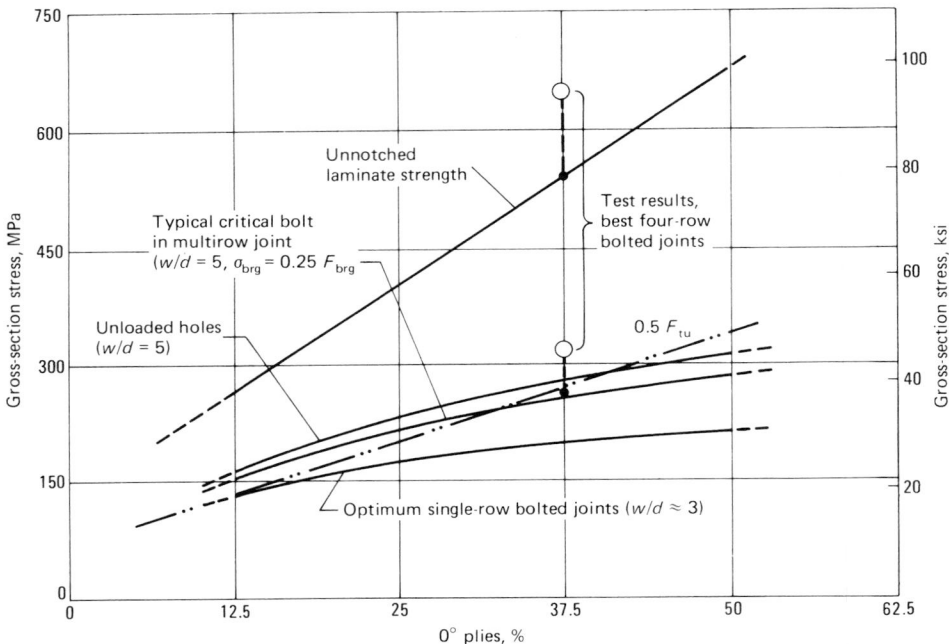

Fig. 32 Optimum proportions for multirow bolted composite joints

Gross-section stress, MPa (left axis) / Gross-section stress, ksi (right axis) vs. 0° plies, %

Unnotched laminate strength

Typical critical bolt in multirow joint ($w/d = 5$, $\sigma_{brg} = 0.25\,F_{brg}$)

Unloaded holes ($w/d = 5$)

Test results, best four-row bolted joints

$0.5\,F_{tu}$

Optimum single-row bolted joints ($w/d \approx 3$)

Fig. 33 Gross-section design stresses for bolted composite structures (carbon-epoxy laminates). Chart applicable for bolts up to 9.5 mm (37 in.) in diameter. Larger bolts are associated with progressively lower laminate stresses.

29 is the linear interaction between bearing and bypass loads whenever the joint fails in tension through the hole; that is:

$$\sigma_{net}\,K_{tc} + \sigma_{brg}K_{tb} \le F_{tu}, \ (\text{and } \sigma_{brg} \le F_{brg}) \quad (\text{Eq } 6)$$

The existence of this linear interaction has been known since about 1970 and confirmed by other analysis methods. Many curves similar to Fig. 29 have been derived by the BJSFM computer program (such as Ref 40, 41) using the characteristic-offset analysis method, and the linearity has been confirmed by extensive testing reported in numerous documents. (The use of the BJSFM analysis method, with a fixed characteristic offset for all hole sizes, permits an assessment of the hole size effect, which requires varying values of the coefficient C with the present method of analysis.)

Equations 1 to 5 permit the construction of joint efficiency charts of the type shown in Fig. 31, which covers all the intermediate cases for single-row joints between a pure bypass load along the upper envelope and a pure bearing load on the lowest curve. Figure 31 reveals that the only way to improve upon the efficiency of an optimized single-row joint is to move the fasteners farther apart and simultaneously decrease the bearing stress on the critical row of fasteners.

These seemingly mutually contradictory requirements can be met only with the assistance of an accurate load-sharing program. Only one has been developed that covers nonlinear behavior in the fastener load deflection characteristic. That is the A4EJ program derived in Ref 27 and described in Ref 20.

Comparative analyses of multirow bolted composite joints reported in Ref 36 have shown that if the basic structure is to be repairable, the optimum splices must contain uniformly thick skins with tapered splice plates, as shown in Fig. 32. Also, the diameter of the fasteners varies throughout the joint. The innermost bolts, adjacent to where the skins butt together, are largest, at about a w/d ratio of 3 to 1. There is no bypass load there in the skin, so that row is optimized as a single-row joint. Obviously,

that is the worst possible thing to do to the splice plates, because the maximum bearing and bypass loads coincide there. Consequently, the splice plates must be suitably reinforced. The weight penalty associated with an off-optimum splice plate is trivial in comparison with the substantial weight savings associated with maximizing the structural efficiency of the entire basic skin.

The next two rows of fasteners are sized at a w/d ratio of 4 to 1 because there is some bypass load there, in all members. The critical fastener row is the outermost one, near a tip of the splice plate. The smallest, most flexible fasteners are used there ($w/d = 5$), and the tip thickness of the splice plate is limited to prevent those bolts from picking up too much load. It was found that the bearing stress in the skin on that critical row of fasteners could be kept under 25% of the ultimate bearing strength. The resulting high structural efficiency, at a gross-section strain of 0.005, is shown by Fig. 31 to be much higher than can be obtained with optimized single-row bolted composite joints. It is also much higher than can be achieved by nonoptimized multirow bolted composite joints. For example, adding a second row of fasteners in tandem, while retaining the optimum spacing for a single-row joint, would increase the joint strength by only about 10%, a rather small gain for having doubled the number, weight, and cost of fasteners.

The strength increase is even greater with respect to off-optimum joints, with one or more rows of fasteners, proportioned in such a way as to enforce benign failures in bearing. While that is a noble goal, it must be recognized that

there is an associated loss of about one-third of the ultimate strength to obtain that noncatastrophic indication of an overload.

Figure 33 shows how the strength of typical bolted composite structures varies with fiber pattern as well as with the efficiency of the joint design for mechanically fastened composite joints. It is apparent that the most efficient multirow joints are about 50% as strong as the unnotched parent laminate. Moreover, the joint strength is not very sensitive to the fiber pattern, even for a 6.5-mm (0.25-in.) diameter bolt, and becomes even less sensitive for much larger fasteners. The joint strength curves in Fig. 33 would tend to become much flatter and lower for larger bolt sizes for both joints, either with more than one row of fasteners or only one, if it were not that the critical fastener of a multirow joint has to be kept small to raise the structural efficiency by limiting the load it can accept.

Practical Considerations

To attain the high structural efficiency of bolted composite joints, each fastener must accept its correct share of the load. Because composites fail at a very low strain level, such proper load sharing is incompatible with loose-fit holes. On the other hand, the testing of interference fit fasteners, which were either hammered in or twisted in, also left something to be desired. The obvious solution to this dilemma is the use of a loose sleeve that subsequently is expanded to fit properly when the fastener (bolt or rivet) is installed. Much valuable work on this subject has been per-

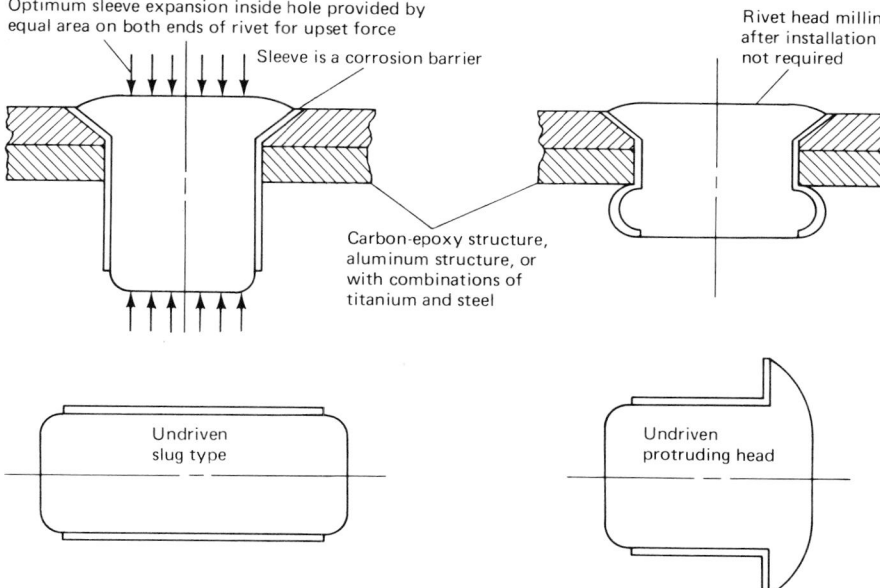

Optimum sleeve expansion inside hole provided by equal area on both ends of rivet for upset force

Sleeve is a corrosion barrier

Rivet head milling after installation not required

Carbon-epoxy structure, aluminum structure, or with combinations of titanium and steel

Undriven slug type

Undriven protruding head

Fig. 34 Sleeved rivet

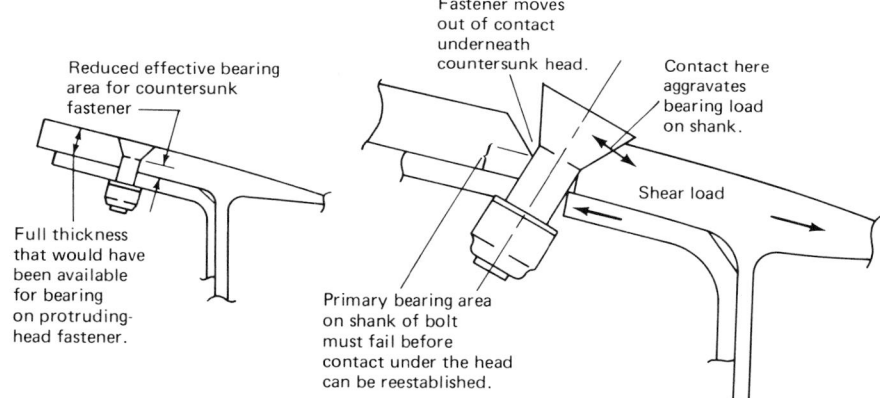

Reduced effective bearing area for countersunk fastener

Full thickness that would have been available for bearing on protruding-head fastener.

Fastener moves out of contact underneath countersunk head.

Contact here aggravates bearing load on shank.

Shear load

Primary bearing area on shank of bolt must fail before contact under the head can be reestablished.

Fig. 35 Problems with using flush fasteners at fittings for composite structures

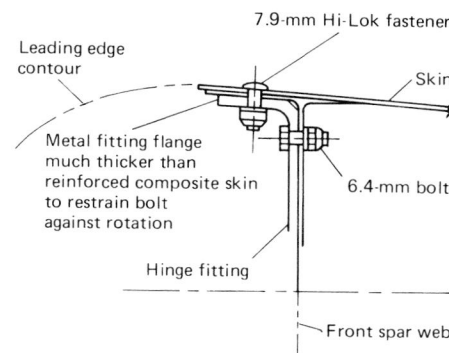

7.9-mm Hi-Lok fastener

Leading edge contour

Skin

Metal fitting flange much thicker than reinforced composite skin to restrain bolt against rotation

6.4-mm bolt

Hinge fitting

Front spar web

Fig. 36 Thickened hinge fitting flanges to minimize bolt bearing stresses

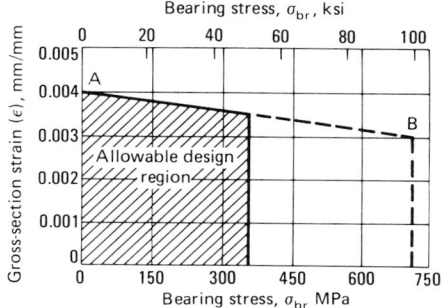

Fig. 37 Design technique for bolted carbon-epoxy structures

formed by E.R. Speakman (Ref 42). A sleeved rivet is illustrated in Fig. 34.

A similar problem exists in composite laminates in regard to the use of countersunk fasteners, particularly single-shear fasteners. As indicated in Fig. 35, as the fastener rotates under low loads, the head moves out of contact with the laminate, and all of the load is reacted by bearing on the shank. For this reason, the author advocates that, in the analysis of countersunk fasteners, the head be totally discounted. Such a procedure means restricting the depth of the countersink to no more than one-half the thickness of the outer member. Only after the shank area has failed in bearing does the head make contact with the laminate again.

Such a design procedure would imply a minimum skin thickness of twice the size of the countersunk head on the fastener, even though

it has become common practice to countersink deeper, in some cases knife-edging the outer composite member. There is no universal analysis method to cover the design of such questionable joints, and there have been failures associated with such concepts, even though some tests show such a practice to be tolerable. It is necessary to confirm the adequacy or inadequacy of such designs by specific testing. Alternatively, because this problem can arise only for relatively thin structures, adhesive bonding might well be considered instead.

A related issue is the installation of fittings in composite control surfaces. Whenever such fittings occur behind cutouts in the leading edge of the surface, it is always permissible to use shallow protruding-head fasteners, as shown in Fig. 36. (The drag from the cutout masks that form the fastener heads.) Not only is there more effective bearing area on the shank than with

countersunk fasteners, but the bolt-bearing allowable strength is also raised by about a factor of 2 because of the through-the-thickness clamp-up on the laminate. This substantial improvement on bearing strength is illustrated in Fig. 27.

A considerable simplification in the design process for mechanically fastened composite structures is afforded by the use of diagrams such as Fig. 37, which was prepared for older carbon fibers. Points A and B in the figure could probably be raised by a strain increment of 0.001 for the newer high-strain fibers. Such a simple chart, which can reasonably be applied when there are orthogonal load components also present, can safely cover all of the relatively lightly loaded fasteners, as at skin-to-spar and skin-to-rib seams on wings, leaving only the few major load transfer splices requiring more detailed analysis of the kind discussed in the section "Multirow Bolted Composite Joints" in this article.

It is not possible to cover all that is known about joints in fibrous composite structures in a single article. Only some of the highlights have been addressed here, and the reader is referred to the references cited and the other copious literature on the subject for further information.

It has been explained here that adhesive bonding is more suitable than mechanical fastening for thin structures (both composite and metal), and that the strength of such joints can, and should, exceed the strength of the members being joined. The issue of the durability of bonded joints is just as important as the short-term static strength, and distinct, albeit simple, design features are needed to ensure that the adhesive bonds will not fail prematurely.

Either mechanical fastening or more complex, bonded stepped-lap joints are needed for thicker, more highly loaded structures; adhesive bonds would only act as weak-link fuses in thick structures. It is never acceptable to design an adhesively bonded joint that is weaker than the members being joined. On the other hand, bolted composite joints can never be as strong as the member being joined, unless one resorts to softening strips and local pad-ups, which leave the basic structure too highly stressed to be repaired. Indeed, designing to structural efficiencies of 50% is quite a challenge. Finally, it should be noted that the design of the joints should always precede the process of filling in the gaps (sizing the laminates) and that due consideration should always be given to designing structures that can be repaired.

REFERENCES

1. L.J. Hart-Smith, "The Design of Repairable Composite Structures," *SAE Trans. 851830*, SAE Aerospace Technology Conference, Society of Automotive Engineers, 1985

2. E.W. Godwin and F.L. Matthews, A Review of the Strength of Joints in Fibre-Reinforced Plastics, Part 1, Mechanically Fastened Joints, *Composites*, Vol 11, 1980, p 155-160

3. F.L. Matthews, P.F. Kilty, and E.W. Godwin, A Review of the Strength of Joints in Fibre-Reinforced Plastics, Part 2, Adhesively Bonded Joints, *Composites*, Vol 13, 1982, p 29-37

4. L.J. Hart-Smith, Effects of Flaws and Porosity on Strength of Adhesive-Bonded Joints, in *Proceedings of 29th SAMPE Annual Symposium and Technical Conference*, Society for the Advancement of Material and Process Engineering, April 1984, p 840-852

5. E.W. Thrall, Jr., Failures in Adhesively Bonded Structures, in *Bonded Joints and Preparation for Bonding*, AGARD-NATO Lecture Series No. 102, Advisory Group for Aerospace Research and Development—North Atlantic Treaty Organization, 1979, p 5-1 to 5-89

6. L.J. Hart-Smith, "Adhesive Bonding of Aircraft Primary Structures," Douglas Paper 6979, *SAE Trans. 801209*, SAE Aerospace Congress and Exhibition, Society of Automotive Engineers, 1980

7. R.W. Shannon *et al.*, "Primary Adhesively Bonded Structure Technology (PABST): General Material Property Data," USAF, AFFDL-TR-77-107, Douglas Aircraft Company, Sept 1978, 1982, 2nd ed.

8. R.J. Schliekelmann, Adhesive Bonding and Composites, in *Progress in Science and Engineering of Composites, Vol 1*, T. Hayashi, K. Kawata, and S. Umekawa, Ed., Fourth International Conference on Composite Materials, North-Holland, 1983, p 63-78

9. L.J. Hart-Smith, R.W. Ochsner, and R.L. Radecky, "Surface Preparation of Fibrous Composites for Adhesive Bonding or Painting," *Douglas Service Magazine*, 1st quarter, 1984, p 12-22

10. A.V. Pocius and R.P. Wenz, Mechanical Surface Preparation of Graphite-Epoxy Composite for Adhesive Bonding, in *Proceedings of 30th National SAMPE Symposium*, Society for the Advancement of Material and Process Engineering, March 1985, p 1073-1087

11. S.H. Myhre, J.D. Labor, and S.C. Aker, Moisture Problems in Advanced Composite Structural Repair, *Composites*, Vol 13, 1982, p 289-297

12. O. Volkersen, The Rivet-Force Distribution in Tension-Stressed Riveted Joints With Constant Sheet Thicknesses, Luftfahrtforschung, Vol 15, 1938, p 4-47

13. M. Goland and E. Reissner, The Stresses in Cemented Joints, *J. Appl. Mech. (Trans. ASME)*, Vol 11, 1944, p A17-A27

14. L.J. Hart-Smith, "Adhesive-Bonded Single-Lap Joints," NASA CR-112236, Douglas Aircraft Company, Jan 1973

15. L.J. Hart-Smith, Stress Analysis: A Continuum Mechanics Approach, in *Developments in Adhesives, 2*, A.J. Kinloch, Ed., Applied Science Publishers, 1981, p 1-43

16. N.A. de Bruyne, "The Strength of Glued Joints," *Aircr. Eng.*, Vol 16, 1944, p 115-118, 140

17. L.J. Hart-Smith, "Analysis and Design of Advanced Composite Bonded Joints," (Douglas Aircraft Company), NASA CR-2218, Jan 1973, reprinted Aug 1974

18. L.J. Hart-Smith, Design and Analysis of Adhesive-Bonded Joints, in *Proceedings of First Air Force Conference on Fibrous Composites in Flight Vehicle Design*, AFFDL-TR-72-130, Air Force Flight Dynamics Laboratory, 1972, p 813-856

19. L.J. Hart-Smith, Advances in the Analysis and Design of Adhesive-Bonded Joints in Composite Aerospace Structures, in *Proceedings of 19th National SAMPE Symposium and Exhibition*, Society for the Advancement of Material and Process Engineering, April 1974, p 722-737

20. L.J. Hart-Smith, Bonded-Bolted Composite Joints, *J. Aircr.*, Vol 22, 1985, p 993-1000

21. L.J. Hart-Smith, Adhesively Bonded Joints for Fibrous Composite Structures, in *Joining Fibre-Reinforced Plastics*, F.L. Matthews, Ed., Elsevier, 1987, p 271-311

22. L.J. Hart-Smith, "Differences Between Adhesive Behavior in Test Coupons and Structural Joints," Paper presented at ASTM Adhesives Committee D-14 Meeting, American Society for Testing and Materials, Phoenix, March 1981

23. L.J. Hart-Smith, "Adhesive Layer Thickness and Porosity Criteria for Bonded Joints," (Douglas Aircraft Company), AFWAL-TR-82-4172, Dec 1982

24. R.B. Krieger, Jr., Analyzing Joint Stresses Using an Extensometer, *Adhesives Age*, Vol 28 (No. 11), Oct 1985, p 26-28

25. L.J. Hart-Smith, "Adhesive-Bonded Double-Lap Joints," (Douglas Aircraft Company), NASA CR-112235, Jan 1973

26. L.J. Hart-Smith, "Adhesive-Bonded Scarf and Stepped-Lap Joints," (Douglas Aircraft Company), NASA CR-112237, Jan 1973

27. L.J. Hart-Smith, "Design Methodology for Bonded-Bolted Composite Joints," (Douglas Aircraft Company), AFWAL-TR-81-3154, Feb 1982

28. L.J. Hart-Smith, Further Developments in the Design and Analysis of Adhesive-Bonded Structural Joints, in *Joining of Composite Materials*, STP 749, K.T. Kedward, Ed., 1981, p 3-31

29. L.J. Hart-Smith, Design and Analysis of Bonded Repairs for Metal Aircraft Structures, in *Proceedings of International Workshop on Defense Applications of Advanced Repair Technology for Metal and Composite Structures*, Naval Research Laboratory, July 1981, p 251-260

30. W.S. Johnson, Ed., *Delamination and Debonding of Materials*, STP 876, American Society for Testing and Materials, 1985

31. L.J. Hart-Smith, "Simplified Estimation of Stiffness and Biaxial Strengths for Design of Carbon-Epoxy Composite Structures," in *Proceedings of Seventh DoD/NASA Conference on Fibrous Composites in Structural Design*, AFWAL-TR-85-3094, 1985, p V(a)-17 to V(a)-52

32. L.J. Hart-Smith, Simplified Estimation of Stiffness and Biaxial Strengths of Woven Carbon-Epoxy Composites, in *Closed-Session Conference Proceedings*, 31st International SAMPE Symposium and Exhibition, Society for the Advancement of Material and Process Engineering, April 1986, p 83-102

33. L.J. Hart-Smith, Mechanically-Fastened Joints for Advanced Composites—Phenomenological Considerations and Simple Analyses, in *Fibrous Composites in Structural Design*, Fourth Conference on Fibrous Composites in Structural Design, E.M. Lenoe, D.W. Oplinger, and J.J. Burke, Ed., Plenum Press, 1980, p 543-574

34. L.J. Hart-Smith, "Bolted Joints in Graphite-Epoxy Laminates," (Douglas Aircraft Company), NASA CR-144899, Jan 1977

35. L.J. Hart-Smith, Design and Analysis of

Bolted and Riveted Joints in Fibrous Composite Structures, in *Joining Fibre-Reinforced Plastics*, J.L. Matthews, Ed., Elsevier, 1987, p 227-269

36. W.D. Nelson, B.L. Bunin, and L.J. Hart-Smith, Critical Joints in Large Composite Aircraft Structure, in *Proceedings of Sixth Conference on Fibrous Composites in Structural Design*, AMMRC MS 83-2, Army Materials and Mechanics Research Center, 1983, p II-1 to II-38

37. J.M. Whitney and R.J. Nuismer, Stress Fracture Criteria for Laminated Composites Containing Stress Concentrations, *J. Compos. Mater.*, Vol 8, 1974, p 253-265

38. R.L. Ramkumar and E.W. Tossavainen, "Bolted Joints in Composite Structures: Design, Analysis and Verification; Task I Test Results—Single Fastener Joints," (Northrop Aircraft Division), AFWAL-TR-84-3047, Aug 1984

39. R.B. Heywood, *Designing by Photoelasticity*, Chapman and Hall, 1952, p 268

40. S.P. Garbo and J.M. Ogonowski, "Effect of Variances and Manufacturing Tolerances on the Design Strength and Life of Mechanically Fastened Composite Joints," (McDonnell Aircraft Company), AFWAL-TR-81-3041, 3 Vol, April 1981

41. S.P. Garbo, "Effects of Bearing/Bypass Load Interaction on Laminate Strength," (McDonnell Aircraft Company), AFWAL-TR-81-3144, Sept 1981

42. E.R. Speakman, Advanced Fastener Technology for Composite and Metallic Joints, in *Fatigue in Mechanically Fastened Composite and Metallic Joints*, STP 927, J.M. Potter, Ed., American Society for Testing and Materials, 1986

SECTION 8

Manufacturing Processes

Chairman: James C. Leslie, Advanced Composite Products & Technology, Inc.
Co-Chairman: Peter Beardmore, Ford Motor Company
Co-Chairman: Hans Borstell, Grumman Aircraft Systems

Introduction

COMPOSITE STRUCTURES are composed of fibrous materials held in place by a matrix system. They derive most of their unique characteristics from the reinforcing fibers. Fabricating a composite part is simply a matter of placing and retaining fibers in the direction and form that is required to provide specified characteristics while the part performs its design function. This Section focuses on the variety of methods and equipment currently being used to cost-effectively manufacture hardware from fibrous reinforcements in organic matrices.

Advanced composites are differentiated from plastics and reinforced plastics because they attain maximum mechanical properties and unique physical characteristics. This is done by obtaining the highest possible fiber-to-resin ratio, using continuous or long-length fibers, and carefully orienting the fibers.

The methods to be used in manufacturing hardware from advanced composite materials must be considered during component design. Unlike metals and many plastics, which exhibit more or less isotropic properties, advanced composites are basically anisotropic. Not all fiber orientations are attainable in a finished component. Certain shapes and configurations are more easily manufactured by one technique than another. The cost of all parts is highly dependent upon the equipment and procedures necessary for their manufacture.

This Section presents all of the basic elements and numerous techniques for manufacturing advanced composite hardware, as well as standard and unique approaches for the cost-effective manufacture of commercial hardware. An article in Section 7 of this Volume, "Design Cost Drivers," focuses on the design process, but should not be overlooked by those working in the manufacturing field.

The development of manufacturing procedures in the 1940s and 1950s came from two areas. Aerospace structures such as rocket motors and satellite components were pushed toward minimum-weight hardware, with manufacturing cost almost being relegated to secondary importance. Conversely, pultrusion and filament winding were quickly and widely adopted for the commercial low-cost production of pipe and other constant cross section hardware.

The aerospace work developed and demonstrated the physical/mechanical capabilities of composites. It also provided the motivational force (and funding) to expand overall manufacturing capability into cost-effective modes that are now being used for commercial purposes.

This Section presents articles on manufacturing methods and other topics that are widely applicable to composite manufacture. These articles are followed by a commercial manufacturing subsection that presents state-of-the-art equipment and processes for minimum-cost high-rate production. It emphasizes historical developments in composite sporting goods manufacture, as well as "now-in-development" (and, in some cases, "in-practice") automotive and transportation hardware production.

Finally, an aerospace subsection provides a comprehensive review of the procedures and equipment used to ensure maximum properties, optimum reproducibility, and the highest possible reliability. The information presented is universally applicable to the manufacture of composite hardware. Although many of these procedures would not be cost effective for high-rate commercial manufacturing, the information on methods and equipment given here is fundamental to the understanding of composite structures fabrication.

Process Modeling and Optimization

Ronald A. Servais, University of Dayton

RESIN MATRIX composite materials typically consist of reinforcing fibers such as glass, carbon, graphite, or metals that are supported by a thermoplastic or thermosetting matrix. In order to tailor the composite shape and its properties, the type and composition of the constituents and the processing procedures are varied. Properties of interest may include density, strength, thermal conductivity, or resistance to environmental exposure. Processing method depends on the type of reinforcement and matrix. Typical thermoplastic processing methods include injection molding, compression molding, and hot and cold stamping. Thermoset processing methods include contact molding, matched die molding, injection molding, pultrusion, filament winding, press molding, and vacuum bag autoclave molding.

Historically, processing procedures have been determined by trial and error, with resulting properties being determined by postprocessing evaluation of the product, as shown in Fig. 1. Although this approach has led to many successful composite material applications, its limitations are severe: processing procedures not represented by available laboratory hardware cannot be considered, determining the effects of systematically varying composition or processing parameters requires many experimental tests, experience accumulated during laboratory processing is difficult to apply to the manufacturing process, and scale-up based on laboratory tests is particularly difficult when processes include chemical reactions. A viable tool for predicting some of these problems and offering potential solutions is process modeling.

Modeling usually consists of representing the process under consideration by mathematical equations and then solving them numerically on a computer in order to predict behavior. These predictions must then be interpreted by an expert. Although models have limitations, which will be described, they can be used for quick, cost-effective investigation of a variety of similar situations and can optimize a given process. In the future, models will be used not only in the development of a process but also as part of the process control package on the manufacturing level.

The history of process modeling is relatively short. Most previous applications of mathematical modeling have been in the fields of heat transfer, fluid flow, and mechanics. Examples include predicting heat flow through complex shapes or sophisticated re-entry thermal protection materials, investigating boundary layer and aerodynamic flow systems, and estimating the stresses in materials under load. The actual computer programs have often been quite large and sophisticated, requiring trained specialists and mainframe computers. These factors have tended to discourage most manufacturing engineers from even exploring the modeling approach. This will rapidly change as some of the models become easier to use and run on microcomputers.

Once a process model has been constructed and verified, it can be used to explore the effects of systematically varying the process variables without resorting to a large number of laboratory experiments. Critical variables can be identified, and the results used to optimize the process. This leads to an improved understanding of the process itself. The model can be used by the materials development engineer to investigate process variables for different product geometries and to consider the changes in processing caused by altering material properties. Different process paths can then be proposed, analyzed, and compared by running the process model. The ability to predict the outcome of a given process path allows the engineer to determine the optimal process path and identify the critical variables that must be monitored during processing to properly control the process.

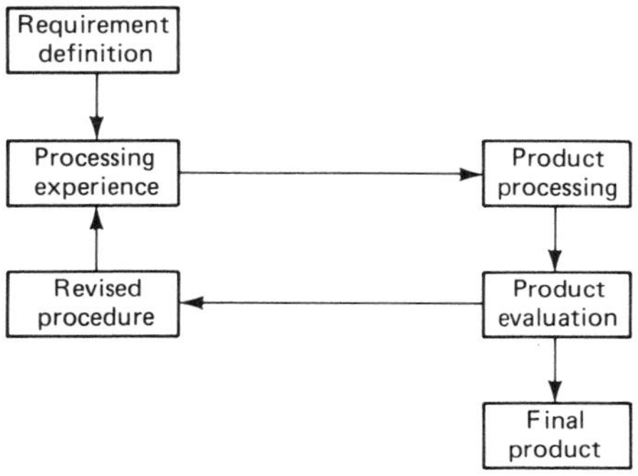

It is important to realize that because of their simplified assumptions, models do not yield exact solutions to problems; by their very nature, modeling solutions are approximate (Ref 1). Therefore, models must be verified by comparison with experimental data. Examples of simplifications include geometry, such as representing a two-dimensional problem with one-dimensional equations; composition, such as treating the material as homogeneous rather than heterogeneous; properties, such as using constant properties instead of temperature-dependent properties; and numerical, such as utilizing a small number of grid points, hence losing information. Some simplifications are made because there is no alternative, such as when specialized property data are not available. Therefore, all models have limited application. It is imperative to recognize these limits in order to draw reasonable conclusions from the predictions. Although comprehensive models can be developed, they are invariably impractical due to the large amount of input data and computer requirements.

Process Model Characteristics

The elements of a process model, shown in Fig. 2, include a mathematical description of the process, property data, a method for its solution, and boundary conditions.

Initially, the purpose of the process model, such as to predict the degree of cure of the final dimensions, is determined. Next, the con-

Fig. 1 Trial-and-error process evolution

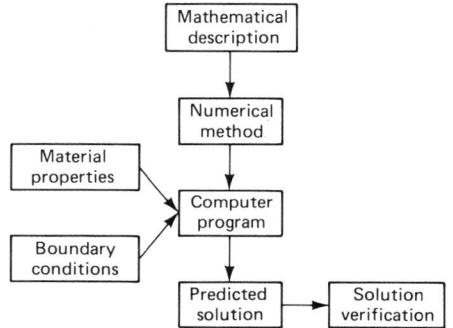

Fig. 2 Process model components

straints on the problem, which might include the maximum heating rate for a press or the minimum temperature in an autoclave, are identified. Assumptions or simplifications, such as assuming constant properties or neglecting pressure effects, can then be reviewed.

Material process modeling involves describing the process with mathematical equations so that changes in the processing variables can be investigated and their effects understood. The relationships available to describe manufacturing processes include mass, momentum, energy balances, and the laws of thermodynamics. By making various assumptions, it may be possible to decouple some of the phenomena and express them in the form of submodels (to be discussed in the next section of this article) that can easily be programmed and modified. Since the processing is inherently dynamic, the balance relations are invariably differential equations (Ref 1, 2).

Property data, such as density, thermal conductivity, specific heat, mechanical strength, and chemical composition, are required inputs for the models and appear as the coefficients in the differential equations. Process models can be used only if the appropriate property data are available. The resulting system of equations usually requires a numerical solution method. To complete the process description, boundary conditions that are external to the composite material are needed and may be in the form of applied pressures, mechanical forces, heat added to the process, specified surface temperatures, or tool shapes.

A comprehensive composite processing description would include heat and fluid flow equations, as well as local mechanical force and moment equations in all three directions. The matrix properties required would include thermodynamic properties, such as density, specific heat, heat of fusion, and heat of reaction; transport properties, such as viscosity and thermal conductivity; and mechanical properties, such as thermal expansion, strength, and elasticity. All of these properties would be dependent upon chemical state, physical state, and temperature. The fiber properties required would include thermodynamic, transport, and mechanical properties that are dependent upon temperature and direction (axial and radial). The chemical state would include history- dependent and temperature-dependent reaction kinetics (Arrhenius coefficients). The physical state would include temperature-dependent phase changes. The solution to the system of equations would yield predicted temperature, pressure, and stress as a function of time and position within the material being processed. This system of equations is not only complicated but impossible to solve analytically. The combination of simplified assumptions (geometry, properties, and magnitude of terms) and the numerical methods used in the solution procedure cause the model to be approximate.

Submodels

The mathematical expressions that can be developed for different material processing phenomena include heat, momentum, material, and force balance. These phenomena are not independent of each other, but are in fact coupled through various laws and conservation equations. Nevertheless, submodels are frequently used to decouple the balance equations in order to investigate a particular aspect of interest or to simplify the manipulations required to solve the equations. It is also convenient to decouple them for discussion purposes.

Heat balance is described by the conservation of energy equation and is expressed as changes in temperature as a function of position and time. Source terms are required to represent chemical kinetics (an exothermic reaction releasing heat within the material) and phase changes (such as heat absorption due to vaporization). The kinetics, in turn, may need to be described by a differential equation if the degree of cure must be calculated (Ref 3, 4). The density and the specific heat of the material are also required data. It can be assumed that heat transfer problems can be adequately described with one-dimensional equations because of the relative dimensions of the material. Sometimes the effect of fiber orientation is important. To accommodate orientation, at least a two-dimensional heat submodel is required. A variety of different heat sources can be used with the heat submodel through boundary conditions; these include conduction (heated platens) or convection (oven or autoclave). Radiation or induction heat sources could be accommodated as well, although they would complicate the boundary condition formulation.

Momentum Balance. Flow models start with the conservation of momentum equation (Ref 5), with pressure, density, viscosity, and local velocity being the primary variables. If properties such as viscosity are dependent upon temperature (Ref 6), the energy balance must also be solved in order to investigate pressure-velocity interactions. Fiber orientation may be included in a sophisticated model but neglected in a simpler model. Alternate schemes may be used to simplify or emphasize special aspects of interest. For example, Darcy's law is sometimes used to investigate vertical flow through a porous media (Ref 7). The resin may be considered to be Newtonian or even viscoelastic in nature (Ref 8). Based on geometric considerations, the number of dimensions in the problems may be reduced.

Material Balance. Modeling the formation and movement of voids (entrapped gas pockets) in composite material processing is of significant interest (Ref 9). Gas formation can be described by thermodynamic principles. Bubble coalescence and movement have been described using empirical correlations and can also be approached statistically. Predicting void location and movement is necessary in order to minimize voids in the cured material by controlling the process variables.

Force Balance. Stresses are formed in the composite during processing because of differences in the coefficients of thermal expansion between the fibers and the matrix, which are due to material compaction during processing and the presence of thermal gradients in the material as it cures. Residual stresses, or those that remain after processing, can lead to reduced mechanical properties in the final product. In some cases, the product will warp substantially because of residual stresses, or the matrix will develop severe microcracks. Relating the stress formation to processing variables is necessary in order to minimize or control residual stresses. Additionally, the submodel should be useful in determining optimum orientation of composite layers in order to control the stresses. Three-dimensional models are usually required to investigate mechanics problems.

Numerical Methods

Most models are expressed in terms of a partial differential equation or a system of partial differential equations, with the solution normally requiring the use of numerical methods. Usually, heat and/or momentum transfer problems are solved using finite-difference methods, while mechanics problems are solved using finite-element methods. An alternate approach that exploits an electrical analogy between the composite problem and elemental electrical circuits leads to a relatively simple set of algebraic equations representing the differential equations of interest. The numerical methods required are described in numerical analysis textbooks.

Finite-difference methods are easier to apply than finite-element methods. Generally, the dimensions of the composite material change during processing, which leads to moving boundaries. Finite-difference methods are particularly complicated for moving-boundary problems. A novel approach involves introducing a moving-grid (body-fitted coordinate) system to the problem (Ref 10). This scheme is compatible with finite-difference methods, while accommodating moving-boundary problems.

Process Model Example

Autoclave curing of a fiber-reinforced thermosetting composite material will be used to illustrate the various components of a process model. The simplified fabrication procedure for thermoset composites consists of impregnating fiber bundles (prepregs) with resin, forming single-bundle-thick (unidirectional) layers, stacking the prepreg layers to a specified orientation and desired thickness on a shaped tool, and then using an autoclave (a large pressurized cooking vessel) to apply heat and pressure according to a predetermined plan (or cure cycle) to form the composite product. The heat initially causes the polymer to melt and flow, triggering an irreversible exothermic reaction and accelerating polymer cross-linking as the temperature increases. The pressure compacts the composite and is useful in minimizing voids. Gases (including air), prepregging volatiles, and water vapor can be entrapped in the composite during the cure cycle, producing voids in the final product. The size and distribution of the voids may make the product unusable. Product characteristics and quality are determined by the properties of its constituents and the cure cycle. The combination of cure-cycle temperatures and pressure histories that leads to an acceptable product is called the processing window. Before developing a procedure that investigates acceptable autoclave temperatures as a function of time (by modeling the energy distribution within the curing material and predicting the internal temperatures and degree of cure), the autoclave temperature history must be assumed. Then, predictions are made and analyzed, a new temperature history is assumed, and the procedure is repeated. Previous experience with cure cycles can also be exploited when establishing the temperature history. For example, if the autoclave temperature is increased too rapidly, the outer layers of the composite will cure prematurely and entrap gases in the central region. If the cycle is too short or the autoclave temperatures too low, the composite will not be cured completely. In addition, if the pressure is too high, too much resin will be squeezed out of the composite, whereas if the pressure is too low, the composite will not be properly compacted.

Assume that the heat flow in only one direction is significant and can be described by the energy balance within the material. The heat transfer is time dependent and includes the effects of the exothermic reaction of the resin. Assume that the effects of resin flow (while in the liquid state) on the energy balance can be neglected. In addition, assume that perfect thermal contact is formed between layers and that the properties are temperature dependent but not directionally dependent. The energy balance can then be written (Ref 2):

$$\left\{\begin{array}{c}\text{Rate of}\\\text{energy}\\\text{accumulated}\end{array}\right\} = \left\{\begin{array}{c}\text{Rate of}\\\text{energy}\\\text{in}\end{array}\right\} - \left\{\begin{array}{c}\text{Rate of}\\\text{energy}\\\text{out}\end{array}\right\} + \left\{\begin{array}{c}\text{Rate of}\\\text{energy}\\\text{generated}\end{array}\right\}$$

$$\rho_c c \frac{\partial T}{\partial t} = \frac{\partial}{\partial x}\left(\frac{K\partial T}{\partial x}\right) + rH_r$$

where ρ_c is the composite density, c is the composite specific heat capacity, T is the local temperature, t is time, x is position, K is the composite thermal conductivity, r is the rate of reaction, and H_r is the heat of reaction. Note that the density, specific heat, thermal conductivity, and rate of reaction must be determined as a function of temperature. The boundary condition for the autoclave curing process is the

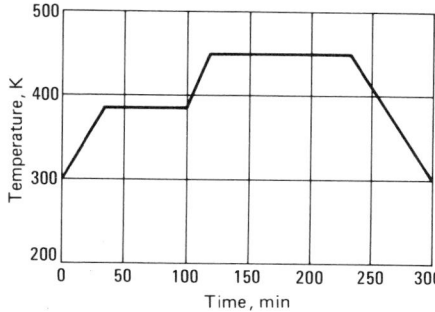

Fig. 3 Typical thermoset cure cycle

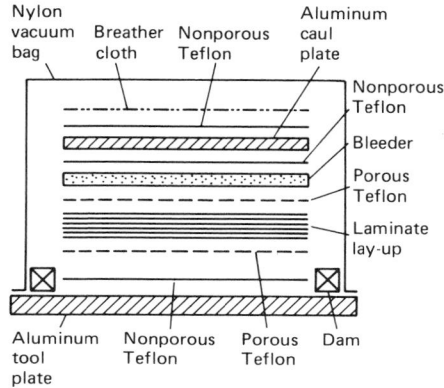

Fig. 4 Autoclave laminate stack lay-up

autoclave air temperature history. A typical temperature cycle is shown in Fig. 3. The energy equation for this process model example can now be solved using a classical finite-difference method. The solution method must be verified by comparison with experimental data, and then the investigation of different temperature boundary conditions can proceed.

Several practical problems can now be identified. Heat transfer through the bagging materials (Fig. 4) as well as the composite itself must be investigated. Data for properties such as density and thermal conductivity are not normally available for the bagging materials. In spite of extensive laboratory investigations of curing cycles, almost no data are available for constituent properties (reinforcement or matrix) as a function of temperature and degree of cure. In addition, virtually no data are available at this time for model verification.

Although this example focuses on the autoclave curing of thermosetting composites, its strategy applies to most manufacturing processes. Several process models described in the literature include a thermoset composite cure model (Ref 3), a winder/cure model (Ref 11), a press-curing model (Ref 12), a carbon-carbon curing model (Ref 13), and a pultrusion model (Ref 14). Two current modeling efforts focus on thermosetting (Ref 15) and thermoplastic composite processing (Ref 16).

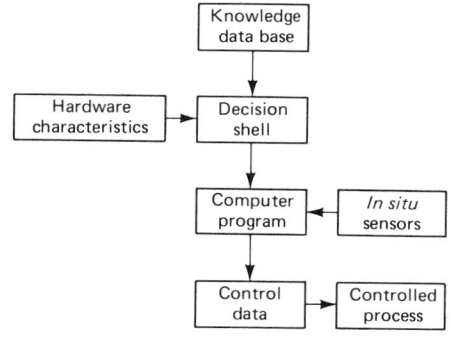

Fig. 5 Expert system components

Future Development of Process Models

As new materials with smaller processing windows (and less room for error) become available, use of a computer process model to predict the behavior of the composite material during the curing cycle, based on its constituent properties, would reduce fabrication costs and ensure product quality.

Although on-line control applications using process models would allow corrections to the cycle, recognizing deviations in initial conditions as well as aberrations in the desired processing conditions, on-line control requires two additional components that have not been addressed. These are *in situ* measurements (sensor feedback) of critical variables and a mathematical definition of acceptable quality as a function of time.

An expert system exploits and combines the information from all of the defined approaches to optimize and control the process on-line. An expert system (see Fig. 5) includes three primary components: a representation of experience (which includes predictions made using a process model), a decision-making shell, and a closed-loop control capability. The expert system approach can be viewed as a new application of process modeling. While the conventional process model remains stagnant, intelligent systems grow as more is learned about the process. More important, this approach ties together various stages of manufacturing, while the conventional model serves only as a tool to the research and development engineer.

REFERENCES

1. J.F. Stanislav, *Mathematical Modeling of Transport Phenomena Processes*, Ann Arbor Science, 1982
2. R.B. Bird, W.E. Stewart, and E.N. Lightfoot, *Transport Phenomena*, John Wiley & Sons, 1960
3. A.C. Loos and G.S. Springer, "Curing of Graphite/Epoxy Composites," AFWAL-TR-83, Air Force Wright Aeronautical Laboratories, 1983

4. K.R. Strother, "Elementary Empirical Process Modeling for Thermoset Resins," Paper presented at the 31st International Symposium, Society for the Advancement of Material and Process Engineering, Las Vegas, 1986

5. J.T. Lindt, "Mechanical Principles of Formation of Fiber Reinforced Materials," Paper presented at the Second International Conference on Reactive Processing of Polymers, Pittsburgh, Nov 1982

6. T.H. Hou, "Chemoviscosity Modeling for Thermosetting Resin," Contract NASA-1600, National Aeronautics and Space Administration, Nov 1984

7. G.S. Springer, Resin Flow During the Cure of Fiber Reinforced Composites, *J. Compos. Mater.*, Vol 16, 1982, p 400-410

8. T.G. Gutowski, S.J. Wineman, and Z. Cai, "Application of the Resin/Fiber Deformation Model," Paper presented at the 31st International Symposium, Society for the Advancement of Material and Process Engineering, Las Vegas, 1986

9. J.L. Kardos, M.P. Dudukovic, E.L. Mckague, and M.W. Lehman, Void Formation and Transport During Composite Laminate Processing: An Initial Model Framework, in *Composite Materials: Quality Assurance and Processing*, STP 797, C.E. Browning, Ed., American Society for Testing and Materials, 1983, p 96-109

10. J.F. Thompson, F.C. Thames, and C.W. Mastin, *Numerical Grid Generation: Foundations and Applications*, North-Holland, 1985

11. E.P. Calius and G.S. Springer, Modeling the Filament Winding Process, in *Proceedings of the Fifth International Conference on Composite Materials*, W.C. Harrigan, Jr., J. Strife, and A.K. Dhingra, Ed., The Metallurgical Society, 1985, p 1071

12. R.A. Servais, J.A. Snide, and P.E. Bern, "A Computer Program for Predicting Temperature Profiles Through a Laminate Stack During a Production Press-Cure Cycle," Paper presented at the Annual Technical Conference, Society of Plastics Engineers, New Orleans, 1984

13. S.R. Soni and N.J. Pagano, "Process Models for 3-D Composites," AFWAL-TR-85-4002, Air Force Wright Aeronautical Laboratories, 1983

14. C.M. Ma, K.Y. Lee, Y.D. Lee, and J.S. Hwang, "The Correlations of Processing Variables for Optimizing the Pultrusion Process," Paper presented at the 31st International Symposium, Society for the Advancement of Material and Process Engineering, Las Vegas, 1986

15. F.C. Campbell, A.R. Mallow, G.A. Blase, and F.R. Muncaster, "Computer-Aided Curing of Composites," Interim Technical Report IR-0355-8, U.S. Air Force Contract No. F33615-83-C-5088, McDonnell Douglas Corporation, April 1986

16. R.M. Smith and F.C. Campbell, "Processing Science of Thermoplastic Composites," Interim Technical Report IR-0384-2, U.S. Air Force Contract No. F33615-85-C-5046, McDonnell Douglas Corporation, March 1986

Filament Winding

S.T. Peters, Westinghouse Electric Corporation
W. Donald Humphrey, Brunswick Corporation

HIGH-SPEED PRECISE LAY-DOWN of continuous reinforcement in predescribed patterns is the basis of the filament winding method. It is a process in which continuous resin-impregnated rovings or tows (gathered strands of fiber) are wound over a rotating male mandrel (Fig. 1). The mandrel can be cylindrical, round, or any shape that does not have reentrant curvature. The reinforcement may be wrapped either in adjacent bands or in repeating bands that are stepped the width of the band and which eventually cover the mandrel surface. The technique has the capacity to vary the winding tension, wind angle, or resin content in each layer of reinforcement until the desired thickness and resin content of the composite are obtained with the required direction of strength.

The most important advantage of filament winding is the cost, which is less than the prepreg cost for most composites (Ref 1). These lower costs are possible in filament winding because a relatively expensive fiber can be combined with an inexpensive resin to yield a relatively inexpensive composite. Also,

cost reductions accrue because of the high speed of fiber lay-down, for example, for large parts such as a missile canister, 45 kg/h (100 lb/h) of low-angle helical, or 320 kg/h (700 lb/h) of hoop windings, as shown in Fig. 2.

The primary advantages of filament winding are:

- The highly repetitive nature of fiber placement (from layer to layer and from part to part)
- The capacity to use continuous fibers over a whole component area (without joints) and to orient fibers easily in load direction
- Elimination of capital expense of autoclave
- Large structures can be built (larger than any autoclave)
- High fiber volume is obtainable
- Lower cost for large numbers of components
- Relatively low material costs because fiber and resin can be used in their lowest cost form rather than as prepreg

The primary disadvantages of filament winding are:

- Shape of component must be such that the mandrel can be removed
- Inability to wind reverse curvature
- Inability to change fiber path easily (in one lamina)
- Need for mandrel, which can be complex and expensive
- Poor external surface, which may hamper aerodynamics

Disadvantages of filament winding have been circumvented. Fabricators of large rocket motors have used plaster mandrels which can be stripped, reduced in size, and passed out through relatively small ports. Reverse curvature can be formed into positive curvature by the addition of oriented fibers or mats. Alternately, if reverse curvature is necessary to the design (such as on an airfoil), it can be obtained by removing the uncured structure from the mandrel and using other means of compaction to form the composite and smooth the external surface. The fiber path can be changed to save material costs or to make a removable portion of a component by winding around pins at the end of a part (Fig. 3) rather than winding over domes (Ref 2). The fiber path can also be altered by pins to avoid slipping or bridging.

Thermoset resins used as the binders for the reinforcements can be applied to the dry roving at the time of winding (wet winding), or can be

Fig. 1 Filament winding of a large graphite-epoxy structure. Courtesy of Hercules Aerospace Company

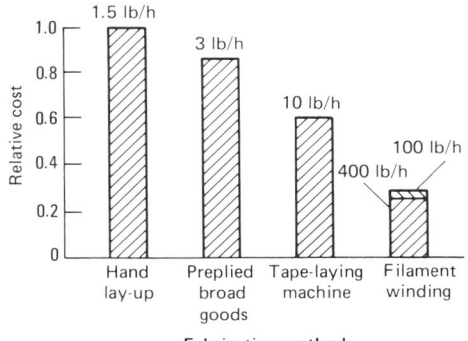

Fig. 2 Relative costs of filament winding and other composite fabrication methods. Source: Ref 1

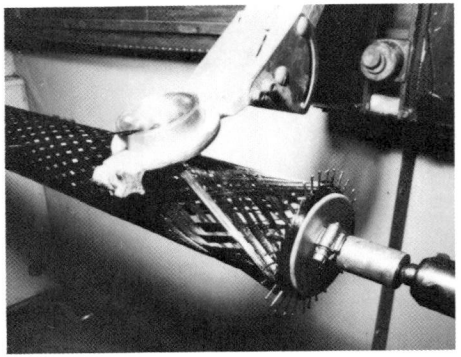

Fig. 3 Typical pin ring winding technique. Courtesy of Westinghouse Electric Company, Marine Division

applied previously and gelled to a B-stage as prepreg. Also, rovings can be impregnated and rerolled without B-staging and either used promptly or refrigerated. The filament-wound composite is usually cured at elevated temperatures, without any additional step for compaction. Mandrel removal, trimming, and other finishing operations complete the process.

Materials

Fibers. The most widely used fiber for filament winding is fiberglass, which has been marketed in several grades in the United States for more than 40 years. Types of glass fibers useful for filament-wound structures are shown in Table 1, which gives the common designation of the fiber, the nominal tensile strength and tensile modulus of the strands, and the maximum number of filaments per strand. The latter is important in selection of a fiber for filament winding because large numbers of filaments can make handling easier. Fiber density is included in the table so that the rule-of-mixtures equations involving fiber volume and resin volume can be used to evaluate void volume and theoretical mechanical values. Fiberglass continues to be useful for filament winding because of low cost, dimensional stability, moderate strength and modulus, and ease of handling.

Aramid fibers, which were initially useful because of their strength- and modulus-to-weight ratios (called specific modulus or specific strength), also show great consistency with a low coefficient of variation, enabling high design allowables. Aramid composites have relatively poor shear and compression properties, which are generally not critical for pressure vessels. A new aramid, type 981, has been developed, that has greater tensile strength (Ref 3) with the same density (hence improved specific tensile strength). Organic fibers are described in Table 2. (The nominal tensile modulus values reflect the greater strain-rate dependence in the oriented polyethylene fibers versus the aramids.)

Table 1 Glass fibers for filament winding (in order of ascending modulus of strand, normalized to 100% fiber volume) (vendor data)

Type	Strand nominal tensile modulus		Strand nominal tensile strength		Maximum number of filaments/strand	Fiber density, g/cm³
	GPa	10⁶ psi	MPa	ksi		
E	72.4	10.5	3447	500	4000	2.60
R	86.2	12.5	2068	300	60	2.49
S	86.9	12.6	4585	665	...	2.55

Table 2 Organic fibers for filament winding (in order of ascending modulus of strand, normalized to 100% fiber volume) (vendor data)

Type	Strand nominal tensile modulus		Strand nominal tensile strength		Maximum number of filaments/strand	Fiber density, g/cm³
	GPa	10⁶ psi	MPa	ksi		
Aramid (medium modulus)	62	9.0	2758	400	1000	1.44
Oriented polyethylene(a)	117	17	2585	375	118	0.97
Aramid(a)	121	17.5	4067	590	...	1.44
Aramid	124	18	3792	550	5000	1.44
Oriented polyethylene(b)	170	24.8	3274	471	...	0.97

(a) Development status. (b) Research and development status

Table 3 Carbon and graphite fibers (in order of ascending modulus of strand, normalized to 100% fiber volume) (vendor data)

Class of fiber	Strand nominal tensile modulus		Strand nominal tensile strength		Maximum number of filaments/strand	Fiber density, g/cm³
	MPa	10⁶ psi	MPa	ksi		
High tensile strength	227	33	3102	450	12 000	1.75
High strain	234	34	4100	594	6 000	1.79
Intermediate modulus	275	40	4295	623	12 000	1.74
High modulus	358	52	2482	360	3000	1.81
High modulus pitch	379	55	2068	300	4000	2.0
Ultra-high modulus	517	75	1816	270	384	1.96
Ultra-high modulus, pitch	517	75	2068	300	2000	2.0
Extreme-high modulus, pitch	689	100	2240	325	2000	2.15

The largest variety of strengths and moduli can be obtained with graphite fibers (Table 3), which have recently been improved in terms of modulus, tensile strength, and strain to failure. Surface finish has also been improved, which facilitates handling for filament winding. Increasing tensile modulus usually lowers tensile strength; the intermediate-modulus fibers have been the only exception. The amount of graphitization increases with increasing modulus, which results in greater thermal and electrical conductivity. Fiber cost also increases, primarily because there is less demand for the high-modulus fibers, and large-scale production economics have not yet been imposed. All fibers, except pitch fibers with a modulus of 517 GPa or greater have been filament wound.

Resin Systems. The resin system in a filament-wound composite serves the same functions as it does in composite structures fabricated by other means, namely:

- Keeping the filaments in the proper position
- Helping to distribute the load
- Protecting the filaments from abrasion (during winding and in the composite)
- Controlling electrical and chemical properties
- Providing the interlaminar shear strength

These are some handling criteria for a wet resin system that are unique to wet filament winding:

- Viscosity should be 2 Pa·s or lower
- Pot life should be as long as possible (preferably more than 6 h)
- Toxicity should be low

One of the important resin properties in the cured structure is adhesive strength to the fiber, which is important for most systems, although rocket motors have been filament-wound with released aramid fibers in the hoop direction (Ref 4). Releasing the aramids increases their performance in a biaxial strain field by eliminating transverse loading in the fiber; aramids have very low transverse strength. Another important property is heat resistance, which is critical. A high heat distortion resin system should be chosen only after a thorough study of the operating environment of the filament-wound component. Also, fatigue strength, chemical resistance, and moisture resistance of the composite are key selection criteria, but should be evaluated only in relation to the required mechanical properties desired in the operating environment. In addition, high strain-to-failure capability of the resin system is important to allow transfer of loads with higher-strength or higher-modulus fibers.

Introduction of the aramid fiber in the early 1970s spurred new investigations of resin systems. Many resin systems which worked well with glass were not appropriate for the advanced fibers, and several key resin systems

Table 4 Representative epoxy formulations for wet filament winding

Component	Curing agent equivalent weight, meq/g	1	2	3	4	5	6	7	8	9	10	11	12	13	14	15	Source
		Long pot life				Toughened				High temperature			High elongation	Anhydride			Source
Resins																	
DER 332	174	100	100	100	100	100	100	100	100								Dow
EA 953A	(a)													100			Hysol
EPON 826	178												100		100		Shell
CIBA 0510	100									100	100						Ciba-Geigy
APCO 2447	(a)											100					Hexcel
EPON 828	188															100	Shell
Diluents, additives																	
ERL 4206	72									20		20					Union Carbide
Kelpoxy 272	340						50.6		25.1								Spencer Kellog
Kelpoxy 293	340					29	50.6	50.1	25.1								Spencer Kellog
5022	130												25				Celanese
EMI-24	55													1			Fike
Heloxy 68	75					84	123.5										Wilmington
Curing agents																	
Jeffamine D230	8.3				34.6												Texaco
Jeffamine T403	6.16	36						51.1	51.1								Texaco
Tonox 60-40	37					45	61				49.8	39.9	30				Uniroyal
APCO 2330	...									43.1							Hexcel
EA 953B	(a)													15			Hysol
MDA	50																Various
DMHDA	72		21.8														Pacific Anchor
MNDA	82			24.8													Rohm & Haas
NMA (CIBA906)	180														90	90	Ciba-Geigy
BDMA	...															1	Various
References		5	6, 7	6, 8	9	10	10	10	10	11	11	11	12, 13	13	14	15	

(a) Proprietary, not disclosed by manufacturer

were withdrawn from the market because of economics or toxicity concerns. New types of resin systems were developed to obtain low-temperature cure, long pot life, high heat distortion, high flexibility or strain-to-failure, and low toxicity (Ref 5-15).

These five categories are somewhat exclusive; that is, a low-temperature cure does not usually result in high heat distortion properties, and high heat distortion generally rules out high strain capability. Representative resin systems and supporting data for several of these categories are shown in Table 4. These resin systems have undergone varying degrees of investigation; they are representative of good starting points that can be further tailored to fit user needs. The systems have been reported so that the starting point is 100 parts of the base resin, and the nominal value is reported if there is a spread; for example, a resin equivalent weight of 180 to 195 is reported as 188. Many of these formulations were developed specifically for filament-wound rocket motors. The proprietary formulations developed for rocket motors (not described here) can be approximated by reference to the reported data—the ingredients are generally listed but not their percentages—and by the use of the formulas that relate the molecular weight and epoxide value to calculate a stoichiometric ratio.

Long pot life systems were developed to allow fabrication of large structures with extended gel time and minimum exotherm. Anhydride-cured systems also usually have long pot lives, coupled with high-temperature cures and high heat-distortion temperatures. Toughened filament-winding resin systems at present depend on the addition of a prereacted epoxy resin/CTBN rubber polymer to achieve toughness without shortening pot life. Toughened, long pot life and flexible formulations have lower heat distortion temperatures (or glass transition temperatures) than do high-temperature resin systems, which usually have higher cure temperatures and lower strain-to-failure rates.

Manufacturing Processes

A rocket motor case manufactured by filament winding will be used as a basis for addressing typical manufacturing processes and concerns. Figure 4 shows a flow diagram of key operations.

Impregnation. Resin and reinforcement are joined in the impregnation process. The types of impregnation processes in common usage are preimpregnated (prepreg) roving (commercial), wet rerolled, and wet winding. The advantages and disadvantages of each are shown in Table 5.

Prepregs offer excellent quality control and reproducibility in resin content, uniformity, and band width control. Many high-performance resins can only be impregnated by special processes, such as the hot-melt process. Many commercial prepregs can be certified to key government specifications; most use solvents or preservatives added to the resin formulation to extend storage life. These can affect the tack of the roving, making it difficult to remove

the roving from the spool during winding. These same solvents can become trapped during B-stage and cure. Trapped volatiles promote void incubation, which decreases mechanical strength (particularly shear and compression) in the finished composite. Often, intermediate compaction and heating operations are required to remove these solvents, particularly in thicker-walled laminates.

With wet rerolled prepreg, a controlled volume of resin is impregnated on a controlled length of fiber reinforcement and then respooled. Quality control can be performed away from the winding operation. Usually no preservatives or solvents are required because the roving is either used immediately or stored in a freezer for future use. This can be a very cost-effective method of obtaining preimpregnated roving.

Wet impregnation of the fiber can be accomplished by pulling the reinforcement either through a resin bath or directly over a roller that contains a metered volume of resin controlled by a doctor blade. This is a low-cost system that is widely used in commercial applications with polyester resins. The resin content is affected by several parameters: resin viscosity, interface pressure at the mandrel surface, winding tension, the number of layers per inch, and the mandrel diameter.

Mandrel Preparation. The tool around which the impregnated roving is wrapped is the mandrel. The principal types of mandrel in common usage in the filament-winding industry are water-soluble sand mandrels; spider/plaster

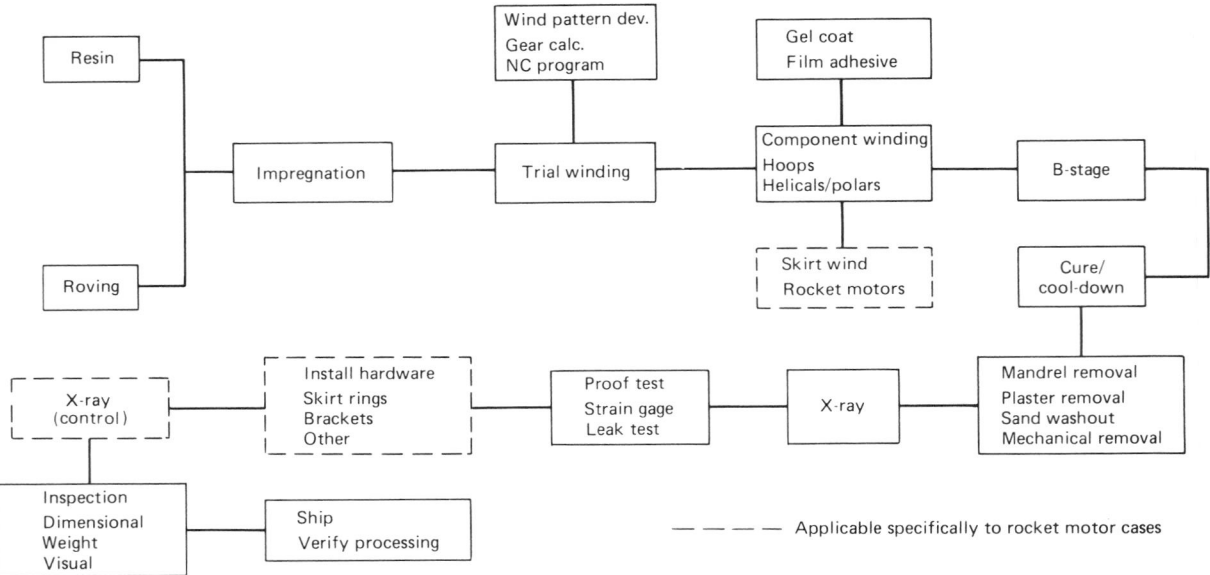

Fig. 4 Typical manufacturing flow diagram for filament winding

Table 5 Comparison of filament-winding impregnation methods

	Prepreg	Wet winding	Wet rerolled
Cleanliness	Best	Worst	Almost equal to prepreg, mess is away from winder
Fiber availability	Poor. Not all fibers are available; many necessitate special order	Best. Any fiber that system will handle	Best. All fibers
Control of resin content	Best. Constant speed and viscosity	Poor. Speed of mandrel varies, viscosity of resin may vary	Better. Process is away from winding and is faster; little viscosity change
Quality assurance.	Highest. Can be done far ahead	Worst. Imposes quality control procedures onto factory floor and can lead to errors	Good. Can be done ahead
Ability to use complex resin systems	Yes. Hot melts available.	Very difficult. Requires complex impregnators to remove solvents or liquify hot melts	Difficult. Still requires complex impregnators
Large data base resin systems	Yes	Commercial resins generally not available as liquids; the wet systems with large data bases may be proprietary	Same as wet winding
Graphite fibers encapsulated (to prevent electronic shortouts)	Yes	No	Graphite fibers not released at winder
Storage	Must be refrigerated and storage records maintained	Easy mix at winder; dry fibers have long shelf life	Must be stored like prepreg, but shorter storage life records must be kept
Fiber damage.	Depends on impregnator; fiber is handled twice	May require special equipment; less damage potential because of less handling	All handling of fiber is under control of user
Cost	Highest	Lowest	Slightly above wet but also requires capital investment for impregnation equipment
Large roving package	Depends on impregnator	Whatever is available dry from fiber manufacturers	Whatever is available dry from fiber manufacturers
Room-temperature cure.	Not possible	Possible	Possible
Simple resin formulation.	Possible	Necessary	Necessary
Winding speed.	Can be highest. Resin throw from fiber is minimized	Lowest speed	Intermediate. Resin can be staged to lower resin throw
Stability on nongeodesic path	Highest possible	Lowest. Wet resin may cause slippage	Intermediate. Resin can be staged to increase tack

mandrels for low-volume products; segmented, collapsible mandrels for continuous production of pipes; tube mandrels; and unremovable liners, such as load-sharing metal liners for pressure vessels.

Water-soluble sand mandrels are used mainly for rocket motor cases, and the insulator is almost always preassembled with the mandrel. Wind axis, polar fittings, and other tooling are preassembled, and a water-soluble sand solution is cast into the mold around the tooling. Following cure of the sand, the two halves of the mandrel are assembled and bonded. The two insulator halves are spliced using uncured rubber, which co-cures with the case. The entire insulation surface is grit-blasted (or rough-sanded) and cleaned with a solvent. Resin gel coat is applied to serve as an adhesive between the insulator and composite overwrap. The use of a film adhesive in place of the resin gel coat is becoming more common. This provides a controlled adhesive thickness with repeatable properties, but the cost can be prohibitive.

Spider/plaster mandrels provide another approach to a high-tolerance mandrel surface through use of a plaster sweep over removable or collapsible tooling. The plaster is cured, then overwrapped with Teflon tape or some other separator film. Following cure, the tooling is removed, the plaster is chipped out, and the release tape is removed, leaving the desired inside contour. In some instances, the rubber insulation is laid up and cured directly over the plaster contour. The process is completed by machining the insulation to the desired contours.

Segmented, collapsible mandrels are specialized and expensive, but the cost is justified for high-production applications because of their reusability and the continuous winding process. Surface preparation before winding consists of an application of a mold release and then an ample gel coat to provide a continuous inside surface. The gel coat in this application is designed to provide a flexible barrier to prevent leakage at low strain levels.

Tube mandrels are used in many applications involving cylindrical metal mandrels in which the cured composite is pushed (or pulled) off the surface after cure; this requires high-quality tooling for trouble-free usage. Chrome plating or hardened and polished surfaces assist in easy mandrel removal. A slight taper along the mandrel length is also beneficial.

Unremovable liners are used for metal-lined pressure vessels, combining the high strength-to-density advantage of composites with a thin, impermeable metal liner. Using this concept, high-pressure low molecular weight gas such as helium or hydrogen can be effectively contained without leakage. The metal liner can be designed to carry a large or small portion of the internal pressure, but in all cases the liner (which initially serves as the winding mandrel) becomes a vital part of the pressure vessel. The mandrel preparation can vary from an adhesive system, where bonding to the composite is desired, to a released system, where independent movement between the liner and structural composite is desired.

Winding Preparation. The manufacturing processes selected for the component are a function of product geometry, weight, and the availability of winding equipment. Most filament winding is still performed using the mechanical gear-driven machines which evolved during the late 1950s. However, many of the winding machines now in use are numerically controlled (NC), providing the latitude to wind nonoptimum shapes where special considerations are required in order to wind the fibers on nongeodesic paths. Analysis techniques have been developed that derive a "slip coefficient" required to prevent fiber slippage from such paths.

Because most reinforcements are packaged on rolls, tension can be introduced at the roll. Tensioning devices include magnetic or friction brakes, electronic rewind, and rotating scissor bars. Because the latter two techniques have the capability to rewind, they allow winding of low-angle patterns around end domes, since the overtravel past the domes can be taken up. The tensioners are often mounted on reels, either remote to the winder or as a part of the carriage that actually travels with the delivery system. The tensioning devices should have variable but controlled tension levels, easy adjustment of tensions, rewind capability to prevent fiber slacking, and uniform tension regardless of roll size.

Component Winding. For rocket motor cases, as for most high-quality filament-wound components, the manufacturing operation is controlled by detailed documentation. The operator follows this documentation, carefully completing patterns, often changing from longitudinal to hoop winding, and verifying quality control. In the prototyping stage of development, the designer's calculations are checked using pi-tape measurements and thickness measurements at the polar bosses. In production, many quality control verifications can be eliminated, particularly in NC winding operations.

Key elements in motor case construction are the skirts at the tangent zones where the domes and cylinders meet (Fig. 5). These are attached by various bonding/winding or riveting methods to convey loads through the motor case assembly. The composite portion of the skirt usually consists of hoops and longitudinal fibers interspersed at approximately a 50:50 ratio. These are wound or laid up using temporary skirt tooling, which is removed following cure. The joint between the skirt and the motor case must transfer the combined loading through shear. Often a shear web consisting of rubber or film adhesive is laid up, overwound, and then co-cured with the rocket motor case body.

B-Stage/Cure. In the first step in the cure cycle, called B-staging, resin viscosity is advanced by means of external heat (lamps or ovens) to the point where cross-linking of the

Fig. 5 Rocket motor case with attached skirts. Courtesy of Brunswick Corporation, Defense Division

epoxide groups is initiated. At this point the resin is still soft to the touch and still exhibits some tack, but will not reflow upon the application of further heat. This operation is performed to allow removal of excess resin before proceeding with cure. This is often done with plastic paddles, wiping off the resin runs periodically until the resin has advanced into the B-stage state. In many rocket motor case applications, the domes are B-staged first, while the cylinder is temporarily covered with insulation to prevent resin advancement. This allows co-curing of the skirt to the cylinder, and guarantees a "run-free" surface for the dome under the skirt tooling.

Fabrication of the composite portion of the assembly is completed by the curing operation. The types of equipment most commonly used are ovens (gas fired or electric), autoclaves, and microwave ovens. Most epoxy resin systems can be cured easily in gas-fired ovens without supplemental pressure, using either air or inert-gas environments. Recently, vacuum bags and bleeder cloths have been used by some manufacturers to produce more compact, void-free laminates. Autoclaves are commonly used with more exotic resins such as bismaleimides and polyimides, where special considerations are required for proper handling of high-volatile contents. Autoclaves are state-of-the-art equipment for nonwound components such as skins and panels for aircraft components, where the interface pressure between the continuous fibers and thermally expanding mandrel does not develop during cure. Microwave cure requires a high initial investment but significantly reduces energy costs and cure times. However, special heating supplements such as induction heaters are required at the composite/metal interfaces.

Mandrel Removal. The water-soluble sand mandrels are the most easily removed; water is added through the wind axis tooling. The sand is washed out and the tooling assemblies removed. Removal is more difficult for mandrels where the tooling is segmented or collapsible. These may require that plaster be chipped out by hand. This operation is laborious and has a high potential for damage to the component.

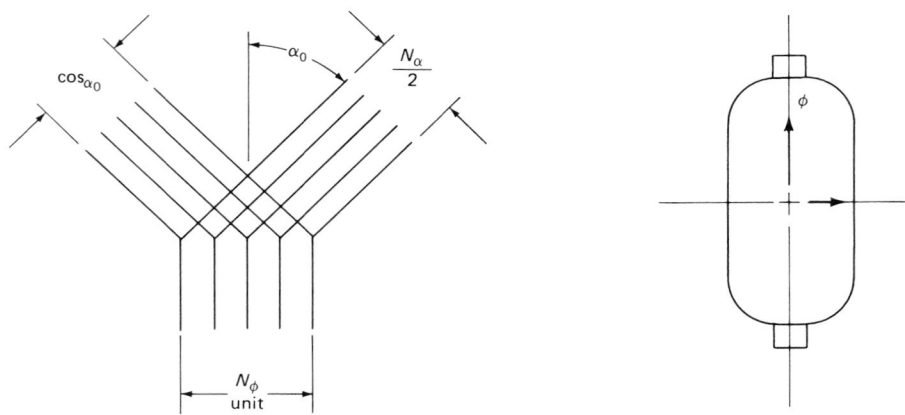

Fig. 6 Free-body diagram for a pressure vessel

Design and Analysis

A rocket motor case application is used here as the typical design problem. The analysis begins with determination of all requirements, including envelope (length and diameter), performance (pressures, weight, and volume), environment (temperature, humidity, and external loads), and interfaces (other stages, nozzles, and ignitors).

With these parameters known, a preliminary design is performed using netting-analysis methods to address the defined loads. Material selection is dictated by performance and environmental constraints. Windability is a major concern, requiring proper selection of wind methods to match the selected geometry with available equipment.

Next, a detailed analysis is performed using the selected wind angles, the calculated fiber thickness, and the resulting composite weight. Polar bosses and skirt thicknesses are selected, and trade studies are performed to optimize the design.

Finally, the detailed stress analysis is performed using finite-element techniques. The thickness of the shear plies at the polar bosses and at the skirt/case interface is analyzed with computer techniques. The skirt stability against buckling is similarly verified.

The design is verified when a full-scale unit is proof-tested and hydroburst. Extensive strain gage and displacement-measuring devices are used along with high-speed video coverage and acoustical measurements.

Netting Analysis. Netting analysis assumes that the fibers provide all the longitudinal stiffness and strength. This assumption is not only conservative but also an excellent basis for quick sizing of composite thickness of a typical rocket motor case or pressure vessel where the liner does not have a significant load share.

In the cylindrical region of a rocket motor case or pressure vessel, the helical fibers carry all of the longitudinal load. A free-body diagram for a pressure vessel is shown in Fig. 6. Summing the running loads in the longitudinal direction, $N\alpha$, the following relationships are obtained:

$$2\left(\frac{N\alpha}{2}\right)\cos^2\alpha - N\phi = 0 \qquad (\text{Eq 1a})$$

$$N\alpha = \frac{N\phi}{\cos^2\alpha} \qquad (\text{Eq 1b})$$

For a pressure vessel with closed ends, the longitudinal running load, $N\phi$, can be defined in terms of the internal pressure, P, on the average wall radius, R:

$$N\phi = \frac{P\bar{R}}{2} \qquad (\text{Eq 2})$$

Substituting Eq 2 into Eq 1b results in:

$$N\alpha = \frac{P\bar{R}}{2\cos^2\alpha} \qquad (\text{Eq 3})$$

The stress in the helical fibers is found by dividing the running load in the direction of the fiber by the helical fiber thickness:

$$\sigma_{\alpha f} = \frac{N\alpha}{t\alpha f} = \frac{P\bar{R}}{2\cos^2\alpha\, t\alpha f} \qquad (\text{Eq 4})$$

The circumferential, or hoop, load in the cylindrical region is carried partly by the hoop fibers. Considering the freebody diagram, it is seen that the contribution of the helical fibers in carrying the hoop load is:

$$N_\Theta' = N\alpha \sin^2\alpha \qquad (\text{Eq 5})$$

The total hoop running load can be defined in terms of the internal pressure and the average wall radius:

$$N\Theta_{\text{tot}} = P\bar{R} \qquad (\text{Eq 6})$$

The running load carried by the hoop fibers is found by subtracting the helical contribution from the total load. Equation 3 can then be substituted into Eq 7:

$$N\Theta = N\Theta_{\text{tot}}\, N\Theta' = P\bar{R} - N\alpha \sin^2\alpha \qquad (\text{Eq 7})$$

$$N\Theta = P\bar{R} - \frac{P\bar{R}\tan^2\alpha}{2} \qquad (\text{Eq 8})$$

The stress in the hoop fibers, $\sigma_{\Theta f}$, is found by dividing the portion of the hoop load, $N\Theta$,

carried in the hoop fibers by the hoop fiber thickness, $t_{\Theta f}$:

$$\sigma_{\Theta f} = \frac{N\Theta}{t_{\Theta f}} = \frac{P\bar{R}}{t_{\Theta f}}\left(1 - \frac{\tan^2\alpha}{2}\right) \qquad (\text{Eq 9})$$

Fiber Thickness Calculation. To solve Eq 4 and 9 in the netting analysis, the fiber thickness of hoop and longitudinal patterns must be determined. The fiber thickness of a ply of reinforcement can be calculated as follows:

$$t_f = $$

$$\frac{\text{number of spools} \times \text{cross-sectional area/roving}}{\text{band width}}$$

$$(\text{Eq 10})$$

To initiate the calculation, the cross-sectional area of the roving must be known. Manufacturers' specifications provide this information; however, if greater accuracy is desired, the cross-sectional area may be determined by weighing a known length of roving and applying the known density.

Band Density. Band density is a term used in the calculation of fiber thickness. It may be noted as ends/inch in glass and aramid applications or as tows/inch or rovings/inch in graphite applications. The inverse of band width is turns per inch (TPI), which is commonly used in filament winding nomenclature. In Eq 10, band density is simply the number of ends or tows divided by band width.

Bulk Factor. Bulk factor is a constant ($K\Theta$ or $K\alpha$) used to convert the calculated fiber thickness ($t_{\Theta f}$ and $t\alpha_f$) into composite thickness and is a measure of the increased volume due to the resin content. It is also the direct inverse of fiber volume (vol%). The calculated fiber thickness is converted to predicted wall thickness by the equation:

$$t_c = t_f K \text{ (per ply)} \qquad (\text{Eq 11})$$

or, more specifically:

$$\text{Hoop layer } t\Theta = t_{\Theta f} \times K\Theta \times (2)^* \qquad (\text{Eq 12})$$

$$\text{Helical or polar layer } t\alpha = t\alpha_f \times K\alpha \times (2)^* \qquad (\text{Eq 13})$$

Typical bulk factors and/or fiber volumes are shown in Table 6. Bulk factors depend on several process and geometric considerations, including resin viscosity, mandrel diameter, winding tension, wind angle, processing time, B-stage temperature, and external pressure during cure. The actual bulk factor derived is calculated using in-processing measurements such as those from pi-tapes.

Composite Density. Composite density may be easily derived, if the bulk factor, fiber volume, or resin volume are known, by using the equations shown in Table 7. These equations assume that there are no voids in the resin (Ref 16).

Winding Patterns. In winding rocket motor cases or pressure vessels, the selection of a

*Typically two plies/wound layer

Table 6 Typical bulk factors and fiber volumes

	Bulk factor	Fiber, vol%
Aramid or glass		
Helical/polar...........	1.82-1.67	0.55-0.60
Hoop.................	1.54-1.43	0.65-0.70
Carbon (graphite)		
Helical/polar...........	2.0-1.8	0.50-0.55
Hoop.................	1.67-1.54	0.60-0.65

Table 7 Methods for determining composite density

Given	Unknown	Equation
% Fiber by volume	% Fiber by weight........................	$\dfrac{va}{va + (1 - v)b}$
Fiber density $= a$ Fiber volume $= v$ Fiber weight $= v \cdot a$ Resin density $= b$ Resin volume $= 1 - v$ Resin weight $= (1 - v) \cdot b$	% Resin by weight........................	$\dfrac{(1 - v)b}{va + (1 - v)b}$
	Composite density	$va + (1 - v)b$
	Bulk factor	$1/v$
% Resin by volume	% Fiber by volume	$\dfrac{(1 - w)/a}{(1 - w)/a + w/b}$
Fiber density $= a$ Fiber weight $= 1 - w$ Fiber volume $= (1 - w)/a$ Resin density $= b$ Resin weight $= w$ Resin volume $= w/b$	% Resin by volume	$\dfrac{w/b}{(1 - w)/a + w/b}$
	Composite density	$\dfrac{1}{(1 - w)/a + w/b}$
	Bulk factor	$1 + \dfrac{wa}{(1 - w)b}$
Bulk factor $= BF$	% Fiber by weight........................	$\dfrac{a}{a + (BF - 1)b}$
Fiber density $= a$ Fiber volume $= 1$ Fiber weight $= a$ Resin density $= b$ Resin volume $= BF - 1$ Resin weight $= (BF - 1)b$	% Resin by weight........................	$\dfrac{(BF - 1)b}{a + (BF - 1)b}$
	% Fiber by volume	$1/BF$
	% Resin by volume	$(BF - 1)/BF$
	Composite density	$\dfrac{a + (BF - 1)b}{BF}$

winding pattern is dictated by the forward and aft opening sizes of each dome and by the cylinder length. The nozzle interface is the primary factor in sizing the aft opening for motor cases. To use standard winding procedures, the opening should be less than 60% of the external diameter (the larger the opening, the more efficient the use of the available envelope). At the forward opening, either the igniter or fabrication tooling (wind axis) usually dictates the dome opening size (here the smaller the opening, the more efficient the use of the available envelope). Length and diameter are, of course, obvious restraints in selecting the winding pattern. Types of winding patterns are shown in Fig. 7.

In polar winding (Fig. 7a), the filament path passes tangent to the polar opening at one end of the chamber, and tangent to the opposite side of the polar opening at the other end. A one-circuit pattern is inherent to the system. The windings are delivered by the arm while describing a great circle, and the laid-down pattern is planar. Although it is the simplest winding method, it is limited to length-to-diameter ratios of <1.8. It is widely used to wind spherical shapes by use of continuous step-outs of the pattern. In applications using different opening sizes, a planar wind angle is generally stable (no slippage) if the difference between the geodesic angle of a dome and the actual winding angle used is small (no more than 12°). The geodesic angle (α_o) of forward and aft domes may be calculated as follows:

$$\alpha_o = \sin^{-1} R_E/R\alpha$$

where R_E is radius to center of winding band adjacent to polar boss and $R\alpha$ is radius to midthickness of longitudinal layers in the cylinder (tangent line). Then the slip parameter, $\Delta\alpha$, is the difference between wind angle α and geodesic angle α_o of each dome. Hence, $\Delta\alpha =$ $\alpha - \alpha_o$. The wind angle through the cylinder (Fig. 7a) is defined as:

$$\bar{\alpha} = \tan^{-1} \frac{R_{EF} + R_{EA}}{L}$$

In helical winding (Fig. 7b), the mandrel rotates more or less continuously while the feed carriage traverses at a speed regulated to generate the desired helical angle. The normal pattern is a multicircuit helical. After the first circuit, the fibers are not adjacent, and a given number of circuits must be traversed before the pattern begins to lay adjacent to previous windings. The helical pattern is characterized by fiber crossovers at certain points along the mandrel. A layer is made up of a two-ply balanced laminate. The mandrel revolutions per circuit will vary with the winding angle, band width, and overall length of the vessel. Helical winding allows winding of longer configurations and provides the designer with a slip-free pattern. However, a constant-angle helical with zero slip potential is limited to applications where both dome openings are equal. The wind angle through the cylinder (Fig. 7b) is defined as:

$$\alpha_o = \sin^{-1} \frac{R_E}{R\alpha}$$

In modified helical winding, the winding angle through the cylinder may be changed as desired to minimize the slippage potential at each dome. This pattern presents the designer with multiple options such as trading off the number of circuits to closure as well as changing wind angles during traverse down the cylinder. Wind angle through the cylinder is arbitrary. The expression for α_o' (Fig. 7c) is:

$$\alpha_o' = \sin^{-1} \frac{R_E \left[1 - C \left(\dfrac{R_N - R_E}{R\alpha - R_E} \right) \right]}{R_N}$$

Hoop Patterns. Also used are hoop patterns, sometimes called girth, 90°, or circumferential winding. Strictly speaking, hoop winding is a high-angle helical winding that approaches an angle of 90°. Hoops are gener-

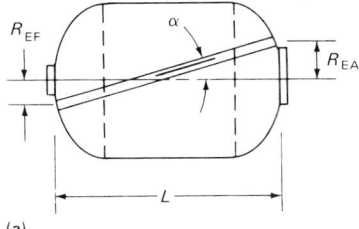

(a)

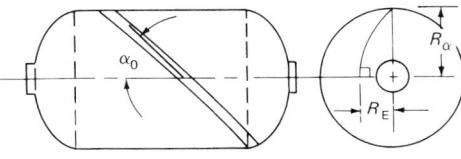

(b)

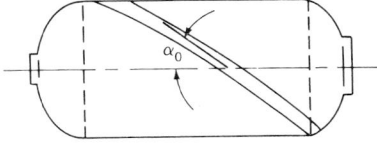

(c)

Fig. 7 Types of filament winding patterns. (a) Polar. (b) Helical. (c) Modified helical

ally confined to the cylindrical portion of the vessel.

Dome Contours. Domes or closures of filament-wound motor cases or pressure vessels have a general surface form of revolution that requires numerical solution using an electronic computer. The two types of dome contours considered are the geodesic (constant stress) and the polar (in plane), each of which requires its own derivation.

The contour of the geodesic dome is derived from the conditions that the filament stress is the same at every point on the dome, and that the principal loads in the surface of revolution are reacted in the direction of the filament. A derivation of the theory is presented in detail in Ref 17. From this derivation, it can be seen that the only geometric variables affecting the dome shape are the radius of the cylindrical portion of the rocket motor case, R_i, and the radius to the center of the filament band at the polar opening, R_E.

For a polar-wound case, the structural elements (domes and cylinder) cannot be considered separately. The polar-dome contour depends not only on the cylindrical radius and the polar opening radius, but also on the polar opening radius of the opposite dome and the overall length of the rocket motor case. Once these geometric variables have been established, the polar winding angle, $\bar{\alpha}$, can be determined. Because the filament path of a polar dome deviates from that of a true geodesic path, additional windings may be required to compensate for increased filament stress. To evaluate the change in filament stress in the dome, the polar filament angle at the dome-to-cylinder tangency (tangent line) must be compared to the stable angle, α_o.

Dome contours are generated by combining netting-analysis equations with membrane shell equations. The result of this combination is a relationship between meridional and circumferential shell curvature:

$$K\phi = K\Theta \ (2 - \tan^2\alpha)$$

where $K\phi$ and $K\Theta$ are curvatures in the meridional and circumferential directions, respectively. The equation is transformed to the variables of the coordinate system and integrated as an initial value problem. A dome generated by this method is called a balanced-stress dome.

Testing

Opportunities and methods for testing to support filament winding use are more varied than those used for testing external pressure cured composites. Any composite test specimen, such as the unidirectional tensile test specimen in ASTM D 3039, can be built by winding unidirectional circumferential impregnated fibers around a large mandrel, cutting and removing the fibers from the mandrel, plying, and then consolidating and curing. In filament winding, however, other types of consolidation are usually avoided because the specimens do

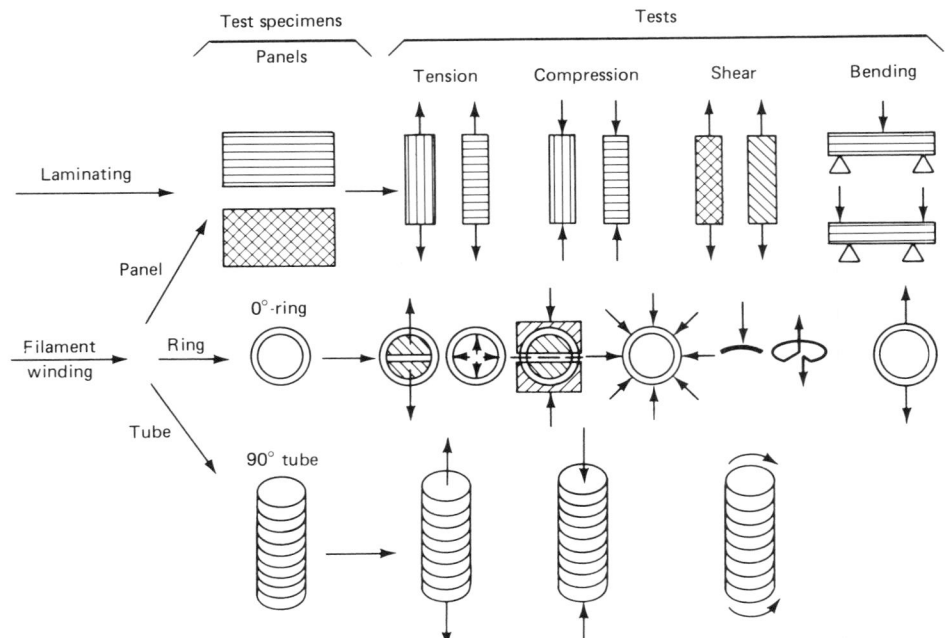

Fig. 8 Methods of material fabrication and respective specimens and test methods. Source: Ref 18

not reflect the properties obtained by filament winding and may be more costly and time consuming to fabricate. It is preferred to filament-wind a tube or a ring and perform all tests on such a specimen. Test specimens and methods are shown in Fig. 8 (Ref 18).

Specimens for ring tests are simple to prepare and test because the ring specimen nearly always consists of only circumferentially wound fiber/composite. Fabrication and testing of a filament-wound ring using split Ds or hydraulic pressure is simpler than testing parallel straight-sided specimens. The filament-wound ring can be wrapped on a parallel fiber reinforced ring (NOL ring) fixture with a lathe. The fiber orientation is automatically correct. The ring, cured without the need for compaction aids such as a vacuum bag or an autoclave, needs no other processing, as opposed to a flat-sided tensile specimen, which needs machining and adhesively bonded end tabs. The flat-sided specimen must be carefully aligned in the test machine; the NOL ring is reasonably self-aligning. The flat-sided test specimen can easily be strain gaged; the NOL ring strain must be determined from test machine deflections or from strain gages mounted on a curved surface, which is less accurate.

If the filament winder uses dry fibers and wet resins, testing is necessary at steps where these are used, as well as for controlling prepreg and finished subcomponent or full-scale parts. Raw materials, resin, and several kinds of fiber can be verified by the standard test methods shown in Table 8. The standard test methods used to control the properties of finished composites and companion-molded test specimens are shown in Table 9. In addition, the test methods used for prepregs suggested by ARP-1617

Table 8 Standard test methods for raw filament winding materials

ASTM No.	Title
C 613	Resin Content of Carbon and Graphite Prepregs by Solvent Extraction
D 1652	Epoxy Content of Epoxy Resins
D 1726	Hydrolyzable Chlorine Content of Liquid Epoxy Resins
D 2587	Acetone Extraction and Ignition of Glass Fiber Strands, Yarns and Roving for Reinforced Plastics
D 3379	Tensile Strength and Young's Modulus for High-Modulus Single-Filament Materials
D 3529	Resin Content of Carbon Fiber-Epoxy Prepreg Tape and Sheet
D 3800	Density of High-Modulus Fibers
D 4018	Tensile Properties of Continuous Filament Carbon and Graphite Yarns, Strands, Rovings and Tows

(Physical-Chemical Characterization Techniques, Epoxy Adhesive and Prepreg Resin Systems, Society of Automotive Engineers) are applicable.

The NOL ring specimen (ASTM D 2290) can be fabricated as an individual ring or by machining rings off a cylinder. The individual rings are preferred and will give higher fiber stress and more consistent results because there are no cut fibers. These test specimens should be wound at a constant but appropriate resin content for the fiber because many tests which are not completely fiber dominated, such as shear, will give inconsistent results if the resin content is not held constant. The tensile test uses a pair of split Ds or a hydraulic pressure apparatus (ASTM D 2290). Neither method gives modulus data.

The NOL ring short-beam shear test is performed in a universal test machine in accor-

Table 9 Standard test methods for filament-wound composites

ASTM No.	Title
D 2290 . . .	Apparent Tensile Strength of Ring or Tubular Plastics and Reinforced Plastics by Split Disk Method
D 2291 . . .	Fabrication of Ring Test Specimens for Glass-Resin Composites
D 2343 . . .	Tensile Properties of Glass Fiber Strands, Yarns and Rovings Used for Reinforced Plastics
D 2344 . . .	Apparent Interlaminar Shear Strength of Parallel Fiber Reinforced Composites by Short Beam Method
D 2584 . . .	Ignition Loss of Cured Reinforced Plastics
D 2585 . . .	Preparation and Tension Testing of Filament-Wound Pressure Vessels
D 2586 . . .	Hydrostatic Compressive Strength of Glass Reinforced Plastic Cylinders
D 2924 . . .	External Pressure Resistance of Reinforced Thermosetting Resin Pipe
D 3039 . . .	Tensile Properties of Fiber-Resin Composites
D 3171 . . .	Fiber Content of Resin Matrix Composites by Matrix Digestion
D 3299 . . .	Filament-Wound Glass Fiber Reinforced Polyester Chemical-Resistant Tanks
D 3379 . . .	Tensile Strength and Young's Modulus for High-Modulus Single-Filament Materials

dance with ASTM D 2344. Data obtained include physical measurements of the test specimen and the ultimate load achieved. These data are used to calculate the apparent shear strength of the mandrel.

Test data obtained from the NOL ring tests are used for comparative purposes only. The data are not suitable for design purposes, but are useful for correlation of pressure vessel performance and NOL ring lab tests and for comparison with published data.

The nonstandard tests that can be used for determining properties of filament-wound composites are shown in Table 10 (Ref 16, 18-26). Many of these tests are similar to the short-beam interlaminar shear test of ASTM D 2344 in that they were not designed to give basic engineering data but are intended to reveal differences in fabrication or materials, such as resin type, fiber type, and sizing; or processing variables, such as fiber tension, resin viscosity or content, and cure schedule. The highly regarded Iosipescu shear test can be used for filament-wound components only if a flat specimen can be made. Although filament winding of a flat specimen is easy to do, it results in a less than optimum specimen because of the lack of compaction across the flat area, leading to an excessive void or resin content unless other methods of compaction are used.

Subscale Test Vessel. In a biaxial strain field, the fiber, resin, and interface are subjected to critical loads. Of all composite test specimens, pressure vessels best serve this purpose. ASTM D 2585 gives the dimensions and test techniques for a standard subscale pressure bottle. However, industry has not reached consensus, and 100-mm (4-in.), 150-mm (6-in.), 170-mm (6.7-in.), and 230-mm (9-in.) subscale pressure bottles are being used in addition to the 146-mm (5.75-in.) inside diameter (ID) bottle recommended by

ASTM. Manufacturers have used their own test vehicles for many years and have developed a large data base, which they are reluctant to give up.

A new subscale test bottle called the Standard Test and Evaluation Bottle (STEB) has been developed under Air Force sponsorship (Ref 27, 28). It is intended to be a standard in the rocket motor case industry for comparing composite materials, processes, and design features. The STEB (Fig. 9) is 250 mm (10 in.) in diameter by 390 mm (15.33 in.) in length, and is modeled as a one-ninth scale of a second-stage ballistic rocket motor case. The cylinder length, however, has been selected to allow either polar or helical winding. The vessel may also be wound with or without stub skirts.

The baseline configuration uses Kevlar 49 fiber in an anhydride epoxy resin and a 75% stress ratio between helical and hoop fibers. Table 11 shows parameters for both helical and polar configurations. Other configurations, including changes in materials, processes, and design features, are easily evaluated using the STEB.

Composite-to-Metal Joints

Although riveting and bolting methods for aircraft composites have undergone a long development and the materials and methods have been well characterized, joining techniques for filament-wound composites have been the sub-

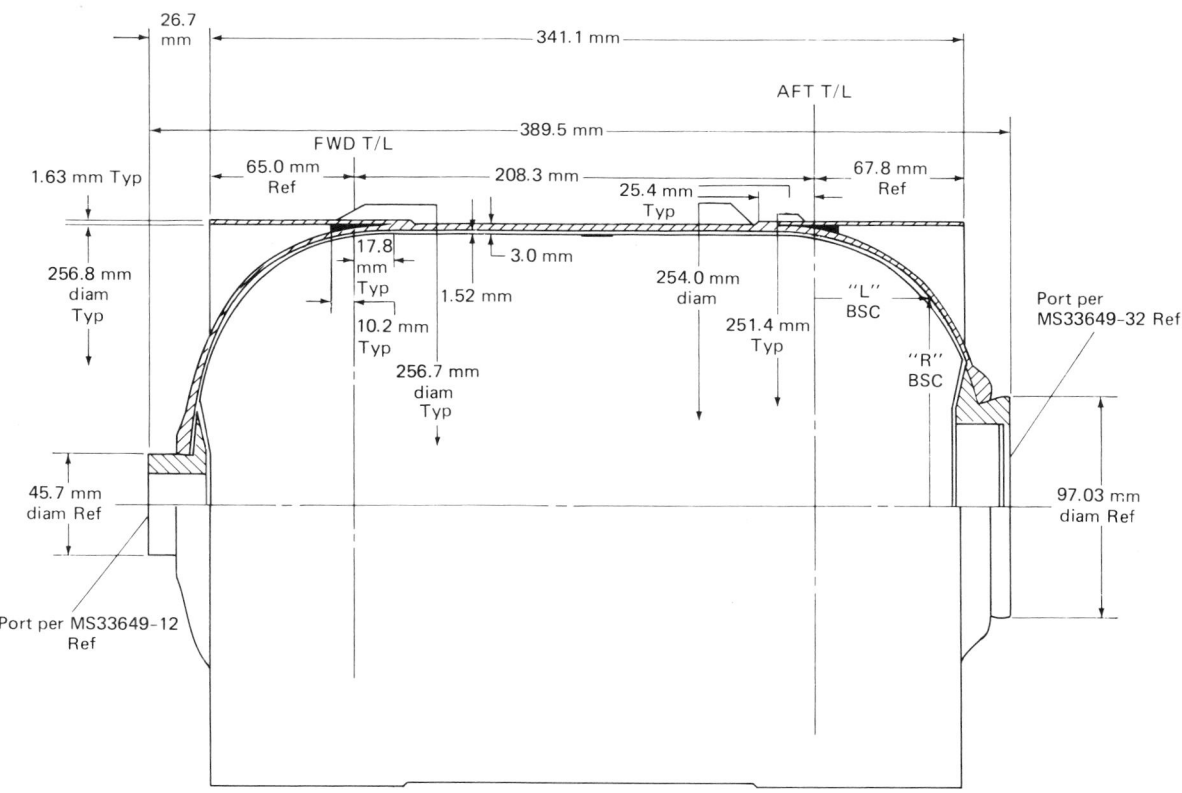

Fig. 9 STEB configuration

Table 10 Nonstandardized test methods for filament winding

No.	Test for	Sample type	Data or fixturing necessary	Test equipment	Comments	Reference
1	Tensile strength Tensile modulus	Elongated NOL ring	Split Ds necessary	Standard tensile machine	Test specimen provides a flat surface for bonded-on strain gages	19
2	Tranverse tensile strength	Cylinder, hoop winding only	Self-aligning end attachments must be bonded on	Standard tensile machine		20
3	Radial compressive strength and modulus	Ring	Compression fixture	Standard compression machine		19
4	Radial compressive strength	Ring	None	Standard compression machine	Useful for determining E, which can then be used to determine G with test No. 8	23
5	Longitudinal compression strength, modulus	Hoop cylinder	None	Standard compression machine		...

(continued)

Table 10   (continued)

No.	Test for	Sample type	Data or fixturing necessary	Test equipment	Comments	References
6	Shear	NOL ring	Compression leveling shear fixture needed	Standard compression machine	Needs thicker test specimen than for normal NOL ring. Used for qualitative comparisons only.	22

7	Shear modulus by torsion of cut ring		U-grips on cut edges of specimen	Standard tensile machine		23

8	Shear modulus by bending intact ring	NOL ring	Must know E, two rods, 25-mm diam, alignment fixture	Standard compression machine		21

9	Shear strength, modulus		Split Ds or internal pressure fixture	Standard tensile machine fixture	Has problems similar to ASTM D 2733	18, 22

10	Shear strength, modulus		Internal pressure fixture	Internal pressure fixture	Same problems as No. 9 above	18, 24

11	Shear strength, modulus by torsion of tube		End caps for cylinder	Torsional test machine	Must use care in manufacture and measurement of ID and OD	25, 26

(continued)

Table 10 (continued)

No.	Test for	Sample type	Data or fixturing necessary	Test equipment	Comments	References
12	Shear strength by compression of a notched tube		None	Standard compression apparatus	Difficult to machine grooves with required accuracy and to ensure failure in gage section	18
13	Bearing	Hoop cylinder	Strain measuring device; fixture to transfer load to pin	Standard compression machine		

Table 11 Baseline STEB design (Kevlar 49)

	Helical	Polar
Helical/polar fibers		
Design strength, MPa (ksi)	2200 (320)	2200 (320)
Wind angle, degrees	18.0	12.5
Fiber thickness, mm (in.)	0.828 (0.0326)	0.853 (0.0336)
Circuits to close pattern	11	1
Circuits per layer	76	76
Band width, mm (in.)	10.1 (0.40)	9.9 (0.39)
Rovings per band	4	4
Band density, ends/mm/ply (ends/in./ply)	1.57 (39.9)	1.62 (41.1)
Number of layers	3	3
Number of dome caps	1	1
Hoop fibers		
Design strength, MPa (ksi)	2950 (425)	2950 (425)
Fiber thickness, mm (in.)	1.10 (0.0433)	1.20 (0.0472)
Band width, mm (in.)	(0.23)	(0.21)
Rovings per band	2	2
Band density, ends/mm/ply (ends/in./ply)	1.39 (35.4)	1.52 (38.5)
Number of plies	9	9
Design burst pressure, MPa (ksi)	26.2 (3.8)	27.6 (4.0)

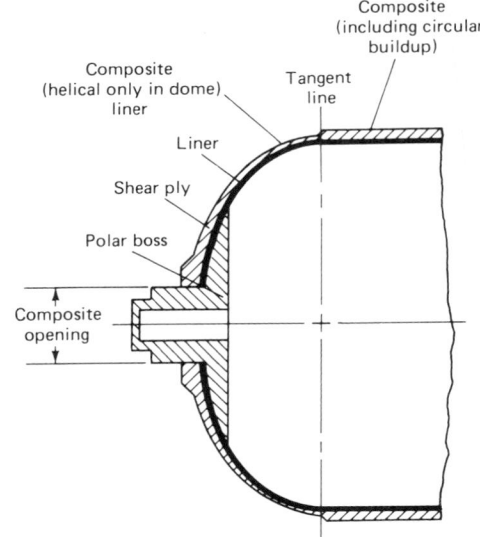

Fig. 10 Dome section of typical cylindrical pressure vessel. Source: Ref 29

ject of relatively little study. The requirements for a filament-wound composite joint are the same as for any other, that is, the joint should be lightweight and noncorrosive, and should not concentrate stresses.

Three types of joints can be made in a filament-wound structure. The first is the concurrent or *in-situ* joint, in which the composite is wound and cured in place while joined to a shear ply or to the metal portion. The *in-situ* joint takes advantage of the filament-winding process and eliminates complex machining, but

has some geometric restrictions. First, the joint must be reasonably smooth and must not have any abrupt changes in contour; if there are changes in elevation, they must be made at a small angle to the mandrel centerline to avoid slipping and bridging. Second, there must not be any protrusions above the composite surface to interfere with the path of the fiber near the joint.

Many rocket motors use the *in-situ* joint on both forward and aft domes, as shown in Fig. 10 (Ref 29). The taper on the boss section serves to

alleviate the stress concentrations, and an adhesive film or rubber shear ply galvanically isolates the composite from the polar boss. The rubber shear ply can be sized in thickness and length to provide reliable bonds in zones of deformation. This type of joint is simple because the joint theoretically reacts only to tensile stress.

Fig. 11 "Viper" motor case, showing "wound-in" holes. Courtesy of Brunswick Corporation, Defense Division

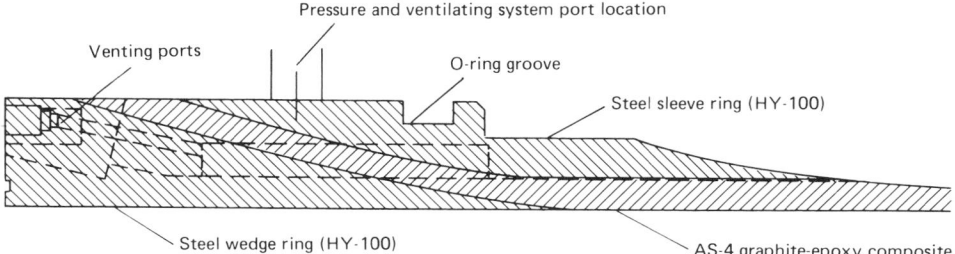

Fig. 12 Clamped composite-to-metal joint

A joint technique using removable pins was developed to allow a 100% opening in a cylindrical rocket motor case (Fig. 11) with the capacity to remove and replace the closure. This technique requires a helical wind pattern and a mandrel that contains removable pegs (Ref 2). An *in-situ* joint between a long composite canister and the metal was developed for ease of manufacture and to resist axial loads and shock and vibration (Fig. 12).

The final type of joint is one that is machined or changed in some way after cure of the composite. Two joint techniques, one single lap and one double lap (Fig. 13), were used on thick MX graphite-epoxy canisters (Ref 30). Another joint, similar in size, was manufactured for proposed space shuttle composite solid rocket motors (Ref 31). Because of the necessity for field assembly and the requirement for commonality with the steel rocket motor sections, a tang and clevis joint was necessary (Fig. 14). All these joints, which involve metallic sleeves or steel clevises, re-

quire drilling and/or machining of both the composite and the metal, which can be time consuming and expensive.

Obviously, it would be advantageous to make a transition to the metal portion without a joint—a contradiction in terms, but possible when a load-sharing metal liner is used. Pressure vessels with load-sharing liners have been used in emergency breathing systems for firemen, escape slide pressurization bottles for aircraft, nitrogen, oxygen, and helium bottles on the space shuttle, and high altitude mountaineering oxygen bottles. The composite and the metallic load-sharing liner must be carefully chosen because the liner carries one-half to one-third of the pressure load and the proof cycle of pressure strains the liner elastically. Subsequent cycles to operating pressure produce loads that can be carried completely within the elastic capabilities of the liner and the reinforcement material.

Environmental Effects

Water is a pervasive substance that is present during the manufacture and service life of any composite. Controlling moisture content throughout the life cycle can be difficult and expensive, but failing to do so can lead to an unqualified product or catastrophic failure. This is particularly true of modern rocket motors, which use organically cross-linked systems susceptible to moisture attack during cure. Even if adequately cured, these materials remain permeable to water throughout their life cycle. It is important to recognize, however, that by nor-

mal standards, these organic materials would be considered adequate moisture barriers; many paints have moisture-permeability values of the same order of magnitude as those of the resin and polymer systems used by the rocket industry. The long service life cycles and the severe consequences of moisture attack on rocket motors force the adoption of a highly efficient moisture barrier system.

A radial section of a typical rocket motor is illustrated in Fig. 15. No single component material is really a continuous, impermeable layer; each is a collection of materials with its own affinity for water and each is held together by some adhesively bonded interface. The potential for leak paths, moisture absorption, diffusion, interface bonding, and so on, is high. Each material is affected differently by moisture. Adhesive bonds are particularly susceptible to degradation. Oxidizer, fuel, fillers, and fibers are bonded to some degree in their respective matrices.

Moisture can be a problem in manufacturing operations, because it affects fibers and resin systems to various degrees. Some levels of humidity can be accepted without jeopardizing the quality of the composite, but occasionally it is necessary to modify the formulations of resin systems and gel coats to negate normal fluctuation in humidity levels.

Moisture can be even more of a problem when the unit is in field service and exposed to temperature and humidity, especially when combined and driven by cyclic changes in environments. The degree of degradation is again a function of the resin and fiber used.

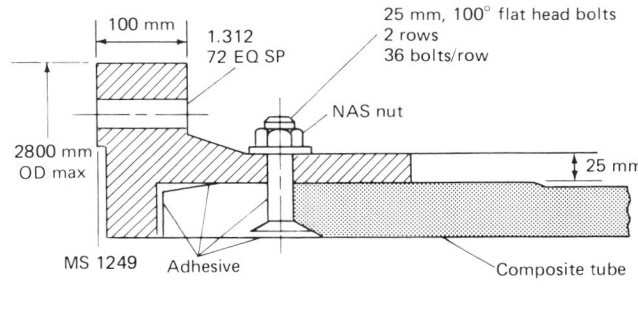

(a)

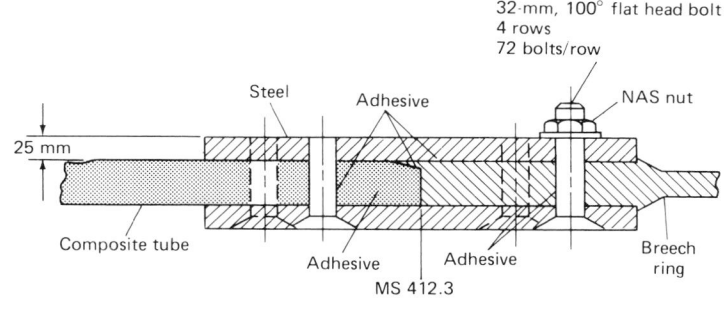

(b)

Fig. 13 Lap-shear metal-to-composite joints. (a) Single lap. (b) Double lap

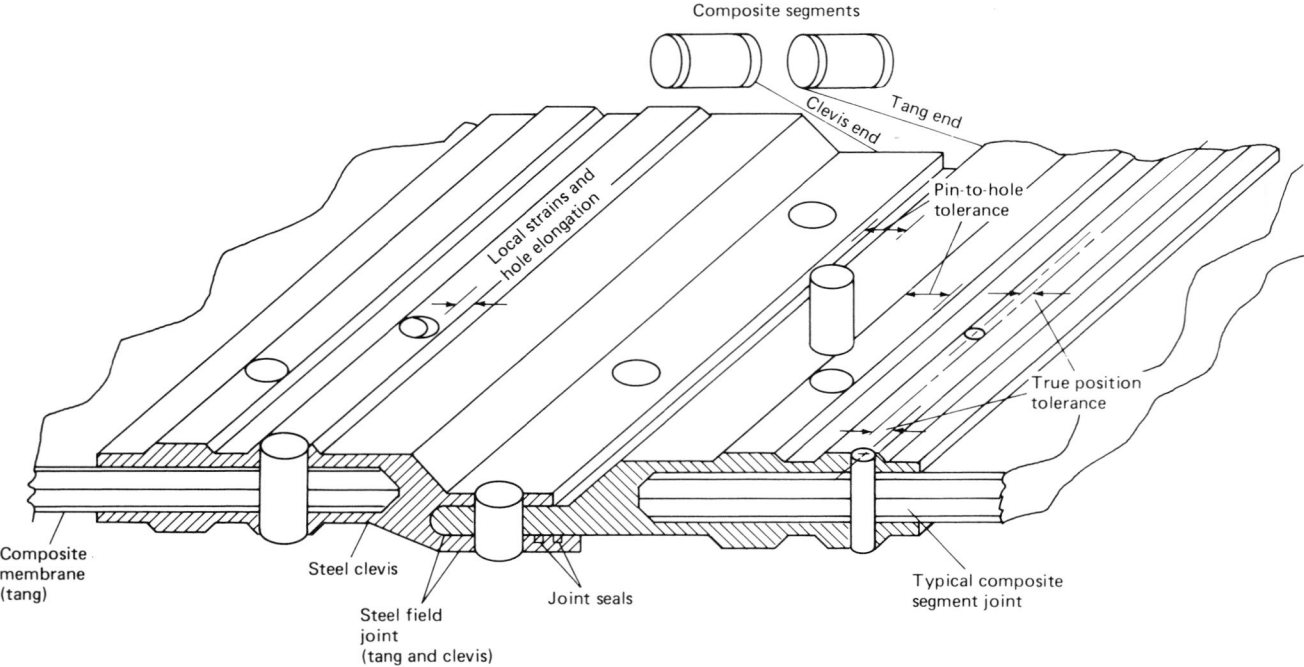

Fig. 14 Filament-wound case joints

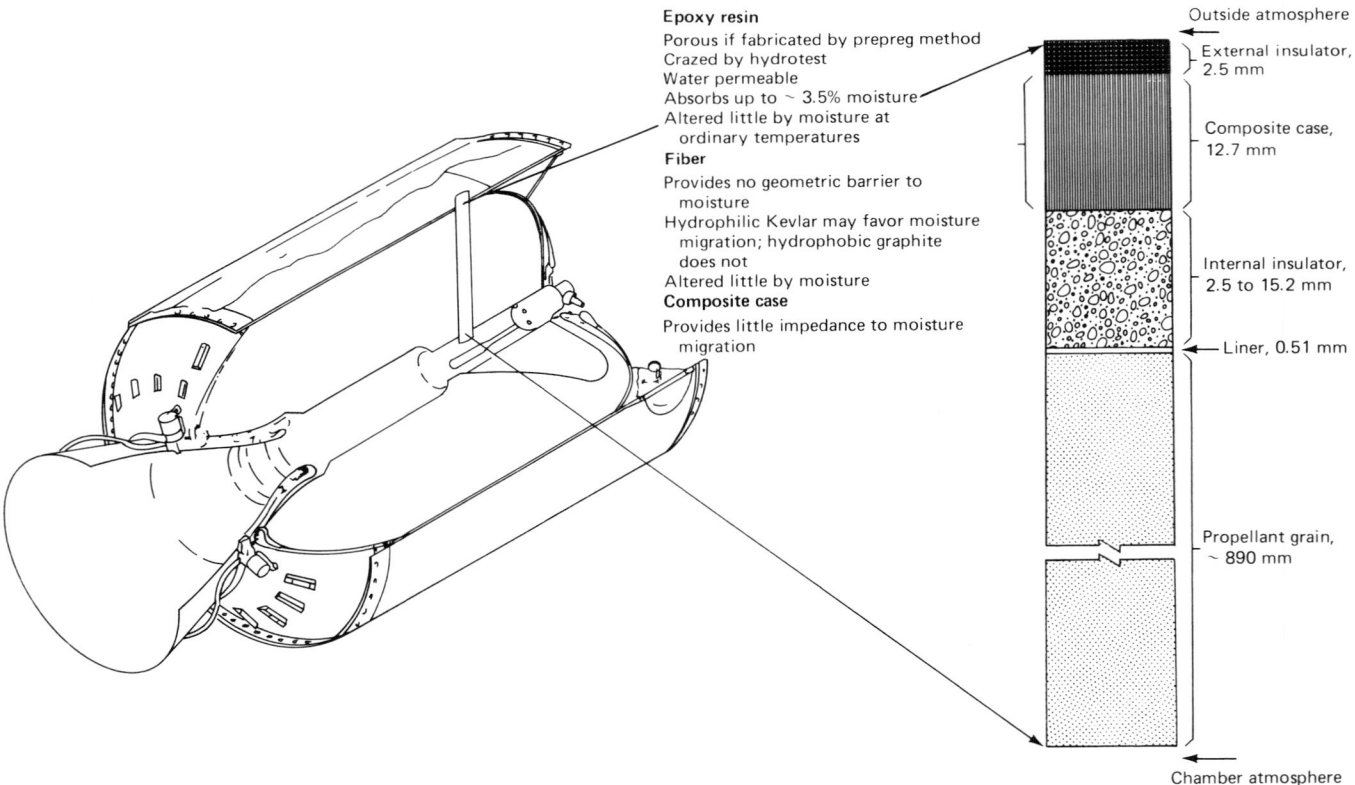

Fig. 15 Radial section of typical rocket motor case. Courtesy of United Technologies Corporation

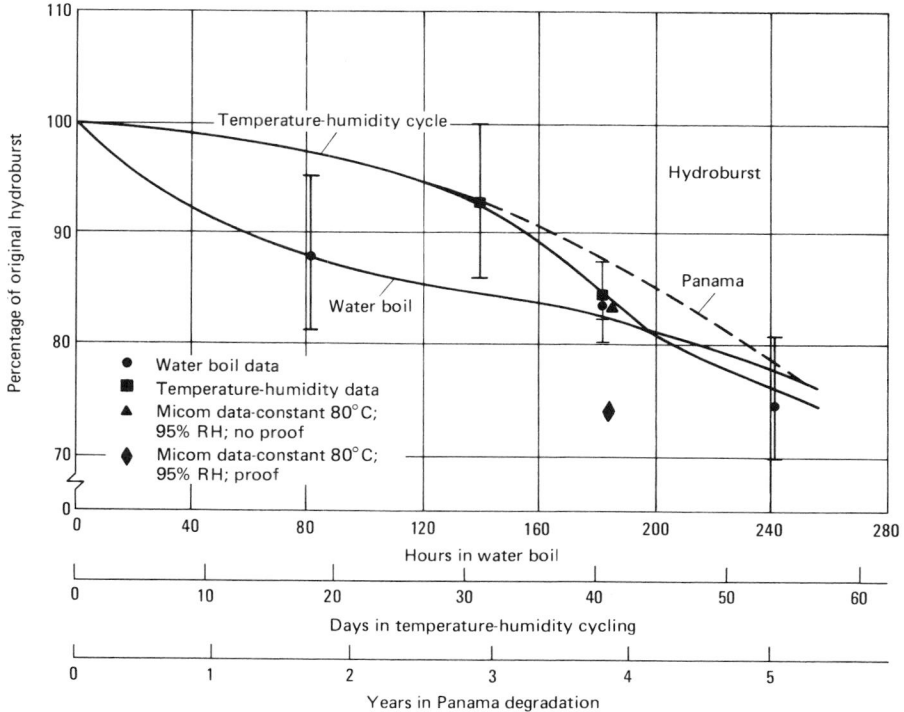

Fig. 16 S-glass-epoxy degradation rates

Resin systems with low cross-link density and low heat distortion temperatures have very little resistance to harsh environments. Other resin systems that have been formulated for high-temperature service, such as bismale-imides and polyimides, offer good moisture resistance, and their use in motor cases is on the increase.

Fiberglass composites degrade as a function of the sizing agent used (Ref 32). However, several studies have shown that, regardless of the sizing, fiberglass composites eventually equilibrate at a certain degradation level regardless of the surface finish used. Other studies (Ref 33, 34) showed that the rate at which degradation occurs in Panama under temperature-humidity cycling is different for various combinations of S-glass and epoxy resins, that the 5-year strength retention is constant for all systems, and that proof testing prior to exposure changes the degradation curve and reduces the ultimate retention strength after 5 years of storage by approximately 12%. To summarize, hydroburst data from various sources show that the ultimate retention strength is about 75% of the original value and that short-term laboratory data correlate with the 5-year Panama exposure data (Fig. 16).

Many studies (Ref 35-37) have shown that absorbed moisture results in significant degradation of the matrix-sensitive mechanical properties of reinforced pressure vessels, particularly at higher temperatures; however, fiber-controlled composite properties are relatively unaffected. Most investigators agree that this degradation occurs because the plasticizing effect of moisture on the resin modulus lowers the glass transition temperature. Water behaves as a plasticizing agent and disrupts the strong hydrogen bond in highly polarized epoxy resin systems. Although reversible to some extent, this reaction can promote a change in failure mode between wet and dry specimens. High heat distortion resin systems are not as significantly affected by moisture exposure at high temperature.

Many epoxies become flexibilized through absorption of moisture; conversely, the same epoxy laminate becomes increasingly brittle as this moisture is removed from the composite. Some pressure vessels designed to fail transverse to the hoops are extremely sensitive to the elongation capability of epoxy resin systems (Ref 38).

Many of the best graphite fibers have not yet been characterized due to the rapid growth of the new fiber sizing technology. However, experience with T300 has indicated that the fiber-resin interface is stable and that the performance of the system is matrix sensitive.

REFERENCES

1. W.T. Freeman, Jr., and B.A. Stein, "Filament Winding; Waking the Sleeping Giant," *Aerospace America*, Oct 1985, p 44-49
2. J. Zitek, "Characterization of Wound-in-Place Holes in Kevlar 49 Composite," Paper 81-0588, American Institute of Aeronautics and Astronautics, 1981
3. G.E. Zahr, An Improved Aramid Fiber for Aerospace Applications, in *Progress in Advanced Materials and Processes: Durability, Reliability and Quality Control*, Elsevier, 1985
4. C.G. Zlomke and A.A. Hughes, "Development of Improved Performance of Organic Fiber Filament Wound Composite for the Second Stage Advanced ICBM Propulsion System, Paper presented at the 1975 JANNAF Propulsion Conference, Joint Army-Navy-NASA-Air Force, Anaheim, CA, Sept 30, 1975
5. T.T. Chiao and R.L. Moore, A Room Temperature-Curable Epoxy for Advance Fiber Composites, in *Proceedings of the 29th Annual Technical Conference*, Reinforced Plastics/Composites Institute, The Society of the Plastics Industry, p 16B-1 to 16B-7
6. R.J. Morgan, C.M. Walkup, F.-M. Kong, and E.T. Mones, Development of Epoxy Matrices for Filament-Wound Graphite Structures, in *30th National SAMPE Symposium*, Society for the Advancement of Material and Process Engineering, March 1985, p 1209-1219
7. J.A. Rinde, I. Chiu, E.T. Mones, and H.A. Newey, 2,5 Dimethyl 2,5 Hexane Diamine: A Promising New Curing Agent for Epoxy Resins, in *11th National SAMPE Technical Conference*, Society for the Advancement of Material and Process Engineering, Nov 1979, p 780-803
8. T.M. Donnellan, "The Curing of a Bisphenol A Type Epoxy Resin With 1, 8 Diamino-P-Menthane," Report NADS 83146-60, Aircraft and Crew Systems Technology Directorate, U.S. Naval Air Development Center, June 1982
9. T.T. Chiao, A.D. Cummins, and R.L. Moore, Fabrication and Testing of Epoxide Resin Tensile Specimens, *Composites*, Jan/Feb 1972, p 10-15
10. Z.N. Sanjana and J.H. Testa, Toughened Epoxy Resins for Filament Winding, in *30th National SAMPE Symposium*, Society for the Advancement of Material and Process Engineering, March 1985
11. J.A. Rinde, E.T. Mones, and H.A. Newey, "Filament Winding Epoxy Resins for Elevated Temperature Service," Report UCRL 52577, Lawrence Livermore Laboratory, Oct 1978
12. F.M. Norton, S.W. Beckwith, and S.R. Swanson, Torsional Shear and Transverse Tensile Evaluation of Carbon Epoxy Composites, in *30th National SAMPE Symposium*, Society of the Advancement of Material and Process Engineering, March 1985, p 443-460
13. R.F. Lark, "Recent Advances in Lightweight, Filament-Wound Composite Pressure Vessel Technology," Paper presented at the 1977 Energy Conference,

American Society of Mechanical Engineering, Houston, Sept 1977; also, NASA TN 73699, National Aeronautics and Space Administration

14. H. Lee and K. Neville, *Handbook of Epoxy Resins*, McGraw-Hill, Rev. ed., 1982

15. "EPON Resin Structural Reference Manual," SC67-81, Shell Chemical Company

16. J.E. Ashton, J.C. Halpin, and P.H. Petit, *Primer on Composite Materials*, Technomic, 1969

17. R.W. Yeager and J.R. Hinchman, "Design Curves for Filament-Wound Rocket Motor Cases," Paper presented at the 20th Technical Conference, Society of Plastic Engineers, 1964

18. Y.M. Tarnopol'skii and T. Kincis, *Static Test Methods for Composites*, Van Nostrand Reinhold, 1985

19. L.L. Clements, R.L. Moore, and T.T. Chiao, Elongated Ring Specimen for Tensile Properties of Filament-Wound Composites, in *Materials Review 1975*, in *Proceedings of the Seventh National SAMPE Technical Conference*, Society for the Advancement of Material and Process Engineering, 1975

20. R.E. Allred, H.K. Street, and R.J. Martinez, Improvement of Transverse Composite Strengths: Test Specimen and Materials Development, *SAMPE National Symposium*, Vol 24, Society for the Advancement of Materials and Process Engineering, 1979, p 31-50

21. L.B. Greszczuk, Application of Four-Point Ring Twist Test for Determining Shear Modulus of Filamentary Composites, in *Test Methods and Design Allowables for Fibrous Composites*, STP 734, C.C. Chamis, Ed., American Society for Testing and Materials, 1981, p 21-33

22. D.V. Rosato and C.S. Grove, Jr., *Filament Winding: Its Development, Manufacture, Applications and Design*, Wiley-Interscience, 1964

23. L.B. Greszczuk, Douglas Ring Test for Shear Modulus Determination of Isotropic and Composite Materials, in *23rd Annual Technical Conference*, Section 17-0, Reinforced Plastics/Composites Division, The Society of the Plastics Industry, 1968

24. N. Fried, Survey of Methods of Test for Parallel Filament Reinforced Plastics, in *Symposium on Standards for Filament Wound Plastics*, STP 327, American Society for Testing and Materials, 1963

25. D. Purslow, "The Shear Properties of Unidirectional Carbon Fibre Reinforced Plastics and Their Experimental Determination," Current Paper 1380, *Aeronautical Research*, Her Majesty's Stationery Office, 1977

26. C.C. Chiao, R.L. Moore, and T.T. Chiao, Measurement of Shear Properties of Fibre Composites, Part 1, Evaluation of Test Methods, *Composites*, Vol 8 (No. 3), 1977

27. N.L. Newhouse and W.D. Humphrey, Development of the Standard Test and Evaluation Bottle, *SAMPE J.*, Vol 22 (No. 2), March/April 1986

28. W.D. Humphrey and N.L. Newhouse, The Standard Test and Evaluation Bottle (STEB) Five Years Later, in *31st International SAMPE Symposium*, Society for the Advancement of Material and Process Engineering, April 1986

29. "Filament-Wound Pressure Vessels," Brunswick Corporation, Oct 1985

30. D.A. MacNab and S.T. Peters, Graphite Epoxy Launch Tube for MX, *SAMPE J.*, Vol 19 (No. 6), Nov/Dec 1983

31. V. Verderaime and M. Rheinfurth, "Identification and Management of Filament-Wound Case Stiffness Parameters," NASA Technical Paper 2117, National Aeronautics and Space Administration, Jan 1983

32. A.E. Reiners, "Characterization of Type 449 Roving," Design and Development Memo 289, Brunswick Corporation, Sept 1978

33. B.L. Lee, R.W. Lewis, and R.E. Sacher, "Environmental Effects on the Mechanical Properties of Glass Fiber/Epoxy Resin Composites," Army Materials and Mechanics Research Center, June 1978

34. A.R. Cederberg, "Development of a Composite Motor Case for the VIPER Antitank Weapon System," Paper presented at the 1979 JANNAF Propulsion Conference, Joint Army-Navy-NASA-Air Force, Anaheim, CA, March 1979

35. W.D. Humphrey, "Degradation Data on Kevlar Pressure Vessels," Publication 563, U.S. Department of Commerce/National Bureau of Standards, Oct 1979

36. C.E. Browning, and J.M. Whitney, The Effects of Moisture on the Properties of High Performance Structural Resins and Composites, in *Fillers and Reinforcements for Plastics*, Advances in Chemistry Series 134, 1974, p 137-148

37. S.S. Tompkins, "Analysis of Moisture Absorption and Diffusion in Fiber Reinforced Polymeric Resin—Matrix Composite Materials," Ph.D thesis, University of Virginia, May 1978

38. W.D. Humphrey, N.L. Newhouse, and N.C. Plass, "Composite Case Technology Program," Final Report AFRPL TR-83-031, Air Force Rocket Propulsion Laboratories, May 1983

Braiding

Frank K. Ko, Fibrous Materials Research Laboratory, Drexel University

BRAIDING IS A TEXTILE PROCESS that is known for its simplicity and versatility. Braided structures are unique in their high level of conformability, torsional stability, and damage resistance. Many intricate material placement techniques can be transferred to and modified for composite prepreg fabrication processes. The extension of two-dimensional braiding to three-dimensional braiding has opened up new opportunities in the near-net shape manufacturing of high damage tolerant structural composites.

In the braiding process, two or more systems of yarns are intertwined in the bias direction to form an integrated structure. Braided material differs from woven and knitted fabrics in the method of yarn introduction into the fabric and in the manner by which the yarns are interlaced. Braided, woven, and knitted fabric are compared in Table 1 and Fig. 1.

Braiding has many similarities to filament winding. Dry or prepreg yarns, tapes, or tow can be braided over a rotating and removable form or mandrel in a controlled manner to assume various shapes, fiber orientations, and fiber volume fractions. Although braiding cannot achieve as high a fiber volume fraction as filament winding, braids can assume more complex shapes (sharper curvatures) than filament-wound preforms. The interlaced nature of braids also provides a higher level of structural integrity, which is essential for ease of handling, joining, and damage resistance. While it is easier to provide hoop (90°) reinforcement by filament winding, longitudinal (0°) reinforcement can be introduced more readily in a triaxial braiding process. In a study performed by McDonnell Douglas, it was found that braided composites can be produced at 56% of the cost of filament-wound composites because of the labor savings in assembly and the simplification of design (Ref 1). By using the three-dimensional braiding process, not only can the intralaminar failure of filament-wound or tape laid-up composites be prevented, but the low interlaminar properties of the laminated composites can also be prevented. A comprehensive treatment of braiding that does not directly relate to composites is provided in Ref 2.

Because of its knot-tying origins, braiding is perhaps one of the oldest textile technologies known to man. From the Kara-Kumi, an Oriental braid for ornamental purposes, to heavy-duty ropes, braids have long been used in many specialized applications. Their modern applications include sutures and high-pressure hose reinforcement. In short, braids have been used wherever a high level of torsional stability, flexibility, and abrasion resistance are required. On the other hand, because of their lack of width and relatively low productivity, braids have not gained as widespread usage in the textile industry as have woven, knitted, and nonwoven fabrics.

As a result of the relatively low use of braids as a textile and clothing material, publications related to braiding are limited. Braids were considered a crafting art in the 1930s (Ref 3); one of the earliest treatments of braids as an engineering structure appeared in an article by W.J. Hamburger in the 1940s (Ref 4) in which the geometric factors related to the performance of braids were examined. The first comprehensive discussion of the formation, geometry, and tensile properties of tubular braids was given by D. Brunnschweiler (Ref 5, 6) in the 1950s. From the machinery and processing point of view, an informative book was written by W.A. Douglass (Ref 7) in the early 1960s. Relating processing parameters to the structure of braids, two articles (Ref 8, 9) reflect the sophistication of the development of braiding technology in Germany. A beautifully illustrated review on the historical development of braiding and its applications and manufacture was published by Ciba-Geigy (Ref 10). Serious consideration of braids as engineering materials did not occur until the later part of the 1970s, when researchers from McDonnell Douglas described the use of braids for composite preforms (Ref 11) to reduce the cost of producing structural shapes. About the same time, the first published article on the structural mechanics of tubular braids by Phoenix appeared (Ref 12), as well as an extended treatment by C.W. Evans of braids and braiding for a pressure hose, which is a flexible composite (Ref 13).

Since the 1980s, most of the published information on braids has been related to composites (Ref 1, 11, 14, 15). A large concentration of articles on three-dimensional braiding has been appearing in the literature. Addressing the delamination problem in state-of-the-art composites and demonstrating the possibility of near-net shape manufacturing, the articles on three-dimensional braiding can be categorized into the areas of applications (Ref 16), processing science and structural geometry (Ref 17), structural analysis (Ref 20), and property characterization (Ref 19). As indicated in this brief review of the literature, braids have gained popularity in the composite industry because of the technological needs of structural composites for the inherent uniqueness of braided structures, as well as the recent progress in hardware and software development for braiding processes.

Coupled with the fully integrated nature and the unique capability for near-net shape manufacturing, the current trend in braiding technology is to expand to large-diameter braiding; develop more sophisticated techniques for braiding over complex-shaped mandrels, multi-directional braiding or near-net shapes; and the extensive use of computer-aided design and manufacturing (CAD/CAM).

Table 1 A comparison of fabric formation techniques

	Braiding	Weaving	Knitting
Basic direction of yarn introduction	One (machine direction)	Two (0°/90°) (warp and fill)	One (0° or 90°) (warp or fill)
Basic formation technique	Intertwining (position displacement)	Interlacing (by selective insertion of 90° yarns into 0° yarn system)	Interlooping (by drawing loops of yarns over previous loops)

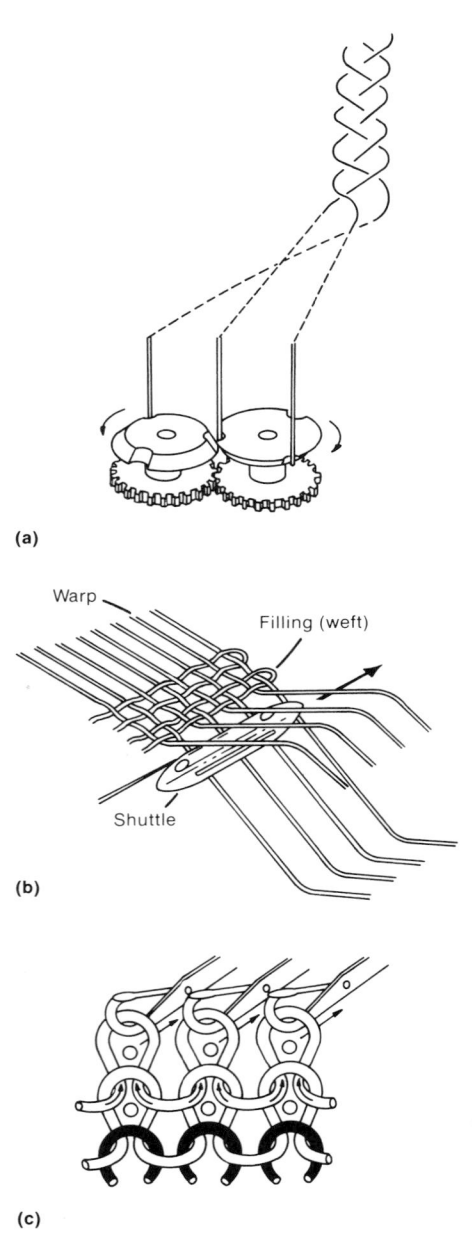

Fig. 1 Fabric techniques. (a) Braided. (b) Woven. (c) Knitted

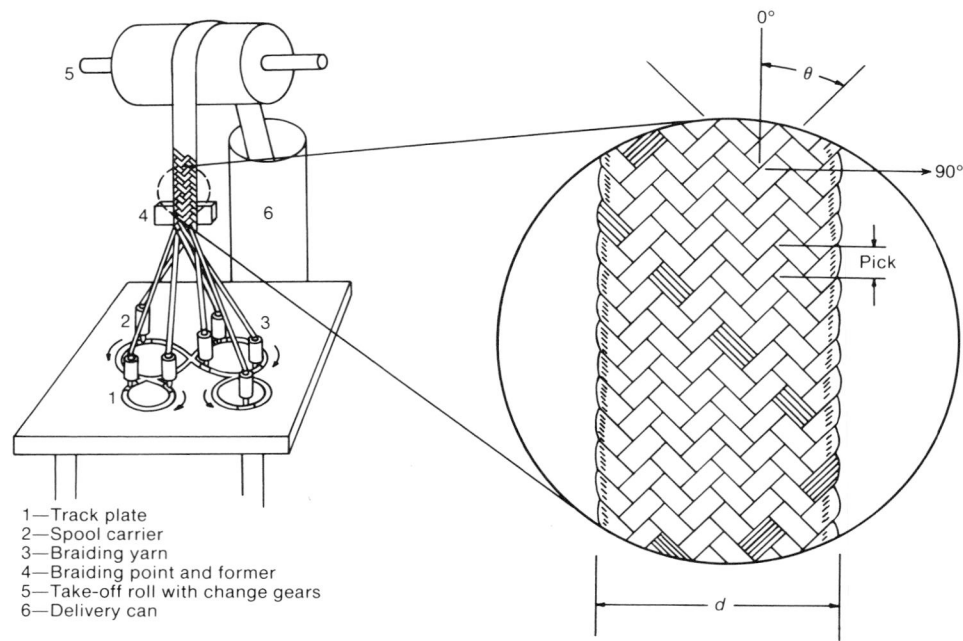

1—Track plate
2—Spool carrier
3—Braiding yarn
4—Braiding point and former
5—Take-off roll with change gears
6—Delivery can

Fig. 2 Flat braider and braid

Table 2 Braiding classifications

Parameter	Levels		
Yarn axes .... Biaxial	Triaxial	Multiaxial
Dimension of braid .... Two-dimensional	Three-dimensional	Three-dimensional
Shaping .... Formed shape		Net shape
Direction of braiding .... Horizontal	Vertical	Inverted vertical
Construction of braid 1/1	2/2	3/3
Control mechanism for carrier motion................... Positive		Jacquard
Braiding type .... Circular	Flat	Jacquard, special

This article describes basic terminology; braiding classifications; and the formation, structure, and properties of the braided structures, with specific attention to composites.

Braiding Classifications

One of the most attractive features of braiding is its simplicity. A typical braiding machine (Fig. 2) essentially consists of a track plate, spool carrier, former, and a take-up device. In some cases, a reversing ring is used to ensure uniform tension on the braiding yarns. The resulting braid geometry is defined by the braiding angle, θ, which is half the angle of the interlacing between yarn systems, with respect to the braiding (or machine) direction. The tightness of the braided structure is reflected in the frequency of interlacings. The distance between interlacing points is known as pick spacing. The width, or diameter, of the braid (flat or tubular) is represented as d.

The track plate supports the carriers, which travel along the path of the tracks. The movement of the carriers can be provided by devices such as horn gears, which propel the carriers around in a maypole fashion. The carriers are devices that carry the yarn packages around the tracks and control the tension of the braiding yarns. At the point of braiding, a former is often used to control the dimension and shape of the braid. The braid is then delivered through the take-up roll at a predetermined rate. If the number of carriers and take-up speed are properly selected, the orientation of the yarn (braiding angle) and the diameter of the braid can be controlled. The direction of braiding is an area of flexibility, because it can be horizontal, vertical from bottom to top, or inverted. When braiding over large mandrels, horizontal braiding is required.

When longitudinal reinforcement is required, a third system of yarns can be inserted between the braiding yarns to produce a triaxial braid with $0° \pm θ°$ fiber orientation. If there is a need for structures having more than three yarn thicknesses, several layers (plies) of fabric can be braided over each other to produce the required thickness. For a higher level of through-thickness reinforcement, multiple track braiding, pin braiding, or three-dimensional braiding, can be used to fabricate structures in an integrated manner. The movement of the carriers can follow a serpentine track pattern or orthogonal track pattern by means of a positive guiding mechanism and/or Jacquard-controlled mechanism (lace braiding). Jacquard braiding uses a mechanism that enables connected groups of yarns to braid different patterns simultaneously. Various criteria and braiding classifications are shown in Table 2. For simplicity, and to be consistent with the literature in the composite community, the dimensions of braided structures are used as the criteria for categorizing braiding. Specifically, a braided structure having two braiding-yarn systems with or without a third laid-in yarn is considered two-dimensional braiding. When three or

Table 3 U.S. braid manufacturers

Atkins and Pearce	Kentucky
Albany International Research	Massachusetts
Bently Harris	Pennsylvania
Fiber Innovations	Massachusetts
McDonnell Douglas	Missouri
Newport Composites	California
Polygon	California
Santa Fe Textiles	California
Cortland Cable Company	New York

Table 4 Applications of braided fabrics and composites

Aircraft interiors	Jet engine spinner	Rocket motor casing
Aircraft propellers	Lightweight bridge structures	Rolling ferel drum
Artificial limbs, tendons, bone	Lightweight submersibles	Rotor blades
Automotive parts	Machine parts	Ski poles
Boats	Military equipment	Skis
Boat masts	Model aircraft	Space struts
Bridge components	Net shape rigid armor	Spar and blades
Chemical containers	Personal armor	Sport cars
Drive shafts	Pressure vessels	Squash rackets
Elbow fittings	Racing canoes	Stiffened panels
Fishing rods	Racing cars (structural panels)	Stocks for high jumping
Frame of airplane seats	Racing sculls and catamarans	Surfboats
Glider	Radar dishes	Tennis rackets
Glider airplanes	Radomes	Wind generator propellers and
Golf clubs	Record brushes	D-spars
Hang-glider frames	Robot arms and fingers	X-ray tables
Hockey and ice hockey sticks	Rocket launcher	

more systems of braiding yarns are involved to form an integrally braided structure, it is known as three-dimensional braiding.

Two-Dimensional Braiding

The equipment for two-dimensional braiding is well established worldwide, but especially in West Germany. One of the oldest braiding machine manufacturers in the U.S. is Mossberg Industry (also known by its former name, New England Butt and now called Wardwell Braiding Machine Company), which manufactures braiders ranging from three-carrier to 144-carrier models. There are a number of braid manufacturers actively producing braided preforms and/or developing braided composites. A sample list of these companies is given in Table 3. A wide range of applications has been reported by these companies, including medical, recreational, military, and aerospace uses, as defined in Table 4.

Figure 3 illustrates a 144-carrier horizontal braider that is capable of biaxial or triaxial braiding. The versatility of braiding for forming complex structural shapes is illustrated in Fig. 4, which shows a fiberglass preform for a composite coupling shaft being formed in the Fibrous Materials Research Laboratory at Drexel University, using a 144-carrier braiding machine. Using a similar braiding machine, a racing car chassis has also been fabricated (Fig. 5) by that laboratory, in conjunction with Fiber Innovations.

Governing Equations. The mechanical behavior of a composite depends upon fiber orientation, fiber properties, fiber volume fraction, and matrix properties. To conduct an intelligent design and selection process for using braids in composites, an understanding of fiber volume fraction and geometry as a function of processing parameters is necessary. The fiber volume fraction is related to the machine in terms of the number of yarns and the orientation of those yarns. The fiber geometry is related to the machine by orientation of the fibers and final shape.

The two-dimensional braider facilitates fabrication of net shape composite preforms. Using a mandrel, the shape is formed, and the fiber volume fraction can readily be determined by the orientation and amount of fiber used. The total material area of yarn in a given cross section of a composite preform can be determined by:

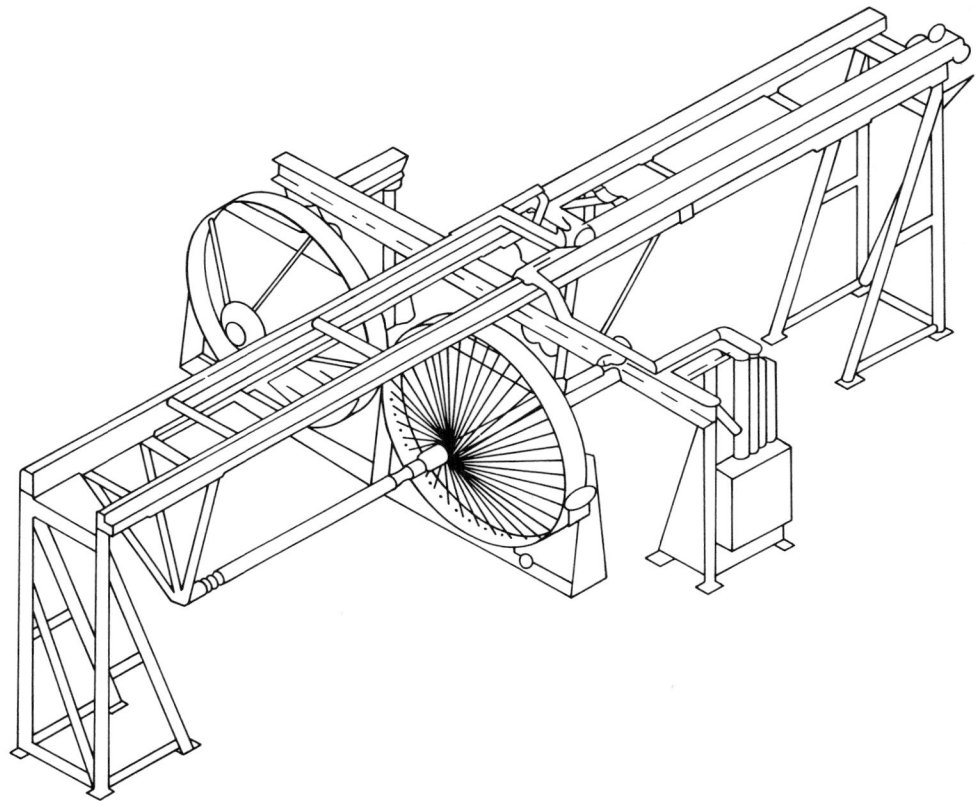

Fig. 3 Braiding machine, 144-carrier horizontal model

$$A_m = A_y \times N_y/\cos(\theta)$$

where A_m is the area of material in the cross section, A_y is the cross-sectional area of the yarn, N_y is the number of yarns on the machine ($M \times$ number of plies, where M is the number of carriers on machine), and θ is the orientation of the yarns with respect to the mandrel axis. Thus, once the composite dimensions are known, the fiber volume fraction can be expressed as:

$$V_f = A_m/A_c$$

where V_f is the fiber volume fraction, A_c is the

cross-sectional area of the composite, and A_m is the cross-sectional area of material in the composite, as above.

If a composite of a given cross-sectional area and a particular yarn and fiber volume fraction are required, the fabric can be designed based on the number of plies and the orientation of the yarns. The analytic relation for this analysis is:

$$\cos(\theta) = M N_{ply}A_y/(V_fA_c)$$

where N_{ply} is the number of plies per bobbin. Thus, the design is determined for some numbers of plies. Figure 6 shows the relationship of θ to V_f for a fixed A_y and A_c at various plies.

Fig. 4 Formation of fiberglass preform for composite coupling shaft

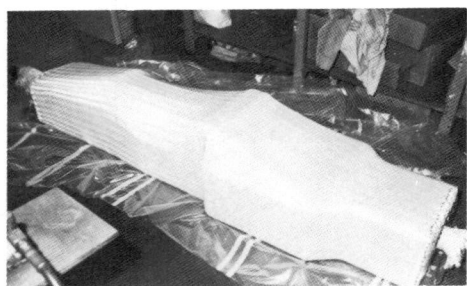

Fig. 5 Braided fiberglass car chassis

In summary, the braiding parameters for braided composites can be presented in the following equations:

$$d_o = M N_{ply} A_y / (\pi T V_f \cos(\theta)) + T$$

$$d_i = M N_{ply} A_y / (\pi T V_f \cos(\theta)) - T$$

where d_o is the outside braid diameter, d_i is the inside braid diameter, and t is the composite (fabric) thickness. With this equation, the effect of braiding angle, fiber volume fraction, and the number of plies on the number of carriers required to produce a given braid diameter for a specific composite can be calculated. As an example, for the following specifications:

- Yarn cross-sectional area, $A_y = 48.1 \cdot 10^{-4}$ cm^2 (7.45 · 10^{-4} in.^{2})
- Composite thickness, $t = 0.318$ mm (0.125 in.)
- Fiber volume fraction, $V_f = 0.50$
- Number of ply/yarn, $N_{ply} = 2$
- Braiding angle, $\theta = 45°$

Figures 7 to 9 illustrate the interrelationship of these parameters. It is quite conceivable that a 5-cm (2-in.) diam braid can be produced on a wide range of braiding machines ranging from

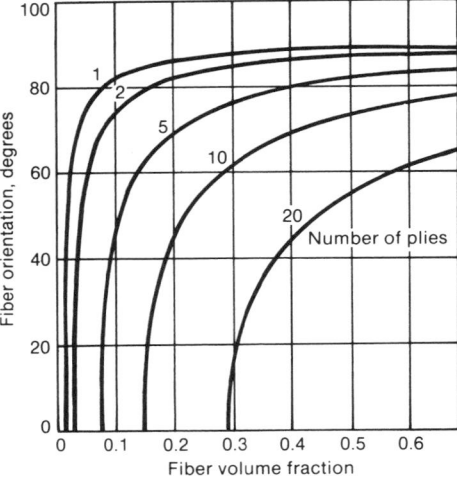

Fig. 6 Relationship of fiber orientation to fiber volume fraction for varying number of plies; 24 carrier, 12K graphite, 80 × 3 mm (3.125 × 0.125 in.)

24 to 144 carriers. However, the resulting braid angle and fabric thickness would vary. On the other hand, if the braiding angle is the key requirement, then one can vary the number of plies per yarn in order to produce a 5-cm (2-in.) diam braid on various braiding machines. It is of interest to note that the higher the number of carriers, the wider the range of diameters that can be produced. Because of the diversity of applications of braided composites, the current trend in the braiding industry is toward larger-diameter braiders.

It should be noted that in addition to the static relationships between braiding angle, θ, braid diameter, d, and number of carriers, M, a dynamic relationship can be established between machine processing parameters, θ, and d. For horizontal braiding over a mandrel, this is particularly meaningful.

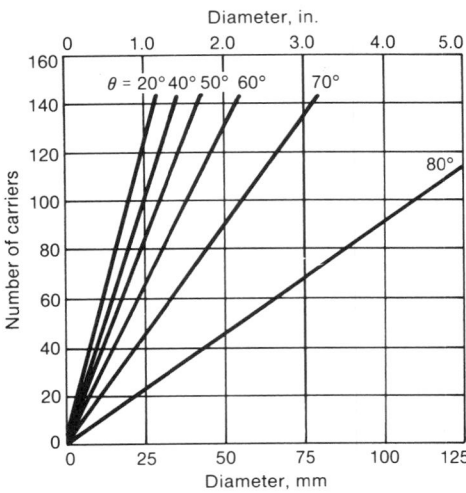

Fig. 7 Effect of braid angle on processing parameters to form a graphite tubular fabric preform with 50 vol% fiber; two-ply, 12K graphite, 3.2 mm (0.125 in.) thick

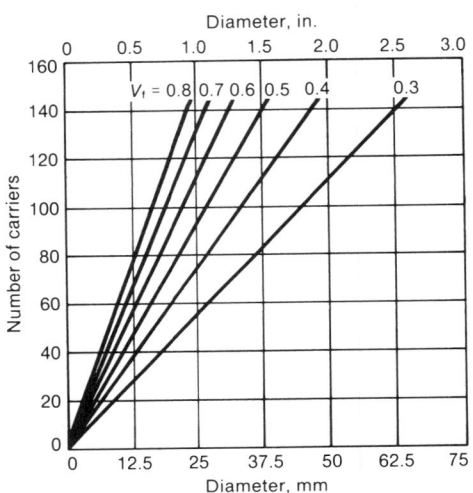

Fig. 8 Effect of fiber volume fraction on processing parameters to form a graphite braided tubular fabric; two-ply, 12K graphite, 3.2 mm (0.125 in.) thick, 45° braid angle

Computer-Aided Manufacturing of Complex-Shaped Structures. The controlling parameter in this case is the take-up/rotation ratio, R. This number indicates the distance the mandrel traverses for one rotation of the carriers. Thus, for a given mandrel diameter, d, the relation between θ and R is:

$$d = R \tan(\theta)/\pi$$

Thus, in order to maintain the proper fiber orientation (and thus the desired fiber volume fraction), the machine should be set for a take-up/rotation ratio of R. Figure 10 illustrates the relationship between mandrel diameter and the take-up/rotation ratio. If the mandrel is of irregular shape, then by using microprocessor control, R can be monitored and modified accordingly along the length of the mandrel.

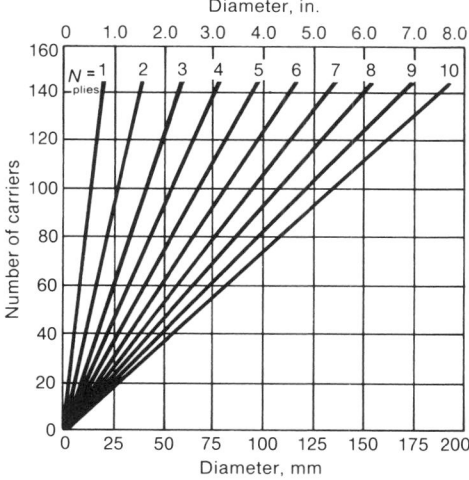

Fig. 9 Effect of number of plies on processing parameters of braided graphite tubular fabric with 50 vol% fiber; two-ply, 12K graphite, 3.2 mm (0.125 in.) thick, 45° braid angle

Knowing the radius and speed of take-up, the controller can output new R values when required. Figure 11 shows the R value required for a truncated conical mandrel as a function of distance along the length of the mandrel. In this way, the operation of the braid can be controlled to provide a constant volume fraction along any arbitrarily complex braidable surface.

Three-Dimensional Braiding

Three-dimensional braiding is an extension of two-dimensional braiding technology in which the fabric is constructed by the intertwining or orthogonal interlacing of two or more yarn systems to form an integral structure. Well-known examples of three-dimensional braids are the diagonal, or packing, braids that are produced by the intertwining of three or more groups of yarns in a square arrangement of horn gears, as shown in Fig. 12. Serious consideration of three-dimensional braid for composites started in the late 1960s in the search for multidirectionally reinforced composites, such as rocket motor components, for aerospace applications. The Omniweave by General Electric (Ref 20) and SCOUDID by Société Européene de Propulsion (Ref 21) are examples of new developments. The mechanism of these braiding methods differs from traditional braiding methods only in the way the carriers are displaced to create the final braid geometry. Instead of moving in a continuous maypole fashion, as does the square braider, these three-dimensional braiding methods invariably move the carrier in a sequential, discrete manner, which is quite suitable for adaptation to computer control.

The track and column machines used by Drexel University and Atlantic Research Cor-

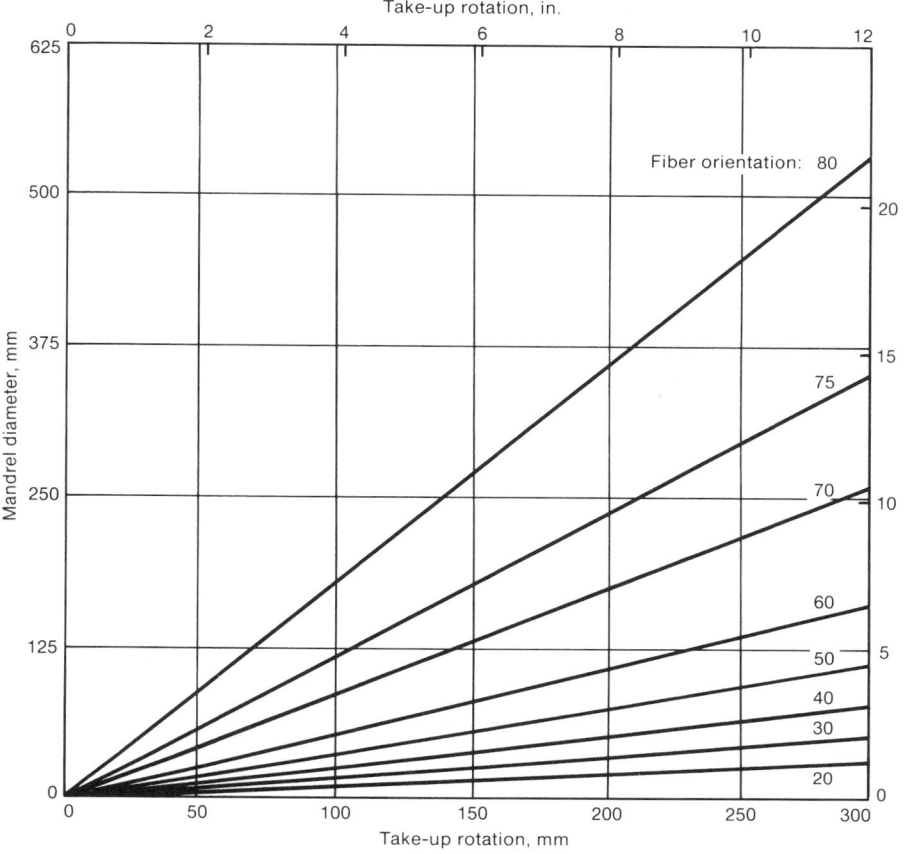

Fig. 10 Relationship between mandrel diameter and take-up rotation ratio

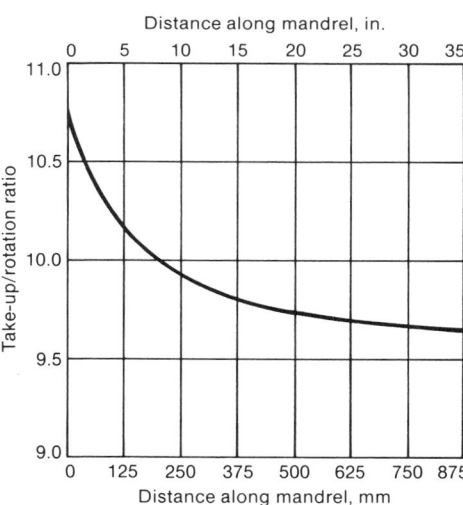

Fig. 11 Take-up rotation ratio as a function of length along a truncated conical mandrel to maintain constant fiber orientation and fiber volume fraction

poration are quite similar to the Maistre method. Similar to the illustration by R.A. Florentine (Ref 22), Fig. 13 shows two basic loom set-ups in circular and rectangular con-

Fig. 12 Horn gear set-up for square braiding

figurations. The basic braiding motion includes the alternate X and Y (or r and θ) displacement of yarn carriers, followed by a compacting motion. The formation of shapes is accomplished by the proper positioning of the carriers and the joining of various rectangular or annular groups through selected carrier movements.

The three-dimensional braiding system can produce thin and thick structures in a wide

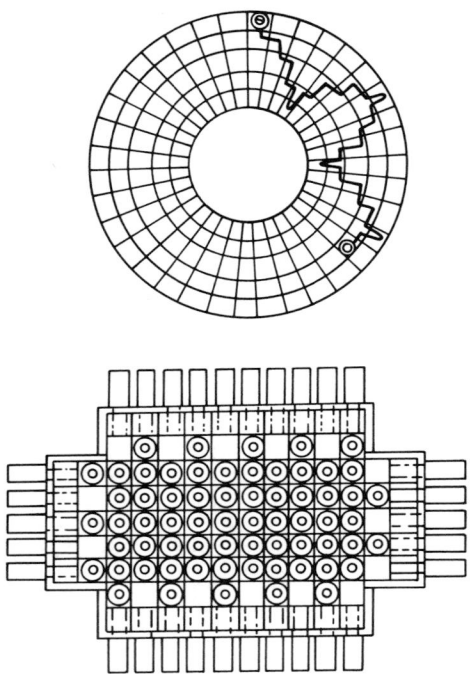

Fig. 13 Circular and rectangular three-dimensional braiding machines

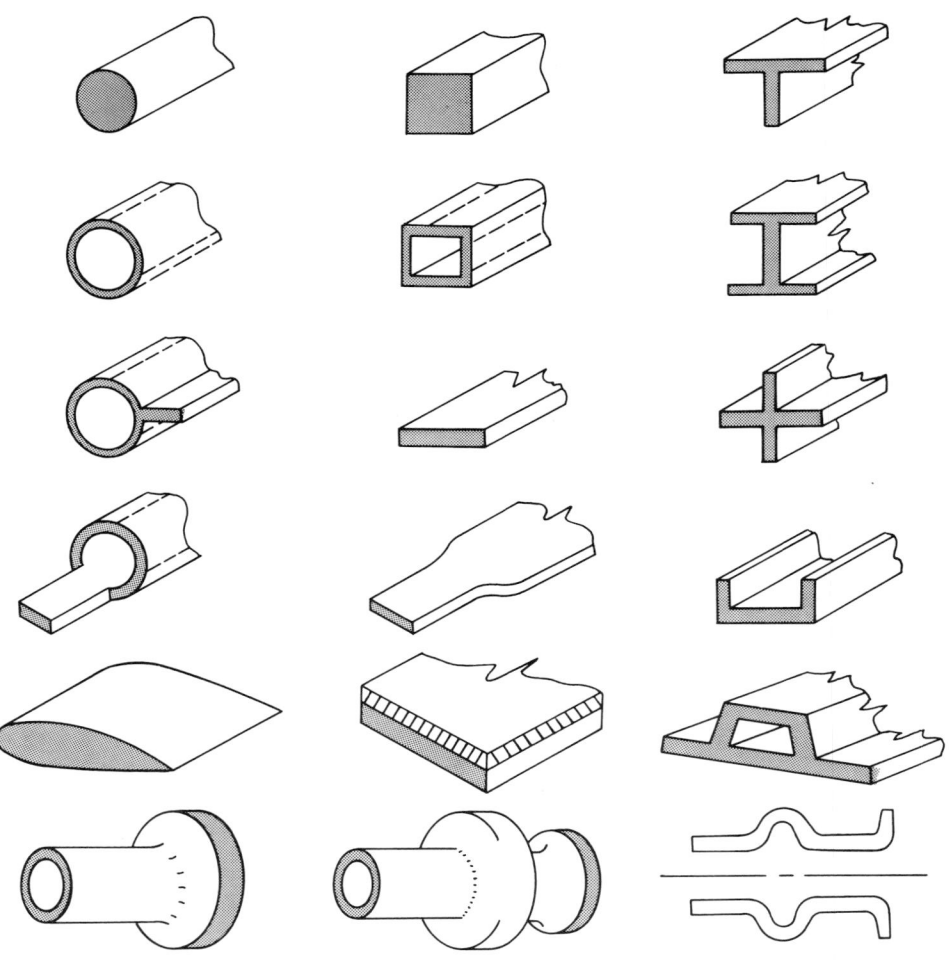

Fig. 14 Net shape structures produced by three-dimensional braiding

variety of complex shapes, as shown in Fig. 14. By proper selection of the yarn bundle sizes, the dimension of these structures can be as thick as desired. Fiber orientation can be chosen, and 0° longitudinal reinforcements can be added as desired. Although this system is not yet fully automated, extensive analytical research has been done in this area, and comprehensive models have been developed relating final shape to manufacturing processes. Considering the potential for near-net shape formation of high damage resistant composites, extensive developmental programs have been carried out in the Drexel laboratory and by Atlantic Research Corporation. Examples of these structures are I-beams, hat sections, rocket motor exit cones, and marine propellers.

Governing Equations. The development of a processing science base for three-dimensional braided fabrics for composites consists of two basic components: quantification of fabric geometry, and determination of fiber volume fraction. With these components, and knowledge of fiber and matrix properties, a fabric can be formed to specification. The mechanical analysis of a composite depends upon the fabric properties that can be quantified using the properties, architecture, orientation, and volume fraction of the fiber. A description of the development of constitutive equations to relate these parameters to the actual fabrication of the fibrous network follows.

Quantification of Fabric Geometry. To establish a geometric model and method for analyzing the properties of the three-dimensional braid, it is necessary to identify the orientation of the yarns in the structure. This is accomplished by identifying a macroscopic unit cell.

The height, width, and thickness of the unit cell can be represented by the parameters W, V, and U, respectively. From trigonometric relationships of the braid, W can be given as:

$$W = (U^2 + V^2)^{1/2}/\tan(\theta)$$

where θ is the angle of inclination of the body diagonally oriented fiber (yarn), and W is the height of the unit cell, which is the distance between picks in the fabric. The above relationship will determine the pick spacing necessary to produce a desired fiber orientation, θ.

Fiber Volume Fraction. Because composite structures are usually made according to a predetermined fiber orientation and volume fraction, the ability to predict the fiber volume fraction in a three-dimensional braided composite structure would be ideal. In making a three-dimensional braid with a given fiber volume fraction, the volume fraction of fiber can be defined as:

$$V_f = V_y/V_c$$

where V_y is the volume of the yarn and V_c is the volume of the composite.

This can be rewritten as:

$$V_f = N_y L_y A_y/(L_c A_c)$$

where N_y is the total number of yarns in the fabric, L_y is the length of each yarn, A_y is the nominal cross-sectional area of each yarn, L_c is the length of the composite, and A_c is the cross-sectional area of the composite. From the definition of denier (linear density), the cross-sectional area (cm²) of a yarn is given as:

$$A_y = D_y/(9 \times 10^5 \rho)$$

where ρ is the density of the fiber (g/cm³) and D_y is the linear density of the yarn (denier).

From trigonometry we can establish the following relationship:

$$W = V/\tan(\theta')$$

where θ' is the surface angle of the yarn. Combining this equation with the previous definition of θ,

$$\tan(\theta) = (1 + k^2)^{1/2} \tan(\theta')/k$$

where k is the ratio of track movement to column movement. Suitable manipulation of this equation gives:

$$\theta = \tan^{-1} ((1 + k^2)^{1/2} \tan(\theta')/k)$$

Table 5 Properties of two-dimensional braided S-2 fiberglass-epoxy composites

Braid angle, degree	Tensile strength				Compressive strength				In-plane shear	
	Hoop		Long		Hoop		Long			
	MPa	ksi	MPa	ksi	MPa	ksi	MPa	ksi	MPa	ksi
89	1320	192	21	3	700	102	220	32	55	8
86.75	1250	182	83	12	380	55	100	14	75	11
82.50	1030	149	. . .	. . .	330	48	. . .	. . .	. . .	. . .
78	730	106	. . .	. . .	275	40	. . .	. . .	. . .	. . .

Table 6 Properties of triaxial braided graphite-epoxy composites

Braid angle, degree	V_f, %	E_{LT}		E_{LC}		E_{HT}		ν_{LHT}	ν_{LHC}	ν_{HLT}
		GPa	10^6 psi	GPa	10^6 psi	GPa	10^6 psi			
45	33.8	61.4	8.9	62.7	9.1	6.8	0.98	0.56	0.64	0.044
63	29.3	49.0	7.1	49.6	7.2	15.2	2.20	0.43	0.45	0.088
80	56.3	. . .	. . .	52.4	7.6	43.6	6.32	. . .	0.13	0.110

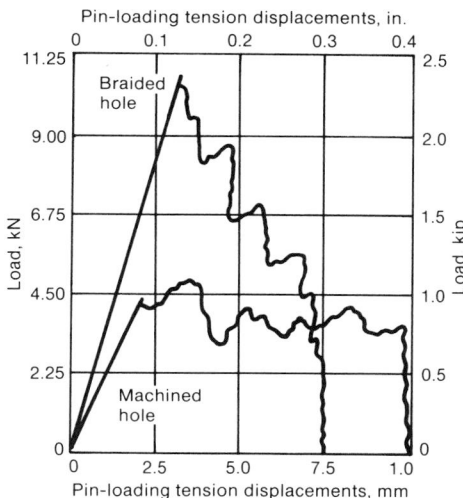

Fig. 15 Strength improvement of braided hole

Incorporating these identities into our original relationship, one can determine the total number of yarns required to make a three-dimensional fabric with a given fiber volume fraction:

$$N_y = V_f A_c \, \rho \, 9 \times 10^5 \cos(\theta)/D_y$$

where A_c is the cross-sectional area of finished composite (cm^2), ρ is the density of fiber (g/cm^3), D_y is the linear density of fiber (denier), θ is the interior angle, and N_y is the total number of yarns in the fabric.

Using this equation, one can easily determine the total number of yarns required to make a fabric with a given fiber volume fraction and cross-sectional area if the parameters of fiber density, yarn linear density, and yarn surface angle are known.

The volume fraction attainable with a given construction has a maximum that is dependent on the fiber architecture. With fibers of circular cross section, the maximum attainable value of V_f in a uniaxially aligned fibrous network is $\pi/2\sqrt{3} = 0.906$. However, with other yarn cross-sectional shapes, larger maximum values of V_f can be achieved. But, when placement of yarns in a planar or three-dimensional system is analyzed, the maximum volume fraction is markedly reduced. It can be shown that the maximum fiber volume fraction of a three-dimensional braid unit cell is $\pi\sqrt{3}/8 = 0.6801$. This means that the maximum volume fraction attainable without distortion is about 68%.

Properties of Braided Composites

The properties of braided composites are not as well characterized as those for unidirectional tape or woven ply laid-up laminated composites. For two-dimensional braided composites, most of the studies have been on tubular braids. For three-dimensional braid, a data base is beginning to be accumulated in academia and government laboratories. In addition to the near-net shape formability, the most outstanding properties noted for two-dimensional and three-dimensional braid composites are their

damage tolerance and their ability to limit impact damage area.

Two-Dimensional Braid Composites. In a study by D.E. Flinchbaugh (Ref 23) on tubular braided S-2 fiberglass-epoxy composites, it was reported that the tensile strength of the braided composites is comparable to that of mild steel at a much lower density. Table 5 summarizes these results. The composite had a density of 1.66 g/cm^3 and a fiber volume fraction of 75%.

The properties of triaxial braided graphite-epoxy composites was demonstrated by T. Tsiang et al. (Ref 24). As shown in Table 6, the hoop modulus was quite sensitive to the braiding angle. In the longitudinal direction, because of the 0° yarn introduced in the triaxial braiding process, the modulus was less sensitive to the braiding angle. It was also shown that the addition of longitudinal yarns can address the concern for the lack of compressive resistance in braids.

In another study by D. Brookstein and T. Tsiang (Ref 25) it was demonstrated, as shown in Fig. 15, that the capability for the formation of holes in the braiding process revealed the superiority of open hole and pin hole strength over that of machined holes.

Three-Dimensional Braid Composites. Since 1983, an intensive effort has been devoted to studying three-dimensional braid composites. Mostly funded by the government, a rather extensive data base is being generated in U.S. government laboratories (with the majority in the Naval labs) and in academia (Drexel and Delaware). The preforms used in these studies are primarily supplied by Drexel and Atlantic Research Corporation. Although research work on three-dimensional braid composites has been carried out on polymer, metal, and ceramic matrix composites, as well as on carbon-carbon composites, the largest data base by far is in polymeric matrix composites. Therefore, for illustration purposes, only their properties are described below.

General Mechanical Properties. The most comprehensive mechanical characterization of

three-dimensional braid composite properties to date has been carried out by A.B. Macander et al. (Ref 26). In this study the effect of cut-edge bundle size and braid construction were examined through tensile, compressive, flexural, and shear tests. It was found that the test specimens were sensitive to cut edges. As shown in Table 7, the tensile strength of a graphite-epoxy (T300/5208) composite was reduced by approximately 60%. When longitudinal yarns (0°) were added, the strength reduction was less than 50%. Accordingly, care should be exercised in the preparation of braided composites to ensure that the yarns on the surfaces are not destroyed. In the same table, one can also see the effect of braid construction, and thus the resulting surface fiber orientation. From a 1 × 1 construction to a 1 × 3 construction, the surface fiber orientation was reduced from 20° to 12°, which resulted in an increase in tensile strength from 661.9 MPa (96 ksi) to 965.3 MPa (140 ksi).

In Table 8, the effect of yarn bundle size is illustrated. It was found that the tensile strength and modulus of the three-dimensional braid composites tend to increase as fiber bundle size increases. This is apparently related to the dependence of fiber orientation on yarn bundle size. A larger yarn bundle size produced lower crimp (fiber angles) and thus higher strength and modulus. From both Tables 7 and 8, one will notice that although the strength and modulus of the braided composites were significantly higher than those of the 0°/90° woven laminates, the Poisson's ratio (or specific Poisson's ratio) of the braided composites were exceedingly high, from 0.67 to 1.36. To address the instability characteristics in the transverse direction, it was found in the Drexel laboratory that by adding 10 vol% transverse (90°) yarns, the Poisson's ratio of the braided composites can be reduced to 0.27 at a reduction of strength and modulus from 1250 MPa

Table 7 Three-dimensional braided graphite-epoxy composite property data

$1 \times 1.3 \times 1$ and $1 \times 1 \times 11$-braid patterns with uncut and cut edges.

Property(a)	T300(b), 1 × 1 (uncut)	T300, 1 × 1 (cut)	T300, 3 × 1 (uncut)	T300, 3 × 1 (cut)	T300, 1 × 1 × 1/2 fixed (uncut)	T300, 1 × 1 × 1/2 fixed (cut)
V_f, %	68	68	68	68	68	68
Tensile strength, MPa (ksi)	665.6 (96.5)	228.7 (33.2)	970.5 (140.8)	363.7 (52.7)	790.6 (114.7)	405.7 (68.9)
Elastic modulus, GPa (10^6 psi)	97.8 (14.2)	50.5 (7.3)	126.4 (18.3)	76.4 (11.1)	117.4 (17.0)	82.4 (12.0)
Compressive strength, MPa (ksi)	... ...	179.5 (26.0)	... ...	226.4 (32.8)	... ...	385.4 (55.9)
Compressive modulus, GPa (10^6 psi)	... ...	38.7 (5.6)	... ...	56.6 (8.2)	... ...	80.8 (11.7)
Flexural strength, MPa (ksi)	813.5 (118.0)	465.2 (67.5)	647.2 (93.9)	508.1 (73.3)	816.0 (118.3)	632.7 (91.8)
Flexural modulus, GPa (10^6 psi)	77.5 (11.2)	34.1 (4.9)	85.4 (12.4)	54.9 (8.0)	86.4 (12.5)	60.8 (8.8)
Poisson's ratio	0.875	1.36	0.566	0.806	0.986	0.667
Apparent fiber angle	±20°	±20°	±12°	±12°	±15°	±12°

(a) Tension and compression specimens were tabbed at grip ends. (b) T300 graphite yarn, 30 000 tow

Table 8 Three-dimensional braided graphite-epoxy composite properties as a function of braid pattern

Uncut specimens, 25.4 mm (1 in.) wide including comparative data for a laminated fabric composite. Tensile specimens were tabbed with 1.6 mm (1/16 in.) thick, 25.4 mm (1 in.) × 63.5 mm (2-1/2 in.) glass reinforced plastic tapered tabs at grip ends. Celion 6K and 12K specimens had cut edges for the short-beam shear tests only.

Property	AS-4, 3K 1 × 1	AS-4, 6K 1 × 1	Celion, 6K 1 × 1	AS-4, 12K 1 × 1	Celion, 12K 1 × 1	T300, 30K 1 × 1	T300, Eight harness satin fabric
V_f, %	68	68	56	68	68	68	65
Tensile strength, MPa (ksi)	736.8 (106.8)	841.4 (122.0)	857.7 (124.4)	1067.2 (154 790)	1219.8 (176 910)	665.6 (96 530)	517.1 (75 000)
Elastic modulus, GPa (10^6 psi)	83.5 (12.1)	119.3 (17.3)	87.8 (12.7)	114.7 (16.6)	113.1 (16.4)	97.8 (14.2)	73.8 (10.7)
Short beam shear, MPa (ksi)	114.8 (16.6)	126.0 (18.2)	71.4 (10.3)	121.4 (17 600)	71.4 (10 350)	... ...	69.0 (10 000)
Poisson's ratio	0.945	1.051	0.968	0.980	0.874	0.875	0.045
Flexural strength, MPa (ksi)	885.3 (128.4)	739.8 (107.3)	... ...	1063.8 (154 210)	... ...	813.5 (117 990)	689.5 (100 000)
Flexural modulus, GPa (10^6 psi)	84.5 (12.3)	95.2 (13.8)	... ...	1385.2 (20.1)	... ...	77.5 (11.2)	65.5 (9.5)
Apparent fiber angle	±19°	±15°	±15°	±13°	±17.5°	±20°	0°

Table 9 Static test summary

	C12K/3501 1 × 1 braid Mean	C.V.	24-ply AS/3501 (42/50/8) Mean	C.V.	C12K/3501 (1 × 1) 1/2 fixed Mean	C.V.
Longitudinal ultimate tensile strength, F_1^{tu}, MPa (ksi)	667.9 (96.8)	9.3%	910.1 (132.0)	7.4%	749.5 (108.7)	6.1%
Transverse ultimate tensile strength, F_2^{tu}, MPa (ksi)	34.5 (5.0)	10.0%	416.5 (60.4)	9.6%	22.8 (3.3)	19.5%
Longitudinal ultimate compressive strength, F_1^{cu} MPa (ksi)	428.2 (62.1)	14.5%	420.0 (60.9)	16.0%	473.0 (68.6)	17.6%
Longitudinal tensile modulus, E_{11}^t, GPa (10^6 psi)	90.3 (13.1)	19.5%	65.5 (9.5)	2.8%	106.2 (15.4)	12.3%
Transverse tensile modulus, E_{22}^t, GPa (10^6 psi)	10.3 (1.5)	9.7%	31.0 (4.5)	13.6%	9.7 (1.4)	9.7%
Longitudinal compressive modulus, E_{11}^c, GPa (10^6 psi)	75.8 (11.0)	21.8%	60.7 (8.8)	5.8%	93.1 (13.5)	19.8%
Longitudinal Poisson's ratio, ν_{12}	1.06	51%	0.42	–	0.81	21.2%
Transverse Poisson's ratio, ν_{21}	0.067	6.7%	0.225	2.8%	0.04	44.9%
Ultimate longitudinal strain, ϵ_1^{tu}	0.00773	13.8%	0.01393	7.9%	0.00733	10.8%
Ultimate compressive strain, ϵ_1^{cu}	0.00640	10.2%	0.00711	20.4%	0.00533	15.7%
Ultimate transverse strain, ϵ_t^{tu}	0.00324	9.7%	0.01474	5.5%	0.00249	21.3%
Longitudinal ultimate tensile strength, F_1^{tu}, MPa (ksi) ($D = 0.25$) G	660.5 (95.8)	11.7%	444.7 (64.5)	2.3%	646.8 (93.8)	9.7%
Longitudinal ultimate tensile strength, F_1^{tu}, MPa (ksi) ($D = 0.25$) N	881.2 (127.8)	11.7%	593.0 (86.0)	2.3%	862.6 (125.1)	9.2%
Longitudinal ultimate compressive strength, F_1^{cu}, MPa (ksi) ($D = 0.25$) G	313.7 (45.5)	12.2%	402.7 (58.4)	6.1%	316.5 (45.9)	11.6%
Longitudinal ultimate compressive strength, F_1^{cu}, MPa (ksi) ($D = 0.25$) N	417.8 (60.6)	12.2%	536.4 (77.8)	6.1%	422.0 (61.2)	11.6%
Compressive bearing strength, F_{br}^c, MPa (ksi) ($D = 0.25$)	335.1 (48.6)	3.8%	577.1 (83.7)	9.5%	362.0 (52.5)	15.3%
Tensile bearing strength, F_{br}^t, MPa (ksi) ($D = 0.25$)	182.7 (26.5)	6.7%	677.1 (98.2)	5.5%	282.7 (41.0)	21.0%

$\epsilon/D = 2.5$

Note: G, gross stress. N, net stress

(180 ksi) and 100 GPa (15×10^6 psi) to 10 MPa (155 ksi) and 90 GPa (13×10^6 psi), respectively.

Damage Tolerance. The first indication of the damage tolerance capability of the three-dimensional braid composites was observed by L.W. Gause *et al.* (Ref 27). In the drill hole test performed on three-dimensional braided Celion 12K/3501 composites and quasi-isotropic composites, it was found that the braided composites were quite insensitive to the drill hole (retaining over 90% of the strength), as shown in Table 9. In the case of the quasi-isotropic

composites, a 50% reduction in strength was observed. In the same study, it was also found that although the braided composites did not increase the damage threshold, they did successfully limit the extent of impact damage of graphite-epoxy, compared to that of conventional laminated constructions. Similar observations were also made by F. Ko *et al.* on glass-epoxy composites (Ref 28) as well as on carbon-PEEK composites (Ref 29). As shown in Table 10, the three-dimensional braid glass-epoxy required significantly higher levels of energy to initiate and propagate damage than

did the laminated composites under drop weight impact test. In the study of three-dimensional braid comingled Celion 3K-PEEK thermoplastic composites, it was found, as shown in Fig. 16, that the compression-after-impact-strength of the three-dimensional composites was less sensitive than for the state-of-the-art, unidirectional tape laid-up graphite-PEEK composites. The most drastic difference, however, was the impact damage area of the three-dimensional braid composite, compared to that of the laminated composites. As shown in Fig. 17, an order of magnitude lower damage area was

Table 10 Instrumented impact properties of E-glass fabric reinforced composites

Sample	E_i		E_p		Ductility index	E_m		σ_m		ρ	t	
	J	ft · lbf	J	ft · lbf		J	ft · lbf	kPa	psi	g/cm³	mm	in.
105 1 × 1 × 1...............	237.4	175.2	649.0	478.7	2.732	237.5	175.21	1589	230.4	1.93	12.7	0.5
106 1 × 1 × 1 (L)(a).........	162.1	119.6	739.3	545.3	4.558	162.2	119.64	2146	311.2	1.96	13.1	0.516
107 1 × 1 × 1 (L)...........	290.8	214.5	575.9	424.8	1.981	290.8	214.48	2725	395.2	1.99	17.1	0.675
108 *XYZ*....................	358.9	264.9	508.0	374.7	1.415	359.2	264.9	2559	371.2	2.10	12.7	0.5
201 *XYZ*....................	235.9	174.1	385.6	284.4	1.634	621.6	458.46	1704	247.2	1.99	11.1	0.437
202 Satin weave	50.4	37.2	384.9	283.9	7.631	387.5	285.82	3464	502.4	2.10	12.7	0.5
203 Plain weave	38.4	28.3	466.7	344.2	12.161	391.0	288.4	3166	459.2	2.05	12.7	0.5

(a) Sample split in two.

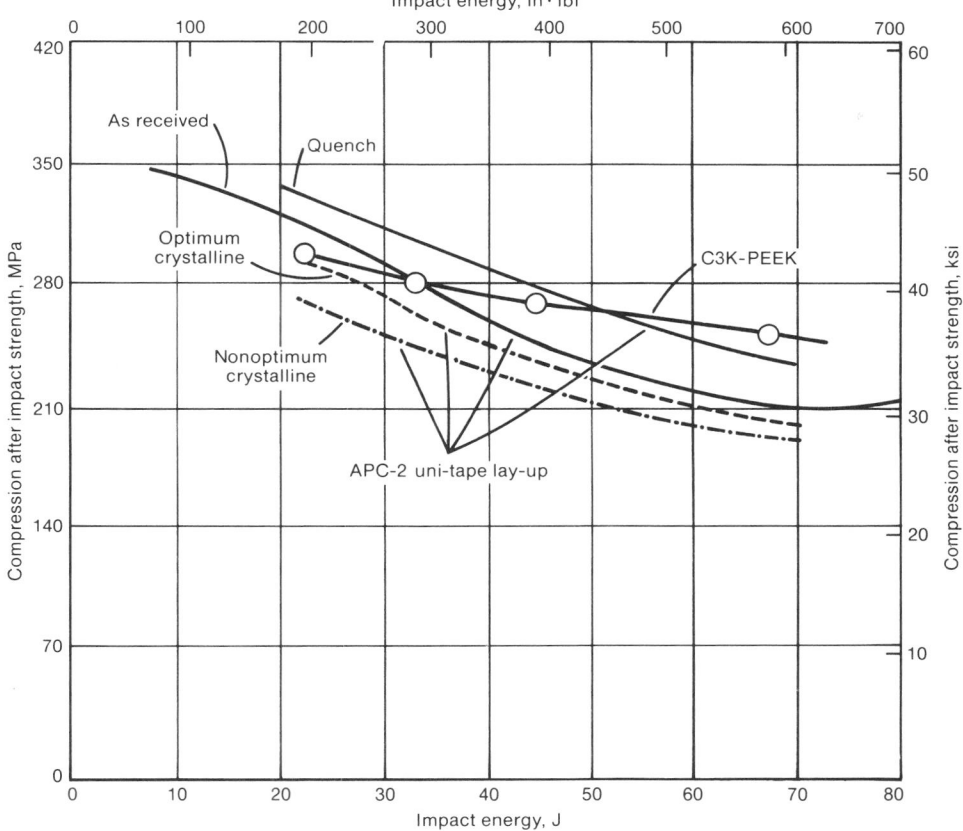

Fig. 16 Effect of impact energy level on compression after impact strength for three-dimensional braid comingled and laminated carbon-PEEK composites

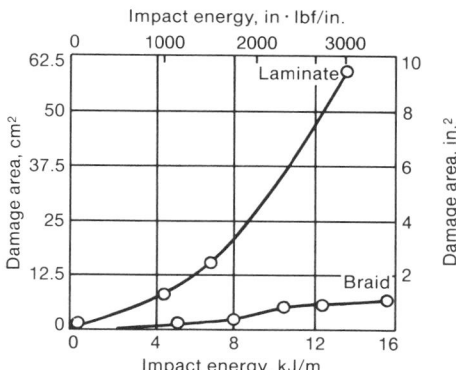

Fig. 17 Effect of impact energy on damage area of three-dimensional braid comingled and laminated carbon-PEEK composites

flexibility and the structural properties of the three-dimensional braid net-shape composite, a study was carried out by S.S. Yau, T.W. Chu, and F.K. Ko on three-dimensional braided E-glass-polyester I-beams (Ref 30). It was demonstrated that mechanical properties of the net-shape composites can be tailored by the strategic placement of materials in the braiding process. For instance, in Table 11, it can be seen that the addition of longitudinal glass yarns in the flanges of the I-beam led to a more than 50% increase in tensile and compressive moduli. Instead of fiberglass, the addition of unidirectional carbon yarns in the flanges of the I-beam produced as much as a three-fold increase in compressive resistance. Furthermore, the delamination failure found in laminated composites was not observed in any of the I-beams as a result of the high degree of through-thickness strength in the three-dimensional braided composites.

ACKNOWLEDGMENTS

This article is the result of contributions by many of the author's students over the past 10 years. The assistance of Christopher Pastore in the preparation of this article is greatly appreciated. Much of the work presented here has been supported by the Office of Naval Research and the Army Research Office.

Table 11 Properties of three-dimensional braid glass-polyester I-beams

	I-beam 1	I-beam 2	I-beam 3
Geometry...Braid		Braid/lay-in	Braid/lay-in
Fiber...Glass		Glass/glass	Glass/carbon
Fiber volume fraction.................................... 50		60	65
Length, mm (in.)..................................... 452 (17.8)		460 (18.1)	447 (17.6)
Test span, mm (in.)................................... 305 (12.0)		305 (12.0)	305 (12.0)
Width, mm (in.).. 31 (1.2)		31 (1.2)	32 (1.25)
Height, mm (in.)... 32 (1.25)		33 (1.3)	33 (1.3)
Tensile modulus, GPa (10⁶ psi)......................... 18.34 (2.66)		30.54 (4.43)	44.82 (6.5)
Compressive modulus, GPa (10⁶ psi)................. 21.10 (3.1)		30.54 (4.43)	68.26 (9.9)
Flexural strength, MPa (ksi)......................... 150.5 (21.8)		237.9 (34.50)	292.0 (42.3)
Compressive modulus, GPa (10⁶ psi)................. 20.62 (3.0)		29.44 (4.27)	68.67 (9.96)
Compressive strength, MPa (ksi).................... 145.1 (21.0)		176.4 (25.58)	175.9 (25.51)

attained with the braided composites, compared to the laminated composites.

Properties of Three-Dimensional Braid Composite I-Beams. To illustrate the design

REFERENCES

1. L.R. Sanders, Braiding—A Mechanical Means of Composite Fabrication, *SAMPE Q.*, 1977, p. 38-44
2. F.K. Ko, *Atkins and Pearce Handbook of Industrial Braids*, to be published
3. C.A. Belash, "Braiding and Knotting for Amateurs," The Beacon Handicraft Series, The Beacon Press, 1936
4. W.J. Hamburger, Effect of Yarn Elongations on Parachute Fabric Strength, *Rayon Textile Monthly*, March and May, 1942
5. D. Brunnschweiler, Braids and Braiding, *J. Textile Ind.*, Vol 44, 1953, p 666
6. D. Brunnschweiler, The Structure and Tensile Properties of Braids, *J. Textile Ind.*, Vol 45, T55-87, 1954
7. W.A. Douglass, *Braiding and Braiding Machinery*, Centrex Publishing, 1964
8. F. Goseberg, The Construction of Braided Goods, *Band-und Flechtindustrie*, No. 2, 1969, p 65-72
9. F. Goseberg, *Training Material Instructional Aid—Textile Technology—Machine Braids*, All Textile Employers Association, 1981
10. W. Weber, The Calculation of Round Braid, *Band-und Flechtindustrie*, No. 1 Part I, p 17-31; No. 3, Part II, p 109-119, 1969
11. R.J. Post, Braiding Composites—Adapting the Process for the Mass Production of Aerospace Components, in *Proceedings of the 22nd National SAMPE Symposium and Exhibition*, Society for the Advancement of Material and Process Engineering, 1977, p 486-503
12. S.L. Phoenix, Mechanical Response of a Tubular Braided Cable with Elastic Core, *Textile Res. J.*, 1977, p 81-91
13. C.W. Evans, Hose Technology, 2nd ed., Applied Science, 1979
14. J.B. Carter, "Fabrication Techniques of Tubular Structures from Braided Pre-impregnated Rovings," Paper EM85-100 presented at Composites in Manufacturing 4, Society of Mechanical Engineers, 1985
15. B.D. Haggard and D.E Flinchbaughy, "Braided Structures for Launchers and Rocket Motor Cases," Paper presented at JANNAF S and MBS/CMCS Subcommittee Meeting, MDAC/Titusville, Nov 1984
16. R.A. Florentine, "Magnaswirl's Integrally Woven Marine Propeller—The Magnaweave Process Extended to Circular Parts," in *Proceedings of the 38th Annual Conference*, Society of the Plastics Industry, Feb 1981
17. F.K. Ko, and C.M. Pastore, "Structure and Properties of an Integrated 3-D Fabric for Structural Composites," Special Technical Testing Publication 864, American Society for Testing and Materials, 1985, p 428-439
18. A. Majidi, J.M. Yang, and T.W. Chou, "Mechanical Behavior of Three Dimensional Woven Fiber Composites, in *Proceedings of the International Conference on Composite Materials V*, 1985
19. C. Croon, Braided Fabrics: Properties and Applications, in *19th National SAMPE Symposium*, Society for the Advancement of Material and Process Engineering, March 1984
20. E.R. Stover, W.C. Mark, I. Marfowitz, and W. Mueller, "Preparation of an Omniweave-Reinforced Carbon-Carbon Cylinder as a Candidate for Evaluation in the Advanced Heat Shield Screening Program," AFML-TR-70-283, March 1971
21. M.A. Maistre, "Construction of a Three Dimensional Structure, German Patent P23016968, 1973
22. R.A. Florentine, Apparatus for Weaving a Three Dimensional Article, U.S. Patent 4, 312, 261, Jan 1982
23. D.E. Flinchbaugh, "Braided Composite Structures," Paper presented at the Composites Material Conference, Dover, Aug 1985
24. T. Tsiang, D. Brookstein, and J. Dent, Mechanical Characterization of Braided Graphite/Epoxy Cylinders, in *Proceedings of the 29th National SAMPE Symposium*, Society for the Advancement of Material and Process Engineering, 1984, p 880
25. D. Brookstein and T. Tsiang, Load-Deformation Behavior of Composite Cylinders with Integrally Formed Braided and Machined Holes, *J. Compos. Mater.*, Vol 19, 1985, p 477
26. A.B. Macander, R.M. Crane, and E.T. Camponeschi, Fabrication and Mechanical Properties of Multidimensionally (X-D) Braided Composite Materials, in *Composite Materials: Testing and Design (Seventh Conference)*, STP 893, J.M. Whitney, Ed., American Society for Testing and Materials, 1986, p 422-443
27. L.W. Gause and J. Alper, "Mechanical Characterization of Magnaweave Braided Composites," Paper presented at the Mechanics of Composites Review, Air Force Materials Laboratory, Oct 1983
28. F. Ko and D. Hartman, Impact Behavior of 2-D and 3-D Glass/Epoxy Composites, *SAMPE J.*, July/August 1986, p 26-29
29. F.K. Ko, H. Chou, and E. Ying, Damage Tolerance of 3-D Braided Comingled Carbon/PEEK Composites, in *Proceedings of the Advanced Composites Conference*, 1986
30. S.S. Yau, T.W. Chu, and F.K. Ko, Flexural and Axial Compressive Failures of Three Dimensionally Braided Composite I-Beams, *Composites*, Vol 17 (No. 3), July 1986

Fiber Preforms and Resin Injection

Thomas W. DeMint, General Dynamics Corporation
H. Gary Van Schooneveld, Xerkon, Inc.

A FIBER PREFORM and *in-situ* resin impregnation approach to manufacturing conventional and advanced composite structures can offer significant advantages over traditional, time-consuming methods of hand lay-up. Of course, this depends on application requirements such as contour, thickness, and durability.

This approach, which emerged from the textile industry, involves the assembly of dry, unimpregnated continuous fiber preforms for subsequent injection or infusion of the matrix resin. It was the increasing use of continuous fiber reinforced composite materials in aerospace, automobile, and other weight-critical applications that generated interest in improving both their durability (for more demanding applications) and cost effectiveness, compared to conventional metallics. This article describes the manufacturing processes for assembling and impregnating oriented, unidirectional graphite fiber preforms by the resin film infusion/pressure molding and resin transfer molding (RTM) processes. The article "Multidirectionally Reinforced Fabrics and Preforms" in this Volume gives further information on fiber preforms.

Preform Materials and Assembly

Materials. Fabric is the starting point for most preforms because it positions the reinforcing fibers into principal directions. Preform fabric can be any one of the conventional fabrics, such as plain weave, five- or eight-harness-satin, unidirectional or multidirectional knitted, or bias weave. Fabric selection is dependent on the final properties desired in the component as well as on the configuration of the manufactured part (simple versus complex contour). As the structural performance required of a component increases, the use of "noncrimped" fabric (that is, knitted and woven unidirectional fabrics) becomes more desirable because of the increase in tensile and compression performance (Fig. 1). Increased performance is gained by eliminating the kinks within fiber bundles that result from the over-

and-under construction of the woven materials (Ref 1).

Predominant starting materials for preform fabrics are fiberglass, carbon, and aramid fibers. Silicon carbide, aluminum oxide, boron, borsic (boron fibers with a silicon carbide coating), quartz, and others are also used in specific applications. As with all composite structures, selecting the reinforcing fiber for the preform depends on the end-use characteristics desired in the composite structure.

The reinforcing thread material used to hold the preform together can be virtually any fiber that will endure the sewing process. The fibers with higher tenacity, such as aramid and polyester, will obviously cause fewer complications than will the more brittle higher-performance fibers, such as graphite, glass, or ceramic.

The operation that requires the most thread strength is the actual stitching of the dry laminate, when the needle penetrates the preform to complete the stitch. A brittle thread will often shear at the needle eye, a tendency that may depend on needle eye diameter, feed rate, and type of stitch used, such as the lock or chain

stitch, shown in Fig. 2. Depending on the application, a thread material with maximum tenacity should be chosen to minimize complications in the stitching process.

Assembly. The desired mechanical and thermal performance of the composite structure is established during the preform design process. Preform assembly converts individual fabrics into the multilayered configuration specified by the composite designer. The fabric layers are assembled into the final configuration in a process analogous to a prepreg lamination operation. Fabric layers are placed in a predetermined orientation by rotating the principal axes of the fabric layers or by using a multidirectional fabric. The assembly process can be automated by using broad goods spreaders and robotic placement (Fig. 3).

In most applications, particularly in aerospace composite structures in which weight is critical, build-ups are used in strategic locations to handle high, localized loads. The build-up is usually an individual preform manufactured in a separate operation. Build-ups generally consist of additional layers of fabrics that are sandwiched within the principal preform layers during the assembly process. Fixturing ensures proper build-up location and stitching in place prevents shifting during impregnation and cure.

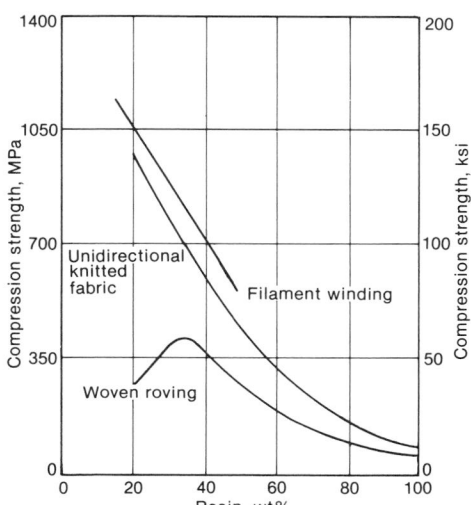

Fig. 1 Comparison of tensile properties for various fiberglass-epoxy fabric types

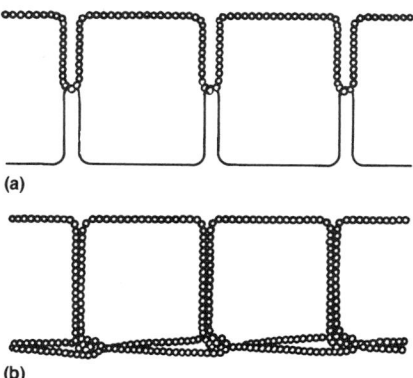

Fig. 2 Stitches used in preform assembly. (a) Lock stitch. (b) Chain stitch

Fig. 3 Broad goods spreader used for automated fabric placement

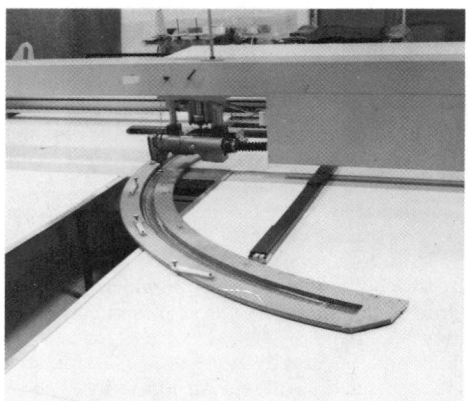

Fig. 4 Computer-controlled large-area sewing equipment

Fig. 5 Resin infusion vacuum chamber

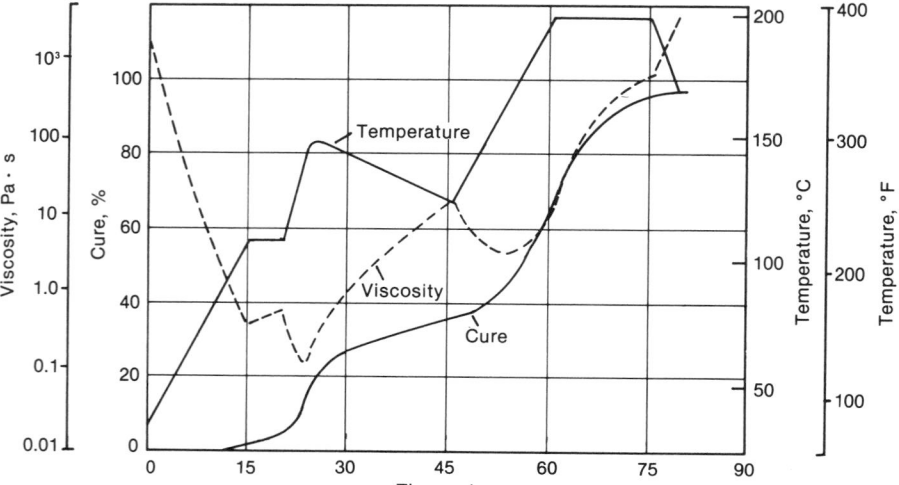

Fig. 6 Viscosity, degree of cure, and temperature versus time for Hercules 3501-6 epoxy resin system

Critical material parameters, such as reinforcement areal weight and fiber orientation, are maintained by using assembly fixturing to lock the fabric in place mechanically prior to the stitching operation and to minimize fiber distortion during subsequent handling operations.

The final operation in preform assembly is the stitching process. Stitching mechanically fixes the final shape of the preform and constrains fiber movement during resin impregnation. In addition to serving as an assembly material, stitching also may be incorporated as a reinforcement in the Z-axis direction. Stitching with structural thread (fiberglass, aramid, or carbon) has been shown to improve compression after impact by 80% and mode I critical strain energy release rate (G_{Ic}) by a factor of 30 (Ref 2). Actual improvement in Z-direction properties depends on the spacing, pitch (the amount, or volume), and type of fiber used as the stitch thread.

Depending on component configuration and type of stitch thread, equipment can range from standard industrial sewing machines to fully computerized equipment with a contour capability (Fig. 4). Stitching with high-modulus low-elongation fibers such as carbon requires specially modified equipment to prevent fiber breakage during the stitching process.

Whenever a stitching operation is to be performed, it is advantageous to work with a fiber preform that is free of resin or heavy binders because their presence limits the mobility of the fiber during needle penetration and results in a large number of fractured preform and thread fibers.

Three-dimensional preforms are being used increasingly. Three-dimensional weaving, tubular braiding, and rectangular braiding are the most common methods currently employed. These processes form a network of fibers that result in some quantity of fiber being oriented through the thickness. Three-dimensional preforms are generally used on parts having a constant cross section that must handle a significant out-of-plane load and/or that require a high degree of damage resistance (Ref 3). Currently, three-dimensional preforms are predominantly used in carbon-carbon composite structures.

Resin Injection

Resin film infusion is a technique for impregnating preforms with hot-melt resin systems, which are resins that are solid at room temperature. These resins include typical aerospace-grade epoxies, bismaleimides, and polyimides.

Preforms are loaded into a holding fixture to stabilize the preform during the infusion process. Resin, in film form, is positioned uniformly onto the preform. The starting ratio of fiber to resin weight is nominally held to 50% resin, 50% fiber. This may vary depending on preform thickness, contour, and subsequent

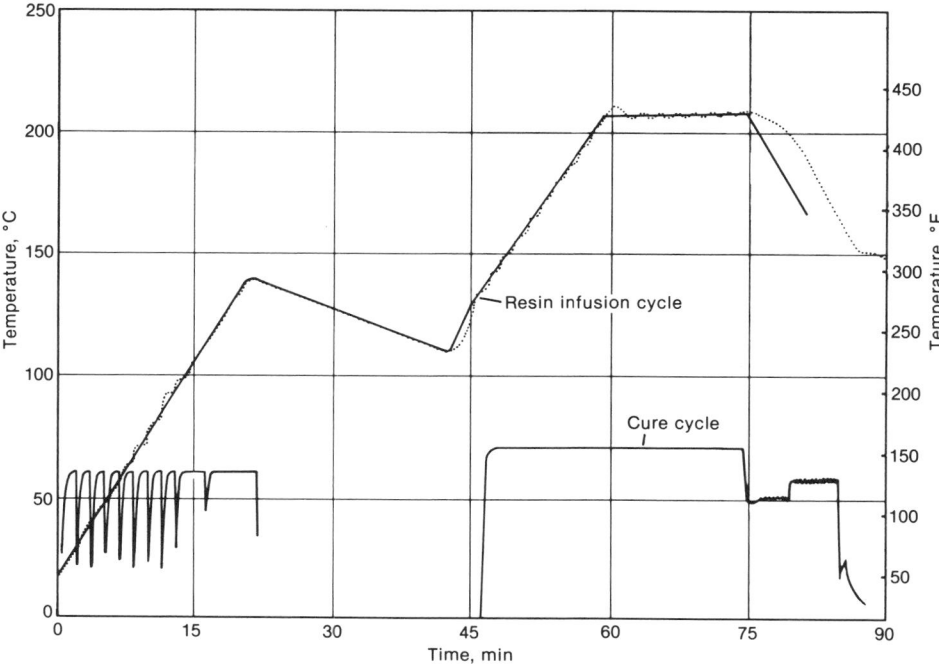

Fig. 7 Typical resin film infusion cycle

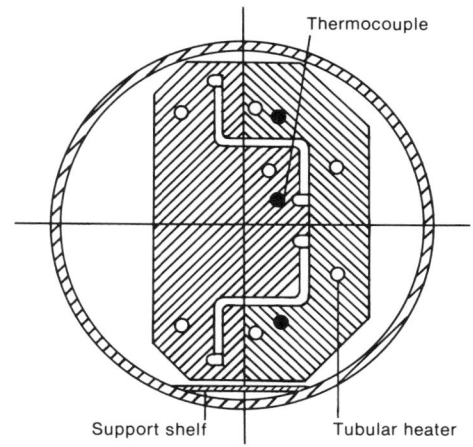

Fig. 8 Cross section of a pressure molding tool

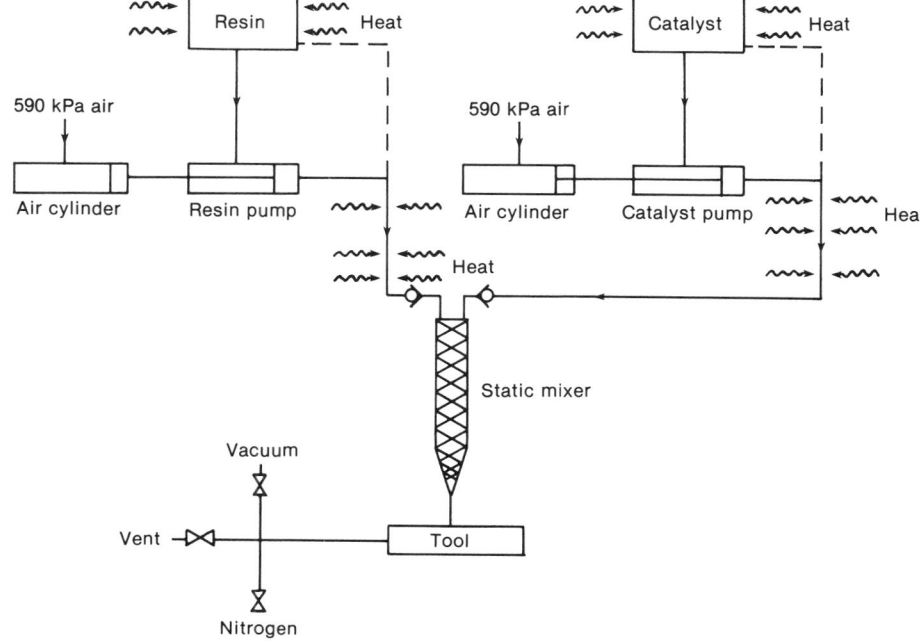

Fig. 9 Resin transfer molding

processing method. During the molding operation, the final ratio of fiber to resin weight for fabrics is established at 32 to 35% resin, 68 to 65% fiber for aircraft applications.

Actual impregnation occurs in a heated vacuum chamber, shown in Fig. 5. The preform, with resin applied, is positioned in the chamber. Heat is then transmitted to the preform and resin by means of infrared radiation. A vacuum is applied during the impregnation cycle to remove entrained air and volatiles from the filmed resin. Preform impregnation occurs through capillary wetting of the preform as the viscosity of the resin decreases (Fig. 6). During impregnation, the vacuum is cycled to prevent excessive resin bubbling and provide a mechanical pumping action to complement capillary wetting. A typical resin film infusion cycle is shown in Fig. 7. At the end of this cycle, the resin is not cured, but it is in an advanced B stage.

In the resin film infusion/pressure molding process, the final sequence is the molding operation itself, which establishes the final shape of the composite structure, fixes the proportions of fiber and resin, cures the resin, and provides the composite structure with its designed mechanical and thermal properties.

The pressure molding process uses integrally heated matched surface tooling in conjunction with an Autocomp vessel to perform the final cure. Positive-stop surfaces within the tool establish the final part thickness. Resin-to-fiber ratio is maintained by controlling the amount of fiber in the cavity and the final thickness of the part. Excess resin is allowed to flow into resin traps incorporated around the perimeter of the part (Fig. 8).

The manufacturing steps for the molding operation are as follows: The tool is prepared by removing any residual material from previous processing and applying a release coat to the molding surface to prevent bonding of the part to the tool. The impregnated preform is placed on the tool, and the remaining tool pieces are installed. The entire tool is inserted into the vessel, which has a permanent, reusable, internal vacuum bag that envelopes the tool as it is inserted. Vacuum is drawn at the beginning of the cure cycle to remove air from the mold cavity and the impregnated preform. Pressure is applied (590 to 690 kPa, or 85 to 100 psi) as the cure progresses and is maintained throughout the cycle. Once the desired level of cure is achieved, the part is cooled under pressure and removed from the tooling. A typical cure cycle by means of this technique is shown in Fig. 7. Additional information on the pressure molding process is presented in Ref 4 and 5.

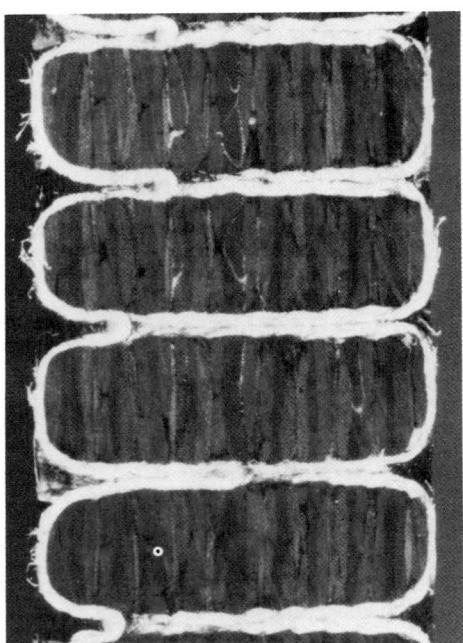

Fig. 10 Piecewise photomicrograph of a stitch in a laminate

Resin transfer molding is a process by which a thermosetting resin is injected into a cavity having the shape of the desired part; the cavity is filled with a dry fiber preform. The preform can include closed-cell cores and metal inserts, in addition to the fabric/mat materials. The process is typically used with low-viscosity fast-curing resins, such as polyester and epoxy, and chopped-strand or continuous mat reinforcement at a low fiber volume (~30% or less). However, the process has been demonstrated with higher fiber volumes (50 to 60%) of oriented material and higher-performance matrices, such as polyimides.

As shown in Fig. 9, a catalyst, or curing agent, is kept separate from the resin to allow the use of highly reactive resin-catalyst systems for fast cures and to facilitate storage of the materials. Using heated flow tubes if necessary, the resin and catalyst are pumped simultaneously through a static mixer and then into the tool, which is maintained at a temperature that allows sufficient time to fill the mold before the resin gels. One or more ports are located on the tool to bleed the air out as the resin fills the cavity and to indicate when the cavity is filled. Vacuum, vent, and nitrogen back pressure can be maintained at each port, if desired. Vacuum is used to evacuate the tool as the cavity is being filled and to facilitate resin flow into the tool. The nitrogen back pressure is applied after the cavity has been filled to stop the resin flow through the tool, keeping the matrix under pressure and preventing void growth in the laminate.

The critical factors in selecting a resin are minimum viscosity and the length of time and temperature the resin will remain at this minimum viscosity (pot life). In general, a maximum viscosity of 1 to 3 P (100 to 300 cP) is desired for the matrix to impregnate the reinforcement properly. This may vary with fiber volume. The time required for the resin to remain at this viscosity depends on several factors: fiber volume, fabric type, wet-out area (or area of reinforcement to be impregnated), and tooling configuration (number and location of injection and vent ports). The resin may be capable of maintaining the required minimum viscosity at room temperature, or it may require a slight application of heat to achieve adequate flow. Applying heat will also accelerate the cure and reduce the time at minimum viscosity.

Inspection

Nondestructive inspection of a cured composite laminate is commonly performed using an ultrasonic signal to detect voids or delamination. This can be accomplished manually by passing the signal through a transducer, a fluid couplant, and the laminate, and then detecting a reflection off the back surface of the laminate (A-scan). Other methods use water as the medium for the ultrasonic signal. Although these methods may be used to inspect an impregnated, stitched laminate, a discontinuity will appear at each stitch in the form of an attenuated, or unreflected, signal. This occurs because the fibers in the out-of-plane direction absorb the signal and do not reflect it as the surrounding area of the laminate does. The photomicrograph of a laminate cross section (Fig. 10) reveals its integrity and that of the stitching.

REFERENCES

1. P. Taft and J. Crowell, The Effect of Weave Distortion on the Physical Properties of RP Structure, in *Proceedings from the 38th Annual Conference, Reinforced Plastics/Composites Institute*, The Society of Plastics Industry Inc., Feb 1983, p 2C1-2C6
2. H.B. Dexter and J.G. Funk, "Impact Resistance and Interlaminar Fracture Toughness of Through-the-Thickness Reinforced Graphite/Epoxy," AIAA Paper 86-1020-CP presented at AIAA/ASME/AHS 27th Structures, Structural Dynamics and Materials Conference, San Antonio, May 1986
3. F.K. Ko, Developments of High Damage Tolerant, Net Shape Composite Through Textile Structure Design, in *Proceedings From the Fifth International Conference on Composite Materials Five*, 1985, p 1201-1210
4. N.C. Olsen, Advanced Manufacturing Technology for Structural Aircraft/Aerospace Components, in *Proceedings From the 31st International SAMPE Symposium*, Society for the Advancement of Process Engineering, April 1986, p 387-395
5. N.C. Olsen, "Advanced Pressure Molding (Autocomp) and Fiber Form Manufacturing Technology for Composite Aircraft/Aerospace Components," EM86-104-1 to EM86-104-13 conference papers presented at Composites in Manufacturing 5, Los Angeles, Jan 1986

Pultrusion

Jeffrey D. Martin and Joseph E. Sumerak, Pultrusion Technology, Inc.

PULTRUSION is an automated process for manufacturing composite materials into continuous, constant cross-sectional profiles. It is probably one of the most versatile composite processes, but it is still one of the least understood.

The term pultrusion refers to the final product and to the process. Most simply, it refers to a nonhomogeneous compilation of materials pulled through a die. In virtually every case, a continuous reinforcing fiber is integral to the process and the finished product. The matrix used is typically a thermosetting resin, which chemically reacts when heat is introduced to create an exothermic reaction. The resulting profile is shaped to the point at which it cannot be reshaped or otherwise altered within its operating temperature range, unlike thermoplastics. In contrast, the extrusion of aluminum and thermoplastic materials involves homogeneous materials that are heated and pushed through a die, then allowed to cool into the final solid shape. Because the material is initially heated and then cooled, it can be heated again and re-formed into another shape.

The pultrusion process has developed relatively slowly compared to other composite processes. There is a significant amount of art in combining the continuous reinforcements and resins in a continuous operation, and developing the science from the art has taken time. During the 1980s, there has been a dramatic increase in market acceptance, technology development, and pultrusion industry sophistication. Today, the number of technically competent personnel in pultrusion is sufficient to provide the base from which dramatic growth can occur. This, coupled with continual increases in cost-competitive advantages, will enable pultruded composites to become a traditional material alongside steel, wood, and aluminum before the end of the 20th century.

Process Description. Of the six key elements in the pultrusion process, the three that precede machine operation are a reinforcement handling system (referred to as creels), a resin impregnation station, and the material forming area. The machine consists of component equipment that heats, consolidates, continuously pulls, and cuts the profiles to a desired length. Although machines can produce profiles that range from 25 mm (1 in.) to 3 to 5 m (10 to 15 ft) per minute, typical line speeds are in the range of 0.6 to 1.2 m/min (2 to 4 ft/min) per cavity.

The process begins when reinforcing fibers are pulled from a series of creels. The fibers proceed through a bath, where they are impregnated with formulated resin. The resin-impregnated fibers are preformed to the shape of the profile to be produced. This composite material is placed in a heated steel die that has been precision machined to the final shape of the part to be manufactured. Heat initiates an exothermic reaction in the thermosetting resin matrix. The profile is continuously pulled and exits the mold as a hot, constant cross-sectional profile. The profile cools in ambient or forced air, or assisted by water as it is continuously pulled by a mechanism that simultaneously clamps and pulls. The product emerges from the puller mechanism and is cut to the desired length by an automatic, flying cutoff saw.

There are two categories of pultrusion products. The first category consists of solid rod and bar stock produced from axial fiberglass reinforcements and polyester resins; these are used to make fishing rods and electrical insulator rods, which require high axial tensile strength. The second category is structural profiles, which use a combination of axial fibers and multidirectional fiber mats to create a set of properties that meet the requirements of the application in the transverse and longitudinal directions.

More than 90% of all pultruded products are fiberglass-reinforced polyester. When better corrosion resistance is required, vinyl ester resins are used. When a combination of superior mechanical and electrical properties is required, epoxy resin is used. Higher temperature resistance and superior mechanical properties generally dictate the use of epoxy resins reinforced with aramid or carbon fibers.

Process Advantages. Pultruded composites exhibit all of the features produced by other composite processes, such as high strength-to-weight ratio, corrosion resistance, electrical insulation, and dimensional stability. Additional advantages are inherent in this process. One is that any transportable length can be produced, because of its axial nature, including small-diameter fiber optic cable core that is 2.2 km (1.4 miles) long, which can be wound on a spool after pultrusion.

Another advantage is that complex, thin-wall shapes, such as those extruded in aluminum, or polyvinyl chloride (PVC), are now possible because of recent process technology advances. Hollow sections can be produced by using cantilevered steel mandrels.

A third advantage is that wire, wood, or foam inserts can be encapsulated on a continuous basis in pultruded products. In addition to symmetrical walls, which are always easier to pultrude, variable wall thicknesses in a constant cross section can be pultruded.

A fourth advantage of the process, which is less obvious, is its ability to use a wide variety of reinforcement types, forms, and styles with many thermosetting resins and fillers. Virtually no other composite process offers as much versatility as pultrusion. Reinforcements can be placed precisely where they are needed for mechanical strength and can be consistently repeated.

Finally, pultruded shapes can be made as large as required because equipment can be built in any size. A corollary advantage of larger equipment is its ability to produce multiple cavities of the same or different profiles, which enables pultrusion to compete with traditional materials because of a relatively low labor cost. The cost of dies for pultruded shapes is also low compared to other composite processes.

Applications. The versatility of the process has enabled pultrusions to penetrate such market areas as land transportation, construction, marine, corrosion-resistant equipment, electrical/electronic, consumer appliances/business equipment, aircraft, and specialties. With the newest forming technology, pultrusion can produce nearly any constant cross-sectional shape that can be extruded. Aluminum extrusions account for approximately 15% of all of the aluminum consumed, while pultrusions account for only 5% of all reinforced plastics produced; thus, pultrusion has much growth potential.

There are several markets in which pultruded profiles have made a significant penetration that is not tied to just price competitiveness. Pultruded products are successful because they provide corollary advantages that are not available with traditional competitive materials, as described in the section "Process Advantages" in this article.

The highest-volume application of pultrusion is the fabrication of nonconductive ladder rails for in-plant and communication utility use. Corrosion-resistant fiberglass sucker rods have replaced steel in the extraction of oil. Semiflexible highway delineator posts that deflect without permanent deformation are used instead of rigid cold-rolled steel posts with plastic reflectors.

In highly corrosive environments, pultruded grating systems have become the standard because of their durability, replacing steel, aluminum, and even stainless steel systems. They are also used in elevated walkways and on steps where the supports are structural profiles, such as I-beams, channels, angles, and tubular shapes, that are made to the same dimensions as steel or aluminum supports. Cable trays of steel are being replaced by pultruded composite cable trays because of their superior corrosion resistance and better electrical insulation values.

Pultruded solid rectangular and square bars are being used in transformers to separate the windings and to permit air circulation. Utility market applications include guy strain insulators, stand-off insulators, hot-line maintenance tools, and the booms for electrical bucket trucks. Other electrical applications include tool handles, bus bar insulator supports, fuse tubes, and lighting poles.

Dunnage bars that separate and isolate loads in trucks and railcars have been made from pultruded lineals for many years. The back doors that roll into the roof of the truck are now also pultruded, as are the structural Z sections between the inner and outer walls of a refrigerated truck trailer.

In many buses, the luggage rack is pultruded. Hollow sections within the rack allow air to be passed for heating or cooling. Because of its continuous nature, the process produces the rack in one piece to span the length of the bus. In other rapid transit applications, continuous lengths of coverboard are pultruded in one piece to cover the current-carrying third rail on rapid transit systems. Because of the design flexibility of composite profiles, the shape is designed to snap over the rail and yet support a load dropped from above.

Technical Process

The basic elements of all pultrusion machines are very similar, but there are differences in the selection of heating components, drive trains, clamping devices, and cutoff saws. A pultruder strives for a common denominator in establishing the processing system and its corresponding process control. Several commercial suppliers are available to provide a full range of hardware as well as process control features. These features help to bridge the gap between the art and experience of the established pultruders and the new companies entering the field.

Material In-Feed. Reinforcements are provided in packages designed for the best continuous run-out of its material form. Continuous glass rovings are provided in center-pull packages that weigh 15 to 25 kg (30 to 50 lb) and are designed for a bookshelf-style creel. Creels of 100 or more packages are common and may be stationary or mobile. The glass roving is usually drawn from the package through a series of ceramic textile thread guides or drilled carding plates of steel or plastic. This allows them to be pulled to the front of the creel while maintaining alignment and minimizing fiber breakage. Some creel designs allow multiple guide eyes (bushings) or guide bars to tailor the tension to each roving. The ease of servicing or replacing roving packages must be considered in selecting a creel design and package capacity.

Some fiberglass rovings are available as center-pull twistless; that is, the natural twist as a consequence of winding has been offset by a built-in reverse twist. Continuous fibers of glass, carbon, and organic polymers can also be supplied on packages with cardboard cores designed for outside pay-out to avoid twist. This style of package dictates the use of a multiple spindle creel design in which the packages are oriented horizontally. Multiple bushings are again used to guide and protect fibers as they are delivered to the front of the creel. An additional consideration of package rotation and the resulting tension necessitates the use of a spindle bearing to provide uniform tension regardless of package size. Because packages in this configuration are usually smaller, a greater number of packages may be found on such a creel design.

The continuous fiber creels are usually the first station on a process line. Directly after the roving creels is a creel designed to accommodate rolls of mat, fabric, or veil. The roll materials are usually supplied in diameters between 305 and 610 mm (12 and 24 in.) with cores of 75 or 100 mm (3 or 4 in.) in inside diameter. The creel must be able to accommodate both the size of the roll and the inside diameter of the core, along with appropriate spindle spacing and core bushings. The ability to lock the position of the roll in the desired location will ensure proper delivery of the material to the desired location. In some cases, it is also necessary to provide for the pay-out of web material in a vertical, rather than a horizontal, format. This requires independent stands in a "lazy susan" type of configuration.

As materials travel forward toward the impregnation area, it is necessary to control the alignment to prevent twisting, knotting, and damage to the reinforcements. This can be accomplished by using creel cards that have predefined specific locations for each material. In some cases, these cards can be used for only one profile. In other cases, a general format for roving and web locations can be easily adapted for a variety of common profiles.

Resin Impregnation/Material Forming. The impregnation of reinforcements with liquid resin is basic to nearly every pultrusion process. The point at which resin is supplied and the manner in which it is delivered can have many different forms. A dip bath is most commonly used. In this process, fibers are passed over and under wet-out bars, which causes the fiber bundles to spread and accept resin. This is suitable for products that are of all-roving construction or for products that are easily formed from the resulting flat ply that exits the wet-out bath. In cases in which it is impractical to dip materials into a bath, such as when vertical mats are required or hollow profiles are made, materials can pass directly into a tailored resin bath through bath walls and plates that have been machined and positioned to accommodate the necessary preform shape and alignment. This alternative method provides the necessary impregnation without the need to move the reinforcements outside of their intended forming path.

Forming is usually accomplished after impregnation, although some initial steps can be carried out during the impregnation process. Forming guides are usually attached to the pultrusion die to ensure positive alignment of the formed materials with the cavity. In the case of tubular pultruded products, a mandrel support is necessary to extend the mandrel in a cantilevered fashion through the pultrusion die while resisting the forward drag on the mandrel. Materials must form sequentially around the mandrel in an alternating fashion to prevent weak areas due to ply overlap joints. Sizing of the forming guide slots, holes, and clearances must be done to prevent excess tension on the relatively weak and wet materials, but must allow sufficient resin removal to prevent too high of a hydrostatic force at the die entrance.

An alternative impregnating and forming method consists of injecting the resin directly into the forming guide or die after the dry materials have been formed. Although this technique minimizes the problems associated with the wet-out bath systems, some limitations exist in the areas of wet-out, air entrapment, and maximum filler content. A combination of techniques may be the answer for a specific profile, depending on its complexity.

The materials commonly used for forming guides include Teflon, ultrahigh molecular weight polyethylene, chromium-plated steel, and various sheet steel alloys. The pultrusion processor who employs a craftsman capable of converting sheet metal and plastic stock into forming guides with precise control would be most successful in processing complex shapes.

Die Heating. A number of different methods can be used to position and anchor the pultrusion die and to apply the heat necessary to initiate the reaction. The use of a stationary die frame with a yoke arrangement that allows the die to be fastened to the frame is the simplest arrangement. In all die-holding designs, the thrust that develops as material is pulled through the die must be transferred to the frame without allowing movement of the die or deflection of the frame. With this yoke arrangement, heating jackets that use hot oil or electrical resistance strip heaters are positioned around the die at desired locations. Thermocouples are placed in the die to control the level of heat applied. Multiple, but individually controlled, zones can be configured in this manner. This approach is well suited to single-cavity set-ups, but it becomes more complex when the number of dies used simultaneously increases because each die requires a heat source and thermocouple feedback provision. Standard heating jackets can be used, and heating plates can be designed to accommodate multiple dies to help alleviate this limitation.

Another popular die station uses heated platens that have fixed zones of heating control with thermocouple feedback from within the platen. The advantage of this method is that all dies can be heated uniformly with reduced-temperature cycling, because changes in temperature are detected early at the source of heat rather than at the load. In the same respect, however, a temperature offset will be common between the platen set point and the actual die temperature. With knowledge of the differential, an appropriate set point can be established. When provided with the means to separate the platens automatically, the advantage of quick set-up and replacement of dies can lead to increased productivity through reduced downtime. One machinery supplier also uses the bottom platen height adjustment feature to exactly align the die centerline with the pulling mechanism to eliminate any product distortion associated with misalignment.

A source of cooling water or air is essential in the front of the die at start-up and during temporary shutdown periods to prevent premature gelation of the resin at the tapered or radiused die entrance. This can be accomplished by using either a jacket or a self-contained zone within the heating platen. Alternatively, the first section of the die can be unheated, and cooling can be accomplished through convection. The most critical pultrusion process control parameter is the die heating profile because it determines the rate of reaction, the position of reaction within the die, and the magnitude of the peak exotherm. Improperly cured materials will exhibit poor physical and mechanical properties, yet may appear identical to adequately cured products. Excess heat input may result in products with thermal cracks or crazes, which destroy the electrical, corrosion resistance, and mechanical properties of the composite. Heat-sinking zones at the end of the die or auxiliary cooling may be necessary to remove heat prior to the exit of the product from the die.

To increase process rates and to reduce temperature differentials that contribute to thermal cracking in large mass products, it is desirable to deliver heat to the material before it enters the die. This is accomplished by radio frequency preheating, induction heating, or conventional conductive heating. Such heating devices are available as either integral units or stand-alone devices, which can be positioned before the die entrance.

One supplier has developed a process optimization instrument that allows tracking in a convenient graphic format of external die temperature profiles and internal product temperatures as a function of die position during the curing process. The data collected at a specific process speed become essentially a photograph of steady-state process conditions to be used for quality control, process engineering, and quality assurance documentation. Further process control developments of this nature will provide improved process capability and production efficiency.

Clamping/Pulling Provisions. A physical separation of 3 m (10 ft) or more between the die exit and the pulling device is provided in order to allow the hot, pultruded product to cool in the atmosphere or in a forced water or air cooling stream. This allows the product to develop adequate strength to resist the clamping forces required to grip the product and pull it through the die. The pulling mechanisms are varied in design among the hundreds of machines built by entrepreneurs or supplied by commercial machinery firms. Three general categories of pulling mechanisms that are used to distinguish pultrusion machines are the intermittent-pull reciprocating clamp, continuous-pull reciprocating clamp, and continuous belt or cleated chain.

The earliest pultrusion machines used a single clamp, which was hydraulically operated to grip the part between contoured pads. A carriage containing this clamping unit was then pulled by a continuous chain, which was driven by a variable-speed reversible drive train for a stroke of 3 to 4 m (10 to 12 ft). At the end of the stroke, the clamp released, and the clamping carriage returned to its starting point. During this return interval, the product remained stationary until the clamping and pulling cycle could be reinitiated. Because of this pull-pause sequence, this style became known as an intermittent-pull machine. Variations of this design are still found in the industry, including multiple clamping heads for multiple-cavity production.

The continuous-pull reciprocating clamp machine, which has become the most popular style, takes this concept one step farther. Its clamping, extension, and retraction cycles are synchronized between two pullers to provide a continuous pulling motion to the product. The value of using the intermittent-pull cycle with slow-cure materials or for purging die buildups is reflected in the fact that commercial reciprocating clamp machines now have intermittent-pull sequences. Subtle variations exist in the use of such drive methods as direct-acting hydraulic cylinders, hydraulic motor chain drives, or recirculating ball screws. Methods of clamping can be hydraulic, pneumatic, or a mechanical wedge action. The basic prerequisite is that sufficient clamping pressure be available on a relatively short ($<$ 460 mm, or 18 in.), contoured puller block that is held within the clamping envelope. In addition, sufficient thrust must be provided to the clamping unit to overcome the die resistance and to maintain a uniform pulling speed. An advantage of the reciprocating clamp system is its need for only two matched puller pads to attain a continuous pulling motion. These pads are easily changed and are generally of durable urethane-coated steel for long life.

Continuous-belt pullers have evolved from extrusion take-off pullers, but they have been modified for higher loads. These pullers are suitable for single-cavity or multiple-cavity production when they are all of the same physical size. Even with this restriction, uneven belt wear can result in slippage of adjacent cavities. On a positive note, the contact area of the belted puller is generally longer than that found with the reciprocating clamp pullers, which allows lower unit pressures on the pultrusion. A more flexible version of the continuous-belt machine is the cleated chain (or caterpillar) puller, which has many individually contoured puller pads attached to chain ears along the chain length. This modification allows the production of complex shapes and multiple cavities. Machine controls are used to ensure that even pressure is maintained between opposing chain pullers. The number of individually contoured puller pads can vary widely, depending on the complexity of the part. For the average part, the number of pads will vary between 12 and 60.

Cutoff Station. Every continuous pultrusion line requires a means of cutting product to length. Many systems employ manual radial arm saws or pivot saws on a table that moves downstream with the product flow. More sophisticated automatic cutoff saws are found on commercial machines; this eliminates the need for operator attention. Both dry-cut and wet-cut saws are available, but regardless of design, a continuous-grit carbide- or diamond-edged blade is used to cut pultruded products. Aramid-reinforced products present a special cutoff problem because of the toughness of the fiber. The use of conventional blades results in jagged edges and delamination. A suitable alternative is still being sought for these composites.

Key Technology Areas. Of the three relevant technology areas, the most critically important is resin formulation. Resin selection itself is important because it governs mechanical characteristics, operating temperature range, electrical insulation, corrosion resistance, and the flame and smoke properties of the profile. In addition, it governs process speed by its reactivity and can significantly control product aesthetics and tolerance capability. The resin matrix can be altered by chemical additives and fillers, which enhance its ability to handle higher temperatures, provide better electrical insulation and corrosion resistance, and lessen flame and smoke propagation. It is essential to the success of any pultruded profile that the polymer matrix be correctly engineered to meet the desired end-use properties and still account for those processing characteristics that are necessary to fulfill the economic goals of the application. This balance requires significant interaction between the end user and the pultruder.

The second key technology is material forming. Reinforcements are selected to meet desired mechanical properties. The electrical insulation and the corrosion resistance of the composite are affected by the amount and type of reinforcement used. In many applications, compromise is required to meet all of the desired properties. Pultrusion is a composite process that allows selective reinforcement of profile sections using different materials. This can be accomplished on a continuous basis with good consistency and repeatability. The only disadvantage is that it must be done continuously so that every foot of profile is identically reinforced.

The material forming area usually comes after the reinforcements have been impregnated with resin. Reinforcements are precisely positioned using porcelain bushings, metal forms, or plastic guides that have been machined to allow axial and multidirectional reinforcements to pass from one to the next before entering the heated die. These guides, if properly designed, can ensure the consistent placement of the different reinforcement types. These forming guides can become very complex and interactive on large profiles and hollow sections. This necessitates engineering the reinforcements, their path, and their forming sequence. However, even with a fair understanding of materials, the techniques required to produce these forming systems can take years to develop. Even with experience, considerable in-process engineering is required to fine-tune the material delivery and forming systems. This technology is fundamental to the consistency and optimization of mechanical properties and tolerances, yet it is too frequently discounted by novices to the process.

The third key technology is the control of product temperature. With thermoset resins, it is important to control the rate and level at which heat is added to the matrix or removed from the profile. If too little heat is put into the die, the composite material will not reach its desired exotherm temperature, resulting in an incomplete cure and less than optimally designed properties in the finished profile. Too high a temperature can induce thermal stress cracking as the profile cools, resulting in a material that exhibits poor electrical insulation or corrosion resistance properties.

Pultrusion tooling has two areas of consideration that are inseparable, yet distinct:

- The primary die, which is the precision-machined component that yields the final profile dimensions
- The fixtures required to align and form the input materials prior to entering the die as well as those required to clamp and draw the product through the die

All the elements needed to produce a specific profile are known as secondary tooling.

The primary die is typically made from tool steel that can be annealed or prehardened to 30 Rockwell C. Dies are usually multiple pieces that are machined and bolted together to form the desired profile cavity. Each piece must be squared, rough machined, and then stress relieved to form the approximate size. The die component is then ground on a linear profile grinder to the finished dimension, with an as-ground finish of 0.50 to 0.65 μm (20 to 25 μin.). Each die component is then aligned to form the finished cavity, and the pieces are doweled for permanent alignment. Alternatively, an integral alignment groove can be provided along the length of the die at the parting line for a separate or integral key. The pieces are then drilled and tapped for bolts that are used to hold the die together against the internal pressure developed in the process.

After positive alignment and fastening, the exterior of the die is again ground flat. The cavity surfaces are polished to a 0.25-μm (10-μin.) finish with buffing wheels and polishing compounds. The die entrances are tapered or radiused to relieve pressures on the material as it enters the die and to decrease associated wear. The finished die is then plated with hard chromium to 0.025 to 0.038 mm (0.001 to 0.0015 in.) and 68 to 72 Rockwell C hardness. Depending on profile complexity, additional grinding may be desirable after chromium plating.

An alternative to the construction described above is the use of air- or oil-hardened steel, which can be heat treated to provide a 60 Rockwell C hardness and does not require chromium plating. With these steels, an allowance must be made for distortion upon heat treating prior to finish grinding. It is more difficult to modify a tool after heat treating if additional tapped holes or grinding is required. The hardened tool is more prone to fracture at areas of stress or impact damage. On the positive side, hardened-steel dies do not require chromium plating and will provide long wear with minimal maintenance. On the other hand, chromium-plated dies must be inspected frequently for chromium coverage, because wear will proceed rapidly on the softer substrate steel once the chromium is removed.

It is difficult to predict die life, but a production rate of 15 200 to 30 500 m (50 000 to 100 000 ft) is not uncommon for chromium-plated dies, which may be replated as necessary to double or triple their service lives. An equivalent service life can be expected with hardened-steel dies without chromium, if their design minimizes stress cracking. Case hardening, ion implantation, and ceramic-base die alternatives are being explored in an attempt to improve die life. Die lengths typically range from 610 to 1500 mm (24 to 60 in.), depending on the size, complexity, and tolerances required.

Clamping fixtures depend on the style of puller used. The continuous-belt cleat-style puller requires numerous custom urethane pads to conform to a complex profile. Reciprocating clamp-style pullers also require contoured puller blocks, but only for two clamping units. In either case, the design of clamping pads is such that adequate pressure can be applied to the product surface without cracking the profile. Consideration of shrinkage, angularity, and heat dissipation is often necessary in designing the clamping hardware.

Forming technology is an art that develops over years of experience and is the most guarded area of processing technology. Improper execution will cause problems with chipping, cracking, poor reinforcement distribution, and frequent breakouts, all of which make continuous, economical processing very difficult. Properly executed, the constant, controlled delivery of material will in turn ensure reproducible quality and profitable production.

Various methods are used, such as step-sequence forms that are machined into steel or plastic plates called carding plates, which gradually collimate, align, and form the input materials. Continuous sheet metal forms can be used as an alternative or in conjunction with carding plates. Individual location guide eyelets and forming rods help to fill areas that are difficult to fill by other means. The cards or guides that are used within the resin pan and on the creels are also considered part of secondary tooling if they are designed specifically for one profile. Although it is relatively easy to find a toolmaker who can provide a pultrusion die, it is nearly impossible to find one with a process insight that allows him to develop secondary tooling. Consequently, the design of secondary tooling is either based on experience or on many forming system optimization trials.

Materials

One of the greatest attributes of the pultrusion process is that a wide range of materials can be used to provide a broad spectrum of composite properties. Given a specific profile geometry, the design engineer has a virtually unlimited supply of material options from which to construct a composite. The engineer must consider the intended function of the finished product, as well as the effects of temperature, atmosphere, environment, and time. Every selection, of course, carries an economic impact, and the optimal cost/performance options can be derived only with a proper understanding of the needs of the application and the available raw materials.

Reinforcements. A composite is, by definition, a combination of reinforcing fibers surrounded by a stress-transferring medium or matrix that allows the development of the full properties of the reinforcing fibers. The level of properties developed within a volume can be described approximately by the rule of mixtures, which, simply stated, predicts the resultant properties displayed in any direction to be proportional to the volume fraction of fibers aligned in that direction. The following factors then become readily apparent in composite property development:

- The type of reinforcement fiber and its ultimate capability
- The form and style of reinforcement as it pertains to orientation
- The proportion of the selected reinforcement relative to the whole

In pultrusion, near-complete versatility is possible with all three of these factors.

Reinforcement Types. The most widely used reinforcement has been and will continue to be glass fibers, because they are readily available and are low in cost. Electrical grade E-glass fibers, the most common, exhibit a tensile strength of approximately 3450 MPa (500 ksi); practical, commercial tensile strengths of 200 to 300 ksi; and a tensile modulus of 70 GPa (10.5×10^6 psi), but they have relatively low elongations of 3 to 4%. These properties result in composites with high strength and elastic limits, but virtually no yield up to ultimate failure—typically brittle fracture. A variety of fiber diameters and yields are available for specific applications. Glass fiber surface sizing chemistry has been developed over many years to provide optimum wet-out and chemical bonding between the fibers and matrix resins, thus ensuring maximum strength development and retention.

Higher tensile strengths can be achieved with S-glass fibers, which were developed for high-performance applications. These fibers exhibit a tensile strength of 4600 MPa (665 ksi) and a tensile modulus of 85 GPa (12.5×10^6 psi). Far greater stiffness can be achieved by using carbon fibers when their conductive nature would not be detrimental to the application.

Carbon fibers are produced from a process of continuous graphitizing and stretching of a textile thread, such as polyacrylonitrile (PAN). The resultant fiber exhibits tensile strength from 2050 to 5500 MPa (300 to 800 ksi) and tensile modulus from 210 to 830 GPa (30 to 120 $\times 10^6$ psi) with elongations of 0.5 to 1.5%. Normally, if high tensile strength is chosen, a lower tensile modulus must be accepted, and conversely. These fibers deliver unique properties, such as electrical conductivity, slightly negative thermal coefficient of expansion, high lubricity, and low specific gravity (1.8 versus 2.60 for E-glass). The price of carbon fibers is often the only limitation to their widespread use.

High-modulus organic fibers, such as the aramids, are an attractive option for providing high tensile strength and modulus of 2750 MPa (400 ksi) and up to 130 GPa (19×10^6 psi), respectively, with elongations of up to 4%. This results in very tough composites that exhibit good flexural and impact strengths, which are well suited to ballistic applications and whenever energy absorption is necessary. The low specific gravity (1.45) of these fibers gives them one of the highest strength-to-weight ratios of any reinforcement available. Deficiencies in compressive strength and interlaminar shear strength are being addressed through improved surface chemistry.

Other organic fibers have recently become available for use in pultrusion. Polyester fibers with appropriate binders have been used as a replacement for glass in applications that would benefit from increased toughness and impact resistance but where tensile and flexural strengths can be sacrificed. These fibers provide a low-modulus capability to composites, thus bridging the gap between thermoplastics and glass-reinforced thermosets. With low specific gravity and only a moderate cost premium over glass, these fibers are well suited for certain commercial and industrial applications. A recently introduced nylon fiber increases the low-cost organic fiber options. A highly oriented polyethylene fiber geared toward higher specific strengths (high properties with low specific gravity) can compete with the aramids in applications requiring stiffness, toughness, and light weight.

Orientation Options

Once the fiber type has been selected, the next most important consideration is the ability to orient it in the desired direction to utilize its properties more advantageously. A common denominator for all materials used in pultrusion is that they must be available in a continuous form to provide reasonably long pay-out periods without defects, splices, or changes in cross-sectional volume.

The most common and lowest-cost form of continuous reinforcement is roving, which consists of continuous axial filaments in single- and multiple-strand configurations. Glass rovings are designated by their yield, which is defined by the number of yards per pound of material or by the European designation, TEX, in grams per kilometer. The two most commonly used yields are 112 yd/lb and 56 or 62 yd/lb, which is the larger tow of the two. The rovings are typically supplied in 20-kg (40-lb) hollow cylindrical packages with a center pay-out that allows them to be stacked on a multiple-shelf (bookshelf) creel configuration. A similar package is available for the organic fibers previously described. Carbon fiber rovings, however, are provided in sizes designated by the number of filaments per tow, with the most common being 3K, 6K, and 12K filaments. The tow sizes are considerably smaller than the glass roving tow,

and the package weights are 0.9 to 2 kg (2 to 5 lb), with an outside pay-out designed for a spindle-style creel system. This difference in material form, although of no consequence in the end product, does present some limitations to the processor with regard to creel style and capacity, splicing frequency, and product size limitations. The recently developed larger tow options (40K, 160K, and 320K) will be of benefit to pultruders.

The roving form allows maximum packing of fibers within a volume to yield the highest possible properties along the product axis, referred to as the longitudinal or machine direction. Maximum axial property development may be diminished by such factors as incomplete wet-out, improper fiber alignment, fiber damage due to creel or forming fixtures, and catenary (uneven fiber-to-fiber tension resulting in loops or nonparallel strands). Given near-perfect alignment, fiber fractions of 65 vol% are achieved (80 wt% for glass fibers). In such a product, there are no fibers oriented in the transverse direction (90° to the longitudinal axis); the strengths exhibited in this direction reflect the strength of the matrix resin only.

To overcome this transverse strength deficiency, reinforcing fibers aligned in the transverse direction must be provided. Continuous-strand mat, which has fibers oriented randomly in all directions, is most commonly used. These fibers are then held together with a thermoset resin binder, which allows the mat to have sufficient tensile strength for processing.

Several mat styles can be used, but the most common is an E-glass mat that has fairly coarse fibers in an open or porous construction but provides high structural efficiency. This mat is used on exterior surfaces and as a center ply to build a laminate with substantially improved transverse physical properties. The porous construction, however, results in a potential for composite surface porosity and provides a very noticeable fiber pattern. When this is unacceptable, a fine-filament A-glass mat (or veil) is used as a surfacing ply to bring more resin to the surface and to achieve a dense, aesthetic surface appearance. Recent strength improvements in fine-filament mats have allowed their use throughout the composite. Regardless of mat style, the processor depends on the mat manufacturer to provide control over fiber distribution, binder content and distribution, and defects that have a serious impact on processing efficiency.

Random-fiber mats are generally used in weights of 0.15 to 0.60 kg/m² (0.5 to 2.0 oz/ft²). To accommodate the volume required by this lower bulk density type of reinforcement, it is necessary to remove fibers from the longitudinal direction. The resultant composite will have a slightly lower overall fiber content by volume because more resin is consumed to fill the open-structure mat described. The resultant increase in transverse and off-axis strengths is accompanied by a decrease in longitudinal properties. It is here that the engineer can exercise control over fiber proportions to achieve the properties necessary for the specific application. One restriction dictated by the nature of composites is the need to provide a symmetrical composite ply structure relative to the centerline of the thickness in order to prevent dimensional problems resulting from differential shrinkages of the plies of different reinforcement forms and styles.

Within the limits of design, mats and rovings are the most common structural composite constructions, with typically 50 wt% fiber. By exercising control of the proportions, however, the overall fiber weight content can be varied between 35 and 65%. Mat construction is also available with carbon fibers having a fine-filament construction. The use of carbon fiber mat in advanced composites, however, is not as common as the use of alternative fabric reinforcements.

Although the random-fiber orientation of a mat provides fiber orientation in all directions, a specific volume of fibers can be oriented transverse to the axis by means of several biaxial fabric reinforcement styles. The traditional material used had been woven roving; however, problems associated with weave stability, wet-out, and ply edge fiber retention limited the effectiveness of this material. The introduction of nonwoven biaxial fabrics employing fibers stitched or knitted together at the interstices has provided an effective solution to the problems mentioned above. The stitched fabrics can be supplied with any proportion of longitudinal to transverse orientation—even to the point of having a 100% transverse fiber continuous ply with only longitudinal stitch fibers. These materials are available from a number of suppliers employing various techniques to ensure fiber directional stability, ply integrity, and thickness and weight control.

The biaxial fabrics are generally introduced as internal plies and used in conjunction with a mat as exterior plies. Although not impossible, the use of these materials as exterior plies is somewhat limited because of the tendency for the transverse fibers to be dragged back by friction at the die surface. Additional problems of splicing and fabric skewing make the use of these materials more of a challenge to the processor.

Multiaxis fabrics typically employing 0, 90, and ±45° orientations have become available and provide additional directional strength as well as an improved level of fabric stability. Double-bias fabrics of ±45° without some 0° (or axial) fiber orientation for pulling strength and stability are impractical for pultrusion. However, several suppliers have overcome this limitation by stitching fibers to a carrier ply of polyester veil or continuous mat.

In all of these multidirectional fabric styles, the supplier has the versatility of using every fiber type previously described in any or all of the directions possible or in an alternating fashion to provide hybrid composites with tailored properties. The challenge to the design engineer becomes the identification of the most cost-effective style of reinforcement to use in an application not satisfied by the conventional mat/roving construction.

Because of the tendency toward ultraviolet degradation of the surfaces of composites used outdoors, a condition known as fiber bloom will occur over time. This term refers to the exposure of fibers at the surface, which can be an irritant to human contact as well as an aesthetic detraction. To resolve this problem, surfacing fabrics of organic fiber composition, primarily polyester and nylon, are being used. The surfacing fabrics are available in a variety of weights and constructions and provide a corollary advantage of contributing to the impact strength and toughness of the composite. They also help the processor from the standpoint of providing a tough material to assist in carrying materials through the die while protecting the die wall from the abrasive glass. The ability to provide a smooth resin-rich surface appearance without fiber patterns and the ability to be screen or roll printed with company logos, information, or wood grains makes this material family an important element in many pultrusion composite designs.

Matrix Choices. Although the fiber type, form, and style determine the ultimate strength potential, the matrix resin determines the actual level of properties realized through effective coupling and stress transfer efficiency. There are composite properties, however, that are determined exclusively by the properties of the matrix resin. Among these are high-temperature performance, corrosion resistance, dielectric properties, flammability, and thermal conductivity. Selection of the resin, as well as formulation chemistry, becomes an important consideration that must be addressed very early in the design process.

Unsaturated polyester resins are most commonly used in pultrusion. Orthophthalic, isophthalic, and teraphthalic acids or anhydrides, in combination with maleic anhydride and various glycols, are the basic elements. The isophthalic polyester has dominated the supply by virtue of its well-rounded physical properties and economics. By adjusting the ratios of backbone chemical units, a variety of strength and elongation characteristics, as well as reactivity levels, can be achieved. A necessary characteristic of a pultrusion polyester is the ability to gel and cure rapidly to form the strong gel structure needed for adequate release at the die wall. Viscosities of 0.5 Pa·s (500 cP) are typical for fully extended resins, or higher-viscosity low-reactive monomer versions can be blended with additional styrene to suit the processing need. The styrene level must be properly maintained to achieve good cross-link structure without leaving residual (unreacted) styrene in the finished composite. The final structure of the cured polyester resin is thermoset, resulting in the associated stability of the finished product to many adverse conditions.

The properties of polyester resins will differ, based on their backbone compositions. However, certain generalizations can be made, as follows.

First, polyester resins exhibit good corrosion resistance to environments of aliphatic hydrocarbons, water, salts, and dilute acids and bases. They do not perform well when exposed to aromatic hydrocarbons, ketones, and some concentrated acids. Corrosion resistance guides provided by the supplier should be consulted for information on suitability to a specific chemical exposure.

Second, because of the high unsaturation of the polyester chain and the resultant high cross-link density, polyesters exhibit shrinkage up to 7%. This level can be reduced by using fillers and low-profile additives, which also help to control microcrack development and shrinkage-related sink marks.

Third, polyester resins have glass transition temperatures ranging from 80 to 120 °C (180 to 250 °F). Exceeding this temperature threshold is accompanied by a rapid reduction in mechanical properties because of molecular chain movement. The actual continuous-use temperature must be defined in the context of the desired performance characteristics. Composites based on polyesters, for example, can be approved for continuous use at 150, 180, and even 200 °C (300, 355, and 390 °F) while retaining a high percentage of their electrical insulation properties. Chemical decomposition of polyesters will begin to occur at temperatures above 250 °C (480 °F).

Fourth, polyester resins can be brittle or tough, depending on the backbone chemistry and the cross-link density. However, maximum elongations are in the 5% range; therefore, impact properties will depend greatly on fiber orientation and on impact-modifying additives.

Fifth, polyesters will support combustion without modification. Through the use of additives or backbone bromination, the flammability and smoke generation properties of polyesters can be greatly improved in order to satisfy most flammability codes.

Sixth, the electrical properties of polyesters make them suitable for use as primary insulators in many high-voltage applications. Secondary insulation applications abound for polyester-base composites. Retention of electrical properties at elevated temperatures has made polyester insulators the material of choice in many applications.

Lastly, the weatherability of polyesters is fair to good, depending on backbone structure. Additional protection is usually sought through a variety of ultraviolet absorption additives, the use of polyester surface veils, and painting (done after pultrusion).

Vinyl Ester Resins. When additional performance is required in the areas of corrosion resistance and elevated-temperature mechanical properties, vinyl esters are available as an alternative to polyesters. The chemical structure of vinyl ester resins is such that the reaction sites are at the end of each polymer chain rather than along the chain length, as with polyesters. This structure results in a thermoset resin that has a lower cross-link density. These resins display greater toughness properties, such as interlaminar shear and impact strength. In addition, there are fewer sites available for chemical decomposition along the vinyl ester molecule. The use of rigid segments along the polymer backbone results in the high-temperature capability of these materials.

In addition to being approximately 75% more expensive than polyesters, an added cost due to lower process speed is characteristic of vinyl esters. This results from the slower reaction rate associated with its lower cross-link density. In every other respect, the processing of vinyl esters is very similar to that of polyester. Both pultrusion-type resins require high-temperature initiation of catalysts to begin the reaction. The room-temperature pot life of these resins is usually 24 to 48 h.

Epoxy resins are used when physical properties of the highest level, as well as elevated-temperature property retention, are required, which is often the case in military and aerospace applications. Although they have increased continuous-use temperatures to about 150 °C (300 °F), composites made with epoxy resin are known for the poor toughness that is due to their rigid structure. Much of this deficiency can be overcome with proper selection of reinforcements to provide impact strength. Epoxy resins do provide increased flexural strengths and shear strengths over polyester and vinyl ester systems. In addition, their excellent electrical properties and corrosion resistance also qualify them for use in many commercial applications requiring superior performance at elevated temperatures.

Epoxy resins are expensive materials in a number of respects. The resins are three to six times more costly than polyesters and have a number of process-related costs not found with polyesters. Because they are formed (cured) by a stepwise reaction rather than an addition reaction, as with polyester resins, their reaction rate is very slow. The gelation of epoxy resins occurs at a later stage of reaction, and it is critical that the exotherm developed be contained within the die. This dictates a slow process rate, which results in a high labor and burden allocation to the product being produced. Because the epoxy resin begins to react slowly as soon as it is mixed, the pot life is short. The resin scrap rate is potentially higher if viscosity buildup affects wet-out to the extent that the bath must be recharged. The die temperature profiles used for epoxy are typically hotter than for polyesters, and the drip-off at the die entrance must often be discarded rather than recirculated to the bath. Because of the tendency for the epoxy resin to bond strongly to the die wall, epoxy products often display surface defects, such as exposed fibers, chipping, or loss of dimension, all of which increase finished-product scrap rates. These ad-

ditional costs place epoxy products in a class in which the end-use requirements must justify the high price.

Much effort is being directed at more processible epoxy systems by using viscosity-modifying fillers, internal mold releases, and hybrid epoxy structures. Some processors use slip sheets between the epoxy resin and the die wall to prevent bonding and to improve surface quality. Ultrasonic vibration of the die is one experimental method recently directed at reducing the adhesion tendency of epoxy resins. Other improved methods, including selection of die materials and designs, will be necessary before epoxy processing is considered routine and therefore more cost competitive.

Other Resins. A variety of resin alternatives beyond the traditional three mentioned above have been in development in the last several years. Resins based on methyl methacrylate are quite promising because they offer the advantages of higher physical properties, high filler loading due to low viscosities, rapid processing speeds, smooth low-profile surfaces, and improved flame-retardant and weathering characteristics. Although they cost twice as much as polyesters, these resins may find use in applications that can exploit their special properties. Higher output offers the processor reduced processing costs; thus, the net cost premium may be moderated, which would make these systems competitive with polyesters in many applications.

Developmental work has been reported on phenolic resins that are suitable for pultrusion but are far from commercialization. These systems will reportedly offer the high heat resistance and flame-retardant/low-smoke characteristic of phenolics. Process rates are expected to be somewhat slower than those of polyesters.

A great deal of interest has developed in recent years in using thermoplastic resins as the matrix for pultruded profiles. The major driving force comes from the desire for improved toughness and postprocessing formability. In addition, several of the engineering thermoplastic resins provide heat distortion properties that are superior to those of the epoxy systems currently used, making them very attractive for advanced composite applications. The technology for impregnating fibers with thermoplastic resins is currently under development by a number of sources using a variety of methods, including hot-melt application, solvent solution impregnation, and fluidized bed particle application. Most of the success reported to date has been limited to the preparation of thin tapes or small-diameter rods, which are then used as molding materials in subsequent processing methods, such as compression molding. The interest in this area is growing, however, and the advantages of the materials will undoubtedly result in the development of a viable pultrusion process.

Additives. All liquid resin systems can be tailored to provide specific performance by using additives. Although the thermoplastic

processor often receives precompounded materials, the pultruder develops his own formulation and prepares it on site for his own use. Resin suppliers have developed recommended formulations that are available to the general market, but the resin formulation is a well-guarded secret of the pultruder and is often considered to be his competitive edge in cost or performance.

Fillers often constitute the greatest proportion of a formulation, second to the base resin. The most commonly used fillers are calcium carbonate, alumina silicate (clay), and alumina trihydrate. Calcium carbonate is primarily used as a volume extender to provide the lowest-cost resin formulation in areas in which performance is not critical. The clay fillers are used where improved corrosion resistance and electrical properties are required. They also impart a superior surface finish to the pultruded product. Alumina trihydrate is a filler that is used for its ability to suppress flame and smoke generation and electrical arc and track development. It is used in many applications to satisfy consumer or governmental flammability codes for product safety. Other fillers, such as mica, talc, calcium sulfate, and various glass beads and bubbles, are offered to the industry for their specific property modification qualities, although they represent a small portion of total use.

Fillers can be incorporated into the resins in quantities up to 50% of the total resin formulation by weight (100 parts filler per 100 parts resin). The usual volume limitation is based on the development of usable viscosity, which depends on the particle size and the characteristics of the resin. Wetting agents have been developed that offer the incorporation of a greater filler volume without increasing formula viscosity. These wetting agents can be added to the filler by the supplier or as an additive by the formulator. Air release agents are added in the same respect, that is, to provide more efficient packing by reducing entrapped air in the liquid resin and void content in the finished product.

Special-purpose additives would include ultraviolet radiation screens for improved weatherability, antimony oxide for flame retardance (used in combination with halogenated resins), pigments for coloration, and low-profile agents for surface smoothness and crack suppression characteristics. A variety of options are available in each of these material categories.

An important category of materials is associated with the curing of thermoset resins. The polyester, vinyl ester, and methacrylate systems are cured by the high-temperature decomposition of organic peroxide catalysts. A variety of catalysts that provide different levels of initiation temperature are available. Pultrusion formulations typically employ a multiple-catalyst system to provide rapid low-temperature initiation, followed by midrange accelerator and high-temperature completion. This combination, which delivers the fastest processing speed, can also reduce resin pot life, especially

Table 1 General properties of fiberglass-reinforced pultruded products

Material	Specific gravity	Tensile strength MPa	ksi	Tensile modulus GPa	10^6 psi	Flexural strength MPa	ksi	Compressive strength, axial MPa	ksi	Dielectric strength, parallel kV/cm	kV/in.	Thermal conductivity W/m · k	Btu · in./h · ft² · °F	Coefficient of thermal expansion, 10^{-6}/K	Water absorption, wt%
Solid rod and bar, 70% unidirectional reinforcement........2.00		690	100	41.4	6.0	690	100	410	60	23.60	60	0.288	2.0	5.4	0.3
Profiles, 50% multidirectional reinforcement........1.80		207	30	17.2	2.5	207	30	276	40	9.84	25	0.144	1.0	9.0	0.5

(a) The axial properties given are typical of those attainable through the pultrusion process. Variations in glass content and orientation, as well as modifications to the resin system, can improve these properties. Composite systems can be developed to meet many customized physical, mechanical, chemical, electrical, flame-resistant, and environmental properties.

Table 2 Effect of fiber type on selected ultimate properties

Fiber type	Specific gravity	Tensile strength MPa	ksi	Tensile modulus GPa	10^6 psi	Compressive strength MPa	ksi	Thermal conductivity W/m · k	Btu · in./h · ft². °F
Glass(a).....	2.0	690	100	40	6	410	60	0.30	2
Carbon(b) ...	1.65	1000-1500	150-220	100-140	15-20	620-970	90-140	0.85-1.4	6-10
Aramid(c) ...	1.28	1400	200	80	12	280	40	0.15	1

(a) E-glass unidirectional rovings. (b) Type AS graphite fibers. (c) DuPont Kevlar 49 fibers. Source: Ref 1

in high ambient temperatures. The use of low-temperature inhibitors and the selection of catalyst type and quantity become important considerations in controlling resin scrap due to premature gelation. Catalysts for epoxy systems are amine or anhydride compounds, which are usually supplied as a component that is matched in specific proportion to the epoxy resin system obtained from the supplier.

Mold releases are important in the development of adequate release from the die wall to provide smooth surfaces and low processing friction. The release cannot leave a surface residual that interferes with bonding if subsequent painting or fabrication is necessary. These releases are added to the resin and are typically metallic stearates or organic phosphate esters suitable for high-temperature use.

Properties

Mechanical Properties. The great amount of latitude that exists in selecting reinforcement type, form, style, and proportion allows a broad spectrum of mechanical properties. The directionality of strength in a pultruded composite can be greatly influenced by substituting longitudinal reinforcement for random mat or directional fabrics. A product with only longitudinal reinforcement will typically exhibit mechanical properties that are at least ten times greater than the same property measured 90° from the longitudinal fibers. In this type of composite, the properties of the fiber dominate the axial properties, but the properties of the resin dominate the transverse properties. As the volume fraction of fibers in the off-axis direction is increased, the longitudinal volume fraction, by necessity, must decrease; thus, the transverse properties are increased at the expense of longitudinal properties. This substitution method can be used to move the directionality of strength toward a one-to-one ratio, or even to the extent of achieving higher transverse properties when using some of the weft transverse orientation fabrics available.

The absolute value of the specific property desired will depend on the fiber type chosen: glass, carbon, aramid, or organic fibers. The magnitudes of typical properties for glass-reinforced pultrusions are given in Table 1 to illustrate the effect of orientation on property levels. Table 2 indicates the effect on selected ultimate properties in the axial direction at an equal volume fraction of alternative continuous fibers. The specific contributions in fiber type become very apparent in the properties selected

Table 3 Property comparison by process

Range of values reflects transverse and axial testing directions as well as percent reinforcement

Process	Reinforcement, wt%	Tensile strength MPa	ksi	Tensile modulus GPa	10^6 psi	Flexural strength MPa	ksi	Compressive strength MPa	ksi	Impact strength J/m	ft · lbf/ft	Thermal conductivity W/m · k	Btu · in./ h · ft² · °F	Heat distortion at 1.8 MPa °C	°F	Dielectric strength kV/cm	kV/in.
Spray(a)	30-50 glass-polyester	60-120	9-18	5.5-12	0.8-1.8	110-190	16-28	100-170	15-25	210-640	48-144	0.17-0.23	1.2-1.6	175-205	350-400	80-160	200-400
Compression(a) .	15-30 glass-SMC	55-140	8-20	11-17	1.6-2.5	120-210	18-30	100-210	15-30	430-1150	96-264	0.19-0.25	1.3-1.7	205-260	400-500	120-180	300-450
Compression(a) .	25-50 glass mat-polyester	170-210	25-30	6.2-14	0.9-2.0	70-280	10-40	100-210	15-30	530-1050	120-240	0.19-0.26	1.3-1.8	175-205	350-400	120-240	300-600
Filament winding(a) ...	30-80 glass-epoxy	550-1700	80-250	28-62	4.0-9.0	690-1850	100-270	310-480	45-70	2150-3200	480-720	0.27-0.33	1.9-2.3	175-205	350-400	120-160	300-400
Pultrusion(b) ...	40-80 glass mat-polyester	410-1050	60-150	28-41	4.0-6.0	690-1050	100-150	210-480	30-70	2400-3200	540-720	0.27-0.33	1.9-2.3	205-260	400-500	80-160	200-400
Pultrusion(b) ...	30-50 glass mat-polyester	80-210	12-30	6.9-17	1.0-2.5	170-210	25-30	210-340	30-50	530-1350	120-300	0.22-0.27	1.5-1.85	95-150	200-300	80-120	200-300
Pultrusion(c) ...	30-55 glass mat and roving-vinyl ester resin	70-280	10-40	6.9-21	1.0-3.0	100-280	15-40	140-340	20-50	270-1600	60-360	0.22-0.33	1.5-2.3	175-230	350-450	80-130	200-325
Pultrusion(c)....	30-55 glass mat and roving-polyester resin	50-240	7-35	5.5-17	0.8-2.5	70-210	10-30	100-280	15-40	210-1350	48-300	0.22-0.33	1.5-2.3	175-205	350-400	80-120	200-300

(a) Ref 2. (b) Ref 3. (c) Ref 4

to illustrate this point. Because so many options exist, even among the unidirectional reinforcements, it is difficult to provide an all-inclusive list of properties. The designer must learn to exploit the specific fiber characteristics to achieve the desired performance characteristics, along with exercising the orientation options available from raw material suppliers.

The finished-product geometry most often dictates the process to be used. A further comparison of selected pultrusion process properties is shown in Table 3.

Frequently, the decision to be made relates to composite selection as an alternative to such traditional materials as steel, aluminum, or wood. Selected relative properties of common alternative materials are listed in Table 4. Although Table 4 compares various materials in terms of absolute strength or stiffness, it is often desirable to determine the thickness of fiberglass-reinforced plastic necessary to achieve strength or rigidity equivalent to that of the traditional material. Such an analysis is shown in Table 5, using the data of Table 4 and contributions of section geometry. When the application does not require equivalency, the product thickness can be reduced accordingly to yield savings in material and processing costs. A similar analysis can be conducted for the alternative fiber types to determine equivalent thickness factors for the high-strength high-modulus materials. The relative costs per unit volume can then be readily determined.

Physical Properties. The thermal conductivity of composites reflects both matrix and fiber characteristics. Generally, the fiberglass-reinforced composites are excellent insulators for thermal and electrical environments. This is also true of the organic fiber composites, regardless of the matrix employed. The use of conductive carbon fibers, however, results in composites that exhibit greater thermal and electrical conductivity; this reduces their effectiveness as insulators (although they are still efficient relative to metals) but creates opportunities because of their static and heat dissipation characteristics.

Although fiberglass-reinforced composites display a modest positive coefficient of thermal expansion, both aramid- and carbon-reinforced composites display a slightly negative coefficient of thermal expansion. This characteristic can be used to advantage in aerospace structures and in producing very tight tolerance parts. The same characteristics can result in molded-in stresses in multidirectional reinforcement systems, particularly if dissimilar

fiber types (hybrids) are used. Carbon-reinforced composites are also noted for their lubricity and wear resistance, which makes them suitable for use as bearing materials. Their capacity for heat dissipation is advantageous in this application.

Specific gravity is a key consideration when strength-to-weight ratios are important, as in aircraft and aerospace applications. Carbon- and aramid-reinforced composites excel because of their low specific gravities and high strength and stiffness characteristics. High-modulus polyethylene fibers, with their very low specific gravity and high strength and stiffness properties at room temperature, have the potential to provide the highest specific strength available.

The impact resistance of organic fiber reinforced composites is quite high, making them suitable for energy absorption applications, such as ballistic shielding. The impact resistance of carbon fibers is generally low, and advanced composites that use carbon rely on the tough resin matrix for impact properties. Fiberglass-reinforced composites are also relatively poor in impact performance compared to the organic fibers, but are superior to carbon-reinforced composites. The fatigue resistance of graphite- and aramid-reinforced composites is superior to that of the fiberglass-reinforced composites, particularly when used with epoxy resins.

Chemical and corrosion resistance characteristics of pultruded composites are predominantly attributed to the resin matrix. In considering a particular resin system for an intended environment, the degree of exposure,

the concentration of the corrosive element, and the temperature of the environment must be known. Resin companies servicing the corrosion markets generally publish a list of chemical environments tested at various concentrations and temperatures, as well as recommendations for the use of their resins. The areas of suitability for polyester, vinyl ester, and epoxy resins are described in the section "Matrix Choices" in this article.

Chemical and corrosion attack can occur at the product surface or end. The presence of a resin-rich barrier layer on the surface can provide a greater degree of corrosion resistance. Although this is readily achievable in the wet lay-up or filament-winding process, the nature of the pultrusion process requires a high fiber volume and significant molding pressure to fill the die volume and to minimize porosity. To achieve a resin-rich surface, a synthetic veil or mat that is typically of polyester fiber is used on the surface of the product when molded. The layer can range from 0.15 to 1.0 mm (0.005 to 0.040 in.) thick, depending on the thickness of the material used.

The end cut of the profile is particularly vulnerable to corrosion because fibers are exposed to the environment. Any matrix crazing or resin-fiber debonding area will promote wicking of the chemical along the fiber surface. Therefore, it is a common practice to dip-coat the end cuts of pultruded profiles to seal them from corrosive attack. If this is not done, the corrosion resistance of the fiber itself becomes an important consideration because the resin does not effectively protect the fiber from attack along the fiber-resin interface.

Table 4 Material properties comparison

	Tensile strength		Rigidity		Flexural strength	
Material	MPa	ksi	GPa	10^6 psi	MPa	ksi
Wood						
Maple	100	15	12.4	1.80	55	8
Pine	60	9	12.1	1.75	35	5
Thermoplastics						
Reinforced (typical)	55	8	3.4	0.5	55	8
Glass reinforced (typical)	100	15	6.9	1.0	140	20
FRP pultrusions						
50% mat and roving	280	40	21	3.0	210	30
70% roving only	690	100	41	6.0	550	80
Metals						
Aluminum	280	40	70	10	280	40
Steel	690	100	210	30	690	100

Table 5 Equivalent thickness factor of FRP pultrusion relative to traditional structural materials

Pultrusion construction	Specific gravity	Steel			Aluminum			Wood		
		Tensile strength	Rigidity	Flexural strength	Tensile strength	Rigidity	Flexural strength	Tensile strength	Rigidity	Flexural strength
50% mat and roving	1.85	2.5	2.15	1.82	1.0	1.49	1.16	0.25	0.79	0.45
70% roving only	2.00	1.0	1.71	1.12	0.4	1.19	0.71	0.10	0.63	0.27

Table 6 Pultrusion product characteristics

Size	Forming guide system and equipment pulling capacity influence size limitations.
Shape	Straight, constant cross sections; some curved sections possible
Length	No limit
Reinforcements	Fiberglass, aramid fiber, carbon fiber, and thermoplastic
Resin systems	Polyester, vinyl ester, and epoxy
Reinforcement, wt% . . .	All roving, 40-80%; mat and roving, 25-50%; 55% woven roving or biaxial materials and mat, 40-70%
Mechanical strength. . . .	Medium to high, primarily unidirectional, approaching isotropic
Labor intensity	Low to medium
Mold cost	Low to medium
Production rate	Shape and thickness related

Table 7 Pultrusion design guidelines

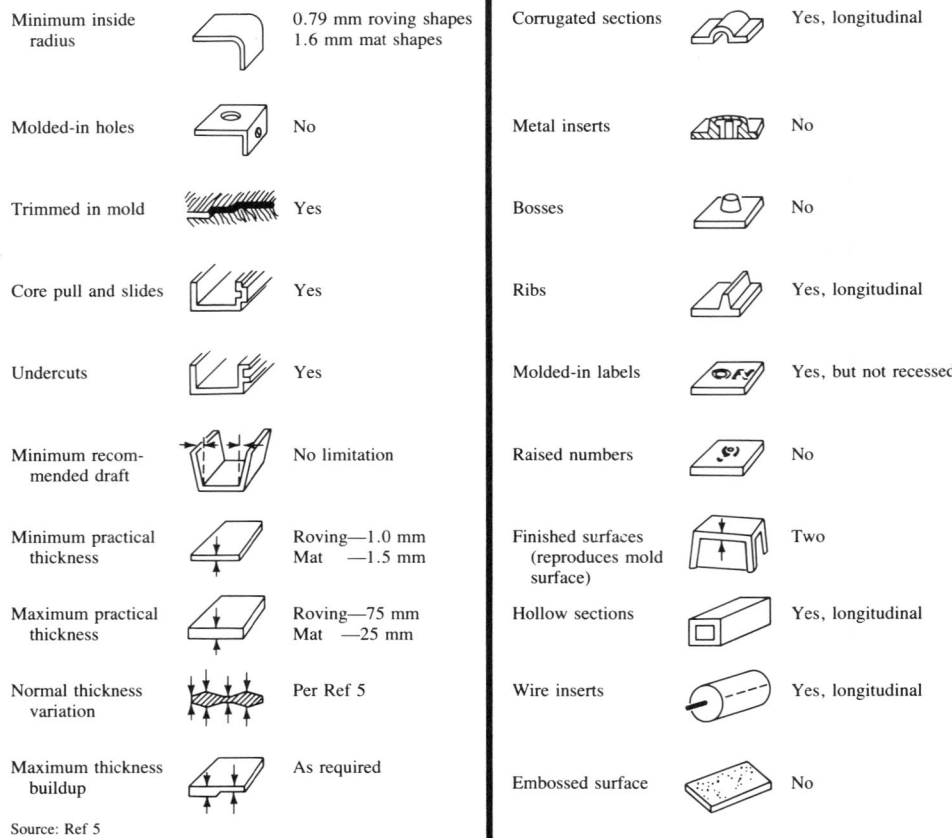

Minimum inside radius		0.79 mm roving shapes 1.6 mm mat shapes
Molded-in holes		No
Trimmed in mold		Yes
Core pull and slides		Yes
Undercuts		Yes
Minimum recommended draft		No limitation
Minimum practical thickness		Roving—1.0 mm Mat —1.5 mm
Maximum practical thickness		Roving—75 mm Mat —25 mm
Normal thickness variation		Per Ref 5
Maximum thickness buildup		As required
Corrugated sections		Yes, longitudinal
Metal inserts		No
Bosses		No
Ribs		Yes, longitudinal
Molded-in labels		Yes, but not recessed
Raised numbers		No
Finished surfaces (reproduces mold surface)		Two
Hollow sections		Yes, longitudinal
Wire inserts		Yes, longitudinal
Embossed surface		No

Source: Ref 5

Of the glass fibers available, C-glass is by far the best in all-around chemical resistance, while ECR glass (modified E-glass) is excellent in most acids. Standard E-glass is relatively inferior in acidic and alkaline environments, and A-glass is poor in water resistance. Standard E-glass is the most popular and is available in pultrusion rovings and mats; some C-glass veil is used for surface coverage. Aramid fibers are resistant to fuels, solvents, and lubricants and are superior to glass fibers in many strong acids and bases. Carbon/graphite fibers are resistant to alkaline and salt solutions, but are subject to attack by strong oxidizing agents and halogenated chemicals, particularly at elevated temperatures. It is important, therefore, to consider the chemical resistance aspects of both the reinforcement and the matrix when considering the application of any composite material in corrosive environments.

Pultrusion Design Guidelines

The design of any pultruded profile requires an adequate understanding of the material and processing contributions that affect the product. Although many material options exist for use in pultrusion, some limit the configuration of the resulting profile.

The primary characteristics of a pultruded product are shown in Table 6. Table 7 shows many of the guidelines for designing pultruded products.

There is no limit to the size of profile that can be pultruded. Standard equipment is available in product envelope sizes of 200 mm (8 in.) wide by 100 mm (4 in.) high to 760 mm (30 in.) wide by 200 mm (8 in.) high. However, several machines can produce parts up to 150 cm (60 in.) wide; others, up to 60 cm (24 in.) high. More than 90% of all profiles can be produced on machines that have envelopes 60 cm (24 in.) wide and 20 cm (8 in.) high. Larger machines

must have more power to pull the raw material through the process. Power is especially important when the resin matrix begins its gel transition to final curing and the resultant part shrinkage occurs. The larger the surface area of the part and the lower the shrinkage, the more power is needed in the equipment. Standard pulling forces are in the range of 5 to 7×10^3 kg (6 to 8 tons), but some machines are capable of up to 18×10^3 kg (20 tons) of power.

One of the limitations to shape is related to the forming area upstream of the die. When forming the raw material for hollow shapes and especially for multicell hollow shapes, the size of the forming area can increase. However, raw material must be formed around fixed steel mandrels, which limit the amount of operating space. As wall thickness increases, this problem intensifies because there is more mat and roving material to deal with.

The pultrusion process can produce virtually any shape that can be extruded. The part must be consistent in cross section over its length. Tapered shapes cannot be produced. Some curved shapes have been reported, but generally, the equipment required for this activity differs from a standard pultrusion machine. Any length can be produced that can be transported.

Machines of any width can be built to meet a market need, although continuous-strand mat is available only to 300 cm (120 in.) in width. On the narrow side, parts having total product dimensions (perimeter) less than 25.4 to 31.8 mm (1 to 1¼ in.) are difficult with mat reinforcement, because cutting a continuous-strand mat less than 100 mm (4 in.) wide eliminates the continuous nature of the strands in the product. Axially reinforced products can be produced down to 0.76 mm (0.030 in.) in diameter.

Because thermoset composites are exothermic in reaction, it is difficult to produce thick-wall products. When mat and roving are used, wall thicknesses are limited to approximately 25 mm (1 in.). Using radio frequency preheating, all roving-reinforced solid rods up to 75 mm (3 in.) or more in diameter can be pultruded. If not properly processed, these thicker parts will crack or delaminate because of lack of control of exothermic temperatures.

Any fiber reinforcement can be used in the process, although fiberglass, carbon, aramid, and thermoplastic fibers are most frequently selected. All of these materials are available in axial, biaxial, and woven fabrics, and in random-oriented mats. In addition, fine-fiber mats can be used at the surface of any part produced. Except for the differing thermal conductivity rates that should be considered, any of these materials can be used together in hybrid composites.

The pultrusion process does not impose any limits on the draft angles designed into a part. The pultruder should consider material shrinkage, especially at corners when calculating draft angles in a die versus the final profile. Undercuts can be incorporated into the design of a part by using multipiece steel dies.

Longitudinal ribs and corrugations are possible, but care should be taken in any composite part to provide sufficient radius where one wall or rib section makes a transition to another. This permits a transfer of load without a concentrated stress in a sharp corner. Another reason for using a radius is related to pultrusion process considerations. Generally, all pultrusion dies are chromium plated. It is difficult to obtain chromium that will not chip in a square corner. Additionally, any square corner or parting line will wear significantly when fiberglass materials pass over them. It is good practice to radius mat-reinforced profiles by a minimum of 1.6 mm (¹⁄₁₆ in.) and all axially reinforced rods and bars by a minimum of 0.79 mm (¹⁄₃₂ in.).

Isolated bosses cannot be produced unless they are longitudinal in nature, in which case they would be a rib. Inserts, like holes, cannot be molded in. Longitudinal inserts, such as antenna wire, plywood, or urethane foam, can be accommodated, but not without a great deal of work by the pultruder.

It is desirable to maintain a uniform wall thickness when designing a profile. However, variations in thickness within a part are possible and even common. The pultruder must produce the profile at the speed required to cure the thickest section. With good communication and planning, the end user and pultruder can design a part that will meet the functional and performance requirements of the end user and give the pultruder the best opportunity to design an efficient pultrusion set-up.

Once a composite is produced, it can be fabricated in the same manner as aluminum or steel. Because of the abrasive nature of fiberglass, continuous-rim diamond tools are recommended. Wet cutting with diamond tools reduces heat buildup and degradation of the composite material. When composites incorporating aramid fibers are fabricated, water-jet cutting is highly recommended and is the preferred technique to provide clean edge cuts.

Joining techniques include mechanical fastening and adhesive bonding. When using adhesive bonding, care must be taken to sand or otherwise prepare the surface before applying the adhesive. The guidelines provided by the adhesive manufacturer should be followed with regard to clamping time and the temperature of the adhesive under clamp pressure. Adhesive bonding, when properly done, is generally more reliable than mechanical fastening.

REFERENCES

1. Compiled data, *Handbook of Fillers and Reinforcements for Plastics*, Van Nostrand Reinhold, 1978
2. Data Sheets, OCF Corporation
3. Data Sheets, Glastic Corporation
4. Data Sheets, Morrison Molded Fiber Glass
5. "Standard Specification for Dimensional Tolerance of Thermosetting Glass-Reinforced Plastic Pultruded Shapes," D 3917, *Annual Book of ASTM Standards*, American Society for Testing and Materials

SELECTED REFERENCES

- M. Colangelo and H. Naitove, Ed., Pultrusion Process Technology: Beyond Infancy, Not Yet Mature, *Plast. Technol.*, Aug 1983
- K.J. Elias and D.K. Watkins, "EPA Requirements for Hazardous Waste and Emission Controls in a Pultrusion Facility," Paper presented at the 42nd Annual SPI Conference, Society of the Plastics Industry, Cincinnati, Feb 1987
- C. Lodge, Ed., Pultruders Pace Composites Growth, *Plast. World*, Dec 1986
- J.D. Martin, Pultrusion: The Other Process, *Plast. Eng.*, Vol 35 (No. 3), March 1979
- J.D. Martin, Pultrusion, in *Plastics Products Design Handbook—Part B, Processes and Design for Processes*, E. Miller, Ed., Marcel Dekker, 1983
- J.D. Martin, Pultrusion and Pulforming, in *Modern Plastics Encyclopedia*, McGraw-Hill, 1986
- J.D. Martin and J.E. Sumerak, "A Review of the Market for Pultruded Applications and Factors Affecting Its Growth," Paper presented at the 38th Annual SPI Conference, Society of the Plastics Industry, Houston, Feb 1983
- J.D. Martin and J.E. Sumerak, "Pultruded Composites—The Case Against Aluminum Extrusions," Paper presented at the 39th Annual SPI Conference, Society of the Plastics Industry, Houston, Jan 1984
- J.D. Martin and J.E. Sumerak, Ed., *Pultrusion Process Seminar Handbook*, 1986
- J.D. Martin and J.E. Sumerak, "The Pultrusion Process and Design Guide," Paper presented at the 42nd Annual Conference, Composites Institute, Society of the Plastics Industry, Cincinnati, Feb 1987
- N. Pennington, Ed., Reinforced Plastic Lineals Challenge Aluminum Extrusions, *Mod. Met.*, Aug 1984
- Pultruders Can Cut Deep into Aluminum Extrusion Market, *Plast. World*, June 1984
- "Pultrusion Process and Design Guide," Pultrusion Technology, Inc., 1987
- J.E. Sumerak, Understanding Pultrusion Process Variables, *Mod. Plast.*, March 1985
- J.E. Sumerak, Moving to Pultruding Advanced Composites, in *Advanced Composites*, Harcourt Brace Jovanovich, Nov/Dec 1986
- J.E. Sumerak, "Understanding Pultrusion Process Variables for the First Time," Paper presented at the 40th Annual Conference of the Reinforced Plastics/Composites Institute, Atlanta, Jan 1987
- J.E. Sumerak and J.D. Martin, "Pultrusion Process Variables and Their Effect Upon Manufacturing Capability," Paper presented at the 39th Annual SPI Conference, Society of the Plastics Industry, Houston, Jan 1984
- J.E. Sumerak and J.D. Martin, It's Time We Really Understood Pultrusion Process Variables, *Plast. Technol.*, Feb 1984
- J.E. Sumerak and J.D. Martin, "Applying Internal Temperature Measurement Data to Pultrusion Process Control," Paper presented at the 41st Annual SPI Conference, Society of the Plastics Industry, Atlanta, Jan 1986
- J.E. Sumerak and J.D. Martin, "Practical Thermal Analysis Applied to Pultruded Tubular Products," Paper presented at the 42nd Annual SPI Conference, Society of the Plastics Industry, Cincinnati, Feb 1987

Thermoplastic Matrix Processing

Martin T. Harvey, ICI Americas Inc.

THERMOPLASTIC ADVANCED COMPOSITES originally generated interest because of their perceived property advantages, which have now been established. Practical application of thermoplastic technology, however, depends on the availability of fabrication processes, joining and adhesive systems, and, most important, economical routes for making structural parts and assemblies.

Processes that are actively being developed and the corresponding economic options are discussed in detail in the sections ''Fabrication'' and ''Process Economics'' in this article. These processes, all of which have been demonstrated under laboratory conditions and many of which have been scaled up to developmental levels, have tremendous potential. It is anticipated that they will increase the rate of development of equipment, materials, and specific processing methods that are required before large production programs for high-quality thermoplastic composite parts can begin.

Although processing is a critical aspect of thermoplastic technology, a description of the resin systems being used and the forms in which they are available has been provided for those seeking an overview. It precedes the discussion of fabrication processes and economics.

Thermoplastic Resin Systems

Most thermoplastics can be used as matrices in composites, from simple filled injection molding compounds, through reinforced thermoplastic molding and stamping materials, to advanced structural composites. This article focuses only on thermoplastic advanced structural composites, which are defined as materials having a relatively high proportion of reinforcing fibers (typically 60 vol%) that are normally in a continuous unidirectional or woven form in a high-performance thermoplastic resin matrix.

Early attempts to produce carbon fiber reinforced thermoplastic composites failed, either because of poor environmental resistance of the resins then available, or from a lack of impregnation technology, especially with those materials that had good solvent resistance and therefore could not form suitable solutions for prepregging.

Since the early 1980s, technology advances in the area of resin chemistry have resulted in new families of thermoplastic advanced composites with high-performance resin systems, such as polyether etherketone (PEEK), polyphenylene sulfide (PPS), and the polyimide (Avimid K-III) and polyamideimide (Torlon) systems, having properties especially tailored to application needs of the aerospace industry. This industry continues to drive the development of advanced composites. In addition, the advanced technology impregnation systems needed to process these new high-viscosity matrix resins satisfactorily have now been developed.

The property performance of these new materials has been extensively evaluated and reported (see the Sections ''Properties of Constituent Materials'' and ''Forms and Properties of Composite Materials'' in this Volume), and the original perception of improved properties is now well accepted. In particular, the thermoplastics have established new levels of improved impact resistance, fracture toughness, and enhanced environmental resistance, without compromising other properties.

Target Properties. To meet the exacting demands of the major industries that provide advanced composites applications, that is, aerospace, transportation, industrial/chemical, and sporting goods, the thermoplastic resin matrix must meet certain target properties, identified in Table 1. These obviously vary according to individual application requirements, but the baseline is dictated by major subsonic aerospace primary and secondary structural applications. Property benefits important to this industry include damage and environmental resistance, as well as high-temperature capability. The performance of selected resins measured against these criteria is illustrated in Fig. 1 to 5.

The minimum solvent resistance and glass transition temperature requirements are difficult to specify exactly. When solvent resistance and temperature are not critical, some resins, such

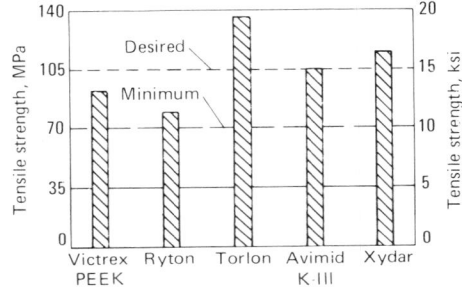

Fig. 1 Tensile strength values of selected thermoplastics

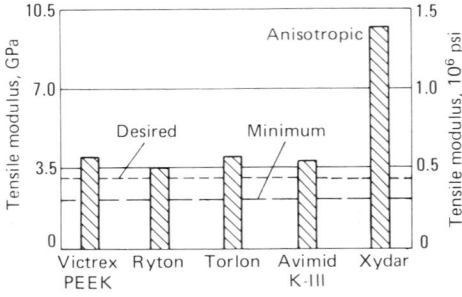

Fig. 2 Tensile modulus values of selected thermoplastics

Table 1 Target neat resin properties

Property	Minimum	Desired	Typical 350 °F (175 °C) epoxy
Tensile strength, MPa (ksi)	70 (10)	>100 (>15)	
Tensile modulus, GPa (10⁶ psi)	2.0 (0.3)	>3.0 (>0.45)	3.8 (0.55)
Ultimate strain, %	5	>10	1-2
Interlaminar fracture toughness, kJ/m² (in. · lb/in.² × 10⁻⁶)	3 (17)	>5-1.8 (>28-10)	0.1 (0.6)
Glass transition temperature, °C (°F)	121 (250)	>177 (>350)	121 (250)
Solvent resistance	...	Equivalent to epoxy	Excellent

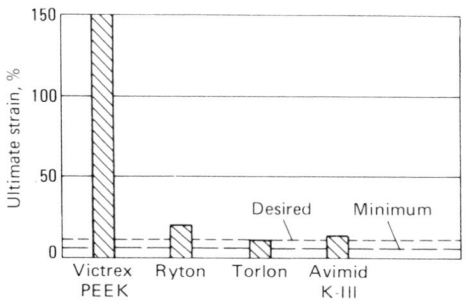

Fig. 3 Ultimate strain values of selected thermoplastics

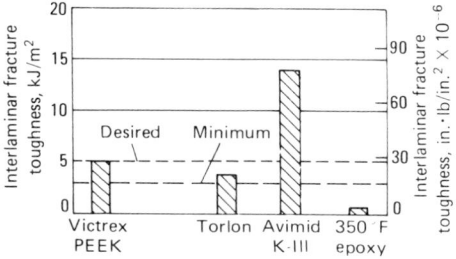

Fig. 4 Interlaminar fracture toughness (mode I) values of selected thermoplastics

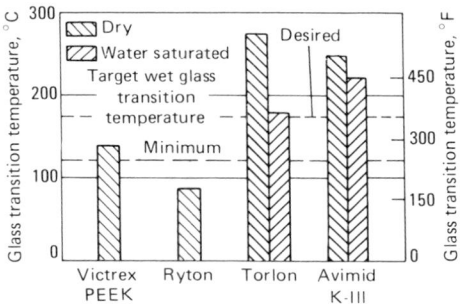

Fig. 5 Glass transition temperature values of selected thermoplastics

Table 2 Candidate matrix resins for thermoplastic advanced composites

Name	Company	Polymer	Glass transition temperature °C	°F	Melt temperature °C	°F
Semicrystalline						
Victrex PEEK.......	ICI	Polyether etherketone	143	290	343	650
Ryton	Phillips	Polyphenylene sulfide	88	190	290	555
HTX	ICI	Poly aromatic ketone	205	400	358	675
Ultra Pek	BASF	Polyether ketone				
		Etherketone ketone	172	340	372	700
Liquid crystal						
Xydar	Dartco	Polyester			415	780
Vectra	Celanese	Polyester			415	780
Amorphous thermoplastic						
Udel	Amoco	Polysulfone	190	375	...	...
Victrex PES.......	ICI	Polyether sulfone	230	445	...	...
Ultem	GE	Polyetherimide	215	420	...	...
Various	Various	Polycarbonate	140-150	285-300	...	...
PASII	Phillips	Poly aromatic sulfide	204	400	...	...
Pseudothermoplastics						
Avimid KIII........	DuPont	Polyimide	255	490	...	...
Torlon...........	AMOCO	Polyamideimide	275	525	...	...
LARC-TPI.........	NASA	Polyimide	255	490	...	...

ling volatiles to achieve target physical properties. Lower initial viscosity allows easier impregnation, although the need to complete chemistry during processing may complicate fabrication or limit subsequent forming or processing operations.

The choice of material from one of these categories is dictated as much by the design of the structure being manufactured, and thus by the fabrication process needed to achieve that shape, as by the intrinsic properties of the composite itself.

Table 2 identifies candidate matrix resins, some of which are still in the early stages of commercial development. The semicrystalline and liquid crystal polymers listed are insoluble in all but the most exotic solvents and are generally hot-melt impregnated. Processes used include either preimpregnation or "postimpregnation" (that is, mixing resin and reinforcement in, for example, cowoven fibers for subsequent melting and consolidation in the fabrication process). The need to keep viscosity (and therefore molecular weight) low enough for impregnation may necessitate a compromise in properties, especially mechanical and impact properties, although some novel processes have been developed in relation to preimpregnated forms.

Amorphous polymers may be hot-melt impregnated or solution impregnated, which is more advantageous, although removing final traces of solvent can be difficult in some processes. Obviously, the environmental resistance of this class to potential solvents is limited.

The pseudothermoplastics generally have too high a viscosity in their final polymeric form for melt impregnation, and their solvent resistance precludes solution impregnation. Monomeric solution or prepolymer solution impregnation processes are normally used, with the final chemistry (polymerization, chain extension, imidization, and/or solvent removal) taking place during the fabrication process.

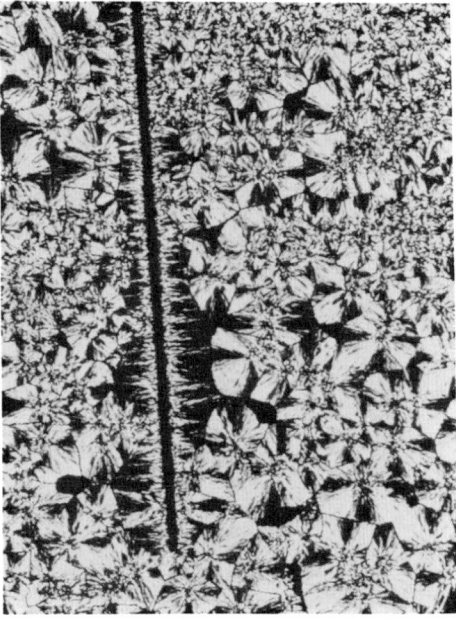

Fig. 6 Single graphite fiber in semicrystalline PEEK matrix

Interface. The major task of the thermoplastic resin, which is to transfer stress from fiber to fiber and hold the composite together, depends not only on the resin and fiber themselves, but also on the interface between them. This involves a very complex interaction of such factors as surface energy, wetting, chemical bonding, mechanical bonding, surface treatment, sizing, and molecular order. These are interactions within the resin phase, in the reinforcing phase, and in an "interphase," which is an area of considerable synergy between those two main phases. This is illustrated by, but not limited to, the "transcrystallinity," or epitaxial, or surface crystallinization of matrix resin, nucleated by the surface molecular order in some graphite fibers, as shown in Fig. 6.

as polycarbonate and polysulfone, that meet all other requirements, have found limited applicability, especially in less-demanding applications (for example, sporting goods).

Resin Categories. The main resin systems being used today fall into two broad categories. One is conventional thermoplastics, in which the chemistry is complete and processing (impregnation and fabrication) involves reversible physical change, such as heating, melting, and cooling to solidify. These thermoplastics are difficult to impregnate because of high viscosity, but are easy to fabricate by means of the heating, melting, consolidating, shaping, and cooling process. Conventional thermoplastics can be further divided into semicrystalline, liquid crystal, and amorphous systems, each of which has benefits and drawbacks.

The other category is pseudothermoplastics, in which chemistry continues during processing, increasing molecular weight and/or expel-

Table 3 Selected process factors for carbon-PEEK composite

	Area/width limits	Weight/thickness	Properties	Economics	Advantages/disadvantages
Preimpregnated					
Woven fabric (APC-2 woven)	460 mm (18 in.) only Wider very difficult	216 g/m² max	~APC-2	Difficult to prepreg	Thin skins, good props, limited widths, no cold drape
Woven prepreg tow (APC-2 tow 12K)	1780 mm (70 in.)	300 g/m²	APC-2	Simple weaving	Large area, thin skins, good props, easy consolidation, no cold drape
Postimpregnated					
Film stacked (PEEK film/woven carbon fiber fabric)	Large area limited only by weaving loom and consolidation press size	200-300 g/m²	Poorer ↑ 90% push and pull 70% bend 50% impact ↑	Simple weaving and film	Straightforward, large area, difficult impregnation, flat sheet only, no drape
Slit film (cowoven slit PEEK film)				Slow weaving Expensive slit tape	Large area, limited drape, difficult impregnation
Cowoven (PEEK monofill or multifill)				Simple weaving	Large area, limited drape with monofill, difficult impregnation
Comingled and woven (PEEK multifill)				Four-step process	Large area, good drape, improved impregnation
Powder impregnated (fine PEEK powder trapped in tow or fabric		200-500 g/m² (0.7-1.6 oz/ft²)	Poor		Better impregnation; poor polymer weight, function, and control
3D woven preform (knitted or Z-stitched thick fabric or true 3D)		Very thick or true 3D		Expensive	May be only way of producing, for example, pultrusion

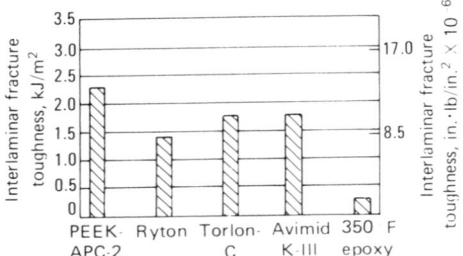

Fig. 7 Effect of polymer type on interlaminar fracture toughness, mode I

Process Factors. The performance of the final composite is also influenced considerably by the impregnation process and subsequent fabrication. Thus, even the same resin and fiber combination, if put together by different processes, can exhibit widely different properties. This is most noticeable in the high-viscosity, nonsolvatable thermoplastics and is probably related to processing dynamics, in which shear mixing might be necessary to effect complete homogeneity and fiber wetting. Table 3 illustrates such a trend for a carbon-PEEK composite, though there are other factors, such as sizing, weaving damage, and so forth, contained in those results.

Thus, there are many factors and compromises involved in choosing a thermoplastic advanced structural composite material for a particular application, including properties, performance, product form availability, fabrication processes, and overall economics.

Although the neat resin properties serve as a guide, they alone cannot be accepted as definitive of final properties, as can be seen by comparing the fracture toughness performance of neat resin and composite forms (Fig. 4 and 7).

Product Forms

As already described, product forms are dictated as much by the resin systems and the processes they lend themselves to as by the form that is required for the subsequent fabrication process and part configuration.

Thus, all the product forms used in thermosetting systems have similar, though not identical, counterparts in the thermoplastic field. Unidirectional tape and tow, woven forms, and three-dimensional knitted preforms that use carbon, glass, quartz, aluminum silicate, or organic fibers are all necessary to meet application requirements.

Product forms can be classified as preimpregnated, in which the fibers are completely wetted and fully impregnated by the resin, forming a continuous phase, or postimpregnated, in which the fibers and resin are merely in close physical juxtaposition and can move interdependently until fused during later processing.

Examples of preimpregnated forms include unidirectional tow and tape, woven fabric, and woven prepreg tow, while postimpregnated forms include film-stacked, cowoven carbon-thermoplastic fibers, comingled carbon-thermoplastic fibers, and thermoplastic powder impregnated woven cloth.

Resins that can be converted to film, powder, or fibrous forms find application in both preimpregnated and postimpregnated systems, whereas others, such as the pseudothermoplastics, are limited to preimpregnation. The intrinsic properties of the resin, such as melt viscosity and solubility, are a major factor in dictating the type of prepreg process: Those resins that are soluble lend themselves to solution impregnation, while those with low melt viscosity are suited to melt impregnation. Others will require monomeric or prepolymer impregnation.

Prepreg Forms

Unidirectional Tows and Tapes. Almost all thermoplastic resin systems are available in carbon fiber single-tow form (3K, 6K, and 12K tows) and in unidirectional tape in 150-mm (6-in.) or 300-mm (12-in.) widths, in an areal weight range from 80 g/m² to 190 g/m². The typical industry standard is 145 g/m² and 0.125 mm (0.005 in.) thick. Lengths range up to 1000 m (3000 ft) for single tow and are typically 50 m (150 ft) for tape.

The conventional thermoplastics produce a prepreg with a "boardy," tack-free feel (like thin cardboard), while the pseudothermoplastics are more pliable and usually have some tack imparted by the solvent still present at this stage.

Pseudothermoplastics and the amorphous, solvent impregnable resins are available in wider widths up to 1.5 cm (60 in.), although a paper/polyethylene carrier system or "tied unidirectional" form may be required to maintain the integrity of the pseudothermoplastic systems during handling. The boardy thermoplastic prepreg tape can be butt-welded side by side in a single, off-line operation to provide any desired width in multiples of the original tape width.

Most of the commercially available prepreg is presently based on standard carbon fiber, although intermediate modulus (IM) fiber systems are becoming available, and experimental quantities of high modulus (HM) material have been made.

Glass fiber forms are available, although their ultimate performance currently seems to be limited by interfacial problems associated with the hydrolytic stability of the glass surface. Experimental samples of quartz, aluminum silicate, and other inorganic fibers are available in many systems. Aramid fibers have also been prepregged successfully in pseudothermoplastic systems and by solution impregnation; however, the high temperature required in melt impregnation of a resin such as PEEK causes some thermal degradation of the aramid.

Preimpregnated Fabrics. Wide fabric forms using carbon, glass, and aramid fibers are available in pseudothermoplastic and amorphous thermoplastic solvent impregnated systems in which viscosity and impregnation temperature are both low.

However, the high-viscosity high-temperature regime of melt impregnation required for semicrystalline and liquid crystal systems makes it difficult to impregnate fabrics directly by this route, especially at the widths required. Narrow material has been produced on an experimental basis, but difficulties in scaling up are considered prohibitive.

A novel route to creating wide fabrics for semicrystalline and liquid crystal based systems is provided by weaving prepreg unidirectional tow to give a closely woven rattanlike product with some limited drapability and other wide-fabric properties, along with the property advantages of preimpregnation.

Postimpregnated Forms

There appears to be a natural hierarchy in terms of ease of postimpregnation and translation of properties, based on the physical intimacy of the two phases. For example, film stacking, in which alternate layers of polymeric film and (usually) woven fabric are fused and consolidated, requires long times and high pressures above the melting point, typically 2 h at 3.5 to 7 MPa (0.5 to 1 ksi), to penetrate the average 30- to 60-filament-deep fiber bundle, and the wetting and properties achieved are less than perfect.

Slitting the film to narrow tapes and coweaving with the reinforcing fiber provides some improvement, although tape shrinkage during final consolidation can cause fiber distortion unless the tape is constrained by a mold. Monofilament or multifilament coweaving can provide further improvements, especially in fine-weave constructions. Comingling fine multifilaments (20 μm, or 790 μin.) with the reinforcing fibers can give almost perfect wetting and yield a translation of fiber properties that is close to that of preimpregnated forms. Powder impregnation processes, in which the powder is trapped between the filaments of individual tows in a fluidized-bed technique by electrostatic charge, provide good results, es-

pecially when very fine powders (micron diameters) are available.

With the exception of film stacking, the postimpregnated forms described above lend themselves to weaving, braiding, and knitting processes, which allow a wide range of fabric constructions, widths, thicknesses, and shapes for subsequent fabrication into parts.

Drapability for laying up complex contoured parts is the primary benefit, although it has to be balanced against a generally more complex fabrication process.

Although most current work has involved carbon fiber reinforcement, glass and aramid fibers can be used, where temperatures allow, although the higher-modulus and more brittle fibers may suffer damage during weaving.

Fabrication

During the early development of the plastics industry, the economic benefits of thermoplastic resins were slow to emerge and were almost entirely derived from the use of new manufacturing processes, such as injection molding and extrusion, that used the rapid heat-to-melt—shape—cool-to-solid automated process that the absence of a chemical curing process allowed.

Similarly, in the development of thermoplastic advanced composites, automated processes must exist before perceived economic benefits can become a reality.

However, the fact that advanced composites have high volume fractions of continuous, inelastic reinforcing fibers presents a challenge to shaping processes that requires the development of some novel techniques. In addition, it is this particular manufacturing stage that will dictate overall economics.

For purposes of discussing fabrication processes, thermoplastic composites again have to be considered in two separate categories: true thermoplastics, in which there is no chemical change during fabrication, and pseudothermoplastics, in which chemistry continues during fabrication to increase molecular weight and/or expel volatiles.

Because forming and fabrication of most of the pseudothermoplastics require careful control of heating rates, temperature, pressure, and time to achieve the necessary chemistry, manufacturing processes are similar to thermosetting fabrication techniques, such as bagging and autoclaving. However, recent developments are bringing pseudothermoplastic forms closer to true thermoplastic behavior; therefore, many of the processes, or modifications of them, which are described in this article, could become applicable to thermoplastics per se.

The true thermoplastics, in which chemistry is complete, such as PEEK and PPS, lend themselves to rapid, automated processing because only physical changes—melting, consolidating, shaping, and solidifying—are required, and these can be achieved very rapidly (in a matter of milliseconds in some processes).

Influence of Product Form. The fabrication process is usually dictated by both the raw material product form and the required part configuration. The differentiation between pre- and postimpregnated forms is one of the basic determining factors in both process requirements and overall economics.

Preimpregnated forms, in which the carbon fiber is completely wetted and fully impregnated with the matrix resin, lend themselves to the rapid, automated manufacturing processes often perceived to be the key to thermoplastic composite competitiveness. Fusing plies to make a finished part requires simply heating to the melting point, applying contact pressure, and cooling to solidify, which is a rapid, low-energy, and inexpensive tooling concept. However, these product forms are stiff and boardy, lacking the drapability of thermosets, which limits cold lay-up to either flat or mildly (single-) contoured parts. Therefore, manufacturing processes generally have to start from flat, unconsolidated lay-ups or preconsolidated sheet and be capable of creating the desired shape and consolidation pressure at the forming temperature of the thermoplastic composite.

By contrast, postimpregnated forms have the drapability and flexibility expected of fabrics, but require much longer times above the melting point and higher pressures to effect impregnation during the shaping process. Because the viscosity of thermoplastics is orders of magnitude higher than that of a B-stage thermoset, impregnation of the many very fine carbon fibers is slow and gradual. Care must be taken in processing these materials to prevent both fiber damage and poor fiber wet-out, which can reduce mechanical properties. Fiber damage is often associated with weaving and high-pressure processing. Poor fiber wet-out can occur because of the absence of the dynamics of shear and flow found in melt pultrusion impregnation processes (see Table 3). Because of time/temperature/pressure requirements, the postimpregnated forms generally have to be processed by matched metal molding press or pultrusion techniques to provide the appropriate impregnation environment.

Manufacturing processes using prepreg forms fall into two main categories: those processes that use preconsolidated sheet feedstock, such as hydro rubber forming, press forming, roll forming, and matched-die molding; and those that use prepreg tape and tow feedstock, such as filament winding, tape laying, braiding, pultrusion, autoclave and vacuum bag molding, and diaphragm forming.

One of the ground rules that covers all advanced composites is that waste has to be minimized when expensive raw materials are used. This encourages the use of additive processes in which parts are built up layer by layer, piece by piece, with the addition of pad-ups and stiffeners where required. Additive processes contrast with the traditional subtractive manufacturing techniques used with metals that re-

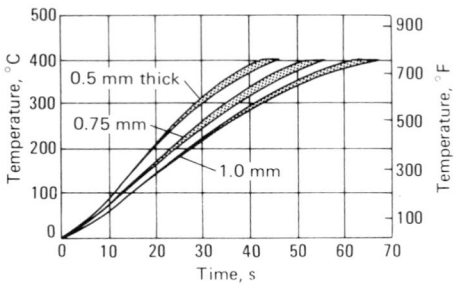

Fig. 8 Typical thermoplastic heating curves for various blank thicknesses

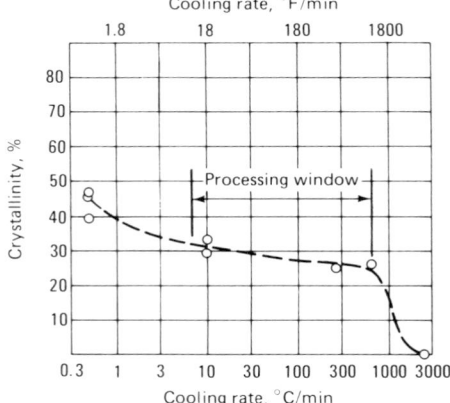

Fig. 9 Cooling-rate processing window for thermoplastic composite

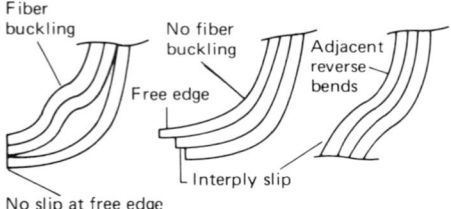

Fig. 10 Interply slippage

quire machining down from ingots or chemical etching.

Heating Methods. The carbon fiber in a thermoplastic composite is a good black-body absorber and can be heated rapidly in a number of ways. Radiant heating using quartz infrared heaters is the most effective method for preconsolidated sheet. Ideally, heating should be applied simultaneously from both sides, using 75 kW/m² (7.5 kW/ft²) energy density. To effect this, the sheet can be supported on a fine wire shelf midway between two banks of quartz infrared heaters placed approximately 300 mm (12 in.) from the material surface. With such an arrangement, heating takes approximately 1 min (see Fig. 8).

Cooling/Morphology. Cooling rates for shaped parts also require consideration, espe-

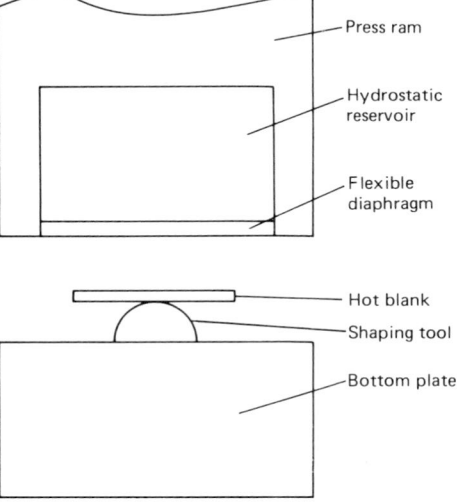

Fig. 11 Hydroforming

cially in the semicrystalline materials, such as PEEK and PPS, in which cooling rate dictates the morphology of the matrix, which in turn determines such properties as toughness and solvent resistance. Too-rapid cooling can result in amorphous material and reduce solvency resistance, whereas long periods at temperatures above the maximum crystallization rate (T_c) can cause overcrystallization and reduce toughness. Fortunately, there is usually a large cooling-rate window that can easily be accommodated in most processes (Fig. 9). Otherwise, postannealing to achieve the desired morphology may be required.

Shaping Mechanisms. Because the continuous carbon fiber reinforcement is essentially inextensible, shaping processes must be capable of accommodating fiber movement. One of the main mechanisms in sheet shaping is interply slippage (Fig. 10), which only occurs above the matrix melting point.

Hydroforming and rubber block forming are similar processes in which sheet material is formed to the shape of a single-sided mold by the application of pressure, either by a hydraulically pressurized rubber diaphragm (hydroforming) or by a mechanically pressurized and deformed rubber block.

Hydroforming (Fig. 11) is a well-known metal-forming process that is widely used in the aircraft industry, and has been used to successfully shape thermoplastic composites using preheated blanks and a standard process.

The advantages of this process are that it can, for a limited-depth draw, accommodate the interply slip that is required to avoid fiber wrinkling in areas being stretched, and it allows some thickening by fiber bunching in those areas (such as flanges) being compressed. It also applies a more gradual folding and feeding action during shaping than can be achieved in matched-die molding. The compliant pressurizing medium applies a relatively even pressure

over the surface of the molding, minimizing fiber damage in high spots and sharp corners.

Infrared heating has proved to be useful for heating the blanks to forming temperatures of up to 380 °C (720 °F), but any means of heating that is capable of reaching this temperature could be used. Because the blank must remain hot until the moment of forming, it must be transferred rapidly from the oven to the mold, typically in less than 10 to 15 s.

The same tooling is used for hydroforming composites as is used for conventional sheet metal forming, although because of the much lower forming pressures required, molds made from aluminum and aluminum-filled epoxy have also been used. Cold aluminum alloy tools can be used; however, depending on the nature of the formed item and other process details, the use of molds heated to 80 to 150 °C (176 to 300 °F) may be a better choice for reducing premature chilling of the blank before forming. This is especially true for thin laminates.

The mold should have a well-polished surface to allow blank slippage during forming, and the use of suitably positioned vent holes of 0.75-mm (0.030-in.) diam is recommended, with adequate channels leading away from the vents. This ensures evacuation of all the air trapped between the blank and the mold and will decrease the number of surface blisters caused by entrapped air.

Roll Forming. Roll forming is a widely used metal-forming process that can produce long, structural sections such as top-hat or Z stiffeners. Thermoplastic composites have been successfully roll formed at production speeds of up to 15 m/min (50 ft/min) using a standard cold-rolling process with preheated blank.

Strip feedstock must be heated to shaping temperature immediately prior to forming. This can be achieved either by using conveying tunnel ovens with infrared elements or by hot-air heating. Tunnel length is dictated by

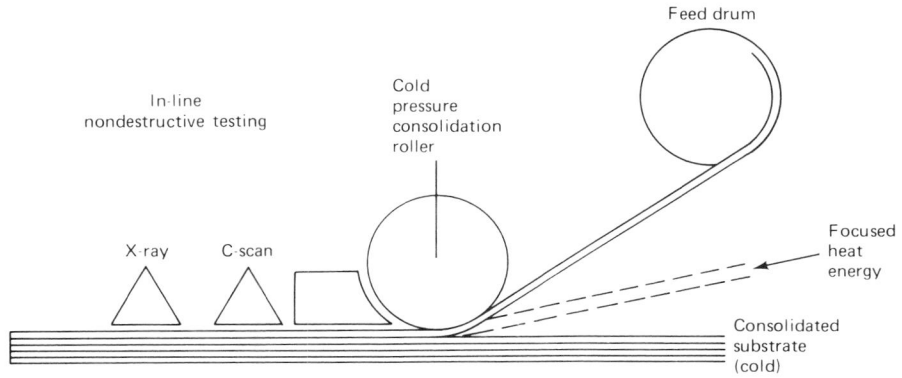

Fig. 12 Automated tape-laying process

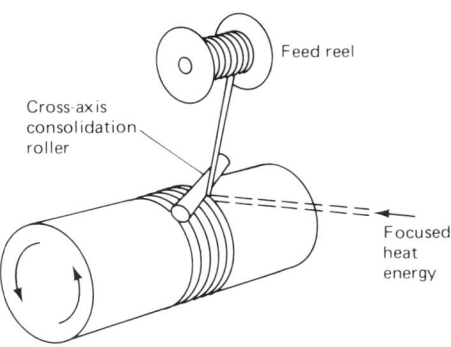

Fig. 13 Thermoplastic filament winding with continuous consolidation

strip thickness, rolling speed, and residence time. Selected masking of the blank is helpful when using infrared heating to produce cold (hence, stiff) areas to aid alignment and feeding to the forming rolls.

A minimum of five shaping rolls is recommended (four to form and one to repeat and cool the full section) at intervals of 150 mm (6 in.). It is desirable to be able to drive top and bottom rolls separately to reduce unwanted interlayer slip within the hot sheet. Alignment and support devices are necessary for the preheated blank before the first stand, as well as between the first and second stands.

No roll lubrication or surface release agent is required during roll forming because a smooth, polished cold roll will not encourage adhesion. The setting of the gaps between the mating faces of the rolls, which is dependent on the blank composition and the section formed, is of critical importance. Typical starting values for the roll gap are 0.075 mm (0.003 in.) less than the average thickness of the cold blank. Extra measures are necessary to maintain side compaction on vertical flanges. Rolled sections have been circled to a 4-m (12-ft) diam by the use of a final stripper plate.

Pultrusion through a hot die can be used to reform thermoplastic composite prepreg tape into other cross sections. The die design must permit heating while gradually changing the cross section as the tape proceeds through the die. Because the carbon fiber is incompressible and there is little excess resin to bleed out, thermoplastic composites have almost no compliance; therefore some compliance may have to be built into the die to maintain a constant pressure and to provide consolidation.

To avoid fiber distortion caused by die wipe, the carbon fiber in the surface layer should run in the direction of pultrusion, with the ±45° and 90° layers in the core. Braided prepreg tow should provide a good basis for the core.

Unless long residence times (slow speed, long dies) can be used, postimpregnated product forms will probably not be suitable for anything but the most simple cross sections.

Autoclave and Vacuum Bag Molding. Ther-moplastic prepreg tape requires only contact pressure for consolidation. Therefore, relatively low pressures can be used if applied in a constant, compliant manner over the entire surface area. Unlike rigid platen pressing, which requires high pressures to press out the high spots and fill in the troughs to mold a surface, with temperature-resistant compliant membranes, pressures as low as 70 kPa (10 psi) can be used for consolidation. Higher pressures may be required for some material forms, especially if shapes are complex.

The standard autoclaving and vacuum bagging techniques, which are well developed for the production of thermoset composite components, can be adapted for thermoplastics. The principal change is in the use of higher-temperature bagging and sealing materials capable of operating at up to 400 °C (750 °F). Temperatures must reach the matrix melting point during the autoclave cycle, and pressures should be maintained down to a point at which the matrix solidifies to ensure good consolidation.

Tape Laying. Large-area planar parts can be produced by a tape-laying technique that involves ironing consecutive layers of prepreg tape onto a mold surface. In this continuous process, prepreg tape is heated as it is ironed or rolled down onto a mold surface, and consolidation pressure is maintained for a few seconds while the tape cools and solidifies. Subsequent plies are welded to previously consolidated layers by further passes of the tape-laying head. With "hot-head" shoe-type tape layers, speeds of approximately 2.5 m/min (100 in./min) can be achieved. The temperature of both the prepreg tape surface and the surface of the previous layer must be above the melting point of the matrix to obtain effective welding and consolidation. However, heat capacity of the thin tape being laid is small compared to that of the mold and previous layers, and cooling to solidification occurs rapidly.

Virtually void-free consolidation has been achieved under laboratory conditions. Further development of prepreg tape quality and hot-head design is required for scale-up. Current commercial tape-laying units designed to pro-duce an intermediate sheet feedstock achieve a 5 to 8% void level, which is reduced to less than 1% on subsequent forming of the sheet to shape.

Because tape laying uses preimpregnated unidirectional tape, which is probably the most attractive material form in terms of cost and properties, it allows the designer to maximize part performance by fully using the anisotropic properties of unidirectional carbon fiber, with placement of each layer so as to provide the maximum mechanical benefit. A fully automated process, using a focused heat source such as a laser, would continuously place, weld, and consolidate parts layer by layer, at speeds up to 25 m/min (1000 in./min) (Fig. 12), with 100% quality assurance.

Suitable lasers can heat 0.13-mm (0.005-in.) thick prepreg to consolidation temperature in 60 ms, and a cold-pressure consolidation roll will cool to solidify in 900 ms for a total cycle of less than 1 s. Such a process has been demonstrated on a laboratory scale and is currently being scaled up. Ply lay-up sequence should be carefully determined to minimize panel warping due to unbalanced stresses during the tape-laying sequence.

Filament Winding. Like the tape-laying process, filament winding can use single-tow prepreg of 3K, 6K, or 12K construction. Flat tow tapes with widths of 3 to 6 mm (⅛ to ¼ in.) and lengths of 1000 m (3000 ft) are available for this purpose. The tow is heated to softening point immediately before being wound onto a mandrel. The mandrel can be heated to assist consolidation and provide a release mechanism by differential contraction, upon cooling. Consolidation pressure can be applied by either filament-winding tension or pressure from rollers (Fig. 13). Again, both surfaces must be above the melting point of the matrix at the point of contact, but cooling to solidify is nonetheless rapid because of the relatively low thermal capacity of the heated area. Proprietary machinery is being developed from existing filament-winding equipment.

Diaphragm Forming. In the diaphragm-forming process, a flat, unconsolidated lay-up is placed between two plastically deformable dia-

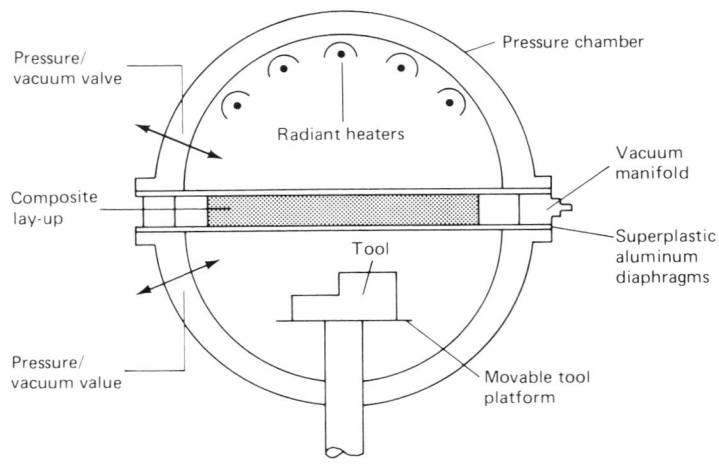

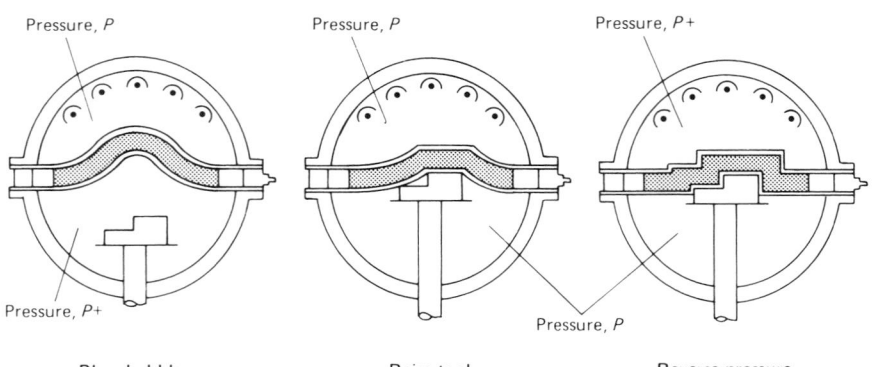

Fig. 14 Diaphragm forming

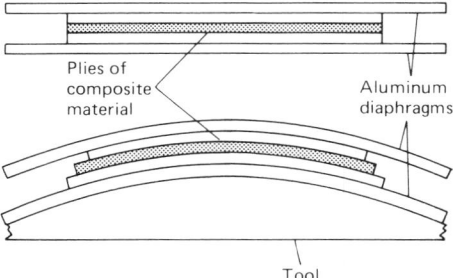

Fig. 15 Detail of diaphragm action

phragms. The entire sandwich is then simultaneously consolidated and formed to the desired shape by heating to soften the composite and by applying hydrostatic pressure, first on either side of the sandwich to consolidate, and then differentially to force the sandwich against a mold of the desired shape (Fig. 14).

This process is derived from a combination of the simple vacuum forming of thermoplastic sheet and the superplastic forming of aluminum or titanium. In those processes the material being shaped is heated and stretched elastically to form the shape and thins in the process because of stretching. The inelastic carbon fiber reinforcement in advanced composites prevents such stretching; consequently, while the diaphragms are clamped around the periphery and stretch to deform, the composite lay-up stops short of the clamping area and is free to float between the diaphragms. Hence, it is formed simultaneously with the diaphragm, but by a process of interply slip and intraply fiber reorganization of the lay-up, rather than by stretching. At the shaping temperature, the stiff and boardy prepreg becomes as soft, pliable, and drapable as a thermoset.

The sliding and stretching action of the diaphragms on the composite constrains the fibers and keeps them under a light tension from the resin viscosity and diaphragm clamping force. This prevents wrinkling and splitting of the material, as well as buckling of the fibers, providing the diaphragm itself does not wrinkle. The diaphragms stretch and smooth over all parts of the surface area simultaneously and uniformly (Fig. 15), providing a steady, overall reorganization of fibers, almost on a fiber-by-fiber basis, and virtually minimizing splits, wrinkles, and thin spots that are caused by gross material movement in some forming processes. This controlled reorganization virtually overcomes the laps-and-gaps problem of laminating tape over compound contours.

This process has been demonstrated using aluminum diaphragm superplastic forming in presses and press claves (Fig. 16). The sandwich must be heated to the resin melting point, typically 370 °C (700 °F); pressures in the range of 350 to 700 kPa (50 to 100 psi) seem adequate. Cycle times of approximately 30 to 40 min have been achieved on nonoptimized equipment; times of 10 to 20 min should be attainable on fully developed equipment. This process, of course, requires that both the diaphragm and the thermoplastic composite be capable of being shaped at the same tempera-

ture and that the diaphragm be capable of being shaped to the desired contour.

High-temperature plastic films can also be utilized as diaphragms in this process, reducing the consumable cost element and providing better surface finish. Similarly, purposely built vacuum-forming fixtures for use in existing autoclaves would minimize capital equipment costs and enable production of the larger parts that are required in the aerospace industry (Fig. 17). With relatively low pressures being applied hydrostatically, and with no requirement for tool heating (because the sandwich can be heated in isolation), the cost of tooling materials and construction could be minimized and perhaps carbon-carbon or ceramic tooling concepts could be employed.

Thus, diaphragm forming produces fully consolidated shaped parts with optional ply drop-off/pad-up and cobonded stiffeners. It uses the simplest, most economical product form—unidirectional tape—and has the potential to reduce or eliminate laps, gaps, splits, and wrinkles. It has a relatively short cycle; low costs for labor, tooling, and consumables; moderate capital expenditure; and versatility.

This process is undergoing active development at a number of equipment and fabrication companies. It has been scaled up to produce parts 3 m (10 ft) in length and is about to be used in a production application.

Automated Lay-Up. One of the benefits of stiff, boardy thermoplastic tape is that it can be continuously butt-welded edge to edge, building from an original tape width of 300 mm (12 in.) to almost any desired width, in multiples of the original, and in continuous lengths (Fig. 18). Entire plies could be automatically cut from such a wide, unidirectional product, picked and placed in the desired lay-up—a simple operation as only flat lay-ups would be needed, and the problems of contouring at this point could be avoided. Each ply is simply tack-welded to the previous one in a few spots to give the lay-up some stability.

However, cutting shapes from wide tape can be very wasteful, especially when shapes are complex, ply orientation and size vary, or drop-offs are involved. A variation of simple butt-welding has been developed in the United Kingdom that involves spiral winding onto adjustable mandrels to provide made-to-order

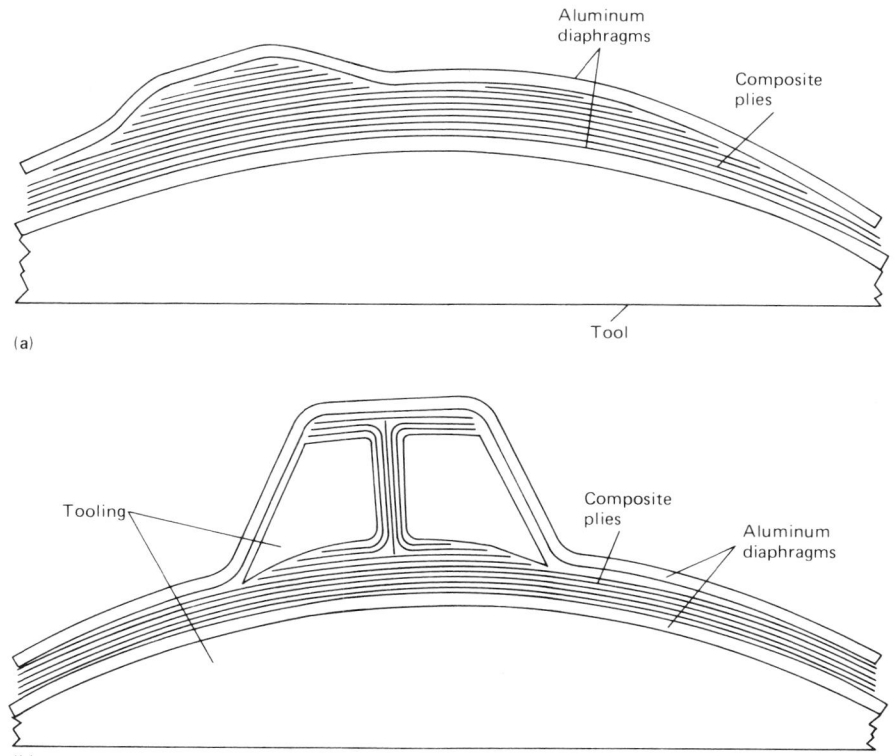

(a)

(b)

Fig. 16 Diaphragm forming. (a) Ply build-up and drop-off. (b) Diffusion-bonded stiffeners (bottom)

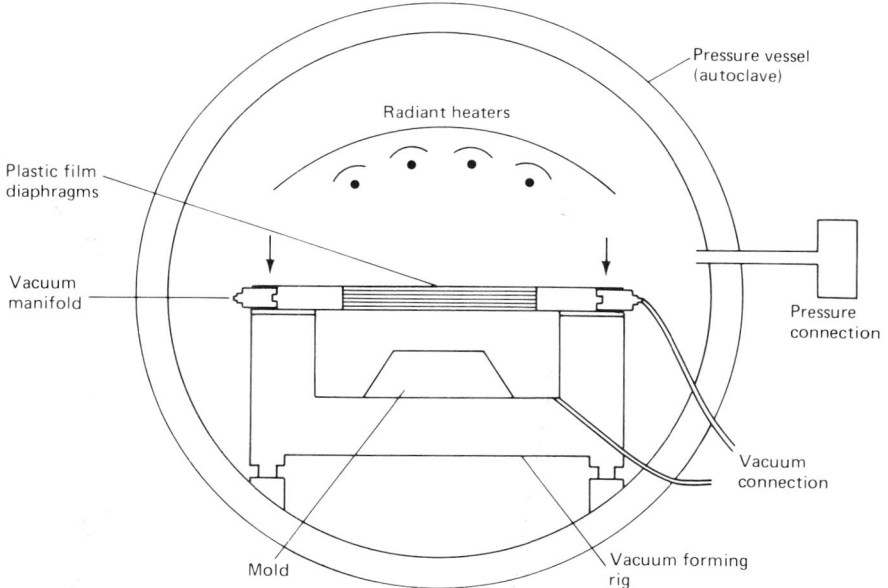

Fig. 17 Vacuum-forming fixture design for use in autoclaves

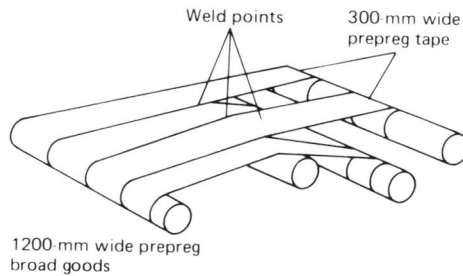

Fig. 18 Simple butt welding of thermoplastic tape

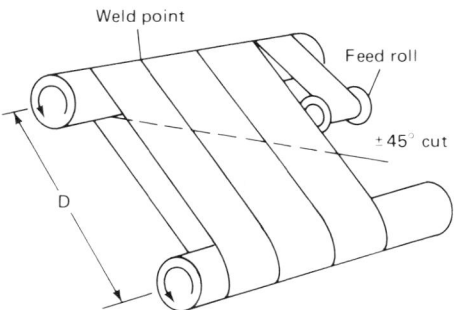

Fig. 19 Butt welding of made-to-order plies by spiral winding onto adjustable mandrels

plies that are close to the required size and orientation (Fig. 19). Standard stock rolls of prepreg tape are used, and the tape is wound in spiral fashion between two parallel rotating mandrels. The tape is continuously butt-welded as it is wound next to the previous turn, to provide a large-diameter, single-ply tube of any desired length. By cutting axially, or at 45° to the axis, 90° and ±45° plies, respectively, can be produced. By adjusting the distance between the mandrels and the number of turns applied, a wide range of made-to-measure sizes and ori- entations can be obtained with minimal waste. This provides a rapid, automated, low energy, low labor, low waste process with no con- sumables, from moderately priced equipment, to integrate with pick-and-place robotics for automated lay-up and feed to the diaphragm- forming process.

Processes for Postimpregnated Forms. Postimpregnated product forms, especially the comingled fibers and woven forms, offer cold drapability, which can provide significant labor savings when making some complex shapes. When starting from postimpregnated product forms, some additional factors must be consid- ered. These generally relate to the need to provide the conditions (temperature, pressure, time, flow) necessary to impregnate and con- solidate during the shaping cycle. The actual conditions required depend on the shape, the process being used, and the fiber-matrix inti- macy in the material form, which dictates the flow required.

Thus, film stacking (in which a film of the thermoplastic resin must be melted and must flow through an average of 30 to 60 fiber layers) requires higher pressures and longer times than comingled fibers, in which fine (<20 μm, or 800 μin.) thermoplastic fibers and the reinforcing fibers are in intimate juxtaposi- tion.

Process and shape are also important consid- erations. Matched-die molding is excellent for relatively planar parts, but may not apply suf- ficient pressure on vertical surfaces of more complex shapes. A compliant-pressure process such as autoclaving or diaphragm forming applies pressure evenly, but may not have the rigidity needed to drive resin flow.

Process temperature must obviously be above the melting point of the resin throughout the shaping cycle. In general, pressures in the range of 1400 to 7000 kPa (200 to 1000 psi) are needed to drive the resin flow, and times in the range of 30 to 120 min provide the best property results. A gradual application of pressure minimizes gross flow and air entrapment.

Final properties of parts manufactured from postimpregnated product forms depend on the effectiveness of the in-process impregnation. At best, they can approach the properties of preimpregnated forms (Table 3).

Adhesive Bonding. Thermoplastic composites for aerospace use are chosen for their environmental resistance and particularly for their resistance to solvents. Thus, some matrix materials do not lend themselves to adhesive bonding because their inert nature limits the solvent or chemical attack usually required in this process.

Although it is possible to bond with traditional epoxy adhesives and obtain reasonable lap-shear strengths, peel strength is generally low. Bonding can be improved by mechanically abrading the composite surface or etching away the surface polymer with chromic or sulfuric acid to expose the carbon fiber surface. Chemical modification of the polymer surface by corona discharge techniques or plasma etching also provides an improvement in some cases.

Fusion Bonding. Fortunately, most of the thermoplastics used in composites are also superb hot-melt adhesives, and fusion bonding techniques work well.

For instance, PEEK can be bonded to itself (and to other materials, including metals) simply by heating the two surfaces to be mated to 400 °C (750 °F) and applying contact pressure while cooling to 200 °C (390 °F) or below. This requires, of course, that the two surfaces mate closely to provide overall contact. The inclusion of a thin PEEK film (200 μm, or 800 μin.) between the surfaces to give a degree of resin richness may help to mate slightly uneven surfaces.

The thermoplastic composite surface can be heated in a number of ways. First, high-intensity infrared radiant heating to melt the surfaces of the areas to be bonded has been found to be satisfactory. This process needs to be done rapidly to avoid delaminating subsurface plies. Also, the mating surfaces must be brought together before the temperature falls below the melting point of the matrix.

Second, prepreg tape can be used as an adhesive layer between two parts. The interface is heated by applying an electric current through the carbon fibers of the prepreg interlayer. The filaments act as resistance heaters and melt the matrix *in situ*, welding the two parts together. A wire mesh can be used as an alternative to carbon fibers if additional thermoplastic resin film is added to the interface. Leaving the mesh in the interface must be

acceptable if this alternative is used. The prepreg tape must extend beyond the bonding area. The exposed ends should be treated to remove the matrix, which otherwise acts as an insulator. This can be done by burning in an oxidizing flame or by chemical etching. Electrodes should be fixed to the exposed carbon fiber. A current of approximately 6 A/cm (15 A/in.) of prepreg width at 30 V dc has been found to provide sufficient heating effect within approximately 1 min. Contact pressure should be maintained during cooling.

Another method involves induction heating of a fine wire gauze at the interface. A thin film of the matrix resin (200 μm or 800 μin.) is placed on either side of the wire gauze, and the sandwich is then placed between the composite parts to be bonded. The entire assembly is held in an induction field, where the gauze heats up and fuses the film at the interface. Moderate pressure is required to obtain strike-through of the gauze and a void-free bond line. To date, this technology has been limited to small laboratory specimens in structural composites. Carbon fibers can also be heated inductively and should be considered in the overall approach to induction joining or bonding.

Ultrasonic heating can be used both to spot-weld thermoplastic composite and to weld it continuously. The correct horn geometry must be used to suit the configuration of the weld to be made. Large bond areas are more difficult, but can be obtained by stepping the ultrasonic horn over the surface of the part for continuous welds.

Finally, hot-plate welding, often used to join large automotive components, can be used to join thermoplastic structures.

Lap-shear joints have been produced by all of the preceding thermobonding processes. Mechanical testing has demonstrated that the strength of the bonds closely approaches the shear strength of the matrix itself. Tooling approaches can become quite complex for these processes, especially if components have to be heated and moved together quickly, as is required for infrared and hot-plate welding.

Machining. A wide variety of metal-machining processes can be applied to most thermoplastic composite sheet. However, tools must be very sharp (usually diamond tipped) and cutting speeds must be high, using slow feed rates and a coolant to wash away swarf and eliminate heat buildup that could melt the thermoplastic. Because of the nonhomogeneous nature of the composite material, a rough finish that is due to fiber pick-out will occur unless these precautions are observed. The desired effect should be that of a high-speed accurate grinding operation rather than the cutting or paring action of conventional metal-machining tools.

Cutting and Sawing. Thermoplastic sheet can be cut on standard band saws and circular saws; the best results are achieved when diamond-coated blades are used. Use of coolant may be required to prevent overheating. Water-jet cut-

ting has been successfully used on thermoplastic sheet when satisfactory back-surface support is provided to prevent fiber pick-out on the rear face. Abrasive water-jet cutting has produced better results than nonabrasive methods. Circular shapes can be cut by trepanning using diamond-tipped tools.

Drilling and Countersinking. Holes can be drilled in thermoplastic sheet using diamond-tipped trepanning drills at a high cutting speed and slow feed rate. High-speed steel, carbide, or cobalt twist drills can be used, but several holes of increasing size should be drilled to achieve the final diameter. The paring action of a twist drill tends to leave a rough surface on the inside of the hole. High-speed steel is dulled quickly by the abrasive fibers, resulting in poor-quality holes.

Countersinking is best achieved by using a high-speed grinding action at slow feed rates. Threads can be tapped into thermoplastic composite sheet using conventional tapping techniques; very high bolt pull-out strengths can be obtained.

Guillotining and Stamping. Because of the toughness of thermoplastic composite, thin sections (of up to 1 mm, or 0.04 in.) can be guillotined or stamped into intricate shapes.

Process Economics

It must be emphasized that practical application of thermoplastic composites depends on economical routes to making both structural parts and assemblies. The economic factors are complex and reach well beyond simple material substitution into the areas of process costs, fabrication, assembly, and design considerations. An integrated design optimization, from raw materials to structure, will be required to achieve the full benefits of using the thermoplastic option.

The first consideration is product form costs because current market prices for thermoplastic composites reflect the development phase of the technology, rather than the cost and profit price that exists for an established product.

Most thermoplastic product forms consist of approximately 68% carbon fiber and 32% thermoplastic resin, and these represent the basic raw-material cost element. The cost structures for the various product forms differ on the basis of the number and complexity of the process steps required to produce a specific product form and on the labor, capital, energy, and efficiency associated with each step.

Industrial experience indicates that most significant manufacturing process steps cost between 1.5 and 2.5 times the raw-material input to the process. Thus, factors can be allocated, admittedly in a fairly arbitrary way, to demonstrate process scenarios.

The most basic and lowest-cost forms of fiber and resin are unsized 12K carbon fiber tow and thermoplastic resin powder, respectively. In addition, there is no intrinsic reason why high-performance thermoplastic resins

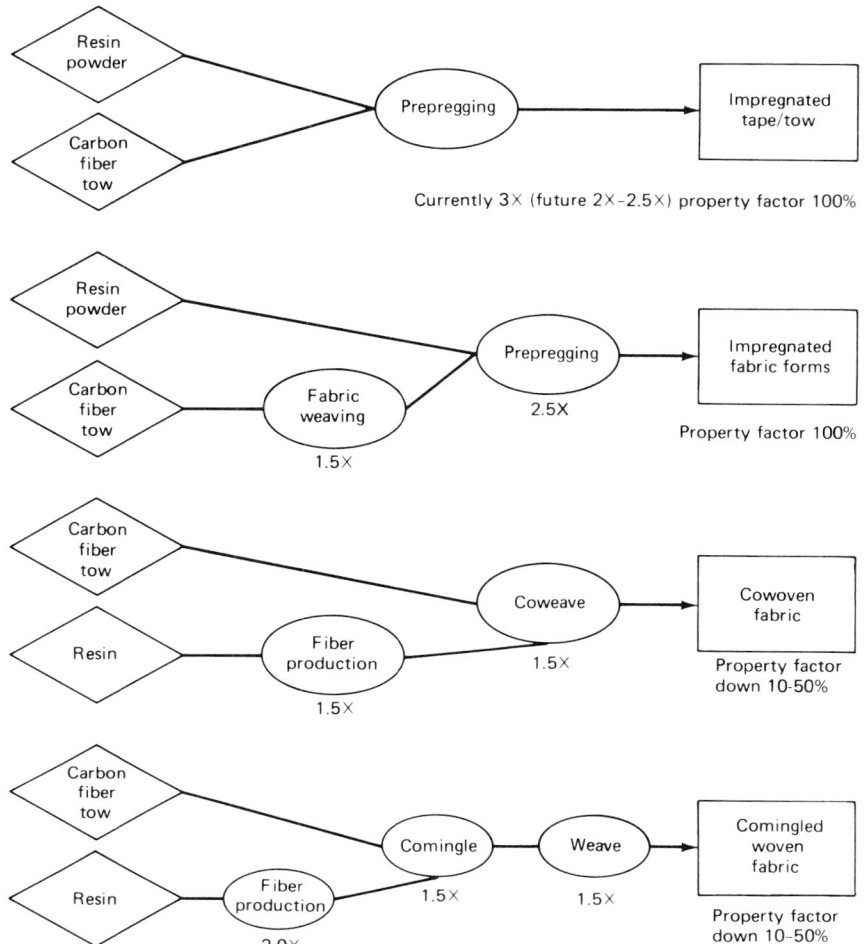

Currently 3× (future 2×-2.5×) property factor 100%

Property factor 100%

Property factor down 10-50%

Property factor down 10-50%

Fig. 20 Process steps and costs for prepreg forms

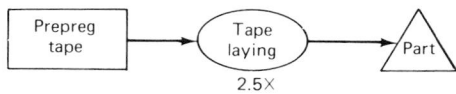

- Single-step process
- Automated
- Maximizes material usage/performance
- High capital cost
- Expensive tooling
- No laps and gaps

Fig. 21 Manufacturing process steps/costs for tape laying

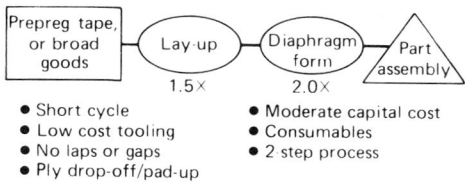

- Short cycle
- Low cost tooling
- No laps or gaps
- Ply drop-off/pad-up
- Moderate capital cost
- Consumables
- 2 step process

Fig. 22 Manufacturing process steps/costs for diaphragm forming

should be more expensive than thermosets in the long run, once economies of scale have been achieved.

With this knowledge, some rough economics models can be established in which process steps are defined and cost factors are allocated to the steps. One example is shown in Fig. 20. Adding a product form/property factor allows the classification of product forms in a cost/property performance relationship. Finally, part fabrication costs involve the consideration of material use, size and complexity of the part, labor requirements, process time and energy (heat), tooling costs, and capital equipment amortization.

Shown in Fig. 21 is a scenario of process steps and costs for tape laying. Figure 22 defines the process steps and costs for diaphragm forming. Although process diagrams can never provide a complete answer, they do show the way the number of process steps and

the complexity of each can have a major effect on economics. The key to achieving economies is to start with the simplest and least costly raw material form; minimize the number of process steps; and contain capital, energy, and labor costs. The heat-to melt—impregnate—shape—cool-to-solid process should be conducted as rapidly and efficiently as possible.

Design and assembly are also very important. However, savings in these areas are difficult to quantify. It is strongly believed that the higher fracture toughness of thermoplastics should allow material and weight savings if they are applied to structures that have previously been designed and meet a damage tolerance standard. A recent U.S. Air Force study on stiffened wing skins indicated that the structure still exceeded previous design requirements even if stiffener load pads were eliminated, thereby providing a 6% savings in material and a significant savings in labor.

Other design allowable improvements in areas such as open-hole compression/tension and fatigue could have similar results. In addition, producing monolithic assemblies by co-consolidation would reduce the number of individual parts used. Fusion bonding processes should allow easier, quicker, and more reliable assembly, eliminating the design-allowable knockdown and the cost of adhesive assembly or fasteners.

Finally, it should be noted that the (thermoset) advanced composite industry has been built upon special materials for special applications, which has led to a myriad of material specifications and qualifications. In turn, this has affected the efficiency of prepreg operations, which, combined with the burden of quality control, is having a significant effect upon costs.

Efforts toward standardizing have resulted in the National Aeronautic and Space Administration (NASA) standardized test methods and the Suppliers of Advanced Composite Materials Association (SACMA) standard specification. Adoption of these standards will be a long and difficult task, especially with thermoset materials, because of the established status quo.

Standardization is not a new idea. Because no specifications or qualifications have yet been established for thermoplastic composites, the opportunity still exists to achieve the ideal of a standardized industry specification and basic qualification, along with their inherent operational savings and resulting cost benefits.

Manufacturing Processes: Consumer Products

Co-chairman: Peter Beardmore, Ford Motor Company

Introduction

IN ANY CONSUMER PRODUCT INDUSTRY, the rate of production is a critical factor in the economics of the business. The axiom that only those manufacturing processes capable of high-speed production are considered is as true for composites as for any other type of material. In a consumer product industry, the product is designed within the constraints of high-speed manufacturing systems to minimize costs, in contrast to an industry such as aerospace, in which the product design is optimized before a manufacturing process is selected, with cost being a secondary consideration.

The use of composites for consumer products is rapidly increasing. Within the sporting goods industry, for example, the manufacture of tennis rackets, fishing rods, golf clubs, and archery bows is being revolutionized by the use of composites. Tennis rackets have undergone a remarkable transition—from wood to metal and then to a completely composite product.

Corresponding to this increased use of composites in consumer industries has been an accelerated development of appropriate manufacturing processes. Some processes are already well developed, such as filament winding, tube rolling, and pultrusion; others, such as resin transfer molding, are still being developed.

Consumer products need a variety of composite forms to satisfy requirements ranging from decorative or low performance, which require minimal mechanical properties, to high performance, which require optimized mechanical properties. Typical examples of low-performance composites are household utensils and automotive-trim components. High-performance products include automobile leaf springs, sporting goods, and artesian wells.

Composites of all types (chopped fibers, continuous fibers, glass fibers, and carbon fibers) can be fabricated into a variety of final products, ranging from the highly oriented fiber systems with high material costs used in automobile leaf springs and tennis rackets to the randomly oriented chopped-fiber systems used in sheet molding compound products such as automobile skin panels, marine engine covers, and firemen's helmets. The more demanding the product requirements, the more precise fiber control must be. This consideration is critical in choosing a fabrication process. The articles that follow focus on the most important areas of composites fabrication relative to consumer products.

Injection molding is a well-developed high-speed process with many variations, all based on the injection of a fluid plastic material into a closed mold, usually of the multicavity type. This injection process is used in a wide spectrum of mass production industries, ranging from toy production to automobile bumper fabrication.

Structural compression molding also represents a mature process for nonstructural and semistructural applications, such as skin panels for automobiles; however, because this fabrication technique is also capable of generating fully structural parts and components, such as automobile floor pans, attention is being given to this emerging aspect of the technology.

Resin transfer molding is widely used in the fabrication of boats, specialty vehicles, and a myriad of other products, all of which are in low-volume production industries. To use this procedure in high-volume applications requires process developments. Developments are underway to automate the fiber preform process required for mold insertion before resin injec-

tion. This and other developments to achieve high-production capability in this promising area will be described in the articles that follow. The extensive interest in this particular process is due to the inherent ability of the technique to control fiber placement precisely, and thus control orientation, resulting in optimization of properties and economics.

Rolled tube manufacture of sporting goods is a highly specialized area that is economically important but has somewhat different constraints than the other segments of consumer products. Leisure-time equipment is not subjected to the same competitive price comparisons as utilitarian products and plays a greater psychological role in purchasing motivation; thus, production economics share an element of the aerospace approach in that performance and aesthetics are probably more important than cost minimization. These factors make the sporting-goods segment of the consumer product market somewhat unique. This is reflected in the extensive use of tube rolling as a basic process in the production of golf club shafts and fishing rods. Rolled tube production gives excellent fiber control and optimized properties, and though it still requires a significant amount of hand labor, the unique economics of the leisure sports industry make the process economically viable.

The following descriptions of these processes are intended to be fairly complete and to provide the reader with a good background knowledge in each area. Obviously, the use of any technique for any given product will require in-depth development, which cannot be supplied here. However, the articles do indicate process advantages and disadvantages, allowing the reader some basis to assess the potential for application.

Injection Molding

Allan D. Murray, Ford Motor Company

INJECTION MOLDING refers to a variety of processes that generally involve forcing or injecting a fluid plastic material into a closed mold (Ref 1, 2). It is differentiated from compression molding, in which plastic materials in a soft but not fluid condition are formed by transferring them into an open mold, which is then forcibly closed. This latter process is fully discussed in the article "Compression Molding" in this Volume. The injection molding process generally has the advantages, first, of being more readily automated and, second, of permitting finer part detail, in contrast to compression molding. The part and mold often can be designed so that no subsequent trimming or machining operations are required. However, not all plastic materials can be injection molded successfully; for example, there is a limit to the amount and types of fibrous reinforcement that can be incorporated in an injection molded part.

There are two basic categories of plastic injection molding: thermoplastic and thermoset. In the former, a thermoplastic material is melted and forced through an orifice, or "gate," into a relatively cool mold in which the material solidifies and from which it can then be removed. In thermoset injection molding, a reacting material is forced into a generally warm mold in which the material further polymerizes or cross-links into a solid part. The article "Injection Molding Compounds" in this Volume describes the materials used in these processes.

Thermoplastic Injection Molding

John and Isaiah Hyatt received a patent in 1872 for a plunger-type injection molding machine for molding plasticized cellulose nitrate (Celluloid) (Ref 3). Injection molding came into significant commercial use following the development of plastic screws. This development started in the 1930s, and culminated in 1952 with the invention of the reciprocating screw plasticizing injection unit by W.H. Willert (Ref 4). The reciprocating screw type of injection machine has largely supplanted the early plunger type of injection machine.

Reciprocating Screw Injection Molding Machine

Although there are many variations in thermoplastic injection molding machines, the reciprocating screw machine has become the most common type, and the following description of its operation illustrates the basic principles of all types. The functions performed by the machine include heating the plastic until it is able to flow readily under pressure, pressurizing this melt to inject it into a closed mold, holding the mold closed both during the injection and while the material is solidifying in the mold, and opening the mold to allow removal of the solid part.

Machine Description. As depicted in Fig. 1, the main components of the machine are the hopper, heated extruder barrel, reciprocating screw, nozzle, and mold clamp. The hopper feeds the unmelted thermoplastic, usually in pellet form, into the barrel. The hopper is often equipped with a desiccant-type drying system to remove moisture from plastics that are susceptible to moisture degradation. A magnet is placed in the hopper throat to remove any iron that has accidentally entered the feedstock.

In the heated extruder barrel, which contains the reciprocating screw, the thermoplastic is gradually melted by a combination of shear heating (caused by the mechanical working of the material as it is conveyed down the barrel) and heat conduction from the barrel. The thermoplastic is gradually conveyed by the rotating screw from the rear of the barrel to the front. The barrel is usually heated by band-type electrical resistance heaters fitted around its periphery. Because the amount of heat required varies with position along the barrel, the band heaters are controlled in several zones along the barrel. Some barrels are equipped about half-way down with a vent to remove gases from the melting feedstock. A vacuum is usually applied to the vent, and a specially designed screw is required to depressurize the melt in the region of the vent to keep it from extruding through the vent.

The reciprocating screw usually has three successive zones, each with specific functions: The feed zone of relatively deep screw flights conveys the unmelted plastic from the hopper throat into the heated barrel, where it begins to melt; the compression zone of decreasing flight depth, and thus volume, provides gas removal and melt densification, along with a material

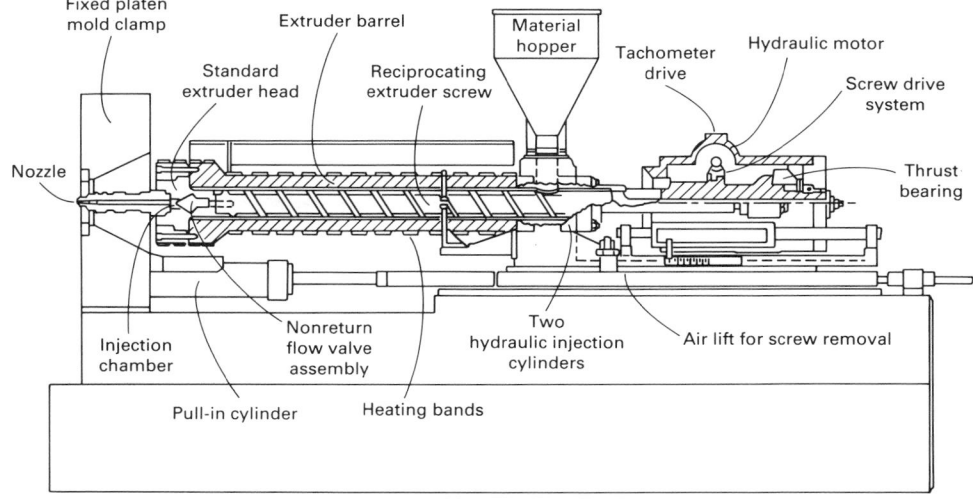

Fig. 1 The injection end of a reciprocating screw injection molding machine. Source: Ref 2

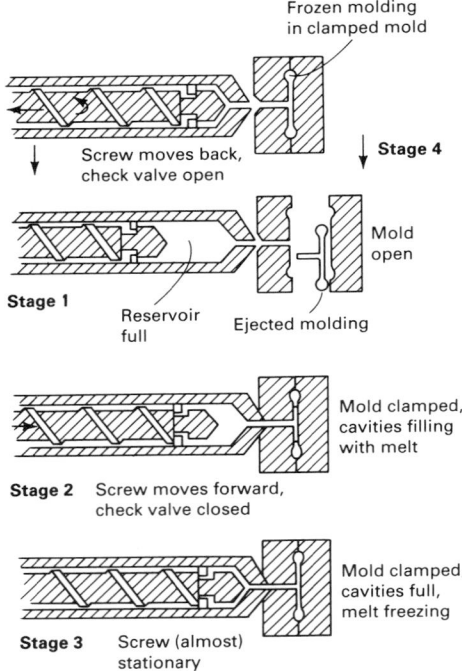

Fig. 2 Stages in the operation of a reciprocating screw injection molding machine

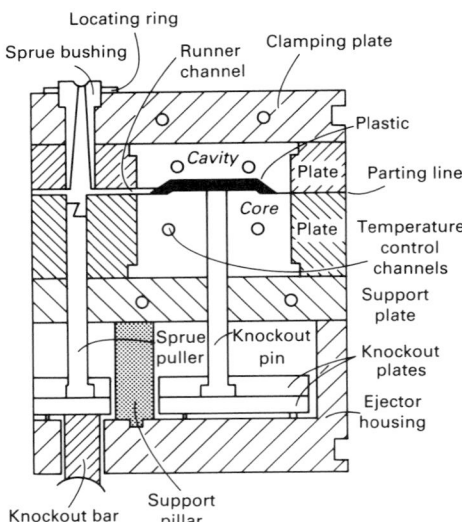

Fig. 3 Two-plate injection mold. Source: Ref 2

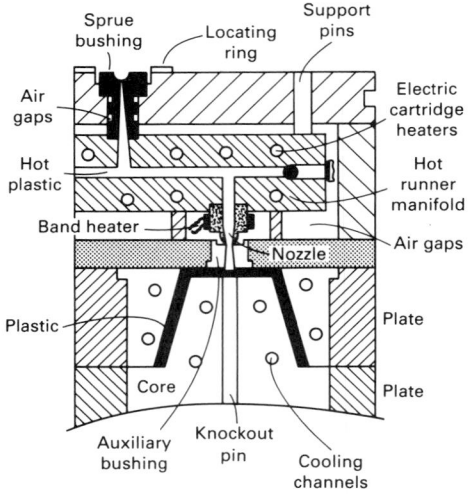

Fig. 4 Hot runner injection mold. Source: Ref 2

mixing action for better material and temperature uniformity; and the metering zone, generally of constant, shallow-flight depth, provides the final shear heating and mixing of the melt. After the melt passes through the metering section and a check valve at the end of the screw, it joins the melt pool in front of the screw. As the volume of melt in front of the screw increases, it forces the screw to the rear of the barrel against an adjustable "back pressure." This back pressure is applied hydraulically to the back end of the screw. Increasing it increases the amount of mechanical working of the feedstock. Screw rotation and feedstock melting continue until a sufficient amount of melt is available in front of the screw to fill the mold. At this stage, the screw rotation is stopped, and the machine is ready for injection. The stages in the operation of a reciprocating screw are shown in Fig. 2.

The melted thermoplastic is injected into the mold through the nozzle, under high pressure (typically 70 to 205 MPa, or 10 to 30 ksi, depending on the mold-filling resistance). Injection occurs as the screw is hydraulically forced forward in the barrel. The hydraulic cylinder is located at the rear of the screw and barrel. A check valve at the tip of the screw keeps the melt from flowing back along the screw as the screw is pushed forward. In contemporary machines, the injection rate (determined by the forward velocity of the screw) and injection pressure are closely controlled throughout the mold-filling stage.

The mold clamp, which holds the halves of the mold closed against the injection pressure of the melt, opens the mold to allow part removal after the thermoplastic has solidified. During the cooling and solidification period, the screw begins to rotate and melt new material for the next shot.

The clamping system consists of a fixed platen and a movable platen, each of which has half of the mold attached. The fixed platen (shown in Fig. 1) has a hole in its center to allow the injection cylinder nozzle to be placed in contact with the sprue bushing of the mold. A movable platen is moved along tie bars by either hydraulic or mechanical means, or a combination of both. The amount of clamping force the machine can apply, rated in tons, generally determines the size of the part that the machine can process. It is this clamping force that overcomes the injection pressure of the melt in the mold and keeps the mold halves together. The clamping force, or "tonnage," required is determined by the amount of injection pressure required to fill the mold and the projected area of the part (that is, the area perpendicular to the axis of the machine). If the machine has inadequate clamping force, the two halves of the tool will begin to separate during injection, causing the melt to squirt out, or "flash," at the mold parting line, potentially causing an incomplete mold-fill, or "short shot."

Contemporary molding machines have two or more closing speeds: a high-speed closing, requiring only a low force, followed by a slow-speed closing stage, which generates the high clamping force to close the mold firmly (stretching the tie bars) prior to injection. The clamp usually opens slowly at first, followed by a rapid traverse. These high-speed motions reduce the time of the overall process cycle.

Injection molds, in their simplest form, consist of two halves, often called the core and cavity (Fig. 3) (Ref 5, 6). A hole or sprue bushing conducts the melt from the injection nozzle through the sprue and through a gate into the mold cavity.

Gates and Runners. Commonly, more than one gate is used to deliver material into the mold, and each gate is fed by a "runner" channel leading from the sprue to the gate. After the mold is filled and the melt has solidified, the material in the sprue and runners also solidifies and must be removed with the part before the next shot. The sprue and runner material are usually reground in a granulator machine and fed back, along with virgin material, into the injection cylinder for reuse. To reduce the amount of material that must be recycled, some molds are equipped with a hot runner manifold (Fig. 4), which keeps the material in the sprue and runners molten to become a part of the next shot.

With larger and more complex parts, it is common to use more than one gate for injecting material into the mold. The configuration of the runners and gates determines the way the mold is filled. This can change the amount of injection pressure required, the location and condition of "knit" lines, the orientation of material flow (which can affect mechanical properties), and the ability to vent air from the mold. Knit lines are surfaces along which the flow fronts meet, and are generally weaker (Ref 7). Non-uniform mold filling, caused by poor gate locations, can result in overpacking the mold in the regions which fill first, causing residual stresses and possible warpage in the part. Computer programs are now available to assist in mold design and the layout of runners and gates for effective mold filling at minimum pressure.

Mold Cooling. With thermoplastic injection molding, a method of cooling is usually incorporated into the mold to speed the solidification of the plastic. This usually consists of holes bored in each half of the mold through which a heat exchange fluid, usually, water, can circulate. The mold temperature is usually controlled above room temperature. The optimum mold temperature depends on the type of plastic being molded, but typical mold temperatures for the more common thermoplastics vary from 40 to 120 °C (100 to 250 °F).

Material Shrinkage. All thermoplastic solidification is accompanied by a volumetric shrinkage. For crystalline plastics, the shrinkage is associated with crystallization. For amorphous plastics, the shrinkage is generally less and is associated with the glass transition. In both cases, the amount of shrinkage depends on various processing parameters, including the mold temperature and rate of cooling of the melt. This shrinkage can continue for a period of time after the part is removed from the mold. It is important that the shrinkage be repeatable, so the mold can be appropriately sized. For parts requiring tight dimensional control or optimum mechanical properties, uniform cooling is essential. If the part does not solidify uniformly in the mold, residual stresses will occur as a result of differential shrinkage. Computer programs are available to assist in the optimum layout of cooling channels for uniform part cooling.

Part Removal. The mold is usually designed such that the part remains on the moving half of the mold when it is opened. Ejector pins are then actuated to separate the part from the mold. These ejector, or knockout, pins are activated either as a direct result of the moving platen (mechanical knockouts), or hydraulically.

The geometry of the mold must be such that the part can be readily removed after the material has solidified, sometimes by means of a mechanical part remover. This requires careful part design and selection of the ''parting surface'' across which the two halves of the mold separate. Neither half of the mold can have undercuts or ''die locks'' that would trap the part or keep it from being ejected. If the part cannot be designed without undercuts, then moving cores, slides, or lifters must be incorporated to eliminate the trapped condition when the mold is opened.

Mold Venting. Air must be removed as the mold fills with plastic. This is usually accomplished by grinding channels, or vents, into the parting line of the mold. These vents must be narrow enough to keep molten plastic from escaping, while allowing air to escape. They should be located adjacent to the last region of the mold to fill. Sometimes undersized ejector pins are also used to assist in venting the mold. Insufficient venting can result in an incompletely filled mold, or in cosmetic defects called diesel burns, which are caused by the heating of the air as it is compressed.

Mold Materials. For high-volume production applications, the mold cavity and core are usually machined from special mold-making steels. Steel is chosen for its wear resistance and durability. For low-volume production or prototype applications, it is common to cast molds from low melting point metals, such as aluminum and zinc. These cast molds are usually less expensive and faster to build, but are not as durable and often do not have as good a surface finish as do machined steel molds.

Molding process controls can be particularly important for parts with tight dimension or performance requirements. Some thermoplastic molding materials require tighter control over these parameters than others.

Predrying. In a molten state, some thermoplastics, such as polyesters and polycarbonates, are very sensitive to the presence of water. Under such conditions, these polymers will degrade or depolymerize, reducing the molecular weight, which in turn reduces impact strength and may cause cosmetic defects on the surface of the part that usually radiate outward from the gates. Known as splay, these marks are caused by microscopic water vapor bubbles at the surface of the part. Predrying is usually achieved by adequate residence time (3 to 4 h) in a hopper drier at a temperature just below the softening point of the material. Too low a temperature will result in inadequate water removal, but too high a temperature may result in fusion of the pellets.

Controlling Shear. Excessive shear can be caused by restrictive check valve assemblies (Ref 8), shut-off nozzles, gates, and even by the part geometry itself. Excessive shear stress during the molding process can cause molecular breakdown or chain scission in thermoplastic materials. Some materials are more susceptible than others and need to be processed with great care. Control of shear stress is particularly critical if mechanical properties, such as impact strength, are important. Shear can be minimized by carefully selecting the injection screw profile (Ref 8), by minimizing the back pressure on the material in the barrel, or by reducing flow restrictions.

Overheating. Thermoplastics are subject to degradation when they are overheated. Some, such as polyvinylchloride (PVC) and acetal, are far more sensitive than others. Degradation can result from chain scission, oxidation, hydrolytic degradation if moisture is present, or chemical interactions, such as transesterification. Material suppliers commonly add stabilizers to those materials that are particularly susceptible to degradation. It is desirable to mold the material at the lowest temperature and shear possible, and to check for hot spots in the machine.

Other Types of Injection Molding Machines

Instead of having a reciprocating screw, a ram injection machine has a simple plunger (ram) that pushes the unmelted plastic down the heated barrel. A spreader or torpedo at the end of the barrel can increase the shear and compression on the material to assist in heating and melting. Because the plunger does not mix and homogenize the melt as well as a reciprocating screw does, the ram injection machine is not often used.

In a two-stage injection or accumulator-type machine, the melting function is separated from the melt accumulation and injection functions. Either a stationary screw in a heated barrel or a plunger in a heated barrel performs the melting function. The melt is then fed into an accumulator cylinder before injection. Once sufficient melt has been accumulated for the next part, the ram forces the melt through the nozzle into the mold. A check valve keeps the melt from being forced back through the melting cylinder. This type of unit is generally not used because of the amount of degradation on the engineering materials.

Thermoplastic Injection Molding Variants

Five types of injection molding modifications are described below.

Thermoplastic structural foam injection uses both a process and machinery that are very similar to conventional injection molding, except that a gas phase is incorporated with the resin. The gas is frequently introduced by a blowing agent that is compounded in with the thermoplastic. The blowing agent decomposes to produce a gas as the plastic melts and causes the plastic to expand as it enters the mold. Molding pressures are substantially less than those for conventional injection molding, reducing the clamping force required to mold a part. Also, the stiffness-to-weight ratio of the part is improved. Generally, the surface finish is marred by a characteristic swirl pattern caused by bubbles breaking at the surface during mold filling.

In sandwich injection molding, two thermoplastic materials are essentially injected simultaneously into the mold. One material forms the skin of the part, while the other forms the core. The core material is frequently foamed, as described above, or filled with a reinforcement. The skin material provides a smooth surface.

Hollow injection molding introduces a gas (N_2) directly into the melt steam during injection. A gas channel is created through the thickest region of the part, such as along a rib, where the resin is most fluid. As with structural foam, the gas reduces the required molding pressure, but because the gas is confined to the center of the part, the surface finish does not deteriorate.

Injection/compression molding involves injecting thermoplastic into a partially open mold and then closing it to fill and pack out the part. Because the melt is injected into a relatively thick tool, the required injection pressure is reduced. Also, dimensional stability of the part is improved as a result of reduced flow orientation and lower residual stresses.

Layer injection molding involves injecting two or more materials sequentially, overlaying the first material with a second.

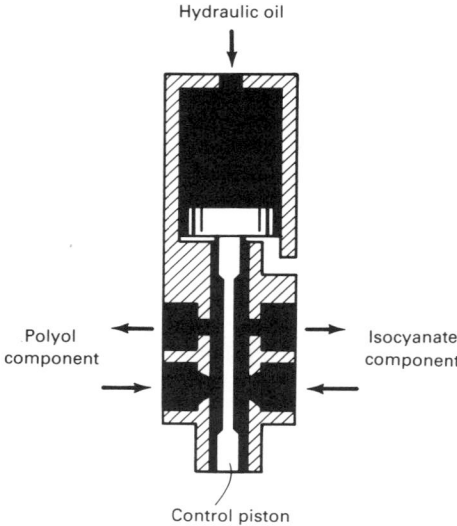

(a)

(b)

Fig. 5 RIM impingement mixing head. (a) Recirculation position just before and just after injection. (b) Shot position during mixing and injection

Thermoset Injection Molding

In thermoset injection molding, a reacting fluid is forced into a generally warm mold, in which the material further polymerizes or cross links into a solid part. The injected material can be either a one-component material or it can be two reacting liquids that are mixed just before injection into the mold, as in polyurethane reaction injection molding.

One-Component Thermoset Injection Molding. Among the materials that are injection molded are phenolics, diallyl phthalate, melamines, and ureas (Ref 9, 10), as well as polyester. These are referred to as one-component materials because only one feedstock material is delivered to the machine. The material, usually a blend of several ingredients that are nonreactive in the solid state at ambient temperature, is processed in reciprocating-screw injection molding equipment, similar to that used for thermoplastic injection molding. The molding machine heats the material in the barrel to a temperature sufficient to permit it to flow readily under pressure, but low enough to keep it from reacting while still in the barrel. With thermoset polyester, twin piston units are used, with no need for external heating. The material is then injected into a heated mold, where it cross links and solidifies. The hot, but solid, part is then ejected. The machine and molds are very similar to those used for thermoplastic injection molding, with the following exceptions discussed below.

First, shear on the material in the barrel is kept to a minimum to prevent overheating and thus precuring. The screw compression ratio is kept very low, with screw flights at a constant depth for most of the length of the screw. Where pistons are used for polyester, no shear is created, and no precuring takes place.

Second, barrel temperature is carefully controlled to keep the material from prematurely reacting. This may actually require removing heat from the barrel and nozzle area, because some shear heating will occur as the material is conveyed down the barrel.

Finally, mold temperature is generally higher than with thermoplastics. Hot oil heaters are often used.

Reaction Injection Molding (RIM). In this process, two reacting liquids (such as a polyol and an isocyanate, the precursors for polyurethanes) are delivered separately to the machine. The machine mixes the liquids just before injection into a closed mold. The liquids react in the mold to form a cross-linked solid part. The two liquid streams, under high pressure, are intimately and instantaneously mixed by impinging on each other in an impingement mixing head, shown in Fig. 5, just before the combined stream flows into the mold. The mixing head plunger eliminates all the mixed material from the head after injection, which precludes the need to flush out the head with solvent after each shot to keep the mixer from clogging, as is required with mechanical-type mixers.

Just after mixing, the liquids have a low enough viscosity to allow large, complex molds to be filled using low injection pressure. Thus, a relatively low clamping force is required,

which makes this process particularly attractive for large parts such as automotive bumper covers and body panels.

Fillers, such as short glass fibers or flakes, can be incorporated in one or both of the liquid streams to provide a part that is somewhat stiffer, and more dimensionally stable. This process is known as reinforced reaction injection molding (RRIM). Special mixing tanks are required to keep the filler in suspension in the liquid. Also, abrasion-resistant equipment is required.

Another modification of the RIM process consists of placing glass reinforcement in the mold before injecting the reacting liquid. The liquid wets the glass and reacts to form a glass-reinforced solid part. This process, which is either called structural RIM, resin transfer molding, or resin injection molding, is fully discussed in the article "Resin Transfer Molding" in this Volume.

REFERENCES

1. I.I. Rubin, *Injection Molding—Theory and Practice*, John Wiley & Sons, 1972
2. I.I. Rubin, Injection Molding, chapter 10 in *Introduction to Polymer Science and Technology: An SPE Textbook*, H.S. Kaufman and J.J. Falcetta, Ed., Society of Plastics Engineers
3. I.I. Rubin, Injection Molding, chapter 10 in *Introduction to Polymer Science and Technology: An SPE Textbook*, H.S. Kaufman and J.J. Falcetta, Ed., Society of Plastics Engineers, p 495
4. J.H. DuBois, *Plastics History U.S.A.*, Cahners Books, 1972
5. A.B. Glanvill and E.N. Denton, *Injection Mould Design Fundamentals*, Industrial Press, 1965
6. J.H. DuBois and W.I. Pribble, Ed., *Plastics Mold Engineering Handbook*, 3rd ed., Van Nostrand Reinhold, 1978
7. R.M. Criens and H.G. Mosle, "The Influence of Knit-Lines on the Tensile Properties of Injection Molded Parts," *Polym. Eng. Sci.*, Vol 23, July 1983, p 591
8. A.J. Keeney, "Free Flow Check Ring and Zero Metering Screw," Paper presented at the 37th Annual Conference, Reinforced Plastics/Composites Institute, 1982
9. J.M. McDonagh, Polymer Fabrication Processes, in *Introduction to Polymer Science and Technology: An SPE Textbook*, H.S. Kaufman and Joseph J. Falcetta, Ed., Society of Plastics Engineers, p 585
10. Injection Molding—A New Day for Thermosets, *Mod. Plast.*, Vol 44, Dec 1966, p 96

Compression Molding

Carl F. Johnson, Ford Motor Company

COMPRESSION MOLDING is a term that actually encompasses several different technologies. Sheet molding compounds (SMC), structural molding compounds, and thermoplastic materials can all be compression molded on various specialized equipment. One attribute all compression processes have in common is the use of a premanufactured material or "charge" that is pressed to shape and cured during the molding operation.

The compression molding process is presently the most technologically developed and versatile way to incorporate either continuous or random chopped fibers into a structural composite. The process is more rapid and complex than the labor-intensive hand lay-up or liquid matched die molding methods that it often replaces, but it has a trade-off with respect to fiber alignment. Depending on component shape and charge pattern, compression molding may involve regions of high resin flow that tend to orient fibers in the flow direction. Random orientation is desirable for chopped-fiber sheet molding compound and long-fiber thermoplastic materials. If directional fibers are desired, flow patterns can be developed.

The compression molding process is most commonly called the SMC process in reference to the precursor sheet molding compound material it uses. The components of the compression molding process are depicted in Fig. 1. The primary application of this technology in the automotive industry has been for grille opening panels and, on selected vehicles, exterior panels. Tailgates (Fig. 2), and hoods (Fig. 3) are two other examples of vehicular use of the SMC process. The entire cab of selected heavy trucks (Fig. 4) is also produced using this process.

A typical process cycle consists of placing sheets of SMC, which consist of 13-mm (0.5-in.) chopped glass fibers in chemically thickened thermoset resin that has a leathery consistency, into a heated mold (typically 150 °C, or 300 °F). The mold is closed under pressures of 7 to 14 MPa (1 to 2 ksi) for about 1 to 3 min to cure the material. Approximately 30 to 80% of the mold surface is covered by the SMC charge, and the material flows to fill the remaining cavity as the mold closes.

Materials for Sheet Manufacture

Fiber utilization in compression molding materials is very versatile. Continuous, cut length, or chopped glass fibers may be used, as well as random continuous glass mats. Carbon or aramid fibers may also be used. Sheet molding compounds can be made in various compositions and by various processes. Continuous, unidirectional molding compounds for structural components generally have 40 to 60 wt% glass fiber reinforcement. Normally, SMC for nonstructural automotive trim and body applications is 27 to 30 wt% glass fiber. Fillers are often used to minimize resin cost and lower thermal expansion of the product.

Resin chemistry has a major influence on the strength and reliability of the final composite component. Although the resin constitutes only 16 to 25 wt% of a typical SMC composite, it

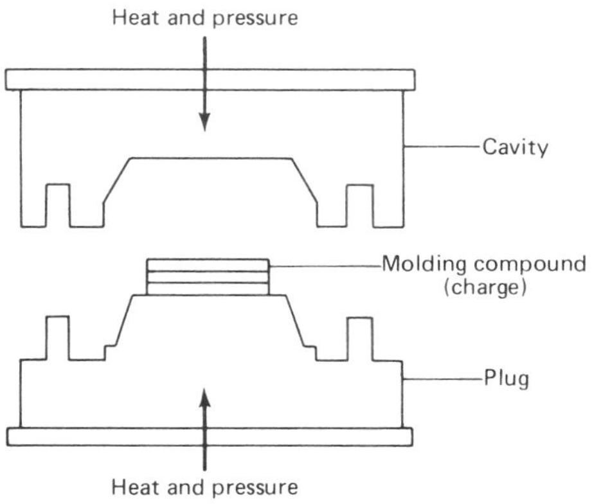

Fig. 3 SMC vehicle hood

Fig. 1 SMC molding operation

Fig. 2 SMC vehicle tailgate

Fig. 4 SMC structural truck cab

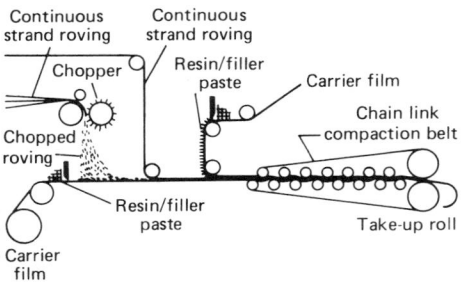

Fig. 5 SMC sheet manufacture

controls flow and moldability. Strength and corrosion resistance must be optimized by the resin selection. However, low-viscosity high-acid resins are often desired for their good wetting and thickening characteristics; for example, magnesium oxide added to the resin system thickens it to a leathery consistency. The reaction takes place over a period of up to 5 days. Although epoxy resins are used in aerospace SMCs, vinyl ester and polyester resins are generally used in automotive applications because of their faster cure time and lower cost. Many SMC materials use a high-reactivity, isophthalic polyester resin (for example, E980) with magnesium oxide for thickening. Ground limestone is generally used if a filler is required.

To form a sheet of SMC material, fibers and resin are distributed on a thin sheet of nonporous material (typically nylon or nylon/polyethylene coextruded film), as shown in Fig. 5. The resin, which may include ingredients such as catalyst, filler, and mold release, mixed with magnesium oxide, is first applied to the carrier sheet as a thin, uniform layer. Fibers are then added in random or continuous form, arranged in the desired orientation. Additional resin paste is applied to a thin cover sheet of nonporous material, and the resulting sandwich structure passes through a series of compaction rollers to thoroughly mix the fibers and resin. Completed SMC is rolled for storage and aged for up to 5 days to allow the thickening reaction to take place. The SMC should be used within a moderately short period of time because of the continually changing chemistry and resulting change in moldability, although a good formulation has a useful window of several weeks.

The production rate for a 30% glass reinforced sheet (SMC-R) is up to 12 m/min (40 ft/min). Material optimization has spurred on the development of new materials and corresponding manufacturing techniques. Currently available types of SMC include SMC-R, SMC-C, SMC-C/R, SMC-D, and XMC.

SMC-R is a sheet with random fiber reinforcement. Fibers that are 75 mm (3 in.) long or less are dispersed in a two-dimensional array. Chopped roving or mat is used as the reinforcement in an amount ranging from 30 to 70 wt%. SMC-R is suited for molding parts of varying cross section, including ribs and bosses. Parts having deeper ribs require shorter

fibers to obtain fill-out and good fiber distribution without resin-rich areas. SMC-R is more or less isotropic in its plane before molding, but becomes anisotropic to varying degrees when molded. Maximizing the charge coverage to keep flow distances to a minimum reduces anisotropy.

SMC-C uses continuous glass fibers with unidirectional orientation. This material gives high unidirectional strength and is highly anisotropic. Flow limitations require almost 100% coverage in the fiber direction, with ribs, bosses, or three-dimensional shapes not being advisable. Note that this is unlike XMC, discussed below.

SMC-C/R combines random chopped fibers and unidirectional, continuous glass fibers in a single SMC sheet. Physical properties are anisotropic, but to a lesser degree than with SMC-C. Strengths are higher in the transverse direction than is the case with SMC-C. The addition of the chopped fiber improves molding properties by allowing small ribs and bosses, limited three-dimensional areas, and more charge pattern flexibility.

SMC-D incorporates lengths of directional but discontinuous fiber as a reinforcement. Fibers are 100 mm (4 in.) or longer in a unidirectional pattern. SMC-D has better moldability than does SMC-C/R, but is lower in strength.

XMC is another material currently used in moderate volume for compression molding. It is a proprietary product (Ref 1) that uses directionally oriented, continuous glass fibers arranged in an X pattern and incorporated in a thickenable thermosetting resin. This material is made into sheets by filament winding on a steel drum. A large-diameter drum is preferred in order to minimize filament length difference, which must be taken into account when the material sheet is cut, removed from the drum, and flattened out. The XMC form is highly anisotropic but is more conformable to complex geometries than are SMC-C materials because of its X construction. As the material flows, the X pattern opens to a larger angle, but does not separate, leaving reinforcement-deficient areas. In the compound XMC-3, random chopped glass fiber replaces about one-third of the continuous glass. XMC-3 has four to five times the transverse strength but only 75 to 80% the longitudinal strength of XMC. Fillers are not normally used.

Although SMC technology is currently the most highly developed composite fabrication process for automotive applications, there are still areas requiring active development if the SMC process is to gain increased penetration into high-volume applications. These areas include optimization of part design to reduce cycle time and fully use material properties; determination of effects related to charge placement, optimization of temperature, time, and pressure relationships; and improvement of chemical and physical consistency of material.

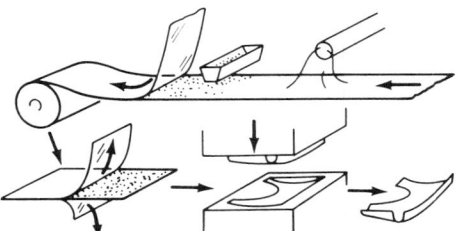

Fig. 6 Molding an SMC component

Processing and Manufacture of Components

To form a component, sheets of SMC material are cut to a desired shape, the carrier film is removed, and the SMC sheets are stacked as a series in a charge pattern. For example, in molding an automotive wheel (Ref 2), nine sheets of SMC are stacked; the charge pattern is then placed between clean, preheated dies; and upon closing, the press spreads out the charge to fill the die cavity. Normal die temperatures are 130 to 160 °C (270 to 320 °F). Typical cavity pressure varies from 4 to 21 MPa (2 to 3 ksi), depending on resin viscosity, glass fiber content, charge pattern, and mold complexity. A schematic of a typical molding operation is shown in Fig. 6.

Cycle time ranges from 1 to 4 min, depending on component complexity, thickness, and the die clean-up required. Cure time is very critical: If the resin cure exotherm is not properly controlled, cracking, blistering, or warping can occur. After all the other steps in the SMC process are automated, the exotherm may well be the rate-controlling factor for thick parts of more than 10 mm (0.4 in.).

Advanced compression mold designs and molding systems are presently being developed to reduce SMC cycle time (Ref 3). Current minimum cycle time is about 1 min, button to button, die closed time. Tooling must be of high quality to maintain these fast molding rates and must be hardened in critical wear areas. It is also necessary in some cases to use compression molding heat transfer analysis techniques to maximize the heat transfer characteristics of the tool (Ref 4). Extra sets of tooling are expensive and introduce variance in part tolerance when numbers of parts must be assembled from moldings of several tools. It is therefore advantageous from several perspectives to produce parts as rapidly as possible using a minimum number of tools.

Material cost for compression-molded SMC was estimated to be approximately 3.5 times the cost of steel by weight, in a study by Volkswagen (Ref 5). To compete with steel, polyester composites must integrate several parts into one, to save assembly, floor space, and storage costs. The Volkswagen study concluded that if suitable part integration could be accomplished, polyester SMC would compete with steel for up to 227 000 units per year. To compete beyond this volume, the SMC cycle time must be reduced. Halving the cycle time would double the level at which composite components would compete with steel. The study concluded that research must be done to speed up the process; eliminate SMC storage by making it on-site; and immediately recycle rejected, uncured SMC to bulk molding compound (BMC).

Many supply companies are conducting research to obtain more reproducible SMC products as well as faster cycle times. Further control of the SMC process has been investigated by the Plastic Process Development Department of General Motors (Ref 6). The rate of pressure application and parallelism of the molding die are being closely controlled by a microcomputer-controlled flow system. Eight hydraulic cylinders, two at each corner of the press, are mounted to oppose the closing action of the press by pressing upward on the platen. Four LVDTs, one on each corner, monitor platen location. As the press closes, before contact of the die by the charge, the counterpressure hydraulic cylinders contact the platen, level it to within 0.25 mm (0.010 in.), and begin controlling the speed and pressure of closing. In molding trials, this system was reported to reduce wall thickness variation on a deck lid by 50%. Tests at General Motors Technical Center on a 1.8×10^6 kg (2000-ton) press demonstrated thickness control of 0.127 mm (0.005 in.) over a piece that was 127 cm (50 in.) long and 2.54 m (100 in.) wide. Parts as thin as 1.5 mm (0.060 in.) have been fabricated on this system.

Many SMC applications are for visible components requiring high-quality surface finishes. Using low-profile resins and highly polished dies improves the as-molded surface. Control of material flow and tooling surface (parallelism) also reduces waviness and improves appearance. A technique for obtaining better surface finish that is currently being used in many automotive applications is called in-mold coating.

To apply an in-mold coating, a thermoset resin is injected into the mold after the SMC component is partially cured. To provide a space for the coating, the press is opened slightly. This opening can be conveniently achieved by using a counteracting force system or, as in the case of the recently developed high-pressure in-mold coating systems (Ref 7, 8), by injecting the coating with a slight yet sufficient pressure to compress the partially cured SMC. The thermoset resin is injected into the die, and the die is reclamped. The press forces the coating to surround and impregnate any surface voids in the SMC in a uniform manner. Urethane is a common in-mold coat. The in-mold coating can eliminate some paint priming operations and significantly reduce the hand finishing required for SMCs. Class A surfaces can be obtained

with a minimum number of paint pops after baking. Research is currently being done to apply in-mold coating to deep sections, such as fender extensions. Present technology extends to shapes such as Corvette hood outer panels and similar large, rather simple geometries. This process does not address problems that exist in areas where subsequent machining will expose new, uncoated surfaces, such as trimmed edges.

The other system in use today to improve SMC surface quality relies on a vacuum to decrease the amount of trapped air and gasses in the molded component. As the mold is closed on the SMC charge, a seal closes around the entire mold and the mold area is evacuated. As the material flows to fill the die, the vacuum enhances the natural expulsion of air and styrene vapor from the SMC material, resulting in higher surface quality and less tendency for the occurrence of subsequent defects during painting.

Process Advantages. The SMC method has several advantages over methods such as hand lay-up or spray-up. Having the liquid resin, catalyst, and glass fiber precombined into a unit can allow better quality control over chemistry, mix, and distribution prior to forming. Using matched metal dies is a closed-mold process that gives better dimensional control and stability than open-mold processes. Higher pressures require more expensive tooling than is necessary for hand lay-up or spray-up but less expensive tooling than is needed for stamping or injection molding.

Higher pressures during compression molding may reduce the blistering, splitting, and paint popping often encountered during paint-baking cycles in moldings made at lower pressure. The short flow lengths that are possible in compression molding tend to minimize fiber movement, and reduce stresses and tool wear, in contrast to injection molding. The process does not constrain the mold designer with sprue and runner lay-outs. Absence of runners and sprues reduces resin degradation due to shear heating and eliminates reinforcement length reduction, which, in the case of injection molding, severely limits the length of reinforcement fibers. Venting tends to be uniform by force, and the dimensional stability of the formed components is better than it is for components formed by higher-flow processes, such as injection molding, or by processes such as hand lay-up.

Process Limitations. There are disadvantages associated with both the SMC material and the compression molding process that must be recognized. A high capital investment is required for sheet forming and molding equipment. The cost of SMC material is also high because of the postfinishing labor and equipment required after molding. The SMC material must be stored properly to prevent thermal and moisture degradation. Cracking and warpage of the final part may result from using a degraded SMC sheet. Also, the SMC

sheet-forming line must be properly protected to prevent toxic vapor emissions.

During the compression molding cycle, most problems are flow related. Finite-element numerical techniques are showing promise in predicting flow of filled-polymer melts of the type encountered in composite fabrication; however, these are not expected to be operational for several years (Ref 9, 10). The problem areas that exist presently are residual stresses, warpage, weld lines, flow orientation of fibers, fiber kinks and folds, delamination, and breakdown of the resin paste and glass interface. While all these problems can be minimized by varying the charge pattern, cycle parameters, or SMC chemistry, the existence of this large number of variables adds complexity and cost to the process.

Structural Compression Molding

The preceding description of the SMC process delineates material primarily used for semistructural applications, rather than high load bearing segments of a structure that must satisfy severe stress load and durability requirements. To sustain more stringent structural demands, it is normally necessary to incorporate appreciable amounts of continuous fibers in predesignated locations and orientations. The same basic SMC operation can be used to incorporate such material modifications, either by formulating the material to include the continuous fibers along with the chopped fibers, or by using separate charge patterns of two different types of material. The complexity of shape and degree of flow possible are governed by the amount and location of the continuous glass material. Careful charge pattern development is necessary for components of complex geometry. A typical example of a prototype rear floor pan fabricated by this technique (Ref 11, 12) is shown in Fig. 7.

The limitations for using compression molding of SMC-type materials in truly structural applications are not yet well established. Assuming that continuous fiber is strategically incorporated, these materials appear capable of providing high structural integrity and may well prove to be the pioneering composite fabrication procedure in high load bearing, high-volume applications. The advanced state of commercialization of this process relative to other evolving techniques provides lead time for optimizing compression molding for higher-performance parts.

Although compression molding of SMC-type materials is an economically viable, high production rate process in current usage, there are some limitations inherent in the process which may, in the long run, restrict applications in structures and tend to favor other processes. For example, the degree of flow required to optimize the mechanical properties results in a variation in mechanical properties that is due to imprecise control of fiber location and orienta-

Fig. 7 SMC structural floor pan

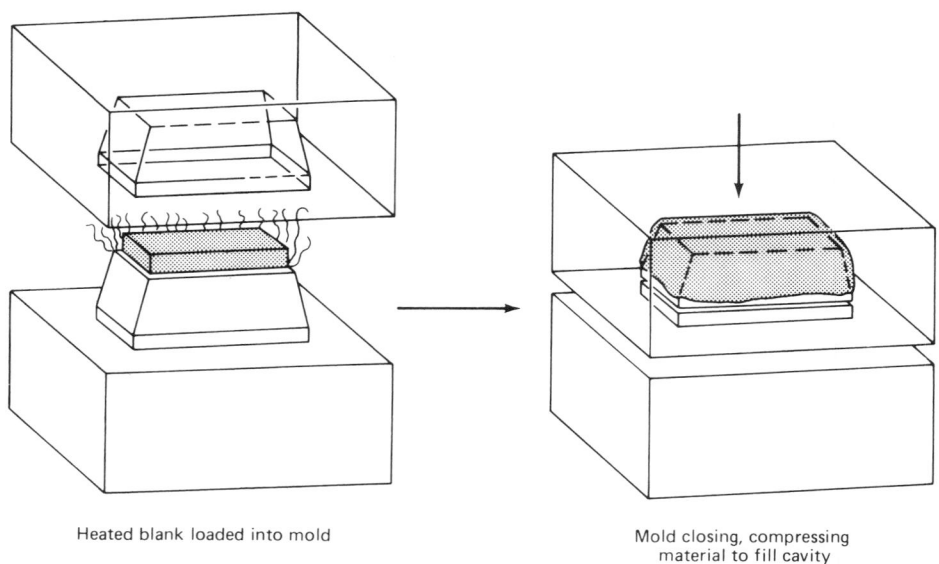

Heated blank loaded into mold

Mold closing, compressing material to fill cavity

Fig. 8 Thermoplastic compression molding. (a) Heated blank loaded into mold. (b) Mold closing, compressing material to fill cavity

tion (Ref 13, 14). Typically, a factor of two variation in mechanical properties throughout the component is not unusual, based on an initial charge pattern coverage of approximately 30 to 80%. Such uncertainty in properties, while tolerated in nonstructural and semistructural applications, introduces reliability issues and conservative design allowables, yielding a heavier than necessary component or structure. Currently, extensive research efforts are underway to develop SMC-type materials that will allow 100% charge pattern coverage and will attain high, uniform mechanical properties with minimal flow. These materials will also be able to be molded at lower pressures on smaller-capacity presses. Material developments such as these may well make the next generation of SMC much more applicable to highly loaded structures than has yet been envisaged.

Another potential limitation of structural compression molding that may turn out to be the main restriction in structural usage is the degree of part integration attainable. The basic strategy in composite application is to integrate as many individual (steel) pieces as possible to minimize fabrication and assembly costs (which offsets increased material costs), and to minimize joints (which increases "effective" stiffness). Compression molding requires fairly high molding pressures (about 7 to 14 MPa, or 1 to 2 ksi), and thus limits potential structures in areal size and complexity (particularly in three-dimensional geometries requiring foam cores). Consequently, while compression molding is likely to play a key role in the development of composites in structural automotive applications in the next decade, the process is ultimately unlikely to provide optimum structural efficiency and weight characteristics. This statement will prove true only if the alternate,

more optimal processes are developed further because, currently, compression molding is the only commercial structural composite process capable of satisfying the economic constraints of a mass-production industry.

Thermoplastic Compression Molding

The process of thermoplastic compression molding (sometimes referred to as thermoplastic stamping) is attractive to the automotive industry because of the rapid cycle time and the potential use of some existing stamping equipment. Thermoplastic compression molding, at its current level of development, achieves cycle times of 1 min for large components. Figure 8 illustrates the process. Typically, a sheet of premanufactured reinforced thermoplastic is preheated above the melting point of the matrix material and then rapidly transferred to the mold (Fig. 8a). The mold is quickly closed until the point at which the material contacts the mold, and then the closing rate is slowed (Fig. 8b). The material is formed to shape and flows to fill the mold cavity, in the same manner as thermoset SMC. After the material is cooled in the mold for a short period, the mold is opened and the component is removed.

Thermoplastic compression molding is currently used in automobiles to form low-cost semistructural components such as bumper back-up beams, seats, and load floors. Other commercial applications range from office seating to large machinery housings. Commercially available materials include wood-filled polypropylene and discontinuous glass fiber filled polypropylene with relatively low physical properties. Other materials, based on oriented reinforcements and such resins as polyether

etherketone (PEEK) and polyphenylene sulfide (PPS), are used in the aerospace industry. These materials are expensive and are limited in their conformability to complex shapes, but, as high-performance materials, are finding increased use in the aircraft industry.

Higher levels of strength and stiffness must be developed in low-cost materials before they can be used in truly structural applications. Attempts have been made to improve the properties of materials through the use of separate, preimpregnated, unidirectionally reinforced tapes. These materials are added to the heated charge at critical locations to improve strength and stiffness locally. Use of these tapes adds to the cost of the material and slightly increases cycle times. Although using preimpregnated tapes is effective in simple configurations, the location of the oriented reinforcement and the reproducibility of that location are problems in complex parts. To be cost effective, this type of reinforcement will ultimately either have to be made part of the premanufactured sheet or applied robotically. Research on the use of oriented reinforcement in critical areas is in process. To be used in structures effectively, these materials will have to retain greater geometric flexibility (that is, ability to form complex shapes with the reinforcement in the correct location) than is currently exhibited in moldings made with today's lower-performance commercial materials.

Part integration is a major benefit in the use of thermoplastic compression molding, but the required high pressures (7 to 21 MPa, or 1 to 3 ksi) limit the size of components that can be manufactured on conventional presses. The process has limited capability for incorporating the complex three-dimensional cores required for some integrated components. If large structures are required, thermoplastic compression molding may not be the best process to use from an economic viewpoint. Ongoing, long-range research in the areas of low-pressure systems and incorporation of foam cores in stampings could significantly alter this outlook.

REFERENCES

1. R.H. Ackley and E.P. Carley, XMC-3 Composite Material—Structural Molding Compound, in *Proceedings of the 34th SPI Annual Technical Conference*, Society of the Plastics Industry, 1979
2. J.A. Woelfel, Fiber Reinforced Composite Wheels—Status Update, Paper 2E, in *Proceedings of the 39th SPI Annual Technical Conference*, Society of the Plastics Industry, 1984
3. D.H. Anderson and G.A. Landsettle, Witchcraft and Wizardry in Composite Molding, in *Proceedings of the First ASM Advanced Composites Conference*, American Society for Metals, 1985, p 139-145
4. M.R. Barrone and D.A. Calk, Optimal Thermal Design of Molds for Chopped-Fiber Composites, *Polym. Eng. Sci.*, Vol 21 (No. 17), Dec 1981, p 1139-1148
5. H. Habblitzel, "Reinforced Polyester in the European Automotive Industry," SAE 790169, Society of Automotive Engineers, 1979
6. G.Y. Yee, "Innovative Sheet Molding Compound Processes," SAE 790170, Society of Automotive Engineers, 1979
7. R.E. Ongena, The Mechanics of Molded Coating for Compression Molded Reinforced Plastics Parts, in *Proceedings of the 33rd SPI Annual Technical Conference*, Society of the Plastics Industry, 1978
8. I. Kocur and J. Ziehm, Process Improvement for In-Mold Coating, Paper 12-D, in *Proceedings of the 41st SPI Annual Technical Conference*, Society of the Plastics Industry, 1986
9. M.R. Barone and D.A. Caulk, Kinematics of Flow in Sheet Molding Compounds, *Polym. Compos.*, Vol 6 (No. 2), April 1985, p 105-109
10. S. Kenig, Fiber Orientation Development in Molding of Polymer Composites, *Polym. Compos.*, Vol 7 (No. 1), Feb 1986, p 50-55
11. C.F. Johnson and F.E. Noggle, A Composite Floor Pan Study, *Body Eng.*, Vol 11 (No. 2), 1983, p 45-58
12. N.G. Chavka and C.F. Johnson, A Composite Rear Floor Pan, Paper 14-D, in *Proceedings of the 40th SPI Annual Technical Conference*, Society of the Plastics Industry, 1985
13. P.K. Mallick, Effect of Misorientation on the Tensile Strength of Compression Molded Continuous Fiber Composites, *Polym. Compos.*, Vol 7 (No. 1), Feb 1986, p 14-18
14. S.Y. Oh and C.D. Han, Processing-Property Relationships in Compression Molding of Sheet Molding Compounds, *Polym. Compos.*, Vol 6 (No. 1), Jan 1985, p 13-19

SELECTED REFERENCES

- D.L. Denton, "The Mechanical Properties of an SMC-R50 Composite," Owens-Corning Fiberglas Corporation, 1979
- J.H. Enos, R.L. Erratt, E. Francis, and R.E. Thomas, Structural Performance of Vinyl Ester Resin Compression Molded High Strength Composites, Paper 11-E, in *Proceedings of the 34th SPI Annual Technical Conference*, Society of the Plastics Industry, 1979
- D.L. Evans and T.J. Nemeth, Improved SMC for Structural Electrical Utility Applications, in *Proceedings of the 34th SPI Annual Technical Conference*, Society of the Plastics Industry, 1979
- R.B. Jutte, "Structural SMC—Material, Process and Performance Review," SAE 780355, Society of Automotive Engineers, 1978
- R.S. Rapp, Sheet Molding Compounds: The Effects of Isophtalic Polyester Processing, in *Proceedings of the 34th SPI Annual Technical Conference*, Society of the Plastics Industry, 1979
- R.T. Silvanieto, B.C. Fisher, and A.W. Birley, Predicting the Flow for Unsaturated Polyester Resin SCM, *Proceedings of the 34th SPI Annual Technical Conference*, Society of the Plastics Industry, 1979
- N.D. Simons, Compression Molding, *Modern Plastics Encyclopedia*, Vol 54 (No. 10A), 1978, p 258
- L. Suter, The First Production SMC Door, in *Proceedings of the 34th SPI Annual Technical Conference*, Society of the Plastics Industry, 1979
- B.M. Walker, chapter in *Handbook of Fillers and Reinforcements for Plastics*, H.S. Katz and J.V. Mikulec, Ed., Van Nostrand Reinhold, 1978, pp 619-632
- E.L. Wood, L. Rockwood, J.E. Foster, Jr., and G. Nelson, Increased Automotive Reinforced Plastic Production and Quality Through Automation of the SMC Process, in *Proceedings of the 34th SPI Annual Technical Conference*, Society of the Plastics Industry, 1979

Resin Transfer Molding

Carl F. Johnson, Ford Motor Company

RESIN TRANSFER MOLDING (RTM) is a closed-mold low-pressure process that allows the fabrication of composites ranging in complexity from simple, low-performance to complex, high-performance articles and in size from small to very large. The process is differentiated from other molding processes in that the dry reinforcement and the resin are combined within the mold to form the composite component. The fiber reinforcement, which may be preshaped, is placed into a tool cavity, which is then closed. A tube connects the closed tool cavity with a supply of liquid resin, which is pumped or transferred into the tool to impregnate the reinforcement, which is subsequently cured. Several similar composite fabrication processes fall into the resin transfer molding category, although there are distinct variants.

Process Variants

Vacuum-Assisted Resin Injection. The most common use of the term RTM describes a process typified by the vacuum-assisted resin injection (VARI) manufacturing process (Ref 1). A mold is constructed of low-cost materials, such as epoxy. Reinforcement is then cut to fit the required geometric pattern and is arranged by hand in the mold. Pieces of reinforcement may be placed in the mold one at a time or preassembled and then placed in the mold as a unit of preform, after which the mold is closed and clamped with bolts or bars. A vacuum can then be applied to the mold to extract the air, and resin is injected at very low pressures, often below atmospheric pressure. Because of the low-cost materials used in mold construction, mold pressures must be low, resulting in slow fill times and limited glass contents. The inability of the mold to tolerate elevated temperatures, coupled with its poor heat transfer, restrict the resin chemistry to slow cure times with minimum exotherm to prevent resin degradation or tool damage. Cycle times of this process are measured in hours and days for large, complex parts. The major benefit of the process is the ability to fabricate large, complex structures with maximum part integration at a low cost.

Preform molding is another process that is sometimes referred to as resin transfer molding (Ref 2). A preform of reinforcement is prepared before it is placed in the mold. The preform is usually made by spraying chopped reinforcement onto a perforated screen. A vacuum applied to the rear of the screen holds the reinforcement in place until the binder, which is sprayed along with the reinforcement, has time to cure. After a postcure, the preform becomes an easily handled, three-dimensional sheet of reinforcement. Cores and changes in thickness are possible but not generally used in this process. Preform materials and techniques are further discussed in the section "Resin Transfer Molding Materials" in this article. Tooling is normally made of steel or zinc alloy, but could also be made of epoxy. The preform is placed in the mold, and a measured quantity of resin is poured or pumped into the open tool. The tool is then closed and compressed to approximately 690 kPa (100 psi), causing the resin to flow and "wet out" the preform. Heated tools are usually used, and the cycle time for large parts of uniform thickness is often 3 min or less. Parts having significant complexity and integration can be made economically using preform molding. Equipment is inexpensive because of the low pressures required. The physical properties of the molded components tend to be very consistent with uniform reinforcement content, even at part edges. Unfortunately, this is at the cost of a relatively high waste factor around the part perimeter. Also, these components do not have the optimum possible level of performance for the amount of reinforcement because of the absence of reinforcement orientation and, to a lesser extent, the presence of a binder.

The structural reaction injection molding process (structural RIM) is similar to the RTM process (Ref 3). It, too, uses a preform that is placed in the tool before the introduction of the resin system. Tooling can be made from various materials, but is typically a metallic shell in order to facilitate heat transfer. The resin is highly reactive and is contained in two separate holding tanks. Resin from each tank is injected under high pressure into an impingement mixing chamber and then directly into the tool. Although the mixing pressure is high, the overall pressure of the resin, once in the tool, is only about 340 to 690 kPa (50 to 100 psi). The resin flows into the tool and wets out the preform as the curing reaction is occurring. A suitable resin system has a viscosity plateau of 0.1 to 0.5 Pa·s (1 to 5 P) for around 20 s, followed by a rapid cure, resulting in cycle times of about 1 min. Reinforcement levels used in this process to date tend to be from 5 to 55 wt%. Resin must enter the tool at a fast rate if flow distances are long because of the fast cure times. These high flow rates restrict the choice of reinforcement to those which are resistant to washing (movement caused by resin flow). Because of the rapid onset of resin cure, flow distances are limited in this process; when flow distances exceed 610 mm (24 in.), multiple inlet ports are desirable.

High-speed resin transfer molding (HSRTM) encompasses portions of all three of the aforementioned processes (Ref 4). This conceptual process, shown in Fig. 1, uses a three-dimensional preform with foam cores and attachment inserts. Glass content is in the 35 to 60 wt% range and can be a mixture of continuous and random material. Tooling for high production volumes should be made of steel in order to contain moderate molding pressures (690 to 3450 kPa, or 100 to 500 psi) and exhibit good heat transfer characteristics. For limited production, aluminum or zinc tooling would be acceptable. Molding is carried out at elevated temperature to minimize the cure time and expand the number of potential resin systems. The preform is placed in the mold and, following minimal hand arrangement of the preform, the mold is closed and resin is injected. At higher reinforcement levels, the mold may be left open slightly during resin injection and then closed completely to promote rapid filling of the mold. Cure should be accomplished in the mold such that the final part will require no postcure and will have acceptable dimensional stability. For some complex components, or components having critical tolerance requirements, a fixtured postcure is required to yield dimensional stability. Cycle times range from 1 min for small components to 8 to 12 min or longer for large, complex structures.

The relationship of RTM to other processes is depicted in Fig. 2. As the degree of complexity, level of performance, and size of composite components increase, labor-inten-

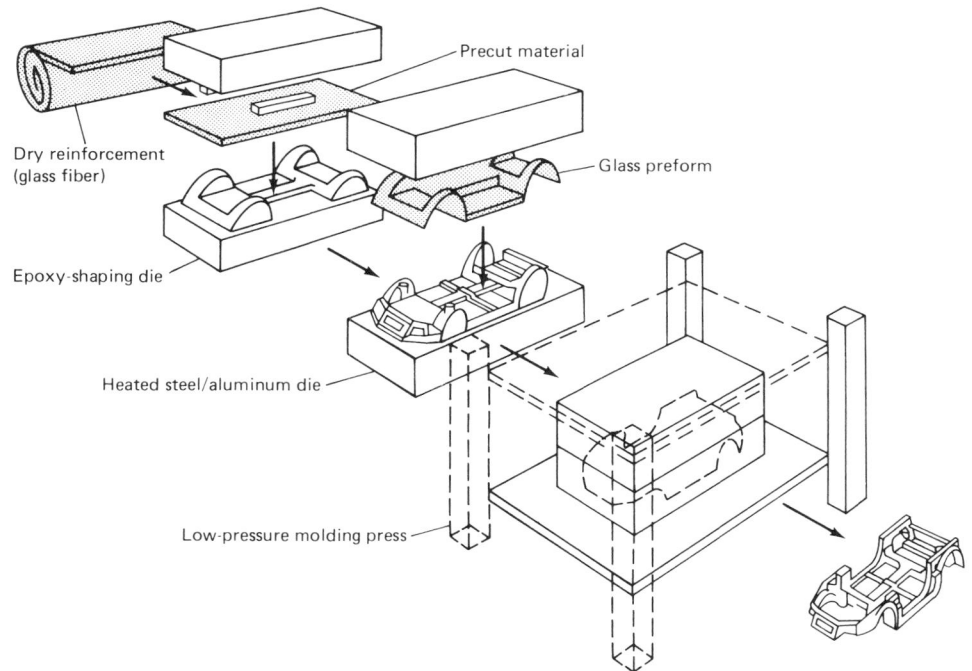

Fig. 1 High-speed resin transfer molding

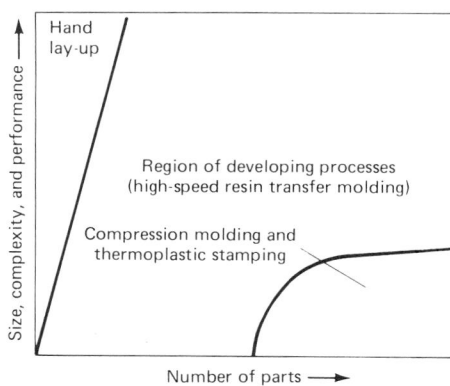

Fig. 2 Process development for automotive composite parts. Steep slope of the hand lay-up process reflects the high-cost and low-volume characteristics of a labor-intensive process

sive processes, such as hand lay-up, may increasingly be chosen to make a number of components. Because of its cost, it is unlikely that this technique can be used to make a significant quantity of parts. At the other end of the spectrum are the compression molding and thermoplastic stamping processes. The equipment required to produce parts with these technologies is expensive, thus there is a threshold of number of parts required before these processes become financially viable.

The level at which this occurs, when the automotive industry considers the replacement of metal components with composites, depends on the cost of the component being replaced. As the cost of the steel component or system being replaced increases, the number of parts to be manufactured will decrease. There is, however, an upper limit to the amount of replacement possible using these processes. Because of the limited mechanical properties of materials used with these processes and the inability to make large, integrated structures without assembling multiple small components, these processes may only economically replace steel in areas where the degree of part integration possible is already limited by function. Examples of such areas include grille opening panels, hoods, deck lids, and doors. The high pressures required (7 to 17 MPa, or 1 to 2.5 ksi) also limit the size of the components fabricated by compression molding and thermoplastic stamping.

The center and upper right-hand area of Fig. 2 represents the areas where the maximum potential exists for producing economical, high-volume composite structures. This area represents high-performance, large, highly in-

tegrated structures produced in medium to high volumes. The RTM process allows placement of preforms of variable thickness containing a variety of fiber types in the mold cavity with minimal subsequent movement of reinforcement during further processing. This contributes to optimum performance at minimum weight. The low pressure required for the process allows the use of less-expensive presses and may slightly reduce the cost of high-volume production tooling, compared to compression molding or thermoplastic stamping. There should, of course, still be the large reduction in tooling expense, when compared to steel components, given a sufficient degree of part integration. The low pressures will also allow much larger structures to be molded. Current compression molding processes are limited by the availability of very large presses. The RTM process allows the incorporation of cores and inserts in the component design. This ability to produce three-dimensional structures with deep sections and cores at low pressures allows the fabrication of large, highly integrated structures. In the automotive industry, these are not surface-quality parts but, rather, unseen structures. A superior surface quality currently requires a high-pressure process.

Status of the RTM Industry. Automotive molders throughout the United States and Europe currently produce composite components and, in several cases, large vehicle structures, using RTM processes (Ref 5-7). Because the processes are slow, using little automation, typical cycle times for small components approach 3 min. For complex or large compo-

nents, a cycle time of 1 to 2 h is typical. Limitations to cycle time arise from two areas: Tooling is typically epoxy, which dictates very low filling pressures and slow resin cure to avoid exotherms that would damage the tooling, and reinforcements are generally cut and placed in the mold by hand for each molding, which increases cycle time considerably for complex parts. For example, one company uses RTM to mold upper and lower body tubs using epoxy tooling and considerable hand reinforcement fitting for each part. Cycle times are approximately 24 h per part. Another company used similar tooling and RTM technology with fiberglass-wrapped foam preforms, which were dropped into the mold, reducing the cycle time to 90 min. Current RTM molding technology, however, does demonstrate the flexibility of the process. Reinforcement fibers are placed in precisely delineated positions, and cores are incorporated, if required, to produce large, complex structures.

Unresolved issues still exist in preforming, tooling, faster resin chemistry, machinery, complex foam cores, resin flow, and faster cycle times to adapt the flexible RTM technology to appropriate production rates. The areas in which these issues exist do not require invention of new technology but rather a systematic development of existing technology. All of the major areas of technology exist today, but due to the fragmented nature of the composites industry and the previous lack of a definition of the required process, the full potential of the RTM process has not been realized.

Equipment

The presses and resin control equipment required for resin transfer molding are readily available from a number of suppliers. Lower-tonnage presses can be used because of the lower pressures required for RTM, compared to processes such as compression molding. Computer control is desirable to sequence the press closing and resin management for applications in which high-speed resin injection and cure are required.

Presses. When required, the presses used in RTM primarily control parallelism of the die set during die opening and closing and also hold the die closed during injection. Because RTM uses lower pressures than does compression molding or thermoplastic stamping, the clamping force required will be considerably lower. Though expensive, injection pressure can be the main factor determining the press size. Basing the size on the average pressure within the tool is a less-expensive approach.

Computer simulations are being developed to allow the calculation of the pressure required to fill a component (Ref 8). One point to consider when sizing a press is the force required to compact the lofted reinforcement preform. For glass preforms in the range of 50 wt%, with a majority of the glass being random material, a closing pressure as high as 690 kPa (100 psi) can be required to close the die.

Tooling. In the case of a large, highly integrated, composite automotive structure, the cost differential between low-cost tooling and standard production steel molding tools may not be significant. There should, however, be large savings in tooling cost over current tooling for fabricating steel structures because of the reduced number of required tools resulting from part consolidation. To realize short cycle times, tooling must be capable of being uniformly heated to about 90 to 150 °C (200 to 300 °F). It must also be rigid enough to compress the lofted reinforcement of the preform as the mold closes, without tool distortion. Hardened shear edges to trim excess reinforcement from the preform in pinch-off areas as the mold is closed will reduce postmolding finishing time and provide a good seal to contain the resin. Because of the abrasive nature of the reinforcement, and the likelihood of molding a large number of components, tooling surfaces should be chrome plated. All of these requirements limit tool choices for high-volume production to materials such as cast aluminum, chemical vapor deposited nickel, and steel. For a production tool, steel will most likely be the best material based on its durability and ability to be easily modified. For low-volume and prototype production, however, lower-cost epoxy and zinc alloy cast tools are an acceptable alternative. Longer cycle times, reduced dimensional accuracy due to mold compliance, and reduced tool life will likely result from this tooling choice.

Resin Transfer Molding Materials

Cost and processability often determine the choice of a resin system. Thermoset materials have been used exclusively to date, but as high-performance thermoplastic materials become available in RIM-type formulations, they may also be used.

Of the several resin systems that perform adequately in an RTM process, polyester is most often used because of its low cost. Epoxy has been used in both aerospace and consumer products and has been demonstrated to provide high physical properties, but at a premium price. Vinyl ester resin has also been used in a number of RTM products and provides properties between those of polyester and epoxy at a moderate price. Other resin systems, such as the acrylamate resin system family and methylmethacrylate vinyl ester, are newer systems that are proving to be very processable with RTM techniques.

In any resin system, a low-viscosity plateau is required, during which the resin can provide constant flow throughout the mold, followed by a fast cure. The resin must not only gel rapidly to be acceptable for rapid RTM cycles, but must provide sufficient Barcol hardness to allow the component to be demolded without distortion. Some resins will yield lower levels of physical properties if cured rapidly, and the processing-property relationships of any resin system must be evaluated for each component before it is produced. The speed of resin cure for many available resin systems is already adequate to achieve rapid cycle times. Systems that yield cycle times of 1 min or less, if desired, are available for use with small components. Developments to increase the time available for mold filling and to improve reinforcement wet-out after the reactive resin components are mixed together are still needed.

Preforms are a critical aspect of the successful implementation of any high-volume RTM project. Development of preforming techniques and characterization of materials resulting from various preforming techniques should be carried out before finalizing component design. An optimized preform process that gives design and fabrication cost equal consideration must be developed. A structure that is optimized to use the ultimate design capabilities of composite materials, while being highly weight efficient, will likely be expensive and is therefore not appropriate in all applications.

The optimal baseline material for consumer products is typically random E-glass. This material is currently being used in many conventional RTM products, including the Lotus, Avanti, and Matra vehicles. The preforming technique most often used with random glass mat is to shape a flat sheet of this material at the time of molding. In the Lotus process, a sheet of mat is cut and formed to fit in the mold. Overlaps of several inches are made at the end of any layer of material. Foam cores are wrapped with sheets of mat before being placed in the mold. This technique is also used in many other resin-transferred components, ranging from small automotive trim to large waste treatment plant components.

A process in which the flat sheets of glass mat are preshaped before insertion into the mold represents the first level of preforming sophistication. This process uses a flat sheet of

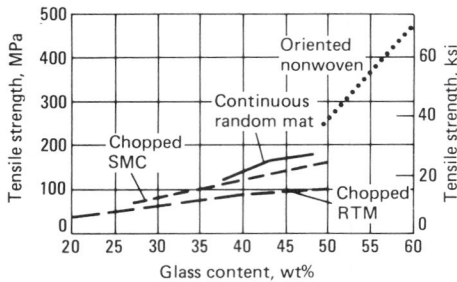

Fig. 3 Tensile strength of various RTM materials with E-glass reinforcement

random mat with a small amount (2 to 5%) of thermoplastic binder applied. The binder allows the sheet to retain a shape when it is heated and pressed in a forming die, imparting a gentle, three-dimensional shape to it without cutting and piecing, as in the previously mentioned process. This process, while faster, cannot achieve radical three-dimensional shape changes, such as deep draws. It is possible, however, to include continuous reinforcement selectively in the preform to improve its physical properties.

The most versatile and widely used preforming technique for creating three-dimensional preforms with complex shapes is the spray-up process. Glass rovings are chopped and sprayed on a rotating screen. A small amount of resin is introduced into the stream of chopped glass, and when the glass accumulates on the screen to the proper weight, the resin is heat-cured, causing the preform to retain its shape. The vacuum applied to the back of the screen not only holds the glass on the screen as it accumulates but also helps maintain uniformity. As the holes in the screen become covered by glass, the open areas tend to attract more glass, imparting a self-leveling action. This process can be fully automated, and the spray-up of a large preform, such as an automobile floor pan, would take 2 min. This process, while yielding complex shapes, tends to produce resin-transferred components at the low end of the potential physical property spectrum. The binders sprayed with the chopped reinforcement tend to cover and seal off the fiber bundles, resulting in incomplete resin impregnation of the reinforcement. Typical physical properties of an E-glass-polyester plaque made by using a sprayed-up preform are presented in Fig. 3 and 4. Those materials with only random reinforcement represent the lower limit of structural composite performance. It is likely that an optimum design will require oriented reinforcements in selected areas. There are currently several techniques for making preforms containing oriented materials.

Continuous material can be added locally to a sprayed-up preform. In fact, continuous glass is added to the rail areas of an Avanti floor pan. In this case, a piece of oriented glass fabric is placed on top of the sprayed-up preform prior to

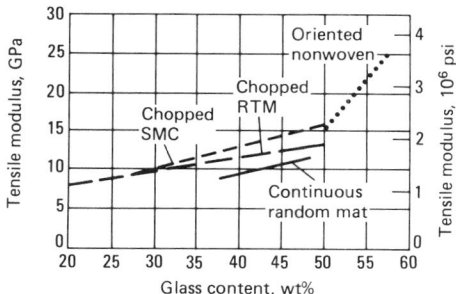

Fig. 4 Tensile modulus of various RTM materials with E-glass reinforcement

Fig. 5 Two-piece front structure made by high-speed resin transfer molding process replaced 90-piece steel structure

mold closing and resin additions. Filament winding, or other robotic fiber placement, is one of the possible techniques for accurately positioning reinforcement where it is needed in a preform.

An approach to preforming that is currently used in the aerospace industry for high-performance components is referred to as the engineered fiberform process. When units of laminates, precut to shape, are stitched together into a three-dimensional structure, they become what is referred to as a fiberform. Fibers used are E-glass, S-glass, carbon, and aramid. The preforming process uses uniaxial fiber rovings that are formed into sheets of varying configurations. Some materials are reinforced solely in the 0° direction, while others are reinforced in the 90° direction. Fiber rovings can also be oriented at any intermediate angle. These materials are not woven, as is cloth, but rather are stitched together with a thread of polyester or aramid. Elimination of the weaving yields improved physical properties and better resin wet-out in the final part. The oriented nonwoven data in Fig. 3 and 4 typify the physical properties of a vinyl ester composite made from this type of E-glass preform.

Fiberforms can be stitched in critical areas with reinforcements such as aramid to increase joint integrity or increase interlaminar shear strength. The fiberform process is similar to the assembly of a suit coat or other garment from fabric building blocks. Because the process can be carried out on high-speed, high-volume, fabric-processing equipment, the cost of large volumes of fiberforms may be low. Assembly of the fiberform may also be accomplished close to the point of use, allowing shipment of flat precut laminate sheets. Some producers are currently supplying fiberforms to the aerospace industry for fabrication trials and limited production using RTM. Layers of the oriented reinforcement sheets can be stacked and further stitched together into an engineered laminate structure, the structural efficiency of which is high because of exact reinforcement placement and orientation. For some simply configured parts, these laminates can be used as RTM preforms.

It is unlikely that any optimum preform could be made from using one of these processes

exclusively. Portions of each process could be used, but because the processes and the design are interdependent, they must be considered in parallel.

Process Advantages, Limitations, and Applications

The many potential advantages of RTM can be summarized as the capability of rapid manufacture of large, complex, high-performance structures. The low pressure of the process allows very large components to be manufactured by means of low-tonnage presses. It also permits the use of foam cores to yield fully three-dimensional parts. The ability to preplace the reinforcement where desired, and have it remain in that location, gives increased design flexibility and a subsequent optimized structure. While composites in general offer this advantage over steel, other composite molding processes are either slow and labor intensive or have some degree of movement of the reinforcement, which results in variations in physical properties throughout the finished part. In applications in which composite materials are competing with steel or other materials on a cost basis, a large amount of part integration is required to offset the increased material cost. Reduced assembly cost, higher quality, and improved functionality at lower material weights are all possible advantages if a sufficient degree of integration can be achieved. Resin transfer molding provides the capability to integrate large numbers of components into one part (Ref 9), and can provide an inexpensive means of obtaining prototype parts or low production volumes. It can also be capable of rapid production of larger volumes when high-quality tooling is used. Resin transfer molding is a closed-mold process, which has advantages

over open-mold processes, including low vapor emissions. The resulting components have both inner and outer surfaces that are dimensionally controlled, thereby requiring a minimum of hand trimming if tooling is properly designed.

Currently, the majority of RTM process limitations stem from the undeveloped nature of the higher-speed versions of the process. While resin systems continue to improve, there are still problems with filling large parts with high glass content at low injection pressures. Preform fabrication for high-volume components also currently has limitations. The absence of reinforcement at part edges may be a limitation if ribs and bosses are required in a design. Ribs and bosses must be loaded individually in the tool cavity, and maintaining reinforcement at the part edge and avoiding resin richness at corners of the part can be difficult. Scrap losses also may be more costly as component integration increases, and if a large component fails, replacement cost to the consumer can be significant.

There are applications of resin-transferred components in the automotive, consumer products, and industrial products industries. Ford Motor Company recently completed a concept study in which the entire 90-piece steel front structure of an Escort automobile was replaced by a 2-piece resin-transferred front structure (Ref 9). The process used was HSRTM. Although only prototype tooling has been used to date, production cycle times were projected to be in the range of 6 to 9 min. The final structure was stiffer and stronger than the steel structure, at 66% of the original weight. Figure 5 illustrates the structure in such a vehicle. The preform used to make the component is shown placed in one half of the prototype tool in Fig. 6.

Examples of consumer product applications of RTM are given in Ref 10-12. Industrial applications are identified in Ref 13-15.

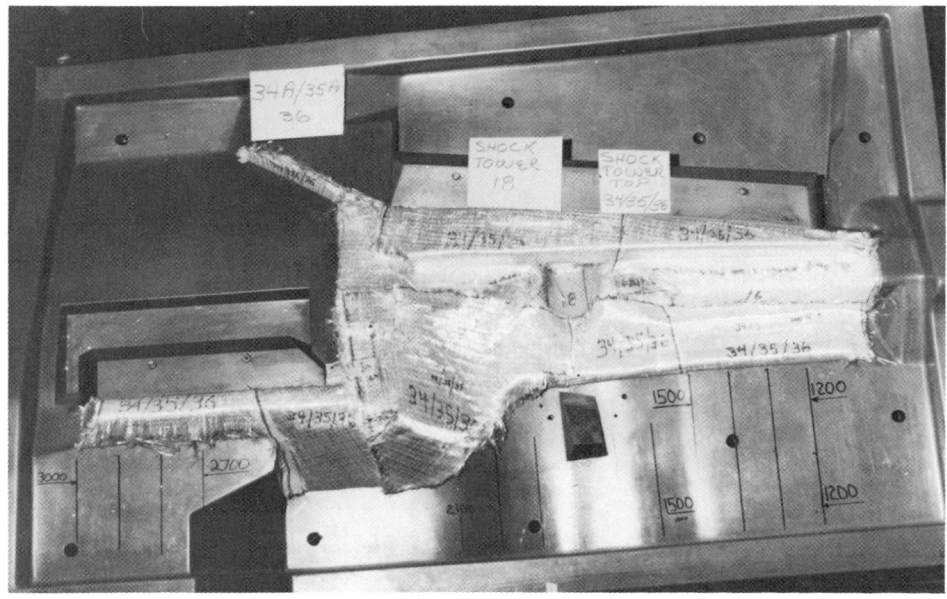

Fig. 6 Preform used to make two-piece structure shown in Fig. 5

REFERENCES

1. A New Look At Lotus, *Autocar*, Vol 150 (No. 4289), Jan 1979, p sup 1-sup 18
2. R.S. Morrison, J. Jensen, and R. Riddell, Meet Your Part Half Way With Preforms, *Plast. Eng.*, May 1980, p 61-64
3. R.D. Farris, R.H. Overchashier, and W.G. Gottenberg, Structural Parts From Epoxy RIM Using Preplaced Reinforcement, Paper 5-E, in Proceedings of the 37th SPI Annual Conference, Society of the Plastics Industry, Jan 1982
4. F.F. Johnson, Rapid Manufacturing Processes for Integrated Automotive Structures, in *Advanced Composites Conference Proceedings*, American Society for Metals, 1985, p 95-99
5. M. Hartung, Resin Transfer Molding: How It's Done On A Large Scale, *Plast. Technol.*, Vol 25 (No. 3), March 1979, p 73-77
6. M.K. McCann, Toward the Ultimate Composite Cab, *Truck and Off-Highway Industries*, Vol 4 (No. 2), March-April 1983, p 36-38
7. I. Sayama, I. Nomura, K. Tabei, and S. Gotoh, New Applications of Resin Injection Molding in Japan, Paper 15-E, in *Proceedings of the 36th SPI Annual Conference*, Society of the Plastics Industry, Feb 1981
8. G.Q. Martin and J.S. Son, Fluid Mechanics of Mold Filling for Fiber Reinforced Plastics, in *Proceedings of the Second Conference on Advanced Composites*, Nov 18-10, 1986, p 149-157

9. C.F. Johnson, N.G. Chavka, and D.Q. Houston, Resin Transfer Molding of Complex Automotive Structures, Paper 12-A, in *Proceedings of the 41st SPI Annual Conference*, Society of the Plastics Industry, Jan 1986
10. J. Raymer and D. Clarke, Evaluation of Economic and Design Consideration for Production of Business Machine Cabinetry With the Resin Injection (RTM) Process, Paper 15-A, in *Proceedings of the 36th SPI Annual Conference*, Society of the Plastics Industry, Feb 1981
11. W.C. Dolan and W.T. Suffern, Functional and Process Considerations in the Design of an RTM Product, Paper 24-F, in *Proceedings of the 38th SPI Annual Conference*, Society of the Plastics Industry, Feb 1983
12. Low Cost Tooling Gets RP Parts to Market—Fast, *Plast. Technol.*, Vol 26 (No. 3), March 1980, p 85-88
13. H.N. Marsh, Jr., T.E. Griffith, and J.V. Spitale, Jr., RTM Tooling and Molding for Corrosion Resistant Applications, Paper 3A, in *Proceedings of the 34th SPI Reinforced Plastics Composites Institute Annual Conference*, Jan/Feb 1979
14. P.W. Vaccarella, Fabrication of FRP Chemical Resistant Equipment by the Resin Transfer Technique, Paper 3B, in *Proceedings of the 34th SPI Reinforced Plastics Composites Institute Annual Conference*, Jan/Feb 1979
15. R.M. Riddell, Resin Transfer Molding—An Update, in *Proceedings of the SAE Earth Moving Conference*, Society of Automotive Engineers, April 1983

SELECTED REFERENCES

- P. Beardmore and C.F. Johnson, The Potential for Composites in Structural Automotive Applications, *Compos. Sci. Technol.*, Vol 26 (No. 4), 1986, p 251-282
- S.G. Dunbar, T.E. Griffith, Reinforcement Selection for Resin Transfer Molding, Paper 15-C, in *Proceedings of the 36th SPI Annual Conference*, Society of the Plastics Industry, Feb 1981
- P. Emrich and R.M. Riddell, Resin Reactivity and Its Effect on Physical Properties of RTM Molded Parts, Paper 24-D, in *Proceedings of the 38th SPI Annual Conference*, Society of the Plastics Industry, Feb 1983
- G.C. Grigoropoulo, Technoeconomical Criteria for Selecting Pressure Molding Processes, Paper 18-C, in *Proceedings of the 40th SPI Annual Conference*, Society of the Plastics Industry, Jan/Feb 1985
- R.D. Howard, D.R. Sayers, and R.A. Mears, The Development of Low Profile Methacrylate Resins for Use in RTM and Cold Press Molding, Paper 19-A, in *Proceedings of the 41st SPI Annual Conference*, Society of the Plastics Industry, Jan 1986
- R.R. Lacovara and J.T. Woehr, Feasibility of Structural Coring in the RTM Process, Paper 15-C, in *Proceedings of the 36th SPI Annual Conference*, Society of the Plastics Industry, Feb 1981
- N.E. Michaels, J. Laven, and J. Bauer, RTM—The Right Choice for the 80's, Paper 15-B, in *Proceedings of the 37th SPI Annual Conference*, Society of the Plastics Industry, Jan 1982
- R.S. Morrison, Resin Transfer Molding of Fiber Glass Preform Reinforced Polyester Resin, Paper 15-D, in *Proceedings of the 36th SPI Annual Conference*, Society of the Plastics Industry, Feb 1981
- R. Riddell, RTM—A New Tool for the Matched Metal Die Molder, Paper 15-D, in *Proceedings of the 37th SPI Annual Conference*, Society of the Plastics Industry, Jan 1982
- D.R. Sayers and R.D. Howard, The Potential for Mass Production With Resin Transfer Molding Using New Methacrylate Based Resins, Paper 18-B, in *Proceedings of the 40th SPI Annual Conference*, Society of the Plastics Industry, Jan/Feb 1985
- K. Tabei, H. Kittaka, S. Yoshimura, and M. Hori, Resin Injection Process—Its Productivity and Cost Effectiveness, Paper 24-B, in *Proceedings of the 38th SPI Annual Conference*, Society of the Plastics Industry, Feb 1983
- P. Vaccarella, RTM: A Proven Molding Process, Paper 24-A, in *Proceedings of the 38th SPI Annual Conference*, Society of the Plastics Industry, Feb 1983

Tube Rolling

Paul A. Roy, Ferro Corporation

FIBER REINFORCED COMPOSITE TUB-ING production methods range from simple hand rolling to continuous pultrusion. To date, most tubing for primary structural application is either rolled or filament wound, although the technology in pultrusion is advancing rapidly.

This article focuses on those methods that are used to roll or wrap finite-length tubing and rod and are particularly applicable to small-diameter cylindrical or tapered tubes, of lengths up to 6 m (20 ft). Tubing diameters that can be rolled most effectively with these methods range from small tubes measuring in tenths of inches, such as fishing rod tips, to those measuring about 150 mm (6 in.) in diameter. Quality tubes with larger diameters can be produced, but production rates diminish rapidly as diameter increases.

Though the approach may vary, the common element in most tube rolling methods is that the material to be wrapped is laid out on a flat surface with no tension applied. Compaction and densification of the material during the wrapping process is accomplished by contact pressure between the mandrel and the material, as shown in Fig. 1. To facilitate a tight wrap, the material to be rolled should have sufficient tack to adhere to the mandrel and to itself. In some instances, it may be necessary to apply a tacking agent to the mandrel or to heat the mandrel. Conventional filament winding/braiding or tape-wrapping techniques will not be addressed, although these methods are used quite extensively and very effectively to manufacture tubing and small- and large-diameter cylinders.

Material Forms

Preimpregnated (prepreg) materials are most suitable for tube rolling when their form is either a fabric or unidirectional tape.

Typically, fabrics are used when a 0°/90° fiber orientation is required. Tubes made with this fiber orientation have good longitudinal and circumferential properties, but poor torsional values. Variants in longitudinal and circumferential mechanical properties depend on the fabric style selected. For example, the standard 181-style glass fabric will produce a tube with nearly equal properties in the two directions, whereas a 1543 weave style will produce a tube with a property imbalance of 8 to 1. Therefore, caution should be exercised when rolling fabrics to ensure that the warp and fill yarns are always wrapped in the proper orientation with respect to the tube axis.

Unidirectional tapes are used for tubes that require a higher degree of refinement in structural properties. It is possible to achieve any combination of axial, hoop, and torsional mechanical properties with unidirectional tapes by using a predetermined combination of wraps. Unidirectional tape also provides better fiber translation than do fabrics and yields a more weight-efficient tube.

Both material forms require cutting of patterns in preparation for rolling. Prepreg pattern design is determined by the product being fabricated, the material form employed, and the equipment being used for fabrication. Generally, tubes made from prepreg fabrics are convolute wrapped, whereas tube based on unidirectional prepreg tapes can be either convolute or spiral wrapped.

Convolute Wrapping of Cylindrical Tubes

Convolute wrapping of constant-diameter tubes is by far the fastest and simplest method of rolling. A rectangular sheet of material with a length equal to the desired number of circumferential wraps is continuously wrapped around the mandrel. For thin-walled shapes, a tube can be wrapped in one operation. When making thick-walled tubing or when using materials with a high bulk factor, it is desirable to wrap the tube in multiple wrap increments, debulking in between.

If too many wraps are attempted in one pass during hand or table rolling, the prepreg material tends to gather ahead of the mandrel, which could cause wrinkles in the finished part. A choice must then be made either to make multiple wraps with splices or to allow wrinkles; the former is usually chosen.

Fabrics, as noted earlier, offer a simple and extremely rapid method of rolling tubes that have a 0°/90° fiber orientation. Patterns are sized to the tube length, and an integral number of circumferential wraps are placed on the rolling surface. The mandrel is tacked in place and then rolled over the material, wrapping the material around it as it advances. To ensure a well-compacted wrap, uniform pressure should be applied to the mandrel throughout the wrapping process, and the mandrel should roll in a straight line.

Unidirectional prepreg tape is the preferred material when a combination of longitudinal and helical wraps are desired. Longitudinal wraps are easily accomplished by cutting sheets to the desired tube length and splicing them together to form one wide sheet for the desired number of wraps. Splicing can be done by overlapping or butting the plies and taping them together on either the prepreg or release paper (back) side. Overlapping is less desirable because it could produce a bump in the tube. To butt-splice, thin strips of supported film adhesive can be used to tape the patterns together on the prepreg surface, but a more efficient method is to tape the sheets together on the release paper side with good quality adhesive tape. The advantages of taping on the release paper side are that the tape is easier to apply because it is not affected by material tack, and the tape facilitates release paper removal. With longitudinal plies, the latter advantage is not as apparent, but it becomes much easier to remove release paper from angle plies, possibly in one operation, if they have all been taped together.

Figure 2 illustrates the procedure for rolling longitudinal (0°) plies. The release paper is removed, the mandrel is tacked in place, and the pattern is simply rolled up. It is generally preferable to lay the pattern on the table with the release paper side up, because it usually has

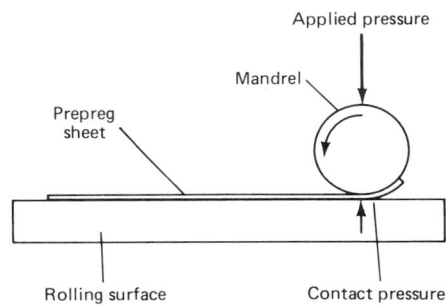

Fig. 1 Tube rolling

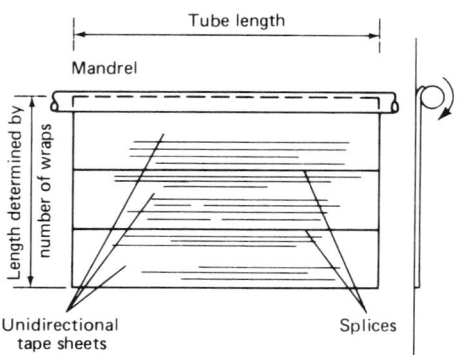

Fig. 2 Unidirectional tape, longitudinal ply rolling

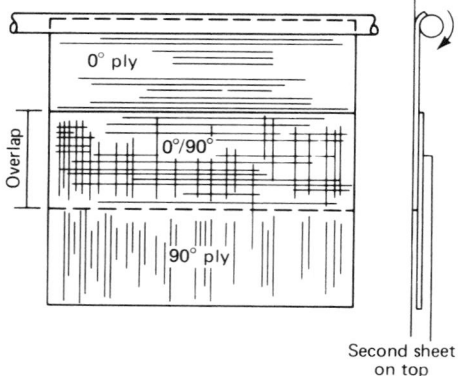

Fig. 3 Unidirectional tape, 0°/90° rolling

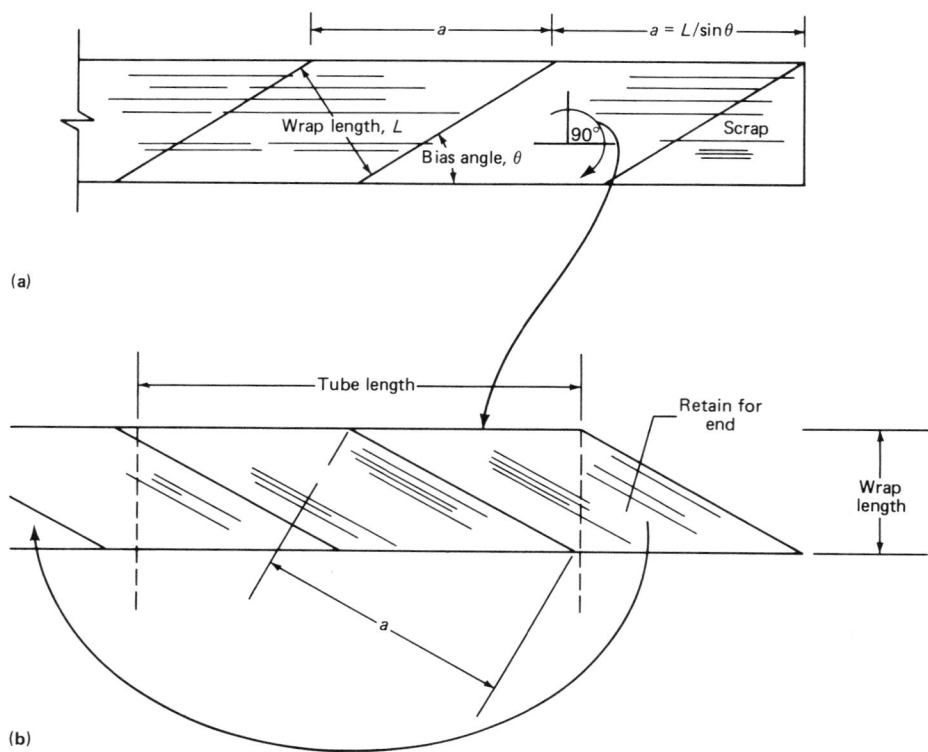

Fig. 4 Generating bias tape from unidirectional tape. (a) Unidirectional tape, bias cuts. (b) Bias tape lay-up

more tack, which makes it easier to start the mandrel.

If circumferential patterns are required, preparing the pattern is slightly more difficult. Typically, because tube length is longer than the width of standard materials, patterns must be spliced together. Although overlap splices are normally out of the question, thin glass tape may be acceptable, or the patterns can be buffed, although they usually do not stay together long enough to be rolled.

A workable solution is to combine the hoop patterns with other patterns to form an integral unit that facilitates alignment and release paper removal. For example, if the tube design dictates a 0°/90° construction, the two patterns can be combined as depicted in Fig. 3. The best way to accomplish this is to lay one pattern face up and apply the opposing pattern face down on top of it. The overlap zone should be firmly pressed together, using heat if necessary, to form a good bond. For volume production these patterns can be stacked for later use, at which time the release paper can be removed. Because the overlap is accomplished with the second pattern on top of the first, rolling is easier. If the overlap were on the underside, it could peel as it winds around the mandrel. The amount of overlap to use is at the designer's discretion. If the tube design indicates a precise ply-stacking

sequence allowing no comingling of plies, the most exacting designer would not want to allow any, in which case spiral wrapping should be considered. In convolute wrapping, at least one circumferential wrap overlap should be used.

With thin prepregs, a full overlap (double ply) could be used, although it is usually easier to start wrapping with a one-ply thickness before moving into the two-ply-thick material. Good design practice dictates that the overlap be made in equal circumferential wrap increments. If overlapping is used, the pattern size should be adjusted to accommodate the change in thickness.

Angle patterns are another matter. First, the unidirectional tape must be converted into a bias tape (Fig. 4). Unidirectional tape is cut at the desired bias angle, with the length of each sheet being determined by the number of wraps desired. The bias sheets are then rotated 90° and spliced together to create a continuous bias tape that can then be cut into the desired patterns. Taping together the sides with the release paper is preferable. By exercising care in the cutting operation, precise angle control can be maintained.

Once the patterns are cut, it is usually best to overlap the opposing angle patterns before rolling. This facilitates release paper removal and helps maintain pattern integrity. The amount of overlap is a matter of choice, but integral multiples of the circumference are desirable.

Although overlapping is considered undesirable by some because of the comingling of

plies, it is usually superior to rolling individual plies. Overlapping helps to align the plies, remove release paper, and hold the pieces together during handling and rolling. These benefits should outweigh any detrimental effects of ply comingling.

Convolute Wrapping of Tapered Tubes

This method is similar to rolling constant-diameter tubes, except that tapered patterns are used and provision must be made to allow the mandrel to roll along a curvilinear path during rolling, as shown in Fig. 5. Ideally, the patterns should be cut on an arc at the ends, as shown by the dashed line, because they should resemble the development for a truncated cone. This luxury is usually omitted, however, because the tube ends are normally trimmed after cure.

If precise fiber alignment is required, multiple convolute wrapping can present a problem. This is best illustrated by considering a multiple wrap using unidirectional tape. Figure 6a presents the usual approach of cutting a symmetrical pattern from the tape without regard to scrap. A close examination of the pattern shows that not all fibers are continuous, which could be a problem, although overlapping during wrapping can provide good fiber translation.

An additional problem is that starting the mandrel so that it is parallel to the fibers and then rolling it straight on turns out to be impossible, as illustrated in Fig. 6b. Because

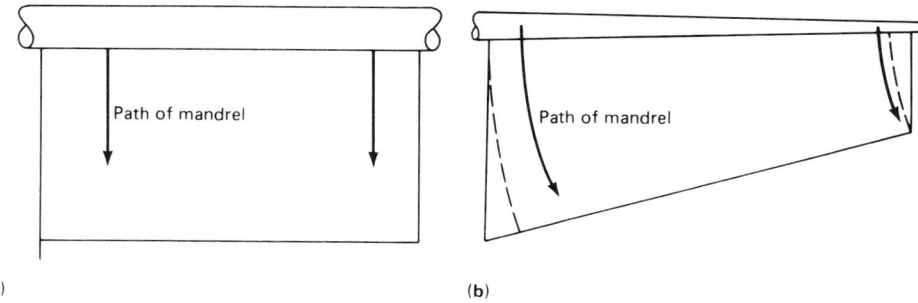

(a)

(b)

Fig. 5 Cylindrical versus tapered tube rolling. (a) Cylindrical mandrel. (b) Tapered mandrel

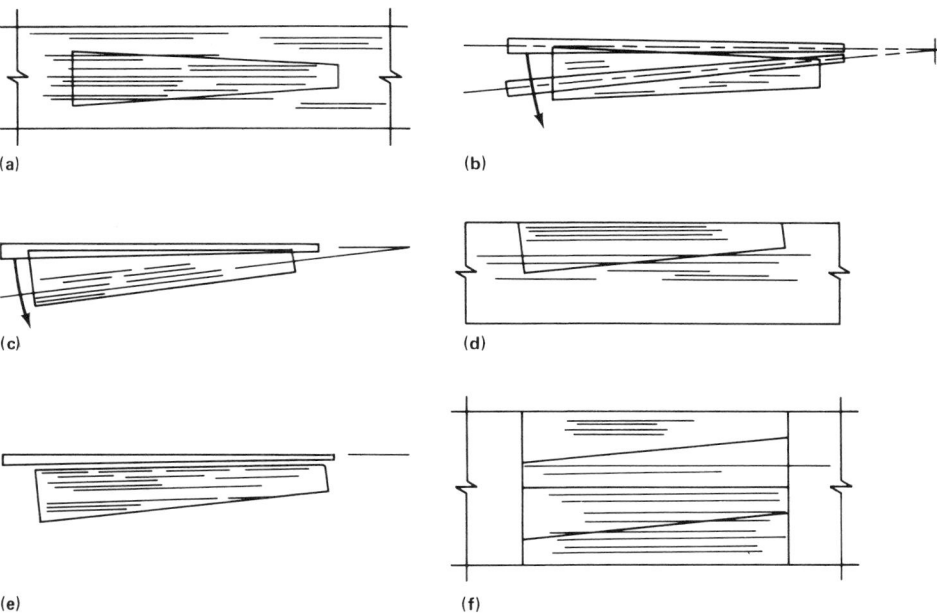

(a)

(b)

(c)

(d)

(e)

(f)

Fig. 6 Tapered tube pattern cutting and rolling

Fig. 7 Large-diameter shallow-angle wrap

Fig. 8 Small-diameter high-angle wrap

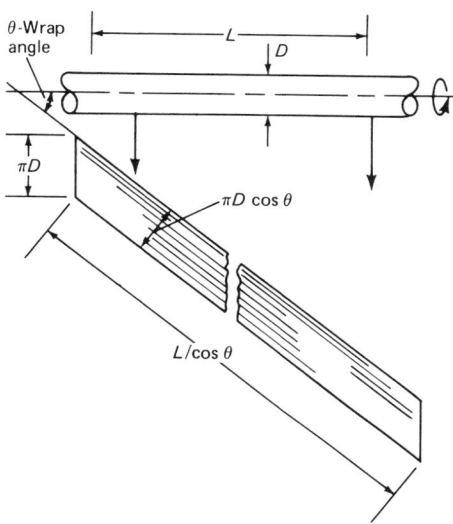

Fig. 9 Spiral wrap pattern, cylindrical tube

the mandrel is tapered, it must roll along an arc, which causes the fibers to be off-axis. Also, an uneven build-up on the mandrel would make it difficult to roll or to obtain good compaction. Starting the mandrel parallel to an edge, as shown in Fig. 6c, improves the quality of rolling but does not alter the fiber misalignment problem. Moreover, starting the rolling with short discontinuous fibers is sometimes difficult. Accepting the fact that one must live with some fiber misalignment when convolute-wrapping taper tubes, Fig. 6d presents an alternative, in which the pattern is cut with one side parallel to the fiber direction. With the continuous fibers along the starting edge, the rolling process is easier to start, as shown in Fig. 6e.

Figure 6f presents the preferred no-scrap method for cutting multiple tapered patterns. A sheet is sized for an even number of patterns and a simple flip-flop cutting procedure will quickly and easily produce a large number of patterns.

Patterns that are cut as in Fig. 6a tend to compensate with a ± angle variation in increments of three layers, whereas wrapping with plies that are cut as in 6d and 6f tends to result in the accumulation of wrap in one angular orientation. This unbalanced angular wrap could cause problems with part stiffness and/or straightness if the angular deviation becomes too great. A greater than 5° angle change should be avoided. One solution is to wrap with single-layer patterns. This slow process is easily accomplished with longitudinal wraps, but is much more difficult for high-angle wraps. A more desirable approach is to use spiral wrapping.

Spiral Wrapping

An alternate method for making tubular shapes from composite materials is spiral wrapping. Hand or table rolling using spiral wrapping is slower than convolute wrapping because only one ply at a time is rolled. With automated wrapping equipment, however, spiral wrapping can be very competitive with convolute wrapping.

Spiral wrapping should be considered for thin-walled tubes to avoid overlaps or gaps in plies that could cause variations in circumferential stiffness (spine), or whenever precise dimensional control is required. Spine problems are common in fishing rods, which is why the best rods are usually spiral wound.

Single-ply angle-wrap applications also favor spiral wrapping because the fibers can be wrapped continuously from end to end with no overlaps or butts. Using ultrahigh-modulus fibers could also dictate the need for a spiral wrap design. Application of pressure to a tube during cure has caused breakage of high-modulus fibers in convolute-wrapped tubes at overlap or gap locations.

In the strictest sense, sprial wrapping can be used with fabrics, unidirectional materials, or even paper. In fact, most cardboard cores are spiral wrapped. The following discussion is restricted to structural tubing in which fiber continuity from end to end is required. Therefore, only unidirectional tape will be considered.

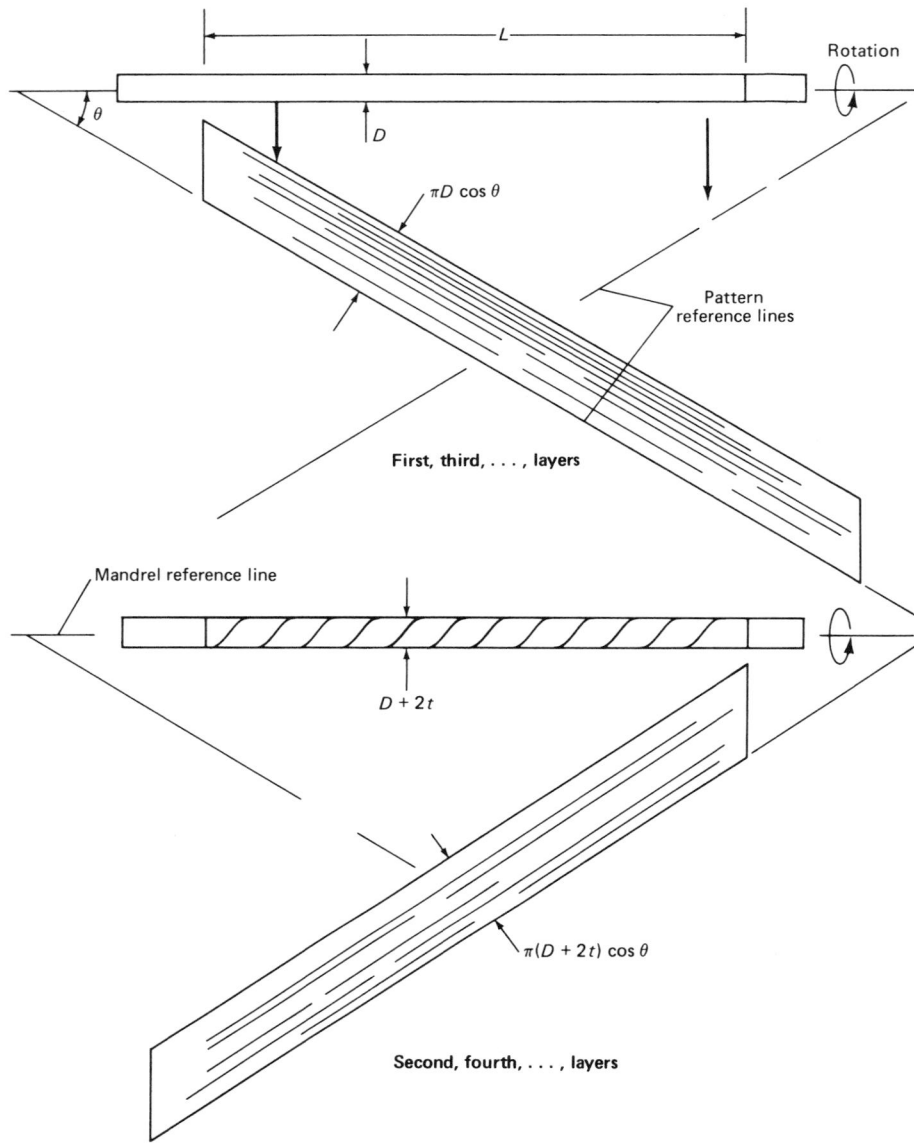

Fig. 10 Multiple spiral wraps

Cylindrical Tubes. The significance of spiral wrapping to structural tubing is in the application of angle plies. Longitudinal plies in spiral or convolute wrapping are equivalent. With angle plies, however, spiral wrapping enables the designer to use the full capability of the composite by having continuous helical fibers with no overlaps or gaps. Spiral wrapping involves the use of long strips of material of a width determined by the tube diameter and the wrap angle. For short tubes or shallow angles, the wrap may not make one complete circumferential circuit around the mandrel, as shown in Fig. 7, whereas for long tubes or high angles, several spiral circuits around the mandrel will be needed, as illustrated in Fig. 8.

Figure 9 illustrates the pattern configuration and rolling procedure for spiral wrapping an angle pattern on a cylindrical mandrel. The pattern shown is a parallelogram with continuous fibers, and its width is a function of the mandrel diameter and the wrap angle. The pattern length is determined by the tube length and the fiber wrap angle. A few simple calculations show that as the wrap angle increases, the pattern width decreases and the pattern length increases. For very steep angles, long, thin patterns are needed.

The difficulty encountered in spiral wrapping multiple angle-plied tubes is that the width of each successive ply must be slightly larger to account for the diameter change introduced by the preceding layer, if a gap-free lay-up is to be achieved. Additionally, the mandrel should al-

ways be rolled in one direction, as in convolute wrapping. Reversing the rolling direction will loosen the ply previously wrapped, resulting in a poorly wrapped tube. With respect to spiral wrapping, this means that each successive ply should be started from opposite ends of the mandrel, as depicted in Fig. 10. The first layer starts at the left end of the mandrel, with the pattern trailing off to the right. The second, opposing angle layer is rolled from the opposite end of the mandrel, maintaining consistency in the number of mandrel rotations. Also shown are reference lines, which are usually inscribed on the rolling table to facilitate mandrel and pattern alignment. With correct sizing of the patterns and proper alignment, the pattern should wrap gap-free around the mandrel.

Tapered Tubes. Spiral wrapping tubes with radical tapers is not recommended because of the complexity of the pattern. Unlike patterns for the constant-diameter tube, the patterns for spiral-wrapped tapered tubes should be cut in true spiral form. Moreover, because the patterns must be cut entirely on a taper from end to end, fiber continuity is lost.

For small-diameter shallow-angle tapered tubes, a common approach is to cut straight-sided tapered patterns and force the edges to butt together by stretching and pulling the material into place as it is wrapped. This distorts the wrap to some degree, but it has been found to be superior to convolute wrapping when overlaps could cause a spine problem. Additionally, with angle plies it is usually easier to spiral wrap very small diameter tubes than to convolute wrap them.

For tubes with large diameters or steep tapers, spiral wrapping becomes less desirable because of the pattern shapes required. Not only are the patterns tapered, but because of the taper, the edges are curved (Fig. 11), and the advantage of end-to-end fiber continuity is lost. In addition, the rolling process becomes rather cumbersome because the roller has to try to maintain a linear rolling direction with a tapered mandrel. A more convenient approach is to rotate the mandrel in a suitable turning fixture, such as a lathe, and apply the pattern in a tape-wrapping mode (Fig. 12). To facilitate wrapping, it is usually desirable to start at the small end and work up the taper. This can be accomplished by starting the first ply on the under surface of the mandrel and the second layer on the top, and alternating for each successive layer (Fig. 12). In this way, the mandrel rotation direction is better maintained and compaction is achieved.

Production Methods

For small- and intermediate-diameter tubing, tube rolling can offer a fast and extremely efficient method for high-volume production. Production equipment is available for all aspects of the process, from pattern cutting through final finishing.

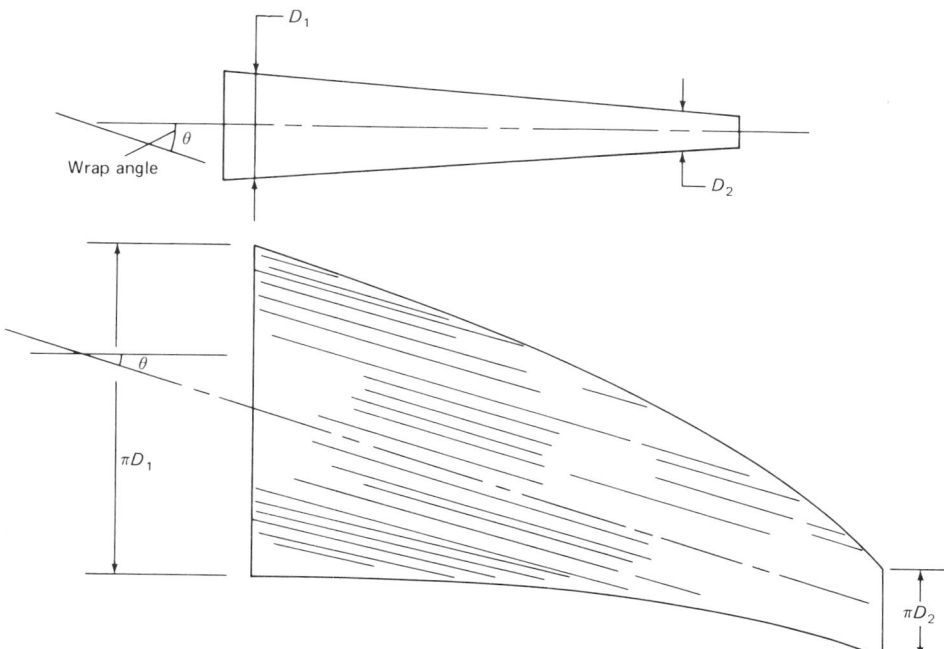

Fig. 11 Curved and tapered patterns for spiral wrapping large-diameter or steeply tapered tubes

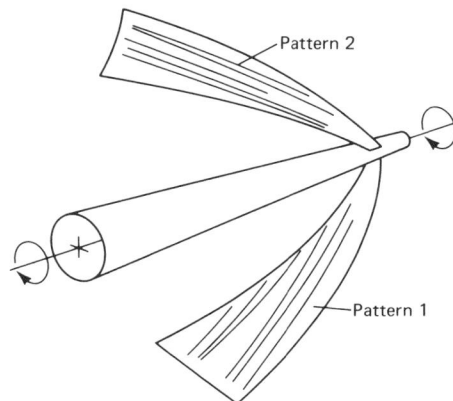

Fig. 12 Spiral wrapping of tapered mandrel

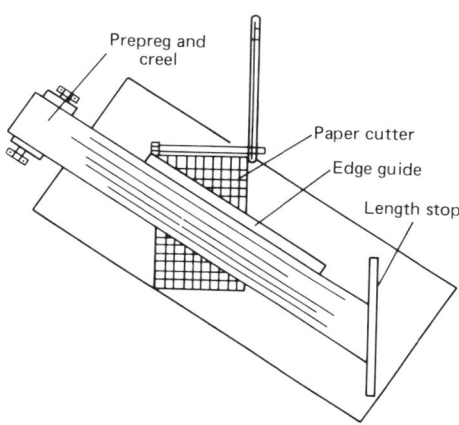

Fig. 13 Paper cutter for angle sheet cutting

Mandrels. A variety of materials have been used for tube rolling mandrels. To produce small- to intermediate-diameter tubing in volume, steel is commonly used because of its strength and stiffness. Steel mandrels are usually hardened and polished or chromed for ease of removal. Because the thermal expansion of steel is relatively low, heavy-duty mandrel pullers are required to remove the tube from the mandrel after cure.

Aluminum, which also has been used as a mandrel, is desirable for larger tubing because of its lower weight. Moreover, for a graphite tube that is greater than 50 mm (2 in.) in diameter, no puller is required because the differential expansion is sufficient to allow part removal after it has cooled. For fiberglass and aramid tubing, however, a mandrel puller is still required. In most instances, mandrel pulling is best attempted when the part is still hot because cold parts are usually more difficult to pull.

Metal mandrels are used when external pressure is applied during cure. For thick-walled tubing where external pressure causes fiber wrinkling, or where tube geometry changes require internal pressure during cure, alternative mandrel materials are available. Female molds are usually used when internal pressure is required during cure.

Both Teflon and solid silicone rubber have been used as materials for mandrels undergoing internal pressure molding. The high thermal expansion coefficient of these materials can be used to exert exceptionally high pressures when properly confined. The amount of pressure applied is dependent upon the cure temperature, mandrel volume, and restrictions imposed.

If more precise pressure control is required, a thin silicone rubber sleeve or a similar bladder material is placed over the mandrel and the part is rolled over the sleeve. The sleeve can then be removed from the mandrel, with the uncured tube in place. The tube and rubber liner are then placed in a closed mold and the bladder is inflated during cure. This method is particularly useful for fabricating irregularly shaped tubes. For round tubes, the mandrel can be left in place and air can be applied through a port in the mandrel.

Sheeting. All rolling operations must start with sheet goods that are subsequently cut into patterns. Sheet cutting can be as simple as using a straight edge and a razor knife. For narrow sheets up to 460 mm (18 in.) wide, standard paper cutters fitted with hardened steel blades for longer life have been used. Outfitted with a creel for material feed, paper cutters can be a very efficient method of cutting large quantities rapidly. A method for cutting angle sheets, shown in Fig. 13, has been used for many years to cut huge quantities of angle sheets.

In the area of automated equipment, high-speed sheet cutters are available. The typical fully automatic prepreg sheeter, (Fig. 14), which digitally controls length of cut, runs at speeds of up to 0.60 m/sec (2 ft/sec). Though shown in a straight cutting mode, it can be modified (at extra cost) to cut angle sheets.

Pattern cutting, like sheet cutting, can be done with a straight edge and razor knife, or with computer-controlled knives or water-jet cutters. Razor knives and templates, though primitive, are still used to cut large quantities of material. The trick is to stack multiple sheets of material and cut several patterns at one time.

Stacks of up to 20 sheets of graphite-epoxy tape can be cut this way. When complex or extremely long patterns are required, or when higher-speed cutting is necessary, the options range from steel-ruled dies and roller presses (Fig. 15) to computer-controlled knives and water-jet cutters.

Tube Rolling. Tube rolling can be accomplished by hand or with rolling equipment. Rolling tables such as that shown in Fig. 16 are widely used to make tubular products such as golf shafts, fishing rods, and other moderate-length tubes. The mandrel, with pattern tacked in place, is placed on the lower platen of the table. With the push of a button, the pneumatically controlled top platen drops to contact the mandrel, after which the movable lower platen slides back to totally envelop the pattern. The top platen is articulated to accommodate rolling either cylindrical or tapered tubes.

Shrink Tape Debulking and Cure. Heat-shrinkable tapes such as cellophane, polypropylene, or polyvinyl fluoride are used to incrementally debulk composite tubing during rolling. They are also used to apply the pressure during cure of noncritical parts such as golf shafts and fishing rods. The amount of actual

Fig. 14 Automatic prepreg sheeter cutting machine. Courtesy of Century Design

Fig. 15 Roller press with steel-rule dies. Courtesy of Century Design

Fig. 16 Tube rolling table. Courtesy of Century Design

Fig. 17 Automatic cellophane wrapping machine. Courtesy of Century Design

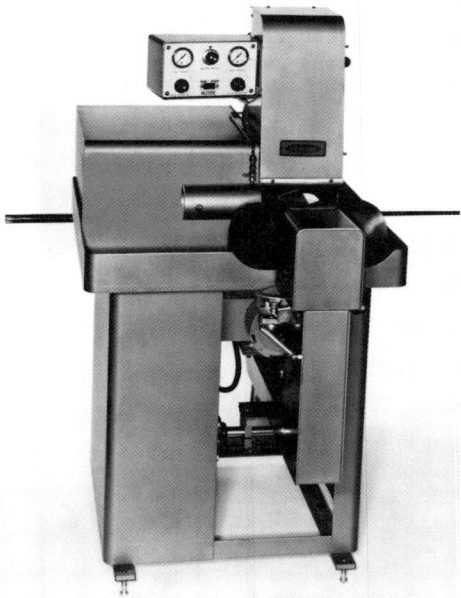

Fig. 18 Centerless sander. Courtesy of Century Design

pressure applied during cure depends upon the type of tape used, its thickness, the number of layers applied, and the tension at which it is applied. Cellophane provides the highest compactor pressure and bonds well to most prepreg resins. To facilitate cellophane removal, it is advisable to apply a base wrap of polyvinyl fluoride or polypropylene and then overwrap with cellophane. Because neither the polyvinyl fluoride nor the polypropylene bond to the resin, they can be easily stripped after cure. Polypropylene should not be used above 150 °C (300 °F) because it begins to degrade. If only cellophane is used, it can be removed by soaking the cured part in water and then stripping it off, which is a very messy job.

Shrink tapes can be applied with modified lathe set-ups where the part is chucked in a lathe and the tape is applied under tension. Care should be taken to ensure a uniform lead, because nonuniform leads can result in undesirable bumps or large resin blisters during cure.

Automatic equipment is also available for small- to intermediate-diameter tubing. One machine that is used extensively in the fishing rod and golf shaft industries is shown in Fig. 17. It can accommodate tubes of up to 75 mm (3 in.) in diameter and control the tape tension and lead. Its feed rate is approximately 0.30 m/sec (1 ft/sec).

Finishing. Tubular products lend themselves to high-speed finishing methods. After cure, some amount of sanding or grinding is usually required. Sanding produces a good quality surface for cosmetic purposes, whereas centerless grinding is used when precise diameter control is required. Centerless sanders are widely used for high production rate items such as golf shafts and fishing rods. The machine shown in Fig. 18 is designed to provide uniform material removal for both straight and tapered parts at rates of up to 0.30 m/sec (1 ft/sec). Material removal is controlled by pressure application and feed rate.

Manufacturing Processes: Aerospace

Chairman: Hans Borstell, Grumman Aircraft Systems

Overview and Basic Operations

AEROSPACE COMPOSITE STRUCTURE FABRICATION has progressed from a laboratory curiosity to world-wide production status in less than 25 years. Despite constantly increasing design requirements (Table 1), composite structures have become more affordable as each operation in the manufacturing sequence has been improved. The limiting factor in the growth of the industry is the cost of the completed parts. Thus, the intent of this Subsection is to present a comprehensive review of the currently available technology in the areas of tooling, ply generation, ply stacking, and curing.

The first high-performance fiber produced in the United States, the 100-μm (4-mil) boron filament was the result of an undesired reaction that occurred during the development of experimental, lightweight rocket fuels. Tungsten heating coils were used to heat gaseous compounds to produce a chemical reaction, but the coils were attacked by the gaseous compounds and covered with a deposit of elemental boron in the form of stiff, strong filaments. Because these filaments were found to have excellent properties, the emphasis was shifted from the rocket fuel to the filaments. In 1965, enough filaments were tediously produced in the laboratory to fabricate five small demonstration components. Concurrently, researchers in the United Kingdom were seeking to improve the mechanical properties of carbon-based fibers. Initially, the properties were not reproducible, and premium-grade carbon fiber was in extremely short supply and very expensive. Therefore, the initial manufacturing efforts were conducted under very closely controlled conditions to maximize the structural results and thus guarantee program success. Demonstrating the excellent

structural properties of advanced composite materials was mandatory in order to continue receiving the funding needed to explore the technology. Cost was not a primary concern at the start. In 1968, boron composites were selected for the empennage of two new fighter aircraft to provide significant weight savings. However, the earlier developmental structure required manufacturing improvements to reduce costs.

The pioneering manufacturing technology was based on the aerospace community's experience with autoclave molding of composite structures made with fiberglass fabric materials. This experience base had to be adapted to unidirectional tapes, which were used with the high-performance fibers in order to realize the best structural properties. This was a nonautomated labor-intensive operation.

The basic operations required to fabricate a part are: production of the individual plies, placement of the plies on the mold, preparation of the lay-up for cure, cure of the matrix in the autoclave under controlled conditions, often requiring additional heating in an oven (postcuring) to complete the resin polymerization, and trim and final inspection. These steps appear simple until the actual requirements defined (Table 2), are considered. The complexity of the process is also illustrated by the shape of some of the plies (Fig. 1), which had to be generated and accurately placed in a curing tool.

The initial test components were produced using manual lay-up techniques in which strips of unidirectional tape cut from a roll were placed on the mold to engineering requirements using various processes. The labor-intensive lay-up (Fig. 2) was the predominant factor in the high cost of this operation, and dictated

Table 1 Design requirements imposed on manufacturing

Year	Requirement
1965	Consider candidates
1966	Demonstrate maximum weight savings
1968	Provide protection against lightning strike
1972	Trade off some weight savings to reduce costs
1973	Save weight at a comparable cost to metal over the life of the vehicle
1975	Compensate for moisture effects
1978	Provide enhanced damage tolerance
1980	Provide reduced radar cross section

Table 2 Typical process requirements

Lay-up	1966 mm	1966 in.	Now mm	Now in.
Gaps between fibers	0.76	0.030	2.3	0.090
Overlaps of fibers allowed on complex contours	Not allowed		2.3	0.090
Size tolerance of partial plies	0.76	0.030	1.5	0.060
Location tolerance of partial plies	0.76	0.030	1.5	0.060
Control of fiber orientation-tape plies		±1°	±2° for flat surface ±7° for complex surface	
Splices in tape	Not allowed		Allowed within limits	
Shape of partial plies	Exotic shapes to save weight		Matched to capability of automated systems	
Cured layer thickness	Nominal ± 5% controlled by resin bleeding		Some fabricators use no bleed process	

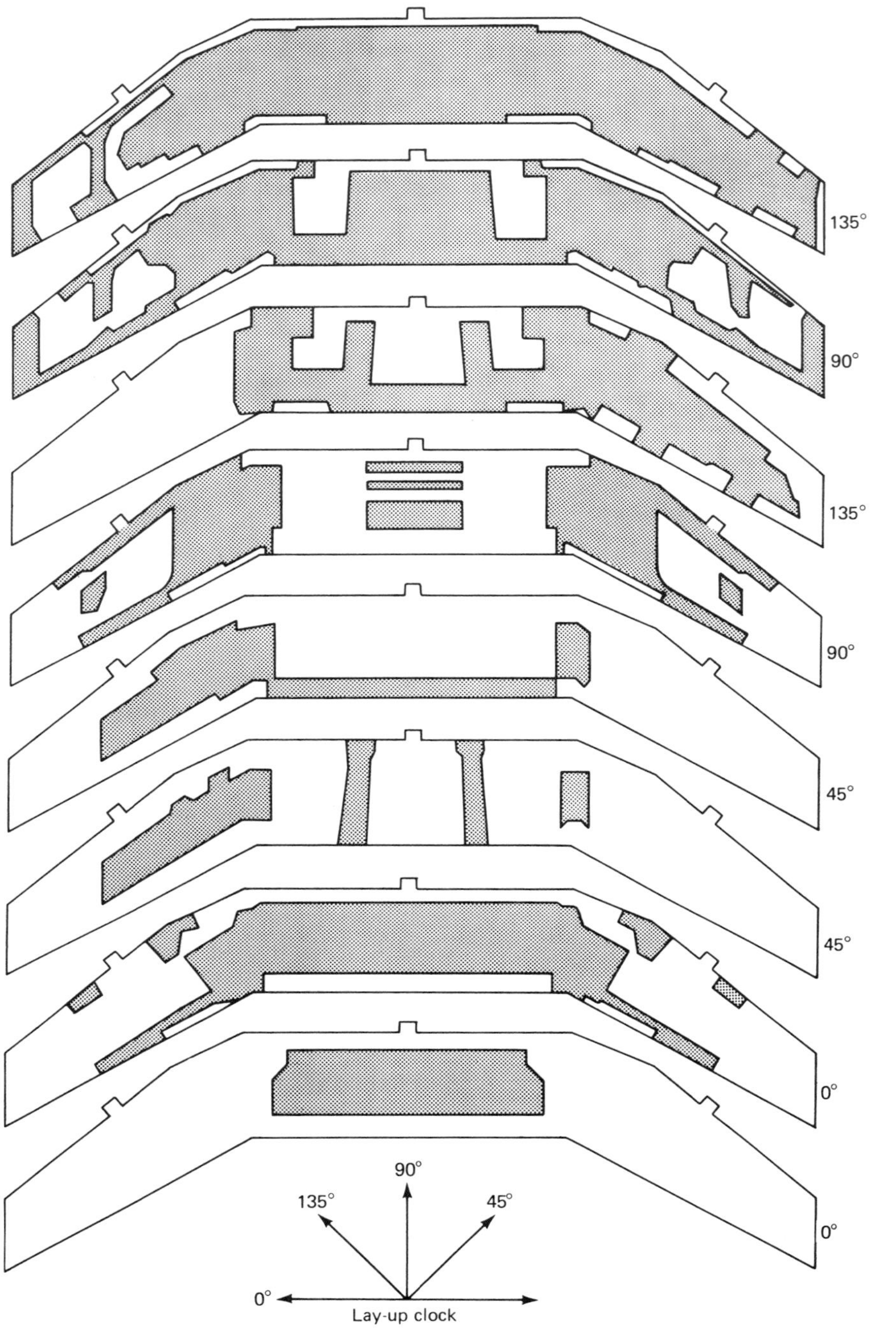

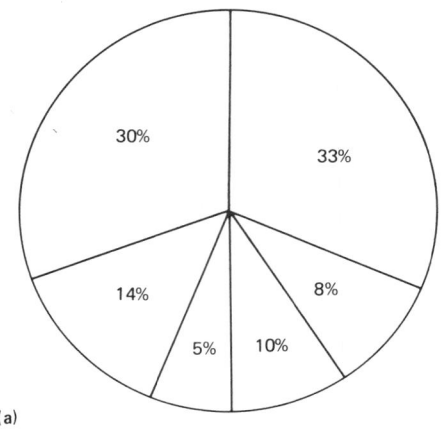

Fig. 1 Typical partial plies in the X-Z9A wing covers

(a)

Sine wave spar fabrication

Tasks	Man-hours, %
Lay-up Mylars	30
Stack plies and close tool	33
Cure/postcure	10
Bleeders	8
Remove part from tool/trim	5
Others(a)	14
	100

Others: Clean mold; vacuum bag; quality control panel; bond washers for fasteners

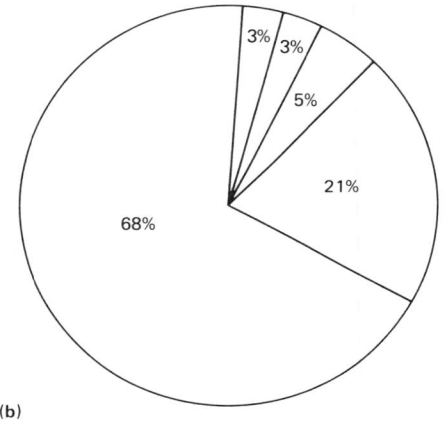

(b)

Cover fabrication

Tasks	Man-hours, %
Lay-up Mylars	68
Stack on MOF	21
Apply bleeder	5
Trim	3
Others(a)	3
	100

Others: Vacuum bag; cure/postcure; remove from tool; apply edge protector; quality control panel

Fig. 2 Man-hour breakdown of molding using manual processes. (a) Sine wave spar fabrication. (b) Cover fabrication

the continuing development of automated/mechanized systems.

The matrix of operations and processes shown in Table 3 identifies the diversity of available options to reduce the cost of the lay-up operation. Initially, the choice of method was a matter of corporate philosophy and part geometry, but now the larger fabricators have installed systems using various types of processes. Just as several types of equipment are needed to machine metal parts of various configurations, the composite factory of the future must have systems dedicated to processing parts of various configurations. The creation of an automated system to produce substructural shapes or fuselage panels is still in the development stage, despite the large number of composite parts being required for each aircraft. The issue of reducing lay-up costs is so basic to manufacturing cost control that as many concepts as possible are presented in the following articles. However, as of 1986, the use of automated systems is limited, and the specific reasons are:

- Structural design must be compatible with automated processing

Table 3 Hand, mechanized, and fully automated lay-up processes

Process	Ply generation	Placement on tool	Forming of prepreg to tool shape
Hand lay-up	Lay tape on tool or on lay-up templates Cut fabric to shape using templates	Place manually using tooling to coordinate the partial plies to the tool: 　Lay-up templates 　Bank rails at edges	Use vacuum bag and localized heating to soften the prepreg
Mechanized	Cut from wide goods with the Gerber cutter Cut from wide goods with Clicker press and steel rule dies	Stack on tool manually as above, but use an optical/layer system to locate plies on the mold	Use vacuum bag plus localized heat as above. Use mechanical devices to seat plies on the webs of sine wave spars (rollers) Use devices to apply vacuum/pressure to seat plies on male tools by means of a diaphragm Use devices to heat the prepreg so it can be formed easily by hand
	Lay-up plies in the flat with a tape layer	Use robotics or other mechanized system to transfer plies to the tool	
Fully automated	For wing and tail skins and parts of similar gentle contour, use contour tape layer to lay up the plies directly on the mold. Applications limited by the capability of the tape layer		
	For automated laminating cell, cut the plies with a Gerber cutter Use a robot to pick up the plies from the Gerber table and place them on the tool		
	Braiding of structural shapes, filament winding of surface of revolution parts, and pultrusion of structural shapes are described elsewhere.		

- Development of systems by the private sector is limited to those having an identifiable commercial market
- Some excellent systems developed with government funding have not been duplicated, and others developed by aircraft manufacturers are proprietary
- Many aerospace contractors have chosen to invest their funds in other manufacturing technologies

Currently, there are approximately 20 automated tape-laying machines supplied by four domestic producers that are being used to lay-up wing and empennage covers. Multiaxis filament winders that are based on commercial fiberglass technology are being used to produce nacelles, rocket motors, tankage, and prototype fuselages. The numerically controlled Gerber cutter has been adapted to cut almost every shape of composite plies at high rates.

After the plies are stacked on the tool, the part is prepared for autoclave cure. In the early days of this technology, all the lay-ups were covered with dry materials to bleed off excess resin, to reduce part weight and improve the mechanical properties of the layers. (Some fabricators now omit this operation, especially for thin parts, which only offer small weight savings.) Then, a nylon film vacuum bag is applied to isolate the part from the autoclave environment. To simplify this critical operation, reinforced rubber caul plates have been developed to eliminate the wrinkles produced by the bag. Reusable rubber bags have been developed to accelerate the bagging operation and reduce risk, but their use has been limited because they require storage.

The autoclave curing process has been made more cost effective by equipment design improvements that reduce energy consumption and by the development of computer controls to significantly reduce manual involvement. Cost reductions arising from improved methods for trimming, assembling, and inspecting cured composite parts are described in subsequent articles.

As composite processing technology has improved, so has tooling technology. Machined metal and laminated fiberglass tools have been supplemented by nickel electroform tools for complex shapes, with savings in acquisition and operating costs. Graphite-epoxy tools have been developed to reduce the thermal expansion mismatches which exist between the composite parts and the metal tools. Silicone rubber tools are sometimes used under the vacuum bag to apply curing pressure more uniformly to minimize voids. Fundamental tooling concepts also have changed to reflect the fact that the layer thickness of composite parts is not constant. Tooling to produce lightly stressed honeycomb panels in one operation has been developed to eliminate the cost of fitting skins to the core.

The molds used for mechanically fastened or secondarily bonded structures are most often made to an inner mold line dimension to simplify fit-up. The most promising emerging technology is integrally molding (co-curing) stiffeners and skins to reduce or eliminate secondary bonding or fastening. Thus, for each assembly operation, a higher initial investment in tooling is traded off against a savings, producing significant cost reduction, in a production program.

Despite the advances in composite manufacturing technology, parts with defects are still produced. It should soon be possible to establish allowable defect criteria which distinguish the benign defects from the dangerous ones. This advancement would permit the use of simplified processes because the penalty for not producing a perfect part would be reduced.

Tooling for Autoclave Molding

Hans Borstell and K.T. Turner, Grumman Aircraft Systems

IF COMPOSITE PARTS were always flat, had constant thickness, and were cured at room temperature, they would fulfill the fantasy of all tool designers. Properly designed tools that produce acceptable parts on a reproducible basis are a must when fabricating composite structures. In fact, their design requires the consideration of as many factors as are studied in the design of the part itself.

The autoclave molding process generally uses a vacuum bag to impose a pressure differential onto the lay-up. However, the mold itself is totally surrounded by the autoclave atmosphere, so that the forces imposed on the tool are small. This type of tool could theoretically be a piece of aluminum foil, if the lay-up and foil were enclosed in an envelope-type vacuum bag. Because the part must have a definite contour specified by an engineering drawing, the mold has to maintain its shape while supporting the weight of the part, being exposed to multiple heating and cooling cycles, and being moved around the plant. These requirements would favor a massive tool. However, the tool must also be heated to a specified temperature at a specified rate under controlled conditions in the autoclave, and these requirements would favor lighter tools of a uniform thermal response throughout.

Although computer control can be used to heat up most tool combinations acceptably, the autoclave cycle times are increased if a massive tool is used concurrently with thinner tools, as shown in Table 1. In addition, more facilities for moving and lifting tools are needed as the tool mass increases. These facilities include cranes and forklift trucks for lifting molds and electrically powered tugs to move them between the lay-up and autoclave areas. Thus, a balance is needed between making a tool that is rigid enough to maintain contour and making one that is overdesigned and therefore difficult to use.

Unlike press molding with matched dies, the tooling for autoclave molding controls the contour of only one surface of the part. The part thickness varies considerably because of tolerances in both the material and process. Supplemental tools, known as caul plates, can be used to improve the contour of the face of the part away from the mold, as discussed below. The decision as to which surface of a part to place on the mold is often a complex one based on factors beyond the control of the tool designer.

Because most process specifications permit part thickness to vary $\pm 5\%$, fit-up of joints is a problem, but it can be reduced by molding on faying surface tools. For mechanically fastened structures, the covers are frequently produced on molds having configurations that match the surface contacting the substructure (inner mold line, or IML). These tools have a more complex surface geometry than air passage (outer mold line, or OML) tools. They are also more expensive and are harder to design. However, the surface mating to the substructure is reproducibly controlled. Thus, if the tolerances in the substructure are reduced, an acceptable fit can be obtained without shimming. Precured covers for highly stressed honeycomb panels are also molded on IML tools so that the contour of the mating surface of the skin and honeycomb is always the same. Typically, OML tooling is used for honeycomb panels made by the co-curing process in one molding operation.

Nose radomes are required to be molded in OML tools to enhance the aerodynamic smoothness of the aircraft. Integrally stiffened panels are usually molded on OML tooling because of the difficulties that would be encountered in the application of curing pressure to the stiffeners if IML tools were used. In contrast, composite parts in the shape of angles, channels, or I-beams are usually cured on male (IML) tools because the lay-up is easier. Thus, the selection of which part surface to create a tool for is complex and must reflect the needs of the entire manufacturing sequence rather than the preference of the tool designer.

Tooling Configurations

Basically, tooling for the autoclave molding process involves two basic configurations: a face sheet supported by stiffeners mounted on a base (the eggcrate concept), and machined blocks or plates. Figure 1 shows a typical mold for a panel-type part, such as a leading edge, that is designed using the eggcrate configuration. A contoured face sheet is supported by ribs to maintain the contour. The ribs are attached to base members, which provide hard points for ease of handling. The attachment of face sheet to rib can be accomplished either by welding or with mechanically fastened clips. (Less common alternatives are hooks that are stud welded to the face sheet and loops that are bolted to the ribs.) The face sheet is typically 6.4-mm (0.25-in.) aluminum or 4.75-mm (0.187-in.) steel plate formed to contour in as many pieces as are required by limitations of stock size or the forming process. The joints are welded and then blended to contour.

The expense and difficulty of producing complex-contour mold surfaces are overcome by use of the process described in the article "Electroformed Nickel Tooling" in this Section of the Volume. In cases of complex contour, it may be desirable to produce the face sheet and ribs concurrently by the casting process. It sometimes becomes necessary to split the mold so that the part can be removed from it, as shown in Fig. 2. Accurate machining of

Table 1 Thermal characteristics of tooling materials

Tool material	Tool thickness		Tool rise time(a) (ambient to 110 °C, or 230 °F) min	Overshoot (degrees over ambient)	
	mm	in.		°C	°F
Aluminum	6.4	0.25	41.6	13	23
Steel	6.4	0.25	(b)	9	15
Carbon-epoxy	6.4	0.25	45.8	14	24
Aluminum	25.4	1.00	48.0	(b)	(b)
Steel	12.7	0.50	48.6	7	12
Carbon-epoxy	12.7	0.50	51.7	13	23
Steel	38.1	1.50	73.8	2	3
Aluminum	101.6	4.00	86.3	0	0

(a) In response to 2 °C/min (4 °F/min) ramp from ambient to 240 °F. (b) Testing difficulties prevented accurate determination of values.

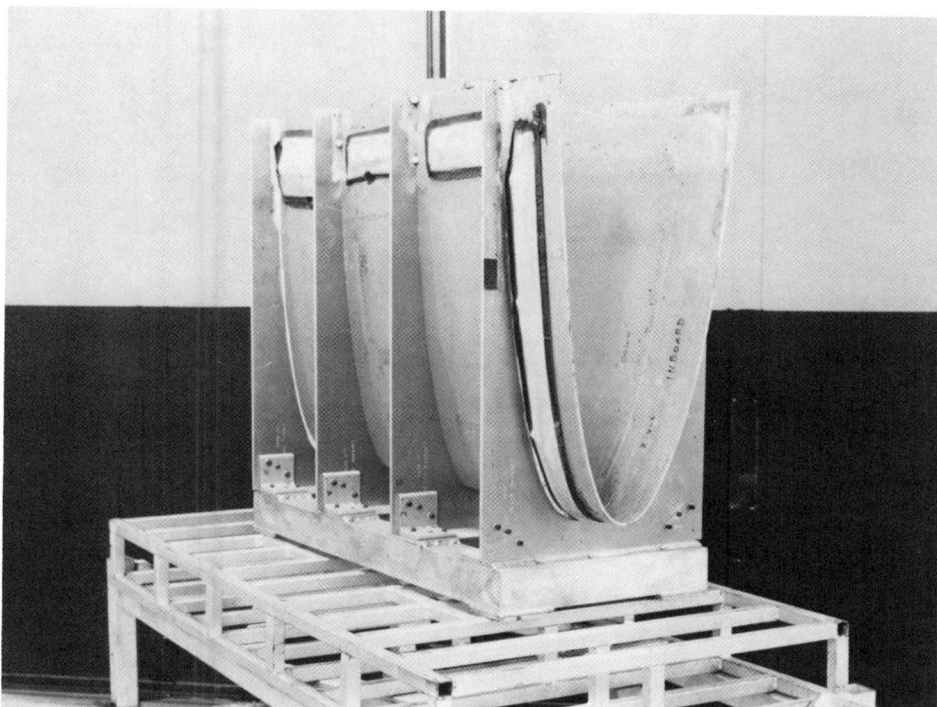

Fig. 1 Typical eggcrate design mold mounted on handling rack

Fig. 2 Split mold produced by casting

fasteners. Adhesive bonding has also been used on limited occasions.

Several design features of the substructure have a significant effect on mold performance. The metal in the face sheet and substructure should be of similar alloys to avoid distortion of the face sheet due to thermal-expansion mismatches. If the base members are closed cross sections, such as square tubing or pipes, they should be sealed at the ends to prevent localized overheating due to the passage of hot air through the member while in the autoclave. If the base members heat up more rapidly than the face sheet, a thermal strain is induced in the face sheet, which will deform the contour of the mold and subsequently, of the part. The substructure must be designed to maximize the flow of air under the face sheet to increase the heat-up rate of the tool and reduce thermal gradients through thick parts. Ideally, the ribs are parallel to the length of the mold. Often, this is not possible if the rib-to-face-sheet joints are used to help seal welded joints in the face sheet. If transverse ribs are used, large cut-outs also should be used to maximize the flow of air under the face sheet.

Once the design concept for the mold is selected, provision for evacuating the material under the vacuum bag must be addressed. Usually, vacuum from the autoclave is introduced to the mold by means of hoses that have quick-disconnect fittings. Piping under the face sheet leads to fittings (vacuum ports) that penetrate the face sheet. For large tools, the piping to the vacuum ports should be designed to match the location of the vacuum sources in the autoclave, to simplify the hook-up. Typically, the vacuum ports are located 50 to 75 mm (2 to 3 in.) beyond the edge of the part to provide space for the use of a porous medium to connect the vacuum ports. This medium, often a chain or fabric rolls of felt or glass, distributes the vacuum completely around the perimeter of the part to accelerate the evacuation of materials under the vacuum bag. A minimum of two vacuum ports are required: one to apply vacuum from the autoclave and one to be connected to a gage to monitor the vacuum under the bag during cure. As the part size increases, more vacuum ports are needed. The number of ports is a trade-off between the cost of hooking/unhooking the hoses and more rapid evacuation of the bag. Typically, the ports are spaced 2 to 3 m (6 to 10 ft) apart.

The vacuum bag is attached to the mold surface by means of a "tacky tape" (bag sealant), a frame that provides clamping pressure, or interlocking rubber extrusions. Because space must be provided for this joint, about 125 to 150 mm (5 to 6 in.) is usually needed between the edge of the part and edge of the mold. A diagram of the relationship between the vacuum system and lay-up is shown in the article "Preparation for Cure" in this Section.

To prevent surface irregularities on the bag side (untooled surface) of the parts, an elas-

deep pockets or steps in the face sheet is difficult and significantly increases the thickness of the face sheet. Thus, stepped build-ups in the face sheet are often produced by adding prefabricated, formed plates locally. These attachments are usually added by welding or

Fig. 3 Male mold fitted with caul plates to mold a composite channel with an integral flange

Table 2   Data for calculating thermal corrections for tool designs

Thermal correction = engineering dimension × (CTE_{part} − CTE_{tool}) × (gel temperature − room temperature).

Material	CTE 10^{-6}/K
Structural composite material	
Boron-epoxy	3.6-10.8
Aramid-epoxy(a)	−2.0-5.8
Graphite-epoxy(a)	1.8-9.0
Fiberglass-epoxy(a)	7.2-9.0
Tooling material	
Graphite-epoxy(a)	4.1-9.0
Cast ceramic.	0.81
Tool steel	11.3
Iron (electroformed)	11.9
Nickel (electroformed).	12.6
High-temperature cast epoxy	19.8
Aluminum	23.2
Silicone rubber.	81-360

(a) Varies as a function of ply orientation. Values shown represent typical parts.

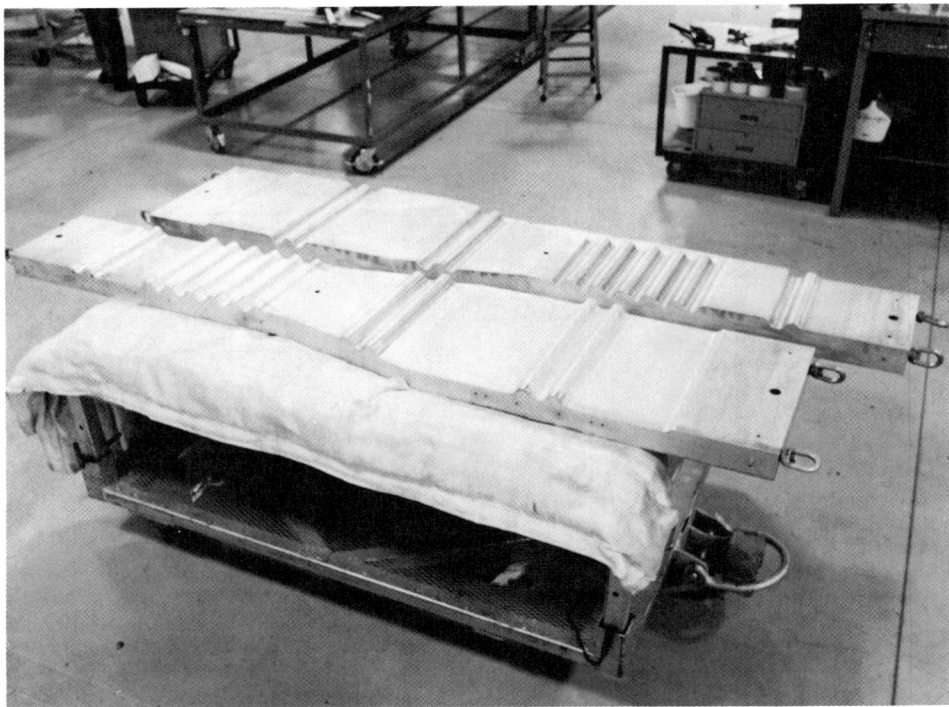

Fig. 4 Mating blocks used to mold an I-beam with a corrugated web

tomeric, flexible caul plate is often used. A caul plate can either be a sheet of thermally stable rubber, such as silicone or fluoroelastomers, or a molded caul plate, as described in the article "Elastomeric Tooling" in this Section. The sole purpose of caul plates is to improve the visual appearance of the parts. They do not control part thickness.

A large number of parts, such as angles, channels, or I-beams, have cross sections. These parts are typically molded on male blocks, as shown in Fig. 3 and 4. If the cross section of the molds for angles and channels is relatively small, acceptable handling and thermal response are obtained from the solid blocks. Small male tools are typically cured on rolling tables fitted with vacuum ports or the large platens usually provided as part of the autoclave. Larger tools may be permanently mounted on a base plate fitted with vacuum ports, lift points, and attachments for wheels to move the tool. Hollow blocks may be used provided that they are sufficiently rigid to

prevent distortion during cure and that all seams are within the vacuum bag.

The mating blocks used to mold I-beams (Fig. 4) are pinned together at the ends beyond the part trim to maintain the alignment needed to control the contour and part thickness specified on the engineering drawing. These tools are usually cured in a vertical position using an envelope vacuum bag and removable vacuum ports.

There are a small number of parts that have complex surfaces and very little contour. If this is the case, a simple alternative to the use of the eggcrate design is to machine the desired contour into a plate and mount the tool on a rolling table.

Regardless of the configuration selected for the tool, there are three issues that remain to be addressed, namely, correcting thermal expansion coefficients, coordinating the location of partial plies, and using rigid caul plates to accurately control the thickness of local areas of the parts.

Thermal expansion correction is required because composites have lower coefficients of thermal expansion (CTE) than do the most commonly used tooling materials, as shown in Table 2. In the autoclave, the temperature at which the resin solidifies is the gel temperature. At that specific temperature, the part is the same size as the thermally expanded mold. At temperatures above the gel temperature, the tool expands more than the partially cured part introducing a thermal strain. As the part and tool cool down from the gel temperature, the

tool usually shrinks more than the part. Because graphite-epoxy molds have coefficients of thermal expansion that are very similar to graphite-epoxy parts, the thermal factors in these cases can be ignored. This feature is a major advantage of graphite-epoxy molds.

As shown in Table 2, the gel temperature and the CTE of the mold and part must be known in order to calculate the dimensional correction required in the tool design. The gel temperature can be easily determined by use of the rheometric spectrometer, as described in the subsequent articles on curing. The coefficients for metallic tooling materials are known. However, the CTE of composite laminates vary significantly with fiber orientation. In the case of a graphite-epoxy laminate, the coefficient varies from almost zero in the fiber direction to 19×10^{-6}/K perpendicular to the fiber.

Although some data has been published, not all composite materials have been measured. One empirical method is to cure a representative panel on a plate of the specified tooling material using the specified cure cycle. Scribe marks are made in the mold at a known distance. The difference in the distance between the scribe marks molded into the part shows the correction needed. One method recommended by a major fabricator is to correct steel or nickel tools by making the tool 0.999% of the engineering dimension, that is, a 2540-mm (100-in.) dimension is tooled to be 2537 mm (99.9 in.). A correction of 0.998 is used for aluminum tools. These corrections are needed

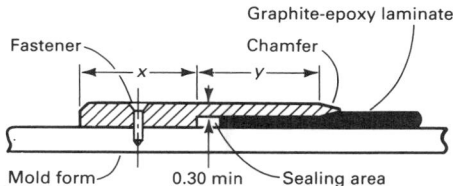

Fig. 5 Caul plate to control the edge of a panel. $x \geq y$

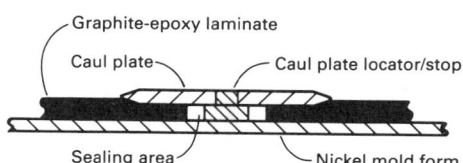

Fig. 6 Caul plate for an access hole

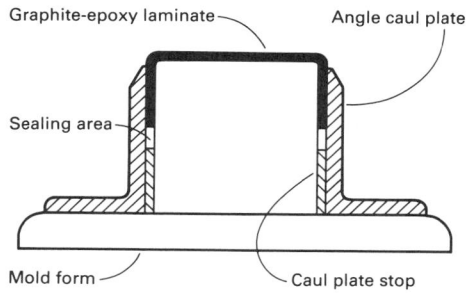

Fig. 7 Caul plates for channel flanges

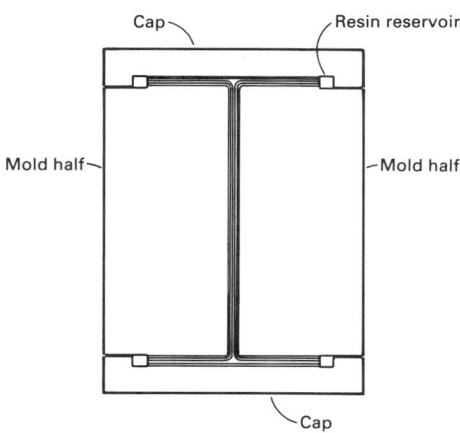

Fig. 8 Caul plates for I-beam caps

to ensure an acceptable fit of mating composite parts.

Location of Partial Plies. Most parts contain partial plies to accommodate local areas of increased stress. The location of these partial plies must be controlled both in making the detail part and in subsequent assembly operations. Separate tooling must be provided for both operations. The techniques for controlling the location of partial plies are given below.

Ply on Polyester Film. The plies are laid up on polyester film templates that are coordinated to the mold by pins on the tool and holes in the templates. The pins should be of different sizes, and located away from the centerline of the part so that the mylars cannot be mislocated on the tool.

Slotted Templates to Locate Precut Plies. A contoured replica of the mold surface is produced either by laying up a fiberglass laminate or forming a thermoplastic replica. The template is coordinated with the mold by pins or stops. The plies are laid out on the template and slots are cut to outline the plies. The mechanic installs the template on the tool, marks the outline of the plies with a colored pencil, and then removes the template. Templates for gently contoured parts can often be scribed by a plotter. The scribe lines are then filled with paint for increased visibility.

Rails and Banking Surfaces to Locate Precut Plies. Metal rails are attached to the mold at the location of oversized part trim. Scribe lines representing the ends of the partial plies are added to the rails in the specified locations. Thus, partial edge plies can be laid up using the rails and scribe marks to control their location. Frequently, access holes requiring local reinforcement are located within the interior of large panels. These partial plies can be located using a block (plug) attached to the mold. The plug should be of some asymmetric configuration to ensure that the precut plies cannot be laid up incorrectly.

Once the part has been cured, tooling holes located in a known relationship to the partial

plies are used in subsequent assembly operations. The location of the tooling holes in gently contoured parts is controlled by a drill plate pinned to the mold. The hole is drilled before the part is broken loose from the mold. In the case of channels and I-beams, an acceptable method is to incorporate the bushing in a trim tool that fits snugly within the flanges of the part. The location of these tooling holes must be coordinated between the molds and assembly tooling. If this coordination is not done correctly, expensive errors will be found when the parts are placed in the assembly or bonding tooling.

Rigid Caul Plates. The last major decision is whether or not to use rigid metal caul plates to control the thickness of composite parts to a precise dimension in local areas. The caul plates are needed to overcome variations in the fiber-resin content of the uncured part and in the tolerances in the process. Typically, these caul plates are used in four applications, namely, to control:

- The edge of a panel, especially when a joggle is used for the attachment of mating structures
- The contour of an access hole fitted with a cover
- The flanges of channels molded on male tools
- The caps of I-beams

The design of these metal caul plates must take into account the fact that the matrix resin melts in the autoclave to a very low viscosity. In many cases, a viscosity similar to lightweight motor oil is maintained for a reasonable length of time during autoclave cure. The caul plate performs by pushing excess resin sideways. A piece of sheet metal laid over the panel does not control part thickness/contour in a reproducible manner because the amount of resin that moves sideways varies. Thus, the rigid metal caul plates used to control part thickness must have:

- High rigidity, so that they do not deflect under autoclave pressure at curing temperature
- A large base beyond part trim, to prevent tipping due to cantilever action
- Tapered edges where the caul plate comes in contact with the composite, to smooth

the path of the vacuum bag and thereby prevent wrinkles
- A sealing area, to prevent a significant loss of resin during cure

The thickness of the caul plates can be calculated by use of the equations for unsupported bending beam analysis. The caul plate deflections should be limited to half the tolerance permitted in the part.

Concepts for the caul plate applications described above are shown in Fig. 5 to 8. Note that the caul plates for the channels shown in Fig. 3 are acceptable for small parts. The concept shown in Fig. 7 is preferred for larger parts because the flanged caul plate is much more resistant to tipping.

Future Trends

An emerging trend is the molding of integrally stiffened panels by a co-curing process. This concept trades off an investment in complex tooling against lower assembly costs for each panel. The stiffeners are formed over mandrels, which must be removed from the cured parts. For open-section stiffeners such as blades, J's and I's, metal mandrels can be used. The mandrel dimensions are thermally corrected. The mandrels may be split to facilitate removal and are located relative to the skin by jig points in the mold.

The use of closed-section stiffeners, such as hat sections, requires the use of Teflon, washout ceramic, rubber, or foam mandrels. Ideally, the thickness of the lay-up under the hat forming mandrels is constant to permit use of Teflon or rubber mandrels that are reusable. Foam or washout-ceramic mandrels are expensive because they must be produced for every part.

The basic concepts and trade-offs in the design of tools for autoclave molding have been discussed as they apply to aluminum and steel molds. The additional data presented in the subsequent articles in this Section also pertains to metal molds.

Electroformed Nickel Tooling

Don L. Sheldon, Electronics Metal Finishing Corporation

MOLD GEOMETRY often becomes more complex as the use of composite structures increases and the creation of new parts is attempted. In the current cost-conscious manufacturing environment, tooling economics must be considered, in terms of acquisition cost, thermal efficiency, maintenance cost, and service life.

Typically, large molds are fabricated using steel or aluminum weldments with a formed face sheet machined to final contour. This expensive process must be repeated to make the duplicate tools needed for high-volume production. An alternative is to fabricate a fiberglass-epoxy-laminated tool using a model of the part surface and building up the mold thickness, layer by layer.

The electroformed nickel tool consists of a 4.6 to 6.4 mm (0.18 to 0.25 in.) thick electrodeposited mold surface that is supported by a simple steel substructure. The mold surface is produced by the electroplating process described later.

Advantages of Electroformed Tooling

The electroformed tooling concept offers these numerous advantages:

- Size of mold restricted only by size of the electroforming tank
- Low cost to produce duplicate tools
- Very smooth and scratch-resistant mold surface
- Reproducible coefficient of thermal expansion (CTE) (13.5×10^{-6}/K) that is approximately 40% less than aluminum
- Rapid heat-up and cool-down rates with excellent thermal uniformity during autoclave cure of composite parts
- Ease of handling and transport because of light weight
- Outstanding durability because mold surface resists cutting or impact damage and is not thermally degraded; some tools have been used for more than 15 years without major repair
- Ease of repair by welding, soldering, silver-soldering, or selective plating, if damaged
- Excellent vacuum integrity at temperatures up to 537 °C (1000 °F)

Fig. 1 Backside of mandrel for aircraft wing

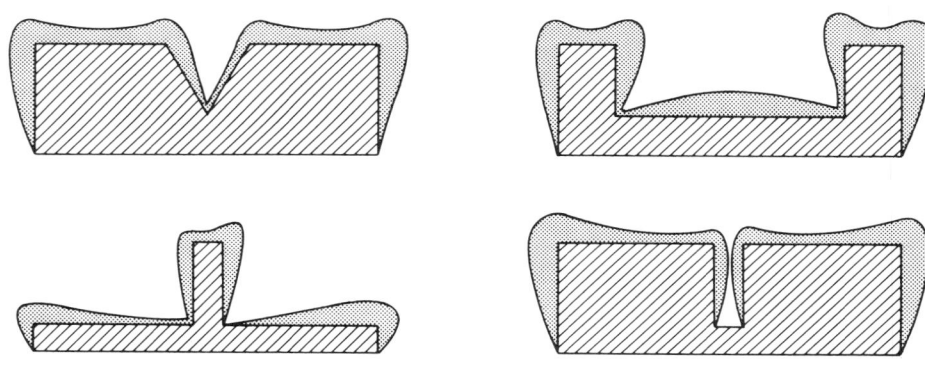

Fig. 2 Extremely difficult mandrel contours

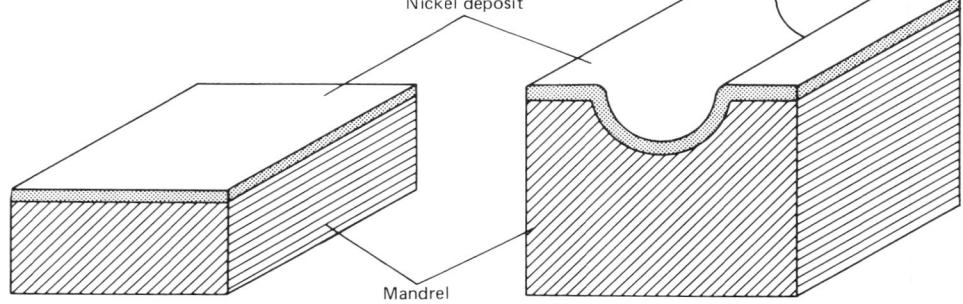

Fig. 3 Easy mandrel shapes

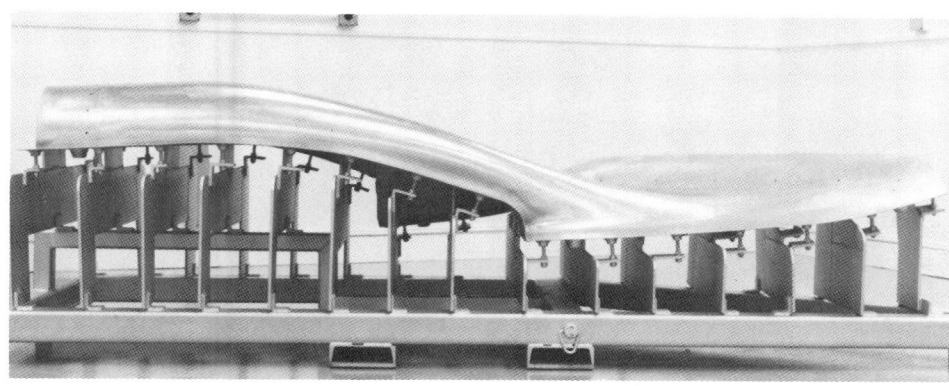

Fig. 4 Lay-up tool for bonding

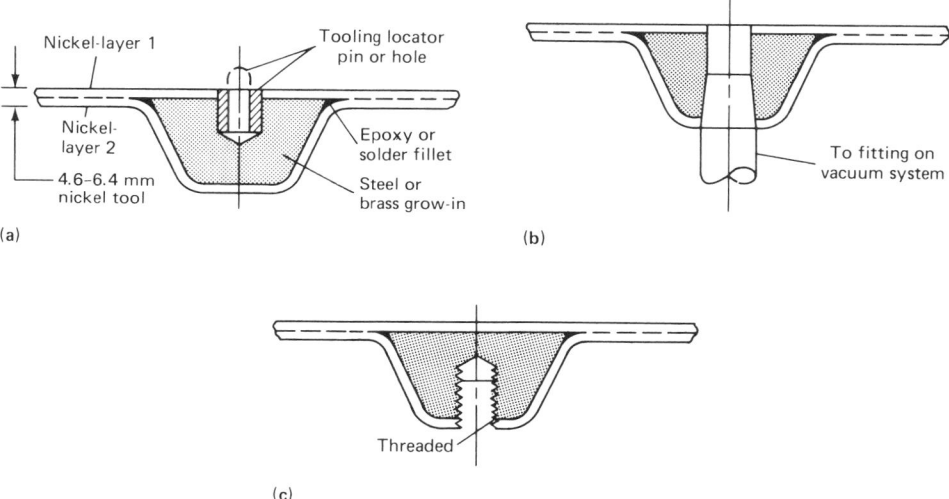

Fig. 5 Types of grow-ins. (a) Grow-in for applications requiring alignment of parts. (b) Vacuum port grown into nickel tool. (c) Grow-in threaded for mounting to welded support

- Complex contours possible without expensive machining
- Good release properties with most resin systems

Mandrel Use in Electroforming Process

Electroforming is the process of producing an article by electrodeposition of a metal onto a conductive mandrel (cathode) surface. An anode suspended in an aqueous electrolyte is connected to the positive pole of a dc electric source, and the mandrel is connected to its negative pole. The flow of electricity or electrons results in the oxidation of a nickel anode to nickel ions and the reduction of nickel ions to nickel metal at the cathode. As the atoms deposit on the mandrel, the electroform thickness begins to grow; the typical rate of growth is approximately 0.013 to 0.025 mm (0.0005 to 0.001 in.) per hour. When the electroform is removed from the mandrel, its surface is a mirror image of the surface of the mandrel. A typical mandrel is shown in Fig. 1.

As in some other types of tooling, constructing a model of the part surface is the first step in creating an electroformed mold. The models are the same net dimensions as the required nickel mold. Compensation may be required when the CTE of the composite part differs greatly from that of the nickel mold. Models are made from plaster, epoxy-faced plaster, fiberglass, fiberglass-epoxy, wood, or other materials. From the model a reverse mandrel or "mother" is generally fabricated from epoxy-faced fiberglass or plaster. The mandrel to be used in electroforming is then copied from the mother, although the model can be used as the mandrel if it is prepared correctly. This elimination of two fabrication steps greatly reduces cost. Because the electroforming process copies with extraordinary accuracy, the slightest pin hole or blemish on the mandrel will be reproduced. An additional hour of mandrel preparation will provide the electroform with a greatly improved finish.

The corners of the mandrel should be designed to have radii in excess of 0.76 mm

Fig. 6 Lay-up tool for side panel

Fig. 7 Aircraft lay-up tool

(a)

(b)

Fig. 8 Mold for radome tool. (a) Side view. (b) End view

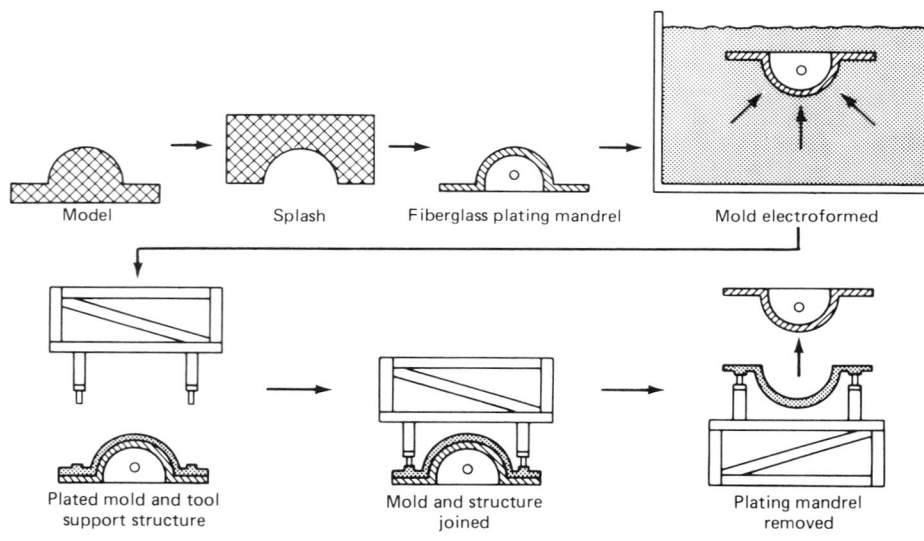

Fig. 9 Electroforming process sequence

(0.030 in.) to avoid thin spots in the deposit. Draft and taper should be designed into the mandrel to facilitate its removal from the electroform. Sharp corners should be avoided if possible. Narrow, deep grooves also should be avoided and can be replaced with shallow, gradual grooves with radii of 0.76 mm (0.030 in.) at the bottom. Figures 2 and 3 depict difficult and easy mandrel shapes, respectively. In all cases, before designing any mandrel, the electroforming vendor should be consulted for design information and specifications. The mandrel can be fabricated from epoxy-faced fiberglass, rubber, or other materials. The back of the mandrel must be reinforced with an egg crate structure of fiberglass, steel, or another suitable material. The reinforcement must be stiff enough to keep the mandrel from distorting during the electroforming process.

Electroforming the Mold

Preparation. When the mandrel is received in the electroforming shop, a buss bar is attached around its perimeter. The buss bar must extend above the solution level of the plating tank for attachment to the cathode from a dc rectifier, and must be large enough to carry the required current.

A natural physical characteristic of electrodeposition is that electric current will tend to localize the deposit on all edges and corners, causing an uneven thickness on the electroform. To overcome this problem, a nonconductive shield is used to alter the lines of current by forcing the ions to flow around it, which results in a more even deposit. Another method that is occasionally used involves placing an additional piece of metal ("thief") near a high current density area to draw off the excess current and the metal. Also, an auxiliary anode may be placed in low current density areas to supply more ions directly to the surface.

The art of controlling the thickness of the deposit on the mandrel is learned through experience only. However, one should not expect a 100% uniform electroform thickness. The thickness should be planned to exceed the minimum specified limits, in the thin areas. The electroforming vendor can give specific recommendations for a specific tool configuration.

After the mandrel has been designed and fabricated, the next step is to clean it thoroughly to remove all moled release material, dirt, or grease. Vigorous use of a nylon brush with either solvents or detergent will accomplish this. When the mandrel is clean, a conductive coating of silver is applied to its surface so that it will carry a current. It is then ready to place into the electroforming tank.

Equipment. The largest tank currently available is 10.36 m long, 2.44 m wide, and 4.57 m deep (34 ft × 8 ft × 15 ft) and contains 115 821 L (306 000 gal) of nickel solution. The mandrel for the tool shown in Fig. 4 was plated at the rate of 161 A/m² (15 A/ft²) with a thickness build-up of 0.19 mm (0.00075 in.) per hour.

Plugs. After the nickel is plated to a thickness of approximately 2.54 mm (0.100 in.), the mold can be removed from the tank. At this time, threaded plugs can be cemented to the nickel electroform with high-temperature epoxy paste adhesive. Other plugs for tooling holes, vacuum attachment, and thermocouples can be applied at this time. Figure 5 depicts several types of grow-ins. The process permits plating-in as many plugs as are needed for the tool design. Up to 162 plugs have been used on a single mandrel. The mold is then returned to the tank and plated to a minimum thickness of 4.6 mm (0.180 in.).

Steel Back-Up Structure. After the desired mold thickness is obtained, the mold is removed from the tank and the steel back-up structure is attached to the threaded plugs. This substructure is designed to hold the nickel electroform mold firmly at all times.

When the steel substructure has been completed, the mandrel can be removed from the nickel mold. The nickel mold can then be polished to the required finish with a minimum amount of effort. The steel frame is then painted for protection from corrosion. The mold is checked for vacuum integrity and other quality assurance requirements before being shipped to the customer.

Typical molds are shown in Fig. 6 and 7 to demonstrate the versatility of nickel electroforming. Figure 8 shows the largest electroform produced yet, with a length of 9.1 m (30 ft) and a total surface area of 24.7 m² (266 ft²). Figure 9 summarizes the sequence of the electroforming process.

Graphite-Epoxy Tooling

Billy D. Harmon, Fiberite Corporation

AS THE USE OF COMPOSITE MATERIALS has increased in the past decade, part size and complexity have grown considerably. Consequently, the selection of the proper tooling material has become critical.

The tooling materials currently used are metals such as aluminum or steel, which are fabricated by standard metalworking techniques; electroformed nickel; and composite tooling materials laid up and cured over a model to produce a mold. The composite tooling materials may be either a wet lay-up system, for which the resin is mixed in the shop and applied to dry fabric by the shop mechanics, or a prepreg system, for which resins that have superior properties, but are difficult to mix, are precisely applied to the fabric by the material supplier.

Advantages of Composite Tools

Composite tools have definite advantages over metal molds for large or highly contoured parts made from graphite fibers. Within the family of composite tools, molds made from tooling-grade prepreg materials are superior to those made by the wet lay-up process.

Low Coefficient of Thermal Expansion. The relationship between the mold and the part during cure at elevated temperature goes through several changes:

- *Initial heat-up.* The tool increases in size because of its thermal expansion. The prepreg is still deformable and moves with the tool
- *Resin gel (vitrification) temperature.* The thermally expanded tool and the part are the same size. However, the part has very low properties because the matrix is only partially cured (gelled)
- *Heat-up to final cure temperature.* The tool and semicured part both expand at their own rates. Due to thermal mismatches, stresses are molded into the part
- *Hold at final cure temperature.* The part receives additional cure to develop higher mechanical properties while it is held in a deformed condition by the tool
- *Cool-down from final cure temperature to end of process (60 to 65 °C, or 140 to 150 °F).* The tool and the part shrink at their own rates of thermal expansion. Because the part is still under pressure, additional distortion may be introduced

Thus, one of the most critical parameters in tool design and performance is the coefficient of thermal expansion (CTE). By using graphite tooling to build graphite parts, and matching the orientation of the plies in the part and tool, the thermal-expansion characteristics of the composite tools can be matched to those of the composite parts. Graphite tools have the lowest CTE of any of the current options (Table 1), allowing tool designers to build parts to tight tolerances because smaller corrections can be made in the tool dimensions. It should be noted that the mold must be shorter and narrower than the engineering call-out to account for thermal expansion.

With different CTE rates for the part and tool, stresses that can cause inaccurate dimensions and warpage are induced in the graphite part, especially in large or complicated parts. However, if the part and the tool are matched thermally, they will expand and contract at the same rates and will therefore be dimensionally accurate. A graphite tool will expand considerably less than an aluminum tool and is therefore more dimensionally accurate. An aluminum tool expands considerably more than a graphite tool and needs five times as much thermal correction (Table 1). Also, by using graphite instead of aluminum, flange springback and radius springback produced by the induced stresses are reduced by 50% and 75%, respectively. In addition, the amount of warpage in a large graphite part is 75% less than with aluminum tooling and 70% less than with wet lay-up tooling (Table 2).

Ease of Preparation. A major advantage of composite tools is that they can be built rapidly and economically using a plaster model. This technology is well known, inexpensive, and fast. Thus, a composite tool can usually be placed into production more rapidly than a metal tool.

The old lay-up systems were limited to those that could be mixed in the shop. They were then applied to dry fabric without much control of the fabric-resin ratio, which affects the CTE. Also, these old systems contained carcinogens and were messy to use. Improved resins have now been developed that suppliers prepare by using specialized equipment in their plants. They then add the resin accurately to the dry fabric using prepreg techniques. Prepregs are not messy to use, contain no chemicals disallowed by the Environmental Protection Agency, and can be precisely controlled.

Low Density. Composite tools are much lighter than metal tools of the same thickness; the density of aluminum is almost twice that of graphite, and the density of steel is nearly five times that of graphite. Extremely heavy metal tools can cause a nonuniform temperature distribution during the cure of a graphite part, which induces stresses that adversely affect tool accuracy (that is, causes part distortion). Graphite-epoxy tools have a relatively uniform temperature distribution, which allows the part to heat up evenly and prevents build-up of internal stresses. Also, their low density makes composite tools easier to handle than metal tools.

Table 1 Performance of tooling materials

| Material | Coefficient of thermal expansion 10^{-6}/K | Thermal correction required | | | | Springback of square corners, degrees (b) |
| | | 0.9-m (3-ft) part | | 9-m (30-ft) part | | |
		mm	in.	mm	in.	
Graphite-epoxy(a)	3.6	0.28	0.011	2.79	0.110	1°15′
Glass-epoxy(a)	11.7-13.1	0.76	0.030	9.14	0.360	...
Steel	12.1	0.76	0.030	9.14	0.360	1°30′
Nickel electroform	13.3	0.76	0.030	8.89	0.350	...
Fiberglass wet lay-up	14.4-18.0	...	...	...	...	...
Aluminum	22.5	1.52	0.060	13.97	0.550	2°0′

(a) MXG-7620 resin. (b) Angle on tool must be larger than engineering call-out.

Table 2 Part warpage with various materials

Tooling material	Warpage at center of 1500- × 2000-mm (60- × 80-in.) tool	
	mm	in.
Aluminum	64	2.5
Fiberglass-epoxy, wet	48	1.9
Fiberglass-epoxy (MXB-7620)	30	1.2
Graphite-epoxy (MXG-7620)	15	0.6

Table 3 Properties of typical tooling prepregs (MXG-7620 graphite prepreg)

Graphite fabric style number	Weight/area		Thickness		Cured-ply thickness, vacuum bag	
	g/m²	oz/yd²	μm	mil	μm	mil
2534	230	9.7	241	9.5	178	7.0
2548	457	19.3	432	17.0	356	14.0
2577	758	32.0	737	29.0	635	25.0

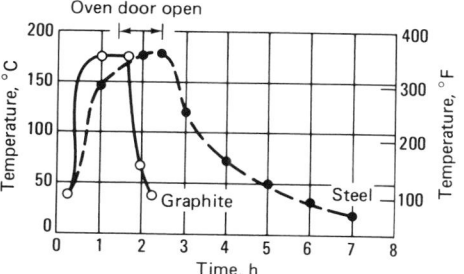

Fig. 1 Heat-up rates of tooling materials

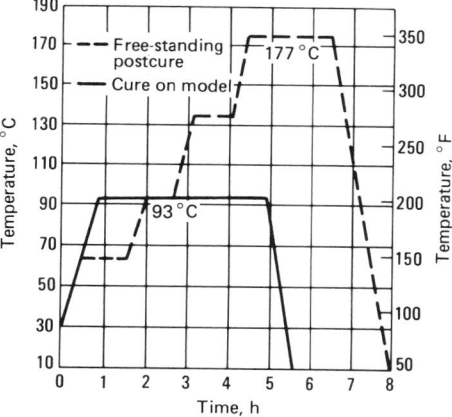

Fig. 2 Two-stage cure cycle of MXG-7620 system

Fig. 3 Sealed plaster model ready for lay-up of composite tool

Increased Productivity. Graphite-epoxy tools provide good thermal conductivity and require less energy than metal tools, to cycle graphite parts through all temperatures with maximum consistency. In part production, thermal conductivity, thermal capacity, and specific heat translate into both time and energy savings, resulting in greater productivity (Fig. 1). Steel requires almost three times as much energy per cubic foot than graphite-epoxy tools, and aluminum requires twice as much. Other factors contributing to higher productivity include reduced support equipment (such as cranes and lifts), reduced handling time, and increased safety due to lower weight.

Thermal Stability. One company has developed a resin system that has high thermal stability and can withstand hundreds of elevated-temperature cure cycles. This thermal stability, along with low void content, high fiber volume, low CTE, uniform thickness, and excellent oxidative stability, provides long service lives for the graphite-epoxy tools that use this resin system.

These thermally stable resins provide the basis for a resin system that can be cured at 95 °C (200 °F), with enough green strength to be demolded and then post cured while free standing at 175 °C (350 °F) to achieve a service temperature of 205 °C (400 °F) (Fig. 2). This system, called MXG-7620, has been used since 1982 to manufacture low-temperature curing, high-temperature service graphite-epoxy tools for the aerospace industry. The glass transition temperature of MXG-7620 resin is approximately 215 °C (420 °F) and allows the material to be used up to 205 °C (400 °F), high enough to cure many of the materials now being used to build composite parts.

Manufacture of Graphite Tools from Prepregs

Many different types of materials are used in building models, including hardwood masters, high-temperature graphite-epoxy tooling aids, plasters, and plastic-faced plasters. With the improved materials made possible by the new resin system, lay-up directly on plasters, woods, and plastic-faced plasters is feasible. Plasters are usually the least expensive materials available. High-quality hardwood models can also be used. Plaster and wood models require proper drying, sealing, and coating with mold release.

Drying and Sealing. Drying times for plasters and woods vary according to their size and mass. For a tool of average size, 72 h at 65 °C (150 °F) is usually sufficient. At this point, it is appropriate to seal and release these masters by applying several coats of a quick-drying lacquer sealer, allowing each coat to dry, and sanding until smooth, before applying the next coat. This operation is continued until the masters are completely sealed, whereupon a coat of hard mold release wax is applied, allowed to dry, and then buffed off. Next, a polyvinyl chloride moisture barrier coat is applied and allowed to dry, and finally a coat of mold release wax is applied and buffed off. These masters are now ready for use.

If plastic-faced plaster is used, the same drying procedure applies, followed by application of three light coats of Freekote 700, allowing each to dry and then buffing each coat. This is followed by application of three coats of a hard mold release wax, with each coat being buffed until glossy. Lay-up can be done directly on the plaster, wood, or plastic-faced plaster master (Fig. 3).

Gel-Coat Application. Liquid gel coats are required to obtain an excellent surface on tools cured by the vacuum bag process, which does not generate enough pressure to ensure a void-free surface. Gel coats are not required, however, on tools cured by the autoclave process, which does provide sufficient positive pressure to achieve a smooth surface. The use of a prepreg film gel coat is required when an extremely smooth surface is required, and to eliminate any fabric print-through.

Graphite Tooling Prepreg Materials. Three different types of graphite prepregs are used in fabricating the tool (Table 3). Light-

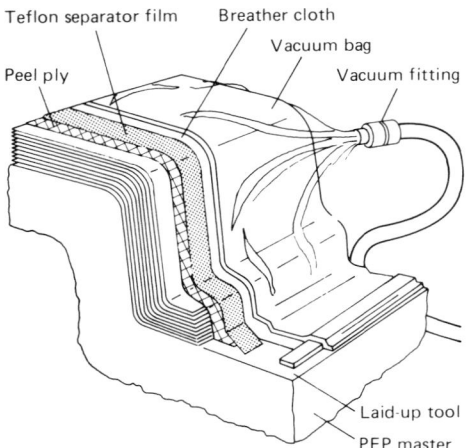

Fig. 4 Vacuum bag installation over tool lay-up

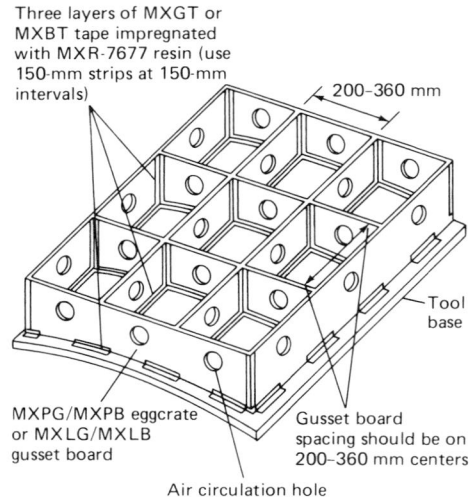

Fig. 5 Support structure details

weight fabrics are used directly against the tool surface, with medium-weight fabrics used behind them, acting as a transition to the heavier fabrics, which build up the thickness. During this lay-up process, careful attention is paid to the isotropic and symmetrical construction of the prepreg layers. The prepregs are cut into rectangular patterns to avoid warpage, leakage, and uncontrolled internal stresses, as well as to facilitate handling of the prepreg lay-up.

Lay-up Procedure. For vacuum bag molded graphite-epoxy tools, first a 0.20-mm (0.008-in.) thick layer of gel coat is applied, after being mixed thoroughly. A good estimate of the amount of gel coat to weigh out is approximately 30 g/m² (1 oz/ft²) of tool. This should be allowed to advance to the stage at which a fingerprint can be left but the surface is no longer sticky. A second layer of gel coat, approximately 0.1 mm (0.005 in.) thick, should then be applied, and prepreg can then be laid up into it. The tool must be laid up and debulked within 10 h of application of the gel coat to ensure good bonding.

For autoclave-molded graphite-epoxy tools, a gel coat is not required, but a film gel coat may be used to ensure an ultrasmooth surface. This is usually applied to the released master. A vacuum bag is then applied, holding 94.5-kPa (28-in. Hg) vacuum for 60 min to compact the gel coat. The bag is then carefully removed to avoid lifting the film from the master.

Application of Prepreg. Three plies of MXG-7620/2534 graphite-epoxy prepreg are applied to the mold, one at a time, oriented at 0°, +45°, and 90°. During the lay-up, care should be taken to work each ply into all radii and corners and to remove all entrapped air. All plies are butt-spliced together to avoid air entrapment and uneven tool thickness.

Four plies of MXG-7620/2548, oriented at 0°, +45°, −45°, and 90°, are applied onto the master, again with extreme care being taken to remove all entrapped air and to work the

material into all corners and radii. All plies are butt-spliced to avoid air entrapment and uneven tool thickness.

Debulking of Lay-Ups. For debulking a vacuum bag tool, the master must be bagged as shown in Fig. 4, with one or more thermocouples between the center plies of the lay-up, a peel ply applied to the lay-up and seated snugly into the corners of the mold. This is followed by application of a solid Teflon separator film, perforated at approximately one hole per square foot. Two plies of breather cloth should then be applied, and a minimum of two vacuum ports placed in the lay-up. Next, a nylon bag is applied and checked for leakage. The maximum leak rate allowed is 17 kPa (5 in. Hg) within 5 min. The lay-up is then debulked under 85 kPa (25 in. Hg) at a part temperature of 65 °C (150 °F) for 1 h. It is cooled to below 38 °C (100 °F) before the vacuum and the bagging materials are removed, with care being taken to avoid lifting the laminate. The lay-up should still be tacky (not cured). This operation should be done within 10 h of the MXR-7676 application.

When curing by the autoclave method, a heat debulk at 38 °C (100 °F) with vacuum pressure is not necessary, but it can be done. An alternative to the heat debulk is a pressure debulk. After the lay-up is bagged as in Fig. 4, the tool is placed in an autoclave for 1 h at 410 to 690 kPa (60 to 100 psi) and full vacuum to compact the material, remove all entrapped air, and consolidate the plies. The bagging materials are then removed as before, and the lay-up is continued to completion.

Final Lay-Up Procedure. With ply orientations of 0°, +45°, 90°, −45°, −45°, 90°, +45°, and 0°, the remaining MXG-7620/2577 plies are applied to the lay-up, again with care being taken to work each ply into all radii and corners and to remove all entrapped air. Again, all plies are butt-spliced. The bagging materials

are applied as before (Fig. 4), being sure not to bleed resin from the lay-up. To do this, the solid separator film is extended beyond the edge of the peel ply (~25 mm, or 1 in., beyond). It should be noted that this is not an isotropic lay-up, but is adequate for most tool configurations.

Cure Cycle. For vacuum bag tools, a vacuum of 85 kPa (25 in. Hg) is applied. The lay-up is heated at a rate of 2 to 4 °C (3 to 8 °F) per minute to a part temperature of 95 to 99 °C (200 to 210 °F), and held a minimum of 4 h at that temperature. Start of all holds should be based on the lagging thermocouple. The lay-up should be cooled below 38 °C (100 °F) under vacuum, and the bagging materials should then be removed.

For autoclave tools, a vacuum of 85 kPa (25 in. Hg) is applied. An autoclave pressure of 690 kPa (100 psi) is applied to the lay-up. The vacuum is vented to the atmosphere at 140 kPa (20 psi) to prevent it from sucking the resin toward the vacuum ports. The lay-up is heated at a rate of 2 to 4 °C (3 to 8 °F) per minute to a part temperature of 95 to 99 °C (200 to 210 °F) and held a minimum of 4 h at this temperature. Start of all holds should be based on the lagging thermocouple. The tool should be cooled to below 38 °C (100 °F) before releasing the pressure, and the bagging materials should be removed from the tool.

Application of Support Structure. With the tool still on the model, the support structure—either a solid laminate or an "egg-crate" panel (aluminum honeycomb with graphite skins)—is attached to it by means of locally applied fabrics, room-temperature curing, and high-temperature resistant resins (Fig. 5). Special care should be taken to leave a gap of 1.5 to 3.0 mm (0.06 to 0.125 in.) between the laminate and the support structure to prevent flat spots and thermal stresses when the support structure expands and is in contact with the laminate. Also, the gap should not be allowed to fill with resin when applying the impregnated tape, and air-circulation holes should be cut into the support structure. Once the support structure is cured to the laminate shell, it is removed from the master. Care should be taken to avoid damaging either the tool or the master.

Postcure of Graphite-Epoxy Tool. The supported or unsupported tool is placed in an oven, and one or more thermocouples are attached. The tool is heated to 66 °C (150 °F) at a rate of 2 to 4 °C (3 to 8 °F) per minute and held at this temperature for 1 h. Next, the temperature should be raised to 93 °C (200 °F) at the same rate, and held for 1 h. The tool should then be heated to 135 °C (275 °F) at the same rate and held for 1 h. Finally, the temperature should be raised to 177 to 191 °C (350 to 375 °F) at the same rate and held for a 2-h minimum. The tool should be cooled to 38 °C (100 °F) at a rate not exceeding 6 °C (10 °F) per minute. The tool is now ready for autoclave use to 205 °C (400 °F).

Fig. 6 Upper and lower wing skin tools and fuse-lage tools for a business jet

Fig. 7 Wing skin built on tools shown in Fig. 7

Fig. 8 Two-part tool for engine intake duct

Fig. 9 Two halves of engine intake duct tool

Fig. 11 Tool for helicopter leading edge

Graphite-Epoxy Tooling Materials in Production

Composite tools are being used successfully throughout the aerospace industry to produce parts that are structurally reliable, reproducible, and dimensionally accurate. For example, upper and lower wing skin tools and the fuselage tools for a business jet are shown in Fig. 6. These tools were autoclave cured and did not use a gel coat; the surface quality is excellent. Figure 7 shows the 18-m (60-ft) all-composite wing skin built on the tools shown in Fig. 6.

The two-part graphite tool shown in Fig. 8 is used to manufacture an asymmetric engine intake part. The built-in bushings are for guide pins that precisely align the halves. Figure 9 shows the two halves mated.

Figure 10 shows the tool for an escape slide component, which has a complex contour. This tool configuration has high deflection resistance, even without the support structure shown.

The leading edge for a helicopter is shown in Fig. 11. The 6.4-mm (0.25-in.) consistent wall thickness with a simple support structure allows for easy air access to facilitate heat-up during cure.

Fig. 10 Tool for escape slide component

SELECTED REFERENCES

- W.P. Benjamin, *Plastic Tooling: Techniques and Applications*, McGraw-Hill, 1972
- J. DelMonte, *Technology of Carbon and Graphite Fiber Composites*, Litton Educational Publications, 1981
- A.B. Kerr, "Innovative Tooling for Advanced Composites," Paper MF 85-507, Society of Manufacturing Engineers, 1985
- W. Lengen and J. Müller, *Moulds Made of Fiber-Reinforced Resin Prepregs for Manufacturing Composite Components by the Autoclave or Vacuum Compression Process*, Kunststoffe, Vol 75 (No. 4), 1985, p 203-209
- K.P. Ralph, "Toolrite Tooling Material System (Composite Tooling)," Paper EM 85-108, Society of Manufacturing Engineers, 1985

Elastomeric Tooling

Marvin Foston and R.C. Adams, Lockheed-Georgia Company

ELASTOMERIC TOOLING, a process that uses rubber tooling details to either generate molding pressure or act as a pressure intensifier to fabricate composite parts, is not a new concept. External intensifier cauls, sheets, rubber tool pads, and removable elastomeric mandrels have been used for years as auxiliary materials in combination with conventional molding materials and molding processes. Thermal expansion molding, however, is a recently developed process in which elastomeric tooling details are constrained within a rigid frame to generate consolidation pressure by thermal expansion during the curing cycle.

Thermal Expansion Molding Methods

Two basic methods employ the principles of thermal expansion molding: the trapped or fixed-volume rubber method and the variable-volume rubber method.

The fixed-volume method (Fig. 1a) exploits the large difference between the coefficient of thermal expansion (CTE) of the elastomer and the CTE of metals. The elastomer is confined within a closed metal tool cavity; when heated, it expands into the cavity, exerting the pressure required to compact a composite laminate. One of the chief attractions is that it allows the manufacturing engineer to ensure adequate pressure with or without a vacuum bag or autoclave.

Elastomeric mandrel compounds have been formulated to meet a wide range of pressure requirements. Compounds are available that can generate controllable molding pressure ranging from less than 7 to greater than 13 800 kPa (less than 1 to greater than 2000 psi) at cure temperature.

The variable-volume method (Fig. 1b) offers more flexibility than the fixed-volume method because a precisely calculated volume of rubber is not normally required. In most applications, the rubber is simply "set back" to allow for the bulk factor of the molding material during assembly of the tooling details. If excess pressure is generated during vacuum bag molding, it is vented against the vacuum bag. In press molding, the excess pressure causes the floating plate to retract the movable press platen.

In an airframe, thermal expansion molding is used primarily in boxlike structures such as rudders, vertical stabilizers, wing boxes, spoilers, and ailerons. Elastomeric tooling provides the means for fabricating integrally stiffened skins with a co-cured substructure in a single curing operation. Secondary bonding is thereby eliminated, substantially reducing direct manufacturing costs.

Applications. The basic mechanism of thermal expansion molding is the transfer of uniform compaction pressure to an uncured composite laminate by an elastomeric tooling detail.

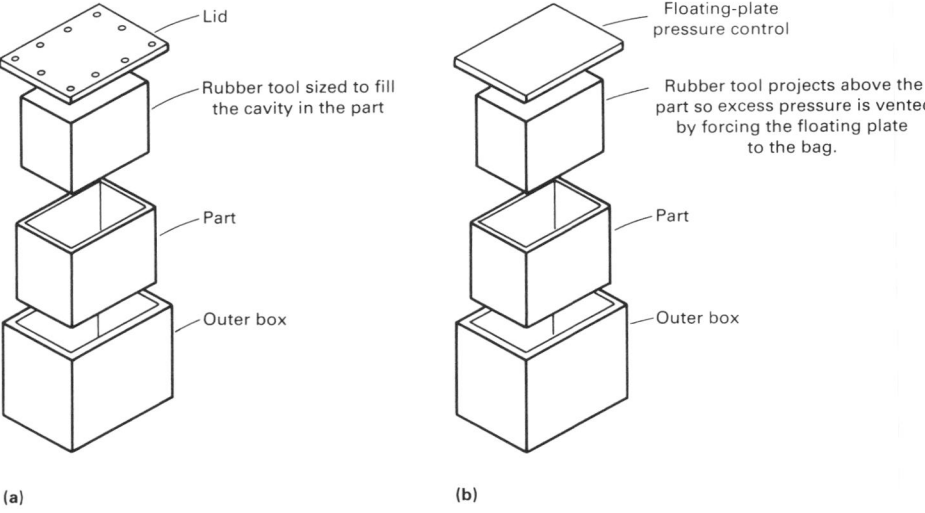

(a) (b)

Fig. 1 Thermal expansion molding methods. (a) Fixed-volume method. (b) Variable-volume method

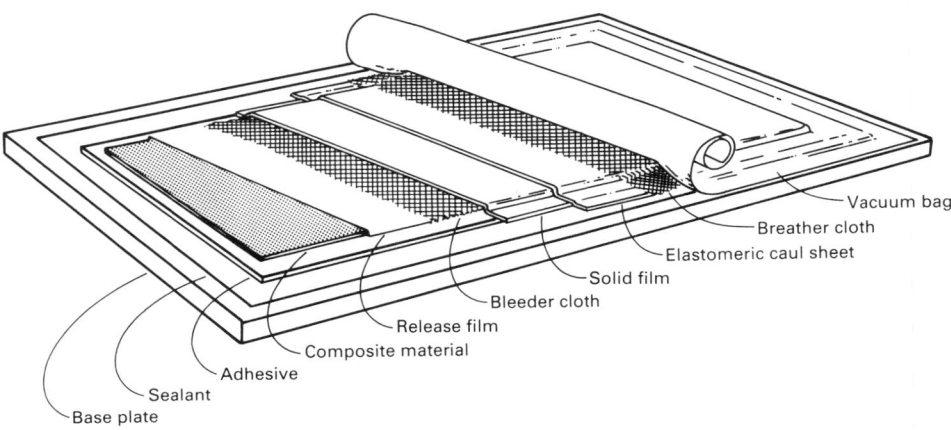

Fig. 2 External pressure intensifier caul

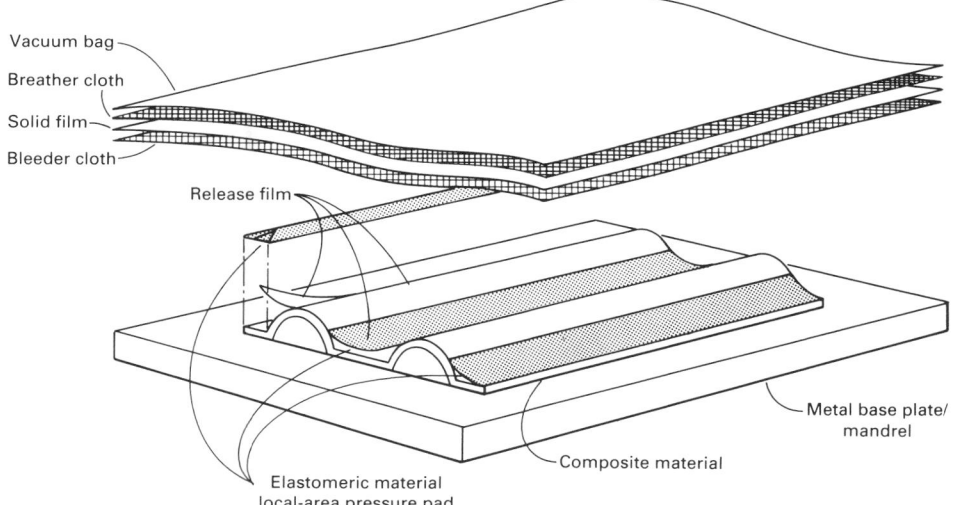

Fig. 3 Local-area pressure pad

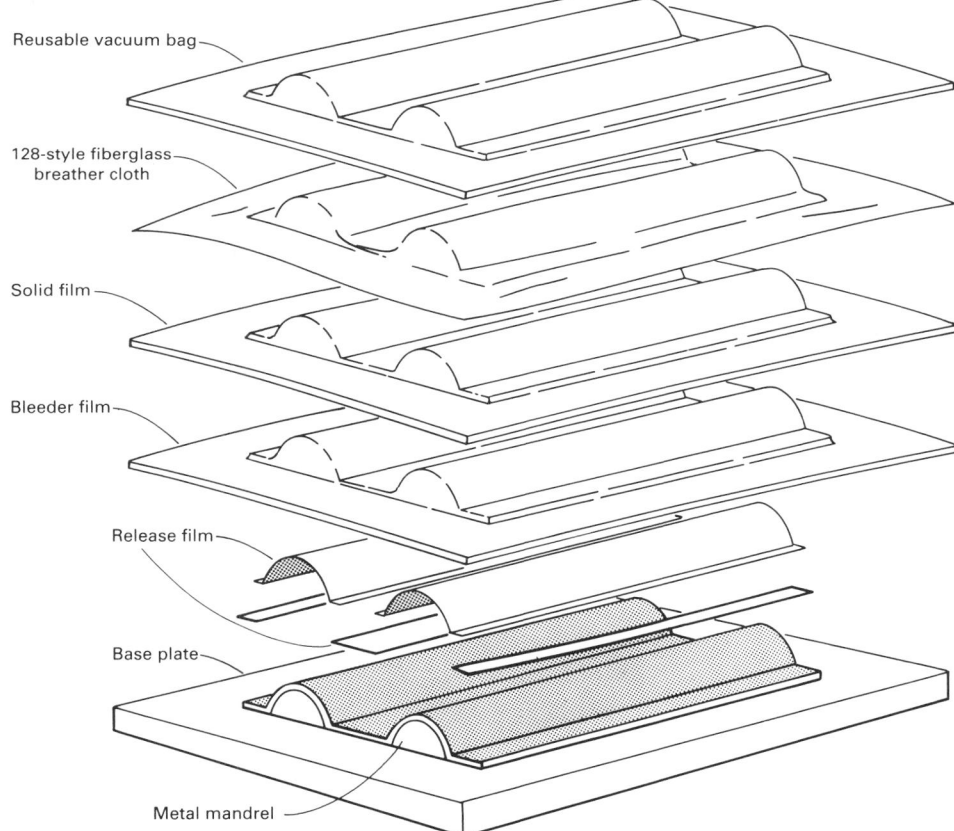

Fig. 4 Reusable vacuum bag

External pressure intensifier cauls are fabricated with cast or calendered elastomeric materials and are described fully in the section "Fabricating Elastomeric Tooling Details" in this article. When heated under autoclave pressure, the caul behaves much like a high-viscosity fluid, transferring the ambient pressure uniformly to all surfaces of the part. A widely used and simple application is a caul sheet, located between the bleeder cloth and vacuum bag, as shown in Fig. 2. The elastomeric caul sheet prevents wrinkles from forming in the part, and blends in any surface irregularities. Thus, a one-time investment in tooling eliminates the recurrent costs of carefully bagging complex parts and reworking surface wrinkles.

Another application is local-area pressure pads that are cast to match the exposed surface (opposite the tool surface) in concave or recessed areas of parts. Using such pads reduces the time needed to bag individual parts, eliminates bag wrinkles on part surfaces, lowers the cost of cleaning parts, and prevents radius thinning when molding radii on male tools. In addition to these benefits, local-area pressure pads (Fig. 3) reduce the chance that a lay-up will not fit the tool snugly in a radius area during cure and cause bag failure. However, this problem can also be eliminated with form-fitting reusable vacuum bags fabricated from calendered elastomeric material, as shown in Fig. 4.

The fabrication processes for the caul and local-area pressure pads are described in the section "Fabricating Elastomeric Tooling Details" in this article. Elastomeric tooling material is also used as a pressure pad (mandrel) for autoclave molding (Fig. 5). Such pads are inherently damage resistant and can be cast to match the outer mold line of the part. The resulting parts, though cured in an autoclave, will have a uniformity and quality approaching that of matched-die molding. This application is further discussed in the section "Design/Fabrication of Elastomeric Mandrels" in this article.

Molding Stages. The main stages of thermal expansion molding for a simplified, one-dimensional case, using the fixed-volume rubber method, are illustrated in Fig. 6. As shown, the rubber block freely expands across a gap and contacts the composite prepreg material, which is compacted under increasing pressure until the resin gels. At the gel point, the laminate has been compacted to its final molding thickness under pressure, P_{gel}. The laminate is cured by dwelling at T_{gel} for the time determined by rheometric data. The cured laminate is then removed from the tooling and postcured unrestrained in an oven.

Curing the laminate at T_{gel}, followed by postcuring, eliminates the excessive pressure that would be generated if the temperature were raised to the final cure temperature, which could cause damage to parts containing stiffeners or ribs. These stages of the thermal expansion molding process are also presented as a pressure-temperature plot, as shown in Fig. 7.

Volumetric Analysis

The significant properties that are required to characterize elastomeric tooling material are CTE, bulk modulus, hardness, and dimensional stability under repeated cycles of pressure and temperature. The nature of the thermal expansion process with an elastomeric material can be obtained by putting pressure transducers on three faces of a rectangular mold to permit the

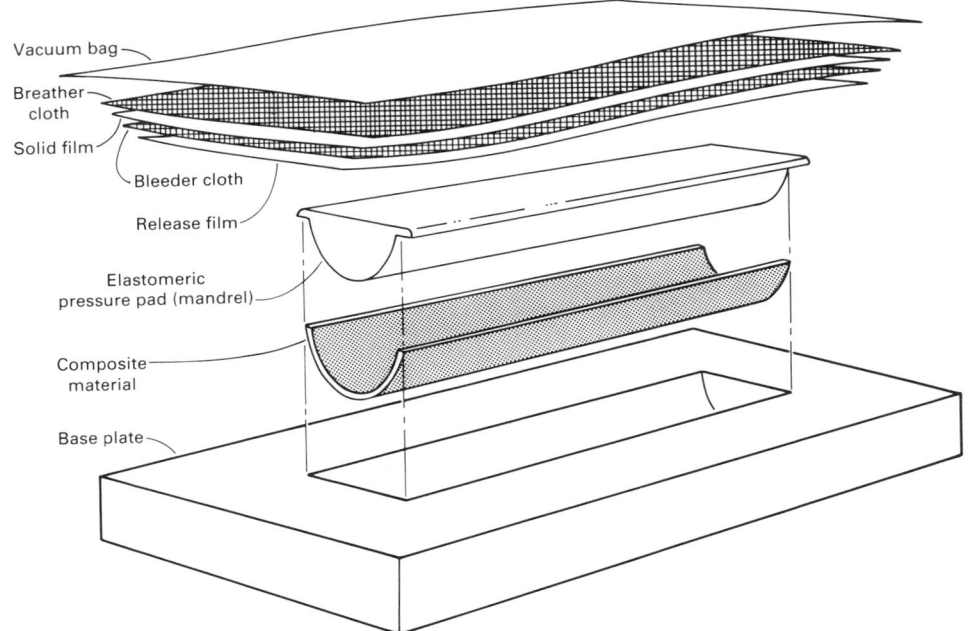

Fig. 5 Autoclave molding pressure pad

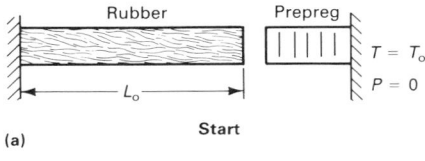

(a) **Start**

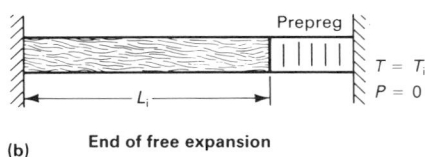

(b) **End of free expansion**

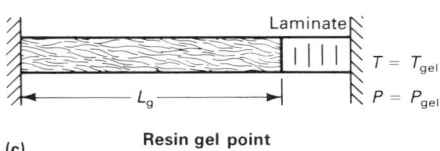

(c) **Resin gel point**

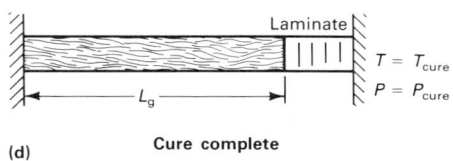

(d) **Cure complete**

Fig. 6 Stages of thermal expansion molding. *L*, length. *T*, temperature. *P*, pressure

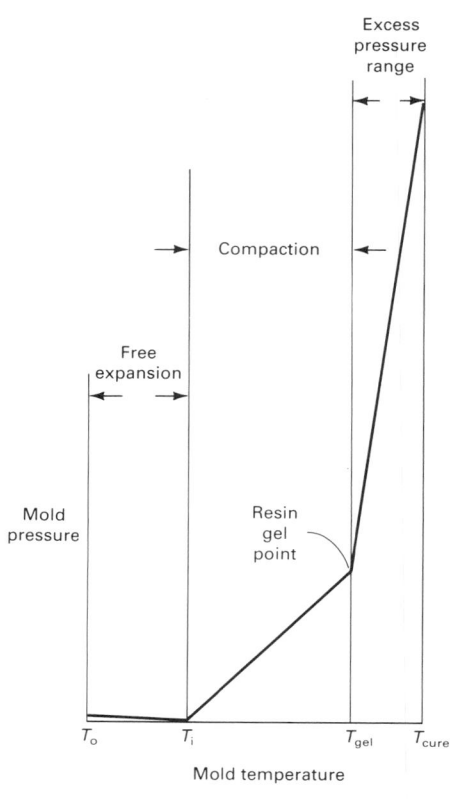

Fig. 7 Conceptual pressure-temperature plot

monitoring of pressure exerted by a rectangular brick in three directions. Using data from the rectangular mold model, volumetric relationships can be derived that are valid for complex as well as for simple tooling configurations.

The controlling equations for pressure in the fixed-volume rubber method of thermal expansion molding can be reduced to a single equation when two characteristics of the elastomeric material being used are determined through accurate tests. These characteristics are the percentage of volume by which the elastomer shrinks between being poured and being cured, and the pressure generated per degree of increase in temperature of a confined elastomeric specimen. Once these properties are determined, the volume lost to shrinkage can be restored during use by adding a suitable ΔT to the nominal positive gradient. This differential is known as the shrinkage compensation factor, or Δts.

Numerically, the Δts is obtained by dividing the percentage of shrinkage by the volumetric expansion factor of the rubber material:

$$\Delta ts = \frac{(X)\%}{B_r} \qquad \text{(Eq 1)}$$

where X is rubber shrinkage and B_r is the volumetric expansion coefficient of rubber.

The pressure compensation factor may also be offset by determining the amount of temperature increase required to provide the desired pressure:

$$\Delta tp = \frac{P}{P/d} \qquad \text{(Eq 2)}$$

where P is the desired pressure and P/d is the pressure generated per degree increase in temperature.

The basic equation controlling the thermal expansion molding process is:

$$V_{r2} = V_{r1} + V_{r1}B_r\Delta T_e \qquad \text{(Eq 3)}$$

where V_{r2} is the volume of rubber at the cure temperature at which pressure is equal to 0, V_{r1} is the volume of rubber at room temperature (RT), and ΔT_e is the effective change in temperature, which is cure temperature $-\Delta ts$ $-\Delta tp$ − RT, where Δts and Δtp refer to Eq 1 and 2. Note that Eq 3 is accurate to less than $\pm$ 1% error, which occurs because of the reduction in the mathematical expansion of:

$$(1 + \alpha t)^3 = (1 + 3\alpha t + 3\alpha^2 t^2 + \alpha^3 t^3)$$

where, for practical purposes, the terms $3\,\alpha^2 t^2$ and $\alpha^3 t^3$ can be dropped because of the infinitesimal value of α.

The initial starting point in tool design for the thermal expansion molding process is at cure temperature. The geometry of the part at temperature is surrounded by the tool inner mold lines, and a cavity is then designed to hold the rubber detail.

The initial value for the volume of rubber at the cure temperature, V_{r2}, is obtained by subtracting the volume of the part at temperature, V_{p2}, from the volume of the total tool cavity at temperature, with pressure equal to 0 psi:

$$V_{r2} = V_{tc2} - V_{p2}$$

where

$$V_{tc2} = V_{tc1} + V_{tc1}\, B_t\, \Delta T$$
$$V_{p2} = V_{p1} + V_{p1}\, B_p\, \Delta T$$

B_t = volumetric expansion coefficient of the metal details

B_p = volumetric expansion coefficient of the material of the part

and

$$\Delta T = \Delta T_2 - \Delta T_1$$

where

ΔT_2 = cure temperature (desired pressure application point) and

ΔT_1 = room temperature

The clearance between the laid-up bulk thickness of the uncured part and the rubber at its room-temperature size should then be determined. This is to ensure that there is enough clearance for tool assembly. If the clearance is inadequate, the depth of the tool cavity must be increased. This larger cavity, in turn, will call for a commensurately thicker rubber mandrel. Generally, the differences between the expansion factors of the hard tool material and the rubber is adequate for tool assembly.

After the proper tool cavity and rubber sizes are obtained, the rubber can be poured directly into the fabricated tool cavity. When this technique is used, the part configuration and set-back size are represented by wax. The volume of wax, V_w, required for set-back is equal to the difference in rubber volumes:

$$V_{r2} - V_{r1} = V_w$$

The shrinkage can also be compensated for at any time after the rubber volume has been determined. One simply increases the physical dimensions by the percentage of shrinkage. Note that this volume of wax does not include the cured-part dimensions.

To ensure accurate geometry of the cured part, the tool designer must take into account the geometries of all three critical surfaces when the set-back wax volume is distributed in the mold. The three geometries can be quite different when a strictly mathematical interpretation is taken. However, by using a two- or three-dimensional computer-aided design, maintaining the geometric configuration becomes straightforward.

Fabricating Elastomeric Tooling Details

By properly combining metal and elastomeric tooling details, the manufacturing engineer can draw on a variety of useful techniques for curing composite structures.

External Pressure Intensifier Caul. To prevent wrinkles and other surface irregularities from forming, an elastomeric caul sheet can be used, as shown in Fig. 2. It can be fabricated either before the first run or during the run. Fabricating the caul from uncured calendered rubber follows this sequence:

- Drape the uncured calendered rubber on a properly contoured dummy part, coated with mold release
- Cover with a ply of release film
- Cover with a one-ply bleeder cloth
- Cover with one ply of solid film
- Cover with breather cloth
- Encase in a vacuum bag
- Cure per vendor instructions
- Postcure as required

To cure the elastomeric intensifier during the first run, place it atop the solid Teflon film that is used to "dress out" the part for cure; then proceed with the above sequence.

Local-Area Pressure Pads. The function of a molding tool is to form uncured material into a structural shape that satisfies certain dimensional requirements. Unfortunately, many tools are designed without considering the processing requirement of the tool/prepreg system as a whole during cure. As a result, prepreg is pushed into a tool that it has problems conforming to. Bridging or unequal buildups cause pressure gradients, which can cause the part to warp as it cures. They also can cause porosity and wrinkles in the finished structure.

Ideally, pressure on the prepreg should be equalized in all directions. The recommended solution is to equalize the pressure on all surfaces with pressure pads. Pressure pads can be fabricated from calendered elastomeric material, using vendor's instructions and the steps described in this article for fabricating cauls.

Design/Fabrication of Elastomeric Mandrels

The sequence for fabricating a composite molding fixture with castable/calendered elastomeric material using the fixed-volume rubber method is:

- Design elastomeric tooling components by taking the part configuration from the engineering drawing, as shown in Fig. 8
- Determine the part thickness
- Determine the per-ply thickness of the cured part
- Determine where in the cavity rubber is needed, as shown in Fig. 9
- Provide no rubber set-back in the lengthwise direction of the elastomeric mandrels
- Cure the part in steps. First, cure at the gel temperature, using the pressure level and time duration dictated by rheometric data. Then postcure unrestrained in an oven per vendor data
- Determine as accurately as possible the volume of the tool cavity at room temperature (Fig. 9)
- Determine the volume of the finished part at the cure temperature (Fig. 8)

For the part shown (a section of the L-1011 vertical fin spar) and the steel tool configuration, the volume of the tool cavity in the area where elastomeric tooling is to be used is 4862.32 cm³ (296.717 in.³) at room temperature (25 °C, or 77 °F).

- Determine the volume of the part at room temperature, which is 161.445 cm³ (9.852 in.³), based on engineering drawing dimensions
- Determine the volume of the part at 143.3 °C (290 °F), which is:

$$V_{p2} = V_{p1}\,(1 + B_p\, \Delta T)$$
$$V_{p2} = (161.445\ \text{cm}^3)\,(9.72 \times 10^{-6}/\text{K})\,(143.3\ ^\circ\text{C}) - (25\ ^\circ\text{C})$$
$$V_{p2} = (161.445\ \text{cm}^3)\,(16.406)\ \text{or}\ (9.852\ \text{in.}^3)\,(1.00115)$$
$$V_{p2} = 161.626\ \text{cm}^3\ (9.836\ \text{in.}^3)$$

where

V_{p2} = volume at 143.3 °C (290 °F)

V_{p1} = volume at 25 °C (77 °F)

B_p = 9.72×10^{-6}/K

The total volume of rubber at 143.3 °C (290 °F) and 0 kPa (0 psi) is:

$$V_{r2} = V_{tc2} - V_{p2}$$

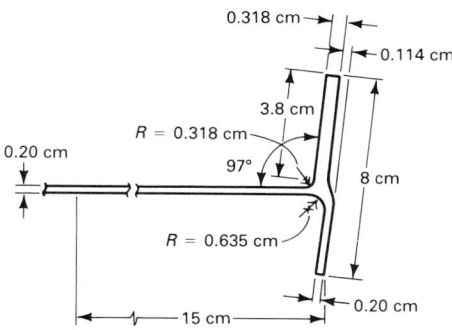

Fig. 8 Section of the part per engineering drawing

0.318 cm
0.114 cm
3.8 cm
R = 0.318 cm
97°
0.20 cm
8 cm
R = 0.635 cm
15 cm
0.20 cm

Determine volume at 15 cm (6 in.) length of web

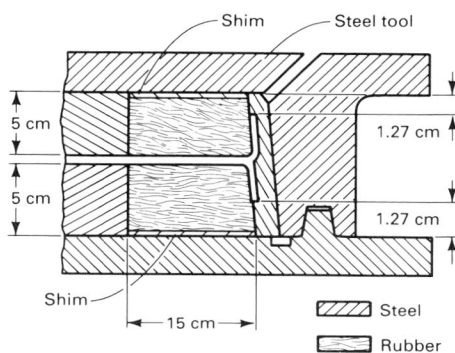

Fig. 9 Tooling concept

Shim
Steel tool
5 cm
5 cm
1.27 cm
1.27 cm
Shim
15 cm
Steel
Rubber

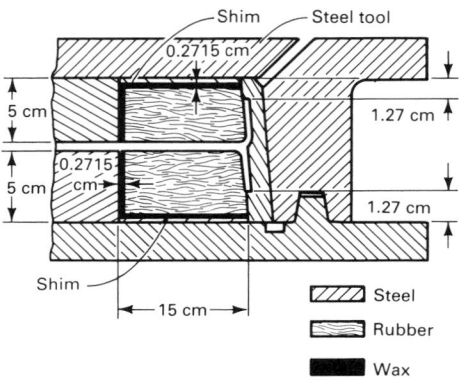

Fig. 10 Waxing tool cavity

where

$$V_{tc2} = V_{tc1} (1 + B_t \Delta T)$$
$$V_{tc2} = 4862.32 \text{ cm}^3 [(1 + 37.8 \times 10^{-6}/\text{K})$$
$$(143.3 \text{ °C}) - (25 \text{ °C})]$$
$$V_{tc2} = 4862.32 \text{ cm}^3 (1.004473)$$
$$V_{tc2} = 4884.066 \text{ cm}^3 (298.044 \text{ in.}^3)$$

B_t = $37.8 \times 10^{-6}/\text{K}$ for steel
V_{tc1} = volume tool cavity at 25 °C (77 °F)
V_{tc2} = volume tool cavity at 143.3 °C (290 °F)

Then

$$V_{r2} = V_{tc2} - V_{p2}$$
$$V_{r2} = 4884.047 \text{ cm}^3 - 161.625 \text{ cm}^3$$
$$(298.044 \text{ in.}^3 - 9.863 \text{ in.}^3)$$
$$V_{r2} = 4722.440 \text{ cm}^3 (288.181 \text{ in.}^3)$$

The room-temperature volume of rubber that will furnish 586 kPa (85 psi) at 143.3 °C (290 °F) is calculated as follows:

$$V_{r1} = \frac{V_{r2}}{(1 + B_r \Delta T_e)}$$

where

V_{r1} = volume rubber at room temperature

V_{r2} = volume rubber at 143.3 °C (290 °F) and 0 kPa (0 psi)

B_r = cubical CTE of rubber

ΔT_e = effective change in temperature

ΔT_e = cure temperature − Δts − Δtp − room temperature

where

Δts = shrinkage compensation factor

Δts = $\dfrac{\text{shrinkage of rubber}}{B_R}$

Δtp = pressure compensation factor

Δtp = desired pressure of 140 kPa/°C (20 psi/°F)

$$\Delta T_e = \left[143.3 \text{ °C} - \left(\frac{0.003}{688 \times 10^{-6}/\text{K}} \right) - \right.$$
$$\left. \left(\frac{586 \text{ kPa}}{140 \text{ kPa/°C}} \right) - 25 \text{ °C} \right]$$

$$\Delta T_e = 93.832 \text{ °C} (200.897 \text{ °F})$$

Then, substituting into the equation for V_{r1}:

$$V_{r1} = \frac{4722.422 \text{ cm}^3}{[1 + (688 \times 10^{-6}/\text{K})(93.832 \text{ °C})]}$$

$$V_{r1} = 4385.849 \text{ cm}^3 (267.642 \text{ in.}^3)$$

The volume of wax, V_w, required for pouring the rubber mandrels must be determined to allow for distribution and pouring of the rubber mandrels. The total wax volume is:

$$V_w = V_{r2} - V_{r1}$$
$$V_w = 4722.422 \text{ cm}^3 - 4385.849 \text{ cm}^3$$
$$(288.181 \text{ in.}^3 - 267.642 \text{ in.}^3)$$
$$V_w = 336.574 \text{ cm}^3 (20.539 \text{ in.}^3)$$

The total wax volume is distributed on two sides in the tool cavity, as shown in Fig. 10. The distribution of the wax on one or all sides of the tool cavity is determined by the tool engineer.

The periphery distance of the two sides selected in the tool cavity for wax application is 40 cm (16 in.), as shown in Fig. 10. The length of the tool cavity is 30 cm (12 in.). Therefore, the thickness of wax required is:

$$T_w = V_w \div S_a$$
$$T_w = 336.573 \text{ cm}^3 \div 1239 \text{ cm}^2 (20.539 \text{ in.}^3 \div 192 \text{ in.}^2)$$
$$T_w = 0.1069 \text{ in.}$$

where

T_w = wax thickness

V_w = volume of wax calculated

S_a = surface area where wax is applied

Materials. Dapco number 1 blue rubber from "D" Aircraft Products Company was the tooling rubber material used to study the process parameters and derive the pressure equations for the thermal expansion molding process. Freeman Thermo Stable Wax number 266 was used to wax-up part thickness and rubber set-back areas.

Elastomeric Tooling Application

C.W. Schneider and H.E. Carroll, Lockheed-Georgia Company

CONTROL SURFACE STRUCTURES demand a high order of structural efficiency. To meet high strength and light weight requirements, the structures must incorporate tailored materials constructed in relatively thin-gage sections. Composite materials provide the capability for optimized structural tailoring to meet strength and stiffness requirements with minimum weight. Composite control surfaces are cured as integral structures to ensure their structural integrity and maximize cost effectiveness. This reduces the number of detail parts fabricated and the need for bonding and mechanical fastening on assembly, thereby effecting significant reductions in production time and cost.

Curing control surfaces as integral structure requires more sophisticated tooling than is needed for curing detail parts. The tooling must ensure that all parts are cured to attain effective bonding of the structure. It must be designed and constructed to accommodate consistent pressure and temperature application throughout the structure. Tooling also must ensure that the interfaces between individual components are capable of eliminating distortion or deformation of the structure.

One effective method for meeting all these requirements is thermal expansion (elastomeric) molding, the theory of which is described in the article "Elastomeric Tooling" in this Section of the Volume. Elastomeric tooling is simple and cost effective, and can eliminate the need for expensive autoclave processing or matched-mold tooling.

Control Surface Construction

Composite control surfaces, such as rudders, ailerons, elevators, and flaps, are currently being produced for many types of aircraft. Formerly laboratory and experimental prototypes, they are now being designed and manufactured for the full service life of the vehicle. Examples being used on commercial aircraft are the Gulfstream III rudder, Gulfstream IV spoilers, and the Boeing 757/767 spoilers. In addition, there are a number of military applications currently in production or retrofit.

Control surface designs may vary in detail, but they are generally constructed in three basic sections: a leading edge, a box, and a trailing edge. Figure 1 shows the breakdown of a typical commercial aircraft aileron. These major sections are generally fabricated separately and assembled later to form the complete control surface.

The preferable manner for manufacturing each section is to form each as one piece and cure it as a monolithic structure, that is, in one piece. This is not always possible, but variations, with large sections being cured together, are still desirable.

As with a conventional metallic structure, a composite structure must be designed for the lowest cost at the outset to ensure cost-effective production. Some of the same basic design philosophies apply, such as minimizing part count, fasteners, and complex machining, all of which minimize production labor, tooling, planning, inspection, and associated overhead, such as inventory storage and detail-part accountability. In composite design, these labor-minimizing approaches to production design are even more important because of the material form. Composite materials are generally considerably more expensive in raw material form than are their conventional counterparts, for example, $7/kg ($3/lb) for aluminum compared with $90/kg ($40/lb) for graphite-epoxy tape. As a result, even higher emphasis must be placed on reducing production labor, and an awareness is necessary of the way the overall mix of costs, in percentages of the total, changes with composite design.

Figure 2 illustrates the change in production cost drivers for composite structures when compared with conventional metallic structures. For the metallic structure, labor represents 47% of the total production cost. For the composite part cost to be equal, the cost driver mix must change considerably. Material cost is a given and represents 35% of the total production cost (as opposed to 20% for metal). Quality assurance is 19% rather than 13%. This leaves the composite structure with 46% for labor and other costs. Holding other costs to 20%, labor cannot exceed 26% in order for the composite structure to compete with a metallic structure

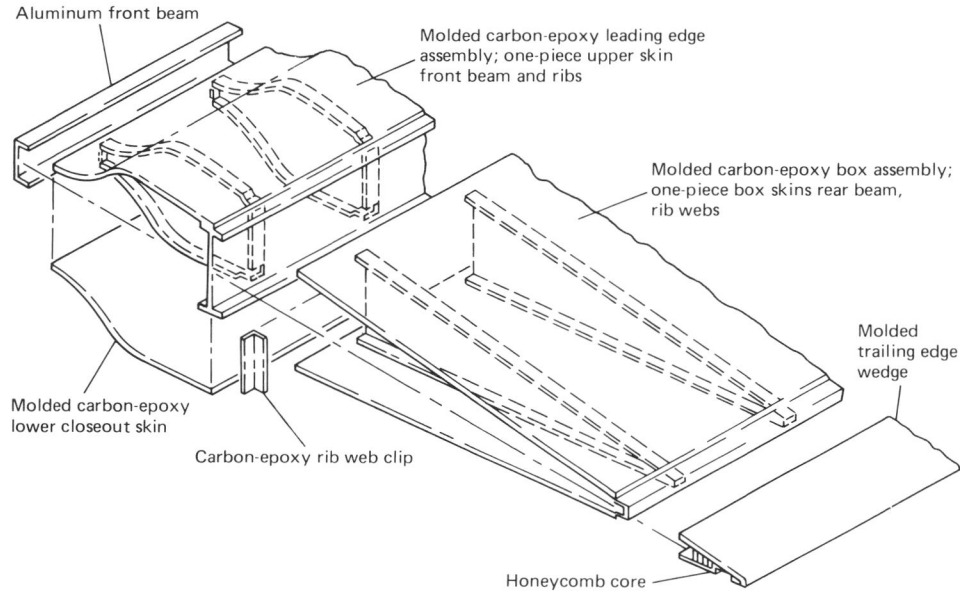

Aluminum front beam

Molded carbon-epoxy leading edge assembly; one-piece upper skin front beam and ribs

Molded carbon-epoxy box assembly; one-piece box skins rear beam, rib webs

Molded trailing edge wedge

Molded carbon-epoxy lower closeout skin

Carbon-epoxy rib web clip

Honeycomb core

Fig. 1 Typical control surface

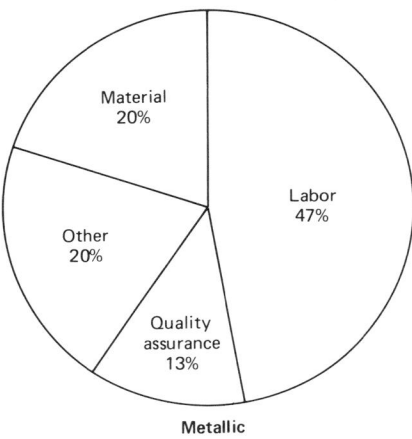

Fig. 2 Production cost elements

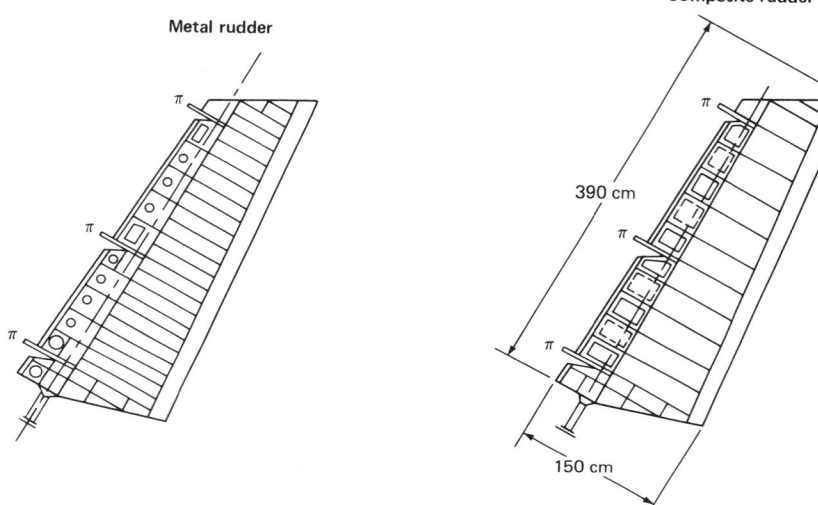

Fig. 3 Comparison of metal and composite rudders. Source: Ref 1

on an equal cost basis. The only way to reduce labor is to reduce the operations required and the time per operation. This can be achieved by taking advantage of the moldability of composite materials and producing assemblies rather than detail parts. Having fewer parts to track and handle and avoiding labor-intensive bonding and fastening operations will effect significant cost savings. Thus, cost-effective components can be obtained by manufacturing them as an integral structure. The key to manufacturing a successful integral structure is tooling.

Integral Structure Tooling

The same design and manufacturing considerations that are used for composite part manufacture also apply to the design and fabrication of composite tooling. Integrally molded composite components require tooling approaches

and materials that provide dimensional stability at cure temperatures, are light weight to simplify handling and minimize heat-up time, and have long tool life.

Design procedures and materials used for composite tools must consider the amount of contouring in the part, the total production quantity, dimensional stability, heat-up rate, thermal coefficient of expansion (CTE) of both the part and the tool surface finish, and cost. Tooling materials can be categorized as metallic (nickel, steel, or aluminum); nonmetallic (fiberglass, graphite, or aramid with thermoset or thermoplastic resins, either in prepreg or wet lay-up forms); and ceramic. Tools for composite parts can be fabricated using any of these materials if the part geometry and contour are simple. Complex parts, such as an integrally molded structure, require tooling materials that thermally match the material in the part to achieve the required

final dimensional tolerances. Thermal coefficient mismatch between the part and tool can also induce undesirable thermal strains during cool-down from the cure temperature. In extreme cases, these thermal strains can cause resin cracking of the part.

Curing large structural parts in a single integral component requires an exterior mold surface that thermally matches the part material. Pressure for curing the laminates can be provided by a combination of autoclave pressure and thermal expansion of internal mandrels. These mandrels can be made of metal or thermally expanding rubber, or combinations thereof. While metal may be preferred from a durability standpoint, the expansion obtained from metal is limited, and may not be sufficient to offset the material bulk factor with thick materials such as fabric.

Integral Structure Design and Fabrication

The tailorability of design permitted by composite materials provides as many approaches to tooling as it does for the composite part. The design, tooling, and fabrication of a typical control surface structure, the Gulfstream III (G-III) rudder, exemplifies a combination of successful tooling approaches.

The design developed for the composite rudder provided an increase in the service life of the rudder at reduced weight when compared with the metal rudder. The hybrid mold tools developed for this program proved to be effective in controlling the thermal expansion of critical points in the structure, while minimizing tool costs and maximizing tool life.

Design. The G-III rudder, shown in Fig. 3, is a sizable structure. The figure highlights two differences between the metal and composite designs: deletion in the latter of 50% of the box ribs, and enlargement of the access holes in the leading-edge structure to provide better accessibility. Other less-obvious features, such as the one-piece skins and the replacement of the honeycomb trailing-edge wedge with a single-laminate wedge, will be discussed in more detail.

The original rudder was redesigned from a metal to a composite structure, with the composite design retaining the same basic major assembly breakdown as the metal counterpart (leading-edge, box, and trailing-edge assemblies, as shown in Fig. 4). To maintain fail-safety, the three existing flight-qualified metal hinge assemblies were retained (with two metal hinge plates per assembly), as well as the metal torque lug box support structure. The entire structural assembly was designed to be fabricated of unidirectional carbon-epoxy prepreg tape. An external ply of fiberglass-epoxy 120-style fabric was used in selected areas for galvanic-corrosion and impact damage protection, as well as to reduce fiber breakout on the back side of the drilled holes.

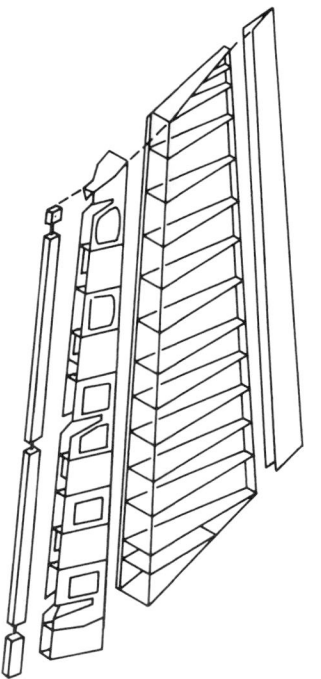

Fig. 4 Rudder assembly breakdown. Source: Ref 1

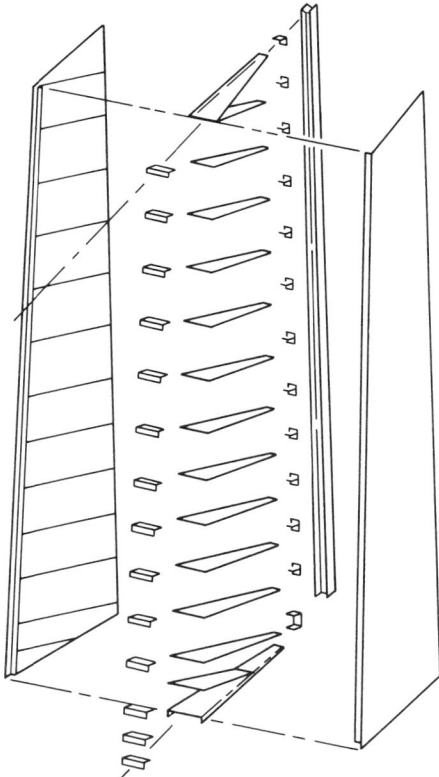

Fig. 5 Box structure breakdown. Source: Ref 1

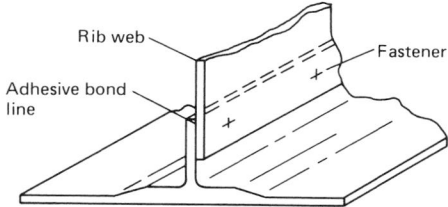

Typical box rib skin and integrally molded rib cap

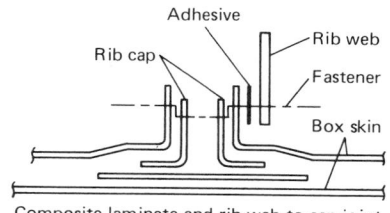

Composite laminate and rib web to cap joint

Fig. 6 Box rib cap configuration. Source: Ref 1

The box assembly (Fig. 5) consists of two integrally stiffened skin covers, an I-section rear spar, and flat laminate front spar and rib webs. The box covers contain co-cured rib caps that are embedded in the center of the skin laminate (Fig. 6). This configuration had been considered on previous designs, but tooling constraints prevented its use. A unique tooling approach, discussed in the next section, permitted this failure-resistant design to be used here. The outer half of the skin is continuous over the entire surface of the rudder box, while the inner half of the skin is discontinuous at the rib caps. The front spar and rib cap material is embedded between the two skin halves, and the upright leg of the cap is sandwiched between adjacent inner skin segments. After curing the spar, skin, and web laminates, the box is assembled using a combination of adhesive bonding and titanium or stainless fasteners, in a fastener-bond process developed for this program. The fasteners are installed on approximately 100-mm (4-in.) centers after application of a paste adhesive containing glass microbeads. The microbeads control the bondline thickness, while the cure clamp-up pressure is provided by the fasteners.

The entire assembly is heated in an oven at 120 °C (250 °F) for 1 h to cure the adhesive. After cure, the fasteners remain in the component as a fail-safe backup in the event of bondline failure, but are not load bearing because the adhesive bondline is capable of withstanding design ultimate load.

The leading-edge design is similar to that of the box, in that the rib caps are embedded

within the skin laminate. As shown in Fig. 7, each skin contains large access holes, which are located over each of the leading-edge bays, on alternating sides, to provide access to the front spar and hinge ribs during fabrication, and to provide inspection access during service. The large cutout area was designed to unload the curved skin segment and prevent buckling. A fiberglass-epoxy doubler was embedded within the skin around the access holes, so that the doors would be mated only to fiberglass to permit use of aluminum fasteners for the door installation. The fiberglass-epoxy doors are installed with faying surface sealant and with loose-fit aluminum screws and nut plates. The side skins are joined together by flat rib webs and adhesively bonded using the fastener-bond process. Two separately cured carbon-epoxy flanged hinge support ribs are used at each of the hinge locations; aluminum hinge plates are bolted to these ribs during final assembly. A metal torque box assembly, identical to that used for the metal rudder, is installed at the lower end of the leading edge to support and rotate the rudder. An aluminum front beam is located at the leading edge to close out the skins and to provide for attachment of the steel balance weights.

The trailing edge is a simple V-shaped laminate without internal stiffening. This replaces an aluminum honeycomb wedge used on the metal rudder. Flanged closeout ribs are attached

to the upper and lower ends to close out and seal the trailing edge.

The individual major assemblies—leading edge, trailing edge, and box—are assembled separately and then joined using the fastener-bond process. The final step in fabrication and assembly is the installation and alignment of the hinge plates on an interchangeability fixture.

This unique design configuration has been shown to be highly resistant to sonic fatigue, which was the controlling criterion for rudder skin designs. The composite rudder was tested to over 200 000 takeoffs/landings with no failures, well beyond the design life of 30 000 takeoffs/landings. In addition, the rib cap design permits the skin to operate in the postbuckled range without danger of skin-stiffener separation, a typical failure mode for most postbuckled design configurations. Cyclic load tests on a box skin segment verified the capability of the design to withstand repeated loads (with visually detectable impact damage) without propagation of the impact delaminations. A box segment successfully withstood ultimate loading conditions with several areas of impact damage. The full-scale rudder successfully completed static loadings up to ultimate load in the ''as-built'' condition; this was followed by a limit load demonstration with initially detectable input damage. The flight flutter performance of the composite rudder was substantiated during a brief flight test program.

A manufacturing plan was developed to establish the manufacturing sequence and to determine the type and quantity of tools needed to produce the rudder. The major tools designed and fabricated to produce rudder composite parts included:

Mold Bond Fixtures

● Box skins, left- and right-hand

- Leading-edge skins, left- and right-hand
- Trailing-edge wedge
- Rear spar
- Leading-edge hinge ribs (six)
- Leading-edge upper and lower closeout ribs (two)
- Box upper and lower closeout ribs (two)
- Trailing-edge upper and lower closeout ribs (two)

Assembly Tools

- Leading edge: assembly of leading-edge skins, rib webs, torque box, and hinge ribs
- Box: assembly of box skins, rear spar, rib webs, and front spar web
- Final assembly: leading edge, box, trailing-edge wedge, and closeout ribs
- Interchangeability fixture: final installation and alignment of hinges and actuator

In addition to these tools, ply templates, lay-up blocks, and other shop-aid trim and drill tools/templates were fabricated to simplify the fabrication process.

Tooling and Fabrication. The design concept for the leading edges and box skins presented a unique challenge to the tool designers, in that each of the box and leading-edge rib stations was required to be precisely located with respect to each other and to matching rib stations on the opposite skin panel. Alternative aluminum and carbon-epoxy tool concepts were evaluated to satisfy the design requirements. The concept selected was hybrid aluminum-carbon tools, to take advantage of the low cost and durability of the aluminum and the low thermal expansion of the carbon-epoxy to control critical rib cap locations. The key to the tooling concept was the selective use of carbon tooling materials to overcome the thermal expansion problems associated with the sole use of aluminum tools at the elevated cure temperature. All carbon-epoxy parts were cured in an autoclave at a temperature of 175 °C (350 °F) and pressure of 585 kPa (85 psi).

This new tooling concept, developed to accommodate the design of integrally molded stiffeners with embedded rib caps, proved to be successful for the rudder leading-edge and box skins. The tool concept has since been applied to other, more complex, co-cured designs.

Leading-Edge Skins. As indicated in Fig. 7, the leading-edge skin assembly consisted of a carbon-epoxy laminate with co-cured rib caps embedded in the skin. The rib cap locations were required to be precisely located with respect to each other. Furthermore, the rib caps on the left-hand and right-hand skin assemblies had to align with each other and with prelocated clips on the metal front beam. A metal tool was designed to control the exterior surface outside mold line (OML), and carbon tooling bars were used to control tolerances on the rib stations during cure.

The leading-edge mold fixture design consisted of a 12.7-mm (½-in.) contoured alumi-

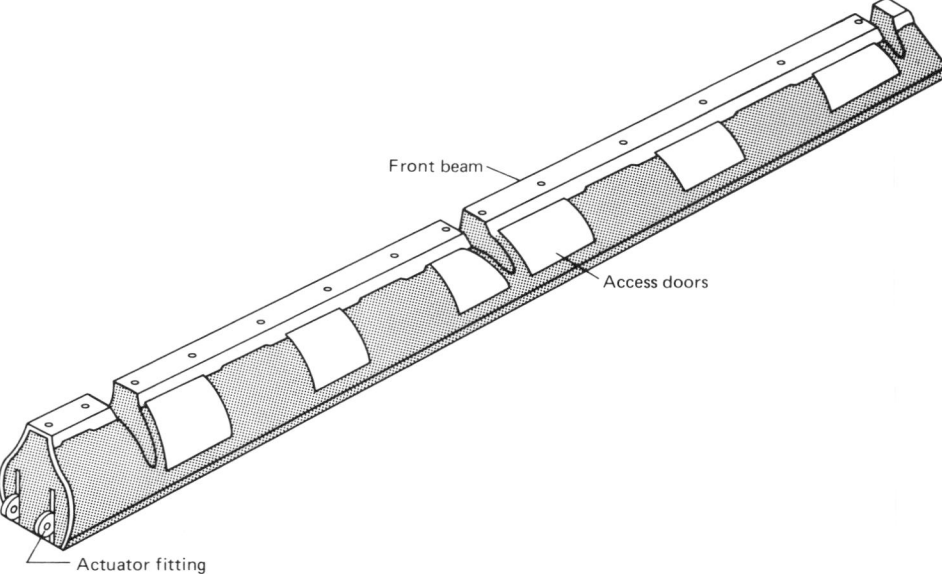

Fig. 7 Leading edge. Source: Ref 1

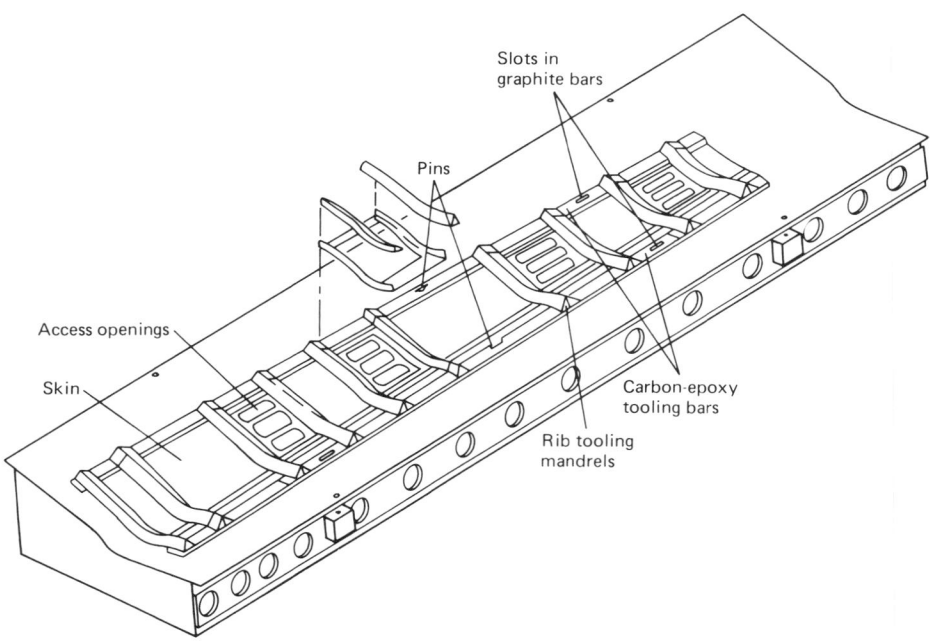

Fig. 8 Leading-edge skin tool. Source: Ref 1

num plate fastened on the back surface to numerical control (NC) machined aluminum headers. The mold surface of the tool was then NC-machined to the OML of the leading-edge skin. Carbon-epoxy tooling bars approximately 12.7 mm (½ in.) thick by 50 mm (2 in.) wide were located on the upper surface of the skin tool in the spanwise direction, as shown in Fig. 8. The carbon tooling bars were pinned to the aluminum tool base at the center of each bar to permit expansion of the aluminum tool base relative to the carbon bar. These carbon bars

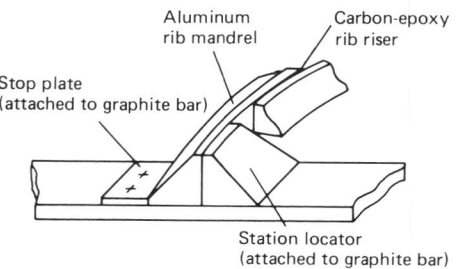

Fig. 9 Rib mandrel detail. Source: Ref 1

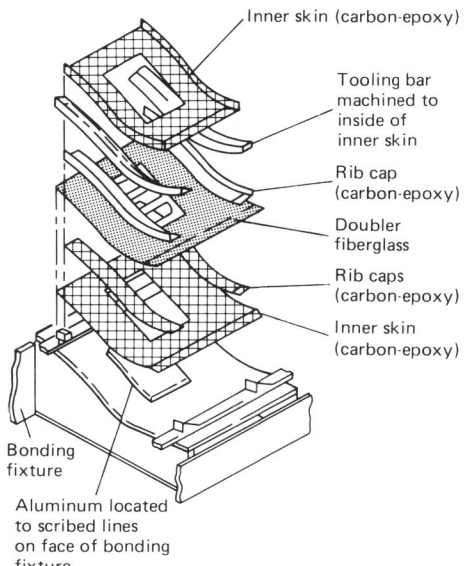

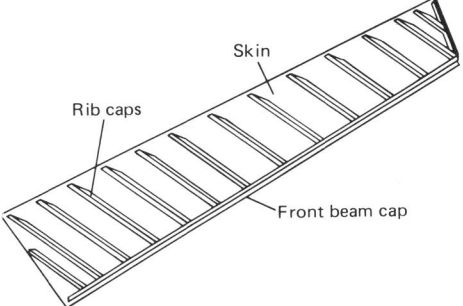

Fig. 11 Box skin (inside view). Source: Ref 1

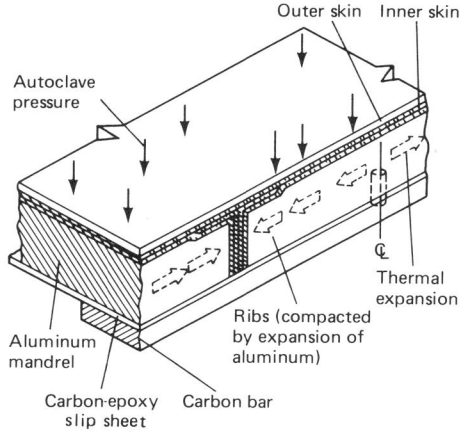

Fig. 13 Box skin aluminum mandrel. Source: Ref 1

Fig. 10 Leading-edge lay-up sequence. Source: Ref 1

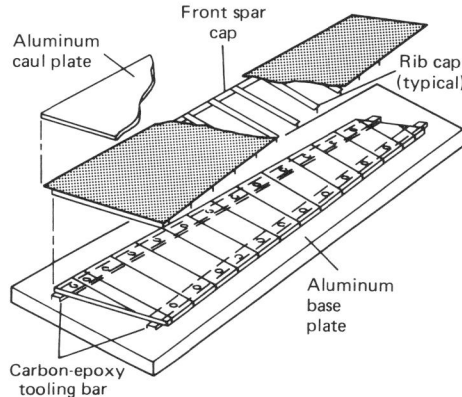

Fig. 12 Box skin fabrication tool. Source: Ref 1

contained stops at each leading-edge rib station to place and control the rib cap location positively. Aluminum rib mandrels, two per rib station, were NC-machined to the inner mold line (IML) of the rib cap flange surface. These rib mandrels apply lateral pressure to the rib cap riser during the cure cycle, as shown in Fig. 9. This pressure is provided by autoclave pressure, because one of the aluminum mandrels is free to move in the lateral direction and both are chamfered so that the pressure bag forces the two mandrels together.

The advantages of this tooling concept are: lower initial cost due to lower cost of aluminum base and rib mandrels; longer tool life due to durability of aluminum tools; and excellent dimensional control of required "hard" points in the design, such as the rib stations, by use of the carbon tooling bars.

The leading-edge lay-up sequence, illustrated in Fig. 10, was:

- The complete exterior skin (one-half of the skin thickness) was laid up on the tool base. Aluminum plates were positioned on the door cutout to mold the door edges
- Rib cap unidirectional reinforcement plies were laid on the inside of the exterior skin
- Each of the inner skin segments was laid up on a fiberglass lay-up block, starting with the inner half of the skin. This was followed by addition of the rib cap plies on each edge
- The inner skin segment was removed from the lay-up block and positioned on the outer skin, and the aluminum rib mandrels were then positioned on both sides of each rib riser
- The completed assembly was then bagged and cured in an autoclave

The box skin design concept is similar to that of the leading edge, but more complex since the front spar cap is included, as shown in Fig. 11. The tooling concept used for the leading-edge skin could not be used for the box skin, because there were no holes through the outer skin through which to attach the aluminum mandrels to the carbon-epoxy tooling bars. Instead, the tool concept was reversed, with the tooling mandrels on the inside of the tool and the carbon-epoxy part exterior skin on the upper side.

The tool concept is illustrated in Fig. 12. While the concept is similar to that of the leading-edge tool, compaction of the rib cap webs is provided by expansion of the aluminum mandrels against each other rather than by autoclave pressure. Again, aluminum was used for its durability, low-cost, and thermal-expansion advantages. Carbon-epoxy was used where necessary to maintain the required accuracy of the station location for each rib cap.

The tool consisted of a flat aluminum base plate, machined from 50-mm (2-in.) thick aluminum plate, with recesses in the lower side to reduce the thickness in most areas to 19.1 mm (¾ in.) to maximize heat transfer to the composite part. Two spanwise recesses were machined into the upper surface to permit installation of carbon-epoxy tooling bars flush with

the upper surface of the plate. These bars were 50 mm (2 in.) wide by 12.7 mm (½ in.) thick and extended the length of the composite box skin. The bars were pinned at the center to the aluminum base plate, and were free to slide in the recess as the aluminum tool expanded or contracted with changing cure temperatures.

Aluminum mandrels were used as lay-up blocks and to apply lateral pressure to the rib cap risers during cure. The individual mandrels were pinned to a fixed round bushing at the front spar carbon bar, and were pinned to a slotted bushing at the rear spar carbon bar. This permitted the mandrel to expand in the fore/aft direction while maintaining the front spar cap plane location.

The aluminum mandrels rested on the carbon-epoxy tooling bars, and because the center of each mandrel was pinned to the bar, each mandrel expanded outward from its center, applying lateral pressure against the rib cap riser. A carbon-epoxy slip sheet approximately 1.3 mm (0.050 in.) thick was used between the aluminum base plate and the aluminum mandrels to reduce frictional forces and expansion strains on the carbon bars, since the base plate could move relative to each mandrel, and to seal the gaps between mandrels. A 12.7-mm (½-in.) thick aluminum caul plate was used on the outside of the box skin to obtain a smooth exterior skin surface.

Figure 13 illustrates the method of pressure application on the rib cap risers. The aluminum mandrels were spaced on the carbon bars such that the gap between adjacent mandrels at 155 °C (310 °F) reached the normal riser thickness. Thermal expansion of the aluminum mandrels provided approximately 585 kPa (85 psi) lateral pressure at the 175 °C (350 °F) cure temperature.

The box skin lay-up procedure was:

- Inner skin segments were laid up on each mandrel, with the laminate wrapped around the front and side edges to form the rib and front spar cap risers

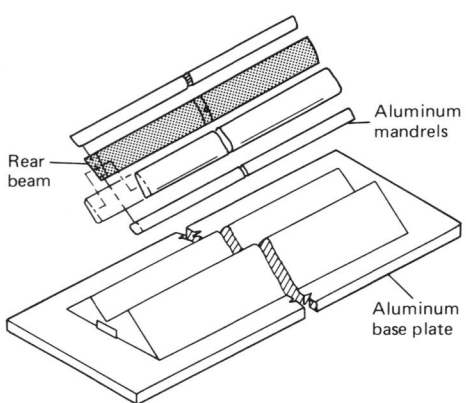

Fig. 14 Rear beam mold tool. Source: Ref 1

- Rib and front spar cap reinforcement plies were laid onto the edges of each mandrel, to complete one-half of each cap
- The mandrels were placed on the tool base, laminate side up, and pinned to the carbon bars. The mandrels were stacked side by side, starting from the center and working toward each end
- The outer skin laminate was placed over the completed inner skin/rib cap lay-up. This skin was continuous over the length and width of the box skin
- The final step consisted of placing the caul plate over the exterior skin surface, followed by application of the pressure bag and autoclave cure at 175 °C (350 °F) and 585 kPa (85 psi)

Rear Spar and Trailing Edge. The rear spar, an I-section co-cured beam, was fabricated using tooling similar to that used previously to fabricate other similar types of stiffening members. The aluminum tooling concept is shown in Fig. 14. The entire tool is bagged to apply autoclave pressure directly to the laminate during cure.

The trailing-edge wedge is a single unstiffened laminate, formed in a tool with a V-shaped interior aluminum wedge. A precured fiberglass arrowhead is located in the center of the laminate, where it bends around the aft edge of the V-insert. The laminate is wrapped around the V-insert, and then placed on a flat tool base with a dam at the aft edge. A caul plate is placed over the other surface, and the entire tool is bagged for autoclave cure.

Flat Laminates and Miscellaneous Parts. The flat rib and spar laminates were fabricated on flat aluminum plates, by laying up a large, uniformly thick panel. After curing, this panel was cut into the appropriately sized components for the rib and spar webs.

Parts such as the hinge ribs, closeout ribs, and small attachment angles were made on aluminum tools. The hinge ribs were made in segmented aluminum female-type tools, mounted to a carbon plate. The carbon plate

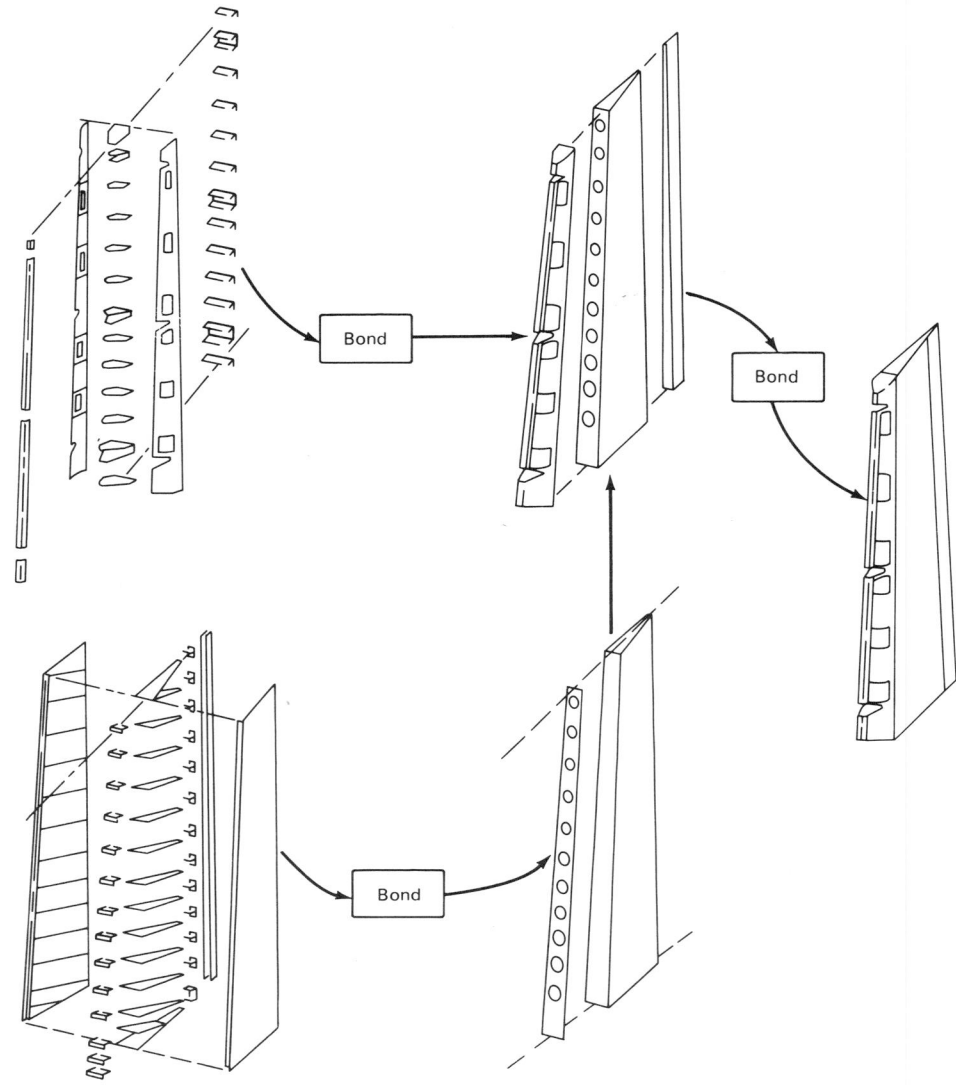

Fig. 15 Rudder assembly flow. Source: Ref 1

provided dimensional control for the external surface of the rib flange, which was a bond surface. Elastomeric rubber was used inside the tool to provide lateral pressure on the rib flange during cure.

Assembly. The components were fabricated, as discussed in the preceding sections, to form the leading and trailing edges and box parts. These components were then assembled in a series of steps outlined in Fig. 15, to form leading-edge and box assemblies and, finally, to form the rudder assembly.

The assembly tools and sequences used were relatively conventional, with the exception of the leading-edge assembly tool. This tool underwent a 120 °C (250 °F) adhesive cure cycle while controlling the hinge and torque tube attachment locations within prescribed tolerances. This was accomplished by using two carbon-epoxy tooling bars similar to those

used on the leading-edge bond fixture. The leading-edge skins were supported by these bars through the hinge attachment points. The carbon bars controlled the distance between hinge points during the adhesive cure cycle.

The box was assembled by drilling the fastener holes on one skin and then the other. After cleanup of the fastener holes, adhesive was applied to all bond surfaces, and the fasteners were installed. The box assembly, consisting of two skins, rear spar, and rib webs, was then cured in an oven at 120 °C (250 °F) for 1 h. The front spar web was not installed during the sequence, to provide access to the box interior for inspection after the adhesive cure cycle. The predrilled front spar web was bonded in a separate cure cycle after inspection of the box.

Upon completion of the leading-edge and box assemblies, they were joined to the trailing edge in a final assembly tool, using the fastener-bond process. A final cure cycle in the oven was accomplished to cure these joints.

The final steps in the assembly sequence involved attaching the hinge plates in an interchangeability fixture, to ensure interchangeability of the rudder between aircraft. The rudder was then painted with a coat of antistatic primer and prepared for shipment to the customer, who applied the final exterior paint finish.

Inspection. Detailed visual, dimensional, and nondestructive inspections were performed on the materials, processes, and finished parts as they progressed through the production cycle. Nondestructive inspections consisted of ultrasonic examination of the laminates, using both A-scan and C-scan techniques. Critical areas of the structure were defined, and received 100% ultrasonic inspection throughout the production program. Less-critical areas received ultrasonic inspections at every tenth article. An automated ultrasonic "squirter" system was used to inspect the flat laminate areas, while the rib caps and other less-accessible areas were inspected with hand-held A-scan probes.

REFERENCE

1. C.W. Schneider and W.T. Terrell, "Composite Control Surface Fabrication," Paper presented at SME Composites in Manufacturing 4 Conference and Exposition, Society of Mechanical Engineers, Anaheim, CA, 1985

SELECTED REFERENCES

● W.S. Cremens, "Manufacturing Technology for Thermal Expansion Molding of Advanced Composite Aircraft Structure," Final Report, Contract F33615-74-C-5150, Aug 1975

● W.E. Harvill, Jr. and A.M. James, "Advanced Composites," Horizons, Feb 1985, p 31-43

● L.W. Lassiter, "Advanced Composite Materials," Paper presented at the 1985 Paris Air Show, Paris, May 1985

● P.J. Marra, "Trapped Rubber Processing for a Composite Rudder," Army/Navy Composites Fabrication Program Review, Marietta, GA, Sept 1977

● C.W. Schneider and D.C. Gibson, "Design and Certification of a Composite Control Surface," Paper 850888 presented at the 1985 SAE Business Aircraft Meeting and Exposition, Society of Automotive Engineers, Wichita, April 1985

Manual Lay-up

Richard A. Brand, General Dynamics Convair Division

WHEN HIGH-MODULUS FIBERS became available, the aerospace industry already had considerable experience with the hand lay-up of fiberglass parts. In this process, woven glass fabric impregnated with resin is manually forced onto a tool to eliminate air bubbles and to squeeze out extra resin; the location of partial plies and the fiber orientation are not tightly controlled. Foilowing the introduction of the high-modulus fiber prepregs, automated lay-up techniques were developed to meet the higher tolerances required. However, hand lay-up remains viable for small-quantity production of test panels, prototype parts, or parts of complex contour. The parts can be laid up from either woven fabric or unidirectional tape. These materials are compared in Table 1.

Ply Flipping. The forms of these materials differ in two respects that affect the lay-up process. Tape prepreg can be spliced parallel to the fiber direction by placing strips next to each other with a small gap, whereas fabric plies are usually spliced with a 13-mm (0.5-in.) overlap. In addition, because woven fabrics are not always symmetrical in weave, problems can arise when plies are stacked without regard for the weave style. A good practice is to flip the plies to obtain alternate nesting. This procedure balances an otherwise asymmetrical stack and prevents warpage of the cured part.

Ply flipping is usually applied to crowfoot and satin weaves because they are asymmetrical from one face to the other. Placing the fabric warp-to-warp and fill-to-fill allows symmetrical stacking of the plies and produces a good interlayer nesting of laminae; this results in better interlaminar shear because fibers from both layers cross interface boundaries. The nesting procedure consists of cutting patterns out of the broad goods prepreg and mating

Table 1 Comparison of tape and fabric prepregs for manual lay-up

Advantages	Disadvantages
Tape	
Better strength/ stiffness control	More plies required
Lower resin content	Longer time to cut patterns and lay up
Can be spliced parallel to fibers	…
Lower coefficient of thermal expansion	…
Fabric	
Fewer plies required	Lower mechanical strength (due to higher resin content)
Less time to lay up; easier to form over large curved areas	Difficulties in splicing large parts
Lower cost	Higher void content

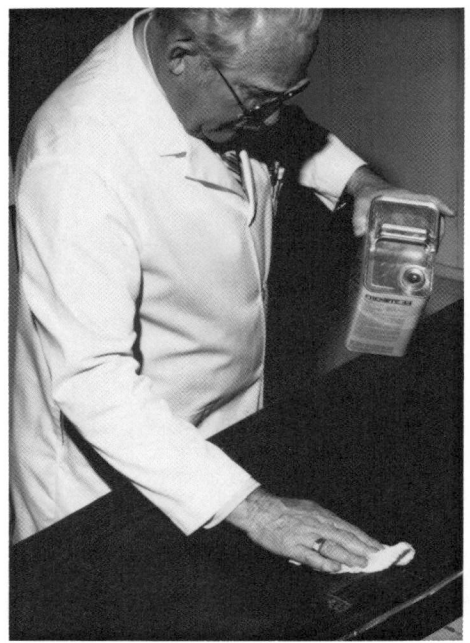

Fig. 1 Typical release agent being applied to tool

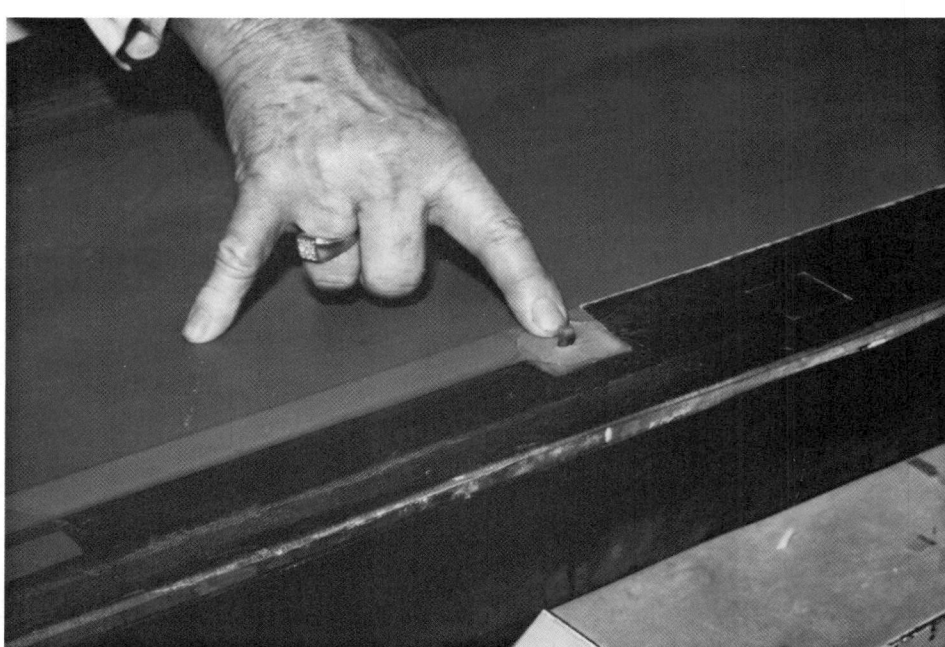

Fig. 2 Locating tab of cured skin on tool surface

Fig. 3 Orientation of precut broad goods prepreg on tool surface

Fig. 4 Preplying with vacuum bag

the sides without release paper (warp-to-warp, fill-to-fill). The release paper is then removed from additional modules, which are again nested to complete the stacking operation. When nesting is performed in this manner, the symmetry of the stack is ensured. Tape prepreg does not present this problem, because it is the same on both faces.

Materials consisting of several plies stitched together became available in 1987. These materials permit more rapid lay-up.

Mold Release. The first step in any lay-up process is to apply a release agent to the mold (Fig. 1) so that the cured part can be removed from the mold without damage. The most popular mold releases use fluorocarbon in a solvent or carrier system that evaporates, leaving a thin film of fluorocarbon release material

on the tool surface. Silicones are less desirable because they leave silicone residues on the cured composite surface that prevent subsequent bonding, priming, or painting. Some polytetrafluoroethylene coatings can be applied to tool surfaces and bonded more securely with a high-temperature curing operation. These coatings are more durable than wipe-on releases, but must be applied before the mold surface has been contaminated with wipe-on releases.

The use of tabs on composite parts beyond the end-of-part boundary (Fig. 2) facilitates removal of the cured part from the mold with wedges or release tools. Any damage due to removal will not be on the tooling surface or will be trimmed from the composite part along with the tabs. The tabs should be provided with

index holes to coordinate the composite part to trim templates, trim shells, router or drill fixtures, and assembly fixtures so that the location of the partial plies relative to the part edge can be controlled throughout the entire manufacturing sequence. In the lay-up phase, the fiber orientation of each ply, the size and location of partial plies, and the sequence in which the plies are stacked on the tool must be controlled. Typical aerospace specifications specify an angular accuracy of $\pm 2°$ in gentle-contour parts and $\pm 7°$ for complex contours.

Ply count is sometimes overlooked and may cause problems. As each ply is laid down, it should be checked off on an accounting sheet to avoid the mistake of including extra plies or shorting the ply count. Composite laminates must be balanced through the centerline of the laminate; otherwise, warpage is likely to occur, especially with thinner laminates. Thick laminates can mask the effect of a missing or extra ply because of the large tolerances permitted in layer thickness.

Another common error is the inadvertent retention of release film or paper on the prepreg material and its incorporation into the laminate stack. Pieces of release paper, tape, or other debris act as delamination areas in the laminate, reducing mechanical properties.

Orientation Accuracy. One of the simplest methods of maintaining lay-up orientation accuracy is to have the angles necessary for the lay-up marked on the tool beyond the trim lines. For gentle contours, the ply-on-Mylar technique is frequently used. The shape of the ply and the fiber orientation are drawn by computer-aided design on dimensionally stable Mylar film. The Mylar templates are coordinated to the mold by index pins in the mold and holes in the Mylars. The Mylars are numbered to reflect the stacking sequence. This prepreg is placed on the template as specified, the assembly is placed on the mold, and the template is removed. The process is accurate and almost foolproof. However, one Mylar is usually made for each of the plies, and their inspection, sorting, storage, and maintenance are costly.

For parts of complex shapes, the ply location and fiber orientation are controlled by a slotted tooling laminate made on the specific mold. After the tooling aid is coordinated to the mold, segments of the outline of the partial plies are traced with a colored pencil on previously laid-up prepreg. The tooling aid is removed, and the plies are stacked, using the marks as a reference. The marking operation is repeated as often as necessary. Again, however, the cost and the maintenance of the tooling aids are unacceptable.

After the mold release has been applied, the first ply on the tool is often a nylon peel ply. It is used either to provide a bondable surface after the part is completely fabricated or to act as a reservoir for trapping volatiles that migrate to the tool surface during cure.

The prepreg is cut according to predetermined patterns, and the tape (or fabric) is oriented along the axis required for that particular ply (Fig. 3). Subsequent adjacent tape patterns are laid in abutting the oriented piece parallel to it in order to retain the angular accuracy of the ply. Care must be taken to ensure that there are no air pockets in the lay-up and that no slippage occurs.

Compacting. As the plies are being stacked, it is desirable to seat them snugly on the tool and to remove some of the air from the lay-up. This process—called compacting, debulking, or preplying—involves use of a vacuum bag made with a plastic bagging film, such as nylon or polyvinylalcohol (Fig. 4). Once the vacuum is established, the prepreg can be worked to remove air pockets and to smooth the surface, if necessary. The vacuum bag is then removed, and additional plies are laid up in the same manner. Several plies (three to ten, if the contour is slight) can be laid up before another compacting sequence is initiated. The tooling often dictates the frequency of compacting. For example, lay-ups on cylinders or curved surfaces require frequent compacting to prevent wrinkling when heat and pressure are applied during a prebleed or precompaction cycle or during cure.

Resin Removal. Many specifications require fiber volume ratios in the cured part that are unobtainable without removing excess resin from the lay-up. The amount of resin that can be removed during final cure by the use of expendable bleeder plies is limited. If the excess resin does not move into the bleeder plies as planned, the parts will be too thick and wavy. They may also be porous. To prevent this from occurring, the lay-ups are often prebled at moderate temperatures by use of the vacuum bag to remove the bulk of the excess resin. The final cure can then be performed with less risk.

Mechanically Assisted Lay-Up

David G. Hess, Northrop Advanced Systems Division

MECHANICALLY ASSISTED LAY-UP reduces labor costs by using machines to replace some of the hand work. It is a compromise between a totally manual process and a fully automated manufacturing sequence. Lay-up system design is determined by material form, part shape, production requirements, and costs. The processes of die, ultrasonic, and computer-controlled ply cutting; flat and contoured tape laying; ply handling; and automated ply lamination are each described in the articles that follow.

The development of systems to lay tape in the flat began about the same time that programs were first initiated to evaluate advanced fibers using tape prepregs. The original concept was to prepare groups of plies in the flat and manually place them on the mold. However, the shape of the individual plies was limited by the capability of the cutters in the tape head. Therefore, recognizing a potential new market, the machine tool industry became involved in tape laying on the basis of experience with machining metal to complex shapes under computer control. In laying tape on contoured molds, gaps and overlaps occur. The machine tool industry contributed to the solution of this problem through application of linear programming to obtain a "best fit" between the tape courses, eliminating overlaps and keeping gaps within permitted ranges. Another major contribution was the use of the expertise of the machine tool industry to improve mechanical designs and computer controls in tape-laying processes.

Tape-Laying Machines. By 1987, tape-laying machines had been developed that offered several advantages for use in mechanically assisted lay-up systems:

- High output of large plies
- Good material utilization
- Capability to lay up moderately contoured parts entirely on the mold
- Capability to produce multiple-ply stock for cutting to shape off-site

However, these machines still had several disadvantages:

- Inefficient operation with plies using short lengths of tape
- Limited capability of the tape layer for ply shape
- No capability for highly contoured parts

Ply Cutting. As advanced composite fibers became increasingly available, they were woven into fabrics, establishing a need for systems that could cut plies from fabric prepreg and wide-tape broad goods. The commercial tech-

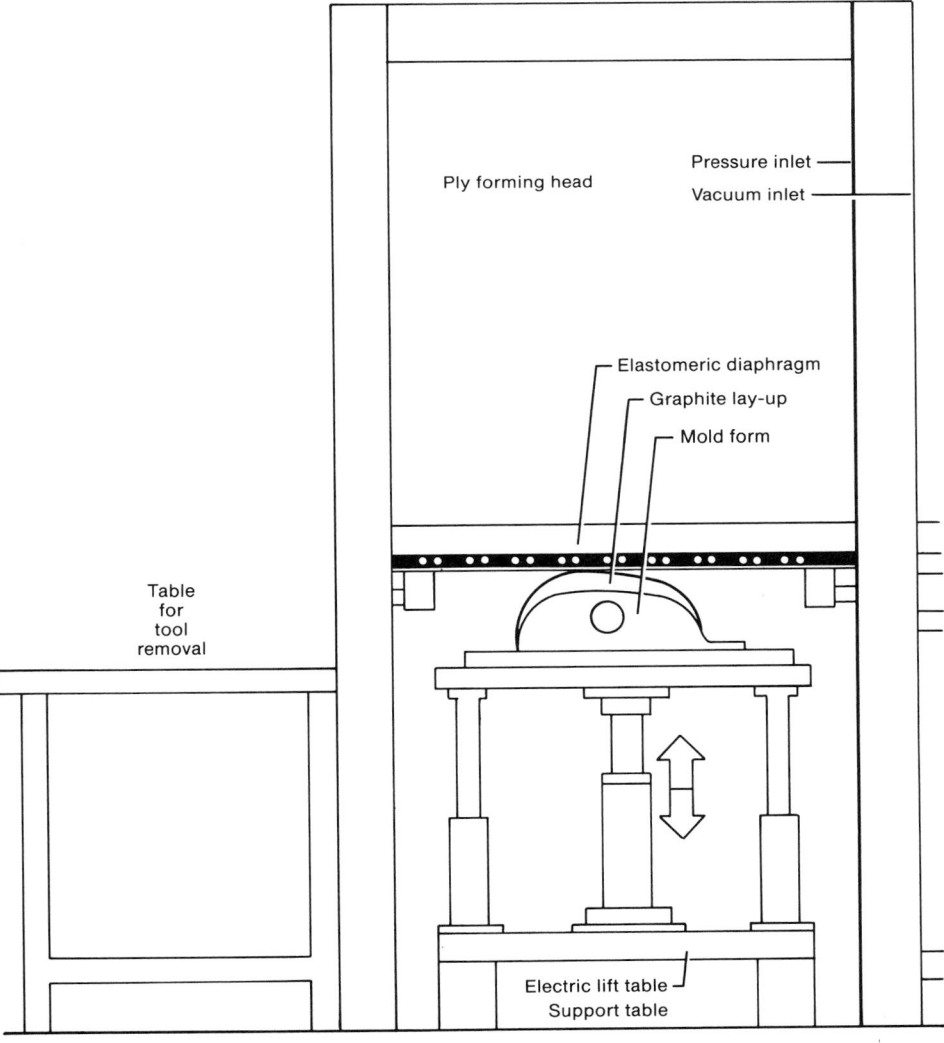

Fig. 1 Mechanized ply-forming station. Draw vacuum in box; push tool into box and release vacuum in box; pressurize box to force diaphragm around tool; move tool down so it is below the diaphragm and can be slid sideways to the tool removal table.

nology provided two solutions: die cutting, and computer-controlled ply-cutting and labeling systems. Die cutting is geared to the production of large quantities of plies, but the number of shapes is limited because a cutting edge is needed in the die to generate the plies. The ply-cutting and labeling systems, however, use a computer-controlled knife to cut the plies and can therefore produce an unlimited number of shapes. One such system can cut up to five layers of fabrics and nine layers of tape concurrently. Advantages of these systems are:

- Ability to process either fabric or wide-tape prepregs
- No significant restrictions on ply shape
- Large output of precision-cut plies

Disadvantages include:

- Need to sort the plies for a part into kits and subsequently place them on the tool
- Greater scrap generation than in the lay-up of narrow tapes (but less than when mechanics cut material off the rolls without measuring)

The cutting patterns, called nests, are designed for maximum material utilization. Often, plies for several parts are included in one nest; sometimes the plies for one part are included in several nests. Complete freedom is possible in large production because the nests are cut frequently.

Ply Sorting and Stacking. Plies can be sorted into kits manually by the cutting-system operators. Also, two mechanized approaches have been investigated using jointed-arm robots equipped with a ply-handling device. This device typically uses vacuum to pick up the plies and mechanical pressure to join them. The simplest approach is to sort the plies into kits for subsequent manual placement on the tool.

A more advanced approach is to pick up the plies and stack them in the sequence and location specified on the engineering drawing. Some attempts have been made to stack them on the tool, but this complicated the system. Therefore, the plies are usually stacked in the flat to form a lay-up, which is then joined to shape on the mold.

A problem with stacking plies in an automated system is that the paper or plastic film interliners used to prevent the prepreg from sticking together must be removed before stacking. A mechanized system to convert plies in a nest to fully stacked lay-ups is being developed (Fig. 1). Steps in the operation of this system are as follows:

- Vacuum is drawn in the box
- The tool is pushed into the box, and the vacuum is released
- The box is pressurized to force the diaphragm around the tool
- The tool is moved down so that it is below the diaphragm and can be slid sideways to the tool-removal table

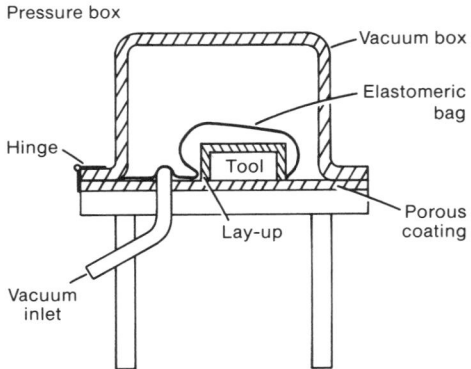

Fig. 2 Prepreg forming table

Locating Partial Plies. When precut plies are collected into kits, partial plies must be precisely located on the tool. The precut prepreg can be laid up on the templates before the ply is stacked on the mold, but this involves handling the prepreg twice. Use of removable templates to mark the location of partial plies interrupts the stacking process.

An approach that eliminates the need for a lay-up tooling aid is the use of structured light, that is, an optical system mounted above the work area to project the shape of the plies onto the tool. Ordinary light systems control the shape of the pattern by passing the beam through several precut templates. Multiple templates are used to control diffraction, which can cause fuzzy edges. A disadvantage of such a system is that the shape to be projected is not easily altered. Also, a new set of templates must be made for each pattern to be projected. Once made, however, the templates can be remotely manipulated, much like the slides in a home projector.

A recently emerging technology is the use of laser light, which offers several advantages over conventional optical methods. The depth of focus of laser beams is often in the range of 2 m (6 ft), so that focusing on the surface of moderately complex contour tools is not usually a problem. Within this range, however, the beam diameter may vary slightly. If maximum precision is needed in scanning the periphery of the ply, the beam can be dynamically focused.

The beam emitted by a helium-neon laser is a brilliant ruby red that stands out on nearly every background, even when applied to the black-on-black contrast of graphite reinforced plies. The optical-contrast ratio of these devices is often as high as 500:1. These systems are composed of a laser projection unit, an operator work station, a tool-locating module, and an engineering file interface. The system reads the ply dimensions from the engineering files and instructs the laser to pass a beam into an optical scanning device. This device takes the single beam and sweeps it

Fig. 3 Use of elastomeric bag to seat plies on complex-contour tool. Courtesy of Grumman Aircraft Systems

Fig. 4 Roller device for seating plies on corrugated tool. Courtesy of Grumman Aircraft Systems

around the periphery of the ply fast enough to be visible as a uniform line.

Complex-Shape Seating Devices. Although mechanical systems are available to produce plies and stack groups of plies in the flat, devices to seat these lay-ups on complex-shaped tools need further development. Composite prepregs vary in tack, which assists in holding plies on the mold surface. The fibers also introduce a spring effect, which makes some fibers more difficult to process than others and usually increases the difficulty in forming tightly woven fabrics. The optimum ply-forming processes require applied pressure and sometimes heat.

A unique approach to this problem, which will increase productivity significantly, is under development. A metal box capable of withstanding vacuum and positive pressure is faced with an elastomeric bag sealed to the lower edge (Fig. 2). The box is placed over the tool/lay-up assembly, and a vacuum is introduced between the bag and the box, pulling the bag out of the way of the tool, facilitating installation. Then, on releasing the vacuum and introducing pressure to the box, the bag conforms to the tool and forces the plies to the shape of the mold. As an alternative, the bag in the box could be sealed to a table fitted with vacuum ports. The space under the bag is evacuated to produce a pressure differential of 100 kPa (14 psi) on the surface of the tool.

Also, devices consisting of a lift table and a diaphragm-faced container have been used to seat plies on the mold. The flat lay-up is placed on the mold and lifted enough to distort the bag sufficiently to form the plies. Use of an elastomeric bag to seat plies on a complex-contour tool is illustrated in Fig. 3.

In addition, devices using mechanical force to shape flat lay-ups on the mold have been developed. These devices use rollers (Fig. 4), wiping blades, downward acting feet, and similar mechanisms. They sometimes are less effective than systems using bladders or bags because the applied load cannot be maintained for long periods.

Ply Die Cutting*

Steven Singer, Ontario Die Company of America

DIE CUTTING is a technology currently in widespread use. Typical applications include commercial seat structures, missile components, radomes, ballistic pads for helicopters, combat helmets, and jet engine parts. The wide variety of composite materials used include graphite, aramid, fiberglass, polyamide film, silica phenolic, boron, swirl mat, woven rovings, and sheet molding compound, in both unidirectional and woven formats, with or without resin impregnation. Cutting dies are also being used to blank a variety of aluminized and phenolic honeycomb core materials. The die-cutting process is extremely efficient for producing large numbers of plies of the same configuration.

Most die-cutting knowledge currently applied to composites originated in the automotive sector, where soft-trim fabricators have always faced the problem of daily producing enough accurately cut pieces from expensive roll good to meet tight sewing-line schedules. Die cutting offers these trim factories, as well as composite fabricators, the benefits of:

- A reliable, economical, easily maintained technology that is not prone to break down. The process is faster by more than 300% than computer-controlled laser, water-jet, or reciprocating-knife cutting equipment.
- Very high throughput rates that do not vary with part shape, material, or ply thickness. This high productivity allows "just-in-time" (JIT) cutting to minimize in-process inventories and part obsolescence
- Consistently accurate parts that minimize scrap rates and sewing errors and are ideal for subsequent robotic picking and handling
- Accurate control of fiber orientation or weave pattern within each ply profile
- Versatility in terms of accommodating a wide variety of materials
- High material yields, achieved by means of efficient nesting of individual die cutters within the total die board

At first glance, automotive cutting technology may not seem applicable to aerospace composites where many fewer units are being produced. Although a relatively small percentage (by weight) of commercial and military aircraft are presently composite, a number of very large clean room cutting installations are already in place. The reason for this is the ply "explosion" that typically occurs when the number of pieces actually required for one finished composite ship set are computed. In one instance, a missile vane, of which there were four per ship set, required 800 intricate pieces, or 3200 total pieces per set. In a second example, a single commercial aircraft engine needed 22 composite vane pads, each made with 28 plies of graphite, for a total of 616 pieces per engine. Thus, as composite volumes rapidly increase, efficient die-cutting technology becomes as essential to the aerospace industry as it already is in the automotive sector.

Die-Cutting System

Any die-cutting system consists of a cutting press, a plastic cutting pad, and cutting dies. Other aspects of the system are material handling, die storage and retrieval, and a trained operator. A typical installation incorporating these elements is shown in Fig. 1. As with any production system, each element is critical to the ultimate success of the system. In the past, companies designing die-cutting systems tended to concentrate most of their attention on the cutting press, to the detriment of the other elements, especially cutting dies and die storage and retrieval. For example, inexpensive steel rule cutting dies designed to cut labels and cardboard boxes will not successfully cut eight to ten plies of graphite or aramid; the result would be excessive die breakdowns and incomplete piece cuts. In addition, a lack of well-designed die storage racks and die-handling devices to facilitate the change of dies would create excessive operator fatigue, die maintenance, and production inflexibility.

The composite material cutting system shown in Fig. 1 is only one of many design possibilities. Design depends on specific production volumes and investment budget. In this particular system, after the material is cut into sheets of proper die length by an automatic cut-off shear, the sheets are aligned on top of the steel rule die and indexed through the cutting press.

The die is then shuttled to an unloading station where the parts are removed, marked, kitted, and either sent directly to the lay-up area or stored in the freezer because of the limited room-temperature shelf life of a preimpregnated composite material. The die is then either recycled through the press or returned to die storage and another die is retrieved.

This relatively simple and inexpensive die-cutting system would give a composite fabricator tremendous cutting capacity in a limited floor area of about 3 × 10 m (10 × 40 ft). Assuming an average die layout size of 1 × 3 m (4 × 10 ft), the system could feed, cut, and pick up an eight-ply lay-up (30 m², or 320 ft²) every 10 min, or about 1400 m² (15 000 ft²) per 8-h shift. Average cutting time for this 3 m (10 ft) die would be about 20 s. Assuming approximately 25 m (83 ft) of rule in the die, the cutting time translates into a production rate of 75 m/min (250 ft/min).

Below is a brief description of each of the six key elements in a successful die-cutting system.

Cutting Press. As shown in Fig. 2, four basic press types are available to composite die cutters. The first two, swing beams and travelling head machines, are inexpensive, space intensive, manually operated units suitable for small parts, gaskets, prototypes, and test panels. However, this article mainly focuses on the more productive wide-area hydraulic beam and mechanical roller presses, both of which cut the full width of the roll goods simultaneously by utilizing incremental feeding of the die (Fig. 3). This type of feed system accommodates any size part, from small fairings to long spars or radome segments, using relatively low-tonnage presses.

Beam presses for composite die cutting range from 65 × 10³ kg to 180 × 10³ kg (70 to 200 tons), with platen or head sizes ranging from 510 mm (920 in.) deep by 1500 mm (60 in.) wide to 1000 mm (40 in.) deep by 2500 mm (98 in.) wide. The size of the press depends on the

*Adapted with permission from Technical Paper EM86-713, in *Fabricating Composites '86 Conference Proceedings*, Society of Manufacturing Engineers, 1986

COMPOSITE DIE CUTTING SYSTEM

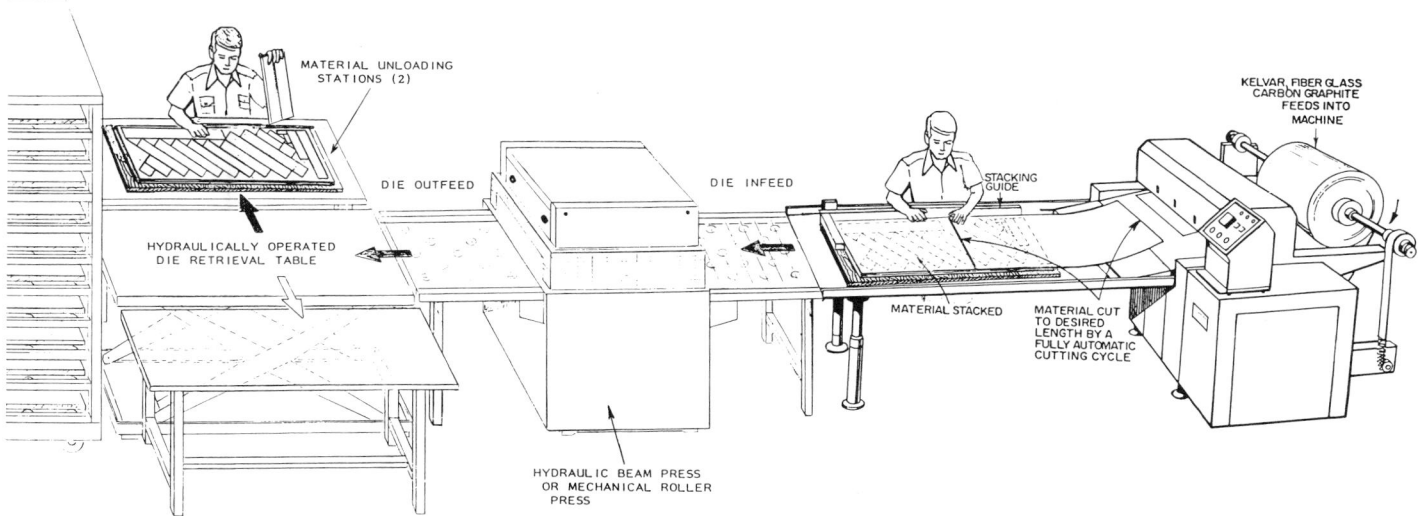

Fig. 1 Composite die-cutting system

type of material and the part size and shape. With smaller and more intricate shapes a large area of the knife can come in contact with the press platen, necessitating higher-tonnage presses. A very general parameter for gaging press capacity or rating tonnage is: 100 linear mm (4 in.) of rule per hydraulic ton.

The press should incorporate parallel platens, excess tonnage capacity, minimum head deflection during cutting, and four guide posts, rather than two. Beam presses normally have 25-mm (1-in.) thick nylon or polypropylene pads, either mounted on the top platen in a fixed position or on a pad shifter that automatically rotates the pad after each press stroke.

Die indexing with beam presses can either be manual or automatic. Manual indexing is done by the operator by means of roller conveyer tracks. Automatic indexing uses an incremental feed system. Board lengths of any size ranging from 0.6 to 10 m (2 to 35 ft) can be accommodated.

Roller presses cut by means of a mechanical pincer action. The die and material are passed between two steel rollers. The top roller is normally adjusted to 0.254 to 0.508 mm (0.010 to 0.20 in.) below the height of the rule. There is no tonnage rating on roller presses because they apply unlimited tonnage at a given point. Roller presses vary in width from 711 to 2502 mm (28 to 98.5 in.), with a variety of sizes to accommodate common material widths.

Roller press widths	
mm	in.
737	29
1041	41
1372	54
1524	60
1600	63

Some roller presses have a retractable top roller to allow the die to return automatically to the starting point without double cutting. This feature allows two teams of operators to work simultaneously from both sides of the press. Roller presses work with 6.4- to 9.5-mm (¼- to ⅜-in.) thick plastic cutting pads placed on top of the material and the die before the press cycle begins. Pad length will vary with the die layout, but can be up to 3 to 3.7 m (10 to 12 ft). Roller presses should have large-diameter (80 mm, or 7 in. or more) hardened steel rollers to minimize roll deflection, and should have an accurate adjustment control. Because of their unlimited tonnage feature, roller presses are particularly suited to cutting masses of small parts where excessive beam press tonnage would be required. Due to the shearing action required to cut on a roller press, dies should be canted or angled slightly (5°) when entering the press. This will prevent incomplete cutting on long sections of rule that run parallel to the rollers.

Both beam and roller presses are widely used by composite fabricators, the choice depending on specific requirements. Features are identified in Table 1.

Purchase price and die life are the two key differences between these press types and often governs their selection. In many cases, fabricators start with an inexpensive roller press and move up to a beam press once they have proved the pay-back of die cutting in their operation.

The cutting pad, through which the die blade penetrates after first passing through the material stack, is important in terms of its type and the maintenance it requires. As a general rule, the pad must be as hard or harder than the material being cut to achieve cleanly cut parts. For most composite fabrics, stress-relieved

white polypropylene in the 75-78 Shore D hardness range will do an excellent job. For other fabrics, such as aramid cloth, a harder, 84 Shore D cast nylon should be employed. Aramid is a particularly difficult material to cut. In many instances a sacrificial layer of aramid fabric is employed in a multistack of material because the knife will cut all but the last layer. This sacrificial layer can be glued down to the plastic cutting surface and replaced periodically when it becomes excessively ragged.

Pad Maintenance. Cutting pads should be inspected periodically by the operator for excessive wear and deformation. Die cutting against pad surfaces that are uneven or not level will result in incomplete cuts, die damage (from driving the die edge too far into the pad), and higher-than-normal press loads. Ideally, the press should be set to "kiss-cut" the die against the pad, allowing for about 0.40-mm (⅟₆₄-in.) pad penetration. Composite cutting should never be done against steel, aluminum, or rubber surfaces. Beam press pads, which are normally 25 mm (1 in.) thick, can be laminated to 25-mm (1-in.) wood blocks so they can be periodically planed down to a new surface. Frequency of planing depends on press adjustment, number of cuts, and the die quality. Spare pads are essential to allow for pad replaning on a regular schedule. Brown paper can be laid over the top ply of material to act as a shim against unevenness or dents in the cutting pad. Release film against the cutting pad is essential in order to prevent the top ply of tacky, preimpregnated material from sticking to the pad.

Roll presses use thinner (6.4- to 9.5-mm, or ¼- to ⅜-in. thick), disposable polypropylene pads that should be rotated 180° and inverted daily to maintain their flatness. These pads are normally available from local sheet plastic dis-

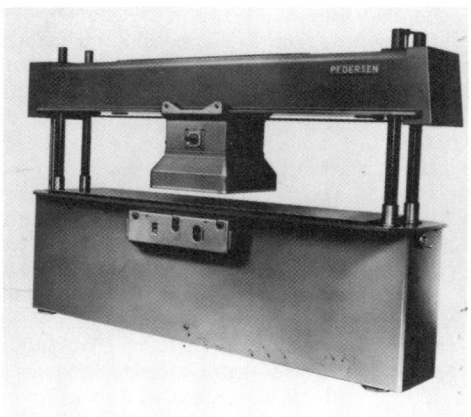

Fig. 2 Types of cutting presses. (a) 18 × 10³ kg (20-ton) swing beam press. Bed size: 760 × 1000 mm (30 × 40 in.), head size: 760 mm (30 in.) wide. Hydraulic operation. Courtesy of Associated Pacific Machine Corporation. (b) 35 × 10³ kg (40-ton) travelling head press. Bed size: 640 × 1600 mm (25 × 63 in.), head size: 610 × 640 mm (24 × 25 in.). Hydraulic operation. Courtesy of Associated Pacific Machine Corporation. (c) 135 × 10³ kg (150-ton) four-pillar beam press. Cutting area: 560 × 1700 mm (22 × 67 in.). Hydraulic operation. Incremental die and feed unit included. Available from 65 × 10³ kg to 180 × 10³ kg (70 to 200 ton) capacity. Courtesy of Emhart, USM Machinery Division. (d) Roller press. Cutting width: 1600 mm (63 in.). Mechanical pincer operation. Incremental die feed. Available in various cutting widths, from 740 to 2500 mm (29 to 98 in.). Courtesy of Associated Pacific Machine Corporation

tributors in sizes up to 2 × 3 m (6.2 × 9.5 ft). Larger sizes are available on a custom basis.

Cutting Die. Figure 4 shows the four main types of cutting dies of interest to composite fabricators. Die 1, with a sharktooth blade, can be used to die cut soft honeycomb core, while die 3, with a 50- to 130-mm high (2- to 5-in.) rule, can be used on aluminum and phenolic honeycomb core materials up to 50 mm (2 in.) thick. Die 4 is a heavy-duty fully heat-treated forged die with wall thicknesses from 6.4 to 11.1 mm (¼ to ⁷⁄₁₆ in.). Forged dies, with their increased depth and wall thickness, are used to cut thick materials and large volumes or high stacks of impregnated or unimpregnated cloth,

including aramid. In some instances, forged dies have been used to trim resin flash off of flat or contoured cured parts, thereby eliminating costly posttrim operations. For aramid cutting, seamless forged dies can be made from tough tool steel for much longer life.

Steel Rule Dies. Die number 2 in Fig. 4 is a steel rule die consisting of a knife-edged strip of steel rule held in place by a slotted wood die board. Steel rule dies are the most commonly used type of composite tooling because they are well suited to jigsawlike arrays to maximize material use. Steel rule dies can vary in length from 0.6 to 10 m (2 to 35 ft) and in width from 0.30 to 1.9 m (1 to 6.2 ft). The term steel rule

is a generic term. It is important that the proper grade be selected for satisfactory composite cutting. Steel rule dies with laser cut boards designed to cut foam packaging, corrugated boxes, paper labels, or gaskets should not be used on composites. These die applications use very thin (0.7-mm, or 0.028-in. thick), soft cutting rules and unwelded fitted joints in order to maintain exacting tolerances (±0.125 mm, or 0.005 in.) and precise levelness for single-ply cutting of nonfibrous materials, often using a steel plate cutting pad. Cutting five to ten plies of tough synthetic fibrous composite materials requires a heavy-duty industrial grade of steel rule die.

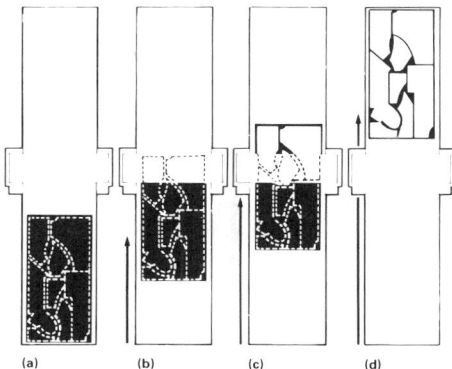

Fig. 3 Incremental work feed type P. (a) Cutting die in position with material laid on top (dotted lines indicate position of dies). (b) Die and material are moved under the head of the press and pause for the first cutting stroke. (c and d) Die and material move forward automatically to next position, and further cutting stroke takes place, until the entire die is complete. Stripping can begin as soon as the cut work appears on the exit side of the press.

The following specifications for composite die construction are recommended when purchasing steel rule composite tooling.

- *Die board:* Densified resin-impregnated beech plywood to hold the cutting rule knife. Low-quality plywood die boards will allow the knife slot to expand in time, resulting in rule loosening and pull-out. One-piece die boards of up to 2 × 3 m (6.2 × 9.5 ft) are obtainable. Piecing small die boards together is not recommended, because the joint areas will be prone to splitting and breaking
- *Steel rule:* A 31.8-mm (1.25-in.) high rule knife with a 2.1-mm (0.084-in.) thickness for maximum durability and cutting capacity. Cutting edge hardness: 50 to 52 RC for maximum life and 40 to 42 RC in the body of the knife to maintain steel flexibility
- *Welding:* All butt joints between the knives should be fully welded at the cutting edges to ensure complete cutting of the fibrous material
- *Rubber ejector pads:* Should be sufficiently dense to eject the cut parts and be locally positioned to minimize material distortion during cutting and minimize double cutting when using the incremental feed system

If these key specifications are followed, composite steel rule dies will normally outlast their planned volume use. Standard die maintenance typically requires periodic removal of resin buildup from the cutting edge as well as honing the die edges. Composite die tolerances are usually ±0.79 mm (±1/32 in.) for parts that require final machine trimming after molding. Net or matched metal molding (closer tolerances of ±0.38 mm, or 0.15 in.) can be achieved for the ply shapes. Steel rule dies can also incorporate internal hole punches from a diameter of 0.75 mm

Table 1 Beam press versus roller press

	Beam	Roller
Tonnage	70 - 200	Unlimited
Cutting speed	125 mm/s (5 in./s)	430 mm/s (17 in./s)
Material ply height	12.7 mm (½ in.) or more, depending on capacity of die and press stroke	3.2-4.8 mm (⅛ - ³⁄₁₆ in.) max
Die maintenance	Up and down hydraulic action, easy on dies	Mechanical pincer action can distort die knives. More die repairs
Press maintenance	Minimal	Minimal
Purchase price	$50 000-$200 000	$10 000-$40 000
Cutting pad	Fixed-in head	Rides on top of die and tends to curl because of the roller action. Disposable pads required
Cuttable materials	All composite unidirectional tapes and fabrics	Same

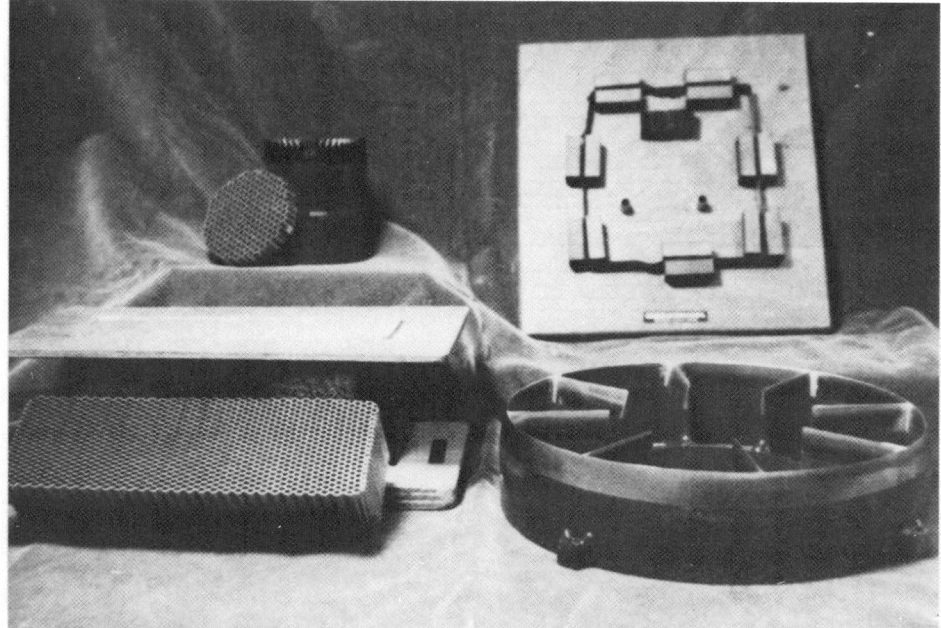

Fig. 4 Principal types of cutting dies, clockwise from left: sharktooth die; steel rule die; high rule die; and heat-treated forged die

(0.030 in.) and up, in addition to ply location notches, slits, or other irregular shapes. Special hardness coatings can be applied to the steel rule edge to increase blade life for long-run or abrasive applications. These coatings can increase surface hardness to the 70-90 HRC range.

Material Handling. *Part Marking.* Marking the cut pieces of material is important in many composite applications in order to correlate the pieces during subsequent ply assembly. Different plies of a composite assembly can look exactly alike, except for the angle they are cut at in relation to the fiber direction (0°, ±45°, 90°). A number of marking techniques can be used with steel rule die cutting.

Self-inking rubber stamps can be installed in each cavity to transfer the alphanumeric markings to the top ply (backing material) of the cut stack. If every ply of the stack must be marked, pointed stab or pin lettering can be inserted to leave a stencilled outline in each ply backing sheet. A paper nesting master, including cavity part and ply designation, in the form of inexpensive paper photocopies, can be produced and interleaved between the material plies before cutting. After the cutting cycle, each cut part will have an individual paper label attached that can be discarded during lay-up on the mold.

Die Nesting. Given the high cost of composite materials, maximizing yield in the cutting operation is essential. Figure 5 illustrates efficient die nestings. Where possible, common-edged lay-outs are used, thereby eliminating scrap between the plies. Parts can be oriented to whatever angle is required in relation to fiber direction, thereby controlling the material lay-up. Long plies can often be separated and butted together later during material lay-up in order to improve the die layout and material use. Grouping all the parts for a particular

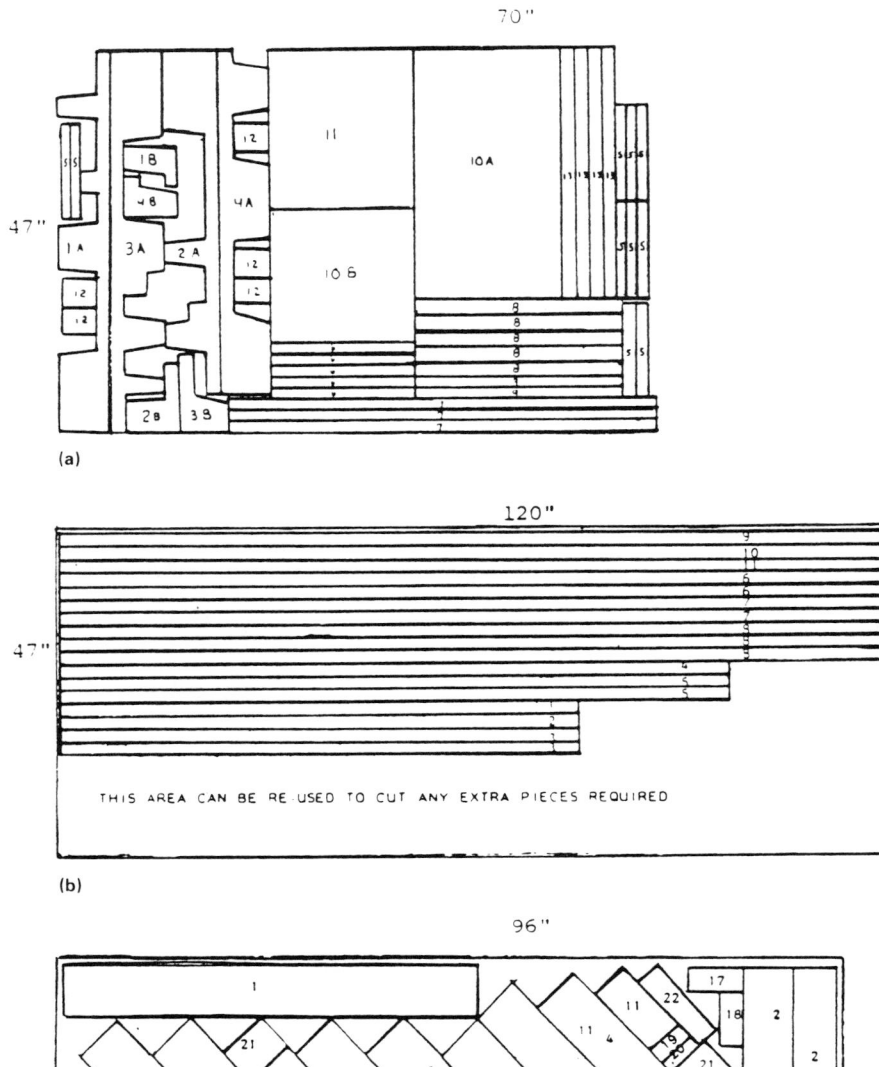

(a)

(b)

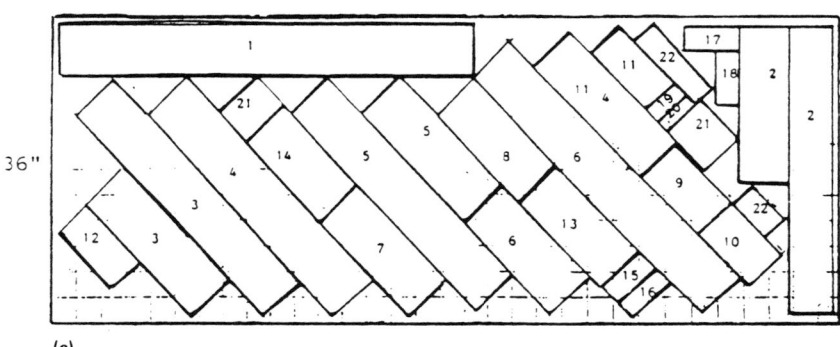

(c)

Fig. 5 Typical steel rule composite die nestings. (a) Aramid fighter door. This die layout illustrates the excellent material yield achievable with an optimum die nesting. All parts were common edged and oriented 0° or 90° as required. Four to six plies of woven aramid prepreg were cut per cycle. The overall layout, 1.8 × 1.2 m (5.8 × 3.9 ft), contained 30 m (107 ft) of steel and cost approximately $2000. (b) Aramid leading-edge channel. This layout was used to cut tapered aramid strips four-ply high to fabricate a leading-edge channel. The strips varied from 1.8 to 2.7 m (5.9 to 8.9 ft) in length and tapered down in width to 0.254 to 2.54 mm (0.010 to 0.100 in.) end to end. These plies were net-molded on a mandrel. All cavities were common edged. The die contained 45 m (149 ft) of steel and cost approximately $2500. (c) Graphite seat doubler. This layout incorporated all 22 plies to produce a graphite doubler. The plies were nested for maximum yield using common edging and 0°, 45°, 90° orientation as required. The overall layout was 1 × 2 m (3 × 8 ft), and the die cut was eight-ply unidirectional material per cycle. This die contained 15 m (56 ft) of rule and cost approximately $1500.

assembly in one die simplifies production control and kitting operations.

While these die nestings can be generated manually using miniature patterns or other time-consuming manual technology, the trend is increasing towards computerized nesting techniques that link up directly with the die making (Fig. 6). Using computers, part shapes for a particular assembly are digitally input and then accessed on a display screen, where they can be arranged for maximum yield. This computer plot can then be output in the form of a full-sized paper or polyester film copy to be used for die making. In more sophisticated systems, the digital nest is transferred by modem to the die maker, who will verify it on the computer and make any changes that will improve material yield or die construction. The nesting tape can then be used to pencil-plot the layout directly onto the die board or to program a laser beam for cutting the knife slot in the die board. Die makers specializing in composites can offer this computerized nesting service to customers who do not have their own equipment.

Material Dispensing. In any die-cutting system, the actual cutting cycle is so rapid that material dispensing and part retrieval normally govern overall productivity. Various dispensing systems can be used, depending on the amount of system throughput desired. The most common method is to lay up one layer at a time from a roll of fabric directly onto the cutting die. More automated material dispensing systems employ cut-off knives in guide tracks to cut off each ply after positioning over the die board. Because of the stability of composite prepreg materials, side and end losses should be limited to about 6.4 mm (¼ in.). Rubber side and end guides can be installed on the steel rule die to aid the operator in accurate ply placement. An automatic sheeter can also be used to shear the plies to precise length. In more mechanized systems, table spreaders like those used in the apparel industry can spread long lays (8 to 16 m, or 25 to 50 ft) onto air flotation tables. Smaller blocks are then cut from these long spreads and transferred on a cushion of air to the cutting die.

Part Retrieval. Die-cut part removal, marking, and kitting should be done off-line to fully use the press. In many incremental feed systems, multiple teams operate from both sides of the press. Up to six teams can be used (three on each side of the press) so that while some teams are laying up material, others are cutting or unloading dies. Multiple teams increase press utilization and, in turn, productivity, reduce press investment, and improve production scheduling and flexibility.

The kitting assembly and transfer of cut pieces into freezer storage is very time consuming in a composites cutting operation. Many fabricators cut only the daily amount required by the lay-up and molding departments. Using this JIT approach reduces in-process inventory, costly freezer space, piece loss, obsolescence, and expensive handling. Die-cutting systems allow speed, flexibility, and minimal set-up, so different dies can easily be shuttled on and off the system as needed, making the JIT approach possible.

Die storage and retrieval systems should be an integral part of an incremental die-cutting system design. Vertical and horizontal die storage racks are commonly used, based upon the factors listed below.

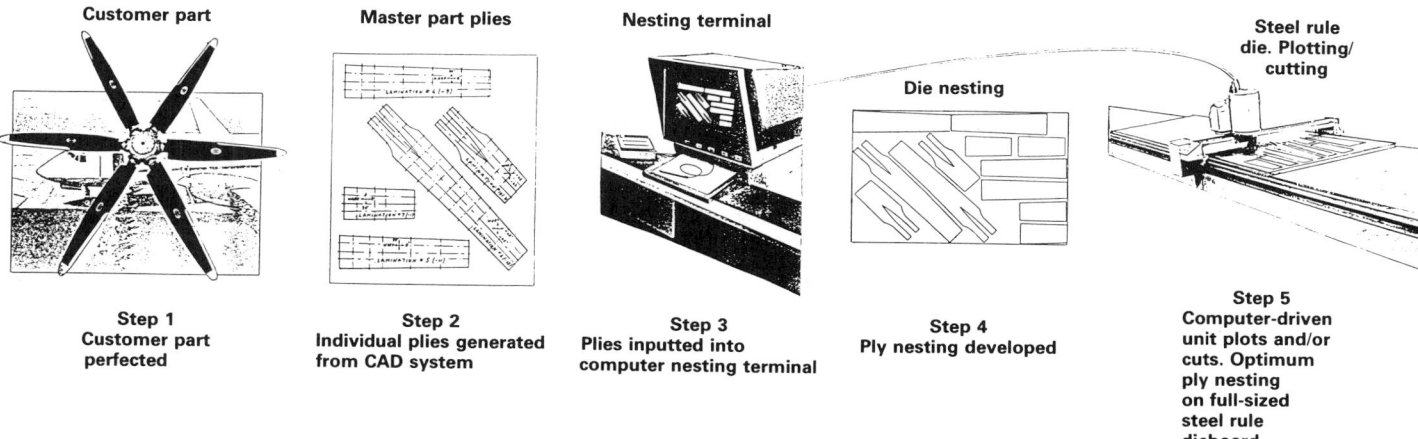

Fig. 6 Composite die nesting, from customer part to steel rule die

Space. A good die storage system minimizes the space required. Vertical racks designed for 31-mm (1.25-in.) high dies can store up to 20 dies in a 2-m (5-ft) wide area.

Board Levelness. Wood die boards will warp if air cannot circulate on both sides or if the die boards are not supported when stored or handled. Horizontal die storage racks should be built with casters or outriggers, not solid bottoms. Die handling devices that move dies to and from the press should fully support the dies. Flexing die boards by manual lifting can cause distortion.

Die Weight. Incremental steel rule dies can be relatively heavy (~70 kg, or 150 lb for a 1- × 2-m, or 4- × 8-ft, die). Moving dies in and out of the racks and onto the press should be done with hydraulic lifting devices or simple trolleys to reduce operator fatigue and simplify die changes. Storage racks should be located close to the press, where possible.

Die Dulling. Only a sharp die will cut composites well. Leaning dies against walls or on top of each other can cause blade damage. Proper die storage will keep tooling sharp and effective.

Standardizing the steel rule board width regardless of the actual cutter design simplifies die storage facilities. Experience has shown that 1- to 1.5-m (4- to 5-ft) board widths are most commonly used for most composite fabrics. Dies should be kept to a manageable length (1.5 to 4.5 m, or 5 to 15 ft). For very long parts, steel rule dies can be made in smaller modules and fastened together to eliminate the problem of handling excessively long boards.

An example of an effective storage system is one that uses a bank of vertical storage compartments with top and bottom guide rollers for the boards. A specially designed die-handling cart is set in the vertical position in front of the correct die storage location. The die is pulled out onto the cart, which is then rotated 90° to a horizontal position, level with the in-feed table of the press system. Another simple die storage system uses layered horizontal racks (ten to a unit) mounted on wheels. This unit can be moved up to the press for loading and unloading as required. Also available are fully automated die storage and handling systems using advanced microprocessor controls and vacuum picking devices to load and unload dies. Regardless of the method, proper die storage and handling techniques are essential to prolonging die life.

The Operator. Because of the simplicity of die-cutting systems, start-up training is minimal, both in terms of operation and maintenance. A new die-cutting system can produce parts within days of its set-up. The key elements in operator training include care in handling cutting dies, accurate material lay-up, proper adjustment of the press, and proper maintenance of the cutting pad.

Automation in Die-cutting Systems

The capital expense and related productivity of die-cutting systems vary greatly. While the basic principles discussed above apply to all systems, the degree of automation designed into a system can result in an investment ranging from $50 000 to $500 000. Tremendous strides in die-cutting automation have occurred in recent years, including automatic handling, storage, and placement of steel rule dies; automated spreading and lay-up systems; and automated die changing. These advanced systems are already finding their way into environments geared for the future in which cutting and lay-up of composites will be fully mechanized and have a minimum of manual control.

Case History. One example of an advanced die-cutting system is a recently installed $400 000 unit at a factory that produces camouflage netting for the U.S. Army. This system can fully incise with small slits and then cut out the pattern shapes in 20 plies of material 2 m (6 ft) wide by 5.5 m (18 ft) long in 3 min/cycle. One operator runs the entire system. The incising die is mounted in the head of a 180×10^3 kg (200 ton) beam press by a robot. Twenty plies of material are automatically dispensed and edge-aligned on a 5.5 m (18 ft) long, 9.5-mm (⅜-in.) thick nylon cutting pad without operator assistance. The loaded cutting pad is shuttled by conveyor and locked into the in-feed of the press. An empty pad takes its place and is loaded with material, while the loaded pad is precisely and progressively indexed under the incising die. After this first pass, the incising die is removed by robot from the press head and a 5.5-m (18-ft) long steel rule pattern die is removed from vertical die storage racks located beside the press, rotated in a special unit to a horizontal position, and vacuum-lifted into precise alignment on top of the incised material. In a second pass, the pad is indexed once again through the beam press, cutting out the pattern shapes required. The fully cut material on the cutting pad is then transported by conveyor to a debulking area in another part of the factory, while the die is vacuum-lifted, either back into storage or into position for a second cycle.

The storage racks hold up to 50 5.5- by 2-m (18- by 6-ft) long steel rule dies. The total system runs on a JIT basis, two shifts a day, with minimal breakdown and maintenance. A similar system currently under development for a robotic composite cutting project illustrates the degree of automation available with die-cutting installations.

If the components of a die-cutting system are properly designed, the economic payback can be substantial. Consistently accurate cut plies speed ply repetition lay-up and reduce the necessity for posttrimming after curing. In addition, the labor and substantial expense associated with manual template cutting are eliminated.

One fabricator gained an average 23% improvement in material yield by changing from manual to incremental steel rule die cutting. Another fabricator documented savings in labor and materials in excess of $1 million over 2 years on an original investment of less than $100 000 by steel rule die cutting the silica-phenolic materials for a missile fin leading edge. In another case, a composite producer reduced his cutting time on a 27-ply net-molded aramid leading edge from 4 h to less than 10 min; another manufacturer reduced his cutting time on a 32-ply, net-molded graphite blade for a hoverdrone from 2 days to less than 10 min.

These examples are not unique and illustrate the effectiveness of die-cutting systems when they are properly designed and used. With the projected increase in the use of composites, die-cutting systems will continue to help manufacturers meet their production goals.

Ultrasonic Ply Cutting

Michael W. Cook and Patrick J. McGill, Technology Marketing Inc.

CUTTING UNCURED COMPOSITE PRE-PREGS using ultrasonic vibratory energy is a relatively new application of ultrasonic technology within the aerospace industry. Ultrasonic vibratory energy was first used in World War II sonar technology applications, and nearly 20 years ago ultrasonic cutting of uncured fiberglass composites and synthetic materials was developed and applied to textile slitting. In the early 1980s, the technology was extended to cutting composite materials. Currently, other industrial applications include nondestructive testing, cleaning, plastic joining, drilling, machining, and biomedical uses. The conventional value of ultrasonic vibrations has been the generation of heat within activated materials for thermoplastic assembly in ultrasonic welding applications.

Theory and Principles of Operation

Ultrasonic energy is simple, mechanical vibratory energy operating in frequencies beyond audible sound, or 18 kHz. During "activation," intense, localized power/energy is imparted to a workpiece without large mechanical displacements or forces. For example, one kilowatt of power supplied to an acoustic tool (horn) is equal to about 4100 kg (9000 lb) of force through a distance of 25 mm (1 in.) in 1 s.

Generating ultrasonic energy starts with converting 50 to 60 Hz of electrical power to 20 kHz of electrical energy using a solid-state power supply. The power supply also can vary the amplitude and frequency of vibration, within limits. High-frequency electrical energy is conducted to a piezoelectric motor (converter) for conversion to unidirectional mechanical vibrations. These vibrations are in turn amplified by an amplitude transformer (booster) and/or horn before being applied to a workpiece (Fig. 1). The maximum available amplitude is determined by the horn design and may be modified by the addition of a booster. The operating amplitude of a stack consisting of the converter, booster, and horn assembly is controlled and regulated by adjustments in the power supply. Ultrasonic vibratory energy is then employed to activate the cutting medium blade at thousands of cycles per second. Ultra-sonic vibrations significantly reduce friction between the blade and the material being cut, which allows faster, easier, and more precise cutting of composite materials than do conventional methods (Fig. 2). Ultrasonic technology has permitted rapid cutting of multiple prepreg plies and a variety of other materials, including thin cured laminates, thick cured aramid laminates, boron prepregs, and honeycomb core materials. One ultrasonic cutting system has cut 25 plies of fiberglass-, aramid-, and graphite-epoxy prepregs in one pass, whereas a razor knife requires 7, 11, and 23 passes to rough-cut the same lay-ups, respectively.

Cutting Medium

Theory. The nature of peak-to-peak cyclic motion characteristic of ultrasonic vibrations is the major factor determining the success or failure of any specific blade length. The point of intersection of a wave with a horizontal axis is known as a nodal point, and the point at which the wave reaches its maximum peak or trough is known as an antinodal point (Fig. 3). With ultrasonic cutting horns, maximum amplitude occurs at the antinodes, and no vibration occurs at the nodes. This physical fact of ultrasonic vibration is visually evident on long blades and has significant impact on cutting ability. Each blade section oscillates at different levels, some in greater degrees than others and some not at all. Nodal points render sections of a blade ineffective as a cutting mechanism because of the absence of vibration. These varying degrees of oscillation within a fixed blade may result in an inability to tune, or frequency-match, a stack. These same oscillations dictate the necessity for a slight sawing motion with blades over 75 mm (3 in.) in length when cutting honeycomb.

Frequencies. The same problems associated with 20 kHz systems are magnified in a 40 kHz system. Because the wavelength of a 40 kHz system is half that of a 20 kHz, there are twice as many nodes, or "dead spots," for a given blade length and twice as much stress within a blade. The most effective frequencies are between 18.5 and 25 kHz, which offer the widest range of design-allowable amplitudes. The most common frequency is 20 kHz, which offers operators the freedom to fluctuate from

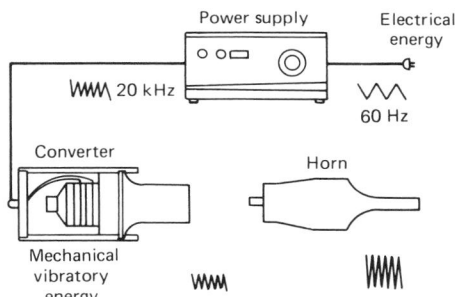

Fig. 1 Basic ultrasonic cutting system components

Fig. 2 Comparison of band saw and ultrasonic cut quality. Band saw cut is shown on left; ultrasonic knife cut is shown on right.

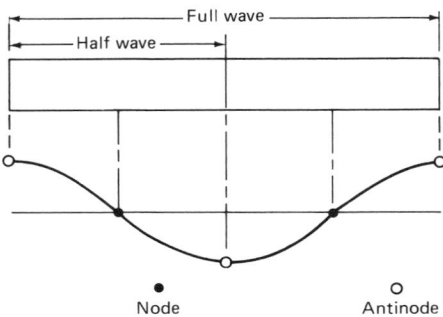

Fig. 3 Nodal and antinodal planes of half- and full-wavelength (250-mm, or 10-in.) horns

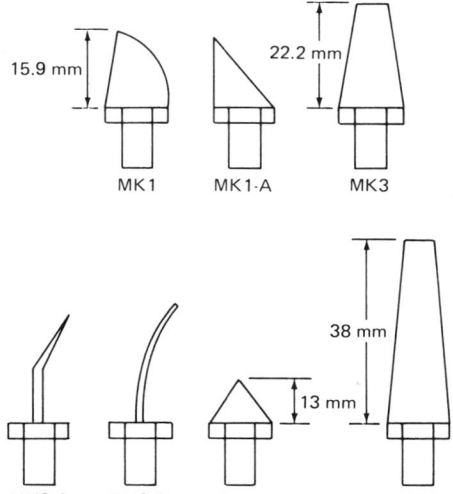

Fig. 4 Various blade types and configurations that have been tested

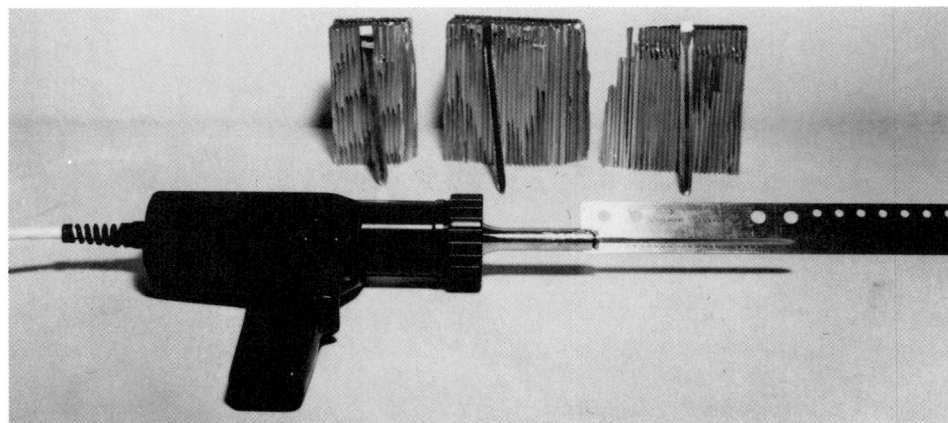

Fig. 5 Experimental blades up to 130 mm (5.25 in.)

an amplitude (blade stroke) of less than 0.025 mm (0.001 in.) for delicate welding or cutting to an amplitude of 0.13 mm (0.005 in.) for difficult, multiple-ply cutting. However, 0.18 mm (0.007 in.) has been demonstrated for experimental purposes.

Blade Technology. Though many blade configurations and lengths have been fabricated and tested (Fig. 4), and a variety may yet prove to be useful for cutting applications, the industry standard remains at a 19-mm (0.75-in.) depth of cut. Research has revealed that the ability to achieve increased blade lengths successfully is dependent upon amplitude. High amplitude produces better cutting ability, but the greater the amplitude, the more difficult it is to increase usable blade length because of metal fatigue and the flexing characteristic of thin blade materials when ultrasonically activated. Thus, effective extended blade lengths are difficult to develop, and many materials will not tune to a resonant system. Current research has produced experimental blades for up to a 130-mm (5.25-in.) depth of cut for 20 kHz systems (Fig. 5). Effective lengths are limited to 38 mm (1.5 in.), though research is continuing in this area, with a goal of 64 mm (2.5 in.).

The type of blade coating is also important. Blades that are carbide coated by a plasma arc process have an effective cutting life of a lineal 4.6 m (15 ft) in 12 plies of graphite unidirectional tape, while those blades treated by carbide impregnation have an effective life of 230 m (750 ft) in the same material. However, experience has shown that the hard, rough carbide coating that is effective on carbon fibers usually hinders the cutting of aramid fibers. Solid carbide blades were at one time thought to be the answer, but their high mass prevents them from being tuned for high-amplitude operation. (The search for the ideal coating suitable for all materials is ongoing.)

Cutting Variables

Amplitude. The relationship between power and amplitude is the most misunderstood aspect of ultrasonic vibratory energy. Two basic concepts should be accepted to ensure a successful cutting operation using ultrasonic vibratory energy. First, the maximum available amplitude of a given blade or horn is determined by the frequency of operation and the design of the horn to which the blade is attached, and is not affected by the wattage rating of a given power supply. Second, the amount of amplitude delivered to the workpiece is adjusted by the setting of the power/output control on the power supply. Thus, the power/output control is in essence an amplitude control. When properly employed, it allows an operator to select the most efficient operating amplitude for the material being processed. Thinner plies of material require less energy to process, while thicker plies require more.

Whether the application is welding, filament winding, tape wrapping, unidirectional tape manufacture, or cutting composite materials, all require different amplitudes to be efficiently processed. Using variable power/output control, an operator can tailor cutting amplitude to the specific composite material type and thickness and to the cutting speed required.

Materials and Thickness. The most common processing assumption is that all materials will be best cut by a high wattage ultrasonic system operating at maximum power; that is, the higher the power, the better the cut. High power availability, however, does not necessarily increase the effectiveness of a system. The wattage rating of a well-constructed power supply indicates the maximum amount of power available to maintain amplitude under load.

An effective system will cut virtually any prepreg material up to the depth of the blade. Some systems have cut 60 plies of graphite prepreg in a single stroke. The factors that can affect the relative ease of the cutting task are material type, material thickness, available amplitude, sharpness or coating on the blade, and operator skill. The material type and thickness dictate the type of blade to be used and the amplitude best suited for the task. Carbon fiber reinforced prepregs, though easily cut with uncoated and relatively dull blades, require carbide-coated blades for adequate blade life. In other prepreg materials, both weave and resin matrix often impact the speed and ease of cutting. In general, honeycomb, fiberglass-epoxy, and aramid-epoxy prepregs cut well with sharp, uncoated blades, while carbon fiber reinforced metal matrix, glass-phenolic, and graphite fiber reinforced thermoplastic prepregs require carbide-coated blades to achieve adequate blade life. Blades dedicated to aramid prepregs may last several hours depending on cutting surface and operator skill. The more brittle (high-modulus) the material is, the lower the required amplitude to cut it, whereas the softer the material, the higher the required amplitude. Blade movement of less than 0.038 mm (0.0015 in.) is generally ineffective and must be complemented by a sawing motion to compensate for the low amplitude. While material type dictates the basic amplitude requirement, material thickness magnifies that requirement and dictates the amount of power needed to maintain it for a given job.

Fig. 6 Ultrasonic prepreg cutter with 450 W of available ultrasonic energy

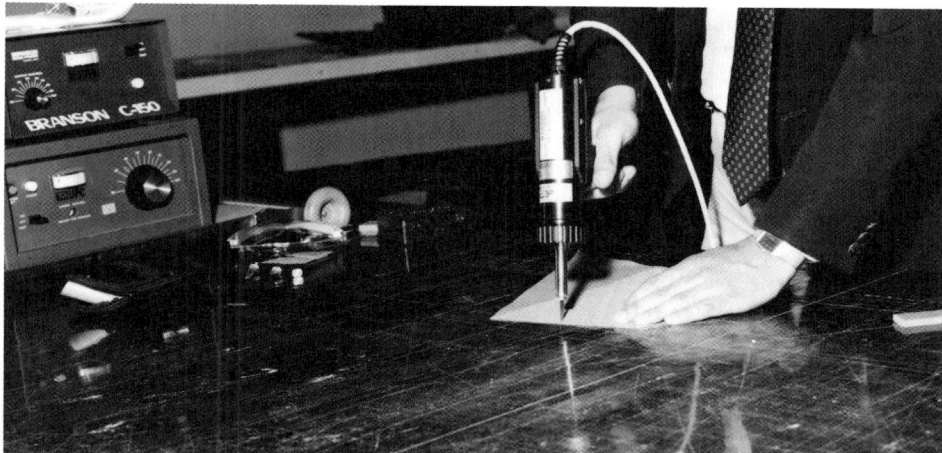

Fig. 7 Ultrasonic trim knife in operation on a mocca-cured urethane table top

Cutting Surfaces. The largest successful hand-held ultrasonic knife (Fig. 6) has a handgun that weighs 2.0 kg (4.5 lb) and is approximately the size of a quality ½-in. electric hand drill. In vertical operation, the weight of the handgun is negated because the blade is literally bouncing on the cutting surface. Imagine the impact force of a 2.0-kg (4.5-lb) mass bouncing on a table top 0.13 mm (0.005 in.) 20 000 times per second. The stress placed upon the end of activated knife blades and the minute portion of a table top with which it comes in contact is tremendous, but can be increased even further by pushing the handgun into the table top. Glass chips, steel, and other metal surfaces will score; cured composites will be cut or deeply scarred; thermoplastics will melt; weak materials will be cut; and blades will quickly dull. The potential problems are almost endless. The type of material needed is a vibration absorbent, cut resistant, stable, and easily cut material that allows an operator to feel the knife blade drag or cut its surface so that plunging the horn through the material being processed can be avoided.

Experience has shown that solid surfaces with a hardness of 80 Shore A are soft enough to attenuate ultrasonic vibrations and allow blades to remain sharp yet hard enough to keep the knife blade from inadvertently embedding in its surface. One such material is mocca-cured urethane, which has a service life of up to 4 years (Fig. 7). When service life is not a significant factor, other cuttable materials may be used, such as fiberglass or aramid laminates, nonmocca urethane, styrofoam sheeting, carpeting, or conveyor belting. In isolated cases, such as certain weaves of aramid fibers and boron prepregs, harder surfaces are desirable for maximum cutting efficiency.

Useful Ultrasonic Cutting System Features

The features that have proved to be useful and/or mandatory for reliable performance of hand-held ultrasonic cutting tools are described below. Most features are equally important to system automation.

Amplitude control. The most important feature for successful cutting operations is a variable power supply (Fig. 8). When properly designed, it provides a correlation between the setting of the power/output control and the operating amplitude of the system. Thus, operators can tailor the amplitude to suit the material being cut.

Automatic Amplitude Compensation. It is important that amplitude remain unchanged during cutting to maintain speed and control. Thus, the power supply must automatically maintain the desired amplitude during any load condition. Without this feature, the operating amplitude would decrease until cutting ability was lost when cutting conditions became difficult or force was increased.

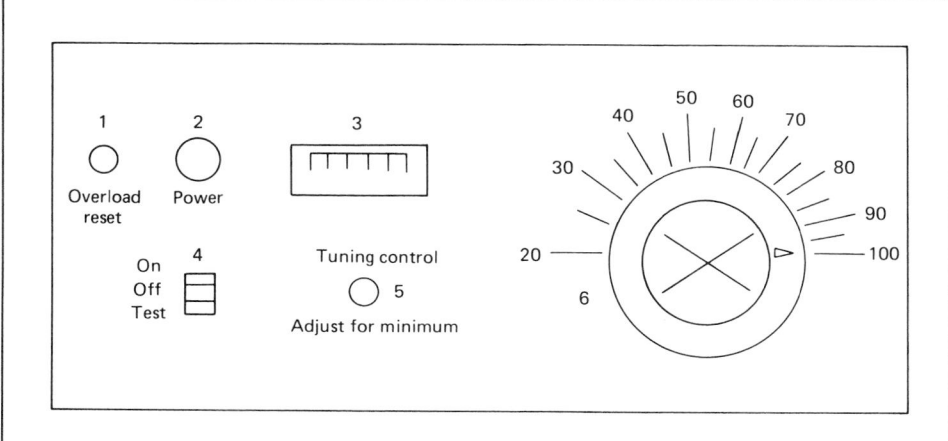

Fig. 8 Variable power supply provides operator control of ultrasonic vibratory energy. (1) Overload reset indicator switch/light indicates whether power supply is overloaded. Resets accessory protection indicator module when installed (S2). (2) Power-on pilot light indicates whether power supply is activated. (3) Loading meter indicates level of ultrasonic power transmitted to horn. (4) On/off/test switch: on position energizes power supply, off position deenergizes power supply, test allows tuning of the power supply (S1). (5) Tuning control adjusts the electrical operating frequency of the power supply to the mechanical resonant frequency of the converter, booster, and horn assembly. (6) Power/output control regulates amplitude of ultrasonic power. Clockwise rotation increases amplitude.

Automatic Frequency Control. The ability of power supply circuits to control frequency variation from operation to operation means that once a system has been tuned for a particular horn/blade type, retuning is unnecessary between horn/blade changes of the same type.

A front panel load meter is necessary to indicate the amount of power, or load, used during system tuning and operation. A numerical load meter is superior to an LED light in that it allows operators to "see" how much power they are using during system operation. With this information, research and development work can be transferred to the production shop by duplicating load meter readings during operation.

A system protection monitor is a circuit that terminates ultrasonic activation to prevent system damage under adverse conditions, such as failed components, improper tuning, blade failure, or excessive power supply loading. This feature is a must for high-output power supplies, such as 450 W (1550 Btu/h) or more.

Trigger-Activated Energy Control. In hand-held cutting operations it is essential for safety that operators have control of the initiation and duration of ultrasonic activation. A properly designed system will have an on/off trigger switch on a comfortable, slip resistant, lightweight handgun to ensure that activation ceases when the trigger is released or the handgun is not in use (Fig. 9).

Converter Compatibility. The availability of interchangeable converters/horns with different design amplitudes can greatly enhance system flexibility. The capabilities of systems offering converter compatibility can be changed extensively, and the systems used for different types of operations can be changed with only minor alterations.

Easy-to-Use Interchangeable Blades. Many blades and blade-holding methods have been tested in the last 3 years: slotted horns with set screws, horns machined as blades, titanium tips machined as blades, and various slotted tips or collets that tighten as they are screwed into tapped horns. However, the most successful blade holder is a simple, slotted screw that is similar in

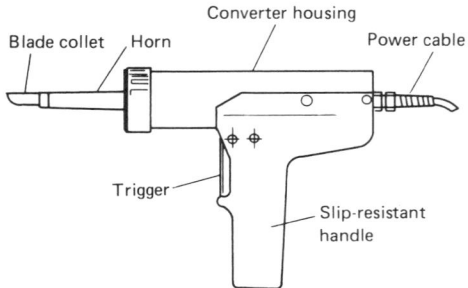

Fig. 9 Lightweight ultrasonic trim knife handgun, with a trigger switch for safety

design to a wood screw. This is a very versatile blade-holding device in production that, though not ideal, allows operators to choose from several interchangeable blades suited to different tasks without changing horns or performing other slow maneuvers required to change blades.

Reliability and Safety

Because system capabilities and reliability vary between manufacturers, the discerning customer will thoroughly evaluate candidate systems and contact current users for performance/reliability reports to ensure suitability for specific production requirements. Although most products perform well during a demonstration, only reliable products perform well under the strain of daily use.

Audible sound is often generated by 20 kHz ultrasonic horns, and although it will not damage human hearing, listening to a monotonous noise for extended periods may be irritating. For this reason, ear protection is recommended. Safety glasses also are recommended because an ultrasonic knife is a power tool that can cause brittle surfaces to chip, creating a potential eye hazard. As with any power tool, the most important safety precautions are operator procedural indoctrination and training, and the use of protective equipment.

Future Developments

Automated ultrasonic cutting is a growing application of ultrasonic technology. Adapting robotic cutting systems to accommodate an ultrasonic knife is relatively easy if the knife carriage and the cutting surface are parallel. Ultrasonic cut quality and speed are superior to those of laser, water-jet, and reciprocating garment cutters.

Ultrasonic cutting systems are commonly used in the biomedical and plastics industries for thinning and mixing liquids and for welding thermoplastics. Extending these applications to hot-melt unidirectional tape manufacture, fiber impregnation, compaction during filament winding, and the tape wrap of ablative materials is natural.

Ultrasonic cutting tools will be an integral part of the efficient, cost-effective, and safe factory of the future. Reduced labor, greater production, more efficient material usage, less material damage, and more precise material cutting and forming are benefits that will be proportionally compounded as the use and application of advanced composites increases.

SELECTED REFERENCES

- "Ultrasonic Plastics Assembly," Branson Sonic Power Company, Branson Ultrasonics Corporation, 1979
- G. Flood, "Ultrasonic Energy, A Proven Process for Laminating and Bonding Nonwoven Web Structures," Branson Sonic Power Company, Branson Ultrasonics Corporation
- W.I. Acton, "The Effects of Airborne Ultrasound and Near-Ultrasound," Wolfson Unit for Noise and Vibration Control, Institute of Sound and Vibration Research, The University, Southampton
- M.W. Cook and C.L. Laurella, "Study for the Development of Ultrasonic Cutting Tools, Final Report," Technology Marketing Inc., June 1985
- "USAF PRAM Program Final Report, Study for the Development of Ultrasonic Cutting Tools," Project No. RA 83-2, Sacramento Air Logistics Center (AFLC), Dec 1985

Computer-Controlled Ply Cutting and Labeling

Robert Forrer, Aerospace Division, Gerber Garment Technology, Inc.

ONE METHOD of producing large numbers of plies from either tape or fabric is to cut them from prepregs up to 150 cm (60 in.) wide under computer control. The most commonly used system employs a reciprocating knife cutter. The prepreg is dispensed over a unique cutting table with the vendor-supplied film having its backing paper facing downward. The table is a box with a surface of rigid plastic bristles, which support the prepreg but move aside as the knife cutter contacts them.

The prepreg is covered with a protective sheet of parchment paper or polyester film. The ply identifications are printed by a magic marker attached to the cutting head, which is mounted on the gantry. A labeling head that is available can identify plies more rapidly. The prepreg is then covered with polyester film, and vacuum is applied to the table to hold the prepreg in place. The plies are then cut, the vacuum is released, the polyester film is removed, and the plies with their protective sheets are sorted into kits. This is usually done manually, but robots with a suction (vacuum) head have been used on a prototype basis. The scrap prepreg is then removed from the table, and the process is repeated.

The primary concern is material utilization, which is a function of how well the various plies can be fitted into the nest (typically, 2 × 14 m, or 6 × 46 ft, maximum). The nests are usually prepared by taking the shape of the plies from the engineering data base and nesting them using special software. Tape plies can be split in the fiber direction as desired to obtain a higher utilization. Fabric plies can be spliced, usually only as directed on the engineering drawing. Some nests contain plies for several parts. Conversely, the plies for a single part may be in several different nests. The trade-off is the material savings that accrues from a better fit-up of plies within the nest as compared to the labor involved in sorting. Because this system can label and cut very rapidly, the sorting of plies into kits is a major limit on throughput.

Automated Ply Pattern Cutting Systems

A cutter designed for aerospace use was developed from a garment cutter in 1979. Its modular features allow a user to begin with a basic single-table cutting system and expand it into a dual-table production system, as shown in Fig. 1. The basic cutting system includes a cutter on a gantry with a magic marker labeling system, a 20-megabyte hard disk computer controller, a 2- × 14-m (6- × 46-ft) cutting table, and a vacuum hold-down system

The capabilities of this single-table S-91-C cutter can be increased by adding more modules so that the system operates at peak efficiency for a larger percentage of time. The options, which can increase productivity up to 40%, include:

- Cutter on one gantry
- Transfer table and a second table to permit labeling/cutting while previously cut plies are removed from the first table
- High-speed labeling system mounted on a separate gantry to perform this operation much more rapidly than the magic marker
- Prepreg dispensing system to unroll the prepreg on the table while the operator is occupied elsewhere

Fig. 1 The dual-table S-91-C cutter

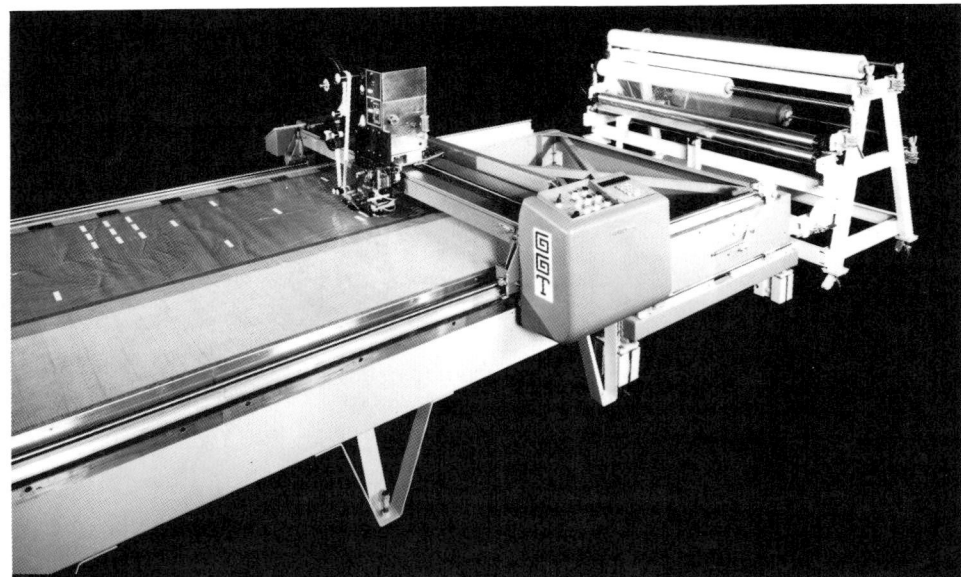

Fig. 2 The S-91-L labeler with material dispensing/spreading system

Fig. 3 The S-91-C-GLH labeling system

- Heavy-duty drill to make tooling holes or other small holes in prepreg materials
- Direct numerical control (DNC) from a host DNC system or computer graphic nesting system
- Spare cutting head
- Magnetic tape reader
- Various manual and automatic material dispenser/spreading systems
- Turnkey installation

The cutter can also be fitted with a conveyor belt to transfer the cut plies to an adjoining area for sorting into kits or stacking into flat lay-ups. The combination of modules beyond the basic system described previously is very flexible. The method of operating these accessories in a more efficient system is discussed below.

Automated Ply Pattern Labeling Systems

Two high-speed pattern labeling systems have been developed. One method is the magic marker system mentioned above. The other class of system, shown in Fig. 2 and 3, is geared to high-volume producers of aircraft structures. The S-91-L system (Fig. 2) has its own separate computer controller, label head and carriage, and full system control that can be switched to the operator control panel on the labeling beam. The system operates in conjunction with one of several types of automated or manual material dispensing/spreading systems to allow high-speed set-up of one of the dual cutting tables while on the other table, previously set-up, prepregs are being cut or patterns removed from the table. Therefore, every ply on every layer of material that is to be cut can be labeled before the cutting operation, which allows maximum output. The options for this system include a spare label head, magnetic tape reader, direct numerical control from a host DNC system or a computer graphic nesting system, various types of automatic or manual material dispensing/spreading systems, and turnkey installation.

The S-91-C-GLH labeling system (Fig. 3) mounts on the cutter head and is controlled from the cutter controller. This system, which can be used on single or dual cutting tables, allows the dispense/spread and label operations to be conducted on one table while cutting is done on the other, and it provides a very effective way to increase productivity through automated ply identification. One aircraft manufacturer intends to use this system in a semi-automated material dispensing/spreading system with a cutting gantry, to which the labeler would be attached. This would allow faster placement of labels during the spread operation for multiple layers or during the label/cut operation for single layers.

Computer Graphic Nesting Systems

A computer graphic nesting system, the Command 1000, was specifically designed and developed for the aerospace industry (Fig. 4). This system drives a cutter/labeler as well as various types of flat-pattern sheet metal routers/flame cutters, water jets, and lasers. Another nesting system, the AM-5, was originally used in the garment industry and was enhanced for the aerospace-oriented cutter/labeler system (Fig. 5). A commonly used program is called Autonest.

The key capabilities of the Command 1000 include interactive and automatic pattern ply nesting, two-way remote job entry, as well as a digitizer for data input. Direct numerical control for cutter/labeler systems is available. The AM-5 nesting system has interactive nesting, remote job entry, and a digitizer for data input. Direct numerical control of cutter/labeler sys-

Fig. 4 The Command 1000 computer graphic nesting system

Fig. 5 The AM-5 computer graphic nesting system

When considering material costs, nesting is the most significant factor in optimizing material utilization. Nesting is the process by which full-scale composite plies are placed within a broad goods material boundary as closely as possible to eliminate expensive scrap and to increase material utilization. The resultant nest will then represent the program to control the material dispensing, identification, and ply cutting processes. Nesting is also flexible, allowing immediate program and geometry changes that will not delay production, as will template or die changes. Another issue is the improvement of material utilization by making all the prepreg laid on the cutting table acceptable. Some companies require that defects be cut from the prepreg and that the residual prepreg be formed by overlap splices. Others are very liberal in accepting defects in the prepregs. Another approach consists of cutting multiple plies concurrently, scrapping defective plies, and recutting additional plies on a best-effort basis in a special nest.

The plies are input into the nesting system by direct connection to an upper-level design network, by manual digitizing, or by pattern design. Direct connection is the fastest, most effective method, and it provides more precise geometry. If a direct connection is not available, the next best approach is digitizing, which allows an operator to trace the shape of a ply from a metal or polyester film template and display the results on the nesting scope. The geometry of the ply can be verified by drawing the results on an automated drafting machine, which is part of the nesting system, and by comparing it to the engineering templates.

Once the plies have been loaded into the nesting system, the operator is ready to establish nesting parameters and to generate a nest. When the nest has been developed, it can be used immediately or stored for future production. It takes approximately 30 min to create a nest, to postprocess the nest into a specific format, and to be ready to begin producing quality ply patterns for the ply lay-up operation. The scope and programming time invested in material-efficient nesting is frequently recovered in several parts because of the savings in material costs.

During the nesting operation, two key parameters are established. The first is the length and width of the material the plies are to be nested within, and the second parameter is the plies to be nested. The plies that are nested will carry the alphanumeric identification within the periphery and the location where the label will be deposited. The cutting surface size will determine the width and length of the nest.

To begin the operation, an operator will load the program into the labeling control unit. Direct numerical control is an effective method because it eliminates computer tape handling and storage. The dispensing gantry grips the width of composite material and pulls it onto the cutting surface to the required length of the nest. The dispensed length is determined by the

tems is also available. This low-cost nesting system works well in the aerospace industry, but does not have all the capabilities of the Command 1000.

Integrated Manufacturing Center (Ref 1)

The integrated manufacturing center shown in Fig. 6 represents the application of state-of-the-art CAD/CAM technology to nest, dispense, identify, and cut composite shapes for the manufacture of composite structures. The integrated manufacturing center approach lends itself to flexible manufacturing techniques by integrating automated systems to eliminate laborious tasks and to enhance the quality of the finished product. The results are a reduction in scrap due to efficient nesting, the elimination of misidentified patterns or plies due to automated material dispensing and identification, and the protection against poorly cut plies through computer numerically controlled (CNC) cutting.

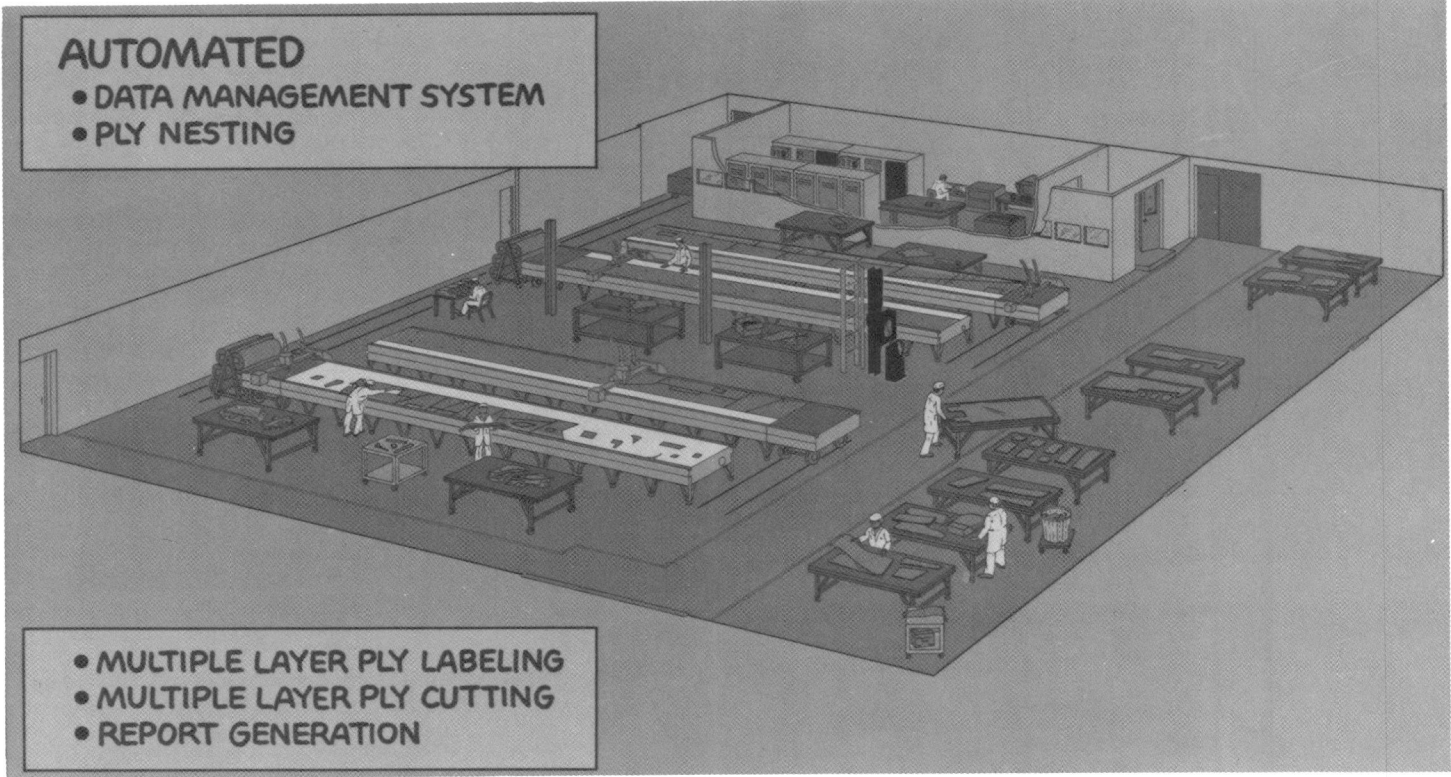

AUTOMATED
• DATA MANAGEMENT SYSTEM
• PLY NESTING

• MULTIPLE LAYER PLY LABELING
• MULTIPLE LAYER PLY CUTTING
• REPORT GENERATION

Fig. 6 Two dual-table cutter/labeler systems with computer graphic nesting form the integrated manufacturing center.

nest program. During the dispensing process, a cover sheet of material is collated onto the exposed side of the composite material to protect it during the kitting and storage process and to provide a medium on which to deposit the adhesive label or the label produced by the magic marker. The cover sheet also enhances the quality of the cut edge on the plies.

Once the first layer of composite material has been dispensed to the programmed length, it is automatically cut off and deposited onto the cutting surface. Traveling with the dispensing gantry is the labeling head. As the dispensing gantry returns to the broad goods carrier, the labeler deposits the labels onto the material. The label will provide up to 30 characters of information. If necessary, it can be rotated to various orientations to fit within a small ply.

After the labels have been deposited, the dispensing gantry is at "home" position. Additional composite layers can now be dispensed and identified; thus, all of the multiple layers will be identified.

After the material is placed on the cutting surface by the labeler, the cutting process can begin. The labeler is removed from the cutting surface and transferred to a parallel cutting surface approximately 4 m (12 ft) away. (This step could be eliminated by using the dispensing gantry to label/cut.) The transfer of the label head and gantry is done on a carriage located at the end of the cutting tables. This allows one

table of material to be dispensed and identified while the other table is being cut.

On a cutting surface that is 2 m (6 ft) wide by 15 m (48 ft) long, the cutting cycle will vary because of the total length of cuts being made, and the complexity of the ply shapes. The CNC reciprocating knife, similar to a jig saw, will operate from the same nest program as the labeler. This nest program standardization eliminates duplicate programs and additional disk storage. The cutter can cut the composites at speeds of up to 15 m/min (600 in./min), depending on the material form and number of layers being cut. A more common speed is 8 m/min (300 in./min), with the common obtained speed of 6 m/min (220 in./min) after corners and acceleration/deceleration are considered. Multiple layers reduce cutting speed somewhat, but total output is still increased.

At the end of the cutting cycle, the cut plies are moved to a kitting area, where they are sorted relative to part number. The completed kits are then sent to the lay-up area, where the plies will be laminated, or placed in freezer storage until needed.

To accommodate the smaller production facilities that require a totally integrated system, a lower-volume lower-cost integrated manufacturing center consisting of the AM-5 graphics nesting system, cutter, and the cutter beam mounted labeler system was created.

The cutter used with this system can be a single- or dual-table unit that operates in a

manner similar to the system discussed above. The automated material dispenser/spreader can work with a single-table system, or a manual material dispenser system can be used on the second table of the dual-table system while the other is being cut and labeled.

System Specifications

The cutter has a cutting accuracy of ±0.76 mm (±0.030 in.) and a tool point repeatability accuracy of ±0.2 mm (±0.007 in.), which represents an improvement created by industry demand. This unit can be retrofitted to systems existing in industry. One labeler can print and place a 30-character label, and the other, a 36-character label, in an average time of 3 s each. These systems can also orient that label in any rotation or 1° increment in 360° to indicate ply orientation.

In terms of labor savings in the area of material cutting, the cutter can cut up to four layers of unidirectional graphite at 6 m/min (250 in./min); up to four layers of woven graphite at 10 m/min (400 in./min), average; up to ten layers of aramid at 15 m/min (600 in./min), average; and up to ten layers of fiberglass at 15 m/min (600 in./min), average. No cut templates are required, nor is trimming required in the cure tool because parts can be cut net. The realistic rate for single-layer manual cutting of composites is 0.5 m/min (20 in./min). This does not include material or template acquisition.

Further labor savings is possible during cutter set-up and dual-table operation by using automated material dispensing, spreading, and ply labeling. The industry standard for labeling a cut ply manually is 1 s per character. If a label contains 30 characters, it will take 30 s per pattern per layer to label a nest that has been cut. If a nest has 60 patterns and is four layers thick, it will take 30 min to label one layer of cut patterns prior to stripping them from the table. This does not include material dispensing, spreading, or alignment, which will take approximately 5 min per layer manually with two people. With automated material dispensing/spreading and labeling systems, it is possible to complete one layer in approximately 7 min. Therefore, for every pattern, up to four layers can be dispensed, spread, and labeled in 30 min or less.

Labor savings during the lay-up process can be attained by using some features of the ply generating process as shop aids to expedite lay-up of the precut plies on the tool. Three examples of these assists for facilitating the lay-up process are as follows:

- If the label from the automated labeling system is placed in the upper right-hand corner of each cut ply in a 0° orientation, a method to check fiber orientation in the lay-up is provided
- A drill unit can provide precisely located tooling holes in the prepreg to coordinate the plies to tooling holes in tabs on the outside edge or trim area, so that the lay-up assembler can quickly locate the pattern in the cure tool. These same tooling tabs/holes could be used to locate the part in a drill fixture for trimming or an assembly/bonding fixture
- Notches cut on the outside scrap areas of full-sized plies can be used as ply locaters on the cure tool for rapid ply placement. Thus, control of the relative location of partial plies can be maintained throughout the entire manufacturing sequence

REFERENCE

1. R.A. Postier, Composite Prepreg Cutting, An Integrated Approach From G.G.T. Aerospace, *Commlink Mag.*, Nov-Dec 1984

Flat Tape Laying

Paul F. Pirrung, Air Force Wright Aeronautical Laboratories

AFTER ADVANCED COMPOSITES were introduced in the early 1960s, it was soon learned that boron fibers were too brittle to be woven without breaking and that their superior strength and high modulus of elasticity were lost when the fibers were chopped into short lengths. It was also learned that graphite fibers abraded when woven into fabric and lost their strength. Thus, unidirectional tapes became the preferred material form for processing boron and graphite fibers.

Originally the advanced composite tapes were laid up by hand (Fig. 1), ply upon ply, to form laminates. This process was labor-intensive, and it was found that irregular placement of the tape caused inconsistency in the quality of cured laminates. Therefore, aerospace companies began to develop tape-laying machines.

Some hand-assisted types, known as Flintstone machines (Fig. 1), had a tape-supply spool and a tape-laying roller, usually mounted on a gantry. After laying a 75-mm (3-in.) course, the laying roller and the spool were moved over in order to lay the next tape course parallel to the one previously laid.

Later, another type of Flintstone machine used a gantry with the supply spool, and a tape-laying roller was hand-moved parallel to the gantry (Fig. 2). This spool could use tape up to 305 mm (12 in.) wide. The tape was usually placed on a mylar lay-up template showing the shape and fiber orientation of the ply. After one course was laid, the mylar was moved the width of the tape, by a mechanized method or by hand, thus placing the second course beside the first and continuing until the entire pattern inscribed on the mylar had been covered. The composite ply was cut at the inscribed pattern lines, forming the first ply in the mold. The remaining plies were laid up like the first, then stacked on the mold in the proper sequence.

Some of the early automated machines used numerically controlled (NC) machine lay-up of all the plies, and the completed lay-up was then placed on the mold (Fig. 1). Other NC machines of this period laid up tape directly on the tool (Fig. 1); a five-axis NC machine of this type is shown in Fig. 3.

Some of the machines developed in the 1960s used the table top as the tape-laying surface. The gantry was completely stationary,

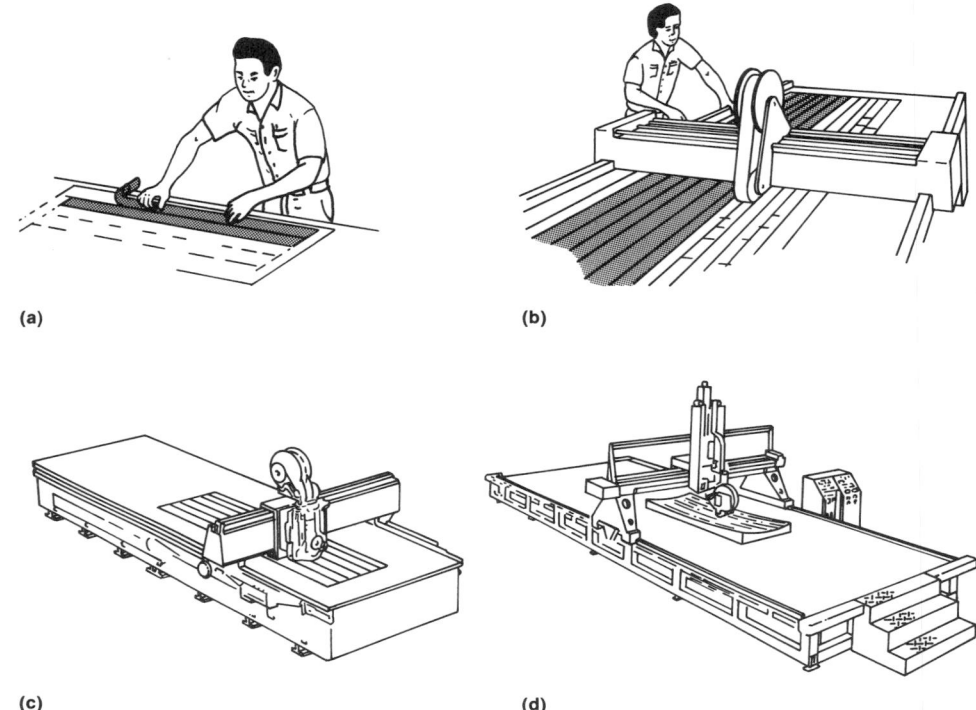

(a)
(b)
(c)
(d)

Fig. 1 Tape lay-up methods (a) Hand lay-up. (b) Hand-assisted Flintstone. (c) Numerically controlled machine lay-up and drape tape material. (d) Numerically controlled machine lay-up directly on tool

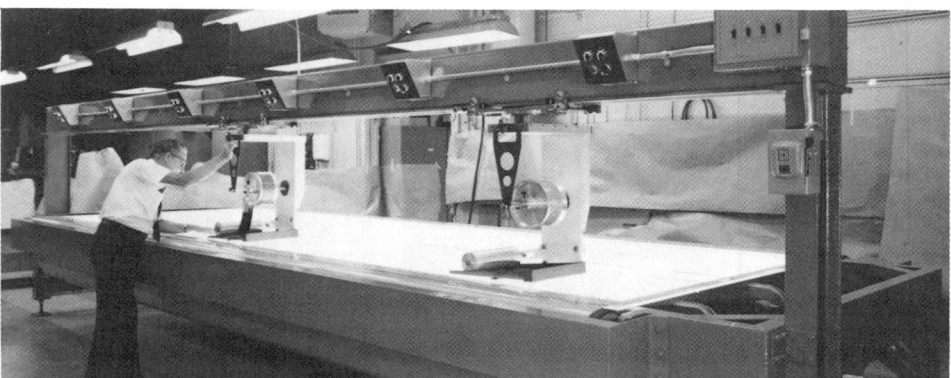

Fig. 2 Manual tape-laying machine. Courtesy of Northrop Manufacturing

Fig. 3 Five-axis NC tape-laying machine. Courtesy of Vought Aeronautics

Fig. 4 First NC tape-laying machine developed under U.S. Air Force contract

and the laying roller moved sideways a distance equal to the width of the tape before starting the next course. The table top moved perpendicular to the gantry.

Machines Developed Under Government Sponsorship

To help the aircraft companies reduce the high cost of hand lay-up, the Manufacturing Technology Division of the Air Force Materials Laboratory (AFML) began sponsoring development of composite tape-laying systems in the late 1960s. The first machine sponsored by AFML (Ref 1) was developed by General Dynamics and the Conrac Corporation (Fig. 4). This machine, with modifications, was the highest-producing tape-laying machine in the late 1970s, at times running three shifts a day, 7 days a week, producing composite parts for the F-16 aircraft. During 1985, it was fabricating F-16 vertical stabilizer skins with an "up" time of greater than 92%. Before being upgraded in 1986, this machine laid 75-mm (3-in.) wide tape, and the maximum gantry movement was 18 m/min (60 ft/min). The bed is 9 m (30 ft) long by 2 m (6 ft) wide. The angle of motion, the length of the tape strips, the shearing mechanism, and the head movements are automatically controlled.

At the start and end of each strip, the machine cutter cuts the tape (without cutting the backing material) at any angle up to 60°. Because the smallest length of tape that this machine could originally lay was 246 mm (9.7 in.), all the machine-laid plies of small composite components for aircraft were laid on picture-frame templates. The overlap of the tape onto the template was cut by hand, on the inner edge of the template. All plies too small to be laid by machine had to be laid by hand.

This type of machine has at least the three basic axes (x, y, and z) programmed, plus the tape-head turning axis. A roll of prepreg tape (normally about 300 to 460 m, or 1000 to 1500 ft long) with a backing separator (usually a coated paper) is supplied to the machine. The tape is threaded past an automatic cutting device and goes under the lay-down roller (or lay-down foot), whereby the tape strip is placed on the machine bed. Either before or after reaching the roller, the coated paper is stripped from the tape and wound on a take-up reel.

To lay the tape automatically, the head is first moved to the starting position for laying the first ply, then lowered to the surface of the machine bed, where it begins moving to lay up the first strip. It accelerates until reaching run speed, runs at this speed, then decelerates to cutting speed and cuts the tape, repeating this process until all of the first strip has been laid. Subsequently, the head is elevated above the bed surface, rotated 180°, indexed over the tape width, and lowered to the bed surface to

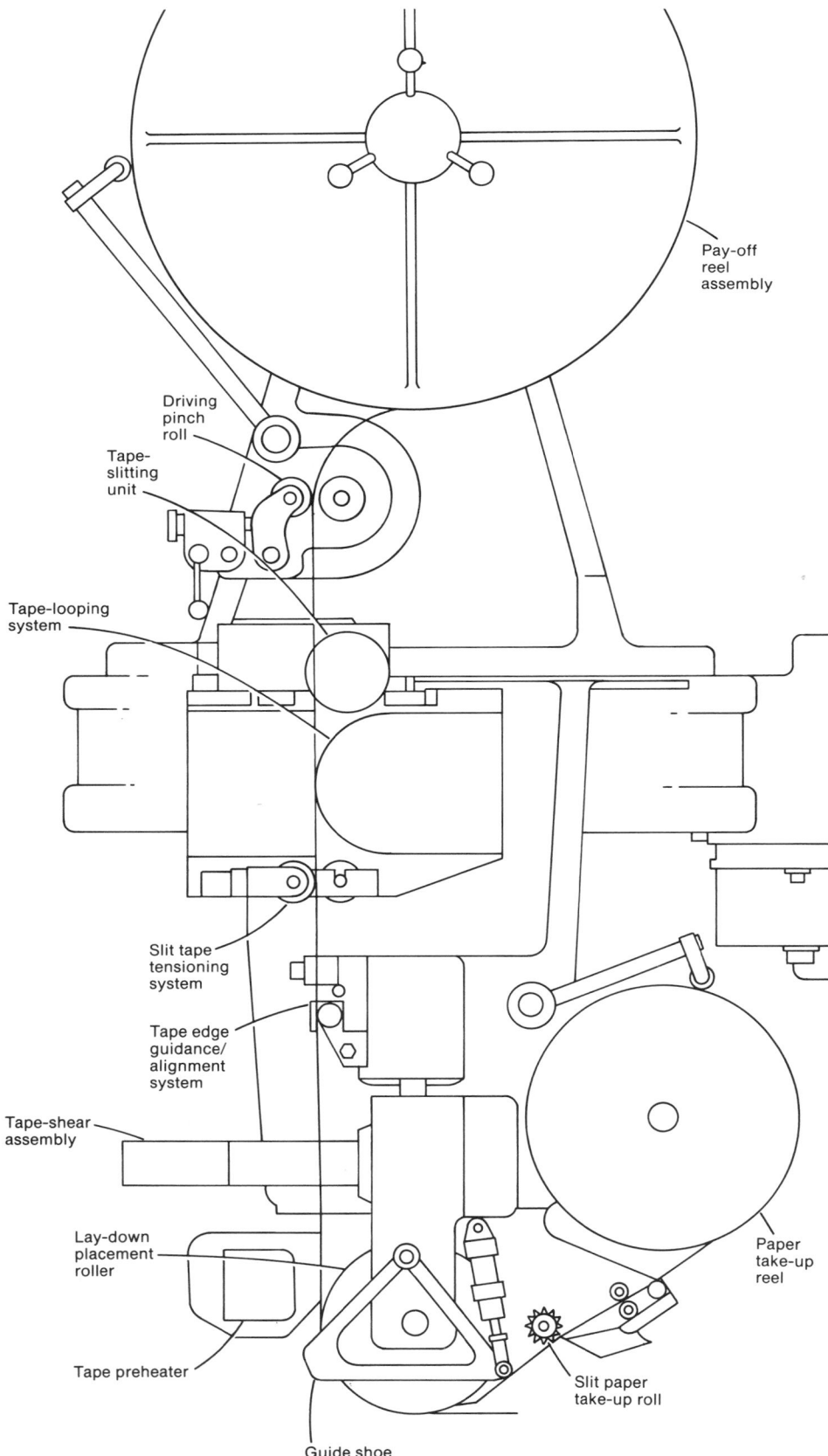

Driving
pinch
roll

Tape-
slitting
unit

Pay-off
reel
assembly

Tape-looping
system

Slit tape
tensioning
system

Tape edge
guidance/
alignment
system

Tape-shear
assembly

Lay-down
placement
roller

Tape preheater

Guide shoe

Paper
take-up
reel

Slit paper
take-up roll

Fig. 5 Atlas tape-laying head

lay the next tape strip, in the opposite direction from the first. This is continued until the pattern for the first ply is complete. Further plies (with their individual patterns) are added directly on top of previous ones until all plies required to fabricate the total component are in place.

This machine was upgraded in early 1986 to increase the maximum tape-laying speed and to incorporate a new head with three 25-mm (1-in.) wide tape cutters, allowing tape angles up to 79° to be cut in a stair-step fashion. It can also lay down short lengths of tape. With these features and a new control system, handwork is no longer required.

Ideas obtained from other AFML-sponsored tape-laying machine development contracts with General Dynamics in the 1970s and 1980s were utilized in the design of two Ingersoll Milling Machine Company machines purchased by General Dynamics. The first of these, which was put into production by General Dynamics in 1980 (Ref 2, 3), can lay 150-mm (6-in.) wide tape and uses computer numerical control/direct numerical control (CNC/DNC). It has a milling-machine-type gantry with a guillotine-type cutter and can lay tape at a speed of 25 m/min (83 ft/min). The machine bed is 9 m (30 ft) long by 3 m (10 ft) wide.

The second machine developed by Ingersoll, which was brought on line by General Dynamics in 1983 (Ref 2), is 6.7 m (22 ft) wide by 20 m (60 ft) long. This CNC/DNC machine has two overhead gantries, one with a head for laying 150-mm (6-in.) wide tape, and the other with a head for laying 25-mm (1-in.) wide tape. The head for the wide tape has six 25-mm (1-in.) wide cutters, enabling it to cut large tape angles in a stair-step fashion. The other head is used for laying small parts. Each 25-mm (1-in.) segment of graphite tape is automatically slit to ensure complete cutting of every graphite fiber. The machine can lay tape at the rate of 25 m/min (83 ft/min), and can cut without slowing down or stopping.

In the early 1970s, an automatic tape lay-up system known as the Atlas (Ref 4) was designed and developed by Goldsworthy Engineering, Inc., under the sponsorship of the United States Army Aviation System Command. The first machine with a gantry made from composite material, the Atlas can use 12.7 to 75 mm (1/2 to 3-in.) wide tapes in up to 300-m (1000-ft) rolls. A drawing of the Atlas head is shown in Fig. 5.

Commercial Tape-Laying Machines

Recent commercial machines include a seven-axis and a ten-axis computer-controlled machine made by Cincinnati Milacron. Designed to provide flexible automation, these machines can automatically lay 3-, 6-, or 12-in. wide tape at speeds of 30 m/min. The tape is auto-

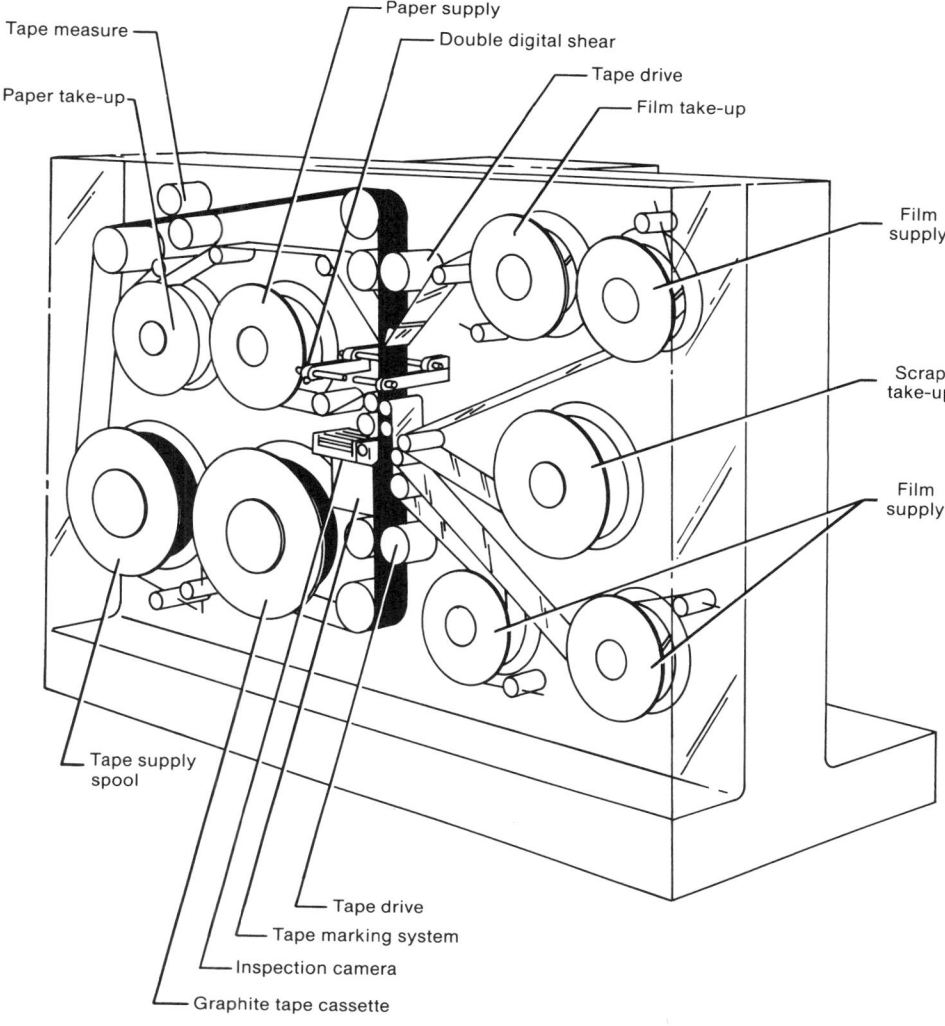

Tape measure

Paper take-up

Paper supply

Double digital shear

Tape drive

Film take-up

Film supply

Scrap take-up

Film supply

Tape supply spool

Tape drive

Tape marking system

Inspection camera

Graphite tape cassette

Fig. 6 Access tape preparation machine

matically debulked, cut, and laid, ply on ply, in directions needed for optimum component strength. These machines can also lay plies on a contour.

Goldsworthy Engineering has developed a two-stage composite tape placement system consisting of a tape preparation machine called Access (Fig. 6) and a tape placement machine called Atlas II (Fig. 7). The Access machine processes prepreg tape from the original supply spool by continuously removing it from backing paper, cutting each strip to the desired length and angle, repositioning the tape on new backing paper, and rewinding it into a cassette for use on the Atlas II.

Ingersoll Milling Machine Company machines now have speeds up to 1200 rpm and have been built in widths up to 9 m (30 ft) and lengths up to 45 m (150 ft), with tape supply reel capacities of 760 m (2500 ft). Ingersoll has an automatic ''on the fly'' dual-rotary shear and a foreign-particle scanner to detect tape defects before lay-up.

A machine with a rotary table has been developed by Vektronics, Inc. It uses a 2.5-mm (1-in.) wide tape application head mounted on a traveling gantry which traverses the rotary table. The gantry has speeds up to 30 m/min (100 ft/min); the application head has speeds up to 60 m/min (200 ft/min).

Advantages of Tape-Laying Machines

A major advantage of tape-laying machines is that they provide a labor savings of up to 86% over hand lay-up (Ref 2). Other advantages of machines include maximum utilization of the tape material. With hand lay-up, waste occasionally occurs at the end of each strip because most operators start each strip from the same side of the lay-up, which causes waste at the ends of every tape cut, with the exception of 90° cuts. Most tape-laying machines, however, turn around 180° at the end of a strip and lay the next strip adjacent to the previous one (parallel

to it), but in the opposite direction. Because the two strip ends on one edge of the ply have the same angle, there is no waste material with this procedure.

In addition, automatic tape-laying machines can help to obtain improved structural properties for composite components. The tape-laying roller, or foot, can apply any desired pressure, which provides even pressure during the total lay-up. When ply upon ply is laid up with pressure on each strip, this uniform pressure gives a high-quality, compacted composite lay-up with less entrapped air to cause porosity in the cured part.

Another advantage is that the results of a composite lay-up with an automatic machine are repeatable because the machine can lay tape in the same place each time, with greater control of the gaps between the tape strip. Therefore, the overall quality of parts fabricated with automatic machines is better than that of those made by hand lay-up, and the scrap rate with machines is lower.

Operating Parameters

One of the important parameters in the laying-up of composite tape strips (made with unidirectional fibers) with an automatic machine is the relation between ply shape and orientation and tape lay-down speed. A composite aircraft component is fabricated from many individual plies, which may differ in shape, size, and fiber orientation. Fiber orientation is usually 0°, −45°, +45°, or 90°. Table 1 shows the effect of fiber orientation on machine operating time. The number of strips to be cut and laid, as well as their lengths, determine the total time required, 12.07 h, which compares with a total time of 119.5 h for hand lay-up of the same wing skin demonstration article.

Table 1 illustrates that for the same area covered, the longer the average strip laid, the faster that area can be covered. Therefore, more weight of tape can be laid in a given time on larger-area plies with longer average strip lengths. This also illustrates the need for a higher rate of acceleration and deceleration at the start and end of each strip. Some machines require that the strip end be cut a short distance away from the lay-down roller but cannot cut the tape at running speed, necessitating decelerating or even stopping during tape cutting. Thus, tape lay-down time is reduced by some modern machines that can cut the strip ends while running at full speed.

Another parameter affecting the efficiency of composite tape laying is the movement at the end of each strip. Some machines will take as long as 20 s for the four movements involved (lift up, rotate, index, and drop the head). However, if the 180° turn and the index movement are done at the same time, if the turn time is decreased to 1 to 2 s, and if the head lift and head set-down time are each decreased to 1 s or less, the time of the four movements at the end

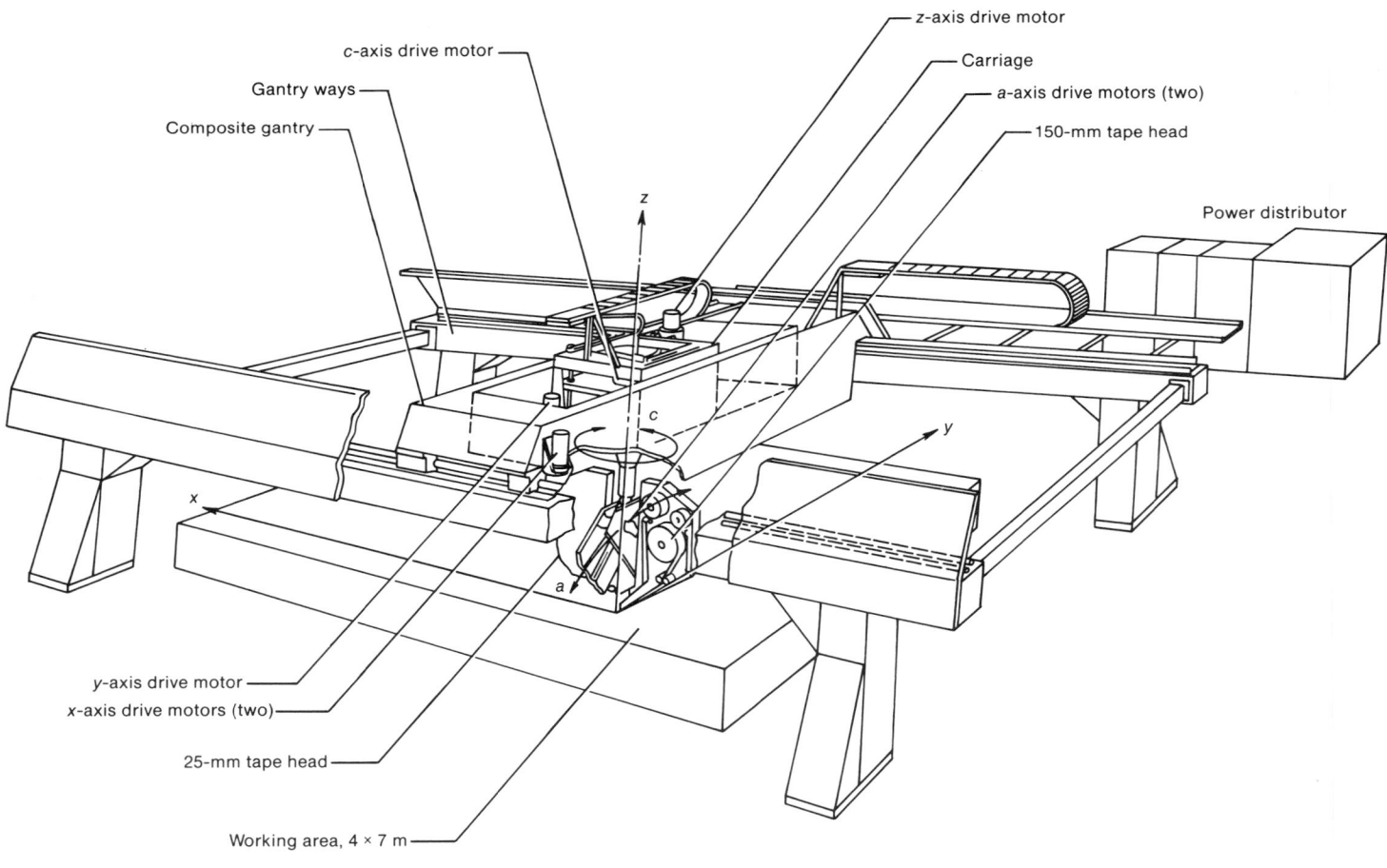

Fig. 7 "Atlas II" tape placement machine

of each strip can be decreased to about 4 s total per strip.

Another possibility is the use of two tape-laying heads. One head lays tape in one direction; then, without turning around, the other head lays tape in the opposite direction. As the first head is lifted from the surface, the other head is lowered. After laying the double strip, the tape head indexes across it to start the third strip. This could change the time of the four movements at the end of each strip to an average of about 1.5 s per strip. Also, with a double head, two rolls of tape could be emptied before a new tape supply would be required.

Differences in characteristics for lay-down angles of 0°, +45°, and −45° are illustrated in Fig. 8 (Ref 5), which also shows how strip orientation affects tape lay-down. For the same outline, the 0° orientation requires 18 strips, the +45° orientation requires 38, and the −45° orientation requires 28.

A detailed time record of a tape machine laying the −45° ply is shown in Fig. 9. The run time includes acceleration time to run speed, time at run speed, deceleration, time to cutting speed, cut, acceleration time to run speed, and

deceleration time to end of strip. With some machines, many of these times can be decreased drastically.

The possibility of waste at the end of each strip is illustrated in Fig. 10. At the end of a 300-mm (12-in.) wide tape strip (Fig. 10a), if the strip is cut at a 90° angle, when the strip is laid at a +45° angle; 0.05 m² (72 in.²) of material is wasted at the end of each 300-mm (12-in.) wide strip. With four 75-mm (3-in.) wide strips (Fig. 10b) cut at a 90° angle, only 0.01 m² (18 in.²) of material is wasted. However, lay-up time is greater for the thinner strips. Although the total waste is less for tape strips that are narrower, some waste almost always occurs at both ends of the strip when the angle of the strip is not perpendicular to the edge of the ply, either when using a dualhead tape layer or a single-head tape layer that is not turned 180° but is returned to the same side of the ply to start each strip of tape. This type of waste can be eliminated by using a single tape-laying head, laying a strip of tape, cutting the tape at the exact angle of the straight ply edge, turning the laying head 180°, and continuing to lay the strip next

to the previous strip but in the opposite direction.

The foregoing considerations are basic to all tape-laying machines. The machines described in this article, however, are intended primarily to lay on the flat. Additional complexity is required to provide the capability to lay on gentle contours.

REFERENCES

1. W.O. Sunafrank, W.H. Drebing, and H.L. Eaton, "Development of Composite Tape-Laying Process for Advanced Fibrous Reinforced Composite Structures," AFML-TR-71-71, Air Force Materials Laboratory, March 1971
2. FZM6982, Composites Laminating Center, General Dynamics, June 1981
3. "Automated Tape-Laying," Production Technology Bulletin 85, General Dynamics, June 1985
4. E.E. Hardesty, Automating the Tape Winding of Composite Shapes, *Mod. Plast.*, Sept 1973
5. Vought Aeronautics, private communication

Table 1   Time required for machine lay-up of wing skin demonstration article

Ply no.	Ply orientation	Total no. of strips per ply	Total strip length		Time used to lay up ply, min
			m	ft	
P1	0°	18	33.2	109	14.48
P2	0°	17	13.7	45	9.92
P3	−45°	28	31.7	104	17.15
P4	+45°	37	31.1	102	21.5
P5	0°	18	33.2	109	14.72
P6	−45°	13	11.6	38	7.19
P7	+45°	22	10.7	35	11.55
P8	0°	18	9.4	31	8.6
P9	−45°	26	28.3	93	14.7
P10	+45°	31	25.6	84	18.45
P11	0°	18	18.3	60	10.94
P12	−45°	15	14.3	47	8.1
P13	+45°	25	15.8	52	13.7
P14	0°	18	33.2	109	14.0
P15	+45°	22	10.1	33	11.5
P16	−45°	8	6.4	21	4.22
P17	0°	17	8.5	28	8.4
P18	−45°	13	11.9	39	7.0
P19	+45°	23	12.8	42	12.4
P20	0°	16	10.4	34	8.6
P21	−45°	28	31.7	104	17.10
P22	+45°	38	31.1	102	21.89
P23	0°	17	5.5	18	7.6
P24	−45°	9	7.9	26	4.9
P25	+45°	21	7.0	23	10.6
P26	0°	17	20.7	68	11.2
P27	−45°	18	18.3	60	10.0
P28	+45°	28	21.0	69	15.7
P29	0°	17	8.2	27	8.1
P30	−45°	8	6.7	22	4.3
P31	0°	16	8.8	29	8.0
P32	0°	17	24.6	81	12.36
P33	0°	16	4.6	15	6.83
P34	+45°	21	5.8	19	10.8
P35	0°	17	8.8	29	8.5
P36	+45°	26	19.2	63	14.8
P37	−45°	20	21.9	72	12.1
P38	0°	17	21.3	70	11.2
P39	+45°	20	6.7	22	8.8
P40	−45°	9	7.6	25	5
P41	0°	17	5.5	18	7.37
P42	+45°	37	31.1	102	21.7
P43	−45°	28	31.7	104	17.1
P44	0°	17	12.5	41	9.2
P45	+45°	22	11.6	38	11.3
P46	−45°	14	12.8	42	7.9
P47	0°	16	4.6	15	6.75
P48	−45°	11	9.8	32	6.1
P49	+45°	20	9.4	31	9.9
P50	0°	16	32.3	106	13.5
P51	+45°	23	13.1	43	12.3
P52	−45°	16	17.1	56	9.5
P53	0°	16	18.9	62	10.2
P54	+45°	33	28.3	93	19.6
P55	−45°	24	27.4	90	14.7
P56	0°	16	4.3	14	6.7
P57	+45°	20	10.4	34	10.2
P58	−45°	12	10.7	35	6.6
P59	0°	16	32.3	106	13.5
P60	+45°	37	31.1	102	21.5
P61	−45°	28	31.4	103	17.1
P62	0°	16	15.5	51	10.1
P63	0°	16	32.6	107	15
Totals		**1254**	**1092.0**	**3584**	**724.72 (12.07 h)**

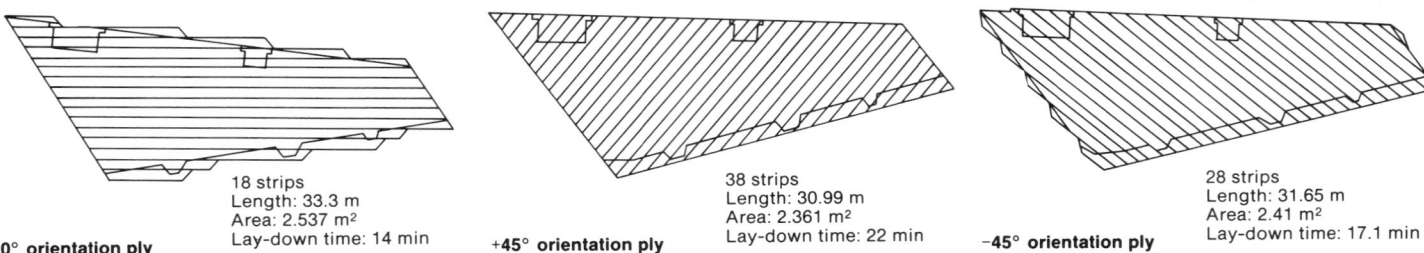

0° orientation ply

18 strips
Length: 33.3 m
Area: 2.537 m²
Lay-down time: 14 min

+45° orientation ply

38 strips
Length: 30.99 m
Area: 2.361 m²
Lay-down time: 22 min

−45° orientation ply

28 strips
Length: 31.65 m
Area: 2.41 m²
Lay-down time: 17.1 min

Fig. 8 Three full plies of composite tape laid up in 0°, +45°, and −45° orientations

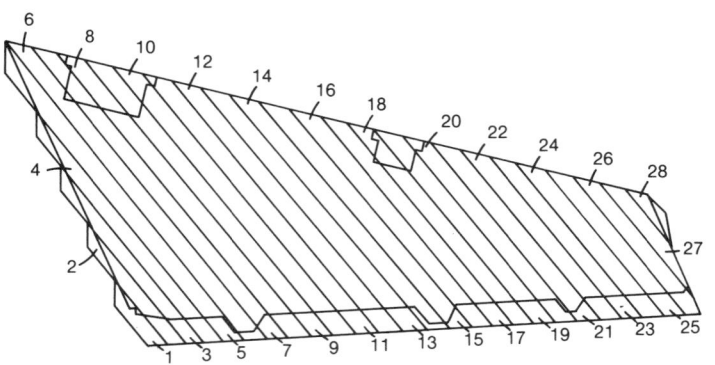

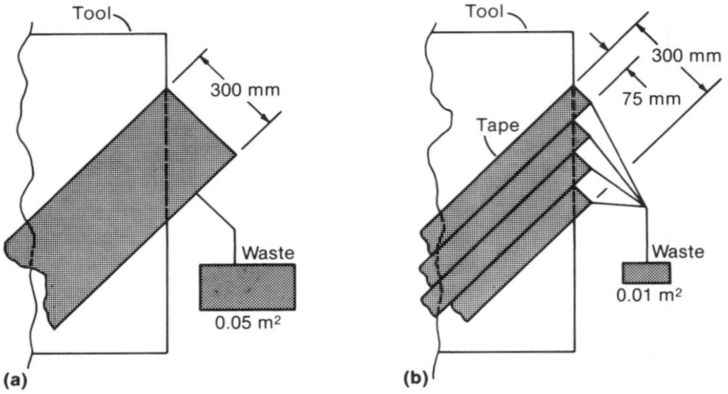

Fig. 10 Waste material as a function of tape width. (a) 300-mm (12-in.) wide tape strip. (b) 75-mm (3-in.) wide tape strip

Strip no.	Strip length m	Strip length ft	Head down	Run time including chop	Head up	Turn around and index 75 mm	Total elapsed time
1	0.3	0.9	7	5	5	9	26
2	0.6	2.1	7	7.3	5	9	28.3
3	1.0	3.3	7	11.4	5	9	32.4
4	1.4	4.5	7	15.8	5	9	36.8
5	1.7	5.7	7	20	5	9	41
6	1.7	5.5	7	19.4	5	9	40.4
7	1.6	5.4	7	19.1	5	9	40.1
8	1.6	5.3	7	18.5	5	9	39.5
9	1.5	5.1	7	18	5	9	39
10	1.5	4.9	7	18	5	9	39
11	1.5	4.8	7	17	5	9	38
12	1.4	4.6	7	16	5	9	37
13	1.4	4.5	7	15	5	9	36
14	1.3	4.3	7	15	5	9	36
15	1.3	4.2	7	14	5	9	35
16	1.2	4.0	7	14	5	9	35
17	1.2	3.8	7	13	5	9	34
18	1.1	3.7	7	13	5	9	34
19	1.1	3.6	7	13	5	9	34
20	1.0	3.4	7	13	5	9	34
21	1.0	3.3	7	12.5	5	9	33.5
22	0.9	3.1	7	12.5	5	9	33.5
23	0.9	2.9	7	11.5	5	9	32.5
24	0.9	2.8	7	12	5	9	33
25	0.8	2.7	7	11	5	9	32
26	0.8	2.5	7	11	5	9	32
27	0.7	2.3	7	4 chops 70	5	9	91
28	0.3	0.9	7	5	5	9	26
Totals	**31.7**	**104.1**	**196**	**441**	**140**	**252**	**1029**
				Total time, min			**17.15**

Fig. 9 Machine-run data sheet (ply No. P3, orientation −45°)

Contoured Tape Laying

Lynn A. Williams III, The Ingersoll Milling Machine Company

THE NEED FOR AUTOMATED PRE-PREG TAPE-LAYING MACHINES was recognized in 1966 as a requirement for cutting manufacturing costs in the aircraft industry. Early machines were made by aircraft companies, machine shops under the direction of material suppliers, small companies willing to try experimental machines, and by modification of existing numerically controlled metal cutting machines by installing a tape-laying head. Today, sophisticated machines are produced by machine tool builders.

Automated Machine Development

In the course of developing the early tape-laying machines, several problems were encountered: The tape, as manufactured, was not straight, so parallel strips wandered; early process specifications required that the tape not overlap, but allowed a narrow gap of up to 0.75 mm (0.03 in.); the ply shapes must be modified to accommodate the capability of the cutter, which contributed an increase to part weight; tape often varied in tack (stickiness); and the tape head had limitations in speed and rotation.

Even with these problems, several developmental machines produced in the 1970s were capable of laying patterns in the flat. Extensive development was funded by the Air Force Materials Laboratory (AFML), and several machine tool companies decided to compete within the viable automated tape layer market. Modern, second-generation machines were designed to cut tape strip ends in a greater variety of shapes to reduce part weight. Because the aircraft industry had learned to accept gaps when tape strips passed over a contour, they accepted a best-fit lay-up concept called linear programming (LP), in which the gap varied within specified limits, so that the best path for the tape on the contoured tool could be obtained. This LP concept complicates the automation program and equipment design, but permits the lay-up of complex plies in the gentle contours found in airfoil shapes. Thus, the part that is laid up by machine can proceed to the autoclave with minimal handling.

The tape-laying machines are more material efficient than the Gerber cutter and significantly increase the output per man-hour of prepreg placed in the tool compared to hand lay-up. The lb/h rate varies with part size and complexity of the plies. Even so, throughput rates of 5 kg/h (10 lb/h) are not unusual. Despite the multimillion dollar expenditure to get a machine running, the savings in material and labor are soon recaptured on production programs. The known cost savings are prompting one fabricator to acquire 26 machines, while a subcontractor plans to add six machines to his existing units. The size limit of gentle-contour parts is limited only by the width of the machine, which is commonly 5 m (15 ft).

The initial development of machinery capable of tape laying, though much was left to be desired in speed, control, tape cutting, and programming, showed a potential for reducing costs by mechanizing a labor-intensive process. Early in the tape-laying development cycle, use of polar robots was tried but eventually discarded because they did not possess the requisite accuracy and rigidity. Prepreg tape is often inflexibly stiff; placed in contact with the surface on which it is to be laid, it can only follow in a line normal to the starting line. If the course to be laid is 45°, the tape needs to be placed at precisely 45° or it will not match the course at which the machine moves. If the tape skids or skews off the lay-down shoe or roller, the process quickly stops. Robots could not provide the appropriate level of accuracy.

The requirement for high machine accuracy seems contradictory to the tolerance of course-to-course accuracy, which is 1.52 to 2.54 mm (0.060 to 0.100 in.). As a practical matter, 150-mm (6-in.) wide unidirectional prepreg tape can be "steered" approximately 2.5 mm (0.1 in.) every 3 m (10 ft) of course length. It is easier to produce parts that have relatively short course lengths of 1.5 to 3 m (5 to 10 ft) than it is to make parts that are 6 to 15 m (20 to 50 ft) long, because the tapes, even as produced by their manufacturers, are not straight enough.

Computer-directed tape-layers are more costly to purchase than ply-cutting systems. However, as tape-laying techniques improve, it seems that they will prove to be the preferred way of producing large aircraft components. The material utilization factor is significantly better, higher strength-to-weight ratios can be realized from tapes than from fabric prepregs, and there is a tremendous potential for reducing labor costs. These issues are the driving forces behind the development of modern tape-laying machines, such as the one shown in Fig. 1.

Machine Features

The machine design aspects and programming of an automated contoured tape laying machine can best be understood visually. Figure 2 depicts the process in which tape is handled by an advanced tape-laying machine. At the upper right of Fig. 2 is the supply reel, which can contain 240 to 370 m (800 to 1200 ft) of 150-mm or 75-mm (6-in. or 3-in.) wide graphite prepreg tape. From the supply reel, the tape moves past a detector that scans the underside of the tape for foreign particles or flaws that would not be visible after the tape is laid

Fig. 1 A large-contour tape-laying machine at work on a contoured tool

down. The tape then moves up through a photoelectric sensing device that reads the edges of the tape and adjusts the tape supply reel laterally, to center the tape across the lay-down roller, which is shown at the bottom of Fig. 2. Then, the tape goes through dual cutting shears so that it may be cut at any angle, "on the fly," to match the ply shape requirement of the engineering drawing for this particular tape strip.

Figure 3 shows two rotary shears. One cutter must operate at a 30° angle to give the cutter enough time to get across the tape, if the cut is to be square to the direction of travel. Another important capability of the cutter is to cut cleanly through the graphite fibers, but not through the paper carrier. The head must stop for rethreading if the carrier is cut accidently. Fully automated lay-up requires that the carrier be wound on the take-up reel as a continuous strip. This was a problem with early tape-laying machines that was solved by the development of a rotary shear. Below the tape shears, and as close to the lay-down roller as possible, is a second electronic guide indicator. This unit works in concert with the first. It tells the computer whether the tape is actually centered on the lay-down roller. It is possible for the tape to have a slight drift or waviness in it due to errors in making prepreg, and it is also possible for the lay-down roller to operate at an angle that is slightly skewed from the course it is supposed to travel. When these conditions occur, the tape slides sideward, off the roller. If permitted to continue, this condition will halt the tape-laying process. Thus, the first tape position sensor reacts by moving the supply reel laterally with respect to the lay-down roller; the second senses that the tracking accuracy (or tape accuracy) of the machine is somehow at a

slight variance to the tape or the mold. It reacts by adjusting the c-axis (rotary) and turning the head into the direction in which the tape is skewed. The amount of steering is very small, on the order of 0.5° or less, but it is capable of midcourse corrections that permit the machine to lay tape to tight tolerances on gently contoured surfaces.

Just below the lower sensor shown in Fig. 4 are the segmented lay-down rollers, which are used to apply a preset mechanical force to seat the prepreg on the tool so that it stays firmly in place. Compaction also forces entrapped air from the lay-up. The lay-down roller is made from 12.7-mm (½-in.) wide segments, each operated by its own controller, so that it can be adjusted to push down either 25-, 75-, or 150-mm (1-, 3-, or 6-in.) wide tape, depending upon the number of rollers that are extended to touch the tape. The segmented lay-down roller also can be programmed to retract sequentially as the machine comes to the end of a lay-down course that is at some angle to the finished edge. When this occurs, the tail of the tape becomes narrower and fewer segments are required to touch the tail. This allows the tail of the laid piece to be compacted into position without pressing down the surface adjacent to it. Conversely, as the tape tail becomes wider, more segments are extended.

Behind the lay-down roller is a third photoelectric web guide detection unit that aligns the carrier take-up roller with the backup paper as it comes off the lay-down roller. There is enough variation in the prepreg to require the backing paper to align with the lay-down roller if it is to be taken up smoothly without tearing. If the paper is torn, the head must stop for rethreading before automated lay-up can resume.

Adjacent to the rear web guide device is a television camera that watches the tape as it is being laid. It also shows the carrier paper, to ensure that no graphite material is sticking to the paper other than the trim scrap that are, in fact, programmed to be taken into the take-up paper chute. There is a monitor at the operator's console so that he can watch the lay-down action.

Fig. 2 Mechanical arrangement of a tape head. The tape supply reel is at the upper right. Tape shears and segmented lay-down rollers are at the lower center.

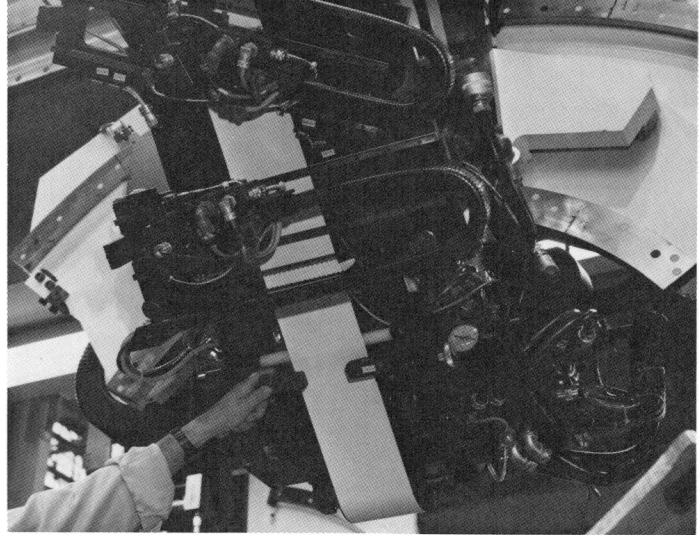

Fig. 3 The dual rotary shears allow a broad range of straight or compound angles to be cut into the tape without disturbing the carrier paper. The lower web guide is below the shears.

Fig. 4 The segmented lay-down roller is required to change tape width, allow for angular tape cuts, and comply to contoured surfaces. The web guide sensors are part of the feedback loop and the control c-axis steering.

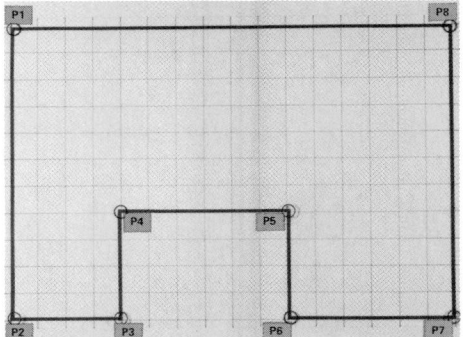

Fig. 5 Step 1 in programming a contoured tape laying machine is to define the boundaries of the part, using a maximum of 20 points and the command PERDEF/P1 (through 20).

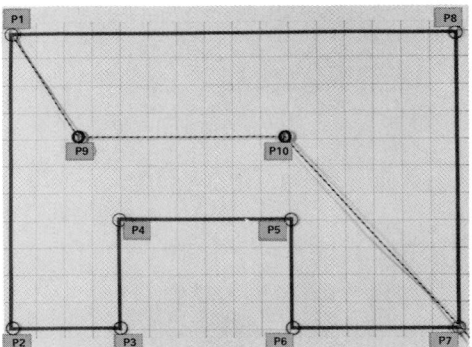

Fig. 6 Step 2 in programming a contoured tape laying machine is to define the control line so that lap/gap has a theoretical value, using a maximum of ten points and the command CTRLIN/P1 (through 10).

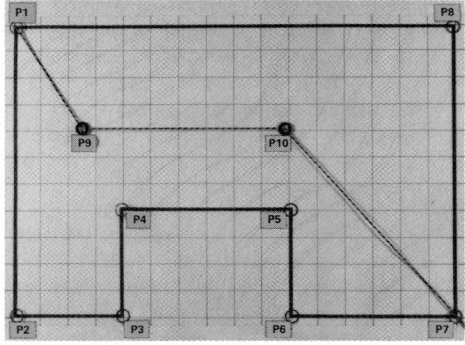

Fig. 7 Step 3 in programming a contoured tape laying machine is to fill the surface with tape using the command PLYFIL/45°.

The tape head is mounted onto a carrier plate that allows it to be infinitely adjusted ±30° from either side of perpendicular. This assembly is mounted on a rotary axis for ±360° of action. The rotary action is carried on the *y*-axis slide, which in turn is carried on the *x*-axis gantry. When operating, there are 11 axes under numeric control (NC), plus NC commands for each of the 12 segments in the lay-down roller.

A unique aspect of automated tape layers is that the machine senses the placement of the tape strip and corrects itself rather than following a rigidly programmed pattern, as is done in metal-working machinery. At the beginning of the laying of a course, the tape is placed down with extreme accuracy. It must be positioned perfectly in the *x, y* coordinates, and its angular position must be set equally well. This is the last time the operator can make a major adjustment; from this point on, the course the machine follows is automatically adjusted to lay the tape in the best path.

The problems identified thus far apply to both flat and contoured tape laying. For contoured tape there are two additional axes of motion that interact with the rest of the system. Also, a programming system has been developed to plot the course the machine must take in following the extremely stiff tape through the contours.

Early in the history of tape laying, many believed that a much simpler machine could be made, working almost wholly through adaptive controls. It was thought that the machine might be made with a lightweight gantry, without high intrinsic accuracy, and that the accuracy requirements might be met by following previously laid tape courses. Some of the early contoured tape laying machines did rely upon adaptive control to guide them across the contoured path. However, this arrangement was found to have limited success, and in the case of laying across compound contours, no success at all. In laying long courses, even on flat tools, the ability of adaptive control was found to be unacceptable. One reason for this was that any error existing in the first course would be

magnified as the machine adapted to the second course. In short, every course would be at least to some degree the product of the previous course. The problem was more difficult when laying a course of tape across a contoured surface. The tape would deviate from a straight line because a course that moved in a straight line through a compound contoured surface required one edge of the tape to become slightly shorter than the other as it moved past one curvature; the opposite occurred as it moved past, or through, the next curve. Therefore, a path had to be plotted through these curves so that, in a sense, the edge of each side of the tape remained the same length. This modified geodesic course was known as "natural path." As the fibers on one edge or the other resisted stretching, if the natural path were not followed, the only other alternative was to buckle the fibers on the short side of the tape, which was unacceptable from an engineering standpoint and difficult to achieve in practice because of the strength of the tape.

When the requirement for natural path was understood, adaptive controls were discarded because it was felt that it was probably impossible to design the extremely complex algorithm required by an adaptive control. For example, if the machine were programmed for a 45° angle, but the natural path of the tape demanded a slightly different angle, the tape would start to move sideways away from the centerline of the machine. Such skewing could be detected through the web guide system, and when that system was incapable of supplying the amount of correction required, it would be next picked up on the *c*-axis steering. However, skewing does not give a clear picture of necessary corrective action to maintain the machine on the centerline of the tape over a longer distance. It was unclear whether tape skewing was due to an angular change in the direction of the tape, which would require an *x* and *y* correction, or whether it was due to a change in angle on the surface, which would require an *a*-axis correction. Sometimes, inaccuracies are due to imperfections in the tape itself. And, even if the specific cause of tape skewing could

have been determined, it would have been difficult to establish the magnitude of correction necessary. Finally, like all adaptive controls, a correction must be instituted after the effect is noticed. Trying to catch up with events that have already occurred precludes the ability to stay ahead of the problem.

Machine Programming

A special program known as CPG-85, developed at Ingersoll, generates the theoretical natural path of the tape, which becomes the input to the NC program for the machine.

For example, the input to CPG-85 are the *x, y,* and *z* coordinates of points on the lay-up surface. No vectors or normals are required. The points need to be in an orderly array, and the rows and columns need to be numbered. Typically, a 25-mm (1-in.) grid is used as input. However, this is a programmable parameter, and other grids can be used. There are several ways in which the data for this grid may be obtained. Frequently, the surface information is taken from the engineering calculations that were used to establish the shape of the particular part, which might be a wing or a tail surface. Sometimes the surface data are obtained by digitizing an existing part, a model, or a lay-up mandrel. In either case, it is not necessary that the original points being calculated or digitized be put on a 25-mm (1-in.) grid. Usually, they are on a much coarser grid. The command FMILL is used to create the intermediate points to a tolerance that is sufficiently accurate.

The next step is to load the surface data into a computer, such as an IBM 3033 or an NSC 9060. Then, the periphery of the part to be laid up is defined. The periphery can be described by as many as 20 points connected by straight lines. The lines can be at angles greater than 180° to each other, and a part can be laid up that looks like a U or a W. Each point has to be defined by its *x, y,* and *z* coordinates. The points need to be in order, counterclockwise, as shown in Fig. 5. The word PERDEF, short for periphery definition, is followed by the name of the point: P1, P2, and so forth.

The next step is to define the control or reference line (CTRLIN), which is the line at which the tape will have its correct spacing and its correct angle, as shown in Fig. 6. Often, this is the line where the heaviest stress occurs in the component to be produced. Every course of a tape must intersect this control line. There can be as many as ten control lines. The lines must be straight and they must be connected. In the case of a U- or W-shaped part, the control line can go across the open spaces.

The final command is PLYFIL, shown in Fig. 7, which is followed by a command that directs the machine to lay tape either in one direction and rapidly return for the next course, or in two directions, as shown in the illustration, whereby the machine goes back and forth as it lays the tape. Two-directional tape laying is more efficient if courses are more than 0.6 or 0.9 m (2 or 3 ft) long. On short courses, perhaps under 0.9 m (3 ft), the time needed to index the tape head around is greater than the rapid return rate, so it is often faster to lay tape in one direction only. It is also true that laying tape in two directions requires the machine to be very accurate because lack of rigidity or misalignment doubles the error as the tape head is indexed 180° for the return course. Figure 7 also shows that there is a need for two other commands, one of which should be GAPLAP, to define the acceptable gap and maximum lap (usually zero for lap). The other command should be WIDTH, defining the nominal width of the tape. These commands are the most likely to be changed by the designer when he detects undesirable gap or lap conditions.

After the total ply lay-up programming effort has been completed, the normal procedure is to plot the results, as shown in Fig. 8. The line going from the upper left-hand corner to the lower right-hand corner is the control line. It has no gaps or laps because it represents the line at which the tape lay-up is theoretically perfect. Because of the contour of the surface, there are fairly large gaps near the edges of each ply. The circled area shows the extent of such a gap. In the drawing that the programmer receives it is scaled and its dimension is accurately known. The programmer, or the cognizant engineer, then decides whether this gap is acceptable. If it is not, his options are to change the angle or to decrease the gap on the control line slightly.

After the programmer or engineer is satisfied with the result, the output of CPG-85 becomes the input to the machine NC program. At this point, the CPG-85 output is postprocessed by the central computer before it is sent to the machine. Normally, this is done by digital NC, although data can be transported on a floppy disk or other medium.

The data is unloaded into an HP-1000, which precedes the standard NC unit. The front-end computer contains software that translates the CPG-85 data into NC instructions. For example, in order to cut the tape to the proper length while laying it, the cutting action has to take place "on the fly" some time and distance

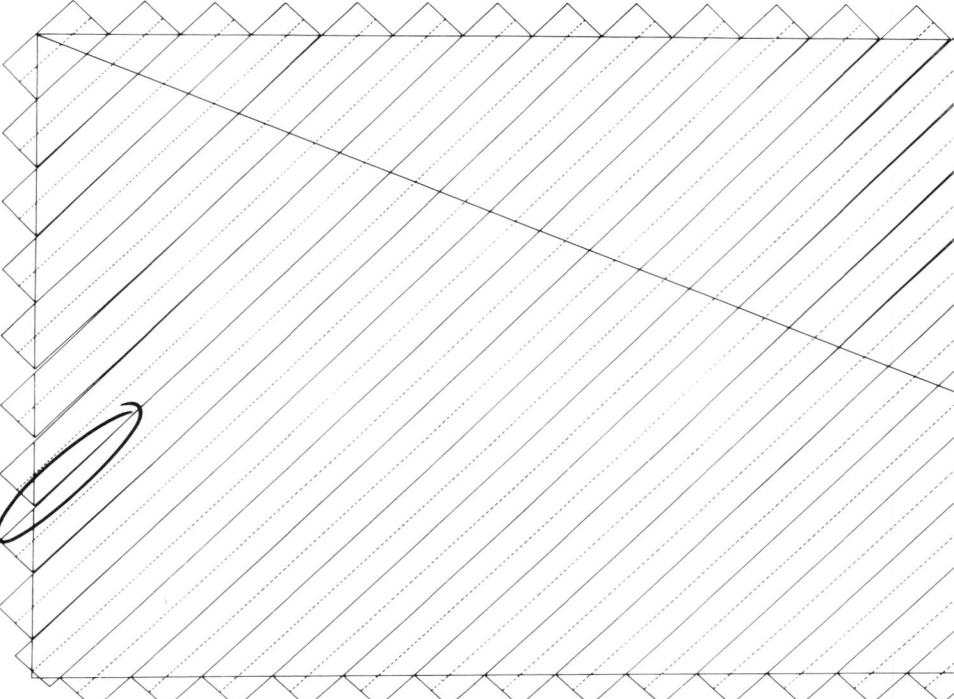

Fig. 8 The CPG-85 program shows, in two dimensions, the gaps (or laps) that will be produced when the "infinitely" stiff tape is laid through a complex contour. The line running diagonally from upper left to lower right is the control line and it may be placed anywhere on the part. The plies are perfect along this line and deviate from theoretical perfection as the contour dictates.

before the cut actually is placed onto the workpiece. This distance is stored in the front-end computer in order to activate the shear at the proper time. After the program has passed through the front-end computer, it is transmitted directly to the machine NC controller in a format that the machine controller can accept. All of this occurs simultaneously and automatically. There is no manual intervention or delay.

In spite of the importance of the NC program, there is still a need for some adaptive control in contoured tape laying. The a-axis, c-axis, and z-axis all respond to some adaptive feedback. However, the adaptive feedback is only intended to make up for minor differences between the theoretical description of the workpiece and the lay-up mandrel. It is also possible to lay two or three plies on top of one another with the same program, rather than having to reprogram the z-axis dimension for every course.

It is easy to appreciate that in addition to the need for accurate programming and the need to move the machine accurately through a contoured path, there is a need for a high degree of accuracy in the lay-up tool, as well. These molds were found to have greater inaccuracy than intended, which was probably due in part to the difficulty of machining large-contour work surfaces, and the difficulty of checking them to tight tolerances. Yet another difficulty probably lies in moving molds with inadequate

stiffness from one place to another. Thermal change also contributes to inaccuracy in the molds. In any case, it is absolutely necessary to have a machine control program that matches the lay-up mandrel; the number of instances in which mold inaccuracy causes poor (or no) results is surprising.

Future Technology

Although a great amount of engineering effort has gone into the development of the current tape-laying machines, there will be significant changes in the future. For instance, current machines are limited to angular displacements of the head of ±30°; machines are on the drawing board whose capabilities are ±45°. But this angular adjustment is somewhat misleading in that the rate of change is limited to something on the order of 1 in 15. This effectively precludes the possibility of laying tape over abrupt surface changes of the type found in honeycomb or other cored composite structures. It tends to limit the usefulness of contoured tape laying machines to large, relatively flat surfaces such as wing skins, ailerons, spoilers, or other aircraft surfaces with relatively shallow contours. The machinery is not currently useful for laying up normal-sized fuselages or components having relatively high curvatures, or other complexly contoured aircraft structures directly on the mold. It is fair to say that tape laying, for both flat and contoured surfaces, is a technique that will be useful in aircraft in the future, as one of many kinds of machines required in aircraft production, just as skin mills are used today to produce wing skins, spar mills for the wing spars, and machining centers for the small parts that compose the aircraft substructure.

Future machines will be simpler than those of the past by virtue of significant improvements to the straightness and control of thickness of the materials (tapes) to be laid or placed into molds. The bulk of carbon fiber tape now produced is used in low-production prototype parts that are laid up by hand. As the need for production increases, so does the need for

machine lay-up, and this will drive tighter controls into tape production.

It also may be possible for materials to have a greater degree of steerability. It is already possible to lay tapes without using carrier paper, which is expected to simplify the process, although it may require modifications in the resin system. This particular work is developing more slowly because the aircraft industry requires the building of a comprehensive and expensive data base to ensure that the materials will perform as expected.

If materials can be made to be somewhat more forgiving, and if developments now underway with respect to fiber and resin formulation

should come to pass, then the required machinery can become significantly simpler. Virtually all of the complexity in the present machinery is used to overcome problems inherent in the material that the machine uses. Simpler machines would translate to lower cost and broader use of composite tape materials.

ACKNOWLEDGMENT

The paragraphs describing NC programming and contoured tape laying are the work of H.W. Lewis, Vice President, The Ingersoll Milling Machine Company, to whom the author is grateful.

Automated Integrated Manufacturing System

A.L. Flescher, Grumman Corporation

AS COMPOSITES GAINED ACCEP-TANCE in the late 1970s, applications were extended to more complex structures. A need arose for mechanized equipment to fabricate severely contoured, integrally stiffened, complex structures for high-performance aircraft. The prime material considered was the easier-to-handle and less costly graphite-epoxy tape in widths up to 30 cm (12 in.) and in woven broad goods. Accordingly, Grumman proposed to the U.S. Air Force that the capabilities of the integrated laminating center (ILC) shown in Fig. 1 be expanded to handle highly contoured plies. Three modules (ply handler, broad goods dispenser, and translaminar stitcher) were subsequently developed, integrated, and validated under contract. The operation of these modules is described later in this article.

Integrated Laminating Center

The ILC is a modular system capable of tape laying, ply trimming, ply handling, and stacking for a variety of aircraft cover or skin configurations. As shown in Fig. 1, the ILC incorporates minicomputer-controlled laydown of composite material in a preprogrammed pattern by a mechanized tape head at the tape-laying station. A transfer station uses its vacuum-holding system to lift and transport the ply to the laser trim station, simultaneously drawing backup material for the next ply across the lay-down station, where it is held by vacuum. When the transfer station has returned to its park position, the laser gantry positions itself for ply trimming. Upon completion of the trim operation, scrap is removed by the operator, the ply is inspected, and the table rotates about one edge (vertical flip), placing the ply on the gently contoured mold form.

Modules of Operation

The ply handler (Fig. 2) automatically locates, drapes, and forms large-area plies to highly contoured tools at a rate of 1300 cm² (200 in.²)/s. This module interfaces with the ILC ply transfer station to automatically receive large-area plies supported on deformable carrier film. Plies are formed to contour by the wiping action of rotary brushes in a programmed cycle that controls the amount of tool rise, split-table separation, and brush arbor motion. Brush rotational speed, bristle density and contour, and programmed cycle configuring allow flexibility to meet specific requirements. Use of this automated module eliminates manual tailoring of strips to contour and permits direct use of automatically prepared flat plies for contoured configurations. The rotary brush force and wiping action imparted to the carrier film and prepreg cause the flat prepreg to conform to the complex-geometry mold.

The broad goods prepreg dispenser (Fig. 3) automatically dispenses and trims graphite-epoxy, aramid-epoxy and fiberglass-epoxy prepregged fabric broad goods in widths of up to 200 cm (80 in.). The dispenser interfaces with the ILC by means of the ply transfer station, which pulls out lengths of material, and the laser trimming station, which cuts off predetermined lengths. This module features quick-load chucks, roll-away freezer storage, and optional dewrinkler roll action. Use of nested, computerized, geometric data provides efficient laser trimming of contoured details with appropriate flange foldouts and ply orientation control. The dispenser can produce plies with computer-generated contours, flanges, stiffener foldouts, splice strips, and edge build-ups.

The translaminar stitcher (Fig. 4) is a self-digitizing, multiaxis module that can stitch along straight, bowed, twisted, or highly con-

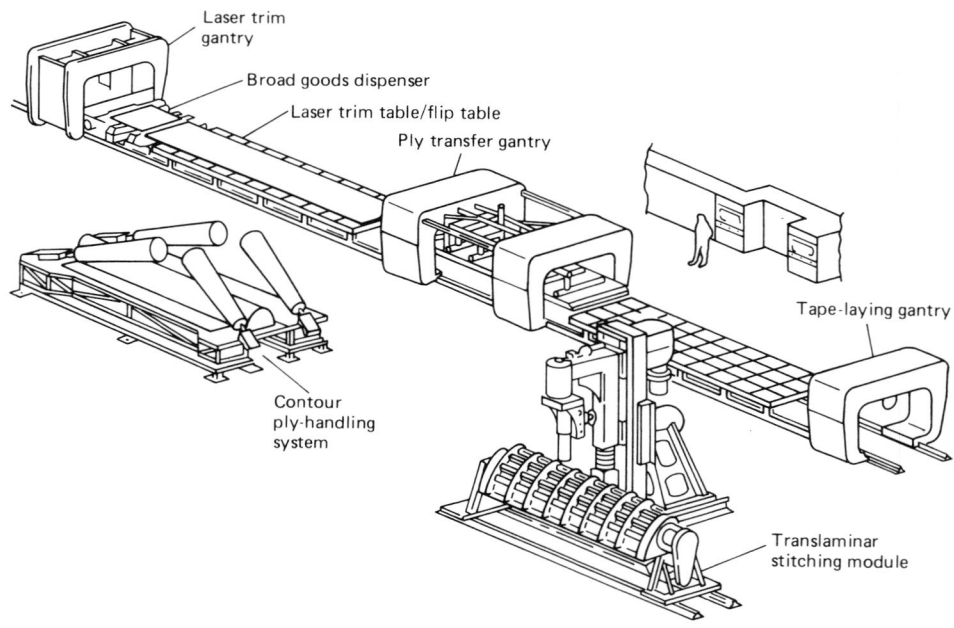

Fig. 1   Integrated laminating center with integrated modules

Fig. 2 Ply handler. (a) Ply handler in position to receive ply. (b) Mold form with ply rising to meet brushes. Carrier film not shown

Fig. 3 Broad goods prepreg dispenser

(a)

(b)

Fig. 4 Translaminar stitcher. (a) Self-digiting multiaxis module. (b) Computer control console

(a)

(b)

Fig. 5 Completed demonstration component. (a) Inside view. (b) Outside view

toured (circumferential) paths. The computer interpolates between selected input points and inserts the required stitch pitch. Use of aramid thread with selected pitch, edge distance, and gage ensures the out-of-plane load-carrying ability of integrally cured, flange-to-skin joints.

Use of these modules in conjunction with the ILC was projected to yield a cost savings of about 50%. The estimated savings were realized and verified by tracking and measuring the work done on a 250-cm (100-in.) long, compound-contour, graphite-epoxy demonstration component (Fig. 5). This component incorporated 113 automatically prepared skin build-ups, nested stiffeners, 162 woven graphite-epoxy frame details, bowed/twisted closed-section hat stiffeners, and stitched and integrally cured stiffener/skin and frame flange/skin joints.

Automated Ply Lamination

Keith Goode, Composite Automation Equipment, Inc.

AUTOMATED PLY LAMINATION is the process of converting rolls of prepreg into flat lay-ups of moderate size, with minimal human intervention, at high production rates. The lay-ups consist of plies of various shapes that are stacked in a specified sequence, with the location of partial plies controlled to tight tolerance. The lay-ups can be automatically placed on gentle-contour molds. Future growth of the system will provide additional devices to seat the flat lay-ups on complex-contour molds.

Most laminating is currently performed by hand, and considerable time is spent locating plies accurately on the mold. It might appear that replacing manual labor with automation would be relatively easy because picking up the precut plies and then manipulating them into a new position before laying them down are all that appear necessary. Several aircraft companies began developing automated lamination cells and have gained considerable knowledge in the past six or seven years. However, it is probably safe to assume that there is no production hardened lamination equipment in use today, which reflects the technical difficulty of its development. These automated systems, though applicable to as many as 98% of the parts required in an aircraft that uses a high percentage of composite structures, are still only in the developmental stage.

The technical problems are largely due to the use of graphite-epoxy prepreg, which is generally supplied on rolls up to 150 cm (60 in.) wide and 910 m (3000 ft) long, weighing over 230 kg (500 lb). The prepreg comes in either woven form or as a unidirectional tape; both have a backing material to separate the layers on the roll. Woven material is approximately 0.028 cm (0.011 in.) thick, excluding its backing material, and unidirectional tape is approximately 0.015 cm (0.005 in.) thick. Also, the tack (stickiness) of the prepreg varies considerably with the resin content of the prepreg, and is subject to tolerances in prepreg manufacture. Tack increases with increasing temperature and humidity levels in the lay-up room, but decreases as the exposure to room temperature increases. With these variables accounted for, conditions must be established such that the prepreg can be easily removed from the backing but maintain enough tack to hold the lay-up together.

The typical automated laminating cell uses a reciprocating knife set-up with a cutting table consisting of numerous bristles. A covering film holds the prepreg on the table under vacuum, and the backing prevents penetration of the bristles in the prepreg. The use of a covering film and vacuum are necessary to hold the material accurately in place while cutting the plies to the desired shapes. Both materials must be removed from the prepreg before the layers are stacked to create the lay-up. Manual removal is currently necessary because an automated system has not yet been qualified for production.

With this cutting process, material use is optimized when large prepreg sheets are cut into plies for several parts. This concept, which imposed restrictions on the device that picks up and stacks the plies, is now workable because those restrictions were resolved. The last technical problem to be solved is to match the capabilities of the system to its engineering design. The processes with a high manual-labor content have permitted complete design freedom to maximize weight savings.

With the introduction of automation, some design restrictions must be identified and implemented to enhance system operation. The few laminating cells that currently exist were built as proof-of-principle demonstrators and consist of standard purchased equipment coupled with specially tailored ply pick-up devices. Typically, computer integration was added after the fact. The various pieces of equipment that form the cells were generally added piecemeal, each one solving a particular problem, but not addressing the laminating process as a whole. References 1 to 3 provide additional information on these cells.

A typical earlier system probably included a commercially available computer-controlled reciprocating knife cutter modified to cut graphite-epoxy prepreg, and a jointed-arm robot manipulating a vacuum head to extract the cut plies from the cutting table and stack them in sequence on a flat lay-up surface (Fig. 1). This system successfully proved the concept and paved the way for the next-generation cell.

Because the first system was limited by the reach of the jointed-arm robot, the next logical development was to produce a variant of the cutting machine with a segmented cutting table. The table segments could then be moved from the cutting machine to the jointed-arm robot (Fig. 2), which allowed a high degree of automation and an unlimited amount of prepreg to be delivered to the robot. The major limitation of this concept was, again, the robot. Its weight-carrying capability restricted the size of the vacuum head and limited the size of the lay-ups that could be produced.

A larger vacuum head was required to increase system capability, which necessitated using a larger robot. Fortunately, the robot industry had started offering gantry-style robots with larger lifting capacities, but a large, flat, vacuum pick-up head would still have been too heavy and cumbersome. Also, the requirement of laying up laminates in contoured tools was emerging, as well as the need to debulk (squeeze out trapped air) during lay-down. Therefore, third-generation lamination equipment consists of the sectional cutting tables; a table transporting device; a large, gantry-style robot; and a computer-controlled, porous, lay-down roller head, constructed like a giant paint roller. The roller head, programmed to roll

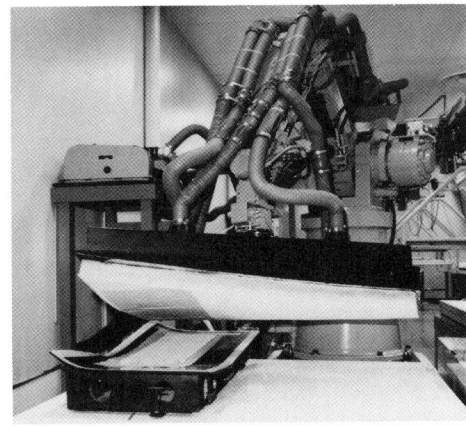

Fig. 1 First-generation laminating cell. Courtesy of Northrop Corporation

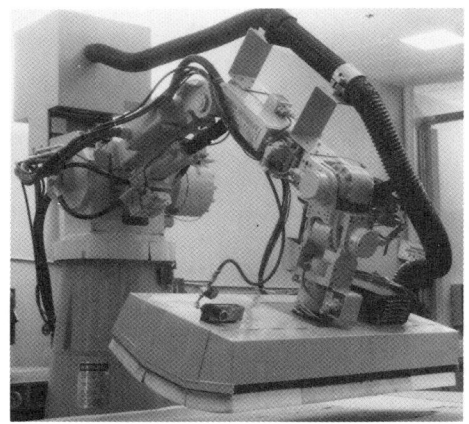

Fig. 2 Second-generation laminating cell. Courtesy of Northrop Corporation

Fig. 3 Third-generation laminating cell. Courtesy of Northrop Corporation

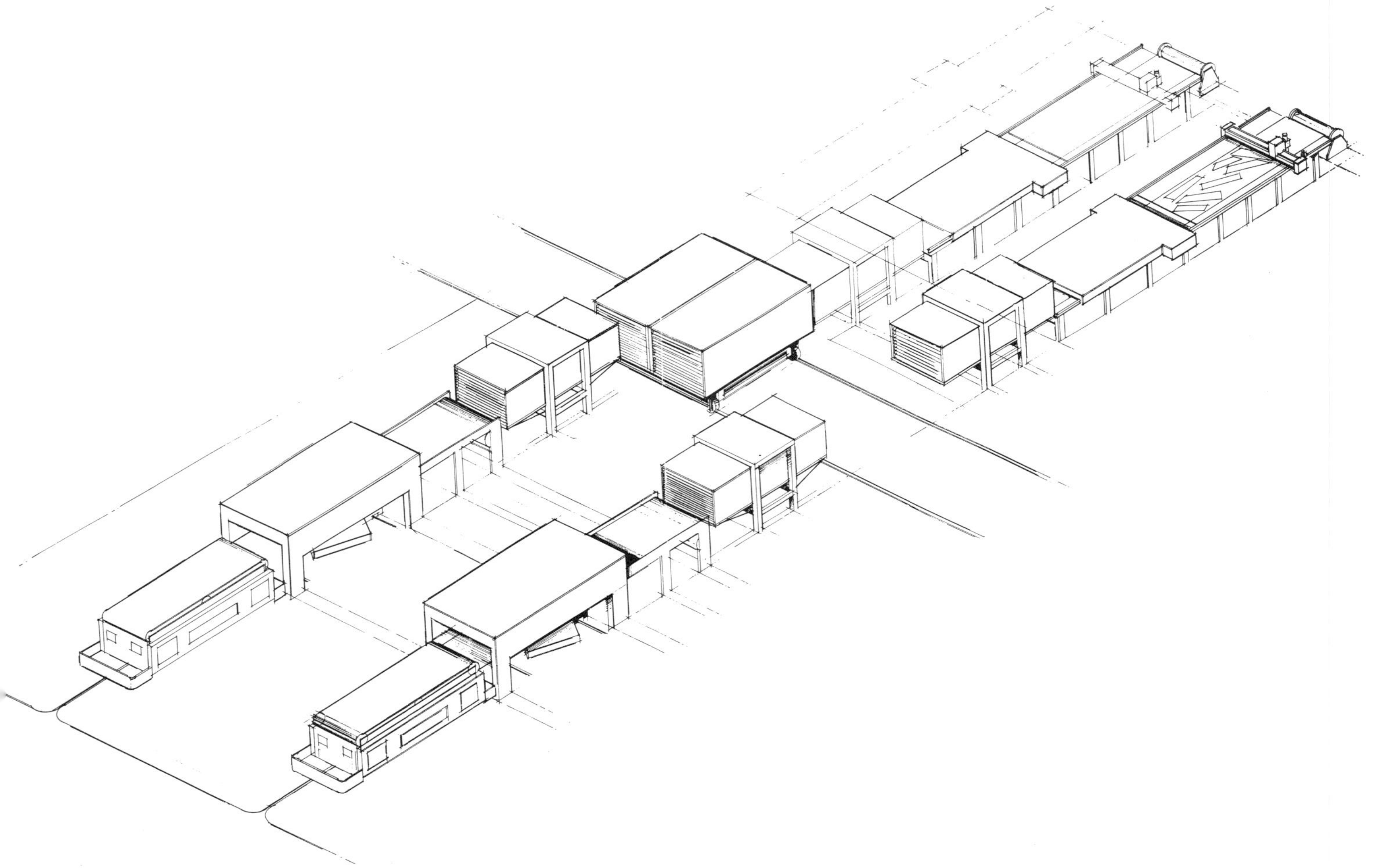

Fig. 4 Next-generation laminating cell

gently over the cut ply nest, automatically introduces vacuum to its surface in specific areas to pick up the desired ply. Once the ply is picked up, the roller head is directed to the lay-down table, where the roller rotates and presses the ply onto the lay-down surface (Fig.

3). Thus, plies can be removed from a larger nest to conserve expensive material, and a lay-up is produced that can be molded into a void-free part.

Currently, this third-generation cell is the most advanced broad goods lamination cell in

use. Operations are controlled from a central computer, which downloads programs to the individual pieces of equipment. The central computer is updated when tasks are completed. In a typical operation, graphite is automatically dispensed onto the table section in the cutting position, the ply nest is downloaded from the cell controller, and the nest is cut. Then, the table section is elevated, picked up by the monorail transportation system, and carried to either a jointed-arm robot or an automatically guided vehicle station, which transports the table into the roller head lamination area. The robot, with its pick-up head or roller head, begins to extract plies in the specified sequence from the cut nest and to make layers. The emptied tables are then transported to a prepreg scrap removal station and recycled to the cutting machine.

Manual intervention is still required at the lamination stations to remove the graphite backing paper and covering film. Even so, throughput times approach 2 min/ply. This third-generation laminating cell produces significant cost savings, compared to kitting precut plies and manual stacking. However, future production requirements demand eight plies/min (or 8s/ply). Thus, an improved system with greater throughput is needed to avoid the expense of owning multiple systems.

The next-generation laminating system needs to overcome the shortcomings of the earlier systems and produce lay-ups containing more than one material. Also, the labor content must be reduced by eliminating the covering film and automatically removing the backing paper. The automated ply-laminating system shown in Fig. 4 represents a continuous flow, just-in-time (JIT) operation. It consists of several individual computer-controlled machines that are integrated into a system. A supervisory computer directs the activities within the system. A schedule for the laminates that are to be produced is generated by the fabricator's production control department and used as the input to drive the system. Each laminate has its own process plan or routing, which is scheduled by the computer that controls the system.

In this automated laminating system, plies are cut on one of the two cutting machines. Cutting is one ply deep, and the cutter penetrates only the graphite, leaving the backing paper intact. The backing paper is removed just before the plies leave the cutting machine. There is no covering film because vacuum is not used during the cutting process. Once cut, each ply is tracked by the cell computer, which uses a cell map in its data base for this purpose. Tracking is initiated at the cutting nest. After the cutting, the nest of plies is conveyed to the next machine, known as a ply plucker, and plies are removed from the cut nest individually on a first-in, first-out basis. Because only one ply is handled at a time, tracking ply location is greatly simplified. Each time a ply moves from one location to the next, the cell map is updated. As plies are picked out of the nest, they are conveyed into a magazine-type storage cabinet with the capacity for at least 40 plies. When the magazine is loaded with the correct number of plies, it is automatically moved to either a staging area or another cutting machine to receive different material or directly to the next machine in the process. The intent is to assemble packs within the magazine that constitute a finished laminate, regardless of the material mix.

The next machine in the process is the laminator. The plies are conveyed from the magazine(s) one at a time and into the laminator in the correct sequence. As they enter, a linear array camera inspects each ply for position. Data concerning ply size and approximate location are known, because each ply has been tracked through the system by the cell map. The measured dimensions for each ply are compared to the programmed data in the data base. This generates offsets (for the X, Y, and A axes of the laminator) that are used in the lay-down process to ensure correct placement of each ply. Each ply is picked up by the laminator and transported to the lay-up table. During this move, each ply is inspected again to ensure correct location before lay-down.

This operation is repeated for every ply until the laminate is complete, at which time it is automatically conveyed onto a wire-guided vehicle. This automated system, as described, has a production capacity of 8 s/ply. Plies can be as small as 25×75 mm (1.0×3.0 in.) to as large as 1.5×4 m (5×12 ft).

REFERENCES

1. M.M. Schwartz, *Composite Materials Handbook*, McGraw-Hill, 1984, p 5.23-5.35
2. J.J. Haggerty, Automation in Aerospace Manufacturing, in *Aerospace*, Aerospace Industries Association, Vol 23 (No. 3), Summer 1985
3. G.F. Lubin, Ed., *Handbook of Composites*, Van Nostrand Reinhold, 1982, p 500-504

Preparation for Cure

Timothy W. McGann and Eugene R. Crilly, Rockwell International Corporation

PROCESSING MATERIALS must be added to a composite ply lay-up before autoclave curing. These materials control the resin content of the cured part and ensure proper application of autoclave pressure to the lay-up.

In selecting materials for use in preparing a laminate for curing, cure temperatures and pressures must be considered, as well as compatibility of the processing materials with the matrix system. The importance of proper techniques and materials in lay-up preparation cannot be overemphasized. To ensure consistent, high-quality parts, it is mandatory that proper materials be used by specially trained personnel.

Material Types and Functions

The materials usually used in preparing a lay-up for autoclave curing are peel ply (optional), separator, bleeder, barrier, breather, dam (depending on laminate thickness and tooling), and vacuum bag. The materials shown in Fig. 1 represent a simple lay-up; examples of more complex lay-ups are shown in Fig. 2 and 3. A generality may be made concerning these materials: Each should be prevented from becoming a possible source of contamination to the composite laminate. Contamination can result in poor adhesion in subsequent bonding or painting operations; also, volatile contaminants can enter the laminate prior to gelation, resulting in porosity and consequent poor matrix-dominant properties. The materials must also be compatible with the maximum cure temperature and pressures required for the matrix system being cured. Each material is discussed below. The typical commercial designations and sources for each material category are listed in Table 1. Maximum usage temperatures can be determined by consultation with suppliers.

The peel ply, if used, is placed immediately on top of or under the composite laminate. It is ultimately removed just before bonding or painting operations so that a clean, bondable surface is available. It is usually a woven fabric and may be either nylon, polyester, or fiberglass. The fabric is treated with a release agent that must not transfer to the laminate; otherwise, subsequent bonding or painting opera-

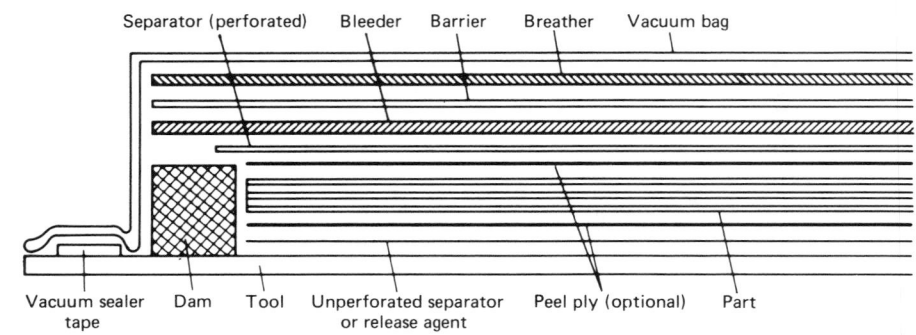

Fig. 1 Simple lay-up

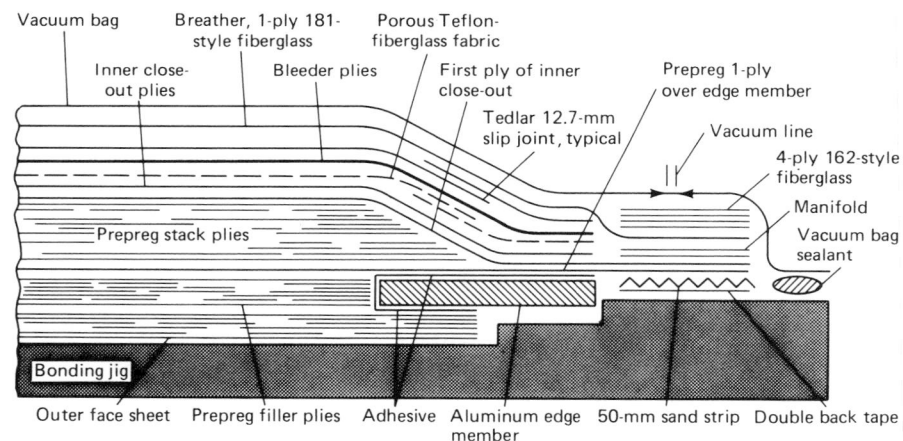

Fig. 2 Complex lay-up, including metal closeout

tions may not be satisfactory. Nylon, polyester, or fiberglass peel ply can be used with most matrix systems, but nylon will not release from phenolics and is not satisfactory for high temperature curing matrices, such as polyimides or bismaleimides, because its upper usage temperature is about 177 °C (350 °F). Nylon can be used if the initial cure temperature is low and if it is removed before postcure. If there is any doubt about using a particular peel ply material for a given system, a test laminate should be made and cured to check for compatibility. A standard adhesion test will determine

whether a given release agent on the peel ply will transfer to part during cure, which, as noted above, would be detrimental to later bonding or painting. A peel ply can also be used between the mold and a thick laminate to accept entrapped volatiles, thereby preventing porosity in the external composite plies.

A separator (release material) is placed on top of or under the laminate and peel ply (if peel ply is used). It allows volatiles and air to escape from the laminate and excess resin to be bled from the laminate into the bleeder plies during cure. It will also give the cured part a smooth

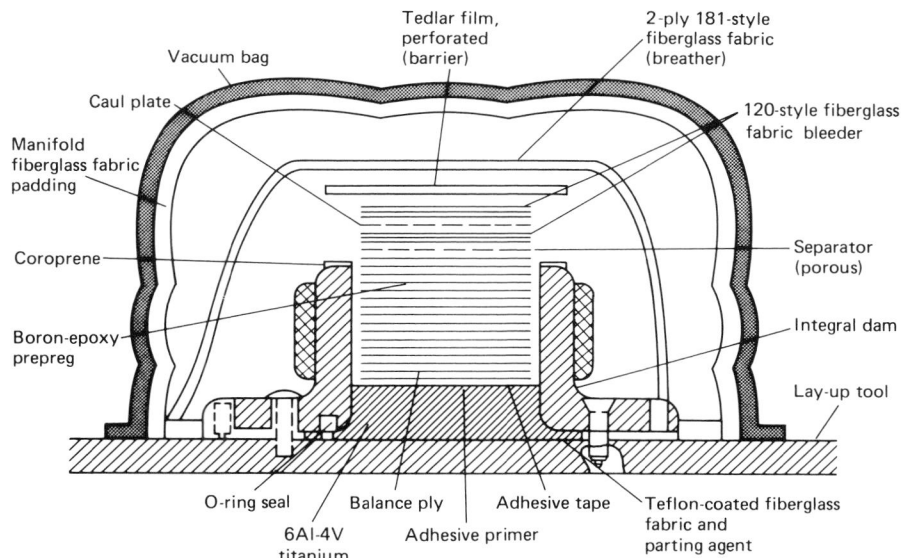

Fig. 3 Complex lay-up showing tool with integral dam. Shown before application of vacuum for clarity. After vacuum is applied, the vacuum bag and fiberglass padding conform to the shape of the tool/lay-up assembly. Note: Lay-up to be co-cured to titanium strap.

surface, except for porous Teflon, which gives a slightly textured surface. After cure, the release material must be easily freed from the cured laminate without causing damage. Most separator materials are porous or perforated and contain fluorocarbon polymers. The size and spacing of perforations or the porosity of the material determines the amount of resin flow from the surface of the laminate, thereby lowering resin content (increasing the fiber volume) of the cured part. With high porosity or large, closely spaced perforations, a large amount of resin will be bled from the laminate

during cure. If little resin is to be removed from the lay-up, the separator material obviously must have small, widely spaced perforations, and if no resin removal during cure is desired, totally unperforated separator film should be used. A prepreg with high flow will not be retarded by porous Teflon, which acts as a mini-bleeder ply and soaks up a small amount of resin.

Bleeder. The purpose of the bleeder material is to absorb excess resin from the lay-up during cure, thereby producing the desired fiber volume. Fiberglass fabric or other absorbent

materials or fabrics are used for this purpose. The amount of bleeder used is a function of its absorbency, the fiber volume desired in the part, and the resin content of the prepreg material used in the lay-up. In advanced composites, essentially all excess resin is bled from the surface of the laminate, with edge bleeding being minimized by properly damming the lay-up edges. Personnel should establish tables giving plies of bleeder per ply of lay-up for each prepreg material. These tables are often set up for a particular bleeder material, and then equivalencies are given for other materials. For instance, 1 ply of 181-style fiberglass is considered equivalent to 2.5 plies of 120-style fiberglass fabric. Table 2 gives an example of a 120-style fiberglass fabric bleeder to prepreg ply ratio to yield a nominal fiber volume of 62% in the cured part, with tape having a nominal areal weight of 0.145 kg/m^2 (0.0296 lb/ft^2).

Table 2 Bleeder ply factors at 62% fiber volume target

Graphite-epoxy prepreg resin content, wt%	Bleeder factor (plies of 120-style fiberglass per ply of graphite-epoxy prepreg)
31	0.04
32	0.09
33	0.15
34	0.21
35	0.27
36	0.33
37	0.39
38	0.46
39	0.53
40	0.60
41	0.67
42	0.75
43	0.83
44	0.91

Table 1 Commercial designations and sources of processing materials

Material	Purpose	Commercial designation	Source	Comments
Peel ply	Provides a bondable surface	Burlease 51789 (nylon-6)	Burlington Mills	Not suitable for phenolics
		Burlease 60 000 (polyester)	Burlington Mills	Suitable for phenolics
		Bleeder-lease E (fiberglass with release agent)	Airtech International	
Separator (release)	Separates cured laminates from other process materials without damage	A5000 Teflon FEP film	Richmond Technology	
		A4000 halogen release film	Airtech International	
		Wrightlon 4500 (halocarbon film)	Airtech International	
		104 TFE Teflon-coated fiberglass-style fabric	Various	
Bleeder	Absorbs excess resin	120-style fiberglass fabric	Commercial	
		181-style fiberglass fabric	Commercial	
		Organic fiber felts	Commercial	
Barrier	Limits resin flow to only bleeder plies	See Separator		
		Unperforated TFE film		
Breather	Evacuates the vacuum bag so that the desired autoclave pressure is applied	Airweave N-10, Airweave N-4	Airtech International	Organic fiber, stretchable felt
		A-3000	Richmond Technology	Organic fiber, stretchable felt
		Airweave HP	Airtech International	1.4-mm (0.055-in.) thick fiberglass fabric
		Burflo 4819	Burlington Mills	Polyester nonwoven fabric
Dam	Prevents resin flow from edges	Rubber neoprene cork, Rubatex 886	Groendyke	Tape with pressure-sensitive adhesive
Vacuum bag	Applies autoclave pressure	Vac-Pak	Richmond Technology	Various types for different cure temperatures
		Wrightlon, IPPLON, Wrightcast, Vacalloy	Airtech International	Various types for different cure temperatures
		Silicone rubbers	Various	Semipermanent

To determine the correct number of plies to use:

● Determine resin content of prepreg from receiving inspection data or vendor certification data and round off to the nearest low whole number
● Obtain bleeder/prepreg ratio from the above table for the given resin content
● Multiply number of plies in the lay-up by the bleeder/ply ratio
● Round off to the nearest low whole number; this is the number of plies of 120-style fiberglass fabric to be used as a bleeder

For example, to determine the number of plies of 120-style fiberglass fabric required for a 24-ply laminate in which the graphite-epoxy has a resin content of 35.7%:

● Round off 35.7% to 35%
● According to the table, ratio of bleeder/ply of graphite-epoxy for 35% is 0.27
● 24 × 0.27 = 6.48
● Round off to 6; this is the number of bleeder plies

Mochburg CW-1850 and 181-style (1581 or 7581) fiberglass may be substituted in these ratios: 1 ply Mochburg CW-1850 = 2 plies 120-style fiberglass; 1 ply 181 = 2.5 plies 120-style fiberglass.

If the user is willing to tolerate a lower fiber volume with lower cured mechanical properties, bleeding may be eliminated. However, a significant weight increase may occur if the prepreg used in the no-bleed process contains more resin than remains in a cured part subjected to bleeding. An alternative is to obtain a prepreg with an existing fiber volume close to that of the desired final part. Either method eliminates the cost of the bleeder, which is an expendable material.

Barrier. A nonadhering material, called the barrier, or barrier film, is commonly placed between the bleeder plies and breather plies. In the case of epoxy resins, it is frequently an unperforated film, or barrier film, so resin removal from the part can be controlled. For resins that produce volatile by-products during cure, a film with small perforations and large spacing is used to prevent the breather material from becoming clogged with resin and unable to perform its function. Often, a low-cost material such as Tedlar film is used instead of nonporous Teflon because of the expense of the latter. Thermocouples are installed at the edge of the part between the barrier film and breather plies to monitor part temperature. They also provide the information needed to allow control of the heating medium operation in computer-controlled autoclaves.

The breather is a material placed on top of the barrier film to allow uniform application of vacuum pressure over the lay-up and removal of entrapped air or volatiles during cure. It may be a drapable, loosely woven fabric, or felt. Care must be exercised in using coarse, open-weave fabrics, because if bridging of the vacuum bag occurs beyond the elongation properties of the bag material, bag failure will occur, which could result in loss of the part. Some lay-ups use an edge breather consisting of a single fiberglass tow laid along the edges of the composite and draped over the dams.

A dam is sometimes located peripherally to minimize edge bleeding. It may be an integral part of the tool or built in position using materials such as rubber neoprene cork pressure-sensitive tape, silicone rubber, or Teflon or metal bars. Dam height should be approximately the same as the lay-up thickness, including release and bleeder plies, to prevent rounding off of the part edge by the action of the vacuum bag.

The vacuum bag is used to contain any vacuum pressure applied to the lay-up before and during cure and to transmit external autoclave pressure to the part. It prevents any gaseous pressurizing medium used in the autoclave (air or inert gas) from permeating the part and causing porosity and poor or unacceptable part quality. Commonly, an expendable material such as nylon is used for this purpose. Because of the physical and chemical properties of the bag, nylon-6 cannot be used above 160 °C (325 °F), and nylon-6,6 cannot be used above 205 to 215 °C (400 to 420 °F). Neither of these materials is compatible with phenolic resins or certain other matrices. Kapton can be used with polyimides and other materials requiring high-temperature cures. For long runs and highly contoured parts, semipermanent bags may be used. Such bags are commonly made of molded-to-shape rubber. To provide a seal between bag and tool, the bag is commonly secured to the tool with a tape material (bag sealant) that adheres to both the tool and the bag.

The application of the vacuum bag is extremely critical. Bag perforation by the sharp edges of the tool and leakage due to improper sealing at the tool edges may result in a porous part. The complex contours of most aircraft parts often require folds in the bag to take up excess bag material. If these folds are not properly made, or if large wrinkles are left in the bag, undesirable wrinkles may develop in the cured part. Thus, this operation is probably the most critical single step to part quality and must be performed carefully by skilled mechanics.

Autoclave Cure Systems

Todd Taricco, Thermal Equipment Corporation

AN AUTOCLAVE SYSTEM allows a complex chemical reaction to occur inside a pressure vessel according to a specified schedule in order to process a variety of materials. The evolution of materials and processes has taken autoclave operating conditions from 120 °C (250 °F) and 275 kPa (40 psi) to well over 760 °C (1400 °F) and 69 000 kPa (10 000 psi). The materials processed in autoclaves range from metal bonding adhesives; reinforced epoxy laminates; thermoplastic laminates; metal, ceramic, and carbon matrix materials; to many other aerospace and electronic components. Although the autoclave system is tailored to specific process requirements, the basic design and subsystems described here are standard for most autoclaves. Shown in Fig. 1 is an 8-m (25-ft) diam by 18-m (60-ft) long autoclave system, one of the world's largest in diameter.

The major elements of an autoclave system (and their functions) are: a vessel to contain pressure, sources to heat the gas stream and circulate it uniformly within the vessel, a subsystem to pressurize the gas stream, a subsystem to apply vacuum to parts covered by a vacuum bag, a subsystem to control operating parameters, and a subsystem to load the molds into the autoclave.

Pressure Vessel

The pressure vessel shell provides the means to retain pressure inside the work space. The pressure vessel typically is fabricated from pressure vessel quality carbon steel. Plates up to 150 mm (6 in.) thick are rolled to shape and joined by arc welding. The dome-shaped heads or ends are fabricated from similar material and either press-formed or spun.

The most critical portion of the vessel is the closure, or breech lock, which is fabricated from three distinct rings fitted with lugs. One ring is welded to the vessel, one is welded to the door, and one rotates. When the door is closed, the locking ring is rotated, and the lugs of the closure engage with the lugs of the head. There are also matched wedges on the lugs to clamp the door surfaces against the seal area. A silicone or fluorocarbon rubber material is normally used in this area to allow good sealing without requiring metal-to-metal contact at the door face. The door is usually hinged or carried to one side by a crane. All hinge structures should be adjustable in all directions of motion to allow realignment, as needed.

It is mandatory to design and fabricate all pressure vessels to American Society of Mechanical Engineers (ASME) requirements. These include material specifications and design stress levels allowable in the vessel, but do not cover any other mechanisms or subsystems within the autoclave. If ASME requirements are met, the vessel will be so stamped after hydrostatic testing. National Board Registration is strongly suggested for future traceability, and is mandatory in some states.

The autoclave owner must have an ASME-approved organization modify or maintain the pressure vessel; any improperly made modification or repair may compromise the integrity of the vessel and subsequent safety. All autoclave vessels and closures should be inspected annually by a recognized inspection agency. Normally, this will reveal any potentially dangerous situations. If properly maintained and inspected, the life of a pressure vessel should exceed 50 years.

Internal Structure. The vessel, upon completion and subsequent testing, is prepared for installation of the internal structure, which provides insulation, duct work, and support for all components to be mounted in the autoclave.

The insulation is used to reduce energy costs by preventing heat transfer to the vessel. It is typically a ceramic wool material, in sheet or blanket form, attached to the vessel with pins. The insulation should not rely on internal duct work for support nor should it be in direct contact with the autoclave atmosphere. The insulation is covered with sheet metal, normally 16- and 18-gage aluminized steel or stainless steel, which is attached in such a way as to allow thermal expansion and protect the insulation from the gas stream in the autoclave.

The internal insulation is very important to the prevention of excessive shell temperature and energy loss. The maximum shell temperature on the autoclave surface should not exceed 60 °C (140 °F) at maximum operating temperature, and external insulation should never be required on a modern autoclave.

The annular duct is attached after insulation, providing a channel for the gas to be circulated in the autoclave. Care must be used in the design to allow for thermal expansion. Because the vessel shell does not get hot, the differential expansion between the interior and exterior may exceed 75 mm (3 in.) on large, high-temperature autoclaves. Provisions for tracks for carts to bring the molds into the autoclave and for structural attachments for internal systems are made at this time.

Gas Stream Heating and Circulation Sources

Currently, several heating methods are available for autoclave systems. The most common method for large autoclaves is indirect gas firing, in which products of combustion from an external chamber are passed through an internal, stainless alloy coil. This system is reliable and, if properly engineered, can be controlled to allow thermal cycling. The materials selected and the design technique must be adequate for the service temperatures and thermal cycling. The gas-fired systems provide substantial operating savings over electrical systems when direct energy and demand charges are considered. Gas heating is regularly used in autoclaves with maximum operating temperatures of 450 to 540 °C (850 to 1000 °F).

In the past, hot oil was used as a heating medium in gas-fired systems in which the oil was circulated from an external heater to an internal coil. These systems are now nearly obsolete because of the potential for contami-

Fig. 1 Autoclave system. Courtesy of Beech Aircraft

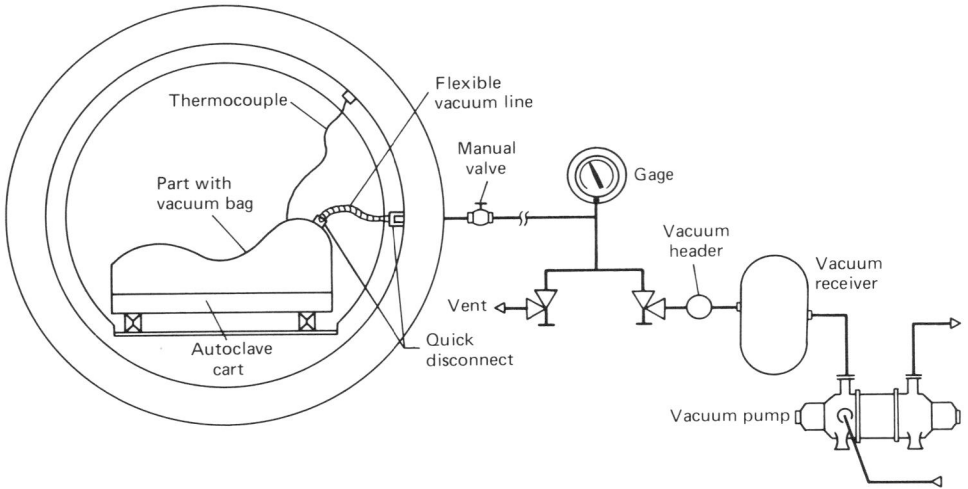

Fig. 2 Conventional vacuum system

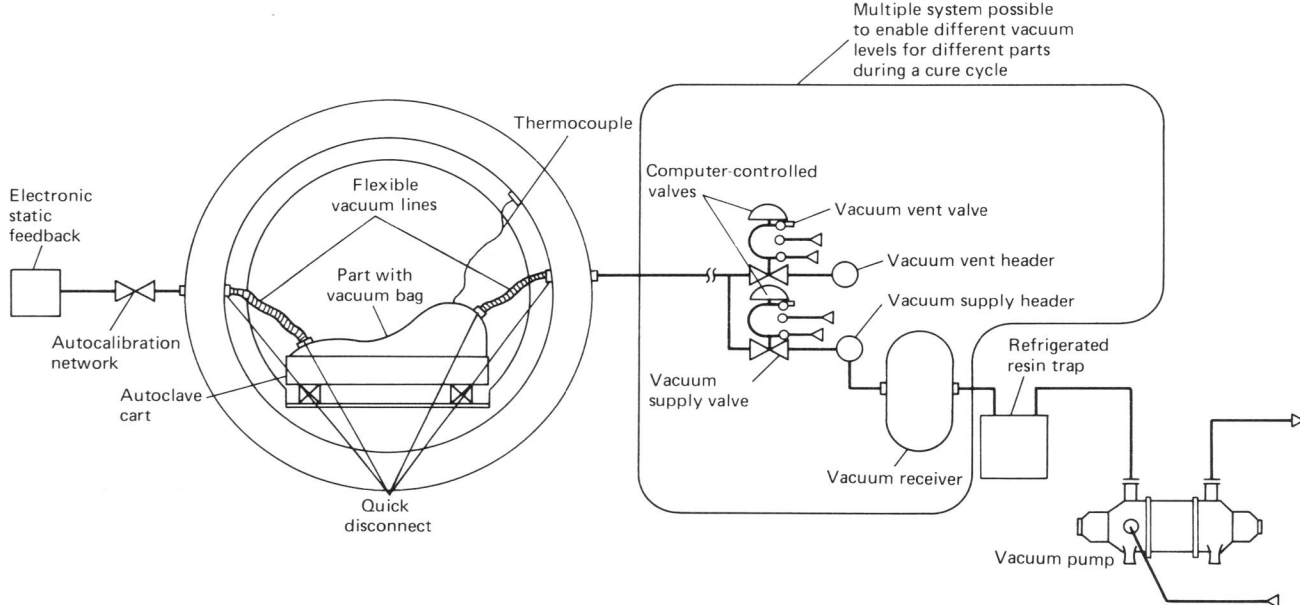

Fig. 3 Advanced vacuum system

nation of the bonding area by PCB, MCB, or silicone-based oils. Area contamination can be disastrous to materials in process. All available heat-transfer oils are combustible and have a maximum operating bulk temperature of 425 °C (800 °F) for silicones and 400 °C (750 °F) for phenyl-based fluids. The potential for fire is significant, and appropriate safeguards must be provided.

Steam heating is often used for autoclaves operating in the 150 to 175 °C (300 to 350 °F) range. The superheated steam is passed through a coil in the autoclave to heat circulating gas. Few steam-heated autoclaves are currently manufactured for composite bonding because of their low operating temperatures.

Most small autoclaves (under 2 m, or 6 ft in diameter) are electrically heated. Electric heating elements are mounted in the circulating gas stream and configured not to radiate onto the workload in the autoclave. Heater elements are made either from open nichrome wire or alloy tubes. Heaters can be either contactor (on-off controlled) or SCR controlled. The major drawback of electrically heating the autoclave is the operating and demand costs, which can be prohibitive for large autoclave systems.

Gas circulation within the autoclave is essential to provide mass flow for temperature uniformity and heat transfer to the part load. This is accomplished with a blower mounted in

the rear of the autoclave. The gas is drawn into the blower through the cooling coil and heater, then returned down the length of the autoclave through the annular duct to the door, where it is then directed through the work space. In the modern autoclave system, the fan motor is mounted in a pressurized housing at the rear. This eliminates the need for pressure seals and external bearings, which have caused maintenance and reliability problems on external motors.

The air circulation should be from 1 to 3 m/s (250 to 500 ft/min) in the work space. Circulation any higher than 3 m/s (500 ft/min) may cause problems with the vacuum bags over the parts if they are not properly attached. Variable-speed fan systems are available for improved part heating performance.

Gas Stream Pressurizing Systems

Three pressurization gases are typically used for autoclaves: air, nitrogen, and carbon dioxide. Gases are introduced into the vessel through solenoids or proportional control valves, the latter being far superior. Proportional inlet and vent valves allow autoclave pressures to be controlled and varied precisely. It is important that this very hot gas being introduced into the pressure vessel not impinge on the part load, as the gas stream could cause part damage due to thermal or mechanical shock. The vessel is vented through similar valves, by means of a silencer to the atmosphere.

Air is relatively inexpensive when supplied in the 690 to 1030 kPa (100 to 150 psi) range, and is acceptable for most 120 °C (250 °F) cures. The main disadvantage of air is that it sustains combustion, and thus may be hazardous at temperatures above 150 °C (300 °F). Nitrogen is the gas most commonly used in autoclaves. The liquid nitrogen is stored in cryogenic form and then vaporized at approximately 1400 to 1550 kPa (200 to 225 psi). Higher pressure tanks and systems are available. Nitrogen suppresses combustion and diffuses well into the air when the autoclave is opened. However, nitrogen costs can be significant if many autoclaves in a plant are using nitrogen and if the autoclaves are large and operating at high pressure. Carbon dioxide is the second most commonly used gas. It is stored as refrigerated liquid at approximately 2050 kPa (300 psi). Its primary disadvantages are high density, hazards to personnel, and physical flow related problems. When using any nonlife-sustaining gas, care should be taken not to enter any vessel without ensuring that adequate oxygen is present.

Vacuum Systems

A very important subsystem is the part vacuum supply. Most parts processed in autoclaves are covered with a vacuum bag, which is used primarily for compaction of laminates (hydrostatic pressure will not compact) and to provide vacuum for removal of volatiles. The bag allows the part to be subjected to differential pressure in the autoclave without being directly exposed to the autoclave atmosphere. The vacuum bag is also used to apply varying levels of vacuum to the part.

New production methods have brought increasing complexity to autoclave vacuum systems. Originally the vacuum systems consisted of a three-way valve which allowed application of vacuum to the part bag or venting of the bag to the atmosphere after pressure application. A gage read the supply, which was nonvariable. This proved to be adequate for simple laminates

Fig. 4 Computer-controlled autoclave system console. Courtesy of Naval Air Rework Facility, San Diego

and metal bonding, but as the resin systems became more sensitive and quality control became more stringent, advanced vacuum systems were developed (see Fig. 2 and 3).

The purpose of these systems is to provide fully computer-controlled manipulation and monitoring of part pressure, not just supply pressure. The ability to provide pressure on the part under the bag by means of vacuum has reduced void content by keeping the dissolved volatiles and water in solution in the resin system itself. This capability is being included in newer systems.

The modern vacuum system can manipulate vacuum levels from the two independently modulated supply headers to the parts. The transfers between controllable vacuum supply levels are computer-controlled as functions of the autoclave pressure or part temperature. These systems are essential in processing advanced high-temperature resin systems.

Vacuum pumps should, as a rule, be located approximately 0.00045 m^3/s (1 ft^3/min) for every 0.23 m^2 (2.5 ft^2) of bag area. The water-seal-type pumps have proved to be the most reliable, as they are not damaged by volatile by-products of the curing reaction, unlike oil-flooded pumps. Refrigerated or water-cooled condensing systems are installed on many systems to trap toxic volatiles produced by some resin systems.

Control Systems

Recent developments in computer technology have greatly increased the ability to monitor and control cure cycles. Progress in the past

5 years rivals that in the preceding 20 years. The objectives have been to increase the reproducibility of repetitive cycles, improve throughput by optimizing process parameters, and reduce labor costs associated with manual control. Currently, microprocessor-based autoclave controls are being replaced in favor of interactive computer systems. A computer-controlled autoclave console is shown in Fig. 4.

The cure cycle is controlled by feedback from thermocouples, transducers, and advanced dielectric and ultrasonic sensors. The software is growing in complexity, with features varying between suppliers of computer systems. When selecting a computer-controlled system, process engineers and users should test it to determine its appropriateness for the application. Computer-controlled systems should always be installed with backup conventional microprocessor controllers to serve as a fail-safe means in case of computer failure.

When a computer system is being considered for existing equipment, the user should realize that a new computer does not solve existing mechanical and electrical problems in an autoclave system. Existing system problems should be remedied before computerization.

Loading Systems

Loading systems are probably the most perplexing aspect of production confronting the autoclave user. Because of the circular configuration of the vessel and the relatively small size of the components, carts must be designed to distribute parts horizontally and vertically in the autoclave. Other considerations are:

- The loaded parts should be accessible to enable repair of bag leaks
- All vacuum source and vacuum sensor lines must be connected to the part when loaded on the cart, and vacuum must be maintained as the parts are introduced into the autoclave
- The cart must be easily rolled or transported into the autoclave

Designing a cart system to meet all the desired criteria is a challenge. The essential loading-system requirements must be determined, because including one feature will often preclude the inclusion of another. Typically, carts of various configurations are used, based on mold size and shape.

Modified Autoclaves for Specialized Applications

The development of thermoplastic matrix composites has necessitated the design of specialized equipment for use within autoclaves to form thermoplastic materials. This need has been met with retrofittable processing fixtures that enable the processing of high-temperature materials in existing low-temperature autoclaves. Often, electrically heated molds are

Fig. 5 Vacuum pressure chamber. Courtesy of Naval Air Rework Facility, San Diego

used. Autoclaves should be manufactured to incorporate design features that make transitions to new materials relatively simple.

The polyimides and related resin matrix systems have created a new set of problems in the design of autoclave systems. Primarily, the vacuum control systems have had to be reengineered to accept the large amount of volatiles (resin solvent) released during cure. Also, additional computer interaction is required to process these materials properly.

Metal matrix composites have been in service for many years and require systems operating from 480 to 700 °C (900 to 1300 °F) at 6900 to 69 000 kPa (1000 to 10 000 psi) to diffusion-bond the matrix into a homogeneous mass. Because of the high-pressure requirement, autoclave systems for curing these composites are usually smaller than others.

Modified autoclaves are currently being used for processing carbon-carbon composites, both for impregnation and carbonization. These complex autoclaves operate at temperatures up to 815 °C (1500 °F).

Phenolic parts have been processed as exit cones and throats for rocket motors in hydroclaves for many years. A hydroclave is similar to an autoclave, except that it is water flooded and pressurized. Hydroclaves are typically run at 6900 kPa (1000 psi) and 150 to 175 °C (300 to 350 °F).

The vacuum pressure chamber (Fig. 5) is a dedicated flexible process center designed for field repair and remanufacturing of composite parts. Future advances will necessitate more dedicated equipment that is directed to specific applications, yet functionally similar to the autoclave. These machines will take the form of bladder presses, versatile process centers, and cavity presses, among others.

Safety and Installation

It is usually standard to have redundant safety features on any autoclave because of the potential seriousness of any malfunction. Overpressure conditions are usually prevented by three different methods: a separate overpressure sensor and shutdown control, rupture disks designed to rupture at pressures above the operating pressure of the autoclave, and pop-off safety valves with the same function. A standard production autoclave has all three. Vessels are usually proof tested to high margins of safety, but the danger posed by the possibility of a burst vessel cannot be overemphasized.

Overtemperature protection is not as critical an issue from an injury standpoint, but overtemperature conditions could damage the interior systems of the autoclave; therefore, overtemperature controls are usually provided. The vessel shell, because of the internal insulation, may not be rated to the maximum operating temperature of the autoclave. As previously noted, the external surface of the autoclave should not exceed 60 °C (140 °F), except at penetrations.

Installation is also an important consideration. Even small autoclaves are not designed to just plug in and run. It is important to consider site preparation before or during design of the system. Important considerations include foundation, cooling water supply, electrical supply, gas (if used for heating), and pressurization medium supply and exhaust arrangements. Even though installation may be provided by the manufacturer (and this is quite valuable), careful site preparation by the user can prevent many future problems.

Computerized Autoclave Cure Control

Richard J. Hinrichs, Applied Polymer Technology, Inc.

AUTOCLAVE CURING OF COMPOSITES attempts to induce specific chemical reactions within polymers that result in predictable engineering properties. Accordingly, control of the curing process should be based on chemical engineering and fluid dynamic principles. This article describes a computerized approach to the simultaneous control of materials reaction behavior and consolidation dynamics, using an autoclave as the reaction vessel.

The primary objectives of computer control of the autoclave process are to improve cured-product quality and reduce fabrication costs by providing:

- Process optimization
- Reduced process inconsistencies and product rejections
- Accurate, real-time quality assurance with rapid error detection and correction
- Verification of process reaction behavior kinetics
- Nondestructive verification of cured properties
- Accurate, permanent process documentation
- Flexibility in adapting to new or modified processes

Fabrication, as used in the present article, includes all operations responsible for the transformation of the prepreg raw material into the final, cured structure. Aircraft structures are designed on the basis of allowables, which in turn are based on the testing of coupons cured by a specified procedure, which includes time, temperature, heat-up rate, pressure, and vacuum. In production, the parts are cured on tools that vary in geometry, materials, and mass and therefore can result in different thermal cycles. This variance can affect the quality and performance of the laminate. This results in a distorted engineering allowables data base.

Originally, the approach was to have an operator control part temperature by adjusting the autoclave air temperature based on part thermocouple readings. Pressure and vacuum were monitored by gage readings and adjusted manually. Because the operator had to perform multiple complex functions and use intuition to control part temperatures, mistakes were made

and engineering evaluation of the affected parts was necessary to ensure that the cure was adequate. Also, the process took longer than necessary because of the reluctance of operators to risk increasing the air temperature beyond the cure temperature. In addition, quality control of the autoclave process was based on reading the charts of the various recorders.

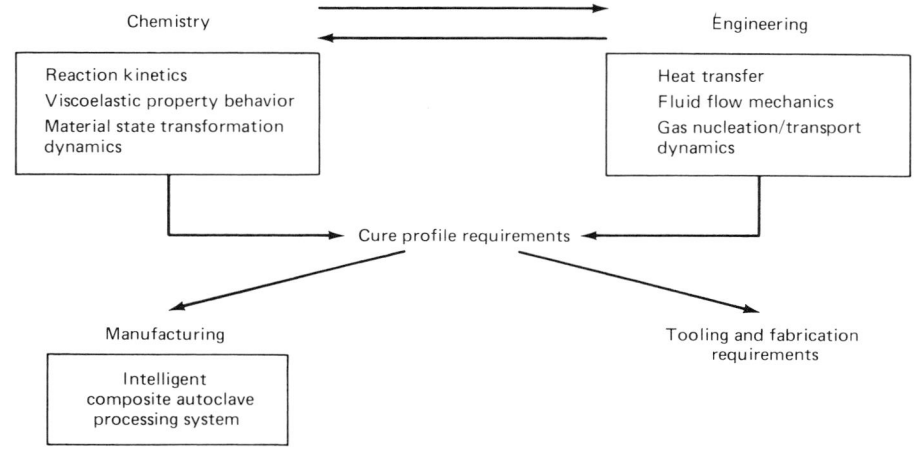

Fig. 1 Composite cure management synergisms

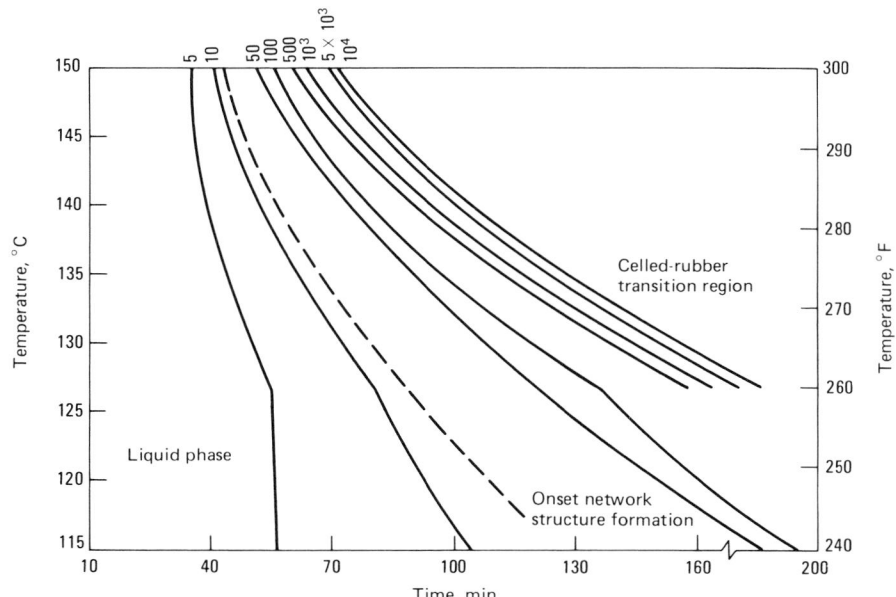

Fig. 2 Effect of temperature and time on material viscosity during cure

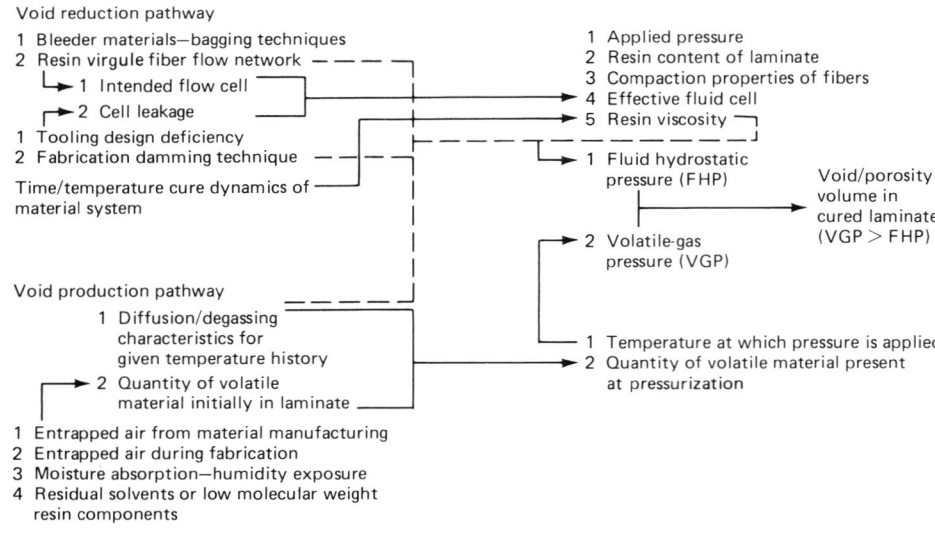

Void reduction pathway
1 Bleeder materials—bagging techniques
2 Resin virgule fiber flow network
    1 Intended flow cell
    2 Cell leakage
1 Tooling design deficiency
2 Fabrication damming technique
Time/temperature cure dynamics of material system

1 Applied pressure
2 Resin content of laminate
3 Compaction properties of fibers
4 Effective fluid cell
5 Resin viscosity
1 Fluid hydrostatic pressure (FHP)
2 Volatile-gas pressure (VGP)
1 Temperature at which pressure is applied
2 Quantity of volatile material present at pressurization

Void/porosity volume in cured laminate (VGP > FHP)

Void production pathway
1 Diffusion/degassing characteristics for given temperature history
2 Quantity of volatile material initially in laminate
1 Entrapped air from material manufacturing
2 Entrapped air during fabrication
3 Moisture absorption—humidity exposure
4 Residual solvents or low molecular weight resin components

Fig. 3 Model for process behavior

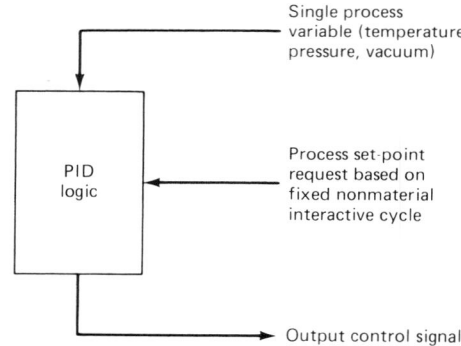

Single process variable (temperature, pressure, vacuum)

PID logic

Process set-point request based on fixed nonmaterial interactive cycle

Output control signal

Fig. 4 Standard control-loop process diagram (vessel, air chamber, and vacuum manifolds). P, proportional band; I, integral element (or reset); D, derivative element (or rate of change)

Occasionally the charts were lost and inspection was not possible.

The computer, however, is programmed to cure the parts according to an algorithm based on the specifications. If any parts cannot be cured properly because of the mass of the tool, the computer identifies them from its calculation and records the discrepancy in its memory. Thus, parts are cured properly and efficiently, and a proper quality control record is maintained.

Control Dynamics

Composite cure control begins with a basic understanding of the reaction process relationships (Fig. 1). The primary regulating parameter for these reactions is temperature. More accurately, the thermal history of a material determines its kinetics, viscoelasticity, morphology, phase precipitation, cross-link density, glass transition properties, and polymer network structure (Ref 1-6). These are the factors that affect both the process behavior (flow consolidation consistency) and the structural-engineering properties evolved from the cured prepreg.

In addition, each material has different thermal reaction sensitivity characteristics, based on its composition, which define the most tolerant and effective thermal-cure profile. It is the total thermal-cure history that governs the consistency, uniformity, and quality of the chemical curing process, as illustrated in Fig. 2, which shows that a shift of 6 to 9 °C (10 to 15 °F) in temperature can affect the viscosity state a material exhibits during cure. This information, when combined with kinetic data, defines the thermal-control requirements of the process (heat rates, allowable thermal gradients across the entire load, and optimum pressurization time).

The chemistry data are also used to define heat transfer, fluid, and gas transport requirements for the specific part configuration being built. This process is shown in Fig. 3, where the interactions among process cycle, tooling, fabrication techniques, viscosity, and volatile-gas formation are illustrated in the consolida-

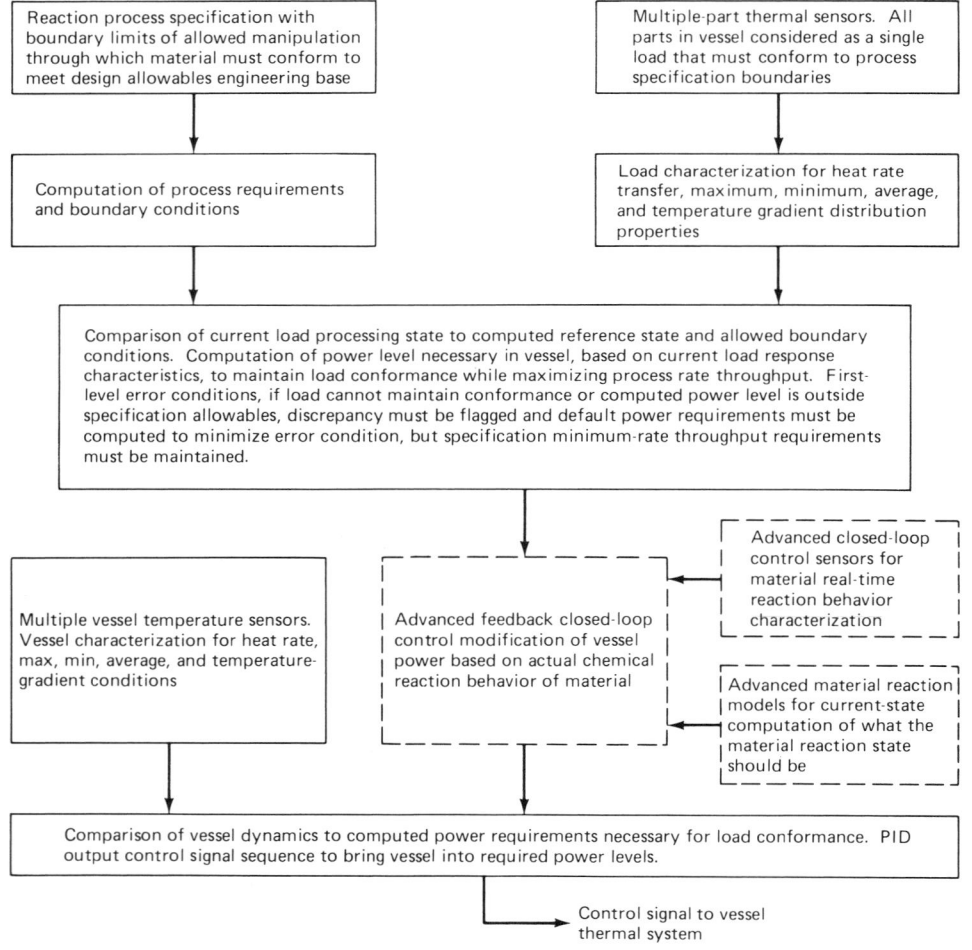

Reaction process specification with boundary limits of allowed manipulation through which material must conform to meet design allowables engineering base

Multiple-part thermal sensors. All parts in vessel considered as a single load that must conform to process specification boundaries

Computation of process requirements and boundary conditions

Load characterization for heat rate transfer, maximum, minimum, average, and temperature gradient distribution properties

Comparison of current load processing state to computed reference state and allowed boundary conditions. Computation of power level necessary in vessel, based on current load response characteristics, to maintain load conformance while maximizing process rate throughput. First-level error conditions, if load cannot maintain conformance or computed power level is outside specification allowables, discrepancy must be flagged and default power requirements must be computed to minimize error condition, but specification minimum-rate throughput requirements must be maintained.

Advanced closed-loop control sensors for material real-time reaction behavior characterization

Multiple vessel temperature sensors. Vessel characterization for heat rate, max, min, average, and temperature-gradient conditions

Advanced feedback closed-loop control modification of vessel power based on actual chemical reaction behavior of material

Advanced material reaction models for current-state computation of what the material reaction state should be

Comparison of vessel dynamics to computed power requirements necessary for load conformance. PID output control signal sequence to bring vessel into required power levels.

Control signal to vessel thermal system

Fig. 5 Temperature-control segment of the composite control-loop diagram

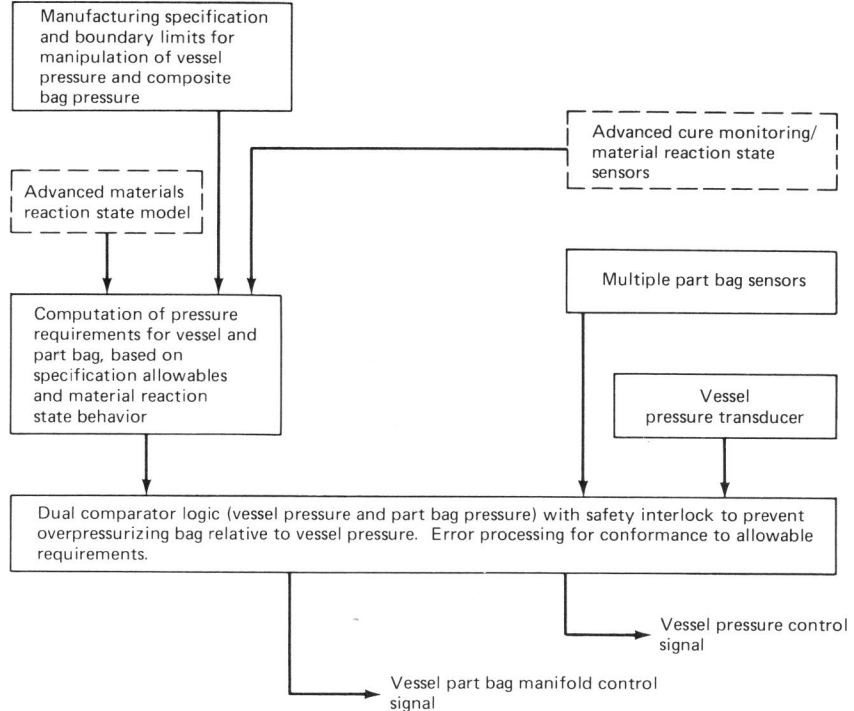

Fig. 6 Pressure/vacuum control segment of the composite control-loop process program

tion process. Resin viscosity and kinetics interact with gas diffusion processes and fluid hydrostatic pressure through the time-temperature cure cycle. The laminate quality depends on how well these interactions are controlled to favor void reduction rather than generation.

The autoclave vessel dynamics must also be recognized. The vessel represents a chamber in which air or inert gas is circulated. The vessel can heat or cool this air stream and increase the chamber pressure. Initially, the main reason for the use of autoclaves in composite manufacturing was to achieve uniform pressurization around complex shapes. Although autoclaves accomplish their function well, autoclave air streams moving across part tooling surfaces create nonuniform air flow disturbance patterns inside the vessel. As a result, heat transfer dynamics to the part become variable and, when coupled with tooling mass variations, part thermal conformance to process requirements is not met. The challenge is to develop logic that recognizes the thermal-variation effects and can control interactively to maintain part conformance.

Standard Control Logic. The dynamics of a control system for chemically reacting materials must contain provisions for precise thermal control and also be sufficiently versatile for use on a variety of materials. Precision control of the time-temperature profile of a material during cure appears to be an insignificant problem. Thus, most commercial control systems (motorized cam or electronic thermal programmers) measure a single thermocouple and output power accordingly (Fig. 4). The process variable (usually a single sensor in a fixed location in the vessel) is used to measure the condition inside the vessel. This signal feeds into a processor logic loop, which compares the single-point information to that requested by the set point. The set point is usually computed using a fixed time based cycle and does not analyze material load-performance characteristics. The resulting output signal is sent back to the vessel-regulating circuits.

This type of control logic is appropriate for manipulative control of the vessel functions and can accurately regulate vessel process variables (temperature, pressure, vacuum, and the like) at the sensor location. Unfortunately, neither parts nor the ambient air inside a manufacturing vessel are uniform or consistent from one load to another. Thus, this level of logic ignores the engineering problems encountered in real manufacturing situations (such as mixed part loads, variations in tooling thermal conductivity, air stream heat deflection anomalies, and so on) that result in nonuniform part thermal profiles that are in turn responsible for variations in part process behavior.

Advanced Control Logic. The logic diagrams and concepts presented in the following sections were developed and patented by Applied Polymer Technology, Inc. and are embodied in the company's commercial control system, CAPS. Similar systems are available from other sources.

The most critical concept is that it is the entire part that must be processed in confor-

mance to the manufacturing specifications; further, all the parts that make up the autoclave load must meet the process requirements. Design allowables and certification procedures do not permit the operator to select the thermocouples or sections of a part to be controlled. Nonetheless, investigation of composite anomalies has shown that selective placement of thermocouples, which made it appear that parts were properly controlled, was contributing to part failures. Use of the proper reaction control logic system, however, ensures that the material is chemically transformed into a primary structural component that conforms to engineering requirements.

Temperature Control Logic. The approach to temperature control logic for chemically reactive material (Fig. 5) uses multiple thermocouple sensors to provide sufficient information for accurate computation of the critical characteristics of the load. All components in the autoclave load are considered as a single unique entity whose thermal characteristics must be kept in conformance to the process requirements. The critical parameters that the system must compute are heat rate transfer and part temperature gradients. The heat rate transfer characteristics of the load determine the efficiency of the translation of the power level of the vessel into part temperatures. It is important to realize that this parameter is not constant during a run or repeatable between autoclave runs. It depends on tooling-mass differentials, autoclave loading geometry (air stream deflection and turbulence), vessel temperature, and vessel pressure. Ideally, the load should have uniform and equal heat transfer characteristics, which would allow for uniform temperature distribution and reaction behavior in the curing materials. Conversely, variation in thermal transfer characteristics of the load results in temperature gradients in the parts and variant reaction behavior. The general chemical rule of thumb is that the reaction rates double for every 10 °C (18 °F). The computer calculates the maximum, average, and minimum load heat transfer characteristics dynamically during the process. This information is coupled with vessel temperatures that will optimize uniformity and process rates.

Part temperature gradients result from variable heat transfer characteristics and must be controlled to achieve uniform properties in the composite structure. The composite laminate undergoes the transitions from liquid to gel to glass (vitrification) as a function of the effect of temperature on kinetic behavior. Thus, allowing significant temperature gradients to exist at these transition functions can cause laminate cure stresses, nonuniform laminate consolidation, and trapped volatiles, all of which degrade performance properties.

The initial objectives are to collect sufficient information to characterize the thermal dynamics of the load in real time and to control the temperature gradient distributions in the part (and, ultimately, the consistency of reaction

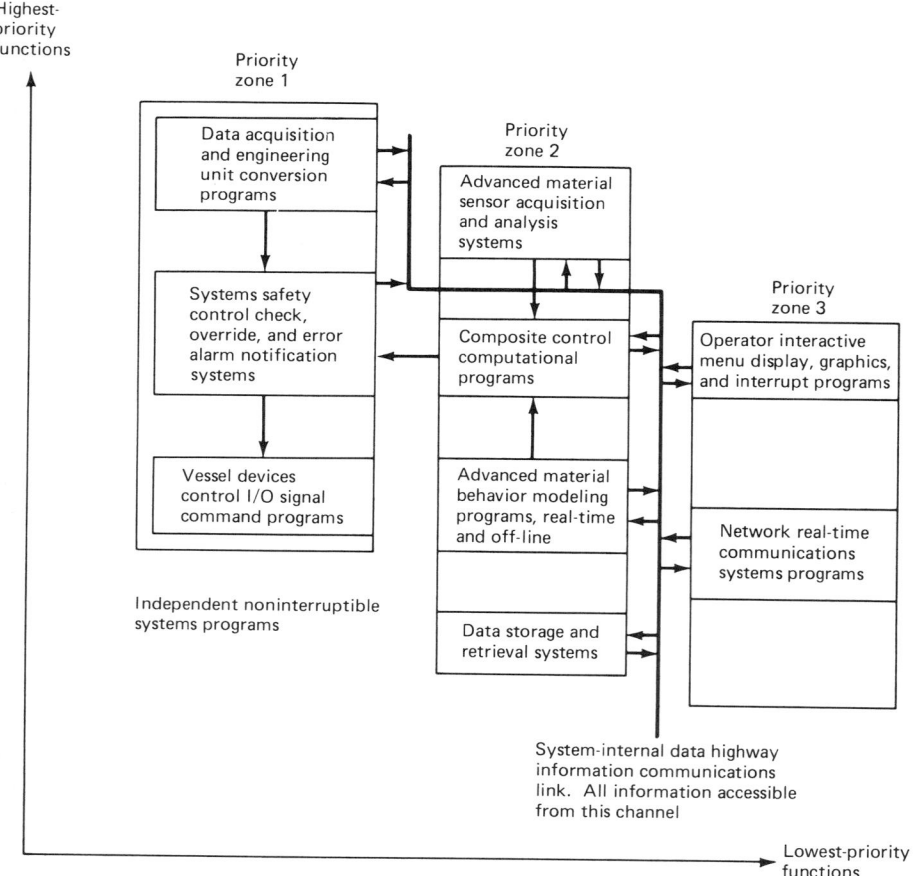

Fig. 7 Simultaneous multitask operating system

cure behavior). The heat transfer characteristics are used to compute effective power levels for the vessel, as illustrated in Fig. 5.

Unfortunately, computing power levels and load characteristics alone does not yield sufficient control over the process. The allowable process boundary conditions must also be computed so that power calculations can be modified to maintain and document run performance. There are four main entities of process temperature control requirements.

The first is the ideal cycle for the load to follow, consisting of desired heat rates, temperature end-point definitions, and amount of time at any hold or dwell point, that is, the thermal cycle that the controller will attempt to achieve. The major requirement relates to the second heat rate boundary limits placed on the process by the specifications, that is, the maximum and minimum heat rate limits that the load can tolerate and still maintain design-allowable performance properties, or boundaries within which the computer can modify load heat rates to maintain uniformity. These boundaries also define quality assurance error detection flags for documentation purposes. The third and most critical entity is the allowable temperature differential gradient for the material at various process temperatures, which ultimately defines

the maximum rate compatible with maintaining the allowed part temperature gradient. The forth entity, which adds a safety boundary check and limitation on the power levels, is the maximum vessel temperature to part temperature overshoot parameters, that is, the final limitation on the rate at which a load can be processed. This results from the fact that it is the differential temperature that drives heat rate transfer characteristics. Thus, the maximum rate condition occurs at the maximum allowed vessel overshoot condition.

This level of logic control and error detection yields optimum vessel power level requirements at any given point in the process cycle, based on predescribed engineering information (process profile specifications) derived from the material reaction chemistry. Advanced feedback control logic (interactive modification of the computed power and pressurization requirements based on material cure monitor sensors) detects material reaction state conditions during cure and uses this information in the control logic algorithms (dashed-line boxes in Fig. 5). This logic, for maximum effectiveness, requires both a sensor system capable of detecting material state changes (viscoelastic properties, degree of cure, resin content/fiber volume, and so forth) and a reaction kinetics model to pre-

dict the correct material state expected at this process point. These concepts enhance quality assurance by documenting actual reaction behavior of the material during cure.

The final logic step in the temperature control loop uses multiple-sensor characterization to define vessel performance. Use of PID regulating logic allows manipulation of the power circuits so as to bring the vessel into conformance with the computed power level requirements.

Pressure and Vacuum Control Logic. Figure 6 presents the logic flow diagram for control of pressure and vacuum. Vacuum, actually a subset of pressure, refers to the level of pressure in the part bag envelope relative to ambient atmospheric pressure. It is usually expressed as negative inches of mercury when pressure inside the bag is less than atmospheric, and as psig when pressure is positive. This is a poor way to specify a control requirement because it depends on local barometric-pressure conditions; therefore, requirements would be better defined in absolute pressure (psia) to avoid significant process variations and confusion.

Pressure control follows a logic pathway that is similar to but less sophisticated than that of temperature. This approach is unique in that it allows the system to control pressure in the vessel and the part bag envelope and to control the relative vessel-to-bag pressure differential as well. This technique (referred to as internal bag pressurization) allows for diffusion control of volatiles, which can reduce the potential for laminate porosity. The process specification defines the vessel and bag pressurization requirements, based on cycle times and part temperature. These requirements are modified by the advanced sensor feedback programs, based on sensing material reaction state, as described in the section "Temperature Control Logic" in this article. A dual logic comparator is used to ensure that pressurization is within safe limits and that the internal bag pressure never exceeds the vessel pressure (otherwise, bag rupture would occur). The resultant control output signals are sent to the vessel and bag manifold regulating circuits, respectively.

Systems Programming Dynamics

A definite hierarchy of programming functions and priorities must exist for efficient use of computer resources and for safe operations. As shown in Fig. 7, these functions are organized into three priority zones in a multitask real-time operating system for computerized control of autoclave curing of composites.

Priority zone 1 contains programs necessary to conduct data acquisition, safety control checks, alarm-error detection, and output control signal functions. These programs are structured to be stand-alone, isolated programs. They have the highest priority in system function and a minimum of interaction with external devices, displays, or human operators. Thus, a problem or delay in other program functions will not affect the ability of the system to ensure control safety and error-alarm notification.

Priority zone 2 contains programs with advanced logic and interface to support peripherals (cure-monitoring sensors, disk mass storage unit, modeling programs, and so forth). They are structured to supply current control requirements to zone 1 programs and minimize any functions with slow input/output response characteristics.

Zone 3 contains the functions of least significance to accuracy of control (that is, those computations which, if requiring a few seconds longer to complete, would not impair control decisions). Also, slow human interactive responses are coordinated into the control system in this zone, along with functions such as graphics/text display updates, operator interrupt requests, and telecommunication to host or other control units for data transfer. These functions are required in composite manufacturing, but their execution does not require high-priority time in the system.

Utility Support-Program Concepts

The programs, logic, and priority functions described in this article constitute the basic autoclave computer control system. However, there are some functions that are not required for control of the vessel but that greatly improve the usability of the system. Referred to as utility support programs, these functions include:

- Quality assurance graphics report and data transfer
- Automatic calibration and computation of linearization coefficients
- Electronic self-diagnosis
- Network line and data exchange to other controllers and heat systems
- Vessel maintenance diagnosis
- Disk backup and file transfer manipulation routines
- Engineering material reaction behavior models for specification development

REFERENCES

1. *Processing Science Program*, AFWAL F33615-80-C-5121, General Dynamics, Convair Division
2. *Processing Science Program*, AFWAL F33615-80-C-5059, General Dynamics, Convair Division
3. *Computer-Aided Curing of Composites*, AFWAL F33615-83-C-5088, McDonnell-Douglas Corporation
4. R.J. Hinrichs and J. Thuen, Environmental Effects on the Control of Advanced Composite Materials Processing, Paper presented at the National SAMPE Symposium, 1979
5. R.J. Hinrichs and J. Thuen, "Structural-Adhesives Rheological Behavior Response to Process-Environmental Variations," Paper presented at the National SAMPE Symposium, 1980
6. R.J. Hinrichs, B. Wade, and S. Hoier, "Modeling Rheological Data for Use in Engineering Process Development," Paper presented at the National SAMPE Symposium, 1984

Curing Epoxy Resins

David J. Boll and John C. Weidner, Hercules Aerospace Company

MANUFACTURING A REINFORCED THERMOSET COMPOSITE involves many steps, of which the cure is the most critical, because it is the point at which the composite lay-up is transformed from a soft, multilayered mixture of resin and fiber to a hard structural component. In essence, the raw materials become the semifinished or finished part.

Epoxy resins used for composite structures can be classified by cure temperature:

- Room temperature cures are mainly used in commercial products to either obtain better properties than would be possible with polyesters or repair structures requiring patches
- 120 °C (250 °F) cures are used for subsonic aircraft and helicopter parts and large commercial parts
- 175 °C (350 °F) cures are mainly used for high-performance aircraft

The room temperature curing systems have a relatively short working life (2-h maximum) and require weighing and mixing on the shop floor. In order to both improve the reproducibility of the parts and permit the use of systems that are difficult to mix, the fabricators of high-performance parts have adapted preimpregnated materials that typically cure at either 120 °C (250 °F) or 175 °C (350 °F). The prepregs are prepared by material suppliers who have considerable expertise in epoxy resin chemistry, specialized equipment, and extensive quality control facilities.

The epoxy curing process requires simultaneous application of heat and pressure, which is normally accomplished through press cure, autoclave cure, or oven cure. In the case of a press or autoclave cure, temperature and pressure are supplied by the equipment. In an oven cure, heat is supplied by the oven, but pressure must come from another source, typically, a vacuum bag, mandrel expansion, shrink film, expandable rubber, or mechanical clamps.

For epoxies, cure refers to the chemical reaction of an epoxide group in the resin component of the matrix with either an amine or anhydride hardener. The reaction is exothermic in nature, and results in increased molecular weight of the matrix, which is a function of the extent of the curing reaction. The measured viscosity is related to the size of the molecules present. As the resin and hardener molecules react, larger molecules are formed, resulting in a significant increase in viscosity at moderate temperatures.

The epoxy resin changes from a liquid to a rubbery state, and eventually to a solid at the gel point. Beyond the gel point, there is little resin movement or flow. Typically, the cure is 40 to 60% completed, and the final shape and thickness of the composite is fixed. Thus, the

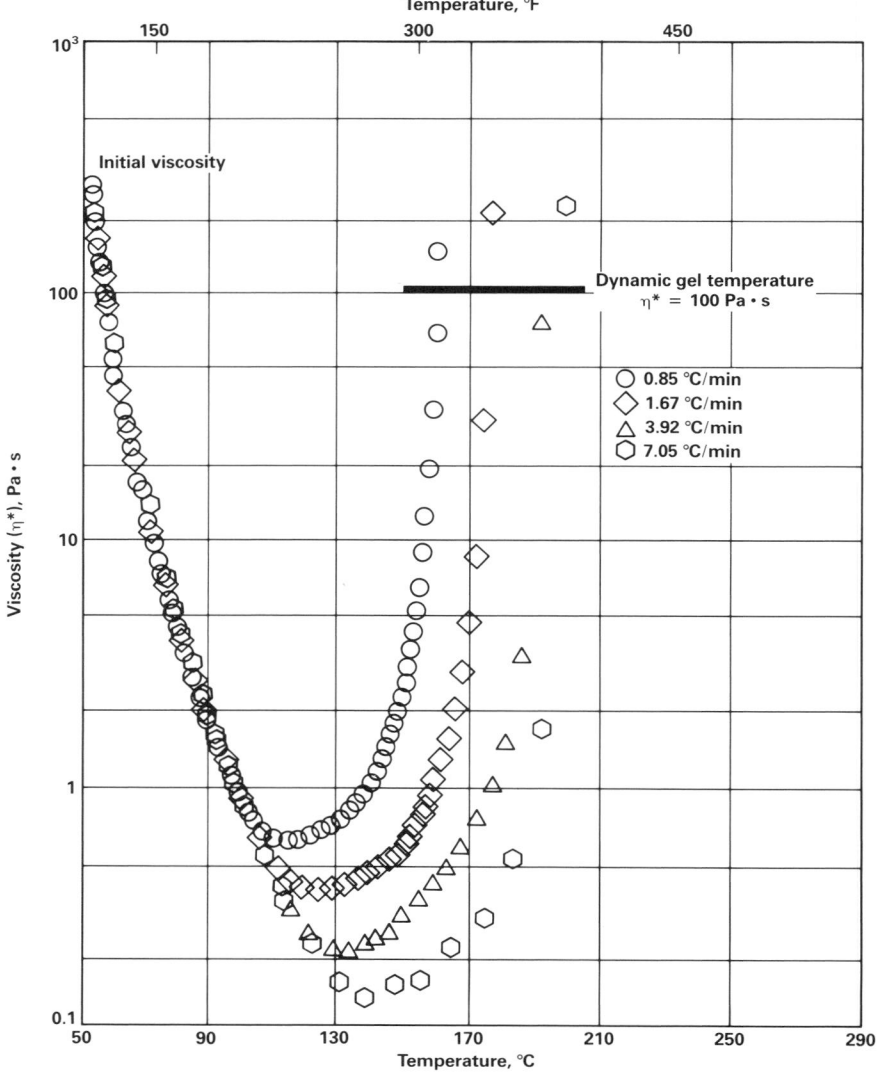

Fig. 1 Dynamic mechanical spectroscopy scans at different heating rates

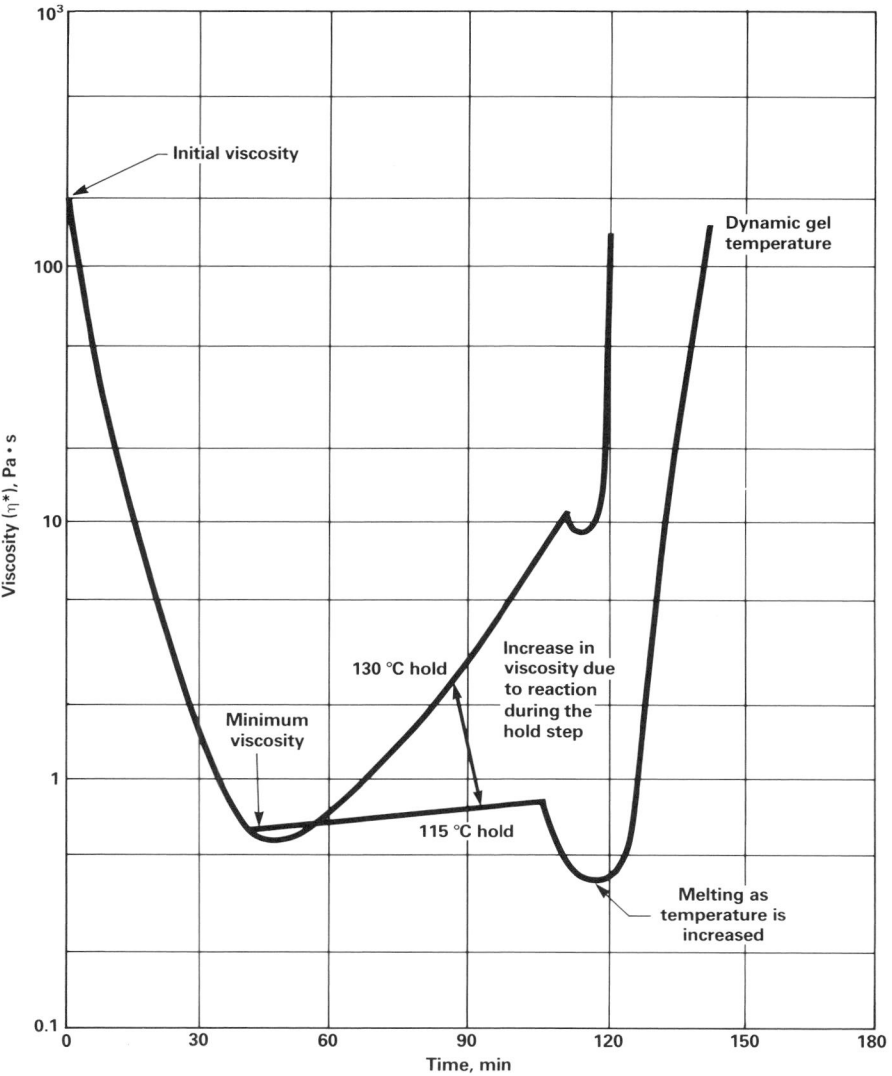

Fig. 2 Dynamic mechanical spectroscopy scans of cure-hold-cure cycles. (Heating rates and hold times are identical.)

- *Resin removal:* Typically, there is 10 to 20% excess resin in the starting materials. By taking advantage of the viscosity change during the cure cycle, this excess resin can be removed. To reduce costs, many parts are now molded using a no-bleed process.

These five items should emphasize the criticality of understanding the rheological properties of the resin. These properties can now be studied in the laboratory and then applied to the specific curing process selected for composite manufacture. This technology has only been available since 1978, and many older curing processes are being revised.

There are various laboratory instruments on the market to measure the rheological properties of fluids. Most instruments either measure point viscosity (one temperature setting, any number of readings), or they only measure very viscous materials (thermoplastic melt flow rheology). In general, these instruments are not practical for modeling the flow rheology of curing thermoset resin systems. Dynamic mechanical rheometers capable of continuously measuring the rheological behavior of the resin are necessary. One such instrument, the Rheometrics Inc. RDS 7700 dynamic mechanical spectrometer (DMS), can measure resin properties on linear heating ramps, or cure cycles having hold steps.

The DMS has become the preferred method for determining thermosetting matrix resin rheological behavior because its data can be used in determining the change in viscosity as a function of temperature at different heating rates (including temperature hold steps) and strain rates. In fact, before a part is worked on, the behavior of the matrix for any cure cycle can be determined and the optimum process can be selected on the basis of realistic data.

The information obtained from a DMS scan can contribute to the process engineer's ability to determine the point in the cure cycle at which to release vacuum and apply pressure or the point at which to have a hold step for resin removal to the part. In addition to providing these critical times in the cure, the DMS curve can provide the valuable dynamic gel temperature, DGT (Ref 1), which is the point in the cure cycle at which the resin gels, or changes from a fluid to a rubbery state. The following discussion will clarify the location of these points on the DMS curve and explain the way to use that data. The first point on the typical DMS curve (Fig. 1) is the initial viscosity at some predetermined temperature. This indicates the onset of matrix melting. The initial viscosity, shown on a very steep portion of the curve, is very sensitive to resin advancement (reaction while mixing the resin system at elevated temperature) and moisture absorption by the resin. As the resin on the prepreg sits at room temperature, the resin advances, because there is a slight reaction. Thus the viscosity will start increasing. If the material is left in a humid environment, moisture will be absorbed by the

reaction beyond the gel point is irreversible. However, at the early stages of the cure, increases in viscosity produced by heating at a fixed temperature are often lost as the part temperature is increased by melting of the higher molecular weight species formed in the early stages of the cure sequence.

During the cure cycle, the process must meet these other requirements to produce a structurally acceptable part:

- *Void elimination:* Many prepregs contain a considerable amount of air when they are supplied to the fabricator. Additional air is introduced as the plies are stacked on each other. For example, graphite-epoxy tapes laid up by machine may contain up to 12 vol%. Also, graphite-epoxy prepregs contain up to 0.5 wt% water, which can boil during cure. Thus, the curing process must prevent unacceptable porosity levels from

these sources by compressing the entrapped air and preventing boiling of the water.
- *Solvent removal:* Some resins are dissolved in solvents before being applied to the fiber. If allowed to remain in the composite during the later portion of the cure, the solvent will create voids and affect final performance of the part. Fortunately, the solvents used for epoxy resins have a low boiling point and evaporate during lay-up.
- *Fiber wet-out:* To achieve optimum performance of the composite, it is critical that the resin coat every fiber. A low viscosity during cure aids fiber wet-out and reduces the porosity in a cured part.
- *Consolidation:* In the case of prepreg materials, the ply thickness of the starting material is usually 20% greater than the desired final thickness. The combination of low resin viscosity and pressure decreases the thickness.

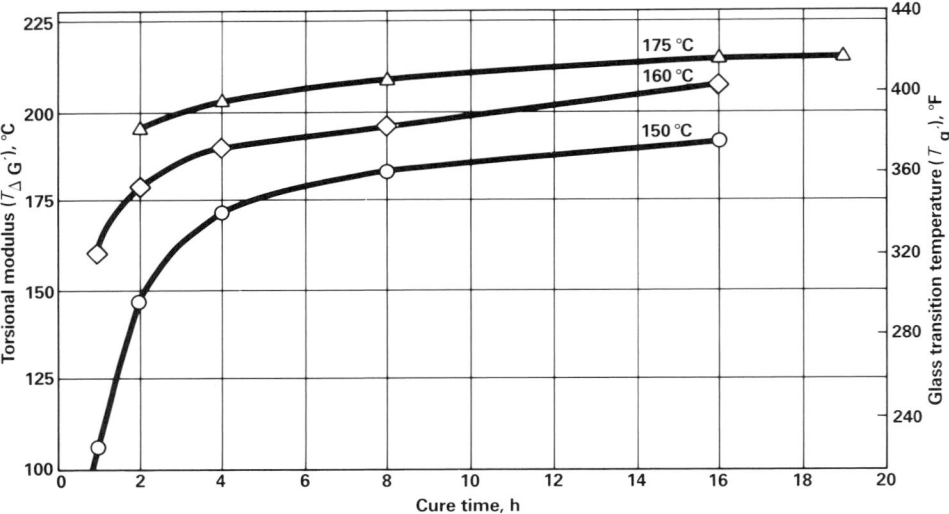

Fig. 3 Cure time/temperature versus T_g by torsional modulus, $T_{\Delta G'}$

matrix and, depending on the resin system, the water will act as a plasticizer and cause a corresponding drop in the resin viscosity. As Fig. 1 shows, the initial portion of the viscosity curve is independent of heating rate. As the resin is heated, two opposing phenomena occur. The temperature increase produces more reaction. However, the predominant effect is melting of the partially cured resin. Thus, the viscosity falls to a minimum, which is a function of the heat-up rate (Ref 2). This data is useful in flow control of the laid-up part. The temperature of minimum viscosity is important in determining when to release vacuum and apply pressure, and it can be useful in establishing when to have hold steps. As stated, the temperature of minimum viscosity is heating-rate dependent; the DMS curve establishes this temperature very effectively. If a hold step is desired, the DMS can be used to monitor the degree of advancement obtained during the hold (Fig. 2). Changing the hold temperature will cause a shift in the slope of the viscosity line during the hold. The DMS will also determine whether there will be appreciable flow after the hold has ended and the next heating ramp has started. The combination of heating

rates, and the use or nonuse of a hold, will affect the DGT.

The DGT, at which the resin gels given a specified heating rate and hold step, becomes very critical when a complex part is being made. If the part is wound or laid-up around an expanding metal mandrel, or has metal inclusions, the temperature rise should be halted at a temperature below the DGT and the resin allowed to gel before continuing the cure to prevent damage to the metal. Because the DGT can be obtained by computer modeling (Ref 1), an infinite number of DMS scans do not have to be run. Also, the DGT can be used to adjust the tools for thermal expansion. After the resin gels, pressure is no longer required because the shape of the part has been fixed. Thus, process failures beyond the gel temperature do not produce a scrap part. In fact, many parts are only slightly cured at a point slightly beyond the gel point under pressure. The chemical reaction is completed by heating the part in an oven.

The DMS also can be used to monitor the completion of cure in a composite system. For a particular cure time/temperature, the G′ loss modulus line can be monitored using torsion

loading of a rectangular specimen to establish the cure advancement. The intersection of the initial straight-line portion and the straight-line portion of the transition from glass to rubber of the G′ line is used to describe the transition temperature of the G′ line, or $T_{\Delta G'}$. As the cure advances at a specified hold temperature, curves can be made to describe the degree of cure at any point in time (Fig. 3). These curves can be used to determine the minimum time required at a specified temperature to achieve full cure, as evidenced by no increase in the $T_{\Delta G'}$. In many cases, adequate structural properties can be obtained without a full cure.

It should be evident that the DMS is a very powerful tool when used in either monitoring or developing cure cycles, using parallel plates for resin or solid rectangular laminate specimens in the torsional cure mode (Ref 3-5). Its data permits tailoring the time, temperature, and pressure parameters of the curing process to repeatedly achieve parts that meet the structural requirements of the engineering design.

REFERENCES

1. R.E. Hoffman and D.J. Boll, The Use of Dynamic Gel Temperatures to Develop Cure Cycles, in *The 29th National SAMPE Symposium*, Vol 29, Society for the Advancement of Material and Process Engineering, 1984, p 1411
2. D.J. Boll, B. Motiee, and W.D. Bascom, Characterizing Matrix Resin Flow in Carbon Fiber Prepreg, *ASTM J. Compos. Technol. Res.*, Vol 8 (No. 2), 1986
3. J.L. Kardos, J.P. Dudukovic, E.L. McKague, and N.W. Lehman, in *Composite Materials: Quality Assurance and Processing*, STP 797, C.E. Browning, Ed., American Society for Testing and Materials, 1983, p 96
4. A.C. Loos and G.S. Springer, Curing of Epoxy Matrix Composites, *J. Compos. Mater.*, Vol 17, 1983, p 134
5. M.G. Maximovich and R.M. Galeos, Rheological Charcterization of Advanced Composite Prepreg Materials, in *Proceedings of the 28th National SAMPE Symposium*, Society for the Advancement of Material and Process Engineering, Vol 28, 1983, p 568

Curing BMI Resins

Mark M. Konarski, U.S. Polymeric, HITCO Materials Division

BISMALEIMIDE (BMI) RESINS are required for graphite composites used at temperatures above 125 °C (260 °F) in the earth environment, because such temperatures severely reduce the properties of epoxy resins. Although autoclave curing of BMI resins is more complex than that of epoxies, it is nevertheless possible to use them for high-quality parts in production quantities, if the user has a good, basic understanding of the effects of curing process parameters on part quality.

Any autoclave cure is expected to produce a high-quality low-void composite within a specified fiber volume range (Ref 1). Lot-to-lot prepreg variations that are manifested as mechanical property variances may be traced to slight differences in raw materials, resin or prepreg processing, fiber and fiber finish, or absorbed material. Because these differences are often very slight and fall within the range of normal quality assurance (QA) tolerances, they must be accepted. Often, the variations that are noticed in the laminate mechanical properties are more a function of resin and void contents in the cured panel (which stem from physical differences of the resin/prepreg lots) rather than actual chemical variations in the resin. When prepreg manufacturers' autoclave cure schedules for BMIs are used unchanged, they can produce high-quality low-void laminates for one lot of material and poor-quality parts for the next. Thus, design allowables must be lowered in order to gain the necessary confidence level for the variation in mechanical properties that results from these laminate quality differences (Ref 2). Resin starvation can result in low compression and shear properties, because the amount of resin becomes insufficient to transfer the load from fiber to fiber. Also, the shear strength drops dramatically as the void content increases past 2% (Ref 1). Often, a composite of higher quality could be produced with only slight changes in the lay-up, bagging, or autoclave cure schedule (Ref 3, 4).

Steady-state temperature, heating and cooling rates, vacuum, and pressure are the parameters that determine the success or failure of any autoclave cure. An oven postcure off the mold is required to complete the chemical reaction. Autoclave cure temperature is usually dictated by resin chemistry. BMI resins usually require

a minimum cure temperature of greater than 150 °C (300 °F) in the autoclave to develop significant strength. Although many resins will gel at significantly lower temperatures, the chemical reaction involved in chain extension, with its associated strengthening, does not occur. Inadequately cured BMI resins are often so weak they cannot be removed from the lay-up tool without damage. Furthermore, after unrestrained postcure, the laminates usually have significant warpage due to softening of the semicured matrix.

The heating rate is most often a function of autoclave capability or thickness of the laminate or tool. Heat capacity of the tool, whether aluminum, steel, or composite, affects the maximum heating rate; however, as long as the heating rate is slow enough to ensure that no part of the laminate is significantly cooler or hotter than the rest, results will usually be satisfactory. Likewise, it is important to control BMI cooldown rate to prevent thermal shock and possible microcracking, which can affect composite mechanical properties. However, the most critical parameters in a successful autoclave cure are the application of vacuum and pressure at the optimum times.

Property Optimization

BMI resins have a processing window in which pressure and vacuum are important factors in creating a part with low void content. Depending on the particular system, there may be too much flow (resin movement) or too little, if processing parameters are not correct. If pressure and vacuum are applied while the flow is too great, areas of porosity due to resin starvation can result, especially in prepregs with low resin content. However, a resin that does not flow well may require holds in the heatup cycle at periods of low viscosity to allow the resin to coalesce into a homogeneous mass between and throughout the plies.

This process is particularly relevant to thick laminates, where heat transfer also is a consideration. For a thick part, there must be a period during which the entire laminate reaches a state of flow, allowing any entrapped air or volatiles to diffuse from the composite. If the top and/or bottom of the laminate increases in viscosity before the center, porosity and poor quality will normally result. Evolution of moisture vapor during gel may also cause high void content.

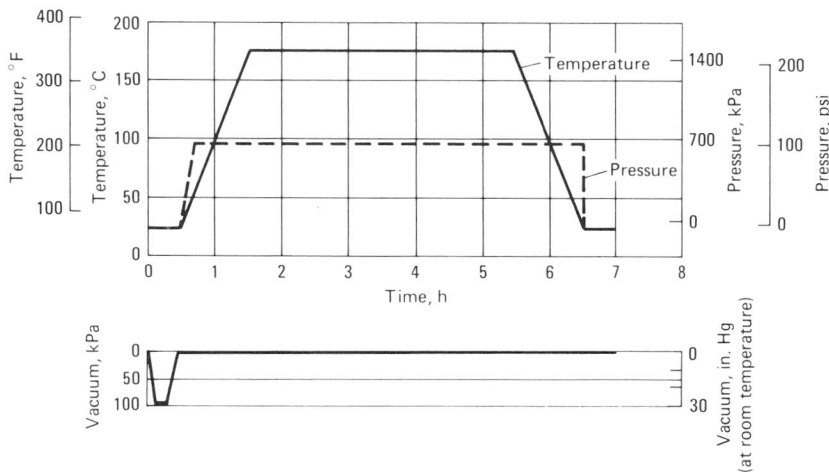

Fig. 1 Standard V-378A cure cycle. Event sequence: apply full vacuum; apply 690 kPa (100 psi) pressure, vent vacuum after 97 kPa (14 psi); heat to 175 °C (350 °F) at ≤3 °C (5 °F)/min; hold at 175 °C (350 °F) for 4 h; cool at ≥3 °C (5 °F)/min to room temperature, then release pressure; postcure at 245 °C (475 °F) for 4 h

Microcracking is another consideration in BMI systems. Heatup and cooldown rates, as well as cure temperature, can determine whether or not a BMI composite will exhibit microcracking. (Ply orientation also affects microcracking.) Generally, most BMI resins do not fully cure until they have received a moderate temperature postcure of 205 to 260 °C (400 to 500 °F). Postcure is usually performed in an oven rather than the autoclave to conserve autoclave time and because an autoclave cure at these temperatures exceeds the capabilities of common bagging and tooling materials. Because BMIs are in their weakest and most fragile state during and after the initial autoclave cure, care must be taken to avoid placing the freshly cured laminate under undue thermal or mechanical stress. It is interesting to note that the optimum cure temperature varies from resin to resin, depending on the chemistry involved in the system. BMI resins with very active catalysts or reactants will cure best (to their toughest state) with a longer cure at lower temperatures. Resins with mild cure catalysts and reactants generally will cure best at the highest temperatures (those closest to the postcure) achievable within the limiting factors of the autoclave, lay-up materials, and tooling.

Cure Cycles

The recommended cure cycles of BMI prepregs are often not the technically optimum cycles to use. Most were developed to duplicate as closely as possible cure cycles developed for epoxy resins, because the demand from the composite industry was for BMI prepregs that process "just like epoxies." For the most part, these compromise cure cycles will work very well, because the resins involved are "extremely processable"; curing under a fairly wide range of conditions will still produce good-quality laminates. Two of the five resin systems described in this section have an alternate cure cycle in addition to the normal "epoxylike" cycle.

The first resin system, U.S. Polymeric V-378A, was one of the first commercially available BMI prepregs. The V-378A resin cures easily with a typical 175-°C (350-°F) cycle (Fig. 1). The alternate cure (Fig. 2) incorporates a hold period at a temperature of low viscosity to enhance removal of volatiles and diffusion of the matrix within the laminate.

The second system is U.S. Polymeric V-388D, which was formulated to exhibit good toughness and excellent processability. The cure cycle is fairly simple and "epoxylike" (Fig. 3).

The third system, U.S. Polymeric V-388E, is similar to V-388D, but has extended high-temperature performance. A simple 160-°C (320-°F) cure cycle (Fig. 4) and a more involved alternate (Fig. 5) are shown.

The fourth system, U.S. Polymeric V-398G, is a very high-performance BMI, exhibiting outstanding toughness and high-temperature ca-

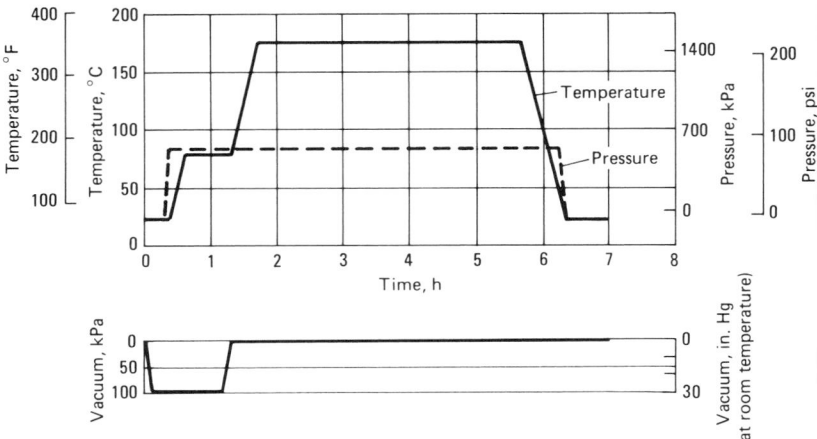

Fig. 2 Alternate cure cycle for V-378A. Event sequence: apply full vacuum and 586 kPa (85 psi) pressure; heat to 80 °C (180 °F) at ≤3 °C (5 °F)/min; hold at 80 °C (180 °F) for 35 min, then vent; heat to 175 °C (350 °F) at ≤3 °C (5 °F)/min; hold at 175 °C (350 °F) for 4 h; cool at ≤3 °C (5 °F)/min to room temperature, then release pressure; postcure at 245 °C (475 °F) for 4 h

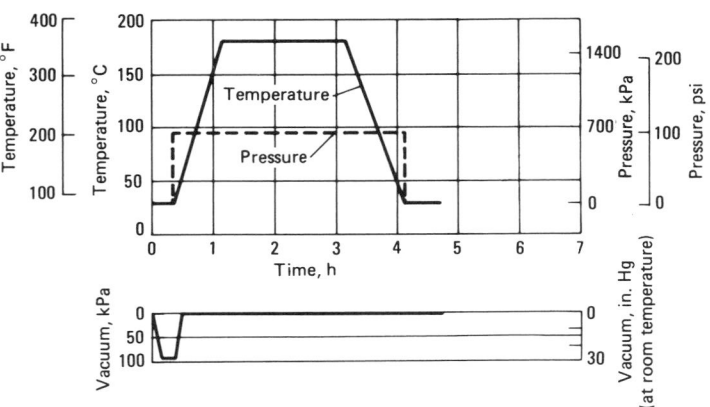

Fig. 3 Cure cycle for V-388D. Event sequence: apply full vacuum; apply 690 kPa (100 psi) pressure and vent after 97 kPa (14 psi); heat to 175 °C (350 °F) at ≤3 °C (5 °F)/min; hold at 175 °C (350 °F) for 2 h; cool at ≤3 °C (5 °F)/min to room temperature, then release pressure; postcure at 205 °C (400 °F) for 2 h, plus 225 °C (440 °F) for 2 h

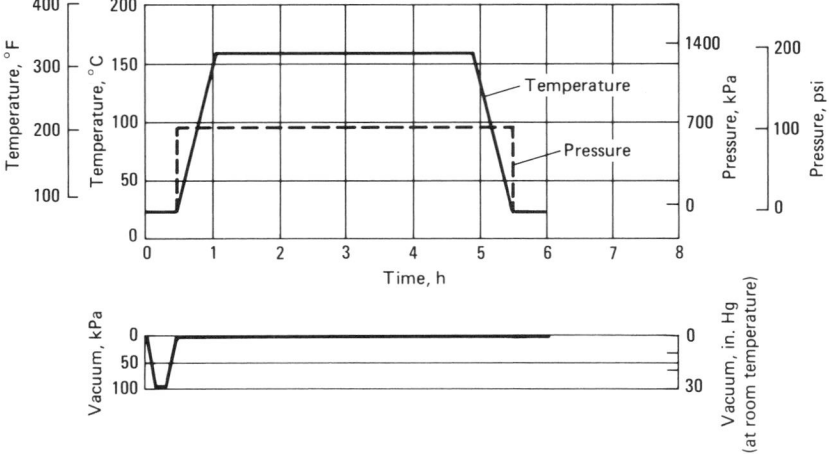

Fig. 4 Standard cure cycle for V-388E. Event sequence: apply full vacuum; apply 690 kPa (100 psi) pressure, vent after 97 kPa (14 psi); heat to 160 °C (320 °F) at ≤3 °C (5 °F)/min; hold at 160 °C (320 °F) for 4 h; cool at ≤3 °C (5 °F)/min to room temperature, then release pressure; postcure at 205 °C (400 °F) for 2 h, plus 245 °C (475 °F) for 4 h

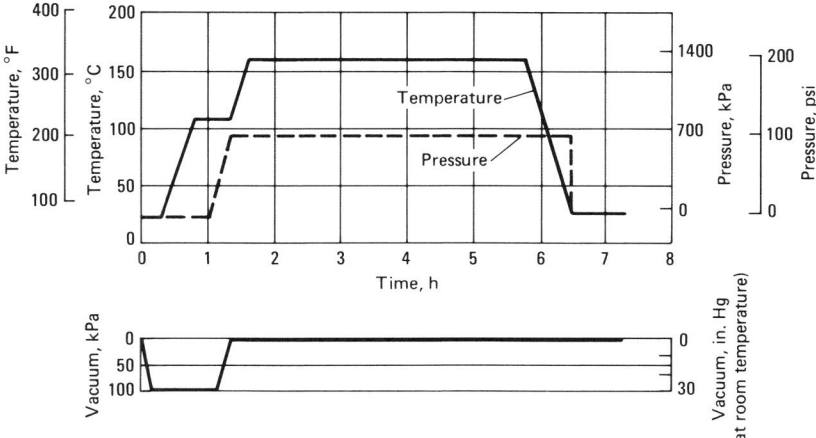

Fig. 5 Alternate cure cycle for V-388E. Event sequence: apply full vacuum; heat to 110 °C (230 °F) at ≤3 °C (5 °F)/min; hold at 110 °C (230 °F) for 30 min; apply 690 kPa (100 psi) pressure, vent after 97 kPa (14 psi); heat to 160 °C (320 °F) at ≤3 °C (5 °F)/min; hold at 160 °C (320 °F) for 4 h; cool at ≤3 °C (5 °F)/min to room temperature, then release pressure; postcure at 205 °C (400 °F) for 2 h, plus 245 °C (475 °F) for 4 h

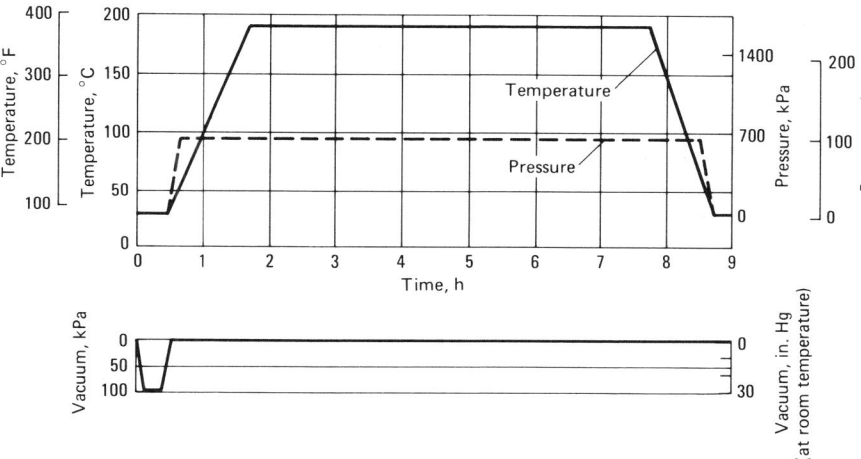

Fig. 6 Cure cycle for V-398G. Event sequence: apply full vacuum; apply 690 ° kPa (100 psi) pressure, vent after 97 kPa (14 psi); heat to 190 °C (375 °F) at ≤3 °C (5 °F)/min; hold at 190 °C (375 °F) for 6 h; cool at ≤3 °C (5 °F)/min to room temperature, then release pressure; postcure at 225 °C (440 °F) for 4 h, plus 245 °C (475 °F) for 4 h

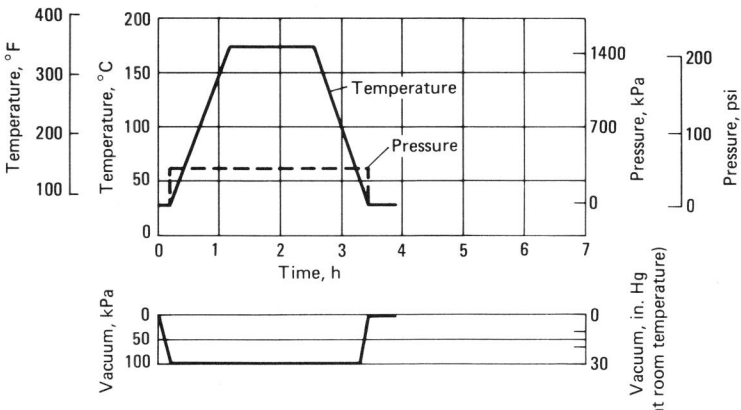

Fig. 7 Cure cycle for CPI-2278. Event sequence: apply full vacuum and 345 kPa (50 psi) pressure; heat to 175 °C (350 °F) at 2-3 °C (3-5 °F)/min; hold at 175 °C (350 °F) for 90 min; cool to room temperature; postcure at 230 °C (450 °F) for 4 h

pability. The cure cycle (Fig. 6) is a simple 190-°C (375-°F) cure.

Also presented is Ferro's CPI-2278. The cycle for this high-temperature BMI (Fig. 7) is a simple 175-°C (350-°F) cure.

Viscosities of BMI Resins During Processing. The rheometrics dynamic spectrometer (RDS) is a useful tool for understanding the behavior of matrices during autoclave cure. The viscosity of a resin sample is measured while subjected to the autoclave cycle. Melting of the polymer is evidenced by a decrease in viscosity. An abrupt increase is evidence that the resin is solidifying. Typical viscosity/temperature curves are presented in Fig. 8 to 11 for V-378A, V-388D, V-388E, and V-398G, respectively. Figure 12 shows a plot of viscosity versus time for CPI-2278 when held for 3 h at 80 °C (180 °F). These figures may be used as guidelines when developing an alternative cure cycle for these materials.

A particular lot of V-388E was found to produce high void contents in laminates cured by the standard cure cycle (Fig. 4), particularly for thick laminates. The problem was attributed to insufficient resin flow during cure. Figure 10 indicates that although the viscosity minimum was low (0.097 Pa · s, or 0.002 lbf · s/ft^2, at 140 °C, or 285 °F), there was a particularly steep rise after that point. It was reasoned that the center of the laminate had not arrived at a high flow state before the surfaces gelled. Air and other volatiles were therefore trapped within the laminates, and voids were produced. The cure cycle was subsequently modified to include a hold at 110 °C (230 °F) (minimum viscosity as determined by RDS). Vacuum is prolonged throughout this hold. Autoclave pressure is not applied until after the intermediate hold, allowing the entrapped air and volatiles to diffuse from the laminate. This cure cycle has been used with good success on both thick and thin panels.

Cure Time/Temperature Relationships for BMIs. The ultimate objective in the autoclave curing of BMI resins is to allow the resin to react to such an extent that reasonable structural integrity is achieved and retained at least up to the postcure temperature. BMIs are affected by cure time and temperature relationships to their glass transition temperature (T_g) in much the same fashion as epoxies. A longer cure at a lower temperature gives the same T_g as a shorter one at a higher temperature. This is true only to a certain degree, however, and is particularly dependent on the individual resin formulation. In general, most BMI systems have multiple cure reaction schemes, or processes, that are involved in the cure. Each of these processes has a minimum activation energy below which the reaction will not occur. These energies manifest themselves in the form of minimum cure temperatures. Once the resin has gone beyond the threshold temperature, additional energy in the form of increased temperature will accelerate the reaction. Conversely, lower temperature might halt the reac-

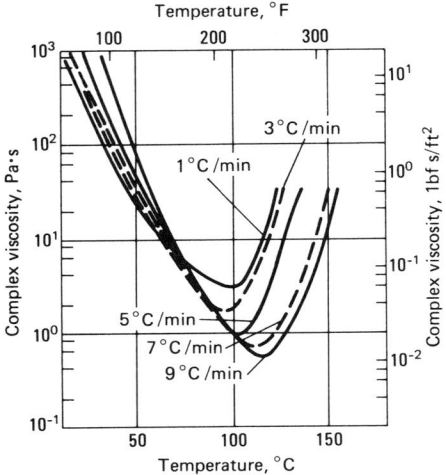

Fig. 8 V-378A resin viscosity curves as a function of RDS heatup rate

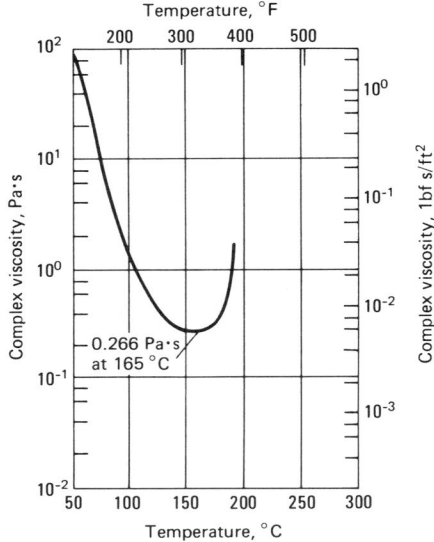

Fig. 9 V-388D resin viscosity profile. 5 °C (9 °F)/min RDS heatup rate

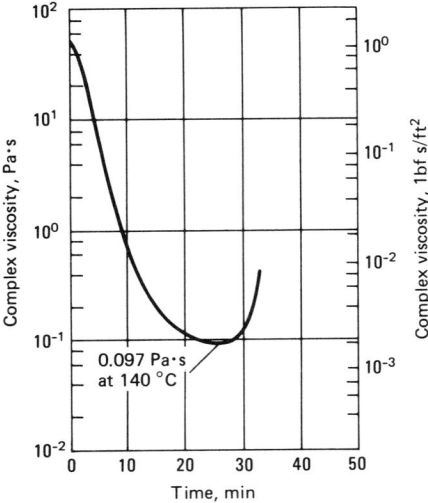

Fig. 10 V-388E resin viscosity profile. 5 °C (9 °F)/min RDS heatup rate

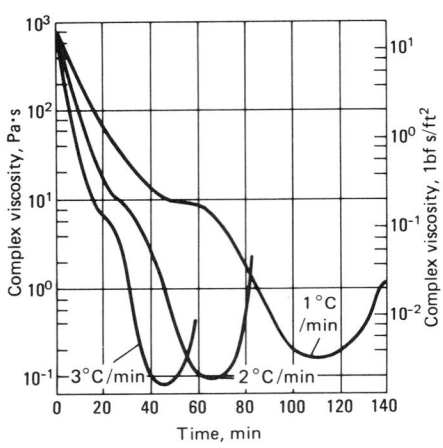

Fig. 11 V-398G resin viscosity curve as a function of RDS heatup rate

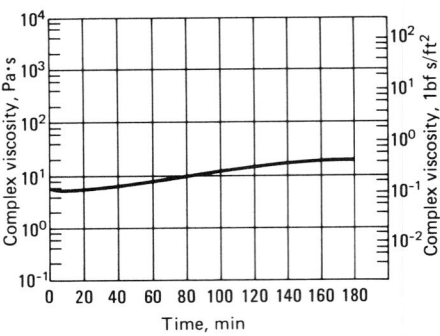

Fig. 12 Isothermal viscosity curve of CPI-2278 at 80 °C (180 °F)

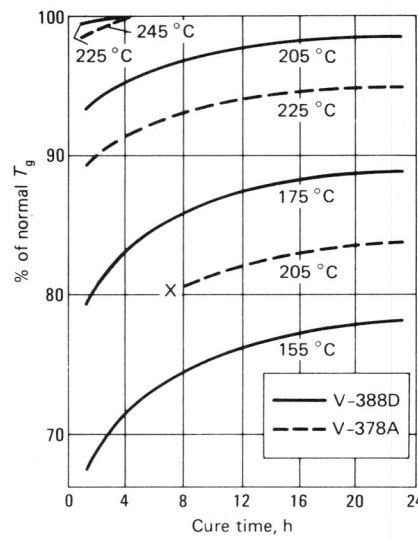

Fig. 13 Cure time/temperature relationship for V-388D and V-378A. X, samples below this time/temperature profile were too fragile to test.

tion completely. Some resins contain very reactive BMI coreactants and will develop reasonable mechanical properties after an extended cure at 150 to 205 °C (300 to 400 °F). However, most systems will require a postcure of 205 to 260 °C (400 to 500 °F). Figure 13 is a comparison of the cure time/temperature relationship for two BMI resins.

V-388D resin is a system with relatively reactive ingredients. This resin develops reasonable T_g's and mechanical properties at cure temperatures as low as 155 °C (310 °F). On the other hand, the cure chemistry of V-378A resin dictates that T_g and mechanical properties will not develop until this resin has been cured above a minimum of 225 °C (440 °F). At all temperatures there is a leveling off in the time/T_g relationship at about 16 to 20 h. Cures prolonged beyond that length of time do not significantly improve T_g's.

Troubleshooting

Although many things can and do go wrong with autoclave cures of BMI composites, the causes can usually be traced to processing anomalies. In any bad cure, the investigator should first consider equipment malfunctions or lay-up errors before blaming the chemistry and cure cycle. Most current BMI resins are fairly forgiving and are formulated to provide ease and latitude in processing.

Resin Starvation. Many BMI prepregs are supplied as net-resin or no-bleed systems because of the good flow and fiber wettability of these materials. This allows more cost-effective processing, since bleeder materials are not required (Ref 5). However, it could also be a source for cure difficulties if excessive bleeding were to occur. Excessive bleeding can produce areas or entire laminates with high void contents, poor mechanical properties, and/or poor visual appearance. Figure 14 shows a resin-starved panel. Resin starvation can be avoided by efficient damming before cure. If the application or tooling concept precludes this "fix," the cure cycle can be changed to minimize the problem. Pressurization at a higher temperature is the easiest answer. Examination of the appropriate viscosity profile to determine a suitable pressurization point past the minimum viscosity would be in order.

Excessive void formation is by far the most common problem encountered in autoclave curing of BMI composites. Although industry specifications generally allow no more than 2% voids, a good cure for BMI composites usually results in void contents substantially less than 1%. A cross section of a panel with excessive porosity is shown in Fig. 15. Voids appear as scratches or pits in the polished surface. There are several possible causes of

Fig. 14 Resin-starved laminate

Fig. 15 Cross section of a BMI composite with high void content. 10X

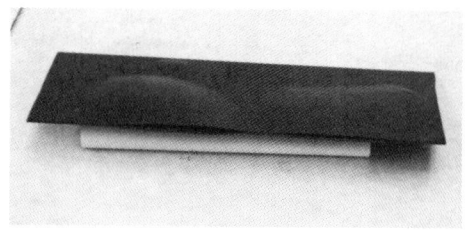

Fig. 16 Large composite blisters

excessive void formation, such as excessive bleeding and resin starvation, the solutions to which appear above.

Voids may also result from volatiles remaining in the resin at the point of gellation. Although BMIs do not evolve volatiles during cure, small amounts of volatile solvent may remain from the processing of the resins or prepregs. A slower heatup rate or an intermediate hold at a time of low viscosity permits these volatiles to diffuse out of the composite. The vacuum should be held throughout this period with minimal pressure until after this hold. In very tacky prepregs, air bubbles can become entrapped during lay-up and act as void-causing volatiles. Treatment is the same, as air will diffuse from the composite in the same manner as residual solvent. The vacuum should be vented immediately after this period of flow because it will tend to magnify any voids as the resin approaches gellation.

Voids can also be caused by waiting too long to apply pressure during the cure cycle. As the resin advances, viscosity increases and flow decreases. If the resin has advanced too far, the application of pressure will not give the compaction necessary to produce a good part. The remedy is simply to apply pressure at an earlier point during the cure. These problems can be resolved by evaluating the viscosity curves and adjusting the autoclave cycle accordingly.

Blistering in a BMI laminate indicates that several problems occurred. First, substantial amounts of volatiles were left in the composites. Second, the resin went through a significant softening (usually during the postcure), indicating that the T_g after cure was not high enough. The first problem can be remedied as outlined above, while the second indicates that either a higher temperature or a longer cure is required. The particular resin lot may be suspect if problems are encountered when it has previously been cured successfully. Another cause of blisters is moisture absorption due to storing improperly packaged material in the freezer; all prepreg packages must be sealed to prevent the ingress of moisture. Figure 16 shows a severely blistered panel. The blisters, which appeared during the postcure, were found to be the result of an abnormally high volatile content in that particular batch of resin.

REFERENCES

1. J.B. Johnson and C.N. Owston, The Effect of Cure Cycle on the Mechanical Properties of Carbon-Fibre/Epoxide Resin, *Composites*, Vol 4 (No. 3), May 1973, p 111
2. G.D. Peddie and D.W. Mayberry, "Quartz Radome and Dielectric Monitoring Development," Report GDC-AKM69-003, General Dynamics Convair, Dec 1969
3. M.J. Yokota, In-Process Controlled Curing of Resin Matrix Composites, in *Proceedings of Symposium on Diversity-Technology Explosion*, Vol 22, Society for the Advancement of Material and Process Engineering, April 1977
4. J. Chottiner and Z.N. Sanjana, Monitoring Cure of Large Autoclave Molded Parts by Dielectric Analysis, in *Proceedings of Symposium on Material and Process Applications: Land, Sea, Air, Space*, Vol 24, Society for the Advancement of Material and Process Engineering, April 1981
5. S.W. Street, V-378A, A New Modified Bismaleimide Matrix Resin for High Modulus Graphite, in *Polyimides*, 1st ed., Vol 1, Plenum Press, 1984, p 77

Curing Polyimide Resins

Wesley C. Mace, W.C. Mace & Associates

POLYIMIDE RESIN MATERIALS were first used in structural composite components in the early 1960s. The impetus behind their development was the supersonic transport program. All of these early resins were the condensation reaction-type, needed a one-stage cure, and gave off copious amounts of water. The solvent used to put the powdered resin in solution was N-methyl-2-pyrrolidone, which was extremely difficult to remove, and the catalyst was an arsenic compound that is considered unacceptable in the current environment under the Occupational Safety and Health Administration. These resins were easily cured, requiring only 340 kPa (50 psi) and full vacuum in an autoclave, and 177 °C (350 °F) curing temperatures. They reached their high-temperature capability (280 °C, or 535 °F) by means of an extended postcure at 288 °C (550 °F). Unfortunately, they produced laminates that were extremely high in void content (approximately 9 vol%) and consequently susceptible to moisture pick-up. Also, if the solvent was not removed before the resin solidified, the resin would precipitate to yield laminates with very low strength. Aircraft companies have since used this tendency toward high void content to produce the sound suppression panels around jet engines. Condensation reaction-type polyimides continue to be used in various acoustic applications.

Two newer polyimide resins that should interest the structural composites engineer have been developed by the National Aeronautics and Space Administration: PMR-15, developed at the NASA Lewis Research Center; and LARC-160, developed at the NASA Langley Research Center. Both polyimides are cured in two stages. The first consists of an imidization reaction, which is a condensation reaction evolving moisture, and the second consists of polymerization reaction that is similar to a common epoxy-type reaction, but is carried out at a higher temperature. These polyimide types are the focus of this article.

Preparation for Autoclave Cure. A debulking process must be performed before autoclave curing for most hardware applications and, in certain cases, for press molding cures. Debulking is required at least every three plies for radical contours. Because a certain amount of heat and pressure are applied during the process, the final cure is affected. In the debulking process, the lay-up is exposed to 82 °C (180 °F) air under a vacuum of 75 kPa (22 in. Hg). This requires that the lay-up be vacuum-bagged on the tool and that the entire assembly then be moved into an oven unless the tool contains its own integral heating mechanism. Obviously, performing this process a multitude of times will affect the lower temperature end of the cure cycle. A majority of the alcohol solvent and some of the absorbed water will be removed this way.

Final preparation for autoclave cure begins after the debulking process is completed. The low-temperature bagging material is removed, and the part is prepared for cure by applying the separator, bleeder, barrier, breather materials, and a Kapton film bag that is sealed with high-temperature bag sealant. Vacuum bagging with Kapton film requires more care and some bagging technique changes than those methods used with epoxy resin systems. Compared to polyimide resins, epoxy resins represent low-temperature curing systems. Nylon films having 300 to 500% elongation are used when processing epoxy-impregnated materials, but these materials cannot be used at the higher temperatures needed to cure polyimides. Because Kapton film has little or no elongation, changes in vacuum-bagging techniques are required. The basic approach to using Kapton film for vacuum bags is to employ excessive material (folds) to offset the lack of elongation. It is imperative that folds are employed where bag bridging may occur, such as at concave or convex surfaces. The Kapton film should never be stressed in any manner. Unlike nylon, Kapton is not tough and damages easily. As PMR-15 resin has a tendency to flow copiously, the bleeder system must be minimized to prevent laminate over-bleeding. For most lay-ups to 50 plies, one layer of 181, 1581, or 7581 fiberglass cloth is sufficient to ensure the release of volatiles, and yet to retain the proper amount of resin in the laminate, that is, 40 to 45 vol%.

If the process has progressed successfully and the vacuum bag passes a leak check, then the cure is considered complete, as shown in Fig. 1. It is very important to note that contrary to some specifications, full vacuum is maintained throughout the cure. No theoretical justification for this exists; rather, it is a result of experience gained during fabrication of thousands of pounds of PMR-15.

Physical Phenomena During Cure. In addition to water that evolves during the first stage of the cure, both PMR-15 and LARC-160 contain a considerable amount of volatile products, particularly alcohol, that were present in the preimpregnated cloth or tape and that must be physically removed, or structures will contain an unacceptable amount of voids. The mechanism used to remove these volatiles is a vacuum source. Although the recommended amount of vacuum to apply varies from source to source, the range is from 15 kPa (5 in. Hg) to the maximum attainable at the altitude involved. The author has had excellent results using the maximum attainable from the onset of cure until cool-down starts.

Another important physical phenomenon is "puffing," or expansion of the laminate during imidization, caused by the evolution of water. This poses no particular problem on thin, relatively flat parts, but it becomes acutely necessary in the case of thick-walled cylinders laid up on male tools to restrain the laminate during "puffing" in order to avoid the formation of large wrinkles.

Alternate Curing Methods. To reduce expensive autoclave time, cure cycles that

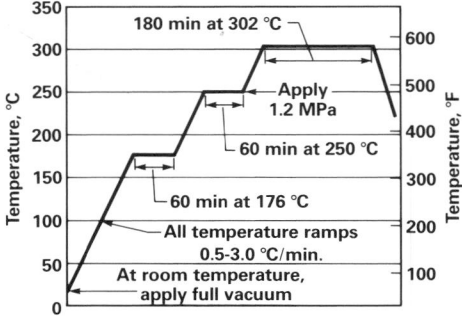

Note: Autoclave cure cycle must be coordinated with part temperature.

Fig. 1 PMR-15 autoclave cure cycle

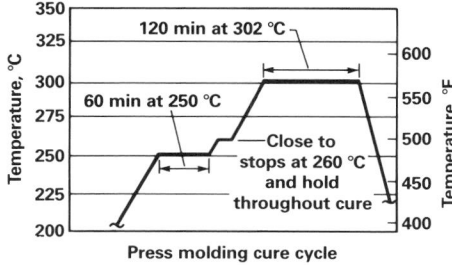

Fig. 2 PMR-15 press-molding cure cycle

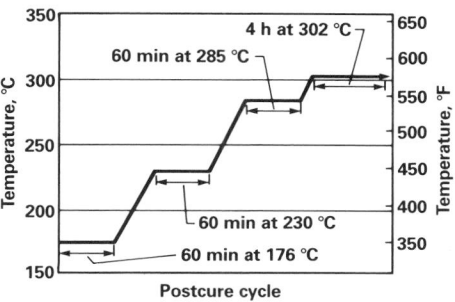

Fig. 3 Postcure cycle

use oven imidization before final cure in the autoclave have been developed. Parts are imidized under vacuum at temperatures from 177 °C (350 °F) to 230 °C (450 °F) and subsequently polymerized in the autoclave or press. This technique is valuable for those parts with tools that can be heated rapidly in the autoclave beyond the imidization temperature and is especially valuable for press-molded parts that do not require the matched die to be cooled below 232 °C (450 °F).

Press Molding. Where geometries permit and the tooling cost can be justified, compression molding of polyimide components is suggested. The relatively simple cure cycle is shown in Fig. 2. Because oven imidization removes the necessity to cool down the mold in the press, cure time is shortened to 3 or 4 h. It should be noted that the number of tool designers and mold fabricators familiar with 316 °C (600 °F) molding temperatures is limited, a fact which must be considered when choosing a tooling source.

Because polyimide resin has low viscosity during the polymerization cycle, care must be taken not to remove excessive amounts of resin. Proper mold design is the best approach, but if for one reason or another, the mold allows too much resin to escape, viscosity can be controlled by adjusting the point at which pressure is applied. One of the problems associated with this tactic is that the variability of resin formulations makes it necessary to determine the pressure application point for each batch.

Postcure. Freestanding postcures in an air-circulating oven are used to increase the glass transition temperature of the completed component. One of the more common postcure cycles used by jet engine component fabricators is illustrated in Fig. 3. Theoretical kinetic calculations would indicate that this postcure time is excessive. However, because oven postcures are relatively inexpensive, many engineering departments hesitate to reduce the time at temperature regardless of theoretical calculations.

It should be noted that the peak temperature allowed is 307 °C (585 °F). Experience has shown that PMR-15 is not a true 316 °C (600 °F) resin system and manufacturing losses increase if postcures are attempted above the peak temperature.

SECTION 9

Machining, Assembly, and Assembly Forms

Chairman: L.E. Roy Meade, Lockheed-Georgia Company

Introduction

EVEN THOUGH USE OF STRUCTURAL COMPOSITE materials can achieve part count reductions up to 60% over conventional metal assemblies, structures made of them often must still be trimmed to net size and/or shape and attached to the final assembly. For this reason, the peculiar machining, drilling, fastening, and assembly requirements of cured composite materials must be addressed.

Cutting and drilling of cured composites can be accomplished using either traditional solid-tool methods (with special cutters) or newer machining technology. In each case, special considerations for application to composite materials must be taken into account. Due to their generally abrasive nature, and to other properties of their constituents, solid-tool cutting and drilling of composites require appropriate tooling. Tool materials, cutter shapes, and feeds and speeds differ for different types of composites, just as they do for different metals. Precautions must be taken during machining operations to avoid damage to the workpiece (delamination, fiber fraying, drill break-through, and so forth) and premature dulling of the tool. The newer machining methods, such as water-jet and laser cutting, while offering some relief in these problem areas, introduce their own complications in terms of equipment size and other inherent limitations when used on composite structures.

Solid-tool, water-jet, abrasive water-jet, and laser cutting machining methods, as applied to some of the more commonly encountered structural composites, are discussed in varying degrees of detail in articles in this Section. One cutting system not included, the ultrasonic knife, which is most often used for cutting uncured prepregs and other forms of reinforcing fibers but has minor application on thin-section cured composites, is covered in the article "Ultrasonic Ply Cutting," in the Section "Manufacturing Processes" in this Volume.

Incorporation of a structural composite into the final assembly for which it is designed usually involves fastening, whether by adhesive means or by mechanical fasteners. Adhesive bonding, the most common approach, is covered here with articles on surface preparation, adhesives selection and specifications, and bonding cure considerations. Other articles cover aerospace practice in bolt fastening. Rivet fastening, important in assembling metal structures, is not widely used with composites, except with automated equipment capable of closely controlling the grip force, and for that reason coverage of this method has been omitted here. Also omitted is rivet bonding, which utilizes widely spaced rivets to hold the structures in place while an adhesive bond is cured. Although this latter approach has been adopted in some applications with a measure of success, the decision to omit it was based on its developmental status and proprietary nature at the time of this writing.

Other important assembly/joining issues are addressed in this Section in articles on dissimilar materials separation and faying surface sealing.

For additional information addressing these fastening and joining related topics from a design viewpoint, see the article "Joints" in the Section "Composite Structures Analysis and Design," in this Volume.

Other approaches to building composite structures, which to some extent offer the possibility of circumventing some of the problem areas inherent in the fastening and joining techniques discussed above, are described in the article "Honeycomb Structure" in this Section, and in the articles "Braiding" and "Fiber Preforms and Resin Injection" in the Section "Manufacturing Processes."

Solid-Tool Machining and Drilling

J.A. Boldt and J.P. Chanani, Northrop Corporation

THOUGH SUPERIOR TO CONVENTIONAL MATERIALS in physical properties, composites pose certain difficulties in attaining acceptable machined edges and drilled holes. This article describes the results of various research programs aimed at optimizing machining and cutting methods. Machining and drilling success is fiber related and totally independent of composite material matrix. Therefore the cutting tools, feed rates, and rotational speeds to be discussed are applicable to thermoplastic, bismaleimide, polyimide, and epoxy matrices, as well as to hybrids of composites and metals. Additional information pertinent to drilling carbon fiber composites for fastener installation appears in the article "Fastener Hole Considerations" in this Volume.

Machining and Drilling Problems

Composite materials used in manufacturing structural assemblies for high-performance fighter aircraft have reduced weight and improved performance. Two representative materials, graphite and aramid composites, have excellent performance properties but when machined or drilled tend to develop the following flaws:

- *Surface delamination:* Separation of plies where the cutter enters and exits the material (Fig. 1)
- *Internal delamination:* Separation that develops between plies as a result of improper machining and drilling (Fig. 2)
- *Fiber/resin pull-out:* Tearing away of fiber or resin from the wall of the machined edge or drilled hole (Fig. 3)

In addition to these manufacturing problems, graphite causes excessive wear on cutting tools because it is abrasive. To minimize this effect, cutting tools made of C-2 tungsten carbide or diamond can be used but require a precise coordination of feed rates and speed to achieve high productivity and quality parts.

Machining Techniques

Most composite structures are fabricated to near-net shape. The most useful machining technique to describe is the trimming operation, including the method, cutting tools, and speeds to obtain the final shape and configuration of cured composite parts. Research and development of other machining techniques and procedures for composites is ongoing but limited.

Cutting tools used to trim composites are of three basic types: circular saws, router cutters, and abrasive tools. Diamond-coated circular saw blades used in portable equipment are useful for straight-line cuts and provide long tool life. Carbide router bits with a diamond-shaped chisel cut (Fig. 4) and diamond-coated router tools are very effective in producing good finishes and may be used to perform any trimming operation. The former are highly recommended because of their lower cost and the surface finish obtained, which is better than 3 μm (125 μin.), in roughness average. Abrasive tools are also used for trimming and in the final finishing operation. Sanding drums, discs, belts, and abrasive cloth are examples of abrasive tools. Table 1 provides recommended tool types and speeds for specific operations.

Drilling Techniques

Hole generation is a major activity in the manufacture of structural assemblies of aircraft. The quality of holes is critical to the life of fastened joints. Hole characteristics, such as waviness/roughness or lack of axial straightness and roundness, can cause stress concentration at the fastener assemblies, leading to premature failure. To avoid this problem, several research projects have established the

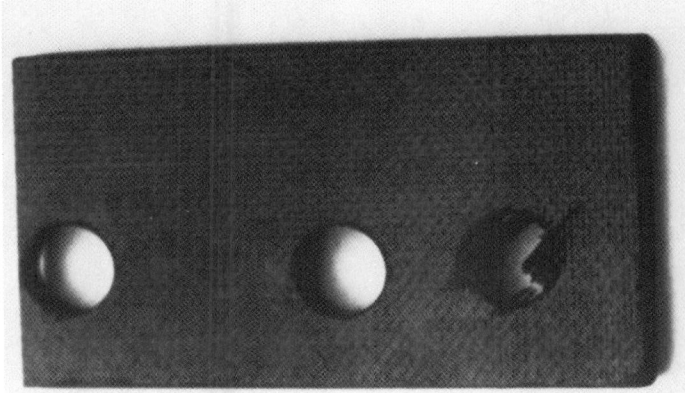

(a)

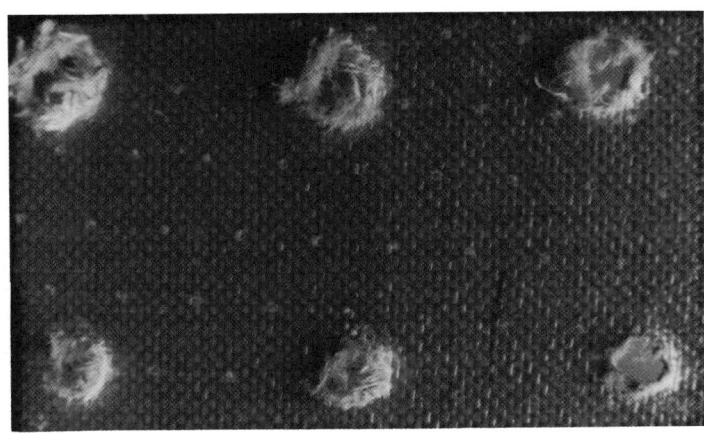

(b)

Fig. 1 Surface delamination characterized (a) by splintering, in graphite composites. (b) By shedding in aramid composites

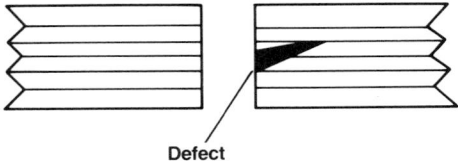

Fig. 2 Internal delamination

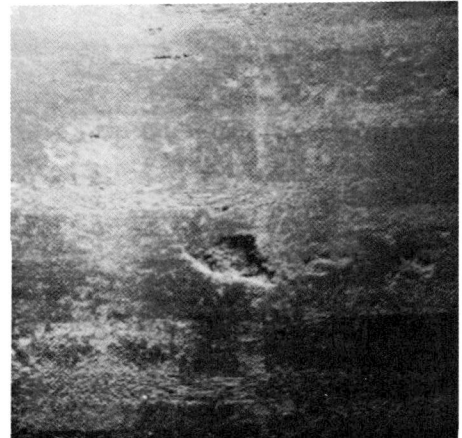

(a)

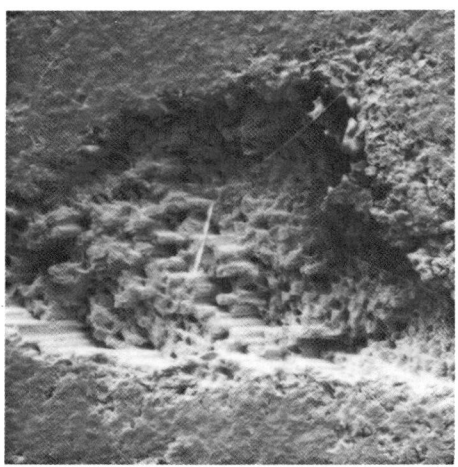

(b)

Fig. 3 Fiber/resin pull-out. (a) 30°; 20×. (b) ±45°; 170×

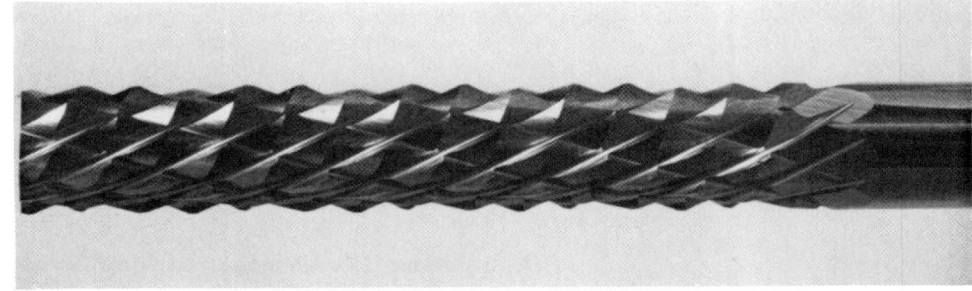

Fig. 4 Carbide router bit with diamond-shaped chisel cut

Table 1 Parameters for composite trimming

Operation	Equipment	Cutter type	Speed m/s	Speed ft/min
Straight-line cuts	Pneumatic saw	Diamond-coated circular saw(a)	60-90	12 000-18 000
	Hand router	Diamond router(b) Carbide router(c)	10-15	2000-3000
Irregular outline	Hand router	Diamond router(b) Carbide router(c)	10-15	2000-3000
Chamfer, deburr	Hand	Abrasive drum(d)	NA	NA
Finish operations	Hand drill motor	Abrasive disc(d)	5-60	1000-12 000
	Hand	Abrasive cloth(d)	NA	NA

(a) Diamond circular saw 0.050 kerf, 36/44 grit. (b) Diamond routers, 36/44 grit roughing, 80/100 grit finishing. (c) Carbide diamond-shaped, chisel-cut routers. (d) 80 grit (rough), 220 grit (finish)

optimum methods and recommendations for ensuring a consistent quality level in composite hole generation.

Graphite Composite Drilling. Because holes in graphite composites tend to exhibit splintering, backups for laminates are generally used. However, using backup material is costly, and it does not completely eliminate splintering when drilling with conventional drill bits. Some holes require reworking, thereby increasing cost and delaying scheduled completion dates. Therefore, a unique drill point geometry having an integral countersink was designed, fabricated, and evaluated to eliminate the need for a material backup. The drill, known as a spade drill/countersink, is shown in Fig. 5, while the point geometry is shown in Fig. 6.

Experiments with this drill on a 12.7-mm (0.500-in.) thick laminate were conducted at an optimum speed of 2800 revolutions per minute (rpm) and a feed rate of 0.040 mm (0.0015 in.) per revolution using a Spacematic-type power feed motor and a light spray of Boelube 70106 coolant at a high pressure (550 kPa, or 80 psi minimum). Test results showed that up to 40 quality holes could be generated (with no splintering and with smoothly cut edges) per tool life cycle, without use of backup material. A comparison between holes drilled with a conventional drill and those drilled with this developed drill is shown in Fig. 7. Recommended speeds and feed rates for different diameters are found in Table 2.

Although this drill was designed for use with a Spacematic-type power feed motor, there are circumstances in which the power feed tool cannot be used and manual drilling becomes necessary. For a manual drilling/reaming operation, the recommended cutting tool is shown in Fig. 8, and the point geometry is shown in Fig. 9. When manual countersinking is required, a carbide countersink with an integral pilot having a proper radius is recommended. A rake angle of 10° (±2°) is preferred.

Aramid composite drilling frequently causes surface delamination, specifically, shredding. The shredded fibers have to be manually removed to obtain a clean hole, which is a time-consuming and costly operation.

(a)

(b)

Fig. 5 Spade drill/countersink. (a) Side view. (b) Front view

Therefore, research sought to develop a method of producing a flawless hole at the time of drilling, precluding any follow-up finishing operation.

Obtaining both good diametric tolerance and no shredding requires that drilling proceed in such a way that aramid fibers are preloaded by tensile stress and then cut by shearing motion. For a rotating tool, this requirement dictates that the cutting edge of the drill be C-shaped to cut from outside to the center. The drill that was designed and fabricated for this purpose incorporated both the conventional drill point and the raidal C-shaped point. The conventional drill point helps prevent "walking" and recentering of the hole, and reduces the

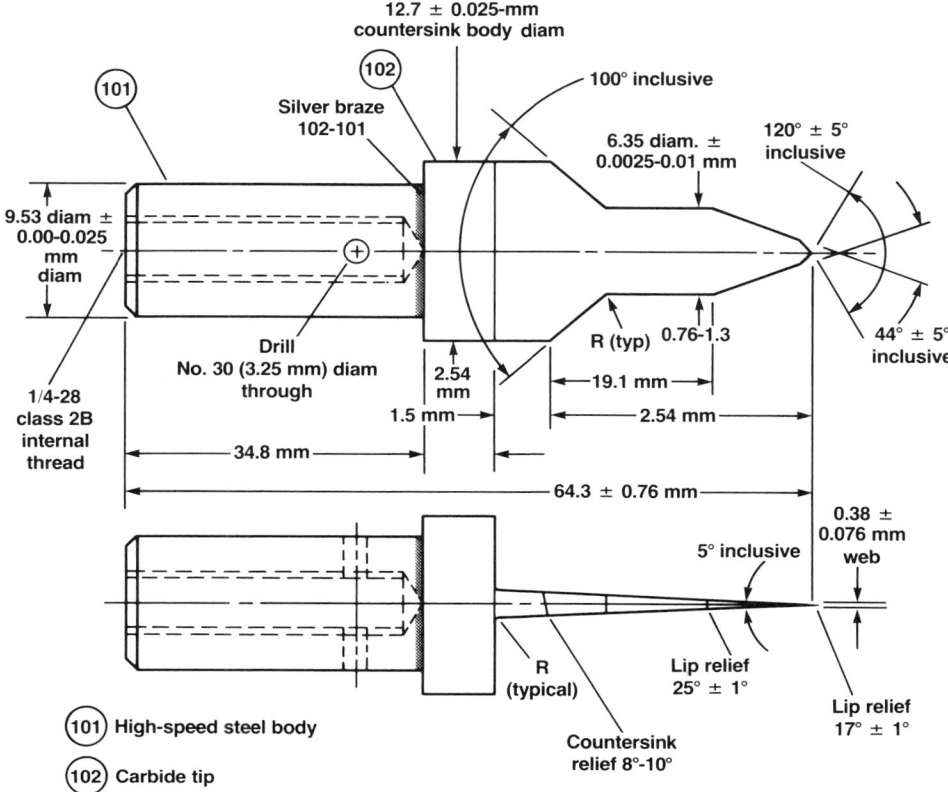

Fig. 6 Point geometry of spade drill/countersink

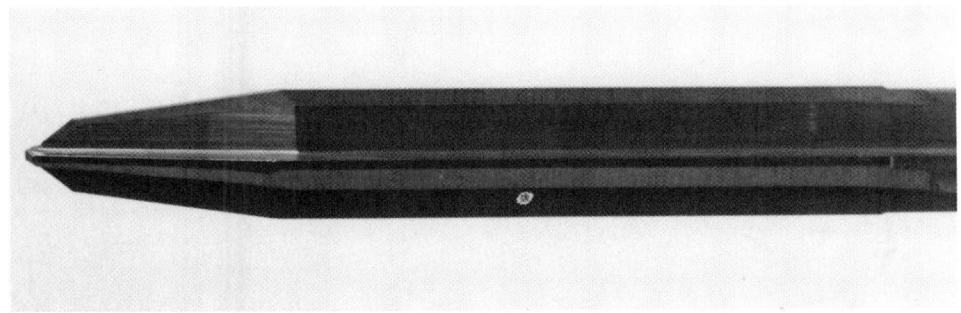

Fig. 8 Drilling/reaming cutting tool

Table 2 Parameters for graphite drilling

| Hole diam, max | | | Feed rate per revolution | |
mm	in.	Speed, rpm	mm	in.
3.967	0.1562	2800	0.025-0.040	0.0010-0.0015
4.763	0.1875	2800	0.025-0.040	0.0010-0.0015
6.350	0.2500	2800	0.025-0.040	0.0010-0.0015
7.938	0.3125	1800	0.045-0.055	0.0017-0.0022
9.525	0.3750	1800	0.045-0.055	0.0017-0.0022

likelihood of chipping the sharp radial C-shaped drill point.

Tests were conducted with a conventional air motor with and without a feed control device on a 3.18-mm (0.125-in.) thick laminate at different speeds and feed rates. The drill diameter used for this test was 6.35 mm (0.250 in.). Test results showed that 5000 rpm and 0.030 mm (0.0012 in.) per revolution were the best speed and feed rate, respectively. The drill tip is shown in Fig. 10, and the holes it produced are shown in Fig. 11.

The shredding problem is also encountered during the countersinking operation. To elimi-

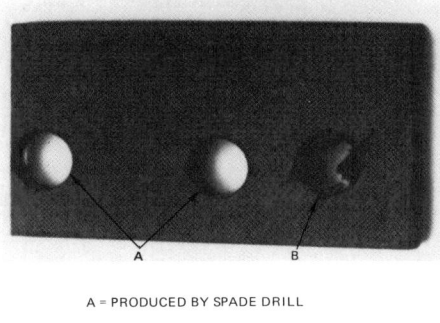

A = PRODUCED BY SPADE DRILL

B = PRODUCED BY CONVENTIONAL DRILL

Fig. 7 Breakout comparison of holes produced by spade drill and conventional drill

nate it, numerous tool configurations were evaluated. Test results show that the countersink that incorporates S-shaped positive-rake cutting flutes produces a fuzz-free countersink edge. This configuration is shown in Fig. 12.

Dissimilar Material Drilling. Hole drilling in hybrid structures, such as a graphite and metal sandwich, is a production reality. Extensive work is still in progress to identify methods to provide quality holes in these structures at a reasonable cost. Presently, several methods can be used to produce holes in these structures. One of the two most useful methods, peck drilling, uses hard-tooling and a lock-in type of bushing equipment, such as Quackenbush; two or more steps are usually required to produce a hole. The other method uses soft-tooling (template) and a Spacematic-type locating guide to drill and countersink in one step. However, this method is not suitable for drilling structures that are thicker than 25 mm (1.000 in.) or that have titanium or steel with a thickness greater than 5 mm (0.200 in.). Both methods employ positive power feed equipment, although power drilling may not be used in all situations because of lack of accessibility, in which case manual drilling may be used.

To prevent the problems associated with entrance and exit surface delamination or shredding in dissimilar material stack-up, backup materials must be used. They can be a form of peel or release ply or integrally cured fiberglass and will ensure a clean hole edge.

Peck Drilling. When Northrop Aircraft Division attempted to drill fastener holes in a 38-mm (1.5-in.) thick dissimilar material (titanium-graphite-titanium) stack-up, conventional mechanical power feed drilling failed to produce a cost-effective, quality hole. Conventional drilling processes utilizing mist coolant caused the titanium chips and graphite sludge (mixture of graphite dust and coolant) to impact into the drill flutes. This condition caused oversized holes in the graphite and eventual drill breakage. To avoid these problems, a peck drill method was developed and produced excellent results: Hole quality was outstanding, and the cost per hole was lower than by the conventional method.

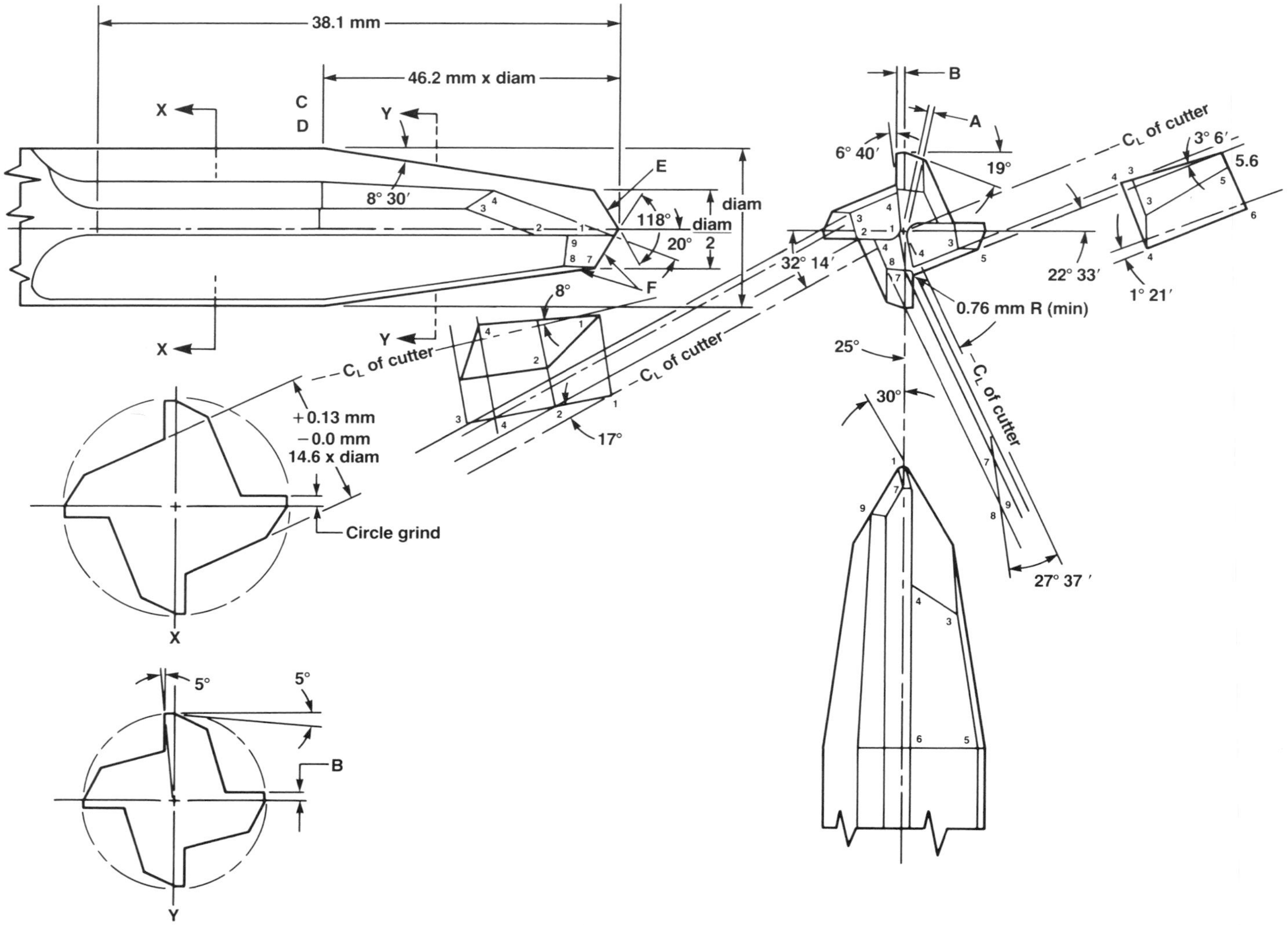

Fig. 9 Point geometry of drill/reamer cutting tool. A, Web thickness at the point to be 1.5 mm times drill radius. B, Margin width to be 2.2 ± 0.05 times drill radius. C, All drill/reamers shall have 0.005 to 0.013 mm/mm uniform back taper starting from intersection of tapered cutting edges with straight edge and continuing for entire flute length. D, Full core area starts here and continues back to point at which the flute washes out on the shank. E, Primary clearance at tip is 10°. F, Total height difference between cutting lips must not exceed 0.025 mm; the same applies to the tapered edge.

Fig. 10 Optimum drill for aramid composites

Peck drilling is a process of drilling deep holes without coolant. The drill advances to a certain depth that can be adjusted by the user and then withdraws from the hole to clear the chips and dissipate heat. It then reenters the hole to drill deeper and repeats the ''woodpecking'' action until the desired depth is achieved (Fig. 13).

Air-fed peck drill motors are automatic, with a rapid-advance capability and automatic retract. The pecking stroke frequency is infinitely variable. The tested equipment had a controlled feed rate ranging from 0.0025 to 0.305 mm (0.0001 to 0.0120 in.) per revolution. To prevent drill bit damage from the pecking motion, the rapid advance stops just prior to contacting the workpiece (~1.5 mm, or 0.060 in.).

A series of tests was conducted to optimize the drilling parameters and peck cycles for titanium, graphite, and aluminum, and for combinations of their stack-up. The tests were conducted using a Gardner-Denver MM8 series motor, model 83660DB5. The rpm availability in this equipment dictated that all tests be carried out at a speed of 550 rpm. When the material stack-up does not contain any hard material (titanium or steel) and higher rpm is available, using a higher rpm is recommended to enhance productivity. All the holes were

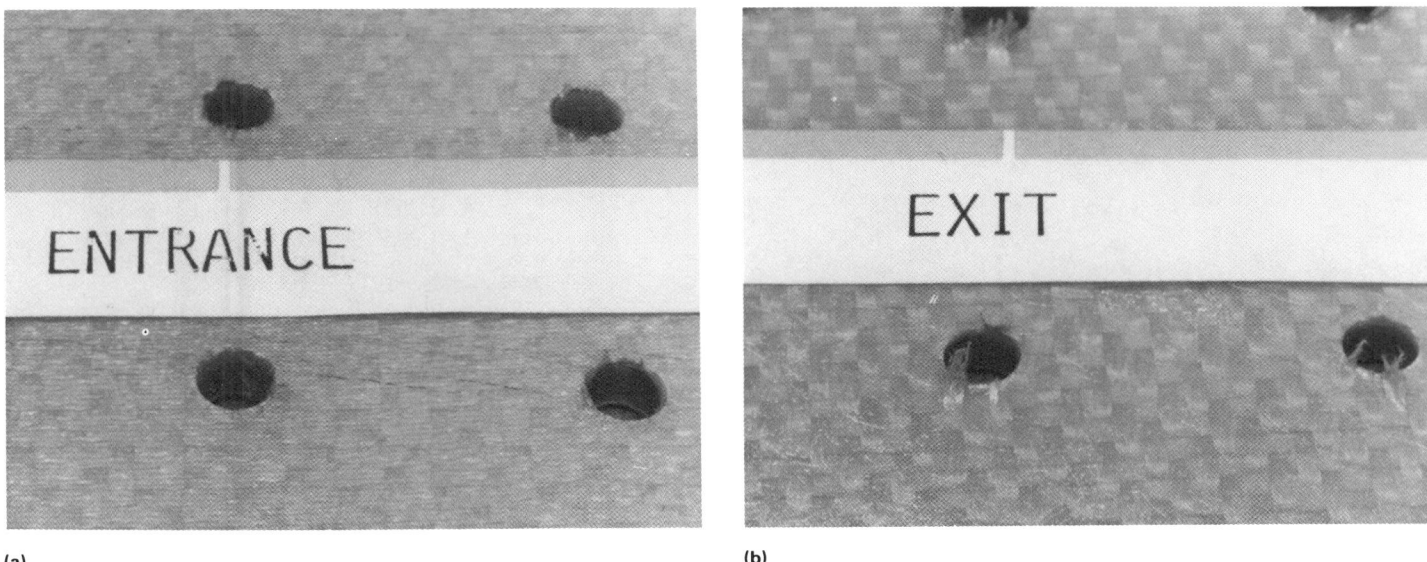

(a)

(b)

Fig. 11 Holes produced in aramid composite by the optimum drill. (a) Entrance. (b) Exit

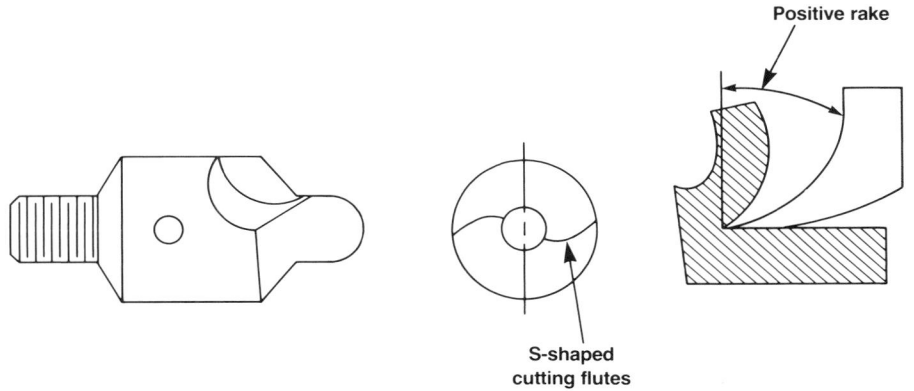

Fig. 12 Optimum countersink drill for aramid composites

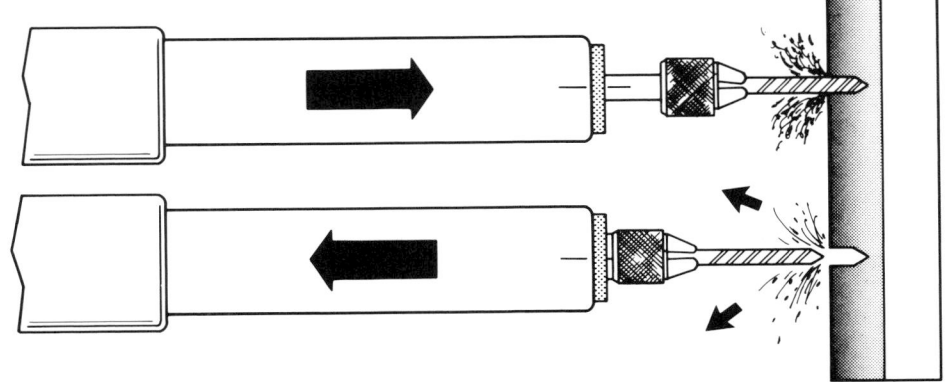

Fig. 13 Peck drilling, in which drill advances into material, retracts, cools, and clears chips before repeating the process

Table 3   Parameters for peck drilling

Material	Speed, rpm	Feed rate per revolution mm	in.	Peck cycle, in.
Titanium	550	0.050	0.002	60 min
Graphite	550	0.10	0.004	30 min
Aluminum	550	0.10	0.004	30 min

produced dry, using carbide-tipped drills having a standard NAS-907 P3 split-point geometry. The recommended parameters for hole sizes up to 9.53 mm (0.375 in.) are listed in Table 3.

The peck drilling process must be followed by the standard reaming operation using coolant, described in the section "Graphite Composite Drilling" in this article. A carbide-tipped reamer having NAS-897-type C geometry is recommended.

The one-step drilling/countersinking method is more cost effective than the peck drilling method. However, it is suitable only for drilling structures that are 25 mm (1.000 in.) thick or less and have hard material in the stack-up that does not exceed 5 mm (0.200 in.). This procedure uses Spacematic-type equipment that drills, reams, and countersinks in one operation. The drill motor is an air-operated, hydraulically controlled, feed-rate-type that is clamped to the work surface by an expandable collet. Using a template foot pad with the collet ensures perpendicularity between the cutter axis and the work material. The operator simply inserts the collet into a previously drilled hole while holding the foot pad firmly against the material surface. Clamping and drilling start when the motor is started.

Table 4 Parameters for one-step drilling/countersinking

Materials	Hole diam, max		Speed, rpm	Feed rate per revolution	
	mm	in.		mm	in.
Graphite and aluminum	3.967	0.1562	2800	0.025-0.040	0.0010-0.0015
	4.763	0.1875	2800	0.025-0.040	0.0010-0.0015
	6.350	0.2500	2800	0.025-0.040	0.0010-0.0015
	7.938	0.3125	1800	0.045-0.055	0.0017-0.0022
	9.525	0.3750	1800	0.045-0.055	0.0017-0.0022
Graphite and titanium or steel	3.967	0.1562	400	0.10-0.15	0.0040-0.0050
	4.763	0.1875	400	0.10-0.15	0.0040-0.0050
	6.350	0.2500	400	0.10-0.15	0.0040-0.0050
	7.938	0.3125	400	0.10-0.15	0.0040-0.0050
	9.525	0.3750	400	0.10-0.15	0.0040-0.0050

Tests were conducted using Dresser-10SC and Deutsch-1000 equipment on graphite-aluminum, graphite-titanium, and graphite-steel stack-ups using coolant. The cutting tool was a carbide-tipped drill/countersink with NAS-907 P3 split-point geometry. The recommended parameters for hole sizes up to 9.53 mm (0.375 in.) are listed in Table 4.

SELECTED REFERENCES

- J.P. Chanani and J.A. Boldt, Manufacturing Methods for Composite Hole Generation, *Trans. SAE*, 1982
- J.A. Boldt, "Machining and Drilling of Composite and Composite/Metallic Structures," Process Specification MA-113, Northrop Corporation, 1985
- J.P. Chanani, "Optimal Methods for Generating Holes in Advanced Composites and Hybrids," American Society for Metals, WESTEC—1985, CA, March 1985

Water-Jet and Abrasive Water-Jet Cutting

Jim Korican, Sundstrand Corporation

WATER-JET CUTTING is relatively new and in many situations offers an advantage over plasma, laser, or conventional cutting methods. Abrasive water-jet cutting is especially suited to nonhomogeneous materials that are abrasive in nature and damaging to conventional cutting tools, or materials that produce dust or toxic fumes during cutting, or to those alloys or materials with properties that are sensitive to high temperatures or the work-hardening effects of other cutting methods. There is the added advantage of intricate shape-cutting capabilities for these difficult materials. Nonabrasive water-jet cutting lends itself to nonmetallic materials such as leather, rubber, urethane, softer plastics, and thin composites, with the advantages of fast, distortion-free, accurate, and clean cutting.

The principle behind water-jet cutting is that a hydraulically driven intensifier unit pumps a fluid, typically filtered and conditioned water, at pressures up to 410 MPa (60 ksi). The fluid is filtered to 0.5 μm (20 μin.) and pumped in relatively low volumes, typically 4 to 8 L/min (1 to 2 gal/min). The high-pressure fluid is expelled through an orifice to form a jet stream. Orifices are typically made of sapphire with typical diameters ranging from 0.8 to 7.6 mm (0.003 to 0.30 in.). The coherent jet of water is propelled at speeds up to approximately 850 m/s (2800 ft/s). Abrasive water-jet cutting uses an abrasive material that is introduced into the water jet after the primary jet is formed. The abrasive is entrained into the water stream in a mixing chamber and then enters a nozzle where it is accelerated by the water jet. A typical abrasive used in this process is garnet, in grit sizes ranging from 16 to 150 mesh.

Specialized nozzle configurations are used in both water-jet and abrasive water-jet cutting. Figure 1 shows the principle of operation of the latter. The process of abrasive water-jet cutting is more complex than nonabrasive water-jet cutting because of the addition of the abrasive feed system.

Several benefits can be realized using a water-jet cutting process. One benefit is that cuts can be initiated at any point on a workpiece if some caution is exercised. Because composite materials can delaminate in the area surrounding the initial penetration, it is recommended that the water jet be activated away from the finished edge of the workpiece and then guided into the desired edge.

Another benefit is that the kerf is not generally affected when the water jet is allowed to dwell. Because the kerf (cut) is relatively narrow, its width being dependent on exit nozzle bore size, material can be saved.

No individual hard tooling is required to cut flat-sheet metal parts if a computer-controlled X-Y machine is used. Because very low forces are imposed on the workpiece, simple clamping (with weights) will often hold the workpiece in position.

A low cutting temperature, with no noticeable heat-affected zone, is another benefit of water-jet and abrasive water-jet cutting, as is the very low dust level produced by the process, reducing the risk of health and fire hazards.

Several problems are also caused by water-jet and abrasive water-jet cutting. High noise levels are generated by air coupling into a large high-velocity volume of air/water. Because the noise level exceeds maximum levels allowed by Occupational Safety and Health Administration, ear protection is required. Commercially available mechanical catcher systems can significantly reduce the noise level. It also can be diminished in some operations by running the tip of the nozzle under the water surface of the catcher tank.

A lesser concern associated with this process is that the cutting fluid must be properly filtered and conditioned to reduce wear and increase life of the jewel orifice. Basic conditioning typically involves use of a water softener, or a chlorinator and activated charcoal filter. Conditioning is followed by the fluid passing through a filter system that removes particulates down to 0.5 μm (20 μin.) in size. Finally, when cutting composite materials, the water jet will delaminate layers if internal pressures are generated by a loss of abrasive or insufficient velocity. Figure 2 shows how this can occur.

Other concerns that are neither benefits nor problems also need to be described. When using abrasive water-jet cutting, the jet stream tends to angle away from the direction of the cutting, similar to flame and laser cutting. This "trailback" effect becomes more pronounced as the thickness of the workpiece increases and/or the nozzle feed rate is increased. Figure 3 clarifies this effect. Cutting nozzle stand-off, depicted in Fig. 3, affects both the capability to cut and the results of the cut edge when the stand-off distance from the workpiece is varied.

Usually, during abrasive water-jet cutting, the kerf is wider at the water-jet entrance than it is at the exit side of the workpiece. This varies with feed rate, and in some materials the angle may reverse, producing a larger kerf at the exit

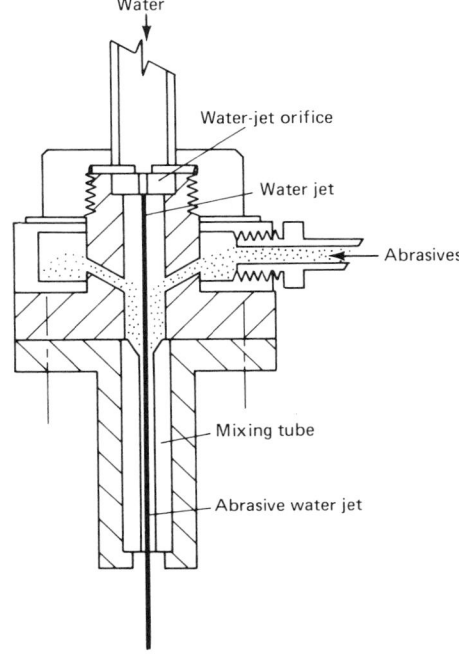

Fig. 1 Water-jet nozzle principle of operation

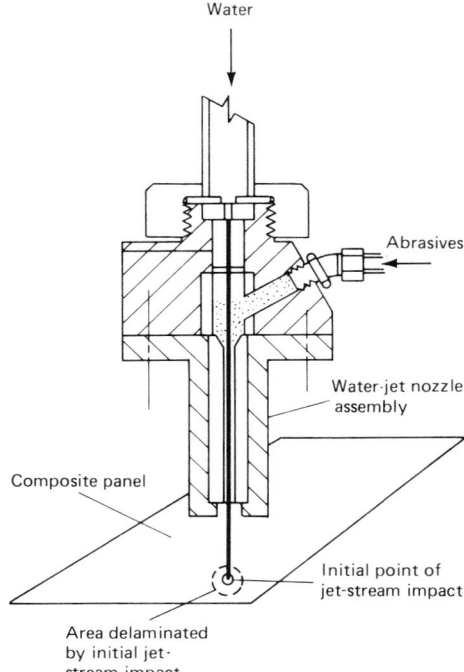

Fig. 2 Process mode for potential damage to composites

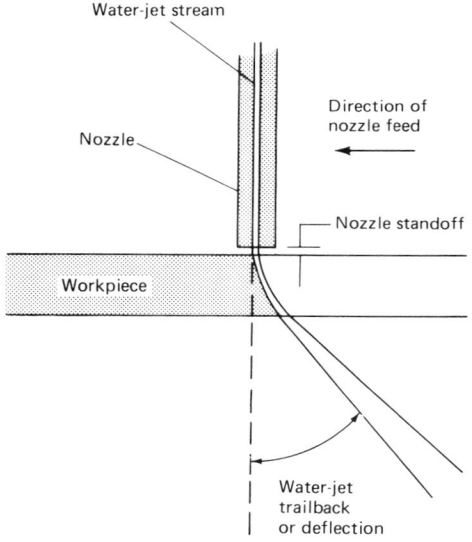

Fig. 3 Jet stream trailback effect

side. Whether this effect is a problem is application-dependent.

In contrast to a saw blade or milling cutter, there is very low horizontal cutting force against the workpiece. There is a need, with lightweight parts, to secure them so that they do not move even under the relatively low cutting force. This is more applicable to parts being cut from flat-sheet metal materials in a nested condition.

Table 1 Cutting speeds versus materials and thicknesses

Material	Thickness		Cutting speed	
	mm	in.	m/min	in./min
Abrasive cutting at 240 MPa (35 ksi), 100-grit garnet, 20 hp				
Resin-impregnated graphite, aramid,				
glass fibers	3.2	0.125	1.60	63
	6.4	0.250	0.75	30
	12.7	0.500	0.46	18
	19.1	0.750	0.30	12
	25.4	1.000	0.13	5
Ferrous metal, stainless, Hastelloy,				
Inconel, mild steel, high carbon,				
R_c60	1.6	0.063	0.50	20
	3.2	0.125	0.28	11
	6.4	0.250	0.18	7
	12.7	0.500	0.075	3
	25.4	1.000	0.019	0.75
Aluminum, magnesium,				
titanium, brass alloys	1.6	0.063	1.47	58
	3.2	0.125	0.74	29
	6.4	0.250	0.41	16
	12.7	0.500	0.25	10
	25.4	1.000	0.075	3
Nonabrasive cutting at 360 MPa (52 ksi), 20 hp				
Balsa wood, light foam rubber,				
styrofoam.	1.6	0.063	15	600 +
	12.7	0.500	12	450
	25.4	1.000	7	275
Polyurethane, rubber compounds,				
polyethylene (a) (30 + durameter)	6.4	0.250	6	225
	12.7	0.500	3	100
	25.4	1.000	1	40
Paper, fabric, corrugated board,				
rubber (a) (30 − durameter).	0.13	0.005	15	600 +
	0.38	0.015	8	300 +
	0.81	0.032	8	300 +
	1.6	0.063	8	300 +
	3.2	0.125	8	300 +
	6.4	0.250	8	300 +

(a) These materials may require additional support on a table.

Cutting Characteristics

Material that is removed by the water-jet or abrasive water-jet process is expelled with the jet stream in small particles. Cutting is basically accomplished through an erosion or shear process. In the case of abrasive water-jet cutting, the sharp edges of each grain of abrasive are worn down by impact with the workpiece. This reduces its cutting efficiency, and it is both impractical and uneconomical to recover and reuse it.

Feed rates for the cutting nozzle are dependent on both the type and thickness of the material being cut. Other factors affecting cutting ability are orifice size, nozzle stand-off distance from the workpiece, and fluid pressure of the jet stream. For the abrasive water-jet process, the amount of abrasive flow also has to be considered. Generally, lower abrasive flow rates result in lower nozzle feed rates.

Another consideration in abrasive water-jet cutting is surface finish of the cut edge. Abrasive grit size is an important factor in determining surface finish. A finer mesh (150-grit) will typically generate a smoother edge finish. Surface finish is also dependent on the type of material being processed. Table 1 shows approximate cutting speeds for different materials of various thicknesses.

Water-jet and abrasive water-jet cutting can result in a kerf width that is about 0.025 mm (0.001 in.) larger than the nozzle orifice diameter for stand-off distance not exceeding 3.2 mm (⅛ in.). In all applications, it is recommended that individual testing be conducted to establish cutting parameters and acceptable results.

Applications

Water-jet and abrasive water-jet cutting need to be discussed separately in terms of applications, although there is some commonality when discussing very thin, soft metals and some composites. However, each is best suited to specific ranges of material types.

Nonabrasive water-jet cutting is best applied to materials with a yield of 80 MPa (12 ksi) or less. Table 2 is a partial list of materials best cut by this process. Material thickness is sometimes a limiting factor in the ability of the water jet to cut some material types effectively. Testing can determine the appropriateness of the application.

An abrasive water jet has the ability to cut almost any type of material, but is best applied to denser and thicker materials, specifically, almost any metallic. Table 3 is a partial list of materials that have been cut with the abrasive

Table 2   Partial list of materials cut with water jet

Acrylic plastics	Nylon (brushes, carpets)
Aluminum	Oak
Aluminum honeycomb	Paper-corrugated boards; filters;
Bakery products	impregnated, coated, laminated newsprint
Boron-epoxy composite	Plastics
Boron-polyester composite	Plexiglass
Box board	Plywood (up to 25-mm or 1 in. thick)
Brass	Polyester fabrics and carpets
Carpets	Polyethylene sheets
Ceramic foam	Polypropylene tubes
Concrete	Polystyrene
Copper	Polyurethane foam (25 mm, or 1 in. thick)
Cork	Poplar
Epoxy-aramid	PVC
Fabrics, automotive	Rocks
Fiberglass-epoxy composite	Roofing (composition, asbestos, rolls,
Fiberglass insulation (13 mm, or 0.5 in. thick)	shingles)
Food (frozen meats, vegetables)	Rubber (foams, reinforced shoe soling,
Furniture components	tire tubes)
Glass fiber reinforced polyester	Shoe leather
Glass-polyimide	Steel (mild and thin)
Graphite-epoxy composite	Styrofoam
(15.9 mm, or 0.625 in. thick)	Synthetic rubber
Gypsum boards	Teflon (9.5 mm, or 0.37 in. thick)
Laminated plastics	Urethane foams
Leathers	Vinyl sheets
Lucite plate (13 mm, or 0.5 in. thick)	Wood (fiber tiles, fiber padding,
Maple	particle board, paneling veneer)
Mild steel	
Neoprene rubber, 50 Shore hardness	
(50 mm, or 2 in. thick)	

Table 3   Partial list of materials cut with abrasive water jet

Metals

718 Inconel
625 Inconel
6AL-4V titanium (3.2 mm, or 0.125 in. thick)
Commercially pure titanium
Hastelloy
321 CRES (75 mm, or 3 in. thick)
15-7 PH CRES
301 half hard CRES
301 full hard CRES
Chromoloy
ESCO 49M-high nickel/high chrome alloy
   (170 mm, or 6.75 in. thick)
Mild steels
Glass (25 mm, or 1 in. thick)
Aluminum (140 mm, or 5.5 in. thick)
Peel shim stock
304L CRES (13 mm, or 0.5 in. thick)

Composites

Metal matrix
Graphite
Aramid
Glass
Laminated glass
Ceramics

water-jet process. Like water-jet cutting, the specific application being considered for abrasive water-jet cutting should be tested to determine suitability.

Equipment and Tools

The requirements for water-jet and abrasive water-jet cutting equipment will vary depending on the selected supplier, but basic system installation is similar. The heart of the process is the hydraulic pump and intensifier unit. A variety of pump capacities provide different water-jet pressures to tailor the process to specific applications. Another component of the process, the nozzle assembly, can involve different nozzle configurations, depending on the supplier and whether the system is non-abrasive water jet or abrasive water jet. If abrasive water-jet cutting is used, an additional system is required to feed the abrasive to the water jet.

Another fundamental part of the process is a water-jet collector and drain. For abrasive water-jet cutting, a settling tank and sump pump may be required. The spent abrasive must be disposed of and will contain traces of the types of material it has cut.

Other major components of the process are a worktable and a device to hold and manipulate the nozzle assembly. The equipment can be as simple as a flat worktable with an overhead gantry having *x-y* movement, or as complex as an articulated robot with multiple-axis cutting capability. Control can be exercised by numerical control (NC), computer numerical control (CNC), or direct numerical control (DNC), depending on the desired application.

Future Trends

Trends in water-jet and abrasive water-jet cutting will focus on continued improvements in the catcher mechanisms used to trap the water-jet stream after it exits the workpiece. Feedback devices will probably be developed to monitor abrasive flow and signal the need for adjustments or repairs to maintain proper flow.

Systems will also be developed to monitor the wear on nozzles and other components of the water-jet system and either signal the operator or automatically change the tools for continued cutting time.

Other types of fluids and higher jet-stream pressures may one day also eliminate the need for abrasives and provide additional flexibility in the application of this process.

Laser Cutting

Leonard R. Migliore, Spectra-Physics

FOCUSED LASER BEAMS to cut composite materials are now being used to cut a wide range of substrates. While most laser cutting has been performed on sheet steel, laser beams facilitate the shaping of many otherwise intractable composites.

Laser cutting is a noncontact, thermal process (Ref 1). These two characteristics are fundamental to its utility. The absence of contact allows intricate cutting of fragile workpieces, and simplifies fixturing. Thermal cutting is independent of the strength or hardness of composite constituents. The thermal nature of laser cutting, however, limits its use when charring or thermal degradation are unacceptable. It is essential, therefore, to understand all elements of laser cutting to apply it properly.

The term laser is actually an acronym for "light amplification by stimulated emission of radiation." Its three essential components are an active medium for light emission; an excitation source for delivering energy to the active medium; and an optical resonator, which is a set of mirrors that supplies feedback to the active medium.

There are many laser types, but only a few can be made powerful enough for use in material processing. Currently, these lasers, which are generally named for their active medium, are known as ruby, neodymium-glass, neodymium-yttrium aluminum garnet (YAG), and carbon dioxide (CO_2) lasers. Excimer lasers, such as xenon chloride and argon fluoride, are currently being developed, and should soon join this group.

A continuous cutting process requires a beam that is on continuously, or at least pulses rapidly enough to behave as if it were. This requirement reduces the suitable laser types to two: neodymium-YAG and CO_2.

The neodymium-YAG laser, referred to as YAG, uses an active medium of neodymium, which is dissolved in a matrix of yttrium aluminum garnet. The resulting crystal is formed into a rod, which is pumped, or excited, by flash lamps. Light is emitted from the rod at a wavelengh of 1.06×10^3 nm (1.06×10^4 Å) in the infrared range. Such lasers used for cutting emit pulses of light at frequencies up to approximately 200 Hz. The peak power for each pulse can be several kilowatts, while the average power does not exceed 500 W. The optical resonator, or head, of a YAG laser is relatively small, at approximately 900 mm (35 in.) × 350 mm (15 in.) wide × 170 mm (7 in.) high and weighs 50 kg (110 lb), for a 500-W unit, which allows it to be carried on a motion system.

The CO_2 laser operates by electrical excitation of CO_2 gas molecules, emitting light in the middle infrared range at 1.06×10^4 nm (1.06×10^5 Å). The output of these lasers may either be continuous or pulsed, at rates up to several kilohertz. Power outputs range up to 9 kW for industrial units, although 500 to 1500 W is a more typical range for laser cutting equipment. Pulse power in CO_2 lasers varies with design, but is less than is available in YAG lasers. A 500- to 600-W CO_2 laser of modern design has a head that is approximately 2 m (7 ft) long × 0.6 m² (6.5 ft²), and weighs 300 kg (660 lb). It is not usually carried on a motion system. For many common composites the CO_2 laser offers a method of generating cuts at high speed and with superior quality. For certain metal matrix composites, CO_2 or YAG lasers are an alternative to diamond tools or grinding.

Laser Cutting

Laser cutting is considered to be a thermal process because when laser light impinges on a surface, a fraction of it is absorbed. The energy associated with this fraction raises the temperature of the material. When enough light is concentrated on the workpiece, the resulting temperature increase causes melting, vaporization, and decomposition.

Lasers are useful light sources for cutting for two reasons: They produce large amounts of power in the form of light and they produce this light in parallel beams that can be focused to small spots.

Focusing Laser Beams. A CO_2 laser with an output of 1500 W generally emits a beam about 20 mm (0.8 in.) in diameter. The power density in this unfocused beam varies across the diameter, and reaches a peak of about 4.0×10^7 W/m² (3.7×10^6 W/ft²). While this level will decompose most organics, and eventually melt some metals, it is not enough for useful cutting. Efficient cutting is accomplished by using a lens or mirror to concentrate the energy. Because laser light is highly monochromatic and parallel, it can be focused to a spot size that is limited primarily by diffraction and focusing optics (Ref 2). The 1500-W CO_2 laser may be focused into a circle 0.15 mm (0.006 in.) in diameter, resulting in a power density of over 1.0×10^{11} W/m² (9.3×10^9 W/ft²), which will melt metal in microseconds (Ref 2) and vaporize any organic material. The small spot size also produces a very narrow kerf that is as small as 0.1 mm (0.004 in.) when the focused beam is traversed over the surface of a sheet. This allows laser cutting to be accurate to ±0.05 mm (0.002 in.), but cutting accuracy is primarily dependent on the motion mechanism. Equally important, the steep gradient in power density preserves material properties adjacent to the edge of the cut.

Some compromises are required when focusing laser beams. Minimum spot sizes, obtained by using short focal length lenses, are achieved at the expense of depth of field: The smallest spots are usable only on the thinnest materials, and the lens-to-workpiece distance must be controlled more accurately (Fig. 1). Typical values are shown in Table 1.

Gas-Assisted Cutting. In most cases, the laser light is not the sole agent for material removal. A gas stream from a nozzle that is coaxial with the focused beam removes material from the kerf and protects the focusing lens from debris (Fig. 2). The gas used to cut most organic materials is air, while oxygen is used for ferrous materials. Argon is used for titanium and other oxidation-sensitive metals to avoid the formation of a hard oxide next to the kerf. Gas flow, nozzle-to-workpiece distance, and nozzle design all affect cutting speed and quality (Ref 3).

Gas-assisted laser cutting operates best when it completely penetrates a sheet of material. The gas then flows through the kerf, sweeping molten or decomposed material away from the interaction region. Thick, organic materials produce high concentrations of particulate matter and gaseous hydrocarbons in the kerf. These fumes absorb and reradiate the laser light, increasing the kerf width in the center of the material, where the particle density is greatest (Fig. 3). The most uniform kerfs in thick,

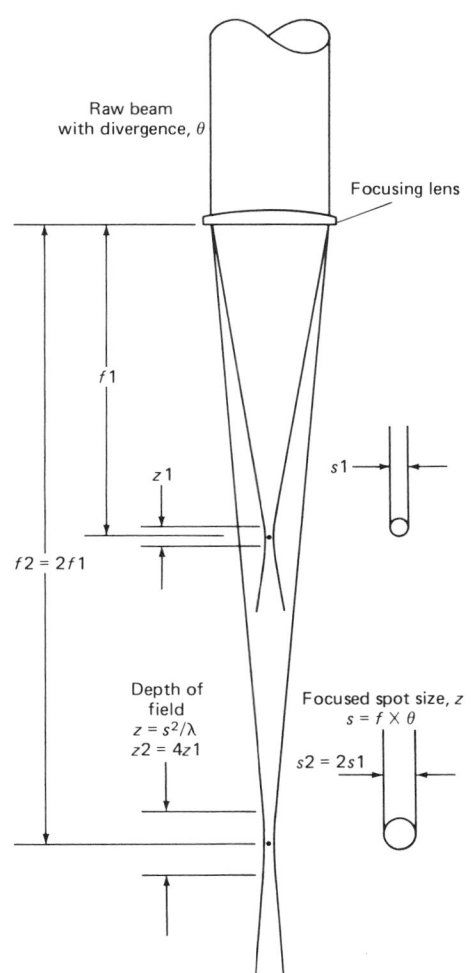

Fig. 1 Variation of laser cutting parameters with focal length. *S*, spot size. *F*, depth of field

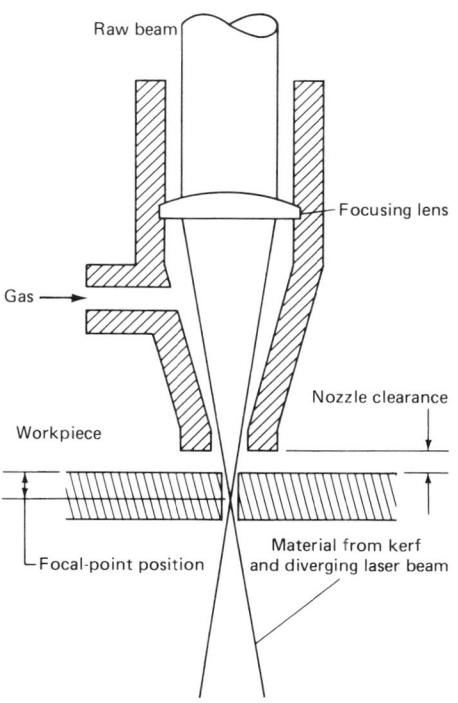

Fig. 2 Gas-assisted laser cutting

Table 1 Typical values for a 20-mm (0.8-in.) raw beam

Lens focal length		Maximum material thickness		Focus tolerance	
mm	in.	mm	in.	mm	in.
63	2.50	3	0.12	±0.15	0.006
127	5.00	10	0.39	±0.6	0.024
190	7.50	25	0.98	±1.5	0.060
254	10.4	50	1.97	±2.5	0.100

Fig. 3 Laser-cut 19-mm (0.75-in.) thick birch plywood showing internal kerf enlargement that is due to short focal length and high travel speed

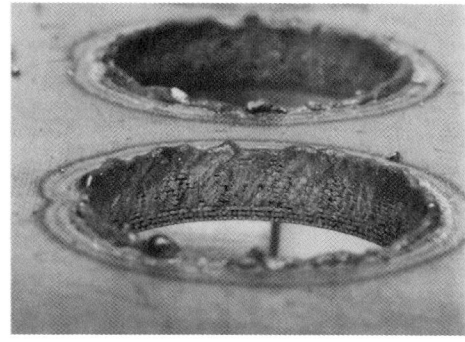

Fig. 4 Laser-cut 3.2-mm (0.13-in.) SiC-Ti composite, where Ti matrix material has flowed away from the SiC fibers and formed a recast layer at the exit (top) side of the cut

Fig. 5 Gas-assisted laser cutting of holes in 19-mm (0.75-in.) thick birch plywood with fixed cutting head, using trepanning technique. The nozzle, which appears to be touching the workpiece, is actually 1 mm (0.04 in.) above it.

organic materials are achieved by using slow travel speeds, which reduce the rate of particle evolution, and long focal-length lenses, which widen the kerf, allowing evolved fumes to escape more easily. The gas-assisted flow must be sufficient to keep the beam path clear of material.

Metallic materials, including metal matrix composites (MMCs) do not exhibit the same behavior as do thick, organic materials because they do not break down into gaseous components during laser cutting. The material just melts and must be physically ejected by the gas-assisted flow. Recast material often adheres to the exit side of the cut (Fig. 4) and may require subsequent operations to remove it. Recast material can also adhere to the walls of the cut, forming a degraded recast layer. Recast material is minimized by using a high-pressure gas-assisted flow, and by using reactive gases if they are metallurgically acceptable.

Failure to penetrate a workpiece completely results in the kerf material being ejected upward toward the laser source. This material generally damages the cut edge, producing a ragged incision that is surrounded by debris. Cutting holes in sheet material, however, requires that the process start on an unbroken surface, unless pilot holes have been provided. Because pilot holes are uneconomical, the standard method of generating high-quality holes is to have the laser perforate the workpiece inside the region to be removed, and then travel out to the periphery of the cut. This technique, called trepanning, is shown in Fig. 5.

Cutting Systems

In order to cut, a laser must be integrated with the means to deliver the beam and handle the workpiece, and an enclosure to ensure safety for personnel. This assembly constitutes a laser-cutting system or laser work station. There are significant variables in the YAG and CO_2 systems. YAG lasers are compact and produce high peak powers that vaporize metal effectively and are thus suitable for MMCs.

The low repetition rate of YAG lasers limits their maximum cutting speed to about 2 m/min (7 ft/min). The wavelength of YAG light is absorbed well by metals, but not by organic composites, which limits the utility of the YAG laser. The CO_2 lasers can produce high average power and generate both continuous and pulsed beams. Continuous wave (CW) operation allows them to cut at very high speeds when the application allows it, such as 75 m/min (250 ft/min) on a 0.1-mm (0.004-in.) graphite-epoxy composite, using a 1500-W unit. In addition, the 1.06×10^4 nm (1.06×10^5 Å) light is strongly absorbed by all organic materials. Most cutting systems use CO_2 lasers because of their higher speed and greater flexibility.

Beam delivery and workpiece handling are two areas of great variation. In the laser cutting operation, the focused beam passes through a nozzle that emits a gas stream, and traverses the part. Either the beam, the part, or both may move. Maintenance of the focal position can be passive or active. Three-dimensional parts require more complex motion than do flat sheets. Descriptions of common systems are given below.

Two-axis, moving workpiece, the simplest and most common laser cutting system, has fixed optics to simplify laser beam delivery. It is only usable on flat sheet material, and takes up at least four times as much floor area as the largest sheet it can handle.

Two-Axis Moving Beam. As workpieces get larger, the expense of making a moving-beam delivery system is less than that of moving the workpiece. Fixturing is simplified with a static workpiece. The size of the workpiece is limited by the optical characteristics of the laser. A CO_2 laser can be designed to send its beam 12 m (40 ft), allowing 6 × 3 m (20 × 10 ft) tables.

A five-axis, moving workpiece is a type of system useful for cutting required by three-dimensional parts. For small parts that are cut at low speed, a milling machine configuration of three orthogonal axes of motion and two axes of rotation allows laser cutting over the entire part. One problem associated with cutting three-dimensional organic materials is that the beam that exits the zone being cut will continue until it hits something, which could be another section of the workpiece. Organic materials will burn under this exposure. Careful programming or metal shielding is often required to prevent damage.

A five-axis gantry is the system on which large, three-dimensional workpieces are best handled. The laser beam is sent by mirrors through three axes of translation and two of rotation to cover any shape within the working volume of the system.

Great flexibility can be achieved by arranging a laser beam so that it is manipulated by a five-axis robot. While much work has been done on this, severe problems remain (Ref 4), specifically laser beam delivery and robot ac-

Fig. 6 Capacitive probe with integral autofocus mechanism

curacy. The most versatile tool for laser cutting composites is a five-axis gantry machine with a CO_2 laser. Its lack of autofocusing mechanisms that are suitable for organic composites in these machines can probably be solved, leaving cost as the only obstacle to their adoption.

Focusing Heads. In all cutting systems, the beam must be focused by an optic device, which is almost always a lens, that must be maintained a set distance from the workpiece. This can be accomplished by one of three types of heads. A fixed cutting head, such as the simple nozzle around a focusing lens shown in Fig. 5, works only if the workpiece position can be accurately maintained. A gravity head, used with essentially flat material, can be attached to a collar and allowed to travel vertically. The collar has ball bearings that roll on the part surface, which maintains lens-to-workpiece distance but distorts thin material and mars finishes. A servo-controlled head can actively sense workpiece position via capacitive, contact, or optical probes that provide feedback to servo drives to maintain focus. Capacitive probes (Fig. 6) work well in all positions and do not protrude from the cutting head, but work only on conductive materials such as metal or a graphite-epoxy composite. Contact probes (Fig. 7) work on any material (although holes represent a problem) and operate best in the vertical direction. Optical probes (Ref 5) are conceptually superior, but have not been proved in industrial use.

Enclosures. Laser system manufacturers are required to certify compliance to federal safety

Fig. 7 Servo-controlled cutting head with contact probe sensing soft organic laminate prior to cutting. (Courtesy of Laserlab Australia)

standards (Ref 6). An interlocked enclosure is required for all material-working laser systems to prevent personnel from receiving either direct or scattered laser light exposure. Two-axis sheet cutters often enclose just the laser beam, while the workpiece is out in the open. Five-axis gantries or robots must be very carefully enclosed because they can send the raw laser beam in all directions.

When cutting composites, the primary hazard is often not the laser beam, but the fumes created by the laser-vaporized material. Thermal breakdown of organic resins and fibers yields irritating, toxic, and carcinogenic gases (Ref 7, 8). Ventilation of the enclosure must be designed to send these gases into filters that absorb them, because it is not enough to just duct them into the atmosphere.

Cutting Composites

It is somewhat misleading to specify the performance of lasers in cutting composites. Lasers are changing rapidly, while composites exhibit sufficient diversity to confound rational classification. In addition, both quality requirements for the cut and part geometry have a strong influence on process parameters. The absence of aerospace specifications for laser cutting of composites reflects this. General guidelines, combined with examples of current capabilities must suffice.

Organic-organic composites all exhibit the same behavior when exposed to CO_2 laser

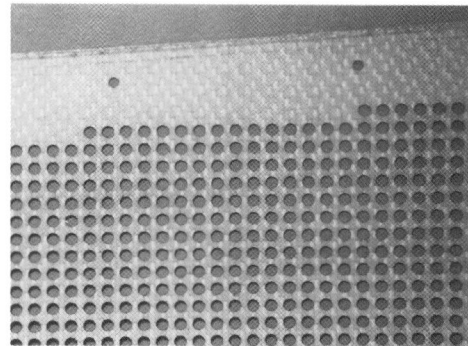

Fig. 8 3.2-mm (0.13-in.) holes cut in 2-mm (0.08-in.) thick aramid-epoxy aerospace component; note clean edges and absence of fraying. (Courtesy Laser Fare, Inc.)

Fig. 9 Small-tab laser cut in 1.5-mm (0.059-in.) thick fiberglass-reinforced polyester. The circular web between the inner hole and the outer U is 1-mm (0.04-in.) thick. Edge quality is good, with slight charring. In electrical applications, the conduction path created by this charring is often undesirable, so light vapor blasting is required to eliminate it.

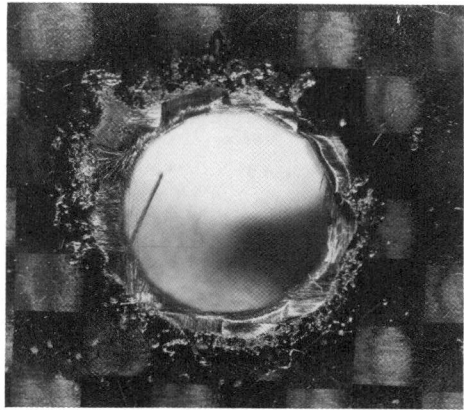

Fig. 10 Laser-cut 6-mm (0.235-in.) hole through a 3.2-mm (0.13-in.) thick graphite-epoxy laminate. Note matrix breakdown caused by heat conducted along the graphite fibers.

Fig. 11 Laser-cut 13-mm (0.50-in.) holes cut through 1-mm (0.04-in.) thick boron-aluminum. Recast aluminum forms a burr on the exit (top) side of the cut.

energy; therefore, their specific compositions are nearly irrelevant to the cutting process. Good results can be obtained with numerous organic-organic materials. For example, a standard laser application is to slit birch plywood for steel rule dies. The slits, produced through 19-mm (0.75-in.) thick wood, must have a kerf width of 0.7 ± 0.02 mm (0.03 ± 0.0008 in.) at the top and bottom of the board. Using a 500-W CO_2 laser, this cut is produced at a linear speed of 0.5 m/min (1.6 ft/min). A 1500-W CO_2 laser operates at 1.5 m/min (4.9 ft/min).

An aramid-epoxy composite, which is an otherwise intractable material, is easily cut by a CO_2 laser in thicknesses up to 10 mm (0.4 in.) (Fig. 8). A cut speed of 6 m/min (20 ft/min) can be achieved using 400 W for 3.2-mm (0.13-in.) material, or 1500 W for 6.35-mm (0.250-in.) material. Charring that is present on the cut edge can be removed by a vapor blast, if necessary (Ref 9). The high speeds required for clean cutting can be a liability, however, because motion systems cannot generate intricate patterns with high accuracy at a rate of several meters per minute. Lowering CW power results in excess burning, so pulsing is used for fine work. It is possible, for example, to cut a clean 2.5-mm (0.1-in.) diam hole through a 6.35-mm (0.250-in.) thickness of aramid-epoxy at a travel speed of 0.2 m/min (0.65 ft/min) with a CO_2 laser using 1500-W pulses that are 700 μs long, at a repetition rate of 220 Hz.

Inorganic-Organic Composites. Inorganic fibers require much more energy to break down than does an organic matrix. Laser cutting of these composites is essentially a process of melting the fibers; the matrix goes along for the ride. Thermal damage to the matrix is a problem because of heat conducted along the fibers. The best results are achieved with high power densities applied to thin materials.

When dealing with fiberglass-reinforced plastics (FRP), a G10 electrical grade material may be laser cut to a thickness of at least 10 mm (0.4 in.). Greater thicknesses can be penetrated, but charring becomes a problem.

Using a 1500-W CO_2 laser with CW power, a 6.35-mm (0.250-in.) material may be cut at 2 m/min (7 ft/min). The more common 1.5-mm (0.059-in.) thickness may be cut at 15 m/min (50 ft/min) using 1000 W. As with aramids, cutting at lower speeds is best done with a pulsed CO_2 beam (Fig. 9). An unusual laser application uses a 5-kW CO_2 laser to cut 6.35-mm (0.250-in.) FR4 at speeds up to 7.5 m/min (25 ft/min). This material is a fiberglass-reinforced polyester containing 20% calcium carbonate filler for fire resistance. The substantial edge charring produced by these parameters is not considered a problem. The fume generation, however, does cause difficulties for personnel and equipment.

Single-layer graphite-epoxy prepregs can be laser cut at a high speed. For 0.1-mm (0.004-in.) thick material, a 1500-W CO_2 laser can cut at 80 m/min (260 ft/min). However, this speed presents severe material-handling problems, especially because prepregs are rather limp. A proposed cutting system rests the prepreg sheet on a table with a rolling slot to handle very large parts. A problem with graphite-reinforced composites is that the graphite fibers are good thermal conductors, and cause the resin to cure near the edge of the cut, which is undesirable.

Cured graphite-epoxy laminates have the same problem in a more severe form. The fibers must be heated to 3600 °C (6500 °F) to vaporize. Energy is conducted along the fibers and radiated from the kerf in sufficient amounts to destroy the epoxy matrix for some distance from the cut edge. Using a 1500-W CO_2 laser, graphite-epoxy that is 3.2 mm (0.13 in.) thick can be cut at 2 m/min (7 ft/min). There will be a 1-mm (0.4-in.) thick zone next to the cut edge where the matrix is destroyed (Fig. 10).

Inorganic-Inorganic Composites. Although the constituents of inorganic composites may vary in their resistance to heat, there is less tendency to damage in the lower-melting species than in organic matrices. Materials that

resist conventional means of machining can be laser cut with good results.

For example, using a 63-mm (2.5-in.) focal-length lens to produce a minimum spot size, a 1500-W CO_2 laser can cut a 1-mm (0.04-in.) thick boron-aluminum composite at 8 m/min (25 ft/min). Cut quality using an air-assisted flow is good, with 0.2 mm (0.008 in.) of slag on the exit side of the kerf (Fig. 11). Pulsing allows good quality cuts at a lower speed. A YAG laser provides a slightly better cut quality than a CO_2 laser applied to a boron-aluminum composite because of its high pulse power. Typical parameters would be 0.3 m/min (1.0 ft/min), with an 80-mm (3-in.) focal-length lens on a pulsed 400-W unit.

Although it is resistant to heat, a silicon carbide-titanium (SiC-Ti) composite may be laser cut in thin layers. An argon-assisted flow should be used to avoid oxidation of the titanium matrix. Using a 1500-W CO_2 laser, 0.8-mm (0.03-in.) thick SiC-Ti was cut at 8 m/min (25 ft/min). Improved quality was achieved by reducing the cutting speed to 0.6 m/min (2.0 ft/min) and using 700-μs pulses of 1000 W at 220 Hz. Cutting 3.2-mm (0.13-in.)

thick SiC-Ti was accomplished with a CO_2 beam of 1.4-ms pulses of 1500 W at 220 Hz at 0.25 m/min (0.82 ft/min). The titanium matrix was observed to have melted back from the edge of the cut for a distance of about 0.1 mm (0.004 in.) (Fig. 4). The resulting slag adhered to the exit side of the cut to a height of 0.5 mm (0.02 in.). Again, a YAG laser provides slightly better quality than a CO_2 laser when used on this material. Typical parameters would be 0.15 m/min (0.50 ft/min) with an 80-mm (3-in.) focal-length lens on a pulsed 400-W unit.

ACKNOWLEDGMENTS

The author wishes to thank Mark Mello of Laser Fare Ltd. for his assistance in the preparation of this article.

REFERENCES

1. J. Darchuk and L. Migliore, Guidelines for Laser Cutting, *Lasers and Applications*, Sept 1985, p 91-97
2. *Guide for Material Processing by Lasers*, Laser Institute of America, Paul M. Harrod Company, 1977, Section 4, p 5; Section 5; Section 8, p 2
3. B. Ward, Supersonic Characteristics of Nozzles Used With Lasers for Cutting, in *Proceedings of the ICALEO 1984*, Laser Institute of America, Vol 44, 1985, p 94-101
4. D. Belforte, Robotic Manipulation of Laser Beams, Paper MS 84-500, Society of Mechanical Engineers, 1984
5. K. Sibayama and K. Itani, Three Dimensional Laser Cutting Machine, Paper MS 85-0495, Society of Mechanical Engineers, 1985
6. NCDRH 21CFR, Parts 1000 and 1040, Center for Devices and Radiological Health, 1985
7. D. Doyle and J. Kokosa, Hazardous By-Products of Plastics Processing With CO_2 Lasers, in *Laser Welding, Machining and Materials Processing*, C. Albright, Ed., Springer-Verlag, 1986
8. D. Doyle and J. Kokosa, in *The Changing Frontiers of Laser Materials Processing*, C.M. Banas and G.L. Whitney, Ed., Springer-Verlag, 1987
9. R.A. VanCleave, "Laser Cutting of Kevlar Laminates," BDX-613-1877, Bendix Corporation, Sept 1977

Adhesive Bonding Surface Preparation

Theodore J. Reinhart, Air Force Wright Aeronautical Laboratories

SURFACE PREPARATION of a material prior to coating or bonding is the keystone upon which the adhesive bond is formed. Extensive field service experience with paint coatings and structural adhesive bonds has repeatedly demonstrated that adhesive durability and longevity depend on the stability and bondability of the adherent surface. Combinations of heat, moisture, and stress have been shown to be particularly effective in discriminating among the various surface preparations used in the prebond conditioning or treatment of both metallic and composite surfaces.

Extensive research and development activities have been directed toward developing prebond surface treatments for metal alloys. The patent and technical literature abound with chemical treatments, all of which claim to provide the best bonding surface possible for these alloys.

Very little published information is available on prebond surface preparation for composite materials. However, Ref 1 provides quantitative data on the use of peel ply to improve composite-to-composite bond strength (which is also covered in this article) and the effect of silicone in degrading composite-to-composite bond strength.

Although it should be understood that the satisfactory performance of the bonded joint is the primary objective, new technology has made it possible to characterize in detail the chemical and physical properties of surface oxide layers on metals. Instrumental techniques, such as scanning electron microscopy, transmission electron microscopy, and surface analysis techniques, such as Auger and ion microprobe analysis, can be used to gain an intimate knowledge of the influence of surface preparation variables on the oxide that is produced. Therefore, with the availability of these powerful analytical tools, a significantly increased understanding of the required chemical and physical characteristics of a metal prebond surface can be expected. These instrumental analysis techniques are also expected to contribute to control of the reliability and reproducibility of bonded composite structures.

In the past, surface treatment evaluation techniques were limited to the stressed exposure of lap shear specimens or, perhaps, hot/wet peel testing. These tests, however, did not readily discriminate among surface treatments of varying durability. It was not until the recent development of the wedge-opening (double-cantilever beam) test by B. Bethune that a test for the discrimination of surface preparations became available.

In general, high-performance structural adhesive bonding requires that great care be exercised throughout the bonding process to ensure the quality of the bonded product. Chemical composition control of the polymeric adhesives; strict control of surface preparation materials and process parameters; and control of the adhesive lay-up, part fit-up, tooling, and the curing process are all required to produce, for example, airworthy structural assemblies. Of course, this is in contrast to the mechanical joining of components, which requires a much lower level of technology to obtain satisfactory performance. Given this situation and the inherent advantages of adhesive bonding compared to mechanical attachment, surface preparations are required that provide optimum adhesion and maximum environmental resistance, at least for critical aerospace equipment.

The formation of a stable oxide on the surfaces of metals to be bonded has been the subject of intense and continuing research. Work on aluminum alloys in recent years has declined with the introduction of the phosphoric acid anodize (PAA) treatment developed by B. Bethune. Although the data show that the PAA treatment, in combination with a corrosion-inhibiting primer, is superior in almost all aspects to any other method of surface preparation for aluminum alloys, it is still not universally accepted by the aerospace industry. The situation with composite materials is not quite as clear cut as that with metallic materials, and much basic and applied work remains to be done to characterize the desired surface characteristics for coating or structural adhesive bonding.

Needless to say, when producing adhesive-bonded flight hardware, whether a primary or secondary structure, great care is taken to obtain the best product possible with state-of-the-art technology. This usually requires the development of and strict adherence to detailed and comprehensive materials and process specifications, as well as end-item nondestructive inspection.

Because the strength properties of composite matrix resins and adhesive materials are very dependent on time, temperature, environment, and stress factors, it is important to know the service conditions of bonded composite joints. All of the factors mentioned above combine to influence the properties of composites and adhesives in ways that are not yet fully known or quantified, but can and do have detrimental effects in many cases. Therefore, materials and process selections should be fully validated by environmentally exposing bonded specimens and prototypes to the actual use conditions anticipated. In addition, such factors as cure cycle, cure pressure, surface preparation, machining, and fabrication will each have influences that must be understood. It is therefore advisable to characterize the specific material design properties and their statistical accuracy before designing with them.

Composite Surface Preparation

The preceding information is general in nature and applies to composites and noncomposites alike. Unlike isotropic materials (such as metals), orthotropic materials (such as advanced composites) may undergo severe damage and weakening when they are cut or machined and thus become more susceptible to interlaminar shear within the substrate. Therefore, bonded rather than mechanically fastened joints are more frequently used in advanced composite joint design. More specifically, bonded joint configurations, such as single and double laps or single- and double-strap joints, are preferred wherever feasible to avoid machining of the composites, even though a weight penalty may be imposed. Exceptions to this practice include beveling or tapering the ends of a lap or strap to alleviate the stress concentrations at the ends. This practice does not damage any fibers in the bond area.

Basic design practice for adhesive-bonded composite joints should include making certain that the surface fibers in a joint are parallel to the direction of pull to minimize interlaminar shear, or failure of the bonded substrate layer.

In designs in which joint areas have been machined to a step-lap configuration, for example, it is possible to have a joint interface composed of fibers at an orientation other than the optimal 0° orientation to the load direction. This tends to induce substrate failure more readily than would otherwise occur. Again, it cannot be over-emphasized that the condition of the surface to be bonded is the most critical parameter in producing reliable, reproducible, and durable adhesive-bonded joints.

Numerous surface preparation techniques are currently used prior to the adhesive bonding of composites. The success of any technique depends on establishing comprehensive material, process, and quality control specifications and adhering to them strictly. One method of preparing composite substrates prior to bonding consists of solvent wiping (or vapor degreasing) to remove surface oil and solvent-soluble contaminants, followed by water rinsing or alkaline degreasing to remove water-soluble materials and surface grime. The surface is then abraded by hand or machine. The first two steps remove solvent- and water-soluble surface contaminants, including waxes and silicones, and tend to remove loosely adhered boundary layer soil. The third step, abrasion, increases the surface energy of the surfaces to be bonded, and surface roughening increases bond area and mechanical interlocking. The abrading operation should be conducted with care, however, to avoid exposing or rupturing the reinforcing fibers of the surface.

Many surface-roughening approaches have been tried, and all have some merit. One method that has gained wide acceptance is the use of peel ply. In this technique, a closely woven nylon or polyester cloth is used as the outer layer of the composite during lay-up; this ply is torn or peeled away just before bonding or painting. The tearing or peeling process fractures the resin matrix coating and exposes a clean, virgin, roughened surface for the bonding process. The surface roughness attained can, to some extent, be determined by the weave characteristics of the peel ply. This method is fast and eliminates the need for solvent cleaning and the removal of abraded dust residue from mechanical abrasion surface treatment methods. In cases in which peel ply is not used, some type of light abrasion is required to break the glazed finish on the matrix resin surface. This can take the form of light sanding, grit honing, or vapor honing. Another approach is to use such products as Scotchbrite material. As a surface-conditioning treatment, this technique does not actually remove significant amounts of material from the composite surface, but it is nevertheless very effective in producing highly bondable surfaces.

Surface-conditioning techniques can be performed either wet or dry and can be automated for use in high-production situations. All surface treatments should have the following principles in common. The surface should be thoroughly cleaned prior to abrasion to avoid smearing the contamination into the surface. The glaze on the matrix surface should be roughened without damaging the reinforcing fibers or forming subsurface cracks in the resin matrix. Any residue should be removed from the abraded surface, which should be dried, if necessary. The prepared surface should be bonded as soon as possible. One such method of surface preparation is outlined as follows:

- Scrub entire surface of panel with a stiff brush and 2% aqueous solution of detergent at 40 to 50 °C (104 to 122 °F)
- Thoroughly rinse in tap water, then immerse in distilled water
- Dry in forced-air oven at 65 °C (150 °F) for 30 min
- Abrade bonding area plus 13 mm (0.5 in.) in margin with 240-grit sandpaper
- Vacuum abraded area
- Swab abraded area with acetone and wet, clean gauze
- Rinse entire panel liberally in distilled water
- Dry in forced-air oven at 100 °C (212 °F) for 60 min, then bond parts immediately

The prebond surface treatment of composite materials for subsequent coating, such as painting, or for structural adhesive bonding is neither standardized nor well understood, and there is no general agreement within the industry as to the best treatment for a specific application. Considering the importance of surface preparation in bond performance, it is advisable to understand the ways in which the bonding process affects joint performance.

In cases in which the composite formulation itself contains internal plasticizers, lubricants, and/or mold release agents, it may be necessary to remove these materials from the surface to be bonded to achieve optimal bond strength and durability. There are many approaches to removing these materials. Their success depends on exact knowledge of the ingredients. These substances can include one or more of the following classes of materials: stearates (metallic soaps), polyethylene (waxy material) oils, and polymeric plasticizers, but, it is hoped, not silicones. Solvent- or water-base detergents and mild acid washes can be effective in removing these materials. In the presence of these materials, wash primers may be of some value in bonding.

Some form of surface preparation is mandatory for the satisfactory structural adhesive bonding of fiber-reinforced resin matrix composites. Adhesives, surface preparations, bonding processes, and joint designs must be thoroughly tested in geometries and under service conditions that closely match the actual service environment. Accelerated or compressed-time testing may be required for expediency, but care should be taken so that potential failure modes are not hidden or masked by the testing. Lastly, because structural adhesive bonding is a high-technology process, strict adherence to detailed materials and processing specifications is mandatory, as is control of the chemical composition of all materials used.

REFERENCE

1. L.J. Matienzo, J.D. Venables, J.D. Fudge, and J.J. Velton, Surface Preparation for Bonding Advanced Composites, in *Proceedings of the 30th National SAMPE Symposium*, Society for the Advancement of Material and Process Engineering, March 1985, p 302-314

SELECTED REFERENCES

- N.M. Bikales, Ed., *Adhesion & Bonding*, Interscience, 1974
- E.J. Bruno, Ed., *Adhesives in Modern Manufacturing*, Society of Manufacturing Engineers, 1970
- C.V. Cagle, *Adhesive Bonding*, McGraw-Hill, 1968
- N. DeLollis, *Metalbonding*, Reinhold, 1980
- R. Houwink and G. Solomon, *Int. J. Adhesion Adhesives*, Vol 2, 2nd ed., Elsevier, p 65
- T.J. Rienhart, "Evolution of Structural Adhesives," Paper presented at SAMPE conference on Adhesive Bonding, Los Angeles, CA, Society for the Advancement of Material and Process Engineering, 1982
- Leeves and Natatojan, Systems Approach to Bonding, *Int. J. Adhesion Adhesives*, Vol 8, p 16

Adhesives Selection

John Williams, Naval Air Development Center
Weldon Scardino, Air Force Wright Aeronautical Laboratories

COMPOSITE COMPONENTS are frequently assembled into larger structures through adhesive bonding. Adhesives can be used to jig components temporarily before drilling and bolting or to bond components together permanently. Finished components that are damaged during assembly or service can also be repaired with bonding techniques.

Adhesive-bonding techniques are used only if subsequent disassembly of the subcomponents is unlikely. Bonded joints may be preferred if the joint is to be highly stressed, particularly if cyclic loads are anticipated or if thin composite sections are to be joined when bearing stresses in bolted joints would be unacceptably high.

Bonded joints may also be preferred if mechanical joints would require extensive drilling and bolting operations. Forming bonded joints by automated techniques is less expensive than mechanically fastening joints, particularly if hot-melt or fast-cure adhesives are used.

Adhesive Joint Design

In a structural adhesive joint, the load in one component must be transferred through the adhesive layers to other components. The efficiency with which this can be done depends on the design of the adhesive joint, the characteristics of the adhesive, and the adhesive/substrate interface. Because the adhesive joint is not reinforced with continuous fiber, the strength of the bulk adhesive in the joint is normally lower than that of the composite substrate. In order to transfer stress through the adhesive, the bonded area usually must be larger than the cross-sectional area of the substrates. This is achieved by overlapping substrates as far as possible. Figure 1 shows typical joint designs for adhesively bonded joints of composite to metal substrate.

The stress distribution in joints such as these has been predicted by using finite-element techniques (Ref 1). Figure 2 shows a typical stress distribution for a double lap shear joint.

The adhesive is required to withstand relatively high local loads near changes in joint section. Therefore, adhesives designed to carry high loads tend to be relatively soft and have lower moduli than the matrix resins in the composite substrate. Adhesives are frequently rubber modified; this practice sacrifices modulus in order to improve fracture toughness and fatigue life. However, in well-designed, toughened adhesives (second-phase toughening), only moderate reduction of the continuous-phase modulus occurs.

The criteria for selecting an adhesive must be considered in view of the joint design. The joint design often must ensure that the adhesive is loaded in shear as far as possible. Cleavage and peel loading (Fig. 3) should be avoided when using adhesives.

Selection Criteria

The first criterion for selecting an adhesive is the capacity to withstand the required stresses in the full range of environmental conditions to which the component may be exposed. The class of adhesive may be dictated by the loading conditions. For example, if high peel loads are unavoidable (such as when a flexible facing is to be bonded to a rigid composite substrate), a highly flexible rubber-base adhesive would probably be more suitable than a rigid epoxy system.

In-service environmental conditions, independent of the required stresses, should also be considered as a selection criterion because, for example, flexible polyurethane adhesives are more sensitive to moisture degradation than the equivalent polychloroprene (neoprene) rubber formulations.

Long-term stress conditions can also affect adhesive selection. Thermoplastic adhesives and highly flexibilized epoxy systems may be susceptible to creep under sustained load. Rigid, glassy adhesives may have limited resistance to cyclic load conditions.

The practicality of the application process must also be considered, specifically, the mode of surface preparation; the restriction on handling prepared surfaces before bonding; the method of mixing (if necessary) and applying the adhesive to the joint; the procedure for

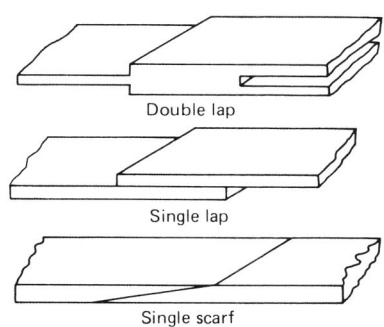

Fig. 1 Typical joint designs

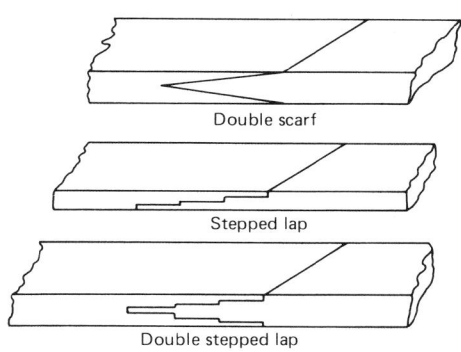

Fig. 2 Typical adhesive shear stresses in bonded joints, measured at room temperature

Table 1 Typical characteristics of adhesive types

Type	Form	Cure temperature, °C (°F)	Maximum use temperature, °C (°F)	Advantages	Disadvantages
Epoxy	Two-part paste	Room or accelerated at 93-178 (200-350)	Generally below 82 (180)	Ease of storage at room temperature; ease of mixing and use; long shelf life; gap filling when filled	Not generally as strong or environmentally resistant as typical heat-cured epoxies
	One-part film	121 (250)	To 82 (180)	Covers large areas; bondline thickness control; wide variety of formulas; higher-temperature curing materials; better environmental properties	Store at 18 °C (0 °F); short shelf life; high-temperature cure; brittle and low peel strength
		149 (300) 178 (350)	149-177 (300-350)		
Acrylic	Two-part liquid or pastes	Room to 100 (212)	105 (221)	Fast setting; easy to mix and use; good moisture resistance; tolerant of surface contamination	Strong, objectionable odor; limited pot life
Polyurethane	One or two parts	Room or heat cure	. . .	Good peel; good for cryogenic use	Moisture sensitive before and after cure
Silicone.	One- and two-part pastes	Room to 260 (500)	To 260 (500)	High peel and impact resistance; easy to use; good heat and moisture resistance	High cost; low strength
Hot melt	One-part	Melt at 190-232 (375-450)	18-171 (120-340)	Rapid application; fast setting; low cost; indefinite shelf life; nontoxic; no mixing	Poor heat resistance; special equipment required; poor creep resistance; low strength; high melt temperature
Bismaleimide (BMI).	One-part paste or film	>178 (350) and 246 (475) postcure	232 (450)	Structural bonds with bismaleimide composites; higher temperature than epoxies; no volatiles; good shelf life	Brittle and low peel; limited formulas available
Polyimide.	Thermoplastic liquids; one- and two-part pastes	260 (500) and postcure	204-260 (400-500)	High-temperature resistance; structural strength	High cost; low peel strength; high cure and postcure temperatures; volatiles for some forms
Phenolic-based	One-part films	163-177 (325-350)	To 177 (350)	High-temperature use	Low peel strength

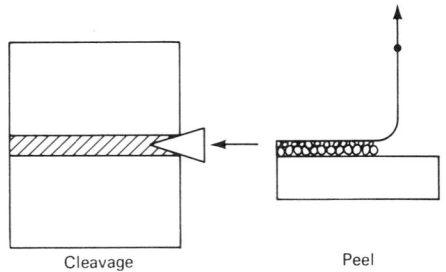

Cleavage   Peel

Fig. 3 Phenomena to avoid in adhesive use

Table 2 Use-temperature guide to structural adhesives

Peel: L, low; M, medium; H, high. Lap shear: P, poor; Mod, moderate; G, good; V, very good; E, excellent. Peel is indicated first, followed by lap shear: peel/lap shear.

Adhesive	Use temperature, °C (°F)								
	−253 (−423)	−196 (−320)	−73 (−100)	−54 (−65)	Room	82 (180)	149 (300)	216 (420)	260 (500)
Epoxy-nitrile modified	L/V	L/V	L/E	L/E	H/E	M/V	L/Mod	. . .	. . .
Epoxy-nylon.	L/E	L/E	L/E	L/E	H/V	L/G	L/Mod	. . .	. . .
Epoxy-phenolic	. . .	. . .	L/V	L/V	L/V	L/V	L/G	G	G
Vinyl-phenolic	. . .	. . .	. . .	L/V	M/E	H/E	L/Mod	. . .	. . .
Nitrile-phenolic	Mod	E	E	L/E	H/V	M/G	L/Mod	. . .	. . .
Bismaleimides	. . .	. . .	. . .	Mod	L/G	L/G	L/G	L/V	. . .
Polyimides	. . .	. . .	. . .	L/V	L/G	L/G	L/G	L/G	L/G
Polyurethanes.	H/V	H/V	H/V	H/G	H/G	H/Mod	H/P	. . .	. . .
Acrylics.	. . .	. . .	. . .	L/P	H/E	M/G	L/P	. . .	. . .

melting or curing the adhesive; the jigging necessary for support of the joint during bonding; and the methods for cleaning the joint after bonding, if necessary, and cleaning application equipment. The elapsed time required by the adhesive-curing process may often be overlapped with other thermal treatments, such as curing or postcuring composite subcomponents or annealing reinforced thermoplastic subcomponents. Adhesive characteristics are listed in Tables 1 and 2. Adhesive selection and use criteria are summarized below:

- The adhesive must be compatible with the adherends and able to retain its required strength when exposed to in-service stresses and environmental factors
- Selection tests for structural adhesives should include stressed durability testing for heat, humidity (and/or fluids), and stress, simultaneously
- Proper joint design should be used, avoiding peel loading whenever possible
- When received, the adhesive should be tested for compliance with the purchase specification. Preferably, this should include both physical and chemical tests, such as infrared, moisture content, resin content, base resin, secondary resins, curing agent, and accelerator

- The adhesive should be stored at the lowest available and recommended temperature
- Cold adhesive should always be warmed to room temperature in a sealed container
- Paste mixes should be degassed, if possible
- Adhesives with volatile evolution should be avoided
- The humidity in the lay-up area should be below 40% relative humidity for most formulations. Lay-up room humidity is adsorbed by the adhesive and is released later during heat cure as steam, yielding porous bondlines and interfering with the cure chemistry

- The recommended pressure and the proper alignment fixtures should be used
- Heat curing is almost always preferred, because it yields bonds that have greater strength, heat, and humidity resistance
- When curing for a second time, such as during repairs, the temperature should be at least 28 °C (50 °F) below the earlier cure temperature. If this is not possible, then a proper and accurate bond form must be used to maintain all parts in proper alignment during the second cure cycle
- Traveler coupons should always be made for testing. These are test coupons that duplicate the adherends to be bonded in material and joint design. The coupon surfaces are prepared by the same method and at the same time as the basic bond. Coupons are also bonded together at the same time with the same adhesive (mix lot, and so on) of the basic joint and subjected to the same curing process simultaneously with the basic bond. Ideally, traveler coupons are cut from the basic part, on which extensions have been provided
- Peel forces should be avoided. Low-modulus (nonbrittle) adhesives having high peel strength should be used where this is a factor
- Tapered ends should be used on lap joints to feather out the edge-of-joint stresses
- Surface preparation should be conducted carefully, avoiding contamination of the bondline with moisture, oil, and so on
- The exposed edge of the bond joint should be protected with an appropriate sealer, such as an elastomeric sealant or paint. Honeycomb assemblies should be hermetically sealed
- For aerospace structural bonding, a full description of good practices can be found in Ref 2

Epoxy Adhesives

Epoxy adhesives are the most commonly used adhesives for assembling and repairing composite structures. A wide range of one-part and two-part systems is available. Some systems cure at room temperature, while others require elevated temperatures.

The advantages of epoxy adhesives include high-strength bonds, particularly to rigid substrates such as composites and metals; good resistance to hostile environments; and good retention of strength over long periods of sustained or alternating load. The disadvantages include sensitivity to surface treatment, especially for metal substrates, and a relatively slow cure process for production applications. Two-part epoxy adhesives usually require mixing in precise proportions to avoid significant loss of cured properties and environmental stability.

Chemistry. Epoxy resins for adhesive use are generally supplied as liquids or low melting temperature solids. They are most commonly bifunctional in epoxy groups, although higher functionalities are available. They can be cured by a variety of curing procedures, including admixture with the stoichiometric proportion of polyfunctional primary amine or acid anhydride. The amine or anhydride groups react with the epoxy groups by a simple addition reaction to give a densely cross-linked product. Some epoxy compositions can be cured through a homopolymerization reaction initiated by strong organic bases (or rarely, acids). These compositions are less sensitive to the mix ratio, but are seldom encountered as two-part systems. The rate of the reaction may be adjusted by adding accelerators in the initial formulation or by increasing the temperature. Because the reaction is also dependent on the mobility of the reaction species, relatively little reaction is possible once a highly cross-linked structure is formed. To improve structural properties, particularly at elevated temperatures, it is common to expose the partially cured material to a postcure at temperatures close to (or preferably over) the maximum use temperature for the structure.

Epoxy resin systems are usually modified by a wide range of additives that control particular properties. These additives include accelerators, viscosity modifiers and other flow control additives, fillers and pigments, flexibilizers, and toughening agents. The versatility of these materials has led to the development of a wide range of epoxy adhesives for specific applications.

Two-Part Room-Temperature Curing Epoxy Liquids and Pastes. The most widely known epoxy adhesives are supplied as two-part systems that cure when mixed in the appropriate proportions at room temperature. These are available as clear liquids or as filled pastes with a consistency ranging from easily mixed thixotropic liquids to heavy-duty putties. The cure time for room-temperature curing systems may be as short as 5 min, but structural applications usually require 8 to 12 h. Full strength for such systems is usually not developed until after 2 to 7 days at room temperature. It is common to postcure room-temperature cured systems for limited periods, such as 1 h at 100 °C (212 °F) or 4 to 6 h at 60 °C (140 °F). The properties of fast-cure epoxy systems are not generally improved by postcuring, and these materials tend to have low strength and tend to be susceptible to attack by such environmental agents as water and common organic liquids. They are also very limited in physical strength, particularly at slightly elevated temperatures (> 50 °C, or 122 °F).

The curing agent used in room-temperature curing epoxy systems often contains aliphatic amines. These materials are primary skin sensitizers, and some caution is necessary to avoid direct contact.

In general, room-temperature cured epoxy systems are relatively brittle glasses. Many toughened systems are commercially available and are used in large volume in structural applications. Their upper use temperature is associated with a transition from glassy to rubbery, generally in the range of 50 to 105 °C (122 to 221 °F), depending on the curing agent and the postcure schedule. Shear properties are very good, but peel properties tend to be low. The formulation of this class of adhesives allows a trade-off of shear strength for improved peel strength (or, preferably, toughness) by using flexibilizers. This type of additive reduces the elastic modulus of the system, which reduces system sensitivity to local regions of high stress concentration, resulting in an apparent increase in toughness.

Heat-Cured Two-Part Epoxy Adhesives. The use of primary amines with reduced basicity allows the formulation of two-part systems, with significant storage life of the mixed system at room temperature. Curing agents based on cycloaliphatic amines can be formulated for use up to 120 °C (248 °F); curing agents based on aromatic amines, up to 160 °C (320 °F). Anhydride-cured systems can be used at temperatures to 250 °C (482 °F). Some reaction is still detectable at room temperature. The pot life is defined as the period between the time of mixing the resin and curing agent and the time at which the viscosity has increased to the point when the adhesive can no longer be successfully applied as an adhesive. This time may range from 2 h to several days, depending on the system and the quantity mixed. The physical properties of heat-cured systems are normally superior to those of room-temperature cured systems. The cured materials are rigid glasses with high elastic moduli and with glass transition temperatures, T_g, of 100 to 200 °C (212 to 392 °F). Adhesives based on these materials generally have low peel strengths and are not recommended for applications where high peel loads are expected or for the bonding of dissimilar materials whose coefficients of thermal expansion are markedly different. It is possible to flexibilize these materials, but this results in a serious degradation of elevated-temperature properties by reducing T_g.

Special formulating techniques are used to improve the fracture toughness of these materials. The usual approach is to incorporate a semicompatible reactive rubber into the adhesive system. During cure, phase separation occurs, leaving particles of cross-linked epoxy-modified rubber distributed through a matrix of slightly flexibilized, rigid glass. The rubber particles operate by several mechanisms to increase fracture toughness markedly, with only a slight degradation in high-temperature performance.

Rubber-toughened epoxy systems are very common in the structural bonding of composites in the aircraft industry. The slight flexibilizing action of the rubber reduces the upper operating temperature to the region of 120 to 160 °C (248 to 320 °F).

One-Part Solid Epoxy Adhesives. The use of low melting temperature solid epoxy resins and high-temperature curing agents allows the formulation of solid epoxy adhesives with indefinite shelf lives at room temperature. Catalytic curing agents, particularly dicyandiamide, are common in these applications. Formulating variables are very restricted for these systems, which must be prepared as melts and cooled rapidly to give cast rods that may subsequently be ground to powder form. Cure is generally at a relatively high temperature, and the product has a very high modulus and a high glass transition temperature. There is, however, virtually no useful way to flexibilize these systems, which consequently have very low peel strengths. These systems should never be used when peel forces are possible during normal use. A tough, one-part paste that is storage stable and easily applied can also be made.

Epoxy Film Adhesives. The most common compromise between the shelf life of one-part systems and formulating flexibility is the use of mixes that require refrigerated storage. Structural adhesives for aerospace applications are generally supplied as thin films supported on a release paper and stored under refrigerated conditions (-18 °C or 0 °F). Film adhesives are available using high-temperature aromatic amine or catalytic curing agents with a wide range of flexibilizing and toughening agents. Rubber-toughened epoxy film adhesives are widely used in the aircraft industry. The upper temperature limit of 118 to 178 °C (250 to 350 °F) is usually dictated by the degree of toughening required and by the overall choice of resins and curing agents. Film materials are frequently supported by fibers that serve to improve handling of the films prior to cure, control adhesive flow during bonding, and assist in bondline thickness. Fibers can be incorporated as short-fiber mats with random orientation or as woven cloth. Commonly encountered fibers are polyesters, polyamides (nylon), and glass. Adhesives containing woven cloth may have slightly degraded environmental properties because of wicking of water by the fiber. Random mat scrim cloth is not as efficient for controlling film thickness as woven cloth, because the unrestricted fibers move during bonding, although new spun-bonded nonwoven scrims do not move and are therefore widely used. In manufacturing composite components, techniques that allow co-curing of matrix and adhesive are common when bonding composites to metals.

Acrylic Adhesives

Acrylic adhesives are derived from substituted acrylic acid esters, which readily homopolymerize if exposed to a source of free radicals. Peroxides and related active oxygen compounds are usually used to initiate polymerization. The polymerization of acrylics results in a large cure shrinkage that would cause severe cure stresses if the acrylic were used directly as an adhesive. To overcome this problem, it is common to dissolve polymerized acrylic in the monomer to give a viscous solution that has a reduced volume change after polymerization. Second-generation acrylics have dissolved high polymer reactant components to help generate cure, reduce shrinkage, and provide toughness. The use of substituted acrylic derivatives and prepolymers allows considerable formulation versatility in developing acrylic adhesives. For example, incorporating a small proportion of polyfunctional acrylic derivatives results in the formation of cross-linked networks.

Acrylics with long-chain ester substituents (for example, butyl methacrylate) are relatively soft, while methyl methacrylates have higher modulus and strength values. Blending methyl methacrylate and butyl methacrylate allows the formulation of adhesives with a wide range of flexibility. Highly reactive substituted acrylates, such as cyanoacrylates, allow the formulation of adhesives that can cure within seconds of application.

Cure Chemistry. Acrylic esters can be polymerized through free radical or anionic mechanisms. Free radical polymerization can be initiated by organic peroxides and some azo compounds; benzoyl peroxide is an example. Normally, the polymerization of such monomers as methyl methacrylate with benzoyl peroxide proceeds at a significant rate only at elevated temperatures (60 to 100 °C, or 140 to 212 °F). To obtain polymerization at lower temperatures, combinations of activators and catalysts are used. Activators induce peroxide initiation at lower temperatures, and catalysts increase the rate of polymerization. Typical activators are reducing agents, and typical accelerators are tertiary amines and some polyvalent metal salts.

Anionic polymerization is commonly used only with highly activated acrylate derivatives. In particular, cyanoacrylic esters, such as ethyl cyanoacrylate, will cure rapidly if mixed with trace amounts of water or amines, which catalyze the reaction. Sufficient water is normally absorbed onto substrate surfaces to allow the cure of thin bondlines in a few seconds. If a gap-filling adhesive is required (because of poor fit of the adherends), an amine catalyst can be used to coat one surface of the joint prior to bonding.

Some acrylic adhesive systems are formulated to take advantage of the fact that atmospheric oxygen can inhibit polymerization. These anaerobic adhesives are supplied in small, partially filled containers that maximize contact of the one-part adhesive with the atmosphere; that is, the ratio of exposed surface to volume is high. Once these materials are trapped in an adhesive joint, inhibition by oxygen is no longer possible, and rapid polymerization occurs.

It is characteristic of acrylic adhesives that polymerization, once initiated, occurs virtually completely and develops an adhesive of high molecular weight and T_g, independent of cure temperature. The T_g and maximum operating temperature are primarily determined by the selection of monomers and prepolymers, not by the cure initiator system or cure conditions Maximum operating temperatures are usually limited to about 105 °C (221 °F), which is the T_g of polymethyl methacrylate. Prolonged exposure to higher temperatures may cause thermal depolymerization of many polyacrylates, which reduces long-term thermal stability.

A second characteristic of acrylic adhesives is their ability to dissolve grease and thus wet-out surfaces that have been contaminated with oils and greases. Acrylic adhesives are very insensitive to the surface preparation of substrates as compared to other glassy adhesives, such as epoxies. Acrylic adhesives are not widely used in structural bonding of aerospace composites.

Rubber-base Adhesives

Rubber-base adhesives generally have low shear strengths as compared to glassy systems, but have a high peel strength. They are often used in applications in which high bond strengths are not required, such as when very large bonded areas are used. They are strongly recommended when peel forces are high, such as when a flexible panel is to be bonded to a rigid composite substrate. Rubber adhesives are usually applied from solution or as hot-melt adhesives, which will be covered in the section ''Hot-Melt Adhesives'' in this article.

Rubber solutions are based on natural and synthetic rubbers of high molecular weight, which are dissolved in hydrocarbon or chlorinated hydrocarbon solvents. Resins (such as terpene resins) that improve tack and additives that protect the rubber from degradation by atmospheric oxygen (antioxidants) are usually incorporated into the solution. The solution is painted onto both adherends, which are allowed to flash dry before their surfaces are pressed together. The last traces of solvent are lost through diffusion. If one of the surfaces is sufficiently porous (for example, a textile), the flash dry period can be reduced or eliminated. Adhesives of this type are often referred to as contact adhesives and are usually used with composites only when decorative films are to be bonded onto exterior surfaces. Some rubber adhesives may be cross linked or vulcanized through unsaturated species in the rubber. Typical unsaturated rubbers include natural rubber and polychloroprene. Elementary sulfur or peroxides are typical cross-linking agents for rubber, and they usually require heat to complete cure. This reaction cross links the preexisting macromolecules. Cross-linked rubber adhesives have improved shear strength and creep resistance as compared to simple solution adhesives, but lose some peel strength. Rubber adhesives usually have poor resistance to solvents and fuels and to atmospheric conditions for long periods, especially if not cross-linked.

Reactive Rubber Adhesives. Some structural adhesive systems are based on monomers that polymerize *in situ* to produce synthetic rubbers. Included in this class are polyurethane, silicone, and polysulfide adhesives.

Polyurethane adhesives exhibit a good compromise between shear strength and peel strength. These adhesives are based on a reacting mixture containing polyisocyanate and either hydroxy-terminated polyethers or polyesters. The isocyanate used is usually derived from aromatic diisocyanate, and the polyhydroxy compound is normally aliphatic. The aromatic components give rise to rigid polymer segments, while the aliphatic material gives rise to flexible polymer segments. A wide range of cured adhesive properties can be obtained by controlling the monomer or prepolymer units. The two-phase attributes of polyurethane systems allow the formulation of very tough adhesives with good shear strengths. Because the adhesive is already in the rubbery phase, the mechanical properties are not dramatically affected by exposure to elevated temperatures. Shear strength tends to fall monotonically as temperature increases, while peel strength may either fall slowly or show an indistinct maximum. The upper operation temperature is then poorly defined. The problems associated with urethane adhesives include the toxicity of the isocyanate component and its sensitivity to attack by water.

Silicone adhesives are based on polyfunctional siloxanes and are available as one-part and two-part systems. One-part systems are generally based on blocked catalysts that are hydrolyzed by atmospheric moisture to liberate the active catalyst. Two-part systems are based on complex cure chemistry involving acids and polyvalent metal salts. The cure rate may vary from minutes to hours depending on the activity of the catalyst. Some systems, particularly one-part systems, are associated with the formation of acidic reacting systems, which may cause corrosion of metallic components.

Silicone adhesives generally have poor shear strength and excellent peel strength, and they can be used to high temperatures (up to 220 °C, or 428 °F). Silicone adhesives do not readily wet-out many surfaces and usually require a primer on the substrate for good adhesion.

Polysulfide adhesives are generally encountered in composite structures as sealants rather than adhesives. Polysulfides are based on sulfur-containing prepolymers that are cross linked through inorganic oxidants and polyvalent metal compounds. Polysulfide systems can provide reasonable shear and peel strengths and have excellent environmental durability, particularly in the presence of fuel and water.

Hot-Melt Adhesives

Hot-melt adhesives are a major class of adhesive used for bonding composite structures. These materials are thermoplastic solids that are applied to a substrate at a temperature well above their liquification temperatures. The structure is assembled, and the polymer is allowed to cool and solidify. The advantage of hot-melt adhesives is the speed with which the bond can be formed under production conditions. The adhesive is supplied as powder, granules, blocks, or film. A wide range of equipment is available to dispense the adhesive, heat it and the substrate to the high temperatures needed for bonding, and clamp the structure until it has cooled sufficiently to allow handling and subsequent assembly stresses. Hot-melt systems are used in large-scale manufacturing operations, such as the assembly of composite and metal automotive components.

The requirements for using hot-melt adhesives in structural applications are resistance to stress under operational environments and the ability to liquefy and wet-out the substrate in a reasonable time at a reasonable temperature without excessive thermal degradation. To ensure that adequate time is available for joint assembly, hot-melt adhesives are usually heated well above their melting points. The melt viscosity should be low but not so low as to cause excessive flow during bonding.

The polymers used for structural applications include the semicrystalline engineering thermoplastics, the most common of which are polyamides and polyesters. Polysulfone is used for high-temperature applications. Polyamides are particularly useful because they melt rapidly to a low-viscosity fluid that has reasonable thermal stability. However, the time between application to the substrate and final assembly (open time) is very short, because the thermal stability of the melt is low and processing temperatures are close to the melting point. Polyesters require high temperatures to achieve a viscosity that is low enough for wetting. Most additives for these adhesives are designed to improve heat resistance in the melt (antioxidants) and tack in the wet state and to decrease melt viscosity.

The advantages of hot-melt adhesives are speed of application and convenience. Mechanical properties include good shear strength and moderately good peel strength. Resistance to solvents and fuels is good. For high-peel applications, the use of a separate class of hot-melt adhesives based on thermoplastic rubbers would be preferable. Temperature resistance is directly related to application temperatures.

Other Considerations

Alternatives to adhesive bonding include redesigning to integrate separate parts into larger components and using mechanical attachments such as fasteners and bolts. Another attractive alternative is the use of adhesive with mechanical fasteners, particularly to aid in assembling the adhesive joint and to reduce peel loads near the edge of lap joints, or to provide a safety factor if the adhesive debonds during use.

Surface Preparation. The most common problem associated with adhesives is surface preparation. For composites, very little surface preparation is needed to ensure an adequate bond with most adhesives. However, some classes of adhesives, particularly epoxies, are sensitive to surface contamination from greases, oils, and moisture after surface preparation. A common method of avoiding surface contamination is to use a peel ply on the composite. This is a sacrificial layer (usually of tightly woven, treated nylon or polyester fabric) that is laminated onto the surface during lay-up. This layer can be peeled off immediately before bonding to allow the adhesive to interact with a clean, freshly roughened surface.

Surface preparation faults are usually encountered with metal substrates, especially if the metal is prone to forming a loose oxide coat. Magnesium, aluminum, and steel are particularly sensitive to faults of this type when bonded with rigid, structural adhesives. Careful control of the adherend surface is necessary to avoid this problem. A full description of surface preparation is provided in the article "Adhesive Bonding Surface Preparation" in this Section of this Volume.

Inadequate Fit of Adherends. Some adhesives are not suitable for joining surfaces with poor fit. The acrylic adhesives that rely on either adsorbed moisture on one adherend or the catalyst application to one adherend are particularly sensitive. These systems will reliably fill gaps only up to 0.25 mm (0.010 in.). If fit tolerance is greater than this, special precautions should be taken, or another adhesive should be selected. Highly fluid adhesives may show excessive flow in joints with poor fit when surface forces are not adequate to prevent adhesive runoff. In this case, thixotropic, filled adhesives would be a better choice.

Void Formation During Bonding. Void formation is a problem that is common to all adhesives. Air becomes entrapped during mixing of two-part adhesives and during assembly of joints. Entrapped air will usually be preferentially expelled during the pressurization of mated surfaces unless the pressure is inadequate for flow or the adhesive is too fluid. If the surface is very rough or poorly fitted, air may be entrapped in resin-rich areas. The most common source of voids in heat-cured adhesives is the volatilization of traces of volatile materials in the adhesive. The most common contaminant that causes voids is water. In adhesive systems cured under conditions in which water can boil, it is very important that moisture does not contaminate the adhesive or that the adhesive does not gel below the boiling point of water under bonding conditions.

Joint Distortion. If dissimilar adherends are bonded with heat-cured systems, considerable thermal stresses may develop and result in premature failure or distortion of the assembly upon cooling. This problem can be prevented through careful system design. It can be minimized by reducing the temperature at which the adhesive gels, but a subsequent postcure will be needed to raise the T_g of the adhesive, as required.

Specifications and Standards. There are two types of specifications and standards: those issued by the government and those issued by industry associations, such as the American Society for Testing and Materials (ASTM), and Society of Automotive Engineers (SAE). Additional information is available in the article "Adhesives Specifications" in this Section. Industry documents can be purchased from the appropriate agency; government documents are free. Generally, the government specifications and standards and the SAE documents refer to material properties and recommended practices or processing methods, while the ASTM specifications refer to test methods. The individual indices of ASTM and SAE should be consulted. The *Department of Defense Index of Specifications and Standards* lists all government specifications and standards and many industry standards that have been adopted in lieu of government documents.

REFERENCES

1. Structural Adhesives, in *Standard Handbook of Plant Engineering*, McGraw-Hill, 1983
2. "Adhesive Bonding (Structural) for Aerospace and Other Systems, Requirements for," MIL-A-83377, Military Specification

SELECTED REFERENCES

- *Adhesive Bonded Aerospace Structures, Standardized Repair Handbook*, MIL-HNDBK 337, U.S. Government Printing Office
- *Adhesives Desk-Top Data Bank*, 3rd ed., The International Plastics Selector, Inc., 1980-1981
- *Adhesives in Modern Manufacturing*, Society of Manufacturing Engineers (SME), 1970
- *Adhesives Red Book, Directory of the Adhesives Industry*, Communication Channels, Inc., 1982
- D. Brewis and J. Comyn, Ed., Advances in Adhesive Application, Materials and Safety, Warwick Publishing, 1983
- J.S. Przemiemecki, Ed., *Composite Materials for Aircraft Structures*, American Institute of Aeronautics and Astronautics, 1986

Adhesives Specifications

Weldon M. Scardino, Consultant

SPECIFICATIONS, including standards, recommended practices, and handbooks, constitute the laws and bylaws that enable designers, manufacturers, and users to communicate in a rational and orderly manner. The various types of specifications available include those generated by individual users or manufacturers as well as the public specifications that will be discussed in detail in this article. Some specifications are simple, such as a data sheet from an adhesive manufacturer. Others are more complex, such as those established by aerospace manufacturers or federal and military specifications. Definitions of specifications and related terms are given in the Appendix to this article.

The Armed Services Procurement Regulations or Defense Acquisition Regulations establish the following mandatory list of specifications for use by the Department of Defense (DoD) in the procurement of supplies and services covered by such specifications:

- Federal specifications, unless determined by the DoD to be inapplicable
- Military specifications approved by the DoD
- Industry documents adopted by the DoD as listed in the *Department of Defense Index of Specifications and Standards*

The lists and descriptions that follow include the principal adhesive-related specifications used by the aircraft industry for metal and composite bonding. Many more specifications that cover special properties or situations can be found in the *Department of Defense Index of Specifications and Standards* and in indices published by the American Society for Testing and Materials (ASTM) and the Society for Automotive Engineers (SAE). It should be noted that no attempt has been made to reproduce the specifications in detail; the purpose of this article is to provide an overview of appropriate specifications and some insight into their content and use.

Although this article relates mainly to composites, it also contains many references to metal surface preparation methods and various metal specimen test methods. This is necessary for two reasons. First, metals are frequently used with composites in composite assemblies,

and, second, there are few adhesive-related composite test methods. Therefore, many investigators are adapting typical metal-to-metal test methods to their composite test needs. The easiest way to accomplish this is to tailor the compliance of composite specimens to match that of metal specimens in typical test methods.

Processing Specifications

ARP 1524 (15 April 78) "Surface Preparation and Priming of Aluminum Alloy Parts for High-Durability Structural Adhesive Bonding, Phosphoric Acid Anodizing." Prepared by SAE.

Scope. This aerospace recommended practice (ARP) describes the processing system and techniques for the surface preparation and priming of aluminum alloy parts through the use of phosphoric acid anodizing (PAA) to achieve optimal durability. The specification was developed for 2024, 7075, and 7475 alloys in the hardened condition.

Content. Discussed is the entire PAA process for aluminum alloy parts where the highest-strength, most durable bonds are necessary. Steps in the process include solvent cleaning, alkaline cleaning, deoxidizing, phosphoric acid anodizing, applying the adhesive, assembling the parts, and curing the assembly. The topics covered in detail are racking procedures, handling precautions, parts storage, equipment requirements (including materials and construction), quality control, process and equipment monitoring, cleaning and etching solutions, voltages, and all process steps.

Notes. This process was primarily recommended for all structural bonding of aluminum alloys using 121 °C (250 °F) and lower curing adhesives and the normal 2000- and 6000-series alloys. Recent data show that the process may also be preferred for 177 °C (350 °F) curing adhesives. The superior environmental durability of adhesively bonded aluminum alloy joints using this process has been demonstrated by many researchers. The process is patented.

ARP 1575 (15 Oct 79) "Surface Preparation and Priming of Aluminum Alloy Parts for High-Durability Structural Adhesive Bonding, Hand-Applied Phosphoric Acid Anodizing." Prepared by SAE.

Scope. This aerospace recommended practice describes a hand-applied, non-tank, PAA method for the surface preparation of aluminum alloys for structural adhesive bonding to achieve optimum bondline durability.

Content. A step-by-step description of the process for hand-applied PAA of aluminum parts prior to bonding is given. The document also contains data on the handling of parts, repair atmosphere environment, safety precautions, and quality assurance provisions. The basic items of equipment for this process are a dc power source (6-V battery is acceptable if large enough to supply the current requirements), stainless steel screen for a cathode, jelled 10 to 12% phosphoric acid, and cheesecloth to serve as an insulator between the screen and the aluminum surface to be anodized.

Use. This anodizing method is intended for use in preparing relatively small areas of aluminum surfaces for repair bonding, either at the bench level or on an aircraft.

Notes. This process is not patented and has been described in the open literature. It can be used on aircraft surfaces, even upside down. Vertical surfaces present the most problems. Additional details and data describing the work leading to this document are given in Ref 1.

MIL-A-9067C (16 March 61) "Adhesive Bonding, Process and Inspection Requirements for—Cancelled; use MIL-A-83377 instead." Prepared by the Bureau of Naval Weapons.

Scope. This specification provides detailed guidance for the preparation of contractors' documents for the processing and inspection of adhesive-bonded parts.

Content. The specification contains requirements for contractors' process specifications, engineering reports, and so on, including guidance on materials handling, part processing, sampling and inspection of adhesive-bonded assemblies, and recommended cleaning solutions.

Notes. Issued in 1961, this specification is out of date and does not reflect the state-of-the-art. It is included here only because many older contracts still specify it. It has been replaced for all military services after 23 Feb 1979, by MIL-A-83377 (see the following).

MIL-A-83377B (6 Oct 78) "Adhesive Bonding (Structural) for Aerospace and Other

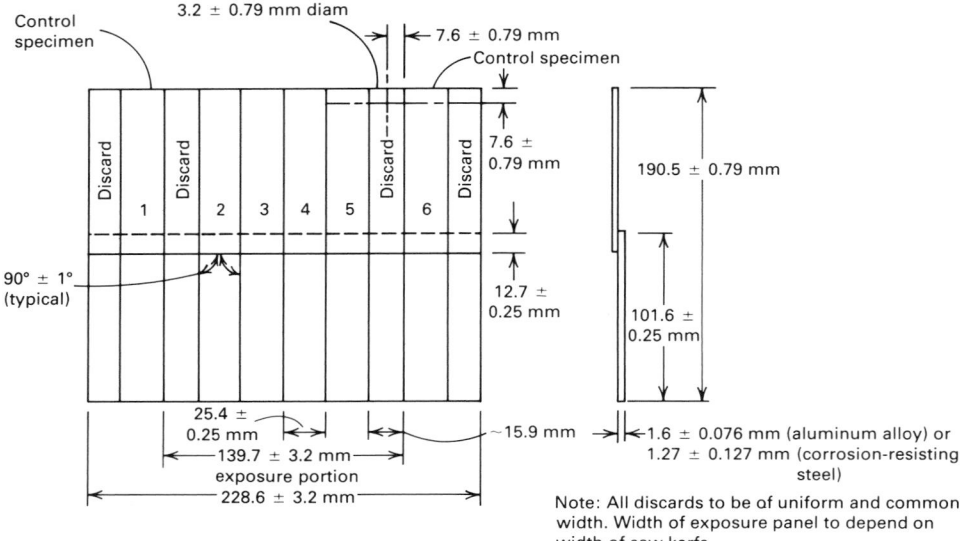

Fig. 1 MMM-A-132 exposure test panel

Note: All discards to be of uniform and common width. Width of exposure panel to depend on width of saw kerfs.

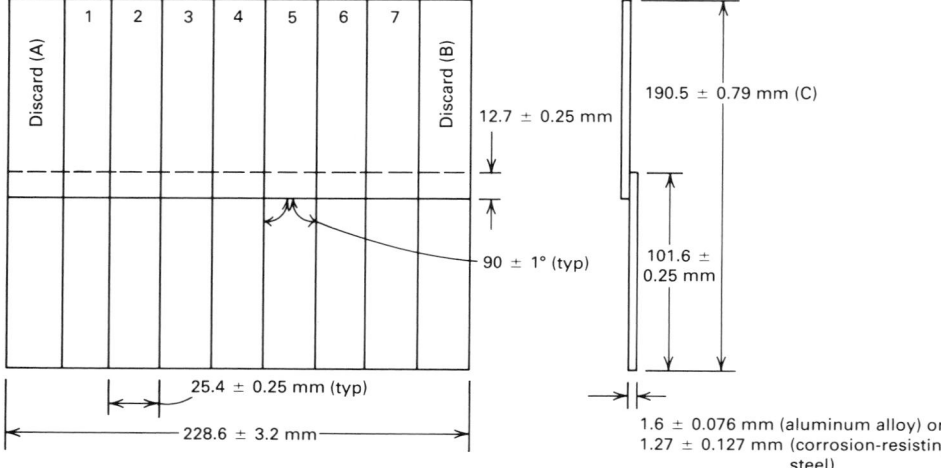

Fig. 2 MMM-A-132 standard lap shear test panel. A, All discards to be of common width as determined by the width of the cutting tool; B, 9.53 ± 0.25 mm (0.375 ± 0.010 in.) for fatigue specimen panel; C, 193.7 ± 0.79 mm (7.625 ± 0.030 in.) for fatigue specimen panel

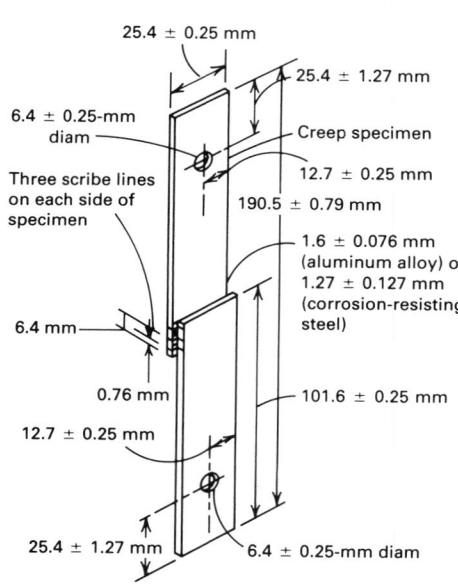

Fig. 3 MMM-A-132 creep rupture test specimen

Material Property Specifications

MMM-A-132A(3) (12 April 84) "Adhesives, Heat-Resistant, Airframe Structural, Metal to Metal." Prepared by the Naval Air Systems Command.

Scope. Discussed are the requirements for heat-resistant adhesives for use in bonding primary and secondary structural and external metallic airframe parts that will be exposed to temperatures from −55 to 260 °C (−67 to 500 °F).

Use. These adhesives are intended for the bonding of metal-to-metal areas of airframe structural components.

Content. This is the primary specification for adhesives used in metal-to-metal aircraft structural bonding. In the absence of any adhesive specification oriented solely toward composites, this is the most applicable specification for composite bonding as well as metal-to-metal bonding. Types I through IV (long-time and short-time exposures at various temperatures) are covered. Materials can be in liquid or film form. Requirements are included for application; pot life; curing time, temperature, pressure, and postcure; storage life; preparation of test panels (Fig. 1 to 6); and sampling, packaging, and so on. Product qualification is required, and a Qualified Products List (QPL) is maintained in conjunction with the specification.

Notes. This is a federal specification. It is not intended for use in honeycomb sandwich construction unless the adhesive also meets MIL-A-25463.

QPL-MMM-A-132-8(3) (3 April 84) "List of Products Qualified Under Federal Specification MMM-A-132." Prepared by the Naval Air Systems Command.

Systems, Requirements for." Prepared by the Aeronautical Systems Division, Wright-Patterson Air Force Base.

Scope. The requirements applicable to the structural adhesive bonding of metal, composites, and core material in any combination are addressed in this specification.

Content. Contained in this publication are a classification of structures, documentary requirements for contractors, precautions on bonding to clad aluminum, material and process selection criteria, underwater leak checking of honeycomb assemblies, recommended surface preparation method, corrosion protection precautions, bonding personnel qualifications, process area physical requirements, part-

processing steps, quality assurance measures, and nondestructive inspection data.

Use. This specification is mandatory for use by contractors to ensure the reliability of adhesive-bonded structural components used in aerospace and other systems.

Notes. Many requirements reflecting the state-of-the-art in structural adhesive bonding are included. This document has evolved over several years and has been discussed and commented on in detail by practitioners in the aerospace industry. The document is under revision to enlarge and clarify the composites-bonding aspects and to reduce the allowable relative humidity in the bond lay-up room from 60 to (probably) below 40%.

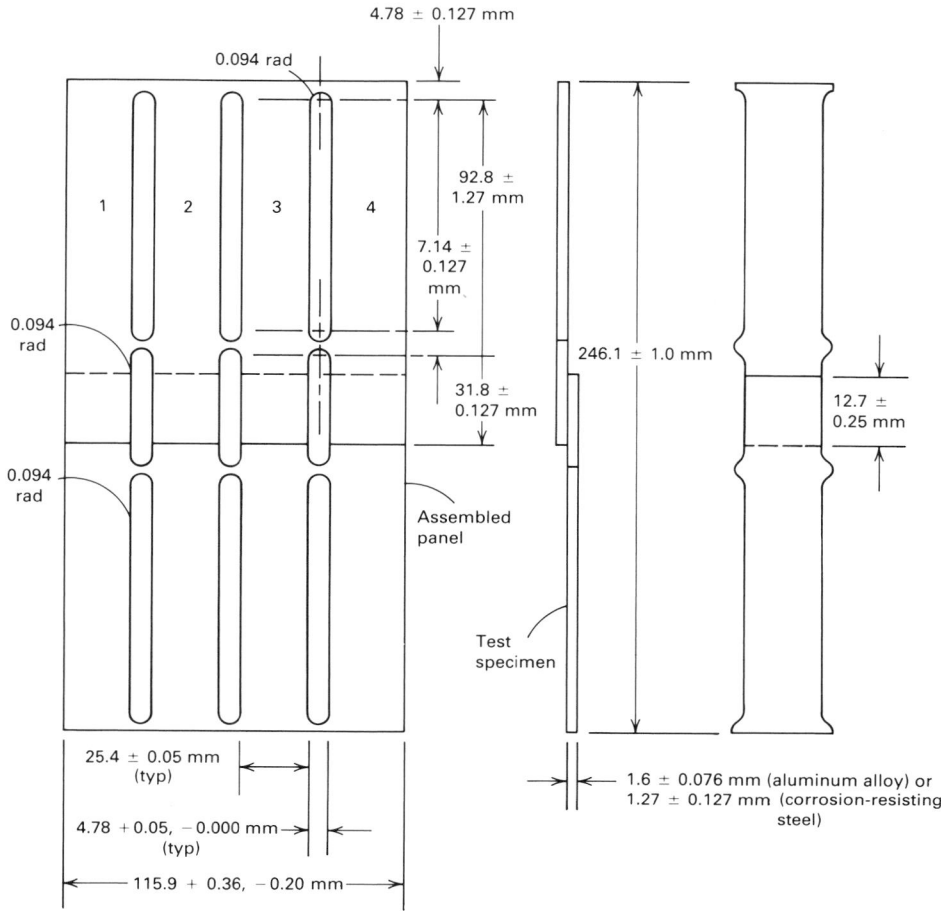

Fig. 4 MMM-A-132 optional lap shear panel

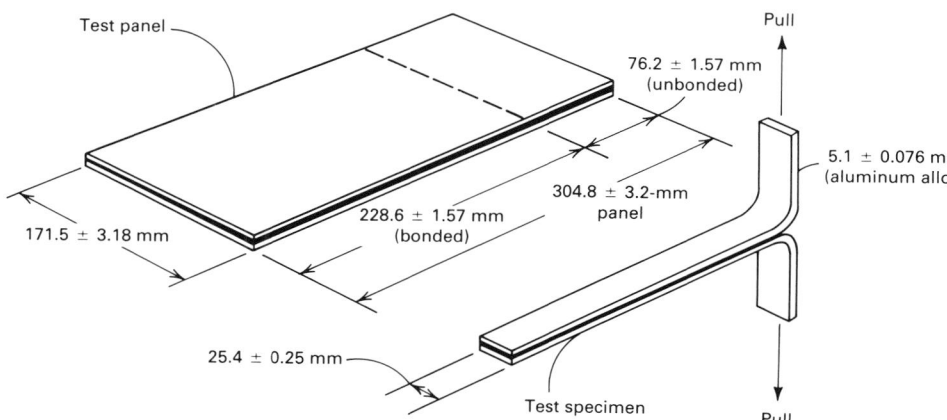

Fig. 5 MMM-A-132 T-peel test specimen

Content. This qualified products list (QPL) contains a list of adhesives and their primers (where applicable) that have been determined by the qualifying agency to meet the requirements of Specification MMM-A-132.

MIL-HDBK-691B (12 March 87) "Military Standardization Handbook, Adhesive Bonding (replaces MIL-HDBK 691A, 17 May 65)." Prepared by the Army Materials Technology Laboratory, Watertown Arsenal.

Scope. Fundamental guidelines are provided for engineers and designers concerned with adhesive bonding.

Content. This handbook contains discussions on adhesives in engineering design; general design data for adhesive joints; designs for specific materials; adhesive selection, bonding conditions, and preparation for bonding; application, assembly, and curing; reliability; process control; and inspection.

Notes. This handbook has recently been updated.

AMS 3107 (1 April 83) "Primer, Adhesive, Corrosion-Inhibiting, for High-Durability Structural Adhesive Bonding," Prepared by SAE.

Scope. This aerospace material specification (AMS) supersedes AMS 3106A (1 March 74). Along with its four supplementary specifications, it covers corrosion-inhibiting, modified-epoxy primers in the form of ready-to-use sprayable liquids.

Use. The adhesive primers covered in this specification and its supplements are intended primarily for use on metal surfaces in preparation for the high-durability structural adhesive bonding of sandwich panels and metal-to-metal attachments.

Content. Test data and requirements for uncured-primer color, solids content, inhibitor content, weight per volume, viscosity, sprayability, and pot life are included. The requirements for cured film on panels include adhesion, flexibility, impact resistance, hardness, fluid resistance, corrosion resistance, heat resistance, low-temperature shock, compatibility with sealant, and compatibility with top coat. The requirements for the cured primer used with its specified adhesive include room-temperature lap shear and metal-to-metal peel.

AMS 3107/1 (1 April 83) "Primer, Adhesive, Corrosion Inhibiting, High-Durability Epoxy, −55 to 95 °C (−65 to 200 °F)." This material must meet the requirements of the basic specification and must be compatible with AMS 3695/1 epoxy film adhesive and MIL-C-83286 polyurethane coating.

AMS 3107/2 (1 April 83) "Primer, Adhesive, Corrosion-Inhibiting, High-Durability Epoxy, −55 to 120 °C (−65 to 250 °F)." This material must meet the requirements of the basic specification and must be compatible with AMS 3695/2 epoxy film adhesive and MIL-C-83286 polyurethane coating.

AMS 3107/3 (1 April 83) "Primer, Adhesive, Corrosion-Inhibiting, High-Durability Epoxy, −55 to 175 °C (−65 to 350 °F)." This material must meet the requirements of the basic specification and must be compatible with AMS 3695/3 epoxy film adhesive and silicone resin-base top coat.

AMS 3107/4 (1 April 83) "Primer, Adhesive, Corrosion-Inhibiting, High-Durability Epoxy, −55 to 215 °C (−65 to 420 °F)." This material must meet the requirements of the basic specification and must be compatible with AMS 3695/4 epoxy film adhesive and silicone resin-base top coat.

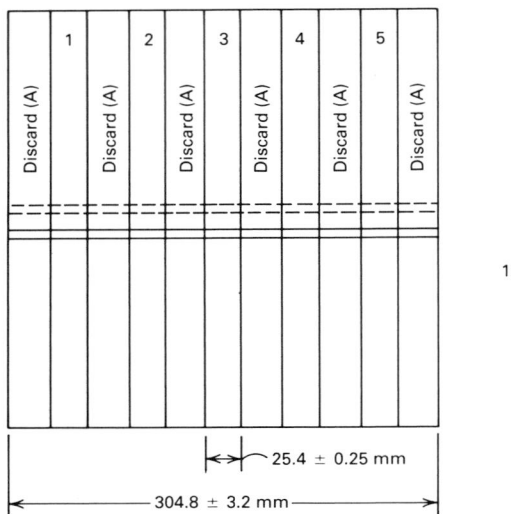

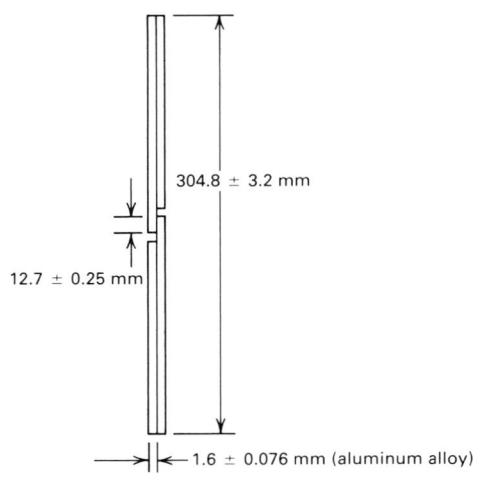

Fig. 6 MMM-A-132 blister detection test panel. A, All discards to be of common width as determined by width of cutting tool

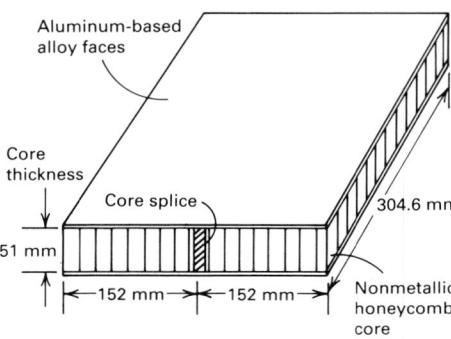

Fig. 7 AMS-3688 peak exotherm test specimen

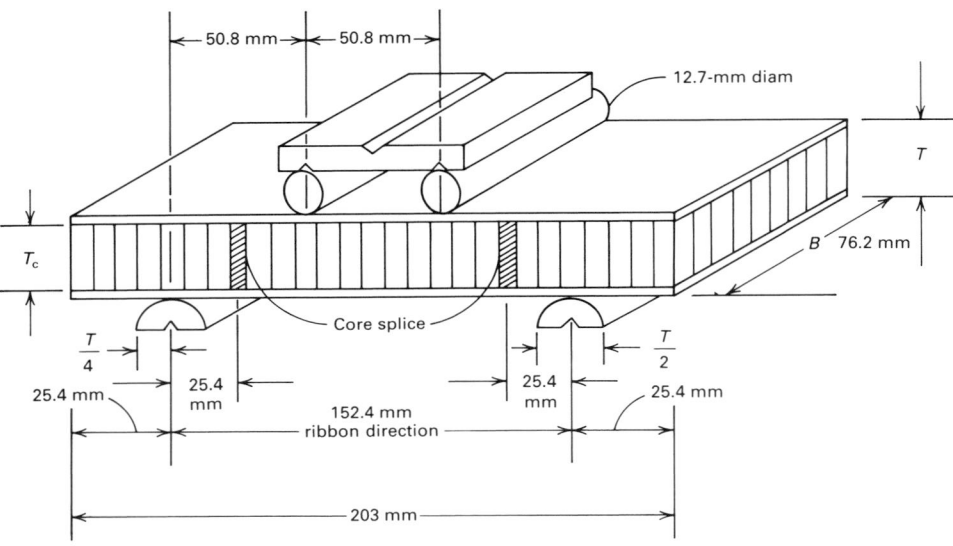

Fig. 8 AMS 3688 beam shear test specimen (core splice). *B*, specimen width (76.2 mm); *T*, specimen thickness; T_c, core thickness

AMS 3686 (15 Sept 75) "Adhesive Polyimide Resin, Film and Paste, High-Temperature Resistant, 315 °C or 600 °F." Prepared by SAE.

Scope. A high-temperature, electrical-grade polyimide resin adhesive (film or paste) is discussed.

Use. This adhesive is intended for the bonding of polyimide laminate faced sandwich structures in radome assemblies for use between −55 to 315 °C (−67 to 600 °F).

Content. Included are test data and properties on uncured adhesive; quality; size and tolerances for film adhesive; and tests for solids content of paste adhesives, volatile content of film adhesives, weight of film adhesive, tensile shear strength, and flatwise tensile strength. Typical properties range from 16.5 MPa (2400 psi) lap shear at room temperature to 8.3 MPa (1200 psi) at 315 °C (599 °F) after 100 h at 315 °C (599 °F). Flatwise tensile properties for honeycomb core at 25, 260, and −55 °C (77, 500, and −67 °F) are also given.

AMS-3688A (1 Oct 81) "Adhesive, Foaming Honeycomb Core Splice, Structural, −55 to 82 °C (−67 to 180 °F)." Prepared by SAE.

Scope. This specification covers a foaming-type, heat-curing, resin-base adhesive in paste or sheet form, nominally 121 °C (250 °F) curing.

Use. This adhesive is used for splicing aluminum or nonmetallic honeycomb core, gap filling, and providing a shear tie between core edges and inserts or edge members in honeycomb assemblies.

Content. The requirements covered include those for storage life, working life, curing properties (sag, expansion, and exotherm) (Fig. 7), cured properties (density and beam shear) (Fig. 8), quality, dimensional sheet tolerance, and quality assurance.

AMS 3689 (1 Oct 81) "Adhesive, Foaming, Honeycomb Core Splice, Structural, −55 to 177 °C (−67 to 350 °F)." Prepared by SAE.

Scope. A foaming-type, heat-curing, resin-base adhesive in paste or sheet form, nominally 177 °C (250 °F) curing, is covered in this specification.

Use. This adhesive is used for splicing aluminum or nonmetallic honeycomb core, gap filling, and providing a shear tie between core edges and inserts or edge members in honeycomb assemblies.

Content. Included are requirements for storage life, working life, curing properties (sag, expansion, and exotherm), cured properties (density and beam shear), quality, dimensional sheet tolerance, and quality assurance.

Notes. A companion specification, AMS 3688, covers nominal 121 °C (250 °F) curing core splice adhesive.

AMS 3693B (1 July 83) "Adhesive, Modified Epoxy, Moderate Heat-Resistant, 120 °C (250 °F) Curing, Film Type." Prepared by SAE.

Scope. Covered in this specification is a modified epoxy adhesive in the form of supported film to be supplied in rolls or sheets. This specification is similar to MMM-A-132, Type 1, Class 2, and MIL-A-25463, Type 1, Class 2.

Use. This supported-film adhesive is primarily for the structural bonding of metallic alloys and rigid nonmetallic surfaces to themselves and to each other and for the bonding of internal and external structural honeycomb components used in the operating range of −55 to 80 °C (−65 to 180 °F).

Content. The resin shall conform to MMM-A-132, Type 1, Class 2, and to MIL-A-25463, Type 1, Class 2. Shelf life must be at least 6 months at −18 °C (0 °F). Cured-product requirements cover tensile shear at three different temperatures to 80 °C (180 °F) and after salt-spray exposure, humidity exposure, and fuel immersion; flatwise tensile strength and flexural strength at various temperatures; T-peel at 24 °C (75 °F); sandwich peel at three different temperatures to 80 °C (180 °F); creep; and corrosivity.

AMS 3695 (1 April 83) "Adhesive Film, Epoxy Base for High-Durability Structural Adhesive Bonding." Prepared by SAE.

Scope. Along with supplementary specifications, this specification covers film adhesives compounded from modified epoxy resins in the form of ready-to-use sheet, either supported by mat or woven monofilaments or unsupported.

Use. This material is primarily intended for bonding metal-to-metal or aluminum honeycomb sandwich assemblies for service over the temperature range specified in the applicable detail specifications. The adhesive is to be used with AMS 3107 corrosion-inhibiting primer and cured in accordance with the applicable detail specification, thus forming an adhesive-primer system. The adhesive must not have a deleterious effect on the surface of materials being bonded.

Content. Uncured-adhesive requirements include color, solids content, weight, thickness, and working life. Cured-adhesive requirements (using metal-to-metal specimens) include tensile shear (dry and after exposure to humidity, salt spray, aromatic fuel, JP-4 fuel, phosphate ester fluid, hydraulic fluid, anti-icing fluid, diester lubricating oil, and polyol ester), fatigue, creep rupture, climbing-drum peel (dry, after humidity exposure, and after salt-spray exposure), crack extension test, and sustained stress loading. Requirements for the cured adhesive for metal-to-honeycomb use include climbing drum peel (dry) and flatwise-tensile.

AMS 3695/1 (1 April 83) "Adhesive Film, Epoxy Base, High Durability, for 95 °C (200 °F) Service."

Use. This adhesive is intended for use with the AMS 3107/1 corrosion-inhibiting primer to form an adhesive-primer system meeting the requirements of the basic AMS 3695 specification.

AMS 3695/2 (1 April 83) "Adhesive Film, Epoxy Base, High Durability, for 120 °C (250 °F) Service."

Use. This adhesive is intended for use with AMS 3107/2 corrosion-inhibiting primer to form an adhesive-primer system meeting AMS 3695 requirements.

AMS 3695/3 (1 April 83) "Adhesive Film, Epoxy Base, High Durability, for 175 °C (350 °F) Service."

Use. This adhesive is intended for use with AMS 3107/3 corrosion-inhibiting primer to form an adhesive-primer system meeting AMS 3695 requirements.

AMS 3695/4 (1 April 83) "Adhesive Film, Epoxy Base, High Durability, for 215 °C (420 °F) Service."

Use. This adhesive is intended for use with AMS 3107/4 corrosion-inhibiting primer to form an adhesive-primer system meeting AMS 3695 requirements.

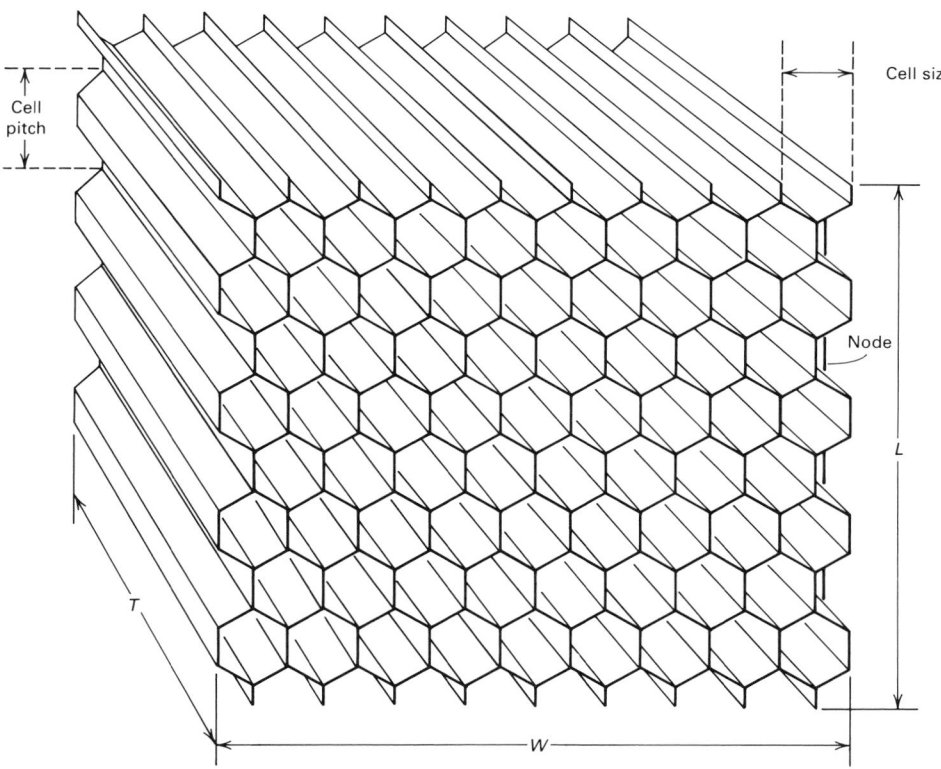

Fig. 9 MIL-C-7438F core reference areas. L, longitudinal direction (parallel to core ribbons); W, transverse direction (perpendicular to core ribbons); T, thickness

MIL-C-7438F (Amendment 1) (1 Feb 77) "Core Material, Aluminum, for Sandwich Construction." Prepared by the Naval Air Engineering Center, Engineering Specifications and Standards Department.

Scope. This specification covers aluminum core material for structural sandwich construction.

Use. This material is intended for structural use in a sandwich construction in airframes and other applications. Specification values should not be used for design allowables.

Content. Included are two grades and one class: Grade B for exposure to 177 °C (350 °F), Grade C for exposure to 221 °C (430 °F), and Class 2—Treated. Also contained are requirements for alloy, finish, trade name and code number, configuration (Fig. 9), perforation (lack of), form, physical properties, color coding, quality assurance, and so on. Alloys covered are 5052, 5056, and 2024. Designated core foil ranges from 0.02 to 0.15 mm (0.0007 to 0.0060 in.). Cell sizes include the range from 4 to 9.5 mm ($^5/_{32}$ to $^3/_8$ in.) and densities from 0.016 to 0.192 g/cm³. Corresponding compressive strengths range from 138 kPa to 15.65 MPa (20 to 2270 psi). Longitudinal shear strength varies from 221 kPa to 9.48 MPa (32 to 1375 psi), and transverse shear strength varies from 138 kPa to 5.65 MPa (20 to 820 psi). Tests are described for core density, cell size, cell pitch, foil thickness, core flatwise compressive

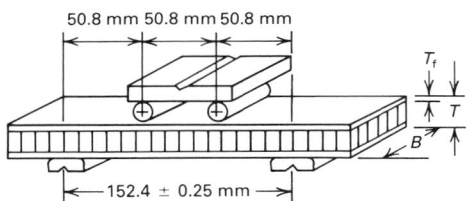

Fig. 10 MIL-C-7438F beam shear test for high-density cores. End support plates are 25.4 × 76.2 × 6.4 mm (1.0 × 3.0 × 0.25 in.) with grooves for alignment. Loaded edges are rounded to 1.5-mm (0.060-in.) radius. Load bars are 12.7 mm (0.50 in.) round. Tolerances as indicated. B: specimen width, 76.2 mm (3.0 in.); T: specimen thickness; T_f: facing thickness. Specimen length: 203 mm (8.0 in.)

strength, shear strength (Fig. 10), delamination strength, and corrosion resistance.

Notes. This fully coordinated specification originally covered two classes: treated core or corrosion-resistant and untreated or noncorrosion resistant. The untreated class was deleted in Amendment 1, which states that treated material may be used where Class 1 (untreated) was previously specified. The intent was to eliminate the use of untreated aluminum core in honeycomb structural sandwich construction.

MIL-C-8073D (5 Dec 79) "Core Material, Plastic Honeycomb, Laminated Glass-Fabric Base, for Aircraft Structural Application." Prepared by the Naval Air Systems Command, Naval Air Engineering Center.

Scope. This specification covers the requirements for glass fabric base plastic honeycomb core materials for aircraft structural applications, including aircraft exterior parts, such as radio and radar antenna housings.

Use. The core material can be used for radar antenna housings and for general aircraft structural parts, depending on class of material.

Content. Types I, II, and III and Classes I and II are covered.

MIL-A-25463B (31 March 82) "Adhesive Film Form, Metallic Structural Sandwich Construction." Prepared by the Naval Air Engineering Center, Engineering Specifications and Standards Department.

Scope. This specification covers film adhesives for bonding metal faces to metal cores and to metal components of sandwich panels for use in primary and secondary structural airframe parts that may be exposed to temperatures to 260 °C (500 °F).

Use. This material is intended for use in bonding sandwich constructions of various materials, provided the specification tests have been made with the material combinations of interest. The test values of the specification are obtained with aluminum facings and aluminum core.

Content. Types I through IV (long-time and short-time exposure at various temperatures) and classes 1 and 3 (bonding metal facings to various components) are covered. The adhesive must meet the properties of the applicable types of MMM-A-132 plus the properties of this specification. Requirements for qualification; working characteristics, including application and curing parameters; storage life; mechanical properties (Fig. 11); and quality assurance provisions are included. Four cure temperature groups are covered.

QPL-25463-30 (22 Oct 84) "Products Qualified Under MIL-A-25463." Prepared by the Naval Air Systems Command.

Content. This document is a list of the adhesives (and their primers, where applicable) that have been determined by the qualifying agency to meet the specification requirements.

MIL-P-47276 (MI) (1) (17 Aug 76) "Primer, Bonding." Prepared by the U.S. Army Missile Command, Redstone Arsenal.

Scope. One type of liquid organic nitrile-phenolic resin primer is discussed.

Use. This primer is intended for use with nitrile-phenolic film adhesives that will be pressed between two metal surfaces.

Content. Requirements are provided for toxicity-flammability identification, quality assurance, and such physical properties as solids, viscosity, density, and shear strength.

Notes. This specification is a limited coordination document that does not contain provisions for peel or wedge testing or for environmental durability.

MIL-C-81986 (AS) (1) (19 Nov 74) "Core Material, Plastic Honeycomb, Nylon Paper Base; for Aircraft Structural Applications." Prepared by the Naval Air Engineering Center.

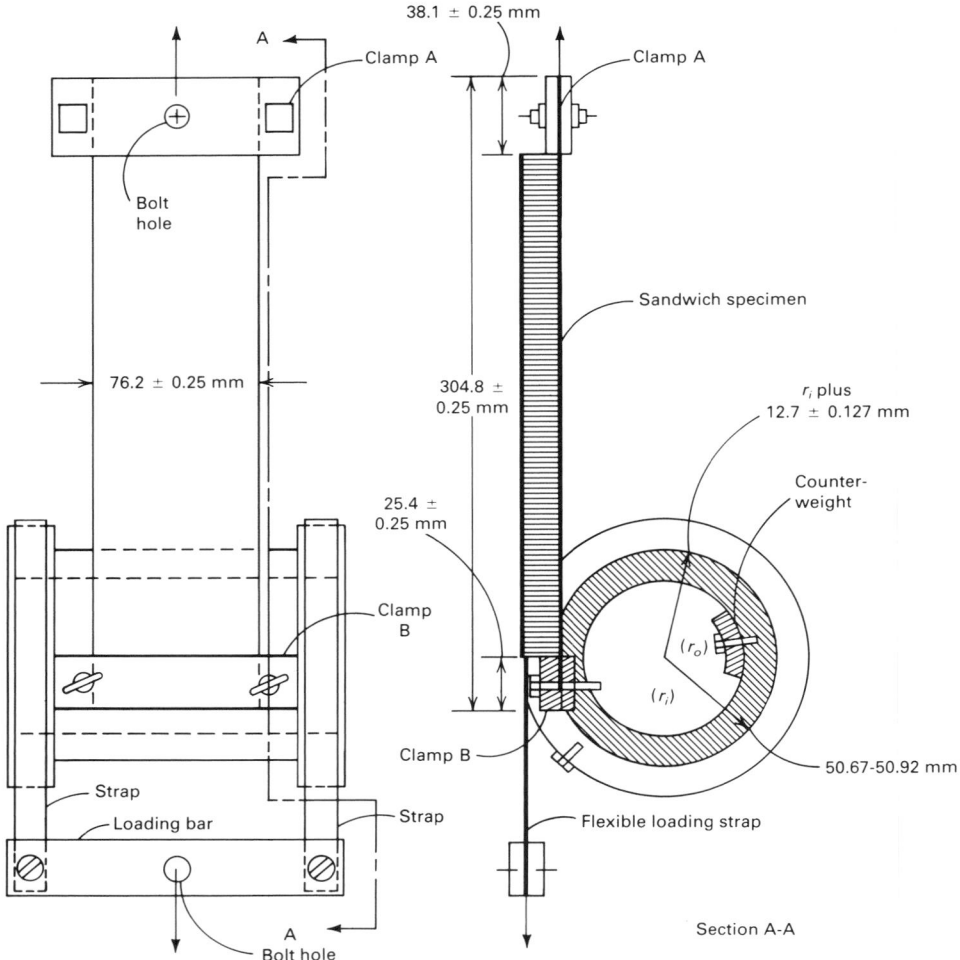

Fig. 11   MIL-A-25463A climbing drum peel test apparatus. Not completely dimensioned. r_i: radius of drum to middepth of specimen facing; r_o: radius of mid-depth of strap

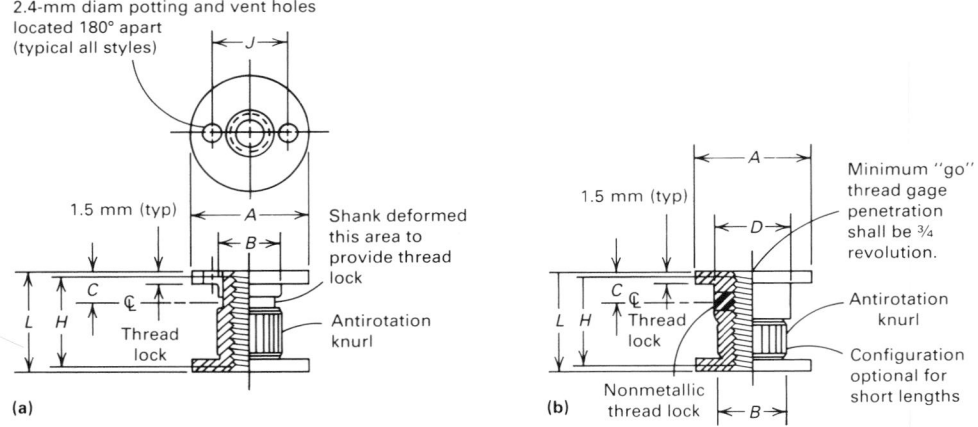

Fig. 12   NAS 1833 potted inserts for threaded sandwich panels. (a) All steel and corrosion-resistant steel and nonself-locking aluminum style. (b) Self-locking aluminum style

Scope. The requirements for nonperforated nylon paper-base plastic-honeycomb core material for aircraft structural applications are covered in this specification.

Use. This material is intended for use in aircraft radar antenna housings and for general aircraft structural and other applications in which temperatures do not exceed 177 °C (350 °F).

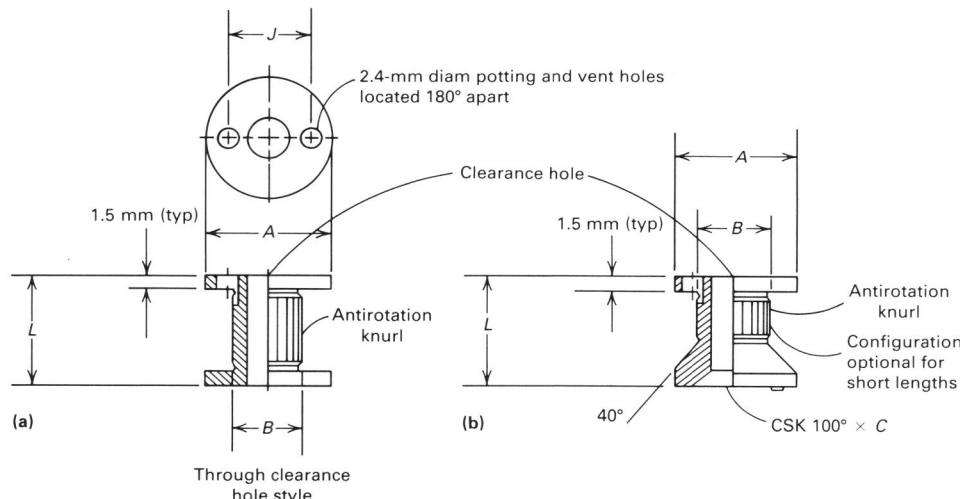

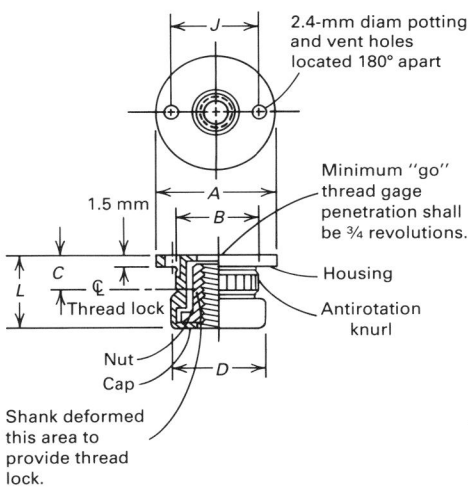

Fig. 13 NAS 1834 potted inserts for smooth clearance hole sandwich panels. (a) Through clearance hole style. (b) Countersunk clearance hole style

Fig. 14 NAS 1835 potted inserts for floating-type sandwich panels

Content. Requirements are addressed for preproduction sampling, material properties, configuration, sizes, slicing resistance, core properties, electrical properties, density, quality, and so on.

Auxiliary Materials And Products

NAS 1833-81 "Insert, Molded-In, Through-Threaded, Self-Locking, Nonself-Locking, Sandwich Panel." Prepared by the Aerospace Industries Association of America (AIAA).

Scope. This national aerospace standard (NAS) includes dimensional drawings covering threaded, self-locking, and regular carbon steel, corrosion-resistant steel, or aluminum alloy inserts with flanges on both ends.

Use. This standard governs a threaded insert to be potted into place in sandwich assemblies to allow bolts to be installed therein or through.

Content. Inserts from 6.4 to 9.6 mm (¼ to ⅜ in.) thick for use with threaded fasteners from No. 6 to 9.6 mm (⅜ in.) (Fig. 12) are discussed.

NAS 1834-81 "Insert, Molded-In, CSK and Through Clearance Hole, Sandwich Panel." Prepared by AIAA.

Scope. Included are dimensional drawings covering carbon steel, corrosion-resistant steel, or aluminum alloy inserts with clearance holes, both through clearance and countersunk clearance holes, flanged on both sides (Fig. 13).

Use. This standard governs a smooth clearance hole insert to be potted into place in sandwich assemblies to allow pins or bolts to penetrate sandwich assemblies.

Content. Inserts from 6.4 to 9.6 mm (¼ to ⅜ in.) thick and from 3.6 to 9.6 mm (0.140 to 0.377 in.) in diameter (Fig. 13) are covered in this standard.

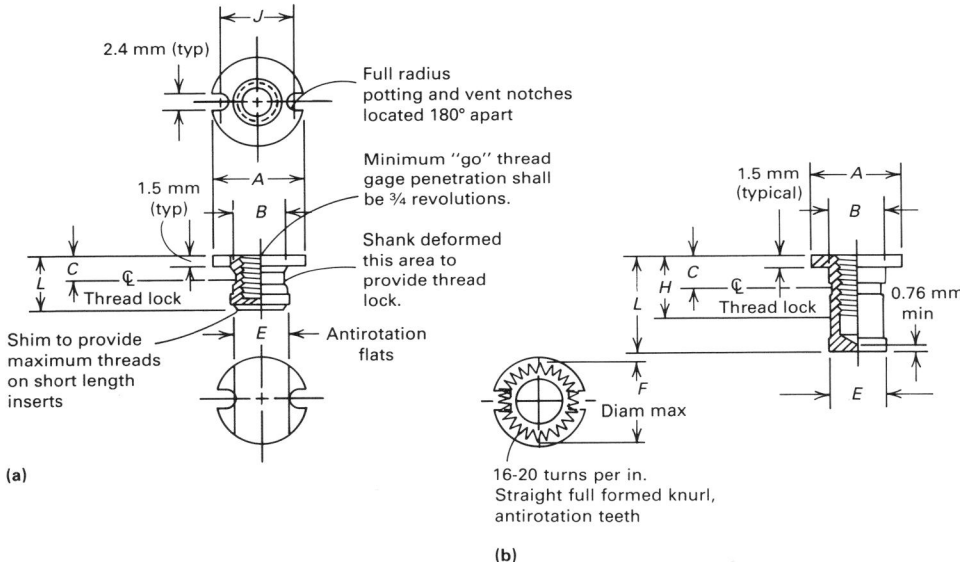

Fig. 15 NAS 1836 potted inserts for sandwich panels—threaded with blind bottom. (a) Shimmed style for short lengths. (b) Blind tapped style for long lengths

NAS 1835-81 "Insert, Molded-In, Blind-Threaded, Self-Locking, Nonself-Locking, Floating, Sandwich Panel." Prepared by AIAA.

Scope. Dimensional and descriptive drawings covering inserts to be potted into place in sandwich assemblies are included. Inserts consist of two parts: a steel, aluminum, or corrosion-resistant steel blind bottom housing and a captured, floating steel or corrosion-resistant steel nut. The housing is flanged on the open side only.

Use. This standard governs a threaded insert to be used in sandwich assemblies to allow bolts to be installed. In these assemblies, slight misplacement of bolts and holes must be taken into consideration (Fig. 14).

Content. Inserts from 9.4 to 20.6 mm (0.37 to 0.81 in.) thick for use with No. 8 to 9.6-mm (⅜-in.) bolts are covered.

NAS 1836-81 "Insert, Molded-In, Blind-Threaded, Self-Locking, Lightweight, Sandwich Panel." Prepared by AIAA.

Scope. Included are dimensional and descriptive drawings covering blind inserts to be potted into sandwich assemblies to allow bolts to be installed. Inserts are flanged one side and may be self-locking or nonself-locking.

Use. This standard governs inserts for use with No. 6 to 6.4-mm (¼-in.) bolts. Inserts can be any length in multiples of 7.9 mm (0.031 in.) above defined minimum lengths (Fig. 15).

AMS 3696 (8 Dec 77) "Aerodynamic Fairing Compound, −55 to 85 °C (−67 to 185 °F)." Prepared by SAE.

Scope. A two-component, air-curing, aluminum-filled, epoxy-base paste or putty is discussed in this specification.

Use. This compound is intended for the filling of small holes, crevices, or gaps and for the aerodynamic smoothing of surfaces for service from −55 to 85 °C (−67 to 185 °F).

Content. Covered in this specification are the base material properties, such as catalyst, appearance, odor and toxicity, storage life, infrared (IR) identification, and nonvolatile content of unmixed compound; the properties of mixed but uncured compounds, such as fineness of grind, color, viscosity, weight per unit volume, pot life, and cure time; and the properties of cured compounds, such as hardness, adhesion, impact resistance, resistance to paint strippers, finish system compatibility, and quality.

AMS 3697 (21 Dec 77) "Aerodynamic Fairing Compound, −55 to 150 °C (−67 to 302 °F)." Prepared by SAE.

Scope. A two-component, air-curing, aluminum-filled epoxy-base paste or putty is discussed.

Use. This compound is intended for the filling of small holes, crevices, or gaps and for the aerodynamic smoothing of surfaces for service from −55 to 150 °C (−67 to 302 °F).

Content is the same as that of AMS 3696, except heat resistance has no effect at 150 °C (302 °F) and fluid resistance has no effect at 25 to 150 °C (77 to 302 °F).

Test Methods

Federal Test Method Standard 175B (1 Sept 83) "Adhesives: Methods of Testing." Prepared by the Army Materials Technology Laboratory.

Scope. This standard consists of only four test methods; most of the many test methods in earlier issues have been replaced by ASTM methods.

Content. The current four methods are Method 1081 "Flexibility of Adhesives," which provides a means for determining the flexibility (elasticity) of single films or systems of films of adhesives; Method 4032.1 "Ash Content of Adhesives"; Method 4041.1 "Grit, Lumps, or Undissolved Matter in Adhesives"; and Method 4051.1 "Odor Test for Adhesives."

MIL-STD-401B (26 Sept 67) "Sandwich Constructions and Core Materials: General Test Methods." Prepared by the Weapons Engineering Standardization Office, Naval Air Engineering Center.

Scope. This standard covers the general requirements and methods for testing sandwich core materials and for testing sandwich construction of the types used primarily in aircraft structures. References to similar nongovernment documents (not necessarily as restrictive) are provided.

Content. The following test methods are discussed (similar nongovernment methods are given in parentheses):

- Core density and specific gravity (ASTM C 271)
- Core moisture sorption (ASTM C 272)
- Core thermal conductivity (ASTM C 177)
- Core compression (ASTM C 365), flatwise and edgewise (Fig. 16)
- Core shear (ASTM C 273) (Fig. 17)
- Core tension (Fig. 18)
- Core water migration
- Node delamination of honeycomb core (ASTM C 363) (Fig. 19)
- Sandwich compression
- Sandwich shear (Fig. 17)
- Sandwich tension (ASTM C 297) (Fig. 18)
- Sandwich flexure (ASTM C 393) (Fig. 20)
- Sandwich thermal conductivity (ASTM C 236)
- Sandwich peel (ASTM D 1781)

General notes only are given for fatigue tests and creep characteristics; no test methods are described.

ASTM D 907-82 (1985) "Standard Definition of Terms Relating to Adhesives."

Content. This specification contains definitions of many terms, such as A-stage, adherend, plasticizer, retrogradation, tack range, and viscosity.

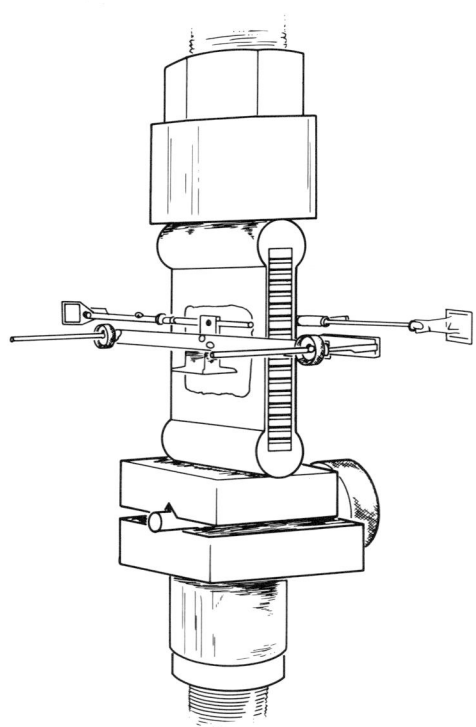

Fig. 16 MIL-STD-401B edgewise compression specimen

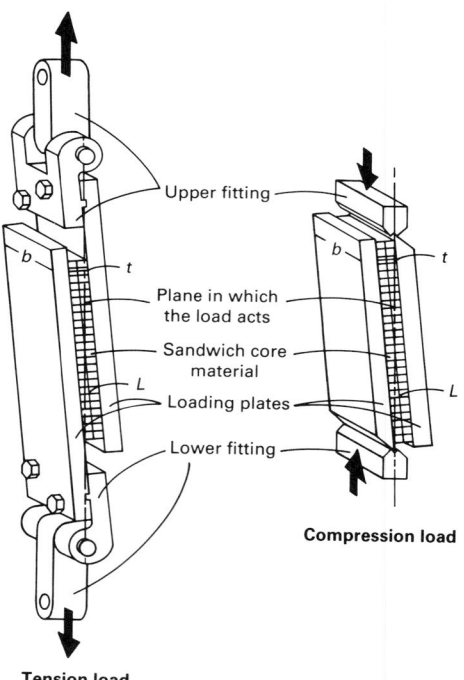

Fig. 17 MIL-STD-401B core and sandwich specimens

ASTM D 1002-72 (1983) "Standard Test Method for Strength Properties of Adhesives in Shear by Tension Loading (Metal-to-Metal)."

Scope. The determination of the shear strength of adhesives for bonding metals (under the specified conditions of preparation and testing) is discussed.

Content. This document contains the complete description for preparing standard specimens from two types of test panels. For the standard panel (Fig. 21), two pieces of aluminum are bonded together, and the test panel is cut into five specimens. The end pieces are discarded. The optional panel (Fig. 22) has premachined spaces; therefore, very little cutting is needed to separate the individual specimens from the bonded test panel. No cutting is necessary through the bond overlap area.

Notes. This test method has been approved for use by the DoD to replace methods 1032 and 1032.1T of FTMS No. 175B. The specimens are virtually identical to those in MMM-A-132.

ASTM D 1151-84 "Standard Test Method for Effect of Moisture and Temperature on Adhesive Bonds."

Scope. This document defines conditions for determining the performance of adhesive bonds when subjected to continuous exposure at specified conditions of moisture and temperature. Results are expressed as the percentage of original strength retained after exposure.

Content. Twenty-one different temperature and humidity exposure conditions are covered.

Notes. This test method has been approved by the DoD to replace methods 2052-T and 2031 of FTMS No. 175B.

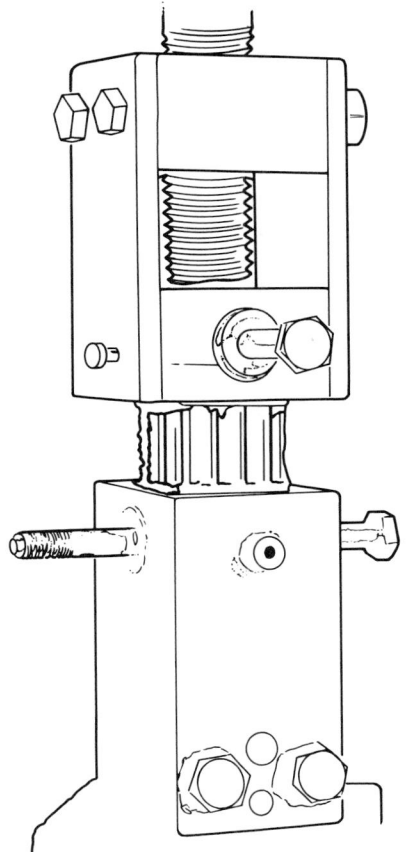

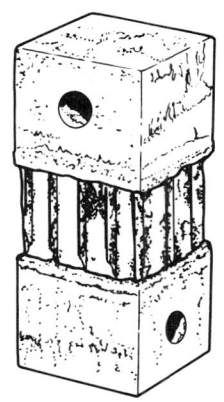

Fig. 18 MIL-STD-401B core and sandwich tension (flatwise) specimens

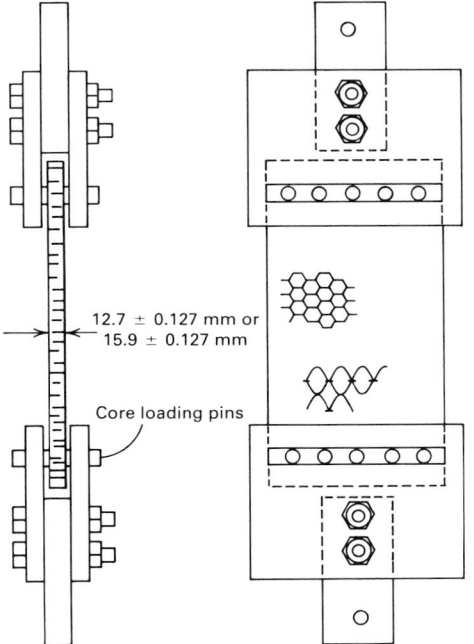

Fig. 19 MIL-STD-401B node strength test

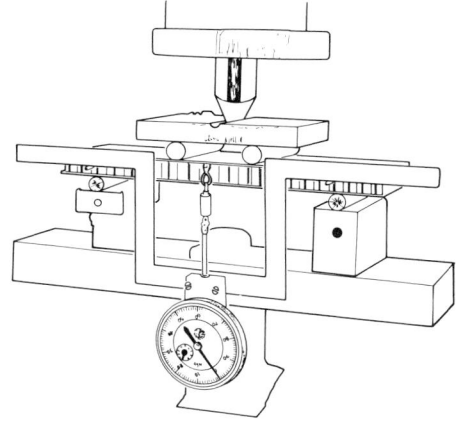

Fig. 20 MIL-STD-401B sandwich flexure

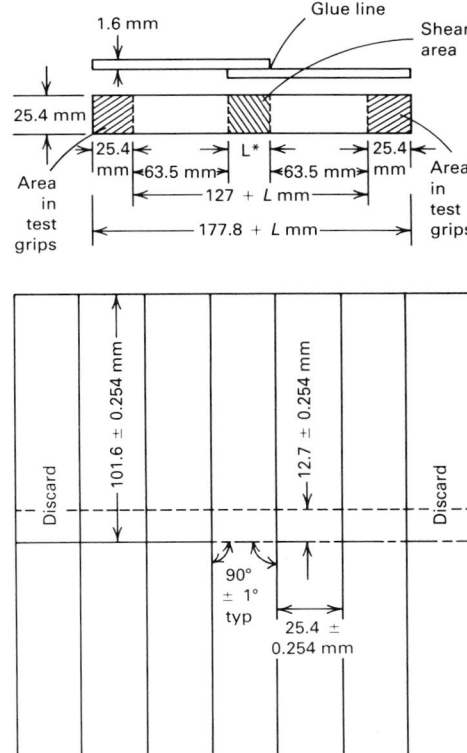

Fig. 21 ASTM D 1002, standard lap shear test panel

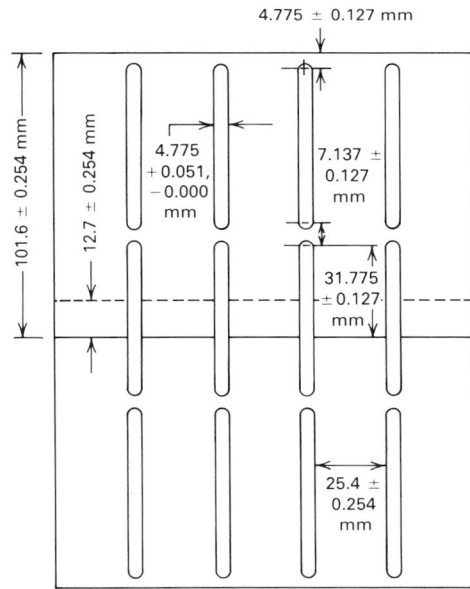

Fig. 22 ASTM D 1002, optional lap shear test panel

ASTM D 1780-72 (1983) "Standard Recommended Practice for Conducting Creep Tests of Metal-to-Metal Adhesives."

Scope. This document addresses the determination of the amount of creep of metal-to-metal adhesive bonds due to the combined effects of temperature, tensile shear stress, and time.

Content. Specimens are cut from two sheets that are overlap bonded to yield a 241- × 305-mm (9.5- × 12-in.) panel (Fig. 23). Three scribe lines are made in the edge of the overlap area on both sides of each specimen, and their divergence is noted during the test and at the end of the loaded test period.

Notes. This practice has been approved by the DoD for use as part of FTMS No. 175B.

ASTM D 1781-76 (1981) "Standard Method for Climbing Drum Peel Test for Adhesives."

Scope. This document covers the determination of the peel strength of adhesive bonds between a flexible adherend and a rigid adherend for metal-to-metal type joints and between the flexible facing of a sandwich structure and its core.

Content. The standard test specimen for laminated assemblies is 25.4 mm (1 in.) wide × at least 25.4 cm (10 in.) long. The standard honeycomb test panel is 76 mm (3 in.) wide ×

30.5 cm (12 in.) long. The extending end of one face for the honeycomb specimen or the base side of the laminated assembly is gripped in a clamp at the top. A 51-mm (2-in.) diam drum is clamped to the bottom of the other face or side of the honeycomb or laminated specimen and is wound up the specimen in a "sardine-can-opener fashion" by the downward pull of straps around the end flanges of the drum (Fig. 11).

Notes. This method has been approved by the DoD to replace method 1042-T of FTMS No. 175B.

ASTM D 1876-72 (1983) "Standard Test Method for Peel Resistance of Adhesives (T-Peel Test)."

Scope. This document concerns the determination of the relative peel resistance of adhesive bonds between flexible adherends.

Content. The standard test specimen is T-shaped and measures 25.4 mm (1 in.) wide × 30.5 cm (12 in.) long. The specimens are cut from a test panel of thin adherends bonded together over a 15.2- × 22.9-cm (6- × 9-in.) area. The bent, unbonded ends are clamped in a tensile test machine capable of autographic recording of load versus head movement or load versus distance peeled. The test specimen is identical to the specimen in MMM-A-132 (Fig. 5) except that a 15.2-cm (6-in.) width is specified and the thickness is not specified.

Notes. This was the most common peel test for many years, but it has recently been largely replaced by the Floating Roller (Bell) Peel Test (ASTM D 3167-76), which gives less data scatter, and by the Climbing Drum Peel Test (ASTM D 1780-72).

ASTM D 2651-79 (1984) "Standard Recommended Practice for Preparation of Metal Surfaces for Adhesive Bonding."

Scope. Various recommended surface preparation methods for metals prior to adhesive bonding are discussed.

Content. This document contains methods for cleaning aluminum alloys, stainless steel, carbon steel, titanium alloy, magnesium alloy, and copper and copper alloys.

Notes. The Forest Products Laboratory etch has recently been updated to reflect the optimized parameters developed for common aircraft alloys. This is described in paragraph 5.7, Method G. Method G is currently being updated by the members of the ASTM D-14 Committee (Adhesives).

ASTM D 2674-72 (1984) "Standard Methods for Analysis of Sulfochromate Etch Solution Used in Surface Preparation of Aluminum."

Scope. This document contains methods for controlling the chemical concentration of the most commonly used aluminum deoxidizer and/or surface treatment solution (sulfuric acid/sodium dichromate).

Content. Procedures for determining sulfuric acid content and its replenishment needs, hexavalent chromium as sodium dichromate, trivalent chromium as sodium dichromate, and replenishment of sodium dichromate amount are covered.

ASTM D 2919-84 "Standard Recommended Practice for Determining Durability of Adhesive Joints Stressed in Shear by Tension Loading."

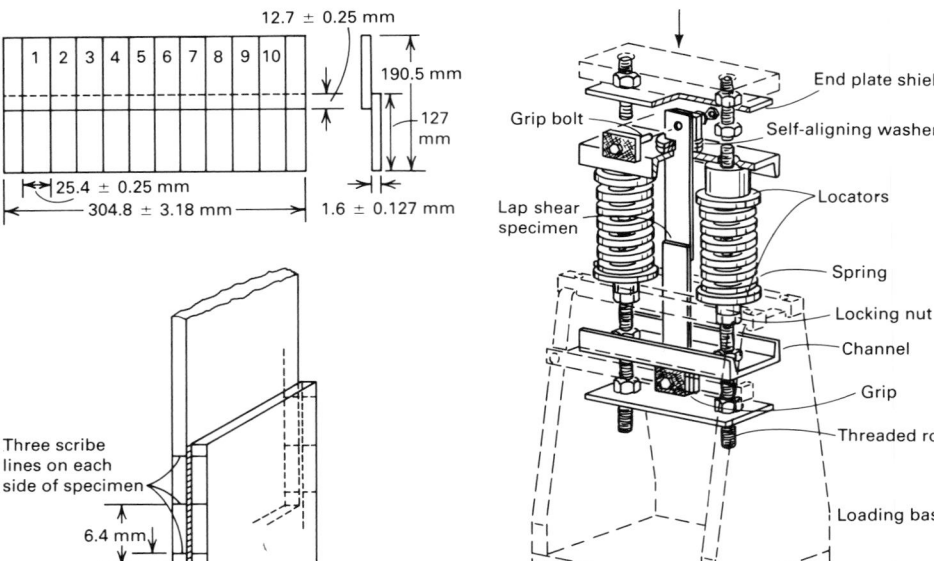

Fig. 25 ASTM D 2919, durability test apparatus

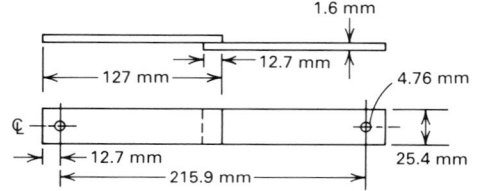

Fig. 23 ASTM D 1780, creep test panel

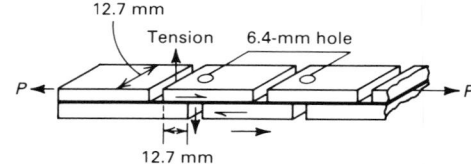

Fig. 26 Raab specimen

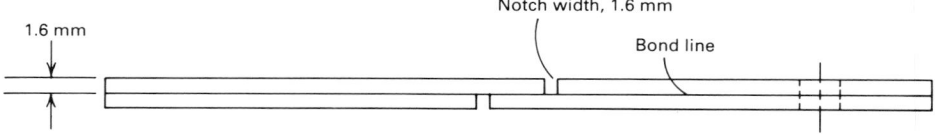

Fig. 24 ASTM D 2919, durability lap shear specimen

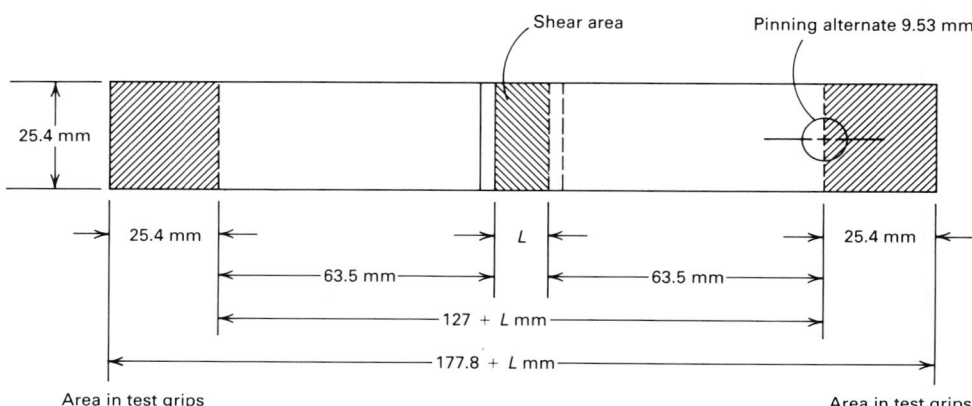

Fig. 27 ASTM D 3165, machined lap shear specimen. L, length of test area, which can be varied. Recommended lap length is 12.7 ± 0.3 mm (0.50 ± 0.01 in.).

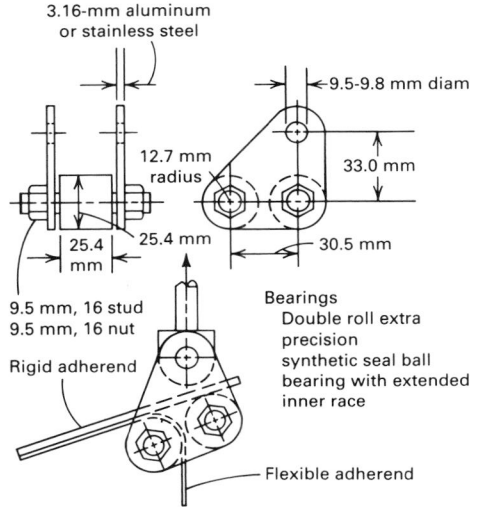

Fig. 28 ASTM D 3167, floating roller bell peel test

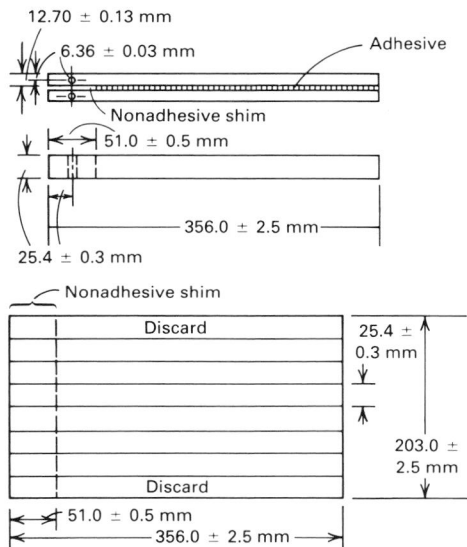

Fig. 29 ASTM D 3433 flat adherend DCB specimen and panel

h	a
12.7	48.62
15.2	64.06
17.8	80.82
20.3	98.86
22.9	118.01
25.4	138.35
27.9	173.41
30.5	182.04
31.8	193.57

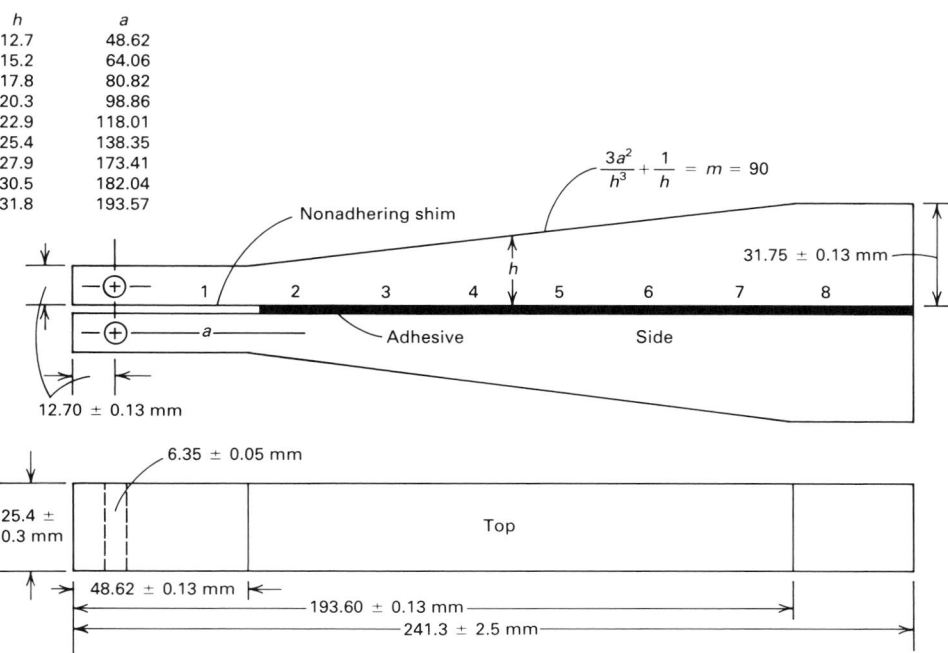

$$\frac{3a^2}{h^3} + \frac{1}{h} = m = 90$$

Fig. 30 ASTM D 3433, contoured DCB specimen

Scope. This document covers the procedure for determining the durability of an adhesive by means of lap shear coupons under stress while in a severe environment.

Content. The test specimens are shear specimens with a hole in each end (Fig. 24). They are placed in a special apparatus in which they are clamped under stress maintained by two coil springs. A special loading base is used to hold the fixture while the load is being established in a tensile machine. After clamp-up, the stress is maintained by the springs (Fig. 25).

Notes. This is a very valuable test, and it has generated much knowledge concerning the durability of adhesives because it allows them to be easily exposed to stress, heat, humidity, or other severe environments simultaneously.

Many adhesives and surface preparation methods yield good results under normal conditions, but are unsatisfactory when placed under load in a severe environment. The stressed durability of adhesives and surface preparation methods is now believed to be their most important property. Other stressed durability tests using multiple string-type specimens have been developed; these are simply long chains of overlap specimens made by making cuts through alternate sides of a laminated bonded assembly. When holes are drilled in each bonded area (Fig. 26), the specimen is known as a Raab specimen. Although no specification exists for these latter two durability specimens, they are included here because of their wide use. A detailed discussion of durability test methods is provided in Ref 2.

ASTM D 3165-73 (1979) "Standard Test Method for Strength Properties of Adhesives in Shear by Tension Loading of Laminated Assemblies."

Scope. This specification concerns the determination of the shear strength of adhesives in large-area joints.

Content. The standard test joint is prepared by bonding two plates together (Fig. 27). Test specimens are usually cut from the central area of this panel. Notches are cut in opposite sides of the test specimens to give an overlap bond.

Notes. This specimen is very similar to the blister detection specimen of MMM-A-132. This type of specimen centralizes the bond area in the test grips, giving less peel effect, and also detects voids in large-area bonds caused by a lack of volatile escape.

ASTM D 3166-73 (1979) "Standard Test Method for Fatigue Properties of Adhesives in Shear by Tension Loading (Metal/Metal)."

Scope. This document covers the standard method for testing adhesives in shear by tension loading in a lap shear joint.

Content. The standard specimen is a typical single-overlap specimen with a free space between the grip area and the overlap bond area.

Notes. This specimen is virtually identical to that in method 1061.1 of FTMS 175B (superseded by ASTM D 1062), but it differs slightly from that in MMM-A-132.

ASTM D 3167-76 (1981) "Standard Test Method for Floating Roller Peel Resistance of Adhesives."

Scope. The determination of the peel strength of adhesive bonds between a rigid and a flexible adherend is covered in this document.

Content. The standard specimens, with flexible and rigid members, are cut from a laminated panel. The flexible adherend is pulled downward over a roller in the test apparatus, which allows the rigid adherend to float upwards at an angle over a second roller (Fig. 28).

Notes. This method, frequently referred to as the bell peel method, gives less data scatter than the T-peel test.

ASTM D 3433-75 (1985) "Standard Recommended Practice for Fracture Strength in Cleavage of Adhesives in Bonded Joints."

Scope. This document covers the determination of the fracture strength in cleavage of adhesives when tested on standard specimens under specified conditions.

Content. Two specimen types are defined: the flat, constant cross section, double cantilever beam (DCB) specimen and the contoured DCB specimen (Fig. 29 and 30). Holes drilled into the unbonded end area of each adherend for both types of DCB specimen are used for grip anchoring. The specimens are pulled to invoke a cleavage force on the bondline. Load versus load displacement across the bondline is recorded autographically. From the data, computations are made for values of interest.

Notes. These specimens can be used to develop design parameters for bonded assemblies.

ASTM D 3528-76 (1981) "Standard Test Method for Strength Properties of Double Lap Shear Adhesive Joints by Tension Loading."

Scope. The determination of the tensile shear strength of adhesives in an essentially peel-free specimen is covered in this specification.

Content. Two types of specimens are used: Type B (Fig. 31), which is the most common, and Type A (Fig. 32).

ASTM D 3762-79 (1983) "Standard Test Method for Adhesive-Bonded Surface Durability of Aluminum (Wedge Test)."

Scope. This test method simulates the forces and effects on an adhesive bond joint at the metal and adhesive/primer interface.

Content. A 25.4- × 25.4- × 3.2-mm (1- × 1- × ⅛-in.) wedge is driven into the end of the specimen. The initial crack length is measured and then monitored for any change or crack propagation during exposure to some severe environment, such as salt spray or 49 to 60 °C (120 to 140 °F) and condensing humidity (Fig. 33).

Notes. This test method is qualitative in nature, but has proved to be very discriminating in determining variations in adherend surface parameters and adhesive environmental durability.

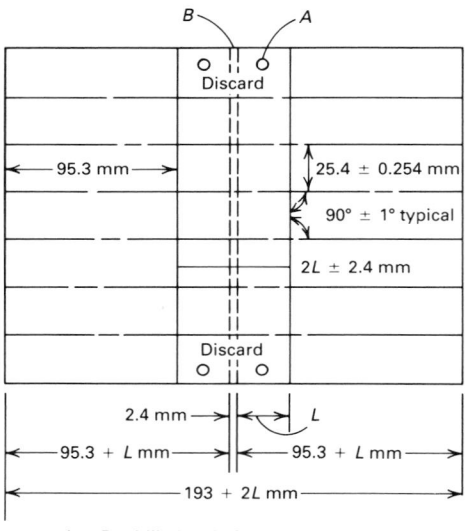

A = Predrilled pinholes (four required, 2.4-mm diam)
B = TFE-fluorocarbon filler suggested for adhesive flash control

(a)

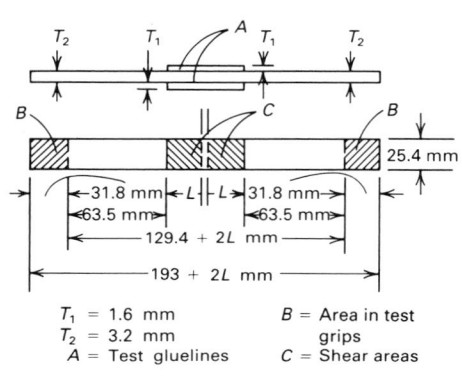

T_1 = 1.6 mm
T_2 = 3.2 mm
A = Test gluelines

B = Area in test grips
C = Shear areas

(b)

Fig. 31 ASTM D 3528, double lap shear specimen, type B. (a) Form and dimensions of standard test panel for type B specimens. (b) Form and dimensions of specimen type B

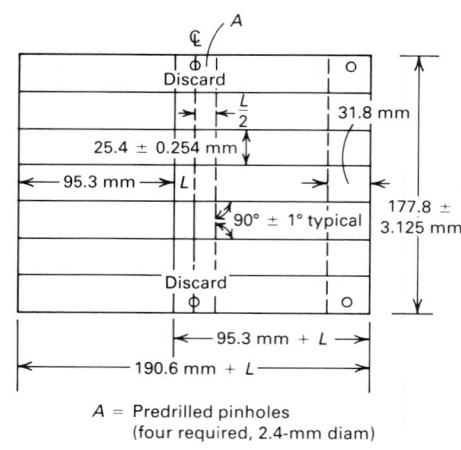

A = Predrilled pinholes (four required, 2.4-mm diam)

(a)

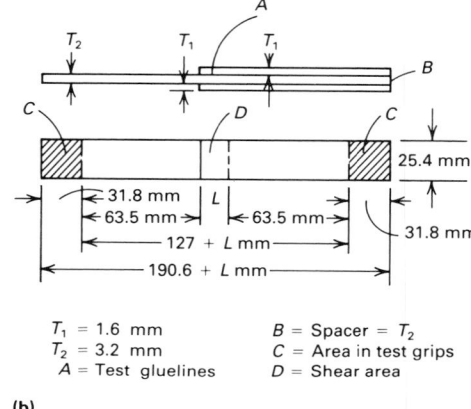

T_1 = 1.6 mm
T_2 = 3.2 mm
A = Test gluelines

B = Spacer = T_2
C = Area in test grips
D = Shear area

(b)

Fig. 32 ASTM D 3528, double lap shear specimen, type A. (a) Form and dimensions of standard test panel for type A specimens. (b) Form and dimensions of specimen type A

*Appendix: Terms and Definitions**

Specifications. A specification is a document intended primarily for use in procurement that clearly and accurately describes the essential and technical requirements for items, materials, or services, including the procedures used to determine if the requirements have been met. Specifications for items and materials may also contain preservation, packing, and marking requirements.

Federal specifications are developed for materials, products, or services that are used or potentially used by two or more Federal agencies, at least one of which is an agency other than the DoD.

*The definitions provided in this Appendix were abstracted from DoD Manual 4120.3-M, MIL-STD-961, MIL-STD-962, and Industry Sources.

Military specifications cover weapons, systems, subsystems, items, or services that are intrinsically military in character; commercial items with features that meet special requirements of the military; or commercial items with no present or known potential use by Federal agencies other than military. Military specifications are issued as either coordinated or limited-coordination documents. Coordinated military specifications are issued to cover items or services required by more than one military department. Limited-coordination military specifications cover items or services of interest to a single department or activity to meet an immediate procurement need where urgency does not permit coordination. The use of limited-coordination specifications is not restricted to the department that prepared the document.

A *use-in-lieu-of limited-coordination military specification* is a revision of a coordinated specification required by a military department to meet a need where time does not permit the preparation of a coordinated revision. Only one use-in-lieu-of specification shall be outstanding per department for each coordinated specification.

Standards. Standards are documents that establish engineering and technical requirements for processes, procedures, practices, and methods that have been adopted as standard. They are primarily for designers, and their procurement use is through reference in specifications.

Federal standards are developed to meet the needs of two or more Federal agencies, at least one of which is an agency other than the DoD.

Military standards are issued as coordinated and limited-coordination documents in sheet, book, and unit-page form.

Handbooks. Handbooks are reference documents that bring together procedural and

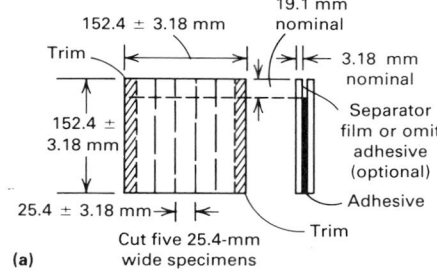

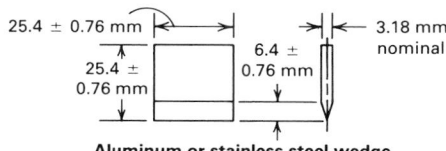

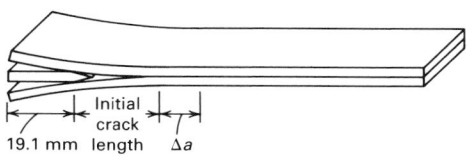

Aluminum or stainless steel wedge

(b)   **Wedged crack extension specimen**

Fig. 33 ASTM D 3762, wedge test specimen. (a) Assembly. (b) Configuration. Δa, crack growth after exposure

technical design information related to commodities, processes, and services. Handbooks may serve as supplements to specifications or standards to provide general design and engineering data. Two valuable handbooks that concern adhesives are military handbook MIL-HDBK-691B (12 March 87) and the *Adhesives Technology Handbook* (1985).

Department of Defense Index of Specifications and Standards (DoDISS). This publication lists Federal and military specifications, standards, and related standardization documents that are used by the military services. Also listed are industry standardization documents selected for use by the military services. The DoDISS is the main source of information for all armed services procurement; the latest supplement should always be checked first. Supplements are issued quarterly; basics, once a year. The DoDISS is issued in three parts: alphabetical, numerical, and Federal Supply Classification, which is actually a companion document to, rather than a part of, the DoDISS.

Sources for Specifications and Standards. The specifications and standards discussed in this article can be obtained from the following organizations:

● DoDISS and Federal and military specifications:
Naval Publications and Forms Center
5801 Tabor Avenue
Philadelphia, PA 19120
(215) 697-2179

Specifications, standards, and so on, are available free.

● SAE (AMS) documents:
Society of Automotive Engineers
400 Commonwealth Drive
Warrendale, PA 15096
(412) 776-4841

● ASTM documents:
American Society for Testing and Materials
1916 Race Street
Philadelphia, PA 19103
(215) 299-5400

● NAS Standards:
National Standards Association, Inc.
1321 Fourteenth St., N.W.
Washington, DC 20005
(301) 951-1389

Where ASTM and SAE (AMS) standards are DoD-adopted, copies may be obtained at no charge from the Naval Publications and Forms Center.

REFERENCES

1. M. Locke, R. Horton, and J. McCarty, "Anodize Optimization and Adhesive Evaluation for Repair Applications," AFML-TR-78-104, Air Force Materials Laboratory, July 1978
2. A. Marceau and W. Scardino, "Durability of Adhesive Bonded Joints," AFML-TR-75-3, Air Force Materials Laboratory, Feb 1975

Bonding Cure Considerations

Robert E. Sanders, Bell Helicopter Textron, Inc.
Saad Taha, Rockwell International Corporation, North American Aircraft Operation

THE COMPOSITE CURING PROCESS transforms relatively soft and flexible resin-impregnated fibers into a stiff, usable structural material. The basic curing process involves the systematic application of heat and pressure for predetermined time periods. Each material system has a slightly different cure cycle, which, in many cases, requires further modification because of the parts being produced. Structural parts have become larger, thicker, and more complex in contour in recent years; the curing process has therefore become more important. However, all cure processes involve:

- Compacting the part to remove entrapped air, volatiles, and excess resin
- Contouring the stacked materials to the specified tool configuration
- Developing the resin chemistry to create a structural material by increasing the polymer chain length and cross linking

Although most cure processes are performed in an autoclave, it is possible to use ovens; heated platen presses; integrally heated, mechanically pressurized tools; or even curing at room temperature with a vacuum bag. The pressurized, heated autoclave can be fully loaded with 30 to 40 parts of different configurations to achieve high production rates (Fig. 1).

Role of Cure Viscosity

In recent years, cure processes have been based primarily on resin viscosity changes. This trend has been accelerated by the advent of the precision-plate rheometric unit, which measures the viscosity changes of the resin as it progresses from room temperature through the dwell at the cure temperature. At room temperature, the impregnating resins are compounded and partially reacted, have very high viscosities, and are therefore applied to the tape or fabric as a solution or a hot-melt coating. Consequently, the resin in the prepreg is a sticky, high-viscosity semisolid, which facilitates handling during storage and lay-up. As the resin is heated during the initial stages of the cure cycle, its viscosity drops significantly to the point at which the fluidity is almost equiv-

alent to salad oil and viscosity is approximately 1 to 2 Pa · s (10 to 20 P). It is in this low-viscosity stage that initial dwell cycles are built into the cure cycle to allow the laminate to consolidate and the volatiles to be removed, along with any excess resin present. Pressure is applied to the lay-up near the end of the low-viscosity stage, preferably before viscosity exceeds 20 to 30 Pa · s (200 to 300 P). Cure pressure continues throughout the cure cycle, while viscosity increases with time and temperature. When viscosity increases beyond 100 Pa · s (1000 P), the practical limit of measurement, the resin is considered to be gelled or converted to a rigid semisolid. This occurs just before the laminate becomes a hard solid. Figure 2 shows a typical change in viscosity for a 3501-6 resin as it progresses through a heat cycle.

Tooling Characteristics

The primary purpose of tooling is to provide the contour configuration that the part assumes when it is finally cured. Transferring heat to the part lay-up is another function of the tool. Tools are generally made from the durable materials listed in Table 1, although single parts can be made on a plaster tool that is usually destroyed by the heat cycle. It is highly desirable to have the coefficient of thermal expansion of the tool closely match the thermal expansion of the part being made; this is the basis of the trend toward using steel or high-temperature carbon-epoxy tools for composites. For small parts, particularly if they are flat, aluminum or steel plate is frequently used as the tool material. For larger, more complex parts, electroless, nickel-plated tools or high-

Fig. 1 Fully loaded autoclave rack with composite parts ready for cure in a 4-m (14-ft) diam autoclave

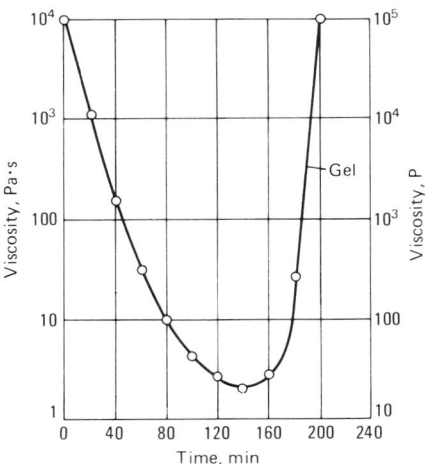

Fig. 2 Viscosity of AS4/3501-6 resin heated at 0.56 °C/min (1 °F/min)

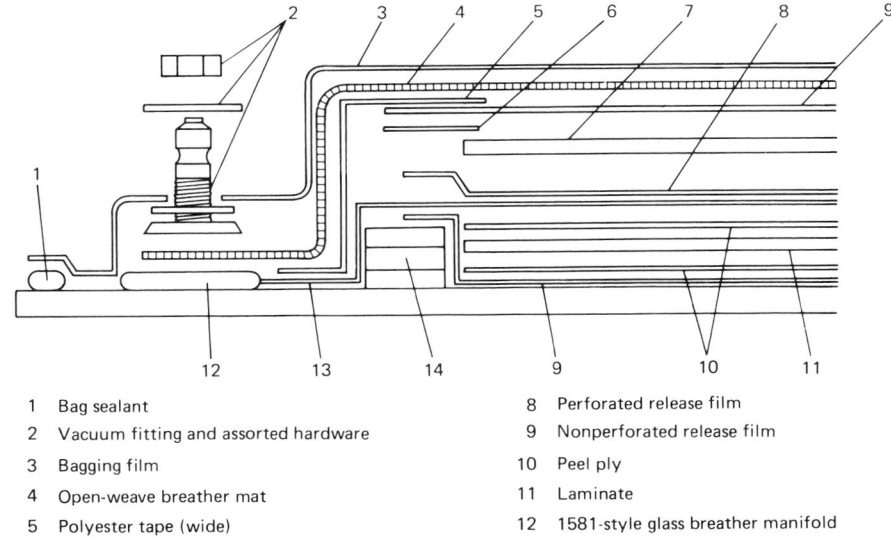

1 Bag sealant
2 Vacuum fitting and assorted hardware
3 Bagging film
4 Open-weave breather mat
5 Polyester tape (wide)
6 Polyester tape (narrow)
7 Caul sheet

8 Perforated release film
9 Nonperforated release film
10 Peel ply
11 Laminate
12 1581-style glass breather manifold
13 1581-style glass bleeder ply
14 Stacked silicon edge dam

Fig. 3 Vacuum bag lay-up sequence

Table 1 Properties of typical composite tooling materials

Material	Coefficient of thermal expansion, 10^{-6}/K	Thermal conductivity		Approximate fabricated cost(a), $/ft^2$
		w/m · K	Btu · in./h · ft² · °F	
Fiberglass-epoxy	7.9			195-292
Graphite-epoxy	−0.9	0.022	0.15	240-360
Aluminum	23.0	0.221	1.53	400-600
Steel	13.9	0.048	0.33	300-500
Electroless nickel	13.3	0.035	0.24	400-600

(a) Includes base or substructure

temperature epoxy with carbon reinforcement is used.

Tools may also contain special inserts, stops, or guides to facilitate part construction. Built-in vacuum and static ports, as well as permanent thermocouples, have also been built into tools. These features are high-maintenance items and tend to increase tooling cost, which must be offset by the gain in efficiency over the long-term use of the tools for several cure cycles. Tooling substructure has a significant impact on the air flow under the tool and therefore on the heat transfer into the part during the cure cycle. The current trend is to incorporate large holes in the tool substructure and to align the substructure supports parallel to the air flow to improve heat transfer.

Tooling is generally heated in an autoclave with convective air flow; ovens are rarely used, because higher pressure is usually necessary for part consolidation. Electrically heated tools have been used infrequently, because the cost of the system is high and there is a potential shock hazard. Tools heated with hot, circulating oil have also been tried, but the potential for fire and operator burns has generally discouraged their use.

Bagging Operation

The term bagging usually refers to the application of a pressure membrane over the part lay-up to provide a transfer medium that allows the autoclave pressure (or vacuum/air pressure for oven cures) to react on the part. The pressure membrane is normally a thin plastic film and is discarded after each use. Permanent rubber bags have been used by some fabricators, but their high initial cost and handling/ storage requirements result in limited and infrequent use. The plastic film is significantly oversized and has numerous folds so that bridging, or stretching, of the film over an angular area or a sharp corner is eliminated. If bridging occurs, the bag will rupture at high temperature

and pressure, causing the loss of the part. The film is adhered to the tool with a temperature-resistant flexible sealing compound.

A typical bagging lay-up is shown in Fig. 3. An edge dam is used with no-bleed or low-bleed resin prepreg systems, but is normally omitted on systems with higher resin content. Nonstick release films are used in addition to the mold release to prevent the parts from sticking to the tools. Peel plies made of thin fiberglass or nylon films are applied directly onto the laminates and are ultimately removed or peeled away before bonding, painting, or applying other finishes to the laminate. Several bleeder plies of fiberglass cloth, nonwoven nylon cloth, or other absorbent materials are also incorporated into the lay-up to absorb any excess resin. Additional plies of bleeder material are placed over corners or sharp angles on the part to prevent bag rupture during curing. A folded and flattened roll of bleeder cloth is placed around the total lay-up just inside the sealing compound to act as a manifold to allow total air evacuation of the cavity under the bag.

At this point, thermocouples are placed in the trim area at the edge of the part, they are pressed into the compound, and additional compound is applied. The thermocouple junc-

tions are positioned at areas of the tool and part determined to be thin (low mass) and thick (high mass). The bag is positioned over the lay-up and pressed into the sealing compound at the edge of the part. Generous folds or gathers in the bag are positioned around the part. Where necessary, these folds are sealed with additional compound. Vacuum connections and connections for static lines are placed in the bag and positioned over the perimeter bleeder manifold, away from areas of possible resin bleed to avoid blocking them. There are no hard and fast rules for the number of connectors, but usually one static line connection is used and at least one vacuum connection for every 0.93 m² (10 ft²) of part. With tools having built-in vacuum/ static line connections, it is not necessary to place the connectors in the bag.

After the bagging operation is completed, full vacuum at room temperature is applied to the bag, except in cases in which parts include pressure-sensitive components, such as a lightweight core. In these cases, a proportionately lower vacuum is applied. The vacuum source is then shut off, and leakage is checked by reading the pressure drop on a gage attached to the static line. In production processes, a pressure drop of more than 16.9 kPa (5 in. Hg) in 5 min is considered excessive leakage. When this occurs the bag must be checked for leaks and resealed or replaced. Several sonic leak detectors are available for locating small bag leaks. Once the leaks are sealed, the lay-up is held under vacuum until it is ready for curing. This prevents movement of the bag and possible bridging. If the vacuum is lost during this period, it should carefully be reapplied and the assembly checked for bridging and leaks.

Autoclave Heat, Pressure, and Control Systems

Most production applications use a pressurized autoclave to apply heat and pressure, but an oven cure with vacuum pressure is sometimes sufficient for small, simply contoured parts. As the part size increases, however, the cost of tooling just to support the curing pressures involved becomes almost exorbitant. This is frequently the case in building self-contained tools. A pressurized autoclave is generally chosen for curing a variety of parts in production quantities. A gas-fired heat exchanger system is frequently the source of heat, although electrically heated autoclaves are used where economical. Most autoclaves are built to operate at 205 °C (400 °F), which will process the 120- to 175-°C (250- to 350-°F) curing material systems commonly used. Many newer autoclaves are being built to operate at maximum temperatures of 315 to 425 °C (600 to 800 °F) in order to process the high-temperature polyimide and thermoplastic systems. These higher temperatures significantly increase the cost of building the autoclave system.

The autoclaves are usually pressurized with nitrogen or carbon dioxide from a liquid storage tank, and vaporized. Early autoclaves were often pressurized with plant air, but fires frequently resulted. Small laboratory autoclaves are set up to operate with nitrogen bottles. Most autoclaves operate at 0.69 MPa (100 psig), with a few produced for operation at 1.4 MPa (200 psig). Operation at these pressures requires that the units be proof pressure tested at least 50% over the standard operating pressure at the scheduled operating temperature (depending upon local and/or state safety requirements).

The control systems on newer autoclaves are generally computer controlled, but have manual backup systems. After complete cure control cycles are entered into the computer, control becomes automatic once the parts are loaded and the proper cure number is requested. The computer controls the autoclave vacuum system, heating rate, and pressurization levels, and it records temperature data from each thermocouple, as well as the autoclave system operating parameters. These recorded data can later be printed and used as a quality control report. In the event of deviations from the programmed control cycle, error deviation messages are printed and recorded, and alarm signals prompt the autoclave operator to correct the deviations or to continue the cycle. Most deviations in

production control runs are related to fast or slow heat-up rates, which generally occur well into the cure cycle. The deviating parts were reviewed individually after the end of the run. For subsequent runs with problem tools, changing their placement or orientation in the autoclave or adding insulation can allow the parts to heat at the proper rate.

The addition of computer-controlled systems to production autoclaves has created a drive for better cure cycles and better understanding of what happens internally to the autoclave during cure. For example, the existence of hot and cold spots in autoclaves has been suspected for many years. Recent studies have verified that autoclave air flow is not uniform (Ref 1). This lack of uniformity in air flow and its effect on heating and cooling rates is suspected of becoming more critical when an autoclave is fully loaded. It is anticipated that air flow studies with loaded autoclaves will continue and that system improvements will allow better heating and cooling and therefore shorter cure cycles.

Specific Cure Cycles

The recommendations of the prepreg material manufacturer are usually effective for developing specific cure cycles for thin laminates of nominal size. However, the specific cure cycles that must be developed for thicker and larger laminates would benefit from two analysis procedures: rheometric dynamic scanning (RDS) and differential scanning calorimetry (DSC). Rheometric dynamic scanning provides information on the changes in viscosity that take place throughout a cure, while DSC analysis determines the amount of heat generated during the cure. These data provide the basis for developing logical cure cycles.

The first step in developing a cure cycle is deciding what heat-up rate to use. A rapid heat-up rate is desirable to reduce processing time, but it must be slow enough so that gross temperature differences do not cause uneven curing and related part warpage or cracking. Although some specifications list heat-up rates as high as 3 °C/min (5 °F/min), this is probably not realistic for curing large parts or for an autoclave load of production parts. A more realistic rate is 0.56 to 2 °C/min (1 to 3 °F/min). Rates lower than 0.56 °C/min (1 °F/min), especially in the early stages of the cycle, will significantly increase costs. This heat-up rate is also compatible with most of the production autoclaves currently in use.

Once the heat-up rate is bracketed, RDS analysis is used to determine the viscosity characteristics. The RDS curves are conducted on the specific resin system using neat resin, which is either taken from a resin sample before impregnation or rendered from the prepreg.

According to the RDS viscosity scan shown in Fig. 2, the viscosity of AS4/3501-6 resin is initially very high; it decreases to a low of about 2 Pa·s (20 P) at 140 min and then gradually increases to a very high level. The material is

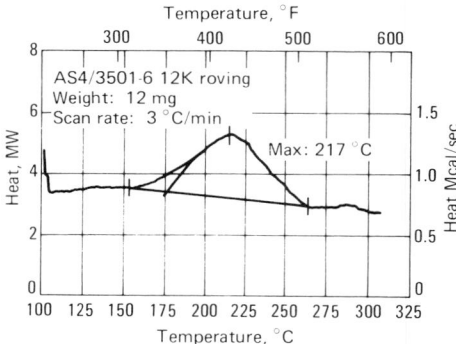

Fig. 4 Differential scanning calorimetry analysis output

considered essentially gelled when the viscosity reaches beyond 100 Pa·s (1000 P). At the point of initial low viscosity or slightly thereafter, all, or at least most, of the autoclave pressure should be applied. It is also desirable at this point to create a dwell period and to vent the vacuum to atmosphere. The dwell period allows the material to consolidate and the curing reaction to advance, which generates heat. The dwell period allows the heat to dissipate, which tends to prevent uncontrolled exotherms. If the viscosity remains low after the initial dwell and if thick laminates are being fabricated, it is usually advisable to implement additional dwells, adjusting their temperatures and times to complete most of the chemical reaction.

The rate of chemical reaction and associated heat generated with the cure of a resin system is determined by conducting a DSC analysis on a small sample of the resin. The DSC scans are run at a heat-up rate of 0.56 to 3 °C/min (1 to 5 °F/min). The typical curve illustrated in Fig. 4 indicates the amount of heat generated during the cure process; some degree of judgment should be used in applying this information to the cure cycle.

Theoretical Aspects of Curing

Numerous theoretical studies conducted in recent years have yielded considerable insight into the curing of epoxy resins. The original work of G.S. Springer and A.C Loos (Ref 2) was probably the first attempt to establish a comprehensive cure model that addressed the following curing aspects:

- Heat transfer to and from the tool and heat source
- Resin flow
- Rate of reaction and heat generation
- Void development
- Development of curing stresses

This model was written in FORTRAN and was programmed to run on a mainframe computer, but portions have been restructured for limited development with a personal computer (Ref 3, 4). Other models, such as ROAST (Ref 5), and CURESIM (Ref 6), are newer. A cure model, WIND, has been developed for filament-wound structures (Ref 7). A less theoretical, but more practical approach to cure modeling is discussed in Ref 8.

Future Development

The concept of placing all pertinent information in computer files and allowing the computer to develop the best cure cycle may become more feasible in light of recent advances in instrumental analysis and autoclave or press control with computers, as well as a significant increase in computing power and a better understanding of computer program development. The following information should be included in the computer files:

- Resin-fiber prepreg characteristics
- Individual tool characteristics and numbers of tools
- Autoclave characteristics

The computer would decide on tool placement and would provide the operator with a graphic representation of the autoclave loading. Once loaded according to the plan, the computer would take over to control the heating and pressurizing of the autoclave and to introduce intermediate and final cure dwell cycles and subsequent cooling to provide an optimum cycle. These concepts are outlined in Ref 9. This intelligent processing, or expert system cure, is perhaps 5 to 10 years away from production installation.

REFERENCES

1. F.C. Campbell *et al.*, ''Computer-Aided Curing of Composites,'' Report IR-0355-1, Contract F33615-83-C-508 for Air Force Wright Aeronautical Laboratories, McDonnell Aircraft Company
2. G.S. Springer and A.C. Loos, ''Curing of Graphite/Epoxy Composites,'' AFWAL-TR-83-4040, Air Force Wright Aeronautical Laboratories, 1983
3. Program available through C.W. Lee, University of Dayton
4. Program available through R.K. Strother, Technology Pipeline
5. A.O. Kays, ''Exploratory Development on Processing Science of Thick Section Composites,'' AFWAL-TR-85-4090, Air Force Wright Aeronautical Laboratories, 1985
6. F.C. Campbell, *et al.*, ''Computer-Aided Curing of Composites,'' Report IR-0355-5, Contract F33615-83-C-508 for Air Force Wright Aeronautical Laboratories, McDonnell Aircraft Company, July 1985
7. E. Calius and G.S. Springer, Selecting the Process Variables for Filament Winding, in *Proceedings of the 31st International SAMPE Symposium*, Society for the Advancement of Material and Process Engineering, April 1986, p 891-899
8. C.G. Ford *et al.*, Advanced Composite Production Cure Cycles Based on Prepreg Rheology, in *Proceedings of the 31st International SAMPE Symposium*, Society for the Advancement of Material and Process Engineering, April 1986, 1300-1312
9. R.A. Servais *et al.*, Intelligent Processing of Composite Materials, in *Proceedings of the 31st International SAMPE Symposium*, Society for the Advancement of Material and Process Engineering, April 1986, p 764-775

Mechanical Fastener Selection

Robert T. Parker, Boeing Commercial Airplane Company

DIFFERENT FASTENER REQUIREMENTS exist for joining carbon fiber reinforced composite structures and must be considered when selecting fasteners during the design process. Fastener selection considerations include corrosion compatibility, fastener materials and strength, head configurations, importance of clamp-up, interference fit, and lightning protection.

When carbon fiber reinforced composite materials began to be used in aerospace and aircraft structures, the inadequacy of conventional fasteners became apparent. Alloy steel was not compatible, cadmium plating quickly corroded away, reduced-shear heads pulled through, and rivets crushed the composite or expanded and delaminated the fibers. A new set of requirements had to be established to determine which characteristics were optimal, and which were detrimental.

Corrosion Compatibility

Although neither fiberglass nor aramid fibers are a problem when used with most fastener materials, composites reinforced with carbon fibers are quite cathodic when used with materials such as aluminum and cadmium, the latter of which is a common plating used on fasteners.

Titanium and its alloys appear to be the fastener materials most compatible with carbon fibers. Fortunately, titanium alloys have the most desirable strength/weight ratio. Austenitic stainless steels, superalloys such as A286, multiphase alloys such as MP135 or MP159, and Inconel 718 also appear to be very compatible with carbon fiber composites, although pitting corrosion has been noted in A286. Copperbearing alloys such as copper-Ni or Monel have a tendency to generate heavy corrosion products, although damage or loss of strength appears to be minimal.

Fastener Materials and Strength Considerations

Titanium alloy 6Al-4V is the most common alloy for fasteners used with carbon fiber reinforced composite structures. Ultimate tensile and shear strengths for Ti-6Al-4V are 1100 MPa (160 ksi) and 660 MPa (95 ksi), respectively. When higher strength is required, materials such as A286 or Inconel 718 can be strengthened by cold-working. Cold-worked A286 fasteners can be obtained with an ultimate tensile strength of 1400 MPa (200 ksi) and an ultimate shear strength of 760 MPa (110 ksi). The respective values for Inconel 718 are 1500 MPa (220 ksi) and 860 MPa (125 ksi). Multiphase alloys can be obtained with an ultimate tensile strength up to 1800 MPa (260 ksi); however, they are presently being used only as fastener components, such as a core bolt or stem of a blind-type fastener, where size is limited and very high strength is required to function.

Head Configuration Selection

Many fastener head styles exist in industry today: 100° reduced shear, 100° shear, 100° tension, AN 509, low-profile protruding, 12-point fatigue rated, and so forth. Because of their viscoelastic properties, carbon fiber reinforced composites are more sensitive to high bearing loads than are metals. This means fastener heads (as well as nuts and collars) should be designed with as much bearing surface as is practicable.

The 130° reduced-height shear head originated during the development of a suitable blind fastener specifically designed for composite structures and introduced to the industry in 1980 (Ref 1). By examining a single fastener lap shear specimen loaded in uniaxial tension, as shown in Fig. 1, it can be seen that as the load is transferred through the fastener, an eccentricity begins to develop and fastener tipping, or "cocking," occurs. This specimen was used to optimize flush head configurations and measure the contribution of the fastener to the stability of the specimen by delaying cocking and pull-through. Figure 2 shows the reaction loads. Assuming that the specimen loads act through the centroids of the fastener, and taking moments about A, it can be shown that very little change is noted by changing the angle of the head; the big difference is in the area of bearing that supports the reaction load. From this analysis, the 130° flush head was developed. The maximum area for a reduced-height shear head was obtained by using the tension head diameter. This resulted in 130°, as shown in Fig. 3. As the distance between the centroids becomes greater, however, the reaction loads go up in value, as demonstrated in Fig. 4. Therefore, the 130° head is limited to thin structures, as shown in Fig. 5. For thicker structures, the tension head optimizes the head bearing area and is recommended for both tension and shear applications where the countersink depth does not exceed 70% of the top sheet thickness. This analysis reduces down to two flush head styles for composite structures: 100° tension head and 130° reduced-height shear head. Flush head fasteners for composite structures should use the maximum bearing area available in the design of the fastener head. A 100° tension head appears to be optimum if the head height does not exceed 70% of the top sheet. Several protruding head configurations are available, but are not as sensitive to performance. B. Cole (Ref 2) used different testing and analyses, including fatigue, and arrived at the same conclusions.

Clamp-Up

When clearance fit holes are used, high clamp-up appears to be beneficial for joint strength and fatigue life. The clamping forces, however, must be spread out over a sufficient area so that the compressive strength of the resin system is not exceeded and the composite

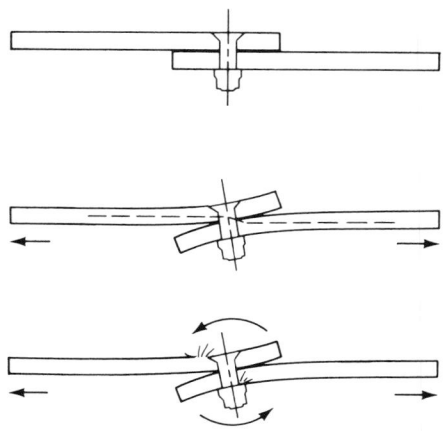

Fig. 1 Single fastener lap shear specimen in uniaxial tension

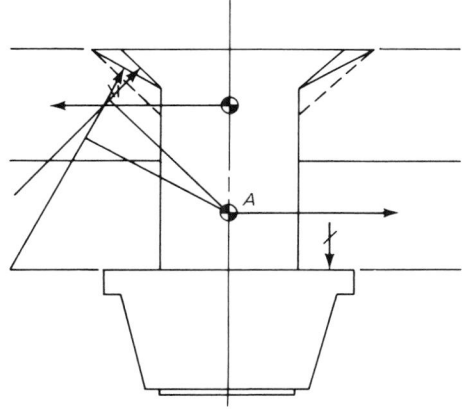

Fig. 2 Reaction loads

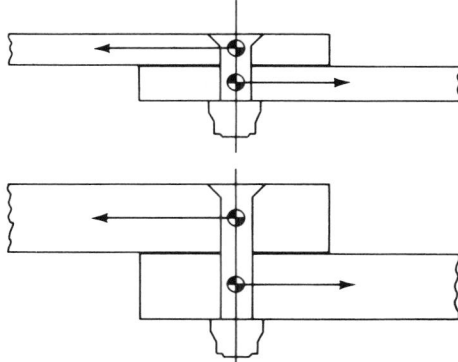

Fig. 4 Load distribution

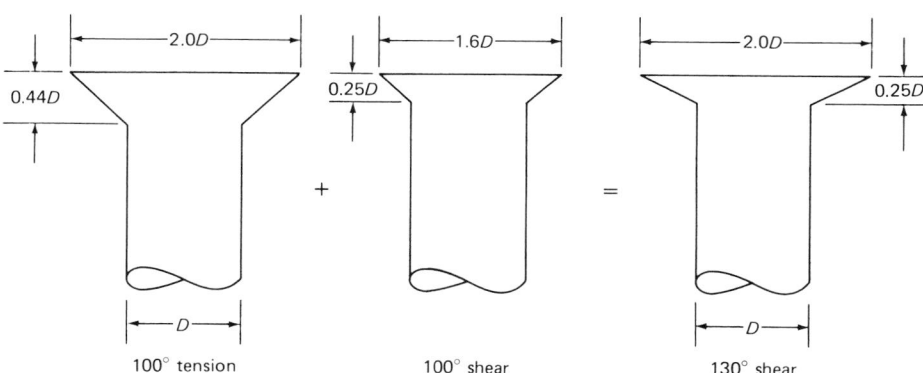

Fig. 3 Development of 130° head for shear

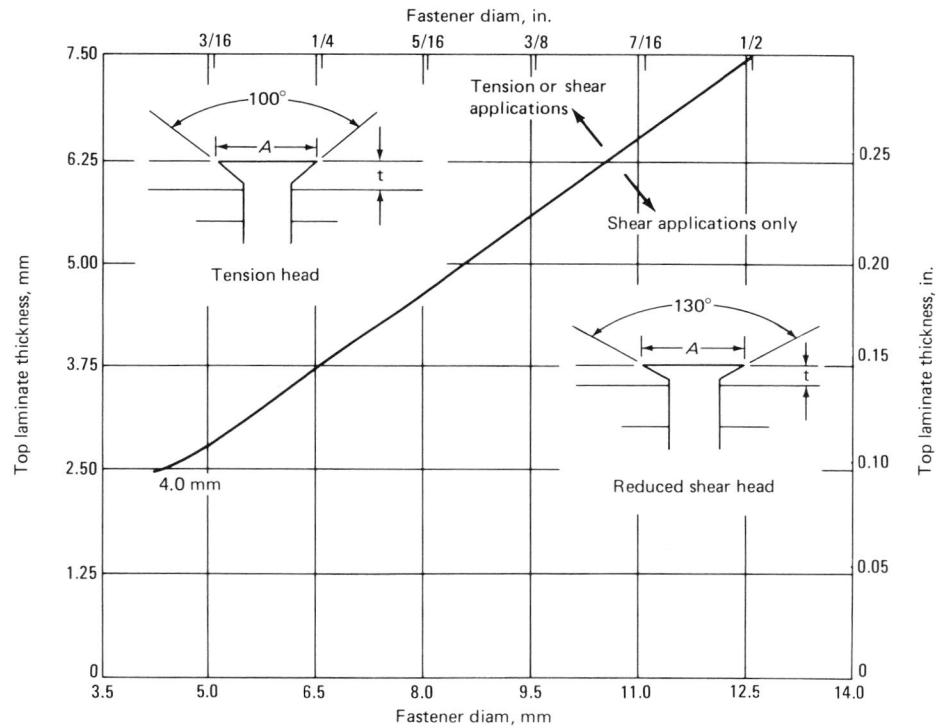

Fig. 5 Flush head applications for composites

crushed. The high clamp-up delays cocking of the fastener in a loaded joint and reduces ratcheting during cyclic loading. J.D. Pratt (Ref 3) also found that high fastener preload, as well as large head bearing areas, were key factors influencing high lap shear joint strengths in composite structures. J. Phillips (Ref 4) found that load transfer fatigue testing with graphite composites shows that gross load and stress-concentration factors (K_t) at holes are not critical. What is critical is bearing stress, because most failures are bearing-type failures caused by the fastener cocking under load and producing a high, localized bearing stress. High clamp-up also delays slippage of the sheets of composite, which contributes to the high concentration of bearing stresses, as illustrated in Fig. 6.

Interference Fit Fasteners

Another method of delaying cocking of the fastener under joint loading is to support the fastener for its full length. A net fit would be ideal, although impractical. In the last few years, fastener manufacturers, working together with airframe companies, have developed methods of accomplishing an interference fit without detriment to joint performance. However, its advantages must be evaluated with regard to cost

and weight increases. In aluminum structures, bolts are frequently installed in "transition" fits, where, because of tolerance overlap, the bolt may be larger than the hole some percentage of the time. The bolts are pressed in or driven in with a rivet gun, with interference fits up to 0.0760 mm (0.003 in.). Not only does this interference fit in aluminum structures lock up the structure, but the hoop tensile stresses generated reduce the amplitude of the cyclic stresses next to the hole and improve the fatigue life of the joint. With composite structures, however, driving the same bolt into interference causes high shearing forces on the reinforcing fibers and bends them down, breaking the matrix resin, as illustrated in Fig. 7(a) and (b). Work by D.M. Shoe (Ref 5) showed damage at as little as

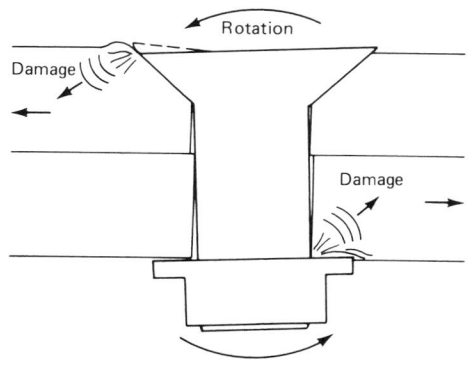

Fig. 6 Clearance in the hole

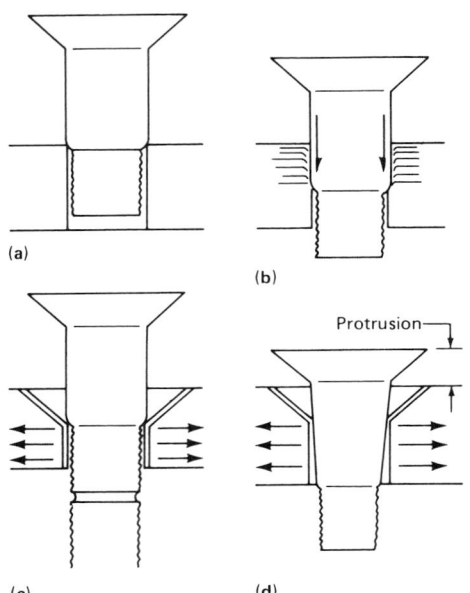

Fig. 7 Interference fit

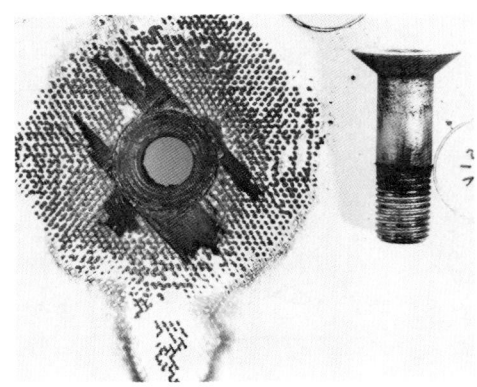

Fig. 8 Fastener with corrosion protection finish, struck by 100 000 A. Heavy damage to composite

Fig. 9 Bare fastener struck by 100 000 A. No damage to composite

0.01780 mm (0.0007 in.) of interference by this method. Because composite fibers will accommodate much more compression, a controlled expansion of a sleeve that remains statically in contact with the fibers (as shown in Fig. 7(c) and (d)) can produce interferences up to 0.1525 mm (0.006 in.). This method is presently being tested on major prototype hardware with excellent results. Unlike aluminum structures, however, the greater interference is of no benefit to composite structures. All that is desired is a "net" fit. The interference is used to absorb the tolerances on the hole and the fastener. Therefore, the interference limit need only be the total accumulation of tolerances to always ensure at least a net fit as a minimum. The major advantages of the net/interference fit are lower joint deflection, equal fastener load sharing, reduction of relative fastener flexibility that causes localized high bearing stresses, and reduction or delay of hole growth/degradation (which can become excessive). An additional advantage is lightning strike protection, which is a design consideration that is covered in the next section of this article. This author believes that major composite structures, such as a wing or large stabilizer, should have a certain percentage of sleeve-type interference fit fasteners to lock up the structure. Fasteners in all clearance holes can ratchet to one side when the structure is loaded and possibly cause aileron reversal. This phenomenon was attributed to a flight test problem on a prototype jet bomber in the late 1940s. Present-day aluminum structures with interference rivets and transition fit bolts do not have this problem, but the problem could surface in composite structures unless a certain number of interference fit fasteners are incorporated. They could be strategically lo-

cated to aid electrical continuity or lightning strike protection.

Lightning Strike Protection

An aluminum airplane is quite conductive and is able to dissipate the high currents resulting from a lightning strike. A carbon fiber reinforced composite airplane would be an anisotropic conductor, that is, not as conductive in all directions, because carbon fibers are 1000 times more resistive than aluminum to current flow, and epoxy resin is 1 000 000 times more resistive (perpendicular to the skin). If the lightning strike attaches to a fastener, the current must be dissipated through the fibers perpendicular to the fastener hole. Intimate contact of a bare fastener with the carbon fibers through an interference fit is the best combination found to date for current dissipation. A swept-stroke lightning strike (defined as a zone 2 strike) attaching to a fastener hole can produce currents as high as 100 000 A. Of course, this current is conducted for a very short period of time (~0.050 s maximum dwell time), but then it must be dissipated in a short period of time to minimize damage. R.O. Brick (Ref 6) has developed a fastener area-current relationship to predict whether a fastener has a large enough countersink and diameter to dissipate lightning strike currents without arc plasma blowby. Although R.O. Brick's paper was directed more toward a fuel environment, the method can also be used to minimize damage to the composite from a lightning strike. Investigators R.O. Brick and J.R. Gozinsky (Ref 7) have determined that large countersinks (tension heads) with high clamping forces or an interference fit are important in the choice of fasteners in a potential lightning strike area. Fastener finish is also critical. Fasteners with corrosion protection coatings, such as NAS 4006 aluminum-phenolic coating or equivalent, are not recommended in potential lightning strike areas. Figure 8 shows a fastener with a corrosion protection finish that has been artificially struck with approximately

100 000 A. Because the protective finish was an insulator, the current avoided the fastener and attempted to dissipate into the top fibers of the composite in a high-concentration gradient. The composite was heavily damaged several laminates deep. In Fig. 9, a bare fastener that was struck with identical current shows more damage to the fastener but no damage to the composite. It is much easier to replace the fastener than it is to repair the composite. The grid pattern shown in Fig. 8 and 9 is a copper mesh overlay used to help dissipate the current on the surface. A bare finish is preferred on the fastener for maximum conductivity; however, depending on the fastener material, a phosphate fluoride or a passivated finish appears to be acceptable and is usually required for paint adhesion.

REFERENCES

1. "Qualification Test Results of Big Foot Blind Fasteners for Boeing Aircraft Co.," Report 518, Monogram Aerospace Fasteners, Division of Monogram Industries, Oct 1980
2. B. Cole, "Special Fastener Development for Composite Structures Program," Paper presented at the National Aerospace Standards Committee Standardization Meeting, May 1982
3. J.D. Pratt, "Blind Fastening of Advanced Composites," Paper presented at CogSME, Fastening Advanced Composites Conference, Renton, WA, Oct 1986
4. J. Phillips, "Fastening Composite Structures with HUCK Fasteners," Technical Paper, Huck Manufacturing Company, 1984
5. D.M. Shoe, Internal report, Boeing Commercial Airplane Company, 1981
6. R.O. Brick, "Multipath Lightning Protection for Composite Structure Integral Fuel Tank Design," Paper presented at the Tenth International Aerospace and Ground Conference on Lightning and Static Electricity, Paris, 1985
7. R.O. Brick and J.R. Gozinsky, Internal report, Boeing Commercial Airplane Company, 1986

Blind Fastening*

John D. Pratt, Monogram Aerospace Fasteners

WHILE THE USE OF ADVANCED COMPOSITES in the aerospace industry has increased at a steady rate, the total number of fasteners required has declined. This is a direct result of the ability to fabricate large, complex components from composite materials with fewer joints. The fasteners that are used, however, are of necessity more sophisticated than those normally used in metallic structures.

Designers must consider several important factors when fastening advanced composite materials. Chief among these are:

- Galvanic compatibility between the fastener and materials being joined
- Installation effects on joined materials
- Joint strength
- Sheet take-up (clamp-up)
- Sensitivity to hole quality
- Robotics or automated assembly of components

Certain types of so-called conventional fasteners, such as Lockbolts and Hi-Loks, are easily adapted for use in composites. Only minor modifications in the clamp-up, head and collar bearing areas, and in selection of compatible materials are required. Other types of conventional blind fasteners are not as easily adapted. Many tend to be installation formed; that is, they undergo uncontrolled shank expansion or form blind side upsets that damage the laminate. Blind fastening has the potential to lower assembly costs through one-man one-side installation and automated assembly. The designer, however, must be fully cognizant of the peculiarities of fastening these materials. A thorough evaluation of the candidate fasteners and actual composite joints under consideration must be conducted early in the design stage. This evaluation can eliminate the burden of future redesign or costly rework on the finished piece.

This article describes those factors that must be considered when using blind fasteners in fiber-reinforced composite materials.

Galvanic Compatibility

Aramid and glass fiber reinforced composites may be fastened with almost any fastener material without fear of galvanic corrosion. Carbon fiber reinforced composites (CFRP), however, cause conventional fastener materials such as aluminum and alloy steel to corrode very rapidly. Salt spray chamber testing has demonstrated the need for fastener materials that are used to join CFRP to be in close proximity to carbon on the electromotive scale. These tests have confirmed that the most compatible fastener materials are titanium alloys, austenitic stainless steels, and certain multiphase and Inconel alloys.

Fasteners used in hybrid joints should be compatible with all materials in those joints. For example, a bare titanium fastener may be used to join CFRP skins to a titanium substructure without corrosion worries. When fastening CFRP skins to an aluminum substructure, however, the fastener should be coated with NAS4006 aluminum coating (or similar material) to provide galvanic protection to the aluminum structure in the vicinity of the titanium fastener. Faying surface sealant should also be used in these cases.

Table 1 lists fastener materials approved for use by a task group of MIL-STD-1515 (currently being converted into MIL-HDBK-1515). Note that the use of a barrier coating is recommended to further protect the fasteners against the corrosive effects of CFRP. The drawback of barrier coatings is that they inhibit electrical continuity between the components being joined, which may be needed for protection against lightning.

Table 1   Approved fastener materials

Materials being joined	Fastener material		
	Preferred	Recommended with barrier coatings	Not recommended
Graphite-epoxy(a) to aluminum(b)	Titanium pin with aluminum collars or nuts bearing on aluminum	Multiphase alloys, Inco 718, or austenitic stainless steel, with aluminum or stainless steel collars or nuts bearing on aluminum, or titanium-columbium systems	Copper, brass, aluminum, low-alloy steel, martensitic stainless steels, or cadmium-plated fastener systems
Graphite-epoxy to titanium, A286, austenitic stainless steel, or graphite-epoxy	Multiphase alloys, Monel fastening systems, Inconel, or titanium, with stainless steel, Inconel, or titanium collars or nuts	Stainless steel pin or screw with aluminum collars or nuts bearing on stainless steel or titanium. Stainless steel collars or nuts bearing on graphite-epoxy, titanium or stainless steel, or titanium-columbium systems	Aluminum collars or nuts bearing on graphite-epoxy, copper, brass, aluminum, or low-alloy steel elements, or cadmium-plated fastener systems

(a) Graphite is used to mean either graphite or carbon reinforcement. (b) These materials are incompatible at the faying surface without a barrier coating.

*Adapted with permission from Technical Paper AD86-795, presented at Fastening Advanced Composites Conference, Society of Manufacturing Engineers, 1986

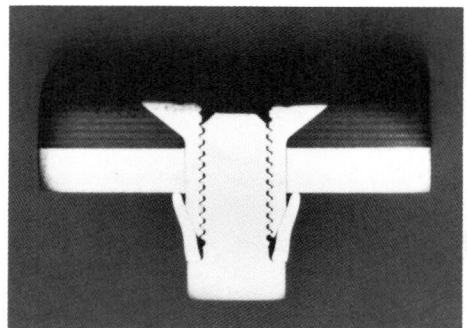

Fig. 1 Conventional blind bolt installed in hybrid joint

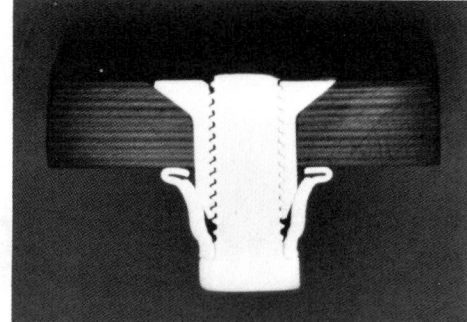

Fig. 2 Composi-Lok blind bolt installed in CFRP joint

Installation Effects

Certain types of conventional blind fasteners may be used to fasten advanced composites to metallic components provided the metallic component constitutes the blind side surface (Fig. 1). Caution should be exercised, however, when considering hole-filling blind fasteners, particularly when the composite is highly directional. Overexpansion of the hole can result in delamination, splitting, or other damage to the laminate.

Joints with a fiber-reinforced composite on the blind side require fasteners specifically designed for this task (Fig. 2). These specialty fasteners are designed to form large, blind-side upsets against the composite material without damaging the laminate. A large-diameter footprint is required to reduce bearing stresses due to flight loads. The manner in which conventional blind fasteners form their blind heads can cause splitting and delamination of the composite material.

It is unlikely that any particular type of blind fastener will satisfy all needs of the designer. Considerations such as allowable joint preload, installed weight, blind-side protrusion (before and after installation), and joint strength requirements all weigh heavily in the selection process. For example, fasteners that impart high preload may be an excellent choice for fastening solid laminates, but these same fasteners may tend to crush fragile sandwich panels in through-fastening applications.

Joint Strength

Tests of tensile pull-through in thin sheets, static joint lap shear strength, and lap shear fatigue have become the three most important means for evaluating new fastener designs. While most aircraft companies insist on this testing, few can agree on standardized test methods. The Fastener Testing and Development Group (FTDG) of MIL-STD-1312 is attempting to standardize test methods of fasteners for advanced composites, but completion is not expected in the near future. An attempt to initiate development of a standard single lap shear static test for establishing design al-

lowables in advanced composites has been started by the MIL-HDBK-5 Committee.

Tensile pull-through in thin sheets is sometimes performed in a fixture similar to that shown in Fig. 3. With the fastener installed in two laminates, an autographic recording of load versus deflection is made while the sheets are separated and the fastener blind head is pulled through the laminate. Obviously, the larger the blind head, the higher the load required to pull the fastener through the sheet.

Static joint lap shear tests are usually conducted in a specimen such as that depicted in Fig. 4. An extensometer is attached across the joint, and load is applied in a tensile testing machine. An autographic recording is generated during the test to assist in the determination of joint yield load (yield is defined as the load at which the specimen would have taken a permanent deformation at 4% of nominal fastener diameter, following removal of load).

Key factors influencing high lap shear joint strengths are good hole quality, close fit between hole and fastener shank, large manufactured and blind head diameters, blind head stiffness, manufactured head stiffness, high fastener preload and clamp-up capability within allowable limits, and class of fit between elements of multiple-component fasteners.

All of the above factors tend to restrict fastener rotation in the joint, thereby minimizing joint elongation as load is applied. For example, an increase of 27% was discovered in lap joint yield strength for CFRP from a 100° shear head to a 130° shear head style. The fastener heads are identical in head height, but the 130° head is larger in diameter. This increase in joint strength has prompted one commercial airplane company to consider the 130° head style for all fasteners used in thin CFRP.

Lap shear fatigue tests are also performed on the specimens depicted in Fig. 4 at several load levels in order to develop S-N curves. It has been demonstrated that fastener preload is the most influential factor affecting results because the higher the preload, the greater the number of cycles to failure. Of course, if preload is too high, crushing of the laminate can occur. High fastener preload provides resistance to movement between the two sheets,

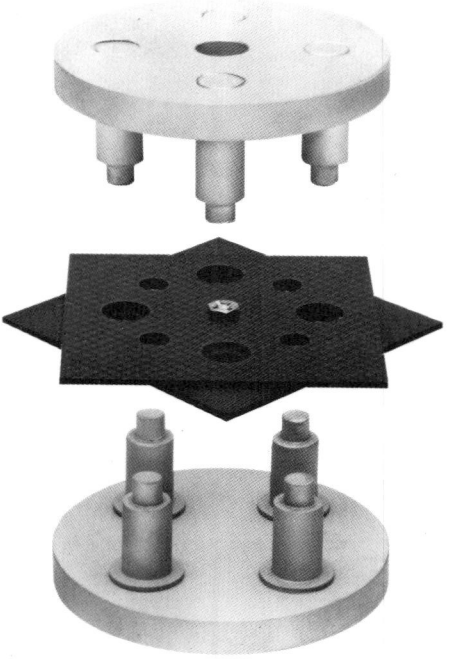

Fig. 3 Thin-sheet pull-through specimen

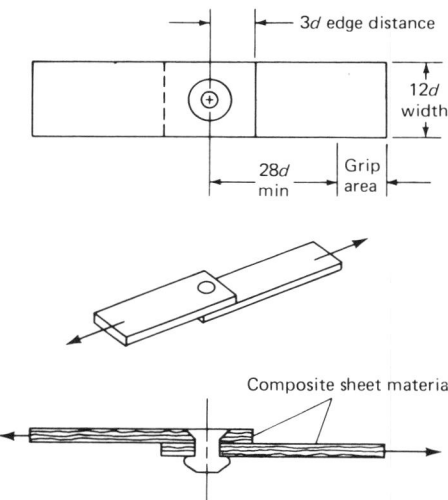

Fig. 4 Lap shear coupon

thereby minimizing heat buildup and fretting of the composite.

Sheet Take-Up

In order to ensure structural integrity, all gaps between the articles to be joined must be removed before installation of the permanent fasteners. The force exerted to close these gaps varies, depending upon the stiffness and cross section of the composites and also on whether sealant is used on the faying surfaces.

The most popular method of temporarily positioning pieces of structure and closing

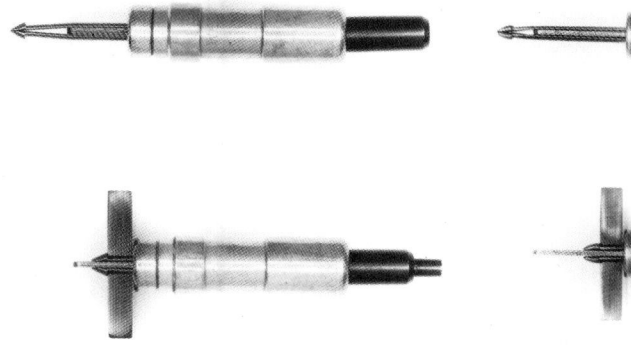

Fig. 5 Blind hole clamp for composite materials

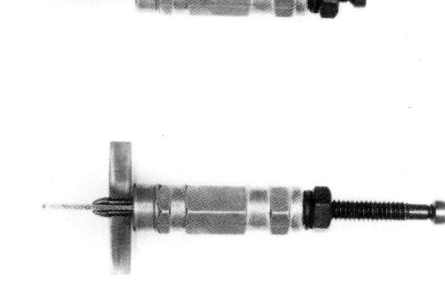

Fig. 6 Blind hole clamp for metallic materials

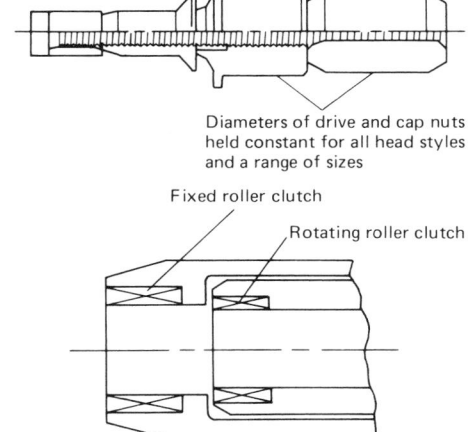

Drive nut Cap nut

Diameters of drive and cap nuts held constant for all head styles and a range of sizes

Fixed roller clutch

Rotating roller clutch

End effector

Fig. 7 Composi-Lok II fastener designed for robotics installation in advanced composite materials

the gaps between them is to install a Wedgelock (or Cleco) type of blind hole clamp (Fig. 5). These fasteners can close gaps requiring up to 140-kg (300-lb) closing force. Steps must be taken to ensure that the force exerted by the clamp is not high enough to damage the composite blind side. The old type of Wedgelock fasteners normally used for metallic structures (Fig. 6) should not be used in advanced composites without extreme caution.

In those applications in which all of the gap cannot be removed with clamps, the permanent fastener must be capable of removing the balance. One of the most common field problems associated with blind fasteners is installation in an out-of-grip condition. The actual grip to be fastened must take into account any gaps between the pieces to be joined.

Sensitivity to Hole Quality

Several diverse techniques exist for producing quality holes in advanced composites. Carbon, aramid, and boron fiber reinforced materials each require different drilling methods and tools. Refer to the article "Fastener Hole Considerations" in this Section of the Volume for a thorough discussion of hole preparation. This topic is briefly addressed in this article in terms of hole quality and its effect on fastener and joint performance.

Fastener holes should be straight and round within limits specified on engineering drawings. Normal hole tolerance is 0.075 mm (0.003 in.). Holes should be drilled perpendicular to the sheet surface, within 1°. It is not usually necessary to provide fillet radius clearance with a chamfer on the edge of the hole in composite material.

Two of the most common problems associated with hole preparation are overheating of the matrix material and delamination of the last one or two plies. Damage due to overheating is not always visible to the mechanic. Installation of certain types of blind fasteners will render this damage visible by distorting the surface of the blind side. Joints in which the matrix material has been overheated will permit greater joint elongation and correspondingly lower strength than those with undamaged holes.

Blind-side delamination caused by poor drilling techniques is not as detrimental to joint performance as are overheated holes, but again, some decrease in joint strength can be expected. Delamination may not always be visible until after installation of the fastener, if then.

If overheating or delamination are suspected, then appropriate inspection by any number of NDT techniques should be employed.

Robotics Installation Capability

Because blind fasteners require access to only one side of a structure, they are better suited than are nonblind fasteners to installation by a robot in both advanced composites and metallic structures. In fact, as automation is applied to existing aircraft components, a number of cases are being witnessed in which conventional fasteners such as Hi-Loks and Lockbolts are being replaced by blind fasteners.

One of the key factors in selecting blind fasteners for automatic installation is the weight and size of the end effector tooling. The end effector consists of a lightweight air motor with a fastener transfer and load mechanism. Generally, the fastener is brought to the end effector from a feeder bowl through a blow tube. The threaded-type blind fastener tooling is lightweight and generally weighs less than 10 kg (25 lb). The tooling used for pull-type blind fasteners is usually very heavy and bulky and may not be suitable for robotic applications.

Another key factor is the ease of automatically orienting and feeding the fastener from a feeder bowl. Easy and reliable engagement of the fastener wrenching surfaces by the installation tool is a must. The ability of a single end effector to install several different diameters and grip lengths, as well as different head-style fasteners, is, in many cases, an absolute must. An example of one fastening system that permits this is the double drive nut Composi-Lok system shown in Fig. 7.

Expendable fastener components, such as pintails and drive nuts, should be selectively ejected or collected from the drive tool into a receptacle in order to prevent foreign-object damage from reaching the work area floor. This is accomplished with the Composi-Lok double drive nut tool by a kick-out piston that is activated by exhaust air from the drive tool.

Fastener Hole Considerations

Joseph L. Phillips, Huck Manufacturing Company
Robert T. Parker, Boeing Commercial Airplane Company

TECHNIQUES AND TOOLS for hole generation by means of drilling, reaming, and countersinking operations on carbon fiber reinforced laminates are the focus of this article. Detailed recommendations for fiber-resin and metal combinations are presented. Because all information is based on the authors' specific experiences, it should be used only as a guide; experimentation for specific applications is encouraged. Additional information on drilling techniques for composite materials can be found in the article "Solid-Tool Machining and Drilling" in this Section of the Volume.

To understand the different techniques and tools used with carbon fiber reinforced laminates, specific characteristics of these materials relative to hole generation need to be understood. For example, graphite material has these characteristics:

- Tolerates high speeds
- Abrasive; requires carbide or polycrystalline diamond (PCD) cutting tools
- Fibers break or cut cleanly
- Limited drill "grabbing"
- Fiber breakout problem
- Subject to delamination
- Cutting fluids are of limited help
- Dust problem

Generally, graphite can be treated similarly to fiberglass, with some exceptions in respect to tool selection. Graphite fibers can be provided in unidirectional-fiber tape or woven-cloth forms. When unidirectional tape is laid up on a part surface, the graphite fibers are extremely prone to fiber breakout when holes are drilled, especially on the drill exit side and particularly if there is some resin starvation on the surface. Unidirectional fibers provide better strength, are more amenable to automated lay-up, have less waste than does cloth, and provide a smoother aerodynamic external surface than does cloth, without much filling, smoothing, and sanding. Methods exist for drilling holes in unidirectional lay-ups without excessive fiber breakout problems, but on drill exit sides it is desirable to have cloth material on the exit surface in localized fastening strips or zones, if possible. Of course, graphite structures or skins attached to metallic substructures will have an automatic backup for drilling and will not have this fiber breakout problem. To date, most military fighter use of graphite has been of this latter design, but there may be a trend toward graphite skins and substructures as a way of reducing the radar image of the structure.

In addition to cutting characteristics, dimensional and tolerance requirements, as well as defect allowances, must be known. Generally, fasteners in composite structures have been installed in clearance holes. Interference fits for fasteners were not possible in the past because they caused serious delamination around the fastener hole when the fastener was installed. This included relatively low levels of interference. The incorporation of a sleeve and special fastener designs have alleviated this problem. However, because a clearance hole was required, and any benefits of an interference fit were basically lost, a practical hole tolerance of 0.075 mm (0.003 in.) on the diameter has generally been used. This usually allows use of one-shot drilling within tolerance without reaming. Thus, hole sizes are generally 0.025 to 0.10 mm (0.001 to 0.004 in.) over maximum fastener diameter, although 0.0127 mm (0.0005 in.) has been reported. Experimental tests have demonstrated that the size of the hole, rather than its quality, affects joint fatigue strength.

The authors know of no exhaustive research on allowable hole anomalies in composites that has been successfully concluded. Generally, allowable anomalies have been arbitrarily defined, and when no connected problems occur with great amounts of test hardware using these parameters, they have been considered acceptable. A value of 0.75 mm (0.030 in.) from the hole edge is often used for acceptable anomalies, such as fiber breakout, delamination, or resin erosion. Because frequent occurrences of such levels of defects might be cosmetically unacceptable to customers, engineers, and inspectors, processes are usually defined to avoid them. Therefore, the conclusion that arbitrarily defined values are acceptable, based on the assumption of difficulties, is somewhat flawed.

One investigation of hole breakout on the exit side showed very little effect on joint fatigue strength. The test evaluated carbon fiber reinforced epoxy tape materials with and without backup during drilling. Figure 1 shows typical backside breakout for a 4.8 mm (3/16 in.) diam drill, when drilling through tape. Delamination is usually limited to one or two laminates. Figure 2 shows the elimination of hole exit breakout when a proper backup is incorporated. Improperly repairing the breakout for cosmetic reasons can actually reduce the strength slightly for this type of joint, as shown by specimens 7 to 9 in Table 1. The cured resin was not sanded flush and did not allow the fastener to seat against the composite surface. This resin-rich (pure resin) area allowed early cocking of the fastener and failure at lower values. Although this test was not in-depth by any means, it does show that breakout may be more of a cosmetic than a strength concern, and care must be taken in use of repair techniques.

To explore the "acceptable defect" subject further, it is necessary to understand how graphite composite joints fail in fatigue. Fatigue failures are not a function of gross stresses amplified by K_t factors and localized strain at holes, but rather are a function of bearing stresses and fasteners tipping in clearance holes to produce highly localized bearing stresses. Therefore, minor fiber breakout on the surface (although cosmetically displeasing) may be relatively unimportant, while delamination may be significant. Delamination of fibers below the first one or two laminates is a problem because,

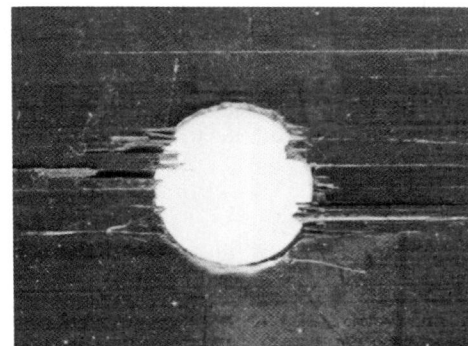

Fig. 1 Typical hole exit breakout damage from drilling without a backup

Fig. 2 Typical hole exit condition when using a backup

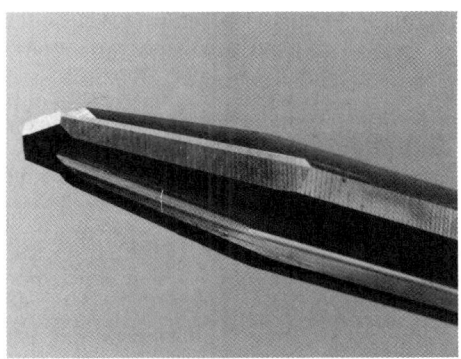

Fig. 3 Four-flute, tapered drill for graphite composites

Table 1   Composite fiber breakout versus strength

Specimen	Lap shear strength		Condition
	N	lbf	
1. 7400		1660	No backup during drilling; no repair
2. 7553		1698 (Average = 7490 or 1684)	
3. 7539		1695	
4. 6934		1559	No backup during drilling; repair breakout with
5. 7264		1633 (Average = 7241 or 1628)	Epocast 50-A/946; wet and install fastener
6. 7526		1692	
7. 6694		1505	No backup during drilling; repair breakout with
8. 7370		1657 (Average = 6952 or 1563)	Epocast 50-A/946; cure 12 h and install
9. 6792		1527	
10. 7597		1708	Used backup during drilling; no breakout
11. 7557		1699 (Average = 7544 or 1696)	
12. 7482		1682	

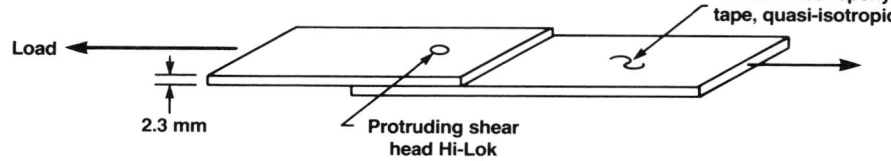

other than using x-rays, it is difficult to determine the extent of the damage. Delamination can be caused by poor drilling techniques, incorrectly installed fasteners, or excessive clamp force. Following is an example of a typical specification for allowable anomalies:

● There shall be no delamination to the external ply
● Countersinks shall be concentric to the hole within 0.075 mm (0.003 in.). CSK axis shall be parallel to the hole within 2°
● There shall be no evidence of delamination or material scorching on the surface of the hole or countersink, except as noted in the last item in this list
● Burrs, dust, or fiber particles that prevent seating of the fastener or intimate contact of structures must be removed
● Breakout damage shall not exceed specified limits. Breakout damage is defined as chipping, splintering, or delamination
● Breakout or chipping of the external ply shall not exceed 0.75 mm (0.030 in.) from the edge of the drilled holes. When the external plies are unidirectional graphite-epoxy tape, the breakout allowance is 2.54 mm (0.100 in.) from the edge of the drilled holes

With graphite materials, the cutters for drilling and countersinking have mostly been developed by the major users. Consequently, there are many cutter configurations in use. This article covers only those cutters that the authors are familiar with or those generally considered best for the operation.

Material Versus Technique/Tool Selection

Technique and tool selection change with fiber type, fiber presentation, fiber combinations, fiber-metal combinations, and resin type. The examples covered in this article are: graphite-epoxy unidirectional tape structure, graphite-epoxy cloth structure, graphite-thermoplastic resin structure, graphite-epoxy with aluminum or titanium substructure, and aramid-graphite-epoxy hybrid structure.

A graphite-epoxy unidirectional tape structure is the most difficult material to drill without encountering surface fiber breakout problems. The abrasiveness of graphite fibers also mandates the use of solid carbide or carbide insert cutters for drills with certain diameters. Because carbide is relatively brittle and very small diameter drills are easily broken when used in manual drill motors, high-speed steel (HSS) drills are often used in pilot-sized diameters, such as number 40. These sizes have acceptable lives for their relatively low costs and relatively low peripheral cutting distances. Above number 40, carbide is essential; HSS drills will produce less than one hole before becoming dull, whereas carbide will produce from 100 to 200 holes. The use of titanium nitride (TiN) coated carbide drills and/or countersinks for graphite applications has been explored and found to be of no value.

Separate countersinking operations using a piloted countersink in a microstop produce rapid wear at the bottom of the countersink such that a maximum of 50 holes can be pro-

duced before resharpening is required. Because of this problem and because an integral drill/countersink tool is required for practical automation, a drill/countersink tool with a removable countersink was developed. Subsequent investigation, however, has shown that an integral drill/countersink tool has uniform wear on the drill tip and countersink and can be used for 200 holes before resharpening is required. The difference is attributed to graphite cutting debris trapped in the countersink by the pilot, in one case, and the lack of trapped cutting debris in the other case (because of the flutes in the drill). This is especially important for practical fastening automation in graphite composites.

If unidirectional tape parts have a layer of cloth on the exit side of the attachment part and a countersink on the entry side, acceptable holes can be drilled rapidly using a high-speed manual method, such as a four-flute, tapered, solid carbide drill (developed by the Boeing Commercial Airplane Company and shown in Fig. 3) in an 18 000 to 20 000 rpm hand drill motor. For this type operation, hole locations are best located by spotting with a small, HSS pilot drill, using a simple location template. When using the four-flute drill, the drill point is first located on the part surface with no drill rotation. Some pressure is applied, and then the trigger is pulled. Because the hole is drilled extremely rapidly with this method, pressure should be removed from the trigger as soon as the hole is drilled. Hole dimensions will generally be 0.025 to 0.050 mm (0.001 to 0.002 in.) over drill size. Material stack-ups of 1.5 mm (0.060 in.) or less should not be drilled with this method because insufficient stability of the drill will produce out-of-round holes. For safety purposes, overall lengths of drills at this rpm rate should be less than 100 mm (4 in.). At this high rpm rate, drill motors must also have

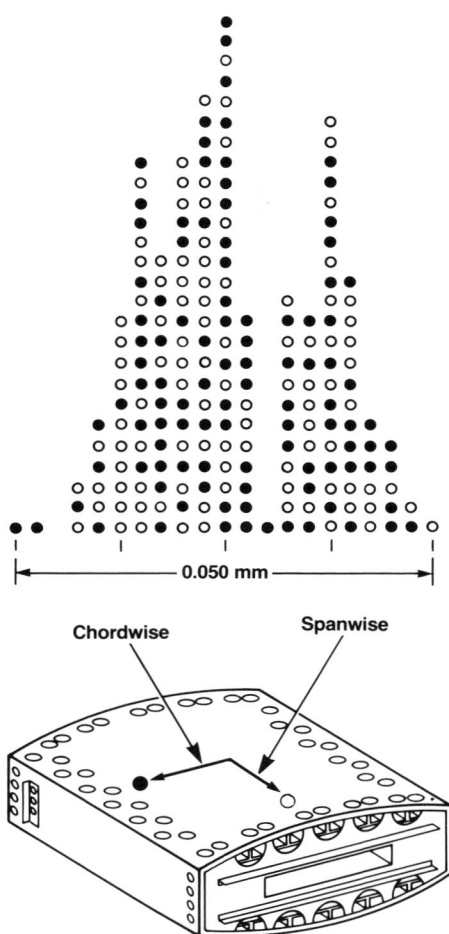

Fig. 4 Hole size population (skin/spar and skin/rib) for JVX (V-22 wing box). (b) Task ID: JVX (V-22 wing box). Hole-size population (skin/spar and skin/rib)

enough torque to do the job properly. This requires 620 kPa (90 psi) air, an air hose with large enough diameter, and a drill motor, such as the 18 000 rpm Rockwell 41D. The 20 000 rpm drill, however, produces a hole of inferior quality compared to a Spacematic-generated hole, but it is less costly and can be used efficiently in nonfatigue-critical locations.

A recent investigation of the capability to produce consistent holes in carbon fiber reinforced epoxy tape composites was performed on a 12-m (40-ft) test wing for the U.S. Navy V-22 Osprey. Approximately 100 holes sampled at random were measured and recorded. Figure 4 shows the population of hole variance. The survey showed all the holes in this diameter (6.4 mm, or 0.25 in.) to be within a 0.050-mm (0.002-in.) range. The holes were drilled manually with a four-flute, tapered, solid carbide drill. Hole tolerances for production hardware of 0.000 to 0.075 mm (0.003 in.) are not unreasonable with the state-of-the-art tooling currently available.

Another drill that can be used with cloth on the backside is a dagger drill, developed by the

McDonnell Aircraft Company and available from the Metal Removal Cutting Tool Division of the Federal Mogul Corporation (FMC) under license. This drill should be used at 2000 to 6000 rpm in manual or fixed-feed operations. It is a lower cost tool, but more fragile.

For a structure that is all unidirectional-tape material, the key word is feed control and the best drill configuration is a solid carbide twist drill with an eight-faceted point, as developed by FMC. The feed rate with hand drill motors must be controlled with hydraulic dashpots. Hand- or robot-held air over oil feed drill motors are also acceptable. Air over oil feed, self-colleting drill motors (Spacematic-type), such as a Winslow drill motor, would not be satisfactory without a metal structure on the backside to avoid colleting damage to the structure. Recommended drill speeds with this drill are 2000 to 3000 rpm. Feed rate should be adjusted to the maximum rate that will not cause exit damage. Using an adjustable collet clamp-up device, as on some Spacematic-type drill motors, and reducing the force would permit drilling without a metal substructure.

Cutting fluid is not normally used or recommended for these drilling operations in graphite. Special graphite dust pickups can be developed to pick up the dust at the point of hole generation. The so-called graphite dust is not really dust, but broken fibers still encapsulated in resin so that they do not freely float in air. Common practice has been to vacuum structures and local areas after drilling or to hold a vacuum hose near the point of drilling. However, one aerospace company does recommend water for all drilling operations because it may avoid temperatures that could exceed resin curing temperatures, and reduces the tendency of graphite dust to float in the air.

Standard carbide insert or solid carbide countersinks are acceptable. Countersinking can be a beneficial operation because it can be used to clean up drill entry damage in unidirectional tape parts. To avoid fiber breakout problems at the edge of a countersink in tape parts, a micarta mask can be used on the front of a microstop to contain the fibers. Countersinking is done through the mask and should be done at only 300 to 500 rpm. If a combined drill/countersink tool is used, a long-throw microstop is required, and the rpm should be defined by the countersinking operation.

Depth control for countersinking in composites requires the use of special techniques. In both graphite and aramid composites, there is some pull-down or localized compression by the fastener when fully clamped. Depth settings for countersinks should be established using a fully installed fastener in sample or scrap material equivalent to that being fastened. Using a loose fastener is not adequate. In comparison to aluminum, countersinks in graphite composites should be 0.10 to 0.125 mm (0.004 to 0.005 in.) shallower.

Because piloted manual countersink tools wear rapidly (even though made of carbide),

the wear will affect countersink apparent depth because the cutter wears more rapidly at the bottom than at the top. Once a sharp countersink tool has been properly set for depth, it should not be readjusted for apparent changes in depth using a loose fastener. Cutters should normally be taken out of service and resharpened after 50 holes. If fasteners do not install flush because of countersink wear, the countersinks can be cleaned up with a properly set, sharp countersink.

With a graphite-epoxy cloth structure, the main difference is the lack of fiber breakout problems when using the right tools and techniques. For this material, the authors recommend using a four-flute, tapered drill at high rpm's with hand-held drills (no feed constraint). The main restraints are parts that are too thin or too thick, the latter of which must be defined by trial and error in sample material. The eight-faceted twist drill should be used for parts that are too thick or thin, with controlled feed, as defined in the section in this article on unidirectional tape.

Again, conventional solid carbide or carbide insert countersinks can be used, but no masks are required to control fiber breakout. Depth control procedures are as defined previously.

Graphite cloth- or tape-thermoplastic structures, as well as ones with high-temperature resin systems, are being considered for the operations described in this article. However, the authors have little familiarity with these systems because of the predominance of epoxy resins. Thermoplastic resin systems obviously will not tolerate a tapered cutter and high rpm's without some melting difficulties. Therefore, it would appear that the eight-faceted twist drill at 2000 to 3000 rpm with controlled feed would be best for this material. Preliminary testing with this material also indicates that less dust and more of a chip is formed when drilling.

Graphite-Epoxy with Aluminum or Titanium Substructures. Drills that work well in graphite-only structures, such as the four-flute, tapered drill, do not cut well in metal. A metal structure requires a more conventional, chip-cutting tool. The graphite still imposes a requirement for carbide cutters, and the titanium imposes a maximum cutting speed restriction of about 300 rpm for most common fastener hole sizes.

The metallic substructure provides an automatic backup that prevents any exit side fiber breakout or delamination in the graphite portion of the structure if it is well clamped. Because it also provides a backing for self-colleting drill units, these tools can be safely used in this type of structure.

The best drill configuration for this type of structure would be the eight-faceted twist drill, and the best equipment would be a Winslow self-colleting drill motor. Air over oil feed, noncolleting drill motors held in a drill jig or by a robot arm would also be satisfactory. Drill speed should be about 2000 rpm with aluminum, and about 300 rpm with titanium. No cutting fluid is recommended in either case.

Pulling aluminum chips through the graphite structure with the drill should cause no problem, but titanium chips could cause some erosion in the graphite, allowing the hole to get oversized. Errosion can also occur to any cast-in-place plastic shim that may be used at the faying surface. If erosion proves to be a problem, the holes have to be drilled undersized and reamed to size with a straight-flute carbide insert reamer. Peck drilling is an option. Considering the reaming characteristics of titanium, one would expect the final hole to be about 0.025 mm (0.001 in.) smaller in diameter in the titanium than in the graphite; the reverse, however, appears to be true, with the hole in the graphite being smaller than the one in the titanium.

When an interference fit fastener is required in the aluminum substructure for fatigue improvement, first the correct-sized hole for the aluminum should be drilled through both the graphite and the aluminum, and then the graphite should be removed and the hole reamed up to its correct clearance size. Recommendations for countersinking in this combination of materials is no different from those covered for graphite-epoxy unidirectional tape or cloth.

Aramid-Graphite-Epoxy Hybrid Structures. Technique and tool selection is dependent upon the type of surface material. An aramid cloth can be cut cleanly with conventional drills if it is not on the surface and is layered with the graphite. It can also be cut cleanly, however, if it is on the surface of the hybrid structure with a layer of 120-style fiberglass on it.

Graphite tape on the surface poses more of a problem than does graphite cloth. The percentage of graphite and aramid materials also affects the degree of restraint relative to self-feeding. The aramid cloth also dictates a slower cutting speed, and the graphite dictates the use of carbide.

Recommendations would be to use the solid carbide eight-faceted twist drill with feed restraint at 2000 rpm except when aramid cloth is on the surface with no fiberglass protection. This situation would dictate the use of the solid carbide fishtail point with the same speed and feed constraints.

Countersinking cutters and techniques would be based on previous recommendations relative to what material is on the surface. Carbide is required. Recommendations for handling aramid-graphite-epoxy hybrids with an aluminum substructure parallel those just described.

Dissimilar Material Separation

Robert H. Stone, Lockheed-California Company

GRAPHITE COMPOSITES present a special problem for assemblies that are bolted and bonded to metal structures because graphite, along with metallic alloys, is in the electromotive series of alloys commonly used in aircraft structures (Fig. 1). A galvanic cell can thus be formed in the presence of moisture or other electrolytes between a graphite composite and any of these metals that it contacts. Graphite, which is at the cathodic end of the series and acts as a noble metal, is impervious to corrosion itself but will accelerate corrosion in the adjacent less noble metal. The accelerated corrosion rates of metals, such as aluminum, that contact graphite composites have been observed and quantified in various test programs (Ref 2).

Graphite Electrochemical and Thermal Expansion Characteristics

As shown in Fig. 1, there is considerably more galvanic potential difference between graphite and aluminum than between graphite and titanium or graphite and stainless steels. Likewise, there is considerably more potential difference between graphite and nonstainless steels than between graphite and stainless steels. In addition, clad alloys and certain aluminum alloys have more galvanic potential than others. Figure 1 provides a guide for materials selection based on the criterion of galvanic corrosion. However, uncoupled corrosion rates for the anodic metal material and the polarization characteristics of the metal are also factors in determining the severity of galvanic corrosion.

Graphite-to-aluminum joints definitely represent a potentially serious corrosion problem. The assembly of graphite to aluminum is not prohibited, except in critical or high-risk applications, but requires special procedures. Using titanium, stainless steel, or graphite composites in place of aluminum is always desirable from the standpoint of galvanic corrosion. Matching materials with similar thermal expansion coefficients is a consideration in any joint design. A comparison of thermal expansion coefficients is given in Table 1. The combination of a graphite composite with aluminum would result in the

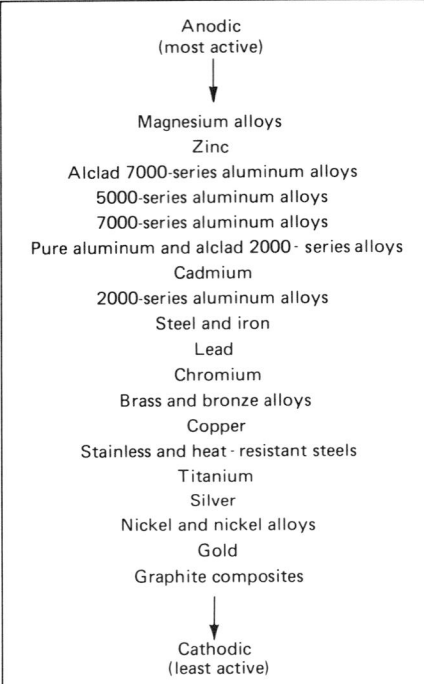

Fig. 1 Electromotive series of aircraft alloys, in descending order of tendency to corrode. Electrolyte is seawater. Source: Ref 1

Anodic (most active)
Magnesium alloys
Zinc
Alclad 7000-series aluminum alloys
5000-series aluminum alloys
7000-series aluminum alloys
Pure aluminum and alclad 2000- series alloys
Cadmium
2000-series aluminum alloys
Steel and iron
Lead
Chromium
Brass and bronze alloys
Copper
Stainless and heat-resistant steels
Titanium
Silver
Nickel and nickel alloys
Gold
Graphite composites
Cathodic (least active)

greatest thermal expansion mismatch, which would increase loads on the fasteners, adhesive, or sealant as the joint was subjected to in-service temperature variations. A thermal expansion mismatch also results in stressed bonded joints when bonding or co-curing is accomplished at elevated temperatures. The design of a graphite-to-metal joint thus involves finding a balance between the lower cost, lighter weight, availability, and ease of machining of aluminum and the better match of properties attainable with titanium or stainless steel. The best galvanic and thermal match, of course, is with graphite itself, and this should be a consideration in materials selection. Graphite-to-graphite bolted joints still present a galvanic problem, however, because metallic fasteners will be in direct contact with the graphite composite.

Table 1 Thermal coefficient of expansion for composite and metallic materials

Material	Coefficient of thermal expansion, 10^{-6}/K Longitudinal	Transverse
Graphite-epoxy (0°)	0.43	29.2
Graphite-epoxy (0°/±45°/90°)	3.4	3.4
Graphite-epoxy fabric (24 × 23 − 8HS)	2.7	4.0
E-glass-epoxy (0°)	8.6	...
E-glass-epoxy (181-style weave)	9.9	12.1
Aramid-epoxy (0°)	−5.4	...
Aramid-epoxy (181-style weave)	−1.8	−1.8
Aluminum alloys	23.4	
Steel	10.8	
Titanium	10.1	
Stainless steel	18	

Values for composite laminates may be taken as typical; however, actual values for specific materials, especially in the graphite-epoxy systems, can vary widely from these values. Source: Ref 3

The most critical requirements for graphite-to-metal joints are using suitable measures for separating the two faying surfaces being joined, finishing the composite and the metal, and preventing moisture entry. Bolted and bonded joints each require special approaches, outlined below.

Bolted Joints

For bolted joints, the initial consideration is fastener selection. The use of aluminum fasteners, collars, and nut plates is never recommended in any joint with graphite composites. Nor are cadmium-plated fastener system elements recommended in any graphite composite bolted joint, because the cadmium coating will be severely attacked. Titanium and stainless steel are the recommended materials for composite fastener systems.

Studies of fastener materials (Ref 4) have established an order of preference for fasteners used in graphite composite joints, as shown in Table 2. Even with compatible fastener materials, proper sealing procedures are required to

Table 2 Galvanic compatibility of fastener materials

Order of preference	Material	Compatibility with graphite-epoxy
1............	Titanium, Ti alloys, Ti-Nb	Compatible
2...........	MP-35N, Inco 600	Compatible
3...........	A-286, PH13-8Mo	Marginally acceptable
4..........	Monel	Marginally acceptable
5...........	Low-alloy steel, martensitic stainless steel	Not compatible

Source: Ref 5

protect the metal in the fastener bore. Wet fastener installation and use of a faying surface sealant, described in the article "Faying Surface Sealing" in this Volume, are both required procedures for graphite-to-metal joints. Wet fastener installation involves placing a bead of sealant around the shank of the fastener so that the sealant forms a complete barrier between the fastener and the material into which it is being installed. The effectiveness of these procedures has been demonstrated by a 13-month real-time exposure in a seacoast environment (Ref 6). In this study, no aluminum component or fastener corrosion occurred when proper sealing measures were employed. This study also developed a ranking of acceptable fastener material selection, in this order: (1) titanium, (2) MP-35N (AMS 5758) or Inco 600 (AMS 5687, and (3) A-286 stainless steel (AMS 5731, 5737).

For graphite-to-graphite assemblies, faying surface sealing and wet fastener installation are not mandatory if acceptable fastener materials are used, but are recommended as an added precaution against moisture entering the joint.

For the typical case of a graphite-to-aluminum bolted joint, the following isolation and protection scheme is strongly recommended. The aluminum component should be surface treated using a suitable corrosion protection procedure, such as chromic or sulfuric acid anodizing. A protective conversion coating (chemical film treatment) can be used, but it is less effective than anodizing and could be a problem because it is conductive. A finish system, such as epoxy primer and polyurethane topcoat, is then applied to the aluminum component before assembly.

A ply of fiberglass or aramid should be co-cured on the surface of the graphite faying with the aluminum component. Both of these materials are electrochemically inert, and using them on the surface provides additional barrier material (primarily the additional resin which encapsulates this outer ply) between the graphite fibers and the metal. Although this layer is most effectively applied as a co-cured ply, it could also be secondarily bonded in the form of either a prepreg or dry fabric impregnated with a wet lay-up resin.

Application of an epoxy primer or primer plus polyurethane topcoat to the fiberglass is an additional protective measure that is strongly recommended. On the composite, the surface treatments and finishes, including the co-cured fiberglass or aramid, should extend at least 25 mm (1 in.) beyond the edges of the joint. In some cases, an extension of at least 102 mm (4 in.) has been found necessary.

The isolation of the composite cut edges in proximity to an aluminum structure requires special attention. The cut edges are very galvanically active because of the presence of graphite fibers with no resin coating. As cut edges are susceptible to abrasion and impact during part handling and installation, the applied isolation must be durable. The easiest form of isolation, primer plus enamel, is the least durable of the alternatives, being relatively susceptible to being chipped off on impact. A more durable, easily applied alternative is brushable sealant, which is flexible and relatively tough. In especially critical applications, it has been found necessary to secondarily bond a ply of fiberglass around the cut edges.

Surface treatment and finishing as described above are applied to detail components. As indicated previously, both proper faying surface sealant with no gaps and proper wet fastener installation that provides a complete barrier against moisture entry through the fastener hole are essential assembly procedures for galvanic corrosion protection. Comparable surface treatment procedures can be used for graphite joints with titanium and stainless steel members, but in most cases are not essential.

Bonded Joints

The fiberglass or aramid protective measure described in the section "Bolted Joints" in this article is also applicable to bonded joints. Particularly with co-cured surface plies, this approach provides a bonded surface that is essentially the same as the basic graphite, which can be prepared for bonding in the same manner. The effectiveness of this approach has been verified by in-house Lockheed test programs (Ref 7), in which specimens had a 7-month seacoast exposure. There was no corrosion of aluminum wire mesh co-cured to graphite composite surfaces with fiberglass and aramid interlayers.

The primary protection against galvanic corrosion in graphite-to-metal bonded joints is the adhesive itself. It performs the same function as a layer of fiberglass or aramid and provides the same protection as a faying surface sealant functioning as a moisture-impermeable barrier. In some cases, such as an epoxy film adhesive with a nylon scrim fabric support, the tendency of the epoxy to absorb moisture and the nylon filaments to wick moisture may result in a less effective barrier than sealant over a long-term service exposure. The selection of film adhe-

sives with less tendency to wick moisture is advisable.

When aluminum honeycomb core is used, the sharp, cut edges of the core may penetrate the adhesive layer and contact the graphite substrate. This is discussed in more detail below. Considerable service experience has been reported for honeycomb sandwich components with graphite skins bonded to an aluminum core. For example, over 6 years of flight service experience have been documented for the Boeing 737 graphite composite spoiler (Ref 8), and over 12 years of experience have been accumulated to date. No core corrosion in the absence of penetration damage has been reported. Other service data have reported varying degrees of effectiveness for several commercially available epoxy film adhesives for preventing galvanic corrosion (Ref 9). Thus, selection of a specific adhesive for a particularly severe corrosive environment might require an evaluation of galvanic protection as well as other structural/environmental test data.

Other considerations for bonded sandwich components include selection of a corrosion-resistant core. Nonmetallic core is obviously the best choice for corrosion resistance. If aluminum core must be used, the 5000-series cores have more intrinsic corrosion resistance than 2024 core, with 5056 preferred over 5052 core. However, as Fig. 1 shows, even the 5000-series alloys have substantial galvanic potential difference with graphite, indicating that caution is needed in all cases to ensure the design is corrosion resistant. In all cases, coated cores should be used. A suggested approach is to dip the cores, after slicing to net thickness and forming, in a corrosion protection coating. Core manufacturers can perform this process, which protects the cut edges of the core that directly contact the skin. A co-cured layer of fiberglass on the side of the skin to which the core is to be bonded ensures galvanic isolation of the graphite from the core.

Other Reinforcements

Composite materials with glass or aramid fiber reinforcements do not present a galvanic corrosion problem. Boron filaments can produce a galvanic cell, however, by contact between the tungsten core of the fiber and the adjacent metal, particularly through a metal fastener. Precautions similar to those described for graphite are thus advisable for boron-to-metal joints.

Testing

When it has been determined that a graphite-to-aluminum design is feasible for a certain application, corrosion testing should be performed to determine the extent of isolation required. A control specimen with service-verified materials and finishes should be run alongside the test specimens as the standard for comparison. Realistic damage and omissions to the isolation systems should be made before testing.

REFERENCES

1. E.L. Riggs and W. Krupp, "Corrosion Principle and Protection," Paper presented at Technical Symposia for China, Peking, Oct 1978
2. M.S. Rosenfeld, The Effect of a Corrosive Environment on the Strength and Life of Graphite/Epoxy Mechanically Fastened Joints, in *Advances in Joining Technology*, Proceedings of the Fourth Army Materials Technology Conference, Naval Air Development Center, Sept 1975, p 569
3. *Design Handbook*, Lockheed-California Company, Nov 1983
4. R.T. Cole, E.J. Bateh, and J. Potter, Fasteners for Composite Structures, *Composites*, Vol 13 (No. 3), July 1982, p 233
5. Data compiled by Aerospace Industries Association of America, Inc., National Aerospace Standards Committee
6. B. Silverman and A. Norrbom, Use of Polysulfide Sealants in Aircraft Composite Structures, *Adhesives Age*, Products Research and Chemical Corporation, June 1983
7. J.L. Wanamaker, "Corrosion Characteristics of Graphite Composite Lightening Protection Systems," Report LR 28947, Lockheed-California Company, Jan 1979
8. D.J. Hoffman, "The 737 Graphite Composite Flight Spoiler Flight Service Evaluation, Annual Report, May 1979—April 1980," NASA CR-159362, Boeing Commercial Airplane Company, Nov 1980
9. D.R. Askins and H.S. Schwartz, Durability of Adhesive Bonded Honeycomb Sandwich in Accelerated Adverse Environments, in *Materials and Processes—In-Service Performance*, Proceedings of the Ninth Technical Conference, Society for the Advancement of Material and Process Engineering, Oct 1977

Faying Surface Sealing

Robert H. Stone, Lockheed-California Company

FAYING SURFACE SEALANT is used for bolted joints in which graphite composites are joined to dissimilar materials. This use is mandatory to avoid galvanic reaction damage (see the article "Dissimilar Material Separation" in this Volume). Wet fastener installation is also mandatory for the same reason, and this article will cover both operations. Fillet or brush sealing of exposed joint edges after assembly is highly recommended for graphite composite bolted and bonded joints in corrosion-prone areas. Aerodynamic sealing is dictated by other considerations and is not covered here.

Sealants Available for Graphite-Composite Assemblies

The sealant materials used for graphite-composite bolted joints are the same as those used for metal-to-metal joints. For normal temperature applications (−50 to 160 °C, or −65 to 325 °F), polysulfide sealants should be used. These sealants are available in a wide variety of types, including those specifically formulated for faying surface sealing and wet fastener installation. Other types are specifically formulated for fillet and aerodynamic sealing, fuel tank sealing, low-density low-adhesion sealing for removable assemblies, and other special conditions. Within each of these categories, there are long application times (pot life), fast cure, minimum-viscosity or low-viscosity (spray or brush) application, and thixotropic (spatula or extrusion gun) application versions.

The sealants listed in Table 1 are identified by specification and usage. Most of these specifications have several types, grades, and classes that identify the application characteristics discussed above. No attempt has been made to list all types within these specifications, and the specifications and/or suppliers should be consulted for specific selections. It should be noted that polysulfide sealants vary somewhat in their maximum use temperature, with maximum recommended temperatures ranging from 80 to 175 °C (180 to 350 °F) (continuous to peak). Some sealants have been formulated to withstand typical adhesive bond cycles of 1 to 2 h at 120 to 175 °C (250 to 350 °F). For use

temperatures exceeding 90 °C (200 °F), care should be taken to select a sealant system capable of operating under the required environmental conditions. For extreme temperature applications (−60 to over 260 °C, or −80 to over 500 °F), silicone sealants are the appropriate choice.

Table 2 gives a sampling of available sealants and illustrates the range of service temperatures, application times, cure cycles, and available forms. Inclusion of products does not constitute a recommendation of them over other comparable products.

Sealant Usage in Graphite Assemblies

The polysulfide sealants listed in Table 1 are generally usable either as faying surface sealants or as fillet/brush sealants. There are some instances where the curing system is anaerobic, that is, cure will take place only in the absence of air. These sealants would be suited only for faying surface sealing or wet fastener installation. There is one major precaution to take in using sealants for faying surfaces and fastener installation: Since some types, especially low-viscosity sprayable and brushable sealants, have significant amounts of solvents, sufficient time after application (typ-

ically, 15 min) must be allowed for the solvent to flash off before assembling the joint or installing the fasteners.

Sealing Procedures for Graphite Assemblies

Sealing procedures are similar to those used in the bolted assembly of metal components. One difference is the surface treatment of the graphite components to promote adhesion of the sealant. The recommended approach is solvent wiping and hand sanding. Surface conditions are not as critical for sealing as for adhesive bonding. Surface treatment and finishing of metal components to protect against galvanic effects is described in the article "Dissimilar Material Separation" in this Volume. Associated primers are recommended by the supplier for use with some sealants, and usage should follow supplier or specification instructions.

Figure 1 shows a typical configuration for a graphite composite joint, with both permanent and removable fastener installations. The methods of application are similar to those used in metal assemblies and include spatula, extrusion gun, brush, and spray applications. Most sealants come in varying types and grades suited for each application: minimum viscosity, spray-

Table 1 Faying surface sealant categories

Material type	Use	Specification(a)
Polysulfide(b)	General-purpose corrosion-inhibiting sealant	MIL-S-81733C(c) (Ref 1)
	Fuel tank sealant (also for general-purpose nonfuel areas)	MIL-S-8802E (z) (Ref 2)
	General-purpose low-adhesion sealant	MIL-S-8784B (Ref 3)
	Low-adhesion sealant for fuel tank areas	AMS 3267(d) (Ref 4)
	High-temperature sealant (to 180 °C, or 360 °F, peak)	MIL-S-83430A (Ref 5)
Silicone(e)	High temperature sealing appplications (−60 to 205 °C , or −80 to 400 °F)(f)	AMS 3373 (Ref 6)

(a) Some of these specifications have several types, grades, and classes. A careful review of these categories and the recommended use for each must be made before sealant selection. (b) Manufacturers of polysulfide sealant include Products Research and Chemical Corporation, Goal Chemical Sealants Corporation, and Chem-Seal Corporation. (c) Material conforming to this specification is recommended for permanent graphite composite assemblies because of the corrosion-inhibiting formulation and also because of the long pot life, which permits a long assembly time for complex structures. (d) Material conforming to this specification is recommended for removable graphite composite assemblies because of the corrosion-inhibiting formulation. (e) Manufacturers of silicone sealant include General Electric Company, Silicone Division, and Dow-Corning Corporation. (f) A two-part silicone sealant, with catalytic curing agents, must be used. One-part silicones require moisture (from exposure to air) for cure and thus cannot be used on a faying surface with more than a 25-mm (1-in.) width.

Table 2   Application characteristics of selected sealant materials

Product and specification	Supplier	Continuous service temperature °C	°F	Application time, h	Cure cycle	Form
PR 1422B MIL-S-8802	Products Research and Chemical Corp.	−50-120	−65-250	0.25-4	24 h at room temperature or 6 h at 55 °C (130 °F)	Paste or liquid (2 parts)
PR 1436 MIL-S-81733	Products Research and Chemical Corp.	−50-120	−65-250	0.25-4	30 h at room temperature or 6 h at 55 °C (130 °F)	Paste or liquid (2 parts)
PR 1750 MIL-S-83430	Products Research and Chemical Corp.	−50-160	−65-325	0.25-4	30 h at room temperature or 6 h at 55 °C (130 °F)	Paste or liquid (2 parts)
Pro-Seal 899 MIL-S-83430	Products Research and Chemical Corp.	−50-160	−65-325	0.50-1	72 h at room temperature	Paste or liquid (2 parts)
Pro-Seal 890 MIL-S-8802	Products Research and Chemical Corp.	−50-120	−65-250	0.25-4	72 h at room temperature	Paste or liquid (2 parts)
RTV-60	General Electric Co.	−60-270	−75-525	0.50-1	24 h at room temperature	Paste (2 parts)

Source: Ref 7

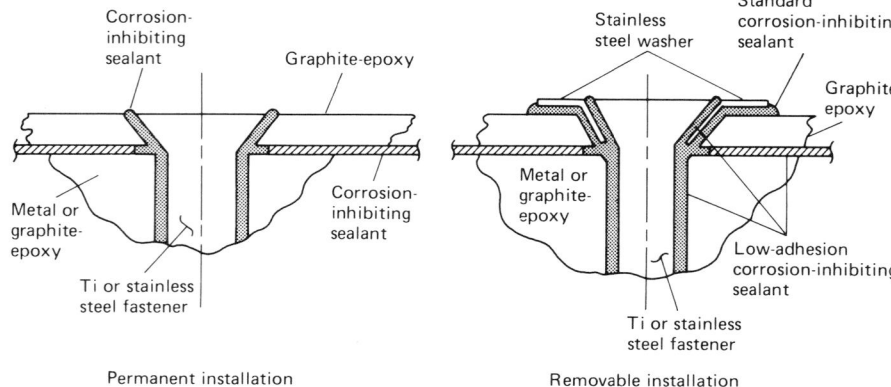

Fig. 1   Sealing arrangement for graphite composite assemblies. Source: Ref 8

able; low viscosity, brushable; thixotropic, spatula or extrusion. Sealants applied by extrusion gun are typically supplied in the form of frozen cartridges (or cartridges containing both sealant and catalyst, separated by a breakable membrane). The sealant may be applied to both faying surfaces, but application to only one surface is recommended by some fabricators to avoid excessive surface separation, which may prevent proper fastener seating. A lightweight scrim cloth can be laid onto one of the surfaces for thickness control, but this may result in a leak path along the fibers in the scrim.

Inspection of Assemblies

The most critical concern in sealant application for faying surfaces and fastener installation is to achieve full coverage. After assembly, it is important to inspect the joint edges for a continuous bead of sealant around the edges as well as each fastener for a continuous sealant bead around the fastener head. Absence of continuous beads indicates incomplete sealing and gaps, which represent a potential corrosion hazard as well as a potential moisture leak path in graphite composite assemblies.

As mentioned previously, applying a fillet of sealant around exposed joint edges is highly recommended for both bolted and bonded graphite composite joints. The fillet seal in fact constitutes the primary seal (backed up by the faying surface sealant) in many joint designs.

Use of O-Rings and Gaskets

Graphite composite assemblies are often designed with a fastener installation that uses an O-ring under the fastener head. Using a sealant in fastener installation with O-rings is not permitted, since the sealant might prevent the fastener from seating properly. An O-ring does not provide an effective barrier against leakage; therefore, this approach should be avoided in corrosion-prone assemblies, such as graphite-to-aluminum.

In other cases, a gasket is used in the faying surface of composite assemblies. This can be considered an acceptable alternative to faying surface sealing, but for a corrosion-sensitive graphite-to-aluminum assembly, this approach should be selected only if the following conditions are met: the selected gasket material has

low compression set, the faying surface members are rigid, the bolt spacing is suitable for holding the gasket firmly in place over the entire faying surface, and the faying surfaces are in the same plane.

Other Reinforcements

The sealant procedures defined in this article are not mandatory for aramid or fiberglass composite assemblies, except as dictated by the metal component. Sealing is typically required for any aluminum assembly, but not for titanium or stainless steel assemblies. A boron composite presents a galvanic problem similar to that of graphite because of its tungsten core.

REFERENCES

1. "Sealing and Coating Compound, Corrosion Inhibitive," MIL-S-81733C, Department of Defense
2. "Sealing Compound, Temperature-Resistant, Integral Fuel Tanks and Fuel Cell Cavities, High Adhesion," MIL-S-8802E(2), Department of Defense
3. "Sealing Compound, Aluminum Integral Fuel Tanks and Fuel Cell Cavities, Low Adhesion, Accelerator Required," MIL-S-8784B, Department of Defense
4. "Low Adhesion, Access Door, Corrosion Inhibiting Sealant (For Fuel Areas)," AMS 3267, Society of Automotive Engineers, April 1984
5. "Integral Fuel Tank Sealant (High Temperature)," MIL-S-83430A, Department of Defense
6. "Compound, Silicone Rubber, Insulating and Sealing," AMS 3373, Society of Automotive Engineers, Jan 1981, p 35-55
7. Data sheets from Products Research and Chemical Corporation and General Electric, Silicone Division
8. *Design Handbook*, Lockheed-California Company, Nov 1983

Honeycomb Structure

John Corden, Hexcel Corporation, Structural Products Division

HONEYCOMB, A PRODUCT consisting of very thin sheets attached in such a manner as to form connecting cells, closely resembles the honeycomb made by bees. Although the Chinese made paper honeycomb about 2000 years ago, honeycomb as a structural product (see Fig. 1) appeared just after 1940 in aircraft sandwich panel construction.

Most honeycomb used today is adhesively bonded core that is subsequently bonded to facings to form a sandwich panel. Other common types of honeycomb cores presently being produced include metallic cores of corrosion-resistant steel, titanium, and nickel-based alloys fabricated by resistance welding. These cores are primarily used for elevated-temperature applications. The facings are usually attached by brazing or diffusion bonding.

A typical sandwich panel, formed by adhesively bonding thin skins to the honeycomb core, is shown in Fig. 2. The sandwich construction is an extremely lightweight structure that exhibits high stiffness and strength-to-weight ratios.

In addition to use in sandwich panels, honeycomb is used for energy absorption, radio frequency shielding, light diffusion, and to direct air flow.

Like most disciplines, the honeycomb industry has its own terminology to define the various aspects of honeycomb core. Figure 3 and the Glossary of Terms in this Volume define those terms.

Manufacturing Methods

The two basic techniques used to manufacture honeycomb are the expansion and corrugated methods. Although other techniques do exist, they are primarily for nonstructural applications and will not be discussed in this article.

The expansion method, shown in Fig. 4, is used for both metallic and nonmetallic core fabrication. The majority of honeycomb core is produced by this method, which consists of printing an adhesive on ribbon sheets, stacking the sheets, and curing the stack in a press at an elevated temperature to form a block of honeycomb. The adhesive lines can be printed "cross-line" or "in-line," as illustrated in Fig.

5. The cross-line method results in a core of fixed thickness, T, (width of ribbon), with a variable L direction. The T direction is parallel with the cell walls, whereas L is the direction of the continuous sheets. This is an advantage in applications that require honeycomb core with a large L dimension. Conversely, the in-line method produces core of variable thickness but of fixed L dimension (width of ribbon).

When aluminum alloy sheet or strip is used in the expansion method, it is cleaned and coated with a corrosion-resistant treatment be-

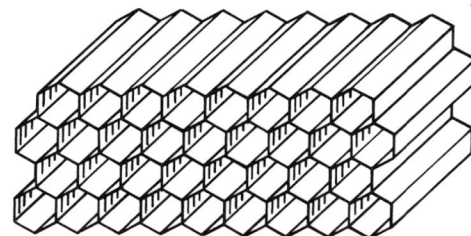

Fig. 1 Hexagonal cell honeycomb

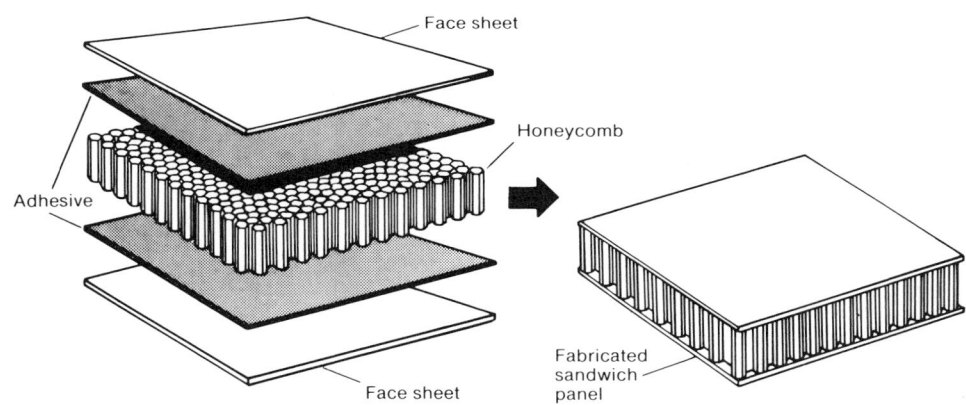

Fig. 2 Example of a bonded sandwich assembly

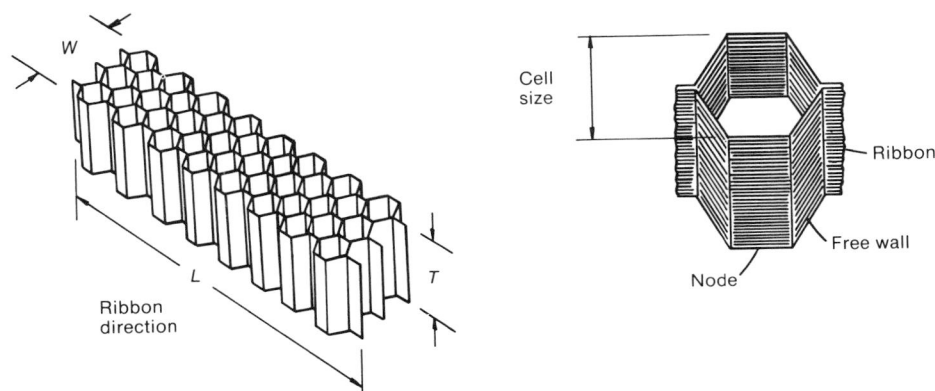

Fig. 3 Honeycomb terminology

fore the adhesive is printed. Once the block of honeycomb is removed from the press, it is ready for expansion. Aluminum blocks are usually cut into slices of the desired thickness before expansion. When the slices are expanded, the individual ribbon sheets yield plastically at the node/free wall joints and retain the expanded shape. The same process can be used for other metallic core materials.

Nonmetallic ribbon sheets are prepared in a similar manner, but with some important differences. Nonmetallic materials do not require the corrosion-resistant treatment before adhesive printing. Some materials may require resin preimpregnation, however. Unlike metallic materials, the nonmetallics do not retain their shape after expanding. They must be racked and heat-set in an oven to hold their expanded shape. The expanded heat-set block is then dipped in liquid resin and oven cured to complete fabrication of the core. This dip-cure cycle is repeated until the block reaches the required density. Slices are then cut from these blocks to the desired thickness.

The corrugated method is most commonly used for high density, high temperature core materials, or when relatively thick ribbon material is used. In this method (Fig. 6) ribbon sheets are corrugated to the desired shape, adhesive is applied to the nodes, the formed ribbon sheets are stacked, and the corrugated block is cured at an elevated temperature. Slices are then cut from the corrugated block to the desired thickness. For some elevated-temperature metallic core materials, the corrugated sheets are resistance welded or brazed rather than adhesively bonded.

Core Characteristics

Honeycomb can be manufactured from virtually any thin-sheet material. Common metallic-core materials are aluminum, corrosion-resistant steel, titanium, and nickel-based alloys. The most common nonmetallic core materials are Nomex, fiberglass, and kraft paper. Nonmetallic core is normally dipped in liquid phenolic, polyester, or polyimide resin to achieve the final density, although other resin systems can be used. Ideally, the resin content should be approximately 50%. Therefore, a range of ribbon thicknesses must be available to make core of various densities, while maintaining the resin content as close to 50% as possible.

To date, honeycomb has been produced from over 500 different materials, the most recent being graphite, aramid, and ceramic.

There are three basic cell configurations: overexpanded, hexagonal, and flexible. These and some of the less common configurations (square, reinforced, and tube cores) are shown in Fig. 7.

The overexpanded cell configuration is obtained by expanding the hexagonal cell until it forms a rectangle. The primary advantage of this configuration is that it is easily formed in the L direction. The hexagonal cell configura-

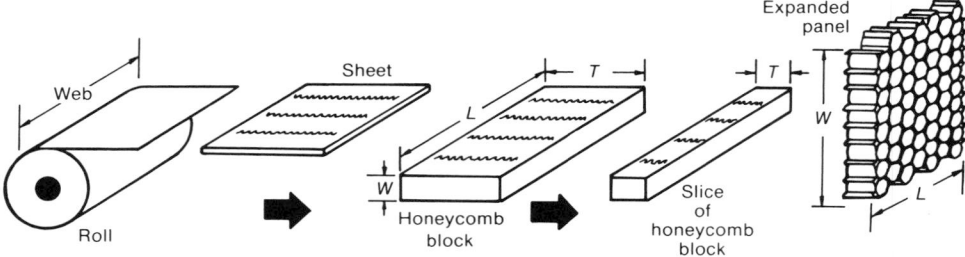

Fig. 4 Expansion method of honeycomb core fabrication

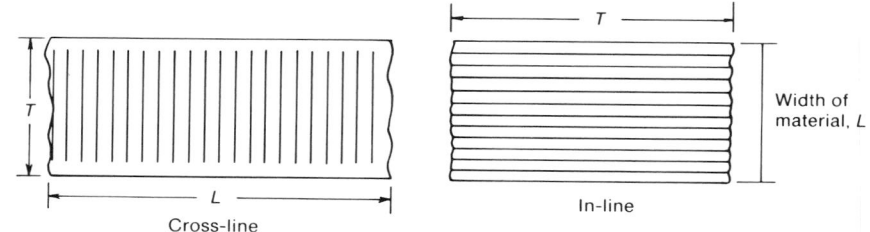

Fig. 5 Adhesive line printing configurations

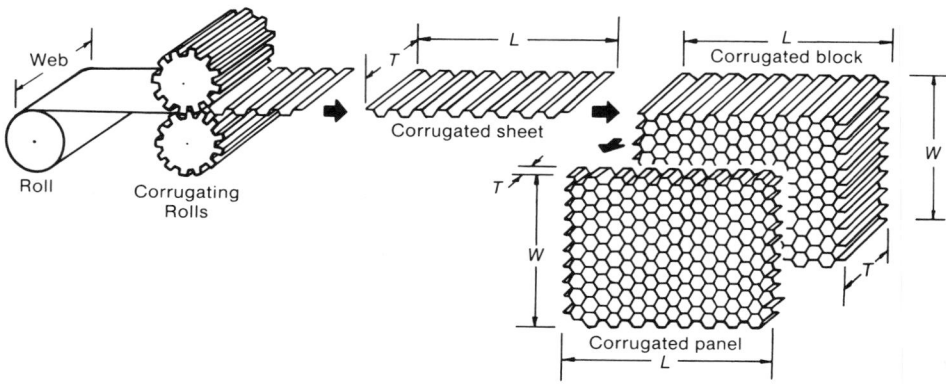

Fig. 6 Corrugation method of honeycomb core fabrication

tion can only be formed in this direction by roll-forming or heat-forming. Also, the W (direction of expansion) shear properties of overexpanded core tend to increase over those of hexagonal core of the same density, while the L shear properties decrease slightly.

The flex-core configuration offers added formability over other types of honeycomb. It forms easily in both the L and W directions, and is ideally suited for applications that require the core to be formed into compound curvatures. Flex-core can be formed into compound curvatures without buckling the cell walls and exhibits greatly reduced anticlastic curvature. When formed over tight radii, it can exhibit higher shear strengths than can hexagonal cell core of equivalent density. Figure 8 shows a typical example of the formability of flex-core.

The square cell configuration is used predominantly for resistance-welded corrugated core. It has a very narrow node, and the free walls are often corrugated to increase their resistance to buckling.

The reinforced cell configuration has a flat sheet placed between the nodes. This results in both increased density and mechanical properties. Aluminum corrugated core has been made in this manner with a density as high as 0.88 g/cm^3.

Tube core is manufactured by spirally wrapping a corrugated sheet with adhesive applied to the nodes and a flat sheet around a mandrel. The inside and outside diameters are variable. This core configuration is used exclusively for energy absorption applications.

The various standard core types available are:

- *5052 aluminum alloy:* Specification grade 5052 H39 aluminum alloy with a corrosion-resistant coating applied for general-purpose applications, particularly in the aerospace industry
- *5056 aluminum alloy:* Specification grade 5056 H39 aluminum alloy with a corrosion-resistant coating. Offers slightly higher mechanical properties than 5052 alloy honeycomb
- *2024 aluminum alloy:* Heat-treatable 2024 aluminum alloy; it combines high room-temperature mechanical properties with good strength retention at elevated temperatures. A corrosion-resistant coating is applied in the same manner as is done for the other aluminum alloy honeycombs
- *Commercial grade aluminum:* A low-cost aluminum honeycomb (usually 3000 series aluminum alloy) with a corrosion-resistant coating applied. It is not used for military applications
- *Naval grade aluminum:* A commercial-grade 5052 alloy honeycomb with a corrosion-resistant coating applied. Primarily for use in U.S. Navy applications (bulkheads, decks, and joiner panels)
- *Fiberglass:* A glass fabric reinforced plastic honeycomb in which heat-resistant phenolic resin is used for both initial impregnation and dip coats. Cores are available with the fabric oriented at either a 0°/90°, or ±45° with respect to the core thickness. The ±45°, or bias weave, orientation offers increased shear properties over the 0°/90° orientation. In addition to the phenolic resin system, fiberglass honeycomb is also available with a polyimide impregnation, node adhesive, and dip. This product exhibits excellent elevated-temperature properties and low dielectric constants
- *Nomex:* Aramid fiber paper dipped in phenolic resin. High strength and toughness, with low density. Also available with a polyimide resin dip (rather than the phenolic). This product has excellent dielectric and loss tangent properties
- *Kraft paper:* A kraft paper core dipped in phenolic resin. Offers low cost and high strength. Used extensively in military air transportable shelters
- *Graphite:* A graphite fiber reinforced plastic honeycomb available in a variety of resin systems. Highest strength nonmetallic core available with properties approaching those of 5052 aluminum alloy core
- *Kevlar:* A Kevlar fabric reinforced honeycomb with epoxy resin used for initial impregnation and dip coats. Used primarily for space-type applications that require ex-

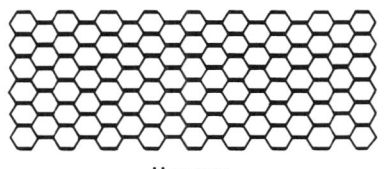

Hexagon

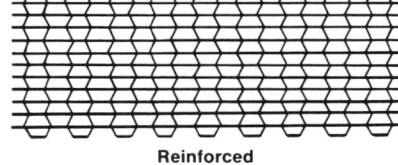

Reinforced

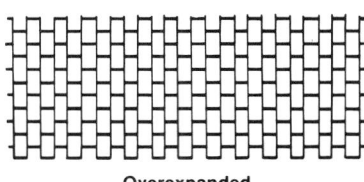

Overexpanded

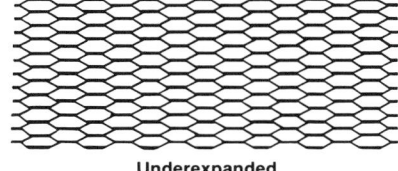

Underexpanded

Tube core

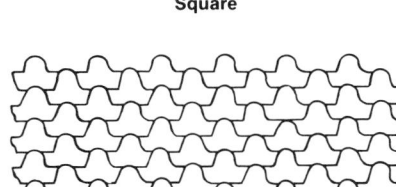

Square

...

Flex-core

Fig. 7 Honeycomb cell configurations. Source: Ref 2, 3

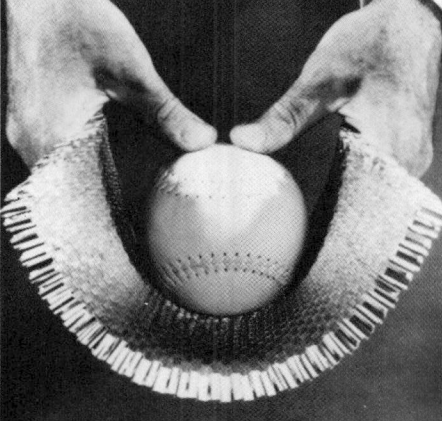

Fig. 8 Formability demonstration of flex-core honeycomb

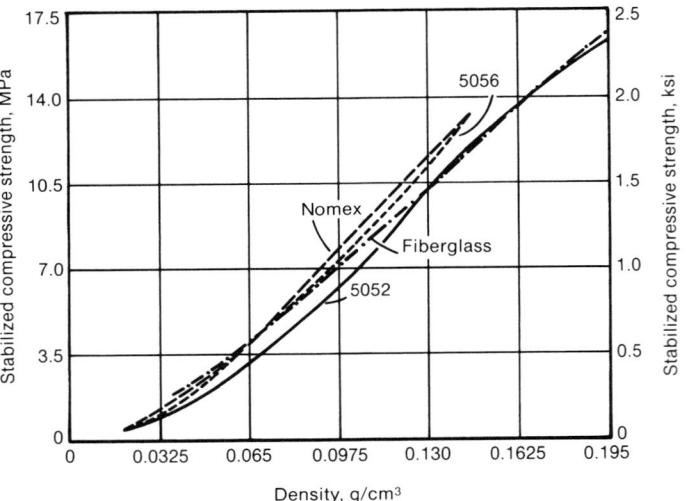

Fig. 9 Typical stabilized compression strength. Source: Ref 3

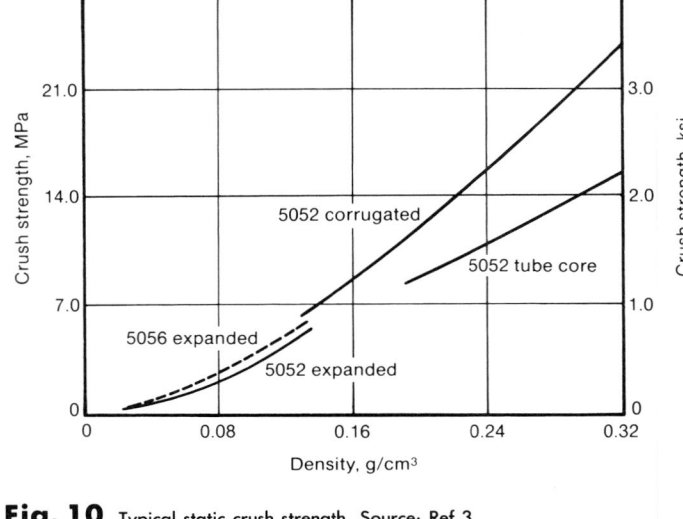

Fig. 10 Typical static crush strength. Source: Ref 3

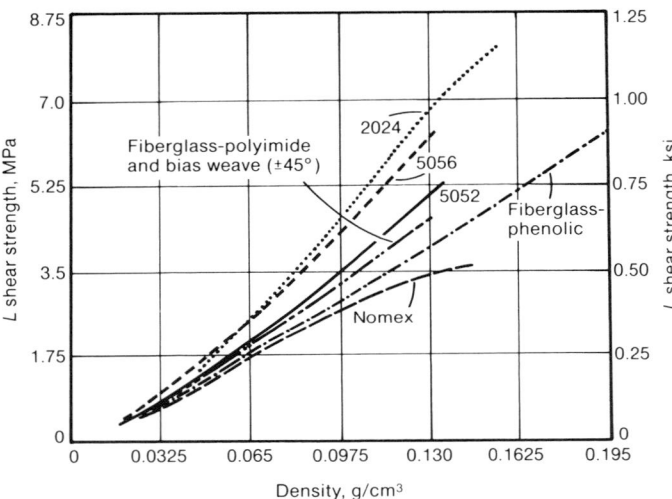

Fig. 11 Typical *L* shear strength. Source: Ref 3

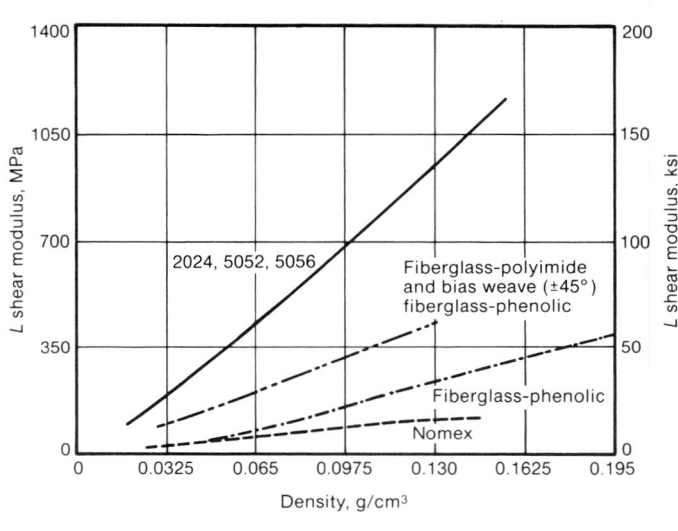

Fig. 12 Typical *L* shear modulus. Source: Ref 3

tremely low coefficient of thermal expansion properties

Honeycomb Properties. The basic mechanical properties used to quantify core are: bare compression strength, stabilized compression strength and modulus, *L* and *W* shear strengths and moduli, and crush strength. Figures 9 to 14 compare hexagonal core materials and their mechanical properties. In general, bare compression strength is roughly 90% of the stabilized strength; crush strength is 50%.

Hexagonal cell honeycomb is very orthotropic, with the *L* (ribbon) direction exhibiting shear strength and modulus values approximately twice those of the *W* direction. When hexagonal core is overexpanded, the *L* and *W* shear strengths are more nearly equal.

Figure 15 shows the mechanical property reduction for various core materials when exposed and tested at elevated temperatures.

Some corrosion-resistant steel and nickel alloy cores can be used at temperatures as high as 650 °C (1200 °F) while retaining 75% of their room-temperature properties.

Test Methods

Honeycomb test methods currently in use are contained in MIL-STD-401B and the ASTM standards listed in Table 1. Specific honeycomb requirements are contained in MIL-C-8073D and MIL-C-81986(AS) for fiberglass and Nomex cores, respectively, in MIL-C-7438G for aluminum alloys, and in MIL-C-21275C for metallic, heat-resistant core materials.

Honeycomb Testing. In the above-mentioned specifications, metallic cores usually are tested at 15.9-mm (0.625-in.) thickness, and nonmetallic cores are tested at 12.7-mm (0.500-in.) thickness.

Table 1   Honeycomb test methods

Test	MIL-STD-401B paragraph No.	ASTM test method
Density	5.1.1	C271
Specific gravity	5.1.1	C271
Water absorption	5.1.2	C272
Thermal conductivity	5.1.3	C177
Bare compression	5.1.4	C365
Stabilized compression	5.1.4	C365
Plate shear	5.1.5	C273
Flatwise tension	5.1.6	C297
Water migration	5.1.7	. . .
Core delamination	5.1.8	C363

Before testing, the density, as well as cell size and pitch, are determined in both the *L* and *W* directions to ensure that core is of the proper configuration.

The most common mechanical property tests

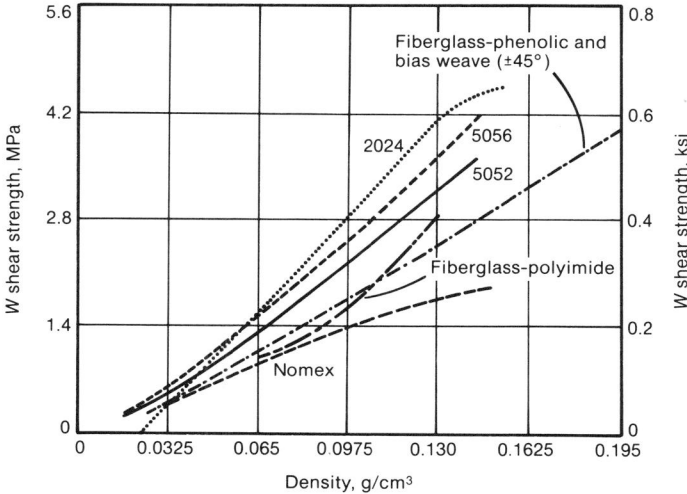

Fig. 13 Typical *W* shear strength. Source: Ref 3

Fig. 14 Typical *W* shear modulus. Source: Ref 3

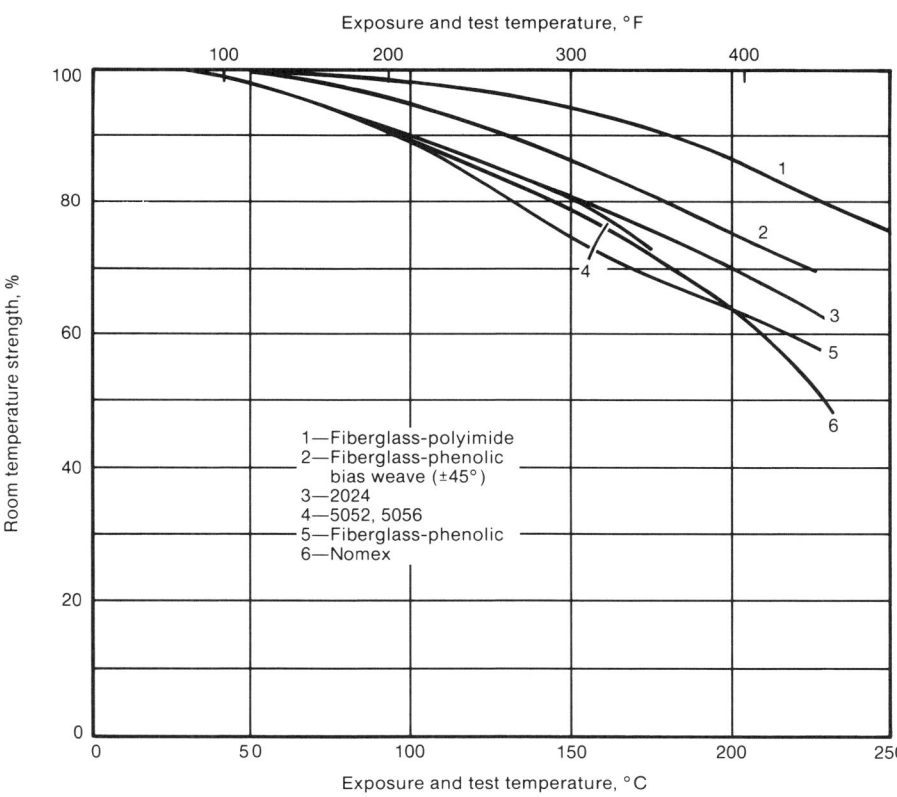

1—Fiberglass-polyimide
2—Fiberglass-phenolic
   bias weave (±45°)
3—2024
4—5052, 5056
5—Fiberglass-phenolic
6—Nomex

Fig. 15 Temperature effects at 30-min exposures. Source: Ref 3

Fig. 16 Heat-formed fiberglass honeycomb nose radome

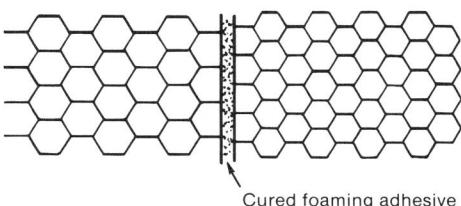

Cured foaming adhesive

Fig. 17 Core splicing

for honeycomb are compression and plate shear. Bare compression tests (no facings) are used as a quick quality control test to determine the compression strength only. Stabilized compression tests are used to determine both the compression strength and modulus. Facings, typically 0.51-mm (0.020-in.) aluminum, are bonded to the core sample for stabilized com-

pression testing. The normal specimen size for all compression tests is 76 × 76 mm (3.0 × 3.0 in.).

The plate shear test determines the core shear strength and modulus and is performed for both the *L* and *W* directions. The test can be run in either tension or compression. Specimen size is typically 51 × 190 mm (2.0 × 7.5 in.) or 51

× 152 mm (2.0 × 6.0 in.) with the bare honeycomb core bonded directly to 12.7-mm (0.5-in.) thick steel plates.

Sandwich panel flexure tests are sometimes used to determine the shear properties of high-density or high-strength core materials.

Sandwich panel testing is accomplished on specimens obtained from actual sandwich

Fig. 18 Machined honeycomb parts

Table 2 Sandwich panel test methods

Test	MIL-STD-401B paragraph No.	ASTM test method
Flatwise compression	5.2.1	C365
Edgewise compression	5.2.1	C364
Plate shear	5.2.2	C273
Flatwise tension	5.2.3	C297
Beam flexure	5.2.4	C393
Thermal conductivity	5.2.5	C236
Climbing drum peel	5.2.6	D1781
Fatigue	5.3	...
Creep	5.4	C480
Laboratory aging	...	C481

panels. Test methods are contained in MIL-STD-401B and the ASTM standards listed in Table 2.

Special Processing

Honeycomb special processing is normally defined as custom shaping of core to fit a customer's specific needs. This shaping includes: perimeter trim, doubler relief routing, chamfering, roll forming, heat forming, core splicing, cell filling (fiberglass, foam, and so forth), and contouring. In general terms, honeycomb special process products are modified from flat core slices into more complex shapes.

Trimming. The four primary tools used to cut honeycomb to plan dimensions are serrated knife, razor blade knife, band saw, and a die. The serrated and razor edge knives and die cutter are used on light-density cores, while heavy-density cores and complex-shape cores are usually cut with a band saw.

Forming. Metallic, hexagonal honeycomb can be roll- or brake-formed into curved parts. The brake-forming method will crush the cell walls and densify the inner radius. Overexpanded honeycomb can be formed to a cylindrical shape on assembly. Flex-core usually can be shaped to compound curvatures on assembly.

Nonmetallic honeycomb can be heat-formed to obtain curved parts. Usually the core slice is placed in an oven at high temperature for a short period of time. The heat softens the resin and allows the cell walls to deform more easily. Upon removal from the oven, the core is quickly placed on a shaped tool and held there until it cools. Figure 16 shows an aircraft nose radome fabricated from heat-formed fiberglass honeycomb.

Splicing. When large pieces of core are required, or when complex shapes dictate, smaller pieces can be spliced together to form the finished part. This is usually accomplished with a foaming adhesive, as shown in Fig. 17. Different core types, cell sizes, or densities can be easily interconnected in this manner.

Machining. In many sandwich panel applications, such as air foils, honeycomb must have its T dimension machined to some contour (Fig. 18). This normally is accomplished using valve-stem-type cutters on expanded core. Occasionally, the solid honeycomb block is ma-

Table 3 Honeycomb sandwich panel structural efficiency

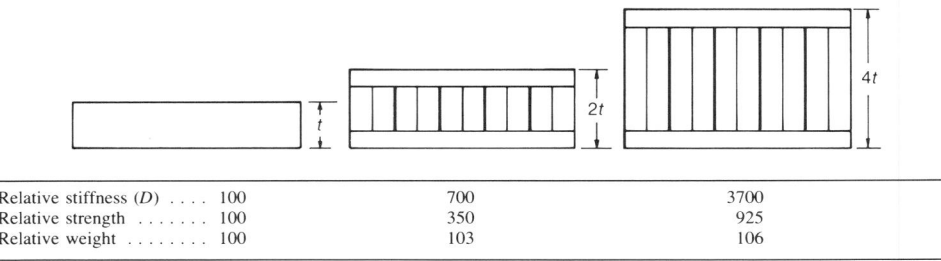

Relative stiffness (D)	100	700	3700
Relative strength	100	350	925
Relative weight	100	103	106

chined using milling cutters. Typical machines used for contour machining (carving) are gantry, apex, three-dimensional tracer, or numerically controlled (NC) five-axis. With five-axis NC machining, the cutting head is controlled by computer tapes, and almost any surface that can be described by x, y, and z coordinates can be produced. These machines can carve honeycomb at speeds of up to 76.2 m/min (3000 in./min) with extreme accuracy. A standard contour tolerance of an NC machine is ±0.13 mm (0.005 in.).

Sandwich Structures

A typical sandwich consists of two thin, high-strength facings bonded to a thick, lightweight core. Each component by itself is relatively weak and flexible, but when combined in a sandwich panel they produce a structure that is stiff, strong, and lightweight. Brazed and diffusion-bonded metallic sandwich panels will not be discussed here, although, in general, the basic sandwich concepts apply in these cases also.

Facing materials that are normally used are aluminum, fiberglass, graphite, and aramid. The facing thickness ranges from approximately 0.25 mm to 12.7 mm (0.010 in. to 0.500 in.). Core densities range from 1 pcf to as high as 55 pcf.

Various forms of adhesives, such as films, liquids, and pastes, can be used to bond the facings to the core. Most structural adhesives require heat and pressure during cure, although there are some two-part systems that cure at room temperature. The primary considerations for selecting an adhesive system are strength

requirements, service temperature range, and ability to form a fillet at the cell wall ends. Sandwich panels are typically used for their structural, electrical, and energy absorption characteristics, or a combination thereof.

Sandwich Concept. The basic concept of a sandwich panel is that the facings carry the bending loads and the honeycomb core carries the shear loads. In most cases the facing stresses are uniformly distributed. The honeycomb core offers no resistance to bending. In other words, the bending modulus, E', of the core is assumed to be zero. This assumption leads to a uniform shear stress throughout the core thickness.

Deflection, in all structures, consists of two components, bending deflection and shear deformation. In nonsandwich structures, such as steel plate, the shear deformation often is negligible and therefore is neglected. In a sandwich structure, on the other hand, the shear deformation can be significant. In most cases it accounts for about 1% of the bending deflection, although it can be much greater for thick panels or short spans.

The honeycomb concept produces extremely stiff and strong structures at minimum weight. Table 3 is a generic example of honeycomb effectiveness: A 0.81-mm (0.032-in.) thick piece of aluminum is compared to two sandwiches made by halving the aluminum into two facings and bonding honeycomb core between them. In bending, the thickest of the two sandwich panels is 37 times stiffer and more than 9 times stronger than the aluminum sheet, with a weight increase of only 6%.

Material Selection Considerations. Honeycomb sandwich construction lends itself

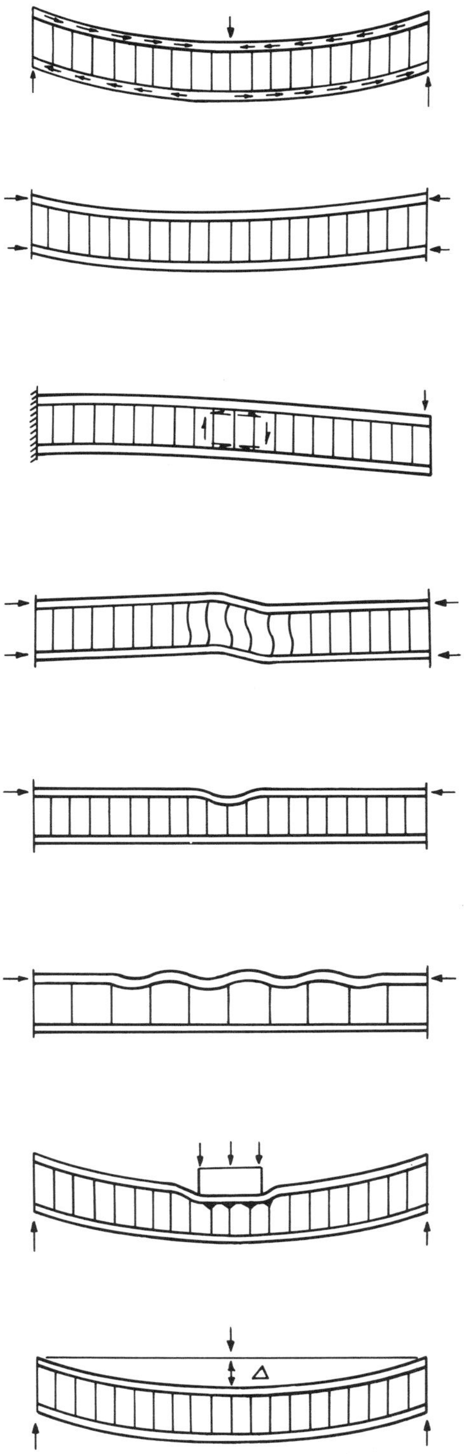

1. The facings should be thick enough to withstand the tensile, compressive, and shear stresses induced by the design load.

2. The core should have sufficient strength to withstand the shear stresses induced by the design loads. Adhesive must have sufficient strength to carry shear stress into core.

3. The core should be thick enough and have sufficient shear modulus to prevent overall buckling of the sandwich under load, and to prevent crimping.

4. Compressive modulus of the core and the compressive modulus of the facings should be sufficient to prevent wrinkling of the faces under design load.

5. The core cells should be small enough to prevent intracell dimpling of the facings under design load.

6. The core should have sufficient compressive strength to resist crushing by design loads acting normal to the panel facings or by compressive stresses induced through flexure.

7. The sandwich structure should have sufficient flexural and shear rigidity to prevent excessive deflections under design load.

Fig. 19 Basic sandwich structural criteria

to an unlimited variety of materials and panel configurations. Because a bonded sandwich is a composite structure, material selection can be tailored to meet specific performance requirements at minimal weight. Certain basics should be addressed when selecting cores and facings.

First, service temperature plays an important role in material selection. Bonded honeycomb structures are generally categorized into four maximum service temperatures: 82 °C (180 °F), 121 °C (250 °F), 232 °C (450 °F), and 288 °C (550 °F). With some of the more exotic systems, short-term temperature capabilities of up to 427 °C (800 °F) are possible. Most of the systems also are suitable for low-temperature environments.

The flammability of materials used in bonded sandwich panels can be another important consideration. Materials are grouped into one of three categories: materials that will not burn to any degree; self-extinguishing materials that will burn when placed in an ignition source but will self-extinguish upon removal; and flammable materials that, after being ignited, will continue to burn upon removal of the ignition source. The degree of flammability is measured by the flame spread rate under controlled conditions. Aerospace and transportation applications usually require nonflammable or self-extinguishing materials.

The degree of heat transfer through a sandwich panel is dependent on convection, conduction, and radiation. Metallic cores and facings tend to maximize heat transfer, while nonmetallic cores and facings tend to minimize heat transfer. Foam-filled core reduces heat transfer further by eliminating the convection and radiation components within the cells.

Bonded structural sandwich panels also can offer excellent acoustical absorption properties. In some cases the core cells are filled with fiberglass or foam, and the facings are perforated.

Metallic materials applied with corrosion-resistant treatments, as well as some nonmetallics, offer superb resistance to moisture and humidity. Further resistance can be imparted to a sandwich panel by using special foam-filling materials that preclude water intrusion, and by proper design of edge close-outs.

Adhesive Bonding. The adhesives must attach the facings rigidly to the core to allow loads to be transmitted from one facing to the other. A low-modulus adhesive such as rubber contact cement would not be acceptable for structural applications. Adhesives for structural sandwich panels must have high strength and modulus. Adhesives that meet or approximate MIL-A-25463 are generally acceptable. In addition, good toughness and peel strength are important, particularly in very lightweight structures that are likely to be subjected to damage in storage, handling, or service.

Design Guidelines. Sandwich structures should be designed to meet the basic structural criteria listed in Fig. 19 (when these criteria apply to the type of loading under consideration), as well as cost effectiveness and durability requirements in the service environment.

The core should be oriented properly in the panel to take advantage of the orthotropic shear properties of honeycomb. If stiffness and deflection are critical design parameters, it is necessary to take into consideration the low shear moduli of most core materials. Shear deflections must be considered in addition to bending deflections.

Potentially, the most critical and difficult problem associated with sandwich panel design is the fitting attachment or panel close-out. Fittings or close-outs serve many purposes and usually have special requirements. Some of the primary functions of fittings or panel close-outs are to:

- Act as a structural tie to adjoining or supporting structure
- Join adjacent panels
- Incorporate hinges or enable detachment
- Provide edge protection
- Seal the panel to prevent moisture intrusion
- Provide a smooth, continuous extension of the facings.

Close-outs can be included when the sandwich panel is assembled or added in a secondary operation.

Close-outs with raised surfaces can be very difficult to process. Special tooling and fixtures are often required to keep close-outs aligned and within dimensional tolerances. For this reason, subsequently assembled close-outs are frequently used. Some facing materials, however, such as fiberglass prepreg, may be molded or shaped to form a close-out in a single curing operation, when the panel is assembled.

Selection of fasteners for sandwich panels is governed by several considerations, such as panel thickness, loading, and environmental exposure. Particular care must be taken to preclude galvanic corrosion due to contact between dissimilar materials. Molded-in fasteners may be preferred because the potting compound bonds the fasteners to both facings and core. A disadvantage is added weight. Grommet-type fasteners consist of a plug and sleeve installed on opposite sides of the panel. They usually are not as strong as molded-in fasteners because the load is only transferred to one or both of the facings. Both types of fasteners are available with these characteristics:

- Flush, raised, or countersunk heads
- Clearance hole or threaded through
- Blind and threaded, with or without self-locking device
- Floating nuts to simplify alignment
- Materials that include aluminum, corrosion-resistant steel, alloy steel, titanium, and nickel alloys

More detailed design information can be found in MIL-HDBK-23A, *Structural Sandwich Composites* and Ref 1.

Air Flow Directionality

Honeycomb is ideally suited for straightening and directing fluid flow in the free atmosphere and in ducts. The thin cell walls of honeycomb result in very low pressure drops and turbulence. The typical open area for honeycomb is 95 to 99%. Available cell sizes range from 1.6 to 25.4 mm (0.06 to 1.00 in.).

The movement of air by a propeller-type fan in open atmosphere results in a conical air flow, with vortices about the periphery recirculating

Table 4   Cell size versus frequency limit

Cell size		Upper frequency limit,
mm	in.	GHz
3.2	1/8	23.6
4.8	3/16	15.7
6.4	1/4	11.8
9.5	3/8	7.9

into the fan. Honeycomb placed directly in front of the fan results in straightened air flow of greater velocity and reduced vortices.

A similar effect is realized in ducted fans. Although the air flow is contained within the duct, there is turbulence downstream from the fan. The addition of honeycomb within the duct results in straightened air flow. For this reason, honeycomb is used in several ducted air flow metering devices.

Turbulence and pulsations in a duct can cause noise, particularly at high air velocities. Straightening the air flow results in reduced power requirements and less noise.

Radio Frequency Shielding

Honeycomb is a unique material that provides a high degree of radio frequency (RF) shielding efficiency in openings used for heating, ventilation, and lighting. Honeycomb can be compared to a set of parallel wave guides, which, if properly designed for cell size and depth, will attenuate a required decibel level through a wide frequency range.

When designing a honeycomb RF shield, the material must be carefully selected. Aluminum honeycomb is the preferred material unless it is prohibited by environmental conditions and/or low-frequency attenuation requirements. Alloy steel and corrosion-resistant steel honeycombs are well suited for corrosive environments, high temperatures, and low-frequency attenuation.

The largest cell size should be selected, consistent with the highest frequency that must be attenuated (see Table 4). Attenuation levels increase with increasing honeycomb thickness for a given cell size.

Low-frequency attenuation performance can be improved by increasing the honeycomb thickness, using a smaller cell size, increasing the foil thickness, or using steel rather than aluminum.

Tooling

Using aluminum honeycomb as the major support structure for composite tooling can result in improved dimensional stability and reduced tool weight. The dimensional stability of honeycomb is, in general, better than that of solid or built-up systems of composites or plaster master tooling, and approaches the stability of welded steel tubing. This stability allows improved control of tolerances in master

die models and check fixtures. In addition, greater accuracy is achieved in male-to-female duplication of a series of tools.

Fiberglass fabric reinforced epoxy resins or aluminum combined with honeycomb form the basis for a class of lightweight, quickly manufactured tools. Tooling that incorporates honeycomb is generally less costly and can be manufactured in a shorter period of time. In a typical case, a master tool constructed of plaster, aluminum, and steel weighing 9091 kg (20 000 lb) was replaced with a honeycomb tool (aluminum core and composite skins) weighing 1136 kg (2500 lb). The cost savings were significant, and tool durability was improved substantially over that of the conventional plaster tool.

Energy Absorption

Honeycomb has proved to be one of the most reliable and efficient methods of providing energy absorption. The action of crushing the honeycomb develops a nearly uniform level of loading that is usually desired for energy absorption.

Aluminum, corrosion-resistant steel, alloy steel, reinforced plastic, and paper honeycombs have all been used to solve energy absorption problems. The hexagonal cell configuration is used for most energy absorption problems, although tubular core (for thin-wall annular column applications) and other configurations also have been used.

Reference 4 is recommended for a detailed discussion on selecting honeycomb for solving energy absorption problems.

REFERENCES

1. "The Basics on Bonded Sandwich Construction," TSB 124, Hexcel Corporation
2. N. Bitzer, "Honeycomb Sandwich Design and Testing," PhD thesis, Century University, Los Angeles, 1980
3. "Mechanical Properties of Hexcel Honeycomb Materials," TSB 120, Hexcel Corporation
4. "Design Data for the Preliminary Selection of Honeycomb Energy Absorption Systems," TSB 122, Hexcel Corporation

SELECTED REFERENCES

- "Aluminum Honeycomb in Tooling Applications," TSB 116, Hexcel Corporation
- "Design Data for the Preliminary Selection of Honeycomb Energy Absorption Systems," TSB 122, Hexcel Corporation
- "Honeycomb in Air Directionalizing Applications," TSB 102, Hexcel Corporation
- "Radio Frequency Shielding Properties of Hexcel Metallic Honeycomb," TSB 113, Hexcel Corporation

Quality Control

Chairman: Richard J. Hinrichs, Applied Polymer Technology Inc.

Introduction

QUALITY ASSURANCE for composites has traditionally centered on techniques for validating the physical/mechanical properties of a cured laminate. This end-result approach has culminated in standard industry reference publications of testing procedures, such as those of the American Society for Testing and Materials and the American Society of Mechanical Engineers. These organizations are engaged in the documentation of methods of testing or referencing the specific properties of a material. In addition, aerospace companies maintain internal reference specification manuals that detail how to test a material to validate its properties.

Real quality assurance, however, begins long before the end-result testing philosophy. One of the purposes of this Volume is to help the engineer understand what constitutes a quality laminate; therefore, a logical approach to quality control follows the fundamentals of laminate production:

- Raw material validation and reaction control
- Material/prepreg characterization
- Fabrication/handling/tooling effects
- Cure process control dynamics and documentation

Article topics range from analyses of fiber-resin consideration to cure dynamics, and from tooling to real-time property monitoring, with the intent of identifying the important engineering aspects of handling materials and administering processes to produce consistent high-quality composite structures.

One of the most difficult concepts to instill into a composite-manufacturing organization is that it has become a chemical material producer. The organization is no longer a material user that can purchase specific material alloys and then simply machine and fasten them into a structure. The metals industry underwent this same transition when the importance of heat-treating and work-hardening effects on chemistry became known. Unfortunately, those associated with material procurement have mistakenly accepted the notion that the quality of a material is defined by the procurement specifications. Although this can be true for materials such as metals that do not undergo chemical changes at the manufacturing facility, it is definitely not true of composite materials. A specification for a composite material is not an effective quality tool. In fact, a current concern is the assumed responsibility/liability for proper materials transformation into the engineering structure (chemical transformation into the engineering polymer) that now resides with

Table 1 Manufacturing transition

Material user	Material producer
Existing materials are bought already containing all engineering properties.	Materials must undergo a chemical reaction process before they achieve engineering properties.
Manufacturing operations are primarily numerically controlled machine shaping and forming	High-performance aerodynamic shapes are formed simultaneously with the cure formation process.
Mechanical robotics processing, routine precision operation, X-Y-Z drilling/positioning.	Chemical process control automation required. Dependent on skilled craftsmen
Very tolerant of manufacturing errors. Functions (mechanical fastening, shaping, or surface treatment) do not change intrinsic material properties.	Structural quality totally dependent on process. Material properties not tolerant of manufacturing errors. Requires a management system able to produce consistent craftsmanship conformance to engineering specifications.
Preceding items limit design and flexibility, cause weight penalties, and lower performance efficiency. Approach is rapidly becoming obsolete.	Flexibility to compete in future aerospace structures design and production. Approach expands customer base and engineering design capability.

fabricators. In other words, the composite manufacturer has become a material producer instead of just a material user. This concept is defined in Table 1.

Process and quality control checks are intended to validate laminate conformance to the manufacturing specification. Theoretically, this should result in an acceptable composite structure of known engineering properties. The critical assumption here is that these checks somehow relate to intrinsic laminate characteristics responsible for the desired properties. The real question, then, is: What defines a good-quality composite laminate? The answer is simplified by identifying the four basic characteristics that structures should meet and upon which all laminate properties are dependent:

- *Void or porosity content:* The volume amount of bubbles or air spaces within a laminate must be less than that allowed by the given structural design (ideally, free of all voids or porosity defects, but generally less than 1 to 2%)
- *Proper level of laminate consolidation:* Resin content, fiber volume, and associated distribution gradients
- *Proper degree of cure:* Formation of the polymeric resin and structure
- *Proper fiber orientation (laminate geometry):* Conforming to the design requirements

These four characteristics govern the resultant mechanical, physical, and chemical behavior of the laminate. It is interesting to point out that the first three are the result of the chemical reaction behavior of the material and the associated interaction with those lay-up techniques related to bagging, involving fluid flow mechanics and hydrostatic pressure generation. The fourth characteristic is dependent on fabrication lay-up techniques. These relationships are described in detail in those articles in this Section dealing with curing considerations.

Given the four basic laminate properties desired, it must be determined whether or not a given prepreg material will process correctly. This involves the age-old battle of assessing whether a materials or a fabrication process problem exists. Actually, the concern can be reduced to determining what defines the quality of a prepreg material. This can also be categorized by four basic characteristics:

- *Formulation verification:* Validation that resin components are what they are supposed to be. The primary tools are high-pressure liquid chromatography (reverse phase), gel-permeation chromatography, and Fourier transform infrared spectroscopy
- *Reaction kinetic behavior:* Verification that the material reacts in accordance with its characteristic data base. The principle tool is rheological evaluation under programmed thermal cycles. Rheological characterization also yields the viscoelastic characteristics critical to the flow consolidation dynamics required to produce a high-quality laminate. Differential scanning calorimetry is also used to validate reaction behavior. The important factors here are that both the kinetic reactor response and the viscoelastic behavior of the material conform to qualification material characteristics
- *Physical form:* Verification that the prepreg material contains the correct resin content, fiber weight, volatile content, and so on
- *Fiber type:* Verification (often ignored) of correct fiber type and coating or surface treatment given to the fiber to assist in weaving or impregnating. Microscopic and mechanical tensile strength are the primary evaluation techniques. Finish concentrations are evaluated by acetone-tetrahydrofuran extraction and evaporations on an infrared sodium chloride crystal for spectrum analysis

If these four properties are consistent with the data base of the material and problems are still encountered, there is a high probability that the handling or fabrication processes are creating the difficulty. These considerations are further developed in the discussions of curing in this Section.

The articles in this Section are based on years of hands-on experience. Many of them attempt to condense the essence of achieving composite quality by focusing on the behavior of the material during cure.

Fiber Properties Analysis

R.H. Ericksen and *R.E. Allred,
Sandia National Laboratories

QUALITY CONTROL, as applied to the mechanical properties of a reinforcing fiber, involves testing the incoming product and monitoring its properties throughout the fabrication process. The main sources of mechanical property data for fibers include tests on single fibers, multifilament yarns, and impregnated strands. Each of these fiber forms poses different problems in sampling, sample preparation, and testing and provides a different type of information. The most common fiber properties used in composite quality control are longitudinal tensile strength, elastic modulus, and failure strain. Creep, fatigue, bending, and other properties are also of interest for more specialized applications. Other properties, such as axial compressive strength or those in the radial direction, are important to composite structure design, but are difficult to measure directly on fibers. These properties are generally not included in quality control testing or specifications, and will not be covered in this article. In many cases, the incoming material is specified as a textile fabric rather than a yarn; therefore, a section on fabric testing is included.

The fibers of most interest are those with high specific strength and high specific modulus for composite reinforcement. These materials include glass, graphite, boron, ceramic, and metallic fibers, as well as organic fibers such as polyaramids, polyesters, and polyethylene.

Depending upon the material type, these fibers may be obtained as monofilaments, multifilament yarns or rovings, or as a woven fabric, braid, mat, or other textile form. The main emphasis of this article is on the testing of continuous fiber or fabric reinforcements, although many of the tests carried out on reinforcements used in discontinuously reinforced composites would be subject to similar considerations. The way mechanical properties are defined may also depend on material type or the general industry convention. In most engineering applications, material strengths are measured as breaking load and reported as stress (defined as load/cross-sectional area). Because it is difficult to measure the cross-sectional area of textile yarns, stress is usually expressed as a specific quantity. The textile industry normalizes, based on linear density. This measure of strength, called tenacity, is an easily measured yarn property. Definitions and conversion factors for the different systems of units that may be encountered in the textile industry are found in Ref 1-3. The relationship between tenacity, T expressed in g/den, and strength, S, in psi, is $S = 1.28 \times 10^4 \times T \times \rho$, where ρ is density (in g/cm^3).

Even fiber samples from a carefully prepared lot will contain flaws or impurities. The type, distribution, and size of flaws have two general effects on strength (Ref 4): a decrease in mean strength with increasing test sample length, and a scatter at fixed length. In addition, the mechanical properties of incoming material may differ from package to package or lot to lot because of variations in processing parameters, handling, and environmental factors.

The strength of a fiber bundle cannot be accurately predicted by simple averaging of the strengths of the individual fibers in the bundle (Ref 5). In the case of parallel multifilament assemblies, rovings, or tows, the length variations (slack) become important. When multifilament bundles, that is, yarns, contain twist, load is differentially shared among surviving fibers during loading. In continuous fiber reinforced composites or strands, the load of a failed fiber is redistributed locally onto surviving neighbors. It is generally observed that composite specimens exhibit less variability in tensile strength than their constituent fibers.

A voluminous amount of literature now treats the statistical nature of filaments, yarns, and strands relative to strength. Some probabilistic concepts, summarized in the section "Statistical Considerations" in this article, provide a basis for understanding the differences in materials, material form, and testing variables. Some manufacturers of composite aircraft components have used these statistical concepts to specify a narrow distribution of graphite fiber strength in yarns in preference to high mean strength (Ref 6).

Mechanical Test Methods

The type of test used to measure strength and modulus depends on the fiber form. For example, the test sample of boron would be a monofilament. In filament winding, yarns or impregnated strands would be tested. In the case of fabric lay-ups, the incoming material would be a fabric.

Before describing each method, it is necessary to address some aspects of sampling briefly. Detailed discussions of sampling plans can be found in sources such as *Quality Control Handbook* (Ref 7). The fibers, yarns, or fabrics of concern here are usually available in a large number of independent packages; the problem with sampling is having to select a small number of packages. A standard test method is available in ASTM D 2258 (Ref 8). McMahon (Ref 9) has reported that well-defined statistical sampling procedures based on random numbers have been demonstrated to produce more consistent single graphite fiber test results between operators.

Single fibers or monofilaments are not usually tested individually if the incoming material is in the form of yarn or a roving, but monofilaments such as boron would be tested individually. S.L. Phoenix (Ref 10) and F.K. Rose, and J.L. Stokes (Ref 11) have described procedures for testing boron fibers. The general problems involved in testing are similar for all single fibers, although certain filaments present specific difficulties. For example, large-diameter filaments tend to be brittle and therefore difficult to handle. Fine fibers taken from yarns are more flexible but are hard to handle and position because of their small diameter. Techniques for evaluating tensile properties of single graphite fiber are reviewed in Ref 9.

The preparation, mounting, and testing of fibers having moduli over 21×10^9 Pa (3×10^6 psi) is addressed in ASTM D 3379 (Ref 12), which includes most fibers used for structural composites. ASTM D 3822 (Ref 13), ASTM D 2101 (Ref 14), and ASTM D 540 (Ref 15) provide methods suitable for most other single fibers. However, these three specifications are not recommended by the American Society for Testing and Materials for acceptance testing of commercial shipments because the between-laboratory precision is not known.

Four areas of difficulty encountered in testing single fibers are discussed below.

*Now with PDA Engineering

Brittleness. High-modulus fibers, such as boron, that have low failure strains and relatively large diameters are susceptible to breakage during handling, damage during pressure clamping, and premature failure due to misalignment. The tab mount method, specified in ASTM D 3379 (Ref 12) mitigates these problems. More details concerning this technique are described in Ref 16 and 17. A theoretical treatment of clamp effects in fiber testing has been reported by S.L. Phoenix and R.G. Sexsmith (Ref 18).

Small Diameter. Single fibers of glass, graphite, or aramid, for example, are fine (about 10 μm, or 390 μin. or less in diameter), and are difficult to see and handle. They can be tested with clamps lined with rubber or other elastomers or with small capstans as alternatives to bonding to cardboard tabs. These methods involve technician dexterity and patience to ensure proper positioning and alignment of the small fibers. The final results, therefore, are more dependent on human factors than macroscopic test results would be.

Cross-Section Area Variation. For quality assurance purposes, the average fiber diameter may be sufficient. An average fiber diameter is used in ASTM D 3379 (Ref 12). If strength distributions are being evaluated using statistical analysis, highly accurate diameter or area measurements must be made on individual test samples. Individual fibers can vary in diameter along their length (Ref 19) with periods shorter than most gage lengths employed in testing (Ref 20). Likewise, all fibers in a yarn may not have exactly the same diameter (Ref 21). Individual fiber diameters large enough for optical resolution can be measured using attachments, such as the Watson image shearing eyepiece (Ref 17), or by projection microscopy (Ref 22). Small-diameter fibers can be measured with scanning electron microscopy. Laser diffraction has also been used (Ref 23). Careful calibration is required, and numerous measurements must be made to obtain an accurate average. An important limitation of optical methods arises when the fiber does not have a circular cross section (Ref 17). An alternate technique measures both fiber weight per unit length and density. When individual fibers are being tested as part of a process control scheme during manufacture, the fiber density may be assumed to remain constant, and weighing techniques as well as periodic density checks can be used (Ref 17). Some density measuring techniques used for carbon fibers are described by J. Donnet and R.C. Bansal (Ref 19).

Strain Measurements. It is difficult to measure strain on a single fiber directly and continuously without the use of highly specialized techniques. Direct clamping of an extensometer on boron filaments is described in Ref 11. Optical techniques can provide a direct measure, but they usually require manual readings. Indirect methods require calibration for compliance and particular care to avoid grip slippage. Increased sensitivity can be attained by increas-ing sample length to achieve more elongation for a given amount of strain. However, if the same samples are used for strength measurements, the average failure stress will be reduced because of the gage length effect.

Testing multifilament yarns in the as-received condition is usually referred to simply as a yarn test. The amount of twist in a yarn is an important parameter affecting strength. The lateral forces between fibers caused by twist allow some transfer of load from a broken fiber to its neighbor. When a multifilament sample is impregnated with an organic resin, the sample is a composite material and the test is referred to as a strand test.

Yarn test methods include ASTM D 2256 (Ref 24), ASTM D 885 (Ref 25), and ASTM D 2970 (Ref 26). The strength of a yarn or strand is statistically related to that of its component fibers and is usually the most important consideration for quality assurance. Testing parameters that must be considered are test environment, sample preconditioning, loading rate, gripping method, and method of elongation measurement. The relative importance of each depends on the material being tested. Some considerations are discussed below.

First, the importance of the clamp arises only in its ability to hold the sample without slipping and without creating bending moments or other stress concentrations that induce premature failure. The clamping method also must be compatible with the technique used to measure sample elongation. One criterion for a satisfactory gripping method is that most failures should occur away from the clamp interface, which can pose evaluation difficulties when the failed region extends over a relatively large fraction of the gage length, or if multiple separations occur during the failure event because of elastic rebound. If preliminary tests that compare the failure loads of samples establish that failure occurred in the grips, near the grips, or clearly in the center gage region, this can be used to establish a more definite criterion of questionable grip methods or of conditions legitimately associated with a grip failure. These factors must be interpreted in terms of the applicable specifications, because consistent test methods can be as important in the quality control function as establishing the highest strength value for a given number of material samples.

Second, the ease of measuring yarn cross-section area or linear density makes textile conventions the preferred method of verifying yarn tenacity (strength). This is reflected in the specifications referenced above.

The tenacity of continuous filament yarns initially increases with twist, and then, at higher twist levels, falls off. Normally, the greatest quality assurance concern is with yarns that are either untwisted (rovings or tows) or have small amounts of twist (producer's twist). Handling is the main problem in testing untwisted yarns because neither slack nor nonuniform fiber lengths can be introduced into the yarn without resulting in reduced measured tenacity. This problem can be avoided by introducing a small amount of twist before testing, a procedure described in ASTM D 2255 (Ref 24).

Finally, in addition to the above considerations for strength determinations, the measurement of elongation (strain and modulus) on yarns is subject to the same difficulties that were described for single fibers. Because loads are higher than for single fibers, slippage in flat clamps is more likely, and when capstans are used, the definition of sample gage length is subject to more uncertainty.

Measured yarn strength reflects some combination of mean strength, scatter, and difficulties inherent in gripping; careful standardization and correct interpretation are required for quality control (Ref 27).

Strand Tests. Impregnated strands have four inherent advantages: the fiber/matrix interaction is taken into consideration; the effect of fiber length on strength is reduced; the coefficient of variation is smaller than that for single filaments; and testing is easier. Strand test methods include ASTM D 2343 (Ref 28) and ASTM D 4018 (Ref 29).

The disadvantages encountered in testing impregnated strands are that more sample preparation steps are required than in simple yarn tests, and the resin content becomes a test variable. Typical resin-impregnating fixtures and procedures are described in the above-referenced specifications. A simple strand test for routine fiber strength and modulus evaluation has been reported (Ref 30). It has also been found that passing the impregnated strand through a circular die reduces the variability in resin content, which in turn improves the strength measurements (Ref 31). Flat-faced grips are suitable for low-strength strands. As strand strength increases, pull-out becomes more of a problem. Cast-resin end tabs are frequently used for samples containing high-strength heavy yarns. An example of tab fabrication is given in ASTM D 4018 (Ref 29).

With strand tests, it is possible to clamp extensometers directly on the sample to obtain strain data. The techniques required are similar to those described for attaching extensometers to boron monofilaments (Ref 11). Although a valid modulus can be obtained from the initial part of the load-elongation curve, it is not always possible to measure failure strain directly, because the samples may fail prematurely at stress concentrations where the extensometer is clamped. Extensometers are often removed before failure to prevent their being damaged. If it can be determined that the extensometer has not reduced the failure stress, it is possible to estimate failure strains by extrapolation using the last part of the loading curve.

Fabric strength tests are outlined in most descriptions of textile test methods, such as Ref 32 and 33 and ASTM D 1682 (Ref 34) and ASTM D 579 (Ref 35). The tensile strength of a fabric is frequently measured by means of the strip test or the grab test. For particular test results, it is important to know which method was used, because the results are not usually the same for the two tests. The grab test is used for control testing, whereas the strip test is preferred for accurate work (Ref 33).

In the strip test, the specimen used is usually a 25- to 50-mm (1- to 2-in.) wide strip of fabric that is prepared by cutting an oversized piece and then ravelling the edge yarns to reduce the width of warp yarns to 50 mm (2 in.). If the fabric is difficult to fray, the sample can be cut to the desired width. The data is reported as breaking load per inch of sample width.

In the grab test, the sample width is larger than the jaws of the grip, and some of the material on either side of the jaws contributes to the measured strength. Grab tests are as precise as ravelled-strip tests, and the specimens require much less time to prepare, although they require more fabric per specimen. There is no simple relationship between grab tests and strip tests, because the amount of additional strength contributed by yarns outside the grip width in the grab test depends on the fabric construction.

For these two tests, the warp yarns are usually aligned in the direction of the tensile loading. Any fabric nonuniformity or misalignment in the grips will lower the measured strength.

The tensile strength of a fabric reflects both yarn strength and the textile construction. The fabric strength can be compared to that of the component yarn by taking the ratio of fabric strength per warp yarn/single warp yarn strength. This is expressed as a percent and called the fabric strength translation efficiency. The grab test cannot be used because the number of load-bearing yarns cannot be defined; therefore the strip test must be used for a meaningful calculation in measuring this parameter. Factors that can influence strength translation efficiency are type of yarn, yarn construction, fabric construction and finish, and weaving parameters such as warp tension uniformity and weaving defects. Fabric strength translation efficiency is usually less than 100%. It is possible to increase the fabric strength above the yarn strength in special cases in which the compacting forces in the fabric prevent fiber slippage within the yarn itself, thereby effectively increasing the yarn strength (Ref 36). Plain-weave fabrics with the highest crimp will generally have the lowest strength translation efficiency (Ref 36). Crimp is the difference between the straightened yarn length and the distance between the ends of the yarn while in the cloth. The extent of deformation depends on the fabric construction.

For lightweight fabric samples, flat-faced grips are often used, although care must be taken when tightening the faces in order to maintain uniform loading across the width of the specimen. A useful technique that sometimes eliminates problems with pull-out from flat-faced grips is to double back the sample in the grip with a pin inserted where the material is folded (this is illustrated in Ref 34). Persistent failure in the grips or nonuniform tear failures usually indicate that the true fabric strength is not being obtained. Capstans are also frequently used for gripping textiles, especially for high-strength narrow samples where normal flat-faced wedge or compression grips usually cause failure at the grip or persistent sample pull-out. The most common problems encountered with aramid fabrics have been described as incomplete failure of warp yarns, nonsimultaneous failure of warp yarns, and failure at the jaw (Ref 37). Gripping procedures for glass fabrics involve impregnating the gripped regions. Procedures are described in Ref 35.

Because the stress-strain curves for the individual yarns of interest here are nearly linearly elastic, the initial portion in the fabric load-deflection curve mainly represents a deformation due to crimp removal or interchange. Following crimp removal, the fabric stiffens, and the slope of the stress-strain curve primarily represents deformation of the yarns themselves. Depending on the type of yarn and the fabric construction, it is possible to alter the total elongation considerably by the method of handling and loading the sample into the grips. If a clip-on type of extensometer is used, some preload is required to straighten the sample and establish a zero strain reference point. These sources of strain variation make the initial part of a fabric stress-strain curve difficult to reproduce, particularly in the case of low-elongation fibers used in composite applications.

Alternate Methods of Measuring Modulus

Sonic modulus can be measured by determining the wavelength of standing waves in a sample mounted under constant tension and driven at one end by a crystal recording head connected to an audio oscillator (Ref 16). Other techniques involve measuring the transit time of a sonic pulse between two transducers in contact with the sample (Ref 38, 39). Dynamic modulus has been measured with a vibrating reed technique that determines resonant frequency of vibration of a fiber (Ref 16). Rolls-Royce Ltd. has developed a method of modulus measurement for carbon fibers that is based on the relationship between tensile modulus and electrical resistivity (Ref 40).

Statistical Considerations

The statistical aspects of comparing test results from a number of samples tested individually under identical conditions (including length) are addressed in the statistics and quality control literature (one example is the statistical methods chapter in Ref 2). Reference 41 presents a useful article, "An Engineer's Guide to Books on Statistics and Data Analysis." Some elementary statistical concepts with textile testing examples are given in Ref 33. Data on the coefficient of variation, level of accuracy, and consistency of properties of carbon fibers are presented in Ref 17. Data for aramid fibers and strands are reported by S.L. Phoenix and E.M. Wu (Ref 42).

A very general overview of the statistical treatment of fracture is presented by L.J. Broutman and R.H. Krock (Ref 27). Detailed analysis of high-modulus fiber strength behavior is difficult. In particular, the relationships between filament, yarn, and strand data for predicting the behavior of composites fabricated from these materials are not well understood.

Single-fiber strengths are often modeled by a two-parameter Weibull distribution (Ref 43, 44) in which the cumulative probability of failure of a fiber loaded to stress level, σ, is:

$$F_l(\sigma) = 1 - \exp\left[-l\left(\frac{\sigma}{\sigma_o}\right)\rho\right] \quad \sigma \geq 0 \qquad \text{(Eq 1)}$$

where F_l is the probability of failure, l is the dimensionless length, σ_o is the scale parameter of unit length, and ρ is the shape parameter. The actual Weibull scale parameter is:

$$\sigma_l = \sigma_o\, l^{-(1/\rho)} \qquad \text{(Eq 2)}$$

The mean, $\bar{\sigma}$, for the above distribution is:

$$\bar{\sigma} = l^{-(1/\rho)}\, \sigma_o\, \Gamma\left(1 + \frac{1}{\rho}\right) \qquad \text{(Eq 3)}$$

where Γ represents the gamma function, and the coefficient of variation is approximately $1.2/\rho$ (Ref 43).

Probability plotting is described by G.J. Hahn and S.S. Shapiro (Ref 45). In the case of the Weibull distribution, Eq 1 can be rearranged into a linearized form:

$$\ln - \ln\left(\frac{1}{1 - F_l(\sigma)}\right) = \rho \ln \sigma + \rho \ln\left[\frac{l^{1/\rho}}{\sigma_o}\right] \quad \text{(Eq 4)}$$

so that a plot of ln-ln [1/1 − $F_l(\sigma)$] versus ln σ will be a straight line whose slope and intercept yield ρ and σ_o. If the assumed model is correct, the plotted points will tend to fall in a straight line. If the model is inadequate, the plot will not be linear. When testing fibers at short gage lengths, a marked effect of the clamps is expected (Ref 18). Some common methods for estimating the Weibull shape parameter are discussed in Ref 46.

An example of a Weibull distribution that accurately models the strength data for aramid filaments tested at a single gage length is presented in Ref 42. H.D. Wagner and S.L. Phoenix (Ref 21) have studied the effect of filament length for aramid fibers and found a discrepancy between the theory and experimental results; the ln-ln dependence of mean strength on length was not linear, reflecting a possible decrease in the Weibull shape parameter, ρ, as length increased.

If the experimental data does not obey Eq 4, a breakdown of the assumptions inherent in Eq 1 is indicated (Ref 44). The most important assumption is that of a single type of defect.

Multiple modes of failure in carbon fibers have been studied by C.P. Beetz (Ref 44, 47). Other deficiencies in using the Weibull distribution to represent fiber strength have been discussed by D.G. Harlow and S.L. Phoenix (Ref 48), who suggest a double Weibull distribution as a more suitable form. For quality assurance purposes, it is not always necessary to know the detailed origin of departures from the Weibull distribution. The utility of these types of analyses and associated statistical testing is that they provide more information that can be used to compare the strength behavior of various sets of fiber, yarn, or strand data in terms of differences between lots, or to evaluate processing and fabrication steps that might be degrading reinforcement strength.

S.L. Phoenix and coworkers have discussed much of the work relating the probability distribution of the strength of multifiber assemblies or composites to the distribution of fiber strengths (Ref 5, 10, 42, 43, 48-51). Other publications summarize additional work in this area (Ref 52-55). As theoretical and experimental efforts continue, a better understanding of the effects of flaws on fiber strengths will lead to improved fibers. Developments in statistical treatments of composite strength should likewise result in more reliable composite structures. Both of these areas suggest that parallel developments in the use of statistical concepts in fiber testing for quality control will play an important role in composite development.

REFERENCES

1. "Standard Terminology Relating to Textiles," D 123, *Annual Book of ASTM Standards*, American Society for Testing and Materials
2. B.C. Goswami, J.G. Martindale, and F.L. Scardino, *Textile Yarns: Technology, Structure, and Applications*, John Wiley & Sons, 1977
3. W.E. Morton and J.W.S. Hearle, *Physical Properties of Textile Fibres*, John Wiley & Sons, 1975
4. C. Zweben, W.S. Smith, and M.W. Wardle, Test Methods for Fiber Tensile Strength, Composite Flexural Modulus, and Properties of Fabric-Reinforced Laminates, in *Composite Materials: Testing and Design (Fifth Conference)*, STP 674, S.W. Tsai, Ed., American Society for Testing and Materials, 1979, p 228-262
5. S.L. Phoenix, Statistical Aspects of Failure of Fibrous Materials, in *Composite Materials: Testing and Design (Fifth Conference)*, STP 674, S.W. Tsai, Ed., American Society for Testing and Materials, p 455-483
6. J.F. McCarthy, Jr., and O. Orringer, Some Approaches to Assessing Failure Probabilities of Redundant Structures, in *Composite Reliability*, STP 580, E.M. Wu, Ed., American Society for Testing and Materials, 1975, p 5-31
7. *Quality Control Handbook*, J.M. Juran, Ed., McGraw-Hill, 1962
8. "Standard Practice for Sampling Yarn for Testing," D 2258, *Annual Book of ASTM Standards*, American Society for Testing and Materials
9. P.E. McMahon, Graphite Fiber Tensile Property Evaluation, in *Analysis of the Test Methods for High Modulus Fibers and Composites*, STP 521, American Society for Testing and Materials, 1973, 367-389
10. S.L. Phoenix, Statistical Analysis of Flaw Strength Spectra of High-Modulus Fibers, in *Composite Reliability*, STP 580, E.M. Wu, Ed., American Society for Testing and Materials, 1975, p 77-89
11. F.K. Rose and J.L. Stokes, "Advanced Methods to Test Thin Gage Materials," AFML-TR-68-64, Air Force Materials Laboratory, 1968
12. "Standard Test Method for Tensile Strength and Young's Modulus for High-Modulus Single-Filament Materials," D 3379, *Annual Book of ASTM Standards*, American Society for Testing and Materials
13. "Standard Test Method for Tensile Properties of Single Textile Fibers," D 3822, *Annual Book of ASTM Standards*, American Society for Testing and Materials
14. "Standard Test Method for Shrinkage of Textile Fibers," D 2102, *Annual Book of ASTM Standards*, American Society for Testing and Materials
15. "Standard Test Method for Staple Length of Man-Made Fibers, Average and Distribution (Fiber Array Method)," D 3660, *Annual Book of ASTM Standards*, American Society for Testing and Materials
16. "Evaluation Techniques for Fibers and Yarns used by the Fibrous Materials Branch, Nonmetallic Materials Division, Air Force Materials Laboratory," AFML-TR-67-159, Air Force Materials Laboratory, 1967
17. R.M. Gill, *Carbon Fibres in Composite Materials*, Iliffe Books, 1972
18. S.L. Phoenix and R.G. Sexsmith, Clamp Effects in Fiber Testing, *J. Compos. Mater.*, Vol 6, 1972, p 322-337
19. J. Donnet and R.C. Bansal, *Carbon Fibers*, Marcel Dekker, 1984
20. W.S. Knoff, E.I. Du Pont de Nemours & Company, Inc., private communication, 1985
21. H.D. Wagner, S.L. Phoenix, and P. Schwartz, A Study of the Statistical Variability in the Strength of Single Aramid Filaments, *J. Compos. Mater.*, Vol 18, 1984, p 297-309
22. H.W.M. Lunney, Random Errors of Observation in the Measurement of Fiber Diameter by Projection-Microscope Methods, *Text. Res. J.*, Vol 50 (No. 12), 1980, p 728-731
23. A.J. Perry, B. Ineichen, and B. Eliasson,
Fibre Diameter Measurement by Laser Diffraction, *J. Mater. Sci. Lett.*, Vol 9, 1974, p 1376-1378
24. "Standard Test Method for Breaking Load (Strength) and Elongation of Yarn by the Single-Strand Method," D 2256, *Annual Book of ASTM Standards*, American Society for Testing and Materials
25. "Standard Methods of Testing Tire Cords, Tire Cord Fabrics, and Industrial Filament Yarns Made From Man-Made Organic-Base Fibers," D 885, *Annual Book of ASTM Standards*, American Society for Testing and Materials
26. "Standard Methods of Testing Tire Cords, Tire Cord Fabrics, and Industrial Yarns Made From Glass Filaments," D 2970, *Annual Book of ASTM Standards*, American Society for Testing and Materials
27. L.L. Broutman and R.H. Krock, *Modern Composite Materials*, Addison-Wesley, 1967
28. "Standard Test Method for Tensile Properties of Glass Fiber Strands, Yarns, and Rovings Used in Reinforced Plastics," D 2343, *Annual Book of ASTM Standards*, American Society for Testing and Materials
29. "Standard Test Methods for Tensile Properties of Continuous Filament Carbon and Graphite Yarns, Strands, Rovings, and Tows," D 4018, *Annual Book of ASTM Standards*, American Society for Testing and Materials
30. H. Morley, A Simple Strand Test for Routine Fibre Strength and Modulus Evaluation, *Composites*, Vol 13 (No. 1), 1982, p 21-23
31. S. Kulkarni, Lawrence Livermore National Laboratory, private communication, 1985
32. "Textile Test Methods," Federal Specification CCC-T-191, U.S. Government Printing Office, 1951
33. J.E. Booth, *Principles of Textile Testing*, Heywood Books, 1968
34. "Standard Test Methods for Breaking Load and Elongation of Textile Fabrics," D 1682, *Annual Book of ASTM Standards*, American Society for Testing and Materials
35. "Standard Specification for Greige Woven Glass Fabrics," D 579, *Annual Book of ASTM Standards*, American Society for Testing and Materials
36. P.R. Lord and M.H. Mohamed, *Weaving: Conversion of Yarn to Fabric*, Merrow Publishing, 1973
37. S.L. Goodwin, N.J. Abbott, "Kevlar Properties Investigation, Development of Kevlar Tensile Test Methods," AFFDL-TR-79-3019, Air Force Flight Dynamics Laboratory, 1979
38. W.N. Reynolds, The Structure and Mechanical Properties of Carbon Fibres, *Third Conference, Industrial Carbons and Graphite*, Society of the Chemical Institute, 1970

39. G. Hinrichsen, S.M. Sadat-Darbandi, and A. Al-Irobaidi, Sonic-Velocity Measurements as a Suitable Tool for the Observation of Structural Changes in Polymeric Materials, *Polym. Bul.*, Vol 13, 1965, p 15-20

40. French Patent 1590257, May 1970

41. G.J. Hahn and W.Q. Meeker, Jr., An Engineer's Guide to Books on Statistics and Data Analysis, *J. Qual. Technol.*, Vol 16 (No. 4), 1984, p 196-218

42. S.L. Phoenix and E.M. Wu, Statistics for the Time Dependent Failure of Kevlar-49/Epoxy Composites: Micromechanical Modeling and Data Interpretation, in *Mechanics of Composite Materials, Recent Advances*, Z. Hashin and C.T. Herakovich, Ed., Pergamon, 1983, p 135-163

43. S.L. Phoenix, Statistics for the Strength of Bundles of Fibers in a Matrix, in *Encyclopedia of Materials Science and Engineering*, M.B. Beaver, Ed., Pergamon Press, 1983

44. C.P. Beetz, Jr., The Analysis of Carbon Fibre Strength Distributions Exhibiting Multiple Modes of Failure, *Fibre Sci. Technol.*, Vol 16, 1982, p 45-59

45. G.J. Hahn and S.S. Shapiro, *Statistical Models in Engineering*, John Wiley & Sons, 1967

46. K. Trustrum and A. De S. Jayatilaka, On Estimating the Weibull Modulus for a Brittle Material, *J. Mater. Sci.*, Vol 14, 1979, p 1080-1084

47. C.P. Beetz, A Self-Consistent Weibull Analysis of Carbon Fiber Strength Distributions, *Fibre Sci. Technol.*, Vol 16, 1982, p 81-94

48. D.G. Harlow and S.L. Phoenix, Probability Distributions for the Strength of Composite Materials I: Two-Level Bounds, *Int. J. Fract.*, Vol 17 (No. 4), 1981, p 347-372

49. S.L. Phoenix, Probabilistic Concepts in Modeling the Tensile Strength Behavior of Fiber Bundles and Unidirectional Fiber/ Matrix Composites, in *Composite Materials: Testing and Design (Third Conference)*, STP 546, C.A. Berg *et al.*, Ed., American Society for Testing and Materials, 1974, p 130-151

50. D.G. Harlow and S.L. Phoenix, Probability Distributions for the Strength of Composite Materials II: A Convergent Sequence of Tight Bounds, *Int. J. Fract.*, Vol 17 (No. 6), 1981, p 601-630

51. D.G. Harlow and S.L. Phoenix, Bounds on the Probability of Failure of Composite Materials, *Int. J. Fract.*, Vol 15 (No. 4), 1979, p 321-336

52. P.W. Manders and T. Chou, Variability of Carbon and Glass Fibers, and the Strength of Aligned Composites, *J. Reinf. Plast. Compos.*, Vol 2, 1983, p 43-59

53. S.B. Batdorf, Tensile Strength of Unidirectionally Reinforced Composites-I, *J. Reinf. Plast. Compos.*, Vol 1, 1982, p 153-164

54. B. Bergman, On the Probability of Failure in the Chain-of-Bundles Model, *J. Compos. Mater.*, Vol 15, 1981, p 92-98

55. P.W. Menders, M.G. Bader, and T. Chou, Monte Carlo Simulation of the Strength of Composite Fibre Bundles, *Fibre Sci. Technol.*, Vol 17, 1982, p 183-204

Resin Properties Analysis

Gerald L. Sauer, BASF Structural Materials Inc.

MOST RESINS USED IN COMPOSITES are blends or systems of ingredients, each of which is added to impart some desired properties to the prepreg or cured composite. Tests are typically performed on the individual component material, on mixtures of several ingredients that later become part of the final blend, and/or on the blend in its final composition. These tests are performed on the uncured resin system.

Polymers with chemically different functions are used to categorize resin families. The earlier polymers were resole-type phenolics and polyesters based on styrene condensation. Epoxy-based polymers are the most common today, although newer systems using bismaleimide functional groups are becoming increasingly popular. Other systems, including polyimides, polysulfides, and vinyls, are used for specialized applications. This article focuses on epoxy-based polymers because they are the most widespread and because they are representative of the quality control measures that are necessary in most resin systems.

Although most of the tests to be described are oriented toward the chemical properties of resin systems or component materials, an evaluation of physical and mechanical properties must be performed to ensure that control is adequate.

Component Material Tests

Because every resin system component material contributes to the final properties of the prepreg or cured composite, it is essential to identify the individual component properties, define acceptable measurement ranges, and assign or develop appropriate test methods. A typical resin system will consist of one or more epoxy resins, a curing agent (usually an amine or anhydride), and a catalyst to control the rate of reaction. In addition, fillers or additional modifiers may be used to alter specific properties. Typical chemical property tests used on component materials are described below.

Epoxy resin tests include the following.

Epoxy per equivalent weight (EEW). The epoxy resin is chemically reacted to determine the epoxide content per unit weight of resin by wet chemical titration. Epoxide content is a measure of potential cross-link density.

Hydrolyzable Chloride. Epoxy resins are frequently made from chlorinated compounds. Any residual hydrolyzable chloride compound can affect the reactivity and therefore the cured resin properties. The chloride content is measured by titration of a soluble sample.

Moisture. Because epoxy resins are hygroscopic materials and absorbed moisture affects reactivity, moisture content is measured by drying or by titration using Karl Fisher apparatus.

Melting Point. The temperature at which heat melts the resin system should be measured.

Softening Point. Because the viscosity of the resin affects system processibility and can be a measure of average molecular weight, the point at which the resin softens should be measured.

Viscosity. Because viscosity affects processibility and, indirectly, the reactivity of the resin system, it is measured by parallel, cone-and-plate, or Brookfield viscometers.

Infrared (IR) Spectroscopy. This test is a rapid method for identifying functionality, which is useful in differentiating epoxies with similar EEWs. A thin film of resin is placed on a crystal, and IR light is passed through the crystal and film at varying frequencies. Functional groups absorb at certain defined frequencies. The test is usually quantitative, but because absorption is proportional to quantity and intensity, qualitative results are also possible.

High-Performance Liquid Chromatography (HPLC). This test is performed by dissolving the resin in a solvent and depositing it on the packing of a column. It is then eluted selectively using a polar-nonpolar solvent mixture. The selective elution occurs as the equipment changes the ratio of solvents in a closely programmed system. This provides a highly reproducible chromatogram. The detector is usually either the ultraviolet or refractive-index technique. Separation is also possible by diverting selected fractions during elution.

Gel Permeation Chromatography (GPC). This test is similar to the HPLC test except that it displays separation by molecular weight groups rather than functional groups.

Other sophisticated techniques such as gas chromatography, mass spectrometry, and nuclear magnetic resonance (NMR) can be used in organic synthesis and to identify unknown organic mixtures, but are not routinely used for quality control.

Curing agents can be liquids or powders. The particle sizes of powdered curing agents are normally the same as those described in the section "Fillers and Modifiers" in this article. An amine equivalent test should be performed in which liquid or solid amine curing agents are dissolved in a liquid media and titrated to determine the number of reactive hydrogens per unit weight. Techniques are available to determine primary, secondary, or tertiary amines. Reactivity decreases as the available hydrogen atoms move from primary to secondary. Only ammonia has an available third hydrogen. However, amines form what is known as quaternary salts with an anion, which can change the behavior of the material.

Other analytical and chemical tests, such as infrared spectroscopy, ultraviolet analysis, gel permeation chromatography, high-performance liquid chromatography, and melting-point determinations may be used to determine the quality of the curing agents. The tests were described in the preceding section on epoxy resins.

Catalysts. Many catalysts or epoxy reactions are molecular complexes with a Lewis acid which is usually BF_3 and a tertiary amine. The complex is a quaternary salt, with the BF_3 acting as the anion, and the amine acting as the cation. Tests include atomic absorption, to quantitatively determine boron concentration; moisture, to assess water content, which tends to degrade the complex; and IR spectroscopy, to identify the amine form.

Fillers and Modifiers. Fillers are often used in the form of ground or pulverized inorganic materials to improve dispersion and reactivity or provide increased surface area for flow control. Screens of increasing size are stacked to separate the particles by size. Finer particles can be evaluated by particle analysis, whereby the sample is dropped in a liquid, and the rate and quantities of falling particles are measured optically. Very fine particles can also be analyzed in a column of controlled dry gas, using a sensitive balance to plot weight gain versus time.

Typical modifiers include:

- Flow control agents, such as finely divided silicon dioxide, clays, and high molecular weight polymers
- Tack enhancers, such as low molecular weight resin or polysulfide, a naturally tacky resin
- Adhesive ingredients, such as soluble rubbers, polysulfides, and dienes
- Flame retardants, such as halogenated resins with cobalts, phosphates, and other inorganic materials
- Color agents, such as pigments and dyes

Mixed Resin System Tests

After the components are mixed, tests are performed to document the presence of each ingredient and to ensure that the proper chemical reaction state has been obtained.

Chemical property tests include:

Gel Time. Several methods are used to determine gel time. In the "probe" method the resin is placed in a test tube, which is then placed in an oil bath at the desired temperature. The resin is probed until gel is reached. Gel is defined as the point at which the resin strings break sharply when probed and strung. Other methods use a modified melting point apparatus called Fisher-John. Some applications use an isothermal hold on a rheometer-type apparatus.

Moisture tests are performed on neat resin using a titration method known as Karl Fisher. A moisture meter with P_2O_5 is also available to measure moisture.

HPLC and IR Spectroscopy. Both chromatographs and spectrograms are used to "fingerprint" the resin. These methods were described in the section "Epoxy Resin" in this article.

Physical property tests include:

Viscosity. Of the various viscometers available, a Brookfield with thermocel or a cone-and-plate type is most often used. Recently, a rheometer-type using oscillating parallel plates held at the desired temperature has become popular. Gel is usually defined by a predetermined viscosity at which the resin will neither flow nor wet-out the dry or precured surfaces. A complete viscosity profile is useful for predicting prepreg curing performance.

Solids Level. When the resin is in a solution impregnation, the solids level is measured by evaporation in an explosion-proof oven.

GPC. A chromatograph is used to define and fingerprint the molecular weight distribution of the resin. Most specifications list GPC under chemical tests because of its convenient analytical nature and the type of equipment used to determine this property.

Prepreg Tests

The prepreg that is formed by combining fiber and resin is also known as an impregnated fiber-reinforced plastic (FRP). While plastic generally refers to thermoplastic resins, the term is also used for thermoset resins.

Chemical property tests for prepreg materials include the following.

Gel Time. This test is measured at a predetermined temperature. Its usefulness is primarily for quality control purposes, but it also can help predict shop life and cure characteristics.

IR Spectroscopy. This test is repeated on the resin after impregnation to ensure that no major change has occurred. Although the test is usually qualitative, techniques can be developed to obtain quantitative results.

HPLC. This test provides a signature of the components, separated by chemical functionality. Both quantitative and qualitative results can be obtained.

Atomic Absorption. When inorganic ingredients are used, they can be measured at low levels by atomic absorption.

Physical property tests are typically run by the manufacturer and also the user to ensure that the prepreg material contains the proper amount of acceptable component materials to meet the user's specifications.

Resin content is the amount of active resin, including the volatiles. The test ensures that the material has the ability to form suitable composite components. The quantity of active resin is expressed as a percentage of the total weight.

Resin solids content is the amount of resin, less the amount of volatiles. In the case of epoxy systems, these two are almost the same because the volatile level is usually less than 1%. Some resins, such as phenolics and polyimides, contain substantial volatiles, in the form of water, formed in the reaction of the prepreg resins during resin cure.

Areal or dry fiber weight is the weight of dry reinforcing fiber in the prepreg. The fiber weight, in conjunction with the resin solids content, allows the manufacturer and user to predict the theoretical ply thickness and the resultant fabricated-part thickness. This property is obtained from the same sample as the resin content.

Resin flow. Samples of plied product are exposed to pressure and heat (sometimes with bleeder) to measure the resin lost because of the flow of the resin at a specified temperature and pressure condition.

Tack. This test evaluates a subjective sticky characteristic of the prepreg composite. The most widely used test requires that a ply of prepreg be "tacked" to a tool and subsequent plies be tacked to the first ply. The normal requirement is that the second ply shall be capable of being removed and repositioned if necessary. Tack is affected by many factors, including resin type, resin content, and degree of resin advancement; temperature when evaluated; and humidity (most epoxies are hygroscopic, and absorbed moisture increases tack). Therefore, it is important to specify the conditions and methods of evaluation. Tack causes more problems for both the manufacturer and user than any other prepreg characteristic. Many attempts to automate this test have had limited success.

Drape. Like tack, drape is subjective. It is also greatly influenced by tack. Drape is the ability of a prepreg to be formed around defined radii and stay tacked to the tool for specified periods of time.

Room temperature out-time is the period of time at ambient temperature for which a prepreg can maintain enough tack and drape to be used to make components in the production environment.

GPC is used by the prepreg manufacturer for resin control after prolonged out-time or storage to determine whether the resin has advanced beyond a usable level.

Resin and Prepreg Mechanical Properties

Mechanical properties that are typically measured during the development and use of an epoxy resin system are shown with the appropriate test methods in Table 1. Because these properties form the basis of subsequent composite properties, the tests are often performed on composite samples as well as on neat resin samples. These properties are also frequently measured after exposure to moisture saturation, and certain resin-dominated tests are performed after chemical and thermal exposures.

Many test procedures are defined by ASTM methods. Moisture, chemical, and thermal conditioning parameters are typically defined by the ultimate end-use to which the composite will be subjected in actual service and are specified by the user.

Table 1 Typical resin and prepreg mechanical property tests

Test	Test method
Tensile strength, modulus, and strain	ASTM D 651
Poisson's ratio	ASTM E 132
Compressive strength, modulus, and strain	ASTM D 695
Shear strength	ASTM D 2344
Shear modulus	ASTM E 143
Flexural strength and modulus	ASTM D 790
Impact toughness	ASTM D 3998
Fatigue	ASTM D 256
	User
Density	ASTM D 1895
Thermal properties	User
Glass transition	
Coefficient of thermal expansion	
Weight loss at temperature	
Chemical exposures	User
Acetone	
Chlorinated solvents	
Oils	
Hydraulic fluids	
Body fluids	
Alcohol	
Moisture effects	User
Flammability	User

Tooling Quality Control

Steven F. Hanson, Composite Tooling International, Inc.

ADVANCED COMPOSITES have moved into considerably more complex and tighter-tolerance applications during the past decade. When increased demands are placed on the tool designer, planner, and fabricator, quality control requirements must also increase. As in all organizations, the primary purposes of the quality control department are to provide assurance that uniformity does exist in the fabrication process and to verify that all intended operations have been performed according to the specified guidelines.

When the primary use of composites in the aircraft/aerospace industry was limited to non-structural applications, the composite part was always more flexible than the adjoining structure. Therefore, if the part did not fit exactly, it was flexible enough to be pulled into contour when it was attached to the supporting structure. However, current tolerance requirements in primary structures are much closer, because the composite part is now the part that determines the shape of the structure. In the past, if a wing-to-body fairing was out of contour by 3.2 mm (0.125 in.), it could be used in most cases. Today, when dealing with such parts as wing spars and ribs, the tolerance for contour error must be kept to 0.125 mm (0.005 in.) or less.

Formerly, the quality control function ensured that the first part off of the tool would indeed fit the assembly. Today, quality control must also verify that the correct materials are being used, that the proper procedures are being followed, and that the tool is dimensionally correct.

Quality control documentation of the tool fabrication process is mostly confined to the tool planning sheet or tool work order. In the case of large or complex tools, a tooling log book is used in conjunction with the tool planning sheet or work order. The tooling log book contains the complete work history of the tool, including changes that occurred after the tool was fabricated. It should remain with the tool or should be filed in the tool shop office. The log book is a good resource when trying to determine what has happened during the life of that particular tool. A third form of quality control documentation is a separate inspection buy-off sheet, which is used when a detailed planning sheet or work order is not provided to build the tool. In the fabrication of a composite tool, whether graphite-epoxy pre-preg or a wet lay-up is used, certain check points can be universally used as inspection verification points through the fabrication of the master model, the pattern taken from it, and the composite tool made from the pattern.

Hand-Faired Master Models

When building a master model from plaster or hand-faired epoxy, the primary areas of concern to the inspector should include the following:

- *Base Plate:* Should be flat and without twist
- *Templates:* Each should be within tolerance on its contour
- *Reference system:* If scribed on the base plate, it should be correct and within tolerance. If an optical reference system is used, it must be verified, and the date of optical equipment calibration must be checked
- *Rigging:* Should be checked to verify that the rigged templates are located properly and are vertical (not bowed or twisted)
- *Faired surface:* Should be free of flat spots, dips, and irregularities
- *Scribe lines:* Trim and reference lines should be verified and checked for sharpness
- *Master documentation and identification:* These are the final items to be checked and bought
- *Final contour check:* Can be verified in its entirety by a coordinate-measuring machine

Machined Master Models

The inspection process is streamlined considerably when inspecting master models that have been machined on a computer numerically controlled lathe or mill. The inspector basically checks to confirm that there have not been any gross programming errors, such as the slip of a decimal point, and determines whether any items should be added to the tool after machining, such as scribe lines and identification. On smaller and simple machined master models, hand-held precision instruments, such as dial calipers and height gages, can be used to check dimensions. Templates are used to show any discrepancies in contours. A coordinate-measuring machine, if available, can be used to check contours, hole locations, and scribe lines.

On large or complex machined master models, the contours can be checked with templates or with a coordinate-measuring machine. Templates, a coordinate-measuring machine, or optics can be used to verify the proper location of tooling holes, scribe lines, and reference points.

Second-Generation Patterns

When a pattern is to be taken from the master model, the inspection is similar to that for a plaster splash, plastic-faced plaster, or epoxy laminate. This is the case before and during fabrication of the pattern.

Before releasing the master model in order to make the pattern, the inspector should verify that:

- The surface of the master model is in an acceptable condition. It should be free of chips, cracks, or other surface blemishes that will affect the surface of the pattern
- The required reference lines are clean and complete
- The master reflects the latest engineering changes. If the master model does not incorporate the latest engineering changes, it may be futile to fabricate any tooling from it

The inspector should also verify that the toolmaker has the correct location on the master model taped off for his pattern and that it is large enough to make a tool, as determined by the tool planning sheet. During pattern fabrication, the inspector should be concerned with each of the areas described below.

Releasing. The master model should be released properly because it is critical that its surface be preserved for future tooling. The various steps of the release procedure can easily be confused or forgotten if the toolmaker is interrupted several times during the release process.

Fabrication. If standard shop practices allow, the inspector can monitor pattern fabrication if it is to be plastic-faced plaster or epoxy laminate. This practice may vary among companies. If a company has a formal fabrication procedure and requires that all patterns be fabricated according to the procedure, then the services of the inspection department will be necessary. If the company is fabricating tooling on a subcontract for a customer, it may be necessary to have an inspector verify that the fabrication procedures of the customer are followed.

Preremoval Inspection. Before the pattern is removed from the master model, an inspector should verify that the pattern is tight to the surface of the master model. This indicates that no warping or twisting of the pattern occurred during cure. If it is difficult to see the edge where the surface of the master model and the pattern meet, the inspector can use a 0.05-mm (0.002-in.) feeler gage to determine if the pattern has lifted off the master.

Cleanup and Scribe Lines. After the pattern has been pulled from the master model, its surface should be cleaned of any release residue, and the scribe lines, which are now high male lines, will have to be reversed back into the surface of the pattern. This reversing procedure is critical and should be monitored to verify that the scribe lines are within the tolerance of the original lines. This can be accomplished by leaving 6.4-mm (¼-in.) long male witness lines every 75 to 100 mm (3 to 4 in.). After verification, these lines are then reversed.

Composite Tools

After the pattern is fabricated and passes inspection, it is ready to be used in composite tool fabrication. The composite tool can be a prepreg or a wet lay-up.

Prepreg tool fabrication involves inspection of the following areas.

Vacuum Integrity. The pattern should have a vacuum bag put on it and be checked for vacuum integrity.

Autoclave Integrity. If the pattern has good integrity at room temperature and pressure, it should then be put into an autoclave, heated to the curing temperature of the prepreg system being used, and pressurized to the curing pressure that will be used. If plastic-faced patterns are used, temperatures less than 110 °C (225 °F) are recommended. This ensures that there are no small voids immediately beneath the surface of the pattern (whether it is plastic-faced plaster or epoxy laminate) that require repair prior to tool lay-up.

Release Procedure. Depending on shop practice, the inspector may need to verify pattern release for coverage and for type.

First-Ply Orientation. When the first ply of the laminate has been laid down on the tool, it should be checked for proper ply orientation and proper butt splicing of individual segments.

First-Ply Debulk. On most prepreg lay-ups, a vacuum debulk is performed after the first ply is laid down in order to pull the ply tightly onto the surface of the pattern. After debulking is performed, the inspector should verify that the proper materials were used, that the bag held the required inches of vacuum, that there was no bridging in the bag, and that the vacuum was held for the required length of time.

Ply Orientation. As the lay-up progresses, the inspector should verify the ply orientation and the splicing of each ply.

Debulking. When additional debulk operations are required, the inspector should again verify the materials, bag integrity, bridging, and time under vacuum. If a heat debulk is used, it will also be necessary to check that the temperature was within the proper parameters for the proper length of time.

Final Bagging. After the lay-up has been completed, the inspector must confirm that the proper layers of peel ply, bleeder, barrier film, breathers, and bagging were used, that there was no bridging in the bag, and that the bag passes the minimum requirement for vacuum integrity.

Cure Cycle. After the tool has been cured in the autoclave, the inspector must verify, from autoclave run charts, that the tool was cured properly.

Postcure. If there is an oven postcure, the inspector must check the chart recorder on the oven to verify that the tool reached the proper curing temperature for the required length of time.

Cleanup and Trimming. After the tool has been popped from the pattern, trimmed, and cleaned up, the inspector must verify that the scribe lines have been reversed properly.

Final Buy-Off. When the tool is submitted to inspection for the final buy-off, the inspector should check that the tool is identified properly and has all of the necessary paperwork with it, that the contour has not changed, and that vacuum integrity exists. If a coordinate-measuring machine is not available, it may be necessary to check the contours with templates or plaster splashes taken from the master model. This final check of the contour is necessary because the tool will tend to spring in on angles and deep contours. Although most of this tendency should be compensated for in the tool design, it is still necessary to confirm where the contour lies in relation to the allowed tolerance. To check the tool for vacuum integrity, it is necessary to bag the part area of the tool and to load the tool into either an oven and then heat it to its operating temperature or, preferably, an autoclave, and check it during a normal autoclave cure cycle. When this has been done, the tool is then ready to release to production fabrication or tool tryout.

Wet lay-up fabrication inspection, although somewhat similar to that for a prepreg tool, is simpler, involving the following areas.

Vacuum Integrity. If the pattern is to be surface bagged instead of envelope bagged, the bag should be checked for vacuum integrity.

Release Procedure. If it is consistent with standard shop practice, the inspector may need to verify the application of the release agent for coverage and for type.

Gel Coat Application. The inspector should verify that the gel coat is of the proper thickness and that it is the specific type of gel coat listed on the tool work order.

Ply Orientation. The orientation of each ply should be verified by an inspector.

Vacuum Debulk. After the first two to four plies have been laid down, there is usually a vacuum debulk operation, which can require 30 min to 12 h, depending on the type of resin being used. The inspector should verify that the materials used are correct, that there is no bridging in the bag or bagging materials, that the bag has vacuum integrity, and that the tool has remained under debulk for the proper amount of time.

Ply Orientation. As the balance of the tool is being laid up, the orientation of each ply should be verified, and any other debulking operations should be conducted.

Final Bagging. When the lay-up is completed, the inspector should verify that the proper layers of peel ply, bleeder, barrier film, breathers, and bagging were used. The bag should be checked for bridging and vacuum integrity.

Cure Cycle. If the tool was cured in an oven, the inspector must examine the charts from the chart recorder and attach them to the tool work order. If the tool was cured in an autoclave (to obtain better compaction), the charts from it will have to be checked.

Scribe Lines. After the tool has been popped from the pattern, trimmed, and cleaned, the inspector must verify that the scribe lines that denote the edge of the part, reference lines, and so on, have been reversed properly.

Final Buy-Off. When the tool is submitted to inspection for the final buy-off, the inspector must verify that the contour is correct, that vacuum integrity exists, and that all paperwork is correct. The contour can be checked with contour templates, a plaster splash taken from the master model, or a coordinate-measuring machine. The inspector should first check the vacuum integrity at room temperature and then during an autoclave run with only a bag and breather on the tool. After the tool has passed both of these checks, all that remains is to ensure that the paperwork is in order. With this process completed, the tool is then ready to be released for production fabrication or tool tryout.

There are other types of composite tooling, such as bond, trim, drill, and assembly fixtures. The fabrication and inspection requirements for these have been well established in the fabrication of metal aircraft components and do not require elaboration.

Reinforcing Material Lay-Up Quality Control

Lawrence A. Lang, Composites Manufacturing Engineering Inc.

IN-PROCESS INSPECTION during composite material lay-up and cure is essential if the structural, dimensional, and environmental performance designed into a part are to be consistently achieved. The inspection techniques are usually based on federal or military standards, company processing specifications, quality assurance requirements, and what are generally described as good shop practices. Although these practices vary from one company or facility to another depending upon prescribed contractual requirements, available equipment, level of personnel training, and documentation systems used, they all fulfill the following criteria.

First, the materials to be used in part fabrication must conform to drawing and material specifications, must have been stored and handled properly, and must be within current shelf life and processing out-time criteria as specified in the applicable processing documents. Indirect material processing aids must be correct for the manufacturing methods to be employed, and tooling and shop aids must conform to the requirements of the processing specifications and manufacturing work orders. The supplier manufacturing the material must have extemely tight process controls to maintain consistency from batch to batch or roll to roll. Variations in prepreg can cause processing problems when fabricating composite details.

Lay-up procedures must be consistent with the drawings for orientation control, and processing specifications must be consistent for handling, debulking, and other operations.

The processing environment must conform to and continue to meet the requirements throughout the lay-up and cure phases. Temperature in bond rooms is extremely critical, because a few degrees can cause material workability problems.

Cure process controls must remain within specified limits; any excursions from specified limits are documented for subsequent disposition and corrective action. Manufacturing should not be handicapped by numerous cure cycles. Straight-up cures should be used whenever possible, and step-up cures should be used for parts approximately 5 mm (0.200 in.) thick or thicker.

Equipment must be within specified calibration requirements. In-process test coupons, trim tabs, or other specified test specimens must be fabricated, traced, and tested in accordance with specification requirements.

Facilities and Equipment

The facilities employed to fabricate composite hardware usually must meet requirements for cleanliness, positive pressure, and humidity and temperature control. The selection of the actual requirements is based upon a set of conditions detailed in a contract or by a process specification in which materials and process engineers have examined the sensitivity of the uncured material to the work environment and determined temperature limits that maximize the working life of the material while retaining acceptable handling characteristics and level of cleanliness. Material suppliers can furnish much information, such as studies of out-time, tack, drape, workability versus temperature and humidity, hot/wet properties, and moisture absorption.

There are many opinions as to what these control limits should be, because the sensitivity of each prepreg or adhesive varies slightly from that of every other. Tolerable limits are usually a matter of the level of sophistication of processing control available and the experience of the manufacturing personnel in dealing with less than desirable conditions.

Recent developments in evaluating the influence of absorbed moisture and its effects on cure kinetic and volatile gas evolution have allowed the implementation of temperature profiles and pressure controls to minimize void formation. Prepreg hot-melt systems (100% solids) can work a lot better than solution-coated systems (70% solids). Hot-melt systems can be a major factor in acquiring better parts with fewer cure kinetic, volatile gas, and moisture pick-up problems.

Dust particles must be limited because they become the nucleating sites for void formation from moisture and entrained air. Aerosols must be controlled as they may act as releasing interfaces between plies, resulting in crack formation under low loading conditions, or they may completely prevent adjacent plies of material from curing together.

The typical environmental conditions employed in industry generally conform to MIL-A-83377B (Ref 1), which provides for a temperature range of 20 to 30 °C (65 to 85 °F), a humidity of 60% RH maximum, and a filtered air supply. This specification, as well as other industrial standards, provide for the establishment of controlled bond room standards at each vendor facility, accounting for all environmental conditions known to influence materials processing. Although military specifications should be used as a guideline whenever possible, many have not been revised to account for advanced technology and current composite materials.

The responsibility of quality control personnel is to ensure that these conditions are met and to verify maintenance of these standards by periodic inspections. This is usually performed at specified time intervals, and the instrument readings, equipment maintenance records, and calibration records are kept in a log book. It is desirable to have automatic recording devices for temperature, humidity, and positive pressure.

A separation of facilities must also be maintained to provide acceptable fabrication conditions. The tool preparation area should be remote from lay-up or bond/assembly areas so that cleaning material and release agents will not migrate into these more sensitive areas. It is highly advisable to establish a formal procedure for cleaning and preparing tools and to verify this process before a tool is allowed to enter another controlled-environment area. A specific concern is the verification of those cures necessary to cure or bake release agents. In addition, any operation that generates dust or aerosols, such as trimming, sanding, grinding, or drilling of materials, should be isolated in a separate facility with its own dust collection system to prevent contamination of lay-up materials and cleaned tools. Most fabricators will also establish controls for using air tools, isolating them from any lay-up or bonding/clean rooms, and/or providing a dry (-20 °C, or 0 °F condensation point), oil-free air or a vacuum exhaust system whenever they are needed in a clean room area.

Processing equipment must also be able to meet established processing standards and verify that all control and recording instrumentation is within current calibration requirements. Standard practice usually conforms to government specifications that specify minimum acceptable practices for calibrating measuring equipment and maintaining traceability of the standards used in the calibration process, such as MIL-STD-45662. Unless all control and recording equipment is in current calibration, verification of the process will be in doubt. The calibration systems used must be defined, and calibration procedures must be in written form.

Standard practices for instrumentation hookups, thermocouple attachment, ultrasonic and dielectric cure monitor use, and cure process verification should be established and followed faithfully. What may appear to be minor variations in placement of thermocouples into or onto a part surface can have very large effects on the temperature to be sensed. Thermocouple lags and variations must be controlled manually or with computer programs to hold all thermocouples within established limits. Autoclave air temperature, tool temperature, and part temperature, during various stages of processing, can vary by 55 °C (100 °F) or more during initial heating up to the processing temperature. A thermocouple that does not provide reliable data can cause the autoclave operator or a computer process controller to make errors which could cause the loss of the entire load. Proper attention to insulation and isolation of the thermocouple is essential. Quality control personnel should verify correct placement and installation of all sensors.

Because equipment is subject to failure, it is advisable to have a back-up control system or recording system available during the cure cycle, even if it is just a certified equipment operator with a clock and a thermocouple reader to record performance of the equipment and adjust the processing cycle.

Material Control

One of the most significant areas in which in-process quality control personnel can be effective in preventing production problems is material control. All aircraft and aerospace companies must have strict, in-place controls to provide material traceability, conformance to shelf life and out-time specifications, and other procedures to ensure that prepreg materials and adhesives remain suitable for the intended purpose. In general, material control systems conform to MIL-I-45208, MIL-Q-9858A or NASA NHB-5300.4. The fabricating facility usually issues a formal quality control manual containing details on the procedure for maintaining and transferring traceability records as materials are consumed in the lay-up of parts. It is also usual procedure for manufacturing work orders (MWOs) to require the verification of materials before their use in a lay-up.

For receiving inspection personnel, this verification involves documenting the source of the material; the specification it was procured to, and the fact that it was tested and accepted to specification requirements; whether it is still within current shelf life and out-time limits; and whether it was maintained at the required storage temperature during transit. Additionally, the material must be of the appropriate type, class, and grade, as called out in the part drawing. Materials not in conformance are not to be used. The use of technical source inspection also could be very cost effective for maintaining quality levels without extensive receiving inspection testing.

As part of in-process surveillance, periodic verification should be made and recorded of the storage conditions of the material. A log book is usually kept to record these inspections. Most prepregs require storage at -20 °C (0 °F) in a sealed polyethylene bag and have a shelf life of 6 months to 2 years under such conditions. A few materials may be tolerant of storage temperatures of 5 °C (40 °F), with a shelf life of 3 months to 1 year. The material specification is usually the controlling document in establishing storage conditions. Storage life is the time during which the material continues to meet all specification criteria. Most material specifications have provisions for retesting to extend the shelf life of materials if they test acceptably after exceeding the first shelf life period. A typical extension period would be half of the original shelf life, with only two extensions allowed. The working life or out-time of a material is the time during which it can be used in a controlled clean room environment and still remain suitable for processing. This is usually judged by its level of tack and drapeability and by the mechanical properties still remaining within material specification limits. The material or process specifications usually contain this requirement for an out-time limit.

Another item of quality control verification is that the material was properly handled in the lay-up room. Good practice dictates that the material be allowed to warm to room temperature in the original sealed package before opening, to prevent surface moisture condensation. Moisture condensed on the surface of prepregs and adhesives is an excellent source for void formation. There are suitable techniques for removing moisture from lay-ups before curing, but each situation should be evaluated by a materials and process engineer to ensure that the most effective practice is used in each instance. As part of routine surveillance, quality control personnel should be on the alert for operations that deviate from good shop practice and should note these on the MWO or other buy-off record for the part. Problem areas include sealing bags damaged during handling, material rolls allowed to rest on adjacent rolls of material (they must be suspended by the ends), and disregard for first-in, first-out practices of inventory control.

Maintaining Traceability. Prepreg materials and film adhesives may be precut to specified ply patterns before being placed in the lay-up mold. In some instances, it may be necessary to package and store kits of prepreg for later use. Quality control personnel must transfer traceability of the material from the original source identification tag to the ply pattern piece. In addition to the requirement for original material information, there is also a need to record the drawing number, ply number, and sequence number, if multiple pieces are required to make up a full ply. This is so the fabricators will know precisely where to place a piece in the lay-up when the kit of cut prepreg is subsequently removed from storage for use. Ultrasonic cutting knives or other computer numerically controlled (CNC) cutting devices are frequently used to generate proper patterns for lay-ups, but manual cutting operations are still used in many shops for short production runs or shops making prototypes. To maintain traceability, the system identification tag should contain the following information:

- Material
- Vendor and vendor identification
- Lot/batch
- Date of manufacture
- Part number
- Ply number, pattern number, and sequence number
- Quality control verification and date

As already indicated, all data relevant to material traceability must be transferred to the MWO or buy-off record for the part, so that the trail of traceability from receiving inspection to its ultimate consumption in a fabricated piece of hardware is complete.

It is essential that proper support and handling practices be used if kits of prepreg have been cut, identified, and stored. Generally, ply pattern pieces are stored on a flat supporting structure so that they do not become distorted during storage and handling. When removed from storage, the pattern pieces are warmed to room temperature before the bag is opened, to prevent moisture condensation. Apart from these considerations, they are handled as any cut piece of prepreg.

The size of a package of prepreg, whether uncut on a roll or precut to pattern pieces, must be kept consistent with the handling equipment available. A great amount of damage to fabric prepregs can occur when a roll is allowed to stand on end or when a ply pattern is twisted in handling to place it in a lay-up mold. A distorted fiber pattern creates such problems as off-axis fiber orientation, misfit ply patterns, and loss of structural properties in the finished part.

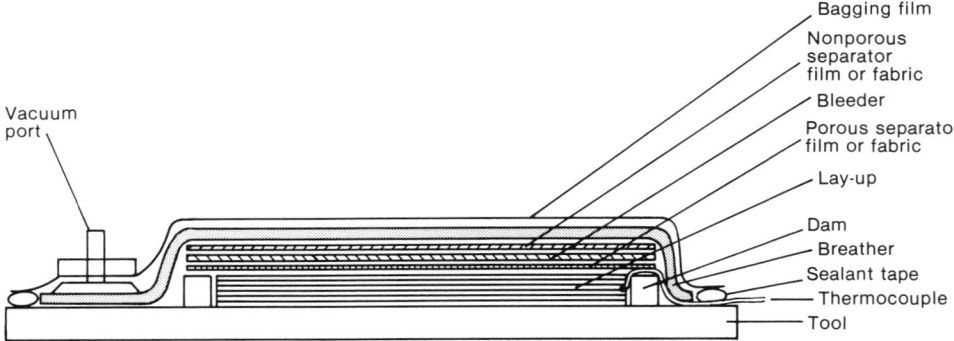

Fig. 1 Typical bagging sequence for lay-ups. Allow one vacuum port for each 0.9 m² (10 ft²) of part surface. Install one thermocouple for each 0.9 m² (10 ft²) of part surface. For a vacuum bag integrity check, draw maximum vacuum on part, disconnect from vacuum source, and measure leak rate for more than 30 min. Leak rate should not exceed 1690 Pa (0.5 in. Hg) per minute.

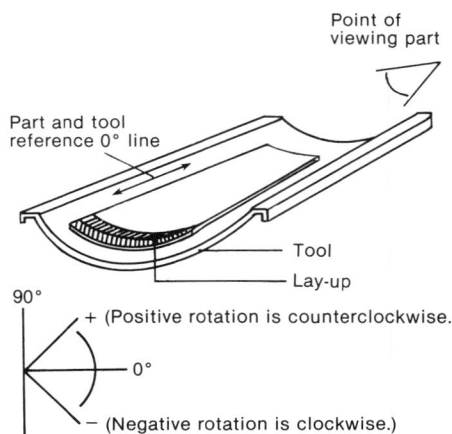

Fig. 2 Both the designer and the fabricator are instructed to view the part from the surface to be tooled inward through its thickness as viewed from the inside. A table is frequently supplied to describe ply number, orientation, and prepreg material.

Additional aspects of material control that could affect part quality if not properly managed are the use of controlled (specification certified) and uncontrolled (nonspecification) materials in the same area and the use of indirect material (discussed below) in processing the lay-up. Prepreg materials, especially when examined only casually, look remarkably alike, and unless strict adherence to material traceability is maintained, mix-ups will occur. Prepregs of 210, 280, or 340 GPa (30, 40, or 50×10^6 psi) modulus fiber cannot be visually distinguished. Likewise, weave styles are easily mistaken. It is advisable to completely segregate the use of controlled materials from uncontrolled materials so that these errors are minimized. Common practices are to require contrasting color backing film for each prepreg and adhesive system and to segregate prototype and production operations.

Indirect processing materials are used to control resin, fiber, and void content of the finished laminate. A typical laminate bagged for processing is illustrated in Fig. 1. The substitution of one bleeder material for another could cause a very large difference in the fiber volume of the finished part. These indirect materials need to be precisely the same as those allowed in processing and recorded in the processing documents, and need to be used in the quantities specified to achieve the desired results. To illustrate this more fully, 120-style glass fabric, when used as a bleeder material, removes resin at approximately 95 g/m² (3 oz/yd²) at full saturation, whereas 181-style glass fabric removes approximately 160 g/m² (5 oz/yd²) under the same conditions. Nonwoven polyester materials frequently used for bleeder show similar variations.

There are two consequences to incorrect selection of material for bleeder or use of the wrong number of plies of bleeder. In the case of an insufficient number of plies of bleeder, high resin content and possible saturation of the breather material can result in loss of vacuum throughout the laminate with entrapment of air and other volatiles within the lay-up during cure. No escape path for volatiles and gases is available, which can result in highly porous parts. In the case of an excessive number of plies of bleeder, failure to achieve a fluid hydrostatic head on the resin during processing, which is necessary for laminate consolidations, can result in a high void content within the laminate. This second condition is frequently described as being "over bled." However, after analysis by resin digestion, the resulting fiber volume is frequently found to be near the nominal range for processing the prepreg in high-quality laminates; only the void content is out of range. Unless a hydrostatic head that is greater than the volatile gas pressure can be developed in the liquid resin during cure, bubble formation will occur, resulting in unacceptable laminates with high void content. Attention to these details in the use of indirect materials is just as important as controlling the direct materials.

Pattern Orientation Control

Orientation of prepreg fiber or fabric patterns in the lay-up precedes the design stage of a composite part. Designers; stress analysts; materials and process engineers; production planners; and tooling, fabrication, and quality control personnel must all have agreed on a convention for interpreting ply orientation.

Typically, the designer and stress analyst will design a part from its outer material limit inward through its thickness. A tool designer views the part from the surface against the mold and fabricates the mold outward from that surface. Additionally, the tool designer must match the coefficient of thermal expansion of the part and tool to achieve the desired dimensions in the part. The planner and fabricator consider the part as it progresses from the tool surface through its thickness during lay-up. This means that two or three reversals in point of view can occur from idea conception to fabrication. Unless the person building the lay-up is interpreting the part as the designer intended, errors in ply lay-up orientation and structural performance are likely to occur. They are less significant in simple, balanced laminates, but become substantial in complex or nonsymmetrical laminates.

The part surface that will be tooled, or, in the event of co-curing, the precured interface surface is the most common place to start. Figure 2 illustrates one possible orientation system, but many others can be used. Once a convention has been selected, it must be remembered that a 0° reference must be defined for each part. This 0° reference must be shown on the part drawing and must also be visually present in the tool so that the lay-up can be properly fabricated.

Manufacturing methods for any particular prepreg material differ very little from one company to another. Processing dynamics must be the same if one is to achieve the same results as any other facility. It is essential that all personnel performing hands-on fabrication work and those inspecting this work be trained in the shop practices required to produce the desired results. Testing and certifying fabricators and inspectors is recommended in order to prove their capabilities to read and interpret drawings, perform lay-up work to prescribed drawings and process specifications, and complete the documentation required to verify compliance with requirements. Obviously, these functions require different levels of proficiency. The use of certified fabricators and inspectors will minimize errors and subsequent part reworking or scrapping.

A formal system to allow quality control personnel to perform in-process surveillance inspection must be instituted. Such a system needs to specify the frequency with which an inspector is to examine and verify ply orientation; material traceability and quality; and the continued use of good shop practice in lay-up, debulking, prebleeding, bagging, and curing. Each specified inspection must be permanently recorded on an MWO, buy-off record, or other specified record. It is common practice to assign some of the responsibility for verifying material use and ply orientation to the lead person or shop floor supervisor, with quality control personnel verifying only the maintenance of these fabrication records. This practice may expedite some work but has adverse consequences if it is misused. The preferred system is to have inspectors who are independent of the production process verify the work.

Inspection Techniques

The precise inspection technique for examining lay-up and ply orientation control is influenced by the manufacturing methods and practices employed in the shop. Most ply orientation patterns are established on the cutting table. When a ply is brought to and positioned on the lay-up mold, the fiber or fabric orientation is aligned with the 0° reference line of the tool and draped into it. The warp yarn, in the case of fabric prepregs, or fiber direction, in the case of tape prepregs, is matched to the 0° reference of the tool and part. Unless otherwise specified, it is usual practice to let the pattern run out naturally and let the angle of orientation change naturally along the part contour (Fig. 3). A part drawing is needed to indicate the place at which orientation control is most critical, and the ply should be positioned accordingly. Most shops allow a placement tolerance of ±5° from the specified angle. This is easily measured by transparent templates or protractors. Ply locator templates are used to identify the location of individual plies or repeating patterns of plies.

If cutting and splicing are necessary to hold orientation, then the drawing or process specification must define the allowed cutting and splicing practice to avoid compromising structural integrity. Splicing techniques usually require an overlap of 40 to 50 mm (1.5 to 2 in.) along the cut fiber, or a tape strap splice that is 40- to 50-mm (1.5- to 2-in.) wide over cut plies that are butted together, as shown in Fig. 4. In these circumstances, inspectors must verify proper orientation and splicing technique, the location of cuts and splices, and their conformance to specified practices. Most splicing practices do not allow stacking one splice on top of another. Each splice is required to be off-set from the previous splice by a certain minimum dimension. Operations that cause fabric weave pattern distortion or fiber buckling, especially in tape lay-ups, are usually prohibited. Likewise, lay-up practices that result in fiber bridging the radii are not allowed. All radii should be examined for bridging, and the ply should be repositioned to eliminate this before covering with the next ply or performing a debulking operation. Once any consolidation of a lay-up is performed, elimination of bridging is very difficult.

When multiple patterns and discontinuous plies are used in a lay-up, a certain ply may exist only at a specific rib, spar, or stringer station and

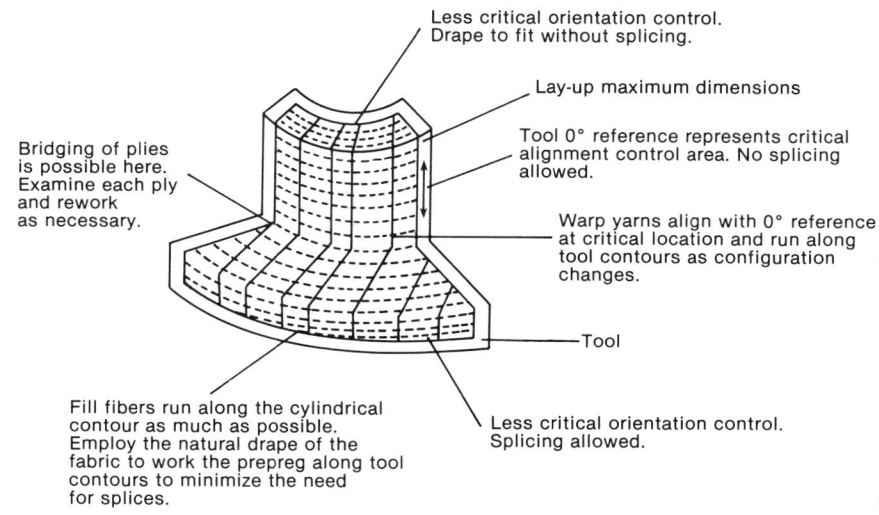

Fig. 3 Natural run-out of fiber orientation

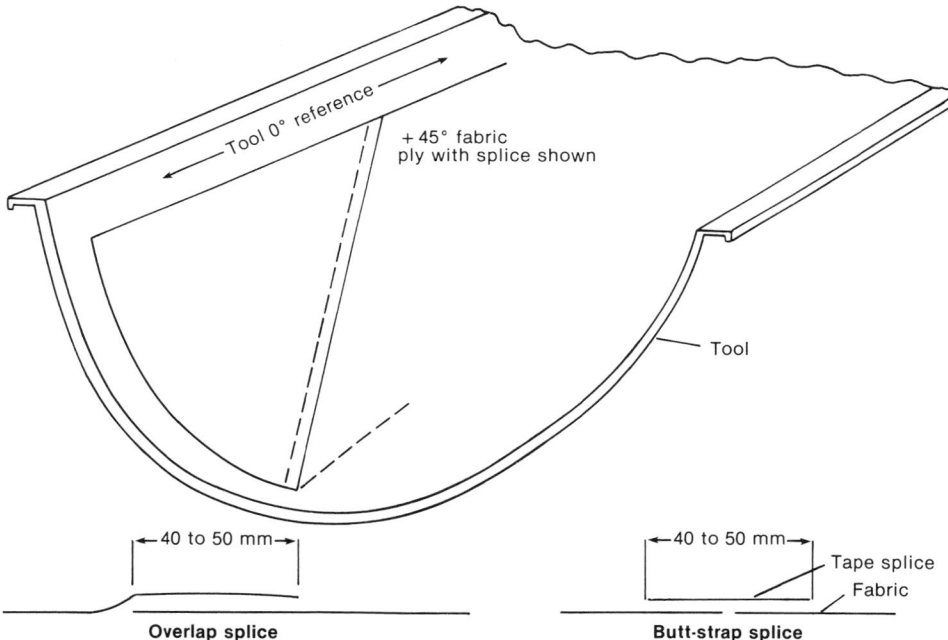

Fig. 4 Splicing of fabric and tape prepregs. Overlap splice is used for fabric at all ply orientations, and for tape in a 0° direction. Butt-strap splice is used for fabric in which only warp or fill fibers have been cut. Tape prepreg is aligned parallel to the cut fibers.

not be continuous throughout the lay-up, particularly on large parts such as a wing skin panel or fuselage skin panel. An inspector must be available to verify that each piece is placed at the prescribed location at the precise angle orientation, and to verify patterns after they are placed on the lay-up tool. It is very difficult to pull up misplaced plies without risking damage and distortion of all plies in the lay-up. The inspector must be on site to verify the work as it is performed during these critical operations. This is even true for automated lay-up operations. Not only angles, but overlap and gap tolerance between segments of tape must also be verified as having been performed correctly.

There is an opportunity for after-the-fact verification of correct ply orientation and number of plies placed in a lay-up if a metal wire tracer has been used in the tape or fabric. An x-ray of the lay-up or cured laminate allows determination of angles from the shadows cast by the wire, and the number of tracers counted at any one angle can be used to verify the ply count. This obviously has limitations but is better than thickness measurement as the only indication of ply count.

Using tracers is one of the best techniques for ensuring correct orientation of fabric or tape prepreg in a lay-up. A common practice is to use aramid yarn tracers in graphite or carbon fiber fabric. The warp surface tracers are usually placed at 50-mm (2-in.) intervals, and the fill surface tracers are placed at 150-mm (6-in.) intervals. Carbon, dyed nylon, glass, or polyester yarns can also be used in aramid or glass materials. In selecting the tracer, its possible shrinkage during the cure cycle, which may result in unacceptable distortion of the material, needs to be considered.

In some lay-ups, the practice of offsetting each ply by a small increment (3.2 to 6.4 mm, or ⅛ to ¼ in.) allows a visual check of the number of plies and the orientation of each. However, small plies that are fully covered by the next ply cannot be examined by this method. There is no good substitute for frequent in-process inspections.

The final check on correct ply placement in a lay-up is made upon examination of the cured laminate. The inspector should be alert to any unusual distortion patterns in the part. A balanced, symmetrical laminate should exhibit little twisting or distortion. A minor amount of distortion may be expected from residual thermal stresses if the laminate experienced some temperature differences throughout its length and thickness during its cure. However, if a part is noticeably distorted and cannot easily be returned to its original mold contour with the application of a minor amount of force, then incorrect ply placement should be suspected. It is possible that gross temperature differences may have occurred during its cure, although a variance in gross temperature sufficient to cause unacceptable warpage is highly unusual.

In-Process Control of Material During Cure

Over the years several methods have been devised, improved, or discarded and replaced with new methods to verify that composite prepregs are properly processed. Each of these methods relies upon a receiving inspection function to verify first that the material conforms to a specification and that by means of the intended processing will result in parts of the desired structural properties. Once the lay-up has been completed and bagged, it is usual practice to verify vacuum integrity of the system. The interior bag pressure should be monitored during cure. Traditionally, the bag is evacuated, disconnected from the vacuum source, and the leak rate measured over a specified period of time. A leak rate of 1690 Pa (0.5 in. Hg) over a 20-min interval or less is a common standard of performance. Each facility must establish a similar standard. In-process test specimen panels and part trim tabs are commonly used to verify processing. Some shops subject the material to additional testing for tack, flow, and gel time before use, to ensure acceptable performance after storage.

In-process test coupon panels do not typically simulate the construction of all parts in an autoclave, press, or oven run because they are usually 2.54 mm (0.100 in.) thick regardless of part configuration. The only similarity is in the materials used. At best, this method only verifies that the material has the potential for being processed into acceptable hardware. More processing specifications are requiring in-process panels to be under the same bag as the part or connected by a vacuum jumper line to another bag. The concerns in using an in-process test coupon panel are:

- It represents only a small quantity of the material in the autoclave
- It usually is made by someone specializing in fabrication of test panels and therefore does not represent the skills used to fabricate the actual part
- It usually has not experienced the same out-time and processing conditions as any of the lay-ups in the batch and certainly does not experience cure conditions representative of any other part in the autoclave
- It creates extra work in preparation, testing analysis, and explaining deviations from specification requirements. These conditions may or may not have been experienced by other parts being processed
- The usual measurements of flexural strength and modulus, short beam shear strength, fiber volume, void content, and degree of cure are very difficult to relate to part performance because test panels are not totally representative of parts during the cure cycle

The use of trim tabs is a significant improvement on the in-process coupon method. A trim tab should be designed as a piece of the part being processed, specifically as an extension of its length or width, preferably at a point of maximum or typical thickness. The trim tab, by virtue of being in the same bag with the part, experiences most of the processing conditions of the part, although obviously, with large parts, it still does not represent the piece of hardware throughout its extremities. Testing trim tabs gives a fair measure of the fiber volume void content, degree of cure, and glass transition temperature if differential scanning calorimetry measurements are performed. These are better than in-process coupons, yet still not as good as using common sense, good instrumentation for the parts being processed, an excellent closed-loop feedback control system, and an understanding of the cure kinetics of the system. Because some part configurations do not allow the use of trim tabs, other techniques are needed. The test limitations of tabs are:

- Tools must afford extra space for them
- Tabs must be cut off; testing cannot begin until trimming is completed
- Irregular edge surfaces around tabs complicate bagging operation
- Mechanical values are usually based on a nominal 2.54-mm (0.100-in.) thickness

Improved Process Verification

If a part has been provided with sufficient thermocouples or other sensors, such as dielectric or ultrasonic transducers, and a cure model has been made for the resin which takes into consideration time, temperature, and viscosity transformations, and coordinates these with the dielectric or ultrasonic sensory data, then part temperature, dielectric, or ultrasonic measurements can be made during the cure cycle and compared to the model. The processing parameters can be adjusted through the closed-loop feedback control system, including management of part bag vacuum and autoclave pressure at optimum times. There can be no greater assurance of proper processing. A visual inspection and nondestructive examination after cure would rapidly verify part quality. The records from the instruments would be sufficient proof of success in achieving the desired results. Because not every shop has this equipment, other methods that are not as capital intensive will continue to be used.

Instrumentation and control systems must be kept within a state of current calibration. In-process verification of the cure cycle is only as good as the instruments and control systems on which it is based. This is becoming increasingly important as new computer-aided process control systems are being installed at many plants throughout the advanced composites industry. Most process specifications will have prescribed limits of temperature variation at any point in the heating cycle, a ramp rate in degrees/minute, a time limit for any hold temperature in the cure cycle, and specified times and/or temperatures at which the vacuum or pressure is to be adjusted. These conditions need to be recorded throughout the cure cycle to verify compliance of the cure conditions to the specification. Time/temperature transformations of the material, part processing profiles, and tool processing profiles need to be understood to know how the material, the tool, and the lay-up will respond in a given processing environment. This knowledge is needed to devise an adequate cure profile for any part from any prepreg. This is not to say that composite parts cannot be produced without a closed-loop feedback control system managed by a computer, because composite hardware has been and will continue to be processed successfully by employing manual control systems and relying on operator experience to achieve the desired results. However, materials science and engineering have come of age, and better, more reliable process controls are available that depend less on art and more heavily on known, manageable engineering principles to provide the best in-process quality control.

REFERENCE

1. "Requirements for Adhesive Bonding (Structural) for Aerospace and Other Systems," MIL-A-83377B

Cure Quality Control

Richard W. Roberts, General Dynamics, Convair Division

THE USE OF COMPOSITES by any manufacturing organization or facility causes that organization to shift its emphasis from being a material user to being a material producer. This is a fundamental shift in the philosophy of manufacturing. The manufacturer, by becoming a material producer, is much more responsible for the properties of the final product than he was as a user. As a user of "classical" materials—iron, steel, aluminum, and other metals in general—the manufacturer is responsible within a single order of magnitude for material properties. For example, changes in modulus from 68.2 GPa (9.9×10^6 psi) in an untreated state to 71 GPa (10.3×10^6 psi) in a treated state usually result from processes such as heat treating, shot peening, and other manipulative methods. As a manufacturer of composite materials, however, that same manufacturer is now responsible for a change of up to six orders of magnitude in the material properties. For example, in the transverse direction of a 0° oriented laminate, the tensile modulus of an uncured composite is 7 kPa (1 psi). After processing and curing, however, the modulus is on the order of 19.9 GPa (2.89×10^6 psi). Most, if not all, of this change occurs during the cure stage of the manufacturing process. Taken in this context, the cure of composites becomes the most important phase in developing acceptable strengths and properties in final parts. Properly cured, the composite has strengths and properties far superior to those of classical materials. Improperly cured, the composite is of no use for structural applications and has no realistic function.

Since the cure of composites is so important for the desired properties, industry has taken a very conservative approach toward the parameters on, and the cure of, these materials. The reasoning behind the current industry approach and the newer approaches being studied will be presented within this article. Some insights into process failures that occasionally occur and ways to prevent them will also be discussed. This article will use epoxy resin systems as the basis for discussion, although the cure considerations of phenolics, polyimides, bismaleimides, and thermoplastics will also be mentioned, where they are similar. The primary

process equipment that will be discussed is the autoclave, although the concepts presented are essentially universal to all kinds of curing devices. Any technical differences will be noted as appropriate.

Curing will be examined by observing the effect of heat transfer on resin cure properties, the effects of heat and age on chemical kinetics of the resin, the effects of chemical kinetics and heat transfer on flow, the effect of flow on hydrostatic pressure and volatile-gas bubble nucleation, and ways to "trick" the resin into showing the effect of increased fluid hydrostatic pressure by diffusion control techniques. Every facet of the chemical reaction of the resin is intricately involved in producing adequately cured resin.

Appropriate tests must be run to ensure that a particular piece of preimpregnated composite

material yields acceptable parts. Traditional tests involve making a panel and performing various mechanical tests, such as tensile, flexure, short-beam shear, and so forth, as well as measuring inherent properties such as tack, resin content, flow, and volatile content. Chemical tests, such as high-pressure liquid chromatography (HPLC), thermogravimetric analysis (TGA), differential scanning calorimetry (DSC), rheology, infrared spectroscopy, and so on, are still considered risky for production use. These tests show the chemical and physical state of the uncured resin. They can indicate the chemical advancement of the resin system, whether the resin is at the proper stage for cure, the ratio of reactants to reaction products, at what temperature the volatiles will evolve, the viscosity, and whether the supplier has sent the correct material. Each of these

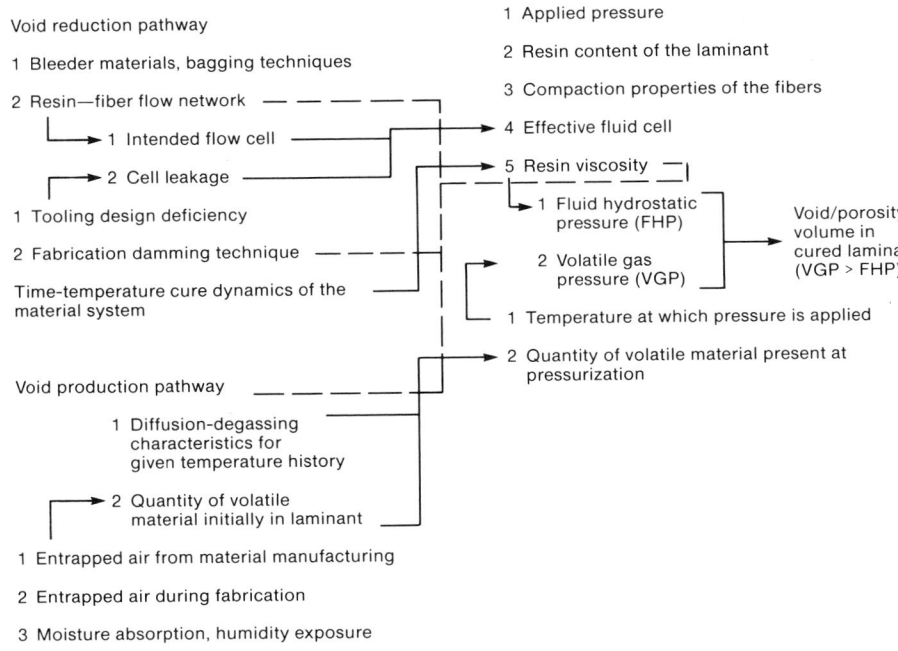

Fig. 1 Model of laminate process behavior. Source: Applied Polymer Technology, Inc.

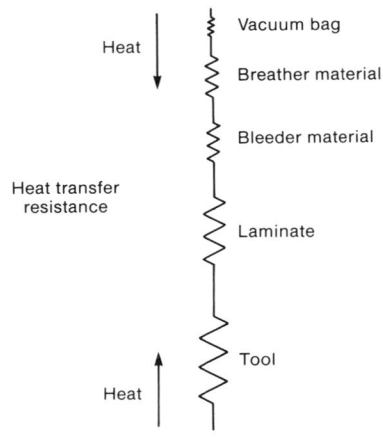

Fig. 2 Heat transfer resistance. Source: General Dynamics, Convair Division

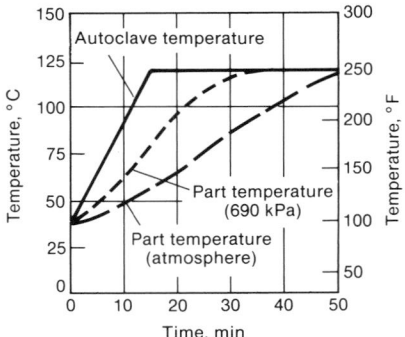

Fig. 3 Effect of pressure on heat rate. Autoclave heated at 6 °C/min (11 °F/min). Atmospheric and 690 kPa (100 psig) pressure. Source: General Dynamics, Convair Division

properties will be shown to have significant impact on the success of the completed part.

A model of laminate process behavior is shown in Fig. 1. It is presented at this point to provide a guide through the chemical cure process that is being considered. It will also help to decide what makes a good laminate. A good laminate has these characteristics: correct ply orientation, correct fiber-to-resin ratio (resin content), correct compaction of the fibers, low void and porosity volume, and correct degree of cure.

Correct ply orientation is a function of the lay-up process, and nothing in the cure or chemical processes can possibly affect a change in ply orientation. Correct resin content is a function of both viscosity and flow during the chemical reaction and the design and placement of bleeder materials. Low void/porosity volume is a result of flow properties, volatile content, viscosity, and fluid hydrostatic pressure during cure. Correct compaction of the fibers is a function of flow, viscosity, applied pressure, and fluid hydrostatic pressure. Correct degree of cure is the function of resin viscosity, temperature, time, and chemical reactions required to cause polymerization of the resin. Three of these characteristics are directly related to the chemical reactions that occur during the cure of the resin.

In terms of the correct degree of cure, cure has been defined as the overall transformation from a low molecular weight resin/hardener system to a polymer network cross-linked through chemical reactions. A correct cure would be one that, during this reaction, progresses to the required state of polymerization and cross-linking to obtain the proper and/or anticipated material properties. Careful consideration of the cure cycle is necessary to obtain the desired properties in the composite.

Heat Transfer Theory

Many heat transfer processes occur during the thermal portion of a cure cycle. The term heat will be used to include not only heating but also cooling, as it is the same process but the flow is in the opposite direction. Heat transfer includes the transfer of heat from the heating medium, generally gas in an autoclave or platen surfaces in presses, to the tool and bag; from the tool, bleeder, breather, and bag to the laminate; and from the outside of the laminate to the inner portions of the laminate. Heat transfer is often considered to be similar to a network of resistors in an electrical circuit. Each boundary and material provides a resistance to the migration of heat throughout the complete network. The concept is depicted in Fig. 2.

An increase or decrease in the resistance of the layers, bleeders, breathers, laminate, tool, and so forth, can greatly affect the ability of the laminate to be heated uniformly. Each of these heat transfers has an important effect on the chemical reaction rates of the resin. Polymeric composite resins follow the relationship:

$$k = A_e^{-(E_a/RT)} \qquad \text{(Eq 1)}$$

where k is the rate constant, A is a constant, E_a is the activation energy, R is the gas constant, and T is the temperature in absolute units (Kelvin or Rankine). This equation shows that as temperature increases, the rate of reaction increases as an exponential function of that temperature. Reactivity will be discussed in detail subsequently, but this concept shows the reason heat transfer theory is so critical.

The first instance of heat transfer in the laminate is the transfer of heat from the heating medium to the tool and bag. The rate of this transfer is related to several factors, such as delta temperature, pressure, autoclave loading, part/tool mass, and air flow. Heat transfer in autoclaves follows classical heat transfer calculations for convection and conductive heating. A sample equation, shown for reference, is:

$$q = (h_c + h_r)A(T_{air} - T_{part}) \qquad \text{(Eq 2)}$$

where q is the rate of heat flow, h_c and h_r are the coefficients for convective and conductive flow respectively, A is the affected area, and ΔT is the difference between the temperature of the air and the temperature of the part. Rigorous

solving of this equation is very difficult and is unnecessary for this application, but it illustrates the area and temperature dependence of the process.

The autoclave provides a high rate of air or gas flow over the laminate and tools. This aids in the heat transfer process by providing contact between the high-temperature air and the parts and results in convective heat transfer. As the pressure within the autoclave is raised, the density of the autoclave atmosphere increases accordingly. This increase in density significantly increases the heat transfer capability of the heating medium and follows the ideal equation:

$$\text{density} = \frac{\text{mass}}{\text{volume}} : \text{density} = \frac{n(MW)}{V} = \frac{P(MW)}{RT}$$
$$\text{(Eq 3)}$$

The increase of pressure (P), with the volume (V) and temperature (T) held constant, results in a proportional increase in the number of moles of the gas (n). MW is the molecular weight of the gas. A corresponding increase in heat capacity due to increased molecules and thereby density is also evidenced (see Fig. 3).

The difference between the part temperature and that of the heating medium, air, platen, and so on, is known as ΔT. The greater ΔT, the faster the heat-up rate will proceed (Eq 2). The following example of this effect is provided. Consider two identical autoclaves with identical parts within them at 20 °C (70 °F). Autoclave 1 is heated at 8 °C/min (15 °F/min) to 120 °C (250 °F). Autoclave 2 is heated at 3 °C/min (5 °F/min) to 120 °C (250 °F). The part in autoclave 1 will heat up much faster than the part in autoclave 2 (Fig. 4). The effect of ΔT is most evident toward the end of the heat-up ramp when the part temperature comes within approximately 11 to 14 °C (20 to 25 °F) of the autoclave ambient temperature. At this time, the heat-up rate of the part slows dramatically. The slowing effect can be so severe that it can lower the average heat-up rate below specification limits.

The effects of ΔT are most easily explained by a proportionality formula. Assuming part mass is constant:

$$\frac{T_{autoclave}}{T_{parts}} \approx HRR \qquad \text{(Eq 4)}$$

The greater the value of HRR (heat-rate ratio), the greater the heat-up rate. This is not an exact formula, but it illustrates the relationship of ΔT to heat-up rate.

This heat-up slowing can be minimized by a technique known as overshooting. The overshoot principle allows the autoclave temperature to be raised above that of the desired part temperature in an effort to maximize ΔT. As the parts approach their required temperature set point, the autoclave temperature is reduced to

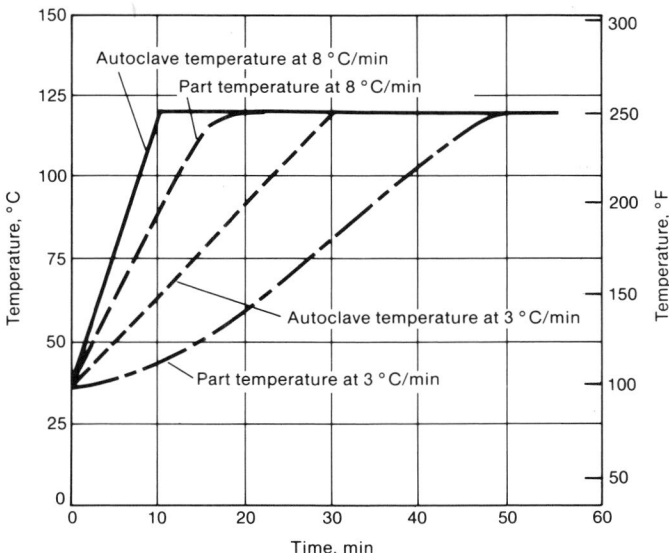

Fig. 4 Effect of autoclave heat rate on part heat rate. Source: General Dynamics, Convair Division

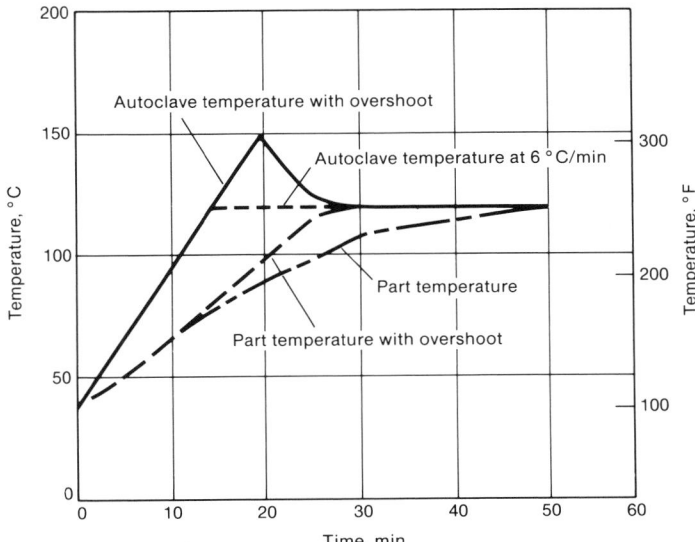

Fig. 5 Effect of autoclave overshoot on part heat rate. Source: General Dynamics, Convair Division

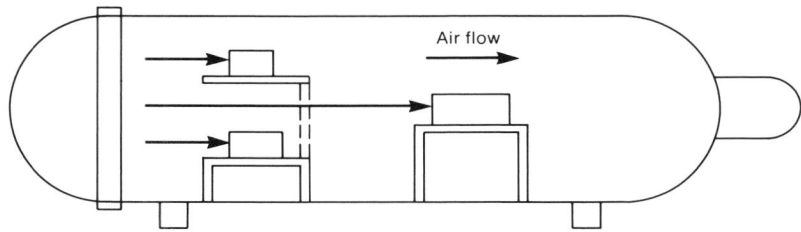

Fig. 6 Autoclave loading. Source: General Dynamics, Convair Division

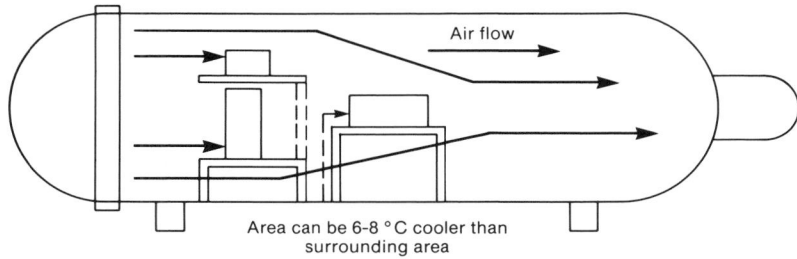

Fig. 7 Effect of shadowing. Source: General Dynamics, Convair Division

within the specification limits. This reduces ΔT and slows the heat rate at a temperature much closer to the required temperature (Fig. 5).

Although overshooting is an effective technique, it has limitations. Extreme care must be exercised to ensure that the autoclave temperature is reduced correctly to maximize the heat-up rate and ensure that the parts remain within temperature specification limits. The overshoot temperature must be selected carefully so that bagging materials and thin sections of the parts are not subjected to degradation temperatures. The reaction time of the autoclave must be well

known so that the overshoot and the subsequent reduction in temperature can be calculated. These are difficult parameters for all but the best operators to follow manually, making computer control of the autoclave desirable.

Part mass is another factor that greatly influences heat-up rate. The definition of a British thermal unit (Btu) is one of the easiest ways to explain this effect. A Btu is defined as the energy required to heat 0.45 kg (1 lb) of water 0.56 °C (1 °F) in 1 min at 16 °C (60 °F). If the mass is doubled, the energy required to heat at the same rate must also be doubled. This leads

to another important relationship: the ratio of the tool surface area to the mass of the tool/part combination. The only way for more energy to enter the tool is through the exposed area of the tool. This is a direct relationship. If the surface-area-to-mass ratio for a particular tool is the same as it is for that of another tool of different configuration but same material, they will show very similar heat rates.

Tool compatibility is also important. When tools are designed, the surface-area-to-mass ratio of the tools and the tooling material must be similar. Another way to state this requirement is that the thermal mass of the tools must be similar. Tool materials can be different as long as the thermal mass is similar. Otherwise, incompatible autoclave loads will result when dissimilar thermal mass tools are run. This occurs when one tool heats quickly while another tool in the same autoclave load heats slowly. Because of the difference in surface-to-mass ratios, it is impossible to heat both tools at a compatible heat-up rate. This has very important ramifications in the chemical reactivity, which will be discussed later.

Often, all production tools cannot be made with the same thermal mass or surface-area-to-mass ratio, such as when making graphite and metal tools. It then becomes necessary to create two different compatibility ranges for the tools, and it is important to keep the similar tools compatible among themselves, and to cure only compatible tools in the same cure load.

All of the foregoing discussion assumes that the part is exposed to uninhibited gas flow, that is, that there are no obstructions to hinder the circulation of the air or gas (Fig. 6). However, this ideal circumstance is not always met during production autoclave loads. Autoclaves should be completely full to maximize energy effi-

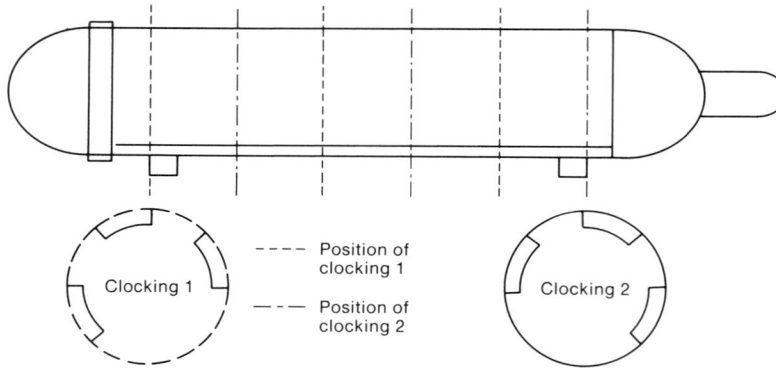

Fig. 8 Deflector arrangement. Source: General Dynamics, Convair Division

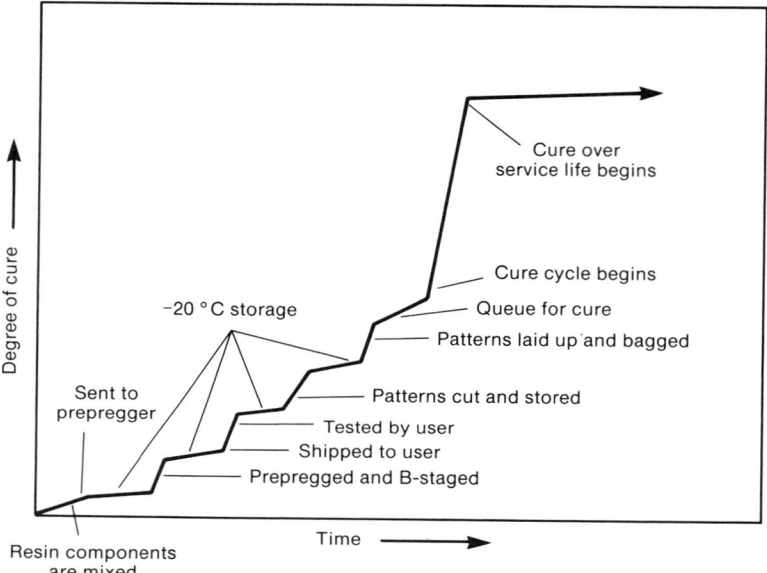

Fig. 9 Resin life cycle. Source: Applied Polymer Technology, Inc.

ciency, but this can cause impediments to circulation. Therefore, the load must be spread evenly within the autoclave. This necessitates the use of multilevel carts in the autoclave for small parts. By spreading the parts equally among the shelves and end-to-end in the autoclave, adequate air flow contact can usually be obtained. When load situations require tight packing on the shelves, the tools should be arranged so that no tool is completely shadowed (hidden) from the flow. Shadowing will result in poor heat transfer to the shadowed tool (Fig. 7).

Generally, autoclave loading should be optimized to include the greatest number of tools possible, thereby significantly reducing part cure cost. Any high-quality autoclave with an air flow in the range of 7 to 8 m/s (1350 to 1500 ft/min) and a 5.5 °C/min (10 °F/min) heat-up capability at operating pressure can be very heavily loaded. The use of air flow deflectors to

break up the laminar flow near the autoclave walls can improve heating uniformity under heavy load conditions (Fig. 8).

Most of the curing facilities for composites are autoclaves, but there are many other methods of applying heat and pressure to cure composites. These methods, such as use of platen presses, cavity or bladder presses, integrally heated tools, ovens, and so forth, while not as versatile as use of an autoclave, can be considerably more cost effective to operate. These pieces of equipment often have the advantage of improved control of heat transfer due to direct heat transfer to tooling or parts. However, they may be limited to a particular part configuration (special tool), shape (presses), or strength requirements (ovens). The same curing rules apply for any of this equipment as apply for autoclaves: The part must heat up evenly over its entire surface, the heat-up must be controllable, and the heat-up

rate must be compatible with other tools in the same load.

Chemical Kinetics

The reason that heat transfer and heat in general are so important to the cure of composites is that all currently used structural composites are heat cured, either at elevated or room temperature. The effect of heat on chemical reactions is best characterized by Eq 1, which showed that the rate of reaction is exponentially proportional to the temperature. This is most evident if one considers that this means that the rate of reaction essentially doubles for each 10 °C (18 °F) temperature rise. (Temperature affects the viscosity as well as the chemical reactivity of the resin; this effect will be more fully described in the section "Flow" in this article.) The degree of cure of a typical resin can be characterized by a time line (Fig. 9), which illustrates the relationship between the quantity of the products of reaction over time. Each steep slope indicates a time when the resin is exposed to temperatures above −20 °C (0 °F). During storage at this temperature, the chemical reaction between resin components is significantly slowed. During processing and room-temperature exposure, the reactants are more reactive but are not forming the polymer matrix desired during cure; that is, they are reacting in an undesired side reaction. During these handling and storage times, the products of reaction usually manifest themselves by a loss of prepreg tack and drapability and increased boardiness (stiffness). Extended room-temperature exposure time causes low-temperature reactions such that no significant desired properties can be obtained from the resin even after a proper high-temperature "full" cure. This chemical advancement problem is the reason for the shelf life and out-time requirements for composite prepregs.

Moisture exposure also has a significant effect on chemical reactivity and kinetics. High moisture content slows the formation of the polymer network, and increases void content and porosity. These are some of the reasons behind the moisture/exposure limit requirements applied to prepregs and resins. Much analytical work has been done to quantify the effect of moisture on epoxy resins, for example, the work at General Dynamics, Convair Division on the USAF Processing Science Program (F33615-80-C-5059) and the work of Hinrichs and Thuen, published in the SAMPE Journal, 1980.

Since chemical reactivity is related exponentially to the temperature, the heat-up rate and ΔT within the laminate are very important to obtaining acceptable composites. If a high heat-up rate is encountered by most resin systems, the reaction will not proceed in a controlled polymerization process. At an uncontrolled rate, the reaction will generally exotherm violently, which will further perpetuate the uncontrolled reaction. The end result is a laminate

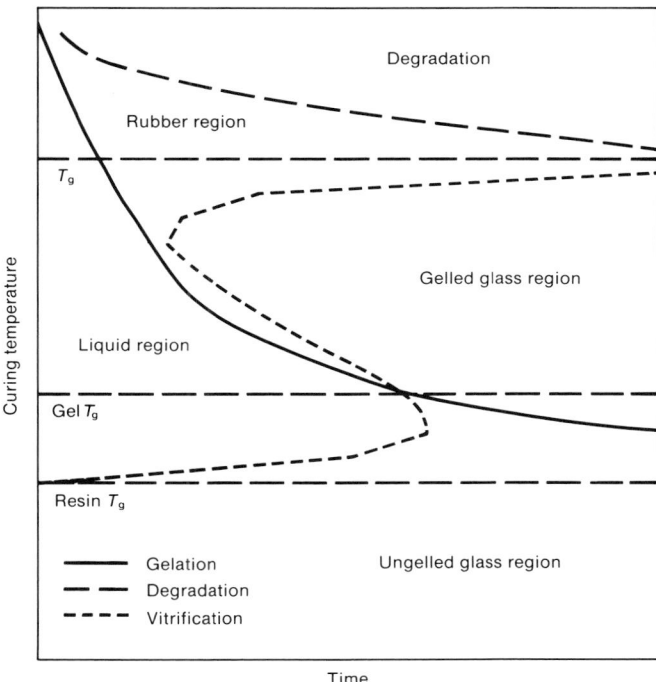

Fig. 10 Isothermal time-temperature transformation. Courtesy of John Gilham, Princeton University

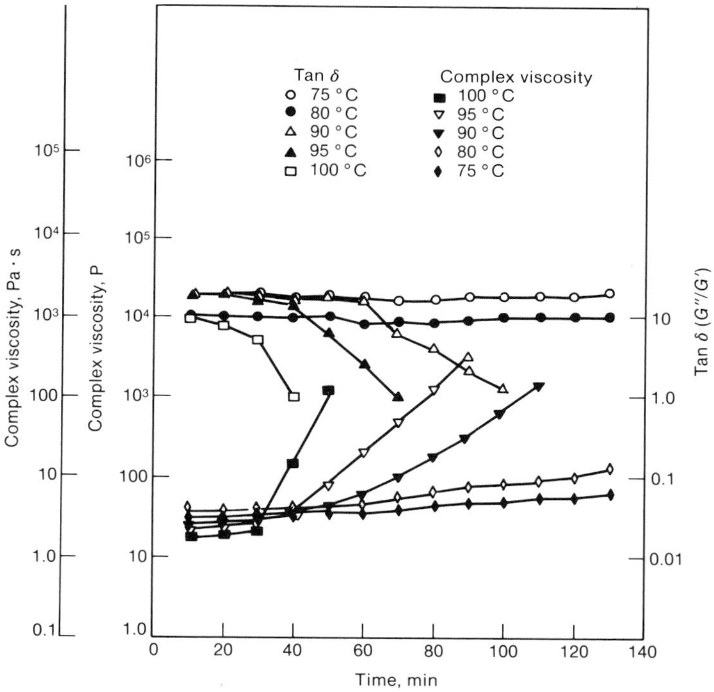

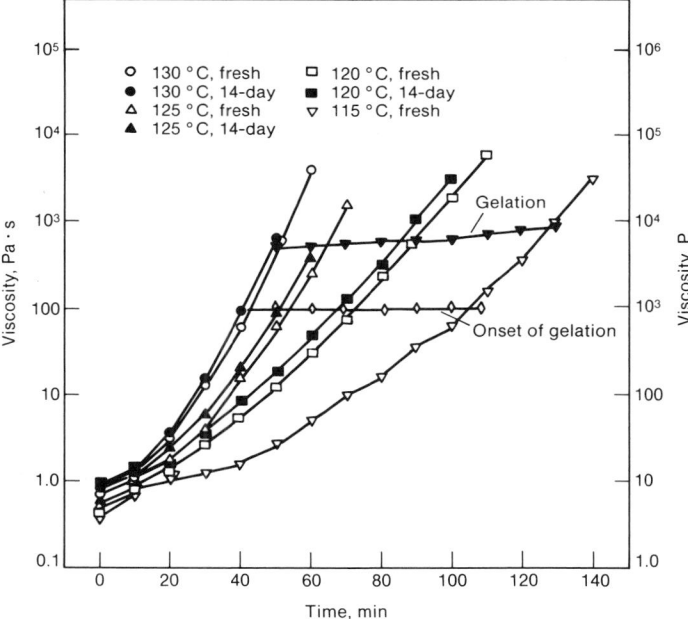

Fig. 12 Time-temperature transformation, isothermal holds. Source: Applied Polymer Technology, Inc.

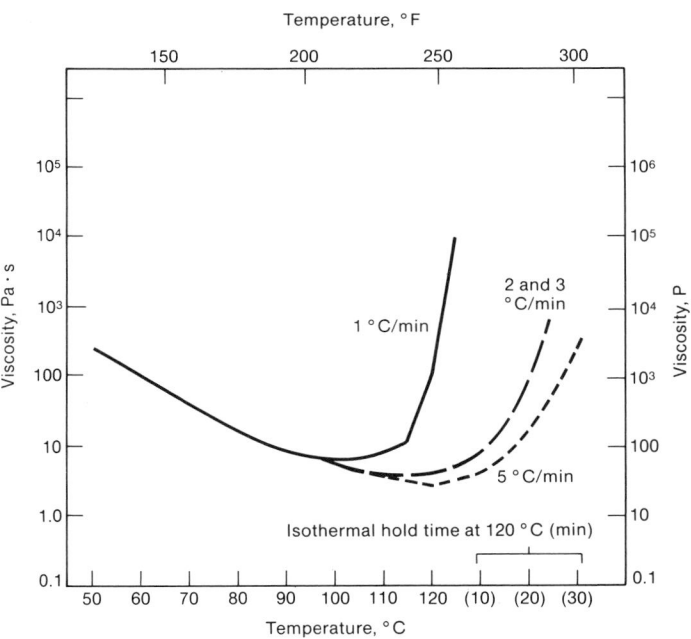

Fig. 11 Isothermal kinetic behavior, accelerated resin system. Source: Applied Polymer Technology, Inc.

Fig. 13 Effect of varying heat rates on viscosity. Source: Applied Polymer Technology, Inc.

that will not meet the properties required. On the other hand, if the heat rate is too slow, the same chemical advancement problem can result that occurs during extended handling and storage; it is thus unlikely that the correct polymerization process will occur. There are many

competing side reactions. The correct heat rate will favor the desired reactions.

An uneven temperature gradient within the laminate causes portions of the part to react at different rates. This can cause porosity, residual stresses (warping/bending), and generally

poor-quality laminates. This becomes especially important when multiple parts are run in a single autoclave. If the tools (and therefore parts) have incompatible heat rates, the parts will not have the same general state of chemical reaction at the same time during the cure cycle.

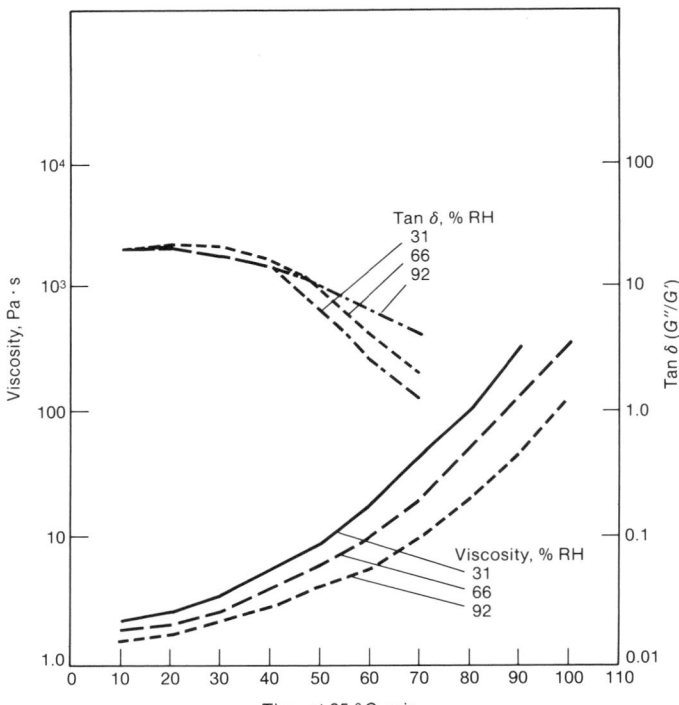

Fig. 14 Moisture effects on an accelerated epoxy resin system. Source: Applied Polymer Technology, Inc.

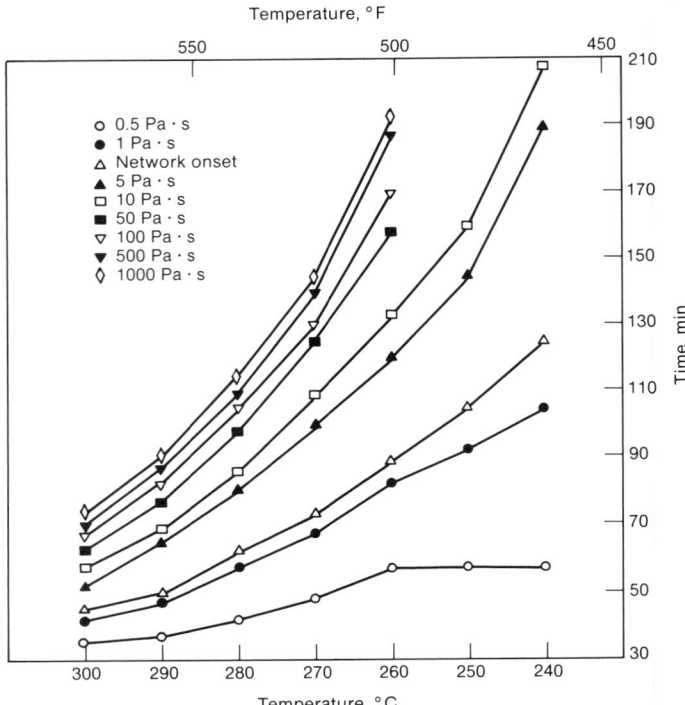

Fig. 15 Engineering transformation phase diagram. Source: Applied Polymer Technology, Inc.

This will cause problems when the pressure is applied at a point late in the cure cycle. If the application of the pressure is delayed until all parts have reached the correct temperature, some of the laminates will have already gelled. This phenomenon will be further explained in the section "Flow" in this article. This is another indication of the necessity for all tools to have similar thermal characteristics.

Chemical reactivity and kinetics are the driving factors for a successful polymerization of the composite resin matrix. Most of the cure cycle parameters are established to ensure correct chemical reactions of the resin. Heat is applied in a controlled manner to allow the resin to polymerize correctly. The application of pressure is designed to take advantage of the viscosity state of the resin that is caused by the chemical and thermal reactions. The cure cycle must be carefully regulated to obtain the desired results from the resin.

The lessons that can be learned from examining the chemical reactivity considerations of curing can be summarized as follows. First, the heat-up rate of the resin must be consistent within the same cure load in order to maintain a minimum part ΔT. Second, the heat-up rate is important to ensure that proper polymerization processes occur. Third, the chemical advancement of the resin must be controlled and understood so that the resin will be exposed to the proper cure conditions established by current industry standards. Specified cure cycles are based on an assumed maximum state of advancement that is dependent on out-times and

expiration dates. Until the advanced curing concepts discussed in the section "Advanced Cure Concepts" in this article are implemented, maintaining control of material within the time, temperature, and humidity bounds established for that material are essential for correct curing.

To further understand the chemical kinetics of resins, several analytical methods are employed. One is the use of rheology to characterize the flow in various resin systems. Rheology is defined as the reaction of material to an applied strain. Rheology measurements are generally expressed in terms of the resin viscosity at various thermal conditions. Viscosity is used to examine resin reactions because of its dependence on the degree of polymerization and its ease of measurement in the laboratory environment. Viscosity also yields two important measures of moduli, G' and G'', the elastic and storage or loss moduli, respectively. These values are important to the ability to determine gel point quantitatively, which is defined as the irreversible change that occurs at the point at which a drastic increase in the viscosity of the resin indicates the initial network formation of the polymer. After gelation occurs, no further flow occurs in the resin system; however, the cross-linking is still incomplete, and the system is reactive. It is important to determine the point of gel (T_{gel}) so that the pressure application and other processing parameters can be accomplished during the fluid portion of the cycle. This correct timing of the pressure application results in

correct compaction and void reduction pathways.

One of the mechanisms for understanding the chemical kinetics of a resin system is the experimental determination of its time-temperature transformation (TTT) diagrams. These diagrams are made by subjecting the resin to a known set of conditions and measuring its response, usually by rheometric methods. The resin is subjected to different heat rates, isothermal holds, degrees of moisture, and age exposures. The results are usually plotted as viscosity versus time at various temperatures (Fig. 10 to 15).

Figure 10 represents a TTT diagram for the entire cure process at isothermal temperatures (as this is an illustrative graph only, there is no scale in either direction). The major points illustrated are the relationships between the liquid, ungelled glass, gelled glass, rubber, and degradation regions of a resin. The vitrification line is the demarcation line between the liquid/rubber regions and the gelled (solid) regions. This graph is important because it represents several areas in the life of a resin. The resin T_g, the glass transition temperature, or temperature at which the resin is an ungelled solid, represents the storage temperature of the resin. In this region very little reaction occurs. However, if the storage time is extremely long, the resin will not enter the liquid region, which is necessary to initiate cure.

The ideal region for cure is when the resin is in the liquid state early in the cure process. In this area on the TTT graph, the molecules of the resin have approximately 10^{11} collisions per second. This is the most reactive region represented on the graph. As the temperature is held within the liquid region for a controlled length of time, the resin approaches gelation. If the temperature of the resin rises above T_g, the resin enters a rubbery region in which the molecules are still reactive but are significantly slower. No matter what state the resin is in, at some point it undergoes vitrification. This is the point at which the liquid resin transforms to a gelled glass, and the number of collisions per second falls to approximately 10^{-3}. A significant drop in reaction rate always follows vitrification, because of a change in the reaction mechanism after gel. It is important to avoid vitrification early in the cure cycle because of this significant reduction in resin reactivity. Vitrification can be avoided by heating into the liquid region early in the cure cycle. Two other significant points shown by this graph are the point at which the resin will never gel at an isothermal, $T_{g\infty}$ (infinite), and the area above the rubber region at which the resin begins to degrade.

Figure 11 illustrates the isothermal kinetics behavior for an accelerated epoxy system. The left axis is a measure of viscosity, and the right axis is a measure of network formation where the smaller the number, the more advanced the network (both scales are logarithmic). This chart shows the different reaction rates for isothermal holds at 75, 80, 90, 95, and 100 °C (170, 180, 190, 200, and 210 °F). The fastest reaction and gelation occurs at 100 °C (210 °F) and 40 min and becomes progressively (exponentially) slower as the temperature drops. This is an excellent representation of the effect of temperature on kinetics.

Figure 12 also illustrates the effect of temperature on kinetics. This time the isothermic holds are at 130, 125, 120, and 115 °C (270, 260, 250, and 240 °F). This chart also shows the effects of 14-day aging of a specimen at constant conditions. The aged specimen shows a reduction in time to onset of gelation (the point at which network formation is significantly growing) and, then, gelation. This is an illustration of the importance of carefully controlling resin out-time to ensure a sufficient cure.

Figure 13 shows the effect of various heat-up rates on the same epoxy resin specimen. At low heating rates, the resin never reaches the lowest viscosity point that the higher rates reach, because of the increased chemical activity that occurs before the heat-induced flow can occur. The effect can be poor flow and porosity remaining within the laminate. This is another example of the necessity for compatible part heat-up rates in cure cycles. These differences in reaction rates of the specimen can cause part of the load to enter gel before another part enters its area of minimum viscosity.

Figure 14 shows the effect of moisture on an accelerated epoxy resin system. The higher the moisture content, the slower the reaction of the

resin. The moisture acts as an inhibitor to the reaction sites on the accelerator for the resin system. This is an example of the reason it is crucial to limit the amount of moisture exposure of a prepreg.

Figure 15 shows the engineering transformation (ET) phase diagram for a current resin system, formulated using the data from TTT charts. The ET phase diagram is useful in the determination of cure cycles. The use of isothermal charts to determine actual cure cycle is difficult because there is no way to attain an isothermal condition in the production environment. However, these isothermal charts do provide important information on the system kinetics that is used to create the ET phase diagram. By using the isothermal lines on the graph, processing windows can be found that will determine optimal processing of the subject materials, and information is made available to determine the maximum ΔT condition that should be allowed to exist within a load or part.

Figure 15 is a good representation of the reason current epoxy cures have low-temperature holds. For example, the size of the processing window at 125 to 130 °C (260 to 270 °F) from 1 to 100 Pa · s (10 to 1000 P) is much larger than the window at 140 to 150 °C (290 to 300 °F) and the same viscosity. This leaves a much larger margin of safety to cure parts using current methods. This use of the safer cure method to ensure proper resin cure is indicative of the lack of faith that industry has in the ability to cure composites. By introducing a higher level of control and implementing more aggressive cure parameters, industry could realize significant cost savings.

A typical epoxy cure cycle (for a 175 °C, or 350 °F curing epoxy) is shown for reference in Fig. 16. The foregoing graphs of chemical kinetic studies are useful in the prediction of resin behavior in the production environment. They also provide valuable insight into the need for proper control of temperature in the curing process. The kinetics of the resin chemical reaction is very complex and difficult to understand. The most important feature to remember about the implementation of chemical kinetics data in a production atmosphere is the absolute dependence of the reaction on temperature. All other features and characteristics are outgrowths of this relationship and should be considered in that light.

Sample Chemical Reactions

Because there are no typical reactions to represent all resin systems, a typical reaction will be shown for epoxies, polyimides, polyesters, bismaleimides, and phenolics. First, reviewing resin and polymer chemistry briefly, there are three types of polymer structures: linear, branched, and networked or cross-linked (Fig. 17 to 19).

Linear polymers are formed by the reaction of difunctional monomers, such as ethylene,

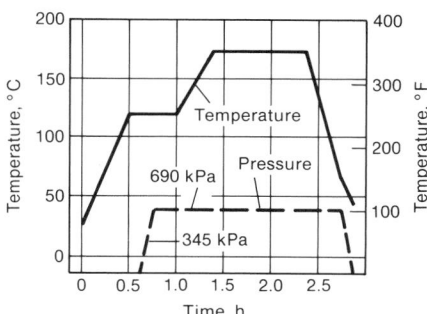

Fig. 16 Typical 175 °C (350 °F) resin cure cycle. Source: General Dynamics, Convair Division

$$-C-C-C-C-C-C-C-C-C-$$

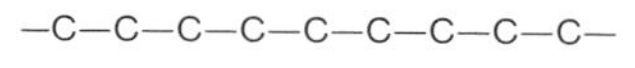

Fig. 17 Linear polymer. Source: General Dynamics, Convair Division

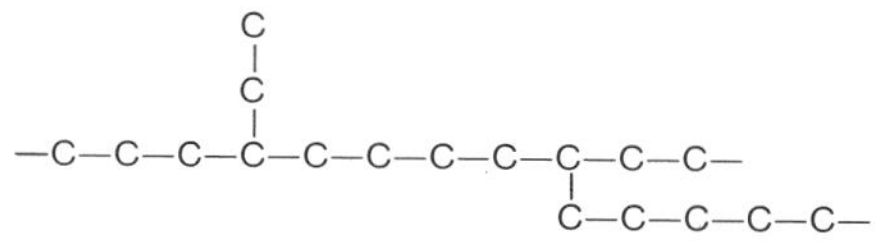

Fig. 18 Branched polymer. Source: General Dynamics, Convair Division

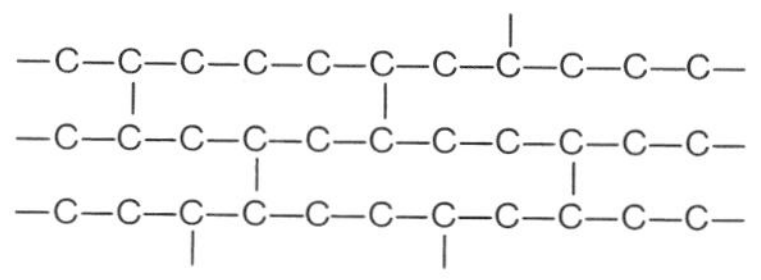

Fig. 19 Networked polymer. Source: General Dynamics, Convair Division

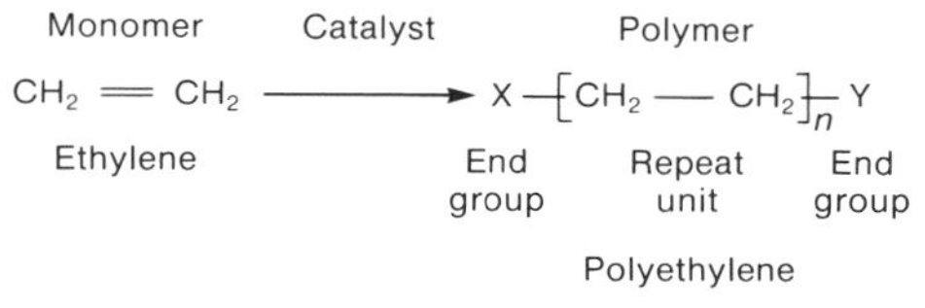

Fig. 20 Linear polymer production. Source: General Dynamics, Convair Division

forming polyethylene or a diamine, and so forth. A typical linear polymer reaction is shown in Fig. 20. Linear polymers are not usually completely linear, because side reactions and chain transfer reactions also occur. Linear polymers are all thermoplastic in form and flow with the application of heat. They also exhibit solubility in various organic and inorganic solvents.

Branched chains are formed by the reaction of three or more functional monomers or by a process known as chain transfer, which causes chains of varying lengths and numbers. This variability in the size of the chains affects the effective molecular weight distribution, melt

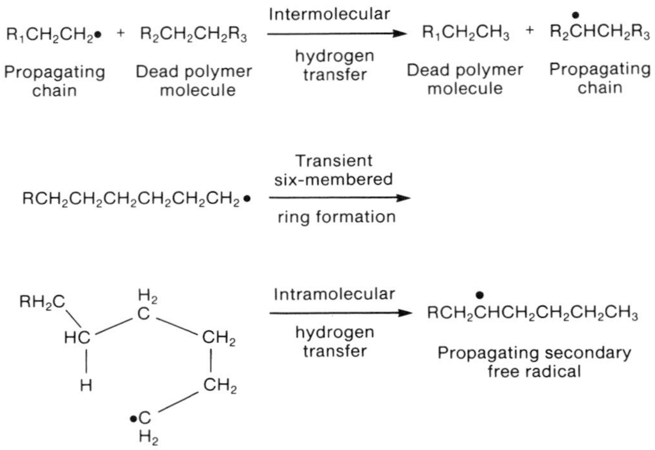

Fig. 21 Branched chain polymerization. The creation of the secondary free radical allows another free radical to join with it at a secondary site and create branched polymers. Source: General Dynamics, Convair Division

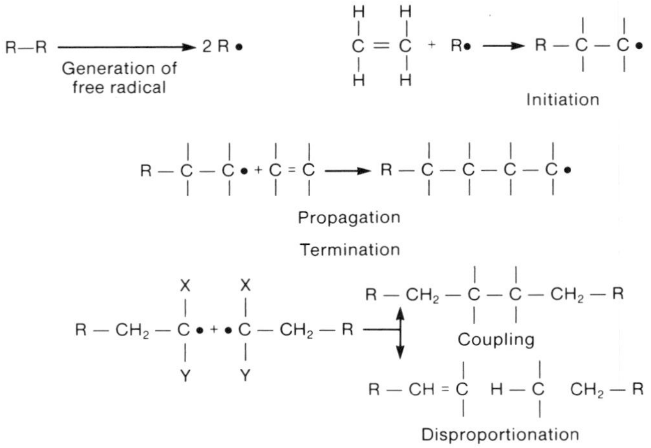

Fig. 22 Addition polymerization mechanism. Source: General Dynamics, Convair Division

viscosity, and crystallinity of the final product. These materials are thermoplastic and are produced by typical reactions, as shown in Fig. 21.

Most current composite production uses the third type of polymer structure, the cross-linked or networked polymer. In these molecules, essentially all chains connect into massive networks. They are produced by the polymerization of multifunctional monomers and are formed by the interconnection of preformed chains. The resins are characterized by being thermosetting or nonreversible structures. They will not dissolve in solvents (although solvents will affect them), and they will not flow at high temperatures after being cured. These materials are classified into two major categories by their cross links per unit molecular weight. Polyester, epoxies, and other hard systems have approximately one cross link per 100 to 500 MW. Elastomers, rubbery type materials, have approximately one cross link per 5000 MW.

There are two major types of polymerization reactions, addition and condensation. The addition reaction, used commercially most frequently, is characterized by the creation of a polymer chain with the same atomic makeup as the reaction monomers; that is, no portion of the monomer is evolved during the reaction. The most important characteristic of the reaction is the generation of free radicals as the reaction proceeds. This process is made up of four general steps: generation of the free radical; initiation; propagation; and termination, either by disproportionation or coupling (Fig. 22). The addition polymerization is formed by the reaction of the monomer with an active center. The chain grows very quickly, as evidenced by the rapid change in viscosity at the gelation point. The high molecular weight is also attained quickly and does not change with increased reaction time. The monomer is consumed by the reaction; thus it is present in decreasing concentrations throughout the reaction. Epoxy resin systems are addition reactions.

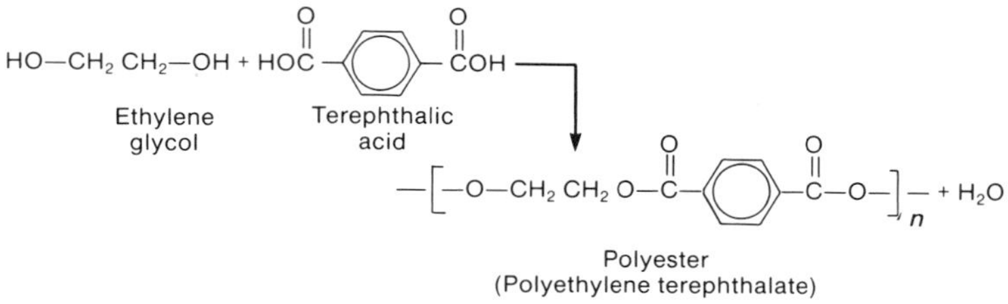

Fig. 23 Condensation polymerization. Source: General Dynamics, Convair Division

In the addition reaction, the gelation point is the point at which the linear chains are formed. No significant dwell time is required by the reaction, as shown previously. The long thermal-soak time compensates for the inability of the operator to predict accurately the onset of gelation in a multipart environment and allows time for proper bleed and compaction prior to gel. Much research is underway to predict or measure resin viscosity in the autoclave during cure. The thermal soak simply allows an adequate window to ensure that resins in different parts that had similar precure environments attain gel within that window. The second full-temperature soak is necessary to provide time for the cross-linking mechanism to occur. Without this second soak, the resin would never attain any significant properties. The establishment of the cross-linking network is essential to meeting the strength requirements of the finished structure. Additional research is underway to establish predictive methods for the cross-linking structure formation so that the cure process can be tailored to individual parts instead of processing windows. This work has the significant possibility of cure cycle time reduction and associated cost savings.

The second major reaction mechanism, the condensation reaction (also called the step reaction), is typical of polyesters, polyimides, polycarbonates, polyurethanes, and others. It is defined as a step-wise reaction of polyfunc-

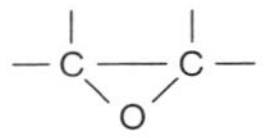

Fig. 24 Epoxide group. Source: General Dynamics, Convair Division

tional monomers to form a polymer with simultaneous elimination of a small molecule. This reaction type is illustrated by the reaction of ethylene glycol and terephthalic acid to form the polyester known as polyethylene terephthalate (Fig. 23).

These reactions require that the functional group reaction be quantitative (stoichiometric) and fast (as is shown in Fig. 23). The polymer must be stable under reaction conditions because the reaction occurs between any two molecules, and the monomer disappears early in the reaction (low molecular weight oligomers form). The polymer molecular weight increases gradually during the reaction and reaches a maximum only at very high monomer conversions (>99%). Condensation reactions introduce the problem of removing volatiles to the curing process. Removal of volatiles is generally done during the early stages of the cure by maintaining a vacuum on the parts long into the pressure application part of the cure cycle. By maintaining this vacuum on the part, the gases are drawn out of the laminate and into the vacuum system. The introduction of these gases

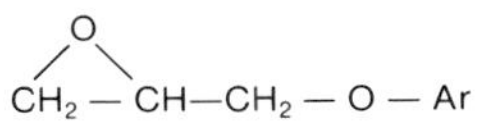

Fig. 25 Glycidyl ether. Source: General Dynamics, Convair Division

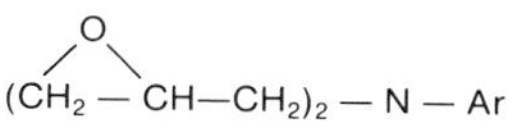

Fig. 26 Glycidyl amine. Source: General Dynamics, Convair Division

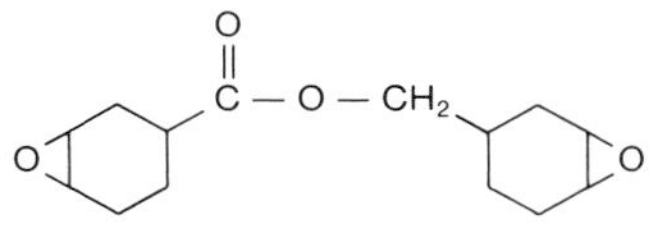

Fig. 27 Cycloaliphatic. Source: General Dynamics, Convair Division

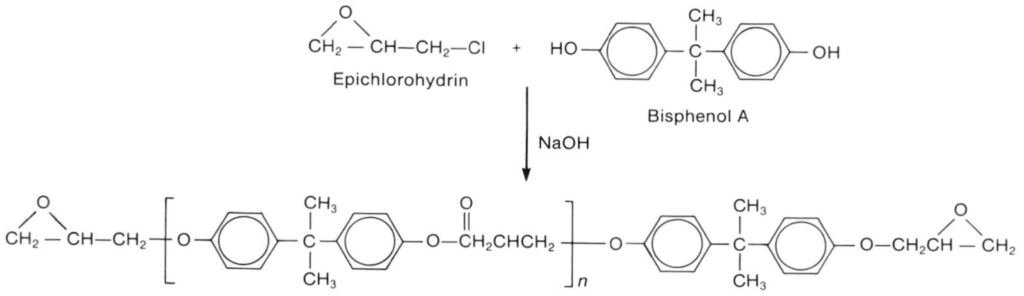

Fig. 28 Manufacture of diglycidyl ether of bisphenol A (DGEBA). Source: General Dynamics, Convair Division

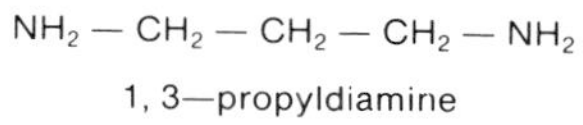

Fig. 29 Aliphatic amine curing agent. Source: General Dynamics, Convair Division

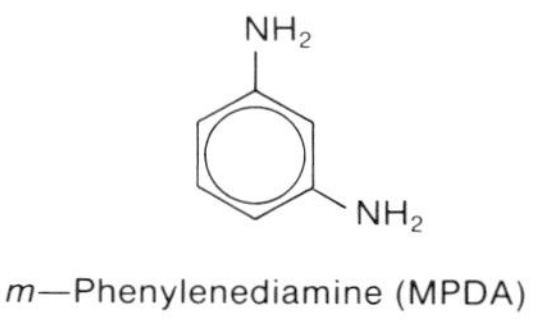

Fig. 30 Aromatic amine curing agent. Source: General Dynamics, Convair Division

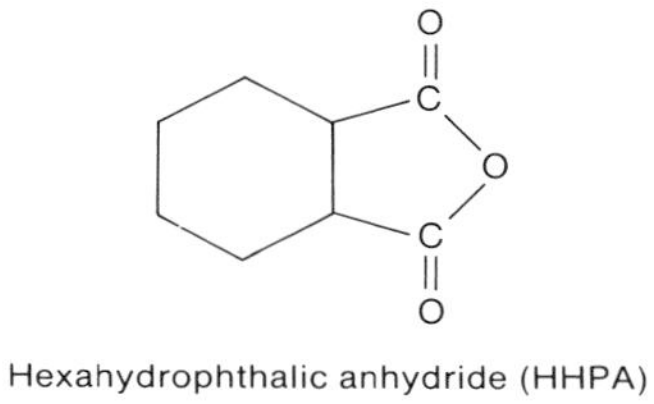

Fig. 31 Anhydride curing agent. Source: General Dynamics, Convair Division

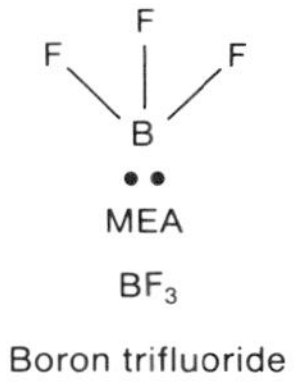

Fig. 32 Catalytic curing agent. Source: General Dynamics, Convair Division

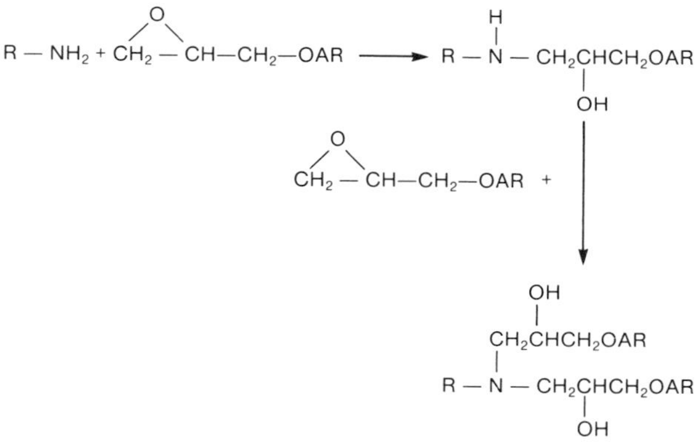

Fig. 33 Basic amine-epoxy reaction. Source: General Dynamics, Convair Division

and volatiles into the vacuum system necessitates the installation of condensate traps on all autoclave vacuum lines. Another very important concept in making a high-quality laminate with these resins is the establishment of an excellent fluid cell to reduce the interlaminar porosity that these volatile products can form. Diffusion control techniques are effective in reducing this porosity.

A sample reaction of an addition-reaction-based resin system is helpful in understanding the chemistry of resins in general. Because epoxy resins represent most structural composite work, a more detailed description of a typical epoxy system is appropriate. Typical of all epoxy systems is the presence of the 1,2-epoxide group shown in Fig. 24. The epoxide group is usually presented as a glycidyl ether, glycidyl amine, or part of an aliphatic ring system (Fig. 25 to 27).

The most widely used epoxide is the diglycidyl ether of bisphenol A (DGEBA), which is formed by the reaction of epichlorohydrin and bisphenol A (Fig. 28). The number n determines the properties and uses of the

system. Numerous commercial varieties are available; every major resin manufacturer makes a resin of this type. The completion of the resin system occurs when the resin is combined with a curing agent, of which there are three basic types: amines, anhydrides, and catalysts. Each curing agent gives different characteristics to the curing procedure and to the cured resin as a whole. Aliphatic amine curing agents are generally used in room-temperature curing applications and give short pot lives and low-temperature service. Aromatic amines, anhydrides, and catalytic curing agents generally require elevated temperature curing and yield higher service temperature resins. An example of each type is shown in Fig. 29 to 32. A sample reaction of an amine-epoxy system is shown in Fig. 33.

Flow

The flow of the resin system generates many of the pathways that are desired during the cure cycle. The application of heat in the early

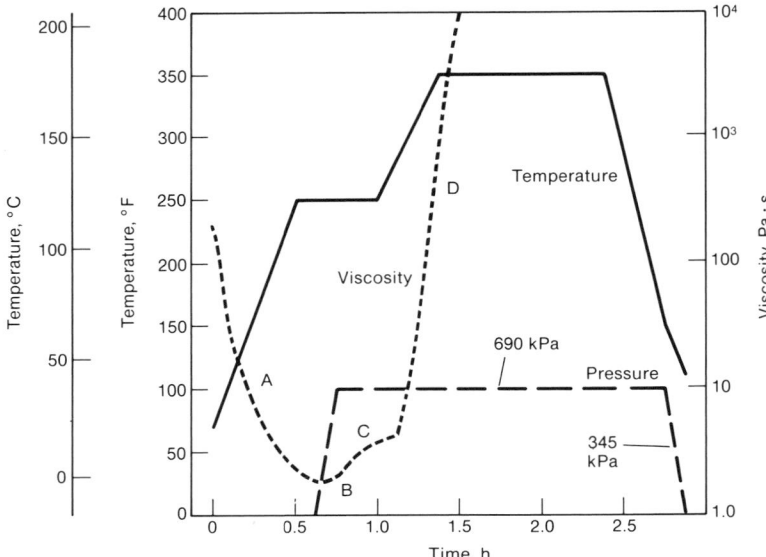

Fig. 34 Typical 175 °C (350 °F) resin cure cycle and viscosity curve. Source: General Dynamics, Convair Division

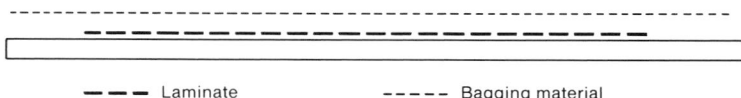

——— Laminate   ----- Bagging material

Fig. 35 Poor fluid cell provided by tool. Source: General Dynamics, Convair Division

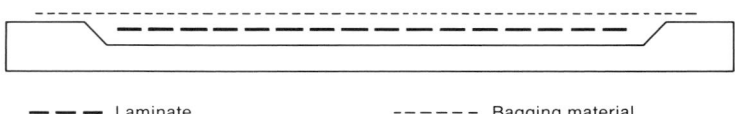

——— Laminate   ------ Bagging material

Fig. 36 Fluid cell provided by tool. Source: General Dynamics, Convair Division

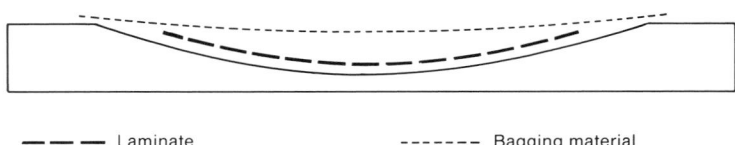

——— Laminate   ------- Bagging material

Fig. 37 Fluid cell provided by part shape. Source: General Dynamics, Convair Division

——— Laminate   ----- Bagging film
—-— Nonporous release film
········· Breather material   ⬭ Bag sealant tape
——— Vacuum line   —··— Static line

Fig. 38 Bagging for no-edge-bleed control and a net resin system. Source: General Dynamics, Convair Division

stages of the cure cycle causes the viscosity of the resin to decrease in proportion to the increase in temperature. This relationship is shown by:

$$v = v_0 + e^{BT} + \int e^{kT} + (H_2O \text{ or another diluent)} \quad \text{(Eq 5)}$$

where v is the viscosity; v_0 is the initial viscosity; e^{BT} is the change of viscosity over temperature; the integral is the function of viscosity over time, temperature, and degree of cure; and T is the absolute temperature (in Kelvin).

This change in viscosity is fundamental to the processing of the resin system; it allows the chemical reactants to move freely in the liquid state and collide with more reactive sites. It also allows the volatiles to nucleate and be drawn out of the laminate. The flow of the resin also enables the application of external pressure to compact the fibers properly. Therefore, the final resin content of the laminate is dependent on the flow characteristics of the resin and the correct application of bleeder material.

The chemical kinetics of a resin are influenced by the flow state of the resin. A typical cure cycle and viscosity curve for an epoxy resin are shown in Fig. 34. The viscosity curve starts out at a relatively high point, 1 Pa · s (10 P) (approximately the consistency of warm tar); and as temperature increases, the viscosity decreases to a level roughly equivalent to that of lightweight oil (0.005 Pa · s, or 0.05 P) (Point A). At this point the resin begins to react vigorously, and the viscosity increases as the polymer network begins to form Point B. The temperature is then raised to increase the reaction rate; and as the temperature increases, the rate of change of the viscosity begins to slow, as indicated by the slope of the line becoming less drastic. This effect is not indicative of the slowing or reversal of the reaction; it is the effect of the increased temperature reducing the viscosity of the resin and the reaction products (Point C). During this period, the resin viscosity rapidly increases to a point of essentially infinite viscosity. This is known as the gel point, or the point at which gelation occurs (Point D). Gelation represents the end of the flow stage of the cure cycle, but does not end the cure cycle. At the gelation point, the resin becomes a solid but is not fully cured. The primary network structure has been formed, but the cross-linking of the networked chains is not complete. The final elevated-temperature soak is provided to allow this cross-linking to occur. This long-term soak has the effect of raising the glass transition point (often used as an indicator of the degree of advancement of the cross-linking process) as a function of time at temperature.

As evidenced by the reaction path, any processing that requires the liquid state must occur prior to the gelation of the resin. This period of time before gelation is when pressure should be

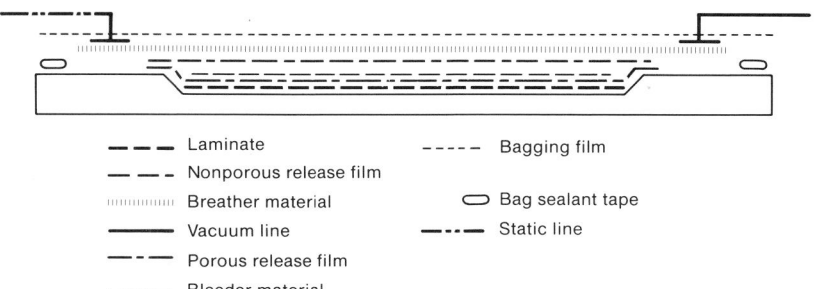

--- Laminate
--- Nonporous release film
||||||||| Breather material
— Vacuum line
--- Porous release film
--- Bleeder material

----- Bagging film
⬭ Bag sealant tape
-·-· Static line

Fig. 39 Bagging for no-edge-bleed control and a bleed resin system. Source: General Dynamics, Convair Division

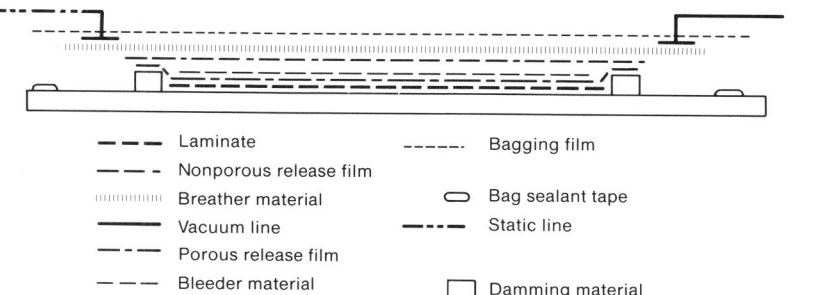

--- Laminate
--- Nonporous release film
||||||||| Breather material
— Vacuum line
--- Porous release film
--- Bleeder material

------ Bagging film
⬭ Bag sealant tape
-·-· Static line
☐ Damming material

Fig. 40 Bagging with damming techniques and a bleed resin system. Source: General Dynamics, Convair Division

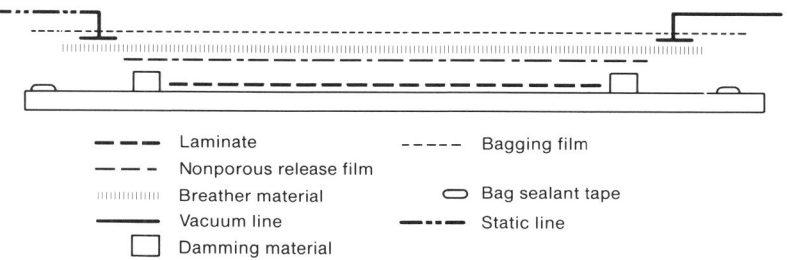

--- Laminate
--- Nonporous release film
||||||||| Breather material
— Vacuum line
☐ Damming material

----- Bagging film
⬭ Bag sealant tape
-·-· Static line

Fig. 41 Bagging with damming techniques and a net resin system. Source: General Dynamics, Convair Division

applied. The reason pressure is applied in the low-viscosity range is to establish fully the maximum fluid hydrostatic pressure, which is important to the compaction of the fibers, establishment of the correct resin content, and evolution of the volatile gases for transport or diffusion back into solution.

Fluid Hydrostatic Pressure

The establishment of an effective fluid hydrostatic cell (FHC) and thereby of an effective fluid hydrostatic pressure (FHP) is essential to the void-free curing of laminates. Tooling concepts are critically important to establishing the proper FHC for curing laminates. The tools must provide for the control of the resin flow during the very low viscosity condition. This has several manifestations in tooling concepts. First, the tool must tolerate early pressure

application by containing the resin in the laminate and bleeders and not allowing uncontrolled edge bleeding of the resin. Second, the tool must allow the FHP to reach its maximum. Third, the tool must allow for the complete compaction of the laminate fibers. A tool that does not provide a good fluid cell is shown in Fig. 35, and some representative design concepts are shown in Fig. 36 and 37 to illustrate the meaning of effective fluid-cell designs. In Fig. 36, the tool is made to create a fluid cell, whereas in Fig. 37 the tool makes use of the part design to create the fluid cell.

An ideal tool and net resin (no bleed) system would allow the part to be bagged without any need to control edge bleeding. The only requirement for the bagging of this ideal part for autoclave curing would be the application of nonporous release films, breather material, and the final bag (Fig. 38). Unfortunately, this ideal is rarely achieved. Net resin systems are more

difficult to control for porosity content, and many older resin systems are not available in the net resin condition. Consequently, many parts require the bleeding of excess resin from the part to obtain the proper resin content. This is accomplished by the use of layered bleeder material in the lay-up of the final bag. After the part is completely laid up, the part surface is covered with a porous release material, and the bleeder is applied on top of the release material, followed by a nonporous release film, the breather, and the final bag (Fig. 39).

The previous two examples provide an adequate FHC, assuming that the tool is designed to control resin flow. This can be done either by creating pockets on the surface of the tool to trap the resin in a designed trap, or by the inherent characteristics of the part configuration. In the event that the tool does not adequately contain the resin flow, a technique known as damming is used, which involves applying a dam around the periphery of the part. The dam can be made of several materials, such as bag sealant tape, nonporous cork strips, and so forth. The idea is to create a barrier to resin flow next to the edge of the part. Dams can be used with either of the foregoing bagging methods as long as the final release film is in intimate contact with the dam surface. The FHC includes all the bleeder material and, as a consequence, the bleeder must not extend over the dam. Figures 40 and 41 show the relationship of the dam to the bagging materials.

Bleeding of the resin system is very difficult to perfect on the first cure of a part, because of the interrelationships of bleeding with other facets of the curing process. Bleeding is dependent not only on resin content but also on part configuration. It is also dependent on material age, because of the effect of age on the complex viscosity of the resin.

A formula is presented in Eq 6 to give a first approximation of a bleeder schedule for a laminate. The required values are fiber areal weight (FAW), in grams, of the material to be bled; the resin content (RC) of the prepreg material to be bled; the desired cured resin content (CRC); and the number of plies (N). The method to obtain the grams of resin weight (RW) to be removed is:

$$RW = \left(\frac{FAW}{100 - RC} + RC \right) -$$

$$\left(\frac{FAW}{100 - CRC} + CRC \right) + N \qquad (Eq\ 6)$$

Three common bleeder materials are 120-style glass cloth, 181-style glass cloth (also called 7781), and Airweave SS. These materials are capable of removing 59, 129, and 218 g (2, 5, and 8 oz) of resin per layer, respectively. This calculation does not have area as a factor in order to simplify calculations. It assumes that the bleeder will be the same size and configuration as the part. To determine the required bleeder, divide the resin weight to be removed by the appropriate factor. For example, in $RW = 580$, the number of 120-style glass bleeder plies is 580/59, or 9.8 (round to 10). The number of 181-style glass bleeder plies is 580/129, or 4.49 (round to 4). The number of Airweave SS bleeder plies is 580/218, or 2.66 (round to 3). Obviously, none of these alternatives yields the best possible bleeder schedule. A combination of bleeders would be best: Divide 580 by 218 to yield 2 with a remainder of 144, then divide this remainder by 129 to yield 1 with a remainder of 15. Also try dividing the remainder by 59 to yield 2 with a remainder of 26. This method indicates the best starting bleeder would be one layer each of Airweave SS and 181-style glass over the whole part.

This method does require proofing on each part, but it provides a good approximation for bleeder schedules. Table 1 shows a BASIC computer program for calculating general bleeder schedules for 1- to 25-ply laminates.

The term breather material is often used interchangeably with the term bleeder. This is a mistake with serious consequences because each material has a distinct and separate function. A bleeder, as previously discussed, is required to remove excess resin from the laminate. A breather is required to provide adequate air flow over the part and to enhance the integrity of the vacuum bag. A breather is required even when a bleeder is not. When a bleeder is present, it must be separated from the breather so that the breather is not impregnated with resin during the cure cycle. Impregnation of the breather nearly always results in resin-starved parts due to removal of excess resin from the part. Breather is also important to

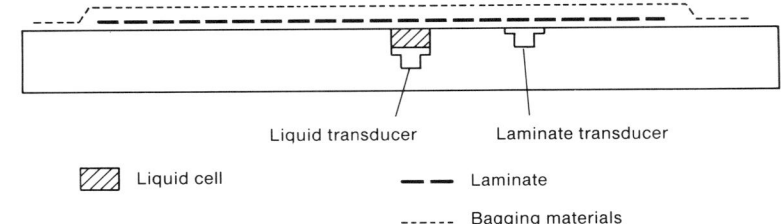

Fig. 42 Test tool for laminate pressure studies. Source: Applied Polymer Technology, Inc./U.S. Air Force Processing Science Program

provide a clear pathway for volatiles to escape from under the bag, which ensures that the entire surface of the part receives the same pressure. If the breather does not permit good air flow, local pockets of lower pressure will occur in areas in which the air under the bag cannot escape.

The FHP on the part is dependent on several factors, as described previously; however, an attempt to determine the actual pressure of the cell would probably result in error. Several experiments to determine cell pressure were run on a special tool with two pressure transducers to measure both laminate and fluid pressure (Fig. 42). The results, shown in Fig. 43, give unexpected results. First, the laminate pressure never actually reached the autoclave pressure. Second, the liquid resin pressure attained only about 18% of the autoclave pressure. Third, the laminate pressure took much longer to build to its final pressure than did the autoclave pressure. Fourth, the pressure within the liquid resin reached a maximum and then tapered off. The explanation for each of these results is relatively easy to deduce. The

laminate never reached the autoclave pressure because of the pressure required to compact the fibers of the laminate, the bleeder/breathers, and the bagging materials. This is also illustrated by the reduction of the laminate pressure as the number of plies is raised, indicating increased compaction requirements. The low liquid pressure is attributable to the fact that the compaction of the fibers limits the amount of pressure the liquid can exhibit. This is a result of the limit on the size of the fluid hydrostatic cell and thereby the amount of pressure applied to the liquid. The corresponding increase between the pressure of the liquid and the number of plies is attributable to the additional resin available to pressurize. This result also reinforces the explanation of fiber compaction requirements that limit the pressure.

The laminate pressure ramp rate is slower than that of the autoclave due to the requirement to compact the fibers at the onset of the pressure application. This is evidenced by the reduction in ramp rate as the number of plies is increased.

Table 1   BASIC computer program for resin bleeder schedule

```
10 REM PROGRAM TO DEVELOP RESIN BLEED SCHEDULES
15 CLEAR 1000
20 READ A
30 FOR X = 1 TO A
40 READ B$,RS,RR,AW,RC
50 FOR Y = RS - RR TO RS + RR
55 LPRINT CHR$(12);" ";CHR$(27);"X";B$;CHR$(27); "Y":LPRINT
   TAB(40);"UNCURED RESIN CONTENT PERCENTAGE:";
56 LPRINT CHR$(27);"X";Y;CHR$(27);"Y":LPRINT:REM NOTE CHR$(27);"X"
   AND "Y" TURNS UNDERLINE ON AND OFF MAY BE DIFFERENT FOR YOU
57 LPRINT:LPRINT " NOMINAL FIBER AREAL WEIGHT";AW;TAB(40);"CURED
   RESIN CONTENT";RC
58 LPRINT:LPRINT " PLIES OF";TAB(30);"BLEEDER REQUIREMENTS"
59 LPRINT " PREPREG"; CHR$(27);CHR$(34):LPRINT
60 RW = ((AW/(100 - Y))*Y) - ((AW/(100 - RC)*RC))
65 B1 = RW/59:B2 = RW/129:B3 = RW/218
70 FOR Z = 1 TO 25
80 C1 = B1*Z:C2 = B2*Z:C3 = B3*Z
90 IF C3 < 1 THEN 100 ELSE 300
100 IF C2 < 1 THEN 110 ELSE 400
110 C1 = INT(C1):C2 = 0:C3 = 0
120 GOTO 500
150 REM PRINT OUT
160 LPRINT TAB(5);Z;TAB(15);Q$
165 IF Z/5 = INT(Z/5) THEN LPRINT
167 GOTO 600
300 C4 = INT(C3):IF C3 - C4 > .25 THEN 310 ELSE C3 = C4:C2 = 0:C1 = 0:GOTO 350
310 IF C3 - C4 > .52 THEN 320 ELSE C3 = C4:C2 = 1:C1 = 0:GOTO 350
320 IF C3 - C4 > .58 THEN 330 ELSE C3 = C4:C2 = 0:C1 = 2:GOTO 350
330 IF C3 - C4 > .788 THEN 340 ELSE C3 = C4:C2 = 1:C1 = 0:GOTO 350
340 IF C3 - C4 > .86 THEN 345 ELSE C3 = C4:C2 = 1:C1 = 1:GOTO 350
345 C3 = C4:C2 = 1:C1 = 2
350 GOTO 500
400 C4 = INT(C2):IF C2 - C4 > .43 THEN 410 ELSE C2 = C4:C1 = 1:C3 = 0:GOTO 450
410 IF C2 - C4 > .88 THEN 420 ELSE C2 = C4:C1 = 1:C3 = 0:GOTO 450
420 C2 = C4:C1 = 2:C3 = 0
450 GOTO 500
500 Q$ = ""
510 IF C3 = 1 THEN Q$ = " 1 PLY AIRWEAVE SS ":GOTO 530
520 IF C3 > 1 THEN Q$ = STR$(C3) + " PLIES AIRWEAVE SS "
530 IF C2 = 1 THEN Q$ = Q$ + " 1 PLY 181 FIBERGLASS ":GOTO 550
540 IF C2 > 1 THEN Q$ = Q$ + STR$(C2) + " PLIES 181 FIBERGLASS "
550 IF C1 = 1 THEN Q$ = Q$ + " 1 PLY 120 FIBERGLASS "
560 IF C1 > 1 THEN Q$ = Q$ + STR$(C1) + " PLIES 120 FIBERGLASS"
565 IF C1 = 0 AND C2 = 0 AND C3 = 0 THEN Q$ = " 1 PLY OF ARMALON OR PEEL
    PLY ONLY"
567 IF C1 < 0 THEN Q$ = " NO BLEEDER REQUIRED"
570 GOTO 150
600 NEXT Z: NEXT Y: NEXT X
8999 REM ONE NUMBER INDICATING NUMBER MATERIALS (DATA LINES)
9000 REM DATA FOR EACH MATERIAL STARTING WITH A STRING TO
9001 REM IDENTIFY MATERIAL, RAW MATERIAL RESIN CONTENT, +/-
     RANGE
9002 REM ON RAW MATERIAL RESIN CONTENT, NOMINAL FIBER AREAL
     WEIGHT,
9003 REM HIGHEST ALLOWABLE CURED RESIN CONTENT.
9004 REM SAMPLE DATA FOLLOWS
9010 REM DATA 0-06166-2 ARAMID EPOXY PER 0-73829-1,50,3,170,43
9999 END
```

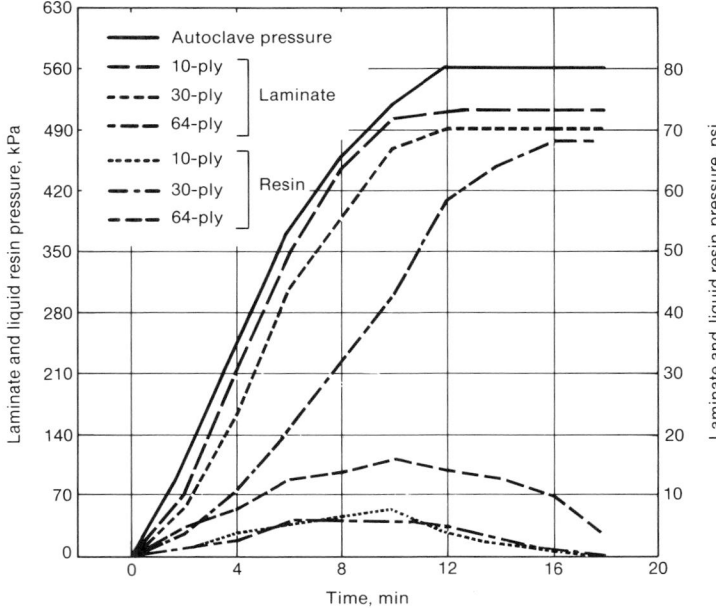

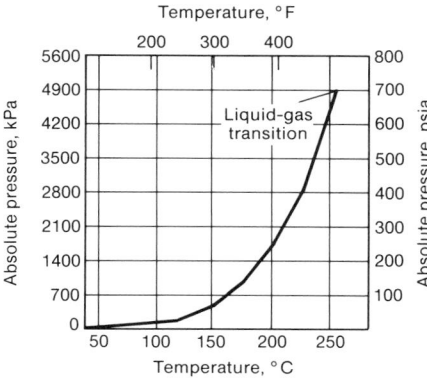

Fig. 43 Results of laminate pressure gradient study. Source: Applied Polymer Technology, Inc./U.S. Air Force Processing Science Program

Fig. 44 Diffusion control pressure requirements. Source: Applied Polymer Technology, Inc./U.S. Air Force Processing Science Program

The reason the liquid pressure lowers after it reaches a maximum point seems to be because of the chemical reaction occurring within the resin, which causes the resin to transfer much of its energy directly to the fibers and not to the liquid. This is evidenced by the steadily decreasing pressure.

The foregoing results reaffirm the need to ensure that the FHC provides the highest integrity during the cure cycle to provide adequate FHP to consolidate the voids and porosity in the resin. This will ensure that the resulting laminates are of the high quality required for structural applications in high-stress areas, where voids and porosity can cause fatigue problems.

Advanced Cure Concepts

Advanced cure concepts include diffusion control and cure modeling.

Diffusion control is a technique used to reduce gas bubble nucleation in the curing of composite resins and therefore in the laminate. This technique, also known as internal bag pressurization (IBP), works by inducing FHP in the resin during the volatile evolution phase of the cure cycle.

Diffusion control is contrary to most current thought about vacuum bag curing. Most cure specifications do not allow any significant pressure to accumulate under the bag. Generally, 35 to 80 kPa (5 to 12 psig) is the specification limit for pressure under the bag during cure.

This technique was discovered somewhat by accident. During experimentation on void and porosity content of laminates, a bag on one of the coupons blew early in the cure cycle; during subsequent testing, the coupon showed no detectable porosity. Further testing was done by replicating the pressure inside the bag by pressurizing the vacuum line during cure. The resulting coupons showed that the internal effects of this technique were predictable and followed the standard steam pressure tables to generate the effect. Additional studies into diffusion control brought to light several explanations for the characteristics of resin behavior, especially the effects of moisture on the void/porosity content of the cured laminate. It was discovered that the higher the moisture content of the resin, the higher the internal bag pressure required to suppress volatile-gas formation. This makes sense because most other volatiles have lower vapor pressures than does water. The relationship of pressure to temperature to limit gas bubble nucleation is shown in Fig. 44.

After this discovery, it was realized that the internal bag pressure actually causes the FHP to increase by the amount of added pressure, thereby greatly magnifying the available FHP from the cure pressure. Additional information derived from this study showed that some voids produced by lay-up anomalies (excess entrapped air) could be reduced but not entirely eliminated by internal bag pressure. This factor reduces some of the risk of voids in composite parts that results from manual fabrication. Also, the study showed that the internal bag pressure required to keep gas bubbles from forming is only approximately one-third of the pressure required to drive the bubbles back into solution after formation.

Another conclusion was that vacuum degassing the resin at impregnation and vacuum debulking the laminate incrementally during lay-up reduces the likelihood of voids and porosity, thereby showing vacuum debulking to be an effective method of reducing void content. Incremental pressure debulking also reduces void content, but it costs more to perform.

A further significant finding was that curing with a vented bag (0 kPa, or 0 psig under the bag) should be avoided from a void/porosity viewpoint. There is sufficient air to create gas bubbles even under a 0% volatile content in the resin. The amount of pressure required to correct this problem is very low; 35 to 70 kPa (5 to 10 psig) will suppress bubble nucleation up to 0.5% moisture content.

The actual moisture content within the resin with its effect on void content actually improved the resin chemical reaction. It solubilizes the resin system so that hardener/epoxy particulate liquid transition occurs at a lower temperature. It also increases the reaction kinetics during cure of a nonaccelerated system, and does not appear to impair the modulus network properties. These concepts, discovered during the research of diffusion control techniques, deviate significantly from those of the accepted state-of-the-art composite cures. However, the results are impressive, and the adaptation of these techniques to current cure methodology is an easy way to ensure that part void and porosity content are maintained at an acceptable level.

The implementation of this technique into production requires several things. First, the autoclave (or other curing device) must have a vacuum system that includes not only vacuum but also pressure capability to each line. These lines must be individually controllable to provide either vacuum or atmospheric (vent) or positive pressure. The installation of proportional control valves on one header to control vacuum and pressure for that header would be adequate to meet this requirement. This would allow the entire line to be vented slowly (which is important when curing parts with a honeycomb core), and then the pressure could be raised to the desired pressure, or the part could be switched to a vent line.

Second, it is important to have a link between the pressure source on the header line

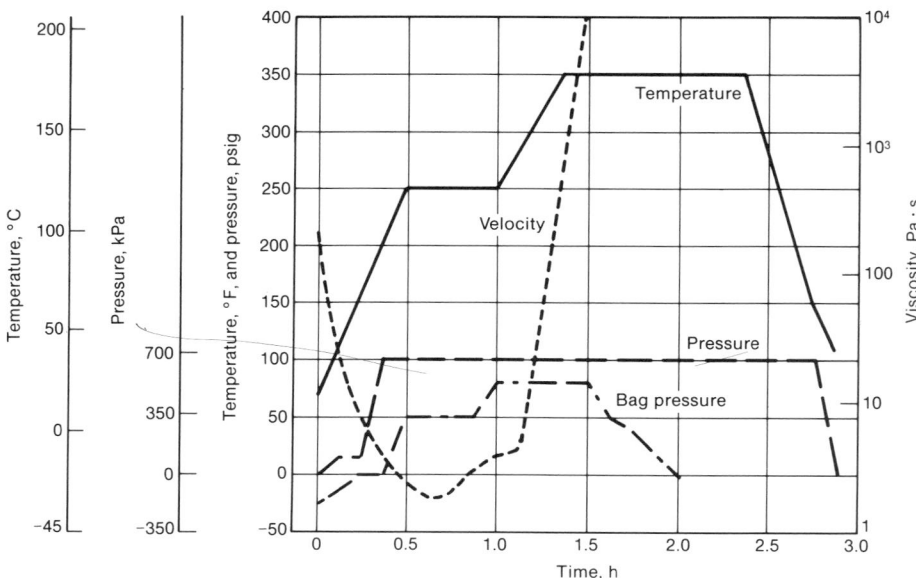

Fig. 45 Resin cure cycle with diffusion control, at 175 °C (350 °F). Source: General Dynamics, Convair Division

and the autoclave pressure, to ensure that the bag pressure never exceeds the autoclave pressure during a run. A pressure that is 70 kPa (10 psig) lower than the autoclave pressure will supply sufficient pressure to the part and guarantee that there will always be a 70 kPa (10 psig) differential on the vacuum bag. If a differential is not maintained, the vacuum bag will literally be blown off the part being cured.

Third, the tooling must include good control of the fluid hydrostatic cell. This is very important, because the resin will be exposed to full autoclave pressure during the low-viscosity region of its viscosity curve.

Fourth, diffusion control is a concept that runs counter to the established curing process for composites, and operators need to adjust to this concept. The benefits are easily recognized, and retraining is well worth the effort because of the increased yield of void-free and porosity-free parts.

A typical cure for a 175 °C (350 °F) curing system that includes diffusion control is shown in Fig. 45. (This cure cycle could be different for differing resin systems.) Steps in this cure cycle are as follows: Load the parts into the autoclave and connect them to normal vacuum lines, static (dead-end) lines, and thermocouples. Start the run and apply heat and vacuum to the parts. When the temperature of the part approaches 65 °C (150 °F), apply 100 kPa (15 psig) autoclave pressure and vent the bags to atmosphere. Heat to approximately 80 °C (180 °F) and apply the full autoclave pressure of 690 kPa (100 psig). When full pressure is reached, or before the part reaches 100 °C (210 °F), apply 345 kPa (50 psig) to the vacuum line. When the parts reach 120 ± 6 °C (250 ± 10 °F), start the

soak time for the parts. After 30 min at 120 ± 6 °C (250 ± 10 °F), raise the internal bag pressure to 550 kPa (80 psig) and the temperature to 175 °C (350 °F) at 1 to 3 °C/min (2 to 6 °F/min). Then, enter a 1-h soak time; at 10 min into the soak, the internal bag pressure can be released, as gelation will have occurred and no further gas bubbles will form. Cool down under pressure after completion of the 1-h soak. At 65 °C (150 °F), release the pressure.

Each phase of this cycle has a distinct purpose. The early venting of the vacuum bag prevents the growth of voids caused by entrapped air. The pressure application prior to the lowest viscosity of the resin ensures good compaction of the laminate by exposing it to maximum autoclave pressure for as long as possible. The application of the internal bag pressure prior to 105 °C (220 °F) ensures that no significant water vapor has evolved before applying the internal pressure. This application of the internal bag pressure is important, as it takes three times more pressure to drive a bubble back into solution than it takes to prevent it from forming.

The resin is left to soak at 120 °C (250 °F) to allow it to gel. The internal bag pressure is raised to 550 kPa (80 psig) before the application of additional pressure to compensate for the increase in vapor pressure from any absorbed water that results as the temperature is raised to 175 °C (350 °F). During the second heat and the first few minutes of the 175 °C (350 °F) soak, the resin finishes gelation. After this point, the internal bag pressure can be slowly reduced to zero. This is because the resin can no longer form gas bubbles, because it is a solid.

The bags are depressurized before the autoclave is, to allow all the pressure in the bags to bleed, which avoids popping the bags upon autoclave venting. The cool-down under pressure is controlled to minimize the effects of the thermal stresses, warpage, and the possibility of forming interlaminar cracks within the planes of resin and reinforcement.

The addition of diffusion control to an autoclave cure cycle is an effective way to increase cure cycle yields by reducing void and porosity content in completed parts. The technique is not simple, and its implementation is another reason for using computers to assist and control the autoclave during operation.

Cure modeling is a new concept for cure control that is currently being developed in which the chemical kinetics and rheology of the resin system are mathematically modeled. A computer controls the cure cycle based on temperature and sensor feedback from the parts and applies that information to the algorithm to determine the current degree of cure. By controlling the degree of cure, the temperature profile can be optimized to obtain the most efficient, complete cure.

The development of cure models is underway on several different resin systems. The model development depends on the ability to monitor the cure kinetics in a way that will relate to the physical characteristics of the part. This is being done by using either dielectric or ultrasonic cure monitors. The development of the ability to monitor an appropriate physical parameter is essential to the accuracy of the model.

Dielectric monitors are used to monitor the change of dielectric properties during the cure. Dielectrics monitor the change in dipole moment and relaxation as the resin viscosity changes. The viscosity of the resin can then be correlated to the dielectric changes. Standard dielectric is unusable beyond the point of gelation because after the resin becomes solid, the dielectric measurements are very difficult to make. A new technology, microdielectrics, has been developed to make measurements into the solid region. However, the accuracy of this data is questionable because of difficulty in reproducing it. Dielectrics in general are extremely sensitive to moisture content in the resin. A very small change in moisture content (±0.1%) will cause a major change in the dielectric characteristics of the resin.

Another monitor under development is based on ultrasonic measurements of the resin during cure. The work with these monitors has shown a very close relationship between the measurements of the resin during cure and the viscoelastic properties (G', G'') of the resin. The data can also be collected during the solid stage of the resin. The readings are currently being correlated to the degree of cure (known as alpha). Being able to read alpha essentially from a part during cure makes the development and validation of cure models much easier.

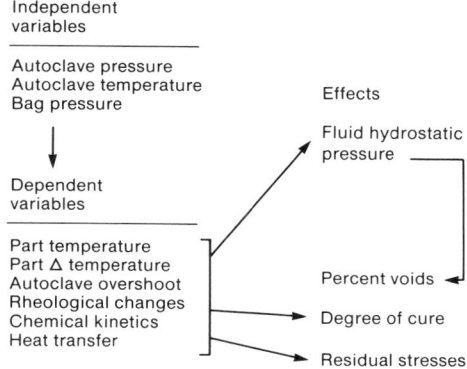

Fig. 46 Curing variables. Source: General Dynamics, Convair Division

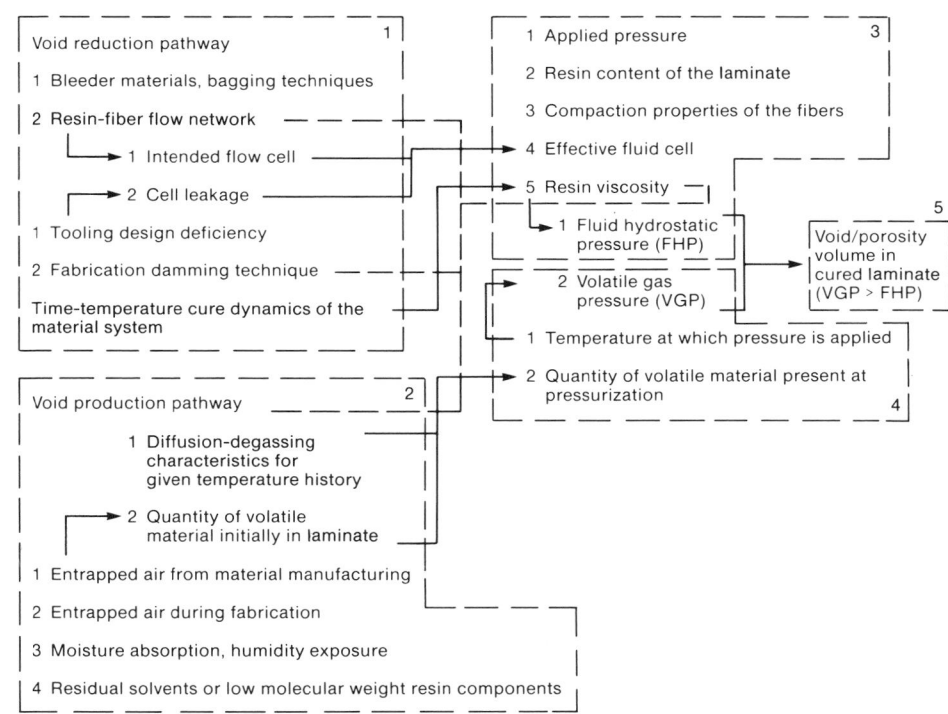

Fig. 47 Model of laminate process behavior. Source: Applied Polymer Technology, Inc.

Cure modeling holds a greater promise for cure cycle quality control than any other emerging technology. When the cure model is validated, both in the lab and in production, the results that it provides on the degree of completeness of cure can be used to accept cure load without any further quality tests. Obviously, this will require a significant change in both thinking and processing in the composites industry.

The cure model is developed using time- and moisture-aged samples of a representative resin and testing them in various ways to determine their reaction pathways and chemical kinetics. These tests include DSC, TGA, and rheometric analysis. The tests characterize the resin and allow its activity to be predicted.

After the model has been developed, software will be written to control cure cycles based on the predictions and feedback from sensors. In the long term, the sensors will be dropped, and the cure model algorithm will be the only control. This new technology has the potential to shorten cure times and thereby reduce costs. Current curing technology cures everything to a standard cycle, which may not be neither optimum nor necessary to cure the resin system fully. In addition, current quality control mechanical tests will be eliminated because they are used mainly to test the degree of cure, which will be verified during the cure itself. Much research is being done on these modeling techniques. The techniques and methods differ slightly, but all use computer-based "expert" systems or artificial intelligence as the method of implementation.

Curing Variables

As shown in the model of laminate process behavior (Fig. 1), the most important entity in the cure cycle appears to be the control of voids and porosity. Theoretically, if heating considerations are taken into account during the design and loading of the autoclave, the only two remaining parameters are the correct heat and pressure profile and the control of voids. The only directly controllable functions during the cure cycle are the heat, pressure, and vacuum bag pressure or vacuum. From the direct control of these functions, it is indirectly possible to control the part ΔT and the autoclave overshoot and therefore influence the rheology, kinetics, and heat transfer of the system. These, then, determine the fluid hydrostatic pressure and degree of cure of the laminate, which influence percent voids, residual stresses, and final cure of the laminate (Fig. 46).

The heat profile itself is a major contributor to the void reduction pathway. Maintaining the minimum practicable part ΔT is imperative for the correct curing of the resin and to the reduction of voids (see Fig. 47). Each area in the chart in Fig. 47 has a reference number for discussion purposes and each area will be summarized to show that the chart represents the entire curing process.

The first area of the chart (area 1) represents the void reduction pathway, that is, all the factors influencing the reduction in void/porosity content of the finished laminate. The three major areas of this pathway are intended flow cell, cell leakage, and time-temperature cure dynamics of the material system. In this pathway, these entities make up the cure requirements for creating an effective fluid cell (item 4 in area 3). The effectiveness of the fluid cell is defined by what the intended flow cell is and what leakage occurs outside of that cell. The intended flow cell is made up of the fiber-resin flow network (the laminate) and the bleeder material and bag. Cell leakage is caused by deficient tool design and the technique for damming (or control of leakage).

The last part of the void reduction pathway is the time-temperature cure dynamics of the material system. This area was discussed in the section "Chemical Kinetics" in this article, in which it was shown that the relationship of time, temperature, viscosity, and cure kinetics is essential to the proper curing of the resin and the control of laminate porosity. Cure pressure must be manipulated effectively to take advantage of the resin-viscosity curve.

The second area of the chart, the void production pathway (area 2), is influenced by two factors, diffusion-degassing characteristics for a given temperature history, and the quantity of volatile material initially within the laminate. The first is related to material properties and the time-temperature characteristics of the material, which can include condensation chemical reactions and the evolution of solvents used in the manufacturing process of the prepreg. This is why thermogravimetric analysis is an important tool in analyzing resins for cure. This test shows how volatiles evolve during the cure process and allows the cure cycle to be tailored to respond to the evolution of these gases. Failure to control these gases results in higher void/porosity content in the completed laminate.

The second factor of the void production pathway is the quantity of volatile material initially in the laminate. These volatiles come from the air entrapped during the material manufacturing process and the laminate manufacturing (lay-up) process, as well as from the absorption of moisture due to humidity exposure during the precured state. There are also residual solvents and low molecular weight components from the resin manufacturing process that become volatiles during cure.

The third area of the chart, the FHP area (area 3), is made up of five major components: applied pressure (from the curing device), resin content of the laminate, compaction properties of the fibers, effective fluid cell (described earlier), and resin viscosity. The only one that is not a function of the cure cycle is the material compaction properties of the fibers.

The fourth area of the chart, the volatile gas pressure region (area 4), is made up of the temperature at which pressure is applied and the quantity of volatile material present at pressurization. The effect of temperature at pressure application is important for two reasons. First, if the temperature exceeds the boiling point of water, then the volatile pressure will be greatly enhanced by the formation of steam from absorbed moisture. Second, the relationship of time and temperature to the viscosity must be accounted for in the proper application of pressure; too early an application of pressure on the resin viscosity curve, and the fluid cell may not be effective; too late, and the resin will have gelled.

The quantity of volatile material present at pressurization is important in order to determine the amount of FHP required to counteract the formation of volatile bubbles. If a resin has minimal moisture content, the amount of required FHP is significantly reduced. This effect translates as a reduced sensitivity to the FHP requirement during cure.

The last area of the chart, void porosity volume in the cured laminate (area 5), is the end result of the composite cure. If the volatile-gas pressure is greater than the FHP, then voids are able to form within the laminate. If the FHP is greater, then an essentially void-free laminate is produced.

This chart summarizes the whole curing process into the void/porosity volume argument. This is based on the assumption that if the time-temperature cure dynamics are correctly followed, a void-free laminate is ensured if all other pathways are followed. Failure to maintain a correct part ΔT relationship across a load or part will most likely manifest itself in the production of porous laminates. Attaining the correct degree of cure within the laminate while following a low void production pathway is the only way to ensure a high-quality laminate.

ACKNOWLEDGMENTS

The author gratefully acknowledges the assistance of these individuals for their help in the preparation of this article: Richard J. Hinrichs, Applied Polymer Technology, Inc.; Wilburn S. Smith, DuPont; Susan K. Ferer, General Dynamics, Convair Division; Amy Bryant, General Dynamics, Convair Division; and Richard Warnock, McClellan Air Force Base.

Closed-Loop Cure

Richard J. Hinrichs, Applied Polymer Technology Inc.

CLOSED-LOOP CURE OF COMPOSITES involves altering the cure based upon thermal and rheological changes taking place in the resin. By continuously making these changes, the resin is "guided" to its optimum quality and cure state, irrespective of its initial state. In contrast, open-loop cure does not have the benefit of feedback from the composite being processed. Rather, process parameters such as temperature and pressure are set by some predetermined schedule that frequently is based on time. Although changes are not made within a cure cycle, they may be made to subsequent cycles once the composite has been cured and evaluated. This article, however, will discuss the principles and techniques of closed-loop cure, including a general description of appropriate instruments and equipment.

Sensors are used to relay data during cure. The real goal of cure processing sensors has been to achieve a signal that defines the critical material properties responsible for the quality of the manufacturing process being evaluated. More important than just the sensing of those properties, however, is the premise that measuring those properties can contribute to intelligent control decisions that ensure the quality of the process. The concept of control by sensor data adds a subconcept: We must also know what state the material is supposed to be in, or where we have requested it to be, so that intelligent decisions to correct the processing environment can be made. Finally, both the real performance data obtained from the sensor and the desired requirements provide a vehicle to produce a quality report simultaneous to the process, which greatly improves the cost efficiency and accuracy of the production system. The principal cure parameters that the sensors need to measure for process control are listed below.

Degree of Cure, α. As the property that defines cure state (0 to 100%), rate of reaction ($d\alpha/dt$), exothermic release function (dq/dt), state transition (liquid-gel-vitrification), length of time to continue cure process, and mechanical properties, the degree of cure is clearly a critical parameter to the cure control/quality assurance (QA) process.

Complex Viscosity, Storage Modulus, Loss Modulus, and Tan Delta. These are the rheological viscoelastic properties of a material, represented by η^*, G', G'', and tan δ, respectively. They are principally responsible for the flow, compaction, and gas transport or diffusion problems that are encountered in laminate consolidation. They also offer a cross check on the reaction kinetic parameters and degree of cure and allow more than one sensor/data source to be used in order to cross check the measured properties to enhance accuracy.

Fiber Volume, Cure Ply Thickness, and Resin Content. A direct measurement of these properties is important to laminate QA. All design properties are based on a projected relationship of fiber volume to mechanical properties, which allows direct QA without time-consuming, secondary postcure operations in a laboratory digestion process.

Material Modulus Property Data. Generating mechanical strength data for QA simultaneous to the cure is a QA element with a cost reduction incentive. During the cure the material is already at elevated temperature, so property versus temperature data can be derived.

This data provides more realistic performance characteristics of the material, because it involves a real part in real time, than does a test coupon.

Cure Stress Residuals. Although this parameter has been greatly ignored, it has been responsible for many delaminations and cracking in space structure applications and carbon-carbon reentry rocket cone work. This parameter should be measured in order to compute the interlaminar cure stresses as a function of the cure process.

The benefits of using an approach based on these parameters are high-yield production loads with greatly reduced risk and cost, reduced costs through higher quality throughput during the manufacturing process, and a reduced need for secondary inspection procedures.

Principles of Closed-Loop Cure

Closed-loop cure may not be the best or most economical mode for processing composites.

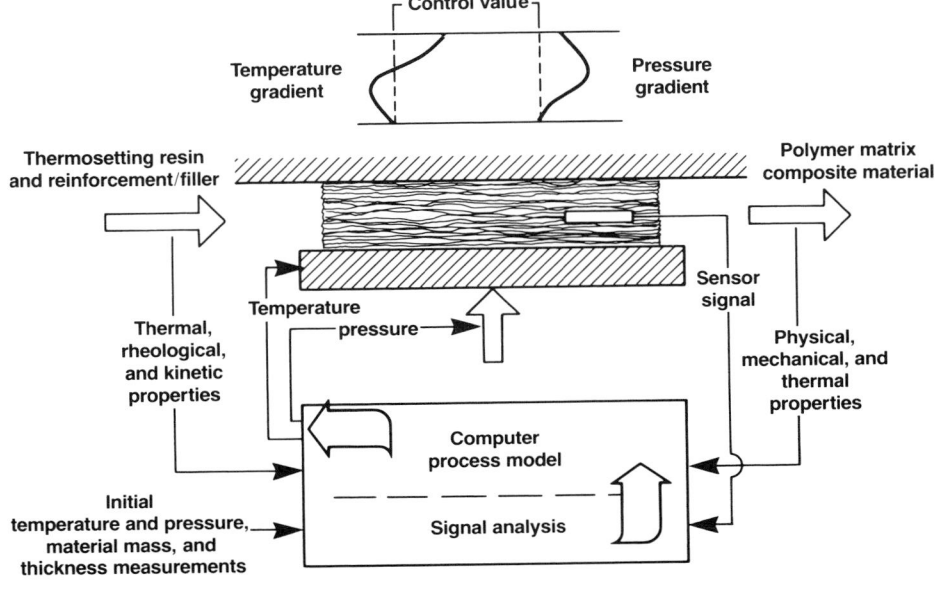

Fig. 1 Closed-loop cure process

For example, if labor and material costs are low, multiple trial processes with parameter changes being made from cycle to cycle may be adequate. Or, if composite performance is not critical (perhaps it just has to have a pleasing appearance), then the cost and complexity of closed-loop cure cannot be justified. However, for performance-driven composites of expensive materials with costly processing, closed-loop cure offers several advantages: controlled and quantified processes, a basis for systematic rather than intuitive process changes, and a procedure for repeatedly obtaining high-quality composites.

The basic elements of closed-loop cure are shown in Fig. 1, which illustrates a process in which thermosetting resin, combined with a reinforcement/filler, is changed by temperature and pressure into a composite material part (in many instances, the material and part are made at the same time). During the process, sensors detect physical and chemical changes in the resin. Sensor signals are monitored many times during the process at intervals that are short relative to the overall process time, and analyzed and fed into a process model. A comparison is made between actual and model values of important process parameters. Based on this comparison, process control signals are issued to reduce the difference between actual and model values. Four elements—sensors, monitoring, model, and control—are important to the closed-loop cure process.

Sensors are devices that are placed in, on, or near the material being processed to detect changes in the resin. All sensors use either mechanical means or electromagnetic radiation to detect these changes. One mechanical technique uses acoustic pulses in either the through-transmission or the pulse echo mode. In through-transmission, the time required for the pulse to travel through the material (from which the sonic speed is inferred) and the attenuation of the pulse by the material are indicators of the resin state. As the resin cures, the speed increases and the attenuation decreases, approaching limiting values characteristic of the composite. In the pulse echo mode, the sensor sends a pulse and then receives the resulting echo. As the resin cures, the echo return time (or time of flight) decreases relative to the original uncured time.

Monitoring is the process by which sensor signals are acquired and analyzed. Signal acquisition frequently is a function of electronic instruments and is beyond the scope of this article. However, signal analysis is very important to the successful implementation of closed-loop cure. The sensor response is usually compared to a reference value, and it is the ratio for the magnitude of this difference that is utilized in characterizing the material. For example, dielectric measurements will show the capacitance and conductivity of the resin as it cures. A more useful form of the information is obtained by converting these measurements to the storage and loss components of permittivity (in-

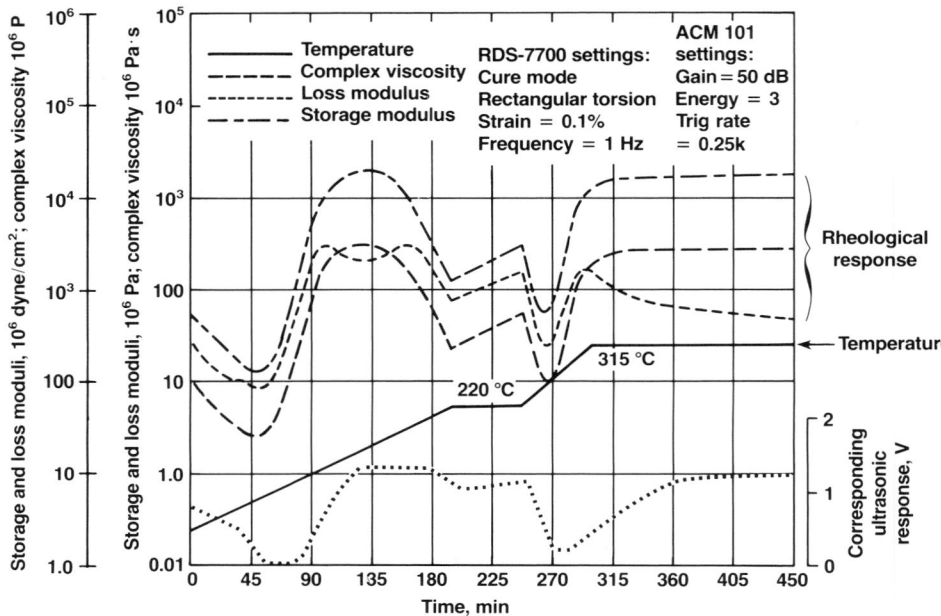

Fig. 2 PMR-15 cure cycle using Advanced Cure Monitor (ACM) 101 instrumentation for ultrasonic response. PMR-15B/Celion-3K-8HSW 42-in. B/G Lot G06197 USP 2130. Resin content: 42.0%. Flow: 16.3%. Volatile content: 9.7%

volving frequency and original capacitance). The ratio of storage and loss components, referred to as the loss tangent, also contains other useful information. These quantities can be used in monitoring because they are intensive variables that are independent of the amount of material and because they change as the resin cures. Whatever the form, the sensor signals must be analyzed in real time to obtain information needed for the next important step in closed-loop cure.

Modeling is the process of preparing a mathematical or numerical construct to indicate the output or cure state of the reacting resin system for a given input or sensor signal. Empirical models are based on experimental correlations between inputs and outputs. Mechanistic models are based on mechanisms or submodels that describe heat transfer, flow, and reaction within the resin and fiber. Mechanistic models require a great deal of information about the materials, such as thermal diffusivity, kinetic constants, and a relationship between reaction and viscosity. Such information rarely is available in the quantity and quality needed for mechanistic models. Consequently, practical processing models are likely to be a combination of empirical and mechanistic models.

All polymer composite processing models must account for heat and pressure transfer from the mold wall or boundary into the material interior. The heat reduces resin viscosity, thus facilitating flow, and initiates the typically exothermic chemical reaction of the resin. This reaction is necessary in order to convert liquid resin of low molecular weight into the higher molecular weight, cross-linked, solid matrix of a composite. The pressure compacts and densi-

fies the material and aids in heat transfer. It also eliminates voids, which are an inherent part of any precursor material. All models must strike a balance between flow and reaction. Pressure must be applied at times and levels that will eliminate voids. In addition, temperature must be increased rapidly to the highest possible values to minimize process time. The model provides a path through these competing mechanisms by which the resin can reach a state of cure that provides high-quality composites.

Control involves changing process parameters, such as temperature and pressure, to reduce the difference between actual and model cure state. Like sensor signal acquisition, control is partly a function of electronic instruments. However, significant nonelectronic delays, filtering, and amplification can have an effect on the accuracy of closed-loop cure. For example, most composite processing relies on mold wall temperature at a limited number of discrete points on the wall. However, these points may not always be representative of the overall temperature distribution. Depending on the distributed mass of the mold tooling and the method of mold heating, some sections of the curing material may be at significantly higher or lower temperatures than those measured. In addition, very thick sections can develop significant temperature gradients through the thickness. The temperature at the center of the curing material will lag behind that of the surface during rapid temperature rise, but may very well surpass the surface temperature once the exothermic resin reaction gets underway. Optimum temperature sensing points must be determined to minimize curing inaccuracies.

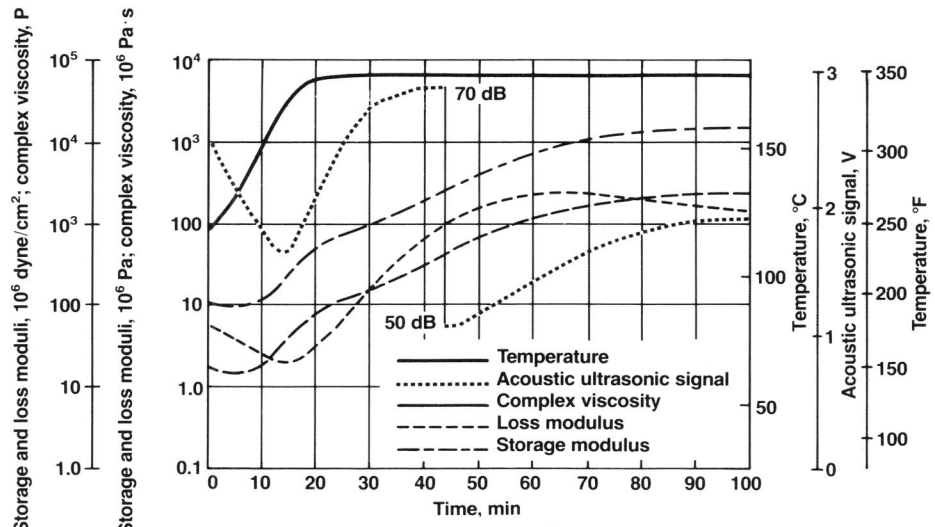

Fig. 3 Acoustic ultrasonic signal with rheometric data in gel-cure portion of cure cycle

Pressure also is applied at the mold wall, the boundary of the reacting system. The pressure aids compaction and heat transfer and suppresses voids near the surface. However, the pressure also will decay deeper in the material and may not be adequate for good compaction and void suppression.

The effect of these physical delays and filters is that the control system may be "mushy"; that is, while the rest of the closed-loop cure system may be working well, the resulting composite may be unsatisfactory unless these gradients and differences are taken into account. A submodel can be appended to the process model to account for material thickness and mold tooling mass. This submodel will modify the commands of the process model based on the ideal situation. In this way, the closed-loop cure system can maintain control over the process.

An example of this concept is illustrated in Fig. 2 and 3. Here the correlation between the viscoelastic properties of the material responsible for flow consolidation behavior of the laminate, or rheological properties, is directly reflected by the Advanced Cure Monitor 101 system. This system employs ultrasonic waveform evaluation of the material during cure. After comparing the signal response to the model calculations for optimum pressurization consolidation, the material can be processed for consistent results. This technique also allows quality control verification of appropriate reaction behavior of the material. Sensor data should conform to known reference reaction behavior for that material. Any significant deviation would indicate abnormal reaction behavior and can be readily identified as an inspection criterion.

SELECTED REFERENCES

- R.J. Ferris and C. Lee, Determination of Time Independent Component of the Complex Modulus During Cure of Thermosetting Systems, *Poly. Eng. Sci.*, Vol 23, July 1983, p 586-590
- R.J. Hinrichs and J.N. Thuen, Control of Composite Cure Processes in *Role of Polymeric Matrix in Processing and Structural Properties of Composite Materials*, J.C. Seferis and L. Nicolais, Ed., Plenum Press, 1983, p 147-158
- D. Kranbuehl, *et al.*, Dynamic Dielectric Analysis: Monitoring the Chemical and Physical Changes During Cure of Thermosets and Thermoplastics, in *Polym. Mater. Sci. Eng.*, Vol 54, 1986, p 535-540
- R.L. Levy, New Fiber-Optic Sensor for Monitoring the Composite-Curing Process, in *Polym. Mater. Sci. Eng.*, Vol 54, 1986, p 321-324
- A.M. Lindrose, Ultrasonic Wave and Moduli Changes in a Curing Epoxy Resin, *Exp. Mech.*, June 1978, p 227-232
- H.L. Price, Phenomenological Model of Isobaric Processing of Composite Materials, in *Proceedings of the Second International Conference, Reactive Processing of Polymers*, Nov 1982, University of Pittsburgh, p 202-216
- S.D. Senturia, *et al.*, *In-situ* Measurement of the Properties of Curing Systems With Microdielectrometry, *J. Adhesion*, Vol 15, 1982, p 69-90
- A. Sofer, *et al.*, Cure of Phenol-Formaldehyde Resin, *Ind. Eng. Chem.*, Vol 45, 1953, p 2743-2748

SECTION 11

Failure Analysis

Chairman: John E. Masters, American Cyanamid Company

Introduction

FRACTURES AND FRACTURE SURFACES have been examined and studied for centuries as a facet of the art of metallurgy. As a result, failure analysis procedures for metallic materials have been well established for some time. However, analogous procedures for composite materials are not as well defined.

The rapid advancement of these materials over the past two decades has outstripped the development of appropriate failure analysis techniques and requisite fractographic data. This is particularly true of the continuous fiber composite material systems used in primary structural applications in the aircraft industry. Although some of the knowledge gained over the years in performing failure analysis on metals is applicable to composites, the fundamentally different nature of the two materials prohibits the widespread transfer of information.

Investigators are now beginning to modify failure analysis techniques developed for metals so that these techniques can be applied to composite materials. Much recent progress has also been made in developing fractographic data for these materials. The objective of this Section, therefore, is to present a complete review of this topic as it is currently practiced. Because this is an area of ongoing study, the procedures employed may be expected to change as the experience of the investigators increases. Similarly, the catalog of fractographic features characteristic of various failure modes of materials is also expected to increase with experience.

As a first step in developing these failure analysis procedures, the U.S. Air Force has, over the past few years, sponsored investigations to analyze failures of continuous fiber reinforced systems. Several reports summarizing the work completed to date have been issued (Ref 1-4). In addition, two follow-up investigations are currently underway to extend these failure analysis procedures further (Ref 5, 6).

The procedures developed in the programs mentioned above involve five investigative operations:

- Collection and review of component history information
- Nondestructive inspection
- Evaluation of component conformity to specifications
- Fractographic examination
- Stress analysis

The sequence in which these operations are employed is illustrated in Fig. 1 (Ref 7).

The discussion of composite failure analysis given in this Section is organized around the operations listed above. Most of the information presented focuses on continuous fiber reinforced composite systems. However, the general procedure outlined is equally applicable to discontinuous fiber reinforced composite systems. Similarly, the general causes of failure, as outlined in Ref 3, are common to both types of material. (See the article "Failure Causes" in this Section.)

The five steps listed above are, in general terms, similar to those employed for other material systems (see the article "Basic Failure Modes of Continuous Fiber Composites" in this Section). However, specific differences exist that are due to material anisotropy and heterogeneity. These differences are apparent in the areas of nondestructive examination, materials verification, and fractography. Therefore, discussions of the use of thermal analysis techniques and destructive and nondestructive test techniques in composites failure analysis are included in this Section. The information presented is applicable to both continuous and discontinuous fiber composites. However, the microstructures of these two types of composites dictate that they will fail in different manners. Therefore, the fractography and failure modes of these two types are discussed separately.

As shown in Fig. 1, component background and service history information is collected and reviewed during the initial stage of the investigation. This includes, when available, data regarding component fabrication, loading, and exposure to environmental conditions. Nondestructive evaluation and thermal analysis procedures are applied after this initial step, which includes a detailed visual inspection and photographic documentation.

A number of nondestructive test methods that have been modified and adapted to monitor damage progression in composite materials will be described. These techniques include ultrasonics, acoustic emission, thermography, stiffness measurement, replication, and radiography. In addition, the two destructive techniques that have proved useful in failure analysis—matrix digestion and deply—will be reviewed. The deply technique is used to obtain information on the nature of fiber fracture in the interior of a composite laminate. Matrix digestion techniques, which determine fiber and void volume fractions, are used to confirm that the material is fabricated according to specification.

In addition to matrix digestion, optical microscopy and thermal analysis techniques are also employed as material/configuration verification procedures. Optical microscopy is used to determine ply orientation and to confirm laminate stacking sequence. It is also used to identify resin-starved or resin-rich regions and to locate misaligned fibers. Thermal analysis techniques are used to probe the matrix resin to determine its glass transition temperature and degree of cure. They are also used to fingerprint the resin by determining its properties over a wide temperature range. The application of these techniques to failure analysis will also be discussed.

Fractography is the detailed, microscopic analysis of the surface of a fracture to determine its cause and the relationship between the fracture mode and the microstructure of the material. It is used to identify the origin of the crack and the crack propagation direction. It can also be used to determine the type of loading that caused the crack to initiate and propagate. Fractography can, therefore, play a key role in failure analysis.

The fractography of composite materials is complicated by the fact that these materials exhibit failure modes that are not normally encountered in metallic materials. The failure types and modes that can develop depend on the direction of applied load and on the orientation of the fibers. Despite the complexity of these failure modes, progress has been made in recent years in identifying fractographic features that are characteristic of certain failure modes. Detailed discussions of these results will be presented for both continuous and discontinuous fiber reinforced composite systems.

The failure analysis of composite materials is an area of active research that is only now beginning to be fully defined. It is hoped, however, that the information presented in this Section will aid investigators initiating failure analyses of composite materials.

REFERENCES

1. K.M. Liechti, J.E. Masters. D.A. Ulman, and M.W. Lehman, "SEM/TEM Fractography of Composite Materials," AFWAL-TR-82-4085, Air Force Wright Aeronautical Laboratories, Sept 1982

2. B.W. Smith, R. Grove, and T. Munns, "Failure Analysis of Composite Structure Materials," AFWAL-TR-86-4033, Air Force Wright Aeronautical Laboratories, May 1986

3. R. Grove and B.W. Smith, "Compendium of Post Failure Analysis Techniques for Composite Materials," AFWAL-TR-86-4137, Air Force Wright Aeronautical Laboratories, 1986

4. B.W. Smith, R. Grove, and T. Munns, "Failure Analysis of Composite Structure Materials," AFWAL-TR-87-4011, Air Force Wright Aeronautical Laboratories, 1987

5. Composite Failure Analysis Handbook, Contract F33615-86-C-5071, U.S. Air Force, to be published

6. Composite Failure Analysis Manual, Contract F33615-87-C-5212, U.S. Air Force, to be published

7. S. Witzler, Composite Failure Analysis, Advanced Composites, Nov/Dec 1986, p 28-33

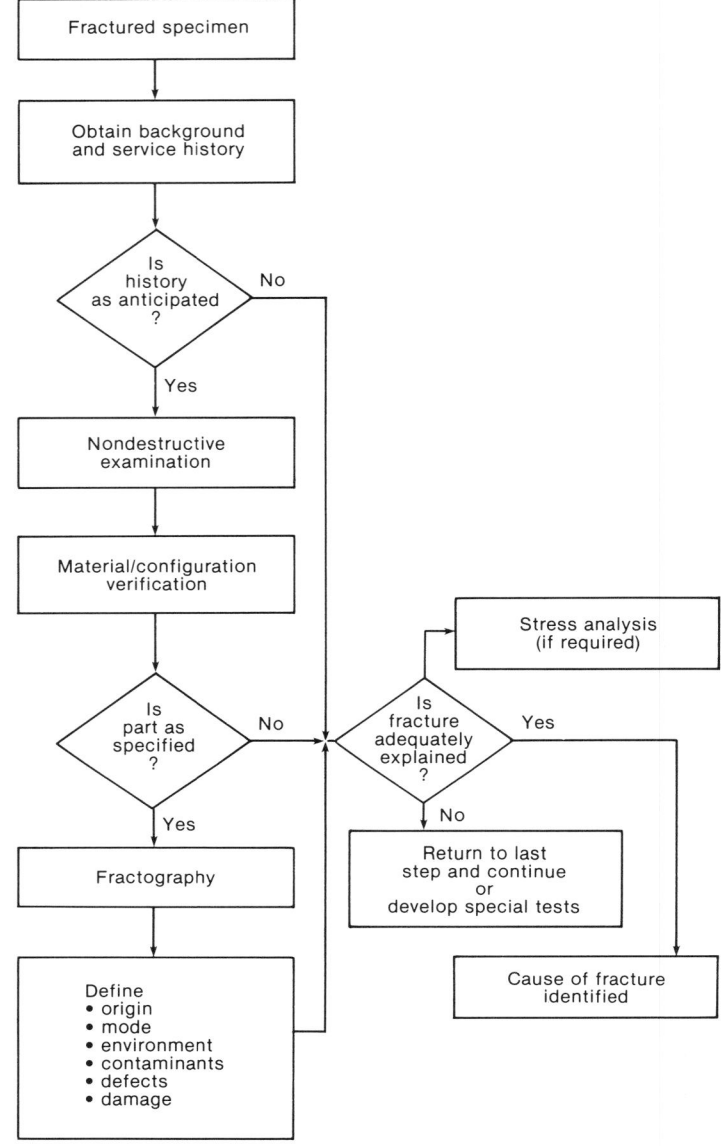

Fig. 1 Failure Analysis Logic Network (FALN) for determining sources of failure in composite parts. Original FALN was developed by Boeing. Source: Ref 7

Failure Causes

Ray A. Grove, Boeing Commercial Airplane Company

NEW CAUSES OF FAILURE IN COMPOSITE MATERIALS are still being uncovered as service experience is gained. Current knowledge, however, indicates that many of the basic sources of failure that occur in metals are likely to be observed in composites. These sources include three basic categories of causes: errors in design, errors in fabrication and processing, and anomalous service conditions. Given the construction, properties, and sensitivities of composite materials, the specific causes that may occur and should be considered during an analysis are worth reviewing.

Design Errors. Composite materials are somewhat unique in that both the fundamental properties of the material and the configuration of the component to be fabricated are subject to design. Correspondingly, design errors can be made at both the material and structural levels of design. Engineering errors related to the material may include a variety of problems. The more common of these include errors in analyzing the effect of individual ply anisotropies or the inadequate assessment of material damage and environmental sensitivities.

Because the level of stress carried by each ply in a uniformly strained laminate depends upon its modulus, large stress gradients and internal shear stresses can exist between plies oriented at significant angles to one another. Such stress gradients can lead to premature fracture, particularly where the magnitude of these gradients is large. Such large gradients are particularly common where groups of adjacent plies with the same orientation are oriented at 90° to another group of adjacent plies.

The highly anisotropic coefficient of thermal expansion of composite materials represents another area where design errors can be made at a material level. Many composite materials exhibit significantly large differences in thermal-expansion coefficients, depending upon their fiber orientation. As a result, changes in temperature, that is, temperatures that differ significantly from the curing temperature, can induce internal stress gradients where plies are oriented at significant angles to one another. These internal stress gradients are analogous to those generated under applied mechanical loads. Because many high-performance composites are cured or formed at elevated temperatures, the cooling of these parts during processing to ambient conditions can induce these internal stresses in the as-fabricated condition. For many designs, the magnitude of these stresses and those generated by additional temperature variations may be relatively inconsequential. However, high internal stress levels may develop in laminates with groups of adjacent plies oriented at large angles to one another or in such structures as space vehicles, in which extreme variations in temperature occur.

On a more general level, errors in material design can include many of the same problems encountered in metals. Failures can be caused by inadequate understanding of environmental sensitivities, the effect of damage, or the fatigue sensitivity of the material used. Because the properties of composites depend upon their ply or fiber orientation and stacking sequences, the sensitivity of each design may vary significantly, posing a potential problem for design. In addition, synergistic effects may exist between these factors, giving rise to further sensitivities not considered during normal design practices.

Design errors relating to the component itself are likely to include unconsidered load sources, stress concentrations, and unanticipated buckling instabilities or modes. As with their metal counterparts, such failures may occur as a result of oversight. In most cases, thorough testing during design uncovers most of these errors. However, those related to fatigue or rarely attained load conditions may not become apparent until well into the life of the part.

Errors in Fabrication and Processing. Typically, the occurrence of defective or anomalous conditions is controlled and prevented by manufacturing controls and material inspection testing imposed during the fabrication process. However, because absolute control and inspection is generally economically infeasible and because human errors do occur, these control and inspection methods sometimes allow occasional errors.

Continuous fiber reinforced composites are usually fabricated by laminating together and curing multiple plies impregnated with unreacted matrix resin. Within this fabrication operation, a number of errors can occur. Because each of the individual plies involved in a laminated composite has highly anisotropic properties, its placement and orientation can be critical in achieving the desired engineering properties. This is particularly true for composites in which each individual ply constitutes a significant percentage of the total laminate, that is, thin gage structures. For example, in a unidirectional laminate, a variation in overall fiber orientation of 15° can generate up to a 50% reduction in ultimate strength.

For thermosetting matrices, reacting the matrix resin represents one of the more critical steps. Either improper amounts of the two resin components or the inadequate application of heat during curing can produce conditions of undercure. Such conditions, when extensive, can significantly degrade the properties of the matrix and its resistance to chemical or environmental exposure. Similarly, inadequate compaction during the lamination process can result in extensive porosity and reductions in material strength and durability.

Anomalous Service Conditions. Particular service anomalies include improper operation or use, faulty maintenance and repair, overloads due to failure of a related part, and environmental- or service-incurred damage beyond that reasonably anticipated. Many of these causes are not unique to composite materials. However, because of their construction, composites are particularly affected by some conditions more than other materials.

The engineering properties of composites can be significantly reduced by variations in temperature, foreign object impact damage, and, with some resin systems, chemical attack. With thermosetting matrices, the effect of temperature can become quite detrimental, particularly if moisture has been absorbed into the resin system. Property reductions due to foreign object damage can be equally extreme. Studies have shown that moderate levels of impact energy can reduce material strength up to 60% (Ref 1).

REFERENCE

1. M.B. Rhodes, Damage Tolerance Research on Composite Panels, in *Selected NASA Research in Composite Material and Structures*, NASA CP-2142, National Aeronautics and Space Administration, Aug 1980

General Considerations for Continuous Fiber Composites

Ray A. Grove, Boeing Commercial Airplane Company

CONTINUOUS FIBER REINFORCED COMPOSITES are seeing significantly expanded levels of use in hardware where reduced weight is critical. This is primarily a result of their tailorability as well as their high strength and modulus-to-density ratios. In recent years, these materials have seen applications ranging from mass-produced tennis rackets to relatively complex structures, such as the wings of the AV-8B Harrier aircraft. As the use of these materials expands, so also does the likelihood of eventual failures. As with their metal counterparts, the occurrence of failure is likely to represent a relatively rare event that is not encountered with most hardware usage. However, when such failures occur, the ability to determine their origin and cause constitutes a critical step that is necessary in providing valuable engineering feedback and ensuring the continued integrity of the components during service.

This article will review the basic methods required to carry out a rudimentary analysis into the cause of failure for continuous fiber reinforced materials. Because these materials are significantly different from their metal counterparts, this article will deal with several considerations unique to composite materials. Specific topics that will be examined include potential failure causes, typical failure modes, and analysis methods for the identification of the origin and direction of fracture. However, because composite materials failure analysis studies have only recently been initiated, it is important to recognize that this article contains many procedures and data that are under development or are not wholly understood for all the possible composite structures and material systems that may be encountered.

Types of Composites

Composites, by definition, are those materials made up of an amalgamation of separate parts or microstructural elements. This amalgamation combines the attributes of each of the separate materials involved, typically resulting in improved properties for the system as a whole. As with metallic alloys, a wide range of discrete material phases and microstructural arrangements is available for composite materials. A variety of composite material microstructures are shown in the article "Fiber Composite Materials" in Volume 9 of the 9th Edition of *Metals Handbook* (Ref 1). The complexity presented is further complicated because composite materials can be easily tailored during fabrication, with the resulting structure and properties dependent upon the process and material forms used. Composite materials can be divided into two basic categories, depending upon their type of reinforcement: continuous fiber reinforced and particulate/short fiber reinforced composites. The first of these two materials typically uses a continuous array of oriented fibers, whereas the second category of material uses randomly dispersed particulates or chopped fibers. In this

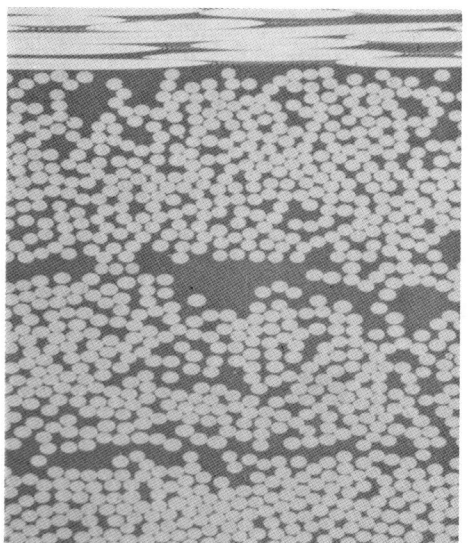 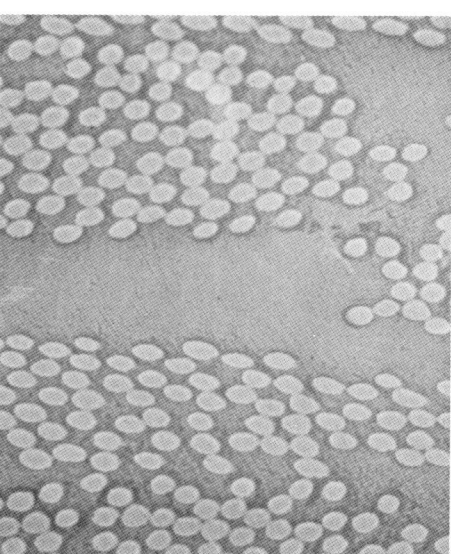 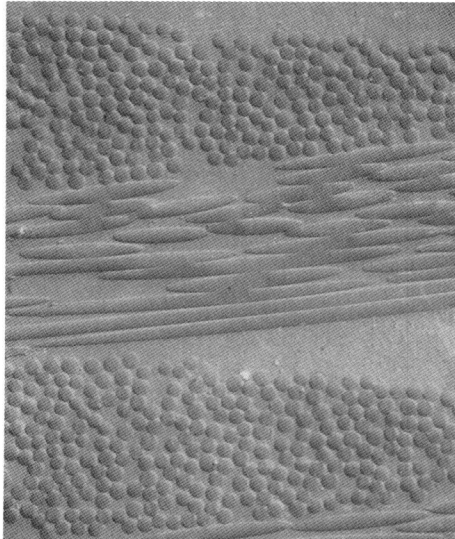

Fig. 1 Microstructures of common composite materials. (a) Graphite-epoxy tape. (b) Aramid-epoxy fabric. (c) Fiberglass-epoxy fabric. All at 200×

article, only continuous fiber reinforced composites will be considered.

Continuous Fiber Composites. Materials in this group are typically made up of 3- to 30-μm (120- to 1180-μin.) diam fibers that are oriented and surrounded in a supportive matrix material. Generally, the fibers used in these material systems are several orders of magnitude stiffer and stronger than the surrounding matrix. Fibers typically used for such applications include graphite, aramid, and fiberglass (Fig. 1). Encompassed within each of these generic fiber types are a number of specialized fibers with specifically tailored properties, for example, high-modulus or high-strength graphite. The high stiffness and strength of these fibers control the characteristic engineering properties, such as tensile, compressive, and shear moduli and strength.

The continuous matrix phase functions as a supportive element to the fibers. In this capacity, fiber orientation and alignment is maintained, load is transferred between fibers, and strength is provided in nonreinforced directions. A wide variety of matrix materials are available for use with each fiber type. By far the most extensive matrices in current use are those employing either elevated-temperature curing epoxies or room-temperature curing vinyl esters. Both of these matrix materials represent a broad class of polymers known as thermosets. These polymers are reacted during processing to form a stable matrix system. Generally, the choice of one thermosetting system over another depends upon a number of variables, including the environment under which usage is likely to occur. Typically, systems cured at elevated temperature exhibit better strengths under elevated temperature, absorbed moisture, and solvent attack than those cured at room temperature. More recently, thermally moldable matrices—thermo-

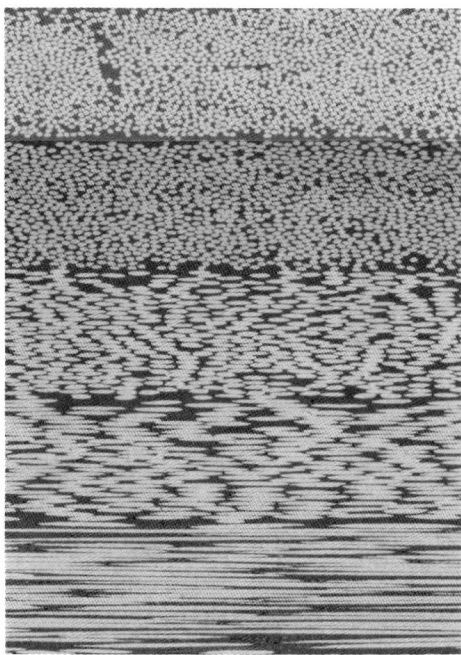

Fig. 2 Cross-sectional optical micrograph illustrating laminated construction typical of continuous fiber composite materials. 50 ×

plastics—have begun to be used for continuous fiber composite applications. These matrices differ from thermosets in that their polymer network can be repeatedly molded and reformed with the application of heat and pressure.

While the principal engineering properties of a composite tend to reflect the constituent fiber used, it is important to recognize that the matrix material contributes significantly to these properties. In distributing fiber-to-fiber loads and supporting the fiber under axial compression, matrix properties directly control compression strength, damage resistance, and residual strength. As a result, both the matrix and the fiber must be considered when carrying out a failure analysis investigation.

One of the unique attributes of continuous fiber reinforced composites is their tailorability. With continuous fibers, control can be exercised over both the orientation and amount of load-carrying fiber reinforcement. This control is generally achieved during manufacture by using fibers arranged in either a fabric cloth or unidirectional tape. By stacking thin layers, a wide variety of thicknesses and fiber orientations can be produced.

Design. Because of their continuity and oriented structure, fiber-reinforced composites have highly anisotropic properties. For example, unidirectional tapes commonly exhibit moduli in the fiber direction 20 times greater than that of the transverse direction. In most designs, this anisotropy is tailored by arranging plies of material (either tape or fabric) at a variety of angles. This structuring produces a stacked, laminated construction (Fig. 2). In the aerospace industry, the angle of plies used in the laminated stack are oriented at fixed angles to the direction of major load. The most common fixed angles are 0, 45, −45, and 90° to the major load axis. By selecting the proper number of plies at each of these angles, composite properties and their anisotropies can be designed for strength, modulus, or even the degree of thermal expansion along each principal direction.

REFERENCE

1. Fiber Composite Materials, in *Metallography and Microstructures*, Vol 9, 9th ed., *Metals Handbook*, American Society for Metals, 1985

Failure Analysis Procedures

Brian W. Smith, Boeing Commercial Airplane Company

FAILURE ANALYSIS PROCEDURES FOR COMPOSITES are generally similar to those used for other material systems. As with any thorough analysis, prudent investigation should follow a basic sequence that at the very least includes a review of available in-service records, a preliminary nondestructive examination of the part, verification of use of the intended material, and the fractographic determination of the origin and mode of fracture. In many cases, evaluation of the stress and fracture mechanics involved in failure will also be necessary to determine the cause adequately. Because this sequence of steps is basically the same for metals, the discussions in the article "General Practice in Failure Analysis" in *Metals Handbook*, are a good source of more detailed information (Ref 1). Most of the methods described in the aforementioned article can be directly applied to composites. However, because of their laminated construction and chemistry, some specific differences exist in the areas of nondestructive examination, materials verification, and fractography.

Nondestructive Examination

Because of their laminated construction, conditions of interlaminar fracture (delaminations) represent a relatively common mode of failure. Because these planes of fracture occur within the laminate, they are not generally visible during routine visual examinations. As a result, nondestructive examinations represent a particularly critical step required to define both the extent and type of damage involved in component failure.

The most successful methods of nondestructive inspection include through-transmission ultrasonics (TTU), pulse-echo ultrasonics, and dye-penetrant enhanced x-ray radiography. In the first two methods, internal damage is detected by the attenuation or reflection of 1- to 25-MHz acoustic waves transmitted into the panel through a transducer. The detection of damage by x-ray radiography is carried out in the same basic way as for metals, except a radiopaque penetrant is wetted into the crack surface to enhance damage visibility. In general, x-ray equipment operable in the 10- to 50-kV, 5-mA range with diiodobutane (DIB) penetrant yields the best results.

Use of any of these nondestructive evaluation techniques requires that some forethought be given to the type of failure mode likely to be encountered. With both ultrasonic methods, liquids are generally used to carry acoustic waves into the panel. Similarly, the radiopaque penetrant used in radiography is also a liquid. The penetration of these liquids into the fracture surface can constitute a form of chemical contamination. In those cases where chemical surface-analysis techniques, such as x-ray photoelectron spectroscopy, are to be used, the presence of such contaminants should be eliminated or prevented from entering critical areas of failure by sealing areas of exposed fracture.

Materials Characterization

As indicated earlier in this article, errors in material lay-up, ply orientation, and degree of matrix curing can lead to or, at the very least, contribute to premature component failures. As a result, the analysis of these basic material and processing errors should be considered a standard operation in most failure analysis cases.

Because the matrices employed in most composites are organic in nature, the conventional test methods used for the evaluation of metals, such as hardness testing, do not translate well to composites. For these materials, assurance of material cure is generally best obtained by measuring both the matrix resin glass transition temperature, T_g, and residual heat of reaction, ΔH. Both of these measurements use standard thermodynamic instrumentation.

The temperature at which the matrix undergoes a transition from a glassy to a rubbery state is measured by monitoring the rate of change of one of several material properties (such as thermal expansion or stiffness) versus temperature. By comparing against known standards, a judgment as to the degree of material cure can be made. In those cases where this value appears lower than normal, an additional assessment of the degree of cure can be obtained by measuring the residual heat of reaction of the material. This measurement is made by using a calorimeter in which heat evolution as a function of temperature is measured. The magnitude of this measured value characteristically corresponds with the degree of material undercure.

In much the same manner as with metals, the number and orientation of plies within the laminate can be examined through simple metallographic cross-sectioning procedures. Polishing is best accomplished with wheels covered with cloths without appreciable nap, such as nylon or silk. Etching is generally not required, because of the distinct optical differences that exist between most fibers and matrix systems. An exception to this is fiberglass, for which a dilute hydrofluoric acid etch may be necessary to enhance fiber visibility. Because most plies are separated by a resin-rich region and because each fiber exhibits a different profile (round, elliptical, and so on) depending upon the cross-section angle, the identification of the number of plies and their orientation is relatively simple. Detailed information on metallographic procedures germane to composites can be found in Ref 2.

Fractography

As pointed out in this article, failures in composites can produce complex fractures exhibiting interlaminar, intralaminar, and translaminar planes of separation. Because the occurrence of such conditions as interlaminar failure can often be quite extensive and can occur on multiple planes, the decision as to which surface to examine and how to examine it often represents one of the most difficult tasks involved in failure analysis.

Compression failures generally exhibit extensive interlaminar fracture, along with through-thickness translaminar-compression failure. Because translaminar-compression fracture surfaces generally exhibit significant postcompression damage, examination tends to be the least informative. On the other hand, interlaminar fracture surfaces for this load state are often undamaged. As a result, the direction and origin of these failures can often be established. Such interlaminar failures may occur secondarily to translaminar fracture. However, their direction of propagation can often be used to infer the origin of primary failure because their occurrence is related to the progression of compressive failure.

In contrast to compression failures, tensile fractures exhibit very little surface damage and lesser amounts of interlaminar failure. As a result, interlaminar, intralaminar, and translaminar separation surfaces can all be examined. For these failures, the selection of which surface to be examined should depend upon the predominant failure mode exhibited and the amount of time available. Where possible, examinations should be carried out on each mode of separation exhibited.

Because the features used to map the direction and mode of failure in composite materials are relatively fine, the examination of large areas of fracture surface can represent a sizable undertaking. As a result, it is important to recognize that many of the features described for interlaminar failures, for example, river patterns and hackles, can be identified using medium-power optical microscopy (200 to $600 \times$). For such large-area failures, the use of optical microscopy before electron microscopy allows several orders of magnitude more area to be examined (details on the principles and applicability of optical microscopy can be found in Ref 3). The ability to survey a large area of fracture is particularly important because extreme variations in the direction of crack growth and failure mode can occur on a local scale.

Scanning electron microscopy (SEM) of composites usually requires gold or gold-palladium coating of the fracture surface to prevent charging and to improve secondary electron yields (The use of conductive coatings in SEM analysis is described in Ref 4). Alternatively, many of the newer scanning electron microscopes can be operated at low electron voltages, eliminating the need for such coatings. Once within the microscope, the analysis of these materials can be carried out in the manner described in the article "Fractography for Continuous Fiber Composites" in this Volume.

Transmission electron microscopy (TEM) of composite fracture surfaces has seen relatively little application to date. While this technique should be translatable to these materials, replication of the fracture surface may prove damaging (TEM replication procedures are described in Ref 5). This may be the case for delicate structures, such as hackles, or for those matrix systems susceptible to attack by replica solvents. For this reason, adequate optical and scanning electron microscopy should be carried out where possible before fracture-surface replication to prevent premature destruction of fracture evidence.

Stress Analysis

In most cases, an accurate understanding of the loads and stress levels in component operation is one of the critical ingredients involved in defining the source of failure. Although other methods of analysis may identify the origin and mode of crack propagation, stress analysis most often provides a quantitative explanation for the cause of failure. Through this analysis step, engineers involved in future or corrective redesigns are provided with direct feedback regarding the actual loads experienced by the part, poor design practices and configurations, and the effectiveness of the analysis methods used in design.

Stress analysis procedures for composite materials can be relatively complex, because of several factors. Because composite materials are fabricated by the lamination of highly anisotropic plies, a nearly infinite variety of directional moduli and strengths can be achieved. Because of this tailorability, a different set of material properties must be considered for each failure case being examined. As a further complication, because of their laminated anisotropic construction, significant variations in stress can exist within the laminate itself. As a result, consideration must be given to failure at both the microstructural (individual ply) and the overall laminate levels. Reflective of this latter problem, the analysis of failures in composite materials has taken two basic approaches. Failures are generally analyzed at the individual ply level (lamination theories) or at the gross laminate level.

Typically, analyses at the individual ply level provide the greatest level of detailed information. This method predicts basic material properties by modeling the composite as a plate built up from individual plies, each with its own discrete lamina properties, that is, lamination theory (Ref 6). These lamina properties may involve considerable detail—for example, moduli, hygrothermal expansion coefficients, moisture diffusivity, and density—but can in general be estimated based upon individual fiber and resin properties using micromechanics equations. Through this method, the stiffness characteristics of the overall laminate can be determined for the general case of in-plane and/or flexural loads. Based upon this stiffness, the gross stress generated for any given load can be calculated.

However, as noted above, consideration must also be given to the stresses carried by each individual ply. These stresses depend on the modulus and orientation of each ply and can be determined using basic lamination theory. In most cases, individual ply stresses are generally different from the stress state of the laminate as a whole. This is due to their orientation, the constraint of neighboring plies, and the residual stresses from temperature or absorbed moisture. For example, uniaxial loading of a composite laminate that consists of plies with different orientations will result in a biaxial state of stress that varies from ply to ply. One of the chief advantages of lamination theory is its ability to quantify these stresses and to examine them on an individual ply level.

Although the lamination theory provides a powerful tool with which to examine the stress state operating within a composite, there are several disadvantages. The most notable of these is in its application in predicting the onset of failure. Such predictions are achieved by evaluating the stress within each individual ply against semiempirical models that define the maximum stress envelope to which a ply can be exposed without failure. However, predictions of the onset of interlaminar and intralaminar failure using such criteria have not been entirely successful. One of the causes of this lack of success centers around the fact that standard lamination theory is unable to predict the three-dimensional state of stress that exists in laminates near free edges and other discontinuities, for example, holes, cut-outs, and ply drop-offs.

Various approaches can be used, however, to predict the three-dimensional state of stress within a laminate. Typically, these methods tend to be rather specialized, including finite-element analysis, for example, generalized plane strain or three-dimensional elements, and approximate strength-of-materials methods (Ref 7). Using these methods, failure criteria based on principles of fracture mechanics (Ref 8) have been somewhat successful in predicting the onset and growth of interlaminar and intralaminar matrix damage, including fatigue effects.

The prediction of catastrophic failure at the laminate level has followed two distinctly different approaches. The first, which may be referred to as a statistical approach, uses coupon test data to determine the maximum allowable strains for each laminate design. Using a statistical approach, traditional methods of stress analysis can be applied (load versus area) to determine the occurrence of failure, without regard to individual ply or interlaminar stresses. In this case, failure occurs when the statistically determined maximum allowable strain is exceeded. The second approach to predicting translaminar failure involves a more mechanistic approach in which the load redistribution due to intralaminar and interlaminar matrix damage is taken into account. In its simplest form, this approach can be applied directly with lamination theory and has been referred to as the global ply discount method (Ref 9). Applications of this method rely on a predetermined definition of characteristic intralaminar and interlaminar damage. Using the lamination theory, the state of stress surrounding this damage is defined, and the magnitude of the applied load necessary to create failure is determined.

Translaminar failure near holes, bolts, and through-thickness cracks has been successfully modeled using modifications to fracture criterion that were used with metals. These modifications involve the determination of semiempirical correction factors that are referred to as characteristic dimensions. These characteristic dimensions describe the distance from a stress concentration for which complete laminate failure will occur if the ultimate strength of the material is reached at this point. A variety of characteristic dimension fracture criterion have

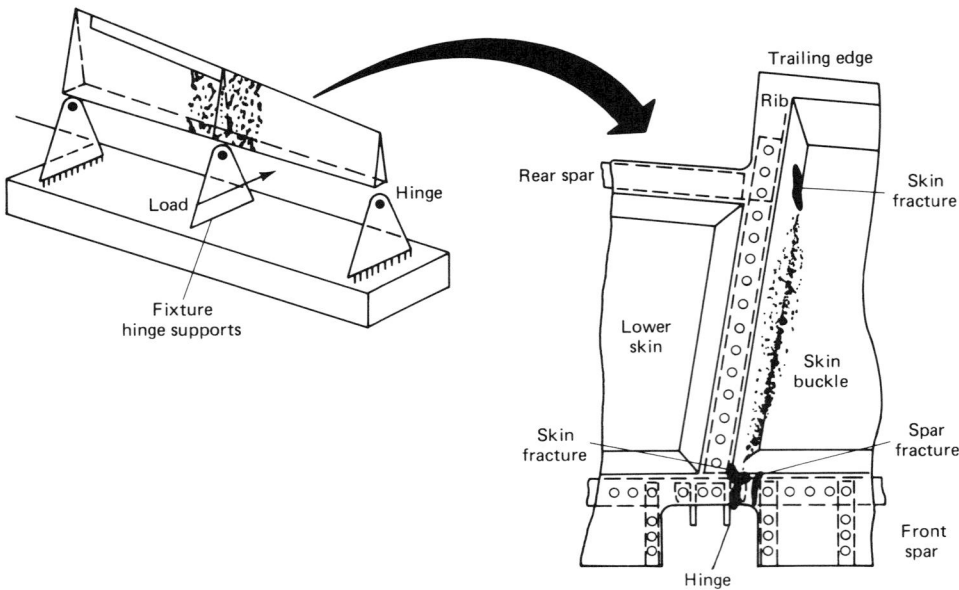

Fig. 1 Schematic of fractured component showing orientation and direction of applied load and approximate fracture location. Source: Ref 10

Buckling Instability. Figure 1 illustrates a portion of a graphite-epoxy tapered-box structure that fractured during testing. This graphite-epoxy box consisted of two honeycomb skin panels fastened to a spanwise spar with intermediate chordwise ribs.

Investigation. A review of the test history revealed that premature fracture occurred during hingeline deflection of the front spar (Fig. 1). Initial nondestructive visual inspection of the fractured box revealed through-thickness cracks in the forward and trailing edges of the compression-loaded skin panel. Upon further examination, some localized buckling of the skin panel was evident between each of these through-thickness fractures. To define areas of nonvisible damage, that is, delamination, a nondestructive evaluation was performed with TTU. This analysis revealed a roughly 10-cm (4-in.) wide band of delamination between the areas of through-thickness skin fracture at the front and rear spars.

Following the definition of the type and extent of fracture, tests were performed to determine if any major material discrepancies existed in either fabrication or processing. Accordingly, sections of the skin, spar, and rib panels were examined to verify the lay-up and to determine the overall panel quality. In addition, thermomechanical analyses (TMA) were performed to verify the extent of cure. Dimensions of panel, spar, and rib details were also measured and checked against required dimensions and tolerances. For each of these analyses, the spar, ribs, and skin panels were found to be in compliance with the drawing requirements.

Because no discrepancies were identified, fractographic examinations were selected as the

been found to be functionally equivalent for engineering applications (Ref 10, 11). Although these correction factors have been proposed for both tension and compression loads, most of the work discussed in the literature has addressed the former.

Analysis of composite panel or column stability is similar to that used for metals. A finite-element approach is used for postbuckling analysis of stiffened panels or box structures. Laminate stacking sequence and ply orientation are important to composite plate-stability. Residual strength after impact damage has been found to be a controlling design parameter for composite laminates. This is particularly true for compression loads in which localized ply buckling and interlaminar damage growth can lead to overall panel failure at load levels that are as low as 40% of the undamaged strength. Methods to predict residual strength after impact have ranged from empirical to an approach involving three-dimensional finite-element models of impact damage and fracture mechanics (Ref 11).

Example 1: Compression Fracture of a Graphite-Epoxy Test Structure Due to a

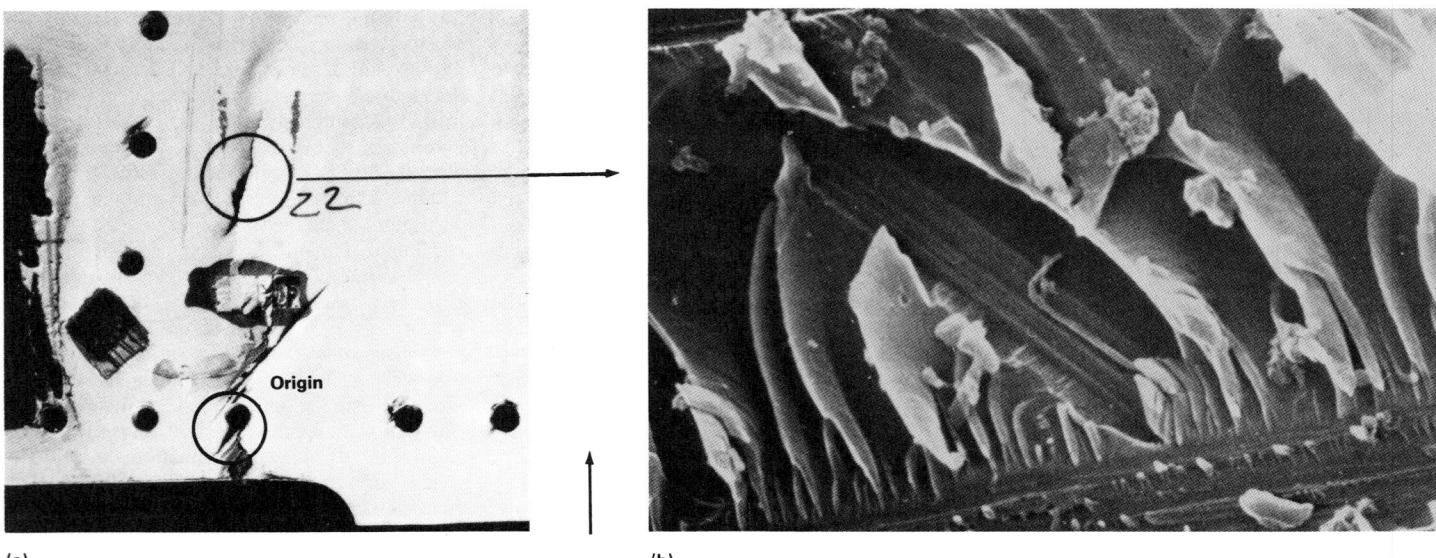

(a)

(b)

Crack direction

Fig. 2 Crack propagation direction and origin identified on fracture test component. (a) Optical micrograph. 0.4×. Crack origin occurred at indicated fastener hole. (b) Scanning electron micrograph of the circled area shown in (a). 5000×. Note river marks that coalesce in the direction of overall crack growth. Source: Ref 10

next investigative operation. Primary emphasis was placed on identifying the direction of crack propagation, origin, and any anomalous conditions that could be associated with fractures. To facilitate examination, fractured areas of the panel were sectioned into roughly 15- × 15-cm (6- × 6-in.) squares and examined optically. These optical examinations were performed at 400×, which provided a rapid and effective means of identifying characteristic fracture features. Scanning electron microscopy was performed on selected areas of interest to examine and document specific fracture-surface features. The orientation of river patterns and resin microflow (Fig. 2) observed on the fracture surface generated a map of the local directions of crack propagation over the fracture surface.

Conclusions and Recommendations. By reconstructing the fracture process, it was discovered that crack initiation occurred at the periphery of a fastener hole located at the front spar. Subsequently, propagation occurred chordwise across the compression-loaded skin panel. Inasmuch as no anomalies were identified at the origin area that might explain premature fracture, detailed stress analyses of this area were initiated. Both the basic in-plane strains and the buckling stability of the origin area were evaluated. These analyses revealed that premature skin buckling occurred because of a relatively large fastener spacing. As a result, further attention was paid to this design

detail, and fastener spacing was reduced to prevent the buckling mode that precipitated fracture.

REFERENCES

1. General Practice in Failure Analysis, in *Failure Analysis and Prevention*, Vol 11, 9th ed., *Metals Handbook*, American Society for Metals, 1986
2. Fiber Composite Materials, in *Metallography and Microstructures*, Vol 9, 9th ed., *Metals Handbook*, American Society for Metals, 1985
3. Optical Microscopy, in *Metallography and Microstructures*, Vol 9, 9th ed., *Metals Handbook*, American Society for Metals, 1985
4. Scanning Electron Microscopy, in *Metallography and Microstructures*, Vol 9, 9th ed., *Metals Handbook*, American Society for Metals, 1985
5. Transmission Electron Microscopy, in *Metallography and Microstructures*, Vol 9, 9th ed., *Metals Handbook*, American Society for Metals, 1985
6. S.W. Tsai and H.T. Hahn, *Introduction to Composite Materials*, Technomic Publishing, 1980
7. W.S. Johnson, Ed., *Delamination and Debonding of Materials*, STP 876, American Society for Testing and Materials, 1985
8. A.S.D. Wang, Fracture Mechanics of Sub-laminate Cracks in Composite Laminates in *AGARD Conference on Characterization, Analysis and Significance of Defects in Composite Materials*, AGARD-CP-355, Advisory Group for Aerospace Research and Development, 1983
9. K.L. Reifsnider, K. Schultz, and J.C. Duke, Long Term Fatigue Behavior of Composite Materials, in *STP 813*, American Society for Testing and Materials, 1983, p 136-159
10. J. Awerbuch and M.S. Madhukar, Notched Strength of Composite Laminates: Predictions and Experiments—A Review, *J. Reinf. Plast. Compos.*, Vol 4 (No. 1), 1985
11. A.A. Baker, R. Jones, and R.J. Callinan, Damage Tolerance of Graphite/Epoxy Composites, *Compos. Struct.*, Vol 4, 1985, p 15-44
12. B.W. Smith *et al.*, Composite Post-Fracture Analysis Experience as Related to a Developing and Evolving CPFA Methodology, in *Proceedings; International Conference: Post-Failure Analysis Techniques for Fiber Reinforced Composites*, Air Force Wright Aeronautical Laboratories, MLSE, Wright Patterson Air Force Base, July 1985

Destructive and Nondestructive Tests

Edmund G. Henneke II, Virginia Polytechnic Institute and State University

MECHANICAL AND ENVIRONMENTAL LOADINGS cause a variety of failure modes in composites, including matrix cracking, fiber-matrix debonding, delamination between plies, and fiber breakage. A cumulative state of damage for a composite includes a percentage of each of these loading-related failure modes as well as material defects originating from processing, which include voids, fiber-rich or matrix-rich regions, fiber misalignment, and laminate stacking errors. The integrated effect of the cumulative damage state of the composite results in the final failure mode. Therefore, to analyze and understand a composite failure, one must recognize the complexity and extent of damage that might have been present before final failure.

Many experimental destructive and nondestructive test (NDT) techniques have been developed to measure damage levels in composites. Destructive techniques are quite useful in the laboratory for evaluating material response, but obviously have limited use for in-service applications. NDT techniques may be categorized as field (bulk) methods or detail (micro) methods, depending on the type of information they yield. Field methods provide test parameters that are a function of the volume of material interrogated by the NDT technique, but do not provide microscopic information on the actual type of damage present. Ultrasonic velocity measurement and attenuation measurement fall into the category of field methods. Detail methods provide information on the microscopic damage state. X-ray stereo-radiography, for example, can provide a three-dimensional picture of the matrix cracking and delamination patterns through the thickness of a composite laminate. As a general rule, two or more NDT methods should be used to provide complementary information on the state of damage to composites.

Destructive Test Techniques

Destructive test techniques are useful for studying damage in composite materials when the objective is to develop fundamental knowledge of material response in the research laboratory, or to analyze a material failure. Most of these techniques are similar to those used to evaluate homogeneous metal and alloy systems. Therefore, only those two methods peculiar to studying composite failures are described here.

Matrix digestion techniques are used to determine fiber and void volume fractions in both metallic and organic matrix composites. The volume fraction of the fiber present is an important aspect of failure analysis because it is used to determine the apparent strength and modulus of the reinforcing fibers in the composite. Void content is one indication of the quality of the fabrication process and resulting material.

Standard test methods (Ref 1, 2) have been established for both metallic and organic matrix composites. The test procedure requires a sample of the material to be weighed and its volume determined by a fluid displacement technique so that the density of the composite can be calculated. The sample is then placed in a hot, liquid medium that can specifically digest the matrix. The mixture is filtered, and the residue is dried and weighed to determine the amount of fiber present. If the density of the fiber is known, the volume fraction of fiber, V_f, present in the composite sample is:

$$V_f = \frac{(W_f/\rho_f)}{(W_c/\rho_c)} \qquad \text{(Eq 1)}$$

where W_f is the fiber weight, W_c is the composite weight, ρ_f is the fiber density, and ρ_c is the composite density. If, in addition, the matrix density, ρ_m, is also known, the same measurements can be used to calculate the volume fraction of void content, V_v, present in the sample:

$$V_v = 1 - \left\{ \frac{[W_f/\rho_f + (W_c - W_f)/\rho_m]}{(W_c + \rho_c)} \right\} \qquad \text{(Eq 2)}$$

It is sometimes necessary to make corrections for loss of fiber material during the digestion process if the fiber material has some degree of solubility in or reactivity to the digestion mixture.

The deply technique developed by S.M. Freeman (Ref 3) provides information on the discrete nature of fiber fracture in the interior of a layered composite and is presently used with organic matrix composites. A sample is heated for a predetermined time period at a temperature sufficient to partially pyrolyze the resin matrix. This process diminishes the interlaminar bond strength to the point that the individual plies can be easily separated by simply applying a piece of adhesive tape to the top lamina, peeling it off, and continuing in the same manner through the laminate, ply by ply. Each lamina can then be examined in an optical or scanning electron microscope (in the latter case, after applying a graphitic or metallic surface coat using a sputtering technique, if fibers are electrically nonconducting). Individual broken fibers can then be easily detected.

Freeman also observed that an image-enhancing agent could be used to delineate in each ply the regions adjacent to delaminations or matrix cracks. This is accomplished by applying a gold chloride solution to the composite before pyrolysis to penetrate cracks and delaminations. During pyrolysis, the solution breaks down and leaves a residue of gold, which demarcates any crack or delamination. When viewed under the microscope, the relationship between broken fibers and cracking/delamination is easily deduced (Ref 4).

Nondestructive Test Techniques

Determining failure modes in a composite is often better performed before final catastrophic failure. Incrementally studying the mechanical state of the composite that is a result of previous loading is useful even though no definitive knowledge exists as to how the final failure event itself would have been related to the damage state prior to failure. Nondestructive tests are performed prior to final failure either in real time—by continuously monitoring the composite during service (or testing in the laboratory)—or at selected service intervals by removing the composite from the load environment to perform the tests. The lack of definition as to what constitutes ''important'' damage necessitates searching for all indications of damage. Because NDT methods have different sensitivities to different types of damage, it is usually necessary to apply several NDT techniques to the same specimen to obtain complementary information on the damage state. In addition to matrix cracking, fiber-matrix debonding, delamination between plies, and fiber breakage, all of which are loading-related damages, other important types of damage resulting from the manufacturing process are voids, foreign objects, misalignment of fibers, and either bunching or lack of fibers.

To monitor damage progression in composites, a number of NDT methods used with homogeneous materials have been modified and adapted. These techniques include edge replication, stiffness measurement, radiography, ultrasonics, acoustic emission, and thermography.

Surface or edge replication is a technique that has found wide application by material scientists studying surface topography. Because replication tape is typically used to record the surface topography along the edge of a composite laminate, the method is known as edge replication. The technique only requires application of: cellulose-acetate tape (or solution) to the surface, a solvent such as acetone to one side of the tape to soften and partially dissolve it, and a small amount of pressure to the back side of the softened tape so that it will conform to the surface of the material. When the tape hardens, it can be removed from the surface and will carry with it a record of the surface topography at the time the replica was made (Fig. 1).

This technique was first applied by D.O. Stalnaker and W.W. Stinchcomb (Ref 5) to the study of composites. Since their early work, edge replication has been used by a large number of investigators and is now recognized as a standard technique for following the development of damage in composites. J.E. Masters and K.L. Reifsnider (Ref 6) applied the technique extensively to composite laminates and found that it provided a very good record of damage progression. Information on damage at the specimen edge provides an accurate record of the transverse cracks occurring in the specimen. Evidence suggests that, at least for laboratory-sized specimens, the cracks observable along the edge run through the entire width of the specimen.

Edge replication can be used to ascertain the degree to which the ''characteristic damage state'' has been achieved. This concept, developed by K.L. Reifsnider *et al.* (Ref 7), describes the state at which a saturated number of transverse cracks develop in each off-axis ply of a composite laminate as a result of tensile loading. Attaining the characteristic damage state corresponds to the knee of the bilinear stress-strain curve in a number of laminates (Ref 7).

Stiffness. The elastic modulus of a composite laminate can be predicted using classical laminate theory (Ref 8) based on the concept of ideal laminate behavior; that is, it assumes that each lamina has its respective elastic properties and is bonded to its neighbor without delaminations, and that no damage exists in the laminae to affect their individual elastic properties. As a

Fig. 1 Edge replica of graphite-epoxy specimen

laminate is quasi-statically loaded to high stress levels or fatigue loaded, damage (such as transverse cracking) develops in the plies. This damage affects laminae stiffnesses and results in as much as 20 to 30% stiffness degradation of the entire laminate, depending on the laminate stacking sequence.

Stiffness degradation in a graphite-epoxy laminate resulting from fatigue loading is shown in Fig. 2. Each of the tensorial stiffness components is affected to some degree (Ref 9). Therefore, stiffness measurement is an acceptable technique for monitoring the degree of damage development in a material. Following the work of T.K. O'Brien and K.L. Reifsnider (Ref 9), stiffness measurements can be made for each of the in-plane tensor elastic moduli. In general, a better correlation between stiffness degradation and damage development is obtained when measurements are made over a

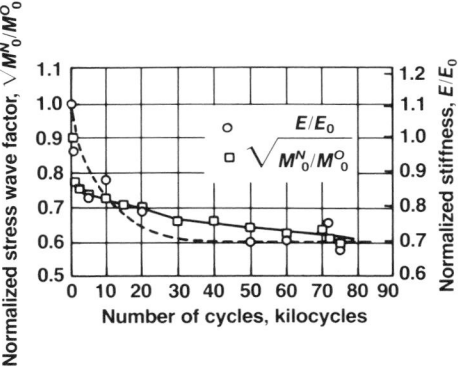

Fig. 2 Change of acousto-ultrasonic stress wave parameter during fatigue test of graphite-epoxy specimen

relatively large specimen length. For example, when using extensometers, a 5-cm (2-in.) gage length will give better results than a 2.5-cm (1-in.) gage length because the greater the length integrated, the more inhomogeneities in the damage state can be averaged out.

Radiography is an NDT technique that can provide extensive and detailed information on the state of damage. A variety of penetrating particles and rays, including neutron, gamma, and x-ray, are used to study composites. For example, W.E. Dance and J.B. Middlebrook (Ref 10) found neutron radiography to be effective in detecting bondline voids/defects in several different composite structures. However, x-ray radiography has been the much preferred technique because it is relatively easy to use and equipment is readily obtainable.

Very low-energy sources of about 15 to 25 kV must be used with x-ray radiography of graphite-epoxy composites. Both the matrix and the fiber are very nearly transparent to even these low-energy x-rays. Hence, to obtain information on interior damage, a good opaque penetrant must be used, thereby limiting the detected damage to that which is open at the surface or adjacent to surface-opening cracks.

Several different image-enhancing penetrants have been used, including tetrabromoethane (TBE), diiodobutane (DIB), and a zinc iodide water solution. Zinc iodide is preferred over TBE because TBE is highly toxic to humans. These substances have been found to have very good penetrating properties, being able to penetrate very small cracks deep within a material and thus provide very good delineation of crack structure. Applying the penetrant while a small, crack-opening load is applied to the material often enables the penetrant to enter surface-opening cracks and branch off and run large distances into adjacent interior cracks without further use of direct surface openings. There is always a question, of course, as to whether some damage has escaped detection because of lack of penetration by the opaque enhancer.

Especially useful for the analysis of damage in thin composite laminates is the technique of x-ray stereo-radiography (Ref 11). This technique employs two x-ray radiographs of the same specimen, taken such that the angle between the beam and the specimen differs by approximately 15° between the two radiographs. The two radiographs are then viewed stereo-optically, one radiograph being viewed by each eye. With a little practice, the human viewer is able to see a full, three-dimensional picture of the internal damage in the laminate. This technique identifies damage progression in composites through the stages of transverse cracking, longitudinal cracking, and local delamination (Ref 12).

X-ray radiography is unable to detect fiber breaks in graphite fiber reinforced or aramid fiber reinforced specimens, although it has been used to find fiber breakage in boron fiber reinforced composites. Figure 3 exemplifies the transverse and longitudinal cracking and local delaminations detected by x-ray radiography of a graphite-epoxy laminate, using an image-enhancing penetrant.

Ultrasonic techniques are most frequently used for nondestructive inspection of composites. Various material properties, including stiffness (through the measurement of elastic wave speeds), volume fraction of voids, and attenuation, in addition to various damage modes, including delamination and matrix cracking, are monitored by different ultrasonic methods. Velocity measurement, attenuation measurement, C-scan, and acousto-ultrasonics are ultrasonic techniques that are related, but are distinguished differently, depending on how the data are obtained, analyzed, and presented.

As a material parameter that can change significantly because of damage development under loading, stiffness can be determined ultrasonically by measuring the velocity of ultrasonic wave propagation. If the material density is known, elastic moduli can be immediately calculated from relationships giving the wave speed in terms of the moduli and density. Ultrasonic measurements have been used to determine all nine tensor elastic moduli (Ref 13) of graphite-epoxy laminates. However, for this determination it is necessary to have thick, unidirectional specimens with the appropriate geometry (Ref 13). Usually, the composite materials are thin, laminated plates, and, in the past, ultrasonic wave speeds were measured only through the thickness, providing a through-the-thickness laminate stiffness calculation that was not generally of importance to the in-plane stresses. Recently, R.C. Stiffler (Ref 14) has suggested using in-plane Lamb waves to determine in-plane laminate stiffnesses. Lamb wave speeds are dependent on the frequency of the wave as well as on stiffness. Wave behavior is presented in terms of a dispersion curve of frequency versus wave number. Because the dispersion curve varies with stiffness, measuring the dispersion provides the apparent laminate stiffness for well-characterized laminate materials.

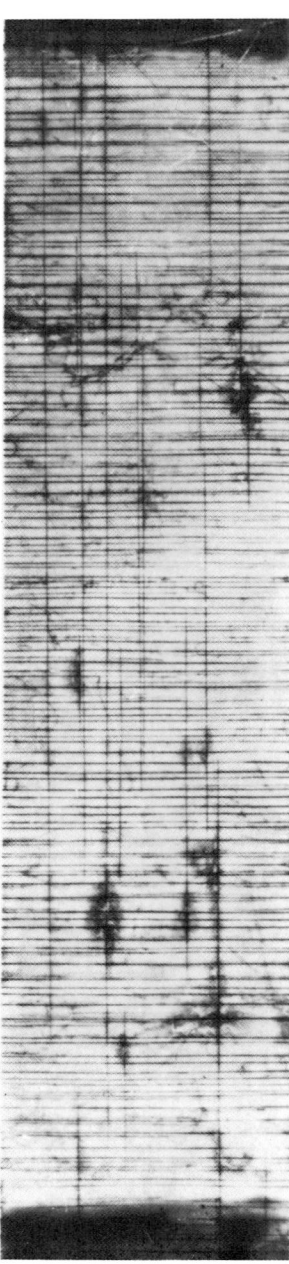

Fig. 3 X-ray radiograph of graphite-epoxy laminate

Attenuation is a general term that describes the rate of energy loss from an ultrasonic wave as it propagates through a material. Applied to an ultrasonic technique, the term refers to the measurement of the energy loss. Energy is irreversibly transformed from mechanical vibrational energy associated with the wave into thermal energy by many loss mechanisms inside a material. In a composite, these loss mechanisms include scattering (by fibers, cracks, delaminations, or voids), thermoelastic effects, and viscoelastic effects. Although these phenomena have not yet been described very well in terms of composite materials, numerous

Fig. 4 Ultrasonic C-scan of graphite-epoxy laminate

experimental papers (Ref 15, 16) have reported correlations of ultrasonic attenuation variations with damage initiation and progression in composites.

The C-scan technique incorporates the effects of ultrasonic attenuation. This technique refers to a method for displaying the relative attenuation of ultrasonic waves across the surface (plan view) of a structural component. An ultrasonic transducer is used to scan the surface of a material mechanically in an x-y raster scan mode while generating and receiving waves. Either the material is immersed in a water bath or columns of water are provided between the transducer and the material as a medium for ultrasonic energy transmissions. The received wave signals are electronically conditioned and measured to determine relative energy losses of the wave as it progresses through the material at each particular location on the specimen. The relative attenuation is compared to an operator-set level and displayed as being either higher or lower than this level. As shown in Fig. 4, the resulting display is a black and white representation of the signal level compared to the set level. By setting a number of comparative levels, colored pictures or gray scales can be displayed. Ultrasonic C-scan has been used extensively to determine both the initial integrity of a manufactured part and the void content, and to follow the initiation and progression of damage resulting from environmental loading.

Acousto-ultrasonics, a technique developed by A. Vary *et al.* (Ref 17-19), is basically a pitch-catch method that uses two transducers, one to generate a wave and one to receive it after it has progressed along the material. It differs from the normal pitch-catch ultrasonic technique in that the two transducers are not placed in sight of one another to receive body waves. Rather, the two transducers are placed

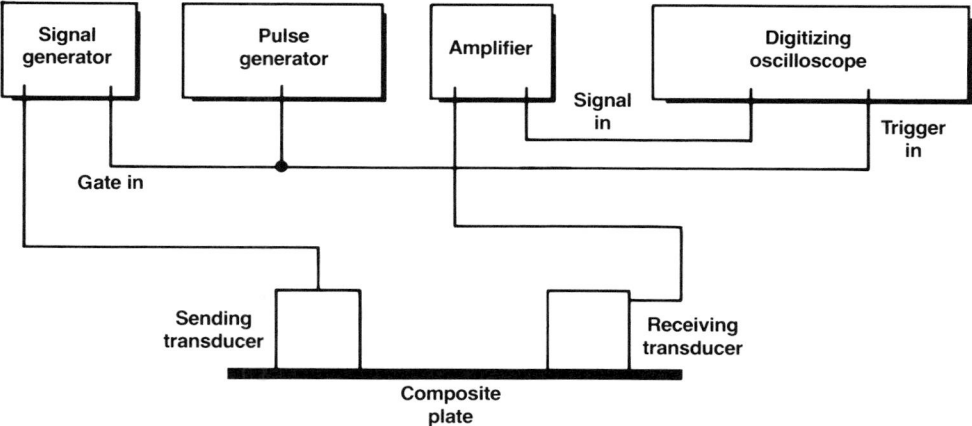

Fig. 5 Acoustic-ultrasonic technique experimental set-up

on the same side of a plate, shown in Fig. 5, and the receiver responds to plate (Lamb) waves that are generated by the transmitter. The transmitter simultaneously generates body waves that are either partially converted into plate waves or reflected back and forth between the front and back faces of the plate under the transmitting transducer until all of their energy is thermally dissipated. The signal received by the second transducer has been found to correlate well with the location of the final catastrophic failure site and with stiffness changes during fatigue (Ref 20, 21). Various signal analysis procedures are sometimes used to obtain more reproducible results from the technique (Ref 22). Figure 2 presents typical results comparing changes in the root mean square of the energy of an acousto-ultrasonic signal with stiffness during a fatigue test of a graphite-epoxy specimen.

Acoustic emission is a phenomenon familiar to metallurgists; mechanical twinning in tin and tin alloys is known as "tin cry." Composite materials are equally noisy under load, particularly during quasi-static tension testing; therefore, acoustic emission (AE) has received significant attention as a technique for monitoring damage development in composites. A 1986 article by M.A. Hamstad (Ref 23) reviews use of AE as a tool for composite material studies, including these studies, which are pertinent to damage analysis: time-dependent composite properties, impact, relationship of AE-detected damage to other measures of damage, interface, and environmental effects.

Many researchers have attempted to develop AE testing to the point that different failure modes could be distinguished by their respective acoustic emissions. However, no definitive conclusions can be made at present. Several authors have suggested that failure modes might be distinguished by grouping received signals according to energy levels; Ref 24 indicates that there are three distinct energy-level groupings that indicate the differing amounts of energy emitted from fiber breakage, matrix cracking, and delamination. Additional

studies of AE in composites need to be performed.

Despite the lack of a firm, fundamental understanding of acoustic emission from composites, the AE technique has been used as the basis for several industry standards for qualifying the initial condition of composite material structures. One such standard, proposed by the Committee for Acoustic Emission from Reinforced Plastics (CARP) for qualifying filament-wound pressure vessels, provides an indirect measure of the process control used in manufacture, and therefore is an indirect measure of product integrity.

Thermography, or surface mapping of isothermal contour lines, can be performed through a variety of techniques, but use of a video-thermographic camera is the most efficient and offers the highest resolution.

Because matter emits infrared radiation with an intensity dependent upon its absolute temperature and surface emissivity, infrared energy sensitive sensors can be used indirectly to measure the temperature of the emitting surface. Other material parameters, specifically surface emissivity, affect the intensity of the emitted radiation and thus the value of the measured temperature. However, as the only concern is relative temperature differences (as in nondestructive inspection for damage); this is of minor concern as long as care is taken to condition the examined surface to ensure that the emissivity is as uniform as possible across the entire surface.

Thermography has been used by a number of investigators to detect and analyze damage in composites (Ref 24-29). Thermal gradients are generated using either a passive or active method to delineate defects or damage. In the passive method, an external heat source is applied to the material. Heat may be conducted either into or away from the examined object by an external heat source of either higher or lower temperature. In the active method, energy is applied in a form other than thermal, such as electrical or mechanical energy. Heat is generated by internal mechanisms that transform the

applied energy into thermal energy. In either method, differences in the local properties of the material affect heat conduction or heat generation through the region surrounding the damage, thus delineating areas of damage. Both the passive and active methods have been used to examine composites (Ref 28, 29).

In vibrothermography, which is one type of active technique, mechanical vibrational energy is applied to the examined object and is preferentially transformed into heat at damaged regions. This energy is applied either in the form of high-amplitude/low-frequency or low-amplitude/high-frequency oscillating loads. High-amplitude/low-frequency loads are applied during fatigue testing. Vibrothermography has been used as an adjunct to fatigue testing to monitor damage initiation and progression resulting from fatigue (Ref 30). It can provide an early warning against damage that could eventually develop into catastrophic failure. It can also promote the study of damage development patterns, such as regions of stress concentration around circular notches (Ref 31).

In thermography tests using low-amplitude/high-frequency loads, a high-frequency shaker is attached to the examined object to vibrate it and thus load it only inertially. This technique, which easily detects delaminated areas in composite plates, applies loads of quite low magnitude, so that the object does not suffer additional damage development as a result of loading. The development of thermal gradients and the degree of heating are dependent on the size of the delamination, extent of the damage, and frequency of the applied mechanical vibrational energy (Ref 28).

REFERENCES

1. "Standard Test Method for Fiber Content of Resin-Matrix Composites by Matrix Digestion," D 3171, *Annual Book of ASTM Standards*, American Society for Testing and Materials
2. "Standard Test Method for Fiber Content by Digestion of Reinforced Metal Matrix Composites," D 3553, *Annual Book of ASTM Standards*, American Society for Testing and Materials
3. S.M. Freeman, Characterization of Lamina and Interlaminar Damage in Graphite-Epoxy Composites by the Deply Technique, in *Composite Materials: Testing and Design (Sixth Conference)*, STP 787, I.M. Daniel, Ed., American Society for Testing and Materials, 1982, p 50-62
4. R.D. Jamison, "Advanced Fatigue Damage Development in Graphite Epoxy Laminates," Ph.D. thesis, Virginia Polytechnic Institute and State University, 1982
5. D.O. Stalnaker and W.W. Stinchcomb, Load History—Edge Damage Studies in Two Quasi-Isotropic Graphite Epoxy Laminates, in *Composite Materials: Testing and Design (Fifth Conference)*, STP 674,

S.W. Tsai, Ed., American Society for Testing and Materials, 1979, p 620-641

6. J.E. Masters and K.L. Reifsnider, An Investigation of Cumulative Damage Development in Quasi-Isotropic Graphite/Epoxy Laminates, in *Damage in Composite Materials*, STP 775, K.L. Reifsnider, Ed., American Society for Testing and Materials, 1982, p 40-62

7. K.L. Reifsnider, E.G. Henneke, and W.W. Stinchcomb, "Defect-Property Relationships in Composite Laminate," AFML Final Report 76-81, Air Force Materials Laboratory, June 1979

8. R.M. Jones, *Mechanics of Composite Materials*, McGraw-Hill, 1975

9. T.K. O'Brien and K.L. Reifsnider, Fatigue Damage: Stiffness/Strength Comparisons for Composite Materials, *J. Test. Eval.*, Vol 5 (No. 5), 1977, p 384-393

10. W.E. Dance and J.B. Middlebrook, Neutron Radiographic Nondestructive Inspection for Bonded Composite Structures, in *Nondestructive Evaluation and Flaw Criticality for Composite Materials*, STP 696, B. Pipes, Ed., American Society for Testing and Materials, 1979, p 5-25

11. W.D. Rummel, T. Tedrow, and H.D. Brinkerhoff, "Enhanced X-Ray Stereoscopic NDE of Composite Materials," AFWAL-TR-80-3053, Air Force Wright Aeronautical Laboratories, June 1980

12. R.D. Jamison, K. Schulte, K.L. Reifsnider, and W.W. Stinchcomb, Characterization and Analysis of Damage Mechanisms in Tension-Tension Fatigue of Graphite/Epoxy Laminates, in *Effects of Defects in Composite Materials*, STP 836, D. Wilkins, Ed., American Society for Testing and Materials, 1984, p 21-55

13. R.D. Kriz and W.W. Stinchcomb, Elastic Moduli of Transversely Isotropic Graphite Fibers and Their Composites, *Exp. Mech.*, Vol 19 (No. 2), 1979, p 41-50

14. R.C. Stiffler, "Wave Propagation in Composite Plates," Ph.D. thesis, Virginia Polytechnic Institute and State University, 1986

15. D.T. Hayford and E.G. Henneke, A Model for Correlating Damage and Ultrasonic Attenuation in Composites, in *Composite Materials Testing and Design*, STP 674, S.W. Tsai, Ed., American Society for Testing and Materials, 1979, p 41-50

16. K.L. Reifsnider, E.G. Henneke, and W.W. Stinchcomb, "Defect-Property Relationships in Composite Materials, Part II," AFML-TR-76-81-Pt 2, Air Force Materials Laboratory, June 1977

17. A. Vary and K.J. Bowles, "Ultrasonic Evaluation of the Strength of Unidirectional Graphite-Polyimide Composites, NASA TM-X-73646," National Aeronautics and Space Administration, 1977

18. A. Vary and K.J. Bowles, "Use of an Ultrasonic-Acoustic Technique for Nondestructive Evaluation of Fiber Composite Strength," NASA TM-73813, National Aeronautics and Space Administration, 1978

19. A. Vary and R.F. Clark, "Correlation of Fiber Composite Tensile Strength With the Ultrasonic Stress Wave Factor," NASA TM-X-78846, National Aeronautics and Space Administration, 1977

20. E.G. Henneke, J.C. Duke, W.W. Stinchcomb, A. Govada, and A. Lemascon, "A Study of the Stress Wave Factor Technique for the Characterization of Composite Materials," Contractor Report 3670, National Aeronautics and Space Administration, Feb 1983

21. A.K. Govada, J.C. Duke, E.G. Henneke, and W.W. Stinchcomb, "A Study of the Stress Wave Factor Technique for the Characterization of Composite Materials," Contractor Report 174870, National Aeronautics and Space Administration, Feb 1985

22. R. Talreja, A. Govada, and E.G. Henneke, Quantitative Assessment of Damage Growth in Graphite Epoxy Laminates by Acousto-Ultrasonic Measurements, in *Review of Progress in Quantitative Nondestructive Evaluation*, Vol 3B, D.O. Thompson and D.E. Chimenti, Ed., Plenum Press, 1984, p 1099-1106

23. M.A. Hamstad, A Review: Acoustic Emission, a Tool for Composite Materials Studies, *Exp. Mech.*, Vol 26 (No. 1), March 1986, p 7-13

24. J. Awerbuch, M.R. Gorman, and M. Madhudar, Monitoring Acoustic Emission During Quasi-Static Loading-Unloading Cycles of Filament-Wound Graphite-Epoxy Laminate Coupons, *Mater. Eval.*, Vol 43 (No. 6), 1985, p 754-764

25. R.T. Schaum, "Development of a Nondestructive Inspection Technique for Advanced Composite Materials Using Cholesteric Liquid Crystals," AD-A032322, National Technical Information Service, Sept 1976

26. J.D. Whitcomb, Thermographic Measurement of Fatigue Damage, in *Composite Materials: Testing and Design*, STP 674, S.W. Tsai, Ed., American Society for Testing and Materials, 1979, p 502-516

27. E.G. Henneke and T.S. Jones, Detection of Damage in Composite Materials by Vibrothermography, in *Nondestructive Evaluation Flaw Criticality*, STP 696, R.B. Pipes, Ed., American Society for Testing and Materials, 1979, p 83-95

28. S.S. Russell and E.G. Henneke, Dynamic Effects During Vibrothermographic NDE of Composites, *NDT Int.*, Vol 17 (No. 1), Feb 1984, p 19-25

29. P.V. McLaughlin, E.V. McAssey, and R.C. Deitrich, Nondestructive Examination of Fiber Composite Structures by Thermal Field Techniques, *NDT Int.*, Vol 13 (No. 2), p 56-62

30. L.A. Marcus and W.W. Stinchcomb, Measurement of Fatigue Damage in Composite Materials, *Exp. Mech.*, Vol 15 (No. 1), 1975, p 55-60

31. W.W. Stinchcomb, K.L. Reifsnider, P. Yeung, and T.K. O'Brien, "Investigation and Characterization of Constraint Effects on Flaw Growth During Fatigue Loading of Composite Materials," Second Semi-Annual Status Report, NASA Grant NSG-1364, National Aeronautics and Space Administration, Jan 1978

Thermal Analysis

Jeanne L. Courter, American Cyanamid Company

FAILURE OF A COMPOSITE STRUC-TURE can be caused by a number of contributing factors. Some failure processes are dominated by fiber properties, such as tensile failures. Others are especially susceptible to fiber-resin interface or interphase problems, such as failures in 90° plies. For such failures, thermal analysis is of limited use. In failure modes sensitive to the properties of the matrix resin, however, thermal analysis can help determine whether the matrix resin of a failed composite has the appropriate glass transition temperature (T_g). Some thermal analysis techniques, especially dynamic mechanical analysis (DMA), can also be used to "fingerprint" resins by determining properties over a wide temperature range.

Composite Failure Modes Affected by Matrix Resin

If failure occurs under high-temperature conditions, then the T_g of the resin should be checked by a thermal analysis technique. Even if failure occurs at low temperatures, T_g should be checked, but this time for indications of overcure, because an excessively high T_g can lead to microcracking. In addition, if the resin is a multiphase resin, for example, elastomeric or thermoplastic modified, thermal analysis techniques can determine the T_g of the modifying second phase. If such a fingerprint differs to a significant degree from a control sample, there may be a problem with the composition of the modifier, the resin preparation (for example, improper prereaction), the formulation (too little or too much modifier), or the cure cycle. Such changes in fingerprint could indicate that the modifiers are not acting to toughen the resin effectively and could lead to low interlaminar toughness, for example.

Test Program

Usually, it is neither possible nor informative to compare the T_g of a laminate to a literature or brochure value. It is therefore necessary to run a control specimen with a test specimen from the failed laminate. Ideally, the control laminate should be made from a batch of prepreg that is within composition specifications and has quality control data available from both the manufacturer and the end user. The control laminate lay-up should be of the same configuration as the failed laminate and should be done under controlled, or at least monitored, humidity. It should then be cured under the same cycle as was the test specimen. Then a thermal analysis can be performed, and the T_g and other transitions for both specimens can be compared.

Reasons for Variations in T_g. There are several reasons that T_g may be lower in the test laminate than in the control laminate. The test laminate may have been undercured because an incorrect cure cycle was used. This sometimes occurs if the operator controls the cycle by tool temperature rather than part temperature when the specified cycle is based on part temperature. The T_g also can be lower if the prepreg were cut and laid up under high relative humidity (RH). For example, T_g can be lowered by as much as 20 °C (36 °F) if it is exposed to 80% RH at 30 °C (85 °F) for 4 h (compared to dry prepreg). Incorrect formulation can also lead to a low T_g.

The T_g of the failed laminate could be higher than the control laminate if an incorrect cure or postcure were used. In addition, if temperatures higher than cure or postcure temperatures were experienced in service, T_g could be increased. Incorrect formulation can also raise T_g.

Thermal Analysis Techniques

Dynamic mechanical analysis (DMA) is probably the most useful thermal analysis technique because it can clearly identify T_g's and other transitions in the polymer matrix. All DMA instruments either apply a sinusoidally varying load and measure the sinusoidally varying displacement, or vice versa, and some instruments can do both. Some instruments hold the frequency of the oscillation fixed, and in others, called natural frequency devices, the frequency changes throughout the analysis.

The ratio of the amplitude of the stress and strain sine waves determines the "storage" or elastic modulus of the specimen, while the phase angle, ϕ, between the two waves is called the loss angle and is a function of the internal friction of the material (Ref 1).

Commonly, the glass transition is determined by measuring the specimen storage modulus and loss (using either tangent (tan) δ or the so-called loss modulus) as a function of temperature. As the specimen goes through the glass transition, the modulus drops by several orders of magnitude, and loss goes through a peak. The temperature at which the loss is at a maximum is defined as the glass transition temperature, T_g.

Regarding the two loss parameters commonly used to define T_g (tan δ and loss modulus) tan δ is equal to the ratio of the loss modulus to the storage modulus. No definition has been agreed upon by the thermal analysis community as to which parameter should be used to define T_g; therefore, it is critical when reporting T_g to state which parameter has been used.

There are many instruments available to carry out DMA, but Fig. 1 shows the geometry of the DuPont Model 982 dynamic mechanical analyzer, a natural frequency device. The sample is clamped between pivoting arms, and the oscillation is maintained at a fixed amplitude. The frequency of the oscillation is related to the storage modulus of the sample, and the power needed to maintain its oscillation is related to the loss in the sample. Other examples of DMA instruments suitable for composites testing include the Rheometrics mechanical and dynamic spectrometers and the Dynastat mechanical spectrometer.

Though specimen size varies from one instrument to another, generally samples for DMA are approximately 13 mm (0.5 in.) wide, over 31.8 mm (1.25 in.) long, and up to 3.8 mm (0.15 in.) thick. When performing any DMA of composites, it is very important that the specimen be loaded in a resin-dominated direction. In fiber-dominated directions, the fiber stiffness will control composite stiffness and will mask stiffness and loss changes in the resin.

Thermomechanical analysis (TMA) measures changes in length or thickness of a specimen as a function of temperature. This

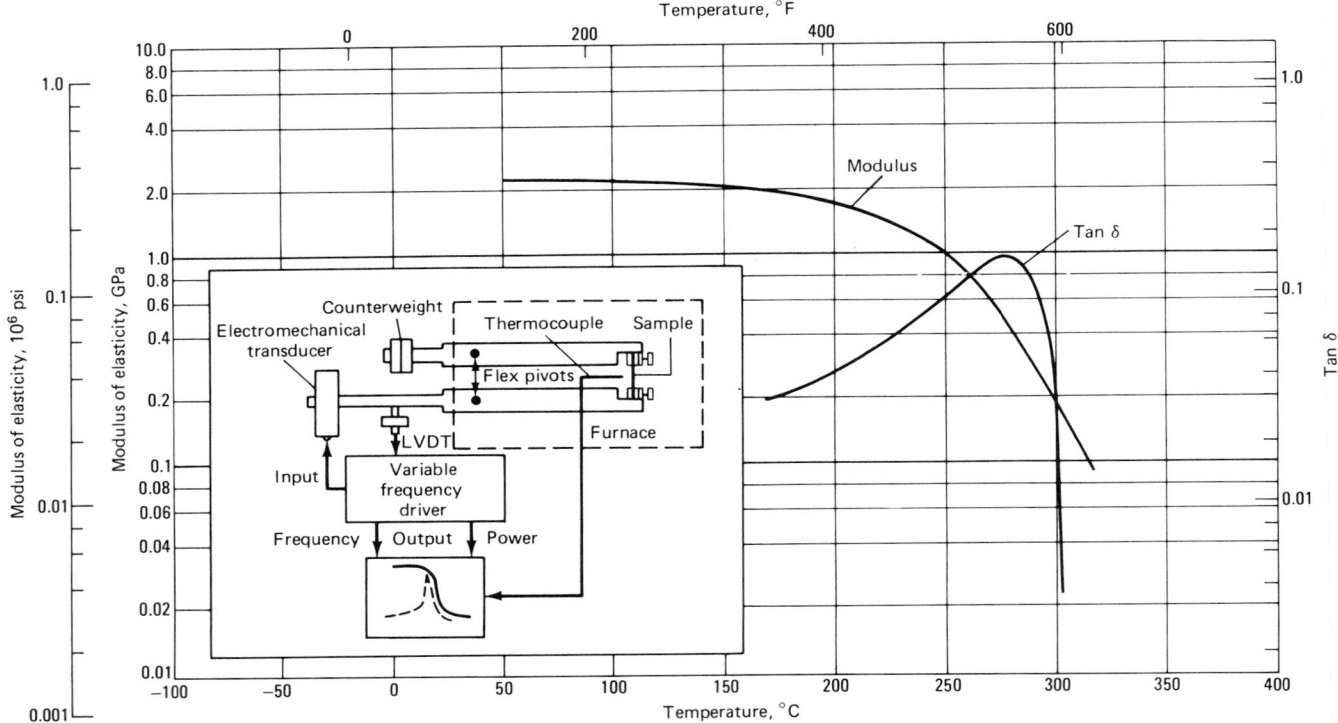

Fig. 1 Typical output for DMA, modulus of elasticity, and tan δ versus temperature curves for an epoxy resin. Inset shows a schematic of the DuPont Model 982 dynamic mechanical analyzer.

commonly used technique has the advantage that small specimens (6.4 mm × 6.4 mm, or 0.25 in. × 0.25 in.) can be used. Its disadvantage is that T_g is measured by determining where the slope of the thickness-temperature curve changes, that is, where the coefficient of thermal expansion changes (changes of a factor of two are common in polymers). This is not as easy as measuring a peak temperature, as is done in DMA. The task becomes even more difficult when moisture exposure has occurred because it broadens the glass transition on the low-temperature side. This makes it difficult to determine where the slope begins to change.

Differential scanning calorimetry (DSC) measures the rate of heat absorption or emission from a specimen as its temperature is raised. The glass transition appears as a step

change in the heat absorption versus temperature curve. This step change indicates the difference in the heat capacity of the glassy and the rubbery state, which is commonly about 10% of the value. The advantage of DSC is that it uses very small samples of about 25 mg (0.39 grain). The disadvantage of DSC is that T_g is difficult or impossible to determine for some of the highest T_g matrices because the difference in heat capacity below and above T_g is very small. This is often due to the very high cross-link densities involved.

The DSC technique can also measure residual cure in cured laminates. Heat given off during any additional cure will appear as an exothermic peak that occurs after T_g. However, reliable comparisons of residual cure heat are difficult because large variations in fiber content can occur in very small samples. There-

fore, any estimate of heat per gram is so strongly affected by fiber content that accurate comparisons are difficult.

REFERENCES

1. W.N. Findley, J.S. Lai, and K. Onaran, *Creep and Relaxation of Nonlinear Viscoelastic Materials*, North-Holland, 1976, p 91

SELECTED REFERENCES

● E.A. Turi, Ed., *Thermal Characterization of Polymeric Materials*, Academic Press, 1981
● I.M. Ward, chapters 5-7, *Mechanical Properties of Solid Polymers*, Wiley-Interscience, 1971

Basic Failure Modes of Continuous Fiber Composites

John E. Masters, American Cyanamid Company

COMPOSITE MATERIAL FAILURE MODE characterization is complicated by the number and complexity of fracture mechanisms exhibited by composites. By definition, composite materials are heterogeneous on a macroscopic scale. Furthermore, the individual lamina that constitute a laminated, continuous fiber reinforced composite are anisotropic. Therefore, the fracture of a composite component is quite different from the fracture of a metallic component. There is no single, self-similar propagating crack as in metals, but rather, the damage zone is characterized by matrix cracking, fiber breakage, and delamination, which ultimately combine to yield component failure.

This article discusses the basic failure modes exhibited by composite laminates. To begin, the loading conditions most generally encountered by these materials should be considered. Because of the overwhelming dominance of the fibers in structural behavior, fibrous composite laminates are highly efficient in carrying membrane loads (Ref 1). Consequently, they are most often designed to carry in-plane tension, compression, or shear loads. Loadings that put the fibers in tension bring out the best features of fibrous composites. The fibers tend to straighten and approach their theoretical stiffness and strength potential. In fatigue, tension-loaded composites are superior to virtually any material of equal weight. In-plane compression of laminates is greatly complicated by the tendency for the fibers to bend or buckle. Because the fibers within a ply are not perfectly straight, axial compression produces shear components of load between the fiber and matrix. These out-of-plane components can lead to tension loads in the matrix that may cause premature matrix failure.

Complex internal stress states develop when lamina are cured to form a laminate and when load is applied to that laminate. These stress states include both in-plane (σ_x, σ_y, and τ_{xy}) and out-of-plane stresses, interlaminar normal, σ_z, and shear stresses (τ_{xz}, τ_{yz}). Because interlaminar stresses are a function of the laminate stacking sequence, small changes in the stacking sequence can significantly alter the internal stress distribution. These different internal stress distributions will in turn alter the pattern of damage development in the laminates and affect the final failure events. Therefore, the types and modes of failure that are encountered depend upon both the direction of applied load and the orientation of the fibers making up the composite laminate. Variations in either of these can produce strikingly different fracture appearances on a macroscopic scale. It is, therefore, important to define the stress state accurately to predict laminate response and failure.

One approach to classifying composite material failure modes is to separate them into two major types: in-plane failure (failure that is due to in-plane stresses) and out-of-plane failure (delamination that is prompted by out-of-plane or interlaminar stresses) (Ref 2). While there is no doubt that these two mechanisms interact, each mode is discussed separately for the purpose of simplification.

The distinction between in-plane and out-of-plane failure is based on the initiation of failure. The examples discussed in this article feature coupon-type specimens. Strictly speaking, out-of-plane forces cannot cause the ultimate fracture of these coupons; that is, they cannot break the specimen into two pieces (Ref 2). In-plane mechanisms must ultimately cause fiber and matrix failure in the plane of loading for this to occur. In the context of this article, in-plane

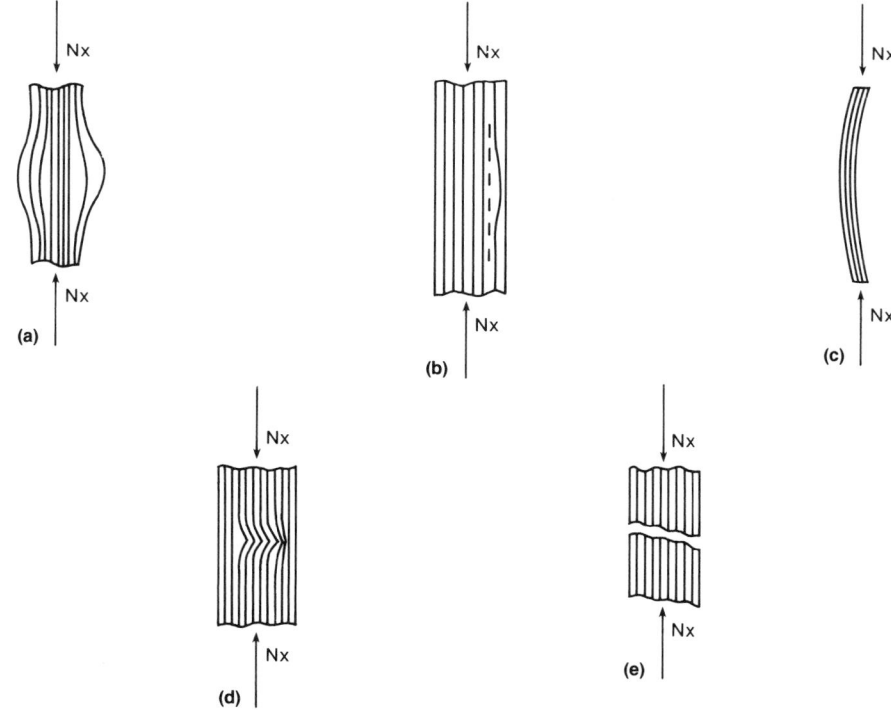

Fig. 1 In-plane compression failure modes. (a) Induced transverse tensile failure. (b) Compressive delamination failure. (c) Evler failure. (d) Microbuckling. (e) Strength failure. Source: Ref 3

and out-of-plane failure refer to the mechanisms that initiate or promote failure. In-plane compression, in-plane tension, and delamination will be discussed. In-plane shear will be assumed to be resolvable into tension and compression components, and will therefore not be specifically addressed.

In-Plane Failure Modes

Compression. Although tension plays a predominant role in fracture mechanics of metals, in the case of laminated composites, in-plane compression loading is an even more severe and limiting test of the material. M. Ashizawa illustrated five distinct failure modes that have been observed in laminated composite structures at the practical fiber volume fractions of 50-65 vol% (Ref 3). These failure modes, which are illustrated in Fig. 1, are influenced by such factors as geometry, matrix and fiber type, interfacial condition, defect size and shape, environmental condition, and resin toughness. Analytical models defining these failure modes have been developed by several authors, including L.B. Greszczuk (Ref 4, 5) and B.W. Rosen (Ref 6), on strength failure and microbuckling; L.J. Broutman (Ref 7), on transverse tensile; and M. Ashizawa (Ref 8), on transverse tension and compressive delamination.

H.T. Hahn and J.G. Williams (Ref 9) gave an excellent review of the compression failure of unidirectional laminates. The following discussion was excerpted from their paper. In that work it was noted that because of the weakness of the matrix and the fiber-matrix interface, compared with the strength of the fibers, unidirectional composites can fracture along the fibers when loaded in compression by a transverse tensile failure mode, as shown in Fig. 1(a). Transverse tensile stresses can develop in the matrix because of Poisson's ratio differences between the matrix and fiber. Additionally, stress concentrations caused by voids can initiate transverse tensile failure either in the matrix or in the fiber-matrix interface (Ref 10).

If a fiber buckles, the fiber-matrix interface may fracture in shear and lead to ultimate failure, as shown in Fig. 1(b). However, if the matrix is ductile and the interface is strong, the fiber can bend without matrix failure and eventually fracture in bending, as shown in Fig. 1(c). The eccentricity introduced by such fiber fracture may lead to longitudinal splitting with continued compression loading.

A more likely failure mode of unidirectional composite laminates associated with fiber microbuckling and fiber kinking, as shown in Fig. 1(d), is shear crippling. Macroscopically, shear crippling looks like a shear failure on a plane at an angle to the direction of loading. Microscopic inspection, however, indicates that shear crippling is frequently the result of kink-band formation.

In graphite fiber reinforced composites, fiber breaks are usually observed at the kink-band boundaries (Ref 11, 12); fiber kinking and extreme fiber bending without fiber fracture are typical of aramid and glass composites, respectively (Ref 13, 14). Kink-bands are most clearly observed when failure is gradual and when longitudinal splitting is prevented by application of hydrostatic pressure (Ref 12). Gradual failure is also observed when there is a gradient in the stress field.

The final failure mode exhibited in unidirectional composites is associated with pure compression failure of the fibers, as in Fig. 1(e). In this case, the fracture surface is likely to be at an angle to the loading direction, usually about 45°. Postfailure examination of fracture surfaces of graphite-epoxy composites alone is usually inadequate to distinguish between fiber kinking and fiber compression failure because the broken fiber segments resulting from kinking failure are randomly displaced during the catastrophic failure event (Ref 15).

H.T. Hahn and J.G. Williams also noted that the data available in the literature suggest that the most likely failure mode in graphite-epoxy composites is shear crippling involving fiber kinking. They suggest that compression failure of a composite starts with kinking of a few fibers. These kinked fibers disrupt the stability of the neighboring fibers such that the neighboring fibers also fail in the kinking mode. This damage propagation process continues until the composite fails completely. In some cases, fiber kinking may be initiated at several different locations and proceed to converge. The transverse tensile stress in the region in which the two advancing kink-bands meet may be sufficiently high to cause longitudinal splitting.

Fiber kinking in 0° plies is also judged to be the mechanism that governs failure initiation in compressively loaded, quasi-isotropic graphite-epoxy laminates (Ref 16). Fiber kinking in these specimens is believed to start at the specimen edge and grow inward toward the specimen center. The broken fiber segments found within the kink-band are subject to both in-plane and out-of-plane rotations. Quasi-isotropic laminates are seen to have considerably higher failure strains than do unidirectional laminates fabricated from the same fiber-matrix system. This is due to the better lateral support provided to the 0° plies by the adjacent off-axis plies in the quasi-isotropic laminates. In-plane kinking is made more difficult in quasi-isotropic laminates because of the higher stiffness provided by the off-axis plies in the direction normal to the loading. Similarly, out-of-plane kinking is also retarded by the off-axis fibers that bridge the kink-band.

The authors postulated that the failure of a 0° ply not only transfers load to other 0° plies but also results in uneven loading of the remaining plies because the specimen loses stiffness on the side with the failed ply. This causes sequential rather than random failure of the remaining plies, starting with the one closest to the failed ply and propagating outward.

The sequence of events subsequent to the initial kinking failure of the 0° ply is, however, affected by resin toughness. A 0° ply with fiber kinking may move relative to the neighboring angle plies. In brittle matrix systems this brings about delamination between the failed 0° ply and the angle plies. Because of their reduced cross sections, the individual sublaminates are more susceptible to buckling than is the original laminate. The global buckling of the sublaminates leads to final failure. In contrast to the brittle resin systems, fiber kinking is well contained in laminates featuring tough resin systems.

This latter observation was also made by M. Ashizawa (Ref 3), who noted that increasing resin toughness helped to increase composite strength when the mode of failure was induced transverse tension. A brittle resin system will fail by this mode, also known as brooming, but a tough resin will tend to prevent the composite from failing in this lower-strength mode. Instead, the tough resin systems will fail by the ultimate strength mode at a much higher compressive load.

The effect of resin toughness on quasi-isotropic laminate compression failure strength has also been observed in specimens that have suffered internal damage because of low-velocity impact (Ref 17, 18) and in notched (open-hole) specimens (Ref 19, 20). Two damage propagation modes, delamination and shear crippling, are again observed in these coupons. Despite the initial impact damage and strain concentrations found in these specimens, damage in brittle resin systems characteristically propagated in a delamination mode, while the more damage-tolerant material systems suppressed delamination and failed at higher loads in a transverse shear crippling mode. Delamination is again seen to permit the local buckling of thin sublaminates, which results in high tension peel stresses at the delamination boundary. This results in sublaminate brooming at final failure. The transverse shear failure mode is caused by shear instability in which filaments buckle locally (0° fiber kinking).

Tension. Because the ultimate in-plane tensile failure of a quasi-isotropic laminate is controlled by the fracture of the 0° lamina, this discussion of laminate tensile failure modes first considers the failure modes exhibited by unidirectional laminates subjected to longitudinal tension, that is, tensile load applied parallel to the fibers.

The simplest failure analysis for this type of loading assumes that a uniform strain exists throughout the composite and that fracture occurs at the failure strain of the fibers alone. The failure strain of the fibers of interest is not, however, a unique quantity. In general, the high-strength high-modulus fibers that are used in composite materials are brittle, having tensile strengths that must be characterized statistically (Ref 21). The individual fibers in a unidirectional laminate therefore fail at various stress levels as the applied tensile load increases

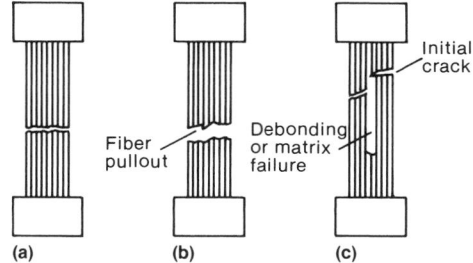

Fig. 2 Failure modes of unidirectional composite subjected to longitudinal tensile load. (a) Brittle failure. (b) Brittle failure with fiber pull-out. (c) Brittle failure with debonding and/or matrix failure. Source: Ref 23

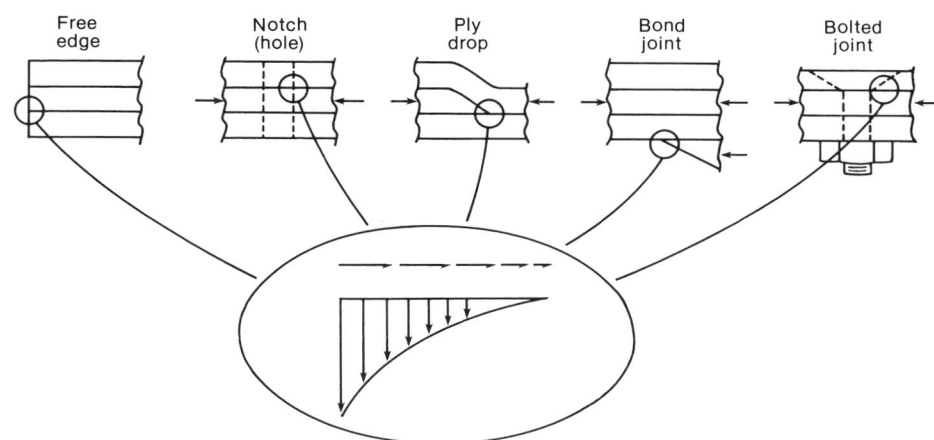

Fig. 3 Sources of out-of-plane loads in typical aircraft design. Source: Ref 1

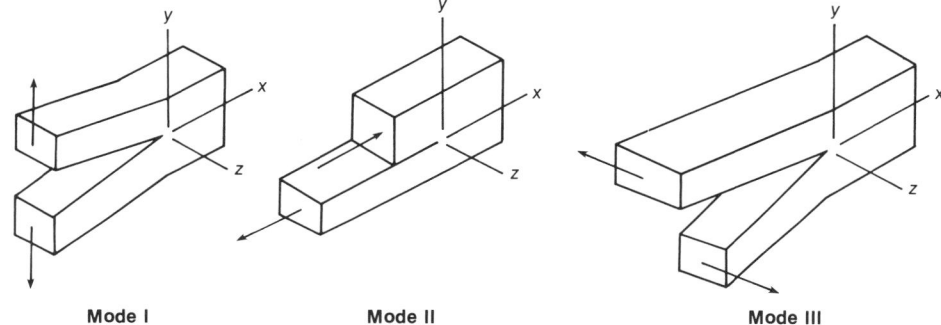

Fig. 4 Basic modes of loading involving different crack surface displacements. Source: Ref 28

(Ref 22). The load on the broken fiber is assumed to be distributed equally among the remaining unbroken fibers in a cross section. Breaking of the fibers is a completely random process. As the number of broken fibers increases, some cross section of the composite will become too weak to support an increased load, thus causing complete rupture of the laminate.

Three failure modes, illustrated in Fig. 2, may develop in these specimens under this longitudinal tensile loading. Stress concentrations created at the broken fiber ends will lead to specimen separation at a given cross section, that is, brittle failure (Fig. 2a). Variation in bond strength and local load transfer mechanisms from matrix to fiber can lead to the pull-out of the fibers from the matrix at fracture (Fig. 2b). Finally, in other cases, cracks at different cross sections of the laminate may join together at fracture through fiber-matrix debonding or by shear failure of the matrix (Fig. 2c). This interfiber matrix shear failure and fiber-matrix debonding can occur either independently or in combination; that is, portions of the failure path may exhibit debonding, while matrix shear failure is evident in other regions (Ref 23).

Fiberglass composites having low fiber volume fractions (<0.40 vol%) predominantly exhibit the brittle-type failure mode. Composites with intermediate fiber volume fractions (0.40 to 0.65 vol%) commonly exhibit brittle failure with pull-out. Composites with high fiber volume fraction (>0.65 vol%) usually exhibit brittle failure with fiber pull-out and debonding of matrix shear failure. These ranges are applicable if the void content in the composite is negligible. Graphite fiber-reinforced composites generally fail by the modes shown in Fig. 2(a) and 2(b) (Ref 23).

The off-axis tensile response of unidirectional lamina has also been the subject of several investigations. C.C. Chamis and J.H. Sinclair (Ref 24, 25) studied the failure surfaces of specimens loaded at several loading angles: 0°, 5°, 10°, 15°, 30°, 45°, 60°, 75°, and 90°. Loading angles were measured with respect to the fiber axial direction. Fracture modes associated with approximate load angle

ranges were identified. The fracture modes identified were: longitudinal tensile fracture, characterized by irregular, tiered failure surfaces dominated by fiber fracture and fiber pull-out; intralaminar shear stress fracture, characterized by regular and level failure surfaces containing extensive matrix shear failure; transverse tensile fracture, again characterized by regular or level failure surfaces, but containing extensive matrix cleavage features; and, finally, mixed mode (transverse tensile and intralaminar shear), characterized by level fracture surfaces containing both resin cleavage and shear failure features.

Two laminate failure modes develop under transverse tensile loading (loading perpendicular to the fibers axis): matrix tensile failure and constituent debonding, that is, failure of the interfacial bonds between fiber and matrix. The failure surface of unidirectional transversely loaded laminates commonly exhibits both of these features. A third possible failure mode, transverse tensile failure of the fibers, is not common in graphite fiber reinforced systems. It may be seen, however, if the fibers are highly oriented and weak in the transverse direction, for example, in boron fiber reinforced composites.

Out-of-Plane Delamination Failure

A major asset of composite laminates is the ability to orient fibers to achieve directional strength and stiffness properties that match the expected loading environment of the structural elements. Composite materials are uniquely suited to this objective because the principal material directions of each layer can be oriented according to need. However, interlaminar stresses develop in laminated composite structures because of the mismatch in engineering properties between individual lamina within the laminate. These stresses are instrumental in delamination initiation and propagation.

Delamination by definition is matrix crack development at the lamina interfaces. The causes of delamination can be attributed, in general terms, to the existence of out-of-plane (interlaminar normal and shear) stresses that develop at structural discontinuities. As discussed earlier, out-of-plane stresses can develop as a result of 0° ply failure and the subsequent internal load redistribution under in-plane compression loading. These stresses are, however, also induced in laminates prior to lamina failure. In coupon specimens, these through-the-thickness forces develop along the traction-free edges of the specimens as load is applied.

More important, in structural applications, out-of-plane stresses are also induced at a variety of common design features involving local discontinuities. Figure 3 depicts five of the most common design details incorporated in structures. Even under in-plane loading, interlaminar shear and normal stresses develop in the gradient stress fields that form at these locations because of local discontinuities. These indirect loads, either alone or in combination with direct out-of-plane loads (from fuel pressures, air pressures, or structural mismatches) can form delaminations.

Delaminations exhibit slow, stable, subcritical flaw growth in structural applications (Ref 26) and may severely reduce the compression residual strength of the structure. Many investigators feel that delamination growth is the fundamental issue in the evaluation of laminated composite structures for durability and damage tolerance. D.J. Wilkins (Ref 27) states, "... when test conditions are extended to explore failure mechanisms, delamination is observed to be the most prevalent life-limiting growth mode."

A number of analysts have addressed the delamination problem. One of the more promising techniques for characterizing delamination growth considers the delamination process as a stable crack growth mechanism and applies linear elastic fracture mechanics (LEFM) to describe the process. In LEFM, the stress fields surrounding a crack tip can be divided into three major modes of loading that involve different crack surface displacements. These three basic modes are shown in Fig. 4. Under mode I, opening or tensile loading, the crack surfaces move directly apart. In mode II, sliding or in-plane shear, the crack surfaces slide over one another in a direction perpendicular to the leading edge of the crack. In Mode III, tearing or antiplane shear, the crack surfaces move relative to one another and parallel to the leading edge of the crack.

In metallic structures, mode I loading is encountered in the overwhelming majority of actual engineering situations involving cracked components. Mode II is found less frequently and is of little engineering importance. Even under mixed mode I and II conditions, analytical methods indicate that the mode I contribution dominates the crack tip stress field (Ref 28). However, this is not necessarily true for composite materials. Delamination of composite laminates generally involves other modes of fracture. The forward shear mode (mode II) and the antiplane shear mode (mode III) are often present in addition to mode I.

While fractures of composite structures are not likely to occur under pure mode I or mode II loading, these two load states provide a logical framework for understanding the characteristics of interlaminar fractures in composite laminates. To this end, experimenters have

developed test specimens that provide one-dimensional fracture and crack growth information. Specimens that exhibit pure mode I and mode II delamination growth have been developed to date. Investigators have not yet succeeded in developing a coupon for mode III crack growth.

In an early study of delamination growth in graphite-epoxy laminates, D.J. Wilkins (Ref 27) characterized interfacial mode I and mode II crack growth in T300/5208 specimens under static and fatigue loading. Among other conclusions, he noted that for this material system (typical of first-generation graphite-epoxy systems) the mode I strain energy release rate, G_{Ic}, for crack growth at 0°-0° ply interface is at least ten times lower than that of structural adhesives. The growth rate exponent for mode I delamination in these specimens was quite high. The applied cyclic load must be nearly equal to the critical static load to obtain observable growth in the tensile, opening mode. Delaminations operating at high percentages of the mode I strain energy release rate will grow rapidly to failure under cyclic loading; delaminations operating below 50% of G_{Ic} will grow very slowly. For this reason, D.J. Wilkins concluded that mode I delaminations can be thought of as a static design issue rather than as a potential fatigue problem. The mode II strain energy release rate measured in this study was about twice the mode I strain energy release rate. The growth rate exponent for mode II delamination in this system was analogous to the growth-exponent measured for mode I crack growth in aluminum, suggesting that mode II delamination growth must be considered a fatigue design issue.

On a microscopic level, delamination or interlaminar fracture of fiber-reinforced composite materials is characterized by fiber-matrix debonding, resin deformation and fracture, and fiber pull-out. Under mode I loading, areas of fiber-matrix separation were found to be relatively smooth and featureless. Conditions of cohesive resin fracture dominate the overall fracture surface topography. As will be discussed in other articles, such areas are flat and exhibit typical cleavage fracture features. Fractures produced under mode II loading exhibit large amounts of fiber-matrix separation and small, finely spaced areas of cohesive matrix fracture. In contrast to the mode I failure surfaces, the regions of matrix fracture found in these specimens exhibited a relatively rough topography (Ref 29). As several investigators have noted, detailed inspection of these areas reveal numerous inclined platelets of fractured resin. Full descriptions and illustrations of these surfaces will be provided in the following articles in this Section of the Volume.

These differences in fracture surface topography can be accounted for by applying the concepts of brittle fracture to the matrix resin (Ref 30). This assumption is valued at the dry,

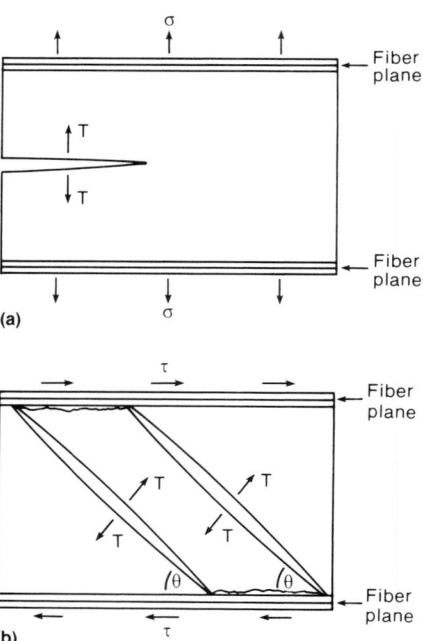

Fig. 5 Schematic of formation of cleavage and hackle features. (a) Local tensile stress field. (b) Local shear stress field. T, maximum principal stress direction. Source: Ref 30

room-temperature test conditions and the relatively high loading rates under which these specimens are usually tested.

The fracture plane in any loading condition in a brittle material is perpendicular to the maximum resolved tensile direction (Fig. 5). When the fibers lie perpendicular to the maximum tensile direction, a single cleavage plane parallel to the fibers results (Fig. 5a). The river pattern and feather pattern features found on mode I failure surfaces fall into this category. However, when the fibers are not perpendicular to the maximum tensile direction (for example, in a situation in which shear is locally predominant), the fracture plane intersects the plane of the fibers. The fiber plane generally restricts any further growth of the crack, with the result that a series of parallel fissures occur in the region between the fiber planes (Fig. 5b). Such conditions give rise to the hackle feature commonly observed in mode II specimens. The occurrence of all three features on some occasions indicates that the local stress state should be considered in conjunction with the globally applied loading when inspecting a failure surface.

REFERENCES

1. D.J. Wilkins, The Engineering Significance of Defects in Composite Structures, in *AGARD Conference Proceedings No. 355, Characterization, Analysis and Significance of Defects in Composite Materials*, April, 1983

2. P.A. Lagace, Postmortem Determination of Delamination Failures, in *Proceedings of International Conference: Post Failure Analysis Techniques for Fiber Reinforced Composites*, July 1985
3. M. Ashizawa, Improving Damage Tolerance of Laminated Composites Through the Use of New Tough Resins, in *Proceedings of Sixth Conference on Fibrous Composites in Structural Design*, Jan 1983
4. L.B. Greszczuk, Microbuckling of Lamina Reinforced Composites, in *Composite Materials: Testing and Design (Third Conference)*, STP 546, American Society for Testing and Materials, 1974
5. L.B. Greszczuk, Microbuckling Failure of Circular Fiber Reinforced Composites, *AIAA J.*, Vol 13 (No. 10), Oct 1975
6. B.W. Rosen, chapter 3 in *Fiber Composite Materials*, American Society for Metals, 1965
7. L.J. Broutman, Failure Mechanisms for Filament Reinforced Plastics, *Mod. Plast.*, April 1965
8. M. Ashizawa, Fast Interlaminar Fracture of a Compressively Loaded Composite Containing a Defect, in *Proceedings of the Fifth NASA/DoD Conference on Fibrous Composites in Structural Design*, Jan 1981
9. H.T. Hahn and J.G. Williams, "Compression Failure Mechanisms in Unidirectional Composites," NASA Technical Memorandum 85834, National Aeronautics and Space Administration, Aug 1984
10. L.B. Greszczuk, On Failure Modes of Unidirectional Composites Under Compression Loading, in *Proceedings of the Second USA-USSR Symposium on Fracture of Composite Materials*, March 1981
11. A.G. Evans, and W.F. Adler, Kinking as a Mode of Structural Degradation in Carbon Fiber Composites, *Acta Metall.*, Vol 26, 1978

12. C.W. Weaver and J.G. Williams, Deformation of a Carbon-Epoxy Composite Under Hydrostatic Pressure, *J. Mater. Sci.*, Vol 10, 1975
13. S.V. Kulkarni, J.R. Rice, and B.W. Rosen, An Investigation of the Compressive Strength of Kevlar 49/Epoxy Composites, *Composites*, Vol 6, 1975
14. C.R. Chaplin, Compressive Fracture in Unidirectional Glass-Reinforced Plastics, *J. Mater. Sci.*, Vol 12, 1977
15. N. Hancox, The Compression Strength of Unidirectional Carbon Fiber Reinforced Plastic, *J. Mater. Sci.*, Vol 10, 1975
16. M.M. Sohi, H.T. Hahn, and J.G. Williams, "The Effect of Resin Toughness and Modulus on Compressive Failure Modes of Quasi-Isotropic Graphite/Epoxy Laminates," to be published as a NASA Contractor Report
17. J.G. Williams and M.D. Rhodes, "The Effect of Resin on The Impact Damage Tolerance of Graphite-Epoxy Laminates," NASA Technical Memorandum 83216, National Aeronautics and Space Administration, Oct 1981
18. B.A. Byers, "Behavior of Damaged Graphite/Epoxy Laminates Under Compression Loading," NASA Contractor Report 159293, National Aeronautics and Space Administration, Aug 1980
19. J.G. Williams, "Effect of Impact Damage and Open Holes on the Compression Strength of Tough Resin/High Strain Fiber Laminates," NASA Technical Memorandum 85756, National Aeronautics and Space Administration, Feb 1984
20. J.H. Starnes and J.G. Williams, "Failure Characteristics of Graphite-Epoxy Structural Components Loaded in Compression," NASA Technical Memorandum 84552, National Aeronautics and Space Administration, Sept 1982
21. C. Zweben, Tensile Failure of Fiber Composites, *AIAA J.*, Vol 6 (No. 12),

Dec 1968
22. B.W. Rosen, Tensile Failure of Fibrous Composites, *AIAA J.*, Vol 2 (No. 11), Nov 1954
23. B.D. Agarwal and L.J. Broutman, *Analysis and Performance of Fiber Composites*, John Wiley & Sons, 1980
24. J.H. Sinclair and C.C. Chamis, "Mechanical Behavior and Fracture Characteristics of Off-Axis Fiber Composites. 1—Experimental Investigation," NASA Technical Paper 1081, National Aeronautics and Space Administration, 1977
25. J.H. Sinclair and C.C. Chamis, "Mechanical Behavior and Fracture Characteristics of Off-Axis Fiber Composites. 2—Theory and Comparisons," NASA Technical Paper 1082, National Aeronautics and Space Administration, 1978
26. D.J. Wilkins, "A Preliminary Damage Tolerance Methodology for Composite Structures," Workshop on Failure Analysis and Mechanisms of Failure of Fibrous Composites, NASA Langley Research Center, 1982
27. D.J. Wilkins, *et al.*, Characterizing Delamination Growth in Graphite-Epoxy, in *Damage in Composite Materials*, STP 775, K. Reifsnider, Ed., American Society for Testing and Materials, 1982
28. R.W. Hertzberg, *Deformation and Fracture Mechanics of Engineering Materials*, John Wiley & Sons, 1976
29. B.W. Smith, R. Grove, and T. Munns, Fractographic Analysis of Interlaminar Fractures in Graphite-Epoxy Material Structures, in *Proceedings of the International Conference: Post Failure Analysis Techniques for Fiber Reinforced Composites*, 1985
30. K.M. Liechti, J.E. Masters, D.A. Ulman, and M.W. Lehman, "SEM/TEM Fractography of Composite Materials," AFWAL-TR-82-4085, Air Force Wright Aeronautical Laboratories, Sept 1982

Fractography for Continuous Fiber Composites

Brian W. Smith, Boeing Commercial Airplane Company

FRACTURES IN COMPOSITES can occur in a number of complex ways because of their laminated anisotropic construction. The types and modes of failure that can be encountered depend upon both the direction of applied load and the orientation of fibers (plies) making up the composite material. Figures 1 and 2 show that variations in either of these can produce strikingly different fracture appearances on a macroscopic scale. This range of diversity precludes the ability to assign well-defined macroscopic fracture types for most applications. The definition of fracture modes on a microscopic scale, however, provides a relatively useful means of classifying failure modes and fracture types in much the same way as with metals.

Fractures in continuous fiber reinforced composites can be divided into three basic fracture types: interlaminar, intralaminar, and translaminar (Fig. 3). As with the intergranular and transgranular terminology commonly used with metals, each of these classifications describes the plane of fracture with respect to the microstructural constituents of the material. Translaminar fractures are those oriented transverse to the laminated plane in which conditions of fiber fracture are generated. Interlaminar fracture, on the other hand, describes failures oriented between plies, whereas intralaminar fractures are those located internally within a ply. Translaminar fractures involve significant fiber fracture, while interlaminar or intralaminar fractures occur in the laminate plane and therefore break few if any fibers.

Using this convention, failures in composites can be described in terms of the failure mechanism exhibited on these differing fracture surfaces. These failure mechanisms reflect the type of load under which microscopic separation occurs—tension, shear, or compression. The following sections discuss these failure modes and their microscopic fracture features.

Interlaminar and Intralaminar Fractures

When considered on a microscale, interlaminar and intralaminar fracture types can be similarly described. In both cases, fracture occurs on a plane parallel to that of the fiber reinforcement. In a similar manner to that described for metals, fracture of either type can occur under mode I tension, mode II in-plane shear, mode III antiplane shear, or any combination of these load conditions. Figure 4 illustrates load states I and II. All three failure modes are still being investigated. As a result, such conditions as mode III antiplane shear have not been thoroughly studied. However, for mode I tension and mode II in-plane shear, enough data exists to model their mechanisms of separation and to describe their fracture characteristics.

Because interlaminar and intralaminar failures occur in the same plane as their fiber reinforcement, their fracture mechanism and appearance tend to be dominated by matrix fracture and fiber-to-matrix separation. In general, separation of the fiber from the matrix occurs at the interface for either mode I tension or mode II in-plane shear-loading conditions. As a result, when cohesive resins are involved, very little fracture occurs along the fiber. Fracture of the matrix resin between fibers exhibits pronounced cohesive fracture characteristics under both mode I tension and mode II in-plane shear loading.

For the majority of thermosetting matrices currently in use, cohesive-matrix failure occurs in a brittle manner. Common to brittle failure in metals and unreinforced polymers, cohesive-resin fracture characteristically exhibits relatively flat fracture planes with very little evidence of material deformation. The plane of such brittle failures is nearly always oriented normal to the direction of locally resolved tension. As shown in Fig. 4, separation under

Fig. 1 Effect of fiber orientation on flexural failure modes in various graphite-epoxy lay-ups. (a) quasi-isotropic. (b) 0°. (c) ± 45°. (d) 0/90°

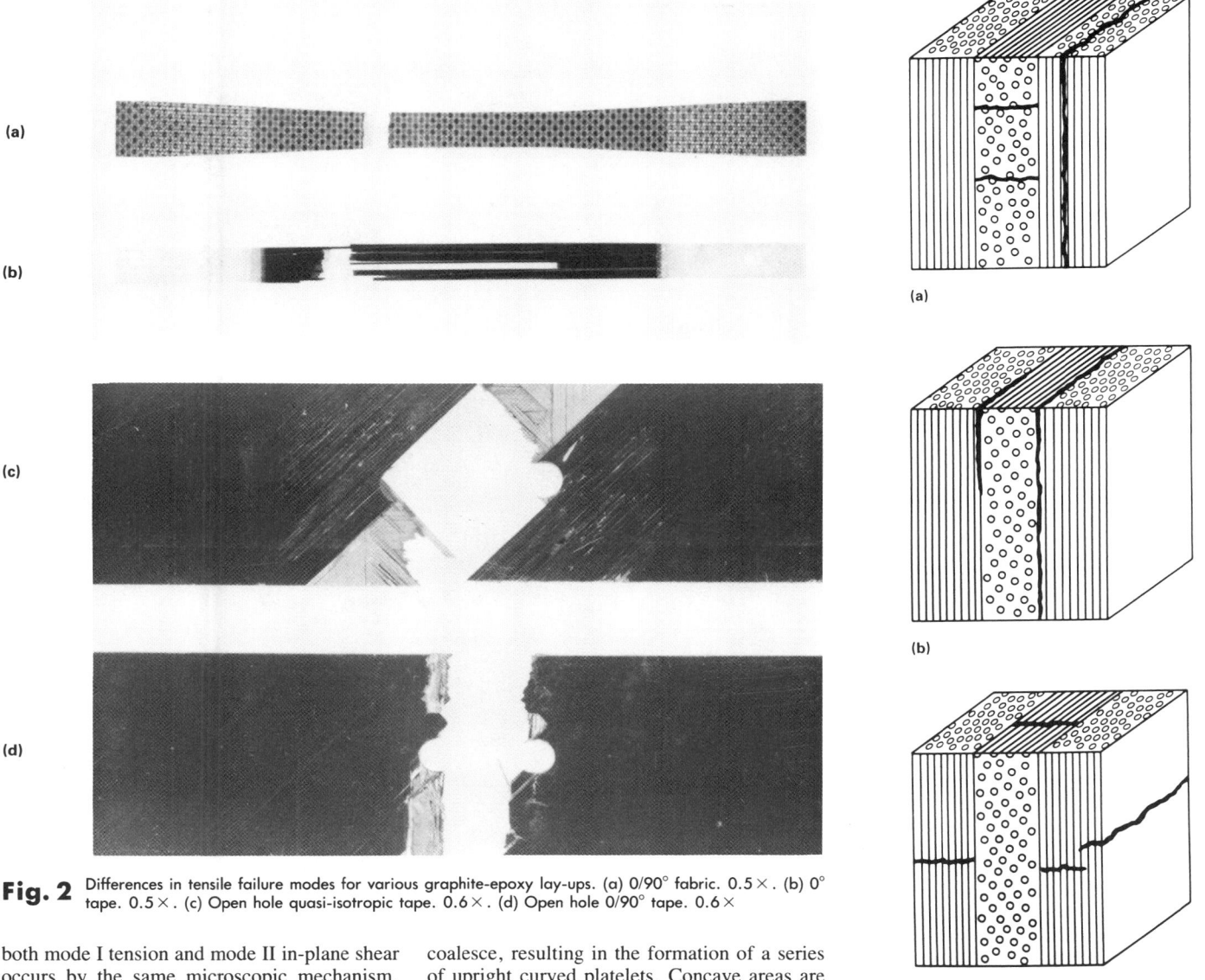

Fig. 2 Differences in tensile failure modes for various graphite-epoxy lay-ups. (a) 0/90° fabric. 0.5×. (b) 0° tape. 0.5×. (c) Open hole quasi-isotropic tape. 0.6×. (d) Open hole 0/90° tape. 0.6×

Fig. 3 Different planes of separation in continuous fiber reinforced fracture. (a) Intralaminar fracture. (b) Interlaminar fracture. (c) Translaminar fracture

both mode I tension and mode II in-plane shear occurs by the same microscopic mechanism, that is, brittle tension. The only difference between these two modes is the orientation of principal tension stress under which microscopic failure occurs.

In the case of mode I tension, the maximum principal tensile stress lies perpendicular to the plane of failure. As a result, brittle cleavage results in distinctly flat areas of cohesive-resin fracture. Failures produced under conditions of mode II in-plane shear, while also occurring by brittle-tensile separation, exhibit an appearance that is distinctly different from mode I tensile failures. Under this load condition, the laminate planes formed on either side of the crack are laterally displaced with respect to one another. As described by Mohr's circle, the principal tensile stress for applied shear is oriented at 45° to the plane of fracture. Because brittle-matrix fracture occurs normal to this resolved tensile stress, a series of distinct, inclined microcracks (Fig. 5) is formed. During the fracture process, these microcracks

coalesce, resulting in the formation of a series of upright curved platelets. Concave areas are found on the mating fracture surface, opposite to where platelet separation has occurred. Several terms have been used to describe each of these features, including lacerations or hackles for the upright platelets and scallops for concave areas. The more common use of hackles to describe platelets and scallops to describe depressed, concave areas will be used in this article.

Crack Directions and Fiber Orientations. The basic failure modes illustrated in Fig. 3 represent a simplified case of interlaminar or intralaminar failure in which fracture progressed parallel with and between plies and in the same direction as the fiber reinforcement. In most actual applications, however, the direction of reinforcement is likely to be oriented at a variety of angles in order to achieve the specific properties desired. For such structures, the direction of crack propagation is likely to occur at a variety of angles to the fiber orientation, depending upon the direction of imposed

stress and material anisotropy. As a result, it is important to understand how these factors alter the basic features illustrated in Fig. 4 and how they can be used to determine the direction of crack propagation.

Mode I Tension Fractures. The easiest mode to consider first is that generated under interlaminar tension. Fracture that is produced parallel to the direction of fiber reinforcement results in flat areas of brittle-matrix failure (Fig. 4a). Such areas typically exhibit distinct river marks. As with metals, these river marks correspond to fracture ridges formed by minutely displaced failure planes. As crack progression occurs, these planes link up, resulting in coalescence of this ridge structure to form a river-

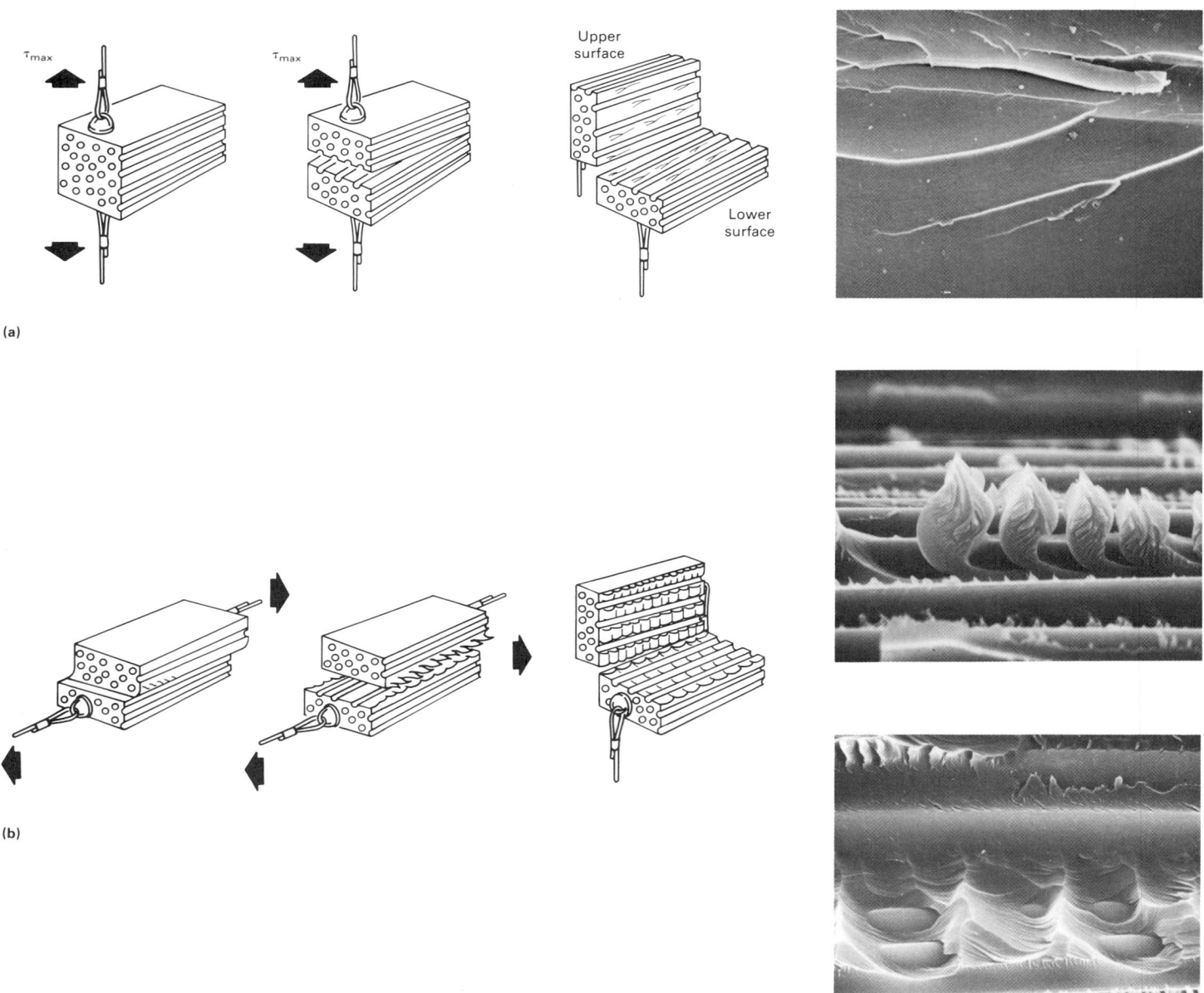

Fig. 4 Failure modes in fiber-reinforced composites. (a) Schematic of mode I interlaminar-tension failure and the resulting fracture surface appearance. (b) Schematic of mode II interlaminar-shear failure and two possible fracture surface morphologies

like pattern. As with metals, the direction of this coalescence can be used to define the direction of crack propagation.

Areas of fractured matrix also exhibit a distinctly textured morphology (Fig. 6). This appearance is discernible only at higher magnifications and in most cases requires tilting of the scanning electron microscope specimen or transmission electron microscope replica for it to become visible. In many ways, this textured appearance is identical to the cleavage feathers characteristic of metallic brittle fractures. These feathers exhibit a distinctive chevron-type appearance, with the pointed end of the chevron oriented toward the origin of propagation.

Mode I tension fractures produced at various angles to the direction of fiber reinforcement typically exhibit both the river markings and feathering noted above. Significant alterations in the general fracture topography can occur, depending upon the number of fibers exposed by fracture and their orientation (Fig. 7). In some cases, the variations can produce relatively large areas of flat-resin fracture with distinct river marks oriented in a consistent direction. Alternatively, extensive amounts of fiber exposure can lead to the existence of extremely localized microscopic areas of fracture. This latter condition often results in river marks and cleavage feathers oriented in a vari-

ety of angles across the fracture surface. Variations in the direction of microscopic crack propagation can in most cases be averaged together to obtain an estimate of the overall direction of crack growth (Fig. 8).

Variations in the direction of microscopic crack growth depend upon several factors. The two most notable factors that must be considered are the formation of localized zones due to fiber intrusion and the magnitude of stress concentration involved in fracture. In the first case, fiber intrusion divides the crack tip into numerous microscopic zones. In general, these zones will exhibit differing growth rates and slightly displaced planes of fracture. Conse-

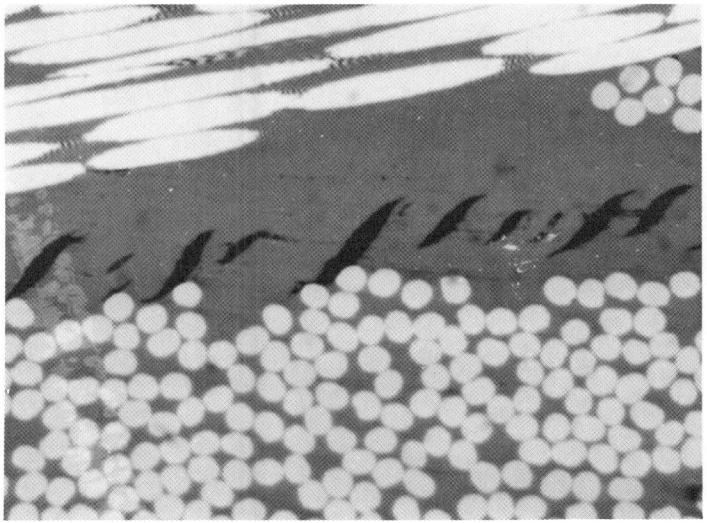

Fig. 5 Inclined microcracks found in short-beam shear specimen tested at 132 °C (270 °F). 500×. Source: Ref 1, 2

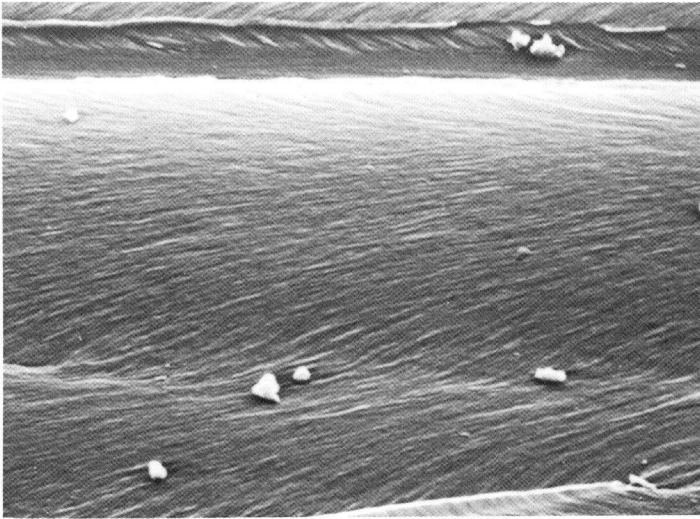

Fig. 6 Matrix feathering produced under interlaminar mode I tension. 5000×. Source: Ref 1

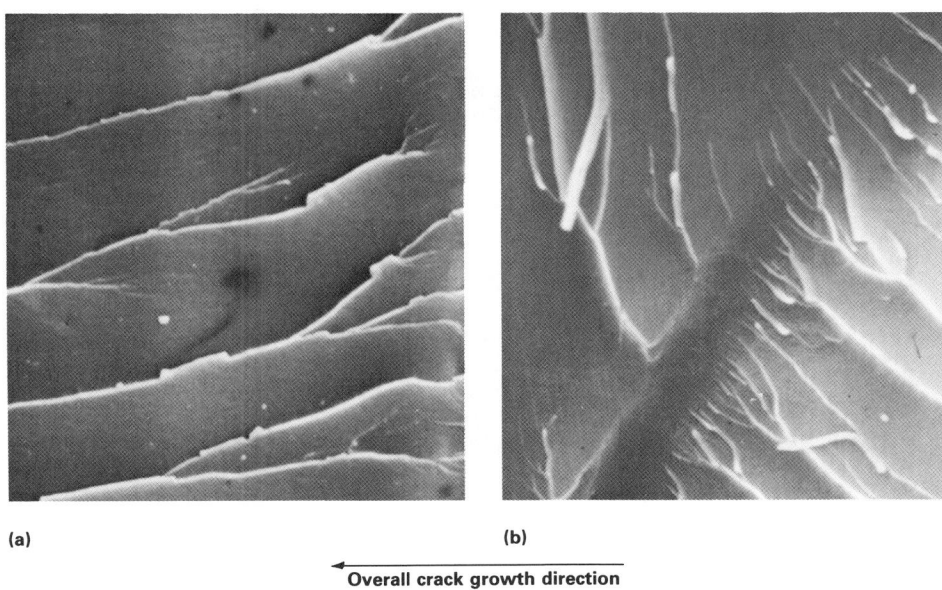

(a)

(b)

← **Overall crack growth direction**

Fig. 7 Mode I tension interlaminar fractures that propagated at various angles to the direction of fiber reinforcement. (a) Fracture between adjacent 0 and 90° plies. (b) Fracture between 45 and −45° plies. Both 2000×. Source: Ref 1

quently, the resultant crack front formed by these zones can be highly irregular, with fingers of advanced growth extending beyond the main crack tip. This crack front profile can produce locally divergent crack directions as zones of unfractured material located between the advancing fingers or behind the main crack tip grow toward immediately adjacent zones of existing fracture.

The second major condition that can lead to crack direction divergence takes place when failure occurs without the formation of any appreciable stress concentration. This condition

is often associated with the initiation of ultimate material failure in unnotched test specimens or components (Fig. 9). Material separation occurs in this case by the initiation of multiple origins without any preferred crack direction. As with ductile rupture in metals, failure occurs when these multiple fracture planes intersect. Such fractures are characterized by the formation of an extensive number of fracture planes throughout the laminate with extreme variations in the direction of crack propagation such that no overall direction of crack propagation often exists.

Mode II In-Plane Shear Fractures. The direction of crack growth for interlaminar-shear failures cannot be established with the same confidence that is possible for interlaminar-tensile fractures. Separation under shear loading occurs by the coalescence of numerous tensile microcracks under continued shear displacement. Growth of each microcrack and subsequent coalescence can occur in one of two principal directions (Fig. 10). Microcrack growth can occur either coincident with or opposite to the direction of overall separation (toward or away from the crack tip). As a result, platelets formed under shear loading and their microstructural details, such as river marks and feathering, can be oriented either toward or away from the direction of overall growth. Consequently, a single direction of crack growth cannot be absolutely defined for shear fractures. However, these features can be used to estimate the direction parallel to which actual crack growth occurred.

The orientation of fiber reinforcement at or adjacent to the delamination plane can have a significant effect on the morphology of both hackles (platelets) and scallops (concave resin fracture areas) that must be considered when determining the direction of crack propagation. Fibers intersecting the fracture surface parallel to the direction of crack growth tend to form roughly orthogonal hackles and scallops, whereas fibers intersecting the fracture surface at an angle to the direction of growth tend to form roughly triangular, asymmetric hackles and scallops (Fig. 11). In the first case, a distinct branched structure exists on both sides of the hackles and scallops where they intersect adjoining areas of fiber-to-matrix separation. Because this symmetry and the roughly orthogonal shape of these features correspond to propagation parallel to the direction of exposed

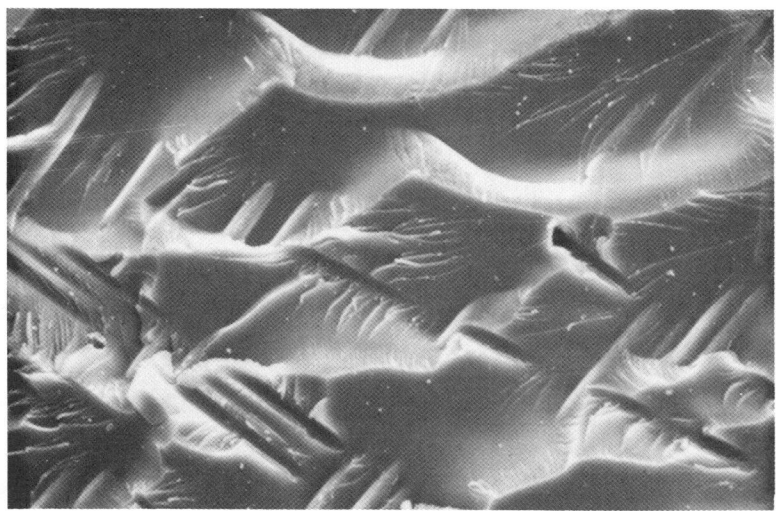 

Overall crack growth direction

Fig. 8 Variations in the crack growth of a graphite-epoxy ($\pm 45°$) composite. The arrows shown in the fractograph on the right-hand side map the crack growth, which enables the overall crack growth direction (shown below fractographs) to be determined. 400×. Source: Ref 1

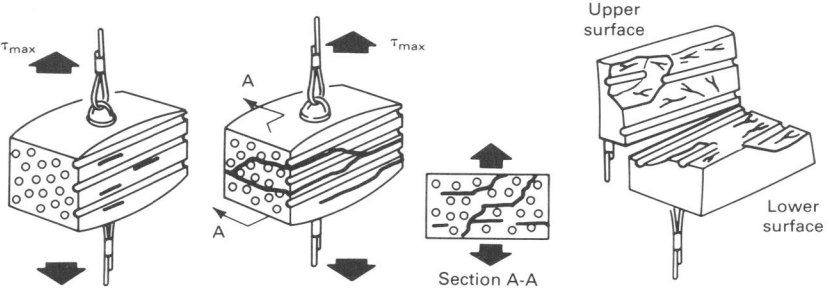

Fig. 9 Schematic of interlaminar-tensile failure without a stress concentration. Variations are produced in microscopic crack direction and multiple failure planes.

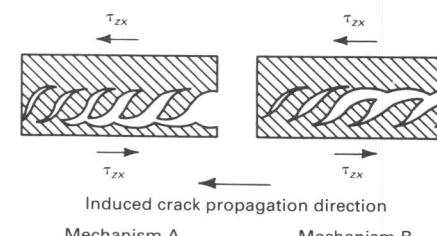

Induced crack propagation direction

Mechanism A   Mechanism B

Fig. 10 Schematic of possible hackle separation mechanisms. Mechanism A illustrates hackle formation coincident with the direction of crack propagation, whereas mechanism B illustrates the formation of hackles opposite to the direction of crack propagation. Source: Ref 1

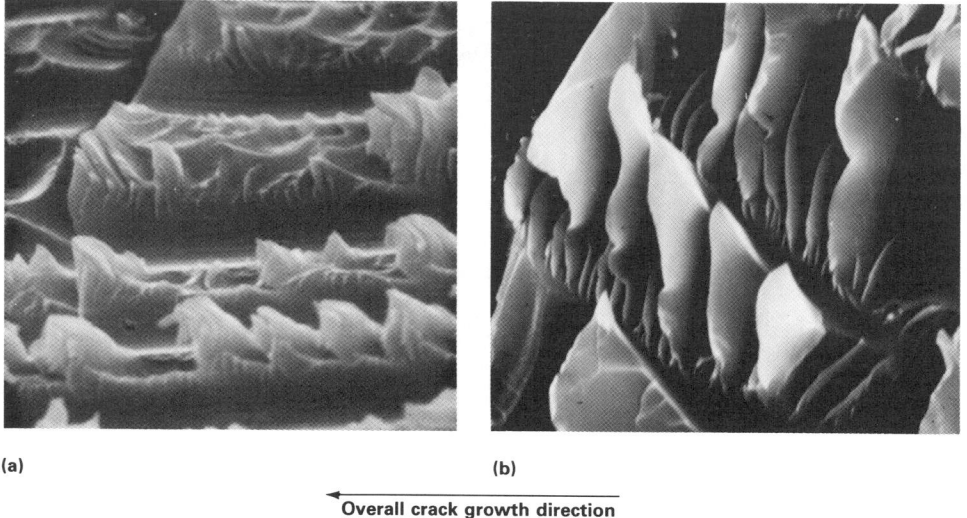

(a)   (b)

Overall crack growth direction

Fig. 11 Interlaminar mode II shear fractures that propagated at an angle to the direction of fiber reinforcement. (a) Delamination between 0 and 90° plies. 5000×. (b) Fracture between 45 and −45° plies. 2000×. Source: Ref 1

reinforcement, these features provide a relatively rapid and easy means of identifying the direction parallel to which crack growth occurred.

When crack propagation occurs at an angle to the direction of exposed reinforcement, the asymmetry of the features produced and the orientation of hackle tilting can be used to define the direction parallel to which fracture occurred. Figure 11 shows that hackles and scallops produced under this condition exhibit a roughly triangular appearance, with distinct branched markings located at the apex of this triangle. For this condition, the crack propagation direction is parallel to the direction of inclined hackle tilting.

Translaminar Fractures

While interlaminar fractures tend to be relatively planar, translaminar fractures are generally identified by their rough fiber-dominated morphologies. Because of the high translaminar

(a)

(b)

(c)

(d)

Fig. 12 Examples of translaminar-tension fractures. (a) Translaminar-tension fracture in graphite-epoxy composite. Note fiber bundles and individual fiber pull-out. 400×. Source: Ref 2. (b) Translaminar-tension failure with localized area of flat fracture. 2000×. Source: Ref 2. (c) Radial fracture topography of an individual graphite-fiber failure under translaminar tension. 10 000×. Source: Ref 2. (d) Variations in fiber fracture mapped to determine overall crack growth direction. 2000×. Source: Ref 3

fracture toughness of composite laminates, a considerable amount of gross damage typically occurs. As with fractures in metallic structures, macroscopic features generally reflect the condition of imposed load at failure. In contrast to interlaminar fractures, load states can be identified by visually examining the displacement between mating fracture surfaces and/or the general morphology of the fracture surfaces. Two distinct translaminar failure mechanisms that occur in laminates are tensile and compression microbuckling. One of these

two separation mechanisms, or a combination, is operative in all translaminar fracture conditions. The following sections discuss the relationships between load states, fracture mechanisms, and characteristic fractographic features.

Translaminar-Tension Fractures. Macroscopically, translaminar-tension fractures exhibit an extremely rough fracture surface, with large amounts of fibers protruding out of the major fracture plane (Fig. 12a). Little or no delamination is evident at or near the fracture

surface. Brittle-tensile failure of individual fibers is the primary operative failure mechanism, with fracture of the surrounding matrix considered secondary. Fibers fracture in groups (bundles) where the fibers within each bundle have a relatively flat, common fracture plane (Fig. 12b). Fiber pull-out, fiber end fracture, and matrix fracture are the characteristic fractographic features of translaminar-tension failures. Figure 12(c) shows the typical radial topography found on broken fiber ends. This radial morphology is analogous to the chevron

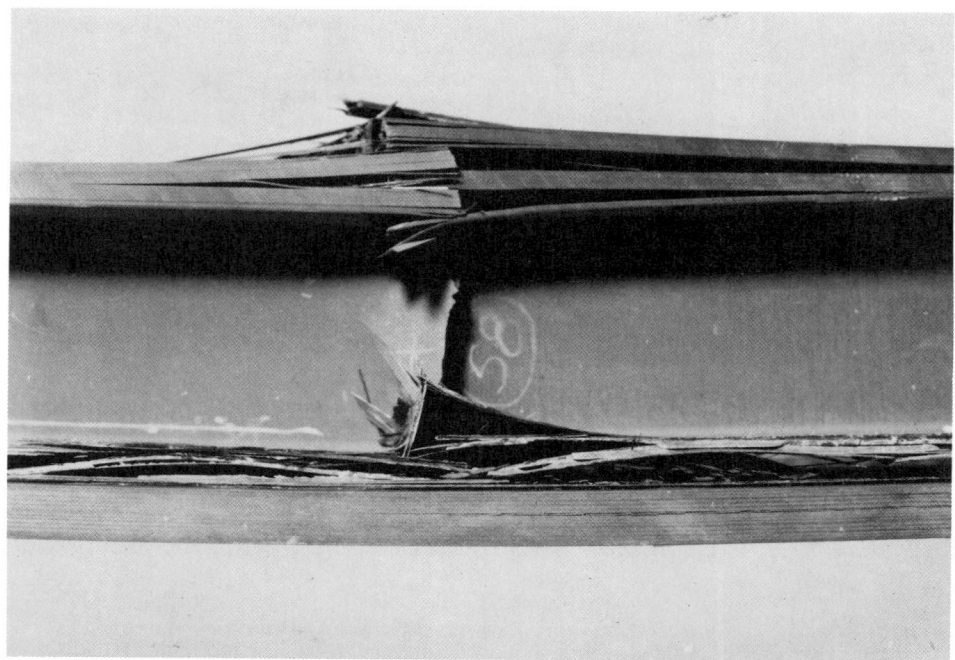

Fig. 13 Side view of a compression fracture with extensive delamination and interlocking. 0.8 ×

The neutral axis lines are commonly found parallel to one another in a given region, indicating that microbuckling occurs on a local scale in a concerted manner and in a unified direction (Fig. 14d). Preliminary controlled crack growth studies have shown that the neutral axis lines are often biased at an angle parallel to the direction of induced crack propagation. This indicates that flexural collapse and fracture propagation on a local scale often occur transverse to the gross crack direction; therefore, neutral axis lines cannot be used to determine the direction of crack propagation.

REFERENCES

1. B.W. Smith *et al.*, Fractographic Analysis of Interlaminar Fractures in Graphite-Epoxy Material Structures, in *Proceedings; International Conference: Post Failure Analysis Techniques for Fiber Reinforced Composites*, Air Force Wright Aeronautical Laboratories, MLSE, July 1985
2. A.G. Miller *et al.*, Fracture Surface Characterization of Commercial Graphite/Epoxy Systems, in *Nondestructive Evaluation and Flaw Criticality for Composite Materials*, STP 696, American Society for Testing and Materials, p 223-273
3. S.W. Tsai and H.T. Hahn, *Introduction to Composite Materials*, Technomic Publishing, 1980

SELECTED REFERENCES

- S.L. Donaldson, Fracture Toughness Testing of Graphite/Epoxy and Graphite/PEEK Composites, *Composites*, Vol 6 (No. 2), April 1985
- T. Johannesson, P. Sjoblom, and R. Seldon, The Detailed Structure of Delamination Fracture Surfaces in Graphite/Epoxy Laminates, *J. Mater. Sci.*, Vol 19, 1984, p 1171-1177
- R.A. Kline and F.H. Chang, Composite Failure Surface Analysis, *J. Compos. Mater.*, Vol 14, Oct 1980, p 315-324
- K.M. Liechti *et al.*, *SEM/TEM Fractography of Composite Materials*, AFWAL-TR-82-4085, Air Force Wright Aeronautical Laboratories, MLSE, Sept 1982
- D. Purslow, Some Fundamental Aspects of Composites Fractography, *Composites*, Vol 12 (No. 4), Oct 1981
- J.H. Sinclair, Fracture Modes in High Modulus Graphite/Epoxy Angleplied Laminates Subjected to Off-Axis Tensile Loads in *Rising to the Challenge of the 80's; 35th Annual Conference and Exhibit*, Society of the Plastics Industry, Inc., 1980, p 12-C1 to 12-C8

patterns in tension failures of metals, particularly for rod or bar forms. The faint lines radiate from the point of fiber fracture initiation and thus indicate the direction of crack propagation for each individual fiber. Consistent with brittle failures, the fiber origins are primarily located at flaws or notches in the crenelated fiber surfaces, although some initiate at internal flaws, such as voids.

Tension failure does not progress by a well-defined crack front. Because of flaw sensitivity, the fracture process involves multiple initiation zones in which all the fiber breaks originate at a single source. Thus, the crack front actually consists of several isolated fracture zones—at different axial planes—that coalesce and propagate in the overall growth direction. Fiber ends within each zone tend to fracture in a variety of directions, although they are often noticeably biased in a single direction. Through extensive mapping of the fiber ends (Fig. 12d), the macroscopic crack growth direction can be determined. Extreme caution must be exercised in the evaluation of isolated microstructural details; therefore, a coherent approach is required to ensure accurate and unbiased crack mapping.

Translaminar-Compression Fractures. Macroscopically, fractures produced under uniaxial compression exhibit gross buckling, extensive delamination, and interlocking of the delamination planes (Fig. 13). An end-on view of the broken fiber ends reveals a distinct, flat fracture surface with extensive postfracture

damage (Fig. 14a). This obliteration of fracture-surface details is due to relative postfailure motion between the fractured surfaces in contact. The surface is much flatter than that in translaminar-tension fractures and is virtually devoid of pulled-out fibers. Fiber buckling, fiber end fracture, resin matrix fracture, and postfracture damage are the characteristic fractographic features of translaminar-compression fractures.

Compression microbuckling is the primary operative failure mechanism; it involves local buckling of individual fibers at a point where a maximum localized lateral instability exists. Under compressive flexure, kinking of each fiber causes fracture at two locations (Fig. 14b), with each fracture separated by 5 to 10 fiber diameters. Short sections of fibers with this length can often be seen on the fracture surface. Figure 14(c) illustrates the typical flexural fracture morphology found on the fiber ends. The radial topography represents the tensile portion of fracture, while the smooth topography represents the compressive fracture. The distinct line shown intersecting the fiber end in Fig. 14(c) is the neutral axis line. For each fiber, the direction of flexure and failure occurs normal to the neutral axis line. For a given fiber, the compression and tension morphologies are reversed when comparing the two breaks. Thus, a singular crack direction cannot be deduced, although the individual fiber fracture propagates perpendicularly to the neutral axis line.

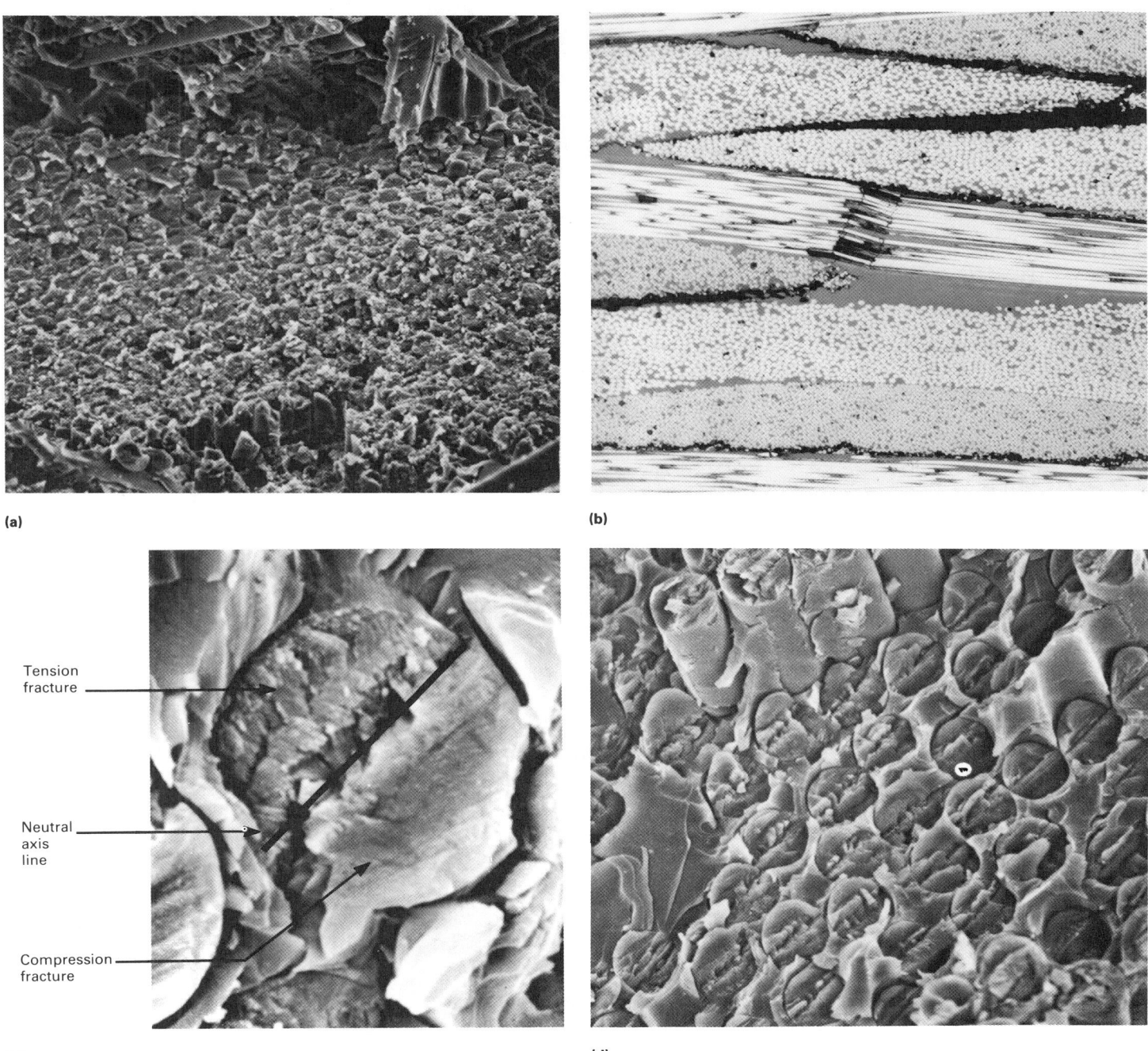

(a)

(b)

Tension fracture

Neutral axis line

Compression fracture

(c)

(d)

Fig. 14 Examples of translaminar-compression fractures. (a) Translaminar-compression fracture with extensive postfailure damage to fiber ends. 750×. (b) Translaminar-compression generated fiber kink in graphite-epoxy fabric. 100×. (c) Flexural fracture characteristics on fiber ends of compression specimen. 10 000×. (d) Translaminar-compression fracture illustrating parallel neutral axis lines representative of unified crack growth. 2000×

Discontinuous Fiber Composites

Richard E. Robertson, The University of Michigan
Viorica E. Mindroiu, Ford Motor Company

FAILURE MODES in discontinuous fiber composite materials, which have a rather complex structure, are best understood in terms of the arrangement of reinforcing fibers and matrix resin.

Structure of Discontinuous Fiber Composites

Sheet molding compounds (SMC) have a typical chopped-fiber construction. Sheet molding compounds are made from glass fiber bundles, each of which contains several thousand filaments that are cut or chopped into lengths of 25 mm (1 in.) or longer. The chopped rovings are arranged more or less in parallel planes, with the fiber axes randomly oriented in these planes (Ref 1). Although the planar texture is retained during molding, the flatness of the planes can be lost because of the ability of sheet molding compounds to be molded into shapes of varying thickness, which is one of their advantages. Therefore, parts of an SMC sheet can be shrunk laterally to fill space, which bunches the fibers together and causes the fiber planes to become wavy.

The fiber bundles in high glass density parts usually end up being flattened, becoming 2 to 2.5 mm wide (0.08 to 0.098 in.), 0.1 mm (0.004 in.) thick, and 25 to 50 mm (1 to 2 in.) long. The flattened fiber bundles are visible in the cross section through a molded SMC sheet, shown in Fig. 1. Designated SMC-R62, it contains 62 wt% glass fiber, or about 42 vol%. The bundles often have a lenticular shape, being thicker in the center and tapering toward the edges. The glass filaments are about 10 μm (390 μin.) in diameter. The heights of the filaments shown in Fig. 1 are all about the same, indicating that the fibers lie in parallel planes. However, the widths of the images vary from layer to layer, indicating different orientations within those layers, with the wider images arising from fibers crossing the section plane at increasingly oblique angles from the perpendicular.

A contrasting view of the same molded SMC sheet is seen in Fig. 2. This is a top view of a section made in the plane of the sheet and is from a planar cut made about 0.5 mm (0.02 in.) below the upper surface of a 3-mm (0.10-in.) thick flat sheet. This photomacrograph was point illuminated from a direction toward the upper left corner, which illuminated only those fibers that were more or less perpendicular to it. Therefore, the most illuminated fibers are those that are parallel to the diagonal across the macrograph from lower left to upper right. The fibers in the other directions are relatively dark. This shows the general tendency toward a lack of strict planarity of the fiber bundles, even for a flat sheet. As indicated by the irregular shape of the light-colored zones, and especially the appearance of circular images that result from sectioning through domelike structures, the flattened bundles are rarely planar but usually have compound curvature, even if it is of slight amplitude.

A view of the arrangement of fibers and resin that permits the space-filling role of the resin to be seen is obtained by cleaving parallel with the plane of the material. The result is shown in

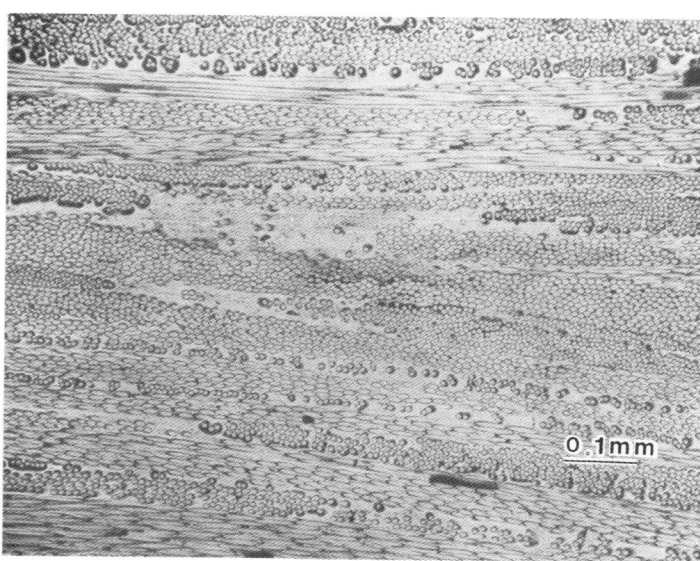

Fig. 1 Light photomicrograph of the cross section through a molded flat sheet of SMC-R62-polyester sheet molding compound

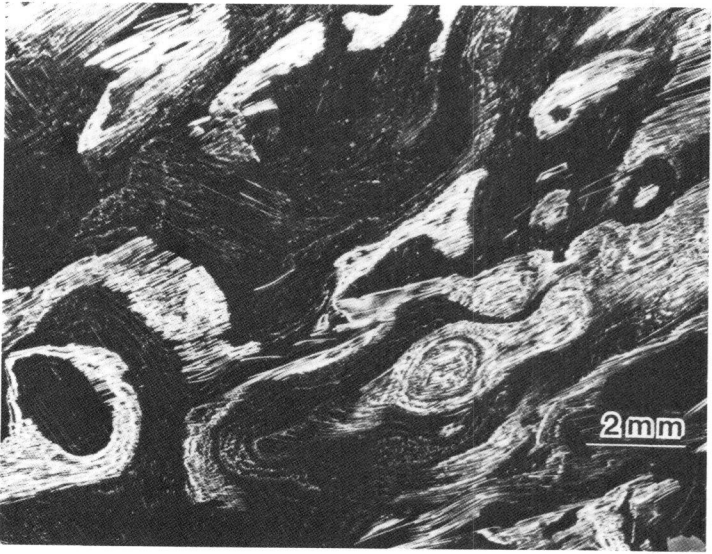

Fig. 2 Light photomacrograph of a section parallel with the plane of the sheet of the same specimen as in Fig. 1. Incident light was directed onto the specimen from a position above the upper left corner.

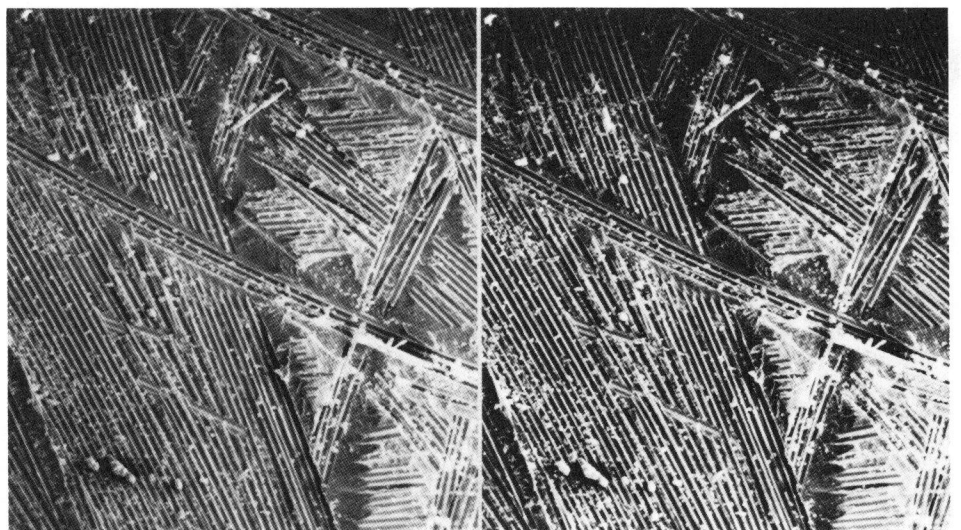

Fig. 3 Stereo pair of SEM photomicrographs of a surface produced by cleaving a flat molded sheet of SMC-R62 more or less parallel with the plane of the sheet. The fiber diameter is 10 μm (390 μin.).

Fig. 5 SEM photomicrograph of a tensile failure near a cut edge in specimens cut from the same SMC-R62 molded sheet as in Fig. 1 to 4. The tensile stress was applied along the horizontal axis. The fiber diameter is 10 μm (390 μin.).

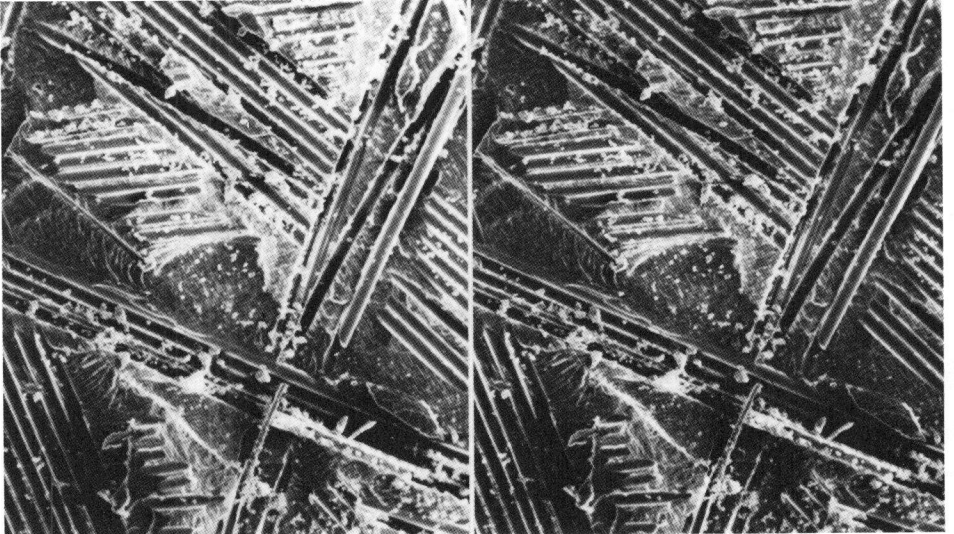

Fig. 4 Stereo pair at higher magnification of the image in Fig. 3. The fiber diameter is 10 μm (390 μin.).

Many other arrangements that are usually less desirable often occur, including the wavy fiber bundles mentioned above, kinked fiber bundles, large resin-rich regions, and parallel fiber bundles. The failure modes in chopped-fiber systems will reflect all of these structures.

Tensile Failure

Tensile failure usually occurs with some separation of fiber bundles along the parallel planes. As with cleavage, from which the fractures shown in Fig. 3 and 4 were obtained, the separation is not very clean. Therefore, one often sees fiber bundle separation and fiber breakage, as shown in Fig. 5 to 7. The tensile stress direction that produced the failures illustrated in Fig. 5 to 7 was horizontal. In general, tensile failure results in much fiber breakage, intrabundle fiber separation, and separation of fiber bundles on planes parallel to the sheet surface (and perpendicular to the nominal fracture plane). Therefore, it is difficult to obtain simple photomicrographs of typical failures, although Fig. 5 to 7, while representative, were selected for their relative simplicity. This simplicity was obtained by recording only the first few layers of the composite. The outer surface of the sheet is visible or nearly visible on the left side of each of these specimens.

The specimen shown in Fig. 5 had the cleanest fracture, owing in part to the edge of the specimen adjacent to the failure. The cut edges exert considerable influence on the failure of chopped-fiber specimens, and the results of mechanical tests, especially on narrow specimens, must be interpreted accordingly. Nonetheless, a significant amount of fiber breakage is apparent in Fig. 5.

Fig. 3 and 4. These photomicrographs should be viewed stereoscopically to see the variation in elevation.

Toward the left in Fig. 3 is the relatively flat, parallel array of grooves left by fibers in a single fiber bundle. Toward the right is the resin acting more or less as a space filler, which occurs because fiber bundles of different orientation cannot lie in the same plane. An enlargement of this is shown in Fig. 4. The resin used was a polyester that produced a large amount of debris during cleavage, some of which is seen adhering to the cleaved surfaces in Fig. 3 and 4. In contrast to this brittle behavior is the surprising ductility of the resin, suggested by the drawn fiber of resin that is seen extending over

a crevice at the lower center in Fig. 4. Although brittle in simple tension, the resin does not necessarily behave in a brittle manner under more complex stress fields that include a superimposed compressive stress. The characteristics seen in Fig. 4, such as the broken glass fiber and the removal of fibers from many different planes, indicate that cleavage parallel to the plane of the SMC sheet is not as simple a failure mode as might be predicted, considering that the fibers lie in parallel planes. This more complex failure mode results from a slight, but effective interweaving of the fiber bundles, as suggested in Fig. 2.

The structure described above is that of chopped-fiber systems generally at their best.

Fig. 6 Photomicrograph of tensile failure in the same SMC-R62 molded sheet, away from a cut edge. The tensile stress was applied along the horizontal axis. The fiber diameter is 10 μm (390 μin.).

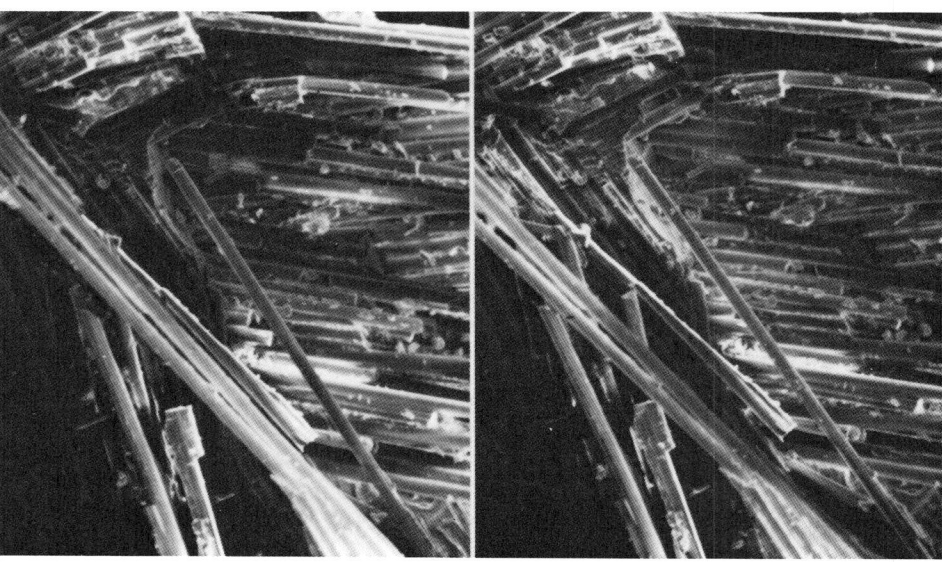

Fig. 7 Stereo pair of SEM photomicrographs of tensile failure showing the separation of fiber bundles around a region filled with resin. The tensile stress was applied along the horizontal axis. The fiber diameter is 10 μm (390 μin.).

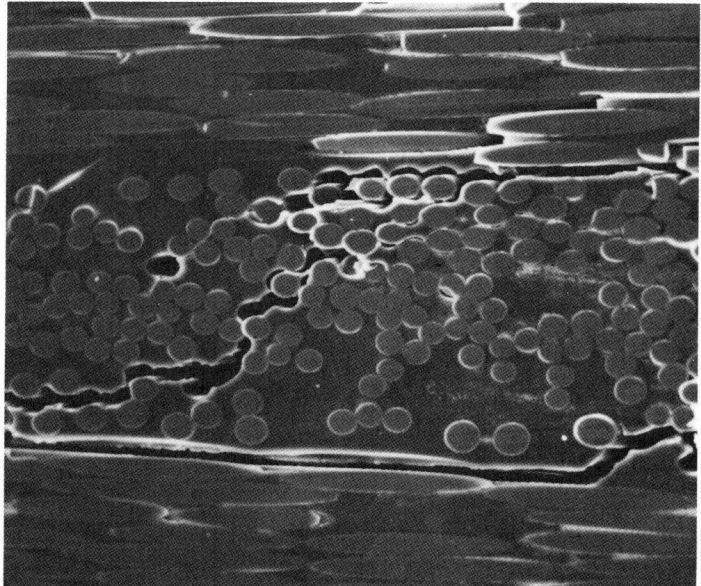

Fig. 8 Shear failure in SMC-C50-R25-polyester. A tensile crack is seen along the diagonal, and a shear crack is seen along the horizontal. The fiber diameter is 10 μm (390 μin.).

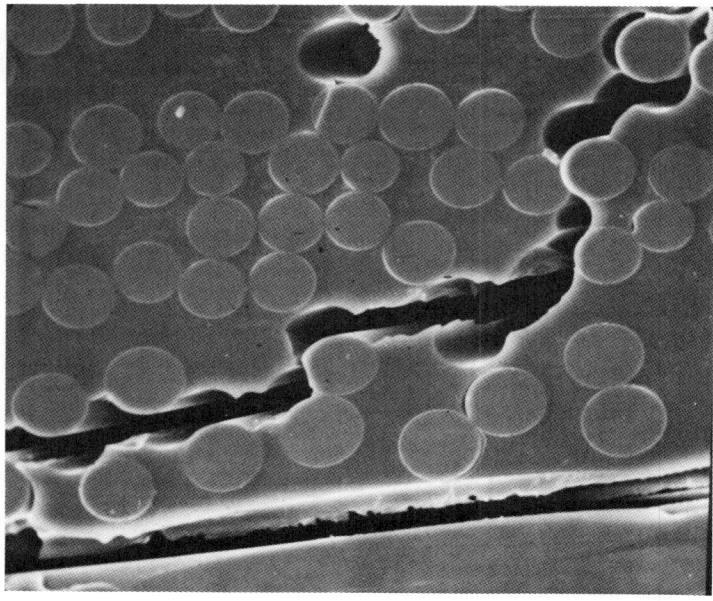

Fig. 9 Higher-magnification view of shear failure shown in Fig. 8. The fiber diameter is 10 μm (390 μin.).

Figure 6 shows a more typical tensile failure. This failure is characterized by separated fiber layers to which pieces of matrix resin that previously separated from adjacent fiber bundles are attached. The separation of fiber bundles arrayed about a pocket of matrix resin is shown in Fig. 7. Because fiber bundles several layers deep have been removed, the relative elevation of the different separated layers is easily visible in Fig. 7. Failure in these chopped E-glass and thermosetting polyester composite materials has resulted in fairly complete de-bonding between fibers and matrix, although matrix residue is visible on many of the fibers. As with the cleaved materials shown in Fig. 3 and 4, a certain amount of resin dust is visible on the fibers.

Compressive Failure

Compressive failures of discontinuous fiber composites are often similar to tensile failures in that separation between the fiber planes occurs. For a compressive stress in the plane of the sheet, failure begins with delamination, followed quickly by buckling of the resulting thin layers. Therefore, this type of failure is characterized by the sharp kinking of the fiber bundles with occasional nonkinked protruding filaments. When the fibers have received support from a rigid matrix before failure, failure usually occurs at high stresses with considerable delamination. With softer resins that offer less support to the fibers, however, the fibers may kink and break locally at relatively low loads and with relatively little delamination.

Shear Failure

Shear failures are most easily seen in specimens having the cut edges polished before failure. Polished edges also reduce adventitious failures. Shear failures are often found to be tensile failures; that is, failure is induced by the tensile component of the stress. This is illustrated in Fig. 8 and 9, in which the sense of the applied shear stress was such that the upper edge of the specimen had been pushed to the left and the lower edge to the right. The principal tensile component is therefore at roughly 45°, along the diagonal from upper left to lower right. This caused fracture perpendicular to it, along the diagonal from lower left to upper right. Also visible in Fig. 8 and 9 is a true shear crack at the bottom of the central fiber bundle. However, such shear cracks are usually not the first cracks to appear. They typically occur upon further loading after the tensile cracks have appeared.

Several other important features are apparent in Fig. 8 and 9. First is the tendency for failure to occur in fiber bundles oriented perpendicularly to the applied shear stress, as is the case in the failed bundle shown in Fig. 8 and 9. The specimen illustrated was constructed of layers of continuous and discontinuous fibers, with the continuous fibers arranged as a ± 15° angle ply. In such materials, the shear failure occurs in the chopped-glass layers, in or about the chopped-fiber bundles oriented perpendicularly to the tensile stress component.

A second feature is the relative motion of the fibers in a bundle. Fiber bundles sometimes come apart during fabrication; therefore, when the material is deformed, the fibers do not all respond the same. As is evident in Fig. 8, and even more so in Fig. 9, a single fiber at the upper center was pulled into the material as it and the neighboring bundle moved in different directions during deformation. Without the freedom of the cut edge to pull away from, this fiber probably would have fractured, even under relatively low loads. A subsidiary result of this is that if the stress is reversed, the fibers usually cannot return to their original positions. During cyclic deformation, then, a considerable amount of working, or fretting, of the material occurs, along with the production of much resin dust.

Another feature that can be seen in Fig. 9 is plastic deformation of the resin. Toward the bottom of the bundle, just above the true shear crack, gaps are visible between the fiber and the matrix for a pair of fibers. Because of their relative locations, these gaps are clear evidence of plastic deformation of the resin. During shearing, the upper right fiber was moved toward the left, and the lower left fiber toward the right, thus debonding the matrix from the fiber and deforming the resin plastically. Upon release of the stress, the resin remained deformed, and the gaps between fibers and matrix shown in Fig. 9 appeared.

Also significant in Fig. 9 is the local fracture path. The crack tends to follow the periphery of the fibers, entering the matrix only as necessary to reach the next fiber. This behavior is consistent with the photomicrographs shown in Fig. 3 to 7 and results in relatively little energy being absorbed during the shear failure of this specimen.

A final feature is fiber breakage. Although rare, fiber breakage sometimes occurs in shear failures. The broken fibers visible in Fig. 8 and 9 usually result from the incomplete removal, during polishing, of the damage that was done to the fibers during specimen cutting.

Fatigue Failure

In general, fatigue results in failures at final separation that are similar to those produced by monotonic stresses, except that great quantities of resin debris or dust are usually produced during fatigue. As mentioned previously, resin dust arises from the relative motions of fiber bundles and of fibers within the same bundle; this fretting can greatly reduce the fatigue life of the material (Ref 2).

When both tensile and compressive loads are applied, as in fully reversed cyclic loading, the failed specimens must be examined to determine which stress caused the failure (Ref 3). The compressive stress causes the greater amount of delamination; that is, compressive fatigue failures tend to be distributed over a larger region than tensile failures.

However, failure in fatigue often occurs long before separation. This is because the utility of a composite structure often depends on its stiffness. Therefore, failure is often defined as the loss of 10% of the modulus, for example. The softening arises from two main processes (Ref 4). These are debonding, which is the earliest to occur, and matrix resin cracking. Both of these phenomena manifest themselves as a whitening of the composite and can be studied in materials with low-to-modest levels of debonding and cracking by transmitted light microscopy.

REFERENCES

1. D.L. Denton, *The Mechanical Properties of an SMC-R50 Composite*, Owens-Corning Fiberglas Corporation, 1979
2. K. Friedrich, Fretting Fatigue Failure of Polyester Resin and Its Glass Fiber Mat Composites, *J. Mater. Sci.*, Vol 21, 1986, p 1700-1706
3. A. Conle and J.P. Ingall, Effect of Mean Stress on the Fatigue of Composite Materials, *Compos. Tech. Rev.*, Spring 1985, p 3-11
4. A.F. Johnson, *Engineering Design Properties of GRP*, The British Plastic Federation, 1979, p 71-75

SECTION 12

Applications and Experience

Chairman: John T. Quinlivan,
Boeing Commercial Airplane Company

Introduction

NEARLY 50 YEARS have passed since plastic resins reinforced with glass fibers were first used for their strength, light weight, low cost, and corrosion resistance. These characteristics not only ensured the widespread use of this material, but also catalyzed the search for advanced materials with even greater performance characteristics. As a result, the age of composite materials was born and today represents the fastest growing of all advanced materials technologies.

Advanced composite materials found their first applications in aerospace structures and sports and leisure equipment, fields which require high specific strength and specific rigidity. The applications being developed today range from artificial joints and organs to reinforced building materials for common construction, and from ship hulls to precision audio equipment. The range of applications appears boundless because man, for the first time, can design a material for the application rather than limit a design to the available materials.

This Section covers selected facets of current advanced composite material applications that will be useful for developing new applications. The aerospace applications of carbon and aramid fibers are nearing 20 years of successful design, development, and service. Advanced composite materials are finding new marine applications not previously possible. Sports and leisure goods continue to be directed by new applications of composite materials to achieve unprecedented performance gains. After many years of development, and even some false starts, the automotive market appears ready for selected composite production commitment. The future for advanced composites remains bright, and an ever-increasing application to new designs is predicted.

Aircraft Applications

Jeanne M. Anglin, Boeing Commercial Airplane Company

FIBER-REINFORCED COMPOSITES have become an increasingly attractive alternative to metal for many aircraft components. Composites are strong, durable, and damage tolerant. They meet design and certification requirements and offer significant weight advantages. Because they readily adapt to innovative manufacturing techniques, composites also can provide significant cost reductions.

The composite materials used in the aircraft industry are generally reinforcing fibers or filaments embedded in a resin matrix. The most common fibers are carbon, aramid, and fiberglass, used alone or in hybrid combinations. Carbon fiber is replacing fiberglass as the most widely used reinforcement. The resin matrix is usually an epoxy-based system requiring curing temperatures between 120 and 175 °C (250 and 350 °F). Both past and current applications of fiber-reinforced composites in commercial and military aircraft are described in this article.

Early Commercial Applications

Contracts awarded in the early 1970s by the U.S. Air Force and the National Aeronautics and Space Administration (NASA) to investigate and encourage the use of composites have resulted in the design, test, certification, and in-service use of a large number of composite components on commercial aircraft. Table 1 summarizes these applications.

Table 1 Composite components in aircraft applications for various types of aircraft

Composite component	F-14	F-15	F-16	F-18	B-1	AV-8B	DC-10 Demo	L-1011 Demo	737 Demo	727	757	767	Lear Fan
Doors	✓			✓	✓	✓					✓	✓	✓
Rudder		✓				✓	✓				✓	✓	✓
Elevator										✓	✓	✓	✓
Vertical tail		✓	✓	✓	✓	✓	✓	✓					✓
Horizontal tail	✓	✓	✓	✓	✓	✓			✓				✓
Aileron						✓		✓			✓	✓	
Spoiler									✓		✓	✓	
Flap					✓	✓					✓		✓
Wing box				✓		✓							✓
Body						✓							✓
Miscellaneous	Fairings	Speed brake		Speed brake, fairings	Slats, inlet	Fairings		Fairings			Fairings	Fairings	Propeller blades

One example of these government-sponsored programs, the composite spoiler for the Boeing 737 jetliner, was funded by the NASA Langley Research Center. Its design featured carbon fiber reinforced skins and end closure ribs made of fiberglass. The rest of the spoiler, including the center hinge fitting, front spar sections, and honeycomb core, were existing aluminum parts. The composite spoiler was completely interchangeable with the standard aluminum version. By July 1973, the Boeing Commercial Airplane Company had placed 111 composite spoilers in regular airline service for evaluation; these were periodically removed and tested in laboratories. Construction details of the part are illustrated in Fig. 1.

Under a NASA contract, Lockheed-California Company manufactured three advanced composite fins for the L-1011 airliner. Shown in Fig. 2, the fin box configuration had a span of 760 cm (300 in.), a root chord of 270 cm (107 in.), and a tip chord of 120 cm (46 in.), for a surface area of 15 m² (150 ft²). The box had 2 covers, 2 spars, and 11 ribs. The skins were co-cured in a single piece from a 175 °C (350 °F) curing tape material.

The National Aeronautics and Space Administration also funded a program for the McDonnell Douglas DC-10 upper aft rudder that resulted in the fabrication of 20 carbon fiber reinforced components for flight service evaluation. Furthermore, NASA sponsored the design and fabrication of elevators and horizontal stabilizer boxes for the empennage of Boeing 727 and 737 jetliners. The 727 elevator shown in Fig. 3 and 4 is composed of one-piece upper and lower honeycomb skin panels, laminated front and rear spars, and honeycomb ribs. The program provided for fabrication of five ship sets for flight service evaluation, all of which have been certified and are currently deployed on commercial aircraft. The 737 horizontal stabilizer box shown in Fig. 5 was the first application of a composite primary structure on a Boeing airplane. Five and one-half ship sets of stabilizer boxes were fabricated. All components were certified and are in commercial service. The first flight of the stabilizers was in 1980. The stabilizer box shown in Fig. 6 consists of I-stiffened laminated lower and upper skins mechanically fastened to laminated front and rear spars. The ribs are of a honeycomb sandwich-type construction and are very similar to those fabricated for the 737 carbon fiber reinforced epoxy elevator. A 22% weight savings (55 kg, or 116 lb) was achieved on the stabilizer box.

Current Production Aircraft

Composite components are used extensively on current commercial production aircraft such as the Boeing 757 and 767, and the Airbus A310, which employ about 1350 kg (3000 lb) each, while smaller planes such as the 737-300 use approximately 680 kg (1500 lb). The composite components of the Boeing 757 are listed in Table 2.

On the 737-300, composites are used for control surfaces, fairings, and nacelle components, as detailed in Fig. 7, for about 3% of the total structural weight of the aircraft. Individual composite parts are 20 to 30% lighter than their conventional counterparts.

Recently designed small general aviation airplanes make extensive use of advanced composites. The Lear Fan 2100 uses carbon, glass, and aramid fiber materials totaling approximately 820 kg (1800 lb) per aircraft; the Beech Starship I will include approximately 1360 kg (3000 lb) of composites.

With the exception of small, detail parts, most composite components for commercial

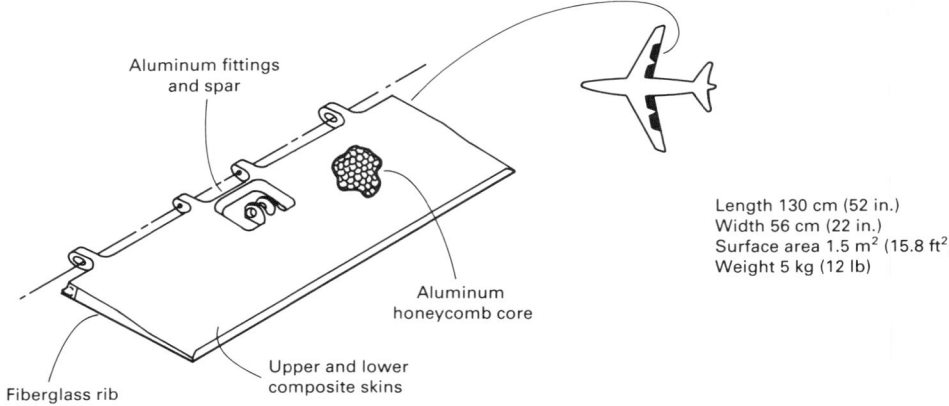

Length 130 cm (52 in.)
Width 56 cm (22 in.)
Surface area 1.5 m² (15.8 ft²)
Weight 5 kg (12 lb)

Fig. 1 Construction of Boeing 737 composite spoiler. Length, 130 cm (52 in.); width, 56 cm (22 in.); surface area, 1.5 m² (15.8 ft²); weight, 5 kg (12 lb)

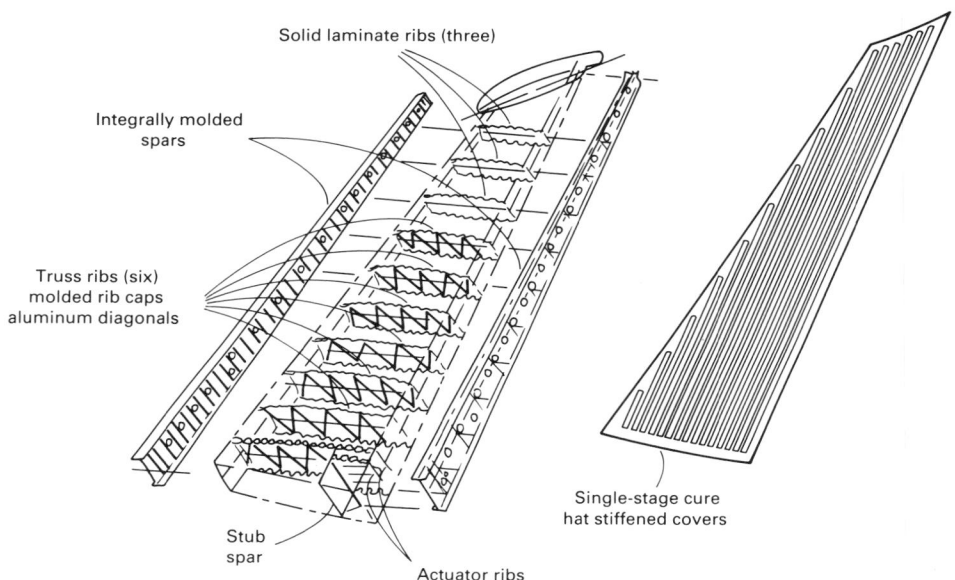

Fig. 2 L-1011 advanced composite vertical fin configuration

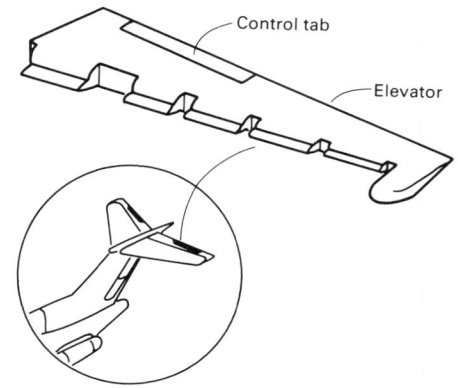

Fig. 3 Boeing 727 elevator

Table 2   Composite components on the Boeing 757

Description	Material	Size T × W × L cm	in.
Body .... Aft wing/body fairing	Aramid, carbon	33 × 50 × 193	13 × 20 × 76
Forward wing/body fairing	Aramid, carbon	38 × 60 × 360	15 × 24 × 142
Forward wing/body fairing	Aramid, carbon	177 × 185 × 188	70 × 73 × 74
Forward wing/body fairing	Aramid, carbon	170 × 198 × 188	67 × 78 × 74
Main landing gear doors	Carbon	81 × 170 × 284	32 × 67 × 112
Nose landing gear doors	Aramid, carbon	64 × 203	25 × 80
Wing .... Flap track fairing	Aramid, carbon	35 × 50 × 295	14 × 20 × 116
Aft outboard trailing edge flaps	Carbon	8 × 20 × 650	3 × 27 × 256
Aft inboard trailing edge flaps	Carbon	13 × 69 × 368	5 × 27 × 145
Aileron	Carbon	15 × 76 × 460	6 × 30 × 180
Inboard spoiler	Carbon	8 × 69 × 190	3 × 27 × 75
Outboard spoiler	Carbon	8 × 58 × 163	3 × 23 × 64
Lower wing inboard fixed leading edge panels	Aramid	46 × 71 × 160	18 × 28 × 63
Wing inboard fixed trailing edge panels	Aramid, carbon	152 × 229	60 × 90
Wing outboard fixed trailing edge panels	Aramid, carbon	36 × 430	14 × 170
Main landing gear trunnion fairing	Aramid, carbon	51 × 51 × 180	20 × 20 × 72
Engine strut fairing - forward	Aramid	41 × 61 × 163	16 × 24 × 64
Engine strut fairing - upper	Aramid	15 × 66 × 160	6 × 26 × 63
Engine strut fairing - U/W forward	Aramid	15 × 51 × 66	6 × 20 × 26
Engine strut fairing - aft	Aramid	3 × 58 × 163	1 × 23 × 64
Engine cowls	Carbon	135 × 135 × 117	53 × 53 × 46
Empennage .... Rudder	Carbon	38 × 234 × 902	15 × 92 × 355
Vertical tip fairing	Aramid	56 × 157	22 × 62
Vertical fixed trailing edge panels	Aramid, carbon	56 × 254	22 × 100
Elevators	Carbon	23 × 135 × 711	9 × 53 × 280
Horizontal tip fairings	Aramid	23 × 117	9 × 46

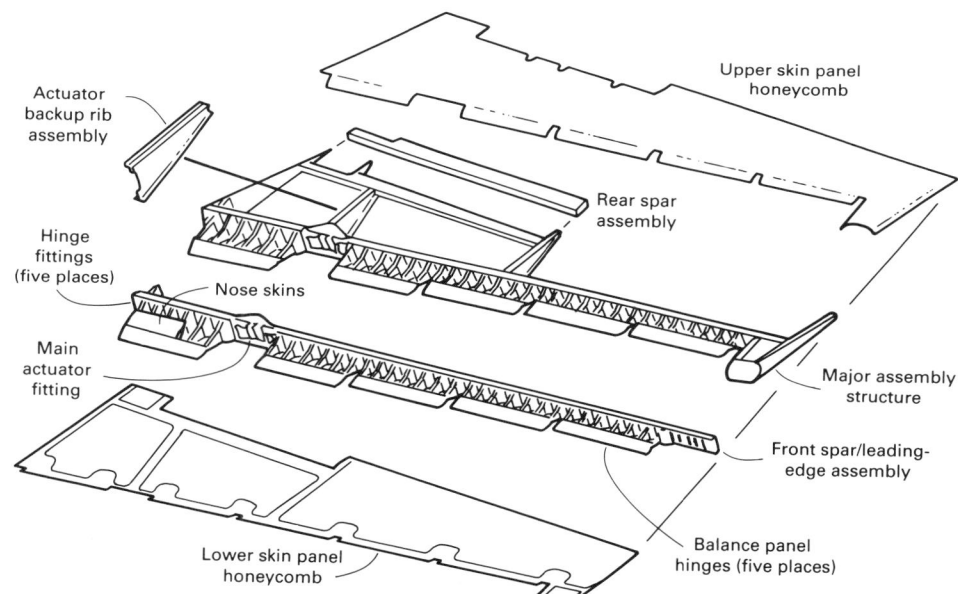

Fig. 4 727 elevator structural arrangement

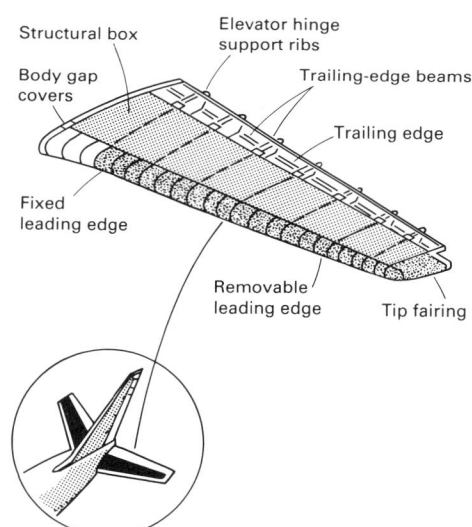

Fig. 5 737 composite stabilizer box

airplanes are of honeycomb sandwich construction. These may be either full-depth designs, such as the 767 outboard aileron shown in Fig. 8, or structures built of separate panels, such as the 767 rudder in Fig. 9. The 767 rudder depicted in Fig. 10 is the largest commercial composite component in service. It is approximately 11 m (36 ft) long and 3 m (8 ft) in chord width at the root. Both examples use 175 °C (350 °F) curing materials. This elevated-temperature cure provides the lightest weight, most

environmentally durable composite, particularly in its strength and modulus retention after moisture exposure.

Figure 11 shows a Boeing 757 as a typical example of composites used for an engine nacelle. Because the structure is in close proximity to the power plant, 175 °C (350 °F) curing materials are generally employed.

Structures such as fairings, fixed wing, and empennage trailing edge panels are generally fabricated as a sandwich. Face sheets for these

panels are made of carbon fiber or carbon fiber combined with aramid or fiberglass fabric. (A typical panel structure is shown in Fig. 12.) Such panels most often employ 120 °C (250 °F) curing systems, and are made either of tape or fabric materials, or with a layer of adhesive for bonding to the honeycomb core. Phenolic-coated fiberglass or honeycomb core is used. The panels are fabricated in a single-stage curing process that provides significant cost advantages in addition to weight savings.

Composites are also widely used in the interiors of commercial aircraft. In addition to meeting mechanical property and processibility

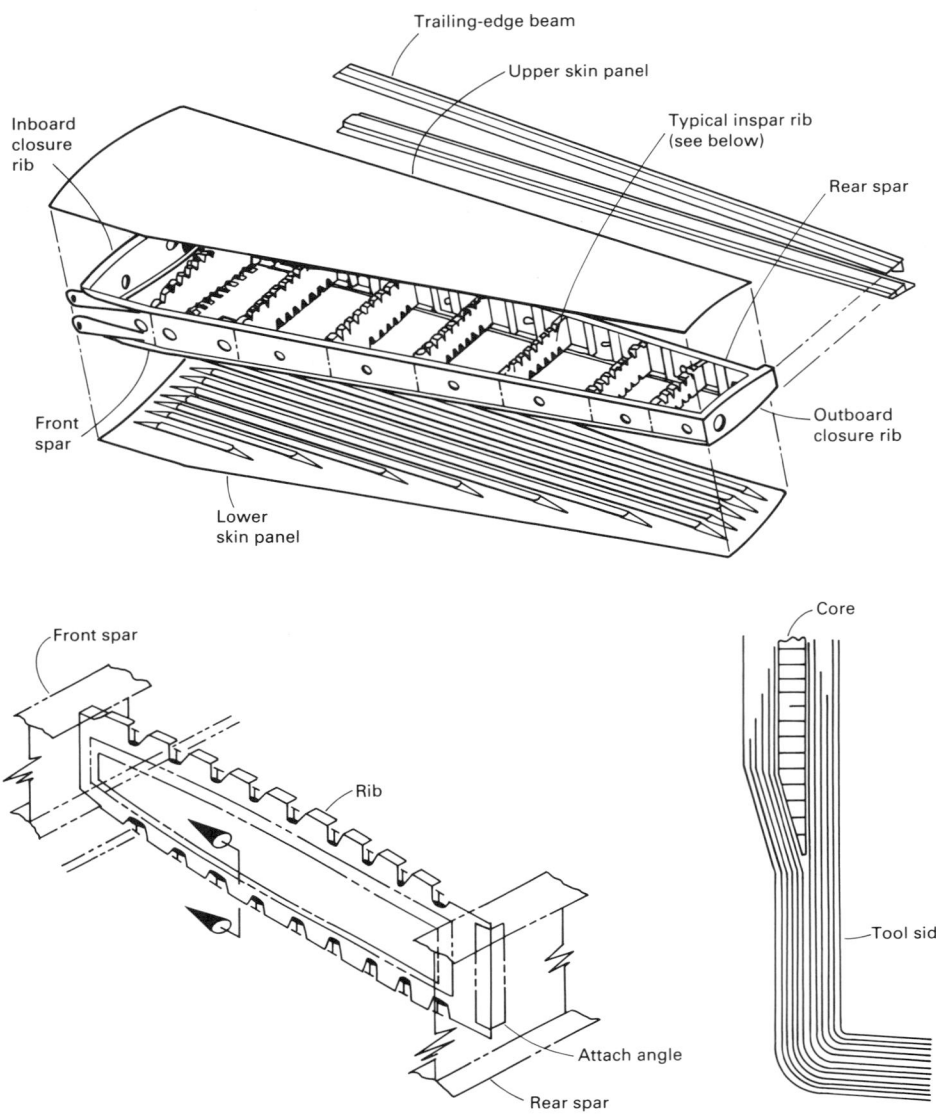

Fig. 6 737 horizontal stabilizer assembly

requirements, all materials used within the pressurized portion of the aircraft must meet both the flammability resistance requirements defined by regulatory agencies and, if applicable, smoke and toxic-gas emission guidelines of the airframe manufacturers. Additionally, visible portions of interior components must meet stringent aesthetic requirements to satisfy the airlines and their customers.

Interior parts such as overhead luggage compartments, sidewalls, ceilings, floors, galleys, lavatories, partitions, cargo liners, and bulkheads are routinely made of composite components (Fig. 13). In general, these are fiber-reinforced epoxy or phenolic resin honeycomb sandwich constructions. The phenolic resin system is used because of its excellent fire-resistant properties, including low flammability and low smoke and toxic gas emissions. The predominant design considerations for interior components are impact resistance, stiffness, and surface smoothness.

The choice of fiber depends not only on structural requirements but on part contour and fabrication method. For relatively flat parts, unidirectional or woven fabrics can be used. For compound contours, stretchable, knitted fabrics are often necessary. The predominant fiber used in interior composites is fiberglass; however, carbon fiber use is increasing as structural applications increase. For example, a filament-wound door spring is employed on the 767. Using unidirectional carbon fibers in an epoxy matrix, the springs are only one-third as heavy as comparable steel springs and only half the weight of state-of-the-art titanium springs.

Military Applications

The major U.S. aerospace industry users of carbon fiber prepreg materials include McDonnell Douglas, Boeing, General Dynamics, and Northrop. The continued rise in sales from about $12 million in 1973 to about $170 million in 1985 indicates the wide acceptance of composites, particularly carbon fiber reinforced composites. The largest application by far of composite material is for military programs, which constitute more than 40% of the aerospace total. For example, in 1985, 181 500 kg (400 000 lb) of composites were used by McDonnell Douglas in St. Louis for the F-15, F-18, and AV-8B fighter aircraft.

About 26% of the structural weight of the U.S. Navy's AV-8B is carbon fiber reinforced composites. Components include the wing box, forward fuselage, horizontal stabilizer, elevators, rudder and other control surfaces, and over-wing fairings. The wing skins are one-piece tip-to-tip laminate, mechanically fastened to a multispar composite substructure; the design of the horizontal stabilizer is similar to that of the wing. Approximately 590 kg (1300 lb) of carbon fiber epoxy is used on the AV-8B, providing a weight reduction of almost 225 kg (500 lb).

On the F-18 aircraft, carbon fiber reinforced composites make up approximately 10% of the structural weight and more than 50% of the surface area, as illustrated in Fig. 14. They are used in the wing skins, the horizontal and vertical tail boxes, the wing and tail control surfaces, the speed brake, the leading edge extension, and various doors. The F-18 composite wing skins are solid laminate; their thickness varies from root to tip, with a minimum thickness of about 2 mm (0.08 in.). The tail primary structure is similar in construction.

The B-1B bomber employs a number of composite structural components. Shown in Fig. 15, these include the dorsal longeron, weapons bay doors, aft equipment bay doors, and flaps. All of the materials, including adhesives, are 175 °C (350 °F) curing systems. The structures include laminate, full-depth honeycomb reinforced panels, and composite face sheets bonded to aluminum core. The bay doors shown in Fig. 16 employ carbon fiber reinforced tape face sheets, aluminum honeycomb core, and titanium fittings. Because the doors are in a position that is particularly vulnerable to foreign object damage, an aramid fiber reinforced phenolic outer layer provides penetration resistance. At a production rate of four aircraft per month, the B-1B uses 127 000 kg (280 000 lb) per year of composite structure—3040 kg (6700 lb) per aircraft—resulting in weight savings of approximately 1360 kg (3000 lb) on each bomber.

Grumman Aerospace Corporation fabricates F-14A horizontal stabilizers from a boron fiber reinforced composite material. The stabilizers are moving surfaces that pivot about shafts that protrude from the fuselage; each has an area of

6.5 m² (70 ft²), with a thickness chord ratio of 5% at the root and 3% at the tip. The stabilizer consists of the main structural box, leading and trailing edge sections, and tip. The latter three components have conventional aluminum skins over a full-depth aluminum honeycomb core. The main box is a boron-fiber reinforced composite structure consisting of the root rib, two intercostals, outer bearing, outboard rib, front and rear beams, tip rib, honeycomb core, and two covers. Figure 17 shows the front and rear beams and the root tip, which are of fiberglass construction. The stabilizer was designed so that there are no mechanical fasteners through the boron. In regions of high shear transfer between the substructure and cover, titanium is carried over the areas to distribute load, and the stabilizer is mechanically fastened. In regions of reduced loads, bonded joints are employed. A unique feature of this structure is a bonded splice at the pivot region shown in Fig. 18. The total weight of the stabilizer is 350 kg (778 lb), a saving of approximately 20%.

General Dynamics of Fort Worth employs a carbon fiber reinforced epoxy horizontal stabilizer, vertical stabilizer, leading edge, and rudder in the empennage of the F-16 fighter, which it manufactures. Figure 19 shows the vertical stabilizer structural box, which is a multispar, multirib, laminate skin design. The horizontal stabilizer in Fig. 20 has composite skins with aluminum honeycomb core.

Two other programs that employ considerable amounts of composite material were in progress as of 1986: the A-6 wing replacement program and the Navy's V-22. In the A-6 program, high-flight-time metal wings were being replaced by lighter composite wings with improved fatigue characteristics and much greater resistance to corrosion. The A-6 wing was being designed and built by Boeing Military Airplane Company. The V-22 aircraft is an innovative design that combines the advantages of the vertical takeoff and landing of a helicopter with the smooth, high-speed cruise and extended range of a fixed-wing airplane. The V-22 engines and propellers are vertically oriented for takeoff and landing, then pivot forward for cruise, with conventional wing surfaces providing aerodynamic lift. The V-22 will be built jointly by Boeing Vertol and Bell Helicopter Textron.

The A-6 replacement program requires a wing structural box made of carbon fiber reinforced epoxy. Figure 21 shows the design configuration of laminate skin fastened to composite intermediate ribs and spars, with titanium front and rear spars, and inboard and outboard tank end ribs. The skin panels are manufactured in a single piece using a 175 °C (350 °F) curing tape material.

The V-22 wing panels have integrally stiffened laminate skins. Figure 22 gives an inside view of a wing test box fabricated in 1985, and Fig. 23 shows an exploded view of the wing structure. The fuselage of the V-22 shown in Fig. 24 is also made of composite materials that

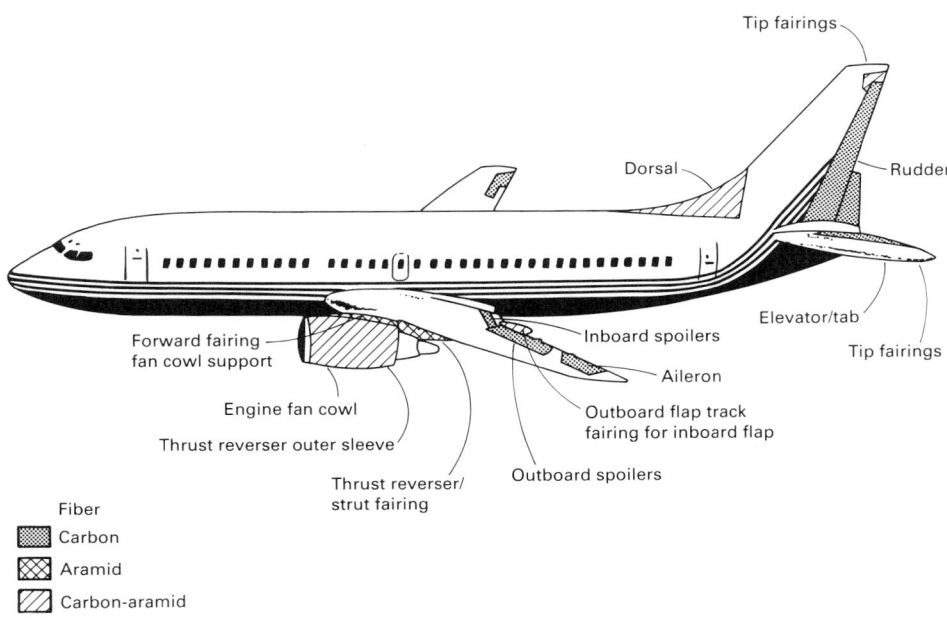

Fiber
◪ Carbon
⊠ Aramid
▨ Carbon-aramid

Fig. 7 Composite applications on the Boeing 737-300

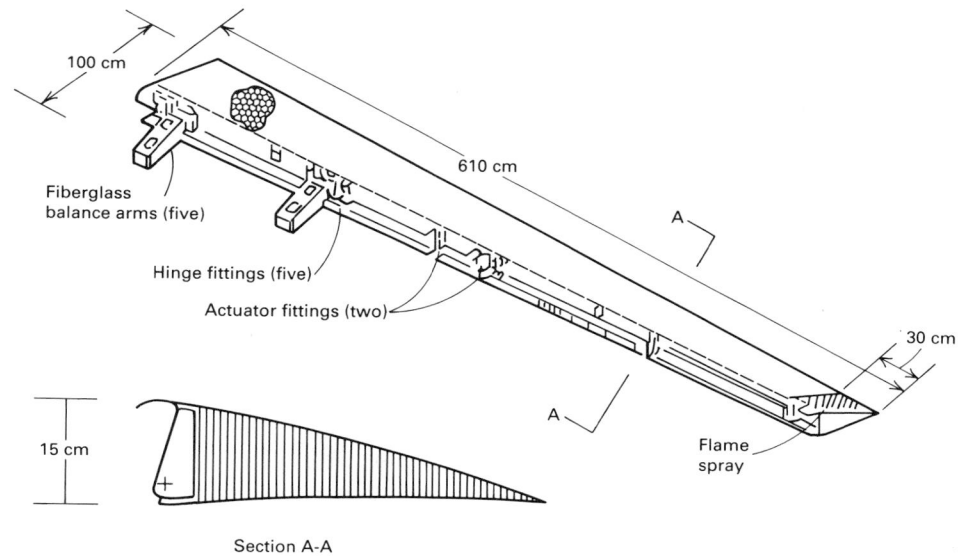

Fig. 8 767 outboard aileron

make up 50% of its structural weight in its current design configuration.

Trends clearly indicate that the use of composite structures will continue to grow in both commercial and military aircraft. New materials, with both improved properties and increased suitability to automated processing, will permit weight savings and cost competitiveness with traditional metal structures. Figure 25 depicts the rapid increase of composites in aircraft and an implied projection of increase in use.

However, factors that may inhibit wider use of composites include cost, schedule, capital investment, and inspectability. Furthermore, inherent variations in raw materials and poorly defined cost data, as well as a lack of uniform, industry-wide specifications, test techniques, standards, and design allowables all combine to create a degree of conservatism at this time. Because of these factors, the check-and-balance approach will result in the selection of composites for applications for which they are technically and economically most appropriate.

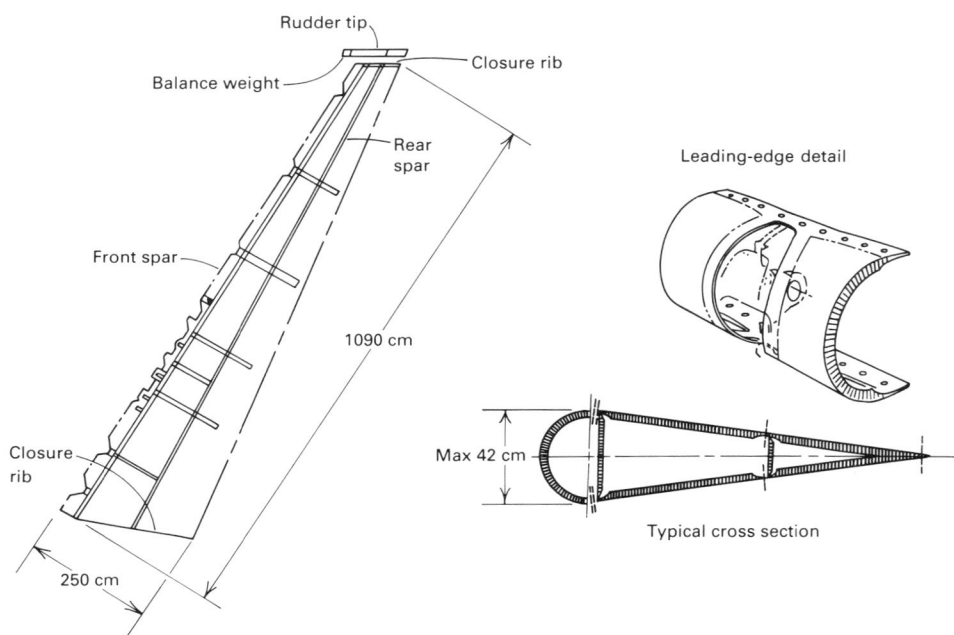

Fig. 9 767 carbon fiber reinforced epoxy rudder

Fig. 10 767 composite rudder

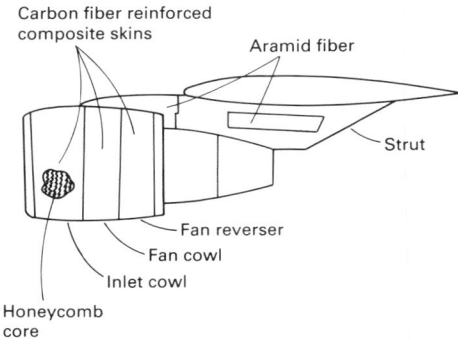

Fig. 11 757 engine strut applications

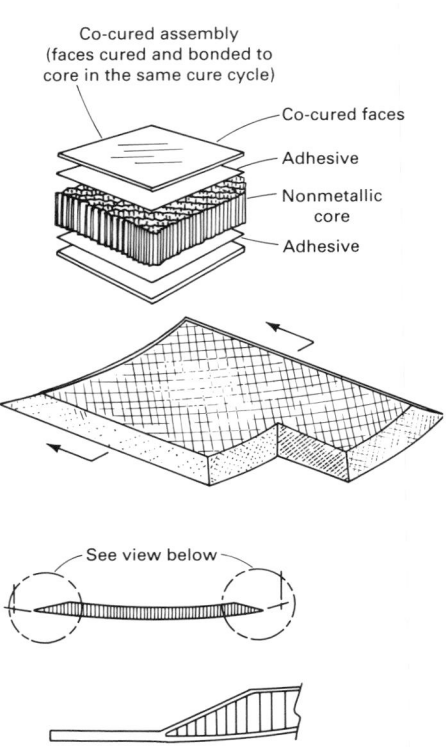

Fig. 12 Typical fairing panel construction

SELECTED REFERENCES

● *DoD/NASA Advanced Composites Design Guide*, Volume III, U.S. Air Force Flight Dynamics Laboratory, July 1983
● R.N. Hadcock, ''Status and Viability of Composite Materials in Structures of High Performance Aircraft,'' Paper presented to the National Research Council, Aeronautics and Space Engineering Board, Naval Postgraduate School, Monterey, CA, Feb 1986
● L.G. Hansen, D. Lossee, and W.L. O'Brien, Advanced Composites Applica-

Fig. 13 Commercial aircraft interior

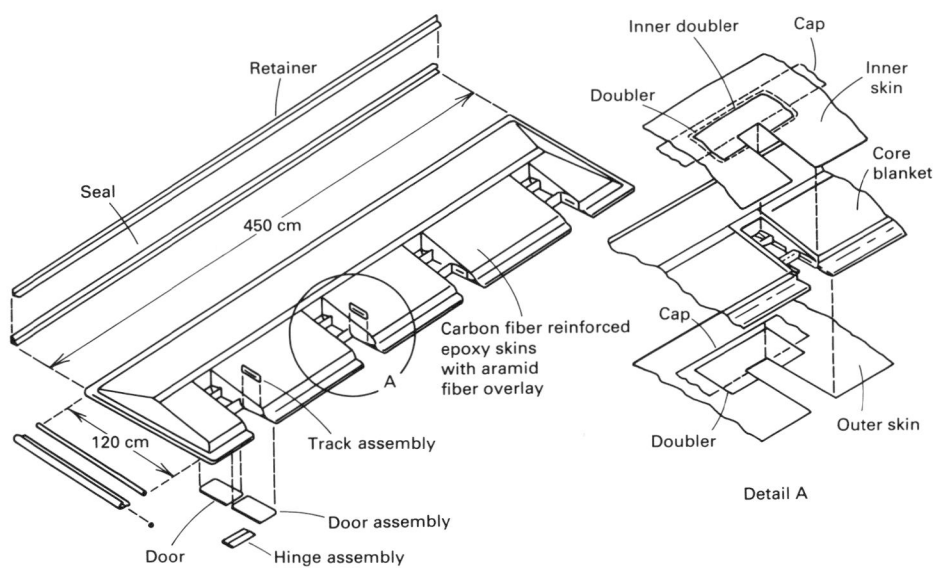

Fig. 16 B-1B weapons bay door

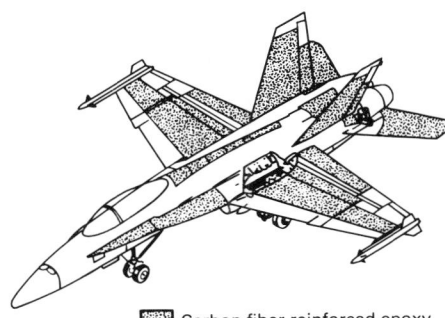

Carbon fiber reinforced epoxy, 10.3% structural weight

Fig. 14 F-18 composites

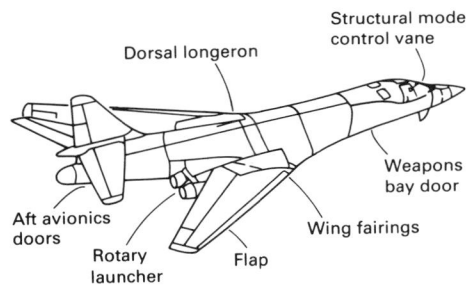

Fig. 15 B-1B composite applications

tions for the B-1B Bomber—An Overview, in *Proceedings of the 31st International SAMPE Symposium*, Society for the Advancement of Material and Process Engineering, 1986

- J.K. Kuno, Growth of the Advanced Composites Industry in the 1980's, in *Proceedings of the 31st International SAMPE Symposium*, Society for the Advancement of Material and Process Engineering, 1986
- J.C. McMillan and J.T. Quinlivan, Commercial Aircraft Applications, in *Advanced Thermoset Composites*, J.M. Margolis, Ed., Van Nostrand Reinhold, 1986
- S.A. Zervas-Berg, "Composites for Applications in Aircraft," *Chem. Eng. Prog.*, June 1986

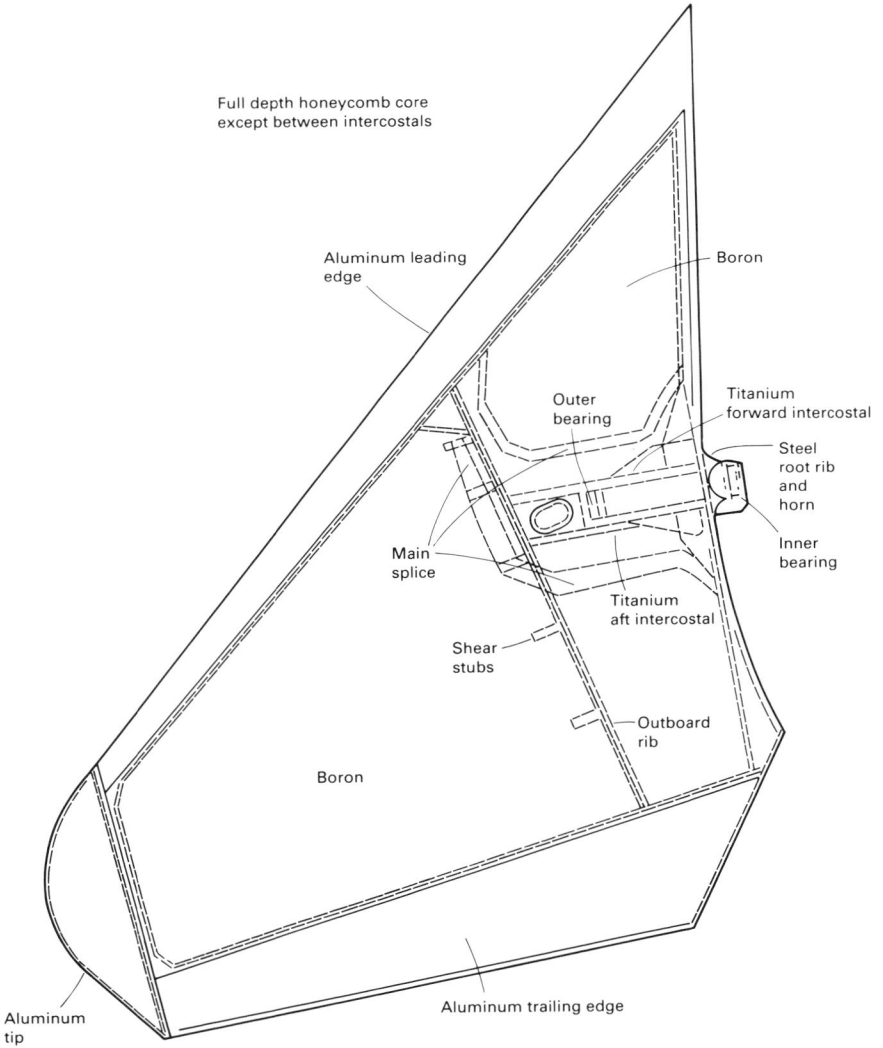

Fig. 17 F-14A boron-epoxy stabilizer, plan view

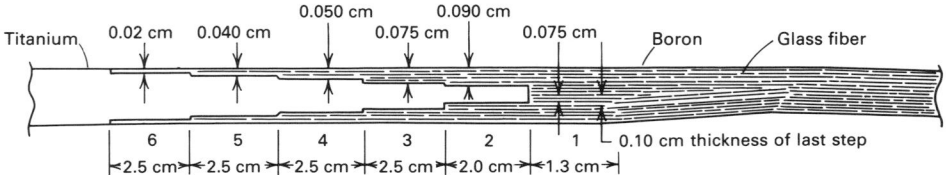

Fig. 18 Section through main splice

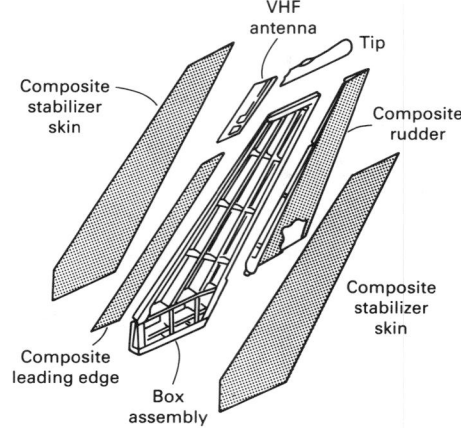

Fig. 19 F-16 vertical stabilizer

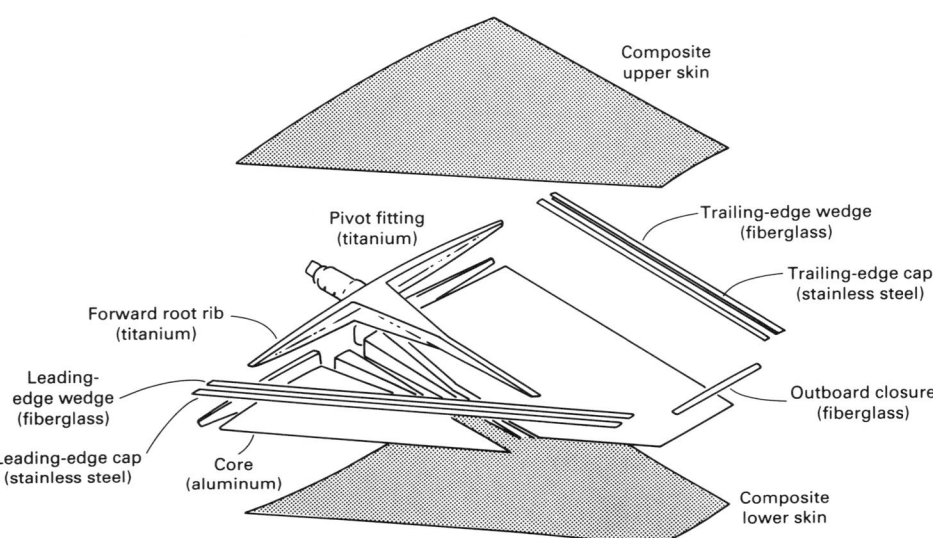

Fig. 20 F-16 composite horizontal stabilizer

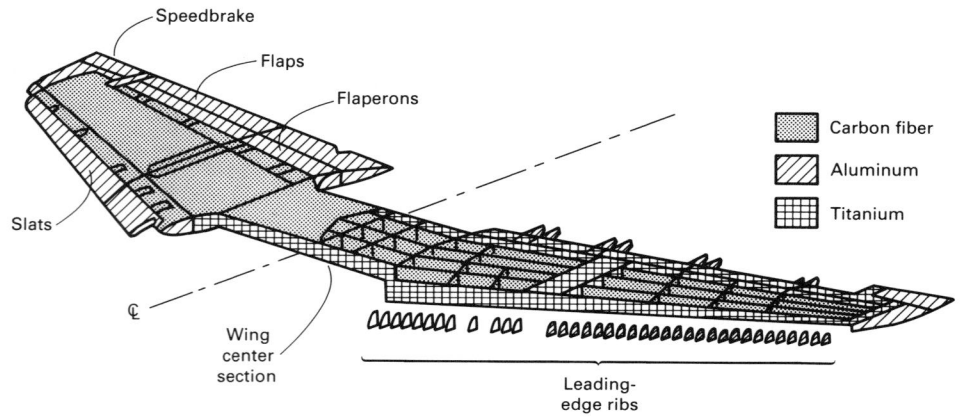

Fig. 21 A-6 composite wing

Fig. 22 V-22 test box

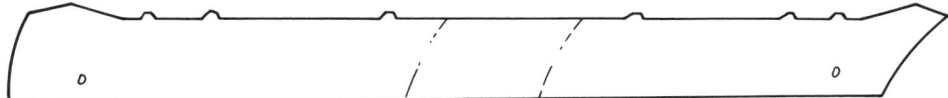

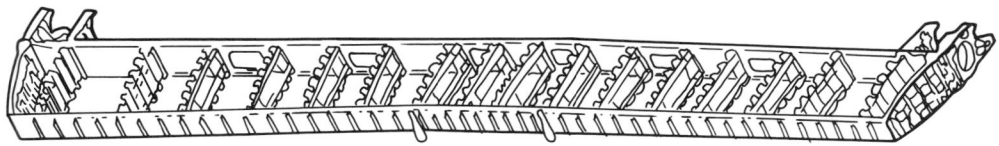

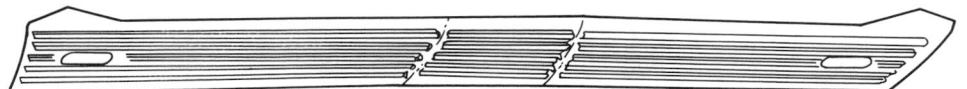

Fig. 23 V-22 composite wing

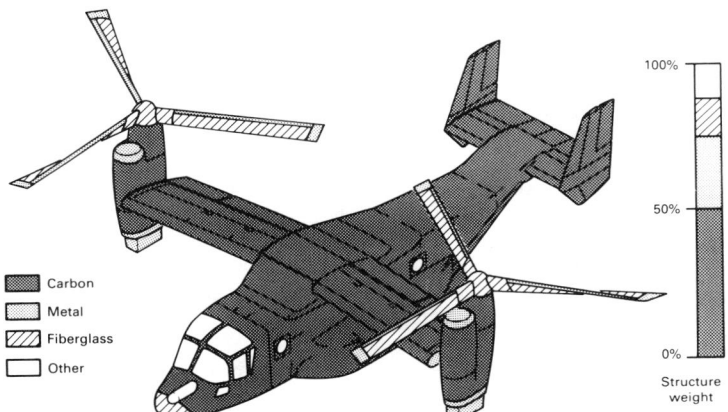

Fig. 24 V-22 material applications. Structure (wing, fuselage, empennage, nacelle, rotor): 6120 kg (13 496 lb). Carbon-epoxy: 3100 kg (6856 lb)

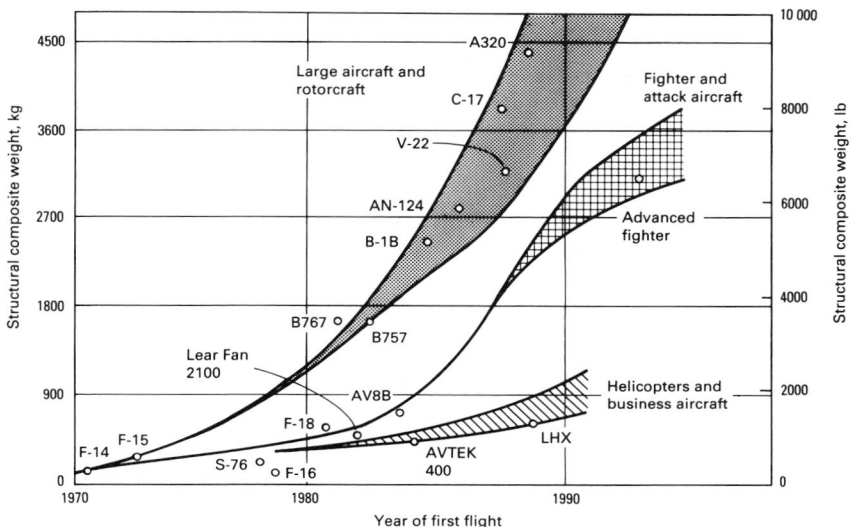

Fig. 25 Composite aircraft structure by weight

High-Temperature Applications

Tito T. Serafini, NASA Lewis Research Center*

HIGH TEMPERATURE RESISTANT POLYMERS are used in aerospace, electrical, electronic, and other applications that demand outstanding elevated-temperature physical and mechanical properties. Major product forms of these polymers include adhesives, coatings, fibers, film, foam, insulation paper, laminating resin solutions, molding powders, and wire enamels. A wide range of literature is available on their chemistry, processing characteristics, properties, and applications.

Until recently, the application of high-temperature polymers as matrix resins in fiber-reinforced composites was limited to nonstructural (nonload-bearing) components. However, following the development of PMR (which stands for *in situ* polymerization of monomer reactants) polyimides at the NASA Lewis Research Center (Ref 1-3) and their commercialization by prepreg suppliers, the fiber-reinforced PMR polyimide based on PMR-15 has found increased acceptance as an engineering material for high-performance structural applications. Because it has become the most widely used high-temperature polymer, most of the applications discussed in this article are representative examples of the use of fiber-reinforced PMR-15 polyimide.

General Characteristics

For a material to function as a viable matrix in a fiber-reinforced composite, the matrix must bind the constituents, transfer loads, provide environmental stability, and perhaps most important, provide processibility. Most organic polymers soften or melt below 204 °C (400 °F). The key to preparing high-temperature polymers is to incorporate highly stable structural units in the polymer chain, such as aromatic and/or heterocyclic rings. These structural units are able to absorb thermal energy and contain a minimum of oxidizable hydrogen atoms.

Commencing in the mid 1950s and continuing into the early 1970s, scores of high-temperature polymers were synthesized by using various types of molecular structural units, either individually or in various combinations. These polymers exhibited high glass transition temperatures and thermal stabilities, as as-

sessed by thermogravimetric analysis (TGA), which led early investigators to proclaim use temperature limits as high as 500 °C (932 °F). However, TGA is a dynamic test that uses high heating rates (5 to 50 °C/min, or 9 to 90 °F/min), and when used by itself, leads to completely erroneous conclusions about continuous-use temperature. Furthermore, the molecular structural units that are responsible for the thermal stability of high-temperature polymers are also responsible for their inherent insolubility and infusibility, commonly referred to as intractability.

In addition to being resistant to elevated temperatures, high-temperature polymers are also resistant to being processed into useful structural components. The poor processing characteristics of early high-temperature polymers prevented them from performing the necessary functions of a matrix material. Composites fabricated with these polymers exhibited high void contents, poor fiber translation efficiencies, inferior elevated-temperature mechanical properties, and inferior thermo-oxidative stabilities. Little progress was made in developing polymers with improved processibility because the major thrust of polymer synthesis research up to 1970 was in the area of improved thermal stability, without any concern for processibility.

Condensation-Type Polyimides

High-temperature laminating resins of the class known as condensation-type polyimides were first commercialized in 1962 and 1965, and emerged as the sole contenders for resin matrix applications. Although these materials presented processing difficulties, they had good high-temperature properties along with cost and availability advantages. Because polyimide laminating resins generally used the same monomers as those used in versions for other product forms such as film and wire enamels, it was a relatively easy matter to make the laminating resins commercially available at a reasonable cost.

The major problems in processing condensation-type polyimide prepreg materials can be attributed both to the inherent nature of condensation reactions (by-product evolution) and to

the use of high boiling point aprotic solvents, leading to void content problems. During thermal processing of the prepreg to effect chain growth and solvent/by-product removal, appreciable imidization also occurs, converting the resin to an intractable state. Reduced resin flow prevents the removal of the last traces of solvent and by-products. The entrapment of these volatile materials results in composites having void contents in the range of 5 to 10 vol%, particularly for thick sections (> 1 mm, or 0.04 in.). The presence of these voids adversely affects mechanical and thermo-oxidative stabilities of the composite.

Applications of Condensation-Type Polyimides. Epoxy resins are the most widely used matrix resins because of their excellent processibility and physical and mechanical properties. Thermal stability and glass transition temperature limitations restrict the upper continuous-use temperature of fiber-reinforced epoxies to about 177 °C (350 °F). To exploit the nearly twofold continuous-use temperature increase provided by condensation-type polyimides, composite fabricators developed complicated, time-consuming cure and postcure schedules. By using bleeder fabrics, vacuum bags, and a "bumping" technique, structural components having void contents below 5 vol% could be fabricated.

The earliest applications of fiber-reinforced condensation-type polyimides, in 1972, were for radomes on advanced aircraft (Ref 4) and for sound-suppression panels in the engine nacelles of subsonic commercial transports. Both applications can be considered secondary structural applications. In fact, the high void content of condensation polyimides is desirable for the sound-suppression panels. The formidable processing problems of these materials were solved by using dynamic dielectric analysis (DDA). More than 200 fiberglass-polyimide radomes measuring 4.6 × 0.5 × 0.6 m (15 × 1.6 × 2.0 ft) have been produced for EA-6B aircraft.

PMR Polyimides

A major advance in high-temperature resins was the development of polyimides that cured by an addition reaction (Ref 5 and 6). Low molecular weight amide-acid prepolymers

---

*Presently Assistant Chief Engineer for Materials, TRW Inc.

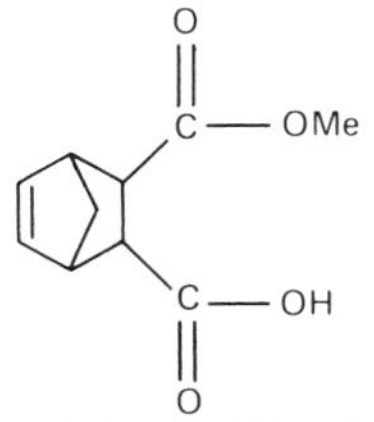

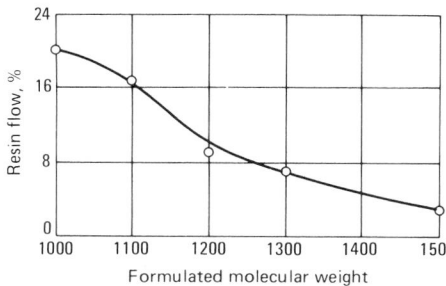

Fig. 1 Structures of monomers used in PMR-15

Monomethyl ester of 5-norbornene-2,3-dicarboxylic acid (NE)

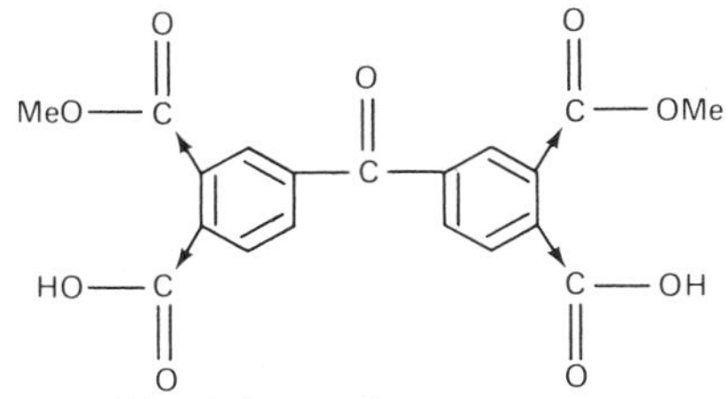

Dimethyl ester of 3,3',4,4'-benzophenonetetracarboxylic acid (BTDE)

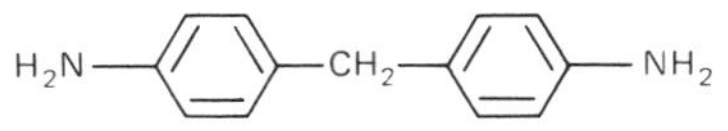

4,4'-methylenedianiline (MDA)

whose chain ends were terminated, or end-capped, were synthesized. In the early 1970s, attempts to commercialize a resin, designated P13N, met with limited success because of several shortcomings.

It was to circumvent the shortcomings of condensation-type and addition-type prepolymer polyimides that investigators at NASA Lewis developed the novel PMR polyimides. In the PMR approach, the reinforcing fibers are impregnated with a monomer reactant mixture dissolved in a low boiling point alkyl alcohol such as methanol or ethanol. *In situ* polymerization of the monomer reactants occurs when the impregnated fibers are heated, followed by final polymerization to a void-free composite. Further details on the polymer reactions are available in Ref 1 and Ref 7-10. Prepreg materials based on PMR-15 are commercially available in the United States from the major prepreg suppliers. The structures of the monomers used in PMR-15 are shown in Fig. 1.

Solutions of PMR-15 containing a dissolved-solids content of 40 to 60 wt% have viscosities in the range of 0.02 to 0.06 Pa · s (0.2 to 0.6 P), which ensures complete wetting of fibers during impregnation. The shelf life at ambient conditions (time to precipitation) of a 50 wt% solids solution is about 2 weeks. Following impregnation, evaporation of the solvent is achieved by mild heating (< 50 °C, or 120 °F), by applying a reduced pressure (84 kPa, or 25 in. Hg), or by allowing the prepreg to remain at room temperature for several hours. Heating the prepreg at temperatures in the range of 121 to 232 °C (250 to 450 °F) results in the *in situ* formation of low molecular weight norbornenyl

end-capped oligomers. The condensation by-products and any residual solvent are completely removed without converting the resin to an intractable state. At still higher temperatures (275 to 350 °C, or 525 to 660 °F), the norbornenyl groups undergo an addition crosslinking reaction without the evolution of undesirable volatile materials.

The early studies (Ref 1, 7) conducted at NASA Lewis also clearly demonstrated the efficacy and versatility of the PMR approach. By varying the chemical nature of either the dialkyl ester acid or the aromatic diamine, or both, and the monomer reactant stoichiometry, PMR matrices having a broad range of processing characteristics and properties could easily be synthesized. A modified PMR-15, called LARC-160, was developed by substituting an aromatic polyamine for 4,4'-methylenedianiline (MDA) (Ref 11). Both of these resins are discussed in the article "Curing Polyimide Resins" in this Volume. Other studies (Ref 12) have shown that the PMR approach has excellent potential for tailoring matrix resins with specific properties. Figure 2 shows that significantly higher resin flow for HTS graphite-PMR composites can be achieved by reducing the formulated molecular weight. However, the PMR compositions that exhibited increased resin flow were found to be less thermo-oxidatively stable at 288 °C (550 °F). The lower resin flow and increased thermo-oxidative stability in going from PMR-10 to PMR-15 clearly show the sensitivity of these properties to imide ring and alicyclic contents. The reduction in resin flow with increased formulated molecular weight also serves to account qualitatively for the intractable nature of condensation-type polyimides at an early stage of their process cycle.

Second-generation graphite fiber reinforced PMR polyimide composites, known as PMR II, were found to exhibit significantly improved thermo-oxidative stability and retention of mechanical properties at 316 °C (600 °F), compared to PMR-15 composites (Ref 13). PMR II prepreg materials have not been commercially available, however, because until recently no commercial source for one of the components has existed.

Fig. 2 Effect of formulated molecular weight on resin flow of HTS graphite fiber-PMR composites

High-pressure (compression) and low-pressure (autoclave) molding cycles have been developed for fabricating fiber-reinforced PMR composites. Although the thermally induced addition cure reaction of the norbornenyl group occurs at temperatures in the range of 275 to 350 °C (527 to 662 °F), nearly all the processes developed use a maximum cure temperature of 316 °C (600 °F). Cure times of 1 to 2 h followed by a free-standing postcure in air at 316 °C (600 °F) for 4 to 16 h are also normally used. Compression molding cycles generally use high heating rates (5 to 10 °C/min, or 9 to 18 °F/min) and pressures in the range of 3.5 to 6.9 MPa (0.5 to 1.0 ksi). Vacuum bag autoclave processes at low heating rates (2 to 4 °C/min, or 3.6 to 7.2 °F/min) and pressures of 1.38 MPa (0.200 ksi) or less have been used successfully to fabricate void-free composites. Autoclave processing methodology can be successfully applied to PMR polyimides because of the presence of a thermal transition, termed melt flow, which occurs over a fairly broad temperature range (Ref 14). The lower limit of the melt-flow temperature range depends on a number of factors, including the chemical nature and stoichiometry of the monomer reactant mixture and the thermal history of the PMR prepreg.

Differential scanning calorimetry analysis has shown that four thermal transitions occur during the overall cure of a PMR polyimide. The first, second, and third transitions are endothermic and are related to the following: melting of the monomer reactant mixture below 100 °C (212 °F), *in situ* reaction of the monomers at 140 °C (285 °F), and melting of the norbornenyl-terminated prepolymers in the (melt-flow) range of 175 °C to 250 °C (350 to 480 °F). The fourth transition, centered near 340 °C (645 °F), is exothermic and is related to the addition crosslinking reaction (Ref 15). To a large extent, the excellent processing characteristics of PMR polyimides can be attributed to these widely separated, chemically distinct thermal transitions.

During the first and second thermal transitions, solvent and condensation by-products can be volatilized without converting the polymer to an intractable state. During the third thermal transition, and with the application of pressure, fusion occurs. At still higher temperatures, and while maintaining the applied pressure, the addition cure reaction occurs. The high-temperature postcure mentioned previously must be used to obtain optimum thermo-oxidative stability and mechanical properties retention at elevated temperatures.

The cure temperature requirements of both first- and second-generation PMR polyimides exceed the capabilities of many existing autoclave facilities and impose a severe strain on the temperature capabilities of current bagging and sealant materials. Various alternatives in the PMR formulation can result in lower cure temperatures, but they entail drawbacks such as poorer thermo-oxidative stability, lower glass

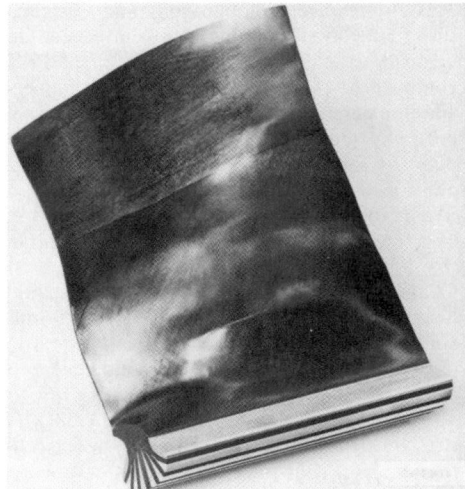

Fig. 3 Graphite fiber-PMR-15 polyimide fan blade

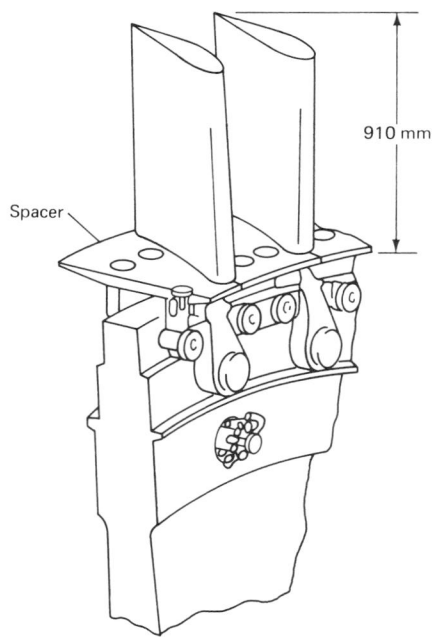

Fig. 4 Application of PMR-15 in supersonic wind tunnel

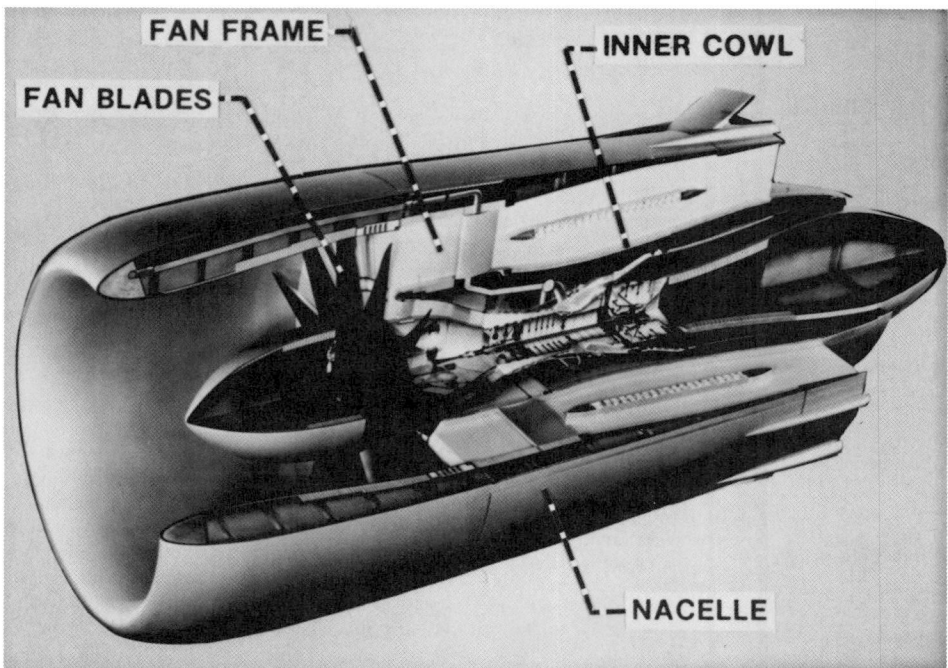

Fig. 5 Application of composites on QCSEE

transition temperature, and higher molding pressure requirements (Ref 16-18).

Many investigations have been conducted to determine the effects of various hostile environments on the physical and mechanical properties of PMR-15 composites. The thermo-oxidative stability and retention of mechanical properties at elevated temperatures have been found to be far in excess of what would be predicted on the basis of polymer molecular structure criteria for polymer thermal stability. Graphite-PMR-15 composites have been reported (Ref 19) to be suitable for use in air at 288 °C (550 °F) for at least 5000 h. At 316 °C

(600 °F), the useful life of graphite-PMR-15 composites is in the range of 1200 to 1400 h.

Various groups have identified and investigated many modified versions of PMR-15. However, the composition of the resin used for commercial products is essentially unchanged from the original composition identified and developed at NASA Lewis in the 1970s.

Fiber-reinforced PMR-15 polyimide composites are finding increased acceptance as engineering materials for the design and fabrication of aerospace structural components, particularly in aeropropulsion. Components being fabricated range from small, compression-molded bearings to large, autoclave-molded aircraft engine cowls and ducts. Processing technology and baseline materials data are also being developed for the application of PMR-15 composites in weapon systems. Firms involved in manufacturing small, compression-molded bearings made from particulate or chopped-fiber PMR molding compounds have found it convenient to become captive producers of PMR-15 resin. In contrast, firms involved in fabricating larger components made from tape or fabric materials rely on traditional sources of composite materials for PMR-15 prepreg. Materials based on PMR-15 are commercially available from the major suppliers of composite materials in the form of solutions, molding compounds, prepregs, and adhesives.

Applications of PMR-15

The blade illustrated in Fig. 3 was the first structural component fabricated as a PMR-15 composite material. The reinforcement is HTS

graphite fiber. The blade was designed and fabricated for an ultrahigh-speed fan stage (Ref 20, 21). The blade span is 279 mm (11 in.), the chord is 203 mm (8 in.), and the thickness ranges from about 13 mm (0.5 in.) just above the mid-point of the wedge-shaped root to 0.51 mm (0.02 in.) at the leading edge. At its thickest section, the composite structure consists of 77 plies of material arranged in varying fiber orientations. The line of demarcation visible in Fig. 3 at approximately one-third the distance from the blade tip resulted from a change in fiber orientation from 40° in the lower region to 75° in the upper region, in order to meet torsional stiffness requirements. Ultrasonic and radiographic examination of the compression-molded blades indicated that they were defect-free. Although some minor internal defects were induced in the blade during low-cycle and high-cycle fatigue testing, the successful fabrication of these highly complex blades established PMR-15 as a processible matrix resin.

Another early application of PMR-15 composites was for the fourth-stage compressor blade shells and spacers in a supersonic wind tunnel (Fig. 4). More than 10 000 kg (12 tons) of fiberglass-PMR-15 were processed for this application (Ref 22). The total quantity of material was about equally divided for 360 blade shells and 600 blade spacers.

Figure 5 shows a cross-sectional view of a quiet, clean, short-haul experimental engine known as QCSEE. To meet the program goals for low noise, clean burn, and improved efficiency, composites were used extensively in the fan blades, frame, nacelle, and inner cowl (Ref 23). All the components except the inner cowl

Fig. 6 Graphite fiber-PMR-15 polyimide QCSEE inner cowl

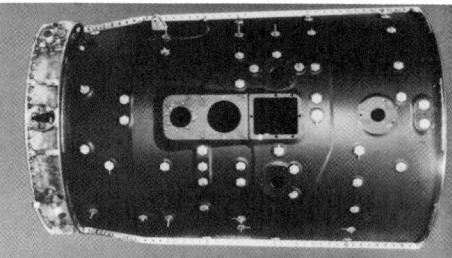

Fig. 7 Graphite fiber-PMR-15 polyimide outer duct for F404 engine

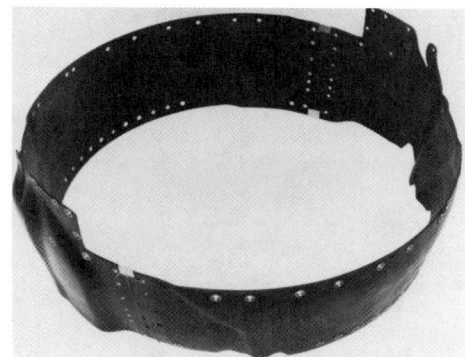

Fig. 8 Graphite fiber-PMR-15 inner duct for F110 fighter engine

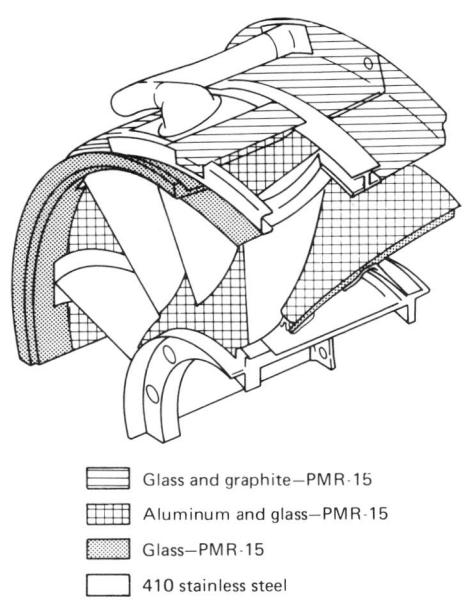

▬	Glass and graphite—PMR-15
▦	Aluminum and glass—PMR-15
▨	Glass—PMR-15
▢	410 stainless steel

Fig. 9 Metal and composite swirl frame for T700 engine

used fiber-reinforced epoxy. Because the temperature requirements of the inner cowl (which defines the inner boundary of the fan air flowpath from the fan frame to the engine core nozzle) exceeded the thermal stability of epoxy-based composites, the cowl was fabricated from a graphite-PMR-15 composite.

Figure 6 shows the inner cowl installed on the QCSEE. The cowl has a maximum diameter of about 889 mm (35 in.) and is primarily of honeycomb sandwich construction. Fiberglass-polyimide honeycomb was used as the core material. The honeycomb core was bonded to the inner surface of the premolded outer skin with a condensation-type polyimide adhesive. The inner skin was then co-cured and bonded to the core surface of the honeycomb-core outer-skin assembly. Complete details of the cowl fabrication process are given in Ref 24. The cowl did not exhibit any degradation on the QCSEE after more than 300 h of ground testing (Ref 25). The maximum temperature experienced by the cowl during testing was 260 °C (500 °F). The successful autoclave fabrication and ground engine testing of the QCSEE inner cowl established the feasibility of using PMR-15 composite materials for large-engine, static structures.

Under a jointly sponsored U.S. Navy/NASA Lewis program (NAS3-21854), a T300 graphite fabric-PMR-15 composite outer duct is being developed to replace the titanium duct presently used on the F404 engine for the Navy F18 strike fighter. The titanium duct is a sophisticated part made by forming and machining titanium plates followed by chemical milling to reduce weight. A preliminary cost benefit study indicated that significant cost and weight savings (Ref 26) could be achieved by replacing the titanium duct with a composite duct. The F404 composite duct differs from the QCSEE inner cowl in several important aspects; namely, it is a monolithic composite structure that needs to withstand fairly high loads and, perhaps most important, it is a production component, not a one-of-a-kind demonstration component.

A full-scale, production-quality composite duct 762 mm (30 in.) in diameter by 1016 mm (40 in.) in length by 2.3 mm (0.09 in.) in wall thickness has been autoclave-fabricated (Fig. 7). The sequence of the major operations included autoclave fabrication of the composite shell, ultrasonic inspection, drilling of the buildups, and attachment of the split line stiffeners. The duct has successfully undergone more than 2500 h of ground testing on an F404 engine. Another duct has been flight-tested for more than 700 h. The fully flight-qualified T300 graphite-PMR-15 composite outer duct provides a weight savings of 3.2 kg (7 lb) per engine and significant total cost savings (estimated to be more than $20 million) compared to the titanium duct. This duct is the first structural component on a production aircraft engine made from a high-temperature polymer matrix composite.

Figure 8 shows a full-scale composite forward inner duct fabricated for the F110 engine. The approximate dimensions of the duct, which was autoclave-fabricated from T300 graphite fabric-PMR-15, are 1020-mm (40-in.) diam × 380-mm (15-in.) length × 1.5-mm (.06-in.) wall thickness. Engine testing demonstrated that the composite duct was fully functional and met engine performance requirements. The T300 graphite fabric-PMR-15 composite inner duct is in the current bill of materials for the F110 engine. Splitter panels for the F110 have also been fabricated from graphite fiber-PMR-15 (Ref 27).

The inlet particle-separator swirl frame on the T700 engine is an all-metal part that involves machining, shape forming, welding, and brazing. Design studies conducted under U.S. Army contract DDAK51-79-C0018 indicated that fabrication of a metal and composite swirl frame could result in a cost and weight savings of about 30%. Figure 9 shows a section of the metal and composite swirl frame that was fabricated from type 410 stainless steel and various kinds of PMR-15 composite materials. The outer casing used stainless steel in the flow path area to meet anti-icing temperature requirements and a T300 graphite and glass fabric-PMR-15 hybrid composite to meet structural requirements. The T300 and glass hybrid composite was selected on the basis of both cost and structural considerations. An aluminum-coated glass fabric-PMR-15 composite material was utilized in the inner-hub flow path to meet heat transfer requirements for anti-icing. The glass fabric-PMR-15 composite used for the front-edge and front-inner surfaces was selected because of cost as well as temperature considerations (Ref 28). A full-scale (outside diameter of ~510 mm, or 20 in.) metal and composite swirl frame, when subjected to sand erosion and ice-ball impact tests, provided improved particle separation and successfully met the impact test requirements. Fabrication feasibility has been demonstrated; if the metal and composite swirl frame successfully meets all performance requirements, it may be used on a growth version of the T700.

Figure 10 shows "committed" and "possible" applications of graphite-PMR-15 composite materials on the PW1120 turbojet under development. A committed application is one for which a metal backup component is not being developed. The only committed applications for graphite-PMR-15 composites are the external nozzle flaps and the airframe interface

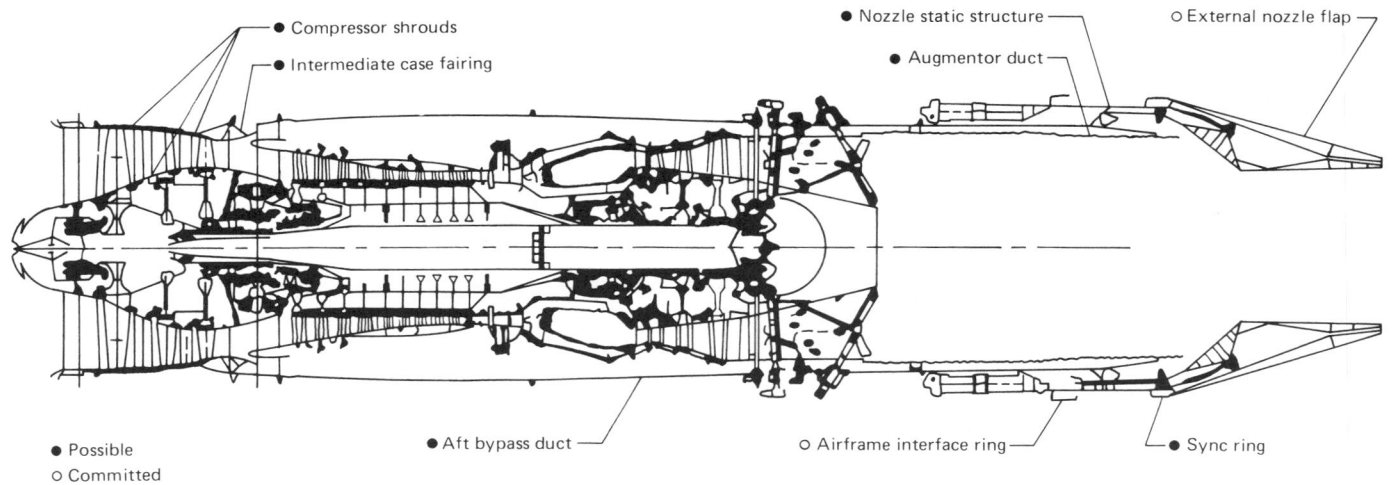

- Compressor shrouds
- Intermediate case fairing
- Nozzle static structure
- External nozzle flap
- Augmentor duct
- Possible
- Committed
- Aft bypass duct
- Airframe interface ring
- Sync ring

Fig. 10 Possible and committed applications of graphite fiber-PMR-15 polyimide composites on PW1120 engine

INTERFACE FAIRING

NOSE CONE

EXTERNAL NOZZLE FLAP

FIRST STAGE VANE CLUSTER

Fig. 11 Graphite fiber-PMR-15 composites on PW-1120 and 1130 engines

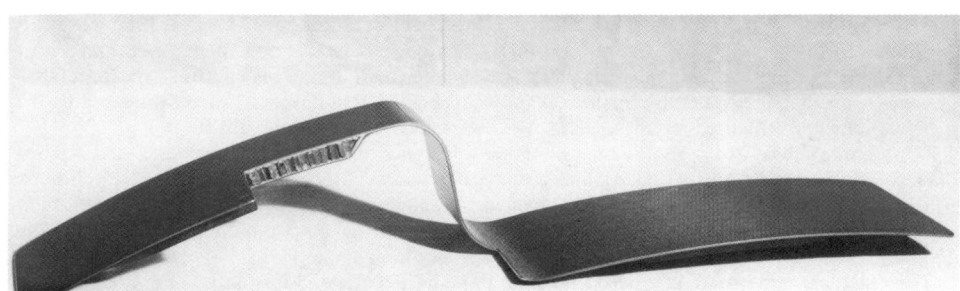

Fig. 12 Graphite fiber-PMR-15 sandwich structure

ring. Many of the possible applications are expected to become committed if engine test schedules can be met. Graphite-PMR-15 external nozzle flaps have been committed for production for the PW1130 turbofan engine. Prepregs made from unidirectionally woven fabrics and PMR-15 are being evaluated for fabrication of the nozzle flaps on both the PW1120 and PW1130 engines. Figure 11 shows the different components fabricated for these two engines using graphite fiber-PMR-15 (Ref 29).

Figure 12 shows a PMR-15 sandwich structure for which the skins were precured and then bonded to the glass-polyimide honeycomb using a commercially available adhesive that was based on PMR-15 (Ref 30). The complexity of this structural component demonstrates the efficacy of the PMR approach for fabricating complex structures. Interest has also been shown in applying PMR-15-based composites in weapon systems. Because many of the intended applications are for structures that are ideal for filament winding, efforts are underway to develop filament-winding technology for PMR-15. Although PMR-15 was not developed as a filament-winding resin, studies performed to date have met with considerable success (Ref 31).

REFERENCES

1. T.T. Serafini, P. Delvigs, and G.R. Lightsey, *J. Appl. Polym. Sci.*, Vol 16, 1972, p 905
2. T.T. Serafini, P. Delvigs, and G.R. Lightsey, U.S. Patent 3,745,149, 1973
3. T.T. Serafini, in *International Conference on Composite Materials*, E. Scala, Ed., Vol 1, American Institute of Mining, Metallurgical, and Petroleum Engineers, 1976, p 202
4. L.M. Poveromo, in *High Temperature Polymer Matrix Composites*, NASA

CP2385, National Aeronautics and Space Administration, 1983, p 339

5. H.R. Lubowitz, "Polyimide Polymers," U.S. Patent 3,528,950, Sept 1970

6. E.A. Burns, H.R. Lubowitz, and J.F. Jones, "Investigation of Resin Systems for Improved Ablative Materials," NASA CR-72460, TRW-05937-6019-RO-00, TRW Systems Group, Oct 1968

7. P. Delvigs, T.T. Serafini, and G.R. Lightsey, NASA TN D-6877, National Aeronautics and Space Administration, 1972

8. F.I. Hurwitz, NASA TM-81580, National Aeronautics and Space Administration, 1980

9. R.J. Jones, R.W. Vaughan, and E.A. Burns, "Thermally Stable Laminating Resins," NASA CR-72984, TRW-16402-6012-RO-00, TRW Systems Group, Feb 1972

10. A.C. Wong and W.M. Ritchey, *Macromolecules*, Vol 14 (No. 3), 1981, p 825

11. T.L. St. Clair and R.A. Jewell, NASA TM-74944, National Aeronautics and Space Administration, 1978

12. T.T. Serafini and R.D. Vannucci, in *Reinforced Plastics—Milestone 30*, Society of the Plastics Industry, 1975, p 14-E1

13. T.T. Serafini, R.D. Vannucci, and W.B. Alston, NASA TMX-71894, National Aeronautics and Space Administration, 1976

14. R.D. Vannucci, in *Materials and Processes—In-Service Performance*, Society for the Advancement of Material and Process Engineering, 1977, p 171

15. R.W. Lauver, *J. Polym. Sci.*, Polymer Chemical Ed., Vol 17, 1979, p 2529

16. T.T. Serafini, P. Delvigs, and R.D. Vannucci, NASA TMX-81705, National Aeronautics and Space Administration, 1981

17. P. Delvigs, NASA TM-82958, National Aeronautics and Space Administration, 1982

18. R.H. Pater, NASA TM-82733, National Aeronautics and Space Administration, 1981

19. C.H. Sheppard and D. McLaren, in *High Temperature Polymer Matrix Composites*, NASA CP 2385, National Aeronautics and Space Administration, p 329

20. J.E. Halle, E.D. Burger, and R.E. Dundas, NASA CR-135122, PWA-5487, Pratt and Whitney Aircraft, 1977

21. P.J. Cavano, NASA CR-134727, TRW-ER-7677-F, TRW Equipment Laboratories, 1974

22. L.A. Lottridge, Hamilton Standard Division, United Technologies, private communication

23. A.P. Adamson, in *Quiet Powered-Lift Propulsion*, NASA CP-2077, National Aeronautics and Space Administration, 1979, p 17

24. C.L. Ruggles, NASA CR-135279, R78AEG206, General Elecric Company, 1978

25. C.L. Stotler, in *Quiet Powered-Lift Propulsion*, NASA CP-2077, National Aeronautics and Space Administration, 1979, p 83

26. C.L. Stotler, The 1980's—Payoff Decade for Advanced Materials, *SAMPE Proc.*, 1980, p 176

27. A.J. Wilson, Aircraft Engine Business Group, General Electric Company, private communication

28. S.C. Harrier, S. Mitchell, and J.A. Saunders, USAAVRADCOM-TR-83-D20A and 20B, U.S. Army, 1983

29. T.E. Schmid, Pratt and Whitney Aircraft Group, private communication

30. R.A. Buchanan, Brunswick Corporation, Defense Division, private communication

31. W.D. Humphrey, Brunswick Corporation, Defense Division, private communication

Space and Missile Systems

Frederick J. Policelli and Albert A. Vicario, Hercules Aerospace Company

USE OF COMPOSITE MATERIALS in primary structures of major aerospace vehicles was based on the successful use of composites in missiles powered by solid-propellant rocket stages. This first application was the rocket motor case in the final propulsion stage of the Vanguard missile, which was used extensively in upper-atmosphere, near-space, and deep-space exploration as well as in data-gathering flights. The X248 Altair rocket motor was the first of a long series of different final propulsion stages that used composite cases and, in many instances, followed the payload to its destination. The Altair all-composite motor casing was constructed of filament-wound glass fiber and epoxy resin.

This particular application eventually became predominant. Major advances were made in composite manufacturing methods, design technology, and materials in the course of extending the high strength and light weight characteristics of composite motor cases. These advancements contributed substantially to the greater use of composites in principal aircraft structures, and the extension of this technology to space vehicle structures.

Thus, the evolution of composite technology in these three major areas of application has resulted in the present level of composite maturity. This level, which has lead to the broad use of composites in principal aircraft structures requiring a high level of design and materials knowledge, has also led to man-rated use of composites in missile and space hardware.

The space shuttle represents one of the first production applications of metal matrix composites. It has 242 unidirectional boron-aluminum circular tubes used in the main-frame and rib-truss struts, frame-stabilizing braces, and nose landing gear drag-brace struts. The boron-aluminum tubes that support its fuselage frame weigh 150 kg (330 lb), reducing weight by 44% over aluminum extrusions. In addition to weight savings, the lower thermal conductivity of the composite tubes results in lower heat flow into the equipment and payload bays, reducing thermal insulation requirements. The smaller tube diameter also permits greater vehicle access.

Design Considerations for Space and Missile Hardware

Space structures are usually required to have low weight, high stiffness, low coefficient of thermal expansion, and dimensional stability during their operational lifetime. Other requirements are that they be resistant to handling damage and unaffected by the space environment.

The major structural components used in space can be grouped into the following categories:

- Trusses
- Platforms
- Pressure vessels and tanks
- Shells

Truss and platform structures usually consist of an assemblage of tubes and flat panel bulkheads. The tubes are designed to have very high axial and bending stiffness and low CTE. Mechanical loads are generally low; however, the tube-to-end fitting joint must be designed to withstand thermal stresses caused by thermal cycling and should be stiffness-compatible to minimize the stresses caused by any imposed loads. If these structures are required to operate for long periods of time in low earth orbit, they must be protected from atomic-oxygen attack and degradation of material properties due to radiation. Materials must be selected that meet current requirements for vacuum outgassing.

Pressure vessels and tanks are required to contain a wide variety of gases and fluids. Because most composite materials are porous, pressure vessels and tanks made from composites must contain some kind of liner. Therefore, the major design consideration for composite pressure vessels and tanks should be centered around the load-sharing and strain compatibility of the liner and composite under pressure/depressurization cycling, the thermal-strain compatibility of the liner and composite under thermal cycling, the stress-rupture and creep capability of the composite under long-term pressurized loads, and the leak-before-burst capability of the liner and composite system.

Compatibility of materials to the storage media and protection of the composite against the exposed space environment must also be considered when designing pressure vessels and tanks for specific applications.

Shell structures can be used in several applications. Like truss structures and pressure vessels, shells must be designed to withstand space environments and to meet the stiffness requirements imposed by the mission. The most critical design consideration for shells is the manner in which the shells are attached to each other or to adjoining structures.

The major structural components of most missile systems are the rocket motor case, nozzle, skirts and interstage structures, control surfaces, and guidance and control structural components. The design, material selection, and fabrication process for these components should result in the most cost-effective structure that satisfies all the mission objectives while maintaining all the imposed constraints.

Tactical missile system components are usually lightweight and small. They must be able to withstand high acceleration and vibration loads, severe temperature excursions, and severe environmental conditions (such as extreme humidity, sand, salt, and chemicals). The rocket motor cases must operate at high pressure and have high axial stiffness. For these reasons, most components for tactical missile systems are currently metal, but composite materials are being evaluated as replacements because of their superior specific strength and stiffness.

Strategic missile system components are usually very large. Their flight loads are low, and they are usually maintained and operated in highly controlled environments. Strategic missiles are not usually required to operate under severe temperature levels. Rocket motor cases usually operate at low pressure levels for relatively long times. Because of the large sizes, the missile components for strategic missiles are usually made of filament-wound composite materials in order to minimize weight.

Defensive missile system components must be lightweight and must withstand very high flight loads. In addition, they are subjected to high launch, acceleration, shock, shipping, and handling loads. They are usually required to withstand moderate environment and temperature extremes. One of the most severe requirements for defensive missiles is that they must be capable of withstanding long exposure to nuclear radiation and must be structurally and aerodynamically adequate when subjected to the large external pressures caused by nuclear-blast loads. Rocket motor cases must also be capable of operating at very high pressures.

Space motor components must be highly reliable and lightweight structures. They vary in size depending on the mission and payload. They may be required to withstand space environments such as electromagnetic and elementary-particle radiation, meteoroid impact, and vacuum conditions. Depending on the mission, they may be required to withstand severe temperature extremes.

Composite Characteristics

One of the major advantages of composite materials is that the material can be varied to meet specific design requirements. This is accomplished by orienting fibers in the main direction of expected load, thereby allowing the very high strength of the fibers to resist the load directly and minimize undesirable secondary load paths. From a design standpoint, the unidirectional properties of the fiber-matrix system are used as input to standard laminate analysis techniques to obtain specific material properties and structural performance. Once the structure has been designed, laminates representing the structure can be characterized for the specific requirements of the structural components by physical testing of coupon specimens.

Because the design of the material begins with selection of the fibers and the matrix, performance characteristics of the more popular fibers and matrix systems are reviewed and compared to the most widely used aerospace metals.

Fibers that are widely used in space and missile composite structures include E-glass, S-glass, aramid, and carbon/graphite. E-glass (electrical) is the oldest and most commonly used glass fiber. It provides excellent electrical resistance and is particularly suited for electrical insulation applications. S-glass (high tensile strength) has both a higher modulus and strength than E-glass and has been used in numerous rocket motor cases. Aramid fiber, introduced in the early 1970s, has a higher specific modulus and higher specific strength than glass fibers and is currently being used in several motor cases for strategic and tactical missiles, as well as for pressure vessels for the space shuttle and several satellite systems. Carbon fibers, also introduced in the early 1970s, are available in a range of CTE, moduli, and strength. Because of the wide range of properties, carbon fiber is used extensively in space structures when dimensional stability and high specific stiffness are required. Because it possesses excellent specific strength, carbon fiber

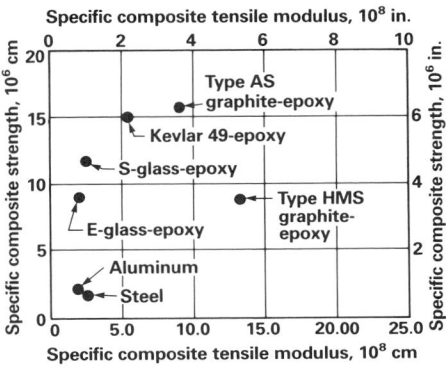

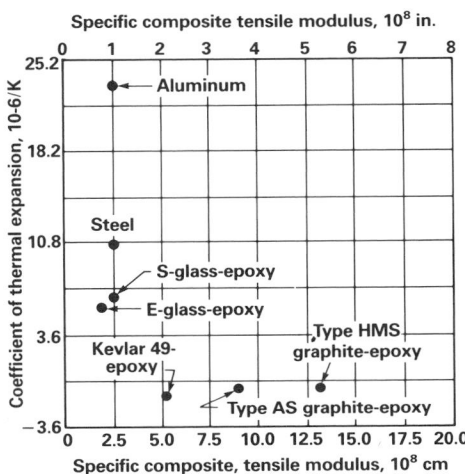

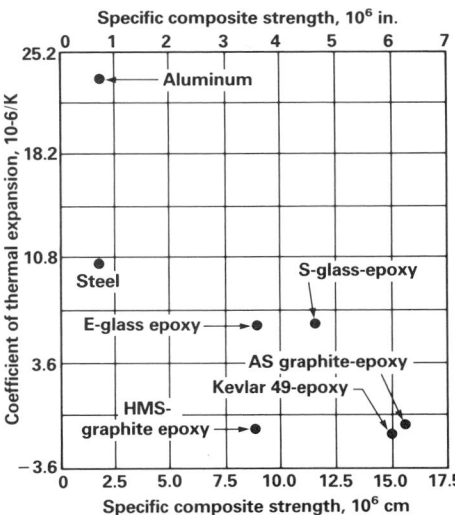

Fig. 1 Specific modulus, specific strength, and coefficient of thermal expansion for various materials. Strength, modulus, and coefficient of thermal expansion are fiber composite properties.

is used in the filament-wound case for the space shuttle rocket booster, in the structure and case of the Trident II strategic missile, and in several new tactical missile systems.

Figure 1 compares the specific modulus, specific strength, and coefficient of thermal expansion of unidirectional fiber reinforced

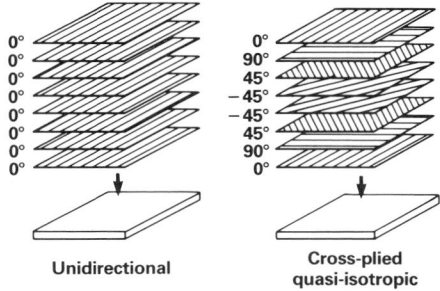

Fig. 2 Typical unidirectional and quasi-isotropic laminate configurations

resin composite materials with aerospace-grade steel and aluminum.

Resins must provide adequate physical and mechanical properties in the cured composite, as well as desired processing properties in the interval between mixing and cure. Such processing properties as initial viscosity, viscosity as a function of time and temperature, pot life, cure conditions that yield the desired composite performance, and the presence or lack of a B-stage and its duration as a function of temperature all affect the performance of the end-item structural component.

Epoxy resins are the most widely used for missile and space applications. Multifunctional epoxies achieve service temperatures from 120 to 180 °C (250 to 350 °F) as well as low-temperature extremes between −73 to 93 °C (−100 to 200 °F). One of the concerns with epoxies for these applications is damage tolerance and toughness. Currently, efforts are being made to improve the toughness of epoxies, and several epoxies are available that provide fracture toughness and compression after impact several times that of standard epoxies.

Other resin systems under development are polyimide and thermoplastic systems. Polyimide resins have performance potential in the 200 to 260 °C (400 to 500 °F) range and are being considered for tactical missile applications where aerodynamic heating is a concern. Polyimides are very difficult to process, have a higher cost than epoxies, and lack toughness. All of these issues are under investigation.

Thermoplastics are new to the aerospace industry. They are very tough and offer the potential for low-cost fabrication of structural components.

Fiber-resin laminates can be tailored to meet specific performance requirements. To compare the performance of specific composite materials with metals, a specific laminate must be designed to meet the given performance requirements. Since there are an infinite number of possible laminate designs, performance comparisons will be presented for only two specific laminates.

The first is a unidirectional laminate in which all the fibers are oriented in one direction. While this is not a practical laminate because the properties in the direction transverse to the

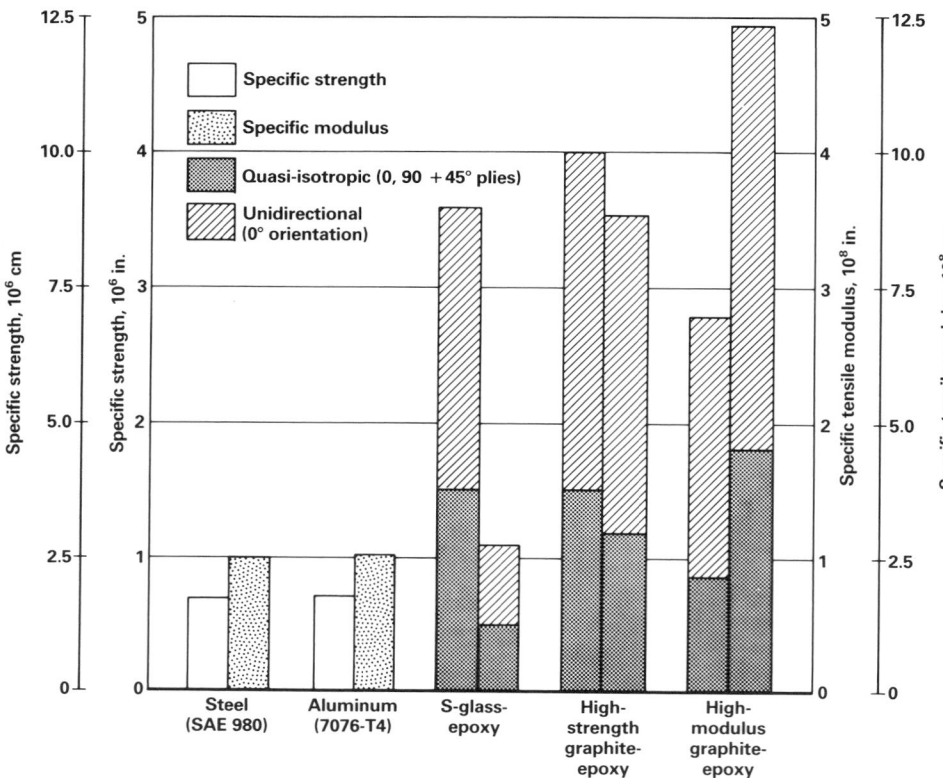

Fig. 3 Comparison of various materials for specific strength and specific modulus

Table 1 Coefficient of thermal expansion and thermal conductivity properties

Type of material	Coefficient of thermal expansion, 10^{-6}/K		Thermal conductivity, W/m · K (Btu · ft/h · ft² · °F)	
	0°	Quasi-isotropic	0°	Quasi-isotropic
AS graphite-epoxy	−0.1	0.6	7-9 (4-5)	5-7 (3-4)
HMS graphite-epoxy	−0.2	0.3	50-55 (30-32)	28-30 (16-18)
Kevlar 49-epoxy	−1.1	−0.63	0.17 (0.101)	------
S-glass-epoxy	1.9	3.3	3.5 (2)	0.35 (0.2)
Aluminum	7.2		140-220 (80-125)	
Steel	3.9		15-110 (9-27)	

fibers are low, many stiffness-critical applications require laminates with most of the fiber oriented in one direction and a small number of fibers oriented to satisfy other requirements. Thus, the unidirectional laminate represents the highest modulus and strength and lowest CTE possible in one direction.

The second laminate is the quasi-isotropic laminate in which unidirectional lamina are oriented in 0° and 90° (orthogonal) and ±45° directions to the geometric axes of the structure. This laminate represents the most conservative orientation and is directly comparable to metals, which are usually considered to be isotropic. Figure 2 presents the two laminate configurations discussed here.

Figure 3 presents the specific strength and stiffness for aluminum, steel, glass, aramid, and several graphite composite unidirectional and quasi-isotropic laminates. Note that in all cases, except for the very high modulus graphite, the strengths exceed those of steel and aluminum.

Table 1 is a summary of the CTE and thermal conductivity of the materials presented in Fig. 3. Again, one may note the wide range of thermal properties available with composites, particularly the graphite composites.

Hardware Applications

Space Applications. Composites were adapted earlier and have been used for a much longer time in space structures than in other structures because of their unique performance characteristics and ability to realize weight savings. The following sections describe specific applications of composite materials for the general categories of structure discussed previously.

Truss Structures. One of the first composite structures to be placed into space was the truss structure for the applications technology satellite (ATS) (Fig. 4). This truss, a stability-critical structure, consisted of eight 89-mm (3.5-in.) diam by 4.4-m (14.5-ft) long tubes arranged with a centrally located stabilizer ring. The end fitting on each tube was titanium, and the tubes were constructed from high-modulus graphite-epoxy. The truss structure total weight was approximately 37 kg (81 lb), which was a 50% weight savings over a similar metallic design. The ATS was launched in May 1974 and operated flawlessly. Many other satellites have since used graphite-epoxy for the main truss structure.

Platforms. Space platforms must have very tightly controlled dimensional stability in addition to high specific stiffness. An example of such a platform using graphite-epoxy composites is the bench structure for the space telescope high-resolution spectrograph (Fig. 5). The requirement for this structure was a CTE of ±0.1 cm/cm/°C (±0.2 in./in./°F). To meet this requirement, tube and flat plate bulkhead components were designed using a hybrid of very high stiffness/low CTE pitch-based graphite fiber, combined with high- and intermediate-modulus PAN-based graphite fiber. The design and material selection were tailored to meet the CTE requirement. In addition, all the components were joined using graphite end fittings. This structure was fabricated in the early 1980s and has met all the performance requirements to date.

Pressure Vessels and Tanks. Because of the high specific strength and stiffness of advanced composites, they are very attractive for use in pressure vessels and tanks. During the late 1970s, small filament-wound tanks with metallic liners were developed for use on the space shuttle and several satellites. These vessels consisted of a steel, aluminum, or titanium metallic liner overwrapped with Kevlar 49-epoxy composite.

With the higher specific modulus and higher specific strength graphite fibers now available, more efficient pressure vessel and tank designs are possible. Tank configurations ranging from spherical and cylindrical to noncircular cross sections are being considered for various space applications. Liner and composite overwrap designs that optimize the load sharing of the liner and overwrap can be achieved by using the wide range of materials available.

Shells. Perhaps the most efficient use of composites from both a cost and performance standpoint is in the design and fabrication of shells for various space applications. Composite shells are being considered for use on very large launch vehicle structures, for future space station module structures, and for various platform applications for which both cost and weight are critical.

Using the design and manufacturing technology currently being developed for large aircraft structures, along with technology that has al-

Fig. 4 Applications technology satellite truss. Courtesy of Hercules, Inc.

Fig. 5 Optical bench for space telescope. Courtesy of Hercules, Inc.

Table 2   Rocket motor cases manufactured with filament-wound composites

Missile category	Solid rocket motor name	Missile stage application	Composite fiber
Space	Altair	4	Glass
	Antares	3	Glass
	BE-3	(Various)	Glass
	IVS/SRM-1	(Various)	Glass
	IVS/SRM-2	(Various)	Glass
	Space shuttle FWC	1	Graphite
Strategic	Polaris A-2	2	Glass
	Minuteman	3	Glass
	Polaris A-3	1, 2	Glass
	Poseidon	1, 2	Kevlar
	Trident I	1, 2, 3	Kevlar
	Peacekeeper	1, 2, 3	Kevlar
	Trident II	1, 2, 3	Graphite
Defense	Sprint	1, 2	Glass
Tactical	Viper	1	Glass
	ADATS	1	Glass
	ASW/SOW	1	Kevlar
	Pershing	1, 2	Kevlar
	Hypervelocity missile	1	Graphite

ready been demonstrated on large filament-wound shells for the space shuttle booster cases, a space station module shell structure of the type shown in Fig. 6 can be completely manufactured using composite materials. This shell would consist of a reinforced skin with built-in meteoroid/space-debris impact protection. It could be produced at a cost and weight savings of up to 30% over conventional metallic construction.

Missile Applications

A large number of solid-propellant missiles use composites for major structural elements, including the rocket motor case. This element of the structure greatly influences missile performance in terms of range and weight of payload. The solid-propellant-type motor burns its fuel inside the case, generating moderately high pressures, which are contained by the large case wall. High tensile strength, together with the low weight of composites, is the most important advantage. Conversely, the large tanks of a liquid pro-

pulsion system contain fuel at a very low pressure, because pumps pressurize the fuel to combustion pressures in the relatively small thrust chamber. The large tanks therefore benefit from higher modulus as opposed to higher strength. This fundamental difference between solid and liquid missiles explains the early exclusive use of composites for the solid-type fuel containment systems. The relatively recent development of much higher-modulus and moderate-cost fibers is beginning to show an advantage on new liquid-propulsion tanks.

Solid-Propellant Rocket Motors. Solid-propellant missiles having one or more composite motor cases are listed in Table 2. All of these cases were manufactured by the filament-winding method, and most contain evolutionary design and manufacturing improvements over their chronological predecessors. Several are described below in approximate order of their development.

The Altair motor case (Fig. 7) was used as the fourth stage on the Vangard missile and on many other missiles for the exploration of

space, beginning in the late 1950s. It represents the initial step in the use of composites in both missile structures and space. Its construction was elemental compared to later composite motor cases, but nonetheless was considerably advanced over alternative metal case designs. The cylindrical body and the hemispherical domes were fabricated in a continuous filament winding operation using a ±45° filament wrap angle. Metal skirt connections were bonded to the exterior.

Polaris. Because of its considerably greater flight complexity, the A-2 second-stage motor case (Fig. 8) used on the strategic Polaris missile contained four nozzle openings and a dome shape of higher performance than the hemisphere. The winding pattern consisted of a low-angled helix, together with circumferential wraps, resulting in a 54-¾° net angle for maximum shell efficiency under internal pressure.

Minuteman. The third stage of the strategic Minuteman missile (Fig. 9) had even greater flight operation requirements, which led to the incorporation of openings in the cylindrical body of the case for thrust termination, integral lightweight thrust skirts, and a winding pattern which resisted high bending loads in addition to

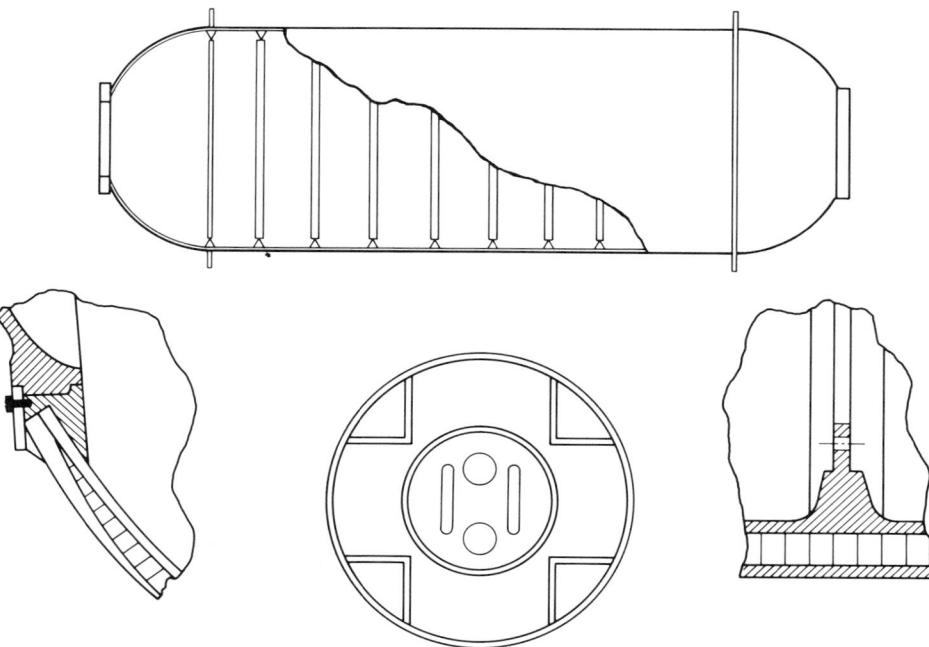

Fig. 6 Typical composite shell construction

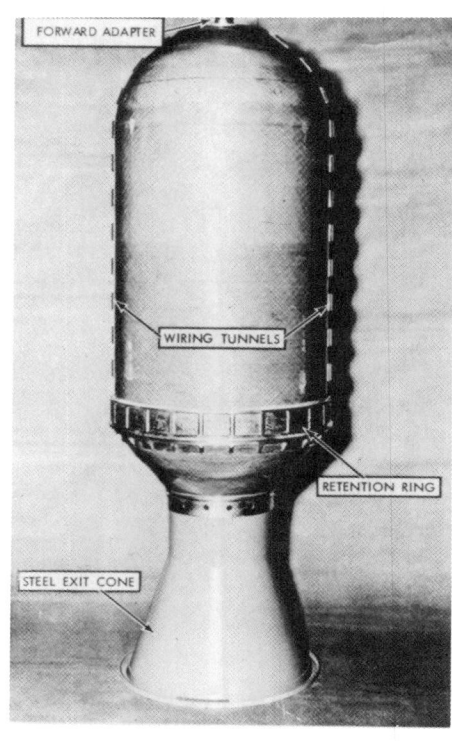

Fig. 7 X248 Altair rocket motor case. Courtesy of Hercules, Inc.

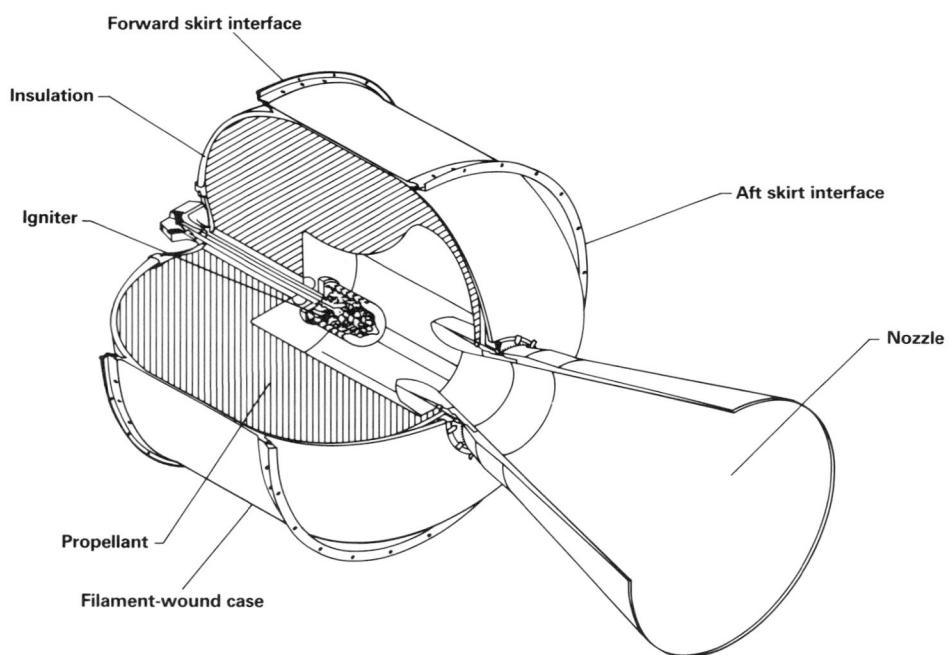

Fig. 10 BE-3 rocket motor case. Courtesy of Hercules, Inc.

Fig. 8 A-2 Polaris second-stage rocket motor. Courtesy of Hercules, Inc.

Fig. 9 Minuteman third-stage rocket motor. Courtesy of Hercules, Inc.

internal pressure. Considerable technology was developed on this program in the area of composite-to-metal joints and fastening.

The BE-3 motor is a very high performance rocket motor that has been used as both a retro rocket and a forward rocket in many space operations. Its high mass fraction results primarily from the use of composites in both the case and the nozzle and secondarily from the optimization of its shape. The near-spherical shape is made up of forward and aft geodesic (approximately elliptical) domes connected by an integrally wound cylindrical section (Fig. 10). The resulting near-sphere is close to optimum for both the propellant and the case, representing maximum performance trade-off design.

Fig. 11 Trident motor cases (left to right: third, second, and first stages)

Fig. 12 Graphite pressure vessel and aluminum liner. Courtesy of Hercules, Inc.

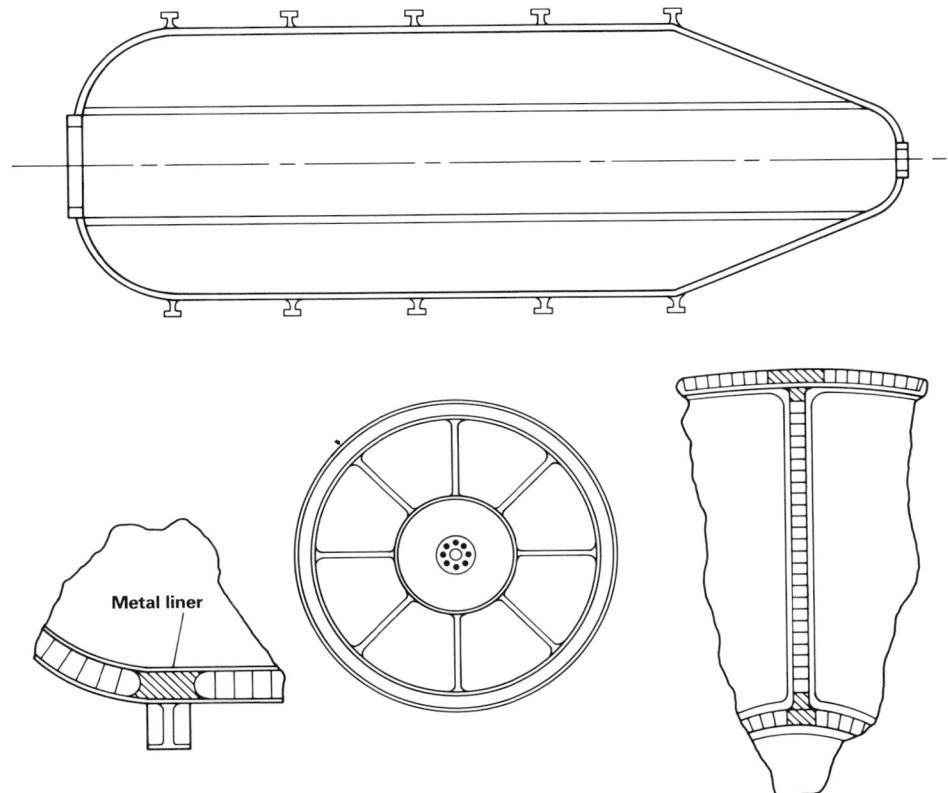

Metal liner

Fig. 13 Concept of multicell fuel tank construction

Table 3   Typical properties of three-dimensionally reinforced carbon-carbon nose tip materials

Property	Property direction	Property value
Bulk density, g/cm^3		2.01
Tensile strength at RT,		
MPa (ksi)	X and Y	90 (13)
	Z	200 (30)
Tensile modulus at RT,		
GPa (10^6 psi)	X and Y	69 (10)
	Z	90 (13)
Compressive strength,		
MPa (ksi)	Z	95 (14)
Compressive modulus,		
GPa (10^6 psi).	Z	110 (16)
Thermal expansion		
at 2100 °C (3800 °F)	Z	0.30%
at 2500 °C (4500 °F)		0.45%

Trident I. Three composite filament-wound motor cases form the primary structure of the Trident I solid-propellant missile. The cases (Fig. 11) use aramid fiber in the pressure vessel body and graphite fiber in the highly loaded thrust skirt. Nozzle attachment flanges are used to attach the large composite single-nozzle thrust cone.

A *filament-wound case* was developed for the space shuttle to replace steel booster cases. The four cylindrical segments that compose each booster case are filament wound with high-strength graphite fiber and epoxy resin. Forge-machined steel rings are pinned to both ends of each segment and are used to join and seal the segments together after propellant loading. The weight reduction of the composite case is approximately 12 500 kg (14 tons) or a 25 000 kg (28 ton) reduction for the set of two boosters used in launching. This results in an appreciable payload performance gain.

Liquid-Propellant Rocket Tanks. The relatively new and rapidly growing potential for application of composites to "liquid missiles" takes advantage of two previously unrelated technology developments. The first is the emerging development of large aircraft fuselages made of composite materials in which integral stiffness, wall thickness distribution, integral reinforcement, and other features are being automated for very large structures. The second is the continued development of very lightweight sealing systems for composite tank walls. Thin metal-lined composite tanks (Fig. 12) have been studied by National Aeronautics and Space Administration for such applications since the late 1970s. The marriage of these technologies will lead to the development of lightweight cost-effective tanks for large rockets.

Multicell Fuel Tankage. Small metal-lined composite pressure vessels have been used for many years in spacecraft and missiles. Much larger vessels having greater structural complexity are needed in future liquid-fuel missiles and aeropropulsion systems. The need for greater complexity comes from structural requirements for bulkheads, stiffeners, fuel separation to assist fuel management and center-of-gravity control, and large-sized vehicle attachment. In Fig. 13, typical construction of a filament-wound fuel tank is shown in which the fuel separation cells are integrally wound and co-cured with an outer sandwich skin shell. External attachment rings or bulkheads are also co-cured to the shell at core-densified areas, and all internal surfaces are lined with metal.

Reentry Vehicles. Three-dimensionally reinforced carbon-carbon composites were first developed for use as nose tips for ballistic missile reentry vehicles. Monolithic bulk graphite had previously been used, but it imposed design and trajectory limitations because of its relatively high ablation rate and vulnerability to thermal-shock failure. Also, it was unsuitable for maneuvering reentry bodies because its low mechanical properties were inadequate for side loading during maneuvering.

The current three-dimensional carbon-carbon composite nose tip materials offer low ablation rates because of their high bulk density of over 2.0 g/cm^3. Also, their superior mechanical properties (Table 3) allow design freedom. In the future, because of even more severe mission requirements, the use of three-dimensional reinforced composites will probably be required for reentry heat shields and structures, as well as nose tips.

SELECTED REFERENCES

- J.C. Bittence *et al.*, Ed., *ASM Guide to Engineered Materials, Metal Progress and Advanced Material & Processes*, June 1986
- F.W. Wendt, H. Liebowitz, and N. Perrone, Mechanics of Composite Materials, in *Proceedings of the Fifth Symposium on Naval Structural Mechanics*, May 1967

Long-Term Environmental Effects and Flight Service Evaluation

H. Benson Dexter, NASA Langley Research Center

THE INFLUENCE of ground and flight environments on the durability of aircraft components fabricated from composite materials is an ongoing concern of aircraft manufacturers and operators. Some of the uncertainties include the effects of moisture absorption, ultraviolet radiation, fuels and fluids, long-term sustained stress, and fatigue loading. Accordingly, in the early 1970s, the NASA Langley Research Center initiated programs to establish the effects of ground and flight environments on several composite material systems. Residual strength and stiffness as a function of exposure time were determined after 10 years of worldwide outdoor exposure. Service performance, maintenance characteristics, and residual strength of numerous composite components installed on commercial and military aircraft and helicopters were determined as a function of flight hours and years in service.

Environmental Effects

Composite structures for commercial and military aircraft must be designed to withstand the great diversity of environments encountered in worldwide operation, such as large variations in temperature and moisture, contact with aircraft fuels and fluids, and lightning strikes. A series of ground-based exposure programs was conducted to establish the effects of various environments on composite materials. Small unpainted test specimens were mounted in outdoor exposure racks and used to measure residual strength and stiffness as a function of exposure location, environment, and time. Specimens were unpainted in order to achieve the maximum effect of the exposure environments. Figure 1 shows the exposure locations and types of specimens. A series of tests was selected that would measure matrix-dependent and fiber-dependent properties. Included were unstressed short-beam shear, compression, and flexure specimens, and sustained-stress tension specimens. Short-beam shear and compression strengths are primarily matrix dependent, whereas flexure and tension properties are more fiber dependent. All the tests were conducted in accordance with recommended procedures of the American Society for Testing and Materials (ASTM).

A variety of exposure sites was selected to represent a broad range of outdoor temperatures and relative humidity. The sites selected had major airport terminals nearby that are used by aircraft with composite components installed. Exposure sites included NASA Langley Research Center, Hampton, Virginia; San Francisco and San Diego, California; Honolulu, Hawaii; Frankfurt, W. Germany; Wellington, New Zealand; and Sao Paulo, Brazil. Unpainted specimens were exposed for up to 10 years, and residual properties were measured after 1, 3, 5, 7, and 10 years. In addition, specimens were exposed to various fuels and fluids in a controlled environment in Seattle, Washington. References 1 to 7 provide a detailed presentation of the test data.

Ten-Year Worldwide Ground-Based Exposure

Triplicate specimens were mounted in exposure racks and placed on rooftops to receive maximum exposure to the environment. The average residual properties (moisture absorption, strength, and modulus of elasticity) were compared to average baseline properties. The unstressed short-beam shear, compression, and flexure specimens were deployed worldwide; however, the stressed and unstressed tension specimens were deployed at NASA Langley and San Francisco only. A variety of composite material systems was selected for the test program, including 121 °C (250 °F) and 177 °C (350 °F) cure materials (Table 1). The material systems selected were also used to fabricate flight service components.

The short-beam shear, flexure, and compression specimens were used to determine changes in mechanical properties. The flexure specimens were also used to measure ultraviolet radiation effects and moisture absorption. The effects of ultraviolet radiation were determined by viewing the exposed surface of the flexure specimens in a scanning electron microscope. Moisture absorption was determined by drying the flexure specimens in an oven after they were tested for residual strength.

Moisture Absorption. The amount of moisture that composite materials absorb is a function of matrix and fiber type, time, specimen or component geometry, temperature, relative humidity, and exposure conditions. Accelerated laboratory tests are often used to saturate composite materials for subsequent mechanical property testing. However, the objective of the tests reported here was to establish the effects of various real-time outdoor environments on the moisture absorption of composite materials. The aramid-epoxy materials absorbed the most moisture during the 10-year exposure period, approximately 2.6%. There was little difference in the amount of moisture absorbed by the Kevlar 49-F-155 and Kevlar 49-F-161 material systems. The moisture absorption of the T300-5209, T300-5208, and AS-3501 graphite-epoxy specimens stabilized at approximately 0.7 to 1.2%, and the T300-2544 graphite-epoxy specimens absorbed up to 2.2%. The Brazil and New Zealand exposures resulted in the highest moisture absorption. These results were not unexpected because the average annual relative humidity in Sao Paulo, Brazil, and Wellington, New Zealand, is about 75 to 80%. The graphite-reinforced composite materials reached moisture equilibrium after about 3 years of exposure, whereas the thicker aramid-reinforced materials reached moisture equilibrium after about 7 years.

Ultraviolet Degradation. Degradation of surface resin on unpainted flexure coupons was noticeable after 3 years of exposure. Typical results for AS-3501 graphite-epoxy and Kevlar 49-F-155 after 7 years are shown in Fig. 2. The scanning electron micrographs on the left of Fig. 2, for specimens with no outdoor exposure, indicate that the surface fibers are fully coated with resin. The micrographs on the right of Fig. 2, however, indicate that the surface

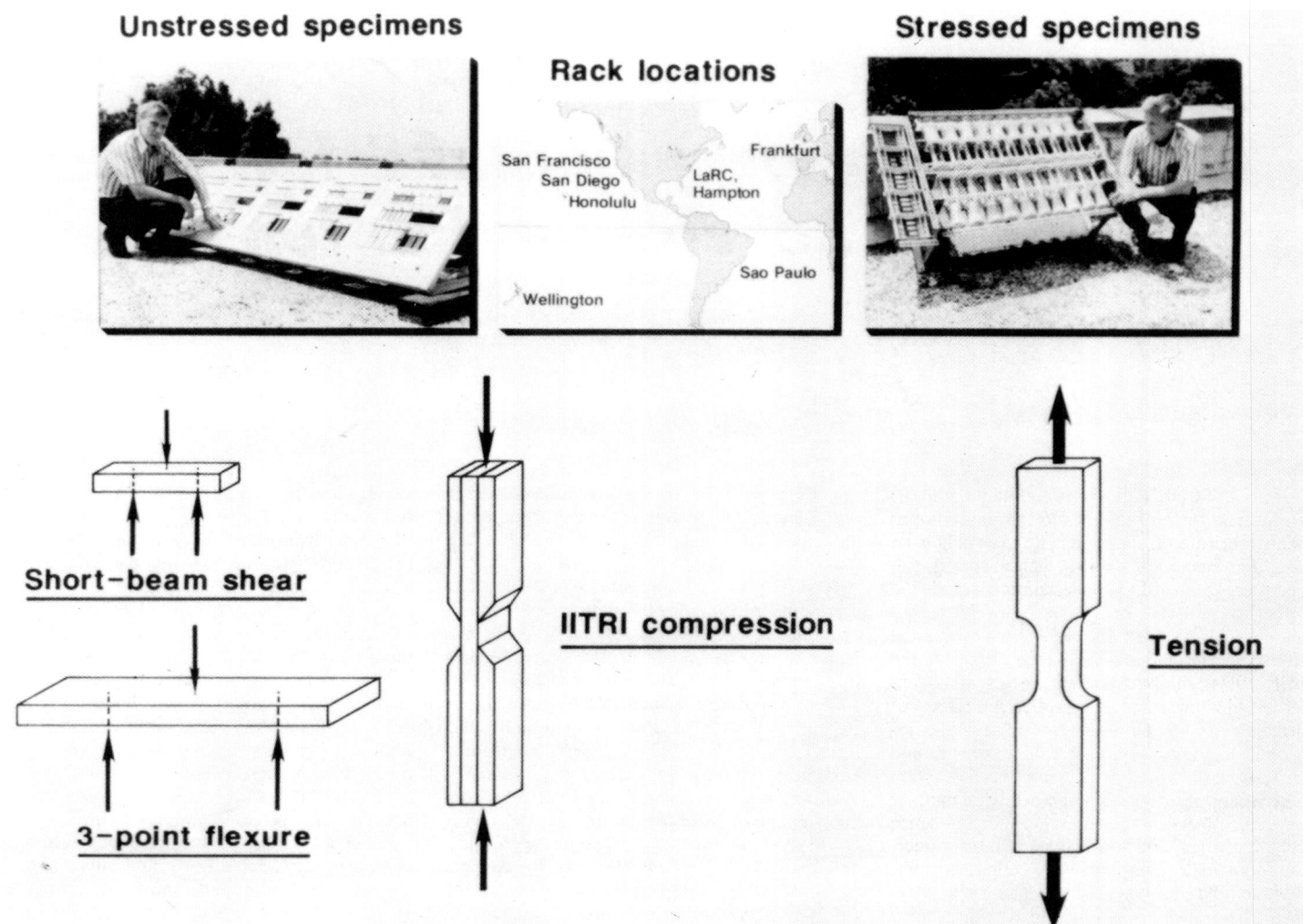

Fig. 1 Exposure locations and specimen types

Table 1 Material systems for test specimens

Material type	Material supplier	Flexure mm	in.	Short-beam shear mm	in.	Compression mm	in.	Fiber lay-up pattern(a)
T300-5209 graphite-epoxy	Amoco Performance Products Narmco Materials	2.06	0.081	3.05	0.120	4.32	0.170	0° tape
T300-2544 graphite-epoxy	Amoco Performance Products	1.78	0.070	2.64	0.104	3.51	0.138	0° tape
AS-3501 graphite-epoxy	Hercules, Inc.	1.75	0.069	2.54	0.100	3.07	0.121	0° tape
K49–F-155 Kevlar-epoxy	Du Pont Hexcel Corp.	3.10	0.122	3.07	0.121	1.27	0.050	(0°/90°) fabric
K49–F-161 Kevlar-epoxy	Du Pont Hexcel Corp.	2.90	0.114	2.95	0.116	1.24	0.049	(0°/90°) fabric
T300-P1700 graphite-polysulfone	Amoco Performance Products U.S. Polymeric	1.93	0.076	2.59	0.102	3.43	0.135	$(+75°, -45°, -75°)_s$ tape
T300-5208 graphite-epoxy	Amoco Performance Products Narmco Materials	2.11	0.083	3.18	0.125	Not tested		0° tape

(a) The 0° fiber direction is oriented along the length of the test specimen.

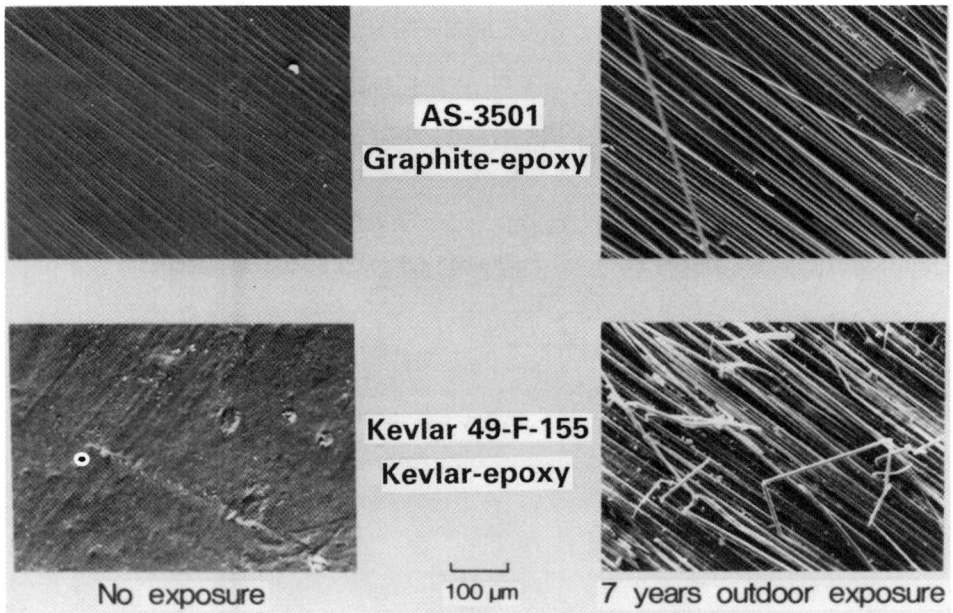

Fig. 2 Effects of ultraviolet radiation on unpainted composite materials

fibers are fully exposed due to ultraviolet degradation of the surface layer of epoxy. The T300-2544 graphite-epoxy and Kevlar 49-F-155 systems were those most affected by ultraviolet radiation. These systems also absorbed the most moisture. These results confirm the need to keep composite aircraft structures painted to prevent degradation of surface resin and exposure of bare fibers. The results of controlled laboratory weatherometer tests (Ref 1) indicate that polyurethane aircraft paint offers substantial protection against ultraviolet radiation.

Residual Flexural Strength. Three-point flexure tests were conducted to assess the effects of outdoor environments on surface fiber strength. The specimens were tested with the exposed surface in compression; in general, however, failure occurred in tension at midspan of the specimens. Although failures were primarily in tension, the test results indicated that specimens with high ultraviolet radiation damage to the compression surface and with the highest moisture absorption also had the highest strength loss, up to 26%. Except for the T300-2544 specimens, most of the graphite-epoxy specimens failed within the baseline strength scatter bands. The aramid-epoxy specimens indicated a steady strength decrease during the 10-year exposure period.

Residual Flexural Modulus. Load-deflection response was recorded for each flexure test, and flexural modulus was calculated using large deflection beam equations. Only three material systems (T300-2544, Kevlar 49-F-155, and Kevlar 49-F-161) indicated any significant loss (up to 28%) in flexural modulus during the 10-year exposure period. The two materials that incurred the most ultraviolet radiation damage also had the highest reduction in flexural

modulus. These results were expected because the unsupported outer fibers could not contribute to bending stiffness.

Residual Short-Beam Shear Strength. Short-beam shear tests were conducted to provide a measure of fiber-to-matrix bond degradation as a function of outdoor exposure time and exposure site. In general, interlaminar failures occurred at midplane in a horizontal shear mode. The T300-P1700 graphite-polysulfone specimens in New Zealand had the most significant shear-strength loss; maximum strength loss of 38% was indicated after 9 years of exposure in New Zealand, versus a 0 to 28% loss for the other systems.

Residual Compression Strength. Compression tests were conducted to determine the effects of outdoor exposure on strength. The test results indicated that most of the data were within the baseline strength scatter band and often above the average baseline strength. However, strength reductions of about 26% occurred for some specimens exposed in Hawaii and New Zealand. These results indicate that the combination of intense ultraviolet radiation and high relative humidity found in New Zealand and Hawaii affected materials more than in the other locations.

Effects of Aircraft Fluids and Fuels on Composite Strength

Aircraft structures are frequently exposed to various combinations of fluids, such as fuel, hydraulic fluid, and water. The effects of combinations of these fluids on composite materials have been evaluated after 5 years of exposure. Short-beam shear and ±45° tension specimens were exposed to six different environmental conditions as follows: ambient air, water, JP-4

fuel, hydraulic fluid, fuel-water mixture, and fuel-air cycling. The water, fuel, and hydraulic fluid were replaced monthly to maintain fresh exposure conditions. Specimens exposed in the fuel-water mixture were positioned with the fuel-water interface at the center of the test specimens. The fuel-air cycling environment consisted of 24 h of fuel immersion followed by 24 h of exposure to air. Residual strengths of T300-5208 graphite-epoxy, T300-5209 graphite-epoxy, and Kevlar 49-5209 fabric were determined after exposure to the six different environments.

The most degrading environment on the T300-5209 and Kevlar 49-5209 materials was the fuel-water combination. The T300-5209 specimens lost about 11% of tensile strength, and the Kevlar 49-5209 specimens lost about 25%. The short-beam shear test is particularly resin sensitive; the largest short-beam shear strength reduction was about 40% for the T300-5209 material system when exposed to the fuel-water combination. These tests were more severe than actual aircraft flight exposures, and the results should represent an upper bound on material property degradation. Additional details on the fluids-and-fuels exposure program can be found in Ref 2.

Effects of Sustained Stress on Strength and Modulus

The effects of sustained stress on tensile strength and modulus were determined during 10-year outdoor exposures in San Francisco and at NASA Langley in Hampton. Quasi-isotropic [0°, ±45°, 90°] T300-5208 tensile specimens were stressed at 40% of baseline ultimate strength. Unstressed specimens were included in the outdoor exposure racks for comparison purposes. Specimens were removed from the racks after 1, 3, 5, 7, and 10 years to measure residual strength and modulus. The tensile strength results indicated that most of the data were within the baseline strength scatter band for both the stressed and unstressed specimens. The residual tensile modulus data indicated a 10 to 15% reduction in modulus after 5 and 7 years of exposure, but no reduction after 10 years.

Effects of Sustained Stress on Strength of Bolted Joints

The effects of sustained stress on the tensile strength of quasi-isotropic graphite-epoxy bolted joints were determined after 7 years of outdoor exposure. Two bolted-joint configurations were tested. The first had a single row of bolts and was fabricated with T300-5208 graphite-epoxy. The second had a double row of bolts and was fabricated with T300-5209 graphite-epoxy. The test specimens were installed in outdoor loading frames at NASA Langley and were subjected to a sustained tensile load of 27% of the design ultimate load for 1, 3, 5, and 7 years. In addition, 0.4 lifetimes of spectrum fatigue loads were applied to the test specimens at the end of each year of exposure.

Test results for the single-row T300-5208 joint indicated a 7.5% decrease in residual strength after 5 years of exposure and two lifetimes of fatigue loading. The double-row T300-5209 joint had a 5.3% decrease in residual strength after 7 years of exposure and 2.8 lifetimes of fatigue loading. These results indicate excellent performance of both graphite-epoxy systems and joint configurations after being subjected to outdoor sustained-load and laboratory spectrum fatigue loading. Additional details on the bolted-joint test program are presented in Ref 3.

Flight Service Evaluation

In 1973 the NASA Langley Research Center initiated a series of programs to evaluate the effects of realistic flight environments on composite components. The objective was to establish confidence in the long-term durability of advanced composites through flight service of numerous composite components on transport aircraft. Emphasis was on commercial aircraft because of their high utilization rates, exposure to worldwide environmental conditions, and systematic maintenance procedures. In 1979 NASA Langley and the U.S. Army initiated joint programs to evaluate composite components on commercial and military helicopters. Although helicopters accumulate fewer flight hours than do transport aircraft, the environments and fatigue loading are sometimes more severe for the helicopter components.

Transport Aircraft Components

The transport aircraft in which composite components were tested in the NASA Langley service evaluation program included the Lockheed L-1011, Boeing 737, Douglas DC-10, and Boeing 727. Eighteen Kevlar 49-epoxy fairings were installed on L-1011 aircraft in 1973, and eight graphite-epoxy ailerons were installed in April 1982. In July 1973, 108 graphite-epoxy spoilers were installed on B-737 aircraft on six different commercial airlines in worldwide service. Ten graphite-epoxy horizontal stabilizers were installed on B-737 aircraft in March 1984. In 1975, 15 graphite-epoxy upper aft rudders and three boron-aluminum aft pylon skin panels were installed on DC-10 aircraft. In 1980, ten graphite-epoxy elevators were installed on B-727 aircraft. In addition to the foregoing commercial aircraft components, two boron-epoxy reinforced aluminum center-wing boxes were installed on U.S. Air Force C-130 transport aircraft in 1974.

L-1011 Kevlar 49-Epoxy Fairings. The L-1011 fairings were fabricated with Kevlar 49 fibers (in fabric form), F-155 and F-161 epoxy resins, and Nomex honeycomb. The configurations of the center-engine fairing, under-wing fillet, and wing-to-body fairing are shown in Fig. 3, along with the various types of damage incurred. During the 10-year service evaluation period, the Kevlar 49-epoxy fairings installed on L-1011 aircraft were inspected annually.

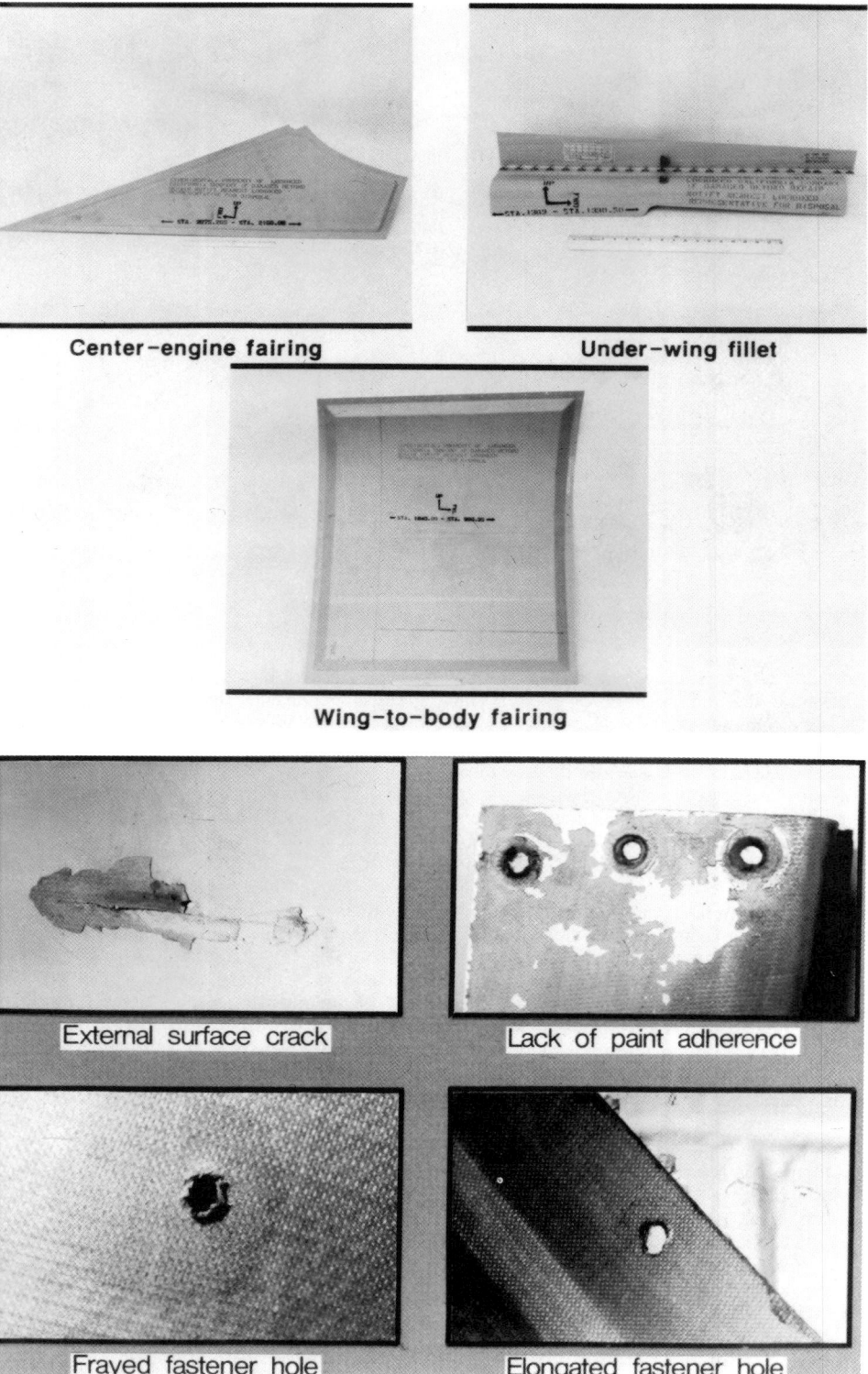

Fig. 3 Configurations of L-1011 Kevlar 49-epoxy fairings (top) and typical in-service conditions (below)

Minor impact damage from equipment and foreign objects was noted on several fairings, primarily the honeycomb sandwich wing-to-body fairings. Surface cracks and indentations were repaired with filler epoxy and, in general, the cracks did not propagate with continued service. Paint adherence was a minor problem, particularly with parts in contact with hydraulic

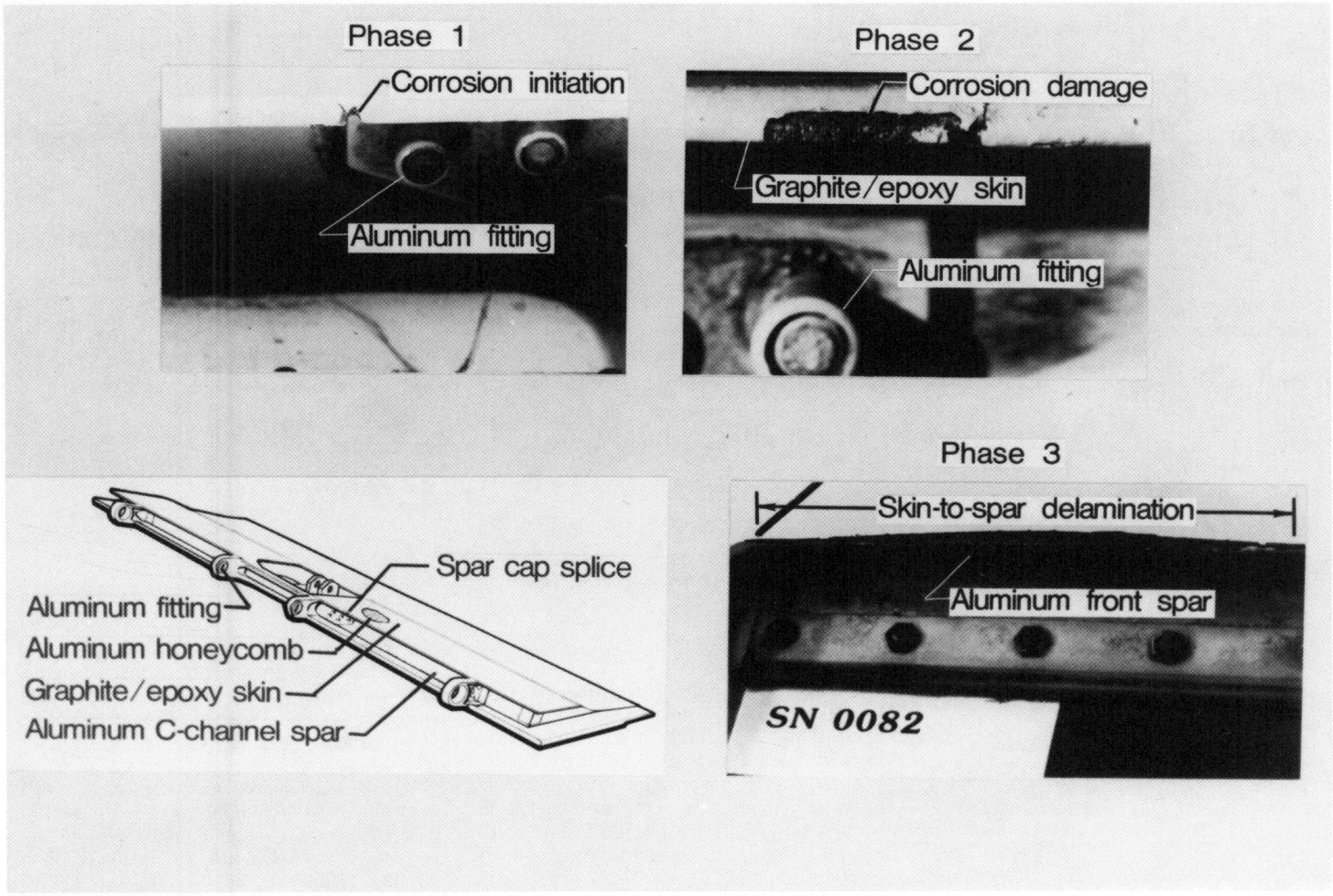

Fig. 4 Corrosion damage to B-737 graphite-epoxy spoiler

fluid. Fastener holes in several fairings were frayed, primarily because of nonoptimum drilling procedures and improper fit. Elongated holes were also noted, primarily caused by improper fit and nonuniform fastener load distribution. There were no moisture intrusion problems with the Kevlar 49-epoxy fairings, and they performed similarly to production fiberglass-epoxy fairings. Additional details on the design, fabrication, and service evaluation of the Kevlar 49-epoxy fairings are presented in Ref 4 and 5.

B-737 Graphite-Epoxy Spoilers. The B-737 spoilers used three different graphite-epoxy unidirectional tape systems: T300-5209, T300-2544, and AS-3501. The spoilers were fabricated with upper and lower graphite-epoxy skins; aluminum fittings, spar, and honeycomb core; and fiberglass-epoxy ribs (Fig. 4).

During the 13-year service evaluation period, several types of damage were encountered, with over 75% of the damage incidents being related to design details. Damage was most often due to actuator rod interference with the

graphite-epoxy skin, which was resolved by redesigning the actuator rod ends. The second most frequent cause of damage was moisture intrusion and corrosion at the spar-to-center hinge fitting splice, which could be resolved by redesigning the splice to prevent disbonds between the skin and spar cap. Miscellaneous cuts and dents related to airline usage were also encountered. Damage from hailstones, bird strikes, and ground handling equipment occurred on several spoilers.

A typical corrosion damage scenario for a spar-to-center hinge fitting splice is shown in Fig. 4. The corrosion damage can be characterized by three phases of development. Phase 1 involves corrosion initiation at an aluminum fitting or at the aluminum spar splice. The corrosion is initiated by moisture intrusion through cracked paint and sealant material. If the corrosion products are not removed and new sealant applied, the damage progresses to phase 2, where moisture penetrates under the graphite-epoxy skin along the aluminum C-channel front spar. Normal service loads, combined

with moisture, contribute to crack growth and subsequent corrosion. If the phase 2 corrosion is not repaired, the damage progresses to phase 3, where extensive skin-to-spar separation takes place. Phase 3 corrosion can result in significant loss of strength and stiffness. Uninterrupted flight service of 2 to 3 years is required before phase 3 separation becomes significant. Residual strength tests were conducted annually to establish the effects of service environments on the graphite-epoxy spoilers. The spoilers were tested with compression load pads on the upper surface to simulate airloads. Trailing-edge tip deflection was measured as a function of applied load for each spoiler tested. The test results were compared with the strength and stiffness of 16 new spoilers. The strength for each spoiler through 12 years of service generally fell within the strength scatter band for the baseline spoilers. However, spoilers with significant corrosion damage that were tested after 7 and 8 years of service, respectively, indicated a strength reduction of up to 35%. The load-deflection response of the spoilers, compared

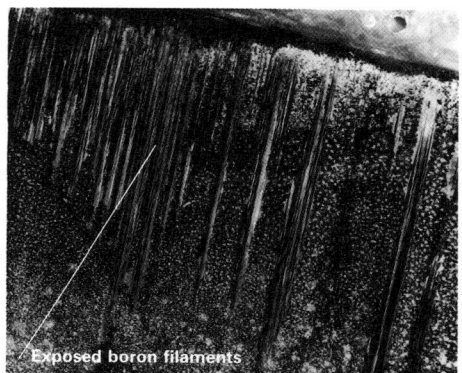

Fig. 5 Interior corrosion damage to DC-10 boron-aluminum aft-pylon skin panel

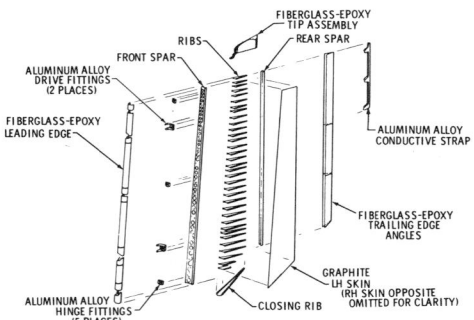

Fig. 6 Configuration of DC-10 graphite-epoxy upper aft rudder

Fig. 7 DC-10 graphite-epoxy rudder damage

with that of the baseline spoilers, followed similar trends.

In addition to structural tests of the spoilers, measurements were made to determine absorbed moisture content of the graphite-epoxy skins. The moisture content was determined from plugs cut near the trailing edge. The plugs consisted of aluminum honeycomb core, two graphite-epoxy facesheets, two layers of epoxy film adhesive, and two exterior coats of polyurethane paint. About 90% of the plug mass was in the composite facesheets, including the paint and adhesive. The moisture content was determined by drying the plugs and recording the mass change.

The data for plugs removed from three spoilers after 9 years of service indicated moisture levels in the graphite-epoxy skins ranging from 0.59 to 0.90%. The moisture levels for the T300-5209 and AS-3501 systems were similar to those for unpainted material coupons exposed to worldwide outdoor environments. However, the moisture content of 0.90% for the T300-2544 plugs was only about half that of the unpainted outdoor-exposure material coupons. Extensive ultraviolet radiation degradation of the T300-2544 unpainted specimens may partially explain the higher moisture absorption. Additional details on the design, fabrication, testing, and service evalu-

ation of the B-737 graphite-epoxy spoilers are presented in Ref 6 and 7.

C-130 Boron-Epoxy Reinforced Center-Wing Boxes. The design of the boron-epoxy reinforced aluminum center-wing boxes installed on the two C-130 aircraft in 1974 included uniaxial strips of boron-epoxy bonded to the aluminum alloy skin panels and to the hat section stringer crown. The objective of this program was to compare the strength and fatigue endurance of the boron-epoxy reinforced boxes and equivalent all-aluminum wing boxes.

The boron-epoxy reinforced aluminum wing boxes performed excellently in service, with no damage or defects reported. No maintenance actions were required during a 12-year service evaluation period. On the basis of results of ground tests on a third wing box, the boron-epoxy reinforced wing boxes are expected to have greater fatigue endurance than the baseline aluminum boxes. Additional details on the design, fabrication, and ground tests of the boron-epoxy reinforced wing boxes can be found in Ref 8-10.

DC-10 Boron-Aluminum Aft-Pylon Skins. The boron-aluminum skin panels installed on the three DC-10 aircraft in 1975 were located above the aft-engine and encountered high acoustic fatigue loading and moderate thermal loads. Two of the skin panels were still in service after 11 years. However, one panel was removed from service after 7 years because of corrosion damage (Fig. 5); the outer layer of boron filaments on the inside of the panel was almost completely exposed. The panel contained a light residue of ester oil similar to turbine engine oil; however, the specific corrodent was not identified. A second panel also had some corrosion damage and a small crack,

but was monitored closely during service to check for crack growth and further corrosion damage. The crack in the panel was probably caused by exterior mechanical damage during removal and reinstallation of the adjacent engine cowling.

It was concluded that the method of corrosion protection used was inadequate. In general, the boron-aluminum panels did not perform as well as similar production titanium panels. Additional details on the DC-10 boron-aluminum skin panels are presented in Ref 11.

DC-10 Graphite-Epoxy Rudders. The configuration of the 15 T300-5208 graphite-epoxy upper aft rudders installed on DC-10 aircraft in 1976 is shown in Fig. 6. The entire structure was co-cured in an oven at the material supplier's recommended cure cycle for T300-5208. Expandable rubber was used inside the rudder to apply pressure against an outside steel tool.

There were seven incidents which required rudder repairs, including three minor disbonds, rib damage due to ground handling, and damage due to lightning. Figure 7 shows minor lightning strike damage to the trailing edge of a rudder and rib damage that occurred while the rudder was off the aircraft for other maintenance. The lightning strike damage was limited to the outer four layers of graphite-epoxy, and a room-temperature repair was performed in accordance with procedures established when the rudders were certified by the FAA. The rib damage was more extensive, and a portion of a rib was removed and rebuilt. A detailed discussion of the repair procedure is given in Ref 12.

More extensive lightning damage was sustained on another graphite-epoxy rudder. Inspection of the rudder revealed that the light-

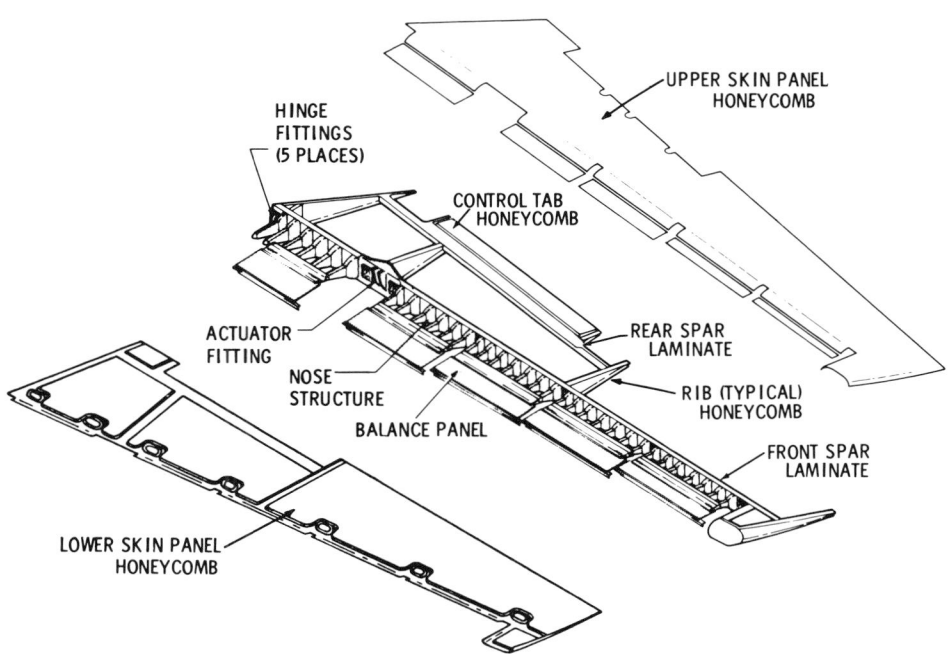

Fig. 8 Configuration of B-727 graphite-epoxy elevator

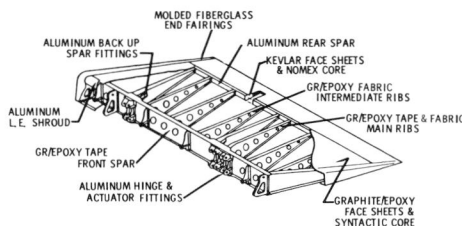

Fig. 10 Configuration of L-1011 graphite-epoxy aileron

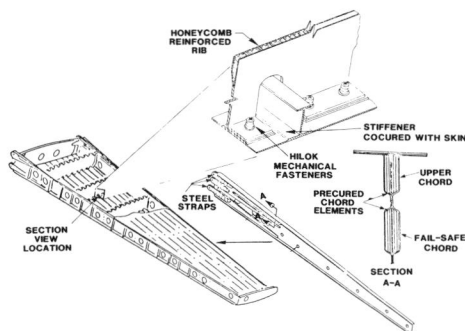

Fig. 11 Configuration of B-737 graphite-epoxy horizontal stabilizer

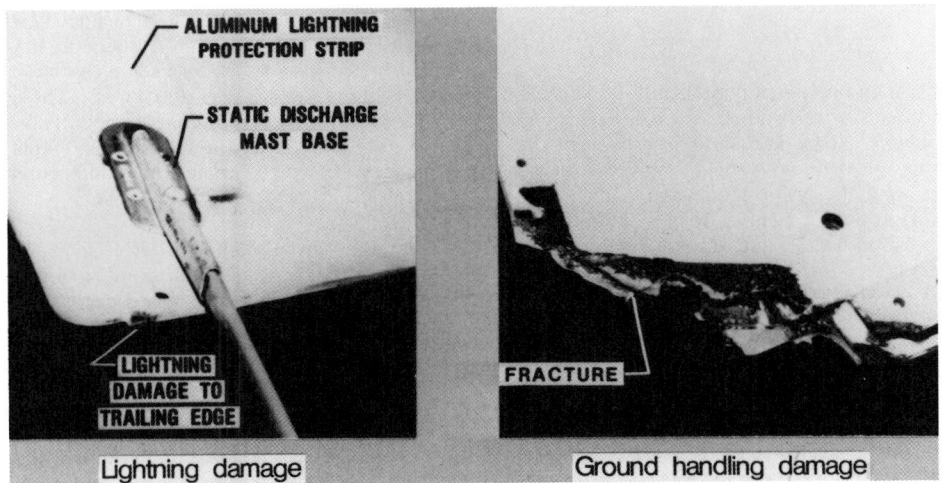

Lightning damage   Ground handling damage

Fig. 9 B-727 graphite-epoxy elevator damage

ning protection strap had been inadvertently left off after the previous maintenance check. The skin in the damaged area was eight plies thick over an eight-ply spar cap. Fiber damage and resin vaporization extended through the skin forward of the spar, and the skin and spar cap aft of the rear spar were completely destroyed. Details of the repair procedures are given in Ref 13.

A graphite-epoxy rudder was removed from service for residual strength testing after 5.7 years and 22 265 flight hours on Air New Zealand. The load-deflection response indicated that this rudder had an initial stiffness higher than that of the baseline rudder, but a similar overall response. The baseline and the 5.7-year tests were stopped at approximately 400% limit load because of instability of the loading apparatus. Although the rudders were designed by stiffness considerations and only one residual strength test was conducted, the overall response of the rudder indicated that no degradation had occurred after 22 265 flight hours.

B-727 Graphite-Epoxy Elevators. The T300-5208 graphite-epoxy elevator (Fig. 8) was constructed with Nomex honeycomb sandwich skins and ribs, and laminated spars. Following initiation of flight service of ten elevators in 1980, three B-727 graphite-epoxy elevators were damaged by minor lightning strikes and two elevators were damaged during ground handling. The most severe damage occurred when the static discharge probe of one B-727 penetrated the elevator of another B-727 during ground handling. Skin panels were punctured, resulting in four holes in the lower surface and one hole in the upper surface, and the lower horizontal flange at the front spar was cut inboard of the outboard hinge. Figure 9 shows typical lightning damage to the trailing edge of an elevator and trailing-edge fracture of another elevator caused by impact from a deicing apparatus. Damage from lightning strikes ranged in severity from scorched paint to skin delamination. All of the elevator repairs were performed by airline maintenance personnel.

The lightning damage was repaired with epoxy filler and milled glass fibers. The skin punctures were repaired with T300-5208 prepreg fabric and Nomex honeycomb core plugs. The front spar was repaired with a machined titanium doubler, which was mechanically fastened to the lower skin flange of the spar chord. The repaired graphite-epoxy elevators were reinstalled on aircraft for continued commercial service. Details of the design and fabrication of the graphite-epoxy elevators are given in Ref 14.

L-1101 Graphite-Epoxy Ailerons. After four shipsets of T300-5208 graphite-epoxy inboard ailerons (Fig. 10) were installed on L-1101 commercial aircraft in 1982, an additional shipset was installed on the L-1101 flight test aircraft. During the 4-year service evaluation period, no damage incidents occurred and no major maintenance actions were required. Minor paint touch-up was performed periodically, and loose fibers around one fastener hole

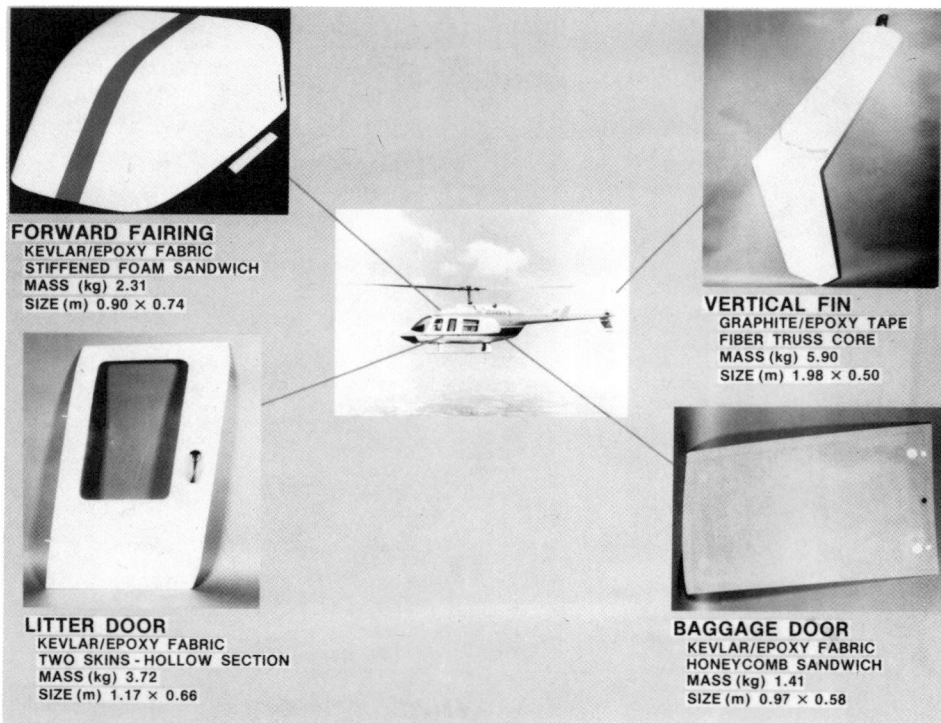

FORWARD FAIRING
KEVLAR/EPOXY FABRIC
STIFFENED FOAM SANDWICH
MASS (kg) 2.31
SIZE (m) 0.90 × 0.74

VERTICAL FIN
GRAPHITE/EPOXY TAPE
FIBER TRUSS CORE
MASS (kg) 5.90
SIZE (m) 1.98 × 0.50

LITTER DOOR
KEVLAR/EPOXY FABRIC
TWO SKINS - HOLLOW SECTION
MASS (kg) 3.72
SIZE (m) 1.17 × 0.66

BAGGAGE DOOR
KEVLAR/EPOXY FABRIC
HONEYCOMB SANDWICH
MASS (kg) 1.41
SIZE (m) 0.97 × 0.58

Fig. 12 Composite components in 206L

on the flight-test aircraft were rebonded with epoxy. Details on the development of the graphite-epoxy ailerons are reported in Ref 15.

B-737 Graphite-Epoxy Horizontal Stabilizers. The configuration of the five shipsets of T300-5208 graphite-epoxy horizontal stabilizers installed on B-737 aircraft beginning in 1984 is shown in Fig. 11. The graphite-epoxy stabilizer features stringer-reinforced skins, laminated spars, and Nomex honeycomb reinforced ribs. This structure was the first certified primary structure for commercial airplanes.

No damage incidents occurred and no maintenance actions were required on any of the stabilizers. Details on the development of the graphite-epoxy stabilizers are reported in Ref 16.

Helicopter Components

The helicopters in which composite components were tested included the Bell 206L and the Sikorsky S-76 and CH-53. Forty shipsets of Kevlar 49-epoxy doors and fairings and graphite-epoxy vertical fins were installed on 206L commercial helicopters for 10 years of service evaluation in Canada, Alaska, and northeast, southwest, and Gulf coast of the United States, with selected components removed periodically from service for residual strength testing. Details on the design, fabrication, and testing of the 206L composite components can be found in Ref 17.

Ten graphite-epoxy tail rotors and four hybrid Kevlar 49-graphite-epoxy horizontal stabilizers were installed on S-76 production heli-

copters and removed periodically to determine the effects of realistic operational service environments. Static and fatigue tests were conducted on the components removed from service, and the results compared with baseline certification test results. Details on the design, fabrication, and testing of the S-76 components are reported in Ref 18.

A Kevlar 49-epoxy cargo ramp skin was installed on a U.S. Marine Corps CH-53D helicopter for service evaluation. Details of its design, fabrication, and installation are reported in Ref 19.

206L Components. The four composite components evaluated on the 206L are shown in Fig. 12. Installation of the 40 shipsets of composite components was initiated in March 1981.

Design-related and normal-usage problems were encountered with some of the components, but service experience with the forward fairing and the vertical fin was excellent. Two graphite-epoxy vertical fins were struck by lightning. One fin was repaired and returned to service, and the second fin was returned to the manufacturer for residual-strength testing. Service experience with the Kevlar 49-epoxy litter door was good, but underdesigned metal hinges caused problems. New hinges were installed on all the litter doors. A design-related thermal distortion problem occurred with plexiglass windows in the litter doors, necessitating redesign of the window attachment to the door. The baggage doors had the poorest service record; the major problem was disbonding of the outer Kevlar 49-epoxy skin from the Nomex honeycomb core,

to which the outer skin had been co-cured with no additional adhesive. Poor resin filleting was the primary cause of the skin-to-core disbonds, which could probably be prevented by using a film adhesive between the skin and core.

Residual-strength tests were conducted on 24 components removed from service in the Gulf of Mexico, eastern Canada, Alaska, and the northeastern United States. The service times ranged from 12 to 34 months, and flight times ranged from 668 to 3387 h. All the flight-service components, except two baggage doors with disbonds, failed at loads higher than the design requirements. Additional details on flight service performance of the 206L components are reported in Ref 20.

S-76 Components. The two composite components evaluated on the S-76 were tail rotors (graphite-epoxy spar, glass-epoxy skin) and horizontal stabilizers (Kevlar 49-epoxy torque tube reinforced with full-depth aluminum honeycomb and graphite-epoxy spar caps, Kevlar 49-epoxy skin, Nomex honeycomb sandwich core). These were baseline designs for the S-76 and are in commercial production.

Three horizontal stabilizers removed from service for residual-strength testing met baseline proof load requirements. Seven tail rotor spars were removed from service for either static or fatigue testing, and test results were compared to the baseline room-temperature dry strength of ten spars tested for Federal Aviation Administration (FAA) certification. These components exhibited minimum strength retention of 93% of FAA certification values, which compares well with strength retention factors projected from laboratory-conditioned specimens (Ref 18).

Coupons cut from spars were tested in short-beam shear, and the results were compared with those from coupons cut from panels exposed in an outdoor exposure rack in Stratford, Connecticut. The average short-beam shear strength of the spar coupons was 5% lower than that of the coupons machined from the outdoor ground-exposure panels. The spar coupons had service times ranging from 37 to 51 months, and the panel coupons had exposure times ranging from 35 to 49 months.

The foregoing results indicate excellent in-service performance for all the S-76 composite components. Additional details on the evaluation program are reported in Ref 21.

CH-53D Cargo-Ramp Skin. A ±45° Kevlar 49 fabric-5143 epoxy composite skin was installed on the aft end of a CH-53D cargo ramp for U.S. Marine Corps service evaluation in 1981 to assess the wear, impact, and damage resistance of Kevlar 49-epoxy in an environment where ground impact occurs frequently. The skin panel was 508 × 2032 mm (20 × 80 in.) and ranged in thickness from 1.0 to 2.0 mm (0.04 to 0.08 in.). Because of the location and thinness of the panel, the potential for damage was significant. However, no damage or service-related problems were reported following annual inspections.

Table 2 NASA composite structures flight service summary

Aircraft component	Total no. of components in service		Start of flight service	Cumulative flight hours	
	Originally	As of March 1987		High-time aircraft	Total no. of components
L-1011 fairing panels................18	15	Jan 1973	39 210	599 210
737 spoiler 108	47	July 1973	38 560	2 380 550
C-130 center wing box 2	2	Oct 1974	8 340	16 430
DC-10 aft pylon skin................3	2	Aug 1975	33 240	83 440
DC-10 upper aft rudder 15	12	April 1976	39 400	350 210
727 elevator 10	8	March 1980	21 980	184 940
L-1011 aileron 8	8	March 1982	16 890	128 060
737 horizontal stabilizer...............10	10	March 1984	8 850	79 040
DC-10 vertical stabilizer...............1	1	Jan 1987	...	...
S-76 tail rotors and horizontal stabilizer 14	3	Feb 1979	5 800	52 100
206L fairing, doors, and vertical fin......................160	84	March 1981	7 300	330 000
CH-53 cargo ramp skin 1	1	May 1981	1 600	1 600
Total	350	193			4 250 610

Overall Flight-Service Results

The NASA Langley flight service program initiated in 1973 included 350 composite components (Table 2). As of March 1987, 193 components were still in service; more than 4 million component flight hours had been accumulated, with the high-time aircraft having more than 39 000 h. Some components were removed from service for residual-strength testing, and others were retired due to damage or other service-related problems previously described.

In general, the composite components performed excellently during the 14-year flight service evaluation. Moisture absorption for the flight components was lower than that for unpainted ground-based exposure specimens. Strength retention for the composite components was as good as or better than that of ground-based exposure specimens with comparable exposure periods. On the basis of component performance in the evaluation program, transport and helicopter manufacturers made production commitments to selected composite components. Additional details on the ground-based environmental exposure program and the flight service program can be found in Ref 22.

REFERENCES

1. H.B. Dexter and A.J. Chapman, NASA Service Experience with Composite Components, in *Proceedings of the 12th National SAMPE Technical Conference*, Society for the Advancement of Materials and Process Engineering, Oct 1980, p 77-99
2. E.Y. Tanimoto, "Effects of Long-Term Exposure to Fuels and Fluids on Behavior of Advanced Composite Materials," NASA CR-165763, National Aeronautics and Space Administration, Aug 1981
3. G.R. Wichorek and J.H. Crews, Jr., "Strength of Graphite/Epoxy Bolted Wing-Skin Splice Specimens Subjected to Outdoor Exposure Under Constant Load and Yearly Fatigue Loading," NASA TP 2542, National Aeronautics and Space Administration, Jan 1986
4. J.H. Wooley, D.R. Paschal, and E.R. Crilly, "Flight Service Evaluation of PRD-49/Epoxy Composite Panels in Wide-Bodied Commercial Transport Aircraft," NASA CR-112250, National Aeronautics and Space Administration, March 1973
5. R.H. Stone, "Flight Service Evaluation of Kevlar-49/Epoxy Composite Panels in Wide-Bodied Commercial Transport Aircraft—Tenth and Final Annual Flight Service Report," NASA CR-172344, National Aeronautics and Space Administration, June 1984
6. R.L. Stoecklin, "A Study of the Effects of Long-Term Ground and Flight Environment Exposure on the Behavior of Graphite/Epoxy Spoilers—Manufacturing and Test," NASA CR-132682, National Aeronautics and Space Administration, June 1975
7. R.L. Coggeshall, "737 Graphite Composite Flight-Spoiler Flight Service Evaluation, Eighth Report," NASA CR-172600, National Aeronautics and Space Administration, May 1985
8. W.E. Harvill, J.J. Duhig, and B.R. Spencer, "Program for Establishing Long-Time Flight Service Performance of Composite Materials in Center Wing Structure of C-130 Aircraft. Phase II—Detailed Design," NASA CR-112272, National Aeronautics and Space Administration, April 1973
9. W.E. Harvill and A.O. Kays, "Program for Establishing Long-Time Flight Service Performance of Composite Materials in Center Wing Structure of C-130 Aircraft. Phase III—Fabrication," NASA CR-132495, National Aeronautics and Space Administration, Sept 1974
10. W.E. Harvill and J.A. Kizer, "Program for Establishing Long-Time Flight Service Performance of Composite Materials in Center Wing Structure of C-130 Aircraft. Phase IV—Ground/Flight Acceptance Tests," NASA CR-145043, National Aeronautics and Space Administration, Sept 1976
11. B.R. Fox, "Flight Service Evaluation of Advanced Metal Matrix Aircraft Structural Component—Flight Service Final Report," MDC Report J3827, McDonnell Douglas Corporation, Aug 1985
12. G.M. Lehman, "Flight-Service Program for Advanced Composite Rudders on Transport Aircraft—Sixth Annual Summary Report," MDC Report J6574, McDonnell Douglas Corporation, Aug 1982
13. B.R. Fox, "Flight-Service Program for Advanced Composite Rudders on Transport Aircraft—Ninth Annual Summary Report," MDC Report J-3871, McDonnell Douglas Corporation, Sept 1985
14. D.V. Chovil *et al.*, "Advanced Composite Elevator for Boeing 727 Aircraft—Volume 2 Final Report," NASA CR-15958, National Aeronautics and Space Administration, Nov 1980
15. C.F. Griffin and E.G. Dunning, "Development of an Advanced Composite Aileron for L-1011 Transport Aircraft," NASA CR-3517, National Aeronautics and Space Administration, Feb 1982
16. J.E. McCarty and D.R. Wilson, Advanced Composite Stabilizer for Boeing 737 Aircraft, in *Proceedings of the Sixth Conference on Fibrous Composites in Structural Design*, AMMRC MS 83-2, Army Materials and Mechanics Research Center, Nov 1983
17. H. Zinberg, "Flight Service Evaluation of Composite Components on Bell Model 206L: Design, Fabrication, and Testing," NASA CR-166002, National Aeronautics and Space Administration, Nov 1982
18. M.J. Rich and D.W. Lowry, "Flight Service Evaluation of Composite Helicopter Components—First Annual Report," NASA CR-165952, National Aeronautics and Space Administration, June 1982
19. D.W. Lowry and M.J. Rich, "Design, Fabrication, Installation, and Flight Service Evaluation of Composite Cargo Ramp Skin on Model CH-53 Helicopter," NASA CR-172126, National Aeronautics and Space Administration, April 1983
20. H. Zinberg, "Flight Service Evaluation of Composite Components on Bell Model 206L—First Annual Flight Service Report," NASA CR-172296, National Aeronautics and Space Administration, March 1984
21. M.J. Rich and D.W. Lowry, "Flight Service Evaluation of Composite Components: Second Annual Report, May 1982 through September 1983," NASA CR-172562, National Aeronautics and Space Administration, April 1985
22. H.B. Dexter, "Long-Term Environmental Effects and Flight Service Evaluation of Composite Materials," NASA TM-89067, National Aeronautics and Space Administration, Jan 1987

Commercial and Automotive Applications

John C. Reindl, General Motors Corporation

THE COMMERCIAL APPLICATION of structural composites has an extensive history in the marine, aerospace, and construction industries. Specialized composites also are making significant impacts in the sporting goods industry. In general, these structural composite applications are produced at relatively low volumes and have long process cycle times, compared with other high-volume structural material applications.

This article addresses the potential for high-volume use of structural composites, mainly in the automotive industry, where low material cost, high productivity, and high quality are necessary for cost-effective operations.

Structural Composites

Composite materials are any combination of two or more distinct and identifiable constituents that differ in form or composition on a macroscale. Composites are often a combination of a fiber phase and a matrix phase; materials commonly used include metals, glasses, nonmetallic crystals, and natural or synthetic polymers. The fiber is generally a strong, stiff material with a high aspect ratio, while the matrix is the material that holds the fibers in place. The composite material generally exhibits many of the properties of the constituents, roughly in proportion to the mix ratio. Structural materials must provide strength and stiffness properties appropriate to the applications for which they are intended. Some applications are strength-critical, some are stiffness-critical, and others have local strength and stiffness requirements that may or may not affect global requirements.

This article primarily focuses on fiber-reinforced plastics, which offer greater design flexibility and higher strength- and stiffness-to-weight ratios than do alternatives such as sheet metal or unreinforced plastics. However, the advantages of structural composites can only be realized if process limitations are understood and if material properties and differing design principles receive proper attention. Engineers who have designed applications using more traditional materials must remember that simple material substitution is not sufficient to ensure a successful product. Simultaneous engineering principles must be followed. An individual design should be made for each material under consideration based on the design criteria for that material. It is also important to document previous materials substitution successes and failures to lay the foundation for product improvements.

Structural Versus Appearance Requirements

Every commercial application has both structural and appearance requirements, which have tended to change with time. Interest in composite automotive body panels can be traced to Henry Ford's demonstration of the impact resistance and durability of a trunk lid made of a soy bean based composite. Today, we need at least the same physical properties, but demand a much higher quality surface appearance.

Automotive applications can be categorized as having mainly structural or mainly appearance requirements. Fenders and fascias are primarily appearance parts, with only local strength and stiffness requirements for attachments and oil-canning resistance. Load floors are structural, with only minor appearance needs. Hoods, roofs, and doors, however, contribute in varying degrees to the overall vehicle structure and must provide the same quality appearance as fenders and fascias.

Structural requirements can be subdivided into primary and secondary requirements. The primary structure reacts to the major loads such as static and dynamic external loads and to component loads, while the secondary structure supplements the primary one by addressing local requirements such as surface appearance, sealing, and carrying minor components. The secondary structure can act to stabilize the primary structure for an optimized overall design.

In the automotive industry, the application of structural composites has been mostly limited to secondary structures and appearance panels. For example, doors contribute to crashworthiness, but are largely secondary structures. Aerospace and aircraft applications are often considerably different from automotive applications in that exterior surfaces are often an integral part of the primary structure of the product. The fact that structural composites have been used extensively for both primary and secondary structures is due to low production rates and an emphasis on reduction of weight as opposed to cost.

Material Descriptions, Properties, and Processes

Descriptions of those composite materials that are widely used commercially, their properties, and generalized process techniques are listed below. Information on two other engineering materials, sheet metal and thermoplastics, is included for comparison.

Sheet molding compounds (SMC) are fiber-reinforced plastics that are fabricated in sheets and compression molded to final shape. They are reinforced in the plane of the sheet, but are not significantly reinforced in the through-the-thickness direction. The reinforcement may be continuous fibers (SMC-C), randomly chopped fibers (SMC-R), or a blend of both (SMC-C/R). The weight percent of fiber is generally indicated after the C or R. Continuous fibers can be placed parallel to get maximum stiffness and strength in one direction, but at the price of very low properties in the direction normal to the fibers. A wide range of intermediate properties can be obtained by combining continuous and random fibers or by laying continuous fibers in an X pattern, oriented at a specific angle to get the desired directional properties (XMC).

Properties of typical appearance-grade (SMC-R25), structural-grade (SMC-R50), mixed continuous/random (SMC-C20/R30), and ±7.5° X-pattern continuous (XMC-3) SMC compounds are shown in Table 1. These properties are from compounds made from typical commercial polyester resins with calcium carbonate filler and glass fibers.

Table 1 Typical SMC properties

	Specific gravity	Glass, wt%	Filler, wt%	Resin, wt%	Tensile strength		Elongation, %	Flexural modulus		Coefficient of linear thermal expansion, 10^{-6}/K
					MPa	ksi		GPa	10^6 psi	
SMC-R25	1.83	25	46	29	82.4	12.0	1.34	11.7	1.70	23.2
SMC-R50	1.87	50	16	34	164	23.8	1.73	15.9	2.30	14.8
SMC-C20/R30	1.81	50	16	34						
longitudinal	...	...	...	...	289	41.9	1.73	25.7	3.73	11.3
lateral	...	...	...	...	84.0	12.2	1.58	5.9	0.86	24.6
XMC-3 ± 7.5°X—										
pattern..................	1.97	75	...	25						
longitudinal	...	...	...	...	561	81.4	1.66	34.1	4.95	8.7
lateral	...	...	...	...	69.9	10.1	1.54	6.8	0.99	28.6

Table 2 Typical RRIM properties

	Specific gravity	Glass, wt%	Resin, wt%	Tensile strength		Elongation, %	Flexural modulus		Coefficient of linear thermal expansion, 10^{-6}/K
				MPa	ksi		GPa	10^6 psi	
RRIM									
polyurethane									
milled glass	1.08	15	85						
longitudinal	...	...	...	19.3	2.80	110	0.538	0.0780	90
lateral	...	...	...	17.9	2.60	140	0.331	0.0480	135
RRIM									
polyurethane									
flake glass	1.15	20	80						
longitudinal	...	...	...	24.1	3.50	25	1.34	0.194	54
lateral	...	...	...	25.2	3.65	25	1.24	0.180	58
RRIM									
polyurea									
flake glass	1.18	20	80						
longitudinal	...	...	...	33.3	4.83	31	1.68	0.244	50
lateral	...	...	...	30.5	4.42	31	1.72	0.250	53
Structural									
RRIM	1.5	37	63	172	24.9	4.2	12.4	1.80	12

The SMC molding process involves placing an SMC charge in a heated mold and compressing the material at pressures of roughly 5.5 to 6.9 MPa (0.80 to 1.0 ksi). Polymerization of the partially reacted matrix is completed in the molding process. A new generation of fast dual-acting presses, with accurately controlled parallelism, closing speed, and pressing force is currently available for production use. These presses have the potential to reduce press cycle time to under 1 min. Molded coating can be applied during the molding cycle to fill surface porosity and uniformly coat the surface for optimum surface appearance and quality.

Complex shapes and refined details that may be impractical with sheet metal are often easy to achieve with SMC. Parts may also be molded with varying thickness for local stiffness requirements. The design must be planned, however, so that all shape and detail is formed in a one-step molding process. While this can be an advantage over the multiple press-forming operations found in sheet metal fabrication, it must be understood that a molded SMC part cannot be reformed or flanged by a restrike operation, as is common with sheet metal parts.

It is critical that SMC parts be designed with sufficient draft to allow easy removal from the mold. Unlike parts made of more flexible polymeric materials, SMC parts cannot be designed with features that require parts to be

snapped or peeled out of the mold. The use of lifters and slides to achieve features that are common in injection molded parts is more restricted in SMC parts because of the requirement for placing the uncured SMC material into the mold before it is closed.

Reaction injection molding (RIM) materials are prepared by polymerizing monomers or prepolymers during the process of molding finished parts. Milled glass fiber can be added to obtain reinforced RIM (RRIM). Milled glass RRIM tends to have directional properties due to fiber orientation during injection. Substituting glass flakes for the milled fibers can greatly reduce directional differences. As in SMC materials, reinforcement is primarily in the local plane of the material, and not through the thickness. The RIM and RRIM materials described above do not have significant structural applications because their strength and stiffness properties are relatively low compared to those of SMC. Strength and stiffness properties similar to those of SMC can be obtained by placing fiber preforms in the mold before injecting and reacting the RIM matrix. This material is referred to as structural RRIM. Properties of typical RRIM and structural RRIM materials—commercial polyurethane and polyurea resins—are shown in Table 2.

The RIM process involves mixing the matrix reactants and injecting them into the mold. The reinforcing material can be injected with the

reactants or can be placed in the mold as a preformed mat before injection. Mold pressures are about 0.7 to 1.0 MPa (0.1 to 0.15 ksi) for RIM and RRIM, increasing to 1.4 to 2.0 MPa (0.20 to 0.30 ksi) for structural RRIM. Fast-reacting polyurea and prepolymer chemicals are being developed that shorten reaction times to 10 s or less. New equipment has been developed that provides precise control of temperatures, mixing ratios, and injection of filled and reinforced materials.

As with SMC, the RIM processes permit complex shapes, refined details, and variations in thickness that would be impractical in sheet metal. Again, this is a one-step process, and restrike operations are not possible. An advantage of the RIM process is that it can use lifters and slides to achieve features common in other injection molding processes. The lower stiffness of RIM and RRIM permit some degree of snapping or peeling the parts out of the mold. On the other hand, the lower stiffness reduces the applicability of these materials to meet many structural requirements unless the operation of fabricating a preform mat is added.

Filament Winding, Tape Laying, and Braiding. Specific fiber orientation is often a structural application requirement. Several combinations of matrix, fiber, and the fiber placement process have shown considerable commercial promise. Epoxy and polyester ma-

Table 3 Typical oriented fiber properties

	Specific gravity	Fiber, wt%	Resin, wt%	Tensile strength MPa	Tensile strength ksi	Flexural modulus GPa	Flexural modulus 10⁶ psi	Coefficient of linear thermal expansion, 10⁻⁶/K
Epoxy 0° S-glass	1.80	70	30					
longitudinal	...	...	...	849	123	40	5.8	...
lateral	...	...	...	47.0	6.82	11	1.6	...
Epoxy 0° AS Carbon	1.26	70	30					
longitudinal	...	...	...	966	140	138	20.0	−0.2
lateral	...	...	...	37.3	5.41	9.7	1.4	13.0
Epoxy ± 45° AS Carbon	1.26	70	30					
longitudinal	...	...	...	89.7	13.0	15.9	2.31	0.8
lateral	...	...	...	89.7	13.0	15.9	2.31	0.8

Table 4 Typical sheet metal properties

	Specific gravity	Tensile strength MPa	Tensile strength ksi	Elongation, %	Flexural modulus GPa	Flexural modulus 10⁶ psi	Coefficient of linear thermal expansion, 10⁻⁶/K
Steel	7.8	345	50.0	35	207	30.0	15
Aluminum	2.7	275	39.9	25	69	10	24

Table 5 Degree of SMC technology implementation

	Front end panels	X-11 hood	Corvette (1983-1984)	Wagon tailgate	Fiero
Mold coating developed	Partial	None	Total	Total	Total
Adhesive developments	None	None	Partial	Partial	Partial
Press controls added	None	None	Partial	Partial	Partial
Material developments	None	None	Partial	Total	Total
Part design criteria	None	None	Partial	Partial	Total
Tooling developments	None	None	None	Total	Total
Process automation	None	None	None	None	None

trices are common, as are glass, aramid, and carbon fibers. Properties for glass and carbon fiber reinforcements, and the effects of fiber orientation are shown in Table 3. Fiber placement techniques include filament winding, tape laying, and braiding.

In general, the processes for each technique are similar in that fibers are precoated with resin by the fiber manufacturer or are coated in the fabrication process just before the fiber placement operation. The parts are then cured, either in a mold or on a supporting surface. The fiber placement techniques are rapidly evolving from their initial low-speed low-volume processes to the point that products are now reaching high-volume commercial markets. An example of a high-volume process is the automated filament-winding machine for production of multiple driveshafts, in which three shafts are wound at one time in about three minutes on removable shuttles.

Complex shapes and varying thickness can be obtained with these processes. However, the design must be planned so that any supporting surface for the fibers can be removed after curing, or can remain as a part of the product. **Resin transfer molding (RTM) and squeeze molding** are low-pressure long cycle time processes that are generally used for low-volume commercial products. They can be thought of as low-productivity, low-pressure versions of the SMC and structural RIM pro-

cesses. They have the capacity for producing parts that approximate the properties of the higher productivity processes, and as such are very useful for producing prototype parts for evaluation.

Thermoplastic resins include any polymers that can be melted and injected under pressure into a mold. These resins include such a wide range of materials that it is difficult to describe them adequately in this forum.

Traditionally thermoplastic resins have not been reinforced with fibers because of processing constraints, although they have been modified with fillers to affect various properties. Thus, the physical properties of the product are more heavily influenced by the polymer properties than is the case with SMC, RRIM, or filament-wound materials. New thermoplastic compounds are being developed that are blends or alloys of polymers with the potential to optimize properties to meet the needs of a specific application.

Processing variables, such as temperature and pressure, are very dependent on the thermoplastic material used. Blends or alloys of thermoplastics can have relatively small processing windows, particularly if the individual constituents have dissimilar processing parameters. A major advantage of most thermoplastic materials is the ability to reuse scrap if it is not contaminated in postmolding operations. Disadvantages include the relatively low heat-distortion temperatures of most commercially

available thermoplastic materials, which also exhibit excessive thermal expansion rates compared to reinforced composites.

Sheet Metal. Steel has long been the preferred material for structural applications, although aluminum is a popular choice because of its light weight and relatively high physical properties. Although steel is more widely used, both steel and aluminum are very acceptable for both secondary structural and appearance applications. Typical sheet metal properties are shown in Table 4.

Sheet metal is generally received in coils, cut into blanks, and then formed to shape in one or more stamping presses. To obtain complex shapes, progressive dies are used to gradually form the final shape and then to remove excess materials. Fast cycle times foster very good productivity and favor high-volume applications. The actual processes for steel and aluminum differ somewhat because of differences in ductility. Disadvantages of sheet metal compared to composites include the relatively high cost of tooling, the difficulty in forming shapes that require excessive draw, inconsistent part shapes due to variations in material properties, the inability to provide significant changes in material thickness within a given stamping, and the need for corrosion protection.

Applications

The degree to which composites have penetrated the commercial market for structural applications has been affected by the rate of technology implementation. Table 5 shows the relative degree of SMC technology implementation in the vehicle program at the General Motors Corporation. Sheet molding compound grill and headlamp mounting panels (front end) have been used in GM products since 1969 and have provided a good method of obtaining a degree of product differentiation at a minimal tooling cost. These panels have pioneered the processing and postprocessing technology for more sophisticated applications.

The Chevrolet X-ll hood, introduced in 1980, was the first major flat panel on a GM car. It provided surface finish and design-for-appearance experience for many subsequent applications.

The current Corvette uses the design and processing criteria established by the front end

Table 6   Relative thickness, weight, and material cost

	Pure membrane resistance ($\alpha = 1$, t^1)			Oil-canning resistance ($\alpha = 2$, t^2)			Flat panel bending ($\alpha = 3$, t^3)		
	t	wt	Cost	t	wt	Cost	t	wt	Cost
Appearance SMC	1.0	1.0	1.0	1.0	1.0	1.0	1.0	1.0	1.0
Structural SMC	0.73	0.77	0.62	0.86	0.91	0.72	0.90	0.95	0.76
RRIM (flake glass)	6.8	4.5	6.0	2.6	1.7	2.3	1.9	1.3	1.8
Structural RRIM	0.94	0.78	0.78	0.97	0.81	0.81	0.98	0.82	0.82
Amorphous nylon	5.6	3.4	9.2	2.4	1.4	3.9	1.8	1.1	2.9
Nonappearance engineering thermoplastic ..	2.1	1.3	1.9	1.5	0.89	1.3	1.3	0.78	1.2
Glass-filled (30%) thermoplastic	1.7	1.3	3.0	1.3	1.0	2.3	1.2	0.92	2.1
Steel	0.06	0.25	0.12	0.24	1.0	0.51	0.38	1.7	0.82
Aluminum	0.17	0.22	0.49	0.41	0.62	1.3	0.55	0.83	1.6

Table 7   Material properties and costs for Table 6

Material	Modulus		Specific gravity	Cost	
	MPa	ksi		$/lb	$/kg
Appearance SMC..................	11.7	1.70	1.8	0.75	0.34
Structural SMC	15.9	2.31	1.9	0.60	0.27
RRIM	1.72	0.249	1.2	1.00	0.454
Structural RRIM..................	12.4	1.80	1.5	0.75	0.34
Amorphous nylon	2.07	0.300	1.1	2.00	0.907
Engineering thermoplastic	5.52	0.801	1.1	1.10	0.499
Glass-filled thermoplastic	6.90	1.00	1.4	1.70	0.771
Steel	207	30.0	7.8	0.38	0.17
Aluminum	69	10.0	2.7	1.44	0.653

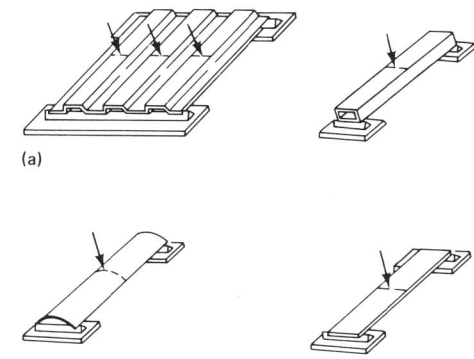

Fig. 1 Panel stiffness criteria. (a) $\alpha = 1$, pure membrane resistance. (b) $\alpha = 2$, oil canning resistance. (c) $\alpha = 3$, flat panel bending

Table 8   Typical values for automotive body panels

	α	Preliminary design requirement	Remark
Floor pan.............	1.0	Stiffness	
Rear compartment pan ...	2.5	Stiffness	
Motor compartment front panel	3.0	Stiffness	
Dash panel............	3.0	Stiffness	
Roof panel............	2.0	Stiffness	Appearance
Rear end panel.........	2.0	Stiffness	Appearance
Quarter inner panel......	3.0	Stiffness	
Quarter outer panel......	2.0	Stiffness	Appearance
Trunk lid inner panel	1.0	Stiffness	
Trunk lid outer panel	2.5	Stiffness	Appearance
Front fenders	2.5	Stiffness	Appearance
Hood assembly..........	2.5	Stiffness	Appearance

panel and X-ll applications. The Corvette now has a much higher surface quality and represents the first major overall secondary structure application using more productive processing.

A tailgate produced for the past several years for the full-sized Buick and Oldsmobile station wagons incorporates advanced design and tooling techniques. This two-piece bonded assembly, weighing 13.5 kg (29.7 lb), replaced a seven-piece steel tailgate weighing 19.5 kg (42.9 lb). It can be opened either as a normal door or drop gate, and meets all design load and deflection criteria for both. All GM General and Astro trucks have three-piece SMC doors that replaced seven-piece steel assemblies and feature modular hardware access.

The Fiero program uses the first composite-skinned vehicle operating at high production line rates. It shows considerable surface quality improvement over the already improved Corvette because of material development. The Fiero has also demonstrated the manufacturer's ability to implement major appearance changes with minimal investment by using many composite parts.

These are just a few of the composite applications that have been put in production. Similar process and material improvements in RIM have been made at GM, from the first RIM fascias, through RRIM fenders on the Sport Omega, to the Fiero. Filament-wound leaf springs were introduced in the 1982 Corvette, and are now being introduced on high-volume car and van programs. Other vehicle manufacturers have had similar experiences with varying degrees of success. All of these developments have shown the absolute need for

simultaneous consideration of material, process, design, and manufacturing techniques.

Design For Simultaneous Engineering

Although the multitude of composite materials and processes now available can make design process decisions difficult, most major companies have developed material selection and design criteria based on continuous improvement and past experience. This section features those points that must be considered when determining which materials are best suited for automotive body panels.

In general, the structural qualities of panels can be categorized by stiffness and strength requirements. The stiffness requirements provide the rigidity necessary to resist deflection, oil-canning, and buckling, and to ensure the desired vibration response. The strength requirements provide the needed failure resistance under service loading. In addition, certain panels contribute to crash resistance and energy management when a vehicle is tested under safety loading conditions.

In the case of body panels the equal stiffness requirement is selected as the only performance constraint. This can be justified for preliminary design evaluation because experience has shown that stiffness requirements are usually the most restrictive. A structural panel can be evaluated for strength after it has been designed for stiffness, and modifications for strength can be added.

Other experience has shown that the secondary structural panels contribute somewhat to the

global stiffness of a vehicle, but that local stiffness requirements predominate. The local stiffness is defined as the stiffness of a support-free region of the panel bounded by stiffeners or edge beams. Therefore, a typical body panel assembly could have one or more support-free regions. In the case of uniform-thickness design, the region with the lowest stiffness is used to determine the panel thickness for the entire panel assembly.

Panel stiffness is proportional to the modulus of elasticity, E, of the material under consideration, and can be approximated as:

$$E \propto t^{\alpha}$$

where t is the panel thickness and α is the stiffness parameter, which is a function of panel curvature and of the loading and boundary conditions imposed on the panel. The value of α for laterally loaded panels can range from $\alpha = 1$ for closed-box sections and corrugated panels governed by membrane stiffness, to $\alpha = 3$ for flat panels governed by pure bending stiffness. In general, all secondary structural applications have an α value between 1 and 3, because they have a combination of bending and membrane requirements. Figure 1 depicts the general conditions found in body panels.

Table 6 lists the relative thickness, weight, and material cost for several materials based on the three general stiffness schemes shown above. In this example the thickness and weight calculations have been normalized to SMC. Approximate material costs (not processing costs) and properties are listed in Table 7. Costs and material properties are approximate and will change with economic, process, and material developments.

To relate the above information to actual applications, Table 8 lists values for typical panel applications. The values can change with size and curvature as well as boundary and loading conditions. It should be noted that using a smaller α value results in a conservative material substitution design.

Preliminary evaluation of alternative materials can be made by comparing Table 6 and 8. After identifying the most likely materials, an in-depth investigation of processing and tooling costs and assembly alternatives should be conducted before finalizing the design and validating the design, process, and manufacturing scheme. In this context, manufacturing includes all postprocessing steps as well, including corrosion protection, joining technique, and final appearance coating.

Marine Applications

John Summerscales, Royal Naval Engineering College

THE MARINE USE of fiber-reinforced materials will be discussed in this article. This includes materials that have been engineered for a specific application or materials that are particularly suited for certain applications. The literature on the design of reinforced plastics for use in marine environments is extensive, and the use of composites in lightly stressed and unloaded components is adequately reviewed elsewhere.

Mine Warfare Vessels

The often tried and proven Royal Navy Ton-class minesweepers were of timber construction, but the escalating cost of this material led to the selection of glass-reinforced plastics (GRP) for a new class of mine warfare vessel. During the 1960s, extensive research was conducted on the use of glass-reinforced plastic for Royal Navy applications. In the late 1960s, the research was extended to include GRP ship construction (Ref 1).

A prototype GRP minehunter was built by Vosper Thornycroft. HMS *Wilton* was similar to the Ton class, but with the aluminum-frame double mahogany hull fabricated of glass-reinforced plastic (Ref 2-5). The design and fabrication of the hull is discussed in detail in Ref 5. The anticipated service life of the hull is 60 years.

As a consequence of the success of HMS *Wilton* and subsequent upon extended shock testing of simulated hull forms, 15 new Hunt-class mine countermeasure vessels (MCMV) were designed (Ref 6, 7). Vosper Thornycroft was contracted as the lead shipbuilder. Yarrow Shipbuilders established a second stream of MCMV building capacity. The experience of Yarrow in MCMV construction was recently published (Ref 8).

In the early 1980s, it was decided that the expensive Hunt-class MCMV should be supplemented by a less expensive vessel dedicated solely to minehunting. Vosper Thornycroft was contracted to design the single role minehunter (SRMH). HMS *Sandown* will be the first ship to enter service with the Royal Navy that has been designed with the aid of a computer.

Design of the ship included the following priorities: very quiet, yet powerful; precise maneuvering at slow speed; high shock resistance; low development risk equipment; and low cost. The hull differs from that of a Hunt-class MCMV in its use of a superior construction technique to eliminate the antishock bolted connections. A 40% weight saving and a 50% cost saving have been achieved for the hull structure (Ref 9-12).

The hull design and production of the Swedish MCMV, built by Karlskronavarvet, were recently described (Ref 13-16). A substantial program to investigate and optimize the strength of the component materials was carried out prior to the choice of a sandwich of 0°/90° woven-roving chopped-strand mat faces (E-glass fiber in isophthalic polyester resin) and a polyvinyl chloride (PVC) foam core. Production methods and economics were studied during the construction of a semiscale minesweeper during 1974.

The Intermarine Shipyard at La Spezia, Italy, was established in 1970 to specialize in GRP naval vessels. Their antishock monocoque GRP minehunter hull design evolved directly from a detailed dynamic analysis of underwater non-contact mine explosion phenomena, not from the analysis of ship structural response behavior under shock loading (Ref 17-19). The main hull girder is fundamentally a simple single-skin monocoque structure without any longitudinal or transverse reinforcements other than the main decks and main bulkheads. The bulkheads are spaced to limit the hull shell panel sizes. The antishock structural solution was intended to optimize the shock resistance of the hull and to ensure a high attenuation of the shock loadings on equipment and crew. Fourteen shock tests were performed on a ⅔-scale fiberglass minehunter hull section, using 40-kg (90-lb) TNT at 9.58 m (3 ft) from the planking of the ship at a 7.31-m (24-ft) depth to give a shock factor of 0.66. The strain gage instrumentation indicated that the tension induced in the section as a result of the repeated underwater explosion was within the elastic limit of the material.

The U.S. Navy recently planned an advanced MCMV fleet by contracting for 17 Cardinal-class 45.7-m (150-ft) minesweeper/hunter surface-effect ships with catamaran hulls (Ref 20-23). The material of construction was to be a sandwich material licensed from Karlskronavarvet. Unofficial estimates were that the vessel would be the largest fiber-reinforced plastic (FRP) structure to be built in the United States and would contain 340×10^3 kg (750 000 lb) of composites per vessel. The contract was cancelled after structural problems in shock testing of a section.

Future Large Ship Hulls: Size and Construction

Glass-reinforced plastic tankers, trawlers, and ferries of up to 80 m (260 ft) may be economically viable, although welded steel construction is likely to persist for vessels over 50 m (165 ft), except for minesweepers or corrosive-cargo carriers where cost is overridden by the special requirements (Ref 24, 25). The construction of GRP ships that are substantially longer than 100 m (30 ft) will become feasible only if (Ref 26):

- Material and construction costs are reduced relative to steel
- Laminate stiffness is substantially increased to use the potentially high specific strength effectively
- Large environmentally controlled laminating sheds can be eliminated through prefabrication

One researcher analyzed the feasibility of filament winding onto a hull-shaped mandrel in such a way as to cover the mandrel with fibers at a variety of angles and to yield a structure conforming to the contours of the mandrel (Ref 27). After suitable fiber placements were found, the feasibility was verified by winding a ¹⁄₄₈-scale hull-shaped model. It was concluded that it is possible to produce 60-m (200-ft) ship hulls by the filament-winding process.

Another investigation examined the problems and possible solutions associated with the filament winding of a 45-m (150-ft) long GRP ship hull (Ref 28). A review was conducted of winding machine and mandrel concepts, as well as potential material and structural requirements. A ⅕-scale model vessel was designed and a mandrel developed.

LeComte Holland BV produces a series of simple, versatile FRP landing craft by using vacuum-assisted injection molding (Ref 29). S-glass, carbon, and aramid fibers can all be used, but the resin system is limited to polyester. The entire hull, including longitudinal, transverse, and side frames; double bottom; engine girders; bulkheads; and tanks, is molded in one piece.

A single 9-m (30-ft) Spear-class patrol boat was constructed with Kevlar 49 aramid fibers instead of the usual glass-reinforced plastic (Ref 30). The specification was calculated on the basis of equal strength and stiffness. The aramid fibers were used in conjunction with chopped-glass strand mat and vinyl ester resin matrix. The structure was 20% lighter than the GRP version, giving an all-up weight saving of over 9%. Trials of the two boats demonstrated several advantages of the aramid boat:

● A 1.7-knot increase in speed at identical horsepower
● Fuel consumption down 20 L/h (5 gal/h) at full throttle
● A two-point decibel reduction in sound levels in the boat

The small waterplane area twin hull (SWATH), or semisubmerged catamaran, is claimed to have the seakeeping abilities of conventional craft that are eight to ten times the SWATH displacement. Twin torpedo-shaped submerged hulls are attached by narrow sidewalls to the main hull/platform so that the cross-sectional area at the water surface is minimized. One U.S. firm has developed methods of constructing these buoyant sections in glass-aramid-carbon laminates for reduced weight and cost (Ref 31).

Sonar Domes

Glass-reinforced plastic has been suggested as a suitable material for Asdic (now known as sonar) domes (Ref 32). These structures are almost constantly immersed in the sea, and are subject to slamming pressures in rough weather. The choice of resin and reinforcement is important, but an additional requirement for minimum void content in the laminate necessitated considerable research into the material and fabrication processes. Woven roving is preferred for strength, but is extremely difficult to manufacture without voids. The air can more easily be worked out of chopped-strand mat, but the materials have insufficient strength. Alternate layers of woven roving and chopped-strand mat could be used, but the performance is influenced by resin variation, monomer content, glass finish, and thixotropic agent (Ref 32).

Table 1 Comparison of sonar dome materials

Characteristic	Metal	Rubber	Glass-reinforced plastic
Attenuation and phase error (3-16 kHz)	Fair on truss-framed single-skin domes, but good on double-skin units	Good	Good, especially on frameless units
Shaping	Double curvature with framed single shells, but single curvature normal on double-skin units	Double curvature within certain limitations	Double curvature easily achievable, including reentrant shapes
Fairing	Single curvature fairing plates common	Single curvature fairing plates common	Double curvature fairing mouldings easily achievable
Installation	Basic, requiring no special considerations	Pressurization system necessary to maintain shape	Basic, requiring no special considerations
Self-noise	Fair, but deteriorates markedly if panels deform between framing or joints loosen	Good up to deformation levels	Good
Repair and maintenance	Single-skin units require no special facilities, but double-skin domes do.	Requires special facilities	Requires special facilities, but these are now available in most dockyards
Durability	Medium	Good	Good
Tooling cost	Moderate	High	Low
Unit cost	Moderate	High	Low
Resistance to damage	Fair	Good	Good

Source: Ref 33

In a recent comparison of the choice of material (steel, GRP, and rubber) for sonar domes, the conclusion was that glass-reinforced plastic is heavily favored, with clear advantages over more traditional materials (Table 1). It is, however, highly likely that the use of carbon and aramid fibers and hybrid (mixed fiber) composites will bring further improvements in the next generation of windows.

Submarine Structures

One of the earliest large GRP structures for marine use was the fairwater of the USS *Halfbeak* submarine (SS-352) (Ref 34). The fairwater was assembled and installed at the Philadelphia Naval Shipyard in late 1953 and placed into service in early 1954. The structure consisted of seven large subassemblies mounted on a large steel coaming. Early in 1965, the original fairwater sail was replaced with one of a more modern design. After 11 years of service, the material was found to possess average properties within the original specification requirements.

Submarine performance at periscope depth is limited by the excessive wake, vibration, and noise produced by periscopes, antennas, and masts. Attempts to streamline fairings have traditionally required a bulky structure to meet the imposed stress level. The U.S. Navy has examined the use of graphite-epoxy composites for submarine masts, utilizing the extreme stiffness and high strength-to-weight ratio of the material (Ref 35).

Submersibles

The possibility of using glass-reinforced plastic for pressure hull and buoyancy structures was investigated (Ref 36, 37). Tests were conducted on fiberglass cylinders and spheres. Hoop-wound reinforcement was recommended to optimize the strength for a given structural weight. The potential for improved designs using sandwich materials and carbon fibers was noted.

Submersibles for commercial operation down to 457 m (1500 ft) have been built using GRP pressure hulls (Ref 38). An earlier model had three seasons of operation in the North Sea, being launched to 366 m (1200 ft) and recovered in seastates up to State 6. This vessel completed over 400 dives with mission times of up to 14 h.

A third-generation remotely operated vehicle, *Solo*, has been developed by Slingsby Engineering Limited (Ref 39). Designed for a variety of inspection and maintenance functions in the offshore industry, it carries a comprehensive array of sophisticated equipment and is designed to operate at a depth of 1524 m (5000 ft) under a hydrostatic pressure of 15.2 MPa (2 ksi). Glass fiber woven roving is used in the construction of the pressure hull, chassis, and fairings. Unidirectional fabric is used in stressed and jointed areas, and chopped-strand mat is used in sections to be machined.

An Italian-built prototype civilian submarine for offshore work has been constructed with an unpressurized aramid-epoxy outer hull in place of glass fiber to allow greater payload and increased operational range and endurance (Ref 40). The new material offers a better combination of low weight with improved stiffness and impact toughness. The operational range at 12 knots has been extended by 2 h. It is anticipated that in future models the inner pressurized steel hull will be fabricated in carbon and aramid fiber composite.

Navigational Aids

The soft-plastic materials (GRP, polyethylene foam/polyurethane elastomer, and syntactic foams) are being used for the progressive replacement of existing steel buoys in the North Sea because of increasing concern of damage to vessels (Ref 41-43). Balmoral Glassfibre produces a comprehensive range of buoys and a light tower made of glass-reinforced plastic that can withstand winds to 56 m/s (125 mph). Buoys are available for a variety of tasks. Anchor mooring buoys supplied to the Egyptian offshore oil industry are believed to be the largest ever produced in glass-reinforced plastic (4-m, or 13 ft, in diameter and 15×10^3 kg, or 16.5 tons, of reserve buoyancy) and are used to anchor tankers of up to 300×10^6 kg (330 600 ton) capacity.

Offshore Engineering

Composite materials already provide a multitude of services in offshore hydrocarbon production, where they usually replace steel because of their lightness, corrosion resistance, and good mechanical performance. It has been proposed that submarine pipelines could be constructed with circumferential carbon fibers for resistance to external pressure and longitudinal glass fibers for lengthwise flexibility (Ref 44).

Drilling risers for use at great water depths are subject to compression and possible failure because the longitudinal resonant period of the disconnected riser is close to that of typical wave periods. The mass reduction on changing from steel to composite materials would significantly reduce the dynamic stress and therefore increase either the working depth or the safety of deep-water drilling. It has been reported that 15-m (50-ft) lines manufactured from carbon and glass fibers have a burst pressure of 168 MPa (25 ksi) (Ref 45). These lines have been effectively subjected to three successive drilling campaigns to 70 MPa (10 psi) from the North Sea rig *Pentagone 84*.

Self-righting, totally enclosed, motor-propelled survival craft and open lifeboats are manufactured in glass-reinforced plastic using fire-retardant resins (Ref 41, 42). The craft range in size from 6.2 m (20 ft) for 21 persons to 8.75 m (29 ft) for 66 persons. As part of the certification trials, the survival craft must withstand 30-m (98-ft) high kerosene flames and temperatures of 1150 °C (2100 °F). Throughout the fire test, the temperature inside the craft never exceeded 27 °C (80 °F).

A new type of rigid-hull, inflatable rescue boat has been introduced by LeComte (Ref 46). The deep V-hull is fabricated in one piece with the deck by vacuum-assisted injection molding of Aramat hybrid aramid-glass fabric around a polyurethane foam core. The boat speeds are above 25 knots, and a 25% weight saving is realized by using hybrids in place of glass fiber.

Hydrofoils

A comparative study of the benefit of foil weight savings was conducted in 1966 for the U.S. Navy experimental patrol craft hydrofoil (PCH-1) *Highpoint* (Ref 47). The study concluded that the achievable gains in performance justified a search for lightweight structural materials. Overall weight savings on the component were 24% for HY 130 steel, 36% for titanium alloy, and 44% for glass-reinforced plastic relative to HY 80 steel.

In the mid-1970s, the U.S. Navy introduced two applications of advanced graphite-epoxy composites: a hydrofoil control flap and a hydrofoil box beam element (Ref 48). The control flap was evaluated as a replacement for an existing steel flap on *Highpoint*. The box beam element was used for laboratory static and fatigue tests in both wet and dry conditions as a prelude to the design of a full-scale foil.

The initial results of a program to consider the use of advanced composites instead of steel in the stabilizing flaps of the Italian RHS 160 hydrofoil were recently reported (Ref 49). It was suggested that the flexural rigidity of the flap could be reduced without reducing the torsional stiffness by careful orientation of the fibers. The principal reinforcement was carbon fibers, but glass fibers were added to impart galvanic corrosion resistance.

Hovercraft

Hovercraft blades of hybrid carbon-glass fiber reinforced plastic have a successful in-service record in the hostile environment of seaspray with sand, which causes erosion and corrosion. These problems and the noise level can be reduced by lowering the propeller tip speed, but the blade length must be increased to retain the same thrust. The composite blade, developed by Dowty-Rotol to replace the 2.7-m (9-ft) SRN6 duralumin blade, consists of a polyurethane foam core covered with a GRP skin. The blade is stiffened with carbon fiber reinforced plastic and a polyurethane strip is used to protect the leading edge. The spars carry both the centrifugal and bending loads (Ref 50-56). The dural and plastic blades are identical aerodynamically and have similar torsional and flexural frequencies. The plastic blade weighs 12.7 kg (28 lb); the metal counterpart weighs 18.4 kg (40 lb). The measured stress margins against fatigue test results are also better with the plastic blade.

The textile reinforcement of elastomers is extensively used in the fabrication of the skirts

that confine the air cushion of hovercraft (Ref 57). Both the fabric and the fiber/rubber bond must resist fatigue and tearing as a result of flexure or abrasion. Bag and finger skirts and loop and segment skirts are the two major types. A third important system is the pericell skirt. The three designs, selection of materials and construction, and skirt deterioration are reviewed in Ref 57.

Passenger Ferries

The Norwegian company Brodrene Aa recently launched the *Hegelandsekspressen*, a 184-passenger commuter ferry. The 27-m (90-ft) catamaran has a beam of 9 m (30 ft) and a draft of 1.05 m (3.5 ft). Construction is PVC foam cored glass-reinforced polyester built on a timber skeleton mold. The structural laminate used for internal bulkheads is given a final coat of colored resin to provide a practical, cleanable, textured surface without adding finishing weight, thus allowing the vessel to achieve a maximum speed of 33.5 knots (Ref 58).

Italcraft has developed a 21-m (70-ft) craft with a new type of aramid/GRP hull with variable geometry to achieve very high speeds (Ref 59). The hull is believed to incorporate a series of integral hydraulically operated steps to optimize the planing surfaces to match the speed. Performance at higher speeds is thus dramatically increased. The prototype, to be fitted out as a motor yacht after trials, has demonstrated its 50-knot speed potential and 40% fuel savings. The concept is being developed for a high-speed passenger ferry able to accommodate 80 passengers at a cruising speed of 43 knots.

Powerboats

The *Aramid Arrow* is a new powerboat whose structural components consist entirely of composite materials based on aramid prepreg and honeycomb in a sandwich laminate (Ref 60). The boat is a copy of a conventional 7.9-m (26-ft) GRP *Nova II* craft, but weighs only 1497 kg (3300 lb), instead of 2000 kg (4400 lb). Fuel consumption is reduced from 60 to 42 L/h (16 to 11 gal/h) at the optimal cruising speed of 32 knots. Top speed is 42 knots.

The 1984 Round Britain Powerboat Race was won by *White Iveco* at an average speed in excess of 52 knots, with an average of 69 knots on one leg (Ref 61). The bottom of the hull consists of three double layers of glass mat/aramid on either side of 19-mm (0.8-in.) Contourkore. The hull sides have two double layers of reinforcement around a 9-mm (0.4-in.) core, and the deck is one double layer around a 12-mm (0.5-in.) core. Total weight of the 13-m (30-ft) hull and deck structure is 2100 kg (4600 lb).

A revolutionary cockpit cell was recently designed that will protect drivers of Grand Prix racing powerboats from impact injury and prevent them from being thrown from the boat (Ref 62). The cell would also keep a driver afloat with his head above water if he were unconscious. The project necessitated the development and supply of carbon and aramid prepreg fabrics, vacuum bagging materials, and autoclave molding facilities.

Racing Yachts

The 23.5-m (77-ft) ketch *Great Britain II* was launched in 1973 (Ref 63-67). The use of a mixed carbon-glass fabric faced Airex core sandwich was claimed to increase hull life because of the reduced vibration, which was achieved with a minimal increase in weight. In the 1973-1974 Whitbread Round-the-World Race, this yacht arrived home first, although it placed sixth overall on handicap. It then won both legs of the 1975-1976 Financial Times Clipper Race. In the 1977-1978 Whitbread Round-the-World race, it set a record of 134½ days for the event.

The yacht was converted from a ketch to a sloop rig for the 1981-1982 Whitbread race, which entered as *United Friendly* and completed in 141 days, 10 h. Having logged approximately 805 000 km (500 000 miles), it entered the 1985-1986 Whitbread race as *Norsk Data GB* and completed the race just 3.5 days outside its own record, after breaking its previous records on each of the first three legs.

The GRP ocean-racing catamaran *British Oxygen*, launched in 1974, had hull and nacelle moldings stiffened with hybrid tape. The 21-m (70-ft) boat won the 1974 Round Britain Race in a record time of 18 days, 4 h, 26 min, almost 37 h inside the previous record, despite encountering Force 8 gales (Ref 67-69).

The 24-m (80-ft) ocean-racing trimaran *Great Britain III* was constructed using 200 kg (440 lb) of pultruded carbon fiber section in the 12-m (40-ft) beams connecting the outriggers to the main hull. In each of the crossbeams, 40 strips of 35- × 5-mm (1.4- × 0.2-in.) pultruded carbon fiber reinforced plastic were used to form the flanges of a double box beam held together by lamination into a glass fiber reinforced polyester (Ref 70, 71).

A later ocean-racing trimaran, *Great Britain IV*, used strips of pultruded carbon fiber composite and a hybrid fabric of carbon and glass fiber to provide additional strength and rigidity in the bow section, deck panel, and dagger board. The yacht had a displacement of 3.6 × 10³ kg (3.9 tons) and an overall length of 15.8 m (50 ft). The crossbeams were fabricated from a unidirectional woven fabric of carbon fiber roving held together with a light glass weft. In its first event—the 1978 Round Britain Race—the boat finished first, despite a split in the fairing to the forward crossbeam, which opened up on the first leg and caused the starboard float to fill with water (Ref 72-74).

The *Brittany Ferries GB*, a 19.8-m (65-ft) trimaran weighing 6.6 × 10³ kg (7.2 tons), was constructed of carbon and aramid composites over an Airex foam core (Ref 75). In the 1981 Two-Handed TransAtlantic Race, it was the

first boat into Newport—17 h ahead of the aluminum-hull *Elf Aquitaine I*—in a record time of 14 days, 13 h, 58 min.

The *Colt Cars GB*, a 20-m (60-ft) trimaran, was launched in 1982 and built of carbon-aramid fiber reinforced epoxy sandwich with an aramid honeycomb core (Ref 76). Before starting the 1984 Single-Handed TransAtlantic Race, it had logged over 24 000 km (15 000 miles) with no structural problems, although the halyards parted in the 1983 Transat en Double. In the 1984 Race, it became dismasted, took on water, and was abandoned.

The *Elf Aquitaine II*, a revolutionary 18.29-m (60-ft) catamaran with a displacement of less than 4540 kg (10 000 lb) was the largest structure ever built in carbon fiber (Ref 76, 77). The materials in this boat were reported to have cost $128 000. Heated molds were used to cure the carbon fiber-epoxy prepreg hulls. The two hulls were connected by an X-beam structure to increase overall vessel torsional stiffness, with a setting wing mast and fully battened sails on a rotating sheeting table that pivots at the center of the crossbeam. The boat took third place on handicap in the 1984 Single-Handed Trans-Atlantic Race.

The world's largest racing catamaran, when launched in September 1983, was *Formule TAG*, sponsored by Techniques Avant Garde (Ref 78). The 24.4- × 12.8-m (80- × 42-ft) boat has hulls manufactured in aramid-epoxy prepreg to give a sailing weight of 8845 kg (20 000 lb). Designed for the Autumn 1984 Quebec-St Malo Race, it is perhaps the fastest sailing yacht in the world, having covered a record 524 nautical miles in 24 h at an average speed of 21.8 knots. It crossed the Atlantic in the 1986 Two-Handed TransAtlantic Race 17 h inside the prerace record, but 14 h behind the winner, *Royale*.

The 11-m (35-ft) contender for the British Admirals Cup, *Summer Wine*, was one of the lightest hulls built, at just 263 kg (580 lb) when lifted from the mold (Ref 79). An aramid-Divynicell sandwich construction was strengthened with unidirectional carbon fibers. The structure, which was built in an environment-controlled tent, will require 900 kg (2000 lb) of ballast to trim the boat to 9.3 m (30.5 ft). A carbon fiber rudder was used.

The 18.3-m (60-ft) trimaran *Apricot* won the 1985 City of Plymouth Round Britain and Ireland Race by a convincing 15.5 h (Ref 80, 81). It fully utilizes modern materials to obtain a faster shape. The weight of the boat is reduced and the hull line narrowed in the search for speed. The rig is a rotating wing mast constructed in aramid on a carbon fiber frame, which allows speeds of 7 knots to be obtained sailing on wing mast alone. Under full sail, *Apricot* is capable of 30 knots. The boat was the undisputed star of the 1985 TAG Round Europe Multihull Race, leading the entire field home 7 h ahead of the next boat. It placed second in the Formula 2 class and sixth overall in the 1986 Trophee des Multicoques and arrived third

overall in the 1986 TransAtlantic Race to win Class II.

The *Colt Cars GB II* was built by Mitsubishi Marine as an entrant for the 1985-1986 Whitbread Race. At the time, it was the world's largest monocoque (no space frame) composite yacht, but weighed 20% less than the contemporary maxis with a displacement of 30 000 kg (66 000 lb). Six skins of aramid unidirectional cloth were laid on a male mold, followed by an aramid honeycomb core and a balanced six skins of aramid cloth. The principal resin was SP1101 epoxy. The hull was vacuum bagged. The deck was fabricated with an aramid-carbon hybrid cloth with carbon fibers preferentially oriented to take the rigging loads. After Mitsubishi withdrew its sponsorship, the hull was sold and fitted out as *Drum*. The yacht capsized in the 1985 Fastnet Race after losing its keel, but was relaunched in less than 1 month for the Whitbread Race. On the first leg, 800 km (500 miles) out of Cape Town, the honeycomb between the skins failed in the port bow. The skin/core interface did not delaminate, but the core failed in shear. Repairs to *Drum* were effected with end-grain balsa glassed in place. In addition, the mast step showed signs of failure, which required fabrication of a steel framework. The repair added 352 kg (800 lb) to the weight of the yacht, but despite this handicap, the boat was placed eighth in the race with a corrected time of 122 days, 6 h, 19 min, 45 min less than the record set in 1985 (Ref 82-84).

The *Enterprise New Zealand* was built using a specially constructed machine to wrap the mold with aramid impregnated with vinyl ester resin so that the fiber was prestressed. The fiber was wound in diagonal layers at 30° to the horizontal, followed by a Divynicell foam core that was vacuum bagged to the mold and overwrapped with an outer aramid skin. The filament-wound hull deck, ring-frames, and keel floors are a single piece without mechanical joints. The 24.25-m (80-ft) boat, constructed for the 1985-1986 Whitbread Race, weighed 4082 kg (9000 lb), with a saving of 771 kg (1700 lb) (Ref 85).

The hull of UBS *Switzerland* was fabricated by placing three layers of aramid prepreg over a male mold to a thickness of 2 mm (0.08 in.). Aramid honeycomb was laid on, followed by seven layers of aramid fabric to produce a hull thickness of 400 mm (16 in.). The primary advantages resulting from the use of these materials are optimal weight distribution, torsional stiffness, excellent damage tolerance, and vibration damping (Ref 86-88). The 24.3-m (80-ft) overall length yacht weighed 2000 kg (4400 lb). UBS *Switzerland* was the first boat to cross the line at the end of the 1985-1986 Whitbread Race.

The 18.3-m (60-ft) Formula 2 trimaran *Paragon*, built using SP Systems materials, had to drop out of the 1985 Round Britain and Ireland Race, but won the 1986 Trophee des Multicoques at La Trinite, not just on handicap but boat-for-boat and three times in succession in the four-race competition (Ref 89, 90). The *Paragon* won the 1986 Round-the-Island (Isle of Wight) Race in a record time of 3 h, 55 min.

Souters at Cowes on the Isle of Wight launched *Ondine VII* in January 1986. This 24.3-m (80-ft) maxiracer was built of E-glass, aramid, and carbon fibers with a Divynicell core (Ref 91). The hull and deck were constructed together without being joined at the gunwhale. The two halves were joined along the deck and hull centerlines to avoid weaknesses at the gunwhale joint. The lightweight rudder was built by Hamble Composites.

A revolutionary 12-m (40-ft) boat, said to be made of composite materials, including graphite, was to be an American contender for the 1987 America's Cup competition in Western Australia. A Cray X-MP supercomputer was extensively used to design the sails and hull. Called *USA*, the boat reportedly featured many technical advances; the world's first front-mounted rudder and canardlike control surfaces were said to be incorporated into the project (Ref 92).

The developments in racing multihulls—particularly with respect to the choice of catamaran or trimaran, hull shapes, parasitic drag, computer-aided design, and rigs—are reviewed in Ref 93. The engineering principles of aircraft design have been developed into an integrated-structure structural analysis technique of multihull design. By using finite-element numerical stress analysis techniques, it is possible to optimize the balance of strength, stiffness, and weight and to avoid major stress concentrations by careful fiber placement. For example, in a design study for the *Seahawk 44*, 20 different possible laminates were considered. The net result was a 900-kg (1-ton) weight saving. For the same cost as a conventional construction, it was possible to use epoxy resin in preference to polyester.

The structural design of multihulls was recently addressed (Ref 94). Simple engineering calculations were performed for the preliminary design to provide a rough estimate of beam cross sections and the required laminates. A general purpose finite-element computer program, suitable for structural analysis of layered anisotropic materials, was used to represent the complex geometry of the structure and to analyze the design loads. Structures were optimized by choice of materials, fiber orientation, and thicknesses, with localized reinforcements added to provide uniform safety factors. The state-of-the-art design process is now limited by the lack of data on real, operational fatigue-load levels.

Pleasure Boats and Luxury Yachts

A mathematical model has been presented that predicts the pressure changes and resulting hull distortion due to temperature variation in a thin, closed shell (Ref 95). The model is specifically applied to the case of a *Laser* 4.3-m (14-ft) racing dinghy constructed with a longitudinally stiffened glass fiber hull and a sandwich construction deck supporting an unstayed mast. The model has been successfully used to analyze the causes of structural failure from launching and beaching such boats in Saudi Arabia.

Laser International has invested $1.5 million in development and tooling costs for a new, bigger boat, the 8.5-m (28-ft) Farr Design Group *Laser 28* (Ref 96). Inner and outer skins of dry aramid fabric are laid into a matched mold on either side of a PVC foam core to produce the deck. A slow-curing liquid resin is then injected through multiple entry ports, starting at the bottom of the mold and working upward.

The *Spectrum 42* cruising catamaran is built using exactly the same structural techniques and materials as the racing catamarans. All of the safety factors are increased to extend the life of the boat. The laminate is varied throughout the boat to achieve optimal strength and fiber orientation. Unidirectional fibers are used to transfer high stresses slowly into lightly loaded parts of the structure, avoiding the stress concentrations at boundaries between areas of high and low loads. Aramid is used throughout the hulls, on the inner skin, in combination with E-glass woven cloth in a triaxial configuration for maximum impact strength and stiffness. The glass is doubled on the bottom of the hulls, and a 300-mm (12-in.) wide strip of aramid is linked through to the outer skin along the keel. In the wide main crossbeam, carbon fibers are used for increased stiffness at minimum weight and are led across the full width of the boat and away into specially reinforced areas of the body. When production began, the boat was estimated to be 30% lighter than any production multihull of equivalent size (Ref 97).

The Guy Couach 2400 luxury motor yacht— 23.5 m (78 ft) long, 35×10^3 kg (34 ton) displacement—was the world's largest yacht made of aramid and glass composites (Ref 98). Hybrid fabric (a balanced warp-weft seven-harness satin weave with equal numbers of glass and aramid fibers stitched to a light chopped-glass strand mat) was used in polyester resin on a balsa wood core. The hybrid composite was specified to fulfill several criteria:

- Weight reduction and therefore lower purchase price and running costs
- Improved safety due to increased toughness
- Increased comfort due to the vibration-damping properties of the composites used
- Higher speed (28 knots)

The engineering of boat hulls is discussed in Ref 99 with a view to the optimal use of carbon fiber reinforcement. Five proposed sandwich laminates of differing lay-ups with S-2 glass fiber, carbon fiber, and core materials were tested in impact and flexural loading. It was suggested that the two-part hull (outer skin to keep water out, inner structure to take rigging loads and provide longitudinal strength and stiffness) is contrary to the established method of boat design and leads to inefficient use of composite materials. The boat should be considered a box section girder in which the skins are stressed and carry the principal loads. The deck and skins should be used as the primary load-carrying structure. A balanced laminate with hybrid carbon-glass faces and an Airex foam core was indicated as the optimal solution.

Composite Masts

The use of composite masts for racing yachts is expanding. The spinnaker pole of the racing yacht *Intrepid* was manufactured in graphite yarn/boron-epoxy/aluminum honeycomb at 18 kg (40 lb) against the all-aluminum design at 38.5 kg (85 lb). Boron fibers were included to improve compressive strength. The main boom of the yacht *Valiant* was fabricated of graphite-epoxy and glass-epoxy inner and outer skins, aluminum honeycomb, and a sail track of extruded aluminum. The composite boom showed a 40% weight saving compared to aluminum (Ref 35).

The *Freedom*-class yachts are instantly recognizable by their unique rig (Ref 66), which requires far fewer sails and sail changes. Instead of a normal stayed mast on which sails are attached to the mast and boom by a groove or tracks, the *Freedom* uses unsupported carbon fiber masts around which the sails are wrapped. The *Freedom 70* was entered in the 1981-1982 Whitbread Race. The unusual rig has led to measurement problems under the International Offshore Rule by which all Whitbread entrants must be rated.

The first wing mast on a racing yacht was fitted to *Elf Aquitaine II*, but the concept gained respectability when *Royale* won the 1984 Quebec-St Malo race. Royale was entered in the 1986 Carlsberg Two-Handed TransAtlantic Race, together with a second French Formula 1 catamaran *Elf Aquitaine III*, both carrying striking wing masts (Ref 100), the rotating booms of which are known in France as Balestrom. The carbon-aramid wing-masted British trimaran *Apricot* was also entered. *Royale* took line honors in a new record time, and *Apricot* won Class 2 and took third place. The huge wing sections are primarily constructed with carbon fiber and rotate on a ball-and-socket type joint. The wing mast allows a cleaner airflow over the sails, which increases the power of the rig. In strong conditions, they can sail under wing mast alone.

Laminated Sailcloths

Laminated sailcloths (Ref 101-105) have been available since 1964, but it is only since 1977 that modern production techniques and advanced design procedures have led to a steady and progressive development in performance. Unlike dinghy sails, windsurfer sails must have a built-in window for safety, they have little scope for fine adjustment of the aerodynamics by skillful use of the sheets, and they are frequently immersed.

The earliest board sails were made of polyester fabric to which a resin finish was applied to impart stiffness and sealing. Such sails sag and lose shape quickly and may even split. Stronger cloth and harder resins performed better, but increased the sail weight and inertia; handling was therefore clumsier. The next stage was the adoption of laminated sailcloth using polyester lightweight weave (open-weave scrim instead of the previous high-sett, close-woven material) bonded to a polyester film. The good results achieved were marred by the disadvantage that an edge notch could easily run across a loaded sail surface.

Another refinement was the use of a sandwich form of construction using a yet more open Terylene weave with the two surface Melinex PET films bonded through the fabric to produce a sail with a weight of just 180 g/m^2 (0.60 oz/ft^2). The penultimate refinement was use of a new scrim of even looser weave with strong 100 D-tex fibers in one direction and 550 D-tex fibers in the other direction. The larger threads are woven at 7 per inch with smaller fibers at 17 per inch to give elongated windows with substantial area for interfilm adhesion. This fabric is used in the highest-performance production sails, and a lightweight aramid-reinforced laminate variant was chosen for *Victory '83* in a recent America's Cup race. The 1986 season saw a further advance with even higher toughness, in which a squarewoven scrim of relatively lightweight thread has heavier threads incorporated at 25-mm (1-in.) spacing in both the warp and weft directions.

Advances in the design of sailboard sails are, of course, easily translated into improved performance for racing yacht sails. The new flexible-composite laminated sailcloths are still in their infancy and will improve considerably over the next decade as the shape and stress distributions are refined. The exploitation of this new technology is an exciting growth area for the application of engineering skills. A number of leading sailmakers have speculated about possible future developments (Ref 101). One of the principal favorites is the extensive use of aramid fibers in laminated yacht sails, especially for the tall high aspect ratio No. 3 genoa and for use when weight saving is significant. A critical factor will be the Offshore Racing Council rules limiting the use of aramid fiber. Hood Sailmakers has been involved in the development of polyethylene fiber for sailmaking to advance the aramid concept further, with higher modulus and ultraviolet light resistance.

REFERENCES

1. D. Henton, Glass Reinforced Plastic in the Royal Navy, *Trans. RINA*, Vol 109 (No. 4), Oct 1967, p 487-510
2. R.F. Beale, Selection of Glass-Reinforced Plastics for Large Marine Structures, *Br. Polym. J.*, Vol 3 (No. 1), Jan 1971, p 1-7
3. R. Dukes and D.L. Griffiths, "Marine Aspects of Carbon Fibre and Glass Fibre-Carbon Fibre Composites," Paper 28, presented at the First International Conference on Carbon Fibres, Plastics Institute, Feb 1971, p 226-231
4. R. Dukes, "Materials Requirements and Validation for Large Marine Structures in Fibre Reinforced Plastics," Paper 2, presented at the Symposium on Reinforced Plastics—Recent Advances in the Marine Field, Plastics Institute Reinforced Plastics Group, Sept 1972
5. R.H. Dixon, B.W. Ramsey, and P.J. Usher, Design and Build of the GRP Hull of HMS Wilton, in *Proceedings of the Symposium on GRP Ship Construction*, Royal Institution of Naval Architects, Oct 1972, p 1-32
6. A.J. Harris, The Hunt Class Mine Countermeasures Vessels, *Trans. RINA*, Vol 122 (No. 6), Nov 1980, p 485-503
7. R.J. Bowen, The "Hunt" Class MCMVs, *Armed Forces*, Vol 3 (No. 8), Aug 1984, p 308-311
8. A.J. Miller, Practical Aspects in the Construction of Large GRP Ships, Proc. Int. Conf. Polymers in a Marine Environment, Institute of Marine Engineers, London, Oct/Nov 1984. *Trans. IMarE Series C*, 1985, 97(Conf 2), Paper 2, p 7-14
9. New British Minehunter Ordered, *Janes Def. Week.*, Vol 1 (No. 2), Jan 1984, p 61
10. $1 Billion to be Spent by Royal Navy on Mine Warfare Resources, *Janes Def. Week.*, Vol 2 (No. 2), July 1984, p 52
11. New Royal Navy Minehunter Contract for Vosper Thornycroft, *Reinf. Plast.*, Vol 29 (No. 10), Oct 1985, p 270
12. Single Role Minehunter Ordered for the Royal Navy, *Combat Craft*, Vol 3 (No. 6), Nov/Dec 1985, p 203
13. C. Rouarch, A New Minehunter for the Swedish Navy, *Int. Def. Rev.*, Vol 17 (No. 9), 1984, p 1277-1279
14. J. Sjogren, C.G. Celsing, K.A. Olsson, C.G. Levander, and S.E. Hellbrat, "Swedish Development of MCMV—Hull Design and Production," Paper 3, presented at the International Symposium on Mine Warfare Vessels and Systems, Royal Institution of Naval Architects, June 1984
15. K.A. Olsson, "GRP-Sandwich Design and Production in Sweden: Developments and Evaluation," Paper M, presented at the International Conference on Marine Applications of Composite Materials, Society of Naval Architects and Marine Engineers, March 1986
16. K.A. Olsson, "GRP-Sandwich Design and Production in Sweden," Paper 3, presented at the International Conference on Polymers in Defence, Plastics and Rubber Institute, March 1987
17. M. Trimming, Marine Applications of Composites, in *Advances in Composite Materials*, G. Piatti, Ed., Applied Science, 1978, p 383-396
18. A. de Marchi, Italian Mine Countermeasures—An Update, *Janes Def. Week.*, Vol 1 (No. 2), Jan 1984, p 74-78
19. Monocoque GRP Minehunters, *Naval Forces*, Vol VII (No. II), special supplement, 1986
20. J.B. Chaplin, "The Application of Air Cushion Technology to Mine Countermeasures in the United States of America," Paper 19, presented at the International Symposium on Mine Warfare Vessels and Systems, Royal Institution of Naval Architects, June 1984
21. Composites Use by the Military to Triple, *Plast. World*, Vol 44, Feb 1986, p 10-11
22. C. Lodge, Anchors Aweigh for Construction of Largest FRP Ship, *Plast. World*, Vol 44, April 1986, p 9-10
23. D. Pike, Design of the Bell Halter Minesweeper Hunter, *Combat Craft*, Vol 3 (No. 3), May/June 1985, p 84-85
24. J. Guiton, "Glass Reinforced Plastic Ship Construction," MSc thesis, Newcastle University, 1968
25. J. Guiton, Production and Design Aspects of GRP Craft Over 90 ft in Length, *Plast. Polym.*, Vol 41 (No. 154), Aug 1973, p 187-199
26. C.S. Smith, Applications of Fibre Reinforced Composites in Marine Technology, in *Proceedings of the Conference on Composites—Standards, Testing and Design*, National Physical Laboratory, April 1974, p 54-69
27. J.L. McLarty, "Feasibility of Filament Winding Large Ship Hulls, McClean-Anderson report J2016, AD A125 771, Dec 1981
28. D.N. Chappelear, T. Aochi, and R.J. Milligan, "Filament Winding of a Ship Hull," Report LMSC-D945 402, AD A134 577, Lockheed Corporation, Oct 1983
29. Injection Moulding for Large Craft, *Ship Boat Int.*, Jan/Feb 1986, p 43-44
30. L.S. Norwood and A. Marchant, "Recent Developments in Polyester Matrices and Reinforcements for Marine Applications, in Particular Polyester/Kevlar Composites," Paper 6, presented at the Port and Coast Services Conference, 1980
31. D. Pike, SWATH Development at Lockheed, *Combat Craft*, Vol 3 (No. 5), Sept/Oct 1985, p 177
32. M.A. Cheetham, Naval Applications of Reinforced Plastics, *Plast. Polym.*, Vol 36 (No. 121), Feb 1968, p 15-20
33. Sonar Domes—The Merits of GRP, *Maritime Def.*, Vol 9 (No. 5), May 1984, p

192-195

34. N. Fried and W.R. Graner, Durability of Reinforced Structural Materials in Marine Service, *Marine Technol.*, Vol 3 (No. 3), July 1966, p 321-327

35. M.M. Schwartz, *Composite Materials Handbook*, McGraw-Hill, 1984, p 7.84

36. K. Hom, Fibre Reinforced Plastics for Hydrospace Applications, in *Mechanics of Composite Materials*, F.W. Wendt, H. Liebowitz, and N. Perrone, Ed., Pergamon Press, 1970, p 455-466

37. W.R. Graner, Reinforced Plastics for Deep-Submergence Application, *Ocean Eng.*, Vol 1, 1969, p 353-372

38. J.S. Tucker, Glass Reinforced Plastic Submersibles, *Trans. NECIES*, Vol 95, Jan 1979, p 49-59

39. Extensive Use of GRP for Tomorrow's Undersea Craft, *Reinf. Plast.*, Vol 27 (No. 9), Sept 1983, p 276

40. Composite Hull Increases Submarine's Range of Action, *Composites*, Vol 14 (No. 3), July 1983, p 314

41. M.J. Seamark, Marine Applications of GRP, in *Energy Business Centre Reference Book and Buyers Guide*, Sterling Publications Ltd, London, 1985, p 183-184

42. M.J. Seamark, Glass Reinforced Plastics (GRP) in Energy Related Industries, in *Diplomatic and Consular Yearbook*, 1985, p 174

43. M.J. Seamark, Plastics Benefitting the Navigational Aids of the Mid-80's, in *Parliamentary Yearbook*, Blakes Houses of Parliament Year Book Ltd, London, 1985, p 278-279

44. G. Bonavent and M. Peniado, Using Composite Materials in Offshore Applications, *Ocean Ind.*, Vol 14 (No. 4), April 1979, p 372-384

45. P. Odru and J.C. Guichard, "Drilling Risers for Great Water Depths: Advantage of Mass Reduction by Means of Composite Materials", SNIAS-852-430-101, N86-18454, Societe National Industrielle Aerospatiale, Oct 1985

46. Seagoing Rescue Boat Has Hybrid Glassfibre/Aramid Reinforced Hull, *Reinf. Plast.*, Vol 30 (No. 6), June 1986, p 173

47. A. Rufolo, Foil Weight Saving and Hydrofoil Performance, *Naval Eng. J.*, Vol 78 (No. 5), Oct 1966, p 905-913

48. A. Macander and A. Silvergleit, The Effect of the Marine Environment on Stressed and Unstressed Graphite/Epoxy Composites, *Naval Eng. J.*, Vol 89 (No. 4), Aug 1977, p 65-72

49. G. Caprino and I. Crivelli-Visconti, Design and Fabrication of Advanced Composite Hydrofoil Flaps, in *Proceedings of the First European Conference on Composite Materials*, European Association for Composite Materials, Sept 1985, p 421-427

50. New Developments From Dowty Rotol, *Eng. Mater. Des.*, Vol 13 (No. 11), Nov 1970, p 1343

51. W.J. Colclough and J.S. Russell, The Development of a Composite Propeller With a Carbon Fibre Reinforced Spar, *J. Roy. Aeronaut. Soc.*, Vol 76 (No. 733), Jan 1972, p 53-57

52. J.G. Russell, "Use of Reinforced Plastics in a Composite Propellor Blade," Paper 16, presented at the Conference on Designing to Avoid Mechanical Failure, Plastics Institute, Jan 1973

53. J.G. Russell, "Propellor and Turbine Engine Fan Blades From Glass and Carbon Reinforced Plastics, Paper presented at the Conference on Reinforced Plastics in Aerospace Applications, Plastics Institute, April 1973

54. G. Arthur, Composite Materials and Their Engineering Application, *Trans. NECIES*, Vol 90 (No. 5), May 1974, p 151-162

55. L.N. Phillips, Carbon Fibre Reinforced Plastics—The First Fifteen Years, *Plast. Rubber Int.*, Vol 3 (No. 6), Nov/Dec 1978, p 239-243

56. R. McCarty, "Fifteen Years Experience in Composite Propeller Blades," Paper 17, presented at the Conference on Advanced Technology in Materials Engineering, Cannes, Society for the Advancement of Material and Process Engineering, Jan 1980

57. E.R. Gardner, Marine Applications, in *Textile Reinforcement of Elastomers*, W.C. Wake and D.B. Wootton, Ed., Applied Science, 1982, p 197-223

58. Fast Ferries a Specialty, *Ship Boat Int.*, March 1986, p 37-39

59. Italcraft M78, *Boat Int.*, No. 7, 1985, p 91

60. Aerospace Composites Technology is Finding New Applications, *Reinf. Plast.*, Vol 27 (No. 8), Aug 1983, p 243-248

61. GRP Proves Its Worth in Tough-Going Powerboat Race, *Reinf. Plast.*, Vol 29 (No. 4), April 1985, p 106-107

62. Advanced Composites Are Helping to Make Power Boat Racing Safer, *Reinf. Plast.*, Vol 29 (No. 8), Aug 1985, p 226

63. Great Britain II, Designed by Alan Gurney, *Yachting World*, Vol 125 (No. 2772), July 1973, p 79-80

64. Carbon Fibres Chosen for Round-the-World Yacht, *Composites*, Vol 4 (No. 5), Sept 1973, p 195-196

65. D.W. Edgell, "Glass Reinforced Plastic Boat Building in the United Kingdom," Paper 21-D, presented at the 35th Annual Reinforced Plastics/Composites Institute Conference, New Orleans, Society of the Plastics Industry, Feb 1980

66. T. Jeffrey, Ed., "Whitbread Round-the-World Race 1981-1982 Official Race Program," IPC Transport Press, Aug 1981

67. S. Holmes, Resources: Materials: The Lightweight Heavyweight, *Insight*, Vol 5 (No. 69), Dec 1981, p 1902-1907

68. Hybrid Carbon Glass Tape, *Reinf. Plast.*, Vol 18 (No. 5), May 1974, p 125

69. "Carboform Hybrid Tape," Publication 72, Fothergill and Harvey Composites, 1976

70. E.M. Trewin, "Carbon Fibres—Performance and Versatility in New Applications," Paper 23, presented at the Tenth Congress, British Plastics Federation Reinforced Plastics Group, Brighton, Nov 1976, p 221-226

71. P.E. Morgan, E.M. Trewin, and I.P. Watson, Some Aspects of the Manufacture and Use of Carbon Fibre Pultrusion, in *Symposium on Fabrication Techniques for Advanced Reinforced Plastics*, IPC S&T Press, 1980, p 69-90

72. GB IV Trimaran for Chay Blyth, *Reinf. Plast.*, Vol 22 (No. 6), June 1978, p 190-191

73. "Official Programme: Royal Western/Observer Round Britain Race 1978," Ocean Publications, June 1978, p 37

74. E.M. Trewin, Carbon Fibres: A Powerful Polymer Reinforcement, *Shell Polym.*, Vol 3 (No. 2), 1979

75. J. Clarke, Partners for the Pond, *Yachting World*, Vol 133 (No. 2866), June 1981, p 70-75

76. "The Observer/Europe I Singlehanded TransAtlantic Race 1984," Sail Publications, 1984

77. Solo Challenge, *Yachting World*, Vol 136 (No. 2902), June 1984, p 56-62

78. Cat With Claws, *Yachting World*, Vol 136 (No. 2896), Dec 1983, p 25

79. T. Jeffrey, First of the Summer Wine, *Yachting World*, Vol 136 (No. 2907), Nov 1984, p 29

80. Racing Logbook: Round-Britain-Race, *Boat Int.*, No. 8, 1985, p 18-22

81. D. Glenn, Playing TAG, *Yachting World*, Vol 137 (No. 2919), Nov. 1985, p 71-75

82. T. Jeffrey, The Colt and the Lion, *Yachting World*, Vol 136 (No. 2900), April 1984, p 98-99

83. R. Fisher, *The Whitbread Round-the-World Yacht Race 1985-1986*, Robertsbridge, 1986

84. T. Jeffrey, Racing Round-Up: Drums Problems, *Yachting World*, Vol 138 (No. 3932), Feb 1986, p 24

85. Whitbread Update, *Yachting World*, Vol 136 (No. 2907), Nov 1984, p 92-95

86. Round-the-World Yacht Race Winner Features Aramid Fibres in Hull, *Reinf. Plast.*, Vol 30 (No. 7), July 1986, p 200

87. The Best May Win, *DuPont Mag.*, (European edition), No. 2, 1985

88. Focus on the Winners, *Yachting World*, Vol 138 (No. 3936), July 1986, p 64-66

89. R. Lean-Vercoe, Trophee des Multicoques, *Boat Int.*, No. 12, 1986, p 84-92

90. Racing in Brief, *Multihull Int.*, Vol 19

(No. 221), June 1986, p 135-136

91. D. Glenn, Best of British, *Yachting World*, Vol 138 (No. 3933), March 1986, p 60-65

92. US Fights to Regain America's Cup, *New Sci.*, Vol 111 (No. 1520), Aug 1986, p 24

93. J. Shuttleworth, Recent Developments in Racing and Cruiser/Racer Designs, *Multihull Int.*, Vol 18 (No. 214), Nov 1985, p 290-295

94. R.P. Reichard, "Structural Design of Multihull Sailboats," Paper P, presented at the International Conference on Marine Applications of Composite Materials, Society of Naval Architects and Marine Engineers, March 1986

95. J.F. Doyle and L. Hadley-Coates, Distortion of Thin Shells Due to Rapid Temperature Changes, *Ocean Eng.*, Vol 13 (No. 2), 1986, p 121-129

96. K.L. Pittman, Breaking the Old Moulds, *Sail*, Jan 1985, p 76-81

97. J. Shuttleworth, Spectrum 42 Production Cat, *Multihull Int.*, Vol 18 (No. 207), April 1985, p 91-94

98. "The World's Largest Yacht in Kevlar," Kevlar Case History, DuPont de Nemours—Geneva

99. E.W. Sponberg, Carbon Fibre Sailboat Hulls: Optimise the Use of an Expensive Material, *Marine Technol.*, Vol 23 (No. 2), April 1986, p 165-174

100. Carlsberg 1986 TransAtlantic Race Official Programme, *Seahorse*, 1986

101. B. Cracknell, J. McWilliam, B. Axford, K. Rose, E. Warden-Owen, and M. Relling, Laminate Sails, *Yachting World*, Vol 136 (No. 2900), April 1984, p 69-71

102. K. Rose, Lighter and Stronger, *Seahorse*, July/Aug 1984, p 23, 26

103. N. Thornton, Yarns From the Trade: Developments in High-Tech Sailcloth, *Yachts Yachting*, 22 March 1985, p 26-28

104. Chain Store Sailmaker, *Yachting World*, Vol 137 (No. 2916), Aug 1985, p 82-83

105. J. Daniels, Putting the Sail in Sailboard, *Plast. Today*, No. 25, Spring 1986, p 15-18

Sports and Recreational Equipment

William J. Spry, Consultant

MODERN COMPOSITE CONSTRUC-TION in recreational equipment and sporting goods has been used for over 100 years. The paper canoe manufactured in 1874 by E. Waters & Sons of Troy, New York, probably the first one ever built, was a very early example of modern composite structure in recreational equipment.

Construction proceeded over a wooden mold that was rabbeted to accept internal framing. Two kinds of paper were used: that made from Manila and that prepared from pure unbleached linen stock. Several sheets were laid on the mold, wetted, and then glued together. An analogy can be drawn directly to modern fiberglass cloth-resin construction. After drying, the hull was removed from the mold, waterproofed, and the woodwork finished. The thickness of the paper layers was controlled to provide adequate strength at various positions on the hull; this was done to make the canoe as light and as strong as possible. This variable-strength feature of composite manufacture provides a unique design freedom that is still used in modern composites.

Unfortunately, Waters & Sons burned down for the last time in 1901. The first paper canoe, the *Maria Theresa*, perished in a 1920 fire at the New York Canoe Club. It had been made famous by Nathaniel Bishop, who paddled it from Troy, New York, to Cedar Keys, Florida. The entire matter is an amazing testimony to the soundness of composite construction in recreational equipment, even with materials that would now be considered inadequate.

Laminated wood construction, or plywood, was later used in various forms for sport and pleasure boating, as well as in such items as tennis rackets and the delicate, glued flyrods constructed of tapered bamboo sections. More recently, the term composite has been associated with such materials as fiberglass, carbon fibers, silicon carbide platelets and whiskers, and aramid fibers. These materials have greatly broadened composite use in sport and recreational equipment by introducing filamentary materials with tensile strength versus weight and stiffness versus weight properties that

greatly exceed those of wood, steel, and other more common materials. All of these new materials are embedded in a resin matrix, such as an epoxy or polyester, for fiber alignment and stable form.

Composite construction permits a designer to vary the mechanical properties almost microscopically in any section of the composite. This is done by controlling the amount of fiber reinforcement inserted, as well as its direction, and by combining layers of different reinforcing fibers of varying elasticity and strength. The end result is a macroscopic body that can have nearly ideal stress-strain and strength properties for the application. This application-efficient construction, along with modern fiber reinforcements, markedly minimizes overall weight in recreational equipment.

The common component of these composites is the resin that binds the mass together. The early non-waterproofed glues were a severe constraint on paper canoes. With the availability of waterproof adhesives, the use of plywood in recreational equipment increased greatly. However, the same development problems remain true today. One of the most recent discoveries is the long-term failure of certain polyester resins used in pleasure craft, which led to severe blistering of the hull surface due to water absorption. The story of composite development in recreational equipment could also be expressed in terms of resin and adhesive development. The combination of modern resins with advanced filaments of great strength and stiffness has created new opportunities for composites in recreational and sporting goods, as well as other applications.

Most applications for new materials in recreational and sports equipment have thrived in less regulated, more individual activities. Organized sports of wide appeal, by their very nature, depend on constant playing conditions and on tradition and comparison with the past. For example, a latter-day Babe Ruth wielding a composite bat might not be accepted as a sports hero. Even in the rather individualistic arena of pleasure boating, fiberglass-resin hulls initially advanced at a slow pace because of the reluc-

tance of boaters to accept a material other than wood. This peculiar parameter of the recreational field is probably more important at the outset of new product introduction than any technical aspect of the problem.

The area in which composites have entered highly organized sports has been auxiliary safety equipment. Modern baseball and hockey helmets are good examples. Their light weight for an achieved impact protection supports the players without altering the competitive terms of the sport.

Fishing. Innovative use of composite materials in a more individual recreational sport can be found in the fishing rod. A good example of a refined manufacturing technique for producing high-quality composite rods is the Howald process used by the Shakespeare Company. Modern high-quality composite rods use a hollow tubular structure to minimize weight and to optimize the strength and sensitivity of the rod in sport fishing. All such tubular fishing rods are created around a removable metal mandrel, which forms the tapered inner diameter of the finished product. The taper and diameter of this cylindrical cone constitute the starting point of the design. On this inner mandrel are placed the various fibers that provide the strength of the fishing rod and the resin that bonds the fibrous structure together.

One application of the Howald construction begins with an inner layer of carbon fiber, wound spirally around the mandrel from the butt end to the tip of the rod. This continuous spiral of fibers dramatically improves the radial integrity, or hoop strength, of the rod blank by distributing pressure over a larger area. It virtually eliminates the collapsing-straw effect that can plague hollow rods constructed with a fiber direction (for example, rolled woven fibers) that does not take into account the actual stress distribution in a bending tube. A second layer of fiberglass fibers is then placed over the initial graphite layer and oriented longitudinally along the axis of the rod, which increases the sensitivity of the rod because the vibrations of the fish are readily transmitted in the direction of the fiber. The final function of these fibers is

Fig. 1 An inner layer of spiral wrapping of carbon filaments being applied to a carbon-fiberglass fishing rod. Courtesy of the Shakespeare Company.

to increase the overall bending strength of the fishing rod. In a premium rod, a third layer of longitudinal carbon fiber is added to the blank for further strength and controlled stiffness. The small, solid tip on such a rod becomes a light, very strong section of 100% carbon fiber embedded in resin.

The end result is a piece of recreational equipment in which material choice, material configuration, and overall shape maximize the strength and sensitivity of the rod at minimum weight, illustrating how modern composite materials and design permit optimization to suit the final purpose of the application. Such design freedom is simply not available with metal or wood.

Figure 1 shows the equipment used to produce high-quality composite rods. The machine starts at the butt of a rod mandrel and uses the maximum number of fiberglass or graphite rovings required by the design. These are then cut out or reduced in number in a preset manner as the rod blank is formed from butt to tip. Consequently, some of the fiber filaments run the full length of the mandrel, and others run only a partial length. Impregnation with resin, usually an epoxy, is carried out simultaneously. By altering the mandrel and changing the amount and locations of the fibers used in construction, fishing rods of different characteristics can be made for all segments of the market on a few basic pieces of manufacturing equipment.

Golfing. The major stress on a golf club occurs when the golfer misses the ball and hits the ground. Both the torsional and bending loads applied to golf club shafts, either in

mishap or in hitting the ball, must be considered in golf club design.

The construction of a golf club is similar to that of a fishing rod; however, the designer must pay more attention to torsional loading, and the requirement of "feel", or control, has an entirely different meaning. To golfers, the control of the club is tied to a "sweetness" upon impact plus the ease and reproducibility of directional control of the ball. The same type of manufacturing equipment used for fishing rods can be employed for golf clubs. This can be a significant advantage of composite technology to a manufacturer of several types of sporting goods. In comparing these two types of construction, the taper of the mandrel and the rate of fiber cutout are the primary variables to be altered.

To combat crushing loads, the manufacturing process begins with a close spiral wrap (to prevent the collapsing-straw effect), followed by a second spiral wrap that places the fibers at a lesser angle relative to the shaft axis. These layers form the effective core that handles torsional and radial impact loads. Here, the material of the wrap and the number of filaments in it control the off-center, springlike response of the golf club head to impact with the ball. The head should return to an appropriately neutral position when the ball leaves contact with the club in order to maintain optimal directional control. The inner windings, with resin applied, are enveloped by a layer of parallel fibers running along the axis of the shaft, which respond to flexural loads along the axis of the golf club. Using this method of construction, adjusting the thicknesses of the

helical and parallel fibers, and using a predetermined inside and outside diameter for the shaft, the designer has enough variables under his control to construct the most efficient golf shaft available. Again, it should be pointed out that material versatility is available by using layers of different reinforcing fibers.

Pole vaulting as a sport has never imposed any particular constraints on the construction or design of the pole. However, there was little change in the design or materials of construction until the advent of fiberglass composites. The poles had been made of bamboo, steel, or wood, but in the 1950s, fiberglass-resin poles were successfully introduced. These poles copied the earlier wood or metal poles without any pronounced use of empty cores or changing of diameters or fiber concentrations to improve the pole in a way comparable to that described for fishing rods or golf clubs.

The engineering problem involves the design of a very light, highly efficient tubular spring that is loaded by impact when the running vaulter places one end of the pole in the planting box beneath the vertical bars of the vault. The kinetic energy of running must be converted into a rotational energy that is sufficient to carry the vaulter to a vertical position and over the measuring rod. The basic vibrational period required for the pole to straighten from its initial loading must equal the time it takes for this quarter revolution of the vaulter to occur. The spring constant must therefore be designed to take the weight of the vaulter into account. Because the vaulter tries to maximize the kinetic energy by sprinting toward the vault, the pole needs to be as light as possible. Above all, it should not break.

This is the type of problem that, once thoroughly understood in engineering terms, makes it impossible to become a successful vaulter. This point may seem facetious, but the customer for recreational equipment must be approached with this matter in mind. The acceptance of an innovative piece of sports equipment is often achieved through performance factors recognizable by the athlete in a framework entirely different from that used by the engineer.

The strength of the composite fiberglass poles per unit weight was greater than earlier materials. Therefore, the poles were made lighter and were more flexible. This required vaulters to develop new techniques for their successful use. The early fiberglass poles were not yet designed for all the desirable requirements from the viewpoint of the vaulter, and this probably slowed their initial acceptance.

The requirements of the vaulting pole have led to a specialized form of fishing rod, rather than a pole that bends uniformly from end to end. The major manufacturers have recently redesigned the poles to be nonlinear springs, which are stiffer at the butt end than at the other. Carbon fiber has also been introduced in an attempt to reduce weight while improving stiffness. Fiberglass, with about five times the strength-to-weight ratio of steel, is roughly equal to steel in specific stiffness. Carbon fibers have about five times the specific strength and about five times the specific stiffness of steel. The resultant optimal pole is therefore likely to be a combination of these two fibrous materials in a suitable resin matrix.

Boating. The use of fiberglass-polyester composites to construct pleasure boats was one of the first extensive applications of modern composite technology in the recreational field. The many advantages of this construction material included its imperviousness to the fungus that produces dry rot in wood. This problem of dry rot had restricted the service lives of wooden boats to a few years, particularly in freshwater, unless they were very carefully maintained.

However, it is not as widely recognized that the relatively large-scale introduction of plywood to boats in the early 1950s occurred at about the same time. This composite construction is also substantially free of dry rot. However, with the exception of cold-molded plywood craft made of narrow, veneerlike strips, the shapes of plywood hulls are constrained to conical sections. The plywood craft are often hard chined; the Lightning day sailer is a well-known example. This design limitation seems to have been a factor that moved boat manufacturers to the current fiberglass mat and cloth construction, which made possible a much greater degree of freedom in hull shape. Another advantage was the ability to carry out major hull construction with unskilled labor, which significantly reduced manufacturing cost.

The use of fiberglass-reinforced plastic hulls also increased the interior volume of the boat.

Wooden hulls require strengthening with numerous interior frames and ribs that steal considerable usable passenger space. The added interior volume in a given length and weight of a fiberglass-resin hull greatly increased the market size and attractiveness of these products. This benefit contributed to the recreational boating boom of the 1970s and 1980s.

Finally, fiberglass-resin construction permits the strength of a hull section to vary in an appropriate and quasi-continuous manner, from very thick and strong near the bottom of the hull, or keel, to relatively thin near the top, or shear. The basic advantage of modern composite construction again emerges with the material designed to fit the stress loads expected on each portion of the overall structure.

In this recreational area, the design of a macroscopic material is perhaps best illustrated by sandwich construction, in which two strong layers of fiberglass cloth and mat are separated by a thick layer of closed-cell foam (usually of polyurethane or polyvinylchloride). This makes a very rigid, strong cross section for a hull. One English firm has carried this approach to the logical point of selling unsinkable keel boats. A typical offering is a 10.59-m (34.75-ft) craft with a 5800-kg (12 800-lb) displacement that is very strong and will not sink when full of water because of the flotation built into the composite hull itself.

A final example of modern composites used in boating is the composite sail. Because a sail is loaded only in tension, sail designers can widen their choice of reinforcing fibers. Aramid fibers, for example, offer excellent strength-to-weight and stiffness-to-weight ra-

tios when loaded in tension, but their properties under compressive loading are not as outstanding. There is also a design problem particular to sails because when in use, the tension may be either in line with or, equally, at an angle to the weave of the cloth. In the latter condition, the cloth stretches by reorienting itself along a bias and distorts the aerodynamic shape of the sail. This reduces the effectiveness of the sail in propelling the boat. In premium racing sails, this problem is minimized by bonding a polyester film to one or both sides of the aramid cloth. Of course, other material combinations are possible.

In short, modern composite construction has been applied in some degree to almost every modern sport. In traditional sports, such as baseball, the applications are directed more toward lightweight and effective safety equipment. In individual sports, such as fishing, the design flexibility and typically lower production costs are utilized to build equipment that provides superior performance. Few people who have used a light, well-designed carbon-fiberglass fishing rod would return to an older steel or bamboo version. Similarly, the wooden boat market, for all of its romance, has succumbed to plastic hulls, spars, and sails reinforced with fiberglass, carbon, or other strong and light fibrous material. The penetration has been so complete that the revolution is almost over in the sport and recreational markets. What remains, however, is significant improvement within the general fiber-resin composite concept as designers learn how to optimize the use of these materials to fit the distinct mechanical loads applied to various parts of sporting equipment.

Metal, Carbon/Graphite, and Ceramic Matrix Composites

Chairman: Walter L. Lachman, Materials International, Inc.
Co-Chairman: Stanley J. Paprocki, Material Concepts, Inc.
Co-Chairman: H. Dean Batha, Fiber Materials, Inc.

Introduction

ADVANCEMENTS IN operational systems and concepts planned by industry and government require performance characteristics that are beyond the established limits of currently available materials. There is every indication that the disparity between materials and design requirements will accelerate. Conventional metals, alloys, and ceramic materials cannot meet the requirements of future systems, and, because of their inherent properties, continued research and development will not render them capable of meeting these future demands. This has spurred the development of a family of new materials that are custom designed to extend the boundaries of performance relative to current limitations; specific properties, fatigue, operational temperature, and thermal stability will receive a dramatic boost in capability.

The most promising emerging materials are the carbon-carbon and metal matrix composite systems, most of which have extremely complex physical interdependencies that inhibit the processing of shapes, the effectivity of nondestructive examinations, and the ability to understand their response to various loads in order to derive the maximum design efficiency. At the present stage of development they are relatively high in cost and, in some instances, difficult to fabricate with a high degree of reliability. However, their continued development is essential because they offer the strongest promise of achieving the properties needed for planned advanced systems.

A renewed interest in ceramic composites has been stimulated by the realization that carbon-carbon composites will be extremely difficult to protect from oxidation (although recent efforts have focused on improving oxidation resistance at high temperatures), and metal matrix composites will have use temperature limitations that are below the level needed for high-performance engine applications. Early ceramic composite progress was discouraging because of the inability of the high-modulus matrix material to transfer the load to fibers. Extensive work with fiber-reinforced glass produced composites in which the lower modulus of the matrix, compared to that of the fibers, provided some reinforcement. Glass matrix composites show promise up to 1000 °C (1830 °F). Higher-temperature whisker-reinforced ceramic composites are being investigated to increase the toughness of the ceramic. The newer materials offer a selection of composites that can be used in turbines, structural components, and insulators when high reliability and performance in oxidizing high-temperature operating conditions preclude the use of conventional ceramics, which fail catastrophically.

Metal matrix composites consist of a metal base that is reinforced with one or more constituents, such as continuous graphite, alumina, silicon carbide, or boron fibers and discontinuous graphite or ceramic materials in particulate or whisker form. In the case of the continuous fiber reinforced composites, the fiber is the dominating constituent, and the metal matrix serves as a vehicle for transmitting the load to the reinforcing fiber. Composites that incorporate discontinuous reinforcement are matrix dominated, forming a pseudo dispersion hardened structure. Optimum properties can be achieved in continuous fiber reinforced composites when the fibers are oriented in one direction. Composites reinforced with whiskers or particulates tend to be isotropic.

Significant improvements are anticipated in the properties of fibers, especially graphite. Cost reduction and process flexibility improvements are also expected in the processing of these materials.

In some cases, the articles included in this Section overlap to a greater or lesser extent in their subject coverage. This is unavoidable, in view of the immaturity of the technology. Individual suppliers have developed different processes to produce similar materials and/or obtain similar properties. At this stage in the development of these materials, it is impossible to predict which approach will prevail. Under these circumstances, the best approach is sometimes to err on the side of redundancy.

Conversely, it is also necessary to recognize the inevitability of gaps, voids, and blind spots in any attempt to synthesize an overview of an emerging technology. Boundaries must be drawn to define the coverage, and this process is by nature arbitrary. (An example is the decision to omit dispersion strengthening and related techniques from the coverage of MMCs, on the basis of their use of microstructural rather than macrostructural reinforcements.) Another arbitrary constraint involving subjective judgment was the decision to include only those materials that had reached a point of demonstrable practical application. Yet another constraint, this one externally imposed, is the fact that information on some of these materials is restricted by proprietary policies and/or the provisions of the Export Control Act and International Traffic-in-Arms Regulations, and for that reason had to be omitted here.

Continuous Boron Fiber MMCs

Michael E. Buck and Raymond J. Suplinskas, Avco Specialty Materials Division, Textron Inc.

BORON, THE FIRST HIGH-STRENGTH, high-modulus reinforcing fiber to be used in metal matrix composite applications, has been investigated for use in both aluminum and titanium matrices. Production of boron-reinforced aluminum composites has been moderately successful, and a large data base has been developed for physical and mechanical properties. In the production of boron-reinforced titanium composites, boron fiber is exposed to severe processing environments that degrade its strength and stiffness. Surface coatings/diffusion barriers have been added to the fiber to solve this problem, but the composite has never achieved the commercial realization of the boron-aluminum composite. For this reason, this article focuses primarily on boron fiber reinforced aluminum matrix composites.

Rather than provide a complete history of boron fiber development, which already exists in the literature (Ref 1, 2), this article describes the chemical vapor deposition process used to manufacture boron fibers, fiber properties, metal matrix preform manufacture, composite properties, and applications.

Boron Fiber Manufacture and Properties

Boron filament is unique among composite reinforcement fibers available in production quantities in that it combines superior tensile, compressive, and flexural strengths, high modulus, and low density in one fiber. It has a tensile strength that exceeds 3.45 GPa (0.500 $\times 10^6$ psi), a compressive strength of about 6.9 GPa (1 $\times 10^6$ psi), a modulus of 400 GPa (58 $\times 10^6$ psi), and a density of only 2.5 g/cm^3.

Boron, which is the fifth element on the periodic table, does not occur freely in nature. Therefore, it must be isolated from boron-containing compounds to be made into a usable fiber form. Several fiber-forming processes have been investigated. These include the thermal decomposition of diborane, the drawing of fibers from molten boron, and the chemical vapor deposition (CVD) of gaseous boron compounds. Of these, the CVD process has been found to be the most economical and technically feasible.

Chemical Vapor Deposition Process. A diagram of the basic deposition unit is shown in

Fig. 1. The reactor module consists of a glass deposition tube that is fitted with gas inlet and outlet ports, two mercury-filled electrodes, a variable dc power supply connected to the two electrodes, a tungsten substrate pay-out system, and a boron filament take-up unit. Tungsten substrate, typically 12.5 μm′ (0.0005 in.) in diameter, is drawn through the reactor and heated through resistance heating by the dc power supply. Before entering the deposition reactor, the tungsten substrate is passed through a short "cleaning" stage (not shown in Fig. 1) in which the substrate is heated to candescence in a hydrogen atmosphere in order to remove surface contaminants and residual lubricants that are used in the drawing of tungsten wire. A stoichiometric mixture of boron trichloride and hydrogen is introduced at the top of the reactor. At about 1300 °C (2370 °F), a mantle of boron is deposited on the tungsten by means of the reaction:

$$BCl_3 + 3/2H_2 \rightarrow B + 3\ HCl \qquad (Eq\ 1)$$

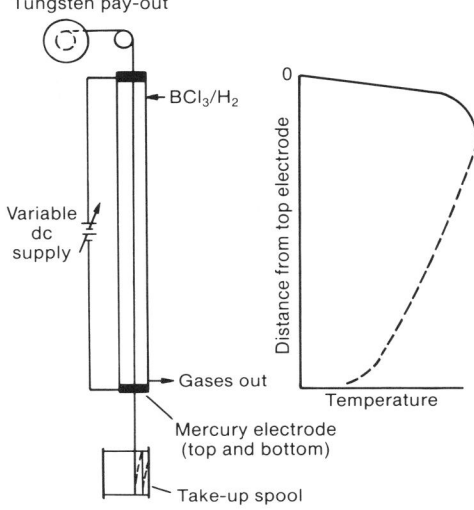

Fig. 1 Boron filament reactor and temperature profile

Fig. 2 Four banks of boron filament reactors

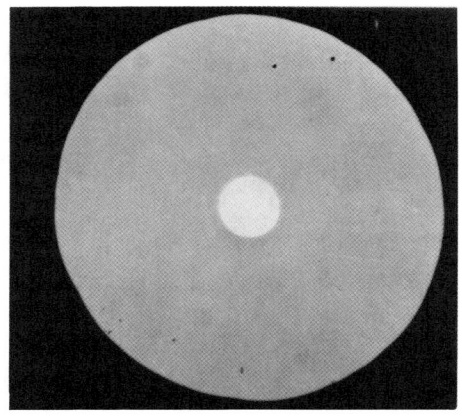

(a)

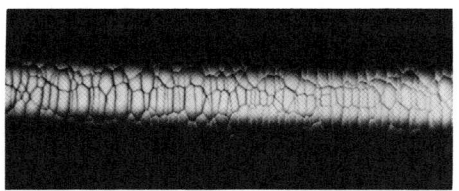

(b)

Fig. 3 A 100-μm (4-mil) boron filament. (a) Cross section. (b) Longitudinal photomicrograph

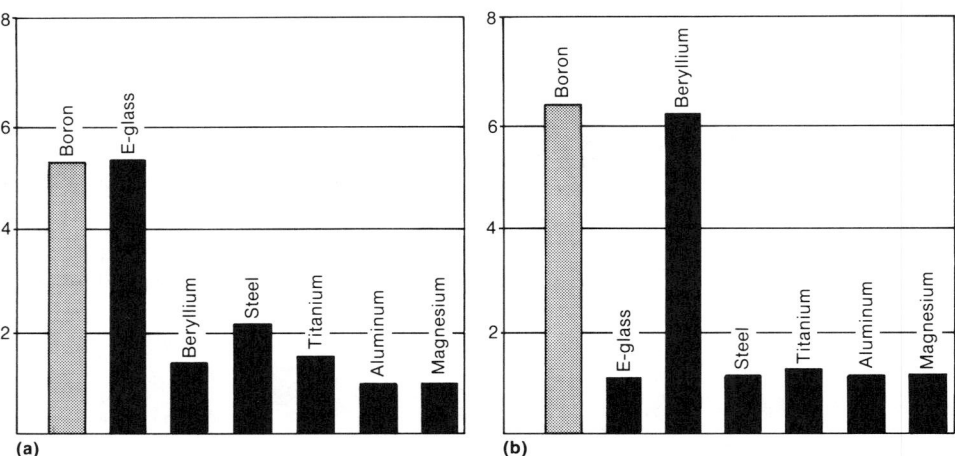

(a)   (b)

Fig. 5 Relative strength/weight ratios (a) and relative stiffness/weight ratios (b) of boron filament

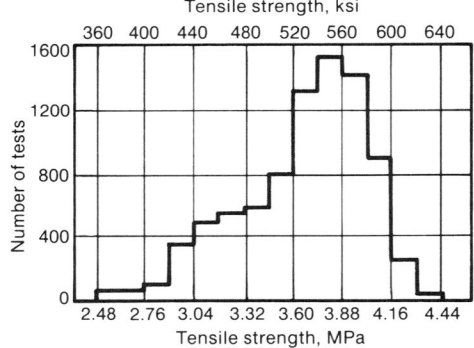

Fig. 4 Typical histogram of boron filament tensile strengths

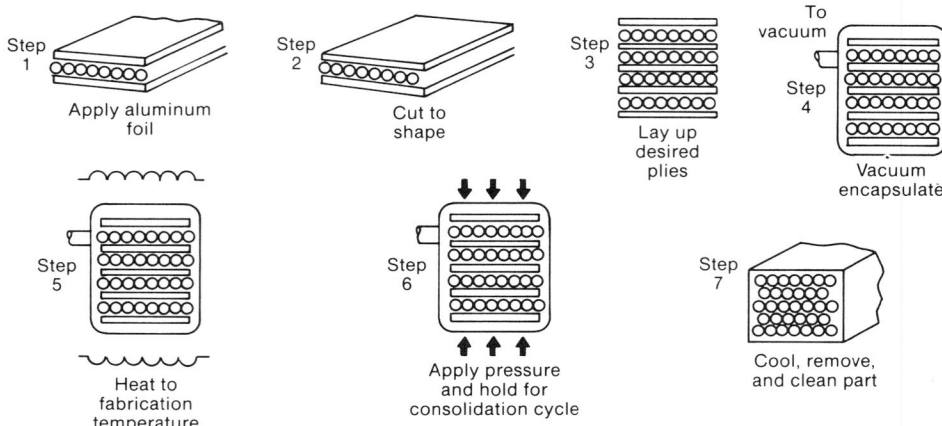

Fig. 6 Typical fabrication process for boron-aluminum composites

Exhaust gases consisting of HCl, intermediate species, and unreacted H_2 and BCl_3 are removed through the outlet port at the bottom of the reactor and are processed to remove and recycle the unreacted BCl_3. A photograph of about 100 boron filament reactors is shown in Fig. 2. Standard diameters of the filament exiting from the reactor are 100 μm (0.004 in.) and 140 μm (0.0056 in.). The filament diameter is changed by varying the drawing rate. Figure 3 shows photomicrographs of cross-sectional and longitudinal views of a 100 μm (0.004 in.) boron filament.

Boron Filament Modifications. Several variations of the standard boron filament have been investigated in an effort to improve temperature resistance, wettability (in metal matri-

ces), and mechanical properties, and to decrease cost. These variations involved adding surface coatings of B_4C or SiC, oxidizing away surface flaws, and replacing the tungsten core with a carbon core.

The B_4C and SiC coatings significantly improved wettability and decreased reactivity of the fiber in the matrix material. Because of the increased cost of this fiber and the development of the potentially lower-cost higher temperature resistant silicon carbide fiber, demand for boron fibers with these surface coatings decreased to such a point that they are no longer commercially available.

J. DiCarlo and T. Wagner (Ref 3) at NASA developed a laboratory postproduction treatment of the boron fiber that significantly reduced the severity of surface flaws on the fiber. They were thereby able to produce fibers with an average ultimate tensile strength of 5.5 GPa (0.800×10^6 psi) and a coefficient of variation of 5%. This postproduction treatment has never been evaluated on a commercial basis.

In an effort to reduce the overall cost of the boron filament, work was conducted at Avco to remove the present high-cost tungsten core and replace it with a lower-cost carbon core. This work was successful, but the inclusion of the larger-diameter carbon core, 33 μm (0.0013 in.), versus 13 μm (0.0005 in.) for the tungsten core, required a subsequent increase in the final boron fiber diameter to maintain an equivalent amount of boron in the cross section. For example, a 107-μm (0.0042-in.) diam carbon core boron fiber was needed to maintain the same boron content in the fiber as an equivalent 100-μm (0.004-in.) diam tungsten core boron fiber. Interest in this new, larger-diameter fiber, though less expensive, was minimal, and work was therefore discontinued.

Fiber Properties. The strength of boron fiber is determined by the statistical distribution of flaws produced during the deposition process. The major flaw types are: voids near the tungsten diboride core/boron mantle inter-

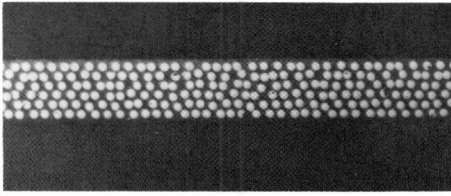

(a)

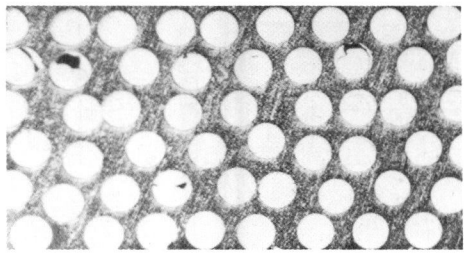

(b)

Fig. 7 Micrographs of cross section of boron-aluminum composite. (a) 8×. (b) 40×

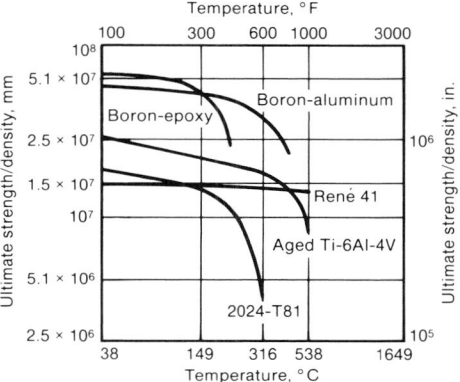

Fig. 8 Strength of axially reinforced boron composites versus metals. Source: Ref 4

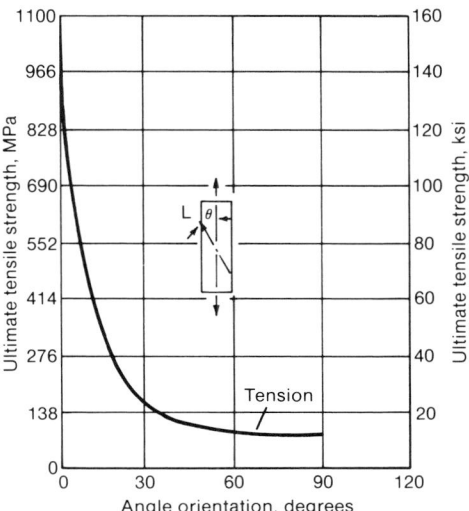

Fig. 9 Tensile strength at room temperature for off-axis loading of boron-aluminum composite. Source: Ref 6

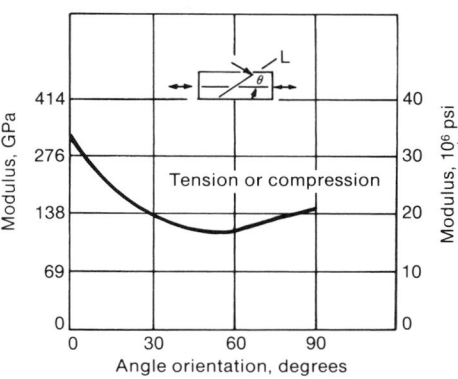

Fig. 10 Elastic modulus at room temperature for off-axis loading on boron-aluminum composite. Source: Ref 6

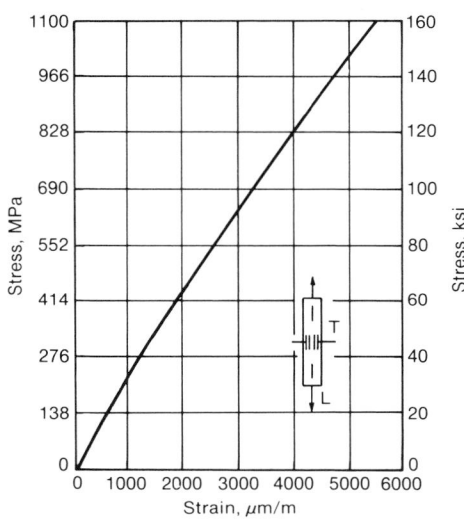

Fig. 11 Typical stress-strain curve for boron-aluminum in longitudinal tension at room temperature. Source: Ref 6

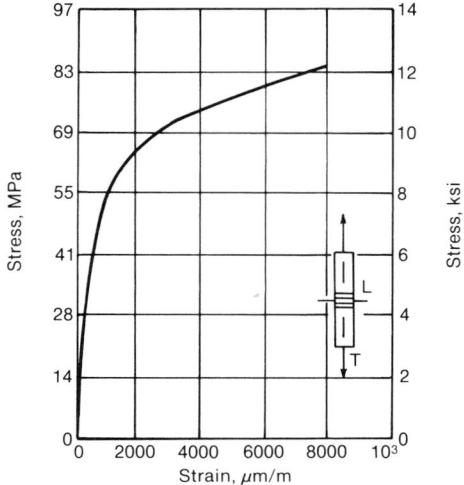

Fig. 12 Typical stress-strain curve for boron-aluminum in transverse tension at room temperature. Source: Ref 6

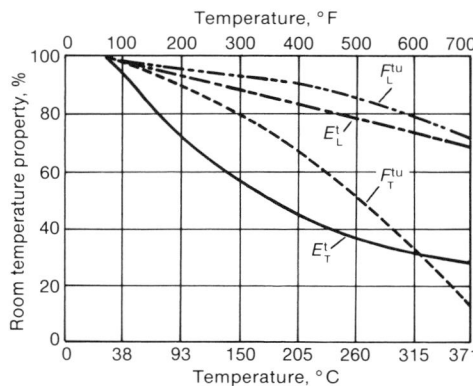

Fig. 13 Strength and modulus versus temperature of boron-aluminum (0°) in longitudinal and transverse tension. Source: Ref 6

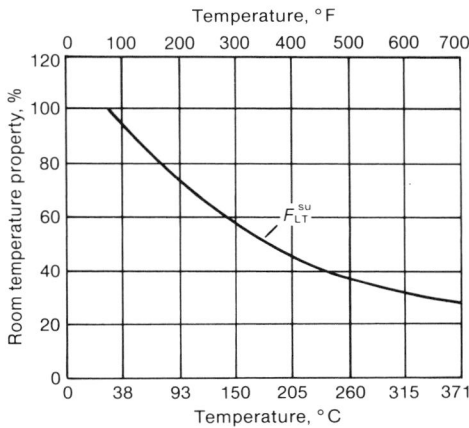

Fig. 14 Strength versus temperature of boron-aluminum (0°) in in-plane shear. Source: Ref 6

face, internal stresses "locked in" during deposition, and surface flaws, principally crystalline or nodular growth (Ref 1). A histogram of fiber strengths does not follow a normal distribution but, rather, is skewed in having a low-strength tail. The distribution is better described by Weibull statistics. Figure 4, a histogram of typical boron fiber strengths, shows an average ultimate tensile strength of about 3.6 GPa (0.520 × 10[6] psi) with a coefficient of variation of about 15%. These values have shown steady improvement over the 15 years that these fibers have been in commercial production. Initial values showed average strengths of 3.1 GPa (0.450 × 10[6] psi) with a 20% coefficient of variation.

Fiber modulus, on the other hand, shows very little variation. It is best regarded as a composite property, depending on the (reacted) substrate and the volume fraction of pure boron in the fiber. For 100-μm (4-mil) boron on tungsten, the value is 400 GPa (58 × 10[6] psi). Figure 5 compares the strength/weight and stiffness/weight ratios of boron fibers to various materials.

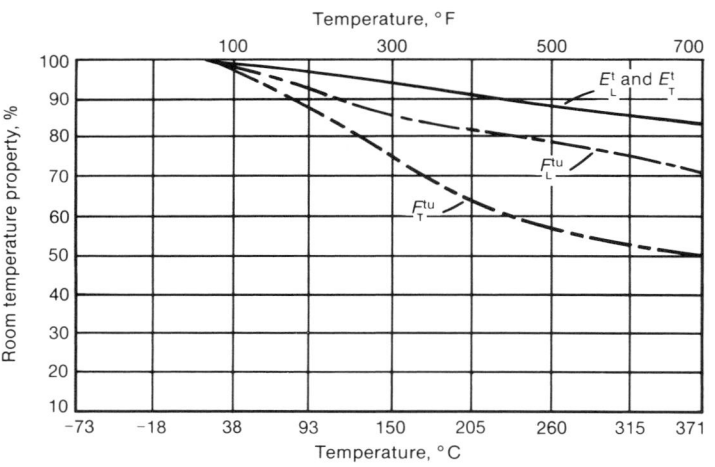

Fig. 15 Strength and modulus versus temperature of boron-aluminum (0°/90°) cross-ply in longitudinal and transverse tension. Source: Ref 6

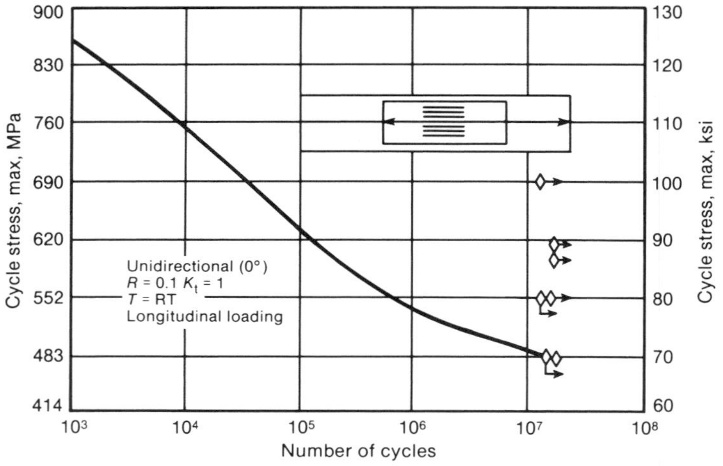

Fig. 16 Constant amplitude fatigue of boron-aluminum in longitudinal loading. Source: Ref 6

Composite Processing

A preform consisting of boron filament and a metal foil is normally used to make a metal matrix composite. The basic process consists of hot-pressing an array of fibers between metal foils. At elevated pressures, the foils deform around the fibers and bond to the fibers and to each other. This preform can then be laid up to form structures. Variations of the above process include plasma spraying of metal on the fiber array to make the preform, and the step-pressing of continuous boron-metal preform tape. In the latter, discrete segments of fiber-foil sandwiches are sequentially diffusion-bonded, producing preform that is continuous in the fiber direction. These preforms are typically fabricated using 140-μm (0.0056-in.) diam fiber. Figure 6 shows a typical fabrication scheme to produce boron-aluminum composites from preform.

Currently, the high-pressure diffusion-bonding-type process is the only commercial fabrication technique used in the manufacture of boron-reinforced MMCs. Other processes, such as casting or powder metallurgy, have been examined. Because of the degradation of mechanical properties of boron at elevated temperatures (uncoated boron fiber will begin to degrade in an aluminum matrix at ~525 °C, or 980 °F), none of these high-temperature low-pressure processes is technically feasible without the expensive B_4C or SiC surface coatings that were previously described. Also, the development of the newer high-temperature fibers has removed the impetus to develop newer, less costly processes.

Boron-reinforced MMCs typically contain approximately 50 vol% filament, although a range from 20 to 60 vol% has been successfully fabricated. Sheet sizes ranging from 1 ply to 100 plies, and from 305 mm × 610 mm (1 ft × 2 ft) to 760 mm × 9200 mm (2.5 ft × 30 ft) have been produced. Figure 7 shows a representative cross section from a six-ply unidirectional boron-aluminum composite plate.

Table 1   Room-temperature properties of boron-aluminum (0°) with 50 vol% filament (Ref 6)

Design strengths			
Longitudinal tensile ultimate	MPa (ksi)	F_L^{tu}	1100 (160.0)
Transverse tensile ultimate	MPa (ksi)	F_T^{tu}	110 (16.0)
Longitudinal compression ultimate	MPa (ksi)	F_L^{cu}	1215 (176.0)
Transverse compression ultimate	MPa (ksi)	F_T^{cu}	159 (23.0)
In-plane shear ultimate	MPa (ksi)	F_{LT}^{su}	69 (10.0)
Interlaminar shear ultimate	MPa (ksi)	F^{isu}	126 (18.3)
Ultimate longitudinal strain	μm/m or μin./in.	ϵ_L^{tu}	5000-6000
Ultimate transverse strain	μm/m or μin./in.	ϵ_T^{tu}	6000-12 000
Plastic properties			
Longitudinal tensile modulus	GPa (10^6 psi)	E_L^t	235 (34.0)
Transverse tensile modulus	GPa (10^6 psi)	E_T^t	138 (20.0)
Longitudinal compression modulus	GPa (10^6 psi)	E_L^c	207 (30.0)
Transverse compression modulus	GPa (10^6 psi)	E_T^c	131 (19.0)
In-plane shear modulus	GPa (10^6 psi)	G_{LT}	66 (9.5)
Longitudinal Poisson's ratio	...	ν_{LT}	0.23
Transverse Poisson's ratio	...	ν_{TL}	0.17
Physical constants			
Density	g/cm³	ρ	2.7
Longitudinal coefficient of thermal expansion	μm/m/°C (μin./in./°F)	α_L	5.8 (3.2)
Transverse coefficient of thermal expansion	μm/m/°C (μin./in./°F)	α_T	19.1 (10.6)

Table 2   Room-temperature properties of cross-ply boron-aluminum (0°/90°) with 50 vol% filament (Ref 6)

Design strengths			
Longitudinal tensile ultimate	MPa (ksi)	F_x^{tu}	483 (70)
Transverse tensile ultimate	MPa (ksi)	F_y^{tu}	483 (70)
Longitudinal compression ultimate	MPa (ksi)	F_x^{cu}	607 (88)
Transverse compression ultimate	MPa (ksi)	F_y^{cu}	607 (88)
In-plane shear ultimate	MPa (ksi)	F_{xy}^{su}	103 (15)
Interlaminar shear ultimate	MPa (ksi)	F^{isu}	96 (10)
Ultimate longitudinal strain	μm/m or μin./in.	ϵ_x^{tu}	6700
Ultimate transverse strain	μm/m or μin./in.	ϵ_y^{tu}	...
Elastic properties			
Longitudinal tensile modulus	GPa (10^6 psi)	E_x^t	145 (21)
Transverse tensile modulus	GPa (10^6 psi)	E_y^t	145 (21)
Longitudinal compression modulus	GPa (10^6 psi)	E_x^c	145 (21)
Transverse compression modulus	GPa (10^6 psi)	E_y^c	145 (21)
In-plane shear modulus	GPa (10^6 psi)	G_{xy}	...
Longitudinal Poisson's ratio	...	ν_{xy}	...
Transverse Poisson's ratio	...	ν_{yx}	...
Physical constants			
Density	g/cm³	ρ	2.7
Longitudinal coefficient of thermal expansion	μm/m/°C (μin./in./°F)	α_x	...
Transverse coefficient of thermal expansion	μm/m/°C (μin./in./°F)	α_y	...

Table 3   Axial tensile strength of 140 μm (5.6 mil) boron-aluminum at various fiber levels (Ref 6)

Matrix	Percent boron	Ultimate tensile strength		Elastic modulus		Strain to failure, %
		MPa	ksi	GPa	10^6 psi	
2024F as fabricated 45	45	1287.5	186.7	202.1	29.3	0.775
	47	1420.7	206.0	222.1	32.2	0.795
	52	1721.0	249.6	. . .	. . .	. . .
	54	1798.6	260.8	. . .	. . .	. . .
	64	1527.6	221.5	275.9	40.0	0.72
	66	1739.2	251.6	. . .	. . .	. . .
	70	1927.6	279.5	. . .	. . .	. . .
2024-T6 46	46	1458.7	211.5	220.7	32.0	0.810
	64	1924.1	279.0	275.9	40.0	0.755
6061F 48	48	1489.7	216.0	. . .	. . .	. . .
	50	1343.4	194.8	217.2	31.5	0.695
6061-T6 51	51	1417.2	205.5	231.7	33.6	0.735

Composite Properties

The primary advantage of a boron-reinforced MMC over its boron-epoxy counterpart is the maximum operating temperature to which the former can be exposed. For example, boron-aluminum offers useful mechanical properties up to 510 °C (950 °F), whereas an equivalent boron-epoxy composite is limited to about 190 °C (375 °F). Figure 8 compares the specific tensile strengths of several materials as a function of temperature.

As with all composites, boron-reinforced MMC mechanical properties depend directly on the lay-up sequence. Properties parallel (longitudinal) to the fiber direction are dominated by the fiber, while those perpendicular (transverse) to the fibers are dominated by the matrix material. In a typical application, both ply orientation and number of plies vary across the final part. For this reason, the mechanical properties of composite materials are anisotropic. Because the fibers dominate in the longitudinal direction, those mechanical properties are quite high, while the mechanical properties in the transverse direction are much lower. As would be expected, the matrix alloy and heat treatment have little effect on the longitudinal properties but can cause a great deal of variation in the transverse direction.

Figures 9 and 10 show the effects on tensile strength of off-axis loading of unidirectional composites. Figures 11 and 12 show typical stress-strain curves for unidirectional boron-aluminum composites in the longitudinal and transverse directions. It is interesting to note the almost perfectly elastic behavior of the composite in the fiber-dominated longitudinal direction, as opposed to the matrix-dominated transverse direction. Tables 1, 2, and 3 list typical mechanical test and design data that are available on boron-aluminum composites. Note the quasi-isotropic properties given for the longitudinal and transverse directions for the cross-ply laminate, shown in Table 2.

The mechanical properties of the composite materials are dependent upon environmental temperature. Figures 13, 14, and 15 show tensile strength, tensile modulus, and shear strength for unidirectional and cross-ply laminates. These figures show that the properties are affected less in the longitudinal fiber direction than the transverse matrix direction. This indicates that for higher-temperature applications, the composite material is typically matrix sensitive.

Because of their anisotropic characteristics, composite materials exhibit very complex fatigue failure mechanisms. Under cyclic loading, MMC materials do not behave like metals. Their fatigue failure is typically characterized by extensive damages throughout the specimen rather than by a predominant, single crack, which is often the failure mechanism observed in most brittle, isotropic materials. The forms of these damages are typically matrix cracking, delamination, fiber breakage, and interfacial debonding. Figures 16 to 19 give some basic fatigue data for boron-aluminum composites. Cross-ply laminates generally show a gradual reduction of strength until failure. This contrasts with the unidirectional laminate, which shows hardly any strength change until immediately before failure. This loss of strength and stiffness in cross-ply composites is due to internal damage and can occur at loads that are well below the endurance limit.

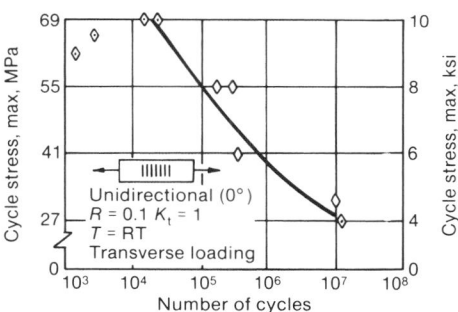

Fig. 17 Constant amplitude fatigue of boron-aluminum in transverse loading. Source: Ref 6

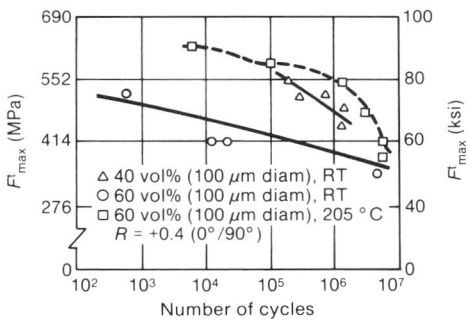

Fig. 18 Longitudinal tension-tension fatigue stress-strain curves of boron-aluminum (0°/90°). Source: Ref 6

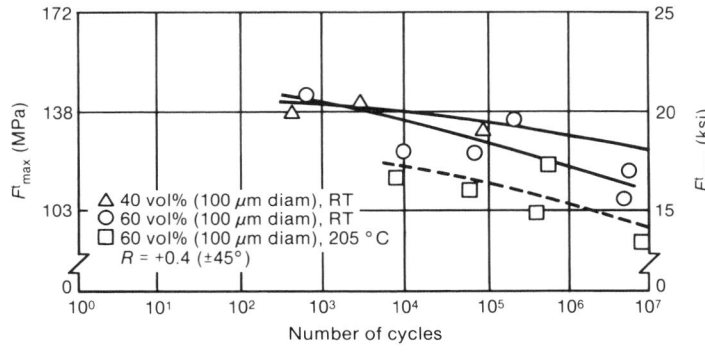

Fig. 19 Longitudinal tension-tension fatigue stress-strain curves of boron-aluminum (±45°). Source: Ref 6

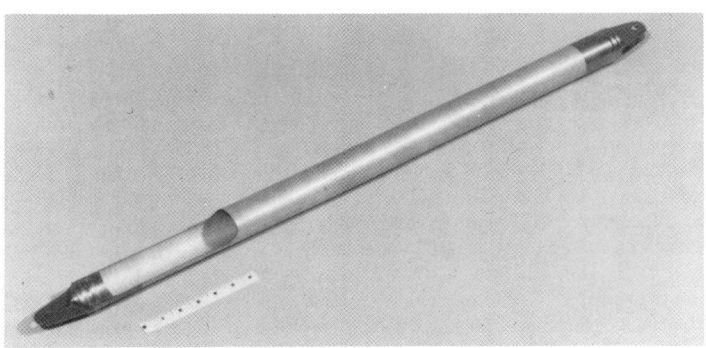

Fig. 20 Boron-aluminum structural tube member used on space shuttle

Fig. 21 Partially completed space shuttle orbiter, showing boron-aluminum tubes in mid-fuselage structure. Source: Ref 4

Fig. 23 Boron-aluminum bicycle frame

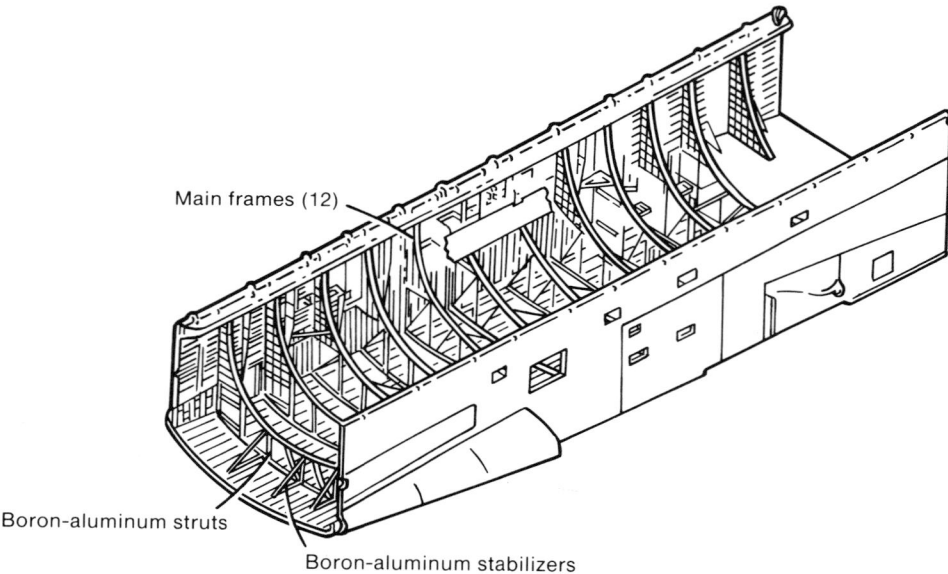

Main frames (12)

Boron-aluminum struts

Boron-aluminum stabilizers

Fig. 22 Schematic of mid-fuselage structure of space shuttle. Source: Ref 4

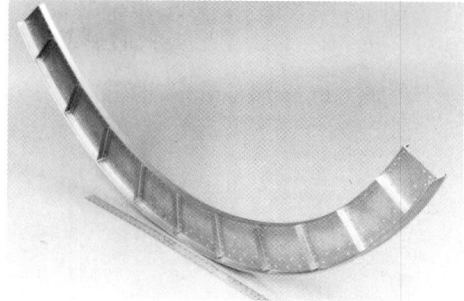

Fig. 24 Curved boron-reinforced aluminum extrusion

Applications

Applications with operating temperatures greater than 175 to 204 °C (350 to 400 °F) have led to the use of boron in metal matrix composites. In contrast to organic matrix composites, boron-reinforced MMCs are characterized by their high strength and stiffness (in tension, compression, and bending), light weight, high thermal conductivity, low coefficient of thermal expansion, and relatively high maximum operating temperature.

The primary boron-aluminum application was for structural tubular struts used as the frame and rib truss members in the mid-fuselage section and as the landing gear drag link of the space shuttle orbiter. In this application, several hundred tube assemblies were fabricated with titanium collars and end fittings for each of the shuttle orbiters. The tube assemblies varied from 25 to 67 mm (1.0 to 2.625 in.) in diameter and from 609 to 1850 mm (24 to 73 in.) in length. The boron-aluminum tube construction resulted in a 44%

weight savings over the original aluminum design. Photographs and drawings of this are shown in Fig. 20 to 22.

Boron-aluminum is also currently being sold as a heat dissipator/cold plate material for multilayer board microchip carriers. This application uses the high thermal conductivity and low thermal expansion properties of boron-aluminum. Heat that is generated by the closely packed semiconductor chips is conducted away by the boron-aluminum composite material. Because boron-aluminum has a coefficient of thermal expansion closer to that of the chips themselves, this places a much lower load on the weld joints (which hold the chips to the substrate material) due to thermal expansion than would a regular copper or aluminum cold plate. This greatly decreases the fatigue of the joints, which dramatically increases their life. Boron-aluminum composites also offer the advantage that their properties may be tailored to fit the application.

A potential application for which boron-aluminum composites are currently being examined is as a neutron shielding material. Elemental boron has a naturally high neutron absorption cross section. Preliminary tests conducted at Brookhaven National Laboratories have shown that high-strength boron composites absorb more neutrons than does an equivalent commercially available material. This will allow designers to reduce the weight and volume of the shielding material, while, at the same time, dramatically increase its strength and stiffness. Some potential applications for this material lie in the areas of spent-fuel transportation casks, portable shielding, spent-fuel storage pools, and control rods.

Other applications for which boron-reinforced MMCs have been evaluated include: jet engine fan blades, aircraft wing skins, structural supports, landing gear components, bicycle frames, and golf club shafts. Figures 23 and 24 illustrate applications for which boron-aluminum has been evaluated.

REFERENCES

1. H.E. Debolt, Boron and Other High Strength, Low Density Filamentary Reinforcing Agents, in *Handbook of Composites*, G. Lubin, Ed., Van Nostrand Reinhold, 1982
2. P. Bracke, H. Schurmans, and J. Verhoest, *Inorganic Fibers and Composite Materials, A Survey of Recent Developments*, Pergamon Press, 1984
3. J. DiCarlo and T. Wagner, "Oxidation-Induced Contraction and Strengthening of Boron Fibers," NASA Technical Memorandum 82599, National Aeronautics and Space Administration, 1981
4. J.D. Forest, *Boron/Aluminum Tube Constructions for Advanced Vehicle Applications*, General Dynamics, Convair Division, March 1975
5. C.J. Hilado, Ed., *Boron Reinforced Aluminum Systems*, Volume 6, Materials Technology Series, Technomic, 1974
6. "Advanced Composites Design Guide," Contract F33615-74-C-5075, Rockwell International, Flight Dynamics Laboratory, Wright Patterson Air Force Base, Sept 1976
7. "*DoD/NASA* Advanced Composites Design Guide," Contract F33615-78-C-3203, Rockwell International, Flight Dynamics Laboratory, Wright Patterson Air Force Base, July 1983

SELECTED REFERENCES

- J.W. Brantley and R.G. Stabrylla, "Fabrication of J79 Boron/Aluminum Compressor Blades," NASA CR-159566, National Aeronautics and Space Administration, 1979
- V.J. Krukonis, "Boron Filaments," chapter 28 in *Handbook of Fillers and Reinforcements for Plastics*, J.V. Milewski and H.S. Katz, Ed., Van Nostrand Reinhold, 1977
- D.L. McDaniels and R. Ravenhall, "Analysis of High-Velocity Ballistic Impact Response of Boron/Aluminum Fan Blades," NASA TM-83498, National Aeronautics and Space Administration, 1983
- C.T. Salemme and S.A. Yokel, "Design of Impact-Resistant Boron/Aluminum Large Fan Blades," NASA CR-135417, National Aeronautics and Space Administration, 1978
- R.J. Suplinskas, "High Strength Boron," NAS3-22187, National Aeronautics and Space Administration, 1984
- R.J. Suplinskas and J.V. Marzik, Boron and Silicon Carbide, in *Handbook of Fillers for Plastics*, H. Katz, Ed., Van Nostrand Reinhold, 1987

Continuous Silicon Carbide Fiber MMCs

John A. McElman, Avco Specialty Materials Division, Textron Inc.

SINCE THE ADVENT of high-strength, high-modulus, low-density boron fiber, the use of fibers produced by chemical vapor deposition (CVD) in high-performance composites has been well established. Although best known for its use as a reinforcement in resin matrix composites (Ref 1, 2), boron fiber has also received considerable attention for its use in metal matrix composites (Ref 3-5). A boron-aluminum composite was used for tube-shaped truss members to reinforce the Space Shuttle Orbiter structure and has been investigated as a fan-blade material for turbofan jet engines.

There are drawbacks, however, to the use of boron in a metal matrix. For example, boron fiber reacts rapidly with molten aluminum (Ref 6), and the mechanical properties of diffusion-bonded boron-aluminum degrade over the long term at temperatures greater than 480 °C (900 °F). Consequently, boron fibers cannot be used either for high-temperature applications or for fabrication methods, such as casting or low-pressure high-temperature pressing, that might be more economically feasible. These drawbacks have led to the development of silicon carbide (SiC) fiber.

SiC Fiber Production Process

Continuous SiC filament is produced in a tubular glass reactor by a two-step CVD process on a resistively heated carbon monofilament substrate (Fig. 1). It should be noted that the process and products described in this article are those of Avco; other companies also produce SiC fibers. During the first step, pyrolytic graphite (PG) approximately 1 μm (39 μin.) thick is deposited to smooth the substrate and enhance electrical conductivity. In the second step, the PG-coated substrate is exposed to silane and hydrogen gases. The former decomposes to form β-SiC continuously on the substrate. Average mechanical and physical properties of the SiC filament are:

- Tensile strength: 3950 MPa (580 ksi)
- Tensile modulus: 400 GPa (60 × 10^6 psi)
- Density: 3.045 g/cm^3
- Coefficient of thermal expansion: 1.5 × 10^{-6}/K
- Diameter: 140 μm (5500 μin.)

Figure 2 is a photomicrograph of the cross section of a fiber surface showing the fiber interior and the carbon monofilament substrate. The various grades of fibers produced include the SCS-2, SCS-6, and SCS-8 grades described in the section "Fiber Variations" in this article; all are based on the β-SiC deposition process in which a crystalline structure is grown onto the carbon substrate. The β-SiC is present as such across all the fiber cross section except for the last few microns at the surface. Here, by altering the gas flows in the bottom of the tubular reactor, the surface composition and structure of the fiber are modified, first by an addition of amorphous carbon that heals the crystalline surface for improved surface strength, then by a modification of the Si-to-C ratio to provide improved bonding with the metal.

Processing Considerations. As in any vapor deposition or vapor transport process, temperature control is critical in producing SiC fiber by the CVD process. A peak deposition temperature of about 1300 °C (2370 °F) is considered optimum. Temperatures significantly higher cause rapid deposition and subsequent grain growth, resulting in a weakening of tensile strength. Temperatures significantly below the optimum cause high internal stresses in the fiber, resulting in a degradation of metal matrix composite properties when machining transverse to the fiber (Ref 7).

Substrate quality is also important in SiC fiber quality. The carbon monofilament substrate, which is melt-spun from coal-tar pitch, has a smooth surface with occasional surface anomalies. If severe enough, a surface anomaly can result in a localized area of irregular deposition of PG and SiC (Fig. 3 and 4), which is a stress-raising region and a strength-limiting flaw in the fiber. The carbon monofilament spinning process is controlled to minimize such local anomalies so as to guarantee routine production of high-strength (>3450 MPa, or >500 ksi) SiC fiber.

Another strength-limiting flaw that can result from an insufficiently controlled CVD process is the PG flaw (Ref 8), which results from irregularities in PG deposition. These can be caused by disruption of the PG layer due to an anomaly in the carbon substrate surface, or mechanical damage to the PG layer before SiC deposition. Examples of both types of flaws are shown in Fig. 4 and 5. Pyrolytic graphite flaws often cause a localized irregularity in SiC deposition, resulting in a bump on the surface (Fig. 5). Poor alignment of the reactor glass can result in mechanical damage to the PG layer by abrasion (Fig. 6a) and a series of PG flaws results in the so-called string of beads phenomenon at the surface of the fiber (Fig. 6b). The

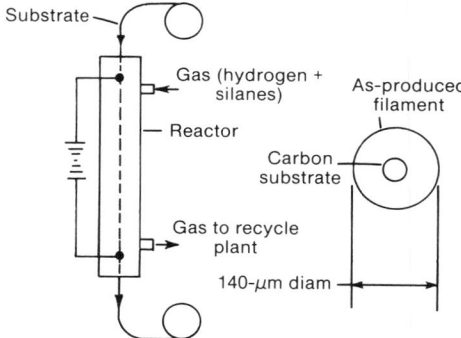

Fig. 1 Fabrication of SiC fiber

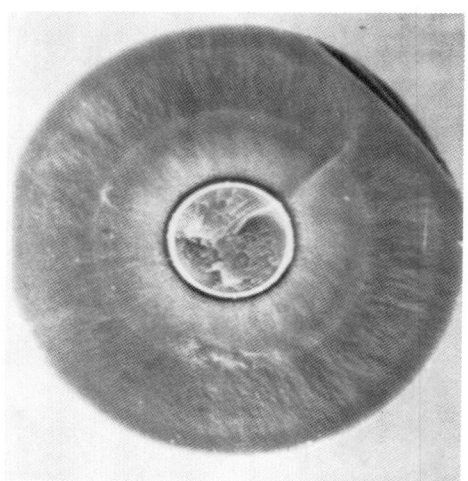

Fig. 2 Cross section of SiC fiber

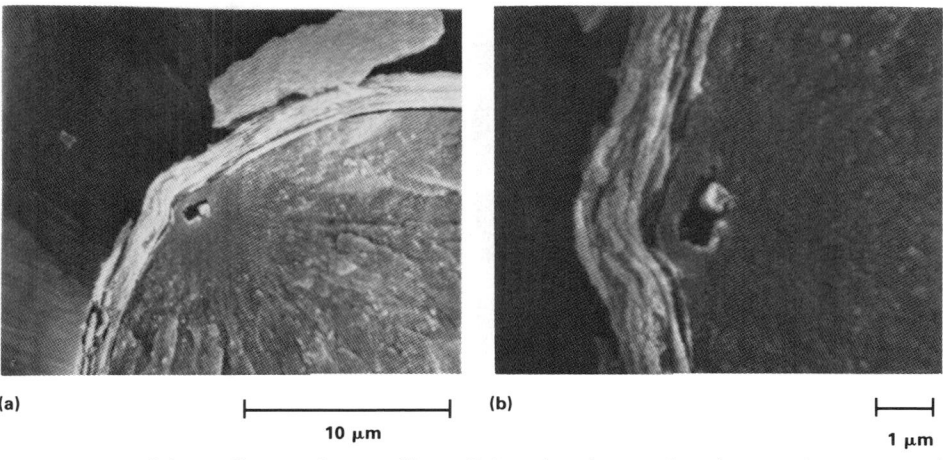

(a)

|— 10 μm —|

(b)

|— 1 μm —|

Fig. 3 Strength-limiting flaw at carbon monofilament/SiC interface due to void at substrate surface (a). Layered morphology of pyrolytic graphite evident in (b)

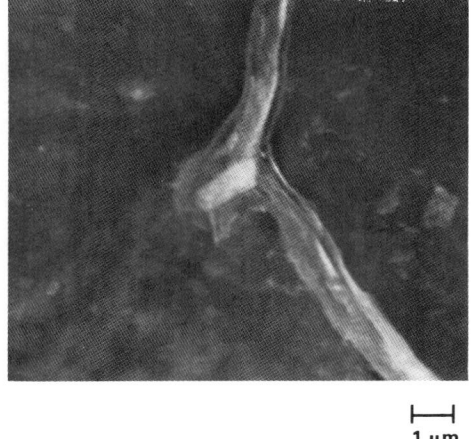

|— 1 μm —|

Fig. 4 Irregularity in PG layer associated with anomaly in substrate surface

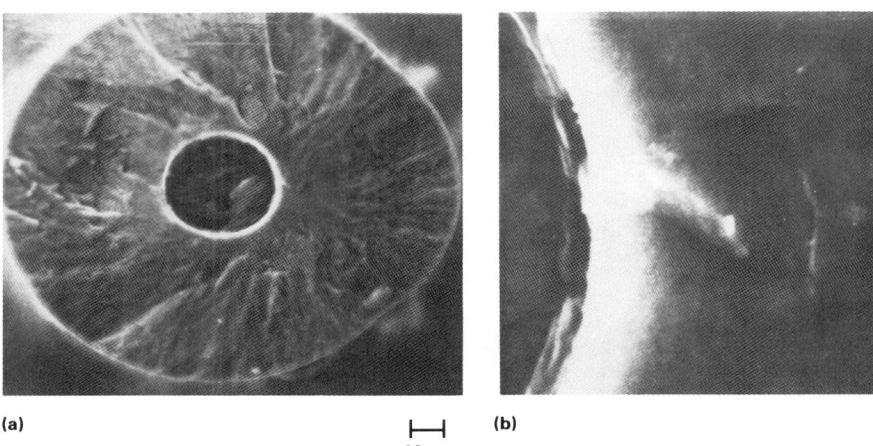

(a)

|— 10 μm —|

(b)

|— 1 μm —|

Fig. 5 Surface bump in PG layer (a) associated with damaged PG layer (b)

mechanical properties of such fibers are severely degraded. These flaws are minimized by careful control of the PG deposition parameters, proper reactor alignment, and minimization of substrate surface anomalies.

The surface region of the CVD SiC fiber is usually carbon rich, which is important in protecting the fiber from surface damage and subsequent strength degradation. An improper surface treatment or mishandling of the fiber that leads to abrasion can result in strength-limiting flaws at the surface. Surface flaws can be identified by optical examination of the fiber fracture surface. These flaws are minimized by proper process control and handling of the fiber (minimizing surface abrasion).

Mechanical properties of CVD SiC fiber usually consist of average tensile strengths of 3790 to 4140 MPa (550 to 600 ksi) and elastic moduli of 400 to 415 GPa (58 to 60×10^6 psi). A typical tensile strength histogram (Fig. 7) shows an average tensile

strength of 4000 MPa (580 ksi) with a coefficient of variation of 15%. Figure 8 shows the strength of SiC fibers after exposure at elevated temperatures in argon, nitrogen, and oxygen.

Fiber Variations. The surface region of the SiC fiber must be tailored to the matrix. Shown in Fig. 9 are diagrams of the surface composition of three fiber types. The fiber SCS-2 has a 1-μm (39-μin.) carbon-rich coating, which increases in silicon content as the outer surface is approached. This fiber has been used extensively to reinforce aluminum. The fiber SCS-6 has a 3-μm (118-μin.) carbon-rich coating in which the silicon content exhibits maxima at the outer surface and at 1.5 μm (59 μin.) from the outer surface. The fiber SCS-6 is primarily used to reinforce titanium.

The fiber SCS-8 was developed to give better mechanical properties than SCS-2 in aluminum composites transverse to the fiber

direction. The SCS-8 fiber consists of 6 μm (236 μin.) of very fine-grained SiC, a carbon-rich region of about 0.5 μm (20 μin.), and a less carbon-rich region of 0.5 μm.

Cost Factors. Silicon carbide is potentially less costly than boron for three reasons: The carbon substrate used for silicon carbide is lower in cost than the tungsten used for boron; raw materials for silicon carbide (chlorosilanes) are less expensive than boron trichloride, the raw material for boron; and deposition rates for silicon carbide are higher than those for boron, allowing more product to be made per unit time.

Composite Processing

Acceptable SiC fiber reinforced metals can be produced easily because the SiC fibers readily bond to the respective metals and resist strength degradation during high-temperature processing. In the past, when boron fibers were evaluated in various aluminum alloys, severe degradation of fiber strength was observed unless complex solid-state (low temperature/high pressure) diffusion-bonding procedures were adopted. Likewise with titanium, unless fabrication times are severely curtailed, fiber-matrix interactions produce brittle intermetallic compounds that again drastically reduce composite strength.

In contrast, the SCS fibers have surfaces that readily bond to the respective metals without destructive reactions. As a result, aluminum composites can be consolidated using less complicated high-temperature processes such as investment casting and low-pressure (hot) molding. Also, with titanium composites, the SCS-6 filament can withstand long exposure at diffusion-bonding temperatures without fiber degradation. As a result, complex shapes with selective composite reinforcement can be fabricated by innovative methods such as super-

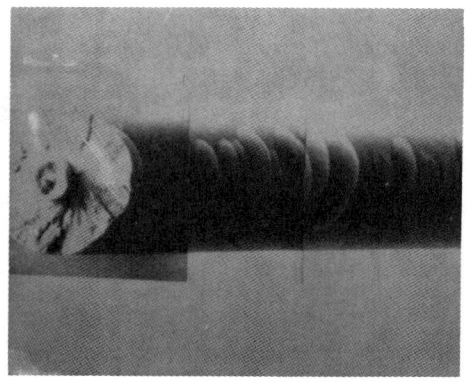

(a)

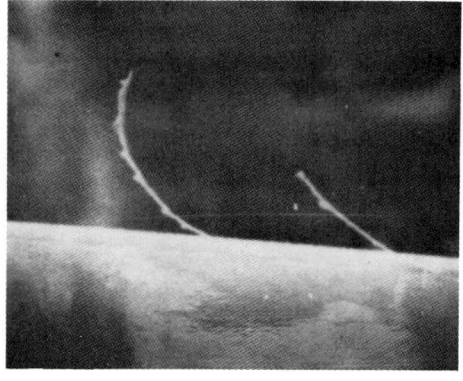

(b)

Fig. 6 Damaged PG layer (a) and associated "string of beads" phenomenon (b)

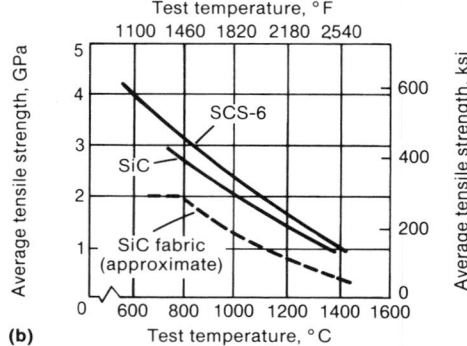

(a)

(b)

Fig. 8 Strength at temperature for SiC fibers in argon or nitrogen (a) and in oxygen (b)

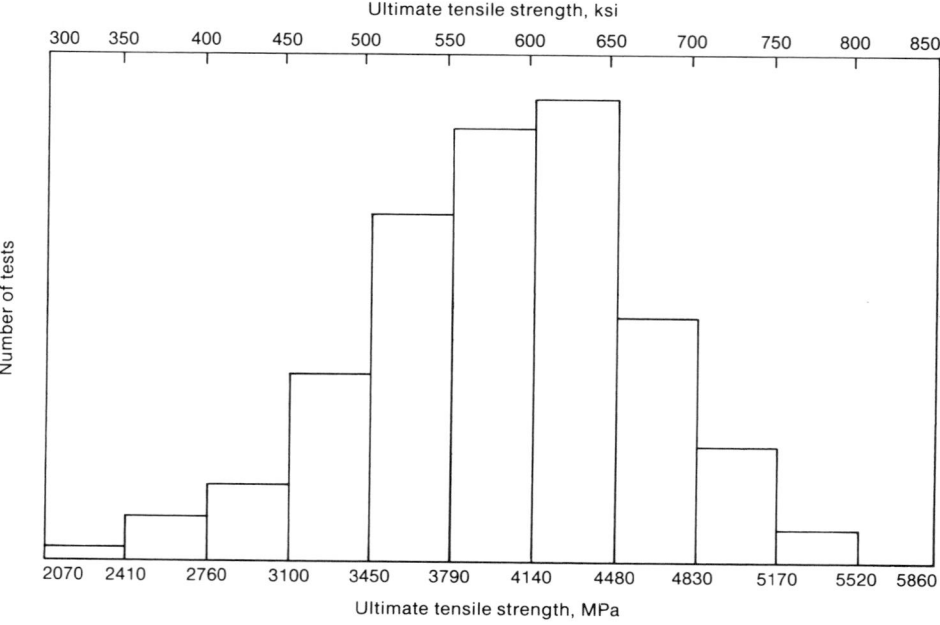

Fig. 7 Histogram of CVD SiC (SCS-2) fiber tensile strength, based on single-reactor run. Mean strength, 400 MPa (580 ksi). Coefficient of variance, 15%

plastic forming/diffusion bonding (SPF/DB) and hot isostatic pressing.

Composite Preforms and Fabrics. Intermediate products such as preforms and fabrics used in component fabrication are produced first to simplify the loading of fibers into a mold and to provide correct alignment and spacing of the fibers. The use of green tape is an old system in which fibers are wound onto a foil-covered rotating drum and oversprayed with resin, followed by cutting of the layer from the drum to provide a flat sheet of prepreg. Prepreg lay-ups can then be prepared sequentially in the mold or tool in required orientations to fabricate laminates. The laminate processing cycle is controlled so as to remove the resin (by heat and vacuum) as volatilization occurs.

Plasma-sprayed aluminum tape is similar to green tape except that the resin binder is replaced with a plasma-sprayed matrix of aluminum. Advantages of this prepreg include absence of contamination from resin residue and faster material-processing times due to elimination of the hold time required to ensure volatilization and removal of the resin binder. As with the green-tape system, lay-ups of the plasma-sprayed preforms are prepared sequentially in the mold as required and pressed to the final shape.

Woven fabric is a universal preform concept that is suitable for many fabrication processes. The fabric is a uniweave system in which relatively large-diameter SiC monofilaments are held straight and parallel, collimated at 100

Fig. 9 Surface-region composition of SiC fibers

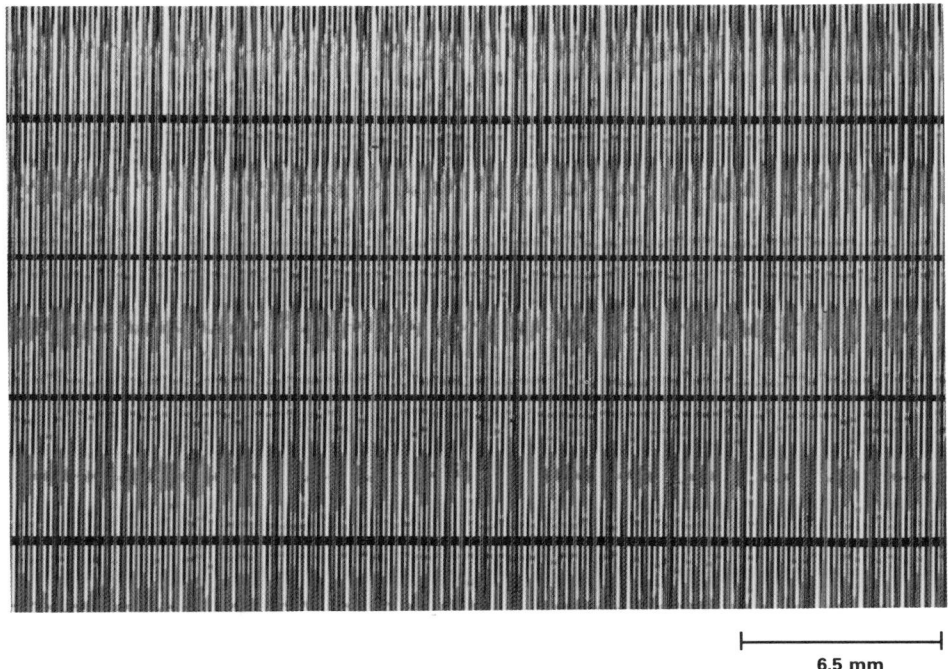

Fig. 10 SiC uniweave fabric with aluminum-ribbon cross weave

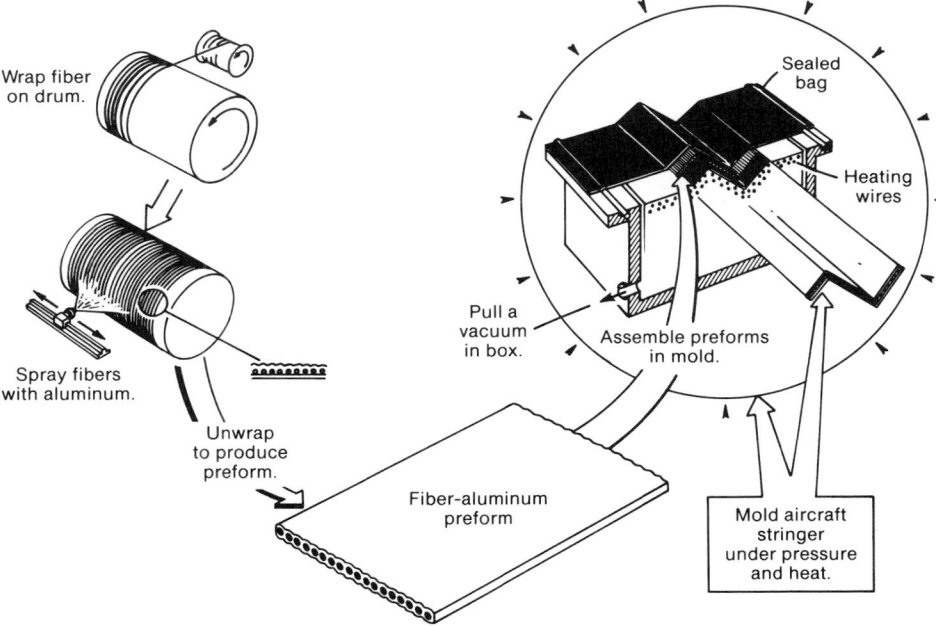

Fig. 11 Hot molding of SiC-Al Z-stiffeners

to 140 filaments per inch, and held together by a cross weave of a low-density yarn or metallic ribbon (Fig. 10). Two types of looms can now be modified to produce uniweave fabric. The first is a single-arm Rapier-type loom capable of producing continuous 1520-mm (60-in.) wide fabric with the SiC filament oriented in the fill (1520-mm) direction. The other is a shuttle-type loom on which the SiC monofilaments are oriented in the continuous direction with the lightweight yarn or metal ribbon in the fill axis. The shuttle loom can weave fabric up to 150 mm (6 in.) wide. Various types of cross-weave materials have been used, such as titanium, aluminum, and ceramic yarns.

Processing methods include investment casting, hot molding, and diffusion bonding. Investment casting has been used for many years and is still universally accepted as a highly cost-effective fabrication technique for producing complex shapes. The aerospace industry has rejected aluminum castings in the past because of the low strengths typically achieved; however, with a material that is now fiber dependent and not predominantly matrix controlled, structural improvements have been significant enough to revive interest in investment casting. Sometimes called the lost wax process, it uses a wax replica of the intended shape to form a porous ceramic shell mold where, upon removal of the wax (by steam heat) from the interior, a cavity for the aluminum is provided.

The SiC fibers are installed in the mold using the fabric previously described, by either placing the fabric into the wax replica or simply splitting open the mold and inserting the fabric into the cavity after the wax has been removed. At present, the latter approach is usually used to avoid contamination and oxidation of the fibers during wax burnout. The necessary techniques for including the fiber in the wax and thereby reducing processing costs will probably be developed eventually.

Hot molding is a low-pressure hot-pressing process designed to fabricate shaped SiC-Al parts at significantly lower cost than is possible with a diffusion-bonding/solid-state process. Because the SCS-2 fibers can withstand molten aluminum for long periods, the molding temperature can be raised into the liquid-plus-solid region of the alloy to ensure aluminum flow and consolidation at low pressure, thereby eliminating the need for high-pressure die molding equipment.

The hot-molding process is analogous to the autoclave molding of graphite-epoxy, in which components are molded in an open-faced tool. The mold in this case is a self-heating slip-cast ceramic tool which contains the profile of the finished part. A plasma-sprayed aluminum preform is laid into the mold, heated to near molten aluminum temperature, and pressure consolidated in an autoclave by a metallic vacuum bag. The process is illustrated in Fig. 11, which shows a stiffener being fabricated in specially prepared molds. These can be profiled as required to produce near-net shape parts including tapered thicknesses and section geometry variations. In Fig. 11, a Z mold tool is being used for manufacture of 0° ± 45° SiC-Al stringers for a stiffened panel.

Diffusion bonding of SiC-Ti is accomplished by hot-pressing (diffusion bonding) technology, using fiber preforms (fabric) that are stacked together between titanium foils for

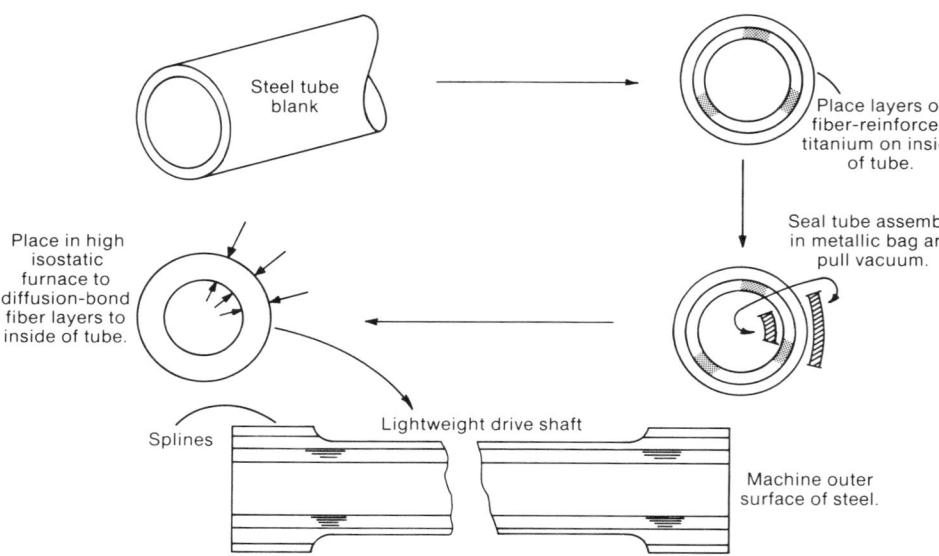

Fig. 12 Hot isostatic pressing of SiC-Ti drive shaft

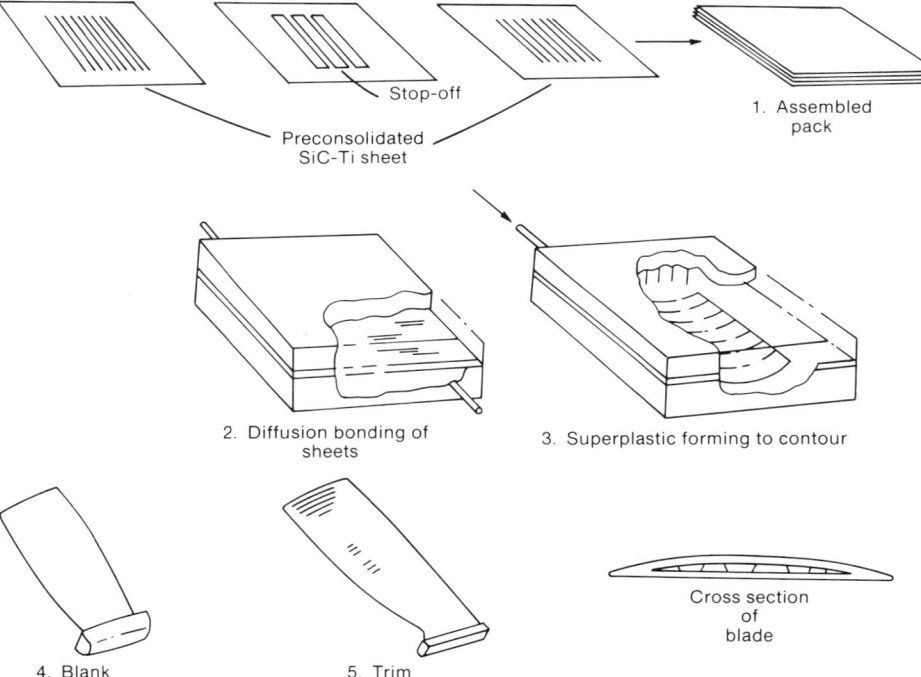

Fig. 13 Superplastic forming/diffusion bonding of SiC-Ti blade

Table 1 Tensile strength of SCS-2-Al
47 vol% fiber

| Fiber orientation | No. of plies | Tensile strength | | Tensile modulus | | Total strain | Poisson's ratio | Coefficient of thermal expansion, 10^{-6}/K |
		MPa	ksi	GPa	10^6 psi			
0°	6,8,12	1462	212	204.1	29.6	0.89	0.268	6.6
90°	6,12,40	86.2	12.5	118.0	17.1	0.08	0.124	21.3
$[0°/90°/0°/90°]_s$	8	673	97.6	136.5	19.8	0.90	. . .	. . .
$[0_2°/90°/0°]_s$	8	1144	166.0	180.0	26.1	0.92	. . .	. . .
$[90_2°/0°/90°]_s$	8	341.3	49.5	96.5	14.0	1.01	. . .	. . .
±45°	8,12,40	309.5	44.9	94.5	13.7	10.6	0.395	. . .
$[0°/±45°/0°]_{s+2s}$	8,16	800.0	116	146.2	21.2	0.86	. . .	. . .
$[0°/±45°/90°]_s$	8	572.3	83.0	127.0	18.4	1.0	. . .	. . .

consolidation. Two methods are being developed by aircraft and engine manufacturers to make complex shapes. One method, based on hot isostatic pressing technology, uses a steel pressure membrane to consolidate components directly from the fiber-metal preform layer. The other method requires the use of previously hot-pressed SCS-Ti laminates that are then diffusion bonded to a titanium substructure during subsequent superplastic forming operations. A typical use of the hot isostatic pressing procedure, for a SCS-Ti engine drive shaft, is illustrated in Fig. 12. The fiber preform is placed onto a titanium foil, which is then spirally wrapped, inserted, and diffusion bonded onto the inner surface of a steel tube using a steel pressure membrane. The steel is subsequently thinned down and machined to form the spline attachment at each end. Shafts are also fabricated without the steel sheath. Figure 13 illustrates the superplastic forming of hollow engine compressor blades. Here the SCS-Ti laminates are first diffusion bonded in a press and then diffusion bonded to form monolithic titanium sheets, with stop-off compounds selectively positioned to preclude bonding in specific areas. Subsequently, the stack-up is sealed into a female die. By pressurizing the interior of the stack-up, the material is blown into the female die to form the desired shape, stretching the monolithic titanium to form the internal corrugations.

The foregoing processes typically require long times at high temperature. All the materials previously used have developed serious matrix-to-fiber interactions that seriously degrade composite strength. The SCS-6 fiber, however, because of its unique surface characteristics, delays intermetallic diffusion and retains its strength up to 7 h in contact with titanium at 925 °C (1700 °F).

Composite Properties

Because continuous SiC reinforced metals have been available for a relatively short period of time, the property data base has been developed sporadically over a brief period, depending on funded applications.

Silicon Carbide-Aluminum. Because hot molding is the most mature of the SiC-Al consolidation approaches, the largest mechanical-property data base developed has been for hot-molded SCS-2-6061 Al. The design data base includes static tension and compression properties, in-plane and interlaminar shear strengths, tension-tension fatigue strengths (S-N curves), flexure strength, notched tension data, and fracture-toughness data. Most of the data have been developed over a temperature range of -55 to $+75$ °C (-65 to $+165$ °F) with static tension test results up to 480 °C (900 °F). The data are summarized in Tables 1 through 4 and Fig. 14 and 15. As can be seen, the inclusion of a high-performance continuous SiC fiber in 6061 Al yields a very high-strength

Table 2 Compression strength of SCS-2-Al
47 vol% fiber

Direction	Plies	Load N	Load lb	Stress MPa	Stress ksi	Tensile modulus GPa	Tensile modulus 10⁶ psi	Poisson's ratio
0°	12	36 000	8 100	2 647	383.9	...	...	...
		38 250	8 600	2 708	392.7	...	...	...
		38 700	8 700	2 739	397.3	...	...	...
		40 500	9 100	2 878	417.4	...	...	...
		48 900	11 000	3 296	478.0	212.4	30.8	0.241
		53 100	11 940	3 689	535.0	222.7	32.3	...
90°	12	4 220	948	294.4	42.7	104.8	15.2	...
		4 380	985	300.6	43.6	116.5	16.9	0.174
		4 270	960	294.4	42.7	...	...	...
		4 230	950	292.3	42.4	113.1	16.4	0.173
		3 960	890	273.0	39.6	115.8	16.8	...
		3 780	850	259.2	37.6	124.1	18.0	...
90°	40	13 480	3 030	293.7	42.6	...	...	...
		14 610	3 285	294.4	42.7	131.7	19.1	0.136
		13 280	2 985	290.0	42.0	102.7	14.9	...
		13 430	3 020	287.5	41.7	108.9	15.8	...
		13 520	3 040	294.4	42.7	115.1	16.7	...
		13 680	3 075	297.2	43.1	142.0	20.6	0.158

Table 3 Shear strength of SCS-2-Al
15° off-axis shear, 47 vol% fiber

Test temperature °C	°F	Failure stress MPa	ksi	Shear strength MPa	ksi	Shear modulus MPa	ksi
Room temperature...........		455.7	66.1	113.8	16.5	42.5	6.17
		452.3	65.6	113.1	16.4	39.5	5.73
		479.2	69.5	120.0	17.4	39.8	5.77
		422.6	61.3	105.5	15.3	40.3	5.85
Average		**452.5**	**65.6**	**113.1**	**16.4**	**40.5**	**5.88**
75 (165).............		437.1	63.4	109.6	15.9	40.2	5.83
		434.4	63.0	108.9	15.8	43.2	6.27
		424.7	61.6	106.2	15.4	41.7	6.05
Average		**432.1**	**62.6**	**108.2**	**15.7**	**41.7**	**6.05**
−55 (−65).............		501.2	72.7	125.5	18.2	44.5	6.46
		482.6	70.0	120.7	17.5	39.6	5.75
		453.0	65.7	113.1	16.4	39.6	5.75
Average		**479.0**	**69.4**	**119.8**	**17.3**	**41.3**	**5.98**

Table 4 Notched strength of SCS-2-Al
12 plies of material, 47 vol% fiber

Specimen	Average gross stress, RT MPa	ksi	Average net stress, RT MPa	ksi	Notch factor RT	Notch factor 75°C (165°F)	Notch factor −55°C (−65°F)
Double-edge notch, 0°	814.8	118.2	1269.5	184.1	0.92	0.85	0.80
Center hole, 1.6-mm (1/16-in.) diam, 0°.........	1125.9	163.3	1163.8	168.8	0.84	...	...
Center hole, 3.2-mm (1/8-in.) diam, 0°	991.7	143.8	1061.6	154.0	0.77	0.75	0.72
	898.0(a)	130.2	956.1(a)	138.7	0.70(a)	...	...
Center hole, 6.4-mm (1/4-in.) diam, 0°	842.1	122.1	966.4	140.2	0.70	...	...
	800.9(a)	116.2	911.3(a)	132.2	0.66(a)	...	...
Center hole, 3.2-mm (1/8-in.) diam, 0°/±45°	728.1	105.6	777.4	112.8	0.90	...	...
Center hole, 6.4-mm (1/4-in.) diam, 0°/±45°	620.5	90.0	710.8	103.1	0.83	...	...
Center hole, 3.2-mm (1/8-in.) diam, 0°/±45°/90° ..	437.1	63.4	467.5	67.8	0.90	...	...
Center hole, 6.4-mm (1/4-in.) diam, 0°/±45°/90° ..	400.6	58.1	460.6	66.8	0.89	...	...
Center hole, 2.4-mm (3/32-in.) diam, ±45°	244.7	35.5	256.6	37.2	0.85	...	...
Center crack, 6.4-mm (1/4-in.) EDM slot, 0°	822.0	119.2	944.2	137.0	0.68	...	...
Center crack, 12.7-mm (1/2-in.) EDM slot, 0°	659.9	95.7	886.9	128.6	0.64	...	...
	621.6(a)	90.20	819.8(a)	118.9	0.60(a)	...	...

RT, room temperature. (a) 40-ply material

(1378 MPa, or 200 ksi) high-modulus (207 GPa, or 30 × 10⁶ psi) anisotropic composite material having a density just slightly greater (2.85 g/cm³) than baseline aluminum. As in organic matrix composites, cross- or angle-plying produces a range of properties useful to designers (Fig. 16).

The property data developed to date for investment-cast SCS-Al have been limited to static tension and compression. Fiber volume fractions (40% max) are lower than those for hot-molded laminates (47% typical) because of volumetric constraints in dry loading the shell molds; however, good rule-of-mixture (ROM) tensile strengths and excellent compression strengths (twice the tensile strength) are being achieved (Table 5).

The use of 6061 Al as the matrix material and the capability of the SiC fiber to withstand molten aluminum have made conventional fusion welding a viable joining technique. Although welded joints would not have continuous fiber across the joint to maintain the very high strengths of the composite, baseline aluminum weld strengths can be obtained. In addition to fusion welding, traditional molten salt bath dip brazing has been demonstrated as an alternative joining method.

Corrosion testing has been performed on SCS-2-6061 Al hot-molded material at the David W. Taylor Naval Ship Research and Development Center (Ref 9) under various conditions (marine atmosphere, ocean splash/spray, alternate tidal immersion, and filtered seawater immersion) for periods of 60 to 365 days. The SCS-Al material performed well in all tests, exhibiting no more pitting damage than baseline 6061 Al alloy.

Silicon Carbide-Titanium. The SCS-6-Ti-6Al-4V composites were originally developed to withstand extended exposure at high temperature. As shown in Table 6, composite strengths remained above 1380 MPa (200 ksi) after extended heat treatment. In a successful program to reinforce the β-Ti alloy 15V-3Sn-3Cr-3Al with SCS fiber (Ref 10), superior composite properties have been achieved (tensile strengths of 1585 to 1930 MPa, or 230 to 280 ksi, see Table 6). Titanium parts have been fabricated by diffusion bonding and by the hot isostatic pressing technique, which has been particularly successful in the forming of shaped reinforced parts (for example, tubes) by the use of woven SiC fabric as a preform. The high-strength high-modulus properties of SCS-6-Ti represent a major improvement over B_4C-β-Ti composites, in which the modulus of the composite is increased relative to the matrix but the tensile strength is not as high as the predicted ROM strength. Modulus and strength at elevated temperature are shown in Fig. 17 and 18.

Silicon Carbide-Magnesium and Silicon Carbide-Copper. The fiber SCS-2 has been successfully cast in magnesium (Ref 11). Properties are listed in Table 7. Under a recent Naval Surface Weapons Center program (Ref 12), development of SiC-reinforced copper has been initiated. At present, about 85% of ROM strengths have been achieved at volume fractions of 20 to 33%. Typical data are presented in Table 8.

Applications

The high strength and stiffness and low density of SiC-reinforced metal matrix composites have generated significant interest in the

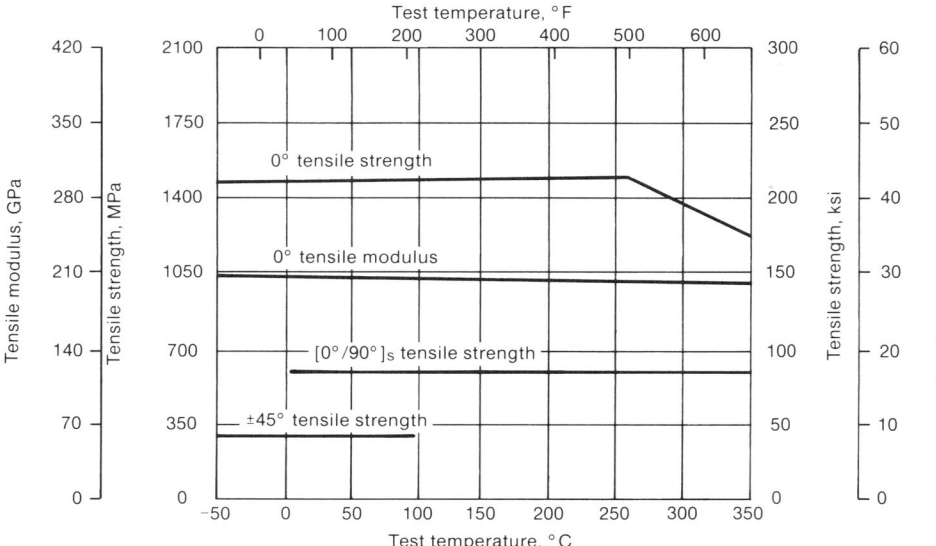

Fig. 14 Composite strength versus temperature, SCS-2-6061 Al

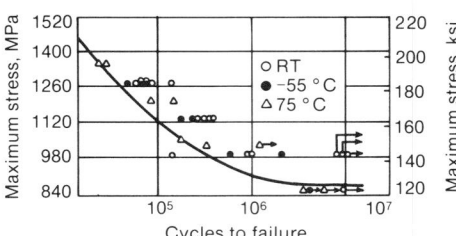

Fig. 15 Tension-tension fatigue for SCS-Al. 0°, 12-ply laminates; 47 vol% fiber; R = min stress/max stress

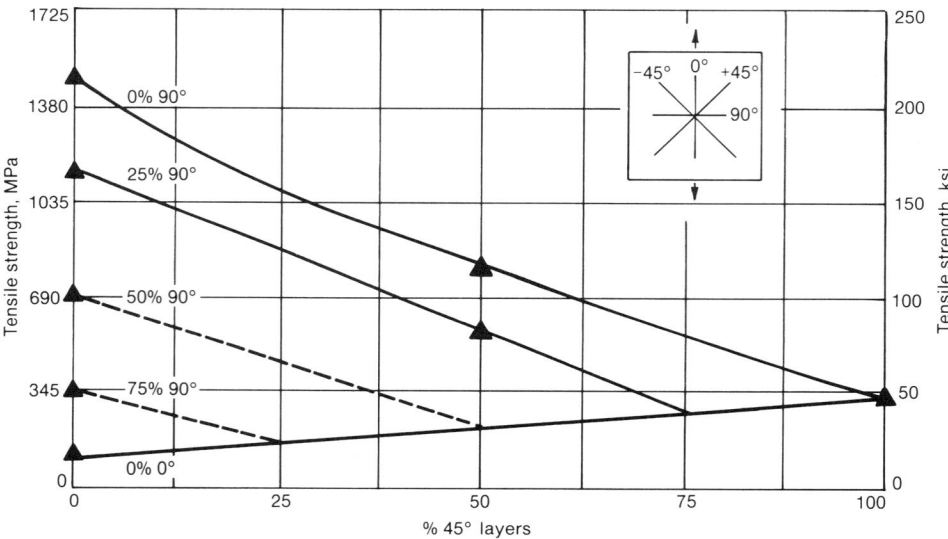

Fig. 16 Composite tensile strength versus ply orientation, SCS-2-6061 Al

Table 5 Data on investment-cast SCS-Al

Fiber orientation	Fiber, vol%	Ultimate tensile strength MPa	ksi	ROM, %	Tensile modulus GPa	10^6 psi	ROM, %	Ultimate compressive strength MPa	ksi	Compressive modulus GPa	10^6 psi
$0_3/90_6/0_3$	33	458.5	66.5	75	122.0	17.7	107	1378.9	200	...	...
$90_3/0_6/90_3$	33	584.0	84.7	95	124.8	18.1	110	1378.9	200	...	...
0°	34	1034.2	150	85	172.4	25	100	1896.1	275	186.2	27.0

aerospace industry, resulting in the initiation of many research and development programs. The principal area of interest is high-performance structures such as aircraft, missiles, and engines. However, as more systems are developing sensitivities to performance and transportation weight, less sophisticated applications for these newer materials are also being considered. The following paragraphs describe a few of these applications.

Silicon carbide-aluminum wing structural elements are being developed. Plans are for 3-m (10-ft) long Z-shaped stiffeners to be hot-molded, then riveted to wing planks for full-scale static and fatigue testing. Experimental results obtained to date on (0° ±45°) SCS-2 fibers in a 6061 alloy have verified material performance and design procedures.

Silicon carbide-aluminum forms are being evaluated as a means of weight reduction in advanced portable bridging. The lower chord, king post, and top compression tubes are some of the components being considered.

Silicon carbide-aluminum internally stiffened cylinders for small, lightweight pressure-vessel applications are being developed using the investment-casting process. A wax replica is first fabricated that incorporates the total shape of the shell, including internal ring stiffeners and the end fittings in two halves of the split ceramic mold (inner and outer). The fabric containing the SiC fibers is then wound onto the inner shell mold, the two halves of the shell are remated and sealed, and the aluminum is then infiltrated.

Silicon carbide-aluminum fins for high-velocity projectiles are being evaluated. Figure 19 shows a SiC-Al missile fin with 30 vol% fiber, designed for low weight and high stiffness to improve flight stability.

Silicon carbide-aluminum missile body casings have been fabricated using a unique variation of filament winding. An aluminum motor case is first produced in the conventional manner but with significantly less wall thickness than is normally required. The casing is then overwrapped with layers of SiC fibers, with each layer sprayed with a plasma of aluminum to build up the matrix thickness. No final consolidation of the 90% dense system is required because the hydrostatic internal pressure on the circular body imposes no (or very low) shear stresses on the matrix. Further development of this technique may permit full consolidation of the matrix by vacuum-bagging the total section and using the hot isostatic pressing technique.

Silicon carbide-titanium drive shafts are being developed and fabricated by the hot isostatic pressing process. These are generally for

Table 6 Data on SCS-6-Ti (sample size, 62 panels)

	Ultimate tensile strength		Elastic modulus		Strain to
	MPa	ksi	GPa	10⁶ psi	failure, %
Mechanical properties of SiC-Ti-6Al-4V (35 vol%)					
As fabricated					
Mean	1690	245	186.2	27.0	0.96
Standard deviation	119.3	17.3	7.58	1.1	0.091
After heating 7 h at 905°C (1660 °F)					
Mean	1434	208	190.3	27.6	0.86
Standard deviation	108.9	15.8	8.3	1.2	0.087
Mechanical properties of SiC-Ti-15V-3Sn-3Cr-3Al (38 to 41 vol%)					
As fabricated					
Mean	1572	228	197.9	28.7	...
Standard deviation	138	20	6.21	0.9	...
After heat treating 16 h at 480 °C (900 °F), 13 samples					
Mean	1951	283	213.0	30.9	...
Standard deviation	96.5	14	4.83	0.7	...

Table 7 Mechanical data on SCS-Mg cast rod
ZE 41 at 675°C, or 1250 °F

Sample No.	Exposure time, min	Ultimate tensile strength MPa	ksi	Strain to failure, %	Elastic modulus GPa	10⁶ psi	Fiber, vol%
VIR 67	5	1000	145	0.83	169.6	24.6	34
VIR 69	10	1524	221	0.88	209.6	30.4	46
VIR 72	10	1331	193	0.78	230.3	33.4	50
VIR 77	10	1379	200	0.95	180.6	26.2	37

Table 8 Mechanical data on SCS-Cu

Panel	Fiber, vol%	Axial ultimate tensile stress MPa	ksi	Axial modulus GPa	10⁶ psi
84-014	0.23	690	100	172.4	25.0
84-153	0.33	965	140	202	29.3
84-377	0.33	900	130	187.5	27.2

sults have been obtained using the higher temperature titanium aluminides as matrix materials. Because the SCS-6 fiber demonstrates high mechanical properties up to 1400 °C (2550 °F), systems such as SiC-nickel aluminides/iron aluminide/superalloys, and so forth, can be projected. On a ROM basis, at least, these systems would have useful properties for engine and hypersonic-vehicle applications. Work required in this area includes diffusion barrier coating development and matrix alloy modifications to facilitate high-temperature fabrication processes. Also required is detailed investigation of any detrimental thermal/mechanical cycling effects that may occur due to mismatch in thermal expansion coefficients between matrix and fiber.

the core of an engine, requiring increased specific stiffness to reduce unsupported length between bearings and to increase critical vibratory speed ranges. Silicon carbide-titanium tubes up to 1.5 m (5 ft) long have been fabricated with a monolithic load transfer section incorporated into their ends for ease of welding to the splined or flanged connections.

Silicon carbide discs for turbine engines currently under development were initially made by winding SiC-Ti monolayer over a mandrel, followed by hydrostatic consolidation. Subsequently, a "doily" approach was developed in which single fibers are hoop-wound between titanium metal ribbons and subsequently pressed together in the axial direction, reducing fiber breakage and simplifying production of tapered cross sections. Figure 20 shows a three-layer doily.

Selectively reinforced SiC-Ti hollow fan blades are being developed. A cross section of a prototype blade shown in Fig. 21 illustrates a high-performance hollow-bending section.

Silicon carbide-copper materials have been fabricated and tested for high-temperature missile applications. Also, SiC-bronze propellers have been cast (Fig. 22) for potential U.S. Navy applications requiring more efficient or quiet propellers.

Future Trends

It has been shown that it is feasible to incorporate SiC fiber in a variety of metal matrix alloys, principally aluminum, magnesium, and titanium. Copper matrix systems are under development, and reasonably good re-

REFERENCES

1. H. DeBolt, Boron and Other Reinforcing Agents, in *Handbook of Composites*, G. Lubin, Ed., Van Nostrand Reinhold, 1982
2. V.J. Krukonis, Boron Filaments, in *Handbook of Fillers and Reinforcements for Plastics*, J.V. Milewski and H.S. Katz, Ed., Van Nostrand Reinhold, 1977
3. D.L. McDanels and R. Ravenhall, "Analysis of High-Velocity Ballistic Impact Response of Boron/Aluminum Fan Blades," NASA TM-834998, National Aeronautics and Space Administration, 1983
4. C.T. Salemme and S.A. Yokel, "Design of Impact-Resistant Boron/Aluminum Large Fan Blades," NASA CR-135417, National Aeronautics and Space Administration, 1978
5. J.W. Brantley and R.G. Stabrylla, "Fabrication of J79 Boron/Aluminum Compres-

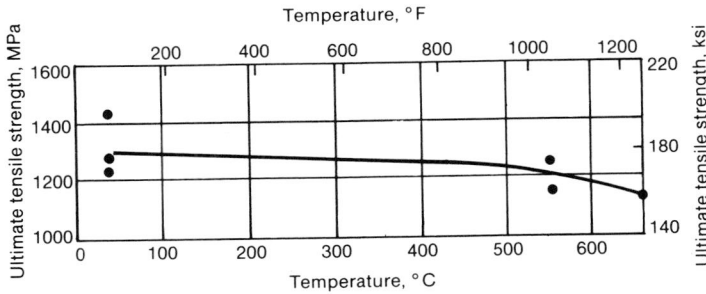

Fig. 17 Ultimate tensile strength versus temperature for SCS-6-Ti

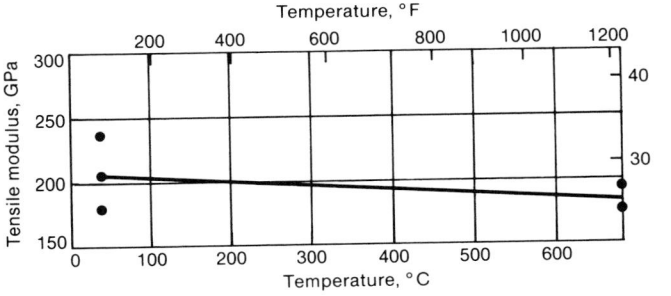

Fig. 18 Tensile modulus versus temperature for SCS-6-Ti

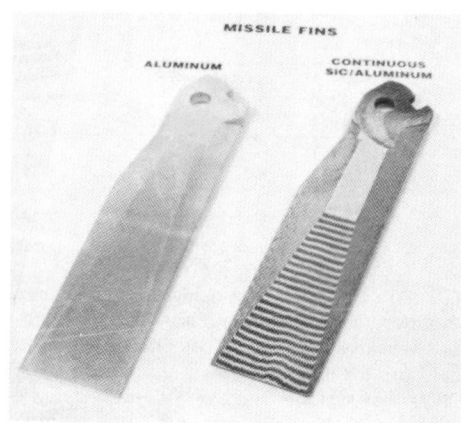

Fig. 19 SiC-Al missile fin

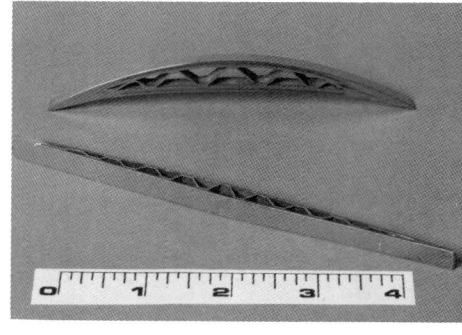

Fig. 21 Hollow SiC-Ti fan-blade section

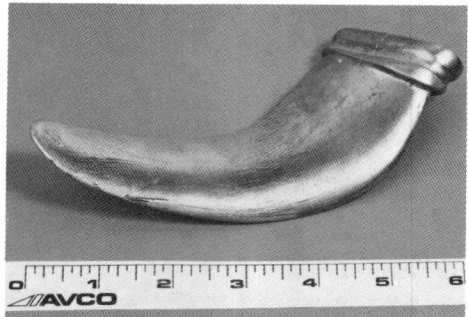

Fig. 22 SiC bronze propeller

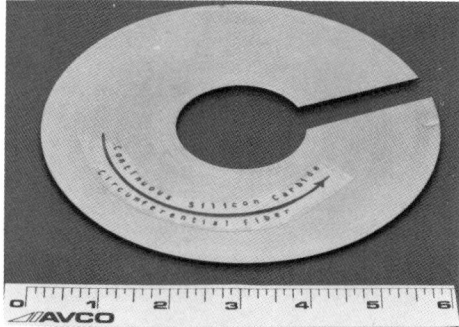

Fig. 20 SiC-Ti turbine disc material

sor Blades," NASA CR-159566, National Aeronautics and Space Administration, 1979

6. E. Wolff, "Boron Filament, Metal Matrix Composite Materials," AF33(615)3164, Air Force Materials Laboratory

7. R.J. Suplinskas, "Manufacturing Technology for Silicon Carbide Fiber," AFWAL-TR-84-4005, Air Force Wright Aeronautical Laboratories, 1983

8. R.J. Suplinskas, "High Strength Boron," NAS3-22187, National Aeronautics and Space Administration, 1984

9. D.M. Aylor, "Assessing the Corrosion Resistance of Metal Matrix Composite Materials in Marine Environments," DTN-SRDC/SMME-83/45, David Taylor Naval Ship Research and Development Center, 1983

10. A.J. Kumnick, R.J. Suplinskas, W.F. Grant, and J.A. Cornie, "Filament Modification to Provide Extended High Temperature Consolidation and Fabrication Capability and to Explore Alternative Consolidation Techniques," N00019-82-C-0282, Naval Surface Weapons Center, 1983

11. J.A. Cornie and Y. Murty, "Evaluation of Silicon Carbide/Magnesium Reinforced Castings," DAAG46-80-C-0076, 1983

12. J.V. Marzik and A.J. Kumnick, "The Development of SCS/Copper Composite Material," N60921-83-C-0183, Naval Surface Weapons Center, 1984

Continuous Graphite Fiber MMCs

David M. Goddard, Patrick D. Burke, and Donald E. Kizer, Material Concepts, Inc.
Roger Bacon, Amoco Performance Products, Inc.
William C. Harrigan, Jr., DWA Composite Specialties, Inc.

CONTINUOUS GRAPHITE FIBER-REINFORCED METAL MATRIX COMPOSITES (MMCs) are a class of extremely high-performance materials. Because the properties of these materials are dictated primarily by the properties of the fibers contained therein, the characteristics of these fibers will be discussed first in this article, and then descriptions of the various methods for fabricating the composite materials will be given. The shapes and material properties produced using each fabrication method also will be described.

Continuous Versus Discontinuous Fibers

A carbon/graphite fiber is basically a long, thin filament composed mainly of carbon with a filament diameter that is normally between 4 and 11 μm (160 and 430 μin.).

Continuous carbon fibers consist of essentially endless filaments grouped together into yarns or tows of 500 to 40 000 or more individual filaments. The yarns may be twisted to provide maximum bundle integrity, but are more commonly untwisted, providing maximum spreadability and filament parallelism. Continuous carbon fibers are used to fabricate composite materials to attain maximum values in properties such as strength or stiffness.

High-modulus graphite fibers have generally been used as continuous reinforcements in MMCs, which yield the utmost in specific mechanical and physical properties. Although discontinuously reinforced MMCs can be isotropic or quasi-isotropic and exhibit other desirable properties, they have lower levels of strength. Discontinuous carbon fibers consist of short filaments ranging in length from 0.2 mm (0.008 in.) to several centimeters. Fiber lengths are uniform if obtained by chopping continuous carbon yarns. Inexpensive discontinuous fibers are sometimes obtained by cutting or grinding random-length fibers. Chopped, high-modulus P55 and P100 fibers (380 and 690 GPa, or 55 and 100 × 10^6 psi, respectively) have been the primary discontinuous reinforcements explored to date with both aluminum and magnesium alloy matrices. These combinations would be expected to yield low-density low-CTE materials with good thermal conductivity.

Manufacturing Process. Nearly all commercial carbon fibers are manufactured by thermal "charring" or "carbonizing" of an organic precursor fiber, followed by a heat treatment to increase carbon content and achieve optimum physical and chemical properties. The final heat-treatment temperature (HTT) is between 1000 and 2000 °C (1830 and 3630 °F) for high-strength and ultrahigh-strength carbon fibers, and between 2000 and 3000 °C (3630 and 5430 °F) for high-modulus and ultrahigh-modulus carbon fibers (Table 1). The term graphite is used to refer to those fibers that have been heat-treated in the higher range of HTT, to distinguish them from carbon fibers treated only to the lower range of HTT. However, the use of the term carbon properly covers both classes.

Several methods of fabricating discontinuous graphite-metal matrix composites have been investigated. E. DiCesare reported success in the casting of the P55-reinforced magnesium (alloy ZE41) by thoroughly mixing the graphite fiber with finely divided titanium prior to the casting operation to promote wetting. The cast samples were subsequently hot-rolled (Ref 1).

Another process, which has been used for fabricating discontinuous graphite-reinforced aluminum, involves the use of liquid metal infiltrated (LMI) precursor wire. The precursor wire was chopped to discrete lengths and then consolidated by diffusion bonding or a solid/liquid consolidation process (Ref 2).

Table 1 shows the normal ranges of properties available in each type of carbon fiber. The key characteristic of the first two fibers listed is high or ultrahigh tensile strength, with values ranging from 2.8 to 5.7 GPa (0.40 to 0.83 × 10^6 psi). The key characteristic of the third and fourth fibers is high or ultrahigh Young's modulus, with values ranging from 350 to 900 GPa (50 to 130 × 10^6 psi). All four fibers are classified as high performance. The fifth type of fiber is called low-modulus carbon fiber and possesses much lower values of both Young's modulus and tensile strength than any of the high-performance fibers. Low-modulus fibers are presently available only in fabric or discontinuous fiber forms.

A typical microstructure of discontinuous 50 vol% P100/6061 graphite-aluminum is shown in Fig. 1. Mechanical property measurements conducted on these materials have shown low values for both modulus and ultimate tensile strength. Modulus values for P100/6061 were generally in the 62 to 73 GPa (9 to 10.6 × 10^6 psi) range. The low strain-to-failure of this material limited the tensile strength values to the 30 to 60 MPa (4 to 9 × 10^3 psi) range. Other available property data are limited to coefficient of thermal expansion (CTE) measurements. Data on two-dimensionally quasi-

Table 1 Properties of carbon fiber types

Fiber type	Density g/cm³	Young's modulus GPa	Young's modulus 10^6 psi	Tensile strength GPa	Tensile strength 10^6 psi	Electric resistivity ohm · m	Thermal conductivity W/m · K	Thermal conductivity BTU · in./h · ft² · °F
High-strength (PAN)	1.7-1.8	230-250	33-36	2.8-4.0	0.41-0.58	12-30	7-10	50-70
Ultrahigh strength (PAN)	1.7-1.8	260-290	38-42	4.1-5.7	0.59-0.83	14-20	7-9	50-60
High-modulus (PAN, mesophase pitch)	1.8-2.0	350-550	50-80	1.7-3.5	0.25-0.50	5-10	60-200	420-1400
Ultrahigh-modulus (mesophase pitch)	2.0-2.2	600-900	90-130	2.1-2.5	0.30-0.36	1-4	400-2500	2800-17 300
Low-modulus (rayon, pitch)	1.3-1.7	40-60	6-9	0.6-1.0	0.085-0.145	30-100	7-28	50-190

isotropic P100/6061 show a fairly low in-plane CTE of 3 to 5 × 10⁻⁶/K, with a CTE perpendicular to the plane of reinforcement of up to 24 × 10⁻⁶/K. While no thermal conductivity data are currently available, estimates suggest that discontinuous graphite-aluminum will exhibit good thermal conductivity. The relatively low strength of these composite materials would appear to preclude their use for structural applications.

Precursors for Carbon Fibers. The precursor fibers from which the various types of carbon fiber are derived are shown in parentheses in the first column of Table 1. High and ultrahigh tensile strength fibers are manufactured from polyacrylonitrile (PAN) precursor fibers. High and ultrahigh Young's modulus fibers are usually prepared from mesophase pitch, a high melting point product of coal or petroleum oil refining. Some high-modulus fibers are prepared from PAN. These latter fibers are usually higher strength than the corresponding fibers from mesophase pitch; however, the PAN-based fibers are also less oxidation resistant, more reactive toward metals, and more expensive. Ultrahigh-modulus fibers are all prepared from mesophase pitch. Low-modulus fibers are prepared from rayon or a low-melting-point nonmesophase pitch.

Carbon Fiber Surfaces. The surface characteristics of carbon fibers fall into two categories, depending upon whether the fibers have been heat treated to a low or a high HTT. The low HTT fibers possess an active surface to which matrix resins and metals tend to bond strongly. The high HTT fibers possess a more inert surface to which most matrix materials bond weakly; these fibers are also more oxidation resistant. The surface activity of either category of carbon fiber may be enhanced by a surface treatment consisting of a chemical or electrochemical oxidation etching. Because this treatment enhances the interlaminar shear strength of the composite, it is sometimes called a shear treatment.

Carbon Fiber Sizings. The bundle integrity and handleability of carbon fibers may be improved by application of a sizing to the yarn. The sizing is a lubricating or protective coating consisting of a waxy or resinous organic substance, applied at a level of ½ to 3 wt%. Sizings are tailored for maximum fiber weavability or spreadability, ease of removal, or compatibility with specific matrix resins or metals. Each manufacturer provides a variety of proprietary sizings.

Precursor to Fabrication

The state-of-the-art method of producing composite precursors for fabricating graphite-aluminum and graphite-magnesium composite products is a liquid metal infiltration (LMI) process. In this process the multifilament graphite tow is first activated by low-temperature (700 °C, or 1290 °F) chemical vapor deposition (CVD) of angstrom layers of titanium boride using titanium tetrachloride and boron trichloride reactants and zinc vapor as a reductant (Ref 3). This surface-activation treatment provides an active yarn tow surface that is wetted by molten aluminum and magnesium alloys to produce a graphite-reinforced composite wire. An eight-line CVD infiltration unit is shown in Fig. 2. A typical metallographic structure of a composite wire is shown in Fig. 3. This structure contains 50 vol% of 10 µm (390 µin.) graphite filaments dispersed in an aluminum alloy matrix.

Composite wire produced in this manner is laid up for final consolidation into composite sheet, tubing, rod, and a variety of other shapes. In addition to providing a composite precursor convenient for composite shape fabrication, the composite wire has several other advantages:

- Near rule of mixtures (ROM) tensile strength and tensile modulus with a variety of graphite tows and alloy matrices
- High-volume production amenable to automation
- Inspectability

- Adaptability to drum winding for low-cost lay-up

In addition to use in the production of composite wire, the metal infiltration process has been modified and utilized successfully to infiltrate woven graphite cloth, producing MMC sheet 305 mm (12 in.) in width and up to 15 m (50 ft) in length. The composite sheet volume loading properties and thickness are dependent on both the characteristics of the yarn used in weaving the cloth and the weave structure. Other applications include the metallization of the surface of monofilament and multifilament tows, including silicon carbide, borsic, niobium nitride, and aluminum oxide.

In the production of graphite-reinforced aluminum and magnesium composite wire, a variety of graphite fiber tows and alloys have been used successfully, including rayon-based T50 and T75; PAN-based T300, GY-70, HM3000, and Celion 6000; pitch-based P55, P75, P100,

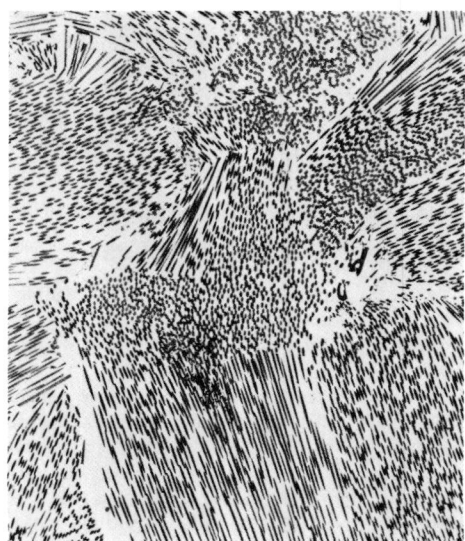

Fig. 1 Typical microstructure of discontinuous 50 vol% P100/6061 graphite-aluminum

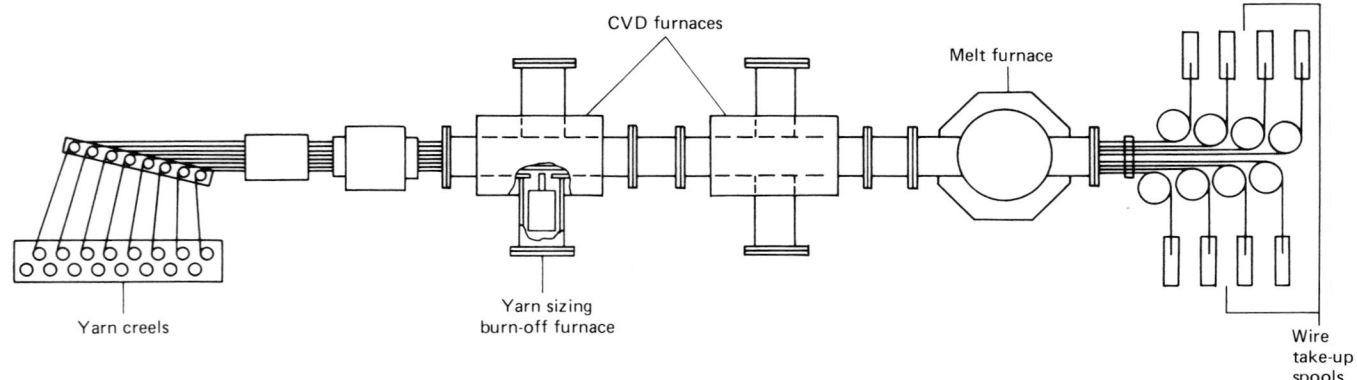

Fig. 2 CVD infiltration unit

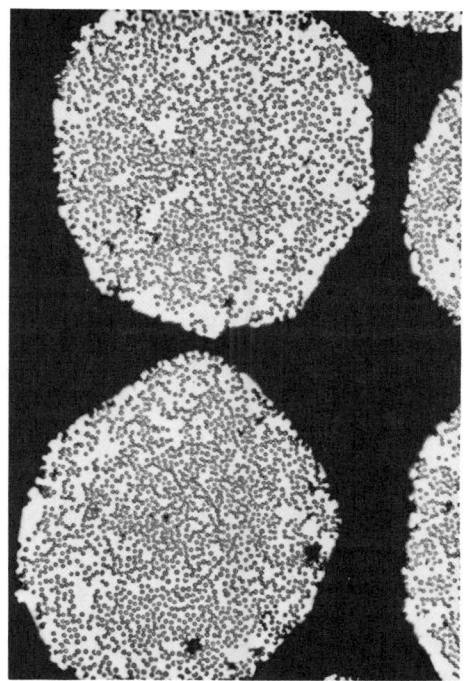

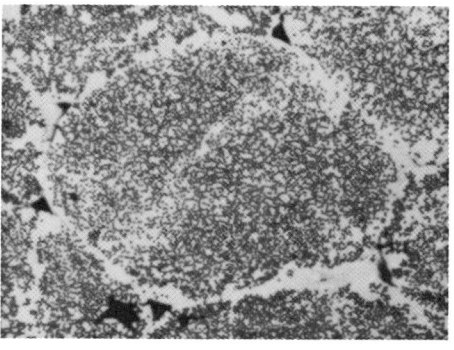

(a)

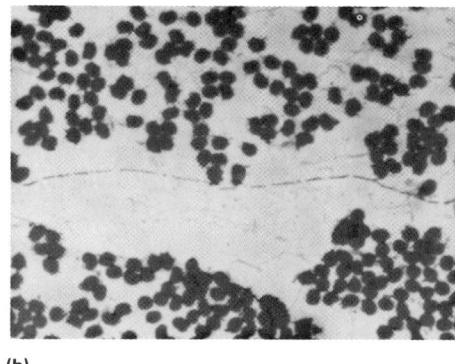

(b)

Fig. 5 Diffusion bonding effects in copper matrix composites. (a) Incomplete consolidation (Celion 6000-copper). (b) Poor oxide break-up (T300-aluminum bronze)

Table 2 Composite wire properties

	6061-Al (%ROM)(a)	AZ91-Mg (%ROM)
Fiber vol%	50.2 ± 2.0(b)	54.3 ± 2.0
Mean tensile strength, MPa (ksi)	1385 ± 110 (201.0 ± 16)(b)(c)	1240 ± 90 (180.0 ± 13.0)(d)(e)
Mean elastic modulus, GPa (10^6 psi)	420 ± 25 (60.7 ± 3.4)(c)(f)	>390 (>57)

(a) Based on the mean properties of virgin yarn. (b) Average of 3200 tests. (c) 100% ROM. (d) Average of 360 tests. (e) >95% ROM. (f) Average of 105 tests

Fig. 3 Photomicrograph of graphite-aluminum MMC wire; 50 vol% fiber. 60×

Fig. 4 Scanning electron micrograph of precursor wire

P120, and P140; graphite; 1100, A201, 2024, 356, 5083, 5154, and 6061 aluminum; and AZ31, AZ91, ZE41, QE22, and EZ33 magnesium.

Because of the applicability of these composite materials for space structures, most extensive work has been carried out using the high-modulus pitch-based fiber P100 (690 GPa, or 100 × 10^6 psi) and 6061 aluminum and AZ91 magnesium alloy. The composite wire properties developed on these systems are presented

in Table 2 to serve as an example of the property translation that can be achieved in production quantities of graphite-reinforced composite wire.

Diffusion Bonding

The first successful method for secondary fabrication of graphite-reinforced MMCs was diffusion bonding. This technique starts with precursor wires, described in the section "Precursor to Fabrication" in this article, and shown in Fig. 4. These wires are collimated and held together in a matt with a binder that thermally decomposes and leaves no deposit. Layers of these matts are stacked together to make the desired number of plies and desired orientation of fibers. Graphite-aluminum composites are normally made in the unidirectional configuration. Thin (0.089 mm, or 0.0035 in.) aluminum foils are placed at the bottom and the top of the wire plies to complete the "green composite." After this green composite is placed in a vacuum retort that is heated to the proper temperature, pressure is applied for a set amount of time, then removed, and the completed composite is removed from the retort. The technique can be applied to ion-plated, sputter-coated, or electroplated fibers as well as LMI precursors.

The diffusion-bonding technique uses the combination of time, temperature, and pressure to deform the precursor wires into intimate contact and to fully knit the wires into a composite by diffusion. This technique has been applied to graphite fiber reinforced aluminum, magnesium, copper, silver, tin, and lead matrices (Ref 3). If an improper combination of

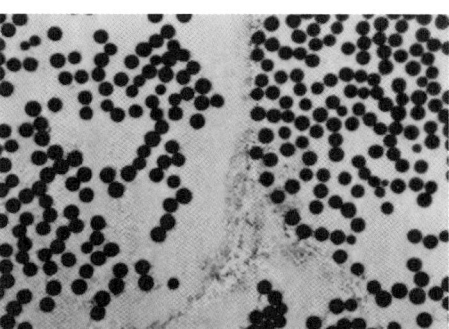

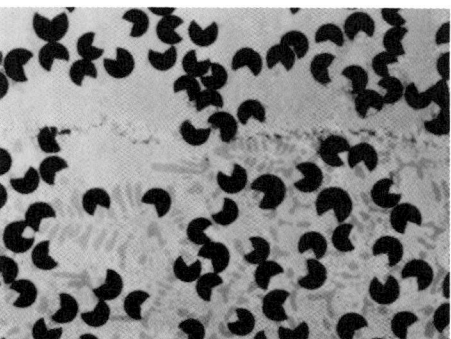

Fig. 6 Diffusion bond zones in silver-copper eutectic matrix composites

time, temperature, and pressure is used, incomplete consolidation can result. In Fig. 5(a), the original precursor wires are clearly visible. The wires did not deform fully to form a fully dense structure. If the wire surfaces are not

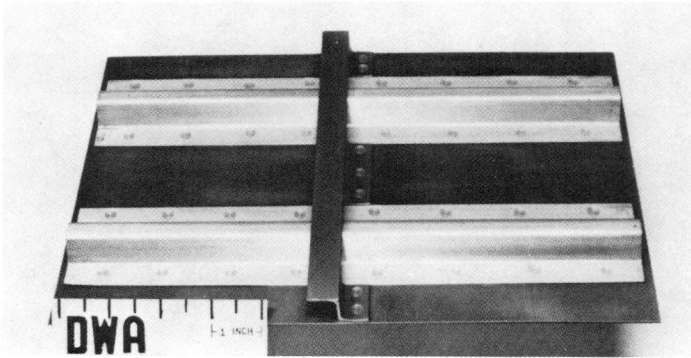

Fig. 7 Spot-welded graphite-aluminum composite stiffened panel (sheet and hat stiffeners made by diffusion bonding)

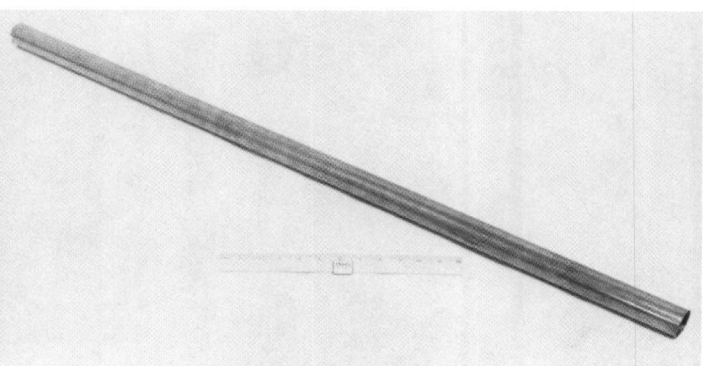

Fig. 8 Thin-walled tube (38-mm, or 1.5-in. diam; 2 m, or 8 ft long)

Fig. 9 High gain antenna mast for Hubbell space telescope, produced with diffusion-bonded P-100 fiber-reinforced aluminum

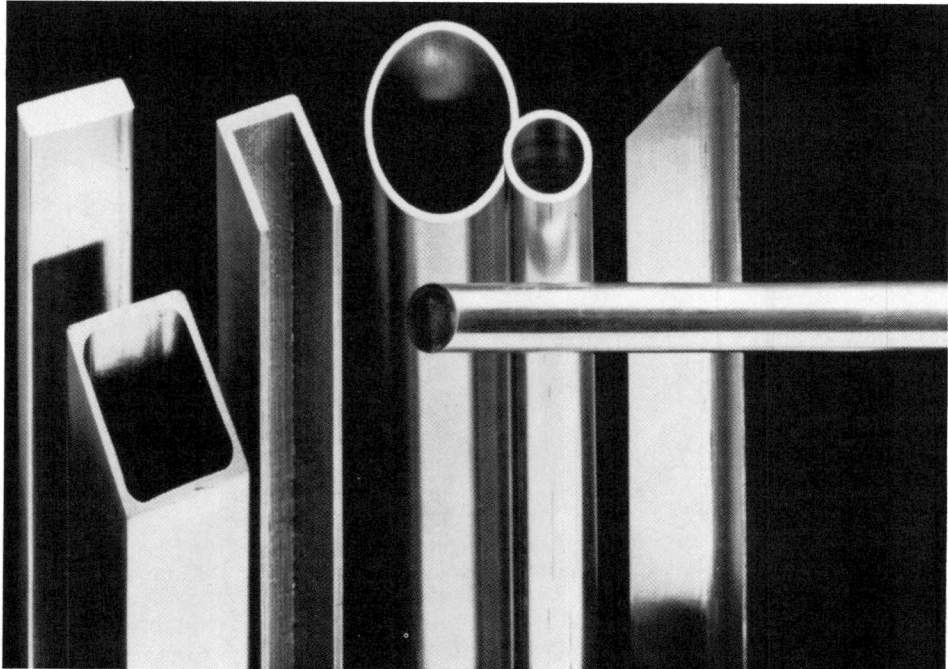

Fig. 10 Various pultruded shapes

cleaned properly prior to bonding, an incomplete bond is made between wires (Fig. 5b). When the precursor is cleaned properly and the correct parameters are used, complete bonds are formed (Fig. 6).

Flat sheet and plate, as well as many shapes, have been made by the diffusion-bonding technique. Examples of some of the shapes are shown in Fig. 7 to 9. Figure 7 is a generic hat-stiffened panel made by spot welding diffusion-bonded hat stiffeners to a sheet of graphite aluminum. Figure 8 is a thin-walled tube 38 mm (1.5 in.) in diameter and 2 m (8 ft) long. Figure 9 is a high-gain antenna boom for the Hubble space telescope made with diffusion-bonded sheet of P100 graphite fibers in 6061 aluminum. This structure is 3.6 m (11.7 ft) long with internal dimension tolerances of ±0.15 mm (±0.005 in.) along the entire length so that the tube can act as a wave guide. This structure also requires the stiffness and low-expansion characteristics of this composite.

Pultrusion

Pultrusion is a hot isothermal drawing process that consolidates and bonds a green shaped lay-up to produce extended lengths of graphite-reinforced MMC. These components have a matrix of either aluminum alloy or magnesium alloy. This low-cost direct fabrication of shapes is suitable for the production of rods, tubing (rectangular and round), C's, T's, channels, and angles having longitudinal reinforcement. Several of these shapes are shown in Fig. 10. A typical microstructure of consolidated two-ply tubing is shown in Fig. 11.

Tubing from 13- to 50-mm (½- to 2-in.) diameters and lengths up to 2 m (6 ft) in one-, two-, and three-ply configurations using LMI wire has been fabricated successfully. Table 3 shows data on tubing fabricated by this process using the 2000-filament P100 pitch-based graphite LMI wire in an aluminum matrix.

In the tubing example cited, the modulus translation of yarn to finished tubing is 90% rule-of-mixture, while the tensile strength translation is nominally 60%. This is com-

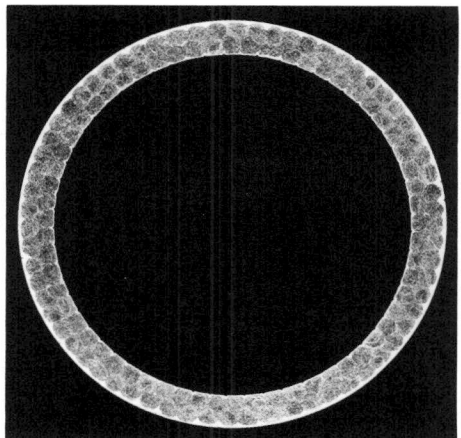

Fig. 11 Typical microstructure of consolidated two-ply pultruded MMC tubing

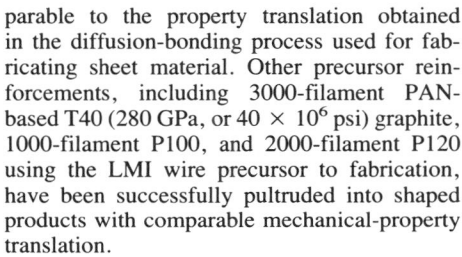

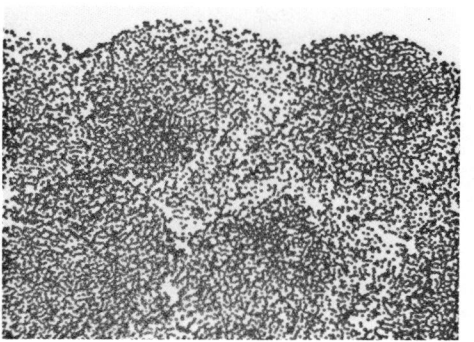

Fig. 12 Typical microstructure of consolidated strip. For 40 vol% graphite: ultimate tensile strength, 660 ± 14 MPa (96 ± 2 ksi); elasticity, 280 ± 20 GPa (41 ± 3 × 10^6 psi). For 41 vol% graphite: ultimate tensile strength, 670 ± 85 MPa (97 ± 12 ksi); elasticity, 340 ± 20 GPa (49 ± 3 × 10^6 psi). Values are the average of three or more duplicate tests.

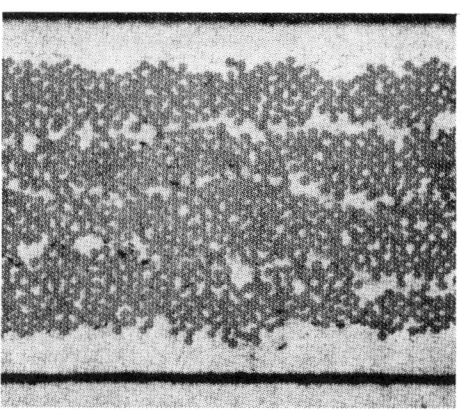

Fig. 13 Typical four-layer 0.41-mm (0.016-in.) thick P75 DWG-Al microsection

parable to the property translation obtained in the diffusion-bonding process used for fabricating sheet material. Other precursor reinforcements, including 3000-filament PAN-based T40 (280 GPa, or 40 × 10^6 psi) graphite, 1000-filament P100, and 2000-filament P120 using the LMI wire precursor to fabrication, have been successfully pultruded into shaped products with comparable mechanical-property translation.

In addition to longitudinal reinforcement components, two-ply tubing with ±17° and ±23° angle-ply construction in 25- and 50-mm (1- and 2-in.) diameters, respectively, have been fabricated to facilitate tailoring of the composite tubing longitudinal CTE for specific applications. The longitudinal CTE of material containing unidirectionally aligned fibers can be predicted based on the following formula:

$$CTE_L = \frac{E_m \cdot \alpha_m \cdot V_m + E_f \cdot \alpha_f \cdot V_f}{E_m \cdot V_m + E_f \cdot V_f}$$

Table 3 Tubing fabricated using LMI wire

Configuration	Ultimate tensile strength		Elastic modulus		Graphite, vol%
	MPa	ksi	GPa	10^6 psi	
19.1 mm (¾-in.) diam, three ply	720	105	345	50	45
	676	98	345	50	45
	710	103	331	48	45
Average	703 + 20	102 + 3	338 + 7	49 + 1	45
1-in. diam, two-ply.	720	105	296	43	40
	703	102	283	41	40
	641	93	310	45	40
Average	690 + 34	100 + 5	296 + 14	43 + 2	40

Table 4 Tensile property summary for DWG-produced composite

Fiber	Elasticity		Ultimate tensile strength, longitudinal		Ultimate tensile strength, transverse	
	GPa	10^6 psi	MPa	ksi	MPa	ksi
P55.	207-221	30-32	520-620	75-90	30-50	4-7
P75.	276-296	40-43	620-720	90-105	30-50	4-7
P100.	379-414	55-60	550-830	80-121	30-50	4-7
P120(a).	469-558	68-81	590-880	85-127	30-50	4-7

(a) Preliminary data

Table 5 Typical properties of graphite-magnesium castings

Fiber type	Fiber content/ orientation	Casting	Fiber preform method	Ultimate tensile strength, 0°		Elasticity, 0°		Ultimate tensile strength, 90°		Elasticity, 90°		CTE 10^{-6}/K
				GPa	10^6 psi	GPa	10^6 psi	GPa	10^6 psi	GPa	10^6 psi	
P55.	40%/0°	Rod	Filament wound	0.72	0.105	172	25	(a)	(a)	(a)	(a)	. . .
P100.	35%/0°	Rod	Filament wound	0.72	0.105	248	36	(a)	(a)	(a)	(a)	. . .
P75.	40%/±16° plus 9%/90°	Hollow cylinder	Filament wound	0.45	0.065	179	26	0.061	0.0089	86	12.5	1.3
P100.	40%/±16°	Hollow cylinder	Filament wound	0.56	0.081	228	33	0.38	0.0055	30	4.4	−0.07
P55.	40%/0°	Plate	Prepreg	0.48(b)	0.070(b)	159	23	0.02	0.003	21	3	2.3
P55.	30%/0° plus 10%/90°	Plate	Prepreg	0.28	0.04	83	12	0.10	0.015	34	5	4.5
P55.	20%/0° plus 20%/90°	Plate	Prepreg	8.45(b)	1.225(b)	90	13	0.24	0.035	90	13	. . .

Note: All materials contain pitch-based fibers. (a) Not determined. (b) Equivalent 0° ultimate tensile strength at 400 °C (750 °F)

Fig. 14 Cast graphite-magnesium simulated rotary engine housing

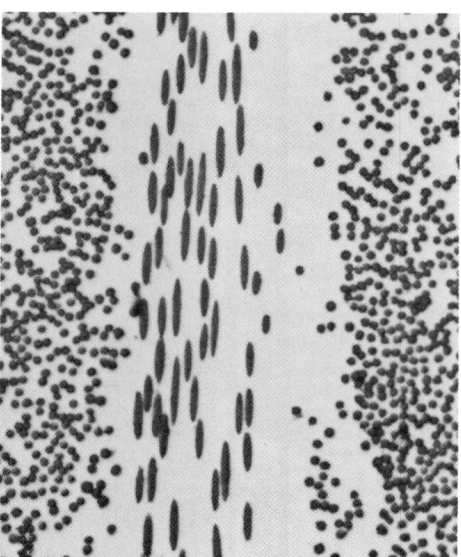

Fig. 15 Microstructure of graphite-magnesium casting containing cross-plied fibers. 100×

where CTE_L is longitudinal CTE, E_m is modulus of the matrix alloy, α_m is CTE of the matrix alloy, V_m is volume percent of the matrix alloy, E_f is longitudinal modulus of the fibers, σ_f is longitudinal CTE of the fibers, and V_f is volume percent of the fibers.

Rapi-Press

The Rapi-Press composite consolidation process is a hot isothermal rolling process that effects consolidation and bonding of a green composite lay-up to produce low-cost composite sheet. This technique has been used successfully to produce composite sheet that is continuously reinforced with graphite in matrix alloys of aluminum, magnesium, and copper, as well as to produce borsic and silicon carbide reinforced aluminum. Various graphite fibers, including the pitch-based P55, P75, P100, and P120, and PAN-based T40 have been used as reinforcement. Precursor to fabrication has included drum-wound silicon carbide and borsic monofilament sandwiched between metallic foils, magnetron sputter-coated and electrocoated graphite fibers, and LMI graphite wire. Both longitudinal and cross-plied sheet have been fabricated in single- and multiple-ply configurations. Sheet thicknesses ranging from 0.15 to 2.5 mm (0.007 to 0.1 in.) have been produced in lengths of 1.2 m (4 ft).

The primary emphasis has been devoted to the fabrication of P100 graphite-aluminum composite sheet using the LMI wire precursor to fabrication. A typical microstructure of a consolidated strip is shown in Fig. 12; resultant mechanical properties are tabulated below the figure.

Direct Metal Infiltration Processing

In 1984, a thin-ply graphite-reinforced aluminum composite designated DWG was introduced. It had ply thicknesses approaching 0.075 mm (0.003 in.) along with a fiber content between 50 and 60 vol%. To date, most of the work done on this system has been with the various pitch-based fibers. A summary of the data derived from this work appears in Table 4. As with other forms of this composite, the modulus values are those predicted by the rule of mixtures, while the strength values are lower than those predicted by it.

A typical microsection for this composite (Fig. 13) shows that the fiber content is high and uniform. This process has been used to make sheets, plates, and structures. Among the structures is a 50-mm (2-in.) diam tube that has a wall thickness of 0.81 mm (0.032 in.) and eight plies aligned at ±15° with respect to the axis for expansion control. This type of struc-

ture becomes very attractive for space structures in which high stiffness and low thermal distortion are critical.

Casting

Casting of continuous graphite fiber MMCs is an especially appropriate fabrication technology when these materials are to be used in complex-shaped parts, such as the joints which connect the load-bearing members of a truss structure or the components of an internal combustion engine. It is also an excellent method for producing these materials in thick sections or in sections containing a variety of fiber orientations, because mechanical pressure is not applied to the material during fabrication, nor are frictional losses a factor in consolidating thick sections, nor is there a problem of damaging cross-plied fibers by forcing them to conform to the small irregularities of the underlying layers. However, casting is not as well suited to the fabrication of thin, flat sheets because of the difficulty of feeding molten metal over long distances through thin cross sections.

In casting, a metal matrix precursor material is not produced. Rather, the fibers are first formed into their desired configuration and placed in a casting mold, and molten metal is added as the final step in forming a near-net shape composite part. In preparing the fiber preform (the arrangement of the fibers into the orientation, packing density, and size desired in the finished part), techniques developed for the resin composite industry (such as filament winding and prepreg formation) are useful. In casting, conventional foundry practice is used to a certain extent. However, in both preform preparation and casting, significant modifications to standard procedures must be made when casting graphite-metal composites.

One major difference between the fiber configurations in resin and those in metal composites is that in metals, conventionally woven fibers are not suitable as reinforcements. The reason is that woven fibers are kinked because of the nature of an over/under weave. For the fibers to provide maximum reinforcement, they must straighten out under load. This is possible in resin composites because of the low elastic modulus of the matrix. However, with metals, the relatively high modulus matrix resists the fiber-straightening process to such an extent that the fibers break before they straighten out. For this reason, fiber-reinforced metal castings use fiber preforms that consist of layers of unkinked fibers, often with the fibers oriented in a variety of directions in successive layers.

Several different casting processes have been used successfully with graphite-reinforced metals. These include permanent mold, split mold plaster, and investment casting. Vacuum or pressure assist is often used in the penetration of the fiber preform by the molten metal. Die casting is generally not used, because the inherent high speed of this process is not compatible with the time required for full penetration of the interfiber channels.

Graphite fibers are not normally wetted by molten metals. Because wettability is a major advantage in the casting process, a proprietary fiber coating has been used which, when applied to the fibers before casting, makes them wettable by molten magnesium alloys. Unlike the previously described titanium boride coating, this coating is air stable; that is, the coating remains wettable even after it is exposed to air. It is this feature that allows filament winding and mold loading procedures to be employed with previously coated fibers. Because suitable air-stable fiber coatings have not yet been identified for metals other than magnesium, the

technology for casting graphite fiber reinforced magnesium is much further advanced than that for other graphite-reinforced metals.

To date, a limited mechanical and physical property data base on graphite fiber reinforced magnesium castings has been generated. Because of the wide variety of fiber types and almost limitless number of fiber orientations that are possible in these castings, a complete data base on all combinations is not realistically feasible. Fortunately, it is also unnecessary, because an increasing number of studies demonstrate that the properties of the composite castings are, in fact, predictable from the properties of unidirectionally reinforced laminates. Some typical properties are shown in Table 5. Note that all materials shown contain pitch-based graphite fibers, because these fibers have received the greatest attention in composite casting development activities. In addition, the tensile strengths achieved in materials produced from prepreg fiber sheets are appreciably lower than those measured in filament-wound materials. The reason for this is believed to be understood; and it is anticipated that in the near future, essentially full theoretical (rule of mixtures) properties will be consistently achieved, no matter what procedure is used to position the fibers.

Graphite-metal composites exhibit good machinability compared with graphite-epoxy composites, which are difficult to machine. Most castings require a certain amount of finish machining in order to achieve close dimensional tolerances or provide particular surface finishes. Unlike many reinforcement materials used in MMCs, such as oxides and carbides, graphite is not an abrasive. While the fibers themselves do not act as a lubricant, neither do they lead to excessive tool wear, and in fact they tend to act to break up chip formation. The only cautionary note on this subject is that deep tool cuts should be avoided, because the rather

weak fiber-to-matrix bond could result in delamination under high shearing forces.

As mentioned previously, casting is an extremely versatile process for producing complex shapes or shapes containing complex fiber orientations. Demonstration castings have ranged from thin-walled 1.3 mm (0.050 in.) tubes containing both low-angle cross-plied and hoop (90° to the axis) fibers to the large simulated rotary engine housing illustrated in Fig. 14. Infiltration of the fibers by the metal is generally excellent, as shown in the microstructure in Fig. 15. In general, although the development time required for a particular shape is longer for graphite fiber reinforced castings than is the case for conventional unreinforced castings (because of the extra difficulties involved with fiber lay-up and metal infiltration), suitable parameter adjustment should allow any part that can be cast in conventional materials to be cast also in graphite-reinforced magnesium and, when suitable fiber coatings become available, other graphite-reinforced metals as well.

REFERENCES

1. E. DiCesare, ''Effect of Rolling Reduction on Discontinuous Graphite Fiber-Reinforced Magnesium Composite,'' MTL TR 85-27, U.S. Army Materials Technology Laboratory

2. P.D. Burke and D.E. Kizer, ''Characterization and Development of Hot Pressed Discontinuous Gr/Al for Low CTE Fittings,'' MCI 85-0640, Contract NAS9-17293, Material Concepts, Inc.

3. W.C. Harrigan, Jr. and R.H. Flowers, Graphite-Metal Composites: Titanium-Boron Vapor Deposit Method of Manufacture, in *Failure Modes in Composites IV*, J.A. Cornie and F.W. Crossman, Ed., TMS-AIME/ASM Joint Composite Materials Committee Symposium, 24-26 Oct 1977

Continuous Aluminum Oxide Fiber MMCs

James C. Romine, E.I. Du Pont de Nemours & Company, Inc.

ALUMINUM OXIDE fiber reinforced metal matrix composites (MMCs) are important materials for weight-sensitive applications operating at elevated temperatures. Under these conditions, MMCs are superior to unreinforced metals in stiffness, strength, fatigue performance, and wear characteristics. These improved features are largely a function of the properties of aluminum oxide fibers.

The chemical and oxidative inertness of aluminum oxide limits fiber degradation during both high-temperature composite fabrication and use. Aluminum oxide has good mechanical properties, with some polycrystalline fibers approaching the theoretical tensile modulus of 400 GPa (58 × 10⁶ psi) (Ref 1). Because strength and stiffness remain high at elevated temperatures, the composites creep less and have significantly better fatigue resistance than unreinforced metals. The hardness of aluminum oxide accounts for the exceptionally good wear properties of the MMC. The chief deficiency of aluminum oxide is its relatively high density (3.2 to 4.0 g/cm³).

Continuous aluminum oxide fibers are now commercially available. They have been evaluated as reinforcing fibers with a number of metal matrices, especially aluminum and magnesium. Composite fabrication is generally accomplished by casting, and many key mechanical properties have been defined.

Constituent Materials

The two continuous aluminum oxide fibers that are currently commercially available are Fiber FP from Du Pont (Ref 2, 3) and Sumika from Sumitomo (Ref 4). These multifilament yarns contain at least 85% aluminum oxide. Saffil (Ref 5), a short staple fiber manufactured by ICI, will also be discussed, because it has been evaluated in many of the same applications as have the continuous fibers (Ref 6). The physical properties of these important aluminum oxide fibers are given in Table 1.

The development of aluminum oxide fiber reinforced MMCs has focused on the low melting point metals, such as aluminum and magnesium and their standard casting alloys. Lead and copper composites have also been produced. Little has been done, however, with steel and superalloy composites because aluminum oxide fibers tend to undergo degrading microstructural changes at the required fabrication and use temperatures. For example, forming operations such as forging or rolling would break the fibers. While composites could be made by casting, the temperature (1600 °C, or 2910 °F) and the aggressive reactivity of molten iron would cause grain growth and chemical attack of the aluminum oxide. In the case of superalloys, composites could be formed by plasma spraying or diffusion bonding, but these alloys are generally used in applications requiring temperatures of 1000 °C (1830 °F) and above. At these use temperatures, most aluminum oxide fibers will eventually lose strength through grain growth and will creep under load.

Fabrication

Continuous fiber MMCs are usually made by casting, which avoids the fiber damage encountered in other metal-forming operations, such as forging and powder metallurgy. Casting also permits production of near-net shape components that require little machining. One of the earliest fabrication developments in this area was vacuum infiltration casting, a process in which molten metal is drawn into an evacuated mold containing a fiber preform (Ref 7). To use the method successfully with aluminum, a small amount of lithium is added to the matrix metal to promote fiber wetting and produce a porosity-free composite. Squeeze casting has also been employed to ensure good infiltration of the metal into a fiber preform (Ref 8). Squeeze casting does not require special alloys and is more amenable to high-production casting operations. Both methods use process optimization to produce superior composite properties. Considerable attention has been given to the role of the rate of solidification in development of matrix microstructures in cast composites (Ref 9). The presence of fibers in these composites also has significant impact on the microstructure, as do chemical reactions between the fibers and matrices (Ref 10, 11). The fibers have been found to control the location and distribution of phases in the matrix and to direct or disrupt dendrite growth. The surface of the fibers can act as nucleation catalysts, for solidification processing, and the fiber-matrix interface is often the site of reactions that occur during processing.

The machining of MMCs presents a unique challenge, because the wear properties are dominated by the aluminum oxide, which is a very abrasive material. One study concluded that the proper choice of machine tool materials and design, combined with slow tool speed and high feed rate, is preferred (Ref 12). Surprisingly, turning and milling of Fiber FP-aluminum at a speed of 30.48 m/min (100 surface ft/min) with C-2 uncoated carbide or ceramic-coated carbide tooling was found to be superior to using more expensive polycrystalline and natural diamond tooling materials. The tool life obtained when machining Fiber FP-aluminum is low, but since the optimum feed rate is the

Table 1 Properties of current aluminum oxide fibers

Fiber	Composition, % Al₂O₃	SiO₂	Crystal form	Filaments/ yarn	Fiber diam μm	μin.	Density, g/cm³	Tensile strength MPa	ksi	Elongation, %	Tensile modulus GPa	10⁶ psi	CTE
Fiber FP	99.5	...	α	200	20	790	3.9	1560	230	0.4	390	55	6.8
Sumika	85	15	γ	380	17	670	3.2	1775	260	0.8	210	30	8.8
Saffil	96-97	3-4	δ	380	3	120	3.3	2000	290	0.7	300	45	...

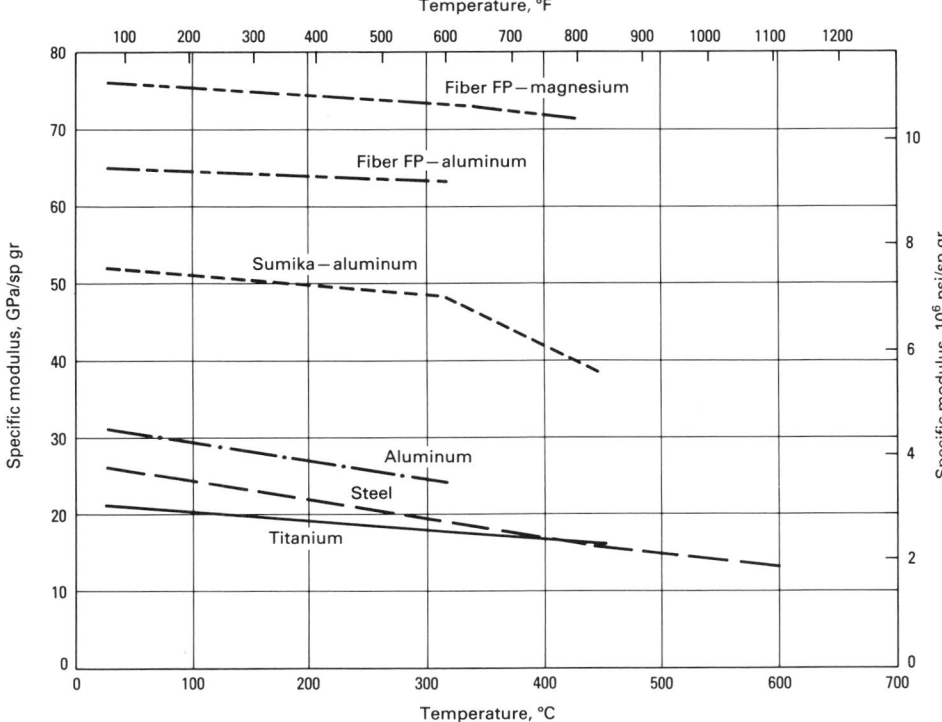

Fig. 1 Specific modulus as a function of temperature

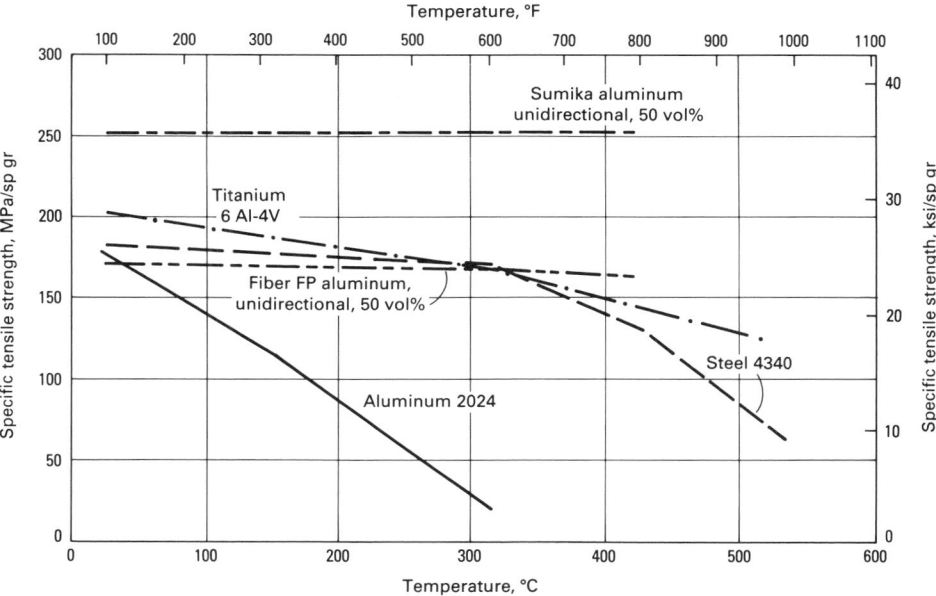

Fig. 2 Specific tensile strength as a function of temperature

developed are the comprehensive data required for design specifications. Recently, a combination of finite-element analysis and mechanical testing has been used to predict and confirm key design data for certain complicated composite applications (Ref 13).

Nondestructive evaluation (NDE) of the composite materials is crucial for quality assurance. Considerable method development has been required to adapt NDE methods to MMC testing, because fibers act as a discrete phase within the matrix and complicate the analysis. Ultrasonic C-scan, x-ray, and acoustic microscopy have been successfully employed. Computer-aided tomography (CAT) is particularly useful for complex shapes and is capable of resolving internal defects in three-dimensional analysis (Ref 14).

Properties

The mechanical properties of aluminum oxide fiber reinforced MMCs have been reported in several comprehensive studies (Ref 15-17). Data have also been generated in connection with the development of specific applications, such as automotive engine components (Ref 18, 19), helicopter transmission housings (Ref 20), and lead composite battery electrodes (Ref 21).

The MMC modulus approximately obeys the rule of mixtures in that a typical unidirectional composite containing 50% fiber by volume will have over twice the tensile modulus of the matrix metal (200 versus 85 GPa, or 29 versus 12×10^6 psi, for Fiber FP reinforced aluminum). Figure 1 shows the specific modulus of several MMCs as a function of testing temperature. Data for unreinforced metals are included for comparison.

The improvement in FP-aluminum composite tensile strength is less dramatic than is that of modulus. On a specific basis, the room temperature strength in the fiber direction for a unidirectional FP-aluminum composite is typically about the same as that of the unreinforced metal. However, as the use temperature increases, the properties of the composite are retained to a greater degree, as shown in Fig. 2.

The fatigue resistance imparted by aluminum oxide fibers to metals such as aluminum and magnesium may allow them to be substituted for dense steel components in weight-sensitive applications such as reciprocating engine parts. Published results show that an aluminum oxide fiber reinforced aluminum composite can provide the required strength, stiffness, and fatigue properties to replace steel connecting rods at half the component weight (Ref 22, 23, 24). Figure 3 shows rotating bending fatigue properties of Fiber FP-aluminum at room temperature and 175 °C (350 °F).

Wear applications seem to be a promising area for using aluminum oxide fiber reinforced metals. In 1982, Toyota Motor Company introduced the first commercial application of an aluminum oxide fiber reinforced aluminum

highest that produces an acceptable surface finish, the production rates are quite high.

Most mechanical testing of aluminum oxide fiber reinforced MMCs has been performed on specimens machined from cast stock. Unidirectional fiber orientation is often used, although some data have been generated on cross-plied, filament-wound, or other more complex fiber geometries. Many standard engineering properties, such as tensile and compressive strengths, flexural behavior, modulus, creep, and mechanical and thermal fatigue have been determined on at least a few representative composite materials. Yet to be

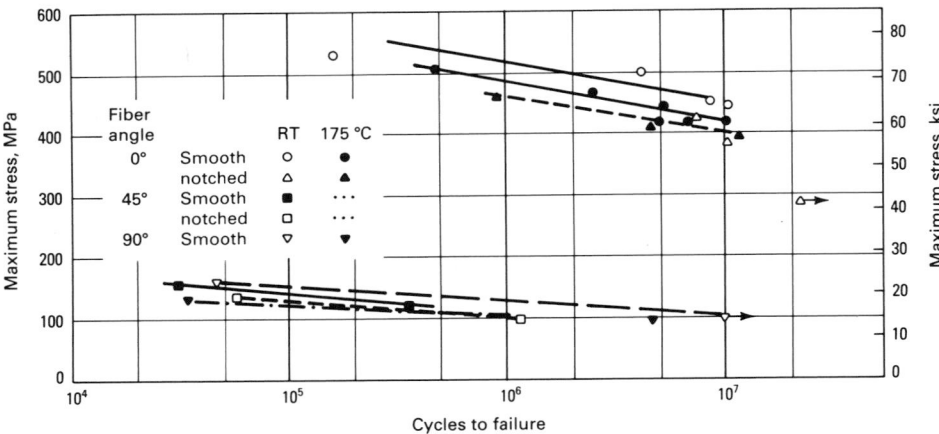

Fig. 3 Rotating bending fatigue performance of 50 vol% Fiber FP aluminum

composite by incorporating Saffil in selectively reinforced diesel engine pistons. The composite piston had about twice the wear and seizure resistance of the incumbent iron piston, as well as lower weight and improved thermal conductivity (120 versus 35 W/m · K, or 70 versus 20 Btu/ft h °F). A compromise between performance and manufacturing considerations was made to optimize the fiber volume fraction to between 3.5 and 5.5% for this application.

Corrosion. Preferential attack by salt water or other chemicals at the fiber-matrix interface and galvanic coupling between conductive fibers and the matrix metal can result in unacceptably high rates of corrosion. A recent review of the topic concludes that aluminum oxide fiber reinforced metals do not have serious corrosion problems. This may promote the use of aluminum oxide fibers in preference to other reinforcing fibers, such as boron and carbon fibers, in applications prone to corrosion (Ref 25).

Applications

In spite of the large aluminum oxide fiber reinforced MMC development effort over the past 15 years, few applications have yet reached commercial production. The Toyota reinforced pistons are evidence that these materials can find value in the marketplace, but further commercial development will depend on economics and technical advances. There has been considerable speculation on the applications most likely to take advantage of the unique properties of aluminum oxide fiber reinforced MMCs (Ref 26). Those applications in which weight savings or high-temperature use can justify the high cost of materials and fabrication are likely to dominate in the near future. The dramatic price drop that is projected for high-volume fiber production should allow aluminum oxide fiber reinforced MMCs to compete in a number of performance-critical applications.

REFERENCES

1. A.K. Dhingra, What Are Fibers Doing in Metal Castings?, *Chemtech*, Oct. 1981, p 600-608
2. A.K. Dhingra, Alumina Fibre FP, *Philos. Trans. R. Soc. (London) A*, Vol 294, 1980, p 411-417
3. A.K. Dhingra, Advances in Inorganic Fiber Developments, in *Contemporary Topics in Polymer Science*, Vol 5, E.J. Vandenberg, Ed., Plenum Press, 1984, p 227-260
4. Y. Abe, K. Fujimura, and S. Horikira, Review: Alumina Fibers and Composite Materials, *J. Japan. Soc. Compos. Mater.*, Vol 6 (No. 3), 1980, p 89-97
5. J.D. Birchall, The Preparation and Properties of Polycrystalline Aluminium Oxide Fibres, in *Fabrication Science*, Vol 3 (No. 33), D. Taylor, Ed., The British Ceramic Society, 1983, p 51-62
6. J. Dinwoodie, E. Moore, C.A.J. Langman, and W.R. Symes, "The Properties and Applications of Short Staple Alumina Fibre Reinforced Aluminum Alloys," in *Conference Proceedings of the Fifth International Conference on Composite Materials*, July/Aug 1985, p 671-685
7. A.K. Dhingra and W.H. Krueger, New Engineering Material-Magnesium Castings Reinforced With Continuous Alumina Fiber FP, in *Proceedings of the 36th World Conference on Magnesium*, International Magnesium Association, 1979
8. M.G. Bader, T.W. Clyne, G.R. Cappleman, and P.A. Hubert, The Fabrication and Properties of Metal-Matrix Composites Based on Aluminum Alloy Infiltrated Alumina Fibre Preforms, *Compos. Sci. Technol.*, Vol 23, 1985, p 287-301
9. A. Mortensen, M.N. Gungor, J.A. Cornie, and M.C. Flemings, Alloy Microstructures in Cast Metal Matrix Composites, *J. Met.*, March 1986, p 30-35
10. C.G. Levi, G.J. Abbaschian, and R. Meh-
rabian, "Interface Interactions During Fabrication of Aluminum Alloy-Alumina Fiber Composites," *Met. Trans. A*, Vol 9A, 1978, p 697-711
11. J.E. Hack, R.A. Page, and R. Sherman, The Influence of Thermal Exposure on Interfacial Reactions and Strength in Aluminum Oxide Fiber Reinforced Magnesium Alloy Composites, *Met. Trans. A*, Vol 16A, 1985, p 2069-2072
12. M.J. McGinty and C.W. Preuss, "Machining Ceramic Fiber Metal Matrix Composites," Paper presented at ASM International Conference on High Productivity Machining, Materials and Processing, American Society for Metals, New Orleans, May 1985
13. F. Folgar, E. Perez, J. Hunt, and D. McCabe, "Finite Element Analysis of Fiber FP/Metal Matrix Composite Connecting Rods," Paper presented at PATRAN User's Conference, Newport Beach, CA, June 1986
14. J.E. Widrig, D.D. McCabe, and R.L. Conner, Nondestructive Evaluation of Fiber FP Reinforced Metal Matrix Composites, in *ASTM Symposium on Testing Technology of Metal Matrix Composites*, Nov 1985, in press
15. A.R. Champion, W.H. Krueger, H.S. Hartmann, and A.K. Dhingra, Fiber FP Reinforced Metal Matrix Composites, in *Proceedings of the Second International Conference on Composite Materials*, April 1978, p 883-904
16. H.R. Shetty and T.-W. Chou, Mechanical Properties and Failure Characteristics of FP/Aluminum and W/Aluminum Composites, *Metal. Trans.*, Vol 16A, 1985, p 853-864
17. M.G. Bader, T.W. Clyne, G.R. Cappleman, and P.A. Hubert, The Fabrication and Properties of Metal-Matrix Composites Based on Aluminium Alloy Infiltrated Alumina Fibre Preforms, *Compos. Sci. Technol.*, Vol 23, 1985, p 287-301
18. T. Donomot, K. Funatani, N. Miura, and N. Miyake, "Ceramic Fiber Reinforced Piston for High Performance Diesel Engines," Paper presented at the SAE International Congress and Exposition, Society of Automotive Engineers, Detroit, 1983, p 1-10
19. W.H. Krueger and A.K. Dhingra, Alumina Fiber Reinforced Metal Composites for Potential Automotive Engine Applications, in *New Composite Materials and Technology*, Symposium Series, American Institute of Chemical Engineers, Vol 78 (No. 217), 1982, p 13-24
20. R.L. Pinckney and J.W. Lenski, Development of Metal-Matrix Helicopter Transmission Cases, in *Metal Matrix Composites II*, NASA TM-82806, Jan 1982, p 55-67
21. H.S. Hartmann and R.A. Sutula, Alumina Fiber FP Reinforced Pure Lead Composites

for Battery Electrodes, *J. Electrochem. Soc.*, Vol 129 (No. 8), 1982, p 1749-1752

22. F. Folgar, W.H. Krueger, and J.G. Goree, Fiber FP/Metal Matrix Composites in Reciprocating Engines, *Metal Matrix, Carbon, and Ceramic Matrix Composites,* NASA Conference Publication 2357, National Aeronautics and Space Administration, 1984, p 43-55

23. J. Nunes, E.S.C. Chin, J.M. Slepetz, and N. Tsangarakis, Tensile and Fatigue Behavior of Alumina Fiber Reinforced Magnesium Composites, in *Proceedings of the Fifth International Conference on Composite Materials*, W.C. Harrigan, Jr., J. Strife, and A.K. Dhingra, Ed., TMS Composite Committee, American Institute of Mining, Metallurgical, and Petroleum Engineers, 1985, p 723-745

24. N. Tsangarakis, J.N. Slepetz, and J. Nunes, Fatigue Behavior of Alumina Fiber Reinforced Aluminum Composites, *Recent Advances in Composites in the United States and Japan*, STP 864, J.R. Vinson and M. Taya, Ed., American Society for Testing and Materials, 1985, p 131-152

25. M. Metzger and S.G. Fishman, Corrosion of Aluminum-Matrix Composites, Status Report, *Ind. Eng. Chem. Prod. Res. Dev.*, Vol 22, 1983, p 296-302

26. C.F. Lewis, The Exciting Promise of Metal-Matrix Composites, *Mater. Eng.*, Vol 107 (No. 5), 1986, p 33-37

Continuous Tungsten Fiber MMCs

Diana M. Essock, TRW Inc.*

THE USE OF CONTINUOUS TUNGSTEN FIBERS to reinforce composites not only provides a strong, stiff addition to matrix material, but imparts an inherently high-temperature capability, good ductility, and high thermal conductivity. Applications of tungsten-reinforced metal matrix composites (MMCs) are particularly appropriate for highly oriented loading structures such as turbine blades, pressure vessels, flywheels, and simply loaded beams. A sizable effort in the area of elevated-temperature MMCs has focused on the development of continuous fiber reinforced superalloys (FRS). The generic designation, FRS, refers to a class of engineering materials in which an oxidation-resistant matrix alloy is reinforced with a strong, stiff, creep-resistant fiber. Although several types of fibers have been investigated for FRS applications, including tungsten and molybdenum alloys, silicon carbide, and aluminum oxide, the major emphasis has been on tungsten and tungsten alloys.

At the present time, the primary contending applications for FRS are in aircraft engines and rocket engine turbopumps. Users of aircraft gas turbines continually demand more efficient engines. Efficiency can be improved by increasing the compressor ratio and the turbine inlet temperature, but engines can achieve these advanced characteristics only if their materials can withstand the increased loads and oxidation burdens caused by the more severe operating conditions. FRS systems enable operation at temperatures that exceed the current limits of monolithic metallic materials.

Research on FRS has been underway to determine the behavior of this composite system under turbine operating conditions. Although a variety of FRS materials have been evaluated, the most commonly studied systems for aircraft gas turbine applications involve tungsten-based wires such as tungsten-thorium dioxide (W-ThO$_2$), tungsten-hafnium-carbon (W-Hf-C), and tungsten-rhenium (W-Re), in matrices of nickel-based and iron-based high-temperature alloys. The properties of these and other high-temperature composite systems will be discussed later in this article.

In addition to superalloy-matrix composite system research on turbine applications, active work with copper, aluminum, and stainless steel matrices has been undertaken for applications involving heat exchangers and similar devices. Tungsten-reinforced copper was originally studied as a model system because adequate fiber-matrix bonding could be achieved without a chemical reaction, since copper and tungsten are mutually insoluble. Because high thermal conductivity is exhibited by both the matrix and fiber, tungsten-reinforced copper composites are now being developed for applications such as cryogenically-cooled thrust chamber liners for rocket engines (Ref 1). Currently, unreinforced liners have limited life because severe cyclic thermal stresses result in thermal fatigue of the copper alloy inner wall.

Table 1 Comparison of fiber-matrix reactions for various matrix materials

Annealing temperature °C	°F	Matrix	No. of compositions investigated	Recrystal-lization	Intermetallic compound	Diffusion penetration	No recrystal-lization	No reaction
1200	2190	Nickel-based	27	93	55	...	7	4
		Cobalt-based	29	10	83	12	90	10
		Iron-based	30	3	30	...	97	70
1300	2370	Nickel-based	27	96	63	...	4	4
		Cobalt-based	19	21	84	...	79	10
		Iron-based	30	20	80	3	80	13

Source: Ref 3, 4

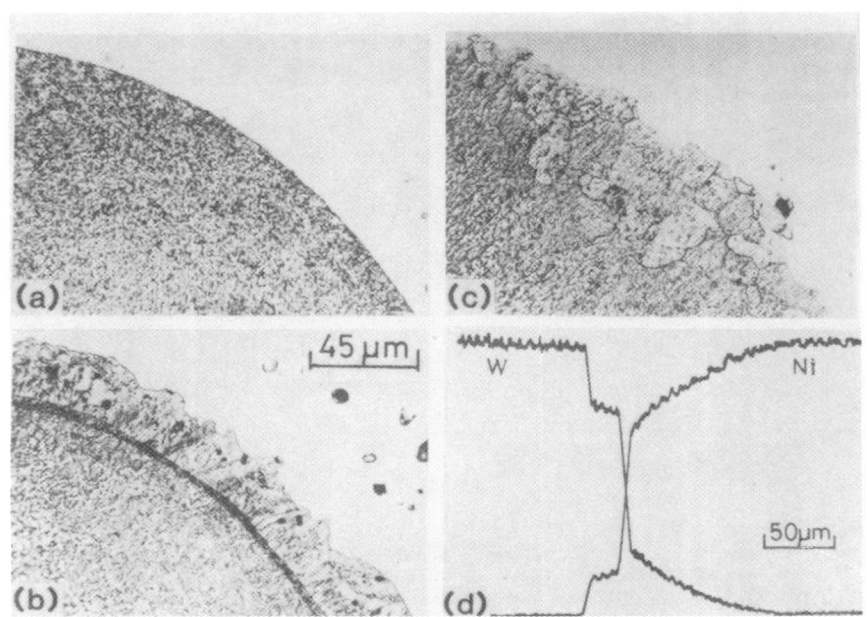

Fig. 1 Microstructures and microprobe line profile at the periphery of uncoated tungsten fibers in nickel matrix. (Ref 8). (a) As hot pressed. (b) 100 h. (c) 200 h. (d) 100 h at 1150 °C (2100 °F)

*Now with General Electric Company

Table 2 Recrystallization temperature of tungsten wires in various matrices

System	Ni content of matrix, wt%	Recrystallization temperature(a) °C	°F	Ref
W-Ni	100	1150-1200	2100-2190	7
W-NiCr	20	1300	2370	7
W-2%ThO$_2$ NiCr	20	1250	2280	7
W-2%ThO$_2$ Ni	100	1080-1130	1980-2065	9
W-2%ThO$_2$ Inconel 718	52	1175	2150	10
W-2%ThO$_2$ Hastelloy X	48	1200	2190	10
W-2%ThO$_2$ Kovar	29.5	1250-1300	2280-2370	9
W-2%ThO$_2$ Stainless steel	10	1435	2615	9
W-2%ThO$_2$ Stainless steel	10	1465	2670	9

(a) 50 μm (1950 μin.) in 1 h. Source: Ref 2

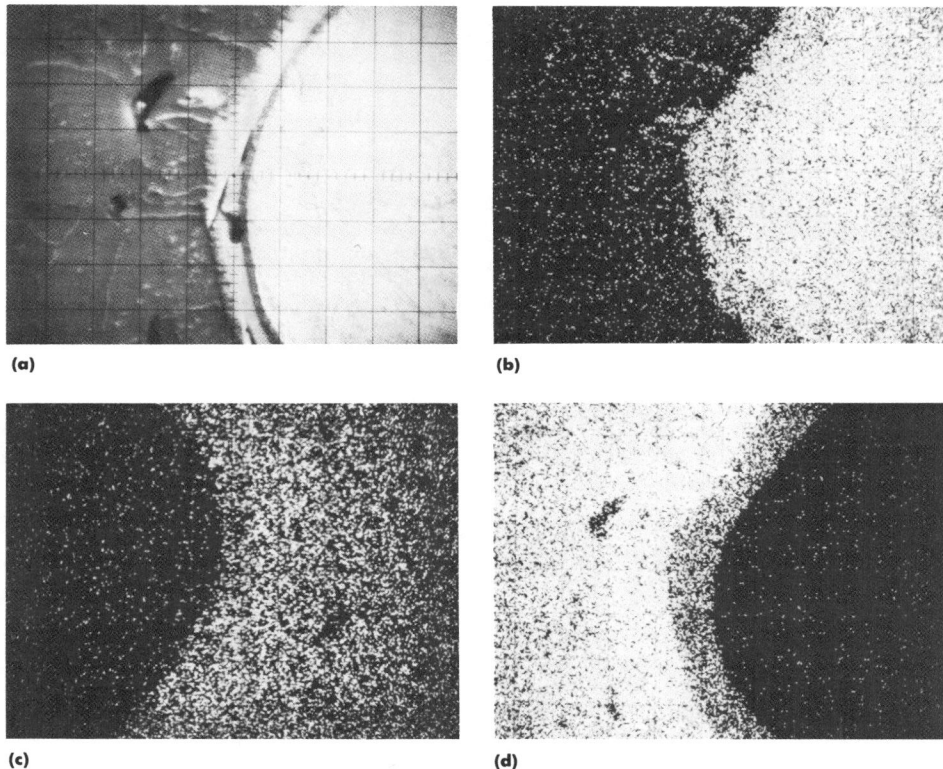

(a)　　　　　　　　　　(b)

(c)　　　　　　　　　　(d)

Fig. 2 Microprobe analysis of interaction zone in W/Fe-Cr-Al-Y specimen thermally exposed 100 h at 1200 °C (2200 °F). (a) Backscatter. (b) Tungsten x-rays. (c) Chromium x-rays. (d) Iron x-rays. 325×. Source: Ref 19

Tungsten fiber reinforcement of the copper is expected to increase the liner wall strength, while retaining efficient heat transfer to the coolant.

The following sections focus on engineering behavior and fabrication processes for continuous tungsten fiber reinforced MMCs.

Fiber-Matrix Compatibility

Without proper selection of fiber-matrix combinations, strength degradation can occur during elevated temperature exposures of MMCs. The observed effects, which result from fiber-matrix interaction, include recrystallization and embrittlement of the fiber, forma-

tion and growth of intermetallic reaction layers, and dissolution of the fiber in the matrix (Ref 2). Screening studies (Ref 3, 4) have been conducted between tungsten fibers and a wide range of binary alloys of iron (Fe), nickel (Ni), cobalt (Co), titanium (Ti), and chromium (Cr), with 5, 10, and 25 wt% additions of aluminum (Al), copper (Cu), silicon (Si), zirconium (Zr), niobium (Nb), molybdenum (Mo), and tungsten (W) (Ref 3). The iron-, nickel-, and cobalt-based alloy samples were annealed at 1200 to 1400 °C (2190 to 2550 °F) for 1 h. A summary of the results is shown in Table 1 (Ref 4).

At 1400 °C (2550 °F), the tungsten fiber recrystallized regardless of the elemental additions. In the nickel-based alloys, tungsten re-

crystallized for almost all ranges of temperature and elemental additions. For the iron-based alloys, additions of 5 to 25 wt% aluminum and 25 wt% titanium resulted in no interaction at 1300 °C (2370 °F), while at 1200 °C (2190 °F) all elemental additions exhibited no interaction, particularly at higher weight percentages. In matrices of cobalt-based alloys, intermetallic compounds were formed at the fiber-matrix interface. The process can be suppressed at 1200 °C (2190 °F) and 1300 °C (2370 °F) only by alloying the cobalt matrix with aluminum.

Numerous studies have been conducted on the reactions between nickel alloys and tungsten fibers. Tungsten recrystallization has been noted in many cases (Ref 5-7) and can be induced at temperatures as low as 950 °C (1740 °F). The recrystallized front advances from the perimeter of the fiber inward and requires a continuous source of nickel for propagation. Figure 1 (Ref 8) illustrates the progression from interface layer formation to tungsten fiber recrystallization in a nickel matrix at 1150 °C (2100 °F). It has also been observed that the recrystallization temperature decreases as the nickel concentration at the interface increases (Ref 9). Data that support this observation are shown in Table 2.

Similar effects have been noted for tungsten fibers in an iron-nickel-based alloy, Incoloy 903 (Ref 11). A reaction zone was noted after a 100-h anneal at 1038 °C (1900 °F) and 1200 °C (2190 °F). Only at the latter temperature was recrystallization of the tungsten fiber seen. The recrystallization temperature can be modified through the presence of metals such as palladium, nickel, aluminum, manganese, platinum, iron, and cobalt (Ref 5). For a nickel-chromium matrix with thoriated tungsten fiber reinforcement, it has been shown that the rate of recrystallization can be significantly affected by the alloying elements in the fiber and matrix, that is, increased by thoria content and decreased by chromium content (Ref 12).

Nickel-based superalloy compositions (Ref 13) that were developed to limit detrimental interactions with tungsten fiber are typified by the alloy Ni-25W-15Cr-2Al-2Ti. The diffusion of nickel into the fibers, hence the rate of recrystallization, is slowed by the formation of titanium (Ti) and aluminum (Al) intermetallic compounds and by the addition of high percentages of tungsten. Conversely, copper alloys with nickel and manganese were developed that encourage the formation of an interaction layer between the matrix and the tungsten reinforcement (Ref 14). It was found that if a reaction layer of 5 to 10 μm (200 to 400 μin.) between the fiber and matrix is produced through alloying and heat treating, the room-temperature tensile strength is improved. However, as the reaction zone increases, the tensile strength decreases rapidly.

An alternative approach to retard fiber-matrix interaction in superalloy matrices is to coat the tungsten fiber with compounds that do not react to the detriment of either the matrix or

Table 3   Representative properties of refractory alloy wires

Alloys	Density, g/cm³	Wire diam		Ultimate tensile strength		Stress for 100-h rupture		Stress/density for 100-h rupture	
		mm	in.	MPa	ksi	MPa	ksi	cm × 10³	in. × 10³
Tungsten alloys, 1093 °C (2000 °F) data									
218CS	19.1	0.20	0.008	869	126	434	63	234	92
W-1%ThO₂	19.1	0.20	0.008	979	142	531	77	282	111
W-2%ThO₂	18.9	0.38	0.015	1193	173	655	95	356	140
W-3%Re	19.4	0.20	0.008	1475	214	476	69	249	98
W-5%Re-2%ThO₂	19.1	0.20	0.008	1213	176	483	70	254	100
W-24%Re-2%ThO₂	19.4	0.20	0.008	1455	211	345	50	183	72
W-Hf-C	19.4	0.38	0.015	1427	207	1110	161	584	230
W-Re-Hf-C	19.4	0.38	0.015	2165	314	1413	205	744	293
Tungsten alloys, 1204 °C (2200 °F) data									
218CS	19.1	0.20	0.008	745	108	317	46	170	67
W-1%ThO₂	19.1	0.20	0.008	841	122	372	54	198	78
W-2%ThO₂	18.9	0.38	0.015	1034	150	483	70	257	101
W-3%Re	19.4	0.20	0.008	1082	157	317	46	168	66
W-5%Re-2%ThO₂	19.1	0.20	0.008	1020	148	303	44	160	63
W-24%Re-2%ThO₂	19.4	0.20	0.008	1014	147	193	28	102	40
W-Hf-C	19.4	0.38	0.015	1386	201	765	111	404	159
W-Re-Hf-C	19.4	0.38	0.015	1937	281	910	132	480	189

Source: Ref 4, 20, 21

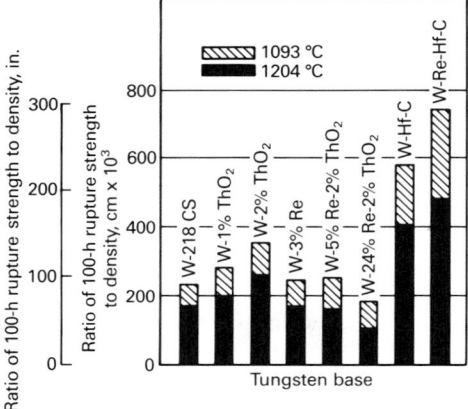

Fig. 3   Ratio of 100 h rupture strength to density for fine-diameter tungsten fibers

the fiber. The key to the future use of such protective coatings is to demonstrate their consistent reliability. Titanium carbide (TiC) and titanium nitride (TiN) coatings have been applied to tungsten fibers through chemical vapor deposition (CVD) (Ref 15) and reactive sputtering (Ref 16). Mechanical testing at 650 °C (1200 °F) and 900 °C (1650 °F) has shown that the coating hinders deleterious effects on strength properties of nickel matrix composites. Additional studies of reactive evaporation-deposited coatings of zirconium carbide (ZrC) and zirconium dioxide (ZrO₂) on tungsten fibers showed that they were effective in controlling interfacial reactions with a nickel matrix at 1150 °C (2100 °F) for 200 h, whereas coatings of hafnium carbide (HfC), hafnium dioxide (HfO₂), tantalum carbide (TaC), aluminum carbide (Al₄C₃), and TiC were somewhat less impervious (Ref 8).

Of all the matrices examined to date, the iron-based alloys appear to show the lowest rate of interaction with tungsten fibers at elevated temperatures. Certain iron-based matrices can be successfully employed with fibers of tungsten and its alloys up to 1300 °C (2370 °F) without inducing recrystallization (Ref 17).

To determine the degree of reactivity of 0.38-mm (0.015-in.) diam W-1%ThO₂ fibers in the iron-based alloy Fe-Cr-Al-Y, test specimens were submitted to isothermal exposures for 10, 100, and 1000 h at 1037, 1093, and 1149 °C (1900, 2000, and 2100 °F) (Ref 18). Results indicated a loss of < 10% of the fiber area after 1000 h at 1093 °C (2000 °F) and 16% after 1000 h at 1149 °C (2100 °F). The rate of growth of the reaction layer is exponential in nature. Microprobe analysis of interaction zones of the same composite samples exposed for 100 h at 1204 °C (2200 °F) are seen in Fig. 2 (Ref 19). The reaction zone appears to consist of W-Fe-Cr ternary compounds. Thus, it can be seen that the choice of a suitable fiber-matrix combination is dependent upon the fabrication temperature, use temperature, and component time at temperature. Consideration of these factors will minimize or avoid recrystallization and dissolution of the tungsten fiber and resultant degradation of composite properties.

Stress Rupture Strength

Stress rupture testing is widely used to indicate use potential of high-temperature material. For tungsten-reinforced MMCs, the composite rupture strength is strongly dependent upon the rupture strength and volume fraction of the fiber. The 100-h rupture strengths for various tungsten alloy fibers at 1093 °C (2000 °F) and 1204 °C (2200 °F) are listed in Table 3 (Ref 4, 20, 21), and are shown in Fig. 3. The superiority of the W-Hf-C and W-Re-Hf-C fibers is evident.

Table 4 presents 100-h rupture strengths and compositions for composites and superalloys at 1100 °C (2010 °F). Because the stresses on turbine components are influenced by the density of the material used, the results presented in Fig. 4 are density compensated for comparison. All of the tungsten-reinforced composites exhibit stress rupture strength that is superior to that of their unreinforced matrices. The potential for a further increase in rupture strength is expected with an increased volume fraction of fiber. Consistent with the fiber rupture properties, the W-Hf-C fiber composites display superior strength. When compared on a density-compensated basis with superalloys, the 40 vol% W-Hf-C composite is almost 2½ times stronger.

Further rupture studies on tungsten-reinforced nickel matrices with varying fiber content (Ref 23) and long application times (Ref 4) indicated that these composites increase their temperature advantage over superalloys with time to rupture, and that stress rupture strengths increase linearly with fiber content. For a tungsten-reinforced iron matrix, thermal cycling before stress rupture testing has shown no evidence of property degradation (Ref 28).

Stress and creep rupture properties were measured for tungsten-reinforced copper alloys at 650 and 815 °C (1200 and 1500 °F) (Ref 29). Equations predicting the creep rate and stress rupture properties of these composites were developed in this study. A comparison of the 100-h rupture stress of several copper-based materials at 815 °C (1500 °F) is shown in Fig. 5. While the dispersion-strengthened copper showed a significant strength increase over unalloyed copper, the tungsten-reinforced composites showed much greater strengths. This strength advantage increases with greater fiber content.

Creep Resistance

Normal creep response showing primary, secondary, and tertiary regions in a strain versus time relationship at constant stress is usually seen for tungsten-reinforced composites. A

Table 4 Rupture strengths and compositions for composites and superalloys

Alloy, wt%	Wire	Wire diam mm	in.	Vol%	Density, g/cm³	100-h rupture strength MPa	ksi	Strength to density for 100-h rupture cm × 10³	in. × 10³	Ref.
100-h rupture strength at 1100 °C (2010 °F)										
Ni-12.5Cr-7W-4.8Mo-5Al-2.5Ti(ZhS6)	VRN Tungsten	0.3-0.5	0.012-0.020	40	12.5	138	20	112.5	44.3	22
Ni-11W-6Al-6Cr-2Mo-1.5Nb(EPD-16)	No reinforcement	...	...	0	8.3	51	7.4	63.5	25.0	23
	Tungsten	0.25	0.010	40	12.7	131	19	104	40.9	
Ni-12.5Cr-2.5Fe-2Nb-4Mo-6Al-1Ti(Nimocast 713C)	No reinforcement	...	...	0	8.0	48	7	61.3	24.1	24
	Tungsten	1.27	0.050	20	10.3	93	13.5	92.7	36.5	
Co-21.5Cr-25W-10Ni-3.5Ta-0.8Ti(Mar-M322E)	No reinforcement	...	...	0	...	48	7	...	...	25
	W-2%ThO₂	0.08	0.003	40	...	207	30	...	...	
Ni-25W-15Cr-2Al-2Ti	No reinforcement	...	...	0	9.15	23	3.3	25.4	10	13
	218CS (Tungsten)	0.38	0.015	40	13.3	138	20	105.8	41.7	
	W-2%ThO₂	0.38	0.015	40	13.0	193	28	151.3	59.6	26
	W-Hf-C	0.38	0.015	40	13.3	324	47	249.1	98.0	27
Fe-24Cr-5Al-1Y	W-1%ThO₂	0.38	0.015	56	12.5	242(a)	35	195.7	76.8	18
Fe-24Cr-5Al-1Y	W-Hf-C	0.38	0.015	35	11.3	242	35	214.7	84.5	28

(a) 831-h rupture strength. Source: Ref 4

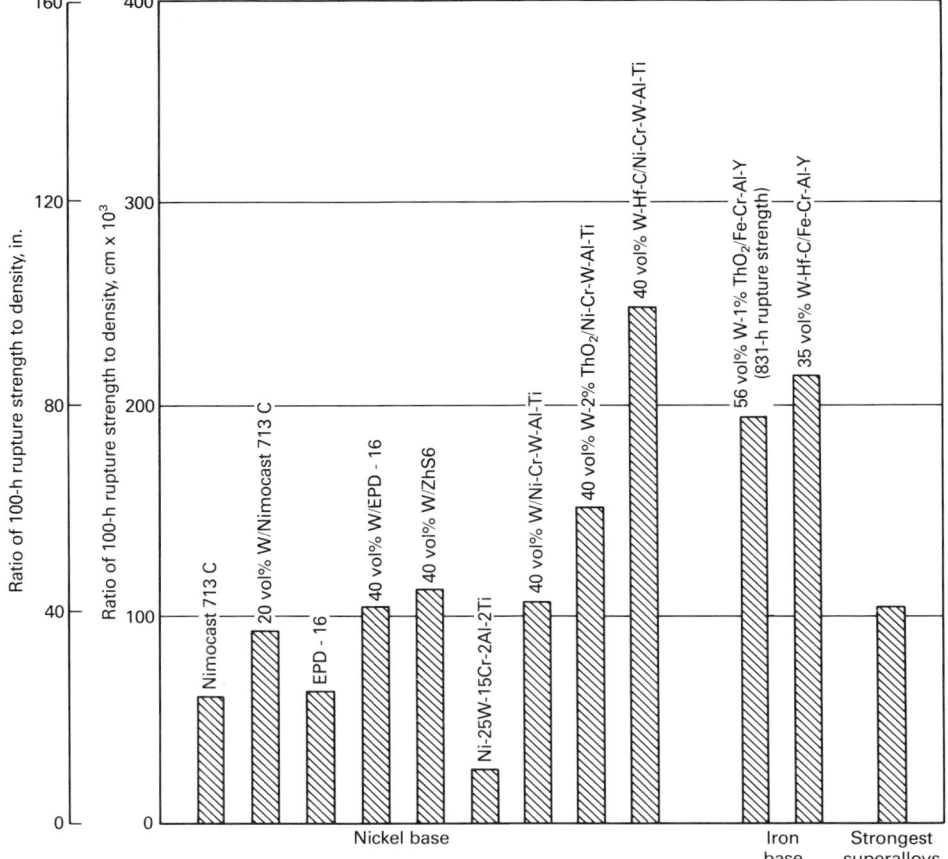

Fig. 4 Comparison of 100 h rupture strength at 1093 °C (2000 °F) for composites and superalloys. Source: Ref 4

series of tests was conducted on 45 vol% 0.38-mm (0.015-in.) diam W-1%ThO₂ fiber in an Fe-Cr-Al-Y matrix (Ref 18). At 1150 °C (2100 °F), the specimens were incrementally loaded to stress levels of 210, 228, and 240 MPa (30, 33, and 35 ksi), as shown in Fig. 6. Because the creep strength for the Fe-Cr-Al-Y matrix is extremely low at these temperatures, the overall creep resistance of the composite is determined by the tungsten fiber. Similar re-

sults are seen for W-5%Re in Nimocast 713C (Ref 24) and W-1%ThO₂ in Hastelloy (Ref 30).

Because of fiber-matrix interface reactions, the creep resistance for tungsten-reinforced composites varies widely with the matrix alloy. For example, W-2%ThO₂ in stainless steel shows an apparent improvement in creep resistance over uncombined wires (Ref 31), while in an Inconel 718 matrix, long-term exposure decreases the creep resistance (Ref 2). Similar results are seen for another tungsten-reinforced nickel matrix (Ref 32). Additional creep rupture testing on tungsten fiber in an Fe-Cr-Al-Y matrix was conducted on test specimens containing fibers of varying diameter in symmetrical, uniaxial configurations (Ref 33). Results showed that the mixed diameter fibers, as well as their stacking sequence, had no adverse effect on composite properties.

Thermal Fatigue

While in service, turbine components are subjected to cyclic mechanical and thermal stresses that result in the initiation and propagation of creep fatigue cracks. In addition to being subjected to cyclic stresses during service, composites have inherent stresses that are generated because of different coefficients of thermal expansion (CTEs) for the individual composite components. In general, the reinforcement CTE is lower than that of the matrix alloy. Typically, the values are 10 to 20 × 10⁻⁶/K for matrix alloys and 5 × 10⁻⁶/K for tungsten alloys.

During the thermal cycle, the matrix is strained in tension upon cooling and strained in compression upon heating, while the reverse occurs for the fiber. Because the tungsten fibers and matrix are significantly ductile at turbine operating temperatures, their tensile and creep properties influence the composite behavior

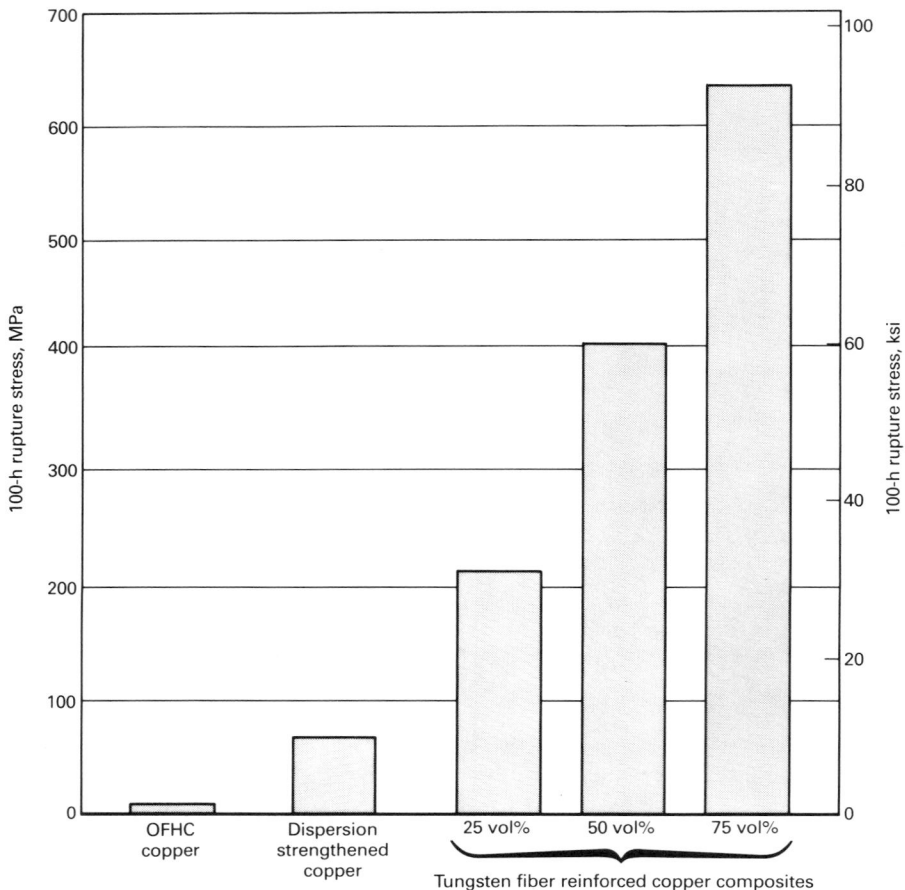

Fig. 5 Comparison of 100 h rupture stress of several copper-based materials at 815 °C (1500 °F). Source: Ref 29

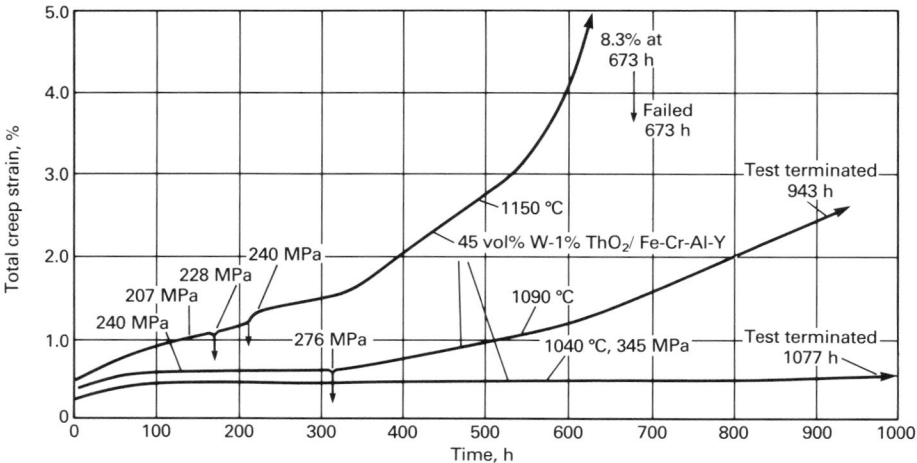

Fig. 6 Creep data for tungsten wire reinforced composites at 1037, 1093, and 1149 °C (1900, 2000, and 2100 °F). Source: Ref 18

during thermal cycling. Theoretical analysis has shown that thermal stresses arising from the composite constituent CTE differences are of such magnitude that damage is possible when the composite is subjected to heating and cooling cycles (Ref 34). These thermal stresses are relieved by one or more mechanisms involving deformation or fracture of the matrix or of the fiber or the fiber-matrix interface, as shown in Table 4 (Ref 2, 4).

Generally, when cycling composites with nickel alloy matrices to temperatures above 900

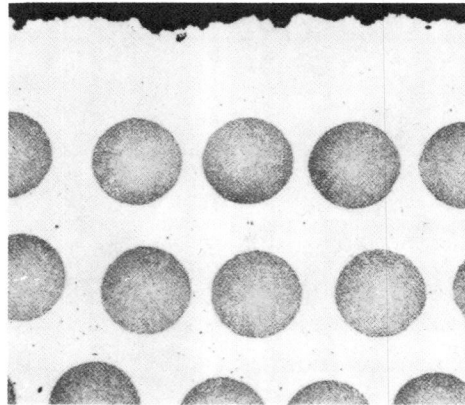

Fig. 7 Surface condition and metallographic cross section of W-1%ThO$_2$/Fe-Cr-Al-Y composite after 1000 cycles, RT to 1204 °C (RT to 2200 °F). 35×. Source: Ref 18

°C (1650 °F) for 1000 cycles, surface cracking of the matrix and distortion are observed. Although this is unsatisfactory behavior for long-term service in an aircraft engine turbine, select systems may be used for shorter life applications, such as ordnance. Distortion was seen in tungsten-reinforced copper composites when cycled from 400 to 800 °C (750 to 1470 °F) for up to 5000 cycles (Ref 40). This dimensional instability was characterized by increasing preferential growth of the matrix along the reinforcing fibers with an increasing number of thermal cycles.

The thermal-cycle response of iron-based matrix composites, in particular Fe-Cr-Al-Y with W-1%ThO$_2$ reinforcement, has shown positive results. After 1000 cycles from room temperature (RT) to 1204 °C (2200 °F), there was a complete absence of internal matrix and fiber cracking, as shown in Fig. 7 (Ref 18). Some roughening of the surface was seen, but there was no significant propagation of surface-initiated cracks. This relatively ductile matrix provides some resistance to crack propagation.

Kovar, an iron-nickel-based alloy with a low CTE, showed a resistance to thermal fatigue up to 500 °C (930 °F) when its CTE is similar to the CTE of the reinforcing tungsten (Ref 36). Reducing the difference in thermal expansion between the fiber and matrix thus offers an avenue for the future development of compatible fiber-matrix systems.

Oxidation and Hot Corrosion

Materials used in air-breathing gas turbines are susceptible to oxidation and hot-corrosion attack. The turbine environment has both an oxidizing atmosphere and fuel combustion products that contain sulfides, sulfates, chlorides, and other by-products. When materials are exposed to the combustion environment, ash or salt are typically deposited on their

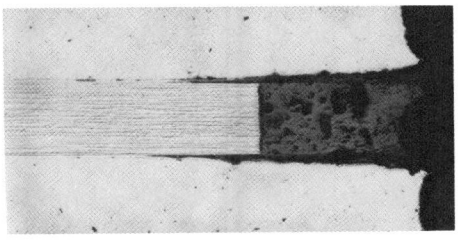
(a)

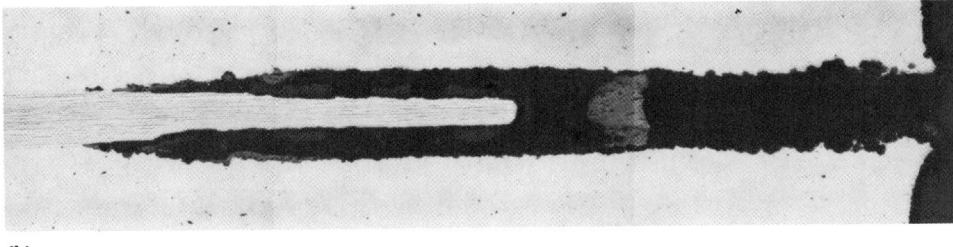
(b)

Fig. 8 The effect of time on oxidation of TaC-coated W fibers in Fe-Cr-Al-Y matrix. Composite subjected to 1204 °C (2200 °F) in air. (a) 1-h exposure. (b) 10-h exposure. 60×. Source: Ref 18

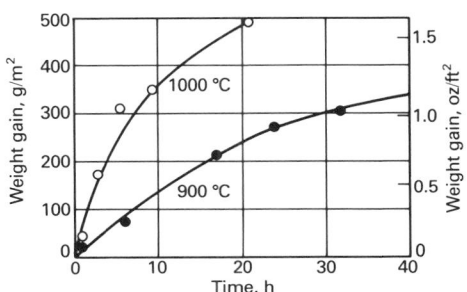

Fig. 9 Oxidation of W-reinforced Ni-20Cr at 900 °C (1650 °F) and 1000 °C (1830 °F). Sources: Ref 4, 42

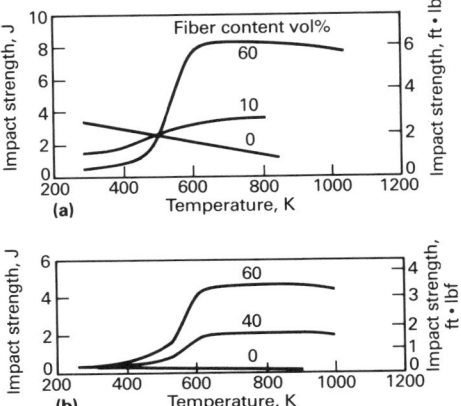

Fig. 10 Impact strength of (a) unnotched and (b) notched as-hot-pressed tungsten-superalloy as a function of temperature and various fiber contents. Source: Ref 43

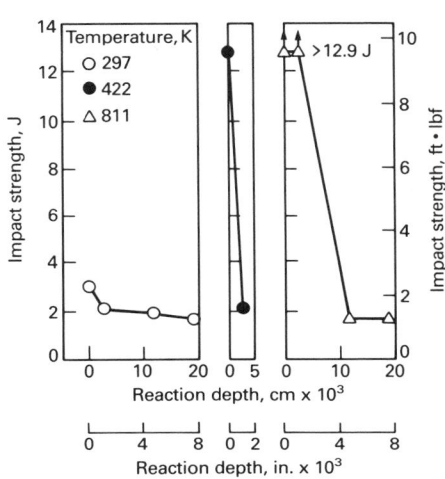

Fig. 11 Impact strength of unnotched tungsten/copper-nickel as a function of fiber-matrix reaction zone depth, approximately 55 vol% fibers. Fiber diameter, 0.38 mm (0.015 in.). Source: Ref 43

surfaces. The nature of surface reactions is substantially different from those reactions occurring in the absence of the deposit. Under such conditions the degradation of the material is referred to as hot corrosion (Ref 41).

The degree of degradation of tungsten-reinforced MMCs in a turbine environment depends on whether the fibers are exposed. When fibers are fully covered with a matrix, the reaction rate is similar to that of an unreinforced matrix. This was demonstrated for composite specimens of tungsten fibers that were fully enclosed in Fe-Cr-Al-Y, Inconel 625, and Inconel 600 matrices, when subjected to 1093 °C (2000 °F) oxidation tests (Ref 28). The specimens were held at temperature for up to 100 h. The weight change data was in agreement with that seen for these matrix materials without reinforcement. When the tungsten-reinforced composites are clad, the oxidation occurs on the outer layer. A nickel-based superalloy reinforced with W-1%ThO$_2$ fibers and clad with Inconel was exposed to 1100 °C (2010 °F) for up to 300 h in air (Ref 26). A coherent oxide formed on the clad surface, and oxidation did not progress into the matrix or fiber layers.

Testing has been performed in which service damage to the external matrix layer of a composite was simulated, exposing the tungsten fiber. Three different tungsten fiber systems (W-1%ThO$_2$, TiC-coated W, and TaC-coated W) in an Fe-Cr-Al-Y matrix were held for 0.1, 1.0, and 10 h in air at 1204 °C (2200 °F), with the exposed fiber ends of the test panels being exposed and oxidized during the test (Ref 18).

Figure 8, a longitudinal cross section of the latter fiber-matrix system, shows a channeling effect where the fiber ends have oxidized, while the adjacent matrix is unaffected. The depth of penetration for 10 h is approximately 2.5 mm (0.1 in.).

Exposed-fiber oxidation and sulfidation tests on 40 vol% tungsten fibers in a Ni-20wt%Cr matrix (Ref 42) exhibited rapid oxidation of the tungsten fiber, followed by considerable distortion and catastrophic degradation of the matrix. Typical weight gain curves for the oxidation of the material at 900 and 1000 °C (1650 and 1830 °F) in 101 kPa (1 atm) oxygen are shown in Fig. 9. For this system, considerable interaction occurred between the oxides in the matrix adjacent to the fibers and the oxide on the fibers themselves. Analysis identified nickel tungsten oxide (NiWO$_4$), nickel chromium oxide, (NiCr$_2$O$_4$), and nickel oxide (NiO). During sulfidation, the matrix was sulfidized, but the fibers did not react. When specimens were presulfidized before oxidation, the matrix tended to oxidize at an accelerated rate, but the fibers were unaffected. Thus, it would seem that hot-corrosion conditions per se are not especially harmful to tungsten-reinforced composites. However, when fibers are exposed, oxidation is clearly an important concern.

Impact

Materials to be used in an aircraft turbine must be able to resist impact failure from ingested foreign objects and from fragments of failed components that pass through the engine. Miniature Izod tests are viewed as reasonable impact screening tests because their results can be closely correlated with composite properties as measured by various ballistic impact studies (Ref 43).

Impact strength was measured for a nickel-based alloy (Ni-25W-15Cr-2Al-2Ti) reinforced with tungsten fibers (Ref 43) and is shown in Fig. 10. In general, impact strength decreased with increasing fiber content at temperatures below the ductile/brittle transition temperature (DBTT) of the fiber, which is about 260 °C (500 °F). Above the DBTT, the plastic deformation of the tungsten fibers significantly increased the composite impact strength; below the DBTT, the impact strength of the composite was a function of matrix toughness. Although notched specimens behaved similarly to unnotched specimens, their impact strengths were much lower. Similar results were seen for Nimocast 713C reinforced with tungsten fiber (Ref 24).

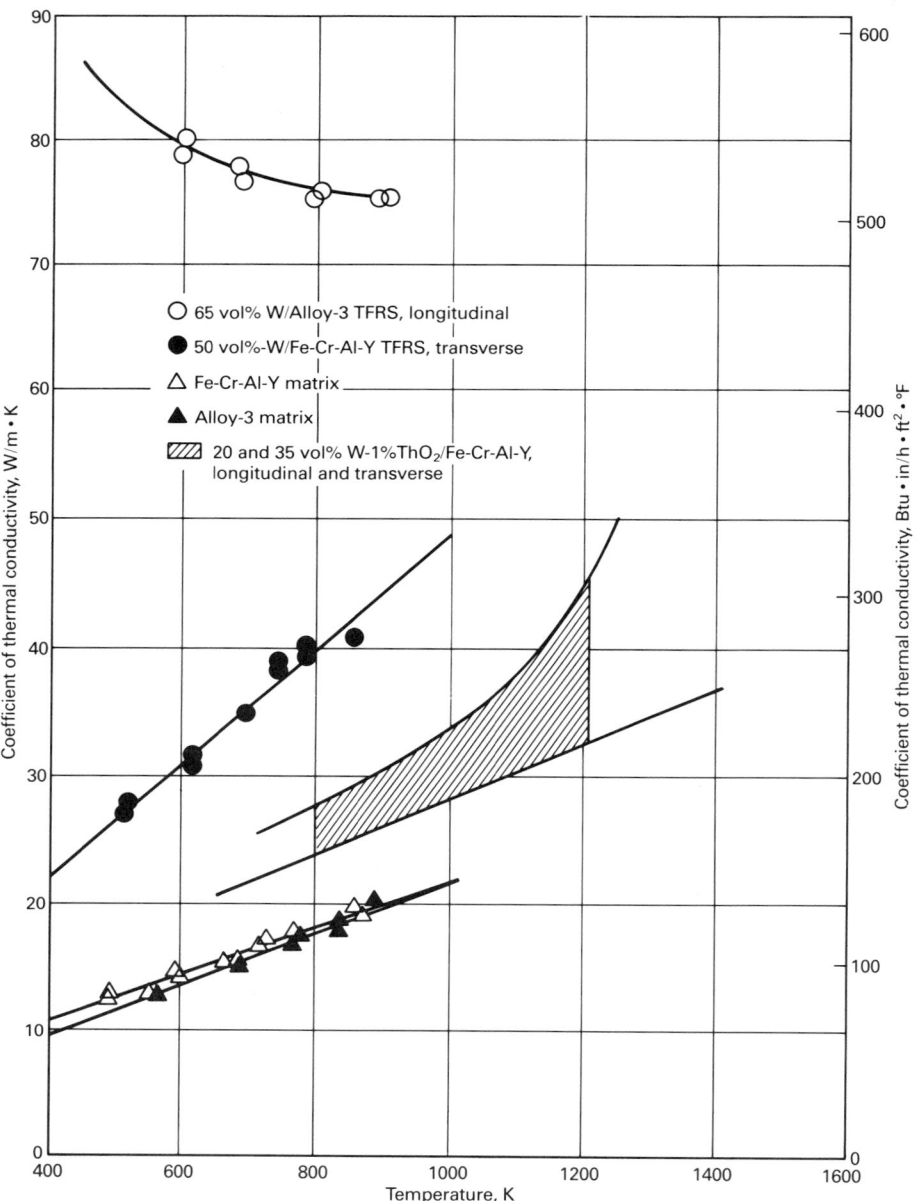

Fig. 12 Thermal conductivity as determined for selected materials over a range of temperatures

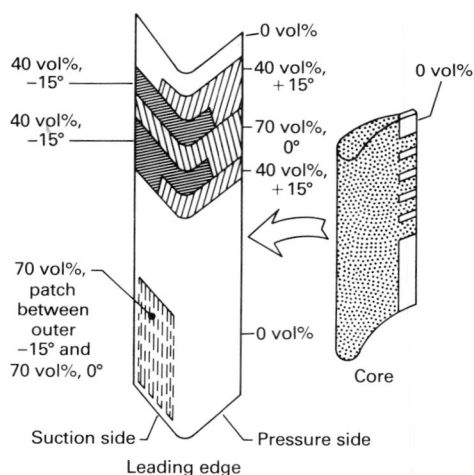

Fig. 13 Reinforcement pattern for demonstration blade. Source: Ref 57

Thermal Conductivity

While in service, turbine components and other high-temperature structures experience localized hot spots on their surfaces that can lead to cracking and distortion. A material with high thermal conductivity tends to reduce these thermally induced strains. The thermal conductivity of tungsten-reinforced Alloy 3 and Fe-Cr-Al-Y was measured, along with that of the unreinforced matrix (Ref 44). The volume fraction of reinforcement was 65 and 50%, respectively. In further studies, the thermal conductivity of 20 and 35 vol% $W-1\%ThO_2$-reinforced Fe-Cr-Al-Y was measured (Ref 45). The results of both studies are presented in Fig. 12. Because the thermal conductivity of the tungsten fiber is higher than that of the matrix, there is an increase in the thermal conductivity of the composite with an increase in the volume fraction reinforcement. Additionally, the thermal conductivity of the composite is greatest in the direction of the fiber axis because there is a continuous path for conduction along the fiber length. Thus, as seen in Fig. 12, even at low levels of reinforcement, the thermal conductivity of tungsten-reinforced composites is superior to that of the superalloys.

Design Considerations

The properties already discussed indicate that continuous tungsten reinforced superalloys represent a material system with desirable characteristics for high-temperature structural applications. Because the anisotropic nature of the composite is adjustable through variations in fiber orientation and volume fraction, it is clear that these composites require close coordination among material engineers, designers, and structure analysts, to achieve maximum material benefits.

Design studies have estimated the effect of different tungsten composite material properties on component function (Ref 44, 46). Because

Improved room-temperature impact strength of tungsten-reinforced nickel composites was noted after heat treatment and hot rolling (Ref 43). Heat treatment nearly doubled the 24 °C (75 °F) impact strength of the 45 vol% composite. Round rolling increased the 24 °C (75 °F) impact strength of a 56 vol% composite by nearly four times.

Additional Izod impact tests on Inconel 600, Inconel 625, and Fe-Cr-Al-Y reinforced with 35 and 50 vol% tungsten (Ref 28) disclosed that at temperatures below the DBTT of the tungsten fiber, the tougher Inconel alloys were superior to the Fe-Cr-Al-Y. Below the DBTT, composite toughness is inversely proportional to fiber content. Above the DBTT, fiber necking and matrix debonding provide the greatest contribution to fracture energy, so composite toughness increased with fiber content.

The effect of fiber-matrix interaction on impact strength was determined for tungsten-reinforced copper-nickel alloy composites, as shown in Fig. 11 (Ref 43). As the reaction zone depth increases, the impact strength of the composite decreases slightly at all test temperatures. In addition, the impact strength of those composites with no reaction layer increases with increasing temperature, while those with reaction layers greater than 0.103 mm (0.004 in.) exhibit no increase of impact strength with temperature.

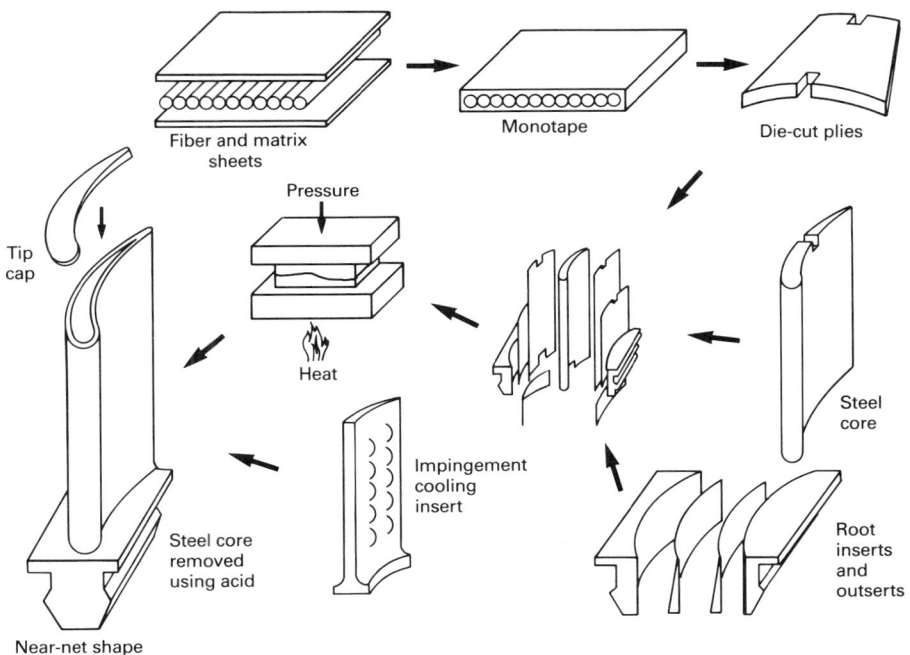

Fig. 14 Blade fabrication process for tungsten fiber reinforced superalloy. Source: Ref 4

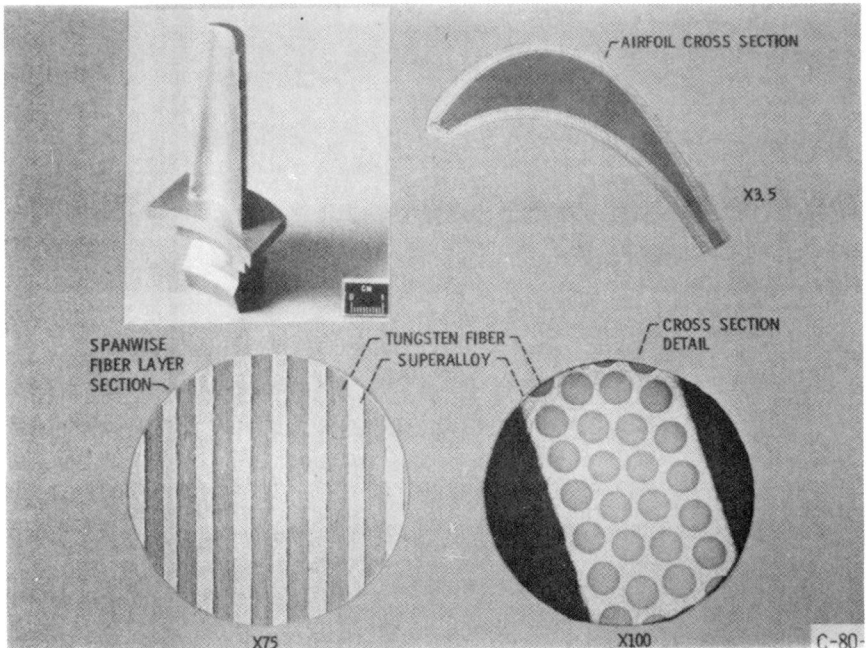

Fig. 15 Tungsten fiber reinforced superalloy composite blade. Source: Ref 4

design capability, and preliminary design and structural analysis (Ref 48). Based on the projection of design properties for composites consisting of 50 vol% W-Re-HfC fiber in a ductile superalloy matrix, it was concluded that using FRS in turbine blades potentially offers significantly improved operating life and higher temperature capability over currently used superalloys.

Fabrication Techniques

The methods developed for the fabrication of continuous tungsten fiber reinforced MMCs are differentiated by the state of the matrix before and during processing. The starting matrix is generally in sheet, powder, or molten metal form. Initial composite fabrication primarily involved casting a molten matrix around reinforcing tungsten fibers. Liquid metal infiltration was successfully used to fabricate test specimens of copper-based alloy matrices (Ref 29, 40). After infiltration, the molten copper was unidirectionally solidified to eliminate macropores caused by freezing shrinkage. Well-consolidated aluminum matrix composites were produced through squeeze casting (Ref 49). No porosity was noted in the test specimens.

Liquid metal infiltration also was used to produce nickel-based alloy composites (Ref 50). The W and W-2%ThO$_2$ fibers were significantly damaged during processing and underwent severe recrystallization through interaction with the molten nickel alloy matrix. Mechanical properties were subsequently improved through mechanical working of the composite.

Investment casting was used to produce turbine blades using nickel (Ref 35, 51) and cobalt-based (Ref 52) matrices reinforced with tungsten fibers. For the nickel-based alloys, extensive interaction with the tungsten fibers was seen. Interaction was also noted for the cobalt-based alloys, and cracking was seen at the fiber-matrix interface.

Solid matrix strips are used in two unique fabrication approaches. The first method consists of cladding the reinforcement fiber in close-fitting fine-diameter tubes (Ref 53). Bundles of the clad wire are sealed into suitable containers and consolidated by hot isostatic pressing. This approach has also been used to fabricate a composite turbine blade (Ref 54). The second method clads the fiber by tightly winding it with a continuous metal matrix strip (Ref 55). The wrapped fibers are then placed in a container and extruded to achieve full consolidation.

Because superalloy sheet is difficult to obtain, powder metallurgy (P/M) approaches have proved to be the favored fabrication method. Tungsten-reinforced nickel-based superalloys, for example, have been produced through a slip-casting process (Ref 26). The powdered nickel alloy is blended with an organic gel in water to form the slip and then poured around

the continuous fiber composites were laminated, many permutations of fiber diameter, volume fraction, and orientation had to be considered, because each variation affects overall composite properties. Composite laminate analysis was applied to the FRS composite system (Ref 47).

Using this approach, aircraft engine turbine blades and vanes were analyzed for tungsten-reinforced composite application. A preliminary design study of the potential of tungsten-reinforced superalloys for the hot section components of rocket engine turbopumps included property evaluation, current structural

Table 5 Thermal cycling of tungsten wire reinforced composites

Composite	Cycle °C	°F	Heating method(a)	No. of cycles	Remarks	Ref.
40 vol% W in						
Nimocast 258	20-1100	68-2010	FB	400	No interfacial cracking	23
13 vol% W in						
Nimocast 713C	20-600	68-1110	FB	200	No cracking	35
	250-1050	480-2100	FB	2-12	Interfacial cracking	35
	20-1050	68-2100	FB	2-25	Interfacial cracking	35
35 vol% W in						
Inconel 600	425-1100	800-2010	SR	1000	No cracks or distortion	28
IN-102	425-1100	800-2010	SR	1000	Surface cracks, distortion	28
Hastelloy X	425-1100	800-2010	SR	1000	Surface cracks, distortion	28
Nimonic 80	425-1100	800-2010	SR	1000	Surface cracks	28
Fe-Ni-Cr-Al-Y	425-1100	800-2010	SR	1000	Surface and interfacial cracking	28
20 vol% W in						
Kovar	100-900	210-1650	PP	1000	No damage	36
Stainless steel	100-900	210-1650	PP		Surface cracks after 50 cycles	36
Inconel 718	100-900	210-1650	PP		Complete fiber debonding after about	36
Hastelloy X	100-900	210-1650	PP		100 cycles	36
30 vol% W in						
Fe-Cr-Al-Y	RT-1204	RT-2200	SR	1000	No damage, some surface roughening, but no cracking	18
20 vol% W in						
Mar-M200	20-1065	68-1950	PP	50	8% length change	37
15, 32 vol% W in						
EI-435	480-700	900-1290	SR	1000	Surface cracks, 15 vol% distortion	38
	500-800	930-1470	SR	1000	Surface cracks, 15 vol% distortion	38
	530-900	990-1650	SR	1000	Surface cracks, 15 vol% distortion	38
	570-1000	1060-1830	SR	1000	Surface cracks, 15 vol% distortion	38
	600-1100	1110-2010	SR	1000	Surface cracks, 15 vol% distortion	38
35, 50 vol% W in						
Ni-W-Cr-Al-Ti	RT-1093	RT-2000	SR	100	35 vol% warpage and shrinkage 50 vol% no damage	39
30, 50 vol% W in						
Ni-Cr-Al-Y	RT-1093	RT-2000	SR	100	35 vol% warpage 50 vol% no damage	39
35, 50 vol% W in						
21DA	RT-1093	RT-2000	SR	100	35 vol% warpage 50 vol% no damage	39
50 vol% W in						
Ni-Cr-Al-Y	427-1093	800-2000	SR	1000	Internal microcracking	39
10, 50 vol% W in						
OFHC Copper	200-800	390-1470	PP	5000	Warpage	40
	400-800	750-1470				

(a) FB, fluidized bed. SR, self-resistance heated. PP, push-pull in and out of preheated furnace. Source: Ref 2, 4

collimated tungsten fibers and vibrated. Excess liquid is siphoned off the top. The casting is then dried, sintered, and hot isostatically pressed to a final, fully dense state. Because consolidation is achieved at temperatures that are lower than any of the casting methods, the reaction with the fiber is less severe.

All of these fabrication methods are best suited to uniaxial reinforcement applications. Composite designs for turbine blades have shown that angle-oriented fibers are desirable in portions of the structure. A fabrication process was developed that used solid-state diffusion bonding in which fiber distribution, alignment, and fiber-matrix reaction could be accurately controlled (Ref 56). The fibers are drum wound with a fugitive binder to produce a collimated mat. The superalloy powder is mixed with an organic binder and either rolled into sheets of powder cloth or sprayed onto a carrier sheet. Alternate layers of powder cloth and fiber mat, or sandwiches of sprayed carrier and fiber mat, are stacked in refractory tooling and consolidated by hot pressing. This technique produces discrete monolayers of composite, which are then cut into any desired shape with various

fiber orientations. These monotapes can be subsequently assembled and diffusion-bonded by hot pressing into a component shape.

This technique was used to demonstrate the fabrication of a hollow JT9D first-stage turbine blade (Ref 57). Figure 13 shows the reinforcement pattern for the demonstration blade. Because it was anticipated that impingement cooling would be used in the composite blade, trailing edge cooling slots were incorporated into the design. The fabrication sequence in Fig. 14 shows that monotapes were produced, cut, and stacked around a steel core. The assembly was then hot-pressed in a refractory die to ensure consolidation and correct contour. The core was then leached out with acid, and the root assembly was brazed onto the airfoil. The tip cap was welded onto the end of the airfoil. The final tungsten-reinforced composite blade shown in Fig. 15 had excellent fiber distribution and alignment.

Although this powder metal/collimated fiber mat approach has produced some excellent products, it has limitations. First, the size of the final component is limited by the size of the bonding dies, which are costly and cumber-

some. Second, the production of monotapes is time consuming and laborious. Finally, the complete removal of the organic binders is always questionable.

An alternate approach, arc spraying, has been developed to produce monotapes (Ref 58). Molten metal is sprayed onto an array of tungsten fibers that have been previously wound onto a large drum. Because this drum assembly is inside an inert atmosphere chamber, no oxide contamination of the matrix occurs during spraying. By varying the spacing of the fibers and the thickness of the deposited layer, monotapes of controlled volume fraction can be produced. Large monotapes without organic impurities are produced in this manner. These monotapes can then be introduced directly into further processing sequences. Diffusion-bonded multi-ply billets produced by means of this process are currently being evaluated. At this stage of development, continuous tungsten-reinforced MMCs are a promising engineering material for devices that must function under conditions of high thermal and mechanical stress. Further efforts to develop higher-strength fibers as well as suitable fiber-matrix combinations are needed to improve the overall composite characteristics.

REFERENCES

1. D.L. McDanels, T.T. Serafini, and J.A. DiCarlo, "Polymer Metal and Ceramic Matrix Composites for Advanced Aircraft Engine Applications," NASA TM-87132, National Aeronautics and Space Administration, 1985

2. R. Warren, The Mechanical Properties of Fiber Reinforced Superalloy Composites, in *Sintered Metal-Ceramic Composites*, G.S. Upadhyaya, Ed., Elsevier, 1984, p 215-237

3. V.S. Mirotvorskii and A.A. Ollshevskii, Interaction of Thoriated Tungsten at 1200-1600 °C with Matrices Based on Various Metals, *Met. Sci. Heat Treat.*, Vol 21, Nov/Dec 1979, p 826-829

4. D.W. Petrasek and R.A. Signorelli, "Tungsten Fiber Reinforced Superalloys—A Status Review," NASA Technical Memorandum 82590, National Aeronautics and Space Administration, 1981

5. T. Montelbano, J. Brett, L. Castleman, and L. Siegle, *Trans. AIME*, Vol 242, 1968, p 1973-1979

6. J. Hoffman, S. Hofmann, and L. Tillman, Recrystallization of Tungsten Fibers in Nickel Matrix Composites, *Z. Metallkd.*, Vol 65, 1974, p 721-726

7. H. Gruenling and G. Hofer, Deferred Recrystallization of Tungsten Wire in Nickel and Nickel-Chromium Matrices, *Z. Werkstofftech.*, Vol 5, 1974, p 69-72

8. C.H. Lee, J. Yamamoto, and S. Umekawa, Effects of Compounds as Diffusion Barrier Coatings Between the Fiber and the Matrix in Tungsten Fiber Reinforced Nickel Matrix Composites, in *International Conference on Composite Materials IV*, Oct 1982

9. R. Warren, L.O. Larsson, and G.H. Andersson, The Effect of Composition on the Microstructure and Properties of W-Wire Reinforced Fe-Alloys, *Dtsch. Ges. Metallkd. Fachber.*, 1981, p 313-324

10. L.O.K. Larsson, "Metal Matrix Composites for Turbine Blades in Aeroengine," Ph.D. Thesis, Chalmers University of Technology, 1981

11. T. Caulfield, R.S. Bellows, and J.K. Tien, "Interdiffusional Effects Between Tungsten Fibers and an Iron-Nickel-Base Alloy," *Metall. Trans. A*, Vol 16A, Nov 1985, p 1961-1968

12. D.M. Karpinos, V.K. Fedorenko, A.L. Burykina, and V.V. Gorskii, "Interfacial Reactions in Composite Materials with 80% Ni-20% Cr Alloy Matrices and Tungsten and Molybdenum-Base Fibers," *Poroshk. Metall.*, Vol 2 (No. 134), Feb 1974, p 64-75

13. D.W. Petrasek, R.A. Signorelli, and J.W. Weeton, "Refractory Metal Fiber Nickel-Base Alloy Composites for Use at High Temperatures," NASA TN-D-4787, National Aeronautics and Space Administration, Sept 1968

14. Y. Umakoshi and T. Yamane, Contribution of the Interfacial Reaction to the Strength of Molybdenum and Tungsten Fiber-Reinforced Composites, *Trans. Jpn. Inst. Met.*, Vol 17, 1976, p 25-34

15. R. Ahlroth, "Effect of TiC and TiN Diffusion Barriers on Tensile and Fatigue Properties of W Fiber-Reinforced Ni Composites," *High Temp. Technol.*, Vol 2 (No. 1), Feb 1984, p 43-47

16. R. Ahlroth and P. Kettunen, Diffusion Barriers Used to Improve Structural Stability and Mechanical Properties of W-Fiber-Ni-Matrix Composites at Elevated Temperatures, in *International Conference on Composite Materials IV*, Vol 1, 1982, p 401-407

17. V.S. Mirotvorskii and A.A. Ollshevskii, "Reaction of Thoriated Tungsten Fibers with Iron-Base Powder Matrices," *Poroshk. Metall.*, Vol 7 (No. 163), July 1976, p 46-52

18. W.D. Brentnall, D.J. Moracz, and I.J. Toth, "Metal Matrix Composites for High Temperature Turbine Blades," TRW ER-7722-F, N00019-74-0122, Naval Air Systems Command, 1975

19. W.D. Brentnall and I.J. Toth, "Metal Matrix Composites for High Temperature Turbine Blades," TRW ER-7634-F, Naval Air Systems Command, April 1974

20. D.W. Petrasek and R.A. Signorelli, "Stress-Rupture and Tensile Properties of Refractory-Metal Wires at 2000° and 2200 °F (1093° and 1204 °C)," NASA TN-D-5139, National Aeronautics and Space Administration, 1969

21. D.W. Petrasek, "High-Temperature Strength of Refractory-Metal Wires and Consideration for Composite Applications," NASA TN-D-6881, National Aeronautics and Space Administration, 1972

22. V.M. Chubarov, Y.V. Levinskii, S.E. Salibekov, A.F. Trefilov, L.V. Grachev, E.M. Rodin, M.K. Levinskaya, and L.V. Dvoichenkova, A Nickel-Base Heat Resistant Composite Material, *Probl. Prochn.*, Vol 3 (No. 7), 1971, p 100-104; *Transl. Strength Mater.*, Vol 3 (No. 7), 1972, p 856-859

23. A.V. Dean, The Reinforcement of Nickel-Base Alloys with High Strength Tungsten Wires, *J. Inst. Met.*, Vol 95, 1967, p 79-86

24. A.W.H. Morris and A. Burwood-Smith, "Some Properties of a Fiber-Reinforced Nickel-Base Alloy," *Fibre Sci. Technol.*, Vol 3 (No. 1), 1970, p 53-78

25. I. Ahmad and J.M. Barranco, "Reinforced Cobalt Alloy Composite for Turbine Blade Applications," *SAMPE Q.*, Vol 8, 1977, p 38-49

26. D.W. Petrasek and R.A. Signorelli, "Preliminary Evaluation of Tungsten-Alloy Fiber Nickel-Base Alloy Composites for Turbojet Engine Applications," NASA TN-D-5575, National Aeronautics and Space Administration, 1970

27. D.W. Petrasek and R.A. Signorelli, "Stress-Rupture Strength and Microstructural Stability of Tungsten-Hafnium-Carbon-Wire Reinforced Superalloy Composites," NASA TN-D-7773, National Aeronautics and Space Administration, 1974

28. G.I. Friedman and J.N. Fleck, "Tungsten Wire-Reinforced Superalloys for 1093 °C (2000 °F) Turbine Blade Applications," NASA CR-159720, National Aeronautics and Space Administration, 1979

29. D.L. McDanels, R.A. Signorelli, and J.W. Weeton, "Analysis of Stress Rupture and Creep Properties of Tungsten Fiber Reinforced Copper Composites," STP 427, American Society for Testing and Materials, 1967, p 124-148

30. R.H. Baskey, "Fiber-Reinforced Metallic Composite Materials," AFML-TR-67-196, Air Force Materials Laboratory, 1967

31. L.O.K. Larsson and R. Warren, "Fiber Reinforced Metals in Turbine Blades," *J. Eng. Power (Trans. ASME)*, Vol 102, July 1980, p 573-578

32. H. Carlsen and H. Linholt, "High Temperature Tensile and Creep Properties of Tungsten Fibre Reinforced Nickel Composites," *Nortemps*, Vol 75 (No. 11), 1975, p 99-114

33. D.M. Essock, "FRS Composites for Advanced Gas Turbine Engine Components," TRW ER-7696-F, NADC 77015-30, Naval Air Defense Command, May 1979

34. L.O.K. Larsson, Thermal Stresses in Metal Matrix Composites, in *International Conference on Composite Materials II*, April 1978, p 805-821

35. A.W.H. Morris and A. Burwood-Smith, "Fiber Strengthened Nickel-Base Alloys," High Temperature Turbines, AGARD-CP-73-71, Jan 1971

36. R. Warren, L.O.K. Larsson, P. Ekstron, and T. Jansson, *Progress in Science and Engineering of Composites*, Japan Society for Composite Materials, 1982, p 1419-1426

37. P.J. Mazzei, G. Van Drunen, and N. Chung, High Temperature Artificial Composite Alloys for Industrial Gas Turbines, in *Proceedings of the Symposium on Fracture Mode and Processing of Composites*, The Metallurgical Society, American Institute of Mining, Metallurgical and Petroleum Engineers, 1977

38. F.P. Banas, A.A. Baranov, and E.V. Yakovleva, "Deformation of Composite Materials During Alternate Heating and Cooling," *Strength Mater.*, Vol 7 (No. 6), 1976, p 744-748

39. W.D. Brentnall and D.J. Moracz, "Tungsten Wire-Nickel-Base Alloy Composite

Development," TRW ER-7849, NASA CR-13502, National Aeronautics and Space Administration, 1976

40. S. Yoda, V. Kurihara, K. Wakashima, and S. Umekawa, "Thermal Cycling Induced Deformation of Fibrous Composites With Particular Reference to the Tungsten-Copper System," *Metall. Trans A*, Vol 9A, Sept 1978, p 1229-1236

41. C.S. Giggins and F.S. Pettit, "Hot Corrosion Degradation of Metals and Alloys—A Unified Theory," PWA Report FR-11545, Contract F44620-76-C-0123, June 1979

42. M.E. El-Dahshan, D.P. Whittle, and J. Stringer, "The Oxidation and Hot Corrosion Behavior of Tungsten Fiber Reinforced Composites," *Oxid. Met.*, Vol 9 (No. 1), 1975, p 45-67

43. E.A. Winsa and D.W. Petrasek, "Factors Affecting Miniature Izod Impact Strength of Tungsten-Fiber-Metal-Matrix Composites," NASA TN-D-7393, National Aeronautics and Space Administration, Oct 1973

44. E.A. Winsa, L.J. Westfall, and D.W. Petrasek, Predicted Inlet Gas Temperatures for Tungsten Fiber Reinforced Superalloy Turbine Blades, in *International Conference on Composite Materials II*, April 1978, p 840-857

45. P. Melnyk and J.N. Fleck, "Fabrication of Tungsten-Wire/FeCrAlY-Matrix Composite Specimens," NAS3-20390, TRW ER-8076, TRW, Inc., July 1979

46. J.W. Weeton, "Design Concepts for Fiber-Metal Matrix Composites for Advanced Gas Turbine Blades," NASA TM-X-5277, National Aeronautics and Space Administration, 1970

47. E.A. Winsa, "Tungsten Fiber Reinforced Superalloy Composite High Temperature Component Design Considerations," NASA TM-82811, National Aeronautics and Space Administration, 1982

48. J.R. Lewis, Design Overview of Fiber-Reinforced Superalloy Composites for the Space Shuttle Main Engine, in *Advanced High Pressure O_2/H_2 Technology*, S.F. Morea and S.T. Wu, Ed., NASA Conference Publication 2372, National Aeronautics and Space Administration, June 1984

49. E. Nakata and Y. Kagawa, "Evaluation of the Toughness of High Volume Fraction W/Al Composites," *J. Met. Sci. Letters*, Vol 3, 1984, p 968-970

50. H.W. Grunling, "Preparation and Properties of Tungsten Wire Reinforced Ni-Cr 80-20," NASA TT F-16, 275, National Aeronautics and Space Administration, 1972

51. R.C. Helmink and T.S. Piwonka, Cast Fiber Reinforced Superalloys, in *Proceedings of the 108th Annual AIME Meeting*, American Institute of Mechanical Engineers, 1979

52. I. Ahmad and J. Barraneo, $W-2\%ThO_2$ Filament Reinforced Cobalt-Base Alloy Composites for High Temperature Application," *Proceedings of the 108th Annual AIME Meeting*, American Institute of Mechanical Engineers, 1979

53. R. Warren, L.O.K. Larsson, and T. Garvare, A Method for the Fabrication of Wire Reinforced Metal Matrix Composites, *Composites*, April 1979, p 126-127

54. L.O.K. Larsson, High Temperature Metal Matrix Composites for Gas Turbines, in *Proceedings of the Fifth International Symposium on Airbreathing Engines*, Feb 1981, p 54-1 to 54-8

55. A.P. Divecha, Method for Fabricating Composite Material Reinforced by Uniformly Spaced Filaments, U.S. Patent 3,828,417, Aug 1974

56. W.D. Brenthall and I.J. Toth, "Fabrication of Tungsten Wire Reinforced Nickel-Base Alloy Composites," NASA CR-134664, TRW ER-7757, TRW, Inc., 1974

57. P. Melnyk and J.N. Fleck, "Tungsten Wire/FeCrAlY Matrix Turbine Blade Fabrication Study," NASA CR-159788, TRW ER-8101, TRW, Inc., 1979

58. L. Westfall, Arc Spray Fabrication of Metal Matrix Composite Monotapes, U.S. Patent 4,518,625, May 1985

Discontinuous Silicon Fiber MMCs

William C. Harrigan, Jr., DWA Composite Specialties, Inc.

DISCONTINUOUSLY REINFORCED metal matrix composites (MMCs) are a class of materials that exhibit a blend of the reinforcement and matrix properties. The reinforcement can be ultrahigh-strength whiskers, short or chopped fibers, or particles. Each reinforcement has property or cost attributes that dictate its use in a given situation. All MMCs have the advantage of being formable by more or less standard metalworking practices. They can be shaped by extrusion, forging, and rolling. The machining, drilling, and grinding operations do not cut or break critical fibers, and therefore do not degrade mechanical properties.

Early work in this area was done by S.S. Brenner (Ref 1, 2) and W.H. Sutton (Ref 3, 4) with α aluminum oxide (Al$_2$O$_3$) whiskers. The cost of the whiskers was high, and the strengths achieved were lower than expected because of bonding difficulties with the alumina whiskers. These difficulties were never overcome. Work by A. Divecha et al. (Ref 5) with β silicon carbide (SiC) whiskers in aluminum demonstrated very good strength, modulus, fatigue, and elevated-temperature properties. However, high cost for these whiskers limited the continued development of this system.

More recently, composites with short, staple, polycrystalline alumina fibers, SiC whiskers made from pyrolyzed rice hulls, and SiC particulates have been investigated. The alumina fibers were first used to reinforce the ring land area of diesel pistons (Ref 6). This development was brought about by the advent of improved grades of the fiber at relatively low cost with high volume availability. These pistons were made by a squeeze casting process that was described in an article by J. Dinwoodie et al. (Ref 7) and shown in Fig. 1. The short fibers did not increase the ultimate strength of the matrix alloy at room temperature; however, strength is retained to temperatures of approximately 300 °C (570 °F) rather than 200 °C (390 °F) for the base alloy (Fig. 2). The elastic modulus of the composite is substantially increased over that of the matrix at all temperatures (Fig. 3). In addition, the incorporation of the fibers decreases the coefficient of thermal expansion (Table 1), and increases the hardness of the metal composite (Table 2). This combination of properties has made the piston a

success. Because of this, other manufacturers have reported producing other pistons reinforced with short fibers (Ref 8).

L. Ackerman et al. reported on composites containing polycrystalline alumina fibers and SiC whiskers in an aluminum casting alloy (Ref 9). These composites were also made by

squeeze casting. In this study, the room-temperature elastic modulus as well as yield strength were improved by the addition of both reinforcements. The SiC whiskers resulted in both higher modulus values and higher yield strengths (Table 3). As in the previous study, strength properties at elevated temperatures

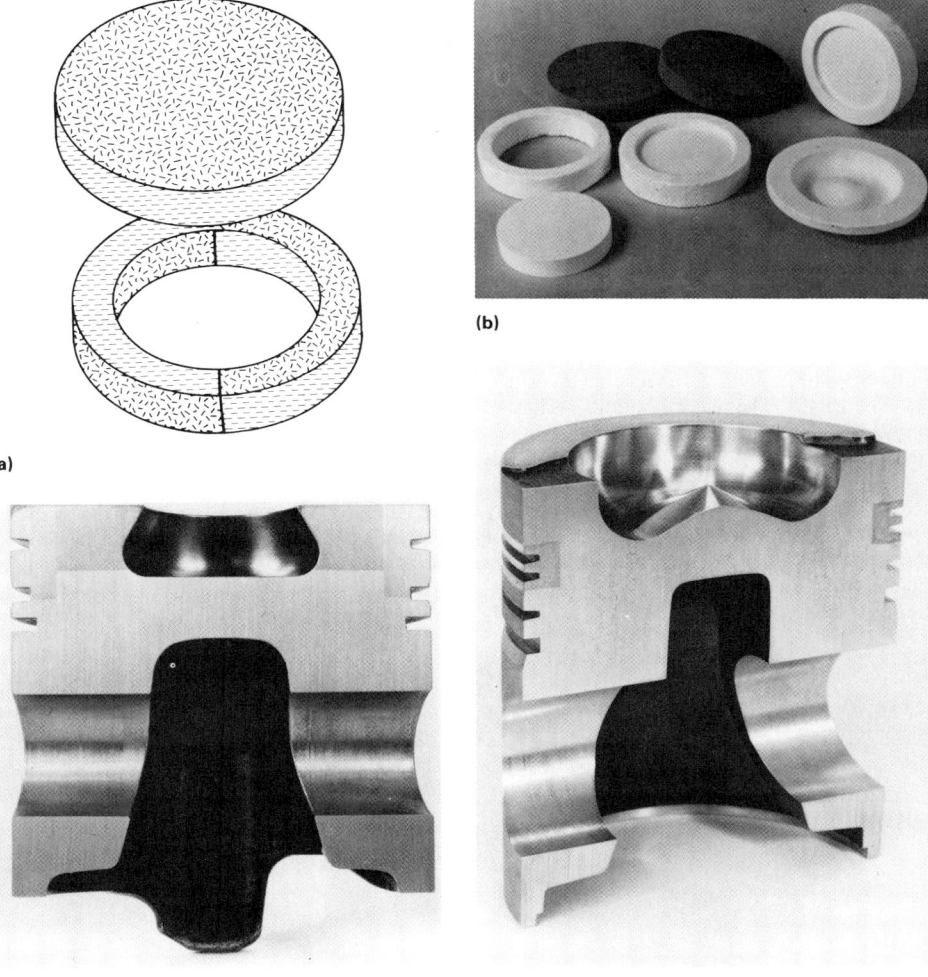

Fig. 1 Drawing and photograph of reinforcement and photographs of reinforced piston sections. Source: Ref 7. (a) Possible orientations in preforms. (b) Simple preform shapes used for piston reinforcement. (c) Reinforced combustion bowl. Courtesy of the AE Group. (d) Ring groove areas of sectioned pistons. Courtesy of Toyota Motor Corporation

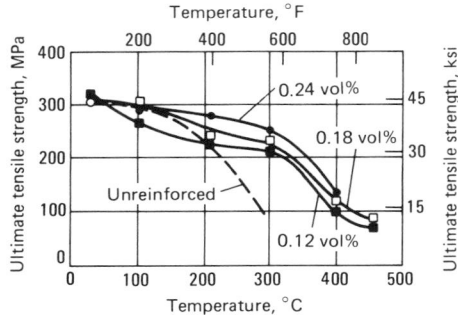

Fig. 2 Effect of temperature on the tensile strength of Al-9Si-3Cu-based composites. Source: Ref 7

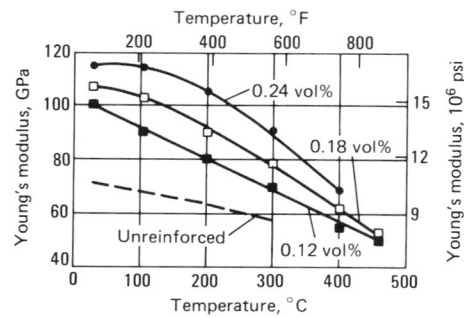

Fig. 3 Effect of temperature on the modulus of Al-9Si-3Cu-based composites. Source: Ref 7

(a)

10 μm

Table 1 Coefficient of thermal expansion, α

Fiber, vol %	α (in-plane), 10^{-6}/K	α (normal), 10^{-6}/K
0 . . . . .2.03		2.03
0.12 . . . . .1.66		1.76
0.18 . . . . .1.54		1.66
0.24 . . . . .1.55		1.57

Source: Ref 7

Table 2 Hardness values at 25 °C (80 °F)

Fiber, vol%	Vickers hardness No., HV10
0 131	
0.12 179	
0.18 190	
0.24 212	

Source: Ref 7

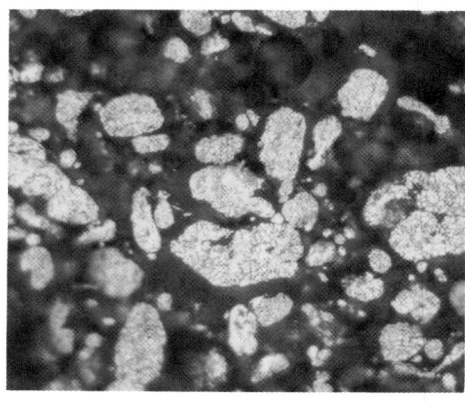

(b)

10 μm

Fig. 4 (a) Scanning electron micrograph of SiC particulate. (b) Optical micrograph of polished and etched 2124 aluminum powder

were retained (Table 4). The strength of either composite containing 20 vol% reinforcement at 350 °C (660 °F) was equal to or greater than the strength of the matrix at 250 °C (480 °F). The fatigue limit at 10^7 cycles was improved from 80 MPa (11.6 ksi) in the matrix alloy to 109 MPa (15.8 ksi) in the 20 vol% polycrystalline alumina system, and 131 MPa (19.0 ksi) in the SiC whisker system.

Composites of SiC particulates in aluminum alloys have been studied by several investigators (Ref 10-12). The majority of the work used composites that were made by blending atomized powders with the SiC reinforcement. The choice of matrix alloys extended from 1100, that is, no alloy additions, to any of the rapidly solidified, high-strength alloys. Examples of atomized aluminum alloy powder and SiC particulate are shown in Fig. 4. The 2124 aluminum powder was metallographically mounted, polished, and etched to show the small grain size found in these powders. The blend was compacted into a billet, which was vacuum hot-pressed and hot-worked into a usable form. The composites behave in a manner similar to that of new, high-strength aluminum alloys made by the powder metallurgy technique (Ref 13); that is, the prior particle oxide skins must be broken up by metalworking operations before the true properties of the composite can be achieved. The most common primary breakdown process has been extrusion. With the particulate reinforcement, extrusion through conical as well as shear face dies is acceptable practice. Composite extrusion is shown in Fig. 5, while a typical extrusion cross section is shown in Fig. 6 and 7. Microstructures of extruded composites are shown in Fig. 8 and 9 for 20 vol% SiC and 30 vol% SiC reinforcement levels in a 6061 aluminum

matrix. These photomicrographs show the uniform distribution of the SiC particulates in longitudinal and transverse planes of the composites.

These composites respond to heat treatment in a manner similar to that of the matrix alloy (Fig. 10). However, the aging times to achieve peak strength are reduced (Ref 14). Therefore, in order to keep the number of discussion

Table 3 Tensile data obtained on polycrystalline alumina and SiC whisker reinforced aluminum alloy

Fiber, vol%	Yield strength (0.2%)				Ultimate tensile strength				Young's modulus			
	MPa	ksi	Standard deviation	Range of measurement	MPa	ksi	Standard deviation	Range of measurement	GPa	10^6 psi	Standard deviation	Range of measurement
Polycrystalline alumina												
0.	210	30.5	3.8	9.5	297	43.1	1.8	3.5	71.9	10.4	4.5	13
0.05	232	33.6	4.2	10.4	282	40.9	6.5	15.1	78.4	11.4	2.3	6
0.12	251.5	36.5	14.6	38.3	273	40.0	19.6	49.6	83.0	12.0	7.8	21
0.20	282.5	41.0	11.3	25.2	312	45.3	16.0	42.3	95.2	13.8	2.7	7
SiC whisker												
0.	210	30.5	3.8	9.5	297	43.1	1.8	3.5	71.9	10.4	4.5	13
0.12	266.5	38.7	4.2	10.6	359	52.1	33.6	85.6	95.3	13.8	1.6	6
0.16	264.5	38.4	0.6	1.6	374	54.2	8.0	23.0	90.0	13.1	3.7	9
0.20	298	43.2	4.0	10.2	383.6	55.6	15.2	38.8	111.0	16.1	5.0	13

Source: Ref 9

Table 4 Yield strength and ultimate tensile strength of aluminum alloy reinforced with polycrystalline alumina and SiC whiskers at different temperatures

Fiber, vol%	350 °C (660 °F) Yield strength MPa	ksi	Ultimate tensile strength MPa	ksi	300 °C (570 °F) Yield strength MPa	ksi	Ultimate tensile strength MPa	ksi	250 °C (480 °F) Yield strength MPa	ksi	Ultimate tensile strength MPa	ksi
Polycrystalline alumina												
0 35	5.1	55	8.0	...	...	70	10.2	70	10.2	115	16.7
0.05 54	7.8	63	9.1	79	11.5	88	12.8	112	16.2	134	19.4
0.12 68	9.9	74	10.7	...	...	...	...	...	...	...	...
0.20 110	16.0	112	16.2	154	22.3	155	22.5	186	27.0	198	28.7
SiC whiskers												
0 35	5.1	55	8.0	...	...	70	10.2	70	10.2	115	16.7
0.12 94	13.6	124	18.0	153	22.2	180	26.1	197	28.6	226	32.8
0.16 120	17.4	147	21.3	...	...	...	...	...	...	...	...
0.20 163	23.6	184	26.3	207	30.0	235	34.1	268	38.9	284	41.2

Source: Ref 9

Fig. 5 Extrusion of 360-mm (14-in.) diam SiC particulate reinforced aluminum composite

Fig. 6 Plate extrusions of SiC particulate reinforced aluminum composites

Fig. 7 Extruded shapes of SiC particulate reinforced aluminum composites

variables to a manageable level, the data for composites in this article will be for the highest-strength heat treatment only. Table 5 contains tensile data for longitudinal samples taken from an extrusion that had area ratios in excess of 15:1. These data demonstrate that the elastic modulus is a function of the amount (vol%) of silicon carbide. The modulus values are quite impressive, being equal to titanium at 25 vol% SiC, and increasing to 140 GPa (20×10^6 psi) at 40 vol%. These composites have stress-strain curves that are similar to those for metals; that is, plastic deformation takes place (Fig. 11).

The yield strength of these composites is defined by offset methods described in ASTM E 8 (Ref 15). The 0.2% yield strength and the ultimate strength are functions of the matrix alloy and the amount of reinforcement. The yield strength of 6061 aluminum is increased from 280 MPa (40 ksi) to 380 MPa (55 ksi) by the addition of 20 vol% SiC, and that of 7091 aluminum is increased from 540 MPa (78 ksi) to 630 MPa (92 ksi) by the addition of 20 vol% SiC. The ductility of these composites is much greater than it is for continuous fiber composites, by a factor of 10 in some cases. The

ductility decreases as the amount of reinforcement increases. In some cases, such as 20 vol% SiC in 6061 aluminum and 2124 aluminum, the ductility is acceptable for standard metal designs requiring at least 5% elongation.

These SiC particulate reinforced aluminum composites have transverse tensile properties that are within 5% of the longitudinal properties. In situations requiring multidirectional reinforcement, these composites can outperform fiber-reinforced composites. Their shear strength is greater than that of the matrix alloy (Table 6). This increased shear strength is

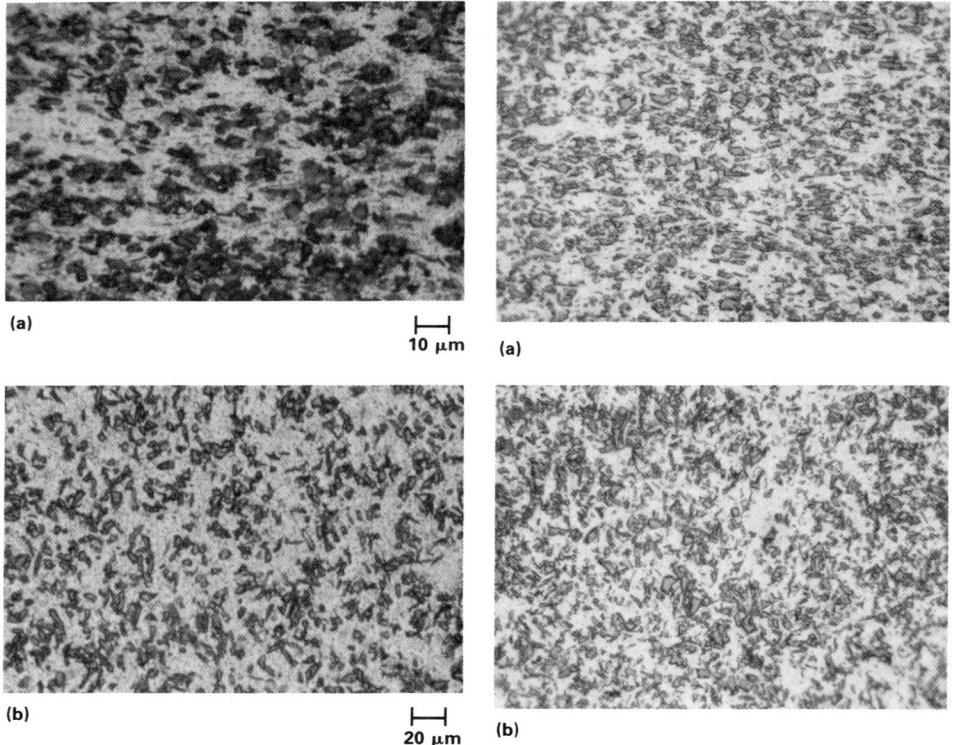

Fig. 8 Optical micrographs of 20 vol% SiC particulate reinforced 6061 aluminum composite after extrusion. (a) Longitudinal. (b) Cross sectional

Fig. 9 Optical micrographs of 30 vol% SiC particulate reinforced 6061 aluminum composite after extrusion. (a) Longitudinal. (b) Cross sectional

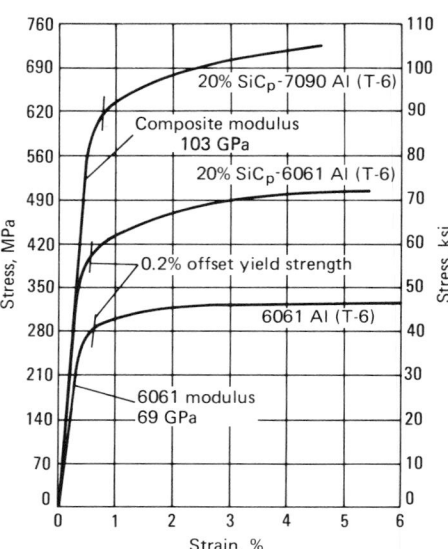

Fig. 11 Stress-strain curves for 6061 aluminum and 20 vol% SiC (particulate) and 7090 and 6061 aluminum matrix composites

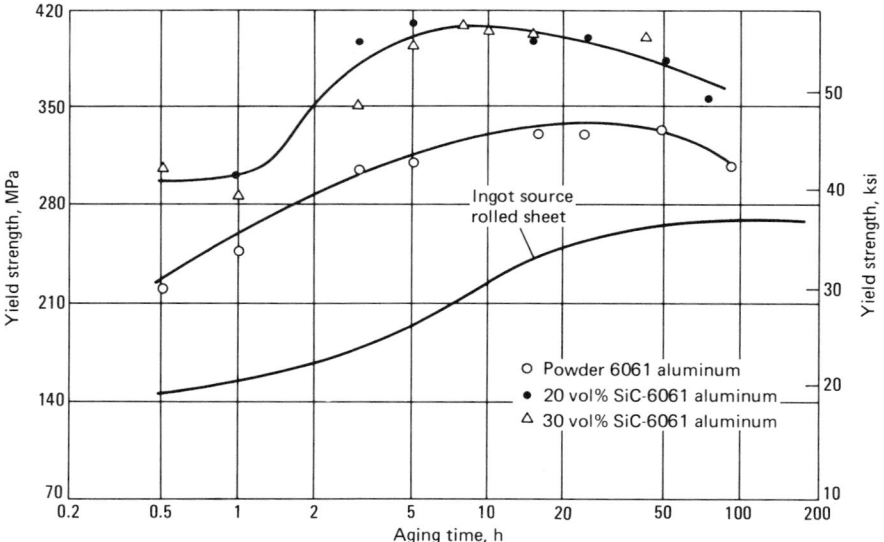

Fig. 10 Aging response for powder and ingot source 6061 aluminum and 20 vol%, 30 vol% SiC-6061 aluminum composites, aged at 160 °C (325 °F). Source: Ref 13

also reflected in an increase in pin bearing strength (Table 7). These bearing strength data are for tests conducted with the center of the pin hole 1.5 and 2 times the pin diameter from the sample edge. Typical edge distances for graphite- or glass-reinforced epoxy composites is 5 times the pin diameter. The SiC aluminum composites are able to save material and weight by reducing the excess edge material at joints.

The expansion coefficient of the aluminum is decreased as SiC is added to the composite (Fig. 12), from 23×10^{-6}/K for aluminum to 11×10^{-6}/K for 40 vol% SiC composites. This expansion behavior is isotropic. A design

of an advanced composite optical system gimbal by General Electric Company (Ref 20) used 40 vol% SiC particulate reinforced 6061 aluminum in low-expansion and joint areas and graphite-epoxy in ultralow-expansion areas. It won first place in the *Materials Engineering* 1985 competition (Fig. 13). The low-expansion behavior of this composite is also being used in place of beryllium for guidance components (Ref 21), as shown in Fig. 14. This composite material is being considered for connecting rods because its expansion is similar to that of steel, and this will reduce the large-end crankshaft clearance problems encountered with aluminum alloys in this application.

The fatigue properties of these composites show very low crack growth rates at low ΔK values. However, as the ΔK levels increase, a rapid increase in growth rate is also observed (Fig. 15). This resistance to crack initiation is reflected in an increase in operating stress levels for S/N-type fatigue tests (Fig. 16). This behavior is for both smooth- and notched-type samples.

The fracture toughness values of the composite are lower than they are for the matrix alloy. Typically, the K_{Ic} values for the composite are one-half to two-thirds that of the matrix alloy. The plasticity of these composites is evident when fracture tests are conducted on thin samples. The fracture conditions change from plane-strain to mixed plane-strain, that is, plane-stress. The fracture toughness value increases as the amount of plasticity increases (Fig. 17). The 1.8-mm (0.070-in.) thick 20 vol% SiC-2124 aluminum samples experienced general yield prior to final failure.

In the late 1970s researchers at the University of Utah developed a process for making SiC whiskers using rice hulls as the starting ingredient. This made SiC whiskers potentially low

Table 5 Typical mechanical properties of SiC particulate reinforced aluminum alloy composites

Alloy and vol%	Modulus of elasticity		Yield strength		Ultimate tensile strength		Ductility, %
	GPa	10^6 psi	MPa	ksi	MPa	ksi	
6061							
Wrought	68.9	10	275.8	40	310.3	45	12
15	96.5	14	400.0	58	455.1	66	7.5
20	103.4	15	413.7	60	496.4	72	5.5
25	113.8	16.5	427.5	62	517.1	75	4.5
30	120.7	17.5	434.3	63	551.6	80	3.0
35	134.5	19.5	455.1	66	551.6	80	2.7
40	144.8	21	448.2	65	586.1	85	2.0
2124							
Wrought	71.0	10.3	420.6	61	455.1	66	9
15	...	...	...	...	...	...	...
20	103.4	15	400.0	58	551.6	80	7.0
25	113.8	16.5	413.7	60	565.4	82	5.6
30	120.7	17.5	441.3	64	593.0	86	4.5
35	...	...	...	...	...	...	...
40	151.7	22	517.1	75	689.5	100	1.1
7090							
Wrought	72.4	10.5	586.1	85	634.3	92	8
15	...	...	...	...	...	...	...
20	103.4	15	655.0	95	724.0	105	2.5
25	115.1	16.7	675.7	98	792.9	115	2.0
30	127.6	18.5	703.3	102	772.2	112	1.2
35	131.0	19	710.2	103	724.0	105	0.90
40	144.8	21	689.5	100	710.2	103	0.90
7091							
Wrought	72.4	10.5	537.8	78	586.1	85	10
15	96.5	14	579.2	84	689.5	100	5.0
20	103.4	15	620.6	90	724.0	105	4.5
25	113.8	16.5	620.6	90	724.0	105	3.0
30	127.6	18.5	675.7	98	765.3	111	2.0
35	...	...	...	...	...	...	...
40	139.3	20.2	620.6	90	655.0	95	1.2

Table 6 Shear strength of SiC particulate reinforced aluminum composites

Composite	Shear strength		Ref
	MPa	ksi	
25 vol% SiC-6061 (T-6)	277.9	40.3	16
30 vol% SiC-6061 (T-6)	289.6	42.0	17
25 vol% SiC-7091 (T-6)	279.2	55.0	18
30 vol% SiC-7090	430.9	62.5	17
25 vol% SiC-2124 (T-4)	344.8	50.0	19

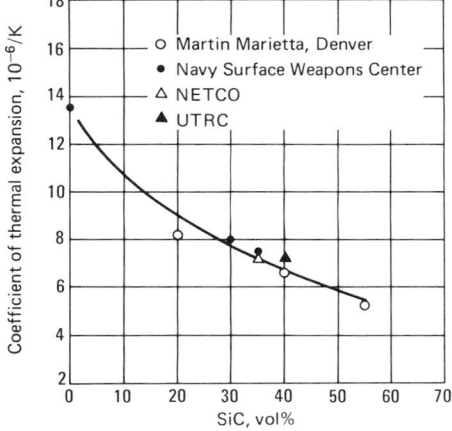

Fig. 12 Coefficient of thermal expansion as a function of reinforcement level, SiC-aluminum composites

Table 7 Pin bearing strengths of SiC particulate reinforced aluminum

Composite	Edge distance (multiplied by pin diameter) from edge	Bearing yield strength		Bearing ultimate strength	
		MPa	ksi	MPa	ksi
20 vol% SiC-7091 (T-6) (Ref 17)	1.5	689.5	100	724.0	105
	2.0	1000.0	145	1310.0	190
	3.0	1000.0	145	1448.0	210
25 vol% SiC-2124 (Ref 19)	2.0	827.4	120	...	...

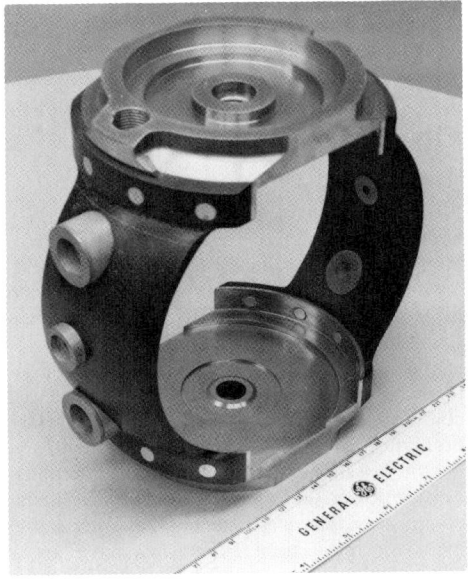

Fig. 13 Advanced composite optical system gimbal

in cost and prompted renewed research with this form of composite (Ref 10-12, 22). These SiC whiskers are approximately 0.1 μm (4 μin.) in diameter and as produced have length-to-diameter ratios of up to 100:1.

Powder metallurgy processing is the most common technique for producing whisker-reinforced aluminum composites. Because of their basis as a powder, these composites must also be metal-worked to develop the best properties. To preserve the whiskers, the extrusion of these composites is restricted to conical or streamlined dies (Ref 23). Even with these precautions, substantial whisker breakage is experienced during metalworking operations. Extrusion of these composites produces alignment of the whiskers and anisotropic mechanical properties. For a given reinforcement level, a given matrix alloy, and comparable metalworking prior to testing, the whisker compos-

ites tend to have higher modulus values in the extrusion (alignment) direction. They also have approximately the same yield strength, a higher ultimate strength, and a lower strain-to-failure rate than do particulate-reinforced composites. The whisker composites also have lower properties in the directions perpendicular to the extrusion. The expansion characteristics of the whisker composites are the same as for the particulate composites. However, the whisker composites are limited to a maximum of 25 vol% because of blending difficulties.

In 1985, D. Schuster presented data for SiC particulate reinforced aluminum composites made by casting techniques (Ref 24). The particulates are coated for protection, injected into a stirred melt of aluminum, and then cast into a mold. The primary reason for using the casting technique is the potential for low-cost billet manufacture and the possibility of replac-

ing standard castings with reinforced castings. Casting to date has been carried out in sand molds, permanent steel molds, and investment molds. There is a limitation on amount of reinforcement because of the increase in viscosity as the particles are added. Reinforcement levels as high as 35 vol% have been attempted;

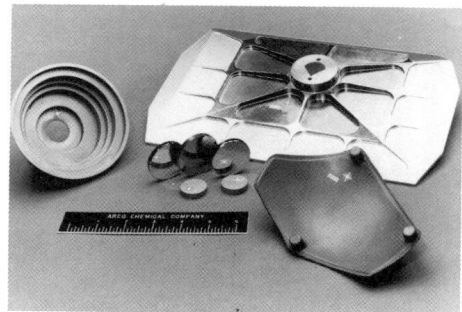

Fig. 14 Prototype SiC particulate reinforced aluminum matrix composite mirrors and precision components. From left, a superplastically formed gyro part, laser mirrors and substrates, an IR tank mirror, and a forged instrument cover. Source: Ref 21

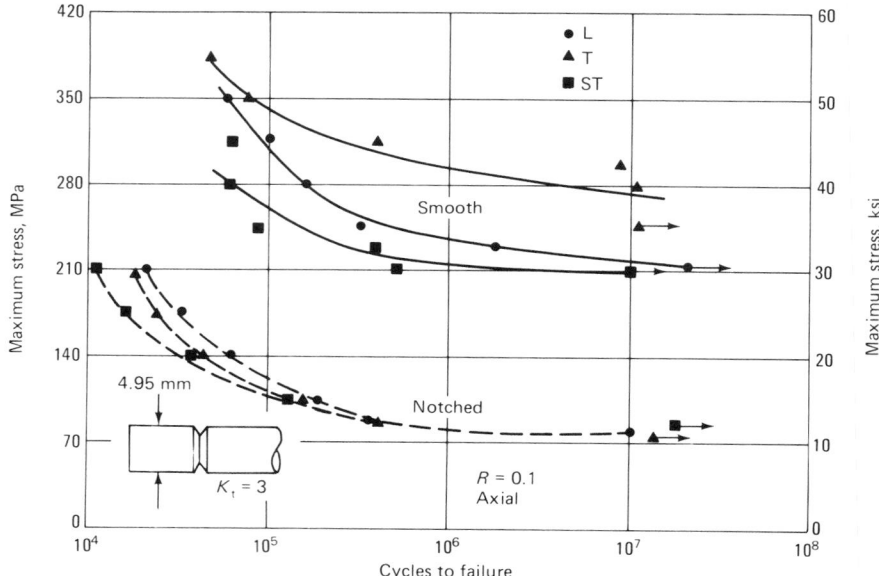

Fig. 16 Axial fatigue properties of smooth and notched specimens taken from plate in different orientations

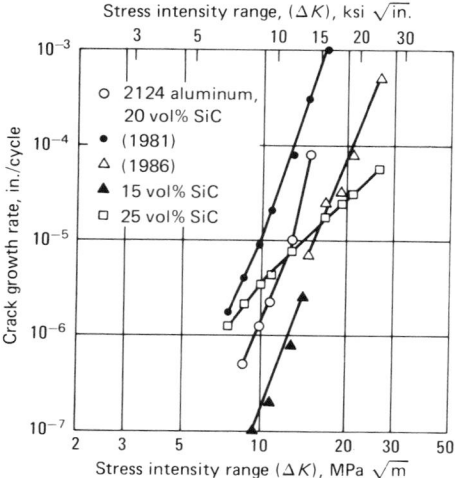

Fig. 15 Fatigue crack growth rate of 2124 aluminum and 2124 matrix composites

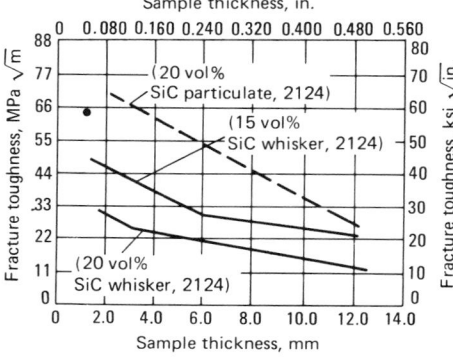

Fig. 17 Fracture toughness as a function of sample thickness

however, the reinforcement level is normally kept to a maximum of 25 vol%. Although these composites do not have prior particle boundaries because of the powder processing, they do have a cast structure that must be broken down by metalworking. These composites have isotropic modulus and yield strength values similar to those of the powder metallurgy processed composites, as well as good fatigue and fracture properties. In general, the ultimate strength and ductility of the cast and metal-worked composites are lower than those of the powder metal composites that have undergone comparable metalworking. These composites have lower expansion properties and increased wear resistance as a consequence of the presence of SiC particles.

Recently, billet sizes for cast composites have increased from nominal 9 kg (20 lb) to 45 kg (100 lb) by increasing the length of the 150-mm (6-in.) diam billet. The cast microstructure of these composites is characterized by the SiC reinforcement being associated with the last liquid to freeze. Even though the particles are wet by the aluminum, they are

rejected by the primary solidifying phase and are segregated to the eutectic regions of the castings. Another consequence of this segregation is the presence of microshrinkage at or near the SiC-metal interface. This microshrinkage casting defect has been overcome almost completely by hot isostatic pressing the castings, as either billets or net castings.

REFERENCES

1. S.S. Brenner, *J. Appl. Phys.*, Vol 33, 1962, p 33
2. S.S. Brenner, *J. Met.*, Vol 14 (No. 11), 1962, p 808
3. W.H. Sutton, Whisker Composite Materials—A Prospectus for the Aerospace Designer, *Astronautics and Aeronautics*, Aug 1966, p 46
4. W.H. Sutton and J. Chorn, *J. Met. Eng. Q.*, Vol 3 (No. 1), 1963, p 44
5. A. Divecha, P. Lare, and H. Hahn, "Silicon Carbide Whisker-Metal Matrix Composites," AFML-TR-69-7, Air Force Materials Laboratory, May 1969
6. T. Donomoto, K. Funatani, N. Miura, and N. Miyake, "Ceramic Fiber Reinforced Piston for High Performance Diesel Engine," Paper 830252, Society of Automotive Engineers, March 1983
7. J. Dinwoodie, E. Moore, C.A.J. Langman, and W.R. Symes, in *Proceedings of the International Conference on Composite Materials V*, W.C. Harrigan, Ed., 1985
8. M. Toaz and M. Smalc, Clevite's Advanced Technology for the Diesel Piston, *Diesel Prog. North Am.*, June 1985, p 56-60
9. L. Ackermann, J. Charbonnier, G. Desplanches, and H. Koslowski, in *Proceedings of the International Conference on Composite Materials V*, W.C. Harrigan, Ed., 1985
10. D.L. McDaniels, *Metall. Trans. A*, Vol 16A, 1985, p 1105
11. R.J. Arsenault, *Mater. Sci. Eng.*, Vol 64, 1984, p 171
12. C.R. Crow, R.A. Gray, and D.F. Hasson, in *Proceedings of the International Conference on Composite Materials V*, W.C. Harrigan, Ed., 1985
13. F.R. Billman, J.C. Kuli, G.J. Hildeman, J.T. Petit, and J.D. Walker, "Processing of P/M Aluminum Alloys 7090 and 7091," Paper presented at The Third Conference on Rapid Solidification Processing: Principles and Technology, National Bureau of Standards, Gaithersburg, MD, Dec 1982
14. W.C. Harrigan, Jr., "Heat Treatment of Silicon Carbide Reinforced 6061 Aluminum Composites," Paper presented at the

AIME Fall Meeting, American Institute of Mining, Metallurgical, and Petroleum Engineers, Oct 1982

15. "Standard Methods of Tension Testing of Metallic Materials," E 8, *Annual Book of ASTM Standards*, American Society for Testing and Materials

16. B. Mosdate, Air Industries, private communication, 1981

17. DWA Composite Specialities Inc., unpublished data, 1985

18. B. Mosdate, Air Industries, private communication, 1982

19. T. Steelman, Rockwell International, unpublished data, 1985

20. 1985 Top Twenty Awards, *Mater. Eng.*, Nov 1985

21. W.R. Mohn, "Dimensionally Stable SXA Engineered Metallic Composites for Guidance Systems and Optics Applications," Paper presented at ASM Materials Week Conference, ASM INTER-NATIONAL, Orlando, Oct 1986

22. A.P. Divecha, S.G. Fishman, and S.D. Karmarkar, *J. Met.*, Vol 9, 1981, p 12

23. H. Gaegle, "Processing of Metal Composites," Paper presented at Westech Conference, Los Angeles, March 1985

24. D. Schuster, "Advanced Ingot-Based SiC Reinforced Aluminum," Paper presented at ASM Materials Week Conference, ASM INTERNATIONAL, Orlando, Oct 1986

Whisker-Reinforced MMCs

Jack L. Cook and Walter R. Mohn, ARCO Chemical Company

WHISKER-REINFORCED metal matrix composites (MMCs) make up a distinct category of advanced engineered materials that provides unique differential advantages over conventional alloys in many high-performance applications. These materials usually consist of a uniform dispersion of needlelike single crystals throughout a homogeneous metallic matrix. Compared to the unreinforced alloy, whisker-reinforced MMCs generally exhibit greater stiffness (Young's modulus), higher strength, and better dimensional stability, without a significant increase in density.

One of the principal advantages of whisker-reinforced MMCs is the opportunity to use conventional metal-forming equipment to texture a preferred orientation of the high-strength single-crystal whiskers in the microstructure. Second, higher directional strengths can often be obtained in finished components through extrusion, rolling, forging, or superplastic forming. Third, the fabrication of complex whisker-reinforced MMC parts in one operation can potentially reduce manufacturing costs and simplify secondary assembly requirements. Finally, such components may be strategically configured with thinner sections than those of unreinforced metal counterparts, with significant weight reduction. Thus, use of whisker-reinforced MMCs can provide opportunities for reducing life-cycle operational costs in many weight-critical applications.

Whisker Reinforcement. Whiskers may best be described as acicular single crystals with a high aspect ratio (L/D): an average length (L) considerably greater than the average maximum diameter (D). Most whiskers have aspect ratios ranging from about 50 to 150. Single-crystal whiskers usually have much greater tensile strength than other types of discontinuous reinforcements, such as polycrystalline flakes, particulates, or chopped fibers. Thus, whisker reinforcement is generally used to produce the highest-strength discontinuous reinforced MMCs.

The superior properties of these materials can be explained in terms of classical load transfer between the whiskers and the metal matrix. The tensile strength, σ_c, of a composite containing aligned short fibers can be calculated by:

$$\sigma_c = \sigma_w f_w\left(1 - \frac{l_c}{2l}\right) + \sigma_m(1 - f_w) \qquad \text{(Eq 1)}$$

where σ_w is the tensile strength of the whiskers, f_w is fractional volume content, l is the length of the whiskers, and σ_m is the stress in the matrix at the onset of fracture of the whiskers (Ref 1). The critical whisker length, l_c, can be calculated by:

$$l_c = d_w\left(\frac{\sigma_w}{\sigma_y}\right) \qquad \text{(Eq 2)}$$

where d_w is the diameter of the whiskers and σ_y is the flow stress of the matrix (Ref 1).

Whiskers have been produced for use in MMCs from over 100 materials (Ref 2). These include metallic whiskers of iron, nickel, and copper, as well as ceramic whiskers of aluminum oxide, graphite, silicon nitride, and silicon carbide. During the past decade, much work has been done in Japan, Europe, the U.S.S.R., and the U.S. to develop whiskers and associated composites.

The characteristics of whiskers produced for use in MMCs vary greatly, depending on manufacturing and processing techniques. Subtleties such as crystal surface chemistry, morphology, flaw sensitivity, and mean aspect ratio of the bulk whiskers ultimately and dramatically affect the properties of the metallic composite material.

The development of silicon carbide (SiC) whisker reinforcement has progressed considerably further than that of other whisker materials. Good quality SiC whiskers can now be produced at relatively low cost by at least two different processes: vapor-liquid-solid (VLS) and rice hull pyrolysis.

Vapor-liquid-solid processed SiC whiskers, shown in Fig. 1, have diameters ranging from 3 to 10 μm (120 to 390 μin.) and lengths as great as 100 mm (4 in.) (Ref 3). SiC whiskers produced by rice hull pyrolysis (Fig. 2) have an average diameter of 0.6 μm (24 μin.) and lengths ranging from 10 to 80 μm (0.40 to 3 mil) (Ref 4). Experimental data indicate that their tensile strength can approach 7000 MPa (about 1000 ksi). Commercial quantities of

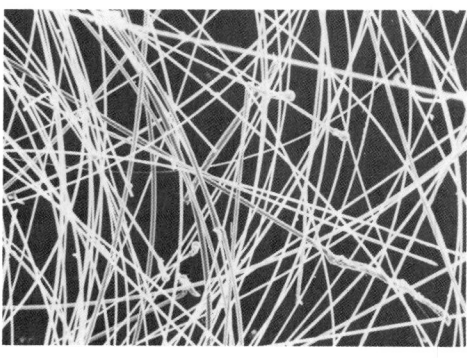

Fig. 1 SiC whiskers produced by the vapor-liquid-solid (VLS) process

rice-hull-derived SiC whiskers are routinely manufactured in the U.S. for use in both MMC and ceramic matrix composite materials and products.

Matrix Metals. Although researchers have experimented with a great variety of matrix metals for use in whisker-reinforced MMCs, most developmental activities have focused on four alloy systems conventionally used in aerospace applications: aluminum, magnesium, titanium, and copper (Ref 5). Selected alloys of these metals collectively exhibit such desirable characteristics as light weight, high specific strength and stiffness, high temperature capability, and high thermal conductivity, when reinforced with 10 to 20 vol% SiC whiskers.

The best properties can be obtained in a composite system when the reinforcing whiskers and matrix alloy are as physically and chemically compatible as possible. Special matrix alloy compositions, in conjunction with unique whisker coatings, have been devised to optimize the performance of certain metallic composites. The mechanisms that affect the matrix-whisker interface and control transfer of stresses from the matrix to the whiskers greatly determine the mechanical behavior of the composite. Because grain structure, precipitation kinetics, and dislocation motions within whisker-reinforced MMCs are complex and not fully understood, many matrix-whisker combinations have been developed empirically.

Fig. 2 SiC whiskers produced by the pyrolysis of rice hulls

10 μm

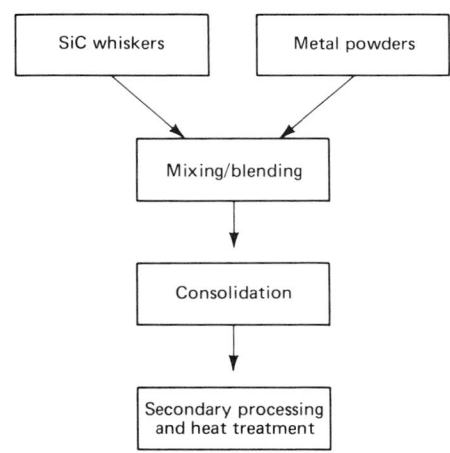

Fig. 3 Powder metallurgy process flow to produce SiC whisker-reinforced MMCs

Fig. 4 Blend of SiC whiskers and metal alloy powders

10 μm

Some of the more successful systems designed for commercial applications use matrices of aluminum or magnesium alloys.

Typical Materials. Advances in technology during the late 1970s demonstrated the feasibility of producing commercial quantities of SiC whiskers economically through rice hull pyrolysis (Ref 6). With increased availability of relatively low-cost SiC whiskers, researchers had more incentive to incorporate them into metals, in order to produce lightweight, high-performance MMCs with high specific strength and stiffness.

During the early 1980s, SiC whisker-reinforced aluminum alloys were the subject of numerous developmental programs funded by both industry and government, with particular emphasis on improving and qualifying whisker-reinforced MMCs for aircraft structural applications. In 1985, these materials were successfully engineered to meet and exceed the design goals established for new-generation tactical aircraft. Recent research and development have focused on low-density metallic composites made up of magnesium alloys reinforced with SiC whiskers, for structural use in advanced flight vehicles and orbital spacecraft.

Commercial products using these new materials were first produced early in 1986. While this discussion focuses on SiC whisker-reinforced aluminum and magnesium alloys, the information presented on processing techniques, material characteristics, and applications is generally relevant to many other whisker-reinforced MMC systems.

Methods of Manufacture

The synthesis of SiC whisker-reinforced MMCs has been successfully accomplished by two distinct processes: a powder metallurgy (P/M) method and a liquid metal (melt) infiltration method. Both processes are described here.

Powder-derived SiC whisker-reinforced MMCs, which require considerable time and care to produce, typically have tensile and fatigue properties superior to those of melt-infiltrated composites. Despite the relative ease of manufacture, the melt-infiltrated materials usually have nonhomogeneous distribution of whiskers in the matrix alloy. They also have flaws in the matrix that occur during solidification. Such anomalies inhibit the effective transfer of stresses from the matrix alloy to the whiskers. Performance and cost considerations must be carefully weighed to determine which property-dependent process is best for producing SiC whisker-reinforced MMCs intended for a given application.

Powder Metallurgy Process. The P/M process begins with mixing and blending prealloyed metallic powder and SiC whiskers, followed by heating and degassing, and, finally, consolidation into intermediate or final product forms (Ref 7). Figure 3 illustrates the P/M manufacture of SiC whisker-reinforced MMCs in a dense, gas-free condition.

Safety Considerations. Safety awareness and good operating practices are paramount in successful P/M production of MMCs. Flammable matrix metals, such as aluminum and magnesium, especially in powder form, require extreme care to prevent accidental fires or explosions. Those who work with these materials should have a thorough knowledge of the potential hazards. Recommendations for proper storage, handling, and processing of flammable metallic powders are available from several excellent sources, including the Aluminum Association, Inc., the National Fire Protection Association, and the U.S. Bureau of Mines.

Mixing and Blending. During the critical stages of production, measured quantities of SiC whiskers and fine-mesh metal alloy powders are thoroughly mixed and blended to establish a high degree of particle intermingling, as shown in Fig. 4. Lubricants and selected additives are usually employed in this kind of metal and ceramic multicomponent powder system to help overcome some of the problems inherent to the mechanics of mixing (Ref 8). The adverse effects of interparticle friction, electrostatic attraction, and density differences must be reduced to facilitate flow during mixing and blending. Mechanically interlocked agglomerates of whiskers also must be separated to establish a statistically random dispersion. This can normally be achieved with high-velocity high-shear blending equipment. However, blending parameters must be carefully selected and controlled to prevent breakage of the high-aspect-ratio SiC whiskers.

Blend Outgassing and Contaminant Control. The production of dense, porosity-free, SiC whisker-reinforced MMCs by a P/M process critically depends on proper treatment of the composite powder blends to remove volatile contaminants effectively. Residual organics, such as lubricants and other mixing and blending additives, must be completely extracted before consolidation. Water vapor and gases adsorbed to the particle surfaces must also be expunged.

Removal of volatile contaminants can be accomplished as the collective result of vaporization, thermal desorption, diffusion, and chemical decomposition (Ref 9). Most volatile residuals can be desorbed from the composite blend by heating at low temperature (below 150 °C, or 300 °F) under vacuum. Continued heating at higher temperatures additionally serves to release chemically bound water through the decomposition of hydrated oxides. Because the characteristics and quantity of hydrated oxides vary with matrix alloy chemistry, separate outgassing procedures usually must be devised for each type of powder composite system. Outgassing techniques are often optimized with information gained from thermogravimetric analysis (TGA) or residual gas analysis (RGA).

Consolidation. Vacuum hot pressing is ideally suited to consolidation, and special tooling and processing conditions can be devised in a vacuum hot press to accomplish outgassing as well. Cylindrical billets and other comparatively simple geometric shapes can be produced by the processing sequence of (1) loading the composite powder blends into the die assembly; (2) placing the loaded die into the press and then evacuating, heating, and outgassing to remove volatile contaminants; (3) heating and pressing to consolidate the blends into a dense form; and (4) cooling the densified form and removing it from the die.

The importance of heating the composite blend to a uniform specified temperature (which depends on matrix alloy) before pressing cannot be overemphasized. Usually a holding time is required to obtain temperature equilibrium within the blend, relative to the surface temperature of the die assembly.

In addition to temperature, chamber vacuum level and ram travel must be checked frequently to ensure normal progression of the hot pressing cycle. Each type of composite blend will have a unique response during the consolidation process.

Once the consolidation sequence for a powder blend has been fully characterized, the process can be automated, using microprocessor control. Use of such a programmed schedule helps reduce manufacturing costs and ensure consistent production of high-quality whisker-reinforced MMCs.

Hot isostatic pressing (HIP) is another processing technique used to consolidate SiC whisker-reinforced composite powder blends into near-net-shape components. Procedures are similar to those used to hot isostatically press conventional unreinforced P/M parts. For additional information, see Ref 10. Particular care must be taken during the encapsulation of composite blends to prevent excessive vibration, which could cause the whiskers to settle and agglomerate.

Melt Infiltration Process. In melt infiltration, a whisker preform is prepared and infiltrated by molten metal (Fig. 5). Subsequent cooling and solidification of the metal produces a whisker-reinforced MMC. A principal advantage of this process is single-step production of net-shape parts. Because of its simplicity, melt infiltration may offer advantages in certain high-volume production applications where lower material and processing costs are more important than high-performance properties. Properties of melt-infiltrated SiC whisker-reinforced composites have typically been inferior to those of P/M-produced counterparts because of the nonhomogeneous distribution of whiskers in the matrix metal, solidification shrinkage, hot tearing, gas porosity, and matrix alloy segregation (Ref 11). In addition, property improvements normally occurring during deformation processing are not attained in net-shape components produced by melt infiltration.

Safety Considerations. Planning and implementing safe operating procedures are vitally important. In addition to the proper handling of molten metals, adequate treatment and drying of the preform and mold assembly are manda-

tory, to prevent dangerous accidents. Violent eruptions of water vapor and gases can occur during melt pouring and whisker preform infiltration if these contaminants are not first removed. Protective shields and clothing should be used during all critical phases of melt infiltration and composite manufacture.

Whisker Preform Preparation. SiC whisker preforms are typically produced either by a process similar to ceramic slip casting or by a pulp molding process similar to that used to make thick paper. Whiskers are first combined with various binders to form a moderately thick slurry. Normally, the binders are organic resins, alumina-silicates, or a complex mixture of several other chemical ingredients. The prepared slurry is poured into slip cavities or vacuum molded and allowed to set up and cure. Mold/preform assemblies are then furnace fired to burn out organic constituents and drive off moisture.

When firing is complete, the delicately bonded and interlocked whiskers define a cohesive, integrally porous preform body. The whisker preforms, which are from 20 to 40% dense, have sufficient ''green strength'' to withstand careful removal from the form or cavity for placement in the infiltration casting mold and final processing.

Metal Casting and Preform Infiltration. Although melt infiltration can be done in air, better results have been achieved in a vacuum. The mold/preform assembly should be preheated to a uniform temperature high enough to promote penetration and free flow of the melt throughout the whisker structure.

To produce parts that are selectively reinforced, such as pistons for reciprocating engines, SiC whisker preforms can be strategically placed in a die and infiltrated by a process known as squeeze casting. The extremely high pressure of this process promotes complete penetration and wetting of the preform by the melt.

Secondary Processing

A major advantage of SiC whisker-reinforced MMCs is that they can be formed and processed using conventional metalworking equipment (Ref 12). Use of readily available production machinery and established manufacturing methods to produce lightweight high-strength whisker-reinforced MMC parts helps keep fabrication costs to a minimum.

Successful use of SiC whisker-reinforced MMCs, however, requires a thorough knowledge of deformation mechanics and an understanding of the important process variables germane to each composite material system. Composite deformation behavior is different from that of unreinforced alloys. Variations in SiC whisker content or in matrix alloy chemistry can radically affect material flow, flow stress, and strain rate sensitivity. In general, these composite types require greater

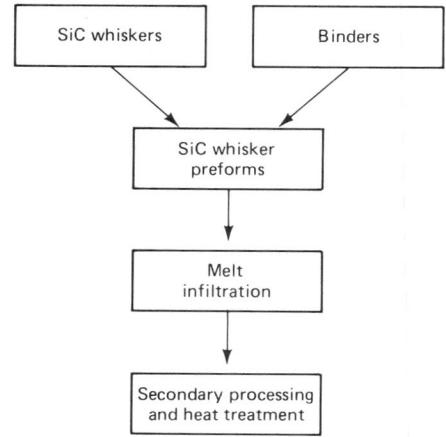

Fig. 5 Melt infiltration process flow to produce whisker-reinforced MMCs

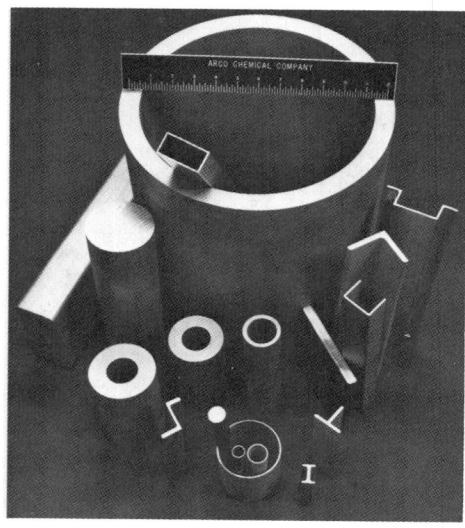

Fig. 6 Cross sections extruded from SiC whisker-reinforced MMCs

control of processing temperature and strain rate than unreinforced metals.

Extrusion. Cylindrical billets of SiC whisker-reinforced aluminum and magnesium alloys have been hot extruded into a wide variety of solid and hollow shapes, including rod, bar, and tube, as shown in Fig. 6. Although the sequence of operations for direct extrusion of these materials is the same as for unreinforced alloys, differences in deformation behavior require changes in process parameters.

Since flow stress of the composite at extrusion temperature is higher than that of the unreinforced alloy, a higher extrusion pressure and lower ram speed are usually required. Load versus ram displacement curves show that more work is needed to initiate deformation and overcome friction and shearing for composites than for the unreinforced alloys. Because additional heat is generated by plastic deformation and friction between the composite material and

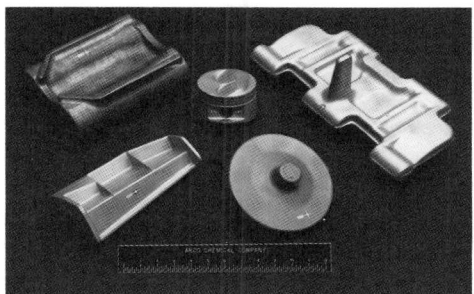

Fig. 7 Component shapes forged from SiC whisker-reinforced MMCs. Clockwise from left of scale: aircraft hinge mount, track shoe for armored vehicle, forged and machined racing piston, tactical tank track shoe, gas turbine compressor wheel

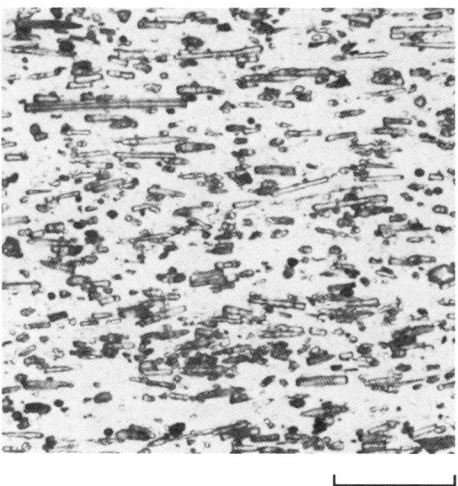

10 µm

Fig. 8 SiC whisker-reinforced aluminum alloy sheet; whiskers aligned in the direction of rolling

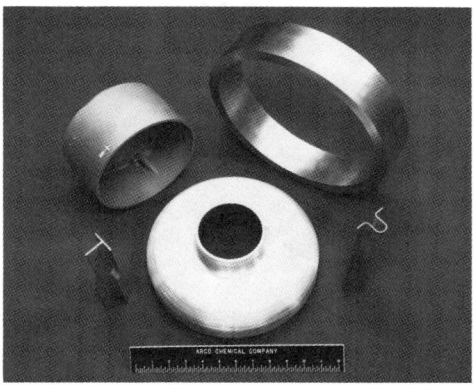

Fig. 9 SiC whisker-reinforced MMC components produced by various deformation processes. Clockwise from left of scale: extruded and jogged structural T-stringer, back-extruded missile guidance bay, ring-rolled and machined missile body stiffener, hot-formed sheet stiffener, spun bell housing

the die, the initial MMC billet temperature must be precisely controlled to prevent incipient melting of the matrix metal.

Composite material extrusions almost always require lubrication to prevent surface defects (Ref 13). Most shapes can be produced using shear face or conical dies. Streamlined dies are required for more complex extrusions. Extrusion ratios as high as 70:1 have been successfully achieved with whisker-reinforced MMCs. Since the whiskers align in the extrusion direction, the highest tensile strengths are established in that direction.

Forging. SiC whisker-reinforced aluminum and magnesium alloy composites can be hot forged into complex shapes with the same forging equipment used for unreinforced metals. Figure 7 shows whisker-reinforced composite products produced by impression-die and closed-die forging.

Single-stroke hydraulic presses are generally preferred for forging because lubrication can be applied easily and pressing speeds can be controlled to obtain the desired strain rates. However, mechanical forge presses are often quite adequate for shaping when the size reduction is low. Strain rate and temperature of the composite workpiece must be carefully controlled in the forging operation to minimize the risk of incipient melting and cracking. Although many forge dies designed for unreinforced metals are satisfactory for forging whisker-reinforced composites, certain shapes require a different die design to allow for the specific flow characteristics of the SiC whisker-reinforced MMCs.

When a specific whisker alignment is desired in a component, multiple forging operations may be required. To establish a radial orientation of whiskers in the MMC turbine disk shown in Fig. 7, an intermediate shape was first produced from an extruded billet using two blocker forging steps. The finished shape was then generated by upsetting the blocker shape through high-energy-rate forging (HERF) (Ref 14).

Rolling of SiC whisker-reinforced aluminum and magnesium alloy composites into

plate, sheet, and foil can be done using a two-high mill with large-diameter heated rolls. Best results are obtained when MMC slabs of appropriate thickness, produced by billet extrusion and forging, are used as starting stock. To minimize the formation of edge cracks, close control of the workpiece temperature, roll temperature, and strain rate is necessary. Heat loss from the workpiece must be kept to a minimum, and reductions should not exceed 10% for each roll pass (Ref 15). With roll surfaces ground to a fine finish, production of 1.8-mm (0.070-in.) thick, aircraft-quality, SiC whisker-reinforced aluminum alloy sheet has become routine.

The large amount of deformation required to produce rolled sheet causes significant alignment of the whiskers in the microstructure (Fig. 8). Cross rolling may be used to establish a more random two-dimensional whisker orientation, if preferred.

The composite sheet can be cold rolled once thickness has been reduced below about 3.2 mm (0.13 in.). As much as 45% cold work can be imparted in multipass rolling to the MMC sheet before edge cracking becomes a problem. With successive anneals between roll passes, SiC whisker-reinforced MMC sheet has been rolled to 50-µm (2000-µin.) thick foils. Thinner foils, 12.7-µm (500-µin.) thick, have been produced for selected applications by chemical milling.

Other Deformation Processes. As the number of applications for SiC whisker-reinforced MMCs continues to increase, many conventional metal-forming techniques are being adapted and developed to produce MMC products economically, including back extrusion, hot forming, superplastic forming, spinning, and ring rolling. Examples of whisker-reinforced MMC parts produced by these forming techniques are shown in Fig. 9. As a general rule, most deformation processes used

successfully with unreinforced aluminum and magnesium alloys can be adjusted to work with whisker-reinforced composites effectively.

The machining characteristics of SiC whisker-reinforced MMCs are considerably different from those of identical unreinforced alloys. The presence of the extremely hard, abrasive, ceramic SiC whiskers within a comparatively soft matrix elevates temperatures between tool and workpiece, resulting in faster tool wear. This abrasive action must be offset by selecting appropriate tool materials and modifying other machining parameters.

Machining practices used with SiC whisker-reinforced aluminum alloy composites are similar to those used for A390 sand cast aluminum (16 to 18% Si). Using polycrystalline diamond cutting tools and adapting standard metal-cutting and finishing techniques developed for other abrasive materials will result in close-tolerance, intricate parts with excellent finish and surface integrity (Ref 16).

Many traditional machining operations, as well as some nontraditional operations (electric discharge machining, electrochemical machining, hydrodynamic abrasive jet machining, chemical milling), have worked well in producing precision MMC components, such as the SiC whisker-reinforced magnesium alloy gears shown in Fig. 10.

Joining Techniques. Metal matrix composites reinforced with SiC whiskers can be joined to similar composites, unreinforced metals, and nonmetals by several methods, including welding, brazing, bonding, and fastening.

Welding processes used successfully are fusion welding, inertia welding, resistance welding, and ultrasonic welding. One drawback of fusion welding is possible formation of unstable aluminum carbides in the weld as a product of the reaction between SiC and molten aluminum (Ref 17). Resistance welding and ultrasonic welding work best when applied to whisker-reinforced MMC sheet materials.

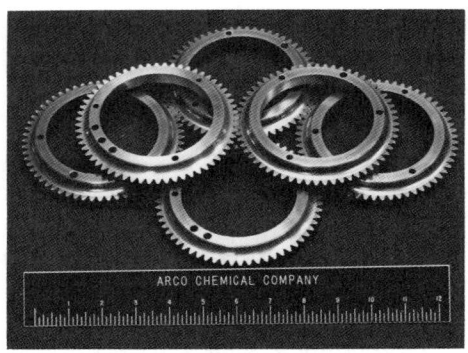

Fig. 10 SiC whisker-reinforced magnesium alloy gears

Table 1 Typical properties of extruded materials showing the effects of SiC whisker reinforcement

Material type	Ultimate tensile strength MPa	ksi	Yield strength(a) MPa	ksi	Elongation %	Young's, modulus, (E), GPa	10^6 psi	Coefficient of thermal expansion (α), 10^{-6}/K	Density (ρ), g/cm^3
2124-T6 482		69.9	448	64.9	7	74.4	10.8	22.5	2.77
2124-T6-20 vol% SiC whisker 856		24	497	72.0	2.4	127	18.4	13.0	2.86
ZK60A-T5. 365		52.9	303	43.9	11	44.8	6.49	24.3	1.83
ZK60A-T5-20 vol% SiC whisker 613		88.9	517	74.9	1.2	96.5	13.9	14.4	2.11

(a) 0.2% offset

Table 2 Typical properties of MMC billet and extruded plate showing the effects of SiC whisker aligment

Material: 2024-T6 reinforced with 20 vol% SiC whiskers

MMC material form	Test specimen orientation	Ultimate tensile strength MPa	ksi	Yield strength(a) MPa	ksi	Coefficient of thermal expansion (α), 10^{-6}/K	Density (ρ), g/cm^3
12-in.-diam cylindrical billet	Longitudinal (axial)	496	71.9	351	50.9	16.1	2.86
½-in. by 5-in. extrusion	Longitudinal	737	107	448	64.9	13.0	2.86
12-in.-diam cylindrical billet	Transverse	503	72.9	358	51.9	16.4	2.86
½-in. by 5-in. extrusion	Transverse (long)	462	67.0	379	54.9	19.6	2.86

(a) 0.2% offset

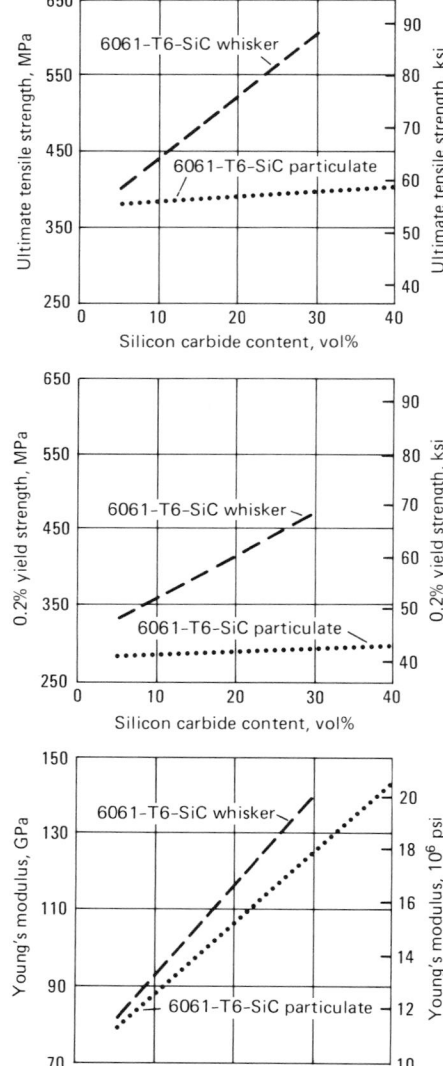

Fig. 11 Effects of SiC content on tensile properties of MMC extrusions

Successful brazing of SiC whisker-reinforced composites normally requires higher braze temperatures and very active fluxes, since the braze metal does not wet the metallic composite material as easily as unreinforced alloy. The composite must be fully degassed to prevent blistering and cracking during the braze cycle.

Both adhesive bonding and mechanical fastening work well in producing prototype hybrid flight vehicle components with SiC whisker-reinforced aluminum alloys. Unlike resin matrix composites, no special fasteners are required for joining MMCs. Conventional aluminum, steel, and titanium fastener systems should be used.

Surface Treatments. Practically all surface metal treatments applied to unreinforced aluminum and magnesium alloys can be applied to SiC whisker-reinforced alloys. Surface processes successfully demonstrated with these composites include shot peening, glass bead blasting, caustic etching, electrochemical etching, plating, cladding, ion vapor deposition, laser glazing, and painting.

Engineered Properties

The mechanical properties of SiC whisker-reinforced MMCs are generally superior to those of unreinforced alloys. Compared to high-strength aluminum and magnesium alloys, SiC whisker-reinforced MMCs have higher strength and stiffness, better dimensional stability, and greater fatigue resistance, especially at elevated temperatures. For many advanced structural applications involving significant aerothermal heating, these materials are preferred over both SiC particulate-reinforced MMCs and polymer matrix composites.

The ability to tailor both their mechanical and physical properties is a unique and important feature. By increasing the content of whiskers in the material, corresponding increases in tensile strength, yield strength, and elastic modulus can be attained. Figure 11 shows the effects on the MMC tensile properties of increasing SiC content in a 6061-T6 aluminum matrix. These comparisons show that the whiskers produce superior properties over particulates at any common volume fraction.

Table 1 compares typical tensile properties of 2124-T6 aluminum and ZK60A-T5 magnesium alloys with their SiC whisker-reinforced counterparts. Enhanced properties, such as tensile strength, yield strength, and elastic modulus, afford reduction of section thickness and structural weight.

Aligning SiC whiskers in the matrix alloy through deformation processing usually produces highest strength in the MMC. Although its properties are virtually isotropic in billet form, its strengths are relatively low. Extrusion of SiC whisker-reinforced MMC billet stock into plate simultaneously results in break-up of oxides at the prior particle boundaries within the billet matrix alloy, relative motion and enhanced bonding between SiC whiskers and the matrix alloy, and alignment of the whiskers in the extrusion direction.

Table 2 shows the tensile properties of SiC whisker-reinforced 2024-T6 MMC before and after extrusion. While the extruded MMC has

Table 3 Tensile properties of a forged SiC whisker-reinforced aluminum alloy turbine wheel

Material: 6061-T6 reinforced with 20 vol% SiC whiskers

Test specimen orientation	Ultimate tensile strength		Yield strength(a)		Elongation (e), %	Young's, modulus, E	
	MPa	ksi	MPa	ksi		GPa	10⁶ psi
Radial	518	75.1	386	55.9	4.7	111	16.1
Transverse (circumferential)	490	71.0	379	54.9	6.1	108	15.7

(a) 0.2% offset

Table 4 Typical properties of SiC whisker-reinforced aluminum alloy sheet

Material: 2124-T6 reinforced with 15 vol% SiC whiskers

Sheet thickness mm	in.	Test specimen orientation	Ultimate tensile strength MPa	ksi	Yield strength(a) MPa	ksi	Elongation (e), %	Young's, modulus, E GPa	10⁶ psi	Fracture toughness, K_c MPa √m	ksi √in.
2.54	0.100	. . . Longitudinal (Along roll direction)	718	104	573	83.1	5.3	114	16.5	55	50
2.54	0.100	. . . Transverse (90° to roll direction)	559	81.0	386	56.4	8.5	95	14	59	54

(a) 0.2% offset

dramatically increased longitudinal tensile strength, relatively little change has occurred in its transverse tensile strength. Alteration of spacing between the aligned whiskers in the extruded plate, in addition to texturing the matrix alloy, has significantly changed both its longitudinal and transverse coefficients of thermal expansion (CTE). The CTE can often be "engineered" to match a desired value by selecting the proper extrusion ratio for a given SiC whisker-reinforcement content.

Like extrusion, forging is another deformation process used to obtain a preferred whisker orientation in selected MMC components. The forged SiC whisker-reinforced 6061-T6 aluminum alloy turbine wheel shown in Fig. 7 was fabricated by forging in such a manner as to align the whiskers radially, thereby enhancing the tensile properties in the desired direction. Table 3 shows both the radial and transverse tensile properties for the wheel.

Silicon carbide whisker-reinforced MMCs can be flattened by means of a carefully planned sequence of extrusion, forging, and rolling to produce MMC sheet with a desired amount of whisker alignment. This control of whisker alignment enables production of SiC whisker-reinforced MMC sheet with directional properties needed for certain high-performance applications. Cross rolling establishes a more random planar whisker alignment, producing MMC sheet with two-dimensional isotropy (Ref 18).

Typical properties of SiC whisker-reinforced MMC sheet are shown in Table 4. The material is characterized by high strength and stiffness, with ductility and toughness sufficient for many advanced aerospace structural applications.

Figures 12 and 13 show the superior properties of the MMC materials for 100-h exposure at elevated temperatures, compared with unreinforced alloys. These composites also perform well during brief thermal exposures up to 370 °C (700 °F); at higher temperatures, the differential advantage of MMCs begins to diminish.

Material Characteristics

The presence of SiC whiskers in an MMC causes it to have different wear and corrosion characteristics from the unreinforced alloy. Most preliminary investigations of these characteristics have involved SiC whisker-reinforced aluminum alloys and have provided only a cursory understanding of their wear and corrosion behavior.

The wear resistance of SiC whisker-reinforced MMCs is superior to that of unreinforced alloys. Generally, the resistance of the composite to wear increases with higher volume content of SiC whiskers, especially when heat treated. Taber abrader tests have shown that wear characteristics of SiC whisker-reinforced aluminum alloy MMCs approach those of 4140 steel (Ref 19).

The electrochemical behavior of SiC whisker-reinforced aluminum alloys has been studied under closely controlled conditions and has provided some important conclusions (Ref 20). First, the SiC reinforcement does not play a part in the galvanic corrosion process. It has also been observed that, as a result of an enhanced general corrosion rate, the SiC whisker-reinforced alloys are more resistant to localized corrosion than conventional wrought-aluminum alloys (Ref 21). Resistance of these MMCs to stress corrosion cracking also appears to be higher.

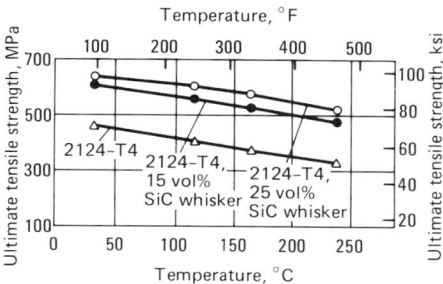

Fig. 12 Effects of SiC whisker reinforcement on the elevated-temperature strength of 2124-T4 aluminum alloy, 100-h exposure

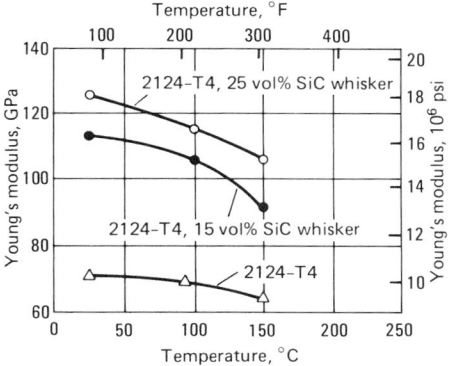

Fig. 13 Effects of SiC whisker reinforcement on the elevated-temperature elastic modulus of 2124-T4 aluminum alloy, 100-h exposure

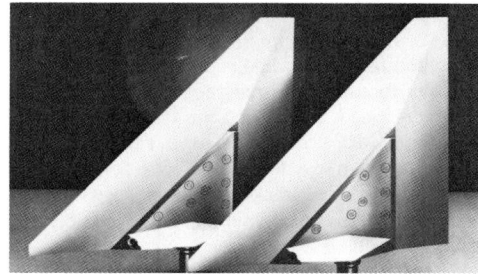

Fig. 14 SiC whisker-reinforced aluminum alloy missile wings

Applications

The differential advantages of SiC whisker-reinforced MMCs have been demonstrated in a number of product forms. For aerospace use, their high compressive and bearing strengths, good fracture toughness, and enhanced elevated-temperature performance translate directly into reduced flight-vehicle weight. This weight reduction may be further translated into increased payload, extended range, or fuel savings, concurrent with savings in life-cycle operational costs.

The prototype SiC whisker-reinforced aluminum MMC missile wings shown in Fig. 14 are an excellent example of high-performance components that can be fabricated from this material. These composites provide the elevated

Fig. 15 SiC whisker-reinforced MMC floor panel being installed in test bed aircraft. Courtesy of Lockheed-Georgia Company

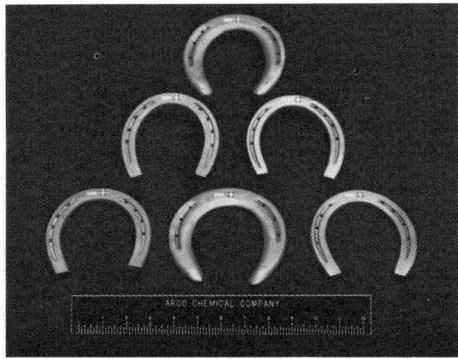

Fig. 16 Superplastically molded SiC whisker-reinforced aluminum alloy horseshoes

temperature strengths and ablation resistance needed to survive the aerothermal heating of supersonic flight. The replacement of heavier conventional titanium missile wings with MMC wings results in significant weight reduction in a tactical missile system.

Figure 15 shows installation of a SiC whisker-reinforced aluminum alloy floor panel in a test bed aircraft. The test panel, which is 1 m (40 in.) long and 0.8 m (31 in.) wide, was incorporated in the cargo floor of the aircraft fuselage. The riveted floor panel assembly contained both whisker-reinforced MMC extruded stiffeners and rolled sheet; successful flight testing of this panel demonstrated the potential of these materials for reducing aircraft weight by 35%.

Special-performance horseshoes, such as those shown in Fig. 16, are superplastically molded from SiC whisker-reinforced aluminum alloys for thoroughbred race horses, jumpers, hunters, and show horses. The lighter and stiffer MMC horseshoes have excellent wear and corrosion resistance, and provide the excellent support and track reliability necessary for endurance in cross-country and steeplechase competition.

Future Directions

Whisker-reinforced MMCs continue to be developed in a variety of reinforcement systems. Much of their future application will be in aerospace structures and engines that must operate at elevated temperatures. Titanium alloys and superalloys reinforced with whiskers of silicon nitride, silicon carbide, or aluminum oxide will probably be good candidates for these systems. The successful solid-state consolidation and secondary processing of whisker-reinforced, rapidly solidified, high-temperature aluminum alloys should also be a focus of considerable developmental work.

REFERENCES

1. A. Kelly and G.J. Davies, The Principles of the Fibre Reinforcement of Metals, *Met. Rev.*, Vol 10 (No. 37), 1965, p 1-77

2. C.T. Lynch and J.P. Kershaw, *Metal Matrix Composites*, CRC Press, 1972
3. J.V. Milewski, F.D. Gac, J.J. Petrovic, and S.R. Skaggs, Growth of Beta-Silicon Carbide Whiskers by the VLS Process, *J. Mater. Sci.*, Vol 20 (No. 4), April 1985
4. "SILAR® Silicon Carbide Whiskers," ARCO Metals Company, 1984
5. C. Zweben, "Metal Matrix Composites Overview," Publication No. 253, Metal Matrix Composites Information Analysis Center, Feb 1985
6. J.-G. Lee and I.B. Cutler, Formation of Silicon Carbide from Rice Hulls, *Am. Ceram. Soc. Bull.*, Vol 54 (No. 2), Feb 1975, p 195-198
7. A.P. Divecha, S.G. Fishman, and J.V. Foltz, Properties of SiC Whisker Reinforced Aluminum Alloys, in *Proceedings of the 24th National SAMPE Symposium*, Vol 24, Society for the Advancement of Material and Process Engineering, May 1979, p 1433-1442
8. P.E. Hood and J.O. Pickens, Silicon Carbide Whisker Composites, U.S. Patent No. 4, 463, 058, July 1984
9. A.I. Litvintsev and L.A. Arbuzova, Kinetics of Degassing of Aluminum Powders, *Poroshk. Metall.*, Vol 49 (No. 1), 1967
10. P.E. Price and S.P. Kohler, Hot Isostatic Pressing of Metal Powders, *Metals Handbook*, Vol 7, 9th ed., American Society for Metals, 1984, p 419-443
11. F. Klovcek and R.F. Singer, Influence of Processing on the Mechanical Properties of SiC Whisker Reinforced Aluminum Composites, in *Proceedings of the 31st International SAMPE Symposium*, Vol 31, Society for the Advancement of Material and Process Engineering, April 1986, p 1701-1712
12. A.P. Divecha, S.G. Fishman, and S.D. Karmarker, Silicon Carbide Reinforced Aluminum—A Formable Composite, *J. Met.*, Sept 1981
13. A.I. Kemppinen, ARCO Chemical Company, Advanced Materials, unpublished research
14. G.A. Gegel, "Fabrication of Turbine Engine Components of SXA® Silicon Carbide Reinforced Aluminum," ARCO Chemical Company, Advanced Materials, 1985
15. P.W. Niskanen, ARCO Chemical Company, Advanced Materials, unpublished research
16. "Guide to Machining SXA® Engineered Materials," ARCO Chemical Company, Advanced Materials, March 1985
17. J.S. Ahearn, C. Cooke, and S.G. Fishman, Fusion Welding of SiC Reinforced Al Composites, *Met. Constr.*, Vol 14 (No. 4), 1982
18. J.A. Walker, ARCO Chemical Company, Advanced Materials, unpublished research
19. "Wear Testing of SXA® Materials," ARCO Chemical Company, Sept 1984
20. S.J. Lenhart, "Electrochemical Characteristics of Silicon Carbide Reinforced Aluminum Alloys," Report No. 84-TA-22, ARCO Metals Company, March 1985
21. D.M. Aylor, David Taylor Naval Ship Research and Development Center, unpublished research

SELECTED REFERENCES

- M.F. Amateau, Progress in the Development of Graphite-Aluminum Composites Using Liquid Infiltration Technology, *J. Compos. Mater.*, Vol 10 (No. 4), Oct 1976, p 279-296
- R.B. Aronson, Metal-Matrix Composites—Materials of the Future, *Mach. Des.*, Aug 1985
- P.L. Blue, Some Ultrasonic Characteristics of Silicon Carbide Whisker and Particulate Reinforced Aluminum Alloy Composites, in *Review of Progress in Quantitative Nondestructive Evaluation*, Vol 5B, D.O. Thompson and D.E. Chimenti, Ed., Plenum Press, 1986, p 1157-1162
- A. Levitt, *Whisker Technology*, J. Wiley & Sons, 1970
- "Magnesium Composite Products—Ceramic Reinforced Magnesium Alloys," ARCO Chemical Co., March 1986
- D.L. McDanels, Analysis of Stress Strain, Fracture and Ductility Behavior of Aluminum Matrix Composites Containing Discontinuous Silicon Carbide Reinforcement, *Metall. Trans.*, Vol 16A, June 1985, p 1105-1115
- J.E. Schoutens, "Introduction to Metal Matrix Composites," Tutorial Series No. 272, Metal Matrix Composites Information Analysis Center, June 1982
- D. Webster, Effect of Lithium on the Mechanical Properties and Microstructure of SiC Whisker Reinforced Aluminum Alloys, *Metall. Trans.*, Vol 13A, Aug 1982, p 1511-1519
- C. Zweben, "Advanced Composites—A Revolution for the Designer," Paper No. AIAA-81-0894, Learn from the Masters Lecture Series, AIAA 50th Anniversary Annual Meeting and Technical Display, American Institute of Aeronautics and Astronautics, May 1981

Discontinuous Ceramic Fiber MMCs

Milton W. Toaz, J.P. Industries, Inc., Engine Products Group*

DISCONTINUOUS FIBER REINFORCED metal matrix composites (MMCs) have been under development since the early 1970s and are now reaching the point at which costs are decreasing and commercial uses are beginning to emerge. Discontinuous fiber reinforced MMCs provide considerable flexibility in terms of composition, fabrication method, and performance. The characteristics of discontinuous fibers allow the use of total or selective reinforcement and the development of isotropic or anisotropic properties in the MMC. Key properties that can be modified by the addition of discontinuous fibers include thermal expansion, thermal conductivity, damping characteristics, modulus, strength, fatigue endurance limit, and wear resistance.

Discontinuous fiber reinforced MMCs are composed of three constituents: the fibers, which alone have no structural value, but generally possess extremely high strength and stiffness; the matrix, which "glues" the fibers together in the correct proportions and orientations and distributes the applied loads; and the interface zone between the fiber and the matrix, which determines wetting, bonding, and load transfer, all of which are critical to composite performance.

Fibers. The compositions of discontinuous fibers that currently predominate as reinforcements of metal matrices are silicon carbide, alumina, and the aluminosilicates. The fiber is sometimes used essentially in its raw form; other times, the raw fiber is processed (for example, by mechanical attrition) before use. It is therefore clear that numerous other materials can be converted for use as discontinuous reinforcements if the application supports the material and processing costs.

The sources of discontinuous fiber are as varied as the compositions that are available. In the case of silicon carbide, the reinforcement may be a fiber (>1 μm, or 40 μin. diam) or a whisker (<1 μm, or 40 μin. diam). The former are produced by the chemical vapor deposition (CVD) process, while the latter are grown from rice hulls or produced by the vapor-liquid-solid (VLS) process. The CVD process is described in detail in the article "Continuous Carbon

Fiber Reinforced Carbon Matrix Composites," while the VLS process is described in "Whisker-Reinforced MMCs," both of which are in this Section of the Volume. Although silicon carbide fibers appear to have been developed initially as reinforcing media, many of the discontinuous fibers presently used as reinforcements were initially developed as insulating and refractory materials. Aluminosilicate, alumina, and zirconia fibers fall into the latter category. Blowing and spinning, the processes typically used to produce these refractory fibers, also produce a high percentage of low-aspect-ratio material, commonly known as shot. The sol-gel process, usually used to produce alumina fibers, appears to provide an initially cleaner material. However, most reinforcing-grade fibers require washing to remove extraneous material.

Important characteristics of several reinforcing grades of discontinuous fibers are given in Table 1. The strength of the ceramic fibers is very high and is retained to the 1500 °C (2730 °F) range. Stiffness is, in most cases, much higher than it is for typical matrix alloys. Of particular importance is the low coefficient of thermal expansion associated with ceramic fibers. The broad spectrum of fiber properties gives the MMC fabricator considerable latitude in creating a composite material to match the engineering requirements of a specific application.

Health safety information on reinforcing grades of discontinuous fiber is relatively sparse because of their brief history. They are currently not regulated in the occupational environment by any state or federal health agency. Nevertheless, the recommended work practices that should be applied when dealing with discontinuous ceramic fiber to minimize personnel exposure are:

- *Loose clothing is recommended.* Long-sleeved shirts that are loose at the neck and wrists, long pants, and caps will protect skin from contacting ceramic fibers, and prevent fibers that do contact the skin from being "ground in" by the clothing
- *Dust should be kept at a minimum.* Mechanical dust collection systems should be used whenever ceramic fiber materials are sawed or machined
- *Skin irritation should be avoided.* If refractory ceramic fibers accumulate on exposed skin areas, the skin should not be rubbed or scratched. The fibers should be removed by gently washing with warm water and mild soap
- *Eyes should be protected.* Safety glasses or face shields should be worn in any situation where ceramic fibers might get into the eyes
- *Respirators are recommended.* Respiratory protection should be worn to protect against breathing contaminated air

A number of medical studies on refractory ceramic fibers are in the initial stages of implementation, and preliminary results are expected by 1992.

Costs of discontinuous fibers vary considerably, as shown in Table 2. Because many of these materials are derivatives of commercial refractory fibers, their base cost is low compared to the cost of current continuous fiber materials ($165 to $330/kg, or $75 to $150/lb). The cost of reinforcing-grade fibers is somewhat higher than that of the base material as a result of the additional processing steps required to eliminate shot. In addition, depending on the MMC fabrication process, other processing costs may be incurred in order to prepare the fibers for incorporation into the matrix metal.

Table 1 Characteristics of discontinuous fibers

Material	Form	Diameter μm	Diameter μin.	Density, g/cm³	Ultimate tensile strength MPa	Ultimate tensile strength ksi	Tensile modulus GPa	Tensile modulus 10⁶ psi	Coefficient of thermal expansion, 10⁻⁶/K
Silicon carbide	Crystalline	0.2	8	3.2	800	120	500	75	4.3
Silicon carbide	Crystalline	120	4700	3.2	...	...	...	...	...
Alumina	Crystalline	3	120	3.3	2000	290	300	45	8.1
Mullite	Crystalline	3	120	3.2	690	100	152	20	5.1
Aluminosilicate	Amorphous	2	80	2.7	1730	250	104	15	...
Zirconia	Crystalline	5	200	5.7	...	...	206	30	10.5

*Formerly Clevite Industries Inc., Engine Parts Division

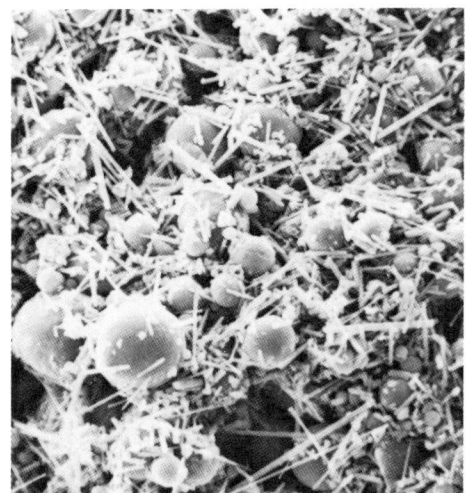

Fig. 1 Precision blended mixture of silicon carbide whiskers with aluminum alloy powder. 950×. Courtesy of Arco Chemical Company

Table 2 Relative costs of discontinuous fibers

Material	$/kg	$/lb
Silicon carbide	90-550	40-250
Zirconia	330-440	150-200
Aluminosilicates	5-45	2-20
Mullite	9-55	4-25
Alumina	35-90	16-40

Matrix Alloys. The predominant matrix alloys being used in discontinuous fiber reinforced MMC are aluminum based. Others include magnesium, titanium, copper, and lead. Aluminum alloy use is relatively evenly divided between casting alloys and wrought alloys. The former class is typically applied by pressure casting techniques, while the latter is usually prepared by powder metallurgy (P/M) techniques. Examples of typical matrix alloys are given in Table 3. It should be noted that there is some overlap in processing; for example, several of the alloys listed under the P/M category have also been processed using casting technology.

Composite Fabrication

Powder Metallurgy. While perhaps best suited to particulate reinforcement, powder metallurgy methods have been very successfully employed in the fabrication of discontinuous fiber composites. The fabrication process begins by blending a relatively fine matrix alloy powder with reinforcing fibers. The parameters of the blending operation are critical because the blending action must be sufficiently vigorous to ensure homogenization of the reinforcing fibers in the matrix alloy, yet not so severe as to damage the fibers seriously. An example of a correctly blended mixture is shown in Fig. 1.

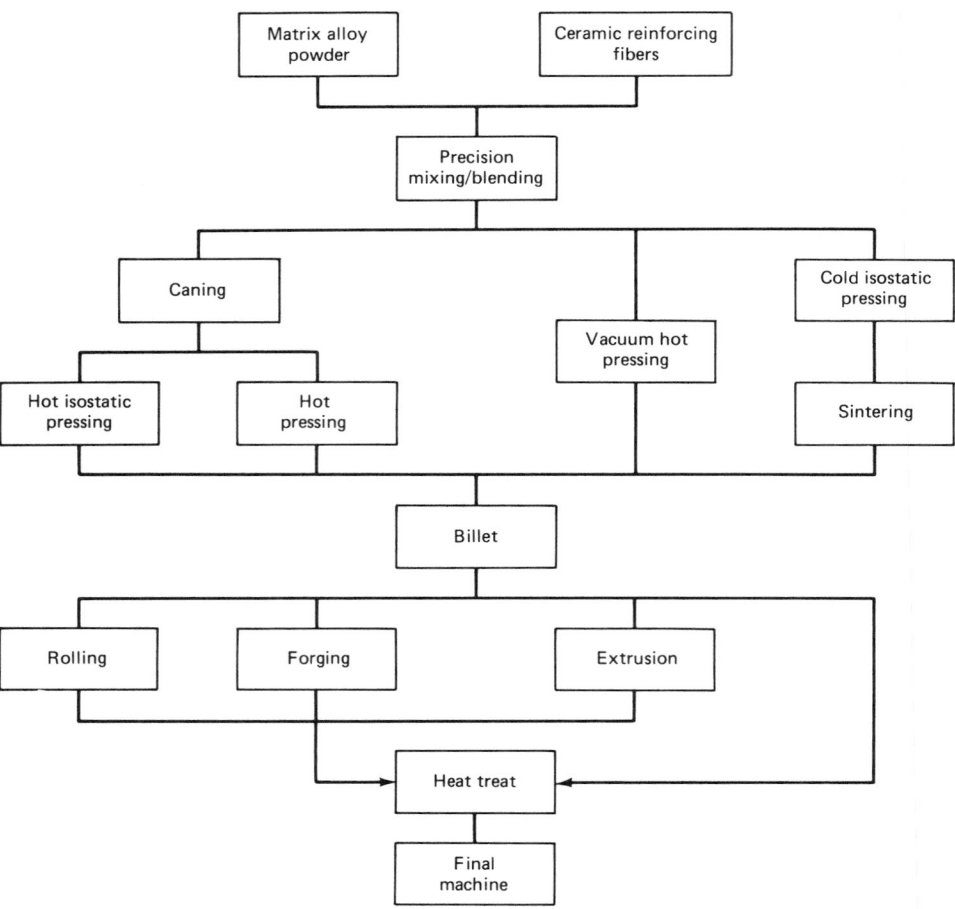

Fig. 2 Process flow chart for P/M fabrication of discontinuous fiber reinforced composites

The blended powder may be cold-compacted and sintered, but it is typically hot-consolidated. Any of the numerous conventional powder metallurgy techniques may be employed, including caning and hot pressing, hot isostatic pressing, or vacuum hot pressing. Figure 2 shows a flow chart of P/M composite fabrication. A criterion of the consolidation process is that the material be raised to a temperature at which the matrix metal is quite plastic before compaction takes place. The reason for this is to avoid diminishing the aspect ratio of the discontinuous fiber reinforcement by fracturing the fibers.

Not uncommonly, as shown in Fig. 2, the consolidated powder billet is fabricated into product by secondary hot-working. Fabrication methods may include forging, rolling, or extrusion. Again, processing temperatures must be high enough to provide sufficient plasticity in the matrix to avoid fiber damage. An example of an extruded structure, fabricated from the blend shown in Fig. 1, is illustrated in Fig. 3. The high degree of fiber alignment as a result of the secondary working is evident.

An alternate P/M fabrication technique that has enjoyed some popularity is spray codeposition. In a sense, the process (Fig. 4) also can be

Fig. 3 Micrograph of an extrusion from the composite blend shown in Fig. 1. Note fiber alignment. 1050×. Courtesy of Arco Chemical Company

considered a casting variant because the MMC is formed when atomized droplets coalesce as they strike the substrate. Advantages include

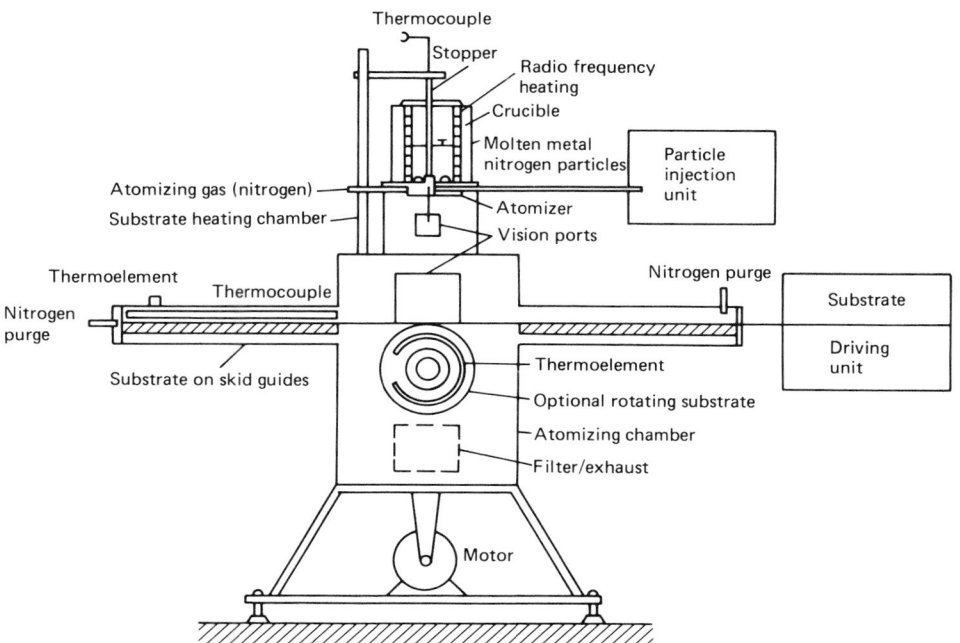

Fig. 4 Equipment for spray codeposition of composite materials

Fig. 5 Examples of vacuum-formed preforms of discontinuous fibers. Parts shown include machined and molded configurations of alumina, aluminosilicate, and silicon carbide fibers. Courtesy of Clevite Industries Inc., Engine Parts Division

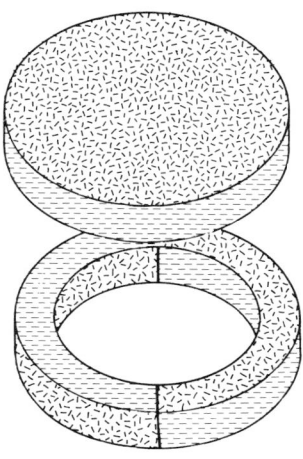

Fig. 6 Two-dimensional random orientation of fibers in vacuum-formed preforms

Table 3 Representative aluminum alloy matrices for discontinuous fiber reinforcement

P/M process	Casting process
2014	242.0
2024	332.0
4032	336.0
6061	339.0
7075	354.0
	356.0
	390.0

Table 4 Influence of alumina and aluminosilicate fiber reinforcement on coefficient of thermal expansion

Matrix	Fiber vol%	Coefficient of thermal expansion, 10^{-6}/K		
		0°		90°
332.0 T5	...	...	24.5	...
	Alumina, 5	23.9	...	23.6
	Mullite, 5	22.3	...	23.8
	Alumina, 15	18.9	...	22.3
339.0 T5	...	...	20.4	...
	Alumina, 10	18.0	...	...
	Alumina, 20	16.4	...	...

near-zero segregation of the reinforcing media when combined with a fine-grained matrix, brought about by the rapid cooling rate. In addition, the rapid cooling rates minimize the development of a significant interfacial phase resulting from chemical reactions between the fiber and the matrix alloy.

Casting has been performed in which the fibers are mixed into the melt and the resulting slurry is poured into the casting mold. One version of this process is known as compocasting, which is a derivation of rheocasting. In this process, the fibers are mechanically entrapped and prevented from settling or agglomerating because the alloy is partially solidified. With continued mixing, the fiber may interact with the matrix alloy to develop adhesion and bonding. After mixing, the slurry is gravity-poured into a mold. The relatively low pour temperatures of compocast composites may result in excessive internal porosity and lack of casting detail. In the case of particulate reinforcements, direct gravity casting of a silicon carbide particulate reinforced aluminum melt into billet and intricate investment cast shapes

has been quite successful. However, gravity-cast silicon carbide whisker reinforced aluminum composites remain experimental.

The most common approach to casting involves preforms that are produced by the vacuum forming process. In this process, the bulk fibers are distributed in an aqueous slurry containing appropriate organic binders, inorganic binders, and deflocculating agents. The fibers are deposited on the surface of a porous mold as vacuum is applied. Subsequently, the preform is removed from the mold, densified if required, dried to eliminate residual moisture, and then fired at high temperature to obtain stability. Examples of preforms are shown in Fig. 5.

This process tends to align fibers in planes parallel to the deposition surface, producing a two-dimensional random orientation, as shown in Fig. 6. The fiber content of these preforms presently ranges between 5 and 30 vol%. The preform is commonly incorporated into the

composite product by means of an infiltration casting technique.

Several approaches have been employed to accomplish infiltration with the molten matrix alloy. Among these are inert gas pressurization, vacuum infiltration, and liquid metal forging (squeeze casting). The latter method will be described because it has enjoyed the greatest degree of success and represents an improvement over powder metallurgy techniques (by a factor of ten) in cost and ease of manufacture. Because the fibers are present as a rigidized body, they can be conveniently placed at any location within the product. This has resulted in selective reinforcement tailored to match application requirements. As shown in Fig. 7, the discontinuous fiber preform is preheated, to avoid subsequently chilling the matrix metal, before being placed in the mold. Molten metal is added to the cavity, and pressure is applied to infiltrate the preform. Normally, after infiltration is achieved, pressure is raised to a high level (>70 MPa, or >10 ksi) which is maintained throughout the solidification process. The structure of a composite pressure casting is shown in Fig. 8, in which the two-dimensional random orientation of the discontinuous fiber is quite evident. The micro-

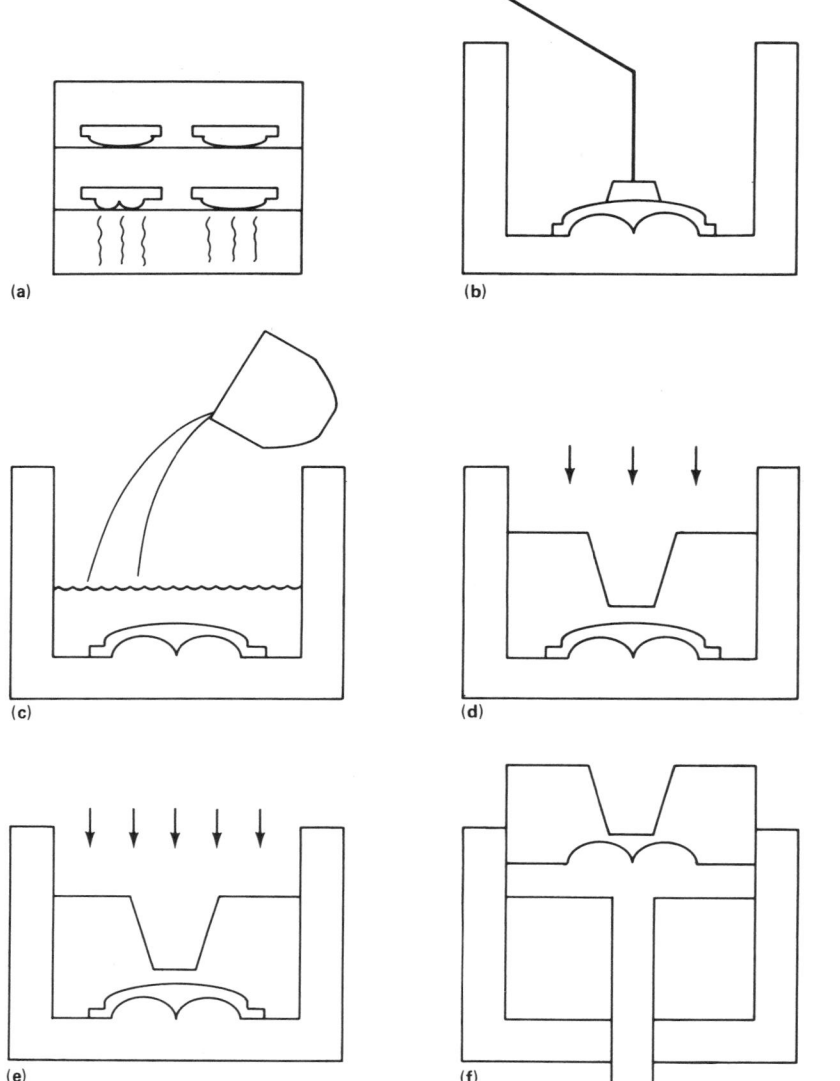

Fig. 7 Liquid metal forging process used to selectively reinforce composite castings selectively. (a) Preheat preforms in oven. (b) Place preform on base of casting tooling. (c) Pour measured quantity of matrix alloy into casting tooling. (d) Close casting cavity with upper ram and apply pressure to infiltrate preform. (e) Increase pressure and hold during metal solidification. (f) Eject near-net shape composite casting

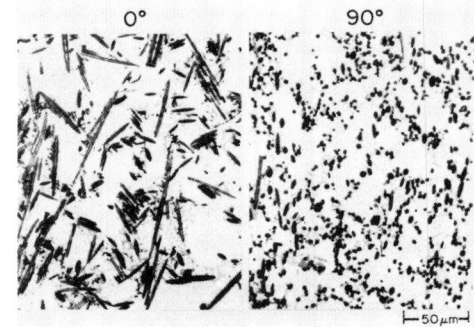

Fig. 8 Optical micrograph of an alumina fiber reinforced hypoeutectic aluminum-silicon alloy. The two-dimensional random orientation of the fibers is similar to the representation in Fig. 6. Courtesy of Clevite Industries Inc., Engine Parts Division

Table 5 Influence of silicon carbide fiber reinforcement on coefficient of thermal expansion

Matrix	Fiber, vol%	Coefficient of thermal expansion, 10^{-6}/K 0°	90°
339.0 T5	...	... 20.4	...
	Silicon carbide, 15	16.9 ...	...
356.0 T5	...	... 21.4	...
	Silicon carbide, 14	16.5 ...	...
2024 T6	...	... 21.1	...
	Silicon carbide, 25	14.9 ...	16.4
	Silicon carbide, 40	13.0 ...	...
6061 T7	...	... 21.6	...
(Extruded)	Silicon carbide, 25	12.1 ...	...

Properties

Physical. The structure of a discontinuous fiber reinforced composite may differ from its continuous fiber reinforced counterpart in that the fibers usually do not exhibit a true uniaxial orientation in the as-fabricated state. In the case of P/M composites, the initial fiber orientation is random in all directions. However, subsequent secondary processing may result in substantial fiber alignment, as was shown in Fig. 3. Cast composites typically exhibit two-dimensional random orientation, a consequence of the preforming process, although a completely random three-dimensional orientation has been achieved on a laboratory basis. Properties of discontinuous fiber reinforced composites are superior in directions corresponding to the length dimension of the fiber (0°), and inferior in the direction transverse to the fiber axis (90°).

The popular processes used to fabricate discontinuous fiber reinforced composites result in complete densification of the matrix metal. The overall density of the product depends on the composition of the fiber and matrix used, according to the rule of mixtures.

Ability to influence the coefficient of thermal expansion (CTE) of a material is particularly important when close tolerances between dis-

structure is devoid of visible porosity, a consequence of the intense casting pressure, and exhibits a close dendrite arm spacing, a result of rapid cooling.

Interfacial Reactions. The high consolidation temperatures and long times required for consolidating a powder metallurgy billet generally produce interactions between the matrix metal and the fiber that result in the formation of an interfacial zone. Similarly, a shallow interfacial zone can develop during the casting process. This is the result of the nonequilibrium status of the composite, which produces a chemical potential gradient at the fiber-matrix interface. The difference in chemical potential provides a driving force for diffusion when the composite material is

exposed to elevated temperatures. One of the key elements in composite strengthening is that the load is effectively transferred from the matrix to the reinforcement. Consequently, the bond integrity and efficiency of load transfer become extremely important factors. The formation of a shallow reaction zone is generally desirable to ensure the development of a strong chemical bond between the fiber and the matrix alloy. However, continued growth of the interfacial zone is usually detrimental to the mechanical properties of the composite. Composite products may be solution heat treated before use, but lengthy exposure to the solution temperature may significantly contribute to growth of the interfacial reaction zone.

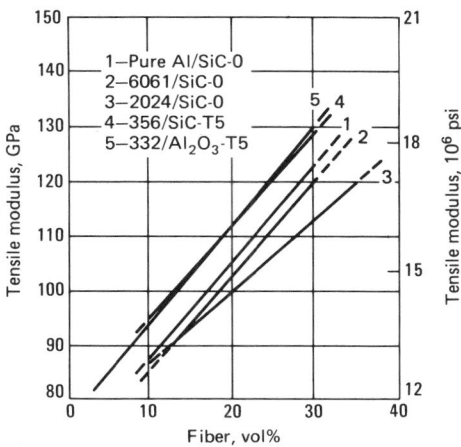

Fig. 9 Elastic modulus of fiber reinforced aluminum matrix composites as a function of fiber content

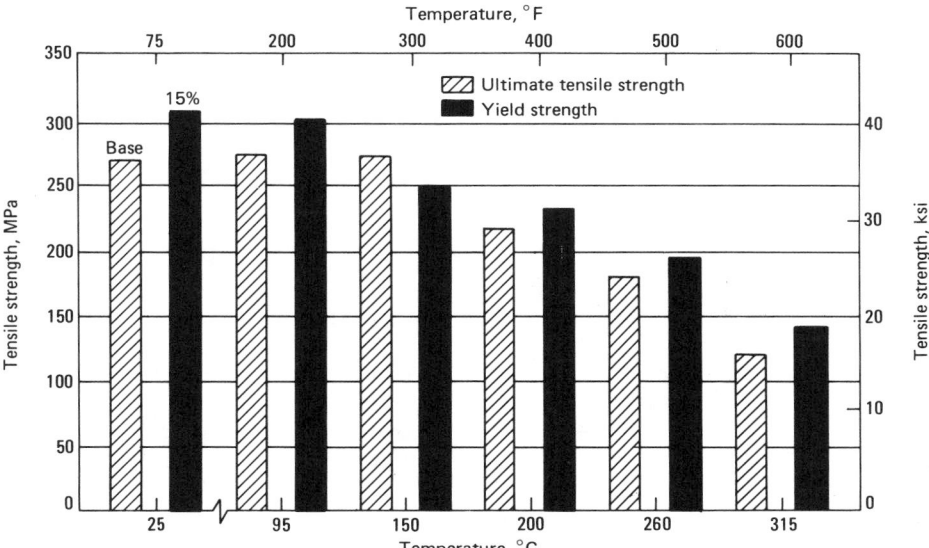

(a)

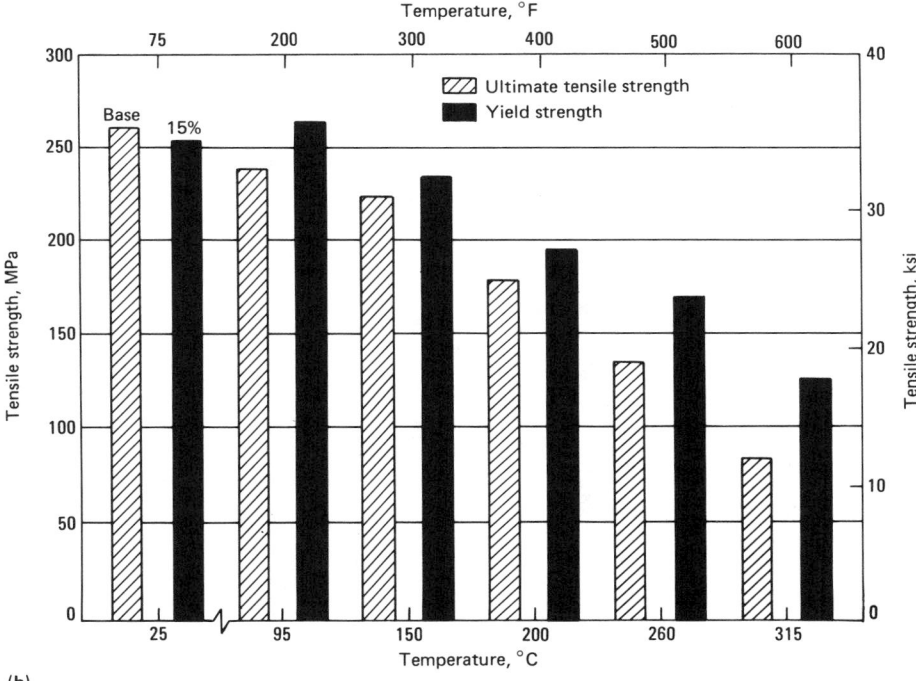

(b)

Fig. 10 Tensile properties as a function of temperature for two discontinuous fiber reinforced casting alloys. (a) 242.0/Al₂O₃ fibers, T5 heat treatment. (b) 339.0/Al₂O₃ fibers, T5 heat treatment

similar metals are required. A practical example is an aluminum piston operating in an iron cylinder. Ceramic fibers exhibit much lower CTE than do metals (Table 1). Consequently, it is possible to alter the expansion characteristics of the discontinuous fiber reinforced composite, particularly in the plane of the fiber axis. The sensitivity of thermal expansion to fiber volume fraction and orientation for a cast alumina fiber reinforced composite is shown in Table 4. Similar data on a silicon carbide reinforced aluminum composite produced by P/M fabrication is given in Table 5.

The thermal conductivity of ceramic fibers is also low relative to most matrix alloys. As one might expect, fiber composition, content, and orientation play a major role in the heat transfer characteristics of discontinuous fiber reinforced composites, as shown in Table 6.

Little information has appeared in the literature regarding the vibrational damping capacity of discontinuous fiber reinforced composites. However, based on damping data reported for both continuous fiber reinforced and particulate reinforced aluminum MMCs, one would expect damping capacity of discontinuous fiber reinforced composites to be improved as a result of the incorporation of ceramic fibers.

Mechanical. Ceramics are considerably harder than metals. Consequently, the addition of ceramic reinforcing fibers raises the hardness of the composite in proportion to the volume fraction of reinforcement. Some ceramic refractory fibers have an elastic modulus that is significantly higher than that of the matrix alloy (Table 1). In any case, the modulus of the composite increases with increasing fiber content according to the rule of mixtures. The high stiffness of some silicon carbide whiskers can result in a significant improvement in stiffness of an aluminum alloy matrix, as shown in Fig. 9.

The influence of discontinuous fiber additions on the strength of an aluminum matrix composite appears to depend on the characteristics of the reinforcing fiber. Several investigators have reported little improvement in room-temperature performance of aluminum alloys reinforced with various volume fractions of alumina fibers. Furthermore, it appears that the strength of an aluminum matrix composite containing aluminosilicate fibers may actually be inferior to the strength of the matrix alloy at low volume fractions (≤10%). Additions of silicon carbide fibers appear to improve room-temperature properties, as shown in

Table 7. The major improvement in strength performance occurs at elevated temperatures, particularly above the aging temperature of the matrix alloy. Examples of two aluminum casting alloys reinforced with alumina fibers are shown in Fig. 10. An example of the performance of a forging alloy reinforced with silicon carbide fibers is shown in Fig. 11.

Most fatigue evaluations on discontinuous fiber reinforced composites have been performed on material fabricated by the P/M method. This appears to be related to the fact

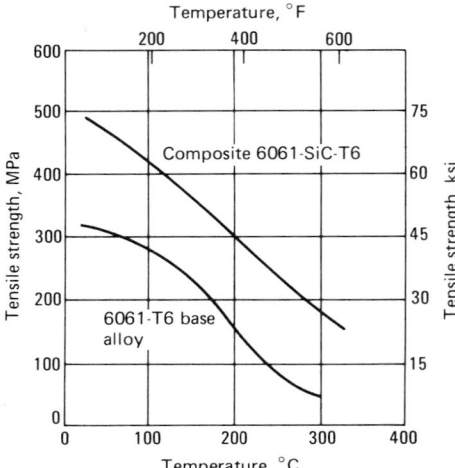

Fig. 11 Tensile strength as a function of temperature for 16 vol% silicon carbide fibers in a 6061-T6 aluminum alloy matrix

Table 6   Influence of fiber reinforcement on thermal conductivity at 200 °C (390 °F)

Matrix	Fiber, vol%	Thermal conductivity, W/m · K (Btu · in./h · ft² · °F)					
		0 °C				90°	
332.0 T5	...	...	...	176	1220	...	...
	Alumina, 5	158	1100	...	...	140	970
	Mullite, 5	152	1050	...	...	128	890
	Alumina, 15	134	930	...	...	98	680
	Zirconia, 19	116	800	...	...	87	600
339.0 T5	...	...	...	144	1000	...	...
	Alumina, 10	112	780	...	...	...	...
	Alumina, 15	99	690	...	...	...	...
	SiC, 15	136	940	...	...	...	...
2024 T6	...	...	...	151	1050	...	...
	SiC, 25	102	710	...	...	...	...

Table 7   Room-temperature tensile strength of silicon carbide-aluminum alloy composites

Material	Fiber, vol%	Ultimate tensile strength			
		Base		Reinforced	
		MPa	ksi	MPa	ksi
Pure aluminum	11	59	8.6	235	34.1
6061-T6	16	300	43.5	441	64.0
2024-T4	20	470	68.2	565	81.9

Table 8   Comparison of a ceramic fiber reinforced ring belt with an austenitic cast iron ring belt in a heavy-duty diesel dynamometer engine

Groove wear				Ring wear			
Ni Resist		Ceramic fiber reinforced ring belt		Ni Resist		Ceramic fiber reinforced ring belt	
mm	mil	mm	mil	mm	mil	mm	mil
0.020	0.80	0.010	0.40(a)	0.000	0.000	0.010	0.40(a)
0.020	0.80	0.020	0.80(b)	0.010	0.40	0.010	0.40(b)
0.020	0.80	0.020	0.80(c)	0.000	0.000	0.023	0.90(c)

(a) 5% alumina fibers. (b) 7% mullite fibers. (c) 10% aluminosilicate fibers

that this material typically employs forging alloy matrices fabricated into structural shapes. These composite materials show a significant improvement over the performance of the matrix alloy (Fig. 12). Fatigue data on casting alloys reinforced with discontinuous fibers is scarce. However, the limited data available suggest that the endurance limit of discontinuous fiber reinforced aluminum alloys is highly dependent on the composition of the matrix alloy. In aluminum-silicon casting alloys, ceramic fiber reinforcement appears to have little impact. In contrast, the addition of discontinuous fiber to aluminum-copper casting alloys results in a substantial improvement in endurance limit.

Wear behavior of discontinuous fiber reinforced composites is particularly interesting. Figure 13 shows that small additions of fibers provide a major improvement in adhesive wear of the composite. Larger additions have a reduced influence. However, it appears that wear of the surface mated to the discontinuous fiber reinforced composite material may be directly related to the fiber content of the composite. Wear comparisons between a Ni-Resist and ceramic fiber reinforced ring belt in a heavy-duty diesel piston are shown in Table 8. The positive influence of discontinuous fiber reinforcement on abrasive wear is shown in Table 9.

Secondary Processing

One of the advantages of the pressure casting process for producing discontinuous fiber reinforced composites is its near-net shape capability. The ability of this process to minimize total fiber consumption (through selective reinforcement) and reduce the machining scrap rate has been a major factor in the use of MMCs in the commercial sector. As a result, there has been little activity devoted to secondary processing of castings except for final machining.

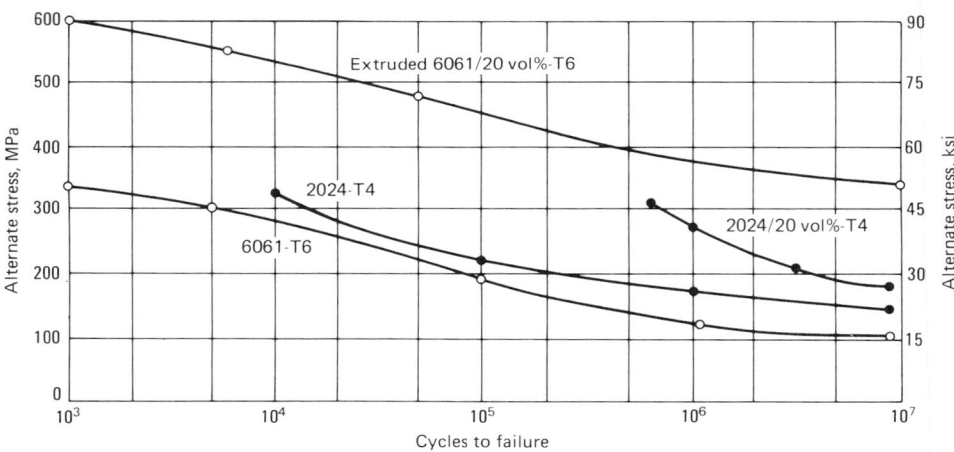

Fig. 12 Fatigue comparison (under rotating bending) of wrought aluminum alloys reinforced with 20 vol% discontinuous silicon carbide fibers

Hot-Working. In contrast to castings, P/M fabricated discontinuous fiber reinforced composites usually receive some secondary fabrication step, as shown in Fig. 2. The P/M process essentially produces billet shapes that can be more efficiently translated into product with secondary operations such as forging, rolling, extrusion, or pultrusion. In addition to bringing the material physically closer to the end shape, properties are modified as a result of densifica-

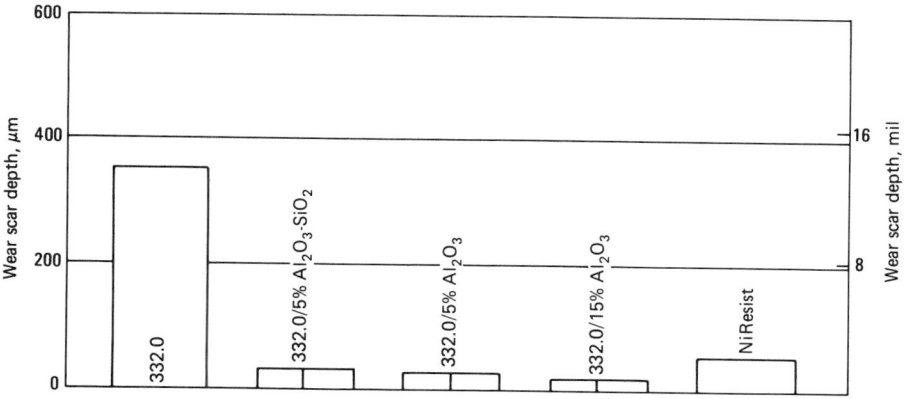

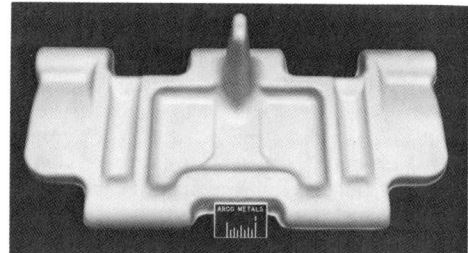

Fig. 14 Prototype track shoe forged from 6061 alloy containing 20 vol% silicon carbide fibers. Courtesy of Arco Chemical Company

Fig. 13 Adhesive wear characteristics of discontinuous fiber reinforced aluminum alloy compared to the matrix and a wear-resistant austenitic cast iron. LFW-1, 190 °C, T5 condition, 10 000 revolutions, nodular iron

Table 9 Abrasive wear testing of silicon carbide reinforced alloys

Technique: Taber-abrader 1000-g load CS-17-type wheel. Alloy: 2024-20 vol% SiC T-6 100 × 100 × 13 mm (4 × 4 × 0.5 in.) plate

	Unreinforced 2024	Silicon carbide-2024
6000 cycles		
Wt loss, g (oz)	0.12 (0.0040)	zero
Scar depth, μm (mil)	15-25 (0.6-1.0)	zero
20 000 cycles		
Wt loss, g (oz)	0.34 (0.0120)	0.06 (0.0020)
Scar depth, μm (mil)	35-60 (1.4-2.4)	1-15 (0.4-0.6)

Fig. 15 Advanced design heavy-duty diesel engine piston with ceramic fiber reinforced ring belt. Courtesy of Clevite Industries Inc., Engine Parts Division

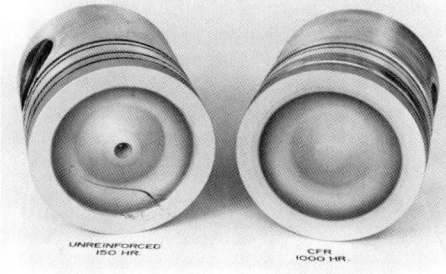

Fig. 16 Comparison of thermal cracking behavior of a standard squeeze cast piston with a ceramic fiber reinforced crown, run under identical conditions in a highly loaded advanced diesel engine. Courtesy of Clevite Industries Inc., Engine Parts Division

tion, texturing, grain refinement, and fiber alignment.

Machining. Because of the abrasive nature of refractory ceramic fibers, discontinuous fiber reinforced composites require the use of special techniques in machining. Specifically, the characteristics of the fibers mandate the use of carbide cutting tools, as a minimum; diamond cutting tools are preferred. The latter can create a surface finish and dimensional detail comparable to those obtained with conventional materials.

Welding. Data on welding discontinuous fiber composites are scarce. Obviously, it would be difficult to maintain structural uniformity in the weld zone between discontinuous fiber reinforced composite sections. Some success has been achieved in friction welding a composite material to an unreinforced structure. Brazing has also been successfully employed to join composite materials.

Applications

Military/Aerospace. A high strength-to-weight ratio, along with the high modulus obtainable from secondarily worked silicon carbide reinforced aluminum composites, have driven this material toward the more demanding applications of the military/aerospace marketplace. An example of a component considered as a high-potential candidate is the forged track shoe (Fig. 14). Another interesting application, relying on the control of thermal expansion coupled with stiffness, is a reflective mirror for optical sensing and transmission.

Automotive. The first commercial application for cast, discontinuous fiber reinforced composites was for automotive pistons. Cost was the driving factor, and these pistons were the first product to use inexpensive fibers commonly used for refractory and insulation purposes. This application has created much interest in the heavy-duty diesel engine market in

which current technological investigation requires substantial improvement in piston design or piston materials. Areas being examined include higher cylinder pressures, increased fuel injection pressures, combustion chamber geometry changes, uncooled engines, and turbocompounding. Operating conditions in the advanced engines being developed to meet these challenges demand the best of the current materials available, particularly from the standpoint of wear resistance and thermal fatigue. Discontinuous fiber reinforced composites show much promise in addressing these challenges. Figure 15 shows a diesel engine piston cross section in which the ring belt has been reinforced to provide excellent wear resistance with no penalty in terms of reciprocating mass. Figure 16 shows a piston for an advanced engine that was run under severe thermal and overload conditions. The application of discontinuous fiber reinforcement to the crown region extended operating life from 150 to 1000 h.

Presently, an exploratory automotive application is the connecting rod. Discontinuous fiber reinforced composites have the potential of maintaining the required strength and stiffness while reducing reciprocating mass, which can be a significant factor in eliminating engine vibration.

Studies are in progress regarding application of discontinuous fiber reinforced composites to brake disks and drums. Previous efforts at fabricating a lightweight brake component have relied on a mechanically bonded ferrous insert to provide wear resistance, but thermal cycling may compromise bond integrity and thwart successful performance. In contrast, a discontinuous fiber reinforced aluminum composite offers the combined advantage of light weight, superior wear resistance, and excellent heat conductivity without the penalty of a ferrous liner. Reportedly, at least one European automobile manufacturer has composite brake rotors in vehicles on its test track. Domestic applications are still in the laboratory stage.

Compressors. A technological advance in compressor design, the scroll compressor, is an interesting candidate for discontinuous fiber reinforced composite material. The scroll itself is a wear-critical component requiring low weight to minimize inertial forces.

SELECTED REFERENCES

- L. Ackerman, J. Charbonnier, G. Desplanches, and H. Kolowski, "Properties of Reinforced Aluminum Foundry Alloys," Paper presented at the International Conference on Composite Materials V, San Diego, July/Aug 1985
- R.J. Arsenault, "Composite Strengthening," DTIC Report ADA 164 838, Defense Technical Information Center, 1985
- R.J. Arsenault and M. Taya, "The Effects and Differences in Thermal Coefficients of Expansion in SiC Whisker 6061 Aluminum Composite," Paper presented at the International Conference on Composite Materials V, San Diego, July/Aug 1985
- M.G. Bader, T.W. Clyne, G.R. Cappleman, and P.A. Hubert, "The Fabrication and Properties of Metal Matrix Composites Based on Aluminum Alloy Infiltrated Alumina Fibre Preforms," University of Surrey, 1985
- D.J. Bray and C.L. Kulick, "Characteristics of Ceramic Fibers for Composite Reinforcement," Paper presented at the American Ceramics Society Symposium, Pittsburgh, March 1986
- G.R. Cappleman, J.F. Watts, and T.W. Clyne, The Interface Region in Squeeze Infiltrated Composites Containing d-Alumina Fibre in Aluminum Matrix, *J. Mater. Sci.*, Vol 20, 1985
- T.W. Chou, A. Kelly, and A. Okura, Fibre-Reinforced Metal Matrix Composites, *Composites*, Vol 16 (No. 3), July 1985
- C.R. Crowe, R.A. Gray, and D.F. Hasson, "Microstructure Controlled Fracture Toughness of SiC/Al Metal Matrix Composites," Paper presented at the International Conference on Composite Materials V, San Diego, July/Aug 1985
- Damping of Metal Matrix Composites, *Mater. Eng.*, April 1986
- "Damping Characteristics of Metal Matrix Composites," DTIC Report ADA 163 569, Defense Technical Information Center, Jan 1986
- A.A. Das and S. Chatterjee, Squeeze Casting of an Aluminium Alloy Containing Small Amounts of Silicon Carbide Whiskers, *Metall. Mater. Technol.*, Jan 1981
- J.A. DiCarlo, "Ceramic Fibers for Metal Matrix Composites," Paper presented at the American Ceramics Society Symposium, Pittsburgh, March 1986
- J. Dinwoodie, E. Moore, C.A.J. Langman, and W.R. Symes, "The Properties and Applications of Short Staple Alumina Fibre Reinforced Aluminum Alloys," Paper presented at the International Conference on Composite Materials V, San Diego, July/Aug 1985
- T. Donomoto, K. Funatani, N. Miura, and N. Miyake, "Ceramic Fiber Piston for High Performance Diesel Engines," Paper 830252 presented at the International Congress, Society of Automotive Engineers, Detroit, Feb/March 1983
- Y. Flom and R.J. Arsenault, Deformation in Aluminum-Silicon Carbide Composites Due to Thermal Stresses, *Mater. Sci. Eng.*, Vol 75, 1985
- Y. Flom and R.J. Arsenault, International Bond Strength in an Aluminum 6061-SiC Composite, *Mater. Sci. Eng.*, Vol 77, 1986
- I.G. Greenfield and R.R. Viguand, Dependence of Abrasive and Sliding Wear in a Metal Matrix Composite, in *Proceedings of the Advanced Composites Conference*, Dec 1985
- B.H. Hamling and R.E. Lattimer, "Zirconia Fibers and Composites in Severe Environments," Paper presented at the American Ceramics Society Symposium, Pittsburgh, March 1986
- "Health and Safety Aspects of Refractory Fibers," Pamphlet, Babcock & Wilcox, Insulating Products Division
- M.A.H. Howes, Ceramic Reinforced Metal Matrix Composites Fabricated by Squeeze Casting, in *Proceedings of the Advanced Composites Conference*, Dec 1985
- G.F. Hurley and J.J. Petrovic, Silicon Carbide Whiskers for Composites—Growth and Properties, in *Proceedings of the Advanced Composites Conference*, Dearborn, MI, Dec 1985
- D.L. McDanels, T.T. Serafini, and J.A. DiCarlo, Polymer, Metal, and Ceramic Matrix Composites for Advanced Aircraft Engine Applications, in *Proceedings of the Advanced Composites Conference*, Dearborn, MI, Dec 1985
- R. Mehrabian, "A Fundamental Study of a New Fabrication Technique for Fiber Reinforced Aluminum Matrix Composites,"
- Report ADA 086 282, National Technical Information Service, April 1980
- Mitsubishi Develops SiC Composite Billet, *Light Met. Age*, Dec 1985
- R. Munro, "The Performance Improvement of Aluminum Alloy Diesel Engine Pistons by Squeeze Casting," Paper 860161 presented at the International Congress, Society of Automotive Engineers, Detroit, MI, Feb 1986
- T. Mura and M. Taya, "Residual Stress in and Around a Short Fiber in Metal Matrix Composite Due to Temperature Change," ASTM Special Technical Publication 864, American Society for Testing and Materials, 1985
- V.C. Nardone and K.M. Prewo, On the Strength of Discontinuous SiC Reinforced Aluminum Composites, *Scr. Metall.*, Vol 20, 1986
- T.G. Nieh, R.A. Rainen, and D.J. Chellman, "Microstructure and Fracture in SiC Whisker Reinforced 2124 Aluminum Composite," Paper presented at the International Conference on Composite Materials V, San Diego, July/Aug 1985
- C.T. Post, Exotic Fibers Teach Old Metals New Tricks, *Iron Age*, July 1985
- "Preparation of Alumina Fiber Preforms for Use in Metal Matrix Composites, e.g. for Reinforcing Piston Crowns," Publication 22419, *Research Disclosure*, Dec 1982
- A. Sakamoto, H. Hasegawa, and Y. Minoda, "Mechanical Properties of SiC Whisker Reinforced Aluminum Composites," Paper presented at the International Conference on Composite Materials V, San Diego, July/Aug 1985
- J.E. Schoutens, Ed., "Discontinuous Silicon Carbide Reinforced Aluminum Metal Matrix Composites Data Review," MMCIAC Report 000461, Metal Matrix Composites Information Analysis Center, Dec 1984
- A.R.E. Singer and S. Ozbek, Metal Matrix Composites Produced by Spray Codeposition, *Powder Metall.*, Vol 28 (No. 2), 1985
- W.G. Spengler and W.B. Young, "Techniques to Upgrade Heavy Duty Aluminum Pistons," Paper 860162 presented at the International Congress, Society of Automotive Engineers, Detroit, Feb 1986
- M.W. Toaz, "Near Net Shape Composite Castings with Tailored Engineering Properties," Paper 8521-008 presented at the Advanced Composites Conference, Dearborn, MI, Dec 1985
- N. Ueda and M. Taya, "A New Model to Predict the Electrical Conductivity of a Misoriented Short Fiber Composite," Paper presented at the International Conference on Composite Materials V, San Diego, July/Aug 1985

Continuous Carbon Fiber Reinforced Carbon Matrix Composites

Russell J. Diefendorf, Rensselaer Polytechnic Institute

CARBON FIBER REINFORCED CARBON MATRIX composites have many of the desirable high-temperature properties of conventional carbons and graphites, including high strength, high modulus, and low creep. In addition, the high thermal conductivity and low coefficient of thermal expansion, coupled with high strength, produce a material with low sensitivity to thermal shock. Also characteristic of carbon-carbon composites are a high fracture toughness and a pseudoplasticity, the latter of which bears a resemblance to fiber-reinforced polymers. These attributes make carbon-carbon composites uniquely useful at temperatures as high as 2800 °C (5070 °F). The major problems with carbon-carbon composites have been high-temperature oxidation and off-fiber-axis properties.

Unidirectional carbon-carbon composites can approach the same strengths and moduli as those achieved with resin matrix composites. Moreover, because their properties are maintained to 2000 °C (3650 °F), they represent the premier material for inert atmosphere or short-time high-temperature applications.

Carbon-Carbon Processes

Carbon-carbon composites were developed to withstand the harsh but different conditions of reentry, rocket motors, and aircraft brakes. Early work on fiber-reinforced polymers used in ablative heat shields indicated that the lowest ablation rate was achieved with polymers that gave high char yields. Still lower ablation rates were achieved by precharring the matrix polymer to produce what was referred to as "burned toast." At the same time, carbon-carbon composites were produced, inadvertently, in the carbon felt insulation in pyrolytic graphite furnaces. Two processes, liquid impregnation and gaseous infiltration, have been developed to produce the present high-performance carbon-carbon composites.

Liquid Impregnation. Carbon fibers can be laid-up uniaxially, in either two-dimensional fabrics or in three-dimensional woven forms before impregnation with a matrix material. (Refer to the article "Multidirectionally Rein-forced Carbon/Graphite Matrix Composites" in this Section of the Volume for detailed information.) Prepregs with phenolic or other high-char-yield resins are used to fabricate the unidirectional laminates. Preliminary processing is much like that for other carbon fiber reinforced resin matrix composites. The laminate is laid up, vacuum bagged, autoclave or press cured, and postcured. Then, the matrix in the composite is carbonized carefully and, often, heat treated to a higher temperature. The matrix material loses mass and densifies during this process. Bulk shrinkage is constrained by the fibers, and extensive matrix microcracking and void formation occur within the composite. The function of subsequent processing is to convert this composite, consisting of fibers loosely coupled by matrix carbon binder bridges, into a strong matrix.

Carbon precursors used in later processing should have low viscosity as well as good wetting to allow thorough impregnation (Fig. 1). Carbon fiber prepregs are fabricated into a desired shape, or alternatively, dry fiber is laid up into a preform and then impregnated with a liquid resin or pitch. The resin is thermoset and then slowly carbonized. During carbonization, the resin microcracks extensively because of weight loss and densification, although resins with low weight loss are used. To achieve a higher density and stronger matrix, the carbon-composite is recycled using resin or pitch impregnation and additional slow carbonization cycles. Graphitization may be included also, to encourage porosity for impregnation and to provide structural stability and better thermal shock resistance.

Ideally, the precursor should be thermosetting, to prevent liquid exudation upon subsequent heating, and it also should have a high carbon yield. Thermosetting phenolic, furfural, or acetylenic resins provide simple processing, but generally give lower density carbon matrices. Thermoplastic impregnants, such as coal tar or petroleum pitch, also can be used. They must be carbonized very slowly and usually under 10 MPa (100 bar) pressure, to prevent gaseous products from exuding the impregnant from the body. However, the density of carbonized pitch is usually higher. Multiple pitch or resin impregnations under pressure, followed by carbonization and high-temperature heat treatments, often with five or more cycles, is the most common sequence used to produce carbon-carbon composites.

The chemical vapor infiltration (CVI) (Ref 1) of carbon uses gaseous hydrocarbons such as methane, propane/propylene, and benzene to deposit a carbon matrix internally in a carbon fiber preform. The process can be performed using three different methods. In the most commonly used technique (Fig. 2a) natu-

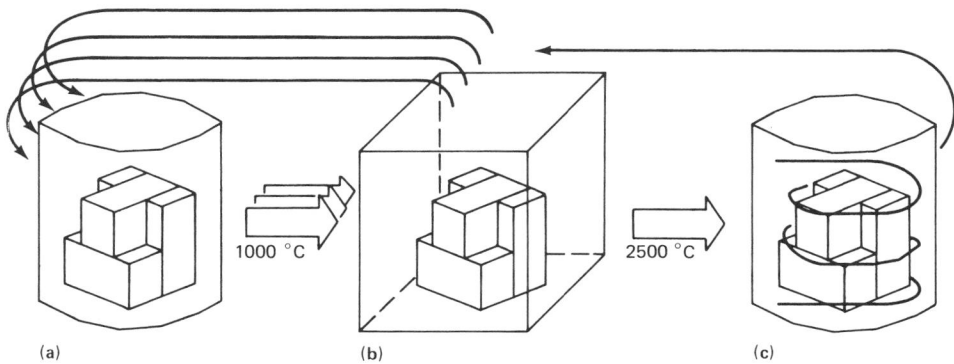

Fig. 1 Composite fabrication of preforms. (a) Impregnation. (b) Carbonization. (c) Graphitization

ral gas or other carbonaceous gases are flowed past and through a carbon fiber preform located in a low-pressure isothermal furnace. Uniform deposition throughout the preform can be achieved by operating at a temperature sufficiently low to permit rapid gaseous diffusion, compared to deposition of carbon. However, the deposition time is usually very long. Mass transfer through the fiber preform can be improved by inducing a pressure gradient through it (Fig. 2b). The deposition rate can be much higher. Enhanced deposition rates can be achieved also by using a temperature gradient deposition process (Fig. 2c).

A major problem with CVI is to achieve the uniform deposition of the carbon matrix throughout a thick preform. Mass transfer from the bulk gas must be sufficiently high in the fiber preform to keep a relatively constant concentration of carbon-containing molecules throughout. Hence, the rate at which carbon is deposited must be slow compared to the mass transfer of carbon into and throughout the preform. Mass transfer of carbon-containing molecules into the preform is usually by diffusion, which slowly increases with temperature. The deposition of carbon is complex, but the overall process has a high temperature coefficient. Hence, the relative rates of the two processes can be varied by adjusting the temperature. A temperature of 1000 to 1100 °C (1830 to 2010 °F) is commonly used, along with a pressure of 500 to 3000 Pa (5 to 30 mbar) to achieve a relatively uniform deposition of carbon throughout a part 10 mm (0.40 in.) thick. More rapid mass transfer can be achieved by placing a pressure drop across the fiber preform (Fig. 2b). However, the deposition rate decreases as the pressure decreases, which produces nonuniform deposition through the preform. A pressure gradient process can be used at the end of a conventional cycle when mass transfer through tiny pores is extremely slow. An alternative is to use a temperature gradient such that carbon is deposited at a moving boundary that sweeps through the thickness (Fig. 2c). The deposition time can be significantly decreased because mass transfer of the deposition gases is mostly through parts of the preform that have not yet been deposited upon. Unfortunately, the technique produces a

variation in microstructure because deposition occurs at different temperatures. A problem with all present CVI processes is closed pore formation caused by the sealing off of bottleneck pores, and, more insidiously, delaminated regions. However, high-temperature heat treatment may be employed to induce microcracks in the matrix to be filled in the subsequent CVI process. Using liquid impregnation to produce relatively uniform open pores, followed by CVI, is another attractive alternative.

Unidirectional Properties

The properties of a carbon-carbon composite depends on the fiber and matrix constituent properties and the coupling between them.

The mechanical behavior of carbon-carbon composites in tensile fracture behavior differs from that of resin matrix composites in two major ways. First, the polymers used in many current applications have much higher strains to failure than do the fiber reinforcement. In carbon-carbon systems, the strain-to-failure of the matrix is usually lower than that of the fibers and, indeed, is frequently microcracked from processing stresses. Hence, instead of the fibers initiating failure, as in a polymeric com-

posite, the carbon matrices fail first. The second major difference is that the moduli of fiber and matrix may be similar. Hence, the load will be shared more evenly between matrix and fiber than for the polymeric system.

Even though these two differences exist between polymeric and carbon matrix systems, the general approach to increasing the strength and fracture toughness is the same. Namely, when fibers or matrix in a composite begin to fail in a statistical sequence, it is advantageous to prevent these local failures from propagating through the rest of the part and causing catastrophic failure.

Fracture Behavior. Generally, the coupling between fibers and matrix that arises from both chemical and mechanical bonding is instrumental in controlling crack propagation and the extent of damage. The fiber type, matrix precursor, and subsequent type of processing determine the strength of the interfacial coupling. Two extreme cases, that of 100% and 0% coupling, describe the limits experimentally observed in carbon-carbon composites.

For a strong interfacial bond, the crack, which forms in the lower strain-to-failure matrix, propagates across the fiber-matrix interface to cause immediate fiber failure (Fig. 3a).

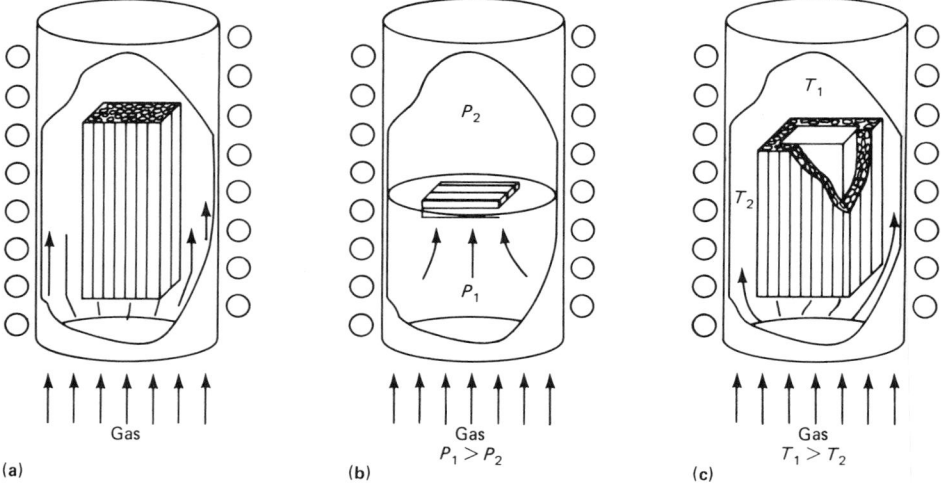

Fig. 2 Chemical vapor infiltration process

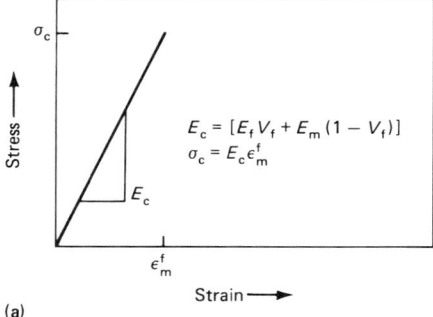

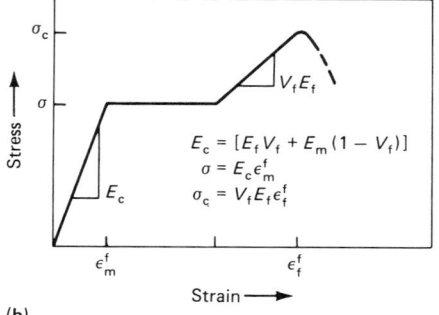

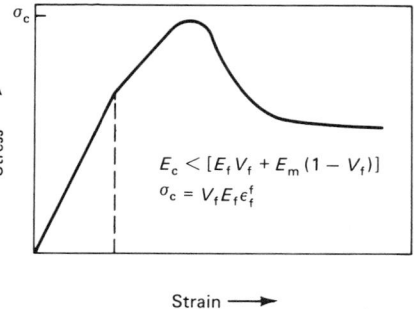

Fig. 3 Stress-strain curves. (a) 100% fiber-matrix coupling. (b) 0° fiber-matrix coupling. (c) Precracked matrix

Table 1 Mechanical properties of unidirectional carbon-carbon composites (~55 vol%)

	Parallel		Perpendicular	
	HTU	HMS	HTU	HMS
Modulus, GPa (10^6psi)				
Tension	125 (20)	220 (30)		
Compression	10 (1.5)	250 (35)	7.5 (1.1)	...
Strength, MPa (ksi)				
Tension	600 (90)	575 (85)	4 (0.60)	5 (0.75)
Compression	285 (40)	380 (55)	25 (4)	50 (7.5)
Bend	1250-1600 (180-230)	825-1000 (120-145)	20 (3)	30 (4.5)
Shear	20 (3)	28 (4)		
Fracture toughness, kJ/m^2 (ft · lbf/ft^2).......	70 (4800)	20 (1370)	0.4 (30)	0.8 (55)

HTU, High tensile untreated surface; HMS, high modulus surface-treated

Table 2 Effect of matrix precursor on composite modulus

	Heat treatment	
	at 1000 °C (1830 °F)	at 2600 °C (4710 °F)
Phenolic resin	110%	140%
Pitch	130%	210%

Note: Fiber stiffening factor assuming all stiffness comes from HM fiber

Therefore, failure is brittle and the strength of the composite is governed by the strain-to-failure of the matrix. Higher-strength composites can be obtained by using higher-modulus fibers, but the strain-to-failure will still remain low (Ref 2). Fibers that are surface treated to promote adhesion and matrices with high carbonization shrinkages both promote this strong coupling between matrix and fiber, in the first case by chemical bonding, and in the second case by compressive stresses normal to the interface (Ref 3, 4).

When poor coupling exists between fiber and matrix, the composite can show a substantial increase in strength and fracture toughness when loaded parallel to the fiber axis. In this case, as the composite is loaded up to the failure strain of the matrix, the matrix cracks cause a debonding between matrix and fiber as the crack approaches the interface. The fiber bridges the crack and maintains load-bearing capability (Fig. 3b). Matrix cracks are often observed to run across the full width of the specimen normal to the stress. Numerous matrix cracks can form, and the effective stiffness of the composite decreases. The exact nature of the stress-strain curve depends on the fiber and matrix moduli and the flaw distributions. Increasing load initiates fiber fracture and failure. If the test is displacement controlled, the load increases to a maximum at which significant fiber fracture occurs, and then decreases slowly as fibers pull out of the matrix (dotted line in Fig. 3b). In many cases with carbon-carbon composites, the matrix is heavily microcracked from processing shrinkages, or from cool-down from a high-temperature heat treatment. Then, the large discontinuity that occurs when the matrix fractures is largely absent (Fig. 3c).

The effect of fiber-matrix coupling on mechanical properties has been documented by studying polyacrylonitrile (PAN)-based carbon fibers of different moduli with and without surface treatments to promote fiber-matrix coupling (Ref 4, 5). The coupling for resin matrix composites is observed to decrease with increasing modulus for both untreated and surface-treated fibers, although the magnitude is higher for surface-treated fibers. C.R. Thomas and E.J. Walker (Ref 5) found that both AU (untreated) and AS (surface-treated) lower modulus (230 GPa, or 35 × 10^6 psi) fibers in carbon-carbon composites failed at the failure strain (0.1%) of the matrix, and only low-strength composites were ever produced. When higher-modulus high-tensile (HT) (265 GPa, or 40 × 10^6 psi) fibers were evaluated, surface-treated fibers still fractured at the matrix failure strain. However, high-tensile untreated surface (HTU) fibers produced much higher strengths and fracture toughness (Table 1). For still higher moduli (400 GPa, or 60 × 10^6 psi) surface-treated (HMS) fibers, the mechanical properties of the composite depended on the heat treatment temperature. Heat treatment at 1000 °C (1830 °F) produced carbon-carbon composites that failed at 250 MPa (35 ksi) or the failure strain of the matrix (0.1 to 0.2%), and had a fracture toughness of 1 kJ/m^2 (70 ft · lbf/ft^2). Heat treatment to a temperature of 2600 °C (6510 °F) caused microcracking on cool-down and decoupling within the carbon-carbon composite. The strength more than doubled and fracture toughness increased by a factor of 20. High-modulus untreated (HMU) surface (400 GPa, or 60 × 10^6 psi) composites produced good toughness and strengths even after carbonization (1000 °C or 1830 °F).

The high-temperature properties of HM composites have been measured from room temperature to 2000 °C (3630 °F). The modulus decreases from 180 GPa (26 × 10^6 psi) at room temperature to 175 GPa (25 × 10^6 psi) at 2000 °C (3630 °F). Tensile strength is also surprisingly constant, varying from 0.95 GPa (0.14 × 10^6 psi) at room temperature to 1.2 GPa (0.17 × 10^6 psi) at 1000 °C (1830 °F) and 1.1 GPa (0.16 × 10^6 psi) at 2000 °C (3630 °F). However, strain-to-failure increases from 0.48% at room temperature to 0.65% at 1000 °C (1830 °F) and 0.73% at 2000 °C (3630 °F) (Ref 5).

Weak fiber-matrix interfaces allow the fiber strength to dominate the composite strength parallel to the fiber axis and to have a high fracture toughness value because of the fiber pullout from the matrix. However, off-axis strengths such as shear and transverse tension are severely degraded (Ref 6). In many applications, it is these off-axis stresses that limit the design.

Moduli. The type of matrix precursor affects the moduli of the composites as well (Table 2) (Ref 4). Resin precursor binders and impreg-nants usually yield low-modulus isotropic carbons, and at low volume fractures do not contribute substantially to composite moduli. By comparison, pitch precursors generally transform to a mesophase (liquid-crystalline) state, which most frequently orders parallel to the fiber axis. A highly oriented graphitic matrix is produced upon graphitization and provides a high-modulus contribution parallel to the fiber axis.

Fatigue. The fatigue of carbon-carbon composites might be expected to be a problem because of the extensive matrix microcracking that is present before even the first loading. Matrix cracks do increase during cycling, and decreasing modulus and dusting of the matrix from the composite have been observed. Transverse tensile strength and shear strength are likely to be degraded during fatigue tests loaded parallel to the fiber axis. However, fatigue has been found to be proportionately as good as with resin matrix composites (Ref 7).

Thermal Expansion. The thermal expansion of unidirectional carbon-carbon composites tends to be dominated by the fibers parallel to the fiber axis. Thermal expansion coefficients vary from being slightly negative at room temperature for high-modulus fibers to being slightly positive for low-modulus fibers. The coefficients become positive at higher temperatures but in all cases remain low. The transverse thermal expansion coefficients depend on matrix, fiber, and voids. The anisotropic nature of the graphite crystal thermal expansion causes microcracking between basal planes upon cool-down because of mechanical constraints. The planar voids, which form upon cool-down, are expanded upon reheating. A lower thermal expansion coefficient than might be expected can result. However, the transverse thermal expansion is several times the axial expansion. This difference in the thermal expansion coefficient will generate sufficiently high residual stresses during cool-down from processing so that laminates, such as 0°/90°$_2$/0°, cannot be manufactured with good integrity. A partial solution has been the use of fabrics, which minimize this problem.

Oxidation of carbon-carbon composites can begin at temperatures as low as 400 °C (750 °F). The rate of oxidation depends on the perfection of the carbon structure and its purity. Highly disordered carbons, such as carbonized resins given low-temperature heat treatments, will oxidize at appreciable rates at 400 °C (750 °F). Highly graphitic structures, such as pitch-based carbon fibers, can be heated as high as 650 °C (1200 °F) before extensive oxidation occurs. At these low temperatures, carbons are very susceptible to catalytic oxidation by alkali metals, such as sodium, and multivalent metals, such as iron and vanadium, at extremely low concentrations. Therefore, the oxidation rate often is determined by the initial purity of the carbon-carbon composite or by in-service contamination. Borates and particularly phosphates have been found to inhibit oxidation up to about 600 °C (1110 °F) (Ref 8). Oxidation at higher temperatures becomes more rapid and by 1300 °C (2370 °F) is completely limited by mass transport of oxygen to the surface and carbon monoxide and dioxide away from it. Oxidation protection at high temperatures is discussed in the article "Oxidation-Resistant Carbon-Carbon Composites" in this Section.

Carbon-carbon composites can be attacked by strongly oxidizing acids, but are inert to most other acids and to all alkalies, salts, and organic solvents.

REFERENCES

1. W.V. Kotlensky, Deposition of Pyrolytic Carbons in Porous Solids, in *Chemistry and Physics of Carbon*, Vol 9, P.L. Walker, Jr. and P.A. Thrower, Ed., Marcel Dekker, 1973, p 173
2. J. Aveston, G. Cooper, and A. Kelly, Single and Multiple Fracture, in *Properties of Fiber Composites*, Paper 2, National Physical Laboratory, Nov 1971
3. E. Fitzer, The Future of Carbon/Carbon Composites, in *Proceedings of the Third Annual Materials Technology Conference, Solid Carbon Materials: Production and Properties*, M.H. Genisio, Ed., Southern Illinois University, 1986, p 4
4. E. Fitzer and W. Hüttner, Structure and Strength of Carbon/Carbon Composites, *J. Phys. D., Appl. Phys.*, Vol 14, 1981, p 347
5. C.R. Thomas and E.J. Walker, Carbon-Carbon Composites as High Strength Refractories, in *High Temperature High Pressure*, Vol 10, 1979, p 79
6. J. Hill, C.R. Thomas, and E.J. Walker, Paper 119, in *Proceedings of the Second International Carbon Fibre Conference*, Plastics Industry, 1974, p 122
7. W. Hüttner, K. Kewscher, and M. Huttinger, in *Ceramics in Surgery*, P. Vincenzini, Ed., Elsevier, 1982, p 225
8. A. Gkogkidis, Ph.D. Thesis, University of Karlsruhe, 1986

Multidirectionally Reinforced Carbon/Graphite Matrix Composites

Lawrence E. McAllister, Allied Bendix Aerospace

MULTIDIRECTIONALLY REINFORCED carbon/graphite matrix composites fall within the class of materials designated as carbon-carbon composites because they contain carbon or graphite fibers in a carbon or graphite matrix. The original carbon-carbon composites developed in the early 1960s were two-directional constructions based on heat-treated carbon fabric-phenolic laminates. The availability of high-strength high-modulus carbon fibers and multidirectional preform substrates in the mid to late 1960s (Ref 1, 2) led to the development of multidirectional carbon-carbon composites (Ref 3).

The multidirectional carbon-carbon composites discussed in this article contain reinforcing fibers in three or more directions. This article also describes multidirectional woven preforms and their processing into carbon-carbon composites, as well as properties and characteristics of these composites. Several review articles on this subject provide additional information (Ref 4-7).

Multidirectional Woven Preforms

The main advantage of multidirectional carbon-carbon composites is the freedom to orient selected fiber types and amounts to accommodate the design loads of the final structural component. Multidirectional fabrication technology provides the means to produce tailored composites.

The simplest type of multidirectional preform is based on a three-directional orthogonal construction, which is normally used to weave rectangular, block-type preforms. As shown in Fig. 1, this preform type consists of multiple yarn bundles located on Cartesian coordinates. Each of the yarn bundles is straight in order to achieve the maximum structural capability of the fiber.

Preforms are described by yarn type, number of yarns per site, spacing between adjacent sites, volume fraction of yarn in each direction, and preform density. The data in Table 1

illustrate the characteristics of selected three-directional carbon fiber preforms.

Several modifications of the basic three-directional orthogonal construction are available in order to achieve more isotropic pre-forms. This is accomplished by introducing yarns in additional directions. For example, a five-directional construction can be achieved by adding two reinforcement directions that are ±45° with respect to the yarns within the X-Y

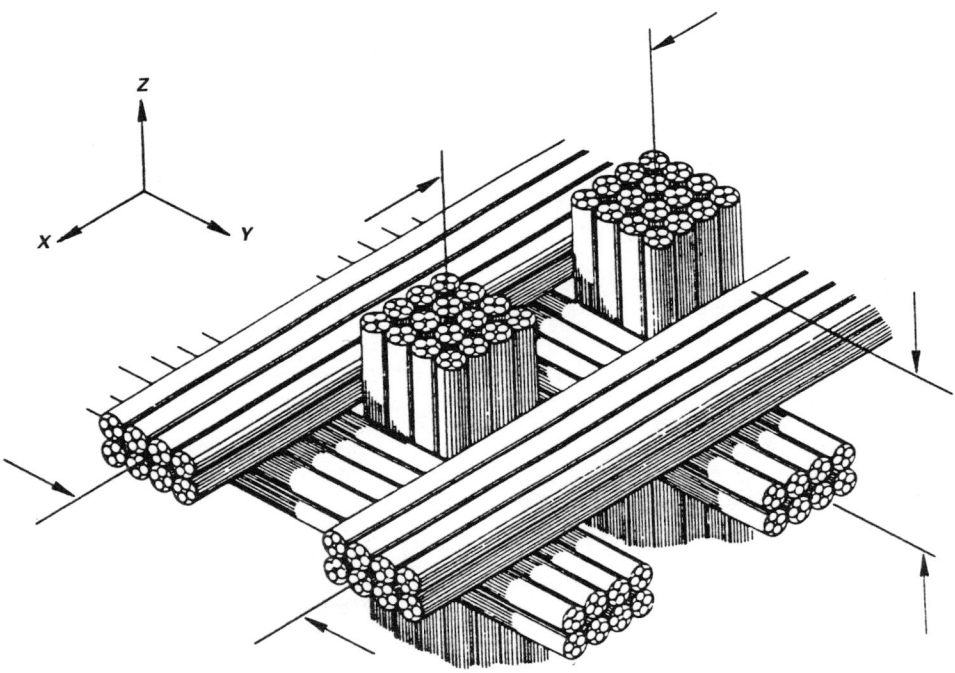

Fig. 1 Three-directional orthogonal preform construction. Source: Ref 8

Table 1 Characteristics of three-directional woven preforms

Material	Bulk density, g/cm³	No. of yarn bundles X	Y	Z	Center-to-center bundle spacing X, Y mm	in.	Z mm	in.	Fiber, vol%(a) X	Y	Z
Thornel 50(b)	0.64	1	1	1	0.56	0.022	0.58	0.023	0.14	0.14	0.13
	0.75	1	1	2	0.71	0.028	0.58	0.023	0.11	0.11	0.23
	0.68	2	2	1	1.02	0.040	0.58	0.023	0.14	0.14	0.12
	0.80	2	2	6	0.69	0.027	1.02	0.040	0.12	0.12	0.24
Thornel 75(b)	0.70	1	1	2	0.56	0.022	0.58	0.023	0.09	0.09	0.17
	0.65	2	2	1	0.84	0.033	0.58	0.023	0.12	0.12	0.09
	0.72	2	2	2	1.07	0.042	0.58	0.023	0.09	0.09	0.18

(a) Volume fraction of total preform volume occupied by fiber in each orthogonal direction. (b) Center-to-center bundle spacing. Source: Ref 8

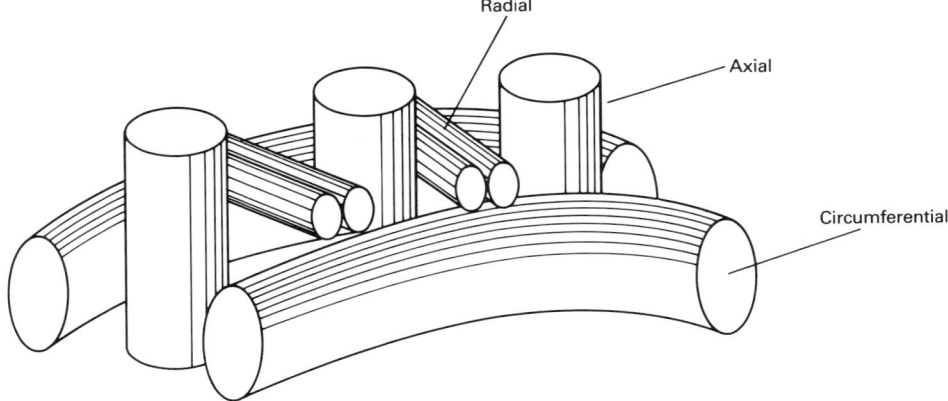

Fig. 2 Three-directional cylindrical preform construction. Source: Ref 8

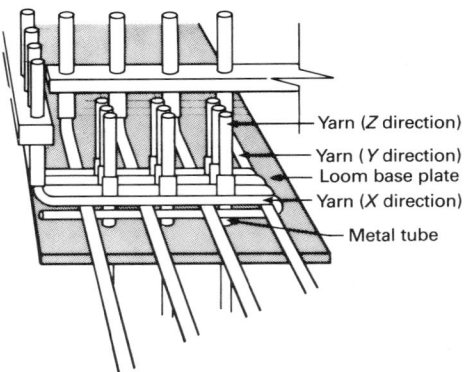

Fig. 3 Three-directional orthogonal weaving. Source: Ref 1

Fig. 4 Three-directional orthogonal block preform. Courtesy of Fiber Materials Inc.

Table 2   Typical properties of coal tar pitch

Softening point, °C (°F)	94-107 (200-225)
Viscosity at 250 °C (482 °F), Pa · s (cP)	0.03-0.05 (30-50)
Benzene insolubles, %	24-28
Quinoline insolubles, %	2-7
Coking value	52-62
Specific gravity	1.28-1.31
Sulfur, %	0.1-0.6
Ash, %	0.2-0.5

Table 3   Typical properties of phenolic resin

Specific gravity	1.08-1.09
Solids content, %	60-62
Viscosity at 25 °C (77 °F), Pa · s (cP)	0.12-0.20 (120-200)
Refractive index	1.518-1.525
Cure time at 165 °C, (329 °F), s	85-105
Free formaldehyde, %	0-0.5
Free phenol, %	11.5-13.5
Trace elements and sodium	< 5 ppm each
Potassium, lithium, iron	< 10 ppm total

Table 4   CVD densification of a multidirectional preform (typical)

Process conditions

Temperature, °C (°F)	1100 (2012)
Pressure, Pa (torr)	1350 (10)
Hydrocarbon	natural gas

Process steps

1. Process three-directional preform to 1.2 g/cm^3 density
2. Machine preform surfaces
3. Process for 650 h to a density of 1.6 g/cm^3

Source: Ref 20

plane of the preform. Another option is to introduce diagonal yarns across the corners and/or across the faces of a rectangular three-directional preform to achieve a nonplanar multidirectional construction (Ref 8).

The type of multidirectional preform construction typically used for cylinders and other shapes of revolution (shown in Fig. 2) is a three-directional construction with yarns oriented on polar coordinates in the radial, axial, and circumferential directions. As with orthogonal block preforms, yarn type, spacing, and volume fraction can be varied in all three directions.

Fiber Selection. Fibers are normally available as yarns or tows containing 1000 to 12 000 filaments per strand. Fibers selected must be compatible with the weaving and densification process, and must provide the physical and structural properties required in the composite.

The highest-modulus fibers have been subjected to the highest heat treatment temperature during manufacture. The properties of these high-modulus fibers are less affected by temperature exposure during carbon-carbon processing than are high-strength intermediate-modulus fibers that have not been previously exposed to graphitizing temperatures.

In most cases, fiber properties are degraded by various handling and processing steps that occur during carbon-carbon composite fabrication. Small amounts of polymeric coatings or finishes are used to reduce handling damage and to improve fiber-matrix compatibility.

Manufacturing. The original multidirectional preforms used precise tooling to locate yarns, but the weaving operations were performed manually. Weaving operations have now been automated, but many details regarding equipment and procedures are proprietary.

Most multidirectional preforms used for carbon-carbon composites are represented by the orthogonal or polar constructions shown in Fig. 1 and 2, respectively, or by some modification of these constructions. The techniques used to manufacture these preforms include weaving dry yarns (Ref 1), piercing fabrics (Ref 9 and 10), assembling resin-rigidized yarns (Ref 11) and filament winding (modified) (Ref 12).

Block Preforms. One method of weaving three-directional orthogonal block preforms involves setting up a precisely spaced rectangular array of thin-walled metal tubes or solid rods

representing the location of each *Z* direction reinforcing yarn (Ref 1 and 13). Alternate *X* and *Y* layers of yarn are built up between the rows of metal tubes, as illustrated in Fig. 3. After the height of the preform has been established by *X-Y* layers, each *Z* direction tube (or rod) is replaced by yarn to establish the *Z* direction of the preform. Figure 4 shows a typical three-directional orthogonal preform that uses a graphite frame to prevent distortion.

A modified three-directional orthogonal block construction is produced by using a two-directional woven fabric instead of *X-Y* yarn layers. These preforms are fabricated by piercing multiple layers of fabric over a precisely spaced rectangular array of metal rods. These metal rods, which represent the *Z* direction of the preform, are replaced with carbon yarns or precured (rigidized) yarn-resin rods as the final step of the process (Ref 10).

Shapes of Revolution. Fully automated computer-controlled equipment for fabricating three-directional cylindrical, conical, and contoured preforms has been developed both in the United States and in France. One version of this type of equipment, shown in Fig. 5, is three-axis computer numerically controlled (CNC) to define the preform configuration accurately and

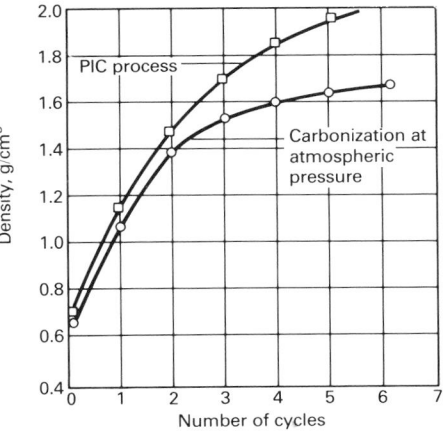

Fig. 7 Densification process comparison - PIC versus atmospheric-pressure carbonization. Source: Ref 8

Fig. 5 Automated three-directional preform fabrication equipment. Courtesy of Fiber Materials Inc.

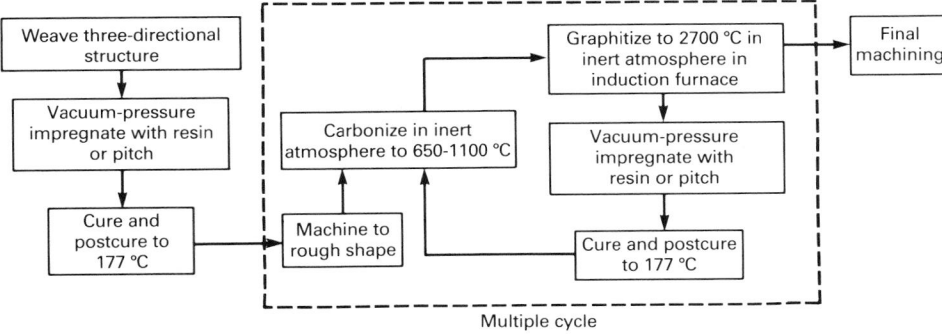

Fig. 6 Typical carbon-carbon process. Source: Ref 3

to place reinforcing fibers in the radial, axial, and circumferential directions (Ref 14). The machine automatically inserts resin-rigidized carbon fiber rods into a mandrel to form the radial reinforcing elements of the preform. The outside surface of the mandrel defines the inside surface of the preform. Alternating axial and circumferential reinforcement in the form of dry yarn is automatically wound onto the mandrel in a pattern established by the radial rod array. This machine can fabricate preforms up to 1400 mm (55 in.) in diameter and 1270 mm (50 in.) in length. The French-developed machine can manufacture preforms up to 900 mm (36 in.) in diameter; a 2000-mm (84-in.) diam capability is under development (Ref 15).

Automated equipment has also been developed to fabricate cylindrical preforms completely from dry yarns (Ref 13, 16). To weave preforms, this type of loom locates yarns in the circumferential and radial directions within an array of axial metal rods. To complete the three-directional preform, the axial metal rods are replaced by dry yarns.

Densification Processing

Generally, the best carbon-carbon composites result from a densification process that fills the open volume of the preform with a dense, well-bonded carbon/graphite matrix. The actual densification process is dictated by the characteristics of the preform (Fig. 6). Carbon fiber type, fiber volume, preform thickness, and void size distribution within the preform are four of the most important factors to consider.

The matrix precursors and densification processes to be discussed here represent examples of typical approaches that have been reported for processing multidirectional carbon-carbon composites. In actual practice, these approaches have to be modified for each particular type of preform being processed.

Matrix Precursor Impregnants. The two general categories of matrix precursors used for carbon-carbon densification are thermosetting resins, such as phenolics and furfurals, and pitches based on coal tar and petroleum. Characteristics of a typical pitch and a typical resin are given in Tables 2 and 3.

The thermosetting resins polymerize to form cross-linked, infusible solids. As a result of pyrolysis, these resins form amorphous (glassy) carbon. The carbon yield at 800 °C (1470 °F) is about 50 to 60 wt%.

The coal tar and petroleum pitches are mixtures of polynuclear aromatic hydrocarbons. From their softening point up to about 400 °C (750 °F), the liquefied pitches undergo various changes, including volatilization of low molecular weight fractions, polymerization, cleav-

Table 5 Typical properties of three-directional orthogonal carbon-carbon composites

Property	Direction	
	Z	X-Y
Density, g/cm³	1.9	1.9
Tensile strength, MPa (ksi)		
at RT	310 (45)	103 (15)
at 1900 °K (2950 °F)	400 (58)	124 (18)
Tensile modulus, GPa (10⁶ psi)		
at RT	152 (22)	62 (90)
at 1900 °K (2950 °F)	159 (23)	83 (120)
Compressive strength, MPa (ksi)		
at RT	159 (23)	117 (17)
at 1900 °K (2950 °F)	196 (28)	166 (24)
Compressive modulus, GPa (10⁶ psi)		
at RT	131 (19)	69 (10)
at 1900 °K (2950 °F)	110 (16)	62 (90)
Thermal conductivity, W/m · K (Btu/ft · h · °F)		
at RT	246 (142)	149 (12)
at 1900 °K (2950 °F)	60 (5)	44 (4)
Coefficient of thermal expansion, 10⁻⁶/K		
at RT	0 (0)	0 (0)
at 1900 °K (2950 °F)	3 (5)	4 (7)
at 3000 °K (4950 °F)	8 (14)	11 (20)

RT, room temperature. Source: Ref 21

age, and rearrangements of the molecular structure. At temperatures above 400 °C (750 °F), mesophase spheres are formed in the isotropic liquid pitch. These mesophase spheres deform, coalesce, and solidify to form regions of extended order. The lamellar arrangement of the molecular structure in these regions favors the formation of a graphitic structure on further heating to above 2000 °C (3630 °F).

Coke yield from coal tar and petroleum pitches is about 50 wt% after pyrolysis at atmospheric pressure. However, pyrolysis of coal tar pitch at 600 °C (1110 °F) under 6.9 MPa (1 ksi) of pressure gives a coke yield of 90% (Ref 17). An increase in pyrolysis pressure does not increase coke yield over the 90% level.

Liquid Impregnation. The general processing technique using organic liquid impregnants as carbon matrix precursors involves multiple cycles of preform impregnation and heat treatment to produce a densified composite. Factors such as impregnant viscosity, and coke yield, density, microstructure, and degree of graphitization must be considered (Ref 7). All of these factors are influenced by the time-temperature-pressure relationships encountered during processing.

A typical densification process is shown in Fig. 7. The process can be modified by performing the carbonization step under pressures ranging from 6.8 to 103 MPa (1 to 15 ksi). This modified process has been designated as the pressure-impregnation-carbonization (PIC) process. Modified hot isostatic press (HIP) equipment is used to impregnate and densify the composite effectively during the melting and coking stages of the carbonization process (Ref 18, 19). Isostatic pressure forces pitch into the small pores that are not filled during initial vacuum impregnation. As the pitch begins to pyrolyze, high isostatic pressure maintains the more volatile fractions of the pitch in a condensed phase. This reduces the amount of liquid forced out of the composite by pitch pyrolysis products.

The curves in Fig. 7 illustrate the advantage of PIC versus atmospheric-pressure carbonization to achieve high-density, multidirectional carbon-carbon composites.

The chemical vapor deposition (CVD) process for carbon-carbon densification involves gas phase pyrolysis of a hydrocarbon such as methane. Carbon is deposited on the external and internal surfaces of the porous multidirectional preform. The carbon fiber preforms are then subjected to radiant heat in a furnace. Hydrocarbon and carrier gasses are introduced into the furnace, usually at reduced pressures (Ref 4). Table 4 summarizes typical processing information for the CVD densification of a multidirectional preform.

Properties

Preform design, fiber type, matrix precursor, and processing will all influence composite properties. There are no standard properties because the number of possible variations is almost limitless. The data in Table 5, presented as typical for three-directional orthogonal carbon-carbon composites, do illustrate these characteristics, which are typical of most carbon-carbon composites:

- Low thermal expansion that increases with temperature
- Strength increase with temperature
- Thermal conductivity decrease with temperature

Applications

The high cost of multidirectional carbon-carbon composites has restricted their use to aerospace and specialty applications. However, the development of cost-effective, automated, three-directional preform manufacturing techniques should lead to new applications.

An early use of multidirectional carbon-carbon composites was for high-performance reentry vehicle nose tip applications (Ref 22). Other applications include integral throat entrance (ITE) components and exit cones for rocket motors (Ref 13, 23). Simplified designs using multidirectional carbon-carbon composites have provided increased reliability and weight savings over earlier multicomponent designs containing tungsten or pyrolytic graphite throat inserts (Ref 23).

Carbon-carbon composites are biocompatible and can be tailored to be structurally compatible with bone for applications such as integral fixation of fractures (Ref 24). Multidirectional carbon-carbon composites are also being studied for potential use in hip joint replacements.

Other reported applications for carbon-carbon composites include friction materials for aircraft and vehicular brake systems, hot pressing molds, thermal protective panels and hot structural components for space transportation systems, and components for advanced gas turbine engines. Multidirectional carbon-carbon composites should have potential use in most of these applications.

REFERENCES

1. R.S. Barton, "A Three Dimensionally Reinforced Material," *SPE J.*, May 1968, Vol 4, p 31-36
2. K.M. Jacobs, A.T. Laskaris, and J.W. Herrick, "Three Dimensionally Reinforced Ablative Rocket Engine Components," Paper 68-598 presented at the AIAA Fourth Propulsion Joint Specialist Conference, American Institute of Aeronautics and Astronautics, Cleveland, 1968
3. L.E. McAllister and A.R. Taverna, "The Development of High Strength Three Dimensionally Reinforced Graphite Composites," Paper presented at the 73rd Annual Meeting, American Ceramic Society, Chicago, 1971
4. H.M. Stoller and E.R. Frye, "Processing of Carbon-Carbon Composites—An Overview," Paper presented at the 73rd Annual Meeting, American Ceramic Society, Chicago, 1971
5. H.M. Stoller, B.L. Butler, J.D. Theis, and M.L. Liberman, "Carbon Fiber Reinforced Carbon-Matrix Composites," Paper presented at the 1971 Fall Meeting of the Metallurgical Society of the American Institute of Mining, Metallurgical, and Petroleum Engineers, Detroit
6. D.L. Schmidt, Carbon-Carbon Composites, *SAMPE J.*, Vol 8 (No. 3), May/June 1972, p 9-19
7. L.E. McAllister and W.L. Lachman, Multidirectional Carbon-Carbon Composites, in *Fabrication of Composites*, Vol 4,

A. Kelly and S.T. Mileiko, Ed., North-Holland, 1983, p 109-175

8. W.L. Lachman, J.A. Crawford, and L.E. McAllister, Multidirectionally Reinforced Carbon-Carbon Composites, in *Proceedings of the International Conference on Composite Materials*, B. Noton, R. Signorelli, K. Street, and L. Phillips, Ed., Metallurgical Society of the American Institute of Mining, Metallurgical, and Petroleum Engineers, 1978, p 1302-1319

9. L.E. McAllister and A.R. Taverna, A Study of Composition-Construction Variations in 3-D Carbon-Carbon Composites, in *Proceedings of the International Conference on Composite Materials*, Vol I, E. Scala, E. Anderson, I. Toth, and B. Noton, Ed., Metallurgical Society of the American Institute of Mining, Metallurgical, and Petroleum Engineers, 1976, p 307-315

10. L.E. McAllister and A.R. Taverna, Development and Evaluation of Mod-3 Carbon-Carbon Composites, in *Proceedings of the 17th National SAMPE Symposium*, Society for the Advancement of Material and Process Engineering, 1972, p III A-3

11. P. Lamicq, "Recent Improvements in 4-D Carbon-Carbon Materials," Paper 77-882 presented at the AIAA/SAE 13th Propulsion Conference, Orlando, 1977

12. C.K. Mullen and P.J. Roy, Fabrication and Properties Description of AVCO 3-D Carbon-Carbon Cylinder Materials, in *Proceedings of the 17th National SAMPE Symposium*, Society for the Advancement of Material and Process Engineering, 1972, p III A-2

13. P.S. Bruno, D.O. Keith, and A.A. Vicario, Jr., Automatically Woven Three-Directional Composite Structures, in *Proceedings of the 31st International SAMPE Symposium*, Society for the Advancement of Material and Process Engineering, 1986, p 103-116

14. H.D. Batha, Fiber Materials Inc., private communication, May 1986

15. P.G. Rolincik, Avco Specialty Materials, Textron Inc., private communication, May 1986

16. Y. Grenie and G. Cahuzac, Automatic Weaving of 3-D Contoured Preforms, in *Proceedings of the 12th National SAMPE Symposium*, Society for the Advancement of Material and Process Engineering, 1980

17. L.E. McAllister and R.L. Burns, Pressure Carbonization of Pitch and Resin Matrix Precursors for Use in Carbon-Carbon Processing, in *16th Biennial Conference on Carbon, Extended Abstracts*, American Carbon Society, 1983, p 478-479

18. R.L. Burns and J.L. Cook, Pressure Carbonization of Petroleum Pitches, in *Petroleum Derived Carbons*, M.L. Deviney and T.M. O'Grady, Ed., ACS Symposium Series 21, American Chemical Society, 1974, p 139

19. W. Chard, M. Conaway, and D. Neisz, Advanced High Pressure Graphite Processing Technology, in *Petroleum Derived Carbons*, M.L. Deviney and T.M. O'Grady, Ed., ACS Symposium Series 21, American Chemical Society, 1974, p 155

20. H. Girard, The Preparation of High Density Carbon-Carbon Composites, in *Proceedings of the Fifth London International Carbon and Graphite Conference*, Vol I, Society of Chemical Industry, 1978, p 483-492

21. A. Levine, "High Pressure Densified Carbon-Carbon Composites, Part II: Testing," Paper FC-21 presented at the 12th Biennial Conference on Carbon, Pittsburgh, 1975

22. V. DiCristina, "Hyperthermal Ablation Performance of Carbon-Carbon Composites," Paper 71-416 presented at the AIAA 6th Thermophysics Conference, American Institute of Mining, Metallurgical, and Petroleum Engineers, Tullahoma, TN, 1971

23. C.W. Hawk and W.C. Kessler, "A Functional Approach to the Application of Carbon-Carbon Composites to Solid Rocket Nozzles," Paper 79-1AF-19 presented at the 30th International Aeronautics Federation Congress, Munich, 1979

24. E.W. Fitzer, L.M. Huttner, L.M. Manocha, and D. Wolter, Carbon Fiber Reinforced Composites for Internal Bone-Plates, in *Proceedings of the Fifth London International Carbon and Graphite Conference*, Vol I, Society of Chemical Industry, 1978, p 454-464

Oxidation-Resistant Carbon-Carbon Composites

James E. Sheehan, GA Technologies Inc.

OXIDATION-RESISTANT carbon fiber reinforced carbon matrix (carbon-carbon) composites are being developed for a variety of military and aerospace applications. This material is currently being used for reentry thermal protection of the nose cap and wing leading edges of the space shuttle vehicles (Ref 1). Future applications include more extensive use on the shuttle vehicles, thrust vectoring exhaust components for fighter aircraft, static and rotating cruise missile turbine engine parts, propulsion components and thermal protection for hypersonic tactical missiles, and the new-generation hypervelocity military and aerospace vehicles (Ref 2-4).

The use of carbon-carbon composites for rocket nozzles and ablative reentry components requires short-duration high-temperature stability and thermal shock resistance, but does not involve sustained operation under significant structural loads. As indicated in Table 1, this is not due to a lack of composite structural capability. Carbon-carbon composites can be fabricated with mechanical properties that are attractive for structural applications at very high temperatures. Substantial reduction of the mechanical properties listed in Table 1 becomes a factor only at temperatures above 2200 °C (4000 °F) (Ref 6, 7). This unique behavior—combined with resistance to catastrophic failure by fiber toughening, low thermal dimensional changes, and densities below 2 g/cm^3—results in a class of materials with the highest potential for thermal structural applications.

The principal uses for high-temperature structural materials are in air-breathing propulsion systems and components that experience aerodynamic heating for extended periods of time. The requirement for oxidation resistance is inherent in these applications and is currently the major barrier to the use of carbon-carbon composites in many important high-temperature systems. Although specific methods for achieving effective oxidation protection are related to particular operating conditions and performance requirements, each approach is based on the common set of considerations discussed below.

General Considerations

All of the methods for protecting carbon-carbon composites from oxidation utilize an external coating as the primary oxygen barrier. The coating must be oxidation resistant, must have low oxygen permeability, and must be compatible with the composite substrate. Much of the current work on coatings for oxidation-resistant carbon-carbon composites is based on past developments for graphite oxidation protection (Ref 8).

The primary coating candidate for oxidation protection at temperatures below 1750 °C (3200 °F) is silicon carbide (SiC), because it is highly oxidation resistant, chemically compatible with carbon, and has a relatively low coefficient of thermal expansion. The temperature limitation for SiC in oxidizing environments is due to the chemical reaction between the SiC and silicon dioxide (SiO_2), which renders the SiO_2 nonprotective, thus resulting in rapid erosion of the SiC coating (Ref 9). Silicon nitride (Si_3N_4) coatings are also being developed to protect carbon-carbon composites. Coatings of Si_3N_4 have a lower coefficient of thermal expansion than SiC, but are generally applied to a thin SiC layer and have essentially the same temperature limitation as SiC.

Although the coefficients of thermal expansion for SiC and Si_3N_4 are low, significant mismatches with high-performance carbon-carbon composites present a problem. The problem results from carbon-carbon having an extremely low thermal expansion rate, so that even coatings with only low thermal expansion rates still result in significant expansion mismatches with the carbon-carbon base material. The tensile cracks that develop in the coatings at temperatures below the coating deposition temperature leave the underlying carbon susceptible to rapid oxidation. One solution to this problem is to use glass to seal the cracks in the outer coating.

Glass sealants must be chemically compatible with carbon and the outer coating, and they must have viscosity and wetting characteristics that promote the formation of continuous adherent layers. The glass former that, upon oxidation, best accomplishes these requirements is boron. Boron has long been known as a key ingredient in oxidation-resistant graphites, and it was used as a surface coating and internal oxidation inhibitor in early carbon-carbon composite development (Ref 10-12). For all but very short-term applications, the use of boric oxide (B_2O_3) or borate glasses to protect carbon materials is restricted to temperatures below about 1400 °C (2550 °F). This limitation is due to B_2O_3 vaporization and chemical reaction with carbon.

Oxidation protection for carbon-carbon composites above 1750 °C (3180 °F) involves the use of either noble metal or highly refractory ceramic coatings as primary oxygen barriers. The most attractive noble metal for carbon protection at very high temperatures is iridium. In addition to a melting point of 2440 °C (4425 °F), iridium has very low oxygen permeability up to 2100 °C (3810 °F), is nonreactive with carbon below 2280 °C (4135 °F), and is a very effective carbon diffusion barrier (Ref 8). The disadvantages of using iridium are erosion by volatile oxide formation, lack of adherence to

Table 1 High-performance carbon-carbon composite properties

Reinforcement	Fiber, vol %	Density, g/cm^3	Flexural strength		Tensile strength		Flexural modulus		Coefficient of thermal expansion (0-1000 °C, or 32-1830 °F), 10^{-6}/K
			MPa	ksi	MPa	ksi	GPa	10^6 psi	
Unidirectional 65		1.7	827	120	690	100	186	30	1.0
Orthogonal 55		1.6	276	40	. . .	. . .	76	10	1.0

Source: Ref 5

carbon, and thermal expansion incompatibility with high-performance carbon-carbon composites. Thermal expansion mismatch is probably the most significant difficulty. Iridium bonding to carbon has been demonstrated with special techniques, and refractory oxide overcoating is a possible solution to the erosion problem.

The use of refractory ceramic coatings for very high-temperature oxidation protection is limited primarily by the high oxygen permeabilities of the refractory oxides. All of the refractory carbides, nitrides, borides, and silicides oxidize rapidly above 1750 °C (3180 °F), and most oxidize at significantly lower temperatures. This is due to the formation of volatile, porous, or nonadherent oxides and to the fact that the stable oxides have inherently high oxygen permeabilities. The disruption or porous nature of the oxide layers is often due to rapid, solid-state oxygen diffusion, resulting in a lack of time to accommodate volatile products or the solid oxide itself. Even the most oxidation-resistant compounds exhibit rapid oxidation at temperatures above 1750 °C (3180 °F). Possible exceptions to this are mixtures of hafnium boride (HfB_2) and SiC, which have the potential as relatively thick coatings to provide effective short-term protection at temperatures as high as 1900 °C (3450 °F) (Ref 13). Based on this discussion, ceramic coating protection of carbon-carbon composites at very high temperatures is currently feasible only for very short periods of time. Long-term protection will require the identification of effective oxygen barriers and the solution of other problems, such as chemical compatibility with carbon and thermal expansion mismatch.

Current Developments

As stated previously, the only oxidation-resistant carbon-carbon composite components currently in use are the nose cap and wing leading edges of the space shuttle vehicles. These components are made from carbon-carbon composites consisting of a low elastic modulus, low-strength carbon fabric and a resin-derived carbon matrix. The coating system that provides reentry protection for short periods of time up to 1300 °C (2370 °F) is composed of an inner SiC conversion layer and an outer layer of glassy silicate filled with SiC powder. The inner coating is formed by converting the surfaces of the carbon-carbon composite components to SiC with a silicon pack cementation process (Ref 14, 15). After pack processing, the pores and cracks present in the SiC conversion layer are filled with SiO_2 by multiple impregnations with tetraethylorthosilicate. The outer SiC-filled silicate coating is a combination of commercial products (Ref 16).

The carbon-carbon thermal protection system described above is effective for the shuttle missions, but is inadequate for such applications as aircraft exhaust components that require structural carbon-carbon composites to operate for extended periods of time at relatively low temperatures with periodic excursions to higher temperatures. This means that a mechanism must be provided for sealing thermal expansion mismatch cracks in the primary coating. As stated above, borate glasses have proved to be ideal for this purpose.

The carbon-carbon materials that are the leading candidates for aircraft exhaust components contain boron both in the composite and as part of the external coating system. These composites employ a dense vapor-deposited SiC or Si_3N_4 outer coating as the primary oxygen barrier, inner coatings that are high in boron, and boron in the carbon matrix of the composite. Oxygen entry through the cracks in the outer coating produces a borate glass that fills the cracks and seals the coating. The presence of boron inside the composite provides additional sealant, protects the carbon constituents from minor oxygen leaks over long periods of time, and guards against catastrophic oxidation if major flaws develop in the outer coating. Materials of this type have exhibited only minor oxidation after hundreds of hours of thermal cycling at temperatures up to 1400 °C (2550 °F).

Oxidation protection at temperatures exceeding 1400 °C (2550 °F) is currently being addressed for missile applications that require lifetimes ranging from minutes to several hours. Both SiC and Si_3N_4 coatings have performed well at temperatures up to 1750 °C (3180 °F). Iridium coatings should be effective up to 2100 °C (3810 °F), but the high cost and unavailability of iridium have hindered this development. High-temperature oxide coatings, such as hafnium oxide (HfO_2) and zirconium oxide (ZrO_2) and their carbide and boride precursors, are being evaluated for extreme temperature ramjet and rocket applications. The inherently high oxygen permeabilities of the refractory oxides limit their potential to only very short-term protection.

REFERENCES

1. H. Goldstein, "Advanced Fibrous Ceramic Heat Shield Materials," Paper presented at the Fibrous Ceramic Materials Technology Seminar, National Aeronautics and Space Administration, March 1983
2. P. Kinnucan, Superfighters, *High Technol.*, April 1984, p 37
3. J.A. Bailie, Application of Composites to Missile Structures, *SAMPE Q.*, Jan 1981, p 1
4. E. Marshall, NASA and Military Press for a Spaceplane, *Science*, Vol 231, Jan 1986, p 105
5. J.E. Hill, E.J. Walker, and C.R. Thomas, High Strength and Modulus Carbon-Carbon Composites (11th Biennial Conference on Carbon), *Extended Abstracts*, June 1973, p 328
6. A. Levine and W. Chard, "High Pressure Densified Carbon-Carbon Composites," Paper presented at the 12th Biennial Conference on Carbon, American Carbon Society, July 1975
7. C.R. Rowe and D.L. Lowe, High Temperature Properties of Carbon Fibers (13th Biennial Conference on Carbon), *Extended Abstracts*, July 1977, p 170
8. "High Temperature Oxidation Resistant Coatings," Publication ISBM 0-309-01769-6, National Academy of Sciences and Engineering, 1970
9. G.H. Schiroky, J.L. Kaae, and J.E. Sheehan, "High Temperature Oxidation of CVD Silicon Carbide," Paper presented at the 87th Annual Meeting, American Ceramic Society, May 1985
10. K.J. Zeitsch, Oxidation-Resistant Graphite-Base Composites, in *Modern Ceramics*, J.E. Hove and W.C. Riley, Ed., John Wiley & Sons, 1967, p 314
11. G.R. Martin, Oxidation Resistant Carbonaceous Bodies and Method of Producing Same, U.S. Patent 3,936,574, 1976
12. L.C. Ehrenreich, Reinforced Carbon and Graphite Articles, U.S. Patent 4,119,189, 1978
13. E.V. Clougherty, R.L. Pober, and L. Kaufman, Synthesis of Oxidation Resistant Metal Diboride Composites, *Trans. AIME*, Vol 242, 1968, p 1077
14. D.C. Rogers, D.M. Shuford, and J.I. Mueller, Formation Mechanism of a Silicon Carbide Coating for a Reinforced Carbon-Carbon Composite, in *Proceedings of the Seventh National SAMPE Technical Conference*, Vol 7, Society for the Advancement of Material and Process Engineering, 1975, p 319
15. D.C. Rogers, R.O. Scott, and D.M. Shuford, Material Development Aspects of an Oxidation Protection System for a Reinforced Carbon-Carbon Composite, in *Proceedings of the Eighth National SAMPE Technical Conference*, Vol 8, Society for the Advancement of Material and Process Engineering, 1976, p 308
16. "Surface Seal for Carbon Parts," NASA Technical Briefs, Vol 6 (No. 2), MSC-18898, National Aeronautics and Space Administration, 1981

Structurally Reinforced Carbon-Carbon Composites

H. Dean Batha, Fiber Materials, Inc.
Charles R. Rowe, Naval Surface Weapons Center

CARBON-CARBON COMPOSITES can be used in numerous applications other than those requiring sustained high-temperature stability. Carbon-carbon composites are biocompatible and can be made with directional properties. The wide range of useful physical properties available to designers permits applications ranging from prosthetics to space structures. Fine-diameter fibers are used to develop turbine rotors and stiff, lightweight space structure components. The frictional characteristics and thermal shock resistance of carbon-carbon composites are exploited in high-performance aircraft brakes. Each of these applications requires a unique combination of fiber, matrix, and composite properties.

Applications

Improved aircraft brakes became necessary as aircraft became heavier and faster. Aborted takeoffs that were terminated at maximum ground speed were uncertain with steel disk brakes. The brakes, if successful in stopping the aircraft, were often destroyed because of warping or melting due to the intense heat generated during the stop. Carbon-carbon composites have a high melting point, are resistant to thermal shock, and have excellent friction and wear characteristics. They can be fabricated in shapes and sizes that are suitable for brake applications. Carbon-carbon brakes, designed to take advantage of these useful properties, can provide superior stopping capability, survive several abortive stops, and last far longer than conventional aircraft brakes. Furthermore, this superior performance is accomplished with a significant weight savings (Ref 1-4).

Disk brakes for aircraft are composed of a number of disks (Fig. 1), half of which are keyed to the nonrotating brake mechanism, and the other half rotate with the wheel to which they are keyed. Braking is accomplished by forcing the disks together, at which time friction is converted to heat, which must be dissipated. This requires a material that is resistant

to thermal shock, stable to very high temperatures, and has low thermal expansion and good thermal conductivity. In addition, the material should have a friction coefficient of about 0.3 to 0.5 for good stopping performance. Carbon-carbon composites have all of these important properties, along with a density of about 1.9 g/cm^3, which provides nearly four times the stopping power of copper or steel brakes. Among the advanced aircraft, the Concorde, Delta 2000, Airbus, and the Falcon 900 require 820 to 1050 kJ/kg (350 to 450 Btu/lb) of the carbon-carbon brake to stop the aircraft. High-performance automobiles require only 300 to 520 kJ/kg (130 to 220 Btu/lb) (Ref 2). Carbon-carbon brakes with fibers in the plane of the flat surface have low wear and good frictional characteristics. Thus, carbon-carbon disks are made with layers of carbon fabric or with random fibers arrayed parallel to the braking surface.

Carbon-carbon brakes are fabricated either by laying up carbon cloth-resin prepreg disks, chopped-cloth prepreg, or random fibers in carbon matrices. In the latter, the matrix is derived from resin or a chemical vapor deposition (CVD) process from methane. A common practice for densifying cloth lay-ups is to use chemical vapor deposition by placing the preform structures in a large furnace that is evacuated, heated, and exposed to methane; the gas pyrolyzes (cracks) in the heated preforms, depositing carbon. The deposition process is continued until the interstices are completely filled with carbon. An alternative method of densification of random-fiber preforms is to impregnate them with resin that is decomposed to form an amorphous carbon matrix. Materials using both systems are manufactured and are in service.

Continued research and development is devoted to reducing the cost of the brakes while retaining or improving performance. Thus, larger CVD furnaces are being used to maximize production in batch processes. The amount of fiber is decreased, and pitch-derived fiber, which is less expensive than poly-

acrylonitrile (PAN) fiber, is used whenever possible. The current high cost of the system precludes the use of these high-performance brakes in land vehicles. However, the payoff for carbon-carbon brakes in aircraft is significant. Automotive use of carbon-carbon brakes at current costs is justifiable only where a marginal performance improvement could provide a competitive edge.

An interesting demonstration of the weight savings and life projections of carbon-carbon over steel-cermet brakes for commercial aircraft is presented in Tables 1 and 2. The significant weight savings, combined with a life of two to four times that of the steel-cermet brakes, is attainable for transport planes and has been demonstrated and accepted on the F-14, F-15, F-16, and F-18, as well as the Concorde and other jet aircraft.

Special modifications to the carbon-carbon brake system include the incorporation of re-

Fig. 1 Carbon-carbon aircraft brake assembly and a stator and rotor disk. Courtesy of Allied Bendix Aerospace

Table 1 Weight savings of carbon over steel-cermet brakes

Aircraft	Total weight savings	
	kg	lb
747	635	1400
767	408	900
757	272	600

Source: Ref 1

Table 2 Life projections for carbon versus steel-cermet brakes

Aircraft	Estimated landings	
	Carbon	Steel-cermet
747	2000	800
757	3000	1500
767	3000	1500
Military	600	150

Source: Ref 1

fractory borides to avoid temporary reduction in the coefficient of friction after a period of idleness. The friction loss is probably caused by water absorption, and it can be overcome operationally by applying brake pressure several times before takeoff. The borides eliminate the need for this preconditioning. Metal caps on the teeth of the rotating disks are used on some systems to reduce wear on the teeth and to increase thermal conductivity.

Prosthetics. The replacement of knee and hip joints with metallic prosthetic devices has provided mobility to many elderly patients suffering from bone diseases. Metal prostheses, anchored with methyl methacrylate cement, have a useful life of 7 to 10 years (Ref 5). Failure of the bond requires a second replacement, but bone resorption resulting from the presence of the device limits the number of transplants possible with each patient to two. Therefore, hip and knee joint replacements are generally limited to patients aged 55 years or older.

Carbon-carbon prosthetic devices have been shown to be compatible with the body. Tissue and bone ingrowth has been demonstrated. Fiber-reinforced composites can be constructed to provide strength and stiffness in the necessary directions. Satisfactory fatigue behavior and toughness, along with the relatively low density of the carbon-carbon composites, combine to make the composite an ideal candidate for prosthetic devices.

The biological compatibility of carbon has been demonstrated and has permitted the use of carbon in the body for a number of nonload-bearing applications. Heart valves have been produced with isotropic pyrolytic carbon, and they have been implanted in more than 200 000 patients (Ref 6). The success of these valves in not triggering blood clotting is attributed to the compatibility of carbon and blood, the high purity, and a polished surface. Silicon-alloyed pyrolytic carbon has also been found to behave similarly.

In a study of the response of bone to carbon-carbon composites, two-dimensional composites were used as implants in the femurs of rats, because carbon has excellent tissue compatibility and the rough surfaces showed a tendency for bone ingrowth (Ref 7). These tests showed that carbon-carbon composites provide better adaption to bone than titanium does. In another study, it was found that carbon fibers were a degradable scaffold upon which ligaments can be regenerated (Ref 8). No allergic, carcinogenic, or biologic problems have been reported

in their use in this application, for which carbon fibers with collagen, sulfone, Teflon, and CVD carbon coatings have been evaluated.

The stresses on the femur in walking have been analyzed (Ref 9, 10). Maximum compressive or tensile loads are approximately 48 to 55 MPa (7.0 to 8.0 ksi) in a fiber direction. Over 10^6 load cycles each year are expected. Carbon-carbon composites typically exhibit tensile strengths of about 138 MPa (20 ksi) and compressive strengths of 207 to 276 MPa (30 to 40 ksi). Shear stress of the composite may be important in this environment, in which the loading is variable and multidirectional. Fatigue experiments with three-dimensional carbon-carbon composites indicate that the degradation of properties is negligible after 10^7 cycles at 90% of the composite stress limit (Ref 11). The femur has a modulus of about 28 GPa (0.29×10^6 psi), a tensile strength of 122 MPa (18 ksi), and a compressive strength of 165 MPa (25 ksi) (Ref 12). Bodies in compression (even nonductile) almost always fail in shear, which makes the carbon-carbon composite an interesting case (Ref 13).

The mechanics of loading of the joint during walking indicate that the stresses are in different directions, depending on the phase of the step. Through careful design of the fiber preform, it is possible to make a prosthetic device with strength in the direction of the applied load. Furthermore, it has been demonstrated that a modulus of 28 GPa (4.0×10^6 psi) and strength of 165 MPa (25 ksi) can be obtained in a carbon-carbon composite.

Although it has been clearly demonstrated that carbon-carbon composites have the physiological and mechanical properties necessary for use in prosthetic devices, the optimum carbon-carbon device has not yet been demonstrated. A carbon-carbon hip for a dog has been designed (Ref 14), but the structure has not been revealed. Bone ingrowth has been indicated in tests with monolithic carbon, but it is clear that the normal surfaces of the dense composites lack the larger (200 μm, or 7900 μin.) pore size conducive to bone growth. Such a requirement can be easily filled by proper design. The interesting biological compatibility of siliconized carbon suggests that a wear surface for the ball and socket interface can be made an integral part of the structure if desired (Ref 15). The benefits that can be expected from the use of carbon-carbon composites for joint replacement will be realized after clinical tests and when close cooperation between the physician and the fabricator results in an optimal structure for resisting the loading of the joint during normal movement of the patient.

Space Structures. Recent initiatives in the exploration and commercialization of space have highlighted the need for new high-performance composites for spacecraft. Many types of composites, including organic resin and metal matrix composites, have applications in this area, but carbon-carbon composites have unique applications when the temperature of

Table 3 Properties of structural metals

Material	Density, g/cm³	Weight ratio to carbon-carbon materials
Carbon-carbon . . .	1.4-2.0	1.0
Aluminum	2.7	1.6
Titanium.	4.5	2.7
Niobium.	8.4	4.9

the structural components exceeds 1000 °C (1830 °F), even for short periods of time.

Carbon-carbon composites have several unique properties that make them attractive for high-temperature space applications. In some cases, they are the only material solution for specific design problems in spacecraft because they are inert in most space environments. Although they do suffer some oxidation from atomic oxygen in low earth orbit, carbon-carbon composites do not degrade or outgas the way organic resin composites do in the high vacuum of deep space. This property is attributed to the high manufacturing temperatures used to pyrolyze the organic precursors such that carbon is the only remaining element.

Carbon-carbon composites are also attractive for space structures because of their relatively low density. Typical densities range from 1.4 g/cm³ for flat panels and tubes to 2.0 g/cm³ for composites reinforced in three directions with high-density fibers, such as pitch-base carbon. This low density (Ref 16) provides a significant weight advantage over such structural metals as aluminum, titanium, and niobium, as evidenced in Table 3.

Thermal expansion is a significant design consideration in space structures. The extent to which a space structure, such as a communications boom, is affected by solar radiation on one side and cold space on the other side is a key design parameter, especially as the sizes of such components increase. Carbon-carbon composites have very low expansion values over the range of −150 to 3000 °C (−240 to 5430 °F). The coefficient of thermal expansion at −150 °C (−240 °F) is typically 10×10^{-6}/K; at 3000 °C (5430 °F), 6×10^{-3}/K. In addition, there is a minimum in the thermal expansion curve between 150 and 300 °C (300 and 570 °F). This is of particular interest because in normal space systems, in which the temperature range is −100 to 100 °C (−150 to 212 °F), the thermal expansion of carbon-carbon composites being considered for spacecraft is nearly 0 (Ref 17).

Another property of interest in applying carbon-carbon composites in spacecraft is high thermal conductivity. It has been shown that the thermal conductivity of carbon fibers is a function of the degree of orientation of the graphite crystallites along the fiber axis. Therefore, carbon-carbon composites fabricated with carbon fibers with highly ordered crystallites, such as pitch-base carbon fibers, will have high thermal conductivity. Table 4 shows the thermal conductivities of metal conductors used in a spacecraft and a prediction of the thermal

Table 4 Thermal conductivity of metallic and composite materials

| Material | Conductivity | | Specific conductivity |
	W/m · k	Btu · in./ h · ft² · °F	W · cm³/g · m · K
Copper	400	2800	45
Silver	390	2700	37
Aluminum	200	1400	74
Carbon-carbon . .	200	1400	115

conductivity of a carbon-carbon composite made with 65 vol% high-modulus pitch-base carbon fiber (Ref 16). The specific conductivity of carbon-carbon exceeds that of metals.

In many spacecraft applications, the mechanical properties of carbon-carbon composites become very important. Component designs that strive for the lowest possible weight result in very thin structural components that depend on materials with very high strength and stiffness. Carbon-carbon composites can meet these needs, especially when the temperature to which the component will be exposed exceeds 1000 °C (1830 °F). Carbon-carbon tubes have been made with a tensile strength greater than 400 MPa (60 ksi), which places it well above the category of advanced carbon-carbon composites (see Ref 18). In addition, the specific elastic modulus is three times that of steel.

The key to applying carbon-carbon composites to spacecraft is in carbon fiber selection. For structural components that are required to be very thin and still have high strength and high modulus, the selection of carbon fibers with filament diameters of about 4 μm (160 μin.) is essential. One such fiber (Apollo 55) has the following properties (Ref 16):

Density . . . . 1.83 g/cm³
Tensile strength . . . . 3.2 GPa (0.46 × 10⁶ psi)
Elastic modulus . . . . 400 GPa (60 × 10⁶ psi)

Also important for spacecraft applications is the availability of high-modulus (and high-conductivity) pitch-base carbon fibers, such as Thornel P-55, P-75, and P-100 fibers.

Manufacture

The methods of manufacturing carbon-carbon composites can be quite varied for flat panels and tubes. The methods can include simple, flat lay-ups; involute lay-ups for tubes and curved surfaces; and braiding for long tubes. In most cases, the fibers are held together through the first steps of the fabrication process with a phenolic resin. The composites are then heat treated to convert the phenolic resin to carbon, usually between 600 and 900 °C (1110 and 1650 °F). After this step, liquid impregnants, such as a resin, a petroleum pitch, or a combination of the two, are typically used to reduce the level of open porosity to 15 to 20%. These impregnants are also converted to carbon at temperatures of 1100 to 2000 °C (2010 to 3650 °F). The final step usually consists of the chemical vapor deposition of carbon from methane-nitrogen gas mixtures at 1100 to 1900 °C (2010 to 3450 °F), which completes the densification of the carbon-carbon composites. Carbon-carbon composites have been manufactured as tubes up to 1 m (3.3 ft) in length, with diameters from 30 to 100 mm (1.2 to 3.9 in.) and wall thicknesses as thin as 1 mm (0.039 in.). Carbon-carbon panels have been fabricated as large as 0.3 × 0.3 m (1.0 × 1.0 ft) and as thin as 0.2 mm (0.0080 in.).

REFERENCES

1. J.P. Ruppe, Today and the Future in Aircraft Wheel and Brake Development, *Can. Aeronaut. Space J.*, Vol 27, 1981, p 212-216
2. L. Heraud and B. Broquere, Carbone Pour le Freinage, Development in the Science and Technology of Composite Materials, *First European Conference on Composite Materials and Exhibition*, A.R. Bunsell, P. Lamicq, and A. Massiah, Ed., Sept 1985, p 440-446
3. P. Turk, Carbon Brakes: The Competition Heats Up, *Interavia*, Sept 1984, p 980-982
4. T. Liu and E. Kartman, "Test Methods for the Evaluation of Carbon-Carbon Composite Friction Material," Paper presented at Conference on Advances in High Performance Composite Technology, Clemson, SC, May 1986
5. H. Brückman and K.J. Hüttinger, Carbon: A Promising Material in Endoprosthetics, Part 1, *Biomater.*, Vol 1, 1980, p 67-72
6. J.C. Bokros, Carbon Biomedical Devices, *Carbon*, Vol 15, Sept 1977, p 355-371
7. D. Adams and D.F. Williams, The Response of Bone to Carbon-Carbon Composites, *Biomater.*, Vol 5, March 1984, p 59-64
8. J.M. Moran, Materials for Ligament Replacement, *SAMPE J.*, May-June 1985, p 10-13
9. E.F. Rybicki, F.A. Simonen, and E.B. Weis, Jr., On the Mathematical Analysis of Stress in the Human Femur, *J. Biomech.*, Vol 5, 1972, p 203-215
10. P. Paul, Load Actions on the Human Femur in Walking and Some Resultant Stresses, *Exp. Mech.*, March 1971, p 121-125
11. J. Pepin, "The Study of Fatigue Mechanisms in 3D Carbon-Carbon Composite," Final Report, NSF Grant ECS-8113723, National Science Foundation, April 1982
12. J.C. Koch, The Laws of Bone Architecture, *Am. J. Anat.*, Vol 21, March 1917, p 177-298
13. J.E. Gordon, *Structures*, Penguin Books, 1978, p 258
14. D.G. Mendes, Haifa Medical Center, private communication, 1984
15. W. Huettner, G. Keuscher, H.-J. Maeurer, and K.J. Huettinger, Experiences With All-Graphite/Siliconcarbide Double Cup Prostheses, in *Clinical Applications of Biomaterials*, A.J.C. Lee, T. Albrektsson, and P.-I. Branemark, Ed., 1982, p 21-29
16. R. Bacon and C.T. Moses, "Carbon Fibers: From Light Bulbs to Outer Space," Paper presented at the American Chemical Society Symposium, April 1986
17. R.A. Meyer, "Matrix Microstructure and Thermal Mechanical Property Behavior of Carbon-Carbon Composites," Paper presented at the Baden-Baden Meeting, German Carbon Society, July 1986
18. A.J. Klein, Carbon-Carbon Composites, *Met. Prog.*, Vol 11, 1986, p 64-68

Structural Ceramic Composites

Thomas Vasilos, Avco Systems Division*

ALTHOUGH MOST CERAMICS ARE COMPOSITES in the sense that they contain more than one discrete crystalline or amorphous, glassy phase, it has been customary to refer to such structures as monolithic because of their phenomenological continuity, particularly in terms of mechanical behavior. The term composite has usually been reserved for two-phase materials expressly fabricated to improve certain properties beyond those expected from conventional or monolithic materials.

Because this article focuses on fibrous ceramic composites for engineering applications, mechanical properties are of primary importance. As an example of the benefits of a composite system, the marked mechanical property improvements of ceramics made possible by fiber reinforcement are indicated in Fig. 1, which compares the stress-strain behavior for glass and carbon fiber reinforced glass (CRG) containing 40 vol% of aligned continuous carbon fibers (Ref 1). The stiffness of the material is increased by a factor of three because the fibers are stiffer than the matrix, and strength is increased by a factor of seven because the fibers are much stronger than the matrix. The most spectacular improvement, however, is in the fracture energy of the material, which is increased by about three orders of magnitude. This results from partly controlled fracture behavior, compared with the catastrophic failure characteristic of unreinforced glass.

Except for stiffness, similar behavior is indicated in Fig. 2, which compares the stress-strain behavior of monolithic reaction-bonded silicon nitride with that of composites containing two varieties of aligned continuous silicon carbide fiber in a reaction-bonded silicon nitride matrix (Ref 2).

Because the development of fiber-reinforced ceramics has followed the development and availability of fibrous materials, much of the early work was concerned with adding metal fibers to ceramic matrices. In general, such composites showed no additional strength, although their thermal-shock resistance was markedly improved, which has been attributed to the role of the metal fiber as a crack stopper

in the brittle matrix. This technique was useful in developing materials for such thermal stress applications as rocket motors and hypersonic vehicles.

Current work often focuses on the development of ceramic composites with improved mechanical properties over the unreinforced ceramic matrix. This effort has taken two different directions. One approach has been to add modest amounts of short reinforcing fibers to a ceramic to produce a composite that largely reflects the properties of the unreinforced matrix but has considerably improved mechanical toughness. The other approach, which has more closely paralleled the development of conventional fiber composites, attempts to fully utilize the high strengths and stiffnesses of carbon and ceramic fibers by producing composites with high volume fractions of continuous fiber reinforcement. In this case the ceramic matrix is of secondary importance to strength, but it does provide a higher temperature capability than can be achieved with polymer or metal matrices. Both approaches have been successful in producing composites with improved properties for use in low to intermediate temperatures. Development of a truly high-temperature composite, particularly for use in oxidizing environments, has been less successful.

Principles of Fiber Reinforcement

The reinforcement phase in a fiber-reinforced composite may be present in a number of forms. Highest strength, stiffness, and toughness are achieved with aligned continuous fibers. Aligning the fibers enables the incorporation of a larger quantity of fiber and permits

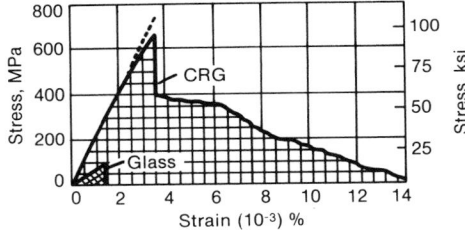

Fig. 1 Stress-strain curves for glass and carbon fiber reinforced glass. Cross-hatched areas represent the fracture energies of the two materials. Source: Ref 1

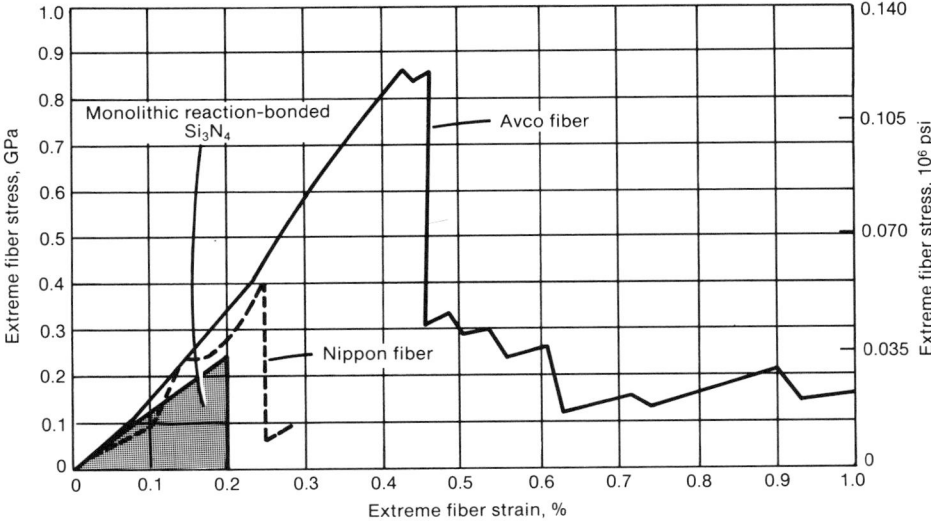

Fig. 2 Norton reaction-bonded silicon nitride (Si₃N₄) composites, load versus deflection. Source: Ref 2

*Now with the University of Lowell

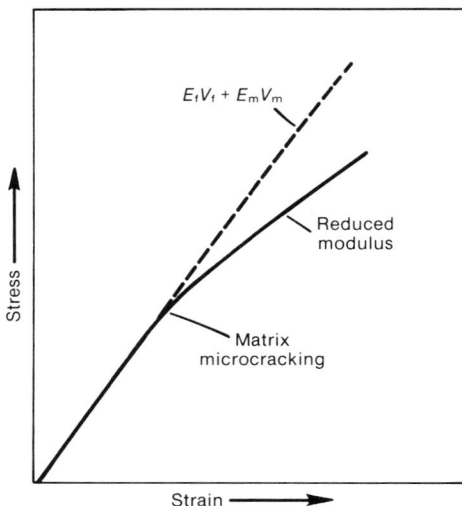

Fig. 3 Stress-strain curves for fibers, matrix, and composite, where E_f and E_m are Young's moduli of fibers and matrix, and V_f and V_m are volume fractions of fibers and matrix

lower porosity in the matrix. In practice, the optimum fiber loading for ceramic matrix composites has been around 50 vol%. Increasing matrix porosity above this level causes a decrease in properties. This has to be contrasted with polymer matrix composites, in which loadings of 60 to 70 vol% are easily achieved. Anisotropy of aligned, continuous fibers can be reduced by laminating plies at angles to one another to obtain useful strengths and stiffness in the required directions. Laminate theory for predicting the properties and laminate design are well developed and may be applicable to ceramic matrix composites.

Ceramic matrix composites are produced by high-temperature fabrication processes; differences between coefficients of thermal expansion (CTE) of fibers and matrix lead to thermal mismatch stresses upon cooling. When the expansion coefficient of the matrix is greater than that of the fibers, the effect can crack the matrix phase severely upon cooling (particularly for dense matrices); when it is lower than that of the fibers, the fibers can shrink away radially from the matrix, diminishing fiber-matrix bonding. The expansion coefficients therefore need to be matched for optimum properties, which rules out many matrix and reinforcing phase combinations. Thermal mismatch cracking is most severe with short, random fibers. Mismatch can be more readily accommodated in aligned, continuous fiber reinforced composites in which cracking is found to be more localized, partly because it is resisted by energy-absorbing mechanisms provided by the reinforcing phase.

For an aligned, continuous fiber reinforced composite that is stressed in the fiber direction, strength and elastic modulus can be calculated to a first approximation by assuming equal strains in fibers and matrix (Ref 3, 4). The

stresses are then distributed between fibers and matrix in a ratio that is determined by the ratio of the moduli of fibers and matrix and the quantity of reinforcement present. This results in the "rule of mixtures," with the elastic modulus, E_c, of the composite expressed as:

$$E_c = E_f V_f + E_m V_m \qquad (\text{Eq 1})$$

where E_f and E_m are Young's moduli of elasticity of fibers and matrix, and V_f and V_m are volume fractions of fibers and matrix. The strength of an aligned, continuous fiber composite $(\sigma_c)_u$ is expressed as:

$$(\sigma_c)_u = \sigma_f V_f + \sigma_m V_m \qquad (\text{Eq 2})$$

where σ_f is the ultimate strength of the fibers, σ_m is the stress in the matrix at that fiber strain, and V_f and V_m are volume fractions of fibers and matrix. Ceramic matrix composites differ significantly from the more usual polymer matrix composites in that the ratio of E_f/E_m is much lower for the ceramic matrix composites. This results in a higher proportion of stress being carried by the matrix, and because the matrix fails at lower strains than does the reinforcement phase, matrix microcracking usually occurs at lower strains. Figure 3 shows the anticipated stress-strain behavior of such a composite. The occurrence of matrix microcracking is important. In practice, it is found that the short ultimate strength, based on fiber strength, is still achieved, and that mechanical fatigue behavior may still be good, although microcracking provides an easy path for oxidation or corrosion and can affect the high-temperature capability of a system.

The ultimate strength of a fiber-reinforced composite depends on a complicated series of events in which weaker fibers break first and then the stresses are redistributed locally among adjacent fibers until final, uncontrolled fracture occurs. A variety of complicated statistical models have been developed to describe these processes. These models are not yet appropriate for designing with real ceramic matrix composite materials, and Eq 2 is adequate for practical purposes. More successful theories have been proposed for matrix microcracking stress, and it has been shown both experimentally and theoretically that microcracking stress is higher than predicted because the presence of fibers inhibits microcracking.

The toughness of a fiber-reinforced composite arises from a number of different mechanisms. The general processes involved are broadly understood, although a detailed understanding of the contribution of each mechanism to specific materials is still lacking. In brittle fiber-reinforced composites, the principal toughening mechanisms are the fiber-matrix debonding energy, the postdebond energy loss, the work done during fiber pull-out, and other related mechanisms (Ref 5). In ceramic matrix composites reinforced with ductile metal wires, the plastic deformation of the wires can be the dominant toughening mechanism. In general,

there is a tendency toward increasing toughness as the fiber-matrix bond strength decreases and the fiber diameter increases. In addition to these processes, in a laminate, both multiple cracking that is parallel to the fibers within a ply and delamination between plies can add significantly to toughness (Ref 6). These processes have not yet been studied in any detail for ceramic matrix composites.

The significance of toughness values for fiber-reinforced ceramics is not as straightforward as it is for unreinforced ceramics because energy can be absorbed by the fiber pull-out process, with energy absorption occurring well behind the tip of the crack as the fracture faces separate. Thus, the total absorbed energy as measured in a fracture test could be much higher than the fracture initiation energy as measured in a linear elastic fracture mechanics (LEFM) test, depending on the fiber pull-out length and size of the resulting process zone behind the crack tip.

Fabrication Processes

Ceramics reinforced with a fibrous phase have been produced by the *in situ* growth of high aspect ratio precipitates in the host matrix. The obtained enhancement in properties, however, has tended to be insignificant because the volume fraction of the fibrous phase has been low, the interface between the phases has been too strong to act as an effective crack deflector, and the small diameter of the fibers gain only a small contribution from the toughening mechanisms listed earlier, all of which show a dependence on fiber diameter. Large improvements in properties over those of unreinforced ceramics have only been obtained by combining previously manufactured fibers with a ceramic matrix.

Nearly the entire range of fabrication processes used for unreinforced ceramics has been employed to produce fiber-reinforced ceramics, with varying degrees of success. The techniques used and some of the systems produced are listed in Table 1 (Ref 7-31).

The most successful composites to date have been produced by hot pressing and, more recently, by chemical vapor infiltration of fibrous preforms. Most of the published work on hot pressing describes the fabrication process and properties of fiber-reinforced ceramics, glass-ceramics, and glasses. Much of the work has been on fiber-reinforced glasses with relatively low softening points at about 600 °C (1110 °F), which precludes their classification as high-temperature materials. Fiber reinforcement of high-silica glasses and glass-ceramics capable of operating at temperatures up to 900 °C (1650 °F) in inert atmospheres extend the load-bearing temperature capability of the systems. Oxide and nitride ceramic matrix systems in inert atmospheres are limited essentially by the temperatures at which creep of the fibers and reaction between fibers and matrix result in degradation.

In principle, glass-ceramic systems offer the possibility of incorporating the matrix as a glass, hot pressing at relatively low temperatures, and then devitrifying at higher temperatures. Early work along these lines was disappointing, and this was attributed to volume changes during devitrification leading to porosity (Ref 15). However, recent publications have indicated that this route can be successful (Ref 24). Cold pressing and sintering have also been unsuccessful because the shrinkage of the matrix phase is resisted by the fibers, resulting in poor consolidation. Chemical vapor infiltration and deposition, which is much used for carbon-carbon composites, has recently been used successfully to develop both silicon carbide fiber-silicon carbide matrix (Ref 31) and carbon fiber-silicon carbide matrix composites.

Preparing materials for hot pressing involves mixing powdered matrix material and fibers or infiltrating matrix powders into fiber tows (or bundles), perhaps with an organic binder to give green strength, burning off the binder, and then hot pressing to consolidate (Ref 15). Composites in which short fibers are randomly aligned in the hot pressing plane can be obtained by simply stirring the powders and fibers in a mixer (Ref 13, 14). A maximum fiber content of about 30 vol% is achievable, above which clustering of fibers and increasing porosity result in poor properties. A degree of alignment can be obtained by techniques that shear the slurry of fibers and powder as it flows, such as extrusion and doctor-blade processes. Alternatively, prealigned fibers in the form of felt and paperlike structures can be impregnated (Ref 19). Continuous fibers can be impregnated by passing a tow through a slurry of powder and organic binder. The coated fibers can then be wound onto a form to produce a preimpregnated sheet (Ref 15). Electrophoretic processes for infiltrating bundles of electrically conducting fibers have also been studied (Ref 26). Higher fiber contents of about 50 vol% are obtainable with aligned, continuous fibers. A method to reduce anisotropy was described in the section "Principles of Fiber Reinforcement" in this article.

Characteristics of Ceramic Composite Systems

In assessing the development of a high-temperature fiber-reinforced ceramic, it is convenient to classify the systems that have been studied into three groups, according to the type of reinforcement: metal fibers and wires, carbon (including graphite) fibers, and ceramic fibers. Although much research has been conducted on glass matrix systems, particularly borosilicate glasses, they cannot properly be classified as high-temperature materials because of their low softening points; however, their study has provided much information that is useful in considering high-temperature combinations. In addition, they could be

Table 1 Manufactured fiber-reinforced ceramic composites and their corresponding processes (Ref 7)

Processing	Composite (fiber-matrix)	Comments
Hot pressing	W-glass, Ni-glass (Ref 8) Mo-thoria (Ref 9,10) Mo-alumina, W-ceramic (Ref 11) Stainless steel-alumina (Ref 12) C-glass (Ref 13-19) C-glass ceramic (Ref 12-15, 20) C-MgO (Ref 13,14) C-Al_2O_3 (Ref 13,14) ZrO_2-MgO (Ref 13) ZrO_2-ZrO_2 (Ref 21) SiC-glass (Ref 22, 23) SiC-glass-ceramics (Ref 24) Al_2O_3-glass (Ref 25) C-Si_3N_2 Ta-Si_3N_4	Fibers and matrix powder are mixed together and hot pressed to produce low-porosity composites, with uncracked matrices, if thermal expansion coefficients are matched. Aligned continuous fiber composites can have very high strengths.
Cold pressing and sintering	C-glass (Ref 15) Metal fiber-ceramic (Ref 11)	Fibers and matrix are mixed, cold pressed, and sintered. Disappointing results because the large shrinkage of the matrix during sintering produces cracked composites.
Devitrification	C-glass-ceramic (Ref 15) SiC-glass-ceramic (Ref 24)	Fibers and glass powder are hot pressed at relatively low temperatures to give a reinforced glass. Further high-temperature heat treatment is used to devitrify the glass to a glass-ceramic. The C-glass system gives disappointingly low strengths, probably because of volume changes during devitrification. The SiC glass-ceramic system is reported to have good properties.
Reaction bonding	Reinforced Si_3N_4 (Ref 26, 27)	Fibers are incorporated into flame-sprayed silicon that is subsequently reaction-sintered in nitrogen.
Slip-casting	Ceramic fiber-fused silica (Ref 28, 29)	Ceramic fibers are incorporated into slips of finely divided fused silica and fired. Increased porosity that is due to the presence of fiber usually results in a degradation of properties.
Plasma-spraying	Mo-Al_2O_3, W-Al_2O_3 (Ref 30)	Alumina powder is plasma-sprayed. Other unpublished work employing plasma-spraying is reported in France and Japan (1983). Processing is believed to be slow.
Chemical vapor infiltration and deposition	SiC fibers in SiC (Ref 31), C fibers in SiC	Fiber integrity can be preserved by lack of mechanical movement and relatively low process temperatures.

useful materials at low to intermediate temperatures.

The following sections highlight the important features of some of the more significant fiber-reinforced ceramic studies, rather than detail all the published work. More comprehensive bibliographies can be obtained in the reviews by D.C. Phillips (Ref 7), I.W. Donald and P. McMillan (Ref 32), and J.J. Krochmal (Ref 33). In particular, Ref 33 provides a review of the thermochemical stability of a range of potential reinforcements with technologically interesting matrix materials, although this is of somewhat academic interest because the primary thermochemical problem has been the oxidation of metal and carbon fibers at elevated temperatures in air.

Metal Reinforcement. The range of metals suitable for reinforcing ceramics is limited by the requirements of chemical compatibility between wires and matrix, mismatching the CTEs, and the need for a high ratio of E_f/E_m to maximize the strains at which matrix microcracking occurs.

All the successful metal-reinforced ceramics have been manufactured by high-temperature processing routes, and the reinforcing phase must be able to retain its integrity at fabrication and use temperatures. This restricts the reinforcement to refractory and other high-temperature metals; most work has been carried out with the metals molybdenum (Mo), tungsten (W), tantalum (Ta), and niobium (Nb). One disadvantage of the refractory metals is their

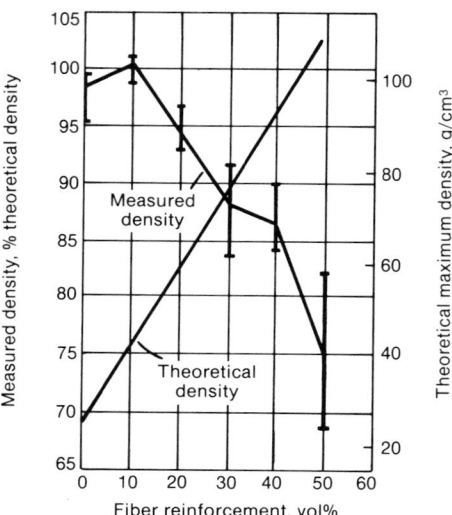

Fig. 4 Percent theoretical density versus vol% fiber reinforcement. Source: Ref 11

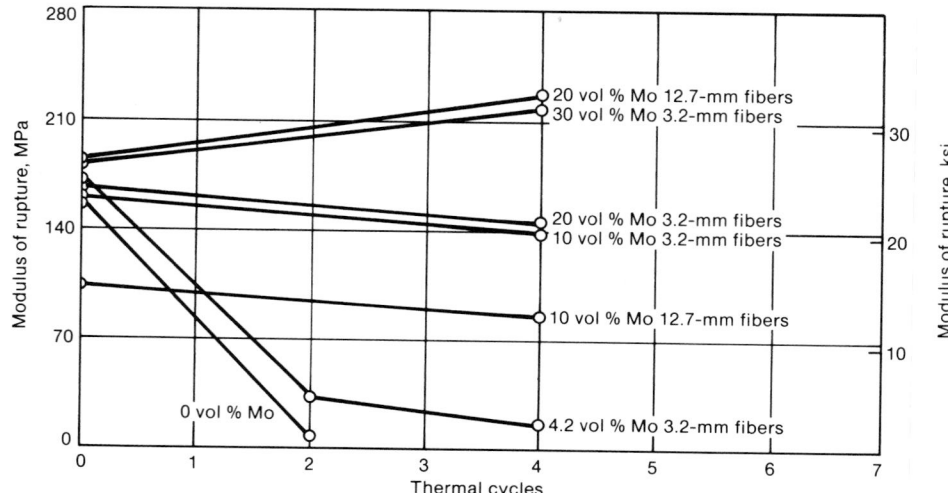

Fig. 6 Modulus of rupture as a function of thermal cycling for the body 712, 0.050-mm (0.002-in.) diam molybdenum fiber system. Source: Ref 11

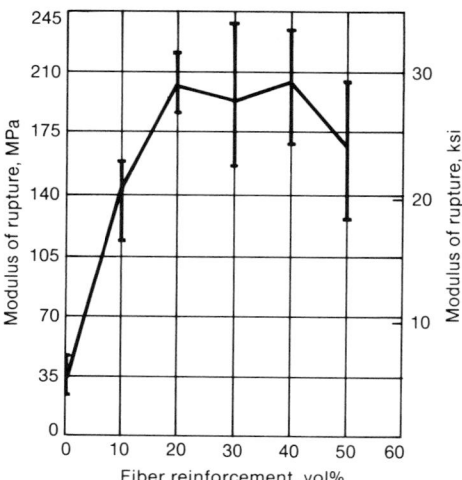

Fig. 5 Modulus of rupture versus vol% fiber reinforcement. Source: Ref 11

susceptibility to oxidation, although in short-term high-aeroheating applications, oxidation is of secondary importance. Silicon nitride (Si_3N_4) composites reinforced with Mo fibers in a three-dimensional array have shown promise.

There should be a match between the CTE of fibers and matrix. When the fibers have a lower CTE than the matrix, matrix cracking is the most severe effect and must be avoided. When the fibers have the greater CTE, the resulting debonding effect could possibly be alleviated either by creating interfacial layers between the fibers and matrix of intermediate CTE or by producing a chemical bond at the interface.

Another disadvantage of refractory metal reinforcement that is severe is its very high density, compared to carbon and ceramic fibers, resulting in composites with relatively poor specific strengths (σ_c/ρ) and stiffnesses (E/ρ). Coupled with poor oxidation resistance, this has resulted in diminished interest in the development of these materials.

Extensive studies of metal-reinforced ceramics carried out in the 1960s are summarized in J.R. Tinklepaugh (Ref 11) and J.J. Krochmal (Ref 33). All the work was carried out on short fibers with typical lengths of 3 mm (0.10 in.), 6 mm (0.25 in.), and 12 mm (6.50 in.) and typical diameters of 50 µm (2 mil) and 250 µm (10 mil) in hot-pressed ceramic matrix composites. Sintering was found to be unsatisfactory because matrix shrinkage led to cracking. As the fiber volume fraction increased, clustering of the fibers occurred; above 20 vol% a matrix density as high as 0.9 of the theoretical (fully compacted) density could not be achieved, as shown in Fig. 4. Accordingly, significant improvements in strength did not occur above about 20 vol% fiber reinforcement (Fig. 5).

A typical system studied was Mo wires in a kaolin/flint/feldspar matrix (designated 712). Recrystallization of the Mo and tungsten (W) wires at hot-pressing temperatures resulted in some loss of wire strength. An Mo/Al_2O_3 composite was cracked at virtually all fiber volume fractions because of thermal expansion mismatch, and strengths were lower than the unreinforced matrix materials. J.R. Tinklepaugh (Ref 11) studied the effects of thermal shock, and Fig. 6 is typical of his results. Specimens were heated at 1200 °C (2190 °F) and then placed on thick steel plates at room temperature. Unreinforced material was unable to survive this thermal cycle, but the fiber-reinforced ceramics, although they developed cracks, retained strength.

Exposure to air at high temperatures led to loss of fiber through oxidation. The effect was much more severe in cracked or porous composites than in the better-quality materials because of the protection afforded by the matrix of the latter. For example, at 1490 °C (2700 °F), uncracked, low-porosity composites containing 15 vol% of W and Mo wires that were protected by a layer of ceramic coating over the surface of the composite lost about 10% of their fibers in 5 h, while cracked or porous composites lost up to 90%. The effects of oxidation, of course, were dependent on the surface area of the wires. For example, exposure of Mo wire composites to 25 h at 1000 °C (1830 °F) led to 10% loss of fiber phase with 50 µm (2 mil) wires but only 1% loss with 150 µm (6 mil) wires (Ref 11).

D.G. Miller *et al.* (Ref 34) studied Mo and W in mullite. They obtained composites with strengths at room temperature that were significantly higher than the unreinforced matrix. However, oxidation resistance was very poor and exposure to air at 1200 °C (2190 °F) for 30 min resulted in severe deterioration. They concluded that the poor oxidation resistance cast doubt on the future of metal wire reinforced ceramics for elevated-temperature use. Y. Baskin *et al.* (Ref 9, 10) produced Mo-reinforced thoria. Thermal mismatch caused extensive cracking, and strengths and moduli were reduced below those of the unreinforced matrix. Severe oxidation occurred at the testing temperature of 1000 °C (1830 °F).

The general conclusion is that metal wire reinforced ceramics can be produced with higher strengths and moduli than unreinforced ceramics, provided that there is a CTE match. However, the matrices gave only limited protection against oxidation. Experiments showed that exposure to temperatures at and above 1000 °C (1830 °F) in air resulted in rapid loss of strength. It can be inferred that prolonged exposure at lower temperatures would have similar effects; therefore, maximum operating temperatures are probably in the range of 500 to 800 °C (930 to 1470 °F). Furthermore, the densities of the composites are very high. Therefore, interest in metal wire reinforcement is now limited to special short-term applications.

Carbon Fiber Reinforcement. Considerable work has been carried out on the development of carbon fiber reinforced glasses and glass-ceramics and, to a lesser extent, ceramics. The advantages of carbon fibers over metal wires include their low density (~1.8 g/cm³), chemical inertness in the absence of oxygen, high strengths and moduli, and ready availability. A major disadvantage is their susceptibility to oxidation, with rapid degradation of exposed fibers occurring at temperatures in excess of around 400 °C (750 °F) in air. This has so far prevented the development of a carbon fiber reinforced ceramic system that can be used at high temperatures in air. Carbon fibers are commercially available in a variety of types. Fibers that are typically about 8 μm (0.31 mil) in diameter are available in continuous lengths, or tows, of several thousand filaments and are anisotropic with low CTE in the axial direction. Typically, the axial CTE varies from -1×10^{-6}/K at ambient temperature to 0 at around 400 °C (750 °F), and becomes positive at higher temperatures, with an average CTE close to 0 in the range of ambient temperature to 1000 °C (1830 °F). The radial CTE is typically around 7×10^{-6}/K.

The earliest work with the development of short-fiber systems (Ref 13) mirrored that of metal wire reinforcement, but the availability of multifilament continuous tows led to the development of continuous impregnation techniques and the production of composites containing aligned, continuous fibers (Ref 14, 15, 20). These composites had much higher strengths and lower densities than did the metal wire composites. Their intrinsic elastic anisotropies could be reduced by laminating the same way that polymer matrix composites were laminated. Strength and stiffness properties of the aligned, continuous fiber composites approached those of the polymer matrix composites.

All the successful carbon fiber reinforced systems have been produced by hot pressing. Because of the low axial CTE of carbon fibers and the requirements for CTE matching, the more successful composites have been produced with low CTE matrices, such as borosilicate glass (CTE = 3.5×10^{-6}/K) and lithium aluminosilicate glass ceramics (CTE = 1.5×10^{-6}/K). Early composites of randomly oriented short carbon fibers in magnesia (CTE = 13.8×10^{-6}/K), alumina (CTE = 8.5×10^{-6}/K) and soda-lime glass (CTE = 8.9×10^{-6}/K) resulted in badly cracked materials (Ref 13).

By matching the CTEs, short-fiber composites with glass and glass-ceramic matrices have been produced in a number of ways. Randomly oriented fiber composites have been produced

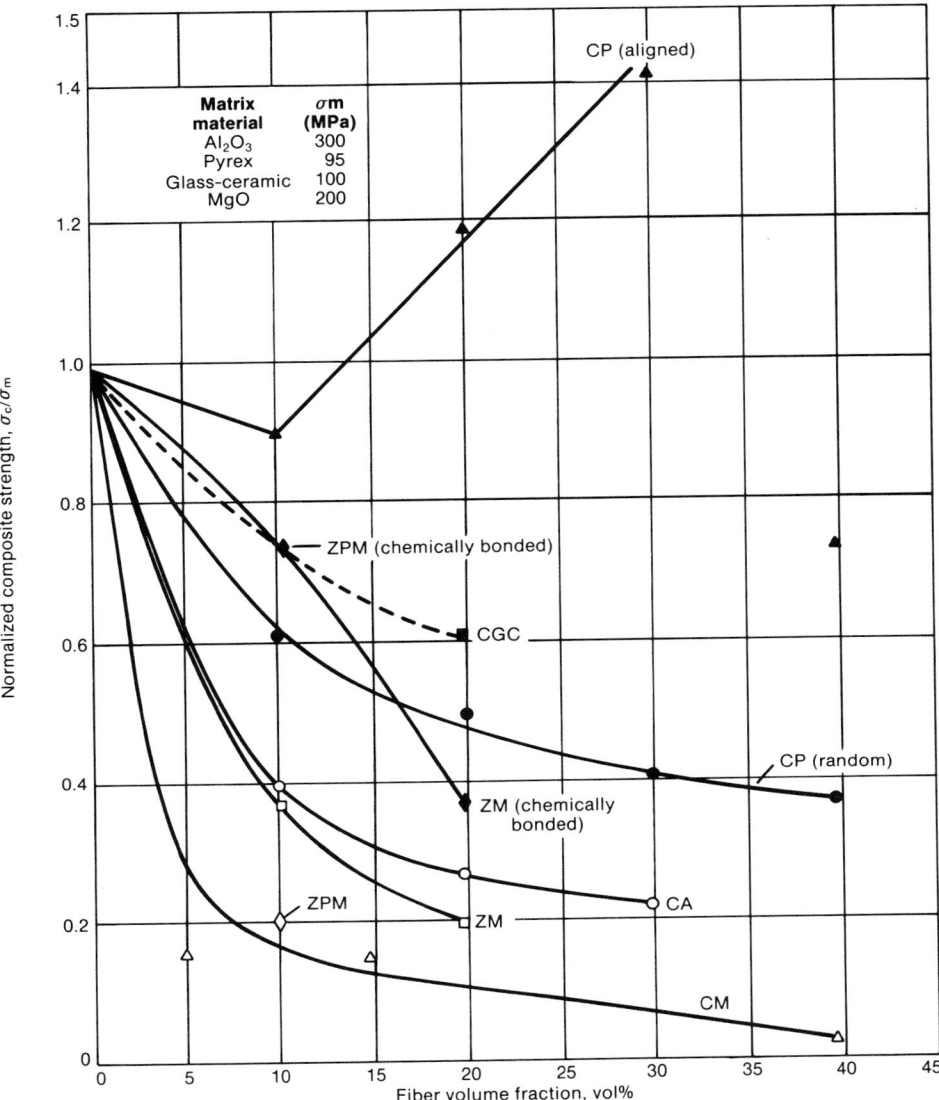

Fig. 7 Flexural strengths of some composites consisting of short carbon fibers in ceramic and glass matrices. Matrices are: CP, borosilicate glass (Pyrex); CGC, lithium aluminosilicate glass-ceramic; CA, alumina; CM, magnesia. ZPM and ZM are magnesia reinforced with zirconia powder and zirconia fibers, respectively. Source: Ref 13

by mixing chopped fibers (~3 mm, or 0.10 in. long) in a slurry of glass or glass-ceramic powder, and then hot pressing. A degree of alignment has been achieved by a doctor-blade process that involves passing the slurry containing an organic binder through a comb to produce a partially aligned, fiber-preimpregnated sheet. This sheet is then cut up into pieces that are stacked together, and then hot pressed. The random-fiber composites produced in this way had lower strengths than the matrix. Aligning the fibers, even partially, resulted in an increase in strength, as shown in Fig. 7, with maximum strength occurring at around 30 vol% of aligned fibers. Substantial increases in toughness were obtained from

approximately 3 J/m² (0.20 ft · lbf/ft²) for unreinforced glass to approximately 400 J/m² (25 ft · lbf/ft²) for composites containing 20 to 30 vol% of fiber (Ref 13). More recently, the availability of graphite fiber paper has resulted in the production of discontinuous fiber composites with interesting properties (Ref 19). The paper was impregnated with glass powder and hot pressed to produce composites with about 30 to 35 vol% reinforcement. The ultimate strengths of these composites were higher, at about 400 to 500 MPa (58 to 73 ksi), than in the earlier work. The onset of cracking occurred at much lower stresses of 50 to 100 MPa (7 to 15 ksi), but the composites retained an increasing load-bearing

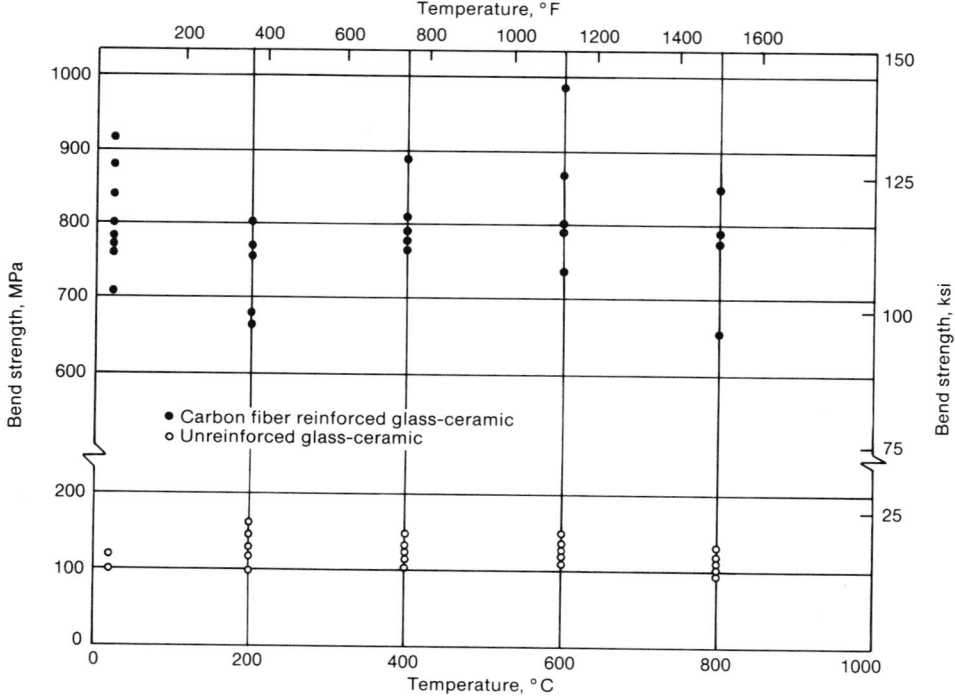

Fig. 8 Flexural strength of unreinforced and carbon fiber reinforced lithia alumina-silica glass-ceramic as a function of temperature in an inert environment. Source: Ref 15

capacity in flexure as strain increased, resulting in a high ultimate failure strain of approximately 1%.

Much higher-strength composites have been produced with aligned, continuous fibers. The fabrication procedure involves passing a tow of carbon fibers through a slurry of powdered matrix material mixed with an organic solvent and binder to impregnate it, and then winding the impregnated tow onto a form. The solvent is then evaporated to leave a handleable sheet that can be stacked in a die for hot pressing (Ref 14, 15, 20). Unidirec-

tional composites produced in this way (Ref 20) had flexural strengths of approximately 900 MPa (130 ksi) and toughnesses of approximately 300 J/m² (20 ft · lbf/ft²). The strengths of these composites vary with fiber volume fraction in agreement with the rule of mixtures, up to around 50 vol%. Because carbon fibers have a higher strain capability than the matrix materials, the composites exhibit matrix microcracking on loading at stresses lower than ultimate fiber strain.

In inert atmospheres, the strengths of the composites are retained to temperatures at

which the matrix becomes softened. Figure 8 shows data for a carbon fiber reinforced glass ceramic. In air, however, oxidation of the fibers begins at about 400 to 500 °C (750 to 930 °F), which causes a rapid loss of strength. Although the matrix provides some measure of protection, the occurrence of matrix micro-cracking under service conditions could lead to early oxidation.

Despite the fact that most work has been on glass and glass-ceramic matrices (Ref 13-20), recent development of a continuous carbon fiber reinforced silicon nitride composite produced by hot pressing has been reported (Ref 24). The relatively high CTE of Si_3N_4 (2.6 × 10^{-6}/K) resulted in much cracking. The strength of these composites at 454 MPa (66 ksi) was lower than those of unreinforced matrix at 473 MPa (69 ksi), but the toughness was much higher.

Carbon fibers produce composites of low specific density with high strengths that are retained to high temperatures in inert atmospheres. In oxidizing atmospheres, however, they degrade rapidly at temperatures above 500 °C (930 °F). Accordingly, for high-temperature applications, attention has begun to focus on oxidation protection for carbon and on ceramic fibers.

Ceramic Fiber Reinforcement. The development and increasing availability of ceramic fibers has improved the possibility of producing high-temperature ceramic matrix composite systems that can be used in air because ceramic fibers have better oxidation resistance in comparison with metal and carbon fibers. Promising fibers that have most often been used in the form of multifilament yarns are silicon carbide (SiC), produced by the Nippon Carbon Company Ltd., chemical vapor deposition SiC produced by Avco; and alumina, produced by E.I. Du Pont de Nemours and Company, Inc. Comparing SiC and alumina, the lower specific density and higher strength of

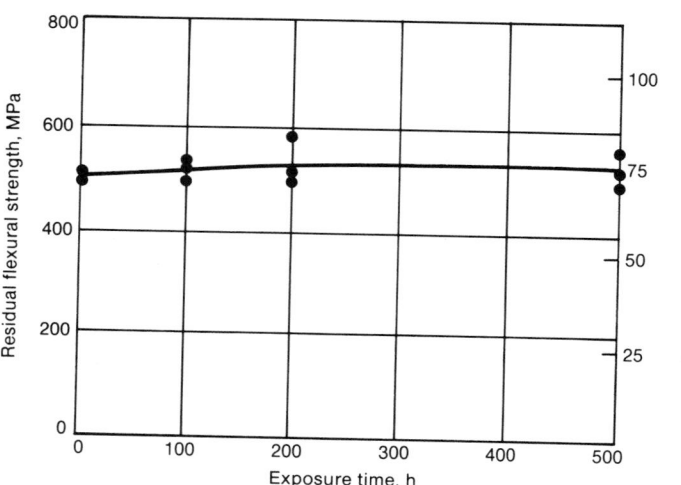

Fig. 9 Composite residual flexural strength at 22 °C (70 °F) after exposure to air at 540 °C (1000 °F). Source: Ref 23

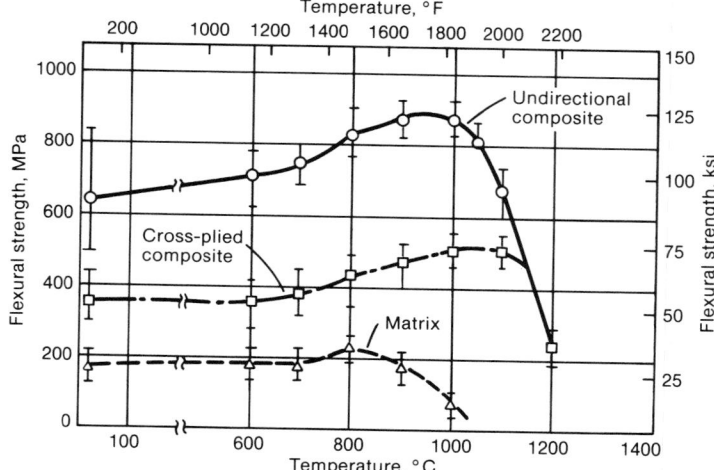

Fig. 10 Flexural strength (three-point) for lithia alumina-silica-SiC yarn composites. Source: Ref 23

SiC should result in composites with higher specific strengths, and its lower CTE should be advantageous in inhibiting matrix microcracking during composite fabrication.

Most development efforts have been with glass and glass-ceramic matrices, but there has been some limited work with true ceramic matrices. In the late 1960s, M.W. Lindley and D.J. Godfrey produced SiC fiber reinforced composites by a number of processes, including flame-spraying silicon powder onto the fibers, and slip-casting silicon powder, followed by nitriding the silicon (Ref 27). Composites produced by the former process had low-porosity matrices, although the processing presumably was slow, while the latter process resulted in higher-porosity matrices, but was presumably faster. Thermal expansion mismatch between the SiC fibers (4.5 to 5.0 × 10⁻⁶/K) and the Si_3N_4 matrix ($2.6 × 10^{-6}$/K) resulted in compressive stresses in the matrix and tensile stresses in the fibers. The SiC fibers available at that time consisted of a tungsten core with a SiC sheath and were relatively dense and expensive. Published data showed that aligned fiber composites with lower fiber volume fractions had modest room-temperature increases in strength over the unreinforced matrix, and a factor of two to three increase in fracture.

In the early 1970s, J. Aveston *et al.* (Ref 35) developed a continuous SiC fiber reinforced glass-ceramic with a strength of approximately 680 MPa (99 ksi) and toughness of approximately $2 × 10^4$ J/m² ($0.15 × 10^4$ ft · lbf/ft²). Creep rupture tests showed a rapid loss of strength at temperatures in excess of 1000 °C (1830 °F). J. Aveston attributed the loss of strength at temperature to thermal degradation of silicon carbide-tungsten fibers.

More recently, K.M Prewo *et al.* (Ref 22-25) have produced a range of glass and glass-ceramic matrix composites reinforced with silicon carbide and alumina fibers. Early strength results were disappointing, but subsequently the strength achieved with silicon carbide fibers began to approach the predictions of the law of mixtures. The typical strength for 50 vol% of SiC fiber at room temperature is approximately 800 MPa (116 ksi). Exposure of such composites to 540 °C (1000 °F) in air for 500 h results in no loss of strength, as shown in Fig. 9. In shorter-term tests, the strengths of glass-ceramic matrix composites increase up to 1000 to 1100 °C (1830 to 2010 °F), and rapidly decrease at higher temperatures (Ref 23, 24), as shown in Fig. 10. The increase in strength with increasing temperature is to be expected, partly because thermal mismatch stresses are relieved and partly because the matrix softens and reduces in brittleness.

The effect of high-temperature exposure on the strength of silicon carbide fibers has been studied by C.H. Anderson and R. Warren (Ref 36) and G. Simon and A.R. Bunsell (Ref 37, 38). Exposure to temperatures in excess of 1000 °C (1830 °F) results in a reduction in strength, partly because the β-SiC recrystallize

and partly because of the reactions of SiO_2 and carbon left in the fiber structure during fabrication.

J.J. Brennen and K.M. Prewo have measured the creep of SiC-glass-ceramic composites at elevated temperatures. They found that no creep occurred at temperatures up to about 900 °C (1650 °F) but that creep occurred at 1000 °C (1830 °F) and increased significantly at higher temperatures (Ref 24). This is partly attributable to creep of the matrix and partly due to creep of the fibers, which are also observed to show significant creep rates at temperatures above 1000 °C (1830 °F).

REFERENCES

1. D.H. Bowen, D.C Phillips, R.A.J. Sambell, and A. Briggs, in *Proceedings of the International Conference on Mechanical Properties of Materials*, Society of Metals, 1972, p 123
2. M. Hasilkorn, Specialty Materials, Textron, Inc., private communication, 1986
3. A. Kelly, *Strong Solids*, 2nd ed., Oxford University Press, 1973
4. D.C. Phillips and B. Harris, The Strength, Toughness and Fatigue Properties of Polymer Composites, in *Polymer Engineering Composites*, M.O.W. Richardson, Ed., Applied Science Publishers, 1977
5. B. Harris, J. Morely, and D.C. Phillips, Fibrous Composites, *J. Mater. Sci.*, Vol 10, 1975, p 2050-2061
6. R.J. Lee and D.C. Phillips, Fracture Toughness Testing of High Performance Laminates, in *Proceedings of the International Conference on Testing, Evaluation and Quality Control of Composites*, T. Feest, Ed., Butterworth Scientific, 1983
7. D.C. Phillips, Fiber Reinforced Ceramics, chapter 10 in *Fabrication of Composites*, Vol 4, *Handbook of Composite Materials*, A. Kelly and S.T. Mileiko, Ed., North Holland Publishing, 1983
8. G. Einmahl, "Metal Fiber Reinforced Glass Composites," M.S. Thesis, University of California, Berkeley, UCRL 16844, 1966
9. Y. Baskin, C.A. Arenburg, and J.H. Handwerk, Thoria Reinforced by Metal Fibers, *Bull. Am. Ceram. Soc.*, Vol 38, 1959, p 345
10. Y. Baskin, Y. Harada, and J.H. Handwerk, *Bull. Am. Ceram. Soc.*, Vol 43, 1960, p 489
11. J.R. Tinklepaugh, Ceramic Metal Fiber Composite Systems, in *Strengthening Mechanisms, Metals and Ceramics*, 12th Sagamore Army Materials Research Conference, J.J. Burke, N.L. Reed, and V. Weiss, Ed., Syracuse University Press, 1965
12. S.A. Bortz and S.L. Blum, A Metal Fiber Reinforced Ceramic System, in *Special Ceramics 3, Proceedings of the Fourth Symposium on Special Ceramics*, BCRA, 1967
13. R.A.J. Sambell, D.H. Bowen, and D.C. Phillips, Carbon Fiber Reinforced Ceramics, *J. Mater. Sci.*, Vol 7, 1972, p 663-675
14. R.A.J. Sambell, A. Briggs, D.C. Phillips, and D.H. Bowen, Fabrication of Fiber Ceramic Composites, *J. Mater. Sci.*, Vol 7, 1972, p 676-681
15. R.A.J. Sambell, D.C Phillips, and D.H. Bowen, The Technology of Carbon Fiber Reinforced Glasses and Ceramics, in *Carbon Fibers, Their Place in Modern Technology, Proceedings of the International Conference*, Plastics Institute, 1974
16. K.M. Prewo and J.F. Bacon, Glass Matrix Composites, 1, Graphite Fiber Reinforced Glass, in *Proceedings of the Second International Conference on Composite Materials*, B. Noton, R. Signorelli, K. Street, and L.N. Phillips, Ed., American Institute of Mining, Metallurgical, and Petroleum Engineers, 1978
17. K.M. Prewo, J.F. Bacon, and D.L. Dicus, Reinforced Glass Composites, *SAMPE Q.*, Vol 10 (No. 4), 1979, p 42-47
18. K.M. Prewo and E.R. Thompson, "Research on Graphite Reinforced Glass Matrix Composites," NASA Contractors Report 165711, May 1981
19. K.M. Prewo, Glass Matrix Composites, *J. Mater. Sci.*, Vol 17, 1982, p 3549-3563
20. S.R. Levitt, Fibrous Composites, *J. Mater. Sci.*, Vol 8, 1973, p 793-806
21. G.A. Graves, C.T. Lynch, and K.S. Mazdiyasni, Alkoxide Derived Zirconia Composites, *Bull. Am. Ceram. Soc.*, Vol 49, 1970, p 797
22. K.M. Prewo and J.J. Brennan, Silicon Carbide Fiber-Glass Matrix Composites, *J. Mater. Sci.*, Vol 15, 1980, p 463-468
23. K.M. Prewo and J.J. Brennan, Properties of Silicon Carbide Fiber-Glass Composites, *J. Mater. Sci.*, Vol 17, 1982, p 1201-1206
24. J.J. Brennan and K.M. Prewo, Carbon Fiber Reinforced Ceramics, *J. Mater. Sci.*, Vol 17, 1982, p 2371-2383
25. J.F. Bacon, K.M Prewo, and R.D. Veltri, Glass Matrix Composites, 2, Alumina Reinforced Glass, in *Proceedings of the Second International Conference on Composite Materials*, B. Noton, R. Signorelli, K. Street, and L.N. Phillips, Ed., American Institute of Mining, Metallurgical, and Petroleum Engineers, 1978
26. R.A.J. Sambell, Reinforced Silicon Nitride, *Composites*, Vol 1, Sept 1970, p 279-285
27. M.W. Lindley and D.J. Godfrey, Reaction Bonded Silicon Nitride, *Nature*, Vol 2296, 15 Jan 1971, p 192-193
28. W.J. Corbett, A.T. Sales, and J.D. Walton, Jr., "Improving the Properties of Slip-Cast Fused Silica by Fibrous Rein-

forcement,'' Special Report 66, Georgia Institute of Technology, 1965

29. W.J. Corbett and J.D. Walton, Jr., Silica Ceramics, in *Proceedings of the Conference on Nuclear Applications of Non-Fissionable Ceramics*, A. Boltax and J.H. Handwerk, Ed., American Nuclear Society Inc., 1966

30. M. Moss, W.L. Cyrus, and B.M. Schuster, Metal Fiber Reinforced Ceramics, *Bull. Am. Ceram. Soc.*, Vol 51, 1972, p 107

31. C.V. Burkland, Amercon Inc., private communication, 1985

32. I.W. Donald and P. McMillan, *J. Mater. Sci.*, Vol 11, 1976, p 949-972

33. J.J. Krochmal, ''Fiber Reinforced Ceramics: A Review and Assessment of Their Potential,'' AFML-TR-67-207, Air Force Materials Laboratory, 1967

34. D.G. Miller, R.H. Singleton, and A.V. Wallace, *Bull. Am. Ceram. Soc.*, Vol 45 (No. 5), 1966, p 513-517

35. J. Aveston, G.A. Cooper, and A. Kelly, Single and Multiple Fracture, in *The Properties of Fiber Composites, Proceedings of a Conference at the National Physical Laboratory*, IPC Science and Technology Press, Nov 1971

36. C.H. Anderson and R. Warren, The Strength and Elastic Modulus of Multifilament SiC Fibers, in *Advances in Composite Materials*, Vol 2, A.R. Bunsell *et al.*, Ed., Pergamon Press, 1980, p 1129

37. G. Simon and A.R. Bunsell, *Composites Rendus des Troisièmes Journées Nationales sur les Composites*, No. 3, Sept 1982, p 195

38. G. Simon and A.R. Bunsell, Elevated Temperature Strength of Silicon Carbide Fibers, *J. Mater. Sci.*, Letters 2, 1983, p 80-82

Multidirectionally Reinforced Ceramics

James P. Brazel, General Electric Company, Re-entry Systems Department

MULTIDIRECTIONALLY REINFORCED ceramic-ceramic composites represent a class of modern engineered materials composed of a three-dimensional preform of continuous ceramic fiber reinforcement that is densified with a ceramic matrix phase. The resulting composite offers significant property improvements in such traditional ceramic deficiencies as failure strain or fracture toughness, while retaining many of the fundamental performance advantages of the ceramic constituents, such as oxidation resistance, high temperature stability, and dielectric quality.

Most of the state-of-the-art materials described in this article have been developed primarily for military applications since the late 1960s, in particular, for hypersonic missile radar window applications. The primary objective of the development work has been to improve fracture toughness for such artificial environments as nuclear weapons effects, and dust and large-particle impact, as well as the natural hyperthermal environment associated with hypersonic flight aerodynamic heating and dynamic pressures, which cause thermal shock and/or aeropressure-induced failure of most conventional ceramics.

Fracture toughness, as measured in various ways to simulate the effects of these environments, has been vastly improved; in some measures, by orders of magnitude. However, mechanical strengths comparable to those of monolithic ceramics or to particulate or discontinuous fiber reinforced ceramic composites of the same chemical composition have not been achieved. This has inhibited the use of (and perhaps development work on) multidirectionally reinforced ceramic composites for high-temperature rotating machinery applications in which high strength and stiffness-to-weight ratios are required.

This article draws on and references only those sources of information not subject to military security or data export restrictions. Specific applications can be shown, however, in which the special properties and flexibility of this composite material offer superior performance advantages.

Multidirectional Continuous Fiber Reinforcement

Multidirectional continuous fiber reinforced composites use a three-dimensional preform array of continuous fiber reinforcement to give the composite its dominant mechanical property potential. The three-dimensional reinforcement has major components in all three mutually perpendicular directions in space, the x, y, and z of familiar Cartesian orthogonal space. This three dimensionality, however, may include multidirectional reinforcement designs other than the three-directional orthogonal. For example, a four-directional reinforcement design might be formed by the four directions along the body diagonals of a cube. Some multidirectional reinforcement designs are shown in Fig. 1. An alternate four-directional reinforcement design (Ref 1) might array three fiber reinforcements in plane, perhaps in an equilateral-triangle arrangement, with the fourth reinforcement aligned perpendicular to the plane of the triangularly arrayed fibers.

Higher order reinforcements include five-directional, one version of which has four fibers along the body diagonals of a rectangular prism, with a fifth arrayed in a direction perpendicular to one of the rectangular prism faces (one of the x, y, or z directions of the three-directional Cartesian orthogonal design). A typ-

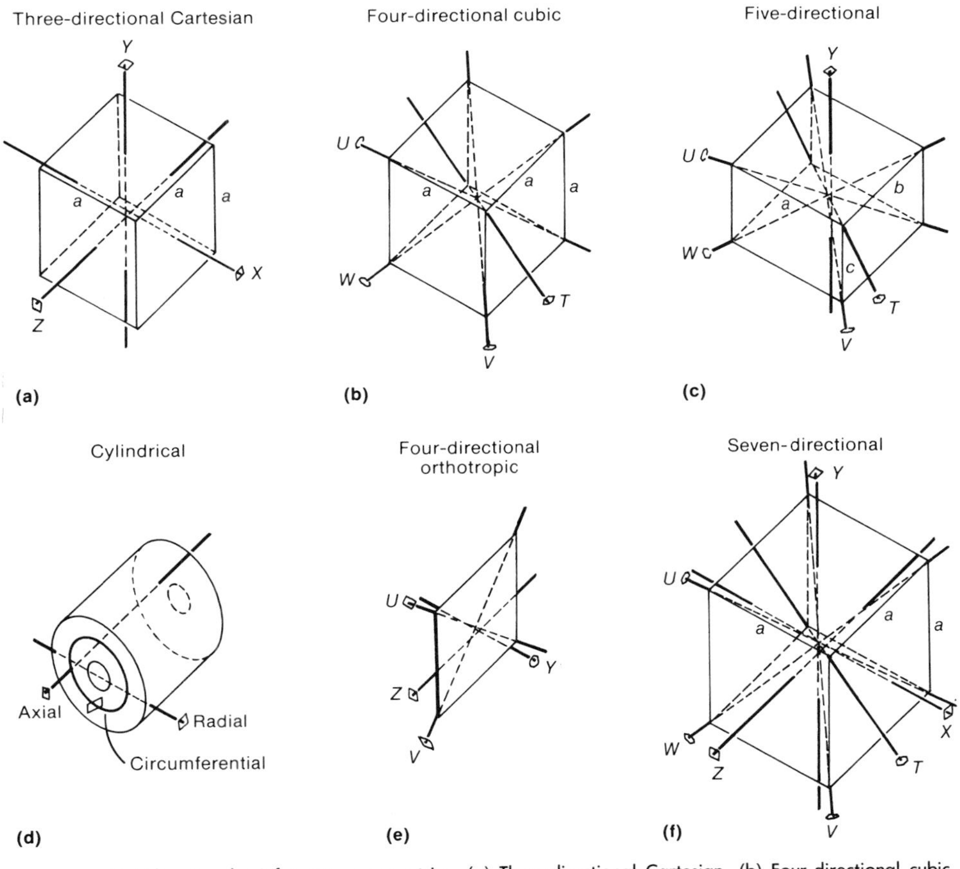

Fig. 1 Multidirectional reinforcement geometries. (a) Three-directional Cartesian. (b) Four-directional cubic. (c) Five-directional. (d) Cylindrical. (e) Four-directional orthotropic. (f) Seven-directional

(a)

(b)

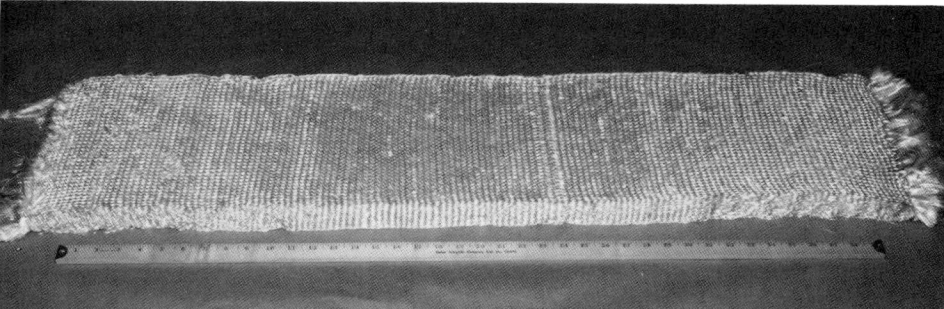

(c)

Fig. 2 Examples of continuous multidirectional fiber preforms in three reinforcement geometries. (a) Three-directional Cartesian orthogonal alumina. (b) Three-directional polar orthogonal hybrid preform of alumina (inside) and silica (outside) fibers. (c) Four-directional cubic silica "omniweave"

combined in layers and pierced by reinforcing interlaminar arrays of individual fibers to form the desired multidirectional design.

The final category is that of braided constructions. An array of fibers is simultaneously and repetitively manipulated to create a multidirectional reinforcement, usually a variation of four-directional cubic design.

These methods may be combined to achieve desired constructions. Some are more suitable than others for various geometries and fiber materials, especially when fiber damage is a concern. A common goal is to achieve a high packing density with few flaws or deviations from design geometry. Figure 2 shows representative multidirectional preforms. Figure 2(a) shows a three-directional orthogonal preform of alumina fibers, while Fig. 2(b) shows a three-directional polar geometry hybrid cylinder preform combining alumina axial and hoop fibers on the inside, silica axial and hoop fibers on the outside, with silica radial fibers running from inside to outside. In Fig. 2(c) a four-directional cubic geometry preform woven by a multidirectional braiding process is shown.

Extensive literature exists for multidirectional preform designs and composite materials based on them. Most of it is for carbon-carbon composites for intercontinental ballistic missile (ICBM) nose tips and rocket nozzle throats, based on continuous graphite fiber preforms. (See the article "Multidirectionally Reinforced Carbon/Graphite Matrix Composites" in this Volume.) These carbon-carbon preform fabrication and composite material design concepts relate well to the subject of this article.

Composite Formulation and Processing

Fiber Choice and Multidirectional Preform Characterization. The choice of ceramic fibers for the multidirectional composites developed to date has been guided primarily by the ruling radar electromagnetic transparency property required by the applications for which development work has been completed. This property is best expressed by loss tangent measurements on densified composites based on the fiber. A benchmark goal of 0.01 across the range of application temperatures has been accepted for this property, but design envelopes considering such application parameters as radome wall thicknesses, maximum temperature, temperature distribution, and radiated power and antenna pattern requirements result in a wide range of allowed deviation from this value (Ref 2). Fiber dielectric data are, however, usually expressed in terms of dc electrical resistivity. Room-temperature values in the range of $10^{10}\ \Omega \cdot cm$ (alumina) to $10^{16}\ \Omega \cdot cm$ (silica) are reported for the fibers used in the radar window materials. Table 1 lists some characteristic properties of these ceramic fibers and preforms woven from them. The continuous filament with which most of the silica-silica composite work has been done has a nominal breaking strength

ical six-directional design would add another of the Cartesian directions. The transition to a seven-directional design can be made with the superposition of three-directional Cartesian orthogonal and the four-directional cubic designs. Multidirectional reinforcement designs containing as many as 13 directions have been reported (Ref 1).

Directionality alone does not specify multidirectional reinforcement design. The relative number of fibers or individual filaments to be arrayed in each direction must be known. In the simplest, or balanced designs, the same number of fibers is used in each direction. A 1-1-1 ratio three-directional orthogonal preform would contain the same reinforcement ratio in each principal reinforcement direction. A cubic four-directional design would have a 1-1-1-1 ratio. A 2-2-3 three-directional orthogonal reinforcement design would have three fibers arrayed in the z direction for every two in the x and y directions.

The absolute values of the arrayed or woven preform fiber dimensions must be added to these considerations of fiber reinforcement directions and/or ratios so that unit cell dimensions and the respective fiber volume fractions can be computed. Fiber center-to-center spac-

ings of from 0.3 to 3 mm (0.010 to 0.10 in.) have been made in preforms with total fiber volume fractions in the 0.30 to 0.55% range and somewhat higher. The fiber reinforcement fraction in any principal reinforcement direction must be calculated from the dimensions of the unit cell. For a three-directional orthogonally balanced 1-1-1 weave of 0.51 vol%, the fiber reinforcement fraction in any of these directions would be 0.17.

Methods of fabricating multidirectional reinforcements have remained highly proprietary, and in some cases, subject to defense security and export control regulations. They may be grouped for purposes of this review into categories. The first is the continuous fiber placement method. Layers of fibers are sequentially arrayed in each reinforcement direction. The fibers may be either dry or rigidized before placement to facilitate handling, but are not moved after placement. The process may include interlocking of fiber, as in conventional two-directional weaving, but more often for multidirectionally reinforced composites, the fibers are arrayed as straight elements with no interlocking.

The second is the pierced fabric method. Previously woven two-directional fabric is

Table 1   Formulation and fabrication parameters of multidirectional continuous fiber reinforced ceramic-ceramic composites

Property/composite	Three-directional silica-silica	Four-directional silica-silica	Three-directional alumina-alumina	Three-directional alumina-silica	Three-directional boron nitride/boron nitride
Preform reinforcement type	Three-directional orthogonal	Four-directional cubic braided	Three-directional orthogonal	Three-directional orthogonal	Three-directional orthogonal
Fiber type	Fused quartz continuous	Fused quartz continuous	Polycrystalline alumina staple	Polycrystalline alumina staple	Continuous polycrystalline boron nitride
Fiber electrical resistivity, ohm · m	1×10^{16} (at 293 K) 2×10^{5} (at 1073 K)	1×10^{16} (at 293 K) 2×10^{5} (at 1073 K)	1×10^{10} (estimated)	1×10^{10} (estimated)	1×10^{8} (at 1075 K)
Fiber strength, GPa (10^6 psi)	0.7 (0.10)	0.7 (0.10)	1.4 (0.20)	1.4 (0.20)	0.35-0.7 (0.05-0.10)
Fiber volume fraction	50%	50%	30%	30%	40%
Matrix/densification	Colloidal silica	Colloidal silica	Colloidal alumina	Collodial silica	Nitrided boric oxide
Process or stabilization temperature, K	923-1013	923	110	923	2073
Composite bulk density, g/cm³	1.6	1.6	1.9	2.0	1.6

of 876 MPa (130 ksi), while one type of staple alumina fiber has three times this strength. In contrast, one type of staple boron nitride fiber has a rather low strength of 414 MPa (60 ksi). Nevertheless, considerable development work has been performed on a three-directional boron nitride-boron nitride composite because of the excellent ablation performance and shock resistance of this material.

State-of-the-art alumina-boria-silica and E-glass dielectric fibers have also been considered in recent years for multidirectional ceramic composite work, but little open literature data is as yet available.

For applications not requiring dielectric behavior, a much wider range of ceramic fibers is available. It is likely that higher-strength multidirectional composites will result from this wider class of compounds. Significant work has been carried out on one-directional and two-directional composites using alumina-boria-silica and various silicon carbide fibers with a dense borosilicate glass matrix (Ref 3). High flexural strengths have been achieved, on the order of 20 times that of the antenna-window-type multidirectional ceramic composites. These material combinations could be extended to multidirectional composites if predicted mechanical properties were found attractive despite the increased cost of the three-dimensional reinforcement component. The increased manufacturing cost-and-complexity factor has undoubtedly slowed adaptation of multidirectional composite technology for structural ceramic applications. However, in arriving at estimates for properties, in particular, mechanical strength, the materials scientist should be aware of two less familiar characteristics of the multidirectional preform technology now available and the implications for designing a ceramic-ceramic composite.

The first is the aspect of interfiber matrix thickness and strength. The crack sensitivity of ceramic matrix composites is a strong function of unreinforced ceramic matrix thickness. For two-directional continuous fiber reinforcement, this is primarily a cross-ply phenomenon. Con-

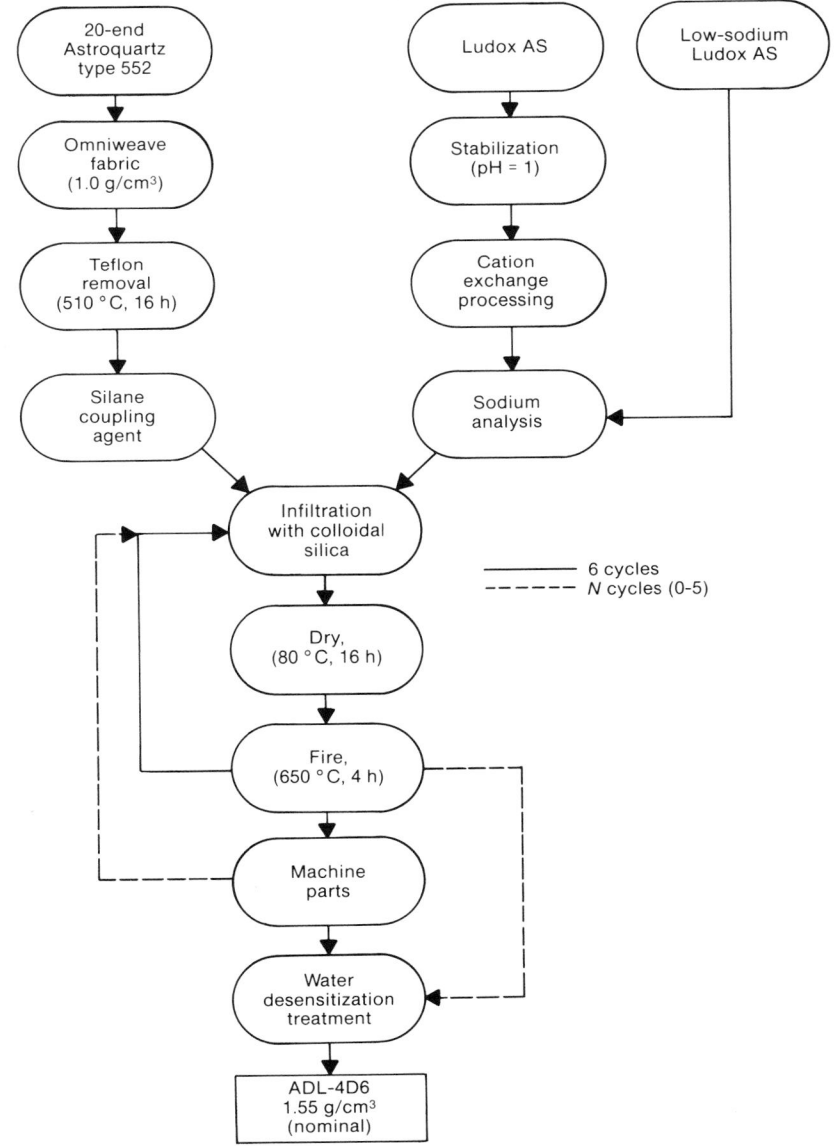

Fig. 3 Densification process for ADL-4D6 four-directional silica-silica composite

tinuous fiber multidirectional reinforcement can, in effect, eliminate this problem on a macrocracking basis, if a suitable densification process is used for introducing the desired ceramic matrix. When a fine-weave texture (close fiber spacing) of high fiber volume fraction is used, the range of macrocrack propagation is further reduced.

The second characteristic is the dominant contribution of preform versus multidirectional composite strength and toughness. For these materials, composite mechanical properties are primarily determined by the preform reinforcement fiber properties. For composite formulation purposes, this means that fiber choice and multidirectional preform design should be the first considerations, and then a compatible densification approach should be chosen to satisfy the remaining performance property criteria, such as melting/softening temperature and dielectric behavior. In particular, the composite mechanical strength may reach a maximum at low values of matrix density and high composite porosity. This contradicts orthodox composite theory and some general theories for ceramic matrix composites (Ref 4) but is common to all highly developed multidirectional composites, including carbon-carbon, for which data bases have been established. The following discussion will show a negative implication, that is, rule-of-mixture composite strengths have not nearly been achieved.

Densification Process. The matrix densification processes that have been developed for the majority of the continuous fiber multidirectionally reinforced ceramic-ceramic composites described in this article have been determined by the primary radar antenna window applications. Starting with the ceramic fiber preform considerations described above, what these processes have in common, in their published final form, is that they:

* Are capable of uniformly infiltrating thick-sectioned preforms (an inch or more) with no practical limitation on lateral extent of preform size
* Require repeated infiltration cycles to reach final density
* Are usually fired at temperatures well below the stable temperature of the fiber component
* Are processed with a goal of maximum strength and/or fracture toughness

A corollary of the last consideration is that other desirable properties, such as dielectric quality (lowest loss tangent) or maximum bulk density for ablation performance, have not been maximized.

The silica-silica composite densification processes use aqueous suspensions of high-purity, colloidal silica. Figure 3 shows the process used to produce a silica-silica composite that is based on a four-directional cubic geometry preform (Table 1).

Teflon-coated silica fiber is used for the Omniweave braiding process to reduce fiber

(a)   **(b)**

Fig. 4 Photomicrographs of polished ADL-4D6 four-directional silica-silica composite flexure bar cross sections. (a) Polished cross section (10 × 10 mm, or 0.4 × 0.4 in.). (b) Scanning electron photomicrograph of fracture surface

damage that is due to the repeated motions of the unbraided remainder of the fiber. Upon completion of weaving, the coating is burned off. In the particular process shown, a silane coupling agent was used to increase colloidal sol adhesion in the first infiltration cycle. The colloidal sol is further purified for removal of alkali and alkaline earth metals. Repeated infiltrations are carried out, followed by drying and firing at 923 K. For the silica-silica composite, six cycles were nominally used. For other versions of four-directional braids and three-directional preforms based on other weaving methods, higher preform densities have been achieved, in some cases, to theoretical packing density. A greater number of infiltrations, perhaps ten, is required for these tighter preforms, but higher bulk density and mechanical properties are also achieved. Along with the details of multidirectional preform weaving methods, one of the more proprietary aspects of these antenna window materials has been the final treatment used to ensure acceptable dielectric properties. The problem arises from atmospheric water absorption by the highly porous and permeable matrix. A survey of alternate variations on colloidal, thin-silica densification processes is described in Ref 5. Other multidirectional composites of this generic silica-silica formulation have been described (Ref 6).

For all of these materials, machining of parts is easily achieved using carbide-tipped tools for blanking or diamond-tipped tools for finer finishing operations. Figure 4(a) shows a polished section of a four-directional silica-silica composite. A scanning electron photomicrograph of a fracture surface taken from this same-sized

specimen is shown in Fig. 4(b). Note the ideal composite fracture surface, with individual filaments pulled out from the matrix, in contrast to the smooth fracture surface found for monolithic ceramics.

An analogous aqueous sol densification process has been developed for a multidirectional composite of high-purity alumina-alumina (Ref 7). The staple fiber alumina is considerably more difficult to weave than is continuous filament silica but development work on various fiber sizings and handling methods has made it possible to produce medium-density preforms without significant fiber damage (Table 1).

A variation of the colloidal silica densification process shown in Fig. 3 has also been used to densify alumina fiber preforms. It will be seen that although high-temperature stiffness and strength are sacrificed, the room- and intermediate-temperature mechanicals are essentially the same as they are for alumina-alumina (the multidirectional preform dominates the mechanical properties). An advantage of the silica matrix is the lower composite loss tangent, at least at this stage of development.

An even greater advantage is in the freedom to densify a hybrid weave preform uniformly. In Ref 8, the design and development of composite structures based on cylindrical and conical frustum-shaped polar weave preforms is described. These use alumina hoop and axial fibers integrally woven inside a silica fiber outside preform of hoops, axials, and radials. A preform of this design is shown as Fig. 2(b). The silica radials tie both zones together in a three-directional hybrid alumina-silica preform

Table 2 **Performance properties of multidirectional continuous fiber reinforced ceramic-ceramic composite materials at 300 K**

Property/material	Three-directional silica-silica (Ref 6, 11)	Four-directional silica-silica (Ref 12)	Three-directional alumina-alumina	Three-directional alumina-silica (Ref 8)	Three-directional boron nitride-boron nitride (Ref 9)
Tensile					
Young's modulus, GPa (10^6 psi)	15.6 (2.26)	9.7-13.1(b) (1.4-1.90)	36.3 (5.26)	33.8 (4.90)	15.4 (2.23)
Ultimate strength, MPa (ksi)	26.7 (3.87)	20.4-26.6(b) (2.96-3.86)	71.1 (10.3)	74.8 (10.8)	24.8 (3.60)
Ultimate strain, %	0.2	1.2	0.2	0.2	0.2
Compressive					
Modulus, GPa (10^6 psi)	21.9 (3.18)	8.6 (1.3)	31.4 (4.55)	. . . . . .	29.2 (4.23)
Ultimate strength, MPa (ksi)	144.8 (21.0)	70.6 (10.2)	224.7 (32.59)	. . . . . .	36.5 (5.29)
Ultimate strain, %	1.6	1.5	>0.6	. . .	0.2
Shear					
Modulus, GPa (10^6 psi)	1.5 (0.22)	4.4 (0.64)	3.4 (0.49)	1.7 (0.25)	. . .
Poisson's ratio	0.09	. . . . . .	. . . . . .	. . . . . .	. . .
Coefficient of thermal expansion, mean, to 600 K, 10^{-6}/K	0.54	0.47	6.4	6.4	2.7
Thermal conductivity 300 K, W/m · K (BTU · in./h · ft^2 · °F)	0.66 (4.6)	0.65 (4.5)	1.62 (11.2)	0.68 (4.7)	9.0 (62.4)
Dielectric constant(a)	3.5	2.9	3.7	3.8(c)	3.0
Loss tangent(a)	0.0009	0.001	0.0045	0.004(c)	0.002

(a) At 8.5 to 9.4 GHz, or as indicated. (b) Lower value for "water desensitized" version corresponding to dielectric properties. (c) At 35 GHz

whose performance properties are described below.

A three-directional boron nitride-boron nitride (BN-BN) composite has been developed for missile antenna window applications requiring the improved intrinsic ablation resistance of BN compared to silica and alumina (Ref 11). Early development work on the continuous polycrystalline fiber produced lots with breaking strengths of 0.35 to 0.70 GPa (0.05 to 0.10 $\times$ 10^6 psi). In subsequent scale-up work, difficulty has been experienced in reproducing this strength level. The working values listed in Table 1 are only 5% of this potential level.

Medium-density preforms have been woven (Table 1). The densification process uses repeated aqueous boric oxide precursor infiltrations, followed by vacuum drying and precipitation, and conversion in anhydrous ammonia to the sintered BN matrix. After densification, the composite is stabilized at 1700 °C (3090 °F) to improve dielectric properties. The composite bulk density is further increased from 1.5 to 1.6 g/cm³ by a hot pressing at 1800 °C (3270 °F), which scissors x and y fibers about the primary z reinforcement direction to maximize ablation resistance performance.

Development work on this promising concept has been reduced, however, by the discovery of high microwave transmission loss characteristics near the sublimation temperatures encountered during ICBM reentry (Ref 10).

Composite Performance Properties

The performance properties of the materials are summarized in Table 2. The sources of the data, where additional processing information, properties, and descriptions of applications may be found, are also listed in Table 2.

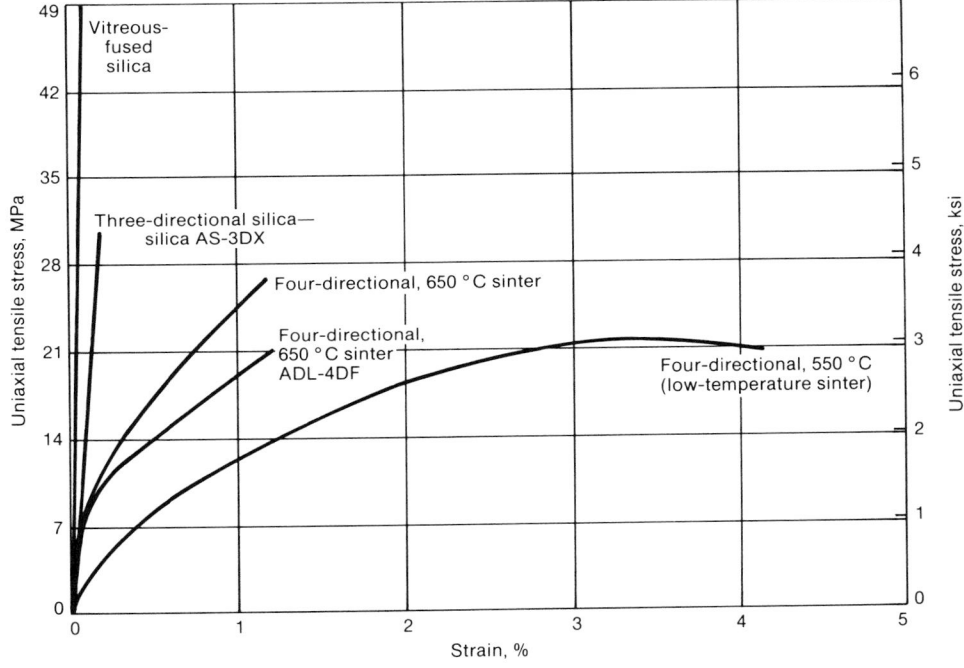

Fig. 5 Tensile stress-strain curve for three-directional and four-directional silica-silica composites

The three-directional and four-directional silica-silica composites are by far the highest developed and characterized of these materials. The various levels of classified and export-controlled documentation contain detailed information on processing, high-temperature performance properties (mechanical, thermal, microwave transmission, and shock/impact). For applications above the process or stabilization temperature noted in Table 1, the mechanical properties in particular are not stable. This is a consequence of the incomplete sintering densification process, which has been found to yield the maximized composite strength and impact resistance properties necessary for missile applications as discussed earlier in this article.

The three-directional and four-directional silica composite mechanical property data shown essentially reflect the behavior of the two reinforcement designs, normalized to similar constituents and densification processes. The three-directional (Cartesian) geometry has slightly higher tensile modulus, but equivalent tensile

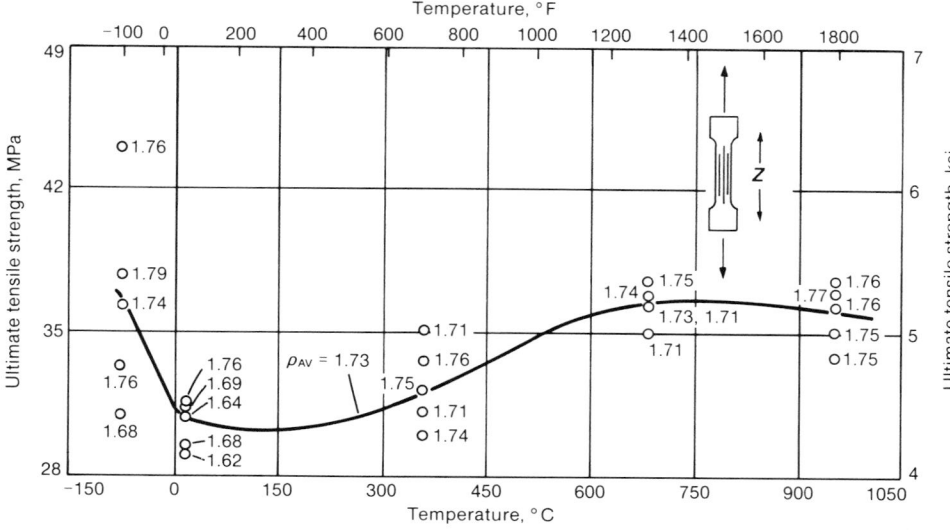

Fig. 6 Temperature dependence of the ultimate tensile strength of AS-3DX three-directional silica-silica (density of specimen blanks indicated as a parameter, measured in the z-direction in g/cm³). Source: Ref 11

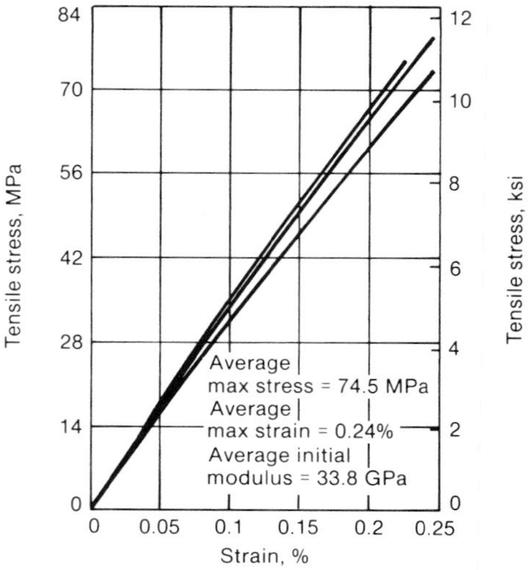

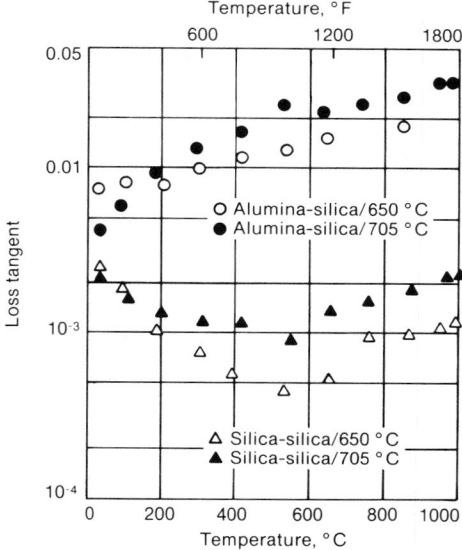

Fig. 7 Tensile stress-strain curves for three-directional alumina fiber-silica matrix composite

strengths of 26.7 MPa (3.9 ksi) are reported. Significant differences occur in strain to failure, with all the three-directional composites failing at nominal 0.2%, while the four-directional goes to a value in excess of 1%. This higher strain behavior for four-directional has also been observed for multidirectional carbon-carbon composites (Ref 2).

The tensile stress-strain curve shown in Fig. 5 illustrates this behavior. Both composites have initial linear elastic behavior. Because of the alignment of a principal fiber reinforcement direction with applied uniaxial stress, the three-directional material shows little departure from elastic behavior. The four-directional material, with all four principal reinforcements aligned as diagonals in the cubic space frame, departs increasingly from linearity as loaded fibers shear through the relatively weak fiber-matrix bonding.

The effect of sintering temperature process variation is also shown in Fig. 5. A four-directional configuration processed at 550 °C (1020 °F) (in contrast to the standard 650 °C, or 1200 °F) has 4% strain to failure, but lower ultimate strength. As shown in Table 2, the three-directional configuration has higher compressive modulus: 21.9 GPa (3.2 × 10⁶ psi) versus 8.6 GPa (1.2 × 10⁶ psi) for the four-directional, which also is lower than the tensile modulus of the four-directional configuration. This behavior is also related to the scissoring action of the four-directional specimen, as stress is applied to the cubic cell face (see Fig. 1). This is a property that has been predicted for various four-directional configurations (Ref 1) and must be traded off in design analyses against other mechanical property advantages, such as strength and modulus of the high principal stress direction.

These are not high-strength composites, even in the context of their use. Accordingly, they

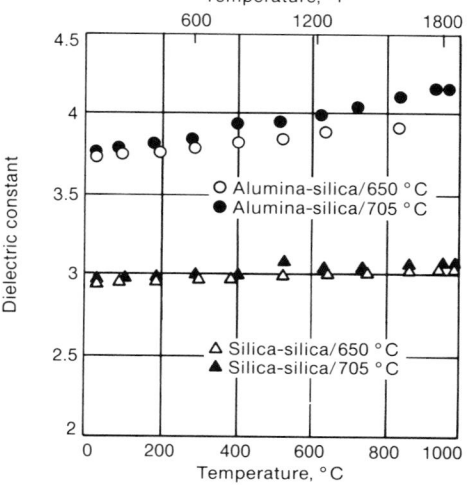

Fig. 8 Temperature dependence of the dielectric constant at 35 GHz of silica-silica and alumina-silica

have been applied to nonstructural but refractory applications where their other performance advantages prevail over the monolithic ceramic predecessors.

In particular, the thermal expansion and thermal conductivity of these silica composites, combined with the improved strain capability, make them even more resistant to thermal shock and efficient as thermal insulators than are the slip-cast and vitreous-fused silica materials they were primarily intended to replace. The temperature dependence of the tensile strength of three-directional silica-silica (Ref 11) is shown in Fig. 6.

The two three-directional alumina-reinforced composites listed in Table 2 show virtually indistinguishable tensile properties. A higher shear modulus (by a factor of two) was mea-

Fig. 9 Temperature dependence of the loss tangent at 35 GHz of silica-silica and alumina-silica

sured for the alumina matrix version: 3.4 GPa (0.5 × 10⁶ psi) versus 1.7 GPa (0.2 × 10⁶ psi). This is probably related to higher fiber-matrix bond strength in the alumina-alumina composite.

These alumina-reinforced multidirectional composites have 2½ to 3 times the tensile modulus of the referenced silica-silica composites, and 3 times the tensile strength. The coefficient of thermal expansion, however, reflects intrinsic alumina ceramic behavior and is an order of magnitude greater than silica. The thermal conductivity is significantly lower than it is for polycrystalline alumina for both ver-

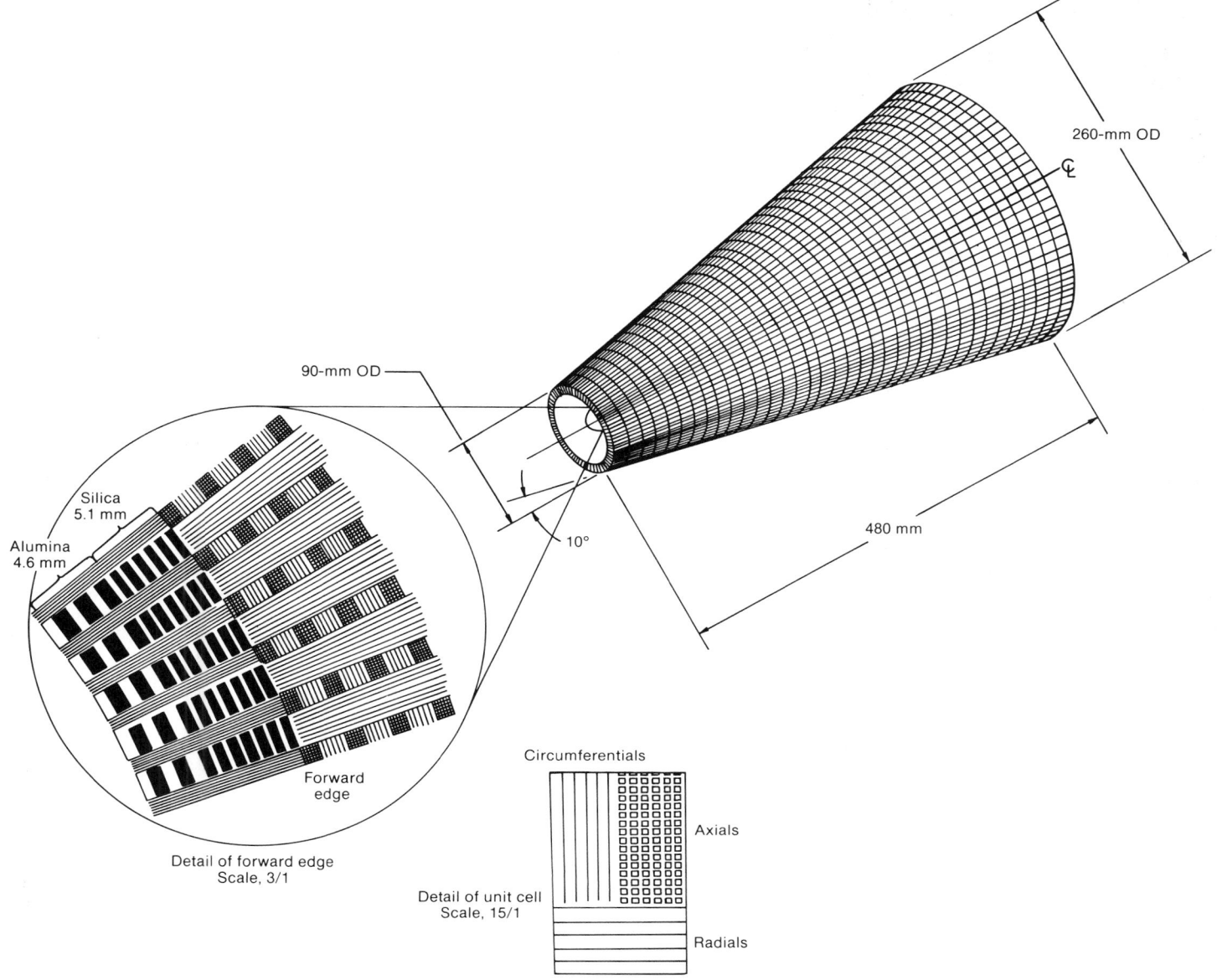

Fig. 10 Design of an alumina-silica hybrid multidirectional composite radome for ballistic missile interceptor seeker radar. OD, outside diameter

sions, and for alumina-silica it is the same as for the silica-silica.

A tensile stress-strain curve for three-directional Cartesian alumina-silica is shown in Fig. 7. This composite used a 1-1-1 balanced weave at a nominal 0.30 fiber volume fraction. The dielectric constant and loss tangent variation with temperature for two versions each of three-directional alumina-silica and silica-silica (Ref 8) are shown in Fig. 8 and 9. Basically, the dielectric constants reflect rule-of-mixture behavior for the constituents, but the loss tangent data shows the effect of atmospheric water absorption by the matrix and the effectiveness of various processes to reduce this effect. Without water desensitization,

room-temperature loss tangents rise above acceptable levels (Ref 6).

Application Examples

A recent design activity illustrates the advantages of multidirectional ceramic composite use and adaptability (Ref 8). A large radome structure in the configuration of a conical frustum shell with a 48-cm (19-in.) length and base outer diameter of 26 cm (10.20 in.), with a cone half-angle of 10°, was required for ballistic missile defense interceptor seeker radar (Fig. 10). The baseline design had called for use of hot-pressed silicon nitride because of the high strength and micro-

wave transparency of this structural ceramic. However, the results of aerothermostructural performance analyses predicted severe thermal stress fracture, especially when Weibull model fracture statistics were applied to impute an allowable tensile stress.

Similar analysis of the performance of the multidirectional silica-silica composites predicted large margins of safety for thermal stress, but unacceptably large radome wall thicknesses would be required to withstand the dynamic aeropressure-induced stress due to the low strength of the silica composite. Alternatively, the alumina-alumina composite originally developed to attain higher strength was predicted to survive the aeropressure-induced

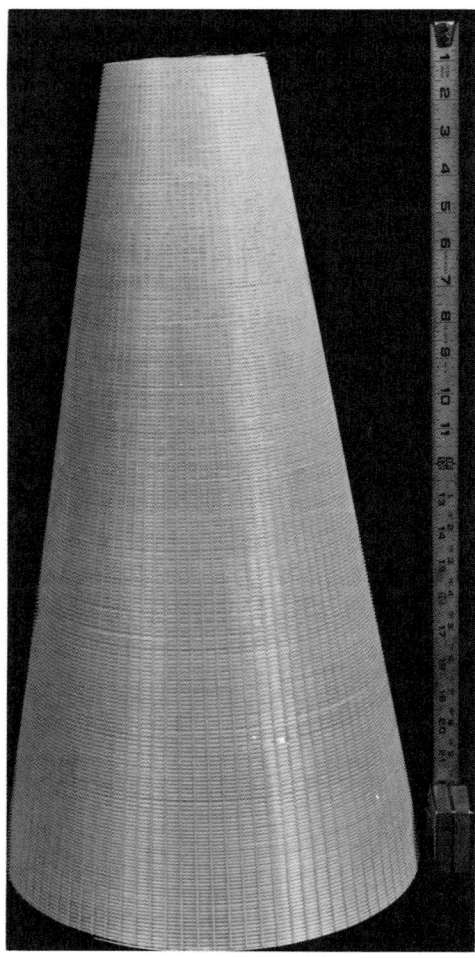

Fig. 11 Full scale alumina-silica hybrid radome multidirectional continuous fiber preform

stress with an acceptable wall thickness, but to fail in thermal stress because of the high alumina thermal expansion.

The hybrid wall design alumina-silica composite radome shown in Fig. 10 was developed as a solution to this problem. The outer 5-mm (0.20-in.) zone of three-directional silica-silica serves as the thermal protection portion of the hybrid wall, with thermal stresses not exceeding 4.8 MPa (0.7 ksi) across a range of tra-

jectory conditions. The alumina 4.5-mm (0.18-in.) thick hoop and axial fiber-silica matrix inner zone remains practically at prelaunch temperature throughout the mission, and its thermal stress rises to a maximum of 18 MPa (2.6 ksi), compared to a 3% statistical probability of failure stress allowable of 69 MPa (9.9 ksi). The higher-modulus and strength alumina-reinforced zone thus provides most of the strength to resist the aeropressure-induced stress throughout the high aerodynamic heating missile trajectory of an approximately 10-s duration. Figure 11 shows the woven hybrid preform (an orthotropic polar geometry) before densification.

Note that this hybrid wall is an application-specific solution. Steady-state high-temperature operation would result in thermal stress fracture of the structure due to the order-of-magnitude difference in thermal expansion of the components. As for composite materials in general, selection of material formulation and component design are best considered together in an integrated development approach.

ACKNOWLEDGMENT

The author wishes to thank his colleague, Richard Fenton, for his assistance in gathering the performance property and processing data for this article.

REFERENCES

1. E.R. Stover, Four Directional Structure for Reinforcement, U.S. Patent 4,400,421, Aug 1983
2. J.D. Walton, *Radome Engineering Handbook, Design and Principles*, Marcel Dekker, 1970
3. K. Prewo, J.J. Brennen, and G.K. Layden, Fiber-Reinforced Glasses and Glass-Ceramics for High Performance Applications, *Am. Ceram. Soc. Bull.*, Vol 65 (No. 2), Jan 1986
4. R.W. Rice, Mechanisms of Toughening in Ceramic Matrix Composites, in *Ceramic Engineering Science Proceedings*, Vol 2, (7-8), 1981, p 661-701
5. J. Brazel and R. Fenton, "Process Optimization Studies on ADL-4D6: A Multidirec-tional-Reinforced Silica-Silica Composite," in *New Horizons—Materials and Processes for the Eighties*, Society for the Advancement of Material and Process Engineering, 1979
6. T.M. Place and D.W. Bridges, Fused Quartz—Reinforced Silica Composites for Radomes, in *Proceedings of the Tenth Symposium on Electromagnetic Windows*, Georgia Institute of Technology, 1970
7. J. Brazel, Alumina-Alumina Composite, U.S. Patent 4,390,583, June 1983
8. J. Brazel, R. Fenton, A. Ross, and J. Dignam, "Design, Development, and Test of an ENNK BMD Millimeter Wave Radome," AIAA paper 85-0703-CP presented at the AIAA/ASME/ASCE/AHS 26th Structures, Dynamics, and Materials Conference, April 1985
9. T.M. Place, Properties of BN-3DX, A 3-Dimensional Reinforced Boron Nitride Composite, in *Proceedings of the 13th Symposium on Electromagnetic Windows*, Georgia Institute of Technology, 1976, p 17-22
10. A.J. Hanawalt, "Plasma ARC Test Technique for Evaluating Antenna Window RF Transmission Performance," AIAA paper 82-0900 presented at AIAA/ASME Third Thermodynamics, Fluids, Plasma, and Heat Transfer Conference, June 1982
11. T. Place, Design Properties for Three Dimensionally Reinforced Silica, in *Proceedings of the 12th Symposium on Electromagnetic Windows*, Georgia Institute of Technology, 1974

SELECTED REFERENCES

● J. Brazel and R. Fenton, ADL-4D6: A Silica-Silica Composite for Hardened Antenna Windows, in *Proceedings of the 13th Symposium on Electromagnetic Windows*, Georgia Institute of Technology, 1976
● J.P. Brazel, R. Fenton, R. Tanzilli, J. Gebhardt, and C. Dulka, "Millimeter Wave Hardened Antenna Window Materials Development," AMMRC TR 81-45, Army Materials and Mechanics Research Center, March 1982

Whisker-Reinforced Ceramics

P.F. Becher and T.N. Tiegs, Oak Ridge National Laboratory

CERAMICS EXHIBIT TENSILE AND FLEXURAL strengths that are extremely sensitive to the size and shape of cracks and defects inherently present in materials. This is due to low fracture toughness values. Typical monolithic ceramics have fracture toughnesses of 1 to 5 $MPa\sqrt{m}$ (0.90 to 4.5 $ksi\sqrt{in.}$), compared to a range of 15 to >150 $MPa\sqrt{m}$ (14 to >140 $ksi\sqrt{in.}$) for metals. There is one group of ceramic materials that exhibits higher toughness (6 to 18 $MPa\sqrt{m}$ or 5.5 to 17 $ksi\sqrt{in.}$) as a result of a stress-induced martensitic transformation. These are known as transformation toughened ceramics and include partially stabilized zirconias, fine-grained tetragonal zirconias, and ceramics containing tetragonal zirconia second-phase particles (Ref 1-4).

The low toughness of ceramics has limited their use as far as tensile and flexural stresses are involved. Ceramics can exhibit very high compressive strengths, but the concern here is only for their tensile and flexural strengths. However, twofold to fivefold increases in the fracture toughness of ceramics can provide considerable improvement in their fracture strength and/or decrease their sensitivity to flaw or defect size. Raising the critical fracture toughness by a factor of two to five substantially increases the achievable fracture strengths for a given flaw size. This also makes it possible to fabricate reliable high-strength ceramics, because higher toughness increases the critical flaw size to a range at which nondestructive analysis techniques can detect flaws of sizes that are larger than are allowed for the desired strength level. As a result, there is substantial interest in processes that can lead to improved fracture toughness in ceramics, such as whisker and fiber reinforcement.

The recent development of fiber- and whisker-reinforced ceramics benefitted considerably from the work carried out in the late 1960s and early 1970s (Ref 5). Substantial advances in ceramic composites occurred with the development of continuous, high-strength, amorphous silicon carbide (SiC) fibers (Ref 6) and their use to form tough and strong glass-ceramic composites (Ref 7-9). As in the case with other composites containing aligned fibers, these ceramic composites exhibit anisotropic properties and low shear strengths. However, such continuous fiber reinforced ceramic composites do exhibit quite stable failure behavior (such as slow loss in load-bearing capacity with increase in strain once failure is initiated) (Ref 10) and thus are very attractive for structural applications.

More recently, SiC whiskers, which are in the form of rod- or needle-shaped single crystals, have been explored as a reinforcing phase for ceramics. These whiskers have several attractive features, such as very high tensile strength (up to 7 GPa, or 1.0×10^6 psi), high Young's modulus (up to 550 GPa, or 79×10^6 psi) (Ref 11), and are microscopic in size (typically 0.1 to 5 μm, or 4 to 200 μin. in diameter and 10 to 100 μm, or ~400 to ~4000 μin. in length) so that they often can be incorporated into ceramics by fairly conventional powder processing techniques.

In addition to SiC, silicon nitride (Si_3N_4) whiskers are also available. Some characteristics of the various whiskers are given in Table 1. The growth of the SiC whiskers can be accomplished by several techniques, including the vapor-liquid-solid method or its variations and by vapor phase processes (Ref 12-16). The phase content and morphology of such whiskers can be influenced by processing parameters, impurities, or additives.

Whisker reinforcement can involve several toughening mechanisms. Bridging of the crack by the whiskers and whisker pull-out demand that the stress transferred to the whisker during matrix fracture be less than the tensile strength of the whiskers. This requires a very high whisker tensile strength, σ_f^w, as already noted. Whisker pull-out requires that the shear strength of the whisker-matrix interface be relatively low to allow maximum whisker pull-out. For a given whisker radius, R, the axial tensile stress, σ_t, generated in the whisker during pull-out will be:

$$\sigma_t = 2\tau_i (L_c/R) \qquad \text{(Eq 1)}$$

where L_c is the critical length of the whisker and σ_t will be less than σ_f^w (Ref 17). Whisker pull-out is favored by an increase in the whisker diameter, as is crack-bridging.

Another toughening process involves crack deflection around the rodlike whiskers. Recent studies indicate that rods are an extremely effective geometry for deflecting cracks, increasing the tortuosity of the crack path (Ref 18, 19). When the crack plane is no longer normal to the applied tensile stress axis, the applied stress must be increased to cause a crack tip stress intensity sufficient to continue crack growth. This deflection therefore causes toughening.

A third process is that of crack pinning by the whiskers in which additional strain energy must be supplied to move the crack past the whiskers. If the crack front is locally pinned at a whisker, this portion of the crack cannot advance until the whisker fails. The toughness then increases as the ratio of the particle or whisker radius to the mean free spacing between whiskers increases (Ref 20, 21). For a fixed radius, the toughness increases with volume fraction of whiskers. In general, each of the above toughening mechanisms leads to increasing toughness with an increase in whisker content.

Properties of SiC Whisker Reinforced Ceramics

While the development of SiC whisker reinforced ceramics is still in its infancy, several

Table 1   Characteristics of ceramic whiskers

Material	Whisker morphology and size	Crystal structure	Properties
SiC	Rod or needle 3-10 μm (120-390 μin.) diam, 10-100 μm (40-3950 μin.) long	Alpha and/or beta phases	> 500 GPa, (>70 $\times 10^6$ psi) Young's modulus; 2-7 GPa (0.3-1.0 $\times 10^6$ psi) tensile strength
Si_3N_4	Rod or needle 0.2-0.5 μm (8-20 μin.) diam, 50-300 μm long	Alpha plus beta phase	390 GPa, (60 $\times 10^6$ psi) Young's modulus; up to 1.5 GPa (0.2 $\times 10^6$ psi) tensile strength

Note: Whiskers are commercially produced by several sources, including Arco Chemical Company; Versar Manufacturing, Inc.; Tateho Chemical Company, Ltd.; and Tokai Carbon Company, Ltd.

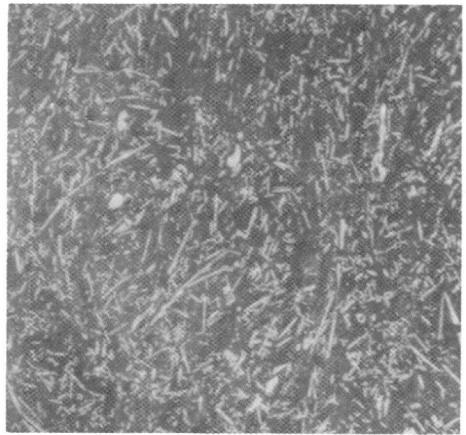

(a)

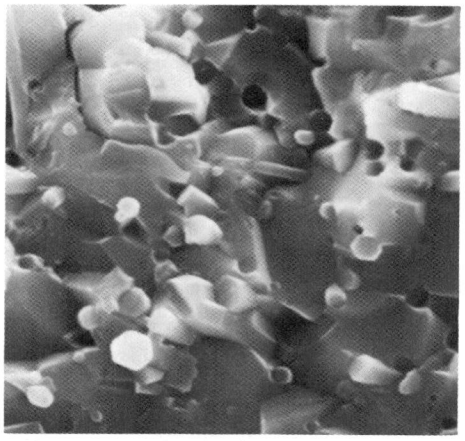

(b)

Fig. 1 Typical microstructures of whisker reinforced alumina ceramic composites containing 20 vol% SiC whiskers. (a) Dispersion of microscopic SiC whiskers (light phase) in alumina matrix is depicted by polished surface (optical micrograph). (b) Rodlike SiC whiskers are observed on fracture surface (scanning electron micrograph).

Table 2 Mechanical properties of SiC whisker reinforced alumina ceramics

Whisker content, vol%(a)	Temperature °C	°F	Fracture strength MPa	ksi	Fracture toughness MPa $\sqrt{m}$	ksi $\sqrt{in.}$
0	22	70	...	...	4.5	4.1
	700	1290	...	...	4.0	3.6
	1000	1830	...	...	3.8	3.5
10	22	70	455 ± 55	65 ± 8	7.1	6.5
	1000	1830	320 ± 36	45 ± 5		
20	22	70	655 ± 135	95 ± 20	7.5-9.0	6.8-8.2
	700	1290	535 ± 35	80 ± 5		
	1000	1830	570 ± 20	85 ± 3	7.0-8.0	6.4-7.3
40	22	70	850 ± 130	120 ± 20	6.0	5.5
	700	1290	740 ± 61	110 ± 9		
	1000	1830	665 ± 88	96 ± 13	6.2	5.6

(a) Hot-pressed mixture of alumina powder and SiC whiskers

Table 3 Mechanical properties of other SiC whisker reinforced ceramics

Composite	Whisker content, vol%	Toughness MPa $\sqrt{m}$	ksi $\sqrt{in.}$	Flexural strength MPa	ksi	Reference
Corning 1723 glass	0	<1	<0.9			24
	25	2.1-3.4	1.9-3.1	200-340	30-50	25
Barium osumilite	25	4.5	4.1	360-400	50-60	25
Si$_3$N$_4$	0	5-7	4.6-6.4	400-650	60-95	26
	10	6.5-9.5	5.9-8.6	400-500	60-75	
	30	7.5-10	6.8-9.1	350-450	50-65	
Spinel	30			415	60	27
Mullite	0	1.8-2.2	1.6-2.0	250	40	28
	20	4.6	4.2	440	65	
ZrO$_2$	0	6.2	5.6	1080	160	29
Toughened alumina	20	8.5-13.5	7.7-12.3	700-880	100-130	
Cordierite	0	2.2	2.0	180	25	29
	20	3.7	3.4	260	40	
MoSi$_2$	0	5.3	4.8	150	20	30
	20	8.2	7.5	310	45	

systems have been explored. The results for various properties obtained to date are summarized below. Most of these composite ceramics have a polycrystalline ceramic matrix containing a dispersion of strong whiskers for reinforcement. Typical examples of the as-polished and fracture surface microstructures are illustrated in Fig. 1.

Fracture Toughness and Flexural Strength. The critical fracture toughness of hot-pressed alumina reinforced with SiC whiskers that are produced by hot pressing increases with higher whisker content (Table 2). Note that the data are representative of fracture in which the crack plane is oriented normal to the plane containing the longitudinal axes of the whiskers. Texture of the whisker axes is induced during hot pressing (or any uniaxial pressing operation) of the ceramic powder-whisker mixture. Thus, there is anisotropy in the toughness and strength properties, with the toughness being greatest for fracture on planes normal to the plane containing the longitudinal axis of the whiskers (Ref 22). One other promising feature of these composites is the insensitivity of the toughness values to increasing test temperature.

The flexural strengths for these same alumina-SiC whisker composites, as measured in four-point bending, are shown as a function of test temperature in Table 2. Note that the flexural-strength data shown here are representive of samples with tensile surfaces prepared by surface grinding with a 220-grit resinoid-bonded diamond wheel. The improved performance of the SiC whisker fine-grained (~2 μm, or 80 μin.) alumina composites is highlighted by the fact that they can be readily machined without the problem of extensive machining damage, chipping, and cracking, which normally occur in fine-grained, fully dense alumina.

As noted by P.F. Becher and G.C. Wei (Ref 22), the flexural strength of such composites is sensitive to the degree of microstructural homogeneity (that is, the uniformity of dispersion of the whiskers in the dense matrix). Also, while

similar composites can be pressureless sintered to achieve comparable toughness values, the fracture strengths are lower as a result of the lower densities currently achieved by this processing route (Ref 23).

Similar improvements in the fracture toughness and strength of other ceramic-based SiC whisker reinforced composites have since been achieved, as noted in Table 3. Increased toughness associated with whisker reinforcement is not always accompanied by an increase in flexural strength. Based on the authors' experience, this is often a result of nonuniform distribution of the whiskers in the composites (Ref 22). Regions of very high or low whisker content often act as the source of mechanical failure and yield lower-strength composites than those achieved with uniform microstructures. One interesting feature observed in carefully prepared whisker-reinforced ceramics is that of improved reliability of flexural strength, as noted by increased Weibull modulus or decreased scatter in flexural strength distributions. For example, K. Ueno and Y. Toibana observed that the distribution of the flexural strength of Si$_3$N$_4$ containing 30 vol% SiC

Table 4   Thermal shock response of SiC whisker reinforced alumina as indicated by flexural strength retained after quenching from elevated temperature into boiling water

	Temperature quenched from		Retained fracture strength	
	°C	°F	MPa	ksi
Alumina-20 vol% SiC	No thermal shock		620	89.9
Single thermal shock cycle	400	750	630	91.4
	600	1110	685	99.3
	800	1470	615	89.2
	1000	1830	710	103.0
Ten thermal shock cycles	400	750	610	88.5
	500	930	570	82.7
	800	1470	540	78.3
	1000	1830	545	79.0
Alumina	No thermal shock		315	45.7
Single thermal shock cycle	300	570	250	36.3
	400	750	225	32.6
	500	930	125	18.1

Table 5   Thermal properties of SiC whisker reinforced ceramics

Composite	Thermal conductivity				Linear coefficient of thermal expansion at 22-1100 °C (70-2010 °F), 10^{-6}/K
	at 22 °C (70 °F)		at 600 °C (1110 °F)		
	W/m · K	Btu · in./ h · ft^2 · °F	W/m · K	Btu · in./ h · ft^2 · °F	
Alumina	36 ± 5	250 ± 35	12 ± 3	85 ± 20	7.8-8.2
Alumina- 20 vol% SiC whiskers	32	220	16	110	7.35
30 vol% SiC whiskers	...	...	...	...	6.70
60 vol% SiC whiskers	...	...	...	...	5.82
SiC	95	660	50	350	4.8
20 vol% SiC whiskers-mullite	7.2	50			5.60

whiskers is quite narrow, and they obtained a Weibull modulus of as high as 20 + with a mean strength of approximately 600 MPa (87 ksi) (Ref 32). Normally, the Weibull modulus of conventional ceramics is in the range of 4 to 8, reflecting a large scatter in strength values that is due to variable flaw size.

Crack Growth Failure at Low Applied Stresses. The resistance of ceramics to slow crack growth at stresses less than those required for instantaneous fracture is of considerable importance. Most ceramics exhibit time-to-failure behavior or loss in strength when subjected to applied stresses that are above 40 to 50% of the stress required for instantaneous fracture as a result of slow crack growth. This results from the fact that the rate of crack growth increases as the applied stress intensity increases. Increasing the applied stress and/or crack size will thus accelerate the rate of crack growth and decrease the time to failure. As discussed by P.F. Becher et al. (Ref 17), the applied stress intensity and, hence, stress or crack size required for slow crack growth are markedly increased in 20 vol% SiC whisker reinforced alumina composites, compared to conventional alumina ceramics. This means that much higher stress intensities and stresses can be applied to these composites while still avoiding time-dependent failure.

Thermal Shock Behavior. The higher toughness and strength of the SiC whisker reinforced alumina can be expected to improve resistance to tensile stresses introduced by thermal gradients imposed by rapid temperature changes or nonuniform heating or cooling. As can be seen in Table 4, the thermal shock resistance of alumina as measured by the strength retained after quenching samples from various elevated temperatures is markedly increased by whisker reinforcement (Ref 23). In fact, the retained strength is only slightly altered after ten thermal shock cycles involving quenching from temperatures up to 1000 °C (1830 °F), compared to either the single-cycle thermal shock or unshocked strength values. Note that the strength of similar fine-grained alumina is severely degraded after a single thermal shock cycle from only 300 °C (570 °F) (Table 4).

Flexural Creep Resistance. Initial results show the creep resistance of alumina can be substantially improved by the addition of the SiC whiskers (Ref 33). At 1550 °C (2800 °F), the creep rate of alumina in four-point flexure decreased by up to two to three orders of magnitude with the addition of 18 vol% of SiC whiskers.

Thermal Conductivity and Thermal Expansion Behavior. The thermal conductivity of the 20 vol% SiC whisker reinforced alumina composite is 32 W/m · K (220 Btu · in./h · ft^2 · °F) at room temperature and decreases to 16 W/m · K (110 Btu · in./h · ft^2 · °F) at 600 °C (1100 °F) (Table 5). The thermal conductivity of such a composite is slightly greater than that of alumina at elevated temperatures as a result of the incorporation of the SiC phase. However, as seen in Table 5, one can modify the thermal properties by selecting other matrix materials (for example, mullite).

As seen in Table 5, the linear thermal expansion coefficients of the SiC whisker reinforced aluminas also exhibit a dependence on the whisker content, decreasing with increasing whisker content. The decrease in expansion coefficients with the increase in SiC content follows the rule-of-mixtures expressions derived for composites (Ref 34). Again, this offers the opportunity to tailor the thermal properties by proper selection of the matrix phase.

Applications

The SiC whisker reinforced aluminas have already demonstrated considerable potential as cutting tools for various metal machining operations (Ref 35). It has been found that the tool life of the SiC whisker reinforced alumina on nickel-based alloys is significantly improved over that of other advanced ceramic cutting tool materials, such as cemented tungsten carbides, and substantially enhanced cutting feed rates and speeds can be achieved. Similar improvements have been obtained in machining various cast iron and steel products.

The current state of SiC whisker reinforced ceramic development indicates their marked potential for application under severe stress and temperature conditions. The enhanced fracture toughness of these composites offers the opportunity to develop composites with improved fracture strengths and narrow strength distributions, which is critical in the fabrication and application of ceramic materials. The findings to date suggest that these composites also exhibit excellent toughness and strength at elevated temperatures and have excellent creep resistance. Their ability to resist time-dependent failure or strength losses, at least at modest temperatures, and to be resistant to thermal shock damage are important attributes. The need now is to continue to explore new applications for such materials.

One must keep in mind that whisker-reinforced ceramic composites are only a recent development and thus have only been explored preliminarily. They have considerable potential, and with the development of new approaches to fabrication and performance optimization, the limits to their use and the factors that must be controlled to further improve these composites can be defined.

ACKNOWLEDGMENTS

The author wishes to thank the following colleagues for providing important contributions to the experimental studies: W.H. Warwick, J.W. Geer, J.C. Ogle, and H. Keating. Research was sponsored by the Advanced Materials Development Program, Office of Transportation Systems, U.S. Department of Energy, under Contract DE-AC05-840R21400 with Martin Marietta Systems, Inc.

REFERENCES

1. A.G. Evans and R.M. Cannon, Toughening of Brittle Solids by Martensitic Transformations, *Acta Metall.*, Vol 34 (No. 5), 1986, p 761-800

2. R. Hannink and M.V. Swain, Magnesia-Partially Stabilized Zirconia: The Influence of Heat Treatment on Thermomechanical Properties, *J. Aust. Ceram. Soc.*, Vol 18 (No. 2), 1983, p 53-62

3. P.F. Becher, Toughening Behavior in Ceramics Associated With the Transformation of Tetragonal ZrO$_2$, *Acta Metall.*, Vol 34 (No. 10), 1986, p 1885-1891; P.F. Becher, M.V. Swain, and M.K. Ferber, Relation of Transformation Temperature to the Fracture Toughness of Transformation Toughened Ceramics, *J. Mater. Sci.*, Vol 22 (No. 1), 1987, p 76-84

4. F.F. Lange, Transformation Toughening—Part 4. Fabrication, Fracture Strength and Toughness of Al$_2$O$_3$-ZrO$_2$ Composites, *J. Mater. Sci.*, Vol 17, 1982, p 247-254

5. A. Kelly, *Strong Solids*, Claredon Press, 1973

6. G. Simon and A.R. Bunsell, Mechanical and Structural Characterization of the Nicalon Silicon Carbide Fibre, *J. Mater. Sci.*, Vol 19, 1984, p 3649-3657

7. R.W. Rice, Mechanisms of Toughening in Ceramic Matrix Composites, *Ceram. Eng. Sci. Proc.*, Vol 2, (7-8), 1981

8. K.M. Prewo and J.J. Brennan, High-Strength Silicon Carbide Fibre-Reinforced Glass-Matrix Composites, *J. Mater. Sci.*, Vol 15, 1980, p 463-468

9. K.M. Prewo and J.J. Brennan, Silicon Carbide Yarn Reinforced Glass Matrix Composites, *J. Mater. Sci.*, Vol 15 (No. 2), 1980, p 463-468

10. D.B. Marshall and A.G. Evans, Failure Mechanisms in Ceramic-Fiber/Ceramic Matrix Composites, *J. Am. Ceram. Soc.*, Vol 68 (No. 5), 1985, p 225-231

11. J.J. Petrovic, J.V. Milewski, D.L. Rohr, and F.D. Gac, Tensile Mechanical Properties of SiC Whiskers, *J. Mater. Sci.*, Vol 20, 1985, p 1167-1177

12. K.S. Mazdiyasni and A. Zangvil, Effect of Impurities on SiC Whisker Morphology, *J. Am. Ceram. Soc.*, Vol 68 (No. 6), 1985, C-142-C-144

13. J.G. Lee and I.B. Cutler, Formation of Silicon Carbide from Rice Hulls, *Am. Ceram. Soc. Bull.*, Vol 54 (No. 2), 1975, p 195-198

14. W.F. Knippenberg and G. Verspui, Growth Mechanisms of Silicon Carbide in Vapour Deposition II, *Silicon Carbide—1972*, R.C. Marshall, J.W. Faust, Jr., and C.E. Ryan, Ed., University of South Carolina Press, 1973, p 108-117

15. C.E. Ryan, I. Berman, R.C. Marshall, D.P. Considine, and J.J. Hawley, Vapor-Liquid-Solid and Melt Growth of Silicon Carbide, *J. Cryst. Growth*, Vol 1, 1967, p 255-262

16. N.K. Sharma, W.S. Williams, and A. Zangvil, Formation and Structure of Silicon Carbide Whiskers From Rice Hulls, *J. Am. Ceram. Soc.*, Vol 67 (No. 11), 1984, p 715-720

17. P.F. Becher, T.N. Tiegs, J.C. Ogle, and W.H. Warwick, Toughening of Ceramics by Whisker Reinforcement, in Fracture Mechanics of Ceramics, R.C. Bradt, A.G. Evans, D.P.H. Hasselman, and F.F. Lange, Ed., Plenum Press, in press

18. K.T. Faber and A.G. Evans, Crack Deflection Processes—I. Theory, *Acta Metall.*, Vol 31 (No. 4), 1983, p 565-576

19. K.T. Faber and A.G. Evans, Crack Deflection Processes—II. Experiment, *Acta Metall.*, Vol 31 (No. 4), 1983, p 577-584

20. F.F. Lange, "The Interaction of a Crack Front With a Second-Phase Dispersion," *Phil. Mag.*, Vol 22 (No. 179), 1970, p 983-992

21. A.G. Evans, "The Strength of Brittle Materials Containing Second Phase Dispersions," *Phil. Mag.*, Vol 26, 1972, p 1327-1344

22. P.F. Becher and G.C. Wei, Toughening Behavior in SiC Whisker-Reinforced Alumina, *J. Am. Ceram. Soc.*, Vol 67 (No. 12), 1984, C-267-C-269

23. T.N. Tiegs and P.F. Becher, Whisker Reinforced Ceramic Composites, in *Tailoring of Multiphase and Composite Ceramics*, R. Tressler and G. Messing, Ed., Plenum Press, in press

24. J.J. Mecholsky, R.W. Rice, and S.W. Freiman, Prediction of Fracture Energy and Flaw Sizes in Glasses From Measurement of Mirror Size, *J. Am. Ceram. Soc.*, Vol 57 (No. 10), 1974, p 440-443

25. K.P. Gadkaree and K. Chyung, Silicon-Carbide-Whisker Reinforced Glass and Glass-Ceramic Composites, *Am. Ceram. Soc. Bull.*, Vol 65 (No. 2), 1986, p 370-376

26. P.D. Shalek, J.J. Petrovic, G.F. Hurley, and F.D. Gac, Hot Pressed SiC Whisker-Si$_3$N$_4$ Matrix Composites, *Am. Ceram. Soc. Bull.*, Vol 65 (No. 2), 1986, p 351-356

27. P.C. Panda and E.R. Seydel, Near-Net-Shape Forming of Magnesia-Alumina Spinel/Silicon Carbide Fiber Composites, *Am. Ceram. Soc. Bull.*, Vol 65 (No. 2), 1986, p 338-341

28. G.C. Wei and P.F. Becher, Development of SiC-Whisker Reinforced Ceramics, *Am. Ceram. Soc. Bull.*, Vol 64 (No. 2), 1985, p 298-304

29. N. Claussen and G. Petzow, Whisker-Reinforced Zirconia Toughened Ceramics, in *Tailoring of Multiphase and Composite Ceramics*, R. Tressler and G. Messing, Ed., Plenum Press, in press

30. F.D. Gac and J.J. Petrovic, Feasibility of a Composite of SiC Whiskers in an MoSi$_2$ Matrix, *J. Am. Ceram. Soc.*, Vol 68 (No. 8), 1985, C-200-C-201

31. F.D. Gac, J.J. Petrovic, J.V. Milewski, and P.D. Shaleh, Performance of Commercial and Research Grade SiC Whiskers in a Borosilicate Glass Matrix, *Ceram. Eng. Sci. Proc.*, Vol 7 (7-8), 1986, p 978-982

32. K. Ueno and Y. Toibana, Mechanical Properties of Silicon Nitride Ceramic Composite Reinforced With Silicon Carbide Whisker, *Yogyo-Kyokai-Shi*, Vol 91 (No. 11), 1983, p 491-497

33. A.H. Chokshi and J.R. Porter, Creep Deformation of an Alumina Matrix Composite Reinforced with Silicon Carbide Whiskers, *J. Am. Ceram. Soc.*, Vol 68 (No. 6), 1985, C-144-C-145

34. D.K. Hale, The Physical Properties of Composite Materials, *J. Mater. Sci.*, Vol 11, 1976, p 2105-2141

35. K.H. Smith, Ceramic Composite Offers Speed, Feed Gains, *Mach. Tool Blue Book*, Vol 81 (No. 1), 1986, p 71-72

Metric Conversion Guide

This Section is intended as a guide for expressing weights and measures in the Système International d'Unités (SI). The purpose of SI units, developed and maintained by the General Conference of Weights and Measures, is to provide a basis for world-wide standardization of units and measure. For more information on metric conversions, the reader should consult the following references:

- "Standard for Metric Practice," E 380, *Annual Book of ASTM Standards,* Vol 14.02, 1987, American Society for Testing and Materials, 1916 Race Street, Philadelphia, PA 19103
- "Metric Practice," ANSI/IEEE 268–1982, American National Standards Institute, 1430 Broadway, New York, NY 10018
- *Metric Practice Guide—Units and Conversion Factors for the Steel Industry,* 1978, American Iron and Steel Institute, 1000 16th Street NW, Washington, DC 20036
- *The International System of Units,* SP 330, 1986, National Bureau of Standards. Order from Superintendent of Documents, U.S. Government Printing Office, Washington, DC 20402-9325
- *Metric Editorial Guide,* 4th ed. (revised), 1985, American National Metric Council, 1010 Vermont Avenue NW, Suite 320, Washington, DC 20005-4960
- *ASME Orientation and Guide for Use of SI (Metric) Units,* ASME Guide SI 1, 9th ed., 1982, The American Society of Mechanical Engineers, 345 East 47th Street, New York, NY 10017

Base, supplementary, and derived SI units

Measure	Unit	Symbol	Measure	Unit	Symbol
			Entropy	joule per kelvin	J/K
Base units			Force	newton	N
			Frequency	hertz	Hz
Amount of substance	mole	mol	Heat capacity	joule per kelvin	J/K
Electric current	ampere	A	Heat flux density	watt per square meter	W/m^2
Length	meter	m	Illuminance	lux	lx
Luminous intensity	candela	cd	Inductance	henry	H
Mass	kilogram	kg	Irradiance	watt per square meter	W/m^2
Thermodynamic temperature	kelvin	K	Luminance	candela per square meter	cd/m^2
Time	second	s	Luminous flux	lumen	lm
			Magnetic field strength	ampere per meter	A/m
			Magnetic flux	weber	Wb
Supplementary units			Magnetic flux density	tesla	T
			Molar energy	joule per mole	J/mol
Plane angle	radian	rad	Molar entropy	joule per mole kelvin	J/mol · K
Solid angle	steradian	sr	Molar heat capacity	joule per mole kelvin	J/mol · K
			Moment of force	newton meter	N · m
Derived units			Permeability	henry per meter	H/m
			Permittivity	farad per meter	F/m
Absorbed dose	gray	Gy	Power, radiant flux	watt	W
Acceleration	meter per second squared	m/s^2	Pressure, stress	pascal	Pa
Activity (of radionuclides)	becquerel	Bq	Quantity of electricity, electric		
Angular acceleration	radian per second squared	rad/s^2	charge	coulomb	C
Angular velocity	radian per second	rad/s	Radiance	watt per square meter steradian	W/m^2 · sr
Area	square meter	m^2			
Capacitance	farad	F	Radiant intensity	watt per steradian	W/sr
Concentration (of amount of			Specific heat capacity	joule per kilogram kelvin	J/kg · K
substance)	mole per cubic meter	mol/m^3	Specific energy	joule per kilogram	J/kg
Conductance	siemens	S	Specific entropy	joule per kilogram kelvin	J/kg · K
Current density	ampere per square meter	A/m^2	Specific volume	cubic meter per kilogram	m^3/kg
Density, mass	kilogram per cubic meter	kg/m^3	Surface tension	newton per meter	N/m
Electric charge density	coulomb per cubic meter	C/m^3	Thermal conductivity	watt per meter kelvin	W/m · K
Electric field strength	volt per meter	V/m	Velocity	meter per second	m/s
Electric flux density	coulomb per square meter	C/m^2	Viscosity, dynamic	pascal second	Pa · s
Electric potential, potential			Viscosity, kinematic	square meter per second	m^2/s
difference, electromotive force	volt	V	Volume	cubic meter	m^3
Electric resistance	ohm	Ω	Wavenumber	1 per meter	1/m
Energy, work, quantity of heat	joule	J			
Energy density	joule per cubic meter	J/m^3			

Conversion factors

To convert from	to	multiply by
Area		
in.^{2}	mm^2	6.451 600 E + 02
in.^{2}	cm^2	6.451 600 E + 00
in.^{2}	m^2	6.451 600 E − 04
ft^2	m^2	9.290 304 E − 02
Bending moment or torque		
lbf · in.	N · m	1.129 848 E − 01
lbf · ft	N · m	1.355 818 E + 00
kgf · m	N · m	9.806 650 E + 00
ozf · in.	N · m	7.061 552 E − 03
Bending moment or torque per unit length		
lbf · in./in.	N · m/m	4.448 222 E + 00
lbf · ft/in.	N · m/m	5.337 866 E + 01
Current density		
A/in.^{2}	A/cm^2	1.550 003 E − 01
A/in.^{2}	A/mm^2	1.550 003 E − 03
A/ft^2	A/m^2	1.076 400 E + 01
Electric field strength		
V/mil	kV/m	3.937 008 E + 01
Electricity and magnetism		
gauss	T	1.000 000 E − 04
maxwell	μWb	1.000 000 E − 02
mho	S	1.000 000 E + 00
Oersted	A/m	7.957 700 E + 01
Ω · cm	Ω · m	1.000 000 E − 02
Ω circular-mil/ft	μΩ · m	1.662 426 E − 03
Energy (impact, other)		
ft · lbf	J	1.355 818 E + 00
Btu (thermochemical)	J	1.054 350 E + 03
cal (thermochemical)	J	4.184 000 E + 00
kW · h	J	3.600 000 E + 06
W · h	J	3.600 000 E + 03
Flow rate		
ft^3/h	L/min	4.719 475 E − 01
ft^3/min	L/min	2.831 000 E + 01
gal/h	L/min	6.309 020 E − 02
gal/min	L/min	3.785 412 E + 00
Force		
lbf	N	4.448 222 E + 00
kip (1000 lbf)	N	4.448 222 E + 03
tonf	kN	8.896 443 E + 00
kgf	N	9.806 650 E + 00
Force per unit length		
lbf/ft	N/m	1.459 390 E + 01
lbf/in.	N/m	1.751 268 E + 02
kip/in.	N/m	1.751 268 E + 05
Fracture toughness		
ksi $\sqrt{\text{in.}}$	MPa$\sqrt{\text{m}}$	1.098 800 E + 00
Heat content		
Btu/lb	kJ/kg	2.326 000 E + 00
cal/g	kJ/kg	4.186 800 E + 00
Heat input		
J/in.	J/m	3.937 008 E + 01
kJ/in.	kJ/m	3.937 008 E + 01

To convert from	to	multiply by
Impact energy per unit area		
ft · lbf/ft^2	J/m^2	1.459 002 E + 01
Length		
Å	nm	1.000 000 E − 01
μin.	μm	2.540 000 E − 02
mil	μm	2.540 000 E + 01
in.	mm	2.540 000 E + 01
in.	cm	2.540 000 E + 00
ft	m	3.048 000 E − 01
yd	m	9.144 000 E − 01
mile	km	1.609 300 E + 00
Length per unit mass		
in./lb	m/kg	5.599 740 E − 02
yd/lb	m/kg	2.015 907 E + 00
Mass		
oz	kg	2.834 952 E − 02
lb	kg	4.535 924 E − 01
ton (short, 2000 lb)	kg	9.071 847 E + 02
ton (short, 2000 lb)	kg × 10^3(a)	9.071 847 E − 01
ton (long, 2240 lb)	kg	1.016 047 E + 03
Mass per unit area		
oz/in.^{2}	kg/m^2	4.395 000 E + 01
oz/ft^2	kg/m^2	3.051 517 E − 01
oz/yd^2	kg/m^2	3.390 575 E − 02
lb/ft^2	kg/m^2	4.882 428 E + 00
Mass per unit length		
lb/ft	kg/m	1.488 164 E + 00
lb/in.	kg/m	1.785 797 E + 01
denier	kg/m	1.111 111 E − 07
tex	kg/m	1.000 000 E − 06
Mass per unit time		
lb/h	kg/s	1.259 979 E − 04
lb/min	kg/s	7.559 873 E − 03
lb/s	kg/s	4.535 924 E − 01
Mass per unit volume (includes density)		
g/cm^3	kg/m^3	1.000 000 E + 03
lb/ft^3	g/cm^3	1.601 846 E − 02
lb/ft^3	kg/m^3	1.601 846 E + 01
lb/in.^{3}	g/cm^3	2.767 990 E + 01
lb/in.^{3}	kg/m^3	2.767 990 E + 04
oz/in.^{3}	kg/m^3	1.729 994 E + 03
Power		
Btu/s	kW	1.055 056 E + 00
Btu/min	kW	1.758 426 E − 02
Btu/h	W	2.928 751 E − 01
erg/s	W	1.000 000 E − 07
ft · lbf/s	W	1.355 818 E + 00
ft · lbf/min	W	2.259 697 E − 02
ft · lbf/h	W	3.766 161 E − 04
hp (550 ft · lbf/s)	kW	7.456 999 E − 01
hp (electric)	kW	7.460 000 E − 01
Power density		
W/in.^{2}	W/m^2	1.550 003 E + 03
Pressure (fluid)		
atm (standard)	Pa	1.013 250 E + 05
bar	Pa	1.000 000 E + 05
in. Hg (32 °F)	Pa	3.386 380 E + 03
in. Hg (60 °F)	Pa	3.376 850 E + 03
lbf/in.^{2} (psi)	Pa	6.894 757 E + 03
torr (mm Hg, 0 °C)	Pa	1.333 220 E + 02

To convert from	to	multiply by
Specific area		
ft^2/lb	m^2/kg	2.048 161 E − 01
Specific energy		
cal/g	J/g	4.186 800 E + 00
Btu/lb	kJ/kg	2.326 000 E + 00
Specific heat capacity		
Btu/lb · °F	J/kg · K	4.186 800 E + 03
cal/g · °C	J/kg · K	4.186 800 E + 03
Stress (force per unit area)		
tonf/in.^{2} (tsi)	MPa	1.378 951 E + 01
kgf/mm^2	MPa	9.806 650 E + 00
ksi	MPa	6.894 757 E + 00
lbf/in.^{2} (psi)	MPa	6.894 757 E − 03
MN/m^2	MPa	1.000 000 E + 00
Temperature		
°F	°C	5/9 · (°F − 32)
°R	K	5/9
°F	K	5/9 · (°F + 459.67)
°C	K	°C + 273.15
Temperature interval		
°F	°C	5/9
Thermal conductivity		
Btu · in./s · ft^2 · °F	W/m · K	5.192 204 E + 02
Btu/ft · h · °F	W/m · K	1.730 735 E + 00
Btu · in./h · ft^2 · °F	W/m · K	1.442 279 E − 01
cal/cm · s · °C	W/m · K	4.184 000 E + 02
Thermal expansion		
μin./in. · °C	10^{-6}/K	1.000 000 E + 00
μin./in. · °F	10^{-6}/K	1.800 000 E + 00
Velocity		
ft/h	m/s	8.466 667 E − 05
ft/min	m/s	5.080 000 E − 03
ft/s	m/s	3.048 000 E − 01
in./s	m/s	2.540 000 E − 02
km/h	m/s	2.777 778 E − 01
mph	km/h	1.609 344 E + 00
Viscosity (dynamic and kinematic)		
poise (P)	Pa · s	1.000 000 E − 01
cP	Pa · s	1.000 000 E − 03
lbf · s/in.^{2}	Pa · s	6.894 757 E + 03
ft^2/s	m^2/s	9.290 304 E − 02
in.^{2}/s	mm^2/s	6.451 600 E + 02
Volume		
in.^{3}	m^3	1.638 706 E − 05
ft^3	m^3	2.831 685 E − 02
fluid oz	m^3	2.957 353 E − 05
gal (U.S. liquid)	m^3	3.785 412 E − 03
Volume per unit time		
ft^3/min	m^3/s	4.719 474 E − 04
ft^3/s	m^3/s	2.831 685 E − 02
in.^{3}/min	m^3/s	2.731 177 E − 07
Wavelength		
Å	nm	1.000 000 E − 01

(a) kg × 10^3 = 1 metric ton

SI prefixes—names and symbols

Exponential expression	Multiplication factor	Prefix	Symbol
10^{18}	1 000 000 000 000 000 000	exa	E
10^{15}	1 000 000 000 000 000	peta	P
10^{12}	1 000 000 000 000	tera	T
10^{9}	1 000 000 000	giga	G
10^{6}	1 000 000	mega	M
10^{3}	1 000	kilo	k
10^{2}	100	hecto(a)	h
10^{1}	10	deka(a)	da
10^{0}	1	BASE UNIT	
10^{-1}	0.1	deci(a)	d
10^{-2}	0.01	centi(a)	c
10^{-3}	0.001	milli	m
10^{-6}	0.000 001	micro	μ
10^{-9}	0.000 000 001	nano	n
10^{-12}	0.000 000 000 001	pico	p
10^{-15}	0.000 000 000 000 001	femto	f
10^{-18}	0.000 000 000 000 000 001	atto	a

(a) Nonpreferred. Prefixes should be selected in steps of 10^3 so that the resultant number before the prefix is between 0.1 and 1000. These prefixes should not be used for units of linear measurement, but may be used for higher order units. For example, the linear measurement, decimeter, is nonpreferred, but square decimeter is acceptable.

Abbreviations and Symbols*

a crack length

A ampere

A area; ratio of the alternating stress amplitude to the mean stress

Å angstrom

ABS acrylonitrile-butadiene-styrene

ac alternating current

ACM advanced cure monitor

ARC accelerated rate calorimeter

at.% atomic percent

AS designation for surface-treated fiber

ATL automated tape layer

atm atmosphere (pressure)

AU designation for untreated fiber

BF_3MEA borontrifluoro-monoethylamine

BGDGE butylene glycol diglycidyl ether

BMC bulk molding compound

BMI bismaleimide (resin)

BT bismaleimide-triazine (resin)

Btu British thermal unit

c composite specific heat (c_p = constant pressure, c_v = constant volume)

C-C carbon-carbon

CAD/CAM computer-aided design/computer-aided manufacturing

CCA composite cylinder assemblage

CAT computer-aided tomography

CFRP carbon fiber reinforced plastic

cm centimeter

cpm cycles per minute

cps cycles per second

CTE coefficient of thermal expansion

CVD chemical vapor deposition

CVN Charpy V-notch (impact test or specimen)

d an operator used in mathematical expressions involving a derivative (denotes rate of change)

d depth; diameter

DADPS diaminodiphenylsulfone

DAIP diallyl isophthalate

DAP diallyl phthalate

DBTT ductile-brittle transition temperature

dc direct current

DDA dynamic dielectric analysis

DGA diglycidyl aniline

DGEBA diglycidyl ether of bisphenol A

DGEBF diglycidyl ether of bisphenol F

DGT dynamic gel temperature

diam diameter

DIB diiodobutane

DMA dynamic mechanical analysis

DSC differential scanning calorimetry

DTA differential thermal analysis

e natural log base, 2.71828

E modulus of elasticity; Young's modulus

Eq equation

et al. and others

ESCA electron spectroscopy for chemical analysis

f fiber

f frequency

F force

FMW formulated molecular weight

FP polycrystalline alumina fiber

FRP fiber-reinforced plastic

FRS fiber-reinforced superalloys

ft foot

FTIR Fourier transform infrared

g gram

G shear modulus

G' storage modulus

G'' loss modulus

gal gallon

GPa gigapascal

GPC gel permeation chromatography

gpd grams per denier

gr grain

G_{xy} in-plane shear modulus of laminate

G_{Ic} interlaminar fracture toughness (mode I, peel; mode II, shear; mode III, scissor shear)

h hour

h plate thickness of composite

H height

HERF high-energy-rate forging

HIP hot isostatic press

HM high modulus

HPLC high-performance liquid chromatography

H_r heat of reaction

HT high tensile

Hz hertz

ID inside diameter

IM intermediate modulus

IR infrared (radiation)

J joule

k notch sensitivity factor

K Kelvin

K coefficient of thermal conductivity; bulk modulus of elasticity

K_I stress-intensity factor

K_c plane-stress fracture toughness

K_{Ic} plane-strain fracture toughness; mode I critical stress-intensity factor

K_{Id} dynamic fracture toughness

K_{Iscc} threshold stress intensity for stress-corrosion cracking

K_t stress-concentration factor

K_t^∞ stress-concentration factor for infinite plate

K_{th} threshold crack tip stress-intensity factor

kg kilogram

km kilometer

kPa kilopascal

ksi kips (1000 lb) per square inch

*Additional abbreviations and symbols and their respective definitions can be found in Tables 1 and 2 in the article ''Continuous Boron Fiber MMCs'' in this Volume.

kV kilovolt

L liter; longitudinal direction

L length

lb pound

LEFM linear-elastic fracture mechanics

ln natural logarithm (base *e*)

m matrix

MDA methylenedianiline

MEKP methyl ethyl ketone peroxide

Mg megagram

min minute; minimum

MJ megajoule

mL milliliter

mm millimeter

MMA methyl methacrylate

MMC metal matrix composite

mol% mole percent

MPa megapascal

mph miles per hour

MVT moisture vapor transmission

N Newton

N fatigue life (number of cycles)

NDE nondestructive evaluation

NDI nondestructive inspection

NDT nondestructive testing

nm nanometer

No. number

OD outside diameter

oz ounce

P applied load; pressure

Pa pascal

PAI polyamideimide

PAN polyacrylonitrile

PAS polyarylsulfone

PBI polybenzimidazole

PBT polybutylene terephthalate

PDCP polydicyclopentadiene

PEEK polyether etherketone

PEI polyetherimide

PES polyether sulfone

PET polyethylene terephthalate

PI polyimide

PIC pressure-impregnation-carbonization

P/M powder metallurgy

PMR *in-situ* polymerization of monomer reactants

ppb parts per billion

ppm parts per million

PPS polyphenylene sulfide

PS polysulfone

psi pounds per square inch

psia pounds per square inch absolute

psid pounds per square inch differential

psig pounds per square inch gage

PTFE polytetrafluoroethylene

PVA polyvinyl alcohol

PVC polyvinyl chloride

R radius; ratio of the minimum stress to the maximum stress

r rate of reaction

RA reduction of area

RDS rheometric dynamic scanning

Ref reference

RGA residual gas analysis

RH relative humidity

RIM reaction injection molding

RMS root mean square

ROM rule of mixtures; rough order of magnitude

rpm revolutions per minute

RRIM reinforced reaction injection molding

RTD room temperature, dry

RTM resin transfer molding

RTW room temperature, wet

RTV room-temperature vulcanizing

RVE representative volume element

RDGE resorcinol diglycidyl ether

s second

SBS short beam shear

SEM scanning electron microscope or microscopy

SMC sheet molding compound

S-N stress-number of cycles

SPF superplastic forming

sp gr specific gravity

t thickness; time

T transverse direction

T temperature; tenacity

TEM transmission electron microscope or microscopy

T_g glass transition temperature

TGA thermogravimetric analysis

TGAP triglycidyl p-laminophenol

TGETPM triglycidyl ether of triphenyl methane

TLC thin-layer chromatography

T_m melting temperature

TMA thermomechanical analysis

TPI turns per inch

tan equal to ratio of the loss modulus to the storage modulus

TTU through-transmission ultrasonics

UDC unidirectional composite

UTS ultimate tensile strength

UV ultraviolet

V_f volume fraction of fiber

V_m volume fraction of matrix

V_v volume fraction of void content

VCDO vinyl cyclohexene diepoxide

vol volume

vol% volume percent

VLS vapor feed gases-liquid catalyst-solid crystalline whisker growth

w whisker

W watt

W width

WPE weight per epoxide

wt% weight percent

XPS x-ray photoelectron spectroscopy

yr year

° angular measure; degree

°C degree Celsius (centigrade)

°F degree Fahrenheit

0° fiber direction

90° perpendicular to fiber direction

α coefficient of thermal expansion

Δ change in quantity; an increment; a range

η viscosity

ε strain

γ shear strain

μin. microinch

μm micrometer (micron)

ν Poisson's ratio

π pi (3.141592)

ψ damping

ρ density

σ tensile stress

τ shear stress

θ angle

⇌ direction of reaction

÷ divided by

= equals

ˆ circumflex

≈ approximately equals

≠ not equal to

≡ identical with

> greater than

≫ much greater than

≥ greater than or equal to

∞ infinity

∝ is proportional to; varies as

∫ integral of

< less than

≪ much less than

≤ less than or equal to

± maximum deviation

− minus; negative ion charge

× diameters (magnification); multiplied by

· multiplied by

Ω ohm

/ per

% percent

+ plus; positive ion charge

√ square root of

~ approximately; similar to

Greek Alphabet

A, α alpha
B, β beta
Γ, γ gamma
Δ, δ delta
E, ε epsilon
Z, ζ zeta
H, η eta
Θ, θ theta

I, ι iota
K, κ kappa
Λ, λ lambda
M, μ mu
N, ν nu
Ξ, ξ xi
O, ο omicron
Π, π pi

P, ρ rho
Σ, σ sigma
T, τ tau
Υ, υ upsilon
Φ, φ phi
X, χ chi
Ψ, ψ psi
Ω, ω omega

Index

F